P9-CBX-699

CIVIL ENGINEER'S REFERENCE BOOK

CIVIL ENGINEER'S REFERENCE BOOK

Edited by
L. S. BLAKE
B.Sc.(Eng.), Ph.D., F.I.C.E., F.I.Struct.E., F.I.H.E.
Director, CIRIA

With specialist contributors

BUTTERWORTHS
LONDON—BOSTON
Sydney—Wellington—Durban—Toronto

First published as 'Civil Engineering Reference Book'
in 1951
Second edition 1961
Third edition (completely revised and reset) 1975
Reprinted 1977, 1979, 1980, 1985, 1986

ISBN 0 408 70475 6

Filmset 'Monophoto' by
Photoprint Plates Ltd, Rayleigh, Essex
Printed and bound in England by
Anchor Brendon Ltd, Tiptree, Essex

PREFACE TO THE THIRD EDITION

'It is becoming increasingly difficult for the civil engineer, who must to some extent specialise, to keep in touch with all modern developments, many of which may possibly be of use in his own branch. More use might be made of proved experience in other fields if the information were more readily available. The aim of this book is therefore to give to the civil engineer, in whatever capacity he may be acting, a concise presentation of the fundamentals of the theory and practice of all branches.'

This aim, stated by E. H. Probst and J. Comrie for the first and second editions of this book, has been applied as far as possible to the third edition. In some other respects, however, this third edition is very different from its predecessors, partly because of the considerable advances in design and construction practices in the past two decades and partly because needs have changed.

First, the overall length of the book has been reduced slightly compared with that of previous editions in order to make the information more accessible and to keep the selling price within the reach of most civil engineers. Secondly, the needs of the different categories of readers have been given careful attention. The student and graduate at the start of his career often needs guidance on practice in the design and construction of various types of civil engineering works, which he will find in Sections 10–42, all of which were written by engineers with considerable experience in their respective fields of work. The civil engineer in mid-career, and later, soon becomes forgetful, or out of date, in some of the basic theories of civil engineering—mathematics, strength of materials, theory of structures, properties of materials, hydraulics, etc. In Sections 1–9 he will find these basic subjects summarised in a form which will enable him to refer to up to date information as and when he requires it, or to revise his own knowledge of a subject. Throughout his career, a civil engineer is frequently faced with an unfamiliar problem or a complete change in occupation, such as a change from building to bridge design or from railway to highway engineering. It is hoped that Sections 18–29 will be of considerable assistance in these cases, as first points of reference into the unfamiliar.

The third edition also includes more—possibly still not enough—information on construction practice, particularly in important areas on which there is a paucity of published information. It is hoped that these Sections, beginning at Section 31 on Contract Management and Control, will be especially useful to most civil engineers.

It would be easy to criticise this book on the grounds that it omits some branches of civil engineering altogether. I am conscious of these omissions, which were dictated by lack of space and, sometimes, by lack of an author of sufficiently high standing. It is impossible within approximately 1700 pages adequately to cover the whole of civil engineering or to deal with any subject in such depth that no further reading is necessary, although, nevertheless, I hope that even a specialist in a particular field will benefit by referring to appropriate Sections. A reference book of this type cannot and should not replace Standards, Codes or standard textbooks. It is a first point of reference which, through its selective lists of references and bibliographies will lead the user to more effective detailed study of a subject.

None of the contributors is a 'professional author'; they are all engineers of considerable standing—designers, contractors, research workers, teachers—who have devoted a considerable proportion of their spare time to setting down their expert knowledge and experience in this book. I am most grateful to them for their hard work, cooperation and help.

Finally, I will welcome any suggestions and criticisms which could help to make future editions of this book as useful as possible to members of the profession.

L. S. BLAKE

Farnham Common
Bucks.

CONTENTS

CONTRIBUTORS

J. ALLEN, D.SC., LL.D., C.Eng., F.I.C.E., F.R.S.E.
Emeritus Professor,
University of Aberdeen (Section 5)

W. H. ARCH, B.SC., C.Eng., M.I.C.E.
Redpath Dorman Long (Contracting) Ltd. (Section 38)

S. C. C. BATE, B.SC., Ph.D., C.Eng., F.I.C.E., F.I.Struct.E.
Building Research Establishment (Section 11)

B. C. BEST, B.SC.
Head of Operational Research Dept.,
Cement and Concrete Association (Section 1)

G. H. CHILD, M.SC., F,G.S.
Soil Mechanics Ltd. (Section 10)

D. H. COOMBS, B.SC., C.Eng., F.I.C.E., M.I.Mech.E.
Barrister-at-Law,
Consultant (formerly Permanent Way Engineer, British Railways Board) (Section 23)

J. B. DWIGHT, M.A., C.Eng., M.I.C.E., M.I.Struct.E.
University of Cambridge (Section 13)

G. R. DARBY, C.Eng., M.I.Mech.E.
Secretary,
Metric Steering Committee, CEGB (Appendix)

N. G. EGGLETON, B.SC., C.Eng., F.I.C.E.
Central Electricity Generating Board (Section 25)

L. B. ESCRITT, A.M.I.C.E., F.R.S.H., F.I.P.H.E., Hon.M.Inst.S.P.
(Section 27)

E. V. FINN, C.Eng., F.I.C.E., F.I.Struct.E., F.R.S.H., M.Soc.C.E.(France), M.Cons.E.
Partner,
Sir Frederick Snow & Partners (Section 22)

J. M. FISHER, B.SC., C.Eng., F.I.C.E., F.I.H.E.
John Laing & Son (Section 36)

P. G. FOOKES, B.SC., Ph.D., M.I.M.M., F.G.S.
Consultant (Section 7)

J. A. FRANKLIN, B.SC.(Eng.), M.SC., Ph.D., D.I.C., C.Eng., M.I.C.E., A.M.I.M.M., F.G.S.
Consultant on Rock Engineering, formerly with Rock Mechanics Ltd. (Section 9)

T. R. GRAVES SMITH, B.SC., Ph.D., C.Eng., M.I.C.E., M.I.Struct.E.
Department of Civil Engineering,
University of Southampton (Section 2)

N. M. GRIEVE, C.Eng., M.I.Struct.E.,
Central Electricity Generating Board (Section 25)

K. P. GRUBB, B.SC.(Eng.), C.Eng., M.I.C.E.
Central Electricity Generating Board (Section 25)

P. H. D. HANCOCK, M.A., C.Eng., F.I.C.E., A.M.B.I.M.
Company Management Development Officer,
George Wimpey & Co. (Section 31)

I. W. HANNA, B.SC., C.Eng., M.I.C.E.
Central Electricity Generating Board (Section 25)

J. S. HARVEY
Edward Lumley & Sons Ltd. (Section 32)

N. B. HOBBS, M.A.(Cantab), D.I.C., C.Eng., M.I.C.E., F.G.S.
Soil Mechanics Ltd. (Section 8)

W. M. JENKINS, B.SC., Ph.D., C.Eng., M.I.C.E., M.I.Struct.E.
Head of Department of Civil and Structural Engineering and Building,
Teesside Polytechnic (Section 3)

W. G. JONES, C.G.I.A., C.Eng., M.I.C.E.
Central Electricity Generating Board (Section 25)

R. W. KIMBER, B.SC.(Eng.), C.Eng., M.I.C.E.
Richard Costain Ltd. (Section 35)

D. J. LEE, B.SC.Tech., D.I.C., C.Eng., F.I.C.E., F.I.Struct.E.
Maunsell & Partners (Section 18)

D. H. LITTLE, O.B.E., B.SC., C.Eng., F.I.C.E.
Water C. Andrews & Partners (Section 24)

T. MALCOLM, B.SC., C.Eng., F.I.C.E.
Richard Costain Ltd. (Section 35)

D. L. McKIE, C.Eng., M.I.C.E.
Central Electricity Generating Board (Section 25)

G. McLEOD, M.SC., C.Eng., F.I.C.E., F.Inst.W.E.
Usk River Authority (Section 28)

A. C. MEIGH, D.SC., C.Eng., F.I.C.E., F.G.S.
Managing Director,
Soil Mechanics Ltd. (Section 8)

A. MEREDITH, C.Eng., M.I.C.E.
Central Electricity Generating Board (Section 25)

A. M. MUIR WOOD, M.A., C.Eng., F.I.C.E., F.G.S.
Partner,
Sir William Halcrow & Partners (Section 30)

D. E. MURCHISON, B.Sc.(Eng.), C.Eng., M.I.C.E.
Department of Civil Engineering,
University of Surrey (Section 6)

F. H. NEEDHAM, B.Sc., A.C.G.I., C.Eng., F.I.C.E., F.I.Struct.E.
British Steel Corporation (Section 12)

I. K. NIXON, A.C.G.I., C.Eng., F.I.C.E., F.G.S.
Director,
Soil Mechanics Ltd. (Section 10)

A. PARRISH, M.B.E., C.Eng., M.I.Mech.E.
Consultant,
Formerly with I.C.I. Ltd. (Appendix)

F. H. POTTER, B.Sc.Tech., C.Eng., M.I.C.E., F.I.W.Sc.
Department of Civil Engineering,
Imperial College of Science & Technology (Section 15)

J. L. PRATT, B.Sc.(Eng.), C.Eng., M.I.E.E., F.Weld.I.
Braithwaite & Co. Engineers Ltd. (Section 37)

R. G. PRICE
Goodman Price Ltd. (Section 42)

D. Y. PRICHARD, C.Eng., M.I.C.E.
Associate Partner,
Harris & Sutherland (Section 14)

P. PULLAR-STRECKER, M.A., C.Eng., M.I.C.E.
Director of Research
C.I.R.I.A. (Section 4)

D. W. QUINION, B.Sc.(Eng.), C.Eng., F.I.C.E., F.I.Struct.E., F.F.B.
Tarmac Construction Ltd. (Section 33)

M. A. RICHARDSON, B.Eng., C.Eng., M.I.C.E.
Rees Construction Ltd. (Section 39)

B. J. RICHMOND, B.Sc., Ph.D., A.C.G.I., D.I.C., C.Eng., M.I.C.E., M.I.Struct.E.
Maunsell & Partners (Section 18)

J. RODIN, B.Sc., C.Eng., F.I.C.E., F.I.Struct.E., M.Cons.E., M.Soc.C.E.(France)
Building Design Partnership (Section 19)

B. H. ROFE, M.A., C.Eng., M.I.C.E., F.I.W.E.
Rofe, Kennard & Lapworth (Section 26)

J. H. SARGENT, C.Eng., F.I.C.E., F.G.S.
Dredging Investigations Ltd. (Section 40)

R. J. M. SUTHERLAND, B.A., C.Eng., F.I.C.E., F.I.Struct.E., M.I.H.E., M.Cons.E.
Harris & Sutherland (Section 14)

F. L. TERRETT, M.Eng., C.Eng., F.I.C.E.
Lewis & Duvivier (Section 29)

A. R. THOMAS, O.B.E., B.Sc.(Eng.), C.Eng., F.I.C.E., F.A.S.C.E.
Chartered Civil Engineer,
formerly of Binnie & Partners (Section 20)

M. J. TOMLINSON, C.Eng., F.I.C.E., F.I.Struct.E.
Director,
Wimpey Laboratories Ltd. (Section 16)

A. D. TOWNEND, B.Sc.(Eng.), C.Eng., A.C.G.I., D.I.C., F.I.C.E.
Partner,
Sir Frederick Snow & Partner (Section 22)

R. C. S. WALTERS, B.Sc., C.Eng., F.I.C.E., P.P.I.W.E., M.Cons.E.
Consultant,
Rofe, Kennard & Lapworth (Section 26)

COMMANDER H. WARDLE, R.N., T.D.
Underwater Engineering Consultant,
Formerly Managing Director,
Strongwork Diving (International) Ltd. (Section 41)

T. D. WILSON, B.Sc., C.Eng., F.I.C.E., F.I.Struct.E., F.I.H.E.
Partner,
Mott. Hay & Anderson (Section 21)

C. J. WILSHERE, B.A., C.Eng., B.A.I., F.I.C.E.
John Laing Design Associates Ltd. (Section 34)

T. A. WYATT, B.Sc.(Eng.), Ph.D., C.Eng., M.I.C.E., M.I.Struct.E.
Department of Civil Engineering,
Imperial College of Science and Technology (Section 17)

1 MATHEMATICS AND STATISTICS

MATHEMATICS AND STATISTICS 1

1 MATHEMATICS AND STATISTICS

B. C. BEST, B.Sc.,
Cement and Concrete Association

MATHEMATICS

ALGEBRA

Powers and roots

The following are true for all values of indices, whether positive, negative or fractional.

$$a^p \times a^q = a^{p+q}$$
$$(a^p)^q = a^{pq}$$
$$(a/b)^p = a^p/b^p$$
$$(ab)^p = a^p b^p$$
$$a^p/a^q = a^{p-q}$$
$$a^{-p} = (1/a)^p = 1/a^p$$
$$\sqrt[p]{a} = a^{1/p}$$
$$a^0 = 1$$
$$0^p = 0$$

Solutions of equations in one unknown

LINEAR EQUATIONS

Generally $ax+b = 0$
of which there is one solution or root $x = -b/a$

QUADRATIC EQUATIONS

Generally $ax^2+bx+c = 0$
of which there are two solutions or roots

$$x = \frac{-b \pm \sqrt{(b^2-4ac)}}{2a}$$

where, if $b^2 > 4ac$, the roots are real and unequal
$b^2 = 4ac$, the roots are real and equal
$b^2 < 4ac$, the roots are conjugate complex.

It is worth attempting to rearrange equations as, often, they can be put into a more familiar form simply by rearrangement.

e.g. $ax^{2m}+bx^m+c = 0$ is a quadratic equation in x^m
while $a/x^2+b/x+c = 0$
is the quadratic $cx^2+bx+a = 0$

CUBIC EQUATIONS

Generally $x^3+bx^2+cx+d=0$
If the substitution: $x=y-b/3$ is made the equation

becomes $$y^3+ey+f+0$$
where $$e=(3c-b^2)/3$$
and $$f=(2b^3-9bc+27d)/27$$
now define

$$A=\sqrt[3]{\left[-\frac{f}{2}+\left(\frac{f^2}{4}+\frac{c^3}{27}\right)\right]}$$

$$B=\sqrt[3]{\left[-\frac{f}{2}-\left(\frac{f^2}{4}+\frac{c^3}{27}\right)\right]}$$

and the three roots, in terms of y are:

$$y_1=[A+B]$$
$$y_{2,3}=[-(A+B)/2\pm\sqrt{-3}(A-B)/2]$$

and in terms of x the three roots are

$$x_{1,2,3}=y_{1,2,3}-\frac{b}{3}$$

EQUATIONS OF HIGHER DEGREE

Equations of degree higher than the second (quadratic equations) are not soluble directly as the method of solving the cubic equation above shows. Generally recourse must be had to either graphical or numerical techniques.

If the equation be of the form:

$$F(x)=0$$

e.g. $$a_nx^n+a_{n-1}x^{n-1}\ldots+a0=0$$

then plot the graph of $y=F(x)$ the values of x at which $y=0$ are the roots or solutions to the equation. Frequently this graphical approach may be used fairly roughly (and therefore quickly) to obtain an estimate of a root. This estimate can then be improved by numerical means. For instance values of $F(x)$ may be calculated for values of x close to that given as a root by the graphical method. The difficulty (which is not serious for hand calculations) is guessing by how much to adjust x to get $F(x)$ nearer to 0.

Newton's method

This is a method of step by step iteration in which an estimate of a root is refined.

Suppose that a_1 is an approximation to a root of an equation then; for small q

$$F(a_1+q)\simeq F(a_1)+qF(a_1)$$

So that if we assume (a_1+q) to be the better solution we are seeking, i.e.

$$F(a_1+q)=0$$

then

$$q = \frac{-F(a_1)}{F'(a_1)}$$

and $a_2 = a_1 + q$ is a second and better approximation.

This is well illustrated by drawing a curve cutting the x-axis, assuming a value a_1 of x near to the intersection to have been found, drawing the ordinate to the curve $x = a_1$ and then constructing the tangent to the curve $y = F(x)$ at the point $x = a_1$.

The point $x = a_2$ where this tangent cuts the axis is plainly a better estimate of the intersection than is a_1.

This technique can be used successfully in automatic calculation on a computer. The problem then becomes that of determining when to stop the iteration process:

$a_1, a_2, a_3 \ldots$

which may be best done by stopping when the change between successive approximations, a_n and a_{n+1} becomes less than some small pre-set amount.

Graphical and numerical methods will generally be required to deal with transcendental equations although in some cases it may be more convenient to find the intersections of two graphs rather than try to compute where a more complicated graph cuts an axis

e.g. $x - \sin x = 0$

is best solved by plotting:

$$y = x$$

and

$$y = \sin x$$

to find the intersection which will give an estimate which can be refined numerically.

Progressions

1 Arithmetic progressions in which the difference between consecutive terms if a constant amount. Thus the terms may be

$$a,\ a+d,\ a+2d,\ a+3d \ldots$$

The nth term is $a+(n-1)d$ and the sum to n terms,

$$S_n = \frac{n}{2}\{2a+(n-1)d\}$$

2 Geometrical progressions in which the ratio between consecutive terms is a constant. Generally terms are:

$$a,\ ar,\ ar^2,\ ar^3 \ldots$$

The nth term is ar^{n-1} and the sum of n terms is

$$S_n = \frac{a(1-r^n)}{1-r}$$

If r is strictly smaller than 1 so $-1 < r < 1$, then r^n tends to zero as n becomes larger so that for such geometric progressions we can find the 'sum to infinity' of the series

$$S_\infty = \frac{a}{1-r}$$

The geometric mean of a set of n numbers is the nth root of their product.

If we limit consideration to non-negative numbers then the arithmetic mean of a set of numbers will be greater than or equal to their geometric mean.

Logarithms

Logarithms, which, short of calculating machinery of some form, are probably the greatest aid to computation are based on the properties of indices.

Thus, if we consider logarithms to base a we have the following results:

$a^x = P$ is equivalent to $\log_a P = x$
$a^1 = a$ is equivalent to $\log_a a = 1$
$a^0 = 1$ is equivalent to $\log_a 1 = 0$

So that using rules for powers given on page 1–2

If
$$a^x = P \text{ and } a^y = Q$$
then
$$PQ = a^{x+y}$$
so
$$\log_a PQ = x+y = \log_a P + \log_a Q$$
Similarly
$$\log_a (P/Q) = \log_a P - \log_a Q$$
Also
$$P^n = a^{nx} \text{ so } \log_a P^n = nx = n \log_a P$$

In computation it is generally convenient to use as base the number 10, i.e. in the expressions given above $a = 10$. However in fundamental work or integration Natural Logarithms (also known as Napierian or hyperbolic logarithms) are generally used. These are logarithms to base e a transcendental number given approximately by

$$e = 2.7182\,8$$

and whose definition can be taken as: 'The value of the solution of the differential equation:

$$dy/dx = y \text{ for } x = 1\text{'}$$

(Note the solution of $dy/dx = y$ is $y = e^x$.)

Permutations and combinations

If, in a sequence of N events, the first can occur in n_1 ways the second in n_2 etc. then the number of ways in which the whole sequence can occur is

$$n_1 n_2 n_3 \ldots n_N$$

PERMUTATIONS

The number of permutations of n different things taken r at a time means the number of ways in which r of these n things can be arranged *in order*. This is denoted by

$$^nPr = n(n-1)(n-2)\ldots(n-r+1) = \frac{n!}{(n-r)!}$$

where $n! = n(n-1)(n-2) \ldots\ldots 3.2.1$ is called factorial n.

It is clear that:

$$^nPn = n!$$

and that

$$^nP_1 = n$$

If, of n things taken r at a time p things, are to occupy fixed positions then the number of permutations is given by

$$^{n-p}Pr-p$$

If in the set of n things, there are g groups each group containing n_1, $n_2 \ldots n_g$ things which are identical then the number of permutations of all n things is

$$\frac{n!}{n_1!n_2!\ldots n_g!}$$

COMBINATIONS

The number of combinations of n different things, into groups of r things at a time is given by

$$^{n}Cr = \frac{n!}{r!(n-r)!} = \frac{^{n}Pr}{r!}$$

It is important to note that, whereas in Permutations the order of the things does matter, in Combinations the order does not matter. From the general expression above, it is clear that

$$^{n}Cn = 1$$

$$^{n}C_1 = n$$

If, of n different things taken r at a time p are always to be taken then the number of combinations is

$$^{n-p}Cr-p.$$

If, of n different things taken r at a time p are never to occur the number of combinations is

$$^{n-p}Cr.$$

Note that combinations from an increasing number of available things are related by

$$^{n+1}Cr = {}^{n}Cr + {}^{n}Cr-1$$

also

$$^{n}Cr = {}^{n}Cn-r$$

The Binomical Theorem

The general form of expansion of $(x+a)^n$ is given by

$$(x+a)^n = {}^{n}C_0x^n + {}^{n}C_1 \times^{n-1}a^r + {}^{n}C_2x^{n-2}a^2 \ldots$$

Alternatively this may be written as

$$(x+a)^n = x^n + nx^{n-1}a + \frac{n(n-1)}{1.2}x^{n-2}a^2 + \frac{n(n-1)(n-2)}{1.2.3}x^{n-3}a^3 \ldots$$

It should be noted that the coefficients of terms equidistant from the end are equal (since $^{n}Cr = {}^{n}Cn-r$).

TRIGONOMETRY

The *trigonometric functions* of the angle α (see Figure 1.1) are defined as follows:

$$\sin\alpha = y/r \qquad \operatorname{cosec}\alpha = r/y$$
$$\cos\alpha = x/r \qquad \sec\alpha = r/x$$
$$\tan\alpha = y/x \qquad \cot\alpha = x/y$$

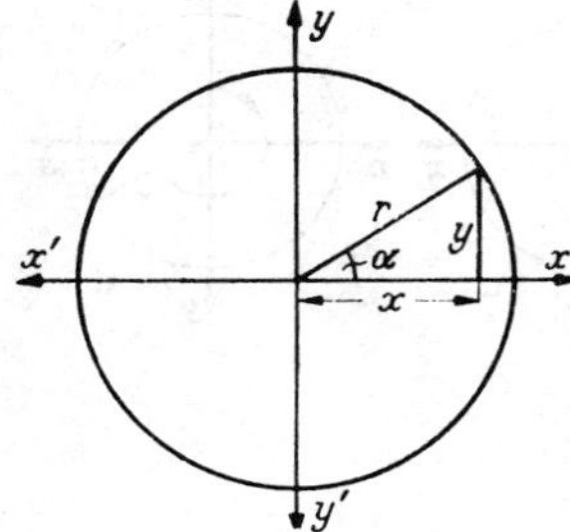

Figure 1.1 Trigonometric functions

These functions satisfy the following identities:

$$\sin^2\alpha + \cos^2\alpha = 1$$
$$1 + \tan^2\alpha = \sec^2\alpha$$
$$1 + \cot^2\alpha = \operatorname{cosec}^2\alpha$$

Positive and negative lines

In trigonometry lines are considered positive or negative according to their location relative to the coordinate axes xOx', yOy', Figure 1.2.

POSITIVE LINES

Radial: any direction.
Horizontal: to right of yOy'.
Vertical: above xOx'.

NEGATIVE LINES

Horizontal: to left of yOy'.
Vertical: below xOx'.

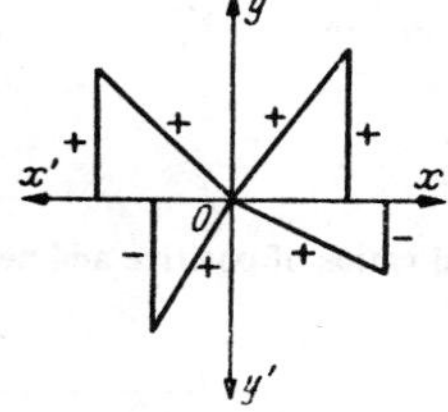

Figure 1.2 Positive and negative lines

Positive and negative angles

Figure 1.3 shows the convention for signs in measuring angles. Angles are positive if the line OP revolves anti-clockwise from Ox as in Figure 1.3a and are negative when OP revolves clockwise from Ox.

Signs of trigonometrical ratios are shown in Figure 1.4 and in Table 1.1.

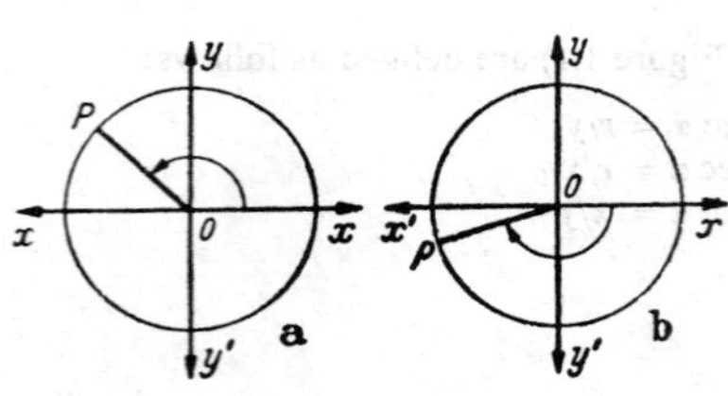

Figure 1.3 (above) (a) Positive, (b) negative angles

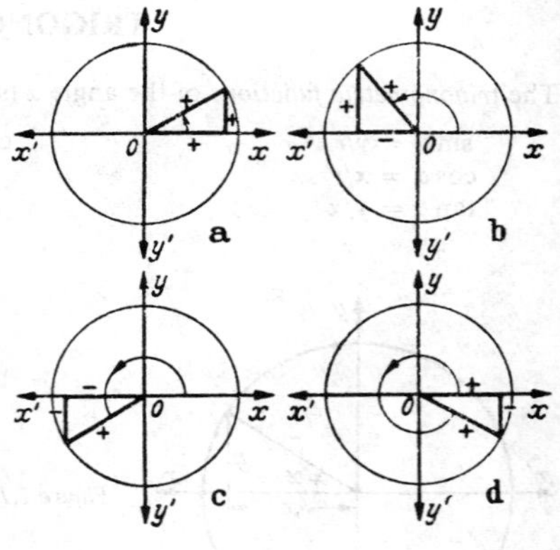

Figure 1.4 (right) (a) Angle in first quadrant; (b) angle in second quadrant; (c) angle in third quadrant; (d) angle in fourth quadrant

Table 1.1

Quadrant	*Sign of ratio*	
	Positive	*Negative*
First	sin cos tan cosec sec cot	
Second	sin cosec	cos sec tan cot
Third	tan cot	sin cosec cos sec
Fourth	cos sec	sin cosec tan cot

Trigonometrical ratios of positive and negative angles

Table 1.2

sin (−α)	= −sin α	tan (−α)	= −tan α	sec (−α)	=	sec α
cos (−α)	= cos α	cot (−α)	= −cot α	cosec (−α)	=	−cosec α
sin (90°−α)	= cos α	tan (90°−α)	= cot α	sec (90°−α)	=	cosec α
cos (90°−α)	= sin α	cot (90°−α)	= tan α	cosec (90°−α)	=	sec α
sin (90°+α)	= cos α	tan (90°+α)	= −cot α	sec (90°+α)	=	−cosec α
cos (90°+α)	= −sin α	cot (90°+α)	= −tan α	cosec (90°+α)	=	sec α
sin (180°−α)	= sin α	tan (180°−α)	= −tan α	sec (180°−α)	=	−sec α
cos (180°−α)	= −cos α	cot (180°−α)	= −cot α	cosec (180°−α)	=	cosec α
sin (180°+α)	= −sin α	tan (180°+α)	= tan α	sec (180°+α)	=	−sec α
cos (180°+α)	= −cos α	cot (180°+α)	= cot α	cosec (180°+α)	=	−cosec α

Measurement of angles

ENGLISH OR SEXAGESIMAL METHOD

1 right angle = 90° (degrees)
1° (degree) = 60′ (minutes)
1′ (minute) = 60″ (seconds)

This convention is universal.

FRENCH OR CENTESIMAL METHOD

This splits angles degrees and minutes into 100th divisions but is not used in practice.

THE RADIAN

This is a constant angular measurement equal to the angle subtended at the centre of any circle by an arc equal in length to the radius of the circle as shown in Figure 1.5.

$$\pi \text{ radians} = 180°$$

$$1 \text{ radian} = \frac{180}{\pi} = \frac{180}{3.141\,6} = 57°\,17'\,44'' \text{ approx.}$$

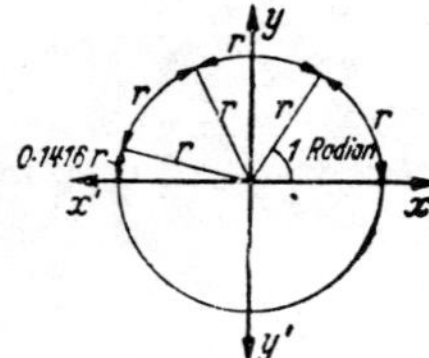

Figure 1.5 The radian

TRIGONOMETRICAL RATIOS EXPRESSED AS SURDS

Table 1.3 gives these ratios for certain angles.

Table 1.3

Angle in radians	0	$\frac{\pi}{6}$	$\frac{\pi}{4}$	$\frac{\pi}{3}$	$\frac{\pi}{2}$
Angle in degrees	0°	30°	45°	60°	90°
sin	0	$\frac{1}{2}$	$\frac{1}{\sqrt{2}}$	$\frac{\sqrt{3}}{2}$	1
cos	1	$\frac{\sqrt{3}}{2}$	$\frac{1}{\sqrt{2}}$	$\frac{1}{2}$	0
tan	0	$\frac{1}{\sqrt{3}}$	1	$\sqrt{3}$	∞

Complementary and supplementary angles

Two angles are complementary when their sum is a right angle; then either is the complement of the other, e.g. the sine of an angle equals the cosine of its complement. Two angles are supplementary when their sum is two right angles.

Graphical interpretation of the trigonometric functions

Figures 1.6 to 1.9 show the variation with α of sin α, cos α, tan α and cosec α respectively. All the trigonometric functions are periodic with period 2π radians (or 360°).

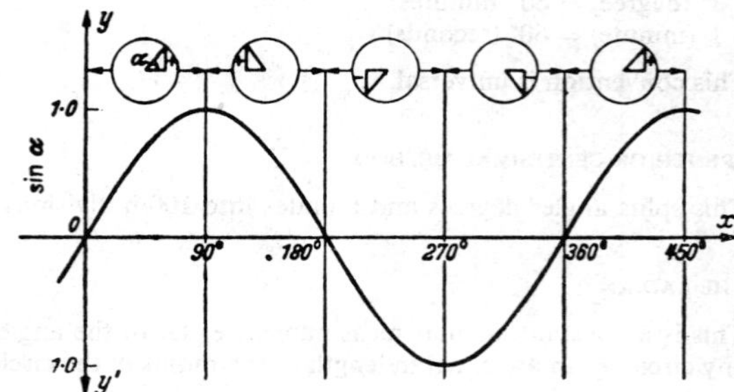

Figure 1.6 Sin α

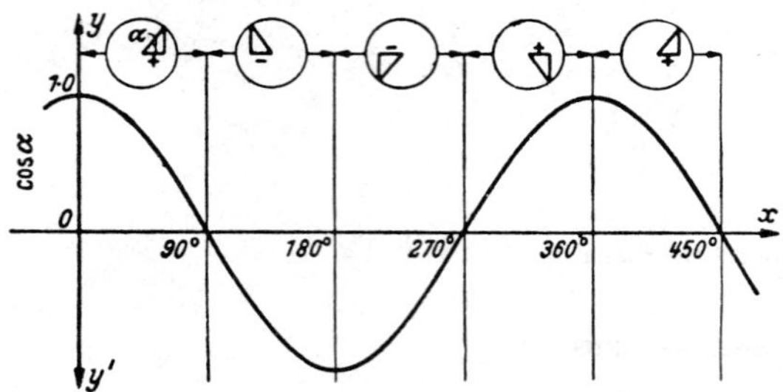

Figure 1.7 Cos α

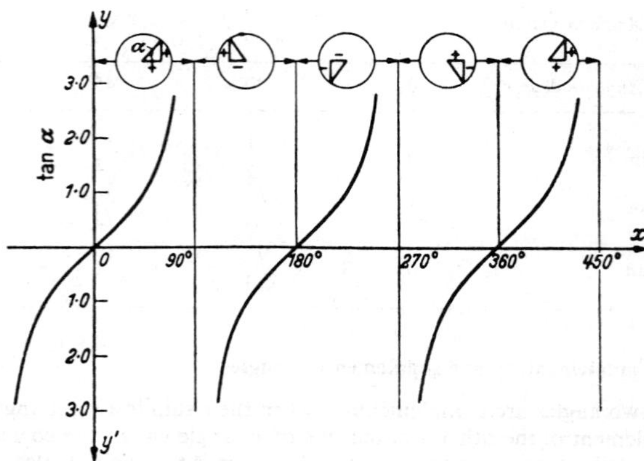

Figure 1.8 Tan α

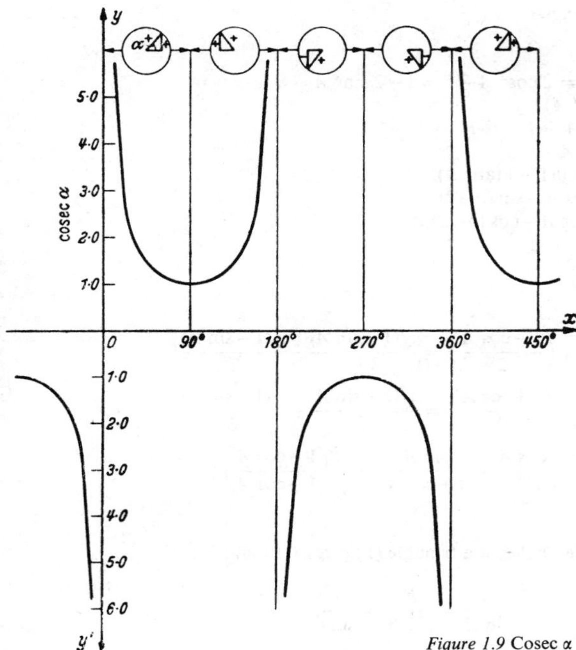

Figure 1.9 Cosec α

Functions of the sum and difference of two angles

$$\sin(A \pm B) = \sin A \cos B \pm \cos A \sin B$$
$$\cos(A \pm B) = \cos A \cos B \pm \sin A \sin B$$
$$\tan(A \pm B) = \frac{\tan A \pm \tan B}{1 \pm \tan A \tan B}$$

Sums and differences of functions

$$\sin A + \sin B = 2 \sin \tfrac{1}{2}(A+B) \cos \tfrac{1}{2}(A-B)$$
$$\sin A - \sin B = 2 \cos \tfrac{1}{2}(A+B) \sin \tfrac{1}{2}(A-B)$$
$$\cos A + \cos B = 2 \cos \tfrac{1}{2}(A+B) \cos \tfrac{1}{2}(A-B)$$
$$\cos A - \cos B = -2 \sin \tfrac{1}{2}(A+B) \sin \tfrac{1}{2}(A-B)$$
$$\sin^2 A - \sin^2 B = \sin(A+B) \sin(A-B)$$
$$\cos^2 A - \cos^2 B = -\sin(A+B) \sin(A-B)$$
$$\cos^2 A - \sin^2 B = \cos(A+B) \cos(A-B)$$

Functions of multiples of angles

$\sin 2A = 2 \sin A \cos A$
$\cos 2A = \cos^2 A - \sin^2 A = 2 \cos^2 A - 1 = 1 - 2 \sin^2 A$
$\tan 2A = 2 \tan A/(1 - \tan^2 A)$
$\sin 3A = 3 \sin A - 4 \sin^3 A$
$\cos 3A = 4 \cos^3 A - 3 \cos A$
$\tan 3A = (3\ \text{tab}\ A - \tan^3 A)/(1 - 3 \tan^2 A)$
$\sin pA = 2 \sin (p-1) A \cos A - \sin (p-2) A$
$\cos pA = 2 \cos (p-1) A \cos A - \cos (p-2) A$

Functions of half angles

$$\sin A/2 = \sqrt{\left(\frac{1-\cos A}{2}\right)} = \frac{\sqrt{(1+\sin A)}}{2} - \frac{\sqrt{(1-\sin A)}}{2}$$

$$\cos A/2 = \sqrt{\left(\frac{1+\cos A}{2}\right)} = \frac{\sqrt{(1+\sin A)}}{2} + \frac{\sqrt{(1-\sin A)}}{2}$$

$$\tan A/2 = \frac{1-\cos A}{\sin A} = \frac{\sin A}{1+\cos A} = \sqrt{\left(\frac{1-\cos A}{1+\cos A}\right)}$$

Relations between sides and angles of a triangle (Figures 1.10 and 1.11)

$$\frac{a}{\sin A} = \frac{b}{\sin B} = \frac{c}{\sin C}$$

$$a = b \cos C + c \cos B$$

$$c^2 = a^2 + b^2 - 2ab \cos C$$

$\sin A = \dfrac{c}{bc}\sqrt{\{s(s-a)(s-b)(s-c)\}}$ where $2s = a+b+c$

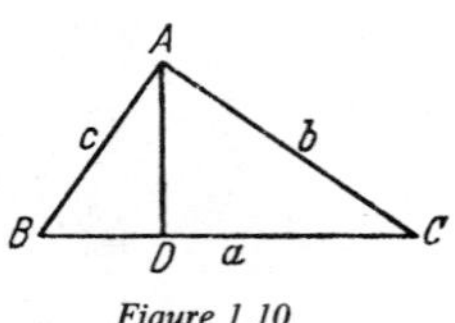

Figure 1.10

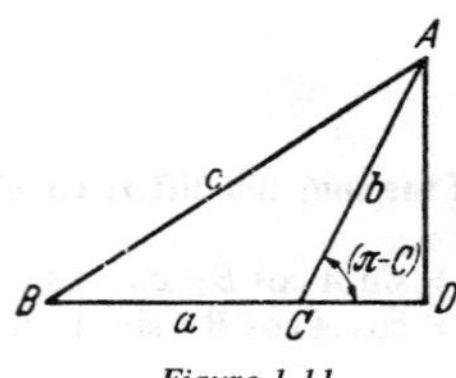

Figure 1.11

Area of triangle $\Delta = \frac{1}{2}ab \sin C = \sqrt{\{s(s-a)(s-b)(s-c)\}}$

$$\tan\frac{A}{2} = \sqrt{\left\{\frac{(s-b)(s-c)}{s(s-a)}\right\}} \qquad \cos\frac{A}{2} = \sqrt{\left\{\frac{s(s-a)}{bc}\right\}}$$

$$\sin\frac{A}{2} = \sqrt{\left\{\frac{(s-b)(s-c)}{bc}\right\}} \qquad \tan\frac{B-C}{2} = \frac{(b-c)}{(b+c)}\cot\frac{A}{2}$$

ANY RIGHT ANGLED TRIANGLE (Figure 1.12)

$$a^2+b^2 = c^2; \quad A+B = 90°; \quad \sin A = \cos B; \quad \cot A = \tan B, \text{etc.}$$

$$\text{Area of } \triangle\, ABC = \tfrac{1}{2}ab = \tfrac{1}{2}bc \sin A = \tfrac{1}{2}ac \sin B$$

ANY EQUILATERAL TRIANGLE (Figure 1.13)

$$a = b = c; \quad A = B = C = 60° = \frac{\pi}{3}$$

$$h = \frac{b\sqrt{3}}{2}; \quad \text{Area} = \frac{b^2\sqrt{3}}{4}$$

Circumscribed circle radius $R = \dfrac{b\sqrt{3}}{3}$

Inscribed circle, radius $r = \dfrac{b\sqrt{3}}{6}$

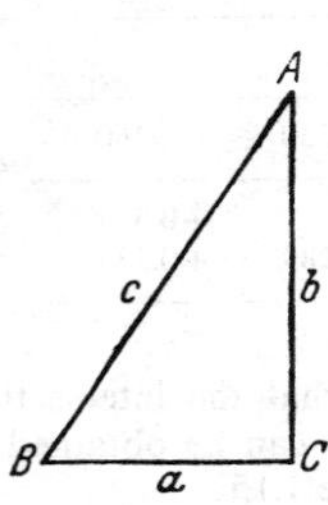

Figure 1.12

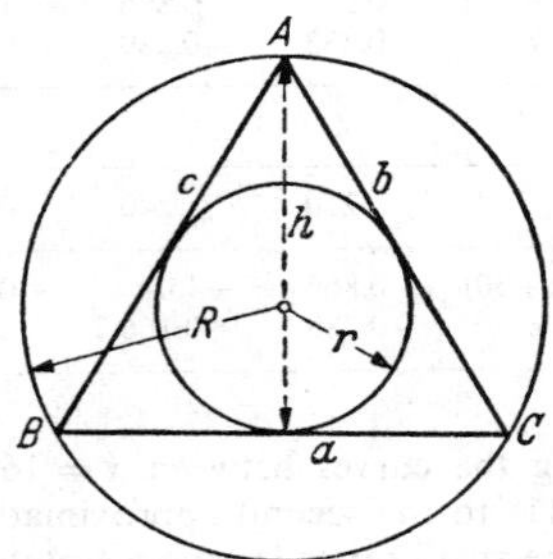

Figure 1.13

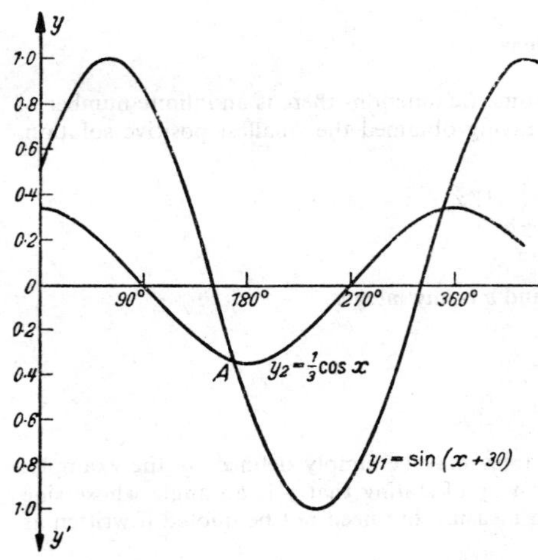

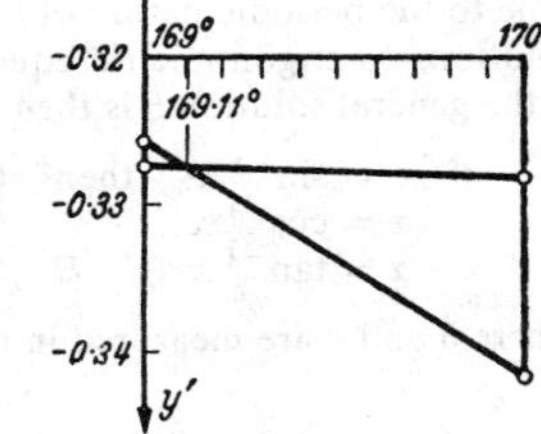

Figure 1.14 (left) Solution of trigonometrical equations

This shows the intersection between $x = 0$ and $x = \pi$ as $x = 169°$ approx.

Figure 1.15 (above) Enlargement at A, Figure 1.14

Solution of trigonometric equations

The method best suited to the solution of trigonometric equations is that described in the section on algebra which deals with the method of solving transcendental equations by means of graphs. The expression to be solved is arranged as two identities and two graphs drawn as shown in Figure 1.14. The points of intersection of the curves projected on to the coordinate axes give the values which will satisfy the trigonometric equation.

Example. Solve $\sin(x+30) = \frac{1}{3}\cos x$ for x between 0 and 2π.

Assigning values to x in Table 1.4 and calculating the corresponding values for $y = \sin(x+30)$ and $y = \frac{1}{3}\cos x$ gives the readings for plotting the curves in Figure 1.14.

Table 1.4

x	0	30	60	90	120	150	180
$y_1 = \sin(x+30)$	0.5	0.866	1.0	0.866	0.5	0	−0.5
$y_2 = \frac{1}{3}\cos x$	0.333	0.289	0.166 7	0	−0.166 7	−0.289	−0.333

x	210	240	270	300	330	360
$y_1 = \sin(x+30)$	−0.866	−1.0	−0.866	−0.5	0	+0.5
$y_2 = \frac{1}{3}\cos x$	−0.289	−0.166 7	0	0.166 7	0.289	+0.333

Plotting the curves between $x = 169°$ and $170°$ shows that the intersection is at $x = 169.11°$ to the second approximation. Greater accuracy can be obtained by continuing the small range large scale plots of the type in Figure 1.15

There is one further value of x between $x = 300°$ and $360°$ which will satisfy the equation as can be seen on Figure 1.14.

General solutions of trigonometric equations

Due to the periodic nature of the trigonometric functions there is an infinite number of solutions to trigonometric equations. Having obtained the smallest positive solution, α, the general solution θ is then given by:

$$
\begin{array}{lll}
\text{if } \alpha = \sin^{-1} x & \text{then} & \theta = n\pi + (-1)^n \alpha \\
\alpha = \cos^{-1} x & & \theta = 2n\pi \pm \alpha \\
\alpha = \tan^{-1} x & H & \theta = n\pi + \alpha
\end{array}
$$

where θ and α are measured in radians and n is any integer.

Inverse trigonometrical functions

Inverse functions of trigonometric variables may be simply defined by the example: $y = \sin^{-1}\frac{1}{2}$ which is merely a symbolic way of stating that y is an angle whose sine is $\frac{1}{2}$ i.e. y is actually 30° or $\pi/6$ in radian measure but need not be quoted if written as $\sin^{-1}\frac{1}{2}$.

SPHERICAL TRIGONOMETRY

Definitions

Referring to Figure 1.16, representing a sphere of radius r:

Small circle. The section of a sphere cut by a plane at a section not on the diameter of the sphere e.g. EFGH.

Great circle. The section of a sphere cut by a plane through any diameter e.g. ACBC′.

Poles. Poles of any circular section of a sphere are the ends of a diameter at right angles to the section e.g. D and D′ are the poles of the great circle ACBC′.

Lunes. The surface areas of that part of the sphere between two great circles; there are two pairs of congruent areas e.g. ACA′C′A; CBC′B′C and ACB′C′A; A′CBC′A′.

Area of lune. If the angle between the planes of two great circles forming the lune is θ (radians), its surface area is equal to $2\theta r^2$.

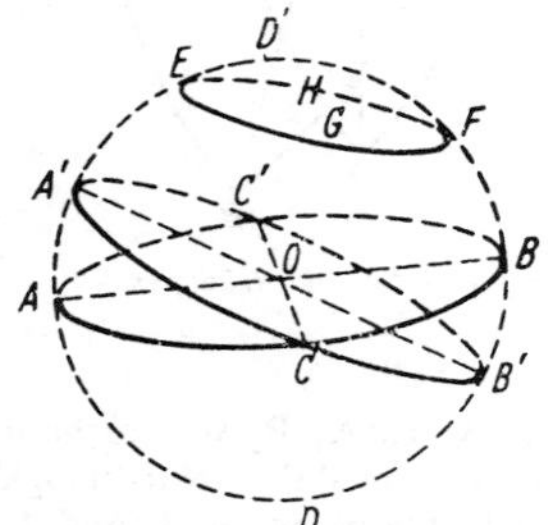

Figure 1.16 Sphere illustrating spherical trigonometry definitions

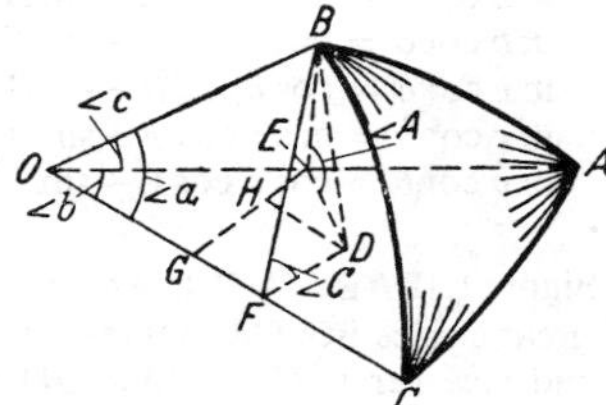

Figure 1.17 Spherical triangles

Spherical triangle. A curved surface included by the arcs of three great circles e.g. CB′B is a spherical triangle formed by one edge BB′ on part of the great circle DB′BA the second edge B′C on great circle B′CA′C′ and edge CB on great circle ACBD′. The angles of a spherical triangle are equal to the angles between the planes of the great circles, or alternatively, the angles between the tangents to the great circles at their points of intersection. They are denoted by the letters C, B', B for the triangle CB′.

Area of spherical triangle. CB′B $= (B'+B+C-\pi)r^2$.

Spherical excess. Comparing a plane triangle with a spherical triangle the sum of the angles of the former is π and the spherical excess E of a spherical triangle is given by $E = B'+B+C-\pi$; hence area of a spherical triangle can be expressed as $(E/4\pi)\times$ surface of sphere.

Spherical polygon. A spherical polygon of n sides can be divided into $(n-2)$ spherical triangles by joining opposite angular points by the arcs of great circles.

$$\text{Area of spherical polygon} = [\text{sum of angles} - (n-2)\pi]r^2$$
$$= \frac{E}{4\pi} \times \text{surface of sphere.}$$

Note that $(n-2)\pi$ is the sum of the angles of a plane polygon of n sides.

Properties of spherical triangles

Let ABC, Figure 1.7, be a spherical triangle; BD is a perpendicular from B on plane OAC and OÊD, OF̂D, OÊB, OF̂B, OĜE, DĤG are right angles; then BÊD $= A$ and BF̂D $= C$ are the angles between the planes OBA, OAC and OBC, OAC respectively. DÊH = CÔA $= b$ also CÔB $= a$, AÔB $= c$, and since OB = OA = OC = radius r of sphere, OF $= r\cos a$, OE $= r\cos c$; then

$$\cos a = \cos b \cos c + \sin b \sin c \cos A$$
$$\cos b = \cos a \cos c + \sin a \sin c \cos B\ H$$
$$\cos c = \cos a \cos b + \sin a \sin b \cos C$$

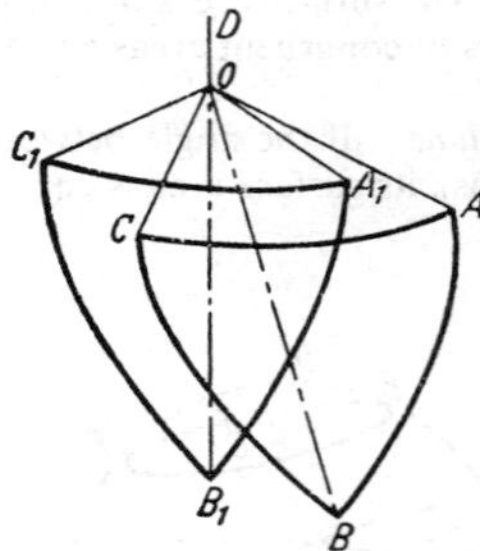

Figure 1.18 Polar triangles

Also the sine formula are:

$$\frac{\sin A}{\sin a} = \frac{\sin B}{\sin b} = \frac{\sin C}{\sin c}$$

and the cotangent formulae are:

$$\sin a \cot c = \cos a \cos B + \sin B \cot C$$
$$\sin b \cot c = \cos b \cos A + \sin A \cot C$$
$$\sin b \cot a = \cos b \cos C + \sin C \cot A$$
$$\sin c \cot a = \cos c \cos B + \sin B \cot A$$
$$\sin c \cot b = \cos c \cos A + \sin A \cot B$$
$$\sin a \cot b = \cos a \cos C + \sin C \cot B$$

In Figure 1.18 ABC, $A_1B_1C_1$ are two spherical triangles in which A_1, B_1, C_1, are the noles of the great circles BC, CA, AB respectively; then $A_1B_1C_1$ is termed the polar triangle of ABC and vice versa. Now OA_1, OD are perpendicular to the planes BOC and AOC respectively; hence A_1ÔD = angle between planes BOC and AOC $= C$. Let sides of triangle $A_1B_1C_1$ be denoted by $a_1b_1c_1$ then $c_1 = A_1\hat{O}B_1 = \pi - C$ also $a_1 = \pi - A$ and $b_1 = \pi - B$; $c = \pi - C_1$; $a = \pi - A_1$; $b = \pi - B_1$ and from these we get

$$\cos b = \frac{\cos B + \cos A \cos C}{\sin A \sin C}$$

$$\cos a = \frac{\cos A + \cos B \cos C}{\sin B \sin C}$$

$$\cos c = \frac{\cos C + \cos A \cos B}{\sin A \sin B}$$

RIGHT ANGLED TRIANGLES

If one angle A of a spherical triangle ABC is 90° then $\cos a = \cos b \cos c = \cot B \cot C$

$$\cos B = \frac{\tan c}{\tan a}: \quad \cos C = \frac{\tan b}{\tan a}: \quad \sin B = \frac{\sin b}{\sin c}: \quad \sin C = \frac{\sin c}{\sin a}$$

$$\tan B = \frac{\tan b}{\sin c}: \quad \tan C = \frac{\tan c}{\sin b}: \quad \cos B = \cos b \sin C; \quad \cos C = \cos c \sin B.$$

HYPERBOLIC TRIGONOMETRY

The hyperbolic functions are related to a rectangular hyperbola in a manner similar to the relationship between the ordinary trigonometric functions and the circle. They are defined by the following exponential equivalents:

$$\sinh\theta = \frac{e^{\theta}-e^{-\theta}}{2} \qquad \operatorname{cosech}\theta = \frac{1}{\sinh\theta}$$

$$\cosh\theta = \frac{e^{\theta}+e^{-\theta}}{2} \qquad \operatorname{sech}\theta = \frac{1}{\cosh\theta}$$

$$\tanh\theta = \frac{\sinh\theta}{\cosh\theta} \qquad \coth\theta = \frac{1}{\tanh\theta}$$

Relation of hyperbolic to circular functions

$$\begin{aligned}
\sin\theta &= -i\sinh i\theta & \sinh\theta &= -i\sin i\theta\\
\cos\theta &= \cosh i\theta & \cosh\theta &= \cos i\theta\\
\tan\theta &= i\tanh i\theta & \tanh\theta &= -i\tan i\theta\\
\operatorname{cosec}\theta &= i\operatorname{cosech} i\theta & \operatorname{cosech}\theta &= i\operatorname{cosec} i\theta\\
\sec\theta &= \operatorname{sech} i\theta & \operatorname{sech}\theta &= \sec i\theta\\
\cot\theta &= i\coth i\theta & \coth\theta &= i\cot i\theta
\end{aligned}$$

Properties of hyperbolic functions

$$\cosh^2\theta-\sinh^2\theta = 1 \qquad \sin 2\theta = 2\sinh\theta\cosh\theta$$

$$\operatorname{sech}^2\theta = 1-\tanh^2\theta \qquad \cosh 2\theta = \cosh^2\theta+\sinh^2\theta$$

$$\operatorname{cosech}^2\theta = \coth^2\theta-1 \qquad \tanh 2\theta = \frac{2\tanh\theta}{1+\tanh^2\theta}$$

$$\sinh(x\pm y) = \sinh x\cosh y\pm\cosh x\sinh y$$

$$\cosh(x\pm y) = \cosh x\cosh y\pm\sinh x\sinh y$$

$$\tanh(x\pm y) = \frac{\tanh x\pm\tanh y}{1\pm\tanh x\tanh y}$$

$$\begin{aligned}
\sinh x+\sinh y &= 2\sinh\tfrac{1}{2}(x+y)\cosh\tfrac{1}{2}(x-y)\\
\sinh x-\sinh y &= 2\cosh\tfrac{1}{2}(x+y)\sinh\tfrac{1}{2}(x-y)\\
\cosh x+\cosh y &= 2\cosh\tfrac{1}{2}(x+y)\cosh\tfrac{1}{2}(x-y)\\
\cosh x-\cosh y &= 2\sinh\tfrac{1}{2}(x+y)\sinh\tfrac{1}{2}(x-y)
\end{aligned}$$

Inverse hyperbolic functions

As with trigonometric functions, we define the inverse hyperbolic functions by $y = \sinh^{-1}x$ where $x = \sinh y$.

$$\therefore\ x = (e^{y}-e^{-y})/2$$

Rearranging and adding x^2 to each side,

$$e^{2y}-2x \,.\, e^{y}+x^2 = x^2+1$$

or,

$$e^{y}-x = \sqrt{(x^2+1)}$$

and

$$\therefore\ y = \sinh^{-1} x = \log_e [x+\sqrt{(x^2+1)}]$$

The other inverse functions may be treated similarly. We find:

$$\sinh^{-1} x = \log [x+\sqrt{(x^2+1)}]; \qquad \operatorname{cosech}^{-1} x = \log \frac{1+\sqrt{(1+x^2)}}{x}$$

$$\cosh^{-1} x = \log [x+\sqrt{(x^2-1)}]; \qquad \operatorname{sech}^{-1} x = \log \frac{1+\sqrt{(1-x^2)}}{x}$$

$$\tanh^{-1} x = \tfrac{1}{2}\log \frac{1+x}{1-x}; \qquad \coth^{-1} x = \tfrac{1}{2}\log \frac{x+1}{x-1}$$

The relationships with the corresponding inverse trigonometric functions are as follows:

$$\sinh^{-1} x = -i \sin^{-1} ix \qquad \sin^{-1} x = -i \sinh^{-1} ix$$
$$\cosh^{-1} x = i \cos^{-1} x \qquad \cos^{-1} x = -i \cosh^{-1} x$$
$$\tanh^{-1} x = -i \tan^{-1} ix \qquad \tan^{-1} x = i \tanh^{-1} ix$$

COORDINATE GEOMETRY

Straight line equations

The equation of a straight line may be expressed as

1.
$$ax+by+c = 0 \text{ or } y = -\frac{a}{b}x-\frac{c}{b} = mx+n$$

where a, b, c, are constants, m is the slope of the line as shown in Figure 1.19.

2.
$$\frac{x}{k}+\frac{y}{l} = 1$$

where k is the intercept on the x axis and l is the intercept on the y axis

3.
$$x \cos \alpha + y \sin \alpha = p$$

where p = length of the perpendicular from the origin to the line and α the inclination of this perpendicular to Ox in Figure 1.20.

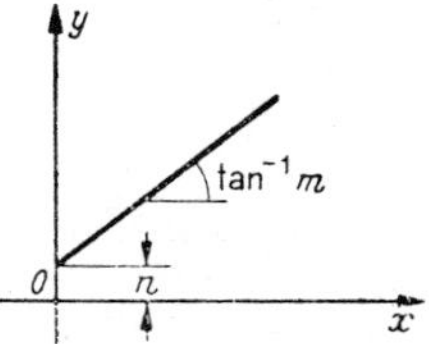

Figure 1.19 Straight line equation $y = mx+n$

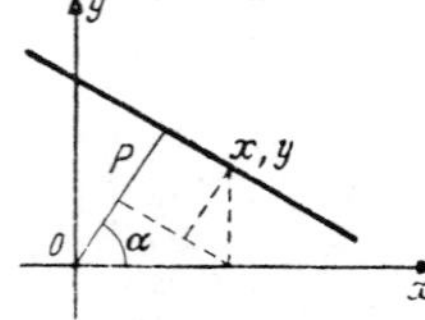

Figure 1.20 Straight line equation $x \cos \alpha + y \sin \alpha = p$

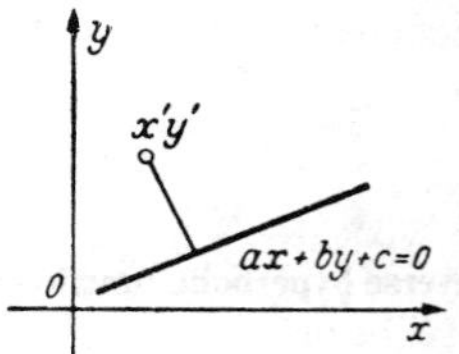

Figure 1.21 Perpendicular to straight line

The length d of a perpendicular (see Figure 1.21) from any point $(x'y')$ to a straight line is given by $ax'+by'+c/[\sqrt{(a^2+b^2)}]$ if the straight line equation is as given in $(x'\cos\alpha+y'\sin\alpha-p)$ if the straight line equation is as given in 3.

The equation of a straight line through one given point $(x'y')$ is $y-y' = m(x-x')$.

The equation of a straight line through two given points (Figure 1.22) $(x_1y_1)(x_2y_2)$ is

$$\frac{y-y_1}{y_2-y_1} = \frac{x-x_1}{x_2-x_1}$$

The angle ψ between two straight lines (Figure 1.23) $y = m_1x+n_1$ and $y = m_2x+n_2$ is given by

$$\tan\psi = \frac{m_1-m_2}{1+m_1m_2}$$

For lines which are parallel $m_1 = m_2$.

For lines at right angles $1+m_1m_2 = 0$.

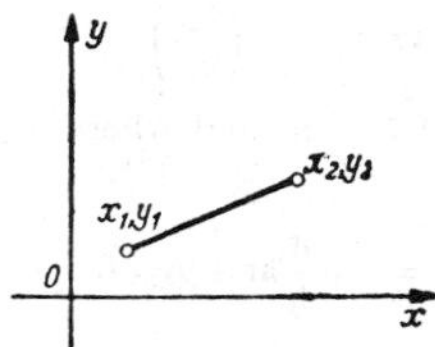

Figure 1.22 Straight line through two points

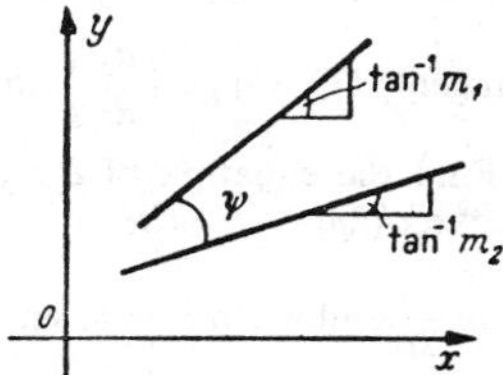

Figure 1.23 Angle ψ between two straight lines

Change of axes

Let the equation of the curve be $y = f(x)$ referred to coordinate axes Ox, Oy; then its equation relative to axes O′x′, O′y′ parallel to Ox, Oy with origin O′ at point (r, s) is given by $y+s = f(x+r)$ in which x and y refer to the new axes.

If the equation of a curve is given by $y = f(x)$ referred to coordinate axes Ox, Oy, then if these axes are each rotated an angle ψ anti-clockwise about O, the equation of the curve referred to the rotated axes is given by $x\sin\psi+y\cos\psi = f(x\cos\psi-y\sin\psi)$.

TANGENT AND NORMAL TO ANY CURVE $y = f(x)$

The tangent PT and the normal PN at any point x_1y_1 on the curve $y = f(x)$ in Figure 1.24 are given by the following equations:

Tangent: $y-y_1 = \dfrac{dy}{dx}(x-x_1)$ where $\dfrac{dy}{dx} = m =$ the slope of the curve at P

Normal: $(y-y_1)\dfrac{dy}{dx}+(x-x_1) = 0$

If ϕ be the angle which the tangent at P makes with the axis of x, then:

$$\tan\phi = \frac{dy}{dx};\quad \cos\phi = \frac{dx}{ds};\quad \sin\phi = \frac{dy}{ds}$$

where s is the distance measured along the curve.

TANGENT AND NORMAL TO ANY CURVE $f(xy) = 0$

The function is implicit in this case so that partial differential coefficients are employed in the equations for the tangent and for the normal at x_1y_1.

$$\text{Tangent: } (y-y_1)\frac{\partial f}{\partial y}+(x-x_1)\frac{\partial f}{\partial x} = 0$$

$$\text{Normal: } \frac{(y-y_1)}{(\partial f/\partial y)} = \frac{(x-x_1)}{(\partial f/\partial x)} \qquad \text{where } \frac{dy}{dx} = -\frac{\partial f}{\partial x}\Big/\frac{\partial f}{\partial y}$$

SUBTANGENT AND SUBNORMAL TO ANY CURVE $y = f(x)$

The subtangent is TQ and the subnormal is QN at any point $P(x_1y_1)$ on the curve $y = f(x)$ in Figure 1.24. Their lengths are given by:

$$\text{Subtangent, TQ} = y_1 \Big/ \left(\frac{dy}{dx}\right)_1 \quad \text{and subnormal, QN} = y_1\left(\frac{dy}{dx}\right)_1$$

Example. Find the equation of the tangent and of the normal where $x = p$ on the curve $y = \cos \pi x/(2p)$

$$\frac{dy}{dx} = -\frac{\pi}{2p}\sin\frac{\pi x}{2p} \text{ and when } x = p, \sin\frac{\pi x}{2p} = 1, \text{ i.e. } \frac{dy}{dx} = -\frac{\pi}{2p} \text{ and } y = 0$$

$$\therefore \text{ the required equation of the tangent is } y = -\frac{\pi}{2p}(x-p)$$

$$\text{and the equation of the normal is } y = \frac{2p}{\pi}(x-p)$$

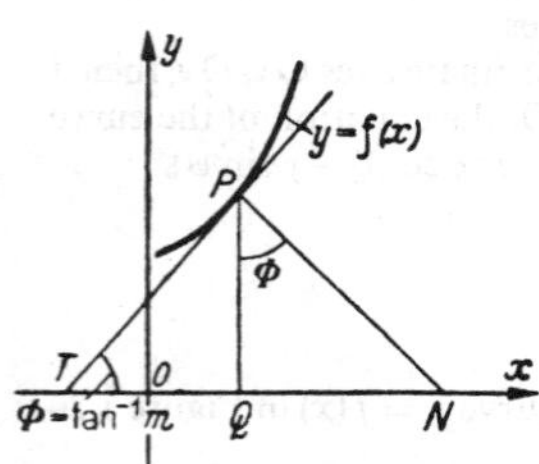

Figure 1.24 Tangent, normal, subtangent and subnormal to a curve

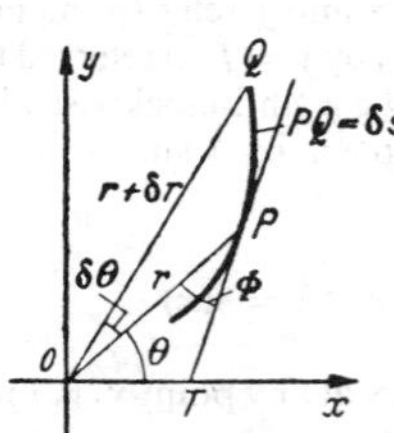

Figure 1.25 Polar coordinates

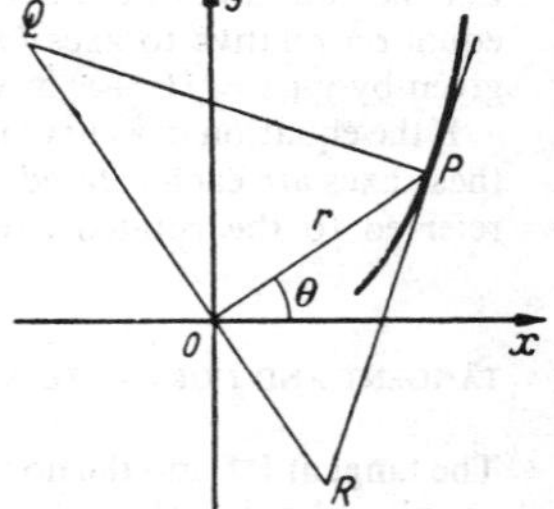

Figure 1.26 Polar subtangent and subnormal

Polar coordinates

The polar coordinates of any point P in a plane are given by r, θ where r is the length of the line joining P to the origin O and θ is the inclination of OP, the radius vector relative to the axis Ox, Figure 1.25.

The relations between the rectangular coordinates x, y and the polar coordinates r, θ are:

$$x = r\cos\theta,\ y = r\sin\theta;\quad r = \sqrt{(x^2+y^2)},\ \theta = \tan^{-1} y/x$$

If PT is a tangent to the curve at point P then

$$\tan\phi = r d\theta/dr;\ \cot\phi = (1/r)(dr/d\theta);\ \sin\phi = r d\theta/ds \text{ and } \cos\phi = dr/ds$$

POLAR SUBTANGENT AND SUBNORMAL

In Figure 1.26 the polar subtangent is OR and the polar subnormal is OQ where QR is perpendicular to OP and their lengths are given by: polar subtangent $= r^2 d\theta/dr$; polar subnormal $= dr/d\theta$.

CURVATURE

Let PQ in Figure 1.27 represent an elemental length δs of a given curve and PS, QT the tangents at the points P, Q then:

Curvature at P $= d\beta/ds$. For a circle centre at C, radius ρ, $ds = \rho\, d\beta$, i.e. curvature $= 1/\rho$

$$\therefore \rho = \frac{ds}{d\beta} = \text{radius of curvature}$$

Putting $\beta = \tan^{-1}\left(\dfrac{dy}{dx}\right)$ and differentiating,

$$\text{curvature} = \frac{d\beta}{ds} = \frac{1}{\rho} = \frac{\dfrac{d^2y}{dx^2}}{\left[1+\left(\dfrac{dy}{dx}\right)^2\right]^{\frac{3}{2}}}$$

$$\text{radius of curvature } \rho = \frac{\left[1+\left(\dfrac{dy}{dx}\right)^2\right]^{\frac{3}{2}}}{\dfrac{d^2y}{dx^2}}.$$

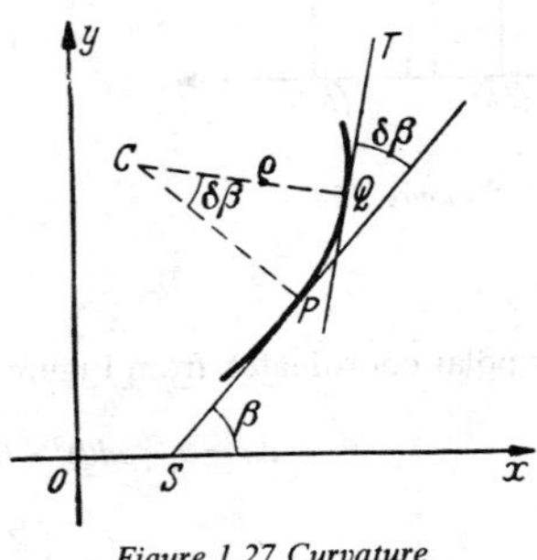

Figure 1.27 Curvature

Where dy/dx is small (as in the bending of beams), the radius of curvature is given by

$$\rho = \frac{1}{d^2y/dx^2}.$$

Example. Find the radius of curvature at any point at the curve $y = a\cos x/a$

$$\frac{dy}{dx} = \sinh\frac{x}{a}\quad \therefore \left[1+\left(\frac{dy}{dx}\right)^2\right]^{\frac{3}{2}} = \left[1+\sinh^2\frac{x}{a}\right]^{\frac{3}{2}} = \left(\cosh^2\frac{x}{a}\right)^{\frac{3}{2}} = \cosh^3\frac{x}{a}$$

$$\frac{d^2y}{dx^2} = \frac{1}{a}\cosh\frac{x}{a}\quad \therefore \rho = \frac{a\cosh^3\dfrac{x}{a}}{\cosh\dfrac{x}{a}} = a\cosh^2\frac{x}{a} = \frac{y^2}{a}$$

Lengths of curves

GENERAL THEORY

From Figure 1.28,

$$ds^2 = dx^2 + dy^2, \text{ hence } ds = \sqrt{\left\{1+\left(\frac{dy}{dx}\right)^2\right\}}\,dx = \sqrt{\left\{1+\left(\frac{dx}{dy}\right)^2\right\}}\,dy$$

$$\therefore s = \int_a^b \sqrt{\left\{1+\left(\frac{dy}{dx}\right)^2\right\}}\,dx$$

or

$$s = \int_c^d \sqrt{\left\{1+\left(\frac{dx}{dy}\right)^2\right\}}\,dy$$

For the evaluation of s for any given continuous function, use the first formula if x is single valued (i.e. if one value of x corresponds to one point only in the function, for example Figure 1.29). If more than one point on the curve corresponds to one value of x, the second formula for a curve of the form shown in Figure 1.30, should be used.

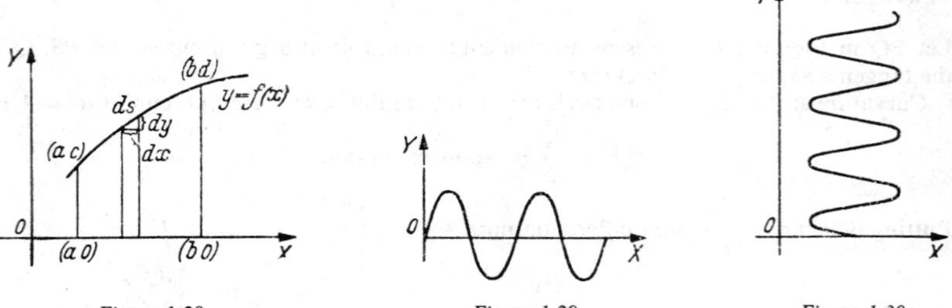

Figure 1.28 Figure 1.29 Figure 1.30

For polar coordinates, from Figure 1.31

$$ds = \sqrt{\{\rho d\theta)^2 + (d\rho)^2\}} = \sqrt{\left\{\rho^2+\left(\frac{d\rho}{d\theta}\right)^2\right\}}\,d\theta$$

$$s = \int_{\theta_1}^{\theta_2} \sqrt{\left\{\rho^2+\left(\frac{d\rho}{d\theta}\right)^2\right\}}\,d\theta$$

or,

$$s = \int_{p_1}^{p_2} \sqrt{\left\{1+\left(\rho\,\frac{d\theta}{d\theta\rho}\right)^2\right\}}\,d\rho$$

Plane areas by integration

See Figures 1.32 and 1.33.

GENERAL THEORY

From Figure 1.32, $A = \int_{x_1}^{x_2} y\,dx = \int_{x_1}^{x_2} f(x)\,dx$

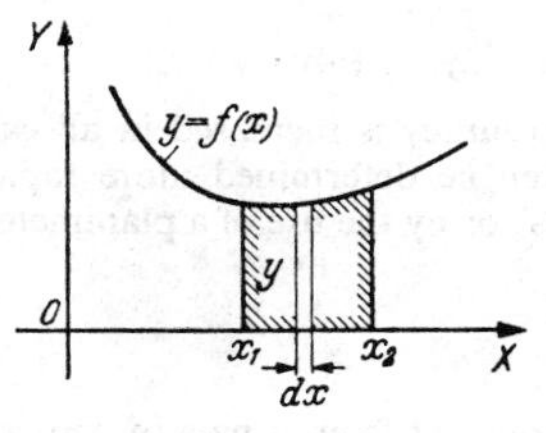

Figure 1.31

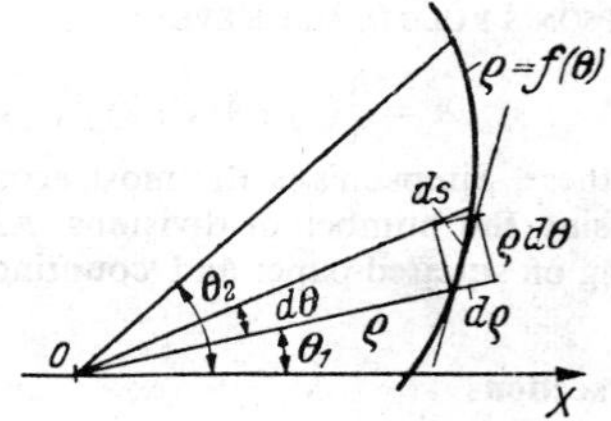

Figure 1.32

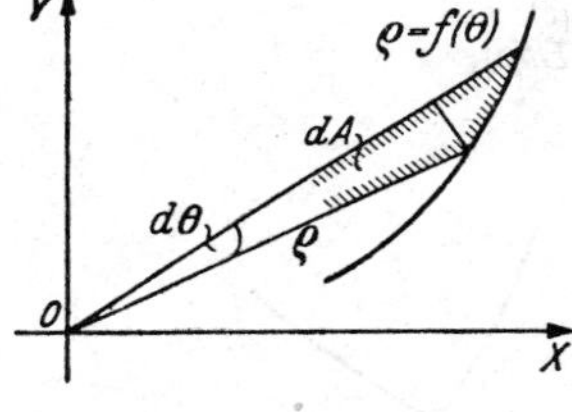

Figure 1.33

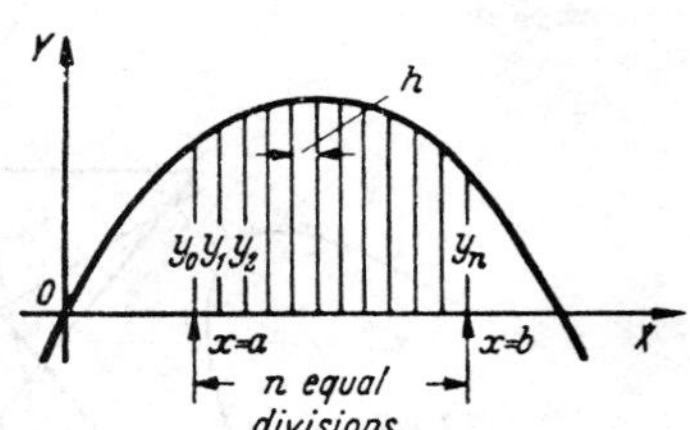

Figure 1.34

POLAR COORDINATES

From Figure 1.33, $dA = \frac{1}{2}\rho^2\, d\theta$ and therefore

$$A = \tfrac{1}{2}\int \rho^2 d\theta = \tfrac{1}{2}\int \{f(\theta)\}^2 d\theta.$$

Note. For curve cutting x axis, equate $f(x)$ to zero, find values of x for $y = 0$ and integrate between these values for the area cut off by the x axis.

When the area lies above and below the x axis integrate the positive and negative areas separately and add algebraically.

Where the area does not extend to the x axis in the case of cartesian coordinates, or to the origin in the case of polar coordinates, then double integration must be used.

Thus $$A = \iint dx\,.\,dy\,.\iint \rho\,.\,dp\,.\,d\theta.$$

Plane area by approximate methods

See Figure 1.34.

1 TRAPEZOIDAL RULE

$$A = \frac{h}{2}\{y_0+2(y_1+y_2+\;\dots\;+y_{n-1})+y_n\}$$

2 DURAND'S RULE

$$A = h(0.4y_0+1.1y_1+y_3+\;\dots\;+y_{n-2}+1.1y_{n-1}+0.4y_n)$$

3 SIMPSON'S RULE (n MADE EVEN)

$$A = \frac{h}{3}(y_0+4y_1+2y_2+4y_3+2y_4+\ \ldots\ +2y_{n-2}+4y_{n-1}+y_n)$$

Of these, Simpson's is the most accurate. The accuracy is increased in all cases by increasing the number of divisions. Areas can often be determined more rapidly by plotting on squared paper and 'counting the squares', or by the use of a planimeter.

Conic sections

Conic sections refer to the various profiles of sections cut from a pair of cones vertex to vertex when intersected by a plane. Figure 1.35 shows a pair of cones generated by

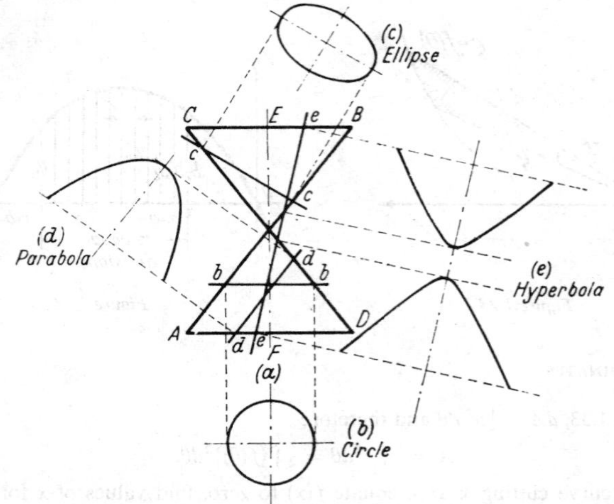

Figure 1.35 Circular cones generated by two intersecting straight lines

two intersecting straight lines AB, CD about the bisector EF of the angle between the lines.

Two straight lines. A section through the axis EF.

Circle. A section bb parallel to the base of a cone.

Ellipse. A section cc not parallel to the base of a cone and intersecting one cone only.

Parabola. A section dd parallel to the side of a cone.

Hyperbola. A section ee inclined to the side of a cone and intersecting both cones.

Properties of conic sections

A conic section is defined as the locus of a point, P, which moves so that its distance from a fixed point, the focus, bears a constant ratio, the eccentricity, to its perpendicular distance from a fixed straight line, the directrix.

Referring to Figure 1.36: the vertex of the curve is at V, the focus of the curve is at F, the directrix of the curve is the line DD parallel to yy'; the latus rectum is the line LR through the focus parallel to DD, FL = FR = l; the eccentricity of the curve is the ratio FP/PQ = FV/VS = e.

Then the curve is a parabola if $e = 1$, an ellipse if $e < 1$; and a hyperbola if $e > 1$. A circle is a particular case of an ellipse in which $e = 0$.

The polar equation of a conic is given by $l = \rho(1 - e\cos\theta)$ where ρ is the radius vector of any point P on the curve, θ the angle the vector makes with VX and l the semi latus rectum.

Parabola ($e = 1$). Figure 1.36.

EQUATIONS

With origin at V and putting a = VS = VF then for P at (x, y): $(x-a)^2 + y^2 = (x+a)^2$, i.e. $y^2 = 4ax$.

TANGENTS

Let PT be a tangent at any point P $(x_1\, y_1)$ then the equation of PT is given by $y - y_1 = m(x - x_1) = (2a/y_1)(x - x_1)$ or $yy_1 = 2a(x + x_1)$ since $d/dx\,(y^2) = 2y\,dy/dx = 4a$, i.e. $m = dy/dx = 2a/v_1$ *at* P $(x_1\, y_1)$.

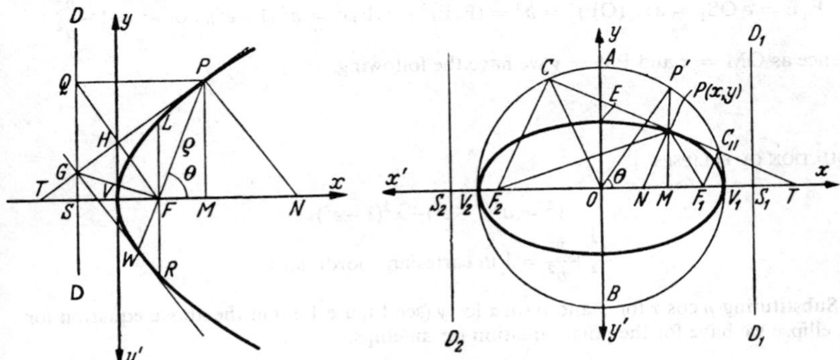

Figure 1.36 Properties of a conic section

Figure 1.37 Ellipse in cartesian coordinates

Alternatively, if any straight line $y = mx + c$ meets the parabola $y^2 = 4ax$ then $(mx+c)^2 = 4ax$ at the points of intersection and this expression will satisfy the condition for tangency if the roots of $m^2x^2 + 2(mc - 2a)x + c^2 = 0$ are equal, i.e. if $4(mc - 2a)^2 = 4m^2c^2$ or $c = a/m$ so that the equation for the tangent may be expressed as $y = mx + a/m$ for all values of m where $m = dy/dx$, and tangency occurs at the point $(a/m^2, 2a/m)$.

NORMAL

Let PN be the normal at any point P $(x_1\, y_1)$; then the equation of PN is given by $y - y_1 = -(y_1/2a)(x - x_1)$.

GENERAL PROPERTIES

Tangents:
1. The tangent PT bisects $F\hat{P}Q$.
2. The tangents PG, GW where PW is a focal chord intersect at G on DD.
3. The tangent PT intersects the axis of the parabola at a point T where TV = VM; TF = SM = PF.
4. The angles GFP, PHQ and PGW are right angles.

Normals: any normal PN intersects VX at N where FT = FN.

Sub-normals: the sub-normal MN is a constant length, i.e. MN = FS = $2a$.

ELLIPSE ($e < 1$)

Referring to Figure 1.37, F_1, F_2 and the foci; D_1D_1, D_2D_2 the directrices.

$$e = \frac{F_1V_1}{S_1V_1} = \frac{F_1V_2}{S_1V_2} = \frac{F_2V_1}{S_2V_1} = \frac{F_1P}{MS_1} = \frac{F_2P}{MS_2} = \frac{OF_1}{OV_1} = \frac{OF_2}{OV_2} = \frac{F_1F_2}{V_1V_2}$$

Let OV_1 the semi-major axis $= a$ and OE the semi-minor axis $= b$,

then $$OF_1 = OF_2 = ae \text{ and } OS_1 = OS_2 = \frac{a}{e}$$

also $$F_1P = a - ex; \quad F_2P = a + ex \therefore F_1P + F_2P = 2a$$

$$F_1E = e\,OS_1 = a; \quad (OE)^2 = b^2 = (F_1E)^2 - (OF_1)^2 = a^2(1-e^2), \quad \text{or } e^2 = 1 - \frac{b^2}{a^2}$$

Hence as OM $= x$ and PM $= y$ we have the following.

EQUATION OF ELLIPSES

$$y^2 = a^2(1-e^2) - x^2(1-e^2)$$

or $$\frac{x^2}{a^2} + \frac{y^2}{b^2} = 1 \text{ in cartesian coordinates.}$$

Substituting $\rho \cos \alpha$ for x and $\rho \sin \alpha$ for y (see Figure 1.38) in the above equation for an ellipse we have for the polar equation for an ellipse:

$$\frac{1}{\rho^2} = \frac{\cos^2 \alpha}{a^2} + \frac{\sin^2 \alpha}{b^2}$$

TANGENT

At any point $P(x_1y_1)$ on the ellipse
Let $f(xy) = 0$ represent the curve, then

$$\frac{dy}{dx} = -\frac{\partial f}{\partial x} \bigg/ \frac{\partial f}{\partial y}$$

$\therefore$ $\partial f/\partial x = 2x/a^2$ and $\partial f/\partial y = 2y/b^2$ so that dy/dx at point $(x_1\, y_1)$ is given by $-b^2x_1/a^2y_1 = m$. Substituting this value of m in the equation of a straight line $(y - y_1) = m(x - x_1)$ we have the equation of tangent PT: $xx_1/a^2 + yy_1/b^2 = 1$.

Alternatively the straight line $y = mx+c$ is a tangent to the ellipse $x^2/a^2+y^2/b^2 = 1$ when the roots of $x^2/a^2+(mx+c)^2/b^2-1 = 0$ are equal, i.e. when $c^2 = a^2m^2+b^2$. Substituting we have for the equation of a tangent at any point P: $y = mx+\sqrt{(a^2m^2+b^2)}$.

The equation to the tangent may also be written in the form

$$\frac{x}{a}\cos\theta+\frac{y}{b}\sin\theta = 1.$$

The coordinates of the point of contact are $(a\cos\theta,\ b\sin\theta)$, θ being known as the eccentric angle (see Figure 1.37).

NORMAL

Substituting the value of m above in the general equation for the normal PN to a curve at point P $(x_1\ y_1)$ given by: $(y-y_1)m+(x-x_1) = 0$ we have as the equation for the normal $(y-y_1)b^2/y_1 = (x-x_1)a^2/x_1$.

GENERAL PROPERTIES

1. The circle AV_2BV_1 is termed the auxiliary circle (Figure 1.37).
2. $OM\times OT = a^2$.
3. $F_2N = e\,F_2P$.
4. $F_1N = e\,F_1P$.
5. PN bisects $\angle F_1PF_2$.
6. The perpendiculars from F_1, F_2 to any tangent meet the tangent on the auxiliary circle.

CIRCLE ($c = 0$)

The circle may be regarded as a particular case of the ellipse (see above). The equation of a circle of radius a with centre at the origin is $x^2+y^2 = a^2$ or, in polar coordinates, $\rho = a$.

The equation of the tangent at the point $(x_1\ y_1)$ is $xx_1+yy_1 = a^2$, or, $y = mx+a\sqrt{(1+m^2)}$. The equation of the normal is $xy_1-yx_1 = 0$.

HYPERBOLA ($e>1$)

This is shown in Figure 1.39 where F_1 F_2 are the foci, D_1 D_1 and D_2 D_2 the directrices and

$$e = \frac{F_1V_1}{S_1V_1} = \frac{F_1P}{MS_1} = \frac{F_2P}{MS_2} = \frac{F_1V_2}{S_1V_2} = \frac{F_2V_1}{S_2V_1} = \frac{V_1V_2}{S_1S_2} = \frac{OV_1}{OS_1} = \frac{OV_2}{OS_2}$$

where O is the origin of the axes x and y.

Putting $OV_1 = OV_2 = a$ then $OF_1 = OF_2 = ea$ and $OS_1 = OS_2 = a/e$; also $F_1P = ex-a$ and $F_2F = ex+a$. Now $(F_1P)^2 = (PM)^2+(F_1M)^2$, so $(ex-a)^2 = y^2+(x-ae)^2$ which becomes $y^2 = (e^2-1)x^2-(e^2-1)a^2$. Putting $(e^2-1)a^2 = b^2$ then $y^2 = (b^2/a^2)x^2-b^2$; therefore the equation of the hyperbola is given by $x^2/a^2-y^2/b^2 = 1$ in Cartesian coordinates, or

$$\frac{1}{\rho^2} = \frac{\cos^2\theta}{a^2}-\frac{\sin^2\theta}{b^2}$$

in polar coordinates.

Rearranging we have $y = b\sqrt{(x^2/a^2-1)}$, i.e. y is imaginary when $x^2<a^2$ and $y = 0$ for $x = \pm a$. y is real when $x>a$ and there are two values for y of opposite sign.

CONJUGATE AXIS

The conjugate axis lies on yy' and is given by CC′ where OC = OC′ = $\pm b$.

TANGENTS

Let the straight line $y = mx+c$ meet the hyperbola $x^2/a^2-y^2/b^2 = 1$; then $x^2/a^2-(mx+c)^2/b^2-1 = 0$ will give the points of intersection. The condition for tangency is that the roots of this equation are equal, i.e. $c = \sqrt{(a^2m^2-b^2)}$ and the equation of the tangent is given by $y = mx+\sqrt{(a^2m^2-b^2)}$ at any point. Alternatively the tangent to the hyperbola at $(x_1\, y_1)$ is given by $xx_1/a^2-yy_1b^2 = 1$.

NORMAL

The equation for the normal at any point $(x_1\, y_1)$ on the curve is given by

$$(y-y_1)b^2/y_1+(x-x_1)a^2/x_1 = 0.$$

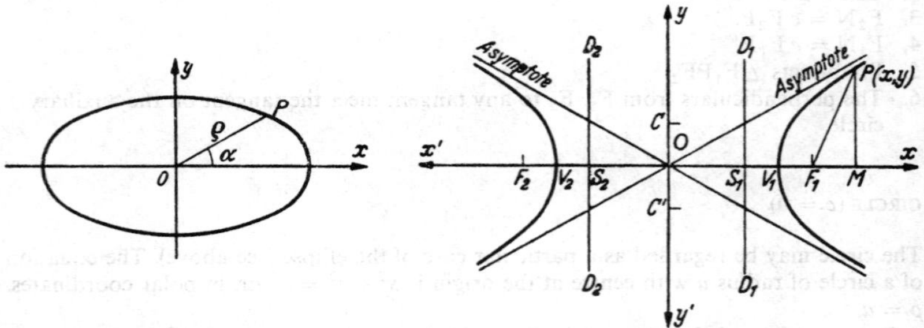

Figure 1.38 Ellipse in polar coordinates *Figure 1.39 Hyperbola*

ASYMPTOTES

The tangent to the hyperbola becomes an asymptote when the roots of the equation $x^2/a^2-(mx+c)^2/b^2-1 = 0$ are both infinite i.e. when $b^2-a^2m^2 = 0$ and $a^2mc = 0$. $\therefore m = \pm b/a$ and $c = 0$. Substituting for m in $y = mx+c$ we have as the equation for an asymptote $y = \pm(b/a)x$. The combined equation for both asymptotes is given by

$$\frac{x^2}{a^2}-\frac{y^2}{b^2} = 0.$$

The equation of the hyperbola referred to its asymptotes as oblique axes is

$$\mathrm{X.\,Y} = \frac{a^2+b^2}{4}.$$

GENERAL PROPERTIES

(i) $\mathrm{F_2P-F_1P} = 2a$.
(ii) The product of the perpendiculars from any point on a hyperbola to its asymptotes is constant and equal to $a^2b^2/(a^2+b^2)$.

RECTANGULAR HYPERBOLA

When the transverse axis V_1V (Figure 1.39) is equal to the conjugate axis CC′ the hyperbola is a rectangular hyperbola, i.e. $a = b$ and the equation for the curve is given by $x^2 - y^2 = a^2$.

The equation for the asymptotes then becomes $y = \pm x$ which represents two straight lines at right angles to each other. The equation of the rectangular hyperbola referred to its asymptotes as axes of coordinates is given by $xy =$ constant.

GENERAL EQUATION OF A CONIC SECTION

The general equation of a conic section has the form:

$$ax^2 + 2hxy + by^2 + 2gx + 2fy + c = 0$$

Let

$$D = \begin{array}{c|c|c} a & h & g \\ h & b & f \\ g & f & c \end{array} \quad \text{and} \quad d = \begin{vmatrix} a & b \\ h & b \end{vmatrix}$$

Then, the general equation represents

1 An ellipse if $d > 0$;
2 A parabola if $d = 0$;
3 A hyperbola if $d < 0$;
4 A circle if $a = b$ and $h = 0$;
5 A rectangular hyperbola if $a + b = 0$;
6 Two straight lines (real or imaginary) if $D = 0$;
7 Two parallel straight lines if $d = 0$ and $D = 0$.

The centre of the conic $(x_0 y_0)$ is determined by the equations: $ax_0 - hy_0 + g = 0$, $hx_0 + by_0 + f = 0$.

THREE-DIMENSIONAL ANALYTICAL GEOMETRY

Sign convention

CARTESIAN COORDINATES

This is shown in Figure 1.40, there being eight compartments formed by the right angled intersection of three planes. The sign of x, y, z follow the convention that these are positive when measured in the directions Ox, Oy, Oz of the coordinate axes and negative when measured in the directions Ox′, Oy′, Oz′ respectively.

POLAR COORDINATES

The location of any point P in space (see Figure 1.41) is fully located by the radius vector ρ and the two angles θ and ϕ thus $(\rho\theta\phi)$. From Figure 1.41

$$\text{OP} = \rho = \sqrt{[(\text{OD})^2 + (\text{OB})^2 + (\text{OC})^2]} = \sqrt{(x^2 + y^2 + z^2)}$$

and

$$x = \rho \sin\theta \cos\phi;\ y = \rho \sin\theta \sin\phi;\ z = \rho \cos\theta.$$

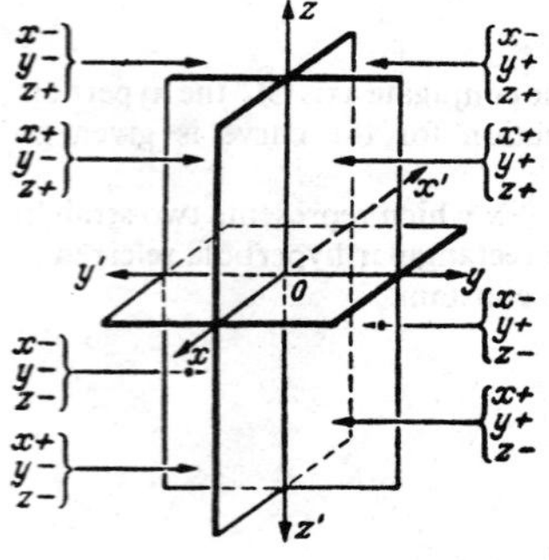

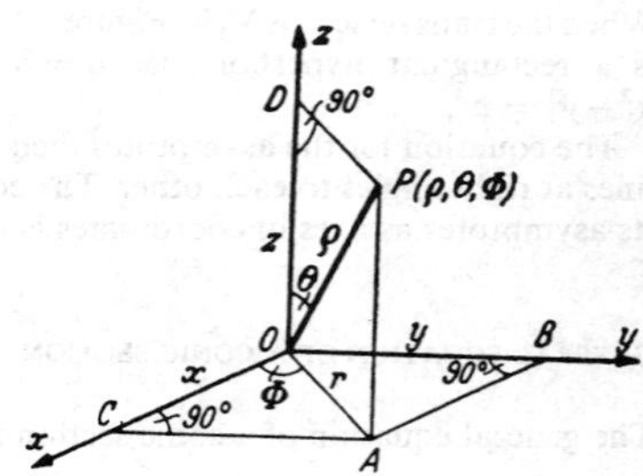

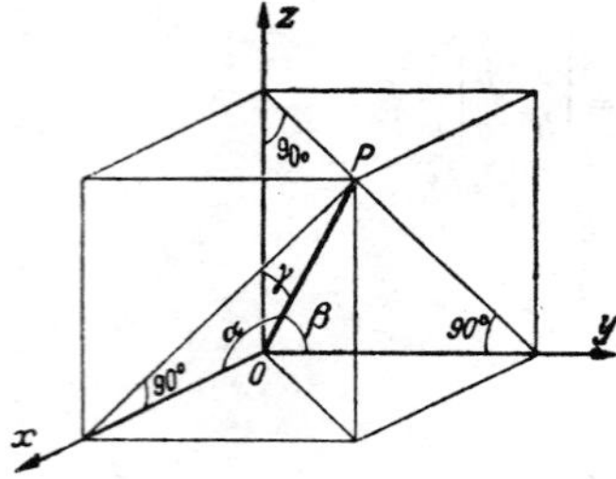

Figure 1.40 (top, left) Sign conventions in analytical solid geometry

Figure 1.41 (above) Polar coordinates in three dimensions

Figure 1.42 (left) Direction-cosines

CYLINDRICAL COORDINATES

In this system the point P (Figure 1.41) is located by its perpendicular distance, z, from the xy plane and the polar coordinates of the foot, A, of that perpendicular in the xy plane. P is the point r, ϕ, z where $OA = r$.

DIRECTION-COSINES OF A STRAIGHT LINE

If the direction of the line OP, Figure 1.42, is determined by α, β, γ then the projections of a unit length of OP on to the axes Ox, Oy, Oz are given by $\cos\alpha$, $\cos\beta$, $\cos\gamma$ respectively, termed direction-cosines. Let $l = \cos\alpha$, $m = \cos\beta$, $n = \cos\gamma$ and $CP = \rho$; then $\rho^2(l^2+m^2+n^2) = x^2+y^2+z^2 = \rho^2$, i.e. $l^2+m^2+n^2 = 1$.

Also $\qquad \sin^2\alpha+\sin^2\beta+\sin^2\gamma = (1-l^2)+(1-m^2)+(1-n^2) = 2.$

Again if $\qquad l:m:n = s:t:u$ then $\dfrac{l^2}{s^2} = \dfrac{m^2}{t^2} = \dfrac{n^2}{u^2} = \dfrac{l^2+m^2+n^2}{s^2+t^2+u^2} = \dfrac{1}{s^2+t^2+u^2}$

i.e. $\qquad l = \dfrac{s}{\sqrt{(s^2+t^2+u^2)}};\quad m = \dfrac{t}{\sqrt{(s^2+t^2+u^2)}};\quad n = \dfrac{u}{\sqrt{(s^2+t^2+u^2)}}$

GENERAL EQUATIONS

The expression $F(xyz) = 0$ represents a surface of some kind and if we put $x = 0$ the resulting equation is for a curve in the yz plane; similarly for $y = 0$ the curve is in the xz plane, etc. In general, any two simultaneous equations, $F(xyz) = 0$, $F'(xyz) = 0$

represent a line (either straight or curved) being the intersection of two surfaces. Any three such simultaneous equations represent a point (or several points).

Equation of a plane

The general equation of a plane is given by the expression $ax+by+cz+d = 0$ ($abcd$ being constants). By putting $y = 0$, $z = 0$ then $x = -d/a = a'$ which is the intercept of the plane on the x axis at a distance a' from the origin. Similarly the intercepts on the y and z axes are b' and c'. Hence $a = -d/a'$; $b = -d/b'$; $c = -d/c'$ and substituting these values in the general equation for the plane we have the intercept equation for a plane as $x/a'+y/b'+z/c' = 1$.

In Figure 1.43 let P be any point on the plane ABC and let OQ of length p be at 90° to the plane ABC; then if l, m, n are the direction cosines of OQ we have $p = lx+my+nz$,

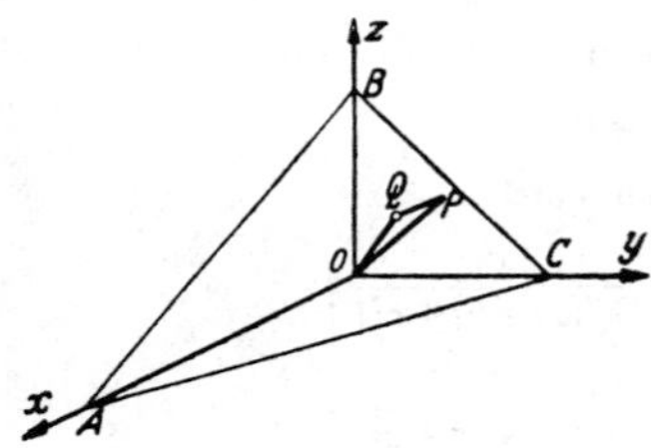

Figure 1.43 Equation of a plane

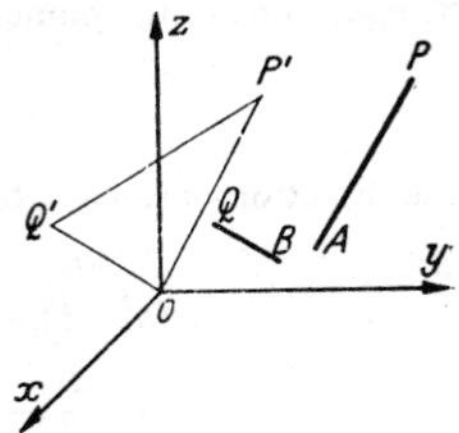

Figure 1.44 Angle between two lines of known direction-cosines

which is the perpendicular form of the equation to a plane. The various forms of the equation to a plane are interchangeable since

$$p = -\frac{d}{\sqrt{(a^2+b^2+c^2)}} = la' = mb' = nc'$$

and

$$\frac{1}{a'^2}+\frac{1}{b'^2}+\frac{1}{c'^2} = \frac{1}{p^2}.$$

Distance between two points in space

Let the two points be $P(x^1\, y_1\, z_1)$; $Q(x_2\, y_2\, z_2)$. Assume origin shifted to P and axes kept parallel to original axes, then coordinates of Q relative to P are (x_2-x_1), (y_2-y_1), (z_2-z_1) and the length PQ $= r$, i.e. $r = \sqrt{[(x_2-x_1)^2+(y_2-y_1)^2+(z_2-z_1)^2]}$ and the locus of Q is a sphere if r is constant.

Equations of a straight line

Using direction cosines for PQ, $l = (x_2-x_1)/r$; $m = (y_2-y_1)/r$; $n = (z_2-z_1)/r$. If Q is taken as any point then the symmetrical equation of a straight line is given by $r = (x-x_1)/l = (y-y_1)/m = (z-z_1)/n$ and the coordinates of any point on the line are given by $x = x_1+rl$; $y = y_1+rm$; $z = z_1+rn$. For a line through the origin this becomes

$$r = \frac{x}{l} = \frac{y}{m} = \frac{z}{n}$$

The equation of the straight line through the points $(x_1 y_1 z_1)$ and $(x_2 y_2 z_2)$ is

$$\frac{x-x_1}{x_2-x_1} = \frac{y-y_1}{y_2-y_1} = \frac{z-z_1}{z_2-z_1}$$

ANGLE BETWEEN TWO LINES OF KNOWN DIRECTION-COSINES

Let PA, QB be any two lines in space (Figure 1.44) and let P′O, Q′O be parallel to PA, QB respectively and having direction cosines $l_1 m_1 n_1$; $l_2 m_2 n_2$ respectively then $\cos\alpha = l_1 l_2 + m_1 m_2 + n_1 n_2$ where $\alpha = \mathrm{P'\hat{O}Q'}$

THE ANGLE BETWEEN TWO PLANES

Let the equations of the planes be

$$a_1x+b_1y+c_1z+d_1 = 0$$

and

$$a_2x+b_2y+c_2z+d_2 = 0$$

then the direction-cosines of the normals to these planes are

$$\frac{a_1}{\sqrt{(a_1^2+b_1^2+c_1^2)}};\quad \frac{b_1}{\sqrt{(a_1^2+b_1^2+c_1^2)}};\quad \frac{c_1}{\sqrt{(a_1^2+b_1^2+c_1^2)}}$$

and

$$\frac{a_2}{\sqrt{(a_2^2+b_2^2+c_2^2)}};\quad \frac{b_2}{\sqrt{(a_2^2+b_2^2+c_2^2)}};\quad \frac{c_2}{\sqrt{(a_2^2+b_2^2+c_2^2)}}$$

If θ is the angle between the planes, this is equal to the angle between the normals to these planes,

i.e.

$$\cos\theta = \frac{a_1a_2+b_1b_2+c_1c_2}{\sqrt{[(a_1^2+b_1^2+c_1^2)(a_2^2+b_2^2+c_2^2)]}}$$

The planes are perpendicular to each other if $a_1a_2+b_1b_2+c_1c_2 = 0$.
They are parallel if $a_1/a_2 = b_1/b_2 = c_1/c_2$.

THE ANGLE BETWEEN A PLANE AND A STRAIGHT LINE

The angle θ between the plane $l_1x+m_1y+n_1z = p$ and the line $(x-x_1)/l_2 = (y-y_1)/m_2 = (z-z_1)/n_2$ is given by $\sin\theta = (l_1l_2+m_1m_2+n_1n_2)$.

LENGTH OF THE PERPENDICULAR FROM A POINT $x_1y_1z_1$ TO A PLANE

1 Where the equation of the plane is the perpendicular form $lx+my+nz = p$ then the equation of a plane containing the point $(x_1 y_1 z_1)$ and parallel to the given plane is given by $lx+my+nz = p'$ where p and p' are the lengths of perpendiculars from the origin. Therefore required length of perpendicular is $p'-p = lx_1+my_1+nz_1-p$, since the point $(x_1 y_1 z_1)$ lies on the second plane.
2 Where the equation of the plane takes the general form $ax+by+cz+d = 0$ the length of perpendicular from point $x_1 y_1 z_1$ is given by

$$\frac{ax_1+by_1+cz_1+d}{\sqrt{(a^2+b^2+c^2)}}$$

In the above the equation of the perpendicular is given by

$$\frac{x-x_1}{l} = \frac{y-y_1}{m} = \frac{z-z_1}{n}$$

CALCULUS

The calculus deals with quantities which vary and with the rate at which this variation takes place.

Variables may be denoted by u, v, w, x, y, z and increments of these variables are denoted by $du, dv \ldots dz$. A simple example concerns the slope of a curve. Suppose that the curve is defined by some function

$$y = f(x)$$

The slope at the point $x = x_1$ may be approximated to as follows. Let x_2 be close in value to x_1 then, provided the curve $f(x)$ is well behaved in the region of x_1, the line joining $f(x_1)$ to $f(x_2)$ is an approximation to the tangent to the curve at $x = x_1$. As x_2 is moved closer to x_1 the approximation becomes better and better, until, in the limit, when x_2 reaches x_1 the tangent (instead of the secant) is obtained and thereby the slope of the curve $y = f(x)$ is found at the point $x = x_1$. This process is known as differentiation.

Differentiation

This process is used to find the value of dy/dx.
For combinations of functions u, and v of x

$$\frac{d}{dx}(uv) = u\frac{dv}{dx} + v\frac{du}{dx}$$

$$\frac{d}{dx}\left(\frac{u}{v}\right) = \frac{v(du/dx) - u(dv/dx)}{v^2}$$

For polynomial functions, $y = ax^n$

$$\frac{dy}{dx} = nax^{n-1}$$

The differentiation process may be carried out more than once. Thus

$$\frac{d}{dx}\frac{(dy)}{dx} = \frac{d^2y}{dx^2} \text{ etc.}$$

As an example, if $y = f(x) = ax^4 + bx^3 + cx^2 + dx + c$

then $$\frac{dy}{dx} = f'(x) = 4ax^3 + 3bx^2 + 2cx + d$$

$$\frac{d^2y}{dx^2} = f''(x) = 12ax^2 + 6bx + 2c$$

$$\frac{d^3y}{dx^3} = f'''(x) = 24ax + 6b$$

$$\frac{d^4y}{dx^4} = f^{iv}(x) = 24a$$

$$\frac{d^5y}{dx^5} = f^{v}(x) = 0$$

It is often convenient, when dealing with long, complicated expressions, to substitute a symbol for a part of a compound expression. Suppose we have

$$y = f(x)$$

a complicated expression and we choose to make a substitution u then the differential, $f'(x)$ can be found from the rule:

$$\frac{dy}{dx} = \frac{dy}{du} \cdot \frac{du}{dx}$$

e.g.

$$y = (x^4 + a^2)^6$$

The substitution:

$$u = x^4 + a^2 \text{ is appropriate}$$

so

$$y = u^6$$

thus

$$\frac{dy}{dx} = 6u^5 \text{ and } \frac{du}{dx} = 4x^3$$

so

$$\frac{dy}{du} = \frac{dy}{du}\,\frac{du}{dx} = 6(x^4 + a^2)^5 4x^3$$

$$= 24x^3(x^4 + a^2)$$

In the case of trigonometric functions the differentiation process can be obtained via the expressions for multiple angles (see section on Trigonometry).
For, suppose

$$y = \sin\theta$$

We let

$$\begin{aligned} y + \delta y &= \sin(\theta + \delta\theta) \\ &= \sin\theta\cos\delta\theta + \cos\theta\sin\delta\theta \\ &= \sin\theta + \cos\theta \,.\, \delta\theta \end{aligned}$$

So

$$\delta y = \cos\theta \,.\, \delta\theta$$

$$\frac{\delta y}{\delta\theta} = \cos\theta \text{ or } \frac{dy}{d\theta} = \cos\theta$$

In cases where inverse trigonometric functions are involved the principle of substitution is employed, for suppose

$$y = \sin^{-1} u$$

then

$$u = \sin y$$

so

$$\frac{du}{dy} = \cos y = (1 - \sin^2 y) = (1 - u^2)$$

so

$$\frac{dy}{du} = 1 \bigg/ \frac{du}{dy} = \frac{1}{(1 - u^2)}$$

In cases where exponentiation is involved the principle of substitution may again be employed,

for suppose

$$y = e^{3x^2/4}$$

we write

$$y = e^u$$

where

$$u = 3x^2/4$$

and $$\frac{dy}{dx} = \frac{dy}{du} \cdot \frac{du}{dx} = e^u 6x/4$$
$$= \frac{3x}{2} e^{3x^2/4}$$

Partial differentiation

The dependent variable u may be a function of more than one independent variable, x and y, and we wish to find the rates of changes of u with respect to u and v separately. These rates of change, the partial differentials with respect to x and y are denoted by

$$\frac{du}{dx} \text{ and } \frac{du}{dy}.$$

In these processes the normal rules of differentiation are followed except that in finding du/dx, y is treated as a constant and in finding du/dy, x is treated as a constant.
The total differential of a function

$$u = f(x, y)$$

where both x and y are functions of t is given by
$$\frac{du}{dt} = \frac{du}{dx}\frac{dx}{dt} + \frac{du}{dy}\frac{dy}{dt}$$

Maxima and minima

Maxima and minima of functions occur when the function has zero slope or first differential. Thus in order to determine a maximum or minimum of a function $y = f(x)$

we set $$\frac{dy}{dx} = f'(x) = 0$$

and solve this equation, say $x = x_1$.
To distinguish between maxima and minima it is necessary to evaluate

$$\frac{d^2y}{dx^2} \text{at the point } x_1$$

For a maximum $$\frac{d^2y}{dx^2} < 0$$

For a minimum $$\frac{d^2y}{dx^2} > 0$$

Integration

Integration is generally the reverse of the process of differentiation. It may also be regarded as equivalent to a process of summing a number of finite quantities but, in the limit the number of quantities becomes infinite and their size becomes infinitesimal.
By the reverse of the differentiation process the integral

$$\int ax^n . dx = \frac{ax^{n+1}}{n+1} + c$$

The c being an arbitrary constant which, for shortness, is frequently not written. This is called an indefinite integral as no range over which the integration is to be performed has been specified. If such a range is specified then we obtain the case of a definite integral,

for example.
If

$$F(x)\,dx = f(x)$$

then

$$\int_b^a F(x)\,dx = f(b) - f(a)$$

In geometrical terms this integral represents the area bounded by the curve $y = F(x)$, the x axis, and the two lines $x = a$, $x = b$.

Successive integration

This is the reverse process from that of successive differentiation, each cycle of operations consisting of the integration of the function resulting from the immediately previous integration. In general terms instructions to carry out successive integration is expressed thus: $y = \int\int\int\int f(x)\,dx, dx, dx, dx$, which means integration is to be successively carried out four times with respect to x. Another form of successive integration is $v = \int\int\int f(x, y, z)\,dx, dy, dz$; referred to as a volume integral. A surface integral would take the form $s = \int\int f(x, y)\,dx, dy$.

Example. Find a general expression for the deflection of a simple span girder of span l loaded uniformly by a load w per unit length of span given that $w = EI\,d^4y/dx^4$ and taking E and I as constant and x as measured from one end.

Load

$$EI\frac{d^4y}{dx^4} = w$$

Shear

$$EI\int\frac{d^4y}{dx^4}dx = EI\frac{d^3y}{dx^3} = wx + c_1 = wx - \frac{wl}{2}; \left(\text{Shear} = -\frac{wl}{2}\text{for } x = 0\right)$$

B.M.

$$EI\int\frac{d^3y}{dx^3}dx = EI\frac{d^2y}{dx^2} = \frac{wx^2}{2} - \frac{wlx}{2} + c_2 = \frac{wx^2}{2} - \frac{wlx}{2}; \text{ (B.M.} = 0 \text{ for } x = 0)$$

Slope

$$EI\int\frac{d^2y}{dx^2}dx = EI\frac{dy}{dx} = \frac{wx^3}{6} - \frac{wlx^2}{4} + c_3 = \frac{wx^3}{6} - \frac{wlx^2}{4} + \frac{wl^3}{24};$$

$$\left(\text{Slope} = 0 \text{ for } x = \frac{l}{2}\right)$$

Deflection

$$EI\int\frac{dy}{dx}dx = EIy = \frac{wx^4}{24} - \frac{wlx^3}{12} + \frac{wl^3x}{24}; \text{ (Deflection} = 0 \text{ for } x = 0)$$

i.e. at any distance x from one end the deflection

$$y = \frac{1}{24}\frac{w}{EI}(x^4 - 2lx^3 + l^3x)$$

$$\int\int\int\int\frac{d^4y}{dx^4}dx\,dx\,dx\,dx = \frac{w}{24EI}(x^4 - 2lx^3 + l^3x)$$

which is in the general form.

For the mid-span deflection the range of integration is from $x = 0$ to $x = \frac{1}{2}l$.

Hence

$$\int\int\int\int_0^{l/2}\frac{d^4y}{dx^4}dx\,.\,dx\,.\,dx\,.\,dx = \frac{wl^4}{24EI}\left(\tfrac{1}{16} - \tfrac{1}{4} + \tfrac{1}{2}\right) = \frac{5wl^4}{384EI}$$

Integration by substitution

The integration of functions can often be simplified by substituting a new variable for a part or the whole of the original function, thereby reducing it to one of the standard forms.

Example. Find the value of $\int\sqrt{(3+x)}\,dx$

Let $3+x = u \therefore dx = du$

so that $\int\sqrt{(3+x)}\,dx = \int u^{\frac{1}{2}}\,du = \frac{2}{3}u^{\frac{3}{2}} = \frac{2}{3}(3+x)^{\frac{3}{2}}$ *or* $\frac{2}{3}\sqrt{(3+x)^3}$

Example. Find the value of

$$2\int\frac{dx}{e^{3x}+c^{-3x}}$$

Let $e^{3x} = v$; then $3e^{3x}\,.\,dx = dv$, or $dx = dv/3v$.

Substituting

$$2\int\frac{dx}{e^{3x}+e^{-3x}} = \frac{2}{3}\int\frac{dv}{v(v+1/v)} = \frac{2}{3}\int\frac{dv}{v^2+1} = \frac{2}{3}\tan^{-1}v = \frac{2}{3}\tan^{-1}e^{3x}.$$

Example. Find the value of $\int\sqrt{(1-x^2)}\,dx$.
Put $x = \sin\theta$; then $\sqrt{(1-x^2)} = \cos\theta$

$$\therefore \int\sqrt{(1-x^2)}\,.\,dx = \int\cos\theta\,.\,d\sin\theta = \int\cos^2\theta\,.\,d\theta = \int\frac{1+\cos 2\theta}{2}\,.\,d\theta$$

$$= \frac{\theta}{2}+\frac{\sin 2\theta}{4} = \frac{1}{2}\{\sin^{-1}x+x\sqrt{(1-x^2)}\}$$

Integration by transformation

The integration of trigonometric functions can often be simplified by transformation into a standard form of integral.

TYPE

$$\int\sin^m\theta\cos^n\theta d\theta$$

Case 1: m = positive odd integer, n = any positive integer.

TRANSFORMATIONS

$$\int\sin^{m-1}\theta\sin\theta\cos^n\theta d\theta$$
$$= \int(1-\cos^2\theta)^{(m-1)/2}\sin\theta\cos^n\theta d\theta$$
$$= -\int(1-\cos^2\theta)^{(m-1)/2}\cos^n\theta d(\cos\theta)$$

Example. Solve $\int\sin^3\theta\cos^2\theta d\theta$

$$\int\sin^3\theta\cos^2\theta d\theta = \int(1-\cos^2\theta)\sin\theta\cos^2\theta d\theta = \int\cos^2\theta\sin\theta d\theta - \int\cos^4\theta\sin\theta d\theta$$

$$= -\frac{\cos^3\theta}{3}+\frac{\cos^5\theta}{5}$$

Case 2: m = any positive integer, n = positive odd integer.

TRANSFORMATION

$\int \sin^{m}\theta \cos^{n-1}\theta \cos\theta d\theta = \int (1-\sin^{2}\theta)^{(n-1)/2} \sin^{m}\theta d(\sin\theta)$.

Example. Solve $\int \sin^{2}\theta \cos^{3}\theta d\theta$

$$\int \sin^{2}\theta \cos^{3}\theta d\theta = \int (1-\sin^{2}\theta) \sin^{2}\theta \cos\theta d\theta = \int \sin^{2}\theta \cos\theta d\theta - \int \sin^{4}\theta \cos\theta d\theta$$

$$= \frac{\sin^{3}\theta}{3} - \frac{\sin^{5}\theta}{5}$$

TYPE

$\int \tan\theta d\theta$ where n is an integer >1.

TRANSFORMATION

$\int \tan^{n-2}\theta \tan^{2}\theta d\theta = \int \tan^{n-2}\theta (\sec^{2}\theta - 1) d\theta = \int \tan^{n-2}\theta \, . \, d\tan\theta - \int \tan^{n-2}\theta \, . \, d\theta$

TYPE

$\int \cot^{n}\theta d\theta$ where n is an integer >1.

TRANSFORMATION

$\int \cot^{n-2}\theta \cot^{2}\theta d\theta = \int \cot^{n-2}\theta (\operatorname{cosec}^{2}\theta - 1) \mathrm{d}\theta = -\int \cot^{n-2}\theta \, . \, d\cot\theta - \int \cot^{n-2}\theta \, . \, d\theta.$

TYPE

$\int \sec^{n}\theta d\theta$ where n is positive and even.

TRANSFORMATION

$\int \sec^{n-2}\theta \sec^{2}\theta d\theta = \int (\tan^{2}\theta + 1)^{(n-2)/2} d\tan\theta.$

TYPE

$\int \operatorname{cosec}^{n}\theta d\theta$ where n is positive and even.

TRANSFORMATION

$\int \operatorname{cosec}^{n-2}\theta \operatorname{cosec}^{2}\theta d\theta = \int -(\cot^{2}\theta + 1)^{(n-2)/2} d\cot\theta$

TYPE

$\int \tan^m \theta \sec^n \theta d\theta$ where n is positive and even.

TRANSFORMATION

$\int \tan^m \theta \sec^{n-2} \theta \sec^2 \theta d\theta = \int \tan^m \theta (\tan^2 \theta + 1)^{(n-2)/2} d \tan \theta.$

TYPE

$\int \cot^m \theta \operatorname{cosec}^n \theta d\theta$ where n is positive and even.

TRANSFORMATION

$\int \cot^m \theta \operatorname{cosec}^{n-2} \theta \operatorname{cosec}^2 \theta d\theta = \int -\cot^m \theta (\cot^2 \theta + 1)^{(n-2)/2} d \cot \theta.$

TYPE

$\int \tan^m \theta \sec^n \theta d\theta$ where m and n are odd.

TRANSFORMATION

$\int \tan^{m-1} \theta \tan \theta \sec^{n-1} \theta \sec \theta d\theta = \int (\sec^2 \theta - 1)^{(m-1)/2} . \sec^{n-1} \theta . d \sec \theta.$

Integration by parts

The integration of functions can often be simplified by breaking up the function into two parts u and dv where u and v are the substituted variables in $\int u . dv = u . v - \int v . du$, the fundamental formula for integration by parts, $\int u . dv$ representing the function to be integrated. In applying this method of integration $\int v . du$ should not be more complex than $\int u . dv$. The integration of logarithmic, exponential, inverse trigonometric and products of algebraic expressions may be simplified by this procedure.

Example. $\int w \sin w\, dw$
Let $u = w$ and $dv = \sin w\, dw$ then $du = dw$ and $v = -\cos w$

$$\therefore \int w \sin w . dw = \int u . dv = -w \cos w + \int \cos w\, dw = -w \cos w + \sin w$$

Example. $\int x e^x\, dx$
Let $u = x$ and $dv = c^x dx$ then $du = dx$ and $v = e^x$

$$\therefore \int x e^x\, dx = \int u . dv = x e^x - \int e^x dx = x e^x - e^x = e^x(x-1)$$

Example. $\int \cos^2 \theta d\theta$
Let $u = \cos \theta$ and $dv = \cos \theta d\theta$ then $du = -\sin \theta d\theta$ and $v = \sin \theta$

$$\therefore \int \cos^2 \theta d\theta = \int u . dv = \cos \theta \sin \theta + \int \sin^2 \theta d\theta = \frac{\sin 2\theta}{2} + \int (1 - \cos^2 \theta)\, d\theta$$

i.e. $$2 \int \cos^2 \theta d\theta = \frac{\sin 2\theta}{2} + \theta \text{ hence } \int \cos^2 \theta d\theta = \frac{\sin 2\theta}{4} + \frac{\theta}{2}$$

Example. $\int \sec^3\theta d\theta$
Let $u = \sec\theta$ and $dv = \sec^2\theta d\theta$ then $du = \sec\theta\tan\theta d\theta$ and $v = \tan\theta$

$$\therefore \int \sec^3\theta d\theta = \sec\theta\tan\theta - \int \tan^2\theta\sec\theta d\theta = \sec\theta\tan\theta - \int\sec^3\theta d\theta + \int\sec\theta d\theta$$

i.e.
$$2\int\sec^3\theta d\theta = \sec\theta\tan\theta + \log_e\left\{\tan\left(\frac{\pi}{4}+\frac{\theta}{2}\right)\right\}$$

$$\therefore \int\sec^3\theta d\theta = \tfrac{1}{2}\left[\sec\theta\tan\theta + \log_e\left\{\tan\left(\frac{\pi}{4}+\frac{\theta}{2}\right)\right\}\right]$$

Integration of fractions

The integration of functions consisting of rational algebraic fractions is best carried out by first splitting the function into partial fractions. It is assumed that the numerator is of lower degree than the denominator; if not, this should first be achieved by dividing out. It may be shown that the prime real factors of any polynomial are either quadratic or linear in form. This leads to four distinct types of partial fraction solutions which are now described.

FRACTIONS TYPE 1

The denominator can be factored into real linear factors all different. The partial fractions are then of the form $a/(bx+c)$.

Example. $\displaystyle\int \frac{2x+3}{x^2-4}dx$

Now
$$\frac{2x+3}{x^2-4} = \frac{A}{x-2}+\frac{B}{x+2}$$

i.e.
$$2x+3 = A(x+2)+B(x-2)$$

i.e.
$$A = \frac{7}{4} \text{ and } B = \frac{1}{4}$$

$$\therefore \int\frac{2x+3}{x^2-4}dx = \frac{7}{4}\int\frac{dx}{x-2}+\frac{1}{4}\int\frac{dx}{x+2} = \frac{7}{4}\log_e(x-2)+\frac{1}{4}\log_e(x+2)$$

FRACTIONS TYPE 2

The prime factors of the denominator include quadratic functions and all factors are different. The partial fractions then include expressions of the form $(ax+b)/(cx^2+dx+e)$.

Example. $\displaystyle\int\frac{7x^2-3}{2x^3-3x^2+4x-6}.dx$

Put
$$\frac{7x^2-3}{2x^3-3x^2+4x-6} = \frac{Ax+B}{x^2+2}+\frac{C}{2x-3}$$

i.e.
$$7x^2-3 = (Ax+B)(2x-3)+C(x^2+2)$$
$$= (2A+C)x^2+(2B-3A)x-(3B-2C)$$

and
$$\therefore A = 2, B = 3, C = 3.$$

$$\therefore \int \frac{7x^2-3}{2x^3-3x^2+4x-6} . dx = \int \frac{2x+3}{x^2+2} . dx + \int \frac{3}{2x-3} dx$$

$$= \int \frac{2x}{x^2+2} . dx + \int \frac{3}{x^2+2} . dx + \int \frac{3}{2x-3} . dx$$

$$= \log_e (x^2+2) + \frac{3}{\sqrt{2}} \tan^{-1} \frac{x}{\sqrt{2}} + \frac{3}{2} \log_e (2x-3)$$

FRACTIONS TYPE 3

The denominator can be factored into real linear factors some of which are repeated. The partial fractions then include expressions of the form $a/(bx+c)^n$.

Example. $\displaystyle\int \frac{3x^2+8x+16}{x^3+3x^2-4} dx = \int \frac{f(x)}{F(x)} dx$

$$F(x) = (x-1)(x+2)^2 \text{ and } \frac{f(x)}{F(x)} = \frac{A}{x-1} + \frac{B}{(x+2)} + \frac{C}{(x+2)^2}$$

Hence $\quad 3x^2+8x+16 = A(x+2)^2 + B(x-1)(x+2) + C(x-1)$

putting $x = 1$ then $A = 3$; $x = -2$ then $C = -4$; substitution gives $B = 0$

$$\therefore \int \frac{f(x)}{F(x)} dx = 3 \int \frac{dx}{x-1} - 4 \int \frac{dx}{(x+2)^2} = 3 \log_e (x-1) + \frac{4}{(x+2)}$$

FRACTIONS TYPE 4

The prime factors of the denominator include quadratic functions some of which are repeated. The partial fractions then include expressions of the form $(ax+b)/(cx^2+dx+e)^n$.

Example. $\displaystyle\int \frac{12x-1}{(x^2+1)^2(x+2)} dx = \int \frac{f(x)}{F(x)} dx$

$$\int \frac{f(x)}{F(x)} dx = \frac{Ax+B}{(x^2+1)^2} + \frac{Cx+D}{(x^2+1)} + \frac{E}{(x+2)}$$

i.e. $\quad 12x-1 = (Ax+B)(x+2) + (Cx+D)(x^2+1)(x+2) + E(x^2+1)^2.$

Put $x = -2$; then $E = -1$.

$$\therefore x^4+2x^2+12x = Cx^4 + (D+2C)x^3 + (A+2D+C)x^2 + (2A+B+D+2C)x + 2(B+D)$$

Equating coefficients, we find $C = 1$, $D = -2$, $B = 2$ and $A = 5$.

$$\therefore \int \frac{f(x)}{F(x)} dx = \int \frac{5x+2}{(x^2+1)^2} dx + \int \frac{x-2}{x^2+1} dx - \int \frac{dx}{x+2}$$

$$= \frac{5}{2} \int \frac{d(x^2+1)}{(x^2+1)^2} + 2 \int \frac{dx}{(x^2+1)^2} + \frac{1}{2} \int \frac{d(x^2+1)}{x^2+1} - 2 \int \frac{dx}{x^2+1} - \int \frac{dx}{x+2}$$

$$= -\frac{5}{2} \int \frac{1}{x^2+1} + \frac{x}{x^2+1} + \int \frac{dx}{x^2+1} + \frac{1}{2} \log_e (x^2+1) - 2 \tan^{-1} x - \log_e (x+2)$$

$$= \frac{2x-5}{2(x^2+1)} + \log_e \frac{\sqrt{(x^2+1)}}{x+2} - \tan^{-1} x$$

MATRIX ALGEBRA

A matrix is an array of *mn* numbers in *m* rows and *n* columns

$$\begin{bmatrix} a_{11} & a_{12} & \cdots & a_{1n} \\ a_{21} & a_{22} & \cdots & a_{2n} \\ \vdots & & & \\ a_{m1} & a_{m2} & \cdots & a_{mn} \end{bmatrix}$$

The element in the *i*th row and *j*th column a_{ij} is called the (i, j)h element and the matrix is often denoted by $[a_{ij}]$ or A. When $m = n$ the matrix is square. An $m \times 1$ matrix is called a column rector or column matrix.

$$\begin{bmatrix} & x_1 \\ & x_2 \\ X = & \vdots \\ & x_m \end{bmatrix}$$

A $1 \times n$ matrix is called a row rector

$$Y = [y_1, y_2 \ldots y_n]$$

Addition of matrices

Two matrices may be added if and only if they are of the same order $m \times n$

Then $$A + B = [a_{ij}] + [b_{ij}] = [(a_{ij} + b_{ij})]$$

(i.e. the sum is formed by adding corresponding elements)

Multiplication of matrices

(1) By a scaler

Any matrix may be multiplied by a scalar

Then $$\lambda A = \lambda[a_{ij}] = [(\lambda a_{ij})] = A\lambda$$

(i.e. all the elements of A are multiplied by λ)

(2) Multiplication of two matrices

Two matrices may be multiplied (A times B in that order) only if the number of columns of A is equal to the number of rows of B. If A is $[a_{ij}]$ of order $m \times n$ and B is $[b_{ij}]$ of order $n \times p$ then

$$AB = [a_{ij}][b_{ij}] = \left[\left(\sum_{k=1}^{n} a_{ik}b_{kj}\right)\right]$$

is of order $m \times p$.

It should be noted that, in general $AB \neq BA$.

The unit matrix

The unit matrix is a square matrix I for which

$$a_{ij} = 0 \text{ for } i \neq j$$
$$a_{ij} = 1 \text{ for } i = j$$

The reciprocal of a matrix

The reciprocal matrix A^{-1} of A exists only if the determinant of A is non zero and is given by

$$AA^{-1} = I = A^{-1}A$$

Determinants

The determinant of a square matrix is defined as

$$1A1 = 11a_{ij}11 = \sum(\pm a_{1\alpha}a_{2\beta}\ldots a_{n\nu})$$

the summation of $n!$ terms being over all the arrangements $(\alpha, \beta, \ldots \nu)$ of the column suffixes and the sign $\pm$ being chosen according to whether the arrangement is even or odd.

In the simplest case

$$\begin{vmatrix} a_{11} & a_{12} \\ a_{21} & a_{22} \end{vmatrix} = a_{11} \quad a_{22} - a_{12} \quad a_{21}$$

and from this can be developed the expressions for the expansion of determinants of higher order than the second. The minor of a_{ij} in A is the determinant of the matrix obtained by deleting the ith row and jth column of A. The co-factor A_{ij} of a_{ij} in A is $(-1)^{i+j} \times$ minor of a_{ij}.

Now the expression for a determinant is given by:

$$|A| = a_{11}|Ai1| + a_{i2}|Ai2| \ldots + a_{in}|A_{in}|$$

The value of a determinant is unaltered by interchanging rows with columns. Interchanging either two rows or two columns changes the sign of a determinant. Thus if either two rows or two columns are identical the determinant is zero.

Simultaneous linear equation

Simultaneous linear equations can be arranged in matrix form and their solution obtained via determinants

$$\begin{matrix} a_{11} & x_1 + a_{12} & x_2 + \ldots + a_{in} & x_n = b_1 \\ \vdots & & & \vdots \quad\; \vdots \\ a_{m1} & x_1 + \ldots & + a_{mn} & x_n = b_m \end{matrix}$$

may be written

$$AX = B$$

and now

$$X = A^{-1}B$$

alternatively

$$x_j = \frac{|Dj|}{|D|}$$

where D denotes the determinant $|a_{ij}|$ and Dj denotes the determinant D with the elements $a_{ij}a_{2j}\ldots a_{mj}$ replaced by $b_1 b_2 \ldots bm$.

COMPUTERS

The advent of computers has had a substantial impact on civil engineers as it has on most branches of engineering. It must be realised that computers are basically very simple machines which can add, subtract, multiply and divide and further, take simple decisions about the equality or inequality of two numbers. The reason that a computer can be a useful tool is that it can carry out these simple operations very quickly and reliably according to a previously determined set of instructions (or programme).

Benefit can be obtained in using a computer if the solution of a problem can be broken down into a (probably fairly large) number of simple steps such as those indicated above. Then a sequence of instructions for carrying out these operations may be written to produce the programme. The programme is useful, however, only if it encourages or uses repetition. Repetition may usefully occur in one of two ways given below. In many cases, programmes involve both forms of repetition.

(i) By virtue of the programme, in its entirety being used frequently on jobs which are similar except for the raw input data. Usually an attempt will be made to allow for this by writing the programme in so general a way that it covers as wide a range of similar problems as can conveniently be arranged. Usually the more general the programme the less efficient it is in dealing with any particular problem.

(ii) By virtue of the programme, within itself, to carry out the same sequence of operation (a loop) as may, for instance, happen in iterative solutions of equations when the process may be one of successive approximation until a sufficiently good answer has been obtained.

Definitions of terms

The field of computers, like any other, is beset with jargon and some of the most frequently used works merit description.

Backing store. Computers are arranged to have an amount of main storage capacity to which access is extremely rapid but which, because of its cost, is limited in size. In addition extra storage is provided which is known as the backing store. This is frequently on magnetic tape or discs, which may be exchanged i.e. stored in a cupboard when not needed and put on-line (i.e. connected) to the computer only when needed.

Card reader/punch. The most common means of communicating with computers is by punched cards in which the presence of holes is sensed by optical means. The peripheral on the computer which does this is the card reader. Most computers which use punched cards will also have card punches for copying and amending sets (or decks) of cards.

Central processor. The heart or control of a computer. This is where the currently active programme is stored while being obeyed, and is where the arithmetic and operations decisions are made. Frequently this may physically be linked to the main store

Core. A form of magnetic storage which is by far the most common form of main storage medium.

Discs. A form of backing store using a two dimensional approach to storing information on circular discs of magnetic material. This is faster and more expensive (per unit of stored information) than magnetic tape and is used for random access applications.

Documentation. A vital aspect of computing often sadly neglected. Ideally there should be:

(i) Notes made by the systems analyst on what is required and how his chosen system will achieve it.
(ii) Notes made by the programmer on what he has done to implement the systems analyst's scheme, and notes of tests failed (and corrective action) and passed by the programme.
(iii) The user manual. Generally only the briefest mention of the techniques used should appear here, i.e. only to indicate limitations of use.

Fast store or main store. Generally core store: see backing store.

Graph plotter. A peripheral which can produce drawings in ink on paper.

Hardware. The computer and its peripherals. This may generally cover the machines or instruments which are in some sense hard or tangible.

Input. Information which is fed into the computer. This may be in the form of either a programme or data.

Line printer. A peripheral used for output by means of which lines of symbols (letters, digits, etc.) are printed on pages of paper.

Magnetic tape. A form of backing store using a one dimensional approach to storing information on long narrow tapes of magnetic material. This is the cheapest form of backing store and is most used in serial applications.

Output. Information produced by a programme, often on a line printer.

Paper tape. An input medium alternative to punched cards.

Peripherals. Units of hardware connected to the central processor, e.g. readers (card and tape), punches (card and tape), disc handling units, magnetic tape handling units, line printer, graph plotter, terminals etc.

Programme. A sequence of instructions, obeyed in order by the computer. The order of obeying may be changed by jumping to some point other than the next in sequence by means of branch instructions possibly as a result of decisions made on equalities or inequalities in tests on the data. Programmes may be written in high level languages such as algol, COBOL or FORTRAN where the programmer can get a large amount of work done in one instruction (which may therefore be quite complicated). Programmes may also be written in a low level language which will often be peculiar to one computer (or range of computers).

Programmer. A person who accepts the systems analyst's instruction covering the general scheme of solution of the problem and then writes and tests the programme instructions.

Software. Programmes, data, test data and results etc., i.e. intangibles (see also Hardware).

Store. That part of the computer where the programme and data may be stored. There are main (fast) stores and backing stores.

Systems analyst. A person who, given the definition of a problem, decides upon the method and general techniques of solution. He may also define certain tests which the programmer must make.

Tape reader (or punch). A peripheral for reading or punching paper tape (see Card reader/punch).

Terminal. An item of hardware allowing access to a multiprogramming (i.e. doing several jobs 'at once') computer. In the minimum form it consists of a typewriter keyboard and typewritten output. It may sometimes have a visual display with the keyboard. In extreme cases a terminal may have input and output peripherals linked to a central processor.

Using a computer

Access to a computer can be achieved in a variety of ways. An organisation may have its own in-house computer, usually in a separate department, or may have terminals to a central computer. Alternatively an organisation may use a bureau service which is operated purely as a business concern.

The most significant uses of computers in the civil engineering field are in structural analysis, where the computer's ability to solve simultaneous linear equations is invaluable, and in network analysis for the control of jobs. New applications are constantly being being found where the computer can give benefit.

STATISTICS

INTRODUCTION

Statistical techniques are used in engineering mainly in connection with the quality control of manufacturing of produced material and with the checking for compliance, of such products, with whatever specifications or clauses are contained in the contracts covering the purchasing and selling of such products. In order to exercise quality control or to check for compliance it is necessary to make measurements of one sort or another. Now it is well established that the result of repeating a measurement (or of repeating an experiment) does not generally repeat the observation or original result. Further repeat measurements will lead to further results and so appears the problem of variability.

Generally speaking the variation in results arises both because the subjects of the measurement are themselves different and also because of errors introduced by the experiment or the measuring technique. Such variation is common experience in the measurement of e.g. the strengths of materials. Often it will be desirable (if only from an economic view) to reduce the variation to as small an amount as can conveniently be arranged. However, it is not generally possible to reduce such variation to an unimportantly small value and so it becomes necessary to deal with the problem posed by the obtaining of different results from apparently identical experiments. It is to deal with the evaluation of such scattered experimental results that statistical techniques have been developed.

It is supposed that, were it possible to continue the experiments indefinitely, the results so obtained would cluster around some fixed value which would be the required value. (It is an implicit assumption that the indefinite series of experiments be conducted under identical conditions.) Since it is not possible to conduct indefinitely long experiments the problem becomes that of trying to determine, from a finite series of experiments, that fixed value (which is presumably the true value) about which the indefinite series of results would cluster. This attempt to determine is known as estimating, and while the use of that particular word does not imply that there has been any guesswork in obtaining it, there is an implication of uncertainty about the result. In statistical methods this uncertainty is calculated and specified in terms of confidence limits. Generally a result obtained after statistical calculations should be given in terms of an estimate surrounded by confidence limits. Of course the more nearly certain we wish to be that the confidence limits contain the true value the wider those limits must be. In cases where the experiment or test is not aimed at the estimating of some particular quantity the form of the estimate and confidence limit changes to one that such and such a result would not have arisen 'by chance' more than on so many percent of occasions in an indefnitely long series of trials.

It is important that statistical results should be properly presented in the form of estimate and confidence limits: having decided upon such a form it is then sensible to use an appropriate precision for the reporting of the values. For example, when estimating the strength of concrete where an estimate might be of the form: 42 ± 5 N/mm^2 (at 95% confidence) there is clearly no point in reporting the result to several decimal places.

When an estimate of some quantity has been obtained, the interval between the confidence limits may be wider than it is desired they should be in which case the interval may be narrowed by accepting a lower confidence. If this is not desirable it will be necessary to:

Take more observations, or

Improve the experimental techniques used to reduce the variability of the results.

It is important that the question of what is required by way of precision should be considered prior to an experiment so that the number of observations necessary to obtain the required precision may be assessed. In making that assessment it will be necessary to have information about the variability of parts of the experiment. This information may be available from previous experience, but if not it must be obtained by a pilot experiment.

It will be clear from the foregoing that any result which is obtained, being subject to error, may cause a wrong decision to be taken. Thus when dealing with, for instance, material to be checked for strength the contract for the supply of the material should indicate a test scheme to determine whether the strength of the material is correct or not.

Such a test scheme will involve experiments, and the possibilities for a wrong decision are:

(i) That the test will, wrongly, show as unsatisfactory material with the correct strength. (This is known as the manufacturer's or supplier's risk.)

(ii) That the test, will wrongly, show as satisfactory material with an incorrect strength. (This is known as the consumer's risk.)

The performance of a test scheme is defined by its power and is represented by a graph showing, on one axis, the true value of the parameter in question (e.g. the strength of the material) plotted against the probability that material will pass the test and so be accepted. The calculation of such graphs is not a simple matter and requires full information about all aspects of the test scheme under consideration. The power curves of two test schemes represent, however, the only way in which the performance of the two schemes may be compared.

In the following sections are presented definitions of some of the terms used in statistical work, descriptions of statistical techniques and tests which may be used as a part of the experimenter's armoury of techniques and a description of central charts as a method of quality control. In the final section the references have, in the first cases been selected for their readibility as well as for their coverage of any particular point. Thus references 1 and 2 are especially recommended as initial reading for anyone interested in statistical problems and techniques.

DEFINITIONS OF ELEMENTARY STATISTICAL CONCEPTS

Statistical unit or item

One of a number of similar articles or parts each of which may possess several different quality characteristics.

Examples. A piece of glass tubing taken from a large number produced in quantity for which the diameter and other characteristics may be measured; a concrete cube for which the strength may be measured.

Observation—observed value

The value of a quality characteristic measured or observed on a unit.

Example. The diameter in mm of a piece of tubing; the strength of a concrete cube.

SAMPLE

A portion of material or a group of units taken from a larger number which is used to obtain estimates of the properties of the larger quantity.

Examples. Forty-eight pieces of tubing sampled from all the pieces produced during a day; the concrete cube made from a batch of concrete.

RANDOM SAMPLE

A sample selected in such a manner that every item has an equal chance of inclusion.

REPRESENTATIVE SAMPLE

A sample whose selection requires planned action to ensure that proportions of it are taken from different sub-portions of the whole.

Examples. The forty-eight pieces of tubing selected two from every hour's production in one day; concrete cubes made, one from every batch, of a lot consisting of several batches.

POPULATION

A large collection of individual units from one source. In particular circumstances this may be:

(a) Output or batch: the bulk of material (concrete) or total collection of units (pieces of tube) produced by a set of machines or a factory in a specified time.

Examples. Pieces of tubing made in a particular factory during a month; the concrete produced by a single plant during one day.

STATISTIC

A statistic is a quantity computed from the observations of a sample.

PARAMETER

A parameter is a quantity computed from the observations made on a sample. Thus the value of a parameter for a population is estimated by the appropriate statistic for the sample.

MEASURES OF LOCATION

ARITHMETIC MEAN

The arithmetic mean, often called the 'mean' or the 'average', is the sum of all the observations divided by the number of observations.

$$\bar{x} = \frac{1}{n} \sum_{i=1}^{n} x_i$$

Example: $\frac{1}{5}(2.540 + 2.538 + 2.547 + 2.544 + 2.541) = 2.542.$

MEDIAN

The value which is greater than one half of the values and less than one half of the values.

Example. The value 2.541 is the median of the above five numbers. (Had there been an even rather than an odd number of numbers the median is the average of the two numbers either side of the median position).

MIDPOINT OR MIDRANGE

The value which lies half way between the extreme values.

Example. Using the numbers above the mid point is

$$\tfrac{1}{2}(2.538 + 2.547) = 2.542\,5$$

MEASURES OF DISPERSION

RANGE

The difference between the largest and the smallest values.

Example. $2.547 - 2.538 = 0.009.$

DEVIATION

The difference between a value and the mean of all the values.

VARIANCE

The variance of a set of values is the mean squared deviation of the individual values and is normally represented by σ^2.

$$\sigma^2 = \frac{1}{n} \sum_{i=1}^{n} (x_i - \mu)^2$$

where μ is the mean value.

A frequently occurring problem is that of estimating the main properties (the mean to describe the location and the variance to describe the dispersion) of a population by measurements (x_i) taken on a sample. From the measurements on the sample we can calculate the sample mean, $\bar{x}$ which is an estimate of the population mean μ. The sum of the squared deviations is smallest about the arithmetic mean; thus for the population an estimate of variance using the sample mean and sample variance will be an under-estimate. In cases where we wish to estimate population parameters from sample observations a correction is made by using $(n-1)$ as divisor instead of n. Thus the estimate of the population variance from observations x_i on a sample is:

$$\sigma^2 = \frac{1}{n-1} \sum_{i=1}^{n} (x_i - \bar{x})^2$$

STANDARD DEVIATION

The standard deviation is the square root of the variance.

$$s = \left[\frac{1}{n-1} \sum_{i=1}^{n} (x_i - \bar{x})^2 \right]^{\frac{1}{2}}$$

As in the case of variance the divisor n is replaced by $(n-1)$ when working with sample observations to estimate a population standard deviation. The standard deviation has the same units as the original observations and their mean, $\bar{x}$.

When carrying out hand calculations the identity:

$$\sum_{i=1}^{n} (x_i - \bar{x})^2 = \sum_{i=1}^{n} (x_i)^2 - n\bar{x}^2$$

frequently saves effort. However, this method is not recommended for use on computers because of the danger of loss of accuracy when n is large and x_i have several significant figures.

COEFFICIENT OF VARIATION

The coefficient of variation is the standard deviation expressed as a percentage of the mean. This is useful for dealing with properties whose standard deviation rises in proportion to the mean, for instance the strengths of concrete as measured by compressive tests on cubes.

STANDARD ERROR

The standard error is the standard deviation of the mean (or of any other statistic). If in repeated samples of size n from a population the sample means are calculated the standard deviation calculated from these means is expected to have a value

$$Sm = \sigma/\sqrt{n}$$

where σ is the standard deviation of the population. An important result is that whatever the distribution of the parent population (Normal or not) the distribution of the sample mean tends rapidly to Normal form as the sample size increases.

REPRESENTATIONS OF SAMPLE AND POPULATION

FREQUENCY

The number of observations having values between two specified limits. It is often convenient to group observations by dividing the range over which they extend into a number of small, equal, intervals. The number of observations falling in each interval is then the frequency for that interval. This allows a convenient representation of the information by means of a histogram.

HISTOGRAM OR BAR CHART

A diagram in which the observations are represented by rectangles or bars with one side equal to the interval over which the observations occurred and the other equal to the frequency of occurrence of observations within that range. (Figure 1.45.)

DISTRIBUTION CURVE

The result of refining a histogram by reducing the size of the intervals and correspondingly increasing the total number of observations. In the limit, when the intervals become infinitesimally small and the number of observations infinitely large the tops of the rectangles of a histogram become a distribution curve. (Figure 1.46.)

NORMAL DISTRIBUTION (OR GAUSSIAN DISTRIBUTION)

A particular type of distribution curve given by:

$$y(x) = \frac{1}{\sigma(2\pi)^{\frac{1}{2}}} \exp\left\{\frac{-\frac{1}{2}(x-\mu)^2}{\sigma^2}\right\}$$

where x is the observational scale value, μ the population mean and σ the population standard deviation. These parameters of the distribution are estimated by the sample mean $\bar{x}$, and standard deviation, s.

It has been found that a great many frequency distribution met with in practice fit quite closely to the Normal distribution. However, one should beware of thinking that there is any law which says this shall be so; it is simply a matter of experience. In circum-

Table 1.5 THE NORMAL DISTRIBUTION

Range	*% of observations within range*
$\mu \pm \sigma$	68.27
$\mu \pm 2\sigma$	95.45
$\mu \pm 3\sigma$	99.73
$\mu \pm 1.96\sigma$	95
$\mu \pm 3.09\sigma$	99.8

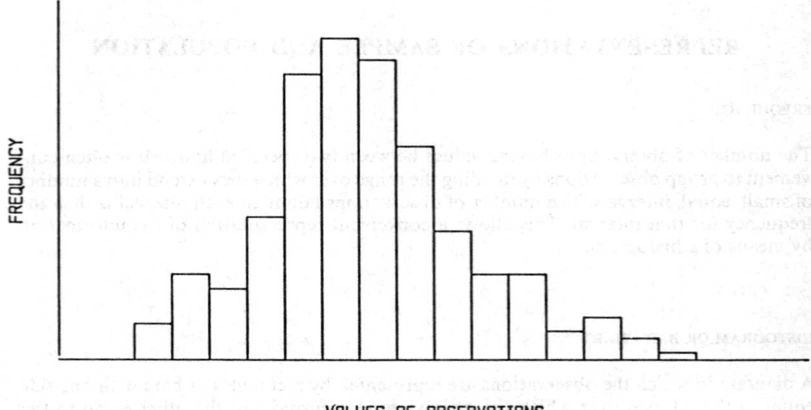

Figure 1.45 A histogram of observations from a sample

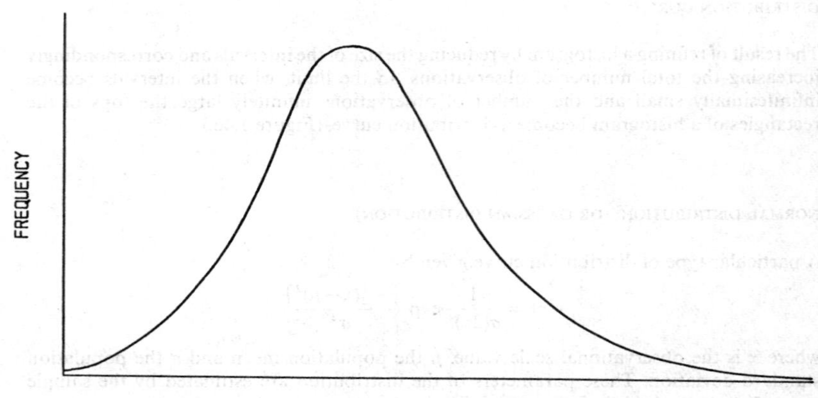

Figure 1.46 A continuous distribution curve

stances where the observed frequency distribution does not appear to be Normal it is often possible to transform the original data (e.g. by taking logarithms, square roots or squares) so that the transformed data is nearly Normal. These two facts explain why so much of the effort in statistical theory has been devoted to treatment of Normal problems.

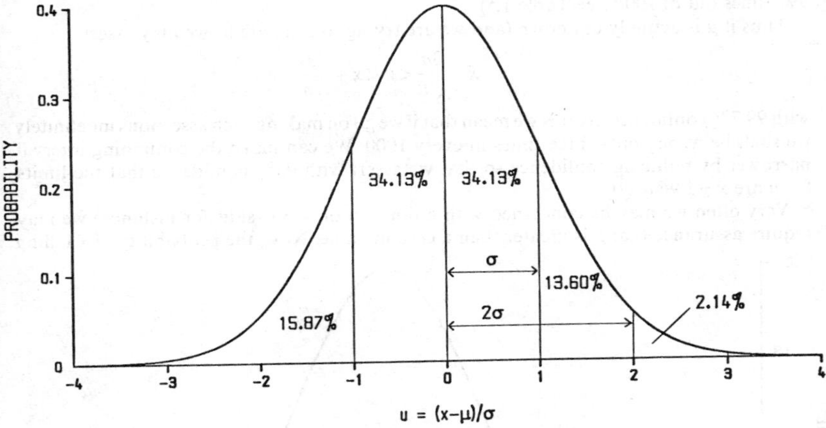

Figure 1.47 The normal distribution with limits

For Normal distributions the percentage of observations (in large samples) lying within certain limits of the observational scale are given in Table 1.5 and Figure 1.47.

THE USE OF STATISTICS IN INDUSTRIAL EXPERIMENTATION

As has been stated, in experimental work units in a sample drawn from a parent population, and the observations made on them, are subject to error and our task for which we use statistics is to make useful statements about the properties of the parent population. To achieve this the most important statistics are the mean and the standard deviation. This section, therefore, considers the obtaining of sample means and standard deviations and confidence limits for them in situations where the parent population is Normally distributed. Tests of significance for comparisons of means and variances are also described. Inevitably only brief summaries are given and a study of the references is advised before using the techniques on any important matters. As an alternative, the help of the statistical expert should be sought. If such assistance is to be obtained it cannot be emphasised too strongly that it should be obtained right at the outset of the problem. It is rarely of much help to anyone (even though it happens only too frequently) for the statistician to be asked: 'Please tell me what these numbers show: they must mean something, I've collected so many, and they cost a great deal to obtain'.

Confidence limits for a mean value

If the form of the distribution were known together with the true mean (μ) and the standard deviation (σ), then it is easy to make statements about the mean of a number of observations. If the population is Normal then the mean ($\bar{x}$) of a sample size n drawn randomly will, on average, satisfy

$$\mu-\frac{3\sigma}{\sqrt{n}}<\bar{x}<\mu+\frac{3\sigma}{\sqrt{n}}$$

997 times out of 1000 (see Table 1.5).

Thus if μ is actually unknown (and we are trying to estimate it) we may assert

$$\bar{x}-\frac{3\sigma}{\sqrt{n}}<\mu<\bar{x}+\frac{3\sigma}{\sqrt{n}}$$

with 99.7% confidence. By this we mean that if we go on making such assertions indefinitely we shall be wrong only three times in every 1000. We can make the containing interval narrower by reducing confidence so that we assert with 95% confidence that the limits for μ are $\bar{x}\pm1.96\,\sigma/\sqrt{n}$.

Very often we may be concerned with a limit on only one side, for instances we may require assurance that μ is greater than a certain value. Now, the probability of $\bar{x}$ falling

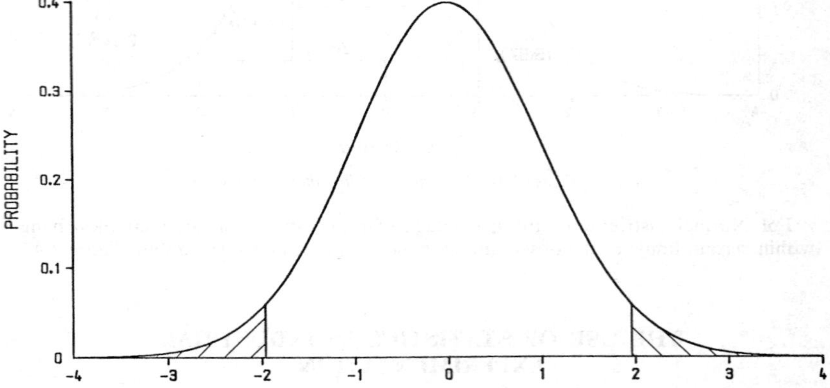

Figure 1.48 Diagrammatic representation of confidence limits (with 95% limits shown)

above $\mu+2\sigma/\sqrt{n}$ is 2.27%. Thus we may assert with 97.73% confidence that μ does not lie below $\bar{x}-2\sigma/\sqrt{n}$.

Generally the proportion of sample means, $\bar{x}$, which exceed $\mu+u_\alpha\sigma/\sqrt{n}$ is equal to α where u_α is the value given in a table of the Normal distribution for a specified probability, say P. Because the distribution is symmetrical, α is also the proportion of sample means which are exceeded by $\mu-u_\alpha\sigma/\sqrt{n}$. Thus the whole range of values which μ may take is divided into three parts and three assertions cak be made, to correspond one with each part:

1. $\mu\geqslant\bar{x}-u_\alpha\sigma/\sqrt{n}$ with confidence $100(1-\alpha)\%$
2. $\mu\leqslant\bar{x}+u_\alpha\sigma/\sqrt{n}$ with confidence $100(1-\alpha)\%$
3. $\bar{x}-u_\alpha\sigma/\sqrt{n}\surd\leqslant\mu\leqslant\bar{x}+u_\alpha\sigma/\sqrt{n}$ with confidence $100(1-2\alpha)\%$

This shows two sorts of statement, the single sided (cases 1 and 2) and the double sided (case 3). When using statistical tables it is important to check whether the tabulation is for single tailed testing or two tailed testing. (This description arises because cases 1 and 2 are, in the practical cases where a useful level of confidence is being used, representable as the two tails of a curve shaped like the Normal distribution curve.)

In the discussion of confidence limits for the mean value, μ, of a population estimated by the mean of the sample $\bar{x}$, above it was assumed that the population standard deviation, was known. Generally this will not be the case and μ will have to be estimated, as s, a sample standard deviation and use s in place of μ in the calculations above. The confidence limits for μ are now wider because of the uncertainty about s and instead of using u from the Normal distribution curve it becomes necessary to use tables of Student's t. The particular value of t to be used depends on how good the estimate s of σ is, which in turn depends upon the number of degrees of freedom in making the estimate. In the case of a standard deviation of n observations the number of degrees of freedom is $(n-1)$. The number of degrees of freedom is generally denoted by ϕ. Some values of t are given in Table 1.6. In using this table the $100(1-2\alpha)\%$ confidence limits are:

Lower limit $\quad \bar{x}-t_\alpha s/\sqrt{n}$
Upper limit $\quad \bar{x}+t_\alpha s/\sqrt{n}$

using the value of t_α for the appropriate number of degrees of freedom.

The difference between two mean values

A problem which arises frequently is that of determining if the difference between two means has occurred by chance because of natural variation or whether there is a real difference. A real difference can only be asserted in the form of a statistical statement that the difference is significant at a certain level, that is to say there is a probability that there is a real difference. This is done by calculating a t statistic from information about the samples and comparing the result with the tabulated t values. The means, standard deviations and number of observations of the two tests are denoted by $\bar{x}_1$, $\bar{x}_2$, s_1, s_2, n_1 and n_2

Calculate
$$t = (\bar{x}_1-\bar{x}_2)/sp$$
where
$$sp = \sqrt{\left[\left(\frac{1}{n_1}+\frac{1}{n_2}\right)\left(\frac{v_1 s_1^2+v_2 s_2^2}{v_1+v_2}\right)\right]}$$
and
$$v_1 = n_1-1, \quad v_2 = n_2-1$$
(Note: sp is the pooled standard deviation for the samples 1 and 2.)

If this calculated value exceeds a tabulated value of t (for $\phi = v_1+v_2$) then the difference is significant at the level determined by the probability heading the column of the t Table.

As an example, consider the comparison of two testing machines for crushing concrete cubes. The machines are to be compared by making a single batch of 12 concrete cubes and testing six cubes on each machine. The results obtained are:

Machine 1	39.2	38.4	44.7	41.0	41.0	44.1
Machine 2	41.1	33.8	42.4	36.8	32.0	40.1

From these observations we can calculate

Sample sizes $\quad n_1 = 6, \quad n_2 = 6$
Sample means $\quad \bar{x}_1 = 41.4, \quad \bar{x} = 37.7$
Sample standard deviations $\quad s_1 = 2.54, \quad s_2 = 4.19$

$$sp = \sqrt{\left\{\left(\frac{1}{6}+\frac{1}{6}\right)\left(\frac{5\times 2.54^2+5\times 4.19^2}{5+5}\right)\right\}} = 2.0$$

so
$$t = (41.4-37.7)/2.0 = 1.85$$

The number of degrees of freedom $\phi = v_1+v_2 = 10$.

From Table 1.6 it is seen that for $\phi = 10$ the single sided 2.5% (or 0.025) point of the t distribution is 2.23 and so the calculated t value is not significant at the $2\times 2.5\%$ level.

Table 1.6 SIGNIFICANCE POINTS OF THE t DISTRIBUTION (SINGLE SIDED)

ϕ	Probability: P				
	0.1	0.05	0.025	0.01	0.005
1	3.08	6.31	12.70	31.80	63.70
2	1.89	2.92	4.30	6.96	9.92
5	1.48	2.01	2.57	3.36	4.03
10	1.37	1.81	2.23	2.76	3.17
20	1.32	1.72	2.09	2.53	2.85
40	1.30	1.68	2.02	2.42	2.70
∞	1.28	1.64	1.96	2.33	2.58

(Note that it is necessary to double the probability value from the table because the question: 'Are the testing machines different?' requires a two sided test to be carried out. By contrast the question: 'Is mean 1 greater than mean 2?' would require a single sided test.)

The ratio between two standard deviations

In a similar way in which it may be desired to compare two means it may be desired to compare two standard deviations. Whereas means are compared by calculating their difference standard deviations are compared by calculating the ratio of the variances (the square of the standard deviation) and comparing the ratio with tabulated values in an F test. In all such calculations the value obtained for the ratio must be greater than unity so that the larger standard deviation (say s_1) must be placed over the smaller (s_2) where s_1 and s_2 are sample standard deviations and so estimates of the population standard deviations. Since we have of necessity $s_1 \geqslant s_2$ when we calculate

$$F = \frac{s_1^2}{s_2^2}$$

the F test is a one sided test.

Values of F for comparison with the calculated value from the observed standard deviations are given in most statistical books.[2] Such tables are presented generally with one table for each specified probability and within such a single table the column headings are the values of v_1 (the number of degrees of freedom, n_1-1, of the larger standard deviation estimate s_1) while the row headings are the values of v_2 (the number of degrees of freedom, n_2-1, of the smaller standard deviation estimate s_2).

By way of illustration consider the example used above for the comparison of means. In that example the observations lead to:

$$n_1 = n_2 = 6$$

So

$$v_1 = v_2 = 5$$

$$s_1 = 2.54 \quad s_2 = 4.19$$

In this example $s_2 > s_1$ so the calculation of F is:

$$F = \left(\frac{4.19}{2.54}\right)^2 = 2.72$$

In the tables (e.g. in reference 2) the tabulated 1% confidence point of F is 10.97 and the 5% point is 5.05 both found for $v_1 = v_2 = 5$. Since the calculated F ratio is not greater than the tabulated values the conclusion to be drawn is that there is not strong evidence that population standard deviations are different.

Analysis of variance

If a manufacturing process or a testing scheme involves a number of independent factors, each of which contributes to the variability of the results, then the variance of the whole system is equal to the sum of the component variances. (Note the variance must be added, not the standard deviation.) This additive property permits the technique of analysis of variance which can take many forms depending on the structure of the process which is being analysed. One of the major difficulties of analysis of variance lies in deciding what form of structure is appropriate to the process being modelled by the analysis of variance. In the majority of cases which are not both simple and short it will be sensible for the arithmetic to be performed by computer. However, in simple and short situations the calculations may reasonably be undertaken by hand.

Probably the most commonly occurring simple situation is that of analysis to determine variance between and within batches. The methods are best described by an example. The example will be one in which concrete cubes are made batches (each of three cubes) and strength tested at (say) 28 days. The first step is to define the statistical model which is being used:

$$Y_{ij} = Y + A_i + E_{ij}$$

where there are I batches each of J cubes,

Y_{ij} is the observed strength of the jth cube in the ith batch,
Y is the average strength (averaged over all tests)
A_i is the difference between Y and the average strength of batch i.
E_{ij} is the difference between the jth cube of bath i and the average strength $Y + A_i$ of that batch.

If the data for four batches of cubes is:

19.8	21.1	19.8	batch 1
21.8	22.0	21.0	batch 2
21.2	21.5	21.2	batch 3
21.4	21.4	21.0	batch 4

it is found that $Y = 21.1$.

From this can now be found the sums of squares of the A_i and E_{ij}. Associated with each sum of squares is a number of degrees of freedom (as usual one less than the number of occurrences) so that dividing the sums of squares by the appropriate number of degrees of freedom gives the mean square. Thus is constructed an analysis of variance table:

Model term	*Sum of squares*	*Degrees of freedom*	*Mean square*
A_i	3.2	3	1.07
E_{ij}	1.9	8	0.23

The method of test is by F ratio so that the larger variance (the average of the sums of squares of errors) is divided by the smaller. Here, 1.07/0.23 = 4.61 with 3 degrees of freedom for the column heading and 8 for the row heading when comparing with the tabulated F values. For $v_1 = 3$, $v_2 = 8$ the tabulated F value at 1% confidence level is 7.59. The observed value does not exceed this and so there is no assertion that can be made at the 1% level. However the tabulated F value at the 5% confidence level is 4.07. The observed value exceeds this and so a result significant at the 5% level has been obtained. Thus although there is not strong evidence there is some evidence of a real difference between batches.

In an example so small as this one the necessary arithmetic (especially if properly organised) may reasonably be tackled by hand. However, as can be deduced from examination of F tables it is not always easy to get significant results with small experiments. Thus the use of the technique will in many cases imply the use of a computer for handling the arithmetic. In such circumstances the engineer is likely to be using an existing computer program and need only concern himself with correctly presenting the data for the program to analyse and then with the interpretation of results and comparisons with tabulated F values. He has no need therefore to develop great skills in short cut arithmetic methods.

Straight line fitting and regression

Experiments may be designed to examine whether two parameters are related. The circumstances may involve the effect on a property of a product of some parameter in the production process. In the experiment the parameter will be controlled or constrained to take a number (n) of prescribed values (x_i) over some range and the consequential observations (y_i) will be paired with them. The question now arises as to the 'best straight line' through the points (x_i, y_i). It is assumed that the x_i values are error free but that the observations y_i are subject to error. The method of obtaining the 'best' straight line in such circumstances is to choose the two parameters m and c of the straight line:

$$y = mx + c$$

in such a way that the sum of the squares of the errors in the y direction is a minimum. This is achieved by making

$$m = \frac{n \Sigma xy - \Sigma x \Sigma y}{n \Sigma x^2 - (\Sigma x)^2}$$

and

$$c = \frac{\Sigma x^2 \Sigma y - \Sigma x \Sigma xy}{n \Sigma x^2 - (\Sigma x)^2}$$

This line is called the line of regression of y on x and one of its properties is that it passes through the centroid ($\bar{\bar{x}}$, $\bar{y}$) of the observed points. The usual statistical question now arises concerning the confidence limits which should be applied to the calculated line which is an estimate of a relationship. To examine this problem the errors or deviations must be calculated. At every observation point (x_1 y_i) which does not actually lie on the calculated line there is an e_i. The variance of y estimated by the regression line is then

$$s_y^2 = \frac{\Sigma e_i^2}{v}$$

where v is the number of degrees of freedom.

Since calculation of m and c impose two restraints the value of v is given by

$$v = n - 2$$

The variance of the mean value, $\bar{y}$ is given by

$$s_{\bar{y}}^2 = \frac{s_y^2}{n}$$

so that the confidence limits for $\bar{y}$ are

$$\bar{y} \pm t s_{\bar{y}}$$

Where, just as for a sample mean the value of t is found from tables using the appropriate number of degrees of freedom.

The variance of the slope m is given by:

$$s_m^2 = \frac{s_y^2}{\Sigma(x-\bar{x})^2}$$

and the confidence band for slope is given by:

$$m \pm ts_m$$

It may be necessary to compare one regression line with another, theoretical one, to see if there is any significant difference between the theoretical slope, m_0, and the observed slope m. This test is performed by calculating a t statistic:

$$t = \frac{m - m_0}{s_b}$$

and comparing with the tabulated values. Just as in the case of comparison of means of samples we can compare the slopes of two observed lines by replacing $m - m_0$ by $m_1 - m_2$ in the equation for t above and using a pooled standard deviation from the variances of the slopes of both lines in place of s_b. The number of degrees of freedom used in the t table will be $n_1 + n_2 - 4$.

TOLERANCE AND QUALITY CONTROL

Material is often manufactured for supply according to a specification which will include compliance clauses for the performance of the product. As an example the Code of Practice for Structural Concrete, CP 110[6] lays down (in section 6.8) certain strength requirements and also suggests a testing plan. The Handbook to that Code[7] discusses the problems of compliance and shows how different forms of testing plan after the operating characteristic of a test plan and so charge the risks run by the producer and by the customer. The customer has, in theory, the opportunity of reducing his risk by adopting a more vigorous testing plan. This, however, is likely to cost more and a customer may well deem this to be not worthwhile. The producer, on the other hand, must expect to have to meet the compliance causes and needs to arrange his production methods so as to make a profit taking account of whatever limits or penalties may be imposed on him by the compliance clauses under which he has to operate. Thus the manufacturer or producer is faced with a problem of how to control his product.

One example of a technique for exercising this control is shown by a system advocated for controlling the strength of ready-mixed concrete[8] by means of the cumulative sum chart which is an improved form of control chart especially developed and adapted to the problems of concrete manufacture.

In the process of manufacture and measurement of some property of the product natural viation will cause the results obtained to be distributed in some way. The problems facing the manufacturer are:

(i) To maintain adequate control over the process so that the variation in results does not become so large that an uneconomic number fall outside the specified tolerances.

(ii) To detect sufficiently early to take useful corrective action any trend for the observations obtained to be moving out of the specified limits.

As usual, samples are taken to estimate the properties of the parent population. To do this comparatively, many samples (25 or more) of comparatively small (but not less than about four and all the same) size are tested and the mean of the means $\bar{\bar{x}}$ used to estimate the population mean. The population standard deviation is estimated from the variance within samples, the average sample standard deviation from the average sample range.

Thus

$$\frac{\bar{\bar{x}} = \bar{x}_1 + \bar{x}_2 \ldots + \bar{x}_k}{k}$$

$$\bar{x} = \frac{s}{\sqrt{n}}$$

for k samples of size n.

Now a chart is drawn with time or sample number in the horizontal axis and observation values on the vertical axis. A line drawn at $\bar{\bar{x}}$ represents the target performance of the process and two surrounding lines at $\bar{\bar{x}} \pm 1.96/(\sqrt{n})s$ represent warning levels for the process while surrounding lines at $\bar{\bar{x}} \pm 3.09/(\sqrt{n})s$ can be regarded as action levels.

The choice of the figures 1.96 and 3.09 have been made on the assumption that the process is functioning in such a way that the specified tolerance limits are reasonable, that is to say they are not so stringent that the chance of the product meeting the requirements is not high while on the other hand the process is not so 'good' (in which case it may be unnecessarily expensive) that all the results obtained lie well within limits.

The design and use of control charts is a valuable use of statistical methods. Generally they are robust in the sense that their usefulness is little affected by factors such as non-Normality of the basic data. However for their efficient use in some area experience of the particular technology is desirable and for a better understanding of the possibilities of the techniques the reader is recommended to references 5 and 8.

BIBLOGRAPHY

The following selection of references is not intended to be exhaustive. The literature of mathematics is vast and that relating to engineers is especially large. Thus the reader should in the first instance turn to works with which he is already familiar as being the quickest way to find the answer to a problem. The list below includes books either of a wide range and general nature and intended for engineers or covering subject matter of recent development, and also some old books on specialist subjects.

BATTERSBY, A., *Network Analysis,* 2nd ed., Macmillan, London (1967)
DOUGLAS, A. H. and TURNER, F. H., *Engineering Mathematics,* Concrete Publications (1964)
HALL, H. S. and KNIGHT, S. R., *Higher Algebra,* 4th ed., Macmillan, London (1892)
KREYSZIG, E., *Advanced Engineering Mathematics,* Wiley, New York (1972)
LAMB, H., *An elementary course of infinitesimal calculus,* 3rd ed., Cambridge (1956)
LONEY, S. L., *The elements of co-ordinate geometry,* Macmillan, London (1922)
MORICE, P. B., *Linear structural analysis,* Thames & Hudson, London (1959)
PIAGGIO, H., *An elementary treatise on differential equations,* Bell, London (1950)
VINE-LOTT, K. M. (Editor), *Computers in civil engineering design,* National Computing Centre, Manchester (1972)

REFERENCES

1. MORONEY, M. J., *Facts from Figures,* Penguin Books, Harmondsworth, 2nd ed. (1953)
2. NEVILLE, A. M. and KENNEDY, J. B., *Basic Statistical Methods for Engineers and Scientists,* International Textbook Company, Scranton (1964)
3. BROWNLEE, K. A., *Industrial Experimentation,* HMSO, London, 4th ed. (1957)
4. COOPER, B. E., *Statistics for Experimentalists,* Pergamon, Oxford (1969)
5. DAVIES, O. L. and GOLDSMITH, P. L., *Statistical Methods in Research and Production,* Oliver and Boyd, Edinburgh, 4th ed. (1972)
6. BRITISH STANDARDS INSTITUTION, Code of Practice for the Structural use of Concrete, CP 110 Pt 1, BSI, London (1972)
7. CEMENT AND CONCRETE ASSOCIATION, *Handbook on the Unified Code for Structural Concrete,* C & CA, London (1972)
8. BRITISH READY MIXED CONCRETE ASSOCIATION, *Authorisation Scheme for Ready Mixed Concrete, BRMCA, Ashford,* 2nd ed. (1972)

2 STRENGTH OF MATERIALS

STRENGTH OF MATERIALS 2

2 STRENGTH OF MATERIALS

T. R. GRAVES SMITH, M.A., Ph.D., C.Eng., M.I.C.E.
Department of Civil Engineering, The University of Southampton

INTRODUCTION

The subject 'Strength of Materials', originates from the earliest attempts to account for the behaviour of structures under load. Thus the problems of particular interest to the first investigators, Galileo and Hooke in the seventeenth century, and Euler and Coulomb in the eighteenth,[1] were the very practical problems associated with the behaviour of beams and columns, whilst at a somewhat later stage, general mathematical investigations of the behaviour of elastic bodies were made by Navier (1821) and Cauchy (1822). The theory of structures has subsequently developed so that it now includes many different and sophisticated fields of interest. Nevertheless, the topic 'Strength of Materials' traditionally covers those aspects of the theory that were the subject of the original research: the theory of bars and the general theory of elasticity. This section, therefore, is essentially a review of the main features of these two somewhat disparate theories, and contains some of the results that are of immediate importance to civil engineers.

THEORY OF ELASTICITY

Internal Stress

Internal stress is the name given to the intensity of the internal forces set up within a body subject to loading. Consider such a body shown in Figure 2.1(a) and an imaginary plane surface within the body passing through a point P. The internal forces exerted between atoms across this surface are represented in the expanded view of Figure 2.1(b). They are described by stress vectors (having the dimensions of force per unit area), and the particular vectors at P give a measure of the intensity of the internal forces at this point. They are denoted by $\boldsymbol{\sigma}$ and called *internal stress vectors*. If they are directed away from the material as in Figure 2.1(c) they are called *tensile*, and if towards the material, *compressive*.

COMPONENTS OF STRESS

The complete state of stress at P is defined in terms of the internal stress vectors acting on three particular surfaces at P called the *positive coordinate surfaces*. (The positive x coordinate surface is the surface parallel to the y–z plane of an x, y, z coordinate system, with the material situated so that a vector directed outwards from the material and normal to the surface is in the positive direction of the x coordinate line as in Figure 2.2.)

These internal stress vectors are distinguished by appropriate subscripts. Thus $\boldsymbol{\sigma}_x$ acts on the positive x coordinate surface, whilst $\boldsymbol{\sigma}_y$ and $\boldsymbol{\sigma}_z$ respectively act on the y and z surfaces. Their scalar components* are then denoted by two subscripts. Thus the com-

*A vector $\boldsymbol{F}$ at P is equal to $F_x\boldsymbol{i}_x + F_y\boldsymbol{i}_y + F_z\boldsymbol{i}_z$, where F_x, F_y and F_z are the scalar components of $\boldsymbol{F}$, and $\boldsymbol{i}_x$, $\boldsymbol{i}_y$ and $\boldsymbol{i}_z$ are unit base vectors parallel respectively to the x, y and z coordinate lines at P.

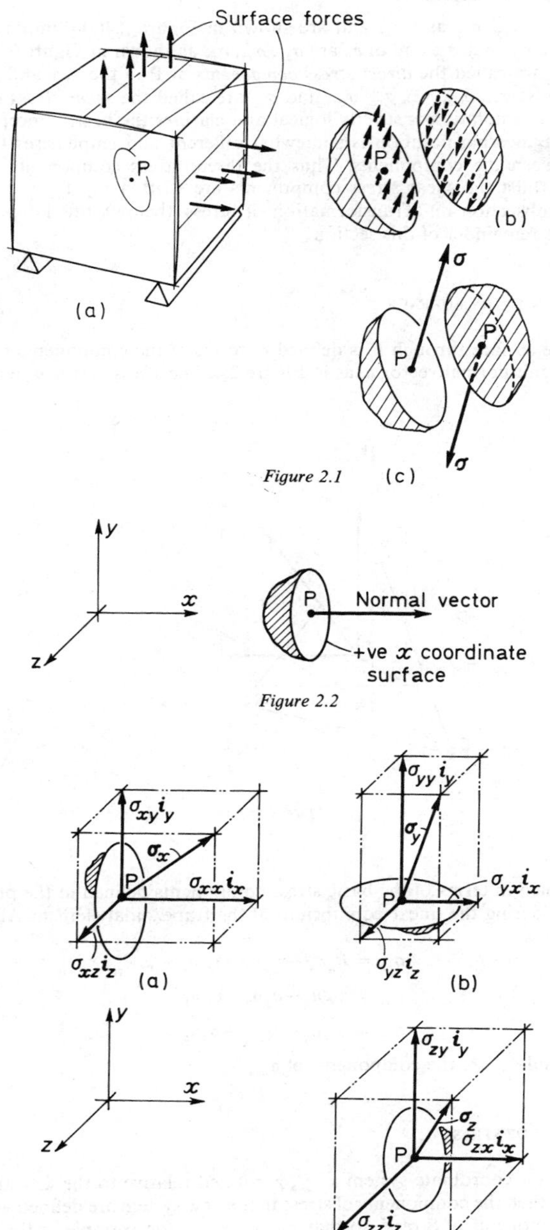

Figure 2.1

Figure 2.2

Figure 2.3

ponents of $\boldsymbol{\sigma}_x$ are σ_{xx}, σ_{xy} and σ_{xz} and are shown in Figure 2.3(a). Similarly the components of $\boldsymbol{\sigma}_y$ are σ_{yx}, σ_{yy}, σ_{yz} and of $\boldsymbol{\sigma}_z$ are σ_{zx}, σ_{zy}, σ_{zz} as shown in Figure 2.3(b) and (c). σ_{xx}, σ_{yy} and σ_{zz} are called the *direct stress components* at P in the x, y and z directions respectively, whilst σ_{xy}, σ_{xz}, σ_{yx}, σ_{yz}, σ_{zx} and σ_{zy} are called the *shear stress components.*

Whilst the above notation is strictly logical and clarifies the basic concepts of stress, conventional engineering notation is somewhat different and emphasises the physical differences between the components. Thus the direct stress components are written σ_x, σ_y and σ_z, whilst the shear stress components are written τ_{xy}, τ_{xz}, τ_{yx}, τ_{yz}, τ_{zx}, τ_{zy}. Except in the subsection on 'Transformation of Stress' (below), this latter notation is employed in the remainder of this section.

STRESS ON AN ARBITRARY SURFACE

Suppose a plane surface through P is defined in terms of the components n_x, n_y and n_z of the outward unit normal vector $\boldsymbol{n}$, as in Figure 2.4. The stress vector $\boldsymbol{\sigma}_n$ acting on this

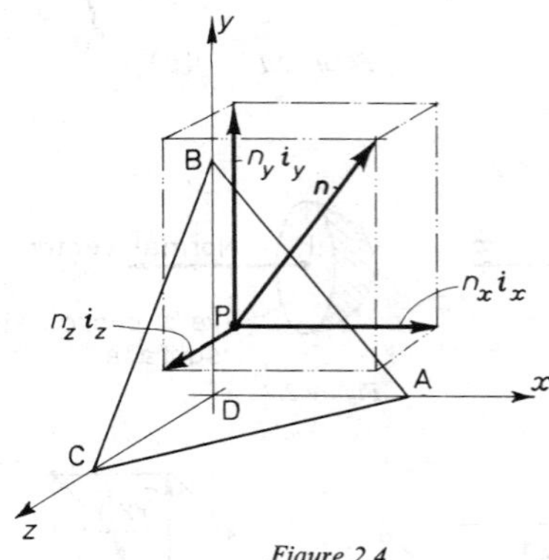

Figure 2.4

surface is obtained in terms of the basic stress components defined in the previous subsection by considering the linear equilibrium of the trapezoidal element ABCD shown in the figure. Thus

$$\sigma_{nx} = \sigma_x n_x + \tau_{yx} n_y + \tau_{zx} n_z \tag{1}$$

$$\sigma_{ny} = \tau_{xy} n_x + \sigma_y n_y + \tau_{zy} n_z \tag{2}$$

$$\sigma_{nz} = \tau_{xz} n_x + \tau_{yz} n_y + \sigma_z n_z \tag{3}$$

where σ_{nx}, σ_{ny} and σ_{nz} are the components of $\boldsymbol{\sigma}_n$.

TRANSFORMATION OF STRESS

Considering a new coordinate system x', y', z' rotated relative to the x, y and z system as in Figure 2.5, then the components of stress in the new system are defined as in the subsection on 'Components of Stress', so that $\tau_{x'y'}$ ($= \sigma_{x'y'}$), for example, is the component in the y' direction of the stress vector acting on the positive x' coordinate surface. The components of stress in the two systems are related by equations of the following type (where for conciseness we employ the original notation of the subsection on 'Components of Stress'):

$$\sigma_{x'y'} = \frac{\partial x}{\partial x'}\frac{\partial x}{\partial y'}\sigma_{xx} + \frac{\partial x}{\partial x'}\frac{\partial y}{\partial y'}\sigma_{xy} + \frac{\partial x}{\partial x'}\frac{\partial z}{\partial y'}\sigma_{xz}$$

$$+\frac{\partial y}{\partial x'}\frac{\partial x}{\partial y'}\sigma_{yx} + \frac{\partial y}{\partial x'}\frac{\partial y}{\partial y'}\sigma_{yy} + \frac{\partial y}{\partial x'}\frac{\partial z}{\partial y'}\sigma_{yz}$$

$$+\frac{\partial z}{\partial x'}\frac{\partial x}{\partial y'}\sigma_{zx} + \frac{\partial z}{\partial x'}\frac{\partial y}{\partial y'}\sigma_{zy} + \frac{\partial z}{\partial x'}\frac{\partial z}{\partial y'}\sigma_{zz} \tag{4}$$

Equation (4) and eight similar equations formed by permuting x', y' and z' are called the *transformation equations of stress.* The partial derivatives in (4) are called direction cosines,

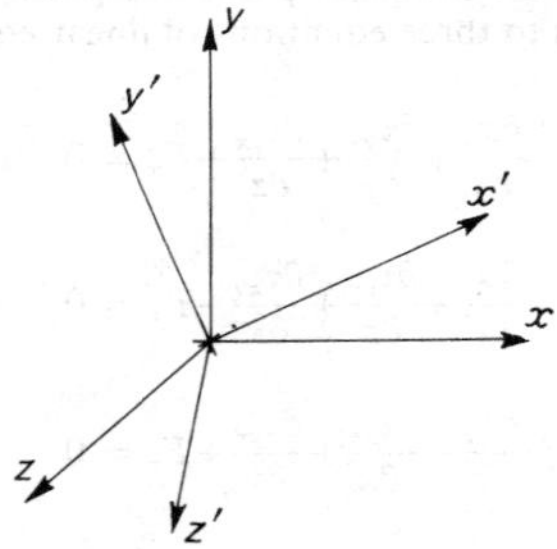

Figure 2.5

since $\partial y/\partial x'$, for example, is equal to the cosine of the angle between the y and x' coordinate lines.

PRINCIPAL STRESSES

For a particular orientation of x', y' and z' it is found that all the shear stress components vanish, that is, that the stress vectors $\boldsymbol{\sigma}_{x'}$, $\boldsymbol{\sigma}_{y'}$ and $\boldsymbol{\sigma}_{z'}$ are directed at right angles to their respective coordinate surfaces. Calling this coordinate system X, Y and Z, the matrix of stress components takes the form

$$\begin{matrix} \sigma_X & 0 & 0 \\ 0 & \sigma_Y & 0 \\ 0 & 0 & \sigma_Z \end{matrix}$$

The direct stresses σ_X, σ_Y and σ_Z are called the *principal stresses* at P, whilst the X, Y and Z coordinate lines are called the *principal directions of stress.*

The values of the principal stresses in terms of the stress components in the x, y and z system are equal to the three roots of the equation

$$\sigma^3 - I_1\sigma^2 + I_2\sigma - I_3 = 0 \tag{5}$$

where

$$I_1 = \sigma_x + \sigma_y + \sigma_z \tag{6}$$
$$I_2 = \sigma_x\sigma_y + \sigma_y\sigma_z + \sigma_z\sigma_x - \tau_{xy}^2 - \tau_{yz}^2 - \tau_{zx}^2 \tag{7}$$
$$I_3 = \sigma_x\sigma_y\sigma_z + 2\tau_{xy}\tau_{yz}\tau_{zx} - \sigma_x\tau_{yz}^2 - \sigma_y\tau_{zx}^2 - \sigma_z\tau_{xy}^2 \tag{8}$$

The direction cosines of the Y coordinate line say, relative to the x, y and z coordinate lines, $(\lambda_{Yx}, \lambda_{Yy}, \lambda_{Yz})$, are found by solving the equations

$$\begin{bmatrix} (\sigma_x-\sigma_Y) & \tau_{xy} & \tau_{xz} \\ \tau_{yx} & (\sigma_y-\sigma_Y) & \tau_{yz} \\ \tau_{zx} & \tau_{zy} & (\sigma_z-\sigma_Y) \end{bmatrix} \begin{bmatrix} \lambda_{Yx} \\ \lambda_{Yy} \\ \lambda_{Yz} \end{bmatrix} = 0 \tag{9}$$

$$(\lambda_{Yx})^2+(\lambda_{Yy})^2+(\lambda_{Yz})^2 = 1 \tag{10}$$

INTERNAL EQUILIBRIUM EQUATIONS

Consideration of the equilibrium of a small parallelepiped element of material surrounding an internal point P, leads to three equations of linear equilibrium

$$\frac{\partial\sigma_x}{\partial x}+\frac{\partial\tau_{yx}}{\partial y}+\frac{\partial\tau_{zx}}{\partial z}+F_x = 0 \tag{11}$$

$$\frac{\partial\sigma_y}{\partial y}+\frac{\partial\tau_{zy}}{\partial z}+\frac{\partial\tau_{xy}}{\partial x}+F_y = 0 \tag{12}$$

$$\frac{\partial\sigma_z}{\partial z}+\frac{\partial\tau_{xz}}{\partial x}+\frac{\partial\tau_{yz}}{\partial y}+F_z = 0 \tag{13}$$

and three equations of rotational equilibrium

$$\tau_{xy} = \tau_{yx} \tag{14}$$

$$\tau_{yz} = \tau_{zy} \tag{15}$$

$$\tau_{zx} = \tau_{xz} \tag{16}$$

In equations (11–13), F_x, F_y and F_z are the components of any body force vector $\boldsymbol{F}$ (units: force per unit volume) acting at P. Note, for example, that a body force vector of magnitude (ρg)/unit volume is exerted by the Earth at all points within a body situated in its gravitational field, ρ being the local density of the body and g being the acceleration due to gravity.

The shear stress components τ_{xy} and τ_{yx} being equal, are called *complementary shear stresses*. It is apparent from (14–16) that if a body is in equilibrium then only six of the nine stress components can take different values at any point.

PLANE STRESS

For structures made of elements whose dimensions in the z direction are much smaller than the dimensions in the x and y directions, such as thin plate girders, slabs, shear walls, etc., the following assumptions can be made:

1 The stress components σ_z, τ_{yz}, τ_{xz} can be ignored.
2 The stress components are uniform across the thickness of the element. That is they are independent of z.

Such a state of stress is called *plane stress*.

For plane stress, the transformation equations (4) take a simple and important form. Suppose the x', y', z' system is formed by a rotation of α^0 anticlockwise about the z axis, as in Figure 2.6, then the transformation equations between σ_x, σ_y, τ_{xy} and $\sigma_{x'}$, $\sigma_{y'}$, $\tau_{x'y'}$ are as follows:

$$\sigma_{x'} = \tfrac{1}{2}(\sigma_x+\sigma_y)+\tfrac{1}{2}(\sigma_x-\sigma_y)\cos(2\alpha)+\tau_{xy}\sin(2\alpha) \tag{17}$$

$$\sigma_{y'} = \tfrac{1}{2}(\sigma_x+\sigma_y)-\tfrac{1}{2}(\sigma_x-\sigma_y)\cos(2\alpha)-\tau_{xy}\sin(2\alpha) \tag{18}$$

$$\tau_{x'y'} = -\tfrac{1}{2}(\sigma_x-\sigma_y)\sin(2\alpha)+\tau_{xy}\cos(2\alpha) \tag{19}$$

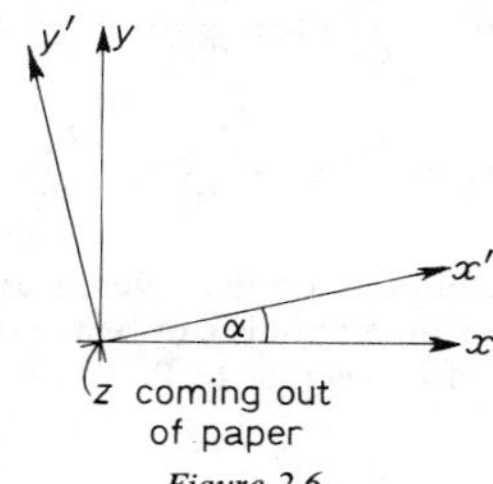

Figure 2.6

These equations can then be represented by the following graphical construction. Two axes are drawn, the vertical representing shear stress and the horizontal, direct stress, and a circle is constructed whose centre is at $(\sigma_x+\sigma_y)/2$ on the direct stress axis, and which passes through the point (σ_x, τ_{xy}) as in Figure 2.7. The line through the centre of the circle at an angle $2\alpha^0$ *clockwise* to the line joining the centre and (σ_x, τ_{xy}) then intersects

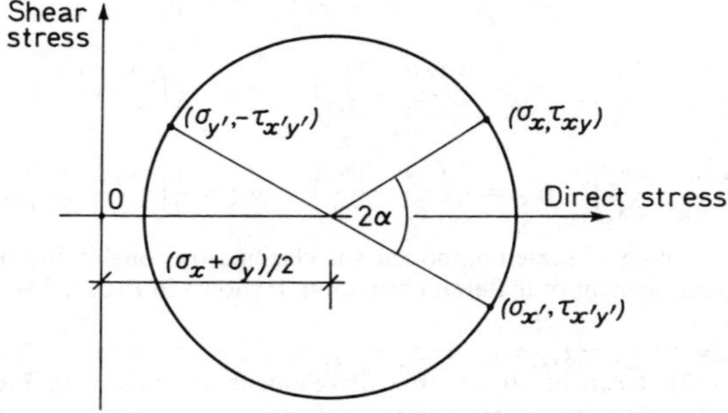

Figure 2.7

the circle at $(\sigma_{x'}, \tau_{x'y'})$. Produced backwards, it intersects the circle at a point whose abscissa is $\sigma_{y'}$. This construction was devised by Otto Mohr in 1882 and the circle is called *Mohr's circle of stress.*

Strain

Strain is the general name given to the deformation of a body subject to loading.

DISPLACEMENTS

A particular point P in a body before loading, occupies its initial position P_i say, and after loading its final position P_f. The line joining P_i to P_f is a vector which is denoted by

$\boldsymbol{u}$ and called the *displacement vector* at P. In general, this vector varies continuously from point to point in the body, and its three components u_x, u_y and u_z are continuous functions of the coordinates of P.†

Consider two neighbouring points P(x, y, z) and P*(x+dx, y+dy, z+dz) in the body. Then

$$du_x = \frac{\partial u_x}{\partial x}dx + \frac{\partial u_x}{\partial y}dy + \frac{\partial u_x}{\partial z}dz \tag{20}$$

$$du_y = \frac{\partial u_y}{\partial x}dx + \frac{\partial u_y}{\partial y}dy + \frac{\partial u_y}{\partial z}dz \tag{21}$$

$$du_z = \frac{\partial u_z}{\partial x}dx + \frac{\partial u_z}{\partial y}dy + \frac{\partial u_z}{\partial z}dz \tag{22}$$

where the differentials du_x, du_y and du_z are the differences between the components of $\boldsymbol{u}$ at the two points. As such, these differentials can be regarded as the components of the vector giving the displacement of P* *relative* to P.

COMPONENTS OF STRAIN

In order to obtain a concise description of the deformation of the material at P it is convenient to define nine components $\varepsilon_{xx}, \varepsilon_{yy}, \varepsilon_{zz}, \varepsilon_{xy}, \varepsilon_{yz}, \varepsilon_{zx}, \omega_{xy}, \omega_{yz}, \omega_{zx}$ by the following equations:

$$\varepsilon_{xx} = \frac{\partial u_x}{\partial x}, \quad \varepsilon_{yy} = \frac{\partial u_y}{\partial y}, \quad \varepsilon_{zz} = \frac{\partial u_z}{\partial z} \tag{23, 24, 25}$$

$$\varepsilon_{xy} = \frac{1}{2}\left(\frac{\partial u_x}{\partial y} + \frac{\partial u_y}{\partial x}\right), \quad \varepsilon_{yz} = \frac{1}{2}\left(\frac{\partial u_y}{\partial z} + \frac{\partial u_z}{\partial y}\right), \quad \varepsilon_{zx} = \frac{1}{2}\left(\frac{\partial u_z}{\partial x} + \frac{\partial u_x}{\partial z}\right) \tag{26, 27, 28}$$

$$\omega_{xy} = \frac{1}{2}\left(\frac{\partial u_x}{\partial y} - \frac{\partial u_y}{\partial x}\right), \quad \omega_{yz} = \frac{1}{2}\left(\frac{\partial u_y}{\partial z} - \frac{\partial u_z}{\partial y}\right), \quad \omega_{zx} = \frac{1}{2}\left(\frac{\partial u_z}{\partial x} - \frac{\partial u_x}{\partial z}\right) \tag{29, 30, 31}$$

The physical meaning of these components is clarified by considering the deformation of the rectangular element of material containing P shown in Figure 2.8(a) for each component in turn.

Thus if $\varepsilon_{xx} \neq 0$, $\varepsilon_{yy} = \varepsilon_{zz} = \varepsilon_{xy} = \varepsilon_{yz} = \varepsilon_{zx} = \omega_{xy} = \omega_{yz} = \omega_{zx} = 0$ then, by using equations (20–22), it can be shown that the element deforms as in Figure 2.8(b). ε_{xx}, corresponding to this type of longitudinal deformation is called the *direct strain component* in the x direction at P. If it is positive it is called *tensile* and the element lengthens and if negative, it is called *compressive* and the element shortens. Similarly, the components ε_{yy} and ε_{zz} corresponding respectively to longitudinal deformation in the y and z directions are called the direct strain components in these directions.

If $\varepsilon_{xy} \neq 0$, $\varepsilon_{xx} = \varepsilon_{yy} = \varepsilon_{zz} = \varepsilon_{yz} = \varepsilon_{zx} = \omega_{xy} = \omega_{yz} = \omega_{zx} = 0$ then $\partial u_x/\partial y = \partial u_y/\partial x = \varepsilon_{xy}$ and again, by using (20–22) it can be shown that the element deforms as in Figure 2.8(c). Deformation of this type is called shear strain, and ε_{xy} is called the ***mathematical shear strain component*** at P. The adjective 'mathematical' is used to distinguish between this and the engineering shear strain at P, which is denoted by γ_{xy} and is equal to the closure in radians of the angle between the x and y coordinate lines. From the geometry of Figure 2.8(c) we have

$$\gamma_{xy} = 2\varepsilon_{xy} \tag{32}$$

*In most cases $\boldsymbol{u}$ is so small that the coordinates of P do not change appreciably during the loading.

Similarly the components ε_{yz} and ε_{zx} correspond to shear strain in the y–z and z–x planes respectively.

Finally, if $\omega_{xy} \neq 0$, $\varepsilon_{xx} = \varepsilon_{yy} = \varepsilon_{zz} = \varepsilon_{xy} = \varepsilon_{yz} = \varepsilon_{zx} = \omega_{yz} = \omega_{zx} = 0$, then $\partial u_x/\partial y = -\partial u_y/\partial x = \omega_{xy}$ and it can be shown that the element rotates without deformation about the z coordinate line as in Figure 2.8(d). ω_{xy} is called the *rotation* at P. Similarly ω_{yz} and ω_{zx} correspond respectively to local rotations about the x and y coordinate lines through P.

As in the case of stresses, the conventional engineering notation for the strain components is somewhat different from the above and the direct strain components are

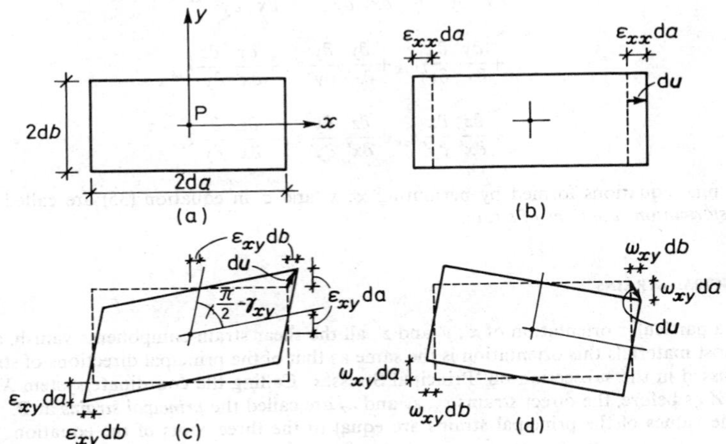

Figure 2.8

written ε_x, ε_y and ε_z. Except in the subsection on 'Transformation of Strain' (below), this latter notation is employed in the remainder of the chapter, and the shear strains are described in terms of γ_{xy}, γ_{yz} and γ_{zx}.

UNIFORM STRAIN

If the displacement components u_x, u_y and u_z are linear functions of the coordinates of P then the corresponding strains given by equations (23–28) are uniform. The overall changes in the geometry of a body are then simply related to the strain components. Thus consider, for example, a line AB in or on the surface of the body which originally coincides with an x coordinate line. If the original length of AB is l and its increase in length is Δl, then

$$\varepsilon_x = \Delta l / l \tag{33}$$

TRANSFORMATION OF STRAIN

Considering again a new coordinate system x', y', z' rotated relative to the x, y and z system as in Figure 2.5, then the components of strain in this new system are defined by strain–displacement relations similar to (23–28). Thus $\gamma_{x'y'}(=2\varepsilon_{x'y'})$ for example is given by

$$\gamma_{x'y'} = \left(\frac{\partial u_{x'}}{\partial y'} + \frac{\partial u_{y'}}{\partial x'}\right) \tag{34}$$

where $u_{x'}$, $u_{y'}$ and $u_{z'}$ are the components of the displacement vector $\boldsymbol{u}$ relative to x', y' and z'. The components of strain in the two systems are related by equations of the same type as (4) (where again for conciseness we employ the original notation of the subsection on 'Components of Strain'). Thus

$$\begin{aligned}\varepsilon_{x'y'} = {} & \frac{\partial x}{\partial x'}\frac{\partial x}{\partial y'}\varepsilon_{xx} + \frac{\partial x}{\partial x'}\frac{\partial y}{\partial y'}\varepsilon_{xy} + \frac{\partial x}{\partial x'}\frac{\partial z}{\partial y'}\varepsilon_{xz} \\ & + \frac{\partial y}{\partial x'}\frac{\partial x}{\partial y'}\varepsilon_{yx} + \frac{\partial y}{\partial x'}\frac{\partial y}{\partial y'}\varepsilon_{yy} + \frac{\partial y}{\partial x'}\frac{\partial z}{\partial y'}\varepsilon_{yz} \\ & + \frac{\partial z}{\partial x'}\frac{\partial x}{\partial y'}\varepsilon_{zx} + \frac{\partial z}{\partial x'}\frac{\partial y}{\partial y'}\varepsilon_{zy} + \frac{\partial z}{\partial x'}\frac{\partial z}{\partial y'}\varepsilon_{zz}\end{aligned} \tag{35}$$

The nine equations formed by permuting x', y' and z' in equation (35) are called the *transformation equations of strain.*

PRINCIPAL STRAINS

For a particular orientation of x', y' and z', all the shear strain components vanish, and in most materials this orientation is the same as that of the principal directions of stress discussed in the subsection on 'Principal Stresses'. Calling the coordinate system X, Y and Z as before, the direct strains ε_X, ε_Y and ε_Z are called the *principal strains* at P.

The values of the principal strains are equal to the three roots of the equation

$$\varepsilon^3 - E_1\varepsilon^2 + E_2\varepsilon - E_3 = 0 \tag{36}$$

where

$$E_1 = \varepsilon_x + \varepsilon_y + \varepsilon_z \tag{37}$$

$$E_2 = \varepsilon_x\varepsilon_y + \varepsilon_y\varepsilon_z + \varepsilon_z\varepsilon_x - \tfrac{1}{4}(\gamma_{xy}^2 + \gamma_{yz}^2 + \gamma_{zx}^2) \tag{38}$$

$$E_3 = \varepsilon_x\varepsilon_y\varepsilon_z + \tfrac{1}{4}(\gamma_{xy}\gamma_{yz}\gamma_{zx} - \varepsilon_x\gamma_{yz}^2 - \varepsilon_y\gamma_{zx}^2 - \varepsilon_z\gamma_{xy}^2) \tag{39}$$

COMPATABILITY EQUATIONS

The three displacement components u_x, u_y and u_z can be eliminated from the six strain–displacement relations (23–28) to produce three equations called the *compatability equations,* which must be satisfied by the strain components. This elimination can be done in different ways to produce different sets of equations. Two such are:

$$\frac{\partial^2\varepsilon_x}{\partial y^2} + \frac{\partial^2\varepsilon_y}{\partial x^2} - \frac{\partial^2\gamma_{xy}}{\partial x\partial y} = 0 \tag{40}$$

$$\frac{\partial^2\varepsilon_y}{\partial z^2} + \frac{\partial^2\varepsilon_z}{\partial y^2} - \frac{\partial^2\gamma_{yz}}{\partial y\partial z} = 0 \tag{41}$$

$$\frac{\partial^2\varepsilon_z}{\partial x^2} + \frac{\partial^2\varepsilon_x}{\partial z^2} - \frac{\partial^2\gamma_{zx}}{\partial z\partial x} = 0 \tag{42}$$

$$\frac{2\partial^2\varepsilon_x}{\partial y\partial z}-\frac{\partial}{\partial x}\left(\frac{\partial\gamma_{xy}}{\partial z}-\frac{\partial\gamma_{yz}}{\partial x}+\frac{\partial\gamma_{zx}}{\partial y}\right)=0 \tag{43}$$

$$\frac{2\partial^2\varepsilon_y}{\partial z\partial x}-\frac{\partial}{\partial y}\left(\frac{\partial\gamma_{yz}}{\partial x}-\frac{\partial\gamma_{zx}}{\partial y}+\frac{\partial\gamma_{xy}}{\partial z}\right)=0 \tag{44}$$

$$\frac{2\partial^2\varepsilon_z}{\partial x\partial y}-\frac{\partial}{\partial z}\left(\frac{\partial\gamma_{zx}}{\partial y}-\frac{\partial\gamma_{xy}}{\partial z}+\frac{\partial\gamma_{yz}}{\partial x}\right)=0 \tag{45}$$

PLANE STRAIN

Plain strain is said to exist when the strain components $\varepsilon_{z,\ yz}$ and ε_{zx} are equal to zero. It occurs when $u_z = 0$ at every point within a region of a body. From symmetry this is the case in the central region of a body which (i) is very long in the z direction, (ii) is of uniform cross section and (iii) is subjected to loading in the z plane that is uniformly

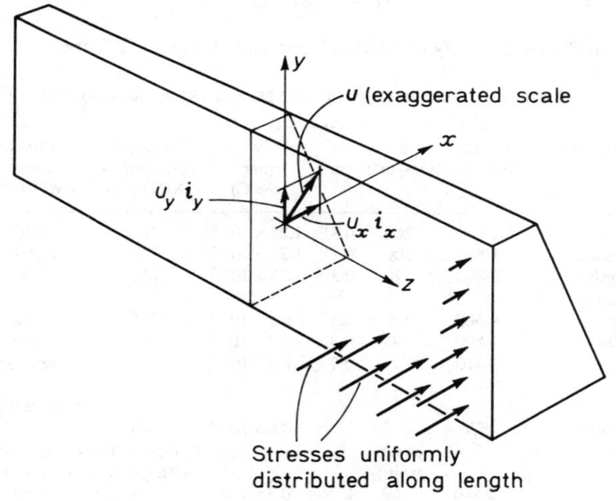

Figure 2.9

distributed along its length (Figure 2.9). It typically occurs in gravity dams, tunnel linings or retaining walls.

Considering again the new coordinate system x', y', z' formed by a rotation of α^0 anticlockwise about the z axis as in Figure 2.6, the transformation equations between ε_x, ε_y, ε_{xy} and $\varepsilon_{x'}$, $\varepsilon_{y'}$, and $\varepsilon_{x'y'}$ take the same form as equations (17–19). These transformation equations are represented by a graphical construction called *Mohr's circle of strain*, whose function is the same as that of Mohr's circle of stress.

The stress–strain relations

The relationship between the stress and strain components at a point in a body is a property of the particular material making up the body. For an isotropic elastic material the stress–strain relations are linear and are independent of the orientation of the x, y, z coordinate system. They take the following form:

$$\varepsilon_x = \frac{1}{E}[\sigma_x - \nu(\sigma_y + \sigma_z)] + \alpha\Delta T \tag{46}$$

$$\varepsilon_y = \frac{1}{E}[\sigma_y - \nu(\sigma_z + \sigma_x)] + \alpha\Delta T \tag{47}$$

$$\varepsilon_z = \frac{1}{E}[\sigma_z - \nu(\sigma_x + \sigma_y)] + \alpha\Delta T \tag{48}$$

$$\gamma_{xy} = \frac{1}{G}\tau_{xy}, \quad \gamma_{yz} = \frac{1}{G}\tau_{yz}, \quad \gamma_{zx} = \frac{1}{G}\tau_{zx} \tag{49, 50, 51}$$

where ΔT is the temperature change from some initial state. E and G are constants having the dimensions of force per unit area and are called *Young's modulus* and the *shear modulus* respectively, ν is a dimensionless constant called *Poisson's ratio* and α is a constant having the dimensions $[\text{deg(C)}]^{-1}$ and is called the *temperature coefficient of expansion.* G in fact is related to E and ν by the following equation:

$$G = E/2(1+\nu) \tag{52}$$

Values of E, ν and α for a variety of practical materials are given in Table 2.1.

Table 2.1 PROPERTIES OF MATERIALS (REPRESENTATIVE)

Property / *Material*	*Density* (kg/m^3)	*E* (GN/m^2)	*u*	α (*per deg* C)	*Limit of proportionality* (MN/m^2)	*Ultimate stress* (MN/m^2)	*Uniform elongation*
Mild steel	7 840	200	0.31	1.25×10^{-5}	280	370	0.30
High-strength steel	7 840	200	0.31	1.25×10^{-5}	770	1 550	0.10
Medium strength aluminium alloy	2 800	70	0.3	2.3×10^{-5}	230	430	0.10
Titanium alloy	4 500	120	0.3	0.9×10^{-5}	385	690	0.15
Magnesium alloy	1 800	45	0.3	2.7×10^{-5}	155	280	0.08
Concrete	2 410	25	0.2	1.2×10^{-5}	—	3 (tension) 30 (compression)	—
Timber (Douglas fir)	576	7 (with grain)		0.6×10^{-5}	43 (compression with grain)	52 (compression with grain)	—
Glass	2 580	60	0.26	0.7×10^{-5}	—	1750	—
Nylon	1 140	2	—	10×10^{-5}	77	90	1.0
Polystyrene (not expanded)	1 050	4	—	10×10^{-5}	46	60	0.03
High-strength glass-fibre composite	2 000	60	—	—	—	1 600	—
Carbon fibre composite	1 600	170	—	—	—	1 400	—

The corresponding inverse stress–strain relations are found by solving equations (46–51) for the stresses and are as follows:

$$\sigma_x = 2\mu\varepsilon_x + \lambda(\varepsilon_x + \varepsilon_y + \varepsilon_z) - (3\lambda + 2\mu)\alpha\Delta T \tag{53}$$

$$\sigma_y = 2\mu\varepsilon_y + \lambda(\varepsilon_x + \varepsilon_y + \varepsilon_z) - (3\lambda + 2\mu)\alpha\Delta T \tag{54}$$

$$\sigma_z = 2\mu\varepsilon_z + \lambda(\varepsilon_x + \varepsilon_y + \varepsilon_z) - (3\lambda + 2\mu)\alpha\Delta T \tag{55}$$

$$\tau_{xy} = \mu\gamma_{xy}, \quad \tau_{yz} = \mu\gamma_{yz}, \quad \tau_{zx} = \mu\gamma_{zx} \tag{56, 57, 58}$$

where for conciseness we employ the *Lamé constants* λ and μ defined in terms of E and v by the equations:

$$\lambda = vE/(1+v)(1-2v) \quad (59)$$

$$\mu = E/2(1+v) \quad (60)$$

The stress–strain relations hold for a wide range of stresses in most practical materials. They become invalid when the interatomic bonds in the materials break down, this process generally being called *yielding*. Yielding can be demonstrated by the tensile test, where a known stress system $\sigma_x \neq 0$, $\sigma_y = \sigma_z = \tau_{xy} = \tau_{yz} = \tau_{zx} = 0$, called *uniaxial stress*, is induced in a specimen and the corresponding strain ε_x is measured. A typical plot of σ_x versus ε_x for a mild steel tensile specimen then takes the form shown in Figure 2.10(a). The initial straight section of the curve of slope equal to E corresponds to equation

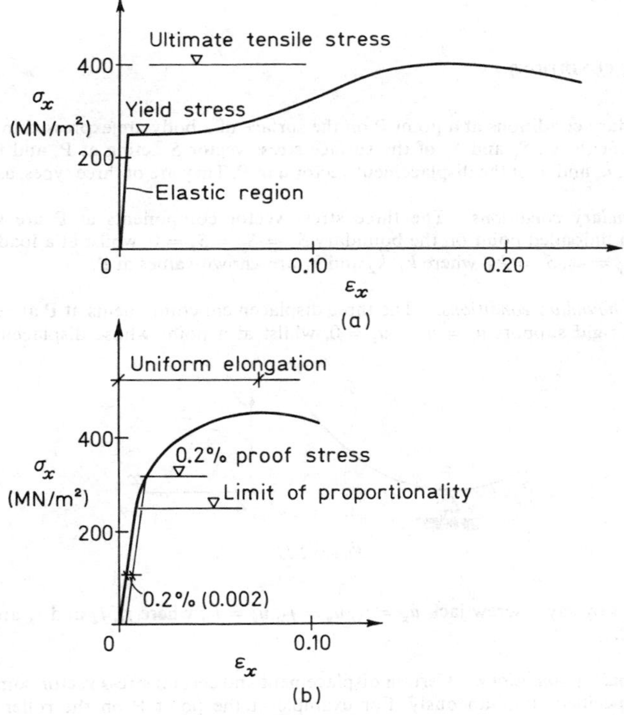

Figure 2.10

(46), but at a certain stress of the order of 250 MN/m^2, the strain increases dramatically with little or no increase of load. This stress is called the *uniaxial yield stress* of mild steel. Subsequently, the stress–strain curve indicates that the specimen supports larger stresses up to a maximum value of the order of 400 MN/m^2 which is called the *ultimate tensile stress*. The uniaxial stress strain curve for an aluminium alloy specimen shown in Figure 2.10(b) does not display a marked yield stress and the material is linear elastic up to a stress called the *limit of proportionality* which again is of the order of 250 MN/m^2.

Two other properties frequently quoted in engineering literature, the 0.2% *proof stress* and the *uniform elongation*, are shown in the figure. Values for the limit of proportionality, ultimate stress and uniform elongation are included in Table 2.1.

For accounts of yield criteria and plastic stress–strain relations corresponding to more general stress systems, see for example, Bisplinghoff *et al*,[2] and Prager and Hodge.[3]

ANALYSIS OF ELASTIC BODIES

The internal equilibrium equations (11–16), strain–displacement relations (23–28) and the stress–strain relations (46–51) are 18 differential equations in the unknowns of the analysis problem, namely the nine stress components, the six strain components and the three displacement components. These equations must be satisfied subject to boundary conditions.

BOUNDARY CONDITIONS

The boundary conditions at a point P on the surface of a body are expressed in terms of the components S_x, S_y and S_z of the surface stress vector $\boldsymbol{S}$ acting at P, and the components u_x, u_y and u_z of the displacement vector $\boldsymbol{u}$ of P. They are of three types, as follows.

Static boundary conditions. The three stress vector components at P are specified. Thus at an unloaded point on the boundary $S_x = S_y = S_z = 0$, whilst at a loaded point $S_x = k_1$, $S_y = k_2$, $S_z = k_3$, where k_1, k_2 and k_3 are known values at P.

Kinematic boundary conditions. The three displacement components at P are specified. Thus at a rigid support $u_x = u_y = u_z = 0$, whilst at a point whose displacements are

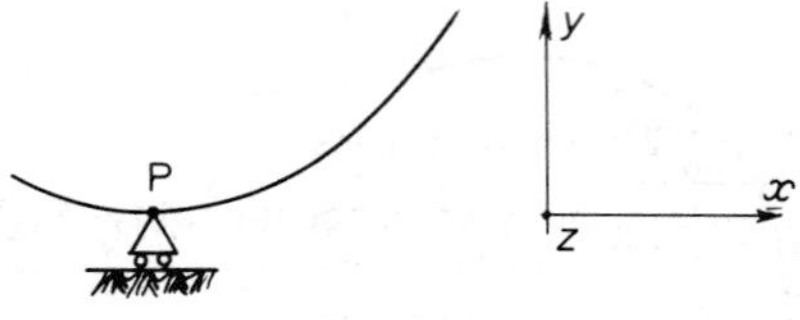

Figure 2.11

constrained by say a screw jack $u_x = j_1$, $u_y = j_2$, $u_z = j_3$, where j_1, j_2 and j_3 are known values at P.

Mixed boundary conditions. Certain displacement and certain stress vector components at P are specified simultaneously. For example at the point P on the roller support shown in Figure 2.11, $S_x = 0$ and $u_y = u_z = 0$.

SOLUTION IN TERMS OF DISPLACEMENTS

A straightforward solution method involves treating the displacement components as the basic unknowns. The three linear equilibrium equations (11–13) are expressed in terms of the displacements by using the stress–strain relations followed by the strain–

displacement relations. The resulting differential equations in u_x, u_y and u_z are called the *Navier equations*. They are as follows:

$$\mu\nabla^2 u_x+(\lambda+\mu)\frac{\partial\Phi}{\partial x}+F_x = 0 \tag{61}$$

$$\mu\nabla^2 u_y+(\lambda+\mu)\frac{\partial\Phi}{\partial y}+F_y = 0 \tag{62}$$

$$\mu\nabla^2 u_z+(\lambda+\mu)\frac{\partial\Phi}{\partial z}+F_z = 0 \tag{63}$$

where

$$\nabla^2 u_x = \frac{\partial^2 u_x}{\partial x^2}+\frac{\partial^2 u_x}{\partial y^2}+\frac{\partial^2 u_x}{\partial z^2} \tag{64}$$

and

$$\Phi = \frac{\partial u_x}{\partial x}+\frac{\partial u_y}{\partial y}+\frac{\partial u_z}{\partial z} \tag{65}$$

In order to solve these equations, the boundary conditions must all be expressed in terms of the displacements of the surface points. In the case of the static boundary conditions, equations for the internal stress components are obtained using equations (1–3), and these are then converted to differential boundary conditions in displacements by again using the stress–strain and the strain–displacement relations. Thus at each internal point and each boundary point there are three simultaneous differential equations in the unknowns u_x, u_y and u_z. In most cases, a direct solution is obtainable only by a numerical procedure such as the finite-difference method.[4]

SOLUTION IN TERMS OF STRESSES

A second solution method involves treating the nine internal stress components as the basic unknowns. Six equations in these unknowns are immediately available from the internal equilibrium equations (11–16). A further three equations are obtained from the compatibility equations (40–42) or (43–45) by using the stress–strain relations to express them in terms of the stress components. The resulting equations are called the *Beltrami–Michell* equations and are as follows:

$$\nabla^2\sigma_x+\frac{1}{(1+\nu)}\frac{\partial^2\Theta}{\partial x^2} = \frac{-\nu}{(1-\nu)}\Psi-\frac{2\partial F_x}{\partial x} \tag{66}$$

$$\nabla^2\sigma_y+\frac{1}{(1+\nu)}\frac{\partial^2\Theta}{\partial y^2} = \frac{-\nu}{(1-\nu)}\Psi-\frac{2\partial F_y}{\partial y} \tag{67}$$

$$\nabla^2\sigma_z+\frac{1}{(1+\nu)}\frac{\partial^2\Theta}{\partial z^2} = \frac{-\nu}{(1-\nu)}\Psi-\frac{2\partial F_z}{\partial z} \tag{68}$$

or

$$\nabla^2\tau_{xy}+\frac{1}{(1+\nu)}\frac{\partial^2\Theta}{\partial x\partial y} = -\left(\frac{\partial F_x}{\partial y}+\frac{\partial F_y}{\partial x}\right) \tag{69}$$

$$\nabla^2\tau_{yz}+\frac{1}{(1+\nu)}\frac{\partial^2\Theta}{\partial y\partial z} = -\left(\frac{\partial F_y}{\partial z}+\frac{\partial F_z}{\partial y}\right) \tag{70}$$

$$\nabla^2\tau_{zx}+\frac{1}{(1+\nu)}\frac{\partial^2\Theta}{\partial z\partial x} = -\left(\frac{\partial F_z}{\partial x}+\frac{\partial F_x}{\partial z}\right) \tag{71}$$

where

$$\Theta = \sigma_x+\sigma_y+\sigma_z \tag{72}$$

$$\Psi = \frac{\partial F_x}{\partial x}+\frac{\partial F_y}{\partial y}+\frac{\partial F_z}{\partial z} \tag{73}$$

The only problems that can be conveniently solved directly in terms of stresses are those in which all the boundary conditions are static boundary conditions. In such problems three equations in the internal stress components are obtained using equations (1–3), and these together with the three equations of rotational equilibrium and the three compatibility equations provide the required nine equations at the boundary. In problems where displacements are specified at various boundary points, the corresponding boundary stresses cannot usually be obtained in advance of the solution except for those special cases where the body is externally statically determinate.

Direct solutions in terms of stresses can in principle be obtained using numerical procedures. However, many solutions, especially to two-dimensional problems,[5,6] have been obtained using stress functions which automatically satisfy the equilibrium equations.

Energy methods

VIRTUAL WORK

Consider a body which is in equilibrium under surface stresses $\boldsymbol{S}$ over part of its surface and body forces $\boldsymbol{F}$. Suppose the corresponding internal stress system is given by σ_x, σ_y, σ_z, τ_{xy}, τ_{yz}, τ_{zx}. This is called an *equilibrium force system.*

Next consider an entirely independent system of displacements $\boldsymbol{u}^*$ which vary continuously from point to point in the body and satisfy the kinematic boundary conditions. Suppose the corresponding strain system is given by ε_x^*, ε_y^*, ε_z^*, γ_{xy}^*, γ_{yz}^*, γ_{zx}^*. This is called a *compatible displacement system.*

The virtual work W_e^* done by the external forces $\boldsymbol{S}$ and $\boldsymbol{F}$, supposing they were to move through $\boldsymbol{u}^*$, is as follows:

$$W_e^* = \int_A (S_xu_x^*+S_yu_y^*+S_zu_z^*)\,\mathrm{d}A+\int_V (F_xu_x^*+F_yu_y^*+F_zu_z^*)\,\mathrm{d}V \tag{74}$$

where $\int_A(\;)\,\mathrm{d}A$ represents an integral taken over the loaded surface of the body, and $\int_V(\;)\,\mathrm{d}V$ represents an integral taken over its volume. By a purely mathematical operation[2] it can be shown that

$$W_e^* = W_i^* \tag{75}$$

where W_i^* is a quantity called the *internal virtual work* and is given by

$$W_i^* = \int_V (\sigma_x\varepsilon_x^*+\sigma_y\varepsilon_y^*+\sigma_z\varepsilon_z^*+\tau_{xy}\gamma_{xy}^*+\tau_{yz}\gamma_{yz}^*+\tau_{zx}\gamma_{zx}^*)\,\mathrm{d}V \tag{76}$$

Equation (75) is called the *equation of virtual work.* Note that its derivation is independent of the nature of the stress–strain relations of the material making up the body.

STRAIN ENERGY

Consider the body in equilibrium under $\boldsymbol{S}$ and $\boldsymbol{F}$ and suppose differential changes in the loading $\mathrm{d}\boldsymbol{S}$ and $\mathrm{d}\boldsymbol{F}$ occur causing corresponding differential changes in the *real* displacements d**u**. d**u** and the strains $d\varepsilon_x$, $d\varepsilon_y$, $d\varepsilon_z$, $d\gamma_{xy}$, $d\gamma_{yz}$, $d\gamma_{zx}$ can be regarded as a compatible system of displacements in (75). The work terms on either side of (75) are then differential quantities of real work caused by the loading change. In particular the internal work is given by

$$dW_i = \int_V (\sigma_x\, d\varepsilon_x + \sigma_y\, d\varepsilon_y + \sigma_z\, d\varepsilon_z + \tau_{xy}\, d\gamma_{xy} + \tau_{yz}\, d\gamma_{yz} + \tau_{zx}\, d\gamma_{zx})\, dV \tag{77}$$

Using the elastic stress–strain relations it is possible to integrate (77) to obtain the total internal work done on an elastic body from the initial state with zero stress to the final state with stresses corresponding to **S** and **F**. This internal work is found to be independent of the loading path to the final state and is called the *elastic strain energy U*. It can be expressed in three forms:

$$\begin{aligned} U &= \int_V \frac{1}{2E}\left[(\sigma_x^2+\sigma_y^2+\sigma_z^2) - 2\nu(\sigma_x\sigma_y+\sigma_y\sigma_z+\sigma_z\sigma_x) + 2(1+\nu)(\tau_{xy}^2+\tau_{yz}^2+\tau_{zx}^2)\right] dV \\ &= \int_V \tfrac{1}{2}(\sigma_x\varepsilon_x+\sigma_y\varepsilon_y+\sigma_z\varepsilon_z+\tau_{xy}\gamma_{xy}+\tau_{yz}\gamma_{yz}+\tau_{zx}\gamma_{zx})\, dV \\ &= \int_V \left[\mu(\varepsilon_x^2+\varepsilon_y^2+\varepsilon_z^2) + \frac{\lambda}{2}(\varepsilon_x+\varepsilon_y+\varepsilon_z)^2 + \frac{\mu}{2}(\gamma_{xy}^2+\gamma_{yz}^2+\gamma_{zx}^2)\right] dV \end{aligned} \tag{78}$$

PRINCIPLE OF MINIMUM TOTAL POTENTIAL ENERGY

The external work done by the loading in the previous subsection is given by

$$dW_e = \int_A (S_x\, du_x + S_y\, du_y + S_z\, du_z)\, dA + \int_V (F_x\, du_x + F_y\, du_y + F_z\, du_z)\, dV \tag{79}$$

If the loading is *conservative*, so that all the loads on the body are independent of the displacements, it is possible to define a function V as follows

$$V = U - \int_A (S_x u_x + S_y u_y + S_z u_z)\, dA - \int_V (F_x u_x + F_y u_y + F_z u_z)\, dV \tag{80}$$

so that the equation of virtual work for the differential change of the body in equilibrium takes the form

$$dV = 0 \tag{81}$$

V is called the *total potential energy* of the system of the body plus loads.

It can further be shown that *for a body in equilibrium, among all the admissible displacements of the body, the actual displacements make the total potential energy an absolute minimum.*[7] This is the most important energy principle, and its method of application for the solution of structures involves expressing all the displacements of the structure in terms of a (usually limited) number of degrees of freedom. (This can be done exactly for frameworks, but only approximately for structures such as slabs.) The total potential energy is then minimised with respect to the degrees of freedom to yield the corresponding exact or approximate displacements of the structure corresponding to equilibrium.

Many other energy principles have been derived and they are fully discussed with their applications by for example Washizu.[7]

Measurement of stress and strain

SURFACE STRAIN

The measurement of strain is usually limited to obtaining direct strains tangential to the surfaces of structures by means of mechanical or electrical strain gauges. If the complete state of tangential strain at a surface point is to be determined, then separate measurements of direct strain have to be obtained in three distinct directions at the point. In interpreting these measurements, we then use the fact that two of the principal directions of stress and strain are tangential to the surface whilst the third is normal to it. Thus using for example, a 45° strain-gauge rosette, producing strain measurements ε_1, ε_2

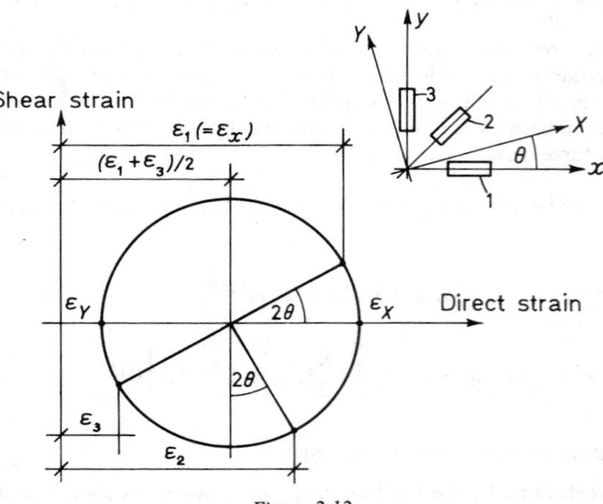

Figure 2.12

and ε_3 as shown in Figure 2.12, it can be shown that the principal direction X is at $\theta°$ anticlockwise to the x coordinate line where

$$\tan(2\theta) = \frac{(2\varepsilon_2 - \varepsilon_1 - \varepsilon_3)}{(\varepsilon_1 - \varepsilon_3)} \tag{82}$$

The two principal surface strains ε_X and ε_Y are then given by

$$\varepsilon_X = \frac{(\varepsilon_1 + \varepsilon_3)}{2} + r \quad \varepsilon_Y = \frac{(\varepsilon_1 + \varepsilon_3)}{2} - r \tag{83, 84}$$

where

$$r = \tfrac{1}{2}[(\varepsilon_1 - \varepsilon_3)^2 + (2\varepsilon_2 - \varepsilon_1 - \varepsilon_3)^2]^{\frac{1}{2}} \tag{85}$$

Example The strains measured by the three gauges of the 45° rosette shown in Figure 2.12 are respectively

$$\varepsilon_1 = -5.0 \times 10^{-4} \quad \varepsilon_2 = +3.0 \times 10^{-4} \quad \varepsilon_3 = +1.0 \times 10^{-4}$$

What are the principal strains at the point and the orientation of the principal direction X, to the x coordinate line?

From equation (85):

$$r = \tfrac{1}{2}[(-5.0 - 1.0)^2 + (2 \times 3.0 + 5.0 - 1.0)^2]^{\frac{1}{2}} \times 10^{-4}$$
$$= 5.8 \times 10^{-4}$$

Thus

$$\varepsilon_X = 3.8 \times 10^{-4} \quad \varepsilon_Y = -7.8 \times 10^{-4}$$

From (82):

$$\tan(2\theta) = -1.667$$

Thus

$$2\theta = -59.0° \quad \text{or} \quad 121.0°$$

The ambiguity in the expression for θ is resolved by examining the position of the strains on the Mohr's circle of strain for the surface plane (Figure 2.13). Thus it is clear that in this example 2θ must be greater than 90°. The X coordinate line is therefore directed at 60.5° anticlockwise to the x coordinate line.

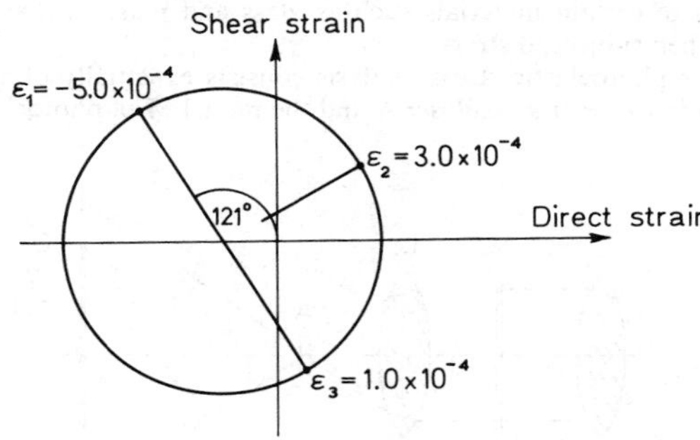

Figure 2.13

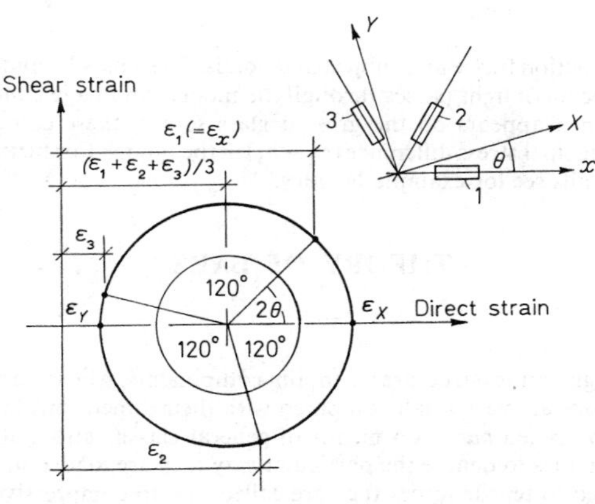

Figure 2.14

Another common layout for strain gauges is the 60° rosette shown in Figure 2.14. The principal direction X is then at θ° anticlockwise to the x' coordinate line where

$$\tan(2\theta) = \sqrt{3}(\varepsilon_2 - \varepsilon_3)/(2\varepsilon_1 - \varepsilon_2 - \varepsilon_3) \tag{86}$$

whilst the principal surface strains ε_X and ε_Y are given by

$$\varepsilon_X = \frac{\varepsilon_1 + \varepsilon_2 + \varepsilon_3}{3} + r \quad \varepsilon_Y = \frac{\varepsilon_1 + \varepsilon_2 + \varepsilon_3}{3} - r \tag{87, 88}$$

where

$$r = \tfrac{2}{3}(\varepsilon_1^2 + \varepsilon_2^2 + \varepsilon_3^2 - \varepsilon_1\varepsilon_2 - \varepsilon_2\varepsilon_3 - \varepsilon_3\varepsilon_1)^{\frac{1}{2}} \tag{89}$$

The complete state of surface stress corresponding to the strains measured above can be found from the stress–strain relations, noting that in the absence of surface loading the state of stress is one of plane stress.

THE PHOTOELASTIC METHOD[8, 9]

A good indication of the internal stresses in model structures can be obtained by making use of the property of certain materials such as glass and plastics, that they become double refracting when subject to stress.

The apparatus for photoelastic stress analysis consists essentially of a light source, L (Figure 2.15), a *polariser* P, and *analyser* A and the model M of photoelastic material,

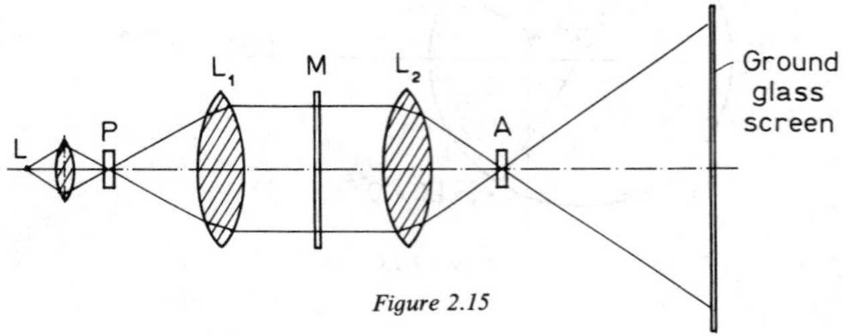

Figure 2.15

which is held in a reaction frame and subjected to loads. The lenses L_1 and L_2 are arranged so that a parallel beam of light passes through the model. An image containing bands of different colours then appears on the ground glass screen, these colours representing regions of equal principal stress difference $(\sigma_X - \sigma_Y)$ in the model. For further experimental and theoretical details see for example Hendry.[9]

THEORY OF BARS

Introduction

A great many engineering structures contain components whose dimensions in two coordinate directions are very small compared with their dimensions in the third. These components can be called *bars* as a means of general classification, although they are often given other names to denote the particular way they are loaded in structures. Thus if they are subjected to tensile forces they are called *ties*, to compressive forces they are called *struts* or *columns*, to lateral forces they are called *beams*, whilst if they are subjected to both compressive and lateral forces they are called *beam-columns*.

Structures completely composed of bars are called *frames*, and are either two-dimensional *plane frames*, or three-dimensional *space frames*.

This section reviews the engineering theory of straight bars of uniform cross section.

Cross-sectional geometry

FIRST MOMENT OF AREA

Consider a bar of some particular cross sectional shape shown in Figure 2.16, and the two orthogonal axes y and z. The first moment of area of the cross section about the y

axis G_y, is defined as the sum of the products obtained by multiplying each element of area dA by its distance z from the y axis. Thus

$$G_y = \int_A z\,dA \tag{90}$$

Similarly

$$G_z = \int_A y\,dA \tag{91}$$

The position of the *centroid* of the cross section is such that the first moment of area about any axis passing through it, is zero. Thus if C is the centroid in Figure 2.16, then

$$G_y = G_z = 0$$

From this it is clear that C must lie on any axis of symmetry of the section. The centroid

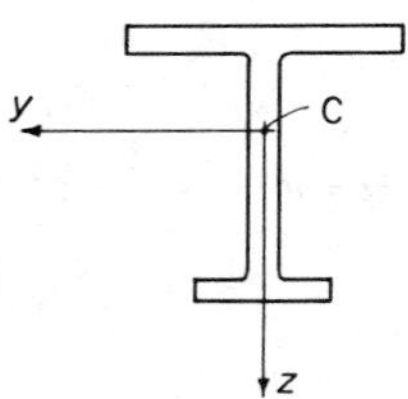

Figure 2.16

can be located in general by selecting any two orthogonal axes y' and z'. The coordinates of the centroid relative to this system, y'_c and z'_c, are then given by

$$y'_c = G_{z'}/A \quad z'_c = G_{y'}/A \tag{92, 93}$$

where A is the area of the cross section. The positions of the centroids of various cross-sectional shapes are shown in Table 2.2.

The *longitudinal axis* of the bar is defined as the line passing through the centroids of its cross sections.

MOMENTS OF INERTIA

The *moment of inertia** of the cross section about the y axis I_y, is defined as the sum of the products obtained by multiplying each element of area dA by the square of its distance z from the y axis. Thus

$$I_y = \int_A z^2\,dA \tag{94}$$

Similarly

$$I_z = \int_A y^2\,dA \tag{95}$$

The *product of inertia*, I_{yz} is defined as

$$I_{yz} = \int_A yz\,dA \tag{96}$$

where y and z are the respective distances of each element of area dA from the z and y axes.

* The term 'moment of inertia' is commonly used in engineering texts because the quantity I_y defined by equation (94) is directly proportional to the mechanical moment of inertia about the y axis, of a thin lamina of the same shape as the cross section. A more precise term for I_y is the 'second moment of area'.

Table 2.2 GEOMETRICAL PROPERTIES OF PLANE SECTIONS

Section	*Area A*	*Position of centroid* C	*Moments of inertia*
1 Rectangle 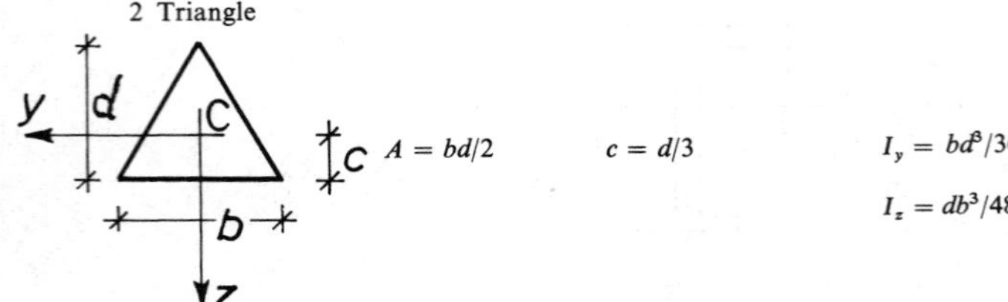	$A = bd$	$c = d/2$	$I_y = bd^3/12$ $I_z = db^3/12$
2 Triangle 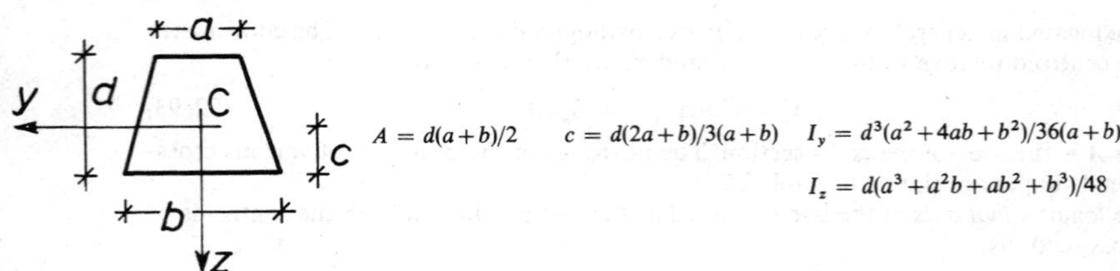	$A = bd/2$	$c = d/3$	$I_y = bd^3/36$ $I_z = db^3/48$
3 Trapezium 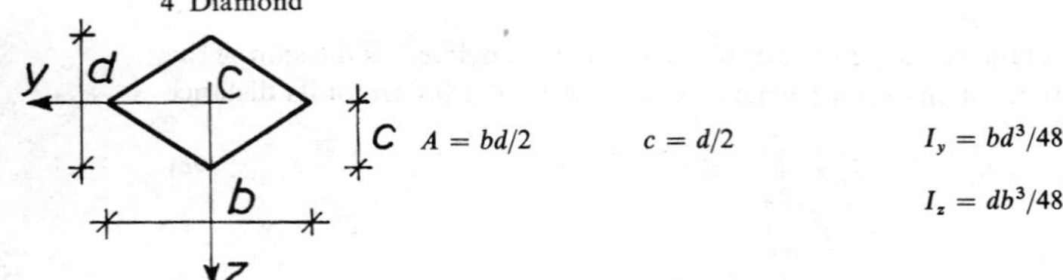	$A = d(a+b)/2$	$c = d(2a+b)/3(a+b)$	$I_y = d^3(a^2+4ab+b^2)/36(a+b)$ $I_z = d(a^3+a^2b+ab^2+b^3)/48$
4 Diamond 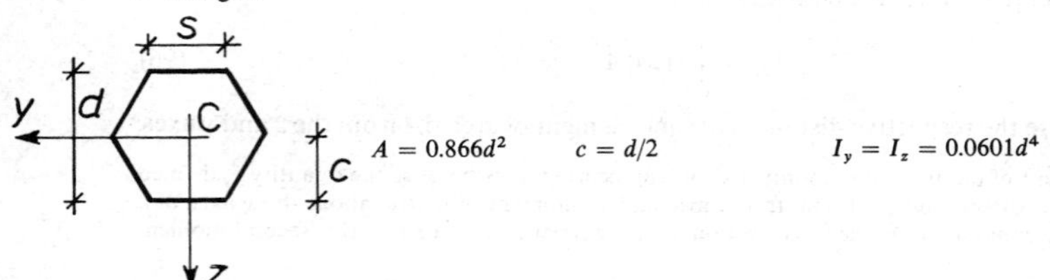	$A = bd/2$	$c = d/2$	$I_y = bd^3/48$ $I_z = db^3/48$
5 Hexagon	$A = 0.866d^2$	$c = d/2$	$I_y = I_z = 0.0601d^4$

Section	*Area A*	*Position of centroid C*	*Moments of inertia*
6 Circle	$A = \pi r^2$ $= 3.1416r^2$	$c = r$	$I_y = I_z = \pi r^4/4$ $= 0.7854r^4$
7 Hollow circle	$A = \pi(r_1^2 - r_2^2)$ $= 3.1416(r_1^2 - r_2^2)$	$c = r_1$	$I_y = I_z = (\pi/4)(r_1^4 - r_2^4)$ $= 0.7854(r_1^4 - r_2^4)$
8 Semicircle	$A = \pi r^2/2$ $= 1.5708r^2$	$c = 0.424r$	$I_y = [(\pi/8) - (8/9\pi)]r^4$ $= 0.1098r^4$ $I_z = \pi r^4/8$ $= 0.3927r^4$
9 Ellipse	$A = \pi ab$	$c = a$	$I_y = (\pi/4)ba^3 = 0.7854ba^3$ $I_z = (\pi/4)ab^3 = 0.7854ab^3$
10 Semiellipse	$A = \pi ab/2$	$c = 0.424a$	$I_y = 0.1098ba^3$ $I_z = 0.3927ab^3$
11 Parabola	$A = 4ab/3$	$c = 2a/5$	$I_y = 0.0914ba^3$ $I_z = 0.2666ab^3$

The *polar moment of inertia* of the cross section about the x axis, I_p, is defined as

$$I_p = \int_A r^2 \, dA \tag{97}$$

where r is the distance of each element of area dA from the x axis. Note that since $r^2 = (y^2+z^2)$

$$I_p = \int_A (y^2+z^2) \, dA = I_z + I_y \tag{98}$$

If y' is an axis parallel to the centroidal axis y and distance c from it, then

$$I_{y'} = I_y + Ac^2 \tag{99}$$

The relationship (99) is known as the *parallel axis theorem.* This theorem facilitates the calculation of the moments of inertia of a complicated cross section, for the section can be divided into separate simpler elements of area A_e say, whose moments of inertia I_{ye} about their *own* centroidal axes are known. If then c_e is the distance of an element centroid from the y axis, we have

$$I_y = \sum_{\text{elements}} (I_{ye} + A_e c_e^2) \tag{100}$$

The moments of inertia about their centroidal axes, of various sectional shapes are given in Table 2.2.

TRANSFORMATION OF MOMENTS OF INERTIA

Consider a new system of centroidal axes, y' and z', formed by a rotation of $\alpha°$ anti-clockwise about the x axis as shown in Figure 2.17. Then the inertias $I_{y'}$, $I_{z'}$ and $I_{y'z'}$,

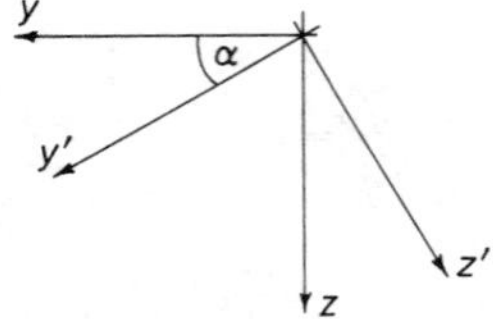

Figure 2.17

being defined in the same way as I_y, I_z and I_{yz} (94–96), are related to I_y, I_z and I_{yz} by the equations

$$I_{y'} = \tfrac{1}{2}(I_y + I_z) + \tfrac{1}{2}(I_y - I_z)\cos(2\alpha) - I_{yz}\sin(2\alpha) \tag{101}$$

$$I_{z'} = \tfrac{1}{2}(I_y + I_z) - \tfrac{1}{2}(I_y - I_z)\cos(2\alpha) + I_{yz}\sin(2\alpha) \tag{102}$$

$$I_{y'z'} = \tfrac{1}{2}(I_y - I_z)\sin(2\alpha) + I_{yz}\cos(2\alpha) \tag{103}$$

Note that these transformation equations are identical in form to the transformation equations of plane stress (17–19).

For a certain orientation of y' and z', the product of inertia $I_{y'z'}$ vanishes. Denoting these coordinates by Y and Z, then I_Y and I_Z are called the *principal moments of inertia* of the cross section, and Y and Z are called the *principal axes.* Concerning their orientation, it can be shown in particular that one of the principal axes always coincides with an axis of symmetry in the section. Values of I_Y and I_Z for standard rolled sections are given in BS 4.[10]

Stress resultants

The stresses acting across a particular cross section of a bar under loads, are conveniently represented by their resultant forces and couples relative to the three coordinate axes x, y and z. Thus the resultants acting on the material of the bar on the negative* side of

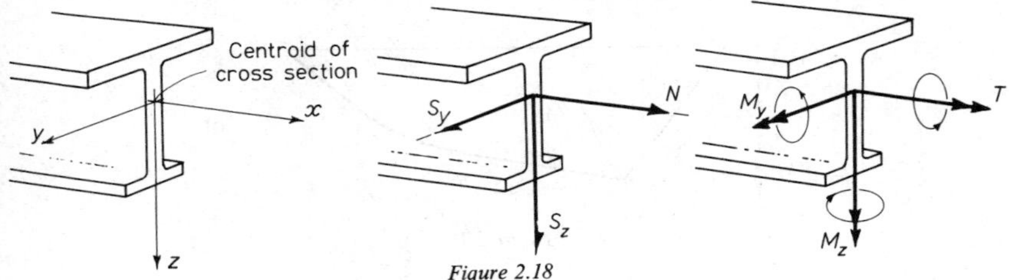

Figure 2.18

the cross section are considered positive when acting in the directions shown in Figure 2.18 and are defined as follows:

Resultant	*Defining equation*	
Axial force N	$N = \int_A \sigma_x \, dA$	(104)
Bending moment about the y axis M_y	$M_y = \int_A \sigma_x z \, dA$	(105)
Bending moment about the z axis M_z	$M_z = -\int_A \sigma_x y \, dA$	(106)
Shear force in the y direction S_y	$S_y = \int_A \tau_{xy} \, dA$	(107)
Shear force in the z direction S_z	$S_z = \int_A \tau_{xz} \, dA$	(108)
Torque T	$T = \int_A (-\tau_{xy}z + \tau_{xz}) \, dA$	(109)

These resultants are in equilibrium with the loads acting on that part of the bar which is on the negative side of the cross section. Thus, if the bar is statically determinate, the resultants can be obtained directly by resolving and taking moments.

A *stress resultant diagram* represents the variation of the stress resultant with x for a specified bar loading. The diagram is drawn positive in the direction of the y and z coordinates. Thus given the beam subject to the vertical forces shown in Figure 2.19(a), the shear force (S_z) diagram and the bending moment (M_y) diagram take the form shown in Figure 2.19(b) and (c) respectively.

* 'negative' means the side in the negative direction of the x axis.

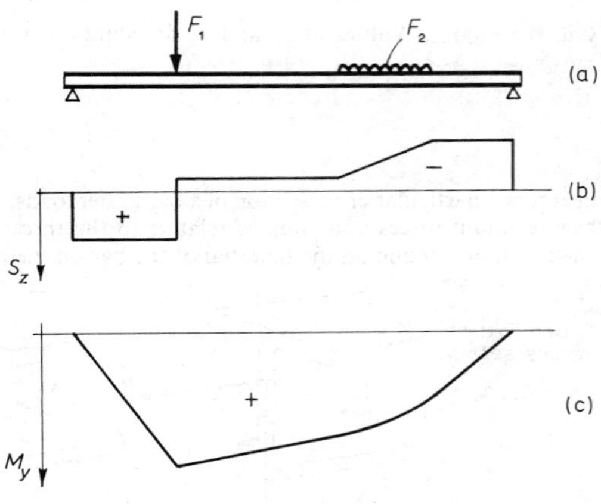

Figure 2.19

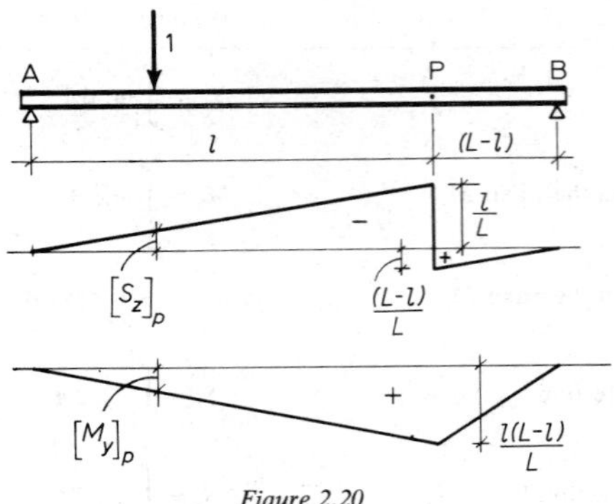

Figure 2.20

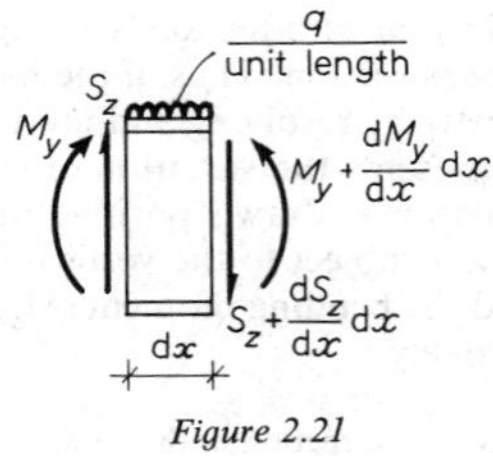

Figure 2.21

It is sometimes of interest in the case of beams to consider the value of a stress resultant (or any other parameter), at a particular point P in the beam, for various positions of a load moving across the beam. If, for example, we consider the bending moment about the y axis at P ($[M_y]_p$), caused by a unit vertical force at point x on the beam, then $[M_y]_p$ is a function of the coordinate x. The plot of $[M_y]_p$ versus x is called the *influence line* of M_y at P. Thus for the beam AB in Figure 2.20 the influence lines for $[S_z]_p$ and $[M_y]_p$ are as shown.

The stress resultants are not all independent of each other. Thus considering the rotational equilibrium about the y axis of a small element of a bar subject to a vertical distributed load q per unit length, as in Figure 2.21, then

$$\mathrm{d}M_y/\mathrm{d}x = S_z \tag{110}$$

Further, considering vertical equilibrium

$$\mathrm{d}S_z/\mathrm{d}x = -\mathrm{q} \tag{111}$$

Whence, combining (110) and (111) gives

$$\mathrm{d}^2M_y/\mathrm{d}x^2 = -q \tag{112}$$

A similar set of equations relates M_z, S_y and the horizontal loading on the bar.

Bars subject to tensile forces (ties)

Consider a bar subject to axial forces N, produced by the loading shown in Figure 2.22.

From the symmetry of the system at some distance from the loading points it can be *deduced* that plane sections originally normal to the longitudinal axis remain plane and

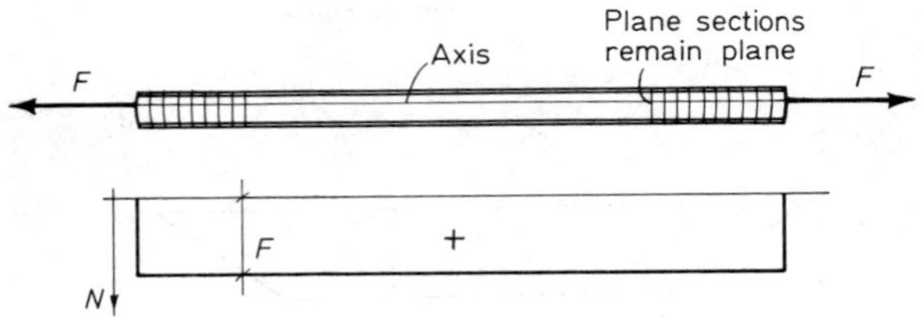

Figure 2.22

normal to the axis after the deformation, whilst from the geometry of the bar, it can be assumed that the only nonzero component of stress is σ_x.[11]

The stress–strain relations corresponding to the uniaxial state of stress take the form:

$$\varepsilon_x = \frac{\sigma_x}{E} + \alpha\Delta T \tag{113}$$

$$\varepsilon_y = \varepsilon_z = -\frac{\nu\sigma_x}{E} + \alpha\Delta T \tag{114}$$

and it follows that at some distance from the loading points

$$\sigma_x = \frac{N}{A} \tag{115}$$

$$\varepsilon_x = \frac{N}{EA} + \alpha \Delta T \tag{116}$$

Beams subject to pure bending

BEAMS SYMMETRIC ABOUT THE VERTICAL PLANE $y = 0$, AND SUBJECT TO VERTICAL LOADING

Consider a beam subject to a uniform bending moment M_y over part of its length, produced for example by the loading shown in Figure 2.23. (Note that (110) implies that a uniform bending moment can only occur in the absence of shear forces.)

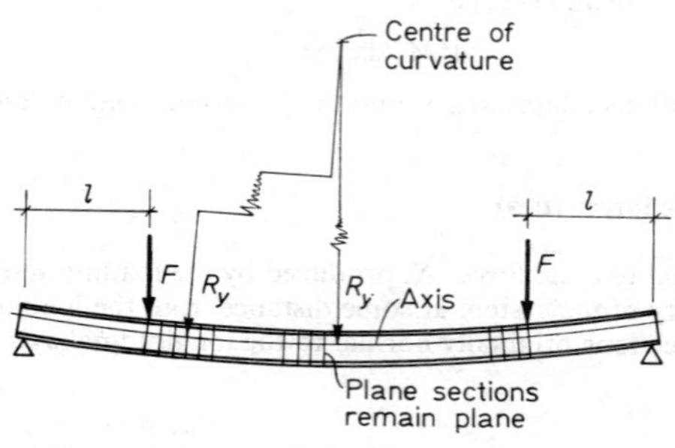

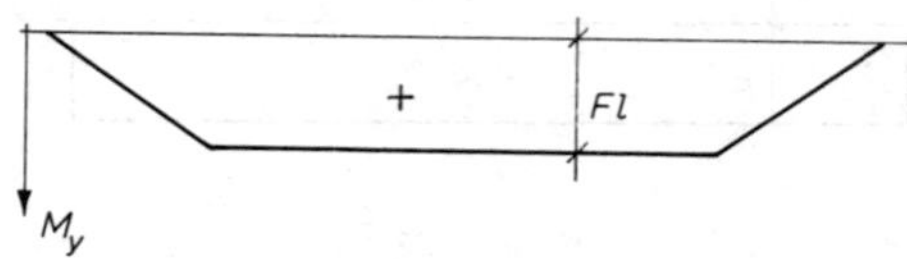

Figure 2.23

From the symmetry of the system it can be *deduced* that

1 the beam deforms in the vertical plane, and straight line generators parallel to the longitudinal axis deform into segments of circles with a common centre
2 planes originally normal to the axis remain plane and normal to the axis after deformation.

It can again be assumed that

3 the only nonzero component of stress is σ_x.

The above three conditions are the fundamental assumptions made in the *engineering theory of the bending of beams.*

The surface containing those points in the beam at which $\varepsilon_x = 0$ is called the *neutral surface.* The intersection of the neutral surface with a cross section produces a line called the *neutral axis.*

From the geometry of the deformation, the uniaxial stress–strain relations (113, 114), and the requirement of axial equilibrium ($N = 0$), it follows that:

1 The neutral axis is given by the equation

$$z = 0 \tag{117}$$

that is, it is a horizontal straight line, coincident with the y coordinate line, and passing through the centroid of the section.

2

$$\sigma_x = \frac{M_y z}{I_y} \tag{118}$$

$$\frac{1}{R_y} = \frac{M_y}{EI_y} \tag{119}$$

where R_y is the vertical radius of curvature of the beam axis.

COMPOSITE BEAMS

Suppose the beam in the previous subsection is made of two materials of Young's modulus E_1 and E_2 respectively comprising areas A_1 and A_2 of the total cross section, as in Figure 2.24.

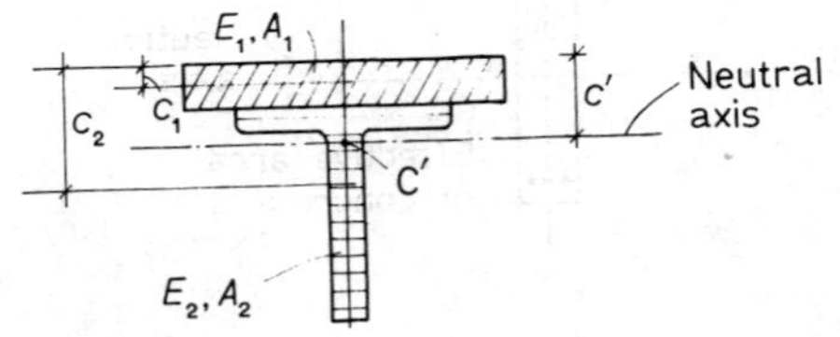

Figure 2.24

The three conditions of the engineering theory of the bending of beams discussed in the previous subsection still apply. It therefore follows that:

1 The neutral axis is a horizontal straight line passing through a point C′ called the *equivalent centroid of the cross section.* This is defined as being such that the first moment of *Young's modulus times area* about any axis passing through it is zero. Thus if c' is the distance of C′ from the upper boundary of the beam and c_1 and c_2 are the distances of the respective centroids of the areas A_1 and A_2 from the upper boundary, then

$$c' = \frac{E_1 A_1 c_1 + E_2 A_2 c_2}{(E_1 A_1 + E_2 A_2)} \tag{120}$$

2

$$[\sigma_x]_{A_1} = \frac{M_y z}{I'_y} \quad [\sigma_x]_{A_2} = \frac{E_2}{E_1}\frac{M_y z}{I'_y} \tag{121, 122}$$

$$\frac{1}{R_y} = \frac{M_y}{E_1 I'_y} \tag{123}$$

where $[\sigma_x]_{A_1}$ represents the axial stress in the area A_1, etc. I'_y is the *equivalent moment of inertia* of the cross section defined as

$$I'_y = \int_{A_1} (z^2)\,\mathrm{d}A_1 + \frac{E_2}{E_1}\int_{A_2} (z^2)\,\mathrm{d}A_2 \tag{124}$$

where $\int_{A_1}(\quad)\,\mathrm{d}A_1$ represents an integral taken over the area A_1, etc. In the above equations, the coordinates are relative to axes y and z passing through the equivalent centroid of the section.

REINFORCED CONCRETE BEAMS

A reinforced concrete beam behaves as a composite beam, except that where the concrete is in tension (that is, below the neutral axis for positive bending about the y axis) its stress bearing capacity is taken to be zero (Figure 2.25). Otherwise the conditions of the engineering theory of the bending of beams still apply.

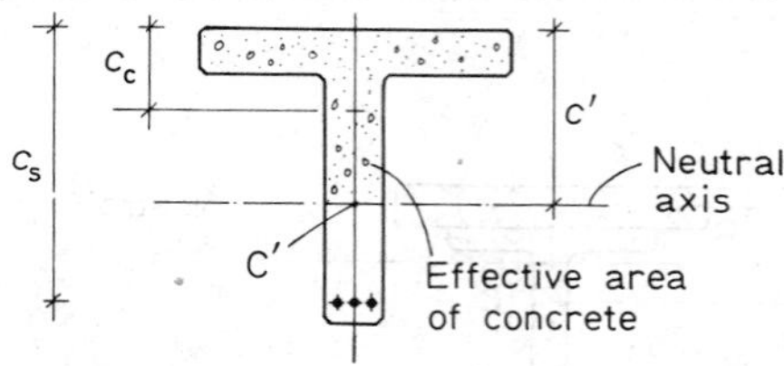

Figure 2.25

Let the subscripts c and s denote parameters associated respectively with the concrete and the steel. It then follows that:

1 The neutral axis is a horizontal straight line passing through the equivalent centroid whose distance c' from the upper boundary of the beam is given by

$$c' = \frac{E_cA_cc_c + E_sA_sc_s}{\quad} \tag{125}$$

(Note that since A_c and c_c are themselves functions of c', (125) is an implicit equation.)

2
$$[\sigma_x]_c = \frac{M_y z}{I'_y} \quad [\sigma_x]_s = \frac{E_s}{E_c}\frac{M_y z}{I'_y} \tag{126, 127}$$

$$\frac{1}{R_y} = \frac{M_y}{E_c I'_y} \tag{128}$$

where

$$I'_y = \int_{A_c} (z)^2\,\mathrm{d}A_c + \frac{E_s}{E_c}\int_{A_s} (z)^2\,\mathrm{d}A_s \tag{129}$$

Example A rectangular reinforced concrete beam with a single layer of reinforcement is shown in Figure 2.26. For this section

$$c' = \frac{E_s}{E_b}\frac{A_s}{b}\left[\left(1+\frac{2E_c}{E_s}\frac{b(d-e)}{A_s}\right)^{\frac{1}{2}}-1\right] \tag{130}$$

$$I'_y = \frac{bc'^3}{3}+\frac{E_s}{E_c}A_s[d-(c'+e)]^2 \tag{131}$$

Note that the ratio E_s/E_c is generally taken to be 15.

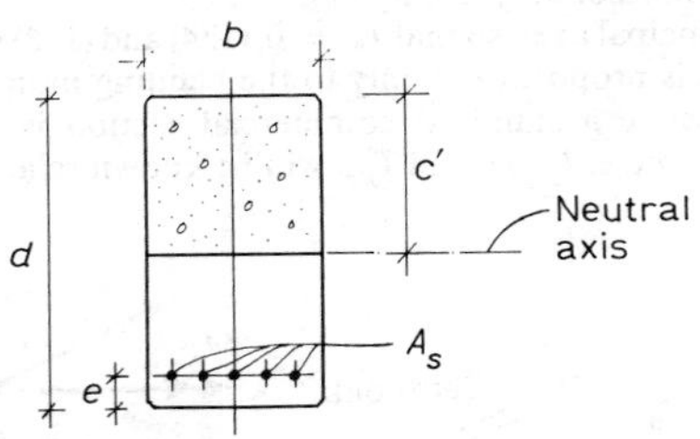

Figure 2.26

BEAMS OF ASYMMETRIC SECTION SUBJECT TO BOTH VERTICAL AND HORIZONTAL LOADING

Consider again a beam of homogeneous material. The general case of pure bending occurs when the beam is of asymmetric section and is subject to uniform bending moments M_y and M_z (Figure 2.27) over part of its length.

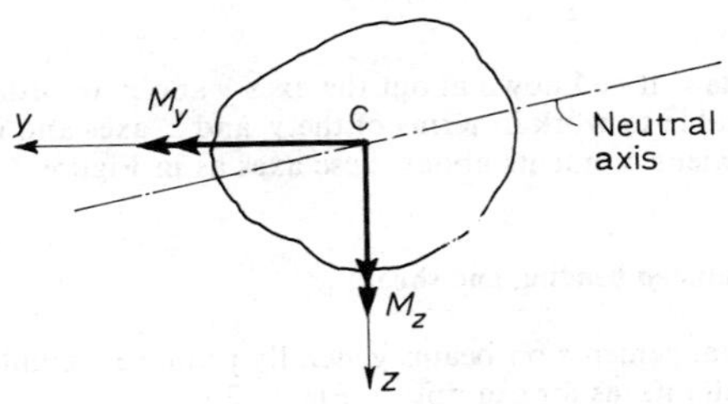

Figure 2.27

From the symmetry of the system is can be *deduced* that straight-line generators parallel to the axis of the beam deform into curves of constant horizontal and vertical curvature. The other conditions discussed on page 2–28 still apply.

From the geometry of the deformation, the uniaxial stress–strain relations and the requirement that $N = 0$, it follows that:

1 The neutral axis is given by the equation

$$(M_yI_z+M_zI_{yz})z-(M_zI_y+M_yI_{yz})y = 0 \tag{132}$$

that is, it is a straight line passing through the centroid of the section, as shown in Figure 2.27.

2

$$\sigma_x = \frac{(M_y I_z + M_z I_{yz})z - (M_z I_y + M_y I_{yz})y}{(I_y I_z - I_{yz}^2)} \tag{133}$$

$$\frac{1}{R_y} = \frac{(M_y I_z + M_z I_{yz})}{E(I_y I_z - I_{yz}^2)} \quad \frac{1}{R_z} = \frac{-(M_z I_y + M_y I_{yz})}{E(I_y I_z - I_{yz}^2)} \tag{134, 135}$$

where R_z is the horizontal radius of curvature of the beam axis.

Note:

1 If the loading is vertical so that $M_z = 0$, equation (135) indicates that the deformed beam is curved horizontally, i.e. $R_z \neq 0$.

2 If y and z are principal axes, so that $I_{yz} = 0$, (134) and (135) indicate that the curvature about each axis is proportional only to the bending moment about that axis.

In some cases, where a standard commercial section is mounted obliquely, as in Figure 2.28(a) for example, $I_{y'}$, $I_{z'}$ and $I_{y'z'}$ will be known relative to the axes y', z', whilst

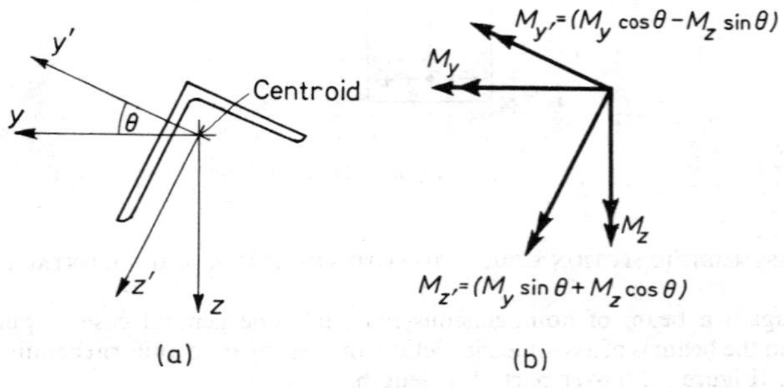

Figure 2.28

the bending moments will be known about the axes y and z. In order to use the results (132–135) it is preferable to work in terms of the y' and z' axes and resolve the bending moments into equivalent moments about these axes as in Figure 2.28(b).

Beams subject to combined bending and shear

Practical loading arrangements on beams generally produce a combination of bending and shear stress resultants, as for example in Figure 2.19.

BEAMS SYMMETRIC ABOUT THE VERTICAL PLANE $y = 0$, AND SUBJECT TO VERTICAL LOADING

Consider a point in a beam at which both M_y and S_z are nonzero. The presence of S_z then implies the existence of the shear stresses τ_{xz} on the cross section and corresponding shear strains γ_{xz}, and much of the symmetry of the deformation of a beam under a uniform bending moment is lost. In particular, plane sections no longer remain plane.

The following approximate analysis of the problem is due to St Venant.[12] It is assumed that the direct stresses σ_x and curvature $(1/R_y)$ are the same as they would be if M_y were acting alone. They are therefore given by (118) and (119). The shear stresses in the beam are then obtained by considering the longitudinal equilibrium of the element of length dx shown shaded in the cross-sectional view of Figure 2.29. Thus employing equations (110) and (118), namely

$$\frac{dM_y}{dx} = S_z \quad \sigma_x = \frac{M_y z}{I_y}$$

it can be shown that the *mean* longitudinal shear stress τ on the surface ABCD is given by

$$\tau = \frac{S_z A_e c_e}{b_e I_y} \tag{136}$$

where A_e is the cross-sectional area of the element, c_e is the distance of its centroid from the neutral axis, and b_e is the length of the curve joining AB (Figure 2.29).

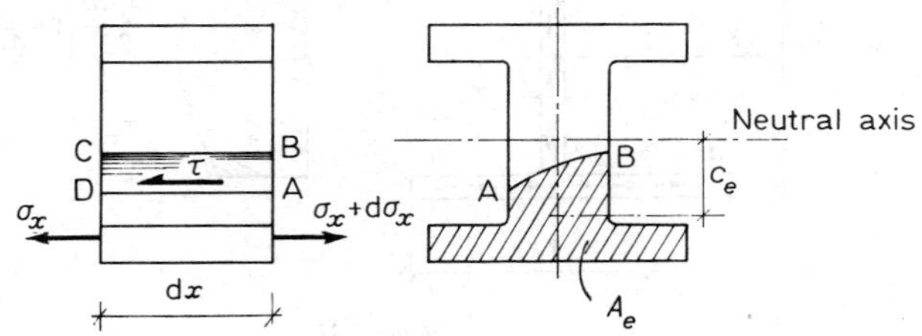

Figure 2.29

τ can then be related to the shear stresses τ_{xy} and τ_{xz} on the cross section as follows. If the cut surface ABCD is a horizontal plane (i.e., it is a z-coordinate surface) then τ is the mean value of the shear stress component τ_{zx} on that surface. Whence, since $\tau_{zx} = \tau_{xz}$, it follows that τ is also the mean value of τ_{xz} on the line AB. For thin sections, we assume that τ_{xz} is uniformly distributed across the width so that

$$\tau_{xz} = \tau \tag{137}$$

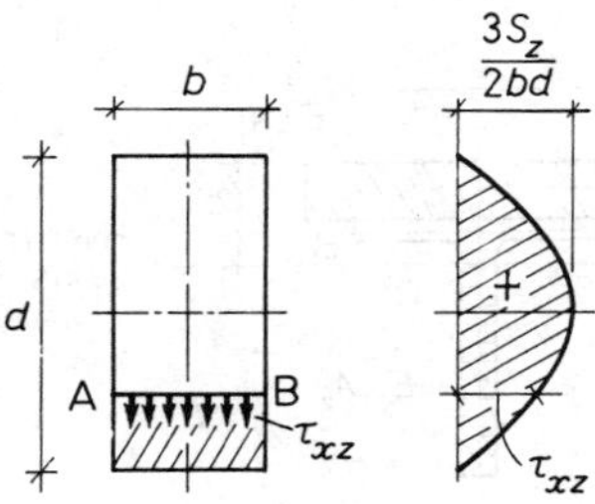

Figure 2.30

Thus for the rectangular section shown in Figure 2.30, (136) gives the following parabolic distribution of shear stress on the cross section:

$$\tau_{xz} = \frac{3S_z}{2bd^3}(d^2 - 4z^2) \tag{138}$$

If the cut surface ABCD is a vertical plane (a y-coordinate surface) then τ is the mean value of the shear stress component τ_{yx} on that surface, or the mean value of τ_{xy} on the line AB. Thus for an I section, the shear stresses in the flanges are as shown in Figure 2.31.

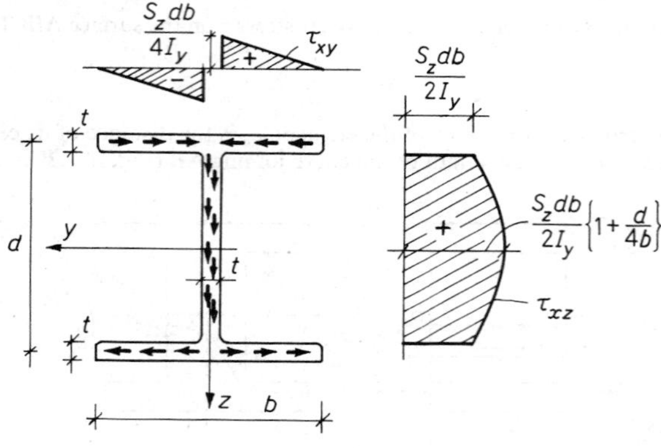

Figure 2.31

COMPOSITE BEAMS

The existence of the longitudinal shear stress τ (Figure 2.29) is of special significance in built-up composite beams, because this stress has to be transmitted between the separate components of the beams by means of suitable bonds such as welds, rivets or shear connectors.

Thus consider a beam composed of two materials of Young's modulus E_1 and E_2 respectively comprising areas A_1 and A_2 of the total cross section (Figure 2.32). The position of the neutral axis and the equivalent moment of inertia of the cross section

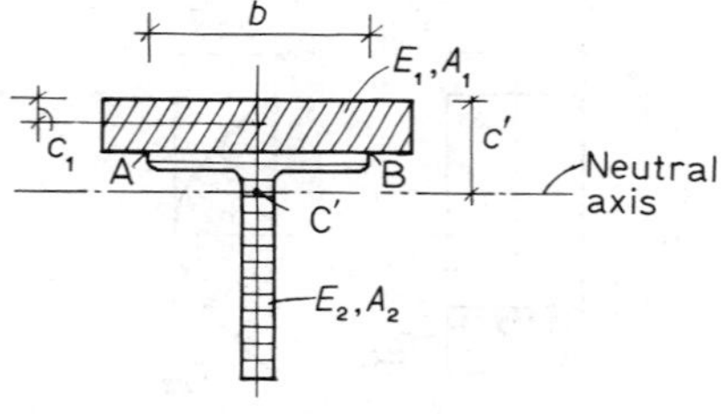

Figure 2.32

are again given by (120) and (124). Whence, employing the assumptions of St Venant's theory, it can be deduced that the mean longitudinal shear stress at the interface AB is given by

$$\tau = \frac{S_z A_1(c' - c_1)}{bI'_y} \tag{139}$$

whilst the corresponding longitudinal shear force/unit length of beam F is given by

$$F = b\tau \tag{140}$$

If the beam were composed say of a concrete slab connected to a steel T-section joist, then F would be transmitted by stud shear connectors of the type shown in Figure 2.33 which would be welded onto the top face of the T section. Supposing that the working shear strength of each connector were known experimentally to be F_s, then the connectors would need to be distributed at a concentration of F/F_s per unit length of beam.

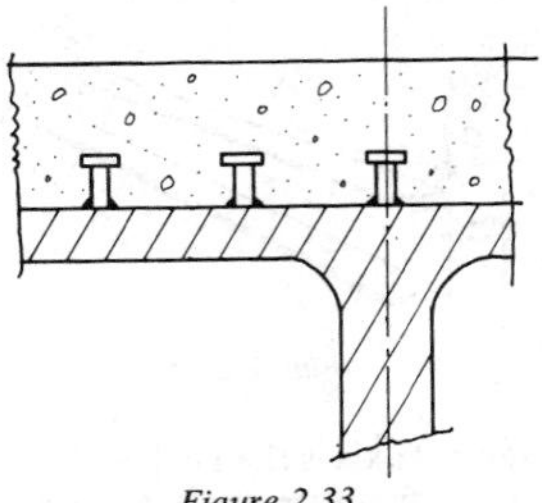

Figure 2.33

THE SHEAR CENTRE (BEAMS ASYMMETRIC ABOUT THE VERTICAL PLANE)

In a beam of asymmetric cross section the shear stresses given by St Venant's theory contribute to a torque T. Consider, for example, the shear stresses produced in the

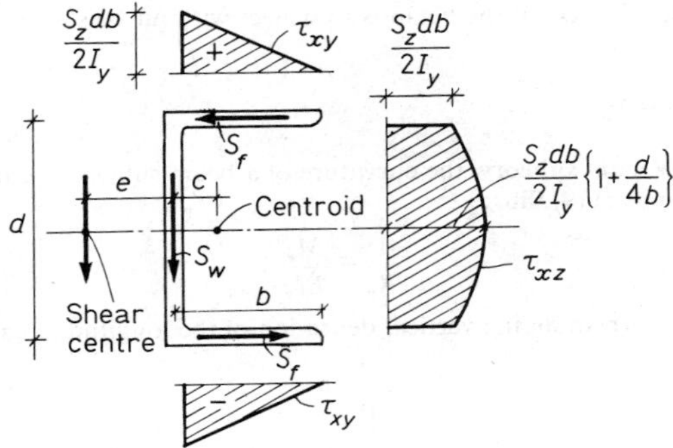

Figure 2.34

channel section shown in Figure 2.34. They are statically equivalent to the stress resultants S_f acting in the two flanges, and S_w in the web, where

$$S_w = S_z \tag{141}$$

$$S_f = \frac{S_z b^2 \, dt}{4I_y} \tag{142}$$

and because of the asymmetry of the section, they produce a torque T acting about the longitudinal axis of the channel given by

$$T = S_z c + \frac{S_z b^2 d^2 t}{4I_y} \tag{143}$$

Now, an important assumption of St Venant's theory is that the beam deflects vertically without twist. Thus it can be deduced that if the loading on the beam is such that it produces the torque T, then twisting does not in fact occur. (If the loading did not produce T then some twisting of the beam would be necessary in order to modify the torque obtained in (143)). T can be applied by positioning the vertical loading so that its resultant at any cross section lies at a suitable distance from the centroid. Thus the torque in the channel can be produced by the loading shown in Figure 2.35. The point at which the

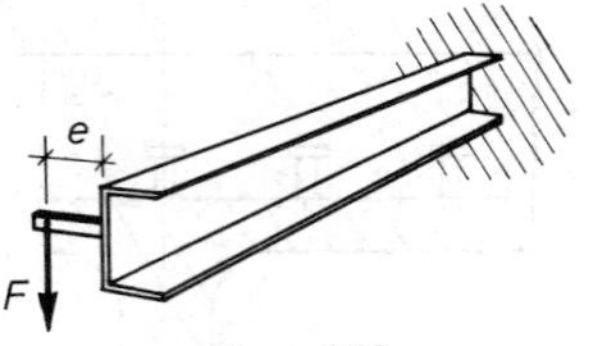

Figure 2.35

vertical resultant crosses the neutral axis is then called the *shear centre*, and for the channel section it is located at a distance e from the web (Figure 2.34) where

$$e = \frac{b^2 d^2 t}{4I_y} \tag{144}$$

The positions of the shear centres of various cross sectional shapes are shown in Table 2.3.

The *shear axis* of the beam is defined as the line passing through the shear centres of its cross sections, and by definition, the resultants of all lateral forces acting on the beam must pass through this axis if the beam is to deflect without twist.

Deflection of beams

According to St Venant's theory, the curvature of a beam subject to combined bending and shear is given by (119) thus:

$$\frac{1}{R_y} = \frac{M_y}{EI_y} \tag{119}$$

Suppose u_z is the corresponding vertical deflection of the longitudinal axis of the beam.

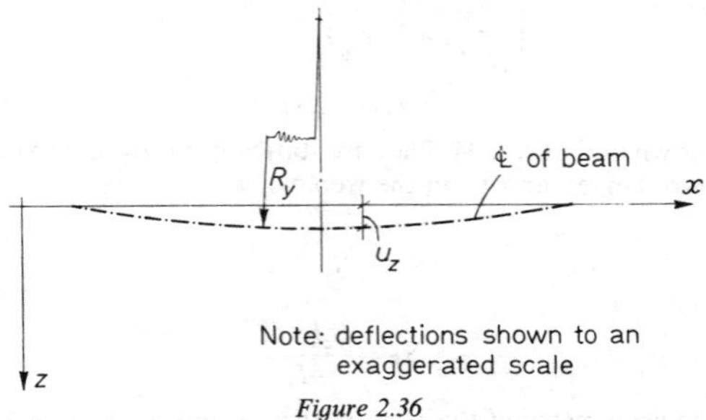

Figure 2.36

Then from the geometry of the deformation (Figure 2.36), it can be shown that

$$\frac{1}{R_y} = -\frac{\mathrm{d}^2 u_z}{\mathrm{d}x^2} \Bigg/ \left(1 + \left(\frac{\mathrm{d}u_z}{\mathrm{d}x}\right)^2\right)^{\frac{3}{2}} \tag{145}$$

In practice, the slopes of beams are extremely small and the denominator of the right-

Table 2.3 SHEAR CENTRES OF THE WALLED SECTIONS

Section	*Position of shear centre* Q

1. Channel

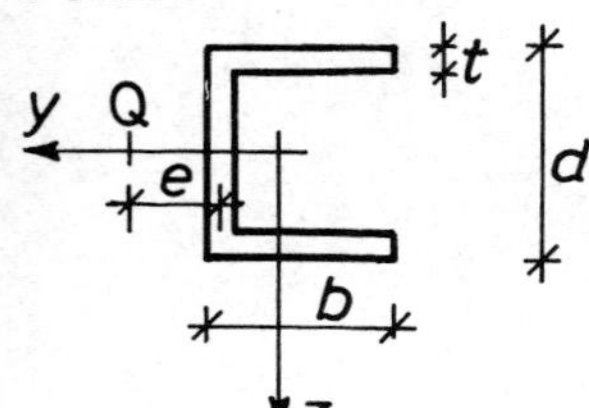

$$e = d\left(\frac{H_{yz}}{I_y}\right)$$

where H_{yz} is the product of inertia of the half section (above the y axis).

If t is uniform

$$e = \frac{b^2 d^2 t}{4I_y}$$

2. Lipped channel.

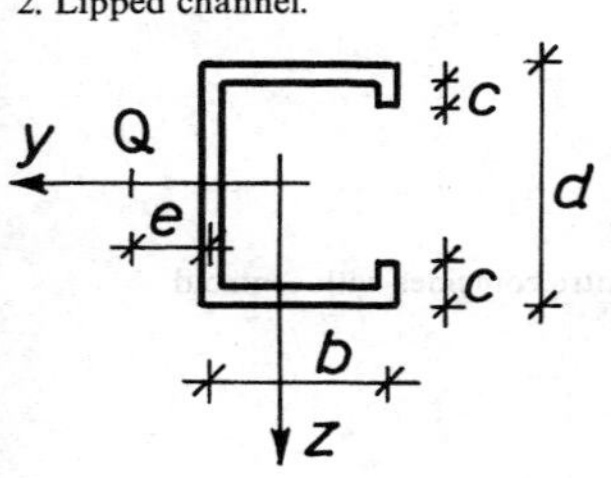

Values of (e/d)

c/d \ b/d	1.0	0.8	0.6	0.4	0.2
0	0.430	0.330	0.236	0.141	0.055
0.1	0.477	0.380	0.280	0.183	0.087
0.2	0.530	0.425	0.325	0.222	0.115
0.3	0.575	0.470	0.365	0.258	0.138
0.4	0.610	0.503	0.394	0.280	0.155
0.5	0.621	0.517	0.405	0.290	0.161

3. Hat section

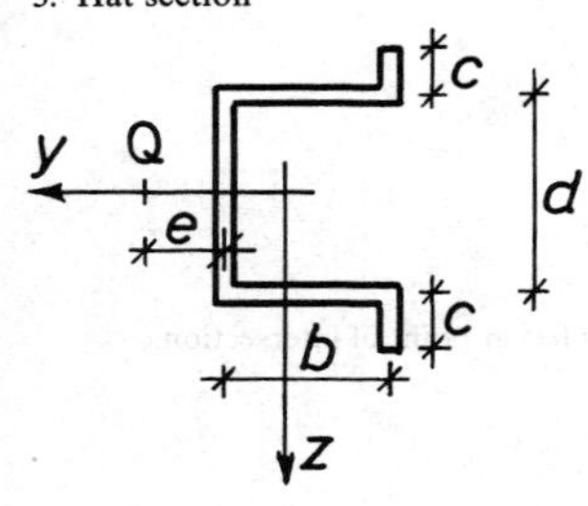

Values of (e/d)

c/d \ b/d	1.0	0.8	0.6	0.4	0.2
0	0.430	0.330	0.236	0.141	0.055
0.1	0.464	0.367	0.270	0.173	0.080
0.2	0.474	0.377	0.280	0.182	0.090
0.3	0.453	0.358	0.265	0.172	0.085
0.4	0.410	0.320	0.235	0.150	0.072
0.5	0.355	0.275	0.196	0.123	0.056
0.6	0.300	0.225	0.155	0.095	0.040

4. I section

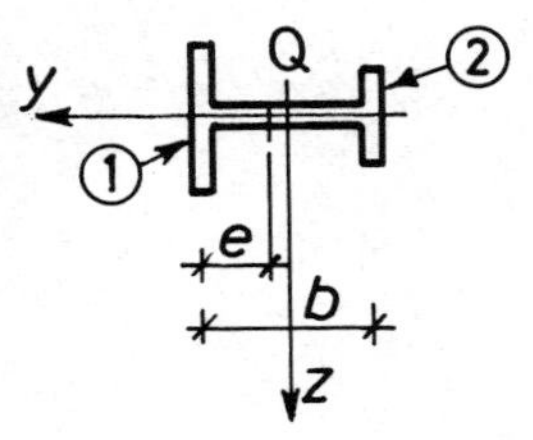

$$e = \frac{bI_2}{I_1 + I_2}$$

where I_1 and I_2 respectively denote the moments of inertia about the y axis of flange 1 and flange 2.

Section	Position of shear centre Q
5. Split circle 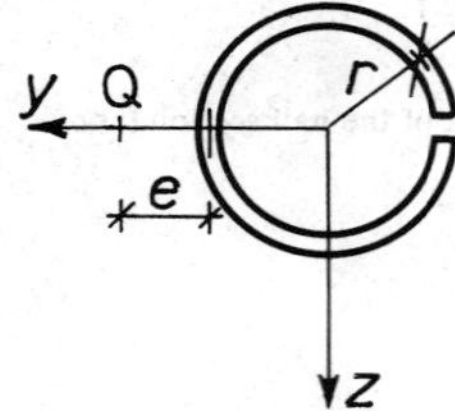	$e = r$
6. Z section 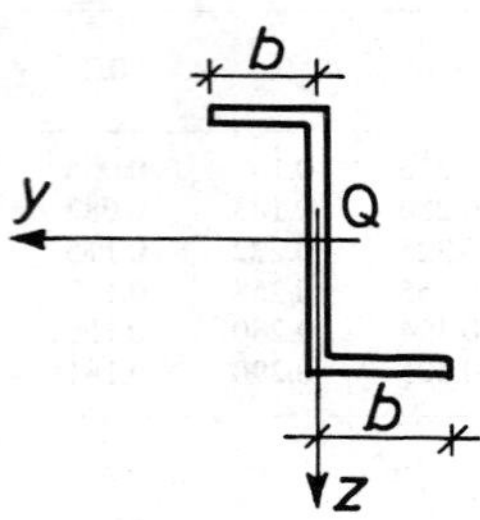	Shear centre coincides with centroid
7. Sections with elements intersecting at a single point 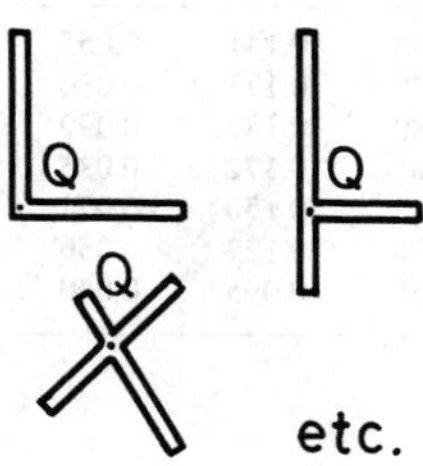	Shear centre lies at point of intersection.

hand side of (145) can be taken to be equal to unity. Whence, combining (119) and (145) gives the following differential equation:

$$\frac{d^2 u_z}{dx^2} + \frac{M_y}{EI_y} = 0 \tag{146}$$

called the *differential equation of beams*. For statically determinate beams, where M_y can be found as a function of x, this second-order equation can be solved subject to boundary conditions by double integration. The solution $u_z(x)$ is then the deflected shape

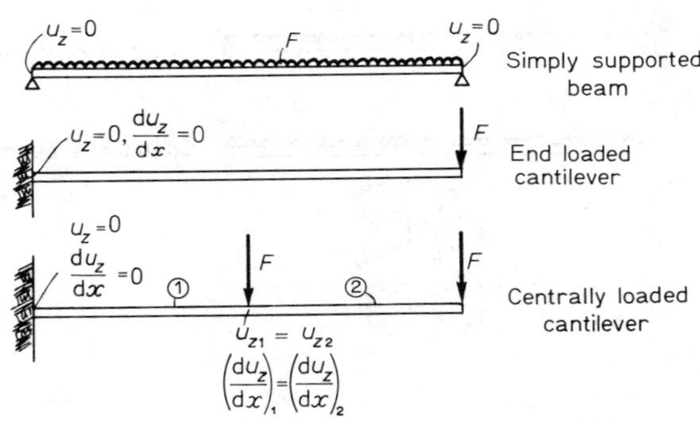

Figure 2.37

of a beam produced by the applied loading. Examples of the boundary conditions for particular cases are shown in Figure 2.37. Special techniques, such as the step function method[13] and the moment-area method[14] have been devised to simplify the analysis.

The differential equation of beams can be expressed in two further forms using the results (110) and (112). Thus from (110) we have

$$\frac{d^3 u_z}{dx^3} + \frac{S_z}{EI_y} = 0 \tag{147}$$

whilst from (112) we have

$$\frac{d^4 u_z}{dx^4} - \frac{q}{EI_y} = 0 \tag{148}$$

Equation (148), expressing the deflections of beams in terms of the lateral loading, is directly equivalent to the three-dimensional Navier equations (61–63), and can be solved if the boundary conditions are expressed in terms of the displacements. The solution of this equation as opposed to (146), is necessary when a beam is statically indeterminate, that is when M_y cannot be found in advance. Examples of the required displacement boundary conditions for particular cases are shown in Figure 2.38.

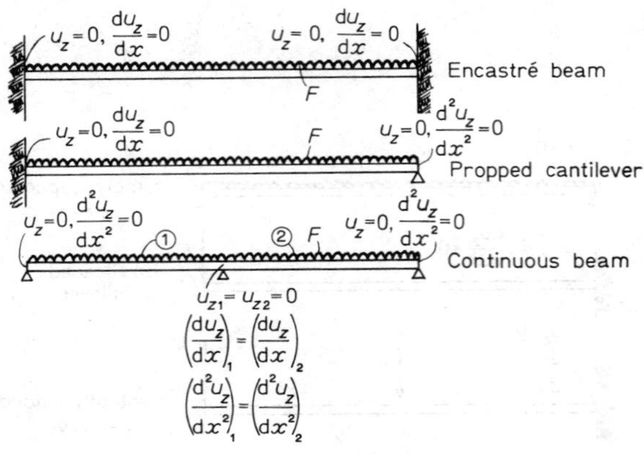

Figure 2.38

An interesting modification of (148) occurs when a beam rests on an elastic foundation. Suppose the stiffness of the foundation if k per unit length of beam. Then in addition to the vertical applied loading q, the foundation resists the deflection of the beam with distributed forces equal to ku_z per unit length. (148) then takes the form

$$\frac{d^4u_z}{dx^4}+ku_z-\frac{q}{EI_y}=0 \tag{149}$$

Examples of the solution of this equation are given by Hetenyi.[15]

Bars subject to a uniform torque

BARS OF CIRCULAR CROSS SECTION

Consider a bar subject to a uniform torque T produced for example by the loading shown in Figure 2.39.

From the symmetry of the system it can be deduced that

1 the bar twists about its longitudinal axis,
2 planes originally normal to the axis remain plane and normal to the axis and rotate like rigid laminae,
3 the rotation θ of any plane is proportional to its distance along the beam.

From the geometry of the deformation and the shear stress–strain relations (49–51), it follows that

$$\tau_{xt} = \frac{Ir}{J} \tag{150}$$

$$\frac{d\theta}{dx} = \frac{T}{GJ} \tag{151}$$

where τ_{xt} is the shear stress on the cross section at a distance r from the axis, and tangential to the circle of radius r (Figure 2.40). J is a sectional constant, equal in this case to the polar moment of inertia I_p about the longitudinal axis.

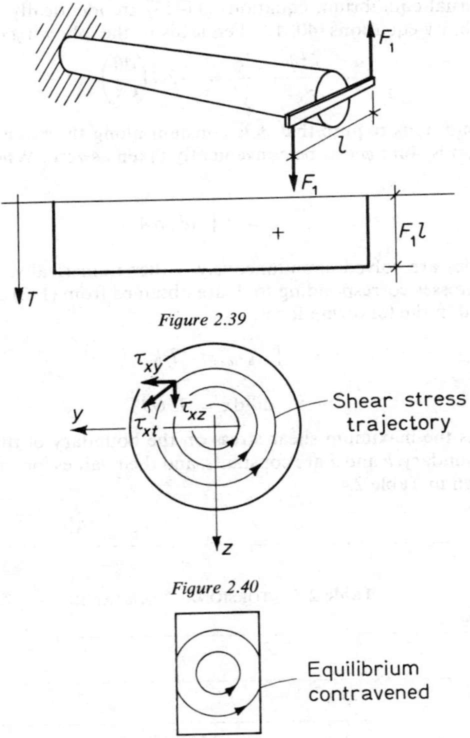

Figure 2.39

Figure 2.40

Figure 2.41

The quantity $d\theta/dx$ being the rate of change of rotation with x is called the *twist* of the bar, and is clearly uniform when the bar is subject to uniform torque.

BARS OF ARBITRARY CROSS SECTION

The three assumptions of the subsection on 'Deflection of Beams' (2–36), can be shown to lead to impossible values of τ_{xt} at the boundaries of an arbitrary section, since in order to satisfy longitudinal equilibrium conditions, τ_{xt} must be tangential to those boundaries (Figure 2.41).

The theory for the analysis of bars of arbitrary section is again due to St Venant.[16] Thus the assumption in the previous subsection that plane sections remain plane is relaxed, and a point P is assumed to have an axial displacement u_x given by

$$u_x = \frac{\mathrm{d}\theta}{\mathrm{d}x}\psi(y, z) \tag{152}$$

u_x is called the *warping* of the section, and is directly proportional to the twist, but is independent of x. The shear stresses τ_{xy} and τ_{xz} are then expressed in terms of a stress function $\phi(y, z)$ by the equations

$$\tau_{xy} = \partial\phi/\partial z \quad \tau_{xz} = -\partial\phi/\partial y \tag{153, 154}$$

so that the internal equilibrium equations (11–13) are identically satisfied. Satisfaction of the compatability equations (40, 42) then leads to the following equation:

$$\frac{\partial^2\phi}{\partial x^2}+\frac{\partial^2\phi}{\partial y^2} = -2G\left(\frac{\mathrm{d}\theta}{\mathrm{d}x}\right) \tag{155}$$

Equilibrium conditions require that ϕ is constant along the boundaries of the section, and if the section is solid ϕ can be conveniently taken as zero. Whence it can be shown that

$$T = 2\int_A (\phi)\,\mathrm{d}A \tag{156}$$

(155) and (156) are solved simultaneously, either numerically, or experimentally,[16] and the shear stresses corresponding to T are obtained from (153) and (154). The results can be expressed in the following form:

$$[\tau_{xb}]_{\max} = T/k \tag{157}$$

$$\mathrm{d}\theta/\mathrm{d}x = T/GJ \tag{158}$$

where $[\tau_{xb}]_{\max}$ is the maximum shear stress on the boundary of the section and is tangential to the boundary. k and J are constants, and their values for various cross-sectional shapes are shown in Table 2.4.

Table 2.4 TORSIONAL CONSTANTS

Section		k	*J*
1 Rectangle	*d/b*		
	1.0	0.208 (b^2d)	0.1406 (b^3d)
	1.2	0.219 (b^2d)	0.166 (b^3d)
	1.5	0.231 (b^2d)	0.196 (b^3d)
	2.0	0.246 (b^2d)	0.229 (b^3d)
	2.5	0.258 (b^2d)	0.249 (b^3d)
	3.0	0.267 (b^2d)	0.263 (b^3d)
	4.0	0.282 (b^2d)	0.281 (b^3d)
	5.0	0.291 (b^2d)	0.291 (b^3d)
	10.0	0.312 (b^2d)	0.312 (b^3d)
	α	$\frac{1}{3}$ (b^2d)	$\frac{1}{3}$ (b^3d)

Section	k	J
2 Equilateral triangle 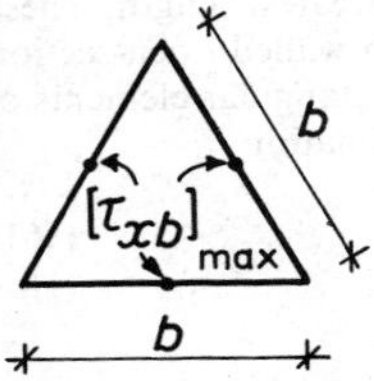	$b^3/20$	$\sqrt{3}b^4/80$
3 Right isosceles triangle 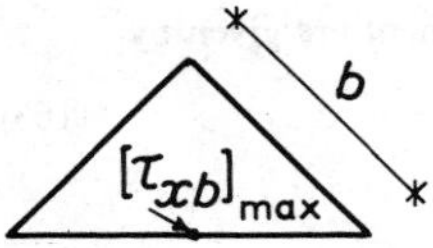	$0.0554\ b^3$	$0.0261\ b^4$
4 Hexagon 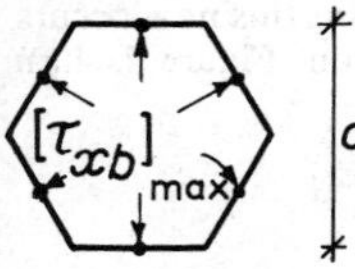	$0.217\ Ad$	$0.133\ Ad^2$
5 Circle 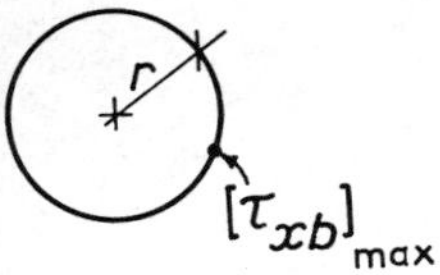	$\pi r^3/2$ $1.5708\ r^3$	$\pi r^4/2$ $1.5708\ r^4$
6 Hollow circle 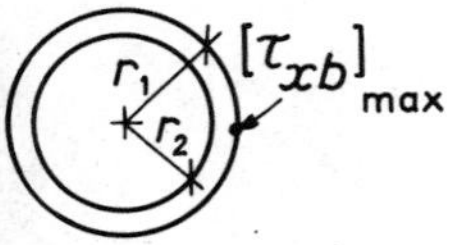	$\frac{\pi}{2}\left(\frac{r_1^4-r_2^4}{r_1}\right)$	$\frac{\pi}{2}(r_1^4-r_2^4)$
7 Ellipse 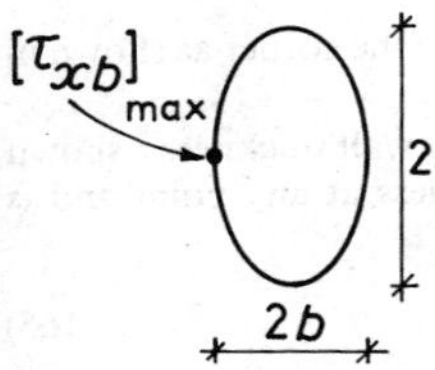	$\pi ab^2/2$	$\pi a^3b^3/(a^2+b^2)$

For the narrow rectangular section shown in Figure 2.42

$$k = t^2d/3 \quad J = t^3d/3 \tag{159, 160}$$

and the maximum shear stress occurs along the boundaries of greatest length. These results can be used to determine the torsional properties of a thin-walled open-section bar, supposing that the cross section can be divided into narrow rectangular elements of thickness t_e and length d_e, for it can be shown that to a first approximation

$$J = \sum_{\text{elements}} \frac{t_e^3 d_e}{3} \tag{161}$$

Thus for the I section shown in Figure 2.43

$$J = \frac{2d_f t_f^3 + d_W t_W^3}{3} \tag{162}$$

The shear stress $[\tau_{xb}]_e$ along the boundaries of a particular element are given by

$$[\tau_{xb}]_e = T/k_e \tag{163}$$

where

$$k_e = J/t_e \tag{164}$$

$[\tau_{xb}]_e$ however, is not the maximum shear stress on the cross section, for this now occurs at the reentrant corners. Thus, in a constant-thickness channel section (Figure 2.44(a))

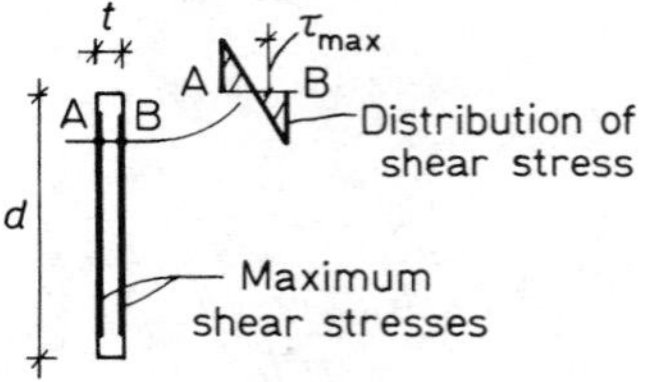

Figure 2.42

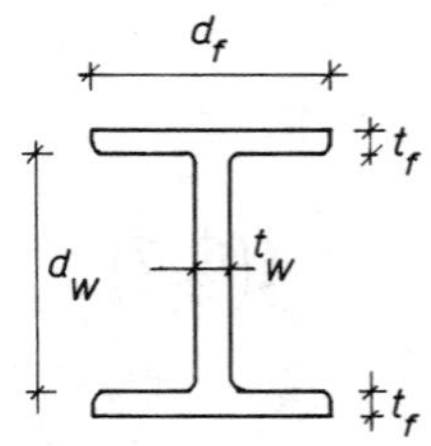

$[\tau_{xb}]_{max}$ occurs at point P, and is related to $[\tau_{xb}]_e$ and the radius of the corner as shown in Figure 2.44(b).[16]

In a thin walled closed-section bar, such as the tube of varying wall thickness t shown in Figure 2.45, the shear stress τ_{xb} is uniform across the thickness at any point and is tangential to the surface of the tube. It is given by

$$\tau_{xb} = T/2At \tag{165}$$

where A is the gross cross-sectional area.
J in (158) is given by

$$J = 4A^2 \Big/ \left(\oint \frac{ds}{t} \right) \tag{166}$$

where ds is an element of length round the tube (Figure 2.45). A further quantity q called the *shear flow* is defined at a point in the tube wall by the equation

$$q = \tau_{xb} t \tag{167}$$

It is then apparent from (165) that the shear flow is independent of t.

In multicell thin-walled bars, as shown in Figure 2.46, the concept of circulatory shear flows q_1, q_2, q_3 is introduced, a concept which automatically satisfied the conditions of

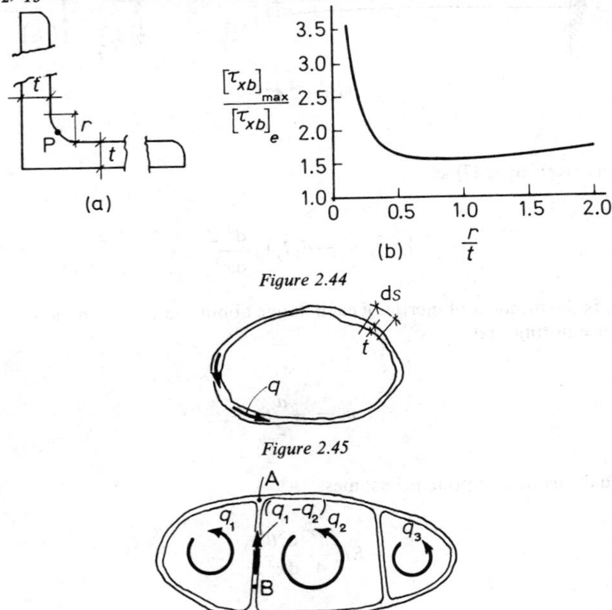

Figure 2.44

Figure 2.45

Figure 2.46

longitudinal equilibrium at junctions such as A. The shear flow at point B for example is then given by $(q_1 - q_2)$. The shear flows and the twist of the bar corresponding to a certain applied torque are calculated from the four simultaneous equations:

$$T = 2q_1A_1 + 2q_2A_2 + 2q_3A_3 \tag{168}$$

$$\frac{d\theta}{dx} = \frac{1}{2A_1G} \oint_1 \frac{q}{t} ds_1 = \frac{1}{2A_2G} \oint_2 \frac{q}{t} ds_2 = \frac{1}{2A_3G} \oint_3 \frac{q}{t} ds_3 \qquad (169, 170, 171)$$

where $\oint_1$ represents the contour integral taken round cell 1, etc.

Nonuniform torsion

Nonuniform torsion in a bar is defined to occur when the twist dθ/dx varies along its length. This situation arises when the warping assumed in St. Venant's theory is restrained at a rigid support, or when the torque exerted by the applied loading is nonuniform.

The nature of the modification necessary to St Venant's theory can be appreciated by considering Figure 2.47. Since $d\theta/dx$ is not constant, the flanges are curved in the z plane. Considering the flanges as subsidiary beams, they contain shear forces $[S_z]_f$ which are related to this curvature. The torque T therefore includes an extra component $[S_z]_f d$.

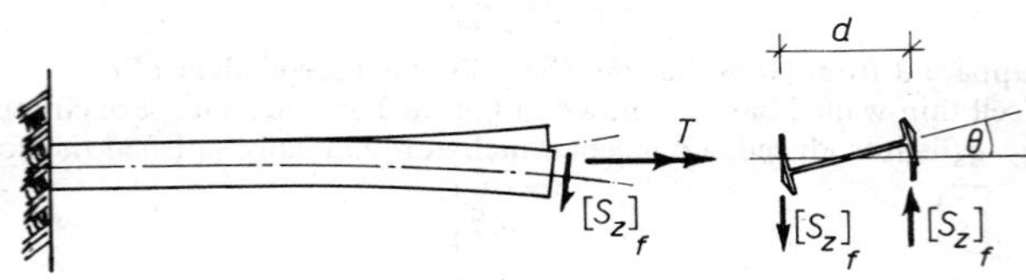

Figure 2.47

Now $[S_z]_f$ is given by (147) as

$$[S_z]_f = -E[I_y]_f \frac{d^3 u_z}{dx^3} \tag{172}$$

where $[I_y]_f$ is the moment of inertia of each flange about the y axis, and u_z is its displacement. Whence noting that

$$u_z = \pm\frac{d}{2}\theta \tag{173}$$

the additional torque component becomes

$$-EI_y \frac{d^2}{4} \frac{d^3\theta}{dx^3}$$

where I_y is the *total* moment of inertia of the cross section about the y axis. Combining this with the torque required for the uniform torsion of the bar, we obtain

$$T = GJ\frac{d\theta}{dx} - EI_y \frac{d^2}{4} \frac{d^3\theta}{dx^3} \tag{174}$$

(174) can be expressed in the more general form

$$T = GJ\frac{d\theta}{dx} - E\Gamma\frac{d^3\theta}{dx^3} \tag{175}$$

where Γ is a constant called the *warping factor*. Its values for various cross-sectional shapes are given in Table 2.5. The differential equation (175) can be solved for various

Table 2.5 WARPING FACTORS

Section	Γ
1 I section 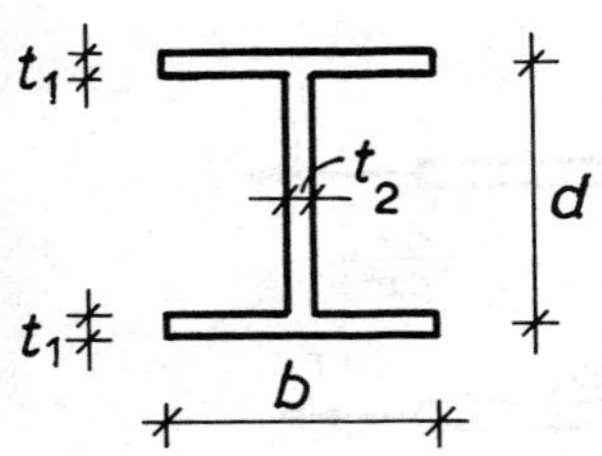	$\frac{1}{24}b^3d^2t_1$
2 Channel section 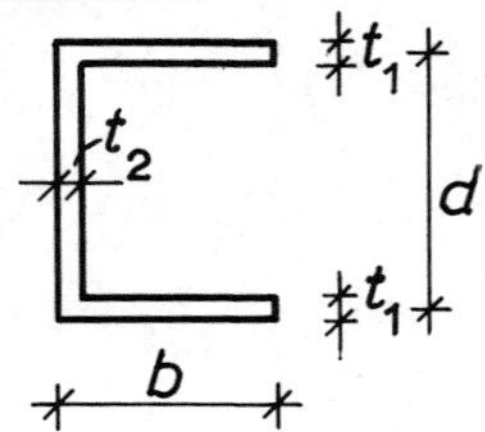	$\frac{1}{12}b^3d^2t_1\left(\frac{3bt_1+2dt_2}{6bt_1+dt_2}\right)$
3 Z section 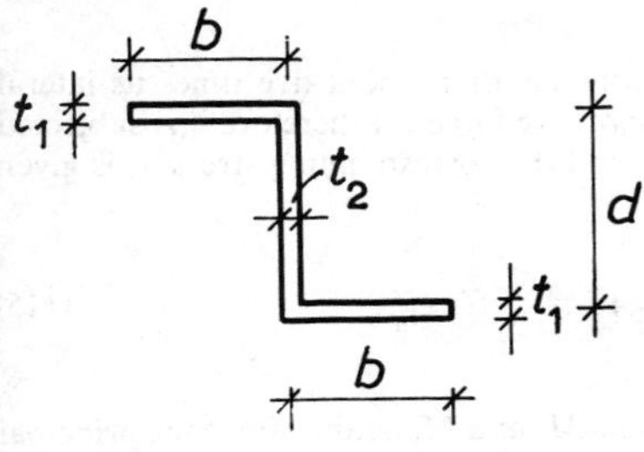	$\frac{1}{12}b^3d^2t_1\left(\frac{bt_1+2dt_2}{2bt_1+dt_2}\right)$
4 Thin-walled sections with elements intersecting at a single point 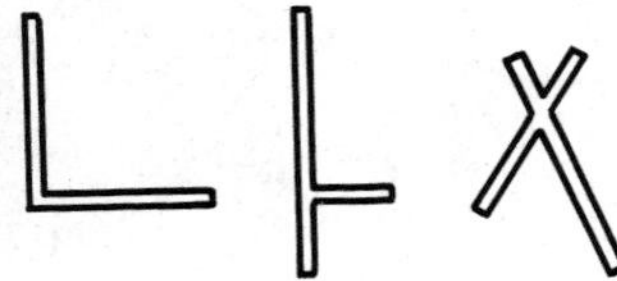	0

values of T applied to the bar, subject to boundary conditions in θ. Examples of these boundary conditions are shown in Figure 2.48.

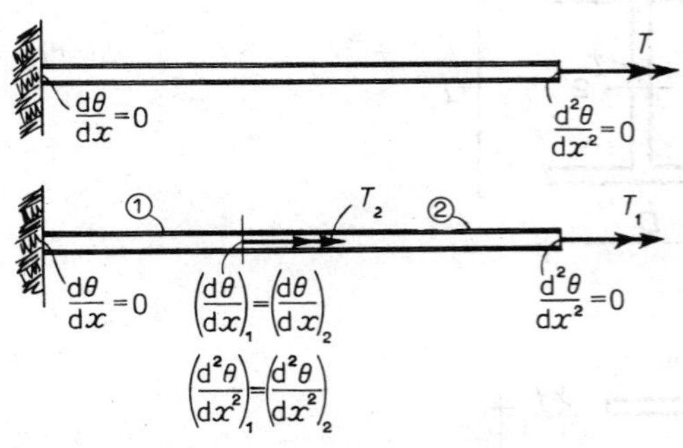

Figure 2.48

Bars subject to compressive forces

SHORT COLUMNS

If the geometry of a bar is such that its length is less than about five times its lateral dimensions, then it is usually stable under compressive forces. If therefore it is subjected to an axial compressive force F, then $N = -F$ and the corresponding stress σ_x is given by (115) as

$$\sigma_x = N/A \tag{115}$$

If further, the bar is subjected to bending moments M_y and M_z acting about the principal axes y and z, then by superposition

$$\sigma_x = \frac{N}{A} + \frac{M_y z}{I_y} - \frac{M_z y}{I_z} \tag{176}$$

and the neutral axis is given by the equation

$$\frac{M_y z}{I_y} - \frac{M_z y}{I_z} + \frac{N}{A} = 0 \tag{177}$$

Combined compressive forces and bending moments occur in the bar if the compressive force F is eccentrically positioned as shown in Figure 2.49. Thus if the resultant due to F passes at a distance n and m from the y and z axes respectively, then

$$N = -F \quad M_y = -Fn \quad M_z = +Fm \tag{178}$$

and

$$\sigma_x = -F\left(\frac{1}{A}+\frac{nz}{I_y}+\frac{my}{I_z}\right) \tag{179}$$

The neutral axis is then given by the equation

$$\frac{nz}{r_y^2}+\frac{my}{r_z^2}+1 = 0 \tag{180}$$

where r_y and r_z are the *radii of gyration* of the cross section defined respectively by the equations

$$r_y^2 = \frac{I_y}{A} \quad r_z^2 = \frac{I_z}{A} \tag{181}$$

Note that if the location of the neutral axis is known, then the maximum and minimum stresses on the section are located at those points which are at the greatest perpendicular distance from this axis. Their positions can easily be found graphically.

It is apparent from (180) that the location of the neutral axis depends only on the coordinates n and m defining the eccentricity of F. If this eccentricity is such that the

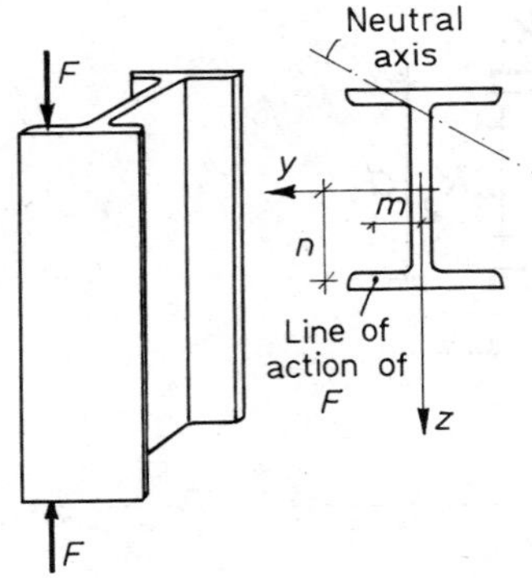

Figure 2.49

neutral axis falls outside the section, then the stress σ_x is negative (or compressive) at all points in the section. This situation arises if the stress resultant lies within an area called the *core* of the section. The dimensions of the cores of regular sections can be found analytically and some examples are given in Table 2.6.

Table 2.6 CORES OF SECTIONS

Section

1 Rectangle

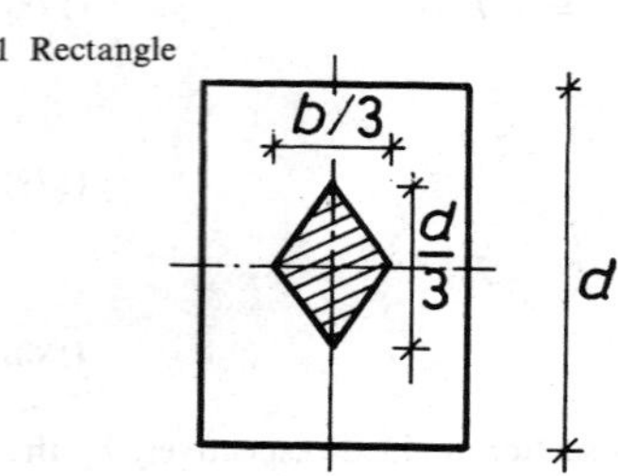

2 Circle

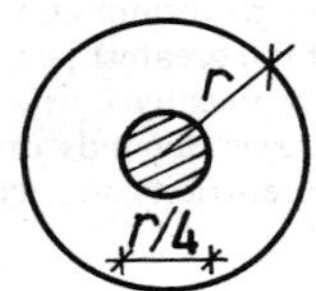

3 I section

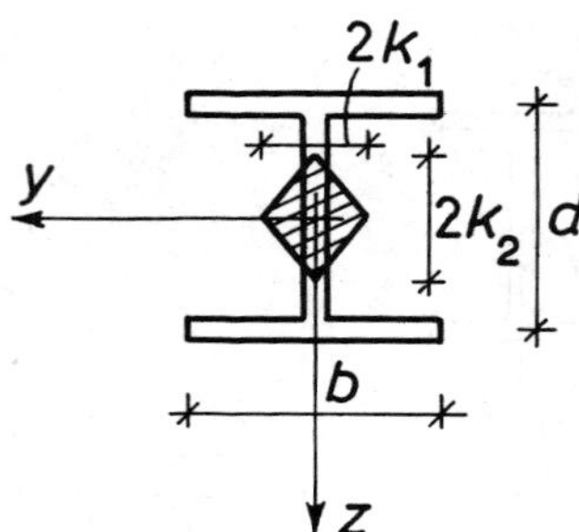

$k_1 = 2r_z^2/b$

$k_2 = 2r_y^2/d$

STRUTS

If the length of a bar is greater than about five times its lateral dimensions, then it can become unstable under compressive forces. Consider for example the pin-ended bar subjected to an axial compressive force F shown in Figure 2.50. If u_z is the lateral displacement in the z direction of a particular cross section, then the moment M_y exerted by F at the section is Fu_z. Thus from (146) we have the differential equation

$$\frac{d^2u_z}{dx^2}+\frac{Fu_z}{EI_y}=0 \tag{182}$$

One solution of (182) is $u_z = 0$, that is, the bar remains straight. However, further nonzero solutions for u_z occur for particular values of F called the *eigenvalues*. The lowest eigenvalue is the critical load F_{cr} of the bar, and can be regarded as the maximum load that can be carried before failure by lateral instability. It can be shown that F_{cr} is given by

$$F_{cr} = \frac{\pi^2 EI_y}{l^2} \tag{183}$$

whilst the corresponding deflected shape of the bar is sinusoidal and of arbitrary amplitude, taking the following form:

$$u_z = A \sin\left(\frac{\pi x}{l}\right) \tag{184}$$

The value for the critical load was first obtained by Euler, and a pin-ended bar subject to axial compression is often called an *Euler strut.*

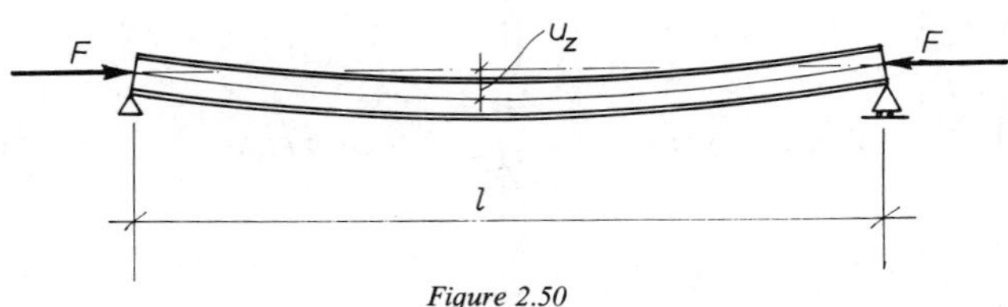

Figure 2.50

Dividing (183) by the area of the bar leads to the following expression for the *critical buckling stress* σ_{cr}:

$$\sigma_{cr} = \frac{\pi^2 E}{(l/r_y)^2} \tag{185}$$

where the quantity (l/r_y) is called the *slenderness ratio.*

When, as in most cases, $I_y \neq I_z$, the strut buckles first about the minor principal axis, that is about that axis for which the moment of inertia of the section is a minimum.

The critical buckling stresses of struts with other than pin-ended boundary conditions are given in Table 2.7. The corresponding *effective lengths* l_e are then defined so that the critical stresses can be given by an equation analogous to (185) namely

$$\sigma_{cr} = \pi^2 E/(l_e/r_y)^2 \tag{186}$$

Values for l_e are included in the table.

The above type of buckling is called *flexural buckling,* and occurs when the cross section of the strut has two axes of symmetry. For unsymmetrical sections, buckling may be accompanied by torsion as well as flexure, producing a correspondingly reduced critical load. Results for such cases are given by Bleich.[17]

DESIGN FORMULAE

The plot of σ_{cr} versus l/r for various strut lengths is the hyperbola shown in Figure 2.51. Clearly when l/r is very small, the critical stress becomes much greater than the yield stress σ_Y of the material, and the failure of the strut is brought about by the yielding of the material rather than by flexural buckling. If the struts were perfectly straight and the axial load had no eccentricity then the ultimate stresses σ_u would be given by the upper curve in Figure 2.51, that is, the elastic buckling hyperbola intersected by the horizontal

Table 2.7 CRITICAL BUCKLING LOADS OF STRUTS

All struts are of length l; $I_z > I_y$

lower end boundary condition	*upper end boundary condition*	*Mode*	P_{cr}	l_e
1 Hinge along y axis	hinge along y axis		$\pi^2 EI_y/l^2$	l
2 Clamped	clamped		$4\pi^2 EI_y/l^2$	$0.5\,l$
3 Clamped	hinge along y axis		$20.19\ EI_y/l^2$	$0.7\,l$
4 Clamped	free		$\pi^2 EI_y/4l^2$	$2.0\,l$
5 Hinge along z axis	hinge along z axis		Smaller of $4\pi^2 EI_y/l^2$ or $\pi^2 EI_z/l^2$	$0.5\,l$
6 Hinge along z axis	hinge along y axis		$20.19 EI_y/l^2$	$0.7\,l$
Special loading cases				
7 Pin-ended strut under end load P_1 and central load P_2			$(P_1 + P_2)_{cr} = \pi^2 EI/(kl)^2$ where $k \simeq 1/(2 - c^{\frac{1}{4}})$, $c = P_1/(P_1 + P_2)$	
8 Cantilever strut under uniformly distributed load q/unit length			$(q_{cr})l = \pi^2 EI/(1.122\,l)^2$	

'squash' line. However, tests show that the strengths of real struts are considerably reduced by initial imperfections when $\sigma_{cr} \simeq \sigma_Y$ as indicated by the lower curve in the figure. The following semi-empirical design formulae have been devised to account for this, giving the ultimate stresses of struts in terms of their geometrical and material properties:

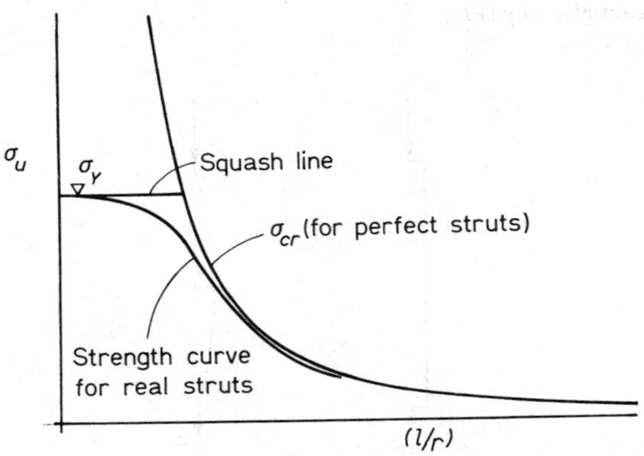

Figure 2.51

The Rankine formula. Some french design regulations are based on the simple interaction formula[18]

$$\frac{1}{\sigma_u} = \frac{1}{\sigma_{cr}} + \frac{1}{\sigma_Y} \tag{187}$$

or

$$\sigma_u = \frac{\sigma_Y}{\left(1 + \frac{\sigma_Y}{\pi^2 E}\left(\frac{l}{r}\right)^2\right)} \tag{188}$$

The Johnson parabola. The design formula recommended by the American Institute of Steel Construction[18, 19] uses the following parabola in the nonelastic range:

$$\sigma_u = \sigma_Y\left(1 - \frac{(l/r)^2 \sigma_Y}{4\pi^2 E}\right) \tag{189}$$

It is tangential to the buckling hyperbola at the point $\sigma_{cr} = 0.5\sigma_Y$.

The secant formula. The well known secant formula is derived assuming that the axial forces on the strut have an initial *eccentricity e* (Figure 2.52(a)). In this case it can be shown that

$$\sigma_u = \frac{\sigma_Y}{[1 + \eta \sec((\pi/2)\sqrt{(\sigma_u/\sigma_{cr})})]} \tag{190}$$

where η is given by

$$\eta = ec/r_y^2 \tag{191}$$

c is the distance from the neutral axis to the extreme fibre of the section.

The Perry–Robertson formula. Assuming that the strut has an initial *curvature* and that its maximum misalignment is e (Figure 2.52(b)), Ayrton and Perry derived the following formula:

$$\sigma_u = \tfrac{1}{2}[\sigma_Y+(1+\eta)\sigma_{cr}]-\left[\left(\frac{\sigma_Y+(1+\eta)\sigma_{cr}}{2}\right)^2-\sigma_Y\sigma_{cr}\right]^{\frac{1}{2}} \tag{192}$$

where η is again given by (191).

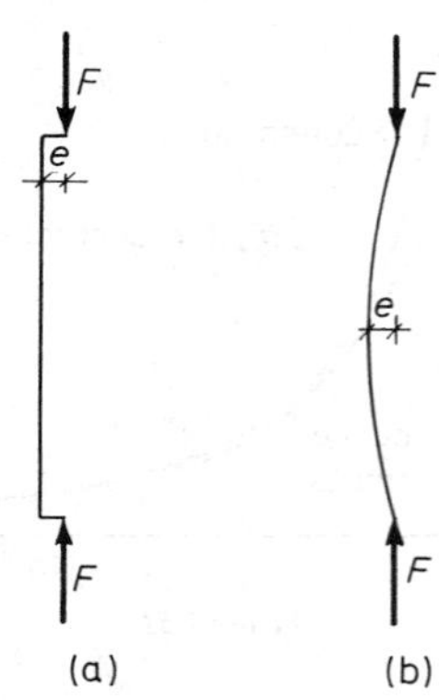

Figure 2.52

Robertson showed by experiment that a good but conservative prediction of the real strengths of struts can be obtained by taking η proportional to (l/r), as follows:

$$\eta = 0.003(l/r) \tag{193}$$

The current British Standard governing the design of steel girder bridges[20] gives strengths based on this assumption. More recent experiments by Dutheil[18] have led to the modified expression

$$\eta = 0.3(l/100r)^2 \tag{194}$$

and this is the basis of the recommendations of the British Standard concerned with the use of structural steel in building.[21]

Virtual work and strain-energy of frameworks

The state of stress and strain at all points in a framework can be expressed in terms of the stress resultants at those points, using the appropriate equations of the previous sections and the stress–strain relations. The internal virtual work done in a framework corresponding to the general expression (76) is then given by

$$W_i^* = \sum_{\text{bars}} \int_0^l \left(\frac{NN^*}{EA}+\frac{M_yM_y^*}{EI_y}+\frac{M_zM_z^*}{EI_z}+\frac{k_yS_yS_y^*}{GA}+\frac{k_zS_zS_z^*}{GA}+\frac{TT^*}{GJ}\right)\mathrm{d}x \tag{195}$$

where k_y and k_z are dimensionless form factors depending on the shape of the bar cross section at each point in the framework. Values of the form factors for some common cross sections are given in Table 2.8.

Similarly the internal strain energy of a framework corresponding to the expression (78) is given by

$$u = \sum_{\text{bars}} \int_0^l \left(\frac{N^2}{2EA} + \frac{M_y^2}{2EI_y} + \frac{M_z^2}{2EI_z} + \frac{k_y S_y^2}{2GA} + \frac{k_z S_z^2}{2GA} + \frac{T^2}{2GJ} \right) dx \tag{196}$$

Table 2.8 FORM FACTORS

Section	k_y		k_z
1 Rectangle		1.2	
2 Circle		1.11	
3 Hollow circle		2.0	
4 I section or hollow rectangle	A/A_{web}	(approx.)	$A/A_{flanges}$

Note on the limitations of the engineering theory of the bending of beams (ETBB)

As noted in the subsection on 'Beams subject to combined bending and shear' (2–32), the basic assumptions of the ETBB, whilst quite correct when the beam is subject to pure bending, become invalid when the beam is also subject to shear. In particular, we can no longer assume that plane sections remain plane.

Some indication of the error involved in using the ETBB is obtained by analysing a thin-walled deep cantilever beam. Treating this as a plane stress problem, a complete solution is possible subject only to certain assumptions regarding the fixity at the encastré end.[22] Thus it can be shown that if the cantilever is loaded by a single vertical load F at its end so that the shear stress resultant is uniform along the length, the direct and shear stresses given by the ETBB are *exact*. However, the deflections u_x and u_z are given by

$$u_x = \frac{F}{2EI_y}(-2lx + x^2)z + \frac{\nu F z^3}{6EI_y} - \frac{F z^3}{6GI_y} \tag{197}$$

$$u_z = \frac{F}{6EI_y}(3lx^2 - x^3) + \frac{\nu F}{2EI_y}(l - x)z^2 + \frac{F d^2 x}{8GI_y} \tag{198}$$

and the corresponding deflected shape of the beam is composed of two components as shown in Figure 2.53. One is the constant curvature shape predicted by the ETBB, whilst the other is a linear vertical displacement due to the shear with the original plane cross section taking up an S shape in side view.

If the cantilever is loaded by a uniformly distributed load F/unit length so that the shear stress resultant varies with x then the *stresses* given by the ETBB are also slightly inaccurate. However, it can be shown that the error is small, provided the span of the beam is large compared with its depth. Further the curvature of the beam is modified from (119) to

$$\frac{1}{R_y} = \left[\frac{M_y}{EI_y} + \frac{F}{EI_y} \frac{d^2}{4} \left(\frac{4}{5} + \frac{\nu}{2} \right) \right] \tag{199}$$

where the second term on the right-hand side represents the effect of the shear forces.

The preceding discussion concerns the behaviour of the webs of beams. However, in the flanges as well, it can be shown that plane sections no longer remain plane when beams are subject to shear. This phenomenon is called *shear lag*. It can be conveniently illustrated by the T section cantilever shown in Figure 2.54(a). According to St Venant's theory (p. 2–33), the forces in the flange are transmitted by longitudinal shear across the section AB, so that the flange can be considered to behave like the cantilever plate shown in Figure 2.54(b) subjected to the uniformly distributed axial load along its centreline. It is then clear that the corresponding displacements u_x and the axial stress σ_x are nonuniform across the width of the flange. The analysis of shear lag for practical cases is complex, and the topic is dealt with at some length by Williams.[23]

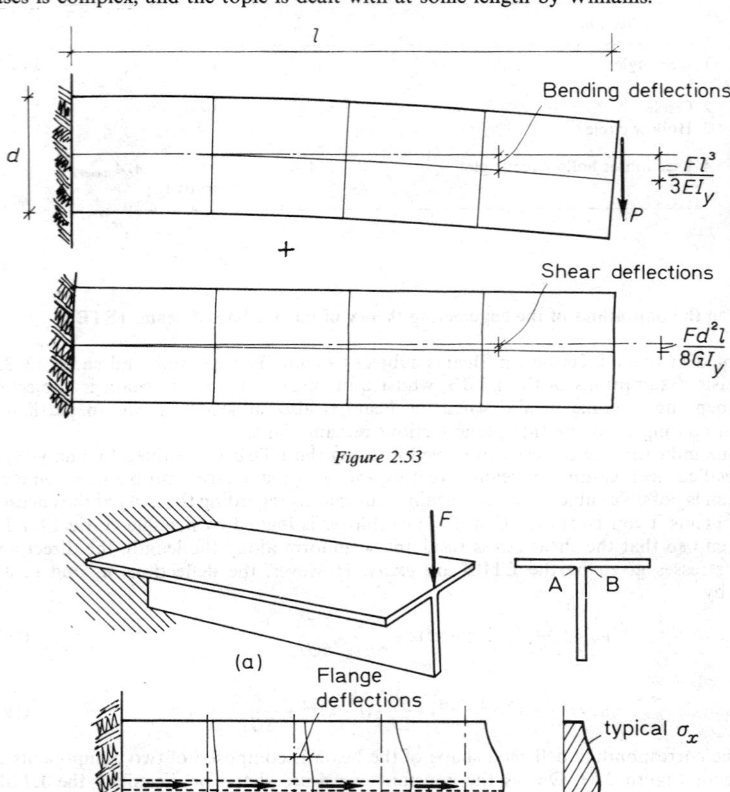

Figure 2.53

Figure 2.54

Further departures from the ETBB occur when beams become geometrically unstable. This instability can take the form of local compressive buckling of the flanges,[24] local shear buckling of the webs[25] and overall torsional buckling.[26]

REFERENCES

1. LOVE, A. E. H., *The mathematical theory of elasticity,* 4th edn, Dover, New York, 1–31 (1944)
2. BISPLINGHOFF, R. L., MAR, J. W., and PIAN, T. H. H., *Statics of deformable solids,* Addison-Wesley, Reading, Mass., 206–230 (1965)
3. PRAGER, W. and HODGE, P. G., *Theory of perfectly plastic solids,* Wiley, New York (1951)
4. DUGDALE, D. S. and RUIZ, C., *Elasticity for engineers,* McGraw-Hill, London, 155–194 (1971)
5. TIMOSHENKO, S. P. and GOODIER, J. N., *Theory of elasticity,* 2nd edn, McGraw-Hill, New York, 29–130 (1951)
6. SOKOLNIKOFF, I. S., *Mathematical theory of elasticity,* 2nd edn, McGraw-Hill, New York, 249–376 (1956)
7. WASHIZU, K., *Variational methods in elasticity and plasticity,* Pergamon, Oxford (1968)
8. FROCHT, M. M., *Photoelasticity,* 2 vols, Wiley, New York (1941)
9. HENDRY, A. W., *Photoelastic Analysis,* Pergamon, Oxford (1966)
10. BS. 4, Part 1, *Structural steel sections,* British Standards Institution, London (1962)
11. TIMOSHENKO, S. P. and GOODIER, J. N., *loc. cit.,* 245
12. TIMOSHENKO, S. P., *Strength of materials,* 3rd edn, Van Nostrand, Princeton, 114 (1955)
13. CASE, J. and CHILVER, A. H., *Strength of materials and structures,* Arnold, London, 225–262 (1971)
14. TIMOSHENKO, S. P., *loc. cit.,* 149–170
15. HETÉNYI, M., *Beams on elastic foundations,* Univ. of Michigan, Ann Arbor (1946)
16. TIMOSHENKO, S. P. and GOODIER, J. N., *loc. cit.,* 258–315
17. BLEICH, F., *Buckling strength of metal structures,* McGraw-Hill, New York 104–147 (1952)
18. GODFREY, G. B., 'The allowable stresses in axially loaded steel struts', *Structural Engineer,* **40,** 97–112 (1962)
19. *Specification for the design, fabrication and erection of structural steel for buildings,* American Institute of Steel Construction, New York, 5–16 (1969)
20. BS. 153, Part 3B, *Steel girder bridges,* British Standards Institution, London, 35 (1958)
21. BS. 449, *The use of structural steel in building.* British Standards Institution, London, 95 (1969)
22. TIMOSHENKO, S. P. and GOODIER, J. N., *loc. cit.,* 35–39
23. WILLIAMS, D., *An introduction to the theory of aircraft structures,* Arnold, London, 233–281 (1960)
24. BLEICH, F., *loc. cit.,* 302–357
25. BLEICH, F., *loc. cit.,* 386–428
26. BLEICH, F., *loc. cit.,* 149–166

BIBLIOGRAPHY

DEN HARTOG, J. P., *Advanced strength of materials,* McGraw-Hill, New York (1952)

DUGDALE, D. S., *Elements of elasticity,* Pergamon, Oxford (1968)

DURELLI, A. J., PHILLIPS, E. A. and TSAO, C. H., *Introduction to the theoretical and experimental analysis of stress and strain,* McGraw-Hill, New York (1958)

GRAVES SMITH, T. R., *Stress and Strain,* Chatto and Windus, London (1974)

GREEN, A. E. and ZERNA, W., *Theoretical elasticity,* 2nd edn, O.U.P., Oxford (1968)

HETENYI, M., *Handbook of experimental stress analysis,* Wiley, London (1950)

HEYWOOD, R. B., *Photoelasticity for designers,* Pergamon, Oxford (1969)

HOFF, N. J., *The analysis of structures,* Wiley, New York (1956)

LANGHAAR, H. L., *Energy methods in applied mechanics,* Wiley, New York (1962)

POPOV, E. P., *Mechanics of materials,* MacDonald, London (1952)

ROARK, R. J., *Formulas for stress and strain,* 3rd edn, McGraw-Hill, London (1954)

SHANLEY, F. R., *Strength of materials,* McGraw-Hill, New York (1957)

SOKOLNIKOFF, I. S., *Mathematical theory of elasticity,* 2nd edn, McGraw-Hill, New York (1956)

TIMOSHENKO, S. P. and GERE, J. M., *Theory of elastic stability,* 2nd edn, McGraw-Hill, New York (1961)

3 THEORY OF STRUCTURES

THEORY OF STRUCTURES 3

3 THEORY OF STRUCTURES

W. M. JENKINS, B.Sc., Ph.D., C.Eng., M.I.C.E., M.I.Struct.E.
Department of Civil and Structural Engineering and Building, Teesside Polytechnic

INTRODUCTION

Basic concepts

The 'Theory of Structures' is concerned with establishing an understanding of the behaviour of structures such as beams, columns, frames, plates and shells, when subjected to applied loads or other actions which have the effect of changing the state of stress and deformation of the structure. The process of 'Structural Analysis' applies the principles established by the Theory of Structures, to analyse a given structure under specified loading and possibly other disturbances such as temperature variation or movement of supports. The drawing of a bending moment diagram for a beam is an act of structural analysis which requires a knowledge of structural theory in order to relate the applied loads, reactive forces and dimensions to actual values of bending moment in the beam. Hence 'theory' and 'analysis' are closely related and in general the term 'theory' is intended to include 'analysis'.

Two aspects of structural behaviour are of paramount importance. If the internal stress distribution in a structural member is examined; it is possible, by integration, to describe the situation in terms of 'stress resultants'. In the general three-dimensional situation, these are six in number; two bending moments, two shear forces, a twisting moment and a thrust. Conversely, it is of course possible to work the other way and convert stress-resultant actions (forces) into stress distributions. The second aspect is that of deformation. It is not usually necessary to describe structural deformation in continuous terms throughout the structure and it is usually sufficient to consider values of displacement at selected discrete points, usually the joints, of the structure.

At certain points in a structure, the continuity of a member, or between members, may be interrupted by a 'release'. This is a device which imposes a zero value on one of the stress resultants. A hinge is a familiar example of a release. Releases may exist as mechanical devices in the real structure or may be introduced, in imagination, in a structure under analysis.

In carrying out a structural analysis it is generally convenient to describe the state of stress or deformation in terms of forces and displacements at selected points, termed 'nodes'. These are usually the ends of members, or the joints and this approach introduces the idea of a structural element such as a beam or column. A knowledge of the forces or displacements at the nodes of a structural element is sufficient to define the complete state of stress or deformation within the element providing the relationships between forces and displacements are established. The establishment of such relationships lies within the province of the theory of structures.

Corresponding to the basic concepts of force and displacement, there are two important physical principles which must be satisfied in a structural analysis. The structure as a whole, and every part of it, must be in equilibrium under the actions of the force system. If, for example, we imagine an element, perhaps a beam, to be removed from a structure by cutting through the ends, the internal stress resultants may now be thought of as external forces and the element must be in equilibrium under the combined action of these forces and any applied loads. In general six independent conditions of equilibrium exist; zero sums of forces in three perpendicular directions, and zero sums of moments about three perpendicular axes. The second principle is termed 'compatibility'. This states that the component parts of a structure must deform in a compatible way, i.e. the parts must fit together without discontinuity at all stages of the loading. Since a release will allow a discontinuity to develop, its introduction will reduce the total number of compatibility conditions by one.

Force-displacement relationships

A simple beam element AB is shown in Figure 3.1. The application of end moments M_A and M_B produces a shear force Q throughout the beam, and end rotations θ_A and θ_B. By the stiffness method (see page 3–18), it may be shown that the end moments and rotations are related as follows,

$$\left.\begin{aligned} M_A &= \frac{4EI\theta_A}{l} + \frac{2EI\theta_B}{l} \\ M_B &= \frac{4EI\theta_B}{l} + \frac{2EI\theta_A}{l} \end{aligned}\right\} \quad (1)$$

Or, in matrix notation,

$$\begin{bmatrix} M_A \\ M_B \end{bmatrix} = \frac{2EI}{l}\begin{bmatrix} 2 & 1 \\ 1 & 2 \end{bmatrix}\begin{bmatrix} \theta_A \\ \theta_B \end{bmatrix}$$

which may be abbreviated to,

$$\mathbf{S} = \mathbf{k}\boldsymbol{\theta} \quad (2)$$

Equation (2) expresses the force-displacement relationships for the beam element of Figure 3.1. The matrices **S** and $\boldsymbol{\theta}$ contain the end 'forces' and displacements respectively.

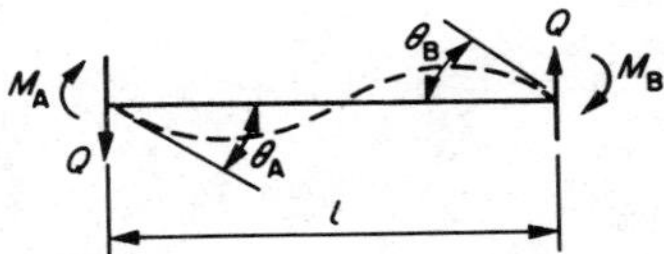

Figure 3.1

The matrix **k** is the *stiffness matrix* of the element since it contains end forces corresponding to *unit* values of the end rotations.

The relationships of equation (2) may be expressed in the inverse form,

$$\begin{bmatrix} \theta_A \\ \theta_B \end{bmatrix} = \frac{l}{6EI}\begin{bmatrix} 2 & -1 \\ -1 & 2 \end{bmatrix}\begin{bmatrix} M_A \\ M_B \end{bmatrix}$$

or

$$\boldsymbol{\theta} = \mathbf{fS} \quad (3)$$

Here the matrix **f** is the *flexibility matrix* of the element since it expresses the end displacements corresponding to *unit* values of the end forces.

It should be noted that an inverse relationship exists between **k** and **f**

i.e.

$$\mathbf{kf} = I$$

or,

$$\mathbf{k} = \mathbf{f}^{-1} \qquad (4)$$

or,

$$\mathbf{f} = \mathbf{k}^{-1}$$

The establishment of force-displacement relationships for structural elements in the form of equations (2) or (3) is an important part of the process of structural analysis since the element properties may then be incorporated in the formulation of a mathematical model of the structure.

Static and kinematic determinacy

If the compatibility conditions for a structure are progressively reduced in number by the introduction of releases, there is reached a state at which the introduction of *one further* release would convert the structure into a mechanism. In this state the structure is *statically determinate* and the nodal forces may be calculated directly from the equilibrium conditions. If the releases are now removed, restoring the structure to its correct condition, nodal forces will be introduced which cannot be determined solely from

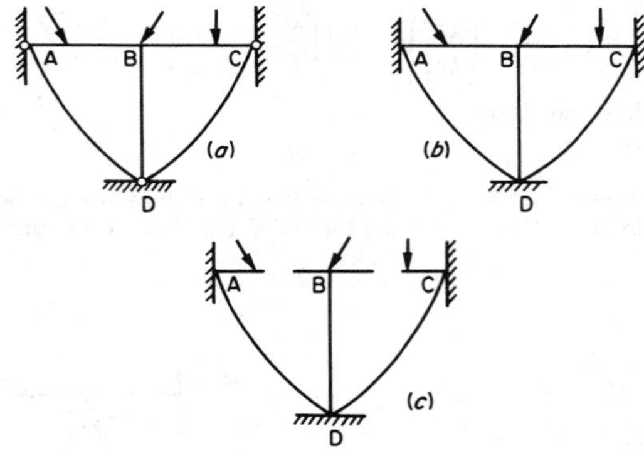

Figure 3.2

equilibrium considerations. The structure is *statically indeterminate* and compatibility conditions are necessary to effect a solution.

The structure shown in Figure 3.2(a) is hinged to rigid foundations at A, C and D. The continuity through the foundations is indicated by the (imaginary) members, AD and CD. If the releases at A, C and D are removed, the structure is as shown in Figure 3.2(b) which is seen to consist of two closed rings. Cutting through the rings as shown in Figure 3.2(c) produces a series of simple cantilevers which are statically determinate. The number of stress resultants released by each cut would be three in the case of a planar structure, six in the case of a space structure. Thus the degree of statical indeterminacy is three or six times the number of rings. It follows that the structure shown in Figure 3.2(b) is six times statically indeterminate whereas the structure of Figure 3.2(a), since releases are introduced at A, C and D, is three times statically indeterminate. A general relationship between the number of members m, number of nodes n, and degree of statical indeterminacy n_s may be obtained[1] as follows

$$n_s = {}^{6}_{3}(m-n+1)-r \tag{5}$$

where r is the number of releases in the actual structure.

Turning now to the question of *kinematical determinacy*; a structure is defined as kinematically determinate if it is possible to obtain the nodal displacements from compatibility conditions without reference to equilibrium conditions. Thus a fixed-end beam is kinematically determinate since the end rotations are known from the compatibility conditions of the supports.

Again consider the structure shown in Figure 3.2(b). The structure is kinematically determinate except for the displacements of joint B. If the members are considered to have infinitely large extensional rigidities then the rotation at B is the only unknown nodal displacement. The degree of kinematical indeterminacy is therefore one. The displacements at B are *constrained* by the assumption of zero vertical and horizontal displacements. A *constraint* is defined as a device which constrains a displacement at a certain node to be the same as the corresponding displacement, usually zero, at another node. Reverting to the structure of Figure 3.2(a), it is seen that three constraints have been removed by the introduction of hinges (releases) at A, C and D. Thus rotational displacements can develop at these nodes and the degree of kinematical indeterminacy is increased from one to four.

A general relationship between the numbers of nodes n, constraints c, releases r, and the degree of kinematical indeterminacy n_k is as follows,

$$n_k = {}^{6}_{3}(n-1)-c+r \tag{6}$$

The coefficient 6 is taken in three-dimensional cases and the coefficient 3 in two-dimensional cases. It should now be apparent that the modern approach to structural theory has developed in a highly organised way. This has been dictated by the development of computer orientated methods which have required a re-assessment of basic principles and their application in the process of analysis. These ideas will be further developed in some of the following sections.

STATICALLY DETERMINATE TRUSS ANALYSIS

Introduction

A structural frame is a system of bars connected by joints. The joints may be, ideally, *pinned* or *rigid,* although in practice the performance of a real joint may lie somewhere between these two extremes. A *truss* is generally considered to be a frame with pinned joints, and if such a frame is loaded only at the joints, then the members carry axial tensions or compressions. *Plane* trusses will resist deformation due to loads acting in the plane of the truss only, whereas *space* trusses can resist loads acting in any direction.

Under load, the members of a truss will change length slightly and the geometry of the frame is thus altered. The effect of such alteration in geometry is generally negligible in the analysis.

The question of statical determinacy has been mentioned in the previous section where a relationship, equation (5), was stated from which the degree of statical indeterminacy could be determined. Although this relationship is of general application, in the case of plane and space trusses, a simpler relationship may be established.

The simplest plane frame is a triangle of three members and three joints. The addition of a fourth joint, in the plane of the triangle, will require two additional members. Thus in a frame having j joints, the number of members is

$$n = 2(j-3)+3 = 2j-3 \tag{7}$$

A truss with this number of members is statically determinate, providing the truss is supported in a statically determinate way. Statically determinate trusses have two important properties. They cannot be altered in shape without altering the length of one or more members, and, secondly, any member may be altered in length without inducing stresses in the truss, i.e. the truss cannot be *self stressed* due to imperfect lengths of members or differential temperature change.

The simplest space truss is in the shape of a tetrahedron with four joints and six members. Each additional joint will require three more members for connection with the tetrahedron, and thus,

$$n = 3(j-4)+6 = 3j-6 \tag{8}$$

A space truss with this number of members is statically determinate, again providing the support system is itself statically determinate. It should be noted that in the assessment of the statical determinacy of a truss, member forces and reactive forces should all be considered when counting the number of unknowns. Since equilibrium conditions will provide two relationships at each joint in a plane truss (three in a space truss), the simplest approach is to find the total number of unknowns, member forces and reactive components, and compare this with two, or three, times the number of joints.

METHODS OF ANALYSIS

Only brief mention will be made here of the methods of statically determinate analysis of trusses. For a more detailed treatment the reader is referred to references 2 and 3.

The *force diagram* method is a graphical solution in which a vector polygon of forces is drawn to scale proceeding from joint to joint. It is necessary to have not more than two unknown forces at any joint, but this requirement can be met with a judicious choice of order. The two conditions of overall equilibrium of the plane structure imply that the force vector polygon will form a closed figure. The method is particularly suitable for trusses with a difficult geometry where it is convenient to work to a scale drawing of the outline of the truss.

The method of *resolution at joints* is suitable for a complete analysis of a truss. The reactions are determined and then, proceeding from joint to joint, the vertical and

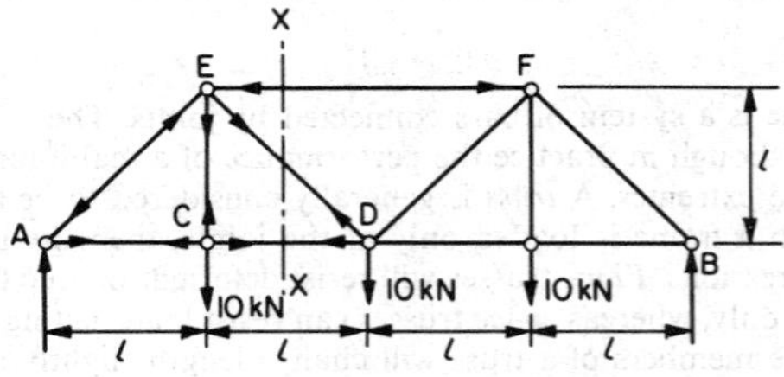

Figure 3.3

horizontal equilibrium conditions are set down in terms of the member forces. Since two equations will result at each joint in a plane truss, it is possible to determine not more than two forces for each pair of equations. As an illustration of the method, consider the plane truss shown in Figure 3.3. The truss is symmetrically loaded and the reactions are clearly 15 kN each.

Consider the equilibrium of joint A,

vertically, $P_{AE}\cos 45° = R_A$; hence $P_{AE} = 15\sqrt{2}$ kN (compression)
horizontally, $P_{AC} = P_{AE}\cos 45°$; hence $P_{AC} =$ 15 kN (tension)

It should be noted that the arrows drawn on the members in Figure 3.3 indicate the directions of forces acting *on the joints.* It is also seen that the directions of the arrows at joint A, for example, are consistent with equilibrium of the joint. Proceeding to joint C it is clear that $P_{CE} = 10$ kN (tension), and that $P_{CD} = P_{AC} = 15$ kN (tension). The remainder of the solution may be obtained by resolving forces at joint E, from which $P_{ED} = 5\sqrt{2}$ kN (tension) and $P_{EF} = 20$ kN (compression).

The *method of sections* is useful when it is required to determine forces in a limited number of the members of a truss. Consider for example the member ED of the truss in Figure 3.3. Imagine a cut to be made along the line XX and consider the vertical equilibrium of the part to the left of XX. The vertical forces acting are, R_A, the 10 kN load at C and the vertical component of the force in ED. The equation of vertical equilibrium is,

$$15-10 = P_{ED}\cos 45° \qquad \text{hence} P_{ED} = 5\sqrt{2}\text{ kN}$$

Since a downwards arrow on the left-hand part of ED is required for equilibrium, it follows that the member is in tension. The method of tension coefficients is particularly suitable for the analysis of space frames and will be outlined in the following section.

Method of tension coefficients

The method is based on the idea of systematic resolution of forces at joints. In Figure 3.4,

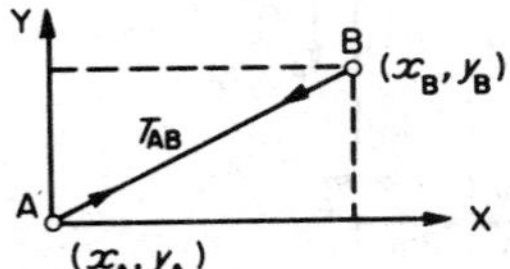

Figure 3.4

let AB be any member in a plane truss, T_{AB} = force in member (tension positive), and L_{AB} = length of member.

We define,

$$T_{AB} = L_{AB}t_{AB} \tag{9}$$

where t_{AB} = tension coefficient.

That is, the tension coefficient is the actual force in the member divided by the length of the member. Now, at A, the component of T_{AB} in the X-direction

$$= T_{AB}\cos \text{BAX}$$

$$= T_{AB}\frac{(x_B-x_A)}{L_{AB}} = t_{AB}(x_B-x_A)$$

Similarly the component of T_{AB} in the Y-direction

$$= t_{AB}(y_B-y_A)$$

At the other end of the member the components are,

$$t_{AB}(x_A-x_B),\ t_{AB}(y_A-y_B)$$

If at A the external forces have components X_A and Y_A, and if there are members AB, AC, AD etc. then the equilibrium conditions for directions X and Y are,

$$\left.\begin{aligned} t_{AB}(x_B-x_A)+t_{AC}(x_C-x_A)+t_{AD}(x_D-x_A)+\ldots\ldots+X_A &= 0 \\ t_{AB}(y_B-y_A)+t_{AC}(y_C-y_A)+t_{AD}(y_D-y_A)+\ldots\ldots+Y_A &= 0 \end{aligned}\right\} \quad (10)$$

Similar equations can be formed at each joint in the truss. Having solved the equations, for the tension coefficients, usually a very simple process, the forces in the members are determined from equation (9).

The extension of the theory to space trusses is straightforward. At each joint we now have three equations of equilibrium, similar to equation (10) with the addition of an equation representing equilibrium in the Z direction,

$$t_{AB}(z_B-z_A)+t_{AC}(z_C-z_A)+\ldots\ldots+Z_A = 0 \quad (11)$$

The method will now be illustrated with an example. The notation is simplified by writing AB in place of t_{AB} etc. A tabular presentation of the work is recommended.

EXAMPLE

A pin-jointed space truss is shown in Figure 3.5. It is required to determine the forces

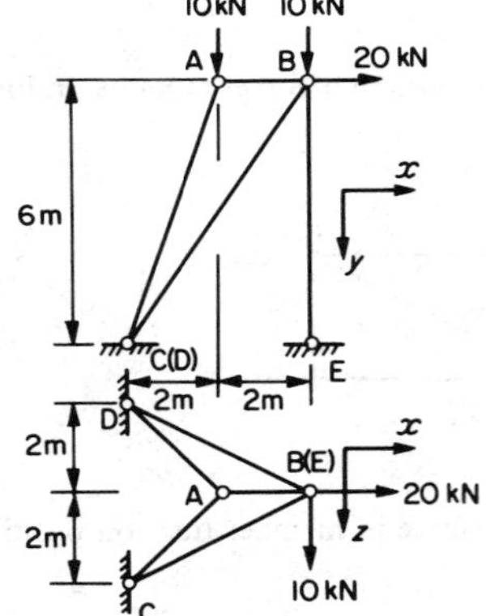

Figure 3.5

in the members using the method of tension coefficients. We first check that the frame is statically determinate as follows,

Number of members = 6
Number of reactions = 9

Total number of unknowns = 15

The number of equations available is three times the number of joints, i.e. 3 × 5 = 15. Hence the truss is statically determinate. In counting the number of reactive components, it should be observed that all components should be included even if the particular geometry of the truss dictates (as in this case at E) that one or more components should be zero.

The solution is set out in Tables 3.1 and 3.2 where it should be noted that, in deriving the equations, the origin of coordinates is taken at the joint being considered. Thus each tension coefficient is multiplied by the projection of the member on the particular axis.

Table 3.1

Joint	*Direction*	*Equations*	*Solutions*
A	x	$-2AC-2AD+2AB = 0$	$AC = AD = -\frac{10}{12}$
	y	$6AC+6AD+10 = 0$	$AB = -\frac{10}{6}$
	z	$2AC-2AD = 0$	$\left.\begin{array}{l}-4BC-4BD+\frac{10}{3}+20 = 0\\ 2BC-2BD+10 = 0\end{array}\right\}$
B	x	$-4BC-4BD-2AB+20 = 0$	$BC = \frac{10}{24}$
	y	$6BC+6BD+6BE+10 = 0$	$BD = \frac{130}{24}$
	z	$-2BD+2BC+10 = 0$	Hence $BE = -\frac{15}{2}$

Table 3.2

Member	*Length (m)*	*Tension coefficient*	*Force (kN) (tension+)*
AB	2	$-\frac{10}{6}$	-3.33
AC	6.62	$-\frac{10}{12}$	-5.52
AD	6.62	$-\frac{10}{12}$	-5.52
BC	7.48	$\frac{10}{24}$	$+3.12$
BD	7.48	$\frac{130}{24}$	$+40.5$
BE	6	$-\frac{15}{2}$	-45.0

THE FLEXIBILITY METHOD

Introduction

The idea of statical determinacy was introduced previously (see page 3–4) and a relationship between the degree of statical indeterminacy and the numbers of members, nodes and releases was stated in equation (5). A statically determinate structure is one for which it is possible to determine the values of forces at all points by the use of equilibrium conditions alone. A statically indeterminate structure, by virtue of the number of members or method of connecting the members together, or the method of support of the structure, has a larger number of forces than can be determined by the application of equilibrium principles alone. In such structures the force analysis requires the use of compatibility conditions. The flexibility method provides a means of analysing statically indeterminate structures.

Consider the propped cantilever shown in Figure 3.6(a). Applying equation (5) the degree of statical indeterminacy is seen to be

$$n_s = 3(2-2+1)-2 = 1$$

(Note that two releases are required at B, one to permit angular rotation and one to permit horizontal sliding, and also that an additional foundation member is inserted connecting A and B. The structure can be made statically determinate by removing the propping force R_B or alternatively by removing the fixing moment at A. We shall proceed by removing the reaction R_B. The structure thus becomes the simple cantilever shown in Figure 3.6(b). The application of the load w produces the deflected shape, shown dotted, and in particular a deflection u at the free end B. Note also that it is now possible to determine the bending moment at A $= wl^2/2$, by simple statical principles. The deflection u may be obtained from elementary beam theory as $wl^4/8EI$. We now remove the applied load w and apply the, unknown, redundant force x at B. It is unnecessary to know the sense of the force x; in this case we have assumed a downwards direction for positive x. The application of the force x produces a displacement at B which we shall call fx; i.e.

a unit value of x would produce a displacement f. The compatibility condition associated with the redundant force x is that the final displacement at B should be zero, i.e.

$$u+fx = 0 \tag{12}$$

and substituting values of u and f,

$$x = -\tfrac{3}{8}wl$$

The process may be regarded as the superposition of the diagrams Figures 3.6(b) and

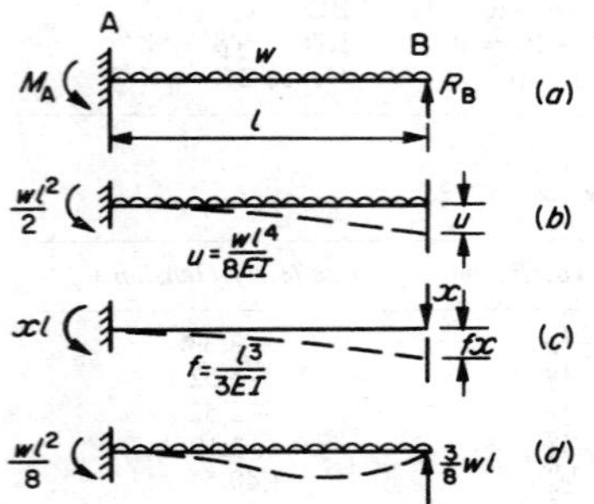

Figure 3.6 Basis of the flexibility method

(c) such that the final displacement at B is zero. The addition of the two systems of forces will also give values of bending moment throughout the beam, for example at A

$$M_A = \frac{wl^2}{2}+xl$$

$$= \frac{wl^2}{2}-\tfrac{3}{8}wl^2 \qquad = \frac{wl^2}{8}$$

The actual values of reactions are as shown in Figure 3.6(d).

The displacement f is called a 'flexibility influence coefficient'. In general f_{rs} is the displacement in direction r in a structure due to unit force in direction s. The subscripts were omitted in the above analysis since the force and displacement considered were at the same position and in the same direction.

Evaluation of flexibility influence coefficients

As seen in the above example, flexibility coefficients are displacements calculated at specified positions, and directions, in a structure due to a prescribed loading condition. The loading condition is that of a single unit load replacing a redundant force in the structure. It should be remembered that at this stage the structure is, or has been made, statically determinate.

For simplicity we restrict our attention to structures in which flexural deformations predominate. The extension to other types of deformation is straightforward.[2] In the case of pure flexural deformation we may evaluate displacements by an application of Castigliano's theorem or use the principal of virtual work.[2] In either case a convenient form is

$$\Delta_i = \int M\partial M/\partial F_i \, \frac{ds}{EI} \tag{13}$$

in which Δ_i is the displacement required, M is a function representing the bending moment distribution and F_i is a force, real or virtual, applied at the position and in the direction designated by i. It follows that $\partial M/\partial F_i$ can be regarded as the bending moment distribution due to unit value of F_i.

Consider the cantilever beam shown in Figure 3.7(a). Forces x_1 and x_2 act on the beam and it is required to determine influence coefficients corresponding to the positions and directions defined by x_1 and x_2. From now on we work with *unit* values of x_1 and x_2 and draw bending moment diagrams, as in Figure 3.7(b) and (c), due to unit values of x_1 and x_2 separately. These are labelled m_1 and m_2. Consider the application of unit force at $x_1(x_2 = 0)$. Displacements will occur in the directions of x_1 and x_2. Applying equation (13) the displacement in the direction of x_1 will be,

$$f_{11} = \int m_1 m_1 \frac{ds}{EI}$$

and in the direction of x_2, (14)

$$f_{21} = \int m_2 m_1 \frac{ds}{EI}$$

Similarly, when we apply $x_2 = 1$, $x_1 = 0$, we obtain,

$$f_{22} = \int m_2 m_2 \frac{ds}{EI}$$

and (15)

$$f_{12} = \int m_1 m_2 \frac{ds}{EI}$$

The general form is,

$$f_{rs} = \int m_r m_s \frac{ds}{EI} \tag{16}$$

The evaluation of equation (16) requires the integration of the product of two bending moment distributions over the complete structure. Such distributions can generally be

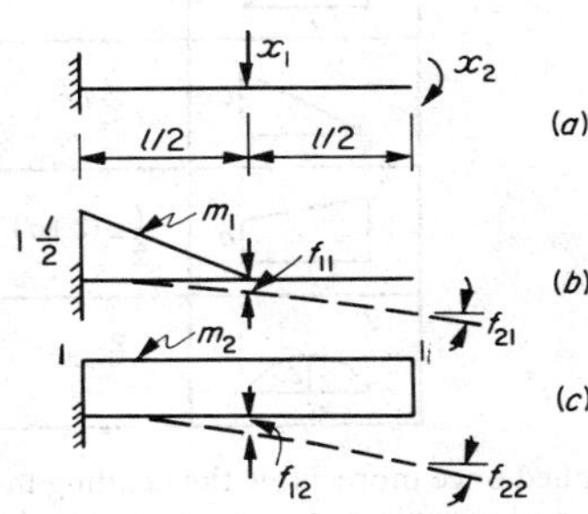

Figure 3.7 Evaluation of flexibility coefficients

represented by simple geometrical figures such as rectangles, triangles and parabolas and standard results can be established in advance. Table 3.3 gives values of product integrals for a range of combinations of diagrams. It should be noted that in applying equation (16) in this way, the flexural rigidity EI is assumed constant over the length of the diagram.

We may now use Table 3.3 to obtain values of the flexibility coefficients for the cantilever beam under consideration. Using equations (14) and (15) with Figures 3.7(b) and (c) we obtain,

$$f_{11} = \frac{1}{3}\cdot\frac{l}{2}\cdot\frac{l}{2}\cdot\frac{l}{2}\cdot\frac{1}{EI} = \frac{l^3}{24EI}$$

$$f_{21} = \frac{1}{2}\cdot\frac{l}{2}\cdot 1\cdot\frac{l}{2}\cdot\frac{1}{EI} = \frac{l^2}{8EI}$$

$$f_{22} = l\cdot 1\cdot 1\cdot\frac{1}{EI} = \frac{l}{EI}$$

$$f_{12} = \frac{1}{2}\cdot\frac{l}{2}\cdot\frac{l}{2}\cdot 1\cdot\frac{1}{EI} = \frac{l^2}{8EI}$$

It is seen that f_{21} and f_{21} are numerically equal, a result which could be established using the Reciprocal Theorem. This is a useful property since in general $f_{rs} = f_{sr}$ and the effect is to reduce the number of separate calculations required. It should be further noted that whilst $f_{21} = f_{12}$; f_{21} is an angular displacement and f_{12} a linear displacement.

The evaluation of the flexibility coefficients f_{rs} provides the displacements at selected points in the structure due to unit values of the associated, redundant, forces. Before the compatibility conditions can be written down, it remains to calculate displacements (u) at corresponding positions due to the actual applied load. The basic equation, (13)

Table 3.3

Product integrals $\int_0^l m_r\, m_s\, ds$ (*EI* uniform)

m_s \ m_r	rectangle, height a, length l	triangle, height a at left, length l	trapezium, heights a and b, length l
rectangle, height c, length l	lac	$\frac{l}{2}ac$	$\frac{l}{2}(a+b)c$
triangle, height c at left, length l	$\frac{l}{2}ac$	$\frac{l}{3}ac$	$\frac{l}{6}(2a+b)c$
triangle, height c at right, length l	$\frac{l}{2}ac$	$\frac{l}{6}ac$	$\frac{l}{6}(a+2b)c$
trapezium, heights c and d, length l	$\frac{l}{2}a(c+d)$	$\frac{l}{6}a(2c+d)$	$\frac{l}{6}\{a(2c+d)+b(2d+c)\}$
parabola, height c, length l	$\frac{2}{3}lac$	$\frac{l}{3}ac$	$\frac{l}{3}(a+b)c$

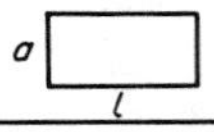

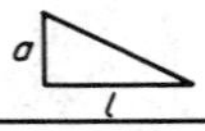

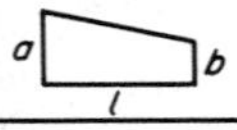

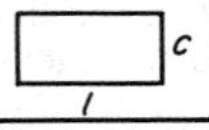

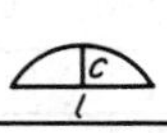

is applied once more. Now the bending moment distribution M is that due to the applied loads and we will re-designate this m_o. As before, $\partial M/\partial F_i = m_i$, and thus,

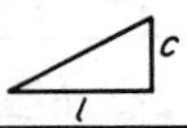

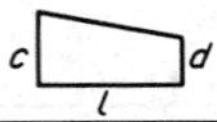

$$u_i = \int m_o m_i \frac{ds}{EI} \tag{17}$$

The table of product integrals, Table 3.3, can be used for evaluating the u_i in the same way as the f_{rs}.

In cases where the bending moment diagrams do not fit the standard values given in Table 3.3 or where a member has a stepped variation in EI, the member may be divided into segments such that the standard results can be applied and the total displacement obtained by addition. In cases where the standard results cannot be applied, for example a continuous variation in EI, then the integration can be carried out conveniently by the use of Simpson's rule,

$$\int m_r m_s \frac{ds}{EI} \simeq \frac{a}{3}(h_1 + 4h_2 + 2h_3 + 4h_4 + \ldots\ldots + h_n)$$

where a = width of strip

$$h_i = \frac{m_r m_s}{EI} \text{ at section } i.$$

In using Simpson's rule it should be remembered that the number of strips must be even, i.e. n must be odd.

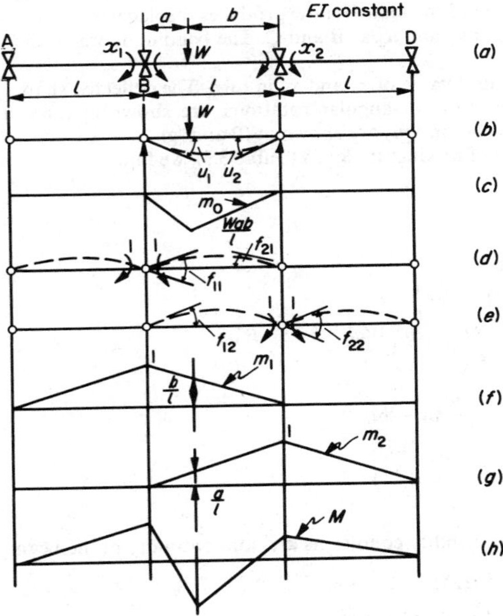

Figure 3.8 Flexibility analysis of continuous beam

SIGN CONVENTION

A flexibility coefficient will be positive if the displacement it represents is in the same sense as the applied, unit, force. The bending moment expressions must carry signs based on the type of curvature developing in the structure. Since the integrand in equation (16) is always the product of two bending moment expressions, it is only the relative sign which is of importance. A useful convention is to draw the diagrams on the tension (convex) sides of the members and then the relative signs of m_r and m_s can readily be seen. In Figure 3.7(b) and (c), both the m_1 and m_2 diagrams are drawn on the top side of the member. Their product is therefore positive. Naturally the product of one diagram and itself will always be positive. This follows from simple physical reasoning since the displacement at a point due to an applied force at the same point will always be in the same sense as the applied force.

Application to beam and rigid frame analysis

The application of the theory will now be illustrated with two examples.

EXAMPLE

Consider the three-span continuous beam shown in Figure 3.8(a). The beam is statically indeterminate to the second degree and we shall choose as redundants the internal bending moments at the interior supports B and C. The beam is made statically determinate by the introduction of moment releases at B and C as in Figure 3.8(b). We note that the application of the load W now produces displacements in span BC only, and in particular rotations u_1 and u_2 at B and C. The bending moment diagram (m_o) is shown in diagram (c).

We now apply unit value of x_1 and x_2 in turn. The deflected shapes and the flexibility coefficients, in the form of angular rotations, are shown at (d) and (e). The bending moment diagrams m_1 and m_2 are shown at (f) and (g).

Using the table of product integrals (Table 3.3), we find,

$$EI f_{11} = \frac{2}{3} l$$

$$EI f_{22} = \frac{2}{3} l$$

$$EI f_{12} = EI f_{21} = \frac{l}{6}$$

$$EI u_1 = -\frac{a}{6}\left(1+\frac{2b}{l}\right)\frac{Wab}{l} - \frac{b}{3}\cdot\frac{b}{l}\cdot\frac{Wab}{l}$$

$$= -\frac{Wab}{6l}(a+2b)$$

and $$EI u_2 = -\frac{Wab}{6l}(b+2a)$$

The required compatibility conditions are, for continuity of the beam,

$$\text{at B,} \quad f_{11}x_1 + f_{12}x_2 + u_1 = 0$$

$$\text{at C,} \quad f_{21}x_1 + f_{22}x_2 + u_2 = 0$$

or, in matrix form,

$$\mathbf{FX} + \mathbf{U} = \mathbf{0} \tag{18}$$

i.e.

$$\frac{l}{6EI}\begin{bmatrix} 4 & 1 \\ 1 & 4 \end{bmatrix}\begin{bmatrix} x_1 \\ x_2 \end{bmatrix} = \frac{Wab}{6EIl}\begin{bmatrix} (a+2b) \\ (b+2a) \end{bmatrix}$$

and the solutions are,

$$\begin{bmatrix} x_1 \\ x_2 \end{bmatrix} = \frac{Wab}{15l^2}\begin{bmatrix} (2a+7b) \\ (2b+7a) \end{bmatrix}$$

The actual bending moment distribution may now be determined by the addition of the three systems, i.e. the applied load and the two redundants. The general expression is,

$$M = m_0 + m_1 x_1 + m_2 x_2 \tag{19}$$

In particular,

$$M_B = x_1 = \frac{Wab}{15l^2}(2a+7b)$$

$$M_C = x_2 = \frac{Wab}{15l^2}(2b+7a)$$

and the bending moment under the load W is

$$M_W = -\frac{Wab}{l} + \frac{b}{l}x_1 + \frac{a}{l}x_2$$

$$= -\frac{2Wab}{15l^3}(4l^2+5ab)$$

The final bending moment diagram is shown in Figure 3.8(h).

EXAMPLE

A portal frame ABCD is shown in Figure 3.9(a). The frame has rigid joints at B and C,

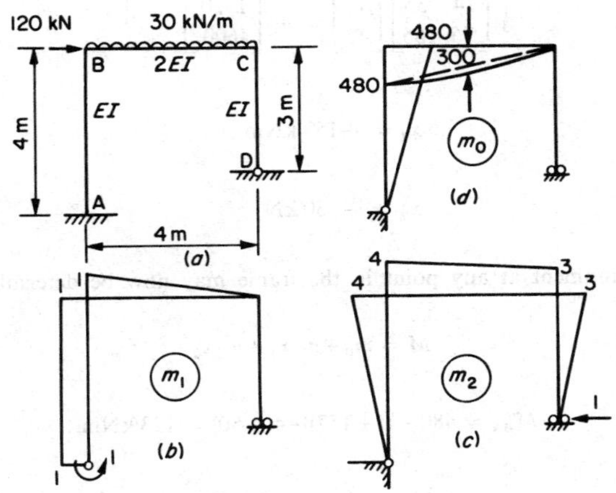

Figure 3.9

a fixed support at A and a hinged support at D. The flexural rigidity of the beam is twice that of the columns.

The frame has two redundancies and these are taken to be the fixing moment at A and the horizontal reaction at D. The bending moment diagrams corresponding to the unit redundancies, m_1 and m_2, and the applied load, m_0, are shown at (b), (c) and (d) in Figure 3.9.

Using the table of product integrals, Table 3.3, we obtain

$$f_{11} = \frac{\int m_1^2 \, ds}{EI} = \frac{14}{3EI}$$

$$f_{22} = \frac{\int m_2^2 \, ds}{EI} = \frac{55}{EI}$$

$$f_{12} = f_{21} = \frac{\int m_1 m_2 \, ds}{EI} = \frac{35}{3EI}$$

$$u_1 = \frac{\int m_0 m_1 \, ds}{EI} = -\frac{1320}{EI}$$

$$u_2 = \int m_0 m_2 \frac{ds}{EI} = -\frac{4600}{EI}$$

Thus the compatibility equations are,

$$\frac{1}{3}\begin{bmatrix} 14 & 35 \\ 35 & 165 \end{bmatrix}\begin{bmatrix} x_1 \\ x_2 \end{bmatrix} = +\begin{bmatrix} 1320 \\ 4600 \end{bmatrix}$$

from which

$$x_1 = +157 \text{ kNm}$$

and

$$x_2 = +\ 50 \text{ kN}$$

The bending moment at any point in the frame may now be determined from the expression

$$M = m_0 + m_1 x_1 + m_2 x_2$$

for example,

$$M_{BA} = 480 - 1(+157) - 4(+50) = 123 \text{ kNm}$$

and

$$M_{CD} = 3x_2 = 150 \text{ kNm}$$

Application to truss analysis

The analysis of statically indeterminate trusses follows closely on that established for rigid frames, however, the problem is simplified due to the fact that for each system of loading investigated, the axial forces are constant within the lengths of the members and thus the integration is considerably simplified. We are now concerned with deformations in the members due to axial forces only and the flexibility coefficients are,

$$f_{rs} = \sum p_r p_s \frac{l}{AE} \tag{20}$$

and,

$$u_i = \sum p_o p_i \frac{l}{AE} \tag{21}$$

in which the p_r system of forces is due to unit tension in the rth redundant member and similarly for p_s and p_i. The p_0 system of forces is that due to the applied load system acting on the statically determinate structure (i.e. with the redundant members omitted). Equations (20) and (21) should be compared with equations (16) and (17) in the flexural case.

EXAMPLE 3

The plane truss shown in Figure 3.10 has two redundancies which we will choose as the forces in members AE and EC. AE is constant for all the members and equal to 1×10^6 kN. The member EC is $l/10\,000$ short in manufacture and has to be forced into position. The

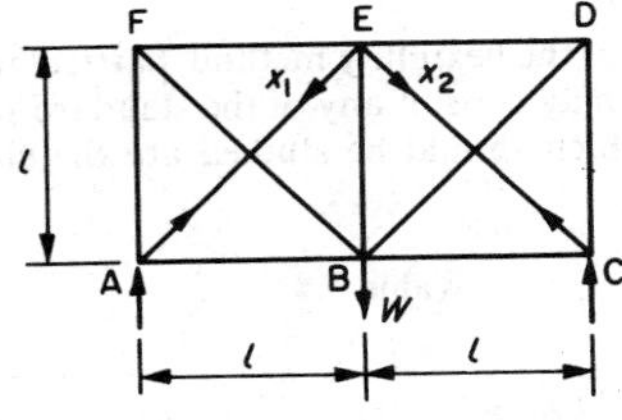

Figure 3.10

member force systems p_0, p_1 and p_2 are found from a simple statical analysis and are listed in Table 3.4.

The flexibility coefficients may now be obtained as follows,

$$f_{11} = \sum p_1 p_1 \frac{l}{AE} = \frac{2l}{AE}(1+\sqrt{2})$$

$$f_{22} = f_{11}$$

$$f_{12} = f_{21} = \sum p_1 p_2 \frac{l}{AE} = \frac{l}{2AE}$$

$$u_1 = \sum p_1 p_0 \frac{l}{AE} = \frac{Wl}{AE}(1+1/\sqrt{2})$$

Ignoring, for the moment, the effect of the shortness in length of member EC, the compatibility equations are,

$$f_{11}x_1 + f_{12}x_2 + u_1 = 0$$

$$f_{21}x_1 + f_{22}x_2 + u_2 = 0$$

Clearly the symmetry will produce $x_1 = x_2$ and thus,

$$x_1 = x_2 = -W\frac{(2+\sqrt{2})}{(5+4\sqrt{2})}$$

The effect of the prestrain caused by the forced fit of member EC may be obtained by putting

$$U = -\begin{bmatrix} 0 \\ 10^{-4}l \end{bmatrix} \tag{22}$$

and then solving **FX**+**U** = **0**
obtaining,

$$x_1 = \frac{-200}{(47+32\sqrt{2})}\,\text{kN}$$

$$x_2 = \frac{800(1+\sqrt{2})}{(47+32\sqrt{2})}\,\text{kN}$$

The forces in the other members may now be obtained from $p = p_0 + p_1x_1 + p_2x_2$.

The sign of the lack of fit in equation (22) should be studied carefully and it should be noted that the convention for the signs of forces is tension positive throughout.

Additional topics in the flexibility method

For a more detailed treatment of the flexibility method, particularly in its more complete matrix formulation the reader may consult any of the standard texts.[1, 2, 3]

Topics of special interest which should be studied are the choice of release system,

Table 3.4

Member	*Length*	p_0/w	p_1	p_2
AB	l	0	$-1/\sqrt{2}$	0
BC	l	0	0	$-1/\sqrt{2}$
CD	l	$-1/2$	0	$-1/\sqrt{2}$
DE	l	$-1/2$	0	$-1/\sqrt{2}$
EF	l	$-1/2$	$-1/\sqrt{2}$	0
AF	l	$-1/2$	$-1/\sqrt{2}$	0
FB	$\sqrt{(2)}l$	$1/\sqrt{2}$	1	0
BE	l	0	$-1/\sqrt{2}$	$-1/\sqrt{2}$
BD	$\sqrt{(2)}l$	$1/\sqrt{2}$	0	1
AE	$\sqrt{(2)}l$	0	1	0
EC	$\sqrt{(2)}l$	0	0	1

treatment of non-prismatic members, multi-action strain energies and the effects of shear forces. It will be found that great care is needed in the use of the statical principles required in the drawing of bending moment diagrams.

THE STIFFNESS METHOD

Introduction

This method has been very extensively developed in recent years and now forms the basis of most structural analysis carried out on digital computers. The method of 'slope-deflection' is an example of the application of the general stiffness method.

Consider the structure shown in Figure 3.11(a) which is fixed at A and C and has a rigid joint at B. The degree of kinematical indeterminacy, from equation (6), is

$$n_k = 3(n-1)-c+r$$
$$= 3(3-1)-5+0$$
$$= 1$$

The five constraints are the zero displacements, three at C and two at B, related to the fixed point A. The single unknown displacement, nodal degree of freedom, is of course the rotation of the joint B.

The procedure is to clamp the joint B so constraining the nodal degree of freedom r. On applying the load W, a constraining force, R, will be required at B to prevent the rotation of the joint. The constraining force R is now applied to the, otherwise unloaded, structure with its sign reversed and the nodal degree of freedom released. The result is a rotation of joint B through angle r. The external moment required to effect this rotation

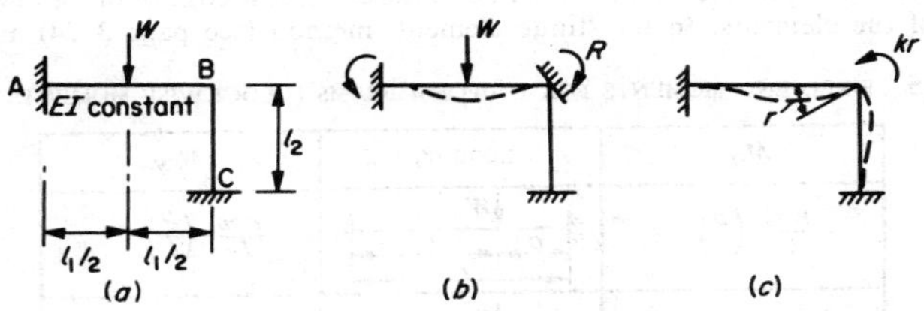

Figure 3.11 Basis of the stiffness method

is kr where k is the stiffness of the structure for this particular displacement. Thus for equilibrium,

$$kr = R \tag{23}$$

From the table of fixed-end moments, Table 3.5,

$$R = \frac{Wl_1}{8},$$

and from the force-displacement relationships of equation (1)

$$k = \frac{4EI}{l_1} + \frac{4EI}{l_2}$$

Thus,

$$4EI\left(\frac{1}{l_1}+\frac{1}{l_2}\right)r = \frac{Wl_1}{8}$$

Hence

$$r = \frac{Wl_1^2 l_2}{32EI(l_1+l_2)}$$

The member forces are now obtained by adding the two systems (b) and (c) in Figure 3.11, for example,

$$M_{BA} = \frac{Wl_1}{8} - \frac{4EI(r)}{l_1} = \frac{Wl_1}{8}\left(1 - \frac{l_2}{l_1 + l_2}\right)$$

$$= \frac{Wl_1^2}{8(l_1 + l_2)}$$

and

$$M_{BC} = -\frac{4EI(r)}{l_2} = -\frac{Wl_1^2}{8(l_1 + l_2)}$$

Note that in the above, clockwise moments are considered positive.

Member stiffness matrix

In the stiffness method, a structure is considered to be an assemblage of discrete elements, beams, columns, plates etc, and the method requires a knowledge of the stiffness characteristics of the elements. In the 'finite element' method (see page 3–24) an artificial

Table 3.5 FIXED-END MOMENTS FOR UNIFORM BEAMS (CLOCKWISE MOMENTS POSITIVE)

M_{FL}	Loading	M_{FR}
$-\frac{Wab}{l}\left(\frac{b}{l}\right)$	W; a; b; l	$\frac{Wab}{l}\left(\frac{a}{l}\right)$
$-\frac{Wab}{2l^2}(a+2b)$	W; a; b	0
$-\frac{wc}{12l^2}\left[12ab^2+c^2(a-2b)\right]$	c; w; a; b	$\frac{wc}{12l^2}\left[12a^2b+c^2(b-2a)\right]$
$-\frac{wl^2}{12}$	w	$\frac{wl^2}{12}$
$-\frac{wl^2}{30}$	w	$\frac{wl^2}{20}$
$-\frac{5}{96}wl^2$	w	$\frac{5}{96}wl^2$
$\frac{Mb}{l^2}(2a-b)$	M; a; b	$\frac{Ma}{l^2}(2b-a)$
$\frac{M}{2l^2}(l^2-3b^2)$	M; a; b	0

discretisation of the structure is adopted. As an example of the determination of stiffness influence coefficients we shall consider the simple beam element shown in Figure 3.12. We neglect any axial deformation.

The expression for the bending moment in the beam with origin at end 1 and deflections y positive downwards is,

$$EId^2y/dx^2 = P_1x - M_1$$

Integrating,

$$EIdy/dx = P_1\frac{x^2}{2} - M_1x + C_1$$

$$= EI\theta_1 \text{ for } x = 0, \text{ hence } C_1 = EI\theta_1$$

$$= EI\theta_2 \text{ for } x = l \text{ hence,}$$

$$EI(\theta_2 - \theta_1) = \frac{P_1l^2}{2} - M_1l \tag{24}$$

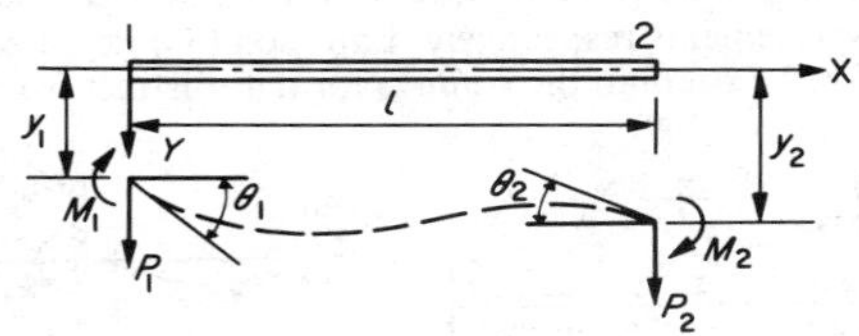

Figure 3.12 Structural beam element

Integrating again,

$$EIy = P_1\frac{x^3}{6} - M_1\frac{x^2}{2} + EI\theta_1x + C_2$$

$$= EIy_1 \text{ for } x = 0, \therefore C_2 = EIy_1$$

$$= EIy_2 \text{ for } x = l$$

hence $$EI(y_2 - y_1) - EI\theta_1 l = P_1\frac{l^3}{6} - M_1\frac{l^2}{2} \tag{25}$$

Solving equations (24) and (25) for M_1 and P_1,

$$M_1 = \frac{4EI\theta_1}{l} + \frac{6EIy_1}{l^2} + \frac{2EI\theta_2}{l} - \frac{6EIy_2}{l^2} \tag{26}$$

and,

$$P_1 = \frac{6EI\theta_1}{l^2} + \frac{12EIy_1}{l^3} + \frac{6EI\theta_2}{l^2} - \frac{12EIy_2}{l^3} \tag{27}$$

Two further relationships between the forces and displacements are obtained from statical equilibrium as follows,

For vertical equilibrium, $P_1 + P_2 = 0$

hence $$P_2 = -P_1 \tag{28}$$

Taking moments about end 1,

$$M_2 = -M_1 - P_2 l$$

$$= \frac{2EI\theta_1}{l} + \frac{6EIy_1}{l^2} + \frac{4EI\theta_2}{l} - \frac{6EIy_2}{l^2} \tag{29}$$

Equations (26), (27), (28) and (29) may be combined in the matrix form,

$$\begin{bmatrix} M_1 \\ P_1 \\ M_2 \\ P_2 \end{bmatrix} = \frac{EI}{l^3} \begin{bmatrix} 4l^2 & 6l & 2l^2 & -6l \\ 6l & 12 & 6l & -12 \\ 2l^2 & 6l & 4l^2 & -6l \\ -6l & -12 & -6l & 12 \end{bmatrix} \begin{bmatrix} \theta_1 \\ y_1 \\ \theta_2 \\ y_2 \end{bmatrix}$$

$$\text{or } \mathbf{S} = \mathbf{k\Delta} \tag{30}$$

The matrix **k** is the stiffness matrix of the beam, and **S** and **Δ** are the matrices of member forces and nodal displacements respectively. Equation (30) expresses the force-displacement relationships for the beam in the *stiffness* form as distinct from the *flexibility* form.

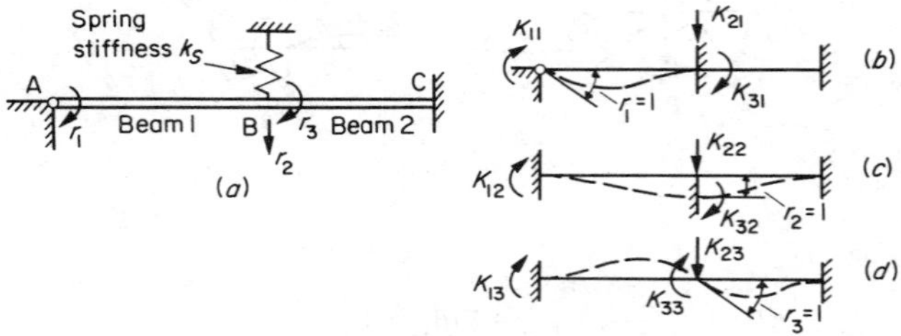

Figure 3.13

The symmetry of the matrix should be noted as consistent with the symmetry exhibited by flexibility coefficients (see page 3–12).

Assembly of Structure Stiffness matrix

The stiffness method involves the solution of a set of linear simultaneous equations, representing equilibrium conditions, which may be expressed in the form,

$$\mathbf{Kr} = \mathbf{R} \tag{31}$$

Equation (31) is similar in form to equation (23) with the important difference that now we are concerned with a multiple degree of freedom system as distinct from a single unknown displacement. **K** is the structure stiffness matrix, **r** is a matrix of nodal displacements and **R** a matrix of applied nodal forces.

The process of assembling the matrix **K** is one of transferring individual element stiffnesses into appropriate positions in the matrix **K**. Naturally this has been the subject of considerable organisation for digital computer analysis and the subject is well documented.[1,4] The basic process will be illustrated with a simple example. Consider the structure shown in Figure 3.13(a). The two beams are rigidly connected together at B where there is a spring support with stiffness k_s. End A is hinged and end C fixed. The structure has three degrees of freedom, rotations r_1 and r_3 at A and B and a vertical displacement r_2 at B. The stiffness matrix for each beam has the form of equation (30) from which **k** may be written in the general form,

$$\mathbf{k} = \begin{bmatrix} k_{11} & k_{12} & k_{13} & k_{14} \\ k_{21} & k_{22} & k_{23} & k_{24} \\ k_{31} & k_{32} & k_{33} & k_{34} \\ k_{41} & k_{42} & k_{43} & k_{44} \end{bmatrix} \tag{32}$$

where $k_{11} = 4EI/l$; $k_{12} = 6EI/l^2$ etc.

We apply unit value of each degree of freedom in turn as shown in Figure 3.13(b), (c) and (d). It should be noted that when $r_1 = 1$ is applied, r_2 and r_3 are constrained at zero value and similarly with $r_2 = 1$ and $r_3 = 1$. The force systems necessary to achieve the unit values of the degrees of freedom are also shown at (b), (c) and (d). The equilibrium conditions are clearly,

$$\begin{aligned} K_{11}r_1 + K_{12}r_2 + K_{13}r_3 &= R_1 \\ K_{21}r_1 + K_{22}r_2 + K_{23}r_3 &= R_2 \\ K_{31}r_1 + K_{32}r_2 + K_{33}r_3 &= R_3 \end{aligned}$$

i.e. $\mathbf{Kr} = \mathbf{R}$

where **R** is the matrix of applied loads. Clearly the forces shown in Figure 3.13(b), (c) and (d) constitute the elements of the stiffness matrix **K** and this may now be assembled by inspection. Using the individual beam elements from equation (30) with the notation of equation (32).

$$\mathbf{K} = \begin{bmatrix} (k_{11})_1 & -(k_{12})_1 & (k_{13})_1 \\ -(k_{12})_1 & (k_{44})_1 + (k_{22})_2 + k_s & (k_{23})_2 - (k_{14})_1 \\ (k_{13})_1 & (k_{23})_2 - (k_{14})_1 & (k_{33})_1 + (k_{11})_2 \end{bmatrix} \tag{33}$$

and more specifically,

$$\mathbf{K} = \begin{bmatrix} 4\left(\frac{EI}{l}\right)_1 & -6\left(\frac{EI}{l^2}\right)_1 & 2\left(\frac{EI}{l}\right)_1 \\ -6\left(\frac{EI}{l^2}\right)_1 & 12\left(\frac{EI}{l^3}\right)_1 + 12\left(\frac{EI}{l^3}\right)_2 + k_s & 6\left(\frac{EI}{l^2}\right)_2 - 6\left(\frac{EI}{l^2}\right)_1 \\ 2\left(\frac{EI}{l}\right)_1 & 6\left(\frac{EI}{l^2}\right)_2 - 6\left(\frac{EI}{l^2}\right)_1 & 4\left(\frac{EI}{l}\right)_1 + 4\left(\frac{EI}{l}\right)_2 \end{bmatrix} \tag{34}$$

Stiffness transformations

The member stiffness matrix **k** in equation (30) is based on a coordinate system which is convenient for the member i.e., origin at one end and X-axis directed along the axis of the beam. Such a coordinate system is termed 'local' as distinct from the 'global' coordinate system which is used for the complete structure. This subject is considered in detail in a number of texts[1, 3, 4] and we shall give only a brief indication of the type of computation required.

Consider a three-dimensional coordinate system $\bar{X}\bar{Y}\bar{Z}$ (global) which is obtained by rotation of the (local) coordinate system XYZ. In the local system the force-displacement relationships for a beam element may be expressed in the partitioned matrix form,

$$\begin{bmatrix} \mathbf{S}_1 \\ \mathbf{S}_2 \end{bmatrix} = \begin{bmatrix} \mathbf{k}_{11} & \mathbf{k}_{12} \\ \mathbf{k}_{21} & \mathbf{k}_{22} \end{bmatrix} \begin{bmatrix} \mathbf{r}_1 \\ \mathbf{r}_2 \end{bmatrix} \tag{35}$$

in which the subscripts refer to ends 1 and 2.

The stiffness expressed in the coordinate system $\bar{X}\bar{Y}\bar{Z}$ may be obtained[1] as follows,

$$\begin{bmatrix} \bar{\mathbf{S}}_1 \\ \bar{\mathbf{S}}_2 \end{bmatrix} = \begin{bmatrix} \boldsymbol{\lambda}\mathbf{k}_{11}\boldsymbol{\lambda}^T & \boldsymbol{\lambda}\mathbf{k}_{12}\boldsymbol{\lambda}^T \\ \boldsymbol{\lambda}\mathbf{k}_{21}\boldsymbol{\lambda}^T & \boldsymbol{\lambda}\mathbf{k}_{22}\boldsymbol{\lambda}^T \end{bmatrix} \begin{bmatrix} \bar{\mathbf{r}}_1 \\ \bar{\mathbf{r}}_2 \end{bmatrix} \tag{36}$$

in which $\boldsymbol{\lambda}$ is a matrix of direction cosines as follows,

$$\boldsymbol{\lambda} = \begin{bmatrix} \lambda_{\bar{x}x} & \lambda_{\bar{x}y} & \lambda_{\bar{x}z} & 0 & 0 & 0 \\ \lambda_{\bar{y}x} & \lambda_{\bar{y}y} & \lambda_{\bar{y}z} & 0 & 0 & 0 \\ \lambda_{\bar{z}x} & \lambda_{\bar{z}y} & \lambda_{\bar{z}z} & 0 & 0 & 0 \\ 0 & 0 & 0 & \lambda_{\bar{x}x} & \lambda_{\bar{x}y} & \lambda_{\bar{x}z} \\ 0 & 0 & 0 & \lambda_{\bar{y}x} & \lambda_{\bar{y}y} & \lambda_{\bar{y}z} \\ 0 & 0 & 0 & \lambda_{\bar{z}x} & \lambda_{\bar{z}y} & \lambda_{\bar{z}z} \end{bmatrix} \tag{37}$$

where, $\lambda_{\bar{x}x} = \cos \bar{X}OX$, etc.

Finite element analysis

This extremely powerful method of analysis has been developed in recent years and is now an established method with wide applications in structural analysis and in other fields. Space permits only the most brief introduction here but the method is extensively documented elsewhere.[1, 3, 5, 6, 7]

We have discussed the application of the stiffness method to framed structures in which the structural elements, beams and columns, have been connected at the nodes and the method observes the correct conditions of displacement compatibility and equilibrium at the nodes. The finite element method was developed, originally, in order to extend the stiffness method to the analysis of elastic continua such as plates and shells and indeed to three-dimensional continua. The first step in the process is to divide the structure into a finite number of discrete parts called 'elements'. The elements may be of any convenient shape, for example a thin plate may be represented by triangular or rectangular elements, and the discretisation may be *coarse*, with a small number of elements, or *fine*, with a large number of elements. The connection between elements now occurs not only at the nodal points but along boundary lines and over boundary faces.

The procedure ensures, as for framed structures, that equilibrium and compatibility conditions are satisfied at the nodes but the regions of connection between nodes are constrained to adopt a chosen form of displacement function. Thus compatibility conditions along the interfaces between elements may not be completely satisfied and a degree of approximation is generally introduced. Once the geometry of the elements has been determined and the displacement function defined, the stiffness matrix of each element, relating nodal forces to nodal displacements can be obtained. The remainder of the structural analysis follows the established procedures similar to those for framed structures. Naturally the best choice of element and discretisation pattern, the precise conditions occurring at the interfaces and the accuracy of the solution are matters which have received a great deal of attention in the literature.

A central stage in the process is the adoption of a suitable displacement function for the element chosen, and the subsequent evaluation of the element stiffnesses. This will be illustrated with one of the simplest possible elements, a triangular plate element for use in a plane stress situation.

TRIANGULAR ELEMENT FOR PLANE STRESS

A triangular element *ijk* is shown in Figure 3.14. Under load, the displacement of any

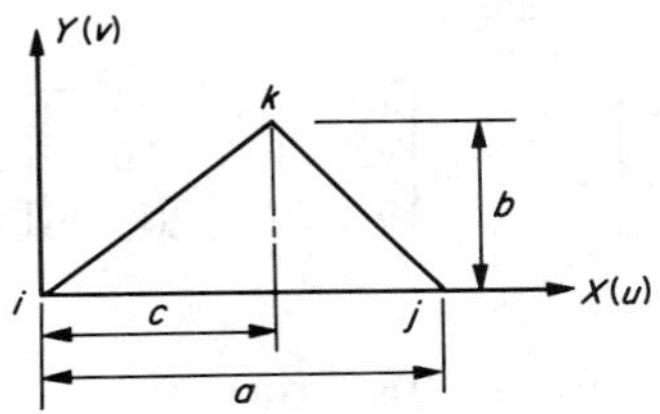

Figure 3.14

point within the element is defined by the displacement components u, v. In particular the nodal displacements are,

$$\mathbf{\Delta} = \{u_i\ u_j\ u_k\ v_i\ v_j\ v_k\} \tag{38}$$

It is now assumed that the displacements u, v are linear functions of x, y as follows,

$$\left.\begin{aligned} u &= \alpha_1 + \alpha_2 x + \alpha_3 y \\ v &= \alpha_4 + \alpha_5 x + \alpha_6 y \end{aligned}\right\} \tag{39}$$

The nodal displacements $\mathbf{\Delta}$ are now expressed in terms of the displacement parameters α, from equations (39) and Figure 3.14,

$$\begin{bmatrix} u_i \\ u_j \\ u_k \\ v_i \\ v_j \\ v_k \end{bmatrix} = \begin{bmatrix} 1 & 0 & 0 & 0 & 0 & 0 \\ 1 & a & 0 & 0 & 0 & 0 \\ 1 & c & b & 0 & 0 & 0 \\ 0 & 0 & 0 & 1 & 0 & 0 \\ 0 & 0 & 0 & 1 & a & 0 \\ 0 & 0 & 0 & 1 & c & b \end{bmatrix} \begin{bmatrix} \alpha_1 \\ \alpha_2 \\ \alpha_3 \\ \alpha_4 \\ \alpha_5 \\ \alpha_6 \end{bmatrix} \tag{40}$$

$$\text{or, } \mathbf{\Delta} = \mathbf{A}\boldsymbol{\alpha}$$

The strains in the element are functions of the derivatives of u and v as follows,

$$\boldsymbol{\varepsilon} = \begin{bmatrix} \varepsilon_x \\ \varepsilon_y \\ \gamma_{xy} \end{bmatrix} = \begin{bmatrix} \partial u/\partial x \\ \partial v/\partial y \\ \partial u/\partial y + \partial v/\partial x \end{bmatrix} \tag{41}$$

$$= \begin{bmatrix} 0 & 1 & 0 & 0 & 0 & 0 \\ 0 & 0 & 0 & 0 & 0 & 1 \\ 0 & 0 & 1 & 0 & 1 & 0 \end{bmatrix} \begin{bmatrix} \alpha_1 \\ \alpha_2 \\ \alpha_3 \\ \alpha_4 \\ \alpha_5 \\ \alpha_6 \end{bmatrix} \tag{42}$$

that is, $\boldsymbol{\varepsilon} = \mathbf{B}\boldsymbol{\alpha} = \mathbf{BA}^{-1}\boldsymbol{\Delta}$, from equation (40), (43)

It should be noted that the matrix **B** in equation (42) contains only constant terms and it follows that the strains are constant within the element.

The stress-strain relationships for plane stress in an isotropic material with Poisson's ratio v and Young's modulus E are,

$$\begin{bmatrix} \sigma_x \\ \sigma_y \\ \tau_{xy} \end{bmatrix} = \frac{E}{(1-v^2)} \begin{bmatrix} 1 & v & 0 \\ v & 1 & 0 \\ 0 & 0 & \frac{1}{2}(1-v) \end{bmatrix} \begin{bmatrix} \varepsilon_x \\ \varepsilon_y \\ \gamma_{xy} \end{bmatrix}$$

that is,

$$\boldsymbol{\sigma} = \mathbf{D}\boldsymbol{\varepsilon} = \mathbf{DBA}^{-1}\boldsymbol{\Delta} \tag{44}$$

Matrix **D** is the 'elasticity' matrix relating stress and strain. To obtain the element stiffness we employ the principle of virtual work and apply arbitrary nodal displacements $\bar{\boldsymbol{\Delta}}$ producing virtual strains in the element,

$$\bar{\boldsymbol{\varepsilon}} = \mathbf{BA}^{-1}\bar{\boldsymbol{\Delta}} \tag{45}$$

The virtual strain energy in the element, from equation (78) of section 2, is

$$\int_{vol} \bar{\boldsymbol{\varepsilon}}^T \boldsymbol{\sigma} \, \mathrm{d}V$$

where V = volume of triangular element = $tab/2$, t = thickness. Substituting for $\bar{\boldsymbol{\varepsilon}}^T$ and $\boldsymbol{\sigma}$ from equations (45) and (44) respectively, the virtual strain energy is,

$$\int_{vol} [\mathbf{BA}^{-1}\bar{\boldsymbol{\Delta}}]^T \mathbf{DBA}^{-1}\boldsymbol{\Delta} \, \mathrm{d}V$$

Now since all the matrices contain constant terms only and are thus independent of x and y, the expression for the virtual strain energy may be written

$$\bar{\boldsymbol{\Delta}}^T \{[\mathbf{A}^{-1}]^T \mathbf{B}^T \mathbf{DBA}^{-1} V\} \boldsymbol{\Delta}$$

The external work is the product of the virtual displacements $\bar{\boldsymbol{\Delta}}$ and the nodal forces **S**, hence equating external virtual work and internal virtual strain energy,

$$\bar{\boldsymbol{\Delta}}^T \mathbf{S} = \bar{\boldsymbol{\Delta}}^T \{[\mathbf{A}^{-1}]^T \mathbf{B}^T \mathbf{DBA}^{-1} V\} \boldsymbol{\Delta}$$

The virtual displacements are quite arbitrary and in particular may be taken to be represented by a unit matrix, thus,

$$\mathbf{S} = \{[\mathbf{A}^{-1}]^T\mathbf{B}^T\mathbf{D}\mathbf{B}\mathbf{A}^{-1}V\}\boldsymbol{\Delta}$$
$$= \mathbf{k}\boldsymbol{\Delta}, \text{ from equation (30),}$$

Thus

$$\mathbf{k} = [\mathbf{A}^{-1}]^T\mathbf{B}^T\mathbf{D}\mathbf{B}\mathbf{A}^{-1}V \tag{46}$$

Before the matrix multiplications required in equation (46) can be performed we need to find $\mathbf{A}^{-1}$. This is easily determined as

$$\mathbf{A}^{-1} = \frac{1}{ab}\begin{bmatrix} ab & 0 & 0 & 0 & 0 & 0 \\ -b & b & 0 & 0 & 0 & 0 \\ (c-a) & -c & a & 0 & 0 & 0 \\ 0 & 0 & 0 & ab & 0 & 0 \\ 0 & 0 & 0 & -b & b & 0 \\ 0 & 0 & 0 & (c-a) & -c & a \end{bmatrix}$$

Hence finally, with $\lambda_1 = \frac{1}{2}(1-\nu)$ and $\lambda_2 = \frac{1}{2}(1+\nu)$ we obtain the stiffness matrix for the plane stress triangular element as,

$$\mathbf{k} = \frac{Et}{2(1-\nu^2)ab}\begin{bmatrix} b^2+\lambda_1(c-a)^2 & & & & & \\ -b^2-\lambda_1 c(c-a) & b^2+\lambda_1 c^2 & & \text{Symmetric} & & \\ \lambda_1 a(c-a) & -\lambda_1 ac & \lambda_1 a^2 & & & \\ -\lambda_2 b(c-a) & \lambda_1 cb+\nu b(c-a) & -\lambda_1 ab & \lambda_1 b^2+(c-a)^2 & & \\ \lambda_1 b(c-a)+\nu cb & -\lambda_2 bc & \lambda_1 ab & -\lambda_1 b^2-c(c-a) & \lambda_1 b^2+c^2 & \\ -\nu ab & \nu ab & 0 & a(c-a) & -ac & a^2 \end{bmatrix} \tag{47}$$

It is neither necessary nor economical to carry out these operations by hand, the computation of the element stiffnesses and, indeed, the entire computational process is easily programmed for the digital computer.

MOMENT DISTRIBUTION

Introduction

Although the stiffness method, described in the previous section has the merit of simplicity, the solution of the equilibrium equations (31), is generally a matter for the digital computer since only for the simplest structures can a hand solution be contemplated. An alternative procedure which is eminently suitable for hand computation, is the method of moment distribution which is essentially an iterative solution of the equations of equilibrium.

As in the general stiffness method, we first imagine all the degrees of freedom, joint rotations and joint translations, to be constrained. We ignore axial effects in members and consider flexure only. The constraints are imagined to be clamps applied to the joints to prevent rotation and translation. The forces required to effect the constraints are applied artificially and in the moment distribution processes these clamping forces are systematically released so as to allow the structure to achieve an equilibrium state. It is important to note that in the method as generally applied, the rotational joint restraints are relaxed by one process and the translational restraints by another. Finally the principle of superposition is used to combine the separate results.

It is necessary to assemble certain standard results before we can consider the actual process.

Distribution factors, carry-over factors and fixed-end moments

For the time being we confine our attention to prismatic members. The treatment of non-uniform section members will be touched on later.

Standard member stiffnesses are required and these are illustrated in Figure 3.15. The member end forces are those required to produce the deflected forms shown. Diagrams (a) and (b) relate to rotation without translation (sway), and diagrams (c) and (d)

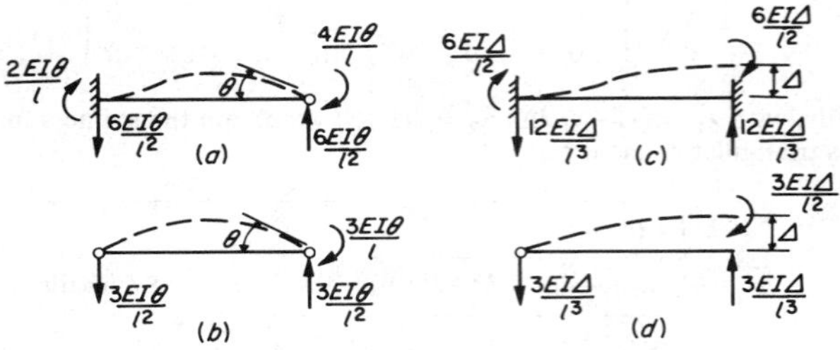

Figure 3.15 Prismatic member stiffnesses

relate to sway without rotation. The results in diagrams (a) and (c) may be deduced from the stiffness matrix in equation (30). The other results may be obtained easily from elementary beam theory, for example in Figure 3.15(b), taking the origin of x at the left-hand end and y positive downwards,

$$EI\mathrm{d}^2y/\mathrm{d}x^2 = \frac{Mx}{l}, \text{ where } M \text{ is the moment, to be determined, at the right-hand end,}$$

$$EI\mathrm{d}y/\mathrm{d}x = \frac{M}{l}\frac{x^2}{2}+C_1$$

$$= EI\theta \text{ for } x = l; \text{ hence } C_1 = EI\theta - M\frac{l}{2}$$

$$EIy = \frac{M}{l}\frac{x^3}{6}+\left(EI\theta - M\frac{l}{2}\right)x+C_2$$

$$= 0 \text{ for } x = 0; \text{ hence } C_2 = 0$$

$$= 0 \text{ for } x = l; \text{ hence, } M = \frac{3EI\theta}{l}.$$

When loads are applied to members which are constrained at the joints, fixed-end moments are required to prevent the end rotations. This is another standard type of result which is required in the moment distribution method. Table 3.5 lists fixed-end moments for a selection of loading cases on uniform section beams. Again, these results may be obtained from elementary beam theory. It should be noted that the sign convention is that a moment is *positive* if tending to produce clockwise rotation of the end of the member at which it acts. This convention is different to, and should not be confused with, the sign convention for constructing bending moment diagrams which must be based on the curvature produced in the member.

As an illustration of the basic process, consider the structure ABC shown in Figure 3.11. This structure was analysed by the stiffness method previously. Joint B is considered to be clamped and thus a system of fixed-end moments is set up in member AB. The end moments in the members are shown in line 1 of Table 3.6. The constraining moment at

Table 3.6

Stage	*Operation*	M_{AB}	M_{BA}	M_{BC}	M_{CD}
1	Fixed-end Moments	$-Wl_1/8$	$+Wl_1/8$	0	0
2	Distribution at B	$-\dfrac{Wl_1}{16}\left(\dfrac{l_2}{l_1+l_2}\right)$	$-\dfrac{Wl_1}{8}\left(\dfrac{l_2}{l_1+l_2}\right)$	$-\dfrac{Wl_1}{8}\left(\dfrac{l_1}{l_1+l_2}\right)$	$-\dfrac{Wl_1}{16}\left(\dfrac{l_1}{l_1+l_2}\right)$
3	Total Moments	$-\dfrac{Wl_1}{16}\left(\dfrac{2l_1+3l_2}{l_1+l_2}\right)$	$\dfrac{Wl_1^2}{8(l_1+l_2)}$	$-\dfrac{Wl_1^2}{8(l_1+l_2)}$	$-\dfrac{Wl_1^2}{16(l_1+l_2)}$

joint B is seen to be $Wl_1/8$ clockwise and we imagine this moment to be removed by the application of a moment $-Wl_1/8$. The subsequent rotation of joint B, anticlockwise through angle θ, will develop moments in both members. Referring to Figure 3.15 the moments induced will be,

$$M_{BA} = -\frac{4EI\theta}{l_1}; \quad M_{AB} = -\frac{2EI\theta}{l_1}$$

$$M_{BC} = -\frac{4EI\theta}{l_2}; \quad M_{CB} = -\frac{2EI\theta}{l_2}$$

For equilibrium of joint B, the applied moment $-Wl_1/8$ must equal the sum of the moments absorbed by the two members meeting at the joint.

Hence
$$\frac{-Wl_1}{8} = -\frac{4EI\theta}{l_1} - \frac{4EI\theta}{l_2} = -4E\theta\left(\frac{I}{l_1}+\frac{I}{l_2}\right)$$

and it is seen that the moment is 'distributed' to the members in proportion to their I/l values.

Thus

$$M_{BA} = \frac{-Wl_1}{8}\,\frac{I/l_1}{(I/l_1+I/l_2)} = \frac{-Wl_1}{8}\left(\frac{l_2}{l_1+l_2}\right)$$

and

$$M_{BC} = \frac{-Wl_1}{8}\,\frac{I/l_2}{(I/l_1+I/l_2)} = \frac{-Wl_1}{8}\left(\frac{l_1}{l_1+l_2}\right)$$

The moments induced at A and C are from Figure 3.15, one-half of those induced at B and the factor of one-half is termed the *carry over* factor. This set of moments is shown in line 2 of Table 3.6.

Joint B is now 'in-balance' and since it was the only joint which was clamped we have reached an equilibrium state and no further distribution of moments is required. The final set of moments is obtained in line 3 of Table 3.6, by the addition of lines 1 and 2.

This result is the same as that obtained from pure stiffness considerations. It should be noted that the zero sum of moments M_{BA} and M_{BC} indicates that joint B is in rotational equilibrium.

Two further points should be noted before we consider the moment distribution process in more detail. Referring to Figure 3.16; of the three members connected at

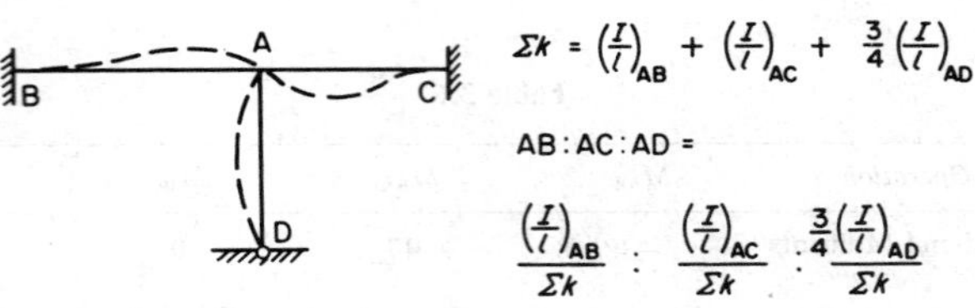

$\Sigma k = \left(\frac{I}{l}\right)_{AB} + \left(\frac{I}{l}\right)_{AC} + \frac{3}{4}\left(\frac{I}{l}\right)_{AD}$
AB : AC : AD =
$\frac{\left(\frac{I}{l}\right)_{AB}}{\Sigma k} : \frac{\left(\frac{I}{l}\right)_{AC}}{\Sigma k} : \frac{\frac{3}{4}\left(\frac{I}{l}\right)_{AD}}{\Sigma k}$
A
B
C
D

Figure 3.16 Distribution factors at typical joint

joint A, member AD is hinged at the end remote from A whereas the other two members are fixed. Since D is hinged no moment can exist there and hence there is no carry-over to D. Furthermore the moment-rotation relationship is different for a member pinned at the remote end, as may be seen by comparing Figures 3.15(a) and (b). In calculating distribution factors this is taken account of by taking $\frac{3}{4}(I/l)$ for such members as compared with I/l for members fixed at the remote end.

Moment distribution without sway

As an example of a structure with two degrees of freedom of joint rotation and no sway, consider the frame shown in Figure 3.17. EI (beams) $= 3 \times EI$ (columns).

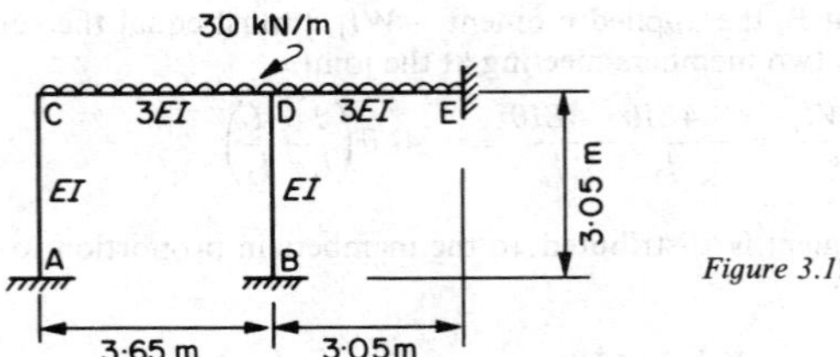

Figure 3.17

The fixed-end moments are, $(wl^2/12)$,

$$M_{FCD} = -30 \times \frac{3.65^2}{12}; \quad M_{FDC} = +30 \times \frac{3.65^2}{12} = 33.3 \text{ kNm}$$

$$F_{FDE} = -30 \times \frac{3.05^2}{12}; \quad M_{FED} = +30 \times \frac{3.05^2}{12} = 23.3 \text{ kNm}$$

and the distribution factors are,

$$\text{at C, CD:CA} = \frac{3/3.65}{(1/3.05)+(3/3.65)} : \frac{1/3.05}{(1/3.05)+(3/3.65)} = 0.715 : 0.285$$

at D, DC: DB: DE =

$$\frac{3/3.65}{(3/3.65)+(1/3.05)+(3/3.05)} : \frac{1/3.05}{(3/3.65)+(1/3.05)+(3/3.05)} : \frac{3/3.05}{(3/3.65)+(1/3.05)+(3/3.05)}$$

$$= 0.386 : 0.154 : 0.460$$

The moment distribution is carried out in Table 3.7. It should be noted that after each distribution at a joint the distributed moments are underlined to indicate that the joint is balanced at that stage. At step 4, the out-of-balance moment to be distributed at D is $+33.3+11.9-23.3 = +21.9$; hence the distributed moments should total -21.9.

Table 3.7
MOMENT DISTRIBUTION FOR FRAME SHOWN IN FIGURE 3.17

Joint	A	C		D			B	E
Distribution Factors		0.285	0.715	0.386	0.154	0.460		
End Moments	AC	CA	CD	DC	DB	DE	BD	ED
1. Fixed-end Moments			−33.3	+33.3		−23.3		+23.3
2. Distribution at C		+9.5	+23.8					
3. Carry-over to A and D	+4.75			+11.9				
4. Distribution at D				−8.45	−3.38	−10.07		
5. Carry-over to C, B and E			−4.23				−1.69	−5.04
6. Distribution at C		+1.20	+3.03					
7. Carry-over to A and D	+0.60			+1.52				
8. Distribution at D				−0.59	−0.23	−0.70		
9. Carry over to C, B and E			−0.30				−0.12	−0.35
10. Distribution at C		+0.09	+0.21					
11. Carry-over to A and D	+0.05			+0.11				
12. Distribution at D				−0.04	−0.02	−0.05		
13. Carry-over to C, B and E				May be neglected				
14. Total moments (kNm)	+5.40	+10.79	−10.79	+37.75	−3.63	−34.12	−1.81	+17.91

Moment distribution with sway

This process will be illustrated with reference to Example 2 (page 3–15) for which the structure is shown in Figure 3.9. We first ignore any horizontal movement (sway) of the joints B and C and carry out a moment distribution.
The fixed-end moments are $wl^2/12 = \pm 40$ kNm;
and the distribution factors are,

$$BA : BC = \tfrac{1}{3} : \tfrac{2}{3}$$
$$CB : CD = \tfrac{2}{3} : \tfrac{1}{3} \text{ (noting } \tfrac{3}{4}I/l \text{ for CD)}$$

This process will be illustrated with reference to Example 2 (page 3–15) for which the line 3 of Table 3.8. We now consider the horizontal equilibrium of the beam BC, Figure 3.18(a), and find that a force F_1 is required to maintain equilibrium. F_1 may be calculated by evaluating the horizontal shear forces at the tops of the columns as follows,

$$F_1 = 120 + \frac{(20+10)}{4} - \frac{20}{3} = 120.8 \text{ kN}$$

This force cannot exist in practice and what happens is that the beam BC deflects to the right and a new set of bending moments is set up with the effect that the out-of-balance horizontal force F_1 is removed. We consider the effect of this sway separately. Referring

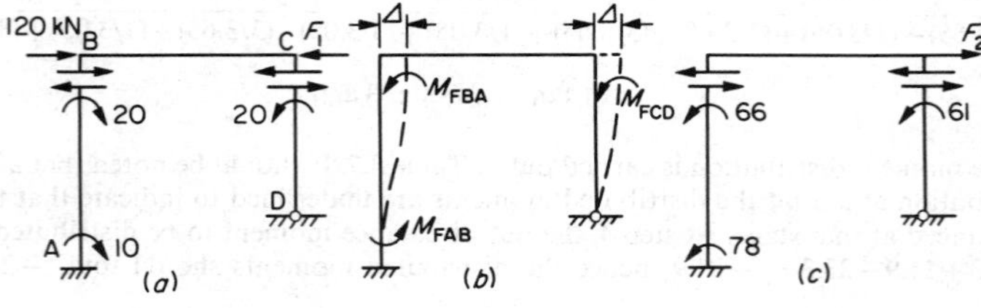

Figure 3.18

to Figure 3.18(b), a movement to the right of Δ, without joint rotation, requires column moments as shown. From Figure 3.15(c) and (d), these column moments are,

$$M_{FBA} = M_{FAB} = -6\left(\frac{EI}{l^2}\right)_{AB}\Delta$$

$$M_{FCD} = -3\left(\frac{EI}{l^2}\right)_{CD}\Delta \qquad \text{(note } M_{FDC} = 0\text{)}$$

We cannot evaluate these moments unless Δ is known but we could proceed with an arbitrary value of Δ, and carry out a distribution to produce rotational equilibrium of the joints B and C. In fact it is seen that any arbitrary values of moments can be used providing these are in the correct proportions between the two columns. The ratio in this example is

$$AB:CD = \left(\frac{I}{l^2}\right)_{AB} : \frac{1}{2}\left(\frac{I}{l^2}\right)_{CD}$$

If we adopt,

$$M_{FBA} = M_{FAB} = -90$$

and

$$M_{FCD} = -80,$$

the moments are in the correct proportion. A second moment distribution is now carried

Table 3.8

Joint	A	B		C	
End moments	AB	BA	BC	CB	CD
1. Arbitrary sway	−78	−66	+66	+61	−61
2. Corrected sway [(1) × λ]	−167	−141	+141	+131	−131
3. No sway moments	+10	+20	−20	+20	−20
4. Final moments [(2)+(3)]	−157	−121	+121	+151	−151

out, using these values of fixed-end moments, and the result is shown in line 1 of Table 3.8. This set of moments is consistent with an applied horizontal force F_2, Figure 3.18(c), and

$$F_2 = \frac{66+78}{4} + \frac{61}{3} = 56.3 \text{ kN}$$

Now F_2 has to be scaled to equal F_1 and the scaling factor is $F_1/F_2 = \lambda = 120.8/56.3 = 2.14$.
The corrected moments are given in line 2 of Table 3.8 and the final moments are in line 4 obtained by adding lines 2 and 3.

Additional topics in moment distribution

Space has permitted only a brief introduction to the method of moment distribution. Additional topics which should be studied by reference to the standard texts,[8, 9, 10] are as follows:

(a) *Frames with multiple degrees of freedom for sway.* These are handled by carrying out an arbitrary sway distribution for each sway in turn. Equilibrium conditions are then used to relate the out-of-balance forces and obtain the correction factors for each sway mode.

(b) *Treatment of symmetry.* In cases of symmetry the moment distribution process can be considerably shortened. Two cases arise and should be studied, systems in which it is known that the final set of moments is symmetrical and systems in which the final moments form an anti-symmetrical system.

(c) *Non-prismatic members.* If the flexural rigidity (EI) of a member varies within its length, then the effect is to change the values of end stiffnesses, carry-over factor and fixed end moments. A suitable general method for handling this situation is to evaluate end flexibilities by the use of Simpson's rule and then convert the flexibilities into stiffnesses. The method is discussed in detail in chapter 5 of reference 1.

INFLUENCE LINES

Introduction and definitions

It is frequently necessary to consider loads which may occupy variable positions on a structure. For example, in bridge design it is important to determine the maximum effects due to the passage of a specified train or system of loads. In other cases the total load on a structure may be comprised of different loads which may be applied in various combinations and this again is a problem of variability of load or load position. The effect of varying a load position may be studied with the help of *influence lines.*

An influence line shows the variation of some resultant action or effect such as bending moment, shear force, deflection etc, at a particular point as a unit load traverses the structure. It is important to observe that the effect considered is at a fixed position, for example *bending moment at* C, and that the independent variable in the influence line diagram is *the load position.* The following is a summary of influence line theory. For a more detailed treatment the reader should consult reference 14.

Influence lines for beams

Consider the simply-supported beam AB, Figure 3.19, carrying a single unit load occupying a variable position distant y from A. We require to obtain influence lines for bending moment and shear force at a fixed point X distant a from A and b from B.

If the unit load lies between X and B,

$$M_x = R_A \,.\, a = 1\,\frac{(l-y)}{l}\,a \tag{48}$$

If the unit load acts between A and X,

$$M_x = R_B \,.\, b = 1 \,.\, y/l \,.\, b \tag{49}$$

Equations (48) and (49) are linear in y and when plotted in the regions to which they relate, form a triangle as shown in Figure 3.19(b). We note that, in both cases, substitution of $y = a$ gives $M_x = 1 \, . \, ab/l$. Thus the influence line for M_x is a triangle with a peak value ab/l at the section X.

Turning now to the influence line for shearing force at X. For unit load between X and B

$$S_x = R_A = \frac{l-y}{l} \tag{50}$$

(and now we have implied a sign convention for shear force namely that S_x is positive if the resultant force to the left of the section is upwards).

$$\text{when } y = a, \; S_x = b/l$$

For unit load between A and X

$$S_x = -R_B = -y/l \tag{51}$$

when $y = a$, $S_x = -a/l$

We note that equations (50) and (51) give different values of S_x for $y = a$ and moreover the signs are opposite. This means that the shear force influence line contains a discontinuity at X as shown in Figure 3.19(c).

In using influence lines with a given system of loads and having determined the locations of the loads on the span, the total effect is evaluated as,

$$\sum(W \times \text{ordinate}), \text{ for concentrated loads,} \tag{52}$$

and,

$$\int whdx = w \text{ (area under influence line),} \tag{53}$$

for distributed loads, (Figure 3.19(d)).

The maximum effect produced at a given position is of interest in the design process. In the case of concentrated loads, from equation (52), this is obtained when

$$\sum(W \times \text{ordinate}) \text{ is a maximum.}$$

The process of locating the loads to produce the maximum value is best done by trial and error. It follows from the straight line nature of a bending moment diagram due to concentrated loads, that the maximum bending moment at a section will be obtained when one of the loads acts at the section. This may be illustrated by reference to the two-load system shown at (e) in Figure 3.19. The shape of the bending moment diagram is as shown at (f) and at (g) is drawn a diagram which shows the maximum value of bending moment at any section in the beam. This is the *maximum bending moment envelope* M_{max} which is seen to consist of two intersecting parabolic curves M_{y1} and M_{y2}.

The curve M_{y1} represents the maximum bending moment at all sections in the beam when this is obtained with load W_1 placed at the section. The curve M_{y2} represents the maximum bending moment at all sections in the beam when this is obtained with load W_2 at the section. It is seen that W_1 should be placed at the section towards the left-hand end of the beam, and W_2 at the section towards the right hand end of the beam.

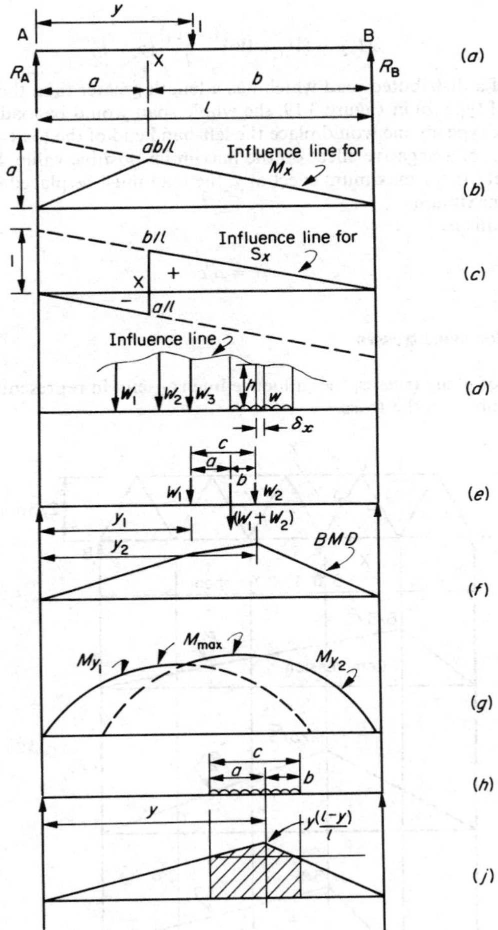

Figure 3.19 Influence lines and related diagrams for simply-supported beams

The expressions for M_{y1} and M_{y2} are as follows,

$$M_{y1} = (W_1 + W_2)\frac{y_1}{l}(l - y_1 - a)$$
$$M_{y2} = (W_1 + W_2)\frac{(l - y_2)}{l}(y_2 - b) \quad (54)$$

In the case of a distributed load which has a length greater than the span, then for an influence line of type (b) in Figure 3.19, the whole span would be loaded whereas for an influence line of type (c) one would place the left-hand end of the load at X thus avoiding the introduction of a negative effect on the maximum positive value. For a short distributed load, as at (h), for maximum effect at y, the load must be placed so that the shaded area in (j) is a maximum.

The rule for this is,

$$y/l = a/c \quad (55)$$

Influence lines for plane trusses

In the analysis of plane trusses, the influence line is useful in representing the variations in forces in members of the truss.

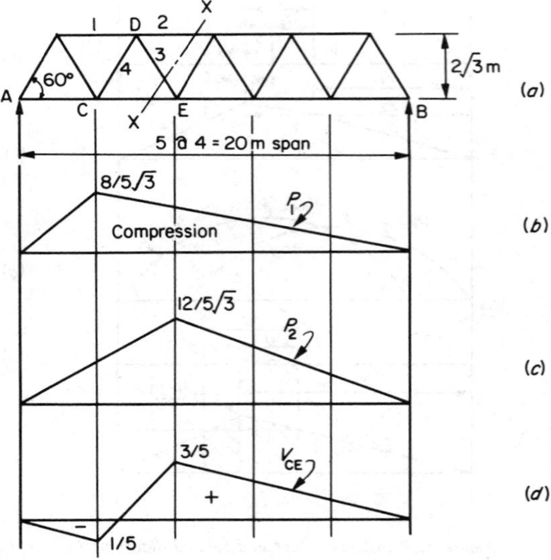

Figure 3.20 Influence lines for plane truss

Figure 3.20(a) shows a Warren girder AB of span 20 m. For the unit acting at any of the lower chord joints, the force in member 1 is

$$P_1 = \frac{4R_A}{2\sqrt{3}}$$

The peak value occurs when the unit load is at C, and thus

$$P_{1\,max} = \frac{2}{\sqrt{3}} \times \frac{4}{5} \times 1 = \frac{8}{5\sqrt{3}}$$

The influence line for P_1 is shown at (b).

For member 2, if the unit load lies between A and E, we take

$$P_2 = \frac{12R_B}{2\sqrt{3}}$$

or, if the unit load lies between E and B we take

$$P_2 = \frac{8R_A}{2\sqrt{3}}$$

The result is a triangle with peak value $12/5\sqrt{3}$ at E, as shown in diagram (c).

It should be noted that both the P_1 and P_2 influence lines indicate compression for all positions of the unit load.

For members 3 and 4 it is useful to note that these members carry the vertical shear force in the panel CE, and we proceed by drawing the influence line for V_{CE} as at (d).

Considering now the force in member 3 and the section XX in diagram (a). It is clear that the relationship is

$$P_3 = \frac{V_{CE}}{\sin 60^\circ}$$

and that P_3 is tensile when V_{CE} is positive and compressive when V_{CE} is negative.

MODEL ANALYSIS

Introduction

In spite of the very sophisticated computing techniques now available for structural analysis, it is sometimes convenient and desirable to use a model of the structure under investigation. There are two kinds of model used in structural analysis and the corresponding methods are termed *direct* and *indirect*.

In a *direct* model analysis, a scale version of the prototype is made and the observations (displacements and strains) under load are related to the prototype through scale factors. It is necessary to define linear scale, elastic modulus scale and load scale and it is further required to use a material with the same value of Poisson's ratio as that of the prototype.

In an *indirect* model analysis the requirements of scale are less exacting. If for example, as is usually the case with this type of model, deformation in the prototype can be considered as due to flexure only, then all that is required is to choose a linear scale for the outline geometry and ensure that the flexural rigidities of model and prototype are proportional. Indirect structural models are used in conjunction with Mueller Breslau's theorem for obtaining influence lines directly in statically indeterminate structures. We shall confine our attentions to a brief summary of the theory of indirect model analysis, for detailed treatment of both types of analysis the reader is referred to the standard texts.[11, 12, 13]

Maxwell's reciprocal theorem

Consider the propped cantilever shown in Figure 3.21 to be subjected to a load W at A, producing displacements f_{11} and f_{21} as shown at (a), and then separately to be subjected to a moment M at B producing displacements f_{12} and f_{22} as at (b). Assuming a linear load-displacement relationship we may use the principle of superposition and obtain the combined effects of W and M by adding (a) and (b). Clearly it will be im-

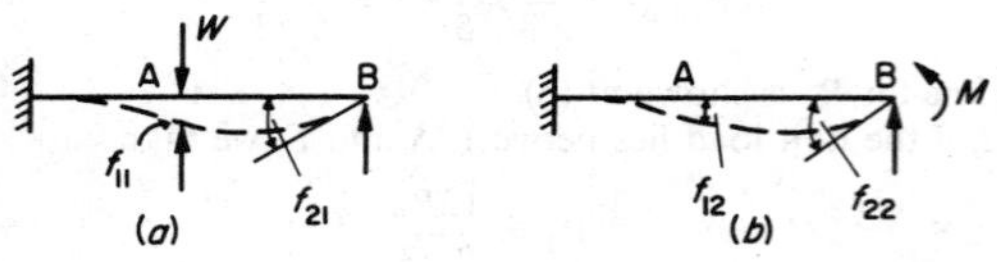

Figure 3.21

material in which order the forces are applied. Applying W first and then M, the work done by the loads will be

$$(\tfrac{1}{2}Wf_{11})+(\tfrac{1}{2}Mf_{22}+Wf_{12}) \tag{56}$$

The first bracket in equation (56) contains the work done during the application of W and the second bracket the work done (by both M and W) during the application of M.

In a similar way, if the order is reversed, the work done is,

$$(\tfrac{1}{2}Mf_{22})+(\tfrac{1}{2}Wf_{11}+Mf_{21}) \tag{57}$$

From equations (56) and (57) it is evident that

$$Wf_{12}=Mf_{21} \tag{58}$$

If the applied actions are taken to have unit values, then equation (58) simplifies to

$$f_{12}=f_{21} \tag{59}$$

Equation (59) is a statement of Maxwell's reciprocal theorem. A more general theorem, of which Maxwell's is a special case, is due to Betti. This latter theorem states that if a system of forces P_i produces displacements p_i at corresponding positions and another set of forces Q_i, at similar positions to P_i, produces displacements q_i, then,

$$P_1q_1+P_2q_2+\ldots+P_nq_n=Q_1p_1+Q_2p_2+\ldots+Q_np_n \tag{60}$$

Mueller Breslau's principle

This principle is the basis of the indirect method of model analysis. It is developed from Maxwell's theorem as follows. Consider the two span continuous beam shown in Figure

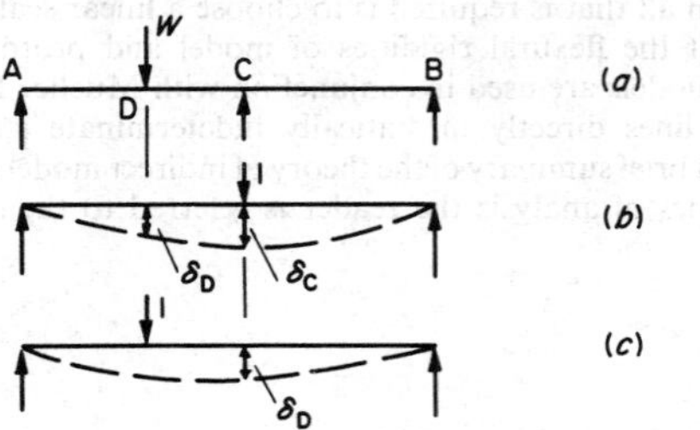

Figure 3.22

3.22(a). On removal of the support at C and the application of a unit load at C, a deflected shape, shown dotted in Figure 3.22(b), is obtained. If a unit load now occupies any arbitrary position D, as at (c), then from Maxwell's theorem the deflection at C will be δ_D. In other words the deflected form (b) is the influence line for deflection of C.

Now the force at C to move C through $\delta_C = 1$

Hence the force at C to move C through $\delta_D = 1 \times \delta_D/\delta_C$.

If a unit load acts at D, producing a deflection δ_D at C, then the upwards force needed to restore C to the level of AB is $1 \times \delta_D/\delta_C$. Hence the reaction at C for unit load at D is $1 \times \delta_D/\delta_C$. Since D is an arbitrary point in the beam then it is seen that the deflected shape due to unit load at C, Figure 3.22(b), is to some scale, the influence line for R_C. The scale of the influence line is determined from the knowledge that the actual ordinate at C should equal unity. Hence the ordinates should all be divided by δ_C.

This result leads to Mueller Breslau's principle which may be stated as follows,

> 'The ordinates of the influence line for a redundant force are equal to those of the deflection curve when a unit load replaces the redundancy, the scale being chosen so that the deflection at the point of application of the redundancy represents unity'.

Application to model analysis

Consider the fixed arch shown in Figure 3.23(a). The arch has three redundancies which may be taken conveniently as H_A, V_A and M_A. We make a simple model of the arch to a chosen linear scale and pin this to a drawing board. End B is fixed in position and direction and the undistorted centre-line is transferred to the drawing paper. We then impose

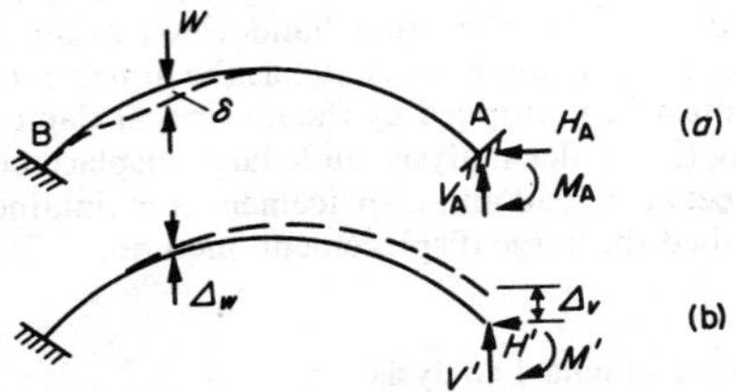

Figure 3.23

a *purely* vertical displacement Δ_v at A and transfer the distorted centre-line to the drawing paper. The distortion produced will require force actions at *A*, *V'*, *H'* and *M'*. Let the displacement of a typical load point be Δ_w. Applying equation (60) to the two systems of forces,

$$V_A(\Delta_v) + H_A(0) + M_A(0) + W(\Delta_w) = V'(0) + H'(0) + M'(0) + 0(\delta)$$

Hence,

$$V_A\Delta_v + W\Delta_w = 0$$

and if

$$W = 1, \qquad V_A = \frac{-\Delta_w}{\Delta_v} \tag{61}$$

Similarly we impose a *purely* horizontal displacement Δ_H and obtain,

$$H_A = \frac{-\Delta'_w}{\Delta_H}, \tag{62}$$

then a pure rotation θ and obtain,

$$M_A = -\frac{\Delta''_w}{\theta} \tag{63}$$

In equations (62) and (63) the displacements Δ'_w and Δ''_w represent the arch displacements due to the imposed horizontal and rotational displacements respectively. In each case the deflected shape, suitably scaled, gives the influence line for the corresponding redundancy.

SIGN CONVENTION

The negative sign in equations (61), (62) and (63) leads to the following convention for signs. On the assumption that a reaction is positive if in the direction of the imposed displacement, then a load W will give a positive value of the reaction if the influence line ordinate at the point of application of the load is opposite to the direction of the load. This is evident in Figure 3.23(b) where the upward deflection Δ_w, being opposed to the direction of the load W, is consistent with a positive (upwards) direction for V_A.

SCALE OF THE MODEL

It should be noted that when using relationships (61) and (62) the ratios Δ_w/Δ_v and Δ'_w/Δ_H are dimensionless and thus the linear scale of the model does not affect the influence line ordinates. On the other hand when using equation (63) in obtaining an influence line for bending moment, Δ_w/θ has the dimensions of length and thus the model displacements must be multiplied by the linear scale factor.

In performing the model analysis, quite large displacements can be used providing the linear relation between load and displacement is maintained. Hence the indirect method is sometimes called the 'large displacement' method.

Example of the use of model analysis

Figure 3.24(a) shows a two-hinged rigid frame of which a model is made to a linear scale of 15. The model is pinned at A and E and the centre lines of the members transferred

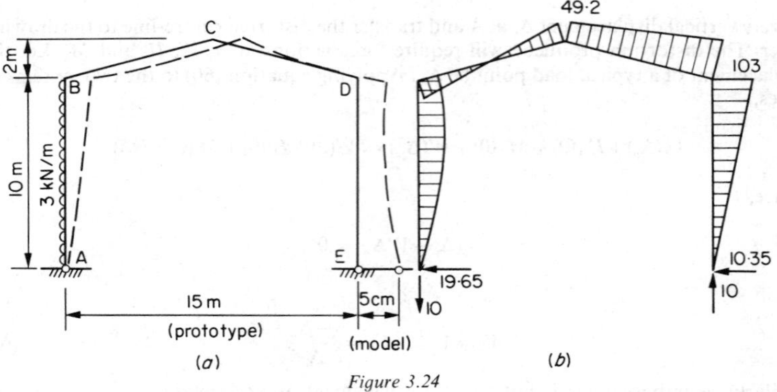

Figure 3.24

to the drawing board. Support E is then moved outwards by 5 cm and the distorted outline of the model plotted. The area between the two diagrams between A and B is measured and found to be 115 cm². Determine H_A and H_E and sketch the bending moment diagram for the prototype frame.

Area under influence line on model = 115/5 = 23 cm².

Area under influence line on prototype = 23 × 15 = 345 cm².

$$\therefore\ H_E = -345 \times \tfrac{3}{100}\ \text{kN} = -10.35\text{kN}$$

The negative sign indicates that H_E acts in the reverse direction to the imposed displacement at E, i.e. from right to left.

Hence $H_A = 3 \times 10 - 10.35 = 19.65$ kN (acting from right to left). Also $V_A = 10$ kN (down) and $V_E = 10$ kN (up).

The bending moment diagram may now be drawn as in Figure 3.24(b).

STRUCTURAL DYNAMICS

Introduction and definitions

Structural vibrations result from the application of *dynamic* loads; i.e. loads which vary with time. Loads applied to structures are often time dependent although in most cases the rate of change of load is slow enough to be neglected and the loads may be regarded as *static*. Certain types of structure may be susceptible to dynamic effects; these include structures designed to carry moving loads, for example bridges and crane girders, and structures required to support machinery. One of the most severe and destructive sources of dynamic disturbance of structures is of course the earthquake.

The dynamic behaviour of structures is generally described in terms of the displacement-time characteristics of the structure, such characteristics being the subject of vibration analysis. Before considering methods of analysis it is helpful to define certain terms used in dynamics.

Amplitude is the maximum displacement from the mean position.

Period is the time for one complete cycle of vibration.

Frequency is the number of vibrations in unit time.

Forced vibration is the vibration caused by a time dependent disturbing force.

Free vibrations are vibrations after the force causing the motion has been removed.

Damping. In structural vibrations, damping is due to, (a) internal molecular friction; (b) loss of energy associated with friction due to slip in joints, and, (c) resistance to motion provided by air or other fluid (drag). The type of damping usually assumed to predominate in structural vibrations is termed *viscous* damping in which the force resisting motion is proportional to the velocity. Viscous damping adequately represents the resistance to motion of the air surrounding a body moving at low speed and also the internal molecular friction.

Degrees of freedom. This is the number of independent displacements or *coordinates* necessary to completely define the deformed state of the structure at any

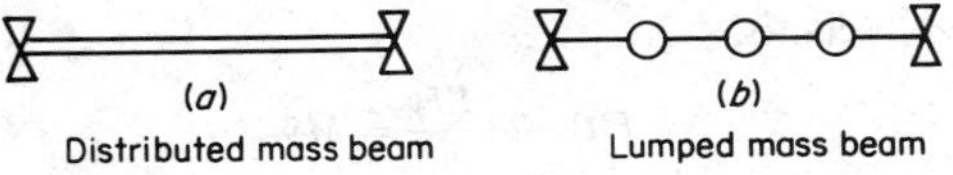

Figure 3.25

instant in time. When a single coordinate is sufficient to define the position of any section of the structure, the structure has a *single degree of freedom.* A continuous structure with a distributed mass, such as a beam, has an infinite number of degrees of freedom. In structural dynamics it is generally satisfactory to transform a structure with an infinite number of degrees of freedom into one with a finite number of freedoms. This is done by adopting a *lumped mass* representation of the structure, as in Figure 3.25. The total mass of the structure is considered to be *lumped* at specified points in the structure and the motion is described in terms of the displacements of the lumped masses. The accuracy of the analysis can be improved by increasing the number of lumped masses. In most cases sufficiently accurate results can be obtained with a comparatively small number of masses.

Single degree of freedom vibrations

The portal frame shown in Figure 3.26 is an example of a structure with a single degree of freedom providing certain assumptions are made. If it is assumed that the entire mass of the structure (M) is located in the girder and that the girder has an infinitely large

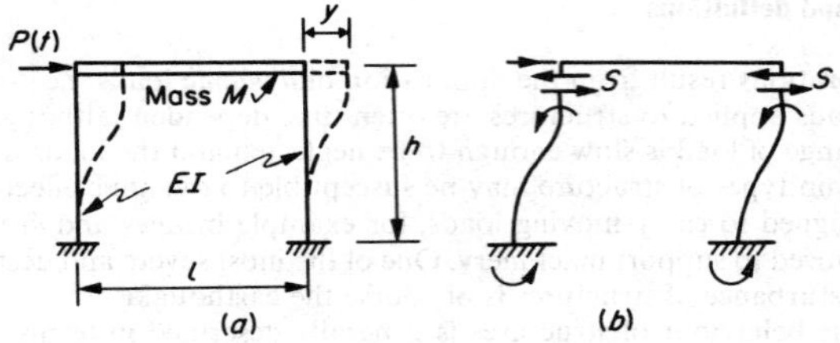

Figure 3.26

flexural rigidity and further, that the columns have infinitely large extensional rigidities; then the displacement of the mass M resulting from the application of an exciting force $P(t)$, is defined by the transverse displacement y. The girder moves in a purely horizontal direction restrained only by the flexure of the columns.

From Newton's second law of motion

$$\text{Force} = \text{mass} \times \text{acceleration}$$

i.e.

$$\sum P = M\ddot{y} \tag{64}$$

Now from Figure 3.26(b), the force resisting motion is

$$2S = 2\left(\frac{12EIy}{h^3}\right)$$

$$= 24\frac{EIy}{h^3} \tag{65}$$

Thus equation (64) becomes,

$$P(t) - 24\frac{EIy}{h^3} = M\ddot{y}$$

or

$$M\ddot{y} + 24\frac{EIy}{h^3} = P(t) \tag{66}$$

If the effect of damping is included then the equation of motion, equation (66) is modified by the inclusion of a term $c\dot{y}$ where c is a constant. It should be noted that since the effect of damping is to resist the motion, then the term $c\dot{y}$ is added to the left-hand side of equation (66). Thus,

$$M\ddot{y}+c\dot{y}+24\frac{EIy}{h^3} = P(t) \tag{67}$$

Equation (67) may be generalised for any single degree of freedom structure by observing that the stiffness of the structure, i.e. force required for unit displacement horizontally, is given by

$$k = 24\frac{EI}{h^3} \tag{68}$$

Combining equations (67) and (68) we obtain the general single degree of freedom equation of motion,

$$M\ddot{y}+c\dot{y}+ky = P(t) \tag{69}$$

If in equation (69), $P(t) = 0$, we have a state of *free vibration* of the structure. The governing equation becomes,

$$M\ddot{y}+c\dot{y}+ky = 0 \tag{70}$$

The situation envisaged by equation (70) would arise if the beam were given a horizontal displacement and then released. The resulting vibrations would depend on the amount of damping present, measured by the coefficient c.

The solution of equation (70) is

$$y = A_1e^{\lambda_1 t}+A_2e^{\lambda_2 t} \tag{71}$$

where A_1 and A_2 are the constants of integration, to be evaluated from initial conditions, and λ_1 and λ_2 are the roots of the auxiliary equation,

$$M\lambda^2+c\lambda+k = 0 \tag{72}$$

or, substituting

$$\left.\begin{aligned} p^2 &= k/M \\ \text{and}\qquad 2n &= c/M \end{aligned}\right\} \tag{73}$$

equation (72) becomes,

$$\lambda^2+2n\lambda+p^2 = 0 \tag{74}$$

Hence,

$$\lambda = -n\pm\sqrt{(n^2-p^2)} \tag{75}$$

Four cases arise:

CASE 1 $\quad p^2<n^2$;

Here (n^2-p^2) is always positive and $<n^2$ and thus λ_1 and λ_2 are real and negative.

Equation (71) takes the form,

$$y = e^{-nt}(A_1e^{\sqrt{(n^2-p^2)}t}+A_2e^{-\sqrt{(n^2-p^2)}t}) \tag{76}$$

The relationship between y and t of equation (76) is shown in Figure 3.27(a) and it is seen that the displacement y gradually returns to zero, no vibrations taking place.

Now, since $n^2>p^2$, then

$$\frac{c^2}{4M^2}>\frac{k}{M}$$

or

$$c>2\sqrt{(Mk)} \tag{77}$$

A structure exhibiting these characteristics is said to be *overdamped*.

CASE 2 $p^2 = n^2$

From equation (75), $\lambda = -n$ (twice)
and hence,

$$y = e^{-nt}(A_1 + A_2 t) \tag{78}$$

Again, no vibrations result and equation (78) has the form shown in Figure 3.27(a).
From equation (73) the value of c for this condition is given by,

$$c_c = 2\sqrt{(Mk)} \tag{79}$$

This is termed *critical damping* and the critical damping coefficient c_c is the value of the damping coefficient at the boundary between vibratory and non-vibratory motion. The critical damping coefficient is a useful measure of the damping capacity of a structure. The damping coefficient of a structure is usually expressed as a percentage of the critical damping coefficient.

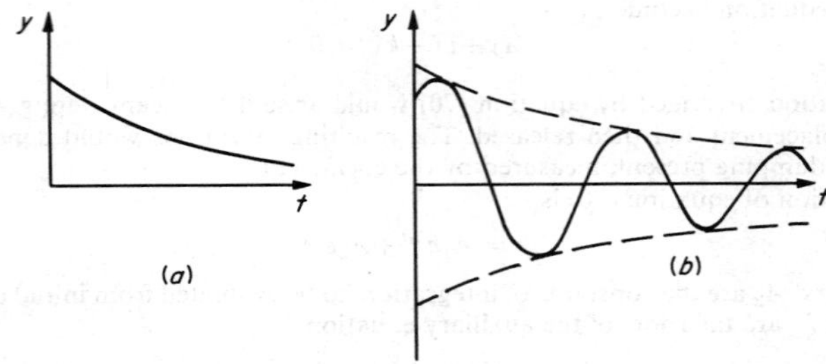

Figure 3.27

CASE 3 $p^2 > n^2$

Here $c < c_c$ and the structure is *underdamped*.
From equation (75), $\lambda = -n \pm i\sqrt{(p^2 - n^2)}$
Hence,

$$y = e^{-nt}(A_1 e^{i\sqrt{(p^2-n^2)}t} + A_2 e^{-i\sqrt{(p^2-n^2)}t})$$

or, putting,

$$(p^2 - n^2) = q^2$$

$$y = e^{-nt}(A_1 e^{iqt} + A_2 e^{-iqt})$$

or

$$y = e^{-nt}(A \cos qt + B \sin qt) \tag{80}$$

A typical displacement-time relationship for this condition is shown in Figure 3.27(b).
An alternative form for equation (80) is

$$y = Ce^{-nt} \sin(qt + \beta) \tag{81}$$

where C and β are new arbitrary constants.

The period $T = \dfrac{2\pi}{q} = \dfrac{2\pi}{p\sqrt{\{1-(n/p)^2\}}}$

The period is constant but the amplitude decreases with time. The decay of amplitude is such that the ratio of amplitudes at intervals equal to the period is constant, i.e.

$$\frac{y_{(t)}}{y_{(t+T)}} = e^{nT}$$

and $\log e^{nT} = nT = \delta$

δ is called the *logarithmic decrement*, and is a useful measure of damping capacity.

The percentage critical damping

$$= 100\frac{c}{c_c}$$

$$= 100\frac{\delta}{pT}$$

This is of the order of 4% for steel frames and 7% for concrete frames.

CASE 4 $c = 0$

In the absence of damping, equation (70) becomes,

$$M\ddot{y}+ky = 0 \qquad (82)$$

The solution of which is,

$$y = A_1e^{\lambda_1 t}+A_2e^{\lambda_2 t}$$

where, from equation (72),

$$\lambda_1 = ip$$

$$\lambda_2 = -ip$$

Thus,

$$y = A\sin pt+B\cos pt \qquad (83)$$

The period is, $T = \dfrac{2\pi}{p}$

where p is the *natural circular frequency*.

The *natural frequency* is $f = \dfrac{1}{T} = \dfrac{p}{2\pi}$

Multi-degree of freedom vibrations

Vibration analysis of systems with many degrees of freedom is a complex subject and only a brief indication of one useful method will be given here. For a more comprehensive and detailed treatment, the reader should consult one of the standard texts.[7, 15, 16]

For a system represented by lumped masses, the governing equations emerge as a set of simultaneous ordinary differential equations equal in number to the number of degrees of freedom. Mathematically the problem is of the *eigenvalue* or *characteristic value* type and the solutions are the *eigenvalues* (frequencies) and the *eigenvectors* (modal shapes). We shall consider the evaluation of mode shapes and fundamental, undamped, frequencies by the process of matrix iteration using the flexibility approach (see page 3–9). The method to be described, leads automatically to the lowest frequency, the fundamental, this being the one of most interest from a practical point of view. The alternative method using a stiffness matrix approach leads to the highest frequency.

Consider the simply-supported, uniform cross-section beam shown in Figure 3.28(a). The mass/unit length is w and we will regard the total mass of the beam to be lumped at the quarter-span points as shown in Figure 3.28(b). We may ignore the end masses $wl/8$

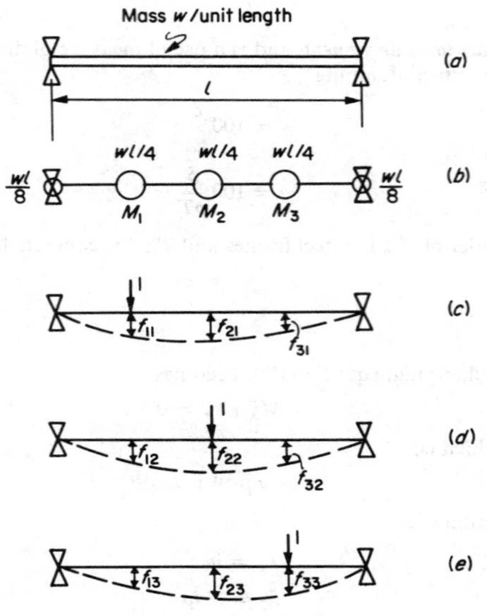

Figure 3.28

since they are not involved in the motion, and consider the three masses $M_1 = M_2 = M_3 = wl/4$.

The appropriate flexibilities, f_{ij}, are shown at (c), (d) and (e).

Using the flexibility method previously described, we may obtain a flexibility matrix as follows,

$$\mathbf{F} = \begin{bmatrix} f_{11} & f_{12} & f_{13} \\ f_{21} & f_{22} & f_{23} \\ f_{31} & f_{32} & f_{33} \end{bmatrix} = \frac{l^3}{256EI}\begin{bmatrix} 3.00 & 3.67 & 2.33 \\ 3.67 & 5.33 & 3.67 \\ 2.33 & 3.67 & 3.00 \end{bmatrix} \tag{84}$$

It should be noted that f_{ij} is the displacement of mass M_i due to unit force acting at mass M_j. Thus, if the forces acting at the positions of the lumped masses are $F_{1,2,3}$ and the corresponding displacements are $y_{1,2,3}$, then,

$$\left.\begin{aligned} y_1 &= f_{11}F_1 + f_{12}F_2 + f_{13}F_3 \\ y_2 &= f_{21}F_1 + f_{22}F_2 + f_{23}F_3 \\ y_3 &= f_{31}F_1 + f_{32}F_2 + f_{33}F_3 \end{aligned}\right\} \tag{85}$$

For free, undamped vibrations F_i is an inertia force $= -M_i\ddot{y}_i$.

Thus,

$$\left.\begin{aligned} y_1+f_{11}M_1\ddot{y}_1+f_{12}M_2\ddot{y}_2+f_{13}M_3\ddot{y}_3 &= 0 \\ y_2+f_{21}M_1\ddot{y}_1+f_{22}M_2\ddot{y}_2+f_{23}M_3\ddot{y}_3 &= 0 \\ y_3+f_{31}M_1\ddot{y}_1+f_{32}M_2\ddot{y}_2+f_{33}M_3\ddot{y}_3 &= 0 \end{aligned}\right\} \tag{86}$$

The solutions take the form

$$y_i = \delta_i \cos(pt+\alpha) \tag{87}$$

hence,

$$\ddot{y}_i = -p^2 y_i \tag{88}$$

Thus, equations (86) become,

$$\left.\begin{aligned} \delta_1-f_{11}M_1p^2\delta_1-f_{12}M_2p^2\delta_2-f_{13}M_3p^2\delta_3 &= 0 \\ \delta_2-f_{21}M_1p^2\delta_1-f_{22}M_2p^2\delta_2-f_{23}M_3p^2\delta_3 &= 0 \\ \delta_3-f_{31}M_1p^2\delta_1-f_{32}M_2p^2\delta_2-f_{33}M_3p^2\delta_3 &= 0 \end{aligned}\right\} \tag{89}$$

or,

$$\mathbf{\Delta} = p^2\mathbf{FM\Delta} \tag{90}$$

where,

$$\mathbf{\Delta} = \begin{bmatrix}\delta_1\\ \delta_2\\ \delta_3\end{bmatrix}; \qquad \mathbf{M} = \begin{bmatrix}M_1 & 0 & 0\\ 0 & M_2 & 0\\ 0 & 0 & M_3\end{bmatrix}$$

The unknowns in equation (90) are the displacement amplitudes δ_i and the frequency p; p has as many values as there are equations in the system, and for every value of p (eigenvalue) there corresponds a set of y (eigenvector).

We adopt an iterative procedure for the solution of equation (90) and first of all re-write the equations in the form,

$$\mathbf{FM\Delta} = \frac{1}{p^2}\mathbf{\Delta} \tag{91}$$

We start with an assumed vector $\mathbf{\Delta}_0$, thus

$$\mathbf{FM\Delta}_0 \simeq \frac{1}{p^2}\mathbf{\Delta}_0$$

Putting $\mathbf{FM\Delta}_0 = \mathbf{\Delta}_1$

then, $\mathbf{\Delta}_1 \simeq \frac{1}{p^2}\mathbf{\Delta}_0$ giving $p^2 \simeq \frac{\mathbf{\Delta}_0}{\mathbf{\Delta}_1}$

We cannot form $\mathbf{\Delta}_0/\mathbf{\Delta}_1$ since each $\mathbf{\Delta}$ is a column matrix, so we take the ratio of corresponding elements in $\mathbf{\Delta}_0$ and $\mathbf{\Delta}_1$ and form the ratio δ_0/δ_1. It is best to use the numerically greatest δ for this purpose.

Continuing the process,

$$\begin{aligned} \mathbf{FM\Delta}_1 &\simeq \frac{1}{p^2}\mathbf{\Delta}_1 \text{ giving } p^2 = \delta_1/\delta_2 \\ &= \mathbf{\Delta}_2 \end{aligned}$$

and again,

$$\begin{aligned} \mathbf{FM\Delta}_2 &\simeq \frac{1}{p^2}\mathbf{\Delta}_2 \\ &= \mathbf{\Delta}_3 \text{ giving } p^2 = \delta_2/\delta_3 \end{aligned}$$

It can be shown that this iterative process converges to the largest value of $1/p^2$ and hence yields the lowest (fundamental mode) frequency.

Applying the iterative scheme to the beam of Figure 3.28, and assuming

$$\mathbf{\Delta}_0 = \begin{bmatrix} 1 \\ 2 \\ 1 \end{bmatrix}$$

then, $\mathbf{\Delta}_1 = \mathbf{FM\Delta}_0$

Where

$$\mathbf{FM} = \frac{l^3}{256EI}\begin{bmatrix} 3.00 & 3.67 & 2.33 \\ 3.67 & 5.33 & 3.67 \\ 2.33 & 3.67 & 3.00 \end{bmatrix}\begin{bmatrix} wl/4 & 0 & 0 \\ 0 & wl/4 & 0 \\ 0 & 0 & wl/4 \end{bmatrix}$$

$$= \frac{wl^4}{1024EI}\begin{bmatrix} 3.00 & 3.67 & 2.33 \\ 3.67 & 5.33 & 3.67 \\ 2.33 & 3.67 & 3.00 \end{bmatrix}$$

Thus,

$$\mathbf{\Delta}_1 = \frac{wl^4}{1024EI}\begin{bmatrix} 12.67 \\ 18.00 \\ 12.67 \end{bmatrix} = \frac{12.67wl^4}{1024EI}\begin{bmatrix} 1.00 \\ 1.42 \\ 1.00 \end{bmatrix}$$

hence,

$$p_1^2 = \frac{\delta_0}{\delta_1} = \frac{2 \times 1024EI}{12.67 \times 1.42wl^4}$$

$$= 114\frac{EI}{wl^4}$$

A second iteration gives,

$$\mathbf{\Delta}_2 = \mathbf{FM\Delta}_1 = \frac{wl^4}{1024EI}\begin{bmatrix} 3.00 & 3.67 & 2.33 \\ 3.67 & 5.33 & 3.67 \\ 2.33 & 3.67 & 3.00 \end{bmatrix}\frac{12.67wl^4}{1024EI}\begin{bmatrix} 1.00 \\ 1.42 \\ 1.00 \end{bmatrix}$$

$$= 12.67\left(\frac{wl^4}{1024EI}\right)^2\begin{bmatrix} 10.54 \\ 14.91 \\ 10.54 \end{bmatrix}$$

hence

$$p_2^2 = \frac{\delta_1}{\delta_2} = \frac{12.67 \times 1.42wl^4}{1024EI} \times \frac{1}{12.67(wl^4/1024EI)^2 \times 14.91}$$

$$= 97.6\frac{EI}{wl^4}$$

This result is very close to that produced by an exact method, i.e. 97.41 EI/wl^4.

PLASTIC ANALYSIS

Introduction

The plastic design of structures is based on the concept of a *load factor* (N), where

$$N = \frac{\text{Collapse load}}{\text{Working load}} = \frac{W_c}{W_w} \tag{92}$$

A structure is considered to be on the point of collapse when finite deformation of at least part of the structure can occur without change in the loads. The simple plastic theory is based on an idealised stress-strain relationship for structural steel as shown in Figure 3.29. A linear, elastic, relationship holds up to a stress σ_y, the *yield stress*, and at this value of stress the material is considered to be in a state of perfect plasticity, capable of

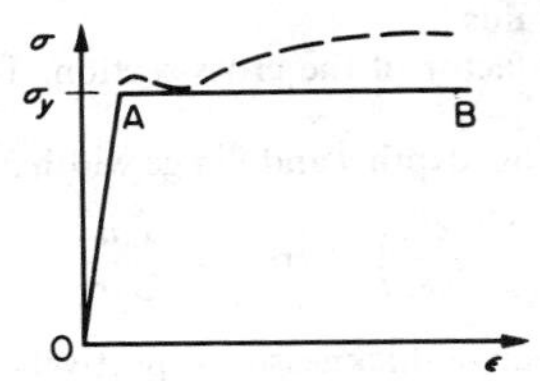

Figure 3.29

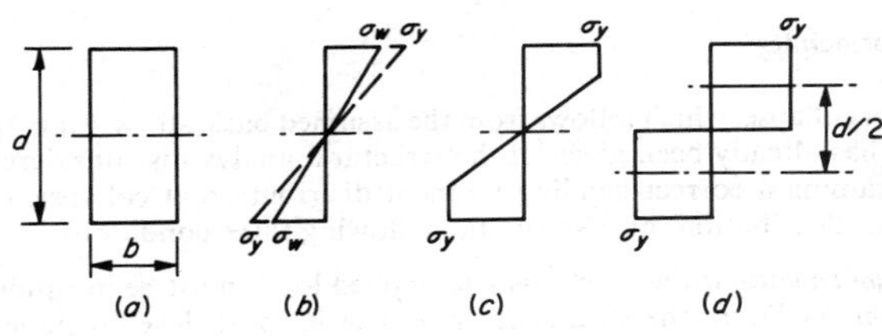

Figure 3.30

infinite strain, represented by the horizontal line AB continued indefinitely to the right. For comparison the dotted line shows the true relationship.

The term '*plastic analysis*' is generally related to steel structures for which the relationship indicated in Figure 3.29 is a good approximation. The equivalent approach when dealing with concrete structures is generally termed 'ultimate load analysis' and requires considerable modification to the method described here.

The stress-strain relationship of Figure 3.29 will now be applied to a simple, rectangular section, beam subjected to an applied bending moment M, Figure 3.30.

Under purely elastic conditions, line OA of Figure 3.29, the stress distribution over the cross-section of the beam will be as shown in Figure 3.30(b) and the limiting condition for elastic behaviour will be reached when the maximum stress reaches the value σ_y. As the applied bending moment is further increased, material within the depth of the section will be subjected to the yield stress σ_y and a condition represented by Figure 3.30(c) will exist in which part of the cross-section of the beam is plastic and part elastic. On further increase of the applied bending moment, ultimately condition (d) will be reached in which the entire cross-section is plastic. It will not be possible to increase the applied bending moment further and any attempt to do so will result in increased curvature, the beam behaving as if hinged at the plastic section. Hence the use of the term *plastic hinge* for a beam section which has become fully plastic.

The moment of resistance of the fully plastic section is, from Figure 3.30(d),

$$\begin{aligned} M_p &= b\frac{d}{2}\sigma_y\frac{d}{2} = \frac{bd^2\sigma_y}{4} \\ &= Z_p\sigma_y \end{aligned} \tag{93}$$

where Z_p = plastic section modulus.

In contrast, the moment of resistance at working stress σ_w is, from Figure 3.30(b),

$$M_w = b\frac{d}{2}\frac{\sigma_w}{2}\frac{2}{3}d = \frac{bd^2}{6}\sigma_w \tag{94}$$

$$= Z_e\sigma_w$$

where Z_e = elastic section modulus.

The ratio Z_p/Z_e is the shape factor of the cross-section. Thus the shape factor for a rectangular cross-section is 1.5.

The shape factor for an I-section, depth d and flange width b, is given approximately by

$$\left(\frac{1+x/2}{1+x/3}\right) \text{ where } x = \frac{t_w d}{2t_f b}$$

and t_w and t_f are the web and flange thicknesses respectively.

Values of plastic section moduli for rolled universal sections are given in steel section tables.

Theorems and principles

The definition of collapse, which follows from the assumed basic stress-strain relationship of Figure 3.29, has already been given. If the structural analysis is considered to be the problem of obtaining a correct bending moment distribution at collapse, then such a bending moment distribution must satisfy the following three conditions,

(a) *Equilibrium condition;* the reactions and applied loads must be in equilibrium
(b) *Mechanism condition;* the structure, or part of it, must develop sufficient plastic hinges to transform it into a mechanism
(c) *Yield condition;* at no point in the structure can the bending moment exceed the full plastic moment of resistance.

In elastic analysis of structures where several loads are acting, for example dead load, superimposed load and wind load, it is permissible to use the principle of superposition and obtain a solution based on the addition of separate analyses for the different loads. In plastic theory the principle of superposition is not applicable and it must be assumed that all the loads bear a constant ratio to one another. This type of loading is called 'proportional loading'. In cases where this assumption cannot be made, a separate plastic analysis must be carried out for each load system considered.

For cases of proportional loading, the uniqueness theorem states that the collapse load factor N_c is uniquely determined if a bending moment distribution can be found which satisfies the three collapse conditions stated.

The collapse load factor N_c may be approached indirectly by adopting a procedure which satisfies two of the conditions but not necessarily the third. There are two approaches of this type,

(i) We may obtain a bending moment distribution which satisfies the equilibrium and mechanism conditions, (a) and (b); in these circumstances it can be shown that the load factor obtained is either greater than or equal to the collapse load factor N_c. This is the 'minimum principle' and a load factor obtained by this approach constitutes an 'upper bound' on the true value.
(ii) We may obtain a bending moment distribution which satisfies the equilibrium and yield conditions, (a) and (c), and in these circumstances it can be shown that the load factor obtained is either less than or equal to the collapse load factor N_c. This is the 'maximum principle' and its application produces a 'lower bound' on the true value.

It should be observed that whilst method (i) is simpler to use in practice, it produces an apparent load factor which is either correct or too high and thus an incorrect solution is on the unsafe side. A most useful approach is to employ both principles in turn and obtain upper and lower bounds which are sufficiently close to form an acceptable practical solution.

Examples of plastic analysis

This section contains some examples of plastic analysis based on the minimum principle. The method employed is termed the 'reactant bending moment diagram method'.

EXAMPLE

The structure is a propped cantilever beam of uniform cross section, carrying a central load W, as shown in Figure 3.31(a). The bending moment distribution under elastic conditions is shown in Figure 3.31(b) and it should be noted that the maximum bending moment occurs at the fixed end A.

As the load W is increased plasticity will develop first at end A. As the load is further increased, end A will eventually become fully plastic with a stress distribution of the type shown in Figure 3.30(d) and the bending moment at A, M_A, will equal M_p the fully plastic

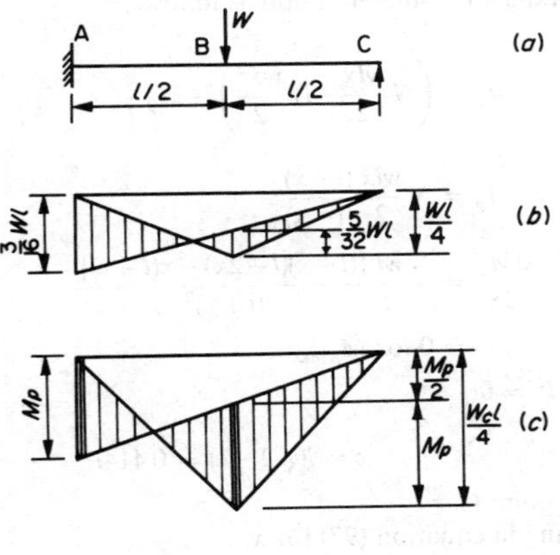

Figure 3.31

moment of the beam. Further increase of load will have no effect on the value of M_A but will increase M_B until it also reaches the value M_p. The resulting bending moment distribution will now be as shown in Figure 3.31(c).

The geometry of the diagram produces a relationship between the load at collapse, W_c, and the plastic moment of resistance of the beam M_p, as follows,

$$\frac{W_c l}{4} = M_p + M_p/2$$

or

$$W_c = 6\frac{M_p}{l} \tag{95}$$

If the working load is W_w then the load factor is given by

$$N = \frac{W_c}{W_w} \tag{96}$$

EXAMPLE 5

This is again a propped cantilever but here the load is uniformly distributed. Figure 3.32(a). At collapse the bending moment diagram will be as shown in Figure 3.32(b) with plastic hinges at A and C. It should be noted that C is not at the centre of the beam.

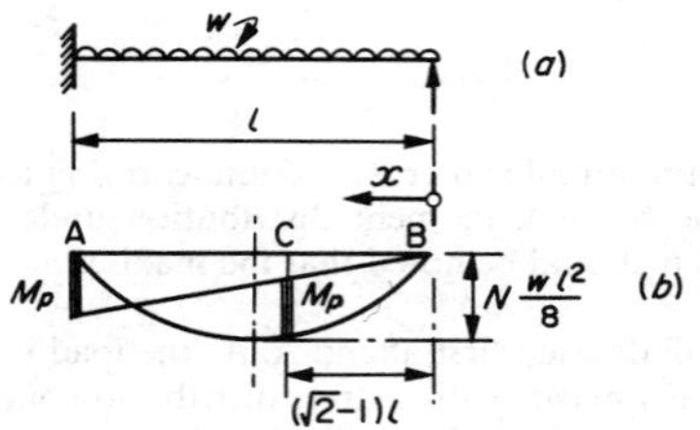

Figure 3.32

The location of the plastic hinge at C and the relationship between the load and the value of M_p may be obtained by differentiation as follows,
At C,

$$M_p = \left(N\frac{wlx}{2} - N\frac{wx^2}{2}\right) - M_p\frac{x}{l}$$

i.e.

$$M_p = N\frac{wlx}{2}\frac{(l-x)}{(l+x)} \tag{97}$$

$$\frac{\mathrm{d}M_p}{\mathrm{d}x} = N\frac{wl}{2}\frac{\{(l+x)(l-2x)-x(l-x)\}}{(l+x)^2}$$

$$= 0 \text{ for } M_{p\,\max}$$

Hence, $x^2+2xl-l^2 = 0$
i.e.

$$x = l(\sqrt{2}-1) = 0.414l$$

which locates the point C.

Also, substituting in equation (97) for x,

$$M_p = \frac{Nwl^2}{2}\frac{(\sqrt{2}-1)}{\sqrt{2}}(2-\sqrt{2})$$

$$= \left(\frac{Nwl^2}{8}\right)4(3-2\sqrt{2})$$

$$= 0.686\left(\frac{Nwl^2}{8}\right)$$

EXAMPLE

A two-span continuous beam is shown in Figure 3.33. The loads shown are maximum working loads and it is required to determine a suitable UB section such that $N = 1.75$ with a yield stress $\sigma_y = 250\ \mathrm{N/mm^2}$. Effects of lateral instability are ignored for the purposes of this example.

With factored loads, the free bending moments are,

$$1.75 \times 30 \times \frac{8^2}{8} = 420 \text{ kNm}$$

$$1.75 \times 30 \times \frac{5^2}{8} + 1.75 \times 40 \times \frac{5}{4} = 252 \text{ kNm}$$

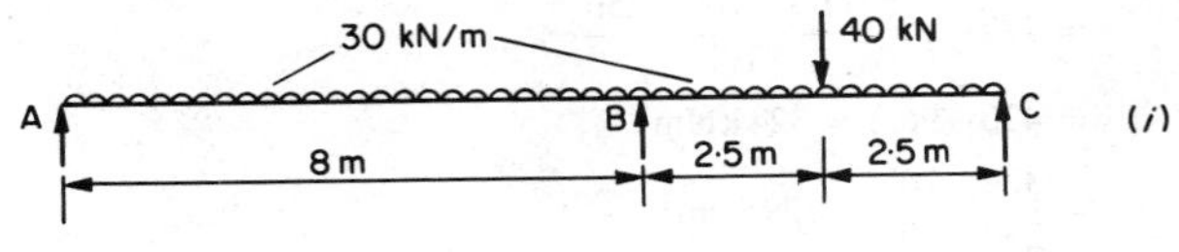

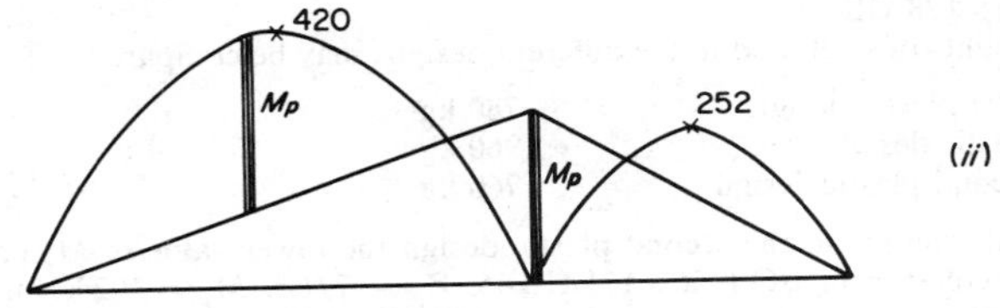

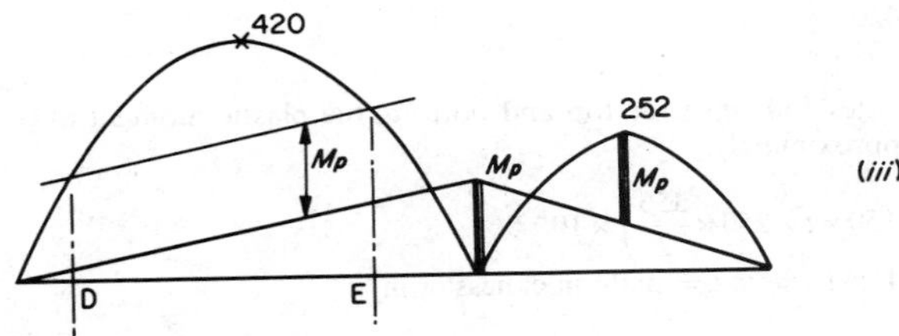

Figure 3.33

For collapse to occur in span AB, Figure 3.33(ii)

$$420 \times 0.686 = M_p = 288 \text{ kNm}$$

For collapse in BC, assuming the span hinge in BC to occur at the centre, Figure 3.33(iii),

$$252 = \frac{3}{2} M_p; \qquad M_p = 168 < 288$$

Hence the beam must be designed for $M_p = 288$ kNm

$$= Z_p \sigma_y$$

Hence,

$$Z_p = \frac{288 \times 10^6}{250 \times 10^3} \text{cm}^3 = 1150 \text{ cm}^3$$

From section tables, select 406 × 178 UB 60 (Z_p = 1195 cm^3).

This design may be compared with elastic theory from which we obtain M_{max} = 198 kNm, Z_e = 1200 cm^3 (using σ_w = 165 N/mm^2). A suitable section would be 406 × 152 UB 74 (Z_e = 1294 cm^3) or, 406 × 178 UB 74 (Z_e = 1322 cm^3).

The plastic design may be improved by choosing different sections for spans AB and BC,

For BC, $M_{PBC} = 168$ giving $Z_p = \frac{168}{250} \times \frac{10^6}{10^3} = 672 \text{ cm}^3$

Select, 356 × 171 UB 45 ($Z_p = 771.5 \text{ cm}^3$)

For AB, $M_{PAB} \simeq 420 - \frac{1}{2}M_{PBC}$

$$= 420 - \tfrac{1}{2} \times \frac{771.5 \times 10^3 \times 250}{10^6}$$

$$= 420 - 96.2 = 324 \text{ kNm}$$

$$\therefore Z_p = \frac{324}{250} \times \frac{10^6}{10^3} = 1295 \text{ cm}^3$$

Select 406 × 178 UB 67.

The weights of steel used in the different designs may be compared,

First plastic design	780 kg
Elastic design	960 kg
Second plastic design	760 kg

As an alternative to the second plastic design the lower value of M_p could be used, based on collapse in BC (356 × 171 UB 45, $Z_p = 771.5$, $M_p = 192$ kNm), and flange plates welded on to the beam in the region DE, Figure 3.33(iii).

The additional M_p required at the plated section

$$= 420 - \tfrac{3}{2} \times 192$$
$$= 132 \text{ kNm}$$

Using plates 150 mm wide top and bottom, the plastic moment of resistance of the plates is approximately,

$$2\left(150 \times t \times 250 \times \frac{356}{2}\right) \times 10^{-6}$$
$$= 13.4\, t \text{ where } t = \text{plate thickness (mm)}$$

Hence

$$t = \frac{132}{13.4} \simeq 10 \text{ mm.}$$

EXAMPLE

Here we consider the plastic analysis of a portal frame type structure as in Figure 3.34(a) and (b). At (a) the frame has pinned supports and at (b) fixed supports. A simple form of loading is used for illustration of the principles.

The frame is made statically determinate by the removal of H_A in both cases, and by the removal of M_A and M_E in case (b). The 'free' bending moment diagram is then as in figure (c) and the reactant bending moment diagrams are as at (d) for H_A and at (e) for M_A and M_E combined. We now seek combinations of the diagrams which will satisfy the conditions of equilibrium, mechanism and yield (see page 3–50). We consider first the case of the two-hinged frame.

Diagram (f)

This is consistent with a pure sideway mode of collapse. From the geometry of the diagram,

$$M_p = \frac{Hh}{2} \tag{98}$$

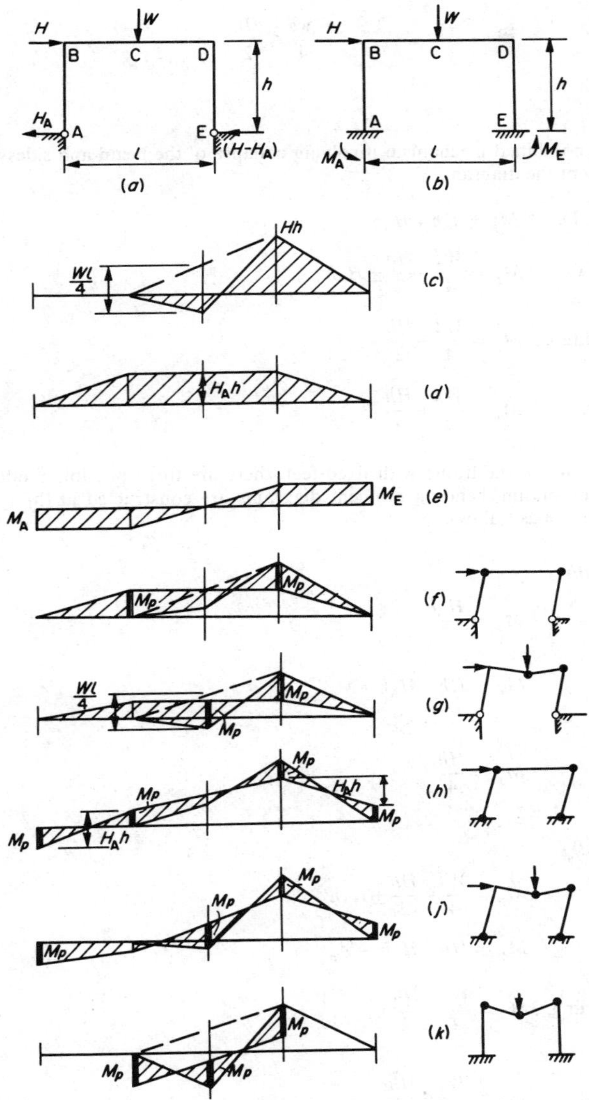
H
W
B
C
D
h
H_A
A
E
$(H-H_A)$
l
(a)
H
W
B
C
D
h
A
E
M_A
l
M_E
(b)
Hh
$\frac{Wl}{4}$
(c)
H_Ah
(d)
M_E
M_A
(e)
M_p
M_p
(f)
$\frac{Wl}{4}$
M_p
M_p
(g)
M_p
M_p
H_Eh
M_p
M_p
H_Ah
(h)
M_p
M_p
M_p
M_p
(j)
M_p
M_p
M_p
(k)

Figure 3.34

The yield condition will be satisfied providing

$$\frac{wl}{4} \leqslant \frac{Hh}{2} \tag{99}$$

Diagram (g)
This is a combined mechanism involving collapse of the beam and sidesway. From the geometry of the diagram,

at D, $M_p = Hh \mp H_A h$

at C, $M_p = \frac{Wl}{4} - \frac{Hh}{2} \pm H_A h$

adding, $2M_p = \frac{Wl}{4} + \frac{Hh}{2}$

or, $$M_p = \frac{Wl}{8} + \frac{Hh}{4} \tag{100}$$

In the case of the frame with fixed feet, there are three possible modes of collapse. The corresponding bending moment diagrams are constructed at (h), (j) and (k) and the results are as follows,

Diagram (h)

$$M_p = \frac{H_A h}{2}$$

$$M_p = Hh - H_A h - M_p$$

Hence

$$M_p = \frac{Hh}{4} \tag{101}$$

Diagram (j)

$$M_p = \frac{Wl}{4} - \frac{Hh}{2} \pm H_A h$$

$$M_p = Hh \mp H_A h - M_p$$

adding, $3M_p = \frac{Wl}{4} + \frac{Hh}{2}$

or, $$M_p = \frac{Wl}{12} + \frac{Hh}{6} \tag{102}$$

Diagram (k)
This mode is the same as the collapse of a fixed end beam; the columns are not involved

in the collapse apart from providing the resisting moment M_p at B and D. From the geometry of the diagram,

$$M_p = \frac{Wl}{8} \tag{103}$$

EXAMPLE

Here we consider a pitched roof frame, a structure which is eminently suitable for design by plastic methods. The frame is shown in Figure 3.35(a). The given loads are already

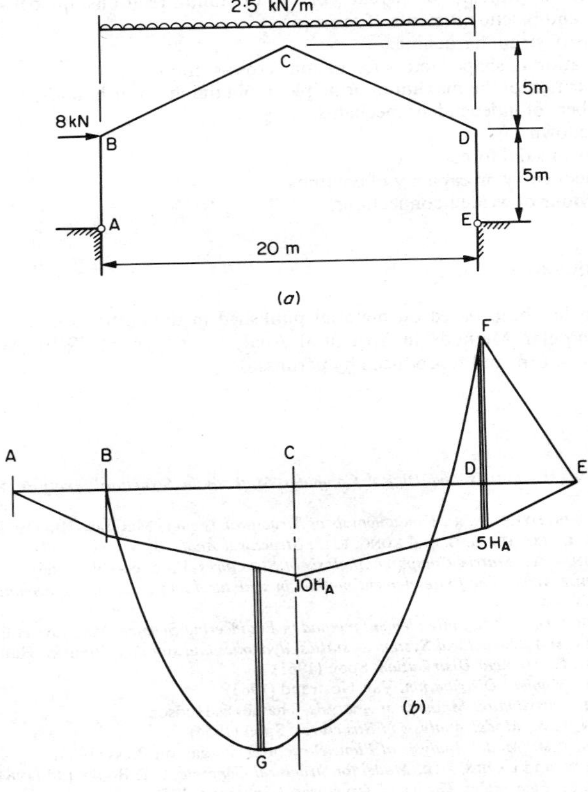

Figure 3.35

factored and we are to find the required section modulus on the basis of a yield stress $\sigma_y = 280$ N/mm^2, neglecting instability tendencies and the reduction in plastic moment of resistance due to axial forces.

The bending moment diagram at collapse is shown in Figure 3.35(b). The free bending moment diagram, EFGB, is drawn to scale after evaluating values of moment at intervals along the rafter members. The reactant line (H_A diagram) is then drawn by trial and error so that the maximum moment in the region BC is equal to the moment at D. This moment is the required M_p for the frame and is found to be

$$M_p = 52 \text{ kNm} = \sigma_y Z_p$$

from which,

$$Z_p = \frac{52 \times 10^3 \times 10^3}{280 \times 10^3} = 186 \text{ cm}^3$$

References 17 and 18 should be consulted for a more detailed study of plastic analysis. Among the topics deserving of further study are,

1. Use of the principle of virtual work in obtaining relationships between applied loads and plastic moments of resistance.
2. Effects of strain hardening.
3. Evaluation of shape factors for various cross-sections.
4. Application of the maximum principle in obtaining lower bounds.
5. Numbers of independent mechanisms.
6. Shakedown.
7. Effects of axial forces.
8. Moment carrying capacity of columns.
9. Behaviour of welded connections.

ACKNOWLEDGEMENT

This section has been based on material published in the author's book 'Matrix and Digital Computer Methods in Structural Analysis' (copyright 1969), McGraw-Hill Publishing Co. Ltd. and reproduced by permission.

REFERENCES

1. JENKINS, W. M., *Matrix and Digital Computer Methods in Structural Analysis,* McGraw-Hill (1969)
2. COULL, A. and DYKES, A. R., *Fundamentals of Structural Theory,* McGraw-Hill (1972)
3. COATES, R. C., COUTIE, M. G. and KONG, F. K., *Structural Analysis, Nelson* (1972)
4. RUBINSTEIN, M. F., *Matrix Computer Analysis of Structures,* Prentice-Hill (1966)
5. ZIENKIEWICZ, O. C., *The finite element method in structural and continuum mechanics,* McGraw-Hill (1967)
6. ZIENKIEWICZ, O. C., *The finite element method in Engineering Science,* McGraw-Hill (1971)
7. RUBINSTEIN, M. F., *Structural Systems—statics, dynamics and stability,* Prentice-Hall (1970)
8. LIGHTFOOT, E., *Moment Distribution,* Spon (1961)
9. GERE, J. M., *Moment Distribution,* Van Nostrand (1963)
10. WRIGHT, J., *Distribution Methods of Analysis,* Thames & Hudson
11. CHARLTON, T. M., *Model Analysis of Structures,* Spon (1954)
12. CHARLTON, T. M., *Model Analysis of Plane Structures,* Pergamon Press (1966)
13. PREECE, B. W. and DAVIES, J. D., *Model for Structural Concrete,* C R Books Ltd (1964)
14. GRASSIE, J. C., *Elementary Theory of Structures,* Longmans (1950)
15. ROGERS, L., *An introduction to the dynamics of framed structures,* Wiley (1959)
16. WARBURTON, G. B., *The dynamical behaviour of structures,* Pergamon (1964)
17. BAKER, J. F. and HEYMAN, J., *Plastic Design of Frames,* Vols 1 and 2, Cambridge University Press (1969)
18. HORNE, M. R., *Plastic Theory of Structures,* Nelson (1971)

4 MATERIALS

MATERIALS 4

4 MATERIALS

P. PULLAR-STRECKER, M.A., M.I.C.E.
Research Manager, CIRIA

A good working knowledge of the materials used in civil engineering is very important to the engineer and in this Reference Book the characteristics and properties of many materials are described appropriately in other sections as indicated below.

Material	*Section*
Soils	8 and 16
Rocks	7 and 9
Reinforcement	11
Steel	12 and 37
Aluminium	13
Bricks and Masonry	14
Timber	15
Asphalt	21 and 22

The extracts from BS 1881: Part 4: 1970, Methods of testing concrete for strength and CP 110, The structural use of concrete, Part 1: 1972 Design, Materials and workmanship are reproduced by permission of the British Standards Institution, 2 Park Street, London W1A 2BS.

This section is concerned with materials which are not covered elsewhere in the Reference Book and therefore considers in detail only:

Concrete	(pages **4**–3 to **4**–30)
Paint	(pages **4**–42 to **4**–47)
Plastics and rubbers	(pages **4**–30 to **4**–42)

In the field of materials especially, the solution of problems often requires a full understanding of technologies outside the engineer's normal experience. Fortunately specialist help is usually readily available (and often without charge), though the enquirer does not always know where to look for it. To help in the search, there follows a selection from the many sources listed in *CIRIA guide to sources of information*.

Aluminium Federation Ltd, Broadway House, Calthorpe Road, Five Ways, Birmingham B15 1TN

Asbestos Information Committee, 10 Wardour Street, London W1V 3HG

Asphalt and Coated Macadam Association Ltd, 25 Lower Belgrave Street, London SW1W 0LS

Association of Bronze and Brass Founders, 69 Harborne Road, Edgbaston, Birmingham B15 3AN

Brick Development Association (1954) Ltd, 19 Grafton Street, London W1X 3LE

British Cast Iron Research Association, Alvechurch, Birmingham B48

British Ceramic Research Association, Beechfields, Queens Road, Penkhull, Stoke-on-Trent ST4 7LQ

British Constructional Steelwork Association, Hancock House, 87 Vincent Square, London SW1P 2PJ

British Glass Industry Research Association, Northumberland Road, Sheffield S10 2UA

British Non-Ferrous Metals Research Association, Euston Street, London NW1 2EU
British Rubber Manufacturers' Research Association Ltd, 9 Whitehall, London SW1A 2DG
British Standards Institution, 2 Park Street, London W1A 2DF
British Steel Corporation, Special Steels Users' Advisory Centre, Swindon Laboratories, Moorgate, Rotherham S60 ZAR
British Stone Federation, Alderman House, 37 Soho Square, London W1V 5DG
British Wood Preserving Association, 71/62 Oxford Street, London W1N 9WD
Building Board Manufacturers' Association of Great Britain Ltd, Plough Place, Fetter Lane, London EC4
Building Centres: Birmingham, Bristol, Cambridge, Coventry, Liverpool, London, Manchester, Nottingham, Southampton, Stoke on Trent, Dublin, Belfast, Glasgow
Building Research Station, Garston, Watford WD2 7JR
Cement and Concrete Association, Wexham Springs, Slough, Bucks SL3 6PL
Clay Pipe Development Association, Drayton House, 30 Gordon Street, London WC1H 0AN
Clay Products Technical Bureau of Great Britain Ltd, Drayton House, 30 Gordon Street, London WC1H 0AN
Concrete Pipe Association, Brenchley, Tonbridge, Kent TN12 7BX
CIRIA—Construction Industry Research and Information Association, 6 Storey's Gate, London SW1P 3AU
CONSTRADO—Constructional Steel Research and Development Organisation, 12 Addiscombe Road, Croydon CR9 3JN
Copper Development Association, Orchard House, Mutton Lane, Potters Bar, Herts
Corporate Engineering Laboratory of British Steel Corporation, 140 Battersea Park Road, London
Finnish Plywood Development Association, Broadmead House, 21 Panton Street, London SW1Y WDR
Flat Glass Association, 6 Mount Row, London W1Y 6DY
Forest Products Research Laboratory, Princes Risborough, Aylesbury, Bucks
Glass Advisory Council, 6 Mount Row, London W1Y 6DY
Institution of Mining and Metallurgy, 44 Portland Place, London W1N 4BR
Lead Development Association, 34 Berkeley Square, London W1X 6AJ
National Physical Laboratory, Teddington, Middlesex TW11 0LW
Pitch Fibre Pipe Association, 35 New Bridge Street, London EC4V 6BH
Plywood Manufacturers of British Columbia, Templar House, 81 High Holborn, London WC1V 6LS
Paint Research Association, Waldegrave Road, Teddington, Middlesex TW11 8LD
Rubber and Plastics Research Association, Shawbury, Shrewsbury, Shropshire SY4 4NR
Timber Research and Development Association, Stocking Lane, Hughenden Valley, High Wycombe, Buckinghamshire HP14 4ND
Zinc Development Association, 34 Berkeley Square, London W1X 6AJ

CONCRETE

Cement

Hydraulic cement, i.e. cement which hardens because of chemical reactions between the cement and water is the main, and often the only binder used in concrete for civil engineering purposes. Portland cement or one of its variants is usually used, but high-alumina cement has advantages for some applications.

ORDINARY PORTLAND CEMENT (OPC)

This is the cheapest and most commonly used form of cement. It is made by heating together raw materials containing alumina and calcium. Clay and chalk or limestone are common sources. During the heating process the materials fuse to form clinker which is subsequently ground to a fine powder, gypsum usually being added at this stage. This powder is the cement in its final form. The fineness of grinding and the raw materials influence the reactivity of the cement, fine cement hardening more quickly than coarse cement of the same composition. Composition, fineness and other properties are governed by BS 12, but the quality of British cement, although varying according to its source, usually exceeds the BS requirements by a considerable margin.

RAPID-HARDENING PORTLAND CEMENT (RHPC)

This is similar to OPC but it is more finely ground. It gains strength more quickly than OPC, though the final strength is only slightly increased. Heat is generated more quickly during the hydration of the cement. This may have advantages in cold weather, but the higher concrete temperature which results may lead to cracking due to subsequent thermal contraction. Composition, etc., are governed by BS 12.

EXTRA RAPID-HARDENING PORTLAND CEMENT

This cement is made by grinding together not less than 35% of OPC clinker with granulated as an accelerator and leads to even greater early strengths and rates of heat evolution than RHPC. The calcium chloride additive increases the danger that reinforcement may corrode if it is not embedded deeply enough in the concrete.

PORTLAND BLAST-FURNACE CEMENT

This cement is made by grinding together not less than 35% of OPC clinker with granulated blast-furnace slag. This cement is less reactive than OPC and gains strength a little more slowly. It has advantages in generating heat less quickly than OPC and in being more resistant than OPC to attack from sulphates. Composition and properties are governed by BS 146. Portland blast-furnace cement is not widely available in the UK except in Scotland (Low-heat portland blast-furnace cement contains more slag but is only manufactured to order in the UK. BS 4246 governs its composition and properties).

LOW-HEAT PORTLAND CEMENT

This cement is less reactive than OPC because it differs in composition, but it is nevertheless more finely ground than OPC. Heat is generated more slowly on hydration and lower concrete temperatures are reached. Early and eventual strengths are less than with OPC and the initial setting time is greater. This cement is only made to order in the UK, its properties and composition being governed by BS 1370.

SULPHATE-RESISTING PORTLAND CEMENT

This cement is similar to OPC, but it is less prone to attack by sulphates. Heat is generated more slowly than with OPC, but a little more quickly than with low-heat portland cement. The composition and properties are governed by BS 4027, but sulphate-resisting cement would also comply with BS 12 for setting time and early and eventual strength. This cement, being generally available, is frequently used as a medium-low heat cement.

SUPERSULPHATED CEMENT

This cement is made from granulated blast-furnace slag, gypsum and not more than 5% of OPC clinker. It is more resistant to sulphate attack than sulphate-resisting cement, and it is not attacked by weak acids. This cement is much finer though less reactive than OPC, but eventual strengths are at least as high. The composition and properties are governed by BS 4248.

COLOURED PORTLAND CEMENT

OPC is made in a variety of colours by the addition of suitable pigments to white or grey cement depending on the shade required. The colours vary with different manufacturers.

ULTRA-HIGH-EARLY-STRENGTH CEMENT

This cement is similar to OPC in composition, but it is very much more finely ground even than RHPC and it contains more gypsum to control the setting time. The cement is very reactive and leads to very early strength development without the use of additives and without reduced setting times.

WATER-REPELLENT CEMENT

This is made from OPC and stearates. It is used to reduce water permeability especially in screeds and rendering. Water-repellent white cement helps white concrete to remain clean.

MASONRY CEMENT

This cement is made by mixing OPC with plasticisers and a fine powder (often whiting). It is used to give plasticity to bricklaying and rendering mortars, especially where the local sand is harsh.

HIGH-ALUMINA CEMENT

HAC differs from ordinary portland cement in its chemical composition and strength development. Although concrete made with this cement initially sets more slowly than OPC concrete it then gains strength very quickly. Strengths of 50 N/mm² can be achieved in HAC concrete of normal mix proportions in 24 hours or less. How strength develops over a longer period depends on the initial water/cement ratio of the concrete and on the condition of exposure. HAC concrete with a w/c ratio of 0.3 or less stored at 18 °C or less gains strength for several years and then remains roughly constant. HAC concrete made with higher w/c ratios than this or any HAC concrete stored at higher temperatures (results are available at 38 °C*) is liable to lose strength substantially as a result of chemical conversion. In extreme cases nearly all of its strength may be lost. This behaviour makes HAC inappropriate for structural use in most cases, except temporary work where only short term strength may be needed.

At temperatures above 700 °C, high-alumina cement forms a ceramic bond with suitable aggregates and it can therefore be used for refactory concrete.

* *Techenné D. C., Longer term research into characteristics of High Alumina Cement Concrete, Magazine of Concrete Research, Vol 27. No. 91, London 1975.*

It is widely believed that high-alumina cement should not be used in contact with even hardened portland cement, and it is established that mixtures of unhardened portland and high-alumina cements lead to very rapid 'flash' setting. This phenomenon has some practical applications where almost instantaneous setting is wanted, but the quality of the resulting concrete will be in most respects inferior to either portland cement concrete or high-alumina cement concrete (see Figure 4.1). The composition and properties of high-alumina cement are governed by BS 915.

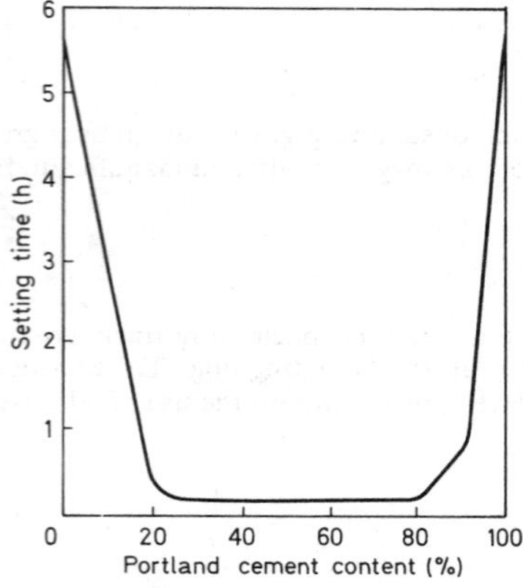

Figure 4.1 Setting times for OPC*–high-alumina mixtures (based on data from reference 1)*

Aggregates

Aggregates form more than three-quarters of the volume of concrete and the selection and proportioning of coarse and fine aggregates greatly influences the properties of both fresh and hardened concrete. The choice of grading, maximum aggregate size and aggregate/cement ratio are subjects for concrete mix design and are dealt with below. In this section the selection of aggregate type will be covered. Broadly, aggregates can be classified according to density as normal, lightweight and heavy aggregates.

NORMAL AGGREGATES

These usually consist of natural materials, hard crushed rock or crushed or natural gravel and their corresponding sands, but artificial materials like crushed brick and blast-furnace slag can also be used. The specific gravity of these materials usually lies between 2.6 and 2.7. Because satisfactory concrete for most purposes can be made with a very wide range of aggregates, local sources of supply usually determine which aggregate will be used. Where very high strength, resistance to skidding, good appearance or other special properties are required, appropriate aggregates will have to be selected, preferably on the basis of previous experience.

For example, the low-speed skidding resistance of concrete roads is affected by the hardness of the sand but only slightly by the polished-stone value of the coarse aggregate.

Thus a hard sand should be chosen for concrete which is to form the surface of a concrete pavement.

Some aggregates have undesirable influences on important concrete properties or are themselves unsound. They should be used with caution if at all. Examples are aggregates with high drying shrinkages, which may lead to poor durability in exposed concrete, aggregates which react with alkalis in the cement paste, aggregates which are readily oxidised, aggregates which can cause surface staining, and aggregates made from weathered partially decomposed rocks.

Other aggregates, although making reasonably satisfactory hardened concrete, for most purposes, may give the fresh concrete poor handling characteristics. Aggregates with flat, flakey, very angular or hollow particles tend to have this effect. In general, aggregates with well rounded particles in the case of gravels, or near-cubical particles in the case of crushed rock, produce concrete with better workability and fewer voids than aggregates with angular particles.

Natural sands have advantages over crushed rock sands because their particles tend to be more rounded and they contain less very fine material (of 150 μm or less), but crushed rock sands may be preferable for example where the grading of locally occurring natural sands is poor, where the colour of natural sands would be unsatisfactory in weathered concrete (many sands weather to a yellowish colour), or where resistance to slipping is important.

The typical properties of some common aggregates are given in Table 4.2.

LIGHTWEIGHT AGGREGATES

These consist of various artificial and natural materials with specific gravities of between 0.1 and 1.2. They are used to make lightweight concrete for structural and insulating applications. In general, concrete made with lightweight aggregates has better fire resistance than dense concrete, but greater shrinkage and moisture movement.

The properties of typical lightweight aggregates are tabulated in Table 4.3.[2]

Examples of lightweight aggregates are given below.

Clinker consists of fused lumps of fuel residues. To be suitable for use as a concreting aggregate it must be low in sulphates and residual fuel. Limits are given in BS 1156.

Foamed blast-furnace slag is made by treating molten blast-furnace slag with water so that the steam which is generated blows the slag. Standards for this material are given in BS 877.

Exfoliated vermiculite is made by heating vermiculite (a mica-like mineral found in Africa and America) to a temperature of about 700 °C when it expands to form a very light material.

Expanded clay, shale, slate and perlite are made by heating suitable grades of these materials to their fusion temperature (about 1000 °C) when they simultaneously fuse and are blown by gases generated within the material.

Pumice is a natural lightweight aggregate consisting of a frothy volcanic glass.

Sintered pulverised fuel ash is made by heating pellets of pulverised fuel ash until they fuse to form hard spherical lumps.

HEAVY AGGREGATES

These consist of natural or artificial materials with specific gravities of 4 or more. They are used to make high-density concrete for radiation shielding.

Examples of heavy aggregates are barytes, which is a naturally occurring rock consisting of 95% barium sulphate (SG about 4.1; density of concrete up to 3700 kg/m^3);

Table 4.1 COMPARATIVE TABLE OF CEMENTS

Cement type	Availability UK	Approx. UK price (% of OPC)	British Standard Specification Requirements: Setting time		Concrete cube strength			Specific surface	BS	Comparative properties of cements
			Initial	Final	3 day	7 day	28 day			
			min	hour	N/mm²			m²/kg		
Ordinary portland cement (OPC)	general	100	$\nless 45$	$\ngtr 10$	8.0	14.0	not specified	225	12	properties of this cement taken as normal
Rapid-hardening portland cement (RHPC)	general	105	$\nless 45$	$\ngtr 10$	12.0	17.0	ca 10% more than OPC	325	12	faster strength development than OPC
Extra-rapid-hardening portland cement	general	130	shorter than OPC				similar to RHPC		none	faster setting than normal; faster strength development than RHPC; may cause reinforcement to corrode
Ultra-high-early-strength cement	delivery from Dunstable	140 ex works	similar to OPC				similar to RHPC	much higher than OPC	none	normal setting; very much faster strength development than RHPC (about twice the strength at 16 h)
Sulphate-resisting portland cement	general	120	$\nless 45$	$\ngtr 10$	8.4		similar to OPC	250	4 027	lower heat evolution than OPC; better resistance to sulphates. (Sulphate resisting cement is often used in place of low-heat portland cement)
Low-heat portland cement	made to order only	depends on quantity	$\nless 60$	$\ngtr 10$	3.5	7.0	14.0	320	1 370	slightly lower heat evolution than sulphate resisting cement; lower final strength and lower strength development than OPC
Portland blast-furnace cement	Scotland only (Glasgow area)	97	$\nless 45$	$\ngtr 10$	8.4	14.0	22.4	225	146	lower heat evolution than OPC; better resistance to sulphates than OPC
Supersulphated cement			$\nless 45$	$\ngtr 10$	8.0	14.0	25.9	400	4 248	more resistant to sulphates than sulphate resisting cement; resists weak acids
White and coloured cement	general (most	170* to	$\nless 45$	$\ngtr 10$	8.0	14.0	similar to OPC	225	12	normal

					OPC			gives better plasticity; richer mixes needed for equivalent strength
High-alumina cement	general	220	$\nless$120 $\ngtr$6 h	$\ngtr$2 after initial		225	915	slow setting but very rapid hardening; full strength reached in 24 h; suitable for refractory concrete resists acids; weakens if kept warm and moist

* The price of coloured cement depends on the colour. Bright colours are more expensive.

Table 4.2 MECHANICAL PROPERTIES OF AGGREGATES

Aggregate	*Specific gravity*			*Crushing strength* (N/mm^2)			*Water absorbtion*	*Thermal expansion*	*Elastic modulus*
	low	*high*	*typical*	*high*	*low*	*typical*	(*% by weight*)	$\times 10^{-6}$ *per* deg C	kN/mm^2
								typical values	
Basalt	2.6	3.0	2.85			200		5 to 9	35
Brick				190	5		5 upwards		
Felsite				520	120	320			
Flint	2.4	2.6	2.55			200	2 to 3	11 to 13	
Gabbro			2.95			200		3 to 10	85
Gneiss				230	95	150		5 to 9	
Granite	2.6	3.0	2.69	260	110	190	0.6	5 to 9	45
Gritstone	2.6	2.9	2.67			220	3.8	4 to 14	up to 35
Hornfels	2.7	3.0	2.88			340		5 to 9	
Limestone	2.5	2.8	2.69	240	90	160	0.5	3 to 9	50
Porphyry	2.6	2.9	2.66			230		5 to 9	
Quartzite	2.6	2.7	2.62	420	120	330	0.8	11 to 13	
Sandstone				240	40	130		4 to 14	up to 35
Schist			2.76	300	90	240			

iron ores such as magnetite, geothite, limonite and ilmenite (SG about 4; density of concrete up to 3900 kg/m^3); iron or steel shot (SG 7.7; concrete density up to 5500 kg/m^3); lead shot (SG 11.4; concrete density up to 7000 kg/m^3) and scrap iron stampings and punchings. Provided that the materials are sound and free from oil, satisfactory concrete of good structural strength can be made, especially if prepared by a method such as prepacked to avoid segregation.

CONTAMINANTS, UNSOUND AGGREGATES AND REACTIVE AGGREGATES

Aggregates may contain impurities which upset the hydration of the cement or coatings which interfere with bond, or the aggregates themselves may be unstable. To some extent impurities and surface coating can be removed by suitable treatments, but aggregates which are unsound or reactive must be avoided.

Organic impurities may or may not delay or prevent the hydration of the cement and it is best to compare the strength of the concrete made with the contaminated aggregate with the strength of concrete made from similar but clean aggregate. Sugar, sugar-like

Table 4.3 PROPERTIES OF LIGHTWEIGHT-AGGREGATE CONCRETES

Aggregate	*Typical range of density* kg/m^3	*Dry density of concrete* kg/m^3	*Compressive strength* MN/m^2	*Drying shrinkage* %	*Thermal conductivity at 5% moisture content* W/m °C
Clinker	720–1 040	1 040–1 520	2.0– 7.0	0.04–0.08	0.35–0.67
Foamed blastfurnace slag	320– 880	960–2 000	2.0–24.0	0.03–0.07	0.24–0.93
Exfoliated vermiculite	65– 200	400– 800	0.7– 3.5	0.25–0.35	0.16–0.26
Expanded perlite	80– 240	400–1 120	0.5– 7.0	0.20–0.30	0.16–0.39
Pumice	500– 880	640–1 440	2.0–14.0	0.04–0.08	0.21–0.60
Expanded clay	320–1 040	720–1 760	2.0–62.0	0.04–0.07	0.24–0.91
Expanded slate	320– 960	560–1 760	1.4–27.5	0.03–0.09	0.19–0.91
Sintered pulverised-fuel ash	640– 960	960–1 760	2.8–55.0	0.04–0.07	0.32–0.91
Gravel or crushed stone	1 300–1 760	2 240–2 480	14.0–70.0	0.03–0.04	1.40–1.80

Reproduced from reference 2.

substances and humic acid are among common contaminants which are known to retard or prevent cement hydration. Products of wood degradation such as 'cellibiose' have a similar effect.

Clay and fine material can contaminate aggregates either as a coating on the coarse aggregate or as a constituent of the fine aggregate. As coatings these materials interfere with bond and therefore reduce concrete strength. As constituents of the mix they are less troublesome unless the quantity is great enough to require the addition of extra water to make the concrete workable. Clay, silt and fine material should not form more than 1% by weight of coarse aggregate, 3% by weight of gravel sand or 15% by weight of crushed rock sand (BS 882).

Salt is usually present in marine aggregates and in small quantities it is harmless. The salt content should however be limited to 1.0% by weight of the cement used, and where metal is to be embedded in the concrete the total of salt and any calcium chloride which might be used must not exceed 1.5% by weight of the cement used. (Calcium chloride is calculated as the anhydrous salt in this case.) Where high-alumina cement is used, or where calcium chloride admixtures would not be permitted, the salt content must not exceed 0.1% by weight of the cement (see CP 110—listed in Bibliography).

Nondurable particles are sometimes found in aggregates which are otherwise satisfactory. Examples of such particles are lumps of clay, shale, wood or coal. Being soft, they are easily eroded and will lead to pitting or spalling of the concrete surface. If more than about 5% of such particles are present in the aggregate they will also cause strength to be reduced. Generally such particles should not form more than 1% of the aggregate by weight.

Reactive particles found in some aggregates may be soluble in or react with water or the hydrating cement paste. Mica and sulphates (for example gypsum) react with cement paste, iron sulphides (for example pyrites and marcasite) react with air and water to form products which then react with the cement paste and cause staining or pop-outs.

Unsound material may form the whole of the aggregate or unsound particles may merely contaminate it. Unsoundness is the property of some aggregates to expand or contract excessively as a result of freezing and thawing, wetting and drying, or temperature changes. Such movements can be large enough to cause the aggregate itself to break down or they may disrupt concrete made with it. Examples of unsound aggregates are rocks with very high water absorbtion, porous cherts, limestones and other sedimentary rocks if they contain laminae of clay, and some shales. Foreknowledge of how such aggregates behave in concrete is the only reliable guide but freezing and thawing tests may give some indication of an aggregate's unsoundness.

Reactive aggregates are those which react chemically with the cement paste, the most common reaction being between reactive silica and alkalis. Reactive silicas occur in opaline and chalcedonic cherts, siliceous limestone, rhyolites, andesite and phyllites. The silica forms a gel with the alkali and this gel expands continuously as it absorbs water, exerting enough force to disrupt the surrounding cement paste in some cases. Fortunately reactive aggregates have not been encountered in the UK.

Admixtures

Relatively small quantities of other materials can be added to concrete to modify its properties. These materials are called admixtures if they are added to the concrete or additives if they are supplied already mixed in with the cement. Calcium chloride in extra-rapid-hardening cement is an example of an additive; there are several useful admixtures which are listed below.

AIR-ENTRAINING AGENTS

These are widely used admixtures, especially for paving concrete. Their most important property is due to the fact that concrete containing a small amount of air in the form of well distributed small bubbles has greater durability than similar concrete made without air-entraining agents, particularly in its resistance to de-icing salts. This increased durability is gained at the expense of strength and it is therefore important to control the amount of entrained air between close limits.

The amount of air that will be entrained with a given addition of air-entraining agent is influenced by the grading of the sand, the workability of the concrete, the type of mixer and the duration of mixing. Trial mixes are essential to establish how much agent is to be added and frequent regular measurements must be made throughout the work to ensure that the correct air content is being maintained (see page **4**–28).

As well as being more resistant to damage from de-icing salts, air-entrained concrete is somewhat more cohesive than concrete made without air-entraining agent and tends to have slightly higher workability, a factor which can be used partly to offset the strength reduction.

The total air content usually required for concrete with 20 mm maximum aggregate is about 5% by volume and this is achieved with a dosage of about 0.05% of agent by weight of the cement. A strength reduction of 10% could be expected for this air content.

WORKABILITY AIDS AND WATER-REDUCING ADMIXTURES

These have the effect of making concrete more workable for a given water content, hence of allowing less water to be used in concrete of a given workability. Workability aids tend to entrain a little air and may retard setting. For these reasons the amount of admixture must be carefully controlled. A typical dosage would be 0.2% by weight of the cement.

ACCELERATORS AND 'ANTIFREEZES'

These are used to hasten the hardening of concrete, particularly in cold weather. The term 'antifreeze' is misleading because these admixtures merely lessen the period when frost damage is likely; they do not prevent concrete from freezing. The active ingredient in accelerators is usually calcium chloride which has the undesirable side effect of promoting corrosion in metals if it is used in excess. The amount of accelerator must be carefully controlled. Up to 1.5% by weight of cement of anhydrous calcium chloride is considered harmless except in cases where concrete is to be pretensioned or if there is minimum cover over the reinforcement, but the use of this additive may increase shrinkage slightly. If the concrete is made with aggregate containing salt the addition of calcium chloride must be reduced by the amount of the salt present (see page **4**–10).

PLASTICISERS

Plasticisers are used to give plasticity or cohesion, mainly to mortars, but sometimes also to concrete. They function by entraining large amounts of air and they therefore reduce strength. This is often acceptable in mortars but not necessarily in concrete.

RETARDERS

These have the effect of delaying the onset of hardening and usually also of reducing the rate of the reaction when it does start. Strengths are unaffected by retardation for several hours but may be reduced if the addition of retarder is excessive. Accidental overdosage may cause retardation of a few days or it may prevent hardening altogether. The fear that this may happen is probably one of the reasons why retarders are seldom used in the UK. Nevertheless retarders can be beneficial where large volumes of concrete have to be poured in one operation or where cracking due to thermal contraction is feared. Trial mixes are essential to determine the dosage at which the retarder is to be used.

POZZOLANAS

Materials containing finely divided silica (pulverised fuel ash is a common example) may be added to concrete to take advantage of the pozzolanic cements formed when silica reacts with free lime. This reaction is a slow one, taking place over many months, but where early strength is not critical it is possible to substitute a pozzolana for part of the cement (30% is not uncommon) and this can have advantages in mass concrete if the generation of heat needs to be reduced.

MIXED ADMIXTURES

Mixed admixtures containing a variety of materials, some to offset the undesirable side effects of others, are available. An example is an admixture which contains air-entraining agent to improve durability, an accelerator to offset the retarding effect, and a water-reducing admixture to compensate for the strength lost due to the entrainment of air.

OTHER ADMIXTURES

These include waterproofers, viscosity modifiers, resin bonding agents, fungicides, etc. They may be useful for specific applications, but the claims made for them should be supported by impartial test results. This applies particularly to the permanence of the effects claimed.

Pigments may be incorporated in concrete mixes as an alternative to the use of coloured cement. If bright or pastel shades are wanted, white cement and light coloured sand must be used for the basic concrete, but low-key colours and dark shades can be obtained with ordinary concrete. The pigments must be stable in cement, fast to light and resistant to being washed out by weathering. The pigment must be uniformly dispersed throughout the mix preferably by thorough mixing with the dry ingredients before the water is added. Contrasting aggregates are best avoided and it should be noted that efflorescence tends to show up more on coloured concrete.

Although a number of organic pigments can be used in concrete, the most commonly used pigments are iron oxides for red, brown, yellow and black, and chromium oxide for green. Synthetic iron oxides have better staining power than natural ones and are available in a greater colour range. Although more expensive than natural oxides, they may be cheaper in use. Carbon black gives a more intense black than ferric oxide, but because it is often greasy it is difficult to disperse and has won the reputation of being easily washed out. Pigment additions vary typically from about 2% to 10% or more by cement weight. With the larger rates of addition some strength reduction should be expected.

Concrete mix design

GENERAL

The purpose of concrete mix design is to choose and proportion the ingredients used in a concrete mix to produce economical concrete which will have the desired properties both when fresh and when hardened. The variables which can be controlled are:

water/cement ratio
maximum aggregate size
aggregate grading
aggregate/cement ratio
use of admixtures

Interactions between the effects of the variables complicate mix design and successive adjustments following trial mixes are usually necessary. The first step in mix design is to determine the properties of the hardened concrete. When this has been done, it is then possible to determine the properties of the fresh concrete.

WATER/CEMENT RATIO

Many of the most important properties of fully compacted hardened concrete, and in particular strength, are for normal concrete virtually decided by the water/cement ratio of the mix. The importance of this parameter is due to the fact that any excess of water

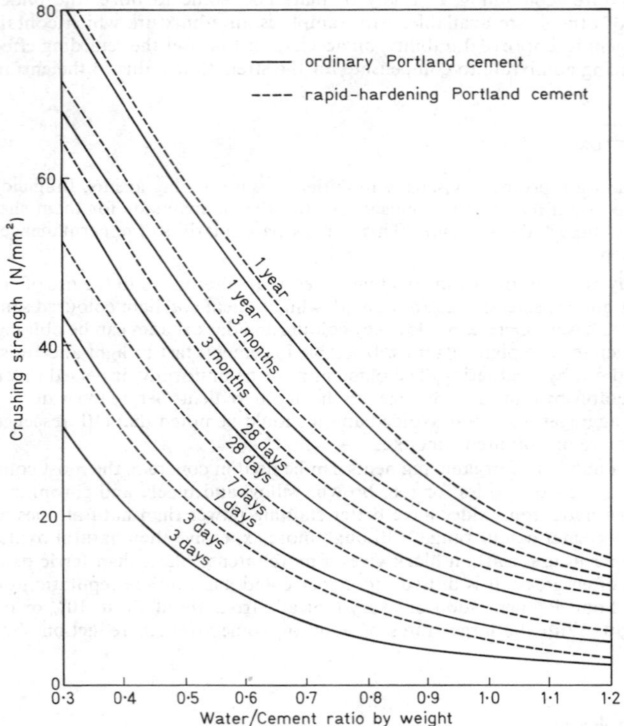

Figure 4.2 Influence of water/cement ratio on strength (based on data from reference 3)

over that needed to hydrate the cement (about 25% by weight) forms voids in the concrete, thus reducing its density. The reduced density leads to reduced compressive, tensile and bond strengths, lower durability, lower resistance to abrasion and greater permeability to water. Excess water cannot be eliminated altogether because it is needed to lubricate the mix and make it workable, but it should be kept to a minimum.

Figure 4.2 shows how strength is influenced by water/cement ratio and the first step in concrete mix design is to fix the water/cement ratio from a knowledge of the strength required.

WORKABILITY

When the concrete is fresh it must be workable or fluid enough to be compacted easily under the conditions in which it will be placed. This is vitally important since loss of density has a very large effect in reducing strength. Table 4.4 gives suggested figures for workability suitable for different circumstances. The cement paste is the lubricant which provides workability, but on grounds of economy as well as for technical reasons such as the limitation of shrinkage and thermal contraction, the amount of cement paste should be as small as possible. The next stages in the mix design process are intended to ensure that the mix will not be richer than is necessary.

MAXIMUM AGGREGATE SIZE

The larger the aggregate the greater the workability for a given richness of mix. Above 40 mm an increase in aggregate size may bring a reduction in strength, but up to this

Table 4.4 SUGGESTED WORKABILITIFS OF CONCRETE MIXES

		Slump (mm)		*Compacting factor*	
Workability	*Suitable use*	19 mm *max. size*	38 mm *max. size*	19 mm *max. size*	38 mm *max. size*
Very low	Vibrated concrete in large sections	0 to 25	0 to 3	0.78	0.75
Low	Mass concrete foundations without vibration. Simple reinforced sections with vibration	25 to 50	3 to 6	0.85	0.83
Medium	Normal reinforced work without vibration, and heavily reinforced sections with vibration	50 to 100	6 to 25	0.92	0.90
High	Sections with congested reinforcements. Not normally suitable for vibration	100 to 180	25 to 100	0.95	0.95

Based on data from reference 3.

figure strength is increased for a given workability and cement content. The largest size of aggregate which can be used is governed by the dimensions of the section being cast and the spacing of the reinforcement. It is unusual to use aggregate with a maximum size of more than 20 mm for reinforced concrete, or with a maximum size greater than 25% of the section thickness for any work.

OVERALL GRADING

Sand and coarse aggregates frequently occur together, e.g. in gravels, but they seldom occur naturally in the best proportions for making concrete. Although all-in aggregates can be used, it is usually more satisfactory and more economical to separate sand from coarse aggregate and then recombine them in the required proportions. Further adjustments could be made by separating the aggregates into smaller size groups and recombining them as required, but it is doubtful if this would repay its cost. Table 4.5 shows the British Standard requirements for aggregate gradings. The criterion for determining what proportions of sand and coarse aggregate shall be used is that the concrete shall be cohesive enough to resist segregation.

Fine sands provide more cohesion than coarse ones so less sand will be needed if it is fine. Very fine sand (zone 4) is not recommended for structural concrete unless tests have shown that concrete made with it is satisfactory. Very coarse sand (zone 1) may cause difficulties in surface finishing if floors or pavements are being constructed. To resist segregation, high workability mixes need more sand than low workability mixes, and the proportion of sand must also be increased as the maximum aggregate size is reduced. Crushed rock coarse aggregates need more sand than rounded gravel aggegates.

Table 4.6 taken from CP 110 gives an indication of sand/coarse-aggregate ratios considered suitable for a range of 'prescribed' mixes.

CEMENT CONTENT

When the water/cement ratio has been fixed the related variable of cement content can be chosen to ensure that there will be enough cement paste to produce a workable mix. The cement content that will be needed to do this depends on the grading and shape of the aggregates, more cement being needed for mixes with a finer overall grading or a more angular coarse aggregate. Trial mixes will usually be needed before the choice of cement content is finally made, but the guidance given in Table 4.7[3] can be used as a starting point. Cement contents considered suitable for a range of prescribed mixes are also included in Table 4.6.

Properties of hardened concrete

COMPRESSIVE STRENGTH

This depends on water/cement ratio, density, the type of cement used and the age of the concrete. To the extent that compressive strength reflects water/cement ratio and density, it is a good indicator of general concrete quality and it is an easy property to measure with reasonable consistency. The cube crushing strength is consequently an important test of both the structural and general quality of the concrete (see page 4–27).

The development of compressive strength with age is greatly influenced by the temperature of the concrete, especially early in its life. Since the hydration of the cement itself generates heat, the temperature of the concrete is influenced not only by its initial temperature and the temperature of the surroundings, but also by the volume and shape of the section. Figure 4.3 indicates how the development of strength is related to the temperature of the concrete itself, and Table 11.4 in Section 11 relates the strength of various grades of concrete to the age at test.

The strength to which a concrete mix is designed depends on structural considerations and the fact that the concrete must be durable. Since there will be some variation in the quality of the concrete made on site and in the results of cube crushing tests, the strength to which the mix is designed must exceed the strength actually needed by a safety factor which will depend on the degree of control which can be exercised over the concrete production process. These matters are discussed in Section 11; the question of the strength is also influenced by the overriding consideration that the concrete must be durable, and this factor often fixes the least cement content which can be used. Table 4.8 gives the requirements of CP 110.

Table 4.5 AGGREGATE GRADING

Description of aggregate	Size or grading zone (mm)	Grading limits defined by British Standards (percentage by weight passing sieve) Sieve aperture size (mm): 150 μm	300 μm	600 μm	1.20	2.40	4.76	9.52	12.70	19.05	38.10	76.20	BS
Fine aggregate (sand)	Zone 1	0 to 10	5 to 20	15 to 34	30 to 70	60 to 95	90 to 100	100					882
	Zone 2	0 to 10	8 to 30	35 to 59	55 to 90	75 to 100	90 to 100	100					882
	Zone 3	0 to 10	12 to 40	60 to 79	75 to 100	85 to 100	90 to 100	100					882
	Zone 4	0 to 15	15 to 50	80 to 100	90 to 100	95 to 100	95 to 100	100					882
Coarse aggregate	5 to 13						0 to 10	40 to 85	90 to 100	100	100		882
	5 to 19						0 to 10	25 to 55		95 to 100	95 to 100	100	882
	5 to 38						0 to 5	10 to 35		30 to 70	100		882
All-in aggregate	19	0 to 6		10 to 35			30 to 50			95 to 100	95 to 100	100	882
	38	0 to 6		8 to 30			25 to 45			45 to 75		100	882
Foamed blast-furnace slag	fine: 3 mm down	5 to 20	10 to 30	20 to 60	45 to 90	70 to 100	90 to 100	100					877
	Coarse: 3 to 13 mm					0 to 10	5 to 40	55 to 100	85 to 100	100			877
Coarse aggregate; blast-furnace slag	13						0 to 5	40 to 85	90 to 100	100			1047
	19						0 to 5	22 to 25		90 to 100	100		1047
	38						0 to 5	0 to 5		30 to 70	95 to 100	100	1047
Granolithic coarse aggregate						0 to 5	0 to 20	85 to 100	100				1201

Granolithic fine aggregate is to comply with zones 1 or 2 (above).

Table 4.6 PRESCRIBED MIXES FOR ORDINARY STRUCTURAL CONCRETE
(Weights of cement and total dry aggregates in kg to produce approximately one cubic metre of fully compacted concrete together with the percentages by weight of fine aggregate in total dry aggregates) (*From CP 110*)

Concrete grade	*Nominal maximum size of aggregate* (mm)	40		20		14		10	
	Workability	*Medium*	*High*	*Medium*	*High*	*Medium*	*High*	*Medium*	*High*
	Limits to slump that may be expected (mm)	*50–100*	*100–150*	*25–75*	*75–125*	*10–50*	*50–100*	*10–25*	*25–50*
7	Cement (kg)	180	200	210	230	—	—	—	—
	Total aggregate (kg)	1 950	1 850	1 900	1 800	—	—	—	—
	Fine aggregate (%)	30–45	30–45	35–50	35–50	—	—	—	—
10	Cement (kg)	210	230	240	260	—	—	—	—
	Total aggregate (kg)	1 900	1 850	1 850	1 800	—	—	—	—
	Fine aggregate (%)	30–45	30–45	35–50	35–50	—	—	—	—
15	Cement (kg)	250	270	280	310	—	—	—	—
	Total aggregate (kg)	1 850	1 800	1 800	1 750	—	—	—	—
	Fine aggregate (%)	30–45	30–45	35–50	35–50	—	—	—	—
20	Cement (kg)	300	320	320	350	340	380	360	410
	Total aggregate (kg)	1 850	1 750	1 800	1 750	1 750	1 700	1 750	1 650
	Sand								
	Zone 1 (%)	35	40	40	45	45	50	50	55
	Zone 2 (%)	30	35	35	40	40	45	45	50
	Zone 3 (%)	30	30	30	35	35	40	40	45
25	Cement (kg)	340	360	360	390	380	420	400	450
	Total aggregate (kg)	1 800	1 750	1 750	1 700	1 700	1 650	1 700	1 600
	Sand								
	Zone 1 (%)	35	40	40	45	45	50	50	55
	Zone 2 (%)	30	35	35	40	40	45	45	50
	Zone 3 (%)	30	30	30	35	35	40	40	45
30	Cement (kg)	370	390	400	430	430	470	460	510
	Total aggregate (kg)	1 750	1 700	1 700	1 650	1 700	1 600	1 650	1 550
	Sand								
	Zone 1 (%)	35	40	40	45	45	50	50	55
	Zone 2 (%)	30	35	35	40	40	45	45	50
	Zone 3 (%)	30	30	30	35	35	40	40	45

TENSILE AND FLEXURAL STRENGTH

The tensile strength of concrete is much smaller than the compressive strength and is in any case usually effectively eliminated by cracking, whether this cracking is visible or not. Consequently the tensile strength of concrete is not usually taken into account for design purposes, though it can be important in as much as it influences the spacing and control of cracks in structures[5] and contributes to the flexural strength of concrete paving.

Tensile strength is measured either directly by testing bobbins or cylinders to failure in tension, or indirectly by the cylinder splitting test or flexural tests on concrete beams. Results from the latter test are referred to as 'modulus of rupture'.

Table 11.2 of Section 11 gives figures for tensile and flexural strengths for several grades of concrete, and testing is discussed on page 4–27.

While tensile and flexural strength both increase with increasing compressive strength, there is no fixed relationship between them. Cylinder splitting tests have shown that the

Table 4.7 TOTAL AGGREGATE/CEMENT RATIO BY WEIGHT REQUIRED TO GIVE DIFFERENT DEGREES OF WORKABILITY WITH AGGREGATES OF DIFFERENT MAXIMUM SIZES (mm)

Workability	*Water/ cement ratio*	*Gravel aggregate*			*Crushed rock aggregate*		
		9.52	19	38	9.52	19	38
Very low	0.4	3.5	4.5	5	3	4	4.5
	0.5	5.5	6.5	7.5	4.5	5.5	6.5
	0.6	7.5	—	—	6	7	—
	0.7	—	—	—	7	—	—
Low	0.4	3	4	4.5	—	3.5	4
Low	0.5	4.5	5.5	6.5	4	5	5.5
	0.6	6	7	7.5	5	6	7
	0.7	7	8	—	6	7	8
Medium	0.4	—	3.5	4	—	3	3.5
	0.5	4	5	5.5	3.5	4	5
	0.6	5	6	7	4.5	5	6
	0.7	6	7	8	5.5	6	7
High	0.4	—	3	3.5	—	3	3
	0.5	3.5	4	5	3	4	4.5
	0.6	4.5	5	6.5	4	4.5	5.5
	0.7	5.5	6	7.5	5	5.5	6.5

Based on data from reference 3.

relationship is influenced by the nature of the aggregate, but some surface characteristic of the aggregate, rather than whether the aggregate is crushed or rounded, is the cause of this influence.[6]

ELASTIC MODULUS

The elastic modulus of concrete is important in designing members to resist deflection, though concrete is not perfectly elastic and does exhibit significant creep behaviour. For design purposes, shrinkage, creep and elastic modulus are often allowed for together by designing on the basis of an 'effective modulus' which takes account of the three factors. This is discussed in Section 11.

The elastic modulus of concrete varies between about 7 and 50 kN/mm^2 depending on the strength of the concrete and the proportion and rigidity of the aggregate. The lowest figure would be applicable to low-strength concrete made with lightweight aggregate while normal structural concrete would have an elastic modulus of 25 to 30 kN/mm^2; some values are given in Table 11.2 of Section 11.

The elastic modulus of concrete can conveniently be measured by vibrating a suitable specimen; the value for the modulus found in this way is termed the 'dynamic' modulus and is considerably higher than the static modulus because no creep occurs under the test condition.

CREEP

Creep is the term given to the tendency for concrete to continue to strain over a period of time when the stress is constant. For design purposes, creep is allowed for by using an 'effective' modulus which takes account of both short- and long-term stress–strain relationships. This is covered in Section 11.

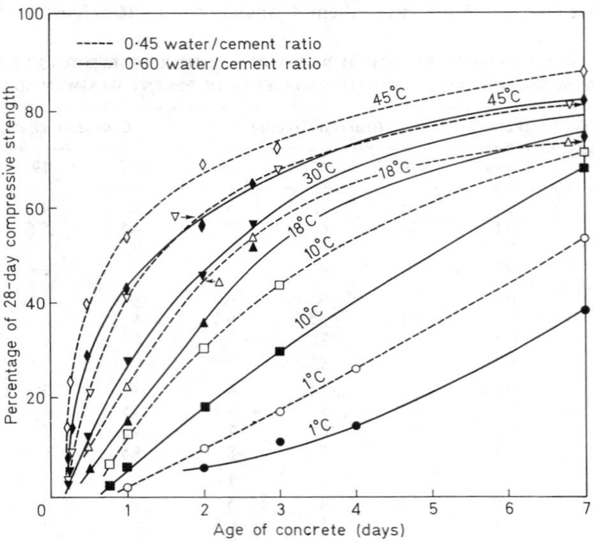

Figure 4.3 Influence of age and temperature on strength (from reference 4)

Factors which tend to increase creep are low strength, low ambient relative humidity, low-modulus aggregates, and high stressing. Methods for calculating creep deflections usually assume that creep is increased by early loading, but recent investigations suggest that this effect may not be significant.[7]

SHRINKAGE AND MOISTURE MOVEMENT

Concrete shrinks when it dries. Part of this shrinkage, usually about 30% but sometimes as much as 60% is reversible and is known also as moisture movement. Shrinkage leads to cracking or distortion in members which are restrained or reinforced, though in this respect it is now considered to be less important than thermal movement (see *Thermal movement* below).

Shrinkage is increased by increasing the richness or the water content of the mix. For these reasons high workability mixes shrink more than low workability mixes of the same strength. Aggregates with high elastic moduli are more effective in restraining shrinkage than low-modulus aggregates, this influence being virtually confined to the coarse aggregate.

Figure 4.4 shows how drying shrinkage is related to the water content of concrete mixes and Figure 4.5 shows the influence of ambient relative humidity on the rate and amount of shrinkage. From the latter it can be seen that shrinkage is a more serious problem in dry countries or inside dry buildings than outside in the UK where the relative humidity usually exceeds 75%: indeed, where concrete remains permanently moist, it increases somewhat in volume.

Lightweight aggregates usually have less effect in restraining shrinkage than normal aggregates, and where aggregate is absent, for example in aerated concrete, products have to be autoclaved to keep the shrinkage and moisture movement within reasonable limits. Autoclaving produces this effect through changes in the structure of the cement paste: curing at atmospheric pressure even with steam does not.

THERMAL MOVEMENT

The linear coefficient of thermal expansion of concrete varies from about 5 to 15 microstrain per deg C, depending on the richness of the mix and the coefficient of expansion of the aggregate. Rich mixes have higher coefficients than lean ones, and siliceous aggregates have higher coefficients than limestone and granite.

Since concrete tends to become heated when the cement hydrates, thermal contraction on cooling and hardening can set up enough stress on restrained members to cause cracking. Even if a reduced coefficient is used for immature concrete (to take creep into account), cooling strains in walls of normal thickness can reach 200 microstrain or more within a few days of the concrete being cast.[5]

DURABILITY

This important property of concrete has already been referred to on page 4–16 where the minimum cement content needed for durability was mentioned in relation to conditions of exposure. Special care must be taken when concrete is exposed to sulphates, acids or salts used for de-icing.

In general, concrete which has low water permeability will be much more durable than concrete which has high permeability and the effect may be so marked that it outweighs the influence of specially resistant cements. Well compacted dense concrete containing sufficient cement and no unnecessary water should always be used where durability is important. Additional measures may also be needed where exposure conditions are severe.

Sulphates in solution can attack cement paste if the concentration is sufficiently high. Sources of sulphate are calcium and magnesium sulphate present in some groundwaters, sulphates contained in seawater and sulphates formed from sulphur dioxide present in the air in urban and industrial areas. Sulphates from the last two sources would be too dilute to attack good quality concrete unless circumstances had allowed them to become concentrated by evaporation. This situation can arise in coastal splash and tidal zones, and on the undersides of units from which contaminated water drips, for example copings on walls. Table 4.9 gives the recommendations of CP 110.

Acids of all kinds attack concrete made with portland cement. Sources of acid are flue gases (if condensation occurs), carbon dioxide dissolved in water (moorland water is frequently acid) and acid formed from sewer gas. Concrete can be protected to some extent by applying acid-resisting coatings, and limestone concrete (curiously) has been found to be more resistant than other concrete possibly because the large area which can

Table 4.8 MINIMUM CEMENT CONTENT REQUIRED IN PORTLAND CEMENT CONCRETE TO ENSURE DURABILITY UNDER SPECIFIED CONDITIONS OF EXPOSURE (*From CP 110*)

Exposure	*Nominal maximum size of aggregate* (mm)				*Nominal maximum size of aggregate* (mm)				*Nominal maximum size of aggregate* (mm)			
	Reinforced concrete				*Prestressed concrete*				*Plain concrete*			
	40	*20*	*14*	*10*	*40*	*20*	*14*	*10*	*40*	*20*	*14*	*10*
	kg/m^3	kg/m^3	kg/m^3	kg/m^3	kg/m^3	kg/m^3	kg/m^3	kg/m^3	kg/m^3	kg/m^3	kg/m^3	kg/m^3
Mild: e.g. completely protected against weather, or aggressive conditions, except for a brief period of exposure to normal weather conditions during construction	220	250	270	290	300	300	300	300	200	220	250	270
Moderate: e.g. sheltered from severe rain and against freezing whilst saturated with water. Buried concrete and concrete continuously under water	260	290	320	340	300	300	320	340	220	250	280	300
Severe: e.g. exposed to sea water, moorland water, driving rain, alternate wetting and drying and to freezing whilst wet. Subject to heavy condensation or corrosive fumes	320	360	390	410	320	360	390	410	270	310	330	360

Exposure	*Reinforced concrete*					*Prestressed concrete*					*Plain concrete*				
	Nominal maximum size of aggregate (mm)				*Maximum free water/cement ratio*	*Nominal maximum size of aggregate* (mm)				*Maximum free water/cement ratio*	*Nominal maximum size of aggregate* (mm)				*Maximum free water/cement ratio*
	40	*20*	*14*	*10*		*40*	*20*	*14*	*10*		*40*	*20*	*14*	*10*	
	kg/m³	kg/m³	kg/m³	kg/m³		kg/m³	kg/m³	kg/m³	kg/m³		kg/m³	kg/m³	kg/m³	kg/m³	
Mild	200	230	250	260	0.65	300	300	300	300	0.65	180	200	220	240	0.70
Moderate	240	260	290	310	0.55	300	300	300	310	0.55	200	230	250	270	0.60
Severe	290	330	350	370	0.45	300	330	350	370	0.45	240	280	300	320	0.50
Salt used for de-icing	240	260	290	310	0.55	300	300	300	310	0.55	220	250	280	300	0.55

be attacked neutralises the acid before much local damage is done to the cement paste alone.

Freezing and thawing cycles attack poor concrete, but good quality concrete is resistant unless de-icing salts are used. Air entrainment (see page 4–11) has been found to provide protection though there are different explanations of the mechanism by which it works. Where no de-icing salt is to be used, but the concrete is liable to become frozen

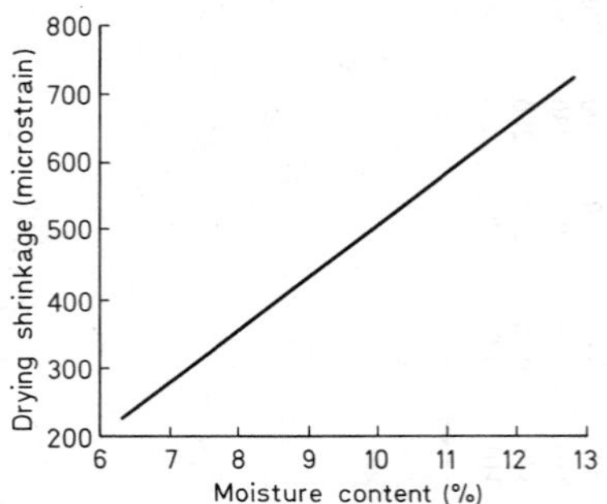

Figure 4.4 Influence of moisture content of fresh concrete on shrinkage (based on data from reference 8)

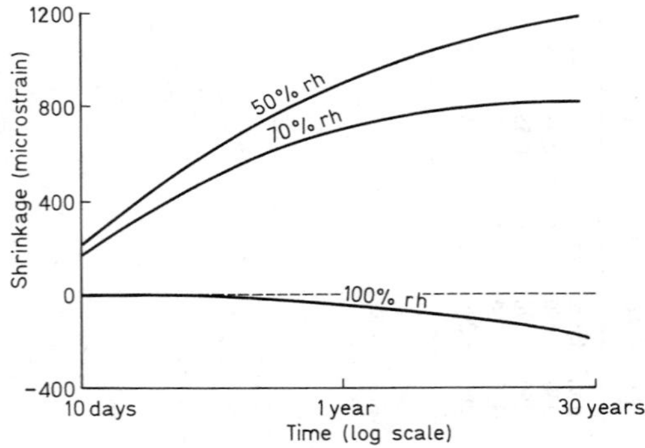

Figure 4.5 Influence of ambient relative humidity on shrinkage (based on data from reference 9)

when wet, air entrainment is not usually specified since concrete with a water/cement ratio of 0.55 : 1 or less should be satisfactory.

The corrosion of reinforcement in concrete is covered in Section 11.

Curing

If newly hardened concrete is to achieve its potential strength and durability, the hydration of the cement must be allowed to continue for several weeks at least. For this purpose

Table 4.9 REQUIREMENTS FOR CONCRETE EXPOSED TO SULPHATE ATTACK (*From CP 110*)

Class	*Concentration of sulphates expressed as SO_3*			Type of cement	*Requirements for dense, fully compacted concrete made with aggregates complying with the requirements of BS 882 or BS 1047*			
	In soil		*In ground water*		*Minimum cement content*			*Maximum free water/cement ratio*
	Total SO_3	*SO_3 in 2:1 water extract*			*Nominal maximum size of aggregate (mm)* 40	20	10	
	%	g/l	Parts per 100 000		kg/m^3	kg/m^3	kg/m^3	
1	Less than 0.2	—	Less than 30	Ordinary Portland or Portland blast-furnace	240	280	330	0.55
2	0.2 to 0.5	—	30 to 120	Ordinary Portland or Portland blast-furnace	290	330	380	0.50
				Sulphate-resisting Portland	240	280	330	0.55
				Supersulphated	270	310	360	0.50
3	0.5 to 1.0	1.9 to 3.1	120 to 250	Sulphate-resisting Portland or supersulphated	290	330	380	0.50
				High alumina	290	330	380	0.45
4	1.0 to 2.0	3.1 to 5.6	250 to 500	Sulphate-resisting Portland or supersulphated	330	370	420	0.45
				High alumina	300	340	410	0.40
5	Over 2	Over 5.6	Over 500	Sulphate-resisting Portland or supersulphated plus adequate protective coatings	330	370	420	0.45
				High alumina	330	370	420	0.35

NOTES

1. This table applies only to concrete made with aggregates complying with the requirements of BS 882 or BS 1047 placed in near-neutral groundwaters of pH 6 to pH 9, containing naturally occurring sulphates but not contaminants such as ammonium salts. Concrete prepared from ordinary or sulphate-resisting Portland cement would not be recommended in acidic conditions (pH 6 or less). High alumina cement can be used down to pH 4.0 and supersulphated cement has given an acceptable life provided that the concrete is dense and prepared with a free water/cement ratio of 0.40 or less, in mineral acids down to pH 3.5.

2. The cement contents given in Class 2 are the minima recommended by the manufacturers. For SO_3 contents near the upper limit of Class 2 cement contents above these minima are advised.

3. Where the total SO_3 in column 2 exceeds 0.5% then a 2:1 water extract may result in a lower site classification if much of the sulphate is present as low solubility calcium sulphate. Reference should be made to BRS Digest 90:1970.

4. For severe conditions, e.g. thin sections, sections under hydrostatic pressure on one side only and sections partly immersed, consideration should be given to a further reduction of water/cement ratio and, if necessary, an increase in cement content to ensure the degree of workability needed for full compaction and thus minimum permeability.

5. The prescribed mixes given in Table 4.6 may be used where appropriate to meet the requirements of this table and for this purpose the maximum free water/cement ratios of grades 20, 25 and 30 shall be taken as 0.55, 0.50 and 0.45 respectively.

an excess of water must be present in the pores of the concrete and the act of ensuring that this is so is curing. The excess water normally present in the concrete is enough to provide curing except in the case of very dry mixes, but near the surface of a member it will escape by evaporation unless this is prevented. Formwork is usually left in place long enough to provide initial curing, and where appearance and durability are not considered important, this may be enough in the UK climate. Where further curing is considered justified, spray applied curing films or other means of preventing evaporation may be used. An alternative is to apply water to the surface for the curing period.[10]

Reinforcement and prestressing steel

These materials are covered in Section 11.

Testing

Most of the tests which are described below have to be carried out on a sample of concrete which will inevitably be very small compared with the work which it is intended to represent. Sampling is therefore of the greatest importance and every care must be taken to ensure that the sample is as representative as possible if the test results are to have any real meaning. BS 1881 part 1 : 1970 gives advice on methods of sampling.

As well as variations in the concrete there will also be variations in the test itself and it is necessary to carry out several tests on concrete which nominally represents the same part of the work. BS 1881 part 3 : 1970 requires that at least three specimens are made for testing at each age. It is frequently necessary to carry out testing on a much larger scale than this especially if the test results show that unacceptable variations are occurring.

WORKABILITY TESTS

These are designed to measure the ease with which concrete can be compacted. Because none of the tests exactly reproduces the conditions under which concrete is compacted on site, each test has some limitations in applicability, though within limits any of the tests is suitable for maintaining uniformity of workability once site use has established what workability will be needed.

SLUMP TEST (For full details of this test see BS 1881 part 2 : 1970)

Application Quick approximate tests for medium and high workability concrete; suitable for site use; simple apparatus.

Special apparatus Mould in shape of inverted cone frustum; flat baseplate.

Method Concrete is compacted into the mould in four approximately equal layers with a 16 mm diameter tamping rod giving 25 tamps per layer. Top surface is struck off and finished with trowel. The mould is then lifted off vertically and the concrete is allowed to slump.

Result Difference in height between moulded and slumped condition measured to nearest 5 mm.
Or if total collapse or shear occur, these facts are recorded.

COMPACTING FACTOR (For full details of this test see BS 1881 part 2)

Application Concrete of all workabilities; suitable for simple site laboratory.

Special apparatus Compacting factor apparatus consisting of two hoppers and a measuring cylinder fixed in vertical alignment; balance to weigh 25 kg to 10 g accuracy.

Method Concrete is filled loosely into top hopper and allowed to fall to next hopper. Concrete from this hopper is then allowed to fall into the measuring cylinder. The surplus is struck off.

Result Ratio of the weight of concrete in the cylinder filled as above to the weight of concrete fully compacted into the cylinder.

'V-B' CONSISTOMETER (For full details of this test see BS 1881 part 2)

Application Concrete of all workabilities; suitable for large site laboratory.

Special apparatus Consistometer consisting of conical mould, cylindrical container, transparent disc kept horizontal by a guide and a vibrating table of specified size, frequency and amplitude; stopwatch.

Method The mould is placed in the cylinder and filled with concrete as for the slump test. The mould is then removed and the disc is allowed to rest on top of the slumped concrete. Vibration is then applied and allowed to continue until the underside of the disc is just covered with grout.

Result Vibration time in seconds to nearest 0.5 s.

BALL PENETRATION TEST (For full details of this test see ASTM C360–55T)

Application Similar to slump test.

Special apparatus Kelly ball consisting of 30 lb, 6 in diameter hemisphere with support frame and graduated scale.

Method The frame is placed on the surface of the concrete (for example in a wheelbarrow) with the bottom of the hemisphere just touching the concrete. When the weight is released, the penetration into the concrete is measured.

Result Penetration in inches.

STRENGTH TESTS

Strength tests are designed to measure the potential strength of concrete: the actual strength in a structural member depends on compaction, curing and uniformity as well

Table 4.10 STRENGTH TESTS FOR CONCRETE SPECIMENS

	Crushing strength	*Flexural strength*	*Indirect tensile strength*
Specimen size (mm)[a]	150 × 150 × 150	150 × 150 × 750	150 dia × 300
Specimen size (mm)[b]	100 × 100 × 100	100 × 100 × 500	100 dia × 200
Rammer size (for hand compaction)[c]	25 × 25 × 380	25 × 25 × 380	25 dia × 480
Blows per layer (smaller specimens)	$\not<$35 ($\not<$25)	$\not<$175 ($\not<$100)	$\not<$30 ($\not<$20)
Rate of loading (smaller specimens)	15 MN/m^2 min (15 MN/m^2 min)	9 000 N/min (4 000 N/min)	1.5 MN/m^2 min (1.5 MN/m^2 min)
Test result	failure stress (f_c)	modulus of rupture (f_b)	tensile strength (f_t)
Calculation	$f_c = \dfrac{\text{load}}{\text{nominal area}}$	$f_b = \dfrac{\text{load} \times \text{outer span}^d}{\text{breadth} \times \text{depth}^2}$ (for failure inside middle third)	$f_t = \dfrac{2 \times \text{load}}{\pi \times \text{dia} \times \text{length}}$
(a = distance in mm from outer support to failure line on underside)	(parts of broken beams may be used for crushing strength tests. The area of the platens is then the nominal area')	$f_b = \dfrac{3 \times \text{load} \times a}{\text{breadth} \times \text{depth}^2}$ (for failure outside middle third)	

[a]Imperial size moulds may be used until 1980 if clearly marked.
[b]The smaller specimen size may be used where the maximum aggregate size is 25 mm or less.
[c]The weight of the rammer is 1.8 kg in each case.
[d]Outer span = 3 × inner span.

as the potential strength. It cannot therefore be measured except by tests on the member (for core tests and nondestructive tests see below).

The main requirements of BS 1881 : 1970 for strength tests on concrete specimens are summarised in Table 4.10. Curing the specimens is to be as follows:

> Until they are strong enough to be demoulded (usually after 24 h), the specimens are kept in their moulds at at least 90% relative humidity and a temperature of 20 ± 2 °C. After being demoulded, they are stored in water at 20 ± 1 °C until tested and are tested while still wet. (Where specimens are cured on site, the temperature tolerances are increased to ±5 °C for the initial period and ±2 °C thereafter.)

Tests on cores cut with diamond-tipped core cutters are described in BS 1881 : 1970. The diameters of the cores should conform with the diameters of cylindrical specimens (see table in BS 1881), but the length cannot usually be chosen. The ends of the cores must be flat and perpendicular to the axis (this can be achieved by capping or grinding). The

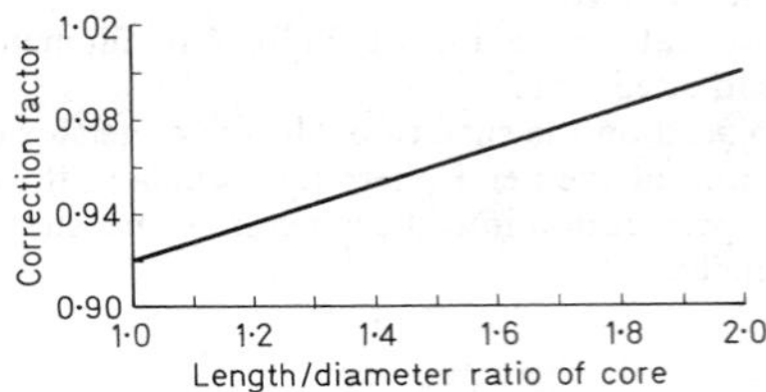

Figure 4.6 Correction factor for compression test on a core (from BS 1881 Part 4 : 1970)

cores must be soaked in water for at least 48 h before being tested and are then tested while still wet. The failure stress is calculated from the load × the correction factor given from Figure 4.6, divided by the average actual area of cross section.

Nondestructive strength tests Some indication of the strength of concrete is given by tests of surface hardness and by ultrasonic pulse velocity tests. The latter can also be used to test the uniformity of the concrete in a structure.

Surface hardness tests measure the rebound of impact hammers (for example the Schmidt hammer) or the depth of indentation. A large number of tests is needed for a satisfactory assessment because the closeness of the aggregate to the test point affects the results. BS 4408 part 4 : 1971 gives details of these tests.

Ultrasonic pulse velocity tests depend on the relationship between transmission time and the density and elastic properties of the concrete, both of which are related to concrete strength. The development of lightweight portable instruments has recently made this test convenient to use on site and research has indicated that it can give useful guidance on the strength of structural members.[11]

TESTS ON AGGREGATES

BS 812 : 1967 gives details of numerous tests on aggregates. A number of these are frequently used in concrete mix design or for quality control. These are briefly summarised in Table 4.11.

MEASUREMENT OF ENTRAINED AIR

It is important to control the air content of air-entrained concrete for the reasons given on page 4–11. Entrained air is measured by compacting a sample of fresh concrete in three layers in a container of known volume (nominally 0.006 m^3). Compaction must be sufficient to remove all entrapped air, but not so prolonged that entrained air is also removed. The container is then clamped to an airtight cover which incorporates a pressure gauge and a graduated sight tube. The space under the cover is filled with water and the vessel is pressurised with an air pump to compress the air contained in the

Table 4.11 TESTS ON AGGREGATES—SUMMARY OF MAIN TESTS IN BS 812 : 1967

Test	*Property measured*	*Principle–apparatus–method*
Sieve analysis	aggregate grading, including clay and fine silt passing 75 microns	Dried sample of aggregate sieved over a number of test sieves conforming with BS 410; weight retained on each is measured.
Sedimentation test	proportion of clay, silt or dust	Fine material in suspension sampled with sedimentation pipette
Field settling test	estimate of clay, silt or dust	Sample of aggregate shaken with salt solution: depth of material which has settled measured
Flakiness test	percentage of flat particles	Sample of aggregate tested in specified slotted gauges
Specific gravity and water absorbtion	specific gravity and percentage water absorbtion of coarse aggregates	Sample of saturated aggregate submerged – loss in weight indicates volume; sample dried to give dry weight and weight of absorbed water
Specific gravity and water absorption	specific gravity and percentage water absorption of coarse aggregates	Sample of saturated aggregate submerged – displaced water indicates volume; remainder as above
Specific gravity and water absorbtion	as above, but for fine aggregates	Volume of sample measured by water displacement in a pycnometer: remainder of test as above
Bulk density	bulk density and void volume of aggregate sample	Weight of aggregate required to fill container of known volume
Oven drying	percentage moisture content	Weighed sample oven dried and reweighed
Siphon can	percentage moisture content	Water volume determined by displacement in siphon can
Aggregate impact value	resistance of aggregate to shock	Percentage of material passing 2.40 mm determined after specified impact test on aggregate sample
Aggregate crushing value	resistance of aggregate to crushing	As above, but specified crushing test instead of impact
10% fines value	resistance of aggregate to crushing	Determination of load required to produce 10% of material passing 2.40 mm in specified crushing test
Aggregate crushing strength	compressive strength of rock	Crushing test on cylinder cut from rock sample
Aggregate abrasion value	resistance of aggregate to surface wear	Determination of percentage loss in weight after specified lapping of aggregate sample

concrete (the air pressure is usually about one atmosphere). The change in volume is indicated by a drop in the water level in the sight tube which is calibrated directly in per cent of entrained air by volume. BS 1881 part 2 : 1970 gives details of this test.

ANALYSIS OF FRESH CONCRETE

This is used to determine the proportions of the constituents and the grading of the aggregates. There are several separate steps in the process which is described in detail in BS 1881 part 2 : 1970.

The fresh analysis (or wet analysis) method is based on the separation of the coarse and fine material in a sample of fresh concrete. This is done by washing the concrete over 5 mm and 150 μm sieves. The weights of material remaining on the sieves are compared with the weights of samples of the aggregates retained on the same sieves. By making this comparison, the proportions of fine and coarse aggregates and of cement can be determined. The moisture content of the mix is determined by oven drying a sample of the concrete.

ANALYSIS OF HARDENED CONCRETE

This is sometimes needed when a failure has occurred, or when, for some other reason the constituents of the hardened concrete have to be determined. Full details of the methods used are given in BS 1881 part 6 : 1971.

Cement content is determined by chemical analysis of a sample ground to a fine powder. If the aggregates contain little or no limestone, reliable estimates can be made quite simply by analysis for calcium oxide using an EDTA titration method. It is assumed that the calcium oxide (which forms 64.5% of OPC) comes from the cement alone. If there is a substantial amount of limestone in the aggregate, analysis becomes more difficult and less certain since estimates have to be made for carbon (liberated as carbon dioxide) and soluble silica.

Aggregate type and grading are determined by breaking down a sample of concrete by heating it to 550 °C for an hour or more. Cement is dissolved from the lumps of fine material with dilute hydrochloric acid, and a sieve analysis is carried out on the insoluble material which remains. Again the method is difficult and rather approximate if the aggregates contain a substantial proportion of limestone.

The original water content is found by saturating a slice (sawn with a diamond saw) with carbon tetrachloride. This fills the pores left by the uncombined water and the volume of the pores is estimated from the weight gained. The combined water is found from the loss in weight on ignition of a sample of the concrete.

PLASTICS AND RUBBERS

Terminology

Plastics and rubbers are organic polymers which are or can become capable of flow. The difference between plastics and rubbers is arbitrary and sometimes unclear, but for practical purposes rubbers can be considered as those materials which recover after they have undergone large strains (say 100% or more).

Table 4.12 lists commonly used plastics and rubbers together with their principle physical properties, typical applications and relevant British Standards and Codes of Practice.

Thermoplastics are those plastics which melt when heated and return to their original state when cooled. They are described as flexible or rigid according to their feel when handled.

Thermosetting plastics are those plastics which do not melt when heated (unless they are heated to temperatures high enough to destroy their properties).

Plasticisers are materials (usually liquids) which are compounded with certain plastics (especially PVC) to make them more flexible.

Curing agents are materials which are added to certain plastics and rubbers to crosslink them and thus form a thermosetting plastic.

Ultraviolet absorbers are materials used to make certain plastics resistant to ultraviolet light (carbon black pigment serves the same purpose where its colour is acceptable).

Physical and chemical properties

FUSIBILITY

Nearly all plastics and rubbers melt or soften when heated, thermoplastics returning to their original state on cooling provided that they have not been degraded by overheating. Some thermoplastics (polystyrene for example) become very fluid when heated and can be used to make castings, others become soft and doughy but do not melt. These compounds (PVC for example) have to be shaped or formed under pressure.

COMBUSTIBILITY

Most plastics and rubbers can be made to burn and some burn very readily. Exceptions include plastics based on chlorides and fluorides, formaldehydes and compounds which have been specially formulated to make them incombustible. Some compounds which contain combustible plasticiser may be combustible even if the basic polymer would not burn. For this reason plasticised PVC is sometimes combustible.

RESISTANCE TO DAYLIGHT AND WEATHERING

Ultraviolet light, present in daylight outside but effectively filtered-out by ordinary window glass, attacks many plastics and rubbers though some (acrylics for example) are unaffected by it. Those materials which are attacked can be made much more resistant by the incorporation of suitable pigments or ultraviolet absorbers in the formulation. The degree of attack naturally depends on the amount of ultraviolet light present, and performance data must relate to the appropriate conditions of exposure. Rain may leach-out constituents of some formulations, and a few plastics and rubbers are not resistant to moisture.

RESISTANCE TO EXTREMES OF TEMPERATURE

The flexibility of plastics compounds increases as the temperature rises and oxidation may degrade plastics and rubbers which are kept at high temperatures for long periods. Thermoplastics particularly are affected by temperature changes, and with some (bitumen is a familiar example), the ambient temperature range is enough to change them from the brittle to the fluid state. With others, polypropylene and thermosetting plastics for example, ambient temperature changes have little effect. These materials are stable at 100 °C or more, and do not become brittle at normal low temperatures.

THERMAL EXPANSION

The coefficient of thermal expansion of polymers tends to be very high compared with conventional construction materials. Formulating compounds with high filler contents

Note: The data given in this table must be considered in relation to the descriptions and guidance given in the text.

Key:

Combustibility

- F flammable; burning continues after ignition
- B combustible but selfextinguishing
- N incombustible
- C burns with clear flame
- S burns with smoky flame

Resistance to attack

- + resistant
- 0 some attack; use with caution if at all
- – little or no resistance; unsuitable for use

Compound or polymer	*Combustibility*	*Specific gravity*	*Fusing temperature (°C)*	*Maximum working temperature (°C)*	*Ultimate tensile strength (N/mm²)*	*Minimum working temperature (°C)*	*Tensile elastic modulus (N/mm²)*	*Tensile strain at failure (%)*	*Coefficient of thermal expansion °C × 10^{-5}*
Acetal copolymers		1.41	160	80 to 120	35 to 80		1 000 to 2 000		8
Acrylic resins	② F C	1.18	100 to 120	80	40 to 70		3 000	5	6
Acrylonitrile-butadiene-styrene (ABS)	③ F S	1.10		80	40		400	1.5	
Butyl rubber	F S	0.92	—	125	5 to 17	– 50		500 to 800	
Chlorosulphonated polyethylene	B	1.10	—		14			300 to 500	
Epoxide resins	F S	1.25 to 1.30	—		55 to 70		5 000	1.6 to 1.8	
Melamine formaldehyde (laminates)	B–	1.45	—	120	95		8 000	0.7	3
Nylon	B S	1.10 to 1.40	220 to 265	80 to 120	50 to 100	– 40	2 000	75	8 to 10

s

1) resistant to aromatic solvents
2) burns with sweet 'gassy' smell
3) very sooty flame, penetrating 'styrene' smell
4) if suitably pigmented or UV stabilised
5) flame-retardant grades are less resistant to weathering
6) burns, drips and smells like candle wax
7) oxidising agents may attack
8) burns with sulphurous fumes
9) burns with sweetish acrid smell

sistance to acids and alkalis			*Resistance to solvents*						
Dil. inorganic acids	*Organic acids*	*Alkalis*	*Petrol*	*Paraffin, diesel and fuel oil*	*Aromatic and chlorinated solvents*	*Ethers, ketones and esters*	*Alcohols*	*Typical applications*	*Relevant British Standards or Codes of Practice*
–	–	+	+	+	① 0	+	+	Plumbing components (e.g. taps) door and window furniture	
+	–	+	+	+	–	–	+	Moulded and shaped lights (e.g. rooflights and domelights); lighting fittings; sanitary ware	
								Waste and drain pipes and fittings; pressure pipes and fittings	BS 3867, BS 3943
+	+	+	–	–	–			Roof coverings; tank linings; damp proof membranes; adhesives and mastics; sealants; bridge bearings	BS 3227, CP 2003 (pt. 1)
			+ to 0	+ to 0				Roof coverings; tank linings	CP 3003 (pt. 1)
+	+	+	+	+	+		+	Adhesives; bedding and jointing mortars and grouts; concrete patching mortars; surface coatings	CP 3003 (pt. 5)
					1			Decorative laminates; cladding; surface coatings	BS 3794
–	–	+	+	+	0	+	+	Door and window furniture; cold water fittings; surface coatings; fairleads; ropes and straps	

Tab

Compound or polymer	*Combustibility*	*Specific gravity*	*Fusing temperature (°C)*	*Maximum working temperature (°C)*	*Ultimate tensile strength (N/mm²)*	*Minimum working temperature (°C)*	*Tensile elastic modulus (N/mm²)*	*Tensile strain at failure (%)*	*Coefficient of thermal expansion °C × 10⁻⁵*
Phenol formaldehyde (figs for laminates)	B–	1.40 (1.40)	—	120 (120)	50 (80)		7 000	0.5	5 (3)
Polychloroprene	B S	1.23	—	120	5 to 20	– 20		600 to 900	
Polyester (figs for laminates)	③ F S	1.1 (1.6)	—	90 (90)	40 to 70 (≯300)		2 000 (10 000)	2 (0.5)	5 to 10 (2)
Polyethylene (low density)	F ⑥	0.91 to 0.94	110 to 125	80	5 to 15	– 60	100	100 to 400	14 to 24
Polyethylene (high density)	F ⑥	0.94 to 0.97		105	20 to 35		1 000	50 to 200	14 to 24
Natural rubber	F S	0.93	—	70	20	– 55	up to 10	500 to 800	
Polypropylene	F ⑥	0.90 to 0.91	165 to 175	120	18 to 30		1 000 to 1 500	5	9.0 to 13.5
Polysulphide	F ⑧	1.34	—	95	4 to 10	– 50		200 to 350	
Polytetrafluoroethylene (PTFE)	N	2.15 to 2.24	325	250	12		400	150	10
Polyvinyl chloride (PVC) rigid	B S ⑨	1.39	75 to 85	65	35 to 55	– 40	5 500	10	6
Polyvinyl chloride (PVC) plasticised	B/F S	1.30	60 to 85	40 to 65	10 to 25		10	200	

ued)

tance to acids and alkalis			*Resistance to solvents*						
Dil. inorganic acids	*Organic acids*	*Alkalis*	*Petrol*	*Paraffin, diesel and fuel oil*	*Aromatic and chlorinated solvents*	*Ethers, ketones and esters*	*Alcohols*	*Typical applications*	*Relevant British Standards or Codes of Practice*
+		–	+	+	+	0 to –	–	Laminates for roofing and walling panels; adhesives for timber; surface coatings. (See Table 4.13 for applications of foamed material)	BS 1203, BS 1204 BS 2572, CP 3003 (pt. 6)
			+ to 0	+	–			Seals, waterstops and gaskets; adhesives; surface coatings; bridge bearings	BS 2752, CP 3003 (pt. 1)
+	+	–	+	+	+	–	+	Surface coatings, as laminated material for: pipes; roofing and cladding; storage tanks; 'architectural' features	BS 3532, BS 3840, BS 4154
+	0	⑦ +	–	0	–	–	0	Damp proof membranes; protective sheeting (temporary); cold water supply pipes; cold water storage tanks; drains	BS 1972, BS 1973, BS 3012, BS 3867, BS 3897, BS 4159
+	0	⑦ +	–	0	–	–	0		BS 3284, BS 3796 BS 3867, BS 4159, BS 4646
+	+	⑦ +	–	–	–			Bridge bearings; adhesives; floor coverings	CP 3003 (pt. 1), BS 1711, CP 203
+	+	⑦ +	0	+	–	0	+	Drain and waste pipes and fittings; hot water tank overflows; valves; containers; pressure pipes; ropes	BS 3867, BS 3943, BS 4159
–	–	–	+	+	①			Flexible sealants	BS 4254
+	+	+	+	+	+	+	+	Bridge bearings; chemically resistant coatings; gaskets	BS 3784, BS 3873 BS 4271
+	–	+	+	+	–	–	+	Cold water supply pipes; drains and wastes; rainwater goods; roofing and cladding; lining panels; ducting	BS 3168, BS 3505, BS 3867, BS 3943, BS 4159, BS 4203, BS 4346, BS 4514, BS 4576, BS 4660, CP 203
+ to –	–	+ to –	+ to –	+ to –	–	–	+ to –	Roof coverings; waterstops; preformed seals; surface coatings; floor coverings	BS 3260, BS 3261, BS 3878, CP 3003, (pt. 4)

reduces this effect, but in design it must always be allowed for, for example by incorporating suitable movement joints. Rigid PVC formulations, such as those used for pipes and claddings, have coefficients of thermal expansion several times those of metals commonly used in construction.

RESISTANCE TO ACIDS AND ALKALIS

Polymers tend to be resistant to attack from acids and alkalis and are generally better than more common construction materials in this respect. The good chemical resistance of many polymers is made use of in formulating protective coatings and linings, but incorporating nonresistant fillers (chalk is a common filler for plastics) in compounds reduces or eliminates their resistance.

RESISTANCE TO OIL AND SOLVENTS

Polymers vary greatly in their resistance to oil and solvents. Many thermoplastics and rubbers are attacked by a variety of solvents; thermosetting plastics tend to be resistant and in some cases they are almost immune from attack. Nylon and PTFE are notable among common thermoplastics for their solvent resistance. The solvent resistance of many polymers is highly specific, and polymers which are unaffected by one solvent may be readily attacked by another. The solubility of many polymers is used in 'solvent welding' and in the formulation of adhesives and coatings. A particular form of solvent attack is the migration of solvents and oils into or out of plasticised compounds. This occurs if the solvent or oil is miscible with the plasticiser in such compounds, even if the basic polymer would be immune from attack. Solvent welding and plasticiser migration are dealt with on pages **4**–38 and **4**–40.

RESISTANCE TO OXIDATION AND OZONE

Some polymers are oxidised to a significant extent at high temperatures (the temperatures at which they fuse for example) and some (rubbers especially) are attacked at ambient temperatures by ozone. Formulation with suitable inhibitors can be used to make rubber and plastics compounds which are resistant to these effects.

RESISTANCE TO BIOLOGICAL ATTACK

Most polymers are immune from biological attack, though attack on ingredients used in the formulations of plastics compounds is not unknown. Casein (a protein) can be attacked though not when it is crosslinked with formaldehyde. Borers, such as woodworm, have been known to make their way into plasticised PVC, but this is unusual and occurs only when the compound is in contact with some more palatable material. Rats sometimes bite through plastics water pipes (as they do through lead) but it is an uncommon hazard.

Mechanical properties

The mechanical properties of rubber and plastics compounds are greatly influenced by both the basic polymer and by the other ingredients used in formulating the compound. The compounding and manufacturing process itself also influences the mechanical properties, especially where molecular orientation occurs.

Data on mechanical properties are thus very difficult to tabulate concisely, also because values vary so much with test conditions such as temperature, duration of loading and method of loading. For such reasons the data given in Table 4.12 are incomplete in some cases, and may appear to be very imprecise in many others.

STRENGTH

Tensile and compressive strengths of plastics compounds vary over a wide range. High tensile strength is a property of polymers such as nylon and some forms (films and fibres) of polyester and polypropylene. The relatively low elastic modulus of many polymers makes the compressive strength more difficult to assess in practical terms. Thermosetting resins tend to have high compressive and tensile strengths, the latter being capable of being greatly increased by the incorporation of reinforcing fibres. Orientation in films and fibres is also a means of increasing strength.

ELASTIC MODULUS

Rubbers and thermosetting polymers behave elastically over a large part of their strain range, but thermoplastic polymers tend to strain irreversibly after a relatively small proportion of their ultimate strain. There are a number of exceptions to this general rule. Among thermoplastics unmodified polystyrene is noted for its exceptionally low strain at failure and it shatters easily. Thermosetting polymers tend to be less flexible than rigid thermoplastics and when broken they often show a brittle fracture.

HARDNESS AND ABRASION RESISTANCE

Rubber and plastics compounds are soft compared with most construction materials, though they are not necessarily easily abraded. Rubbers and flexible thermoplastics are softer than rigid thermoplastics; thermosetting plastics being generally harder still. Abrasion resistance depends on several factors including hardness, elasticity, surface friction and the ability for abrasive particles to become embedded. Factors increasing abrasion resistance for some of these reasons tend to reduce it for others and it is a property which is difficult to predict without tests.

CREEP

Strength and elastic modulus measured at high rates of loading are much higher than those which are obtained at very low rates of loading for most plastics compounds, though rubbers and thermosetting polymers are less prone to creep than thermoplastic polymers. When creep deflection is considered important, care must be taken to choose suitable compounds and to limit stresses to those which will not lead to unacceptable creep. Where loads are to be applied intermittently, creep is unlikely to become a problem as recovery can take place over a relatively long period. Creep in plastics increases greatly with higher temperatures.

Compounding, processing and fabrication

COMPOUNDING

Some of the ingredients which are used in formulating plastics compounds have been mentioned on page **4**–30. Many polymers are used in commercial applications without addition, but the art or science of formulating PVC compounds suitable for particular applications is the converter's most important contribution in the manufacture of plastics articles and compounds based on this polymer. Guidance on formulation cannot be given here, but it is necessary that the engineer should understand that formulation is important.

PROCESSING METHODS

There are many ways of making plastics compounds into useful articles or materials; some of the most usual methods are:

- Extrusion, where the compound is continuously forced through a die;
- Calendering, where the compound is forced between a series of rollers to form a sheet;
- Injection moulding, where the compound is forced into a die or mould;
- Spreading, where the compound (usually PVC) is spread onto a support (often temporary) to form a sheet;
- Casting, where the compound is allowed to flow into a mould under gravity or by centrifugal force;
- Dough moulding, where the compound is shaped under pressure by a die;
- Vacuum forming, where previously made sheet is shaped by being heated and forced onto an evacuated former under air pressure.

INFLUENCE OF PROCESSING METHODS ON PROPERTIES

All processing methods except some used for thermosetting polymers need the polymer or compound to be heated, and many thermoplastic compounds are degraded by prolonged heating. Thus processing methods, like extrusion, which need the compound to be heated for only a short time have inherent technical advantages over methods like calendering where the compound may have to be kept hot over a long period.

Processing methods for compounds which do not become truly fluid on being heated, shape the compound under pressure into a form which it will largely retain on cooling. However some tendency to return to the unformed shape may remain, and 'relaxation' of newly formed shapes (calendered sheet especially) in thermoplastics should be allowed for.

Where thermoplastics compounds are to be used at temperatures which even begin to approach the processing temperature, relaxation can be a severe problem. An example is vacuum formed shapes which have been formed from sheet heated only enough to soften it slightly. Such shapes may relax enough to be considerably distorted even by the temperatures caused by sunshine on a summer's day.

FABRICATION METHODS FOR MATERIALS AND COMPONENTS

Materials made from thermoplastic polymers or compounds can be fabricated by heat or friction welding; and in the case of those which are soluble, by solvent welding. Mechanical methods of fabrication can also be used.

It is usually possible to find a solvent which can be used for solvent welding thermoplastics, though not all the solvents which attack a material are suitable for welding it. Important among thermoplastics which cannot be solvent welded are polyethylene, polypropylene, PTFE and nylon. The welding solvent may be modified by the addition of a separate polymer to make it tacky. This is useful in keeping joined parts in position while the solvent is doing its work. Properly made heat-or solvent-welded joints are usually as strong as the parent material. Materials made from thermosetting polymers and crosslinked rubbers cannot be welded, though many can be glued satisfactorily using thermosetting resin applied before it has crosslinked, or with solvent based adhesives made from other polymers. Glueing is likely to be less strong than welding, but good bond can be obtained with thermosetting polymer adhesives. Because plastics materials are relatively simple to make in almost any shape using one or other of the methods noted above (**4**–38), fabrication can usually be limited to the joining of finished units, and even these can often have mechanical joints built into them during manufacture.

FABRICATION METHODS—DIRECT FABRICATION FROM POLYMERS AND COMPOUNDS: CONTACT MOULDING

Manufacture of the material and fabrication into the required unit can often be combined into one operation. Glass-reinforced thermosetting polymers are often used in this way, and if the polymer can be cured under ambient site conditions fabrication on site is possible. When contemplating on-site fabrication of plastics components or the direct application of compounds as surface coatings, it should be noted that full curing under ambient conditions (which might need to be modified by installing heating) will be needed. It sometimes happens that a compound which will harden under ambient conditions, and look as if it has cured fully, does not in fact cross-link sufficiently to develop desired properties of strength, durability and solvent resistance.

On-site fabrication, or surface coating with thermosetting compounds, usually needs a curing agent to be added to the polymer shortly before fabrication. This is necessary because most compounds which will cure under ambient conditions would also cure during storage and could not be kept ready mixed for more than a few hours. Exceptions include thermosetting compounds which cure through the absorbtion of atmospheric moisture and compounds whose storage life can be extended usefully by storing them under refrigeration.

Where heat can be applied to promote curing, ready mixed thermosetting compounds which can be stored at ambient temperature are often convenient to use, since curing starts when heat is applied, and this time is under the fabricator's control. As well as thermosetting compounds which already contain the curing agent, pre-impregnated glass cloth can be fabricated in this way. This cloth is usually made with glass strand mats and compounds which have a high enough viscosity at ambient temperatures to give a conveniently handled material. Before it is cured, the compound is in a thermoplastic condition, and the material can be shaped easily if it is slightly heated. Prolonged heating, or heating to a higher temperature is then used to cure the compound after shaping and fabrication.

APPLICATION OF PLASTICS MATERIALS AND COMPONENTS

The properties of plastics compounds described in the above sections should give the designer some guide on the virtues and limitations of the materials themselves, but in their application the interaction between plastics and other materials must also be taken into account. Two important limitations are the high coefficient of thermal expansion of plastics materials and the phenomenon of plasticiser migration.

In case of flexible plastics, the high coefficient of thermal expansion causes few problems because the material's tendency to strain with temperature changes does not produce high stresses in the plastics materials or at the interface between plastics materials and the materials to which they are applied.

With rigid plastics materials, however, the stresses produced by restrained thermal expansion can be high enough to produce distortion, failure at the interfaces or even failure of the materials themselves. When designing fixings for rigid plastics components, provision must be made for thermal movement. Fixing through slotted holes or with clips is satisfactory provided that they are not fastened too tightly to allow free movement. Where weatherproofing has to be provided by plastics components the design of joints which will remain weathertight while allowing movement is essential.

Plasticiser migration can be a serious problem when flexible thermoplastics containing plasticisers are bonded with adhesives which contain similar materials. In such cases the plasticiser and constituents of the adhesive diffuse into each other with the result that the plastics material may shrink and become brittle if there is a net loss of plasticiser, or soften excessively if there is a net gain and the adhesive may suffer similarly. The problem is best avoided by the choice of suitable adhesives, but where plasticised materials have to be applied over substrates into which plasticiser can migrate (for example over bituminous materials), a coating or an intermediate layer can be used to provide a barrier to the migration of the plasticiser.

Identification of polymers and plastics compounds

The suitability of any polymer or compound for any particular application will depend greatly on which compound is chosen, and it is therefore helpful to know how different compounds and polymers can be recognised. Although precise identification is often impossible without modern analytical equipment, a useful idea of the nature of the material can be obtained quite easily in many cases. The following is intended as a general guide.

Flexibility Rubbers can be bent without breaking or cracking and they snap back when released.

Flexible thermoplastics can also be bent without breaking or cracking, though usually not as much as rubbers, and they do not snap back. Rigid thermoplastics can usually be bent a little, but continued attempts to bend them result in breaking or cracking. Polystyrene, however, is rigid and brittle unless modified. It cannot be bent. Thermosetting plastics are usually very rigid and break cleanly if an attempt is made to bend them.

Feel is a difficult sensation to describe accurately, but polyethylene and PTFE have a waxy feel which other plastics do not have.

Bounce Most rubbers (but not butyl rubber) bounce.

Density A simple division can be made between polymers which float in water (a minority), and those which do not. (Table 4.12 lists specific gravities.)

Burning Many polymers support combustion and, of those that do, some burn with a smoky flame and others with a clear flame. Table 4.12 indicates behaviour on ignition.

Chemical tests Details of chemical tests are too long to be included here, but engineers who wish to carry out further tests for the identification of polymers and compounds will find that many of the simple tests described by Saunders (see Bibliography) can be carried out with rudimentary chemical knowledge and apparatus.

Foamed and expanded plastics

Thermal insulation is a very important application of plastics when they are in a foamed or expanded form. Very low bulk densities combined with sufficient strength for satisfactory handling and fixing can be obtained with these materials, and some of them have

Table 4.13 PROPERTIES OF SOME FOAMED AND EXPANDED PLASTICS

Expanded or foamed polymer	*Bulk density* kg/m³	*Thermal conductivity* W/m deg C	*Maximum working temperature* °C	*Water absorption at 7 days* % by vol	*Water vapour 'diffusance* g/in² s bar	*Combustibility*	*Typical applications*	*British Standards or Codes of Practice*
Expanded polystyrene	16 to 24	0.033 to 0.035	80	2.5 to 3.0	0.010 to 0.015	(1) F	Lining walls and ceilings; insulating cold water services. Integral wall, floor and roof insulation	BS 3932; 1965 BS 3837: 1965 CP 144
Foamed polystyrene	32 to 40	0.033 to 0.035	80	1.0 to 1.5	0.003	(1) F	Similar to above	No BS
Expanded PVC	24 to 125	0.035 to 0.055	65	3.0 to 4.0	0.002 to 0.004	B	Lining walls and roofs; integral insulation in sandwich panels	BS 3869: 1965
Foamed phenol formaldehyde	32 to 64	0.036	130	high	0.20	N	Roof insulation under hot applied finishes	BS 3927: 1965
Foamed urea formaldehyde	8	0.038	100	high	0.18	N	In situ cavity wall insulation	No BS
Foamed polyurethane	32	0.020 to 0.025	100	2.5	0.01	(1) F	Lining walls and ceilings; integral insulation in sandwich panels; insulating pipes, sprayed on in-situ insulation	

KEY
(1) Flame retardant grades are available.
F Flammable and continues to burn after ignition.
B Combustible but selfextinguishing.
N Noncombustible under normal conditions of use.

the additional advantage for low temperature insulation of low water vapour diffusance. Commonly used insulating materials made from plastics are listed in Table 4.13 together with their most important physical properties, typical applications and relevant British Standards and Codes of Practice.

PAINT FOR STEEL

General principles of painting steelwork

MECHANISM OF RUSTING

Iron and steel rust when moist because electrochemical reactions dissolve iron at anodic areas and the dissolved iron combines with atmospheric oxygen to form hydrated ferric oxide which is the main constituent of rust. Anodic areas (areas which are electrically positive) arise from several causes including variations in the composition of the steel, breaks in millscale and deficiency in oxygen relative to other areas. They cannot be avoided unless the whole of the steel is made cathodic (electrically negative) relative to some separate metal surface which is also in good electrical contact with the moisture immediately surrounding the steel. This arrangement is called cathodic protection.

INFLUENCE OF MOISTURE

Moisture is essential for corrosion and the more moisture comes into contact with the steel surface, the faster it will rust. In theory painting could protect the steel surface completely from moisture, but in practice the paint would have to be very impermeable and be applied very thickly. Even if this were possible, damage to the coating, which is almost unavoidable, would allow moisture to reach the steel. Moisture at the time of painting should also be avoided and features which trap moisture or encourage it to hang in drips should be eliminated by good design.

INFLUENCE OF CONTAMINATION

Dissolved salts make moisture more readily able to dissolve steel as they hasten the electrochemical reaction. Thus dirty steel in a polluted atmosphere rusts more quickly than clean new steel in a clean atmosphere. Steel which is exposed for even a few days in an industrial atmosphere collects enough surface contamination to increase the rate of rusting many times, and the contamination is very difficult to remove completely even if the steel is cleaned to a 'white metal' finish. Ideally steel should be cleaned of millscale and primed at the rolling mill: in practice this is seldom done and primer thick enough to protect the steel during subsequent storage and fabrication is liable to interfere with fabrication processes, especially welding. It is sound economics to clean steel as thoroughly as possible before it is painted. The paint life will be greatly increased if this is done.

INFLUENCE OF PRIMER TYPE

Although good practice will reduce hazards it is inevitable that moisture will penetrate to the steel surface and that the surface will be contaminated with corrosion promoters. Because this has to be accepted, it is usual for the first coat of a painting system to be a primer which will inhibit rusting, either by chemical action in making the dissolved contaminants less aggressive, or by electrical action in providing an anodic layer in electrical contact with the steel surface. Red lead in linseed oil, which forms an insoluble salt with sulphate solutions, is an example of the first type of inhibitor; zinc-rich primer, which provides a layer of metallic (anodic) zinc in contact with the steel, is an example of the second type.

Cathodic protection is sometimes used in addition to painting, especially where exposure is severe, but it should be noted that oil-based primers and finishing paints are liable to be degraded by saponification of the fatty esters which they contain; this phenomenon occurs under alkaline conditions such as those which are created at cathodic areas when cathodic protection is used.

Primers

PRETREATMENT PRIMERS

These primers (also known as wash primers and etch primers), are applied in thin layers to prepared steel surfaces to increase the adhesion and life of the painting system. The normal dry film thickness of about 10 μm is too small to interfere with fabrication methods and gives protection for only a very short period. Etch primers are usually based on phosphoric acid and a water soluble or alcohol-soluble binder with inhibitive pigments.

PREFABRICATION PRIMERS

These are designed to be applied to the steel immediately after it has been cleaned and are supposed to give protection throughout the fabrication period without adversely affecting welding or other fabrication processes. Prefabrication primers are usually made from zinc dust in epoxide resin binder and are applied to give a 10 to 25 μm thick dry film containing 90% or more of metallic zinc. There is doubt about the length of time for which these primers give effective protection, and welders object to the fumes which come from them when they are heated. Nevertheless these primers do give at least some temporary protection and the fume problem can be overcome by efficient ventilation.

RED-LEAD PRIMERS

Red lead, especially if it is dispersed in linseed oil, is a primer which will give protection to steel even if some residual contamination remains, as for example, when a steel surface has been cleaned by wire brushing. Red lead has the important disadvantage of being poisonous, and red lead in linseed oil dries very slowly. Red lead in fast-drying media (alkyd resin for example) is not especially tolerant to poor surface preparation.

METALLIC LEAD PRIMERS

These consist of lead dust and other pigments in oleo-resinous, chlorinated rubber, or epoxide resin media. The lead content is 40% or more and the dry films are hard and resist damage well. These primers are not as tolerant to poor surface preparation as red lead in linseed oil, though metallic lead in oleo-resinous media is fairly tolerant. These primers are considered better than red lead paints in marine environments, and similar paints can be used as finishing coats.

CALCIUM PLUMBATE PRIMERS

These are particularly useful over hot-dip galvanised surfaces because they adhere to them well, but not all finishing paints (especially alkyd based paints) can be used over them. Calcium plumbate primers dry quite quickly but do not tolerate poor surface preparation.

ZINC CHROMATE PRIMERS

Zinc chromate primers are based on mixtures of zinc chromate and other pigments usually in oleo-resinous media. They dry quite quickly but give thinner dry-film thicknesses than lead-based primers and are not as resistant to aggressive industrial atmospheres. In marine atmospheres, zinc chromate primers give good protection, but they are more sensitive to surface preparation than red lead in linseed oil. Zinc chromate primers should contain 40% or more of zinc chromate for best results.

ZINC PHOSPHATE PRIMERS

These are based on zinc phosphate, normally in oleo-resinous or alkyd media. They are relatively tolerant to imperfect surface preparation, are quick drying and nontoxic.

ZINC-RICH PRIMERS

Zinc-rich primers are made from not less than 85% of metallic zinc dust in styrene, chlorinated rubber or epoxide media. The dry film should contain at least 90% of zinc which is a high enough concentration to give electrical contact between the zinc and the steel. Cathodic protection therefore results, though it is believed that this mechanism of protection is soon replaced by the protective action of a dense layer of zinc and its corrosion products which forms through the action of contamination by the atmosphere. Very good surface preparation is essential for these primers to ensure good electrical contact and continued adhesion. The solubility of zinc corrosion products formed on the surface of the primer makes it essential to wash primed surfaces very thoroughly before applying subsequent coats of the painting system.

(It should be noted that aluminium in priming paints does not function in a similar way to zinc because the oxide film formed round the metal particles prevents electrical contact.)

OTHER PRIMERS

Commonly used priming paints have been mentioned above. The information is necessarily brief and does not cover primers which are based on variations of the basic systems, or those based on uncommon systems, though many such primers are produced commercially. Paint suppliers who are able to produce impartial evidence for the quality of their products, or painting consultants can be relied upon to give good advice where the user is in doubt.

Finishing paints

Finishing paints protect the primer from the external environment and in many cases also give the steelwork an attractive appearance.

DRYING OIL, ALKYD AND OLEO-RESINOUS BASED PAINTS

Many finishing paints are based on pigments in drying oils, the pigments sometimes having plate-like (lamellar) particles. Silica graphite, flake aluminium, flake stainless steel and micaceous iron oxide are examples of pigments which have such particles. The overlapping of the pigment particles helps to protect the medium from attack by the weather, especially ultraviolet light. Such pigments tend to limit the colour or gloss (or both) and for this reason they are sometimes used in intermediate coats which are subsequently covered with more decorative though perhaps less durable paints. Alkyd resin or oleo-resinous media are often used as alternatives to drying oils, but alkyd resins cannot be used over calcium plumbate primers and may be unsatisfactory over zinc-rich primers. Dry film thicknesses of up to 125 μm per coat are possible with some formulations.

CHLORINATED RUBBER PAINTS

Solutions of chlorinated rubber and plasticiser in suitable solvents are the media in so called 'chemically resistant' paints of this type. These paints are not degraded by saponification (as are oil-based paints when in alkaline conditions) and they resist attack by atmospheric pollutants, but they are not resistant to solvents and this can create problems when these paints are recoated. Dry film thicknesses of up to 100 μm can be achieved with some formulations.

VINYL RESIN PAINTS

These are based on solutions of polyvinyl chloride or its copolymers in suitable solvents. The paints tend to have low dry film thicknesses but their chemical and water resistance is good. Adhesion can be a problem with these paints unless the surface is carefully prepared, though the addition of maleic anhydride is a means of improving adhesion. These paints are not resistant to solvents.

POLYURETHANE PAINTS

These are based on polyurethane resin in solvent, the most chemically resistant type being a two-pack material in which resin and curing agent have to be mixed together shortly before the paint is applied. One-pack air or moisture cured polyurethane paints are available but they do not have the outstanding chemical resistance of the fully cured two-pack systems. Although moisture is used to initiate the curing of some polyurethane paints, moisture at the time of application degrades these paints and the surface to be painted must always be perfectly dry. Adhesion can be difficult over some surfaces, and intercoat adhesion between a fully cured coating and a subsequent coat can only be achieved with careful surface preparation.

EPOXIDE RESIN PAINTS

Epoxide resin paints may be two-pack materials (solvent-containing or solvent-free) in which resin and curing agent have to be mixed together shortly before the paint is applied, or they may be oil-based paints which contain a proportion of uncured high-viscosity epoxide resin. Two-pack epoxide resin paints, especially those which contain little or no solvent, have excellent resistance to chemicals, most solvents and abrasion, and their adhesion is very good. At low temperatures curing may be very slow or may cease altogether unless a suitable formulation is used. Epoxide resin paints can be formulated to give very high dry film thicknesses, 250 μm in one coat being possible.

COAL-TAR PITCH AND BITUMEN PAINTS

These are based on solutions of these materials in solvents. Pigments and thickening agents have to be added to these solutions if the paints are to give adequate film thicknesses and be resistant to ultraviolet light. High-build paints can be formulated to give dry film thicknesses of up to 250 μm in one coat. These paints are normally black, but a limited range of dark colours can be produced.

PITCH/EPOXIDE PAINTS

Pitch/epoxide paints are similar to coal-tar pitch paints, but have much better chemical and weathering resistance because of the addition of 30% or more of epoxide resin. These paints can be formulated to give dry film thicknesses of 250 μm or more in one coat and they give good protection in very severe exposure conditions. These paints are black or very dark in colour.

PAINT FOR WOOD

General principles of painting and preserving wood

There are some similarities between painting steel to stop it from rusting and painting wood to stop it from rotting. In each case the surface must first be prepared to make protection as simple as possible, an inhibitor to prevent degradation must then be applied, and finally, the surface must be weatherproofed. The surface preparation of wood for painting must aim at providing a continuous surface to which paint can be applied without the paint film having to span open cell cavities. Sanding helps to close the cell cavities in hardwoods but in softwoods the cavities are too large to be much affected by it. Nevertheless the smoother the surface of softwoods the longer the paint will last.

Preservatives

Preservative treatment is essential for the long life of nearly all softwoods and for some hardwoods too, especially if sapwood is present. This is because rotting starts to take place if the moisture content of wood rises above about 20%, and no painting system is good enough to prevent this reliably if the wood is exposed to the weather. Wood preservatives are based on chemicals which are toxic to the microorganisms which cause decay. They are usually prepared as solutions in water or solvent and are applied to the wood by dipping, diffusion, pressure impregnation or surface soaking. Apart from water-repellent preservatives which are intended as a finish for the wood, preservative treatments do not interfere with the adhesion or integrity of subsequently applied coatings. Chemicals on which wood preservatives are commonly based include salts of arsenic, chromium, copper, zinc and tin, and organic derivatives of phenol and naphthalene (see also Section 15).

Primers

Primers are applied to the preservative-treated wood to fill the remaining cell cavities and to even out surface roughness. Wood primers usually consist of fillers in resinous solutions, but emulsion primers based on polyvinyl acetate or acrylic emulsions have some advantages because very high pigment loadings are possible with these paints. Lead-based pigments and aluminium powder are particularly good fillers for wood primers, and aluminium primers are noted for their water resistance.

Undercoats and finishing paints

These are usually based on alkyd resin solutions pigmented with titanium dioxide and other fillers, but polyurethane resins (either one- or two-pack) are sometimes used. For decorative purposes an important function of the undercoat is to provide a surface of the desired colour to which the finishing paint will adhere with minimal further surface preparation. Finishing coats contain relatively more resin and less pigment and their function is to provide weather resistance and gloss if required. It is important that all coats of the painting system, and especially the finishing coats, should remain flexible enough to accommodate the movement of the wood.

REFERENCES

1. LEA, F. M., The chemistry of cement and concrete, 3rd edn, Arnold, Glasgow (1970)
2. *Lightweight aggregate concretes,* Digest 123 (2nd series), Building Research Station, HMSO, London (1970)
3. *An introduction to concrete,* Eb 1, Cement and Concrete Association, London (1958)
4. SADGROVE, B. M., *The early development of strength in concrete,* TN12, CIRIA, London (1970)
5. HUGHES, B. P., *Control of thermal and shrinkage cracking in restrained reinforced concrete walls,* TN 21, CIRIA, London (1971)
6. CHAPMAN, G. P., 'The cylinder splitting test with particular reference to concrete made with natural aggregates', *Concrete,* **2,** No. 2, London (February 1968)
7. SADGROVE, B. M., *The strength and deflection of reinforced concrete beams loaded at an early age,* TN 31, CIRIA, London (1971)
8. SHACKLOCK, B. W. and KEENE, P. W., *The effect of mix proportions and testing conditions on drying shrinkage and moisture movement of concrete,* TRA 266, Cement and Concrete Association, London (1957)
9. TROXELL, G. E., RAPHAEL, J. M. and DAVIS, R. E., 'Long term creep and shrinkage tests of plain and reinforced concrete', *Proc. ASTM,* **58,** 1101–120 (1958)
10. BIRT, J. C., 'Curing concrete: an appraisal of attitudes practice and knowledge', Rep. 43, CIRIA, London (1973)
11. ELVERY, R. H., *An assessment of ultrasonic testing for reinforced concrete beams,* TN 4, CIRIA, London (1969)

BIBLIOGRAPHY

Applications and durability of plastics, Digest 67 (2nd series), Building Research Station

BATE, S. C., *et al., Handbook on the unified code for structural concrete,* Cement and Concrete Association, London (1972)

Cellular plastics for building 2, Digest 94 (2nd series), Building Research Station

DAVEY, A. B. and PAYNE, A. R., *Rubber in engineering practice,* Maclaren

FANCUTT, F., HUDSON, J. C., RUDRAM, A. T. S. and STANNERS, J. F., *Protective painting of structural steel,* Chapman and Hall (1968)

GRANFIELD, E. F., 'Protection of small steel bridges from corrosion', Forestry Commission forest record, No. 81, HMSO (1972)

HANSON, R. P., 'The importance of correct surface preparation and paint application in the protection of marine and industrial steel from corrosion', *Trans. NE Coast Inst of Eng and Shipbuilders,* V88 (1972)

KINNEY, G. F., *Engineering properties and applications of plastics,* Wiley

LEA, F. M., *The chemistry of cement and concrete,* 3rd ed, Arnold, Glasgow (1970)

MCINTOSH, J. D., *Concrete mix design*, 2nd ed, Cement and Concrete Association, London (1966)
NEVILLE, A. M., *Properties of concrete*, Pitman, London (1963)
OGORKIEWICZ, R. M. (Ed), *Engineering properties of thermoplastics*, Wiley (1970)
Painting metals in buildings, BRS Digests 70 and 71 (2nd series), HMSO (1966)
Painting woodwork, BRS Digest 106, HMSO (1969)
Plastics materials guide, Engineering, Chemical and Marine Press Ltd
SAUNDERS, K. J., *The identification of plastics and rubbers*, Science Paperbacks, Chapman and Hall (1966)
'Thermoplastics and mechanical engineering design', Technical Service Note G117 (2nd edn), ICI Plastics Division

LIST OF CODES OF PRACTICE AND STANDARDS PUBLISHED BY THE BRITISH STANDARDS INSTITUTION AND REFERRED TO IN SECTION 4

CP 98
Preservative treatment for constructional timber
CP 110
The structural use of concrete
CP 203
Sheet and tile flooring
CP 231
Painting of buildings
CP 2008
Protection of iron and steel structures from corrosion
CP 3003
Lining of vessels and equipment for chemical purposes
Part 1: Rubber
Part 4: Plasticised PVC sheet
Part 5: Epoxide resins
Part 6: Phenolic resins

In preparation at time of writing
Plastics pipework (thermoplastic material)

BS 12: 1958
Portland cement (ordinary and rapid hardening)
BS 146: 1958
Portland–blast-furnace cement
BS 277: 1936
Ready mixed paints (oil glass) zinc oxide base
BS 282: 1963
Lead chromes and zinc chromes for paints
BS 284: 1952
Black (carbon) pigments for paints
BS 332: 1956
Liquid driers for oil paints
BS 388: 1964
Aluminium flake pigments (powder and paste) for paints
BS 390: 1953
Oil pastes for paints
BS 410: 1969
Test sieves
BS 729: 1971
Zinc coatings on iron and steel articles
BS 812: 1967
Methods for sampling and testing of mineral aggregates, sands and fillers
BS 877: 1967
Foamed or expanded blast-furnace slag lightweight aggregate for concrete

BS 882 and 1201: 1965
Specification for aggregates from natural sources for concrete
BS 913: 1954
Pressure creosoting of timber
BS 915: 1947
High alumina cement
BS 1014: 1961
Pigments for cement, magnesium oxychloride and concrete
BS 1047: 1952
Air cooled blastfurnace slag coarse aggregate for concrete
BS 1070: 1956
Black paint (tar base)
BS 1165: 1957
Clinker aggregate for plain and precast concrete
BS 1203: 1963
Synthetic resin adhesives (phenolic and aminoplastic) for plywood
BS 1204: 1964
Synthetic resin adhesives (phenolic and aminoplastic) for wood
BS 1282: 1959
Classification of wood preservatives and their methods of application
BS 1370: 1958
Low heat Portland cement
BS 1711: 1951
Solid rubber flooring
BS 1795: 1965
Extenders for paints
BS 1881
Methods of testing concrete
Part 1: 1970 Methods of sampling fresh concrete
Part 2: 1970 Methods of testing fresh concrete
Part 3: 1970 Methods of making and curing test specimens
Part 4: 1970 Methods of testing concrete for strength
Part 5: 1970 Methods of testing hardened concrete for other than strength
Part 6: 1971 Analysis of hardened concrete
BS 1972: 1963
Polythene pipe (type 32) for cold water services
BS 1973: 1970
Polythene pipe (type 32) for general purposes including chemical and food industry uses
BS 2015: 1965
Glossary of paint terms
BS 2029: 1953
White oil pastes for paints
BS 2451: 1963
Chilled iron shot and grit
BS 2521: 1966
Lead-based priming paints
BS 2523: 1966
Lead-based priming paints for iron and steel (types A, B and C)
BS 2524: 1966
Red oxide-linseed oil priming paint
BS 2525: 1969
Undercoating and finishing paints for protective purposes (white-lead based)
BS 2572: 1955
Phenolic laminated sheet
BS 2660: 1955
Colours for building and decorative paints
BS 2752: 1956
Vulcanised chloroprene rubber compounds
BS 3012: 1970
Low and intermediate polythene sheet for general purposes
BS 3148: 1959
Tests for water for making concrete

BS 3168: 1959
Rigid PVC extrusion and moulding compounds
BS 3189: 1959
Phosphate treatment of iron and steel for protection against corrosion
BS 3227: 1960
Vulcanised butyl rubber compounds
BS 3260: 1969
PVC (vinyl) asbestos floor tiles
BS 3261: 1960
Flexible PVC flooring
BS 3284: 1967
Polythene pipe (type 50) for cold water services
BS 3412: 1966
Polythene materials for moulding and extrusion
BS 3452: 1962
Copper/chrome water-borne wood preservatives and their application
BS 3453: 1962
Fluoride/arsenate/chromate/dinitrophenol water-borne wood preservatives and their application
BS 3483: 1962
Methods for testing pigments for paints
BS 3505: 1968
Unplasticised PVC pipe for cold water services
BS 3506: 1969
Unplasticised PVC pipe for industrial purposes
BS 3532: 1962
Unsaturated polyester resin systems for low pressure fibre reinforced plastics
BS 3681: 1963
Methods of sampling and testing lightweight aggregates for concrete
BS 3698: 1964
Calcium plumbate priming paints
BS 3699: 1964
Calcium plumbate for paints
BS 3784: 1964
Polytetrafluoroethylene sheet
BS 3796: 1970
Polythene pipe (type 50) for general purposes including chemical and food industry uses
BS 3797: 1964
Lightweight aggregates for concrete
BS 3840: 1965
Polyester dough moulding compounds
BS 3842: 1965
Wood preservative treatment of plywood
BS 3867: 1969
Outside diameters and pressure ratings for pipe of plastics materials
BS 3873: 1965
Polytetrafluoroethylene basic moulded shapes
BS 3878: 1965
Flexible PVC sheeting for hospital use
BS 3892: 1965
Pulverised-fuel ash for use in concrete
BS 3900
Methods of test for paints
BS 3943: 1965
Plastics waste traps
BS 3982
Zinc dust pigment
BS 4027: 1966
Sulphate-resisting Portland cement
BS 4072: 1966
Wood preservation by means of water-borne copper/chrome/arsenic compounds

BS 4154: 1967
Corrugated plastics translucent sheets made from thermosetting polyester resin (glass fibre reinforced)
BS 4159: 1967
Colour markings of plastics pipes to indicate pressure ratings
BS 4203: 1967
Extruded rigid PVC corrugated sheeting
BS 4232: 1967
Surface finish of blast cleaned steel for painting
BS 4246: 1968
Low-heat Portland–blast-furnace cement
BS 4248: 1968
Supersulphated cement
BS 4254: 1967
Two-part polysulphide-based sealing compounds for the building industry
BS 4346: 1969
Joints and fittings for use with unplasticised PVC pressure pipes
BS 4408: Part 4: 1971
Recommendations for non-destructive methods of test for concrete-surface hardness methods
BS 4514: 1969
Unplasticised PVC soil and ventilating pipe, fittings and accessories
BS 4550
Methods of testing cement
Part 2: 1970: Chemical tests
BS 4576: 1970
Unplasticised PVC rainwater goods
BS 4619: 1970
Heavy aggregates for concrete and gypsum plaster
BS 4646: 1970
High density polythene sheet for general purposes
BS 4652: 1971
Metallic zinc-rich priming paint (organic media)
BS 4660: 1971
Unplasticised PVC underground drain and pipe fittings
BS 4725: 1971
Linseed stand oils
BS 4756: 1971
Ready mixed aluminium priming paints for woodwork

In preparation at time of writing
ABS pressure pipes and fittings for industrial applications
Polypropylene pressure pipes for industrial use

5 HYDRAULICS

HYDRAULICS 5

5 HYDRAULICS

J. ALLEN, D.Sc., LL.D., F.I.C.E., F.R.S.E.
Emeritus Professor, University of Aberdeen

PHYSICAL PROPERTIES OF WATER

DENSITY

For most purposes in hydraulic engineering, the density of fresh water may be taken to be 62.4 lb/ft^3 (*c.* 1000 kg/m^3). Correspondingly, the weight of one gallon of water is 10.0 lb or, the weight of 1 litre is approximately 1 kg. In more precise work, usually of a laboratory or experimental nature, the variation of density with temperature may have to be taken into account in accordance with Table 5.1.

Table 5.1 DENSITY OF FRESH WATER AT ATMOSPHERIC PRESSURE

Temperature		*Density*	
°C	°F	lb/ft^3	kg/m^3
0	32	62.42	999.9
4	39.2	62.43	1 000.0
10	50	62.41	999.7
20	68	62.32	998.2
30	86	62.16	995.7
40	104	61.94	992.2
50	122	61.68	988.1
60	140	61.38	983.3
70	158	61.04	977.8
80	176	60.67	971.9
90	194	60.26	965.3
100	212	59.83	958.4

The density of sea water depends on the locality but for general calculations the open sea may be assumed to weigh 64.0 lb/ft^3 (1025 kg/m^3). In a tidal river the density varies appreciably from place to place and time to time; it is influenced by the state of the tide and by the amount of fresh water flowing into the estuary from the higher reaches or from drains and other sources. At any one spot it may also vary through the depth of the water owing to imperfect mixing of the fresh and saline constituents.

VISCOSITY

Let us visualise a layer of a fluid as represented in Figure 5.1. The thickness of the layer is δy and particles in the plane AB are supposed to have a velocity v while those in CD have a different velocity, say $v + \delta v$. The plan area of each plane, AB or CD, is a, say. Now the fluid bounded by AB and CD experiences a resistance to relative motion along

AB analogous to shear resistance in solid mechanics. This force of resistance, divided by the area a, will give a resistance per unit area, or a stress f. Then

$$f = \eta \frac{dv}{dy} \tag{1}$$

as δy tends to zero, or as the layer assumes an infinitesimal thickness, so that dv/dy becomes the velocity gradient, i.e. the rate at which the velocity changes as we proceed

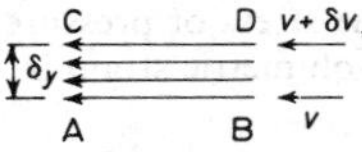

Figure 5.1 Layer of fluid illustrating laminar flow

outwards in a direction normal to the plane AB. η in equation (1) is known as the coefficient of viscosity. If force is defined by force = mass × acceleration then η will have the units of $f/(dv/dy)$, or

$$\frac{[M]\times[L]\times[T^{-2}]\times[L^{-2}]}{[L]\times[T^{-1}]\times[L^{-1}]}$$

i.e. $[ML^{-1}T^{-1}]$, where $[M]$ represents mass, $[L]$ length, and $[T]$ time.

Thus if newtons (i.e. kg m/s^2) are adopted for the force of resistance, metres for length, metres squared for area, and metres per second for velocity, then the coefficient of viscosity takes the units kg/m s. For example, the numerical value of η in the case of water at 10 °C is 0.001 31 kg/m s. This is the same as 0.0131 poise, i.e. 0.0131 g/cm s.

Kinematic viscosity. Kinematic viscosity ν is defined as the ratio of the viscosity η to the density ρ of a fluid, or

$$\nu = \frac{\eta}{\rho} \tag{2}$$

It follows from this definition that if η is in kg/m s and ρ in kg/m^3 then ν will be in m^2/s. Again, considering water at 10 °C, ν is 1.31×10^{-6} m^2/s or 1.31×10^{-2} cm^2/s.

Typical values of η and ν, for both water and air, are given in Table 5.2, from which it will be seen that temperature has quite different effects on these two fluids.

Table 5.2 VISCOSITIES OF WATER AND DRY AIR AT ATMOSPHERIC PRESSURE

Temperature		*Water*		*Air*	
°C	°F	$10^3\eta$ kg/m s	$10^6\nu$ m^2/s	$10^5\eta$ kg/m s	$10^5\nu$ m^2/s
0	32	1.79*	1.79*	1.71*	1.32*
5	41	1.52	1.52	1.73	1.36
10	50	1.31	1.31	1.76	1.41
15	59	1.14	1.14	1.78	1.45
20	68	1.01	1.01	1.81	1.50
25	77	0.894	0.897	1.83	1.55
30	86	0.801	0.804	1.86	1.59
35	95	0.723	0.727	1.88	1.64
40	104	0.656	0.661	1.90	1.69
50	122	0.549	0.556	1.95	1.79
60	140	0.469	0.477	2.00	1.88
80	176	0.357	0.367	2.09	2.09
100	212	0.284	0.296	2.18	2.30

* To avoid any possible misinterpretation of the column headings note that, at 0°C,
η for water is 1.79×10^{-3} kg/m s; ν is 1.79×10^{-6} m^2/s;
η for air is 1.71×10^{-5} kg/m s; ν is 1.32×10^{-5} m^2/s.

COMPRESSIBILITY

In the vast majority of engineering calculations water may be treated as an incompressible fluid. Exceptions arise when large and sudden changes of velocity occur, as in certain problems associated with the rapid opening or closing of a valve. If we imagine a mass of water to have its volume changed from V to $V-\delta V$ by an increase of pressure δp applied uniformly round its surface, then the bulk modulus of compressibility K is defined as the stress or pressure intensity δp divided by the volumetric strain produced by δp. This volumetric strain is $-\delta V/V$. Hence

$$K = -\delta p \bigg/ \frac{\delta V}{V} \tag{3}$$

δV itself being treated as negative.

The value of K depends somewhat on the temperature and absolute pressure of the water, but in round numbers it is usually sufficiently accurate to take it as 300 000 lbf/in^2 (c 2×10^9 N/m^2).

SURFACE TENSION

Surface tension is the property which enables water and other liquids to assume the form of drops, when it appears that the water is bounded by an elastic skin or membrane under tension. Another important manifestation is related to small waves or ripples where the form and motion are restricted or influenced by the tension in the surface. Surface tension depends on the liquid and gas in contact with one another. Suppose a portion of the liquid to have a bounding surface with radii of curvature in two mutually

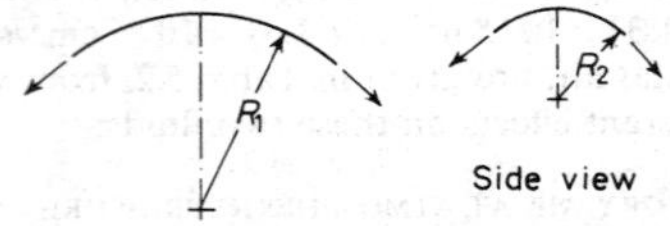

Figure 5.2 Bounding surface of a liquid

perpendicular directions R_1 and R_2 as in Figure 5.2. Then the excess of pressure intensity inside the boundary, over that outside it, is $\gamma(1/R_1+1/R_2)$, where γ is the surface tension, a force per unit length of the line to which it is normal.

Table 5.3 SURFACE TENSION OF WATER IN CONTACT WITH AIR

(°C)	0	20	40	60	80	100
γ (N/m)	0.0756	0.0728	0.0700	0.0671	0.0643	0.0615

The surface tension of mercury in contact with air at 20 °C (68 °F) is approximately 0.51 N/m.

CAPILLARITY

If a vertical tube is placed in a vessel containing water, the water will be drawn up the tube by capillary attraction. The angle α in Figure 5.3 is known as the angle of contact, and approaches the value zero for clean water in contact with a clean glass tube. In that case

$$h = \frac{4 \times 10^6 \gamma}{\rho d g} \text{ mm, approximately, or say } \frac{4 \times 10^5 \gamma}{\rho d} \tag{4}$$

where γ is the surface tension in N/m, ρ the density in kg/m^3, d the bore of the tube in mm, and g is in m/s^2.

At 20 °C (68 °F), therefore, the elevation of the water in the glass tube amounts to about $30/d$ mm, d being measured in mm, or $0.046/d$ in if d is measured in inches. It should be emphasised, however, that capillary attraction depends to a marked degree upon the state of cleanliness of the liquid and the tube.

If mercury is considered instead of water, there is a depression of the liquid in the tube as shown in Figure 5.4. The angle β is approximately 53° and h, at room temperature, is about $9/d$ mm, if d is in mm, or $0.013/d$ in, if d is in inches. Capillary attraction may be important in connection with the technique of measurement. For example, if the level of water in a tank is read, for convenience, on an external gauge as depicted in Figure 5.5, the bottom of the meniscus, or curved surface of the water in the gauge-tube, will stand higher than in the tank. Whether the effect is serious depends, of course, upon the standard

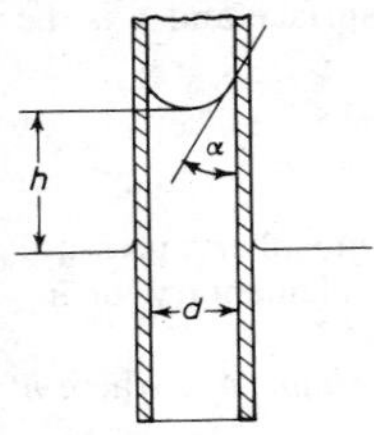

Figure 5.3 Capillary rise of water in a tube

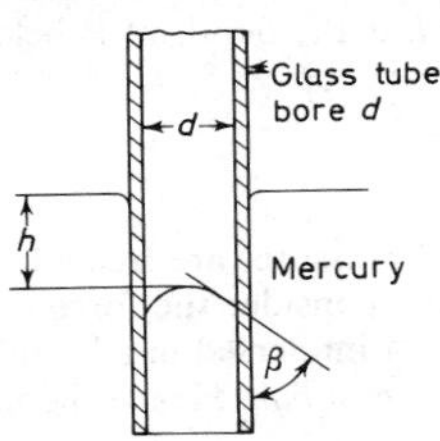

Figure 5.4 Capillary depression of mercury in a tube

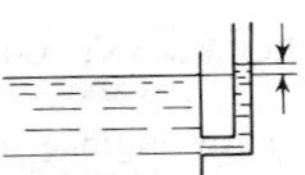

Figure 5.5 Effect of capillary attraction on measurement of height of water in a tank

of accuracy demanded, but it is generally advisable in such a case, or when using a differential gauge having two limbs, to use a tube not smaller than $\frac{3}{8}$ in (9.5 mm) bore, and as uniform as practicable throughout its length.

SOLUBILITY OF GASES IN WATER

At atmospheric pressure, water is capable of dissolving approximately 3, 2 and 1% of its own volume of air at temperatures of 0, 20 and 100 °C respectively. Certain other gases, such as carbon dioxide, are dissolved in much greater volumes, but the presence of air alone, together with the phenomenon of vapour pressure of water, may lead to complications in certain pipelines or machines. To take an example, at the highest point of a siphon the pressure is below atmospheric and it is possible for air to come out of solution there which was originally dissolved in the water at atmospheric pressure. This accumulation of air may ultimately reduce the flow along the siphon very appreciably, or even break it entirely, unless precautions are taken to draw off the air and vapour as it collects. The suction-lift of pumps is also restricted in practice by the difficulty of release of air and generation of vapour so that, in practice, it is usual to limit the suction-lift to about 28 ft (8.5 m) instead of the full height of the water barometer, say 34 ft (10.4 m).

VAPOUR PRESSURE

If a liquid is contained within a closed vessel the space above it becomes saturated with its vapour and the space is subjected to an increase of pressure, which is the vapour pressure of the liquid at the temperature then obtaining.

Table 5.4 VAPOUR PRESSURE p_v (N/m^2) OF WATER AT VARIOUS TEMPERATURES

(°C)	0	5	10	15	20	30	40
10^{-4} p_v	0.0610	0.0875	0.123	0.170	0.235	0.423	0.736
(°C)	50	60	70	80	90	100	
10^{-4} p_v	1.23	2.00	3.09	4.76	7.00	10.1	

To obtain p_v in metres of water at a given temperature, divide p_v (N/m^2) by $g\rho$ where ρ (kg/m^3) is given in Table 5.1. For example, at 100 °C, p_v is $(10.1 \times 10^4)/(9.81 \times 9.58 \times 10^2)$, i.e. 10.8 metres of water.

HYDROSTATICS

1. A fluid at rest exerts a pressure which is everywhere normal to any surface immersed in it.
2. The pressure intensity at a point P in a liquid is equal to that at the free surface of the liquid together with ρgh, where h is the depth of P below the free surface and ρ is the density of the liquid.

FORCE ON ANY AREA

In many engineering problems all pressures are treated relative to atmospheric pressure as a datum. Adopting that system, consider the force exerted on an elementary, or infinitesimal, portion δA of an area A immersed in a liquid, see Figure 5.6.

The pressure intensity on δA is $p = \rho gh$. Hence the force on δA is $\rho gh \,.\, \delta A$, where h is the vertical depth of δA.

The total force on the whole area A of which δA is an element is the arithmetical

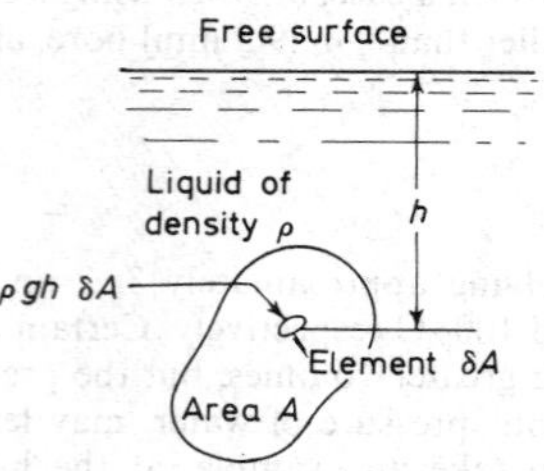

Figure 5.6 Elementary area δA immersed in liquid

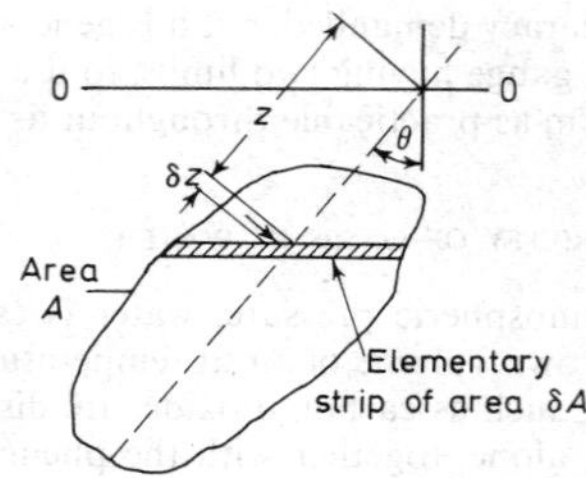

Figure 5.7 Plane area immersed in liquid

sum of the forces on all its constituent elements, $\Sigma\rho gh \,.\, \delta A = \rho g\Sigma h \,.\, \delta A$, assuming ρ to be constant throughout the liquid. Hence

$$\text{the total force} = \rho gH \,.\, A \tag{5}$$

where H is the vertical depth of the centroid of the whole area.

This total force is only equal to the resultant force if the area under consideration is a plane one. If the area is curved, then the forces acting on its elementary portions are not all parallel to one another so their simple arithmetic sum is not the same as their resultant.

FORCE ON PLANE AREAS (FIGURE 5.7)

$$\text{Force on element} = (\rho g z \cos\theta)\delta A$$

$$\begin{aligned}\text{Resultant force} &= \text{total force in this case}\\ &= \Sigma(\rho g z \cos\theta)\delta A\\ &= \rho g \bar{z} A \cos\theta \qquad (6)\end{aligned}$$

where $\bar{z}$ is the inclined depth of centroid of A. The resultant force will act through a point in the immersed area A known as its centre of pressure and such that its inclined depth Z is given by

$$Z = \frac{\int z^2\, dA}{\bar{z}A} = \frac{\text{Second moment of area } A \text{ about } 00}{\bar{z}A}$$

$$= \frac{I_{00}}{A\bar{z}} \qquad (7)$$

or

$$Z = \frac{k_{00}^2}{\bar{z}} \qquad (8)$$

where k_{00} is the radius of gyration about 00 and $k_{00}^2 = k^2 + \bar{z}^2$, where k is the radius of gyration about an axis through the centroid parallel to 00.

Examples. In the following examples C is the centroid, P the centre of pressure.

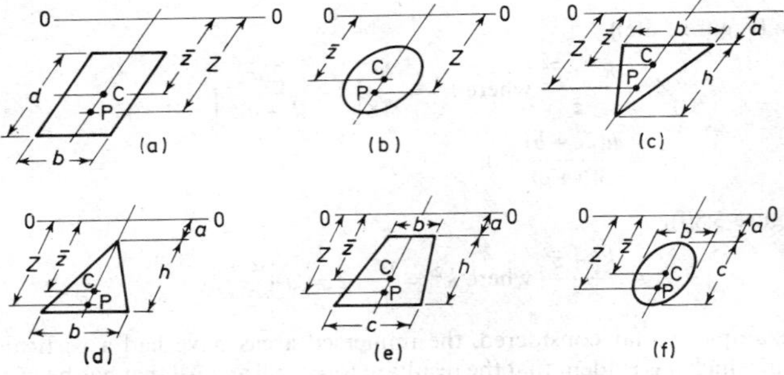

Figure 5.8 (a) Parallelogram; (b) Circular area; (c) Triangular area, apex downwards; (d) Triangular area, apex upwards; (e) Trapezium; (f) Ellipse

Parallelogram (Figure 5.8(a))

$$I_{00} = \frac{bd^3}{12} + bd(\bar{z})^2$$

$$Z = \frac{bd^3/12 + bd(\bar{z}^2)}{bd\bar{z}} = \frac{d^2/12 + \bar{z}^2}{\bar{z}}$$

$$= \tfrac{2}{3}d \text{ if upper edge of parallelogram is in surface.}$$

Circular area, diameter d (Figure 5.8(b)

$$I_{00} = \frac{\pi d^4}{64} + \frac{\pi d^2}{4}\bar{z}^2$$

$$Z = \frac{(\pi d^4/64)+(\pi d^2/4)\bar{z}^2}{(\pi d^2/4)\bar{z}} = \frac{d^2/16+\bar{z}^2}{\bar{z}}$$

$$= \frac{5d}{8} \text{ if circle touches 00.}$$

Triangular area, apex downwards (Figure 5.8(c))

$$I_{00} = \frac{bh^3}{36} + \frac{bh}{2}\bar{z}^2$$

$$Z = \frac{(bh^3/36)+(bh/2)\bar{z}^2}{(bh/2)\bar{z}} = \frac{h^2/18+\bar{z}^2}{\bar{z}} \text{ where } \bar{z} = a+\frac{h}{3}$$

$$Z = \frac{h}{2} \text{ if } a = 0.$$

Triangular area, apex upwards (Figure 5.8(d))

$$I_{00} = \frac{bh^3}{36} + \frac{bh}{2}\bar{z}^2$$

$$Z = \frac{h^2/18+\bar{z}^2}{\bar{z}} \text{ where } \bar{z} = a+\frac{2}{3}h$$

$$Z = \frac{3}{4}h \text{ if } a = 0$$

Trapezium (Figure 5.8(e))

$$Z = \frac{k^2+\bar{z}^2}{\bar{z}} \text{ where } k^2 = \frac{h^2}{18}\left[1+\frac{2bc}{(b+c)^2}\right]$$

$$\bar{z} = \frac{h(2c+b)}{3(b+c)}+a$$

Ellipse (Figure 5.8(f))

$$Z = \frac{k^2+\bar{z}^2}{\bar{z}} \text{ where } k^2 = \frac{c^2}{16};\ \bar{z} = a+\frac{c}{2}$$

In the examples so far considered, the immersed areas have had a vertical plane of symmetry in which it is evident that the resultant force will act. All that has been necessary, therefore, was to determine the position of the resultant force in that plane of symmetry.

Force on an unsymmetrical plane area. Choose any convenient axes OX, OY. OX may be the line of intersection of the plane of the immersed area with the surface of the liquid. The elementary area δA has coordinates x and y relative to the chosen axes (Figure 5.9).

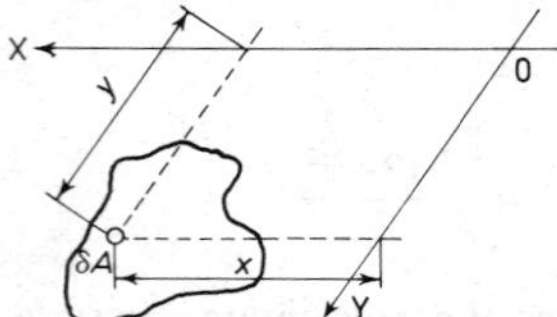

Figure 5.9 Unsymmetrical plane area

Let $\bar{x}$ and $\bar{y}$ be the coordinates of the centre of pressure of the whole area relative to these same axes. Then, by using the principle that the moment of the resultant force is equal to the sum of the moments of the individual elementary forces,

$$\bar{y} = \frac{\Sigma y^2\,\delta A}{\Sigma y\,\delta A} \qquad \bar{x} = \frac{\Sigma xy\,\delta A}{\Sigma y\,\delta A}$$

or

$$\bar{y} = \frac{\iint y^2\,dx\,dy}{\iint y\,dx\,dy} \qquad \bar{x} = \frac{\iint xy\,dx\,dy}{\iint y\,dx\,dy} \tag{9}$$

But

$$\begin{aligned} \Sigma y^2\,\delta A \quad \text{or} \quad \iint y^2\,dx\,dy &= Ak^2 \\ \Sigma y\,\delta A \quad \text{or} \quad \iint y\,dx\,dy &= Ay_0 \end{aligned} \tag{10}$$

where A is the total area, k its radius of gyration about OX and y_0 the ordinate of the centroid of the area.

FORCE ON CURVED AREAS

The following examples will serve to illustrate some useful principles.

Hemispherical bowl, radius r, just full of water (Figure 5.10).

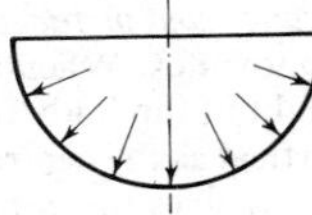

Figure 5.10 Hemispherical bowl just full of water

$$\begin{aligned} \text{Total force} &= \text{arithmetical sum of forces acting on the surface} \\ &= \text{area} \times \text{pressure intensity at centroid} \\ &= 2\pi r^2 \times \text{density of water} \times \text{depth of centroid} \times g \\ &= 2\pi r^2 \times \rho \times \frac{r}{2} g \\ &= \pi r^3 \rho g \end{aligned}$$

But the horizontal components of the corresponding forces on opposite sides of the vertical axis counterbalance one another, and

$$\begin{aligned} \text{Resultant force} &= \text{weight of water contained} \\ &= \text{volume of hemisphere} \times \text{density of water} \times g \\ &= \tfrac{2}{3}\pi r^3 \rho g \end{aligned}$$

Cylindrical vessel with hemispherical end, just full of water.

1. (Figure 5.11(a)). Force on lid due to water $= 0$

Resultant force (vertical) on hemispherical base $= (\pi r^2 h + \frac{2}{3}\pi r^3)\rho g$

2. (Figure 5.11(b)). Resultant force (vertical) on flat base $= \pi r^2 (h + r)\rho g$

Resultant force (vertically upwards) on dome $= \pi r^2(h+r)\rho g - (\pi r^2 h + \frac{2}{3}\pi r^3)\rho g = \frac{1}{3}\pi r^3 \rho g$

3. (Figure 5.11(c)). Horizontal force on either end $= \pi r^2(r\rho)g = \pi r^3 \rho g$

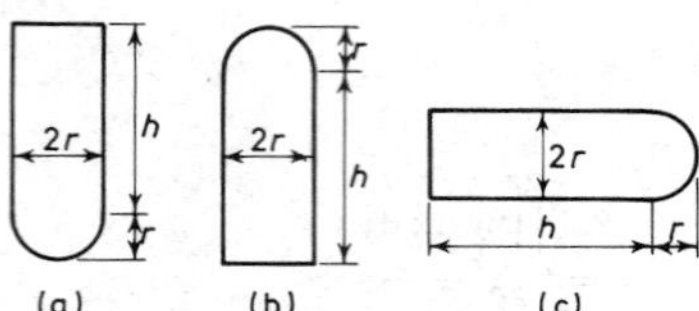

Figure 5.11 (a) *Cylindrical vessel with hemispherical end just full of water;* (b) *The same inverted;* (c) *The same lying with axis horizontal*

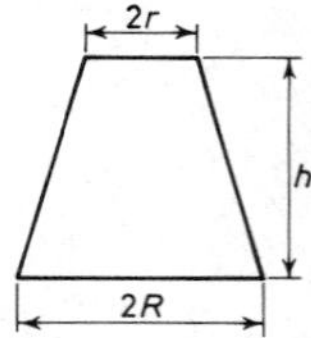

Figure 5.12 Truncated cone just full of water

Truncated cone, just full of water (Figure 5.12).

Resultant force (vertically downwards) on base $= \pi R^2 h\rho g$

Volume of water contained $= \frac{1}{3}\pi h(R^2+Rr+r^2)$

Resultant force (vertically upwards) on curved side $= [\pi R^2 h\rho - \frac{1}{3}\pi h(R^2+Rr+r^2)\rho]g$
$= (\frac{2}{3}\pi R^2 h - \frac{1}{3}\pi Rrh - \frac{1}{3}\pi r^2 h)\rho g$

BUOYANCY

A liquid of density ρ exerts a vertical upwards force $V\rho g$ on an immersed body of volume V (Figure 5.13). If the weight of the body is greater than $V\rho g$ it will sink. If the weight of the body is less than $V\rho g$ the body will float in such a way that the portion immersed has a volume V' which satisfies the following equation:

$$V'\rho g = \text{total weight of body} \tag{11}$$

Centre of buoyancy and metacentre. The centre of gravity of the displaced fluid is called the centre of buoyancy. When a body is floating freely, the weight of the fluid displaced equals the weight of the body itself, and its centre of buoyancy, for equilibrium, must be in the same vertical as the centre of gravity of the body. The degree of stability for angular displacements involves the conception of the metacentre. In Figure 5.14, XX represents

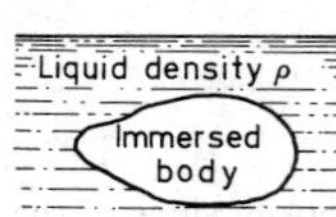

Figure 5.13 Body immersed in liquid

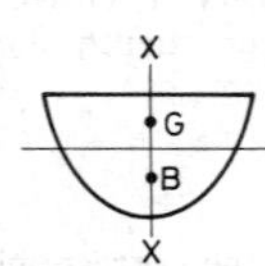

Figure 5.14 Centre of buoyancy

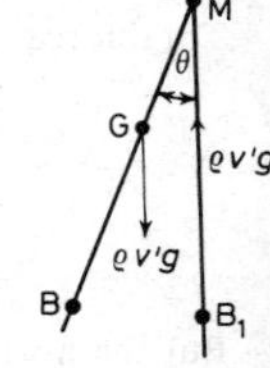

Figure 5.15 Metacentre

the vertical axis of symmetry of a floating body with a centre of gravity, owing to the distribution of its weight, we suppose to be at G. B is the centre of buoyancy.

Let the body be displaced so that B_1 becomes the new centre of buoyancy, i.e. the centre of gravity of the liquid as displaced in the new position (Figure 5.15).

The new force of buoyancy acts vertically upwards through B_1, to intersect the deflected line BG in M, the metacentre. Strictly speaking, the metacentre is the position assumed by M as the angle of displacement θ tends to zero.

If M is above G, there will be a 'righting moment' $\rho V' . GM \sin \theta . g$.

The condition for initial stability, or stability during small displacements, is then that M shall be above G.

The metacentric height may be calculated from the equation

$$GM = \frac{I}{V'} \pm GB \qquad (12)$$

where $I = Ak^2$: the plus sign is used if G is below B and the minus sign if G is above B.

A is the area of the water-line section and k is its radius of gyration about the axis Oy. V' is the volume of the immersed portion of body (shown in Figure 5.16).

HYDRODYNAMICS

ENERGY

A liquid possesses energy by virtue of the pressure under which it exists, its velocity and its height above some datum level of potential energy. These three forms of energy – pressure, kinetic and potential – may be expressed as quantities per unit weight of the liquid concerned. The result is the pressure, kinetic or potential head. Thus, referring to Figure

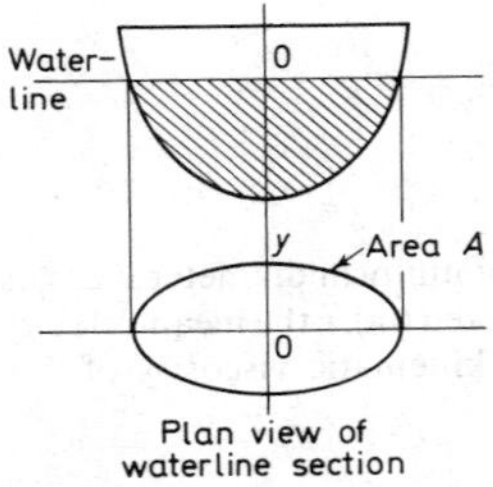

Figure 5.16 Metacentric height

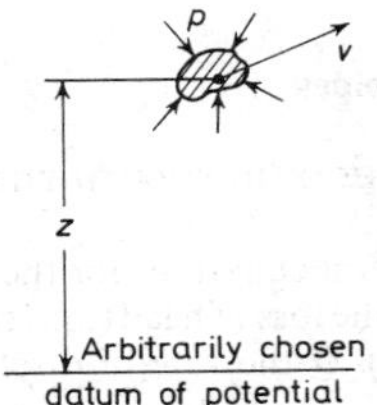

Figure 5.17 Mass of liquid subjected to pressure p N/m² *and moving with velocity v* m/s

5.17, in which a mass of liquid is represented as subjected to a pressure p N/m², moving with a velocity v m/s and having its centre of mass at a height z above a datum of potential energy,

$$\text{the total head} = \frac{p}{\rho g} + \frac{v^2}{2g} + z \qquad (13)$$

in metres, where ρ is the density in kg/m³.

If a gas is considered, then account should be taken of its elasticity and the work done in compressing a given mass of it as it passes from a region of low to higher pressure.

BERNOULLI'S THEOREM

This states that in the streamline motion of an incompressible and inviscid fluid the total head remains constant from section to section along the stream tube, i.e.

$$\frac{p}{\rho g} + \frac{v^2}{2g} + z = \text{constant} \qquad (14)$$

The idealised circumstances envisaged in this statement would seem, at first sight, to render the theorem useless for the solution of problems dealing with natural fluids, which are viscous, and especially when such fluids are not moving in streamlines, i.e. when the motion is turbulent, as it usually is in hydraulics. In fact, however, the theorem forms the basis of the majority of practical calculations if and when appropriate terms are added or coefficients introduced to allow for losses of head arising from various causes. The numerical values of these terms and coefficients are almost always the result of experiment or experience.

STREAMLINE AND TURBULENT MOTION

If we concentrate attention upon one point P in the cross section of a pipe or channel along which a fluid is moving at a constant rate, the motion at P may be called streamline if the velocity there is constant in magnitude and direction. On the other hand, the motion at P will be turbulent if the velocity there varies from time to time in magnitude and/or direction, despite the fact that the general rate of flow along the channel is constant. In this turbulent motion, the instantaneous velocity at P depends upon how the eddies are passing it at the moment under consideration. The eddies which characterise turbulent motion require energy for their creation and maintenance, and the law of resistance in streamline (sometimes called laminar) flow is quite different from that in turbulent flow.

Flow in pipes

LOSS OF HEAD IN SMOOTH PIPES

The general equations for the loss of head in a pipe of uniform diameter d are as follows. Let h be the loss of head (m), l the length of pipe considered (m), v the mean velocity (m/s) $= Q/(\pi d^2/4)$, Q the rate of discharge (m^3/s) and ν the kinematic viscosity of fluid (m^2/s). Then

$$h = K\frac{lv^{2-n}\nu^n}{gd^{3-n}} \quad \text{where } K \text{ is a coefficient} \tag{15}$$

Both K and n depend upon R_e the Reynolds number, vd/ν. If R_e is less than 2100, $K = 32$ and $n = 1$.

These values give the equation for streamline flow, which may be deduced mathematically:

$$h = \frac{32\nu lv}{gd^2} \tag{16}$$

Alternatively we may use the equation* commonly adopted by hydraulic engineers, i.e.

$$h = \frac{flv^2}{2gm} \tag{17}$$

in which m represents the hydraulic mean depth or the ratio of the area of section to the wetted perimeter. In a cylindrical pipe running full, $m = d/4$.

For values of R_e up to 2100:

$$f = 16\left(\frac{vd}{\nu}\right)^{-1} \tag{18}$$

* Some writers prefer to use h $= \lambda lv^2/2gd$, rather than $4flv^2/2gd$, for cylindrical pipes. Their friction factor λ is then $4f$.

For values of R_e between 3000 and 150 000 (Davis and White[1]):

$$f = 0.08\left(\frac{vd}{\nu}\right)^{-0.25} \tag{19}$$

Alternatively, if R_e exceeds 4000, the Prandtl equation may be used:

$$\frac{1}{\sqrt{(4f)}} = 2.0\ \lg\left[R_e\sqrt{(4f)}\right] - 0.8 \tag{20}$$

These relationships apply to smooth pipes of, say, glass, drawn brass, copper or large pipes with a smooth cement finish.

In calculating Reynolds numbers, the units in which v, d and ν are measured should be consistent with one another; for example, v in m/s, d in m, ν in m²/s. Values of f for

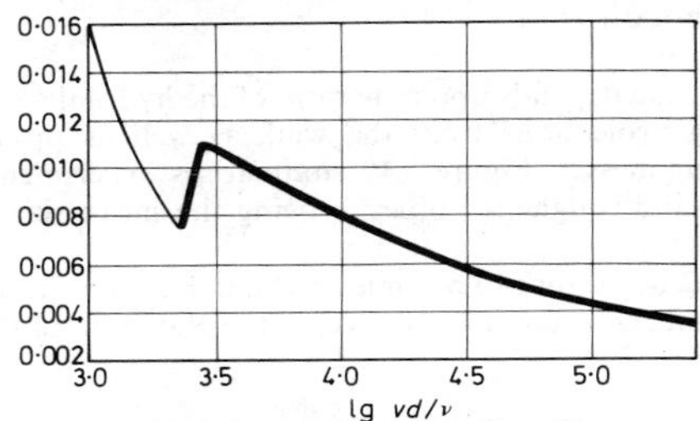

Figure 5.18 Values of f for smooth pipes (Stanton and Pannell[2])

smooth pipes in the equation $h = flv^2/2gm$ are plotted against $\lg vd/\nu$ in Figure 5.18 (after Stanton and Pannell[2]).

At a temperature of 15 °C, ν for water is 1.14×10^{-6} m²/s.

PIPES OF NONCIRCULAR SECTION

There is experimental evidence showing that if the flow is *turbulent*, the value of f for various shapes of section is approximately the same as for a cylindrical pipe at the same value of vm/ν, where m again represents the hydraulic mean depth.

The critical value of vm/ν, below which the motion is normally laminar, does depend to some extent, however, on the shape of the section. For a circular section it is 525 (i.e. $vd/\nu = 2100$). For rectangular sections the critical vm/ν varies with the ratio of the lengths of the sides and has approximate values of 525 for a square section, 590 for a section having one side three times the other, and 730 for a section in which the length of one side is large compared with the other.

During truly *laminar* or *viscous* motion, the loss of head h for various shapes of section is as follows (v being the mean velocity through the section):

Circular section (diameter d): $h = 32\nu lv/gd^2$

Rectangular section (one side $2a$, other side $2b$):

$$h = \frac{3\nu lv}{gb^2\left[1 - \frac{192}{\pi^5}\frac{b}{a}\left(\tanh\frac{\pi a}{2b} + \frac{1}{3^5}\tanh\frac{3\pi a}{2b} + \ldots\right)\right]}$$

Square section (each side $2a$):

$$h = \frac{7.12\nu lv}{ga^2}$$

Rectangular section having *a* large compared with *b*:

$$h \to \frac{3\nu l v}{g b^2}$$

Circular annulus (mean velocity v through space of area $\pi(d_1^2 - d_0^2)/4$):

$$h = \frac{32\nu l v}{g d_1^2 \left[1 + (d_0/d_1)^2 + \dfrac{1-(d_0/d_1)^2}{\ln(d_0/d_1)}\right]}$$

where d_1 is the outside diameter and d_0 the inside diameter.

LOSS OF HEAD IN ROUGH PIPES

Here the value of f also depends upon the ratio of the hydraulic mean depth to the height of the roughening projections from the wall, as well as upon the distribution and shape of these roughnesses. Figure 5.19 summarises experimental results obtained by Nikuradse[3] with sand-roughened pipes, k being the mean size of the grain projecting from the wall.

The general tendency is for f to become constant, for a given rough pipe, at sufficiently high Reynolds numbers. A commonly accepted explanation of this is that first of all the

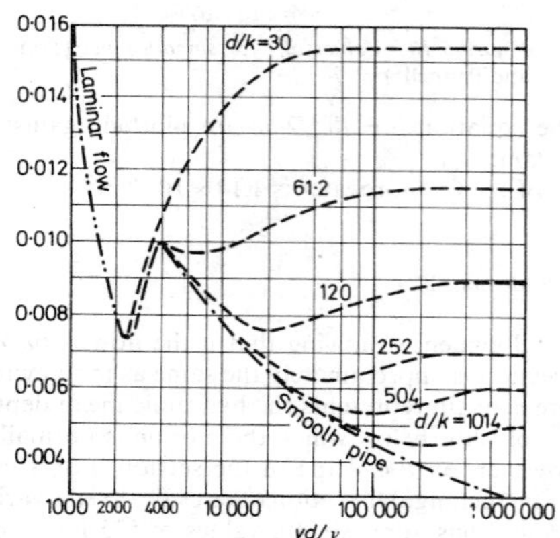

Figure 5.19 Values of f for rough pipes (Nikuradse[3])

surface grains lie inside a very thin viscous layer at the wall of the pipe, even when the main motion is turbulent; at higher values of R_e, however, they begin to project from this layer and to shed eddies for the maintenance of which additional energy is required.

Prandtl and von Kármán[4] have shown that Nikuradse's results may be made to lie within one band by plotting the quantity

$$\frac{1}{\sqrt{(4f)}} - 2\lg\left(\frac{d}{2k}\right)$$

against a new Reynolds number $V_* k/\nu$, in which V_* represents $v\sqrt{(f/2)}$.

Again, if $V_* k/\nu$ exceeds 60, f becomes constant for a given pipe, the flow then being 'fully turbulent' and the resistance proportional to v^2.

Under those conditions ($V_*k/\nu > 60$),

$$\frac{1}{f} = 16\left(\lg \frac{3.7d}{k}\right)^2 \tag{21}$$

A pipe may be regarded as 'hydraulically smooth' if $V_*k/\nu < 3$.

To take a specific example, namely, $d/k = 252$: for values of the original Reynolds number vd/ν less than about 11 500, f is the same as for a smooth pipe (see Figure 5.19), while if vd/ν exceeds 250 000, f assumes a constant value of 0.007. Between the two there is a curve which represents a transition and which covers a wide range of Reynolds numbers (11 500 to 250 000 in the example $d/k = 252$).

A large proportion of the cases which occur in engineering will be found to lie within the zone of transition, for which Colebrook and White[5a, b] have evolved the equation

$$\frac{1}{\sqrt{(4f)}} = -2\lg\left(\frac{k}{3.7d} + \frac{2.51}{R_e\sqrt{(4f)}}\right) \tag{22}$$

FORMULAE FOR CALCULATING PIPE FRICTION (TURBULENT FLOW)

With the velocities commonly encountered in water pipes, the motion is turbulent. These velocities in fact are frequently within the range 1.5 to 3.5 m/s, whereas in general the motion can only be expected to be streamline, considering water at ordinary temperatures, for velocities lower than $2.4/d$ m/s, where d is the pipe diameter (mm).

Manning's formula. Among the many formulae which have been suggested from time to time, that of Manning[6] is much favoured:

$$v = \frac{m^{2/3} i^{1/2}}{n} \quad \text{(m/s)} \tag{23}$$

In this form it applies to pipes and open channels.

m is the hydraulic mean depth (metres), i the virtual slope of the pipe (that is h/l) or the actual slope of the open channel under conditions of uniform flow. n depends upon the material of which the conduit is made.

Alternatively

$$v = \frac{m^{2/3} i^{1/2}}{100n} \quad \text{(m/s)} \tag{23a}$$

if m is expressed in mm.

For cylindrical pipes, the Manning formula (23) gives

$$h = \frac{n^2 l v^2}{m^{4/3}} \tag{24}$$

But

$$m = \frac{d}{4}$$

Hence

$$h = n^2 (4)^{4/3} \frac{l v^2}{d^{4/3}} = 6.35 n^2 \frac{l v^2}{d^{4/3}} \tag{25}$$

If we write this in the form

$$h = A \frac{l v^2}{d^{4/3}}$$

the following are appropriate values of A for new pipes, see Table 5.5.

Table 5.5

Material	n	$A = 6.35\,n^2$
Clean uncoated cast iron	0.013	0.001 1
Clean coated cast iron	0.012	0.000 92
Riveted steel	0.015	0.001 4
Galvanised iron	0.014	0.001 2
Brass, copper, or glass	0.010	0.000 64
Wood-stave	0.012	0.000 92
Smooth concrete	0.012	0.000 92
Cement mortar finish	0.013	0.001 1
Vitrified sewer pipe	0.011	0.000 77

Comparing the formulae

$$h = \frac{4flv^2}{2gd} \text{ and } h = \frac{Alv^2}{d^{4/3}}$$

it appears that

$$f = \frac{4.91}{d^{1/3}} A \tag{26}$$

or

$$f = \frac{31.2n^2}{d^{1/3}} \tag{27}$$

where d is in metres. If d is in mm,

$$f = \frac{49.1}{d^{1/3}} A = \frac{312n^2}{d^{1/3}} \tag{27a}$$

Hydraulic Research Papers, Nos 1 and 2 (Ackers[7]), first published by HM Stationery Office in 1958, contain not only a fascinating review of the resistance of fluids in channels and pipes but also tables and graphs for the use of designers. The results are derived from the formula of Colebrook and White (equation 22) and values of the effective roughness dimension k are quoted for a great variety of commercial pipes, while in addition a supplementary note provides information concerning the *actual* diameters of different classes of pipe in relation to their nominal bores.

DETERIORATION OF PIPES

Pipes deteriorate with age and to allow for this reduction in their carrying capacity Barnes[8] has suggested the following; see Table 5.6.

Table 5.6

Type of pipe	*Discharge for which to design, in terms of required discharge Q*
Uncoated cast iron	1.55 Q
Asphalted cast iron	1.45 Q
Asphalted riveted wrought iron or steel	1.33 Q
Wood-stave	1.08 Q
Neat cement or concrete	1.06 Q

None of these values can be at all precise, since the reduction of carrying capacity must depend not only upon the material but also upon the nature and velocity of the water and upon the diameter of the pipe; an increased roughness due to tuberculation will be more troublesome, proportionately, in small- than in large-diameter pipes.[9]

LOSSES OF HEAD IN PIPES DUE TO CAUSES OTHER THAN FRICTION

Sudden enlargement. With sufficient accuracy for practical purposes, loss due to sudden enlargement (Figure 5.20) = $(v_1 - v_2)^2/2g$ (28)

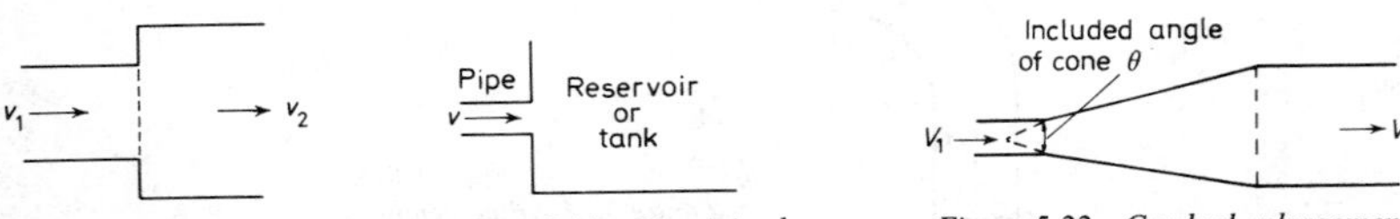

Figure 5.20 Loss of head due to sudden enlargement

Figure 5.21 Pipe joined to tank or reservoir

Figure 5.22 Gradual enlargement of pipe

If the enlarged section is very large, as when a pipe is joined to a tank or reservoir, Figure 5.21,

$$\text{loss} = \frac{v^2}{2g} \tag{29}$$

Gradual enlargement (Figure 5.22). Loss of head (including friction) may be taken as

$$\frac{k(v_1 - v_2)^2}{2g} \tag{30}$$

where k depends upon the angle of divergence θ in the following manner (Gibson[10]).

Table 5.7

θ (degrees)	2	5	10	15	20	40	60	90	120	180
k (circular pipe)	0.20	0.13	0.18	0.27	0.43	0.91	1.12	1.07	1.05	1.00

Sudden contraction. The loss due to sudden contraction is almost entirely due to the subsequent re-enlargement of the contracted stream (Figure 5.23). For practical purposes,

$$\text{loss} = \frac{1}{2}\frac{v^2}{2g} \tag{31}$$

and this may be taken as the immediate loss of head experienced as water flows from a reservoir into a pipe in which it attains a velocity v.

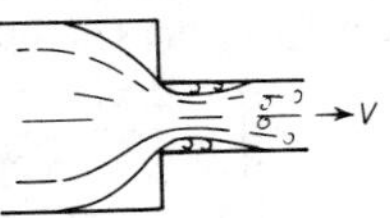

Figure 5.23 Loss of head due to sudden contraction

With a reasonably rounded entrance, the loss may be reduced to about

$$\frac{1}{20}\frac{v^2}{2g}$$

If the pipe has a sharp entrance but projects into the reservoir and forms a re-entrant mouthpiece, the loss of head at the entrance is approximately 0.80 $v^2/2g$ (assuming that the pipe runs full).

LOSSES AT PIPE BENDS

Owing to the many variables involved (e.g. size of pipe, radius of bend, velocity of flow), it is impracticable at present to generalise with any degree of certainty, but the following data may be helpful in ordinary calculations (Figure 5.24)

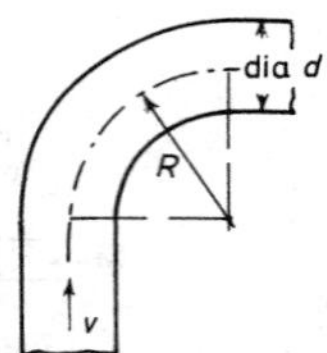

Figure 5.24 Pipe bend

Defining $Kv^2/2g$ as the loss in excess of that which would arise from friction in the same length of straight pipe, then approximate values of K are:

$R/d = 1$; $K = 0.50$ for either 90° or 180° bends;
$R/d = 2$ to 8; $K = 0.30$ for 90° and $K = 0.35$ for 180° bends.

For 90° elbows, $K = 0.75$ and for square, or sharp, elbows, $K = 1.25$.

The motion round the bend tends to take on the characteristics of a free vortex, having a greater velocity at the inside than at the outside. Correspondingly, the pressure at the inside is less than at the outside. Consider a section half-way round the bend and let $d = 2r$. If the discharge (volume per second) is Q, the velocities at the inside and outside of the bend are approximately

$$v_i = \frac{Q}{2\pi(n-1)r^2[n-(n^2-1)^{1/2}]}$$

$$v_o = \frac{Q}{2\pi(n+1)r^2[n-(n^2-1)^{1/2}]}$$

where

$$n = R/r.$$

Correspondingly, the effect of the free vortex itself is to make the pressure head at the outside of the bend exceed that at the inside by an amount $(v_i^2 - v_o^2)/2g$.

LOSSES AT VALVES

These depend, of course, upon the relative size and design, but the order of magnitude involved may be judged from Gibson's experiments[10] for

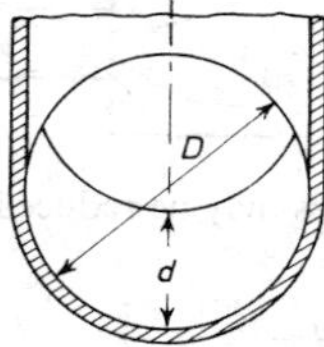

Figure 5.25 Circular sluice gate valve

1. Circular sluice gate valves (Figure 5.25)

$$\text{Loss} = F\frac{v^2}{2g} \text{ where } v = \text{velocity in pipe} \tag{32}$$

Table 5.8

	F for d/D						
	0.2	0.3	0.4	0.5	0.6	0.8	1.0
D = 2 in	30	11	4.2	2.1	0.9	0.22	0
D = 24 in	36	11	3.0	1.6	1.0	—	—

2. Butterfly valve (Figure 5.26)

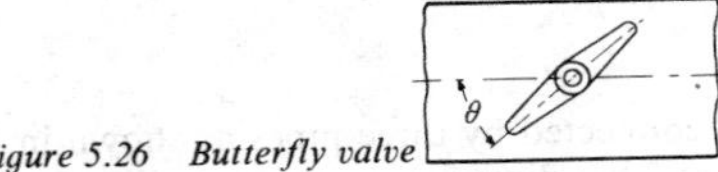

Figure 5.26 Butterfly valve

Table 5.9

θ	5°	10°	20°	30°	40°	50°	60°	70°
F	0.24	0.52	1.54	3.9	10.8	32.6	118	751

GRAPHICAL REPRESENTATION OF PIPE FLOW PROBLEMS

This may be illustrated by the example of a pipeline joining two reservoirs, Figure 5.27, and for the sake of the example we will suppose that the pipe is enlarged somewhere along its length. Consider a particle of water to find its way, in effect, from the surface in the upper reservoir to that in the lower. The velocity of fall of the upper surface, or of rise of the lower one, may be treated as negligible in comparison with the velocity through the pipe.

Accordingly, the particle starting in surface *A* has originally no kinetic head. Taking atmospheric pressure as the datum of reference for pressure energy, it has also no pressure

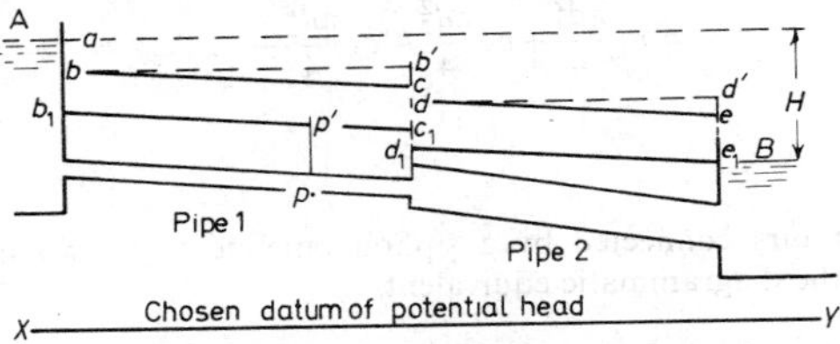

Figure 5.27 Graphical representation of pipe flow

head. Its head is, in fact, entirely potential and may be represented diagrammatically by the height of *A* above any arbitrarily chosen datum of potential, XY. As the water enters the pipe there is a loss *ab* due to the sudden contraction: this is followed by a friction loss *b′c* in pipe 1. Then comes the loss *cd* caused by the enlargement and this is followed by the friction loss *d′e* in the second length of pipe.

The height of the line *abcde* above XY therefore represents the total head.

Now drop *bc* by a distance bb_1, representing the kinetic head $v_1^2/2g$ in pipe 1, and *de* by a distance dd_1, representing the kinetic head $v_2^2/2g$ in pipe 2.

The height of $ab_1c_1d_1e_1$ above XY must then represent total head minus kinetic head, i.e. pressure plus potential head. $b_1c_1d_1e_1$ is then the line of virtual slope or hydraulic gradient of the pipe, and its height above the centreline of the pipe gives the pressure head in the pipe. Thus, Pp' is the pressure head at P. If p' were below P, it would follow that the pressure in the pipe at P was below atmospheric and one of the advantages of this graphical method is to reveal clearly such points of suction.

Further, considering the bottom end of the pipeline, if the pipe 2 has a sharp, or flanged, connection to the reservoir, the whole of the kinetic head $v_2^2/2g$ will be lost in the creation of eddies at the enlargement. The point e_1 will then define the surface level in the lower reservoir, the total head under which flow is taking place being H. With a gradual enlargement, joining in a curved bell mouth to the lower reservoir, a proportion of $v_2^2/2g$, up to possibly $\frac{3}{4}(v_2^2/2g)$ might be regained and the level in B would be correspondingly higher than e_1 by this, say, $\frac{3}{4}(v_2^2/2g)$ amount of reconverted kinetic head.

PIPES IN PARALLEL

Consider two reservoirs connected by three pipes as shown in Figure 5.28. Pipe 1 is of diameter d_1, length l_1. Pipe 2 has diameter d_2, length l_2. Pipe 3 is of diameter d_3, length l_3. H is the loss of head in any one of the pipes. Neglecting end effects

$$H = \frac{4f_1l_1v_1^2}{2gd_1} = \frac{4f_2l_2v_2^2}{2gd_2} = \frac{4f_3l_3v_3^2}{2gd_3}, \text{ etc.} \tag{33}$$

or, in the Manning form,

$$H = \frac{A_1l_1v_1^2}{d_1^{4/3}} = \frac{A_2l_2v_2^2}{d_2^{4/3}} = \frac{A_3l_3v_3^2}{d_3^{4/3}}, \text{ etc} \tag{34}$$

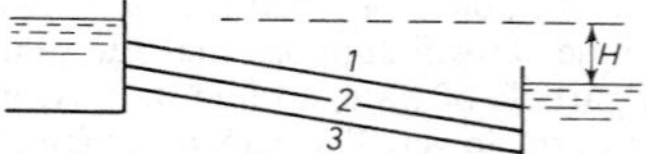

Figure 5.28 Pipes in parallel

The total rate of flow along the pipes

$$= v_1\frac{\pi d_1^2}{4} + v_2\frac{\pi d_2^2}{4} + v_3\frac{\pi d_3^2}{4} + \ldots \tag{35}$$

SIPHONS

Consider two reservoirs connected by a siphon pipeline as shown in Figure 5.29a, Figure 5.29b being the diagrammatic equivalent.

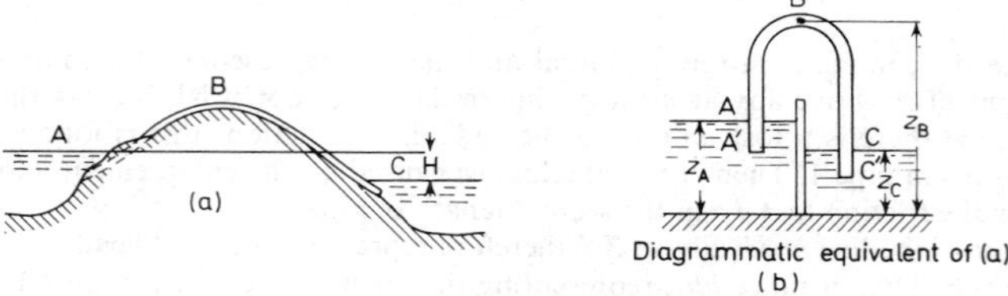

Figure 5.29 The siphon

If v is the velocity through the siphon, then H = loss at entry+friction loss+loss at exit, i.e.

$$H = 0.80\frac{v^2}{2g} + \text{friction loss} + \frac{v^2}{2g}$$

for submerged sharp entrance and exit ends of pipe. The end effects are negligible in a reasonably long pipe.

Let F_1 be the friction loss of head in portion A′B of length l_1. Then pressure head at crown B is

$$\frac{p_B}{\rho g} = z_A - z_B - \left(F_1 + 0.80\frac{v^2}{2g} + \frac{v^2}{2g}\right) \tag{36}$$

assuming a loss of $0.80\frac{v^2}{2g}$ at A′.

Hence p_B is negative, i.e. less than atmospheric.

In practice, to avoid undue difficulty arising from accumulation of air and vapour at B, the numerical value of $p_B/\rho g$ should not exceed 28 ft (8.5 m) of water, an absolute pressure of about 6 ft (*c.* 2.0 m) of water being then left at the crown.

The analysis giving equation (36) assumes that the pressure over the pipe-section at B is uniform. If the curvature of the pipe is pronounced, however (as it frequently is in the case of siphon-spillways), the free-vortex phenomenon mentioned earlier (under the heading of 'Losses at pipe bends') becomes important. The difference of pressure between the inside and the outside of the bend has already been given for a *circular section.* In the case of a *rectangular section* having a breadth b, an inside radius r_i and an outer radius r_0, the pressure head at the outside of the bend exceeds that at the inside by an amount, due to the free vortex alone,

$$\frac{K^2}{2g}\left(\frac{1}{r_1^2} - \frac{1}{r_0^2}\right)$$

where $K = Q/b \ln r_0/r_1$ and Q is the volume flowing per second.

Difficulties may arise, therefore, in bends of severe curvature, through high suction at the inside of the bend.

NOZZLE AT END OF A PIPELINE (FIGURE 5.30)

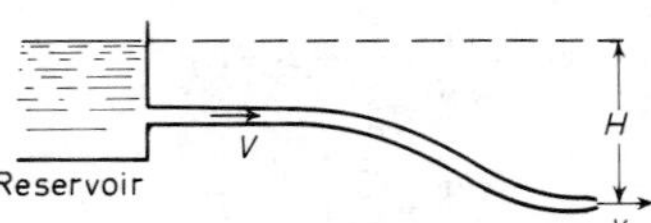

Figure 5.30 Nozzle at end of a pipeline

Let H be the total head (m), V the velocity in the pipe (m/s), v the velocity in the jet (m/s), D the diameter of the pipe (m) and d the diameter of the nozzle end (m).

$$\text{Discharge} = v\frac{\pi d^2}{4}\ \text{m}^3/\text{s}. \qquad \text{Then } v = V\left(\frac{D}{d}\right)^2$$

Effective head behind nozzle = H − loss at entrance to pipe − friction in pipe. Hence $v^2/2g = H$ − friction head in pipe, neglecting entrance effect, or, more precisely,

$$v = C_N[2g(H-F)]^{1/2} \tag{37}$$

where F is the friction loss of head in pipe and C_N the coefficient of the nozzle.

Although C_N depends upon the Reynolds number vd/ν, for practical purposes it is usually sufficiently accurate to give it the value 0.98, either for elaborately streamlined nozzles or for a straight taper form ending in a cylindrical portion of length $d/2$ and diameter d.

Writing F in the form $F = 4flV^2/2gD$,

$$v^2 = \frac{2gC_N^2H}{1+[4flC_N^2/D(D/d)^4]} \tag{38}$$

The energy delivered in the jet $= \rho a v^3/2$ N m/s, where ρ is the density of water ($= 1000$ kg/m^3) and $a = \pi d^2/4$.

This energy is a maximum, for a given H, if

$$d^2 = \frac{D^2}{C_N}\sqrt{\left(\frac{D}{8fl}\right)} \tag{39}$$

The velocities are then such that the friction head F is very nearly equal to $H/3$.

MULTIPLE PIPES

Consider the example shown in Figure 5.31, in which it is desired to calculate the flow along the three pipes.

Height of A, B, C, J above some chosen datum of potential

$= z_A, z_B, z_C, z_J$ respectively.

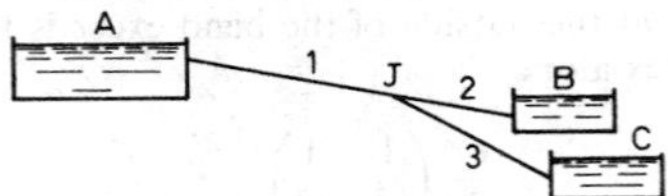

Figure 5.31 Multiple pipes

Neglecting losses other than those due to pipe friction F,

$$z_A = \frac{p_J}{\rho g}+\frac{v_1^2}{2g}+z_J+F_1 \tag{40}$$

$$z_B = \frac{p_J}{\rho g}+\frac{v_2^2}{2g}+z_J-F_2 \tag{41}$$

$$z_C = \frac{p_J}{\rho g}+\frac{v_3^2}{2g}+z_J-F_3 \tag{42}$$

Also

$$v_1\frac{\pi d_1^2}{4} = v_2\frac{\pi d_2^2}{4}+v_3\frac{\pi d_3^2}{4} \tag{43}$$

Calling

$$F_1 = \frac{4f_1l_1v_1^2}{2gd_1};\ F_2 = \frac{4f_2l_2v_2^2}{2gd_2};\ F_3 = \frac{4f_3l_3v_3^2}{2gd_3}$$

and assigning values to f_1, f_2, f_3, equations (40), (41), (42) and (43) are sufficient for the determination of the four unknowns p_J, v_1, v_2, v_3, and if necessary the solutions may be modified by further calculation should the resulting v suggest values of f_1, f_2, f_3 different from those originally assumed.

The solution of (in this example) four simultaneous equations is, however, cumbersome and full of possibilities of arithmetical slips. A simpler and quicker method[11] is as follows.

If *assumed* total head difference between two ends of a pipe = h, then

$$h = \frac{4flv^2}{2gd} = \frac{4fl}{2gd}\left(\frac{Q}{a}\right)^2 = KQ^2 \tag{44}$$

where Q is the rate of flow = va, a is area of section = $\pi d^2/4$, and $K = 2fl/gda^2$.

If the correct values of h and Q are $h+\delta h$ and $Q+\delta Q$, then $\delta h = 2h.\ \delta Q/Q$ to a first approximation.

Now the error δQ for any one pipe is unknown, but the sum of the errors, $\Sigma(\delta Q)$, is known, being equal to the unbalanced flow, and

$$\delta h = \frac{\Sigma(\delta Q)}{\Sigma(Q/2h)} \tag{45}$$

Example. $z_A = 30.50$; $z_J = 12.20$; $z_B = 6.10$; $z_C = 3.05$, all in metres.

(These might be levels with reference to ordnance datum, say).

Pipe 1 3.00 km long, dia. 60.0 cm, $f = 0.007$
Pipe 2 1.50 km long, dia. 30.0 cm, $f = 0.007$
Pipe 3 1.50 km long, dia. 30.0 cm, $f = 0.007$

Let

$$H = \frac{p}{\rho g} + \frac{v^2}{2g} + z$$

Steps in solution

1. Assign some value to the total head at J. Evidently it must be less than that at A, which is 30.5 m. Hence we might first try $H_J = 20$ m, say.
2. It then follows that

$$F_1 = (4 \times 0.007 \times 3000v_1^2)/(2 \times 9.81 \times 0.600) = 30.5 - 20.0 = 10.5$$

Hence

$$v_1 = \sqrt{(1.47)} = 1.21 \text{ m/s}$$
$$Q_1 = 0.342 \text{ m}^3\text{/s}$$

3. Similarly $F_2 = 20.0 - 6.1 = 13.9$
therefore

$$\frac{4 \times 0.007 \times 1500v_2^2}{2 \times 9.81 \times 0.300} = 13.9$$

$$v_2 = 1.40 \text{ m/s}$$
$$Q_2 = 0.099\,0 \text{ m}^3\text{/s}$$

$F_3 = 20.0 - 3.05 = 16.95$
therefore

$$\frac{4 \times 0.007 \times 1500v_3^2}{2 \times 9.81 \times 0.300} = 16.95$$

$$v_3 = 1.54 \text{ m/s}$$
$$Q_3 = 0.109 \text{ m}^3\text{/s}$$

Hence our original assumption (that total head at J = 20 m) has led to an out-of-balance flow of $0.342-(0.099\,0+0.109)$, or $0.134 \text{ m}^3\text{/s}$. Hence

$$\Sigma\delta Q = 0.134$$

4. But $\frac{Q}{2h}$ for pipe 1 $= \frac{0.342}{2 \times 10.5} = 0.016\,3$ therefore $\Sigma\left(\frac{Q}{2h}\right) = 0.023\,1$

But $\frac{Q}{2h}$ for pipe 2 $= \frac{0.099\,0}{27.8} = 0.003\,57$

But $\frac{Q}{2h}$ for pipe 3 $= \frac{0.109}{33.9} = 0.003\,22$ So that $\frac{\Sigma\delta Q}{\Sigma(Q/2h)} = \frac{0.134}{0.023\,1} = 5.80$

5. Now it is evident that we must aim at decreasing our original estimate of Q_1 while increasing those of Q_2 and Q_3.
We now try total head at J: $20.0+5.8 = 25.8$ m

$$\text{Our new estimate of } Q_1 = 0.342 \times \sqrt{\left(\frac{30.5-25.8}{30.5-20.0}\right)} = 0.229 \text{ m}^3/\text{s}$$

$$\text{Our new estimate of } Q_2 = 0.099\,0 \times \sqrt{\left(\frac{19.7}{13.9}\right)} = 0.118 \text{ m}^3/\text{s}$$

$$\text{Our new estimate of } Q_3 = 0.109 \times \sqrt{\left(\frac{22.75}{16.95}\right)} = 0.126 \text{ m}^3/\text{s}$$

and

$$\Sigma\delta Q = 0.118+0.126-0.229 = 0.015$$

$$\frac{Q}{2h} = \frac{0.229}{9.40} = 0.0244 \text{ for pipe 1}$$

$$\frac{Q}{2h} = \frac{0.118}{39.4} = 0.0030 \text{ for pipe 2}$$

$$\frac{Q}{2h} = \frac{0.126}{45.5} = 0.0028 \text{ for pipe 3}$$

Therefore

$$\Sigma\left(\frac{Q}{2h}\right) = 0.030\,2$$

$$\frac{\Sigma\delta Q}{\Sigma(Q/2h)} = \frac{0.015}{0.030\,2} = 0.497$$

6. Now make total head at J: $25.8-0.50 = 25.3$

$$Q_1 \text{ then} = 0.342 \times \sqrt{\left(\frac{5.20}{10.5}\right)} = 0.241$$

$$\left.\begin{aligned} Q_2 \text{ then} &= 0.099\,0 \times \sqrt{\left(\frac{19.2}{13.9}\right)} = 0.116 \\ Q_3 \text{ then} &= 0.109 \times \sqrt{\left(\frac{22.25}{16.95}\right)} = 0.125 \end{aligned}\right\} 0.024\,1$$

The flows are now in balance; the number of steps required in the process of successive approximation depends on the accuracy of the original guess at the total head of J.

Having obtained a sensibly accurate balance, we may if we choose carry out refined calculations based upon more acceptable values of f and including losses due to other causes such as the junction J, any bends, and so forth.

The example considered above is, however, comparatively simple. The more complicated cases frequently encountered in practice are nowadays analysed with the aid of analogue or digital computers.[12, 13, 14, 15]

Flow measurement in pipes

THE VENTURI METER

The proportions shown in Figure 5.32 are not essential, but are fairly representative; the gently tapering divergent portion following the convergent tube (which is the real meter) is intended to minimise the overall loss of head due to eddies and friction between 1 and 3.

$$Q = \text{rate of flow} = va = C\frac{\pi d^2}{4}\left[\frac{2g(h_1-h_2)}{(a_1/a_2)^2-1}\right]^{1/2} \tag{46}$$

where $h_1(=p_1/\rho g)$ and $h_2(=p_2/\rho g)$ are the pressure heads at sections 1 and 2 respectively.

If the throat diameter is too small, the pressure p_2 may be so low as to encourage release of air and vapour which will cause the flow to fluctuate and will introduce an uncertainty.

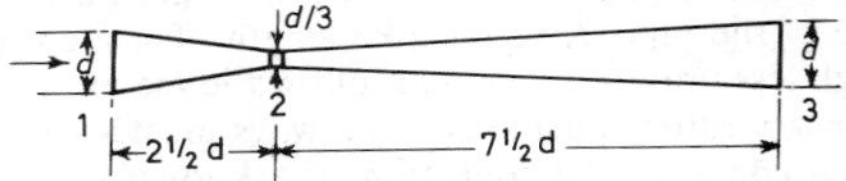

Figure 5.32 The Venturi meter

To avoid this, it is advisable to use proportions which will not cause p_2 to be less than 3 lbf/in^2 (21×10^3 N/m^2); that is h_2 not less than 2.1 m of water (absolute).

Values of C. The value of C, the coefficient of the meter, is influenced by the Reynolds number and for the sort of designs of meter commonly used in practice the following are approximately correct values[16] (Table 5.10).

Table 5.10

Reynolds number vd/ν as measured at throat	*C*
2 000	0.91
6 000	0.94
10 000	0.95
100 000	0.98
1 000 000	0.99

PIPE ORIFICE AS METER

The pipe orifice as a measuring device is conveniently installed at a flange-joint but has the disadvantage of creating an appreciable obstruction and consequent loss of head. In Figure 5.33, a and b are pressure tappings.

$$Q = CA\sqrt{\frac{2gh}{(D/d)^4-1}} \tag{47}$$

where $A=\pi D^2/4$, and h is the pressure head difference between points a, b. C is of the order of 0.61 for values of $(d/D)^2$ between 0.3 and 0.6, but it should be noted that the accuracy of the machining is of great importance, since the quantity $(D/d)^4$ occurs in the

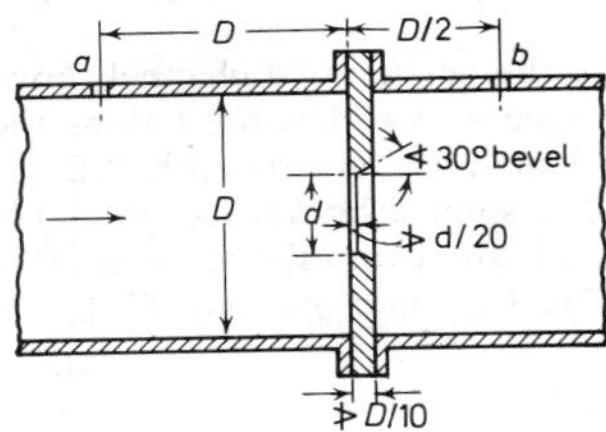

Figure 5.33 The pipe orifice as a meter

formula. Similarly it is important to have a high degree of accuracy in the measurement of the diameter D of the pipe in which the plate is installed.

If a well shaped convergent nozzle is used instead of a sharp edged orifice plate, C has a value of 0.98 to 0.99 if $(d/D)^2$ does not exceed 0.2.

GENERAL NOTES ON METERS

The pressure holes used for meters or other purposes should be finished flush with the inside of the pipe. A reasonable length of straight pipe should precede the meter, and, though less important, should follow the meter.

For laboratory purposes it is always most satisfying to calibrate any meter *in situ* by comparison with the collection of a known weight of water in a measured time, but considerable accuracy may be expected from observance of the recommendations in the British Standard Code,[17] BS 1042, 1943, which covers Venturi tubes, orifice plates and nozzles, and pitot tubes: it deals with gases as well as liquids. The U.S. Standard[18] is *ASME Fluid Meters Report.*

WATER HAMMER IN PIPES

If a valve is closed suddenly, successive masses of the water in the pipe are brought to rest; their kinetic energy is converted to strain energy and the effect is transmitted along the pipe with the velocity of sound waves in water. Some energy is in fact expended in stretching the pipe walls, thus reducing the water hammer pressure, but if this effect is neglected the rise of pressure p at the valve is given by

$$p = v\sqrt{(K\rho)} \qquad (\mathrm{N/m^2}) \tag{48}$$

where v is the velocity of flow before the valve is closed (m/s), K the bulk modulus of compressibility of the water, equal to about 2×10^9 N/m² and ρ is the density of water, 1000 kg/m³.

This formula leads to the result

$$\text{water hammer pressure} = 1.4\times10^6\, v \qquad (\mathrm{N/m^2}) \tag{49}$$

v being the original velocity of flow (m/s).

Pressures of this order of magnitude will result if a valve is closed in a time not exceeding $2l/V_p$ s, where $V_p = \sqrt{(K/\rho)}$ is the velocity of sound waves in the water and where l is the length of pipe (m). In round numbers, V_p may be taken as 1400 m/s.

If the time of closure exceeds $4l/V_p$ the stoppage becomes gradual. Supposing the valve to be then closed in such a manner as to cause a constant retardation α m/s² of the water column, the resulting rise of pressure will be $\rho l\alpha$ N/m², or $l\alpha/g$ m head of water.

Flow in open channels

FORMULAE FOR OPEN CHANNELS

Consider the portion of a channel shown in Figure 5.34. AB represents the surface of a stream: section A is distance l along the channel, section B a distance $l+\delta l$ along. h is the depth at A, $h+\delta h$ the depth at B. v is the mean velocity at A, r the depth of surface at A below some arbitrary datum and $(r+\delta r)$ the depth of surface at B below the same datum. m is the hydraulic mean depth, equal to the ratio of area to wetted perimeter, and f is the friction coefficient. Then

$$\frac{\mathrm{d}r}{\mathrm{d}l} = \frac{v}{g}\frac{\mathrm{d}v}{\mathrm{d}l} + \frac{fv^2}{2gm} \tag{50}$$

If the flow is uniform and the mean velocity constant from section to section along a channel of constant cross section, this assumes the familiar form

$$\begin{aligned} i &= \text{slope of bed} = \text{slope of water surface} \\ &= \text{fall per unit length} \\ &= \frac{fv^2}{2gm} \end{aligned} \tag{51}$$

alternatively, as in the Chézy equation, equation (51) can be written

$$v = c\sqrt{(mi)} \tag{52}$$

where c is known as the Chézy coefficient and is related to the friction coefficient f in the formula:

$$\text{friction head} = flv^2/2gm \quad \text{by} \quad c^2 = \frac{2g}{f} \tag{53}$$

The numerical value of c depends on the units adopted; it has the units of $[L^{\frac{1}{2}}]/[T]$. Consequently, in the ft s system, c has the units of $\text{ft}^{\frac{1}{2}}/\text{s}$, whereas in the m s system it is measured in $\text{m}^{\frac{1}{2}}/\text{s}$. On the other hand, f is dimensionless: it has the same numerical value in either system.

Chézy's c depends upon the nature of the channel and also upon the hydraulic mean depth of a channel of given material. To some extent it also depends upon the mean

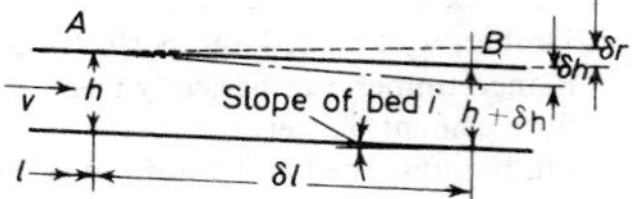

Figure 5.34 Portion of open channel

velocity of the stream, although in most practical examples of open channel flow with which the engineer is concerned, this effect is of minor importance.

Although old fashioned in the sense that it dates back to the last century, a formula due to Bazin[19] is very reliable:

$$c = \frac{158}{1.81 + \text{N}/\sqrt{\text{m}}} \qquad (\text{m}^{\frac{1}{2}}/\text{s}) \tag{54}$$

the hydraulic mean depth m being measured in metres and N having the values given in Table 5.11.

Table 5.11

Class	*N*	*Application*
I	0.109	Smoothed cement or planed wood
II	0.290	Planks, bricks or cut stone
III	0.833	Rubble masonry
IV	1.54	Earth channels of very regular surface, or revetted with stone
V	2.35	Ordinary earth channels
VI	3.17	Exceptionally rough earth channels (bed covered with boulders) or weed-grown sides

As one of the many proposed alternatives to the Chézy-Bazin treatment ($v = c\sqrt{(mi)}$ where $c = [158/(1.81 + \text{N}/\sqrt{\text{m}})]$), the formula due to Manning[6] is much favoured and regarded by many as more convenient, though giving much the same result.

For the classes of channel already described in Table 5.11 in connection with Bazin's N, Manning's n may be taken as given in Table 5.12.

$$v = \frac{m^{2/3} i^{1/2}}{n} \quad \text{(m/s)} \tag{23}$$

where m is in metres.

With rather more precision, the values of n given by Parker[20] are quoted in Table 5.13.

Table 5.12

Class	n
I	0.009 3
II	0.012 9
III	0.018 2
IV	0.022 5
V	0.025 8
VI	0.028 4

Table 5.13

Nature of channel	n
Timber, well planed and perfectly continuous	0.009
Planed timber, not perfectly true	0.010
Pure cement plaster	0.010
Timber unplaned and continuous; new brickwork	0.012
Rubble masonry in cement, in good order	0.017
Earthen channels in faultless condition	0.017
Earthen channels in very good order or heavily silted in the past	0.018
Large earthen channels maintained with care	0.0225
Small earthen channels maintained with care	0.025
Channels in order, below the average	0.0275
Channels in bad order	0.030

Note that by comparing $v = c\sqrt{(mi)}$ with $v = (m^{2/3} i^{1/2})/n$ we may obtain the result

$$c = \frac{m^{1/6}}{n}$$

or

$$f = \frac{2g}{c^2} = \frac{19.6n^2}{m^{1/3}} \tag{55}$$

Incidentally the formula $c = 20.7 + 17.7 \lg m/k$ $\text{m}^{\frac{1}{2}}/\text{s}$ covers a remarkably wide range of both rough pipes and rough open channels.[21] Here k represents the size of roughening excrescences.

FORM OF CHANNEL FOR MAXIMUM V AND Q

Q = rate of discharge (m^3/s) $= vA$, where A is now the area of section (m^2) and v is the mean velocity (m/s).

Adopting the Manning formula

$$Q = vA = \frac{A m^{2/3} i^{1/2}}{n} \tag{23a}$$

for a given material of channel, and a given slope i, v is a max when m, or when A/P, is a maximum, P being the wetted perimeter.

For Q to be a maximum, however, $Am^{2/3}$ must be a maximum. Hence,

$$\left.\begin{aligned} \text{for max } v,\ P\,dA - A\,dP &= 0 \\ \text{for max } Q,\ 5P\,dA - 2A\,dP &= 0 \end{aligned}\right\} \tag{56}$$

Examples. *Rectangular channel* (Figure 5.35). $A = bd$; $P = b + 2d$. If A, n and i are fixed, max v and max Q will occur when $b = 2d$.

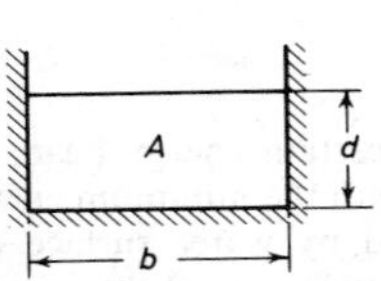

Figure 5.35 Rectangular channel

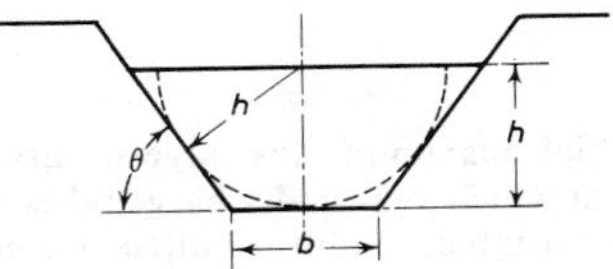

Figure 5.36 Trapezoidal channel

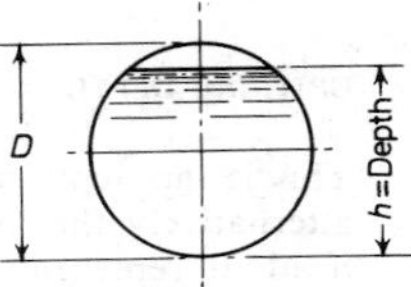

Figure 5.37 Circular channel

Trapezoidal channel (Figure 5.36). For max v and max Q with given A,

$$\sqrt{(1+s^2)} = \frac{b+2sh}{2h}, \quad \text{where} \quad \tan\theta = 1/s$$

This is satisfied if a semicircle can be drawn, centred in the water surface and touching both sides and bottom.

Circular channel (Figure 5.37). For max v, $h = 0.813D$; for max Q, $h = 0.938D$.

RESISTANCE OF NATURAL RIVER CHANNELS

This is complicated by the losses of energy at bends and at relatively sudden changes of cross-sectional area. As these depend on the precise dimensions and shapes, it is quite impossible to generalise, but they are nevertheless important. For example, it has been shown that in an eight mile tortuous stretch of the river Mersey[22] the textural roughness of the bed and sides accounts for only 25 to 50% of the total loss of head depending on the rate of flow, the rest of the resistance being due principally to the bends.

Somewhat similar conclusions have been reached in a study of the river Irwell.[23]

VELOCITY DISTRIBUTION IN OPEN CHANNELS

Side and bottom friction cause the stream to be retarded. The highest velocity in any vertical at a particular section is usually found some distance below the surface; the mean velocity in a vertical line occurs at about 6/10 of the depth, whether the wind is blowing up or downstream. This is the basis of one method of stream-gauging, in which the section is considered divided into strips of equal width and the velocity in these strips, or panels, is measured by current meter at 6/10 of the depth of each individual panel (Figure 5.38). The area of a strip, as found by sounding the bed, multiplied by the velocity so measured

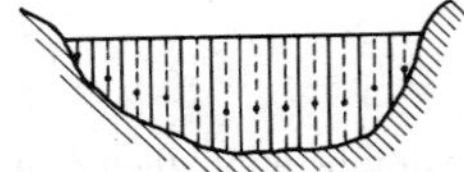

Figure 5.38 Measurement of velocity in open channel

is assumed to give the flow through the strip and the addition for the total number of strips gives the flow through the whole section.

ENERGY OF A STREAM IN AN OPEN CHANNEL, OR 'SPECIFIC ENERGY'

If D is the depth and v the mean velocity, the energy head H_e, taking atmospheric pressure as the datum of pressure and the bottom of the channel as the datum of potential, is $D+v^2/2g$, or $D+Q^2/2gA^2$.

CRITICAL DEPTH

This is the depth at which maximum discharge occurs for a certain energy head, or, alternatively, the depth at which a given discharge takes place with the minimum energy head. It represents an unstable condition, often accompanied by water surface undulations. Under these conditions $D = v^2/g$ in a rectangular channel and Q then equals $(0.544b\sqrt{g})H_e^{3/2}$.

NONUNIFORM FLOW IN OPEN CHANNELS

If h is the depth at a distance l along the channel with slope i

$$\frac{dh}{dl} = \frac{i-(fv^2/2gm)}{1-(v^2/gh)} \tag{57}$$

This condition of varying depth, even in a stream of constant width and constant rate of discharge, as here assumed, may be brought about by obstructions or irregularities.

h is now the depth actually found at a particular section, whereas H may be called the depth which would apply under conditions of uniform flow corresponding to the simple equation (52). In other words, h becomes equal to H if $dh/dl = 0$.

Suppose now, in order to examine general trends, that the width of the stream is large compared with its depth. In such a case the hydraulic mean depth m is approximately the same as h, at any rate in channels having approximately uniform depth across their width. Then

$$\frac{dh}{dl} = \frac{i-(fv^2/2gh)}{1-(v^2/gh)}$$

and $Q = vbh$. Q is also equal to VbH, where V and H are the velocity and depth which would be obtained with uniform flow.

It then follows that

$$\frac{dh}{dl} = \frac{i[1-(H/h)^3]}{1-(2i/f)(H/h)^3} \tag{58}$$

SPECIAL CASES OF NONUNIFORM FLOW

1. *Sluice gate in channel with small slope and/or rough bed*

$2i/f<1$, $h^3<(2i/f)H^3$ (Figure 5.39). dh/dl becomes infinite when

$$h = \left(\frac{2i}{f}\right)^{1/3} H \tag{59}$$

An 'hydraulic jump' then results.

2. *Sluice gate in channel with steep slope and/or smooth bed*
$2i/f > 1$, $h < H$. dh/dl is again positive, but tends to zero, i.e. the depth increases gradually until it reaches that appropriate to uniform flow (Figure 5.40).

3. *Weir or dam in channel with small slope and/or rough bed*
$2i/f < 1$, $h > H$ (Figure 5.41).

$$h_1 - h_2 = i(l_1 - l_2) + H\left(1 - \frac{2i}{f}\right)[\phi(y_1) - \phi(y_2)] \tag{60}$$

where $\phi(y)$ = backwater function $= -\int \frac{dy}{y^3 - 1}$, and $y = h/H$.

For values of the backwater function, see Figure 5.42.

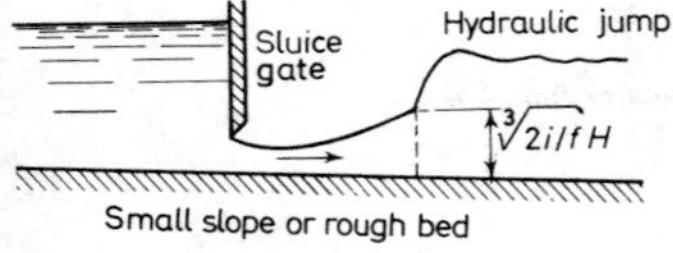

Figure 5.39 Sluice gate in channel, small slope and/or rough bed

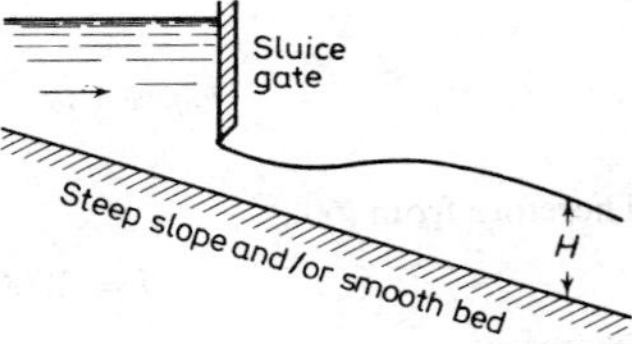

Figure 5.40 Sluice gate in channel, steep slope and/or smooth bed

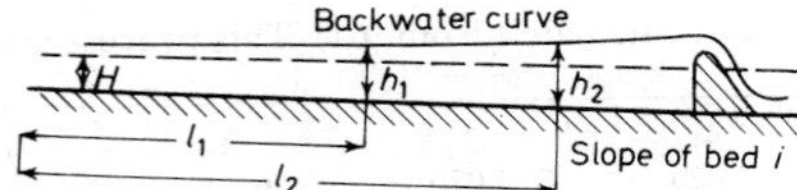

Figure 5.41 Weir or dam in channel, small slope and/or rough bed

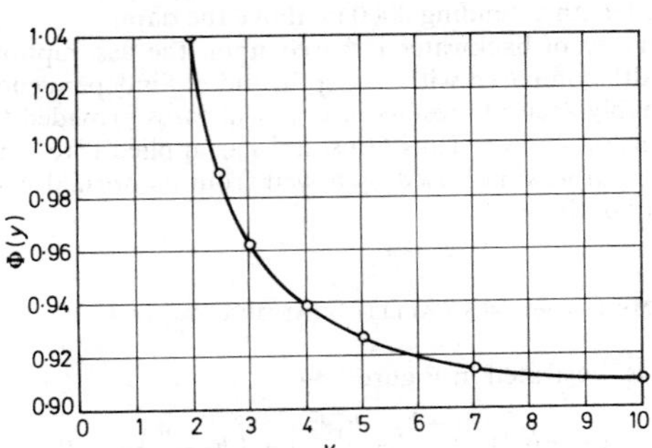

Figure 5.42 Backwater function

Specific values are tabulated below:

y	1.000	1.005	1.010	1.015	1.020	1.050	1.100
$\phi(y)$	∞	2.555	2.326	2.192	2.098	1.803	1.587
y	1.200	1.500	1.800	2.000	2.50	5.00	10.0
$\phi(y)$	1.387	1.162	1.073	1.039	0.989	0.927	0.911

Example. The following example illustrates the application of the backwater function. A dam is built across a stream which was previously flowing with a depth of 1 m. The effect of the dam is to raise the level just behind it by 4 m. The slope of the bed is 1:2000 and $f = 0.01$. What is the effect of the dam on the levels upstream? (Figure 5.43.)

$$i = 1/2000;\ 2i/f = 1/10;\ H = 1.00\ \text{m};\ H(1-2i/f) = 0.900$$
$$h_2 = \text{depth behind dam} = 5.00\ \text{m}$$
$$y_2 = h_2/H = 5.00;\ \phi(y_2) = 0.927$$

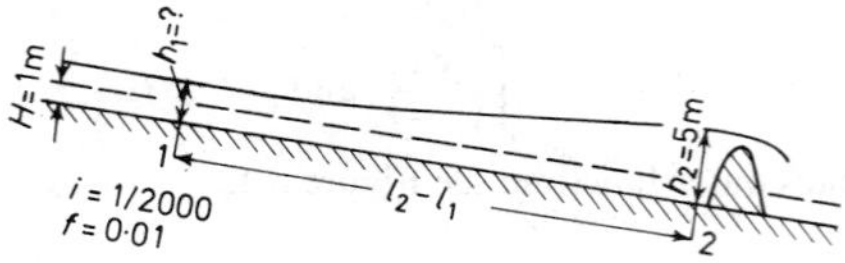

Figure 5.43 Application of backwater function

Therefore from (60),

$$l = 2000[4.166 - h_1 + 0.900\phi(y_1)] \tag{61}$$

where

$$l = l_2 - l_1;\ h_1 = \text{depth at distance } l \text{ from dam};\ y_1 = h_1/H = h_1/1.00.$$

Now assign values to h_1 and calculate l from (61). This process gives:

h_1 (m)	5	4	3	2	1.5	1.2
l (m)	0	2 020	4 070	6 200	7 420	8 430

Inspection of the data given for this example will show that the water surface is virtually horizontal over the length extending 2000 m above the dam.

Although this theory of backwater is based upon the assumption of a rectangular channel of great width compared with its depth, and of f independent of depth, it nevertheless gives reasonably accurate results in practical cases provided that the value of H for the actual channel is known. Thus Gibson[24] has applied it to a circular conduit, in which the central depth was increased by a weir from its normal value H of 1.04 m to 1.67 m in the vicinity of the weir.

THE HYDRAULIC JUMP (SOMETIMES CALLED 'STANDING WAVE')

An hydraulic jump is illustrated in Figure 5.44.

$$(h_1 - h_2)\left(\frac{h_1 + h_2}{2} - \frac{h_1 v_1^2}{h_2 g}\right) + \left(h_2 - \frac{d}{2}\right)d = 0$$

or, in a practically horizontal channel ($d = 0$),

$$h_2 = -\frac{h_1}{2} + \left(\frac{h_1^2}{4} + \frac{2h_1 v_1^2}{g}\right)^{\frac{1}{2}}$$

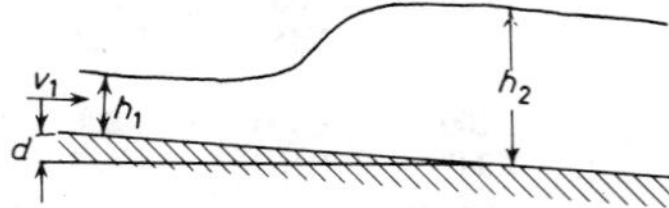

Figure 5.44 Hydraulic jump

and

$$(h_2-h_1) = \left(\frac{h_1^2}{4}+\frac{2h_1v_1^2}{g}\right)^{\frac{1}{2}} - \frac{3}{2}h_1 \tag{62}$$

For information concerning the length of channel covered in forming the jump, see Allen and Hamid.[25]

THE VENTURI FLUME

This device for measurement of rates of flow in an open channel is not so liable to damage as a weir and does not offer the same obstruction to the flow. In order to preserve its surface and its hydraulic characteristics, it is sometimes lined with stainless steel.

It may be formed by inserting 'streamlined' humps on the sides, Figure 5.45a, and/or the bed, Figure 5.45b, of the channel.

If the discharge is 'free' as represented by the broken lines in Figure 5.45 and accompanied by the formation of a standing wave, the rate of discharge depends only upon the

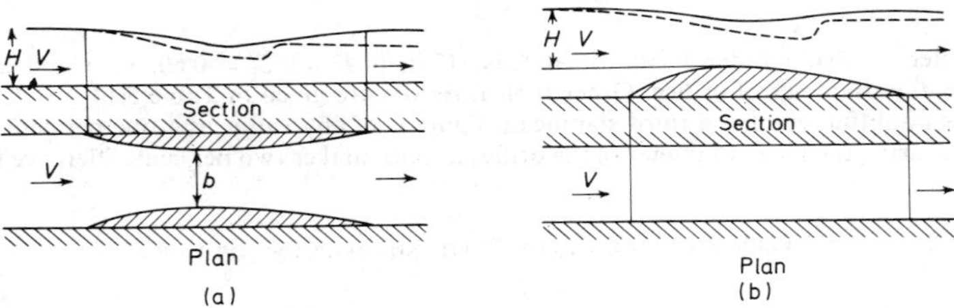

Figure 5.45 Venturi flume

depth upstream of the constriction. With a throat of rectangular section and width b, Q is approximately $0.54bg^{\frac{1}{2}}(H+V^2/2g)^{3/2}$, V itself of course depending upon H.

The general equation is

$$Q = C_D \frac{b_2 d_2}{[1-(b_2 d_2/b_1 d_1)^2]^{\frac{1}{2}}}[2g(d_1-d_2)]^{\frac{1}{2}} \tag{63}$$

where b_2 is the breadth at throat, d_2 the depth at throat, d_1 the depth upstream of the constriction, b_1 the breadth upstream of the constriction, and C_D is the coefficient of discharge.

For particular designs, C_D is best found by scale-model experiments.

Details as to proportions, shapes and types of flow may be found in papers by Engel.[26] See also Elsden.[27]

Orifices

SHARP-EDGED ORIFICE (FIGURE 5.46)

Velocity at *vena contracta* = $C_v\sqrt{(2gh)}$, where C_v is the coefficient of velocity, about 0.985.

Neglecting air resistance

$$x^2 = 4yhC_v^2 = 3.88yh \tag{64}$$

$$\text{Discharge } Q = C_v a_c \sqrt{(2gh)}$$
$$= C_v C_c a\sqrt{(2gh)} \tag{65a}$$

where C_c is the coefficient of contraction or

$$Q = C'a\sqrt{(2gh)} \tag{65b}$$

where a is the area of the orifice and C' is the coefficient of discharge.

Consideration of various published data[28] indicates that for orifices of $2\frac{1}{2}$ in (6.35 cm)

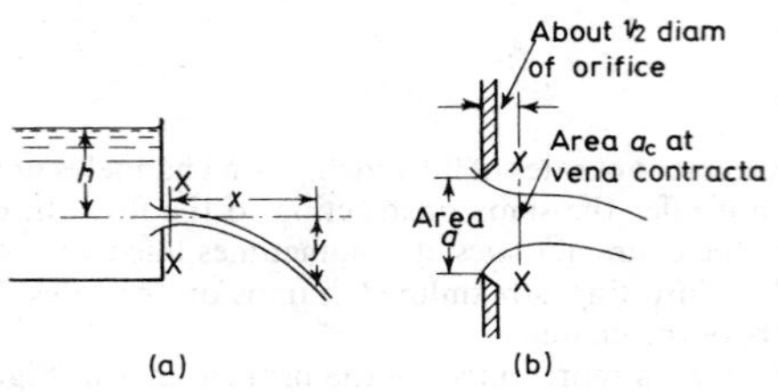

Figure 5.46 Sharp-edged orifice

diameter or over, under heads of at least 17 in (0.43 m) $C' = 0.60$, provided $h \nless 3d$, where d = diameter of orifice. Other typical results are given in Table 5.14.

It is doubtful whether a third significant figure is of any value, as a one per cent error in measuring the mean diameter of the orifice at once makes two per cent difference to the

Table 5.14 VALUES OF C' FOR SHARP-EDGED ORIFICE

Head, h		*d of circular orifice, or side of square orifice*					
(ft)	*(m)*	*(in)* 0.25 *(cm)* 0.64	*(in)* 0.5 *(cm)* 1.27	*(in)* 1.0 *(cm)* 2.54	*(in)* 2.5 *(cm)* 6.35	*(in)* 6 *(cm)* 15.2	*(in)* 12 *(cm)* 30.5
0.4	0.12	—	0.64	0.63	0.61	—	—
0.6	0.18	0.66	0.64	0.62	0.61	0.60	—
1.0	0.31	0.65	0.63	0.62	0.61	0.60	0.60
2.0	0.61	0.63	0.62	0.62	0.61	0.60	0.60
4.0	1.22	0.62	0.61	0.61	0.61	0.60	0.60
10.0	3.05	0.61	0.61	0.61	0.60	0.60	0.60
50.0	15.24	0.60	0.60	0.60	0.60	0.60	0.60
100.0	30.48	0.60	0.60	0.60	0.60	0.60	0.60

computed discharge. The absolute sharpness of the edge must also have some bearing upon the results.

The value of C' for a rectangular orifice appears to be somewhat higher than for a circular or square one of the same area. The difference amounts to about two per cent for rectangles having a 4:1 ratio of sides and four per cent for a ratio of 10:1.

ROUNDED OR BELL-MOUTHED ORIFICE

For a design such as that shown in Figure 5.47, $C_c = 1$ and $C' = C_v$.

$$Q = 0.95\frac{\pi d^2}{4}\sqrt{(2gh)} \quad \text{to} \quad 0.99\frac{\pi d^2}{4}\sqrt{(2gh)} \tag{66}$$

SUBMERGED ORIFICES

For a submerged orifice as shown in Figure 5.48

$$Q = C'a\sqrt{(2gh)} \tag{67}$$

where C' is substantially the same as for free discharge into the atmosphere.

TIME OF DISCHARGE THROUGH AN ORIFICE

The equation for time of discharge through an orifice from a tank, Figure 5.49, without any simultaneous inflow is

$$-\frac{dh}{dt} = C'\frac{a}{A}\sqrt{(2gh)}$$

or

$$t_2 - t_1 = -\frac{1}{C'a\sqrt{(2g)}}\int_{H_1}^{H_2} Ah^{-1/2}\,dh \tag{68}$$

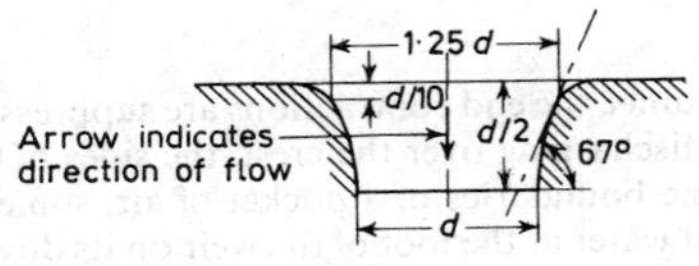

Figure 5.47 Rounded orifice

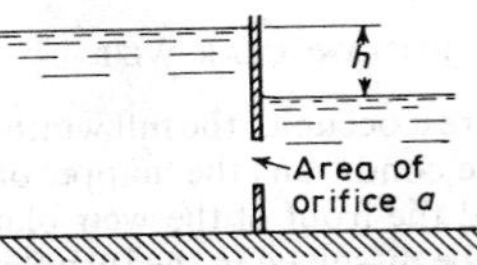

Figure 5.48 Submerged orifice

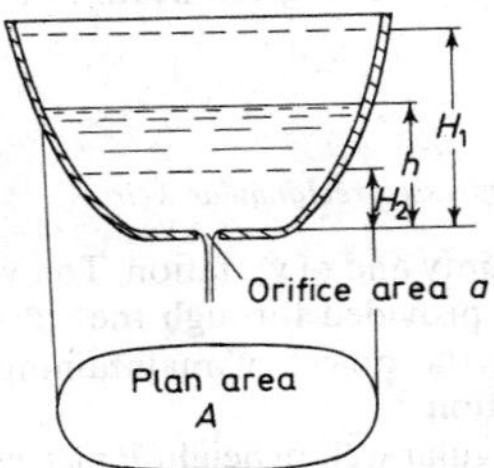

Figure 5.49 Discharge through an orifice

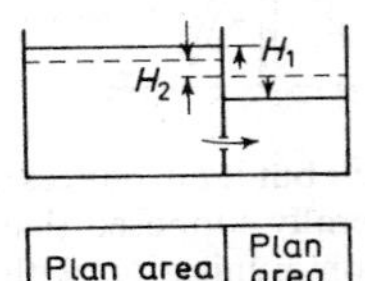

Figure 5.50 Discharge through submerged orifice

This is soluble if A can be expressed in terms of h, the instantaneous head at time t.

If A is constant, or independent of h,

$$\frac{C'a\sqrt{(2g)}}{A}(t_2 - t_1) = 2(\sqrt{H_1} - \sqrt{H_2}) \tag{69}$$

treating C' as independent of h.

The time of discharge through a submerged orifice (Figure 5.50) is given by the equation

$$(t_2 - t_1) = \frac{2}{C'(1/A_1 + 1/A_2)a\sqrt{(2g)}}(\sqrt{H_1} - \sqrt{H_2}) \tag{70}$$

Weirs and notches

The term 'notch' is used for the smaller weirs common in laboratories, as distinct from outdoors.

RECTANGULAR SHARP-CRESTED WEIR

For a weir as shown in Figure 5.51 in which $a \not< 4H$ and $c \not< 3H$, and H is measured at a distance six to ten times H behind the weir, then

$$Q = [0.410(2g)^{\frac{1}{2}}](b - H/10)H^{3/2} \tag{71}$$

This formula is probably accurate within two per cent for all values of H from 0.25 to 2 ft (0.08 to 0.61 m) provided $b/H > 2$ and provided b is $\not< 1$ ft (0.305 m).

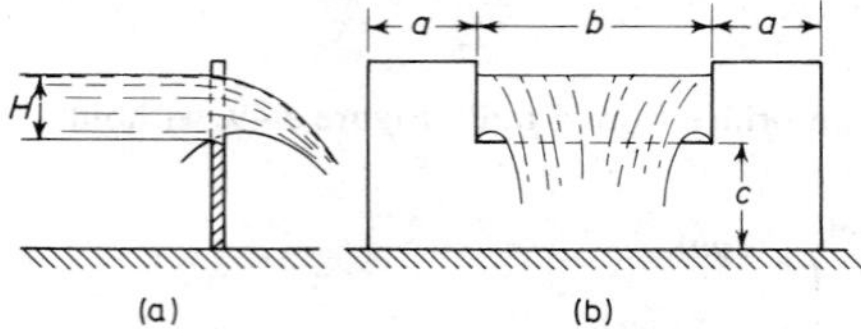

Figure 5.51 Rectangular sharp-crested weir

The effect of the velocity of the approaching stream is automatically allowed for in this formula, as in all others to be presented.

SUPPRESSED RECTANGULAR WEIR

If the weir crest occupies the full width of the channel, the end contractions are suppressed. Under these conditions the 'nappe' or stream discharging over the crest, the sides of the channel and the front of the weir plate form the boundaries of a pocket of air, some of which may be dissolved in the turbulent mass of water at the foot of the weir on its downstream side and carried away (Figure 5.52). The effect of this would be to reduce the pressure below the nappe and hence to increase the discharge for a given head. In itself this

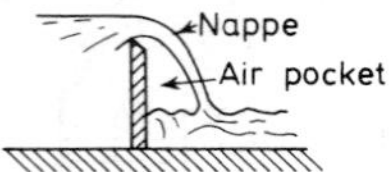

Figure 5.52 Suppressed rectangular weir

is no detriment but it introduces an element of uncertainty and of variation. To overcome this it is generally supposed that air vents should be provided through the sides of the channel in communication with the air-pocket with the object of maintaining atmospheric pressure and preserving a standardised condition.

Under such conditions (ventilated suppressed rectangular weir of height P metres above the bed of the channel), the Rehbock formula[29] is perhaps accurate to within one or two per cent. It reads

$$Q = \frac{2}{3}\sqrt{(2g)}b\left(0.605 + \frac{1}{1050H - 3} + \frac{0.08H}{P}\right)H^{3/2} \qquad (\mathrm{m^3/s}) \tag{72}$$

Writing this as

$$Q = Cb\sqrt{(2g)}H^{3/2} \tag{73}$$

values of C are as given in Table 5.15.*

THE 90-DEGREE VEE-NOTCH (SHARP-EDGED)

Measuring the head with reference to the point v (Figure 5.53)

$$Q = 1.34H^{2.48} \qquad (\mathrm{m^3/s}) \tag{74}$$

over a wide range, H being in metres. (If H is in feet, $Q = 2.48H^{2.48}$ ft³/s.)

* $Q = 4.43CbH^{\frac{3}{2}}$ m³/s if b and H are in metres;
$Q = 8.02CbH^{\frac{3}{2}}$ ft³/s if b and H are in feet.

Table 5.15 VALUES OF C

P		H							
(ft)	(m)	(ft) 0.2 (m) 0.06	(ft) 0.5 (m) 0.15	(ft) 1.0 (m) 0.31	(ft) 1.5 (m) 0.46	(ft) 2.0 (m) 0.61	(ft) 3.0 (m) 0.91	(ft) 4.0 (m) 1.22	(ft) 5.0 (m) 1.52
0.5	0.15	0.436	0.461	0.512	0.565	0.617	0.724	0.831	0.936
1.0	0.31	0.425	0.434	0.459	0.486	0.511	0.564	0.617	0.672
2.0	0.61	0.421	0.422	0.433	0.446	0.458	0.484	0.510	0.537
3.0	0.91	0.418	0.417	0.423	0.431	0.440	0.458	0.475	0.492
5.0	1.52	0.416	0.413	0.416	0.421	0.426	0.436	0.446	0.457
10.0	3.05	0.415	0.411	0.411	0.412	0.416	0.421	0.425	0.431

This notch is more convenient than the rectangular form for the measurement of small quantities but it should be remembered that an error of one per cent in the measurement of H means $2\frac{1}{2}$ per cent in the resulting estimate of Q, whereas with a rectangular notch the corresponding error is $1\frac{1}{2}$ per cent.

THE CIPPOLETTI WEIR (SHARP-CRESTED)

The discharge over this type of weir (Figure 5.54) is

$$Q = 0.420\sqrt{(2g)}b[(H+h)^{3/2} - h^{3/2}] \qquad \text{if } \tan\theta = \tfrac{1}{4} \tag{75}$$

where $h = v^2/2g$, v being the mean velocity in the approach channel.

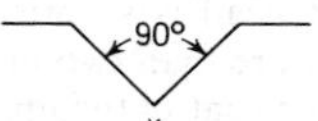

Figure 5.53 Sharp-edged 90° vee-notch

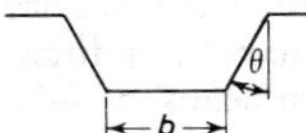

Figure 5.54 Sharp-crested Cippoletti weir

v cannot be allowed for until Q is known. Hence as a first approximation, find Q from $Q = 0.420(2g)^{\frac{1}{2}}bH^{3/2}$. Then calculate $v = Q$/area of section of approach channel, and re-evaluate Q from equation (75).

WEIRS WITHOUT SHARP CRESTS

Some typical examples are shown in Figure 5.55(a), (b), (c) and (d). The discharge $Q = C(2g)^{\frac{1}{2}}bH^{3/2}$.

5.55(a)

H (ft)	H (m)	C
0.5	0.15	0.400
1.0	0.31	0.426
1.5	0.46	0.441
2.0	0.61	0.442
3.0	0.91	0.411
4.0	1.22	0.392

5.55(b)

H (ft)	H (m)	C
0.5	0.15	0.402
1.0	0.31	0.407
1.5	0.46	0.424
2.0	0.61	0.431
3.0	0.91	0.445
4.0	1.22	0.455

5.55(c)

H (ft)	H (m)	C
0.5	0.15	0.392
1.0	0.31	0.426
1.5	0.46	0.439
2.0	0.61	0.450
3.0	0.91	0.456
4.0	1.22	0.456

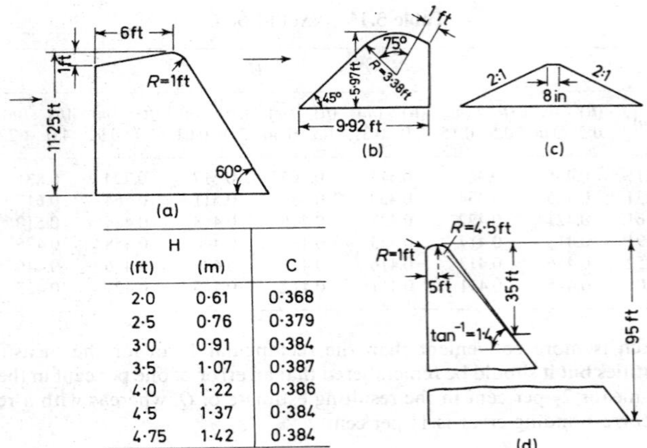

H (ft)	H (m)	C
2·0	0·61	0·368
2·5	0·76	0·379
3·0	0·91	0·384
3·5	1·07	0·387
4·0	1·22	0·386
4·5	1·37	0·384
4·75	1·42	0·384

Figure 5.55 Weirs without sharp crests, with corresponding H, C values

See also *Hydraulics Research,* HMSO Annual Reports, e.g. 1964 p. 15 and 1965 p. 7 (the 'Crump' weir).

SUBMERGED WEIRS

If the downstream level rises above the crest of the weir, a 'drowned weir' results. The effect of this upon the discharge for a given upstream head is surprisingly small: in general, the reduction in discharge will not amount to more than two or three per cent for 'downstream heads' or submergences up to twenty per cent of the upstream head.

TIME OF DISCHARGE OVER A WEIR (FIGURE 5.56)

The time of discharge over a weir or spillway of length b may be calculated as follows.
1. *With no inflow*

$$A\,\delta H = -C(2g)^{\frac{1}{2}} b H^{3/2}\,\delta t$$

therefore

$$\int_{t_1}^{t_2} \mathrm{d}t = -\int_{H_1}^{H_2} \frac{A}{C(2g)^{\frac{1}{2}} b} H^{-3/2}\,\mathrm{d}H$$

Or, time for head to fall from H_1 to H_2 is

$$(t_2 - t_1) = t = -\frac{1}{b}\int_{H_1}^{H_2} \frac{A}{C(2g)^{\frac{1}{2}}} H^{-3/2}\,\mathrm{d}H$$

This may be solved by splitting the change between H_1 and H_2 into stages over which mean values of A and C are applied.

If A and C are treated as constant

$$t = \frac{2A}{bC(2g)^{\frac{1}{2}}}\left(\frac{1}{\sqrt{H_2}} - \frac{1}{\sqrt{H_1}}\right) \tag{76}$$

2. *Reservoir with inflow as well as outflow* (Figure 5.56)

Let A be the surface area of reservoir, Q the inflow and h the instantaneous head over spillway of length b.

Then excess of inflow over outflow $= Q - C(2g)^{\frac{1}{2}}bh^{3/2}$. Hence

$$\frac{dh}{dt} = \frac{Q - C(2g)^{\frac{1}{2}}bh^{3/2}}{A}$$

Let H be the head over spillway which would make the rate of outflow equal to the rate of inflow. Then

$$Q = C(2g)^{\frac{1}{2}}bH^{3/2}$$

Let

$$r = h/H \qquad \text{and} \qquad K_1 = C(2g)^{\frac{1}{2}}b.$$

The time taken for the head to change from h_1 to h_2 is given by

$$t_2 - t_1 = \frac{A}{K_1\sqrt{H}}[\phi(r_2) - \phi(r_1)] \tag{77}$$

In this equation, $r_1 = h_1/H$, $r_2 = h_2/H$ and ϕ represents Gould's function[30] of h/H, i.e. of r.

Detailed values of $\phi(r)$ for use when the time-interval is expressed in the form given

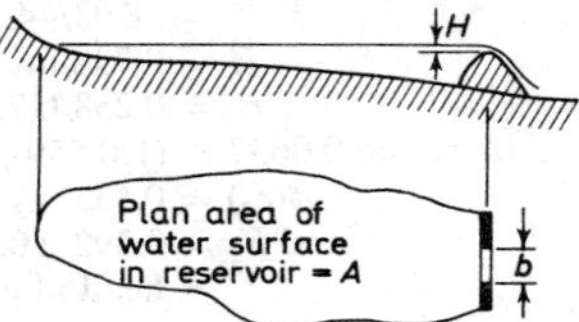

Figure 5.56 Discharge over a weir

by equation (77) (where b as well as $C\sqrt{(2g)}$ is absorbed in K_1) have been calculated by Mathieson.[31]

When $r < 1$,

$$\phi(r) = \frac{2}{3}\left[\ln\left(\frac{(r+\sqrt{r}+1)^{\frac{1}{2}}}{1-\sqrt{r}}\right) - \sqrt{3}\left\{\tan^{-1}\left(\frac{2\sqrt{r}+1}{\sqrt{3}}\right) - \frac{\pi}{6}\right\}\right]$$

When $r > 1$,

$$\phi(r) = \frac{2}{3}\left[\ln\left(\frac{(r+\sqrt{r}+1)^{\frac{1}{2}}}{\sqrt{r}-1}\right) - \sqrt{3}\left\{\tan^{-1}\left(\frac{2\sqrt{r}+1}{\sqrt{3}}\right) - \frac{\pi}{2}\right\}\right]$$

Some of the values given by Mathieson are quoted below.

r	0	0.10	0.20	0.30	0.40	0.50	0.60
ϕ (r)	0	0.1013	0.2076	0.3220	0.4482	0.5920	0.7615
r	0.70	0.80	0.90	0.99	1.01	1.02	1.04
ϕ (r)	0.9729	1.2619	1.7414	3.2925	4.4948	4.0426	3.5795
r	1.06	1.10	1.50	2.00	5	10	∞
ϕ (r)	3.3202	2.9838	1.9708	1.5730	0.9155	0.6376	0

Example. A reservoir has a surface area of 2.50 square kilometres. It is provided with an overflow weir of length 25 m, $C = 0.400$. Initially there is a steady head of 0.250 m over the weir, but superimposed upon the discharge corresponding with this condition, additional flood water enters the reservoir as detailed below.

Time (h)	0	1	2	3	4	5	6	7	8
Additional inflow (m³/s)	0	10.0	35.0	50.0	40.0	20.0	10.0	0	−2.75

Investigate the variation of water level.

$$K_1 = Cb\sqrt{(2g)} = 0.400 \times 25.0 \times 4.43 = 44.3$$

$$A/K_1 = 2.50 \times 10^6/44.3 = 5.64 \times 10^4;\ 3600K_1/A = 0.0637$$

Initial inflow = initial outflow = $44.3(0.250)^{3/2} = 5.53\ \text{m}^3/\text{s}$.

First hour

$$\begin{aligned}
\text{Mean } Q &= 5.53 + 5.00 = 10.53\ \text{m}^3/\text{s} \\
\text{therefore } H^{3/2} &= 10.53/44.3 = 0.238 \\
\text{and } H &= 0.384\ \text{m};\ \sqrt{H} = 0.620 \\
r_1 = h_1/H &= 0.250/0.384 = 0.651;\ \phi(r_1) = 0.863 \\
3600(\text{s}) &= (5.64 \times 10^4/0.620)(\phi(r_2) - 0.863) \\
\phi(r_2) &= 0.903;\ r_2 = 0.670 \\
h_2 &= 0.670 \times 0.384 = 0.258\ \text{m} \\
&= \text{head over spillway at end of 1 hour.}
\end{aligned}$$

Second hour

$$\begin{aligned}
\text{Mean } Q &= 5.53 + 22.5 = 28.03\ \text{m}^3/\text{s} \\
H^{3/2} &= 28.03/44.3 = 0.634 \\
H &= 0.738\ \text{m};\ \sqrt{H} = 0.859 \\
r_1 &= 0.258/0.738 = 0.350;\ \phi(r_1) = 0.384 \\
\text{therefore } 0.0637 &= (1/0.859)(\phi(r_2) - 0.384) \\
\phi(r_2) &= 0.438;\ r_2 = 0.392 \\
\text{and } h_2 &= 0.392 \times 0.738 = 0.289\ \text{m} \\
&= \text{head at end of two hours.}
\end{aligned}$$

Proceeding in this way, we obtain:

Hour	0	1	2	3	4	5	6	7	8
Head (m)	0.250	0.258	0.289	0.348	0.407	0.442	0.453	0.449	0.438

So the maximum head is 0.453 m at 6.10 hours. This head would give a maximum *out*flow of $44.3\ (0.453)^{3/2}$, or $13.5\ \text{m}^3/\text{s}$, as compared with the maximum *in*flow of $55.5\ \text{m}^3/\text{s}$ at 3.10 hours.

Impact of jets on smooth surfaces

SINGLE MOVING VANE

Let v be the absolute velocity of jet, u the absolute velocity of vane, assumed parallel to v, and a the area of section of jet (Figure 5.57).

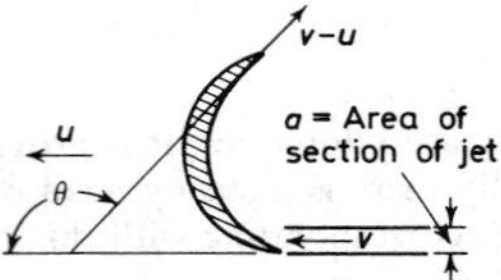

Figure 5.57 Impact of jet on single moving vane or series of moving vanes

Velocity of jet relative to vane $= v - u$.

Therefore mass striking vane $= \rho a(v - u)$.

This is unaltered in flow over the vane. (If roughness is taken into account, the final relative velocity $= k(v - u)$ where $k < 1$.)

Initial momentum/second, of jet $= \rho a(v-u)v \leftarrow$
Final momentum/second, of jet $= \rho a(v-u)\,[u+(v-u)\cos\theta] \leftarrow$
Therefore force exerted on vane $= \rho a(v-u)^2\,(1-\cos\theta) \leftarrow = x$, say
Work done on vane $= \rho a(v-u)^2\,(1-\cos\theta)u$

Initial kinetic energy of jet $= \dfrac{\rho a v^3}{2}$

$$\text{Therefore efficiency} = \frac{2(v-u)^2(1-\cos\theta)u}{v^3} \qquad (78)$$

Initial momentum per second of jet, in ↑ direction = 0
Final momentum per second of jet, in ↑ direction $= \rho a(v-u)^2 \sin\theta$
Therefore force exerted on vane in direction ↓ $= \rho a(v-u)^2\sin\theta = y$, say

$$\text{Resultant force on vane} = \sqrt{(x^2+y^2)} \qquad (79)$$

SERIES OF MOVING VANES

Figure 5.57 also applies. Since successive vanes intercept the jet, the mass of water striking them per second now is $\rho a v$. ρ is again the density of the water.

Force exerted in direction ← $= \rho a v(v-u)\,(1-\cos\theta) = x$, say
Work done on vanes $= \rho a v u(v-u)\,(1-\cos\theta)$
Efficiency $= 2u(v-u)\,(1-\cos\theta)/v^2$ (80)
Force exerted in direction ↓ $= \rho a v(v-u)\sin\theta = y$, say (81)

$$\text{Resultant force} = \sqrt{(x^2+y^2)}$$

CUBICAL BLOCK RESTING ON THE BED OF A STREAM (FIGURE 5.58)

Let ρ' be the density of the material of the block (kg/m³), μ the coefficient of friction between the block and bed of stream and v the velocity of stream at height y above bed (m/s).

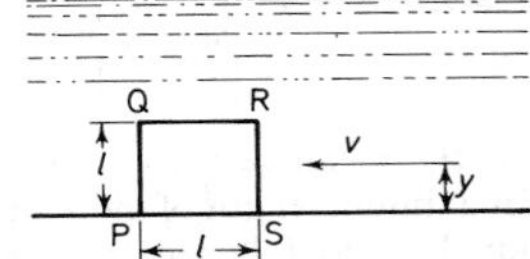

Figure 5.58 Impact of jet on cubical block resting on bed of stream

$$\text{Resistance to overturning about P} = \frac{(\rho'-\rho)l^4}{2}g$$

Resistance to sliding $= \mu(\rho'-\rho)l^3 g$
where μ is the coefficient of sliding friction.

$$\text{Force of impact of stream against face RS} = K'\rho l\int_0^l v^2\,\mathrm{d}y$$

where K' is a coefficient, ~0.70, to allow for the fact that not all the forward momentum of the stream is 'destroyed'.

Therefore, block will overturn if

$$K'\int_0^l v^2 y\,\mathrm{d}y > \frac{g(\rho'-\rho)l^3}{2\rho} \qquad (82)$$

or will slide if

$$K' \int_0^l v^2 \, dy > \frac{\mu g(\rho' - \rho) l^2}{\rho} \tag{83}$$

These results neglect other effects such as reduction of pressure on top and lee faces. They serve to show, however, that the stability of the block is essentially dependent upon the way in which the velocity is distributed through the depth of the stream.

Experiments in laboratory flumes indicate that

$\bar{v}$ = mean velocity of stream for which block overturns (m/s)

= rate of discharge divided by (area of section of channel – area of section of block)

$$= \frac{L}{l}\left(5.52 - \frac{2930}{(32 + h/l)^2}\right)\left[\left(\frac{\rho' - \rho}{\rho}\right) l\right]^{\frac{1}{2}} \tag{84}$$

where L is the length of block measured parallel to direction of current (m), l the depth of block (m), h the depth of water above top of block (m), ρ' the density of material of block (kg/m^3) and ρ the density of water (kg/m^3).

Example. A concrete cube 1 m × 1 m × 1 m of density 2400 kg/m^3.

Total depth D m	*Depth over cube h* m	$\bar{v}$ m/s
2	1	3.34
3	2	3.53
4	3	3.70
6	5	3.99
11	10	4.56

For stability of more than one block, arranged in rows and tiers, or heaped at random, see Allen.[32]

STABILITY OF FLAT BEDS AND MOUNDS OF BROKEN STONE AND SAND

The paper just cited[32] suggests that, over a wide range including materials having 'equivalent cube lengths' of 0.003 84 to 1.03 in (0.098 to 26.2 mm) and specific gravities of 2.016 to 3.89,

$$\bar{v} = k\left(\frac{L+B}{2l}\right)^{0.44} (\sigma' - \sigma)^{0.22}\, l^{0.22}\, M^{-0.22}\, h^{0.06}\, m^{0.22} \tag{85}$$

where $\bar{v}$ is the mean velocity (m/s) of stream flowing over a flat bed, L is the average value of maximum length of individual grains, B the average value of maximum breadth of individual grains, l the length of side of a cube of the same weight as the average piece, σ' the specific gravity of material, σ the specific gravity of water (say unity), M the uniformity modulus of material, h the depth of stream, m the hydraulic mean depth and k is a constant.

k is equal to 0.067 for movement of the first few particles if L, B, l, h and m are all in millimetres.

The uniformity modulus M is defined as indicated in Figure 5.59.

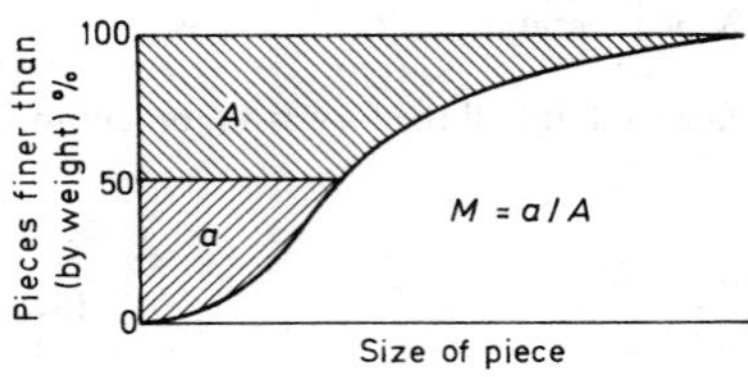

Figure 5.59 Definition of uniformity modulus

Similarly, for mounds,

$$\bar{v} = 0.079(\sigma' - \sigma)^{0.27} l^{0.27} M^{-0.27} h^{0.21} m^{0.02}$$

for initial disturbance. (86)

$$\bar{v} = 0.083 \left(\frac{L+B}{2l}\right)^{0.54} (\sigma' - \sigma)^{0.27} l^{0.27} M^{-0.27} h^{0.21} m^{0.02}$$

for 'flattening the crest of the mound.' (87)

Here $\bar{v}$ is the velocity (mean) of the stream flowing over the top of the mound and h the depth over the top (mm).

Equations (86) and (87) apply to materials of irregular shape (sand or broken stone), but the case of mounds formed of *cubes* laid in random fashion has also been investigated in reference 32.

Examples.
(a) 1 m cubes weighing 2400 kg so that $\sigma' = 2.40$.
(b) Broken stone, equivalent cube length $l = 1$ metre, $\sigma' = 2.64$, $M = 0.90$

Depth over crest of mound laid in random fashion m	$\bar{v}$ *for initial movement (m/s)*	
	1 m *cubes*	*Broken stone* $l = 1$ m $M = 0.90$
1	1.6	2.9
2	1.9	3.4
5	3.1	4.2
10	4.0	5.0

Note that σ', the true specific gravity, of silica or quartz sands is usually 2.64, or very nearly so.

Vortices

FORCED VORTICES

Forced vortices are of the type caused by stirring a liquid in a dish or by rotating the dish (Figure 5.60).

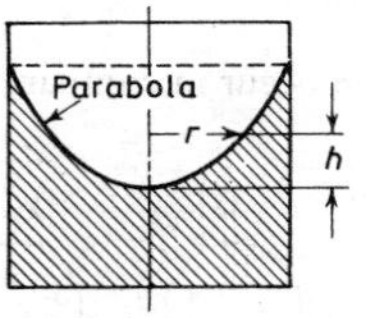

Figure 5.60 Forced vortex

$$h = \frac{\omega^2 r^2}{2g} \tag{88}$$

where ω is the angular velocity of rotation (rad/s).

FREE VORTICES

Free vortices are of the type which results when a liquid flows through a hole in the bottom of a vessel. The water moves in stream lines spirally towards the centre, where an air core tries to form. The coefficient of discharge through the orifice is greatly reduced as compared with its value in the absence of a vortex, i.e. with larger heads.

If the hole is now closed, the vortex motion persists for a time (until damped out by

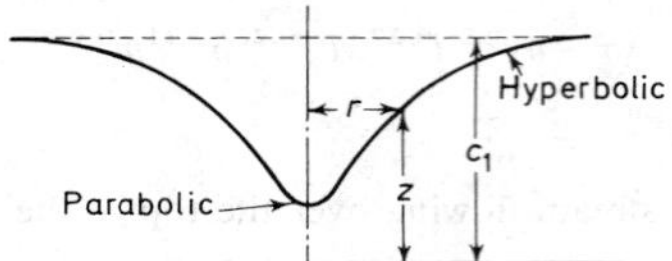

Figure 5.61 Free vortex

viscous resistance), the lower part having the characteristics of a forced vortex and the upper of a free cylindrical vortex (Figure 5.61).

$$c_1 - z = \frac{\text{Constant}}{2gr^2} \tag{89}$$

Waves

WAVES OF TRANSMISSION OR TRANSLATION

Waves of transmission or translation are of the type formed when a sluice gate is suddenly opened to admit water to a channel, or when a tidal bore advances along a river.

Let h be the depth of the stream, v the velocity of the stream moving in the direction opposite to the wave, k the height of the wave crest above the surface of the stream and V the velocity of propagation of the wave. Then

$$V = \left[\frac{2g(h+k)}{1+h/(h+k)}\right]^{\frac{1}{2}} - v \tag{90}$$

If k is small compared with h

$$V = [g(h+k)]^{\frac{1}{2}} - v$$

If k is very small compared with h

$$V = (gh)^{\frac{1}{2}} - v$$

WAVES OF OSCILLATION

Waves of oscillation occur in comparatively deep water (Figure 5.62).

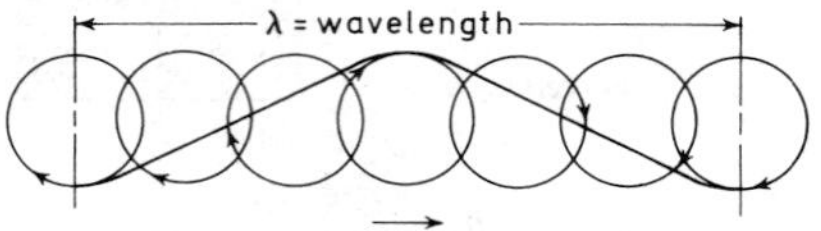

Figure 5.62 Waves of oscillation

Particles in the surface describe circular orbits, giving the appearance of a wave crest advancing with velocity $\sqrt{(g\lambda/2\pi)}$. The period of oscillation is $\sqrt{(2\pi\lambda/g)}$. In shallower water the orbits become distorted into approximate ellipses – a condition intermediate between a wave of oscillation and one of translation – and finally the wave breaks on the shore.

DYNAMIC EFFECTS ON HYDRAULIC STRUCTURES

These are not confined to the direct impact of streams or waves. They may also result from eddies giving rise to periodic forces and may become serious, for example, with long, slender piles in deep tidal waters. An important occurrence of this phenomenon was investigated recently at Immingham Oil Terminal.[33, 34, 35]

Dimensional analysis

Dimensional analysis is a valuable tool in reducing the apparent chaos of experimental results involving many variables and also in the systematic design of experimental procedure or technique.

Consider the resistance R of a certain shape of submerged body. Let l be any representative dimension, such as the length, v the velocity relative to the stream, ρ the density of fluid, μ the viscosity of fluid and g the acceleration due to gravity.

$$\text{Suppose } R \text{ depends upon } l, v, \rho, \mu \text{ and } g \tag{91}$$

Six quantities are involved here, and together they depend upon the three fundamental units or quantities of mass, length and time.

We therefore expect to find (6 – 3) or 3 dimensionless groups of the original quantities R, l, v, ρ, μ and g, related to one another.

Let N be a dimensionless group.

Choose any three quantities from equation (91) such that together they include mass M, length L, and time T.

Suppose we choose R, l, ρ.

R (a force) has the dimensions of mass × acceleration, or $[MLT^{-2}]$

l has the dimensions of length $[L]$

ρ has the dimensions of $[ML^{-3}]$

Now write $N_1 = R^{a_1}l^{b_1}\rho^{c_1}$, the last quantity v being chosen at random from the remaining symbols of equation (91).

Now N_1 is to have no dimensions in mass, length or time. Hence, dimensionally, $0 = [M^{a_1}][L^{a_1}][T^{-2a_1}][L^{b_1}][M^{c_1}][L^{-3c_1}][LT^{-1}]$

Equating indices of $[M]$, $[L]$, $[T]$ in turn

$$\text{for } [M],\ 0 = a_1 + c_1$$
$$\text{for } [L],\ 0 = a_1 + b_1 - 3c_1 + 1$$
$$\text{for } [T],\ 0 = -2a_1 - 1$$

These simultaneous equations give $a_1 = -\frac{1}{2}$, $c_1 = \frac{1}{2}$, $b_1 = 1$.

Hence

$$N_1 = \frac{\rho^{\frac{1}{2}} l v}{R^{\frac{1}{2}}}$$

Similarly, writing $N_2 = R^{a_2}l^{b_2}\rho^{c_2}\mu$ and remembering that the units of μ are $[M][L^{-1}][T^{-1}]$, we should find

$$N_2 = \frac{\mu}{R^{\frac{1}{2}}\rho^{\frac{1}{2}}}$$

Again, writing

$$N_3 = R^{a_3}l^{b_3}\rho^{c_3}g$$

we should get

$$N_3 = R^{-1}l^3\rho g$$

Now write

$$N_1 = \text{a function of } N_2, N_3 = \phi(N_2, N_3)$$

i.e.

$$\frac{\rho^{\frac{1}{2}}lv}{R^{\frac{1}{2}}} = \phi\left(\frac{\mu}{R^{\frac{1}{2}}\rho^{\frac{1}{2}}}, \frac{l^3\rho g}{R}\right)$$

which result is not invalidated if we multiply any one of the dimensionless 'groups' N_1, N_2, N_3, by a function of itself or of one of the others, so long as we do not reduc[e] the total number of such distinctive groups.

Therefore

$$\frac{R}{\rho l^2 v^2} = \phi\left(\frac{vl\rho}{\mu}, \frac{v^2}{lg}\right) \quad (92)$$

In this, $vl\rho/\mu$ is the so-called Reynolds number, and v^2/lg is the so-called Froude numbe[r] although some writers define it as $v/\sqrt{(lg)}$.

For alternative methods of dimensional analysis, see Duncan[36] and Whittington.[3]

APPLICATION TO SCALE MODELS

The principle of dynamical similarity indicates that if a model is operated at a spee[d] truly corresponding with the full size project, then

$$\frac{R_1}{R_2} = \frac{\rho_1 l_1^2 v_1^2}{\rho_2 l_2^2 v_2^2}$$

where the suffix 1 refers to the full size and the suffix 2 to its model.

As far as the resistance of submerged bodies is concerned, therefore, it follows fro[m] equation (92) that the Reynolds number in the model should be the same as in the actua[l] and (at the same time) the Froude numbers should be the same in model and full-siz[e] project.

The one condition then requires

$$v_2 = v_1\left(\frac{l_1}{l_2}\right)\left(\frac{\rho_1}{\rho_2}\right)\left(\frac{\mu_2}{\mu_1}\right) \quad (9$$

and the other requires

$$v_2 = v_1\sqrt{\left(\frac{l_2}{l_1}\right)} \quad (9$$

These two conditions cannot, in general, be satisfied at one and the same time a[nd] one of the features of scale model technique is to choose one or other as the more importa[nt] and then to discover, as a result of experiments on models of different scales, or by co[m]parison of model results with prototype values, what is the inaccuracy or 'scale effe[ct'] caused by neglect of the other requirement. In more complicated examples, more tha[n] two requirements may ideally be necessary.

SHIP MODELS

In testing the resistance of a ship's hull, the model is towed at a speed given by $v_2 =$ $\times (1/S)^{\frac{1}{2}}$, where $1/S$ represents the geometrical scale. Its total resistance R_T is then measur[ed] and from this is subtracted the skin-frictional resistance R_S as estimated from expe[r] ments on long 'boards' of similar surface roughness. The residue $(R_T - R_S)$ represents eddy and wave-making resistance R_E.

$$R_E \text{ multiplied by } \frac{\rho_1 l_1^2 v_1^2}{\rho_2 l_2 v_2^2}, \quad \text{or} \quad \frac{\rho_1}{\rho_2} S^3$$

then gives the eddy and wave-making resistance of the full size ship at the corresponding speed $v_1 = v_2\sqrt{S}$, and to this is added the estimated skin friction of the ship at that speed.

MODELS OF SLUICE-GATES, WEIRS, ETC.

In models of sluice-gates, weirs and spillways, the Froude number is to be adopted, provided the scale is well chosen. For example, a model of an overflow spillway for a dam may usually be relied upon if the lowest head on which deductions are to be based is not less than 0.25 in (6.4 mm) in the model. Under heads smaller than this, surface tension and viscosity are generally responsible for appreciable scale effects. With this reservation, however, if we call h the head observed in the model and q the discharge, then the discharge of the full size spillway under a head of Sh will be $qS^{5/2}$, the model scale being $1:S$.

RIVER MODELS

River models are frequently constructed to different scales horizontally and vertically. Many considerations influence the scales actually to be chosen. If the horizontal scale is $1:x$ and the vertical scale $1:y$, there will be a vertical exaggeration, or distortion, of scale, equal to x/y. This exaggeration is desirable in order to improve the prospects of (a) the flow being turbulent in the model as it is in nature, (b) the water-surface slopes being reproduced, (c) bed material being shifted by the currents available in the model. The smaller the value of x, the smaller must be the exaggeration x/y. The nominal velocity scale for sensibly horizontal stream velocities in such a model is then $1:\sqrt{y}$ and the scale of time $1:x/\sqrt{y}$. If q is the rate at which water is fed to the model in order to simulate a flow Q in nature, then q should be equal to $Q/xy^{3/2}$. Some river models have, however, been operated with such flows and velocities, discovered by trial, as to give the proper gradients and bed movements irrespective of these ideal conditions which should, if practicable, be observed.

Sand or other material used on the bed of river models to give qualitative or approximately quantitative indications of scour is not necessarily reduced to scale. Frequently such materials are of the same order of size as in nature, the feasibility of this depending on the fact that the scouring property of a shallow stream is greater than that of a deep one of the same mean velocity.

HARBOUR MODELS

Models of harbours specifically concerned with surface waves produced by storms have been successful with scales (undistorted) in the region of 1:50, 1:100 or 1:180, and with model waves in the exposed area outside the harbour works about 20 mm or more in height. The velocity scale of such a model having a geometrical scale of 1:100 would be $1:\sqrt{100}$; or 1:10, and its time scale would also be 1:10.

REFERENCES

1. DAVIS, S. J. and WHITE, C. M., 'A Review of Flow in Pipes and Channels', *Engng*, **78,** 71 (1929)
2. STANTON, T. E. and PANNELL, J. R., *Phil. Trans. A,* **214,** 199 (1914); *National Phys. Lab. Coll. Res.,* **11,** 293 (1914)
3. NIKURADSE, J., *Forschungsh. Ver. dtsch. Ing.,* No. 361 (1933)
4. PRANDTL, L. and VON KÁRMÁN, T., *Z. Ver. dtsch. Ing.,* **77,**105 (1933); *Proc., 4th Int. Congr. appl. Mech.,* Cambridge, 1935

5a. COLEBROOK, C. F. and WHITE, C. M., 'Experiments with Fluid Friction in Roughened Pipes', *Proc. R. Soc. A,***161,** 367 (1937)

5b. COLEBROOK, C. F., 'Turbulent Flow in Pipes, with particular reference to the Transition Region between the Smooth and Rough Pipe Laws', *J. Instn civ. Engrs,* **11,** 133 (1939)

6. MANNING, R., 'Flow of Water in open Channels and Pipes', *Trans. Instn civ. Engrs, Ireland,* **20,** 161 (1891); **24,** 179 (1895)
7. ACKERS, P., 'Resistance of Fluids flowing in Channels and Pipes', *Hydraulics Research Papers,* Nos 1 and 2, HMSO (1958)
8. BARNES, A. A., *Hydraulic Flow Reviewed,* Spon, London (1916)
9. COLEBROOK, C. F. and WHITE, C. M., 'The Reduction of Carrying Capacity of Pipes with Age', *J. Instn civ. Engrs,* **7,** 99 (1937)
10. GIBSON, A. H., Loss of Head in Gradual Enlargements, *Proc. R. Soc. A,* **83,** 366 (1910); 'Loss at Enlargements and at Valves', *Trans. R. Soc. Edinburgh,* **48,** (1911)
11. CORNISH, R. J., The Analysis of Flow in Networks of Pipes, *J. Instn civ. Engrs,* **13,** 147 (1939)
12. SKEAT, W. O. and DANGERFIELD, B. J. (Eds), *Manual of British Water Engineering Practice,* 4th edn, Vol. III, *Instn water Engrs,* London 168–174 (1969)
13. STUCKEY, A. T., Methods used for the analysis of Pipe Networks, WWE (Mar. 1969)
14. BARLOW, J. F. and MARKLAND, E., 'Computer analysis of pipe networks', *Proc. Instn civ. Engrs,* **43,** 249 (1969)
15. AL-NASSRI, S. A., 'Flow in Pipes and Pipe Networks', PhD thesis, Univ. of Liverpool (1971)
16. O'BRIEN, M. P. and HICKOX, G. H., *Applied Fluid Mechanics,* McGraw-Hill, New York (1937)
17. *Flow Measurement* BS 1042, British Standards Institution (1943)
18. *Am. Soc. Mech. Engrs Fluid Meters Report,* 4th edn (1931)
19. BAZIN, H. E. *Ann. des Ponts et Chaussées,* **4,** 20 (1897)
20. PARKER, P. à M., *The Control of Water,* Routledge, London (1915)
21. ALLEN, J., 'Roughness Factors in Fluid Motion through Cylindrical Pipes and through Open Channels', *J. Instn civ. Engrs,* **20,** 91 (1943)
22. ALLEN, J., 'The Resistance to Flow of Water along a Tortuous Stretch of River and in a Scale Model of the Same', *J. Instn civ. Engrs,* **11,** 115 (1939)
23. ALLEN, J. and SHAHWAN, A., 'The Resistance to Flow of Water along a Tortuous Stretch of the River Irwell (Lancashire)—an Investigation with the Aid of Scale-Model Experiments', *Proc. Instn civ. Engrs,* Pt III, Vol. 3, No. 1, 144 (1954)
24. GIBSON, A. H., *Hydro-Electric Engineering,* Blackie, London **1,** 67 (1924)
25. ALLEN, J. and HAMID, H. I., The hydraulic jump and other phenomena associated with flow under rectangular sluice-gates', *Proc. Instn civ. Engrs,* **40,** 345 (1968)
26. ENGEL, F. V. A., 'Non-Uniform Flow of Water', *The Engineer,* London, 21, 28 April, 5 May (1933); 'The Venturi Flume', *The Engineer,* London, 3 & 10 August (1934)
27. ELSDEN, O., 'Flow Measurement', Ch. II, Vol. I, *Hydro-Electric Engineering Practice,* 2nd edn, Ed. J. Guthrie Brown, Blackie, London & Glasgow (1964)
28. HAMILTON SMITH, JR., *Hydraulics* (1886)
    BILTON, H. J. I., *Victorian Inst. Engrs* (1908)
    JUDD, H. and KING, R. S., *Am. Assoc. Adv. Sci.* (1906)
    SMITH, E. S. and WALKER, W. H., *Proc. Instn mech. Engrs* (1923)
    BOND, W. N., *An Introduction to Fluid Motion,* Arnold, London
29. REHBOCK, T., *Handbuch der Ingenieur wissenschaften,* Vol. 1, Pt 3/2 (1912); discussion *in Trans. Am. Soc. civ. Engrs* 93 (1929); also details in *Hydraulic Laboratory Practice* (Ed. FREEMAN, J. R.) Am. Soc. Mech. Engrs (1929)
30. GOULD, *Engineering News* (1901)
    HORTON, D. F., *Engineering News Record* (1918)
    GIBSON, A. H., *Hydro-Electric Engineering,* Vol. 1, Blackie, London & Glasgow, 75 (1924)
31. MATHIESON, R., 'The Generalized Gould's Function', *Proc. Instn civ. Engrs,* Pt III, **2,** No. 1, 142 (Apr. 1953)
32. ALLEN, J., 'An Investigation of the Stability of Bed Materials in a Stream of Water', *J. Instn civ. Engrs,* **18,** 1 (1942)

33. SAINSBURY, R. N. and KING, D., 'The flow induced oscillation of marine structures', *Proc. Instn civ. Engrs*, **49,** 269 (1971)
34. Constr. Industry Res. and Inform. Assoc., London, Report, Project 143, p. 21 (1970–1971)
35. *Docks*, **8,** No. 11, British Transport Docks Board, London, 5 (Nov. 1971)
36. DUNCAN, W. J., *Physical Similarity and Dimensional Analysis*, Arnold, London (1953)
37. WHITTINGTON, R. B., 'A simple dimensional method for hydraulic problems', *J. Hydr. Div , Proc. Am. Soc. civ. Engrs.*, **89,** No. HY5, 1 (Sept. 1963)

BIBLIOGRAPHY

ADDISON, H., *A Treatise on Applied Hydraulics*, 5th edn, Chapman and Hall, London (1964); *Hydraulic Measurements*, Chapman and Hall, London (1940)
ALLEN, J., *Scale Models in Hydraulic Engineering*, Longmans, London (1947)
BROWN, J. GUTHRIE, *Hydro-Electric Engineering Practice*, Vol. I, 2nd edn, Blackie, London and Glasgow (1964)
DUNCAN, W. J., THOM, A. S. and YOUNG, A. D., *The Mechanics of Fluids*, Arnold, London (1960)
FRANCIS, J. R. D., *A Textbook of Fluid Mechanics for Engineering Students*, Arnold, London (1958)
GIBSON, A. H., *Hydraulics and its Applications*, 5th edn, Constable, London (1952)
Hydraulics Research, HMSO Annual Reports, London
JAEGER, C., *Engineering Fluid Mechanics*, transl. and Ed. P. O. Wolf, Blackie, London and Glasgow (1956)
JAMESON, A. H., *An Introduction to Fluid Mechanics*, Longmans, London (1937)
KING, H. W., *Handbook of Hydraulics*, 4th revised edn, Ed. E. F. Brater, McGraw-Hill (1954)
LEWITT, E. H., *Hydraulics and the Mechanics of Fluids*, 9th edn, Pitman (1952)
Manual of British Water Engineering Practice (3 Vols) 4th edn, Inst. Water Engrs, London (1969)
O'BRIEN, M. P. and HICKOX, G. H., *Applied Fluid Mechanics*, McGraw-Hill, New York (1937)
PAO, R. H. F., *Fluid Mechanics*, Wiley, New York (1961)
PARKER, P. à M., *The Control of Water*, Routledge, London (1915)
PRANDTL, L., *The Essentials of Fluid Dynamics*, Blackie, London and Glasgow (1952)
ROUSE, H., *Elementary Mechanics of Fluids*, Wiley, New York (1948)
VENNARD, J. K., *Elementary Fluid Mechanics*, 4th edn, Wiley, New York (1962)

6 SURVEYING AND PHOTOGRAMMETRY

SURVEYING AND PHOTOGRAMMETRY 6

6 SURVEYING AND PHOTOGRAMMETRY

D. E. MURCHISON, B.Sc. (Eng.), C.Eng., M.I.C.E.
Department of Civil Engineering, University of Surrey

In the past decade standard surveying instruments have been improved and developed in many ways; they are more precise, more compact, lighter in weight and generally easier to use than the instruments of ten years ago. New electronic distance measuring equipment has been introduced and developed during this time. In addition, with the advent of electronic computers, some of the adjustment techniques applied to surveying networks have been changed entirely to take advantage of these modern technological advances.

Civil engineers are making use of site plans, contours, profiles, cross sections and data in the form of digital ground models obtained by photogrammetric measuring techniques, not only for project planning but also for detailed design work. An attempt is made in this section to describe the basic principles and the surveying techniques involved, giving brief descriptions of a selection of typical surveying and photogrammetric instruments in current use. The work is confined to plane surveying, with some reference to the effects of earth curvature and terrestrial refraction on the determination of levels for control purposes. Angles are expressed in radians or degrees, minutes and seconds as appropriate.

LINEAR MEASUREMENT

Tapes

All site plans drawn to scale which show positions of natural topographical features and man-made construction works are based on linear measurements reduced to the horizontal plane. It is essential that all linear measurements in surveying, regardless of the final accuracy required, are horizontal measurements, and precautions must be taken to account for sloping ground. The most basic piece of equipment in common use is the ordinary steel or glass-fibre-reinforced plastic tape. Simple precautions and attention to detail can avoid many errors which arise frequently in using tapes for site measurements. Initially it is necessary, in planning the work, to assess the final accuracy of measurement required. Three classes of accuracy are considered in using tapes for linear measurement:

(i) Measurement of length and offsets over rough ground to an accuracy of $\pm 1/1000$;
(ii) Precise measurements required for close tolerances when using prefabricated units in the construction industry or taking measurements for planimetric control surveys to an accuracy of $\pm 1/50\,000$;
(iii) Ordnance Survey base-line measurement techniques over long distances to achieve an accuracy of $\pm 1/500\,000$.

For the first category, a 30 m steel tape and a 30 m fibre tape are required, together with the following accessories: good steel arrows for marking the ends of intermediate measurements, ranging rods for keeping a straight alignment by sighting along the length of the line, an Abney level clinometer hand-held for measurement of changes of slope, and a hand level.

The marking arrow is advanced along the length of the slope by a distance equal to the slope correction for every 30 m and the correction is subtracted from the closing measurement at the end of the line as indicated in Figure 6.1. Offsets perpendicular to the line are measured using the fibre tape held horizontally and 'stepping' horizontal measurements to ranging rods held vertically at intervals up or down the transverse slope. It is advisable to use a hand level to ensure that ranging rods are held vertically.

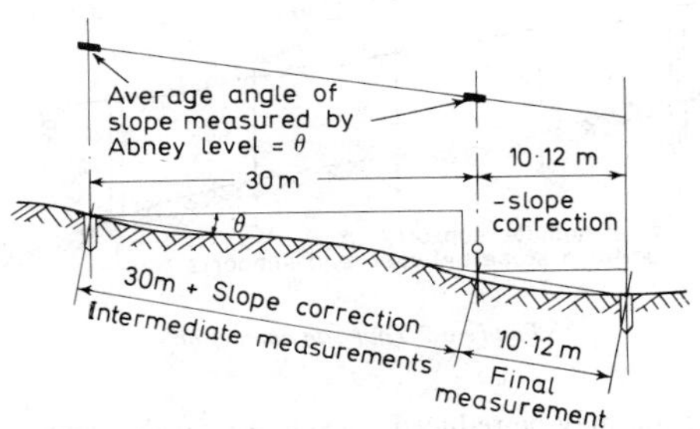

Figure 6.1 Slope corrections

Important points of detail are fixed by taking an intersection of two oblique horizontal measurements to the line using the same procedure.

In the second category of measurement, it is necessary to suspend a steel tape between intermediate pegs aligned by theodolite and to take field measurements of tape tension and atmospheric temperature. The differences in level between the pegs are also measured, using an engineer's level. Care must be taken to provide a good index mark to which measurements are made on the tops of the pegs; a metallic strip which can be scribed with an index mark is ideal for this purpose. The tape tension may be measured with an ordinary spring balance or by using a constant-tension handle developed by the Building Research Station. This constant-tension handle applies a constant tension of 70 N to a 9.5 mm-wide tape and is now available commercially.

In order to achieve the required accuracy of $\pm 1/50\,000$, it is necessary to use a steel tape which has been standardised or to compare the measuring tape with a standard tape. This standardisation may be carried out at the National Physical Laboratory or may be undertaken by certain commercial firms or other organisations who will supply a standardisation certificate for the steel tape. The tape will be supported on a flat surface and compared with a standard reference tape at a given temperature and tension, e.g. 15 °C and 44.5 N. The length of the nominal 0–30 m interval will be given, e.g. 30.0046 m. If required, the coefficient of linear expansion, Young's modulus and the average cross sectional area of the tape will be checked.

The following corrections must be made to the average observed length between pegs.

(1) SLOPE CORRECTION $= -\dfrac{d^2}{2L} - \dfrac{d^4}{8L^3}$

where d is the difference in level between the tops of pegs and L is the observed length.

The first term will give the required accuracy of $\pm 1/50\,000$ for slopes less than 6 degrees.

(2) SAG CORRECTION $= -\dfrac{1}{24}\left(\dfrac{W}{P_F}\right)^2 L$

where W is the weight of the suspended part of the tape, P_F is the field tension and L the observed length.

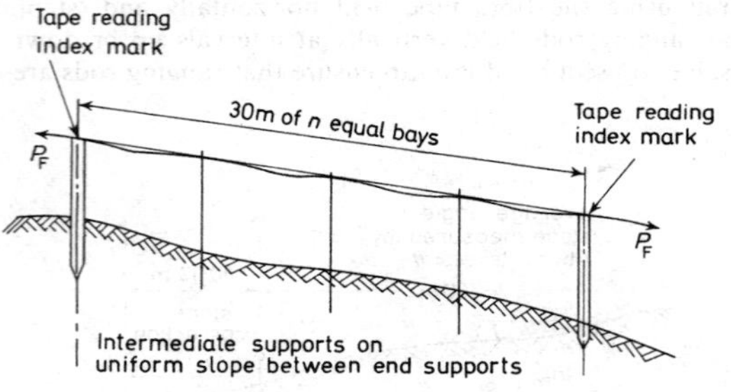

Figure 6.2 Tape sag correction

The sag correction may be reduced considerably if the tape is supported between the end pegs, as shown in Figure 6.2, making n equal bays.

SAG CORRECTION PER BAY $= -\dfrac{1}{24}\left(\dfrac{W/n}{P_F}\right)^2 \dfrac{L}{n}$

$$= -\frac{1}{24}\left(\frac{W}{P_F}\right)^2 \frac{L}{n^3}$$

MODIFIED SAG CORRECTION $= -\dfrac{1}{24}\left(\dfrac{W}{P_F}\right)^2 \dfrac{L}{n^2}$

(3) TENSION CORRECTION $= +\dfrac{(P_F - P_S)L}{AE}$

where P_F is the field tension, P_S the standard tension, A the tape cross-sectional area, E is Young's modulus and L is the observed length. For steel, $E = 200$ kN/mm^2.

(4) TEMPERATURE CORRECTION $= +L\alpha(T_F - T_S)$
where T_F is the field temperature, T_S the standard temperature, α the coefficient of linear expansion and L is the observed length.
For steel, $\alpha = 0.000\,011\,2/°C$.

(5) STANDARD CORRECTION
This is added or subtracted according to the length of the 0–30 m interval given on the standardisation certificate.

Examples of accurate measurements are given in BRS publications.[1, 2]
The procedure for the third category of linear measurement, to achieve a base-line accuracy of 1/500 000, requires special equipment and extreme precautions but the basic

principle is the same as that already described. The tape is freely suspended between straining tripods and the tension applied by weights passing over pulleys and hanging freely. Readings are taken by hinged magnifiers against special index marks carried on measuring tripods. Precautions are taken to prevent tape sway by providing wind shields along the whole length of the tape. Corrections are made to observed lengths for height above mean sea level for changes in gravity with latitude and elevation of the station, in addition to those already described. The site for the base line is carefully chosen and measurements are taken only in the most favourable atmospheric conditions, often at night. In addition, the tape used is manufactured from invar steel which has a very low linear coefficient of thermal expansion. This method of measurement to an accuracy of 1/500 000 is becoming obsolete with the introduction and development of electronic distance measuring equipment.

Electronic measuring equipment

In 1950 the first electronic instrument for distance measurement, called the geodimeter, was introduced in Sweden. A light source was used to produce the measuring signal and this was reflected from a mirror-type reflector at the distant station. Shortly after this the tellurometer was introduced in South Africa. This instrument used a transmitted radio wave for the measuring signal which was received and transmitted by a combined radio receiver and transmitter at the distant station.[6] These two instruments were regarded as complementary; the geodimeter was best suited to the measurement of short-range distance and the tellurometer to long range. In the past ten years many new instruments have been introduced; lasers are used to increase the range and efficiency of modern geodimeters and the tellurometer range includes instruments which emit near infra-red modulated light beams. Manufacturers in many countries have entered this field of surveying equipment. Developments in electronic engineering have made all these instruments and their associated power supplies very compact, light weight, easy to use, versatile and reliable. The principle used for determination of distance is outlined with reference to the basic geodimeter and a brief description of some typical modern instruments is given below.

Figure 6.4 shows a typical assembly of three retrodirective prism reflectors mounted on a tripod for use with a geodimeter. The advantage of using the retrodirective prism reflector is that the reflected beam is parallel to the incident beam within a permissible pointing error from the reflector to the geodimeter of 20 degrees. To take advantage of this, the transmitting and receiving optics of the instrument are placed coaxially. The geodimeter transmits an intensity-modulated light beam to the reflector, this light beam returns to the geodimeter out of phase with reference to the transmitted signal (unless the distance between the geodimeter and reflector happens to be an exact multiple of the wavelength of the modulated beam). The light signal is converted to an electrical signal and the phase difference between the transmitted and received signal is measured at the instrument.

In the basic geodimeter, the modulated light signal is transmitted at three different frequencies and the distance determined as follows:

At frequency 1, the wavelength of the modulated light beam $x_1 = 5$ m. Then with reference to Figure 6.6:

$$D = nx_1 + L_1 \tag{1}$$

where n is the number of full wavelengths in the distance D and L_1 is calculated from the recorded phase difference.

It follows that if the distance can be estimated to the nearest 5 m it can be determined from equation (1). Alternatively, if the modulated beam is transmitted at another frequency f_3 where the wavelength $x_3 = (20/21)x_1$,

$$D = nx_3 + L_3 \tag{2}$$

Figure 6.3 (left) The geodimeter

Figure 6.4 (above) Geodimeter reflector
(Courtesy: Aga Geodimeters)

and n will be the same as that in equation (1) up to $D = 100$ m $= 21x_3 = 20x_1$, with one exception shown by the line AB in Figure 6.6. If the reflector is placed in any such position along the line at approximately 5 m intervals then n_3 is one more than n_1. This is detected by noting that $L_3 < L_1$ and, if so, one whole wavelength $(20/21)x_1$ is added to L_3 and n becomes the same in both equations (1) and (2). The simultaneous solution of these equations and elimination of n gives

$$D = 21(L_3 - L_1) + L_1 \qquad (3)$$

which will resolve distances up to 100 m without ambiguity. An intermediate frequency f_2 where $x_2 = (400/401)x_1$ is introduced so that

$$D = 401(L_2 - L_1) + L_1 \qquad (4)$$

which will resolve distances up to $400x_1 = 401x_2 = 2000$ m without ambiguity. Equations (3) and (4) are used together to determine the measured distance and units of 2000 m are estimated in the final analysis.

The readings will be slightly affected by atmospheric temperature and pressure, instability of the delay line and other factors, so that at each frequency an internal distance known as the instrument constant, given for each individual instrument, is measured by an internal reflection of the transmitted and received signals. The readings taken with reference to this internal distance are called calibration readings and denoted by C, whilst the external reflector readings are denoted by R. Each reading of C and R

is repeated at four difference phases of the transmitted signal and recorded with the appropriate sign. Thus, for one determination of distance, 24 geodimeter readings are recorded. Taking account of signs, it can be shown that

$$L = R - C$$

If $> R$, add the appropriate 1/2 wavelength and change the sign of R.
If signs are opposite, add the appropriate 1/2 wavelength and change the sign again.
Calculate L_1, L_2 and L_3 taking the mean reading of four phases.
If L_3 or $L_2 < L_1$, add the appropriate whole wavelength to L_3 or L_2.
Use equation (3) to find the part distance from 0 to 100 m to nearest 5 m

$$D_1 = 21(L_3 - L_1)$$

Use equation (4) to find the part distance from 100 m to 2000 m to the nearest 100 m

$$D_2 = 401(L_2 - L_1) - D_1$$

Then the approximate distance D', to the nearest 5 m, is

$$D' = D_1 + D_2$$

Add to this the mean value of L_1 obtained at f_1, f_2 and f_3 as follows:

at f_1, L_1 is the direct reading;

at f_2, $L_1 = L_2 - \dfrac{(D_1 + D_2)}{401}$;

at f_3, $L_1 = L_3 - (D_1/21)$.

Then the final slope distance D is given by

$$\begin{aligned} D = {} & D_1 + D_2 + L_{1MEAN} + \text{INSTRUMENT CONSTANT} \\ & + \text{REFLECTOR CONSTANT} + \text{ATMOSPHERIC CORRECTION} \\ & + \text{UNITS OF 2000 m BY ESTIMATION} \end{aligned} \tag{5}$$

The reflector constant arises owing to the mounting eccentricity on the tripod, and to the fact that the speed of light is slower in glass than in air. The correction for the geodimeter reflector is -0.03 m. In addition, the speed of light is constant in a vacuum but in the atmosphere varies with temperature, pressure and, slightly, with humidity. Thus the atmospheric temperature and pressure are recorded and the correction calculated from the equation

$$\text{ATMOSPHERIC CORRECTION} = \left(309.2 - \frac{83\,189.4}{(273.2+t)}\frac{p}{760}\right)\frac{D}{10^6} \tag{6}$$

where t is in °C and P in mm Hg.

The accuracy of measurement using this model 6 geodimeter is from ± 10 mm/km to 2 mm/km. Improvements in later models give distance measurements to an accuracy of from ± 5 mm/km to ± 1 mm/km.

Model MA 100 in the tellurometer range uses a gallium arsenide diode emitting a near-infrared modulated light beam. The beam is reflected from a prism reflector and returned to the instrument. The distance is measured in a similar way to that already described but direct readings are taken on a four-digit automatic display. These readings are selected by means of a switch in five ranges from millimetres to metres × 10, and the final slope distance is obtained by simple addition. If corrections are made for slope, atmospheric variations, instrument and reflector constants, the accuracy of the final measurement varies from ± 1.5 mm at short ranges to about ± 5 mm at 2 km.

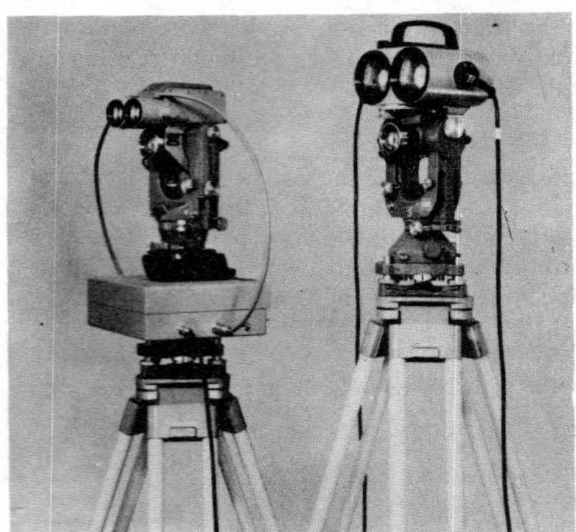

Figure 6.5 (left) Distomat D10 (right) Distomat D13 (Courtesy: Wild Heerlrugg Ltd.)

In the most recent instruments, the cycle of operations is carried out automatically; once set by the operator and after a few seconds delay, the digital display gives a direct reading of the distance required. The following instruments represent a selection of those available at the present time which have many applications in the civil engineering construction industry.

The geodimeter 700 shown in Figure 6.3 is combined with a field computer to give a digital display of distance to an accuracy from ±5 mm/km over a range from 100 m with a small plastic reflector to 5 km with 6 prism reflectors. This instrument uses a gas laser as light source, the slope distance can be automatically corrected to the horizontal for distances up to 500 m and thus the movements of the reflector can be followed on the digital display when setting out. In addition, an automatic vertical circle index and a horizontal circle are provided, so that measured horizontal and vertical angles are displayed as a digital output in degrees, minutes and seconds of arc to the nearest two seconds. An automatic tape punch recording device is available.

The distomat DI 10 shown in Figure 6.5 may be fitted to the top of some standard theodolites to provide a unit for distance and angular measurement combined. The distance is read on a five figure display to the nearest 1 cm. Its range is up to 600 m with one prism reflector to 1 000 m with a six prism reflector attachment. A later, short range, development of this instrument, the DI 3 distomat, is combined with a field computer to read the horizontal distance with an accuracy from ±5 mm to ±10 mm for a range up to 300 m. This instrument, shown on the left in Figure 6.5, gives readings of height difference in addition to distance. Another instrument made in Switzerland, the Kern DM 500, is adapted to slide over the eye-piece of the DKM 2A

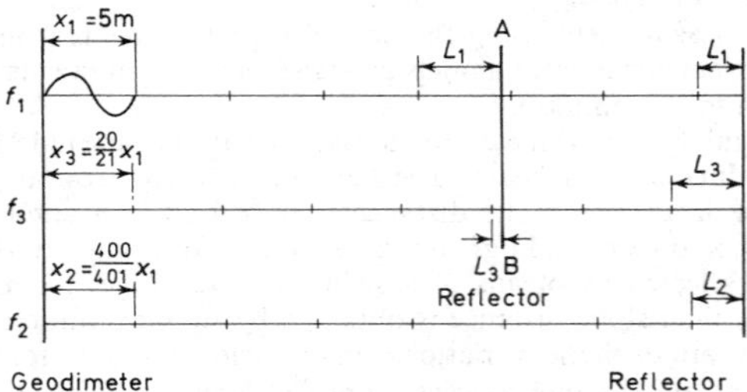

Figure 6.6 Derivation of distance

theodolite (page 6–13). In this instrument the light source is an infra-red beam with a range from 300 m with one reflector to 500 m with three reflectors. Within 15 seconds the slope distance is computed and displayed directly to an accuracy of ±10 mm throughout the range.

Two instruments made in America, the Hewlett Packard distance meter and the Cubitape, have the following similar specifications giving a direct readout of slope distance to an accuracy of ±(5 mm+1 : 100 000) over a range from 1 000 m with a single prism reflector to over 2 000 m with triple reflectors. Similar instruments in the tellurometer range, CD 6, and in the geodimeter range, model 12, are expected to be available in the near future. For long range measurement up to 30 km, the tellurometer microwave instrument CA 1000 consisting of master and remote instruments with a two-way speech channel duplexed on to the radio beam is available. The probable error of a single determination of distance with this instrument is ±(15 mm+5 parts per million).

Finally, a system developed by the National Physical Laboratory may be used for the measurement of vertical and horizontal deflection to an accuracy of ±1 mm in 1 000 m in structures such as dams, bridges, tunnels, piles and towers. The system consists of a laser beam mounted on a rigid support such that the beam passes through a zone plate which is fixed to the structure for which the movements are required to be measured. The zone plate focuses the beam which is projected on to a grid plate mounted on a rigid support on the opposite side of the zone plate. Any movement of the structure is magnified according to distances between the zone plate, laser support and grid plate and is read by means of the position of the spot light on the grid plate.

ANGULAR MEASUREMENT

Bearings and coordinates

The bearing of a line is defined as the horizontal angle between the line and some fixed reference direction.

Whole circle bearing (WCB) is the clockwise horizontal angle from the reference direction to the line.

Reduced bearing (RB)θ is the horizontal angle less than 90 degrees between the line and the reference direction; it is one of the four following directions: Nθ_1E, Sθ_2E, Sθ_3W or Nθ_4W.

Magnetic bearing. When the reference direction is magnetic north, the WCB is a magnetic bearing.

Azimuth. When the reference direction is true north, the WCB is an azimuth. The true north direction is the direction towards the geographical north pole of the meridian passing through the earth's north and south poles.

Magnetic declination is the horizontal angle between true north and magnetic north and is changing continuously at the rate of approximately 10 minutes every year. Its value and rate of change are stated on some Ordnance Survey maps.

The National Grid is superimposed on all Ordnance Survey maps to provide a single reference system for the whole country. It is an orthomorphic or transverse mercator projection with its origin at Longitude 2° W, Latitude 49° N. The central meridian is a straight line with a scale factor of 2499/2500. Thus the scale varies over the whole country plotted on the National Grid network from 0.04% too small at the central meridian to 0.04% too large near the east and west coasts. At about 180 km on each side of the central meridian the scale is exact. These variations have no visible effect upon the representation of topography on the largest scale Ordnance Survey map of 1/1250. The grid is a series of lines drawn parallel and at right angles to the central meridian. Hence the *grid bearing* taken with reference to grid north will differ from both azimuth and magnetic bearing in most cases. The variation between true north and grid north is stated on some

Ordnance Survey maps. Further information is available in three HMSO publications.[3, 4, 5]

Latitudes and departures. In plane surveying, the projection of any line on to the reference direction is called its latitude and the projection on to a direction perpendicular to the reference direction is its departure.

Thus

$$\begin{aligned} \text{LATITUDE N OR S} &= \text{LENGTH} \times \text{COSINE RB} \\ \text{DEPARTURE E OR W} &= \text{LENGTH} \times \text{SINE RB} \end{aligned} \tag{7}$$

The theodolite

There are many variations of the theodolite available commercially, designed for different requirements in accuracy and classes of work. However, the basic functions of this instrument are to measure the horizontal angle between two lines regardless of their difference in gradient, and to measure the vertical angle to any point with reference to a horizontal plane defined by a spirit level attached to the vertical circle index. Provision is made for levelling the instrument, which is mounted on a firm tripod, and for centering over a fixed point on the ground. The modern theodolite is robust, precise, fully protected against adverse weather conditions, easy to use by day or by night, reliable, compact and generally adaptable for many applications in civil engineering where accurate measurement of angles is required. There are too many refinements in the construction and measuring techniques used in modern theodolites to describe fully in this section. A good basic instrument, satisfactory for most applications in civil engineering, other than the most precise work required for primary control survey on a very large scale, is

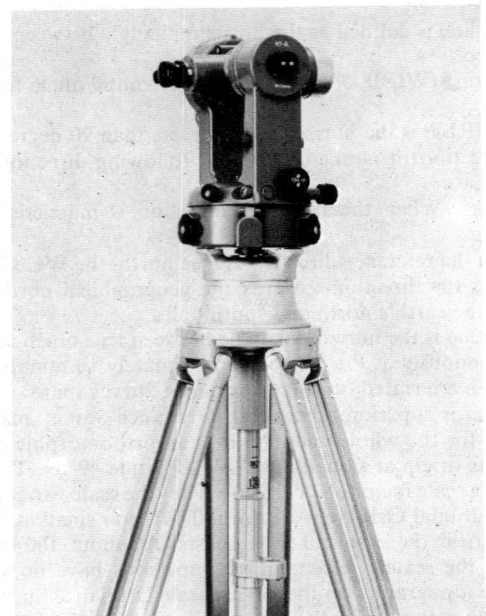

Figure 6.7 Kern theodolite K1A (Courtesy: Kern & Co. Ltd)

one which will read directly to 10 or 20 seconds of arc, fitted with an optical plumb or central levelling rod attachment for precise setting up; it will have glass circles with optical mean reading micrometers. Preferably, the vertical circle will be provided with an automatic compensator to ensure that the plane of reference for measurement of vertical angles is always truly horizontal. Instruments of this class, with some or all of these features, are manufactured by Rank Precision Industries (ST series microptic theodolite), by Wild Heerbrugg (the T1A) and by Kern (the K1A). For more precise work, such as large-scale control surveys for tunnel alignment, requiring direct reading of angles to the nearest one second of arc, one of the most modern theodolites at present available is the Kern DKM2A. Geodetic theodolites are in the class of instruments above this, reading by estimation to 0.1 seconds of arc and have applications in civil engineering work only in special cases such as the measurement of deformation of dams and setting-out work demanding the highest precision. As a typical example of a modern theodolite in current use by large numbers of civil engineers all over the world, a brief description of the K1A is given.

In Figure 6.7 it is seen that the external controls are reduced to the minimum number possible. Horizontal and vertical circle clamp screws are eliminated by a friction coupling device. Only one slow-motion control screw is provided for taking horizontal circle readings and one for vertical circle readings. There is an automatic pendulum compensator, see Figure 6.8, to ensure that the true vertical angle is read, regardless of the level of the horizontal plate which, in Figure 6.8(b), is shown at angle β to the horizontal. Readings of the horizontal circle may be changed or set to zero using an auxiliary control under a safety cover. The main levelling foot screws are placed to the side, so that the instrument can be mounted on the tripod shown in Figures 6.7 and 6.9 using the levelling

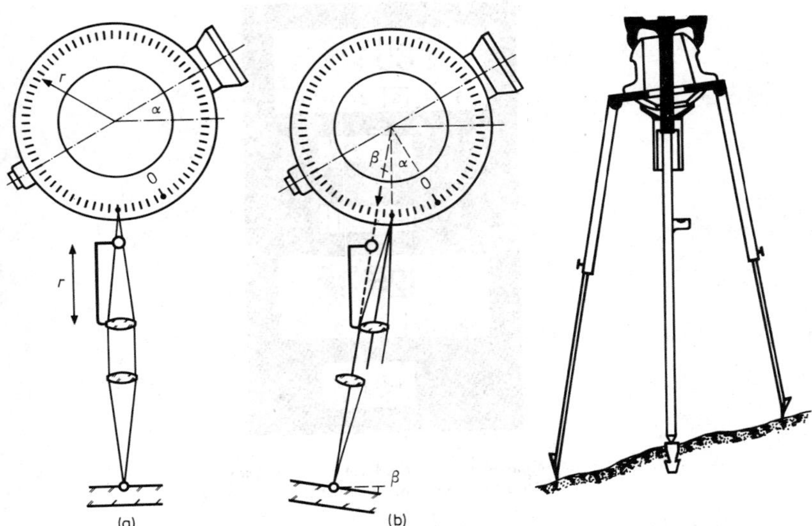

Figure 6.8 Kern vertical circle pendulum compensator (Courtesy: Kern & Co. Ltd)

Figure 6.9 Kern tripod (Courtesy: Kern & Co. Ltd)

rod for centering over the ground station. The point of the rod is first inserted into th peg marking the station and by means of the adjustable tripod legs the central rod made vertical as seen by the circular level bubble attached. The instrument is the clamped to the top of the tripod, a fine adjustment made to the spherical seating required and then the three foot screws are used in the usual way to level the pla bubble. The theodolite is now ready for use and examples of the readings of the vertic and horizontal circles are shown in Figure 6.10, from which it may be seen that resul

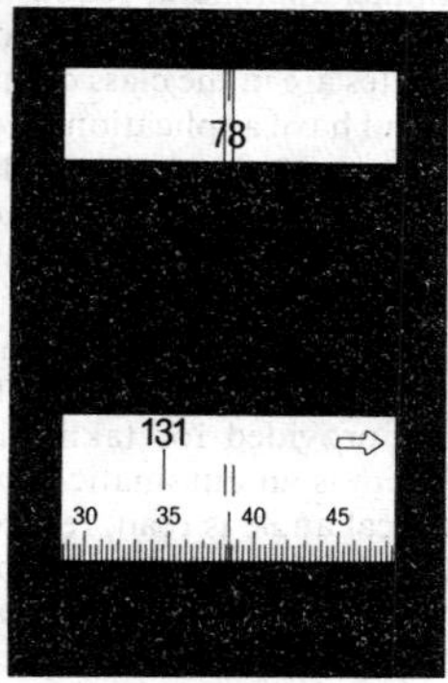

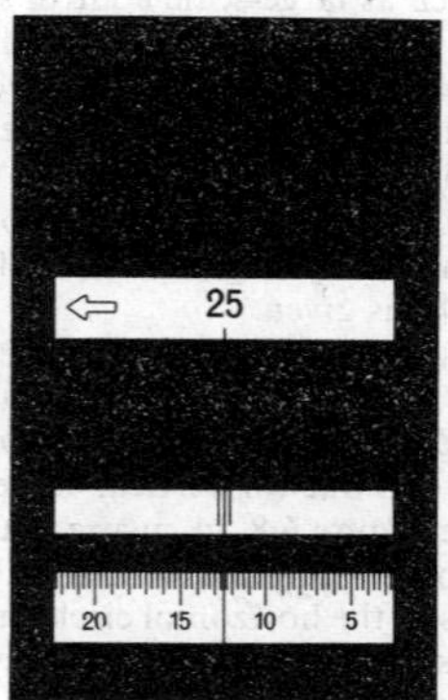

Figure 6.10 Kern K1A circle readings (Courtesy: Kern & Co. Ltd)

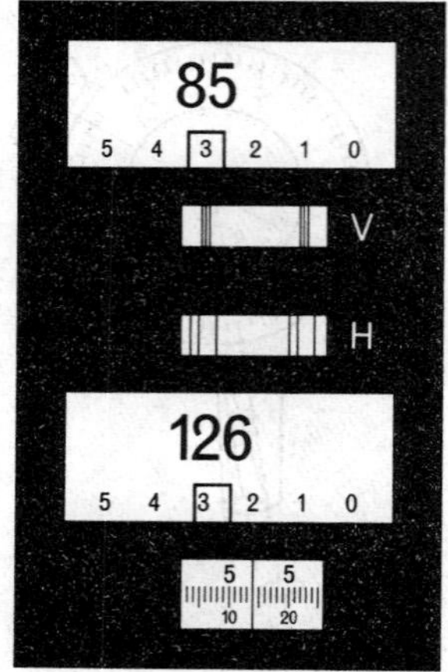

Figure 6.11 Kern DKM2A circle readings
(Courtesy: Kern & Co. Ltd)

estimated to the nearest five seconds are obtained. By operating a switch, the horizonta circle may be read clockwise or anticlockwise in order to facilitate setting out left- o right-hand curves without subtraction from 360°. For rapid angular measurement, th theodolite is made interchangeable with a standard glass target having vertical an horizontal reference marks; the heights of instrument and target are identical and ar read against a metric scale on the tripod centering rod.

The more precise theodolite DKM2A is used in a similar way but when centering closer than ± 5 mm is required, the central rod can be removed from the tripod and the instrument centred using its own optical plumb. An example of the vertical circle reading where the micrometer is set in the 'V' window is shown in Figure 6.11 to be 85° 35′ 14″. The digital circle reading, estimated to the nearest 0.5 seconds of arc, is self-explanatory.

Theodolite field adjustments

The following tests may be carried out in the field but frequent adjustments should not be required.

(1) PLATE BUBBLE Set the bubble central in two directions at right angles, turn through 180° and, if the bubble goes off centre, correct half the error using the main instrument levelling screws. The bubble is now in its true level position and will remain stationary for all directions in azimuth. Use the bubble adjustment screw to place the bubble in its central position with reference to the tube graduations.

(2) HORIZONTAL COLLIMATION Place a levelling staff horizontally at 50 m from, and at the same level as, the axis of the theodolite. In the same straight line as the centre of the staff and the theodolite, at 50 m on the opposite side of the instrument, set up a reference target. Sight the target, transit the telescope and take a reading S_1 on the staff. Turn the horizontal circle through 180° to sight the target again, transit the telescope and take another staff reading S_2.
Then

$$S_1 - S_2 = 4e$$

where e is the horizontal collimation error in 50 m. Correct one quarter of the error by using the adjusting screws for movement of the vertical hair line engraved on the telescope diaphragm.

For further information and specialist requirements such as those for adjustment of pendulum compensators, readers are referred to the appropriate instruction manuals published by the instrument manufacturers.

HEIGHT MEASUREMENT

Datum level

The figure of the earth called a *Geoid* approximates closely to an *Ellipsoid* and for many purposes the surface may be assumed spherical.

A level line as defined by a spirit level is everywhere perpendicular to the direction of gravity. It is therefore a curved line parallel to the mean surface of the earth.

The reduced level of a point is defined as its height above a fixed reference level. In the UK the reference level, established by the Ordnance Survey, is mean sea level at Newlyn, Cornwall, known as the *Ordnance Datum*. Reference levels of points throughout this country are determined by the Ordnance Survey and shown as *Bench Marks* on some O.S. maps. A bench mark list is published by the Ordnance Survey for every 1 km square of the National Grid shown on the 1/2500 O.S. maps. This list gives a description of the bench-mark location with its grid reference, height above mean sea level and the date of levelling. It is common practice in all major civil engineering construction work to refer levels to O.S. datum.

Since the level line is a curved surface, it will be necessary to take this into account when long-distance sights are required, such as across wide rivers or valleys. For this purpose the surface is assumed spherical but, in addition, allowance for terrestrial refraction must be made (see Figure 6.12 where the effect of curvature and refraction is shown).

If D is the length of sight, e_c the error due to curvature, e_r the error due to refraction, R the mean radius of the earth's surface and K is the coefficient of refraction, then

$$e_c = \frac{D^2}{2R} \tag{8}$$

$$e_r = K\frac{D^2}{R} \tag{9}$$

The combined effect

$$e_c - e_r = \frac{D^2}{2R}(1-2K) \tag{10}$$

The values of R and K may be taken as 6370 km and 0.07 respectively.

If D is in metres then

$$e_c - e_r = 67.5\times10^{-6}D^2 \text{ mm} \tag{11}$$

should be added to the apparent level of point B in Figure 6.12.

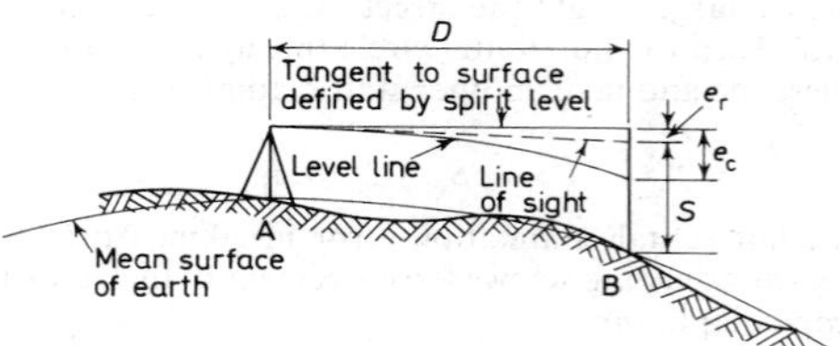

Figure 6.12 Earth curvature and terrestrial refraction

For $D < 172$ m this error is less than 2 mm and negligible for ordinary levelling work in plane surveying. By taking reciprocal observations in long-distance levelling, the effect of these errors is eliminated when the mean difference of level from each end is computed.

The engineer's level

There are four main types of level in common use which may be classified as follows:

(i) Dumpy level
(ii) Tilting level
(iii) Automatic level fitted with a pendulum compensator to ensure that the line of sight is maintained horizontal automatically
(iv) Precise level used in conjunction with a parallel-plate micrometer and special levelling staff.

The dumpy level consists of a telescope with spirit level attached, rigidly connected to a vertical axis by means of a spindle rotating in a bearing allowing free movement in a horizontal plane. Three foot screws are provided for levelling the instrument when mounted on a tripod. Assuming that the spirit level bubble tube axis is perpendicular to the vertical axis and that the collimation line or line of sight is parallel to the bubble tube axis then, once levelled in any position, the dumpy level is ready for use and readings are taken on a levelling staff without any further adjustment.

The tilting level differs from the dumpy level in that the line of sight must be adjusted by means of a tilting screw before each staff reading is taken. In this instrument it is necessary only for the collimation line and bubble tube axis to be parallel, since the line of sight is adjusted with reference to the main spirit level before every reading. The initial setting-up time is less but individual readings will take slightly longer than those using the dumpy level.

An automatic level is shown in Figure 6.13, the Kern GK1A. Its main feature is the automatic compensator, pendulum operated, to ensure that the line of collimation is

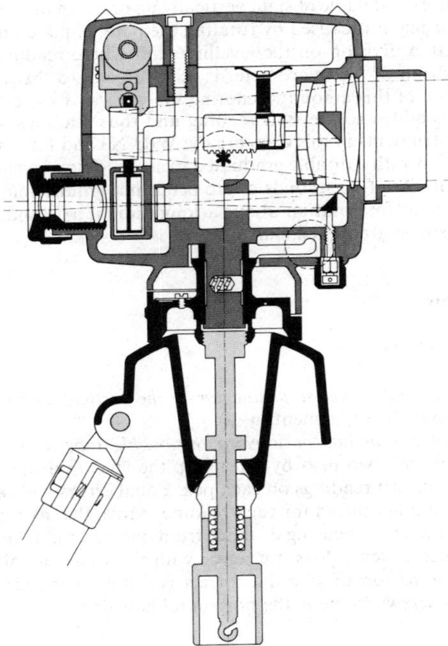

Figure 6.13 Kern automatic level GK1A (Courtesy: Kern & Co. Ltd)

maintained horizontal regardless of a small displacement of the vertical axis. The instrument is mounted on a tripod by means of a spherical seating and clamping screw so that no levelling foot screws are provided. A small circular bubble is used for initial setting up. Selfaligning levels made by Rank Precision Industries, Vickers Instruments, and Wild Heerbrugg are also in common use. Some of these instruments are mounted on a standard three-screw levelling base and the operation of the automatic compensator may be tested quickly by turning a foot screw and noticing that the staff reading returns to its original value automatically. Recently a red warning filter has been introduced into the Kern GKOA automatic level to show the limit of the compensator range.

These three types of instruments are used in conjunction with metric staves graduated in $\frac{1}{2}$ cm or 1 cm units. It is recommended that staff readings are estimated to the nearest 0.2 cm for ordinary levelling work.

For certain classes of work such as the establishment of level control points over a large area, measurement of small deformations or movements due to instability and creep, laboratory applications and setting-out foundations for measuring equipment of high precision, a level is required giving greater reading accuracy than the instruments already described. This is achieved by providing a telescope objective with greater magnification—up to 40 times compared with 20 – 25 times for ordinary engineer's levels—having an increased sensitivity of the main spirit level of 10 seconds of arc per 2 mm graduation compared with 30 seconds, using a parallel-plate micrometer and a special precise levelling staff.

The parallel-plate micrometer is an attachment mounted in front of the telescope objective which displaces the horizontal line of sight vertically up or down but retains its horizontal direction. The displacement is caused by rotating the parallel plate until the line of sight coincides exactly with a division on the levelling staff where readings are taken to the nearest 5 mm. The drum of the micrometer is read directly to the nearest 0.1 mm and estimated to a fraction of this amount, denoting the vertical displacement of the line of sight. This reading is added to the staff reading and thus readings are obtained to the nearest 0.05 mm. Instruments in this class are the Wild N3 and Kern GK23 levels which are used in conjunction with a double graduated invar precision levelling staff. The Wild N2 level with a sensitivity of 30 seconds of arc per 2 mm graduation of the bubble tube has a telescope mounted for rotation about a longitudinal axis and thus two readings may be taken on a single graduated staff.

Level field adjustments

DUMPY LEVEL ADJUSTMENTS

(a) *To make the bubble tube axis perpendicular to the vertical axis of the instrument*—see theodolite plate bubble adjustment, **6**–13.
(b) *To make the collimation line parallel to the bubble tube axis.* Find the correct difference in level between two pegs by setting up the level midway between them and taking the difference of staff readings on each peg. Equal errors will cancel out provided that the bubble is carefully centred for each reading. Move the instrument to a position over one peg or to the nearest focusing distance from one peg and take two staff readings again. If the difference in level does not agree with the first true difference obtained, adjust the line of collimation to give the correct reading on the far staff by using the diaphragm adjusting screws to move the horizontal hair line.

TILTING AND AUTOMATIC LEVEL ADJUSTMENT

Check the line of collimation using the two-peg test in (b) above and refer to the manufacturer's instruction manual for the method of adjustment. This may be by the bubble-tube adjustment screws, by the diaphragm adjustment screws, or other means according to the particular instrument.

INDIRECT MEASUREMENT

Tacheometry

All standard theodolites and levels are provided with a diaphragm consisting of a circular glass plate on to which the image of the staff or target is focused. A vertical line and three horizontal lines are etched on the diaphragm, the vertical line provides an index mark for reading horizontal angles, the centre horizontal line is used for taking levels as already described; the other two horizontal lines are used in conjunction with a levelling staff to determine the distance from the staff to the instrument. This system of indirect measurement of horizontal distance by stadia readings on a levelling staff is called tacheometry and is described fully in reference 8.

There are two methods of tacheometry in common use on civil engineering construction sites:

(i) Vertical Staff Tacheometry using a standard theodolite in conjunction with a normal levelling staff

(ii) Horizontal Staff Tacheometry using a double image selfreducing tacheometer and special horizontal staff.

In method (i) some calculations are required to determine horizontal distance and difference of level but tacheometry tables[9] are available for this purpose. This method is not so accurate as method (ii) where readings of horizontal distance are obtained on a horizontal staff using a special selfreducing tacheometer, such as the Kern DK-RT or Wild RDH, which have built-in compensators in the form of rotating optical wedges to

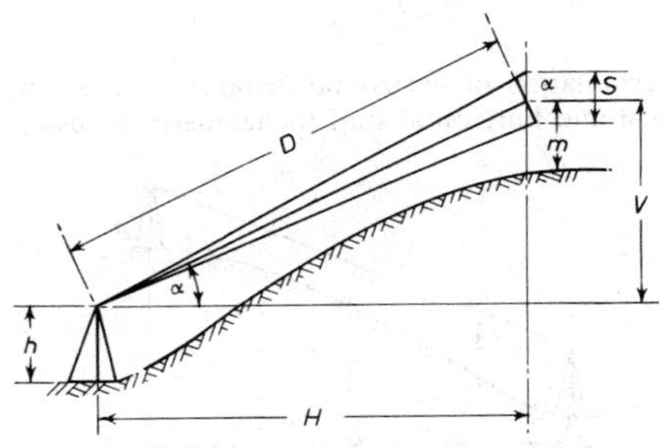

Figure 6.14 Vertical staff tacheometry

allow for inclined sights automatically.[8] The cost of the special instrument is approximately three times that of the standard engineer's theodolite. The order of accuracy in determination of horizontal distance, however, is up to $\pm 1/5000$ for horizontal staff tacheometry, whereas with an ordinary vertical levelling staff graduated in 1 cm units the accuracy may be only $\pm 1/500$.

Figure 6.14 shows the measurements taken in vertical staff tacheometry where the following calculations are required to obtain horizontal distance and difference in level. S is the stadia intercept or difference between upper and lower stadia readings, α is the angle of elevation (+) or depression (−), m is the centre stadia reading and h the height of the instrument above ground level.

For *horizontal sights* it can be shown that

$$H = CS + K \tag{12}$$

where C is a multiplying constant usually 100 and K is an additive constant which varies with H in instruments having internally focusing telescopes. Modern instruments are designed so that for average lengths of sight between 10 m and 100 m the value of K is negligible compared with CS and may be taken as zero, for all practical purposes giving an accuracy from $\pm 1/500$ to $\pm 1/1000$ in H. See *Modern Theodolites and Levels*,[7] pp. 20, 21.

Thus in Figure 6.14

$$D = CS \cos \alpha$$

$$H = CS \cos^2 \alpha \tag{13}$$

$$V = \pm \tfrac{1}{2} CS \sin 2\alpha \tag{14}$$

which reduces to

$$H = 100S\cos^2\alpha \tag{15}$$
$$V = \pm 50S\sin 2\alpha \tag{16}$$

and

$$\text{Difference in level} = \pm V + h - m \tag{17}$$

To facilitate field calculations tables of $100\cos^2\alpha$ and $50\sin 2\alpha$ are available.

In horizontal staff tacheometry, difference of level is obtained using a tangent scale provided on the vertical circle of the tacheometer. The tangent of the angle of elevation or depression is read from the scale and multiplied by the staff reading of H to give V. The heights of instrument and staff are read from the central levelling rod on the tripod and equation (17) is used to find the difference in level between the staff and the instrument station.

Subtense bar

The accuracy of determination of horizontal distance may be improved considerably compared with vertical and horizontal staff tacheometry by using the subtense bar in

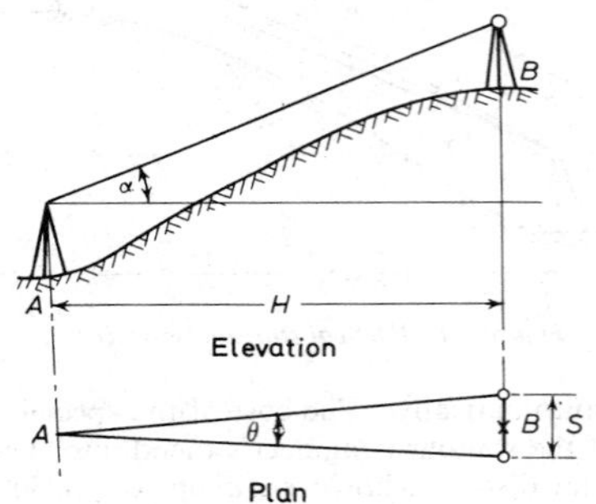

Figure 6.15 Subtense bar

conjunction with a precise standard theodolite, such as the Kern DKM2A already described. The principle of this method is shown in Figure 6.15 where a subtense bar consisting of two targets at a fixed distance S apart is shown mounted on a tripod B perpendicular to the line of sight of the theodolite A. Regardless of the angle of elevation α, the true horizontal angle θ is measured.

Then

$$H = \frac{S}{2}\cot\frac{\theta}{2} \tag{18}$$

For distances up to 100 m where $\theta > 1.5$ deg measurements may be taken in this way. An auxiliary base, as shown in Figure 6.16, is used to avoid measurement of very small angles for distances between 100 m and 1000 m. Angle β is set out at approximately 90° but subsequently measured precisely by taking the mean angle between the subtense bar targets. Angles θ and α are measured to the same degree of precision. It is advisable when setting out the auxiliary base H to choose its length so that angles θ and α are approximately equal. If S is 2 metres,

$$H = (2D)^{\frac{1}{2}} \quad \text{approximately} \tag{19}$$

and D is obtained from equation (20) using the measured values of α, β and θ.

$$D = \frac{\cot(\theta/2)\sin(\alpha+\beta)}{\sin\alpha} \quad \text{metres} \tag{20}$$

Subtense bars manufactured by Rank Precision Industries, Kern, Wild and Zeiss have targets mounted at each end of an invar strip protected by a surrounding aluminium tube

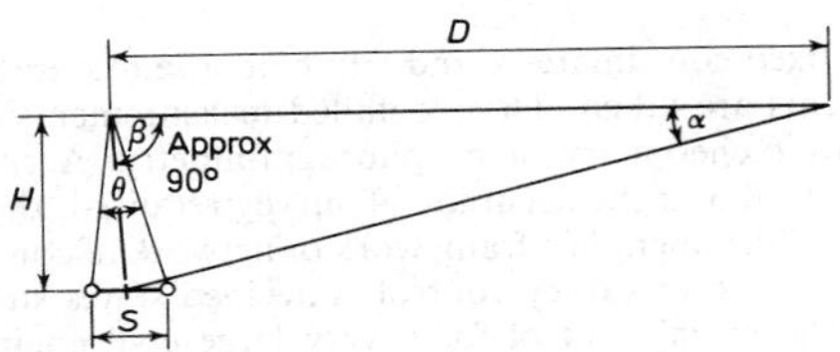

Figure 6.16 Subtense bar auxiliary base

in order to ensure that, for all practical purposes, the length of the bar remains constant at 2 m. Construction details and further information are given in reference 8.

Estimation of accuracy

Where indirect measurements of angles are made to determine horizontal distances and differences of level, the following method is used to estimate the accuracy attained.

If $x = f(I_1 I_2 \ldots)$ is a function of independently observed quantities $I_1 I_2 \ldots$ where the errors in these quantities are $e_1 e_2 \ldots$ respectively, then the error in x is given by

$$e_x = \left[\left(\frac{\partial f}{\partial I_1} e_1\right)^2 + \left(\frac{\partial f}{\partial I_2} e_2\right)^2 + \cdots\right]^{\frac{1}{2}} \tag{21}$$

For example, the distance D is determined using equation (20) from angles measured to the nearest ± 1 second. In order to estimate the probable accuracy in this distance, equation (21) could be applied directly but it is simpler for the purpose of estimation of accuracy to approximate as follows:

With reference to Figure 6.16

$$S/H = \theta \quad \text{and} \quad H/D = \alpha$$

Then

$$D = \frac{S}{\theta\alpha}$$

Assume that there is no error in the length S of the subtense bar. From equation (21):

$$e_D = \left[\left(-\frac{S}{\alpha\theta^2} e_\theta\right)^2 + \left(-\frac{S}{\theta\alpha^2} e_\alpha\right)^2\right]^{\frac{1}{2}}$$

or

$$\frac{e_D}{D} = \left[\left(\frac{e_\theta}{\theta}\right)^2 + \left(\frac{e_\alpha}{\alpha}\right)^2\right]^{\frac{1}{2}} \tag{22}$$

From equation (19), $\theta = \alpha = S/H = (2/D)^{\frac{1}{2}}$

$$e_D/D = \sin 1''(D)^{\frac{1}{2}} \tag{23}$$

If $D = 400$ m then $e_D/D = \pm 1/10\,000$.

ESTABLISHMENT OF CONTROL

Planimetric control

Control points, with fixed coordinates x and y in plan, are required before any detailed surveying measurements are taken. These detailed measurements may be obtained by offsets using tapes, by tacheometry or by photogrammetry. According to the overall size of the area concerned and the accuracy of survey required, there are various techniques available for establishment of a framework or network of control points. The basic principle on which the type of survey control is decided states simply 'work from the whole to the part'. The initial control for a very large civil engineering construction site is carried out by triangulation, trilateration or a combination of both techniques. Usually secondary control is established between the main control points by traversing.

Triangulation is the traditional surveying method where a base line is measured to an accuracy from $\pm 1/50\,000$ to $\pm 1/500\,000$ and from this base line a network of triangles is formed over the whole area. All the angles of all the triangles are measured. The coordinates of the control points are established by solution of the triangles, using a continuous application of the sine rule

$$\frac{a}{\sin A} = \frac{b}{\sin B} = \frac{c}{\sin C}$$

after adjustment of the observed angles to satisfy certain equations of condition. With the introduction and development of electronic distance measuring equipment, triangulation as a method of control survey is becoming obsolete.

Trilateration is a method of establishment of planimetric control where all the sides of a network of triangles are measured directly. In order to improve the rigidity of the network and the consequent accuracy of the final coordinates, it is usual to take both linear and angular measurements. Dual-purpose electronic instruments such as the Aga geodimeter model 700 are available for this work (see **6**–5). The independent measurement of azimuth of some of the survey lines may be carried out using a gyro-theodolite. This instrument seeks true north automatically, to within one or two seconds of arc, giving the azimuth of any line directly.

The adjustment of this type of network in which angles, distances and azimuths are measured, makes full use of matrix operations and electronic digital computers (see **6**–21).

Traversing is the establishing of a framework of control points in the form of a closed polygon. The coordinates of the points are determined from measurements of lengths and interior angles of the polygon. Most small engineering surveys required for schemes such as road improvement, housing-estate development, etc., are based on closed traverses. The traverse also provides an ideal form of linkage between major trilateration control points where the terminal coordinates of the traverse are known. The traverse lines are established close to main features such as roads, rivers and buildings, etc., which will be plotted on the final plan. The method of linear measurement may be vertical or horizontal staff tacheometry, subtense bar, steel tape or electronic distance measuring equipment according to the final accuracy of control point coordinates required.

The fieldwork is planned to provide checks, such as taking face-left and face-right readings of angles and double checking lengths where necessary, so that mistakes in measurement are avoided. Corrections are made to eliminate systematic errors and the remaining closing errors in latitudes and departures are calculated by taking the difference between their sums. With reference to equation (7),

$$\Sigma(N)-\Sigma(S) = e_y = \text{error in latitudes}$$
$$\Sigma(W)-\Sigma(E) = e_x = \text{error in departures}$$

A correction is made to every latitude and departure, adding or subtracting in each case in order to diminish the total error. In this way the closing error is distributed proportionally between all the latitudes and departures.

Correction to latitude

$$L_n = \frac{L_n \times e_y}{\Sigma(N+S)} \tag{24}$$

Correction to departure

$$D_n = \frac{D_n \times e_x}{\Sigma(W+E)} \tag{25}$$

It is advisable to check that the sum of interior angles is equal to $(2n-4)$ right angles, where n is the number of traverse sides, before calculating the latitudes and departures. For normal engineering traverse surveying, an error of approximately $20''\sqrt{n}$ is acceptable No corrections need be applied to the observed angles, since these will be changed when corrections are made to latitudes and departures using equations (24) and (25).

Variations of coordinates

Figure 6.17 shows a small network of control points in which all the lengths and angles are measured. A general method of adjustment of any system of planimetric control points, called 'variation of coordinates', may be used where an electronic digital computer is available. The stages in the computation are as follows:

(i) Assume provisional coordinates for all points. These coordinates may be obtained either graphically or preferably from a preliminary calculation based on the azimuth of one line and a sufficient number of observed quantities of lengths and angles.
(ii) Calculate rigorously the lengths, azimuths and angles of the control network from these provisional coordinates.
(iii) Set up observation equations with the variation of coordinates, δx and δy of each point, as unknowns.
(iv) Form normal equations and solve these linear simultaneous equations for the unknown variations of coordinates.
(v) Add these variations to the original provisional coordinates.
(vi) Repeat this process a sufficient number of times to reduce the variation of coordinates to negligible proportions and thus obtain the most probable values of final coordinates of all the points.

With reference to Figure 6.17 the observation equations are expressed as follows:

$$\delta l_{AB} = -\delta x_A \sin\theta_{AB} - \delta y_A \cos\theta_{AB} + \delta x_B \sin\theta_{AB} + \delta y_B \cos\theta_{AB} \tag{26}$$

$$\delta\theta''_{AB} = \frac{1}{l_{AB}\sin 1''}(-\delta x_A \cos\theta_{AB} + \delta y_A \sin\theta_{AB} + \delta x_B \cos\theta_{AB} - \delta y_B \sin\theta_{AB}) \tag{27}$$

If O_l is the observed length, C_l the computed length from provisional coordinates and δl the change in length expressed in equation (26), then (28)

$$O_l - C_l = \delta l$$

Similarly, for angular measurement α,

$$O_\alpha - C_\alpha = \delta\alpha \tag{29}$$

$\delta\alpha$ is given by the difference in changes of azimuths for the two lines containing angle α, e.g. for the angle BAD,

$$\delta\alpha_{BAD} = \delta\theta_{AD} - \delta\theta_{AB}$$

For the network shown in Figure 6.17 there is one equation for each observed length, similar to equation (28), and one equation for each observed angle, similar to equation (29), making a total of 15 observation equations.

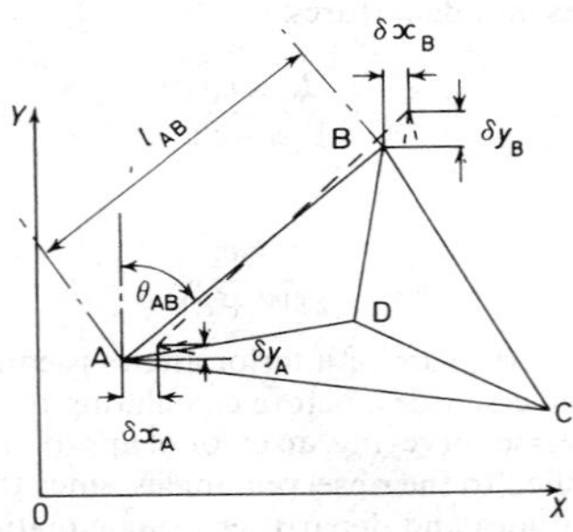

Figure 6.17 Variations of coordinates

If point A and the direction AB are regarded as fixed then

$$\delta x_A = \delta y_A = 0$$

and

$$\delta y_B = \delta x_B \cot \theta_{AB}$$

Hence there are five unknowns

$$\delta x_B, \delta x_C, \delta y_C, \delta x_D \text{ and } \delta y_D.$$

The 15 observation equations are expressed in matrix form thus

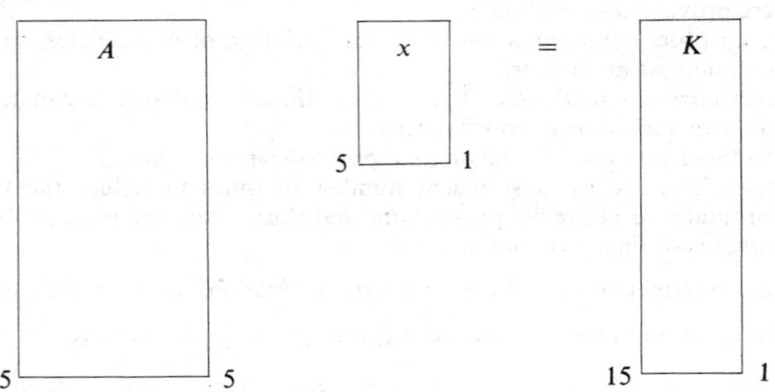

where A is the matrix of coefficients of the unknowns from equations (26), (27) and K the column matrix of constant terms $(O-C)$.

For which the solution is found by forming the normal equations

$$A^T A x = A^T K \tag{30}$$

and taking the inverse of $(A^{T}A)$

$$x = (A^{T}A)^{-1}A^{T}K \tag{31}$$

For further information and methods of weighting, see reference 11.

Level control

Provision of control points for levelling is particularly important for surveys carried out by photogrammetry. Ordinary levelling techniques are used in most cases where the reduced levels of a network of control points are determined using an engineer's level and vertical staff. If closed loops of level controls are formed, then the closing errors must be distributed throughout the network. The usual method of adjustment is to set up condition equations in terms of the unknown corrections to the observed differences in level and to calculate the most probable values of the corrections using the principle of 'least squares'.[12]

Where electronic distance measurement techniques are used over long distances for planimetric control, it is usually more convenient and quicker to observe differences of level by reciprocal trigonometrical levelling. Then the observed slope distances may be reduced to the horizontal immediately.

Figure 6.18 shows how these observations are taken so that errors due to earth curvature and terrestrial refraction evaluated in equations (10) and (11) are eliminated.

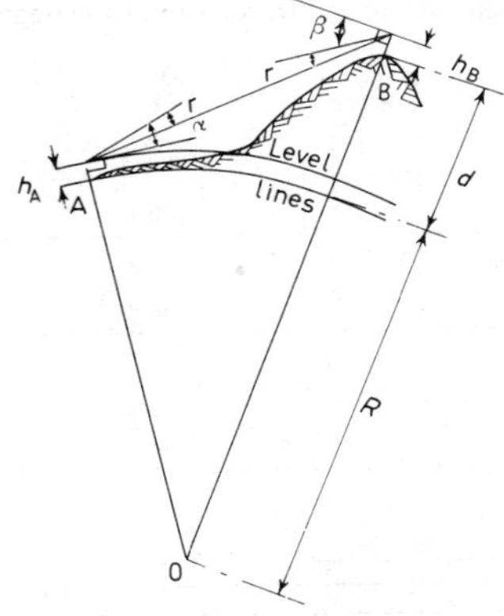

Figure 6.18 Trigonometrical levelling

Modern equipment for trigonometrical levelling has the theodolite and target at the same height above ground level. This height may be read from the scale provided on the tripod central levelling rod. Reciprocal sights are taken from A and B by interchanging the instrument and target without removing the tripods. Angles of elevation α at A and depression β at B are observed together with the heights of instrument above ground level h_A at A and h_B at B. Then the difference in ground levels between A and B is given by

$$d = D\tan\left(\frac{\alpha+\beta}{2}\right)\left[1+\frac{D}{2R}\tan\left(\frac{\alpha+\beta}{2}\right)\right]+h_A-h_B \tag{32}$$

where D is the distance between A and B and R the mean radius of the earth.

The accuracy of the determination of d is estimated using equation (21) taking $e_\alpha = e_\beta = \pm 1''$ and $e_D = \pm 1/50\,000$ approximately.

PHOTOGRAMMETRY

If ground points both planimetric and level control are provided by ground surveyors and identified in overlapping stereo pairs of photographs, then nearly all site plans, contours and data in the form of digital ground models required by civil engineers can be provided by photogrammetry. In the past ten years many technological advances in equipment and improved techniques have been introduced, making it possible to take measurements in the stereoscopic model formed by a pair of overlapping vertical air photographs to a high degree of accuracy.

Basic principles

Photogrammetry is the science of taking measurements from photographs, usually in stereo pairs, to provide surveying data in the form of three-dimensional coordinates, contoured plans, profiles and cross sections drawn to scale. Photographs are taken from an aircraft fitted with an automatic film camera mounted with its axis as near vertical as possible. Exposures, up to 320 in number on one film, are taken at regular time intervals in flight. At each exposure the film is held flat against a register glass by a vacuum pump, ensuring a negative with distortion less than ± 0.01 mm. Collimation marks in the form of crosses at the centre of each side and at the centre of the negative or in the four corners

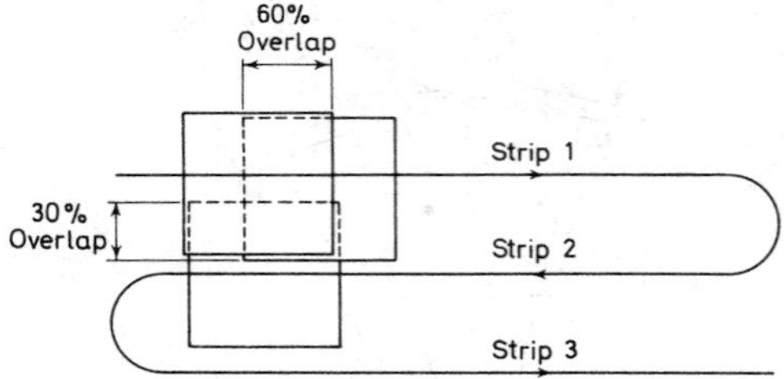

Figure 6.19 Air photography flying pattern

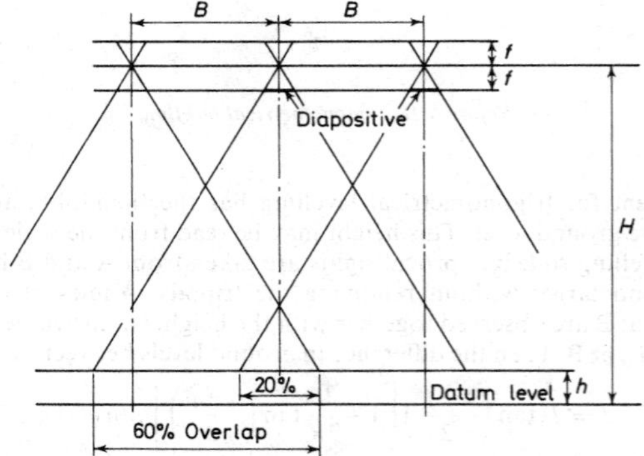

Figure 6.20 Air photography longitudinal overlaps

are provided, together with a data panel on each negative giving such information as flying height, focal length of lens, serial number, etc. Exposures are timed to give a 60% longitudinal overlap between successive images appearing on the final prints or diapositives. The route followed by the aircraft is shown in Figure 6.19 where the flying strips can be seen covering the whole area of the survey with 60% longitudinal overlaps and 30% transverse overlaps between strips. A wide-angle lens with a total field coverage

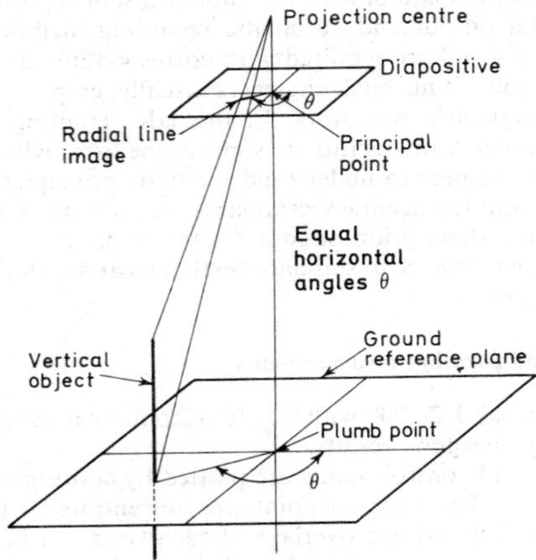

Figure 6.21 Vertical photography central projection

of approximately 93°, equivalent focal length 152 mm and aperture $f/5.6$ giving a negative size of 23 cm × 23 cm, such as the Wild Heerbrugg Aviogon, is used for most applications of photogrammetry to civil engineering surveys.

A longitudinal section showing the ground coverage of three successive photographs is outlined in Figure 6.20. B is the air base, f the focal length of the lens, H the flying height above datum level and h is the mean ground height above datum level.

The approximate photograph contact scale S is given by

$$S = \frac{f}{H-h} \tag{33}$$

There are a number of practical problems to be overcome before accurate survey data can be obtained from these air photographs:

(i) The aircraft cannot maintain perfectly level flight.
(ii) The camera axis is not always truly vertical.
(iii) The height of flight above datum level cannot be measured with precision and varies at each air station.
(iv) The position of air stations at the ends of the air base B cannot be measured in relation to any ground survey control.
(v) A transformation from the photograph central projection to the site plan orthogonal projection is required. In Figure 6.21 it can be seen that, for a truly vertical photograph, the angle between two radial lines from the principal point is a true horizontal angle regardless of the scale of the photography and of any displacement of the image due to vertical relief on the ground. This radial line principle may be used to produce an orthogonal projection of detail on a site plan from pairs of overlapping photographs.

To overcome these practical difficulties the first essential is to provide both planimetric and level control points on the ground which can be identified in the common overlaps of stereo pairs of photographs. These points are surveyed and may be pre-marked before photography for easy identification. Alternatively, conspicuous points of detail identified in the photographs may be surveyed on the ground after photography. Secondly, according to the scale and accuracy of the final contoured plan or survey data required, the stereo pairs of photographs are set up for measurement in various types of photogrammetric laboratory equipment. The capital cost of sophisticated stereo plotting equipment with digital output and automatic recording facilities is initially high. If large-scale plans up to 1/500 are required with corresponding accuracy in coordinates and reduced levels, consulting civil engineers usually employ specialist firms with experienced photogrammetric operators. In the UK, Hunting Surveys Ltd, Fairey Surveys Ltd and Meridian Airmaps Ltd are some of the firms who undertake this work. It is necessary for the engineer to understand the basic principles of these photogrammetric measurements and the accuracy attainable, in order to specify his requirements most efficiently. A brief description of some of the techniques and typical measuring equipment is given with reference to standard textbooks and technical papers for further information where required.

Radial line methods and parallax measurements

For a small-scale plan, say 1/20 000 with a 5–10 m contour interval, simple photogrammetric equipment may give good results.

The planimetric ground control points are plotted by coordinates using a 1 km grid, i.e. 5 cm at 1/20 000 scale. These control points are concentrated in the corners of the area to be surveyed, in the first and last overlaps of each strip and in some of the extreme lateral overlaps. Few points are required in the central area of the survey. All these points are marked on the photographs and the centre points of each photograph, called *principal points,* are transferred to the next photograph of the strip in the common overlap. In addition, points of conspicuous detail not surveyed on the ground but appearing in the common longitudinal and transverse overlaps, called *minor control points,* are marked on all the photographs. Then each photograph has nine points marked as shown in Figure 6.22 which represents the second photograph P_2 in the first strip of photography. Radial lines from the principal points are drawn through these marked

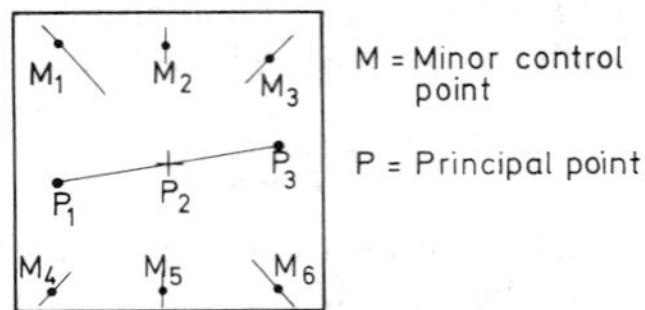

Figure 6.22 Photograph control points

points on a transparent template covering each photograph. A slotted template cutting machine (available from C. F. Casella and Co. Ltd) is used to cut precision slots along the radial lines in the transparent templates. Circular holes are punched at each principal point. Studs are provided to fit the holes and slots. Commencing at the fixed ground control points the templates are assembled over the 1 km grid already plotted. Using studs in the slots which become fixed in plan position through multiple intersections of the radial lines at points in the common overlaps of the photographs, a network of control points is established.[13]

In order to plot details on the plan a simple radial line plotting machine is employed, such as the SB 100 manufactured by Rank Precision Industries Ltd. The photographs are mounted successively in pairs in the plotter which is placed over the control point plan.

There is a small range of scale adjustment provided so that the machine scale is fitted to the ground control. The operator views the stereoscopic model obtained from the pair of photographs and, by moving a point defined by the intersection of two lines in the model, along any ground feature, plots the details automatically in orthogonal projection to a scale of 1/20 000. A full description of this machine, its principle and method of

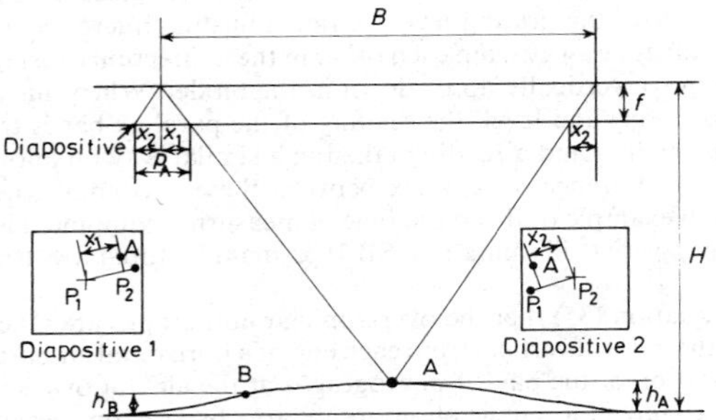

Figure 6.23 x-parallax

operation can be found in a technical publication obtainable from Rank Precision Industries Ltd. This method of plotting will not eliminate entirely the distortions due to relative tilt between the photographs when taken from the air stations. With good flying within 3° of the vertical, provided that ground heights do not exceed about 8% of the flying height, then for small-scale plans the errors in plotting are negligible. It should be noted that the scale of the plan must be near to the average contact scale of the photography. This means that with f = 152 mm, the flying height H is approximately 3000 m for a plan scale of 1/20 000.

The principle of determination of height by measurement of x-parallax is shown in Figure 6.23.

$$x\text{-parallax} = x_1-(-x_2) = p_A$$

The air base is B, the focal length f, the flying height H, the height of point A is h_A and the height of point B is h_B.
By similar triangles

$$p_A = \frac{fB}{H-h_A} \tag{34}$$

It follows that

$$h_A-h_B = \frac{(H-h_A)(p_A-p_B)}{p_B} \tag{35}$$

H is estimated from measurements taken in the common overlap between control points P and Q of known height and distance apart on the ground.

$$H = \frac{fD}{d}+h_m \tag{36}$$

where D is the ground distance PQ, d is the photo distance PQ and h_m is the mean height of P and Q.
The parallax of point A is measured using a travelling microscope taking the difference in x coordinates of this point, see Figure 6.23. Subsequently, differences of parallax between A and any other point are measured in the stereoscopic model using a *parallax bar*. Hence equation (35) gives the approximate height of any point relative to the known

height of point A. The technique of using a parallax bar or stereometer is extremely important as an introduction to the use of every type of stereo-plotting machine because of the idea of the 'floating mark'. It consists of seeing the model formed in the common overlap of a pair of photographs through a stereoscope and placing a pair of measuring marks, consisting of black dots or spots of light, on each of the pair of photographs covering the same point of detail. Then in the model only one measuring mark appears as a 'floating mark' above the ground level. A fine adjusting micrometer screw moves the pair of marks towards or away from each other in the x direction causing the 'floating mark' to appear to move vertically up or down in the model. When this mark appears to the observer to be at ground level, the reading of the parallax bar is taken. The difference between this reading and a reading taken in a similar way on another point is a direct measure of the difference in parallax between these two points, i.e. $(p_A - p_B)$ in equation (35). A good example of this basic type of measuring equipment is the scanning stereoscope with diapositive illumination SB 190, manufactured by Rank Precision Industries Ltd.

It is assumed in equation (35) that the overlapping photographs are taken without tilt and with the axis of the camera vertical from each end of a horizontal air base. A difference in the relative orientation of the pair of photographs at the air stations is inevitable and gives rise to a distorted model when the photographs are observed on a level base through a simple stereoscope such as that described above. The resulting errors in height measurement are very much larger than those obtained in the radial line planimetric control using the same photography. It is essential that corrections are made to height measurements in order to achieve an acceptable accuracy in levels of approximately 1/5000 of the flying height. Differential longitudinal tilt results in a parabolic deformation of the datum plane and differential transverse tilt results in a hyperbolic paraboloid deformation. It is shown in reference 14 that under certain assumptions

$$p' = p + a_0 + a_1 x + a_2 y + a_3 x^2 + a_4 xy \tag{37}$$

where $(p' - p)$ is the correction required to the observed x-parallax of a point in the common overlap with coordinates (x, y), and a_0, a_1, a_2, a_3, a_4 are constant coefficients taking into account the parabolic and hyperbolic paraboloid deformations of the datum plane. These five coefficients are obtained using five ground control points of known height in each overlap. The method is described fully by Professor Thompson in *Heights from Parallax Measurements*[15] and is now published in *Plane Surveying*[16] and *Elementary Air Survey*.[13]

Stereo plotting equipment

In general, there are 12 degrees of freedom between a pair of photographs taken from two air stations. With reference to an X, Y, Z coordinate system, the degrees of freedom are the three displacements x, y, z along the coordinate axes and the three corresponding rotations, ω, ϕ and κ about these axes, for each photograph. It is possible to determine seven of these unknown parameters by measurements referred to ground control points in the procedure called *absolute orientation* and the remaining five parameters by *relative orientation*. These procedures are carried out by mounting the diapositives in a stereo-plotting machine such as the Thompson–Watts Plotter, illustrated in Figures 6.24 and 6.25.

The diapositive plate carriers shown at A and B may be rotated independently about the machine Y and Z coordinate axes. The machine X-axis is common to both plates hence the differential rotation of one plate relative to the other about the X-axis is achieved by rotating one plate only. By viewing the stereoscopic model formed by the overlapping pair of photographs, the machine operator places the two diapositives at A and B (Figure 6.25) in the same relative positions as they occupied at the air stations. This process is called *relative orientation*; the five parameters $\kappa_A \kappa_B$, ϕ_A, ϕ_B and $\delta\omega$ are fixed and a true stereo model, free from distortion, is formed. The plotter is provided

Figure 6.24 Thompson–Watts plotter Model 2[17] **(Courtesy: Rank Precision Industries Ltd)**

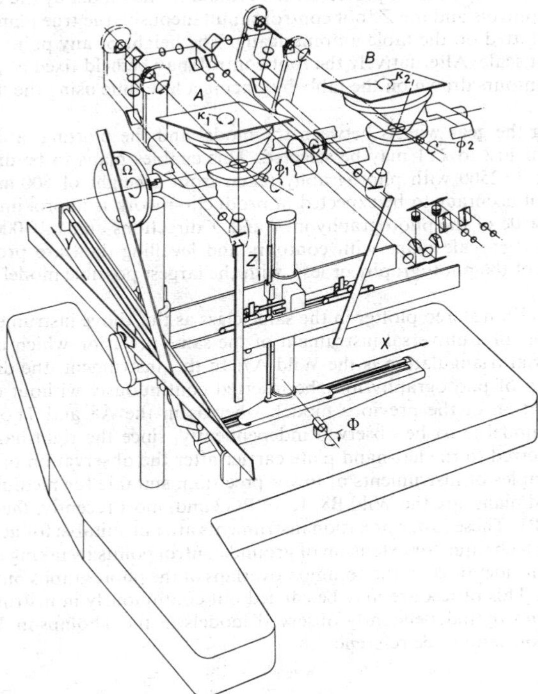

Figure 6.25 Thompson–Watts plotter, schematic representation **(Courtesy: Rank Precision Industries Ltd)**

with micrometer drums for measuring X and Y coordinates directly to 0.02 mm and a scale for Z coordinates reading directly to 0.1 mm in the model.

It is necessary now to refer to ground control points in order to fix the remaining seven degrees of freedom by *absolute orientation*. This process is carried out in two parts, first scaling and second levelling. The XY machine coordinates of two points P and Q in the model are measured and the distance between these points compared with the ground distance obtained from the ground coordinates X'_P, Y'_P, X'_Q, Y'_Q. The scale of the model in the machine is changed by a control provided and thus four of the remaining seven parameters are fixed. The model, considered as a rigid body, is now rotated about the X and Y machine coordinate axes, without upsetting the relative orientation between the two diapositives. The three remaining parameters are fixed with reference to the known heights of at least three ground control points. It is usually more convenient to measure the heights in the model of four points near to the corners of the overlap and to compare the differences of heights with those obtained from the known ground levels of these points. The required rotations of the model about the X and Y machine axes are calculated and rotations Φ about the Y axis and Ω about the X axis are applied using the machine controls provided.

The model is now fitted completely to ground control and plotting is commenced on the coordinatograph table seen in Figure 6.24. Through a mechanical linkage and gear box, a pencil on the table follows the 'floating mark' seen in the stereoscopic model. It is necessary to orient and locate the plotting paper on the table so that the XY ground control system of coordinates coincides with the XY model coordinate system. Then provided the 'floating mark' is placed on the ground in the model by the operator using the XY hand controls and the Z foot control simultaneously, the true plan at the correct scale will be plotted on the table automatically. The height of any point is read on the machine height scale. Alternatively the foot control may be held fixed at predetermined heights and contours drawn on the table by tracing a level line using the 'floating mark' in the model.

By changing the gear wheels between the model and the plotting table a model-to-plan ratio from 1:8 to 2:1 may be selected. This enables plans to be drawn to scales from 1:250 to 1:2500 with photography taken from a height of 500 m for example. The standard of accuracy to be expected in production work is approximately 6 micrometres at the scale of the photography in X and Y directions and 1/10 000 of the flying height in Z. Large-scale plans with contours and levelling data are produced at five times the scale of the photography or less, with the largest possible model scale, for best results.

Another precision stereo plotter in the same class as the above instrument is the Wild A8. An example of a universal instrument of the same precision which is designed for continuous aerial triangulation is the Wild A7. In this instrument, the diapositives for one whole strip of photography may be inserted continuously without disturbing the relative orientation of the previous model, whereas in the A8 and Thompson–Watts plotters each model must be observed independently, since the right-hand diapositive must be transferred to the left-hand plate carrier after the observation of any model of the strip. Examples of instruments of lower precision, suitable for medium- and small-scale contoured plans, are the Wild B8, Kern PG2 and, most recently, the Cartographic Engineering CP1. These lower precision instruments are not suitable for aerial triangulation, which is a technique for extension of ground control points by taking measurements to selected points identified in the common overlaps of the photographs but not surveyed on the ground. This procedure may be carried out continuously in instruments such as the Wild A7 and by independently observed models in the Thompson–Watts plotter. For further information, see reference 18.

Applications

Civil engineers engaged on highway location and design use plans and data prepared by photogrammetry in all stages of their work. For the reconnaissance survey, where no suitable maps are available, small-scale maps are prepared from super-wide-angle lens photography. In most developed countries, existing maps are available and may be supplemented by a photo-mosaic to say, 1/25 000 scale, where contact prints of recent photographs are assembled, approximately matched at the edges of each overlap. It is possible to improve the matching at the edges by rectification of the prints taken from the negatives, in order to eliminate scale and tilt variations, before assembling the mosaic. These controlled mosaics are not equivalent to a map of the area, however, owing to relief displacement caused by height differences in the topography, and for reconnaissance work an uncontrolled mosaic may be quite adequate. An interpreter, trained to study aerial photographs, can obtain useful information about types of soil and vegetation, geological features and land forms, etc., as well as new developments which take place after the existing maps are prepared. Research into the use of colour photography for this purpose and for mapping is producing good results—see *Manual of Color Aerial Photography*[19] and *Use of Colour Photography for Large-scale Mapping*.[20]

The survey for preliminary planning, where the work is concentrated in the area proposed for a new highway is usually carried out by photogrammetry. A plan from 1/5000 to 1/2500 scale, with a contour interval 5 m to 2 m, is prepared from air photographs. Finally, once the route has been selected, new low-level photography is taken from about 500 m flying height in order to prepare site plans for all interchanges, bridges and service areas to a scale of 1/500. Contours at 1/2 m interval can be drawn if required.

A recent development is the use of digital ground models for route selection and final design of motorway interchanges. A digital ground model is a mathematical model of the topography where each point is defined by its *XYZ* coordinate. These models are produced most efficiently and economically on a large scale by photogrammetry. The coordinates are measured in a photogrammetric plotting machine, such as the Thompson–Watts model 2 or Wild A8, provided with an automatic digital recording device. The coordinates are automatically recorded on punched cards or tape to provide data for an electronic computer where the information is stored. The coordinates are transformed to the ground coordinate system in three dimensions and a method of interpolation used to determine the level of any point within the digital model, given its *XY* ground coordinates. In this way, the earthwork quantities of any selected alignment of a proposed road are calculated very rapidly and several trial routes are easily compared. Further information is given in reference 21.

Another photogrammetric technique called orthophotography was introduced commercially for the first time in the UK in 1970 by Fairey Air Surveys Ltd. Photogrammetric equipment developed by Zeiss Jena is used to produce a photograph from the original stereo pair of overlapping diapositives in which the details appear in orthogonal projection free from radial displacement due to topographical relief. The stereo pair of photographs is set up in the machine in the usual way. The 'floating mark' automatically scans the model in the *XY* directions whilst the operator keeps the mark in contact with the ground in the model by continuous adjustment of the *Z* control. The image, free from relief displacement, is projected on to film which is later developed in order to produce a photo-plan of the area. Contours are superimposed afterwards from a drop-line chart, automatically produced for the whole overlap during the process. Other equipment such as the Wild B8 Stereomat is available for the automatic production of orthophotographs. In other countries, United States, Sweden, Germany and France for example, orthophotography is used for regular production of photomaps at scales from 1/25 000 to 1/2000.

Finally, many interesting projects are in progress using photogrammetry in studies of architecture, archeology and structural engineering. A recent example is the study of a three-dimensional, large-scale model of a suspended cable roof structure using stereometric cameras. These cameras are mounted at each end of a rigid base support at a fixed distance apart, from 500 mm to 2 m, and a pair of overlapping photographs are taken. Ground control points for scaling and levelling are used in the same way as that already described for aerial surveying and the diapositives are set up in a plotting machine such as the Thompson–Watts. *XYZ* coordinates of points in the structural model are measured and transformed to the ground control system of coordinates to study the structural form of the model. Loading tests may be carried out and deflections of a large number of points accurately measured very rapidly in the photogrammetric plotting machine. If the stereometric cameras are mounted at a height of 5 or 6 m above the model with a base length of 2 m, the positions of points such as the intersection of suspended cables may be measured to an accuracy of ± 1 mm.

Acknowledgments are due to Aga (UK) Ltd, Kern and Co. Ltd, Survey and General Instrument Co. Ltd, Rank Precision Industries Ltd, Fairey Surveys Ltd, Wild Heerbrugg Ltd, Tellurometer (UK) Ltd and A. Clarkson and Co. Ltd.

REFERENCES

1. MILLER, R. M., *Accuracy of measurement with steel tapes,* Building Research Paper CP 51/69
2. PENMAN, A. D. M. and CHARLES, J. A., *Measurement of movement of engineering structures,* Building Research Paper CP 32/71
3. *An introduction to the projection for Ordnance Survey maps and the National Reference System,* HMSO (1951)
4. *Constants, formulae and methods used in transverse mercator projection,* HMSO (1951)
5. Projection tables for the transverse mercator projection of Great Britain, HMSO (1951)
6. SANDOVER, J. A. and BILL, CDR R., *Measurement of distance by radio waves and its application to survey problems,* Paper No. 6643, ICE (1963)
7. COOPER, M. A. R., *Modern Theodolites and Levels,* Crosby Lockwood (1971)
8. HODGES, D. J. and GREENWOOD, J. B., *Optical Distance Measurement,* Butterworths, London (1971)
9. MUNSEY, D. T. F., *Tacheometric tables for the metric user,* Technical Press (1971)
10. BURNSIDE, C. D., *Electromagnetic Distance Measurement,* Crosby Lockwood (1971)
11. ASHKENAZI, V., 'Solution and error analysis of large geodetic networks', *Survey Rev.,* XIX, No. 146 (1967); No. 147 (1968)
12. RAINSFORD, H. F., *Survey Adjustments and Least Squares,* Constable (1957)
13. KILFORD, W. K., *Elementary Air Survey,* Pitman (1969)
14. THOMPSON, E. H., 'Correction to x-parallaxes', *Photogrammetric Record,* **VI,** No. 32 (1968)
15. THOMPSON, E. H., 'Heights from parallax measurements', *Photogrammetric Record,* **I,** No. 4 (1954)
16. SANDOVER, J. A. and MALING, D. H., *Plane Surveying,* Edward Arnold (1961)
17. THOMPSON, E. H., 'The Thompson-Watts plotter model 2', *Photogrammetric Record,* **IV,** No. 23 (1964)
18. *Manual of Photogrammetry,* American Society of Photogrammetry (1966)
19. *Manual of Color Aerial Photography,* American Society of Photogrammetry (1968)
20. WOODROW, H. C., 'Use of colour photography for large scale mapping', *Photogrammetric Record,* **V,** No. 30 (1967)
21. *Photogrammetry and Engineering,* Planning and Transport Research and Computation Co. Ltd (1968)

BIBLIOGRAPHY

ALLAN, A. L., HOLLWEY, J. R. and MAYNES, J. H. B., *Practical Field Surveying and Computations,* Heinemann (1968)

ASHWORTH, R., *Highway Engineering,* Heinemann (1966)

BANNISTER, A. and RAYMOND, S., *Surveying,* Pitman (1965)

CLARK, D., *Plane and Geodetic Surveying for Engineers,* revised by J. E. Jackson, Constable (1969)

CLENDINNING, J. and OLLIVER, J. G., *Principles and Use of Surveying Instruments,* Blackie (1969)

CRONE, D. R., *Elementary Photogrammetry,* Edward Arnold (1963)

HAYWARD, L. M., *Survey Practice on Construction Sites,* Pitman (1968)

HAZAY, I., *Adjusting Calculations in Surveying,* Akademiai Kiado Budapest (1970)

MEYER, C. F., *Route Surveying and Design,* International Textbook Co. (1969)

MIDDLETON, R. E., CHADWICK, O., *A Treatise in Surveying* Spon (1955)

MOFFIT, F. H., *Photogrammetry,* International Textbook Co. (1967)

RICHARDUS, P. and ALLMAN, J. S., *Project Surveying,* North-Holland (1966)

WHYTE, W. S., *Basic Metric Surveying,* Butterworth, London (1969)

7 GEOLOGY FOR ENGINEERS

GEOLOGY FOR ENGINEERS 7

7 GEOLOGY FOR ENGINEERS

P. G. FOOKES, B.Sc., Ph.D., M.I.M.M., F.G.S.
Consultant

This section introduces civil engineers to some basic geology and outlines the broad concepts of the subject.

Geology is concerned with the science of the Earth and the materials comprising the Earth. This includes *physical geology* or *geomorphology* (the form of the Earth), *palaeontology* (study of fossils), *stratigraphy* (the chronological sequence of rocks), *mineralogy* (study of minerals), *petrology* (study of the composition of rocks) and *structural geology* or *tectonics* (the broad structure of rocks). Together with newer and closely related branches such as geochemistry, geophysics or mathematical geology, and applied and biological aspects, the whole subject is rapidly developing and is now generally being called Earth Science.

Engineering geology is the branch of geology applied to Civil Engineering and, in Britain particularly, is applied to all aspects of foundation and excavation design, construction and performance. The extremes of the subject merge into the practices of Soil Mechanics, Rock Mechanics and some aspects of the Extractive Industries, as sand and gravel or opencast mining (Price[1]).

BASIC GEOLOGY

Introduction

Rock is strictly defined in geology as any natural solid portion of the earth's crust which has recognisable appearance and composition. Some rocks are not necessarily hard and in discussion a geologist may call peat or clay a rock as he would granite or limestone.

There are three major classes of rocks:

(i) *Sedimentary Rocks* formed by the deposition of material at the earth's crust, e.g. sandstone, clay

(ii) *Igneous Rocks* formed from molten rock magma solidifying either at the earth's surface or within the crust, e.g. basalt, granite (*s.l.*).

(iii) *Metamorphic Rocks* produced deep in the earth by the transformation of existing rocks through the action of heat and pressure, e.g. marble, slate

The interrelation and continual recycling of rock over periods of geological time is illustrated in Figure 7.1.

Principles of stratigraphy

Sedimentary rocks cover some 75% of the earth's land surface but form only a discontinuous and relatively thin cover to the underlying igneous and metamorphic rocks of the 'sial'.

The sedimentary layers (*strata*) normally lie one above another in order of decreasing age, but where there has been structural disturbance they are faulted and folded. Study of the strata in a particular area enables their sequence to be recorded, and this can then be compared with other local sequences. From such observations the general succession of sedimentary rocks over a wider area can be established; this has been done, for example, for nearly the whole of the British Isles. A list of strata for England and Wales was compiled by William Smith, 'the father of English Geology'; in 1815 he produced the first simple coloured geological map of the country. As a result of his studies he stated two basic principles of stratigraphy, that 'the same strata are always found in the same order

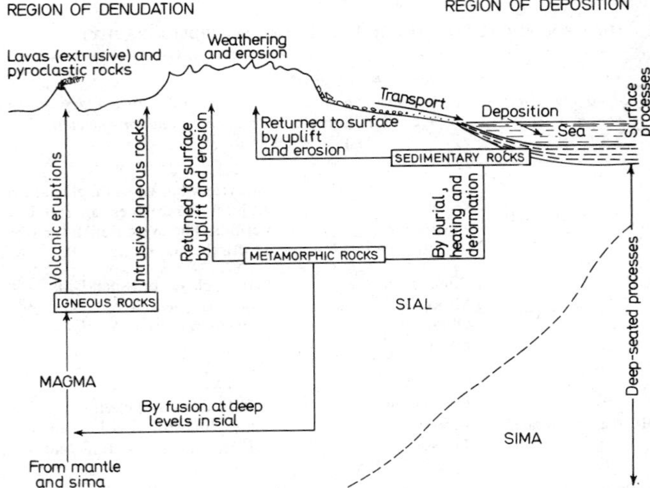

Figure 7.1 Diagrammatic representation of the long-term cycling of rocks (After Read and Watson—see Bibliography)

of superposition, and contain the same peculiar fossils'. These principles are still used to determine the relative ages of strata, i.e. in the order of superposition for an undisturbed series of sedimentary beds the oldest (i.e. the first deposited) is at the bottom, and successively younger beds lie upon it. Sedimentary strata in different localities can usually be correlated by the diagnostic fossil remains they contain. Rapidly evolving fossils act as horizon markers so that a specimen of one of these enables the particular level of the rock outcrop in which it occurs to be identified in the geological column wherever in the world it is found.

The whole sequence of rocks comprising the geological column is broadly divided into the Systems and Groups shown in Table 7.1; this column applies particularly to British strata. The column shows the age of each group relative to the others, and was in use long before any of the recent radiometric methods of determining the absolute age in years was discovered. Names of the geological Systems, and of the larger Groups are of world-wide application; they are also used to express the periods of time during which the rocks of the different Systems were formed, e.g. the Jurassic System and the Jurassic Period, or Mesozoic Group and the Mesozoic Era. The times of major mountain-building episodes (*orogenies*) and of phases of igneous activity in Britain are given in the third column of the table.

There are numerous further subdivisions down to 'zones' and even 'horizons', many of the smaller divisions being based on specific fossils.

In any given area the deposition of sediments was not continuous throughout the geological Periods. There are breaks in the sequence of deposits, marked by *unconformities* which represent intervals of time during which there was no deposition and erosion took place. The sea floors with their sediments were raised and became subject to erosion by wind and water. There were also periods of quiet sedimentation, when the seas covered the land, and intervening episodes of disturbance when uplift and folding took place. This broad pattern of events—the transgression of the sea over the lands, then the regression of the sea, followed by orogenic upheaval—has been repeated many times throughout geological history; see Figure 7.2 which shows the typical simplified borehole sequence of such a chain of events.

Table 7.1 THE GEOLOGICAL COLUMN (AFTER BLYTH—SEE BIBLIOGRAPHY)

Name of geological group or era		*Name of geological system or period* (ages in millions of years)	*General nature of deposits, major orogenies, and igneous activity in Britain*
CAINOZOIC	Quaternary	Recent Pleistocene	Alluvium, blown sand, glacial drifts, etc. At least five ice ages separated by warmer periods. The Weichselian (or Newer Drift), the last ice age
	Tertiary	Pliocene Miocene Oligocene Eocene (70)	Sands, clays, and shell beds *Alpine Orogeny* *Igneous activity in Scotland and Ireland*
MESOZOIC (or Secondary)		Cretaceous Jurassic Triassic (225)	Sands, clays and chalk Clays, limestones, some sands Desert sands, sandstone and marls
PALAEOZOIC (or Primary)	Newer	Permian	Breccias, marls, dolomitic limestone *Hercynian Orogeny* *Igneous activity*
		Carboniferous	Limestones, shales, coals and sandstones
		Devonian (and Old Red Sandstone) (c. 400)	Marine sediments (Lacustrine sands and marls) *Igneous activity* *Caledonian Orogeny*
	Older	Silurian Ordovician Cambrian (c. 600)	Thick shallow-water sediments, shales and sandstones. Volcanic activity in the Ordovician
PRE-CAMBRIAN		*Dalradian*	Schists
		Moinian (740+)	Schists and granulites
		Torridonian	Sandstones and arkoses
		Uriconian	Lavas and tuffs (Shropshire) *Pre-Cambrian Orogenies*
		Lewisian (3 500+)	Orthogneisses, etc.

Unconformities are often marked by beds of pebble gravel, the beach deposits of a sea which gradually inundated the land during its submergence (see Figure 7.3). Examples of this are the pebbly quartzites at the base of the Cambrian, or the rounded flints at the base of the Eocene deposits of south-east England overlying the Chalk, both marking the oncoming of marine transgression. Boulderbeds and hill or mountain screes formed on an old land surface during erosion, after uplift has taken place, may also be preserved as the lowest members of a newer series of rocks resting unconformably on older rocks; an example is the boulders and coarse sands at the base of the Torridonian in north-west

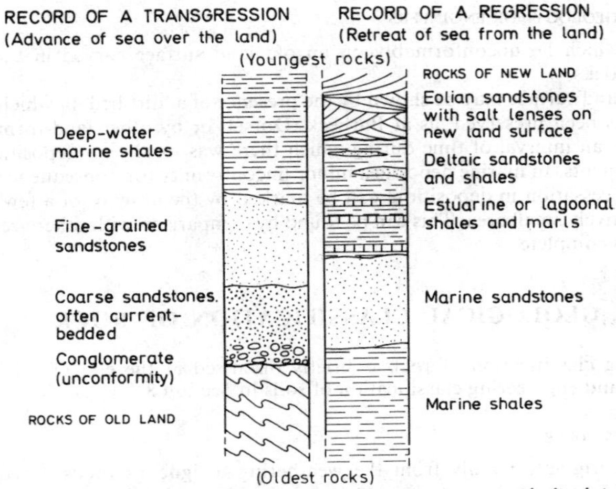

RECORD OF A TRANSGRESSION
(Advance of sea over the land)

RECORD OF A REGRESSION
(Retreat of sea from the land)

(Youngest rocks)

Deep-water marine shales

Fine-grained sandstones

Coarse sandstones, often current-bedded

Conglomerate (unconformity)

ROCKS OF OLD LAND

ROCKS OF NEW LAND

Eolian sandstones with salt lenses on new land surface

Deltaic sandstones and shales

Estuarine or lagoonal shales and marls

Marine sandstones

Marine shales

(Oldest rocks)

Figure 7.2 Marine transgression and regression as seen idealised in borehole core **(After Read and Watson)**

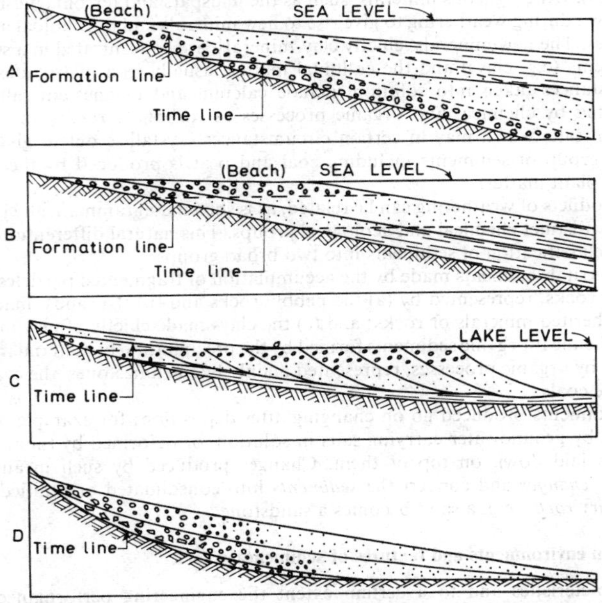

Figure 7.3 Examples of common marine and freshwater transgressions and regressions showing types and geometric distribution of sediment deposited **(After Lahee—see Bibliography).** *Lines parallel to lake and sea floors are time lines as they join sediment deposited contemporaneously. Lines essentially parallel to gravel, sand or clay deposits are formation lines. A, a marine transgression over the land; B, a marine regression from the land; C, a lake regression from the land; lake bottom muds are gradually covered by coarser sediments. Later transgression is shown left of a; D, an alluvial transgression by growth of a cone of river alluvium in mountainous area overlooking a plain*

Scotland which lie unconformably on an old land surface carved in the underlying Lewisian rocks.

An old land surface may be shown by the presence of a 'dirt bed' in which some of the old soil has been preserved, as at Purbeck, Dorset, or by other land-formed deposits. It indicates an interval of time during which there was locally no deposition of water-borne sediments. In marine deposits a minor unconformity (or nonsequence), representing a local cessation in deposition, can be marked by the absence of a few feet of beds over a relatively small area. This can be found by comparison with other areas where the sequence is complete.

GEOLOGICAL CLASSIFICATION OF ROCK

Engineering classification of rock is briefly discussed at the end of this section and Section 9, and engineering classification of soils in Section 8.

Sedimentary rocks

Sediments originate mainly from the weathering of igneous rocks. Certain resistant minerals in igneous rocks such as quartz survive unchanged and are eventually incorporated in the new sediments; often they tend to be concentrated in certain types of sediment (e.g. sands). Other igneous minerals, such as the feldspars and ferromagnesian minerals, break down during weathering to give rise to new minerals and to colloidal and dissolved substances. The new minerals, chiefly clay-minerals, are concentrated in a second group of sediments (e.g. clays) and the colloidal matter, usually iron hydroxides, in a third. The substances taken into solution include calcium and magnesium salts which are precipitated by chemical and organic processes as carbonate rocks, and sodium and potassium salts which may in certain circumstances crystallise out to give evaporites. Another group of sediments including coal and peat is produced by the piling up of decaying plant matter.

The products of weathering can be related, as is shown diagrammatically in Figure 7.4, into fairly distinct chemical and geological groups. This natural differentiation provides a simple classification of sediments into two broad groups:

(i) Detrital sediments made by the accumulation of fragmented particles of minerals or rocks, represented by (a) the pebbly rocks and (b) the sands, made chiefly of inherited minerals or rocks: and (c) the clays made chiefly of new minerals.

(ii) Chemical-organic sediments formed by the precipitation of material from solution or by organic processes, represented mainly by the limestones, the evaporites and the coals.

The sediments produced go on changing after deposition; for example, they may be saturated by groundwater carrying salts in solution, or deformed by the weight of new sediments laid down on top of them. Changes produced by such means are called *diagenetic changes* and convert the *sediments* into consolidated or lithified (hardened) *sedimentary rocks*, e.g. a sand becomes a sandstone.

Deposition environments and textures of sedimentary rock

The characteristics and to a certain extent the engineering performance of Recent sediments can be directly related to the environment occurring at their location of deposition, because the agents of deposition can still be seen in action. In the older sedimentary rocks, the environment of deposition can be reconstructed from the characters of the rocks themselves. The evidence for this reconstruction is provided by the *composition* and *texture* of the rock, the type of bedding, the fossil content and the relationship between any one bed and its neighbours. The sum of all these features decides its sedimentary *facies* and from this it is generally possible to deduce the conditions under which each rock was formed. This is summarised in Table 7.2.

Table 7.2 ENVIRONMENTS OF DEPOSITION OF SEDIMENTARY ROCKS

Environment of deposition			*Common sedimentary rocks produced by the environment*
SEA			
Shallow seas (continental shelf)	Littoral (beaches, sandbanks, tidal flats)		Conglomerate, sandstone, shale
	Neritic	Shelf seas in stable areas	Orthoquartzite, current-bedded sandstone, shale, organic and chemical limestones
		Restricted deep basins	Black shale
Deep seas		Geosynclinal seas in mobile belts Deep seas in stable areas	As for shelf seas with in addition greywackes and other turbidites
Abyssal seas			Calcareous ooze, siliceous ooze, Red Clay
LAND/SEA			
Deltas			Mainly sandstone, shale
Estuaries, lagoons			Shale
LAND			
Flood-plain			Conglomerate, sandstone, shale
Lakes	with outlet to sea		Sandstone, shale, freshwater limestone
	in basins of interior drainage		Sandstone, shale, evaporates
Deserts			Sandstone, conglomerate, breccia
Piedmont (intermontane basins, alluvial fans)			Conglomerate, breccia, arkose, sandstone
Areas of glaciation			Tillite

The most obvious and characteristic feature of sedimentary rocks is *bedding*, i.e. the presence of recognisably different beds or strata in a sedimentary succession, and the presence within any one bed of depositional surfaces which are the bedding planes—see Figure 7.5.

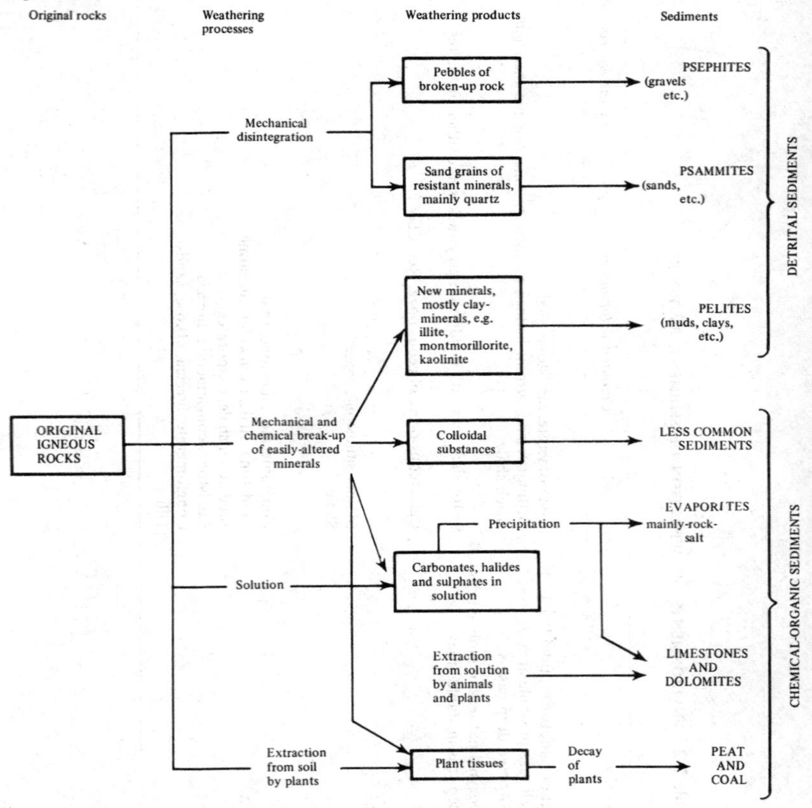

Figure 7.4 Sedimentary differentiation (After Read and Watson)

Although many beds are homogeneous, some show considerable variation, especially *graded beds,* in which there is a passage from coarser to finer particles towards the top; lateral gradation may also be found. Thin laminae or layers, differing somewhat in colour or texture, may be present without causing a bed to lose its individuality. A bed is characterised by all of its lithological features. These indicate that it was laid down in a particular environment, either uniform, or varying systematically. Although it may be arbitrary, some very thin strata may best be regarded as beds rather than as laminae within a bed. For example, in glacial varves each annual deposit of summer silt and winter clay is an individual bed even though its thickness is measured in fractions of an inch, whereas sandy laminae in a graded greywacke are parts of the whole graded unit—see Figure 7.6.

In describing bedding it is necessary to distinguish firstly between bedding planes which are individual structures where each planar surface may be distinguished, and also depositional textures, which result from the parallel orientation of particles through-

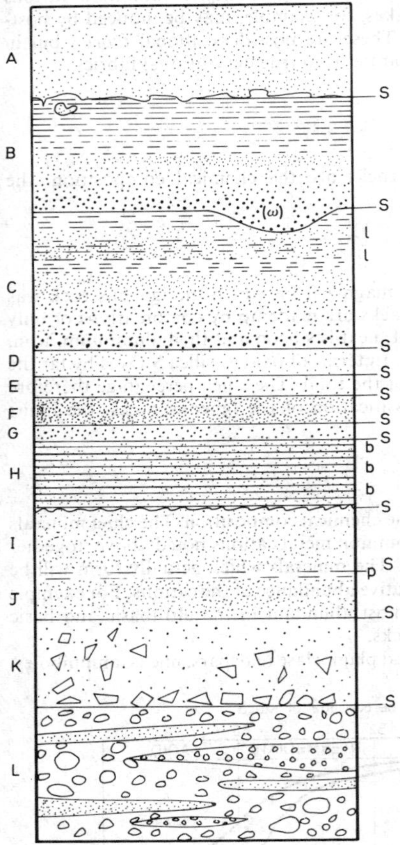

Figure 7.5 Idealised types of sedimentary beds (After Sherbon Hills—see Bibliography). *Beds are bounded by separation planes (S). A, uniform, massive sandstone with bottom structures at its base; B, simple graded bed, with uniform grading from coarse sandstone below to shale above and a washout (w); C, complex graded bed with thin sandstone laminae (l, l) in the shale; D, E, F, G, individual thin beds; H, single sandstone bed with discrete bedding planes (b, b, etc.); I, J, two sandstone beds separated by shale parting (p); K, heterogeneous bed of sandstone containing angular shale fragments; L, heterogeneous bed of conglomerate containing lenses of sand and gravel*

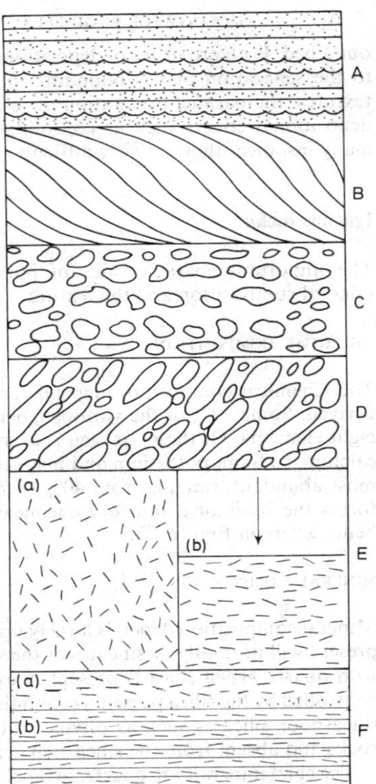

Figure 7.6 Idealised types of sedimentary bedding. A, sandstone with discrete bedding planes parallel to separation planes. Some beds ripple-marked (r); B, sandstone with discrete bedding planes inclined to separation planes (false or cross bedding; an inclined depositional texture); C, conglomerate with long axes of pebbles approximately parallel to separation planes (a parallel depositional texture); D, edgewise conglomerate with long axes of pebbles inclined to separation planes (an inclined depositional texture); E(a), unconsolidated mud with random orientation of mica flakes and clay particles (a random depositional texture); E(b), consolidated clay or lithified mudstone with flaky particles approximately parallel, and parallel with separation planes (a parallel consolidation texture); F(a), mudstone with mica flakes deposited parallel to separation planes, but lacking discrete bedding planes (a parallel depositional texture, cf. C above); F(b), mudstone with mica flakes deposited parallel to separation planes, and showing discrete bedding planes. A thin bed of sandstone lies between the two mudstones

out a bed. Both are primary depositional features, and may be either parallel or inclined to the separation planes bounding individual beds (Figure 7.6). In addition, various textures, the parallel orientation of mica-flakes, for example, may be induced by post-depositional effects such as consolidation. These are post-depositional *fabrics* but in many instances they are very difficult to separate from true depositional fabrics.

Igneous rocks

The important characteristics of igneous rocks are the chemical composition, the mineral composition and the texture.

CHEMICAL COMPOSITION

The Chemical composition depends on the magma from which the igneous rock was derived. Some 99% of the various igneous rocks are made up by combinations of only eight elements. Of these, oxygen is dominant, next is silicon and then aluminium, iron, calcium, sodium, potassium and magnesium. In terms of oxides, silica SiO_2 is by far the most abundant, ranging from 40% to 75% of the total. The silica percentage therefore forms the basis of a fourfold chemical classification of the igneous rocks, the limits being given on Figure 7.7.

MINERAL COMPOSITION

Mineral composition depends largely upon the chemical composition. The chief minerals present will normally be silicates of the six common metal cations noted above, together with quartz, when silica is present in excess. The minerals which actually form will be controlled by the silica percentage and the relative abundance of the cations. For example silica-poor silicates such as olivine will be most abundant in the ultrabasic and basic rocks and absent from the silica-rich acid rocks.

The chief minerals are quartz, orthoclase and plagioclase feldspars, micas, amphiboles,

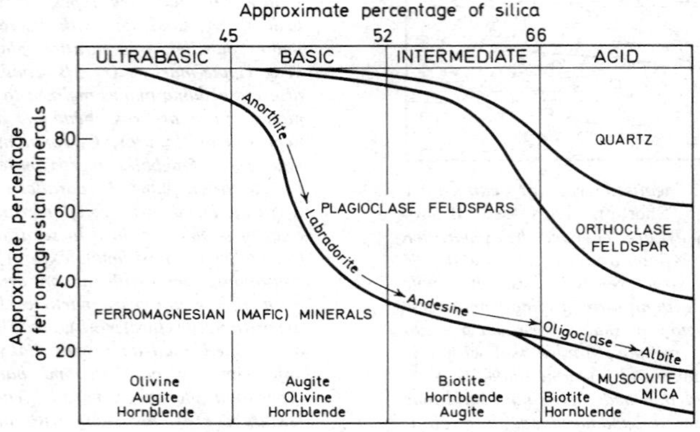

Figure 7.7 A classification of igneous rocks based on silica percentage

pyroxenes and olivines. Their distribution in the four groups established by silica percentage is shown diagrammatically in Figure 7.7. Many of the names given to igneous rocks are defined according to the presence of two or three particular minerals which are the essential minerals for that rock type. Other accessory minerals may also be present in small quantities, e.g. the essential minerals of granite are quartz, feldspar and mica; common accessories are zircon and iron oxide.

The predominant minerals of an igneous rock may determine its general appearance and it is usually possible to get some idea of its composition from its colour and density. Quartz is commonly colourless and transparent, feldspars opaque but pale coloured. Rocks made mostly of these minerals (that is, acid and intermediate rocks) are therefore usually pale in colour and relatively light in weight. The coloured ferromagnesian silicates, olivines, pyroxenes and amphiboles, are abundant in basic and ultrabasic rocks which are usually dark and relatively heavy. Two important exceptions are, very fine-grained or glassy rocks which tend to look dark whatever their composition; and weathering or other alteration which changes the colours of minerals. It is, therefore, usually necessary to look at freshly broken surfaces to diagnose the parent rock type.

TEXTURE

The texture of an igneous rock is shown by the arrangement of the constituent minerals and the relation of each mineral to its neighbours. The main textural character is the

Table 7.3 A CLASSIFICATION OF IGNEOUS ROCKS ON SILICA PERCENTAGE AND GRAIN SIZE

Acid	*Intermediate*	*Basic*
Coarse-grained (plutonic) rocks. Grain size larger than about 5 mm. Liable to be brittle owing to presence of large crystals		
Granite Granodiorite (Widely distributed in the British Isles)	Syenite Diorite (Comparatively rare in the British Isles)	Gabbro Norite (Not very common in the British Isles)
Medium-grained (hypabyssal) rocks. Grain size between about 1 and 5 mm. Very frequently possess intergrown texture: include some of the best roadstones		
Microgranite Granophyre (Fairly common in the British Isles)	Porphyry Porphyrite (Fairly common in the British Isles)	Dolerite Diabase (Widely distributed in the British Isles)
Fine-grained (volcanic) rocks. Grain size below about 1 mm, i.e. below the limit of visible recognition. Similar to medium-grained rocks, but sometimes liable to be brittle and splintery		
Rhyolite Felsite (Not very common in the British Isles)	Trachyte Andesite (Not very common in the British Isles)	Basalt Spilite (Widely distributed in the British Isles)
← Continuous variation in properties →		
Light colour ←	(Due to increase in ferromagnesian minerals)	→ Dark colour
Low specific gravity ← (2.6)	(Due to increase in ferromagnesian minerals)	→ High specific gravity (2.9)

grain size and in a general way this depends on the rate of cooling of the magma. Coarse-grained rocks are the result of slow cooling which allowed time for the growth of large crystals; fine-grained rocks are produced by rapid cooling. By extremely rapid cooling, no time at all is given for crystallisation and glasses are formed. *Holo-crystalline* rocks are entirely crystalline, *hypo-crystalline* are partly crystals, partly glass. A common distinctive texture is the *porphyritic* texture in which crystals of two different sizes occur: large *phenocrysts* are scattered through a finer-grained or glassy groundmass. The texture is an important controlling factor in the engineering performance of the rock.

CLASSIFICATION

Classification of the common igneous rocks is usually made on the basis of grain size and silica percentage as given in Table 7.3. The characteristic minerals of rocks of different compositions are shown in Figure 7.7 which should be studied with Table 7.3.

FORM

A body of magma which is under pressure in the sial may be forced upwards intruding into the upper rocks of the crust. During the process of intrusion it may incorporate some of the rocks with which it comes into contact, by assimilation. In some cases it may also give off mobile fluids which penetrate and change the rocks in its immediate neighbourhood and mineralisation may occur. If the intrusive magma cools at some depth below the surface, the rocks which result are called *plutonic* rocks and are coarsely crystalline; a large mass of this kind constitutes a major intrusion, e.g. a granite batholith. When magma rises and fills fractures or other lines of weakness in the crust, it forms minor intrusions. These include *dykes,* which are steep or vertical wall-like masses, with more or less parallel sides, and *sills,* which are sheets of igneous rock intruded between bedding planes or sedimentary rocks and lying more or less horizontal. Dyke and sill rocks commonly have a fine-grained texture. Veins are smaller and irregular bodies of igneous material, filling cracks which may run in any direction.

Magma which rises to the earth's surface and flows out as a lava, is called extrusive, and under these conditions it loses most of its gas content. These are the volcanic rocks, and since they have cooled comparatively quickly in the atmosphere they are frequently glassy (i.e. noncrystalline), or very fine-grained with some larger crystals.

These forms are summarised in Figure 7.8.

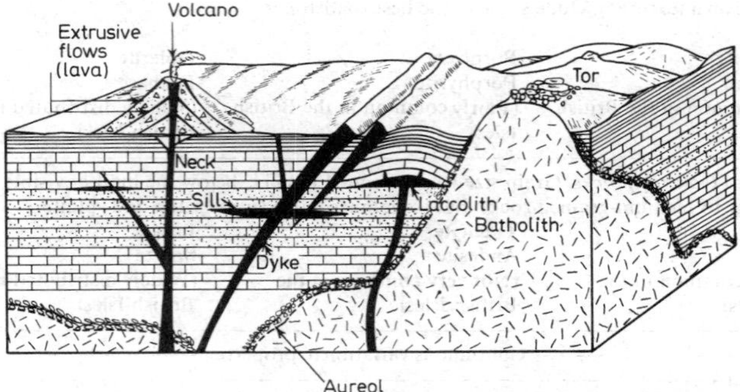

Figure 7.8 Idealised forms of intrusive plutonic rocks (After Vitaliano)

STRUCTURE

The use of the term *structure* is reserved for more pronounced features of a rock than those described by the term texture. In igneous rocks the structure may indicate a relative arrangement of different spatial features of the rock, both small (*microscopic*) and large (*macroscopic*). For example, gas bubble holes in an igneous rock may be characteristic of its structure. A *vesicular* structure is the presence of small holes, or vesicles, throughout the igneous rock, such as are found in pumices and some basalts. Holes larger than vesicles are *vugs* and are generally filled with minerals other than those forming the rock.

An important macroscopic structural feature is jointing of the rock. Joints are fractures and may be open or closed and run in different directions. They usually represent more-or-less regular systems and may tend to break the rock into cubes or other regular blocks. This is an important engineering property and is discussed further later. Fractures or cracks are also macroscopic features and may run in any direction and may intersect each other at any angle. A fracture is usually irregular in contrast to the planar or even surface of a joint.

FABRIC

Fabric is a controversial term which sometimes is considered as a generalisation of the term texture. Here, igneous fabric denotes the spatial pattern of the rock particles which includes grain sizes and their ratios, grain shapes, grain orientation, microfracturing, packing and interlocking of particles, the character of the matrix, and so on, all of which help control the engineering performance of the rock.

Metamorphic rocks

Rocks formed by the complete or incomplete recrystallisation, i.e. the change in crystal shape or in composition, of igneous or sedimentary rocks by high temperatures, high pressures, and/or high shearing stresses are metamorphic rocks. A platy or foliated structure in such rocks indicates that high shearing stresses have been the principal agency in their formation.

Table 7.4 METAMORPHIC ROCK CLASSIFICATION

Structure and texture	*Composition*	*Rock name*
FOLIATED OR PLATY	Various tabular and/or prismatic minerals (generally elongated)	Schist, some serpentines, slate, phyllite
MASSIVE:		
Banded, consisting of alternating lenses	Various tabular, prismatic, and granular minerals (frequently elongated)	Gneiss
Granular, consisting mostly of equidimensional grains	Calcite, dolomite, quartz, in small particles	Marble or quartzite

Foliation is not always visible to the naked eye, but individual grains may exhibit strain lines when seen under the microscope. Metamorphic rocks formed without intense shear action have a massive structure. In Table 7.4 the most common metamorphic rocks are subdivided into two basic classes according to their structure. Foliated rocks usually have directional engineering properties.

Field identification of common rocks

Table 7.5 gives a simple field guide to the identification of the more common rocks. It is after the scheme by Krynine and Judd (see Bibliography) for engineers with little training in geology and has been devised to present those features first seen when picking up a hand specimen. It is based primarily on texture and structure. They consider the scheme fits the most common occurrences of the rock but some variations will occur.

Table 7.5 FIELD IDENTIFICATION OF ROCKS (SPECIMENS SHOULD BE UNWEATHERED AND NOT ALTERED)

[LIGHT-COLOURED]

GRAINS OR CRYSTALS VISIBLE TO NAKED EYE								
Angular particles				*Rounded particles*			*Erratic large grains*	
Large	*Fine to medium*	*Very fine*	*Foliated or banded*	*Large*	*Fine to medium*	*Very fine*	*Rounded*	*Angular*
Pegmatite Granite (+Q, +F)* Granodiorite (+Q, +F)* Monzonite (−Q, +F)* Syenite (−Q, +F)* Marble (reacts with HCl) Arkose (usually bedded)	Tuff (contains glasslike fragments)	Felsite* (rhyolite +Q and trachyte −Q)	Schist (shiny) Gneiss (may have sub-angular particles)	Conglomerate (+10% of grains over 2 mm diameter) Sandstone (bedded) (if it reacts to HCl = calcareous sandstone; if it gets slick when wet = argillaceous sandstone)	Quartzite (not friable and very hard)	Siltstone	Depositional breccia	Volcanic breccia and agglomerate fault breccia (may have clay)

NO GRAINS OR SPARSE CRYSTALS VISIBLE TO NAKED EYE										
Glassy lustre	*Dull lustre*	*Shiny lustre*	*Earthy appearance*		*Laminated*					
Quartzite	Felsites* (rhyolite, trachyte)	Schist (foliated)	*Spongy, light wt*	*Porous moderate wt*	*Slick when wet*	*Not slick*	*Slick when wet*	*Not slick*		
								Reaction to HCl	*No reaction to cold HCl*	
			Pumice Volcanic ash	Chalk (HCl reaction)	Shale	Shale Slate (dull) Phyllite (shiny)	Claystone Mudstone Serpentine (usually greasy and may be banded)	Limestone Chalk (earthy)	Dolomite	

[DARK-COLOURED (DARK GREY OR GREEN TO BLACK)]

GRAINS OR CRYSTALS VISIBLE TO NAKED EYE		
Angular particles		*Rounded to subangular particles*
Fine to medium	*Very fine to glassy*	Graywacke (fine to medium-grained) Dark sandstones
Peridotite (−Q, −B)* Gabbro (−Q, −B)* Diorite (−Q, +B)* Dolerite (−Q, +B)*	Andesite* Basalt (usually vesicular)*	

NO GRAINS OR SPARSE CRYSTALS VISIBLE TO NAKED EYE			
Glassy lustre	*Dull lustre—laminated*		*Dull lustre—not laminated*
	Slick when wet	*Not slick*	
Obsidian	Shale	Shale (flexible) Slate (brittle) (dull) Phyllite (shiny)	Basalt* Serpentine (usually greasy and may be banded)

* Rocks may contain occasional large crystals embedded in a very fine-grained matrix or occasional very large crystals in a medium-grained matrix – in either case the term *porphyry* is appended to the rock name, e.g. syenite porphyry.

(+Q) = contains numerous white or colourless quartz crystals.
(−Q) = contains little or no quartz.
(+F) = contains numerous white to pink feldspar crystals.
(+B) = contains numerous flakes of black mica (biotite).
(−B) = contains little or no black mica.

Textbooks such as the one by Lahee (see Bibliography), give more specialised field identification techniques. Difficult or contentious identification should be carried out by a geologist who may require thin section examination of a slice of the rock or even geochemical methods for complete identification.

GEOLOGICAL ENVIRONMENTS

A geological environment is the sum total of the external conditions which may act upon the situation. For example, a 'shallow marine environment' is all the conditions acting offshore which control the formation of deposits on the sea bed; the water, temperature, light, current action, biological agencies, source of sediment, sea bed chemistry and so on.

The concept of geological environment forms a suitable basis to study systematically the engineering geology of the deposits formed in or influenced by the various environments, as they condition the *in situ* engineering behaviour of the various deposits. A knowledge of the parameters of the environment enables predictions and explanations of the engineering behaviour to be attempted.

PROCESSES ACTING ON THE EARTH SURFACE

A *landform* may be defined as an area of the earth surface differing by its form and other features from the neighbouring areas. Mountains, valleys, plains and even swamps are landforms.

The principal processes that are continually acting on the earth's surface are gradation, diastrophism and vulcanism.

> *Gradation* is the building up or wearing down of existing landforms (including mountains), formation of soil and various deposits. Erosion is a particular case of gradation by the action of water, wind or ice.
>
> *Diastrophism* is the process where solid, and usually the relatively large, portions of the earth move with respect to one another as in faulting or folding.
>
> *Vulcanism* is the action of magma, both on the earth's surface and within the earth.

With the exception of vulcanism and sometimes erosion, these processes may take hundreds and even millions of years to change the face of the earth significantly. The sudden eruption of a volcano, for example, with the ensuing flow of lava or deposition of volcanic ash, can abruptly change land overnight.

Origin of soils. The majority of the soils are formed by the destruction of rocks. The destructive process may be physical, as the disintegration of rock by alternate freezing and thawing or day/night temperature changes. It may also be by chemical decomposition resulting in changes in the mineral constituents of the parent rock and the formation of new ones.

Soils formed by disintegration and chemical decomposition may be subsequently transported by the water, wind or ice before deposition. In this case they are classified as alluvial, aeolian, or glacial soils and are generally called *transported* soils. However, in many parts of the world, the newly formed soils remain in place. These are called *residual* soils.

In addition to the two major categories of transported and residual soils, there exist a number of soils that are not derived from the destruction of rocks. For example, peat is formed by the decomposition of vegetation in swamps; some marly soils are the result of precipitation of dissolved calcium carbonate.

Soil-forming processes. There are very many and varied processes that take place in weathered rock and soils that affect the formation of soil profiles to varying degrees, but the major soil-forming processes are:

organic accumulation, eluviation, leaching, illuviation, precipitation, cheluviation and organic sorting

The soil-forming processes produce an assemblage of soil layers at horizons, called the *soil profile*. In its simplest it is categorised as three layers A, B and C (see CP 2001[2]) but numerous varieties of this and many other soil classifications exist. Probably the most generally accepted one is that based on a geographical approach. This is the zonal scheme thought to reflect zones of climate, vegetation and other factors of the local environment. A map showing the world's soils broadly divided on this basis is given in reference 6.

Engineering geology significance of selected environments

ROCK WEATHERING

A rock can weather by physical breakdown without considerable change of its constituent minerals by disintegration. The residual or transported soil derived from this process

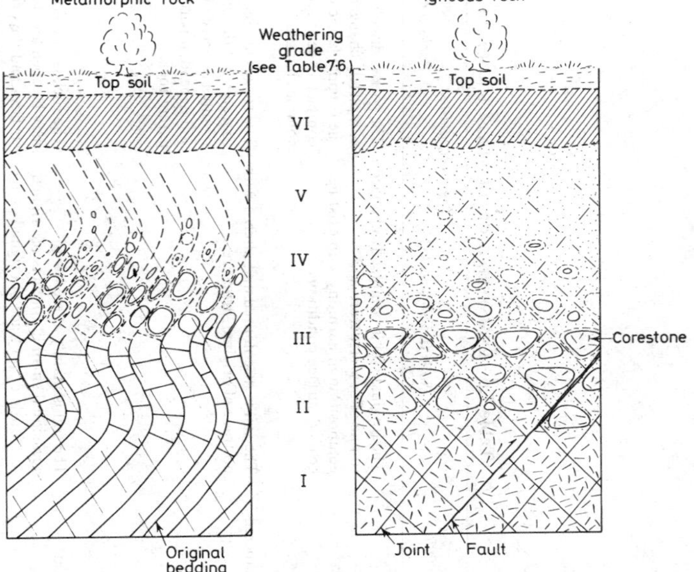

Figure 7.9 Diagrammatic weathering profile of an igneous and a metamorphic rock (After Deere and Patton[5])

consists of an accumulation of mineral and rock fragments virtually unchanged from the original rock. This type of weathering is found mainly in arid or cold climates. Chemical deposition leads to the thorough alteration of a large number of minerals and only a few, among them quartz, may remain unaffected. The greater the percentage of weatherable minerals in the original rock, particularly the ferromagnesian minerals, the more conspicuous is the change from rock to soil. This type of weathering is generally found in

Table 7.6 ENGINEERING GRADE CLASSIFICATION OF WEATHERED ROCK

Grade	*Degree of decomposition*	*Field recognition*		*Engineering properties of rocks*
		Soils (*i.e. soft rocks*)	*Rocks* (*i.e. hard rocks*)	
VI	Soil	The original soil is completely changed to one of new structure and composition in harmony with existing ground surface conditions.	The rock is discoloured and is completely changed to a soil in which the original fabric of the rock is completely destroyed. There is a large volume change.	Unsuitable for important foundations. Unsuitable on slopes when vegetation cover is destroyed, and may erode easily unless a hard cap present. Requires selection before use as fill.
V	Completely weathered	The soil is discoloured and altered with no trace of original structures.	The rock is discoloured and is changed to a soil, but the original fabric is mainly preserved. The properties of the soil depend in part on the nature of the parent rock.	Can be excavated by hand or ripping without use of explosives. Unsuitable for foundations of concrete dams or large structures. May be suitable for foundations of earth dams and for fill. Unstable in high cuttings at steep angles. New joint patterns may have formed. Requires erosion protection.
IV	Highly weathered*	The soil is mainly altered with occasional small lithorelicts of original soil. Little or no trace of original structures.	The rock is discoloured; discontinuities may be open and have discoloured surfaces and the original fabric of the rock near the discontinuities is altered; alteration penetrates deeply inwards, but corestones are still present.	Similar to grade V. Unlikely to be suitable for foundations of concrete dams. Erratic presence of boulders makes it an unreliable foundation for large structures.

III	Moderately weathered*	The soil is composed of large discoloured lithorelicts of original soil separated by altered material. Alteration penetrates inwards from the surfaces of discontinuities.	The rock is discoloured; discontinuities may be open and surfaces will have greater discolouration with the alteration penetrating inwards; the intact rock is noticeably weaker, as determined in the field, than the fresh rock.	Excavated with difficulty without use of explosives. Mostly crushes under bulldozer tracks. Suitable for foundations of small concrete structures and rockfill dams. May be suitable for semipervious fill. Stability in cuttings may depend on structural features, especially joint attitudes.
II	Slightly weathered	The material is composed of angular blocks of fresh soil, which may or may not be discoloured. Some altered material starting to penetrate inwards from discontinuities separating blocks.	The rock may be slightly discoloured; particularly adjacent to discontinuities which may be open and have slightly discoloured surfaces; the intact rock is not noticeably weaker than the fresh rock.	Requires explosives for excavation. Suitable for concrete dam foundations. Highly permeable through open joints. Often more permeable than the zones above or below. Questionable as concrete aggregate.
I	Fresh rock	The parent soil shows no discolouration, loss of strength or other effects due to weathering.	The parent rock shows no discolouration, loss of strength or any other effects due to weathering.	Staining indicates water percolation along joints; individual pieces may be loosened by blasting or stress relief and support may be required in tunnels and shafts.

* The ratio of original soil or rock to altered material should be estimated where possible.

warm and hot climates and can lead to great thickness of weathered rock and soil. Biological weathering is generally of less importance than physical or chemical weathering and is a combination of biochemical and biophysical effects.

The weathering process produces a gradational and often quite irregular change in the rock from fresh some distance below ground surface to more or less completely weathered at the surface; Figure 7.9 shows somewhat schematically two examples of weathered rock profiles.

There are several generalised weathering rock classifications for engineering purposes; the one reproduced here, Table 7.6, is from Fookes, Dearman and Franklin[3] which discusses weathered rock mainly in Britain. For information on the engineering performance of weathered rock elsewhere see also Little[4] and Deere and Patton[5].

RESIDUAL SOILS

A residual soil is the end product (i.e. grade VI) of rock weathering. Different types of residual soils are produced in different environments (see Figure 7.10).

Though broadly speaking the engineering characteristics of each residual soil type will be similar, great differences may exist between each type; some soils will be well cemented (as some laterites), some will have high montmorillonite clay contents (as some

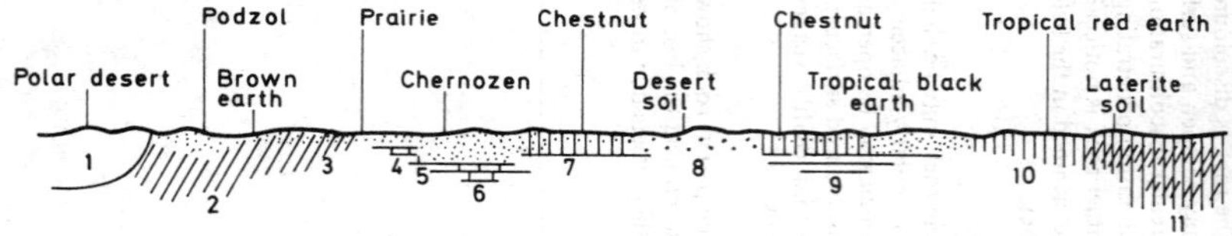

1. Frozen ground
2. Intense leaching and illuviation
3. Less leaching more mixing
4. Carbonate appears
5. Organic matter increases
6. Carbonate prominent
7. Organic matter decreases
8. Gypsum and salt accumulate
9. Organic matter increases
10. Ferrallitisation becomes dominant
11. Ferricrete formation

Figure 7.10 Diagrammatic cross section of zonal soils from pole to equator (After Ollier—see Bibliography)

black cotton soils) and so on. It is therefore difficult and probably dangerous to generalise on the engineering performance of residual soil. Reference should be made to the appropriate literature for the particular soil type, though this itself may be difficult since in many of the published engineering articles the soil type is usually not described sufficiently to characterise it. In addition, a great variety of different residual soils seem to be dubbed with the title of 'laterite' often quite erroneously. See Sanders and Fookes[6] for a general study of foundation conditions related to four principal climate zones, periglacial, temperate, arid and humid tropical; Deere and Patton[5] for extensive treatise on slope stability aspects; Little[4] for 'laterites' and the Proceedings of the Institution of Civil Engineers[7] for a symposium on road and airfield construction in tropical soils.

DESERT SOILS

Desert soils are formed in dry environments where the evaporation exceeds the precipitation and are generally associated with the world's hot deserts. Rainfall is low (say less than 150 mm per annum) and often seasonal. Physical weathering is dominant and the disintegration of the rock mainly results from insolation, but often other factors such as abrasion of windborne particles may contribute.

The products of this type of weathering are mainly of coarser-grained materials near hills or mountains. Parent materials of a high-silica content produce detrital sands and gravels, which, when transported away from high land by wind, give sand-dune deposits and possibly loess or, when transported by water, give alluvial sands and gravels. Calcareous parent materials result in calcareous sands and gravels and evaporite salts are often present throughout the soil profile especially in internally draining areas as playa or salina flats.

Fookes and Knill[8] (Figure 7.11) divided inter-montane desert basins into four sediment deposition zones which may be correlated with the degree of distintegration of the parent material.

Engineering problems provided by desert conditions are principally those related to the grading of the material. Coarse, angular, ill-sorted material generally occurs in zone II (Figure 7.11) and intermittent stream flow and occasional flash floods indicate

Zone	*I*	*II*	*III*	*IV*
Principal engineering soft types		Rock fans	Silty stony desert and sandy stony desert. Some evaporites	Sand-dunes, loess and evaporites
Slope angle of desert surface		2–12 °	$\frac{1}{2}$–2 °	0–$\frac{1}{2}$ °
Principal transporting agent of the environment		Gravity, and as III	Intermittent stream flow and sheet floods	Wind and evaporation
Geotechnical features		Good for foundation and fill	Generally very good foundation and fill material. Saline. May be pervious in foundations	Erratic behaviour to load bearing. Migrating dunes. Metastable loess. Saline. Material

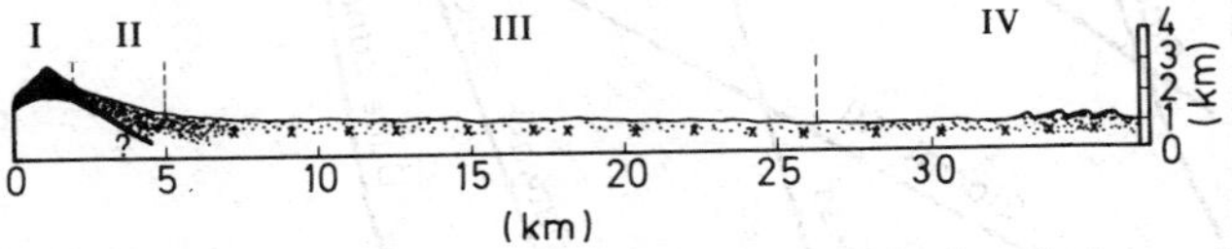

I II III IV
0 5 10 15 20 25 30
(km)
0 1 2 3 4 (km)

Figure 7.11 Simplified cross section of typical inter-montane desert basin normal to mountain range

that adequate drainage and run-off measures are required for engineering works. Better sorted and finer material occurs in zone III and the danger of sheet flood here may be greater. In zone IV, mobile sand dunes may require stabilisation and loess soils can suffer metastable collapse on loading. Evaporite salts in the soil may cause expansive problems under blacktop.

Desert engineering problems in general are discussed in Fookes and Knill[8], metastable loess soils in Holtz and Gibbs[9] and soluble salts in soils in Weinert and Clauss.[10]

GLACIAL SOILS ('DRIFT')

The five major glaciations of the Pleistocene period, begun about two million years ago, constitute the last major episode in the shaping of much of the world's land surface. During each glaciation ice advanced over large areas of the northern and southern hemisphere. Post-glacial changes which have only occurred within the last 15 000 years or so have been relatively limited. They are mostly confined to low ground, where alluvial deposits have tended to accumulate in response to the world-wide rise in sea level caused by the final recession of the ice sheets and glaciers. The deposits of one or more of the Pleistocene ice advances lie at the surface over, very approximately, 50% of Britain's land area, whilst roughly another 10% is covered with post-glacial alluvium sometimes concealing glacial materials at various depths. These deposits are generally known as drift in Britain.

A large proportion of British site investigations therefore encounter glacial materials in one or more of their varying forms, and the property of rapid lateral and vertical change shown by some types of deposit has become notorious since construction has often revealed features undisclosed by the site investigation. It is commonly thought of as random and unpredictable but this is not always true. An understanding of the different facets of the glacial environment, each with its characteristic landforms, erosional processes and assemblages of deposits, can be of great value in predicting not only the range of variation but often also the actual location of anomalous geotechnical features.

Glacial till and outwash. Moving glaciers excavate soils and rocks in their paths which are carried along and released as the ice melts away. The material deposited directly by the glacier as it melts is called 'boulder clay' or much better *till*. If the ice front remains more-or-less stationary for a long period of time, a considerable amount of till may accumulate along the ice front. Tills are also laid down in the form of extensive plains as the glaciers retreat during periods of rapid melting and are characterised by lack of stratification and large range in particle size (Figure 7.12). Many tills deposited by continental glaciers contain substantial amounts of clay-size particles and these are sometimes

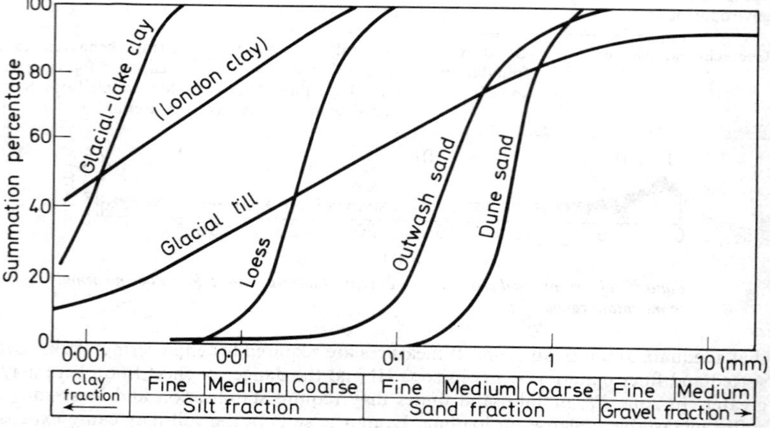

Figure 7.12 Examples of particle-size curves of some glacial soils and a London Clay for comparison

overconsolidated and form fairly stiff clays. Even though a deposit of till may be extensive in size and uniform in texture, its strength may vary considerably from place to place.

Along the front of a glacier, water from the melting of the ice gathers to form large torrential streams which are capable of transporting great quantities of sediments. As the streams spread out over the plains most of the coarse sediment is deposited as *fluvioglacial* alluvium which has the characteristics of braided-stream deposits. This consists of granular soils with lenses of gravel, sand, or silt, which generally occur in front of a till moraine. In some localities extensive areas are covered by fluvio-glacial deposits up to 100 ft (30 m) in thickness. In other areas they exist as thin lenses of limited lateral extent included between layers of till or peat.

Deposits of fluvio-glacial soils may also occur in river valleys (valley trains) that once served as drainage outlets for glacial meltwater or locally in the form of ridges (eskers) and terraces (kames) as a result of various complications in the drainage system around glaciers. These glacio-fluvial soils are composed primarily of silt, sand and gravel. Some of them are unstratified and others may exhibit irregular stratification. Glacial-lake deposits are often varved clays which exhibit characteristic thin stratification of silt and clay.

The principal engineering problem produced by the glacial environment is probably the difficulty of satisfactorily investigating the site. This is variously due to the rapidly changing soil type and engineering properties, large boulders in the till, lenses of clay, silt, sand or open gravel in other materials, concealed and weathered rockhead topography, structural disturbance of glacial deposits and complex groundwater conditions.

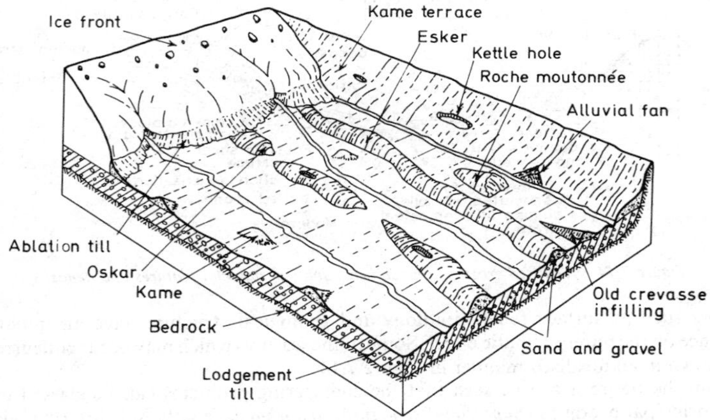

Figure 7.13 Idealised block diagram of a glaciated valley showing some typical glacial deposits and their geometric distribution

Figure 7.13 of an idealised glaciated valley shows some of the features which can be found near the end of a retreating glacier. Differential settlement may occur with heavy bearing structures and for water-retaining structures permeability may be a problem.

Higginbottom, Fookes and Gordon[11] discuss methods of site investigation; Linell and Shea[12] and Anderson and Trigg[13] discuss geotechnical properties of glacial sediments.

PERIGLACIAL SOILS ('DRIFT' AND 'HEAD')

The term periglacial is used to denote conditions under which frost action is the predominant weathering process. Mass transportation, wind action, or both, may occur, but only in association with very cold climatic conditions such as those near the margins of glacial ice. Perennially frozen ground (permafrost) is an important characteristic, but is not essential to the definition of the periglacial zone. The inner boundary of the zone is sharply defined by the current margin of the ice sheet, but the outer edge is gradational and the radial width of the periglacial zone is indefinite.

The distribution of permafrost may be strongly influenced by ground conditions and topographic features. The surface strata must be sufficiently porous or jointed to contain water, and their thickness must be greater than the potential thickness of the *active* layer. Permafrost may be thin or absent under surface features such as large bodies of water but it is still extensive near the northern and southern polar ice caps, in parts of Canada, Siberia and elsewhere.

Most periglacial effects from the Pleistocene glaciations on the topography and surface deposits are well preserved in southern England, between the limits of the last (Weichselian) glaciation and the loess belts of north and central Europe. However, they are not restricted to this area but can be found over the whole of Britain since the periglacial zone moved northward in the wake of the receding ice sheet. In the unglaciated areas, the relationship of frozen-ground features to older or younger glacial drifts often establishes them as independent of specifically glacial processes.

The phenomena associated with the periglacial environment almost defy classification since they are esentially overlapping aspects of a continuously evolving situation.

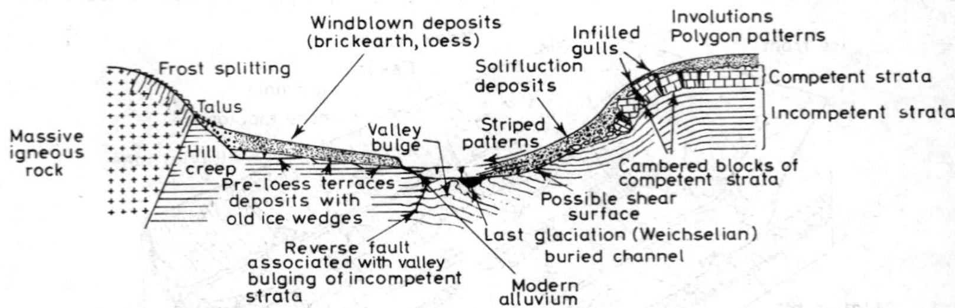

Figure 7.14 Idealised cross section showing some periglacial features and deposits

Factors such as surface relief, lithology and geological structure have an important influence on the purely climatic effect. Some of the features which may occur concurrently are shown in an idealised manner in Figure 7.14.

From the figure it can be seen that the engineering problems can be classed under three principal headings. *Superficial structural disturbance,* e.g. frost shattering, glacial shear, hill creep, ice wedges and involutions or chemical weathering. *Mass movements,* e.g. cambering and valley bulging, landsliding or mudflows. *Periglacial deposits,* e.g. loess, head or solifluxion soils. A full discussion of these problems in Britain is given in Higginbottom and Fookes.[14] Weeks[15] discussed periglacial slope problems and Fookes and Best[16] discuss periglacial metastable soils.

ALLUVIAL SOILS OF RIVERS

Cycle of valley erosion A river flowing in a valley erodes the material of its bed and local surface run-off contributes to the erosion of the walls of the valley. The eroded materials are transported in the form of sediment by the river and are eventually deposited.

Considered simply, rivers and the valleys along which they flow may be youthful, mature or old. At these three stages in the life of a river or valley its longitudinal profile, cross-section, and plan undergo gradual changes. At the *youthful* stage of a valley, its longitudinal profile is irregular and contains rapids, falls and even lakes because of local obstructions, such as hard rock strata, and its cross-section tends to be V-shaped. The plan of a youthful valley or river is somewhat angular or zigzag.

As erosion progresses, the river reaches *maturity*, irregularities gradually disappear and the plan acquires the shape of a smooth sinusoidal curve. The longitudinal profile also becomes reduced in gradient, decreasing gradually towards the mouth of the river. The valley at its mature stage is wide; its slopes are flatter than in its youth and often covered with talus (hillside rock debris).

Periodic floods contribute to the gradual widening of the valley until at its *old* age it becomes a wide peneplain. Between the floods, the old river meanders, changes its plan, but stays within a certain meander belt at the central part of the peneplain. In shifting from one location to another, a meandering river may leave behind oxbow lakes or abandoned oxbow-shaped depressions. An example of a meandering river is the Thames.

Alluvial soils. Eroded soil transported by water and deposited is alluvial (water-laid) soil, or alluvium. Immediately adjacent to the steep portion of the valley, boulders and coarser gravel might be expected and there will be a minimum of fine sizes. At a distance of several miles from the place of original erosion, fines may predominate.

Alluvial deposits are in many respects similar to glacial but are generally more stratified and their properties might be determined from fewer boreholes than under equal conditions in glacial soils. The alluvial deposits are somewhat heterogeneous. It is not unusual, for example, to find a bed of alluvial clay several feet long, although it may be fairly narrow and only a few feet thick. Rather uniform sand and gravel beds of varying dimensions may be found and, although there may be lens-like inclusions of sand in gravel beds and vice versa, these deposits as a whole are fairly continuous.

Besides forming terraces and benches in the valley itself, deposition of alluvium also may occur on river plains and form relatively flat deposits. Large plains are not necessarily continuous but may be interrupted by isolated hills and occasional valleys. The sediment carried by a flow moving across a plain during a flood may be spread if the gradient of the stream decreases gradually and in this case a flood-plain is formed. However, if the gradient decreases abruptly, a larger part of the sediment carried by the stream drops in one place and forms an alluvial fan, a broad cone with the apex at the point where the gradient breaks.

Particular cases of recent alluvium are organic silt and mud. These are fine outwash from hills and mountain ridges, deposited in estuaries and in the rivers flowing into them, especially in the lower reaches of these rivers. The greater part of the organic silt consists of angular fragments of quartz and feldspar, abundant sericite (fine mica), and clayey matter; numerous micro-organisms are also present. In the natural state, organic silt is dark and smells unpleasant; after drying it can become light grey and lose its characteristic odour.

For further reading see the Bibliography on General and Physical Geology, and for engineering behaviour see Section 8.

ROCK DEFORMATION IN NATURE—FRACTURES AND FAULTS

Joints

Joints are fractures without any displacement. They may appear to be somewhat random in direction, but a careful field examination will usually show that they have some relation to the host rock, e.g. with the bedding in sedimentary rock or with flow lines in igneous rock*.

In igneous rocks there are often three regular sets of joints (Figure 7.15). In an ideal situation one set lies approximately horizontal and parallel to the flow lines and is termed flat-lying. Another set, the cross joints, is roughly perpendicular to the flow lines. The third set, the longitudinal joints, dips steeply and strikes parallel to the flow lines if the latter are projected to a plane surface such as a map.

In sedimentary rocks, there are often two systems of mutually perpendicular joints, both perpendicular to the bedding plane.

Joints also may be grouped into strike joints and dip joints. Figure 7.16 illustrates the terms *strike* and *dip* where the rock surface is assumed to be an oblique plane. Strike is the direction of contour lines or lines of equal elevation on the surface of the rock mass,

* Lines showing the flow of the originally liquid magma and indicated by the long axes of crystals

and the dip is the maximum slope of its surface. In Figure 7.16 the dip is the angle α made by the line AB with the horizon. In measurements of dip, it is important to measure the 'true' dip, i.e. the angle located in a plane perpendicular to the strike; otherwise, a misleading *apparent* dip, β in Figure 7.16, is recorded. These terms also apply to beds and other geometric features.

Joints and their orientation with respect to other structures have been widely studied

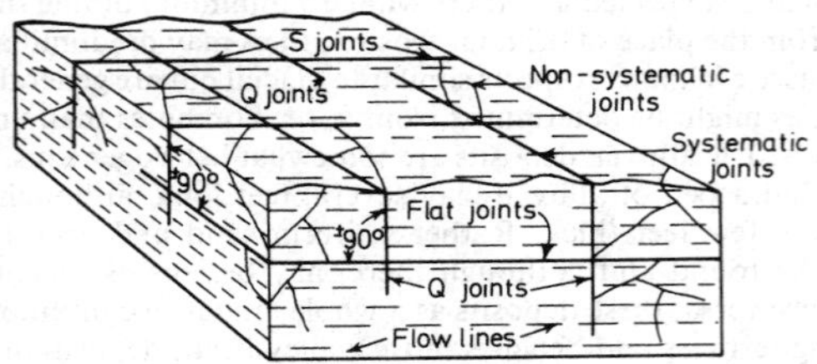

Figure 7.15 Block diagram of simple joint systems in an igneous rock. Systematic S joints are more commonly called longitudinal joints, and Q joints are more commonly called cross joints perpendicular to the flow lines of the original molten rock (After Hills and Cloos)

in the field and it has been established that systematic joints usually show well defined relationships to folds and faults which develop during the same tectonic episode.

The spacing of joints varies considerably and is of importance in engineering. Some rocks, such as sandstones and limestones in which the joints may be widely spaced, yield large blocks and may be suitable for masonry, for example, whereas other rocks may be so closely jointed as to break up into small pieces and may be suitable for aggregate or other purposes. Some joints in sedimentary rocks run only from one bedding plane to the next, but others may cross several bedding planes, and are called *master joints.*

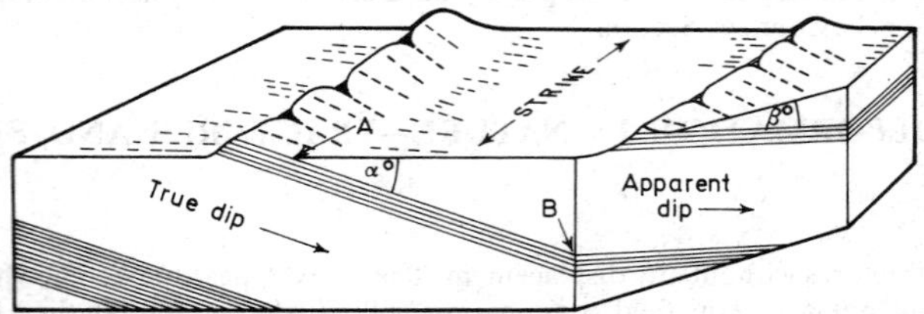

Figure 7.16 Idealised block diagram to show dip and strike relationships

The ease of quarrying, excavating or tunnelling in hard rocks largely depends on the regular or irregular nature of the joints and their surface characteristics, attitude, size, frequency and spacing. Joints and other fractures control ground water flow in otherwise intact rock and therefore help to promote rock weathering.

Faults

Faults are fractures in the crust along which there has been displacement of the rocks on one side relative to those on the other.

The surface on which movement takes place during faulting is the fault plane. It may be vertical, steeply inclined or gently inclined as with thrust faults. The intersection of a fault with the ground surface is known as the fault line or fault trace. The upper side of an inclined fault, and the rock which lies above it, is referred to as the hanging wall. Rock below it is the foot wall; dip faults strike parallel to the local direction of dip of the beds, strike faults are parallel to the strike and oblique faults cut across both strike and dip directions.

Movements on a fault may be in any direction. The displacement or slip is the sum of all the previous effects of movement and is shown by the relative positions on either side of the fault of two originally contiguous features as a bedding plane. The vertical component of the slip, taken by itself, is called the throw of the fault (see Figure 7.17).

Faults can be classified, according to the direction of movement that has taken place on them, into normal faults, reverse faults and transcurrent or strike-slip faults.

Normal faults (originally so called because they are the normal type found in British coalfields) are those in which the hanging-wall rocks have moved down the dip of the fault plane. Small normal faults are extremely common in almost all geological situations and may even occur in Quaternary sediments. Large normal faults, occurring in groups,

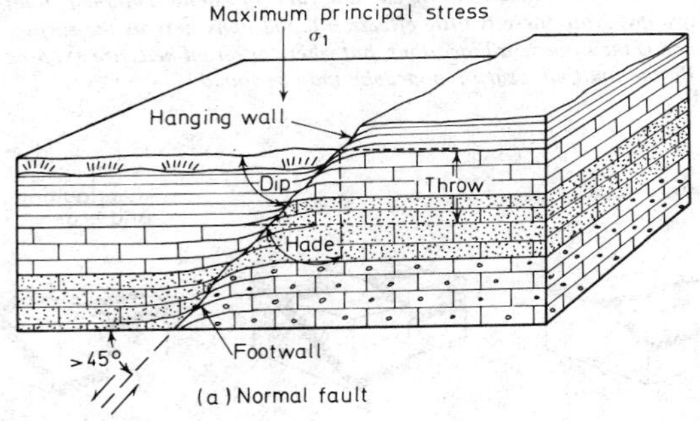

(a) Normal fault

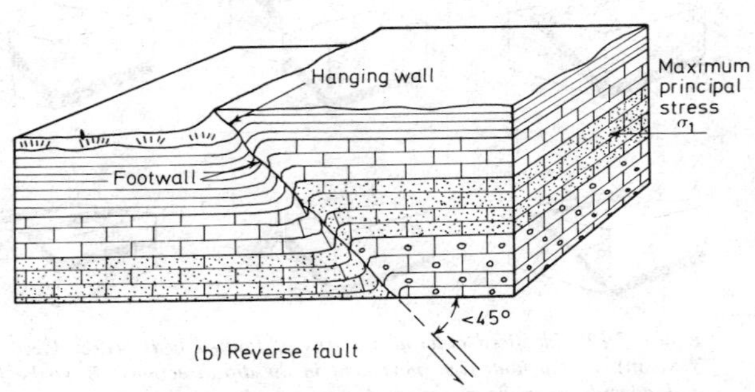

(b) Reverse fault

Figure 7.17 After Blyth

produce a considerable effect of lengthening and are especially common in the more stable areas of the earth's crust. Groups of faults are arranged so that alternate dislocations dip in opposite directions and produce the effect of block faulting illustrated in Figure 7.18; the crust is separated into high blocks or horsts between outward-dipping faults and low blocks, troughs or graben between inward-dipping faults.

Reverse faults are those on which the rocks of the hanging wall move up the dip of the fault plane. They result in shortening across the fault and in duplication of strata; reverse faults with low dips are thrusts.

Transcurrent faults are wrench faults, tear faults or strike-slip faults on which horizontal movement takes place. The fault planes are almost vertical and the effect of faulting

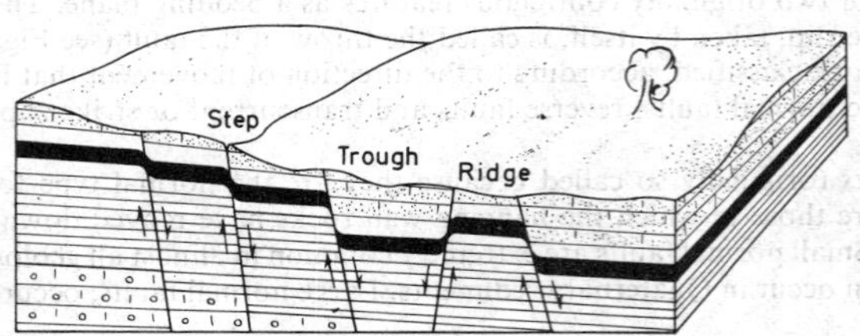

Figure 7.18 Idealised block diagram of some common fault groups. Note there is little effect on topography here as the surface bed is the same in all locations, but where different beds are exposed by the faulting, scarp topography may be found

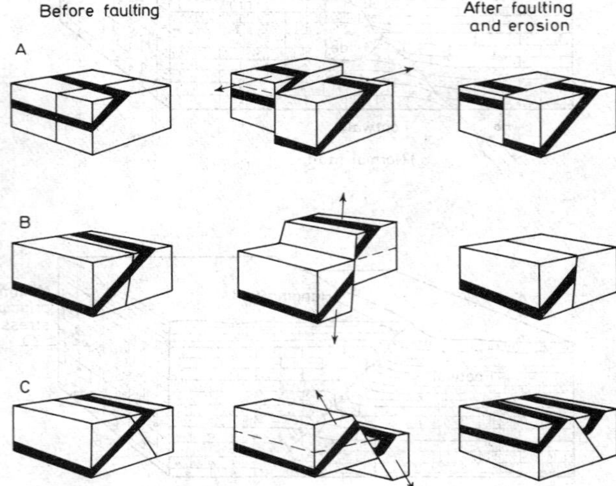

Figure 7.19 Idealised outcrop patterns of faulted beds (After Reed and Watson). *A, dip fault (i.e. movement in the dip direction); B, strike fault with downthrow in dip direction; C, strike fault with downthrow against the dip angle*

when seen on a map is to shift rocks laterally, even for many tens of miles. Examples of mapped outcrop patterns of faults are shown in Figure 7.19.

An example of the relationship between faulting and jointing in one complete episode is shown in Figure 7.21 from the textbook by Price. Techniques and the use of stereographic projection in geology is given in Phillips (see Bibliography).

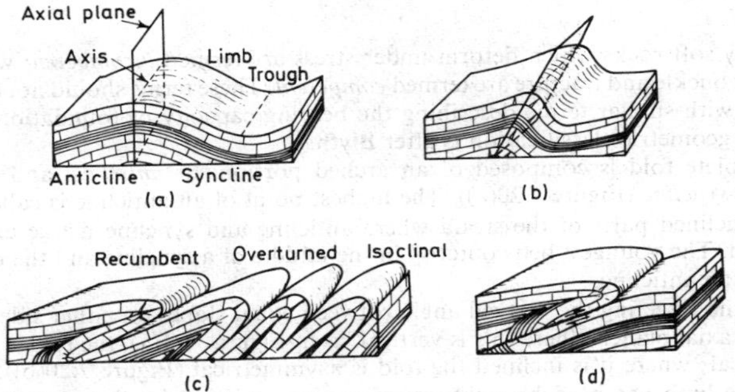

Figure 7.20 Idealised fold types (After Blyth). *(a) Simple or gentle symmetrical; (b) simple or gentle asymmetrical; (c) tight asymmetrical, recumbent, overturned and isoclinal; (d) recumbent passing into a thrust fault*

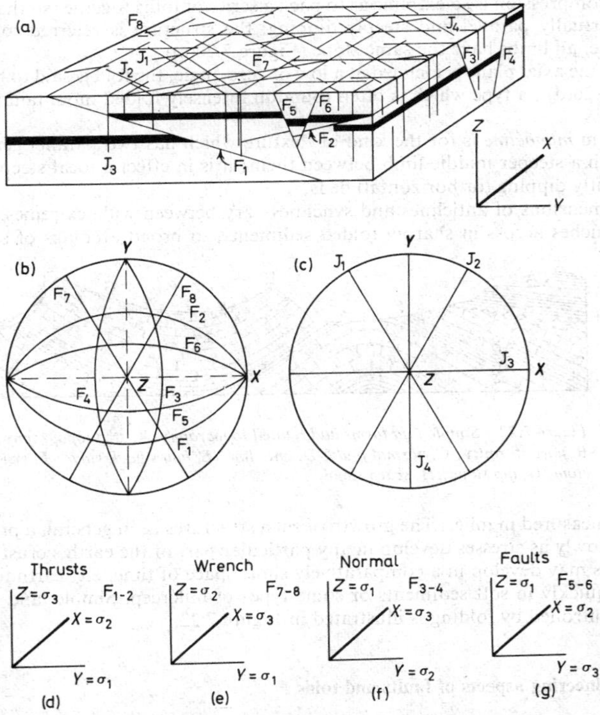

Figure 7.21 (a) Block diagram showing orientation of faults and joints in unfolded rocks which may result from various phases of compression and tension related to one complete tectonic episode; (b) stereogram of fault orientations shown in (a); (c) stereogram of joint orientation shown in (a); (d)–(g) orientation of stress fields when the various groups of faults were initiated. (Redrawn from N. J. Price)

Folds

In geology soft rocks which deform under stress are termed *incompetent* whereas hard rocks that buckle and fracture are termed *competent*. These terms should not be confused, however, with similar terms describing the bearing capacity of foundation rocks. The following geometric classification is after Blyth.

A complete fold is composed of an arched portion, or *anticline*, and a depressed trough or *syncline* (Figure 7.20(a)). The highest point of an anticline is called the crest, and the inclined parts of the strata where anticline and syncline merge are the limbs of the fold. The youngest beds outcrop in the middle of a syncline and the oldest in the middle of an anticline.

The plane bisecting the vertical angle between equal slopes on either side of the crest line is the axial plane. Where this is vertical, as in Figure 7.21(a), the fold is upright and symmetrical; where it is inclined the fold is asymmetrical (Figure 7.20(b)). Sometimes the middle limb has been brought into a vertical position by the compression which buckled the strata, and under still more severe conditions an *overturned fold,* or overfold, is produced (Figure 7.20(c)). Here the middle limb is inclined in the same attitude as the axial plane, and the beds of which it is composed have a reversed dip, i.e. upper beds are now brought to dip steeply beneath lower beds, an inversion of the true sequence.

If the compression is so extreme as to pack a series of folds together so that their limbs are all virtually parallel and steeply dipping, the structure is referred to as *isoclinal folding*, i.e. all limbs have the same slope (Figure 7.20(c)).

Where the axial plane is inclined at a low or zero angle, the fold is said to be *recumbent* (Figure 7.20(d)), a type which is often found in intensely folded mountain regions such as the Alps.

The term *monocline* is for the kind of flexture which has two parallel gently dipping limbs with a steeper middle limb between them; it is in effect a local steepening of the dip in gently dipping (or horizontal) beds.

The dimensions of anticlines and synclines vary between wide extremes, from small puckers inches across in sharply folded sediments, to broad archings of strata whose

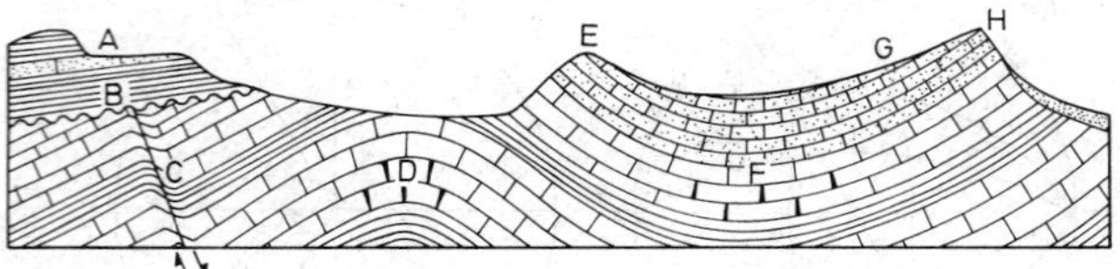

Figure 7.22 Simple fold forms and related topography. A, step topography; B, unconformity; C, normal fault; D, anticline; E, hog's back ridge; F, syncline; G, dip slope; H, scarp slope

extent is measured in miles. The growth of such structures is, in general, a process which goes on slowly as stresses develop in any particular part of the earth's crust; but superficial folds may develop in a comparatively short space of time, e.g. earthquake ripples forming quickly in soft sediments or some types of hillcreep. Simple land topography largely controlled by folding is illustrated in Figure 7.22.

Some engineering aspects of faults and folds

A search for faults is not always effective during site investigation, and significant faults are sometimes not discovered until construction or even afterwards. Stability of hillsides, cut slopes, quarry faces and so on may often be controlled by the geometric arrangement of joints and faults. Also the ground water pattern may be controlled by the condition of the joints and faults whether they are open or closed or filled with debris or gouge.

On large works the determination of whether a fault is *active*, *inactive* or *passive* may be important. Active faults are those in which movements have occurred during the recorded history and along which further movements can be expected any time (such as the San Andreas and some other faults in California). Inactive faults have no recorded history of movement and are assumed to be and probably will remain in a static condition. Unfortunately, it is not possible yet to state definitely if an apparently inactive fault will remain so. The fault may reopen, either because of a new strain accumulation in the locality or from the effect of earthquake vibrations.

From the alteration products of faulting, gouge is probably of the most concern in foundation problems. This is usually a relatively impervious clay-grade material and may hinder or stop the movement of ground water from one side of the fault to the other and so create hydrostatic heads, e.g. if encountered in a tunnel. It may also reduce sliding friction along the fault plane. The presence of soft fault breccia or gouge may cause sudden squeezes in a tunnel that intersects the fault. Arch action of rocks in tunnels may be reduced by the presence of joints and faults. Rock falls on cuts and in tunnels, patterns of rock bolts, grout holes and so on are all controlled to a large extent by the joint and fault pattern.

Dipping beds, which must be part of a fold system, may cause stability problems if the dip is unfavourable into a cut face. Serious water problems may arise in the construction and maintenance of tunnels intersecting synclines containing water-bearing strata. In deep cuts, analogous water problems arise that may create continuous maintenance problems.

In foundations, folds are generally not so critical as faults though they may give stability problems if their geometry is unfavourable. Occasionally, folds may influence the selection of a dam site; for example, when the reservoir is located over a monocline containing previous strata, there may be excessive seepage if the monocline dips downstream. If the monocline were to dip upstream, the reservoir might have little seepage providing the monocline contained some impervious layers such as shale which was not fractured in the folding.

ENGINEERING GEOLOGY CLASSIFICATION OF ROCK

Intact Rock

Intact rock refers to the rock material which is free of the larger-scale structural features such as joints, bedding planes, partings or shear zones. There are many classification schemes in existence. Coates,[17] Coates and Parsons[18] and others have done extensive work on the classification of intact rock on the basis of laboratory values of the mechanical properties. Deere and Miller[19] give a modified version of Miller's earlier work and it is this classification which is described in the following, but attention is drawn to British systems given in the Working Party reports of the Geological Society[20,21] and those in Section 9.

The classification is based on two important engineering properties of the rock—the uniaxial compressive strength and the modulus of elasticity. A tangent modulus taken at a stress level equal to one-half the ultimate strength of the rock is used and the compressive strength is that determined on specimens with a length/diameter ratio of at least 2. The rock is divided into one of six categories of strength as shown in Figure 7.23.

Strength categories follow a geometric progression with the dividing line between A and B categories chosen at 32 000 lbf/in^2 (2250 kgf/cm^2) since this is about the upper limit of strength of most common rocks. Only a few rock types fall in the A category—including quartzite, dolerite and dense basalts. The B category, between 16 000 and 32 000 lbf/in^2 (1125 and 2250 kgf/cm^2), includes the majority of the igneous rocks, the stronger metamorphic rocks and well-cemented sandstones, hard shales and the majority of the limestones and dolomites. The C category, the medium strength rocks in the 8000–16 000 lbf/in^2 (562–1125 kgf/cm^2) range, includes many shales, porous sandstones and limestones and the more schistose varieties of the metamorphic rocks (e.g. chlorite, mica or talc schists). The D and E rocks are of low and very low strength and include porous or low-density rocks as friable sandstone, porous tuff, clay-shales, rock salt and weathered or chemically altered rocks of any lithology.

The second element of the classification system is the ratio of the modulus to the uniaxial compressive strength giving the modulus ratio, as indicated in Figure 7.23. The values of the compressive strength and the modulus are plotted on a logarithmic scale to accommodate a wide range of values, and strength categories are shown across the top. The modulus ratio is obtained from the plotted position with respect to the diagonal lines. The stippled zone is bounded by an upper line with a modulus ratio of 500:1 and by a lower line of 200:1. This zone is referred to as M the zone of average, or medium, modulus ratio. Rocks possessing an interlocking fabric, e.g. the majority of igneous rocks (Figure 7.23), and little or no anisotropy, normally fall in this category.

The summary plot for sedimentary rocks is given in Figure 7.24. Limestone and dolomites fall mostly in strength categories B and C although a few samples plot as A, very high strength rocks, and D, low strength rocks.

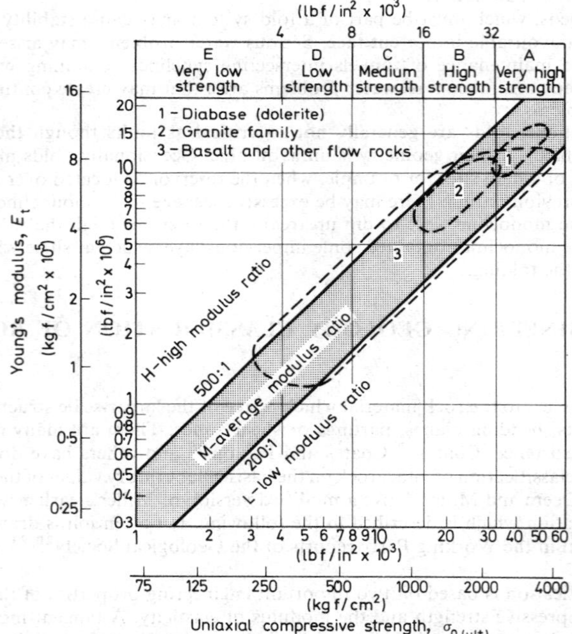

Figure 7.23 Engineering classification of intact rock; summary plot for igneous rocks (176 specimens 75% of points). (Redrawn from Deere and Miller[19])

This position seems to be in response both to the fabric (interlocking) and to the particular mineralogy (calcite and dolomite). The sandstone and shale plots are shown open-ended in their lower portions because several samples failed at strengths of less than 1000 lbf/in² (70.307 kgf/cm². It can be seen that both the sandstone and shale envelopes extend into the zone of low modulus ratio. This position is considered to be the result of anisotropy created by the bedding or laminations and the modulus values are low because almost all samples were tested with the core axis perpendicular to the bedding.

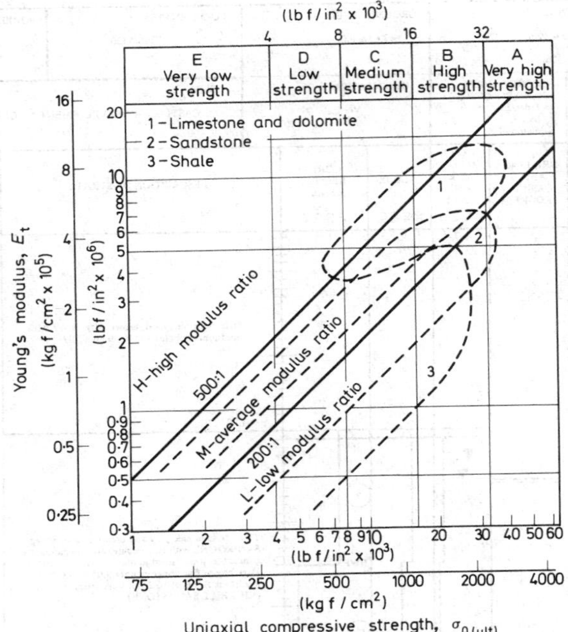

Figure 7.24 Engineering classification for intact rock; summary plot for sedimentary rock (193 specimens 75% of points). (Redrawn from Deere and Miller[19])

This orientation does not affect the strength but gives low modulus values because of deformation caused by closure across incipient bedding planes and the aligned or platy minerals, particularly in the shales.

Structure and fracture in rock

Any geological structure that influences one of the mass properties of the *in situ* rock, such as the strength, modulus of deformation or permeability, is highly significant. The most common structural features of significance are joints, bedding planes and foliation surfaces and 'shears' or faults. Because these are all planar or near-planar discontinuities, they have a strong anisotropic effect on the mass properties.

ENGINEERING MAPPING AND LOGGING OF GEOLOGICAL STRUCTURES

It is important to map carefully all these geological structures by location, orientation (strike and dip) and spacing. The physical and geological characteristics should also be described. Even from rock cores obtained in diamond drilling the tightness and irregularity of the surfaces of discontinuity as well as the kind of filling material between or along the adjacent surfaces can be observed and should be recorded. Terms such as tight or open should be used to describe the tightness; plane, curved or irregular to describe the degree

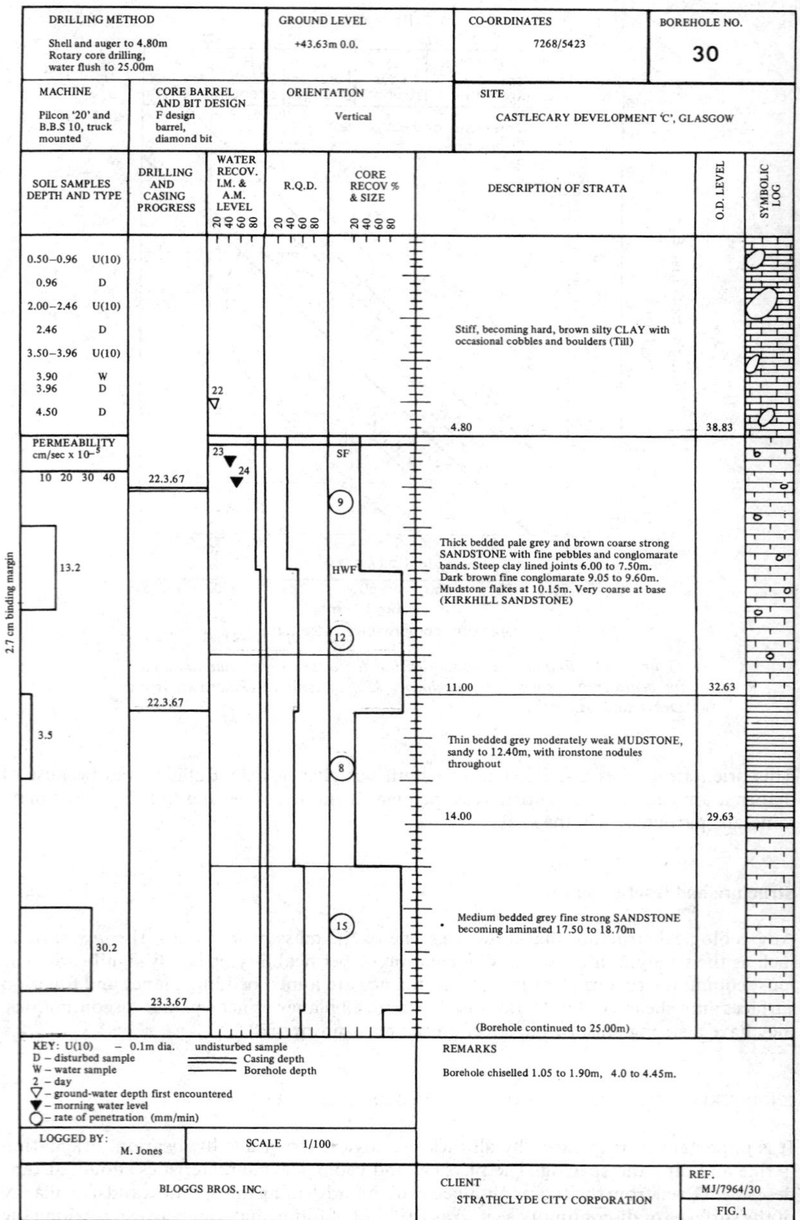

DRILLING METHOD
Shell and auger to 4.80m
Rotary core drilling, water flush to 25.00m
GROUND LEVEL
+43.63m O.D.
CO-ORDINATES
7268/5423
BOREHOLE NO.
30
MACHINE
Pilcon '20' and B.B.S 10, truck mounted
CORE BARREL AND BIT DESIGN
F design barrel, diamond bit
ORIENTATION
Vertical
SITE
CASTLECARY DEVELOPMENT 'C', GLASGOW
SOIL SAMPLES DEPTH AND TYPE
DRILLING AND CASING PROGRESS
WATER RECOV. I.M. & A.M. LEVEL
R.Q.D.
CORE RECOV % & SIZE
DESCRIPTION OF STRATA
O.D. LEVEL
SYMBOLIC LOG
20 40 60 80
0.50–0.96 U(10)
0.96 D
2.00–2.46 U(10)
2.46 D
3.50–3.96 U(10)
3.90 W
3.96 D
4.50 D
22
Stiff, becoming hard, brown silty CLAY with occasional cobbles and boulders (Till)
4.80
38.83
PERMEABILITY cm/sec x 10–5
10 20 30 40
22.3.67
23
24
SF
9
13.2
HWF
Thick bedded pale grey and brown coarse strong SANDSTONE with fine pebbles and conglomarate bands. Steep clay lined joints 6.00 to 7.50m. Dark brown fine conglomarate 9.05 to 9.60m. Mudstone flakes at 10.15m. Very coarse at base (KIRKHILL SANDSTONE)
12
11.00
32.63
22.3.67
3.5
Thin bedded grey moderately weak MUDSTONE, sandy to 12.40m, with ironstone nodules throughout
8
14.00
29.63
15
Medium bedded grey fine strong SANDSTONE becoming laminated 17.50 to 18.70m
30.2
23.3.67
(Borehole continued to 25.00m)
KEY: U(10) – 0.1m dia. undisturbed sample
D – disturbed sample
W – water sample
2 – day
Casing depth
Borehole depth
– ground-water depth first encountered
– morning water level
– rate of penetration (mm/min)
REMARKS
Borehole chiselled 1.05 to 1.90m, 4.0 to 4.45m.
LOGGED BY:
M. Jones
SCALE 1/100
BLOGGS BROS. INC.
CLIENT
STRATHCLYDE CITY CORPORATION
REF.
MJ/7964/30
FIG. 1
2.7 cm binding margin

Figure 7.25 Example of proposed borehole log redrawn from the Geological Society Working Party report on the logging of rock cores for engineering purposes[20]

of planeness; and slick, smooth or rough to describe the smoothness and so on. Numerical values should be obtained by measurement where possible.

Field mapping will often provide useful data concerning the above. However, care must be taken that erroneous assumptions are not drawn from the measurements because the rock outcrops may not be sufficiently numerous to provide a statistically valid sample. The major discontinuities, as faults or shear zones, may not be visible because erosion or deep weathering along them may have obscured them and the rock outcrops may not provide sufficient three-dimensional exposures to allow the true number and spacing of all the discontinuities to be obtained. The discontinuities and rock types at depth may also differ considerably from those exposed at the surface. For those reasons it is often advisable to supplement surface mapping with test shafts, test adits and core borings.

Two principal methods are used for mapping in boreholes. One by use of oriented core, the other by television, colour film or endoscope techniques. The advantage of the photographic methods is that some information may be gained about the width of the structural feature, and about whether it is open or filled.

Detailed recommendations for *core logging* are given in the Geological Society Engineering Group Working Party Report on logging of rock cores.[20] In brief the scheme for logging recommends the following factors in the log for adequate engineering geological description: (see also Figure 7.25)

(i) systematic description
(ii) alteration and weathering state
(iii) structure and discontinuities
(iv) assessment of rock material strength
(v) other features, including stratigraphy.

SYSTEMATIC ROCK DESCRIPTION

The following standard sequence of systematic description is proposed: i—weathered state; ii—structure; iii—colour; iv—grain-size; (C.P. 2001); (iv(a)—subordinate particle size; iv(b)—texture; iv(c)—alteration state; iv(d)—cementation state as relevant); v—rock material strength; vi(a)—mineral type as relevant); vi—ROCK NAME.

It is considered that the qualifications are more important in core descriptions than the actual rock name and, for this reason, the name was placed last. Such a system is appropriate to an engineering description where classification by mechanical properties is more significant than classification by mineralogy and texture. The following examples are provided for illustrative purposes:

Fresh	Foliated	Dark grey	Coarse	Very strong	Hornblende	GNEISS
i	ii	iii	iv	v	vi(a)	vi

Moderately weathered	Thickly bedded	Cream	Medium-grained	Strong	Dolomitic	LIMESTONE
i	ii	iii	iv	v	vi(a)	vi

Completely weathered	Thinly flow-banded	Mid-grey	Very coarse	Porphyritic	Kaolinised	Weak	Tourmaline	GRANITE
i	ii	iii	iv	iv(b)	iv(c)	v	vi(a)	vi

Detailed recommendations for *engineering geology mapping* are given in the Geological Society Engineering Group Working Party Report on Engineering Geology Maps.[21]

The systematic classification of rock for engineering geological mapping purposes is generally similar to that for core logging with the addition of estimates for the mechanical strength of the rock material and estimate of its mass permeability and other terms indicating special engineering characteristics which might be estimated from inspection of exposed rock. Samples of weathered state, structure and estimated field strength are recommended. That for weathering is given in Table 7.6 and for fracture spacing and strength as follows, Tables 7.7 and 7.8.

Table 7.7 BEDDING AND FRACTURE (DISCONTINUITY) SPACING OF ROCK MATERIAL

Bedding spacing term	*Spacing*	*Fracture spacing term*
Very thickly bedded	>2 m	Very widely spaced
Thickly bedded	600 mm–2 m	Widely spaced
Medium bedded	200 mm–600 mm	Moderately widely spaced
Thinly bedded	60 mm–200 mm	Closely spaced
Very thinly bedded	20 mm–60 mm	Very closely spaced
Laminated (sedimentary) Closely (metamorphic and igneous)	6 mm–20 mm	Extremely closely spaced
Thinly laminated (sedimentary) Very closely (metamorphic and igneous)	<6 mm	

Table 7.8 ESTIMATED MECHANICAL STRENGTH OF ROCK MATERIAL

Term	*Uniaxial compressive strength* MN/m^2 1MN/m^2 = 145 lbf/in^2	*Comparable point load strength* kN/m^2 1kN/m^2 = 145 × 10^{-3} lbf/in^2
Extremely strong	>200	>12 000
Very strong	100–200	6 000–12 000
Strong	50–100	3 000–6 000
Moderately strong	12.5–50	750–3 000
Moderately weak	5–12.5	300–750
Weak	1.25–5	75–300
Very weak	<1.25	<75

GEOLOGICAL MAPS

Maps of the *Institute of Geological Sciences* are published in two principal forms. 'Solid' maps show the rock outcrops as they would appear with the overburden removed. 'Drift' maps show overburden, usually with dotted lines to indicate the probable extent of the underlying outcrops. All show outcrop patterns, not just the actual exposure of the rock at the surface. As it is only exposures which can be seen, geological maps are necessarily in part conjectural. Mapping is based on various techniques and the completed map is not simply a survey, but the sum of all the information gathered from various sources. The small-scale map, e.g. usually 10 miles to 1 inch or larger, is useful to obtain a general appreciation of the country over a relatively wide area, whilst the large scale map, 1 mile to 6 inches or 1 mile to 1 inch, is more for detailed information. Regional Guides and detailed Memoirs are also published by the Institute of Geological Sciences as well as Water and Mineral memoirs and so on, to supplement their maps.

One of the first things to be considered on looking at a geological map is to determine whether the rocks shown are igneous, metamorphic or sedimentary, by checking the main outcrops shown against the key provided on the map. Assuming the rocks are sediments, the strike is usually the long axis of the outcrop if it is a small-scale map. A more accurate picture is obtained by comparing the outcrop with the contours. By definition, in dipping beds the strike is the direction in which similar horizons of the strata are at the same elevation, so it follows that where the top (or bottom) boundary of a bed twice crosses the same contour line, the top of the bed will be at the same altitude at those two points and that a line drawn connecting them will show the strike direction.

The dip can be identified by noting the direction of dip arrows if these are shown, otherwise it can be deduced. The key will show which are the older of two successive beds; a boundary line on flat ground indicates that the older beds have emerged from below the newer beds, or in other words that the older beds are dipping towards the newer beds. If the boundaries follow contour lines, the beds are more-or-less horizontal, and if the boundaries cross contour lines at right angles, the beds are vertical. The relation of the outcrops to the contour lines in valleys is particularly helpful in diagnosing the general dip of the bed.

The degree of dip is obtainable by drawing parallel strike lines, at different contour levels, using the same boundary between beds. Thus if a boundary between two beds crosses, say, the 200 ft, 250 ft and 300 ft contours on each side of a valley, and the strike lines are drawn in at these levels, the distance between each strike line on the map is the distance over which the bed has dipped 50 ft. The degree of dip may then be calculated or obtained graphically.

There are many other problems that can be solved by a study of geological maps, and textbooks dealing with this are given in the Bibliography.

REFERENCES

1. PRICE, D. G., 'Engineering geology in the urban environment', *Q. J. Engng Geol.*, **4**, 191–208 (1971)
2. *Site Investigations*, C.P. 2001, British Standards Institution, London (1957)
3. FOOKES, P. G., DEARMAN, W. R. and FRANKLIN, J. A., 'Some engineering aspects of rock weathering with field examples from Dartmoor and elsewhere', *Q. J. Engng Geol.*, **4**, 139–186 (1971)
4. LITTLE, A. L. 'Laterites', *Proc. 3rd Asian Reg. Conf. Soil Mech. Found. Engng*, Haifa, **2**, 61–71 (1967)
5. DEERE, D. U. and PATTON, F. D., 'Stability of slopes in weathered rock', *4th Pan. Am. Conf.*, 87–163 (1971)
6. SANDERS, M. K. and FOOKES, P. G., 'A review of the relationship of rock weathering and climate and its significance to foundation engineering', *Engng Geol*, **4**, 289–325 (1970)
7. Symposium on airfield construction on overseas soils, *Proc. Instn Civ. Engrs*, **8**, 211–292 (Nov. 1957)
8. FOOKES, P. G. and KNILL, J. L., 'The application of engineering geology in the regional development of northern and central Iran', *Engng Geol.*, **3**, 81–120 (1969)
9. HOLTZ, W. G. and GIBBS, H. J., 'Consolidation and related properties of loessial soils', *Am. Soc. Testing Materials*, Spec. Tech. Publ. 126, 9–33 (1952)
10. WEINERT, H. H. and CLAUSS, M. A., 'Soluble salts in road foundations 4th Reg. Conf. for Africa', *Soil Mech. and Found. Engng*, 213–218 (1967)
11. HIGGINBOTTOM, I. E., FOOKES, P. G. and GORDON, D., 'Engineering aspects of glacial features in Britain and their investigation', *Q. J. Engng Geol.* (in the press)
12. LINELL, K. A. and SHEA, H. F., 'Strength and deformation characteristics of various glacial tills in New England', *Am. Soc. Civ. Engng Reg. Conf. on Shear Strength of Cohesive Soils*, 275–314 (1960)
13. ANDERSON, J. G. C. and TRIGG, C. F., 'Geotechnical factors in the redevelopment of South Wales valleys', *Civil Engineering Problems in South Wales Valleys*, Institutions of Civil Engineers, 13–22 (1970)
14. HIGGINBOTTOM, I. E. and FOOKES, P. G., 'Engineering aspects of periglacial features in Britain', *Q. J. Engng Geol.*, **3**, 86–117 (1970)
15. WEEKS, A. G., 'The stability of slopes in south east England as affected by periglacial activity', *Q. J. Engng Geol.*, **2**, 49–62 (1969)
16. FOOKES, P. G. and BEST, R., 'Consolidation characteristics of some late Pleistocene periglacial metastable soils of east Kent', *Q. J. Engng Geol.*, **2**, 103–128 (1968)

17. COATES, D. F., 'Classification of rocks for rock mechanics', *Int. J. Rock Mech. and Mining Sci.*, **1,** 421–429 (1966)
18. COATES, D. F. and PARSONS, R. D., 'Experimental criteria for classification of rock', *Int. J. Rock Mech. and Mining Sci.*, **3,** 181–189 (1966)
19. DEERE, D. U. and MILLER, R. P., 'Engineering classification and index properties for intact rock', *Tech. Rep. No.* AFWL-TR-65-116, Air Force Weapons Lab., Kirtland Air Force Base, New Mexico (1966)
20. 'Geological Society Working Party Report on the logging of rock cores for engineering purposes', *Q. J. Engng Geol.*, **3,** 1–19 (1970)
21. 'Geological Society Working Party Report on the preparation of maps and plans in terms of engineering geology', *Q. J. Engng Geol.* **5,** No. 4, 293–381

BIBLIOGRAPHY

Periodicals

Engineering Geology (Pub. quarterly by Elsevier)
Géotechnique (Pub. quarterly by Institution of Civil Engineers)
International Journal of Rock Mechanics and *Mining Sciences* (Pub. bimonthly by Pergamon)
Quarterly Journal of Engineering Geology (Pub. by Geological Society)
Rock Mechanics and Engineering Geology (Pub. quarterly by Springer-Verlag)

Geology Textbooks

General and Physical Geology

AGER, D. V., *Introducing Geology*, Faber (1961)
BLYTH, F. G. H., *A Geology for engineers*, 5th edn, Arnold (1967)
COTTON, C. A., *Geomorphology*, 7th edn (rev.), Whitcombe & Tombs (1958)
DURY, G. H., *The face of the Earth*, rev. edn, Penguin Books (1966)
FLINT, R. F., *Glacial and Quaternary geology*, Wiley (1971)
GASS, I. G., SMITH, P. J. and WILSON, R. C. L., *Understanding the Earth*, Artemis Press (1971)
HOLMES, A., *Principles of physical geology*, 2nd edn, Nelson (1965)
KING, C. A. M., *Beaches and coasts*, 2nd edn, Arnold (1972)
OLLIER, C. D., *Weathering*, Oliver and Boyd (1969)
READ, H. H. and WATSON, J., *Beginning geology*, Macmillan/Allen & Unwin (1966)
SCHEIDEGGER, A. E., *Theoretical geomorphology*, 2nd edn, Allen & Unwin (1970)
SHEPHARD, F. P., *Submarine geology*, 2nd edn, Harper (1963)
SMALL, R. J., *The study of landforms*, Cambridge (1970)
SPARKS, B. W., *Geomorphology*, Longmans (1960)
WEST, R. G., *Pleistocene geology and biology*, Longmans (1968)
WOOLDRIDGE, S. W. and LINTON, D. L., *Structure, surface and drainage in South East England*, Philip (1955)

Fieldwork and Mapwork

BADGLEY, P. C., *Structural methods for the exploration geologist*, Harper (1959)
BENNISON, G. M., *An introduction to geological structures and maps*, 2nd edn, Arnold (1969)
BLYTH, F. G. H., *Geological maps and their interpretation*, Arnold (1965)
HIMUS, G. W. and SWEETING, G. S., *The elements of field geology*, University Tutorial Press (1951)
LAHEE, F. H., *Field geology*, 6th edn, McGraw-Hill (1961)

Engineering Geology

DUNCAN, N., *Engineering geology and rock mechanics*, 2 Vols, Leonard Hill (1969)
KRYNINE, D. P. and JUDD, W. R., *Principles of engineering geology and geotechnics*, McGraw-Hill (1957)
LEGGET, R. F., *Geology & engineering*, 2nd edn, McGraw-Hill (1962)
STAGG, K. G. and ZIENKIEWICZ, O. C., *Rock mechanics in engineering practice*, Wiley (1968)

Mineralogy and Petrology

GRIM, R. E., *Applied clay mineralogy*, McGraw-Hill (1962)
HATCH, F. H., WELLS, A. K. and WELLS, M. K., *The petrology of igneous rocks*, 12th edn, Murby (1961)
HATCH, F. M. and RASTALL, R. H., *The petrology of the sedimentary rocks*, 5th edn, Revised by J. T. Greensmith, Murby (1971)
KRUMBEIN, W. C. and SLOSS, L. L., *Stratigraphy and sedimentation*, 2nd edn, Freeman (1963)
PETTIJOHN, F. J., *Sedimentary rocks*, 2nd edn, Harper (1957)
READ, H. H., *Rutleys elements of mineralogy*, 26th edn (1970)
TRASK, P. D., *Applied sedimentation*, Wiley (1950)

Stratigraphy and Palaeontology

DAVIES, A. M., *An introduction to palaeontology*, 3rd edn, Murby (1962)
GIGNOUX, M., *Stratigraphic geology*, Freeman (1955)
WELLS A. K. and KIRKCALDY J. F., *Outlines of historical geology*, 6th edn, Murby (1967)
WILLIS, L. J., *A palaeogeographical atlas of the British Isles and adjacent parts of Europe*, Blackie (1952)
WOODS, H., *Palaeontology; invertebrate*, 8th edn, Cambridge (1947)

Structural Geology

BILLINGS, M. P., *Structural geology*, 2nd edn, Prentice-Hall (1954)
HILLS, E. SHERBON, *Elements of structural geology*, 2nd edn, Chapman & Hall (1972)
PHILLIPS, F. C., *The use of stereographic projection in structural geology*, 3rd edn, Arnold (1971)
PRICE, N. J., *Fault and joint development in brittle and semi brittle rock*, Pergamon (1966)
RAMSAY, J. G., *Folding and fracturing of rocks*, McGraw-Hill (1967)

In addition to these general works, the following series are of interest to British engineers:

British Regional Geology: handbooks published by the Institute of Geological Sciences, Geological Museum, London, SW7

British Palaeozoic Fossils, British Mesozoic Fossils, British Cainozoic Fossils } Handbooks published by the British Museum (Natural History) London SW7

Geologists' Association Guides: a series of excursion guides to selected British localities (available from The Scientific Anglian, 30/30A St Benedict's St, Norwich)

Publications of the Geological Society of London, Burlington House, Piccadilly, London W1V OJU : various monographs and authoritative works.

8 SOIL MECHANICS

8 SOIL MECHANICS 8

8 SOIL MECHANICS

A. C. MEIGH, M.Sc (Eng.), F.I.C.E., F.G.S., and N. B. HOBBS, M.A. (Cantab.), D.I.C., M.I.C.E., F.G.S.
Soil Mechanics Ltd.

SOIL PROPERTIES AND SOIL TESTING

Porosity, density, water content

DEFINITIONS

All soils consist of solid particles assembled in a relatively open packing. The voids may be filled completely with water (fully saturated soils) or partly with water and partly with air (partially saturated soils).

The relationships between void space and the volume occupied by the particles are fundamental and are characterised by the following definitions.

Porosity n = volume of voids V_v/total volume of soil V_t

Voids ratio $e = V_v$/volume of soil particles V_s

Hence $e = n/(1-n)$ and $n = e/(1+e)$ (1)

Degree of saturation S_r = volume of water/V_v

Water content w = weight of water/weight of soil particles

Hence, if G_s is the specific gravity of soil particles,

$$w = S_r e/G_s$$

For fully saturated soils $S_r = 1$ and $w = e/G_s$

Bulk density γ = total weight of soil and water W_t/V_t

Hence

$$\gamma = (G_s + S_r e)\gamma_w/(1+e) \qquad (2)$$

where γ_w is the density of water (1 Mg/m^3).

when $S_r = 1$, $\gamma = (G_s + e)\gamma_w/(1+e)$ (3)

Dry density $\gamma_d = W_s/V_t$

Hence $\gamma_d = G_s\gamma_w/(1+e)$ (4)

and $\gamma = \gamma_d(1+w)$ (5)

Also $n = 1-\gamma_d/G_s\gamma_w$ (6)

and $S_r = w/(\gamma_w/\gamma_d - 1/G_s)$ (7)

The submerged density, of soils below water table, is given by

$$\gamma_s = (G_s - 1)\gamma_w/(1+e)$$

$$= (G_s - 1)\gamma / G_s(1 + w) \quad (8)$$

Percentage air voids A = volume of air × 100/V_t

$$A = 1 - \gamma_d(1 + wG_s)/\gamma_w G_s \quad (9)$$

All the foregoing definitions and relationships are in constant use in soil mechanics problems.

DETERMINATION OF WATER CONTENT, SPECIFIC GRAVITY AND BULK DENSITY

(Standard methods are given in BS 1377[1])

A sample of soil is weighed and then placed in a ventilated oven maintained at a temperature of 105 to 110 °C until a constant weight is reached. 24 hours is usually sufficient. Water content = loss in weight/dry weight.

The choice of 105 to 110 °C as a drying temperature range is arbitrary, since at temperatures above this a higher moisture content sometimes results.

The specific gravity of soil particles smaller than 6 mm may be readily determined by the usual pycnometer method, as described in textbooks on elementary physics.

In the laboratory, density is readily determined by measuring the volume of a known weight of soil. If in addition the water content is determined the dry density of the sample can immediately be calculated.

For cohesive soils the simplest method is to cut a specimen by pressing in a thin-walled cylinder and trimming the ends flush. The cylinder is then weighed and from its known weight and volume the density is found by direct calculation.

For sand samples which have been obtained in the compressed air sand sampler (see Section 10), the density is obtained by carefully trimming the ends (inside the sampler), measuring the length of the sample and pushing it out into a vessel, when its weight can be determined. In general, in laboratory tests on sand and gravel, density is calculated by using a known weight of material and measuring the volume which it occupies in the test apparatus.

It is often necessary to measure the density of soil *in situ.* The two most important methods are the 'sand replacement method' and the 'core cutter method'.

In the sand replacement method a hole about 150 mm diameter and 150 mm deep is made in the soil whose density is to be determined and all the material taken from the hole is carefully kept and weighed. The volume of the hole is then determined by filling it with sand in a standard manner from a container holding a known weight of sand. The weight of sand required to fill the hole is obtained by difference, and the volume of this weight of sand is known by previous calibration of the sand and method of filling when used to fill standard containers. The method as described is applicable only to fine soils, but by suitable modification of the equipment it can also be used with coarse materials.

In the core cutter method, which is suitable for cohesive and non-stony soils only, a metal cylinder of known volume is forced into the soil and dug out. The soil is then trimmed off flush with the ends of the cylinder and the density determined as described above.

Recently, nuclear methods have been used for *in situ* measurement of moisture content and density in compacted files and in natural soils, both at the surface (Gardner *et al.*[2]) and at depth in boreholes (Lefevre and Manke[3]).

DETERMINATION OF POROSITY, VOIDS RATIO AND DEGREE OF SATURATION

These properties are rarely measured directly but are deduced from the values for water content, bulk density and specific gravity, by means of the relationships already given.

Classification tests

LIQUID AND PLASTIC LIMIT, PLASTICITY AND LIQUIDITY INDICES

The liquid limit is that water content above which a soil behaves as a fluid. Obviously there will be a range of water content over which any soil will become gradually more fluid with increasing water content. Some point must be arbitrarily chosen as the liquid limit; that chosen in BS 1377,[1] which is in agreement with the practice in most countries, is the water content at which 25 small standard shocks will close a standard groove for a length of 13 mm. For a description of the apparatus and method of test see Akroyd.[4]

The plastic limit is that water content below which a soil ceases to behave in a plastic manner. A sample of the soil is dried by kneading it in the hand until, when rolled out under the palm on a sheet of glass to a thread of 3 mm diameter, the thread breaks up into short lengths. The water content is then determined. A repeat test is carried out and if the two results are close the mean water content is taken as the plastic limit. Although this test appears to be rather arbitrary, consistent results are soon obtained by an intelligent operator.

The difference between the liquid limit (LL) and the plastic limit (PL) is called the plasticity index (PI). The liquidity index (LI) is given by:

$$\text{LI} = (w - \text{PL})/(\text{LL} - \text{PL})$$
$$= (w - \text{PL})/\text{PI}$$

When the natural water content of a soil is equal to its liquid limit the LI is unity, and when the natural water content is equal to the plastic limit the LI is zero.

ACTIVITY

Another index property is the activity (Skempton[5]).

$$\text{Activity} = \text{PI}/\mu$$

where μ is the clay fraction content, i.e. the percentage by weight of material finer than 0.002 mm. Activity is related to the mineralogy and geological history of clays and to the proportion of their shear strength contributed by surface activity between the particles. Experience has shown that difficulty in obtaining undisturbed samples from deep beds of normally consolidated clays is restricted to clays with an activity less than 0.75.

RELATIVE DENSITY, MAXIMUM AND MINIMUM DENSITIES

The classification tests already described apply to predominantly cohesive soils. For granular soils the significant index properties are the maximum and minimum densities. The maximum density, γ_{max}, is the highest dry density at which the soil can exist without a breakdown of the particles, and the minimum density, γ_{min}, is the lowest possible dry density. The relative density, which is an important factor governing the strength of granular soils, is expressed as

$$\text{RD} = (\gamma - \gamma_{min})/(\gamma_{max} - \gamma_{min})$$

Techniques for measurement of maximum and minimum density are discussed by Kolbuszewski.[6]

MECHANICAL ANALYSIS

By mechanical analysis is meant the determination of the sizes of the particles composing a soil and the percentage of particles of a given size present. For particles larger than

0.06 mm, i.e. those particles which are retained by a No. 200 BS sieve (0.07 mm mesh), the sizing is done by sieving through standard sieves. These sizes comprise the sands and gravels. For particles smaller than 0.06 mm, i.e. the silt and clay range, the sizing is done by measuring the settling velocity v of the particles in water and calculating the particle size from Stokes' law, the assumption being made that the particles are spherical. The method is described in detail by Taylor.[7]

Full details for carrying out the test are given in BS 1377.[1] The results are plotted as shown in Figure 8.1.

The uniformity coefficient U is defined as the ratio: amount by weight of D_{60} size divided by the amount by weight of D_{10} size. A low value of U indicates a uniform soil and a high value a well graded soil.

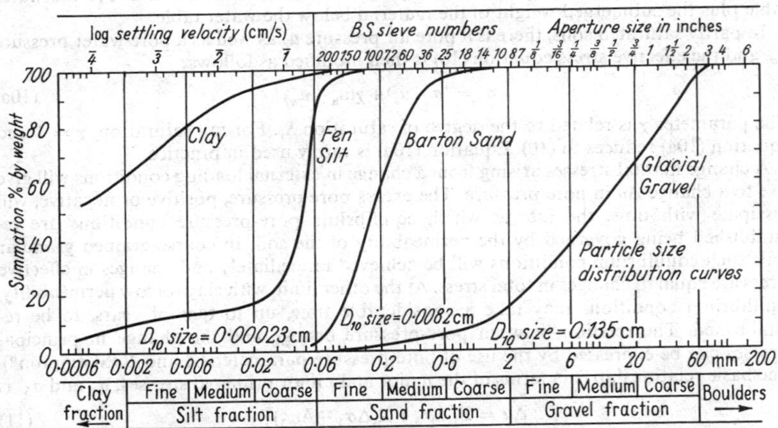

Figure 8.1 Particle size distribution curves (on standard sheet) showing D_{10} size. (The D_{10} size is now given in millimetres)

SHRINKAGE LIMIT

When a saturated clay soil is allowed to dry out in air it shrinks. At first the loss in volume is equal to the volume of water lost by drying, but finally a point is reached at which the volume ceases to diminish although the weight continues to drop. The water content at which this occurs is called the shrinkage limit.

Total pressure, pore-pressure and effective stress

Soil can be considered as a two-phase system consisting of a solid phase, the skeleton of soil particles, and a fluid phase, water plus air in a partially saturated soil and water alone in a saturated soil. It follows that the normal stress across a plane within a soil mass will have two components, an intergranular pressure, known as the effective pressure or effective stress σ' and a fluid pressure, known as the pore pressure or neutral pressure u. The sum of these will constitute the total normal stress. The volume change characteristics and the strength of a soil are controlled by the effective stress, the pore pressure being significant only in so far as it determines the magnitude of the effective stress for a given total stress.

The simplest illustration of pore pressure and effective stress is given by consideration of the vertical stresses acting on a horizontal plane at depth h under equilibrium conditions with a horizontal water table. The total vertical stress σ is given by the weight per unit area of soil and water above the plane

$$\sigma = h\gamma = h(G_s + S_r e)\gamma_w/(1+e)$$

The pore pressure will be the water pressure, and if the plane is at a depth h_w below the water table then $u = h_w\gamma_w$

The effective vertical stress is the difference between these:

$$\sigma' = \sigma - u \tag{10}$$

It should be noted that for positions below the water table the vertical effective stress can also be calculated from the total weight per unit area of material above the water table plus the submerged weight of the material below the water table.

In partly saturated soils, there is a pore air pressure u_a as well as a pore water pressure u_w, and the effective stress equation (10) is then modified as follows:

$$\sigma' = (\sigma - u_a) + \chi(u_a - u_w) \tag{10a}$$

The parameter χ is related to the degree of saturation S_r. For full saturation, $\chi = 1$ and equation (10a) reduces to (10). Equation (10a) is rarely used in practice.

A change in total stresses arising from a change in external loading conditions will give rise to a change Δu in pore pressure. The excess pore pressure, positive or negative, will dissipate with time, the rate at which equilibrium pore-pressure conditions are re-established being governed by the permeability of the soil. In coarse-grained granular soils such equilibrium conditions will be achieved immediately and changes in effective stress are equal to changes in total stress. At the other limit, with clays of low permeability, equilibrium conditions may take a considerable time, up to tens of years, to be re-established. The relation between pore-pressure change and the change in principal stresses can be expressed by the use of pore pressure parameters A and B (Skempton[8]). The basic relationship is in terms of the major and minor principal stresses, σ_1 and σ_3, is

$$\Delta u = B[\Delta\sigma_3 + A(\Delta\sigma_1 - \Delta\sigma_3)] \tag{11}$$

It is also useful to relate the pore pressure change to the change in deviator stress $(\Delta\sigma_1 - \Delta\sigma_3)$ alone and also to the change in the major principal stress $(\Delta\sigma_1)$. For these purposes two further parameters $\bar{A}$ and $\bar{B}$ are used as follows:

$$\Delta u = B\Delta\sigma_3 + \bar{A}(\Delta\sigma_1 - \Delta\sigma_3) \qquad \text{or} \qquad \Delta u = \bar{B}\Delta\sigma_1 \tag{12}$$

If the soil structure behaved in an elastic manner the values of the pore pressures could be established theoretically, e.g. A would have a value of $\frac{1}{3}$. However, soils behave nonelastically and A can have values ranging between $+1.3$ and -0.7 (values at failure in a triaxial compression test). Typical values of the pore pressure parameters are given by Bishop and Henkel[9] For a full discussion of the parameters see Skempton.[8]

Permeability

Permeability is that property of a soil which controls the rate of flow of water through the soil. In soil mechanics permeability is defined by the equation (Darcy's law):

$$v = ki \tag{13}$$

where v is the superficial velocity of flow through the soil, i is the hydraulic gradient and k is the permeability. k therefore has the dimensions of a velocity; it depends chiefly on particle size and grading, and to a lesser extent on the particle packing.

Typical values of permeability for soils range from 1×10^{-5} m/s for a coarse sand to

1×10^{-9} m/s for a clay. A very rough estimate of permeability can be obtained for a relatively uniform sand from Hazen's law,

$$100\,k = D_{10}^2$$

where D_{10} is the 10% size or effective size in mm and k is in m/s.

The 10% size or effective size is the particle size at which the grading curve crosses the 10% line (see Figure 8.1).

KOZENY'S FORMULA AND LOUDON'S FORMULA

Research by Loudon[10] has shown that the permeability (m/s) of clean sand may be computed from simple soil tests, using Kozeny's formula, with an accuracy of ±20%:

$$kS^2 = 1{\cdot}5 \times 10^{-4} \frac{n^3}{(1-n)^2} \text{ (SI units)} \qquad (14)$$

He suggests an alternative formula which is easier to use and of equal accuracy:

$$\log_{10}(kS^2) = 1.365 + 5.15n \qquad (15)$$

In both of these formulae n is porosity and S denotes specific surface (mm^2/mm^3):

$$S = f(x_1 S_1 + x_2 S_2 + \ldots x_n S_n) \qquad (16)$$

where f is an angularity factor, varying between 1.1 for a rounded sand and 1.4 for an angular sand

$x_1, x_2, \ldots$ = fraction in each sieve range

and

$S_1, S_2, \ldots$ = specific surface for each sieve range; from Loudon's tables. (Note that Loudon's tables are given in cm^2/cm^3 units for S. It is therefore necessary to convert Loudon's values for S using $1\ cm^{-1} = 0.1\ mm^{-1}$.)

MEASUREMENT OF PERMEABILITY

Permeability is measured in an apparatus known as a permeameter. Two forms are in general use, the falling-head permeameter and the constant-head permeameter. In general the falling-head type is used for soils with relatively low permeability, and the constant-head type for soils with high permeability.

It should be remembered that in the ground the horizontal permeability may be very different from (generally higher than) the vertical permeability. Lamination, or arrangement of the soil particles, which is almost invisible to the naked eye, may multiply the permeability by ten or more. Laboratory measurements should be used with caution in making calculations and should be looked upon as an estimate giving only the order of the permeability. For methods of carrying out permeability tests, see Taylor.[7]

Since the mass permeability of a clay soil is dominated by the fabric of fissures, silt bands and organic inclusions, it is desirable to test as large a sample as possible.

In situ tests are thus often a more reliable means of determining the permeability of a soil mass (see section 10).

Shear strength

THEORY OF SHEAR STRENGTH

Shear strength of a soil is commonly thought of as having two components, cohesion and frictional resistance. Clays are often described as cohesive soils in which the shear strength or cohesion is independent of applied stresses, and sands and gravels are described as noncohesive or frictional soils in which the shearing resistance along any plane is directly proportional to the normal stress across that plane:

$$s = p \tan \phi$$

The concepts of cohesion and friction were combined in Coulomb's equation for the shear strength of the soil:

$$s = c + p \tan \phi \tag{17}$$

where c is the cohesion and ϕ is the 'angle of internal friction'.

Such simple concepts are, however, inadequate to deal with the complex problem of the shear strength of soils. The early history of the study of shear strength is somewhat confused. Attempts were made to represent the shear strength of a soil by the envelope to a Mohr circle diagram of stress (Taylor[7]), the intercept on the vertical axis being taken as cohesion c, and the slope of the envelope being taken as the friction angle ϕ. It was found that except in sands and gravels the results for a given soil varied considerably depending on the test procedure used, particularly the rate of testing and the conditions of drainage of the specimens during test. However, following the realisation that the strength of a soil is governed by the effective stress, it was possible to achieve a better understanding of the shear strength characteristics of soils. The shear strength can be expressed as

$$\tau_f = c' + (\sigma - u) \tan \phi' \tag{18}$$

where c' and ϕ' are effective stress parameters; c' is the apparent cohesion, ϕ' the angle of shearing resistance and u the excess pore-water pressure.

The Mohr circle diagram can be plotted in terms of effective stress, with c' as the cohesion intercept and ϕ' as the slope of the envelope (Figure 8.2).

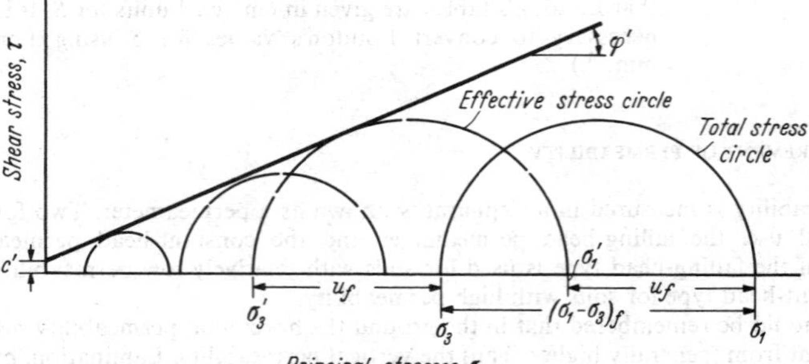

Figure 8.2 Mohr circle diagram

In terms of effective principal stresses in the Mohr diagram the Coulomb failure criterion may be expressed as

$$(\sigma'_1 - \sigma'_3) = \sin \phi'(\sigma'_1 + \sigma'_3) + 2c' \cos \phi' \tag{19}$$

and if c' is zero, then

$$\sigma'_3 = \sigma'_1(1 - \sin \phi')/(1 + \sin \phi') \tag{20}$$

which is the well known Rankine failure condition (Terzaghi and Peck[11]).

TRIAXIAL COMPRESSION TESTS

The apparatus generally used for measuring the shear strength of soils is the triaxial compression apparatus (Figure 8.3).

In this test a cylindrical specimen of soil is enclosed in a watertight rubber membrane. The specimen is contained within a chamber, the triaxial cell, in which a fluid can be placed under pressure. An axial load is applied at a constant rate of strain by means of a plunger passing vertically through the top of the cell. In normal testing procedures the stress applied by the plunger is the deviator stress $(\sigma_1-\sigma_3)$ and the cell pressure is the minor principal stress σ_3 (the intermediate principal stress $\sigma_2=\sigma_3$). The value of $\sigma_1-\sigma_3$ at failure $(\sigma_1-\sigma_3)_f$ is the diameter of the Mohr circle at failure (Figure 8.2). Density and moisture content are also measured as part of the test procedure, and the stress–strain modulus E is obtained from the initial tangent to the stress–strain curve.

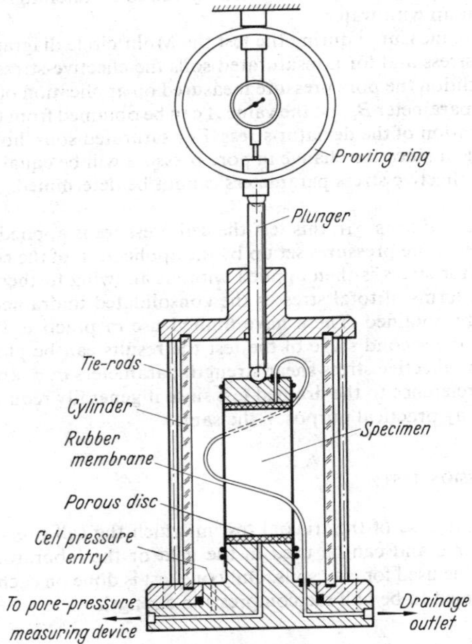

Figure 8.3 Apparatus for triaxial compression test

Connections to the top and bottom of the specimen can be made. One of these is used to control the drainage of the specimen and the other can be connected to the pore pressure measurement device. Two standard sizes of specimen are used, 75 mm long × 38 mm diameter and 200 mm long × 100 mm diameter.

Three standard test procedures, differing in the drainage conditions which apply, are discussed below. For a further description of these tests, and of a number of special tests which can be made in the triaxial apparatus, reference should be made to Bishop and Henkel.[9]

1. *Drained tests* Drainage is allowed throughout the test, and the rate of test is sufficiently slow to ensure that full dissipation of excess pore pressure occurs. The results are plotted in terms of effective stresses (in this case equal to total stresses); the terms c_d and ϕ_d are sometimes used in place of c' and ϕ' in describing the results of these tests.

2. *Undrained tests* No drainage is allowed. The results are plotted in terms of total stresses and the parameters are denoted as c_u and ϕ_u. For saturated soils the undrained

angle of shearing resistance, ϕ_u is zero and the apparent cohesion c_u is equal to half the deviator stress at failure. The undrained shear strength c_u is useful in many practical problems, particularly in estimating bearing capacity of clays and in stability analyses for conditions in which no change in pore pressure can take place. This method of analysis is known as the $\phi = 0$ method.

The results of tests on soft clays often show a large scatter when plotted against depth. This is due to the pore pressures set up in the specimen by the end restraints, and can be largely overcome by carrying out the tests with lubricated end plattens, at a slower rate of strain to allow equalisation of the pore pressures within the specimen (Barden and McDermott[12]). The scatter can also be caused by random softening in boreholes that have not been topped up with water.

If pore pressures are measured during the test the Mohr circle diagram can be plotted in terms of effective stress and for nonsaturated soils the effective-stress parameters can be determined. In addition the pore pressure measured on application of the cell pressure gives the value of the parameter B, and the value A can be obtained from the pore pressure change during application of the deviator stress. For saturated soils, however, the Mohr circles will be coincident since the change in pore pressure will be equal to the change in cell pressure and the effective stress parameters cannot be determined.

3. *Consolidated undrained tests* In this test the cell pressure is applied and drainage is allowed until the excess pore pressures set up by the application of the cell pressure have dissipated. The deviator stress is then applied without allowing further drainage. If the results are plotted in terms of total stresses the consolidated undrained shear strength parameters c_{cu}, ϕ_{cu} are obtained. These have limited use in practice. If pore pressures are measured during the second stage of the test the results can be plotted in terms of effective stress, and the effective stress shear strength parameters c', ϕ' are obtained. This test is often used in preference to the drained test since it generally requires less time and the results are, for many practical purposes, the same.

UNCONFINED COMPRESSION TEST

This test is a particular case of the triaxial test in which the cell pressure is zero. The apparatus is very simple and can be used in the field or the laboratory (Cooling and Golder[13]). It can only be used for clay soils. Only one test is done on each sample and the shear strength is assumed to be half the compression strength.

BOX SHEAR TEST

For determining the angle of shearing resistance of free-draining granular soils a box shear test is often used. The sample is contained in a box split horizontally in which the bottom of the box can be moved relative to the top, thus shearing the sample along a horizontal plane. A vertical load is applied to the sample by means of a weighted hanger and a lever arm system. It is also used to determine the residual strength of clays under drained conditions, extending the strain by reversing the movement a number of times. (See Figure 8.4, and Skempton.[14])

Sufficient tests are carried out under different vertical loads and the results are plotted as shear stress against normal pressure. The cohesion is taken as the intercept of the shear stress axis and the angle of shearing resistance is obtained from the slope of the plot.

TYPICAL SHEAR STRENGTH VALUES

Clays fall into two main groups, normally consolidated clays, those which have not been subjected to loads greater than their present overburden pressure, and overconsolidated

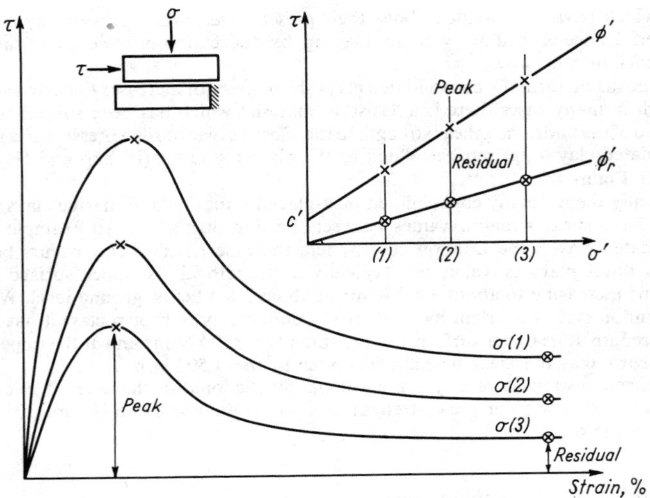

Figure 8.4 Drained shear box test on stiff clay

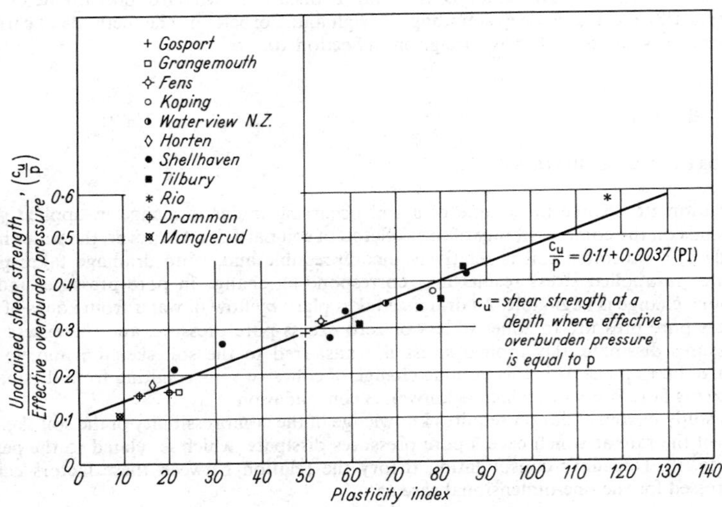

Figure 8.5 The relationship between c_u/p and plasticity index for normally consolidated clays

clays, which have been loaded above their present overburden pressure by the weight of material since eroded away, by ice loading, by desiccation or lowering of the ground water level, or otherwise.

In deposits of normally consolidated clays shear strength increases linearly with depth, although in many cases there is a 'crust' of material which has been subject to drying. The ratio of the undrained shear strength to the effective overburden pressure in a normally consolidated clay (c_u/p) can be related to the plasticity index (Bishop and Henkel[9]) as shown in Figure 8.5.

Typically the normally consolidated post-glacial estuarine and marine clays of Great Britain have shear strength values between 7.5 and 30 kN/m^2. An example of over-consolidated clay is the London clay. Where the blue London clay occurs below the Thames flood plain gravel it has typically a strength at its upper surface of some 75 kN/m^2 increasing to about 300 kN/m^2 at about 30 m below ground level. Where the blue London clay is overlain by weathered London clay (a brown clay) it has a higher shear strength at its upper surface, usually some 125–150 kN/m^2 and at the upper surface of the brown clay the shear strength falls often to about 50 kN/m^2.

The effective strength of clay is predominantly frictional in character. Typical values for London clay for the peak strength ϕ' and residual strength ϕ'_r are 20° and 16° respectively (Skempton[14]).

IN SITU MEASUREMENT OF SHEAR STRENGTH

The preceding discussion has been concrned with laboratory testing to determine shear strength. This can also be obtained from *in situ* tests using a variety of methods. Two of these give the shear strength directly; the *in situ* shear box test carried out on an exposed surface and generally used in overconsolidated clays, and the vane test carried out below the base of a borehole or by direct penetration and used usually in normally consolidated or lightly overconsolidated clays. Indirect methods require the application of a bearing capacity factor determined by the geometry of the test. These include plate-loading tests – either carried out on an exposed surface or in a borehole where it is sometimes called a piston test (Burland, Butler and Dunican[15]) – Menard pressuremeter tests (Menard[16]) and Dutch deep soundings (Meigh and Corbett[17]). The methods of carrying out the tests mentioned above are given in Section 10.

Consolidation

THEORY OF CONSOLIDATION

The ultimate change in volume of a soil occurring under a change in applied stress depends on the compressibility of the skeleton of soil particles. However, the water in the voids of a saturated soil is relatively incompressible and, if no drainage takes place, change in applied stress results in a corresponding change in pore pressure, and the volume change is negligible. As drainage takes place by flow of water from zones of high excess pore pressure to zones of less or zero excess pore pressure, and the excess pore pressures dissipate, the applied stress is transferred to the soil skeleton and volume change takes place. It is this volume change of cohesive soils resulting from dissipation of excess pore pressures which is known as consolidation.

A study of consolidation requires knowledge of the compressibility of the soil skeleton and of the rate at which excess pore pressures dissipate, which is related to the permeability. In Terzaghi's consolidation theory the relation between these factors can be expressed for the one-dimensional case as

$$c_v \partial^2 u/\partial z^2 = \partial u/\partial t \qquad (21)$$

where c_v is the coefficient of consolidation and is given by

$$c_v = k/m_v\gamma_w \tag{22}$$

u is excess pore-water pressure, z the thickness of the stratum, t is time, and m_v is the modulus of volume compressibility, defined as

$$m_v = -(de/dp)\,(l+e)^{-1} \tag{23}$$

where e is the voids ratio and p the effective pressure.

The solution of the consolidation equation has been given by Taylor[7] and values have been tabulated for the degree of consolidation U against the time factor T, where

$$T = c_v t/H^2 \tag{24}$$

and H is the length of the drainage path (see p. 8–56). Values of c_v and m_v are determined by laboratory tests known as oedometer tests or consolidation tests.

CONSOLIDATION TEST

The consolidation test is usually carried out on clays and silts. The samples are usually undisturbed, but cases can arise in which tests on remoulded soils are required (e.g. assessing the consolidation characteristics of fill for an earth dam). The purpose of the test is to obtain the effective pressure/voids ratio curve and the coefficient of consolidation for the soil in question.

Two types of apparatus are used; the cells are shown in Figure 8.6. In the standard

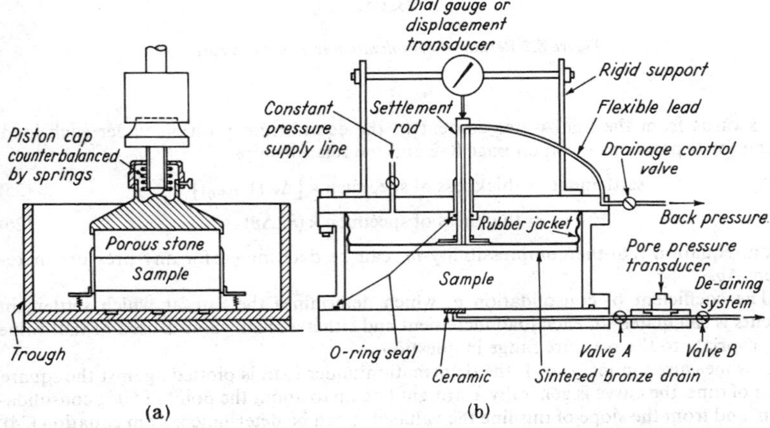

Figure 8.6 Oedometer (consolidation) press cells. (a) Standard oedometer cell; (b) Rowe hydraulic oedometer cell

test a specimen contained in a ring, usually 76 mm diameter and 200 mm long, is loaded axially between two porous discs, the load being applied mechanically by a lever arm (Figure 8.6a). In the special test (Figure 8.6b) the specimen, either 150 mm or 250 mm diameter, with either vertical or radial drainage, is loaded hydraulically with pore pressure measurement at the bottom surface (Rowe and Barden[18]). This Rowe cell test is likely to replace the standard test.

For a given load increment the settlement with time is measured. When movement has ceased, or virtually ceased, the load is increased and settlement with time is again measured. This is repeated several times until the pressures are greater than any likely to be applied to the soil in the practical problem. Finally the water content of the sample is determined. The initial water content is measured on pieces of soil adjacent to the sample and acts as a check on the calculations. In the Rowe cell test the pore-pressure dissipation is also observed for each pressure increment.

The pressure/voids ratio curve (e–log p curve (Figure 8.7) is calculated by working

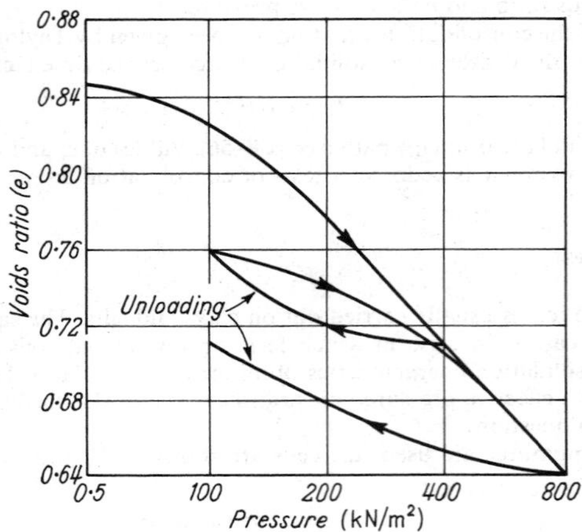

Figure 8.7 Results of consolidation test: e–log p curve

backwards from the final water content to the equilibrium position under each load, using the expressions given on page 8–2 and the relationships

$$\text{settlement} = \text{thickness of specimen} \times [\Delta e/(1+e_0)] \quad (25)$$

$$= \text{thickness of specimen} \times (m_v \Delta p) \quad (26)$$

From equation (26) the compressibility m_v can be determined for any pressure increment Δp.

The coefficient of consolidation c_v which determines the rate at which settlement occurs is calculated for each load increment and either a mean value is used or that value appropriate to the pressure range in question.

c_v is measured in m²/year. If the deformation under load is plotted against the square root of time, the curve is generally a straight line up to about the point of 50% consolidation, and from the slope of this line the value of c_v can be determined, from equation (24), putting a value of 0.848 for T at 90% consolidation,

$$c_v = 0.848H^2/t_{90} \qquad \text{m}^2\text{/years} \quad (27)$$

where t_{90} is the time in years to 90% consolidation, H is the length of drainage path ($= \frac{1}{2}$ thickness of sample in metres).

A construction to determine t_{90} is given in Figure 8.8 (Taylor[7]). In some cases it is preferable to plot time on a log basis, and the construction, in this case to determine t_{100}, is shown in Figure 8.9. c_v can also be determined from the rate of dissipation of pore pressure measured in the Rowe cell, as described below.

Experience has shown that the rates of consolidation under actual structures are many times faster than those predicted in laboratory tests on small specimens. This is due mainly to the influence of soil structure on permeability (Rowe[19]). It is now generally accepted that the best method of estimating the field c_v value is to combine the permeability measured at appropriate depths in a borehole with the m_v values determined in the oedometer on samples from the same depths. Equation (22) is used for this purpose.

A phenomenon known as the 'secondary time effect' must be taken into account in

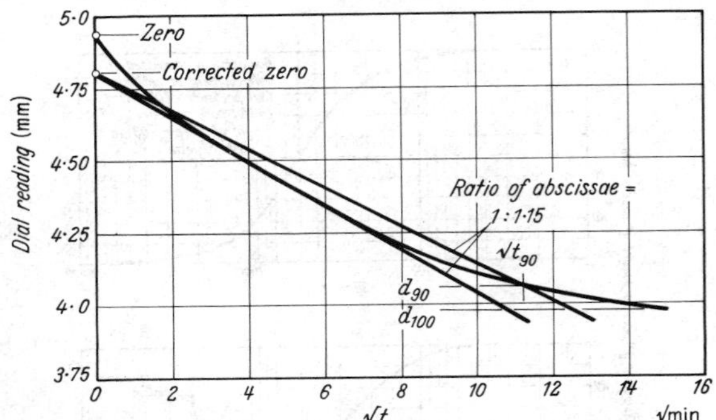

Figure 8.8 Time–settlement curve for consolidation test. Square-root fitting

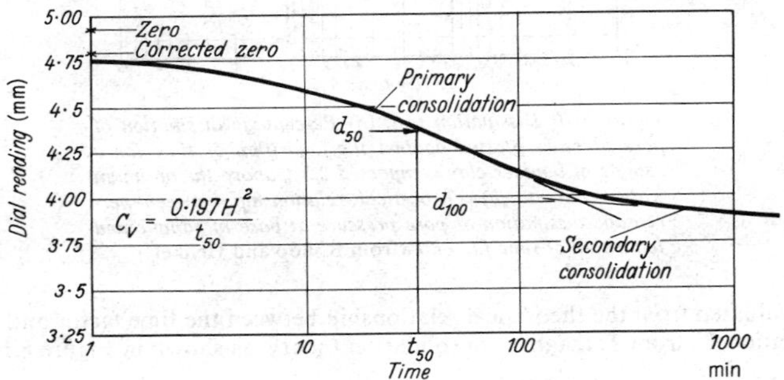

Figure 8.9 Time–settlement curve for consolidation test. Log-fitting method

some cases, but for details of this, reference must be made to a textbook on soil mechanics (Terzaghi and Peck[11]), also to a paper by Bjerrum.[20]

Pore-pressure dissipation

While for settlement problems the indirect measurement of rate of dissipation obtained from the oedometer test may be adequate, for some other problems, however, particularly those involving stability calculations in which strength, and hence the factor of safety, is dependent on the magnitude of the pore pressures, it is important to have a more direct measurement of the pore pressure/time relationship.

In the pore-pressure dissipation test the rate of dissipation is measured directly. A sample is set up as for a triaxial compression test with porous discs at each end. The top disc is connected to an external drainage tube and the bottom disc is connected to a pore-pressure measuring device. For a given cell pressure the pore pressure in the sample is allowed to reach a steady state with the drainage connection shut off. The connection is then opened and the rate of dissipation of the pore pressure is measured. The consolidation coefficient c_v is calculated using equation (24), where H is the drainage path (in this case the length of the specimen) t is the time to a given degree of consolidation (usually taken as 50% consolidation), and T is the time factor. A pore pressure dissipation test can also be conducted in the Rowe cell under either vertical or radial drainage.

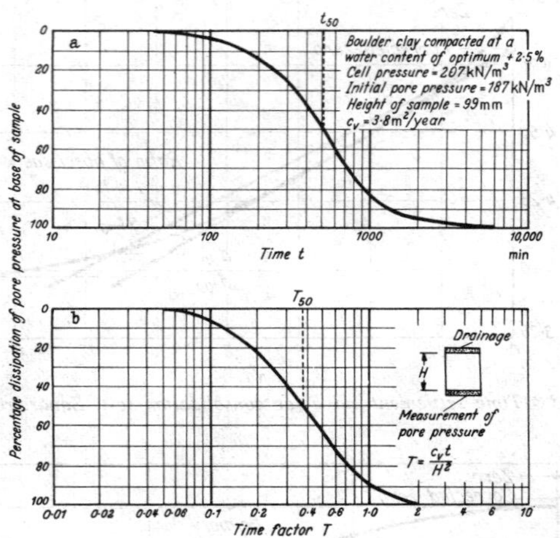

Figure 8.10 Dissipation test. (a) Percentage dissipation of pore pressure plotted against the logarithm of time for a sample of boulder clay compacted 2.5% above the optimum water content. (b) Theoretical relationship between percentage dissipation of pore pressure at base of sample and logarithm of time factor T (from Bishop and Henkel[9])

T is evaluated from the theoretical relationship between the time factor and degree of consolidation U, from Terzaghi's consolidation theory, as shown in Figure 8.10.

Compaction, California bearing ratio, and modulus of subgrade reaction

Certain other soil properties and the corresponding tests must be considered in relation to problems which arise in the construction of earth dams, roads and runways.

COMPACTION

When soil is used as a constructional material in earth dams and the basecourses and subgrades of roads and runways it is important to ensure that it is placed in as compact a condition as possible in order to obtain high strength, and a minimum of softening and settlement.

A standard test originally known as the Proctor test and now known in Great Britain as the British Standard Compaction Test is used to determine the compaction characteristics of soils. This test is fully described in BS 1377.[1] It is carried out by compacting the soil in a mould in the standard manner at several different moisture contents and recording the density obtained at each. The moisture content at which the maximum dry density is obtained is the optimum moisture content. The results are plotted as shown in Figure 8.11.

The optimum moisture content and maximum dry density as given by the standard compaction test are looked upon by some authorities as classification tests rather than values to be used in the field. This is because for a different amount of work used in the

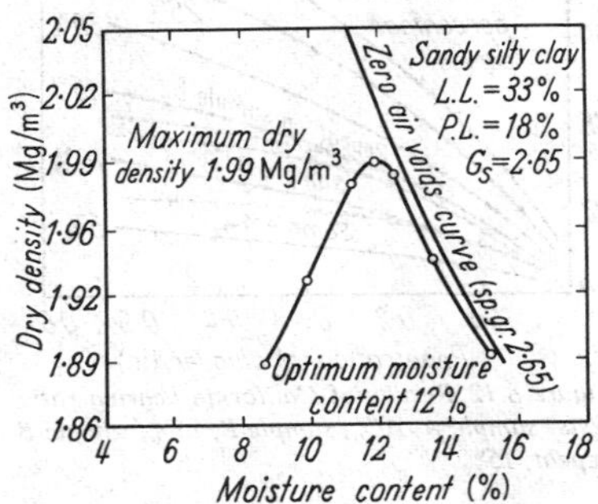

Figure 8.11 Results of standard compaction test

test the values of these properties obtained are different. For heavy compaction, particularly with well graded nonplastic soil, a modified test is sometimes used in which a greater amount of work is employed. This is particularly so in the United States where the AASHO standard is used since the results obtained have been found to agree better with the field results on dry soils using the heavy compacting equipment available in that country. There is no British Standard corresponding to this heavier test.

In practice, the optimum moisture content is best determined in the field for the particular plant which it is proposed to use (Road Research Laboratory[21]).

CALIFORNIA BEARING RATIO (CBR)

The California bearing ratio test is an empirical test described fully in reference 21 for the design of flexible or nontensile road pavements. It has been extended by large scale tests to cover the design of flexible runways and taxi-tracks for aircraft.

Although the test can be carried out on undisturbed soil it is generally done on compacted soil. The soil is compacted into a mould of 150 mm diameter in a manner similar to that used in the British standard compaction test described above, or in the manner used in the AASHO test, but the amount of work is increased in the ratio of the volume of the mould for each of these tests.

After compaction a circular plunger of 1936 mm^2 area is forced into the soil at a steady rate of 1.3 mm/min until a penetration of about 1.3 mm has been obtained. The pressure at various depths of penetration is recorded. The pressure corresponding to a penetration of 2.5 mm is read off and is expressed as a percentage of the standard pressure for a material having a CBR of 100, which is defined as 6.9 MN/m^2. The value for a penetration of 3 mm is also read off and expressed as a percentage of 10.35 Mn/m^2. The larger of these two values is taken as the CBR. A typical result is shown in Figure 8.12.

The test can be carried out either on the soil as compacted or after the soil has been soaked. This latter test is advisable if there is any possibility of the soil on the actual job soaking up water from the water table, or being subjected to heavy rains.

To carry out the soaked test a circular 45 N weight is placed on top of the sample as a surcharge and the mould is then placed under water for four days. The expansion of the surface is measured and this is expressed as a percentage of the height of the sample. After four days' soaking the plunger penetration test is carried out as described above, the plunger operating through a suitable circular hole in the centre of the surcharge weight.

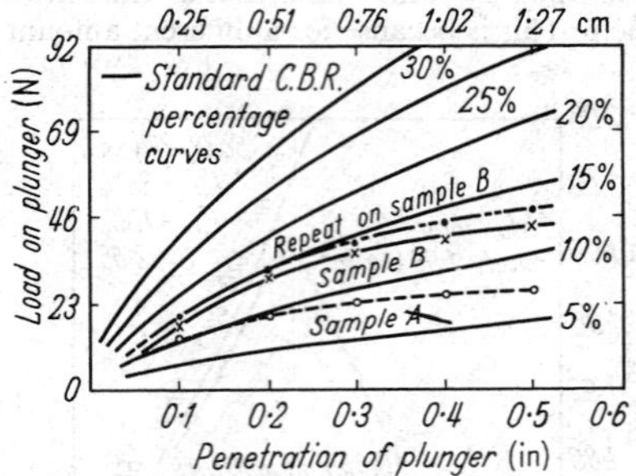

Figure 8.12 Results of California bearing ratio tests: sample A, 10%; sample B, 15%; sample B repeat, 15%

MODULUS OF SUBGRADE REACTION

The design of concrete pavements may be based on Westergaard's analysis (Road Research Laboratory[21]) in which the subgrade reaction is assumed to be proportional to the vertical deflection. The modulus of subgrade reaction is measured by a plate-loading test in the field. If under a pressure p the plate settles an amount ρ then

$$p = k_s\rho \tag{28}$$

where k_s is the modulus of subgrade reaction.

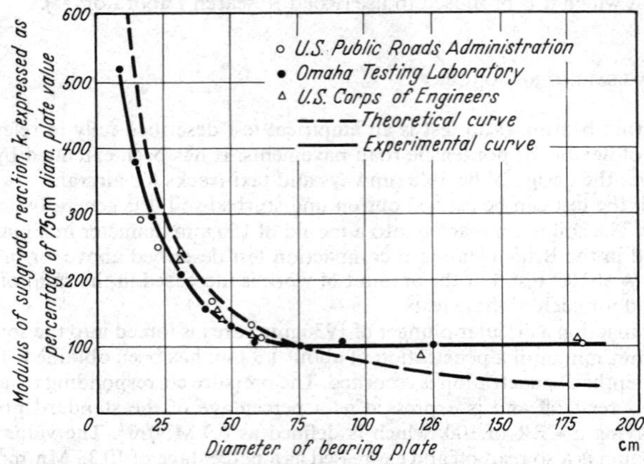

Figure 8.13 Relation between modulus of subgrade reaction and diameter of bearing plate

From elastic theory and field bearing tests it is known that k_s has no unique value but depends on the size of the loaded area.

For roads, 12 in (300 mm) diameter plates are usually used for the test and for runways 30 in (750 mm) diameter plates are used. When it is impossible to use the larger size a correction can be applied to the results of tests on the smaller plate (Road Research Laboratory[21]). A graph giving this correction is shown in Figure 8.13.

The test is carried out on the surface of the soil by bedding the plate carefully, using fine sand or plaster of Paris. The plate is then loaded to 10 lbf/in^2 (69 kN/m^2) and the settlement is measured, k_s is then equal to $69/p$ kN/m^3.

Alternatively, pressure–settlement readings are taken at intervals up to a total settlement of at least 1.8 mm. The pressure causing a settlement of 1.3 mm is then found from the settlement–pressure graph and the value of k_s is obtained from the expression $k_s = p/1.3$ kN/m^2.

To allow for the effect of possible future softening of the subgrade the US Corps of Engineers[22] recommend that the value of k_s should be corrected as follows. Two consolidation tests are carried out on samples of the soil, (i) in the condition in which it was loaded, (ii) after soaking under a surcharge pressure equal to the weight of the slab. A pressure of 69 kN/m is then applied to each sample, and if the compression of the soaked and unsoaked samples after consolidation are s_2 and s_1 respectively, the corrected subgrade modulus is taken to be

$$k_s \text{ (soaked)} = k_s s_1/s_2$$

For a discussion of the factors affecting the evaluation of the subgrade modulus in sand and stiff clays and a review of the practical application of the theories of subgrade reaction, see Terzaghi.[23]

STABILITY OF SLOPES

ANGLE OF REPOSE

The problem of the stability of earth slopes has traditionally been approached by the concept of an 'angle of repose' for the material, in spite of the fact that both experience and some of the earliest work on soil mechanics (Coulomb,[24] 1773, Francais,[25] 1820) showed this to be untrue for cohesive soils. The application of elegant mathematical solutions to problems in the nineteenth century, based on the assumption of an ideal material, led to the assumption that cohesion was nonexistent or should be ignored, and that soils had an angle of repose which was the same as their angle of internal friction (assumed constant) and that at this angle a slope would be stable to any height. For clean dry, or clean submerged sands this conception is true, with the qualification that the angle of repose is equal to the lower limit of the angle of internal friction. For cohesive soils, however, the limiting slope is a function of the height of the bank and of time.

FRICTIONAL SOILS

The angle of repose of a frictional soil can be found by (*a*) observations in the field on existing slopes; (*b*) laboratory tests – the lower limit of the angle of internal friction; (*c*) pouring a heap of dry sand from a funnel on to a level surface and measuring the angle of slope.

If a slope is cut or built in a dry frictional soil it should be stable provided that the inclination is less than the angle of repose. Failure of a bank in frictional soil is generally due to the effects of water. For example, seepage through a bank can cause erosion, pore-water pressures due to seepage pressures or arising from sudden drawdown of external water level, give rise to a reduction in effective stress which may lead to failure, and vibration of a loose fine saturated sand can result in a flow slide. Shallow surface troubles are common in these slopes, deep-seated shear slides being due either to sudden drawdown or to heavy external loads applied at the top of the bank. These conditions can be analysed by the circular arc method using the graphical method of slices (see p. 8–21).

COHESIVE SOILS

In cohesive soils, slope and height are interdependent and can only be determined when the shear characteristics of the material are known. In such soils failure usually occurs along curved surfaces of rupture which approximate to cylindrical surfaces, and one method of analysis used is called the circular arc method.

CIRCULAR ARC METHOD

In principle the method is very simple. An arc is chosen and moments are taken about its centre of the disturbing forces (weight) and the resisting forces (strengths). The factor of safety is equal to the moments of the resisting forces divided by the moments of the disturbing forces. A search is made for the most dangerous circle i.e. the one with the lowest factor of safety. This is the factor of safety of the slope. For steep slopes the worst circle goes through the toe; for flatter slopes it goes below the toe, and tends to go as deep as possible, unless shear strength increases rapidly with depth.

Two main cases arise which are known as $\phi = 0$ and c', ϕ' cases.

$\phi = 0$ case In this case the soil is assumed to have a shear strength c which does not change and the angle of shearing resistance is assumed to be equal to zero. This case applies for 'end of construction' conditions.

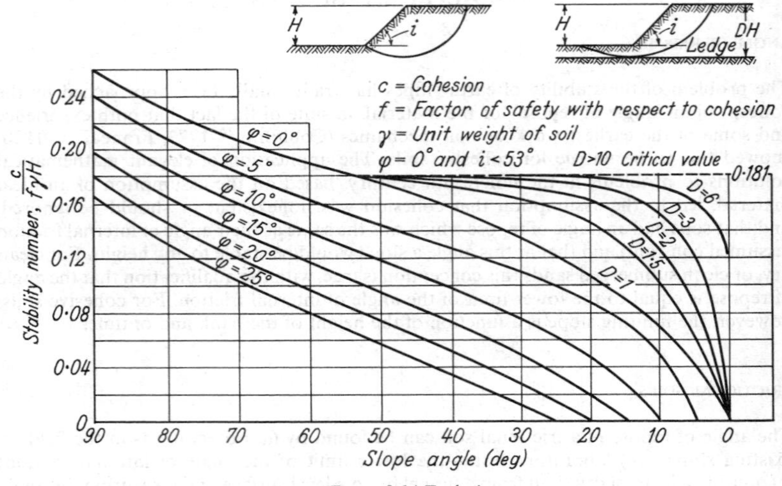

Figure 8.14 Taylor's curves

c', ϕ' case In this case the soil is assumed to have both friction and cohesion and the effective-stress shear-strength parameters, c' and ϕ' are used. This case represents the conditions for long-term stability. The analysis is known as an effective-stress analysis.

For an analytical solution for simple cases see Figure 8.14 (Taylor's[7] curves) including the effect of depth factor. For cases in which irregular slope outlines or external loads occur, the graphical method must be used, a search being made for the worst circle. The exact position of this circle is not critical and the work is not onerous. In Figure 8.15, which shows a typical calculation for the $\phi = 0$ case, the factor of safety F is given by the equation

$$F = slR/Wd \tag{29}$$

where W is the weight of soil, l the length of arc and s the shear strength.

The reactions across the arc, N, are normal to the surface of rupture (since $\phi = 0$) and therefore have no moment about the centre of the circle.

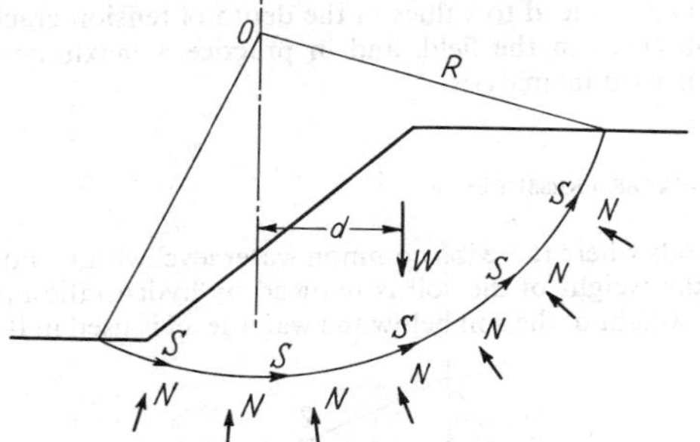

Figure 8.15 Circular arc method of analysis for cohesive soil ($\phi = 0$)

METHOD OF SLICES

The above simple method cannot be applied to an effective-stress analysis since the N forces, although they still pass through the centre of the circle, determine the value of the frictional resistance along the surface of rupture which must be included in the moment equation, and the magnitude and distribution of these N-forces is unknown. The wedge of earth bounded by the arc of rupture is divided into vertical slices and the assumption is made that the normal force across the arc is equal to the weight of the slice resolved normally to the arc. The forces between the slices are neglected. In Bishop's improved method these forces are taken into account (Bishop[26]).

The forces acting on one slice are then as shown in Figure 8.16 where W_{a6} is the weight of slice 6 above water level, W_{b6} is the weight of slice 6 below water level, E_{5-6} are the normal forces between slices 5 and 6, f_{5-6} the tangential force between slices 5 and 6. N is the normal reaction at the base of slice 6, l_6 is the length of arc at tne base of tne slice, d_6 the lever arm for slice 6. For slice 6 the disturbing moment is W_6d_6 and the resisting moment is $(N \tan \phi' + c'l_6)R$. The moments of all these forces about the centre of the arc are calculated and summed for all the slices. The factor of safety is as before:

$$F = \frac{\textit{Moment of resisting forces}}{\textit{Moment of disturbing forces}}$$

and for the whole bank, ignoring E and f forces, and assuming no excess pore pressures exist,

$$F = \frac{\Sigma(N \tan \phi' + c'l)R}{\Sigma Wd} \tag{30}$$

where c' and ϕ' are the effective-stress and shear-strength parameters of the soil. The worst circle must be found by trial and error.

TENSION CRACKS

In a soil possessing cohesion, tension cracks can form and so reduce the length of the arc around which shear strength is resisting the disturbing forces. This should be taken into account in the analysis. The depth z_c to which tension exists in the soil is theoretically given by

$$z_c = 2c/\gamma \text{ in a purely cohesive soil}$$

and

$$z_c = 2c \cos \phi/\gamma(1-\sin \phi) \text{ in a soil with friction as well}$$

The above formulae can lead to values of the depth of tension crack which are much greater than any observed in the field, and in practice a maximum depth of tension crack of 1.5 to 3 m is used in analysis.

WATER LEVEL AND WATER PRESSURE

In using these methods where there is a common water level within and without the slope, as in Figure 8.16, the weight of the soil is reduced by hydrostatic uplift and, provided that the submerged weight of the soil below the water level is used in the calculations, the

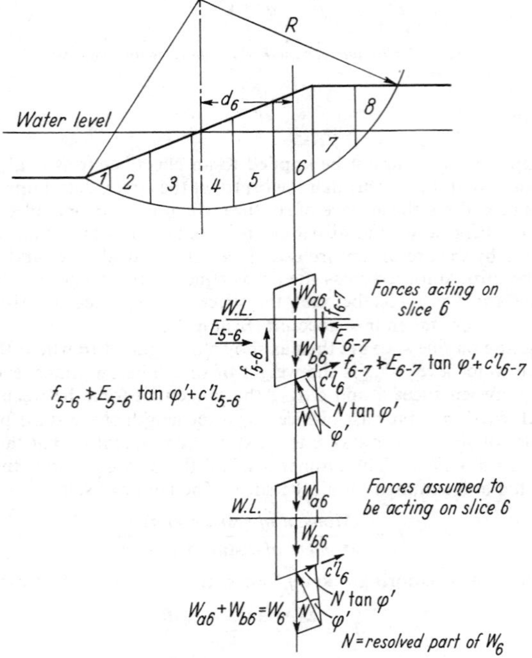

Figure 8.16 Conventional circular arc method (effective stress analysis)

effect of the water outside of the slope is ignored. In practice, however, slopes are frequently charged with water above any external water level and the same rules apply, the effective resistance on the base of the typical slice being

$$(N-ul)\tan\phi'+c'l \tag{31}$$

where the excess pore pressure is $u = h_w \gamma_w$, where h_w is the excess head of water above the external water level, if any, or above the base of the slice, whichever is at a higher level. Analysis in the conventional method then proceeds as in equation (30).

It is convenient to express the excess water pressure, $u = h_w \gamma_w$, as a ratio r_u of the total stress σ_T at the base of the typical slice using submerged densities below external water level,

$$r_u = h_w \gamma_w / \sigma_T \tag{32}$$

since in many practical problems r_u can be given an appropriate average value, and this greatly simplifies stability problems and enables Bishop's[26] more rigorous analysis to be used with economic advantage. Bishop and Morgenstern,[27] observing a closely linear relationship between the factor of safety F and r_u over a wide range of slopes and r_u values, have proposed that

$$F = m - n r_u \tag{33}$$

where m and n are *stability coefficients* which have been calculated for a variety of slope angles and heights and soil parameters γ, ϕ' and c' The results are given in both graphical and tabular form in reference 27. By this means a wide variety of slope angles with varying ground water conditions can be rapidly analysed.

NON-HOMOGENEOUS CASES

While the circular-arc method is appropriate for homogeneous ground conditions, in those cases where zones of weaker material occur the slip surface will tend to depart from the circular form (which permits solid-body movement with minimum of disruption) and to accommodate itself to the path of least resistance (Figure 8.17). As a consequence, noncircular failures are characterised by considerable disruption, and the appearance of the slip topography will often give an indication of the type of failure. Because of this disruption it is essential to take some account of the lateral forces between the arbitrary selected slices in the analysis.

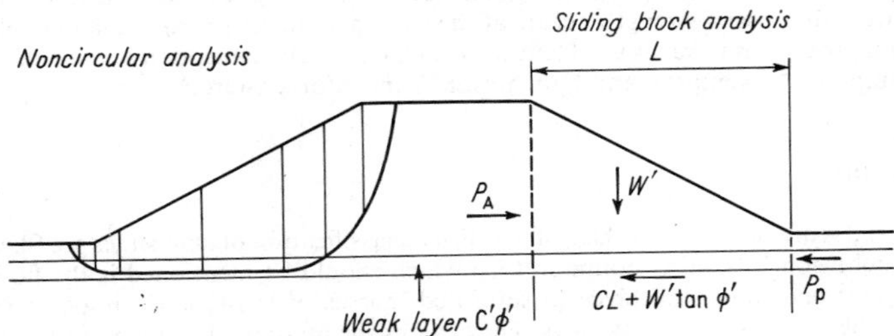

Figure 8.17 Noncircular slips analyses

Analysis proceeds by trial as before but with smoothly flowing slip surfaces adapted to the ground conditions. In this case, however, equation (30) cannot be used and summation of the components of the disturbing forces and the frictional and cohesive resistance is carried out along the slip surface. In the simplified method due to Taylor,[7] the moment equilibrium of the individual slices is ignored and solutions can be rapidly obtained; see Sherard, Woodward, Gizienski and Clevenger[28] for two variations on this method.

In 1965 Morgenstern[29] presented a solution to the noncircular stability problem, taking account of all the necessary conditions of internal equilibrium. As in Bishop's circular method, the equations are non explicit and in addition are considerably more complex. Computation by computer is therefore necessary.

In simple cases where the weak layer is well defined, an approximate solution can be obtained using the sliding block analysis (Taylor[7]). The mechanics of the method are given in Figure 8.17. The active thrusting force P_A and the passive resisting force P_p at the head and toe of the slope respectively are determined by earth pressure theory (see p. 8–28), where the factor of safety F is given by

$$F = \frac{(W' \tan \phi' + c'L) + P_p}{P_A} \tag{34}$$

W' is the effective weight of the slope mass.

Where the weak layer is of low permeability, failure occurs under undrained conditions, $\phi = 0$, and the weight W of the slope does not enter into the calculations, unless the clay layer is sloping in which case the appropriate component is included.

STIFF-FISSURED CLAYS

The above methods of analysis are not directly applicable to stiff-fissured clays. Any excavation in these materials allows the fissures to open. Water then penetrates these fissures, causing softening, which gradually increases in extent. Although the cut is stable when made, the factor of safety drops with time until failure occurs, often many years after the cut was first made. Although for temporary works, cuts can often be left very steep, short-term failures have occurred (Skempton[30]). For permanent work the best measure is to carry out a c', ϕ' analysis using the effective-stress shear-strength parameters in conjunction with effective stresses. This means that the ground-water surface in the clay must be known. It is usually assumed that $c' = 0$ under long term conditions.

Failure often occurs at flatter slopes than analysis using the effective stress parameters indicates to be safe. This is due to the shear strength along fissures being nearer the residual value ϕ'_r than the peak strength ϕ' (see Figure 8.4), leading to a lower average shear strength. Loss of strength occurs at the toe where water pressure may be relatively high, and progresses inwards until failure occurs. For a full description of the mechanism, see Skempton,[14] Skempton and Hutchinson[31] and Morgenstern.[32]

FLOW SLIDES

Not all slides are deep-seated shear slides. For a classification of land slides see Skempton and Hutchinson.[31] A fairly common type which should be mentioned is the flow slide. Flow slides generally take place in saturated masses of loose, fairly impervious soils such as fine sands, silts or silty clays. They can occur on quite flat slopes and can travel long distances, the soil and water flowing as a liquid mass. They are caused by a sudden reduction of the shear strength to zero, or close to it, by the transfer of all the pressure to the water in the voids, a process known as liquefaction. The chief factor here is the relative density of the soil, loose soils having a relative density less than a critical value below which volume reduction occurs on shearing contrary to dense soils which tend to dilate thus throwing a tension on to the pore water. Casagrande considers that sands having a relative density greater than 50% would be safe against liquefaction (Green and Ferguson[34]). Casagrande and others have drawn attention to a more general phenomenon called ‘fluidisation’ where the fluid phase is mainly air. Such phenomena are not necessarily confined to fine-grained soils. They can occur under a variety of circumstances, the chief of which is a sudden disturbance of a loose soild by a heavy shock or vibration. Drainage and compaction are the two main remedial and preventive measures.

SENSITIVE CLAYS

In some cases the existing undisturbed strength is much greater than the strength after remoulding. The ratio of these two strengths is called the sensitivity of the clay. In general, the undisturbed strength is measured *in situ* by means of a vane test (see p. 10–21). Much work on this problem has been done in Norway and Sweden (Skempton and Northey[33]). In England, sensitivities are usually below 10 and in these cases an analysis can be made using the unconfined compression strength and the $\phi = 0$ method.

With clays of high sensitivity, that is over 20 and up to 100, such as occur in southern Norway and Sweden, experience shows that an analysis based on the $\phi = 0$ case and the undisturbed strength of the clay measured by a vane test gives results which do not necessarily agree with practice (Bjerrum[35]). This problem is not yet completely understood and some aspects of it are controversial. When failure takes place in a clay of high sensitivity the result is in many respects similar to a flow slide described in the paragraph above, the clay becoming practically fluid and flowing through quite small apertures down flat valleys or hillsides for a long distance. The formation of these very sensitive clays is believed to be due to the clay being laid down in salt water, then being raised above water level by isostatic readjustment, the salt later being leached out by the percolation of fresh water; thus the liquid limit of the clay is reduced but the moisture content remains high. On disturbance, therefore, the moisture content greatly exceeds the liquid limit and the clay acts as a heavy fluid. For a description of the leading process see Bjerrum.[20]

PROTECTIVE AND REMEDIAL MEASURES

Certain protective and remedial measures can be taken to prevent a slip or to stabilise a slip which has occurred. Of these the most important is drainage. The majority of troubles are due to water; well designed and adequately maintained drainage can prevent the ingress of water to a bank and so stop or considerably delay any softening which may take place and prevent the build-up of high pore pressures.

On sidelong ground a drain should be installed at the top of the slope to catch surface water. Water from the slope itself should be caught in a toe drain and led away from the toe. On a long slope it may be necessary to catch surface water in herring-bone drains or even to lead it to a longitudinal drain on a berm halfway up the slope. It is important to line the inverts of these drains.

A considerable degree of protection can be obtained by grassing the slope and the level area at the top where shrinkage cracks are likely to appear. Cracks once formed can become filled with water the pressure of which exerts a considerable disturbing force on the bank. A good carpet of grass prevents not only drying and cracking but also surface erosion. Bushes and trees with strong root systems can also be of help, but trees which grow rapidly and absorb large amounts of water from the soil should be avoided, e.g. poplars, elms.

Deeper drainage can be achieved by drilling horizontally into the slope, or in some cases adits can be used, either alone or in conjunction with vertical drainage holes. Counterforts are also used to improve drainage. These consist of trenches excavated back into the slope, backfilled with free-draining granular material.

In addition to drainage methods, the stability of deep-seated slides can be improved either by reducing the disturbing forces or by increasing the resisting forces, by one or other of the following methods:

(a) Reducing the disturbing forces by loading the toe, or removing material from the top, or replacing material at the top by lighter material, or altering the bank profile, e.g. introducing berms.

(b) Increasing the resisting forces by increasing the shear strength, by drying or adding frictional counterforts or keys through the slip surface.
(c) It may in some instances be possible to improve stability by introducing 'rigid' elements, such as sheet-piling driven through the toe, with or without ground anchors. However, great caution is required, since the rigid element will attract load and very high bending moments will be developed.

FLOW NETS

A flow net is a graphical representation of the pattern of the seepage or flow of water through a permeable soil. By means of a flow net it is possible amongst other things to calculate the hydrostatic uplift on a structure such as a dam or barrage, the amount of seepage through an earth dam or under a barrage, or to estimate the probability of piping occurring in a cofferdam.

A flow line is the path followed by a particle of water flowing through a soil mass. Flow lines are always smooth even curves as shown in Figure 8.18. An equipotential line is a line joining points at which the hydraulic head is equal; therefore if standpipes are inserted into any two points on an equipotential line the water will rise to the same level in each standpipe. Flow lines and equipotential lines are always at right angles to each other. For a discussion on the theory of flow nets, see Taylor.[7]

CONSTRUCTION OF FLOW NETS

Four methods of constructing flow nets are in general use.

(*a*) *Mathematical* For simple boundary conditions the governing differential equation (Laplace) can be solved mathematically. With the increasing use of electronic computers allied with relaxation methods, quite complicated cases can be tackled (Remson, Hornberger and Molz[36]).
(*b*) *Electrical analogy* The differential equation for flow nets is the same as that for flow of electricity and flow nets can be drawn by using an electrical model and tracing lines of equal potential with a wandering probe. The soil is represented by a suitably shaped card coated with graphite, strips of copper represent water surfaces and an insulating material represents an impervious surface. Once the equipotential lines are drawn, the flow lines can easily be drawn in at right angles to them.

This method assumes that the permeability in both directions, horizontally and vertically, is the same. In the electrical resistance network method a scaled network of resistances of values proportional to the permeability is set up, the electrical potential at each node being a direct measure of the hydraulic potential at that point (Karplus[37]).
(*c*) *Hydraulic models* An obvious approach is to construct a model of the problem in sand behind glass, to allow water to flow through it and to trace the flow lines by inserting a small drop of dye at the soil surface. The flow of the water can be clearly seen by the streak left by the dye as it flows through the soil. This trace is then drawn on the glass with a wax pencil and the procedure repeated from a different point.
(*d*) *Graphical method* After a little practice it is quite possible to sketch a flow net for many problems which is quite accurate enough for most practical purposes.[7] The cross section is drawn and the boundary conditions are clearly marked. The flow net is then tentatively sketched in, bearing in mind that flow lines and equipotential lines are at right angles to each other, that flow lines always start at right angles to a free water surface and equipotential lines start or finish at right angles to an impervious surface. The number of flow and equipotential lines is chosen to divide the seepage area into shapes which are approximately square and which are bounded by two flow lines and two equipotential lines.

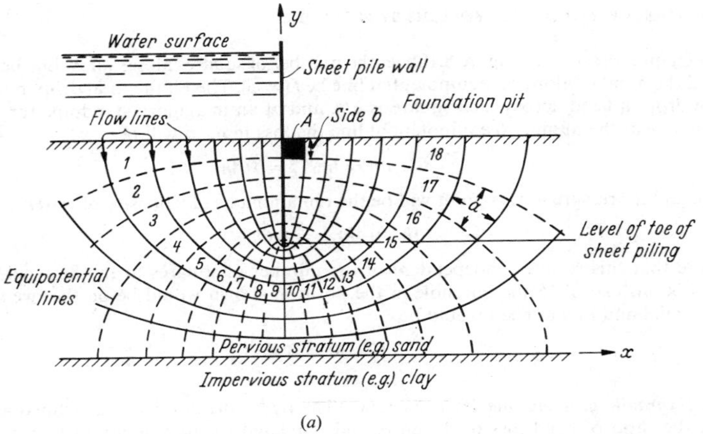

(a)

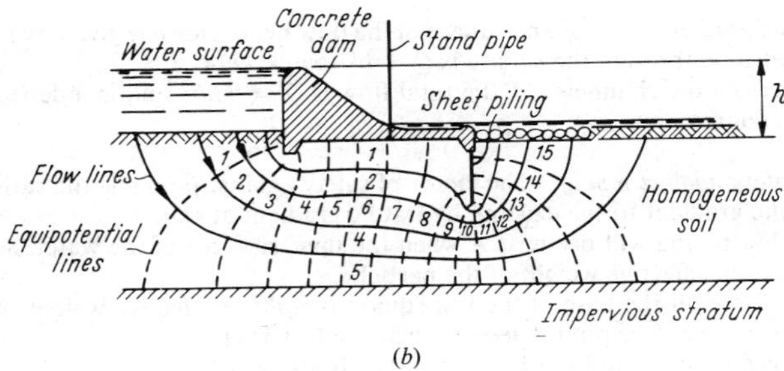

(b)

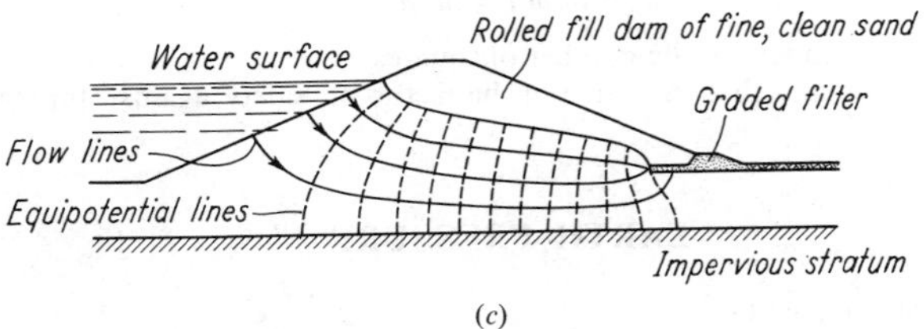

(c)

Figure 8.18 Example of flow nets for simple cases: (a) beneath sheet pile wall; (b) beneath concrete dam on sand with sheet pile cut-off wall; (c) through rolled fill dam with toe drain

SOLUTION OF HYDRAULIC PROBLEMS BY FLOW NETS

(a) Uplift pressure In Figure 8.18b let the number of squares along a flow line be $n(=15)$ and the number along an equipotential line be $f(=5)$. Then if the total drop in head is h, the drop in head across each square is h/n, and at an imaginary standpipe through the concrete at the number 6 equipotential line the loss in head will be

$$6 \times h/n = 6h/15 = 0.4h$$

The uplift pressure at this point will be the remaining head × density of water, that is

$$(h-0.4h)\gamma_w = 0.6h\gamma_w$$

Note that this result is independent of the number of squares in the flow net, since if $n = 30$ instead of 15 the borehole in the position shown would be on the twelfth equipotential and loss in head would be

$$12h/30 = 0.4h$$

(b) Hydraulic gradient and D'Arcy's Law The hydraulic gradient i is defined as $\Delta h/\Delta l$, i.e. the ratio of head loss to distance, and is related to the velocity of flow v and the permeability k by D'Arcy's law,

$$v = ki \tag{35}$$

(c) Amount of seepage The quantity of water Q flowing under the dam in Figure 8.18b is given by $Q = vAt$, where A is the area of flow, t is time; and introducing D'Arcy's law, $Q = Atki$.

The hydraulic gradient i across any square of the flow net of side b is given by $i = h/nb$. The flow in unit time through the square is $Q = bkh/nb = kh/n$.

If the number of flow channels is f, the total flow is $Q = fkh/n$, and is independent of the size of the squares.

(d) Factor of safety against piping The factor of safety against piping is the ratio of the critical hydraulic gradient to the existing hydraulic gradient at exit.

In Figure 8.18a piping will occur at A when the upward force of the water issuing at A is greater than the effective weight of the particles.

The seepage force on the base of the last square is $\gamma_w ib^2$ = effective weight of soil = $b^2\gamma_w(\rho-1)/(1+e)$. Therefore piping occurs when $i = (\rho-1)/(1+e)$.

For sand, ρ is about 2.7, and in the loose state e is about 0.7, giving

$$i = (2.7-1)/(1+0.7) = 1,$$

i.e. the critical hydraulic gradient is about unity.

The exit hydraulic gradient at A, Figure 8.18a is $(h/n)/b$.

Therefore the factor of safety against piping is

$$1/(h/nb) = nb/h$$

Note that nb is independent of the number of squares.

It is generally considered that the value of the factor of safety against piping should be 4 or greater.

EARTH PRESSURE

ACTIVE AND PASSIVE PRESSURE

The problem of earth pressure is the oldest soil mechanics problem. The lowest pressure which a retaining structure must be capable of resisting to prevent a soil mass from collapsing is the active pressure. The highest pressure which a structure can exert on a bank of earth without causing it to move in the direction of the pressure is the passive pressure or passive resistance. Examples of both are given in Figure 8.19.

Below a level ground surface, the horizontal pressure is known as the 'pressure at rest'. This pressure lies between the active and passive pressures and is usually designated by the factor K_0 which is the ratio of the horizontal and vertical pressures at any given depth. The factors K_a and K_p similarly relate the active and passive horizontal pressures to the vertical pressure.

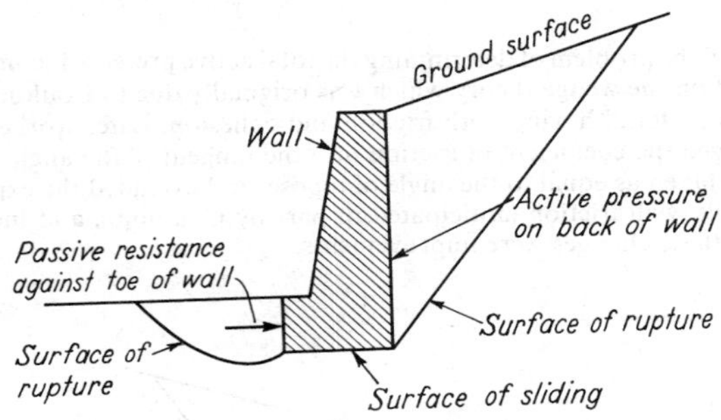

Figure 8.19 Active pressure and passive resistance

The value of K_0 depends on the depositional conditions and stress history of the ground. For loose sands the value of K_0 is about 0.45, falling to about 0.35 in dense sands, following the relationship given by Jáky, $K_0 = (1 - \sin \phi)$. A wider range is encountered in clays, typically 0.4 to 0.7, but in some heavily overconsolidated clays values considerably in excess of unity occur.

Tables 8.1 and 8.2 from the Civil Engineering Code of Practice No. 2,[38] give typical values for K_a and K_p for cohesionless materials, vertical walls with horizontal ground where ϕ is the angle of friction for the soil and δ the angle of wall friction.

Table 8.1

Values of δ	*Values of φ* 25°	30°	35°	40°	45°
	Values of K_a				
0°	0.41	0.33	0.27	0.22	0.17
10°	0.37	0.31	0.25	0.20	0.16
20°	0.34	0.28	0.23	0.19	0.15
30°	—	0.26	0.21	0.17	0.14

Table 8.2

Values of δ	*Values of φ* 25°	30°	35°	40°
	Values of K_p			
0°	2.5	3.0	3.7	4.6
10°	3.1	4.0	4.8	6.5
20°	3.7	4.9	6.0	8.8
30°	—	5.8	7.3	11.4

Recent experimental work by Rowe and Peaker[39] has emphasised the dominant role of the angle of wall friction, but has shown that the code values of K_p can be as much as 50% too high in loose sands and 30% in dense sands.

In order that the lateral pressure may change from the pressure at rest to either the active or the passive value, movement must take place to mobilise shear forces. This generally occurs during the construction of the retaining structure.

Active pressure

For an ideal material the problem of determining the total active pressure is comparatively simple and is based on the wedge theory which was originally due to Coulomb[40] (1773) who solved it for a material having both friction and cohesion. Later workers omitted the cohesion, changed the coefficient of friction into the tangent of the angle of internal friction, which was taken as equal to the angle of repose, and extended the expression to include sloping walls, wall friction (anticipated in part by Coulomb), and inclined surcharges. Not all of these changes were improvements.

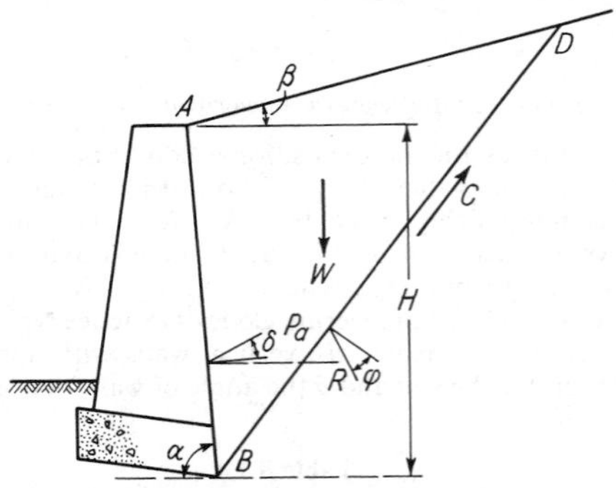

Figure 8.20 Coulomb or general wedge theory

In the wedge theory it is assumed that the pressure on the wall is due to a wedge of earth which tends to slip down an inclined plane as shown in Figure 8.20. The forces acting on the wedge are also shown in the figure. The inclination of the plane *BD* is altered until the position which gives the greatest value for the force P_a is found. This can either be done analytically or graphically.

The general formula for the total pressure over depth *H* exerted by a frictional material having no cohesion is

$$P_a = \tfrac{1}{2}\gamma H^2 \left(\frac{K_a}{\sin\alpha\cos\delta}\right) \tag{36}$$

where the coefficient of active earth pressure

$$K_a = \frac{\sin^2(\alpha+\phi)\cos\delta}{\sin\alpha\sin(\alpha-\delta)\left[1+\left\{\dfrac{\sin(\phi+\delta)\sin(\phi-\beta)}{\sin(\alpha-\delta)\sin(\alpha+\beta)}\right\}^{\frac{1}{2}}\right]^2} \tag{37}$$

where ϕ is the angle of friction, δ the angle of wall friction and α, β and H are as shown in Figure 8.20.

For the case of a vertical wall and horizontal backfill, the values of K_a to be used in equation (36) are as in Table 8.1. For the special case of no wall friction, vertical wall and horizontal backfill, this reduces to the Rankine formula

$$P_a = \tfrac{1}{2}\gamma H^2(1-\sin\phi)/(1+\sin\phi) = \tfrac{1}{2}\gamma H^2 K_a \tag{38}$$

No analytical solution exists for the general case of a soil having both friction and cohesion, but for the special case of a vertical wall, horizontal backfill and no wall friction or adhesion, the total pressure is given by

$$P_a = \tfrac{1}{2}\gamma H^2 K_a - 2cH\sqrt{K_a} \tag{39}$$

For a purely cohesive material, equation (39) reduces to $P_a = \tfrac{1}{2}\gamma H^2 - 2cH$ for the case of no wall adhesion, or if wall adhesion is taken as equal to the cohesion the formula becomes $P_a = \tfrac{1}{2}\gamma H^2 - 2(\sqrt{2})cH$.

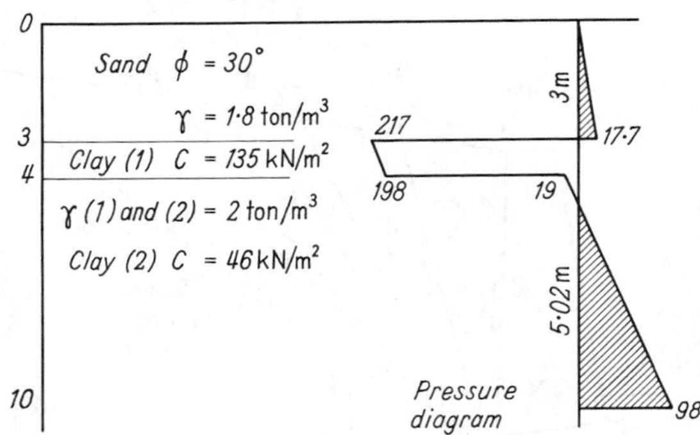

Figure 8.21 Earth pressure in layered system (pressure in kN/m²)

Equation (40), which gives the intensity of active pressure at any level, can be applied to simple cases of layered systems by calculating the pressure at the top and bottom of each layer, treating the layers above as a surcharge.

$$p_a = \gamma z K_a - 2c\sqrt{K_a} \tag{40}$$

Below water, the earth pressure is calculated using the submerged weight of the soil, but the full water pressure is added to this.

Example To find the total pressure for the layered system and soil conditions shown in Figure 8.21.

The pressure p_a at any depth z in soil of weight γ is $\gamma z(1-\sin\phi)/(1+\sin\phi)$ in sand and $\gamma z - 2c$ in clay, where ϕ is the angle of shearing resistance and c the cohesion.

At $z =$ 3 m (sand)	$p_a = 3\times 1.8\times .33(9.81)$	$= 17.7$ kN/m²
= 3 m (clay)	$= 53-270$	$= -217$ kN/m²
= 4 m (clay)	$= [53+1\times 2(9.81)]-270$	$= -197.4$ kN/m²
= 4 m (clay)	$= [53+1\times 2(9.81)]-92$	$= -19.4$ kN/m²
= 10 m (clay)	$= (53+19.6)+[6\times 2(9.81)]-92$	$= +98$ kN/m²

Total area (positive areas only) $= (3\times 17.7/2)+(5.02\times 98/2)$

Total force/metre run $= 274.5$ kN

For a discussion on the practical aspects of retaining-wall problems, reference should be made to Terzagh and Peck[11] and to reference 38.

It must be clearly understood that although the equations (36)–(40) give an approximation to the total pressure which will be exerted on a wall, and that this is given by the total area of the pressure diagram, they do not necessarily give the actual distribution of the pressure.

GRAPHICAL METHOD

A simple graphical solution can be used for walls with irregular outlines, for irregular backfills, external loads on the backfill, and variation in the properties of the backfill. The principle is that of the wedge theory, the most dangerous surface of slip being found by trial.

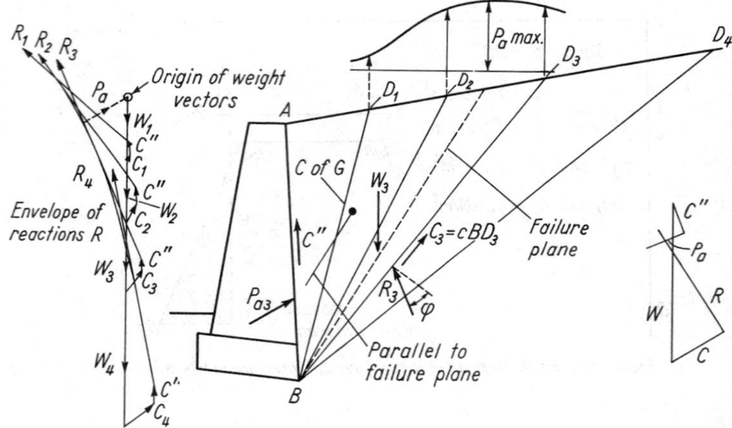

Figure 8.22 Engesser's method for determination of active pressure

The polygons of forces are drawn for a number of wedges ABD_1, ABD_2, etc., as shown in Figure 8.22. The forces acting on each wedge are: its weight W, the reaction R across the plane BD, the cohesion C acting up the plane BD, and the earth pressure P which it is required to find. For limiting equilibrium, R is inclined at ϕ to the normal to the plane BD, and $C = c \times \mathrm{BD}$. If the direction of P is assumed, the polygon of forces can be completed, giving the value of P. It is convenient to plot all the polygons from a common origin of W as shown. The maximum value of P is then given by the point at which the envelope of the R lines cuts the line representing E. This P line can be drawn from the origin in any assumed direction, thus allowing for wall friction. Wall adhesion (cohesion) can be included by subtracting it from the weight of the wedge.

Alternatively, the force polygons can be drawn separately for each wedge and the value of P_a can be plotted above the position of the corresponding slip surface.

SURCHARGE LOADS

The effect of a distributed surcharge of magnitude W_u is to increase the earth pressure over the whole height of the wall by an amount W_uK_a, in the case of granular backfills, and by an amount W_u in the case of cohesive backfills. The effect of a line load W_l can be estimated with sufficient accuracy for most designs by the construction shown in Figure 8.23.

The effect of a point load W_i on the backfill is more difficult to estimate. The problem has been investigated by Gerber[41] and by Spangler.[42] An approximate method which is given in Civil Engineering Code of Practice No. 2, (C.P. 2),[38] is to assume that the load is spread through the backing at an angle of dispersion of 45° on each side of the load. The lateral pressure at any point due to the surcharge is then taken as K_a times the vertical pressure at the point. This method, however, tends to give results on the unsafe side. The following tentative approximate method is suggested. The line of action of the resultant force is obtained by a construction similar to that for a line load (Figure 8.23), the 40°

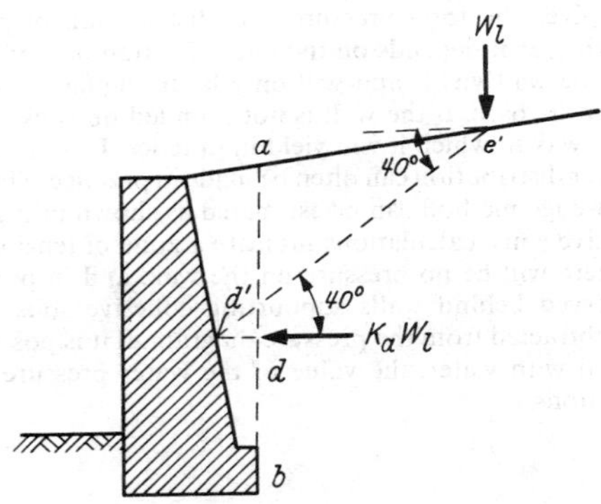

Figure 8.23 Method of estimating magnitude and line of action of pressure due to a line load

line being constructed from the centre of the loaded area. It is assumed that, if the length of the loaded area be L and the distance between the back of the wall and the near edge of that area be x, the resultant lateral thrust will be distributed along a length of wall equal to $L+x$. Then, if W_i be the load on the area, the resultant thrust per unit length of wall will be $K_a W_i(L+x)$.

Passive resistance

In the case of active pressure the assumption of a plane surface of failure gives results which are within 3 or 4% of the true value, and this is close enough for all practical purposes. With passive pressure, however, this may lead to results which differ significantly from the true values if wall friction is taken into account. Only for the case of no wall friction does the wedge theory give the true value. The passive pressure in this case is given by the formula

$$P_p = \tfrac{1}{2}\gamma H^2 K_p + 2cH\sqrt{K_p} \tag{41}$$

Wall friction adds greatly to the passive resistance, but not so much as the wedge theory indicates; it is seldom, therefore, that it can be neglected. The wedge theory does not give the correct answer because the surface of failure is not a plane but is curved, as shown in Figure 8.19.

The effect of curved failure surfaces can be analysed using the log-spiral method, a graphical procedure which is given in detail by Terzaghi and Peck,[11] and which forms one basis for the computation of bearing capacity.

For the simple case of horizontal ground and vertical wall, in cohesionless soil, Table 8.2 gives values of the passive earth-pressure coefficient, K_p for use in the equation

$$P_p = \tfrac{1}{2}\gamma H^2 K_p \sec \delta \qquad (42)$$

Distribution of pressure

It is generally assumed that pressure increases uniformly with depth. This is only true in certain special cases although the assumption is not unreasonable in some other cases for which it is not strictly true. Cases in which the assumption should not be made are dealt with below.

The wedge theory gives the total pressure; the distribution of pressure cannot be obtained from this theory as it depends on the lines of action of the forces involved and on the way in which the wall yields and will only be triangular (hydraulic) if the wall yields by turning about its base. If the wall is not founded on rock and is very rigid in itself this is usually the way in which it will yield in practice. It is for this reason that the assumption of triangular distribution can often be made in practice. The centre of pressure on the wall using the wedge method can be estimated as shown in Figure 8.22.

In the case of cohesive soils, calculations indicate a zone of tension at the top of the wall. In this region there will be no pressure on the wall, and in practice deep tension cracks are often observed behind walls supporting cohesive soils. The value of this tension must not be subtracted from the pressure diagram. If it is possible for the tension cracks to become filled with water, the value of the water pressure must be included in the pressure calculations.

Strutted excavations

For excavations below ground-water level a sheeted excavation is often used in preference to an open excavation with battered sides, particularly where space is limited and where the piles can be driven into a relatively impervious stratum to provide a cut-off against ground water in overlying pervious strata. Such sheeting is usually braced by frames consisting of walings and struts. Calculations of the earth pressures on the sheeting

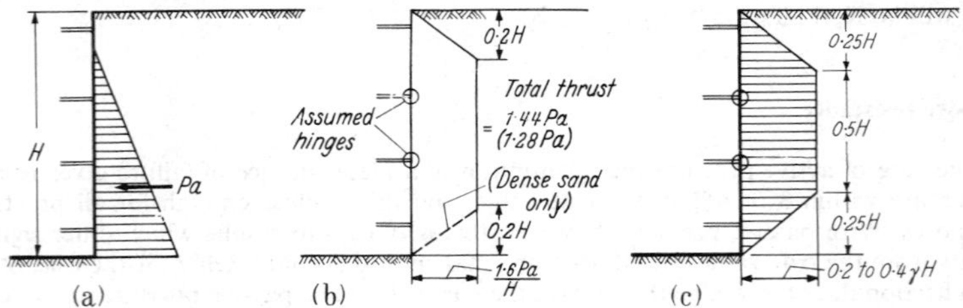

Figure 8.24 Earth pressure in strutted excavations. (a) Calculated earth pressure; (b) earth pressure distribution by British code of practice (CP2), based on Terzaghi's method; (c) stiff fissured clay (Terzaghi and Peck)[11]

follow the same lines as for retaining walls. However, during progressive excavation and placing of frames, deflections of the sheeting occur which lead to frame loads which are not in agreement with those calculated from the earth-pressure diagram, assuming hinge points at all frame levels below the top frame. In general the load in the struts does not increase with depth below about one quarter of the depth of the excavation.

For sands, Terzaghi has proposed an empirical design rule, based on a number of field observations, as shown in Figure 8.24(b); and the same rule, with a modification for use in clay, has been adopted in the Civil Engineering Code of Practice No. 2 (C.P.2).[38] Recently Peck[43] and Tomlinson[44] have discussed this question and that of adjacent associated movements in some detail. Typical pressure distributions are given in Figures 24(b) and (c).

It should be emphasised that the redistributed pressure diagrams described above are in effect design devices for obtaining frame loads and do not necessarily imply any actual redistribution of earth pressure. In fact Skempton and Ward[45] have described the results of strut and waling load measurements in a cofferdam at Shellhaven and have interpreted results to show that the observed frame loads can be accounted for in terms of deflections of the sheet piling prior to placing the struts in position and without the need for assuming any redistribution of earth pressure.

Two useful rules in determining levels at which frames should be placed have been given by Ward[46] as follows:

(a) In a deep deposit of normally consolidated clay the uppermost frame of struts should be placed across a cofferdam before the depth of excavation (H_1) reaches a value given by $H_1 = 2c/\gamma$ and

(b) the second frame of struts should be placed before the depth of excavation reaches a depth H_2 given by $H_2 = 4c/\gamma$.

Recent practice, particularly for larger excavations, is to avoid internal strutting by the use of ground anchors, placed at suitable levels as excavation proceeds, and which with advantage can be prestressed to stipulated loads. Reference should be made to articles by Littlejohn[47] for information on anchor design and construction.

Anchored bulkheads

An anchored bulkhead is usually in the form of a steel sheet-pile wall supported by ties at one level only and by passive pressure against the toe. However, anchored bulkheads may also be constructed with timber, pre-cast reinforced concrete sheet piles, or continuous

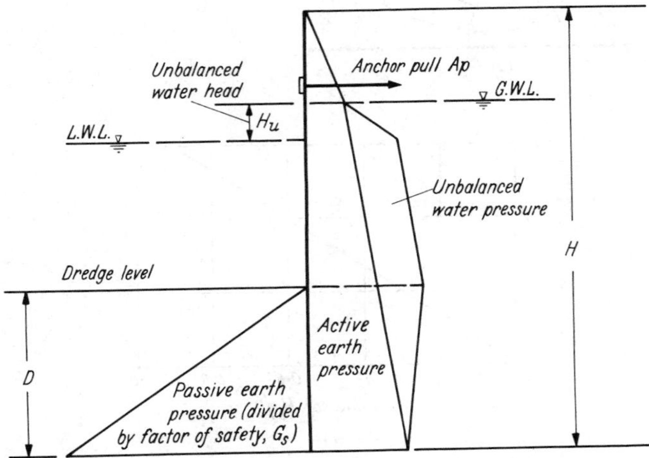

Figure 8.25 Dimensions and forces in anchored bulkhead calculation. Earth pressure diagrams illustrated are for homogeneous cohesionless soil. No pressures from surcharge loads have been shown

bored piles. Calculations of active and passive earth pressures follow the same lines as for retaining walls but analysis of the stability of an anchored bulkhead requires the determination of bending moments in the piling and of the magnitude of the anchor pull.

The magnitude of the maximum bending moment occurring in the bulkhead will be influenced by the conditions of fixity at the toe. If there is no fixity, i.e. with 'free earth support', the maximum bending moment will be at its highest value. With increasing end fixity the maximum bending moment is reduced. It has been shown by Rowe[48] that the maximum bending moment is greatly influenced by the flexibility of the wall and the density of the material into which the wall penetrates.

The various dimensions and forces entering into a bulkhead calculation are indicated in Figure 8.25. A design procedure based on Rowe's method, and similar to that described by Terzaghi[49] is given in (A)–(D) following.

(A) FORCES ACTING ON FACES OF BULKHEAD

The active and passive pressures are first calculated. The active pressure calculations must allow for the maximum possible unbalanced water pressure and for any surcharge in the form of distributed, line or point loads supported directly on the backfill.

Strictly speaking the water pressures should be calculated from a flow net (Figure 8.26(a)) taking into account the effects of stratification in the soils present. However,

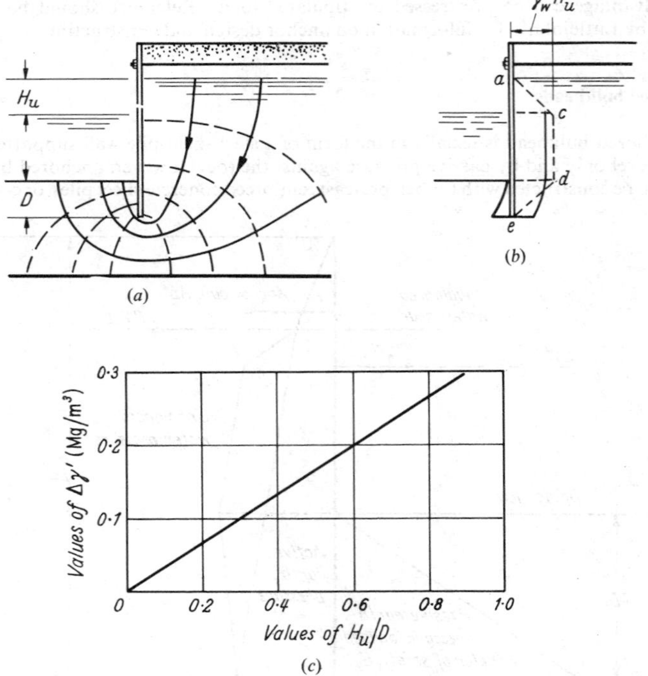

Figure 8.26 Unbalanced water pressure. (a) flow net; (b) distribution of unbalanced water pressure; (c) average reduction of effective unit weight of passive wedge due to seepage pressure exerted by the upward flow of water (from Terzaghi[49])

if the soils do not vary widely in their permeabilities it is sufficient to use the simplified pressure distribution shown in Figure 8.26(b). Allowance must also be made, where the passive pressure is provided by a permeable stratum, for a reduction in passive pressure due to the seepage gradients (Figure 8.26(c)).

(B) COMPUTATION OF DEPTH OF PENETRATION (free-end method)

A diagram of active and passive pressures is drawn as shown in Figure 8.25 for a trial penetration of the piling. The effects of any surcharge loads should also be included, as described on page 8–32. Before passive pressures are plotted, a factor of safety G_s is applied. For granular soils it is applied to the whole of the passive pressure and for cohesive soils it is applied to the shear strength component. The choice of a value for this factor of safety for a given design depends on the accuracy of the data on which the earth pressures have been based, but in general it should not be less than 2.

The earth-pressure diagram is then divided up into a number of convenient areas and the total load on each of these areas and its point of application is estimated. Moments of these loads are then taken about the line of action of the anchor pull. This is repeated for other trial depths of penetration until a depth giving zero total moment is obtained. Alternatively the pressures below dredge level may be expressed in terms of the penetration D and the required depth found analytically. It is usual to increase the calculated depth of penetration by 20% to allow for the possibility of scour or overdredging.

(C) ANCHOR PULL

The anchor pull is determined by resolving horizontally all the forces acting on the bulkhead. There are, however, a number of factors which may lead to an anchor pull somewhat greater than that computed, and conservative design stresses should therefore be adopted.

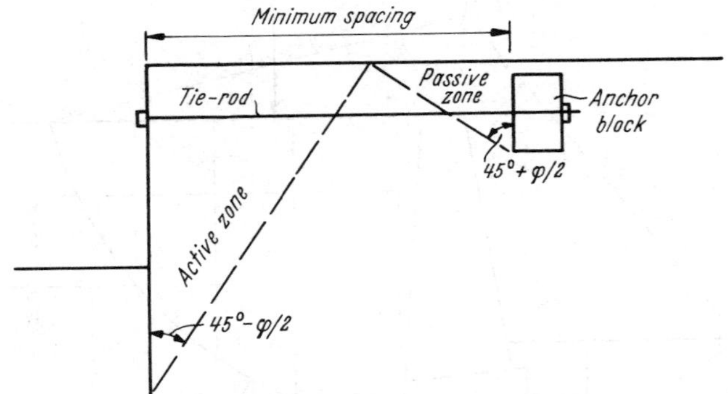

Figure 8.27 Minimum length of anchor ties

Where the ties are taken back to blocks or beams the position of these should be such that no overlapping of active and passive zones occurs, as illustrated in Figure 8.27.

(D) COMPUTATION OF MAXIMUM BENDING MOMENTS

The bending moments are first calculated on the assumption of 'free earth support', and the maximum bending moment is determined. This is a straightforward calculation based on the resultant forces of the areas into which the pressure diagram has been divided, using either analytical or graphical methods.

If the sheet piles are to be driven into a fairly homogeneous stratum of clean sand with a known relative density, the calculated maximum bending moment for free earth support can be reduced on the basis of Rowe's investigations, as illustrated in Figure 8.28.

As a first step the flexibility number is calculated for a trial section of bulkhead. For

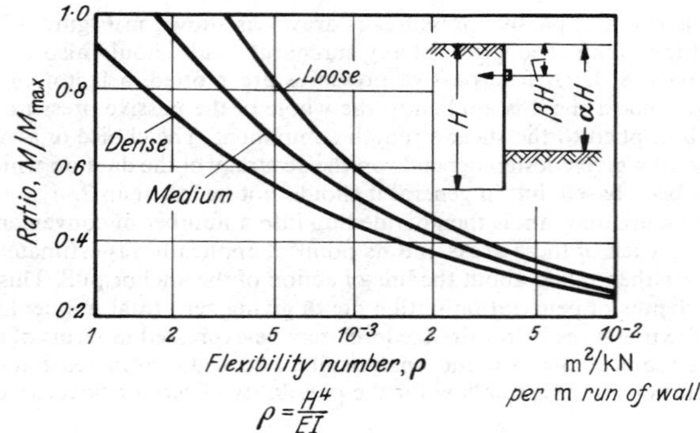

Figure 8.28 Relation between the flexibility number ρ of sheet piles and bending moment ratio M/M_{max} (logarithmic scale) (After : Terzaghi[49])

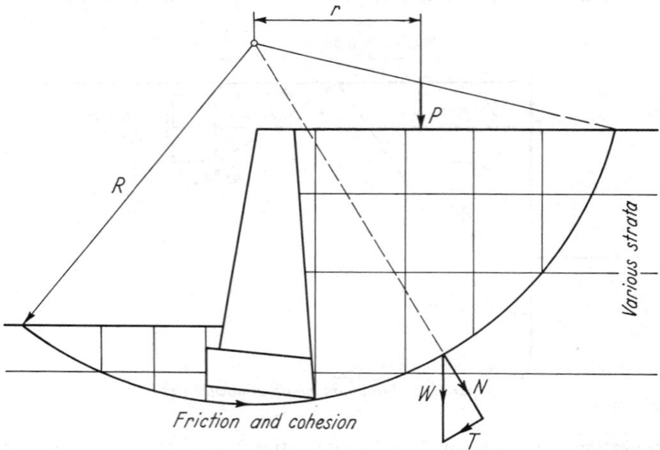

Figure 8.29 Overall stability of a retaining wall

the calculated value of the flexibility number a value of the moment reduction ratio M/M_{max} can be read off, depending on the relative density, and hence the reduced moment can be obtained and compared with the moment of resistance of the piling. The trial is repeated until the most suitable section of the bulkhead is obtained, i.e. until the reduced moment is equal to or just less than the moment of resistance of the section. If required the calculation can be extended to cover alternative construction materials.

If the sheet piles are to be driven into a homogeneous stratum of dense or medium-dense silty sand, the moment reduction curves for medium and loose sand should be used instead of those for dense and medium dense sand. Sheet piles to be driven into loose silty sand should be calculated on the figure for free earth support, since compressibility of such sands may be high. Recent work by Rowe[50] has shown that in some cases moment reduction can be made where piles penetrate clay below dredge level.

Overall stability

The design of a retaining wall, whether a mass wall or a sheet pile wall, should always be considered from the point of view of overall stability, i.e. failure as a bank of earth. The forces are shown in Figure 8.29.

$$\text{Disturbing moment} = \Sigma TR + Pr$$

$$\text{Resisting moment} = (\Sigma N \tan\phi + \Sigma cL)R$$

where L is the length of the arc over which c acts and T and N are the tangential and normal components of W

$$\text{Factor of safety} = \frac{(\Sigma N \tan\phi + \Sigma cL)R}{\Sigma TR + Pr}$$

The methods described in the subsection on 'Slope Stability' are directly applicable.

FOUNDATIONS

Bearing capacity and settlement

There are two ways in which a foundation can fail; by shear failure and by settlement. In the first case a surface of rupture is formed in the soil, the foundation settles considerably and probably tilts to one side, and heaving of the soil occurs on one or both sides of the foundation. In the second case failure of the soil in shear does not occur, but the existing deformations are large enough to cause failure of the structure which the foundation is supporting.

Failure by settlement is therefore a function of the particular structure as well as the underlying soil. Skempton and Macdonald[51] have given a criterion for framed buildings based on angular distortion, which is expressed by the ratio of differential settlement, δ, to the distance, l, between two points, usually the column positions. From a detailed study of field data a limiting value of $\delta/l = 1/300$ has been determined. More flexible structures, oil tanks for example, may undergo considerably greater settlements without sustaining damage. On the other hand some sensitive machinery will tolerate very little settlement.

The ultimate bearing capacity of a foundation is the value of the net loading intensity at which the ground fails in shear. Before discussing bearing capacity, several definitions are necessary:

(1) The gross loading intensity, p, is the pressure due to the applied load and the total weight of foundation, including any backfill above the foundation.

(2) The net loading intensity, p_n, is the gross foundation pressure less the weight of material (soil and water) displaced by the foundation (and by the backfill above the foundation). Alternatively the net pressure can be considered as equal to the gross pressure less the total overburden pressure, $p_n = p - p_0$.

(3) The safe bearing pressure is the ultimate bearing capacity divided by the factor of safety, $q_s = q_u/F$.
The term 'presumed bearing value' has recently been introduced (CP 2004[52]). This is defined as the net loading intensity considered appropriate to the particular type of ground for preliminary design purposes. It is either based on local experience or taken as the safe bearing pressure. Presumed bearing values are given in Table 8.3.

Table 8.3 PRESUMED BEARING VALUES UNDER VERTICAL STATIC LOADING
(Note. These values are for preliminary design purposes only, and may need alteration upwards or downwards. No addition has been made for the depth of embedment of the foundation.) Reference should be made to CP 2004 when using this Table.

Group	*Class*	*Types of rocks and soils*	*Presumed bearing value* (kN/m²*)	(kgf/cm²) *or* (tonf/ft²*)	*Remarks*
I Rocks	1	Hard igneous and gneissic rocks in sound condition	10 000	100	These values are based on the assumption that the foundations are carried down to unweathered rock
	2	Hard limestones and hard sandstones	4 000	40	
	3	Schists and slates	3 000	30	
	4	Hard shales, hard mudstones and soft sandstones	2 000	20	
	5	Soft shales and soft mudstones	600 to 1 000	6 to 10	
	6	Hard sound chalk, soft limestone	600	6	
	7	Thinly bedded limestones, sandstones, shales	To be assessed after inspection		
	8	Heavily shattered rocks			
II Noncohesive soils	9	Compact gravel, or compact sand and gravel	>600	>6	Width of foundation (*B*) not less than 1 m (3ft). Ground-water level assumed to be a depth not less than *B* below the base of the foundation
	10	Medium dense gravel, or medium dense sand and gravel	200 to 600	2 to 6	
	11	Loose gravel, or loose sand and gravel	<200	<2	
	12	Compact sand	>300	>3	
	13	Medium dense sand	100 to 300	1 to 3	
	14	Loose sand	<100	<1	
III Cohesive soils	15	Very stiff boulder clays and hard clays	300 to 600	3 to 6	Group III is susceptible to long-term consolidation settlement
	16	Stiff clays	150 to 300	1.5 to 3	
	17	Firm clays	75 to 150	0.75 to 1.5	
	18	Soft clays and silts	<75	<0.75	
	19	Very soft clays and silts	Not applicable		
IV	20	Peat and organic soils	Not applicable		
V	21	Made ground or fill	Not applicable		

*1 tonf/ft² = 1.094 kgf/cm² = 107.25 kN/m²

(4) The allowable bearing pressure q_a is less than or equal to the safe bearing capacity, depending on the settlements which are expected and which can be tolerated.

There are two groups of methods of determining ultimate bearing capacity: analytical methods and graphical methods, as with earth-pressure problems to which this problem is closely related. The analytical solutions are often easier and quicker to use than graphical methods but they can only be applied to cases in which the soil is fairly uniform. The graphical methods are very flexible and will cover any conditions likely to be found in practice.

The most general formula for net ultimate bearing capacity of a strip footing is that given by Terzaghi,[57] and modified by Skempton.[53]

$$q = cN_c + 0.5\gamma BN_\gamma + \gamma D(N_q - 1) \tag{43}$$

where N_c, N_γ, and N_q are bearing capacity factors depending on the angle of shearing resistance ϕ (Figure 8.30).

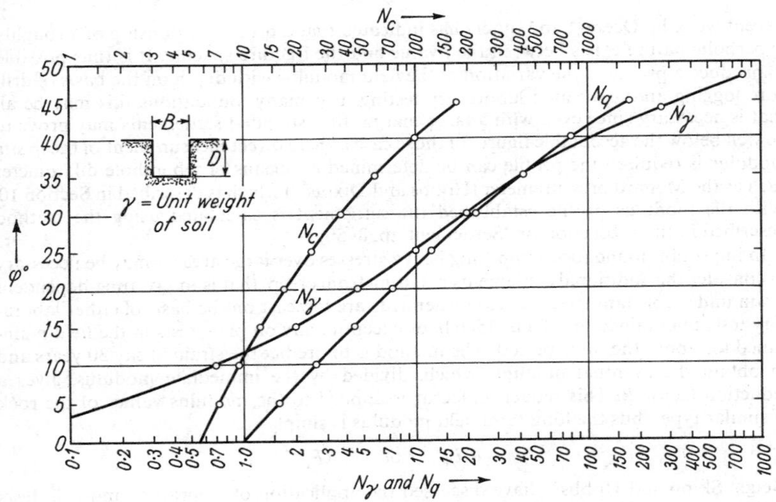

Figure 8.30 Terzaghi's bearing capacity coefficients for shallow footings

For a circular footing of diameter D:

$$q = 1.3cN_c + 0.3\gamma BN_\gamma + \gamma D(N_q - 1) \tag{44}$$

For a square footing of width B:

$$q = 1.3cN_c + 0.4\gamma BN_\gamma + \gamma D(N_q - 1) \tag{45}$$

Although the Terzaghi formula is a useful summary of factors governing ultimate bearing capacity, in its full form it is not used in practice. The methods for calculating both ultimate bearing capacity and settlements are generally related to the type of ground involved under three main groups, rocks, granular soils (sands and gravels) and cohesive soils (clays).

The discussion above has been concerned with net pressures. As indicated, the total overburden pressure should be added to net allowable bearing pressure (without application of a factor of safety to the overburden pressure), where the foundation is fully contained, in order to arrive at a gross allowable bearing pressure.

Foundations on rock

The bearing capacity of rock is not readily determined by tests on specimens and mathematical analysis, since it is greatly dependent on the large-scale structural features of the rock stratum. Some guidance concerning presumed bearing values for rocks is given in Table 8.3 and reference 52.

Generally the modulus of sound unweathered rocks is so high that settlements are not normally considered a factor in the design of foundations on massive rocks except for special structures such as nuclear reactors. The modulus of a rock mass, however, falls rapidly from the intact value appertaining to the rock substance, as measured on small specimens in the laboratory, as the intensity of the natural fractures or discontinuities in the rock mass increase. It is convenient to define a rock mass factor j as the ratio of the modulus of the mass in the field, E_f, determined by an appropriate test to the elastic modulus E_l determined in the laboratory, thus

$$j = E_f/E_l$$

Recent work by Deere[54] and others has indicated that a broad relationship of a roughly hyperbolic nature exists between the j value and the fracture spacing. It is thus possible to produce a profile of the variation of the field modulus with depth on the basis of drill core logging and associated laboratory testing. For many foundations this may be all that is necessary since even with a large margin the estimated settlements may prove to be well below the acceptable figure. In those cases where direct measurement of the *in situ* modulus is required the profile can be determined by means of a borehole dilatometer such as the Menard pressuremeter (Hobbs and Dixon[55]) which is described in Section 10. With the modulus profile established the settlement is calculated using the method described in the subsection on 'Settlement' (p. **8**–50).

In important foundations imposing heavy stresses over large areas it may be necessary to consider the additional settlements resulting from creep, that is to say, time-dependent strain under constant stress. These settlements are forecast on the basis of either laboratory tests maintained for 500 to 1500 h, or exceptionally on plate tests in the field maintained for about the same period. The method is to predict the strain at say 30 years and so obtain the eventual modulus which, divided by the immediate modulus, gives a reduction factor R. This reduction factor is applied to the modulus values of the rock of similar type, thus the long-term field modulus is simply

$$RjE_l \quad \text{or} \quad RE_f$$

Meigh, Skipp and Hobbs[56] have discussed the application of laboratory and field tests to the creep properties of a Bunter sandstone using these methods.

Foundations on sand

For uncomplicated ground conditions an analytical solution is used and this is described in detail below. For complicated conditions graphical methods are necessary (Terzaghi and Peck;[11] Terzaghi[57]).

SHALLOW FOOTINGS ON SAND

In granular soils the cohesion is negligible; putting $c = 0$, the Terzaghi/Skempton formula for a strip footing reduces to

$$q = 0{\cdot}5\gamma BN_\gamma + \gamma D(N_q - 1) \qquad (46)$$

in which the first term represents the effect of the strength of the sand below foundation level and the second term represents the effect of the surcharge of material above foundation level. This is often used as a basis for calculation of the net ultimate bearing capacity of shallow footings on sand or gravel. In order to determine the values of the bearing capacity factors N_γ and N_q it is, however, necessary to determine the angle of shearing resistance, ϕ. With sand and gravels it is unfortunately difficult and expensive to obtain an adequate number of undisturbed samples, except at very shallow depths, on which to carry out the necessary shear testing in the laboratory. Even if 'undisturbed' samples are obtained it is often virtually impossible to prepare an undisturbed test specimen; recompaction of the specimen is usually necessary. For this reason it is common practice to determine the approximate relative density of granular materials by means of dynamic penetration tests (the Standard Penetration Test, see Section 10). From these an approximate value of ϕ can be obtained for use in the bearing capacity formula. Values of the number of blows to drive the standard tool a distance of 12 in (30.48 cm) using the standard equipment and procedure, N values, are determined over a range of levels below the proposed foundation level in a number of boreholes. The average value for

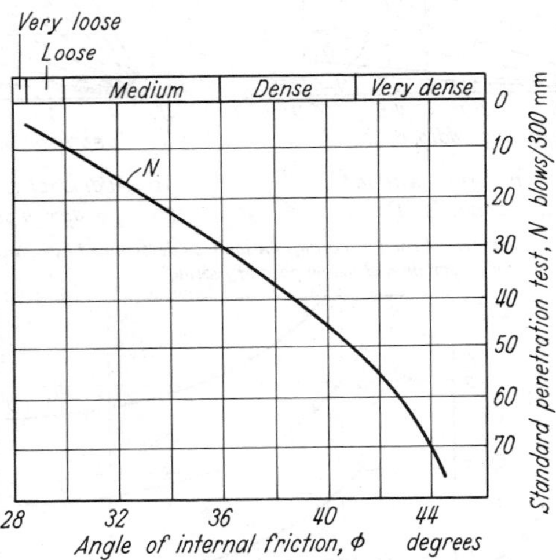

Figure 8.31 Curves showing the relationship between ϕ and values of N from the standard penetration test

each borehole is obtained and the minimum average value is used in the calculation. The empirical correlation between N value, relative density and ϕ is given in Figure 8.31.

However, the procedure has been further simplified by Peck, Hanson and Thorburn[58] who have drawn graphs of safe bearing pressure directly against N values. These assume a bulk density of 100 lb/ft^3 (1602 kg/m^3) and a factor of safety of 3. The values of safe bearing pressure are given in two components corresponding to the two terms in the bearing capacity formula, one relating to strength and hence relative density, and the other to overburden, in Figures 8.32(a) and 8.32(b).

The values are based on a level of the water table at a depth below foundation level at least equal to the width of the footing. If ground-water level is at foundation level the component of safe bearing pressure corresponding to the first term (Figure 8.32(a)) is halved, and for intermediate positions of the water table a linear interpolation is used. For ground-water level at ground level the component of bearing pressure corresponding to the second term (Figure 36(b)) is also halved. Where fine sand exists below the water

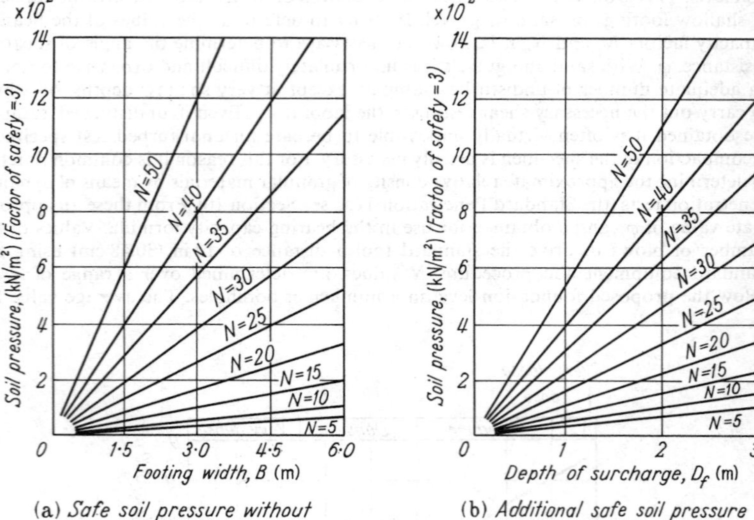

Figure 8.32 Safe soil pressures beneath footings on sand, as determined by bearing capacity. Charts based on water table not closer than B below base of footing

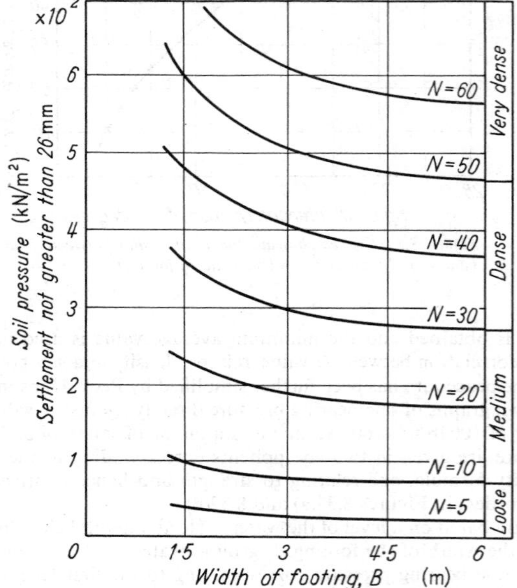

Figure 8.33 Soil pressure corresponding to 25 mm settlement of footings on sand. Chart based on water table not closer than B below base of footing

table the corrected value of N must be used. Where the measured N is greater than 15,

$$N_{corr} = 15 + \tfrac{1}{2}(N - 15) \tag{47}$$

The dominant part of equation (46) is the first term, $0.5\gamma BN_\gamma$ which is directly proportional to the width of the footing. For all but the very narrow footings the safe load is high and the limiting factor is settlement. Peck, Hanson and Thorburn have also prepared graphs of allowable bearing pressure against N values (Figure 8.33). These are based partly on theoretical considerations and partly on a study of field data; the criterion is that of 25 mm settlement being tolerable. Again it is assumed that ground-water level is at a depth below foundation level at least equal to the width of the foundation. If water level is at foundation level the allowable bearing pressure must be halved. Allowable bearing pressures corresponding to other settlement criteria can be obtained by assuming the settlement to be proportional to bearing pressure.

The above procedure is now generally regarded as being unnecessarily conservative. Meyerhof[59] has suggested that the penetration test automatically takes account of the water table and that no reduction of bearing pressure need be made provided the water table remains reasonably constant. It has also been suggested that the allowable pressures derived from Figure 8.33 may be increased by 50%.

Meyerhof[60] in 1956 pointed out a correlation between the standard penetration N value and the cone penetration resistance C_r obtained from static sounding using the Dutch apparatus (De Beer[61]),

$$C_r = K(107)N \tag{48}$$

with C_r in kN/m^2. Meyerhof found that K varied from 2.2 to 6.3 and suggested an average value of 4. This work has been extended by Meigh and Nixon,[62] Rodin,[63] and others, to take account of grain size and grading. Typical values of K are given in Table 8.4 (Simons[64]).

Table 8.4

Soil description	K
Sandy silt	2.5
Fine sand	4
Fine to medium sand	5
Sand with sandy gravel	8
Medium to coarse sand	8
Gravelly sand	8–18
Sandy gravel	12–16

While it is clear that the value of K increases with the particle size, comparison between tests on coarse materials becomes somewhat uncertain once the particle size becomes comparable with the diameter of the instruments, i.e. the penetrometer cone and the standard-penetration test spoon. Difficulties also arise on making comparisons at the other end of the scale owing to the tendency of finer sands to pipe; this results in an underestimate of the N value (see Section 10).

De Beer[61] has suggested that the compressibility C_s of a granular deposit could be related to the cone resistance C_r and the effective overburden pressure P'_0 by the relation

$$C_s = \frac{1.5C_r}{P'_0} \tag{49}$$

the compression in a layer of thickness H being given by the expression

$$\rho = \frac{H}{C_s}\log_e\left(\frac{P'_0 + \Delta p}{P'_0}\right) \tag{50}$$

where Δp is the increase in stress due to the net foundation pressure at the centre of the layer. The settlement is obtained by summing the compression in the various layers comprising the supporting ground, as discussed below in the subsection on clay.

Schmertmann,[65] following a field study and the examination of case records, has proposed that the ground modulus profile in sands may be determined from the static sounding cone resistance profile, that is the variation with depth, by the relationship,

$$E_s = 2C_r,$$

the settlement being obtained by the summation of compressions in the various layers of soil comprising the foundation down to a depth equal to twice the width of the foundation. The stress increments may be determined by the elastic theory, as in Figure 8.39, or from Newmark's influence values as discussed below under the settlement of clay foundations. For a critical study of the settlement of spread footings on granular foundations, see references 65 and 66.

It will be seen from the foregoing discussion that there is a good deal of uncertainty about the estimation of settlements of shallow foundations on cohesionless soils. The following procedure is suggested for use with standard penetration test values (N) or with static cone resistance values (C_r).

Settlement (ρ_N) at a given net bearing pressure is first estimated using the Terzaghi and Peck curves (Figure 8.33), adopting an appropriate value of N and ignoring the effect of the water table. If only deep sounding values are available, the value of N is decided with the help of Table 8.4. The settlement is then corrected to allow for the effects of grain size and grading, as follows:

$$\rho_{\text{corrected}} = \rho_N \times \frac{4N}{C_r} \qquad (51)$$

For example,

$$\text{in gravel } C_r = 8N \text{ and thus } \rho_{\text{corr}} = \tfrac{1}{2}\rho_N$$

$$\text{in silty sands } C_r = 2N \text{ and thus } \rho_{\text{corr}} = 2\rho_N$$

If only N values are available, again Table 8.4 can be used to estimate the corresponding value of C_r.

This is a new proposal not previously published, but a preliminary check against case histories indicates that it gives satisfactory results.

SHALLOW RAFTS ON SAND

For rafts the safe bearing pressure is always very high and settlement becomes the sole criterion. Since the effects of loose pockets is less marked with a raft than with isolated footings it is possible to accept a greater total settlement criterion. It can be seen from Figure 8.33 that for foundation widths greater than about 7 m the allowable bearing pressure is independent of the width of the footing. Hence for rafts it is possible to derive a simple expression [59] relating settlement to N value:

$$\rho = 3q/N \qquad \text{min} \qquad (52)$$

This relationship is sensibly independent of the shape of the foundation. If bedrock is at a depth below foundation level less than the width B, then some reduction in settlement may be allowed.

DEEP FOUNDATIONS IN SAND

The Terzaghi bearing capacity factors have been generally accepted as a basis for the design of shallow foundations, but not for foundations where the depth greatly exceeds the width as in the case of piers and piles. Meyerhof[67] suggested that the shear surface beneath the base would return upwards and inwards to reach the shaft, and produced graphs of a general bearing capacity factor $N_{\gamma q}$ for various values of ϕ and foundation depth ratio D/B. Large-scale experiments by Kerisel[68] have indicated that, in addition to ϕ, both depth and size influence the bearing capacity factor N_q. In a comprehensive study of deep foundations Vesic[69] has observed that there is no evidence of failure surfaces reverting to the shaft.

The design of deep foundations in granular materials is a highly complex matter. Terzaghi's bearing capacity factor N_q is too conservative for higher values of ϕ and is

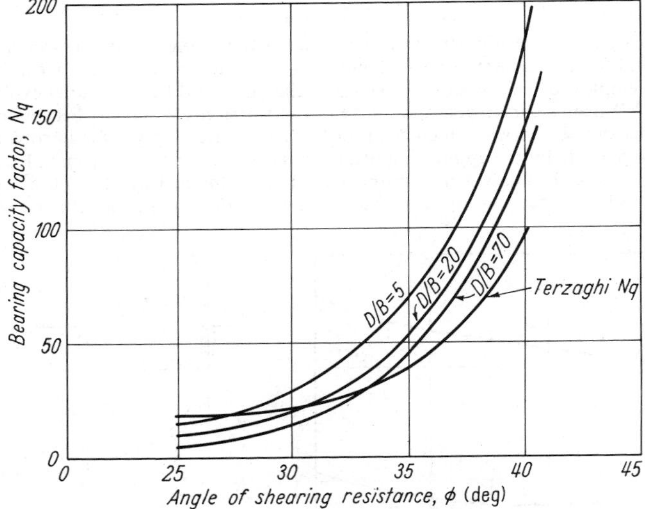

Figure 8.34 Berezantsev's bearing capacity factor N_q

independent of the depth/width ratio D/B. Berezantsev's[70] curves relating N_q to ϕ (Figure 8.34) are now recognised as giving the best representation of the ultimate bearing capacity of deep foundations in terms of N_q and D/B, thus

$$Q_u = A\gamma' D N_q$$

where A is the base area of the pier or pile. See also Tomlinson[71] for a discussion on deep foundations,

Foundations on clay

BEARING CAPACITY

Under the loading imposed by a foundation, clay strata consolidate and undergo an increase in strength. However, except in special cases, e.g. the foundations of an earth-fill dam which is under construction for two or three years, the amount of consolidation which occurs during the construction and first loading period is negligible, so that calculations of ultimate bearing capacity can be more easily made in terms of the total applied stresses, and not the effective stresses, and the angle of shearing resistance with respect to these applied stresses is zero for saturated clays, $\phi = 0$. This is a direct parallel with the $\phi = 0$ method for 'end of construction' stability analysis of slopes; in fact the circular arc method of analysis is applicable, as discussed below.

Putting $\phi = 0$ in the general bearing capacity formula eliminates the terms involving N_γ and N_q, leaving the simple expression

$$q = cN_c \tag{53}$$

where c, the undrained shear strength, is obtained from triaxial compression tests; N_c is a function of the shape and depth of the footing and values of N_c have been obtained by both analytical and graphical methods. Prandtl[72] in 1920 gave a solution for surface strip footings which he developed originally for metals in a plastic state:

$$N_c = (\pi + 2) = 5.14 \tag{54}$$

The analysis has been extended to cover circular footings, both shallow and deep, by Meyerhof,[67] and others (see also Tomlinson[71]). In addition, the circular arc method has been employed. This was introduced by Fellenius[73] and has been fully explained by Wilson.[74] Wilson's curves give the centre of the worst circle and the bearing capacity. The former can be a very useful guide in starting a graphical analysis for a problem which is too complicated for the application of a formula. These curves are given in Figure 8.35.

The results of the analytical solutions have been combined by Skempton[75] into two expressions for the bearing capacity factor N_c in terms of the depth, length and width of a

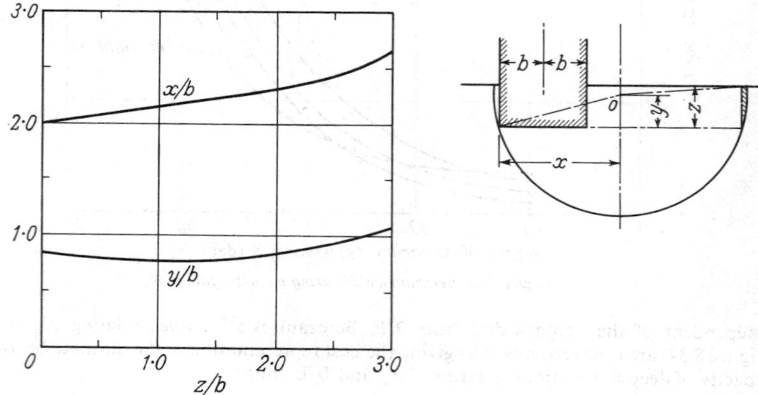

Figure 8.35 Circular arc method (After: Guthlac Wilson)

foundation. He has also shown that the resulting values are in good agreement with field data and with results of laboratory tests.

For a depth/width ratio, D/B, of less than $2\frac{1}{2}$

$$N_c = 5(1 + D/5B)(1 + B/5L) \tag{55}$$

in which the term in the first bracket is the depth factor and the term in the second bracket is the shape factor.

For deep footings, $D/B \geqslant 2\frac{1}{2}$,

$$N_c = 7.5(1 + B/5L) \tag{56}$$

Circular footings have the same N_c values as square footings ($B/L = 1$) and hence for deep circular or square footings, and for piles, $N_c = 9$.

In using the values of ultimate bearing capacity given by equations (55) and (56), it is necessary to apply a factor of safety, F, usually 3, in arriving at the net safe bearing pressure. The value of c taken in the calculation is the average value for a depth below foundation level equal to about $\frac{2}{3}B$, provided that the shear strength within this depth does not vary more than about $\pm 50\%$ of this average value. Overstressing will occur where the net applied pressure p_n is greater than πc_{min}, where c_{min} is the minimum shear strength within a depth equal to $B/2$. Except in special cases, applied pressure should not exceed πc_{min} or $c_{av}N_c/F$ whichever is the less. As mentioned, with complex foundation conditions, graphical analysis may be the more appropriate method.

Settlement (clays)

An idealised representation of settlement of foundations in clays and of the heave on excavation is given in Figure 8.36.

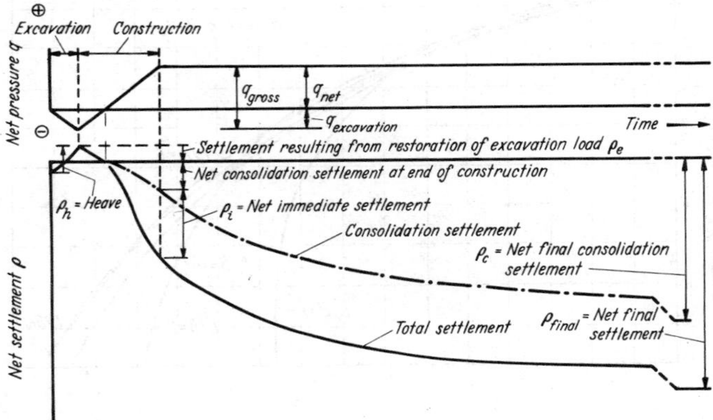

Figure 8.36 Idealised representation of settlement of foundations in clay

On excavation, heave occurs and the majority of this is recovered when the original total overburden pressure p_0 is replaced. This recovery of heave is often neglected in settlement calculations. With further application of load, i.e. with increase in the net applied pressure, and immediate settlement occurs without volume change of the clay, followed by consolidation settlement as the excess pore pressures set up by the applied load dissipate. In practice, the consolidation settlement starts immediately the net pressure is greater than zero, but at a very slow rate, so that it is convenient to ignore the consolidation settlement occurring during construction or, alternatively, to consider it as starting at halfway through the construction period. See Taylor[7] for a fuller treatment of this problem.

The net final settlement is the sum of the immediate settlement and the consolidation settlement:

$$\rho_{final} = \rho_i + \rho_c \tag{57}$$

The net settlement at any time t is given by

$$\rho_t = \rho_i + \bar{U}\rho_c \tag{58}$$

where $\bar{U}$ is the degree of consolidation after time t.

IMMEDIATE SETTLEMENT

The immediate settlement below the corner of a uniformly loaded rectangular area can be calculated from elastic theory using the Steinbrenner[76] equation (see also Terzaghi[57]):

$$\rho_i = \frac{3}{4} q \frac{B}{E} I_\rho \tag{59}$$

where E is Young's modulus for the clay, as measured by the appropriate tangent modulus of stress–strain curves from triaxial tests, and I_ρ is an influence factor which is a function

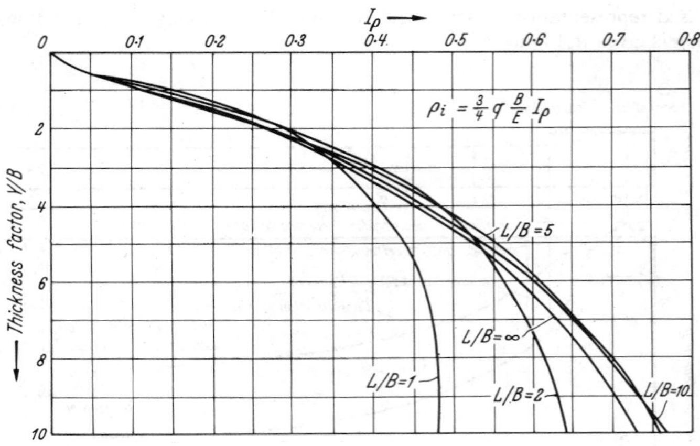

Figure 8.37 Steinbrenner's influence factors for loaded area L × B on compressible stratum of thickness Y

of the length and width of the foundation and the thickness of the compressible layer below foundation level. Values of I_ρ for a Poisson's ratio value of 0.5 are given in Figure 8.37.

Settlements at other points below a rectangular area can be calculated by splitting the area into a number of rectangles and using the principle of superposition.

CONSOLIDATION SETTLEMENT

The oedometer settlement is calculated from the equation

$$\rho_{oed} = \int_0^z m_v \Delta\sigma_1 \, dz \tag{60}$$

In practice this expression is evaluated by dividing the compressible strata into layers and the oedometer settlement of each layer is calculated using the average stress increment $\Delta\sigma_1$ in that layer and the average value of the modulus of compressibility m_v (obtained from an oedometer test) in that layer. The total oedometer settlement is then the sum of the settlements of the layers. The increment of vertical stress in any layer can be calculated from elastic theory. Boussinesq[77] gave the following equation for the vertical stress increase at depth z due to a point load P on the surface of a semi-infinite solid:

$$\sigma_z = \frac{P}{z^2}\frac{3}{2\pi}\left(\frac{1}{\{1+(r/z)^2\}^{\frac{5}{2}}}\right) \qquad (61)$$

where z and r are defined in Figure 8.38. This can be written

$$\sigma_z = PK/z^2 \qquad (62)$$

Values of K are given in Table 8.5.

Table 8.5 VALUES OF COEFFICIENT K IN EQUATION(62)

Ratio $\frac{r}{z}$	Coefficient K	Ratio $\frac{r}{z}$	Coefficient K	Ratio $\frac{r}{z}$	Coefficient K	Ratio $\frac{r}{z}$	Coefficient K	Ratio $\frac{r}{z}$	Coefficient K	Ratio $\frac{r}{z}$	Coefficient K
0.00	0.477 5	0.60	0.221 4	1.20	0.051 3	1.80	0.012 9	2.40	0.004 0	2.84	0.001 9
								45	0.003 7	2.91	0.001 7
0.10	0.465 7	0.70	0.176 2	1.30	0.040 2	1.90	0.010 5	2.50	0.003 4	2.99	0.001 5
0.15	0.451 6	0.75	0.156 5	1.35	0.035 7	1.95	0.009 5	2.55	0.003 1	3.08	0.001 3
0.20	0.432 9	0.80	0.138 6	1.40	0.031 7	2.00	0.008 5	2.60	0.002 9	3.19	0.001 1
0.25	0.410 3	0.85	0.122 6	1.45	0.028 2	2.05	0.007 8	2.65	0.002 6	3.31	0.000 9
0.30	0.384 9	0.90	0.108 3	1.50	0.025 1	2.10	0.007 0	2.70	0.002 4	3.50	0.000 7
0.35	0.357 7	0.95	0.095 6	1.55	0.022 4	2.15	0.006 4	2.72	0.002 3	3.75	0.000 5
0.40	0.329 4	1.00	0.084 4	1.60	0.020 0	2.20	0.005 8	2.74	0.002 3	4.13	0.000 3
0.45	0.301 1	1.05	0.074 4	1.65	0.017 9	2.25	0.005 3	2.76	0.002 2	4.91	0.000 1
0.50	0.273 3	1.10	0.065 8	1.70	0.016 0	2.30	0.004 8	2.78	0.002 1	6.15	0.000 1
0.55	0.246 6	1.15	0.058 1	1.75	0.014 4	2.35	0.004 4	2.80	0.0002 1		

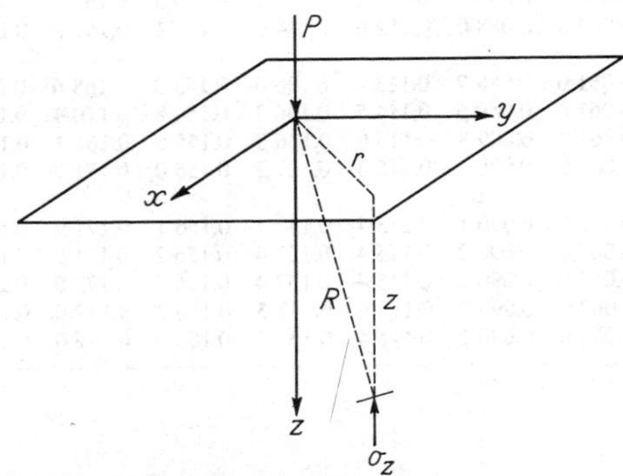

Figure 8.38 Diagram showing z and r in Boussinesq equation for concentrated load

Equation (62) has been integrated and tabulated by Newmark[78] to give the pressure below the corner of a rectangle uniformly loaded at the surface and the values are given in Table 8.6. In order to obtain the pressure below any other point it is necessary to regard that point as the corner of four adjoining rectangles, not necessarily the same shape or size, to calculate the pressure below the corner of each and add these pressures; e.g. the pressure below the centre of a rectangle is four times the pressure beneath the corner of a rectangle whose sides are half the sides of the original rectangle. The principle can be extended to points outside the original rectangle by addition and subtraction of

Table 8.6 VERTICAL PRESSURE σ_z UNDER CORNER OF RECTANGLE $a \times b$ L

α/β	0.1	0.2	0.3	0.4	0.5	0.6	0.7	0.8	0.9	1.0
0.1	0.004 7	0.009 2	0.013 2	0.016 8	0.019 8	0.022 2	0.024 2	0.025 8	0.027 0	0.027 9
0.2	0.009 2	0.017 9	0.025 9	0.032 8	0.038 7	0.043 5	0.047 4	0.050 4	0.052 8	0.054 7
0.3	0.013 2	0.025 9	0.037 4	0.047 4	0.055 9	0.062 9	0.068 6	0.073 1	0.076 6	0.079 4
0.4	0.016 8	0.032 8	0.047 4	0.060 2	0.071 1	0.080 1	0.087 3	0.093 1	0.097 7	0.101 3
0.5	0.019 8	0.038 7	0.055 9	0.071 1	0.084 0	0.094 7	0.103 4	0.110 4	0.115 8	0.120 2
0.6	0.022 2	0.043 5	0.062 9	0.080 1	0.094 7	0.106 9	0.116 8	0.124 7	0.131 1	0.136 1
0.7	0.024 2	0.047 4	0.068 6	0.087 3	0.103 4	0.116 8	0.127 7	0.136 5	0.143 6	0.149 1
0.8	0.025 8	0.050 4	0.073 1	0.093 1	0.110 4	0.124 7	0.136 5	0.146 1	0.153 7	0.159 8
0.9	0.027 0	0.052 8	0.076 6	0.097 7	0.115 8	0.131 1	0.143 6	0.153 7	0.161 9	0.168 4
1.0	0.027 9	0.054 7	0.079 4	0.101 3	0.120 2	0.136 1	0.149 1	0.159 8	0.168 4	0.175 2
1.2	0.029 3	0.057 3	0.083 2	0.106 3	0.126 3	0.143 1	0.157 0	0.168 4	0.177 7	0.185 1
1.4	0.030 1	0.058 9	0.085 6	0.109 4	0.130 0	0.147 5	0.162 0	0.173 9	0.183 6	0.191 4
1.6	0.030 6	0.059 9	0.087 1	0.111 4	0.132 4	0.150 3	0.165 2	0.177 4	0.187 4	0.195 5
1.8	0.030 9	0.060 6	0.088 0	0.112 6	0.134 0	0.152 1	0.167 2	0.179 7	0.189 9	0.198 1
2	0.031 1	0.061 0	0.088 7	0.113 4	0.135 0	0.153 3	0.168 6	0.181 2	0.191 5	0.199 9
2.5	0.031 4	0.061 6	0.089 5	0.114 5	0.136 3	0.154 8	0.170 4	0.183 2	0.193 8	0.202 4
3	0.031 5	0.061 8	0.089 8	0.115 0	0.136 8	0.155 5	0.161 1	0.184 1	0.194 7	0.203 4
4	0.031 6	0.061 9	0.090 1	0.115 3	0.137 2	0.156 0	0.171 7	0.184 7	0.195 4	0.204 2
5	0.031 6	0.062 0	0.090 1	0.115 4	0.137 4	0.156 1	0.171 9	0.184 9	0.195 6	0.204 4
6	0.031 6	0.062 0	0.090 2	0.115 4	0.137 4	0.156 2	0.171 9	0.185 0	0.195 7	0.204 5
8	0.031 6	0.062 0	0.090 2	0.115 4	0.137 4	0.156 2	0.172 0	0.185 0	0.195 7	0.204 6
10	0.031 6	0.062 0	0.090 2	0.115 4	0.137 5	0.156 2	0.172 0	0.185 0	0.195 8	0.204 6
∞	0.031 6	0.062 0	0.090 2	0.115 4	0.137 5	0.156 2	0.172 0	0.185 0	0.195 8	0.204 6

RMLY WITH INTENSITY q. σ_z/q FOR VALUES OF $\alpha = a/z$ AND $\beta = b/z$

	1.6	1.8	2.0	2.5	3.0	4.0	5.0	6.0	8.0	10.0	∞
	0.030 6	0.030 9	0.031 1	0.031 4	0.031 5	0.031 6	0.031 6	0.031 6	0.031 6	0.031 6	0.031 6
	0.059 9	0.060 6	0.061 0	0.061 6	0.061 8	0.061 9	0.062 0	0.062 0	0.062 0	0.062 0	0.062 0
	0.087 1	0.088 0	0.088 7	0.089 5	0.089 8	0.090 1	0.090 1	0.090 2	0.090 2	0.090 2	0.090 2
	0.111 4	0.112 6	0.113 4	0.114 5	0.115 0	0.115 3	0.115 4	0.115 4	0.115 4	0.115 4	0.115 4
	0.132 4	0.134 0	0.135 0	0.136 3	0.136 8	0.137 2	0.137 4	0.137 4	0.137 4	0.137 5	0.137 5
5	0.150 3	0.152 1	0.153 3	0.154 8	0.155 5	0.156 0	0.156 1	0.156 2	0.156 2	0.156 2	0.156 2
0	0.165 2	0.167 2	0.168 6	0.170 4	0.171 1	0.171 7	0.171 9	0.171 9	0.172 0	0.172 0	0.172 0
9	0.177 4	0.179 7	0.181 2	0.183 2	0.184 1	0.184 7	0.184 9	0.185 0	0.185 0	0.185 0	0.185 0
6	0.187 4	0.189 9	0.191 5	0.193 8	0.194 7	0.195 4	0.195 6	0.195 7	0.195 7	0.195 8	0.195 8
4	0.195 5	0.198 1	0.199 9	0.202 4	0.203 4	0.204 2	0.204 4	0.204 5	0.204 6	0.204 6	0.204 6
8	0.207 3	0.210 3	0.212 4	0.215 1	0.216 3	0.217 2	0.217 5	0.217 6	0.217 7	0.217 7	0.217 7
2	0.215 1	0.218 4	0.220 6	0.223 6	0.225 0	0.226 0	0.226 3	0.226 4	0.226 5	0.226 5	0.226 6
1	0.220 3	0.223 7	0.226 1	0.229 4	0.230 9	0.232 0	0.232 4	0.232 5	0.232 6	0.232 6	0.232 6
4	0.223 7	0.227 4	0.229 9	0.233 3	0.235 0	0.236 0	0.236 4	0.236 7	0.236 8	0.236 8	0.236 9
6	0.226 1	0.229 9	0.235 5	0.236 1	0.237 8	0.239 1	0.239 5	0.239 7	0.239 8	0.239 9	0.239 9
6	0.229 4	0.233 3	0.236 2	0.240 4	0.242 0	0.243 4	0.243 9	0.244 1	0.244 3	0.244 3	0.244 3
0	0.230 9	0.235 0	0.237 8	0.242 0	0.243 9	0.245 5	0.246 1	0.246 3	0.246 5	0.246 5	0.246 5
0	0.232 0	0.236 0	0.239 1	0.243 4	0.245 5	0.247 3	0.247 9	0.248 2	0.248 4	0.248 4	0.248 4
3	0.232 4	0.236 4	0.239 5	0.243 9	0.246 1	0.247 9	0.248 6	0.248 9	0.249 1	0.249 1	0.249 2
4	0.232 5	0.236 7	0.239 7	0.244 1	0.246 3	0.248 2	0.248 9	0.249 2	0.249 4	0.249 5	0.249 5
5	0.232 6	0.236 8	0.239 8	0.244 3	0.246 5	0.248 4	0.249 1	0.249 4	0.249 6	0.249 7	0.249 8
5	0.232 6	0.236 8	0.239 9	0.244 3	0.246 5	0.248 4	0.249 1	0.249 5	0.249 7	0.249 8	0.249 9
6	0.232 6	0.236 9	0.239 9	0.244 3	0.246 5	0.248 5	0.249 2	0.249 5	0.249 8	0.249 9	0.250 0

rectangles. It is implied in the above that the pressure is uniformly distributed at the surface of the ground.

Jurgenson[79] has tabulated influence values for grids of points below a variety of flexible loaded areas, with uniform, triangular and 'terrace' loadings. More recently, solutions to problems involving loaded areas of various shapes, nonuniform flexible loads (such as embankments, etc.), layered foundations and finite foundation thicknesses, have been given in graphic and tabular form (Taylor,[7] Terzaghi and Peck,[11] Terzaghi,[57] Tomlinson,[71] Little[80]).

A flexible load is one in which the contact reaction from the ground is identical to the applied pressure at each point. Uniformly flexible loads on relatively thick foundation soils produce a dish-shaped settlement profile. Under rigid foundations the settlement is uniform or planar but the contact pressure varies with soil type. Intermediate cases occur, depending on the degree of relative rigidity of the structure and ground, but present analytical difficulties. In practice, foundations are generally considered as being either flexible (e.g. oil tanks) or rigid (e.g. high buildings on stiff rafts). The pressure distribution for the latter case has been worked out by Fox,[81] who gives a series of curves, Figure 8.39, for the mean vertical stress σ_z at a depth z beneath a rectangular area $a \times b$ uniformly loaded with a pressure q, the rectangle being on the surface.

For more important cases, recent practice has been to use the finite-element method to investigate soil/structure interaction, stress distribution within the structure and the ground, and the associated deformations.

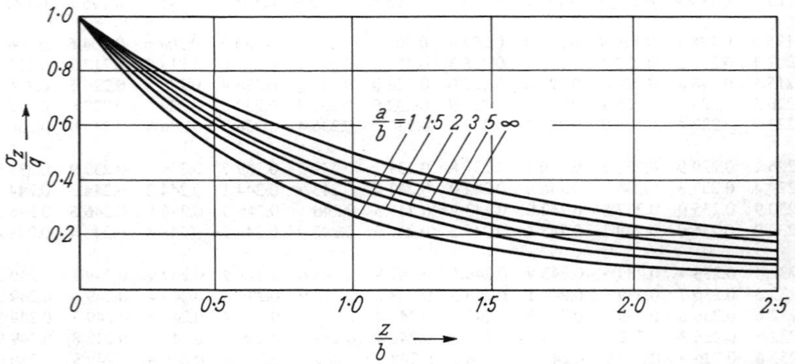

Figure 8.39 Mean pressure under stiff foundation (After: Fox[81])

In view of the geometrical and stress differences between the oedometer and the ground supporting the foundation, it is necessary to make a correction to the settlement, determined from equation (60). Skempton and Bjerrum[82] have evaluated the coefficient μ in the expression $\rho_c = \mu\rho_{oed}$, to take account of these factors and the consolidation history of the clay. The coefficient μ is related to the pore pressure coefficient A by the equation

$$\mu = A + \alpha(1 - A) \tag{63}$$

where α is a coefficient depending on the geometry of the problem. It has been computed for circular and strip footings, with various ratios of the thickness of the clay z to the width of the footing B. The results are given in Figure 8.40, in terms of the consolidation history.

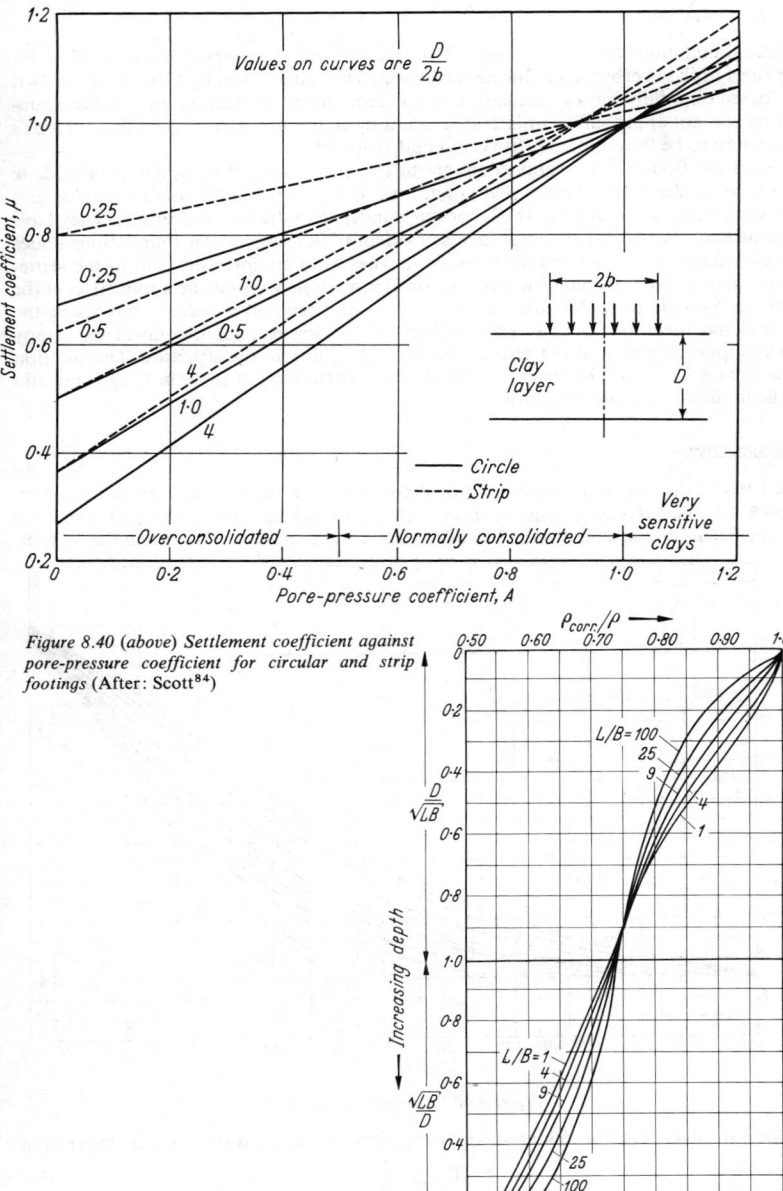

Figure 8.40 (above) Settlement coefficient against pore-pressure coefficient for circular and strip footings (After: Scott[84])

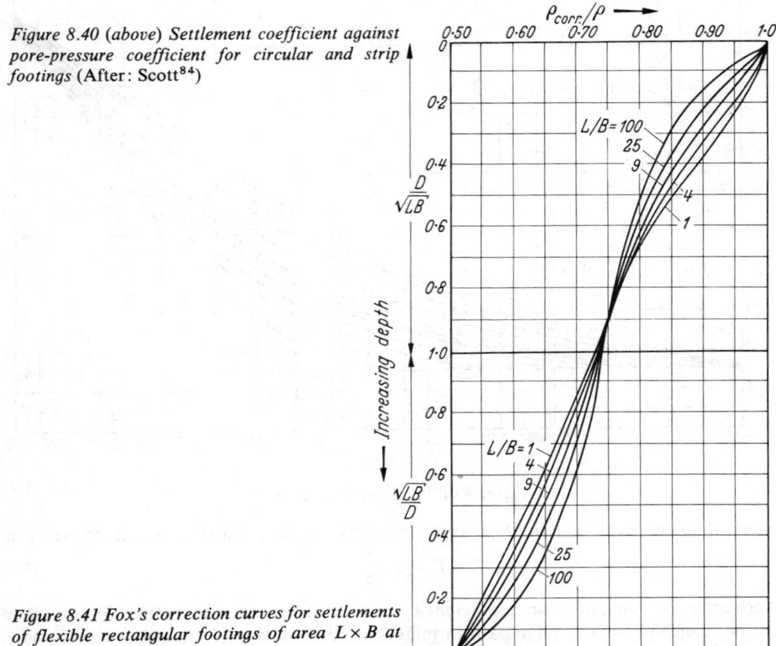

Figure 8.41 Fox's correction curves for settlements of flexible rectangular footings of area L × B at depth D

DEPTH CORRECTION

Where foundations are below ground level a correction is generally made to allow for the effect of the overburden reducing settlement. The values given by Fox[81] (Figure 8.41) are based on a fully buried foundation in an incompressible medium. In practice, foundations do not apply sufficient lateral restraint upon the soil, apart from piles, for Fox's correction to be fully effective (Burkland and Hobbs[83]).

Davis and Poulos[85] have applied the elastic theory to settlement prediction, immediate and final, under three-dimensional conditions, using values of Young's modulus and Poisson's ratio determined in the laboratory under fully drained conditions. They have obtained curves relating the total final settlement of flexible circular foundations under three-dimensional conditions for various thicknesses of compressible soil to the settlement obtained from the conventional oedometer analysis without the application of the factor μ. Specific three-dimensional problems have been solved numerically using the finite-element method (Zienkiewicz and Cheung[86]) taking account not only of anisotropy in the supporting ground but also of the flexural rigidity of the structure. The solution gives the settlement of the structure and also the distribution of stress in the ground and in the foundation of the structure.

TIME SETTLEMENT

The basis of the Terzaghi one-dimensional consolidation theory has been discussed on page 8–12. Once the total consolidation settlement has been determined and a value of the coefficient of consolidation c_v has been selected, it remains to determine the relation

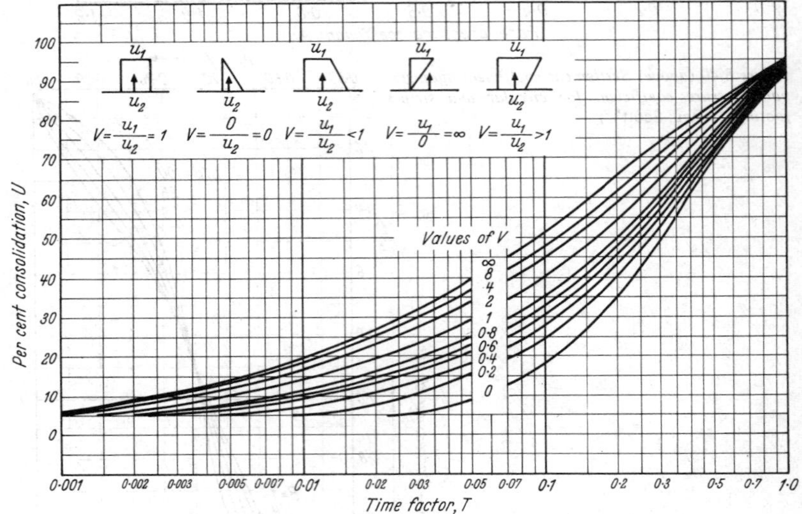

Figure 8.42 Values of time factor T

between the degree of consolidation and time. This can be obtained from the expression

$$T = c_v \frac{t}{H^2} \tag{64}$$

(where c_v is the consolidation coefficient in metres^2 per year, t is the time in years and H is the length of the drainage path in metres), and from the relation between U and t. This relationship is dependent on the initial distribution of excess pore pressure. Figure

8.42 gives a plot of U against t for various ratios of the initial excess pore pressure at the top and bottom of the compressible stratum u_1/u_2. The cases given are all for single drainage.

For double drainage, i.e. where drainage can take place at the top and bottom of the layer, values corresponding to $u_1/u_2 = 1$ can be used for all ratios of initial pore pressure, but it should be noted that in the double drainage case H is taken as only half of the layer thickness.

One-dimensional drainage is seldom fully realised in practice and for important calculations, particularly where the loaded area is small in comparison with the thickness of the compressible stratum, two- or three-dimensional consolidation should be considered (see Gibson and Lumb,[87] and also Davis and Poulos[85]).

Furthermore, in many deposits, lateral permeability can be up to orders greater than vertical owing to the presence of a laminar structure of thin layers and partings of silt and fine sand. This will have a very marked effect on rate of consolidation (Rowe[88]). Problems involving a number of layers having differerent consolidation characteristics have been solved numerically using the finite-difference method (Schiffman and Stein[89]).

Piled foundations

Piles are used to transfer foundation loads to a deeper stratum when the surface soils are too weak or too compressible to carry the load. They can be divided into two main types, driven and bored.

DRIVEN PILES

Driven piles are generally large-displacement piles which can often give rise to problems of ground heave and vibration. They fall into two groups, preformed and cast-in-place. Preformed piles are constructed of timber, concrete (reinforced or prestressed), or of steel in tube, box or H section. H-piles will lead to considerably lower displacements.

Driven cast-in-place piles are formed by driving a closed ended tube, usually of steel, to form a void, the void being filled with concrete as the tube is withdrawn. The tube may be driven at the top or by means of an internal mandrel, or at the toe usually by driving on a plug of green concrete.

BORED PILES

Bored piles are nondisplacement piles which are formed by boring with drop tools, augers, or rotary boring buckets. The hole is lined if necessary to support the surrounding ground over part or whole of the depth. The lining is usually removed as the concrete is placed, although in some cases it is left in the ground. Alternatively, bentonite mud may be used to support the hole in unstable ground. However, this will lead to a reduction in shaft friction. Enlarged toes may be formed in suitable ground by use of a belling tool.

Under certain conditions a combined bored and driven pile called a prebored pile is sometimes used where it is required to reduce displacement. The hole is bored usually with a continuous flight auger to a somewhat smaller cross-sectional area and to a lesser depth than the pile which is subsequently driven into it.

GROUPS OF PILES

A piled foundation generally consists of a group of several piles, the behaviour of which should be considered as an entity.

The piled group will consist of either point-bearing piles which transfer their load to a hard stratum of soil on which their points bear, e.g. piles to rock, or friction piles which transfer their load mainly by friction on the sides of the piles to a firm stratum into which they penetrate. Friction piles into a firm clay stratum will usually penetrate about 20 to 30 times the pile diameter into the clay. Piles driven through soft material into compact sand or gravel will usually penetrate about five times the diameter and will be partly point-bearing and partly frictional.

When friction piles are used in conditions in which the increase in strength of the soil with depth is only gradual, they must be of a length comparable to the size of the building to be of much advantage. This is shown in Figure 8.43 in which the stress distribution

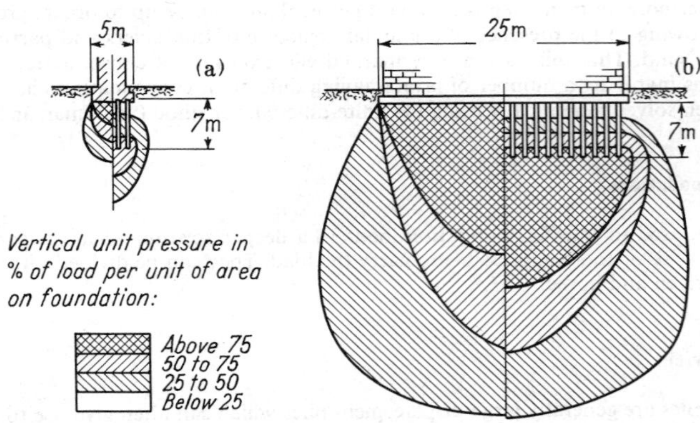

Figure 8.43 Illustrating increase of vertical pressure in soil beneath friction pile foundations having piles of equal lengths carrying equal loads. In (a), width of foundation small compared with pile length; in (b), width of foundation large compared with pile length

with and without piles is shown for two buildings of different widths but with piles of the same length. Unless the use of piles changes the stress pattern radically their use is probably not economic.

The design of a foundation on friction piles should always be checked from the point of view of overall stability, as described in Section 16 'Foundations Design', and assuming that the whole of the load is distributed uniformly over the area of the building and acts at the foot of the piles. The friction round the circumference of the block of soil containing the piles should be subtracted from the foundation load.

If the foundation on friction piles is supported by a bed of clay, consolidation settlements will occur which can be estimated as described for deep foundations above.

SINGLE PILES

Although the carrying capacity of a group of piles is not that of a single pile times the number of piles in the group, it is useful to know the load which a single pile will carry.

PILE BEARING ON ROCK

Where the bedrock is massive and strong, bearing capacity is usually not a problem. However, if the rock surface is steeply sloping, it may be necessary to provide a driven pile with a special point to make sure that it is adequately toed in (Bjerrum[90]). Bored piles into strong rock will only need a small penetration, say half to one pile diameter, in order to develop a working load equal to the maximum allowable working stress in the concrete – which is taken as 5000 kN/m^2. With small penetrations, particular care is needed to get a good contact between the pile toe and the rock.

With weaker and fractured rocks, it is still generally possible to develop the full allowable working stress in the concrete, but this will require greater penetration in order to carry some of the load in shear in the 'rock socket' (Thorburn[91]), e.g. penetrations of some three or four pile diameters into unweathered rock are considered advisable in coal measure shales and mudstones (Meigh[92]).

PILES BEARING IN DEEP DEPOSITS OF SAND AND GRAVEL

The ultimate point bearing capacity of a pile of end area A_B in sand may be estimated directly from the cone resistance C_r of the Dutch deep sounding test (Huizinga[93]), thus

$$Q_{uB} = A_B C_r \tag{65}$$

When using this method, however, it is necessary to take due account of the difference in scale between the cone (end area 10 cm^2) and the pile by ensuring that the pile penetrates the bearing layer some 4 to 6 or more diameters for piles less than 450 mm diameter. For larger piles, greater penetration ratios are necessary. (See Tomlinson,[71] Kerisel,[68] and Vesic[69] for further information on this problem.) The bearing capacity can also be determined from Berezantsev's[70] curves, as discussed above in the subsection on 'Deep Foundations in Sand'. However, results of theoretical calculations may be unreliable, and past experience is a better guide.

In deeply embedded piles it is also necessary to take account of the side friction, a factor which depends upon the horizontal earth pressure coefficient K_s, the overburden pressure $\gamma' D$ and the angle of shearing resistance δ between the ground and the pile; thus the ultimate shaft resistance is given by

$$Q_{us} = A_s \bar{K}_s(\gamma' \bar{D} \tan \delta) \tag{66}$$

where A_s is the shaft area and $\bar{K}_s$ and $\gamma' \bar{D}$ refer to average values of K_s and $\gamma' D$. The horizontal coefficient K_s depends not only upon the relative density and stress history of the deposit, but also upon the method of forming the pile. In bored piles it can be as low as the active earth pressure coefficient, rising to values in excess of unity for driven tapered piles. The value of δ depends upon the value of ϕ and the pile material; δ/ϕ varies from 0.54 to 0.80 for steel and pre-cast concrete respectively. See Tomlinson[71] for a full discussion on the question of shaft resistance.

The frictional resistance of the ground for driven piles can also be determined by use of the friction sleeve adaptation to the Dutch deep-sounding apparatus (Begemann[94]). This gives the term $K_s \gamma' D$, $\tan \delta$ throughout the depth of the deposit directly. (See Section 10.)

The ultimate bearing capacity of the pile is given by the sum of equations (65) and (66):

$$Q_u = Q_{uB} + Q_{us} \tag{67}$$

Frequently, pile bearing capacities have to be based on standard penetration test data. Meyerhof[60] has proposed the following empirical relationships for the components Q_u and Q_{us}

$$Q_u = A_B(107)4N + A_s \frac{\bar{N}}{50}(107) \tag{68}$$

where Q_u is in kilonewtons and A_B and A_s are in m^2, and $\bar{N}$ is the average N value along the pile shaft. An upper limit of 60 kN/m^2 is suggested for the shaft resistance.

In the subsection on 'Deep Foundations' above, attention was drawn to the dependence of the coefficient K in the relationship $C_r = KN$ on the soil type, a range of values being given in Table 8.3. It is suggested that the point bearing capacity term in equation (68) might be written as follows, to take account of the actual soil type into which the pile is driven:

$$Q_{uB} = A_B(107)KN \tag{69}$$

When it is possible to carry out a loading test on a full-scale pile this should be done, the test being carried to failure, i.e. until settlement continues under constant load.

The use of dynamic pile-driving formulae to calculate the ultimate load is not to be recommended for two reasons, namely (a) the information is obtained too late, i.e. at the construction stage instead of the design stage unless previous expensive tests are carried out, in which case loading tests should be included; and (b) a very wide range of answers can be obtained depending on the formula used and the choice of constants in the formula.

Dynamic pile-driving formulae have their uses, however, and records of set and energy should always be kept as they are a guide to the variation in ultimate load over a site on which one or two loading tests have already been carried out. Also engineers of wide experience can make an estimate of the load-carrying capacity from the results of a driving test, provided always that their experience was obtained with conditions similar to those relating to the test. The relation between the true ultimate load and that given by the dynamic formula is empirical and should be recognised as such in spite of the mathematical basis of Newtonian impact mechanics on which such formulae appear to be founded.

PILES BEARING IN CLAY

A pile bearing in clay receives support from the adhesion along the shaft and from the resistance at its base, the relative contribution of these two components depending upon the strength–depth profile of the clay, the ratio of the length to diameter of the pile and the manner of its installation, i.e. whether bored, driven preformed, or cast-in-place. Skempton[95] has given the following expression for the bearing capacity of a bored pile in London clay:

$$Q = 9c_u A_b + \alpha \bar{c}_u A_s \tag{70}$$

where α is an empirical factor which depends on the type of pile, the clay and its condition, and varies between 0.3 and 0.6 for a bored pile in London clay, with a mean value of 0.45 for a well constructed pile in typical unweathered clay; $\bar{c}_u$ is the average shear strength along the pile shaft.

The bearing capacity of a bored pile can be greatly increased by constructing an enlarged base (belling or under-reaming); diameters up to 5 m are possible in good conditions, though 2 to 2.5 m is customary.

The shear strength of the clay in commercial practice is determined on 38 mm specimens cut from 100 mm driven samples, and experience over the last decade has shown that, owing to the fissured nature of London clay, this practice results in considerably higher strengths being used in the assessment of end bearing capacity than obtain *in situ* (see Whitaker and Cooke[96] and also Burland, Butler and Dunican[15]). It is necessary therefore to allow for this factor by introducing an empirical coefficient ω (Whitaker and Cooke[96]) related to the base diameter B of the pile, thus:

$$Q = 9\omega c A_B + \alpha \bar{c}_u A_s \tag{71}$$

where

$$\omega = 0{\cdot}8 \text{ for } B < 1\text{ m}$$
$$0'75 \text{ for } B > 1\text{ m}$$

and αc_u has an approximate limit of 100 kN/m^2.

In the event of the strength being determined on larger specimens or from *in situ* piston tests, upward modification in the value of ω is necessary. Load factors used in the design range from 2 for straight shafted piles to 2.5 for base diameter less than 2 m as these values will generally restrict the short-term settlement of the single pile to less than 10 mm. With diameters larger than 2 m the working load should be checked by calculating the settlement using the curves relating Q_a/Q to settlement shown in Figure 8.44.

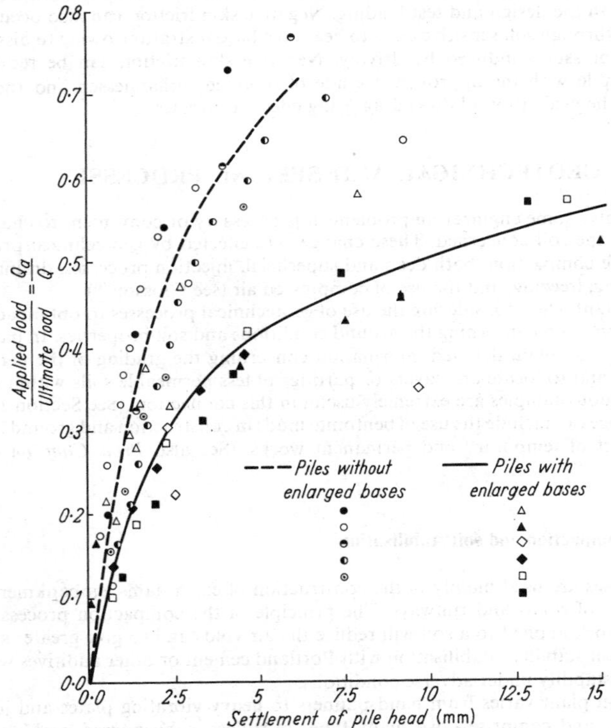

Figure 8.44 Q_a/Q against settlement in loading tests, showing the mean curves for piles with and without enlarged bases (After: Whitaker and Cooke)

The range of α values discussed above is based on experience of London clay. In dealing with other overconsolidated clays, for example, the Lias, Kimmeridge and Oxford clays and in weathered marls, some caution is needed in assigning α values, and where previous experience is not available, loading tests designed to establish α values should be undertaken.

Piles driven into soft sensitive clays result in loss of strength in the clay in contact with the pile, and thus test loading should be delayed for as long as possible after driving, preferably at least a month, to allow thixotropic regain of strength.

Piles driven into stiff clay cause severe disturbance and fracturing ot the clay and experience has shown that the effective cohesion can be extremely erratic, even on the same site (Tomlinson[71]). The lower limit of the cohesion factor, Tomlinson suggests, ranges from about 0.5 for soft clays to 0.2 for stiff and very stiff clays. There appears to be an upper limit of about 40 kN/m^2 for the shaft cohesion of driven preformed piles and there is some evidence that the strength may decrease somewhat with time owing to 'strain-softening effects'.

Loading tests should be carried out whenever possible.

Piles driven into ground which is subject to consolidation from loading, or which is still settling under prior loading, are subject to downdrag forces (negative skin friction) which place an additional load on the toe. These forces can be severe and should be taken into account in the design and test loading. Negative skin friction can also occur, when driving piles through soft sensitive clays to bear on a harder stratum, owing to dissipation of the pore pressures induced by driving. Negative skin friction can be reduced by slicking the pile with the appropriate grade of bitumen. Johannessen and Bjerrum[97] recommend the evaluation of downdrag using effective stresses.

GEOTECHNICAL AND SPECIAL PROCESSES

In order to solve some engineering problems it is necessary or convenient to change the properties of the soil concerned. These changes are effected by geotechnical processes, which include compaction, both deep and superficial, injection processes, drainage and water lowering, freezing, and the use of compressed air (see Glossop[98]).

It is important when considering the use of geotechnical processes to obtain adequate relevant information concerning the ground conditions and soil properties. In particular it is necessary to obtain detailed information concerning the grading of the permeable soils present, and to locate any layers or partings of less permeable soils within the soil mass. Continuous samples are extremely useful in this connection. (See Section 10.)

Special processes include the use of bentonite muds in construction and ground anchors in the support of temporary and permanent works. (See also *Proc. Conf. on Ground Engineering*.)

Superficial compaction and soil stabilisation

These processes are used mainly in the construction of earth dams, embankments, and the subgrades of roads and runways. The principle of the compaction process is that mechanical work applied to a soil will reduce the air voids and so give greater strength and mechanical stability. Stabilisation with Portland cement or other additives will then maintain the stability under adverse conditions.

Compaction plant varies from hand punners to heavy vibrating plates and jumping frog rammers, and comprises also many types of rollers, rubber-tyred, wobbly wheel, sheepsfoot and smooth steel-tyred. Some rollers can be ballasted by water or broken stone, and some incorporate vibrators.

Stabilisation plant includes all the above compaction plant plus pulverising and mixing plant which can vary from simple ploughs, harrows and disc cultivators to impressive single-pass machines, some of which pulverise, moisten, mix, lay and compact the soil in one pass of the machine. (See *Soil Mechanics for Road Engineers.*[21])

It is easier to compact or to stabilise a soil satisfactorily if the grading approximates to that of a naturally stable material (see Figure 8.45). The grading of a soil which departs from this can be corrected by mixing another soil with it, provided always that the soil is such that mechanical mixing is possible, e.g. it is not always possible to mix another soil with some very sticky or very hard clays. But a stiff clay can sometimes be stabilised with cement by using a machine with a high-speed rotor which cuts the clay into flakes which then act as an aggregate and are coated with cement and compacted.

The moisture content of the soil must be close to the optimum for the particular compaction plant and process used. This is often best determined by field trials based on preliminary laboratory tests.

The soil should always be compacted to the maximum possible density. This will not only give it greater immediate strength, but will greatly reduce any future softening which may take place. There is an obvious economic limit to the degree of compaction which can be obtained with the plant available, since the increase in density per pass of the roller drops off rapidly after eight or ten passes.

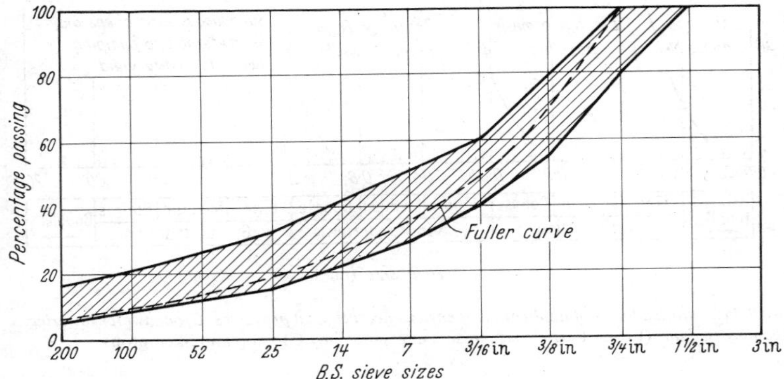

Figure 8.45 Suggested particle-size limits for bases. 1.5 in (38 mm) maximum aggregate size (For description of the Fuller Curve, see Reference 21)

A stabilising agent may be mixed with the soil before compaction. The purpose of this is to waterproof or to cement the soil, or to prevent the pickup of water and thus reduce softening. Various substances can be and have been used as stabilisers, amongst which are cement, resins (sodium rosinate, *Vinsol*), bituminous emulsions, chemicals (calcium chloride, common salt), molasses, oils (Shell stabilising oil) (see references 21 and 99). A recent development is the construction of impervious soil blankets by soil stabilisation methods used to line reservoirs.

Deep compaction

This process is used to increase the density of sandy and gravelly soils before building on them. The method relies on vibration as the compacting agency.

Three methods exist. The first is the simpler and consists of driving piles into the soil. The vibration compacts the soil and the pile occupies the void resulting from the compaction. Driven *in situ* piles can be used; it is unnecessary to add cement to the pile, an inert filler being used since the pile is not required to carry a load. This is an advantage in acid soils.

The second method consists of dropping a large heavy weight (e.g. a 2 m concrete cube) from a height of up to 5 in. This will induce compaction up to depths of about 15 m, and is particularly effective in the more freely draining soils.

The third method, which originated in Germany, is known as vibroflotation. The vibrator is jetted into the ground by means of a water jet. It is then withdrawn with the vibrator motor running and an upward flow of water issuing from the bottom of the machine. Sand is fed into the hole made by the vibrator and fills the void created by the compaction of the soil. Although the machine itself is only about 0.3 m diameter, the effective radius of the vibration is about one metre.

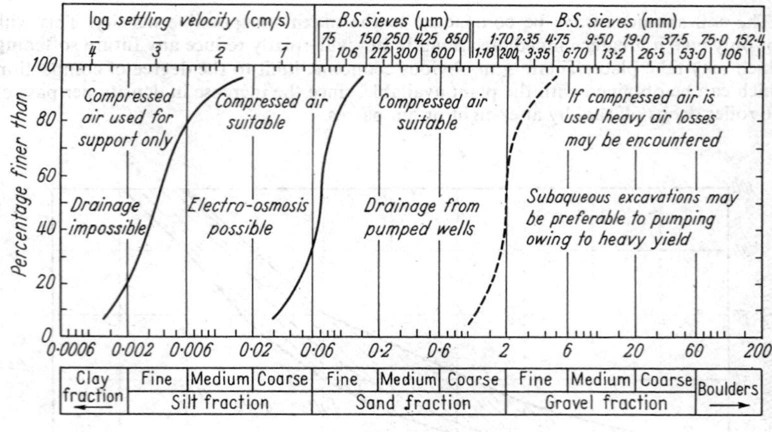

Figure 8.46 Tentative limits of application of various geotechnical processes. Grounawater lowering and compressed air. (From 'Particle size in silts and sands', by Glossop and Skempton)

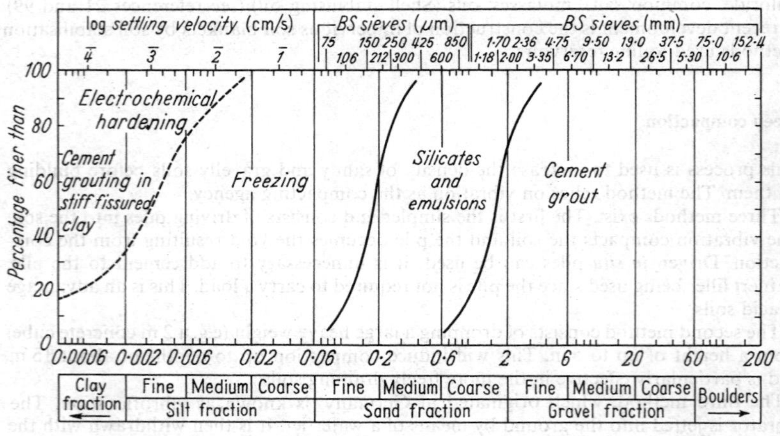

Figure 8.47 Tentative limits of application of various geotechnical processes. Artificial cementing. (From 'Particle size in silts and sands' by Glossop and Skempton)

Injection processes—grouting

PURPOSE

It is sometimes possible to change the properties of the soils encountered in an engineering problem by injecting materials of various sorts into the voids of the soil. These changes include reduction in permeability, increase in strength and decrease in compressibility, or a combination of these.

Cases in which the reduction in permeability is important include the formation of grouted cut-offs under dams, grouting fissured rocks, grouting sand and gravel to reduce air losses during construction work in compressed air, and sealing gaps in sheet piling. The increase in strength is important in underpinning problems and in support of excavation in tunnelling. Injection processes can also be used to lift tanks and pavement slabs by hydraulic pressure, the grout later setting and supporting the structure in the raised position.

SOILS WHICH CAN BE GROUTED

Grouts can be injected into the fissures in a rock. This is probably one of the earliest applications of the process. Rockfill and rubble masonry can also be grouted. Usually cement grouts are used in these cases since the voids are fairly large.

Gravels and sands can be grouted successfully by a variety of different processes as described below, but clays and silts cannot be grouted because their voids are too small and their permeabilities are too low (Glossop and Skempton[100]). An exception to this is the use of the technique called claquage grouting, in which tongues of high-pressure grout penetrate in planes and zones of weakness within the soil body. However, this is used infrequently.

Figures 8.46 and 8.47 show the tentative limits of application, with respect to particle size and grading, of the various geotechnical processes, including grouting (Glossop and Skempton[100]).

MATERIALS USED AS GROUTS

Grouts generally consist of suspensions, emulsions or solutions, or mixtures of these.

SUSPENSIONS

The majority of grouts are suspensions of cement in water. If the voids to be filled are large then sand is also used to reduce the cost and the shrinkage on setting. Fly-ash is now sometimes used when available cheaply. Bentonite, or bentonitic clays, which form a thixotropic gel, are now commonly used, either in conjunction with cement, or alone, in which case they have a low strength and are used only for reduction of permeability.

The particles of cement and clay will not penetrate into the voids of a soil finer than fine gravel or possibly coarse sand, but in a mixed soil the grouting is often accompanied by compression and compaction of the soil due to the penetration of the grout along the coarser layers since most soils of this type are stratified to some extent.

Another group of grouts comprises suspensions in solutions, e.g. in a solution of sodium silicate. An example would be a combined cement–bentonite–silicate grout.

EMULSIONS

Bituminous emulsions can be used as grouts. When used alone, as in the Shell-Perm process, they can be injected into soils down to the fine sand range. They reduce the permeability considerably and are therefore used to form cut-offs under dams and barrages; but they add practically nothing to the strength of the soil and are therefore of no use in underpinning problems.

Bituminous emulsions can also be used mixed with sand for filling large voids under pavement slabs and floors. In this case cut-back emulsions are used which coagulate and set, developing considerable strength.

SOLUTIONS

There are several grouting processes in which solutions of chemicals based on sodium silicate are injected. The chemical processes can be used down to the fine sand range. They divide into the two-solution and the single-solution processes.

In the two-solution processes (Joosten and Guttman processes) the first chemical injected is sodium silicate, and this is followed immediately by calcium chloride or some such salt. The reaction is almost immediate and for this reason the solutions cannot penetrate far from the injection pipes which are therefore spaced at about 2 ft (61 cm) centres. The process gives considerable strength to the soil and also reduces the permeability to a very small fraction of its previous value.

In the single-solution processes two chemicals are mixed before injection, possibly with a third chemical to delay the setting action for some time. The injection pipes can therefore be at greater centres. The processes reduce the permeability but do not give strengths comparable to the two-solution process.

Recent developments include grouts having acrylic, formaldehyde, lignin, and epoxide bases, etc. These grouts have low viscosities and therefore considerable penetration power, and achieve comparatively high strengths. They are, however, more expensive than the common grouts (see Dempsey and Moller[101]).

METHOD OF INJECTION

In nearly all grouting work the injections are made by drilling or driving pipes into the ground and pumping the grout in under pressure through hoses attached to the pipes.

The spacing of the pipes varies widely with the process and the conditions from 2 ft (0.61 m) for the two-solution chemical process in sand, to about 10 ft (3.05 m) for clay injections in alluvium, and up to 20 ft (6.10 m) or more for cement grouts in fissured rocks.

When filling fissures in rock with cement grout it is usual to use piston pumps which will give a pressure up to 7500 kN/m^2. The same pumps can be used for clay injections with alluvial sands and gravels but the pressures must be carefully controlled in relation to the depth and nature of the overburden to avoid undue lifting of the ground surface. If the limitation of ground heaving is important, suitable instrumentation for the monitoring of heave may be necessary. In the Joosten and Guttman two-solution chemical processes the amount of grout required to fill the voids in the soil between injection pipes is pumped in with less regard to the pressure, subject to a maximum pressure of about 3000 kN/m^2. Piston pumps are used.

For very simple grouting jobs a grout pan may be used. The cement grout is mixed in the pan by a paddle driven by hand or by a compressed air motor, an air pressure up to 750 kN/m^2 is then applied to the surface of the grout in the pan and this drives the grout through the hose into the injection pipe and so into the ground. This suffers the limitation that the injection pressure cannot easily be varied to suit the ground conditions.

For complex soils, sleeve grouting is frequently used. The system comprises a PVC tube of about 30 mm bore which is installed into a borehole of about 90 mm diameter and sealed into the ground with a relatively weak bentonite-cement sleeve grout. The tube is equipped at short intervals, normally 300 mm, with rubber sleeves covering perforations. An injection device is located against selected perforations in turn between upper and lower packers. The grout lifts the sleeve, fractures the sleeve grout and enters the ground. With this device it is possible to return to any position and regrout.

In general grouting work is not simple and damage can be caused by the indiscriminate use of high pressures by inexperienced operators. The work should be planned and carried out by engineers and operators experienced in the use of grouting methods (see Dempsey and Moller[101]).

Drainage and water lowering

Ordinary methods of drainage, although correctly classed as a geotechnical process, are so well known that they need not be considered here.

Water lowering can be used in sands and gravels to allow an excavation to be carried out in the dry below the water table, or to reduce the water pressure either on the side or below the bottom of an excavation. It is not normally of use in silts owing to the long time required to drain a silt. It differs from pumping from sumps in that the flow of water is away from the excavation, thus increasing the stability and minimising the danger of 'blows'.

Continuous pumping is required during the lowering period. The spacing of the wells and capacity of the pumps is theoretically calculable provided that the permeability of the ground is known (see Glossop and Collingridge[102] and Cashman and Haws[103]) but a great deal of experience and judgment is also called for. In practice, provision should always be made for more wells than are theoretically needed, and these should be installed for use in the event of a pump failure.

PERMEABILITY AND FILTERS

The determination of the permeability of the ground is part of the site investigation without which no water lowering project should ever be considered (see Section 10). The permeability (transmissibility divided by the depth of the aquifer) can be determined by pumping from a large-diameter well while monitoring the draw-down in a number of adjacent observation wells. The shape of the draw-down/time curve is matched to type curves derived theoretically, and the transmissibility and storage coefficients so deduced. A wide range of available type curves enables varying aquifiers and hydrologic boundary conditions to be considered; the most important of these are summarised by Walton.[104] Alternatively, for a fully penetrating well into an unconfined aquifer, permeability can be estimated from the equilibrium draw-down/distance curve, using equation (72) below. This latter method, which is the older of the two described, requires the establishment of equilibrium conditions, which may take many days; in contrast the time-variant method can be applied within a matter of hours after commencement of pumping.

Pumping tests give the most reliable value for k but an order of permeability can be obtained from grading curves (see Loudon[10]). Undisturbed samples of fine sands are essential in order to see if the material is laminated – a fact which naturally can have a marked effect on the horizontal permeability.

Grading curves are also used to choose suitable sand as a filter medium using Terzaghi's empirical rule which states that the grading curve for the filter material should be the same shape as that for the material to be filtered, and the point at which it crosses the 15% line should lie between four times the 15% size and four times the 85% size for the latter (see Figure 8.48).

PUMPING CAPACITY

The quantity of water to be pumped from a fully penetrating well into an unconfined aquifer can be calculated from the equation

$$Q = \frac{\pi k(H^2 - h^2)}{\log_e R/A} \quad (\text{m}^3/\text{s}) \qquad (72)$$

where k is the permeability (m/s), H the depth from normal water level to the impermeable stratum (m), h the depth from lowered water level to the impermeable stratum (m), R the radius of cone of depression (m) and A the radius of circle of area equal to area surrounded by wells (m). R can be obtained from the empirical relation

$$R = 30(H-h)\sqrt{k} \qquad \text{(m)} \tag{73}$$

R is not critical, and can be assumed where equation (73) gives a value which experience indicates to be unrealistic.

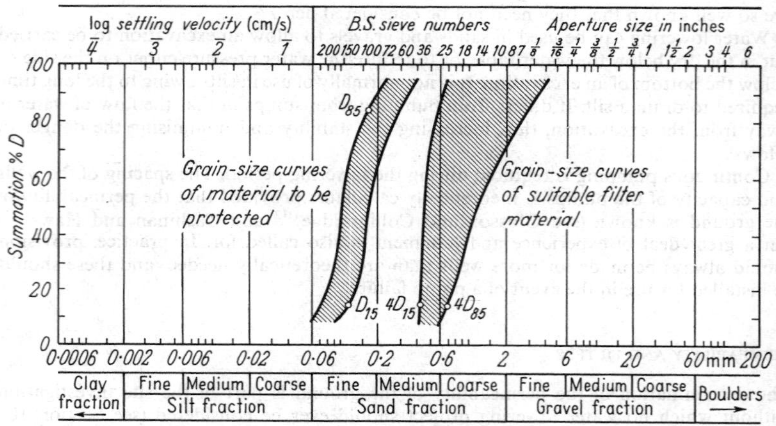

Figure 8.48 Diagram illustrating Terzaghi's rule for filter materials. Left-hand zone encloses particle-size distribution curves of material to be protected; right-hand zone encloses particle size distribution curves of filter material

The number of wells required can be obtained from the empirical relationship

$$Q = 3.63\times10^{-6} r_0 h_0 n k^{\frac{1}{2}} \qquad (\text{m}^3/\text{s}) \tag{74}$$

where r_0 is the radius of a well (m), h_0 the water level outside a well (m), k the permeability (m/s) and, n the number of wells required.

The expression $2\pi r_0 h_0 n$ is the immersed area required. h_0 is generally fixed by the geometry of the problem, but r_0 and n can be varied together at will (see Glossop and Collingridge[102]).

WELL-POINTS, SUCTION WELLS AND DEEP WELLS

Three systems of water lowering are in common use and each has certain advantages and disadvantages.

Well points When the lowering of the water level required is 4.5 m or less a well-point system can be used. If the system operates very efficiently, greater lowering can be obtained, but this should not be assumed at the planning stage.

A well point is a metal tube of about 50 mm diameter carrying a gauze filter about one metre long at its lower end. New types of well points are now on the market with slotted plastic outer covering and metal centre tubes. The well points are jetted into the ground at intervals of one to two metres and then connected to a header main through which the water is extracted by a well-point pump for exhausting the main and well points, and a centrifugal pump to remove the water.

Well points are cheap to install and for a long progressive excavation such as a pipe or sewer trench they are usually the most economical system. If greater lowering than 4.5 m is required a two-stage system can be used.

A special trenching machine is now available for laying a horizontal porous pipe of 100 mm diamter at a depth of up to 5 m below ground level. A pipe up to a maximum length of about 230 m can be installed, and for continuous trench work the pipes are overlapped by about 4.5 m.

Shallow wells These are in principle the same as well points but the wells are bored into the ground. The wells are usually about 600 mm diameter with a 300 mm filter tube and a 75 mm to 100 mm diameter suction pipe. The space outside the 300 mm tube is filled with a gravel filter as the boring tube is withdrawn. Because of their greater diameter the wells can usually be spaced about 10 to 15 m apart, and since there are many fewer connections than in a well-point system the efficiency is greater. The wells are connected to a ring main and a well-point pump.

The cost of pumping is much the same for a given lowering of the water level from either a well-point or a shallow well system, but the cost of installation of the shallow wells is higher. The shallow well installation is often to be preferred for an excavation of rectangular shape (as opposed to a long trench) where pumping must continue for many months, in fine sands where a graded filter is necessary or in laminated soils where a definite vertical connection between the aquifers is required.

As in the case of well-points two-stage systems can be used for a lowering of more than 4.5 m but in general deep wells will prove cheaper.

Deep wells With the deep-well system any required lowering of the water level can be achieved in one stage. This is because the pumps are placed at the bottom of the wells and deliver the water against pressure; there is no suction. The pumps used are electrically driven submersible pumps. They require a 200 mm to 400 mm diameter filter tube to accommodate them and the ring main is 150 mm to 300 mm diameter.

The well cannot be less than 450 mm diameter, which allows a 75 mm thickness of filter gravel, and if a two-stage filter is desired the well will be 600 mm diameter. For this reason the capacity of the wells is much greater than in the case of shallow wells and they can therefore be spaced further apart, in general up to 30 m, but this will vary considerably on different installations.

The cost of a deep-well system is high but it generally gives safe dry excavation and can often reduce the time of construction considerably. For safety, two independent sources of electric power must be provided for the pumps, since once they have started pumping it might be disastrous if they failed.

VACUUM DRAINAGE

In soils of low permeability such as coarse silts, drainage can sometimes be effected by sealing the wells or well points and exhausting the air from them. The pressure of the atmosphere then acts as a surcharge on the soil causing it to consolidate, and water is squeezed out of the soil into the filters of the wells. The amount of water removed is very small but the increase in strength of the silt is marked, and excavation into it is greatly facilitated (see Terzaghi and Peck[11]).

ELECTRO-OSMOSIS

Electro-osmosis is a further drainage process which can be used in silts. It is based on the principle that if a direct electric current is passed through the soil a flow of water takes place from anode to cathode. The cathode is made into a well and the water which reaches it is pumped out; the amount of water removed is small. As with the vacuum method the success of the method depends on the fact that the flow of water is away from the excavation, the free water surface is lowered and the water which remains in the soil above this surface is in tension, and that the capillary tensions add greatly to the strength of a silt. To some extent also the water content of the soil is reduced, thus resulting in an increased strength. Electro-osmosis is an expensive process and should only be considered if more normal methods of construction are inapplicable. In many silts the vacuum method of drainage is probably nearly as effective and much cheaper. (See Casagrande.[105, 106])

SETTLEMENTS CAUSED BY WATER LOWERING

When the water level is lowered the effective weight of the soil between the original and the lowered water levels is increased because the buoyancy effect has been removed. Where the soil concerned is sand and gravel any settlements due to this increase in weight are normally small, but where silt, clay or peat occurs in the zone referred to, settlement will occur with time owing to the consolidation of this material under its own increased weight.

Before installing a ground-water lowering system it is essential therefore to consider what effect such settlements may have on structures within the zone of influence. Important structures will probably be founded below the compressible material, either directly or on piles, and will be unaffected. For structures founded on the compressible strata it is necessary to calculate the probable settlement and to estimate what damage, if any, to the structure would result.

In order to limit the radius of influence of a ground-water lowering system, some of the pumped water can be 'recharged' or fed back into the aquifer by means of infiltration wells sited close to the structure below which it is desired to limit the potential settlement (Cashman and Haws[103]).

SAND DRAINS

Sand drains are used to accelerate the settlement of layers of soft clay or silt under applied loads. In many cases settlement can be tolerated provided it occurs quickly, preferably during the construction period, e.g. road embankments on soft clay.

The rate at which a uniform thin clay layer consolidates is inversely proportional to the square of the drainage path, which is etiher the thickness or half the thickness of the layer depending on the drainage conditions. The principle of the sand drain method is to provide vertical drains of sand in the clay. The drainage path is then reduced to half the spacing of the drains. When the load, for example a fill, is applied, the settlements take place quickly, in say a few months instead of many months or even years. It is important to note that sand drains do not reduce the amount of settlement.

Sand drains are particularly effective in deposits where the horizontal permeability is high compared with the vertical permeability. However, care must be taken to avoid local reduction of horizontal permeability by the process of installation of the drain. In deposits of exceptionally high lateral permeability, sand drains may not be necessary. It is important therefore to investigate the horizontal drainage characteristics with great care. (Rowe;[88] see also the subsections on the consolidation test, p. 8–13, and on time-settlement, p. 8–56.) For a discussion on the design and effectiveness of sand drains, see Richart[107] and Casagrande and Poulos.[108]

The consolidation of the clay under load increases its strength, a fact which can sometimes be made use of by construction, in stages, of a fill which would cause foundation failure if placed in one operation.

Such drains are usually 250 mm to 500 mm diameter and spaced at anything between 2 and 10 m centres depending on the conditions. They are particularly effective in laminated soils, where the horizontal permeability is very large in comparison with the vertical permeability. They are constructed by making a vertical hole in the clay and filling it with sand. Methods of installation of the drains which cause remoulding of sensitive soil should be avoided. A horizontal drainage blanket must be provided at ground level to link the drains together before the fill is placed, unless the fill itself is permeable.

Compressed air

The use of compressed air is well known in underground work. It again is a temporary process used during construction only. It can be used in sands and gravels, silts and clays. The air pressure, acting on the surface of the soil in the excavation, or more correctly on the water surfaces in the voids of the soil, prevents the flow of water through the soil and acts as a support. Theoretically the air pressure must be equal to the water pressure. In practice a pressure somewhat lower than the theoretical is often satisfactory. In gravels the losses of air through the gravel may be serious and these can sometimes be cut down by injections of clay suspensions into the gravel before commencing excavation in order to reduce the permeability. There are short- and long-term hazards to health arising from working in compressed air (Catton,[109], Haxton[110]).

Freezing

Another temporary method of preventing the access of ground water to excavations is by freezing the water. This is normally done by boring vertical holes into the ground, installing pipes in them and circulating brine, cooled to below the freezing point of water, through the pipes. The freezing process is expensive and slow but once the water is frozen excavation can safely take place inside the frozen ring. The freezing must, of course, be continued until the permanent work is completed. One of the disadvantages of brine is that if a leak occurs in the pipes it will escape into the ground water and it may then prove impossible to freeze it. To overcome this objection the Dehottay process was introduced. In this process liquid carbon dioxide is circulated instead of brine; more recently, liquid nitrogen has been used. In general the freezing process is applied to narrow, deep excavations such as mine shafts, but cases are on record of its use in foundation work, see Daxhelhofer[111] and Mussche and Waddington.[112]

In situ concrete walls

Reinforced concrete diaphragm walls can be constructed to considerable depths by placing concrete by tremie in a narrow trench supported by bentonite mud. The trench is excavated in short panels, 2 to 7 m in length, using special cutters with circulating mud, or grabs and stationary mud. Junctions between panels are formed by means of steel tubes acting as end shutters. The mud displaced by the rising concrete is used in subsequent panels provided it is still in good condition. (See Littlejohn, Jack and Sliwinski.[113]) It is possible to construct such walls in all types of ground, though difficulties have been known to occur when concreting in very soft alluvium owing to displacement of the clay under the high head of liquid concrete. In very permeable ground, significant mud losses may occur.

Walls consisting of soldiers of bored piles can also be made with or without the use of mud depending upon the ability of the ground to support itself. The spacing can be varied and in the case of contiguous bored pile walls each pile is bored slightly into the completed adjacent pile while the concrete is still green.

Ground anchors

Diaphragm and pile walls are thin and generally require support which nowadays is provided by anchors, rather than strutting and bracing, as this facilitates excavation. However, for narrow excavations, strutting is often cheaper. With modern boring and injection techniques it is possible to install anchors at reasonably flat angles in all manner of soils.

Rock anchors and bolts have been in regular use since 1934 and their installation and design presents little difficulty.

The method comprises boring a hole using augers in clay and rotary percussion with water flushing and casing support in granular soils. Bar or strand is inserted into the hole and the predetermined anchor length grouted up with neat cement cement under pressure, the free end of the bar being sleeved off. The anchor can be stressed to loads in excess of the working load if necessary to test its capacity. When pulled to failure, special test anchors are installed.

In clays the design procedure is somewhat similar to that for bored piles. In sands a semi-empirical approach is used based upon the density and grain size of the sand, the overburden pressure and the injection pressure. (See Littlejohn.[114])

REFERENCES

1. *Methods of test for soil classification and compaction*, BS 1377, British Standards Institution, London (1948)
2. GARDNER, R. P., *et al.*, 'Optimization of density and moisture content measurements by nuclear methods', Highway Research Board, Nat. Cooperative Highway Res. Programme Rep. 125
3. LEFEVRE, E. W., and MANKE, P. B., 'A tentative calibration procedure for nuclear depth/moisture/density gauges', Highway Res. Board Record No. 248 (1968)
4. AKROYD, T. N. W., 'Laboratory testing in soil engineering', Soil Mechanics Ltd, Bracknell, Berks (1954)
5. SKEMPTON, A. W., "The colloidal 'activity' of clays", *Proc. 3rd Int. Conf. on SM & FE,* Vol. 1, 57, Zurich (1953)
6. KOLBUSZEWSKI, J. J., 'An experimental study of the maximum and minimum porosities of sands', *Proc. 2nd Int. Conf. on SM & FE,* Vol. 1, 158, Rotterdam (1948)
7. TAYLOR, D. W., *Fundamentals of soil mechanics*, NEW YORK (1948)
8. SKEMPTON, A. W., 'The pore-pressure coefficients A and B', *Géotechnique*, **4,** 143 (1954)
9. BISHOP, A. W., and HENKEL, D. J., *The measurement of soil properties in the triaxial test*, Edward Arnold, London (1957)
10. LOUDON, A. G., 'The computation of permeability from simple soil tests', *Géotechnique*, **3,** 165 (1952)
11. TERZAGHI, K., and PECK, R. B., *Soil mechanics in engineering practice*, Wiley, New York (1948)
12. BARDEN, L., and MCDERMOTT, R. J. W., 'Use of free ends in triaxial testing of clays', *ASCE, JSM & FE Div.,* **91,** (SM6), 1–23 (1965)
13. COOLING, L. F., and GOLDER, H. Q., 'A portable apparatus for compression tests on clays soils', *Engineering, Lond.*, **149,** 57 (1940)
14. SKEMPTON, A. W., 'Long term stability of clay slopes', *Géotechnique*, **14,** 77 (1964)
15. BURLAND, J. B., BUTLER, F. G., and DUNICAN, P., 'The behaviour and design of large diameter bored piles in stiff clay', *Prof. Conf. on Large Bored Piles*, Institution of Civil Engineers, London (1966)
16. MENARD, L., 'The determination of the bearing capacity and settlement of foundations from pressuremeter tests', *Proc. 6th Int. Conf. on SM & FE, Montreal,* **II,** 295 (1965)
17. MEIGH, A. C., and CORBETT, B. O., 'A comparison of in-situ measurements in a soft clay with laboratory tests and the settlement of oil tanks', *Proc. Conf. on In-situ investigations in Soils and Rocks*, B.G.S., London (1970)
18. ROWE, P. W., and BARDEN, L., 'A new consolidation cell', *Géotechnique*, **16,** 162–170 (1966)
19. ROWE, P. S., 'The relevance of soil fabric to site investigation practice, *Géotechnique*, **22,** 193–300 (1972)

20. BJERRUM, L., 'Engineering geology of Norwegian normally consolidated marine clays as related to the settlement of buildings', *Géotechnique*, **17**, 81–118 (1967)
21. *Soil Mechanics for Road Engineers*, Road Research Laboratory, HMSO (1968)
22. US Corps of Engineers Engineering Manual, part 12 chap. 3 (1946)
23. TERZAGHI, K., 'Evaluation of coefficients of subgrade reaction', *Géotechnique*, **5**, 297 (1955)
24. COULOMB, C. A., 'Essai sur une application des règles de maximis et minimis à quelques problèmes de statique relatifs á l'architecture', *Mem. Div. Sav. Acad. des Sciences*, **7**, 343 (1776)
25. FRANCAIS, 'Recherches sur la poussée des terres, sur la forme et les dimensions des revêtements et sur les Talus d'excavations', *Mem. de l'Officier due Génie*, **4**, 157 (1820)
26. BISHOP, A. W., 'The use of the slip circle in the stability analysis of slopes', *Géotechnique*, **5**, 7 (1955)
27. BISHOP, A. W., and MORGENSTERN, N. R., 'Stability coefficients for earth slopes', *Géotechnique*, **10**, 129–150 (1960)
28. SHERARD, J. L., WOODWARD, R. J., GIZIENSKI, S. F., and CLEVENGER, W. A., *Earth-rock dams*, Wiley, New York (1965)
29. MORGENSTERN, N. R., 'The analysis of the stability of general slip surfaces', *Géotechnique*, **15**, 79–93 (1965)
30. SKEMPTON, A. W., 'The Bradwell slip: a short term failure in London clay', *Géotechnique*, **15**, 221–242 (1965)
31. SKEMPTON, A. W., and HUTCHINSON, J. N., 'The stability of natural slopes and embankment foundations, State of the Art Report', *7th Int. Conf. on SM & FE, Mexico* (1969)
32. MORGENSTERN, N. R., 'Shear strength of stiff clay. General Report', *Géotechnique, Conf. Oslo* (1967)
33. SKEMPTON, A. W., and NORTHEY, R. D., 'The sensitivity of clays', *Géotechnique*, **3**, 30 (1952)
34. GREEN, P. A., and FERGUSON, P. A. S., 'On liquefaction phenomena', by Prof. A. Casagrande: Report of Lecture. *Géotechnique*, **21**, 197 (1971)
35. BJERRUM, L., 'Engineering properties of normally consolidated clays', Lecture at King's College, London (1972)
36. REMSON, I., HORNBERGER, G. M., and MOLZ, F. J., *Numerical methods in subsurface hydrology*, Wiley-Interscience, London (1971)
37. KARPLUS, W. J., *Analog simulation*, McGraw-Hill, London (1958)
38. *Earth Retaining Structures*, Civil Engineering Code of Practice No. 2
39. ROWE, P. W., and PEAKER, K., 'Passive earth pressure measurements', *Géotechnique*, **15**, 57–78 (1965)
40. COULOMB, C. A., 'Essai sur une application des règles de maximis et minimis à des revêtements et sur les Talus d'excavations', *Mem. de l'Officier du Génie des Sciences*, **7**, 343 (1773)
41. GERBER, E., 'Untersuchungen uber die Druckverteilung im ortlich belasteten Sand', Zurich (1929)
42. SPANGLER, M. G., 'Horizontal pressures on retaining walls due to concentrated surface loads', *Iowa Eng. Expt. Stn. Bull.*, No. 140
43. PECK, R. B., 'Deep excavations and tunnelling in soft ground. State of the Art Report', *7th Int. Conf. on SM & FE Mexico* (1969)
44. TOMLINSON, M. J., 'Lateral support of deep excavations', *Proc. Conf. on Ground Engineering*, Institution of Civil Engineers, 55 (1970)
45. SKEMPTON, A. W., and WARD, W. H., 'Investigations concerning a deep cofferdam in the Thames Estuary Clay at Shellhaven', *Géotechnique*, **3**, 119 (1952)
46. WARD, W. H., 'Experiences with some sheet-pile cofferdams at Tilbury', *Géotechnique*, **5**, 327 (1955)
47. LITTLEJOHN, G. S., 'Soil anchors', *Proc. Symp. on Ground Engineering*, ICE, London (1970)
48. ROWE, P. W., 'Anchored sheet pile walls', *J. Instn civ. Engrs*, **1**, (1) 27 (1952)
49. TERZAGHI, K., 'Anchored bulkheads', *Proc. Am. Soc. civ. Engrs*, **79**, 262 (1953)
50. ROWE, P. W., 'Sheet-pile walls in clay', *J. Instn civ. Engrs*, **7**, 629 (1959)
51. SKEMPTON, A. W., and MACDONALD, D. H., 'The allowable settlements of buildings', J. Instn civ. Engrs, **5**, Pt. III, (3) 727 (1956)
52. *Foundations*, CP 2004, British Standards Institution, London (1972)
53. SKEMPTON, A. W., *Bearing capacity of clays*, Building Research Congress (1951)
54. DEERE, D. U., HENDRON, A. J., PATTON, F. D., and CORDING, E. J., 'Design of surface and near surface construction in rock', *Proc. 8th Conf. on Failure and breakage of rock' Minnesota* (1968)
55. HOBBS, N. B., and DIXON, J. C., 'In-situ testing for bridge foundations in the Devonian Marl', *Proc. Conf. on In-situ investigations in Soils and Rocks*, B.G.S. London (1970)
56. MEIGH, A. C., SKIPP, B. O., and HOBBS, N. B., 'Field and laboratory creep tests on weak rocks', *Proc. 8th Int. Conf. on SM & FE, Moscow* (1973)
57. TERZAGHI, K., *Theoretical soil mechanics*, Chapman Hall, London (1943)

58. PECK, R. B., HANSON, W. E., and THORBURN, T. H., *Foundation Engineering,* Wiley, New York (1953)
59. MEYERHOF, G. G., 'Shallow foundations', *Proc. ASCE,* **91,** (1965)
60. MEYERHOF, G. G., 'Penetration tests and bearing capacity of cohesionless soils', *Proc. ASCE,* **82** (SM1) (1956)
61. DE BEER, E., 'Settlement records on bridges founded in sand', *Proc. 2nd Int. Conf., on SM & FE,* **2,** 111 (1948)
62. MEIGH, A. C., and NIXON, I. K., 'Comparison of in-situ tests for granular soils', *Proc. 5th Int. Conf. on SM & FE* **I,** Paris (1962)
63. RODIN, S., 'Experiences with penetrometers with particular reference to the Standard Penetration Test', *Proc. 5th Int. Conf. on SM & FE.* **I,** Paris (1961)
64. SIMONS, N. E., 'Prediction of settlements of structures on granular soil', *Ground Eng,* **5,** (1) (Jan. 1972)
65. SCHMERTMANN, J. H., 'Static cone to compute settlement over sand', *Proc. ASCE J.S.M. & F. Div.,* **96,** 1011–1043 (1970)
66. D'APPOLONIA, D. J., D'APPOLONIA, E., and BRISSETTE, R. F., 'Settlement of spread footings on sand', *Proc. ASCE J.S.M. & F. Div.,* **54** (SM3), 735–760 (1968)
67. MEYERHOF, G. G., 'The bearing capacity of foundations', *Géotechnique,* **2,** 301 (1951)
68. KERISEL, J., 'Fondations profondes en milieu sableux', *Proc. 5th Int. Conf. on SM & FE,* **II,** Paris, 73 (1961)
69. VESIC, A., *A study of the bearing capacity of deep foundations,* Georgia Inst. of Technology (1967)
70. BEREZANTSEV, V. G., 'Load bearing capacity and deformation of piled foundations', *Proc. 5th Int. Conf. on SM & FE,* **II,** Paris, 11 (1961)
71. TOMLINSON, M. J., *Foundation design and construction,* Pitman (1970)
72. PRANDTL, L., 'Ueber die Härte plastischer Körper', *Nachr. Kgl. Gesell. Wiss, Göttingen, Math-phys, Klasse,* 74 (1920)
73. FELLENIUS, W., 'Jordstatiska Beräkningar för vertikal Belastning pá horisontal Mark under antagande av cirkulär-cylindriska Glidytor', *Tekn. Tidskr,* **59,** 57 (1929)
74. WILSON, G., 'The calculation of the bearing capacity of footings on clay', *J. Instn civ. Engrs,* **17**, 87 (1942)
75. SKEMPTON, A. W., 'The bearing capacity of clays', Building Research Congress Div. I, part III, 180 (1951)
76. STEINBRENNER, W., 'Tafeln sur Setsungsberechnung', *Die Strasse,* **1,** 121 (1934)
77. BOUSSINESQ, J., *Application des potentiels à l'étude de l'équilibre et du mouvement des solides elastiques,* Paris (1885)
78. NEWMARK, N. M., 'Simplified computation of vertical pressures in elastic foundations', *Circ. Univ.* III, *Eng. Expt. Stn. Bull.* No. 24 (1935)
79. JURGENSON, L., 'The application of theories of elasticity and plasticity to foundation problems', *J. Boston Soc. civ. Engrs,* **21, 206** (1934)
80. LITTLE, A. L., *Foundations,* Arnold, London (1961)
81. FOX, E. N., 'The mean elastic settlement of a uniformly loaded area at a depth below the ground surface', *Proc. 2nd Int. Conf. on SM & FE,* **1,** 192 (1948)
82. SKEMPTON, A. W., and BJERRUM, L. A., 'A contribution to the settlement analysis of foundations on clay', *Geotechnique,* **7,** 168 (1957)
83. BURLAND, J., and HOBBS, N. B., 'Discussion on papers in Session A—Properties of Rocks', *Proc. In-situ investigation of Soils and Rocks,* Instituttion of Civil Engineers, London (1970)
84. SCOTT, R. F., *Principles of Soil Mechanics,* Addison-Wesley (1963)
85. DAVIS, E. H., and POULOS, H. G., 'The use of elastic theory for settlement predictions under three-dimensional conditions', *Géotechnique,* **18** (1) (1968)
86. ZIENKIEWICZ, O. C., and CHEUNG, Y. K., *The finite element method in structural and continuum mechanics,* McGraw-Hill, London (1970)
87. GIBSON, R. E., and LUMB, P., 'Numerical solution of some problems in the consolidation of clay', *J. Instn civ. Engrs,* **2** part I, 182 (1953)
88. ROWE, P. W., 'The influence of geological features of clay deposits on the design and performance of sand drains', *Proc. Instn civ. Engrs,* Supplementary Volume (1968)
89. SCHIFFMAN, R. L., and STEIN, J. R., 'A computer programme to calculate the progress of ground settlement', Univ. of Colorado, Rep. 69–9a (1969)
90. BJERRUM, L., 'Norwegian experience with steel piles to rock', *Géotechnique,* **7,** 73 (1957)
91. THORBURN, S., 'Large diameter piles founded on bedrock', *Proc. Conf. Large bored piles,* Institution of Civil Engineers, London (1966)
92. MEIGH, A. C., 'Foundation characteristics of the Upper Carboniferous Rocks', *Q. J. Engg Geol.,* **I** (2) (1968)

93. HUIZINGA, T. K., 'Application of results f deep penetration tests to foundation piles', *Building Res. Congress Div.* 1, Pt III, 173 (1951)
94. BEGEMANN, H. K. S., 'The Dutch static penetration test with the adhesion jacket cone', *LGM Mededelingen Delft,* Deel XII, No 4 (1969)
95. SKEMPTON, A. W., 'Cast in-situ bored piles in London Clay', *Géotechnique,* **9,** (1959)
96. WHITAKER, T., and COOKE, R. W., 'An investigation of the shaft and base resistance of large bored piles in London Clay', *Proc. Conf. on Large Bored Piles,* Institution of Civil Engineers, London (1966)
97. JOHANNESSEN, I. J., and BJERRUM, L., 'Measurement of the compression of a steel pile to rock due to settlement of the surrounding clay', *Proc. 6th Int. Conf. on SM & FE,* II, Montreal, 37 (1965)
98. GLOSSOP, R., 'Classification of Geotechnical Process', *Geotechnique,* **2,** 3 (1950)
99. 'Soil and Soil-Aggregate Stabilization', *Highway Research Board Bull.* 108 (1955)
100. GLOSSOP, R., and SKEMPTON, A. W., 'Particle-size in silts and sands', *J. Instn civ Engrs,* **25,** 81 (1950)
101. DEMPSEY, J. A., and MOLLER, K., 'Grouting in ground engineering', *Proc. Conf. on Ground Engineering,* Institution of Civil Engineers, London (1970)
102. GLOSSOP, R., and COLLINGRIDGE, V. H., 'Notes on groundwater lowering by means of filter wells', *Proc. 2nd Int. Conf. on Soil Mech.,* **2,** 320 (1948)
103. CASHMAN, P. M., and HAWS, E. T., 'Control of ground water by water lowering', *Proc. Conf. on Ground Engineering,* Institution of Civil Engineers, London (1970)
104. WALTON, W. C., 'Selected analytical methods for well and aquifer evaluation'. *Illinois State Water Survey Bull.* 49 (1962)
105. CASAGRANDE, L., 'Electro-osmosis', *Proc. 2nd Int. Conf. on SM & FE,* **1,** 218–223 (1948)
106. CASAGRANDE, L., 'Electro-osmosis in soils', *Géotechnique,* **1,** 159–177 (1949)
107. RICHART, F. E., 'Review of theories for sand drains', *Trans. ASCE,* **124,** (1959)
108. CASAGRANDE, L., and POULOS, S., 'On the effectiveness of sand drains', *Canadian Geotechnical Journal,* **6,** 287 (1969)
109. CATTON, M. J., 'Compressed air working', *Consulting Engineer* (Apr. 1968)
110. HAXTON, A. F., *Compressed air tunnelling—safety, health, welfare,* London Engineers Guild (1960)
111. DAXHELHOFER, J. P., 'Un nouveau procédé de congélation et ses possibilités d'application', *Erdbaukurs der Eidg. Techn. Hochschule Section* 17, Zürich (1938)
112. MUSSCHE, H. E., and WADDINGTON, J. C., 'Applications of the freezing process to Civil Engineering Works', ICE Works Construction Paper No. 5 (1946)
113. LITTLEJOHN, G. S., JACK, B., and SLIWINSKI, Z. 'Anchored diaphragm walls in sand', *Ground Engineering,* **4**(6), 18–21 (1971)
114. LITTLEJOHN, G. S., 'Soil anchors', *Proc. Conf. on Ground Engineering,* Institution of Civil Engineers, London (1970)

9 ROCK MECHANICS

ROCK MECHANICS 9

9 ROCK MECHANICS

J. A. FRANKLIN, B.Sc.(Eng.), M.Sc., Ph.D., C.Eng., M.I.C.E., A.M.I.M.M., F.G.S., D.I.C.
*Rock Mechanics Ltd**

INTRODUCTION

The scope of rock mechanics

Rock mechanics has evolved over the last thirty to forty years from the disciplines of mining and civil engineering, structural geology and engineering geology. An estimated 2000 papers are currently published each year on this and closely related subjects. Some of the relevant journals and textbooks are listed in the Bibliography together with abstract bulletins that relate to rock mechanics topics. The International Society for Rock Mechanics has established commissions to attempt the standardisation of terminology and test techniques, and to investigate various topics including rock mechanics teaching, research and rock classification.

The development of rock mechanics has received the majority of its impetus from major disasters occurring throughout the world, frequently involving considerable loss of life and damage to property.[1] These and more everyday occurrences emphasise how little has been understood of the mechanical behaviour of rock or of the means for controlling its behaviour in engineering works. However, considerable progress has recently been made. An increasing demand for large underground excavations particularly in hydroelectric works has led to improved techniques, for example resin grouted rockbolts, sprayed concrete and other support methods that allow excavation of large unobstructed spans. Blasting and excavating methods have been improved. Reliable instrumentation is now available to monitor and warn of rock movements in both surface and underground works. Design methods, particularly computation techniques, have evolved to the stage where rock conditions can be modelled with some degree of realism in order to predict rock behaviour.

The discontinuous rock mass

The mechanical behaviour of rock is controlled very largely by the presence and characteristics of discontinuities contained within it, for example joints, faults, and cleavage surfaces. A free-standing column of rock without such discontinuities, for example basalt with a uniaxial compressive strength of 70 MPa (1 Pa (pascal) = 1 N/m^2) and a specific gravity of 2.8, in theory would reach a height of over 2500 m before crushing occurred at the base of the column. In reality this would not happen. The rock mass is intersected by sets of fissures, typically spaced at 20–2000 mm, and behaves as a closely packed and interlocking stack of discreet 'rock blocks', each of which is relatively indestructible compared with the strength of the fissures that bound each block. These fissures also play a major role in determining water and stress environments in the mass. The mechanical behaviour of intensely fractured rock can sometimes be approximated to that of a soil. At the other extreme, for example in very deep underground works where the rock is massive and the fractures confined, the rock can some-

* Now a Rock Engineering Consultant.

times be considered as a continuous medium. More often, rock must be regarded as a discontinuum. The mechanical properties of discontinuities are therefore of considerable relevance. *Roughness, tightness* and *filling* in particular control the shear strength and deformability of fractures. Even a thin weathered layer in a joint can considerably reduce the strength afforded by tightly interlocking roughness asperities.[2] Discontinuities that *persist* smoothly and without interruption over extensive areas, for example certain types of fault and bedding plane,[3] offer considerably less resistance to shearing than discontinuities of irregular and interrupted pattern. The *orientation* of fractures relative to the exposed rock surface is also critical in determining rock mass stability. Techniques are available for rapid assessment of fracture orientations using stereophotogrammetric methods for presentation of fracture orientations in statistical and graphical forms and for processing this type of data in rock mechanics calculations.[4] Finally, *fracture spacing* is important, since it determines the size of rock blocks: the 'grain size' of the rock mass. Fracture spacing is relevant to problems of excavation stability and the design of bolted support systems, also in rock excavation and in determining its suitability for use as construction material.

The origin and nature of discontinuities are discussed at length in textbooks on structural geology.[5] Recent developments in rock mechanics have contributed to a better understanding of fault and fold formation, so that there is a continual feedback between geological and engineering aspects of the science.

Voids and water in rock

Intact rock material contains grains and intergranular pores filled with air and water. The relative volumes and weights of these three constituents determine porosity, density, saturation and various related parameters (Table 9.1). Void space is perhaps the most important rock constituent since it controls water flow and pressures,[6, 7] also

Table 9.1 POROSITY AND DENSITY DEFINITIONS AND INTERRELATIONSHIP FORMULAE

The components of a rock sample are defined as follows:

Grains: weight G_w, volume G_v
Pore water, weight W_w and volume W_v
Pore air: zero weight and volume A_v
Pores: volume $P_v = W_v + A_v$
Bulk sample: weight $B_w = G_w + W_w$
Bulk sample: volume $B_v = P_v + G_v$
Density of water: ρ_w = mass of water per unit volume

The physical properties—density, porosity, etc.—may be defined in terms of the above components as follows:

Water content	$w = (W_w/G_w) \times 100\%$	
Degree of saturation	$S_r = (W_v/P_v) \times 100\%$	
Porosity	$n = (P_v/B_v) \times 100\%$	
Void ratio	$e = P_v/G_v$	
Dry density of rock	$\rho_d = G_w/B_v$	(dry specific gravity $d_d = \rho_d/\rho_w$)
Density of rock	$\rho = (G_w + W_w)/B_v$	(specific gravity $d = \rho/\rho_w$)
Saturated density of rock	$\rho_s = (G_w + P_v\rho_w)B_v$	(saturated specific gravity $d_s = \rho_s/\rho_w$)
Grain density	$\rho_g = G_w/G_v$	(Grain specific gravity $d_g = \rho_g/\rho_w$)

Having defined the three properties, water content, porosity and dry density, the remaining properties may be calculated from the following interrelationships:

$$S_r = 100(w\rho_d)/n\rho_w$$
$$e = n/(100-n)$$
$$\rho = (1+w/100)\rho_d$$
$$\rho_g = \rho_d(1-n/100)$$

Table 9.2 PHYSICAL AND MECHANICAL PROPERTIES OF ROCK MATERIALS

Rock type	*Porosity*[a] (%)	*Dry bulk*[b] *density* Mg/m^3	*Lab. permea-bility*[c] m/s	*Uniaxial compressive strength*[d] MPa	*Uniaxial tensile strength*[d] MPa	*Young's modulus (static lab.)* ($\times 10^3$ MPa)	*Poissons ratio*[f] *(lab.)*	c(MPa) *(intact peak lab.)*[g]	tan ϕ *(intact peak lab.)*[g]	tan $\phi, c = 0$ *(smooth joints)*[h]	*P wave velocity*[i] *(lab.)* km/s	*S wave velocity (lab.)* km/s
Crystalline limestones, dolomites, marble, evaporites	0.5–5 (1)	2.5–3.0 (2.7)	10^{-8}–10^{-16} (10^{-11})	40–240 (70)	7–20 (10)	45–90 (70)	0.1–0.3 (0.2)	15–50 (35)	0.3–0.7 (0.5)	0.6–0.8 (0.7)	3.7–6.9 (6.0)	2.5–3.9 (3.6)
Porous tragmental limestones, dolomites	3–20 (10)	2.2–2.7 (2.4)	10^{-6}–10^{-12} (10^{-9})	30–350 (100)	5–25 (10)	15–105 (50)	0.1–0.3 (0.2)	7–60 (30)	0.2–1.0 (0.5)	0.4–0.6 (0.5)	1.7–6.5 (5.0)	2.6–3.7 (2.9)
Chalk	17–53 (35)	1.3–2.2 (1.7)	10^{-5}–10^{-9} (10^{-6})	0.2–7 (6)		0.1–12 (2)	0.05–0.15 (0.1)	0.1–2 (1)	0.2–1.0 (0.8)	0.5–0.7 (0.6)	2.0–3.0 (2.5)	1.0–1.5 (1.2)
Sandstones and siltstones	3–33 (15)	1.9–2.7 (2.2)	10^{-4}–10^{-13} (10^{-6})	7–300 (70)	4–25 (13)	3–80 (10)	0.02–0.13 (0.08)	1–50 (25)	0.3–1.1 (0.7)	0.5–0.8 (0.6)	1.5–5.8 (4.0)	1.0–3.9 (2.7)
Crystalline quartzite	0.1–5 (1)	2.5–2.8 (2.6)		120–600 (290)	10–30 (24)	55–105 (83)		30–50 (42)	0.9–1.4 (1.2)		3.1–6.0 (5.0)	
Indurated, high durability shales and	0.3–5 (1)	2.5–2.8 (2.7)	10^{-8}–10^{-15} (10^{-10})	100–200 (120)	7–20 (15)	20–90 (30)	0.02–0.20 (0.1)	1–60 (30)	0.3–0.9 (0.8)	0.6–1.1 (0.8)	3.0–5.5 (4.0)	1.7–3.1 (2.2)

gabbro and coarse grained igneous rocks	(1)	(2.7–2.9)	(10^{-11})	(200)	(30)	(50)	(0.1)	(60)	(1.1)	(0.6)	(5.2)	(3.2)
Basalts, rhyolites and fine grained igneous rocks	0.1–25 (0.5)	2.2–3.0 (2.6–2.9)		20–400 (250)	10–40 (20)	7–110 (40)	0.15–0.25 (0.17)	20–80 (50)	0.3–1.1 (0.7)	0.5–1.1 (0.7)	2.7–6.4 (5.0)	2.7–3.0 (3.0)
Schists	1–10 (2)	2.4–2.9 (2.7)		10–500 (50–300)		5–80 (20)	0.01–0.20 (0.07)	2–15 (5)	0.5–2.9 (1.3)	0.6–1.2 (1.0)	3.3–5.8 (4.3)	2.3–4.0 (3.0)

These tables are compiled from published test data and should be used only as an approximate guide. Each rock type, e.g. 'Limestone', encompasses a wide variety of materials with differing mechanical properties. The range of published values has been given, together with a value deemed 'typical' for the rock (in parentheses). 1 MPa = 1 MN/m^2 ≃ 145 lbf/in^2.

[a] Porosity is closely related to other properties such as strength. High porosities are associated with most sedimentary noncrystalline rocks, weathered igneous rocks and vesicular volcanics.

[b] Dry bulk density is related to both porosity and mineral specific gravities. Basic igneous rocks will usually have densities high in the stated range, due to an abundance of mafic (dense) minerals; in comparison the acid rocks are less dense.

[c] Only laboratory values of permeability are given. Field values are usually greater by a factor of 10^2–10^6 owing to the important influence of joints and fissures.

[d] The uniaxial and triaxial strengths of schists and other foliated rocks depend on the percent mica in the rock, and also on the orientation of load to the cleavage direction. Published data are scarce, so that a wide range of values has been quoted. Uniaxial compressive strength is given approximately by $24 \times I_s$, the point load strength of the rock.

[e] Field values of Young's modulus are usually less than laboratory values by a factor of 1.5–10 depending on the presence of open fissures, on the number of cycles and on the stress level.

[f] Quoted values for Poisson's ratio are unreliable owing to scarcity of published data. Poisson's ratio is also strongly dependent on stress level.

[g] Peak direct shear strength values obtained from laboratory triaxial tests.

[h] Peak direct shear strength of smooth joints: laboratory shear box tests. (Limited test data, so that values are not necessarily comparable with those in column g).

[i] Laboratory ultrasonic pulse velocity measurements. Sonic velocity values generally increase as the rock becomes saturated; ranges given encompass both dry and wet rocks. Field velocities are usually less than those measured in the laboratory, by a factor of 1.1–1.4, owing to the presence of open fissures in the rock mass.

rock strength[8] and deformability.[9] The porous texture of rocks has been studied with the help of the scanning electron microscope.[10] Igneous and metamorphic rock materials usually have a very low porosity with elongated pores at grain boundaries, although weathering which progresses outward from grain boundary cracks can result in porosities of 50% or more. Sedimentary rocks with the exception of crystalline and indurated types usually have a higher porosity (Table 9.2).

The rock mass contains fissures as well as pores. The volume occupied by fissures is usually much less than that of pores so that the porosity and bulk density of the rock mass and of intact material are usually similar. Fissures however, have a much greater influence on permeability because they afford a less tortuous path for water flow than does a network of pores. Louis[11] has demonstrated that joints separated by 1 m and open by only 10 μm result in an equivalent rock mass permeability of about 10^{-10} m/s. Usually it is possible to ignore the effect of intact material permeability in studies of water flow.

Rock stress

The state of stress existing at a point in the rock mass can be represented by three orthogonal principal stress components with magnitude and direction that generally

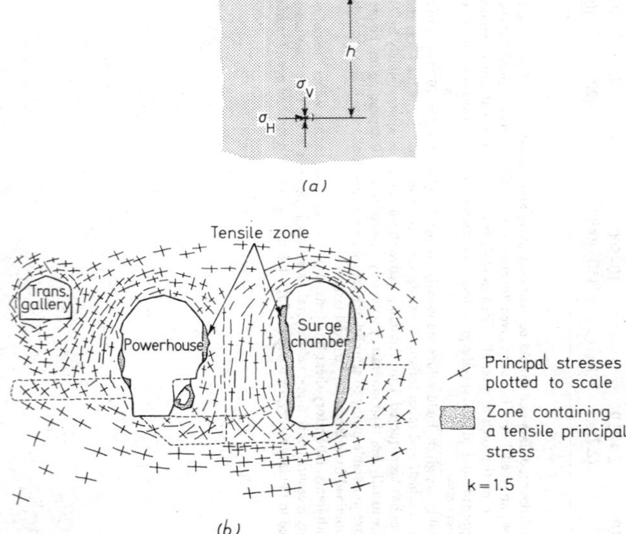

Figure 9.1 In situ stresses. (a) *Prior to excavation;* (b) *Stresses caused by excavation (from reference 19)*

vary from place to place. Experimental and mathematical techniques for analysing the stress 'field' are described in basic texts on continuum mechanics[12–18] and are briefly reviewed later in this Section.

In situ rock is under stress from the gravity load of overlying material and also from the tectonic forces responsible for crustal deformation. Gravitational stresses existing in a linear elastic rock mass with a flat topography are relatively easy to calculate (Figure 9.1(a)).

For a perfect fluid; $\sigma_v = \gamma h$, $\sigma_H = \gamma h$. Alternatively, for a perfectly elastic solid,

$$\sigma_v = \gamma h$$

$$\sigma_H = \frac{\mu}{1-\mu}\gamma h = K\gamma h$$

where h is the depth below horizontal ground surface, γ is the unit weight of rock and μ is Poisson's ratio for rock mass.

These virgin stresses may be very different if tectonic forces are acting, or if a substantial thickness of overburden has been removed by geological erosion.

Gravitational stresses in a real inelastic rock mass can also be computed if the material properties are known. Tectonic stresses, however, are much more difficult to predict, since the geological processes involved can not be evaluated with any precision. Ample evidence is available, however, to show that 'geological' stresses are of major importance, for example in California where sedimentary materials of recent age have been upturned and contorted by compressive forces acting perpendicular to the coastline. The earth's crust comprises old 'craton' blocks that are drifting, impacting and separating, and there is evidence to show the directions of stresses to be expected on this account.[20] On a smaller scale these geological stresses are evidenced by specimens that explode and by quarry floors that buckle and burst.

Erosion processes also have their effect on ground stress. Firstly they control the ground topography and hence the configuration of gravitational loading; the directions of principal stresses tend to follow the topographic surface and this can have a considerable influence on the design of tunnels and other excavations in rugged terrain.[21] Secondly, 'residual stresses' can remain after considerable thicknesses of overburden have been removed by erosion.

The above discussion relates to 'virgin stresses' that exist prior to any excavation. A man-made excavation will considerably modify the pre-existing virgin stress field (Figure 9.1(b)). Design requires a knowledge of the stresses prior to excavation in order to predict subsequent conditions. These stresses are difficult to predict and are usually measured *in situ* using methods described on page **9**–33. Measurement also involves difficulties and approximations but even limited information, such as the directions and relative rather than absolute magnitudes of principal stresses, can often be sufficient for design calculations.

ROCK TESTS AND PROPERTIES

Table 9.3 illustrates three roles of testing. Tests for calssification and characterisation of rock (index tests) are used for rock quality description and mapping and must be quick and cheap. Engineering design tests give data specifically required for design calculations and often may be complex and expensive. The third category, research tests, are required for academic investigations into rock properties and behaviour and fall beyond the scope of this Section. Tests should be selected to suit a specific and well defined requirement. Some of the more commonly used tests are shown in Figures 9.2–9.5.

Tests for density, porosity and water content

These laboratory tests[22] require measurement of the volumes and weights of rock constituents and of the bulk sample. Bulk volume can be measured directly from caliper

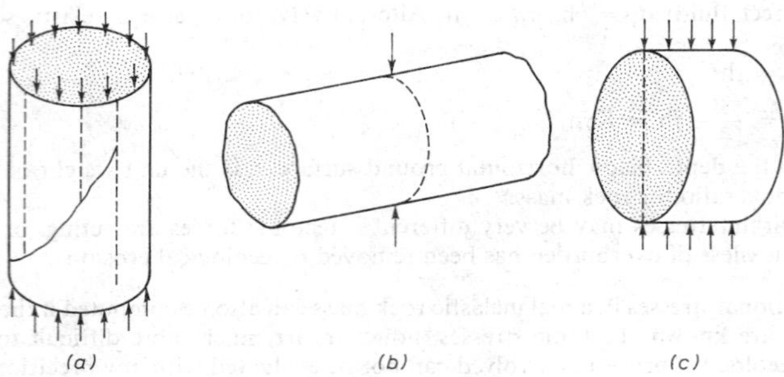

Figure 9.2 Strength index tests showing the loading configuration, specimen shape and failure pattern in the uniaxial compressive strength test (a) *the diametral point-load test* (b) *and the Brazilian test* (c)

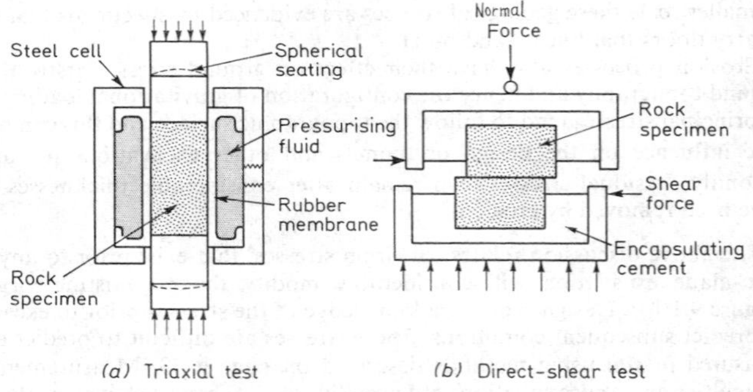

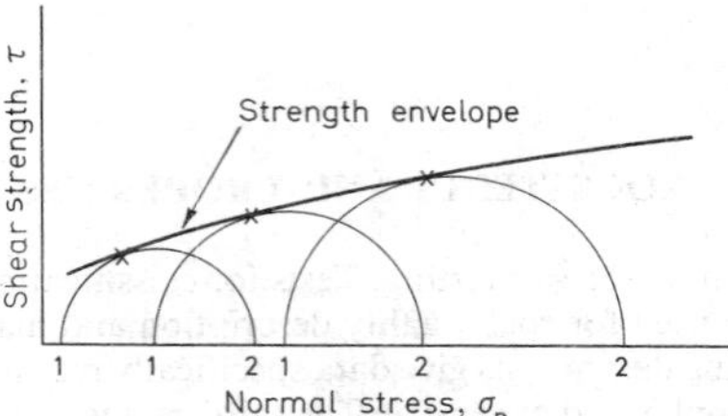

(c) Graphical presentation of strength test data
(Principal stresses 1 and 2 relate to failure conditions in three triaxial strength tests)

Figure 9.3 Triaxial and direct-shear strength tests in the laboratory. The triaxial test (a) *is most often used for intact rock, and the shear test* (b) *for determining the strength of joints or weakness planes. Results from direct-shear tests or triaxial tests may be plotted directly on the Mohr–Coulomb diagram* (c) *to obtain a strength envelope for the rock*

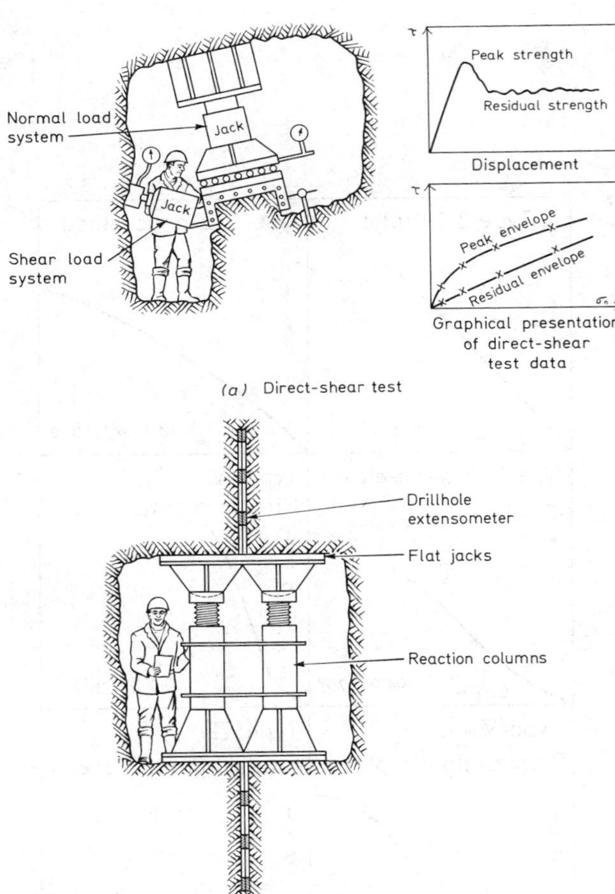

Figure 9.4 Field strength and deformability tests

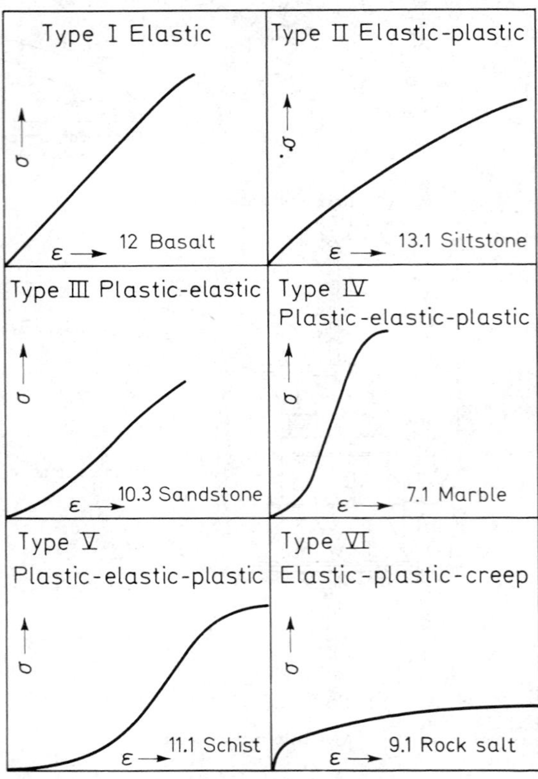

Type I Elastic
σ
ε
12 Basalt
Type II Elastic-plastic
σ
ε
13.1 Siltstone
Type III Plastic-elastic
σ
ε
10.3 Sandstone
Type IV
Plastic-elastic-plastic
σ
ε
7.1 Marble
Type V
Plastic-elastic-plastic
σ
ε
11.1 Schist
Type VI
Elastic-plastic-creep
σ
ε
9.1 Rock salt

(a)

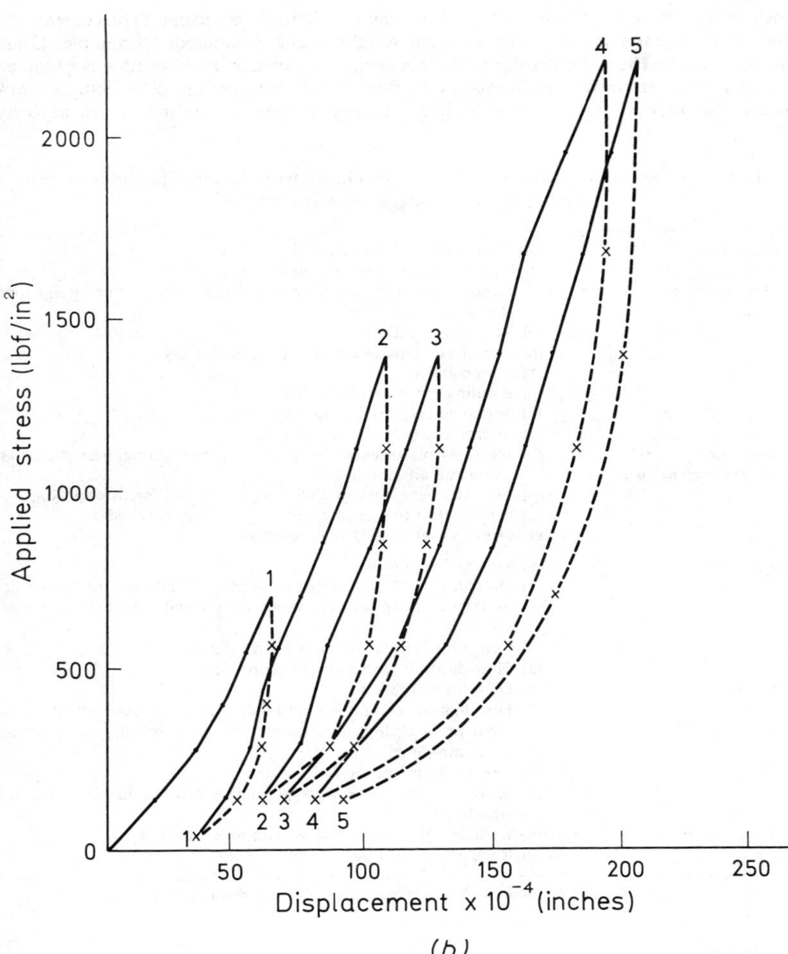

Figure 9.5 Stress–strain curves for laboratory and field deformability tests. (a) Laboratory stress–strain curves after Hendron[23] showing the influence of rock type on the shape of the curve. (b) Result of a field plate loading test on granulite rock at Monar Dam, Scotland, showing the effect of repeated cycling of the applied load. A deformability modulus must be selected to suit the problem, and depends on the stress level, number of cycles, and whether load is increasing or decreasing

measurements on specimens of regular geometry, alternatively using displacement of a fluid such as mercury or water, or from weights using Archimedes' principle. Grain volume can be found by crushing the specimen to a powder. Pore volume is obtained using a water saturation technique or in Boyle's law gas pressure cells. Simple *quick absorption methods* have been described[22] that give approximate indexes to porosity

Table 9.3 ROCK TESTING CATEGORIES (reproduced from International Society for Rock Mechanics suggested test methods[22])

Category I:	*Classification and characterisation*
Rock material (laboratory tests)	(1) Density, water content, porosity, absorption
	(2) Strength and deformability in uniaxial compression; point load strength
	(3) Anisotropy indices
	(4) Hardness, abrasiveness, attrition, drillability
	(5) Permeability
	(6) Swelling and slake-durability
	(7) Sound velocity; pulse and resonance
	(8) Micro-petrographic descriptions
Rock mass (field observations)	(9) Joint systems; orientation, spacing, openness, roughness, geometry, filling and alteration
	(10) Core recovery, rock quality designation and fracture spacing
	(11) Seismic tests for mapping and as a rock quality index
	(12) Geophysical logging of boreholes
Category II:	*Engineering design tests*
Laboratory	(1) Determination of strength envelope and elastic properties (triaxial, biaxial, uniaxial compression and tensile tests; direct-shear tests)
	(2) Strength of joints and planes of weakness
	(3) Time-dependent and plastic properties
In-situ	(4) Deformability tests
	(5) Direct-shear tests (intact material, joints, rock/concrete interface)
	(6) Field permeability, groundwater pressure and flow monitoring, water sampling
	(7) Stress determination
	(8) Monitoring of movements, support pressures, anchor loads and vibrations.
	(9) Uniaxial, biaxial and triaxial compressive strength
	(10) Rock anchor testing

Category III, not shown in the tabulation, comprises special 'research tests'.

and density and may be suitable for rock classification purposes. Microscopic methods may be necessary in order to examine the pore structure in detail.

Swelling, slaking and durability tests

Clay-bearing rocks (shales, mudstone and some weathered igneous rocks) can swell or disintegrate when exposed to atmospheric wetting and drying. *Swelling tests*[22] similar to soil consolidation tests can be used to measure swelling pressure or strain during wetting of the specimen, alternatively the swelling of an unconfined rock cube or cylinder can be measured as an index property. A quick *slake-durability index test* can be used to measure the disintegration of rock subjected to wetting and atmospheric weathering.[24]

Sodium-sulphate soundness and other salt crystallisation tests[25] subject rock specimens to the disruptive action of salt crystals growing from solution in the pore space, and are appropriate in assessing the durability of aggregates and building stones exposed to saline conditions. *Freezing and thawing tests*[26, 27] are available to measure susceptibility to frost damage.

Strength tests on intact rock

The *uniaxial* (*unconfined*) *compression test* is mainly used for strength classification.[22, 28] Rock cylinders are crushed by axial loading between steel platens to give compressive strength, defined as the ratio of failure stress to specimen cross-sectional area (Figure 9.2). The length/diameter ratio for specimens should be between 2.0 and 2.5 to avoid the confining effect due to platen friction, and the end faces of the cylinder should be machined flat and parallel. Alternative crushing tests on cubes have been used for classification of rock aggregates but are not recommended in other applications.

The *point-load strength test* is an alternative for strength classification.[22, 29] Rock in the form of either core or irregular lumps is tested by compressing the specimen between conical platens, and a strength index is obtained as the ratio of failure load to the square of the distance separating the platen contact points. The test does not require machined specimens, employs portable testing equipment, and may be carried out in the field to produce a strength log for rock core. The *Brazilian test*[30] employs line loading of machined discs rather than point loading and is more often used in testing concrete. *Impact tests* employing a pendulum or falling weight[31] are sometimes used for strength classification of rock aggregates but the results correlate closely with those of other, simpler, strength tests that can be used as an alternative. Typical strengths are given in Table 9.2.

Hardness, abrasion, attrition and drillability tests

The concept of hardness[23, 27, 32–34] is usually associated with individual rock minerals rather than with rock itself. Moh's scale of hardness commonly employed in petrographic studies is based on a scratch test applied to minerals. *Micro-indentation* tests are available to measure mineral hardness under the microscope for purposes of mineral identification. The *Shore scleroscope* measures mineral hardness by recording the rebound height of a small diamond-tipped plunger after its impact with the rock surface. The Schmidt rebound hammer works on a similar principle but uses a large plunger so that a measure of 'average rock hardness' is obtained. Care must be taken in standardising the methods for holding the specimen and for preparing the surface at the point of impact. However, the instrument is portable and has frequently been used in the field.

Space does not permit a description of other specialised tests in this category, for example the abrasiveness and drillability tests,[33] the abrasion resistance tests for aggregates[27] and the polishing tests for roadstones.[34] Hardness and related properties are closely linked with the mineral content of the rock; quartz content in particular can be a useful guide.

Geophysical methods

Seismic, resistivity, magnetic and gravimetric techniques[35] were developed for mineral and oil prospecting, but have been successfully adapted to the more detailed surveys required in civil engineering. The techniques are used mainly to map rock structure (this aspect is covered in Section 7) but can also be used to give index values related to the mechanical character of rocks and soils.

Refraction seismic surveys measure the velocity of sound emitted usually by an explosive charge and received by one or more geophones after travelling through the ground. Velocity is usually highest in rocks of low porosity or when the pores are filled with water. Therefore velocity measurements can be used to indicate the spacing and openness of fissures, particularly if the velocity of sound *in situ* is compared with that through unfissured laboratory specimens of the same rock. The method has proved useful in mapping the quality of rock in dam foundations and abutments and in assessing the effectiveness of grouting treatment.[36]

Hammer seismograph equipment, where the explosive source is replaced by a sledge-hammer blow, is convenient for small-scale engineering surveys. Sound velocity can be measured between addits or, using a piezoelectric or a sparker source, can be measured between drillholes or in a single drillhole, to give a more complete and accurate picture of rock quality variation.

Sound-velocity values can be used in deriving dynamic elastic parameters (Young's modulus and Poisson's ratio) for the rock,[37] but the calculated values are usually quite different from the 'static' values measured for example by plate testing. The reason is that the rock strains produced by a travelling stress wave are of smaller amplitude and higher stress rate and gradient than in static loading. Sound velocity can, however, be used as an index to rock deformability if a suitable correlation is established.

Field resistivity, magnetometry and gravimetric surveys are generally only used for structural mapping, and then in relatively few applications in comparison with seismic methods. However, these and other techniques are becoming increasingly used in *downhole geophysical logging*.[38] Instrument probes are lowered to the base of a drillhole on a wireline and then drawn up the hole producing a record or 'log' of various rock properties. Common instruments measure sound velocity, spontaneous electric potentials in drilling fluid and natural or induced nuclear radiation. The drillhole logs can be used to evaluate rock porosity, density, saturation, clay content and other mechanically relevant information. Caliper logging, which shows variations in hole diameter that occur in rocks of differing competence, is usually included.

Permeability tests

The majority of flow through the rock mass usually occurs along fissures rather than pores, so that tests on *laboratory specimens* usually have limited significance. Such tests are, however, useful if the rock is very porous. Often they employ air or an inert gas rather than water as the test fluid, in order to speed up the testing procedure.

Permeability can be measured in the field using a *packer test* where a section of borehole is isolated by pneumatically inflated packers.[39] Tests in rock require packers that are long in relation to the test section and, preferably, also to the fracture spacing in the rock. Borehole permeability tests can be carried out by 'pumping in' under conditions of either constant or falling head, or by 'pumping out'. A graph of flow rate against test pressure is usually linear at lower pressures where laminar flow occurs, but may become nonlinear at higher pressures owing to turbulence or to the effect of water pressures in increasing the width of existing fissures. Analysis of test results and water flow problems in general should account for the influence of both total stress and pore pressure on rock permeability. Anisotropic permeability may require that tests be carried out with holes drilled at different inclinations. In some circumstances permeability in rock can also be evaluated using the *well-point drawdown techniques* common in soil mechanics and discussed in Section 8.

Strength testing for design

Triaxial strength tests[40] are most often required for stress/strength design particularly of deep underground excavations. Cylindrical specimens are tested in the laboratory by applying a confining pressure to the curved surface of the specimen, using a pressurising fluid confined in a triaxial cell and separated from the specimen by a flexible membrane to avoid generation of pore pressures. The confining pressure is radially symmetric and provides the (equal) intermediate and minor principal stresses acting on the specimen. The major principal stress is usually applied with a hydraulic ram acting through steel plattens along the axis of the cylinder. This stress is increased, usually at constant confining pressure, until the specimen fails. The test is repeated for a range of confining pressure in order to define a 'strength criterion' for the material (Figure 9.3).

Direct-shear tests are usually employed to provide data for limit equilibrium analyses, particularly for analysis of the stability of rock slopes, dam foundations and abutments.[41, 42] They are best suited to testing a well defined discontinuity such as a joint, bedding plane or the interface between concrete and rock rather than to testing of intact material. Typically an *in situ* test block with dimensions 700 × 700 × 350 mm is first isolated by sawing or line drilling. Stress is applied normal to the discontinuity to be tested, and a force is then applied to shear the discontinuity (Figure 9.4). To avoid generation of tension at the heel of the test block, the shearing force may be inclined so as to pass through its centre of area. A graph showing shear displacement as a function of shear force is used to find values of 'peak' and 'residual' strength. Further tests at different normal stress values allow strengths to be plotted as a function of normal stress. The shearing resistance of the discontinuity, thus evaluated, may be directly applied in limit equilibrium design calculations.

The strength of a discontinuity in rock is a function of the roughness of the surface and of the strength of interlocking roughness asperities. At low normal stress values, shearing tends to be accompanied by 'riding' of asperities whereas at higher normal stresses the discontinuity is prevented from dilating and the asperities shear through. Laboratory direct-shear tests on discontinuities can be unrealistic since very large specimens may be needed to incorporate a representative 'roughness sample'. If discontinuities are very rough even the tests on field-size blocks may give strengths that are lower than would pertain for the full-size structure, and a correction for the additional strength afforded by large-amplitude roughnesses should be made.[4]

Deformability testing for design

Deformability test data are required for designs involving analysis of stresses and displacements in rock; for example, analysis of foundation settlements on softer rocks, analysis of stresses around underground openings or design of tunnel linings. The most commonly used design methods assume linear elastic behaviour for the rock and require values for the elastic parameters: Young's modulus and Poisson's ratio.

Laboratory deformability tests can be used in situations where the rock material is likely to be much more deformable than the discontinuities, for example in soft rocks and at depths where the discontinuities are tight. A cylindrical specimen is loaded along its axis, and strains occurring in the central third of the specimen are compared with the corresponding levels of stress. Strains are typically measured with electric resistance strain gauges or transducers bonded or clamped to the specimen surface. The stress–strain curves (Figure 9.5) are typically nonlinear, reflecting inelasticity due to the closing of pores at low stress levels and to the generation of failure cracks at higher levels of stress.[23] Elastic parameters can be obtained by approximating the slopes of these curves to straight lines over the restricted ranges of stress that are of relevance to the problem. Hysteresis effects (where the curve for unloading does not correspond to the loading curve) are observed in both laboratory and field tests and are probably due to friction acting on the surfaces of cracks and pores.

Rock fissures and joints are typically more open near the surface owing to weathering and stress relief, and in such materials laboratory tests can indicate that rocks are less deformable than is really the case. Field tests are designed to affect as large a volume of rock as possible in order to fully reflect the contribution of fissures to deformability.

The most economic field deformability tests are carried out in drillholes using a jack or *dilatometer*.[43] This instrument applies a radial pressure through either a rigid split cylinder or a flexible membrane, and the resulting radial displacements may then be analysed in terms of elastic parameters. In softer ground these tests can also give an approximation to the strength of the rock.

Plate-loading tests[19, 44] (Figure 9.4) can be used to obtain deformability values that reflect the behaviour of a larger volume of rock. Tests using smaller-diameter plates in drillholes have the advantage of simplicity when a large number of strata are to be tested. The trend, however, is to employ loaded areas of increasingly large diameter. Tests at diameters in excess of 1 m require *flat jack techniques*[45] in order to ensure uniformity of loading and to facilitate application of the high forces that are required. Flat jacks comprise two thin steel plates circumferentially welded and inflated by hydraulic pressure. Pressure is applied in increments and the corresponding rock displacements are recorded by extensometers either at the surface or at depths beneath the loaded plate or flat jack. Reaction can be provided by the opposite wall of an addit or test chamber, or methods are available where a flat jack is grouted into a slot machined into the rock face. Even larger volumes of rock can be tested by *radial jacking in a test addit* using flat jacks distributed around the circumference of a ring beam, or by applying hydraulic pressure to a section of adit that has been lined and plugged with concrete.

Recent developments allow design based on nonlinear test data.[9] In situations where rock behaviour cannot reasonably be assumed to be elastic, because of viscous effects, the creep or time-dependent behaviour of the rock must be measured.[46] These properties are of greatest relevance to geological processes involving crustal deformations over long periods of time and in engineering their relevance is restricted to certain rock types such as evaporates, salt deposits and clay-bearing rocks under conditions where flow and squeezing of rock is likely.

Rock classifications

Until recently rocks were classified using geological names only. This approach can mislead because often the names are too general and sometimes they depend on properties that are of little engineering significance. For example 'granite' can be a crumbly sand or a broken rubble rather than the monolithic material implied by the name. Shales, mudstones and limestones can also exhibit an extremely broad range of engineering properties. On the other hand there are over 2000 igneous rock names in existence, reflecting minor minerological changes that are usually mechanically insignificant.

Test data have been gradually introduced to supplement the classification. Many alternative types of index test can be used, the longest established being the uniaxial compressive strength test on intact material.[28] The simpler point-load strength test has more recently been used as an alternative.[29] Intact strength is only one aspect of rock quality however, and a more realistic classification requires data on further properties. The classification can be made more realistic by including properties characteristic of *in situ* conditions as well as hand-specimen conditions, for example characteristics of the fractures and discontinuities. Fracture spacing (p. **9**–3) has been used together with intact strength in several of the more commonly used rock classification schemes[47, 48] (Figure 9.6).

Other classifications have been formulated on the basis of Young's modulus and uniaxial compressive strength tests[49] (Figure 9.6), and using porosity, density and crystallinity.[50] Important considerations in selecting a suitable classification are that the required tests should be feasible within the project budget and should be relevant to the engineering problems on site. The ultimate object of a rock classification is to allow

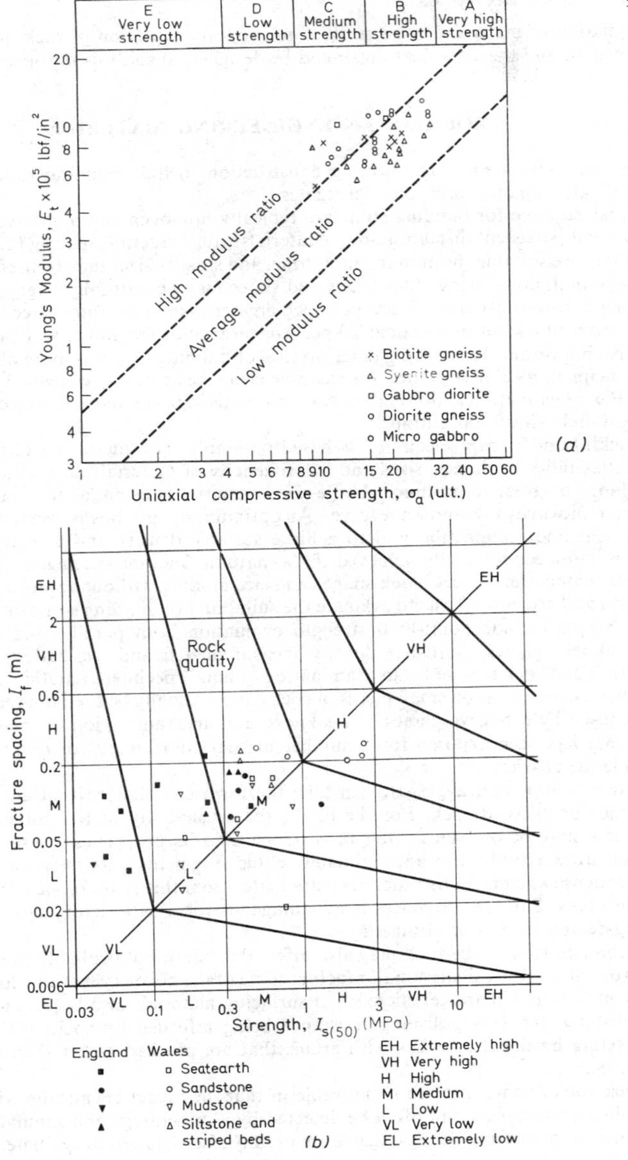

Figure 9.6 Rock classification methods. (a) Rock classification after Deere[49] using laboratory measurements of Young's modulus and uniaxial compressive strength. The plotted results relate to specimens from the Churchill Falls underground powerhouse[19]. (b) Rock classification after Franklin et al.[48] using observations of fracture spacing and point-load strength. The plotted results relate to specimens of Coal Measures rock core

compilation of maps and cross sections to show the variation of rock quality in depth and extent, and the classification should be designed to suit this purpose.

ROCK AS AN ENGINEERING MATERIAL

Rock materials used in engineering construction include building stone, riprap and rockfill, also concrete and road aggregates.

Rocks suitable for building stone are typically homogeneous and have well defined, planar and persistent discontinuities. Materials with a fracture spacing (p. **9**–3) of about 1000 mm are suitable for monumental stone, and smaller sizes may be useful for general-purpose building. Facing stone, flags and slates may be naturally flaggy, with fracture spacing in two orthogonal directions much greater than in the third direction, or may be sawn from blocks of more cubic shape. Appearance of the stone, and ease of dressing are also important. Weathering deterioration of building stones is more often associated with incipient weakness planes, for example cleavage or poorly cemented bedding, than with the intact material, although porous materials in particular are subject to the action of frost and salt crystallisation.[51]

Rockfill and riprap selection[52] is based on similar considerations although there is more flexibility in shapes, sizes and heterogeneity of materials, and visual appearance is seldom of great importance. Large-sized material is essential for marine works to inhibit removal by tide and wave forces. An optimum size grading is essential for adequate placement and compaction and to achieve suitable density and placed permeability. This is most economically achieved if the natural fracture spacing is such as to give approximately the correct block shapes and size grading without appreciable secondary blasting and crushing. Tests to evaluate the suitability of building stone or rockfill might for example include point-load strength evaluation both parallel and perpendicular to weakness planes, porosity, density measurements, and evaluation of cementing materials and rock tecture by an examination of hand specimens and thin sections under the microscope. Tests on small pieces of rock cannot usually be used to predict deterioration caused by extensive planes of weakness, and an examination of the weathering of rock that has been exposed for a number of years in a quarry or rockface can often provide the answer.

Concrete and road aggregate can take the form of either natural gravels, artificial materials or crushed rock. For the latter, the crushability of the potential source of aggregate may be of even greater importance than its properties in use. The material should break readily into approximately cubic fragments without an excess of fines. Brittle, dense and crystalline materials are better from this point of view than porous or friable rocks. Surface roughness is advantageous for a satisfactory bond between the aggregate and cement or bitumen.

The mineralogy of the rock may also affect this bond, but probably to a lesser extent than roughness. Porosity is a major factor; some of the porous yet strong limestones give excellent bonding characteristics. Road surfacing materials also require to be resistant to polishing, the best polishing resistance being afforded by rocks with minerals of contrasting hardness or rocks with grains that are plucked rather than worn smooth by traffic.

Rock constituents that are undesirable in that they react chemically with cement or bituminous substances can often be detected by a mineralogical examination, but tests on concrete made from the aggregates are usually also required to evaluate this hazard.[53]

DESIGN METHODS

The purpose of designing a rock structure, for example a tunnel, open cut or foundation, is to decide the size and shape of the excavation, to determine whether measures to improve the rock conditions such as grouting, anchoring or drainage are needed and, if so, to select and design the rock improvement system. In this subsection some of the available design methods will be reviewed.

The usual procedure is first to select a design that, on the basis of experience, will satisfy the practical and economic considerations, then to check this design for mechanical stability by estimating the rock stresses, displacements or a factor of safety against failure, then to modify the design until a satisfactory compromise is reached between practical and economic requirements and mechanical performance. The performance of the structure and the adequacy of design can then be further checked by instrumentation and monitoring.

Any attempt to analyse a rock structure should start with a complete statement of the factors involved. These usually include the geometry of the structure, the structure of the rock, mechanical properties of intact rock material and the discontinuities, and the nature of water and stress conditions in the mass. It is not usually possible to take each factor into full account, so that analysis is often based on simplified mathematical or physical 'models'. The choice of simplifying assumptions and the errors that these assumptions are likely to introduce are matters for engineering judgement, and it is often advisable to check a design by carrying out more than one type of analysis.

Empirical methods

The oldest approach is design by 'rule of thumb' using design principles that experience has shown to give satisfactory results. Where no rule of thumb is available, one can be established by undertaking a programme of field observations to establish a relationship between 'cause' and 'effect'. This is more likely to succeed if it is preceded by an attempt to establish, using theory and simple methods, those parameters that might prove fundamental to the design, and the trends to be expected.[4, 54] Empirical design rules are usually only safe to apply in the context for which they were originally formulated, and extrapolation can be unreliable particularly if the method has no theoretical basis.

Physical models

Physical as opposed to mathematical models can be used in a laboratory analysis of stresses and displacements in a rock structure, and can sometimes also be applied to the study of fracture and failure.[55]

Elastic models may be used to analyse strains and stresses where the rock may reasonably be assumed to behave elastically. The most common type, a *photoelastic model,* is machined from a stress-birefringent material such as glass or plastic. When loaded and viewed in polarised light the model exhibits coloured fringes (isochromatics) that follow contours of maximum shear stress in the model. Black fringes (isoclinics) visible in plane polarised light indicate the principal stress directions. Models constructed from a highly deformable material such as gelatin develop these photoelastic patterns under their own weight; other materials require superficial loading. Stress-freezing and other methods are available for analysis of problems in three dimensions. *Other types of elastic model* have also been used, for example rock or metal slabs with resistance strain gauges or moiré fringe grids mounted on the exposed surface. Elastic models are being superseded in most applications by computer techniques that afford greater flexibility, can be more reliable, and can also solve inelastic problems. Photoelastic models, however, can still be useful in presenting a visual and easily understood representation of stress distributions, and can cope easily with complex geometrical configurations.

Inelastic physical models are built from materials chosen, according to principles of dimensional similarity, to scale down prototype properties such as density, strength and deformability.[56] Physical models have been used, for example, to investigate the behaviour of rock slopes (Figure 9.7) and of underground excavations at various stages of construction. Their main disadvantage is that they can take some considerable time to construct, and then must be rebuilt for each study. Simple and approximate physical

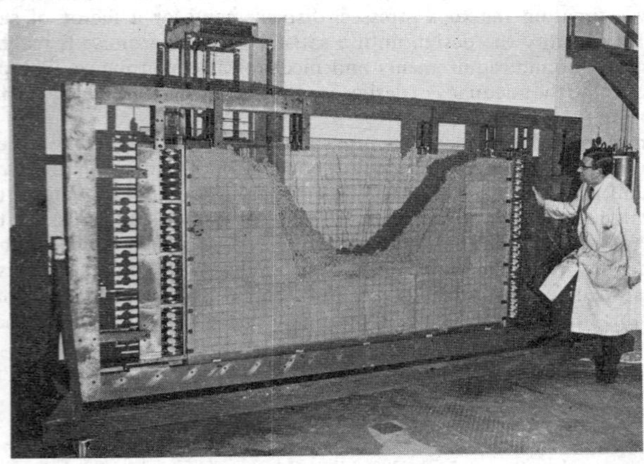

Figure 9.7 Physical and analogue modelling. The photographs show work at the Rock Mechanics Project, Imperial College, and are reproduced with the permission of Professor E. Hoek. (a) Physical model of a rock slope failure; (b) resistance analogue modelling of water flow in rock

models can be valuable at the early stages of analysis in helping to visualise possible mechanisms and in formulating the problem, but care is needed to select appropriate scaling factors.

Analogue methods

The equations governing, for example, electric potential differences and currents, are—with suitable transformations—identical in form (analogous) to the equations governing stress distributions or water flow in the rock mass. This enables stress or water problems to be simulated and solved by electric analogue methods.[57, 58] *The conducting-paper method,* of limited accuracy and flexibility but simple, uses an impregnated paper with probes to monitor surface potential differences. *The resistance network method* uses a grid of interchangeable or variable resistors, can solve anisotropic and heterogeneous problems and is more accurate, but has the disadvantage of restricting measurements to a limited number of elements or nodal points. Three-dimensional problems can be solved by transformation of the problem into two dimensions, or by using an electrolytic tank. Analogues other than electric ones are possible, for example the Hele–Shaw method uses the flow of viscous fluids between closely spaced parallel plates.[59]

Analogue models are in some ways more convenient than digital computation methods in that they are easy to control and to visualise. There are, however, severe restrictions to the problems that can be simulated in this way, and in analysing more complex situations an analogue model can take some considerable time to construct or modify. Simple analogues can nevertheless be most useful.

Numerical and computational methods of analysis

Analysis of stresses and displacements is usually based on principles of continuum mechanics[13] assuming that the material is continuous throughout and that, even locally, conditions of equilibrium and compatibility of displacements are satisfied. The constitutive equation, that is the relationship between stress, strain and time for the rock mass, must be known or assumed in order to formulate the problem. This relationship is in theory established by testing, although in practice tests serve only to measure the parameters in an idealised constitutive equation such as one of linear elasticity. A satisfactory constitutive equation should account for rock behaviour both before, during and after failure of intact material. In most rock structures, zones of fractured rock can be formed which, because they are confined by unfractured material, do not lead to collapse. The presence of fractures or discontinuities does not invalidate the premises of continuum mechanics provided that a constitutive equation can be formulated for an 'element' or test specimen that incorporates a large number of such discontinuities. Soil materials, for example, contain discrete grains bounded by discontinuities but can be tested in this way, so that continuum mechanics can be applied to soil problems.

Having formulated a constitutive equation for the material the next step in design analysis is to solve this equation taking into account the geometry and boundary conditions for the rock structure, and ensuring that conditions of equilibrium and compatability of displacements are satisfied. A wide range of *exact or 'closed form' solutions* are available for solving two-dimensional problems, particularly for linear elastic behaviour, but very few solutions have been formulated for three-dimensional problems. However, Boussinesq and Cerruti give solutions for normal and shear point-loading on a three-dimensional elastic half-space that can be used to build up, by a process of simple superposition, the distribution of stresses and displacements for any system of applied loads.[15, 16] Savin[17] gives closed-form solutions for stress concentrations around holes in an elastic 'plate'; these and other solutions are reviewed by Obert and Duvall.[60]

To solve the many stress analysis problems for which no solution in closed form is available one must resort to *numerical approximation methods* for solving the continuum mechanics equations. The *finite-difference* (*relaxation*) *method*[15] has had a long-established usage in civil engineering. Partial differential equations that define material behaviour and boundary conditions are replaced by finite-difference approximations at a number of discreet points throughout the rock mass. The resulting set of simultaneous equations is then solved, usually with the aid of a digital computer.

The finite-element method[18, 61] is similar in many respects, except that the rock mass is subdivided into a number of structural components or elements that may be of irregular and variable shape (Figure 9.8). A judicious selection of element is critical to the efficiency of computation. The elements are assumed to be interconnected at a discrete number of points on their boundaries, and a function is chosen to define uniquely the state of displacement within each element in terms of nodal displacements at element boundaries.

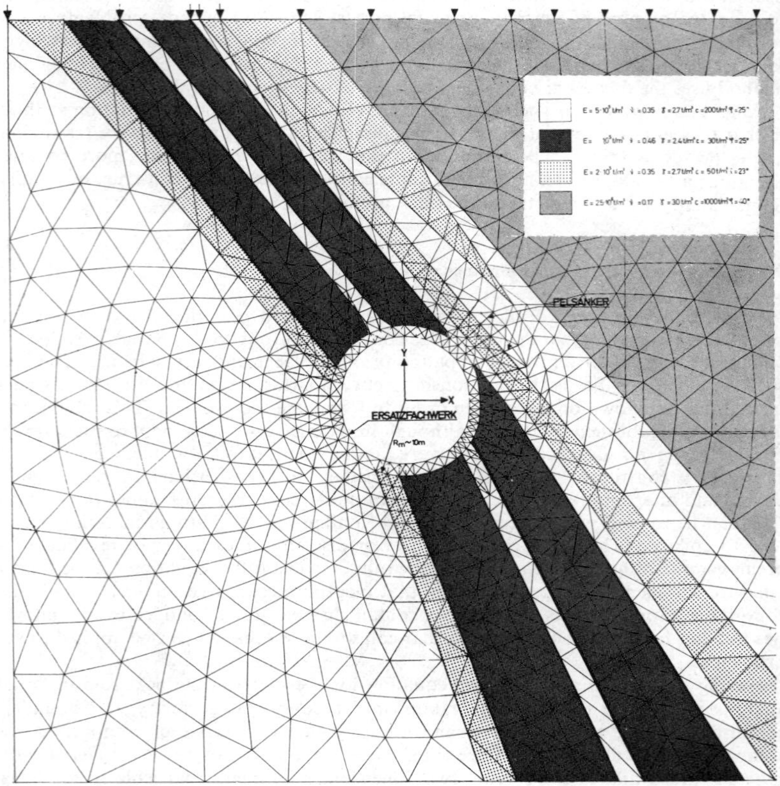

Figure 9.8 Finite-element method. Example showing finite-element mesh for the analysis of stresses acting on a pressure tunnel lining. Elements of varying stiffness have been used to simulate rock zones of varying competence (after Grob *et al., 2nd Int. Conf. on Rock Mechanics, Belgrade,* paper 4–69 (1970))

Strain may then be defined and also stress using the constitutive equation for the material. Nodal forces are determined in such a way as to equilibriate boundary stresses, and the stiffness of the whole model may then be formulated as the sum of contributions from individual elements. The response of the structure to loading may then be computed by the solution of a set of simultaneous equations. Finite-element computer programs have been written for a variety of rock-mechanics problems, to tackle both two- and three-dimensional situations, elastic, plastic, and viscous materials, and to incorporate 'no-tension zones', joints, faults and anisotropic behaviour. The method is also used to solve water flow problems, heat flow problems, and an even wider scope of situations unrelated to rock engineering.

The dynamic relaxation method[62] allows a problem to be formulated assuming rock blocks to be rigid, with deformation and movement occurring only at the joints and fissures so that for this type of analysis no information is needed on the deformability and strength of intact rock. Calculations are based on laws relating forces and displacements between blocks (e.g. laws of elasticity or friction) and on the laws of motion (e.g. creep, viscosity or Newton's first law). Behaviour of the model is constricted to be compatible with boundary force or displacement conditions. Large movements can be modelled—not normally possible with any accuracy using a finite-element technique. The method of computation is ideally suited for considering the development of rock movements incrementally with time (Figure 9.9). The principal advantage of the method, however, is that it requires considerably less computer storage than a finite-element program, so that more complex problems can be solved with greater accuracy. However, the method is demanding in its requirements for computer time. Dynamic relaxation is a comparatively new technique in rock engineering but has been applied to the study of

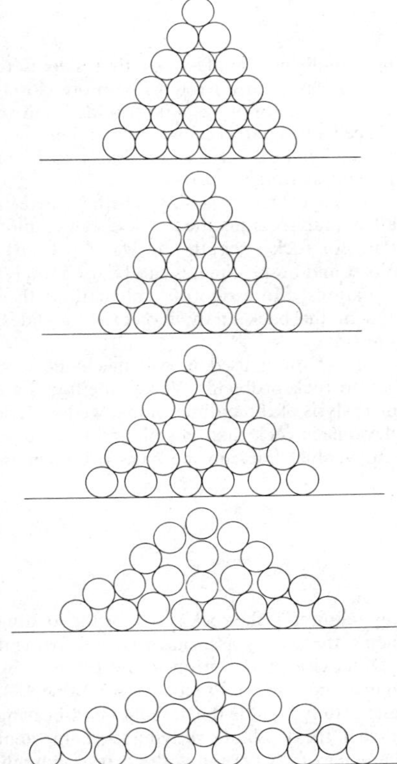

Figure 9.9 Dynamic relaxation method. The figure shows progressive collapse of a stack of cylinders, with displacements computed as a function of time using the dynamic relaxation method (after Cundall[62]). *Similar calculations can be used to show the collapse of rectangular blocks such as comprise a rock mass*

progressive collapse of a rock slope, and to vibration effects from blasting and earthquakes.

In *the limiting equilibrium method*[63] a rock mass is considered under conditions where the mass is on the point of becoming unstable. The method gives no information on magnitudes of displacement or on rock behaviour prior to failure, so that the design calculations cannot readily be checked by instrumentation and monitoring of rock movements.

Equilibrium is examined by relating the shear and normal forces on the sliding surface to the sliding resistance of that surface. Shear tests are necessary to evaluate sliding resistance, but otherwise a constitutive equation for the rock mass is not required. The geometry and position of the sliding surface must be predicted in advance, and for this reason the method has been most commonly applied to slope stability problems where the sliding surface is more readily predicted than in underground situations. Newer developments include three-dimensional wedge analyses, improved design charts that give quick and easy access to limit-equilibrium solutions, and use of the technique in analysing the stability of rock subject to toppling, rather than to a sliding mode of failure.[4]

Slope design

Slope design[4, 64, 65] usually employs limiting equilibrium analysis. A first step is to assess whether the mechanisms of potential slope failure are likely to be more closely approximated by a sliding wedge, rotational slip or a toppling failure model, and to identify the beds, joints or faults that could conceivably control such a failure. Through-going discontinuities such as faults, beds or older failure surfaces are likely to be of considerably greater significance than impersistent or rough features.

Quick and approximate calculations at this stage help to assess whether there is indeed a problem, and whether a more detailed analysis is justified. These can employ hand calculations or design charts[4] using data for rock strength and water pressures estimated after examining the rock *in situ*. Worst and best estimates can be used to give upper and lower bounds in the stability calculations. More rigorous calculations then require *in situ* measurement of shear strength or the back analysis of existing slides, water pressure monitoring and permeability testing.

The two-dimensional methods of analysis most often used in soil mechanics can sometimes prove grossly inaccurate if applied to rock problems. Vector methods[4] are particularly suited to the limiting equilibrium analysis of three-dimensional wedges. The kinematics of stability are also of greater relevance in rock than in soil, and techniques are available for selecting probable from improbable slides on the basis of kinematic considerations[66].

Foundation design

Detailed design calculations are generally only required if the rock is soft or the loading unusually high, and in these cases the problems are usually associated with settlement prediction rather than foundation failure. Otherwise guidance on allowable bearing pressures may be obtained from the Civil Engineering Code of Practice[67] (see Table 9.4).

Typical problems that require a more detailed study include the design of end-bearing piles or caissons carried to rock, particularly when the depth of overlying less competent material is such as to require a minimum diameter of excavation with correspondingly high bearing pressures. Certain types of structure are particularly vulnerable to differential settlements or impose high foundation loads, for example, arch dams and the heavier types of nuclear reactor[68, 69] and in these instances even rock foundations require careful design.

Site exploration is primarily aimed at locating suitable foundation levels, and the relative rather than the absolute competence of strata. Rock quality maps can be useful in making this choice. The depth of rock weathering (Section 7) is often of particular significance. Approximate allowable bearing pressures can be estimated empirically[67] (Section 16) and often at this stage the foundation design can be modified or improved by grouting.

More detailed analyses of foundation behaviour usually employ closed form elastic or plastic solutions or a finite-element approach. These require information on rock deformability that is usually obtained by *in situ* plate-loading tests. Foundations on argillaceous rocks can be subject to viscous flow under high contact stresses and may require a study of long-term (time-dependent) behaviour. Stability analyses in addition

Table 9.4 MAXIMUM SAFE BEARING CAPACITIES FOR HORIZONTAL FOUNDATIONS AT 2 ft DEPTH (0.6 m) BELOW GROUND SURFACE UNDER VERTICAL STATIC LOADING (reproduced from the Civil Engineering Code of Practice[67])

		Types of rocks and soils	*Maximum safe bearing capacity*[a] tonf ft^2 (tonne f m^2)	*Remarks*
I	1	Igneous and gneissic rocks in sound condition	100 (1094)	
Rocks	2	Massively-bedded limestones and hard sandstones	40 (437)	
	3	Schists and slates	30 (328)	
	4	Hard shales, mudstones, and soft sandstones	20 (219)	
	5	Clay shales	10 (109)	
	6	Hard solid chalk	6 (65.6)	
	7	Thinly-bedded limestones and sandstones	—	To be assessed after inspection
	8	Heavily-shattered rocks		

[a] Maximum safe bearing capacity is defined as the maximum intensity of loading that the rock will safely carry without risk of shear failure irrespective of any consolidation settlement that may result. The tabulated values incorporate a factor of safety of about 2 with respect to shear failure. The metric values in parentheses are not included in the original table and are approximate.

to settlement calculations are necessary when designing dam foundations and abutments, and when a foundation is situated above a rock slope. Limiting equilibrium methods are appropriate for these analyses.

Design of underground openings

Although the shape and size of underground excavations are often dictated by economic and functional considerations, their precise location and orientation can often be adjusted to suit ground conditions. Optimum orientation requires a knowledge of rock structure and also of the directions and relative magnitudes of principal stresses in the ground prior to excavation.

A detailed analysis is most often required for large openings of restricted extent, for example, powerhouse excavations, and for design of tunnel linings and supports. Finite-element analyses have been used most frequently for this purpose although photoelastic and electric analogue methods can also be useful[70, 71]. Computation requires information on *in situ* stress, and on rock strength and deformability.

Certain types of civil engineering excavation may give rise to subsidence problems, for example, unsupported chambers for storing water and gas. Furthermore, the civil engineer is often affected in his surface construction operations by mining excavations beneath the site. Subsidence can to some extent be predicted although, since the phenomenon is

Table 9.5 ELECTRO-OPTICAL DISTANCE MEASURING DEVICES

Instrument	*Light source*	*Range* m	*Measuring time* s	*Accuracy* mm	*Weight* kg	*Manufacturer*
Geodimeter 700 (inc. theodolite)	1 mw He–Ne laser reflected and modulated	5 000	15	±5	14.5	AGA
Geodimeter 6BL	1 mw He–Ne laser reflected and modulated	25 000	60	±10	15	
			180	±5		
3800A/B	Gallium arsenide infrared diode	3 000	120	±5	14.2	Hewlett Packard
ME 3000 (Mekometer)	White light	3 000	120	±0.1	17	Kern
DM 1000	Gallium arsenide infrared diode	2 500	15	±2–±8	9.9	
MA 100	Gallium arsenide infrared diode	2 000	30	±1.5	17.3	Tellurometer
MRA 3	Microwave (klystron tube) and receiver	80 000	120	±15	17	
CD6	Gallium arsenide infrared diode	2 000	6	±5	2.5	Tellurometer
DI 10	Gallium arsenide infrared diode	2 000	15	±5–±10	17.8	Wild Heerbrugg
DI 3	Gallium arsenide infrared diode	300	10	±5–±10	9	Wild Heerbrugg
EOK 2000	Gallium arsenide infrared diode	2 000	90	±10	13	VEB Carl Zeiss Jena

essentially time dependent, the analysis is complex and often based on empirical observations[72].

MONITORING OF PERFORMANCE

Instrumentation and monitoring gives a check on the design and its inherent assumptions and simplifications.[73–75] Alternatively it can be used in cases where a detailed analytic design is not justified by the nature of the problem, but when rock stability remains in question. The object of performance monitoring is to give sufficient warning of adverse or unpredicted behaviour to allow timely remedial action, and to assess the effect of remedial works.

Movement monitoring

Conventional survey techniques, for example, precise levelling and triangulation, provide the simplest and perhaps the most reliable control but their accuracy is not always sufficient for detection of movements smaller than a few millimetres.

Electronic distance measuring methods (Table 9.5, see also Section 6) are now capable of this accuracy, with the added advantages of speed and of a range greater than 1 km.

Photogrammetric techniques (ground photogrammetry in particular) are being increasingly used (Table 9.6). Although their accuracy is usually not better than 1 or 2 cm, depending on the object distance, movements can be detected even if they do not coincide

Table 9.6 PHOTOGRAMMETRIC EQUIPMENT

Instrument	*Type*	*Focal length* mm	*Plate size* mm	*Manufacturer*
Askania	1″ theodolite with camera	100	130 × 180	Askania
Wild P30	below telescope	165	100 × 150	Wild Heerbrugg
FTG 1/b	1″ theodolite with camera	155	100 × 150	Galileo-Santoni
Wild P32	above telescope	64	65 × 90	Wild Heerbrugg
Photheo 19/1318	Photo theodolite	190	130 × 180	
UMK 10/1318	Universal photogrammetric camera	100	130 × 180	VEB Carl Zeiss Jena
TMK 6	Terrestrial camera	60	90 × 120	Carl Zeiss
TMK 12	Terrestrial camera	120	90 × 120	Oberkochen/Wuertt
Wild P31	Universal terrestrial camera	100	50 × 117	Wild Heerbrugg

with prelocated survey markers. *Lasers* can be used to detect changes in alignment, and *surface extensometers* (Table 9.7) to monitor the development of tension cracks, tunnel convergence, and the superficial movements of rock slopes.

Movement monitoring devices can also be installed in boreholes to supplement, or sometimes to replace surface monitoring. *Borehole extensometers* (Table 9.8) measure changes in length of the borehole. Multiple position extensometers are available to measure differential movement at as many as ten or fifteen anchors at varying depth. Simple settlement devices often work on a hydraulic or ‘U-tube’ principle. Multiple-position instruments may employ rods or wires and either a mechanical or an electric transducer system for recording movements.

Borehole inclinometers (Tables 9.9 and 9.10) record changes in borehole inclination. Moving-probe inclinometers employ a capsule or probe that travels along the borehole which for this purpose is cased with flexible plastic or aluminium grooved tubing. The

Table 9.7 SURFACE EXTENSOMETERS

Type	*Operating principle*	*Gauge length* mm	*Reading range* mm	*Sensitivity* mm	*Manufacturer*
Dial gauge extensometer	Dial gauge micrometer incorporated in bar and lever system usually portable and locating in targets cemented to rock	To suit 250 (50–2 000 available)	— 5	0.02 0.000 2	Soil Instruments W. H. Mayes
Vibrating wire extensometer	Similar to dial extensometer but employing vibrating wire measuring device in place of micrometer	114–250 92 100	0.1–100 0.08 0.30	0.000 02 0.000 05 0.000 01	Mahiak Geonor Slope Indicator
Tensioned tape or wire	Wire or tape extended between anchor points and incorporating transducer measurements	Up to 20 metres	50 + tape	0.1	Terrametrics Interfels
Disc and stylus	Stylus inscribes movement on acrylic disc	Up to 75	5	0.05	Building Research Station (not marketed)
Photoelastic disc	Disc cemented into portable instrument: records displacement as photoelastic pattern	250	5	0.000 2	Stress Engineering

Table 9.8 DRILLHOLE EXTENSOMETERS

Type		*Maximum number of anchors*	*Type of anchor*	*Drillhole diameter* mm	*Range without reset* mm	*Sensitivity* mm	*Manufacturer*
Multi-wire vernier	MK1	Not stated	2 Hydraulic	48–139	25	0.025	Peter Smith
	MK2		1 Mechanical	40–60	150	0.025	Instrumentation
Multi-wire		8	Mechanical	56	50	0.03	Terrametrics
					25	0.08	
Multi-wire screw micrometer		10	Mechanical	54	75	0.01	ELE (U. of Nottingham)
Multi-wire pointer and scale		10	Mechanical	54	100	0.5	ELE (U. of Nottingham)
Multi-wire electric transducer		8	Mechanical	56	15	0.025	Terrametrics
		6	Grouted	50–100	50	0.05	Slope indicator
Single rod (fixed), dial micrometer		1	Mechanical	42	50	0.025	Slope Indicator
			Grouted or hydraulic	25	20	0.1/0.01	Interfels
Single rod linear potentiometer		1	Mechanical	50	25/75/150	0.025/0.075/0.15	Slope Indicator
			Mechanical			1 (scale readout)	ELE
Single rod (probe), multiple magnetic anchor		Indefinite	Magnetic	100	Infinite	0.02 (micrometer)	Soil Instruments
Idel Sonde radiofrequency transmitter probe		Indefinite	Grouted steel rings or plates	110	Infinite	1	Maihak
Multi-rod dial micrometer		8	Grouted	60	20	0.1/0.01	Interfels
Single- or multi-rod vibrating wire		6	Mechanical or grouted	50–100	100	0.003	Maihak

Table 9.9 DRILLHOLE INCLINOMETERS (FIXED-POSITION TYPE)

Type	*Trade name*	*Drillhole diameter* mm	*Maximum number and type of anchors*	*Range*		*Sensitivity*		*Manufacturer*
				mm/m	minutes	mm/m	seconds	
Anchored chain of rods with transducers at pivots	Lateral deformation indicator/chain deflectometer	116–146 (cased)	Not determined	±10	35	0.1 –0.01	20–2	Eastman Interfels
Pivoted rod and proximity transducer	Multiple-position deflectometer	75–100	Not determined	±12	40	0.03	6	Terrametrics
Flexible steel strip with strain gauges in parallel to monitor strip bending	Strip gauge	75 or larger as required	Continuous	60 mm radius subject to metal thickness	5 400	6.0	1 200	Savage (not yet marketed)
Tiltmeter incorporating pendulum and vibrating-wire measurement for mounting on retaining walls, etc., or rods in drillhole	MDS 81 MDS 81B MDS 82B	If used arbitrary	Arbitrary	±3 ±6 ±12	10 20 40	0.01 –0.002 0.025–0.004 0.050–0.007	2–0.3 5–0.7 10–1.4	Maihak
Flexible breakable strip with resistors in series to detect depth of movement horizon	Shear strips	76 or larger	60 m strip lengths in series	Shear detection movement only		2–50 mm		Terrametrics

Table 9.10 DRILLHOLE INCLINOMETERS (PROBE TYPE)

Type	Trade name	Approximate casing size mm	Casing type	Range mm/m	Range degrees	Sensitivity mm/m	Sensitivity seconds	Manufacturer
Strain gauged pendulum	CRL inclinometer	45 × 45	Square aluminium duct	±88 from vertical	±5	0.075	15	Cementation Research
	Inclinometer	50	Aluminium tubing with keyways	360 from vertical	±20	0.2	36	Soil Instruments
	Borehole clinometer	76 × 76	Square steel tube	±175 from vertical	±10	0.1	20	Structural Behaviour Eng. Lab.
	C-350 slope meter	45 × 45	Square steel tube	±577 from vertical	±30	0.075	15	Soiltest
Pendulum with rheostat	Series 200-B slope indicator	81	Aluminium tubing	±467 ±87 from vertical	±25 ±5	1.0	180	Slope Indicator
2 electrolevels at 90°, servomotor and compass	Slope reader	51	Plastic	±175 from vertical	±10	0.1	20	Eastman
Servo accelerometers	Digitilt	30/70/81	Aluminium/ plastic tube	±577 infinite	±30 ±90	0.1	18	Slope Indicator
Pendulum with vibrating wire, 2 direction, compass or keyway	MDS 83	50 or larger	Aluminium or plastic, keyways optional	±290	±15	0.05	10	Maihak
Pendulum with vibrating wire	68-062 Inclinometer	50	Aluminium alloy	±792	±45	0.15	30	ELE Geonor

probe may comprise, for example, a cantilever pendulum, a short length of pivoted rod, an inertial system or a system where the position of a bubble in a 'spirit level' is monitored electrically. Inclinometers of the fixed-position type usually employ a system of pivoted rods anchored at various positions along the borehole. A simple inclinometer, seldom used in rocks because of its limited sensitivity, employs a rod inserted into or drawn up the casing to detect the depth of changes in curvature. Another device, the *shear strip,* comprises a set of electric resistors connected in parallel at regular intervals along the length of a printed circuit conductor and is used to detect the depth at which the strip, grouted into the borehole, is sheared by ground movements.

Table 9.11 ROCK NOISE MONITORING INSTRUMENTS

Type	*Number of channels*	*Detector*	*Amplification and frequency range*	*Output*	*Manufacturer*
Geomonitor MS1	1	Crystal pick-up in drillhole	1.5×10^6 1.5 Hz–50 kHz	Headphones	Slope Indicator
Geomonitor MS2	Up to 8	Crystal pick-up in drillhole	1.5×10^6 1.5 Hz–50 kHz	Chart recorder	Slope Indicator

Blasting, earthquake and vibration studies require monitoring of *dynamic movements* with an array of geophones located in drillholes or at the rock surface. Seismic arrays can be used either to locate the sources of 'rock noise', for example, in monitoring landslide movements or rockburst phenomena[76] (Table 9.11), or to record the waveforms associated with blast or earthquake tremors, traffic or machinery vibrations.

Water pressure and flow

Piezometers (water-pressure measuring devices) are installed in drillholes to provide information for design analysis, or to record changes in water pressures so as to monitor one of the major causes of instability. They may take the form of *simple standpipes* where pressure is measured as a change in water level using a probe lowered into the standpipe tube. Artesian pressures and local pressure anomalies are common in rock, and the drillhole must usually be sealed with grout over its complete length except for the test sections at which pressures are to be measured. Several standpipes recording pressures at test sections of different elevation, may be installed in a single drillhole.

Standpipe piezometers are only suitable for pressure monitoring in near-vertical drillholes and also have a comparatively slow response to water pressure fluctuations. This response can be improved by use of a *pressure transducer* method. Pressure transducers can be used in drillholes at any inclination and usually employ a flexible diaphragm exposed to water pressure on one side. They incorporate a device for measuring diaphragm deflection that typically uses bonded resistance strain gauges or a vibrating-wire system.

Water-flow monitoring is most often required in connection with reservoir leakage problems, pollution and hydrogeological studies, but a knowledge of flow velocities, directions and the identity of particular fissures carrying the majority of flow is often needed for other types of problem. Flow directions and velocities can be evaluated using radioactive or dye tracer techniques where a concentrated tracer is injected into one drillhole and the time taken for the tracer to appear in nearby holes recorded. Flow can also be measured in a single drillhole by observing the rate of dilution of a tracer or saline solution. Flow velocities and directions can also be logged in the drillhole by a probe incorporating a small propellor or electrically heated wires, the object being to find horizons of greatest permeability prior to installation of piezometers or to permeability testing.

Stress measurement

Rock stress measurements have the object of evaluating the virgin stress field for purposes of design (p. **9**–6) or of monitoring stress changes for purposes of control and warning.[77] Overcoring ('trepanning') methods are more usual in the first application. These make the assumption that rock core is relieved of its *in situ* stress by core drilling and will return to its initial unstressed configuration. The expansion of the rock core is measured by first bonding a stressmeter device into a small-diameter drillhole, recording initial readings, and then drilling a concentric hole of larger diameter to remove an annulus of core with the stressmeter inside. The stress prior to overcoring is then either measured by applying external stress to the core until the stressmeter again registers its initial values, or computed from initial and final strain measurements assuming elastic properties for the rock.

Many types of *borehole stressmeter* are available. The South African doorstopper cell employs a rosette of resistance strain gauges mounted on a rubber plug that is bonded to the end face of the drillhole. The photoelastic glass stressplug utilises a glass cylinder with a centrally drilled hole. Photoelastic fringe patterns (p. **9**–19) are observed before and after overcoring to give stress magnitudes and directions. With these devices more than one drillhole is required for a complete stress-ellipsoid evaluation. The Lisbon stressmeter and others similar incorporate resistance strain gauges in more than one plane, bonded into a resin plug, and so can be used to determine the complete state of stress at a given location.

Borehole stressmeters have the advantage of measuring stresses deep in the rock, away from disturbing influences due to the ground surface or the presence of excavated cavities. However, they need considerable expertise to install reliably, and analysis of results must account for the stress disturbance due to the drillhole itself. They can be installed and left in place without overcoring if the object is to measure stress changes rather than virgin stress.

A further technique for stress measurement is that of *hydrofracturing*. In this method fluid is pumped at high pressure into the drillhole and an accurate record is kept of the relationship between pressure and flow volume. A sudden increase in volume at constant pressure indicates that the rock has been fractured (usually this involves joint or bed separation rather than fracture of intact material) and gives an estimate of the pre-existing rock pressure normal to the fracture plane. The direction of the fracture plane must be known for meaningful interpretation of results.

Flat-jack techniques can also be used to measure rock stresses in close proximity to an excavation. A pattern of displacement measuring points is bonded to the rock surface, and after taking initial readings a slot is cut between the points. A flat jack (p. **9**–16) is then grouted into the slot and inflated until the initial displacement values have been reinstated. The pressure at which this is achieved gives a measure of the rock stress that existed perpendicular to the plane of the slot prior to slot cutting. The Lisbon technique incorporates a displacement-measuring device within the flat jack. Slots at various positions and orientations around the excavations are required, and the evaluation of stresses must depend on assumptions as to the effect of the excavation on the virgin stress field.

Pressure and load

Pressures on retaining walls and tunnel linings are usually monitored with flat jacks (p. **9**–16). The hydraulically inflated flat jack may be connected to a pressure gauge as a sealed system so that changes in pressure are recorded directly, or there may be provision for inflation of the jack prior to taking readings, so that a null displacement condition is achieved.

Load cells (force transducers) can be used to record tension in rockbolts and anchors, compression in ribs, steel sets and prop supports, or can be incorporated in walls and linings as a means of measuring pressure. Essentially they use a 'proving ring' principle with a semi-rigid member to carry the load to be measured, and a means for monitoring the strain in this member. Electric-resistance load cells use one or more metal columns onto which the strain gauges are bonded. Hydraulic load cells employ a sealed hydraulic capsule with measurement of internal fluid pressure. Rockbolt and anchor load cells are available that work on an electric-resistance principle, use a photoelastic glass plug (p. 9–33), or employ a sandwich of rubber between metal plates whose separation is measured with a micrometer. Load cells often incorporate a spherical seating to ensure that the measured load is coaxial with the cell.

ROCK IMPROVEMENT

Unsatisfactory rock conditions can very often be improved by bolting and support, grouting or drainage. The cost of rock improvement is frequently offset by the benefits, for example a rock slope can in some cases be steepened by ten or more degrees if an efficient drainage system is used; in deep rock cuts this appreciably reduces excavation costs.

Rockbolting and anchoring

A rockbolt assembly usually comprises a bar with an anchor at one end and a faceplate assembly (faceplate, nut and wedge or spherical washer) at the other. The anchor may be mechanical or grouted, and the bar may be replaced by a cable to achieve greater lengths and loads.

The *dowel, or fully bonded rockbolt,* comprises a bar installed in a drillhole and bonded to the rock over its full length. A cement-grouted bar can be installed by packing the drillhole with lean quick-set mortar into which the bar is driven. The 'Perfo' system uses a split perforated sleeve to contain the mortar, which is extruded through the perforations as the bar is driven home. Fully bonded resin anchors usually employ a polyester or epoxide resin and catalyst in the form of cartridges that are ruptured and mixed by a bar driven into the drillhole with a small rotary drill. Resin anchors can offer greater strength and rigidity than mechanical or cement systems but can be more expensive.[78]

Point-anchored rockbolts comprise a bar anchored over a comparatively limited length. The slot and wedge system uses a bar with a longitudinally sawn slot. A wedge inserted into the slot expands the slotted section against the rock when the bar is driven home. Sliding wedge anchors employ a pair of wedges drawn over each other when the bolt is rotated. Expansion shell anchors employ two or more wedges or 'feathers' that are expanded by a threaded cone. Explosive anchors have also been used in softer rocks and comprise a split tube, sections of which are driven into the rock by an explosive detonated within the tube. The resin point anchor is identical to the fully bonded resin anchor described above, except that a limited quantity of resin is used to provide an anchor only at the base of the drillhole.

Point-anchored rockbolts must essentially be *tensioned* to work efficiently, since the action of opposing anchor and bearing plate forces effectively tightens the superficial zone of loose rock. This zone can then make a significant contribution to the support of rock at greater depth. Dowels cannot gain from tensioning during installation but are selftensioning when the rock starts to move and dilate. Rockbolts should be installed as soon as feasible after excavation, before rock has started to move with consequent loss of interlocking resistance.

Grouting of point-anchored rockbolts reinforces the anchor, protects against bolt corrosion, and is particularly necessary in softer rocks where point anchorages are seldom reliable in the long term. Grout should be injected at the lowest point in the drillhole, using a bleeder tube to remove air. Hollow-centre rockbolts offer a simpler and more efficient means of introducing grout. Resin-bonded bolts can be arranged so that a fast-setting resin is introduced first in the drillhole, followed by slower setting resin cartridges. The bolt is tensioned when the point anchor provided by the fast-setting cartridge has gained sufficient strength, and the slow-setting cartridges then polymerise to grout the remainder of the bolt length. An interesting development is the use of fully resin-bonded wooden rockbolts. These are cheaper and have the added advantage that

Figure 9.10 Rockbolt testing in granite

they can be readily cut and so can be used for temporary support at a tunnel portal or advancing face.

Rockbolts should be field tested in the rocks in which they are to be installed in order to establish their design performance (Figure 9.10). A bolting pattern may then be designed on the basis of test results, taking into account the rock structure and the size and shape of the slope or underground excavation.[79, 80]

Table 9.12 GUIDELINES FOR SELECTING PRIMARY SUPPORT FOR 7–14 m DIAMETER TUNNELS IN ROCK (from Deere *et al.*[80])

Rock quality	*Construction method*	*Alternative support systems*							
		Steel sets			*Rock bolts*[a] (*Conditional use in poor and very poor rock*)		*Shotcrete*[b] (*Conditional use in poor and very poor rock*)		
		Rock load (*B = tunnel width*)	*Weight of sets*	*Spacing*[c]	*Spacing of pattern bolts*	*Additional requirements and anchorage limitations*[a]	*Total thickness*		*Additional support*
							Crown	*Sides*	
Excellent[d] RQD[e] > 90	Boring machine	(0.0 to 0.2)*B*	Light	None to occasional	None to occasional	Rare	None to occasional local application	None	None
	Drilling and blasting	(0.0 to 0.3)*B*	Light	None to occasional	None to occasional	Rare	None to occasional local application 2 to 3 in	None	None
Good[d] RQD = 73 to 90	Boring machine	(0.0 to 0.4)*B*	Light	Occasional to 5 to 6ft	Occasional to 5 to 6 ft	Occasional mesh and straps	Local application 2 to 3 in	None	None
	Drilling and blasting	(0.3 to 0.6)*B*	Light	5 to 6ft	5 to 6 ft	Occasional mesh or straps	Local application	None	None

RQD = 60	machine		medium						
to 75	Drilling and blasting	(0.6 to 1.3)B	Light to medium	4 to 5 ft	3 to 5 ft	Mesh and straps as required	4 in or more	4 in or more	Provide for rockbolts
Poor RQD = 25 to 50	Boring machine	(1.0 to 1.6)B	Medium circular	3 to 4 ft	3 to 5 ft	Anchorage may be hard to obtain. Considerable mesh and straps required	4 to 6 in	4 to 6 in	Rockbolts as required (~4–6 ft cc)
	Drilling and blasting	(1.3 to 2.0)B	Medium to heavy circular	2 to 4 ft	2 to 4 ft	Anchorage may be hard to obtain. Considerable mesh and straps required	6 in or more	6 in or more	Rockbolts as required (~4–6 ft cc)
Very poor RQD<25 (Excluding squeezing and swelling ground)	Boring machine	(1.6 to 2.2)B	Medium to heavy circular	2 ft	2 to 4 ft	Anchorage may be impossible. 100% mesh and straps required	6 in or more on whole section		Medium sets as required
	Drilling and blasting	(2.0 to 2.8)B	Heavy circular	2 ft	3 ft	Anchorage may be impossible. 100% mesh and straps required	6 in or more on whole section		Medium to heavy sets as required
Very poor, squeezing or swelling ground	Both methods	up to 250B	Very heavy circular	2 ft	2 to 3 ft	Anchorage may be impossible. 100% mesh and straps required	6 in or more on whole section		Heavy sets as required

Note: Table reflects 1969 technology in the United States. Groundwater conditions and the details of jointing and weathering should be considered in conjunction with these guidelines particularly in the poorer quality rock.

[a] Bolt diameter = 1 inch. Length = $\frac{1}{2}$ to $\frac{1}{4}$ tunnel width. It may be difficult or impossible to obtain anchorage with mechanically anchored rockbolts in poor and very poor rock. Grouted anchors may alo be unsatisfactory in very wet tunnels.

[b] Because shotcrete experience is limited, only general guidelines are given for support in the poorer quality rock.

[c] Lagging requirements for steel sets will usually be minimal in excellent rock and will range from up to 25% in good rock to 100% in very poor rock.

[d] In good and excellent quality rock, the support requirement will in general be minimal but will be dependent on joint quality, tunnel diameter, and relative orientations of joints and tunnel.

[e] RQD (Rock Quality Designation) is defined as the percentage of a core run comprising pieces greater than 10 cm length (see reference 49).

Sprayed concrete and other lining methods

A first essential in rock support is to prevent even small quantities of material from ravelling from the rock face, since this can lead to general loosening and progressive failure. The size of rockbolt faceplates is often adjusted to minimise ravelling, and *wire mesh or ribs* can be installed beneath the plates to give added protection. *Sprayed concrete* ('Gunite' or 'Shotcrete') can be used to supplement bolting and mesh, or may under some circumstances provide adequate support on its own[80, 81]; it is a particularly appropriate technique for preventing the slaking deterioration of mudstones and shales. A thin sprayed concrete lining, unlike more rigid methods of support, will crack to reveal zones of instability before these have had chance to develop, allowing the placing of additional local support. Sprayed linings as thin as 5–10 mm have been used effectively in some instances. Several proprietary systems for a more rigid tunnel lining using, for example, interlinked expanded metal sheeting and pumped or sprayed concrete are in use and the methods are described in greater detail in Section 30. Table 9.12 gives a guide to the use of alternative support methods underground.

Grouting

The injection of a grout into the rock mass so that air or water in fissures is replaced by a solid material or gel will inhibit percolation of water and may also provide added strength.[81] Injection into rock normally requires a grout consisting of a mixture of portland cement and water. Sand, clay, or other inert materials may be added to reduce cost provided that these filler grouts can flow, without undue segregation, through the sizes of fissure present in the rock. High grouting pressures can be necessary for adequate grout emplacement, depending on rock mass permeability and on the fluidity of the grout. Grouting at high pressure can result in hydraulic fracturing and lifting of beds. Although this assists emplacement it can be detrimental to the resulting strength of grouted fissures, can result in damage to nearby structures, and in the worst case can itself initiate rock collapse. The efficient grouting of narrow fissures can require grouts of greater than usual fluidity, and in these cases chemical grouting materials can be used. Permeability tests (p. **9**–14) are usually needed to select appropriate grouting pressures and materials. Grouting is commonly used to reduce leakage beneath dams and into tunnels or excavations beneath the water table. It is also used in association with drainage measures to control uplift and pore pressures in dam foundations and abutments, and to consolidate loose rock in foundations or in the vicinity of an excavation. *Consolidation grouting* (distinct from the term consolidation as applied in soil mechanics) serve to improve rock strength but, more important, it considerably reduces rock deformability. Cementitious materials are usually required. *Grout curtains* on the other hand are used to reduce permeability, for example beneath a dam, and may employ lower strength gel-type materials. Temporary control of water flow, and temporary rock mass consolidation, can sometimes be provided by *freezing techniques,* used for example in the driving of shafts through highly fractured rock. The reader is referred to Sections 7 and 30 for further information on the application of grouting techniques. The efficiency of rock grouting is usually assessed by monitoring of the grouting operation itself, by visual inspection, or using sonic velocity techniques (p. **9**–14).

Drainage

Installation of a drainage system in a rock structure can have the immediate advantage of improving working conditions, for example in a tunnel or cut excavation, but its principal objective is usually to reduce water pressures within the rock mass and hence improve its stability.[82, 83] Problems can be due to either excessive discharge or excessive pressure. If a rock mass is very tight, excessive pressures can exist without large discharges. Alternatively if abundant water is available and the rock mass is open, then substantial discharges can occur under quite small pressures. The apparent contradiction between the complementary use of grouting and drainage to control groundwater can be appreciated if it is understood that control of seepage by grout injection usually results in a build-up of water pressures which must be relieved to ensure stability. Sprayed concrete linings, for example, must usually be provided with drain holes if they are to remain intact under conditions of high water pressure.

Design of drainage systems requires a comparison of water pressures and flow paths in the drained and undrained structure. Darcy's law, that the flow velocity is proportional to the change in pressure per unit distance (hydraulic gradient) along the flow path, can

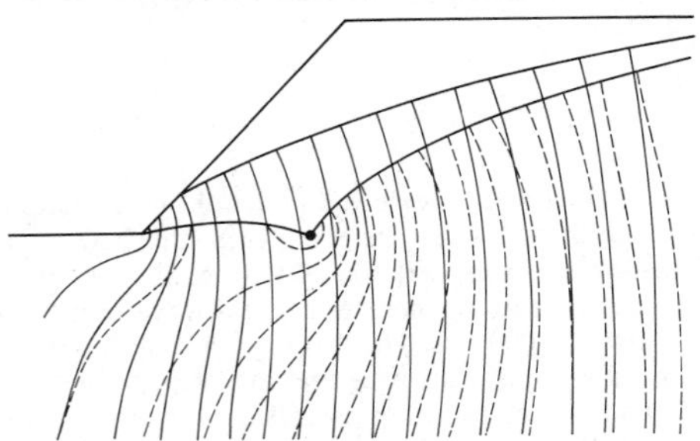

Figure 9.11 Comparison between drawdown curves and equipotential lines for a 45° slope in isotropic material with and without drainage (after Sharp *et al.* [82])

be assumed as valid in most cases. Flow through the rock is expressed in terms of the Laplace equations with appropriate boundary conditions, and these may be solved using graphical methods (flow-net sketching), analogue methods or numerical techniques as discussed on pages **9**–21 and **9**–22. The graphical methods are particularly suitable for an initial examination of the problem, and can be relatively accurate given some practice in flow-net construction and reliable data on rock conditions.[84]

Flow nets (Figure 9.11) are constructed as a grid of equipotential lines (lines of equal potential head) and flow lines (lines giving the directions of flow). The construction is based on certain boundary requirements, also the two sets of lines must meet at right angles and in a homogeneous stratum must form approximate squares. If these conditions are followed, then pairs of adjacent equipotentials have equal head losses, and an equal quantity of seepage flows between any two adjacent flow lines. Flow lines are refracted across a boundary between two materials of differing permeability. Anisotropic permeability conditions can be analysed by a geometric transformation, that is by shrinking the dimensions of a cross section in the direction of greater permeability.

DRILLING, BLASTING, EXCAVATION AND TUNNELLING

Processes of rock fragmentation are known collectively as comminution processes. In spite of a considerable amount of research aimed at improving these techniques[85] the gap between theory and practice is still great and an empirical approach is more often used.

Drilling

Much rock mechanics research has been directed towards understanding the mechanisms of fragmentation during drilling, in order to improve the design of conventional mechanical bits (diamond bits, percussion or rotary drag bits) and to develop new ways of drilling such as by water-jet cutting, flame cutting and pellet impact.[86]

Mechanical drilling techniques suffer from energy losses due to inefficient transfer of energy from the bit to the rock. Ideally the drilling process should produce fragments small enough to be flushed from the drillhole but not so intensely crushed as to absorb a considerable proportion of the input energy.

Obviously, minimum energy consumption is not the only requirement. Large drilling rigs in the petroleum industry may cost about $30 000 a day to operate, of which actual power cost is maybe $1 per hour.[87] Other factors such as drilling rate and quality of core recovery are of primary importance. The reader is referred to Sections 8 and 10 for details on core drilling methods.

Blasting

The theory and practice of blasting are reviewed in detail by Langefors and Kihlstrom[88] and details of techniques and explosives are given in Section 30.

Blast pressure in a drillhole can exceed 100 000 atmospheres. This pressure shatters the area adjacent to the drillhole and generates a stress wave that travels outward from the hole at a velocity of 3000 to 5000 m/s. The leading front of the stress wave is compressive, but is closely followed by tensile stresses that are responsible for the major part of rock fragmentation. When the stress wave is reflected from a nearby exposed rock surface it again gives rise to tensile stresses which may cause scabbing of the superficial rock. The stress wave is the initial cause of fracturing, gas pressures serving to widen and extend the cracks previously generated. Rock mechanics research has been directed towards the modelling and simulation of blasting processes in order to understand the mechanisms involved.

Variables in designing a blasting pattern include the degree of fragmentation required (size of fragments), the explosive used, the diameter, inclination and method of loading the drillhole, the burden (distance from the drillhole to the free face) and the spacing between holes. Also the sequence of firing can be varied, for example with delayed charges, to minimise vibration levels and unwanted rock damage and to give a more efficient pattern of rock removal. A well designed blast gives maximum yield, controls the size and shape of fragments (critical to subsequent crushing processes for aggregate production and to utilisation of material in rockfill constructions), controls the throw and scatter of fragments, and minimises the amount of drilling and explosive required.

Low benches can be blasted with multiple rows of vertically drilled holes, blasting continuously without mucking. This reduces cost and also the throw and scatter of debris. In road cuts blasting is complicated by the continually varying height of bench; heights of more than 10 m are usually blasted in more than one lift. In foundation blasting it is particularly important to control throw and also vibration levels, by means of short-delay multiple-row blasting with small-diameter drillholes and reduced charges.

In tunnel blasting (Section 30) the first holes to be detonated should create an opening towards which the rest of the rock is successively blasted. The holes of this 'cut' are usually arranged in a wedge, fan or cone pattern. The remainder of the round is designed to leave the intended contour undamaged. In long excavations of diameter larger than 8 m the upper section is often removed first, followed by removal of the remaining bench in one operation after installation of roof support. This can give a more economic result and also facilitates both mucking out and the installation of support.

Smooth-wall blasting is a comparatively recent innovation that greatly improves rock stability and at the same time reduces the amount of concrete required for lining the excavation. Often only minor adjustments to conventional procedures are needed. In *pre-split blasting*[89] the cracks for the final contour are created prior to drilling holes for the rest of the pattern. Spacing for the contour holes is typically 10–20 times the hole diameter, and holes are loaded with a charge about one tenth the normal. The contour holes should preferably be ignited simultaneously. The same effect can sometimes be produced with shaped charges.

Excavation and tunnelling

The development of larger and more efficient ripping and tunnelling machines together with the continuing increase in manpower costs has resulted in the increasing use of mechanical excavation as an alternative to blasting. Ripping can be highly competitive particularly in the larger-scale surface mining operations.[90]

The mechanics of rock excavation with a ripper or with the cutting head of a tunnelling machine are in some respects similar to those of drilling, although on a larger scale the fractures and planes of weakness in the rock play an increasingly important role. Also a tunnelling machine, unlike a drill bit, cannot be changed at will to suit rock conditions. Versatile machines are needed, where the spacing, size and type of cutting disc or pick as well as the thrust and speed of cut can be varied.[91]

The considerable capital investment associated with the purchase of ripping and, particularly, tunnelling plant requires a careful study of rock conditions prior to selecting a machine. Rock quality classifications (p. **9**–16) can help in making this choice, particularly if used in association with careful site investigation and geological studies. Fracture spacing is often the most relevant property to note, since it determines the sizes of rock block to be excavated. Intact material strength and abrasiveness are also of relevance. Sonic velocity mapping (pp. **9**–13, **9**–14) has sometimes been used to assist in assessment of the state of fracturing of the ground and hence its rippability.

ACKNOWLEDGEMENTS

The author is indebted to Rock Mechanics Ltd for assistance in preparing the manuscript and to ex-colleagues at the Rock Mechanics Project, Imperial College, for help in reviewing sections of the text.

REFERENCES

1. JAEGER, C., 'The Vajont rock slide', *Water Power,* **16,** 3 (1964)
2. BREKKE, T. L. and SELMER-OLSEN, R., 'A survey of the main factors influencing the stability of underground constructions in Norway', *Proc. 1st Int. Congr. on Rock Mech., Lisbon,* **2,** 257–260 (1966)
3. STIMPSON, B. and WALTON, G., 'Clay Mylonites in English coal measures: Their significance in opencast slope stability', *Proc. 1st. Int. Congr. Int. Assoc. Engng Geol.,* **2,** 1388–1393 (1970)
4. HOEK, E., 'Rock Slope Engineering', Inst. Min. Metal, London (in the press.)
5. RAMSAY, J. G., *Folding and fracturing of rocks,* McGraw-Hill, New York (1967)
6. SERAFIM, J. L., 'Influence of interstitial water on the behaviour of rock masses in Rock Mechanics', Eds Stagg and Zienkiewicz, Wiley, 55–97 (1968)
7. SNOW, D. T., 'The frequency and apertures of fractures in rock', *Int. J. Rock Mech. Min. Sci.,* **7** (1970)
8. MURRELL, S. A. F., 'A criterion for brittle fracture of rocks and concrete under triaxial stress and the effect of pore pressure on the criterion', *Proc. 5th Rock Mech. Symp.,* University of Minnesota, in *Rock Mechanics,* Ed. C. Fairhurst, Pergamon, 563–577 (1963)
9. MORGENSTERN, N. R. and PHUKAN, A. L. T., 'Non linear deformation of a sandstone', *Proc. 1st Int. Congr. Rock Mech., Lisbon,* **1,** 543–548 (1966)
10. WEINBRANDT, R. M. and FATT, I., 'A scanning electron microscope study of the pore structure of a sandstone', *J. Petrol. Techn.,* **21,** 543–548 (1969)
11. LOUIS, C., 'A study of groundwater flow in jointed rock and its influence on the stability of rock masses', Imperial College Rock Mech. Res. Rep. No. 10 (Sept. 1969)
12. HETENYI, M., Ed., *Handbook of experimental stress analysis,* Wiley, New York (1950)
13. FUNG, Y. C., Foundations of solid mechanics, Prentice-Hall, New Jersey (1965)
14. EIRACH, F. R., *Rheology,* Academic, New York (1958)
15. TIMOSHENKO, S. P. and GOODIER, J. N., *Theory of elasticity,* 3rd edn, McGraw-Hill (1951)
16. LYSMER, J. and DUNCAN, J. M., 'Stresses and deflections in foundations and pavements', 4th edn, University of California, Berkeley, Dept of Civil Engineering (1969)
17. SAVIN, G. N., *Stress concentration around holes,* Pergamon (1961)
18. ZIENKIEWICZ, O. C. and CHEUNG, Y. K., *The finite element method in structural and continuum mechanics,* McGraw-Hill, London (1967)

19. BENSON, R. P., MURPHY, D. K. and MCCREATH, D. R., 'Modulus testing of rock at the Churchill Falls underground powerhouse, Labrador', ASTM Special Tech. Pub. 477, *Determination of the in situ modulus of deformation of rock*, 89–116 (1970)
20. HAST, N., 'The state of stresses in the upper part of the earth's crust', *Engng Geol.*, **2**, 5–17 (1967)
21. ELFMAN, S., 'Design of a tunnel in a rockburst zone', *Proc. Int. Symp. on Large Permanent Underground Excavations, Oslo*, 79–85 (1970)
22. Commission on standardisation of laboratory and field tests: 'Suggested methods for determining the slaking, swelling, porosity, density and related rock index properties', International Society for Rock Mechanics, Lisbon (1971, revised 1973); 'Suggested methods for determining the uniaxial compressive strength and point-load strength index of rock materials', International Society for Rock Mechanics, Lisbon (1973)
23. HENDRON, A. J., 'Mechanical properties of intact rock', in *Rock Mechanics*, Ed. Stagg and Zienkiewicz, Wiley, 21–53 (1968)
24. FRANKLIN, J. A. and CHANDRA, R., 'The slake-durability test', *Int. J. Rock Mech. Min. Sci.*, **9**, 325–341 (1972)
25. DE PUY, G. W., 'Petrographic investigations of rock durability and comparisons of various test procedures', *J. Am. Ass. Engng Geol.*, **2**, 31–46 (1965)
26. STINI, J. and PETZNY, H., 'Freezing water and its damage to rock' (in German), *Geologie und Bauwesen*, **23** (1957)
27. *Concrete and mineral aggregates*, Standards Part 10, American Society for Testing and Materials (1969)
28. HAWKES, I. and MELLOR, M., 'Uniaxial testing in rock mechanics laboratories', *Engng Geol.*, **4**, No. 3, 177–285 (1970)
29. BROCH, E. and FRANKLIN, J. A., 'The point load strength test', *Int. J. Rock Mech. Min. Sci.*, **9**, 669–697 (1972)
30. MELLOR, M. and HAWKES, I., 'Measurement of tensile strength by diametral compression of discs and annuli', *Engng Geol.*, **5**, 173–225 (1971)
31. RAMSEY, D. M., 'Factors influencing aggregate impact value in rock aggregate', *Quarry Managers. J.*, **49**, 129–134 (1965)
32. GYSS, E. E. and DAVIS, H. E., 'The hardness and toughness of rocks', *Mining and Metall.*, Am. Inst. Min. Metall. Engrs, **8**, 261 (1927)
33. FURBY, J., 'Tests for rock drillability', *Mine and Quarry Engng*, **30**, 292–298 (1964)
34. GILES, C. G., SABEY, B. E. and CARDEW, K. H. F., 'Development and performance of the portable skid resistance tester', Road Research Laboratory, Tech. Paper TP66 (1964)
35. GRIFFITH, D. H. and KING, R. F., *Applied Geophysics for engineers and geologists*, Pergamon (1965)
36. KNILL, J. L., 'The application of seismic methods in the prediction of grout take in rock', *Proc. Conf. on In Situ Investigations in Soils and Rocks*, 1969, British Geotechnical Society, London, 93–100 (1970)
37. MASUDA, H., 'Utilization of elastic longitudinal wave velocities for determining the elastic properties of dam foundations rock', *Proc. 8th Int. Congr. Large Dams*, **1**, 253–272 (1964)
38. KENNETT, P., 'Geophysical borehole logs as an aid to ground engineering', *Ground Engng*, **4**, No. 5, 30–32 (Sept. 1971)
39. MUIR WOOD, A. M. and CASTE, C., 'In situ testing for the Channel Tunnel', *Proc. Conf. on In situ Testing in Soils and Rocks*, British Geotechnical Society, 109–116 (1970)
40. FRANKLIN, J. A. and HOEK, E., 'Developments in triaxial testing technique', *Rock Mech*, **2**, 223–228 (1970); and *Rock Mech.*, **3**, 86–98 (1971)
41. KUTTER, H. K., 'Stress distribution in direct shear test samples,' *Proc. Int. Symp. on Rock Mech. Rock Fracture, Nancy*, paper 2–6 (1971)
42. LAJTAI, E. Z., 'Shear strength of weakness planes in rock', *Int. J. Rock Mech. Min. Sci.*, **6**, 499 (1969)
43. ROCHA, M., SILVEIRA, A., RODRIGUES, F. P., SILVERIO, A. and FERREIRA, A., 'Characterization of the deformability of rock masses by dilatometer tests', *Proc. 2nd Congr. on Rock Mech., Belgrade*, **1**, 2–32 (1970)
44. COATES, D. F. and GYENGE, M., 'Plate load testing on rock for deformation and strength properties', ASTM Special Tech. Pub. 402, *Testing techniques for rock mechanics*, 19–35 (1966)
45. ROCHA, M., 'New techniques in deformability testing of in situ rock masses', ASTM Special Tech. Pub. 477, *Determination of the in situ modulus of deformation of rock*, 39–57 (1970)
46. MEIGH, A. C., SKIPP, B. O. and HOBBS, N. B., 'Field and laboratory creep tests on weak rocks', *8th Int. Conf. on Soil Mech. and Found. Engng* (1973)
47. WARD, H. H., BURLAND, J. B. and GALLOIS, R. W., 'Geotechnical assessment of a site at Mundford, Norfolk, for a large proton accelerator', *Geotechnique*, **18**, 399–431 (1968)
48. FRANKLIN, J. A., BROCH, E. and WALTON, G., 'Logging the mechanical character of rock', *Trans. Inst. Min. Metall.*, **80**, A1–A10 (1971); discussion, **81**, A43–A51 (1972)

49. DEERE, D. U., 'Geological considerations', in *Rock Mechanics,* Ed. Stagg and Zienkiewicz, Wiley, 1–20 (1968)
50. DUNCAN, N., 'Rock Mechanics and earthworks engineering', *Muck Shifter,* parts 1–8 (June 1966 to Feb. 1967)
51. HONEYBORNE, D. B. and HARRIS, P. B., 'The structure of porous building stone and its relation to weathering behaviour', *Proc. 10th Symp. Colston Res. Soc. Bristol,* Butterworth, London 343–354 (1958)
52. MARACHI, N. D., CHAN, C. K. and SEED, H. B., 'Evaluation and properties of rockfill materials', *J. Soil Mech. Found. Div. ASCE,* **98,** SM1, 95–114 (1972)
53. BLOEM, D. L., 'Concrete aggregates—soundness and deleterious substances', *Am. Soc. Testing and Mater.,* STP 169-A, 497–512 (1966)
54. CORDING, E. J., HENDRON, A. J. and DEERE, D. U., 'Rock engineering for underground caverns', *Proc. ASCE, National Water Resources Meeting, Phoenix* (Jan. 1971)
55. FUMAGALLI, E., 'Model simulation of rock mechanics problems', in *Rock Mechanics,* Eds Stagg and Zienkiewicz, Wiley, 353–384 (1968)
56. STIMPSON, B., 'Modelling materials for engineering rock mechanics', *Int. J. Rock Mech. Min. Sci.,* **7,** No. 1, 77–121 (1970)
57. HERBERT, R. and RUSHTON, K. R., 'Groundwater flow studies by resistance networks', *Geotechnique,* **16,** 53–75 (1966)
58. WILSON, J. W. and MORE-O'FERRALL, R. C., 'Application of the electric resistance analogue to mining operations', *Trans Inst. Min. Metall.,* Sect. A., **79** (1970)
59. SANTING, G., 'The development of groundwater resources with special reference to deltaic areas', UNESCO Water resources series 24, Chap. 16, 85–87 (1963)
60. OBERT, L. and DUVALL, W. I., *Rock mechanics and the design of structures in rock,* Wiley, New York (1967)
61. BREBBIA, C. A. and CONNOR, J. J., *Fundamentals of Finite Element Techniques,* Butterworth, London (1973)
62. CUNDALL, P. A., 'A computer model for simulating progressive large scale movements in blocky rock systems', *Proc. Int. Symp. Rock Mech. Rock Fracture, Nancy,* paper 2–8 (1971)
63. MORGENSTERN, N. R., 'Ultimate behaviour of rock structures', in *Rock Mechanics,* Eds Zienkiewicz and Stagg, Wiley, Chap. 8 (1968)
64. STEWART, R. M. and KENNEDY, B. A., 'The role of slope stability in the economics, design and operation of open pit mines', *Proc. 1st Int. Conf. on stability in Open Pit Mining, Vancouver,* 5–22 (1970)
65. HAMEL, J. V., 'The slide at Brilliant Cut', *Proc. 13th Symp. on Rock Mech., Illinois* (preprint) (1971)
66. HEUZE, F. E. and GOODMAN, R. E., 'A design procedure for high cuts in jointed hard rock—three dimensional solutions', Final Rep. Contract 14-06 D-6990, US Bur. Reclamation. Pub., University of California, Berkeley (1971)
67. *Civil Engineering Code of Practice,* No. 4, *Foundations,* Institution of Civil Engineers (1954)
68. HOBBS, N. B., 'Settlement of foundations on rock', General Rep., *Proc. Conf. on Settlement of Structures, Cambridge,* (1974)
69. ROCHA, M., 'Arch dam design and observations of arch dams in Portugal', *Proc. Am. Soc. Civ. Engrs,* **82,** 997 (1956)
70. HOEK, E., 'Photoelastic technique for determination of potential fracture zones in rock structures', *Proc. 8th Symp. on Rock Mech., Minnesota,* 94–112 (1967)
71. ALEXANDER, L. G., WOROTNIKI, G. and AUBREY, K., 'Stress and deformation in rock support, Tumut 1 and 2 underground power stations', *Proc. 4th Aust. N. Zealand Conf. on Soil Mech. and Found. Engng* (1963)
72. HOOK, I., *Subsidence engineering,* National Coal Board (1966)
73. FRANKLIN, J. A., 'The monitoring of rock slopes', *Proc. Regional Meeting, Engng Group Geol. Soc. London,* Bristol (1972): (*Q. J. Engng Geol.,* in the press)
74. SELLERS, J. B., 'Rock Instrumentation in tunnels', ASCE, *Water Power,* 1–7 (1968)
75. ROUSE, G. C., RICHARDSON, J. T. and MISTEREK, D. L., 'Measurement of rock deformation in foundations for mass concrete dams', ASTM STP 392, *Instruments and Apparatus for Soil and Rock Mechanics,* 98–114 (1965)
76. GOODMAN, R. E. and BLAKE, W., 'An investigation of rock noise in landslides and cut slopes', *Rock Mech. and Engng Geol. Suppl.,* II, 88–93 (1965)
77. ROBERTS, A., 'The measurement of strain and stress in rock masses', in *Rock Mechanics,* Eds Stagg and Zienkiewicz, Wiley, 157–202 (1968)
78. FRANKLIN, J. A. and WOODFIELD, P. F., 'Comparison of a polyester resin and a mechanical rockbolt anchor', *Trans. Inst. Min. Met.,* **80,** A91-A100 (1971)
79. GOODMAN, R. E. and EWOLDSEN, H. M., 'A design approach for rockbolt reinforcement in underground galleries', *Proc. Int. Symp. on Large Permanent Underground Openings, Oslo,* 181–195 (1970)

80. DEERE, D. U., PECK, R. B., *et al.*, 'Design of tunnel support systems', *Highway Res. Record*, **339**, 26–33 (1970)
81. LANCASTER-JONES, P. F. F., 'Methods of improving the properties of rock masses', in *Rock Mechanics*, Eds. Stagg and Zienkiewicz, Wiley, 385–429 (1968)
82. SHARP, J. C., HOEK, E. and BRAWNER, C. O., 'Influence of groundwater on the stability of rock masses: 2-drainage systems for increasing the stability of slopes', *Trans. Inst. Min. Met.*, **81**, A113–A120 (1972)
83. MORGENSTERN, N. R., 'The influence of groundwater on stability', *Proc. 1st Int. Conf. on Stability in Open Pit Mining, Vancouver*, 65–82 (1970)
84. CEDERGREN, H. R., Seepage, drainage and flow nets, Wiley, New York (1967)
85. BAILEY, J. J. and DEAN, R. C., 'Rock Mechanics and the evolution of improved rock cutting methods', in *Failure and Breakage of Rocks*, Ed. C. Fairhurst, Am.Inst. Mining, Metal. and Pet. Engrs. (1967)
86. MAURER, W. C., *Novel drilling techniques*, Pergamon, Oxford (1966)
87. JACKSON, R. A., 'Cost/foot: key to economic selection of rock bits', *World Oil*, **174**, No. 7, 83–86 (1972)
88. LANGEFORS, U. and KIHLSTROM, B., *The modern technique of rock blasting*, Wiley (1963)
89. PAINE, R. S., HOLMES, D. K. and CLARK, G. B., 'Presplit blasting at the Niagra Power Project', *Expl. Engng*, 72–92 (1961)
90. ATKINSON, T., 'Ground preparation by ripping in open pit mining', *Min. Mag.*, **122**, 458–469 (1970)
91. GAYE, F., 'Efficient excavation, with particular reference to cutting head design of hard rock tunneling machines', *Tunnels and Tunneling*, **4**, 39–48, 135–143, 249–259 (1972)

BIBLIOGRAPHY

ABSTRACT JOURNALS

Geotechnical Abstracts (Published by German National Society of Soil Mechanics & Foundation Engineering, Essen, Germany)
Geomechanics Abstracts (Published by Pergamon as part II of *Int. J. Rock Mech. Min. Sci.*, Compiled by Rock Mechanics Information Service, Imperial College, London)

JOURNALS (ENGLISH LANGUAGE)

Bulletin, Association of Engineering Geologists
Engineering Geology (Elsevier)
Geotechnique (Institution of Civil Engineers, London)
International Journal of Rock Mechanics & Mining Sciences (Pergamon)
Journal, ASCE, Soil Mech. Foundation Engng Div.
Quarterly, Colorado School of Mines
Quarterly Journal of Engineering Geology (Geological Society, London)
Rock Mechanics (Springer-Verlag)

TEXTBOOKS

COATES, D. F., 'Rock mechanics principles', Dept. of Mines & Technical Surveys (Canada) Mines Branch Monograph 874, Ottawa (1965)
HOEK, E., 'Rock Slope Engineering', Inst. Min. Metal., London (in the press)
JAEGER, C., *Rock mechanics and engineering*, Cambridge University Press (1972)
JAEGER, J. C. and COOK, N. G. W., *Fundamentals of rock mechanics*, Methuen (1969)
MULLER, L., *Der Felsbau*, F. Enke-Verlag, Stuttgart (1963)
OBERT, L. and DUVALL, W. I., *Rock Mechanics and the design of structures in rock*, Wiley, New York (1967)
STAGG, K. G. and ZIENKIEWICZ, O. C., *Rock Mechanics in engineering practice*, Wiley (1968)
TALOBRE, J., *La méchanique des roches*, Dunod, Paris (1957)

10 SITE INVESTIGATION

SITE INVESTIGATION 10

10 SITE INVESTIGATION

I. K. NIXON, A.C.G.I., M.I.C.E., F.G.S. and G. H. CHILD, M.Sc., F.G.S.
Soil Mechanics

PRELIMINARY CONSIDERATIONS

The primary factors influencing the choice of a site are the economic, social, topographical, administrative and environmental considerations. Thus water supply may strictly control the siting of a nuclear power station, whereas limiting environmental pollution may govern the location of an airport, or preferential government aid a new industrial development. Geology as well as topography will influence the siting of a dam. Each of the primary factors should be considered in sufficient depth to disclose any adverse matter that may be critical before deciding upon feasibility of the site. Guidelines for this study are given below.

LOCATION

At the very beginning the site should be identified on existing maps and any property and building lines established.

ADMINISTRATION

National or local plans for development or redevelopment should be inspected. Restrictions of access, time of working, noise, atmospheric pollution and site rehabilitation should be examined. The existence of mineral rights, ancient monuments, burial grounds and rights of light, support and way, including any easements, should be established.

Proposals for development with outline plans should be submitted to the local planning authority and an application made for approval of use of land before the preparation of a detailed scheme. It may be necessary to present evidence at a public enquiry.

ENVIRONMENT

This refers to the local conditions and resources both natural and those already existing as part of the infrastructure. A site reconnaissance should be carried out at an early stage in order to study the surface features and the surroundings in relation to the proposed works. The principal aspects in this group are given below.

Living amenities. The availability of housing at present and in the future; the extent and standard of the community services either existing or planned.

Public and private services. Availability of a suitable workforce, transportation facilities for access, water supply, sewerage and drainage, disposal of wastes, power and telecommunications. The ownership of roads should be established and the future of any railway.

Geology. The conditions for foundations and earthworks would normally be indicated sufficiently at this stage from the Geological Survey maps, a site visit and an enquiry addressed to the local authority. Check for adverse natural conditions such as unstable ground and underground caverns.

Construction materials. The Geological Survey memoirs refer to natural materials. Also consider use of old tips.

Hydrology. River and tide levels, stream flow, flood levels and groundwater conditions. Periodic occurrence of springs.

Climate. Rainfall, temperature, humidity, prevailing winds and fog.

Other uses of site. Mine workings, past, current or in the future (underground and opencast), tunnels, refilled gravel pits, refuse tips, reclamation, waste and spoil dumps, buried pipelines.

DETAILED EXAMINATION

Once the feasibility of a site for a new project has been established, the next step is to undertake the detailed examination of the relevant aspects of the primary factors required for the design of the project.

TOPOGRAPHICAL SURVEY

The first stage in a detailed examination is to prepare an accurate survey from which plots on any required scale may be made. All levels should be referred to the Ordnance Datum and the site related to the Ordnance Survey grid. Where underground cavities or old workings are accessible, they should be included in this survey.

On large sites it is often convenient to set up a local graticule of coordinates employing permanent beacons. Aerial photography can be advantageous, particularly on extended sites such as motorways. Vertical photographs with stereoscopic overlap permit measurements to be made provided there is ground control. Photographs can be rectified to produce a true map, and are then known as orthophotos. Such photographs can have contours plotted directly on to them. All aerial photographs can be produced as black and white with or without enhanced tones or infrared sensitive. Natural colour or false colour may also be used to identify special features. Existing air-photo cover sufficient for preliminary purposes will probably be available from the Department of the Environment. Also the main air survey companies and a few other organisations have photo libraries. The techniques and their applications in civil engineering have been summarised.[1]

HYDROGRAPHIC SURVEY

For small sites hand sounding may be sufficient with an accuracy of probably about ± 75 mm. For larger works in tidal areas, deep water and high flows, more sophisticated methods may be required. Echo sounding may be employed to provide a surface profile and under the most favourable conditions would be more accurate than hand soundings. Surface profiling may be combined with sub-surface work by employing continuous seismic profiling or side scan sonar systems, although the accuracy would be less. A comprehensive description of hydrographic surveying is given in the Admiralty Manual.[2]

ENVIRONMENTAL SURVEY

The effects of the proposed development on the local, human and natural environment must be taken into consideration. It may be necessary to carry out special studies among which may be included the following:

(i) Preservation of social amenities
(ii) Prevention of pollution
(iii) Disposal or re-use of waste
(iv) Preservation of wildlife
(v) Prevention of siltation and scour
(vi) Preservation of areas of archaeological, historical or other special interest.

GROUND INVESTIGATION

This refers to the collection and interpretation of the data on the ground conditions and groundwater hydrology which may influence the design of structures, the likely methods of construction and their subsequent performance. Occasionally it also includes, where appropriate, investigations of earlier underground works no longer easily accessible, such as old mine workings, possible future changes that might occur in the ground conditions (e.g. development of instability, mining subsidence) and it may be necessary to take account of the likelihood of earthquake damage.

The methods used to obtain this data are described in the following sections.

PRINCIPLES OF GROUND INVESTIGATION

Governing factors and limitations

Sufficient knowledge and experience exist of the difficulties in predicting ground conditions locally and the inherent weaknesses in many soils and rocks to justify the need for ground investigations in order to provide safe and economical designs.[3, 4] However, it is not possible to be precise about the manner and extent of the study, even for a particular case, and judgement plays an important part in deciding the course of any investigation.

The primary object of ground investigation is to determine the stratigraphy and relevant physical properties of the ground appertaining to the project in sufficient detail and extent in order to determine its design, construction and performance. However, the infinite variety in geological conditions and range of design and construction problems to be solved make the subject complex. This has led to the development of a considerable number of different investigation techniques. All have limitations. The basic intent is to collect representative data but, as this is generally obtained indirectly and significant changes in the ground conditions can occur in very short distances within a site, the validity of the data may often be difficult to confirm and reliance must be placed on experience.

Cost

The price of the investigation may sometimes be related to possible savings in the overall cost of the new project, but its value as an insurance against costly delays during construction due to unexpected ground conditions, possible structural damage or even failure should never be overlooked. It varies with the type of structure, its size and the nature and extent of the ground problems. The influence of the last named is such that it is not always possible to make a reliable forecast of the total expenditure, but as a guide the ground investigation may cost about 0.1 to 0.5% of the capital cost of new works and about 0.1 to 2% of earthworks and foundation costs, although exceptionally the cost may be several times these ranges.[5]

Planning the ground investigation

For a sound and economical approach there should be a systematic expansion in knowledge of the ground conditions, taking full advantage initially of the considerable body of information often already available in this country, then investigating its application to the site, and finally verifying it as necessary, according to the limitations of the investigation techniques and the experience available on the geological conditions. As a result it is possible to distinguish three stages in the complete investigation process: preliminary appreciation, main investigation, construction review.

Preliminary appreciation

In order to proceed it is necessary to have some knowledge of the project besides knowing its location. Minimal information initially would be the approximate overall size, layout and purpose of the principal structural units. It needs to be sufficient to establish the main geotechnical problems in relation to the ground and site conditions revealed by the appreciation. However, at an early stage and before deciding upon the detailed plan for the main investigation, it will be essential to have more particulars such as loadings, floor levels and settlement tolerances of buildings, to enable the scope of the testing to be determined.

A considerable amount of information on the geology of the area will normally be available in the form of maps and publications of the Institute of Geological Sciences, from the Geological Society and the Water Resources Boards, as well as mining records (from the NCB in the case of coal). Also much local general information may be obtained from local geological departments of colleges and universities and local government bodies as well as the central advisory bodies such as the Transportation and Road Research Laboratory and the Building Research Station. There may well be previous investigations which have been carried out on or near the site or records of previous construction work which indicate the ground conditions.

All such information should be studied together with the results of a site inspection which for an important project should also include a visit by an experienced geologist. This may reveal significant landform features, outcrops, surface drainage and vegetation which can assist in this preliminary appraisal. Aerial photographs can also be of help, particularly stereopairs where there are changes of level, because the distant view often provides a better means of interpreting likely difficulties such as unstable areas and opportunities for inspecting important geographical and geological features, such as quarries, which may be located outside the actual site area.[6, 25]

Advantage should also be taken of the growing body of soil and rock mechanics information on the more important ground formations present in this country. These offer useful quantitative data for making tentative predictions on the engineering characteristics of the ground as well as suggesting the particular problems pertaining to the formations in question that have been noted from experience and which might not necessarily be revealed by an individual investigation at a site but nevertheless may deserve safeguards in the design. For example, where cavities may possibly be present, as in chalk or limestone formations, whether or not any are revealed, foundations should be reinforced against local collapse and in certain cases permanent precautions taken during the subsequent use of the site against initiating subsidence by restricting the use of hoses for watering gardens and by employing flexible joints on water mains.

Enquiries on other uses of the site can be complicated and extensive, particularly where mining is concerned. Inspections of Ordnance Survey plans (collection at the British Museum) can be helpful with respect to earlier uses. In the mining areas there are a number of consultants with specialist knowledge of particular localities.

Further particulars of preliminary sources of information on ground conditions, drawn up originally for road engineers but which have a much wider application, have been published by the Transportation and Road Research Laboratory of the Department of the Environment.[7]

The preliminary appreciation of the available data, which must be properly listed for future reference with a description of the project, should have the following objectives:

(a) As clear a conception as is possible of the ground structure, the formations present and likely to be affected, the possible alternatives and the probable degree of accuracy of each.
(b) The kinds of solutions indicated for the foundations and earthworks of new projects.
(c) The principal ground engineering problems anticipated in the design and for the construction, e.g. possibility of fill or mine workings; past, present or in the future.
(d) Detailed proposals for the main investigation having regard to the factors outlined in the next section, including the methods to be used and the amount of work necessary in each. The budget cost and possible extent of contingencies that may be required having regard to the probable degree of accuracy of the preliminary information.

Where the available data on a site for new works is found to be inadequate to indicate the course of the main investigation or where important inconsistencies arise at this stage that raise doubts on the best exploratory method to be employed, it may be necessary to make a preliminary survey by carrying out a limited amount of field and laboratory work for the preliminary appreciation. Whilst the amount of such work should be kept to a minimum for economical reasons, it is important that this initial work is of a sufficiently high quality to enable the preliminary appreciation to be soundly based and that it is carried out sufficiently in advance of the main investigation in order to provide adequate time to consider the results and to make the appropriate arrangements.

Main investigation

The object is to develop the initial concept of the ground conditions deduced from the preliminary appreciation, correcting and adjusting it according to what is revealed, to the extent that sufficient details become known to establish a reliable picture of the ground profile (with groundwater data), including its structure and the composition of the individual formations, as well as obtaining the engineering characteristics essential for the detailed design and construction requirements. The former invariably requires fieldwork, generally with mechanical plant, whilst the engineering characteristics of soils and rocks are determined by means of in situ and/or laboratory tests.

Information gathered in the early stages should always be studied without delay by the ground engineer and the designer in case some important discrepancy arises which may necessitate a major change in the scheme for the main investigation or the methods being employed for determining the relevant information.

Whilst the size and importance of the project will obviously have an important bearing on the overall extent of the investigation, it is the geological factors that generally determine the methods for exploring the stratigraphy, whereas the amount and types of testing needed are usually governed more by the requirements of the project, although due regard has also to be taken of the types of material present. For example, a tall block of flats on a stiff clay stratum may require details of bearing capacity and settlement for shallow footings, a raft or piles. In order to determine these the ground would have to be investigated by a pattern of borings with sampling and testing in some detail to a comparatively large depth. An oil tank on the same profile may only require limited data to shallow depth on bearing capacity and settlement characteristics because of the greater tolerances.

Particulars of the various methods used for ground investigations, including the selection and extent of the sampling and testing, are given later in this chapter together with some general remarks on the geological conditions to which they are suited and other factors that determine their choice.

EXTENT OF MAIN INVESTIGATION

The extent of the investigation is determined but not necessarily limited by the plan area of a new project and the complexity of both the design and geological conditions. Basically the field exploration must prove the conditions at least up to the boundary of the area and identify all the required data for the ground within it. By initially carrying out exploratory boreholes actually at the corner boundaries, reliable information is obtained of the limits of the site which can then be extended by further borings within the site so that the variations are disclosed in sufficient detail. There are dangers in assuming that an investigation at a point is representative of the area around it in all directions and that the bedding between formations is horizontal. The limit is reached in the spacing between boreholes when their results are generally consistent with each other and with the geological inferences. In the absence of other determining features, boreholes are normally spaced not greater than 100 m apart, and may be only 20 m apart or less in order to detect significant local variations where structures are situated on changing strata.

The possibility of dangerous conditions originating outside the boundaries of the project should not be overlooked. Normally the topography and geological information will indicate the importance of this aspect which includes such cases as land slips, internal erosion and very occasionally the possibility of weak material immediately adjoining the site.

Sound bedrock in contrast to soil generally has sufficient mechanical strength for engineering design and the testing of it, where important, such as for dams, is generally limited to index measurements. On the other hand rock structures are often more complex than soil due to the faulting and shearing that may have taken place giving rise most commonly to a close pattern of discontinuities and occasionally, but no less important, to local changes in form, e.g. caverns and interbedded soil layers. Consequently, it is of prime importance to obtain adequate data on the structure commensurate with the problems, extending the investigation if necessary by means of inclined drillings (particularly suitable for locating faults, steeply dipping strata and fracture pattern) and the use of adits. Further information on this aspect is given in Section 9, and the methods of exploration (Table 10.1). Weathered rock and the weaker rocks like Keuper Marl need to be treated more like soils as well as taking account of their structure.

DEPTH OF MAIN INVESTIGATION

The investigation should extend to such a depth that it penetrates all the strata that are to be significantly affected by the project. Any excavation work (surface or underground) that is planned should always be adequately supported by investigation beyond its full depth.

Table 10.1 BOREHOLES, PITS, TRENCHES AND ADITS

Method	Geology	Technique	Application in Civil Engineering
Boreholes	Clays, silty clays and peats.	Hand or power auger (single blade or continuous spiral).	Shallow reconnaissance. Power operation fast. Limited to non-caving ground except for hollow continuous augers.
	As above, also silts, sands and gravels.	Wash boring	Inexpensive equipment. Unsatisfactory for precise investigation.
	As above with occasional cobbles and boulders also decomposed rocks.	Percussion cable tool boring with casing.	Standard for soil exploration. Water added below water table to stabilise base of boring. Clean out with auger before core sampling cohesive soils.
	As above and up to moderately weak rocks.	Pneumatic chisel , rotary tricone bit.	Fast, unsatisfactory for precise investigation. Limited to location of hard ground (check for presence of boulders).
	All rocks	Rotary core drilling usually with water flush but mud or air can be used.	Standard for rock exploration. Reliability depends on correct selection of core barrels, bits and flush fluid. Water table observations difficult.
Pits	Clays and peats	Excavation by hand, power-grabs, or augers with support as required.	Direct access gives best opportunity for detailed studies of ground in situ, presence of stratification and thin clay layers. Depth usually limited by problems of ground water lowering.
	Silts, sands and gravels.	Close timbering, ground water lowering essential below water table.	
	Weak to moderately weak rocks.	Hand excavation power grabs or augers with support as required.	Detailed study of local variations; bedding, fissures and joints. Depth usually limited as above.
Trenches	Clays, silts, peats, and sand and gravel above water table.	Excavation usually by machine such as hydraulic powered excavators. Support as required.	Direct access with extended inspection of lateral variations. Exploration of borrow areas.
Adits	All soils and rocks.	Appropriate forms of hand excavation and timbering as for tunnelling.	Established method for detailed exploration of dam abutments and underground structures. Sub-surface exploration of steeply inclined rock strata.

After Glossop. (1)

The normal rule where stresses are to be induced, such as beneath a foundation, is that the investigation should penetrate to a depth at which the net increase in soil stress under the weight of the structure is less than 10% of the average load of the structure or less than 5% of the effective stress in the soil at that depth. Generally this means that borings should be extended to at least one and a half times the breadth of the loaded area and where there are a number of closely spaced areas these must be measured overall. Where piles or caissons are to be employed, the depth should be reckoned from the pile toe. Where deep excavations are proposed, the depth of the investigations should be similarly assessed and in accordance with the changes of soil stress. Where artesian or subartesian conditions are suspected, the depth of exploration will need to be extended to below the aquifer. Relaxation of the general depth rule may be permitted for building foundations when a well known stratum of adequate quality is present within the depth to be significantly stressed. The weathered crust of bedrock should be fully penetrated and careful consideration must be given to the possibility of encountering boulders or overhanging ledges and the maximum thickness that might be present beneath the site. Boulders have been encountered up to more than 10 m in thickness, beneath which were relatively weak soils.

The influence of natural seasonal changes is a limiting factor for some very shallow foundations such as roads, airfields and lightly loaded buildings. The effect of these changes in Britain may extend to depths of about 2 m and overseas in certain cases up to about 5 m.

TYPES OF MAIN INVESTIGATION

Whilst the majority of investigations concern new works, there are occasions when they are required for other purposes. Some of the more important general considerations for each type are given below. Further details are given in the British Standard Code of Practice on *Site Investigations,* CP 2001 (1957).[8]

New works. Careful attention needs to be given to every aspect of the site and the possible effects on adjacent properties. It may be necessary to consider alternative locations for the project. The least favourable conditions should be taken into account for design, and information obtained for the construction as well as for any precautions to safeguard the subsequent performance.

Extensions to existing works. Where an investigation was carried out for the existing works, this should be reviewed and extended by further work to cover the extension. The study should include a review of the method of construction and any problems encountered and the performance of the structure since construction. It will be necessary to consider the effect of the new work on the old as this may influence radically the type of foundation to be employed.

See also remarks under *Safety of existing works.*

Damaged works. In such cases it is generally necessary to establish the causes of the problems that have arisen as well as to obtain the information required for any remedial measures. Although often clearly defined in plan, it generally involves gathering more details on particular aspects than for new works. Measurements should be made where movement is still taking place.

It is inadvisable to proceed with remedial works until the true cause of the problem has been identified, because this is always of considerable assistance when deciding upon a satisfactory solution.

Safety of existing works. Where safety is concerned, the key factor is to establish all the possible problems that may give rise to a critical situation and this involves taking full advantage of experience and expert advice. Particular attention is often needed to assess the effects of changes that have occurred or are likely to occur in the ground conditions.

Some reasons for examining the safety of works could be further development in the vicinity such as extensions, an adjacent excavation, underground works, or changes in the hydrological conditions that may give rise to settlement of the ground surface.

Fill for construction. Pits or trenches are an appropriate and inexpensive means for investigating sites for borrow areas in soils. This method permits examination of local variations and gives some indication of the problems of excavation. Classification of soils generally needs to be more detailed than for foundation investigations.

When exploring bedrock for quarrying purposes, the classification should be on the basis of the properties of the broken rock for its proposed use. Drillings should be planned to determine the structure in some detail, including the jointing to assist in assessing extraction costs. At any new site a blasting trial should also be considered.

INTERPRETATION AND THE REPORT

Interpretation of the field data should be a continuous process leading in the first instance to reliable ground profiles and ultimately to the selection of the appropriate types of foundations for the project. Scheduling the laboratory tests can follow after inspection of the samples but the standard in situ tests are done at the same time as the field work to explore the stratigraphy. Accordingly, an experienced ground engineer should be employed full time on site to supervise important investigations. The ground engineer should check the test results with his descriptions of the samples of the corresponding ground formations. Due regard must be taken of such factors as sample disturbance, size of sample, type of test and ground structure. Judgement and experience both have an important bearing on the interpretation work.

Full records should be kept and included in the report of all the findings of the field and laboratory work. The evidence should be listed in a systematic manner with particulars of the essential identification data and individual measurements in order that an independent assessment is possible. A form of graphical record for borehole data is given in reference 8. More recently a working party of the Engineering Geology Group of the Geological Society of London[9] has made valuable recommendations for logging of rock cores (example in Figure 7.25, Section 7). In the case of trial pits, trenches and adits, sections should show the geological details for each face with measurements and ground descriptions, the location of samples and any tests, also the bearing and the name of the ground engineer or geologist responsible for the observations. Particular attention should be given to recording all the information obtained on groundwater not only because of its significance in the design but the sometimes greater importance for the construction. Useful guidance on the preparation of a standard investigation report has been given by Palmer[10] and more generally by Cooper.[11]

Construction review

The reliability of the results of all investigations must have limitations to some extent by reason of being based on the principles of sampling and there is always the possibility of unusual conditions being present, for instance between two adjacent boreholes. Accordingly during construction full advantage should be taken to verify the results of the main investigation. In simple cases it will consist of comparing the conditions revealed in the excavations for the foundations with the predicted soil profile. Significant differences that arise may require amendment in the design after further investigation. Such differences should be properly recorded for use later when modifications may be introduced or extensions added.

In the case of specialist geotechnical processes, for example piling, grouting, ground anchors, and diaphragm walls, check tests may sometimes be required to compare local conditions with design criteria established from the main investigation. This third stage would also normally include full scale trials made at the commencement of the contract.

The full extent of the bedrock structure can only be seen properly in an excavation and full allowance should be provided in the design for all reasonable eventualities. This also applies particularly to earthworks and dam construction where some modifications in the design to suit the conditions revealed has to be accepted.

Groundwater observations may have to extend into the construction period. Records are also needed when groundwater lowering is being used to note its effect on the excavation work and possibly outside the site where the cone of depression may affect water supply and cause ground settlement.

Instrumentation measurements are often usefully continued after construction to observe the performance of the project. This is particularly desirable in the case of dams. Valuable data are also gained for the advancement of the art—an aspect that can never be stressed too strongly.

In some cases where the interaction of the project and ground conditions is a complex one, the construction review becomes more fundamental in the solution of the problem. This is considered further under observational methods later in this chapter.

METHODS OF GROUND INVESTIGATION

The methods of investigation can be divided into two groups: those that determine the stratigraphy and those that measure the engineering properties. In any complete ground investigation both are required, the relative amounts depending on existing information and the proposed project.

Stratigraphical methods

GEOLOGICAL MAPPING

The structure in depth is inferred from mapping of the surface features. This gives a general indication of the ground conditions and may give a very good indication of the structure where there are numerous surface features. However, it may fail to reveal comparatively minor geological features which have a decisive influence on the project. This method is fully discussed in Section 7.

DIRECT EXPLORATION BY BOREHOLES, PITS, TRENCHES AND ADITS

Geological mapping should be supplemented by exploration by boreholes, pits, trenches or adits in which the ground is exposed for direct examination and representative samples taken for identification and laboratory testing (see *Measurement of engineering properties* on page **10**–20). The various techniques of exploration and their application to the ground conditions are discussed in Table 10.1, and those for sampling in Table 10.2. The most widely used method of exploration is that of boreholes; cable tool boring in superficial deposits and rotary core drilling in rocks. The standard equipment is capable of penetrating to a depth of 50 m and much more, is easily moved by landrover or tractor, can be operated in restricted conditions on land, or from staging or a suitably equipped craft overwater. Wash borings are quick and cheap but only tailings are recovered unless the work is carried out in a large enough size for standard sampling. A good water supply and disposal facilities are necessary. Pits and trenches are very suitable for detailed studies of the ground at comparatively shallow depth. Although

Tab

Source	Geology		S Disturbed
Boreholes	Clays, silty clays and peats	Hand auger	Normally representative of compositio unreliable for examination of structure
		Clay cutter	As above, but liable to more mixing.
	As above, also silts and sands	Shell	Standard for non-cohesive strata to ex composition. Best when whole content shell is emptied into tank and allowed settle before taking representative sam from sediment.
		Powered auger	Liable to considerable disturbance and mixing except when conditions in dept very uniform.
		Water flush	Liable to serious disturbance and mixi
		Standard penetration test sampler	Provides small specimens of both cohe and non-cohesive soils for classificatio purposes but is not normally suitable f retaining structural features.
	Gravel, cobbles and boulders	Shell	Standard for gravel, but grading may unreliable.
		Power auger	Specimens up to gravel size may be re without reliance on source.
	Weak rocks (including hard clays)	Auger	Sample identification generally mislea due to remoulding which produces a v material.
		Air flush (vacuum recovery)	Possible for study of mineral composi above water table.

PLING

que and application in civil engineering

	Undisturbed
drive samplers[8] ratio not ding 30%)	Standard usually 100 mm dia × 450 mm long, occasionally 38 mm dia × 150 mm long. Suitable for local stratagraphical identification and soil mechanics testing excluding pore water pressure measurement on softer materials.
samplers[8]	Less disturbance and better recovery than for open drive samplers. Fixed piston superior to free piston. Non-cohesive strata only retained within mud filled borehole. Improved quality helpful when testing soft recent clays and for effective stress analysis. Reliability aids studies of specific horizons. Sample diameters range from 60 to 250 mm and lengths up to 1 m.
nuous samplers lly commenced ground surface)	(a) Delft 29 mm dia (nylon stocking)[29] rapid method with individual samples up to 18 m long in recent alluvium (Dutch cone resistance below 10 MN/m^2 for stratagraphical identification). (b) Delft 66 mm dia (nylon stocking)[29] as for 29 mm sampler, also all standard soil mechanics testing. (c) Swedish 68 mm dia (steel foils).[32] Individual samples up to about 29 m in soft recent alluvial clays and laboratory strength tests correspond to in situ vane results. Can also be used in silts and sands of medium and low density.
pressed air ler[8] (60 mm	For recovery of silt and sand strata from above or below water table without use of mud, to study laminar structure and composition, density and permeability.
	No common method in use, although injection of chemical grout has been tried[67]
n samplers	Shatter during driving causes serious structural disturbance which can affect results of soil mechanics tests.
ry core barrels	Double and triple tube types, see below, and Pitcher Sampler.[18]

(continued on pages **10**–14 *and* **10**–15*)*

Table 10.2—*continued*

Source	*Geology*	*Sa*	
		Disturbed	
All rocks	Water flush	Rock sludge samples provide opportunity for identification by microscope when conditions are uniform if no core is recovered.	
Pits, trenches and adits	Clays and peats, silts, sands and gravels. Up to moderately strong rock	Hand excavation	For identification purposes particularly useful to study local variations and ano Ensure fresh in situ surface is exposed b sampling.
Groundwater		Bail out borehole or pool and sample after the water ha source. Ensure surface or rain water has not diluted wa	

After Glossop[1]

ue and application in civil engineering

	Undisturbed
y core barrels[31]	Single tube. Simplest type suitable only for massive uniformlý strong rock. Double tube types support and protect core during drilling. Inner tube rigid: least likely to jam but liable to cause serious sample disturbance in variable and broken rock. Inner tube swivel: Internal discharge adversely affects core recovery in variable and broken rock which is minimised when discharge is below core lifter. Face discharge although expensive is considered the best method to minimise losses in variable and broken rock. Triple tube types provide extra split inner tube which assists in removal of core from barrel with least disturbance. Other special barrels include spring loaded inner barrel which extends to protect core in weak layers. Wire line barrels provide facility to withdraw and return inner barrel and core from bottom of hole independently of outer barrel and bit. Water flush is generally used to cool bit and remove cuttings. Air flush requires special equipment to maintain air speeds, can have advantages when coring above the water table. Mud flush can be helpful to reduce erosion of core. Rock cutting is usually with diamond bits but tungsten carbide inserts are applicable for uniform soft rocks. Chilled steel shot is used only for large diameter cores (over 150 mm dia) when some loss is acceptable. Fissures must be grouted to prevent loss of shot. Suitable ancillary equipment as well as skilful operation are essential for good core recovery and the greater the complexity in ground conditions the higher the degree of skill required. The more broken the ground is the shorter each drill run should be to ensure good recovery. The core should be preserved in 'lay-flat' plastic tubing and any length selected for laboratory testing in polyurathene foam as below.
drive and samplers. samples[28, 30]	See notes under boreholes. Offers opportunity for horizontal and inclined as well as vertical tube samples, in silts and sands as well as clays. Ensure fresh surface is exposed before sampling. Hand cut specimens of self supporting soil or weak rock, carefully cut and trimmed in situ to provide undisturbed sample with minimal disturbance. Samples, often 150 mm cube are coated in wax reinforced with muslin as each face is exposed or wrapped in foil and encapsulated in polyurathene foam.[68]

ed to its former level. Rinse the container thoroughly beforehand, preferably using water from test ed.

adits can be used in all soils, they are more appropriate in rocks at depths greater than those possible by pitting or trenching.

The levels at which water is struck should be noted and the rate and extent of any rise in level due to artesian pressure. Such observations may be necessary for a number of water-bearing strata separated by impermeable beds. Observations made during boring,

(*a*) (*b*)

Figure 10.1 The Delft continuous sampling equipment for use in alluvial sands, silts and clays. There are two sizes, 29 mm *and 66* mm *diameter. The illustrations are of the former:* (*a*) *shows the sampler head being attached to a 2* Mg *sounding machine for driving purposes,* (*b*) *is an enlargement showing a split sample of laminated sand and silt, partly air dried to enhance the structure*

however, may be affected by the boring operations and moreover may not be representative. It is preferable, therefore, to use observation wells which also take account of tidal and seasonal variations in level, in order that measurements may be made from time to time to establish the worst conditions. When drilling, levels at which the circulating water fails to return should be noted as this denotes the existence of open fissures.

Sufficient samples of the right size and type should be taken in order to fully represent the ground being investigated. In soils this means that each stratum should be sampled at regular intervals over its whole depth. Samples are either 'disturbed', i.e. taken from the spoil from the borehole, pit, etc., and not therefore representative of the soil structure, or 'undisturbed', i.e. showing the undisturbed soil structure. The latter are obtained, in the case of boreholes, from driving or pushing tubes into the ground below the base and, in the case of pits, by cutting blocks (see Table 10.2). Typical spacings of samples in routine investigations are disturbed samples at 1 m interval in granular soils, and alternating disturbed and undisturbed samples at 1 m interval in cohesive soils.

Figure 10.2 A block sample being taken for shearbox test. Initially an upstanding block is cut with the chain saw, then set with resin in the apparatus frame, after which the block is detached by undercutting

The size of the standard undisturbed sample is normally sufficient for the usual laboratory tests, although larger diameter or block samples are sometimes required.

The size of the disturbed sample should be governed by the nature of the soil and the type and number of tests which are to be made upon it. Typical sizes are as follows:

Purpose of sample	*Type of soil*	*Minimum amount of sample required* kg
Soil identification, natural moisture content and chemical tests	Cohesive soils and sands	1
	Gravelly soils	3
Compaction tests	Cohesive soils and sands	12
	Gravelly soils	25
Comprehensive examinations of construction materials including soil stabilisation	Cohesive soils and sands	25–45
	Gravelly soils	45–90

In very variable formations or where the macrostructure has great engineering significance, continuous sampling may be required. This could consist of a continuous run of undisturbed samples or in the relatively stiffer cohesive soils careful core drilling could be used. In soft alluvial clays and peats the Delft continuous sampler could be employed up to depths of about 17 m (see Table 10.2). Where quality of samples is very important it may be necessary to modify standard boring procedure.[5] Continuous cores are preferred in rock formations in order to study the joints and fissures as well as the material itself. Since any loss is usually from the weakest, and therefore probably the most important materials, every effort should be made to reduce this to a minimum.

Samples of groundwater should be taken in order that it can be checked for aggressiveness to underground structures of metal and concrete. It may be necessary to test its fitness for domestic or industrial purposes. Care should be taken to ensure that the samples are not mixed with water from other sources, i.e. rain or surface water entering the borehole pit etc., or tap water in the sample containers. For considering aggressiveness to concrete see reference 12.

All samples should be labelled as soon as possible and the following information recorded: number of sample, number of pit or borehole, container number, type of sample, contract or site, depth of sample below ground level, and date. Samples should be sealed in airtight containers to prevent drying out. Rock cores may also need to be sealed (see Table 10.2). All samples should be described in accordance with the principles of classification given later in this Section and by trained personnel. Where samples have significant features they should be photographed.[5]

INDIRECT EXPLORATION BY PROBING AND SOUNDING

Probings and soundings are used to extend information on the soil profile obtained by direct methods when there is sufficient contrast in the penetration resistance between formations to differentiate between them. The methods are comparatively cheap and quick, and many types exist.

The simplest form is probing (see Table 10.4). The change in the resistance to driving is the indication of a change of stratum and the method is commonly used to explore the extent of thin beds of peat and soft clay overlying gravel or rock. The maximum depth possible is probably about 10 m. Dynamic sounding is the term used when comparative measurements are made of the number of standard blows per unit of distance penetrated and a plot of these against depth indicates the variations in the soil conditions. The size of the equipment can vary between say 50 mm rod driven by a 50 kg mass with a fall of 1 m up to a full-size pile-driving equipment.

Static sounding consists of pushing rods into the ground using kentledge or ground anchors to provide the reaction. The most widely used is the Dutch deep sounding which consists of a 60° cone, 1000 mm^2 in cross-sectional area, attached to a rod which is in a tube. This enables the point or cone resistance to be isolated from the total friction thereby obtaining greater sensitivity in determining changes in strata. The tube, rod and cone are pushed down together to the required depth and then only the cone is advanced 60 mm in order to measure the cone resistance. The process is repeated every say 1 m depth and cone and total resistance are plotted against depth. Two later developments are an additional short sleeve just above the cone to measure local friction and continuous electric recording. Capacities of the equipment range up to 200 kN enabling soft deposits to be investigated to a depth of 30 m or more. With both dynamic and static soundings, the results can be used in a quantitative way to assess the in situ soil properties. A full discussion of sounding methods is given in reference 13.

GEOPHYSICAL METHODS

The techniques used are in situ methods of measuring contrasts in certain engineering properties of strata and hence determining the stratigraphy. They are based on contrasts

Table 10.3 GEOPHYSICAL METHODS

Method	Principle	Application in Civil Engineering
Electrical resistivity	The form of flow of an induced electric current is affected by variations in ground resistivity, due mainly to the pore or crack water. Current is passed through an outer pair of electrodes whilst the potential drop is measured between the inner pair.	Simplest and least expensive form of geophysical survey. –Exploration of simple geological features, stratigraphy and irregularities in soils, rocks and groundwater (subsurface saline bodies). –'Expanding' electrode technique for changes in depth. –Constant separation' technique for lateral delineation of soil boundaries, e.g. sand and gravel. Location of faults. –Extension of direct measurements of porosity, saturation and permeability. –Analysis is often done by theoretical curve-fitting techniques..
Seismic	The speed of propogation of an induced seismic impulse or wave is affected by the elastic properties and density of the ground.	Most highly developed form of geophysical survey. Can be quite accurate under suitable conditions, particularly for horizontally layered structures. –Also for ground vibration problems and estimation of Young's modulus of elasticity and Poissons Ratio.
	'Refraction' technique with single shots concerns travel times of refracted waves which travel through sub-strata and are rebounded to the surface. Valid only when seismic velocities increase with depth. separate short traverses are used to check this.	Determination of depth to bedrock, including horizontal and inclined surfaces, also buried channels and domes and rippability. –Direct evidence of seismic velocities in refracting strata. –For checking effectiveness of cement grouting of rock.
	'Reflection' technique with single shots concerns the directly reflected impulses from horizons of abrupt increase in seismic velocity.	–Interpretation generally possible only for depths greater than is normally required for civil engineering.
	'Continuous seismic profiling' systems concern the reflected wave trace from a regular series of low frequency acoustic (sonar) impulses of ultra short period for high resolution.	–Submarine exploration of general stratigraphy, also possible beneath lakes and rivers. –Seismic velocities are not normally calculable and the pattern of the records has to be calibrated with drillhole data.
Gravitational	The earth's natural gravitational field is affected by local variations in ground density. Measurements are made of differences between stations in the vertical component of the strength of gravity, which is then corrected for latitude, height and topography to reflect only changes due to sub-surface geology. Careful topographical survey of stations is necessary to obtain reliable results as differences are small.	–The interpretation of regional geology, without depth control, mainly where some geological information is already available. –For distinguishing anomalies such as rock ridges, large domes, faults, intrusions, and steeply inclined strata. Also for positioning buried channels, cavities and old shafts. –Fitting techniques based on simplified structures can be applied for studying anomalies.
Magnetic	Many rocks are weakly magnetic and the strength varies with the rock type depending upon the amount of ferromagnetic minerals present. This modifies the earth's field. Surveys are similar to those for gravity measurements. Although the field work is simpler the interpretation is more difficult.	–Mainly qualitative assessment of regional structures. –For locating the hidden boundaries between different types of crystalline rock and positions of faults, ridges and dykes. Also for positioning buried channels, cavities and old shafts.
Borehole logging	The application of geophysical methods in boreholes.	–Electrical and sonic methods to distinguish between strata especially where core recovery is difficult.

Notes: 1. Quality of the interpretation is very dependent upon the experience used in the analysis and the amount of geological knowledge that is available.
2. Whenever possible carefully link the results with drill hole data or rock outcrops.

After Glossop (1)

in the electrical resistivity of strata, tne velocity of sound transmission (seismic), rock density (gravimetric) and magnetism (magnetic). They can be carried out from the surface except in the case where they are used down a borehole for 'borehole logging', where resistivity, acoustic, magnetic and nuclear measurements can be made. The methods are summarised in Table 10.3. Although there are no great difficulties in carrying out the site measurements, experience and a geological knowledge are essential in interpreting the data correctly. The results should always be checked with some form of direct exploration, such as a rotary core drilling.

These methods have the advantage that the fieldwork for quite large areas can be carried out rapidly in a preliminary survey to the main investigation or as part of the latter to delineate top of bedrock between boreholes. In this way they are particularly useful where there would otherwise have to be a large number of boreholes to give the same coverage.

Usually geophysical methods cannot be applied where there would be interference from surface or near surface structures, e.g. railway lines, underground and overhead cables or pipes, and it is often necessary to work well outside the site boundaries. Also it should be borne in mind that, since these methods depend on contrast in properties of strata, they will only be successful where there is sufficient contrast over a large enough area to be reflected in the results.

A particular application of the seismic work is that to overwater work. Reference has already been made to continuous seismic profiling and side scan sonar systems under hydrographical surveying. By careful choice of techniques these methods can be used to present a continuous visual profile of the stratigraphy below the sea bed. These methods are often attractive when compared with the high cost of boreholes overwater. All methods are discussed in references 14, 15 and 16.

THE OBSERVATIONAL METHOD

Sometimes, because of the complexity of the soil conditions, it is not possible to assess completely the problem after the main investigation. However, provided the project is flexible enough, it should be possible to monitor the construction so that the design can be checked and modified where necessary. A good example of this 'observational method' is the construction of road embankments over soft ground where the preliminary assessment indicates a very low factor of safety. Construction is monitored, the design checked and if necessary side slopes, rate of earthmoving, etc. adjusted.

The method is equally applicable to investigations other than those for new works and especially for investigations into failures. It should be noted that successful application of the method to any project depends upon obtaining reliable relevant field data, which in turn requires the correct field instrumentation (see Table 10.4). Basically, instrumentation is to enable measurements to be made of displacement, earth pressure and pore water pressure. Two important points need always to be borne in mind. Firstly select the simplest form of apparatus consistent with the required accuracy and secondly always make provision for some breakdowns due to the difficulties inherent in the installation and operating environments. The observational method is discussed in reference 18 and some of its limitations in reference 19. Field instrumentation is fully discussed in references 17, 26 and 27.

Measurement of engineering properties

IN SITU TESTING AND INSTRUMENTATION

It is usually necessary to determine the engineering properties of the various strata as well as the stratigraphy. In situ methods may be divided into those which attempt to measure a soil parameter, e.g. the vane test, which determines shear strength of soft clays, and those which are empirical, e.g. the various penetration tests, and in particular the standard penetration test. The latter, which is most useful in the case of noncohesive soils, may need to be correlated against laboratory or other in situ measurements. In situ measurements can be carried out: in pits, e.g. plate loading tests; in boreholes, e.g.

Figure 10.3 The BRE plate bearing test equipment for use in an augered hole to determine bearing capacity at depth. Reaction beam, jack, hydraulic power pack and displacement gauges can be seen. The plate under load is 840 mm *diameter. The whole equipment can be assembled and dismantled easily and quickly for carrying out tests at different depths in one hole*

standard penetration tests; or independent of either, e.g. soundings. In addition to in situ testing it may be necessary to observe a particular feature in which case appropriate instruments need to be installed. The most common feature required is that of the groundwater level or levels. The various techniques of testing and instrumentation and their application to engineering problems are summarised in Table 10.4.

LABORATORY TESTING OF REPRESENTATIVE SAMPLES

The samples obtained from the exploration may need to be tested in a laboratory to assist in the identification of strata and to determine their relevant engineering properties. The various laboratory tests are summarised in Table 10.5 with their application to engineering problems.

Table 10.4 IN SITU TESTING AND FIELD INSTRUMENTATION

Nature of works	*Geology*		*Technique*	*Application in civil engineering*
Foundations for static loads	Soft recent clay	Vane test	Direct penetration from surface and in borehole or pit.[33]	Undrained shear strength, particularly for sensitive clays.
	Soils	Simple probe	Driving usually by drop hammer or pneumatic hammer.	Location of hard ground beneath weak strata. Beware of boulders.
		Dynamic sounding	Probing with standardised dynamic driving procedure.[18]	See remarks for simple probe.
		Static sounding test	Standardised test with shielded rod and constant rate of penetration. Adaptable for small piston sampling.[18]	Bearing value and length of piles in silts and sands. Relative densities. Location of weak zones.
		Standard penetration test	Important always to maintain positive head in borehole. Provides small sample except in gravel when solid cone is used.[33]	Bearing values of non-cohesive soils. Relative densities. Unreliable in gravel. Correction to be applied to tests in fine grained soils.
		Piston loading test	Plate loading test on base of borehole above water table.[44]	In situ bearing value for clays. Rarely used in non-cohesive soils.
		Pile tests	Loading, pulling and lateral as required. (i) Maintained load method.[43, 45] (ii) Constant rate of penetration method (C.R.P.).[46] (iii) Equilibrium load method (E.L.) requires fairly even temperatures and leak proof ram. In all types of tests, end load can be measured separately by load cell.[26]	Pile design. Ratio of settlement in sands between individual test and group suggested by Skempton.[48] M.L. method represents conventional technique. C.R.P. method is very quick for load carrying behaviour. E.L. method is compromise for determining load carrying behaviour quickly. Load increment is applied and load system sealed so that as settlement occurs loading decreases until equilibrium is reached.
		Hydraulic fracturing	In hydraulic piezometers.	Measurement of minor stress.
	Soils and particularly stony clays, weak and weathered rocks	Plate bearing tests	At shallow depth in pits. At greater depth in augered holes. Ensure test load is carried out well clear of plate.[8, 43]	Bearing value for foundation design. Test by boring for softer deposits at depth.
		Borehole loading test	(e.g. Menard pressuremeter.) Radial compression test on wall of borehole	Modulus of deformation, creep limit, shear strength up to 1 MN/m^2 and under good conditions K_0

			predetermined plane, with or without normal loading.†	elasticity.	
		Drillhole loading test	Expansion of segmented cylindrical shell in truly drilled hole.†	Modulus of elasticity of strong rocks; directional pressure.	
		Drillhole dilatometer	With flexible membrane.†	Modulus of elasticity of weak to strong rocks; all round pressure.	
		Plate jacking test.	Usually carried out between two sides of pit or adit.†	Modulus of elasticity and bearing value involving maximum volume of rock to take account of discontinuities.	
		Seismic measurements	Determination of seismic velocities in various modes.†	Dynamic modulii and indirectly strength. Preferably confined to extension of direct measurements. Field delineation of low velocity areas can indicate fractured rock.	
		Stress measurements:†	(i) Overcoring methods. A device is fixed in place (drill hole or surface) and observations are made before and after overcoring. Types include stress plugs, discs or meters; utilising transducers, strain gauges or birefringent elements. (ii) Flat jack methods. Strain gauges are fixed on rock surface and observations are made before and after a slot is cut, and again after a flat jack is grouted into slot and pressurised to restore ground strain to unrelieved state. (iii) Hydraulic fracturing in packer tests. Determines minor principal stress only.[51]	Absolute stresses in rock masses. All methods require measurements in three planes for complete stress ellipsoid.	
Foundations for dynamic loads	Soils and rocks	Static loading test	Extra sensitive plate test cycled over expected stress range to give a modulus of reaction.[43]	'Spring constant' for foundation design.	49, 50
		Dynamic loading test	Small vibrators mounted on soil to give resonance response.	Values of dynamic modulii, Poisson's ratio and damping.	49, 50
		Seismic velocity measurements	In various modes.	Dynamic modulii. (See note above opposite 'Stress measurements'.)	49, 50

Table 10.4—*continued*

Nature of works	*Geology*		*Technique*	*Application in civil engineering*
Earthworks, soil and rock slopes	Soils	In situ shear strength	Normally undrained direct shear test.[52]	Undrained in situ shear strength having regard to structure and orientation of failure plane particularly for slope design.
		Sand replacements	(Also water balloon device.)[43] Calibrate sand at natural humidity.[33]	Bulk density during construction.
		Nuclear devices at surface	Radioactive sources and counting unit.[43, 53]	Bulk density (preferably by attenuation method) and in situ moisture content.
		Nuclear density probe	Usually back-scatter method with radioactive isotopes.[53]	Bulk density measurements above and below water table, with casing if required.
		Proctor needle	In earthwork construction.[43]	Field control and consistency of fine grained soils.
		In situ CBR	In earthwork construction.[20]	Only appropriate in clay soils and subject to climatic changes.
		Piezometers	High air entry value in partially saturated soils.[18, 26]	
		Total pressure cells	Require very careful positioning.[18, 26]	Total earth pressure against substructures and within a soil mass.
		Settlement and heave instruments	Types: water, mercury, magnetic ring, buried plates, rods and notched tubes.[18, 26]	Total and relative settlement.
		Conventional survey methods	Laser, photogrammetry.[26]	Total and relative surface movement.
	Soils and rocks	Inclinometers and deflectometers	Portable and installed.[18, 26]	Creep and slip detection.
		Extensometers[26]		Expansion due to relief of stress and across tensile zones arising from differential settlement.
Groundwater permeability etc	Soils and rocks	Observation wells and piezometers	Use effective filter, test regularly and seal from extraneous infiltration.[18, 26]	Level of water table, artesian and sub-artesian conditions.

	Sands and gravels	Pumping tests	Pump to equilibrium conditions measuring transients during draw-down and recovery. Use at least two lines of observation wells [55]	Best form of test for natural permeability measurement. Transient measurements provide storage coefficient, and indication of vertical variations of permeability.
		Two well pumping test	Established technique.	Estimation of difference between horizontal and vertical permeability.
		Radioactive tracers	Various.[56]	(as above)
		In situ permeability	Careful shelling beforehand. Constant rising and falling head tests possible. Latter two tests to be run jointly. Temperature of injected water to be 5 °C greater than groundwater.[54]	Local measurement of in situ permeability either through base of borehole or after placing coarse filter and withdrawing casing. Treat results with caution. A considerable number of tests are required to compensate for scatter.
		Electrical resistivity	Four electrodes. Wenner or Schlumberger configuration.[14]	Extension of direct measurement of porosity, degree of saturation, and permeability.
	Rocks	Formation tests	Expanding packers isolate zone under test.[58] Keep excess head below effective overburden pressure.	Joint seepage and condition of joints by measuring flow under varying pressures, rising and falling.
Miscellaneous, corrosion, etc.	Soils	Thermocouples and thermistors.[59, 60]		Ground temperature of coal tips on fire, beneath boilers and refrigeration plant.
		Electrical resistivity	Four electrode system. Wenner configuration or two electrode probe.[14]	Electrical resistivity for corrosion survey. [65, 66]
		Corrosion probe	Short circuit current between magnesium iron cell and earth.	Depolarising ability of soil for corrosion survey. [65, 66]
		Stray current measurement.		For corrosive effect. [65, 66]
	Soils and rocks	Periscope calipers and borehole cameras, with video tape recording.[61, 62, 63, 64]		Defining cavities, fractures, etc.
	Rocks	Noise detectors†	Considerable amplification required, and quiet environment.	Incipient ground movement at faults, slopes, tunnels.

†For reference to tests and instrumentation in rocks, see also *Rock Mechanics*, Section 9.
After Glossop[1]

Table 10.5 LABORATORY TESTS

Category	Test (Application)	Foundations: Bearing capacity			Foundations: Settlements			Slopes		Dams and embankments		Earth pressure		Water flow		Roads, runways and construction				Reference and remarks
	Description	C	NC	R	C	NC	R	C	NC	C	NC	C	NC	C	NC	C	NC	R	S	
Identification	Visual inspection																			
	Density	*	*	2	2	2	2	*	*	*	*	*	*	2	2	2	2	2	2	33
	Moisture content	2			2			2		2		2		2		2	2			33
	Atterberg limits	2			2			2		2		2		2		*				33
	Shrinkage limit				2					2										33
	Max and min densities		2			2			2		2		2		2					34
	Particle size distribution		2			*			2	2	2		2	*	*	*	*		*	33
	Particle shape and texture		3						3		3		3				3			35
	Specific gravity of particles			2	*	*	2			*	*							2		33+
	Porosity			2			2													20+
	Organic content	3			3			3		3						3			2	36
Total strength	Unconfined compression	( )		*																33+
	Laboratory vane	*						*		*		*				*				37 soft clays
	Triaxial quick undrained	*		*				*		*		*				*				33
	Triaxial slow undrained	*						*		*		*				*				38 soft clays
	Shearbox		*						*		*		*							34

Effective strength (including residual)	Triaxial undrained with pore pressures and drained		2					*		*		*	*							39
	Shear box drained multireversal							*		*										40
Compressibility and permeability	Oedometer consolidation				*					2										33, 41
	Triaxial consolidation				*		*			2										39
	Young's modulus and Poisson's ratio				*		*													42+
	Triaxial permeability									2	2			*						39
	Permeameter (constant) and falling head)							2	2		2				*					34
Other tests	Compaction									*	*					*	*		*	33
	CBR															*	*		*	33
	Freeze and thaw																	*	*	34+
	Swelling			2														2		+
	Slake durability			2														2		+
	Sodium sulphate soundness																	3		+
	Brazilian tension			*																
	Hardness																	*		+
	Drillability																	*		+
	K_0 triaxial										*	*								39
	$\bar{B}$ triaxial								*											39
	Chemical pH, SO_3	3	3							3						3	3		3	33

After Glossop[1]

Key
C—Cohesive soils
NC—Non-cohesive soils
R—Soft rocks
S—Stabilised soil
*—Fundamental values
2—Other essential data
3—Additional useful data
( )—Alternative tests
+For reference, see Section 9.

GEOPHYSICAL METHODS

Although most geophysical work falls into the category of 'Stratigraphical Methods', some techniques can be used for ascertaining certain engineering properties of rock as given in Table 10.4.

MODEL AND PROTOTYPE TESTS

It may be necessary to carry out full-scale trials in the field or model tests in the laboratory to check the parameters used in the preliminary analyses based on the results of the work carried out under the above methods of measuring engineering properties. For example, trial embankments may be constructed in the field on soft ground to check for stability or settlement, pile loading tests carried out to measure shaft adhesion and/or end bearing, or compaction trials run to test the suitability of fill. In the laboratory, dams, embankments and cuttings can be modelled and tested in a centrifuge and problems of permeability and seepage can be investigated in a flow tank. Model piles and footings can also be tested. Various examples are given in reference 28.

Figure 10.4 A borehole loading test using the Menard pressuremeter probe. The instrument in its slotted sheath is being lowered into the drill hole to the test position. On the left is the control and measurement panel. The diamond drill rig used for making the hole is in the background

Choice of method

Consideration of the foregoing indicates that the selection of the appropriate methods to be employed is governed by the following main factors: project requirements, geological conditions, scope of method.

The interaction of these factors in determining the methods to be employed is illustrated in the following example. A power station is to be built on a site which from a preliminary appreciation is believed to consist of alluvial and glacial deposits over sandstone bedrock which may be faulted and with igneous intrusions.

To clarify bedrock condition with regard to the siting of the project, and hence the location of the main investigation, an initial programme of geophysical work is carried out with a limited number of borings and drillings. This is also used to determine the most appropriate methods for the main investigation. This may consist of a comprehensive programme of borings and drillings, with sophisticated sampling and testing, in the selected areas for the major structures to ensure that these are not sited on a major fault zone or spanning two different ground formations. If the alluvial deposits are shown to be soft clays in places, piston sampling and vane testing may be carried out for say stability of tankage. It may be necessary to investigate the fabric of the alluvial clays to determine rate of settlement. On the other hand, if loose sand is encountered and piling is envisaged under heavy structures, soundings may be more appropriate for pile design.

For sensitive structures settlement would be monitored during and after construction.

CLASSIFICATION OF SOILS AND ROCKS

The need for classification

The purpose of classifying soils is to provide an accepted, concise and reasonably systematic method of designating the various types of materials encountered in order to enable useful conclusions to be drawn from a knowledge of the type of material. The degree and type of classification required on any particular project or soil will depend on its relative importance. Thus, for extensive earthworks for roads, soils will need to be fully classified to determine their suitability for fill and the appropriate type of compaction required. For other problems it may only be necessary to classify soils sufficiently to assist in their identification for stratigraphical purposes.

Classification of soils

The general basis for field identification and simple classification of soils is based upon grain size, strength and structure. The simple nonorganic soil types are divided into six classes in decreasing order of grain size; boulders, cobbles, gravels, sands, silts, and clays. The limits of these classes correspond approximately to important changes in the engineering properties of the soils.

The principal soil types usually occur in nature as siliceous sands and silts and as alumino-siliceous clays, but varieties very different chemically and mineralogically also occur. These may give rise to peculiar mechanical and chemical characteristics which, from the engineering standpoint, may be of sufficient importance to require special consideration. The following are examples. Lateritic weathering may give rise to deposits with unusually low silica contents, which are either soft nodular gravels or clays; but intermediate grades are rare. Volcanic ash may give rise to deposits of very variable composition which may come under any of the principal soil types. Deposits of sand grains may be composed of calcareous material (e.g. shell sand, coral sand) or may contain considerable proportions of mica (where grain shape is important) or glauconite (where softness of individual grains is important). They can be banded or laminated with other sand, silt or clay. Deposits of silt and clay may contain a large proportion of organic matter (organic silts or clays and clays may be calcareous (marls)). As well as being banded or laminated they may be fissured. The importance of the soil fabric is emphasised by Rowe[5] and this should be included in the description of undisturbed samples giving, for example, thicknesses of bands or laminations of, say, silt or sand in samples of, predominantly, clay, and the nature of any fissures.

Table 10.6 GENERAL BASIS FOR

In the order suggested for description.

COLOUR	GRAIN SIZE	DISCONTINUITY SPACING: Structure (Mainly sedimentary rocks)	DISCONTINUITY SPACING: Fracture Spacing (Joints, fissures etc
For example: Grey, Blue, Brown, Green etc.	Very coarse – grained	Very thickly bedded	Very widely spaced
	60mm	2m	
Supplement where necessary with	Coarse – grained	Thickly bedded	Widely spaced
	2mm	600mm	
Light or Dark	Medium – grained	Medium bedded	Moderately widely spa
and: Greyish, Bluish, Brownish, Greenish etc.	0.6mm	200mm	
	Fine – grained	Thinly bedded	Closely spaced
	0.2mm	60mm	
	Very fine – grained	Very thinly bedded	Very closely space
		20mm	
		Laminated closely	Extremely closely spaced.
		6mm	
		Thinly laminated very closely	

1. Some rock names imply a particular grain size. Nevertheless this should be included since the correct rock name cannot always be readily given.

2. Can normally only be assessed from exposures or seve boreholes.

3. Alternative term for mainly igneous and metamorphic e.g. flow-banded and foliated respectively. Structure also be described as massive, cleared, veined etc.

Examples of description: Dark brown, fine to medium grained, moderately widely spaced joints, slightly weathered, contact metamorphosed, DOLERIT

Greenish – grey, fine grained, thickly bedded, closely to very closely jointed, fresh SHALE, moderately strong.

'FICATION AND CLASSIFICATION OF ROCKS

	WEATHERING
h	Parent rock showing no discolouration, loss of strength or any other weathering effects.
tly hered	Rock may be slightly discoloured, indicative of some deterioration of strength, particularly adjacent to discontinuities, but away from discontinuities, which may be open, rock is effectively fresh.
rately hered	Rock is discoloured; discontinuities may be open with discoloured surfaces and alteration, representing significant weakening, starting to penetrate. The rock is everywhere weaker than fresh rock.
y hered	Rock is discoloured; discontinuities may be open with discoloured surfaces and alteration penetrating deeply, but core stones still present.
letely hered	Rock is discoloured and changed to a soil but original fabric is mainly preserved. There may be occasional small core stones. Soil properties largely dependent on parent rock.
ual	Rock is discoloured and completely changed to a soil in which original rock fabric is completely destroyed. There is usually a large change in volume.

scolouration may sometimes be difficult to identify particularly in predominately d rocks.

teration may be described e.g. Kaolinized, mineralized.

TYPICAL ROCK NAMES
Sedimentary:
Conglomerate
Breccia
Sandstone
Siltstone
Chalk
Igneous:
Granite
Diorite
Gabbro
Andesite
Basalt
Tuff
Metamorphic:
Quartzite
Schist
Gneiss
Serpentine

6. For more detailed rock names and typical description of rocks see 'Geology for Engineers' chapter

STRENGTH: Uniaxial compressive strength MN/m^2	Term
	Very weak
1.25	
	Weak
5	
	Moderately Weak
12.5	
	Moderately Strong
50	
	Strong
100	
	Very Strong
200	
	Extremely Strong

Based on report by Geological Society Engineering Group Working Party on the preparation of drawings and plans in terms of Engineering Geology. London 1972.

Table 10.7 GENERAL BASIS FOR FIELD IDENTIFICATION AND CLASSIFICATION OF SOILS

In the order suggested for description.

COLOUR	Class	INSITU STRENGTH: Visual: Term	INSITU STRENGTH: Visual: Field Test	INSITU STRENGTH: Measured	STRUCTURE: Term	STRUCTURE: Field Identification	SOIL NAME: Particle Size mm	SOIL NAME: Basic Types	SOIL NAME: Visual Identification	SOIL NAME: Examples of composite types	OTHER INFORMATION
For example:	Non-cohesive (coarse grained): Cobbles and Boulders	Loose	By inspection of voids and particle packing.	No method	Homogeneous	Deposit consisting essentially of one type.	200	BOULDERS*	Only seen complete in exposures or pits.	Secondary constituents would be included in description as sandy, gravelly, silty, clayey. e.g: boulder GRAVEL	
Pink Red		Dense			Stratified	Alternating layers of varying types, or with bands, lenses or other material.	60	COBBLES*	Often difficult to recover from boreholes.		
Yellow Brown	Non-cohesive (coarse grained): Sands and Gravels	Loose	Can be excavated with spade. 50mm wooden peg can be easily driven.	Relative Density / S.P.T. N values blows/300mm Very Loose 0-4	Hetrogeneous	A mixture of types.	Coarse 20 Medium 6 Fine	GRAVELS*	Easily visible to naked eye. Particle shape can be described. Well or poorly graded	sandy sub-rounded GRAVEL	
Olive Green		Dense	Requires pick for excavating. 50mm wooden peg hard to drive more than a few inches.	Loose 4-10 Med. dense 10-30	Weathered	Particles are weakened and may show concentric layering.	2 Coarse 0.6 Medium 0.2 Fine	SANDS*	Visible to naked eye. Very little or no cohesion when dry. Well or poorly graded	slightly silty fine and medium SAND clayey fine SAND	
White Grey		Slightly cemented	Visual examination. Pick removes soil in lumps which can be abraded with thumb.	Dense 30-50 Very dense >50							
Supplement where necessary with:	Intermediate	Soft or loose	Easily moulded in the fingers.	Coarse material can be tested by SPT and finer material by shear strength measurement.	As for non-cohesive soils		0.06 Coarse 0.02 Medium 0.006 Fine	SILTS	Only coarse silt barely visible to naked eye. A little plasticity and exhibits marked dilatancy. Slightly granular touch. Disintegrates in water. Lumps dry quickly possess cohesion but can be powdered easily in the fingers.	clayey SILT organic SILT slightly sandy SILT	Minor constituents Shells Roots Crystals of gypsum selenite etc.
Light or Dark		Firm or dense	Can be moulded by strong pressure in the fingers.								
and: Pinkish	Cohesive (fine grained)	Very soft	Exudes between fingers when squeezed in fist.	Shear Strength kN/m² Very soft 20	Fissured	Breaks into polyhedral fragments along fissure planes.	0.002	CLAYS	Dry lumps can be broken but not powdered. They also disintegrate under water but more slowly than silt. Smooth touch and plastic, no dilatancy. Sticks to the fingers and dries slowly. Shrinks appreciably on drying usually showing cracks. Lean and fat clays show these properties to a moderate and high degree respectively.	gravelly CLAY sandy CLAY silty CLAY organic CLAY	Minor features: Bands Lenses Pockets Inclusions
Reddish		Soft	Easily moulded in fingers	Soft 20-50	Intact	No fissures					
Yellowish		Firm	Can be moulded by strong pressure in the fingers.	Firm 50-100	Homogeneous	Deposits consisting essentially of one type.					
Brownish		Stiff	Cannot be moulded in fingers.	Stiff 100-150	Stratified	Alternating layers of varying types. If layers are thin the soil may be described as banded or if very thinly laminated.					
Olive		Hard	Brittle or very tough	Hard 150	Weathered	Crumb or columnar structure.					
Greenish Bluish Greyish	Organic	Firm	Fibres compressed together	Shear strength can be measured in some circumstances.	Fibrous	Plant remains recognisable	Varies	PEAT and all vegetable matter	Usually dark brown or black in colour, often with distinctive smell. Light in weight.	clayey PEAT	
		Spongy	Very compressible and open structure.								
		Plastic	Can be moulded in hands and smeared between fingers.		Amorphous	Recognisable plant remains absent.					

After C.P. 2001 (8)

Examples of [illegible] Reddish brown medium dense sandy subangular fine and medium GRAVEL.
Dark grey stiff fissured CLAY with occasional shells.

* 1. Grading would be described according to the presence [illegible]

The basic information required for describing soils is as follows:

(a) Strength (i) cohesive soils—consistency
(ii) granular soils—relative density
(b) Colour
(c) Structure and texture
(d) Secondary constituents
(e) Primary constituents
(f) Minor features.

Table 10.6 gives the general basis for description and classification of soils with examples for typical descriptions of individual samples and is based on table I in CP 2001.[8] For roads and airfields the extended Casagrande Soil Classification was proposed by the then Road Research Laboratory.[20] This is on the basis of particle sizes for coarse soils and Atterberg limits for fine soils. This kind of system is used in the grouping of soils for earthworks materials in the *Specification for Road and Bridge Works*.[21] Dumbleton[22] proposed in 1968 a system based on this with the following modifications:

(a) Class boundaries are more precisely defined
(b) For coarse soils, more precise definition of poorly graded soils and soils containing a high proportion of fines
(c) For fine soils, more precise definition of silt and classes extended to cover very high plasticity clays and an appreciable proportion of coarse particles
(d) More comprehensive classification of organic soils
(e) Classification of boulder- and cobble-sized material.

Classification of rocks

The general basis for the classification of rocks for engineering purposes is rather more complex than that for soils. It is based upon weathering, structure, strength and rock type. Rocks may be igneous, metamorphic or sedimentary in origin with a very wide range of chemical and mineralogical compositions. Their intact strength may vary from little more than a hard clay to stronger than concrete, but their mass strength may be completely controlled by the intensity and direction of fractures, bedding planes and joints and the degree to which the rock is weathered. This latter may be slight, being limited to the surfaces of major discontinuities, or be so far advanced that the rock may be reduced to a soil consistency. For example, many mudstones and shales, when exposed, weather to clay consistency. The rock may consist of two or more distinct types interbedded in a regular or irregular way. It may be interbedded with soil, as in the Lower Lias Clay, which is often stiff clay and limestone interbedded. Any description should attempt to take account of these factors.

The basic information required in the description of rocks is as follows:

(a) Weathering
(b) Discontinuity spacing
(c) Strength
(d) Colour
(e) Grain size
(f) Rock name.

The classification is set out in detail in Table 10.6 and in many respects is similar to that proposed in reference 9, which particularly applies to the logging of rock cores. Where rocks are partly weathered to a soil, it may be necessary to develop a classification on the lines of that proposed for Keuper Marl[23] or Middle Chalk.[24]

Where rocks are to be used as constructional materials, e.g. rockfill, concrete aggregate, rip-rap, it will be necessary to extend the more general classification to include other properties. For example, the suitability of a rock for rockfill will depend on its response to blasting or its rippability by earthmoving plant and its stability under compaction plant. For use as concrete aggregate, shape when crushed and behaviour with cement would be important. This subject is further dealt with in Section 9.

REFERENCES

1. GLOSSOP, R., 'The Rise of Geotechnology and its Influence in Engineering Practice', *Geotechnique,* **18,** No. 2, 107–150 (1968)
2. *Admiralty Manual of Hydrographic Surveying,* Hydrographic Department, Admiralty, London (1967)
3. SZECHY, C., *Foundation Failures,* Concrete Publications Ltd (1961)
4. FELD, J., *Failures in Foundations,* Soiltest Inc., Evanston, USA (1965)
5. ROWE, P. W., 'The Relevance of Soil Fabric to Site Investigation Practice', *Geotechnique,* **22,** No. 2, 193–300 (1972)
6. MOLLARD, J. D., 'Photo Analysis and Interpretation in Engineering Geology Investigations: A Review', from *Reviews in Engineering Geology,* Ed T. Fluhr and R. F. Legget, *Geol. Soc. America,* New York (1962)
7. DUMBLETON, M. J. and WEST, G., 'Preliminary Sources of Information for Site Investigations in Britain', *Department of the Environment* RRL Report LR 403, Crowthorne (1971)
8. *Site Investigations,* CP 2001, British Standards Institution, London (1957)
9. 'Working Party Report on Logging of Rock Cores', Engineering Group, Geological Society of London (1970)
10. PALMER, D. J., *Writing Reports,* Soil Mechanics Ltd, London (1957)
11. COOPER, B. M., *Writing Technical Reports,* Penguin, Harmondsworth (1964)
12. 'Concrete in Sulphate-Bearing Soils and Groundwater', BRS Digest 90, HMSO (1970)
13. SANGLERAT, G., *The Penetrometer and Soil Exploration,* ELSEVIER BOOK DIV., PARIS (1972)
14. GRIFFITHS, D. H. and KING, R. F., *Applied Geophysics for Engineers and Geologists,* Pergamon, London (1969)
15. PARASNIS, D. S., *Principles of Applied Geophysics,* Chapman and Hall, London (1971)
16. SARGENT, G. E. G., 'Review of Acoustic Equipment for Studying Submarine Sediments', *Trans Section B. Inst. Min. and Met.,* **75,** (1966)
17. FRANKLIN, J. A. and DENTON, P. E., 'The Monitoring of Rock Slopes', *Q.J.Engng Geol.* (to be published)
18. TERZAGHI, K. and PECK, R. B., *Soil Mechanics in Engineering Practice,* Wiley, New York (1961)
19. PECK, R. B., 'Advantages and Limitations of the Observational Method in Applied Soil Mechanics', *Geotechnique,* **19,** No. 2, 171–187 (1969)
20. *Soil Mechanics for Road Engineers,* Road Research Laboratory, HMSO (1952)
21. *Specification for Road and Bridge Works,* Ministry of Transport, HMSO (1969)
22. DUMBLETON, M. J., 'The Classification and Description of Soils for Engineering Purposes. A suggested revision of the British system', RRL Rep. LR 182, Road Research Laboratory (1968)
23. CHANDLER, R. J., 'The Effect of Weathering on the Shear Strength Properties of Keuper Marl', *Geotechnique,* **19,** No. 3, 321–334 (1969)
24. WARD, W. H., BURLAND, J. B. and GALLOIS, R. W., 'Geotechnical Assessment of a Site at Mundford, Norfolk, for a Large Proton Accelerator', *Geotechnique,* **18,** No. 4, 399–431 (1968)
25. DUMBLETON, M. J. and WEST, G., 'Air Photograph Interpretation for Road Engineers in Britain', RRL Rep. LR 369, Road Research Laboratory, Crowthorne (1970)
26. HANNA, T. H., 'Foundation Instrumentation', *Trans. Tech. Pub.,* Cleveland (1973)
27. *Field Instrumentation in Geotechnical Engineering* (1973), *Proc. Symp. Brit. Geotech. Soc.,* Butterworths, London (1973)
28. WARD, W. H., MARSLAND, A. and SAMUEL, S. G., 'Properties of the London Clay at the Ashford Common Shaft: in situ and undrained strength tests', *Geotechnique,* **15,** No. 4, 321–344 (1965)
29. 'A New Approach for Taking a Continuous Soil Sample', Laboratorium voor Grondmechanica Paper No. 4, Delft, Holland (1966)
30. EURENIUS, J. and FAGERSTROMH, H., 'Sampling and Testing of Soft Rock with Weak Layers', *Geotechnique,* **19,** No. 1, 133–139 (1969)
31. *Specification for Rotary Core Drilling Equipment,* BS 4109, Part 1, British Standards Institution, London (1966)
32. 'Soil Sampler with metal foils. Device for taking undisturbed samples of very great length', Royal Swedish Geotechnical Institute Proc. No. 1 (1950)
33. *Methods of Testing Soils for Civil Engineering Purposes,* BS 1377, British Standards Institution, London (1967)

34. AKROYD, T. N. W., *Laboratory Testing in Soil Engineering,* Marshall, London (1957)
35. *Sampling and Testing of Mineral Aggregates, Sands and Filters,* BS 812, British Standards Institution, London (1967)
36. SKEMPTON, A. W. and PETLEY, D. J., 'Ignition Loss and other properties of Peats and Clays from Avonmouth, Kings Lynn and Cranberry Moss', *Geotechnique,* **20,** No. 4, 343–356 (1970)
37. 'Symposium on Vane Shear Testing of Soils', ASTM Special Technical Publication No. 193 (1957)
38. ROWE, P. W. and BARDEN, L., 'Importance of Free Ends in Triaxial Testing', *J.S.M.F. Div. ASCE,* **90,** SM1, 1–24 (1964)
39. BISHOP, A. W., and HENKEL, D. J., 'The Measurement of Soil Properties in the Triaxial Test', *Arnold,* London (1962)
40. SKEMPTON, A. W., 'Long-term stability of Clay Slopes', *Geotechnique,* **14,** No. 2, 75–102 (1964)
41. ROWE, P. W. and BARDEN, L., 'A New Consolidation Cell', *Geotechnique,* **16,** 162–170 (1966)
42. LADD, C. C., 'Stress Strain Modulus of Clay in Undrained Shear', *Proc. ASCE,* SMFE Div., **90,** 43 (1964)
43. *Annual Book of ASTM Standards,* Part II, 'Bituminous Materials for Highway Construction. Waterproofing and Roofing. Soils and Rocks. Peats, Moss and Humus Skid Resistance', ASTM (1973)
44. BUTLER, F. G., 'Piston Loading Tests in London Clay', *Proc. Midland Soil Mechanics Symposium,* Birmingham University (1964)
45. TOMLINSON, M. J., *Foundation Design and Construction,* Pitman, London (1969)
46. WHITAKER, T. and COOKE, R. W., 'A New Approach to Pile Testing', *5th Int. Conf. on SM and FE,* **2,** Dunod, Paris, 171–176 (1961)
47. WILUN, Z. and STARZEWSKI, K., *Soil Mechanics in Foundation Engineering,* Intertext Books, London (1972)
48. SKEMPTON, A. W., 'Discussion on Piles and Pile Foundations', *Proc. 3rd Int. Conf. on SM and FE,* Zurich, **3,** 172 (1953)
49. RICHART, F. E., HALL, J. R. and WOODS, R. D., *Vibrations of Soils and Foundations,* Prentice-Hall, New Jersey (1970)
50. GROOTENHUIS, P. and AWOJOBI, A. O., 'The In situ Measurement of the Dynamic Properties of Soils', *Proc. Symp. on Vibration in Civil Engineering,* Butterworths, London, 181–187 (1966)
51. HUBBERT, M. K. and WILLIS, D. G., 'Mechanics of Hydraulic Fracturing', *Trans. Am. Inst. Min. Engrs,* **210,** 153–168 (1957)
52. BISHOP, A. W., 'The Strength of Soils as Engineering Materials', *Geotechnique,* **16,** No. 2, 91–130 (1966)
53. MEIGH, A. C. and SKIPP, B. O., 'Gamma Ray and Neutron Methods of Measuring Soil Density and Moisture', *Geotechnique,* **10,** No. 2, 110–128 (1960)
54. WILKINSON, W. B., 'Constant Head Insitu Permeability Tests in Clay Strata', *Geotechnique,* **18,** No. 2, 172–194 (1968)
55. TODD, D. K., *Ground Water Hydrology,* Wiley, New York (1959)
56. STOUT, G. E., Ed, *Isotope Techniques in the Hydrologic Cycle,* Am. Geophys. Union, Byrd Press, Richmond, Virginia (1967)
57. HVORSLEV, M. J., 'Time Log and Soil Permeability in Groundwater Observations', *Bull. No. 36,* WES CoE, US Army, Vicksburg, Miss. (1951)
58. *Earth Manual,* US Department of the Interior, Bureau of Reclamation, Denver, Colorado, 1–751 (1962)
59. RICHARDS, B. G., 'Pavement Temperatures and their Engineering Significance in Australia', H.R.B. Special Report 103: 'Effects of Temperature and Heat on Engineering Behaviour of Soils', 254–265; Highway Research Board, Washington (1969)
60. CERNI, R. H. and FOSTER, L. E., *Instrumentation for Engineering Measurement,* Wiley, New York (1962)
61. 'Using a Remotely Controlled Borehole Camera', *Ground Engineering,* **3,** No. 5, 20–21 (1970)
62. 'Closed Circuit Systems ease Inspection', *Ground Engineering,* **6,** No. 1, 26 (1973)
63. 'Borehole Caliper', *Ground Engineering,* **6,** No. 1, 54 (1973)
64. KREBS, E., 'Optical Surveying with the Borehole Periscope', *Mining Magazine,* **116,** No. 6, 390–399 (June 1967)
65. *Cathodic Protection,* CP (to be published), British Standards Institution, London
66. SKIPP, B. O., 'Corrosion and Site Investigation', *Corrosion Technology,* Soil Mechanics Ltd (1961)
67. ARTHUR, J. R. F. and SHAMASH, S. J., 'Sampling of Cohesionless Soils without Disturbing the Particle Packing'. *Geotechnique* **20,** 4, 439–440 (1970).
68. STIMPSON, B., METCALFE, R. G. and WALTON, G., 'A New Field Technique for Sealing and Packing Rock and Soil Samples'. *Q.J.E.G.,* **3,** 2, 127–133 (1970).

11 REINFORCED AND PRESTRESSED CONCRETE DESIGN

REINFORCED AND PRESTRESSED CONCRETE DESIGN 11

11 REINFORCED AND PRESTRESSED CONCRETE DESIGN

S. C. C. BATE, B.Sc.(Eng.), Ph.D., F.I.C.E., F.I.Struct.E.
Building Research Establishment

INTRODUCTION

The design of reinforced and prestressed concrete has been increasingly codified during the post-war period. Before the war, recommendations for design of reinforced concrete had been contained in the Code of Practice[1] prepared by the Department of Scientific and Industrial Research, published in 1934, and in the Building Bylaws[2] of the London County Council, issued in 1938.

In 1951, the Institution of Structural Engineers published the First Report on Prestressed Concrete[3] which gave design procedure for prestressed concrete. At the present time BS Codes of Practice exist for reinforced concrete, CP 114,[4] for prestressed concrete, CP 115,[5] and for pre-cast concrete, CP 116,[6] which meet the deemed-to-satisfy provisions of the Building Regulations. In addition a BS Code, CP 2007,[7] deals with concrete liquid-retaining structures, and Standards for the design of bridges, including concrete bridges, and prestressed concrete pressure vessels are nearing completion.

The most important recent innovation however has been the publication in 1972 of a Unified Code[8] for Structural Concrete which will in due course supersede CP 114, CP 115 and CP 116. This code introduces a new approach to design, namely limit state design, which is based on the concept of designing structures for safety, serviceability and economy by taking into account specific risks for the occurrence of failure or unserviceability due to variability of the materials, inaccuracy in design assumptions or in construction, the variability of loading and the incidence of accidental damage. Whilst the approach to design is different, it is still necessary to maintain many existing methods of calculation and analysis but to regard them in a new light. Provision is, however, made for accommodating new data obtained from the statistical appraisal of the factors governing design, construction and performance as these become available. The basis for this approach has been developed mainly by the European Committee for Concrete assisted more recently by the International Federation for Prestressing, and their recommendations[9,10] have helped in the preparation of the new BS Code, CP 110. In applying these new recommendations to the new Code, the object has been to make the principles clear without making substantial differences to the dimensions of structures from those obtained using existing codes. When sufficient experience has been obtained in application, it will be possible to introduce more rational margins of safety in design and substantial economic advantages may then be expected.

In this section, the design philosophy and the procedures given will be directly related to CP 110. (The notation used is that of CP 110.) This Code, apart from being much more comprehensive than previous codes, also provides design charts for dealing with most calculations relating to the strength of sections and has since been amplified by the publication of a Handbook.[11] It is therefore an essential reference for designers and the object here is to give a brief explanation of its main content and background as an aid to its use.

Definitions

This section is concerned with the basic approach to design of reinforced and prestressed concrete. It deals with both cast in-place and pre-cast concrete whether reinforced or prestressed. It includes information on the use of plain or deformed steel reinforcing bars and with tendons which may be either pretensioned or post-tensioned. In this context some definitions and an indication of limitations may be useful.

(a) Reinforcement which is used to provide the tensile component of internal forces in reinforced concrete, generally consists of one of three types of material; plain round mild-steel bar produced by hot-rolling; plain square or plain chamfered square twisted mild-steel bar which has had its yield stress raised by cold-working and which is described as a Type 1 deformed bar when the pitch of twists is not greater than 18 times the nominal size of the bar (the nominal size is diameter of bar with the same cross-sectional area); ribbed bars, which may be hot rolled from steel with high yield stress or cold-worked by twisting from hot-rolled mild-steel, and which are known as Type 2 deformed bars when the transverse ribs have a spacing not greater than 0.8ϕ (when ϕ is their diameter in mm) with an area of projecting rib on a plane transverse to the axis of the bar of not less than 0.15ϕ mm^2 per mm.

Since steel reinforcement can only develop an effective tensile force by extension of the concrete by cracking, there is a limit on the maximum strength of steel that can be used. In general the yield stress should not exceed 500 N/mm^2 although higher strength steels may be used if particular care is taken to avoid excessive cracking or deflection.

(b) Tendons are used to impart a prestress to concrete before service loads are applied which offsets the tensile stresses which will later result from the application of these loads. Tendons are usually comprised of plain, indented or deformed cold-drawn carbon steel wire, of seven-wire or nineteen-wire strand spun from one or two layers respectively of cold-drawn carbon steel wire around a core wire, or of high-tensile alloy steel bar. The strength of steel used must be high enough for it to be extended sufficiently to avoid excessive loss of tension due to elastic contraction, creep and shrinkage of the concrete. In general it is not of lower tensile strength than about 1000 N/mm^2.

(c) In prestressed concrete, prestressing may be effected by pretensioning or post-tensioning the tendons. Pretensioned tendons are stressed before the concrete is cast. They are stretched either between temporary anchorages placed sufficiently far apart for a number of moulds to be assembled in line around the tendons, i.e. the 'long-line' method, or between the ends of specially strong moulds, i.e. the 'individual' mould method; in each case, concrete is then cast and allowed to harden before the tendons are released from their temporary anchorages. The methods are best suited to mass production in the factory and usually use wire or the smaller sizes of strand as tendons.

With post-tensioning, however, the tendons are stressed after the concrete has hardened and are usually accommodated in ducts within the concrete being held at their ends by anchorages, of which there are various proprietary types. Subsequently the ducts are grouted with cement grout to protect the tendons from corrosion. This method is mostly applied to site construction and tends to use tendons of relatively large size.

BEHAVIOUR OF STRUCTURAL CONCRETE

The characteristics of concrete that have conditioned its development as a structural material are its high compressive strength and relatively low tensile strength. In consequence its use for flexural members did not become practicable until it was discovered that steel reinforcement could be cast in the concrete to carry the bending tensile stresses whilst relying on the concrete to carry the bending compressive stresses. Experiment showed that mild steel, when present in the tension zone in relatively small amounts, provided a material with characteristics for deformation and strength which complemented those for concrete and provided a practical form of construction. Early research workers concluded that the presence of the steel increased the extensibility of the concrete. Later experiments showed, however, that this was not so. It then became clear that as the tensile stress in the steel of a beam increased beyond a small amount, which is appreciably less than that developed under service loading, cracks developed in the concrete. These cracks were controlled in width and numbers by the position of the reinforcement relative to the concrete surface and by the size of bars used. Thus with closely spaced bars near the surface, a large number of small cracks would develop, but with large widely spaced bars, the cracks would be fewer in number and much larger for the same stress in the steel. If the stress in the steel were increased the size of the cracks increased and their size was little influenced by the surface roughness of the steel, although at one time it was thought that roughening of the surface resulted in appreciably smaller cracks of larger numbers. It was eventually established that the main benefit of using bars with a roughened surface was in developing good end anchorage.

Because steel needs to extend to develop stress and hence causes cracking and deformation of the concrete, there is a limit to the strength of steel that can be used efficiently for reinforcement, since unsightly cracking, which could lead to severe corrosion in adverse conditions and unacceptable deflections, must be avoided. The use of steel in prestressed concrete, where the stress in the steel is imposed before the concrete member is subjected to external load, avoids this problem, since the initial tensile force is developed without extending the concrete, and so no upper limit is imposed on the strength of steel that can be employed. This was not, however, appreciated in the early development of prestressed concrete. Then, steel of relatively low strength was used with a small initial tension. The experimenters found that, although this was effective at the start, the initial prestress disappeared with time. Eventually, however, it was established that this nonelastic behaviour was limited in extent and that if a sufficiently large elastic extension was imparted to the steel, the nonelastic effects of creep and shrinkage of the concrete did no more than reduce the prestress by an acceptable amount. Although for a time, there was a tendency to underestimate the losses of prestress due to contraction of the concrete and to ignore creep in the steel tendons, recent research has now, however, clearly set the limits on what needs to be considered in design.

The performance of reinforced concrete and prestressed concrete beams under increasing load is characteristically different since cracking develops in different ways in each form of construction. This is illustrated by the results of tests on beams in each form of construction as illustrated in Figures 11.1 and 11.2.

Examined in more detail the deformation of the reinforced concrete beam under load is linear until cracking occurs; thereafter it approximates to a linear relationship until the steel yields as cracking becomes more extensive for beams of normal design. Subsequent deformation leads to the development of a hinge with continued yielding of the steel accompanied by damage to the concrete. This deformation continues at approximately constant moment until a stage is reached where the resistance reduces. The occurrence of this stage is influenced by the amount of transverse shear reinforcement in the section.

The prestressed concrete beam however remains uncracked usually until the service load is exceeded, and in this range its deformation is elastic. Once cracking has occurred deformation increases disproportionately rapidly with increasing load as cracks widen until the maximum load is reached. Subsequently there is a rapid reduction in resistance. Since the prestressed concrete beam is usually uncracked under service conditions its stiffness is greater than that of reinforced concrete beams of the same overall depth.

In continuous construction subjected to applied loads of short duration, deformation of both reinforced concrete and prestressed concrete members is elastic or effectively

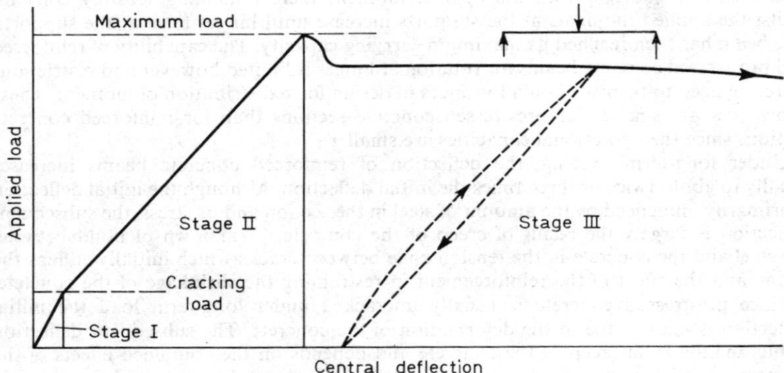

Figure 11.1 Relationship between applied load and deflection for a reinforced concrete beam showing recovery and reloading

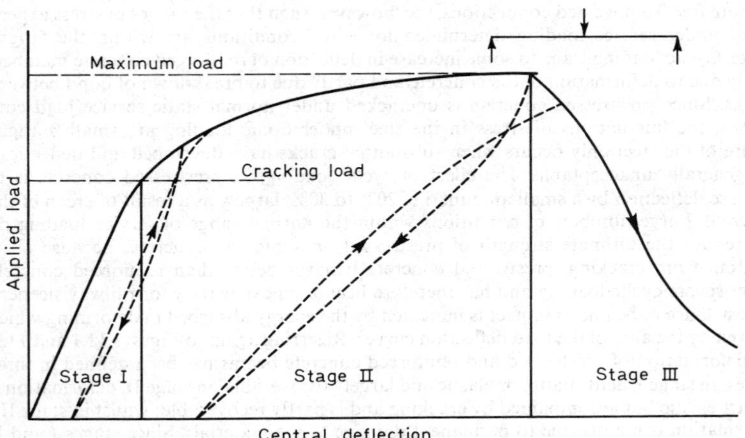

Figure 11.2 Relationship between applied load and deflection for a prestressed concrete beam showing recovery and reloading

elastic until service loads are exceeded. With further loading, as the applied moment at any section approaches the resistance moment at that section, there is a tendency for the moment to be relaxed and redistributed to sections that are less seriously stressed. Thus a loaded beam, built in at each end, may reach its maximum resistance moment at mid-span before the maximum resistance moments at the supports are attained; a hinge then forms at midspan with the applied moment there remaining sensibly constant whilst the applied moments at the supports increase until hinges form at the supports. The beam has then reached its maximum carrying capacity. The capability of reinforced and prestressed concrete beams for rotation at hinges is limited however and restrictions therefore need to be placed on allowances in design for redistribution of moment. These allowances are smaller for prestressed concrete sections than for reinforced concrete sections since their rotational capacities are smaller.

Under long-term loading, the deflection of reinforced concrete beams increases usually to about twice or three times the initial deflection. Although the initial deflection is primarily influenced by the amount of steel in the section and its stress, the subsequent deflection is largely the result of creep of the concrete, breakdown of bond between the steel and the concrete in the tension zone between cracks which initially stiffens the beam, and the effect of the reinforcement in restraining the shrinkage of the conctete.

Since prestressed concrete is usually uncracked under long-term load the initial deflection is mainly due to the deformation of the concrete. The subsequent deflection results mainly from creep of the concrete and depends on the combined effects of the prestress and the stresses due to applied load. The former tend to deform the member in the opposite direction to the latter. In consequence a loaded prestressed concrete member may initially have an upward deflection which can continue to develop upwards or downwards depending on how heavily it is loaded.

Under cyclic loading, reinforced concrete members usually fail in fatigue by fracture or yield of the reinforcement. The properties of most reinforcing steels, provided that they are free from welded connections, are, however, such that the ranges of stress experienced under service loading determined for static conditions are within the fatigue range. Cyclic loading leads to some increase in deflection of reinforced concrete members partly due to deformation of the concrete and partly due to breakdown of bond between cracks. Since prestressed concrete is uncracked under normal static service load conditions, the fluctuations of stress in the steel under cyclic loading are small. Fatigue failure of the steel only occurs when substantial cracks have developed and deflections are generally unacceptable. The effect of cyclic loading on prestressed concrete is to increase deflection by a small amount, i.e. 20% to 30%, largely as a result of creep of the concrete. Large numbers of repetitions within the normal range of service loading do not reduce the ultimate strength of prestressed or reinforced concrete. Because of its freedom from cracking, prestressed concrete behaves better than reinforced concrete under severe cyclic loading and has therefore been used extensively for railway sleepers.

Resistance of beams to impact is indicated by the energy absorbed in deforming which is given by the area of the load deflection curves. Referring again to Figures 11.1 and 11.2, the deformation of prestressed and reinforced concrete beams has been defined in three stages. In stage I, deformation is elastic and largely recoverable; in stage II, deformation is in part elastic but accompanied by cracking and is partly recoverable; whilst in stage III, deformation is mainly due to permanent damage to the materials. Since stages I and II represent the largest amounts of absorbed energy for prestressed concrete, this material has a considerable capacity for recovery after impact. For reinforced concrete, the energy absorbed in stage III is substantially greater than in the other two stages. Thus reinforced concrete does not show much recovery after impact but has a high ultimate impact resistance which is appreciably higher than that for prestressed beams designed for the same static loads. Prestressed concrete beams are, however, better in resisting repetitions of relatively light impacts with little residual damage.

So far performance has been considered mainly in terms of bending conditions, but conditions of direct stress in compression may exist in columns and walls. In such construction, unless high bending moments are also likely to occur, prestressed concrete would be unsuitable and reinforced concrete should be used with the steel acting in compression. For columns, transverse steel in the form of links is essential to contain the longitudinal steel and ensure ultimate resistance to strains in excess of those causing failure of plain concrete. Evidence from long-term tests also shows that the effect of creep of the concrete in a column under load is to raise the stress in the longitudinal steel to its yield stress and hence there is a need to retain it in its correct alignment. Walls when lightly reinforced are slightly weaker than walls without reinforcement and they can therefore only be treated as reinforced when the longitudinal reinforcement exceeds a specific minimum.

Other aspects of behaviour which are of importance are shear and torsion. In each case if these cause failure, the mode of failure tends to be brittle and less ductile than bending failures. Hence in design, the procedure is to avoid such failure by the inclusion of sufficient transverse reinforcement to ensure bending or compression failure in the event of severe overloading.

Members subjected solely to tension are relatively rare. If they are of reinforced concrete, then the role of the concrete is to protect the reinforcement which is designed to take the whole tensile force. In prestressed members, however, the precompressed concrete can sustain the tension until the load exceeds the cracking loading when the behaviour reverts to that of reinforced concrete with the steel carrying the whole of the tension, stiffened to some extent between cracks by the concrete.

For most building structures, the Building Regulations define fire resistance requirements, which are expressed in terms of a required endurance under service load when components are subjected to a standard heating regime. Both reinforced concrete and prestressed concrete are primarily influenced in their behaviour in fire by the behaviour of the steel at high temperature, as its temperature is raised its strength and yield characteristics are reduced. For reinforcing steels the rate of reduction in strength is lower than for steels used in tendons and hence greater amounts of protection are needed for prestressed concrete. This may take the form of concrete cover and the optional addition of insulating material. It is often easier, however, to provide the greater thicknesses of cover needed for tendons without loss of efficiency than that needed for reinforcement, since the positioning of tendons is governed by different requirements.

The need to provide adequate durability also affects the amount of cover required to the reinforcement or tendons. As concrete ages, carbon dioxide in the air causes carbonation of the concrete which, as it progresses, reduces its capability for inhibiting rusting of the steel. For dense concrete the rate of progress is very low but, since defects exist, experience has shown that a greater thickness of concrete is required to prevent spalling of the concrete caused by expansion of the corrosion products on rusting. Cover requirements also affect the width of cracks that are likely to occur and hence needs attention in dealing with serviceability.

These characteristics of the behaviour of both reinforced and prestressed concrete are considered in more detail in presenting design procedures.

PHILOSOPHY OF DESIGN

The early developments of the design of reinforced concrete were crystallised in this country by the issue in 1934 of Recommendations for a Code of Practice[1] prepared by a Committee set up by the Department of Scientific and Industrial Research. It was based on the premise that the stresses in the steel and concrete should not exceed certain permissible values, related to the strengths of the materials by safety factors, when the structure was subjected to the maximum loads that it would need to carry in service. The materials were assumed to behave elastically and compatability of strains between steel and concrete was ensured by assigning a value for the ratio of their moduli of elasticity. Some account was taken of the inelastic effects of creep of concrete by adopting a low value for the modulus of elasticity of concrete in determining the modular ratio for use in the design calculations. No account was taken of the effects of shrinkage and no estimate was made of the ultimate strength of the structure. When the British Standards Institution issued its first Code for Reinforced Concrete, CP 114, in 1948, it followed the same general approach. In the revision in 1957, however, there was an alternative method for design in flexure which limited the stresses to the same permissible values as for elastic design but assumed that they were distributed as at failure and avoided the use of the modular ratio; this was therefore a form of ultimate strength design.

Limitations on the permissible stresses in the steel and on span/depth ratios were imposed to guard against excessive deflection or cracking. Thus it could be argued that CP 114 provided for safety against failure and for the avoidance of unserviceability.

The earliest formal presentation of a design procedure for prestressed concrete was contained in the First Report on Prestressed Concrete[3] published by the Institution of Structural Engineers in 1951. Many of the recommendations in that report found their way into the British Standard Code of Practice for Prestressed Concrete, CP 115, issued in 1959. It conformed with CP 114 in the sense that it was based primarily on the limitation of stresses to permissible values related to the strengths of the materials with the object of preventing cracking and avoiding excessive deflection. It also provided for the calculation of ultimate strength and introduced separate requirements for minimum load factors for the dead and imposed loads.

Thus when the drafting of the unified code commenced in 1964 it had already been demonstrated that there were a number of limiting conditions or limit states which had to be considered by the designer in the overall conception of structural safety and adequacy. These were primarily limits of collapse, deformation and cracking, but other matters such as the effects of vibration, of fatigue, of deterioration with time or as a result of fire, needed attention in the design process.

Until recently the BS Code of Practice defining loading on buildings and other documents giving specific requirements for loading have specified precise maximum values. If the designer felt that higher loads might be imposed on a structure then it was his responsibility to determine how much higher that load should be, his decision being based primarily on his intuition and experience. In the new approach, the object is to state the level of loading which has a defined risk of being exceeded in a certain number of instances only for different specific forms of occupancy in buildings or other types of structure. Intuition and experience are still necessary but more guidance will be given in quantifying the risks in making design decisions.

A first step has already been made in expressing loads for design in a statistical form in CP 3, Chapter V, Part II, *Wind Loading.* The recommendations for the calculation of wind loads are based on the maximum wind speed for a three-second gust likely to occur once in the lifetime of the structure. There is therefore a specifically defined chance that the structure will have to stand a higher level of loading during its service life. Surveys of floor loading are now being made from which recommended values for loadings, related to their being exceeded on a specific proportion of occasions, will be determined. These provisions are, however, still to come in the future. For the present, floor loadings given in CP 3, Chapter V, Part I, will apply as now recognised in Building Regulations.

Since it is the object of limit state design to embrace all design aspects which need to be considered by the designer in preparing a design to satisfy all his clients' requirements, he must also review the possible consequences of accidents to the structure and ensure that unacceptable damage cannot be caused by trivial incidents. Recent occurrences of structural failures have placed emphasis on this aspect and have led to amendment of Building Regulations and to modification of Codes of Practice. So far these changes in requirement have been based mainly on engineering appraisals of structural performance made by experienced engineers. For the future, amplification of this approach is necessary to take account of current research on the incidence of accidental damage and the actual performance of construction to provide a more rational balance between the severity and likelihood of the incidents, and the extent of precautionary measures required.

A further major change that has taken place in the content of the structural codes of practice by the introduction of CP 110 in 1972 is the move from documentation giving guidance on the design of component parts of the structure and some aspects of their interaction to a code setting out the principles on which overall structural safety and serviceability are based together with better coordinated procedures for dealing with the more detailed aspects. Such developments have become necessary partly as a result of evolution of the philosophy in design and partly because the utilisation of materials, determined by the increases in the levels of stress in both steel and concrete under service conditions, has become much more onerous.

Criteria for limit state design

The basis for limit state design is the recognition of the engineer's intentions in the design process to provide, with reasonable certainty safe, serviceable and economic construction with a more rational expression of the margins of safety and of ignorance that are involved. It takes account of the variations that are likely to occur in the loads to which the structures are likely to be subjected, and of the variations in the strengths of the materials of which they are comprised. It covers the inadequacies of construction and methods of analysis and ultimately it will lead to design being based on specific degrees of risk related to particular conditions of unserviceability and failure.

Ideally the variation of load should be expressed in statistical terms which enable the characteristic load to be defined precisely, the characteristic load being the load with a small and acceptable risk of being exceeded in service. It may be expressed as:

$$\text{Characteristic load} = \text{Mean load} + (K \times \text{standard deviation of load})$$

or (1)

$$F_k = F_m + K\sigma_F$$

K is a factor with a value which ensures that the risk is small. For the time being, however, the characteristic loads are those which should be taken into account according to CP 3, Chapter V, Parts 1 and 2, and the Building Regulations, when designing buildings.

The variations in the strengths of materials are treated similarly. The characteristic strength of a material is the level of strength below which only a specific number of test results will lie. Thus:

$$\text{Characteristic strength} = \text{Mean strength} - (K_1 \times \text{standard deviation of strength})$$

or (2)

$$f_k = f_m - k_1\sigma_f$$

K_1 may be taken to have a value of 1.64 which ensures for a normal distribution of data that only 1 in 20 test results will be less than the characteristic strength. This approach to the definition of material strengths has already been accepted for both steel and concrete in the appropriate British Standards.

The characteristic values of the loads take account of normally expected variations in loading but do not allow for:

(i) unforeseen levels of loading
(ii) lack of precision in design calculations
(iii) inadequacy of the methods of analysis
(iv) dimensional errors in construction which alter the assumed positions or directions of loads and their effects, e.g. incorrect positioning of reinforcement and inaccurate alignment of columns in successive storeys.

The magnitude of the load used in design is therefore increased by factors, termed partial safety factors, to cater for these effects and to provide a margin of safety appropriate to the need for ensuring that a particular limit state is not reached. Thus for conditions of failure higher values are used than for those of serviceability. Where more than one form of load is taken as acting, values for the individual sources of load are reduced since their coincidence at high values is less likely. The loads for use in design are therefore the characteristic load $\times \gamma_f$, where γ_f is the partial safety factor appropriate to the limit state and the combination of loads considered. For simplicity, the Structural Code for Concrete restricts the number of different values to a minimum as will be seen later.

The strengths of the materials which are assumed in the design calculations are then defined in the specification for the construction and checked by physical tests. The strengths of the materials in the structure, however, are likely to differ from those determined from test specimens and may deteriorate with time. Partial safety factors for the materials are therefore introduced which allow for these features. The strengths of the materials used in design are therefore taken as the characteristic strength/γ_m, where γ_m has a value depending on the limit state being considered and the extent to which the material in the construction is likely to vary from that specified; it is therefore different for different materials and less for steel than for concrete.

The idealised and simplified situation for a homogeneous material is therefore illustrated in Figure 11.3. The provisions for safety outlined so far then require:

$$(F_m + K\sigma_F)\gamma_f \leqslant \frac{(f_m - K_1\sigma_f)}{\gamma_m} \tag{3}$$

This conforms relatively closely with accepted practice in recent revisions of some codes and is adopted in the new Code for Structural Concrete, CP 110. Current thought, however, accepts the view that a further partial safety factor should be introduced which

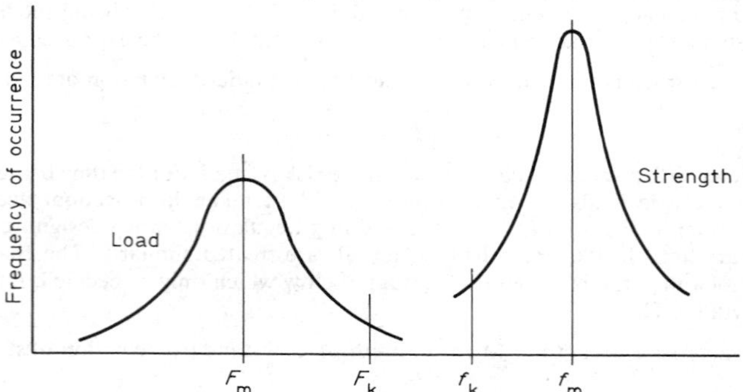

Figure 11.3 Idealised relationship between load and strength for a structure

would take into account the nature of the construction and its behaviour under overload conditions, e.g. whether it is capable of sustaining large deformations and so giving warning of the imminence of collapse, and the seriousness of failure in terms of risk to health, life and property. This factor γ_c might have a value of less than 1.0 for temporary construction which is not occupied by human beings but more than 1.0 for buildings with large spans used for public assemblies. Thus design would then require

$$(F_m + K\sigma_F)\gamma_f\gamma_c \leqslant \frac{(f_m - K_1\sigma_f)}{\gamma_m} \tag{4}$$

For a homogeneous material the global factor of safety relating characteristic loads to characteristic strength is then $\gamma_f\gamma_c\gamma_m$.

If the concept of relating the factors of safety to the nature of construction is not introduced then the global factor is $\gamma_f\gamma_m$. Since reinforced concrete and prestressed concrete are composite materials the value of the global factor for each limit state cannot be expressed as simply as this; it is dependent on the interaction of steel and concrete each of which have different values for γ_m. Also γ_f cannot be given a single value for each limit state since the partial safety factors for dead, imposed, wind and other loads may differ and change with different combinations of loads. Hence only upper and lower values for the global factor can be defined which makes comparison of the new code with earlier codes imprecise. Nevertheless in preparing the new code, the aim has been to avoid substantial changes in the dimensions of the resulting structures whilst at the same time obtaining more consistent levels of safety and leaving room for development on more rational lines in the future.

It is convenient to divide the limit states to be considered in design into two kinds, namely those concerned with collapse and those concerned with serviceability. Limit states of collapse deal with overturning of the complete structure, failure of the whole or a large part of the structure as a result of overstressing of a number of sections or buckling of a number of compression members or as a result of a serious accident; the effects of fire and fatigue may also be included. Deflection, cracking, deterioration, corrosion and vibration are all aspects of serviceability and require limits of acceptability to be set for consideration. In the new Code for Structural Concrete the limit states specifically dealt with are ultimate conditions in general, and deflection and cracking under the heading of serviceability. The criteria defining the serviceability limits are set out in Table 11.1.

The partial safety factors γ_f to be used with the characteristic loads for dead, imposed and wind loads obtained from CP 3, Chapter V, or other appropriate specification, are set out in Table 11.2 with notes on interpretation for ultimate and serviceability limit states. The combinations of loading to be taken are those which create the most severe conditions within the limits specified.

The partial safety factors for materials, γ_m, for the limit states considered are given in Table 11.3, also with notes on their interpretation.

The Code for Structural Concrete has special provisions to satisfy the requirement that, when a building suffers accidental damage, the amount of damage caused shall not be inconsistent with the original cause. It would seem reasonable to apply this same approach to other structures where safety and avoidance of excessive damage are necessary considerations in the event of accidents. To achieve this in buildings, attention should be given to the choice of an appropriate plan form since this may have a large influence on the mode of collapse as a result of an accident. When it is necessary to consider the effects of excessive loads outside those normally likely to be experienced or the residual strength of a structure after accidental damage the value of γ_f can be taken as 1.05 for those loads likely to be experienced. In these circumstances also, the values for γ_m for steel and concrete may be taken as 1.0 and 1.3 respectively. These low values for the factors are acceptable because the loading considered will not be experienced by most buildings and it would therefore be uneconomic to design for it to be sustained without damage.

CHARACTERISTICS OF MATERIALS (see also Chapter 4)

The grades of concrete used for reinforced and prestressed concrete construction in the Structural Concrete Code are expressed as the characteristic strengths determined from 28-days-tests on cubes; they are given in Table 11.4 with their application and properties relevant to design, including the increase in cube strength with age. No data are given for lightweight aggregate concrete since its properties are dependent on density in addition to strength as well as on the type of aggregate. The figures for flexural and indirect tensile strength refer to concretes made with smooth gravel aggregates; for crushed rock aggregates of rough texture, tensile strengths for the same grades of concrete would be somewhat higher. Generally the minimum grade of concrete for reinforced concrete will be grade 25, there are, however, areas in Britain where the natural aggregates are not of high enough quality for concrete to meet this grade even though its cement content is sufficient to conform with requirements for durability. Unless there are special needs, grades stronger than grade 40 are unlikely to be used for reinforced

Table 11.1 CRITERIA FOR LIMITS OF SERVICEABILITY

Limit state	*Reinforced concrete*	*Prestressed concrete*
Cracking	Maximum surface width of cracks ≯ 0·3 mm. Maximum surface width of cracks adjacent to steel for 'severe' aggressive environments ≯ 0.004 times nominal cover	Class I – no tensile stresses Class II – no visible cracking but limited tensile stresses Class III – maximum surface width of cracks ≯ 0.2 mm but for 'severe' aggressive environments ≯ 0.1 mm
Deflection	Final deflection ≯ Span/250 Where partitions or finishes may be cracked deflection ≯ 20 mm or ≯ Span/350 whichever is the less	As for reinforced concrete Upward camber if uniformity of adjacent units cannot be ensured ≯ Span/300

Table 11.2 PARTIAL SAFETY FACTORS γ_f FOR LOADS AND LOAD EFFECTS

	Limit state design loads[a]		
	Ultimate		*Serviceability*
Load combination			
Dead and imposed load	$\begin{cases}1.4\,G_k+1.6\,Q_k\\1.0\,G_k\end{cases}$	See Note (1)	$\begin{cases}1.0\,G_k+1.0\,Q_k\\1.0\,G_k\end{cases}$
Dead and wind load	$\begin{cases}0.9\,G_k+1.4\,W_k\\1.4\,G_k+1.4\,W_k\end{cases}$	See Note (2)	$1.0\,G_k+1.0\,W_k$
Dead imposed and wind load	$1.2\,G_k+1.2\,Q_k+1.2W_k$		$1.0\,G_k+0.8Q_k+0.8\,W_k$

[a] The figures given in the table are the values for the partial safety factors, γ_f.

NOTES

(1) The minimum load for this combination should not be less than $1.0\,G_k$. When alternate spans are considered loaded in the design of continuous beams, for example, the loaded spans should be assumed to carry $1.4\,G_k+1.6\,Q_k$ and the 'unloaded' spans to carry $1.0\,G_k$.

(2) The most serious load condition will usually occur when the design dead load is taken as $0.9\,G_k$ but for certain cantilvered structures, for example, a more serious situation may exist when the design dead load for part of the structure is $1.4\,G_k$.

Table 11.3 PARTIAL SAFETY FACTORS FOR MATERIALS, γ_m

Material	*Limit state values for* γ_m		
	Ultimate	*Serviceability*	
		Deflection	*Cracking*
Concrete	1.5[a]	1.0[c]	1.3[d]
Steel	1.15[b]	1.0	1.0

[a] This value is related to the standards of workmanship and supervision advocated in the code for the production of concrete. If these standards are not applied, a higher value should be used. It relates primarily to compressive strength of concrete.

[b] This value is for reinforcement in tension or tendons. For reinforcement in compression it is increased to $1.15+f_y/2000$.

[c] Calculations of deflection are based on the characteristic strength of the material and therefore the modulus of elasticity of concrete derived for this strength is less than the mean value for the component or structure which strictly speaking would be more relevant. This slightly conservative approach is justified in the interests of simplicity.

[d] This higher value for γ_m is selected for all calculations of stress.

Table 11.4 GRADES AND PROPERTIES OF STRUCTURAL CONCRETE

Grade[a] *(characteristic strength (28 days)* (N/mm^2))	*Cube strength*[a] (N/mm^2) *at the age of:* 7 days	28 days	2 months	3 months	6 months	1 year	*Flexural strength* (N/mm^2) *at 28 days*	*Indirect tensile strength* (N/mm^2) *at 28 days*	*Modulus*[a] *of elasticity* (KN/mm^2)	*Use*[a]
15	—	15	—	—	—	—	—	—	—	Reinforced concrete with lightweight aggregate
20	13.5	20	22	23	24	25	2.3	1.5	25	Reinforced concrete with natural dense aggregates
25	16.5	25	27.5	29	30	31	2.7	1.8	26	
30	20	30	33	35	36	37	3.1	2.1	28	Prestressed concrete for post-tensioning
40	28	40	44	45.5	47.5	50	3.7	2.5	31	Prestressed concrete for pre-tensioning
50	36	50	54	55.5	57.5	60	4.2	2.8	34	
60	45	60	64[b]	65.5[b]	67.5[b]	70[b]	4.6	3.1	36	

[a] Recommendations in the Code of Practice for Structural Concrete.
[b] Subject to demonstration to Engineer.

concrete. When lightweight aggregate is used, a lower grade, grade 15, is acceptable for reinforced concrete but it is preferable to use a higher grade for the lightweight aggregates of higher strength. No upper limit needs to be set on the strength for prestressed concrete and higher grades than grade 60 may therefore be used, but only special circumstances would justify the much greater cost and need for control and supervision.

Calculations for conformity with ultimate and serviceability limit states require the strength and deformation characteristics for concrete to be defined in numerical terms. In particular, data are required on the relationships between stress and strain in compression under short-term loading and on creep and shrinkage when serviceability in the longer term is being considered. These aspects of behaviour are dealt with in Section 4 and are simplified for design later in this section, but it must be recognised that there are substantial variations in the behaviour of concrete, depending on its constituent materials and environment, and that the values given for calculation should only be adopted if more reliable data are not available.

The strength properties of steel reinforcement and steel tendons are defined in British Standards which are summarised in Tables 11.5 and 11.6. For reinforcing bars of hot-rolled steel the characteristic strength is derived from the yield stress, but for cold-worked bars or wire reinforcement, it is derived from the 0.2% proof stress. The characteristic strength of tendons for prestressed concrete however is derived from their ultimate tensile strengths. In each case these are the relevant strengths for calculating ultimate strength for structural concrete members. Also in each case, the conformity with the specified characteristic strength is determined by ensuring that not more than 2 in 40 consecutive results of tests made during the production of the steel falls below the specified value.

The design calculations for serviceability of structural concrete require information on the modulus of elasticity of steel. The values adopted in the new Code are: for reinforcement for all types of loading 200 kN/mm^2, and for short-term loading for wire and strand of small diameter 200 kN/mm^2 and for alloy bars and strand of large diameter 175 kN/mm^2.

In prestressed concrete, considerations of serviceability require allowance not only for the effects of creep and shrinkage of the concrete but also relaxation of the tendons which may modify the prestress conditions substantially. Appropriate requirements are incorporated in the standards which therefore provide guidance on values for relaxation to be used in design.

The stress–strain characteristics for concrete and steel may be needed for calculations of the deformation of structural members under short-term loading or for assessing ultimate strength. These are given in Figure 11.4 for concrete, in Figure 11.5 for reinforcement and in Figures 11.6 and 11.7 for tendons.

In interpreting these curves, the value of γ_m appropriate to the limit state being considered should be obtained from Table 11.3. The values for the modulus of elasticity given in these figures should not be used for estimating the required extension of tendons. This data should be obtained from stress–strain curves for actual material being stressed, which are supplied by the manufacturers.

The creep and shrinkage characteristics of concrete are considered in Section 4. Where it is necessary to calculate long-term deformation, the effects of creep can be conveniently allowed for by adopting an effective modulus:

$$E_{c.eff} = \frac{E_{ci}}{1+\phi_c E_{ci}} \tag{5}$$

where E_{ci} is the short-term modulus of elasticity of concrete and ϕ_c is the creep of concrete under a unit stress of 1 N/mm^2.

The effects of shrinkage may be treated by assuming that the concrete contracts without a change in stress except for that caused by the effect of the change in strain on the stress in the steel. Some readjustment of strains then becomes necessary to balance the forces in the cross section by assuming that the concrete is stressed under this strain in proportion to the effective modulus.

Table 11.5 BRITISH STANDARDS FOR REINFORCING BARS FOR CONCRETE

BS No.	*Title—Specification for:*	*Material*	*Specified characteristic strength* (N/mm^2)	*Other requirements*
BS 4449	Hot-rolled steel bars for the reinforcement of concrete	Plain and deformed mild steel bars and high yield deformed bars	Mild steel – 250 High yield steel – 410 (yield stress)	Tensile strength at least 15% greater than actual yield stress. Elongation on 5.65(area)$^{1/2}$ at least 22% for mild steel and 14% for high yield steel. (Composition, bend and rebend tests.)
BS 4461	Cold-worked steel bars for the reinforcement of concrete	Rolled steel bars other than round bars cold worked to eliminate their yield point	For bars between 6 and 16 mm size – 460 For bars greater than 20 mm size – 425 (0.2% proof stress)	Tensile strength at least 10% greater than actual 0.2% proof stress. Elongation on 5.65(area)$^{1/2}$ at least 12% for bars 6–16 mm size and 14% for bars greater than 20 mm size (composition, bend and rebend tests).
BS 4482	Hard-drawn mild-steel wire for the reinforcement of concrete	Plain, indented or deformed hard-drawn mild steel wire specifically excluding wire for prestressing concrete	485 (0.2% proof stress)	Tensile strength at least 10% greater than actual 0.2% proof stress. (Composition and rebend test.)
BS 4483	Steel fabric for the reinforcement of concrete	Factory-made fabric formed by welding plain round, indented or other deformed wire to BS 4482 or by welding or interweaving cold worked steel bars to BS 4461	See BS 4482 and BS 4461	(Welding and dimensions of fabric.)

Table 11.6 BRITISH STANDARDS FOR PRESTRESSING TENDONS FOR CONCRETE

BS No.	*Title – Specification for:*	*Material*	*Size (mm)*	*Specified characteristic strength*	*Other requirements*
BS 2691	Steel wire for prestressed concrete	Round cold-drawn high-tensile steel wire, plain or deformed: (1) pre-straightened and normal relaxation, (2) prestraightened and low relaxation, or (3) in mild coils	(1) and (2) { 7 5 4 5 4 3.25 3	1 570 N/mm^2 1 570 1 720 1 570 1 720 1 720 1 720	0.2 % proof stress as a ratio of specified characteristic strength ≮ (1) 0.85 (2) 0.90 (3) 0.75 Relaxation – 1000 hour requirements for (1) and (2) in Table 11.18. (Composition, coil size and reverse bend test.)
BS 3617	Seven-wire strand for prestressed concrete	High-tensile steel wire strand which has been given a final heat treatment to produce (1) normal relaxation strand, or (2) low relaxation strand	(1) and (2) (all sizes) 6.4 7.9 9.3 10.9 12.5 15.2	44.5 kN (1 820)[a] 69.0 (1 840) 93.5 (1 790) 125 (1 760) 165 (1 750) 227 (1 640)	0.2 % proof load as a ratio of specified characteristic strength ≮ (1) 0.85 (2) 0.90 Relaxation – 1000 hour requirements in Table 11.18. (Composition, coil size and elongation)
BS 4757	Nineteen-wire steel strand for prestressed concrete	High-tensile steel wire strand: (1) as spun or which has been given a final heat treatment to produce (2) normal relaxation strand or (3) low relaxation strand	(1) { 25.4 28.6 31.8 (2) and (3) } 18.0	659 kN (1 550) 823 (1 540) 979 (1 480) 370 (1 760)	0.2 % proof load as a ratio of specified characteristic strength ≮ (2) 0.85 (3) 0.90 Relaxation – 1000 hour requirements in Table 11.18. (Composition, coil size and elongation)
BS 4496	Cold-worked high tensile alloy steel bars for prestressed concrete	As title of BS	20 25 32 40	325 kN (1 030) 500 (1 020) 800 (995) 1 250 (995)	0.2 % proof load 0.85 times specified characteristic strength. Relaxation – 1000 hour requirement in Table 11.18. Elongation on 5.65(area)$^{1/2}$ ≮ 6 %.

[a] Characteristic strengths N/mm^2 on nominal area.

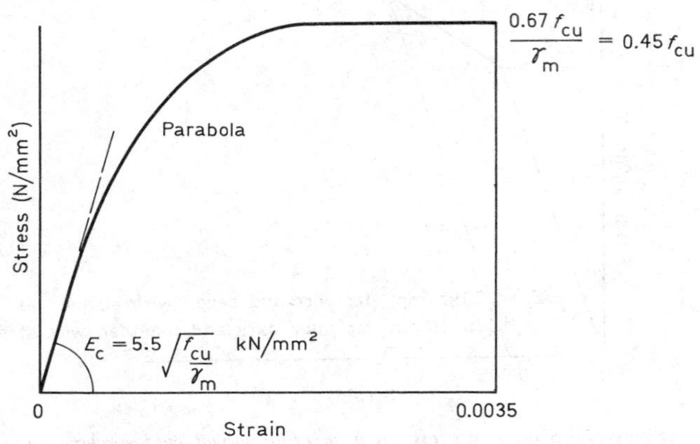

Figure 11.4 Short-term stress–strain curve for concrete

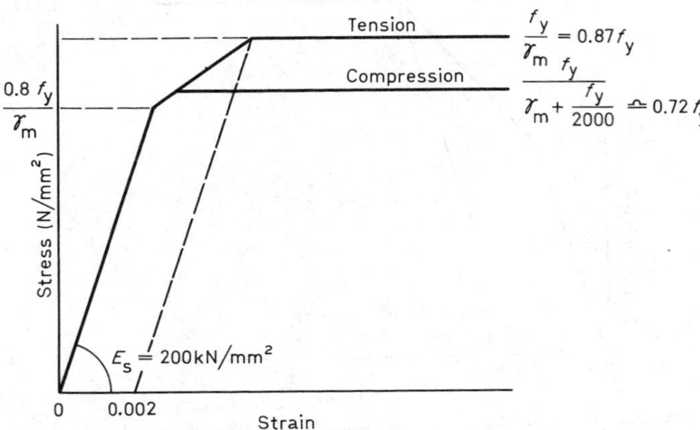

Figure 11.5 Stress–strain curve for reinforcement

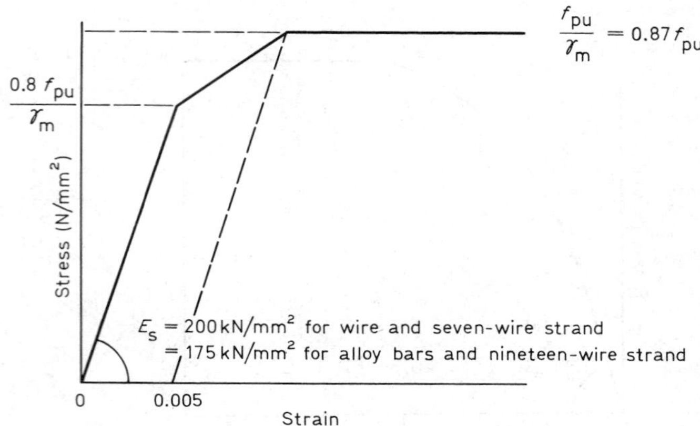

Figure 11.6 Short-term stress–strain curves for normal and low relaxation tendons

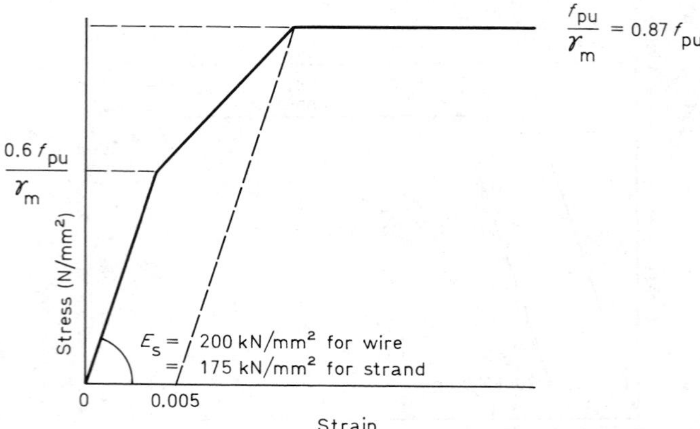

Figure 11.7 Short-term stress–strain curve for 'as-drawn' wire and 'as-spun' strand

ANALYTICAL AND DESIGN PROCEDURES

General

For most forms of concrete construction, with the possible exception of slabs, it is most convenient at the present time to base all design on elastic analysis of the structural system. The analysis would then apply directly to the serviceability limit states of deflection and cracking and, with some limited redistribution of moments and shear forces to the ultimate limit state. For slabs, other than one-way spanning slabs, it will usually be most convenient to use yield line methods or the strip method for ultimate design. For most construction it will usually be preferable to determine conformity with the serviceability limit states by using the arbitrary rules given in the Code for span/depth ratios and reinforcement detailing instead of calculating deflections and widths of cracks.

Other methods of analysis and design, where experimental procedures are used to develop the theoretical approach or to determine performance, are acceptable but will normally only be employed for specially complex structures or where repetition justifies more refinement than is obtained by established methods of calculation. The assessment of stresses in the region of load concentrations or of holes in continuous construction may be determined by photoelastic procedures. Model testing using special materials or scaled concrete has found applications in developing design methods, for example, in the design of concrete box-girder bridges and pressure vessels for nuclear power stations. In pre-cast concrete construction particularly, the behaviour of joints can only be established by tests on full-scale assemblies. It may also be economic to derive the dimensions of pre-cast components for mass production by testing successively refined prototypes to obtain the final form; this approach applies particularly in dealing with the requirements for fire resistance. The interpretation of test data for design requires the special care of experienced engineers since tests cannot embrace all the loads and load effects that may need to be sustained and the circumstances that exist in actual structures cannot necessarily be fully reproduced experimentally. When tests results are applied therefore, there is a need to show convincingly the justification for departures from established practice, especially so, if these lead to less conservative design. If test data are applied in contexts for which they were not originally sought even more caution is necessary.

For the purpose of analysis, the code offers three alternative methods for estimating beam and column stiffnesses:

(i) the concrete section
(ii) the gross section
(iii) the transformed section

The concrete section is the whole concrete section excluding the reinforcement, the gross section is the whole concrete section including the reinforcement allowing for the modular ratio, and the transformed section is the section of concrete in compression together with the reinforcement again allowing for the modular ratio. Generally, the concrete section is most convenient for use in design. For checking existing structures or for design in special circumstances, it would be more appropriate for reinforced concrete to use the transformed section; in construction where flexural cracking has occurred, however, the actual stiffnesses obtained by this assumption will be greater since the concrete exerts some tensile stiffening in the regions between cracks through bond with the reinforcement. The appropriate section for checking the design of existing or special prestressed concrete structures is the gross section since cracking does not usually occur with elastic deformation under service loads even for class 3 prestressed concrete.

Beams and slabs

The effective span (l) of beams or slabs, which are simply supported, is taken as either the distance between the centres of bearings or the clear distance between supports plus the effective depth whichever is the smaller. For continuous members, however, the effective span is the distance between the centres of the supports. Whilst for a cantilever which forms part of a continuous beam or slab, it is to the centre of the support, but for an isolated cantilever the effective span is to the centre of the support plus half the effective depth.

The effective width of a flange to a T beam may be taken as the smaller of the width of the web plus one fifth of the distance between points of zero moment or the actual width. Similarly the effective width of flange for an L beam is taken as the smaller of the width of the web plus one tenth of the distance between points of zero moment; for continuous beams the distance between points of zero moment may be assumed to be 0.7 L.

The lateral stability of beams may need attention, usually by providing for adequate restraints and stiffness. The limits between lateral restraints for simply supported beams or continuous beams should not exceed 60 b_c or 250 b_c^2/d, where d is the effective depth and b_c the breadth of the compression face midway between supports. For cantilevers restrained only at the support, its length should not exceed 25 b_c or 100 b_c^2/d.

Columns

A column is slender according to the Code when the ratio of the effective length to the corresponding breadth with respect to either axis is greater than 12. If each ratio is less than 12, the column is defined as short. The slenderness of a column should not exceed 60 times its minimum breadth.

Differentiation is also made between braced and unbraced columns and a column is described as braced when the lateral stability of the whole structure is ensured by providing walls or bracing to resist all horizontal forces.

The effective height of a column l_e may be assessed from Table 11.7 which indicates values for specific circumstances and is based on CP 114: 1967, or for framed structures from the formulae given below, in each case l_o is the clear height of column between end restraints.

For a braced column, the lesser of

$$l_e = l_o\,[0.7+0.05(\alpha_{c1}+\alpha_{c2})] \leqslant l_o \tag{6}$$

or

$$= l_o(0.85+0.05\alpha_{c\,min}) \leqslant l_o \tag{7}$$

For an unbraced column, the lesser of

$$l_e = l_o[1.0+0.15(\alpha_{c1}+\alpha_{c2})] \tag{8}$$

or

$$= l_o(2.0+0.3\alpha_{c\,min}) \tag{9}$$

where α_{c1} is the ratio of the sum of the column stiffnesses to the sum of the beam stiffnesses at one end of the column, α_{c2} is the ratio of the sum of the column stiffnesses at the other end of the column and $\alpha_{c\,min}$ is the lesser of α_{c1} and α_{c2}.

The values of α_c should be calculated from the stiffnesses of beams and columns in the appropriate plane of bending and should only take account of members which are properly framed into the structure. The stiffness should be taken as the ratio of the second moment of area of the concrete to its actual length.

Walls

The procedures for dealing with walls have much in common with those for columns. Plain or reinforced members are described as walls for design when the greater lateral dimension is at least four times the smaller lateral dimension. Although for plain walls reduction factors are given when the greater lateral dimension is less. To be described as a reinforced wall, the vertical reinforcement should not be less than 0.4% of the cross section (fire resistance requirements are that a reinforced wall should have at least 1.0% of vertical reinforcement); if it is less the wall should be treated as a plain wall. Short walls are defined as for short columns, i.e. the ratio of effective height to thickness should

Table 11.7 EFFECTIVE COLUMN HEIGHT

Type of column	*Effective column height, l_e*
Braced column properly restrained in direction at both ends	$0.75\ l_o$
Braced column imperfectly restrained in direction at one or both ends	A value intermediate between $0.75\ l_o$ and l_o depending upon the efficiency of the directional restraint
Unbraced or partially braced column, properly restrained in direction at one end but improperly restrained in direction at the other	A value intermediate between l_o and $2l_o$ depending upon the efficiency of the directional restraint and bracing

not exceed 12 (10 for lightweight aggregate concrete) otherwise it is treated as slender. A wall is also described as braced when walls or bracing are provided at right angles to withstand all horizontal loading. The overall stability of a building or other structure should not depend on unbraced walls and their use is therefore abnormal and it is not dealt with here although methods for design are given in the Code.

Where reinforced walls are built monolithically with other parts of the structure, their effective height should be determined as for columns. Otherwise when they are assumed to be loaded by simply supported members, reinforced walls should be treated in the same way as plain walls. For braced plain walls the effective height is three-quarters of the distance between lateral supports or twice the distance between a support and a free edge where supports are vertical or horizontal and provide resistance to lateral movement and rotation. If the lateral supports give resistance to lateral movement only, the effective height is the distance between centres of support or two-and-a-half times the distance between a support and a free edge.

The limit for the ratio of the effective height to thickness for a braced reinforced walls is 40 which may be increased to 45 if more than 1% of vertical reinforcement is provided. For unbraced reinforced walls and plain walls, this ratio should not exceed 30.

The axial forces on plain and reinforced concrete walls are determined by assuming that the beams and slabs, through which they are loaded, are simply supported. The moments on monolithically constructed reinforced walls should be determined by elastic analysis following the procedures used for frameworks. For plain walls and for reinforced walls where the construction is assumed to be simply supported, the eccentricity at right angles to the wall used for calculation should allow for the eccentricity of the forces applied to the wall, for tolerances on dimensions and for slender walls the deflection under load. The load from a concrete floor or roof is assumed to act at one-third of the bearing width from the loaded face of the wall. Where special hangers are used the eccentricity may exceed one-half the thickness of the wall. For braced walls, the resultant of the total load at any level immediately above a lateral support may be assumed to have no eccentricity at right angles to the wall.

Eccentricity in the plane of the wall should be calculated by elastic analysis. When horizontal forces need to be resisted by several walls they should be divided in proportion to the relative stiffnesses provided that the resultant eccentricity in each wall is not more than one third of its length.

Frames

The loads to be considered in design with their approrpiate γ_f factors have already been given in Table 11.2. When dealing with the ultimate limit state, these loads should be combined in worst manner possible is assessing the values of the maximum forces and

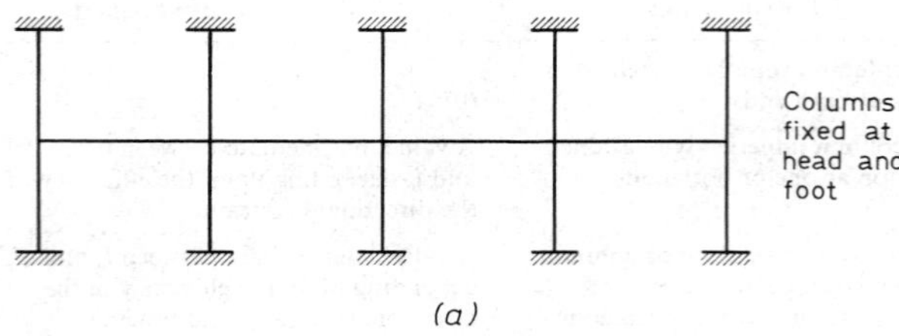

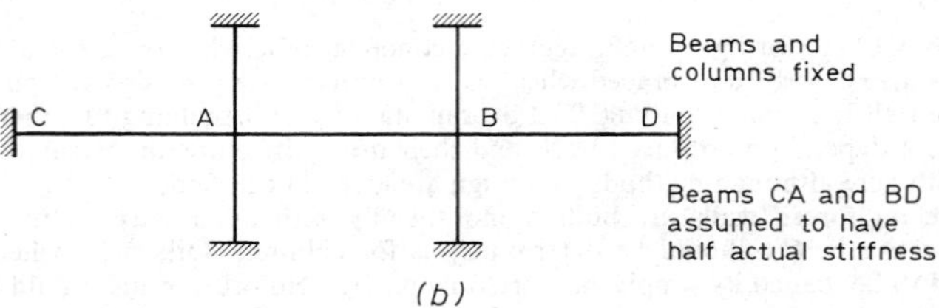

Figure 11.8 Alternative subframes for elastic analysis of braced frames for moments and forces.
(a) For columns and beams—vertical loads only. Loads to be considered
(1) Alternative spans loaded $1.4G_k + 1.6Q_k$
Other spans loaded $1.0G_k$
and
(2) Any two adjacent spans loaded $1.4G_k + 1.6Q_k$
Other spans loaded $1.0G_k$
(b) For beam AB only—vertical loads only. Loads to be considered:
(1) Centre span loaded $1.4G_k + 1.6Q_k$
Other spans loaded $1.0G_k$
and
(2) Any two spans loaded $1.4G_k + 1.6Q_k$
Other spans loaded $1.0G_k$
If the centre beam is the longer of two spans framing into a column, the moments in the column may also be determined from this subframe

moments in the structure. Since these calculations might be unnecessarily tedious for many forms of structure, the Code gives simplified rules for elastic analysis of frameworks by considering subframes which are illustrated in Figure 11.8 (a) and (b), for no-sway frames subjected to vertical loads only. For sway frames subjected to vertical and lateral load, the most severe of the moments and forces should be determined by elastic analysis from consideration, firstly, of the loading conditions given in Figure 11.8 for no-sway frames and, secondly, of a combination of loading conditions comprising:

(i) vertical loads of $1.2\ G_k + 1.2\ Q_k$ applied to all beams in subframes given in Figure 11.8(a), together with

(ii) a lateral load of $1.2\ W_k$ applied to the whole frame assumed to have points of contraflexure at the centre of all beams and mid-height of all columns.

In cases where overturning is a special consideration, for example in cantilevered structures, a loading combination which must be considered is comprised of $0.9\ G_k$ for dead loads counterbalancing overturning moments together with $1.4\ W_k$ for wind loads and $1.4\ G_k$ for dead loads which contribute to overturning.

If elastic analysis is used for frames, either an exact method or the approximate method quoted above from the Code, it is reasonable to allow for redistribution of moments for the ultimate limit state. Certain limitations are, however, necessary:

(i) Internal and external forces should be in equilibrium under each combination of loads considered.

(ii) The resistance moment provided at any section should be less than 70% for reinforced concrete or 80% for prestressed concrete of the moment at that Section determined from the analysis, to avoid due cracking under service loads.

(iii) The reduction in the elastic moment at any section due to a particular combination of loads should not be more than 30% for reinforced concrete or 20% for prestressed concrete of the maximum moment in the member determined elastically for all appropriate loading conditions since the capacity for 'plastic' deformation is limited.

(iv) Since redistribution cannot take place unless there is sufficient capacity for plastic deformation at the section, it is necessary to set an upper limit on the depth of the neutral axis when the section is subjected to the reduced moment to ensure ductility. For reinforced concrete this is

$$(0.6 - \beta_{red})d$$

and for prestressed concrete it is

$$(0.5 - \beta_{red})d$$

where β_{red} is the ratio of the reduction in resistance moment to the maximum moment determined elastically for all appropriate loading conditions.

Since the successive development of hinges in multi-storey frameworks leads to decreasing stiffness and hence to a greater risk of overall frame instability, a limit of 10% is set on redistribution of moments for all frameworks over four storeys in height.

Continuous beams

The distribution of moments and forces in continuous beams should be derived for the loading conditions as given Figure 11.8(a) and allowance should be made for moment redistribution as for frames. Where spans in reinforced continuous beams of three or more spans are within 15% of the largest span, the code allows the use in design of the moments and shears given in Table 11.8.

These moments and shears allow for redistribution of moments and apply to reinforced concrete beams but not prestressed concrete beams.

Continuous and two-way solid slabs

Slabs which are continuous in extent in one or two directions may be designed as simply supported provided that continuous ties that may be required for stability are incorporated. In such cases, cracking will develop in the top surface at their supports and some provision for dealing with this cracking will be needed in applying finishes.

Where slabs are required to span in one direction over a number of supports, they should be designed for the same moments and shear as for continuous beams.

If solid slabs are designed to span in two directions yield line methods of analysis or the strip method of design may be applied. The Code, however, gives simple methods for

Table 11.8 MOMENT AND SHEAR COEFFICIENTS FOR CONTINUOUS BEAMS

Position	*Moment* $-x(1.4\ G_k + 1.6\ Q_k)$	*Shear* $-x(1.4\ G_k + 1.6\ Q_k)$
At outer support	0	0.45
Near middle of end span	0.091	—
At first interior support	−0.111	0.6
At middle of interior spans	0.071	—
At interior supports	−0.100	0.55

the design of rectangular slabs for simply supported two-way slabs and two-way continuous or restrained slabs. Both analytical procedures for deriving moments are given in this section.

For rectangular two-way slabs, simply supported at their edges, the ultimate bending moments are obtained from the following:

Moment in the direction of the short span (l_x),

$$M_{sx} = \alpha_{yx}(1.4\ g_k + 1.6\ g_k)l_x^2 \tag{10}$$

Moment in the direction of the long span (l_y)

$$M_{sy} = \alpha_{sy}(1.4\ g_k + 1.6\ g_k)l_y^2 \tag{11}$$

where α_{sx} and α_{sy} are coefficients given in Table 11.9.

Table 11.9 BENDING MOMENT COEFFICIENTS FOR SLABS SPANNING IN TWO DIRECTIONS AT RIGHT ANGLES, SIMPLY SUPPORTED ON FOUR SIDES

$l_y l_x$	1.0	1.1	1.2	1.3	1.4	1.5	1.75	2.00	2.5	3.0
α_{sx}	0.062	0.074	0.084	0.093	0.099	0.104	0.113	0.118	0.122	0.124
α_{sy}	0.062	0.061	0.059	0.055	0.051	0.046	0.037	0.029	0.020	0.014

The moments obtained in this way tend to be conservative but since no provision is made in the Code for the torsional effects of preventing uplift at corners, this is desirable.

For two-way restrained slabs, the moments given are derived from analyses using yield line methods with provision for moment redistribution. These moments are expressed as follows:

Moment in the direction of the short span, (l_x),

$$M_{sx} = \beta_{sx}(1.4\ g_k + 1.6\ g_k)l_x^2 \tag{12}$$

Moment in the direction of the long span, (l_y),

$$M_{sy} = \beta_{sy}(1.4\ g_k + 1.6\ g_k)l_y^2 \tag{13}$$

where β_{sx} and β_{sy} are coefficients obtained from Table 11.10.

The loads transmitted to beams from two-way spanning slabs are obtained by dividing the slab geometrically as shown in Figure 11.9 and assuming that the loads on these contributory areas are supported by the adjacent beams.

Table 11.10 BENDING MOMENT COEFFICIENTS FOR RECTANGULAR PANELS SUPPORTED ON FOUR SIDES WITH PROVISION FOR TORSION AT CORNERS

Type of panel and moments considered	Short-span coefficients β_{sx}								Long-span coefficients β_{sy} for all values of l_y/l_x
	Values of l_y/l_x								
	1.0	1.1	1.2	1.3	1.4	1.5	1.75	2.0	
Interior panels									
1 Negative moment at continuous edge	0.032	0.037	0.043	0.047	0.051	0.053	0.060	0.065	0.032
Positive moment at midspan	0.024	0.028	0.032	0.036	0.039	0.041	0.045	0.049	0.024
One short edge discontinuous									
2 Negative moment at continuous edge	0.037	0.043	0.048	0.051	0.055	0.057	0.064	0.068	0.037
Positive moment at midspan	0.028	0.032	0.036	0.039	0.041	0.044	0.048	0.052	0.028
One long edge discontinuous									
3 Negative moment at continuous edge	0.037	0.044	0.052	0.057	0.063	0.067	0.077	0.085	0.037
Positive moment at midspan	0.028	0.033	0.039	0.044	0.047	0.051	0.059	0.065	0.028
Two adjacent edges discontinuous									
4 Negative moment at continuous edge	0.047	0.053	0.060	0.065	0.071	0.075	0.084	0.091	0.042
Positive moment at midspan	0.035	0.040	0.045	0.049	0.053	0.056	0.063	0.069	0.035
Two short edges discontinuous									
5 Negative moment at continuous edge	0.045	0.049	0.052	0.056	0.059	0.060	0.065	0.069	—
Positive moment at midspan	0.035	0.037	0.040	0.043	0.044	0.045	0.049	0.052	0.035
Two long edges discontinuous									
6 Negative moment at continuous edge	—	—	—	—	—	—	—	—	0.045
Positive moment at midspan	0.035	0.043	0.051	0.057	0.063	0.068	0.080	0.088	0.035
Three edges discontinuous (one long edge continuous)									
7 Negative moment at continuous edge	0.057	0.064	0.071	0.076	0.080	0.084	0.091	0.097	—
Positive moment at midspan	0.043	0.048	0.053	0.057	0.060	0.064	0.069	0.073	0.043
Three edges discontinuous (one short edge continuous)									
8 Negative moment at continuous edge	—	—	—	—	—	—	—	—	0.057
Positive moment at midspan	0.043	0.051	0.059	0.065	0.071	0.076	0.087	0.076	0.043
Four edges discontinuous									
9 Positive moment at midspan	0.056	0.064	0.072	0.079	0.085	0.089	0.1.00	0.107	0.056

Flat slab construction

Generally flat slab construction consists of a slab usually without supporting beams which spans between columns in two directions. Sometimes the depth of the slab is increased above the columns to provide drops and sometimes the column heads is flared to reduce shear stresses. The Code, CP 110, offers the choice of two methods for determining design moments, either the structure should be divided into two series of continuous frames, one longitudinally and the other transversely, or an empirical method

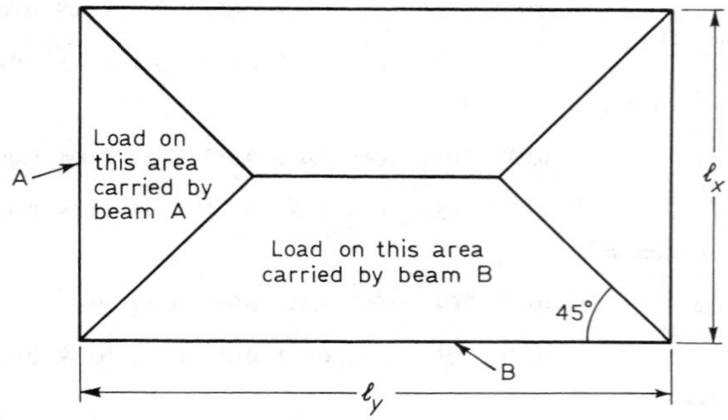

Figure 11.9 Two-way spanning slabs—areas contributing load to adjacent supporting beams

of analysis should be used. The latter method will be described here since it covers most normal requirements although some limitations are set on its application.

For the purpose of design the slab is split for convenience into column strips and middle strips in each direction. When drops are not used the column strip and the middle strip are each half of the transverse span. However, when drops are used, the column strip is the transverse width of the drop, the middle strip being the remainder of the width. The limitations imposed on form are that the panels should be rectangular of nearly uniform thickness with not less than three rows in each direction; the length of each panel should not exceed 4/3 times its width; adjacent panels should not differ in span or width by more than 15% and end widths or spans should not exceed those of adjacent interior panels. Drops should be rectangular in plan with a length of not less than one-third of the panel length in that direction. Lateral resistance of the structure to wind loading or other loads should be provided by shear walls or bracing.

The moment assumed to act in each direction

$$M_{\mathrm{ds}} = (1.4_{g}{}^{\mathrm{k}} + 1.6g_{\mathrm{k}}\frac{l_2}{8}\left(l_1 - \frac{2h_{\mathrm{c}}}{3}\right)^2 \qquad (14)$$

where g_{k} and g_{k} are the dead and imposed characteristic loads per unit area, l_1 is the span longitudinally and l_2 is the span laterally, and h_{c} is the average equivalent diameter of the column heads.

This moment is apportioned between the column strips and middle strips as given in Table 11.11.

If the width of the drop is greater than half the width of the panel, the moment in the column strip should be increased and the moment in the middle strip should be reduced in proportion. The figures in the table for exterior panels assume that columns provide the support, if, however, panels are supported on walls these will not apply and the Code should be consulted to obtain appropriate figures.

Stability

The Building Regulations require that all buildings of more than four storeys should be designed to resist accidental damage. The Regulations give specific requirements that they should be capable of sustaining removal of a structural member with excessive collapse resulting or should be able to sustain a pressure of 34 kN/m². The Code, which it is intended should become a deemed-to-satisfy document, makes certain recommendations to meet the functional requirement implied in the Regulations.

The layout of the structure and its general form should be such that it is not sensitive to accidental damage from any cause. If appropriate, it should be protected from the

Table 11.11 PROPORTION OF MOMENT IN COLUMN STRIP AND MIDDLE STRIP FOR FLAT SLABS

	Apportionment of moment as a percentage of M_{ds}	
	Column strip	*Middle Strip*
Interior panels		
With drops		
Negative moments	50	15
Positive moments	20	15
Without drops		
Negative moments	46	16
Positive moments	22	16
Exterior panels		
With drops		
Exterior negative moments	45	10
Positive moments	25	19
Interior negative moments	50	15
Without drops		
Exterior negative moments	41	10
Positive moments	28	20
Interior negative moments	46	16

impact of vehicles by provision of earth banks or bollards. In particular, the following provisions should be made:

(a) The horizontal wind load taken for ultimate design should not be less than 1½% of the total characteristic dead load above any level.
(b) Effective horizontal ties should be provided in directions at right angles in each storey level for all buildings around the periphery and internally.
(c) In buildings of five or more storeys in height, a continuous vertical tie should be provided in all walls and columns but special provisions should be made for plain load-bearing wall construction.

The reinforcement used for these ties should be located in such a way that the structure is effectively tied together and may be used for other purposes in the structure.

The peripheral steel tie, which should be positioned within 1.2 m of the edge of the building at each floor level and at roof level, should be effectively continuous and should be capable of sustaining a force of F_t when acting at its characteristic strength. $F_t = (20+4n_s)$ in kN, when n_s is the number of storeys but not more than 10, i.e. $F_t \not> 60$ kN.

The internal ties in each direction should be capable of sustaining a force of

$$\frac{F_t(g_k+g_k)}{7.5}\frac{l}{5} \quad \text{kN per metre width}$$

where (g_k+g_k) is the sum of the average characteristic dead and live loads on the floor in kN/m^2, and l, in metres, is the lesser of the greatest span in the direction considered between vertical supports or five times the clear storey height.

The continuous vertical ties in columns and walls should not be less than the minimum amounts of reinforcement given in Table 11.16. External walls and columns should be tied horizontally to the structure using the peripheral or internal ties or additional ties by steel reinforcement capable of developing at its characteristic strength of a force:

(i) For each column or metre width of wall, 2 F_t or $(l_o/2.5)F_t$ in kN, where whichever is lesser, where l_o is the floor-to-ceiling height in metres;

or

(ii) 3% of the total ultimate vertical load for which the floor member has been designed if this is greater.

Corner columns should be tied to the structure in two directions by such ties.

These requirements introduce a new feature into the design of buildings. They relate the magnitude of the ties provided to both the height of the building and the spans of the floors and thus aim to relate the margin of safety to the seriousness of failure if it occurred.

For plain load-bearing concrete walls, whether pre-cast or cast in situ, there are certain additional or alternative requirements. With respect to vertical ties, less steel than that required in reinforced walls may be provided in structures of five or more storeys. If it is less than 0.2% of the cross section of the walls, then the structure should be able to sustain the removal of any length of wall between effective lateral supports in any part of the structure. Effective lateral supports are stiffened sections of wall not less than 1 m in length capable of withstanding a horizontal force of 1.5 F_t kN/m height of the wall or substantial partitions at right angles with a weight of not less than 150 kg/m^2 tied to the wall with a tie capable of sustaining 0.5 F_t kN/m height of wall. The design should then provide for spanning or bridging across the missing component by cantilever or catenary action. Alternatively if the form of the structure makes it impossible to provide a bridging solution for dealing with a missing component, this component should be designed to sustain a lateral load of 34 kN/m^2 applied to the component from any direction together with the effects of this pressure on any subsidiary members attached to the component.

In making these calculations to deal with accidental effects, the value of γ_m for steel may be taken as 1.0, i.e. the tie force is obtained by assuming that the steel is acting at its characteristic strength and again that this is so if catenary action is involved. The value of γ_m for concrete should be taken as 1.3 and therefore if it is necessary to consider anchorage of the steel in the concrete then the anchorage stresses given in Table 11.15 may be increased in the ratio of 1.5 to 1.3. Only those loads likely to be acting simultaneously at the time of any accident or before remedial measures can be introduced need be considered using a γ_f value of 1.05. It would, for example, be appropriate to use a lower gust velocity in dealing with wind to suit the short time that the structure is at risk in a damaged state.

The purpose of these recommendations is to ensure that all structures are robust and insensitive to damage from localised disturbances. It is therefore important in providing for tie forces or for catenary or other action that the arrangements for anchoring reinforcement and for providing lateral supports are based on sound engineering principles.

REINFORCED CONCRETE

General

In the design of reinforced concrete to meet the requirements of the Code for Structural Concrete, it will usually be most appropriate to consider the ultimate limit state first and then check the design against the requirements for cracking and deflection. This might be inappropriate in exceptional circumstances, however, where steels of characteristic strength in excess of 500 N/mm^2 were being used or where spans were exceptionally long; in these cases cracking or deflection might govern design. In the sections that follow, design will be treated on the assumption that normal conditions obtain. For these, the Code gives simplified treatments for dealing with both cracking and deflection, it also gives methods more suited to the exceptional cases to which reference should be made if necessary.

Beams

BENDING

Ultimate resistance in bending is calculated by assuming that:

(a) sections which are plane before bending remain plane after bending
(b) stresses in the concrete may be determined using the stress–strain curve in Figure 11.4 (as assessed in the preparation of the design charts in the Code), or may be taken as uniformly distributed across the concrete in the compression zone with a value of $0.6f_{cu}/\gamma_m$, i.e. $0.4f_{cu}$ for deriving simplified formulae. Ultimate compressive strain in the concrete for analysis of sections is 0.003 5. For beams reinforced in tension only the compression zone should not have a greater depth than $d/2$
(c) the strength of the concrete in tension is ignored
(d) the stress in the steel is derived from the stress–strain relationships in Figure 11.5 with a value not greater than f_y/γ_m, i.e. $0.87f_y$ in tension and not greater than $f_y/(\gamma_m+f_y/2000)$ in compression, which is simplified to $0.72f_y$ for the development of formulae.

The simplified assumptions given may be used to derive design formulae which are shown in Figure 11.10 (a), (b), (c) and (d). The Code allows these formulae to be applied

For beams reinforced in tension only
$$C = 0.4f_{cu}b\,d_c \ngtr 0.2f_{cu}b\,d$$
$$T \ngtr 0.87f_yA_s$$

If $d_c < 0.5d$ then

$$M_u = 0.87f_yA_s\,d\left(1-\frac{1.1f_yA_s}{f_{cu}b\,d}\right) \qquad (15)$$

$$\ngtr 0.15f_{cu}b\,d^2 \qquad (16)$$

If $\dfrac{A_s}{bd} > 0.23\dfrac{f_{cu}}{f_y}$ then stress in the reinforcement $<f_y$.

For beams reinforced in tension and compression
$$C_c = 0.4f_{cu}b\,d_c \ngtr 0.2f_{cu}b\,d$$
$$C_s = 0.0035\left(\frac{d_c-d_1}{d_c}\right)A_s'E_s \ngtr 0.72f_yA_s'$$
$$T = 0.0035\left(\frac{d-d_c}{d}\right)A_sE_s \ngtr 0.87f_yA_s$$

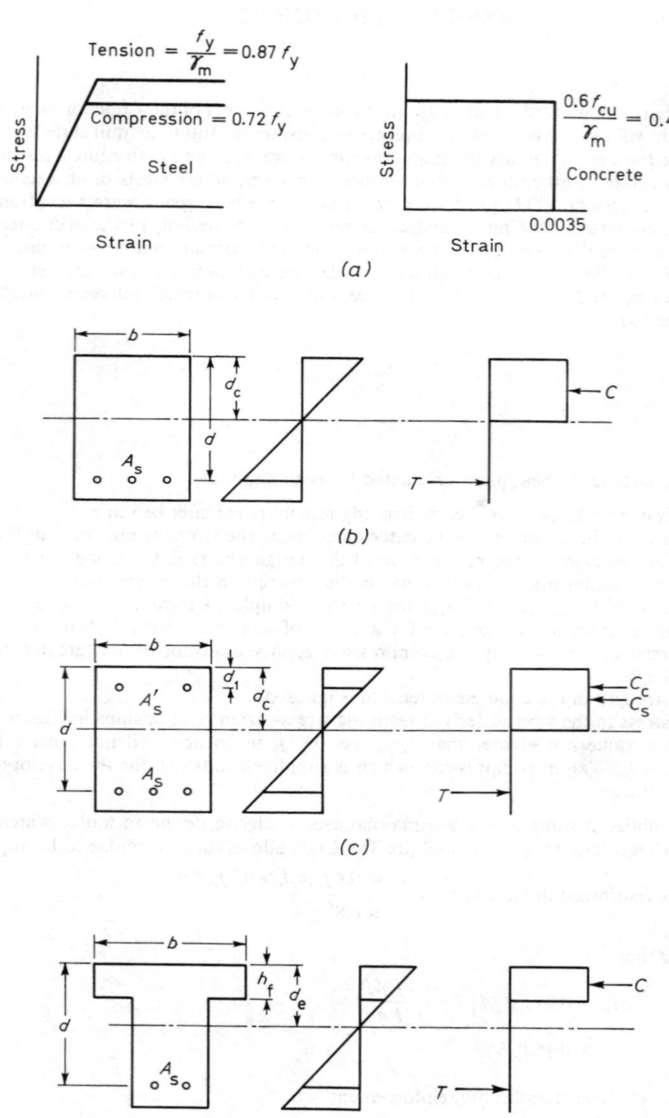

Figure 11.10 Flexural strength of beams—approximate methods. (a) Stress–strain curves assumed; (b) beams reinforced in tension only; (c) beams reinforced in tension and compression; (d) flanged beams

If $d_c \not> 0.5d$ and $d' \not> 0.5d_c$ where

$$d_c = \frac{0.87 f_y A_s - 0.72 f_y A_s'}{0.4 f_{cu} b}$$

then

$$M_u = 0.72 f_y A_s'(d - d_1)$$

For flanged beams $$+(0.87 f_y A_s - 0.72 f_y A_s')d\left(1 - \frac{1.1 f_y A_s - 0.9 f_y A_s'}{f_{cu} b\, d}\right) \tag{17}$$

If $h_f < d_c < d/2$ then

$$C = 0.4 f_{cu} b\, h_f$$

$$T = 0.87 f_y A_s$$

$$M_u = 0.87 f_y A_s(d - h_f/2) \tag{18}$$

$$\not> 0.4 f_{cu} b\, h_f(d - h_f/2) \tag{19}$$

provided that moment redistribution is restricted to not more than 10%. For full moment redistribution, considered earlier, either the more complex stress–strain relationships should be used in the calculations or, more readily, the Code design charts should be employed. This also applies when the form of section cannot readily be dealt with by the simple formulae.

SHEAR

The resistance of beams to shear is calculated for the ultimate limit state and takes account of the contribution from the concrete in addition to that provided by vertical link reinforcement or bent-up bars. All beams except those of little importance should be reinforced with links with a spacing not exceeding 0.75 d which should enclose all the tension reinforcement. The minimum area of both legs of this reinforcement:

$$A_{sv} = 0.001\,2\, s_v \,.\, b_t \text{ for high-yield steel}$$

$$A_{sv} = 0.002\,0\, s_v \,.\, b_t \text{ for mild steel,}$$

where s_v is the spacing of the links and b_t is the breadth of the beam at the level of the tension reinforcement.

The amount of shear that can be carried by the concrete is dependent on the strength of the concrete and the amount of longitudinal tension reinforcement at the section considered. It is determined from the Code as being $v_c bd$, where v_c is obtained from Table 11.12 for natural aggregate concrete and lightweight aggregate concrete and b is the breadth of the beam or of the rib for T beams.

Shear in excess of what can be carried by the concrete should be resisted by links and bent-up bars; the latter may be used as 50% of the shear reinforcement needed provided the minimum requirement for area and spacing of link reinforcement is met. Then

$$A_{sv} = \frac{s_v b(v - v_c)}{0.87 f_{yv}} \tag{20}$$

where $v = V/bd$, V is the shear force at ultimate load and f_{yv} is the characteristic strength of the link reinforcement but not more than 425 N/mm^2. Bent-up bars may be taken as contributing a vertical component in resisting the shear, provided that bearing and anchorage stresses are acceptable.

An overall limit is set on the shear that can be carried by a section irrespective of the amount of reinforcement. This is given by the limits to shear stress for natural and lightweight aggregate concrete given in Table 11.13.

DEFLECTION

The accuracy of any calculation of deflection is dependent on the extent to which the conditions of loading are known both with respect to position and duration, and to which the assumptions made in design conform with the behaviour of the structure in reality. Apart from the dead load on the structure which may be known with reasonable accuracy, the imposed load that is actually applied may be unpredictable. The structure itself may have non-load-bearing components such as floor screeds and partitions which

Table 11.12 ULTIMATE SHEAR STRESS IN CONCRETE v_c (N/mm^2)

Material	$100A_s/bd$	*Concrete grade*				
		15	20	25	30	40 *or more*
Natural aggregate concrete	0.25	—	0.35	0.35	0.35	0.35
	0.5	—	0.45	0.50	0.55	0.55
	1.0	—	0.60	0.65	0.70	0.75
	2.0	—	0.80	0.85	0.90	0.95
	3.0	—	0.85	0.90	0.95	1.00
Lightweight aggregate concrete	0.25	0.15	0.28	0.28	0.28	0.28
	0.5	0.20	0.36	0.40	0.44	0.44
	1.0	0.20	0.48	0.52	0.56	0.60
	2.0	0.25	0.64	0.68	0.72	0.76
	3.0	0.30	0.68	0.72	0.76	0.80

Table 11.13 MAXIMUM SHEAR STRESS IN CONCRETE BEAMS (N/mm^2)

Material	*Concrete grade*				
	15	20	25	30	40 *or more*
Natural aggregate concrete	—	3.35	3.75	4.10	4.75
Lightweight aggregate concrete	2.30	2.68	3.00	3.28	3.80

make a substantial contribution to its stiffness. The characteristics of the concrete may also not be known precisely since these are dependent on the different constituents actually used and provide additional uncertainty.

In most cases therefore it is not practical to calculate long-term deflections and this is recognised in the Code which gives a method of complying with the limit state requirements for deflection which take a number of features into account and define limits for span/depth ratios.

In defining these limits it is assumed that deflection of beams is primarily influenced by the conditions of support, the proportions of tension and compression reinforcement in the section and their levels of stress under service loading. These features are dealt with by introducing modifying factors given with the basic span/depth ratios in Figure 11.11(a), (b), (c), (d) and (e). To determine the limiting maximum value for the span/depth ratio when natural aggregate concretes are used, the appropriate value is obtained from (a); if the span is greater than 10 m for beams other than cantilevers, then it must be decided whether a deflection of $l/250$ is acceptable, if so the values given by the dashed lines should be used, if the limitation is 20 mm deflection then the full line should be used for spans up to 20 m in each case. This value for the span/depth ratio should be multiplied by a factor obtained from (b) which depends on the service stress (f_s) in the tensile reinforcement for particular circumstances which is obtained from (c), this allows for the effects of redistribution of moments and for the effects that result if more reinforcement is provided in the section than is required for ultimate conditions. Finally the basic span/depth ratio should also be multiplied by a factor from (d) if there is compression steel in the section and a factor from (e) if the width of the compression zone is greater than the width of the tension zone.

For lightweight aggregate concrete in reinforced members, the limit on span/depth ratio obtained for natural aggregate concrete should be multiplied by 0.85 to take account of the smaller elastic modulus and slightly greater creep and shrinkage that may be expected.

For special structures and those with spans in excess of 20 m (10 m for cantilevers) more detailed methods of calculation are necessary as indicated in Appendix A to the Code.

CONTROL OF CRACKING

The width of cracks at a particular level in a flexural member is dependent on a large number of parameters of which the following have been found by experimental investigations to be the most important:

(a) the distance from the nearest reinforcing bar spanning the crack;
(b) the distance from the neutral axis of the cross section; and
(c) the mean strain at the level of the section considered.

These investigations, which have shown that the surface characteristics of the bars have only a relatively insignificant effect, have led to the derivation of formulae recommended in the Code for use in special problems. For majority of circumstances, satisfaction of the requirements for the cracking limit state is provided for by meeting the detailed needs for distribution of reinforcement in the concrete section with respect to location and spacing, which is dealt with in the section on Requirements for reinforcement.

For liquid-retaining structures, cracking of the reinforced concrete adjacent to the liquid is restricted in the Code of Practice, CP 2007, by ensuring that the stress in the concrete in tension does not exceed certain values and relating the stress in the steel under the design loads for serviceability to these values using an effective modular ratio of 15 and the gross concrete section. For special mixes, the permissible concrete stresses in N/mm^2 are $f_{cu}/40+0.69$ in direct tension and $1.4\ (f_{cu}/40+0.69)$ in tension in bending.

In all construction and particularly in water-retaining structures, moist curing of the concrete plays an important part in minimising the extent of cracking due to drying shrinkage.

Slabs

The flexural strength of slab sections, including ribbed and flat slab sections, is treated in the same manner as for beam sections dealt with previously, design being based on the moments derived from the form of analysis referred to on page **11**–29. The Code gives specific rules for positioning the reinforcement in slabs.

For rectangular slabs simply supported along all four edges by beams or walls, all the reinforcement in each direction should be taken to within 0.1 of the span from the support and half the reinforcement should extend to the support. For restrained rectangular slabs, the slab is divided in each direction into a middle strip which is three quarters of the width and edge strips which are one eighth of the width of the slab. The maximum moments are taken as applicable without redistribution in the middle strip only. In the edge strips, the minimum amount of reinforcement allowed (see Table 11.16) should be incorporated. Tension reinforcement at midspan should extend in the bottom of the slab to within $0.25\,l$ of a continuous edge or $0.15\,l$ for a discontinuous edge and half of this reinforcement should extend to within $0.15\,l$ of a continuous edge and to within

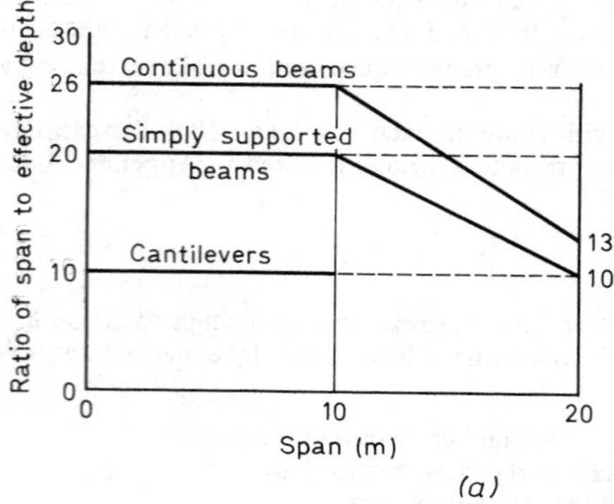

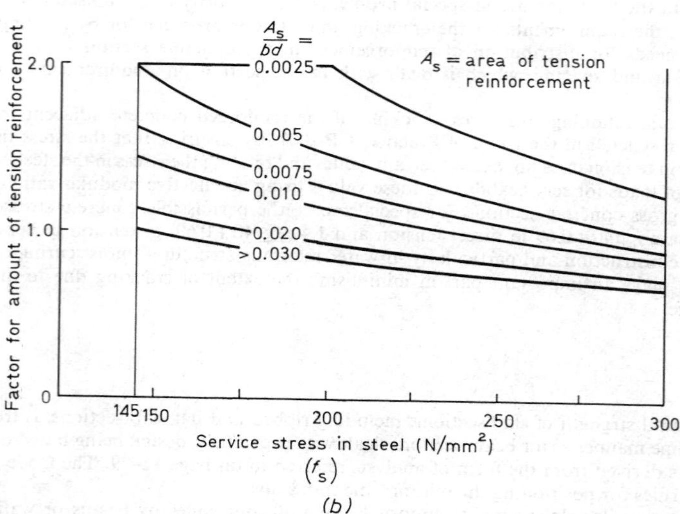

Figure 11.11 Factors for determining limiting span/depth ratios. (a) Basic span/depth ratios; (b) factor for tension reinforcement; (c) factor for determining service stress f_s for use in (b); (d) factor for compression reinforcement; (e) factor for web width

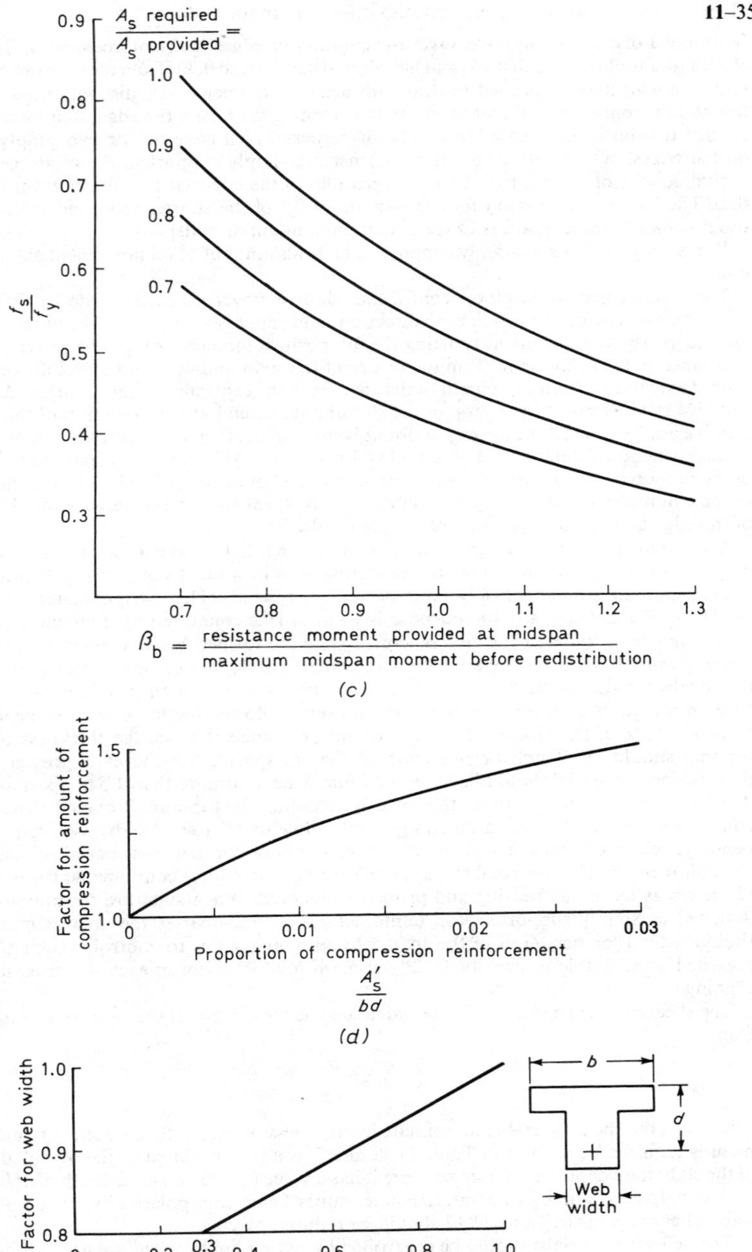
0.9
0.8
0.7
0.6
0.5
0.4
0.3
$\frac{f_s}{f_y}$
$\frac{A_s \text{ required}}{A_s \text{ provided}} =$
1.0
0.9
0.8
0.7
0.7
0.8
0.9
1.0
1.1
1.2
1.3
$\beta_b = \frac{\text{resistance moment provided at midspan}}{\text{maximum midspan moment before redistribution}}$
(c)
Factor for amount of compression reinforcement
1.5
1.0
0
0.01
0.02
0.03
Proportion of compression reinforcement
$\frac{A'_s}{bd}$
(d)
Factor for web width
1.0
0.9
0.8
0
0.2
0.3
0.4
0.6
0.8
1.0
Web width/Effective depth
b
d
Web width
(e)

50 mm of a discontinuous edge. Over the continuous edges the reinforcement in the top of the slab should extend to 0.15 l and half should continue to 0.30 l from the support. Some reinforcement may be needed to deal with negative moments at a discontinuous edge; this should consist of half that at midspan extending 0.1 l from the edge. Reinforcement against torsion is required additionally for corners which have one or two simply supported edges. When both edges at the corner are simply supported, the reinforcement should consist of bars in two directions parallel to the edges at top and bottom of the slab. The bars should extend for a minimum of 0.2 of the shorter span and each layer should contain three quarters of the maximum amount at midspan.

When only one edge is simply supported, half this amount of reinforcement should be used.

The arrangement of reinforcement in flat slabs is governed in the code by different rules. It should be positioned in both directions and should extend across the full width of each strip. The reinforcement resisting the full positive moments should be provided for a distance of 0.3 of the span from the centre of the span and 40% of this reinforcement should extend to within one eighth of the span from the centreline of the columns. All the negative reinforcement in the top of the slab should extend at least one fifth of the span into the adjacent panel the average amount being not less than one quarter of the span. If the equivalent diameter of the column head is less than 0.15 times the span, two-thirds of the reinforcement to resist the negative moments should be confined in the central half of the column strip. At the edges of slabs, reinforcement should extend to within 70 mm of the edge and be anchored by right-angle bends.

A convenient form of design of floor is the ribbed, hollow block or voided floor. It may be formed by special formwork, by casting in hollow blocks of ceramic or concrete with a minimum strength of 14 N/mm^2, or by forming voids of lightweight material which are left in place. Generally the purpose is to form ribs connected by concrete topping of the same quality as in the rib. Where the ribs are formed by permanent blocks the topping must have a thickness which is the greater of 40 mm or one-tenth the spacing of the ribs for the construction to be regarded as a slab rather than individual beams. Certain smaller thicknesses of topping are, however, allowed by the code for some forms of block floor. If the ribs are formed without permanent blocks, the thickness of the topping should be 50 mm or one-tenth of the rib spacing whichever is greater. The dimensions of the ribs should be at least 65 mm wide, not more than 1.5 m apart and in depth not more than four times their width excluding the topping. Provided that these conditions are met the design may in general follow that for solid slabs with respect to bending, deflection and shear. As in other forms of floor construction special rules apply to reinforcement. Half the total tension reinforcement should be continued at the bottom of the rib as far as the bearing and properly anchored. When slabs are continuous but designed as simply supported, top reinforcement of one-quarter the area at midspan should extend for one-tenth of the span into adjoining spans to control cracking. It is regarded as advisable to provide 0.12% of mesh reinforcement in each direction in the topping.

The shear stress in a slab should be calculated as for a beam. If the value of v obtained from

$$v = \frac{V}{b\,d} \qquad (21)$$

where V is the shear force due to ultimate loads, is less than $\xi_s v_c$, then no shear reinforcement is required. v_c is given in Table 11.12 and ξ_s is a factor related to the overall depth of the slab; for a depth of 250 mm or more it has a value of 1.00 and for a depth of 150 mm or less it has a value of 1.20, intermediate values being interpolated linearly. For flat slabs, shear stresses in Table 11.12 should be reduced by 20%.

The deflection of slabs should be controlled by setting limits on the span/depth ratios. For one-way spanning slabs, the same treatment should be adopted as for beams given earlier. The treatment should also be adopted for two-way spanning slabs basing the ratio on the smaller span and its reinforcement.

The control of cracking will be satisfactorily imposed by following the rules for the spacing of bars given on pages **11–44** and **11–45**.

Columns

The Code draws particular attention to the need when commencing the design of columns to consider the dimensional requirements for cover for durability and for cover and minimum dimensions for fire resistance. Minimum amounts of reinforcement are given in Table 11.16.

Moments forces and shears in columns are derived by the analytical and design procedures considered earlier. For most construction with braced columns, i.e. where the structure is fully braced against lateral loading, the ratio of effective height to minimum breadth will not exceed 12 and the columns may be treated as short columns.

For short braced axially loaded columns the ultimate load is derived from the assumptions made for beams and is given by:

$$N = 0.45 f_{cu} A_c + 0.72\, A_{sc} f_y \tag{22}$$

but to allow for inaccuracy in construction it is reduced by about 10% to give the relationship in the Code:

$$N = 0.4 f_{cu} A_c + 0.67\, A_{sc} f_y \tag{23}$$

where f_{cu} is the characteristic strength of concrete, f_y is the characteristic strength of steel in compression, A_c the area of concrete and A_{sc} the area of steel in compression.

If the braced short column has an approximately symmetrical arrangement (i.e. within 15% of span) of uniformly loaded beams then loading may be treated as axial using a reduced value for N to deal with the small moments induced:

$$N = 0.35 f_{cu} A_c + 0.60 f_y A_{sc} \tag{24}$$

This formula should not, however, be used for unsymmetrically loaded columns, e.g. corner columns.

Short columns subjected to moments and axial force should be designed using the same assumptions as those already given for beams. Appropriate formulae are derived in Figure 11.12, and conform with those in the Code.

$$N = 0.4 f_{cu} b\, d_c$$

Nominally reinforced section $\dfrac{M}{N} = e = \dfrac{h}{2} - \dfrac{d_c}{2}$. Therefore $d_c = (h - 2e)$

$$N = 0.4 f_{cu} b (h - 2e) \tag{25}$$

e should not exceed $h/2 - d_1$ to ensure that centre of compression remains within the reinforcing cage.

Fully reinforced section $N = 0.4 f_{cu} b\, d_c + f_1 A'_s + f_2 A_{s2}$ (26)

where $f_1 = 0.72 f_y$ for compression, $f_2 = -0.87 f_y$ for tension.

$$M = Ne = 0.2 f_{cv} b\, d_c (h - d_c) + f_1 A'_s (n/2 - d_1) - f_2 A_{s2} (h/2 - d_2) \tag{27}$$

If $e > (h/2 - d_2)$ then alternatively the section may be designed for $M_a = M + N(h/2 - d_2)$ and A_{s2} may be reduced by $N/0.37 f_y$.

In the use of these design formulae for short columns giving resistance moments and axial forces, an additional moment should be introduced to allow for the effects of tolerances in construction. This moment should be obtained from the axial load acting at an eccentricity of 0.05 times the minimum depth of the cross section. If the design charts in

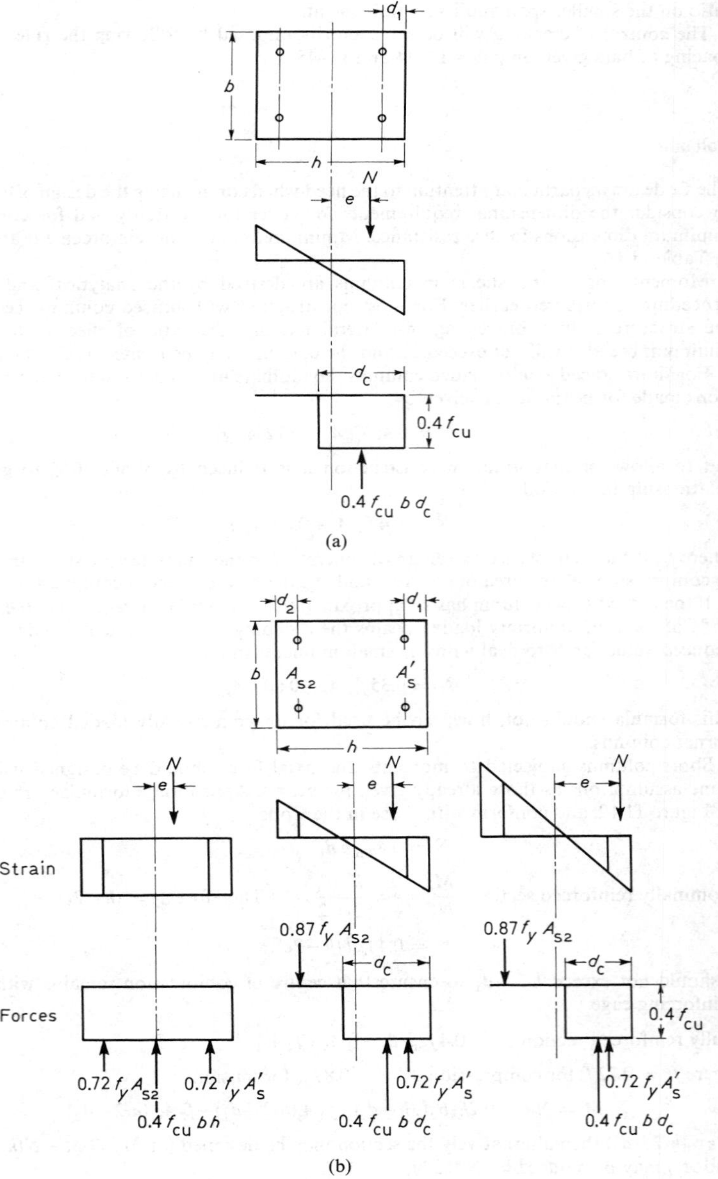

Figure 11.12 Strength of short columns (a) nominally reinforced section; (b) fully reinforced section

Parts 2 and 3 of the Code are used, it should be noted that allowance is made automatically for this additional moment.

Generally only single-axis bending about the critical axis need be taken into account even though significant bending may arise about the other axis. Where it needs attention the following conditions should be satisfied as required by the Code:

$$\left(\frac{M_x}{M_{ux}}\right)^{\alpha_n} + \left(\frac{M_y}{M_{uy}}\right)^{\alpha_n} \leqslant 1.0 \tag{28}$$

Where, M_x and M_y are the moments under load about the major and minor axis respectively and M_{ux} and M_{yx} are the maximum moment capacities for bending about each respective axis only under the ultimate axial load N.

α_n has a value of 1.0 when the ratio $N/N_{uz} \geqslant 0.2$ and 2.0 when $N/N_{uz} \geqslant 0.8$ with linearly interpolated values for intermediate values of N/N_{uz}, $N_{uz} = 0.45 f_{cu} A_c + 0.72 f_y A_s'$.

Short columns in lightweight aggregate concrete should receive the same treatment in design except that the ratio of l_c to thickness h should not exceed 10 instead of 12.

Slender columns

The methods suitable for dimensioning the cross sections of slender columns are the same as those for short columns. In the calculation of the moments, however, account needs to be taken of the additional moments resulting from deflection of the columns under load. Slender uniform columns of rectangular section with symmetrical reinforcement may be designed for N and M_t, where N is the axial ultimate load and $M_t = M_i + Ne_{add}$ in the following way.

For bending about a minor axis:

$$e_{add} = \frac{h}{1750}\left(\frac{l_e}{h}\right)^2\left(1 - 0.003\,5\frac{l_e}{h}\right) \tag{29}$$

For bending about a major axis:

$$e_{add} = \frac{h}{1750}\left(\frac{l_e}{b}\right)^2\left(1 - 0.003\,5\frac{l_e}{b}\right) \tag{30}$$

M_i is the maximum initial moment due to ultimate loads (but not less than 0.05 Nh) calculated from simple elastic analysis; for a braced column under ultimate load, where no transverse loads occur in its height M_i may be reduced to $0.4M_1 + 0.6M_2$ where M_1 is the smaller initial end moment (assumed negative if the column is bent in double curvature) and M_2 is the larger end moment (assumed positive). h is the overall depth of the section in the plane of bending, b is the width of the section at right angles to the plane of bending and l_e is the effective height in whichever plane it is the greater, as determined earlier.

M_t should in no case be less than M_2 nor should M_i be less than $0.4M_2$.

If slender columns are bent about both axes they should be designed for N and for moments M_{tx} about the major axis and M_{ty} about the minor axis, given by:

$$M_{tx} = M_{ix} + Ne_{add\,x} \tag{31}$$

$$M_{ty} = M_{iy} + Ne_{add\,y} \tag{32}$$

where $$e_{add\,x} = \frac{h}{1750}\left(\frac{l_{ex}}{h}\right)^2\left(1 - 0.003\,5\frac{l_{ex}}{h}\right)$$

where $$e_{add\,y} = \frac{b}{1750}\left(\frac{l_{ey}}{b}\right)^2\left(1 - 0.003\,5\frac{l_{ey}}{b}\right)$$

where M_{ix} and M_{iy} are the initial moments due to ultimate loads about the major and minor axis respectively; l_{ex} and l_{ey} are the effective heights in respect of the major and minor axis respectively, as determined from pages **11**–20 and **11**–21, and b and h are the width and depth of the column in respect of the major axis respectively. (These notation are not identical with those for bending about major or minor axis.)

For slender columns in lightweight aggregate concrete, the divisor in the foregoing expressions should be 1200 in place of 1750, otherwise the method of design is unchanged.

DEFLECTION AND CRACKING

In general no check is needed to assess the deflection of columns. The Code suggests that a check is only necessary for unbraced columns with an average l_e/h for all columns of more than 30. This does not apply for structures of single-storey construction with claddings or finishes which are not susceptible to damage due to movement.

Cracks due to bending are only likely to be experienced in columns when the ultimate axial load is less than about half the ultimate compressive capacity of the concrete under axial loading. Thus the reinforcement provided should conform with the requirements given for beams.

Walls

REINFORCED CONCRETE WALLS

The design of short braced reinforced concrete walls follows closely that of columns. Where a wall cannot be subjected to significant moments its ultimate load is given by

$$N = 0.4 f_{cu} A_c + 0.67\, A'_s f_y \tag{33}$$

This is the same as the formula for columns and includes a reduction to allow for the effects of constructional tolerances.

If the spans on either side of a wall do not differ by more than 15% and are uniformly loaded then it may be assumed that

$$N = 0.35 f_{cu} A_c + 0.60\, A'_s f_y \tag{34}$$

Where axial forces and moments need to be resisted, the moment should not be less than $0.05 n_w h$ where n_w is the axial load for ultimate loads per unit length and h is the

Table 11.14 λ_w FOR DIFFERENT TYPES OF CONCRETE AND DIFFERENT LENGTHS OF WALL

Type of concrete	*Ratio* Clear height between supports / Length of wall: 1.5 *or greater*	0.5 *or less*
Fully compacted concrete – grade 25 or above	0.4	0.5
Fully compacted concrete – grades 10, 15 and 20	0.35	0.4
No-fines concrete	0.3	0.35

thickness of the wall. For design of the section, the assumptions made for beams and columns should be applied.

If there is substantial bending in the plane of the wall and at right angles to it, the distribution of load along the length of the wall should be determined and then unit lengths of wall should be considered separately to deal with transverse bending.

Slender walls are treated as slender columns.

PLAIN CONCRETE WALLS

For short braced plain walls, the axial ultimate load per unit length, n_w, should be:

$$n_w = (h - 2e_x)\lambda_w f_{cu} \qquad (35)$$

where e_x is the resultant eccentricity at right right angles to the wall but not less than 0.05 h, h being the thickness of the wall, λ_w is a coefficient with values given in Table 11.14. Where the length of the wall is less than four times the thickness λ_w should be reduced by a factor varying linearly from 1.0 to 0.8 as the length reduces from 4 to 1 times its thickness.

At every section of a braced slender plain concrete wall the following relationship must also be satisfied:

$$n_w = (h - 1.2\,e_x - 2\,e_a) \qquad (36)$$

where e_a is the additional eccentricity and is $l_e^2/2500\,h$, l_e being the effective height of the wall.

Requirements for reinforcement

BOND AND ANCHORAGE

Bond and anchorage as distinct from cracking in reinforced concrete is substantially affected by the surface characteristics of the reinforcement. As already mentioned, the Code recognises three types of bar surface, those corresponding to plain round bars, Type (1) deformed bars which are usually twisted bars of square or chamfered square cross section and Type (2) deformed bars which are usually of circular cross section with transverse ribs.

Where there are rapid changes in the stress in the longitudinal direction over a relatively short length, or there are changes in the depth of the section, the local bond stresses should be checked for the ultimate condition to ensure that the stresses given in Table 11.15 for the different grades of concrete and types of steel are not exceeded. The local bond stress

$$f_{bs} = \frac{V}{\Sigma u_s d} \qquad (37)$$

where, Σu_s is the sum of the perimeters of the bars, V is the shear at the section and d is the effective depth.

If there is an angle θ_s between the tension steel and the compression face of the member and M is the moment at the section, then V should be increased or decreased, depending on the effect of the slope on the stress in the steel, by $(M/d)\tan\theta_s$.

At the end of any bar, a sufficient length should be provided to anchor effectively the tensile or compressive force in that bar. The length required is found by dividing the force in the bar by the product of the effective perimeter of the bar or groups of bars and the ultimate anchorage bond stress which is given in Table 11.15 for the different grades of concrete, types of bar and whether in tension or compression.

The stress values given in Table 11.15 apply for concretes made with natural aggregates, for lightweight weight aggregate concrete they should be multiplied by 0.5 for plain bars and 0.8 for both types of deformed bar.

The effective perimeter of a bar is taken as 3.14 times its nominal size, which is the diameter of a bar of the same cross-sectional area as the actual bar for bars of uniform cross section and for ribbed bars with less than 3% of their weight in the transverse ribs; where the bars have more than this amount of steel in the ribs that in excess of 3 % is ignored in determining nominal size. For a group of bars, the effective perimeter is the sum of the effective perimeters of the individual bars multiplied by 0.8 for two bars, by 0.6 for three bars and by 0.4 for four bars.

Anchorage provided by hooks and bends conforming with BS 4466 is taken as the smaller length of 24 times the bar size or 8 times the internal radius for hooks or 4 times

Table 11.15 ULTIMATE LOCAL BOND STRESSES AND ANCHORAGE BOND STRESSES (N/mm²)

Material	*Condition*	*Concrete grade*				
		15	20	25	30	40 *or more*
Natural aggregate	Local bond					
	plain bars	—	1.7	2.0	2.2	2.7
	square twisted bars	—	2.1	2.5	2.8	3.4
	ribbed bars	—	2.5	3.0	3.4	4.1
	Anchorage bond					
	Tension					
	plain bars	—	1.2	1.4	1.5	1.9
	square twisted bars	—	1.7	1.9	2.2	2.6
	ribbed bars	—	2.2	2.5	2.9	3.4
	Compression					
	plain bars	—	1.5	1.7	1.9	2.3
	square twisted bars	—	2.1	2.4	2.7	3.2
	ribbed bars	—	2.7	3.1	3.5	4.2
Lightweight aggregate concrete	Local bond					
	plain bars	0.7	0.9	1.0	1.1	1.3
	square twisted bars	1.4	1.7	2.0	2.2	2.7
	ribbed bars	1.6	2.0	2.4	2.7	3.3
	Anchorage bond					
	Tension					
	plain bars	0.5	0.6	0.7	0.8	1.0
	square twisted bars	1.1	1.4	1.6	1.8	2.1
	ribbed bars	1.4	1.8	2.0	2.3	2.7
	Compression					
	plain bars	0.6	0.8	0.9	1.0	1.2
	square twisted bars	1.3	1.7	1.9	2.2	2.6
	ribbed bars	1.8	2.2	2.5	2.8	3.4

To conform with the description of square twisted bar and ribbed bar given above the deformation of the bars should satisfy the requirements for Type 1 and Type 2 deformed bars respectively as stated on page **11**–3.

the internal radius for 90° bends. The radius of the bend is limited to twice the bend test radius given in the appropriate British Standard or by the ultimate bearing stress in the concrete. The bearing stress is given by

$$\text{Bearing stress} = \frac{F_{bt}}{r\phi} \text{ and should not exceed } \frac{1.5 f_{cu}}{1+2\phi/a_b} \qquad (38)$$

where, F_{bt} is the tensile force in the reinforcement, r is the internal radius of the bend. ϕ is the size of the bar or size of bar of the same area for a bundle, f_{cu} is the characteristic cube strength and a_b is the distance between bars perpendicular to the bend or the cover $+\phi$ for bars next to a face. For lightweight aggregate concretes, the bearing stress should not exceed 2/3 of the bearing stress for natural aggregate concretes of the same strength.

Anchorage of links may be achieved by passing the link 90° around a longitudinal bar

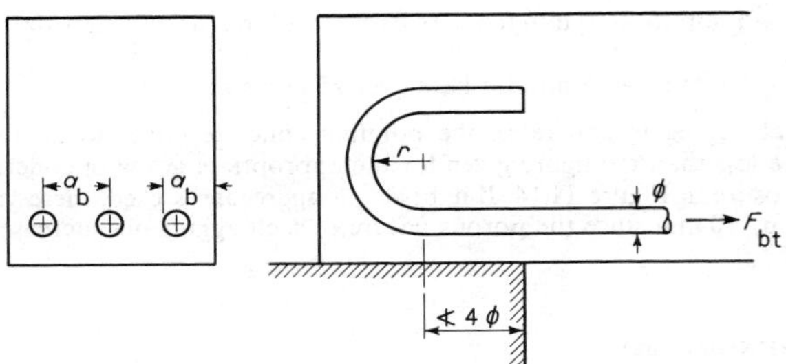

Figure 11.13 Bearing requirements at the end of a beam

not less than its own size continuing for a length not less than 8ϕ or through 180° continuing for a length of not less than 4ϕ. Again the internal radius of the bend should not be less than twice the bend test radius.

COVER

Concrete provides protection to the steel against corrosion and against too-rapid heating in the case of fire. Thus the conditions of exposure and the fire resistance requirements (see pages **11**–69 and **11**–70) have a major influence in determining the cover. Other factors are the dimensions of the reinforcing bars, the nature of the aggregate and the quality of the concrete. The Code refers to nominal cover, which should not be less than

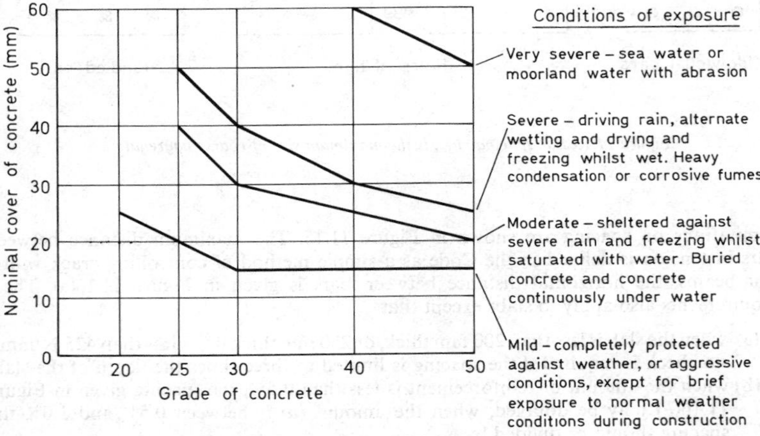

Figure 11.14 Cover to reinforcement and tendons. Notes: Actual cover should not be less than nominal cover less 5 mm*. For lightweight aggregate concrete, the nominal cover should be increased by 10* mm *except for the mild conditions of exposure for which it remains unchanged and for which it should be 25* mm *for Grade 15 concrete*

the size of the bar for single bars or the size of the equivalent single bar for a bundle of bars. The nominal cover is given tolerances of:

-5 mm to $+5$ mm for bars up to 12 mm size

-5 mm to $+10$ mm for bars between 12 mm and 25 mm size

and

-5 mm to $+15$ mm for bars over 25 mm size

For natural aggregate concretes, the nominal concrete cover to all reinforcement should not be less than the figure given for the appropriate grade of concrete and condition of exposure in Figure 11.14. If lightweight aggregate is used, these values should be increased by 10 mm since the porous nature of such aggregates increases the rate of carbonation.

SPACING OF REINFORCEMENTS

The spacing and location of reinforcement must be sufficient to allow compaction of the concrete around the reinforcement but not so great that the control of cracking is insufficient to satisfy the limit-state requirements.

In general the maximum size of the aggregate governs the minimum spacing of bars, but when the size of the largest bars is 5 mm greater than that of the aggregate the spacing should not usually be less than the bar size. Bars or groups of bars should be located in horizontal layers with the gaps between the bars or groups in each layer in line vertically.

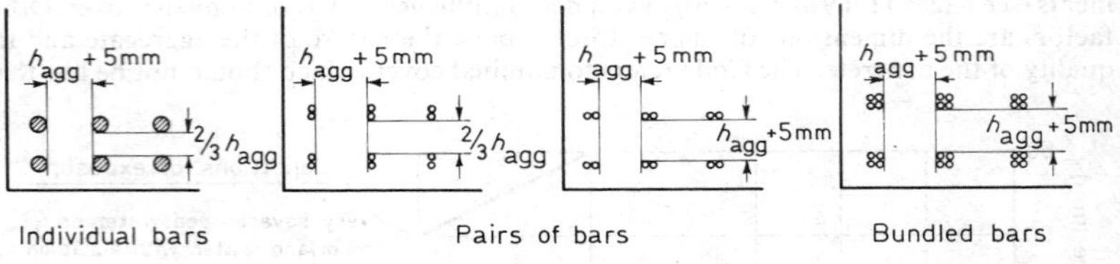

Figure 11.15 Bar spacing. h_{agg} is the maximum size of coarse aggregate

Limitations on spacing are shown in Figure 11.15. The maximum distance between bars in tension is defined in the Code as a simple method of controlling crack width. For beams, the horizontal distance between bars is given in Figure 11.16(a). These requirements also apply to slabs except that

(a) when the slab is less than 200 mm thick, or 250 mm thick if f_y is less than 425 N/mm^2, no check is required if the spacing is limited to three times the depth of the slab;
(b) when the amount of reinforcement is less than 0.5%, the spacing given in Figure 11.16(*a*) may be doubled, when the amount (p) is between 0.5% and 1.0% the spacing should be divided by p.

The amount and spacing of side reinforcement required for beams of greater depth than 750 mm is illustrated in Figure 11.16(b).

These recommendations relate to bars which are not less in size than 0.45 times the largest size of bar in the section, except for side reinforcement, and do not apply for particularly aggressive environments for higher values of f_y than 300 N/mm^2.

LAPS AND JOINTS

Bars can be lapped, welded or joined with mechanical devices to obtain continuity but these joins should be located away from points of maximum stress. Load may be transferred in compression by cutting the ends of the bars square and holding them in direct alignment by a mechanical sleeve.

The length of lap for lapped bars should be not less than the anchorage length for the

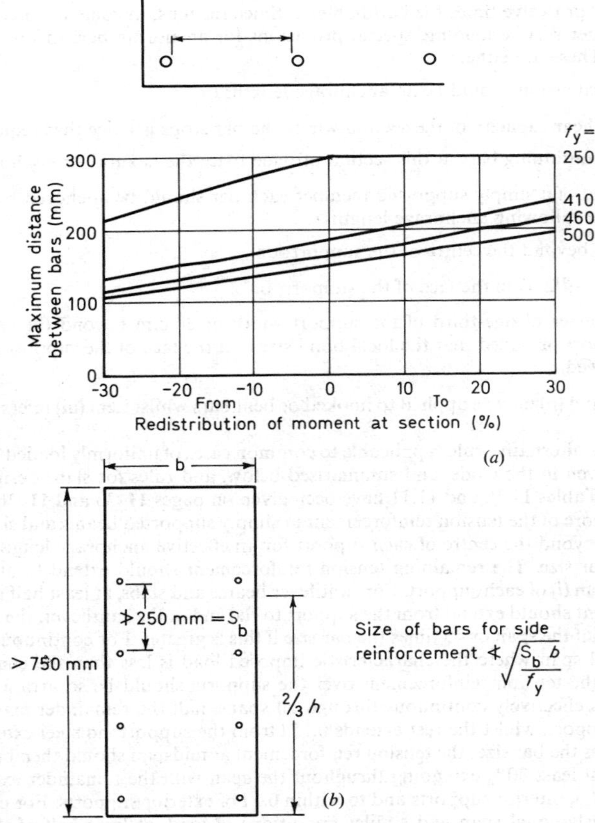

Figure 11.16 Reinforcement for crack control. (a) Maximum distances between bars in beams and slabs; (b) minimum amounts and maximum spacing of side reinforcement for beams of 750 mm depth or more

smaller bar, except that for deformed tension reinforcement the length should be 25% greater. Nevertheless the lap should have a length of not less than 25 times the bar size plus 150 mm for tension steel and of not less than 20 times the bar size plus 150 mm for compression steel. Bars in bundled reinforcement should be lapped individually.

Welding of joints in bars should be avoided if loading is likely to be cyclic. Where the strength of welds has been proved by tests to be as strong as the parent bar, welded bars may be subjected under ultimate load to $f_y/1.15$ for compression joints and $0.8\,f_y/1.15$ for tension joints.

CURTAILMENT AND ANCHORAGE

In principle all reinforcement should extend beyond the point where it is no longer needed. The amount of this extension should not be less than either the effective depth of the member or twelve times the bar diameter. Since the tension zone in concrete may be cracked under service loading, special provisions for anchoring bars in this region are necessary. These are either:

(i) the extension should be an anchorage length; or

(ii) the shear capacity of the section where the bar stops is twice that required; or

(iii) the continuing bars at this section provide twice the flexural strength required.

At the end of a simply supported member each bar should be anchored by provision of one of the following anchorage lengths:

(i) $12\,\phi$ beyond the centre of the support; or

(ii) $12\,\phi + d/2$ from the face of the support; or

(iii) the lesser of one-third of the support width or 30 mm beyond the centre if the support provided that the local bond stress at the face of the support is half that allowed.

Items (i) and (ii) may be applied to hooked or bent bars whilst item (iii) refers to straight bars.

Simplified alternative rules applicable to common cases of uniformly loaded beams and slabs are given in the Code, and summarised below, and rules for slabs designed using the data in Tables 11.10 and 11.11 have been given on pages **11**–33 and **11**–36.

Half or more of the tension reinforcement in simply supported beams and slabs should be carried beyond the centre of each support for an effective anchorage length of twelve times the bar size. The remaining tension reinforcement should extend to at least 0.08 times the span (l) of each support. For cantilever beams and slabs, at least half the tension reinforcement should extend from the support to the end of the cantilever, the remainder extending half the span or 45 times the bar size if this is greater. For continuous beams of nearly equal span where the characteristic imposed load is less than the characteristic dead load, the tension reinforcement over the supports should be so arranged that at least 20% is effectively continuous through all spans, half the remainder extends $0.25\,l$ from the support, whilst the rest extends $0.15\,l$ from the support, no steel extending less than 45 times the bar size; the tension reinforcement at midspan should then be arranged to provide at least 30% extending thrughout the span with the remainder extending to within $0.15\,l$ of interior supports and to within $0.1\,l$ of exterior supports. For continuous slabs of nearly equal span and similar conditions of load, at least half of the tension reinforcement over the support should extend $0.3\,l$ into the span and the remainder a distance of not less than $0.1\,l$ or 45 times the bar size; at least half of the tension reinforcement at midspan should extend to the supports and all bars should extend to within $0.2\,l$ of internal supports and $0.1\,l$ of external supports. These rules for continuous beams and slabs only apply when the distributions of moment and shear given in Table 11.8 are used. Where the end of a continuous slab has a monolithic connection with its supporting

Table 11.16 MAXIMUM AND MINIMUM REQUIREMENTS FOR REINFORCEMENT

Member	*Maximum*	*Minimum*
Beams – Longitudinally	4 % on the gross cross sectional area of the concrete for tension or compression	0.15 % on $b_t \times d$ for high yield steel, 0.25 % on $b_t \times d$ for mild steel, where b_t is the average breadth of the concrete below the flange.
– Transversely (links)	—	(a) 0.12 % on $b_t \times s_v$ for high yield steel, 0.20 % on $b_t \times s_v$ for mild steel, where b_t is the breadth of the beam at the level of the tension reinforcement and s_v is the link spacing (the area given is the area of both legs of each link); (b) where longitudinal compression reinforcement is included in the section, links should have a size $\frac{1}{4}$ times the size of the largest compression bar spaced at 12 times the size of the smallest compression bar.
– Transversely in flanges	—	0.3 % of the longitudinal cross-sectional area of the flange.
Slabs – Longitudinally	—	As for beams.
– Transversely	—	0.12 % of gross concrete section for high yield steel. 0.15 % of gross concrete section for mild steel.
Columns – Longitudinally	6 % for vertically cast columns 8 % for horizontally cast columns but up to 10 % at laps in each case	1 % on the cross-sectional area but 4 bars of 12 mm ϕ for square columns and 6 bars of 12 mm ϕ for circular columns. But for lightly loaded members, area of steel can have a minimum value of $A_{se} = \dfrac{0.15N}{f_y}$ where N is the ultimate axial load.
– Transversely (links)	—	As (b) for beams.
Walls – Vertically	4 % of the gross cross-sectional area of the concrete	0.4 % of gross concrete sections. Where fire resistance is considered 1.0 % (otherwise walls are treated as unreinforced).
– Horizontally	—	For walls with compression reinforcement 0.25 % of gross concrete section for high yield steel and 0.30 % of gross concrete for mild steel links as for (b) in beams may also be required.

beam, it may be assumed that the tension reinforcement at the support should be half that at midspan and should extend not less than 0.1 l or 45 times the bar size into the span.

LIMITS ON THE AMOUNT OF REINFORCEMENT

For a concrete structure to be regarded as properly reinforced it is necessary to have reinforcement crossing all sections which could otherwise develop fracture planes and lead to failure. The Code only exempts certain columns and plain walls from this requirement. Generally it sets out lower limits for structural members which are listed in Table 11.16. These are supplementary to the amounts of steel required to provide stability in the event of partial damage (see pages **11**–27 and **11**–28) and are contributory to these requirements.

Small amounts of vertical steel in walls do not contribute to the strength of the wall and hence the minimum percentage is 0.4% except when fire resistance is required when the minimum is 1.0%. For axially loaded reinforced walls the steel may be placed in one layer and in that case transverse links are not necessary, but if two layers are used the Code required transverse links.

To avoid difficulty in compaction of concrete, upper limits are set on the amounts of steel in the section, and these are also shown in Table 11.16.

PRESTRESSED CONCRETE

General

The primary objective in prestressing is to avoid excessive cracking and deflection whilst at the same time enabling high-strength materials, particularly high-tensile steel, to be used efficiently in construction. The main criteria governing the design of prestressed concrete are therefore characteristics of serviceability rather than ultimate strength.

In setting out the criteria for serviceability of prestressed concrete in Table 11.1, three classes of structure have been identified but no indication was given, nor is it given in the Code, for what purposes these different classes of structure should be used. They nevertheless represent a logical progression from reinforced concrete construction which is likely to be cracked under service loading through Class 3 and Class 2 to Class 1 prestressed concrete construction which is not only completely free from cracking but free from flexural tensile stresses under service conditions.

Where there are particularly adverse conditions of exposure or where cyclic or dynamic loading is severe, it may be appropriate to use Class 1 structures. Where on the other hand these effects do not exist and cracking is acceptable Class 3 structures would be more appropriate. For general purposes, however, including water-retaining structures, Class 2 structures offer most advantages being free from cracking but more economical than Class 1 structures.

Since the serviceability requirements tend to dominate the design process rather than ultimate strength as in reinforced concrete (and in some prestressed concrete Class 3 construction), the procedures for calculating stresses due to the prestress and likely service loadings in relation to serviceability will be given first. The main advantages of prestressing and its most important applications are seen in flexural members and main attention will therefore be given to beams and slabs with secondary attention to ties and columns subjected to bending; little advantage is gained by prestressing members subjected mainly to compression and such columns and walls are not therefore considered.

In dealing with serviceability, several different sets of conditions need to be examined:

(i) During the imposition of the prestress, the stresses in the materials should not exceed certain values determined by the need to avoid failure during the transfer operations or excessive loss of prestress due to creep effects in the materials.

(ii) After the losses of prestress have occurred due to creep and shrinkage of the concrete and relaxation of the steel, the remaining prestress should be sufficient to ensure that the limit state for cracking does not occur under the appropriate design loads.
(iii) During none of these stages should the deflections exceed the limits set.

Additionally, attention has to be paid to the secondary effects that arise in anchoring prestressing tendons and finally to the need to meet the ultimate loading conditions.

Prestress and serviceability

GENERAL

In assessing the conditions of prestress, it is sufficient to assume that the concrete deforms elastically under short-term loading, that creep of concrete can be treated by adopting an effective modulus for the concrete (see page **11**–14), and that shrinkage is uniform across the section.

PRISMATIC MEMBERS

The stress conditions in a uniform member, subjected to external moment m and external force F with a prestressing force P with eccentricity to one axis only of e are shown in Figure 11.17.

Steel is normally used to impose the prestressing force although it can be applied by jacks or by the use of other materials. Such applications are, however, so rare that they are not considered further and this section is therefore concerned only with design for pretensioning and post-tensioning with steel tendons.

For pretensioning, the stress in the steel immediately after transfer, f_{p2}, is less than the initial stress, f_{p1}, by an amount corresponding to the relaxation of the steel at that stage and to the elastic contraction of the concrete adjacent to the steel; the stress is then

$$f_{p2} = f_{p1} - \Delta f_{p1} - \frac{E_s}{E_{c1}}\left[f_{p2}A_{ps}\left(\frac{1}{A_c}+\frac{e^2}{I_c}\right)-\frac{M_g e}{I_c}\right] \tag{39}$$

where, Δf_{p1} is the relaxation of steel before transfer, M_g is the moment due to the proportion of dead load effective at transfer, E_s and E_{c1} are the moduli of elasticity of concrete and steel respectively, A_c and I_c are the area and second moment of area respectively of the concrete section and e is the eccentricity of the tendons.

The final term in the equation is the stress in the concrete adjacent to the steel multiplied by the modular ratio, to give the equivalent change in stress in the steel.

For beams of uniform section with steel at constant depth along the beam, the moment due to dead load, M_g, should be ignored, since the most severe conditions of prestress occur away from midspan near the supports where M_g is small. Where the beams are of nonuniform section, the eccentricity of the steel is normally reduced towards the ends and checks on the stress conditions are required at several sections in the span.

In post-tensioning where a number of tendons are stressed successively, only the first tendon to be stressed contracts by the full amount of the elastic shortening of the concrete and the stress in the steel after transfer is then given by

$$f_{p2} = f_{p1} - \frac{E_s}{2E_{c1}}\left[f_{p2}A_{ps}\left(\frac{1}{A_c}+\frac{e^2}{I_c}\right)-\frac{M_g e}{I_c}\right] \tag{40}$$

Δf_{p1} is not included since there is little relaxation of steel between stressing and anchoring.

In structures with post-tensioned tendons, the dead-load moment effective at transfer may be large and may include dead load from additional superstructure which is temporarily propped but becomes effective on stressing the tendons. This dead-load moment has an important influence on design since it can, if properly manipulated, lead to improvements in efficiency.

The effect of time on the deformation of concrete and steel is taken into account in the following expressions which give the stresses in the tendons. For pretensioning:

$$f_{p3} = f_{p1} - \Delta f_p - SE_s - \left(\frac{E_s}{E_{c1}} + \phi E_s\right)\left[f_{p3}A_{ps}\left(\frac{1}{A_c} + \frac{e^2}{I_c}\right) - \frac{M_g e}{I_c}\right] \qquad (41)$$

For post-tensioning:

$$f_{p3} = f_{p1} - \Delta f_p - SE_s - \left(\frac{E_s}{2E_{c1}} + \phi E_s\right)\left[f_{p3}A_{ps}\left(\frac{1}{A_c} + \frac{e^2}{I_c}\right) - \frac{M_g e}{I_c}\right] \qquad (42)$$

where Δf_p is the total relaxation loss in the steel, S is the shrinkage strain and ϕ is the creep strain for unit stress.

Using these formulae, the value of P is obtained from

$$P = f_p A_{ps} \qquad (43)$$

where f_p is the stress in the tendon at the stage considered. P is then substituted in the appropriate expressions in Figure 11.17 to obtain the required stress conditions immediately after transfer and subsequently under the loads for serviceability limit states.

In members subjected to bending in one direction only it will be normal to locate the centre of the prestressing tendons at an eccentricity which will provide maximum compression at what will become the tension face with a small amount of tension or compression at what will become the compression face under load. For members subjected to loads from any transverse direction, the tendons will be placed concentrically to give a uniform prestress. The formulae given apply to either case.

LOSSES OF PRESTRESS

In making these calculations for the stress conditions in manufacture and service, quantitative allowances must be made for shrinkage and creep of concrete and relaxation of the steel. These characteristics are variable and are much influenced by the nature of the materials used and methods of production. Where data are not available from measurement, values given in the Code should be used. For shrinkage, these are given in Table 11.17.

For creep, it is assumed that the total creep is proportional to applied stress for stresses up to one-third of the cube strength at the time of stressing and may be calculated from a

Table 11.17 SHRINKAGE OF CONCRETE

System	*Shrinkage per unit length*	
	Humid exposure (90% *R.H.*)	*Normal exposure* (70% *R.H.*)
Pretensioning – transfer at between 3 and 5 days after concreting	100×10^{-6}	300×10^{-6}
Post-tensioning – transfer at between 7 and 14 days after concreting	70×10^{-6}	200×10^{-6}

given value of creep for unit stress ϕ. For higher stresses, the creep for unit stress is increased linearly by up to 25% for stresses of half the cube strength at transfer. When the cube strength at transfer is less than 40 N/mm² the creep per unit stress is increased by

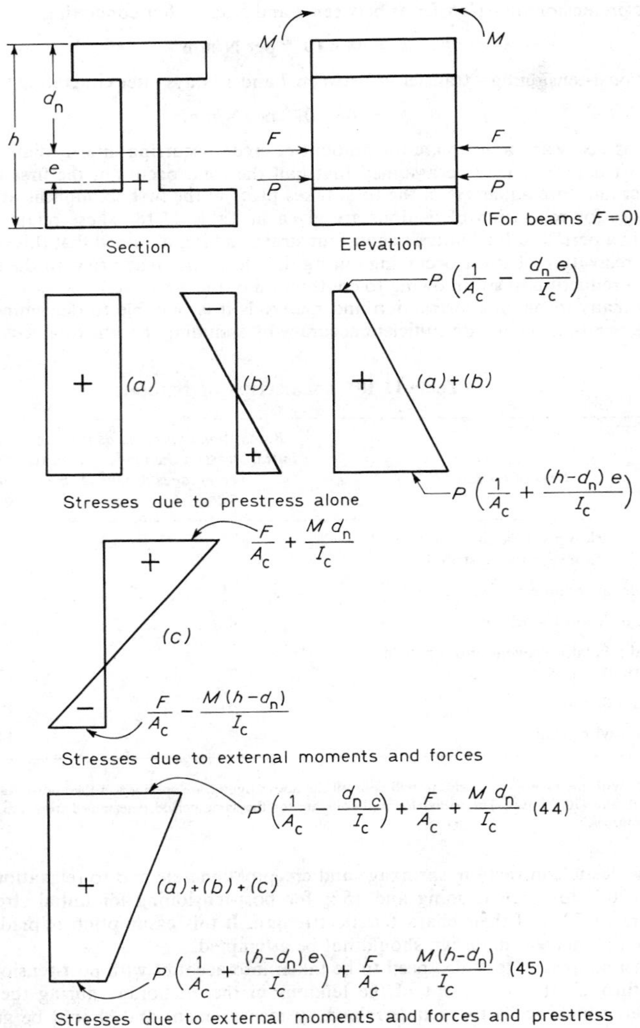

Figure 11.17 Elastic analysis of prestressed concrete sections. M, external moment; F, external force; P, prestressing force; A_c, area of concrete; I_c, second moment of concrete area; A_{ps}, area of tendons.

a factor of 40 divided by the actual cube strength at transfer in N/mm^2. Basic values for creep per unit stress are given below:

For pretensioning – transfer at between 3 and 5 days after concreting

$$\phi = 48 \times 10^{-6} \text{ per N/mm}^2$$

For post-tensioning – transfer at between 7 and 14 days after concreting

$$\phi = 36 \times 10^{-6} \text{ per N/mm}^2$$

If it is necessary to estimate the amount of creep occurring intermediately in the life of the structure, it may be assumed that half the total occurs in the first month after transfer and three-quarters of the total takes place in the first six months after transfer.

Values for relaxation of tendons are given in Table 11.18. These relate to losses of stress in a period of 1000 hours at constant strain and it is assumed that this is equivalent to the relaxation of stress occurring during the life of the structure with the steel undergoing a reduction in strain owing to contraction of the concrete.

For many forms of construction and materials it is possible to determine the losses of prestress that occur with sufficient accuracy by assuming that the total loss of prestress due to elastic contraction, shrinkage and creep of concrete and to relaxation of steel as about 20% for pretensioning and 15% for post-tensioning for initial stresses in the tendons of 70% of their characteristic strength. If this assumption is made, however, detailed refinement in design should not be attempted.

Table 11.18 RELAXATION OF TENDONS

Type of tendon	*Relaxation expressed as a percentage of the initial stress in the tendon for initial stresses as percentages of tensile strength of:* 70%	80%
Low-relaxation pre-straightened wire Low-relaxation seven-wire strand Low-relaxation nineteen-wire strand	$2\frac{1}{2}$	$3\frac{1}{2}$
High-tensile alloy steel bar	4	—
Normal relaxation seven- and nineteen-wire strand	7	12
Wire in mill coils	8	10
Nineteen-wire strand	9	14

Notes: With the exception of wire in mill coils all the above figures are specified in the appropriate standard for relaxation after 1000 hours. The values for mill coils are for the same period determined from tests on wire from several sources.

Other sources of loss also need to be taken into account with post-tensioning. These arise through the movement of the tendons in the anchorage during the process of transfer, which needs to be determined by measurement and should be given by the manufacturer of the system, and through the development of friction between the tendon and its surroundings.

The profiles of the cables or bars may be curved to provide for counteracting variation of moment due to dead and imposed load or due to continuity. As a result, friction develops during stressing between the cables or bars and the inner surfaces of ducts or tendon deflectors. The amount of friction depends on the construction of the cable, the materials in sliding contact and the angular displacement. For long cables, the actual profiles are likely to deviate from their correct position to such an extent that they have an effective additional curvature, which causes considerable frictional effects. Then the force in a tendon, P_x, at a distance x from the jacking point is given by:

$$P_x = P_o \exp - \left(\frac{\mu x}{r_{ps}} + Kx\right) \tag{46}$$

where P_o is the force in the tendon at the jacking end, μ is the coefficient of friction from Table 11.19, r_{ps} is the radius of curvature, x/r_{ps} is the angle of deviation over length x) and K is the constant and the form of tendon and duct which has a usual value of 33×10^{-4} metre but may be reduced to 17×10^{-4} metre for rigid sheaths or rigidly fixed duct formers.

Table 11.19 VALUES FOR COEFFICIENT OF FRICTION μ

Condition	μ
Steel moving on concrete	0.55
Steel moving on steel	0.30
Steel moving on lead	0.25

STRESS LIMITATIONS FOR SERVICEABILITY

Limits need to be set on the stresses in the steel and concrete at transfer to ensure that deformation of these materials is not excessive since this could lead to high losses of prestress, excessive deflection and the possibility of premature cracking. Limits also need to be imposed on the stresses likely to occur in service in order to keep deflections within reasonable bounds and maintain the requirements with respect to tensile stress and cracking set out for Class 1, 2 and 3 structures.

The initial stress in tendons should in general not exceed 70% of their characteristic strength. A stress as high as 80% is, however, allowed when there is reasonable certainty that it will not be exceeded locally in the anchorages or as a result of friction in ducts or at points where tendons are deflected. Experimental evidence should also be available to show that proper provision is made in design for relaxation of the steel at this level of stress.

The allowable levels of compressive stress in the Code in the concrete at transfer are as follows:

For triangular distribution of prestress	0.5 times f_{ci} (the cube strength of the concrete at transfer)
For rectangular distribution of prestress	0.4 times f_{ci}

It should be noted, however, that the use of these maximum values entails the use of higher values for ϕ, the creep coefficient referred to earlier. Limitations on stress under the loads at the serviceability limit state impose a maximum value for compressive stress of $0.33 f_{cu}$, the characteristic strength of the concrete.

Tensile stresses in the concrete at transfer should not exceed 1 N/mm² for Class 1 structures. For Class 2 and 3 structures, the allowable tensile stresses should not exceed those given in Table 11.20.

For Class 2 structures the tendons should be distributed across the section in a uniform manner when pretensioning is used, and nominal secondary reinforcement should be provided when the tendons are post-tensioned. For Class 3 structures the concrete may be allowed to sustain greater strains than those corresponding to these stresses but it must then be assumed to be cracked and the properties of the section calculated on this assumption.

Under the service loads, the allowable compressive stresses in the concrete are given by:

Bending	$0.33 f_{cu}$ which may be increased to $0.40 f_{cu}$ for continuous beams in the negative moment region
Direct stress	$0.25 f_{cu}$

The corresponding tensile stresses for Class 2 and 3 structures are set out in Table 11.21. By definition, no tensile stresses are permitted in Class 1 structures.

Safeguards against cracking in prestressed concrete liquid retaining structures are obtained by imposing restrictions on the tensile stresses in the concrete under conditions corresponding to the limit state of serviceability. These stresses are defined in CP 2007 and are given in Table 11.22.

Table 11.20 TENSILE STRESSES FOR CLASS 2 AND 3 STRUCTURES AT TRANSFER (N/mm^2)

Method of stressing	*Concrete grade*			
	30	40	50	60
Pretensioning	—	2.9	3.2	3.5
Post-tensioning	2.1	2.3	2.55	2.8

BEAMS

Bending Ultimate flexural resistance of prestressed concrete sections is calculated in a similar way to that for reinforced concrete sections except that the effect of prestress on the stress in the steel under ultimate conditions must be allowed for. The assumptions made are:

(a) Sections which are plane before remain plane after bending.
(b) The stresses in the concrete may be determined using the stress–strain curve in Figure 11.4, or may be taken as uniformly distributed across the concrete in the compression zone with a value of f_{cu}/γ_m, i.e. $0.4 f_{cu}$ for deriving simplified formulae. Ultimate compressive strain in the concrete for anlaysis of sections is 0.003 5.
(c) The tensile strength of the concrete is ignored.
(d) The stresses in pretensioned tendons and bonded and post-tensioned tendons at ultimate load may be derived from the appropriate stress–strain curves in Figures 11.6 and 11.7, or may be obtained from the empirical relationships in Figure 11.18(a). These latter relationships give the method of calculation outlined in Figure 11.18(c). They should only be used, however, when the effective prestress in the tendons is greater than $0.45 f_{pu}$. If tendons are located within the compressive zone they should be ignored. Additional mild steel reinforcement close to the tension face of the beam may be assumed to be at its characteristic yield strength; if it is within the compression zone, it should be ignored.
(e) The stress in unbonded tendons (or poorly bonded tendons) at ultimate load may be obtained from the data in Figure 11.19(a). This method only applies for beam sections which are rectangular since it is assumed that the stress distribution is uniform across the compression. It is also assumed that the effective prestress in the tendons is not greater than $0.6 f_{pu}$ and that the position of the tendons for estimating their effective depth is the top of the ducts or underside of guides or diaphragms which position the tendons. It should be noted that for certain situations of ratio of effective span to effective depth of l/d of 10 for $f_{pe}/f_{pu} = 0.6$, ungrouted

Table 11.21 FLEXURAL TENSILE STRESSES FOR THE SERVICEABILITY LIMIT STATE (N/mm²)

Class of structure	Comments	Pretensioning			Post-tensioning			
		Grade of concrete			Grade of concrete			
		40	50	60	30	40	50	60
1	—	0	0	0	0	0	0	0
2	—	2.9	3.2	3.5	2.1	2.3	2.55	2.8
	These stresses may be increased to:	4.6	4.9	5.2	3.8	4.0	4.25	4.5
	It should then be shown that: (i) the enhanced stress is not greater than ¾ of the tensile stress corresponding to the appearance of a crack in a test on a member (ii) the prestress is not less than 10 N/mm² (iii) pretensioned tendons are distributed in the section and nominal reinforcement is provided with post-tensioned tendons							
	These stresses may be further increased to:	6.3	6.6	6.9	5.5	5.7	5.95	6.2
	It should then also be established that these stresses occur only under exceptional conditions and not normally when no tensile stress should occur.							
3	Pretensioned and post-tensioned grouted tendons							
	Crack width (mm) 0.1	4.1	4.8	4.8	3.2	4.1	4.8	4.8
	0.2	5.0	5.8	5.8	3.8	5.0	5.8	5.8
	For pretensioned tendons only when these are distributed in the section and located near the tension face							
	Crack width (mm) 0.1	5.3	6.3	6.3				
	0.2	6.3	7.3	7.3				
	Pretensioned and post-tensioned grouted tendons as above but with 1% additional steel reinforcement							
	Crack width (mm) 0.1	8.1	8.8	8.8	7.2	8.1	8.8	8.8
	0.2	9.0	9.8	9.8	7.8	9.0	9.8	9.8
	For pretensioned tendons only where these are distributed in the section and located near the tension face as above but with 1% additional steel reinforcement[a]							
	Crack width (mm) 0.1	8.3	9.3	9.3				
	0.2	9.3	10.3	10.3				

[a] When less than 1% additional reinforcement is incorporated in the section the additional tensile stress should be interpolated linearly. These stresses for additional reinforcement should be multiplied by the following factors related to depth: less than 200 mm, 1.1, greater than 1000 mm, 0.7, with linear interpolation. No tensile stress should exceed $f_{cu}/4$.

Table 11.22 TENSILE STRENGTH OF CONCRETE FOR ESTIMATING THE RESISTANCE TO CRACKING OF PRESTRESSED CONCRETE LIQUID-RETAINING STRUCTURES

Grade of concrete	*Direct tensile strength* (N/mm²)	*Bending tensile strength* (N/mm²)
35	1.58	3.16
40	1.67	3.34
55	1.89	3.78
70	2.10	4.20

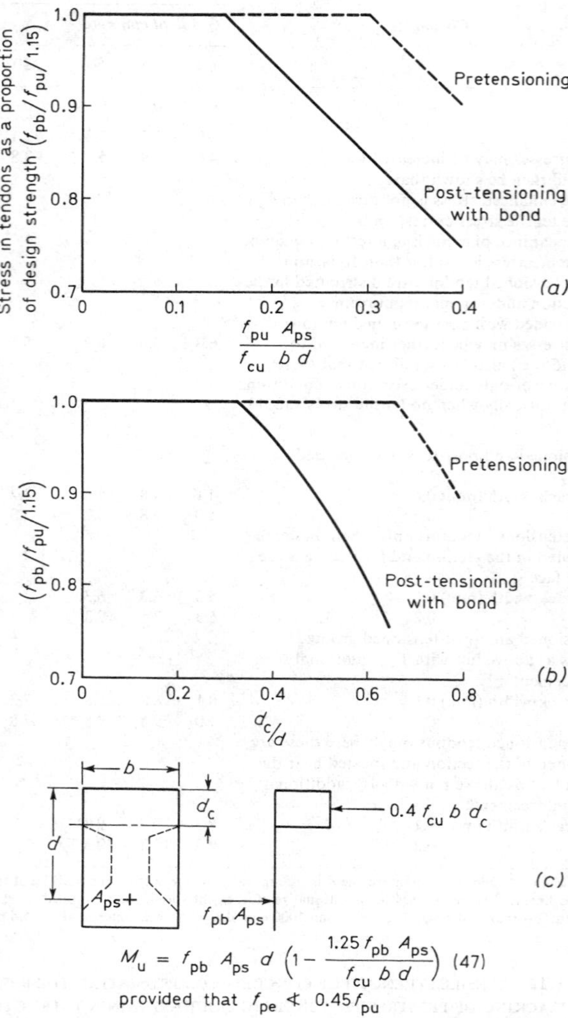

Figure 11.18 Flexural strength of beams—approximate methods (pre-tensioning and post-tensioning with bond)

beams give a higher calculated ultimate strength than if they were grouted. In such cases, the lower value of the two values for strength should be used.

If the compression zone at failure is not rectangular then the ultimate strength should be calculated from first principles using the stress–strain relationships of Figures 11.4, 11.6 and 11.7.

Shear Shear conditions in prestressed concrete beams need only be considered for ultimate conditions. Sections subjected to shear with little bending are uncracked under these conditions but those subjected to both substantial bending and shear will be cracked in flexure. Both situations may need to be dealt with in design and in each case

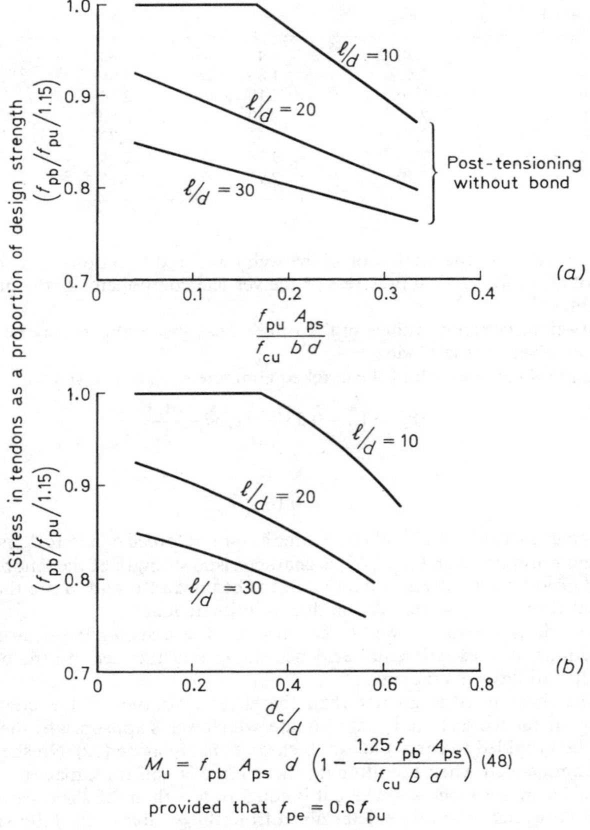

Figure 11.19 Flexural strength of beams—approximate methods (post-tensioning without bond)

this is done by determining the shear resistance of the concrete alone and then adding sufficient shear reinforcement to provide the additional resistance required. Certain types of member also require shear reinforcement because of their importance to the stability of the structure.

For uncracked sections the ultimate resistance of the concrete (V_{co}) is given by

$$V_{co} = v_{co}bh \tag{49}$$

where v_{cs} the ultimate shear stress in the concrete is given in Table 11.23 which is obtained from the Code and has been derived from limitations of principle tensile stress in the concrete, f_{cp} is the compressive stress at the centroid of the section due to the prestress

Table 11.23 ULTIMATE SHEAR STRESS IN CONCRETE FOR UNCRACKED BEAMS (N/mm^2)

Compressive stress at the centroid due to prestress, f_{cp} (N/mm^2)	*Grade of concrete*			
	30	40	50	60
2	1.30	1.45	1.60	1.70
4	1.65	1.80	1.95	2.05
6	1.90	2.10	2.20	2.35
8	2.15	2.30	2.50	2.65
10	2.35	2.55	2.70	2.85
12	2.55	2.75	2.95	3.10
14	2.70	2.95	3.15	3.30

and b is the breadth of the section or of the web for T and I sections. The data in Table 11.23 do not allow for vertical prestress or the vertical components of the prestress from inclined cables.

For lightweight concrete, values of 0.8 times those given above may be appropriate although not given in the Code.

The ultimate shear strength of the cracked concrete section is given by

$$V_{cr} = \left(1-0.55\frac{f_{pe}}{f_{pu}}\right) v_c bd + \frac{M_o V}{M} \tag{50}$$

but

$$V_{cr} \not> 0.1bd\sqrt{f_{cu}}$$

where v_c is given in Table 11.12 and is the same as for reinforced concrete, f_{pe} is the effective prestress and not more than $0.6\,f_{pu}$ (f_{pu} is characteristic strength of the tendons, M_o is the moment to produce zero stress at depth d in the beam, and V and M are the shear force and moment respectively at the section due to ultimate loads.

Inclined tendons are not allowed to be considered as assisting the shear resistance of cracked sections since experimental evidence shows that they lead to the occurrence of more widespread flexural cracking.

Where the shear force is greater than the shear resistance of the cracked section, V_{cr}, or of the uncracked section, V_{co}, at ultimate, whichever is appropriate, shear reinforcement must be provided to carry the excess shear as in equation (52). No shear reinforcement is recommended when the ultimate shear force is half the capacity of the section in shear nor for minor members where it is equal or less than the shear capacity. Otherwise the reinforcement should be determined from the greater of the following:

$$\frac{A_{sv}}{S_v} = \frac{0.4\,b}{0.87\,f_{yv}} \tag{51}$$

or

$$\frac{S_{sv}}{s_v} = \frac{V-V_c}{0.87\,f_{yv}d_t} \tag{52}$$

where f_{yv} is the characteristic strength of the reinforcement which should not exceed 425 N/mm^2, A_{sv} is the cross-sectional area of two legs of the link, S_v is the longitudinal link spacing, d_t is the depth from the compression face to the centroid of the bars or tendons, V is the ultimate shear and V_c is V_{co} or V_{cr}, whichever is appropriate.

Links should pass round longitudinal bars or tendons of larger diameter of the link should be located as near the tension and compression faces as possible and should enclose all tendons. The spacing of links should generally be less than 0.75 d_t of four times the web thickness except when $V > 1.8\ V_c$ when it should be 0.5 d_t; lateral spacing should also be 0.75 d_t.

In no case should the shear stress V/bd on the section exceed the values given in Table 11.24.

Table 11.24 MAXIMUM ULTIMATE SHEAR STRESS (N/mm^2)

Grade of concrete			
30	40	50	60
4.1	4.7	5.3	5.8

These values are consistent with those for reinforced concrete given in Table for natural aggregate concrete. For lightweight aggregate concrete, they should be multiplied by 0.8; this is not, however, referred to in the Code.

Deflection The immediate deflection of prestressed concrete beams is calculated by assuming that the materials deform elastically up to the appearance of first cracks. This method therefore applies to all Class 1 and Class 2 construction and to Class 3 construction provided that the loads do not cause cracking. For long-term loading the effective modulus, given on page **11**–14, should be used instead of the elastic modulus to determine the deflection. Values for the creep coefficient should preferably be those known by experience to be suitable for the type of aggregate and grade of concrete but, in the absence of this information, the value should be obtained from those given for calculating losses.

If Class 3 members are likely to be cracked under permanent load then the span/depth ratios used for reinforced concrete should be used. The Code defines this condition as existing when the permanent load is greater than 25% of the design imposed load.

It should be noted that, for members with an eccentric prestress uniform along their length, the upward deflection may need calculation if the permanent imposed loading is light and the limit on upward deflection must then be met. Even if such members are heavily loaded the regions near their ends will be predominantly influenced by the prestress giving a reverse curvature due to creep from that at midspan. Hence for these members deflection is unlikely to be critical in many cases.

OTHER FORMS OF MEMBER

Slabs The design of slabs in prestressed concrete follows that for reinforced concrete slabs except that the flexural strength is assessed in the same way as that for prestressed concrete beams.

With regard to shear, no shear reinforcement is necessary if the resistance of the concrete to shear is greater than the shear stress.

Columns It will usually only be appropriate to use prestressed concrete columns when the eccentricities of load are high and so require high bending stresses to be resisted. The

Code defines a prestressed column as one in which the mean prestress is 2.5 N/mm^2 or more.

The analysis of columns in prestressed concrete is the same as for reinforced concrete and the sections are designed using the assumptions made for prestressed concrete beams.

Tension members Although prestressed concrete is seldom used for tension members, it is particularly well suited for this purpose.

Ultimate strength should be calculated by assuming that both tendons and any additional reinforcement is acting at 0.87 times its characteristic strength. For the serviceability limit state of cracking, the tensile stresses in the concrete should be limited to about half those permitted for beams for Class 2 or Class 3 structures.

Requirements for tendons and reinforcement

TRANSMISSION LENGTHS FOR PRETENSIONING

With pretensioning it is usual to anchor the tendons by bond. In this case the stress in the steel and the prestress in the concrete builds up along the length of the member near its ends. Consequently in dealing with shear in this region some allowance must be made for this build up of prestress.

Values for this transmission length which have been obtained from measurements in the field and in the laboratory have shown that very substantial variations can occur and that where possible, therefore, advantage should be taken of experience under the conditions of manufacture in determining the appropriate length. The Code nevertheless recommends certain values which are summarised in Table 11.25. These relate to concrete with a strength at transfer of not less than 35 N/mm^2.

Table 11.25 TRANSMISSION LENGTHS FOR TENDONS

Material	*Transmission length*
Plain or indented wire	100 diameters (80% in 70 diameters)
Wire with an offset crimp of 1 mm in 40 mm	65 diameters (80% in 56 diameters)
9.3 mm strand	200 mm
12.5 mm strand	330 mm
18 mm strand	500 mm

Within the anchorage region, the pretensioned tendons should be distributed as uniformly as possible across the section to avoid the development of unnecessary bursting stresses between groups of tendons. Guidance on the amount of reinforcement required can be obtained by considering the needs for reinforcement for end-blocks dealt with in the next paragraph. Such reinforcement should extend for at least the depth of the section.

END-BLOCKS FOR POST-TENSIONING

The anchorage of post-tensioned tendons at the ends of members give rise to high bursting stresses in the concrete. Recommendations for the design of end-blocks may be made by the manufacturers of proprietary systems and if so should be followed.

If they are not made the recommendations in the code should be followed. These give a method of estimating the bursting forces for square blocks which may be applied also to rectangular blocks or combinations of rectangular blocks to cover irregular shapes of anchorage. The basic formula is:

$$\frac{F_{bst}}{P_k} = 0.32 - 0.30\frac{y_{po}}{y} \qquad (53)$$

for

$$0.3 < y_{po}/y < 0.7$$

where F_{bst} is the bursting force, P_k is the maximum prestressing force during jacking when tendons are grouted or, for ungrouted tendons, the force in the tendons at ultimate if this is greater, y_{po} is the half-side of the loaded square end plate when y is the half-side of the square block.

The reinforcement required in each direction should be located transversely in the distance 0.2 y_{po} to 2 y_{po}. For rectangular end blocks, the treatment should be similar giving different amounts of reinforcement in each direction. Circular bearing plates should be treated as square plates of the same area. Provided that the reinforcement has a cover of 50 mm or more it may be stressed to $0.87 f_y$; if the cover is less the stress should also not exceed 200 N/mm^2 to avoid excessive cracking.

PROPORTIONS OF PRESTRESSING STEEL

The characteristics of prestressed concrete of being crack-free under normal conditions and of exhibiting extensive cracking and deflection under overloads have been emphasised as advantages, but these are not necessarily obtained with all prestressed concrete construction. Brittle fracture will occur if there is insufficient prestressing steel in the section to sustain the tensile forces transferred to it on the development of cracking. To prevent this from taking place, it is recommended that the amount of steel in tendons as a percentage of the area given by the product of the overall depth of the beam and its width at the soffit should not be less than 0.15. Brittle failure can also occur by crushing of the concrete in 'over-reinforced' members. It may be experienced with pre-cast members of inverted T section which are designed for incorporation in composite construction but which are susceptible to this form of failure during erection. Difficulty can be avoided by consideration of erection stresses and proper supervision of construction.

COVER AND SPACING FOR TENDONS

In general the requirements for concrete cover to tendons and for their spacing are governed by the same needs as for reinforced concrete.

The cover of concrete necessary to protect the steel against different conditions of exposure is given in Figure 11.14 and applies to both reinforcement and tendons. For concretes of higher grade than Grade 50, there should be no reduction in the thickness of the cover. Since prestressing steels are more sensitive to the effects of heat than reinforcing steels, the requirements for concrete cover for the protection of tendons in fire are more onerous and are dealt with later on pages **11**–69 and **11**–70.

Experience has shown that ends of individual pretensioned tendons do not require concrete cover for protection and may be left cut flush with the end of the member. Where post-tensioned tendons are positioned outside the member, they are normally protected with added concrete to provide the cover; some attention should be given, however, to the way in which this is done since the development of possible paths for penetration of moisture to the steel must be avoided.

There should be sufficient space between tendons to allow proper compaction of the concrete and the rules applied to reinforced concrete should therefore apply. Large ducts, however, provide more difficulties than are experienced with large reinforcing bars and careful attention is needed in detailing. If these ducts are required in thin diaphragms or webs then care must also be given to the avoidance of bursting the concrete and possibly to the provision of additional reinforcement. Additional reinforcement may also be required in members with pretensioned steel if the individual tendons are in separate groups to prevent longitudinal splitting at the ends where transmission of the prestress by bond is developed.

SECONDARY REINFORCEMENT

Reinforcement is required in prestressed concrete to meet requirements for resisting shear, to permit higher tensile stresses in Class 3 structures, to prevent bursting in the region of end anchorages, to reinforce thin webs and to retain concrete cover in place for longer periods of fire resistance. Any longitudinal reinforcement provided for these purposes can be taken as contributing to ultimate strength as can the ties needed to ensure stability in the event of accidental damage.

Reinforcement may also be desirable for members prestressed by post-tensioning to restrict cracking after casting caused by restraint of the formwork due to shrinkage or cooling of the concrete.

PRE-CAST AND COMPOSITE CONSTRUCTION

General

The previous sections have dealt with concrete construction in general and when the concrete is cast in-place. Not uncommonly, however, it is economic or convenient to pre-cast concrete in the factory and use it in construction with site-cast concrete to form composite members or structures. Since the strength of the concrete construction depends on the compressive strength of the concrete and the tensile strength of the reinforcement at all sections, it usually is necessary to provide continuity of reinforcement across joints with the object of attaining some degree of the monolithic action obtained when the whole structure is cast in-place.

This section therefore deals with the problems that arise in the use of pre-cast concrete, which have not already been covered, and with the design of joints. The provision of ties for ensuring the overall stability of the structure including pre-cast and large panel construction has already been considered on pages **11**–27 and **11**–28.

Pre-cast concrete

Since pre-cast members must be handled, transported, stored, and erected sometimes at a very early age, design must take account not only of the resulting stresses that occur but also must ensure that the design of the concrete mix will give sufficient strength to avoid damage. It is therefore not uncommon for pre-cast members to be made of concrete of a much higher strength than would be the case if they were cast in situ; cube strengths in excess of 60 N/mm^2 at 28 days may be obtained since the manufacturing requirement may be for 10 N/mm^2 or more at 24 hours. The joints between members therefore not only provide discontinuities in members of the structure but may also lead to discontinuities in the strength of the structure. Particular care is therefore required in design to ensure that the conditions assumed to obtain at joints can actually be attained in the structure and that the details of the joints are such that they can be formed and inspected without difficulty on the site.

Where continuity of beams, frames and slabs is obtained by the appropriate positioning and anchoring of steel reir 'orcement or tendons, the amount of redistribution of moments allowed for in situ construction may also be adopted for pre-cast construction.

BEARINGS

If a pre-cast concrete unit rests directly on another cast concrete surface, the contact surfaces should be free from irregularities and the bearing stress should not exceed $0.4f_{cu}$ for ultimate conditions, where f_{cu} is the cube strength of the weaker of the two concretes. Such direct bearing of concrete surfaces, even with a flexible padding insert, is not recommended for the bearing of a column on a column or a wall on a wall; in these cases the joint should be made with placed concrete or mortar. For other types of bearing where special provision is made, for example, using binding reinforcement near specially prepared concrete bearing faces with a flexible padding, a higher bearing stress may be used up to $0.8f_{cu}$. If a particular type of bearing is to be used frequently, it is recommended that tests should be made to ensure that the strength and mode of failure are acceptable; experimental evidence shows that in suitable circumstances very high stresses may be used.

The corbel is a common form of support for beams on columns, which the Code defines as a short cantilever beam with the limiting dimensions given in Figure 11.20. The corbel should be designed as an inclined strut with a tie member of reinforcement, equal to at

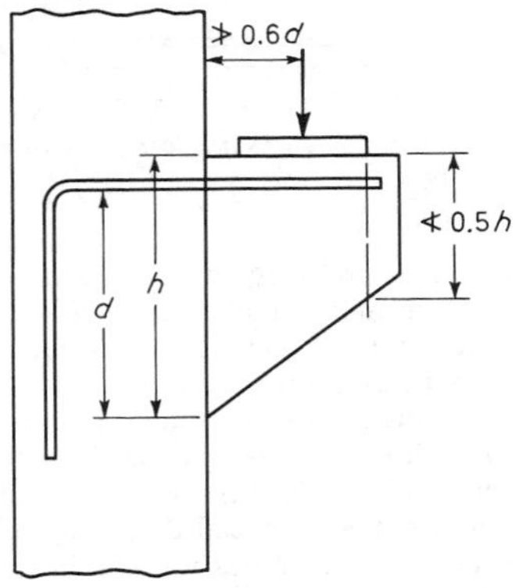

Figure 11.20 Dimensions of corbels

least 0.4% of the cross section of the corbel at the column face, adequately anchored in the beam or column of which it is a part. Reinforcement should also be provided horizontally to resist shear in the upper two-thirds of the corbel. If a horizontal force can develop at the bearing or is required to be resisted, then the whole of this force should be carried by additional horizontal reinforcement welded to the bearing plate and fully anchored in the member.

Where pre-cast floor units are supported by beams, a continuous concrete rib may be required along the length of the beams. These may be designed by assuming that the effective depth and eccentricity of loading are as shown in Figure 11.21. Shear requirements should be met and the vertical leg of additional link reinforcement in the beam should transfer the load carried to the compression zone of the beam.

STRUCTURAL CONNECTIONS BETWEEN UNITS

The first requirement in the design of structural joints is to provide for transmitting the tie forces required for stability. In dealing with the reinforcement that is needed for this purpose it should be ensured that the provisions for anchorage are realistic and can be obtained with normal site workmanship.

When the joint is not required to transmit horizontal forces or moment, the detailing should be such that these effects are not in fact transmitted or if this is not possible that

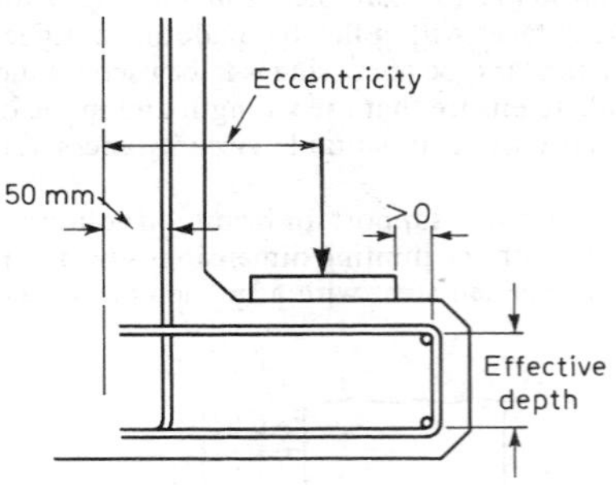

Figure 11.21 Dimensions of continuous nibs

any unintended transmission is not likely to lead to untoward cracking or local damage. It must be recognised that shrinkage and creep effects can lead to the development of substantial restraints in structures if there is no provision for such movements and that these effects can be more serious in pre-cast structures.

In dealing with the transmission for forces and moments at connections, the normal procedures for calculation should be used for reinforced concrete, prestressed concrete or structural steel. Where special difficulties arise, tests should be made to assess both strength and mode of failure. Some special details are given in the Code of Practice, CP 110, for dealing with joints in compression and in shear and with halved joints, to which reference should be made.

Attention needs to be paid to the protection of joints against the weather and corrosion and against fire to make sure that the performance of the joint under these conditions is not inferior to that of the rest of the structure. Emphasis has already been placed on the importance of detailing joints in such a way that they can be made properly on the site and effectively inspected. To achieve this, any projecting bars, ribs or fins should be sufficiently robust to avoid damage in transit and erection and sufficiently accurately located and dimensioned for ease of casting and assembly. The space between members being jointed should be sufficient to allow filling of the joint without difficulty and for subsequent checks on workmanship. Further, to make sure that joints are made properly, written instructions should be given to the site giving full details and sequence for jointing with information on what should be done in the event of misfits. The instructions should also contain information on the making of the joints in relation to progress of construction since stability during erection may well depend on the extent of the completion of joints in other parts of the structure.

Continuity of reinforcement in pre-cast construction may require special consideration. If sockets or slots are left for lapping or anchoring reinforcement they should be sufficiently large and of a form likely to achieve their object with a suitable form of surface for developing bond between the infill concrete, mortar or grout and the pre-cast units. Sleeves or threaded anchors for connecting reinforcement may be used as well as welded connections with structural steel sections. Test data for many of these types of connections are available.

Composite construction

Pre-cast members of reinforced or prestressed concrete are commonly used in composite construction, particularly pre-cast beam units acting with cast in situ slabs for floor construction. Design of composite sections follows the methods developed for sections in reinforced or prestressed concrete.

Generally the ultimate flexural strength of composite sections may be calculated by the methods used for reinforced or prestressed concrete provided that there is adequate connection between the two concretes. The same approach applies to vertical shear when the pre-cast member is of reinforced concrete. When it is of prestressed concrete, it will often be found that the pre-cast section has adequate shear resistance for the composite section. When the pre-cast prestressed unit cannot provide the whole shear resistance and the in situ concrete is placed between the units, the composite concrete section should be used in the calculation with the principal tensile stress in the prestressed unit as not greater than $0.24(f_{cu})^{\frac{1}{2}}$.

With regard to serviceability limit states, the composite reinforced concrete member is treated as an ordinary reinforced concrete member. For prestressed concrete composite members, the limitations of stress for ordinary prestressed concrete members

Table 11.26 ACCEPTABLE HORIZONTAL SHEAR STRESSES UNDER SERVICE DESIGN LOAD (N/mm^2)

Grade of in situ concrete	25	30	40	50	60
Surface					
Type 1 No links. Surface prepared after setting by finely spraying with water or brushing with a stiff brush to expose but not distort coarse aggregate	0.38	0.45	0.54	0.59	0.64
Type 2 Links equivalent to 0.15% of contact surface spaced at 4 times depth of in situ concrete or 600 mm whichever the least. No special treatment of concrete	0.36	0.38	0.42	0.46	0.50
Type 3 Links at (2) surface as (1). Where additional links are provided shear stress may be increased by 0.5 N/mm^2 for each 1% of contact area increase in steel	1.22	1.25	1.32	1.38	1.45

still apply and tensile stresses in the cast in situ concrete are limited to about $0.65(f_{cu})^{\frac{1}{2}}$, these stresses may, however, be increased by up to 50% provided that there is a 50% reduction in the tensile stresses in the prestressed unit. The compressive stresses in prestressed units may be increased by up to 50% at the interface with the cast in situ concrete so long as the failure of the composite member is due to the steel reaching its tensile strength.

Particular attention needs to be given to designing for shear at the interface between the in situ concrete and the pre-cast unit. The Code treats this situation by limiting the horizontal shear stress under the design load for serviceability. The shear stress

$$v_h = \frac{V_d S_v}{I b_c} \tag{54}$$

where V_d is the vertical shear at the point considered under $(G_k + Q_k)$, S_v is the first moment about the neutral axis of the concrete area to one side of the contact surface, I is the second moment of area of the transformed composite section, and b_c is the width of the contact surface.

Values for the acceptable level of shear stress according to the nature of the contact surface for beams are given in Table 11.26.

Shear reinforcement provided to resist vertical shear should be extended into the compression zone and may be used to resist horizontal shear at the interface.

For slabs in which it is usually inappropriate to include vertical reinforcement the stresses that can be used for the Type 1 surface should be not less than 1.2 times the figures in the Table for a Type 1 surface or 0.7 N/mm^2 whichever is less. For a Type 2 surface without links the figures should be 0.8 times those given in the Table for this surface.

Differential shrinkage effects do not in general influence the ultimate strength of composite members but they may affect serviceability and the Code gives details of treatment which may only be needed if the in situ and pre-cast concrete differ in quality by more than one grade, i.e. 10 N/mm^2.

STRUCTURAL TESTING

In design and construction in reinforced and prestressed concrete the testing of components and structures can play important role. It may take the form of:

(a) model or full-scale testing to aid in the evolution of improved analytical and design methods;
(b) development testing of prototypes of components or ancillary equipment for performance or feasability of construction procedure;
(c) check testing of factory production as a control on quality of output; and
(d) investigations of structural adequacy of construction which for some reason may have become suspect.

Model or full-scale testing to obtain a better understanding of structural behaviour has been used for many types of construction, as already mentioned. The tests may relate to solving general structural problems or may deal with the design of specific structures. In either case, testing requires sophisticated backing of experimental facilities and staff and can only be undertaken by universities and technical colleges, government laboratories and established industrial research organisations. The planning of the experiments and the interpretation of the data is largely a matter for those with the expert knowledge and experience.

The development testing of components possibly for mass production in the factory or of ancillary equipment such as bridge bearings, prestressing jacks and structural connections is the direct concern of designers and specialist contractors. The objects may be to produce economic design, to simplify procedures or to ensure that details in design give the performance required. Where concern is with repetitive production or procedures, statistical methods of analysis should be used; where it is with structural performance some care may be necessary to be sure that, in isolating the problem for experimental evaluation, it has not been so simplified that its solution is irrelevant.

Testing as a control procedure has been used very widely by industry for many years and has found application in the production of pre-cast concrete components. It may take the form of nondestructive testing of a proportion of production possibly with a smaller proportion being tested to failure. The types of test and the procedures are covered by British Standards for a large number of products, such as kerbs, paving slabs, pipes and lamp standards.

The Code of Practice, CP 110, deals with some of the aspects of testing concerned with checking concrete quality and refers to the cutting of cores and the use of gamma radiography, of ultrasonic tests, of covermeters and of rebound hammers. These methods provide some check on construction when the quality of the work is in doubt and are used to give guidance on the need for structural tests on components and structures. Such checks may become necessary because of faults in construction, because the structure has been damaged possibly by fire or because of a change of occupancy.

The Code sets out specific forms of test for satisfying the serviceability limit states and the ultimate limit state for both components and structures.

For components the nondestructive test for serviceability consists of supporting the member as it would be used in structure and applying to it a load of $1\frac{1}{4}$ times the characteristic imposed load. The deflection under this load after a period of five minutes should not exceed the value set by the designer before the test. The load is then removed, the recovery after five minutes recorded, and the loading cycle is then repeated. The percentage recovery after the second loading should not be less than that after the first loading nor less than 90%. During the test, the component should not show any signs of distress, damage or faulty construction.

Under the destructive test described in the Code the component must sustain its ultimate design load for 15 minutes without failure and without exceeding a deflection of span/40.

The load to be applied to structures or parts of structures to check their serviceability with respect to deflection and cracking is the characteristic imposed load with any necessary additional load component to bring the dead load up to the value of the characteristic dead load. The structure is considered to pass the serviceability test if, immediately after application of the load, the width of cracks for reinforced concrete members and Class 3 prestressed concrete members is not greater than two-thirds of the values set for the limit state and no cracks occur in Class 1 and 2 prestressed concrete members. The immediate deflection should not exceed 1/500 of the effective span.

Compliance with the ultimate limit-state requirements for structure is more difficult to ascertain since satisfactory structures would be irrevocably damaged if they were loaded near to their ultimate design load. The test procedure given in the Code can only therefore be used to supplement the evidence from other sources on the adequacy of the structure. The test requires the application for 24 hours of a load of $1\frac{1}{4}$ times the characteristic imposed load with any additional load to ensure that the test load includes the full characteristic dead load. The load is then removed and recovery allowed to take place over a further period of 24 hours. If the maximum deflection in millimetres is less than $40l^2/h$ where l is the effective span in metres and h the overall depth in millimetres, then no requirements are placed on the amount of recovery. Reinforced and Class 3 prestressed concrete structures should show a recovery of at least 75%, if this is not obtained, the test should be repeated and the recovery must then be at least 75% of the deflection in the second test. The corresponding recovery for Class 1 and 2 prestressed concrete structures is 85% for each situation.

For prestressed concrete structures any departure from linearity of the load–deflection curve when plotted will give some indication of the level of prestress existing in the structure. This information is particularly useful in case of prestressed concrete damaged by fire since the retention of a high proportion of prestress indicates that the tendons have not lost strength.

FIRE RESISTANCE

The Building Regulations define requirements for health and safety in general and set amongst other matters the provisions that must be made to ensure safety in buildings in the event of fire. The Regulations deal with the prevention of the spread of fire, means of escape and facilitating the fighting of fires. They set out the maximum size of buildings or compartments into which buildings should be divided according to the class of occupancy and height, and then define the required fire endurance for the size of building and occupancy. The endurance is determined by tests conforming with the requirements of BS 476.

Owing to lack of comprehensive information on overall behaviour of structures it is assumed that, if each component part of the structure, walls, floors, columns and beams, is designed for the required fire resistance, the whole structure will have the necessary fire resistance. In general, this assumption would appear to be conservative but for certain types of structure, e.g. unbraced tall framed buildings, it might be conceivable for a fire to become widespread in one storey and so lead to instability of the columns at that level as their stiffness reduces with increasing temperature. Isolated fires would probably not produce this effect since the requirements for ties to prevent excessive damage due to accidents would enable the structure to bridge over any individual failures. The possibilities should, however, be given some consideration in design.

Both the Building Regulations and the Code give requirements for the minimum dimensions of members for various periods of fire resistance for concrete floors, walls, beams and columns. These are based on the results of fire tests in which the members were tested in a furnace heated at a rate laid down as a standard time–temperature curve. Results of these tests show that a number of factors may have an important influence although not all can yet be taken into account in design; these include the effects of size and shape of member, the properties of the different types of steel and concrete that may be used, the protection afforded to the steel, the level of stress that must be sustained by the steel (which is dependent on the proportion of the service load carried at the time of the fire) and the degree of restraint provided by the rest of the structure.

The Code tabulates the requirements for common types of structural component and some of these are summarised in Tables 11.27, 11.28 and 11.29. Full details for other forms of construction should be obtained from the Code.

When the concrete cover for beams of siliceous aggregate concrete, given in Table 11.27 is in excess of 40 mm, reinforcement of a light mesh or expanded metal is required to retain the cover in position; this may not be required for limestone aggregate concrete although the same thicknesses of concrete cover apply; it is not required with lightweight aggregate concrete where cover requirements are smaller.

The fire resistance of floors, given in Table 11.28, may be increased by applying insulating material to the soffit. For example, a gypsum sand or cement sand finish on expanded metal as a suspended ceiling will increase the fire resistance by 3 hours when 25 mm thick, by 2 hours when 20 mm thick and by 1 hour when 10 mm thick. Details for other finishes are given in the Code. If lightweight aggregate concrete is used in floors, smaller dimensions may be appropriate but they must be determined by test. Walls with more than 1 % vertical reinforcement which are exposed to fire on both faces should have a minimum thickness equal to the minimum dimension for columns exposed on all four faces.

Some indication of fire resistance of concrete members may be obtained from the rate of temperature rise in the concrete and the changes that this produces in the properties of the materials. Tests show that, where gravel aggregate concrete is heated following the standard time–temperature curve, the temperature rises to 400 °C at a depth of 20 mm in 30 minutes, of 40 mm in 1 hour, of 80 mm in about 2 hours, and it rises to 600 °C at a depth of 20 mm in 1 hour and of 40 mm in about 2 hours. Lightweight aggregate concrete has a lower thermal conductivity and these thicknesses may be reduced to about 3/4 for similar rises in temperature. This information on temperature in the concrete taken with the deterioration in behaviour of steel can be used to estimate the fire resistance of beams and floors.

Cold-drawn prestressing steels lose about half their strength at 400 °C and most

Table 11.27 FIRE RESISTANCE OF STRUCTURAL CONCRETE BEAMS

Description	*Dimensions* (mm) *for fire resistance in hours of:* 4	3	2	1½	1	½
Siliceous aggregate concrete						
Beam width	280	240	180	140	110	80
Cover of concrete to main reinforcement for reinforced concrete	65[a]	55[a]	45[a]	35	25	15
Average cover of concrete to tendons for prestressed concrete	100[a]	85[a]	65[a]	50[a]	40	25
Lightweight aggregate concrete						
Beam width	250	200	160	130	100	80
Cover of concrete to main reinforcement for reinforced concrete	50	45	35	30	20	15
Average cover of concrete to tendons for prestressed concrete	80	65	50	40	30	20

[a] Supplementary reinforcement, at least 0.5 kg/m², is required to retain the cover in place.

Table 11.28 FIRE RESISTANCE OF STRUCTURAL CONCRETE FLOORS (SILICEOUS AND CALCAREOUS AGGREGATES)

Description	*Dimensions* (mm) *for fire resistance in hours of:* 4	3	2	1½	1	½
(a) Solid slab						
Overall depth (including noncombustible screeds and finishes)	150	150	125	125	100	100[a]
Average cover to reinforcement for reinforced concrete	25	25	20	20	15	15
Average cover to tendons for prestressed concrete	65[b]	50[b]	40	30	25	15
(b) Ribbed floor with hollow infill blocks of clay or inverted T sections with hollow infill blocks of concrete or clay. A floor in which less than 50% of the gross cross section is solid material must be provided with 15 mm of plaster coating on soffit						
Overall depth (including noncombustible screeds and finishes)	190	175	160	140	110	100
Width of rib or beam at soffit	125	100	90	80	70	50
Average cover to reinforcement for reinforced concrete	25	25	20	20	15	15
Average cover to tendons for prestressed concrete	65[b]	50[b]	40	30	25	15

[a] 90 mm for prestressed concrete.
[b] Supplementary reinforcement required in cover where no ceiling protection is provided.

reinforcing steels yield at half their yield stress at about 600 °C. When these temperatures are reached in the steel in prestressed concrete and reinforced concrete respectively, collapse may be expected. For compression members, the concrete makes a major contribution. Test data show that for gravel concrete, its strength is reduced to 90% at 300 °C, to 50% at 500 °C and to 30% at 700 °C; limestone aggregate and lightweight aggregate concretes behave a little better. These results provide some basis for estimating

Table 11.29 FIRE RESISTANCE OF CONCRETE COLUMNS AND WALLS

Description	*Dimensions* (mm) *for fire resistance in hours of:* 4	3	2	$1\frac{1}{2}$	1	$\frac{1}{2}$
Columns (all faces exposed) – minimum dimensions						
Siliceous aggregate concrete	450	400	300	250	200	150
Limestone aggregate concrete or siliceous aggregate concrete with supplementary reinforcement in cover	300	275	225	200	190	150
Lightweight aggregate concrete	300	275	225	200	150	150
Columns (one face exposed) – minimum dimensions						
Siliceous aggregate concrete	180	150	100	100	75	75
Walls (one face exposed) –						
Siliceous aggregate concrete with 1% vertical reinforcement						
minimum thickness	180	150	100	100	75	75
concrete cover	25	25	25	25	15	15
minimum thickness	—	—	—	175	150	—

the strength of walls and columns. Whatever estimates are made, however, tests are necessary to give the information required to satisfy the Regulations.

EXAMPLES OF CALCULATIONS FOR REINFORCED AND PRESTRESSED CONCRETE

It should be noted that these examples illustrate calculations. In the Design Office, it would be more appropriate to use the design charts in Parts 2 and 3 of CP 110: 1972.

Reinforced concrete

1 (a) Design a reinforced concrete beam of rectangular section to support a uniform characteristic imposed load of 20 kN/m on a span of 18 m, using concrete of Grade 40 and steel with $f_y = 250$ N/mm².

$$\text{Assume } h = 1000 \text{ mm}, \quad b = 400 \text{ mm}$$
$$\text{take } \quad d = 930 \text{ mm}$$
$$\text{density of concrete} = 2400 \text{ kg/m}^3$$

$$\text{Dead load, } G_k = 2400 \times 10 \times 0.4 \times 18/10^3 = 173 \text{ kN}$$
$$\text{Live load, } Q_k = 20 \times 18 = 360 \text{ kN}$$

ULTIMATE LIMIT STATE

$$\text{Moment due to } G_k = 173 \times 18 \times 1.4/8 = 540 \text{ kN m}$$
$$\text{Moment due to } Q_k = 360 \times 18 \times 1.6/8 = 1300 \text{ kN m}$$
$$\text{Total} = 1840 \text{ kN m}$$

From equation (15):

$$M_u = \frac{0.87 \times 250 \times A_s \times 930}{10^6}\left(1 - \frac{1.1 \times 250\, A_s}{40 \times 400 \times 930}\right) = 1840$$

whence

$$A_s = 11.6 \times 10^3\ \text{mm}^2$$

use 10/40 mm ϕ bars giving

$$A_s = 10 \times (\pi/4) \times 40^2 = 12.5 \times 10^3\ \text{mm}^2$$

For mild exposure, cover is 15 mm and with bars in two rows with vertical spacing of 14 mm for 20 mm maximum size of aggregate (Figure 11.15) and links of 8 mm dia. gives

$$d = 930\ \text{mm as assumed.}$$

Shear

Minimum reinforcement required at 250 mm spacing (maximum spacing allowed = 0.75 d = 0.70 m) for steel with f_y = 250 N/mm² (see page **11**–31).

$$A_{sv} = 0.002 \times 250 \times 400 = 200\ \text{mm}^2$$

Use 4 legs of 8 mm ϕ bars giving

$$A_{sv} = 4 \times (\pi/4) \times 8^2 = 201\ \text{mm}^2$$

$$\text{Maximum shear } V = \tfrac{1}{2}(173 \times 1.4 + 360 \times 1.6) = 409\ \text{kN}$$
$$v = 409 \times 10^3/400 \times 930 = 1.10\ \text{N/mm}^2$$

From equation (20) and Table 11.12,

$$A_{sv} = \frac{s_v \times 400(1.10 - 1.00)}{0.87 \times 250} = 0.18 s_v$$

If

$$A_{sv} = 201\ \text{mm}^2 \text{ as above}$$
$$s_v = 1100\ \text{mm}$$

Therefore minimum reinforcement provides all that is needed.

SERVICEABILITY LIMIT STATE

Deflection is checked from Figure 11.11(a), (b) and (c). From Figure 11.11(c),

$$\frac{\text{As required}}{\text{As provided}} = \frac{11.6 \times 10^3}{12.5 \times 10^3} = 0.93$$

and

$$\text{since} \qquad \beta_b = 1.00, \quad f_s/f_y = 0.54$$

Therefore $f_s = 0.54 \times 250 = 135$ N/mm².

From Figure 11.11(b) with $A_s/bd = 12.5 \times 10^3/400 \times 930 = 0.034$.

The factor for tension steel may be taken as approximately = 1.0.

From Figure 11.11(a), span/depth ratio may be 12 where deflection is critical, or 20 where deflection of $l/250$ is acceptable. If deflection is critical, then section must be increased in depth, possibly with the addition of compression steel.

Cracking—The maximum distance permitted between bars to control cracking is given in Figure 11.15(a) where, for f_y = 250 N/mm², the spacing should not be greater than 300 mm. Actual spacing is not greater than 40 mm.

Side reinforcement is required as in Figure 11.16(b). For vertical spacing of 250 mm,

$$\text{size of bars} = (250 \times 400/250)^{\frac{1}{2}} = 20 \text{ mm}$$

1 (b) For the same loads and dimensions as in example **1 (a)** but using Grade 20 concrete and compression reinforcement, check requirements for longitudinal reinforcement, and examine effect on deflection.

Use 8 No. 40 mm ϕ bars in two layers of 3 and 5 bars as tension steel giving

$$d = 937 \text{ mm and } A_s = 10.1 \times 10^3 \text{ mm}^2$$

5 No. 40 mm ϕ bars in one layer as compression reinforcement giving

$$d_1 = 43 \text{ mm and } A_s' = 6.3 \times 10^3 \text{ mm}^2.$$

ULTIMATE LIMIT STATE

From equation (17):

$$M_u = 1820 \text{ kN m}$$

But if stress in compression steel is assessed from Figure 11.5 instead of as $0.72 f_y$, i.e. as $250/(1.15+250/2000) = 196$ N/mm² instead of 180 N/mm², then

$$M_u = 1840 \text{ kN m as required.}$$

Since the concrete is weaker, more shear reinforcement would be required.

SERVICEABILITY LIMIT STATE

Deflection—from Figure 11.11(a), (b) and (c), taking $A_s/bd = 0.027$ and $A_s'/bd = 0.017$. Span/depth ratio

where deflection of $l/250$ is acceptable

$$= 20 \times 1.05 \times 1.36 = 28.6$$

where deflection is more critical

$$= 12 \times 1.05 \times 1.36 = 17.1$$

Actual span/depth ratio $= 18/0.937 = 19.2$

Thus, deflection should be less than in example **1 (a)**. No additional steel to that in example **1 (a)** would be needed to control cracking.

1 (c) Design a floor slab to support a characteristic imposed load of 3·0 kN/m² on a simply supported span of 4 m. Finishes are 1.2 kN/m². Use Grade 30 concrete and steel with $f_y = 460$ N/mm².

Dead load	finishes	1.2 kN/m²
	slab (say)	3.6 kN/m²
		4.8 kN/m²

Assume $h = 150$ mm
$d = 130$ mm
10 mm ϕ bars at 150 mm centres

$$A_s = \frac{1000}{150} \times \frac{\pi}{4} \times 10^2 = 520\ \text{mm}^2/\text{m}$$

$$\frac{A_s}{bd} = 0.004.$$

ULTIMATE LIMIT STATE

Moment due to dead load

$$= \frac{4.8 \times 4^2 \times 1.4}{8} = 13.4\ \text{kN m/m}$$

Moment due to imposed load

$$= \frac{3.0 \times 4^2 \times 1.6}{8} = \underline{9.6\ \text{kN m/m}}$$

$$\underline{23.0\ \text{kN m/m}}$$

From equation (15):

$M_u = 25.2$ kN m/m and therefore adequate.

Shear due to dead load $= 4.8 \times 2 \times 1.4 = 13.4$ kN/m
Shear due to imposed load $= 3.0 \times 2 \times 1.6 = \underline{9.6}$ kN/m

Total $\underline{23.0}$ kN/m

From equation (21):

$$v = \frac{23.0 \times 10^3}{1000 \times 130} = 0.18\ \text{N/mm}^2$$

$$\xi_s v_c = 1.20 \times 0.47 = 0.56\ \text{N/mm}^2$$

No provision for shear reinforcement is therefore required.

SERVICEABILITY LIMIT STATE

Deflection Actual span/depth ratio = 4000/130 = 30.7.
From Figure 11.11(a), (b) and (c) with

$$\beta = 25.2/23 = 1.09$$

$$\frac{A_{\text{required}}}{A_{\text{provided}}} = 0.91$$

$$f_s = 460 \times 0.49 = 226\ \text{N/mm}^2$$

$$\frac{A_s}{bd} = 0.004\,0$$

Therefore allowable span/depth ratio = 29.6.

Although this does not strictly comply it would probably be satisfactory. To comply, however, the spacing of the reinforcement should be reduced to 140 mm. Then the allowable span/depth ratio becomes 32.4.

Cracking From Table 11.16(a), spacing should not exceed 160 mm.

From these calculations, it is evident that deflection can control design and that when high tensile reinforcement is used in slabs, cracking may limit design.

1 (d) Determine the eccentricity at which a square column 200 mm × 200 mm reinforced with 3 No. 20 mm ϕ bars at 25 mm from tension and compression faces can sustain an ultimate design load of 240 kN. Grade of concrete = 30, f_y = 250 N/mm².

$$A'_s = A_{2s} = 3 \times (\pi/4) \times 20^2 = 940 \text{ mm}^2$$

From equation (26):

$$240 \times 10^3 = 0.4 \times 30 \times 200 \times d_c + 0.72 \times 250 \times 940 - 0.87 \times 250 \times 940$$

whence $d_c = 115$ mm

From equation (27):

$$240 \times 10^3 \times e = 0.2 \times 30 \times 200 \times 115(85) + 0.72 \times 250 \times 940(100-25) + 0.87 \times 250 \times 940(100-25)$$

whence $e = 165$ mm

Transverse reinforcement (See Table 11.16)

Size of bars $\nless \frac{1}{4} \times 20$, use 6 mm ϕ bars.
Spaced at $\ngtr 12 \times 20 = 240$ mm

Prestressed concrete

Compare examples **1 (a)** and **1 (b)** with **2 (a)** and **2 (b)**, etc.

2 (a) Design a Class 1 prestressed concrete beam of rectangular section with post-tensioned grouted tendons to support a uniformly distributed characteristic imposed load of 20 kN/m on a span of 18 m using concrete Grade 50.

Assume
$$h = 1000 \text{ mm}, \ b = 400 \text{ mm}$$
$$A_c = 0.4 \times 10^6 \text{ mm}^2$$
$$I_c = 33.3 \times 10^9 \text{ mm}^4$$
$$d_n = 500 \text{ mm}$$

Dead load $$G_k = \frac{2400 \times 10 \times 0.4 \times 18}{1000} = 173 \text{ kN}$$

Imposed load $$Q_k = 20 \times 18 = 360 \text{ kN}$$

SERVICEABILITY LIMIT STATE

Moment due to $G_k = 173 \times 18/8 =$ 390 kN m
Moment due to $Q_k = 360 \times 18/8 =$ 810 kN m
Total 1200 kN m

From equations (44) and (45):

Concrete stress due to prestress and dead load, N/mm² (P in N), at top surface:

$$= P\left(\frac{1}{0.4 \times 10^6} - \frac{500e}{33.3 \times 10^9}\right) + \frac{390 \times 500}{33.3 \times 10^9} \times 10^6$$

$$= P \times 10^{-6}(2.50 - 0.015e) + 5.85 \nless 0$$

at bottom surface:

$$= P \times 10^{-6}(2.50+0.015e) - 5.85 \not> 16.7 \text{ (allowable stress)}$$

Therefore at limit $P = 3.33 \times 10^6$ N

$e = 284$ mm

Using 18 mm tendons stressed to 80% of f_{pu},

$$\text{number of tendons} = \frac{3.33 \times 10^6}{370 \times 10^3 \times 0.8} = 11.3$$

(From Table 11.6, Characteristic strength of 18 mm tendon = 370 kN.)

Use 11 tendons,

$$A_{ps} = 11 \times 210 = 2310 \text{ mm}^2$$
$$f_{pu} = 370 \times 10^3/210 = 1760 \text{ N/mm}^2$$

Initial stress = $0.8 \times 1760 = 1410$ N/mm^2

$e = 280$ mm.

From equation (40), stress in the steel after transfer, N/mm^2,

$$f_{p2} = 1410 - \frac{175 \times 10^3}{2 \times 34 \times 10^3}\left[f_{p2} \times 2310\left(\frac{1}{0.4 \times 10^6} + \frac{280^2}{33.3 \times 10^9}\right) - \frac{390 \times 280 \times 10^6}{33.3 \times 10^9}\right]$$

$$f_{p2} = 1370$$

From equations (43), (44) and (45), stress in concrete after transfer, N/mm^2, under prestress and dead load,

$$\text{at top surface} = \frac{1370 \times 2310}{10^6}(2.5-4.2) + 5.85 = 0.5$$

$$\text{at bottom surface} = \frac{1370 \times 2310}{10^6}(2.5+4.2) - 5.85 = 15.3$$

Required strength of concrete at transfer N/mm^2 = 30.6. Take, $\Delta f_p = 3.5\%$, $S = 200 \times 10^{-6}$ and $C = 1.25 \times 36 \times 10^{-6}$/N/mm^2. From equation (41), stress in steel after all losses (N/mm^2)

$$f_{p3} = 1410 - 0.035 \times 1410 - 200 \times 10^{-6} \times 175 \times 10^3$$
$$-\left(\frac{175 \times 10^3}{2 \times 34 \times 10^3} + 175 \times 10^3 \times 45 \times 10^{-6}\right)\left[f_{p3} \times 2310\left(\frac{1}{0.4 \times 10^6} + \frac{280^2}{33.3 \times 10^9}\right)\right.$$
$$\left. - \frac{390 \times 280 \times 10^6}{33.3 \times 10^9}\right]$$

$$f_{p3} = 1210$$

$$\text{Percentage loss of prestress} = \left(\frac{1410 - 1210}{1410}\right) \times 100 = 14$$

From equations (43), (44) and (45), as before, stress in concrete after all losses (N/mm^2) under prestress and dead load,

at top surface = 1.1
at bottom surface = 13.0

under prestress, dead load and imposed load,

at top surface = 13.3
at bottom surface = 0.8

These stress are within allowable limits for serviceability of 0 to 16.7 N/mm^2, and they would not have been exceeded had the serviceability design moment for imposed load been 870 kN m instead of 810 kN m.

ULTIMATE LIMIT STATE

Moment due to $1.4\,G_k + 1.6\,Q_k = 1.4 \times 390 + 1.6 \times 810 = 1840$ kN m

$$\frac{f_{pu}A_{ps}}{f_{cu}bd} = \frac{1760 \times 2310}{50 \times 400 \times 780} = 0.261$$

From Figure 11.18(a)

$$f_{pb} = \frac{0.889}{1.15} \times f_{pu} = 1360 \text{ N/mm}^2$$

From equation (47):

$$M_u = \frac{1360}{1000} \times 2310 \times \frac{780}{1000}\left(1 - \frac{1.25 \times 1360 \times 2310}{50 \times 400 \times 780}\right)$$
$$= 1840 \text{ kN m}$$

Since secondary steel would be provided in a beam of this size, this would be satisfactory.

2 (b) Determine the imposed load that could have been carried had Class 3 tensile stresses been adopted using the same quality of concrete and tendons.

SERVICEABILITY LIMIT STATE

Moments as before.

Concrete stress due to prestress and dead load (N/mm²), (P in N) as before.

at top surface $= P \times 10^{-6}(2.50 - 0.015e) + 5.85 \nless -2.55$ (from Table 11.20)

at bottom surface $= P \times 10^{-6}(2.50 + 0.015e) - 5.85 \ngtr 16.7$

Therefore at limit

$$P = 2.83 \times 10^6 \text{ N}$$
$$e = 365 \text{ mm}$$

Using 18 mm tendons stressed to 80% of f_{pu},

number of tendons $= 2.83 \times 10^6/370 \times 10^3 \times 0.8 = 9.6$

Use 9 tendons,

$$A_{ps} = 9 \times 210 = 1890 \text{ mm}^2$$
$$0.8\,f_{pu} = 0.8 \times 1760 = 1410 \text{ N/mm}^2$$
$$e = 360$$

Stress in concrete after transfer, N/mm², under prestress and dead load,

at top surface $= -1.8$

at bottom surface $= 14.9$

Losses of prestress 13%

Stress in concrete after all losses, N/mm^2, under prestress and dead load,

$$\text{at top surface} = -0.9$$
$$\text{at bottom surface} = 12.5$$

For class 3 structures in normal environments (i.e. cracks of up to 0.2 mm) allowable stresses must be within the limits of 16.7 to $-5.8\ N/mm^2$ if there is no additional reinforcement or 16.7 to $-6.9\ N/mm^2$ for 1% reinforcement (see Table 11.21 and note allowance for depth).

Possible range of stress (N/mm^2), for no extra reinforcement

$$\text{at top surface} = -0.9 \text{ to } 16.7 = 17.6$$
$$\text{at bottom surface} = 12.5 \text{ to } -5.8 = 18.3$$

Imposed load moment for serviceability could therefore be

$$\frac{17.6}{10^6} \times \frac{33.3 \times 10^6}{500} = 1170 \text{ kN m}$$

Since the stress range at the top surface controls, there is no apparent advantage from using additional reinforcement for considerations of serviceability.

ULTIMATE LIMIT STATE

$$\frac{f_{pu}A_{ps}}{f_{cu}bd} = \frac{1760 \times 1890}{50 \times 400 \times 860} = 0.193$$

$$f_{pb} = 1460 \text{ N/mm}^2$$

$$M_u = 1840 \text{ kN m}$$

By coincidence, this is the same as before and therefore the ultimate condition controls. The use of additional reinforcement could increase the ultimate strength.

Assume additional 1% of reinforcement of mild steel, $f_y = 250\ N/mm^2$, $A_s = 4000\ mm^2$.

Assume depth of neutral axis $= 0.5\,d$

Total compression, from Figure 11.18.

$$= 0.4 \times 50 \times 400 \times 860/2 = 3.44 \times 10^6 \text{ N}$$

Stress in tendons $d_c = 0.5\,d = 0.89 \times 1760/1.15$ from Figure 11.18(b)

$$= 1360 \text{ N/mm}^2$$

Total tension

$$= 1360 \times 1890 + 250 \times 4000/1.15 = 3.44 \times 10^6 \text{ N}$$

$$M_u = \frac{1360}{10^6} \times 1890 \times (860 - 0.5 \times 430) + \frac{250 \times 4000}{10^6 \times 1.15}(1000 - 40 - 0.5 \times 430)$$

$$= 2310 \text{ kN m (centre of steel 40 mm above soffit)}$$

This moment corresponds to a moment due to the characteristic imposed load of:

$$(2310 - 1.4 \times 390)/1.6 = 1100 \text{ kN m}$$

Therefore ultimate limit still controls but characteristic imposed load could now be 27 kN/m instead of 20 kN/m if the changed conditions are adopted.

2 (c) Assume a section given below for a Class 1 prestressed concrete beam having the same area and using the same materials as for example **(a)**. Determine the acceptable characteristic imposed load.

$$h = 1000 \text{ mm}, b = 800 \text{ mm}$$
$$A_c = 0.4 \times 10^6 \text{ mm}^2$$
$$I_c = 46.0 \times 10^9 \text{ mm}^4$$
$$d_n = 440 \text{ mm}$$
$$G_k = 173 \text{ kN (as before)}$$

SERVICEABILITY LIMIT STATE

Moment due to $G_k = 390$ kN m

At limit as before $P = 2.68 \times 10^6$ N
$e = 490$ mm

Using 9/18 mm tendons

$$A_{ps} = 1890 \text{ mm}^2$$

Calculations made as for previous examples.

Stress in concrete after transfer (N/mm^2) due to prestress and dead load

at top surface = 0.2
at bottom surface = 15.9

Cube strength at transfer > 31.8 N/mm^2
Total loss of prestress 16%

Stress in concrete after all losses (N/mm^2) due to prestress and dead load

at top surface = 0.9
at bottom surface = 13.1

After all losses, the stresses under prestress, dead load and imposed load must remain

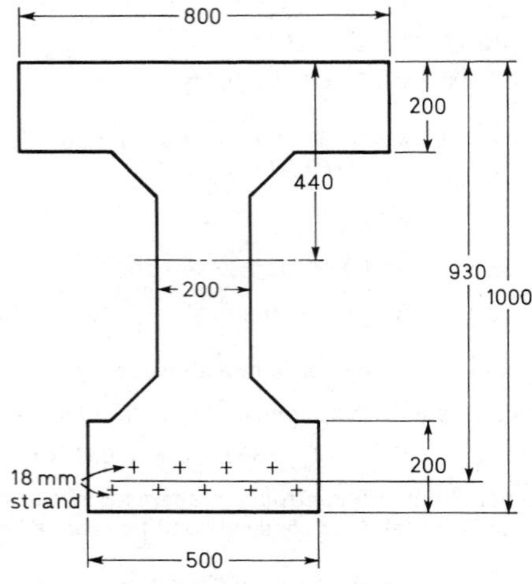

Figure 11.22

within the range 0 to 16.7 N/mm^2.

Therefore, maximum moment due to imposed load

$$= 13.1 \times 46.0 \times 10^9/560 \times 10^6 = 1080 \text{ kN m}$$

ULTIMATE LIMIT STATE

$$M_u \text{ (calculated as before)} = 2430 \text{ kN m}$$

corresponding to a characteristic imposed load-moment of 1170 kN m. Therefore serviceability controls and $Q_k = 27$ kN m.

Thus by changing the shape of the cross section and reducing the number of cables from 11 to 9, the imposed load moment has been increased by $\frac{1}{3}$, comparing example **2 (a)** with **2 (c)**. This is the same increase as obtained by changing from Class 1 to Class 3 stress (comparing example **2 (a)** with **2 (b)**).

It should be noted that it is assumed that the full dead load is operative at the time of stressing for examples **2 (a)**, **(b)** and **(c)**. Stresses have been calculated at midspan to determine the mean depth of the cable. At other points in the span, the dead load moment is smaller and therefore the mean depth of the cable must be smaller. For example **2 (c)**, the mean depth is 930 mm at midspan and must not be greater than 870 mm at $\frac{1}{4}$ points of the span and 700 mm at the supports.

2 (d) Calculate the characteristic imposed load for the section used in example **2 (a)** assuming pretensioning for the same quality of materials.

In this case, the beam is taken as being of uniform section and the tendons as being straight. Hence, the dead load moment cannot be assumed to act at transfer, since sections away from midspan are more critical.

SERVICEABILITY LIMIT STATE

Calculated as before

$$P = 3.33 \times 10^6 \text{ N (identical with (a))}$$
$$e = 167 \text{ mm (284 mm with (a))}$$

Number of 18 mm wire strands

$$= 11.3 \text{ (as for (a))}$$

Use 12 tendons since losses are higher at transfer for pretensioning.

Stress in concrete after transfer (N/mm^2) prestress alone,

at top surface = 0
at bottom surface = 16.4

with dead load also acting (at midspan)

at top surface = 5.8
at bottom surface = 10.6

Taking $S = 300 \times 10^{-6}$ and $\phi = 60 \times 10^{-6}$/N/mm^2

Total losses of prestress = 16%

Stress in concrete after all losses (N/mm^2) under prestress and dead load,

at top surface = 5.8
at bottom surface = 9.1

Moment due to imposed load could be

$$9.1 \times 33.3 \times 10^9/500 \times 10^6 = 610 \text{ kN m}$$

ULTIMATE LIMIT STATE

M_u (calculated as before) = 1615 kN m
Corresponding characteristic imposed load moment = 670 kN m

Serviceability therefore governs and, compared with example **2 (a),** imposed load capacity is 25% less.

The shape of the pretensioned section could be improved to give a better performance by choosing an I section.

2 (e) Taking example **2 (c),** assume that the imposed load is a central point load equivalent to the uniformly distributed load for flexure. Determine what shear reinforcement is required.

ULTIMATE LIMIT STATE

Equivalent point load at midspan,

$$Q_k = 27 \times 18/2 = 240 \text{ kN}$$

Ultimate shear at midspan $= 120 \times 1.6 = 192$ kN

From equation (50):

$$V_{cr} = (1 - 0.55 f_{pe}/f_{pu}) v_c bd + \frac{M_o V}{M}$$

$$\frac{f_{pe}}{f_{pu}} = \frac{0.8(1-0.16) f_{pu}}{f_{pu}} = 0.67 \text{ (16\% loss of prestress)} > 0.6$$

$$v_c \text{ from Table 11.12} = 0.75 \text{ N/mm}^2 \left(\frac{100 A_s}{bd} = 1.02\right)$$

$$V = 192 \text{ kN}$$

$$M = 390 \times 1.4 + 1080 \times 1.6 = 2280 \text{ kN m}$$

$$M_o = 390 + 12.3 \times 46.0 \times 10^9/490 \times 10^6 = 1540 \text{ kN m}$$

(see Figure 11.23)

$$b = 200 \text{ mm}$$
$$d = 930 \text{ mm}$$

Whence $V_{cr} = 223$ kN. But $V_{cr} \not> 0.1 bd\sqrt{fc} = 131$ kN

From equation (52)

$$\frac{A_{sv}}{s_v} = \frac{V - V_{cr}}{0.87 f_{yv} d_t} = 0.30 \text{ mm}$$

From equation (51) the maximum is:

$$\frac{A_{sv}}{s_v} = \frac{0.4b}{0.87 f_{yv}} = 0.37 \text{ mm}$$

If $s_v = 250$ mm, $A_{sv} = 93$ mm^2 for $f_{yv} = 250$ N/mm^2.
Use 8 mm links with $A_{sv} = 100$ mm^2.

Over most of the length of the beam, the section is uncracked and the shear conditions are less critical. It would be appropriate to continue the reinforcement throughout the length of the beam.

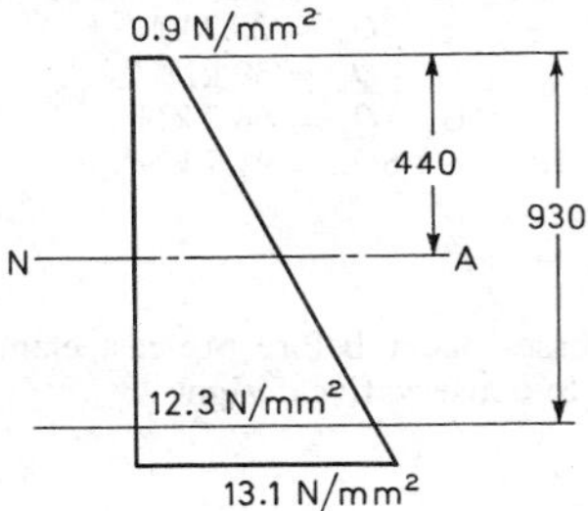

Figure 11.23 Stresses due to prestress and dead load

2 (f) Design a composite section for a ribbed floor spanning 6 m with ribs at 1 m centres. Assume loading due to finishes of 1.5 kN/m² and imposed load of 6.5 kN/m². Take f_{cu} as 60 N/mm² for pre-cast concrete and as 40 N/mm² for in situ concrete. Use 5 mm wire with low relaxation with f_{pu} as 1570 N/mm². Assume Class 2 tensile stresses. The composite section shown below has been obtained by trial.

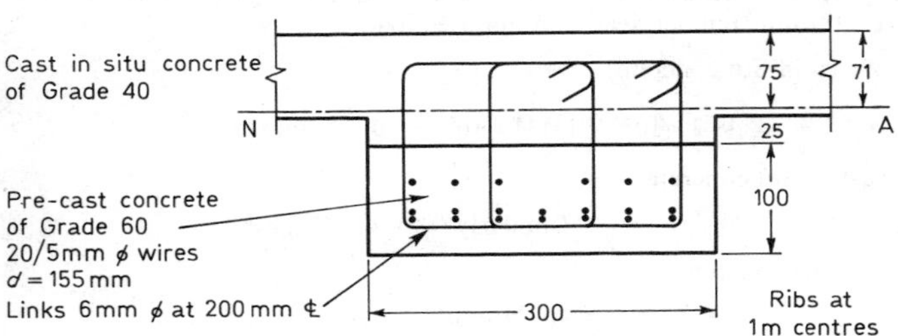

Figure 11.24

Prestressed pre-cast section

$$A_c = 30 \times 10^3 \text{ mm}^2$$
$$I_c = 25 \times 10^6 \text{ mm}^6$$
$$d_n = 50 \text{ mm}$$
$$A_{ps} = 393 \text{ mm}^2$$
$$e = 5 \text{ mm}$$

Composite section

$$A_c = 112.5 \times 10^3 \text{ mm}^2$$
$$I_c = 334 \times 10^6 \text{ mm}^2$$
$$d_n = 71 \text{ mm}$$

No allowance has been made for differences in E_c since the area of steel has also been neglected.

$$G_k = 25.2 \text{ kN}$$
$$Q_k = 39 \text{ kN}$$
$$G_k + Q_k = 64.2 \text{ kN}$$
$$1.4G_k + 1.6Q_k = 97.7 \text{ kN}$$

SERVICEABILITY LIMIT STATE

Assume that all prestress losses occur before pre-cast element is incorporated in the structure; this usually leads to conservative design.

Pre-cast unit:
From equation (41):

$f_{p3} = 940 \text{ N/mm}^2$ after taking $f_{p1} = 1260 \text{ N/mm}^2$
Total loss of prestress $= 25\%$

Stress in concrete (N/mm²) prestress only after all losses,

at top surface $= 8.6 \text{ N/mm}^2$
at bottom surface $= 16.0 \text{ N/mm}^2$

Composite section:
Assume unit is propped during casting of floor moment due to $G_k + Q_k = 48.2$ kN m
Stress in concrete (N/mm²) under dead and imposed load,

at top surface of cast in situ concrete

$$= 48.2 \times 10^6 \times 71/334 \times 10^6 = 10.3 \text{ N/mm}^2$$

at interface in cast in situ concrete

$$\text{(similarly)} = -4.2 \text{ N/mm}^2$$

at interface in pre-cast concrete

$$= -4.2 + 8.6 = 4.4 \text{ N/mm}^2$$

at bottom surface of pre-cast concrete

$$= -(48.2 \times 10^6 \times 129/334 \times 10^6) + 16.0 = -2.6 \text{ N/mm}^2$$

All stresses are therefore within limits prescribed for the Grades of concrete for Class 2 structures.

ULTIMATE LIMIT STATE

Ultimate moment from equation (47)

$$= 74.1 \text{ kN m}$$

Moment of resistance required ($1.4\,G_k + 1.6\,Q_k$)

$$= 73.5 \text{ kN m}$$

Shear Since the prestressed unit is a small part of the total cross section, the beam has been treated as being of reinforced concrete. Also at the ends of the units, the support is likely to be within the transmission length for the wire of 500 mm (100 × 5 mm) and therefore not fully stressed.

From equation (20):

$$
\begin{aligned}
V &= 48.9 \text{ kN} \\
v &= 1.05 \text{ N/mm}^2 \\
100\, A_s/bd &= 0.85 \\
v_c &= 0.69 \text{ N/mm}^2 \text{ for Grade 40 concrete} \\
\text{with } S_v &= 200 \text{ mm} \\
\text{and } f_{yv} &= 250 \text{ N/mm}^2 \\
A_{sv} &= 200 \times 300 \times (1.05-0.69)/0.87 \times 250 = 100 \text{ mm}^2
\end{aligned}
$$

Use double links of 6 mm

$$\text{giving } A_{sv} = 4 \times \pi 6^2/4 = 113 \text{ mm}^2$$

Check shear at interface for serviceability limit state from equation (54).

$$
\begin{aligned}
V_d &= 32.1 \text{ kN} \\
v_h &= 32.1 \times 10^3 \times 30 \times 10^3 \times 79/334 \times 10^6 \times 300 \text{ (section is not cracked)} \\
&= 0.76 \text{ N/mm}^2.
\end{aligned}
$$

Therefore Type 3 surface is required from Table 11.26 with 0.15% reinforcement.

$$
\begin{aligned}
A_{sv} \text{ required} &= (0.15/100) \times 200 \times 300 \\
&= 90 \text{ mm}^2
\end{aligned}
$$

Sufficient is therefore available.

Over the transmission length of 500 mm this reinforcement should be spaced more closely, say at 100 mm centres, and should enclose the tendons.

REFERENCES

1. *Code of Practice for reinforced concrete,* DSIR, HMSO, London (1934)
2. *Construction of buildings in London,* LCC (1938)
3. *First report on prestressed concrete,* Institution of Structural Engineers, London (1951)
4. *The structural use of reinforced concrete,* CP 114, British Standards Institution, London (1969)
5. *The structural use of prestressed concrete,* CP 115, British Standards Institution, London (1965)
6. *The structural use of precast concrete,* CP 116, British Standards Institution, London (1969)
7. *Design and construction of reinforced and prestressed concrete structures for the storage of water and other aqueous liquids,* CP 2007, British Standards Institution, London (1970)
8. *The structural use of concrete,* CP 110, Parts 1, 2 and 3, British Standards Institution, London (1972)
9. *Recommendations for an international code of practice for reinforced concrete,* CEB, Cement and Concrete Association (1964)
10. *International recommendations for the design and construction of concrete structures,* CEB/FIP, Cement and Concrete Association (1970)
11. *Handbook on the unified code for structural concrete* (CP 110: 1972), Cement and Concrete Association (1972)

12 PRACTICAL STEELWORK DESIGN

PRACTICAL STEELWORK DESIGN 12

12 PRACTICAL STEELWORK DESIGN

F. H. NEEDHAM, B.Sc.(Eng.), A.C.G.I., F.I.Struct.E., F.I.C.E.
British Steel Corporation

BRITISH STANDARDS FOR DESIGN

The British Standards pertaining to the design of steelwork are BS 449, 1959–69 *The Use of Steel in Building,* and BS 153, 1958–72 *Steel Girder Bridges.* It has been recognised for some time that both these Standards are in need of revision to bring them into line with the latest developments in philosophy and knowledge gained from research. Accordingly, at the time of writing, British Standards Committees are in being for the purpose of drafting revised documents. The policy is that design standards for composite and non-composite steel bridges will be incorporated into a new document dealing with bridges generally. The design of steelwork for buildings, or rather the design of all structural steelwork other than bridges, will be covered by another document including guidance in the matter of composite action.

The philosophies behind both new documents will differ considerably from those which have gone before. In particular they will require limit state design to be adopted, requiring predictions of collapse loads and limits of serviceability, and the adoption of partial load factors for different classes of loading taking into account, perhaps somewhat crudely, the statistical probabilities of different classes of loading occurring simultaneously.

Thus in this section it would not be appropriate to include numbers of fully worked examples of the design of structural elements in accordance with design rules shortly to be superseded, neither is it yet possible to anticipate the final forms which the new design rules will take. However, this may be a beneficial situation in that it presents the opportunity to take a broad look at the process of steelwork design which will enable various matters to be brought into a better perspective than otherwise would be the case.

STEEL AS A STRUCTURAL MATERIAL

Fundamentals of the steelmaking process

It is axiomatic that the designer of any engineering work ought to have a fair understanding of the nature of the material he proposes to use. This ideal does not always obtain in steelwork or other building materials. To provide some grasp of the varying characteristics of steel, and the origin and nature of possible defects, it is necessary to consider the manufacturing processes by which steel plates, beams and sections are made.

The production of structural plates and sections is a three-stage process, namely, ironmaking, steelmaking and rolling. Ironmaking is performed in a blast furnace and consists of chemically reducing iron ore, using coke and crushed limestone, and is essentially a continuous process. The resulting material, cast iron, is high in carbon, sulphur and phosphorus. Steelmaking, on the other hand, is a batch process, and consists of refining the iron to reduce and control the C, S and P levels and also to add where necessary, depending on the grade of material to be produced, controlled proportions of manganese, chromium, nickel, vanadium, niobium, etc. This process is carried out in an open hearth furnace, an electric furnace or a Bessemer-type converter. The latter method, once very popular and later superseded, is again coming to the fore in the form of the LD Converter with the availability of bulk tonnage oxygen, dramatically affecting the economics of the operation.

Whilst cast iron has been mentioned, in fact comparatively little iron is allowed to solidify, the metal mostly being tapped and transferred directly, in the liquid state, to the steelmaking furnace. The melt, so called, in a steel furnace may well be several hundred tons, economy deriving from bulk production. This in large measure explains why small quantities of specially alloyed steels are expensive.

The steelmaking process may last an hour or more, during which chemical change is taking place. Samples are taken at intervals and analysed for composition in a laboratory. During the minutes which this takes the chemical process continues, and it remains a matter of nice judgement when to stop the process by cutting the oxygen and tapping the

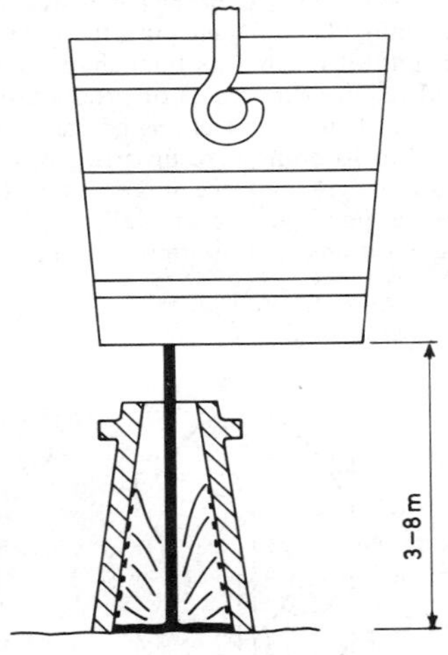

Figure 12.1 Teeming

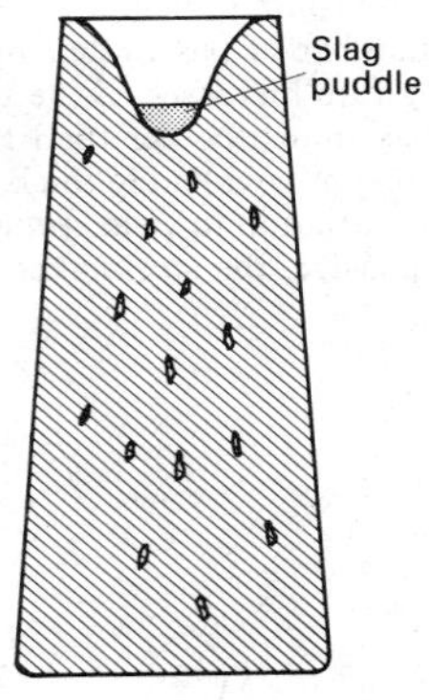

Figure 12.2 Inclusions

melt into a ladle. (Current research is aimed at providing instant spectrographic analysis, permitting greater control). Once tapped into the ladle, a further sample analysis is made, the 'ladle analysis', which is taken as a record of the whole melt, but it must be recognised that there may be some variability in the dispersion throughout the melt which explains why samples of a part of the product may show slight variations from the ladle analysis.

At this stage, the steel is poured, or 'teemed', from the ladle into moulds to form ingots (Fig. 12.1). Not so long ago, a 10 tonne ingot was considered big—today a 40 tonne ingot is not uncommon. Defects that concern the structural engineer may occur at this stage. In the first place, the steel may be poured from as high as 6 or 9 metres into the bottom of the mould, splashing up the sides. Some drops, which freeze instantly on contact with the cold mould, may not re-melt when the surface level rises to encompass them, and may even not fully forge into the bulk of the steel on rolling. This results in surface imperfections, which are mostly of little importance, apart from appearance. More seriously, oxidation inevitably occurs at the free surface of liquid metal, and the melt may also contain some slag. Most of this nonmetallic material floats to the surface of the ingot before solidification, but some may remain in the body of the steel, leading to internal laminations after rolling (Figure 12.2).

Ingots free of slag inclusions can be produced by special processes, such as uphill teeming, but inevitably are more expensive. For the bulk of heavy engineering purposes, therefore, one must expect a certain amount of laminations. As with knots in wood, what matters is where they are, how big they are and whether they render the material unfit for its intended purpose.

After the ingot has solidified and the mould been stripped off, the ingot is transferred to a soaking pit, in which numbers of such hot ingots are stacked, to ensure even distribution of heat. The ingot is then passed to the rolling mill, at which the first operation is the cropping off of the top end containing the slag puddle. The amount to be cropped is a matter of nice judgement on the part of the mill operator, the ingot passing towards him top end first. If he crops too much the yield (i.e. of rollable product) is reduced, if too little, end piping or lamination will be present. Of course, he errs on the safe side but mistakes can occur leading to laminations in the end bar or two of a rolling. The ingot is now rolled down in a cogging or blooming mill by a series of passes to and fro between the rolls, which are closed between each pass, reducing the girth of the metal and elongating it. Passing to the final mill, a universal beam mill, plate mill, etc., it is further reduced in size by rolling through a series of reversals until it reaches its final shape. Thus the total number of passes depends upon the final required thickness of product, and the cooling rate increases as the thickness is reduced. Thus thin sections go through their final pass much cooler than thick sections, and in so doing take up a degree of work hardening not evident in thick sections. Since for any grade of steel it is desirable to have as little variation in yield strength with thickness as possible, and since alloying elements are expensive, the steelmaker aims at keeping additions to a minimum if thin sections

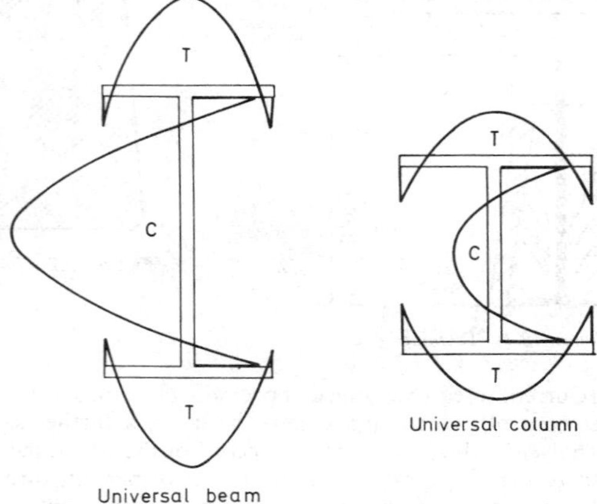

Figure 12.3 Residual stresses in beams

are being rolled. Likewise thick sections are likely to contain alloying elements near to the maximum for the particular specification. Thus the material of which thin sections are made is inherently easier to weld than thick sections, apart from thermal problems which are likely to arise with thick material.

After rolling, the hot bars or plates are sawn to length and transferred to a cooling bank. It is to be noted that at this stage the steel is most unlikely to be straight or flat. The rate of cooling of different parts of the plate or section will vary, depending upon its exposure. For instance the toes of flanges and the centre portion of a deep thin web of an I-section will cool faster than the thicker material at the junction of the web and flanges, and it is this circumstance which creates the pattern of residual stresses (Figure 12.3). When cold the material is cold straightened, sometimes by rolling and sometimes by gag pressing, dependent on the shape, and in turn this affects the level and distribution of residual stresses. The straightening process has a stress-relieving effect and a bar which needs extensive treatment will finish with a low level of residual stress, while the rare bar which remains straight when cooled will have the highest level. The presence of residual stresses is most clearly seen when a member is cut longitudinally, as when an I-member is slit into two Ts, which curve noticeably on division owing to redistribution of longitudinal stresses. Plate also shows similar behaviour to a smaller extent.

Fundamental properties of structural steel

Figure 12.4 indicates the familiar tensile stress–strain curve for mild steel. The properties of usual concern to the design engineer are the value of the yield stress and the gradient of the elastic portion, i.e. the modulus of elasticity. Fortunately or otherwise (some think otherwise) the second property varies little from one grade of steel to another

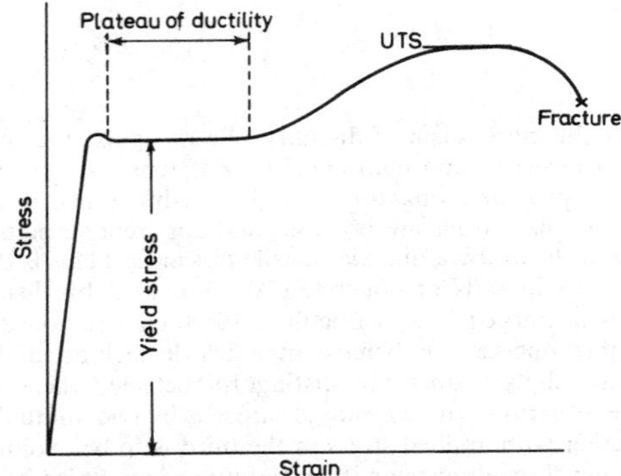

Figure 12.4 Tensile stress–strain diagram (mild steel)

and cannot be controlled. It is to be noted that the finite value of the Ultimate Tensile Strength (UTS) is of little significance in structural design, since a structure can have manifestly failed without the stress exceeding yield stress. The gradient of the line immediately after yield, i.e. the strain hardening modulus, can be of significance in plastic design as it can affect moment/rotation behaviour. But the property which is at least as significant from a structural point of view as the yield strength is the ductility, expressed by the length of the horizontal portion of the curve. It is this plateau of ductility which enables the steel to relieve and redistribute residual stresses arising from cooling and welding.

Another factor to be recognised is that the stress–strain curve usually refers to a specimen cut in the rolling direction. A similar specimen cut transversely to the direction of rolling would show lower characteristics and, for thick plate, through thickness strength can be markedly less. In the case of plates, in an attempt to equalise longitudinal and transverse strengths, a plate may be turned and rolled sideways for one or two passes, if the dimensions permit.

Having referred to residual welding stresses some remarks concerning their mode of origin and magnitude seem appropriate. During welding, the heat put into the Heat Affected Zone (HAZ) causes this to try to expand, but this expansion is prevented by the surrounding cold material. The HAZ therefore yields in compression and becomes effectively shorter and when cooled attempts to contract, causing stress reversal. The weld and the HAZ is then in a state of residual tensile stress, with a corresponding residual compressive stress in the main bulk of the material maintaining equilibrium. It can be assumed that, as welded, all weld metal and HAZ material is in a state of tensile stress equal to the yield stress. These residual stresses can be relieved, by heating in a stress-relieving furnace or by local heating with electric mats or by proof loading to a degree greater than that anticipated in service. It is here that the property of ductility comes into play. For the great bulk of structural purposes, stress relieving is too expensive, and occurs fortuitously on the first application of severe load, during which the 'peaks' of the residuals are cut off so to speak. This explains the partly inelastic initial deflection of a heavily welded member. Recoveries of initial deflection of 80 to 90% have been observed, after which behaviour is fully elastic.

The foregoing underlines the necessity for achieving adequate ductility in the weld metal. Usually the attainment of strength presents no problem, owing to the quenching effect of fairly rapid cooling. This quenching effect can be too severe tending to produce brittleness, as for instance in multi-pass welds in thick material. Therefore preheating and, in extreme cases, post-heating, are needed in order to slow the cooling rate in the interests of achieving ductility.

Notch ductility

Let us now consider the mechanism of ductility. We recognise that uniaxial extension is accompanied by transverse contraction. If biaxial tensions are present, only one dimension remains to provide elongation, that is a reduction in thickness. If triaxial tensions are present no distortions are possible, and apparent strength is much greater than the yield strength shown by a uniaxial tensile test piece. (This is the converse effect of triaxial compressions in soils or concrete.) Accompanied by this increased tensile strength, however, is a marked loss of ductility. We then have the situation where a material demonstrably ductile can behave in a brittle fashion under certain stress conditions. The material itself does not distinguish between the origins of different stresses, so that the situation where residual stresses in two mutually perpendicular directions exist together with applied stress in the third is to be avoided. This situation occurs in details giving three-dimensional continuity accompanied by restraint against contraction.

Stresses at the tip of a crack subjected to tension perpendicular to the plane of the crack are essentially triaxial, which we have seen leads to brittleness. The property of notch ductility or toughness is the ability to resist the propagation of such cracks under tensile stress.

One must recognise that even with the closest of control over an industrial process, some defects will occur. It therefore follows that no weld can ever be perfect and may have slag inclusions, porosity and cracking. For design purposes, therefore, one must assume that welds contain cracks which are undetectable. We come, therefore, to face the fact that adequate notch ductility is vital in any welded structure. The problem, however, is one of measurement.

The generally accepted measure of notch ductility is given by the Charpy impact test. In this a 10 mm square notched test piece 100 mm long is struck and broken by a pendulum and the loss of kinetic energy is measured. The specimen is supported at both ends and struck in the centre opposite the notch, putting the notch tip into tension. The radius at the notch tip is 0.25 mm, that is very much greater than that at a crack tip. The specimen is also quite small relative to actual structures, and of course not subject to residual stresses.

If, however, standard material is tested over a range of temperatures, and energy value is plotted against temperature, a curve of the form shown in Figure 12.5 results. Steel shows the characteristic of high notch ductility at high temperatures and a marked fall in notch ductility at low temperatures. The temperature, or rather range of temperature, over which this transition takes place is known as the 'transition temperature'. The transition temperature can be altered by heat treatment of the steel, quenching tending to raise the level and annealing or normalising tending to lower it. For a given steel, strength can be raised at the expense of notch ductility and transition temperature, and vice versa. For any particular requirement a balance has to be struck.

It is important to recognise, though, that the Charpy test has its limitations. Firstly, the specimen is small and does not reveal the inherent increase of brittleness with thickness due to triaxial effects. Secondly, the notch tip is relatively blunt, and finally the strain

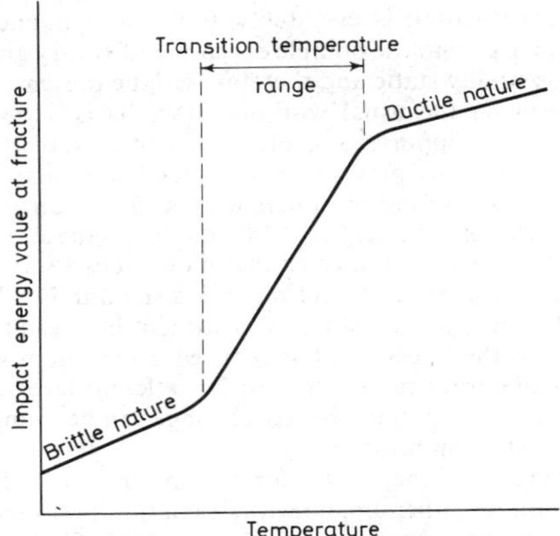

Figure 12.5 Impact–transition temperature curve

rate is high. High strain rate leads to loss of ductility, so the combination of these effects means that a finite value of a Charpy test does not give any accurate guide as to the actual in-service behaviour of a particular structure or detail. What it does provide is a quality control standard at the steelworks by which a batch of steel can be compared with another of known performance.

Other tests giving much closer correlation to in-service performance have been evolved, notably the Wells wide plate test, the Crack Opening Displacement test and the Pellini Drop Weight Tear test.

Factors leading to possible brittleness include three-dimensional continuity or restraint, such as is usually present in welded components of thick material, coupled with loading giving high strain rates. But in addition to these, three features are essential for a brittle fracture to occur, namely, a notch or severe stress concentration, a tensile stress and a service temperature below the transition temperature. If any one of these three features is absent, brittle fracture cannot occur. The task of the designer, therefore, is to make a careful assessment of the particular situation in order to determine the degree of risk. In the majority of structural configurations the risk is slight, but where it is not he must modify such factors as are under his control, namely the detailed design and the material quality, in that order. It is true to say that, of the failures that have occurred, more could have been prevented by good detailed design than by varying the steel quality.

For a fuller treatment of the subject of brittle fracture and testing procedures the reader should refer to the Welding Institute Publication, *Brittle Fracture of Welded Structures.*[1]

Fatigue

Whereas failure from brittle fracture is sudden and temperature dependent and can occur at any time in the life of a structure (though usually early owing to the effect of in-service stress relieving), failure due to fatigue is quite different. It arises principally due to the cumulative effect of pulsating or alternating stresses, and propagates slowly, often over a period of years. As one would expect, initiation occurs at a stress raiser, such as a sharp notch or crack and its rate of growth is dependent both on the magnitude of the stress variation and the total number of cycles of stress. Whilst fatigue crack propagation occurs slowly it must be appreciated that it is non-ductile in nature, that is that no significant deformation takes place, and therefore detection of cracking can be difficult.

In considering the design of a structural element, the designer's first task must be to assess whether the live loading is essentially static or dynamic. In the majority of cases of building steelwork, with the notable exception of gantry girders and other lifting gear, the loading is essentially static and therefore fatigue presents no problems. Design may, therefore, proceed on the usual working stress basis. However, in bridgework, crane structures and their supports, and other structures subject to dynamic loading such as wind-induced vibrations, prevention of fatigue failure may determine the design.

As indicated earlier, stresses can be fluctuating above a certain minimum level or alternating or somewhere between the two. What matters is the stress range, or amplitude and the mean stress. It has been established that if one does a series of tests on identical specimens, varying the stress range S and recording the number N of cycles to failure, and the results are plotted of S against log N, a straight line results, as shown in Figure 12.6. This shows that, as the stress range is reduced, so the number of cycles to failure is increased. A family of curves can be derived for different levels of mean stress. If the proposed life is known, then a permissible stress range can be found and compared with that permitted under static conditions.

If we now study the effect of varying the form of specimen to include on the one hand plain rolled plate and on the other a plate with discontinuities, such as holes, notches or welded attachments, another phenomenon presents itself. That is that the presence of discontinuities significantly reduces the fatigue life, the more severe the discontinuity,

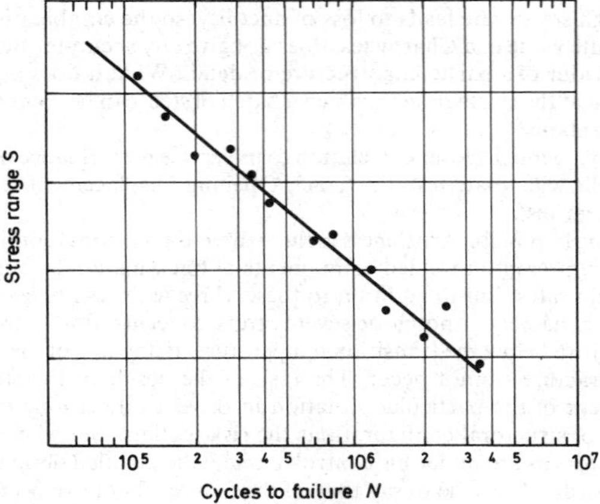

Figure 12.6 Fatigue curve

the greater the effect. A plain rolled plate at a strength of 247 N/mm^2 will give a fatigue life of 2×10^6 cycles in pulsating tension. That is, at 2×10^6 cycles plain mild steel plate has a strength equal to the yield stress. (Note that there is no particular magic about 2×10^6 cycles. It may be a more than adequate life or it may be grossly insufficient in any particular case.)

A detail with a marked discontinuity can show a strength at 2×10^6 cycles, as low as 30 or 40 N/mm^2. As a result of extensive experimental work at the Welding Institute it has been possible to classify various structural details in terms of their severity as regards fatigue. These are clearly presented in the fatigue clauses of BS 153, 1972, to which reference should be made.

It may seem curious that fatigue can be a consideration when the applied loading is wholly compressive. This is because, as explained earlier, the presence of high residual stress due to welding can cause local reversals. Thus a fatigue crack in a zone of high residual tensile stress, not overcome by the applied compressive stress, will propagate until it reaches a zone where the total stress is wholly compressive. Care must also be taken to see that zones of high local stress are not overlooked, as for example the severe stress occurring in the top flange-to-web joint in a gantry girder immediately below a local wheel load.

In considering fatigue one recognises that, in practice, stress variations are mostly random in nature and rarely of a uniform amplitude. If a picture of the stress spectrum can be arrived at, either by prediction or by measurement, cumulative damage can be assessed by applying Miner's Rule.

It should also be emphasised that fatigue behaviour cannot be assessed with any great accuracy, there being a wide scatter evident in test results. This arises partly because the degree of severity at a discontinuity, such as a butt weld or fillet weld, depends upon the weld shape and the presence or otherwise of weld undercutting. Thus the skill of the individual welder enters into the picture, which is of course a variable.

The Welding Institute also publish valuable booklets on *The Fatigue Strength of Welded Structures*,[2] and *The Weldability of Steel*[3] which are worthy of study if fuller information is needed.

Engineering judgement

Summarising the matter of fatigue and brittle fracture one must put it that in practice, in the majority of structural configurations, these factors do not dominate. When they do, however, they can be absolutely critical. The safety and life of the structure thus depends on the engineering judgement of the designer in his assessment of the various factors involved. This cannot be divorced from considerations of the consequences of failure, which may vary from simple inconvenience to catastrophe. Accordingly he must then err on the optimistic or pessimistic side.

AVAILABLE STRUCTURAL STEELS

In 1968 a new British Standard was published, BS 4360, entitled *Weldable Structural Steels*, which superseded all previous standards for structural steel. So important has welding become as a fabricating technique that all steels now marketed can be broadly classified as weldable and no non-welding qualities are produced. All the weldable steels previously covered are included, and a new grade of high-tensile steel with yield strength in the range 402–448 N/mm^2 (depending on thickness) was introduced. Steels of this order of strength had previously been available under various trade names but had not made any serious structural impact.

BS 4360 presents compositions, tensile strengths and Charpy impact properties for four principal grades, i.e. 40, 43, 50 and 55 (these numbers roughly corresponding to Ultimate Tensile Strengths expressed in kgf/mm^2) for plates, rolled sections and tubes. It also shows the nearest equivalent previous British Standard, in each case. The dimensional tolerances and rolling margins for plate material are given, and also details of all mills tests and their frequency. These include tensile tests, bend tests, flattening tests for hollow sections and, where the specified properties require, Charpy impact tests. Other matters dealt with include identification, marking, provision of test certificates, and inspection.

It is understood, at the time of writing, that an amendment is in course of preparation to extend this standard to cover the so-called weathering or slow-rusting steels, whose properties fall close to Grade 50 as at present specified.

One of the principal concerns of the designer is the matter of economy, and this must be considered when specifying the grade of steel to be used. Clear guidance on this issue is somewhat sparse. Not only is the pricing structure of the various grades complex and subject to periodic change, but even if material costs are known, no clear-cut answer can be given in most instances. This is due largely to the fact that more than half of the cost of a finished steel structure derives from fabrication and erection costs. The higher grades of steel require more carefully controlled, and therefore more costly, welding procedures, applied to a reduced total tonnage. Therefore estimated rates per ton can be misleading, total costs being the criteria. In particular, fabricators tend to load tender rates when steels with which they are not familiar are specified. Suffice it to say that in many structural applications the high-tensile steels possess potential which has not as yet been fully exploited. The best attempts so far made to draw realistic cost comparisons between components and structures fabricated from different grades of steel have been by A. A. Doncaster whose articles in *Construction Steelwork and Metals*[4, 5] are worthy of study.

One must add that if an engineer has a need for particular properties which fall outside the properties specified in BS 4360, and the tonnage required is likely to warrant the making of a special cast (say 250 tonnes and upwards), he should explore the possibilities with the steelmakers. Also worthy of mention is the possibility of supplying, for structural purposes, quenched and tempered (QT) steels of tensile properties much higher than Grade 55. Such QT steels are only on the fringe of the structural market and are expensive but, for particular requirements, they can be made available against a special order.

TYPES OF STEEL STRUCTURE

It would be helpful in this section to review the classes of structure in which structural steelwork finds its principal applications. Accurate statistics of usage are elusive, the best available being recorded by the British Steel Corporation. However, the rolling mills produce a much greater tonnage of plates and section than are revealed in construction statistics. Much of this difference is accounted for by sales to stockholders, the ultimate destiny being obscure, and some can be attributed to inevitable waste in fabrication. There remains, though, an inexplicable discrepancy in the figures.

Below is given a schedule of types of structure, roughly in descending order of tonnage usage to put matters in perspective, followed by guidance on some of them which it is hoped will prove of assistance in the early stages of design.

(a) Heavy industrial buildings, e.g. power stations, steelworks mill buildings, etc.
(b) Light industrial buildings, e.g. factories, warehouses, etc.
(c) Industrial plant, e.g. bunkers, hoppers, conveyor structures, tanks, petrochemical plant, etc.
(d) Pressure vessels
(e) Short-and medium-span bridges
(f) Major bridges
(g) Multi-storey buildings
(h) Other miscellaneous structures, e.g. electricity transmission towers, TV masts, railway electrification structures, temporary bridges, space frame roofs, etc.
(i) Light industrialised buildings, e.g. system-built schools, hospitals, etc.

It is to be noted that tubular structures have not been shown as a separate category since circular and rectangular hollow sections (CHS and RHS) are being more widely used of late alongside more traditional sections. Indeed, not until these shapes are accepted throughout the industry as one of the arrows in the quiver will their full versatility have been exploited. The same principle applies to sections cold rolled from strip.

Heavy industrial buildings

In considering this group one must emphasise that great accuracy of calculation and sophisticated techniques of analysis are largely misplaced. Highly competitive designs which give rise to a host of minor troubles to the user are not preferred to relatively crudely designed structures of a high standard of robustness. The principal loading arises from the plant to be accommodated and whilst in an ideal world the plant design should be finalised before the structural engineer starts, in practice this situation never obtains. Instead one can expect loading figures given in the first instance to be subject to almost continuous and sometimes radical change during the development of the project, even up to the erection stage. Accordingly, the wisdom of providing an additional margin of stress in design to accommodate some of these loading changes will be evident. There is also the strong possibility of structural requirements changing during the lifetime of the building, and the designer will not be blessed if a succession of strengthening operations are needed.

In this class of structure there is not a great deal of scope for originality on the part of the designer. For the most part, plant requirements dominate and the structural layout is not decided – it occurs. It is important, nonetheless, to ensure adequate economy by studying actual needs. For example, a steel melting shop and a power station turbine house may demand cranes of similar capacity but, whereas the ladle crane may make a dozen passes a day at maximum load, the turbine house crane only needs its full capacity for the replacement of a turbo-alternator every few years. Clearly, different standards of fatigue strength are required in these instances.

Light industrial buildings

It is in this application that there exists the greatest competition between fabricators, which is reflected in design. There is plenty of scope for new ideas and changes in techniques as indicated by the swing from roof trusses to portal frames. Plastic design has been widely adopted for design of portals, but it is worth pointing out that if absolute economy is desired (and the roof space is not needed for stacking goods, etc.) tied portals show benefit. That is, using a horizontal slung tie between the knees at eaves level. Controversy continues as to whether haunched or uniform section portals are most economic. Whilst material cost tends to increase, so too do labour rates, but at a faster rate. Therefore in today's conditions and for the foreseeable future the amount of fabrication should be minimised – if necessary at the expense of total weight. Tapered members are often used, but in many cases the overall economy achieved is arguable.

Whilst not condemning the pursuit of economy in the design of frames it must be stressed that it is unwise to economise on wind and rafter bracing. There have been too many instances of frames going over before the cladding has been fixed. In this context interesting work has been done recently by Professor E. R. Bryan of Salford University on the stiffening and bracing effect of cladding. If his methods of stressed skin design are contemplated, adequate erection bracing, re-usable, must be included.

It is not often appreciated that the cost of purlins and side cladding rails form a significant portion of the total. Two points should be considered here. Firstly, if the structural designer has freedom in fixing the spacing of trusses or frames it is worth designing the purlins and rails first and stressing them fully by adjusting their spans. Secondly, cold-formed sections show cost advantage over hot-rolled angle purlins. This arises largely from their being symmetrical and therefore more efficient beam sections than angles. If the first procedure is adopted, hot-rolled channel purlins can be used,

Figure 12.7 Light industrial building. New IBM warehouse, Greenock (Courtesy: R. T. James & Partners)

Figure 12.8 Crane gantry structure. Redpath Dorman Long Ltd, Cambuslang

and their main disadvantage – an insufficient range in the smaller sizes – overcome. In addition, purlin cleats can be dispensed with on the frames.

In light factory construction, increasing emphasis is being put on the benefit to productivity of a pleasant environment, and this tends to influence design. Tubular space frames have much to offer here, and should be considered. On the other hand, for high-bay warehouses and warehouses designed round the characteristics of the fork-lift truck, internal aesthetics do not figure and structural capability is all-important. Benefit can be derived by integrating the roof and wall structure with the internal racking.

Industrial plant

In many respects this group is a speciality and indeed has its own trade association. Design problems are likely to arise in many respects, notably in connection with working stresses and whether those specified in BS 449 are applicable and appropriate. Judgment is required, particularly in respect of the possibility of overloading, accidental or deliberate. Clearly, tanks cannot be overloaded, and a slender load factor may be appropriate whereas in other applications normal load factors will result in structures insufficiently robust for their purpose.

In design, fatigue may be a consideration in dynamically loaded structures such as conveyor frames, gantry structures, etc. Reference to the BS 153 rules will indicate whether this is a governing factor.

In such work, maintenance costs to prevent corrosion can be heavy. It is not sufficient to design and detail a steel structure and then ask oneself what means of protection should be applied. It may by then be too late since moisture traps may have been built in, and certain areas may be inaccessible for painting after erection, e.g. back-to-back angles. Care must be taken in detailing to ensure that water cannot collect in puddles. If this situation is unavoidable, drain holes should be provided. The BSC publish a useful booklet *How to Prevent Rusting*[6] giving simple guidance. It should be borne in mind that hot-dip galvanising gives excellent protection at modest cost and the trend is for longer baths enabling greater lengths and areas of steel to be treated.

Pressure vessels

The design of pressure vessels forms the subject of a separate British Standard, No. 1500, Part 1, 1958. The requirements laid down therein are considerably more stringent than for normal steelwork, both as regards material quality and fabrication, and inspection procedures are correspondingly more severe. That this is right is undeniable, having regard to the catastrophic consequences of possible failure, but it does lead to fabrication costs greatly in excess of the usual. Specialist plant is needed, such as that necessary to bend thick plates and spin dished ends.

Short- and medium-span bridges

Competition with concrete here is very severe. It will frequently be found that the most economic solution is arrived at by using steel beams or girders acting compositely with a reinforced concrete deck. In design it is no longer adequate to consider statical transverse distribution. An analysis in accordance with the theories of Hendry and Jaeger[7] or Morice and Little,[8] or a finite-element grillage analysis will demonstrate great economy in the sizes of beams at the expense of a minimum addition to the transverse reinforcement in the slab. If full transverse distribution is assumed it is usual to provide continuous transverse top steel in the slab.

A number of model tests have been carried out on bridges designed to normal stresses on these principles in accordance with CP 117, Part II. Departure from strict linearity occurs at around twice working load, which indicates reasonable economy with adequate safety. Destructive tests, however, give enormous margins of strength at collapse, ultimate load factors of 10 not being uncommon. In one such test which the author witnessed, an ultimate load factor of 7 was achieved with a system in which the transverse top slab steel was discontinuous, i.e. over the beams only, calling in question the necessity for continuous top steel.

Girders may be rolled universal beams, autofab beams or purpose-built plate girders, either with equal flanges or with a top flange smaller and narrower than the bottom. In the latter case, the intermediate design stage must be examined when the laterally unrestrained beam is supporting the weight of the wet concrete. It is rare for such systems to be designed as propped, and thus composite action is assumed for live load only.

Figure 12.9 Plate girder for BSC Anchor project, Scunthorpe, being welded

The theoretical advantage of unequal flanges giving minimum weight of steel will prove largely illusory if a rolled beam could have been used as an alternative. Minimum fabrication leads to minimum total cost.

An interesting variant, finding increasing application, is the proprietary 'Preflex' beam. Such beams are bent elastically in the works, and the tension flange then cased in concrete. After curing and subsequent release, the concrete naturally goes into compression. The moment of inertia of the resulting composite section can then be used in design in the calculation of deflections. This system does not increase the ultimate strength of the girder, but does enable high-tensile steel to be used, with resulting economy, in those cases where deflection considerations would otherwise have precluded its use.

It is in the kind of bridge discussed, where the girders are largely protected from driving rain, that the 'Cor-ten' or weathering steels are likely to become popular. At small extra cost, bridges can be built which will be effectively maintenance-free during their life, being particularly attractive as motorway overbridges for this reason.

Major bridges

It is here that the designer has greatest scope for imagination and inventiveness, as is illustrated by the wide variety of elegant structures which have been built of recent years throughout the world. Steel dominates the scene here, largely due to its strength-to-weight ratio, since the greater part of the load sustained by a large span bridge is its own self-weight. It is, unfortunately, given to few of us to have a hand in the design of such works but, with increasing perplexity, design becomes a matter of team work, and no one name can any longer be attached to one structure.

On the other hand the hazards are great, as events have shown.

Multi-storey buildings

Steelwork for this application faces the severest competition from concrete. So much so that recently it seemed as if it had almost been ousted from the scene, except for very special applications. This has largely come about as a result of fire regulations, with which one would not wish to argue. There are signs, however, which indicate that, with some of the newer concepts, particularly for very high-rise prestige office buildings, steelwork offers economic and practical solutions. Two factors have brought this situation about. Firstly, the fire regulations require the lift and stair well enclosures to be separated and protected from the spread of fire from the occupancy floors. Secondly, the latest wind loadings require larger transverse forces to be taken by the building. What better solution

Figure 12.10 Multi-storey building: P & O building during erection (Courtesy: Scott Wilson Kirkpatrick & Partners)

could there be than to have a reinforced concrete core containing the lifts and services which is itself capable of resisting transverse forces and surrounding it with a simple steel frame supporting vertical forces only. In certain cases the height and plan form of the core may require vertical post-tensioning to enable lateral forces to be resisted. This post-tensioning can be provided by the surrounding office floors if these are suspended from the top of the core. Several examples of such structures have recently been constructed; these have the added advantage, in city centre developments, of keeping all the foundation work within the plan area of the building.

For the less dramatic type of building, other developments have been taking place, notably in the design of frames with rigid connections. Prior to the publication of the first report of the Institute of Welding and the Institution of Structural Engineers Joint Committee on Fully Rigid Welded Multi-Storey Frames,[9] design problems were intractable. Following the publication of this work the writer was associated with full-scale loading tests on rigid jointed frames which culminated in the paper by R. F. Smith and E. H. Roberts in the *Structural Engineer*, November 1971. The tests therein described, which were certainly the largest and most exhaustive ever carried out, have yielded a wealth of information, not only as regards overall frame behaviour, but also behaviour of the joints. The very considerable reserve of strength evident in the columns has led Dr R. H. Wood, of the Building Research Establishment where the tests were conducted, to formulate an entirely new column method based on the concept of 'vanishing stiffness'. Inasmuch as his method predicts column failure with a much greater degree of accuracy than any theory hitherto, when fully developed it offers promise of great economy in column design.

Having dealt shortly with two of the latest developments one must still admit that now, and in the future, the bulk of multi-storey steel frames will be discontinuous column and beam frames designed on simple methods. That is to say that beams are assumed to be simply supported and columns are assumed to carry only such bending moments as arise from the end reaction of the beams acting at some eccentricity from the column centreline. This method is easy and quick to apply and has stood the test of time in that failures in such structures are extremely rare indeed, and when they do occur it is usually due to instability during erection. It is important to realise, though, that the stress values arrived at in calculations bear little relationship to those actually experienced by the frame. This is due primarily to the omission of any consideration of joint stiffness between beam and column. In practice a true simple support is very rare and, even with the simplest connection, some moment transference will take place between them. In the event of both beams and columns being encased in concrete, the degree of joint stiffness will be considerable. Thus the simple design method tends to overestimate beam moments and underestimate column moments. This does not matter, certainly so far as internal columns supporting beams of approximately equal span and loading are concerned, but outer columns, and particularly corner columns, carrying assymmetrical moments deserve some thought, certainly if uncased. Therefore, in a highly competitive design it is quite safe to trim beam design, and that of internal columns, to the bone, or even accept some degree of overstress, leaving the face and corner columns conservatively designed.

The results of tests on full-scale structures (apart from tests on isolated elements) have indicated a very considerable margin of strength in the columns, well in excess of any predictions made by elastic design. Clearly, further research is needed before this can be codified into a simple design method.

Other miscellaneous structures

Within this group one should include electricity transmission towers (themselves something of a speciality), lighting masts, television masts, railway electrification structures, temporary bridges, footbridges, offshore structures, etc.

Frequently, structures are evolved which could not be visualised by the drafting committees responsible for British Standards, for example, railway electrification structures supported only on one side of the track, gallows style. In this instance conductor tension can produce considerable torsion in the column and severe moments arise when the track is on a curve. Accordingly, strict application of BS 449 rules need to be tempered by engineering judgement, and it is arguable whether the British Standard is really applicable in such cases.

Likewise, difficulties have been known to arise in transmission towers. Ever greater line voltages have necessitated taller and taller towers. Some years ago 30 m was a tall tower, today 75 m is not uncommon and for long river crossings even greater heights are needed. It is not sufficient for the designer merely to extrapolate from existing designs. There will always be a tendency for the designer to assume that what is flat on his drawing, will also be planar when erected. This may be a reasonable assumption up to a certain size, but ceases to be accurate to a larger scale. Self-weight deflections of long members of lattice structures can become significant, particularly if the member is intended to act as a strut.

Fortunately, it is often the case that such structures are intended to be built in large numbers, and so it becomes practical and economic to carry out full-scale destructive tests of the proposed structure. It also offers the possibility of effecting economies by trimming the design. The Central Electricity Generating Board insist upon such tests on all new designs of transmission towers and have set up a facility for doing so which is available to outside users on a contract basis. One should beware though of placing too much emphasis on the results of one such test. Preferably one should test duplicates and ideally carry out at least three tests to get some measure of scatter.

Figure 12.11 Electricity transmission tower (Courtesy: CEGB)

Results obtained from scale model tests should be interpreted with some caution. Apart from scale effects, such as that observed above, it is often very difficult to reproduce a steel structure in model form, owing to the fact that the range of rolled sections is intended to be used in full-size structures. Further, it is almost impossible to reproduce models with the same relative standard of tolerance, and scaled-down welds also present their problems. There is also the human factor that people tend, quite rightly, to place more reliance on full-size tests. With these factors in mind, where great repetition occurs, full-scale testing is well worthwhile.

Other structures in this group, however, do not lend themselves to this kind of practical approach, particularly offshore structures. Recent developments in the North Sea have opened up a vast potential market for steel structures in the future. The sizes of many of those built so far beggar description, and the limit has by no means been reached. Again, it is arguable whether BS 449 really applies, and the design engineer is in many ways out on a limb and must work from first principles. To start with, the assessment of loading is far from straightforward and much has yet to be learnt about the effect of wave forces on structures of various shapes. Temperature must also be considered and the principles mentioned on pages **12**–6 and **12**–7 must be observed. From this it follows that a structure satisfactorily designed for use in a tropical sea may not be suitable, indeed may be downright dangerous, in Arctic conditions. The problems, therefore, would seem to warrant a conservative approach, tempered in the light of developing experience. One such rig, the Sea Gem,[10] which failed, has spelt out some fundamental lessons, but no doubt there are other traps awaiting the unwary. As designers tend to overlook stability questions on land-based structures in the part-erected condition, so also is there a danger of failing to recognise the situation when an off-shore structure is being towed, usually on its side, to its final location. Buoyancy considerations arise which warrant more than cursory attention.

Light industrialised buildings

The first, and by far the most successful application of structural steelwork in industrialised building has been in schools designed and built by the various schools consortia such as CLASP, SCOLA, SEAC, etc., which have come into being in recent years. Schools present an almost unique situation inasmuch as accommodation requirements in terms of room sizes, corridors, assembly halls, cloakrooms, storey heights, etc., do not

Figure 12.12 Industrialised building: British Aircraft Corporation Offices, Stevenage (Courtesy: SEAC)

differ with locality. Also buildings are limited to one or at most two storeys. Consequently it has proved possible to develop entirely modular designs, which, coupled with the policy of bulk purchase of components, has enabled such buildings to be competitive with traditional construction. Indeed, over 50% of schools currently under construction are of such industrialised form, all of which utilise light, simply designed steel frames. Most floor and roof beams are of lightweight open web or lattice form ('bar-joists'), diagonal bracing is extensively used and columns are of various forms. One system, for example, uses nothing but 'square' 150 mm columns, consisting of cold-formed inwardly lipped channels, 152 mm × 76 mm hot-rolled channels welded toe-to-toe, 152 mm × 152 mm RHS of varying thickness, or 152 mm × 152 mm Universal Columns of varying weights. In this way a range of load-carrying capacities can be provided without altering the dimensions of the modular infill panels. Many of the beam designs look complicated and expensive but, when great repetition is anticipated, production jigs can be used and fabrication labour thereby reduced.

Attempts have been made from time to time to develop similar systems for other applications, notably light industrial buildings such as small factories, warehouses and farm buildings, with only varying degrees of success. The difficulty is that in most cases the client, usually a private company, expects, and is prepared to pay for, buildings tailor-made to suit the particular site. Whilst for social reasons most schools require a measure of surrounding open space, this does not obtain in industrial premises. The application in the public sector which in many ways parallels the schools' situation is in hospital buildings where common accommodation requirements arise. The long overdue upsurge in hospital building may well generate similar consortia and indeed certain semi-industrialised designs have been developed using steel framing, and these are being marketed.

OVERALL STRUCTURAL BEHAVIOUR

Having referred frequently to British Standards for design, i.e. BS 449 and BS 153, a word of caution would seem to be warranted. These two documents, as all British Standards, were drafted by committees of experienced people drawn from all quarters. They were charged with the task of formulating design rules in the light of the state of knowledge at the time, incorporating the best modern practice and exploiting the latest research. This was no easy task. They recognised that many of those who would use the documents would not understand, indeed would not want to understand, the background theory behind many of the requirements. Inevitably many areas of doubt and uncertainty were encountered and differences of opinion expressed. At the end of the day, in such situations, they rightly tended on the conservative side, though, in one or two instances, subsequent research has indicated that their rules were not sufficiently cautious. However, both documents have stood the test of time and, by and large, when properly interpreted result in structures in which safety and economy are reasonably in balance. But they tended to lay undue emphasis on the determination of exact values of permissible stresses and insufficient attention, certainly in the case of BS 449, to questions of overall structural behaviour and stability, particularly during erection.

The young designer can be forgiven for supposing that if all members and connections in a structure are assessed and designed strictly in accordance with the British Standard then the complete structure will be adequate for its task. In the great majority of cases this will be so, but in rare instances it has been found to be a fatal error. In a number of cases the designer has, unconsciously perhaps, relied on the stiffening effect of the cladding to provide overall stability and a collapse has occurred, sometimes under the weight of an erection crane before the cladding has been fixed. What one does not know, and cannot assess, is how many structures are satisfactorily in service which were within a hair's breadth of collapse at some stage during erection. As an illustration, it may seem an economic solution in the design of a frame for a multi-storey building to have a number of load-carrying plane frames, interconnected with light ties. In the absence of temporary bracing such a structure relies for stability on the stiffness of the joints at the ends of the tie beams. Such joints, being on nominally unloaded or lightly loaded members, are often regarded as being unimportant and are detailed by an office junior. Thus in the unclad state the frame possesses insufficient longitudinal stability. This illustration is but one of many which could have been quoted but it is hoped that the point has been driven home. Overall supervision of the design of both members and connections by a competent engineer, and the intelligent assessment of the problems by the erection supervisor, should be sufficient to prevent such mishaps.

The stiffening effect of cladding, which has been referred to, is very great indeed but, except in the case of simple structures, it is likely to remain impossible to assess in the foreseeable future. Because of this and because of the accidental and unquantifiable composite action which occurs between steel members and other materials, the actual stresses experienced in service are likely to be very different, usually much lower, than those calculated. This fact, coupled with the fact that loadings which are specified, particularly wind loadings, are at best only general estimates, means that great nicety of calculation is largely misplaced. Thus the pursuit of notional absolute economy is a cardinal error if it means that under pressure of time the consideration of overall behaviour is omitted.

Relating to this point, a problem frequently encountered, on which little guidance is to be had, concerns the amount of sway or horizontal deflection permitted at the top of a tall multi-storey building under wind loading. It has been common to limit such sway to 1/500 of the height, calculated on the stiffness of the bare steel frame. This largely begs the question, since most of the wind force would be absent without cladding. With it in place, the deflection must be considerably reduced in consequence of its stiffening effect. What is being considered here is not structural safety, that is not in question, but serviceability. In the absence of any knowledge at all of cases where discomfort has been caused to occupants or damage to finishes through excessive sway, the writer concludes that the above notional limit is conservative. Clearly, long-term observations are called for which may take years to yield worthwhile data. Meanwhile the writer would suggest that if the above limitation is deemed acceptable for dwellings with curtain walling, then relaxation of the limitation would seem appropriate in respect of two factors, namely other uses, and stiffer cladding. Thus a limit of 1/275 might be rational for an office block with brick or masonry cladding. Again, in city centres, the sheltering effect of neighbouring buildings may be taken into account in assessing wind forces. Is one to take into account the possibility of their being demolished sometime in the future? Such considerations need careful assessment, and authoritative advice.

DESIGN OF STRUCTURAL COMPONENTS

As indicated in the preamble, there would seem little point in including worked examples in this section, in view of the forthcoming revisions to the British Standards which will render them obsolete. On the other hand it would be premature to anticipate the contents of the new documents.

Importance of correct loading assessment

Great care should be taken in the assessment of dead and live loading, as mistakes at this stage can make all subsequent calculations abortive and, in extreme cases, result in completed members being scrapped. This has, however, been known to happen, not all that infrequently. Common causes of mistakes in this matter are failing to make adequate allowances for finishes, as for instance in providing screeding to falls in a roof slab to provide adequate drainage; failure to make proper allowance for surges in gantry structures; incorrect interpretation of wind loading requirements; effect on pressure of flow in granular materials etc., etc. Security requires that a full loading schedule for the intended structure should be drawn up and checked independently before the detailed design of members is commenced.

Determination of structural layout

It is all too rarely, in building structures at least, that the structural engineer is summoned sufficiently early in the planning of a project to be able to contribute to the overall efficiency of the building by influencing the choice of layout. More often the steelwork designer is set an almost impossible problem by a form of building already finalised by others with little thought of structural needs. Between these two extremes falls the majority of problems. Thus it frequently comes about that the structural layout merely occurs and the steelwork designer has little room to manoeuvre in such matters as column layout, overall beam depth, pitch of roof, etc. Where he does have such freedom he should explore the effect upon overall economy of various arrangements before committing himself to one particular layout, time permitting.

Calculations

All calculations for a project should be neatly prepared in logical sequence, properly indexed and crossreferenced. Apart from the value to the designer himself as the work proceeds, it should be remembered that in most cases calculations have to be submitted to a Local Authority for approval. Incomprehensible papers invite criticism and do not smooth the road to acceptance. Additionally, mistakes are more easily spotted if the work is clearly presented, and the effect of structural alterations, perhaps at a much later date, can be more readily followed.

Consideration of individual members

ANGLE MEMBERS

Angle members, i.e. purlins, cladding rails, bracing members, truss components, usually act as simple beams, over one or two spans, or pin-ended struts and ties. Within the stress rules laid down, having regard to deduction for holes and eccentricity of connections, etc., design is a matter of trial and error and simple arithmetic. Certain practical points though are worth bearing in mind. Mistakes made by the writer in the past include the transport of double-angle rafters in too long lengths, resulting in extensive damage in transit; and the design of diagonal bracing to resist longitudinal surge in a gantry, which although structurally adequate was much too flimsy. In this latter case, when examined, the bracing looked far too light, and since the gantry served a scrapyard it was obvious that they would be quickly damaged.

SIMPLY SUPPORTED BEAMS

Such members fall into two categories, namely, laterally restrained and laterally unrestrained. In the former case direct design is possible and simple arithmetic will usually arrive at a section modulus required, either elastic or plastic depending on the design method. The necessary lateral support is often provided by the load itself either connecting load bearing beams or a concrete slab floor suitably secured in some way to the beam. In the case of long spans, it is sometimes economic to provide restraint in the form of light tie beams rather than design the main beam as unrestrained.

The laterally unrestrained beam is designed to a lower working stress depending on a number of factors, namely, section shape, slenderness ratio, effective length and torsional restraint at the supports. The designer must first assess the effective length of the compression flange in lateral buckling and BS 449 provides guidance which in turn depends partly on torsional restraint at the supports. A trial section is then selected, the slenderness ratio l/ry calculated, the shape parameter allowed for and a permissible stress derived. Thus design is a trial-and-error method.

It must be understood that the assignment of an effective length is a somewhat arbitrary process and therefore it follows that permissible stresses based on this value are themselves somewhat notional. Again, great accuracy in calculating actual stresses is therefore unnecessary. In illustration, tests carried out some years ago on beams subjected to uniform moment with the supports torsionally restrained gave some interesting results. At the commencement of loading the points of contraflexure of the compression flange were indeed about 0.7 l apart, as the rules supposed. With increasing load, however, they tended to move towards each other, until at failure they were some 0.4 l apart. The load factor at failure was considerably higher than one would have expected, all things being taken into account, partly, no doubt, due to this phenomenon, indicating a measure of conservatism in the current BS 449 rules.

One word of warning: the Safe Load tables for beams refer to beams fully laterally restrained, and therefore beam sizes cannot be selected direct from these tables if there is any doubt whatever on this issue. Examples of unrestrained beams include wall-bearing beams not at a floor level, as for instance at the rear of a lift shaft, beams supporting runways or hoists, and of course the worst possible case, gantry beams, where the load far from restraining the compression flange can impart disturbing forces due to surge and impact.

PLATE GIRDERS

This is a subject on which whole books have been written. At one time most steel bridges were built of plate girders in one form or another, and only comparatively recently have box girders entered the scene in a big way. Plate girders have also themselves been the subject of rapid development and change, owing to the almost complete supersession of riveting in favour of welding. Nowadays most plate girders consist principally of three plates. Variations include assymmetrical girders with larger tension flanges than compression for use in composite action with a concrete deck, and girders which are the opposite, where the top compression flange is wider than the tension flange to resist surge in the case of gantry girders. Girders can also have continuous or curtailed flanges. The wisdom of curtailing flanges depends on the size of the girder, and to an extent the form of the loading or rather the shape of the bending moment diagram or envelope. Generalising, it can be said that if the size of the girder is such that flanges can be supplied in one piece, it is never economic to curtail the flanges by introducing butt welds in them, with all that this entails in terms of machined preparations, possible preheating, welding and final nondestructive testing. Minimum weight does not necessarily mean minimum cost.

It is rare in the case of rolled beams that the web requires stiffening to resist buckling tendencies, and then only at the supports and points of concentrated loads. In plate girders, however, the web thickness is decided by the designer and is not simply presented as one of the dimensions of a section of adequate bending resistance. Thus the web can be much thinner relative to the girder depth than in the case of rolled beams, bringing in its train the necessity to provide web stiffeners, which are spaced apart a maximum distance of $1\frac{1}{2}$ times the web depth. Except in the case of bearing stiffeners, and those under a concentrated load, stiffeners do not have to be of great size. Quite small restraining forces are needed to keep an initially flat plate in that condition. Thus, if appearance requires it, stiffeners can be provided on one side of the web only.

Whilst rolled beams are mostly simply supported, worthwhile economy can be derived in plate girders by making them continuous or semicontinuous (i.e. alternate cantilevers and suspended spans). This creates a situation over the supports where maximum bending moment and maximum shear force act together, a circumstance which requires special consideration of combined stresses.

Plate girders with thin webs often display waves in the webs during fabrication owing to the shortening of the weld and HAZ in the web-to-flange joint. This can make the subsequent fitting of the stiffeners difficult. Where possible, therefore, the stiffeners should be welded to the web beforehand, but this precludes the use of automatic welding, which is a disadvantage. Thus it can come about that it is not necessarily economic to design with the bare minimum web thickness, and additional web area does also add to the bending resistance.

If it is necessary to make a full-strength splice in a plate girder the web and flange welds should coincide. With the flanges butted for welding there should be a $\frac{1}{8}$ inch clearance in the web to allow for shrinkage when welding the flanges. Otherwise the waviness mentioned above will again be evident. The reason the welds should coincide is that there will inevitably be a slight difference between the two web depths, making fit-up difficult. Only by great good luck will the fit be exact.

COLUMNS

Columns can be continuous, as in a multi-storey building, or discontinuous, as in a column-and-truss shed. They may be laterally restrained at intervals by other members, or entirely unrestrained throughout their lengths. These factors are taken into account in design of an axially loaded column by assigning an 'effective length' somewhat arbitrarily as a first step in design. Thereafter a section is selected for trial, its slenderness ratio, i.e. effective length/ry, calculated, and from this a permissible stress found from a column-stress formula. Such a formula is usually presented in the form of tables or curves.

There are many such stress formulae, used in various codes throughout the world. Most of them incorporate some factor to cover for notional initial imperfections, and none of them are exact, since they are all attempts to express mathematically what is in essence a naturally occurring situation. Further, all of them are based on the behaviour, under laboratory tests, of pin-ended struts, and it should be understood that a truly pin-ended strut is a very rare occurrence in structural engineering.

Almost equally rare is a truly axially loaded column, and some bending moment is almost invariably imparted by the connecting members, about one or both axes, at one or both ends. The values of these bending moments may be insignificant or they may be such that the stresses induced thereby dominate the situation and render the axial stress insignificant. In order to take account of this, some form of stress interaction formula is used such as that in the current BS 449, i.e.

$$\frac{f_a}{p_a}+\frac{f_{bc}}{p_{bc}}\leqslant 1$$

where f_a and p_a are calculated and permissible axial stresses and f_{bc} and p_{bc} are calculated and permissible bending stresses. The permissible stress in bending is established in the same manner as for a laterally unrestrained beam. It can be argued that this manoeuvre is not accurate, as indeed it is not, but it has the virtue of simplicity and can be readily applied to a variety of cases. Suffice it to say that it has stood the test of time and results in safe columns but, as indicated earlier, recent test results on full-scale structures have indicated that it is conservative. However, in view of the arbitrary manner in which the notional effective length is decided, it is doubtful if there is any virtue in attempting to replace it with another method more accurate but offering greater complication.

Effective lengths, or lengths between points of contraflexure, have also been seen to vary with the magnitude of the axial load; they also vary with the loading pattern. Thus in a multi-storey frame, variations of loading can induce conditions ranging from full double curvature to full single curvature, i.e. effective lengths of 0.5 *l* and 1.0 *l* (Figure 12.13(a) and (b)). One discounts the possibility of 'chequer board' loading and adopts an intermediate value, say 0.7 *l* or 0.85 *l*.

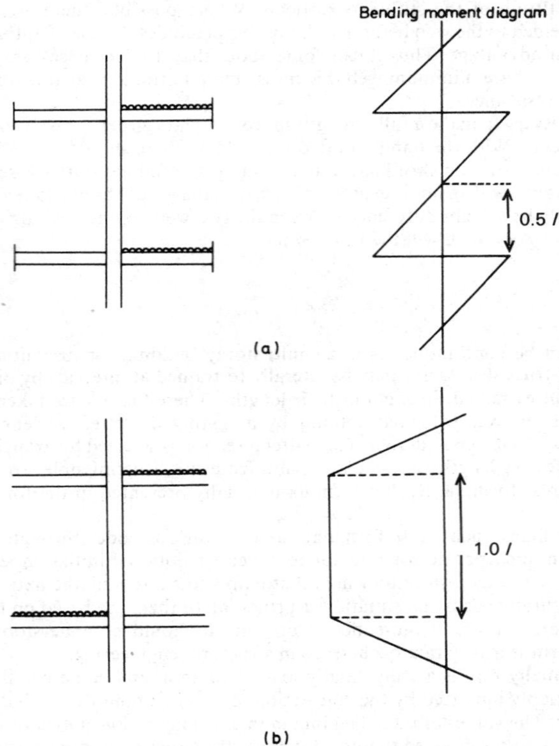

Figure 12.13 Column bending moments

All the foregoing refers to elastic design of columns, and as indicated the accepted methods err on the side of conservatism. Plastic design methods cannot as yet be universally adopted, and form the subject of widespread research. The first practical design method, for columns in portal frames, was evolved by Professor M. R. Horne,[11] and the theories of Dr R. H. Wood have been referred to earlier in connection with rigid-jointed frames. It is in column design that the greatest potential exists for achieving economy through research in pursuit of realistic design methods.

BOX GIRDERS

At the time of writing, nationwide research is in hand at the behest of the Merrison Committee into a number of governing parameters in box-girder design in order to establish new design rules following the recent failures of certain bridges. Much has appeared in the public press which is misleading and some which is downright wrong. It may therefore be helpful to spell out some of the first principles in order to put matters into perspective.

Firstly, box girder construction is not new; some structures built in Victorian times incorporated steel box girders albeit of riveted construction. The particular virtue of box construction is the much greater torsional resistance it offers compared with normal single web girders. For this reason, welded box girders have been used for a number of years in crane construction, particularly heavy EOT cranes subjected to racking and surge forces. Of course, with the complete change-over from riveting to welding since the war has come the greater ease with which one can tailor a member to suit a need. It is worth remembering that many hundreds of welded box-girder bridges have been successfully put into service, on the Continent and elsewhere, and none have failed in service.

It is possible, therefore, with the data already possessed, to design such structures successfully. Were one not under pressure to build the cheapest possible structure (in the eyes of some this wrongly equates to minimum weight) there would be no great problems. It has been in pursuit of maximum economy, particularly in plate thicknesses, that the trouble has arisen.

With the exception of the long suspended span, it is true to say that box-girder construction is almost always more expensive than using discrete single web girders, with or without cross beams. For a straight bridge this is always so—it is only when the structure is subject to high torsional forces, as for instance in a curved bridge, that the particular virtue of high torsional resistance renders box construction economic. For the most part it has been the clean appearance and good corrosion performance of boxes which have determined the choice.

One would not attempt to anticipate the final report of the Merrison Committee. There have been many gloomy forebodings that costs will escalate and design be rendered so complex as not to be worthwhile. The writer does not share this view and indeed it may be that, ultimately, welded box girders will be cheaper than hitherto, if a bit heavier.

Turning to detailed design, one must accept that it has become a speciality of its own. The determination of approximate web and flange sizes to suit the calculated bending moments and shear forces is not of itself difficult, if a bit tedious in the absence of a computer program. But it is in the refinement of the outline design in regard to such matters as support diaphragms, web and flange stiffening, torsional stresses, the effect of shear lag, etc., that the process becomes intricate.

TUBULAR MEMBERS

Structural tubes have been available for many years, supplemented a decade ago by square and rectangular tubes in a large range of sizes. The use of these members is accelerating rapidly. Initially inhibited by a high price, ex-mill, compared with traditional sections, this factor is of decreasing importance having regard to the greater rise in fabrication labour rates.

Although it has long been appreciated that a circular tube is technically the most efficient form of strut, the greater ease with which connections can be effected with the square and rectangular shapes has led to their wide acceptance. For example, to fabricate a lattice girder from circular tubes (CHS) requires the use of a profiling machine to generate the interpenetration curves necessary to give a fit-up adequate for welding, whereas only straight cuts are needed for square or rectangular sections (RHS).

Care must be taken, however, in detailing the joints in such structures. It is unwise to use internal tubes significantly smaller than the main chords, unless some form of stiffening can be used, since the effect of applying a load to a small area of a flat face can lead to premature distortions in the main member. Such mistakes can lead to yielding occurring at or below working load.

Tubes are widely used for space structures, which have aesthetic appeal, but the problems of jointing are considerable. Consequently the cost of jointing dominates. In order to overcome this difficulty, the Tubes Division of the British Steel Corporation has developed and patented a special joint for space frames which they call the 'Nodus' joint.

It is unfortunate that the way that tubular fabrication has developed into a specialisation has tended to lead to structures which are either all tubular or all traditional. Clearly, the greatest potential for both types of section will only be realised when they are fully blended.

The versatility of tubes is demonstrated by the very wide range of applications, from steel furniture to aircraft hangar roofs. On this topic, a notable success was recently achieved with the construction of the 'Jumbo-Jet' hangar at London Airport, which broke new ground on two scores. Firstly it is by far the largest space frame yet built, the fascia girder carrying the doors spanning 130 m. The roof behind carries heavy loads from servicing equipment. Secondly it is the first large structure in this country making use of the new Grade 55 high-tensile steel of 448–463 N/mm^2 yield strength, which was successfully shop and site welded in large thicknesses, notably in the main chords which were 9 ft (2.74 m) in diameter. This structure clearly points the way for others.

COLD-FORMED MEMBERS

As indicated earlier, cold-rolled purlins have already established a wide market. Other shapes are used in the industrialised building field and in proprietary components such as lightweight lattice beams. These forms of structural sections have certain advantages. Firstly, when the loading is light, as in the case of purlins and cladding rails, the section can be structurally more efficient than a hot-rolled section owing to the limited number of smaller sizes of the latter. Secondly, they can be made from tight coated galvanised strip, which with a subsequent paint coat gives excellent corrosion protection.

It is not always appreciated that a cold-forming mill is a relatively cheap piece of equipment, only a fraction of the cost of a hot mill. Thus, if a reasonable market is anticipated the designer has the freedom to design his own tailor-made section to suit his particular requirement. This is not to say that it would be economic to design sections for a particular building, but it would be for a range of standard buildings. There are several manufacturers willing to undertake such work and to give guidance.

In the design of sections, prevention of local buckling is the prime consideration, and the derivation of exact solutions can be most complex. To prevent local buckling the edges of members are frequently lipped, inwards or outwards, and comparatively slender webs are often formed with a ridge or groove, longitudinally. In deriving a section, it is often helpful to ensure that sections will 'nest'. Apart from saving space in transport, it also ensures that the minimum damage will occur in transit, to which cold-formed sections are otherwise somewhat sensitive.

It is worth pointing out that, in the manufacturing process, since it is done cold, a significant amount of work hardening takes place, particularly in those zones bent to a small radius. Thus a section tends to be stronger than calculated on the basis of the yield strength of the flat strip. This effect is not at present taken into account in design and hence one has more margin of stress than one might suppose.

GANTRY GIRDERS

These members are being considered separately since they present unique problems, not found in other members. Firstly one must admit that there is a degree of irrationality in British Standards, as it would seem logical for gantry girders to be designed to the same stresses and safety factors as the crane structures they support. However, this is not the case and crane structures are designed with higher margins than the supporting structure. Of course a line of demarcation has to be drawn somewhere and one can put up a good case for making this between the moving and the static structure, which is the situation which obtains. Nonetheless, this brings in its train certain difficulties. Firstly, for insurance purposes EOT cranes are subjected to a test load 50% higher than the normal crane capacity. If this should be done with the crane midway between columns, with the crab at the end of its cross travel, flange stresses in the gantry girder may well be approaching yield stress. Secondly, because the operatives are aware of the test overload requirement, a blind eye is often turned on the specified Safe Working Load when a particularly heavy load has to be lifted, and again it is often not possible to estimate the weight of a complicated piece of machinery or other load accurately. These factors lead to occasional overloading of the gantries. Finally, in many applications, notably in steelworks, it is not uncommon for a crane to be uprated by retesting, with little thought being given to the supporting structure. In view of all this, it is most unwise to design gantry girders, nor indeed their supporting columns, to fine limits. Prudence suggests that a fair margin of stress be left, consistent with reasonable economy. These comments apply with particular force to the heavier types of crane supported by plate girders, and less so to lighter cranes carried on rolled beams.

Turning to the principles of design, BS 449 lays down certain factors for longitudinal and transverse surge, and vertical impact. These represent a reasonable average and are unlikely to be changed in the new edition although one suspects that the factors tend to be too conservative for heavy cranes and the opposite for light cranes. The treatment of load combinations is given, vertical loading being taken together with surge either along the rail or transverse to it.

In calculating vertical bending moments, one must know not only the load to be lifted but also the crane characteristics in terms of self weight, crab weight, end-carriage wheel spacing, etc. Envelope diagrams of shear and bending moment should be derived. In this context it is worth bearing in mind that two cranes often run on the same track. One therefore has to take into account how closely they can be spaced. Sometimes long buffers are fixed to the cranes in order to prevent the two cranes from running on to one girder. The writer has seen a case where such buffers were removed by the operatives because they were inconvenient!

The worst condition, which usually governs the design, is with maximum vertical bending moment and transverse surge. In the calculation of stresses it is usual to consider that only the top flange resists the transverse surge, this being somewhat conservative. If a rolled beam can be used it will offer by far the cheapest solution. The next best alternative is a rolled beam as a core section with either a flat plate or a toe-down channel welded to the top flange to accommodate surge stresses. If calculations suggest that the bottom flange also should be reinforced, one should then turn to a tailor-made plate girder, since there will be no difference in the number of main weld runs and the section will be lighter. Finally, very heavy gantries are often built in two parts connected at intervals, i.e. main girder to resist vertical loads and horizontal surge girder for transverse loads, this latter often serving also as a maintenance walkway. Final selection of section size is a trial-and-error process which can be very tedious. As a first trial one could assume a span/depth ratio of about 15 and a top flange with 30 to 50% greater cross-sectional area than the bottom. At this stage it is worth examining the shear situation at the supports in order to determine an appropriate web thickness, which will affect the overall moment of resistance. Ideally one should arrive at a section where maximum permissible bending stresses are approached simultaneously in top and bottom flanges. This can only rarely be achieved and in any case it is an illusion to calculate stresses to a greater degree of accuracy than one's real knowledge of the loads.

Finally, certain detail points should be considered. Experience with the earliest welded gantry girders was unfortunate, since little regard had been paid to the local effect of rolling wheel loads. In the absence of accurate web-to-flange fit-up, the fillet welds were required to transmit the whole of the local compressive stresses under the wheel. This was sometimes compounded by a slightly eccentric rail, creating a rocking effect to the top flange. The result was early fatigue failure of the web-to-flange fillet welds. Several solutions have subsequently demonstrated their effectiveness. Firstly, the rail can be mounted on a resilient pad, to give overall rather than point contact on the high spots. Secondly, where possible, the weld position can be moved to a zone of lower local stress by using a T-section top flange, formed by splitting a Universal Beam, the web plate being butt-welded to the stalk; and finally, if this is not possible, the web-to-flange weld should be made by a full-strength double-V butt.

The effect of misalignment should also be considered. With the best of efforts, no foundation can be guaranteed free from some degree of settlement in the lifetime of a building and, if magnified by height, some realignment of the track may be needed. This is greatly facilitated if there is some means provided at the girder supports for transverse adjustment. If this is not provided, the only solution is to realign the rail on the top flange, leading to eccentricity of loading. Some degree of eccentricity is, however, inevitable, owing to lack of fit of the rail, even with a pad, and the wheels not running on the crown of the rail. For this reason, web stiffeners should not be spaced too widely apart and it may be desirable to introduce short intermediate stiffeners supporting the top flange.

Various means have been used for securing the rails on gantries. For light work, bolts through both rail and girder flanges suffice, but these should not be at too wide a spacing otherwise each bolt will in turn suffer the benefit of full transverse surge. Rail clips are more popular, can be purpose made, and offer the possibility of adjustment without the necessity to slot holes. Direct welding of rail to flange has been tried, often unsuccessfully. This is mostly due to the fact that rail steels have a high manganese content to give resistance to wear and require a welding technique foreign to many fabricating shops.

METHODS OF DESIGN

The two current British Standards for the design of steelwork, BS 449 and BS 153, are based on elastic design principles, except that BS 449 has a 'let-out' clause which permits other proven methods of design or design on an experimental basis. This allows plastic design to be adopted where this is possible (e.g. in continuous beams and single-storey portal frames) but does not lay down any design criteria, or even a recommended overall load factor.

Briefly, elastic design is based on the philosophy of permissible working stresses for different situations, these being some proportion of yield stress in the case of tie-members or laterally restrained beams, and some proportion of the critical buckling load in the cases of laterally unrestrained beams and columns. A similar principle applies to shear stresses. Unfortunately, these proportions or so-called 'safety-factors' are not stated but are implicit in the permissible stresses specified. If one delves deeply enough one can establish that the safety factors differ from one member to another, which is irrational to say the least. It is this anomaly which the drafting committees now sitting are anxious to eradicate.

If one considers the basic aims of structural design one comes to realise the fundamentals, that structures should have an adequate margin against collapse and not become unserviceable (due perhaps to excessive deflection) under normally anticipated service loading. The only real purpose in calculating levels of stress assumes that from this knowledge one can predict what margin against failure exists. And here we must come to define failure. The principles of elastic design implicitly define failure as either the attainment of first yield in an extreme fibre in a tension situation, or the attainment of the critical buckling stress in a compression or shear situation. So far this appears to be rational, and it is on such lines that the design of the majority of building steelwork and all bridge steelwork is based. It is argued that if one is assured of a known factor against first yield one is satisfied that a safe structure results, as indeed it does, and that what happens under greater load than that necessary to achieve yield is of no consequence.

Such philosophy has been upset with the gathering of knowledge of actual behaviour gained from practical research. Firstly it is now widely recognised that although members may be elastically designed many parts of a structure may reach yield stress, and indeed physically yield, on the first application of working load. A typical example of such a situation is the traditional seating bracket and top cleat beam-to-column connection. Simple design assumes this to be a pinned connection and the top cleat is simply a stabilising fixing, not intended to transmit a fixed-end-moment. For the design assumption to be realised, some end rotation must take place, causing permanent but local yielding in the top cleat. One can find yield stress attained in many places, particularly if one takes into account residual cooling and welding stresses. But this is not to say that the structure is unsafe.

Secondly, it has been established experimentally that all structures possess a margin of strength considerably greater than that calculated on the elastic basis. That is to say, on yielding, stresses will be redistributed extensively before a structure shows permanent deformation such as to render it unserviceable. What now appears to make the elastic philosophy irrational is that this extra margin of strength differs from one situation to another, as for instance a laterally restrained beam compared to a slender strut. Redefining failure as excessive and unacceptable permanent deformation, ultimate load philosophy (plastic design in the case of steelwork) attempts to quantify this additional margin of strength in different situations, and to utilise it as part of the overall margin against failure, with the aim of achieving economy.

The principles of plastic design were originated by Sir John Baker at Cambridge, developed there by him, with others, and later taken up extensively at Manchester and Lehigh Universities, and elsewhere. As a result, plastic design can now be readily applied to continuous beams and portal frames of various shapes. Now over 50% of the steel portal frames built in this country are designed plastically and, as indicated on page **12**–16, efforts are being made, with some success, to extend the application to rigid-jointed multi-storey frames. The justification, if any is needed, for the extensive research which is proceeding lies in the fact that plastic design is by far the most accurate method yet established of predicting that load at which real structural failure will occur.

As stated, plastic design consists of recognising, and quantifying, the additional margin of strength in bending, beyond the attainment of first yield in an I-section. It has been established experimentally that before a beam or structure can be made to collapse it must be transformed from a structure to a mechanism. This occurs due to the formation of 'plastic hinges' at points of severe moment, which occur only when the whole depth of a member has reached yield stress, compressive on one side of the neutral axis and tensile on the other, i.e. no portion of the member depth remains elastic. The principles of design for continuous beams and portal frames are clearly set forth in BCSA pamphlets Nos21,[12] 28[13] and 29[14] and other works.[15]

Having established that a structure has a known margin against collapse one has satisfied the strength criterion, and the value of stresses in various parts under working load are of little significance provided that overall and local stability requirements are satisfied. Serviceability conditions, that is, elastic deflections under working load, may, however, not have been satisfied, and in effect one has to do an elastic analysis in order to establish the position, which lengthens the design process somewhat. Computer programs now exist for portal frames, which, for given loading conditions, will select a section found plastically and provide values for elastic deflections.

Fully rigid multi-storey frames, however, present considerable problems. Under sway conditions a large frame may require very many hinges to form before a mechanism is created and at any intermediate stage the frame is part elastic and part plastic. Proper understanding and control of stability considerations is essential and we are as yet some way from producing rules of thumb suitable for Codes of Practice.

As an intermediate step it was proposed, many years ago, that frames could be designed as 'semi-rigid'. That is, with certain connection requirements satisfied, part transference of moment from beam to column could be assumed. This never found favour, owing to a complicated design process, but a gross simplification of it is allowed in the current BS 449 whereby beam moments may be reduced by 10% provided that the columns are designed to resist such extra moment and structurally cased. This is a swings-and-roundabouts situation, not leading to any great economy.

PARTIAL LOAD FACTORS

Alongside the development of ultimate load philosophy (i.e. plastic design), has come consideration of appropriate load factors – that is the determination of the right margin necessary against collapse. As indicated earlier, there are inconsistencies in present practice.

Clearly a structure must never closely approach collapse conditions in service, but have some margin. This margin is to allow for a number of uncertainties, including design inaccuracies, variations in material strengths, fabrication errors, lack-of-fit, foundation settlement, residual stresses and errors in assumed loading. Most of these uncertainties will never be quantified, but must be allowed for by a global factor, with two exceptions; these are material strength which can be treated statistically, and secondly errors in assumed loading. As yet, adequate strength data sufficient for a rational statistical treatment are lacking, but are being accumulated. Loading, however, can now be dealt with rather better.

In the design of a building, for instance, one can calculate and control self-weight or dead load quite closely. The magnitude of applied load is very much less certain, and with possible change of use in the lifetime of a building, control is difficult, even by legislation. Further, naturally occurring loading such as wind loading and snow loading entail the consideration of the likely frequency of attainment and the probability of this occurring simultaneously with maximum service loading. One must also consider in continuous or semicontinuous structures that dead load may have a counterbalancing effect to certain live load situations. Since it is possible for self-weight to be overestimated it is therefore necessary to apply minimum as well as maximum dead-load factors.

A rational approach therefore is to adopt maximum and minimum dead-load factors, acting either alone or in combination with different factors for imposed and wind loads, the values of the latter depending on whether they act together or separately. This procedure will be adopted in the new British Standards.

The effect of this will be to complicate the design process somewhat, but the benefit will be much greater consistency in safety margins. When applied it will appear to make little difference to the run-of-the-mill structure, but for the type of structure in which one kind of loading dominates, e.g. dead load or wind load, significant differences will be apparent.

CORROSION PROTECTION

The British Standard document giving guidance on this topic is Code of Practice 2008, *The Protection of Iron and Steel Structures From Corrosion.* Many textbooks have also been written on the subject, to which reference may be made.[16] For specialist advice the British Steel Corporation runs a Corrosion Advice Bureau for dealing with *ad hoc* problems, sometimes free of charge.

Generally, one must admit that, at one time, corrosion protection was given far too scant attention, as regards the effect of both structural detail and protective treatment. In recent years great advances have been made in both directions. Welding has relieved a lot of detail difficulties, and sophisticated surface and paint treatments have been developed which promise long repaint lives.

Whilst major exposed structures such as big bridges, where repainting is expensive, rightly receive 'Rolls-Royce' treatment, it should be remembered that such may not be appropriate in each and every case. It seems to the writer that in some respects the pendulum has swung too far the other way, some engineers specifying treatments and inspection standards not warranted in many cases. Once again judgement enters the picture and in any design one must first assess whether a problem exists at all. In the case of steelwork which one can guarantee will be kept dry, say internal to a heated building, little problem exists. If exposed within the building, painting is carried out for cosmetic reasons.

Steelwork which is exposed to the elements presents an entirely different problem which should receive attention before design commences, let alone detailing. Again, though properly one should take into account the probable life of the structure, its use, atmospheric environment and location before coming to a judgement, structures in the public eye rightly demand the full treatment. Industrial structures of a temporary or semi-temporary nature do not warrant expensive treatment, particularly if they are likely to be subject to accidental damage.

The treatment appropriate to a particular case can thus vary from nothing to the grit-blast, metal spray and four-coat paint system. Further information on methods of treatment is given in Section 4.

DETAILED DESIGN

Ideally, design should only be carried out by those familiar with shop-floor problems and procedures, but this is a counsel of perfection. Where doubt exists an approach to a fabricator is worthwhile and advice will be given without commitment.

What one is referring to here are the difficulties which arise because a designer has interpreted technical information literally without recognising its limitations. The rolled section tables published in British Standard No. 4, and in the various handbooks giving dimensions and web and flange thicknesses, are *average* values only, being subject to rolling margins of $\pm 2\frac{1}{2}\%$. Sections can also be out of square, and the limits of tolerance on shape are given. Whilst it would be unfair to assume that all member sizes and shapes are at the tolerance limits, it is also unfair to suppose that all dimensions are strictly accurate to three significant figures. One must have some regard to the possibility of members being slightly out of true. No bar is ever completely straight, no plate flat, and no flange perpendicular to its web. Fortunately, in most cases the work can be forced into alignment (albeit introducing locked-in stresses) but occasions can sometimes arise where this is not possible. When this occurs it becomes necessary to use packing pieces, sometimes needing costly machining which cannot be charged for. The designer's aim, therefore, must be to eliminate the need for accurate fit-up and to use machining only as a last resort.

Another point frequently overlooked is the matter of accessibility for welding. The easiest, and therefore the best fillet weld results when the electrode, either manual or automatic, can be offered at 45°. The limits for satisfactory work are roughly 30° and 60°, and outside these limits poor welds result since the electrode must be bent, breaking the flux coating and introducing many stops and starts. The designer cannot be expected to foresee all possible difficulties, but having made some attempt he should retain an open mind and be prepared to consider sequences and procedures suggested from the shop floor.

Handling and floor space deserve some thought. For straightforward fabrication one should aim to keep complicated weldments small so that they can be readily handled to enable all welding to be done downhand. A complicated end detail to a long member makes this difficult if not impossible. If design can be effected such that all that long members need is to be cut to length and drilled, probably on an automatic machine, competitive tender sums will be offered. This is at least partly due to the fact that shop-floor space used is kept to a minimum and hence throughput can be high.

Where possible, repetition should be the aim. Money will not be saved if in the pursuit of imaginary economy a great variety of member sizes is used for broadly similar loading conditions. A recent instance occurred in a bridge consisting of 54 girders, all different. Admittedly this was an extreme case but some saving through repetition must have been possible. Building work offers the greatest scope here, which should not be overlooked.

CONNECTIONS

The greater part of the cost of a steel structure is in the connections, whether they be bolted or welded. Simplicity therefore must be the keynote, with the greatest standardisation possible if economy is to result. Typical examples of a great variety of connections are to be found illustrated in the *Steel Designers' Manual.*[17] Suffice it to say here that the trend is towards shop welding and site bolting. Site welding tends to be very expensive and should only be considered if extensive work is in hand, as for instance in a big bridge or long pipeline.

The British Standards for welding of greatest concern to the structural engineer are BS 1856, 1964 *General Requirements for the Metal-Arc Welding of Mild Steel* and BS 2642, 1965 *General Requirements for the Arc Welding of Carbon Manganese Steels.* There are many other current British Standards covering various welding processes, inspection procedures, and welding of special alloy steels and other materials. Indeed, so extensive is the coverage that to the uninitiated great difficulty may be experienced in selecting the appropriate standard or most effective procedure. However, excellent guidance is available from the Welding Institute which publishes a series of booklets to some of which reference has already been made; *The Weldability of Steel*[3] is recommended. Couched in easily understood language it defines the fundamentals and points out pitfalls for the unwary. Armed with such guidance an intelligent attempt can be made to propose details and procedures for a particular case, but an open mind should be retained for ideas and proposals from the welding engineers responsible for carrying out the work.

The various types of bolts in structural use and the respective British Standard to which they are made, or governing their use, are as follows:

Black bolts (BSW)	BS 916, 1953
	BS 325, 1947
High-tensile bolts (BSW)	BS 3692, 1967
Turned and fitted bolts	BS 2708, 1956
Turned barrel bolts	
High-strength friction-grip bolts	BS 4395, Part 1, General grade
	Part 2, Higher grade
	BS 4604, Part 1, Use of general grade
	Part 2, Use of higher grade

By far the greatest majority of bolts used in structures are, and will continue to be, black bolts. This is because, although their efficiency is lower than other types, they are cheap in themselves, readily available and cheap to install, needing no special equipment. Permissible shear and bearing values are given in the design standards, and detail design guidance can be found in the *Steel Designers' Manual.*[17] Black bolts should, however, mostly be used in a shear situation. Where details are such that bolts are required to resist tension, ordinary high-tensile bolts should be used, since bearing values do not enter the picture, and bolt quality tends to be more consistent. Further, since the bolts are tightened manually using the same kind of spanner there is less likelihood of these bolts being damaged by over-tightening.

Turned and fitted bolts are only used where accurate fit-up is essential. This occurs not infrequently in bridgework, but rarely in building work. Having regard to the points put on page **12**–31, efforts should be made to design such that the need for their use is avoided. Apart from the expense of the bolts themselves, the parts to be joined must be fabricated to a standard of accuracy matching that of the bolts themselves, particularly as regards hole centre dimensions. If this is not achieved it becomes impossible to insert the bolts on site without resort to costly and difficult site reamering.

Turned barrel bolts are bolts in which the machined shank is of a larger diameter than the protruding threaded end, being shouldered at the spigot. They are for the situation where it is necessary to secure the bolt effectively without gripping the work and exerting pressure between the plies. Such a situation occurs in an expansion joint where one of the holes is slotted to allow for movement.

High-strength friction-grip (HSFG) bolts are becoming increasingly popular. Both general and higher grade bolts, as the name implies, resist shear through the interface friction arising from bolt tension. The coefficient of friction to be assumed is called the slip factor, and the strength of a bolt is calculated as bolt tension × (slip factor/load factor) × number of interfaces. It is assumed that bolts are tightened up to their proof load in tension and, to ensure this, alternative methods of tightening are specified in BS 4604. These are the part-turn method and the torque control method. In the former method the bolts are brought up hand-spanner tight, to bring the surfaces into contact, the nuts and threads marked, and tightening is then continued by a predetermined amount depending on dimensions. The latter method depends upon the use of either a manual or power-driven tool preset to slip at a particular torque value, which can be adjusted to suit the case. Of the two methods, the former is the more accurate and reliable, but tedious, whilst the latter is much quicker but less accurate owing to the fact that torque and bolt tension do not necessarily relate exactly, being dependent on thread fit.

The slip factor normally adapted is 0.45 for untreated surfaces, but BS 4604 gives details of a slip factor test to determine the value in other cases. The current BS 449 permits of a load factor in design of 1.4 whereas BS 153 calls for higher load factors depending on load combinations. (It is to be hoped that in the new documents these values will be brought more into line.) The reason for the apparently low load factor lies in the fact that a bolt possesses a margin of strength after slip has occurred, when the bolt starts to act in bearing as well as friction.

It might be supposed from the apparently full coverage in British Standards outlined above that all outstanding problems regarding HSFG bolts had been solved. Unfortunately, this is far from being the case. In the first place, post-slip strength is uncertain and clearly depends to some extent on the thickness of plies. Secondly, the test to determine slip factors consists of four bolts in line, two either side of the joint, with double cover plates putting the bolts into double shear. The effect of eccentricity arising in single-shear conditions is not determined. More particularly though, it has been established that very large or long joints do not behave in the simple fashion assumed. For instance, a very long cover plate acting in tension transverse to its length also develops longitudinal tension. The Poisson's ratio effect in this biaxial tension situation brings about a reduction in thickness and thus bolt relaxation. This can reduce bolt strengths by as much as 20%. Similar biaxial tension effects can occur in large joints in lattice girders, particularly when more than three plies are involved. Indeed the effect of a large number of plies is completely unknown.

Another difficulty arises in large joints. That is that, with the best of fabrication, if faces are not machined truly flat an indeterminate amount of bolt tension is used to bring the surfaces in contact. An extreme case would occur in a splice in a box girder. If the overall widths and depths of the two lengths do not exactly coincide, towards the corners, proper contact cannot be made without the use of machined packings. Proper contact would only be made towards the middle of the faces, where biaxial tensions exist. Large joints therefore need careful thought and cannot be treated by applying rule-of-thumb methods.

REFERENCES

1. *Brittle Fracture of Welded Structures,* Welding Institute
2. *The Fatigue Strength of Welded Structures,* Welding Institute
3. *The Weldability of Steel,* Welding Institute
4. DONCASTER, A. A., 'Economics of Design in High Strength Structural Steels, Part 1', *Construction Steelwork & Metals* (Dec. 1970)
5. DONCASTER, A. A., 'Economics of Design in High Strength Structural Steels, Part 2', *Construction Steelwork & Metals* (Mar. 1971)
6. *How to Prevent Rusting,* British Iron & Steel Research Association, and the British Steel Corporation
7. HENDRY, A. W. and JAEGAR, L. G., *The Analysis of Grid Frameworks and Related Structures,* Chatto & Windus (1958)
8. MORICE, P. B. and LITTLE, G., *The Analysis of Right Bridge Decks Subjected to Abnormal Loading,* Publication Db 11, Cement & Concrete Association
9. Institute of Welding/Institution of Structural Engineers Joint Reports on Fully Rigid Welded Multi-Storey Steel Frames (Dec. 1964 and May 1971)
10. *Sea Gem: Enquiry into the Drilling Rig Accident,* Command Paper No. 3409, HMSO (1967)
11. *Plastic Design of Columns,* British Constructional Steelwork Association Publication No. 23 (1964)
12. *Plastic Design in Steel to* BS 968, British Constructional Steelwork Association Publication No. 21 (1963)
13. *Plastic Design,* British Constructional Steelwork Association Publication No. 28 (1965)
14. *Plastic Design of Portal Frames in Steel to* BS 968, British Constructional Steelwork Association Publication No. 29 (1966)
15. NEAL, B. G., *The Plastic Methods of Structural Analysis,* Chapman & Hall (1963)
16. FANCUTT, F., HUDSON, J. C., RUDERAM and STANNERS, J. F., *Protective Painting of Structural Steel,* Chapman & Hall
17. *The Steel Designers' Manual,* Crosby Lockwood

LIST OF CODES OF PRACTICE AND STANDARDS PUBLISHED BY THE BRITISH STANDARDS INSTITUTION AND REFERRED TO IN SECTION 12

CP 117:1967 Part II
Beams of bridges
CP 2008:19
The protection of iron and steel structures from corrosion
BS 153:1958–72
Steel girder bridges
BS 325:1947
Black cup and countersunk bolts and nuts
BS 449:1959–69
The use of structural steel in building
BS 916:1953
Black bolts, screws and nuts (BSW)
BS 968:1962
High yield stress (welding quality) structural steel
BS 1500 : 1958 Part 1
BS 1856 : 1964
General requirements for the metal-arc welding of mild steel
BS 2642 : 1965
General requirements for the arc welding of carbon manganese steels
BS 2708 : 1956
Unified black square and hexagon bolts, screws and nuts (UNC and UNF threads). Normal series
BS 3692:1967
ISO Metric precision hexagon bolts, screws and nuts
BS 4395:1969

High strength friction grip bolts. Part 1, General grade; Part 2, Higher grade
BS 4360:1968–72
Weldable structural steels
BS 4604:1970
The use of high strength friction grip bolts. Part 1, General grade; Part 2, Higher grade

13 ALUMINIUM AND ALUMINIUM ALLOYS

ALUMINIUM AND ALUMINIUM ALLOYS 13

13 ALUMINIUM AND ALUMINIUM ALLOYS

J. B. DWIGHT, M.A., M.Sc., M.I.Struct.E., M.I.Mech.E.
University of Cambridge

INTRODUCTION

Aluminium (or 'Aluminum' in N. America) covers any aluminium based alloy. When the pure metal is specifically intended and not an alloy, this is referred to as Pure Aluminium. Aluminium is the most abundant of the metals in the earth's crust. It is second to steel as the cheapest metal suitable for structural use. The quantity of aluminium used is equal to that of all the other non-ferrous metals put together.

Until 1890 when the modern electrolytic method of smelting was invented, aluminium was ranked as a precious metal. The first strong alloy ('duralumin') was developed in

Table 13.1 RELEVANT BRITISH STANDARDS

DESIGN	
CP 118	Structural use of aluminium
SPECIFICATIONS FOR WROUGHT ALUMINIUM MATERIAL:	
BS 1470	Plate, sheet and strip
BS 1471	Drawn tube
BS 1472	Forgings
BS 1473	Rivet, bolt and screw stock
BS 1474	Extruded sections (solid and hollow)
BS 1475	Wire
BS 4300/14/15	H17 plate, sheet and sections
JOINTS	
BS 275, 641	Rivet dimensions
BS 1974	Large aluminium rivets
BS 3019 (Pt. 1)	TIG welding
BS 3571 (Pt. 1)	MIG welding
BS 2901 (Pt. 4)	Filler rods for welding
BS 3451	Testing welds in aluminium
PROTECTION:	
CP 231	Painting of buildings
BS 1615, 3987	Anodised coatings
BS 3416	Bituminous coatings
OTHER STANDARDS:	
BS 1161	Some standardised structural sections
BS 1490	Aluminium castings
CP 143 (Pt. 7)	Aluminium roof and wall coverings
BS 1494 (Pt. 1)	Fixings for same
BS 4300/1	Aluminium welded tube
BS 1500 (Pt. 3)	Welded aluminium pressure vessels
BS 3989	Aluminium lampposts
BS 3660	Glossary of terms used in aluminium

1905, making possible the structural use of aluminium in the German Zeppelins of World War I. Its use for aircraft construction developed during the 1930's and led to a vast increase in aluminium production capacity during World War II. This was accompanied by a decrease in cost, relative to other metals. When the demand for aircraft suddenly shrank in 1945, there was great pressure to develop other outlets for aluminium and many new markets were found.

Today aluminium is well established in a wide range of industries including: building, road and rail transport, shipbuilding, electrical engineering, chemical engineering. Aircraft accounts for a fairly small part of the total tonnage.

The use of aluminium for large civil engineering structures was pioneered in the USA in the 1930's, and included drag-line jibs, overhead cranes and bridges. Immediately after the war great efforts were made to establish aluminium as a primary structural material in competition with steel. Many successful structures were built in that period including, large span roofs, bascule bridges, footbridges, crane jibs, overhead cranes, transmission towers, large hangar doors. In the 1950's the momentum subsided, however, and today aluminium finds relatively little use in primary structures. It has become established only where its special properties warrant the considerable extra cost over steel as, for example, in military bridges.

By far a greater tonnage is carried in secondary structural applications, such as glazing bars, metal windows, curtain walling, shop-fitting, prefabricated buildings, commercial greenhouses, hand-rails, bridge guard-rails and lamp-posts. Typical of these well-established applications is the use of proprietary systems based on ingenious use of special extruded sections.

An important area of aluminium usage is sheet metal cladding (building sheet). This ranges from ordinary 'corrugated' to specially developed trough profiles for use on large industrial buildings, where its low maintenance costs have led to growing use.

Aluminium sheet is also available in soft form as an alternative to lead or copper for flashings and fully-supported roofing.

The purpose of this section is to give the civil engineer a basic understanding of the properties and technology of aluminium so that he can talk intelligently to the specialist firms who supply him with aluminium products. The final part of the section deals with allowable stresses which is intended to be of assistance in the design of aluminium structures. Reference may also be made to the relevant BS listed in Table 13.1.

BASIC CHARACTERISTICS

The characteristic physical properties of aluminium are summarised in Table 13.2.

The following characteristics are usually advantageous:

(a) *Lightness*. Aluminium has a density one-third that of steel.
(b) *Durability*. Aluminium has a tenacious oxide film and does not rust.
(c) *Extrudability*. Aluminium is ideally suited to the extrusion process. Intricate section shapes can be produced giving flexibility in design.
(d) *Low temperature properties*. Brittle fracture is not a problem with aluminium.
(e) *Electrical conductivity*. Aluminium has a high electrical conductivity (although the rapid forming oxide film necessitates special connection techniques).

The following are usually a disadvantage:

(f) *Modulus*. Aluminium has a low modulus of elasticity, one-third that of steel. This increases deflections and the tendency to buckle (but it reduces impact stresses).
(g) *Strength*. Pure aluminium has low strength, and the range of mechanical properties available with the alloys is narrower than that for steels. Acceptable structural alloys exist, but they tend to be less tough than steels of comparable tensile strength. Fatigue is a factor.

(h) *Melting point.* Aluminium has a melting point of only 660 °C, and its strength falls off rapidly above 200 °C.

(i) *Expansion.* The coefficient of thermal expansion is twice that for steel necessitating wider gaps at expansion joints. (But the temperature stresses are only two-thirds of those for steel, because of the lower modulus.)

(j) *Electrolytic attack.* Despite its basically good durability, aluminium can suffer serious local corrosion from the proximity of certain other materials, including steel and copper. Precautions must be taken against this.

Table 13.2 PHYSICAL PROPERTIES OF ALUMINIUM

Density	2.71 g/cm³
Mass of sections	271A g/m
(where A = section area in cm²)	
Mass of plate and sheet	2 710T g/m²
(where T = thickness in cm)	
Modulus of elasticity	70 kN/mm²
Shear modules	26 kN/mm²
Poisson's ratio	0.33
Melting point	660 °C
Coefficient of linear expansion	24×10^{-6} per °C
Thermal conductivity	220 W/m per °C
Electrical resistivity	2.7 microhm cm

Note. These figures are for commercial purity aluminium. The only significant differences for the alloys are as follows:

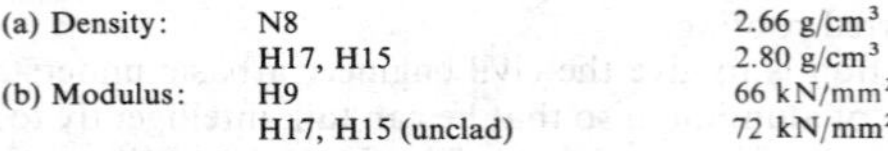

(a) Density:	N8	2.66 g/cm³
	H17, H15	2.80 g/cm³
(b) Modulus:	H9	66 kN/mm²
	H17, H15 (unclad)	72 kN/mm²

(c) Conductivities are lower.

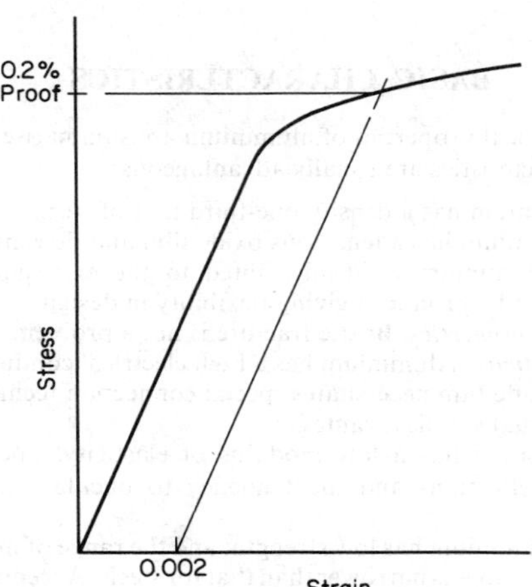

Figure 13.1 Definition of proof stress

The following characteristics may or may not be advantageous.

(k) *Stress-strain curve.* Aluminium in common with most non-ferrous metals has a stress-strain curve which bends over gradually, without a clearly defined yield (Figure 13.1). 'Yield' is represented by specifying the 0.2% proof stress.

(l) *Thermal conductivity.* Aluminium has a high thermal conductivity.

Finally, it must be noted that wrought aluminium material costs three times as much as steel, volume for volume. It occupies a midway position between steel on the one hand (which is cheap), and the rest of the non-ferrous metals on the other (which are expensive).

MANUFACTURE

Primary production

Aluminium is obtained from bauxite, which is a plentiful reddish ore got by open-cast mining. The first process is to extract alumina, a white powder. This is then shipped to a smelter, which is sited for cheap electricity, and which may be thousands of miles from the mine. The smelting is done electrolytically in relatively small furnaces (or 'pots') with big carbon electrodes, the alumina being dissolved in molten cryolite. The product of the smelter is pure aluminium ingot.

The ingot is shipped to secondary plants where it is remelted, alloyed and used to produce wrought or 'semi-fabricated' products, namely: plate, sheet and strip, sections, hollows, rod and wire.

Plate and sheet

Mill practice for flat material is much as for steel. Slabs are hot rolled to produce plate, which may then be cold reduced to make strip, from which sheets are cut. The product has a superior surface finish to steel.

Material is described as plate when it is over 3 mm thick. A product known as cold-rolled plate is supplied, which is held to closer tolerance than the ordinary hot-rolled plate.

Material known as 'clad' sheet and plate is also available. This consists of a high strength alloy core having a thin layer of pure aluminium rolled on to each outer surface, to give improved corrosion resistance.

Building sheet of corrugated or troughed profile is continuously rolled in the mill in long lengths. Other patterns of ribbed sheeting are also available.

Sections

Aluminium sections are produced in horizontal extrusion presses. A preheated cylindrical billet is inserted in a heated container and is forced through an aperture in a die, the emerging material having a cross-section determined by the shape of the aperture.

The die is cheap when compared with the set of rolls needed for a steel section, and the time needed to change it on the press is a fraction of that wanted for a roll change. There is also little limit to the shape and complexity of section that can be extruded. It is therefore common practice to design a special section, or suite of sections, tailor-made to suit the job in hand. Great skill is needed in die cutting, in order to make the section come out reasonably straight, without too much curvature or twist. Some crookedness is inevitable and this becomes worse after quenching (for heat-treated materials). Before final cutting to length, the sections must therefore go through various correction processes. These can consume considerable time (and cost) and are often the critical factor—

rather than the actual extrusion—in deciding whether a proposed section will be a practical proposition.

Cold-rolled sections formed from strip are also produced, but they are used to a far lesser extent than in steel.

Hollows

Hollow aluminium sections can be produced in three ways, namely hollow extrusions, drawn tube, welded tube.

Hollow extrusions can be of complex shape or simple tubular cross-section. They are usually produced by means of a 'bridge die' in which a mandrel, defining the internal shape of the section, is supported on feet within the die. During extrusion, the hot plastic metal has to flow around these feet and reunite, so that the final section contains longitudinal welds. These cannot be detected by eye, and in the vast majority of applications hollow extrusions produced in this way are perfectly acceptable. The cost per kg is more than for a non-hollow, because of the slower extrusion speed. Also the die cost is considerably increased.

An alternative method of making a hollow extrusion is to employ a long mandrel coming right through the middle of the billet, the tip of which gives the internal shape to the section. This technique becomes necessary for those applications where the welds from a bridge die are not acceptable, owing to stringent safety requirements. It tends to be more expensive, and the concentricity of the bore is not so good.

Drawn tubes can be either round or of simple non-circular cross-section, such as rectangle or oval, the wall thickness being nominally constant. The tubes are seamless and are called for when the accuracy required is too high for extrusion, or the wall thickness too low; they cost more than extrusions.

Welded tubes are made from strip and are relatively cheap for large tonnages. A typical application is irrigation pipes.

CONTROL OF MECHANICAL PROPERTIES

Most of the aluminium alloys, unlike structural steel, are unacceptably weak in the annealed state and before they can be used they must be brought up to strength. There are two classes of alloy: 'non-heat-treatable' and 'heat-treatable'. The former can be strengthened by cold work, the latter by heat treatment. Both types can be softened again by annealing.

Cold work

The non-heat-treatable alloys, and also pure aluminium, can only attain improved mechanical properties by means of strain hardening (cold work) applied during manufacture. These alloys are therefore suitable for sheet or drawn tube, which are cold reduced, and also in some cases for plate. They are less suitable for extrusions, which are formed by a hot process.

The mechanical properties of the final product (sheet or tube) are controlled by suitably adjusting the sequence of cold reductions in the mill, and by inter-pass annealing when necessary. Material can be produced that has as high a strength as is possible for the alloy concerned ('fully-hard'), but with limited ductility. Alternatively, half-hard or quarter-hard material can be made, which is weaker but more ductile (and formable). The latter will have a much-reduced proof stress, as compared with the fully-hard condition, but only a slightly lower ultimate stress.

The condition of non-heat-treatable material is specified as H2 to H8, depending on its degree of strain-hardening (and strength). H8 denotes the fully-hard condition.

Heat treatment

The heat-treatable alloys, which tend to be stronger than the non-heat-treatable, derive their strength as a result of heat treatment applied after the material has reached its final dimensions. They are typically chosen for extrusions (which cannot receive cold work during manufacture), and also for sheet, plate and tube in high strength applications.

The basic heat treatment procedure consists of quenching followed by ageing. The quench, known as 'solution treatment', takes place from about 500 °C. This does not have any immediate effect, but with the passage of time the material gradually gets stronger (or 'ages'), and reaches a final strength after a period of some days. Material so produced is described as 'quenched and naturally aged', the ageing having taken place at room temperature.

Alternatively, the ageing process can be speeded up by treating the quenched material in a furnace for some hours. This is called 'precipitation treatment' (or 'artificial ageing'). It leads to a higher final strength, but with some loss of ductility. Material produced in this manner can be described as 'quenched and artificially aged' or 'fully heat-treated'. The precipitation treatment need not be performed immediately after quenching, and can be used to bring up strength of material that has already been allowed to age naturally. The temperature at which it is done is not critical, and lies in the range 150–200 °C.

The quenching operation is normally done in water. In the case of the strongest alloys it is critical and has to take place from an accurately controlled temperature, which involves the use of special furnaces. With some alloys, however, it is possible in the case of extruded material to carry out the solution treatment simply by spraying the section with water as it emerges from the die, thereby saving heat and reducing distortion. There may be some quenching action even when the spray is turned off ('air quenching').

The following letters are used to specify the condition of heat-treated material (in ascending order of strength):

TB Water quenched and naturally aged.
TE Air quenched and artificially aged.
TF Water quenched and artificially aged (i.e. fully heat-treated).

Annealing

Material in both the non-heat-treatable and heat-treatable alloys can be brought back to its fully soft condition by means of annealing. This consists of heating in a furnace to a temperature in the region of 400 °C, and then allowing to cool. For the non-heat-treatable alloys the material is heated up quickly and the holding time limited to 20 min, to prevent undue grain growth. For the heat-treatable alloys a holding time of an hour or more is employed and it is necessary to control the rate of cooling (to prevent any quenching action).

THE ALLOY GROUPS

The newcomer to aluminium is confronted by a baffling array of specifications. There are no simple names, such as 'mild steel', which have a worldwide meaning, and the impression is gained that there is a vast selection of different alloys. In fact there are seven alloys that matter.

The reader is recommended to concentrate on the British Standard (General Engineering) system of alloy nomenclature and to ignore the proprietary names which appear in trade literature.

British numbering system

A typical specification for an item of aluminium, using the British Standard system, would be as follows:

HE30—TF

which conveys four pieces of information:

H shows whether it is a non-heat-treatable (N) or heat-treatable (H) alloy; in this case it is the latter.
E indicates the form; in this case an extruded section.
30 specifies the chemical composition; in this case Al-Mg-Si.
TF indicates the condition of the material; in this case fully heat-treated.

Forms, specified by the letter in the second position (E in the example), include:

Plate and sheet (S)
Clad plate and sheet (C)
Drawn tube (T)
Wire (G)
Solid and hollow extrusions (E)
Rivets (R)
Bolts (B)
Forgings (F)

The following numbers are used to denote the important alloy compositions (30 in the example):

Non-heat-treatable (N)	3, 4, 8
Heat-treatable (H)	9, 15, 17, 30

When it is necessary to mention a particular alloy without specifying its form or condition, it is normal to include the initial N or H along with the composition number. Thus, for example, one would refer in general terms to N8 or H30.

Symbols used after the hyphen to denote the condition of the material (TF in the example) are as follows:

Non-heat-treatable alloys (including pure aluminium)	H2 to H8
Heat-treatable alloys	TB, TE, TF

In the case of a non-heat-treatable alloy the symbol indicates the degree to which the material has been strain hardened, while for a non-heat-treatable alloy it denotes the heat treatment condition. There are also the two following conditions, which can apply to material in any alloy.

Annealed (O)	As-manufactured (M)

The latter is a rather vague condition, and would refer to an as-extruded section or to a rolled plate, without either strain-hardening or formal heat-treatment.

Pure aluminium is referred to as 1, 1A, 1B, or 1C depending on the purity. It is in fact a non-heat-treatable material, but the N is left out. Thus S1-0 means 99.99% pure aluminium sheet in the annealed condition, while TIC-H4 indicates drawn tube of the lowest commercial purity (99.0%) in the half hard condition.

Aircraft materials, which are made to closer limits and which cost more, are controlled by a separate system of specification.

Other numbering systems

The American system of alloy naming has a considerable worldwide usage. A typical designation would be as follows:

6061-T6

where the 6061 specifies the alloy composition and the T6 the condition of the material. The broad correlation with the British system is given in Table 13.3. This table also includes the equivalent ISO alloy numbers.

Table 13.3 FOREIGN EQUIVALENTS OF BRITISH ALLOYS

Britain	*ISO* *ISO R209*	*USA*
N3	Al-Mn 1	3003
N4	Al-Mg 2.5	5052
N5	Al-Mg 3.5	5154
N8	Al-Mg 4.5	5083
H9	Al-MgSi	6063
H20	Al-Mg1SiCu	6061
H30	Al-SilMg	6351
H17	Al-ZnMg 1	—
H15	Al-Cu4SiMg	2014

The alloy groups

The aluminium alloys together with pure aluminium fall into six well defined groups, as follows:

	Main alloying ingredients	*Important alloys in group*
Non-heat-treatable	Al	Pure aluminium
	Al-Mn	N3
	Al-Mg	N4, N8
Heat-treatable	Al-Mg-Si	H9, H30
	Al-Zn-Mg	H17
	Al-Cu+others	H15

Mechanical properties

Table 13.4 lists the mechanical properties of the above alloys, the figures given being specification values. The properties are slightly different for forms and thicknesses other than those covered in the table, and for fuller information reference should be made to the appropriate British Standards (Table 13.1).

Typical properties will be somewhat higher than the specification values.

Table 13.4 MINIMUM MECHANICAL PROPERTIES

Alloy	*Form*	*Thickness* mm	*Specification*	*0.2% proof* N/mm²	*Tensile strength* N/mm²	*Elongn* %
99% pure	Sheet	2	S1C-0	—	70	30
		2	S1C-H4	—	110	5
		2	S1C-H8	—	140	4
N3	Sheet	1	NS3-H8	—	175	3
N4	Sheet	2	NS4-0	60	160	18
		2	NS4-H3	130	200	6
		2	NS4-H6	175	225	5
N8	Plate	6	NS8-0	125	275	16
		6	NS8-H2	235	310	8
H9	Section	3	HE9-TE	110	150	7
		3	HE9-TF	160	185	7
H30	Section	6	HE30-TF	255	295	7
H17	Section	6	HE17-TF	280	340	8
H15	Section	6	HE15-TB	230	370	10
		6	HE15-TF	370	435	6

Notes. The values quoted are the minimum tensile properties as given by BS 1470, BS 1474 or BS 4300/15.
Each set of values refers to the condition form and thickness indicated, these having been selected to represent typical usage. For other forms or thicknesses the properties may be slightly different; refer to the BS for exact details.
The elongation figures are for a 50 mm gauge length.

Characteristics of the principal alloys

Characteristics and applications of the various alloys are summarised in groups as follows:

PURE ALUMINIUM

Pure aluminium is ductile, but too weak to be of much structural value, even in the work-hardened condition. It tends to be selected where corrosion resistance or high conductivity is the major consideration. It is commonly used in the 'commercial purity' grade IC (99.0% minimum purity); higher purity improves the corrosion resistance, conductivity and ductility.

In the form of plate and tube it finds use in welded chemical plant. It is also employed in the form of annealed sheet for flashings and fully supported roofing (in place of lead), often at 99.99% purity (S1-0). Pure aluminium is also used for electrical conductors.

AL-MN ALLOY

The one alloy in the group, N3, contains a nominal 1¼% of manganese. It is a sheet alloy, combining a corrosion resistance close to that of pure aluminium with a slightly improved strength. Its only application, an important one, is in 'building sheet' (corrugated and troughed sheeting) for which it is the standard alloy and for which it is used in the fully hard temper (NS3-H8).

AL-MG ALLOYS

The alloys in this important group also have excellent corrosion resistance, especially in marine environments. In the annealed condition they are ductile, but harden up quickly when worked. They are relatively tough and are readily welded. Extrudability is rather poor.

N4 contains a nominal 2% of magnesium. It is a useful sheet alloy, commonly used in the half-hard temper (NS4-H3). In this form it is reasonably ductile for bending and forming, and at the same time has adequate strength for sheet metal fabrication.

N8 contains $4\frac{1}{2}$% of magnesium and $\frac{3}{4}$% of manganese. It is primarily a plate alloy, used in the as-rolled (NS8-M) or slightly work-hardened (NS8-H2) condition. To some extent it is the aluminium equivalent of structural mild steel, being tough, formable and weldable; but it has a somewhat low proof stress. It is typically used for welded plate work where a high yield strength is unnecessary, such as ship superstructures and welded truck bodies. Extruded sections in NE8-M are used for stiffeners and framing in such structures. N8 is also a standard material for liquid methane tanks, brittle fracture being no problem.

N5 is another alloy in the group, having properties intermediate between N4 and N8. It is used for rivets, and it is popular for boat building.

AL-MG-SI ALLOYS

The alloys in this other important group are noted for extrudability. Not only do they extrude easily, but they are suitable for spray quenching at the die, enabling sections to be produced that are thinner and more intricate than is possible in other alloys. They are normally used in the fully heat-treated (TF) condition, in which their ductility is adequate. They are readily welded, but suffer severe local softening in the region of the weld. Their corrosion resistance is good.

H9, containing some $\frac{3}{4}$% of magnesium and $\frac{1}{2}$% of silicon, is the extrusion alloy par excellence. It is the standard choice for thin intricate architectural sections such as window sections and glazing bars, where stiffness rather than strength is required. In the fully heat-treated condition (HE9-TF), it is only about half as strong as mild steel, and is therefore unsuitable for highly stressed situations. When an exceptionally thin and awkward section is required use can be made of the air-quenching property of this alloy, which helps to cut distortion (HE9-TE). Very good surface finish is possible with H9.

H30 contains about $\frac{3}{4}$% of magnesium and 1% of silicon, with $\frac{3}{4}$% of manganese. It has an extrudability approaching that of H9, but with considerably better properties. The yield strength in the TF condition equals that of mild steel, although the tensile strength and ductility are appreciably lower. It is the alloy most commonly chosen for load-bearing structures, and has been used in a wide range of structural applications, such as bridges, roof structures, lorry bodies, ladders and scaffolding. It is used mostly in the extruded form (HE30-TF), but also as plate or sheet for gussets. Assembly is nearly always by riveting or bolting, not welding.

The alloy H20 is similar to H30, being the version produced in North America (6061).

AL-ZN-MG ALLOYS

This group includes the strongest aluminium alloys available, with tensile strengths up to 600 N/mm^2. Developed for use in military aircraft, these ultra-strong alloys are of no interest to a civil engineer because of their cost and low ductility.

H17 is a medium strength alloy in the same group, with a nominal $4\frac{1}{2}\%$ zinc and $1\frac{1}{4}\%$ magnesium. It has only recently been standardised in Britain, and its range of application is still being established. Fully heat-treated, it is a little stronger than H30. Previously there were some doubts on the score of stress corrosion, but it is believed that in the finally standardised version this possible weakness has been eliminated. H17 extrudes readily, but not as well as H30. Its corrosion resistance is good, but again is slightly inferior to H30.

The great attraction of H17 lies in its good strength in the welded condition. Unlike H30, the local softening close to the weld is only temporary and after a passage of time the heat affected material regains much of its lost strength, by a natural ageing process; after 30 days the tensile strength in this zone is up to over 80% of the fully heat-treated TF value. This gives the same freedom in the siting of welds as with N8, but with the advantages of a much higher proof stress. H17 has been adopted for military bridges, and there is no doubt that other structural applications will follow.

AL-CU ALLOYS

This group often referred to unofficially as 'duralumin' is typified by a high copper content (around 4%) and includes the high strength alloys used for aircraft. The copper content impairs the corrosion resistance, so that the durability is not as good as for the other groups. For this reason clad sheet is often specified. Manufacture and heat-treatment of these alloys has to be carefully controlled and they extrude slowly; they therefore cost more. They are not suitable for welding.

H15 is the standardised commercial version, containing $4\frac{1}{2}\%$ of copper together with some magnesium, silicon and manganese. It can be used either naturally aged (TB) or else fully heat-treated (TF). In the TB condition it is stronger and more ductile than H30-TF, its properties being similar to those of mild steel. In the TF condition it has a much higher proof stress, but a rather low ductility. The TB condition is the one of possible interest to civil engineers, and an equivalent to HE15-TB was in fact used for the early aluminium civil structures—this being before the more corrosion-resistant (and cheaper) Al-Mg-Si alloys came to the fore. The TF condition is of limited value, because of its lower ductility, and would only be selected where a high proof stress is the dominating requirement.

FABRICATION

Cutting

Thin material can be cut by guillotine or shears. For thicker material cold sawing is used, with either a circular or band saw. Thick plate can be cut in this way, but for best results a saw with relatively coarse teeth should be used.

Flame cutting with the oxy-acetylene torch is unsuitable, because it leaves an unacceptably rough edge. Instead it is possible to employ 'plasma arc' cutting, an adaptation of the TIG welding process (see below). This is quick and convenient, but involves the use of special equipment to produce the necessary constricted arc.

Bending

With the non-heat-treatable alloys the practice for bending and forming is much as for steel. Cold bending is employed when possible, the maximum severity of bend being governed by the temper. Reasonable bends are possible except when the material is in the fully hard (H8) condition. Spring back is more than for steel.

When the required bend is too severe to be done cold, heat may be applied locally with a gas flame, the necessary temperature being 450–500 °C. This is far below red heat, and there is no colour change to show when it has been reached. Great care must therefore be taken to ensure that the aluminium is not overheated. One method is to rub a pine stick on to the heated area and see if it leaves a mark. (Alternatively, colour-sensitive crayons may be used.)

Heat-treated material, which is normally used in the TF condition, cannot be formed so readily. Being less ductile, it can only accept a small deformation when bent cold. Heating, on the other hand, disturbs the heat-treatment and produces local weakening.

When severe forming of heat-treated material cannot be avoided, it must be done when the material is in the more ductile TB condition. The part can then be artificially aged later, to bring it up to the full strength TF properties.

Welding

Pure aluminium is readily welded. Of the alloys, the most suitable for welded construction are N4, N8 and H17 which retain most of their strength after welding. H30 can also be readily welded, but serious local weakening in the heat-affected zone brings design limitations. H15 is unsuitable for welding.

Stick welding of aluminium with coated electrodes is possible but unsatisfactory because of the poor appearance and the corrosive nature of the flux.

The standard arc welding process is MIG (metal-inert gas) which is similar to CO_2 welding in steel. This is a semi-automatic gas-shielded d.c. process, in which the filler wire is fed in automatically through the hand held torch, the shielding gas being argon. The MIG process is easy to operate, and is suitable for positional welds. It can be used in material down to about 2 mm thick.

The alternative arc-welding method is TIG (tungsten-inert gas); an a.c. process. In this the arc is struck from a non-expendable tungsten electrode, the filler wire being added separately with the other hand. It requires more skill than MIG; also it is slower and causes more distortion. Its main role is in light-gauge work. The choice of filler wire is important and depends on the parent alloy (see Table 13.5). Weld preparations are generally as for steel, but with greater emphasis on good fit up.

Table 13.5 FILLER WIRE FOR MIG OR TIG WELDING

Parent alloy	*Acceptable filler materials (in order of preference)*
N3	NG3, GIB
N4	NG6, NG61
N8	NG6, NG61
H9, H30	NG21, NG6. NG61
H17	NG61

Notes. NG6 and NG61 are broadly similar in composition to alloy N8, but with a slightly higher magnesium content.
NG21 has the composition Al-5%Si.
Refer to BS 1475 for further information.

Aluminium can be spot-welded, in the same way as steel. Stud-welding is also available, using a special gun.

Riveting

Structures built from H30 or H15 material are nearly always of riveted construction. Hot driven steel rivets were used originally, but these gave corrosion trouble and have been superseded by aluminium rivets.

In H30 structures the rivets are usually in NR5, driven cold. The force required to close these is considerable. Squeeze riveting is therefore used where possible, and a smaller head (the 'small pan') employed than for steel. Rather small clearances are necessary. For large rivets (above about 15 mm) hot-driving is required, involving the use of special rivet furnaces.

An alternative technique is to use HR30-TB rivets which have been held at a low temperature since quenching, to suppress the natural ageing. These are readily driven cold, after which they age-harden in position to attain a final shear strength that is better than NR5. HR15-TB rivets can be used in the same way and would be the logical choice for an H15 structure.

Bolting and screwing

The choice of bolt material for an aluminium structure would normally be HB30-TF, which has a shear strength appropriate for use with N8, H30 or H17 members.

Aluminium bolts are not too good in tension, especially under fatigue conditions, and when better tensile properties are needed, recourse must be had to steel bolts. These may be of mild steel, in which case a suitable coating is necessary, either of zinc or cadmium. Even so they impair the durability of an otherwise durable structure. Alternatively the bolts can be of stainless steel; this is the ideal, but the more expensive solution.

A bolted joint in an aluminium structure must be assumed to act entirely by shear and bearing, with very little friction grip. Therefore, if anything like a stiff joint is to be obtained, it is necessary to use close fitting bolts in reamered holes.

Tapped holes in aluminium are not very satisfactory. For situations where repeated screwing and unscrewing has to take place, patent thread-inserts give good service.

Glueing

Glueing with epoxy resin adhesives is a perfectly practical method of jointing aluminium. It has been successfully used for lamp-posts and other small structures. The epoxies are attractive because of their ability to tolerate poor fit-up, and may be used rather like a cement. They can be either cold or hot-setting, the latter being stronger and quicker.

Shear strengths of the order of 15 N/mm^2 can be developed, but it is important to guard, against the risk of premature failure by peeling. Extruded tongue-and-groove joints are a good way of getting over this.

DURABILITY AND CORROSION

Durability of aluminium

The atmospheric corrosion of aluminium proceeds by pitting, an entirely different form of attack from the rusting of steel. The corrosion products formed at each pit are voluminous and give an exaggerated impression of the actual damage, the amount of metal consumed, even in a badly corroded specimen, being a minute proportion of the total.

The rate of corrosion, defined by the depth of pitting, depends on the alloy and on the environment. Tests have shown that it is relatively rapid in the first two or three years, but is thereafter stifled by the corrosion products and proceeds much more slowly. The corrosion is always less severe on exposed surfaces which are washed by rain.

With the exception of H15, aluminium and its alloys are highly durable and can very often be left unpainted. Any corrosion that occurs tends to affect the ductility (and the appearance), rather than the actual strength.

The extent to which painting may or may not be necessary depends on the Durability Rating of the alloy concerned. The following guide is taken from CP 118.

Durability rating	*Alloys*	
A	N3, N4, N8	Can almost always be left unpainted. N4 and N8 are especially good in marine atmospheres.
B	H9, H30, H17	Painting necessary when exposed to severe industrial environments especially if the metal is under 3 mm thick. Painting also desirable for thin material in a marine situation.
C	H15 (unclad)	Painting necessary unless in a dry unpolluted situation. Note that clad sheet (HC15) has an A rating.

Painting

If a structure is painted, it is important that the priming and subsequent coats should contain no copper, mercury or graphite, and preferably no lead. A zinc chromate priming coat is recommended.

Surface finish

In an unpainted structure exposed to the elements, the surface of the aluminium weathers and grows rough with time. In many situations this does not matter, provided the strength is unimpaired. If, however, it is important to retain a smooth appearance, as for window frames and curtain-wall mullions, regular washing is necessary. A good rule is to wash the aluminium at the same time as the glass.

For best results anodised material may be used. Anodising is a process whereby the inherent oxide film is artificially increased in thickness—to 25μ for 'architectural anodising'. This gives a pleasant satin appearance, which will last for many years if regularly washed. If not washed, exposure to an industrial atmosphere will cause even an anodised surface to break down in time.

Coloured and also black anodised finishes are available.

Contact with other materials

When aluminium is in electrical contact with certain other metals and moisture is present, the adjacent aluminium gets eaten away. This is known as 'electrolytic' or 'galvanic' corrosion. Failure to prevent it is the most common cause of corrosion trouble in aluminium.

Corrosion occurs when aluminium is in contact with steel (other than stainless) and, more severely, with copper, brass and bronze. The effect becomes more pronounced when the area of the sacrificed metal (the aluminium) is small compared with the other, i.e. aluminium bolt-heads on a steel plate. Electrolytic corrosion can be prevented by ensuring that there is no direct contact between the two metals, either by means of bitumastic paint or else with a suitable interposed gasket.

With copper the effect is so strong, that even water dripping off a copper roof on to an aluminium one will quickly perforate the aluminium, because of dissolved copper ions. Such situations must be avoided.

Electrolytic action between aluminium and lead is only slight.

When aluminium and zinc are in contact, it is the zinc that gets consumed. Galvanised bolts in an exposed aluminium structure tend to lose their protection rather quickly.

Aluminium that is to be embedded in concrete should be protected with bitumastic paint. Otherwise it will suffer attack while the concrete is green.

STRUCTURAL DESIGN

Primary structures in aluminium will usually be built in H30-TF (riveted) or H17-TF (welded) which compare with mild steel. They may be in the more ductile, but weaker alloy N8 (welded). In exceptional circumstances H15-TF (riveted) may be used, which is stronger but less ductile and less durable.

The chief differences from steel design are as follows:

(a) The greater need to save weight, at the expense of labour.
(b) The ingenious use of extruded sections.
(c) The use of thinner material.
(d) Deflection is more critical.
(e) Buckling is more critical.

The allowable stresses given in Table 13.6 are based on CP 118. Limit state design has not yet arrived for aluminium.

Available sections

The available types of extruded section range from the conventional structural shapes, such as angles and channels, to special sections combining function with structure. Dies exist for a vast range of the former, but the latter must usually be designed for the job in hand and a new die cut. Of particular interest are planking sections, originally developed for lorry flooring, and since applied in various fields including bridge decks.

In H30 the minimum feasible thickness for a 200 mm wide extrusion would be of the order of 2.5 mm. In H17 it would be slightly more, and in N8 or H15 a lot more. For a smaller section the minimum thickness would be correspondingly less. The maximum size of an extrusion is 350 mm.

Basic permissible stresses

Table 13.6 gives static permissible stresses for the recommended structural alloys (N8, H30, H17, H15). These incorporate a safety factor on the 0.2% proof stress; this factor depends to some extent on the ductility and the proof/ultimate ratio.

Effective area in tension

In riveted or bolted construction the effective area is taken as the gross area minus a deduction for holes—in the normal way. In the case of discontinuous angle ties a further deduction is made as follows per outstanding leg:

	Single angle connected through one leg	*Two angles connected both sides of a gusset*
N8	0.4A	Nil
H30, H17	0.6A	0.2A

where A is the area of the outstanding leg.

Welded members

The strength of N8-M members may be assumed to be unaffected by welding. In H17-TF (slightly) and H30-TF (severely) there is a locally weakened zone in the vicinity of a weld, for which reduced value of permissible stress is applicable (Table 13.6). For a butt-welded member (with a fully-softened cross-section) this value determines the load, that can be carried. In other cases, e.g. members containing longitudinal welds, the load carrying capacity is obtained by combining the strengths of the reduced-strength zones (at the reduced stress) and the rest of the cross-section (at full stress).

Table 13.6 BASIC PERMISSIBLE STRESSES

Alloy	*Condition*	*Axial T, C* N/mm²	*Bending T, C* N/mm²	*Shear* N/mm²	*Bearing* N/mm²
N4	H3	75, 70	85, 80	45	165
	welded	50	50	30	—
N8	O, M	80	90	50	200
	welded	80	90	50	—
H9	TE	60	70	35	115
	TF	85	95	50	140
	welded	30	30	20	—
H30	TF	135	150	80	215
	welded	50	50	30	—
H17	TF	150	160	90	225
	welded	125	125	75	—
H15	TB	135, 125	150, 140	80	240
	TF	150, 200	150, 220	105	270

Notes. The welded figures refer to material within 25 mm of the centre line of a butt weld or root of a fillet weld. The other figures are for non-welded construction or, in welded structures, refer to material outside the above zone.

The table is based on CP 118: 1969 (amended March 1973), and incorporates a safety factor (varying slightly from alloy to alloy) which is comparable to that inherent in BS 449. A 25% increase in stress is allowed under wind loading.

There is a small difference from CP 118, in that CP 118 gives slightly varying permissible stresses depending on the form and the thickness, for each alloy, whereas the above table gives a single, averaged value. The only significant divergence arises for HE30-TF sections above 20 mm thick, for which the table is 10% safer than CP 118.

Where two figures appear in one column, the first is for tension and the second compression. Otherwise axial stresses apply in both tension and compression.

The reduced-strength zones are assumed to extend up to a distance of 25 mm from the centre line of a butt weld or the root of a fillet, irrespective of the thickness.

Struts

Buckling is often a critical factor in aluminium design. Here it is only possible to give an outline treatment, and the reader should consult CP 118 if more refined calculations are needed. Generally speaking it is necessary to consider: (a) overall buckling of the member as a whole, and (b) local buckling.

Column buckling

Overall column buckling is covered by the curves in Figure 13.2, the permissible stress being read off at $\lambda = KL/r$, in the usual way.

With struts of thin angle or channel section it is also necessary to check for torsional buckling, another mode of overall failure. The permissible stress corresponding to this is obtained by entering Figure 13.2 at $\lambda = \lambda_t$, where $\lambda_t = \pi\sqrt{(E/\sigma_t)}$ and σ_t is the elastic critical stress for tensional buckling (see Timoshenko). Formulae are given in CP 118.

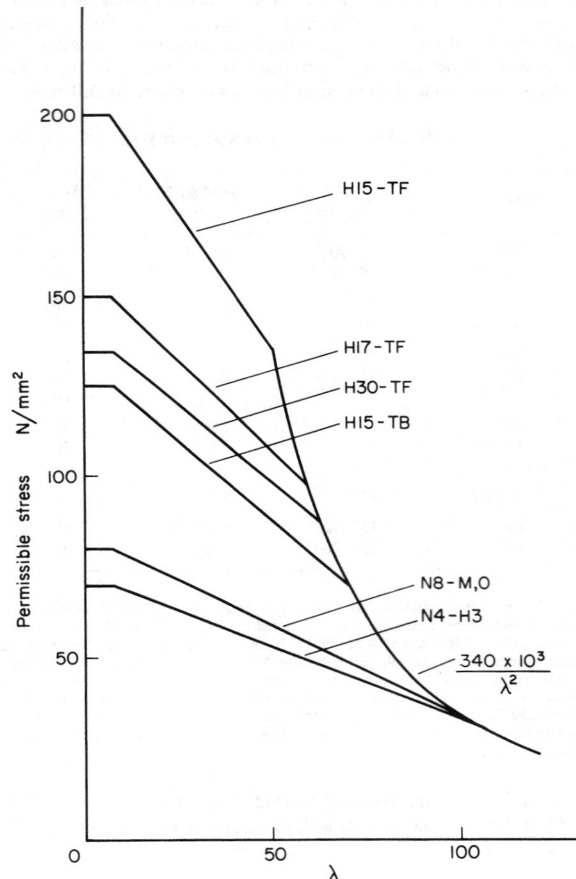

Figure 13.2 Permissible stresses for column buckling and torsional buckling

In the case of a simple angle (no bulbs), it is reasonable to treat torsional buckling as a form of local buckling. Local buckling is covered below.

Beams

Lateral-torsion buckling of beams between points of lateral support is covered by Figure 13.3. The permissible extreme fibre stress is read off at $\lambda = \lambda_{lat}$ where

$\lambda_{lat} = \pi\sqrt{(E/\sigma_{lat})}$ and σ_{lat} is the extreme fibre stress corresponding to the elastical critical moment (see Timoshenko). Approximate rules for finding λ_{lat} appear in CP 118.

For local buckling of the compression flange see below.

Local buckling in compression

Local buckling of the component plate elements is also covered by Figure 13.3. Two types of element are recognised, a 'web', supported along both edges; and an 'outstand' supported along one edge only. The permissible stress is obtained by entering the curve

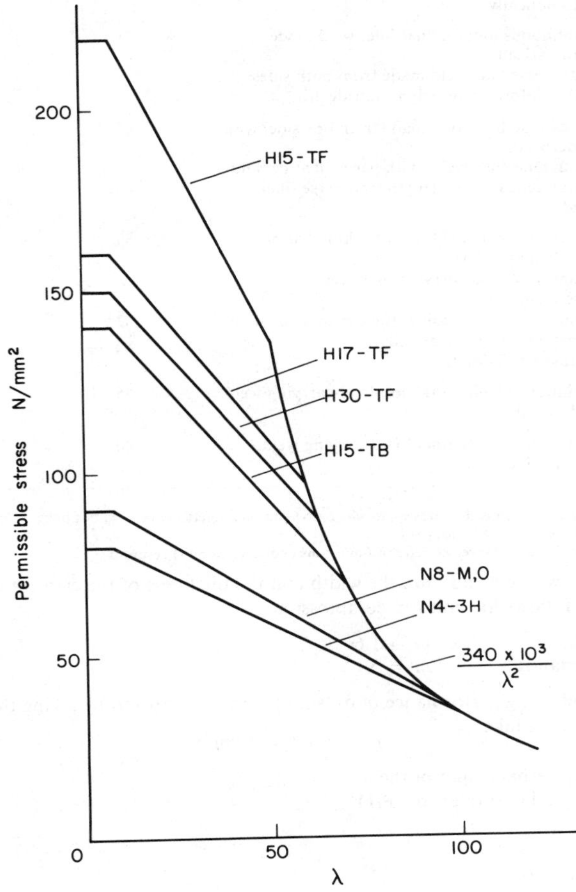

Figure 13.3 Permissible stresses for lateral buckling (beams) and local buckling

Table 13.7 FATIGUE STRENGTH

	Description of joint	*Pulsating tension* N/mm²	*Full reversal* N/mm²
1	As-extruded or machined material (away from joints)	120	82
2	Continuous, full-penetration butt weld (longitudinal or transverse) with reinforcement dressed flush	99	69
3	(i) Joint made with close-fitting bolts or cold-driven rivets (single-lap joints not allowed), (ii) Full penetration transverse butt weld made from both sides, with reinforcement height on each side not exceeding the lesser of $t/5$ and 3.2 mm, (iii) Full-penetration longitudinal butt weld, made automatically.	79	56
4	(i) Continuous longitudinal fillet weld made automatically, (ii) Transverse butt weld made from both sides with reinforcement height outside 3(ii).	71	48
5	(i) Transverse butt weld made from one side, with underbead, (ii) Transverse butt weld, with permanent backing bar attached by full length transverse fillet welds	65	45
6	(i) Transverse butt weld, with backing bar not attached as in 5(i), (ii) Transverse load-carrying fillet weld or cruciform joint.	59	40
7	(i) Continuous longitudinal fillet weld made with interruptions (i.e. manually), (ii) Transverse T-joint.	53	36
8	Discontinuous longitudinal non-load-carrying weld, butt or fillet	45	29
9	Discontinuous longitudinal load-carrying weld, butt or fillet	40	25

Notes. The table gives permissible stresses, to which no additional safety factor need be applied, for 10^5 reversals. The values apply equally to all alloys.
'Pulsating tension' and 'full reversal' refer to f-min/f-max equal to 0 and -1 respectively.

at $\lambda = m(b/t)$, where b and t are the width and the thickness of the element considered. m is taken as follows for uniform compression:

Webs	$m = 1.6$
Outstands	$m = 5.1$

The local buckling performance of outstands can be improved by giving them a small lip. In this case we take:

$$m = 5.1(1-c^2/80t^2)$$

where c is the internal height of the lip.

For further guidance refer to CP 118.

Local buckling in shear

The strength of shear panels can be safely based on elastic buckling theory, assuming simply supported edges.

Table 13.8 PERMISSIBLE STRESSES FOR BOLTS AND RIVETS

Alloy and details	*Shear* N/mm^2	*Tension* N/mm^2
BOLTS H30-TF (closefitting)		
Under 13 mm dia.	60	70
13 mm dia. or over	70	90
Steel		
Under 22 mm dia.	95	95
22 to 29 mm dia.	95	110
RIVETS H30-TB		
Cold driven	60	—
N5−0		
Cold driven	55	—
Hot driven	50	50
Steel		
Power driven (shop)	100	95
Power driven (field)	95	95

Table 13.9 PERMISSIBLE STRESSES FOR WELDS

Alloy	*Butt welds*		*Fillet welds*	
	Tension N/mm^2	*Shear* N/mm^2	*Transverse* N/mm^2	*Longitudinal* N/mm^2
N4	50	30	55	30
N8	80	50	75	45
H9	30	20	55	30
H30	50	30	55	30
H17	125	75	70	45

The above values only apply if the appropriate filler material is used (Table 13.5).

Joints

Riveted and bolted joints are designed for shear and bearing, as in steel. Friction is ignored. Allowable stresses are given in Table 13.8.

Permissible stresses for welded joints are given in Table 13.9. They assume that the correct filler wire has been used, depending on the parent alloy (Table 13.5).

Fatigue

The performance of aluminium in fatigue is poor, and for design purposes can be assumed to be the same for all structural alloys. CP 118 gives comprehensive data on fatigue strength, set out the same way as in the steel codes. Here it will suffice to give some typical values. These are listed in Table 13.7 and are permissible values to which no additional safety factor need be applied.

Permissible deflection

The design of aluminium beams often tends to be governed by deflection, because of the low E. It is therefore advantageous to use continuity when possible. It is also sensible to work to larger permitted deflections than those customary in steel, provided they do not impair the strength, function or appearance of the structure.

CP 118: 1969 suggests the following deflections for buildings:

Beams carrying plaster finish	L/360
Purlins and sheeting rails Dead load only	L/200
Worst combination of all loads	L/100
Curtain wall mullions	L/173
Members carrying glass direct (where L is the span in metres)	$0.36L^2$ mm

BIBLIOGRAPHY

TIMOSHENKO, S. P and GERE, J. M. *'Theory of Elastic Stability'*, McGraw Hill (1961)
'Aluminium in Building' *Proc. of Symposium, Aluminium Federation*, (July 1959)
'Aluminium in Structural Engineering' *Proc. of Symposium, Aluminium Federation*, (1963)

14 LOAD-BEARING MASONRY

LOAD-BEARING MASONRY 14

14 LOAD-BEARING MASONRY

R. J. M. SUTHERLAND, B.A., F.I.C.E., F.I.Struct.E., M.I.H.E., M.Cons.E.
and D. Y. PRICHARD, M.I.C.E.
Harris & Sutherland

INTRODUCTION

The rapid progress over recent years, both in Great Britain and abroad, in the understanding of the materials and the considerable advances in the methods of design have led to the increasing acceptance of load-bearing masonry as a viable structural engineering material.

Up till now masonry has been designed almost universally throughout the world on the basis of actual loads and permissible stresses, the safety factor being incorporated in the stresses. In Britain the Code of Practice CP 111 (published by the British Standards Institution) is being redrafted in limit-state form to bring it into line with the Unified Code for Concrete. It is not possible to write the design sections of this Section to line up with the limit-state version of CP 111 because, at the time of writing, this is still some way off from publication. The only alternative is to discuss structural design within the framework of the present code noting that many references will eventually be out of date. The disadvantages of this course are not so great as they may seem to be at first because the amendments to CP 111 of 11 June 1971 cover many of the useful revisions agreed for the new issue of the complete code.

The existing CP 111 is often criticised because it is confined to load-bearing walls. It specifically excludes panel walls (i.e. walls subject to wind but otherwise unloaded) and gives no guidance on the treatment of masonry beams and slabs nor on earth pressures, explosive forces and overall stability. In its treatment of reinforced masonry it is particularly inadequate compared with the codes issued in the USA and elsewhere. These subjects will all be better covered in the new CP 111, together possibly with a separate code on reinforced masonry. However, as the following will show, it is perfectly possible to design high-performance structures both in brickwork and concrete blockwork—with or without reinforcement—within the limits of the present British codes and regulations.

MATERIALS

Bricks

DEFINITION

A brick is defined as a walling unit not exceeding 337.5 mm length, 225 mm in width or 112.5 mm in height. They may be of clay, calcium silicate or concrete, as specified in BS 3921, BS 187 and BS 1180 respectively. A survey of brick materials is given in CP 121.101.

CLAY BRICKS

Clay bricks are manufactured by pressing or extruding the basic unit and then firing it in a kiln. Although generally dimensionally stable, clay bricks do undergo a slight change in size (see **14**.8).

The bricks are defined in BS 3921 under the following headings: *variety,* which grades the bricks according to their use under three general headings of 'common', 'facing' and 'engineering'; *quality,* which qualifies the durability of a brick under three headings 'internal quality', 'ordinary quality' and 'special quality'; *type,* which distinguishes bricks according to their internal form under four categories 'solid', 'perforated', 'hollow' and 'cellular'.

The high compressive strength of engineering bricks is rarely required in practice, and these bricks are usually only specified where a high resistance to moisture penetration is required, i.e. below damp-proof courses and in manhole construction, etc. A list of manufacturers may be obtained from the British Engineering Brick Association.

CALCIUM SILICATE BRICKS

These bricks are made from mixtures of either sand and/or siliceous gravels with lime, and depending upon the raw material are often referred to as 'sandlime' or 'flintlime'. The mixture of raw materials is heated under steam pressure resulting in the formation of hydrated calcium silicate. These bricks are generally less stable dimensionally than clay bricks (see **14**.8).

They are classified according to their strength and drying shrinkage in table 2 of BS 187. Appendix E of the same Standard gives recommendations on the use of each class in relation to the constructional position and purpose of the brick, and the appropriate mortars for use with them. The compatability of the mortar with the class of brick is particularly important in relation to the drying shrinkage, as it has been found that with bricks of high shrinkage under conditions of restraint (i.e. as infill panels between frames) there is a real risk of cracking the bricks if strong cement mortars are used.

CEMENT BRICKS

These bricks are made from a mixture of cement and inert aggregate (either natural or man-made). They are classified according to their size and strength in BS 1180, with specific limits for drying shrinkage for each category of strength.

Recommendations on the use of concrete bricks is given in Appendix E of BS 1180, including advice on the choice of mortar for various situations and degrees of exposure.

SIZE

Following the change to the metric system, the present format for standard bricks is 225 mm × 112.5 mm × 75 mm (actual: 215 × 102.5 × 65 for clay and 215 × 103 × 65 for calcium silicate) which is 1.6% smaller than the old imperial format 9 in × 4½ in × 3 in (or 228.6 mm × 114.3 mm × 76.2 mm). In deriving the coursing with metric bricks, the mortar joint is 10 mm.

In the 'Foreword' to the metric versions of both BS 3921 and BS 187 the fact that the metric format for bricks does not comply with the moves towards dimensionally co-ordinated sizes is acknowledged, thus both standards were issued as interim measures. BS 1180, on the other hand, publishes tables for both the metric format and modular metric bricks. Dimensionally coordinated (or modular bricks) are manufactured in the sizes given in Table 14.1.

The change to modular bricks requires a new approach to bonding, particularly with the 300 mm bricks which require the use of 'third bond', thus posing problems at piers, returns, etc. The dilemma that a 90 mm-wide brick is less than the 100 mm requirement of certain of the 'Deemed to Satisfy' provisions of the Building Regulations (e.g. sound insulation) has, at the time of publication, to be resolved; in the meantime the designer is obliged to demonstrate that their use is satisfactory.

STRENGTH

The classification of bricks according to their strength is given in Table 14.2.

CHOICE OF BRICK

The choice of a brick will depend on the desired behaviour in use with respect to strength, appearance, durability and the effect that the various properties of bricks will have on its performance (expansion, shrinkage, efflorescence, etc.).

Table 14.1 MODULAR BRICK SIZES

Format mm	*Actual size* mm
200 × 100 × 75	190 × 90 × 65
200 × 100 × 100	190 × 90 × 90
300 × 100 × 75	290 × 90 × 65
300 × 100 × 100	290 × 90 × 90

Table 14.2 CLASSIFICATION OF BRICK STRENGTHS

Average compressive strength (MN/m²)	*Clay to* BS 3921	*Calcium silicate to* BS 187	*Concrete to* BS 1180
69.0	A Engineering		
48.5	B Engineering		
103.5	15 Load-bearing		
69.0	10 Load-bearing		
48.5	7 Load-bearing	7 Load-bearing	Separate
34.5	5 Load-bearing	5 Load-bearing	approach to
27.5	4 Load-bearing	4 Load-bearing	classification,
20.5	3 Load-bearing	3A, 3B Load-bearing	strength range
14.0	2 Load-bearing	2A, 2B Load-bearing	from 7.0 to
7.0	1 Load-bearing	1 Load-bearing	40.0 MN/m²

Guidance on the selection of clay bricks is given in Building Research Station Digests 65 and 66.

A list of manufacturers, classifying the material format, size, strength and variety of bricks produced has been published.[1]

Concrete blocks

DEFINITION

A block is defined as a walling unit exceeding the dimension specified for a brick. In this country load-bearing blocks are generally of pre-cast concrete manufacture, as defined in BS 2028, 1364. A variety of proprietary insulation blocks, manufactured from sintered ashes, etc., is also produced but their use is limited in load-bearing construction because of their relatively low strength.

In BS 2028, 1364, blocks are distinguished according to their internal form (i.e. 'solid', 'hollow', 'cellular', 'aerated'), classified according to their intended use (i.e. type A, B or C), and finally according to their surface treatment. Further information on the materials and manufacture of concrete blocks is given by Gage and Kirkbride[2] and others.

SIZE

Unlike bricks, the change to metric has been readily accepted with blocks, so that dimensionally coordinated block sizes are now specified in the amended table 1 of BS 2028, 1364 (amended January 1970).

STRENGTH, DENSITY AND DRYING SHRINKAGE

The various requirements of BS 2028, 1364, for each type of block in respect of strength, density and drying shrinkage are usefully correlated in table 1 of the Cement and Concrete Association publication, *Concrete Blocks to British Standard Specification.*

CHOICE OF BLOCK

The choice of block will depend upon:

the intended use
the desired strength } determines block type A, B or C
the limit on shrinkage
and the surface characteristics

In considering the intended use it is important to note that both the density and the maximum permitted drying shrinkage are minimum standards, and the designer may consider it necessary for a particular application to lay down higher standards for these characteristics.

The categories of special faces given in clause 10 of BS 2028, 1364 apply more particularly to blocks with a deliberately featured surface. This clause does not cover internal blocks which require a fair face suitable for painting or to be left untreated. Where such a surface is desired, the specification must either be worded to this effect or a known block nominated.

A list of block manufacturers has been published[3] which gives blocks according to type, size, strength, thermal value, etc.

Natural stone and reconstructed stone

The use of natural stone and reconstructed stone is covered by CP 121.201.

Mortar

TYPES

The various types of mortar are:

Lime mortar—a mixture of lime and sand
Cement mortar—a mixture of cement and sand
Cement–lime mortar—a mixture of cement, lime and sand
Air-entrained (plasticised) mortar—a mixture of cement and sand with a plasticiser added to the mixing water
Masonry cement mortar—a mixture of cement and sand, but with the cement mixed at manufacture with a very fine mineral filler and an air-entraining agent (in powder form)

A fuller description of each type is given in Building Research Station Digest 58 (revised 1970), while advice on the properties of different mortars is given in SCP 1.[4]

Mortars of equivalent strength are designated by five grades, but as grade V is not used for load-bearing work, it is not classified with a strength in Table 14.3. A comprehensive list of all mortars and their proportions is given by Gage and Kirkbride.[5]

STRENGTH OF MORTARS

It has been found, notably with weaker units, that an increase in mortar strength does not lead to a proportional increase in wall strength. This fact is reflected in the basic stress given in tables 3A and 3B of CP 111 (i.e. for units of 7.0 MN/m^2 and less the same basic compressive stress is given for the use of Grades 1, 2 and 3 mortars).

As with concrete, mortar strength increases with time, the cement mortars doing so more quickly than the nonhydraulic lime mortars. It has been shown that cement mortars more than double in strength between the first and twelfth months. The strength of a mortar is also affected by the w/c ratio.

Gage and Kirkbride[6] have suggested minimum compressive stresses at 28 days for different grades of mortar, as shown in Table 14.3

Table 14.3 COMPRESSIVE STRESSES FOR MORTARS

Mortar designation	*Proportions*	*Minimum compressive stress at 28 days* (MN/m^2)
I or 1	$1:\frac{1}{4}:3$	11.0
II 2	$1:\frac{1}{2}:4\frac{1}{2}$	5.5
III 3	$1:1:6$	2.75
IV 4	$1:2:9$	1.00

Note: the above proportions are for cement : lime : sand and will be different where either plasticisers or masonry cements are used (see above).

SELECTION OF A MORTAR

General guidance on the choice of mortar is given in tables 1 and 2 of Building Research Station Digest 58 (revised November 1970); while comprehensive recommendations for the choice of mortar with calcium silicate and concrete bricks are referenced on pages **14**–3.

Not only is no advantage to be gained by using a strong mortar with the weaker masonry units, but there is a very definite disadvantage if the mortar is stronger than the units, as this can lead to the cracking of the unit if settlement or shrinkage movements occur. As a general rule, therefore, the mortar should always be weaker than the unit so that any cracks will tend to form in the mortar joints where they are less conspicuous, although the aim of good design is to avoid even these cracks!

Reinforcement

TYPES

There are three principal types of reinforcement used in masonry construction, namely; bar, wire and expanded metal. The first two may be designed to act compositely with the masonry, but expanded metal is only used as a means of controlling cracking.

COVER

Recent tests have shown that the covers given in clause 323 of CP 111 may be inadequate. Where the cover is less than 100 mm, consideration should be given to galvanising reinforcement in masonry exposed to weather or against the ground.

Ties

TIES IN CAVITY WALLS

The requirements on the spacing and use of ties in cavity construction are given in CP 111, and BS 1243 specifies the three types of metal tie to be used in cavity construction.

The purpose of wall ties is to share lateral forces and deflections between the two leaves and to provide a stiffening effect for the support of vertical loads when either one or both leaves are loaded; further information can be obtained in Building Research Station Digest 61 (reprinted 1970).

The strength of ties and other fixings is discussed on page **14**–10.

The question of corrosion and fire resistance should also be examined in the choice of tie.

Damp-proof courses

TYPES

Materials for damp-proof courses are specified in BS 743.

SPECIAL CONDITIONS

Where conditions of high stress occur, it is important that the damp-proof course is capable of withstanding the load, as discussed by Plowman.[7]

PROPERTIES OF MATERIALS

Movement in masonry

GENERAL

Movement in masonry construction varies with the type of material and its mode of manufacture, fluctuations in the ambient conditions, the aspect, the choice of mortar and the influence of adjacent structural elements.

MOVEMENT DUE TO THE PROPERTY OF THE MATERIAL

While kiln-fired clay and shale bricks exhibit a permanent moisture expansion after manufacture, concrete blocks and calcium silicate and concrete bricks exhibit a shrinkage following manufacture.

In the case of kiln-fired bricks, this moisture expansion starts as the bricks commence to cool soon after firing in the kiln. Thereafter the process continues with time and exposure, but an experiment[8] has shown that a significant percentage of this movement takes place within the first 14 days after manufacture and a storage period of at least this time is therefore recommended.

While the shrinkage effect of calcium silicate and concrete bricks and concrete blocks may be reduced by allowing a storage period of at least 14 days after manufacture (in dry conditions) it is still necessary to consider the need for joints in the permanent structure, particularly if the drying shrinkage exceeds 0.02%. A 10 mm joint every 6–9 m for calcium silicate bricks and every 6 m for concrete blocks and bricks is about the order which is generally recommended, although it appears that shape is more critical than size and that cracking is more likely to occur if the length of panel exceeds about $1\frac{1}{2}$ to 2 times its height.

THERMAL MOVEMENT

CP 121.101 quotes the linear expansion of bricks as lying between 3 and 6×10^{-6} per deg F. This is probably a reasonable range for most clay bricks, but the respective values for calcium silicate and concrete bricks and concrete blocks may well be higher. However, with these latter units, the spacing for joints given for shrinkage should cater for expansion. For most clay bricks, the frequently quoted figure of 12 m for joint spacing probably errs on the safe side and greater spacings have been successfully used. This matter is very much one of judgement in which the aspect of the wall (south-facing walls can expand up to 30% more than north-facing ones[9]), the position of openings, and points of restraint (bonding to piers and other walls), all need to be taken into account.

DIFFERENTIAL MOVEMENT

It is important to consider the differing movement characteristics when it is proposed to use dissimilar materials in one wall (e.g. clay bricks and concrete blocks). Rigid bonding should be avoided and cavity construction adopted, see page **14–9**.

JOINT FILLERS

The material for filling expansion joints must be readily compressible and proprietary fibre boards (traditionally used in the joints of concrete construction) should not be used. A number of suitable materials, such as extruded closed-cell rubber or plastics (polyurethane and polythene), are available, but most require the use of a sealant at the external face.

Efflorescence, staining and chemical attack

EFFLORESCENCE AND STAINING

Efflorescence and staining of brickwork is reviewed by Butterworth.[10]

SULPHATE ATTACK

Sulphate attack and its prevention is discussed in Building Research Station Digest 89 (second series).

Thermal insulation

The thermal transmittance of various wall constructions is given by Beard and Dinnie.[11]

Sound insulation

The basic principles are reviewed in Building Research Station Digest 143, and the sound insulation achieved by various types of wall construction is given in appendix D of CP 3, Chapter III.

Fire resistance

Some guidance on the fire resistance of load-bearing walls is given in CP 111 and further in Joint Fire Research Organisation Fire Note No. 6 'Fire resistance of brick and block walls'.

In the case of concrete blocks it is necessary to consider the choice of aggregate; aggregates are divided into two classes, those in the first being more effective than those in the second.

Influence of adjacent structural elements

MASONRY PANELS

Masonry panels which are bounded by structural frames of concrete or steel require special attention in the detailing of the junctions between the two elements, since movements within the frame due to thermal changes, shrinkage and creep can lead to considerable forces being applied to the panels.

This is particularly relevant in multi-storey reinforced-concrete (RC) frame buildings, where there have been several examples of severe cracking, especially where brick slips have been used to mask the horizontal edge of the RC element, leading to the actual spalling of the brickwork. It is apparent that in such cases the vertical shrinkage and creep of the concrete, perhaps coupled with the contrary expansion of the masonry due to thermal and moisture movement, has led to the buckling of the brick skin. Measures to overcome this by means of a compressible joint between the top of each storey of masonry and the underside of the structure are described by Foster.[12]

The possibility of differential movement between the two leaves should be also considered. CP 111 gives some guidance on this matter, laying down certain limits on panel sizes. The choice of tie can also play a part here in that the butterfly type will provide greater facility for differential movement, but will be less effective than the rigid twist tie in sharing loads between the two leaves.

BOUNDARY FIXINGS OF INFILL PANELS

Two situations arise; one in which the masonry panel is located between the structural frame resulting in a shear loading on the fixing; and the other in which the panel is located in front of the frame resulting in a tensile or compressive loading. The results of tests and recommendations for the shear and tensile capacity of various types of fixings are given by Thomas.[13]

ON LOAD-BEARING MASONRY WALLS

Where the movement of RC slabs (notably roofs) and beams supported on masonry walls has not been allowed for, there have been many incidents of cracking. To a large extent, this problem with RC roof slabs can be overcome by the provision of sliding joints between the top of the wall and the underside of the slab around the perimeter (Figure 14.1). In the case of long roofs, consideration should be given to subdividing the roof into a series of 'islands' by means of movement joints (Figure 14.2).

Another frequent trouble spot occurs where lintels are built into blockwork. As far as possible these should be avoided by using storey height openings. If this is not desired, the provision of expanded metal mesh in the bed joints over the opening can alleviate the problem.

STRUCTURAL DESIGN OF UNREINFORCED MASONRY

Codes

For the reasons given in the introduction to this Section, the future limit-state version of CP 111 will be ignored in the following. The existing code is likely to be in use for several years and, since most masonry building in Britain must comply with it, frequent references will be made to its requirements. However, CP 111 is not mandatory in all fields even in Britain and compliance with it does not in itself ensure structural safety. Designers are urged always to consider engineering principles first and codes second.

Vertical (axial) loading

PERMISSIBLE STRESSES

In Britain permissible compressive stresses are derived by applying modification and reduction factors to basic stresses which are dependent on brick or block unit strengths and mortar mixes. The basic stresses for brickwork and concrete blockwork are found in Clause 315 of CP 111. They are plotted in Figure 14.3 in relation to available strengths of brickwork and solid blockwork using the strongest mortar ($1:\frac{1}{4}:3$); the basic stresses will be correspondingly weaker for cement lime or lime mixes.

Instead of just accepting the concept of permissible stresses it is useful if not essential to have some idea of how masonry initially responds to loading and how it will ultimately fail. Failure is seldom caused by crushing of the material but generally by vertical splitting both within the plane of a wall or perpendicular to it, the former more often being the direct cause of collapse. The failure mechanism in its simplest form is well described by Hilsdorf.[14]

Within the permissible compressive stress range which generally gives factors of safety of 5 or more (see **14**–26) all types of masonry (clay, calcium silicate or concrete) can be assumed to behave elastically with a rectangular, trapezoidal or triangular stress distribution across the section, the following factors affecting the strength of a wall or column in addition to unit strength and mortar mix.

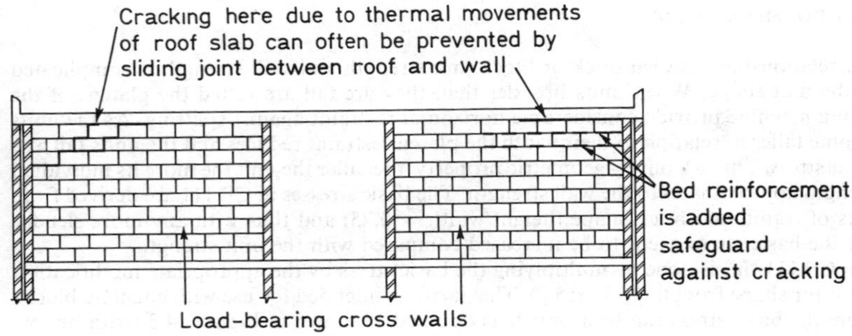

Figure 14.1 Typical cracking below a concrete roof slab

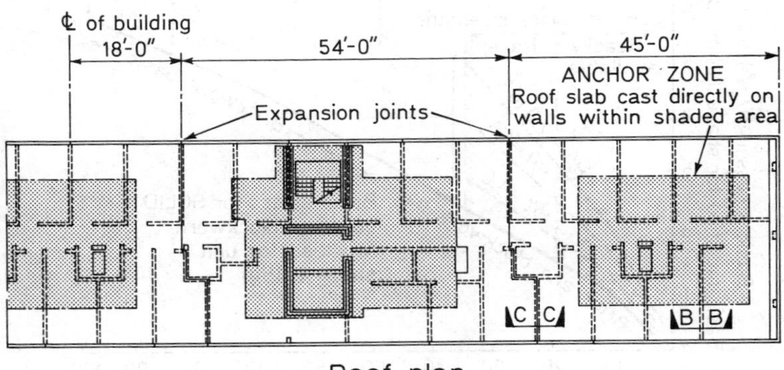

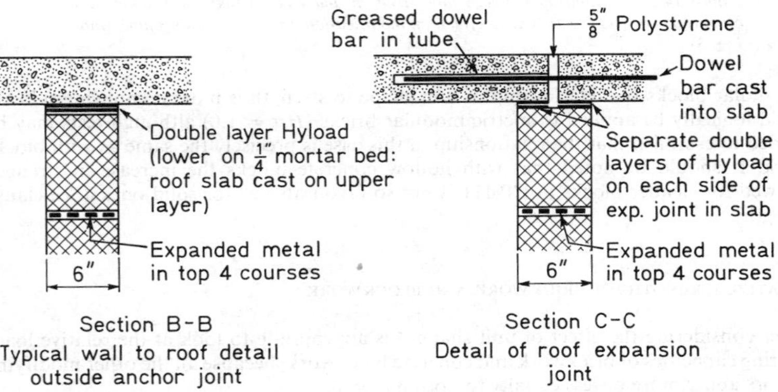

Figure 14.2 Division of reinforced-concrete roof into islands

EFFECT OF SHAPE OF UNIT

The relationship between brick or block unit strength and wall strength is complicated by the unit shape. When units broader than they are tall are tested the platens of the testing machine provide considerable horizontal restraint against splitting. As the units become taller in relation to their width the platen restraint reduces and the units fail at a lower stress. This is a purely geometric property, the taller the unit the more its individual strength approximates to the wall strength. The basic stresses in CP 111 are derived from units of traditional brick shape (height/width = 0.75) and thus with any more slender unit the basic stress needs to be increased compared with the unit strength.

In CP 111 this is done by multiplying the basic stress by the appropriate modification factor for shape from table 5 (315 g). This factor is intended for use with concrete blocks where the basic stress can be as much as doubled as shown in Figure 14.3 (with hollow

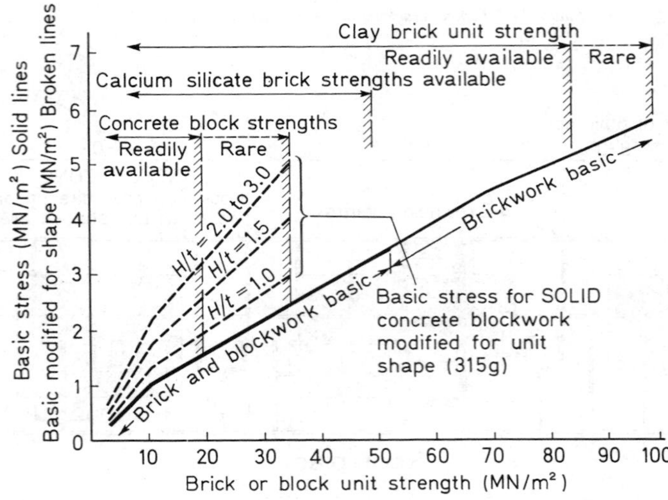

Figure 14.3 Relationship between unit strength and basic stress for brickwork and blockwork (CP111: 1970) and effect of modification factor for unit shape (mortar 1:¼:3)

or cellular blocks the relationship of unit shape to strength is more complicated) but it could logically be applied to metric modular bricks (H/t = 1.0) although tests may be needed to establish that the relationship in this case is precisely the same as with blockwork. It should be noted that with hollow concrete blocks the increase in strength allowed for slender blocks in CP 111 is not so favourable as for solid ones. (See clause 315 g(ii).)

RELATIVE STRENGTHS OF BRICKWORK AND BLOCKWORK

After considering the effect of unit shape it is appropriate to look at the relative load-bearing capacities of brickwork and concrete blockwork, because all the other modifying factors apply more or less equally to both materials.

As Figure 14.3 shows, there are many makes of clay brick with unit strengths about four times as great as those readily available for concrete blocks. However, with concrete blocks the shape factor helps and can as much as double the effective basic stress although on strength clay brick still wins. The question is whether this is significant or not.

Even in tall structures, tension or limiting slenderness tend to rule rather than compressive stress, and as a result brick strengths over 50 MN/m^2 are seldom needed, although bricks which happen to be stronger than this are often chosen for other reasons. With concrete blockwork, strength is more of a limitation but for the vast majority of structures where brick strengths of 20 to 35 MN/m^2 are adequate, concrete blockwork is comparable in performance.

EFFECT OF SIZE OF WALL OR COLUMN

In large walls the presence of a few weak bricks or blocks is of no significance; further there is no way of rejecting these because the relevant British Standards lay down minimum average strengths for test batches rather than absolute minima. As walls or columns become smaller the effect of weak units becomes more serious and to guard against this CP 111 (315b) gives a reduction factor which cuts the basic stress by up to about 22%. This should cover most variations but designers might also be wise to check more closely on the unit strengths where very small independent columns are used.

EFFECT OF SLENDERNESS

In common with all other materials, compressive stresses in masonry need to be reduced (and overall limits set) to guard against buckling as walls and columns become more slender. In CP 111 a table of reduction factors is given (table 4 in 315c) for a range of slenderness ratios up to 27, depending on the type of structure (see clause 309). The slenderness ratio is defined as the effective height (or length) divided by the effective thickness. The terms height, length and thickness in this context are defined in clauses 304 to 307 of CP 111.

EFFECT OF BONDING PATTERN

Tests on brickwork show that for the same brick strength, mortar type and slenderness, single-skin walls fail at higher stress than thicker ones in say Flemish or English bond. This is understandable bearing in mind the normal failure pattern due to splitting in the plane of the wall, but no allowance is made for this fact in CP 111 at present, the basic stresses being appropriate for the worst condition for 'solidly' bonded walls. With concrete blockwork the difference in strength would presumably be the same, but the multiskin bonding condition seldom arises.

With cavity walls in brick or block the slenderness lies between that of a single skin and a solidly bonded wall using the same amount of material (as discussed in **14**–7); CP 111 covers this by making the effective thickness of cavity walls equal to two-thirds of the total material thickness.

Distribution of vertical loading on masonry walls

This is simple enough to calculate with one-way floor slabs of uniform span, but becomes complex with two-way slabs and irregular layouts. The simplest analytical approach is to divide up the slab into triangular or trapezoidal areas and apply the resulting loads to the walls as with beam and slab construction.

Eccentricity of vertical loading

There is a wide gulf between theory and current practice in the treatment both of eccentric loads on masonry walls and of fixity at the junctions of walls and slabs. Floor and roof loadings are generally considered axial unless they are obviously eccentric. Figure 14.4 taken from Building Research Station Digest No. 61 shows assumptions which have been current for some time. Although as noted in the Digest these were intended to apply to two- or three-storey house construction, they have been used repeatedly and successfully in recent years for load-bearing masonry buildings of 15 storeys or more.

In accordance with simple elastic theory these assumptions are not strictly tenable. For instance in diagram (a) of Figure 14.4 the end rotation of the slab due to its deflection —ignoring for the moment any load from the wall above—would move the reaction away from the centre of the wall thus applying a bending moment to it. Further, if the load from the wall above is great enough to clamp the slab and prevent end rotation this moment may well become very much increased. The reason why theory has been ignored so widely without trouble is not just that masonry does not behave wholly elastically but that, even in high-rise building, most applications have been domestic in scale, generally with two-way slabs of modest spans and thus small deflections.

With most lead-bearing masonry there is nothing to be gained from calculating stresses due to fixity, which at best can seldom be done with any accuracy. However, it is always worth taking reasonable precautions to minimise bending moments in masonry due to any cause. The following points are worth considering.

(i) Design floors and roofs for the minimum deflection especially due to dead load. Where economic, prestressed is better than reinforced concrete and two-way spans better than one-way.
(ii) Allow the maximum proportion of the elastic and creep deflections to take place before the slabs are fully bedded and clamped at these ends.
(iii) Avoid building in deep edge beams as any rotation of these is particularly disruptive to masonry walls.
(iv) Do not apply loads eccentrically (on corbels or as in diagram (b) of Figure 14.4) if this can be avoided.

Fixing and eccentricity are normally purely local in their effects and seldom cumulative. Thus in multi-storey construction they just cause a constant ripple on the downward-increasing axial stress in the walls. If the ripple is large, it is likely to be more serious at the upper levels where it can cause cracking than lower down where it may or may not lead to overstressing. CP 111 allows a 25% increase in permissible stress due to eccentricity across the wall but, even if this increase is exceeded in theory, long-term creep will tend to reduce any peak of stress to a level compatible with the materials used.

While experience shows that eccentricity and fixity may cause only minor problems in 'small cell' buildings with two-way spans, they can become much more critical where long single-span roofs or floors rest on tall walls as in warehouses, transit sheds or factories. Here there is not only a danger of cracking due to eccentricity or fixity but even of these causing buckling of the walls. With such structures the rather cavalier approach of ignoring fixity is clearly out of place and a more rigorous analysis of stress limits may be needed. However, in such cases the calculations tend to be simpler and the results mean more than with complex domestic layouts.

Lateral loading (wind)

In masonry structures lateral loads (wind mainly in Britain) are usually transferred to shear walls by floor or roof slabs acting as horizontal plates. Each wall resisting the lateral force must then be checked, both to see that the average shear stress is not excessive and that the variation in loading along the wall is not great enough either to cause tension

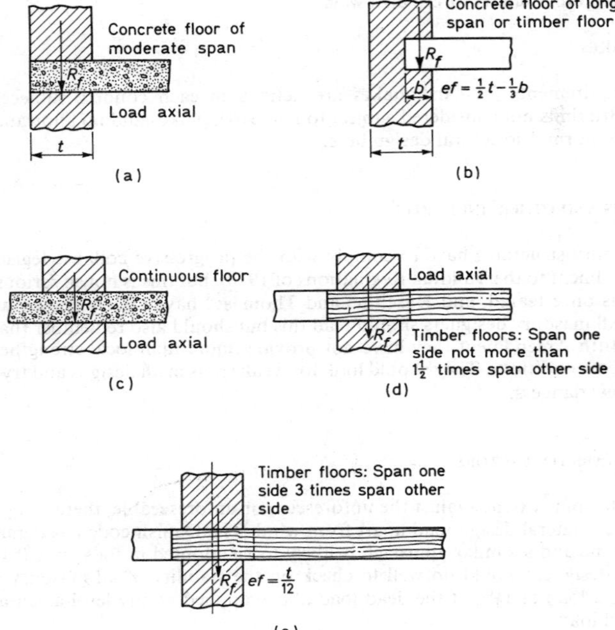

Figure 14.4 Axial and eccentric loading on solid walls. ef is the eccentricity of the reaction of floor load or loads (Reproduced from BRE Digest 61 'Strength of brickwork, blockwork and concrete wall' with the permission of the Controller of Her Majesty's Stationary office)

or bring the compressive stress more than 25% above that permissible for axial loading (see **14**–16).

Calculation of the distribution of lateral loads between shear walls is really outside the scope of this Section. It is perhaps enough to say that in complex structures where the distribution is not obvious the connected cantilever method may be used. In this the wall units are considered as separate vertical cantilevers all connected together by notionally hinged links at each floor level. The lateral load is then distributed between them in proportion to their stiffnesses and the resulting stresses calculated. If this method leads to apparent tensions, a more sophisticated composite action study can be carried out usually by computer. It is, of course, only necessary to find enough resistance to match the lateral forces without distress; if two parallel walls alone will suffice there is no virtue in considering any more.

Wind forces acting on the faces of individual load-bearing walls are seldom serious because the downward load on these is usually ample to ensure stability due to vertical arching alone. It is with unloaded panel walls usually in framed buildings where the wind stability problems arise. Many of these stand up which theoretically should fail and an extensive test programme is now in hand to establish a design procedure for such walls. Pending publication of the results of this, designers are advised to detail panel walls so that they can arch vertically or horizontally (or both) or to reinforce them (see 'Reinforced masonry').

Stability and other lateral forces than wind

EARTHQUAKES

Design requirements for earthquakes are defined in each country subject to major tremors. Britain is not considered subject to real earthquakes and no allowance is made for these in normal structural design here.

EXPLOSIONS AND OTHER 'INCIDENTS'

Most masonry structures have to comply with the progressive collapse legislation (the fifth Amendment to the Building Regulations of 1970), but this is not so serious a restriction as was once feared, and Haseltine and Thomas[15] have clearly shown what to do about it. All masonry designers should read this but should also remember that by satisfying the fifth Amendment they have not provided more than local strengthening for a limited set of conditions. They should look for weak spots in all designs and try to foresee possible new dangers.

OVERALL STABILITY FACTORS

To guard to some extent against the unforeseen, or unforeseeable, there is a strong case for minimum lateral design load apart from wind. The Danish code has demanded this for some time and a similar approach is likely to be adopted in the new CP 111. In the meantime designers would do well to check on the stability of all masonry structures using, say, a load of $1\frac{1}{2}\%$ of the dead load above applied at any level as an alternative to the wind load.

Combination of eccentric and lateral loading

Where eccentricity and lateral loading (whether from wind or any stability factor) are combined the worst combination should cause neither tension nor a compressive stress more than 25% in excess of that permitted for axial loading.

Design procedure for masonry structures

Because of widespread confusion over the interpretation of CP 111 and the unfamiliarity of many engineers with the properties of masonry it has been found that a definite sequence of working is useful in the design office. Such a sequence is given in Table 14.4. Some people experienced in this field may wish to add to it or modify it. However, it is suggested that this or a similar scheme is used for checking, even if not for design.

Particular attention is drawn to the dangers of combining materials with different properties in the same wall, for instance clay and concrete (see **14**–8) and to the avoidance of cracking due to thermal movements and shrinkage.

REINFORCED MASONRY

General

Until recently reinforced masonry has scarcely been taken seriously at all in Britain although in the USA and elsewhere it has been a well established and strong competitor to reinforced concrete for many decades. It is particularly popular in earthquake areas like California and New Zealand. With no earthquakes in Britain it has not generally been necessary to reinforce walls and this lack of need may well explain why reinforced masonry is still a rarely used structural technique here.

In the last few years interest in reinforced masonry has grown. Extensive testing is being carried out both on reinforced brickwork and blockwork and work has started towards producing a new code for reinforced masonry rather on the lines of the American SCPI one but on a limit-state basis and, it is hoped, with appreciably higher stresses in flexure than those permitted in CP 111.

What is not generally realised is that in spite of the inadequacies of CP 111 even within its limits the scope for reinforced masonry is large and so are the economic advantages.

Permissible stresses in reinforced masonry

The range of permissible stresses for different materials and unit strengths according to CP 111 is shown in Table 14.5. From this it will be seen that the effective maximum flexural stress is just over 4.5 MN/m^2 which although appreciable is little more than one-third of that permitted in the USA. Current tests at the Building Research Establishment are confirming that the flexural stresses in CP 111 are in practice considerably too low.

Basis of design

CP 111 states that reinforced masonry should be designed using the principles for elastic design given in CP 114. However, in the case of masonry the modular ratio is taken as varying with the unit strength, ranging from 12 to 23 compared with a constant 15 for dense concrete (see tables 6a and 6b in CP 111).

Suitable masonry units and mortars

The limiting properties of bricks, blocks and mortars suitable for reinforced masonry are specified in clause 321h of CP 111 (amendment No. 1, June 1971).

Reinforcement

Permissible tensile stresses are defined in CP 111 and are virtually the same as those for reinforced concrete. Cover and precautions against corrosion are discussed on page **14**–7.

Direct compression

Logically enough the permissible direct compressive stresses are the same as those derived from the basic stresses for unreinforced masonry. By using steel in compression it is possible to increase the load-bearing capacity of a wall or column, the steel stress being taken as equal to the stress in the surrounding masonry multiplied by the appropriate modular ratio. In practice this gives little real benefit because, economics apart, it is almost impossible to fit in enough steel to give even a 10% increase in strength.

Table 14.4

Action		*Procedure*		
		Axial vertical load	*Eccentric vertical load*	*Lateral load and axial vertical*
Calculate compressive stress in actual wall or pier		Wall thickness, W; $\frac{W}{A}$ Stress (uniform) $= \frac{W}{A}$ Note: If load varies from one end of wall to the other consider conditions at the more heavily loaded end.	Wall thickness, W, e; $\frac{W}{A}+\frac{We}{Z_1}$; $\frac{W}{A}+\frac{We}{Z_1}$	Length of wall unit; Moment due to lateral load M_2; $\frac{W}{A}-\frac{M_2}{Z_2}$; $\frac{W}{A}+\frac{M_2}{Z_2}$ Note: The same procedure applies for a moment M_1 due to lateral forces perpendicular to wall face except that stress $= \frac{W}{A} \pm \frac{M_1}{Z_1}$
			Max. calculated stress $= \frac{W}{A}+\frac{W.e}{Z_1}+\frac{M_1}{Z_1}+\frac{M_2}{Z_2}$ Min. calculated stress $= \frac{W}{A}-\frac{W.e}{Z_1}-\frac{M_1}{Z_1}-\frac{M_2}{Z_2}$	In the *unlikely* case of all occurring at the same point
Calculate permissible compressive stress	(1) Select appropriate basic stress	See tables 3a and 3b in CP111 (315b)		
	(2) Check for shape of unit	Multiply basic stress by modification factor (M_f) for shape of unit (CP111:315g)		

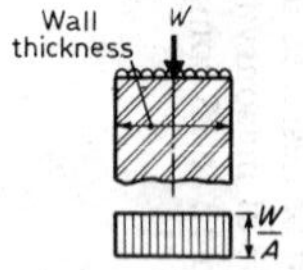

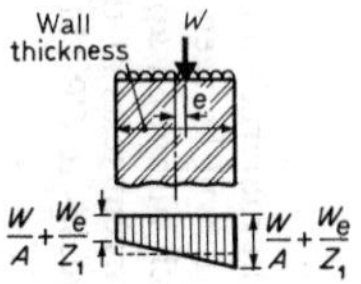

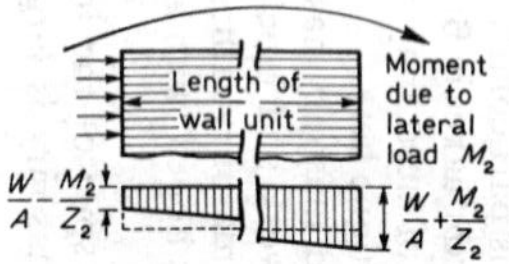

slenderness		Determine effective thickness of wall or pier (CP111:307 or 308) Calculate slenderness ratio $= \dfrac{\text{Effective height (or length)}}{\text{Effective thickness}}$ Check that slenderness ratio is not above the maximum permitted (CP111:309)	
		Multiply basic stress by 'axially loaded' reduction factor for calculated slenderness ratio (R_{f2a}) (CP111:315c)	Multiply basic stress by reduction factor for calculated eccentricity and slenderness ratio (R_{f2e}) (CP111:315d)
Check whether wall or pier is acceptable for compression	(1) Accept if:	Ignoring eccentricity and lateral loads calculated compressive stress $\leqslant$ basic $\times M_f \times R_{f1} \times R_{f2a}$	
	(2) And if:		With worst combination of eccentricity and lateral loading (a) Max. calculated stress $\leqslant 1.25 \times$ basic $\times M_f \times R_{f1} \times R_{f2e}$ (b) Min. calculated stress $\geqslant$ zero (i.e. no tension)
Make the following additional checks	(1) Shear (where applicable)	Divide maximum lateral force by the plan area of the wall and check that the resulting shear stress does not exceed the limits given in CP111:317.	
	(2) Lateral support	Check that each horizontal support to a wall or pier is capable of resisting $2\frac{1}{2}\%$ of the vertical load at that point as well as any horizontal forces (CP111:304a)	
	(3) Point loads	Check that the local stress under any concentrated load (girder bearings, etc.) does not exceed permissible compressive stress as calculated above by more than 50% (CP111:315e) but subject to the allowance given by 315f.	
	(4) Progressive collapse and overall stability	Check that the structure satisfies the Fifth Amendment to the Building Regulations (if applicable: see **14**–16) Try to assess the structure for overall stability: it is recommended that lateral forces used in the calculations above should allow for a minimum stability force (see **14**–16)	
	(5) Shrinkage thermal movement moisture expansion	See **14**–7 to **14**–8. Special care needed where (a) below concrete roof slab (see **14**–10) (b) clay, brick and concrete combined (see **14**–9)	
	(6) Chemical attack efflorescence staining	See **14**–8 to **14**–9. Make sure especially that precautions against sulphate attack are adequate	

Note: The factors R_{f1}, R_{f2} etc. are derived for the purpose of this chapter and are not from CP111.

Table 14.5 PERMISSIBLE FLEXURAL STRESSES (CP111) FOR DIFFERENT FORMS AND STRENGTHS OF MASONRY

Typical form	*Brick or block unit strength* N/mm²	*Permissible flexural stress* N/mm²	*Material*
I Vertically spanning brickwork			BRICKWORK (Flexural stress = basic × 4/3)
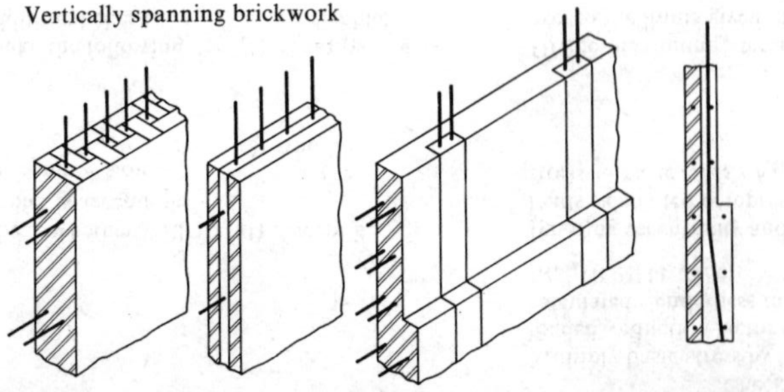 (a) Quetta bond (b) Grouted cavity (c) Pocket type Section through pocket	20.5 (Minimum suitable)	1.65 × 4/3 = 2.20	
	52.0 or above	3.5 × 4/3 = 4.65	

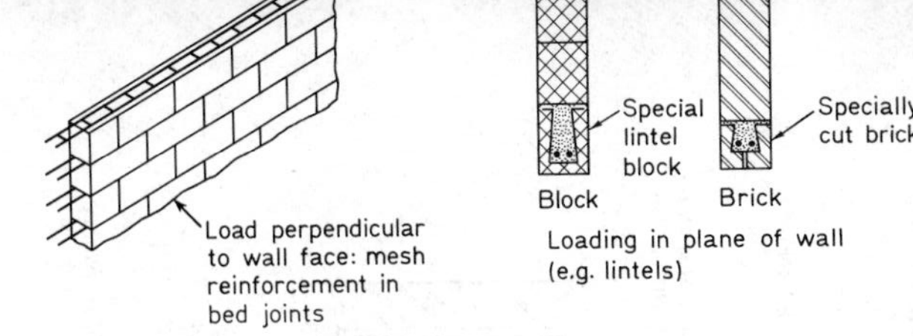

III Vertically spanning solid concrete blockwork

Grouted cavity type of wall as for brickwork (see above) most suitable

IV Vertically spanning hollow concrete blockwork

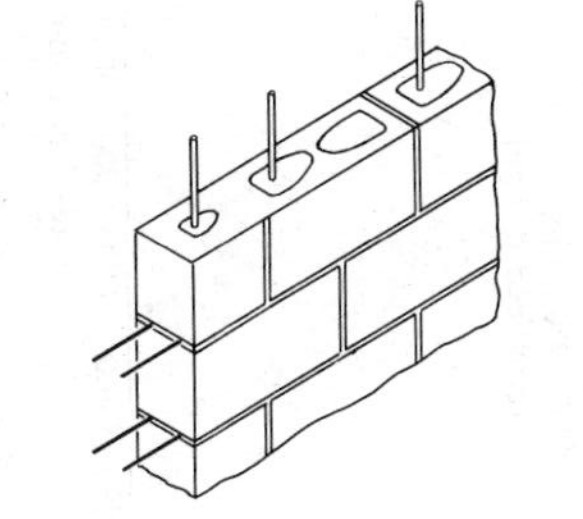

Reinforcement in concrete filled cavities in blocks
Mesh in joints as necessary

SOLID CONCRETE BLOCKWORK (Flexural stress = basic × shape factor × 4/3)

Unit strength		
5.5 (Min. suitable)	$0.55 \times 1.2 \times 4/3$	= 0.88
	$0.55 \times 2.0 \times 4/3$	= 1.46
21.0 (Max. readily available)	$1.7 \times 1.2 \times 4/3$	= 2.71
	$1.7 \times 2.0 \times 4/3$	= 4.52
35.0 (Max. in CP111)	$2.5 \times 1.2 \times 4/3$	= 4.0
	$2.5 \times 2.0 \times 4/3$	= 6.65

HOLLOW CONCRETE BLOCKWORK Flexural stress derived by linear interpolation between shape modified basic for 5.5 N/mm² units and unmodified basic for 21.0 N/mm² units multiplied by 4/3 in each case

Unit strength		
5.5 (Min. suitable)	$0.55 \times 1.2 \times 4/3$	= 0.88
	$0.55 \times 2.0 \times 4/3$	= 1.46
21.0 or above	$1.7 \times 4/3$	= 2.26

Flexure

Figure 14.5 shows the very small size of theoretical compressive stress block for the highest strength of masonry compared with that for a moderate strength of reinforced concrete; with most masonry units this stress block would be even smaller. This figure also shows what happens (again theoretically) if extra steel is added say to halve the tensile stress. With the reinforced concrete section this increases the resistance moment by about 20% while with the reinforced masonry one the area of the stress block is nearly doubled with only a small reduction in lever arm; this gives an increase of some 60% in resistance moment for no change in material or wall thickness. Figure 14.6 shows a typical relationship between steel area, steel stress and resistance moment for a constant

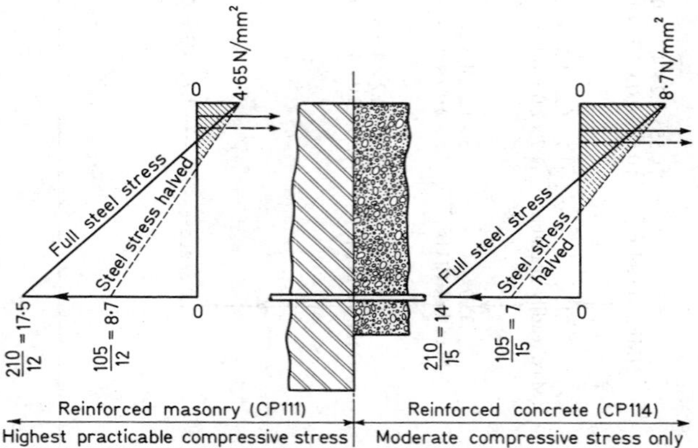

Figure 14.5 Comparison of 'elastic' stress blocks for reinforced masonry and concrete in flexure

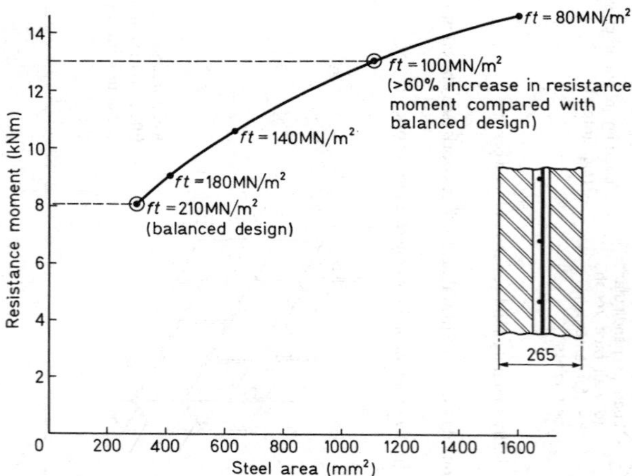

Figure 14.6 Example of relationship of steel area, steel stress and resistance moment with reinforced masonry. ft = steel stress

thickness and compressive stress. Although this method of increasing performance makes little engineering sense and is only really a freak of CP 111, it is useful to remember when designing within the limits of this code; it largely gets over the otherwise coarse steps in strength given by the general need to increase the thickness of masonry walls (other than hollow block ones), in steps of 100 mm or more.

Given more realistic flexural stresses this apparent advantage of masonry would disappear but with higher stresses there would be less need for it.

Flexure and direct compressions

Although not covered in CP 111 it has been common practice for some time to check combined stresses in accordance with the traditional formula:

$$\frac{\text{Actual direct stress}}{\text{Permissible direct stress}} + \frac{\text{Actual flexural stress}}{\text{Permissible flexural stress}} \leqslant 1$$

A more realistic treatment may be possible when limit-state design is accepted.

Shear

The permissible shear stress in CP 111 is 0.1 MN/m^2 increasing to a maximum of 0.5 MN/m^2 in proportion to the precompression across the shear plane, or 'as may be justified by suitably placed reinforcement'. This last phrase is not very helpful to designers and where shear forces are large it is best to consider the problem in terms of principal stresses and reinforce to suit these rather than to treat shear and flexure separately. However, as most reinforced masonry is in the form of walls, shear stresses seldom exceed 0.1 MN/m^2 in practice.

Applications of reinforced masonry

Masonry is essentially a wall material and this goes for reinforced masonry as much as unreinforced. Beams and slabs have been built in reinforced masonry, but with the exception of deep wall beams it is hard to justify them in comparison with reinforced concrete ones. It is in walls subject to bending perpendicular to the wall plane that the real advantage of reinforced masonry lies; it combines flexibility of form with good finish and frequently a large cost saving compared with reinforced concrete largely derived from the elimination of concrete's most expensive element, the shuttering. The following applications of reinforced masonry are worth noting:

(a) In horizontally spanning cladding where it is not possible to prove stability in wind due to arching. With mesh in the horizontal joints 100 mm single skin walls can be made to span 4 m or more with normal wind pressures.
(b) Cantilevering vertically in boundary walls or tall sheds where the walls cannot be restrained at the top. Theoretically a 275 mm grouted cavity wall can be made to cantilver 5 m in most wind conditions but this height to thickness ratio may be considered excessive.
(c) For retaining walls of up to 6 m height; this is one of the most fruitful fields for masonry in civil engineering. Table 14.6 shows what can be done using various types of brick wall and filled hollow blocks, ones with a drained granular fill—all strictly in line with CP 111. Above 4 m height the economy of reinforced brickwork relative to reinforced concrete tends to be lost, but here there may be a case for using brickwork or block as permanent shuttering to a wall designed in reinforced concrete.

Table 14.6 PERFORMANCE TABLE FOR MASONRY RETAINING WALLS

Material	*Type of wall*	*Thickness of wall* (mm)	*Approximate height range possible* (m)	*Notes*
BRICKWORK Units with crushing strength of 55N/mm² or above in Grade I mortar (1 : $\frac{1}{4}$: 3)	Grouted cavity (as I(b) in Table 14.5)	275	2.0–2.5	Wall heights based on drained granular fill without surcharge
	Quetta bond (as I(a) in Table 14.5)	372	2.3–3.0	
	Pocket type (as I(c) in Table 14.5)	215	2.3–3.0	Height range for given type and thickness of wall depends mainly on steel stress. Reinforcement percentages would be high at top of range
		215 increasing to 327 at base	3.0–4.0	
		215 increasing to 440 at base	4.0–5.0	
		215 increasing to 552 at base	5.0–6.0	
SOLID CONCRETE BLOCKWORK 21 N/mm² units in Grade I mortar	Grouted cavity (as I(b) in Table 14.5)	275	2.0–2.5	
	Pocket-type walls and even Quetta bond are possible in solid concrete blockwork but there are bonding problems with the larger dimensions of blocks	220	1.0–1.8	
HOLLOW CONCRETE BLOCKWORK 21 N/mm² units in Grade I mortar with concrete fill	Bars in concrete fill to cavities in blocks (As IV in Table 14.5)	220 increasing to 390 at base	1.8–3.0	

TESTING

Testing of materials

The various tests on masonry materials are summarised under Table 14.7. Before making a final choice of a brick or block and mortar, it is advisable to build a sample panel for all materials to be used as 'facing' work. Having decided on a suitable unit and mortar, it is normally only necessary to carry out occasional compressive tests on the units and mortar during construction.

Testing of masonry

The tests described in Table 14.7 can only be used to verify the properties of the materials and there has been considerable research to establish a control test for the masonry assemblage.

Table 14.7 TESTS FOR MASONRY MATERIALS

Material	*Tests*		*Remarks*
	Source	*Description*	
Clay bricks	BS 3921	Dimensional Water absorption Soluble salts Efflorescence Compressive	Dimensional, water absorption, soluble salts, efflorescence: Determines suitability for face work
Calcium silicate bricks	BS 187	Dimensional Drying shrinkage Compressive	Drying shrinkage: Indicates possible jointing requirement and compatability with other materials
Concrete bricks	BS 1180	Dimensional Drying shrinkage Compressive	Drying shrinkage: Indicates possible jointing requirement and compatability with other materials
Concrete blocks	BS 2028, 1364	Dimensional Density Drying shrinkage Compressive	Indicates suitability for face work Aids assessment of sound insulation Indicates possible jointing requirement and compatability with other materials
Cement	BS 12	Various	
Sand	BS 812	Various	
Mortar	BS 4551	Various	

In Britain the 9 in cube has been widely used in the test for brickwork. The procedure of making and testing cubes is described in Special Publication 56, 'Model Specification for Load-bearing Clay Brickwork', of the British Ceramic Research Association. In a further publication, No. 60, it was concluded that an adequate representation of the wall strength is given by brickwork cubes and a straightline relationship was suggested.

Because of the relative size of blocks, cubes are not appropriate to the testing of blockwork, and at the present time there is no established test.

Quality control

While the testing of materials is an important aspect of quality control, a very high standard of workmanship is also required for structural masonry. A high degree of supervision must be exercised, particularly at the beginning of contracts when it is often necessary to re-educate brick/block layers from certain traditional 'bad habits'.

It is because of the variation in strength of the units and mortar and the need for a very high standard of workmanship that higher load factors are used with structural masonry than with other structural materials, as concrete. In the Special Publication No. 56, referred to above, it is recommended that for most bricks and mortars the compressive stress in the brickwork cubes should be at least five times the basic compressive strength given in CP 111 for the unit and mortar in question. Depending upon the degree of quality control which it is considered it is possible to exercise, the designer may decide on a figure higher than 5.

The effects of various errors in the workmanship of brickwork on the strength of the resulting masonry are reviewed by Hendry.[16]

NEW DEVELOPMENTS

These range from the introduction or extension of techniques already familiar in other countries to the development of entirely new ideas.

High-bond mortars, which are far from new in America, may become popular here if the present tentative swing to prefrabrication continues; in our climate they are more suited to factory production than open-air work. Similarly, prestressing used with masonry in New Zealand largely to resist earthquake forces could become much more attractive when combined with prefabrication if only grouting could be avoided; plastic-coated cables could be the key to this. Composite action between masonry and reinforced concrete is likely to be used more in future; again prestressing is the relatively new ingredient which could make this really popular.

New forms of masonry unit will doubtless be produced from time to time, but to date there has tended to be no demand until the units become available, and few manufacturers are willing to invest in innovation without an assured market; large clay structural units, even up to the size of storey height blocks, made experimentally by the British Ceramic Research Association, could revolutionise masonry construction.

REFERENCES

1. Bricks (Revision 1), 'NBA + Building Commodity File', in *Building* (26 Aug. 1972)
2. GAGE, M., and KIRKBRIDE, T., *Design in Blockwork*
3. Concrete Blocks, 'NBA + Building Commodity File', in *Building* (26 May 1972)
4. 'Calculated Brickwork Data Sheets', SCP 1, Structural Clay Products Ltd, Architectural Press/C & CA, London (1972)
5. GAGE, M., and KIRKBRIDE, T., *Design in Blockwork*
6. GAGE, M., and KIRKBRIDE, T., *Design in Blockwork*
7. PLOWMAN, J. M., 'Damp-proof Coursing in Loadbearing Brickwork', No. 11, *Proc. Brit. Ceram. Soc.* (July 1968)
8. THOMAS, A., 'Moisture Expansion in Brickwork', Vol. 70 No. 1, *Trans. J. Brit. Ceram. Soc.* (1971)
9. BEARD, R., DINNIE, A., and RICHARDS, R., 'Movement in Brickwork', Vol. 68 No. 2, *Trans. Brit. Ceram. Soc.* (March 1969)
10. BUTTERWORTH, B., 'Efflorescence and Staining of Brickwork', reprinted from *The Brick Bulletin* (Dec. 1962)
11. BEARD, R., and DINNIE, A., 'Thermal Transmittance of Wall Constructions', Vol. 2, No. 5, Clay Products Technical Bureau Technical Note (Nov. 1968)
12. FOSTER, D., 'Some Observations on the Design of Brickwork Cladding to Multi-storey RC Framed Structures', Vol. 1, No. 4, Brick Development Association Technical Note (Sept. 1971)

13. THOMAS, K., 'The Strength, Function and other Properties of Wall-ties', Vol. 1, No. 2, Brick Development Association Research Notes (March 1970)
14. HILSDORF, H. K., 'Failure Mechanism of Masonry', State of Art Report No. 2, Committee C7, Conference on Planning and Design of Tall Buildings, Bethlem Pa USA (1972)
15. HASELTINE, B. A., and THOMAS, K., 'Loadbearing Brickwork—Design for the Fifth Amendment', Vol. 1, No. 3, Brick Development Association Technical Note (July 1971)
16. HENDRY, A. W., 'Workmanship Factors in Brickwork Strength', Vol. 1, No. 6, Brick Development Technical Note (Nov. 1972)

LIST OF CODES OF PRACTICE AND STANDARDS PUBLISHED BY THE BRITISH STANDARDS INSTITUTION AND REFERRED TO IN SECTION 14

CP 3	*Code of basic data for the design of buildings*
CP 111 :	*Structural recommendations for loadbearing walls*
: Part 1 : 1964	*Imperial units*
Part 2 : 1970	*Metric units*
Amendment : June 1971	
CP 121.101 : 1951	*Brickwork*
CP 121.201 : 1951	*Masonry walls ashlared with natural stone or with cast stone*
BS 12 : 1958	*Portland cement (ordinary and rapid-hardening)*
BS 187	*Calcium silicate (sandlime and flintlime) bricks*
BS 743 : 1970	*Materials for damp proof courses*
BS 812 : 1967	*Methods for the sampling and testing of mineral aggregates, sands and fillers*
BS 1180 : 1944	*Concrete bricks and fixing bricks*
BS 1243 : 1964	*Metal ties for cavity wall construction*
BS 2028, 1364 : 1968	*Precast concrete blocks*
BS 3921	*Standard special bricks*

15 TIMBER DESIGN

TIMBER DESIGN 15

15 TIMBER DESIGN

F. H. POTTER, B.Sc. Tech., C.Eng., M.I.C.E., F.I.W.Sc.
Imperial College of Science and Technology

INTRODUCTION

Timber is one of the finest structural materials: it has a high specific strength, can be easily worked and jointed and does not inhibit design. Like most other structural materials it suffers attack causing deterioration (corrosion, weathering and biodeterioration) but once the material is known and the causes understood, effective preventative measures can be taken easily and economically.

Design is thus a confluence of specification, structural analysis, detailing and protection, each of which is of equal importance if an effective design is to be achieved.

Nowadays, structural design covers a wider range of components than ever before, for the intense wind loadings in high-rise building coupled with large glazed areas has meant that much window joinery is now subject to structural design. In addition, the effects of wind loadings together with the requirements of the Building Regulations and the newer building shapes has meant that even in low-rise buildings, components which once were built must now be designed.

Timber is thus used for a wide variety of structural purposes, either on its own or in combination with one of the 'heavier' materials. It can take extremely simple forms such as solid beams, joists and purlins or can be used in the more recent forms of glued–laminated construction or plywood panel construction. These latter forms allow the design of exciting and economic structural shapes, the variety of which may be judged from Tables 15.3 and 15.4.

Some of the characteristics of timber may be found in Table 15.2, whilst general properties are given in other publications.[1, 2, 3]

DESIGN BY SPECIFICATION

Essentially this is a prescription of fitness for use under service conditions and requires the choice not only of an appropriate material but also of its condition, use and protection.

The success of the specification will depend upon its interpretation; standard glossaries are available for timber and woodwork,[4] nomenclature of timber[5] and preservative treatment.[6]

Functional and user needs will dictate the choice of material based on the following factors.

Species and use

Very many timbers are structurally useful, whereas usefulness for joinery purposes is often more restrictive. Where timber is used for structural joinery, the combination of requirements may be even more restrictive.

Table 15.2 lists those timbers and their characteristics for which working stresses are given in CP112,[31] whilst BS 1186[7] indicates the joinery use of specific species. A comprehensive guide to West African species and their uses, both structural and joinery is given in pamphlets issued by the United Africa Company.[10]

Flooring, particularly industrial flooring, has particular requirements and recommendations for suitable timbers for these requirements are given in PRL Technical Note 49.[8]

Availability and sectional properties

Availability is equally important, and Table 15.2 indicates the availability of the structural timbers from the viewpoints of supply and length. The geometric properties of sawn and precision timber to be used in design are given in CP 112,[31] whilst the available sizes for hardwoods are given by TRADA Advisory Service Leaflet No. 9.[9]

The movement of timber

Even with dried timber, changes in atmospheric conditions will result in a varying moisture content which will induce fluctuating dimensional changes in the timber, known as 'movement'. The variation can be designed for quite simply but some knowledge of the degree of possible movement is helpful. Some indication can be obtained from Table 15.2, whilst further information can be obtained from the publications of the Princes Risborough Laboratory.[2, 3, 11, 12]

Moisture content and end use

Every species of timber will achieve a fairly steady moisture content for a particular environment—the equilibrium moisture content. The Princes Risborough Laboratory has established moisture contents for various environments.[12] Greater reliability can be achieved by drying timbers to these moisture contents before construction.

Working properties

Ease of fabrication is indicated in Table 15.2 although more detailed information may be found in Princes Risborough Publications.[2, 3]

Natural resistance to attack

Timber has a widely varying resistance to attack by fungi, insects, marine borers and termites. Fungi will not normally attack timber having a moisture content lower than

Table 15.1 DURABILITY CLASSIFICATION OF THE HEARTWOOD OF UNTREATED TIMBERS

Grade of durability	*Approximate life in ground contact* (years)
Very durable	More than 25
Durable	15–25
Moderately durable	10–15
Nondurable	5–10
Perishable	Less than 10

20%, but a timber's ability to resist fungal attack is classified according to Table 15.1.

The natural durability of some structural timbers is given in Table 15.2. Information on further timbers will be found in Technical Note 40[13] and the Handbooks of Softwood and Hardwoods[2, 3] whilst further advice on the control of decay will be found in Technical Note 29.[14]

Termite attack and its prevention are dealt with in Leaflet 38[16] in which the following timbers are mentioned as being generally resistant:

Iroko Opepe Californian redwood Teak

Other timbers are given in the Handbooks on Softwood and Hardwoods.[2, 3]

Marine borers are a hazard in the sea or brackish waters and PRL Leaflet 46[15] gives advice on the protection of timbers against this attack. Highly resistant timbers suitable for marine works are: greenheart, pyinkadò, turpentine, totara, jarrah, basralocus and manbarklak.

Preservative treatment

The sapwood of all timbers is liable to attack by fungi and insects but it is often possible to obtain a more attack-resistant structure by pressure-impregnating nondurable or perishable timbers than by using durable species. Indeed it is sometimes more economic also.

The amenability of timbers to preservative treatment is given in Table 15.2 and is related to the following classification:

Permeable:	easily treated by either pressure or open tank
Moderately resistant:	fairly easy to treat by pressure, penetration 6 to 20 mm in 2 to 3 h
Resistant:	difficult to impregnate, incising often used. Penetration often little more than 6 mm
Extremely resistant:	very little penetration can be achieved even after prolonged treatment

Further information and guidance on satisfactory types and methods of treatment may be found in publications of the BSI and TRADA[18, 19, 20] The economics of timber preservation is discussed in Timberlab 17.[17]

Fire resistance

Although timber ignites spontaneously at about 250 °C, ignition is a function of the external surface area to the total volume of timber and the rate of charring does not significantly increase with a rise of temperature. The rate of charring is generally taken as about 0.5 mm/min (Western red cedar 0.85 mm/min, dense hardwood 0.42 mm/min), but perhaps the most important factor is that the structural properties of uncharred material remain virtually unchanged.

Thus if adequate protection against combustion is provided, timber is one of the safest structural materials in a severe fire. These measures are usually: the provision of sacrificial material, chemical impregnation and protective covering.[21, 22]

STRESS GRADING AND PERMISSIBLE STRESSES

Timber is a natural organic material and therefore is subject to wide variability because of environmental, species and genetic effects. This variability affects both visible quality and strength.

Table 15.2 CHARACTERISTICS AND AVAILABILITY OF SOME STRUCTURAL TIMBERS

Standard name	*Approx. density at* M/C 18% (kg/m³)	*Strength group*		*Natural durability*	*Resistance to preservative treatment*	*Moisture movement*	*Working quality*	*Availability*		*Relative price*
		Wood	*Joint*					*Supply*	*Normal length* (m)	
SOFTWOODS (IMPORTED)										
Douglas fir	590	S1	J2	Moderately	Resistant	Small	Good	Good	4.20–4.80	Medium
W hemlock (unmixed)	540	S2	J3	Not	Resistant	Medium	Good	Good	4.20–4.50	Low
W hemlock (comm)	530	S2	J3	Not	Resistant	Medium	Good	Good	4.20–4.50	Low
Parana pine	560	S2	J3	Not	Moderately	Medium	Good	Good	3.60–3.90	Low
Pitch pine	720	S1	J2	Durable	Resistant	Medium	Good	Good	4.50–9.00	Medium
E redwood	540	S2	J3	Not	Moderately	Medium	Good	Good	1.50–7.00	Low
E whitewood	510	S2	J3	Not	Resistant	Small	Good	Good	1.50–7.00	Low
Canadian spruce	450	S2	J3	Not	Very	Medium	Good	Good	2.40–5.10	Low
W red cedar	380	S3	J4	Durable	Resistant	Small	Good	Good	2.40–7.30	Low
SOFTWOODS (HOME GROWN)										
Douglas fir	560	S1	J2	Moderately	Resistant	Small	Good	Fair	1.80–4.50	Low
Larch (E—Japan)	560	S1	J2	Moderately	Resistant	Medium	Good	Good	1.80–3.60	Medium
Scots pine	540	S2	J3	Not	Moderately	Medium	Good	Good	1.80–3.60	Low
European spruce	380	S3	J4	Not	Resistant	Small	Good	Good	1.80–3.60	Low
Sitka spruce	400	S3	J4	Not	Resistant	Small	Good	Good	1.80–3.60	Medium
HARDWOODS (IMPORTED)										
Abura	590		J2	Perishable	Moderately	Small	Good	Good	1.80–6.00	Low
African mahogany	590		J2	Moderately	Extremely	Small	Medium	Good	1.80–7.30	Medium
Afrormosia	720		J1	Very	Extremely	Small	Medium	Good	2.40–7.30	Med high
Greenheart	1 060		J1	Very	Extremely	Medium	Difficult	Good	4.80–9.00	High
Gurjun/Keruing	720		J1	Moderately	Resistant	Large	Medium	Good	1.80–7.30	Low
Iroko	690		J1	Very	Extremely	Small	Medium	Good	1.80–6.00	Medium
Jarrah	910		J1	Very	Extremely	Medium	Difficult	Good	1.80–8.40	Med high
Karri	930		J1	Durable	Impermeable	Large	Difficult	Good	1.80 up	Med high
Opepe	780		J1	Very	Moderately	Small	Medium	Good	1.80–6.00	Medium
Red meranti	540		J3	Moderately	Resistant	Small	Good	Good	1.80–7.30	Low
Sapele	690		J1	Moderately	Resistant	Medium	Good	Good	1.80 up	Medium
Teak	720		J1	Very	Extremely	Small	Medium	Good	1.80 up	High
HARDWOODS (HOME GROWN)										
European ash	720		J1	Perishable	Moderately	Medium	Good	Fair	1.80 up	Low
European beech	720		J1	Perishable	Permeable	Large	Good	Good	1.80 up	Medium
European oak	720		J1	Durable	Extremely	Medium	Medium	Good	1.80 up	Medium

If for any particular property and species only one design stress were specified this would have to be set so low (to allow for variability) that the material would have a very limited structural application. In consequence, a number of stress grades has been adopted, leading not only to a more economic use of the material but also to a higher yield of structurally useful material.

There are two main methods of stress grading for solid timber, visual grading and mechanical grading; each of which requires a different procedure.

Visual stress grading

In visual grading, data obtained from clear material (straight-grained and free from knots and fissures) are analysed statistically for each species and basic stresses for each property are derived. These basic stresses are then reduced by factors which take account of the strength-reducing effects of the permissible growth characteristics for each stress grade.

At the present time there are two sets of quality requirements used for visual stress grading in this country as follows.

The first set is given in Appendix A of CP 112.[13] This defines the requirements for four grades for solid timber construction (grades 75, 65, 50 and 40) and for three laminating timber grades (LA, LB and LC).

The second set is given in BS 4978.[24] Besides setting the requirements for two grades for solid timber construction (SS and GS), this standard restates the requirements for the laminating timber grades given in Appendix A of CP 112. The standard replaces BS 1860 and it is intended that its requirements will eventually replace those currently given in Appendix A of CP 112.

The stresses for all these grades are tabulated either in CP 112: Part 2 or in CP 112: Addendum No. 1 for both green and dry conditions.

Mechanical stress grading

Mechanical stress grading[23] is a method of non-destructive testing each piece to be graded. The pieces are bent under a constant central load over a constant short span. The strength of the material can then be calculated accurately from the resultant deflection. The same grade stresses as for visually graded timber are specified, but higher moduli of elasticity are given. Further information on the derivation of these grade stresses can be found in Bulletin 47.[26] The grade stresses (M75, M50, MSS and MGS) for both green and dry conditions are tabulated in CP 112: Addendum 1 (1973) and at present are limited to the five softwoods most commonly available.

Glued–laminated timber grades

In glued–laminated members, the presence of strength-reducing characteristics will have a smaller effect than in solid beams, since the probability of identical structural defects appearing in identical positions in adjacent laminations is very small. CP 112 therefore allows higher grade stresses for glued–laminated timbers, these being obtained by applying tabulated factors to the basic stress for each species. The derivation of each grade stress is given in Bulletin 53.[25]

Grouped softwood species

To simplify design, softwoods have been grouped together according to their properties of strength and stiffness, the group classification being given in Table 15.2. The strength properties of each group are necessarily based on the weakest member of the group, but the system allows group timbers to be specified instead of particular species.

Because of the more limited data available on machine graded species, the grouping system is at present restricted to visually graded material. In the same way, hardwoods, being more variable than softwoods are not grouped at present, but it is hoped that grouping may be developed soon.

Permissible stresses

Permissible design stresses for both solid and laminated timber components are governed by the type of component, the conditions of service and the type of loading. They are obtained from grade stresses by applying the appropriate modification factors.

DESIGN—GENERAL

Design in timber is similar to that in any other structural material as long as timber's peculiar qualities are acknowledged, indeed, these qualities can be exploited by resourceful designers. Timber is idealised as an orthotropic material, but in practice, only two directions need be considered: that parallel to the grain (along the trunk) and that perpendicular to the grain. Most strength properties, of both timber and joint fasteners, vary according to these directions and the variation has been found to follow the Hankinson relationship:

$$N = \frac{PQ}{P \sin^2 \theta + Q \cos^2 \theta}$$

where θ is the angle between directions of load and grain, N the stress at θ to the grain, P the stress parallel to the grain and Q the stress perpendicular to the grain,

from which intermediate stress or strength values can be calculated. This is not normally required for solid beams, joists and columns where only the major directions are used, but is often met where members intersect at joints. The stresses given in CP 112 are for permanent loading and increased values are allowed for short- and medium-term loads. This Code of Practice governs general design, but additional information may be found from references 30–39.

In the past, design has been inhibited by the relatively short lengths of timber available (Table 15.2 indicates availability). However, the production of durable resin adhesives has led to new construction techniques and structural forms being developed. Glued–laminated timber in which thin laminae are glued together to form structural components of almost any shape or length is a common reality, whilst structural plywood can be combined with either solid or glued–laminated timber to produce composite components which are lightweight, reliable and pleasing. The design in these forms is more complex than in solid timber but information on a wide variety of structural forms can be found in references 47–72 whilst advice on the selection of a particular form is given in Tables 15.3 and 15.4 General design advice is provided by TRADA.[40]

No matter which form of construction is chosen, with flat-roof construction, water ponding and snow loading should be carefully considered, since these can affect deflection of the roof system, both significantly and progressively.

DESIGN IN SOLID TIMBER

Since permissible stresses are maxima there may be some advantage in using structural hardwoods or the higher-grade-stress softwoods whenever stress governs design.

Table 15.3 ROOF SELECTION

Division	*Subdivision*	*Construction*	*Minimum support conditions*	*Maximum economic spans* (m)	*Fastenings*
Beams		Solid timber		6	None
		Laminated, either vertically or horizontally, depending on size	Vertical support at ends	24	Glue
		I or box sections: flanges solid or laminated. Webs plywood or diagonally boarded		30	Glue and/or nails
Arches		Laminated horizontally	Vertical and horizontal support at ends	46	Glue and/or nails for laminating. Connectors for site joints
		I or box sections: flanges laminated horizontally. Webs diagonally boarded		46	
Portals		Laminated horizontally		24	Glue and/or nails for laminating. Connectors for site joints
		I or box sections: flanges solid or laminated. Webs plywood or diagonally boarded	Vertical and horizontal support at ends	46	
Trusses	Belgian	Solid timber	Vertical support at ends	12	Nails and/or glue
				24	Connectors
	Warren	Solid timber		12	Nails and/or glue
				30	Connectors
	Bowstring	Laminated chords. Solid webs		46	Glue and/or nails for laminating. Connectors at joints

(After: L. G. Booth, *Engineering*, 25 March 1960)

Table 15.4 ROOF SELECTION

Division	*Subdivision*	*Construction*	*Minimum support conditions*	*Maximum economic sizes* (m)	*Fastenings*
Plates	*Flat*	Membrane formed from plywood or layers of diagonal boarding A single-skin structure may have stiffening ribs	Vertical support at corners	12 × 12	Nails and/or glue
	Folded	A double-skin structure will have spacing ribs Edge beams and end diaphragms required	Vertical support at corners	18 × 9	Membrane with nails and/or glue Diaphragms with nails or connectors
Singly curved shells	*Circular cylindrical*	Membrane formed with layers of diagonal boarding. May have stiffening ribs Edge beams required End diaphragms required	Vertical support at corners	30 × 12	Membrane with nails and/or glue Beams (see Table **15.3**) Diaphragms with nails or connectors
Doubly curved shells	*Spherical dome*	Boarded membrane with or without laminated ribs Laminated ring beam	Ring beam to be supported at intervals	30 dia.	Membrane with nails and/or glue Ribs and ring beam glued
	Hyperbolic paraboloid	Boarded membrane with laminated edge beams	Vertical support only at low points, if columns tied together. Otherwise buttresses at low points	21 × 21	Membrane with nails and/or glue Edge beams glued
	Elliptical paraboloid	Boarded membrane with laminated tied arches along edges	Vertical support at corners	24 × 24	Membrane with nails and/or glue Tied arches with glue and connectors
	Conoid	Boarded membrane with laminated tied arches on ends Edge beams required	Vertical support at corners	30 × 9	Membrane with nails and/or glue Beams (see Table **15.3**) Tied arches with glue and connectors

(After: L. G. Booth, *Engineering*, 25 March 1960)

However, if deflection governs, there will be no advantage in using these more expensive materials unless the moduli of elasticity are sufficiently high. A possible exception is keruing (*dipterocarpus* spp) whose current cost is roughly similar to that of softwoods. Some indication of price is given in Table 15.2.

As design in solid timber is limited by the maximum size of timber available, this has led to the development of many types of girder framework: however, where there is sufficient headroom, trussed beams can give an economic solution for heavy loads and long spans.[30, 35, 39]

One of the major advances in CP 112 : Part 2 is that it specified *basic sizes*. The measurement of these sizes is at 20% moisture content and the sizes should be adjusted for moisture contents other than 20%. Of the three types of timber covered: sawn timber; precision timber; processed timber; sawn and precision timber have limited permissible reductions in section. Basic sizes can therefore be taken as actual sizes and no allowances need be made when calculating the geometric properties of the sections. However, the permissible reductions for processed timber are not only greater but also more variable. Therefore minimum sizes are specified and the geometric properties for processed timber tabulated in Appendix A are based on those minimum sizes.

Further reductions in section should be made for notches, mortices and bolt, screw and connector holes. Modification factors may also be required for the length and position of bearing, the shape of a beam and its depth if greater than 300 mm, whilst for compression members, combined factors are given for both slenderness and loading.[31]

Lateral stability is important both for deep beams and for compression members, and in built-up members web-stiffeners should be provided wherever concentrated loads occur.

General design data are available[24, 36, 31, 40] applicable to particular structural forms (references 50, 56–60, 64, 66, 67, 71 and 72) whilst design aids have been published for solid beams, portal frames and trussed rafters.

GLUED—LAMINATED TIMBER ASSEMBLIES

Glued–laminated timber is essentially a built-up section of two or more pieces of timber whose grains are approximately parallel and which are fastened together with glue throughout their length. This enables the properties of timber to be regulated to some degree and provides structural sizes and shapes which would not be possible in solid timber. Variation in section is possible, whilst high-grade material can be placed in zones of high stress and low grade material in zones of low stress. All softwoods glue well and are generally preferred in Britain although occasionally there can be some advantage in using wholly hardwood laminae.

Construction may use either vertical or horizontal laminations.

With vertical laminations, the zones of equal stress are shared between the laminations so that the strength of a beam can be said to the sum of the individual laminations. This load-sharing concept has led to the grade-stress modification factors tabled in CP 112[31] which give higher permissible stresses for the laminated beam.

Horizontally laminated beams have been permitted under CP 112 since 1967 but the philosophy for behaviour is entirely different from that for vertical laminations. A beam will consist of material containing knots whose presence will affect the strength ratio of the beam. Since the knot effect will vary according to the sizes of knots and the number of laminations, CP 112 tables *basic stress* modification factors according to these variables. The basis for determining these grade stresses is discussed by Curry.[25, 41]

Since curved laminated beams are fabricated by bending the individual laminations, fabrication stresses are induced which depend upon the degree of curvature, the thickness of the lamination and the species of timber. CP 112[31] therefore tables modification factors to be applied to the grade bending stresses for different values of t/R.

The production of long laminations depends upon the use of efficient methods of end jointing. Where the efficiency of an end-joint is known the laminations containing them can be included when calculating the section properties, but where efficiencies are not known, the laminations containing the end joints must be omitted when calculating section properties. Efficiency ratings for plain scarf joints are given in CP 112[31]: these are used to modify the basic stresses to give the maximum stresses to which any particular lamination may be subjected. Finger joints are now possible which give efficiencies that are comparable with those for plain scarf joints. Butt joints do not transmit load and should only be used in zones of zero or very low stress.

Apart from the consideration of end joints and curvature, design is similar to that for solid timber and several references are given (25, 41, 42, 51 and 65) whilst design aids are noted for glulam beams.

PLYWOOD ASSEMBLIES

Plywood is a type of glued–laminated construction in which the laminae are formed from thin flat veneers of timber. These veneers are produced by the rotary cutting of logs and are laid alternately at right angles in an odd number of layers. Since both the shrinkage and strength of timber differ according to the grain direction, the type of construction gives greater dimensional stability and tends to equalise the strength properties in both major directions of the plywood sheet.

There are two distinct design philosophies, the North American approach which only considers the 'parallel plies', that is those plies whose grain lies in the direction of the load. The approach is based on the basic stresses and moduli for solid timber. The Finnish and British approach is known as the 'full cross-section' approach in which grade stresses and moduli for the sheet materials have been determined from tests. In CP 112[31] all the grade stresses and moduli are for full cross section, but it is well to remember that many North American design manuals will be based on the 'parallel-plies' approach. Plywood is a strong, durable and lightweight structural material which can be used to produce exciting structural shapes (see references 48, 49, 52–55, 57, 61–63, 68 to 72). Design data are available for a variety of constructions[43–46] whilst design aids are available for stressed skin panels and portal frames.

Perhaps plywood's most useful property is that of providing excellent shear resistance for a small cross section, although it is well to remember that lateral stability constraints may be required.

TIMBER FASTENINGS

Available methods of jointing are perhaps the most important criteria for the design of structural components. This is particularly true in timber for which highly efficient methods of transferring tensile loads have been developed only during the past 50 years. Split-ring and tooth-plate connectors are now available which have efficiencies two to three times those for nailed and bolted joints. A comparative indicator of fastener efficiency and the required member sizes is given in Table 15.5

The strength of mechanical fasteners depends upon member size and thickness and the spacing of the fasteners. CP 112[31] tabulates permissible values of a wide range of variables, whereas some manuals prefer a presentation as a series of design curves (references 30, 33, 34, 36, 37).

However, the major advance in fastening techniques has been in glued joints. Early glues were unreliable, deteriorating quickly, but the present phenolic and resorcinol resins are so durable that the risk of delamination has been almost entirely eliminated even under extreme exposure. Unfortunately gluing still requires controlled conditions and its application to site work is still limited.

Table 15.5 THE RELATIVE STRENGTHS OF TIMBER JOINTS

Fastener unit	*Minimum number of fastener units for a load capacity of:* 10 kN		20 kN		*Joint dimensions for load of* 20 kN *Joint overlap*	*Width*	*Thickness of inner member*	*Load capacity timber*
	//g	⊥g	//g	⊥g	mm	mm	mm	kN
SPLIT RINGS 2/64 mm dia. + 12.7 mm bolt	1	1	1	2	280	88	50	21
SHEAR PLATE 2/67 mm dia. + 19 mm bolt	1	1	1	2	280	88	72	30.4
TOOTH PLATE 2/64 mm dia. + 12.7 mm bolt	1	1	2	3	285	74	100	35.5
BOLTS								
12.7 mm dia.	4	5	8	9	279	89	100	42.7
25.4 mm dia.	1	2	2	3	457	127	100	61.0
NAILS								
9 gauge	18	18	36	36	450	192	22	20.6
3 gauge	5	5	10	10	512	256	62	76.0

Assumptions: (i) three-member tension lap joints, timber to timber, (ii) J2 group timbers

Since the shear strength of adhesives is usually higher than that of timber, a fastener efficiency of 100% can be achieved. Nevertheless it is important to remember that glues seldom have a good tensile strength, so that they should be stressed in shear as much as possible.

Information is available on gluing,[27] the requirements for adhesives[29] and the compatibility between glues and preservatives.[28] The permissible stresses for glued joints are the shear stresses for the timber,[31] however, regard must be paid not only to the variation of that shear strength but also to the possibility of differential shrinkage and stress concentrations in the joint.

The type of fastener chosen will depend upon the skills and equipment available, possible fabrication conditions, relative costs and whether or not it is necessary to take down and reassemble the structural component.

TIMBER-FRAME CONSTRUCTION

It is estimated that the major use of structural timber will be in the housing field.

In high-rise construction, timber will play a supplementary role to the heavy material, being used for partitions, infill panels and floor and roofing systems. In this role, timber's ready adaptability to prefabrication is a great benefit.

In low-rise construction on the other hand, timber is increasingly being used to provide the structural skeleton for the building; indeed at the present time, timber-frame construction constitutes some 20% of all house construction. The method of construction is a simplification and refinement of that which has been used successfully for many centuries, but which is equally well applicable to many other uses besides housing.

The structural form is that of a free-standing skeleton for which standard details have been produced.[56] The basis of the skeleton is the stud-framed panel for which Peek describes the design[59] whilst the application of the principle to four-storey buildings is explained by Johnson and Burgess.[58]

Of especial importance is the structure's ability to withstand lateral loading and particularly that resistance to planar deformation of a wall panel known as its racking resistance. Figure 15.1 shows the deformation of an idealised frame structure under lateral loading. Figure 15.1(a) shows the uniform translation of the walls which occurs when there is complete symmetry of both structure and loading. This hardly ever occurs and this lack of symmetry causes a rotation in addition to the translation (Figure 15.1(b)). A review is available describing the theoretical and experimental attempts to solve the racking problem.[60]

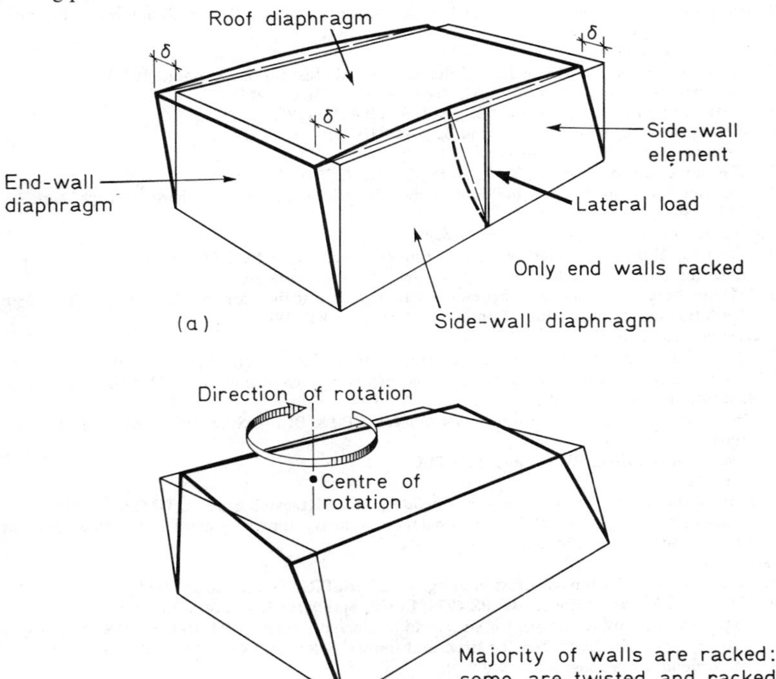

Figure 15.1 Deformation of idealised house structure under later loading. (a) Uniform translation of top of wall parallel to the lateral load; (b) rotation after lateral translation, caused by lack of symmetry

DESIGN AIDS

There are three areas in which design aids can make a valuable contribution to the design process. These are: rapid preliminary design either for comparison or estimation of cost, routine elementary design and complex design processes for which the design time can be drastically reduced.

Nomograms and load-span tables have been used for many years, but the application of computer programming has extended considerably the use of design aids.

The Bibliography indicates some of the design aids which are now available for structural design in timber.

REFERENCES

General properties of timber

1. EVERETT, A., *Mitchells Building Construction Materials*, Batsford, London (1970)
2. *Handbook of hardwoods*, PRL, HMSO (1972)
3. *Handbook of softwoods*, PRL, HMSO (1957)

Specification

Glossaries

4. *Glossary of terms relating to timber and woodwork*, BS 565, British Standards Institution, London (1972)
5. *Nomenclature of commercial timbers*, BS 881 and 589, British Standards Institution, London, (1958)
6. *Glossary of terms relating to timber preservatives*, BS 4261, British Standards Institution, London (1968)

Species, use and availability

7. *Quality of timber*, BS 1186 Part 1, British Standards Institution, London (1971)
8. 'Hardwoods for industrial flooring', Tech. Note 49, PRL (1971)
9. 'Sizes of main building hardwoods', ASL 9, TRADA (1962)
10. *West African hardwoods*, Part 1 and 2, UAC (1971)

Moisture content, moisture movement

11. 'The movement of timbers', Tech. Note 38, PRL (1969)
12. 'Flooring and joinery in new buildings. How to minimize dimensional changes', Tech. Note 12, PRL (1971)

Natural durability and the protection of timber

13. 'The natural durability classification of timber', Tech. Note 40, PRL (1969)
14. 'Ensuring good service life for window joinery', Tech. Note 29, PRL (1968)
15. 'Marine borers and methods of preserving timber against their attack', Leaflet 46, PRL (1950)
16. 'Termites and the protection of timber', Leaflet 38, PRL (1965)

Preservative treatment

17. TACK, C. H., 'The economics of timber preservation', PRL Timberlab 17 (1969)
18. *Classification of wood preservatives and their method of application*, BS 1282, British Standards Institution, London (1959)
19. *Preservative treatments for construction timbers*, CP 98, British Standards Institution, London (1964)
20. *Timber preservation*, BWPA and TRADA

Fire Resistance

21. *Fire and the structural use of timber in buildings*, Fire Research Station, HMSO (1970)
22. WARDLE, T. M., 'Notes on the fire resistance of heavy timber construction', New Zealand Forestry Service Information Series 53 (1966)

Stress grading

23. CURRY, W. T., 'Mechanical stress grading of timber', PRL Timberlab 18 (1969)
24. *Timber grades for structural use*, BS 4978, British Standards Institution, London
25. CURRY, W. T., 'Grade stresses for structural laminated timber', PRL Bull 53, HMSO (1970)
26. SUNLEY, J. G., 'Grade stresses for structural timber', PRL Bull 47, HMSO (1968)

Glues for structural components

27. 'The gluing of wooden components', Tech. Note 4, PRL (1967)
28. 'Gluing preservative treated wood', Tech. Note 31, PRL (1968)
29. KNIGHT, R. A. G., and NEWALL, K. J., 'Requirements and properties of adhesives for wood', PRL Bull 20, HMSO (1971)

Design

Textbooks, etc.

30. *AITC Timber construction manual*, Wiley, New York (1966)
31. *The structural use of timber*, CP 112, British Standards Institution, London (1971)
32. BOOTH, L. G., and REECE, P. O., *The structural use of timber (a commentary on BS CP 112)*, Spoon, London (1967)
33. *Timber construction manual*, CITC, Ottawa (1959)
34. HANSEN, H. J., *Modern timber design*, Wiley, New York (1962)
35. KARLSEN, G. G., *Wooden structures*, Mir, Moscow (1967)
36. LEVIN, E., 'Design data for timber structures', Pub. RSA 18, TRADA (1967)
37. PEARSON, R. G., KLOOT, N. H., and BOYD, J. D., *Timber engineering design handbook*, Jacaranda, Melbourne (1967)

38. REECE, P. O., *The design of timber structures*, Spon, London (1949)
39. *Wood Handbook*, US Department of Agriculture, US Printers Office, Washington DC (1955)

General design data

40. *Design of timber members*, TRADA (1967)
41. CURRY, W. T., 'Laminated beams from two species of timber. Theory of design', PRL Special Rep. No. 10, HMSO (1955)
42. FREAS, A. D., and SELBO, M. L., 'Fabrication and design of glued laminated wood structural members', USDA Technical Bulletin 1069 (1954)
43. *Canadian fir plywood data for designers*, COFI (1972)
44. *Fir plywood design fundamentals and physical properties*, COFI (1969)
45. CURRY, W. T., and HEARMON, R. F. S., 'The strength properties of plywood. Part 2. Effect of geometry of construction', PRL Bull 33, HMSO (1967)
46. *Finnish birch plywood handbook*, FPDA (1964)

Arches and portal frames

47. BURGESS, H. J., *Exploiting geometrical symmetry in timber structures*, TRADA (1970)
48. *Portal frame manual*, COFI (1972)
49. KHARNA, J., and HOOLEY, R. F., 'Design of fir plywood panel arches', COFI Rep. TDD-43 (1965)
50. 'Ridged portals in solid timber', TRADA E/IB/18 (1969)
51. WILSON, T. R. C., 'The glued laminated wood arch', USDA Tech Bull 691 (1939)

Barrel vaults

52. KHARNA, J., 'Design of fir plywood barrel vaults', COFI Rep. TDD-40 (1964)

Folded plates

53. *Fir plywood folded plate design*, COFI (1969)

Formwork

54. *Fir plywood concrete form manual*, COFI (1967).
55. LEE, I. D. G., 'Film faced plywood for concrete formwork', FPDA Reprint (1967)

Housing

56. *Canadian wood frame house construction*, COFI, Central Mortgage and Housing Corporation (1967)
57. 'The uses and application of Finnish birch plywood, blockboard and laminboard in housing', Tech. Pub 14, FPDA (1969)
58. JOHNSON, V. C., and BURGESS, H. J., 'The structural design of the High Wycombe four storey timber framed maisonettes', TRADA (1966)
59. PEEK, J. D., 'Design of timber stud walling', TRADA E/IB/19 (1969)
60. POTTER, F. H., 'The racking resistance of timber framed walls', Paper to the Institute of Wood Science (1968)

Plyweb beams

61. BURGESS, H. J., 'Introduction to the design of ply-web beams', TRADA E/IB/24 (1970)
62. *Nailed fir plywood web beams*, COFI (1963)
63. *Fir plywood web beam design*, COFI (1970)

Shells

64. BOOTH, L. G., 'The model testing, design and construction of a timber conoid roof', *1st Int. Conf. on Timber Engg, Southampton University* (1960)
65. KERESZTCSY, L. O., 'Interconnected, prefabricated laminated timber diamond type shell', *Int. Conf. for Space Structures, Surrey University* (1966)
66. TOTTENHAM, H., 'The analysis of hyperbolic paraboloid shells', TRADA E/RR/5 (1958)
67. TOTTENHAM, H., 'Analysis of orthotropic cylindrical shells', *Civil Engineering* (May 1959)
68. *Fir plywood stressed skin panels*, COFI (1971)
69. 'Design data for stressed skin panels in Finnish birch plywood', Tech. Bull 11 (M), FPDA (1970)
70. WARDLE, T. M., and PEEK, J. D., 'Plywood stressed skin panels. Geometric properties and selected designs', TRADA E/IB/22 (1970)

Trussed rafters

71. BURGESS, H. J., 'A theoretical approach to the design of gussetted trussed rafters', TRADA E/RR/27 (1967)
72. *The structural use of timber. Part 3: Trussed rafters for roofs of dwellings*, CP 112: Part 3, British Standards Institution, London

BIBLIOGRAPHY

BURGESS, H. J., 'Design aids including computer programmes'; Paper WCH/71/5/8. World Work

Consultation Housing, Vancouver (1971). (General appraisal of the development of design aids by TRADA)

'Design for roof structures in Finply', Tech. Pub. 17, FPDA (1972). (Includes standard designs for box and I beams, stressed skin panels, portal frames and gussetted trusses.)

'Design of timber members', TBL 33, TRADA (1967). (Typical calculations for beams and columns together with beam-span tables for a wide range of applications.)

'The structural use of hardwoods', Tech. Note, TRADA (1971). (Span tables for 65 grade Keruing for floor and roof joists, purlins, ply-box beams and two-hinged portals.)

Guide to the use of West African hardwoods, UAC (1972). (Load–span charts and tables for beams, joists, purlins, studs and ridged portal frame members.)

(Universal span charts for any timber, grade and load duration together with simplified tables.)

Solid timber beams and joists

BURGESS, H. J., and PEEK, J. D., 'Span charts for solid timber beams', TRADA TBL 34 (1968).

BURGESS, H. J., 'Further applications of TRADA span charts', TRADA (1971). (Permissible loads and deflections.)

HEARMON, R. F. S., and RIXON, B. E., 'Limiting spans for machine stress-graded Eu. redwood and whitewood', PRL Timberlab 30 (1970). (Span tables.)

Hem-fir, COFI (1973). Load–span tables for floor, roof and ceiling joists for a wide variety of distributed and concentrated loads.)

Glued-laminated timber beams

Span–load tables for glued–laminated softwood beams, BWMA (1967)

Plywood box and web beams

Computer analysis of plywood web beams, COFI (1968). (A Fortran IV programme for the analysis of both symmetrical and unsymmetrical beams.)

Fir plywood web-beam selection manual, COFI (1968). (Tabulates the properties of 4000 standard glued beams.)

Nailed fir plywood web-beams, COFI (1971). (Load span-deflection tables for 24 standard beams.)

WARDLE, T. M., and PEEK, J. D., 'Load tables for ply-box beams', TRADA E/IB/33 (1969). (Load span tables for 72 standard glued beams.)

Portal frames

BURGESS, H. J., *et al.,* 'Span tables for ridged portals in solid timber', TRADA E/IB/17 (1970). (Selection tables for portal member sizes for 5 different timbers.)

Portal frame manual, COFI (1972). (Design and selection manual.)

JOHNSON, V. C., 'Use of portal span tables for farm buildings', TRADA ST 235 (1970)

Stressed skin panels

Fir plywood stressed skin panels, COFI (1971). (Tabulated load-span data.)

'Design data for stressed skin panels', Tech. Pub. 11(M), FPDA (1970). (Tables of data for 900 design cases.)

WARDLE, T. M., and PEEK, J. D., 'Plywood stressed skin panels', TRADA E/IB/22 (1970). (Geometric properties and selected designs.)

Trussed rafters

MAYO, A. P., 'Recommended spans for Fink and Fan trussed rafters made from machine stress-graded timber', PRL Timberlab 32 (1970). (Limiting span tables for M75 timber.)

'Trussed rafters for roofs of dwellings', CP 112: Part 3 (1973), British Standards Institution, London

ABBREVIATIONS AND USEFUL ADDRESSES

AITC	American Institute of Timber Construction Washington DC
ABPVM	Association of British Plywood and Veneer Manufacturers 23–25 City Road, London EC1
BSI	British Standards Institution 2 Park Street, London W1A 2BS
BWMA	British Woodwork Manufacturers Association 26 Store Street, London WC1 E7BT

BWPA	British Wood Preserving Association 62 Oxford Street, London W1N 9WD
COFI	Council of Forest Industries of British Columbia Templar House, 81 High Holborn, London WC1
CITC	Canadian Institute of Timber Construction Ottawa
FPDA	Finnish Plywood Development Association Broadmead House, 21 Panton Street, London SW1
PRL	Princes Risborough Laboratory, (Timberlab) Building Research Establishment, Aylesbury, Bucks
TRADA	Timber Research and Development Association Stocking Lane, Hughenden Valley, High Wycombe, Bucks
UAC	United Africa Company (Timber) Ltd United Africa House, Blackfriars Road, London SE1

16 FOUNDATIONS DESIGN

FOUNDATIONS DESIGN 16

16 FOUNDATIONS DESIGN

M. J. TOMLINSON, F.I.C.E., F.I.Struct.E.,
Wimpey Laboratories Ltd.

GENERAL PRINCIPLES

The function of foundations

A *foundation* is defined in the British Standard Code of Practice for Foundations (CP 2004) as 'that part of the substructure in direct contact with and transmitting loads to the ground'. The *substructure* is defined in the same Code as 'that part of any structure (including building, road, runway, or earthwork) which is below natural or artificial ground level'.

Foundations have the function of spreading the load from the superstructure so that the pressure transmitted to the ground is not of a magnitude such as to cause the ground to fail in shear, or to induce settlement of the ground that will cause distortion and structural failure or unacceptable architectural damage. In fulfilling these functions the foundation, substructure and superstructure should be considered as one unit. The tolerable total and differential settlement must be related to the type and use of the structure and its relationship to the surroundings. Foundations should be designed to be capable of being constructed economically and without risk of protracted delays. The construction stage of foundation work is not infrequently subjected to delays arising from unforeseen ground conditions. The latter cannot always be eliminated even after making detailed site investigations. Thus elaborate and sophisticated designs and construction techniques which depend on an exact foreknowledge of the soil strata should be avoided. Designs should be capable of easy adjustment in depth or lateral extent to allow for variations in ground conditions and should take account of the need for dealing with ground water.

Foundation designs must take into account the effects of construction on adjacent property, and the effects on the environment of such factors as pile-driving vibrations, pumping and discharge of ground water, the disposal of waste materials, and the operation of heavy mechanical plant.

Foundations must be durable to resist attack by aggressive substances in the sea and rivers, in soils and rocks and in ground waters. They must also be designed to resist or to accommodate movement from external causes such as seasonal moisture changes in the soil, frost heave, erosion and seepage, landslides, earthquakes and mining subsidence.

General procedure in foundation design

The various steps which should be followed in the design of foundations are as follows.

(i) A *site investigation* should be undertaken to determine the physical and chemical characteristics of the soils and rocks beneath the site, to observe ground water levels, and to obtain information on all factors relevant to the design of the foundations and their behaviour in service. The general principles and procedures described in Section 10 should be followed.

(ii) *The magnitude and distribution of loading* from the superstructure should be established and placed in the various categories, namely:

Dead loading (permanent structure and self-weight of foundations)
'Permanent' live loading (e.g. materials stored in silos, bunkers or warehouses)
Intermittent live loading (human occupancy of buildings, vehicular traffic, wind pressures)
Dynamic loading (traffic and machinery vibrations, wind gusts, earthquakes).

(iii) The *total and differential settlements* which can be tolerated by the structure should be established. The tolerable limits depend on the allowable stresses in the superstructure, the need to avoid 'architectural' damage to claddings and finishes, and the effects on surrounding works such as damage to piped connections or reversal of fall in drainage outlets. Acceptable differential settlements depend on the type of structure, e.g. framed industrial shedding with pin-jointed steel or pre-cast concrete elements and sheet metal cladding can withstand a much greater degree of differential settlement than a 'prestige' office building with plastered finishes and tiled floors.

(iv) The most suitable *type* of foundation and its *depth* below ground level should be established having regard to the information obtained from the site investigation and taking into consideration the functional requirements of the substructure. For example a basement may be needed for storage purposes or for parking cars.

(v) Preliminary values of the *allowable bearing pressures* (*or pile loadings*) appropriate to the type of foundation should be determined from a knowledge of the ground conditions and the tolerable settlements.

(vi) The *pressure distribution* beneath the foundations should be calculated based on an assessment of foundation widths corresponding to the preliminary bearing pressures or pile loadings, and taking into account eccentric or inclined loading.

(vii) A *settlement analysis* should be made, and from the results the preliminary bearing pressures or foundation depths may need to be adjusted to ensure that total and differential settlements are within acceptable limits. The settlement analysis may be based on simple empirical rules (see Section 8) or a mathematical analysis taking into account the measured compressibility of the soil.

(viii) Approximate *cost estimates* should be made of alternative designs, from which the final design should be selected.

(ix) *Materials* for foundations should be selected and concrete mixes designed taking into account any aggressive substances which may be present in the soil or ground water, or in the overlying water in submerged foundations.

(x) The *structural design* should be prepared.

(xi) The *working drawings* should be made. These should take into account the constructional problems involved and where necessary they should be accompanied by drawings showing the various stages of construction and the design of temporary works such as cofferdams, shoring or underpinning.

Foundation loading

A foundation is required to support the dead load of the superstructure and substructure, the live load resulting from the materials stored in the structure or its occupancy, the weight of any materials used in backfilling above the foundations, and also wind loading.

When considering the factor of safety against shear failure of the soil (see Section 8) the dead loading together with the maximum live load may be either a statutory or code of practice requirement (e.g. the requirements of the British Standard Code of

Practice for Loading, CP 3) or it may be directly calculated if the loads to be applied are known with some precision.

With regard to wind loading, CP 2004 states 'where the foundation loading beneath a structure due to wind is a relatively small proportion of the total loading, it may be permissible to ignore the wind loading in the assessment of allowable bearing pressure, provided the overall factor of safety against shear failure is adequate. For example, where individual foundation loads due to wind are less than 25% of the loadings due to dead and live loads, the wind loads may be neglected in this assessment. Where this ratio exceeds 25%, foundations may be so proportioned that the pressure due to combined dead, live and wind loads does not exceed the allowable bearing pressure by more than 25%.

When considering the long-term settlement of foundations the live load should be taken as the likely realistic applied load over the early years of occupancy of the structure. Consolidation settlements should not necessarily be calculated on the basis of the maximum live load.

Loadings on foundations from machinery are a special case which will be discussed on pp. **16**–52, 53.

The design of foundations to eliminate or reduce total and differential settlements

The amount of differential settlement which is experienced by a structure depends on the variation in compressibility of the ground and the variation in thickness of the compressible material below foundation level. It also depends on the stiffness of the combined foundation and superstructure. Excessive differential settlement results in cracking of claddings and finishes and, in severe cases, to structural damage. Differential settlement can be expressed in terms of angular distortion of the structure. Bjerrum[1] has established danger limits for distortion covering a range of structural conditions (Table 16.1).

Table 16.1 DANGER LIMITS FOR DISTORTION OF STRUCTURES

Angular distortion	*Behaviour of structure*
1/750	Limit where difficulties with machinery sensitive to settlements may be encountered.
1/600	Limit of danger for frames with diagonals.
1/500	Safe limit for buildings where cracking is not permissible.
1/300	Limit where first cracking in panel walls is to be expected.
1/300	Limit where difficulties with overhead cranes are to be expected.
1/250	Limit where tilting of high, rigid buildings might become visible.
1/150	Considerable cracking in panel walls and brick walls.
1/150	Safe limit for flexible brick walls ($h/L < \frac{1}{4}$).
1/150	Limit where general structural damage of buildings is to be feared.

Differential settlement may be eliminated or reduced to a tolerable degree by one or a combination of the following measures:

(a) Provision of a rigid raft either as a thick slab, or with deep beams in two directions, or in cellular construction.
(b) Provision of deep basements or buoyancy rafts to reduce the net bearing pressure on the soil (see pp. **16**–16 and **16**–22).
(c) Transference of foundation loading to deeper and less compressible soil by basements (p. **16**–16), caissons (p. **16**–24), shafts (p. **16**–32) or piles (p. **16**–33).
(d) Provision of jacking pockets within the substructure, or brackets on columns from which to re-level the superstructure by jacking.

(e) Provision of additional loading on lightly loaded areas by ballasting with kentledge or soil.
(f) Ground treatment processes to reduce the compressibility of the soil (p. **16**–14).

Legal requirements and codes of practice

Foundations constructed in the United Kingdom must conform to one of the following sets of building regulations, depending on the location of the site:

The Building Regulations 1972[2]	England, Wales and Northern Ireland
The Building Standards (Scotland) (Consolidation) Regulations 1971[3]	Scotland
London Building Acts 1930–1939 Constructional By-laws[4]	The Inner London Boroughs, i.e. the territory formerly administered by the London County Council.

With the exception of the Inner London Boroughs the building regulations give 'deemed-to-satisfy' provisions for the design and construction of foundations based on Codes of Practice. 'Deemed-to-satisfy' provisions are shown for the following subjects:

FOUNDATIONS

Civil Engineering Code of Practice No. 4 (1954) *Foundations* (now published as CP 2004).

REINFORCED CONCRETE WORK IN FOUNDATIONS

BS Code of Practice 114, Part 2, 1969 *The Structural Use of Reinforced Concrete in Buildings.*

FOUNDATIONS OF BUILDINGS HAVING NOT MORE THAN FOUR STOREYS

BS Code of Practice 101, 1963 *Foundations for substructures for non-industrial buildings of not more than four storeys.*

This 'deemed-to-satisfy' provision does not apply to factory or storage buildings, and in Scotland it does not apply to buildings more than two storeys high. In the Inner London Boroughs the by-laws require the design and construction of the foundations to be to the satisfaction of the District Surveyor. Other British Standard Codes of Practice which are particularly relevant to foundation work are CP 2 *Earth retaining structures,* and CP 2003 *Earthworks.*

SHALLOW FOUNDATIONS

Definitions

CP 2004 defines shallow foundations as those where the depth below finished ground level is less than 3 m and include many strip, pad and raft foundations. The Code states that the choice of 3 m is arbitrary, and shallow foundations where the depth-to-breadth ratio is high may need to be designed as deep foundations.

Pad foundation is an isolated foundation to spread a concentrated load (Figure 16.1).

Strip foundation is a foundation providing a continuous longitudinal bearing (Figure 16.2).

Raft foundation is a foundation continuous in two directions, usually covering an area equal to or greater than the base area of the structure (Figure 16.3).

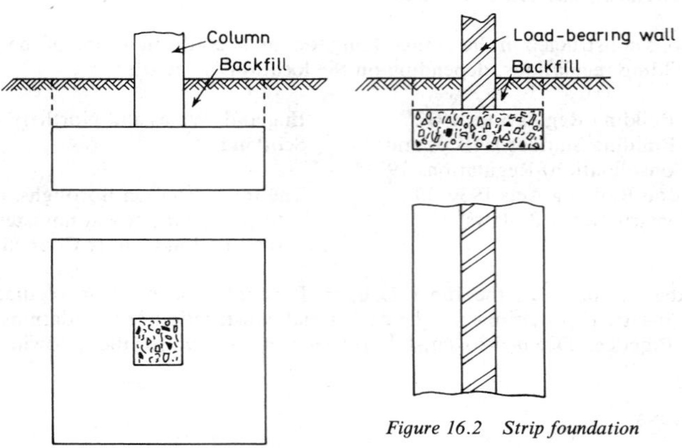

Figure 16.1 Pad foundation

Figure 16.2 Strip foundation

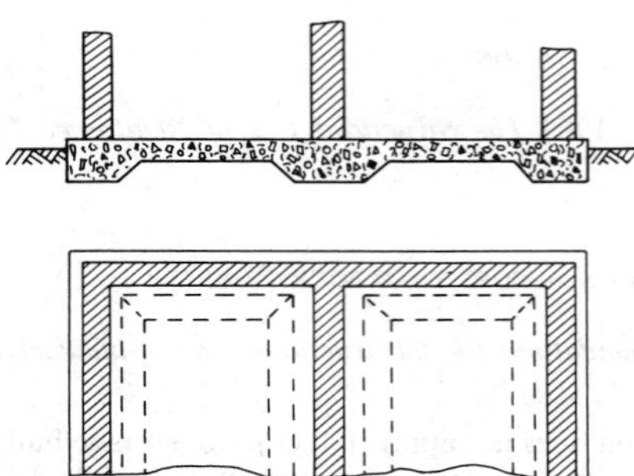

Figure 16.3 Raft foundation

Foundation depths

The Building Regulations[2] merely state that the foundations of a building shall 'be taken down to such a depth, or be so constructed, as to safeguard the building against damage by swelling, shrinking or freezing of the subsoil'.

The first consideration is, of course, that the foundation should be taken down to a depth where the bearing capacity of the soil is adequate to support the foundation loading without failure of the soil in shear or excessive consolidation of the soil. The minimum requirement is thus to take the foundations below loose or disturbed topsoil, or soil liable to erosion by winds or floods. Provided these considerations are met the object should then be to avoid too great a depth to foundation level. A depth greater than 1.5 m will probably require support of the excavation to ensure safe working conditions for operatives fixing reinforcing steel or formwork, which adds to the cost of the work. If at all possible the foundations should be kept above ground water level in order to avoid the costs of pumping, and possible instability of the soil due to seepage of water into the bottom of an excavation. It is usually more economical to adopt wide foundations at a comparatively low bearing pressure, or even to adopt the alternative of piled foundations, than to excavate below ground water level in a water-bearing gravel, sand or silt.

Apart from considerations of allowable bearing pressures, shallow foundations in clay soils are subject to the influences of ground movements caused by swelling and shrinkage (due to seasonal moisture changes or tree root action), in cohesive soils and soft rocks to frost action, and in most ground conditions to the effects of adjacent construction operations such as excavations or pile driving.

It is usual to provide a minimum depth of 500 mm for strip or pad foundations as a safeguard against minor soil erosion, the burrowing of insects or animals, frost heave (in British climatic conditions other than those sites subject to severe frost exposure), and minor local excavations and soil cultivation. This minimum depth is inadequate for foundations on shrinkable clays where swelling and shrinkage of the soil due to seasonal moisture changes may cause appreciable movements of foundations placed at a depth of 1.2 m or less below the ground surface.[5] A depth of 0.9 to 1 m is regarded as a minimum at which some seasonal movement will occur but is unlikely to be of a magnitude sufficient to cause damage to the superstructure or ordinary building finishes.

Movements of clay soils can take place to much greater depths where the soil is affected by the drying action of trees and hedges, and in countries where there is a wide difference between the rainfall in the dry season and wet season. Permafrost (permanently frozen ground) has a considerable influence on foundation depths (see p. **16**–58).

Consideration should be given to the stability of shallow foundations on stepped or sloping ground. Analyses as described in Section 8 should be made to ensure that there is an adequate safety factor against a shear slide due to loading transmitted to the slope from the foundations.

The depth of foundations in relation to mining subsidence problems is discussed on pp. **16**–54 to **16**–58.

Allowable bearing pressures

Allowable bearing pressures (see definition in Section 8) for shallow foundations may be based on experience, or for preliminary design purposes on simple tables of presumed bearing values for a standard range of soil and rock conditions.

Where appropriate, more precise allowable bearing pressures for shallow foundations on cohesionless soils may be obtained from empirical relationships based on the results of *in situ* tests made on the soils (Section 10). In the case of shallow foundations on cohesive soils, the allowable bearing pressures may be obtained by applying an arbitrary safety factor to the ultimate bearing capacity calculated from shear strength determinations on the soil (Section 8). Where settlements are a critical factor in the design of foundations, detailed settlement analyses will be required based on the measured compressibility of the soil (Section 8).

Description of types of shallow foundations

PAD FOUNDATIONS

Pad foundations (Figure 16.1) are suitable to support the columns of framed structures. Pad foundations supporting lightly loaded columns can be constructed using unreinforced concrete, in which case the depth is proportioned so that the angle of spread from the base of the column to the outer edge of the ground bearing does not exceed 1 vertical to

1 horizontal (Figure 16.4). The thickness of the foundation should not be less than the projection from the base of the column to its outer edge, and it should not be less than 150 mm.

Pad foundations to be excavated by a powered rotary auger should be circular in plan, so providing a selfsupporting excavation in firm to stiff cohesive soils and soft rocks. Square or rectangular foundations can be excavated by mechanical grabs or backacters. The designs should not require the bottom to be trimmed by hand to a regular profile (Figure 16.4). This necessitates operatives working at the bottom of excavations in confined conditions, and for safety reasons the sides of excavations deeper than 1.5 m may have to be supported.

Savings in the volume of concrete can be obtained by providing steel reinforcement for pad foundations where heavy column loads are to be carried, and it may be advantageous to save depth of excavation by adopting a relatively thin base slab section (Figure 16.5). Reinforcement is also necessary for foundations carrying eccentric loading

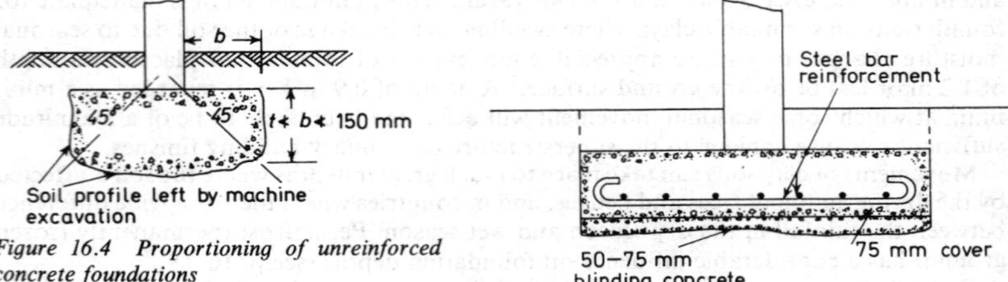

Figure 16.4 Proportioning of unreinforced concrete foundations

Figure 16.5 Reinforced concrete strip foundation

which may induce heavy bending moments and shear forces in the base slab. The procedure for reinforced concrete design is described on pp. **16**–11 to **16**–13.

STRIP FOUNDATIONS

Strip foundations are suitable for supporting load-bearing walls in brickwork or blockwork. The traditional form of strip foundations is shown in Figure 16.6(a). The concrete-filled trench foundation (Figure 16.6(b)) is suitable for stable soils in level ground conditions but should not be used where soil swelling may occur owing, say, to removal of trees or hedges. The swelling is accompanied by horizontal thrust on the foundation followed by movement of the foundation and superstructure. Strip foundations are also an economical method of supporting a row of closely spaced columns (Figure 16.7).

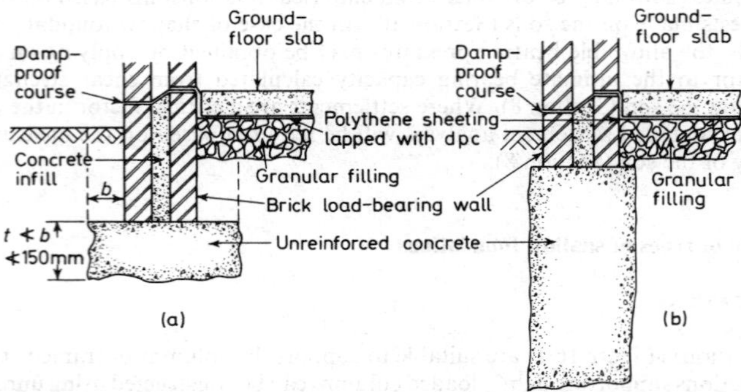

Figure 16.6 Unreinforced concrete strip foundations for load bearing walls. (a) traditional; (b) concrete filled trench

For unreinforced strip foundations the Building Regulations[2] require that the thickness should not be less than the projection from the base of the wall and in no case less than 150 mm. The Regulations also require that where foundations are laid at more

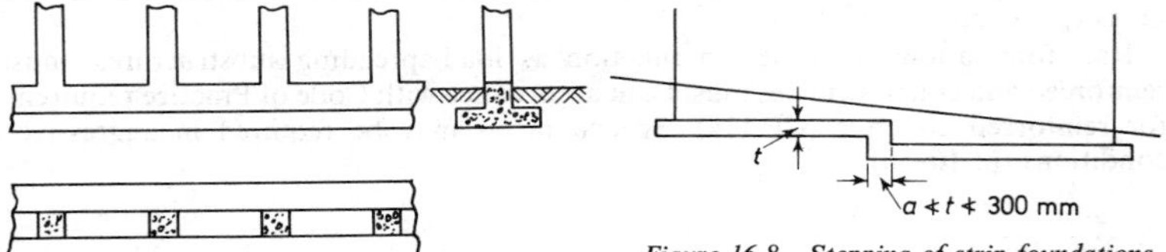

Figure 16.7 Strip foundation for closely spaced columns

Figure 16.8 Stepping of strip foundations

than one level, at each change of level the higher foundation should extend over and unite with the lower one for a distance of not less than the thickness of the foundation and in no case less than 300 mm (Figure 16.8).

The excavations for strip foundations are normally undertaken by a backacter machine, and it is usually possible to trim by the machine bucket to a rectangular bottom profile.

Reinforcement can be provided to strip foundations to enable savings to be made in the volume of concrete and also in foundation depths owing to the lesser required thickness of the base slab. Reinforcement is also necessary to enable the foundations to bridge over weak pockets of soil or to minimise differential settlement due to variable loading conditions, e.g. when a strip foundation is provided to support a row of columns carrying different loads.

The procedure for the design of reinforced concrete foundations is described on p. **16**–13. For unreinforced concrete strip foundations the Building Regulations require the concrete to be composed of 50 kg of cement to not more than 0.1 m^3 of fine aggregate and 0.2 m^3 of coarse aggregate. The design of concrete mixes suitable for aggressive soil conditions is described on p. **16**–60.

RAFT FOUNDATIONS

Raft foundations are a means of spreading foundation loads over a wide area thus minimising bearing pressures and limiting settlement. By stiffening the rafts with beams and providing reinforcement in two directions the differential settlements can be reduced to a minimum.

Edge beams and internal beams can be designed as 'upstand' or 'downstand' projections (Figures 16.9(a) and (b)). Downstand beams save formwork and allow the rafts to

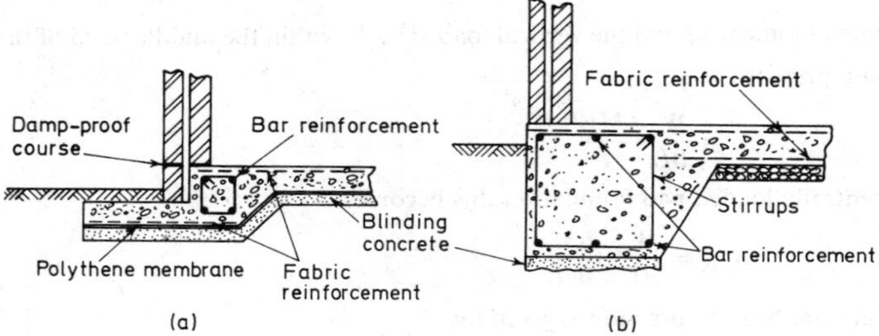

Figure 16.9 Reinforced concrete raft foundations. (a) with upstand beam; (b) with downstand beam

be concreted in one pour. However, the required trench excavations may not be self-supporting in loose soils and there are difficulties in maintaining the required profile in water-bearing ground. Upstand beams are required where rafts are designed to allow horizontal ground movements to take place beneath them, as in mining subsidence areas (p. **16**–56).

Raft foundations, in order to function as load-spreading substructures, must be reinforced and concrete mixes must be in accordance with Code of Practice requirements for reinforced concrete (CP 114). Special mixes may be required in aggressive soil conditions (p. **16**–60).

Shallow foundations carrying eccentric loading

The soil adjacent to the sides of shallow foundations cannot be relied on to provide resistance to overturning moments caused by eccentric loading on the foundations. This is because in clays the soil is likely to shrink away from the foundation in dry weather and, in the case of cohesionless soils, excavation and subsequent backfilling will cause loose conditions around the sides. It is therefore necessary to check that the soil beneath the foundation will not be overstressed or suffer excessive compression under the unequal bearing pressures induced by the eccentric loading.

The pressure distribution beneath an eccentrically loaded foundation is assumed to be linear. For the pad foundation shown in Figure 16.10(a) where the resultant of the

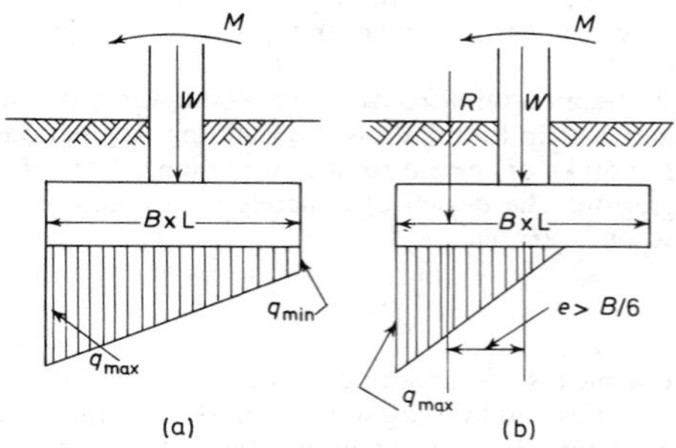

Figure 16.10 Eccentrically loaded foundations. (a) resultant within middle third; (b) resultant outside middle third

overturning moment M and the vertical load W falls within the middle third of the base:

Maximum pressure $= q_{max}$

$$= \frac{W}{BL} + \frac{My}{I} \tag{16.1}$$

For a centrally loaded pad foundation this becomes

$$q_{max} = \frac{W}{BL} + \frac{6M}{B^2L} \tag{16.2}$$

The minimum bearing pressure is given by

$$q_{min} = \frac{W}{BL} - \frac{6M}{B^2L} \tag{16.3}$$

When the resultant of W and M falls outside the middle third of the base, equation (16.3) indicates that tension theoretically occurs beneath the base. However, tension cannot develop and redistribution of bearing pressure will occur as shown in Figure 16.10(b). The maximum bearing pressure is then given by

$$q_{max} = \frac{4W}{3L(B-2e)} \qquad (16.4)$$

In above equations, W is the total axial load on the column, M is the bending moment on the column, y is the distance from the centroid of the pad to the edge, I is the moment of inertia of the plan dimensions of the pad, e is the distance from the centroid of the pad to the line of action of the resultant loading.

The maximum bearing pressure q_{max} should not exceed the allowable bearing pressure appropriate to the depth and width of the foundation, but the effective width for consideration of settlement in cohesionless soils (see Section 8) can be taken as one third of the overall width for the pressure distribution shown in Figure 16.10(b) or for a triangular distribution of pressure.

The structural design of shallow foundations

PAD FOUNDATIONS

The following steps should be taken in the structural design of a pad foundation.

(i) Calculate the base area of the foundation by dividing the total net load by the allowable bearing pressure on the soil, taking into account any eccentric loading.
(ii) Calculate the required overall depth of the base slab at the point of maximum bending moment.
(iii) Decide on either a simple slab base with horizontal upper surface or a sloping upper surface, depending on the economics of construction.
(iv) Check the calculated depth of the slab by computing the beam shear stress at critical sections on the assumption that diagonal shear reinforcement should not be provided.
(v) Design the reinforcement.
(vi) Check the bond stress in the steel.

The following example will illustrate this procedure. A column 600 mm × 600 mm carrying a vertical central load of 2400 kN and a reversible bending moment of 1600 kN m is sited on a soil having an allowable bearing pressure of 180 kN/m^2. Design a suitable foundation for a 1 : 2 : 4 nominal concrete mix as in Table 1 of CP 114.

For the stated concrete mix the other design data are:

Maximum compressive stress in concrete	$= p_{cb} = 7.00$ N/mm^2
Maximum tensile stress in steel	$= p_{st} = 140$ N/mm^2
Modular ratio	$= m = 15$
Maximum shear stress	$= 0.70$ N/mm^2
Maximum average shear bond stress	$= 0.83$ N/mm^2
Maximum local shear bond stress	$= 1.25$ N/mm^2
Lever arm	$= l_a = 0.83 \times$ effective depth
Moment of resistance of most economical section	$= M_r = 1.286\, bd^2$ N

As a trial, assume base dimensions of 5 m × 5 m. For the stated central load and bending moment it can be shown that the resultant force lies within the middle third of the base.

Therefore, from equations (16.2) and (16.3),

$$\text{Maximum edge pressure} = q_{max} = \frac{2400}{5\times 5}+\frac{6\times 1600}{5^2\times 5}$$

$$= 96.00+76.80 = 172.80\ \text{kN/m}^2$$

$$\text{Minimum edge pressure} = q_{min} = 96.00-76.80 = 19.20\ \text{kN/m}^2$$

Thus the maximum pressure is within the allowable value of 180 kN/m² and uplift does not occur beneath the base.

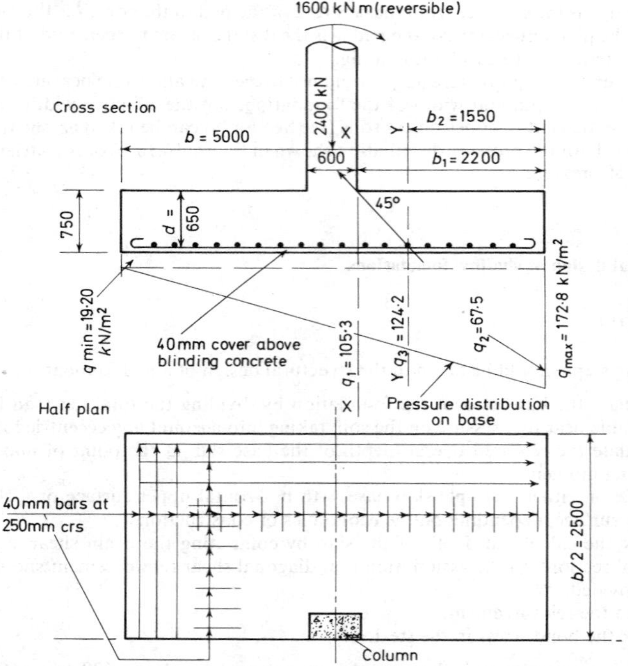

Figure 16.11 Eccentrically loaded reinforced concrete pad foundation

From Figure 16.11

$$\text{Bending moment at face of column} = \frac{q_1\times b_1^2}{2}+\frac{q_2\times b_1}{2}\times\frac{2}{3}\times b_1$$

$$= \frac{105.3\times 2.20^2}{2}+\frac{67.5\times 2.20^2}{3}$$

$$= 364.3\ \text{kN m/m run}$$

$$\text{Required minimum effective depth } d=\left(\frac{364.3\times 1000}{1.286\times 1}\right)^{\frac{1}{2}}$$

$$= 533\ \text{mm}$$

For 40 mm cover and $1\frac{1}{2}\times$ bar diameter, say of 40 mm, the required overall depth is $533+40+1.5\times 40 = 633$ mm.

In order to avoid congestion of close-spaced two-layer reinforcement and to provide a stiff slab it will be desirable to adopt an overall depth of 750 mm, when the effective depth will become 750 − (40 + 60) = 650 mm.

Checking shear stress at critical section Y–Y

$$\text{Shear stress on } bl_a \text{ area} = \frac{\frac{1}{2}(q_3 + q_{max}) \times b_2}{bl_a}$$

$$= \frac{\frac{1}{2}(124.2 + 172.8) \times 1.55 \times 1000}{1000 \times 0.83 \times 650}$$

$$= 0.41 \text{ N/mm}^2 \text{ per metre width (which is within safe limits)}$$

Required area of main steel per metre width in top layer is given by

$$\frac{M}{p_{st} \times 0.83 \times d} = \frac{364.3 \times 1000^2}{140 \times 0.83 \times 650}$$

$$= 4650 \text{ mm}^2$$

Provide 40 mm *bars at* 250 mm *centres* (5026.4 mm²/m) *in top layer and same reinforcement in bottom layer.*

Checking local bond stress at face of column:

$$\text{Bond stress} = \frac{(105.3 \times 2.20 + \frac{1}{2} \times 2.20 \times 67.5) \times 1000}{0.83 \times 650 \times 4 \times \pi \times 40}$$

$$= 1.09 \text{ N/mm}^2 \text{ (which is within safe limits)}$$

If the loading conditions are such as to cause the bending moment to act from any direction it will be necessary to check the bending moments and shears across a diagonal. The arrangement of the reinforcement is shown in Figure 16.11.

STRIP FOUNDATIONS

Strip foundations are designed in a manner similar to that described for pad foundations. For the case of a uniformly loaded strip (Figure 16.12):

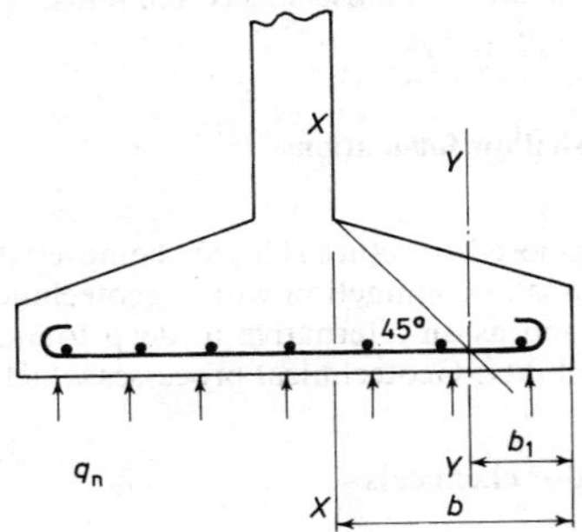

Figure 16.12 Reinforced concrete strip foundation

Bending moment at critical section at face of wall (Section X–X) is

$$\frac{q_n \times b^2}{2} \text{ per unit length of wall}$$

The shear stress should be checked at the critical section Y–Y, and in the case of a steeply sloping upper face and a wide strip it may be necessary to check bending moments and shears at other sections.

$$\text{Shear stress at Section Y–Y} = \frac{q_n \times b_1}{I_a} \text{ per unit length of wall.}$$

RAFT FOUNDATIONS

Rafts are provided on compressible soils, and particularly on soils of variable compressibility. Thus wherever rafts are needed from the aspect of soil compressibility, some settlement is inevitable, either in the form of dishing (on soils of uniform compressibility) or hogging (where the compressibility of the soil or the thickness of the compressible layer varies across the raft) or twisting where the compressibility conditions are irregular.

Distortion of a raft will also occur as a result of variation in the superimposed loading. The magnitude of dishing, hogging or twisting, i.e. the angular distortion of the raft, will depend on the stiffness of the raft and of the superstructure. Only in the case of a uniformly loaded raft on a soil of uniform compressibility can the raft be designed as an inverted floor, either in slab and beam construction or as a stiff slab (Figure 16.3). In all other cases the design is a complex process of redistributing column loads bending moments and shears by the amount calculated from a consideration of the stiffness of the substructure and superstructure and the settlement of the soil. The starting point is always the theoretical total and differential settlements calculated by the soil mechanics engineer on the assumption of a fully flexible foundation. Flexibility of the raft is desirable to keep bending moments and shears to a minimum, but if the raft is too flexible there will be excessive distortion of the superstructure.

A method of design is described by Baker.[6] Analysis of the complex interaction between the raft structure and a subgrade soil undergoing elastic or plastic deformation lends itself to computer methods for solution. Sawko[7] has discussed the problems involved in computer analysis. Where settlements are expected to be fairly small, the complexities of raft design can be avoided by designing the substructure as a series of touching but not interconnected pad or strip foundations. This will greatly reduce the amount of reinforcement required to resist the high bending moments and shears which occur in the short stiff members of a raft with close-spaced columns.

Ground treatment beneath shallow foundations

If the ground beneath a proposed structure is highly compressible it may be economical to adopt shallow foundations in conjunction with a geotechnical process to reduce the compressibility of the ground as an alternative to deep foundations taken down to a stratum of lower compressibility. Geotechnical processes which may be considered are:

(i) Preloading
(ii) Injection of cement or chemicals
(iii) Deep vibration.

Preloading consists of applying a load to the ground equal to or greater than the proposed foundation loading so that settlement of the ground will be complete before the structure is erected. The method is applicable to loose granular soils or granular fills, where the settlement will be rapid. It is generally unsuitable for soft clays where shear failure may occur under rapid application of preload and because of the long-term character of consolidation settlement the preloading would have to be sustained over a long period to be effective. Preloading is most economical over a large area where the granular material such as gravel or colliery waste can be provided in bulk and moved progressively across a site using earth-moving machinery.

The injection of cement or chemicals is suitable for treatment of loose granular soils or fills where the particle size distribution of the materials is suitable for the acceptance of grouts (see Section 8). The effect of the injections is to replace the void spaces by relatively incompressible material, thus greatly reducing the overall compressibility of the ground mass.

Cement or chemicals used for injection are costly and the process is not normally recommended for dealing with large foundation areas or deep compressible strata. The process is usually restricted to small-scale application beneath important structures such as complex machinery installations. It is also employed as a remedial treatment to arrest the excessive settlement of foundations.

Deep vibration methods comprise the insertion of a large vibrating unit into the soil for the full depth required followed by its slow withdrawal. Granular material is fed into the depression surrounding the vibration unit as it is withdrawn, and the unit is re-inserted several times to form a cylinder of densely compacted soil mixed with the imported material. By adopting close-spaced insertions on a grid pattern beneath loaded areas or in single or double rows beneath strip foundations the whole mass of compressible soil can be compacted to a reasonably uniform state, thus reducing the total and differential settlements beneath the applied loading.

In the 'vibroflotation' process the vibratory unit is assisted in its insertion by water jetting. During withdrawal the direction of the jets is reversed to consolidate the added materials. In the 'vibro-replacement' process no water jetting is used, the vibratory unit resembling a large poker vibrator. The process of deep vibration has been described by Greenwood.[8]

The vibroflotation or vibro-replacement processes are not cheap and the cost per metre of depth treated at each insertion point is not very much less than the cost per metre of driven and cast-in-place piling. The process is ineffective in clays or silts, and the depth of treatment is limited to the maximum depth to which the vibratory unit can be inserted which, with the models in current use, is about 11 m. The process has been used to advantage in compacting very loosely placed brick rubble and building debris filling on urban redevelopment sites. Houses can then be built on conventional strip foundations on the fill which has been compacted to a reasonably uniform state of density. The process may not be suitable if the debris contains a high proportion of timber or other organic materials which may decay over a period of years, resulting in further settlement of the fill.

DEEP FOUNDATIONS

Definitions

Deep foundations are required to carry loads from a structure through weak compressible soils or fills on to stronger and less compressible soils or rocks at depth, or for functional reasons. The types of deep foundations in general use are as follows.

Basements are hollow substructures designed to provide working or storage space below ground level. The structural design is governed by their functional requirements rather than from considerations of the most efficient method of resisting external earth and hydrostatic pressures. They are constructed in place in open excavations.

Buoyancy rafts (*hollow box foundations*) are hollow substructures designed to provide a buoyant or semi-buoyant substructure beneath which the net loading on the soil is reduced to the desired low intensity. Buoyancy rafts can be designed to be sunk as caissons (see below): they can also be constructed in place in open excavations.

Caissons are hollow substructures designed to be constructed on or near the surface and then sunk as a single unit to their required level.

Cylinders are small single-cell caissons.

Shaft foundations are constructed within deep excavations supported by lining constructed in place and subsequently filled with concrete or other prefabricated load-bearing units.

Piles are relatively long and slender members constructed by driving preformed units to the desired founding level, or by driving or drilling-in tubes to the required depth—the tubes being filled with concrete before or during withdrawal—or by drilling unlined or wholly or partly lined boreholes which are then filled with concrete. Piles form a large group within the general classification of deep foundations and will be described separately on p. **16**–33 to **16**–47.

The design of basements

GENERAL

Basements are constructed in place in an open excavation. The latter can be excavated with sloping sides, or with ground support in the form of sheeting or sheet piling. The choice of either excavation method depends on the clear space available around the substructure and the need to safeguard existing structures adjacent to the excavation. It may be economical to use the permanent retaining walls as the means of ground support as described on p. **16**–19. A circular shape to a basement can save construction costs where ground support is required, as cross bracing to support the sheeted sides may not be needed. A circular plan should always be considered for structures such as underground pumping stations, or car parks.

The walls of basements are designed as retaining walls subjected to external earth pressure and water pressure. The methods of calculating earth pressure on retaining walls are described in Section 8. If no ground water is encountered in site investigation boreholes it must not be assumed that there will not be any water pressure. For example, where backfill is placed between the walls of a basement and the sides of an excavation in clay soil a reservoir will be formed in which surface water running across the site will collect and a head of water will progressively rise around the walls. Such accumulations of water will not occur in permeable soil or rock formations in which the rate of downward seepage exceeds the inflow from surface water.

The floors of basements are designed to resist the upward earth pressure and any water pressure. The basement slabs span between the external walls or cross walls or between ground beams placed along the lines of the interior columns. Alternatively they can be designed as flat slabs propped at column and wall positions. They act as raft foundations subjected to bending moments and shears induced by differential settlements. The results of the site investigation will normally provide estimates of total and differential settlement on the alternative assumptions of a rigid raft (heavy beam and slab construction) or a fully flexible raft (thin flat slab construction). It is then a matter for the structural designer's judgement to assess the degree of flexibility of the raft and its interaction with the superstructure for the particular design under consideration. The complexities of this assessment have already been discussed on p. **16**–14. Particular points to be taken into consideration with basement floor designs are noted below.

Basements constructed in water-bearing strata may become buoyant if the ground water level in the excavation around the completed (or partly completed) structure is allowed to rise to its normal rest level. At this stage there may not be sufficient loading from the superstructure to prevent uplift occurring. Therefore care should be taken to keep the excavation pumped down until the structural loads have reached the stage when uplift cannot occur.

DESIGN OF BASEMENT FLOORS

Basement floors founded on rock or other relatively incompressible soils will not undergo appreciable downward movement due to elastic or consolidation settlement of the subgrade material. Then differential settlements will be negligible and it will be necessary only to design the floor to resist upward water pressure. If no water table exists or cannot develop in the future then columns and walls can be designed with independent foundations, the floor slab being only of nominal thickness (Figure 16.13).

Where appreciable total and differential settlements of the substructure can occur the basement floor should be designed as a stiff raft, either in slab and beam construction (Figure 16.14(a)) or as a flat slab (Figure 16.14(b)). Design practices are similar to those described on p. **16**–14 for surface rafts.

When basements are supported on piles, and settlements are expected in the pile group, i.e. where the piles terminate on compressible soils, some loading will be transferred to the underside of the floor slab. The magnitude of the pressure which develops

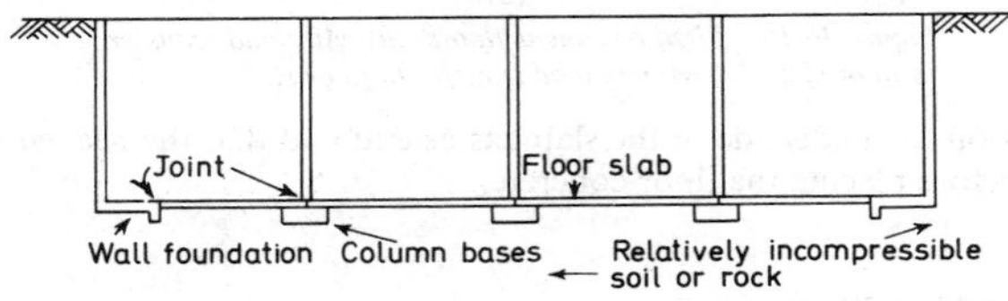

Figure 16.13 Basement floor founded on relatively incompressible stratum

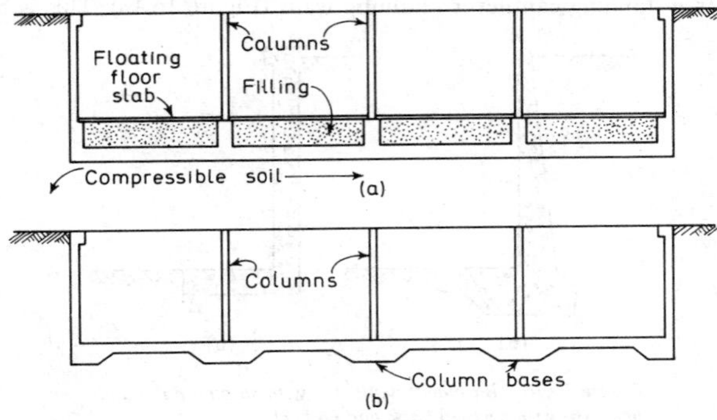

Figure 16.14 Basement floor founded on compressible stratum

will depend on the amount of settlement of the piles, the amount of heave of the base of the excavation due to relief of overburden pressure, the amount of heave and reconsolidation of the soil due to the installation of the piles and the time interval between completion of the excavation (including final trimming and removal of heaved soil) and the time when yielding of the piles commences due to superstructure loading. In all cases where there is potential load transfer to the underside of the floor slab, or where hydrostatic pressure has to be resisted, the piled raft (Figure 16.15(a)) is the appropriate form of construction.

Where the piles are terminated on rock or other relatively incompressible material, and there is no hydrostatic pressure, there will be no load transfer to the floor slab, the latter being only of nominal thickness (Figure 16.15(b)). This assumes that ground heave

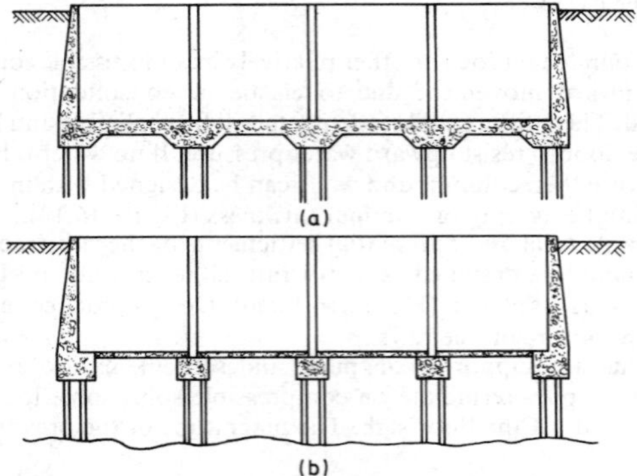

Figure 16.15 Piled basement floors. (a) with load transfer to floor slab; (b) with no load transfer to floor slab

causing uplift on the underside of the slab has ceased and that the heaved soil has been stripped off before placing the floor concrete.

DESIGN OF BASEMENT WALLS

Although the exterior walls of basements are supported by the ground-floor slab of the main structure, and any intermediate subfloors in deep basements, they should be designed as free-standing cantilever retaining walls (Figure 16.16). This is because the

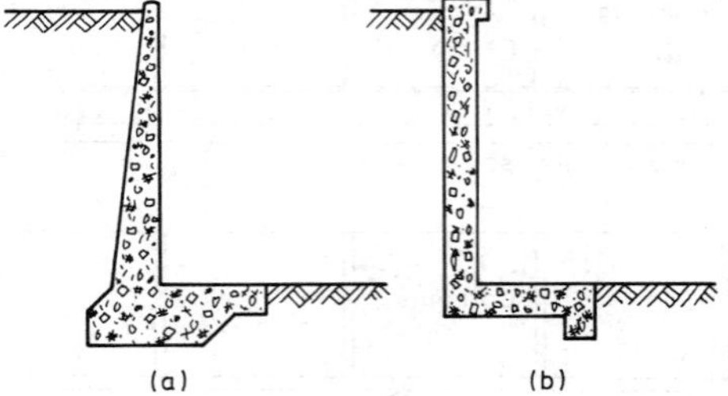

Figure 16.16 Basement walls. (a) with sloping back and heel; (b) with vertical back and no heel

supporting floors are not usually constructed until the final stage of the work (a special method of supporting the external walls of deep basements is shown in Figure 16.22). Similarly the foundation slab of the retaining wall should not be dependent on its connection to the basement floor slab for stability.

The structural form of the retaining wall is governed to some extent by the ground conditions and by the need or otherwise for waterproofing treatment (see below). Thus the sloping back and projecting heel shown in Figure 16.16(a) require additional width of excavation, the cost of which may outweigh the increase in concrete volume required by a wall of uniform thickness (Figure 16.16(b)). In stable ground it may be possible to undercut the excavated face to form the heel enlargement. The wider excavation required for the sloping back wall (Figure 16.16(a)) may be needed in any case to allow room for applying a waterproof asphalt layer, whereas the vertical back requires either an enlarged excavation or the construction of a separate vertical backing wall on which to apply asphalt.

The basement walls can be constructed as diaphragm walls by excavating a narrow trench by mechanical grab using bentonite to support the excavation (Figure 16.17).

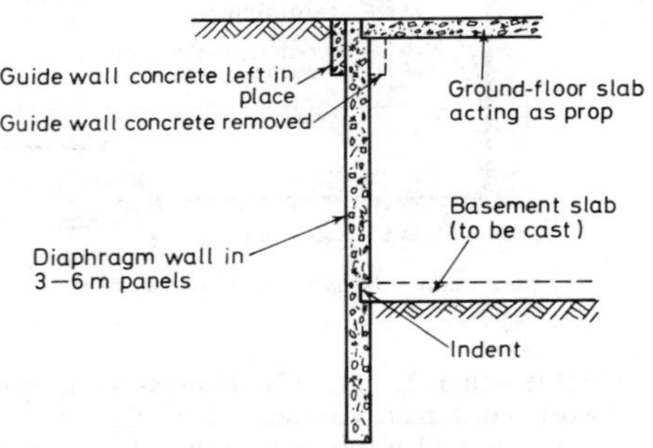

Figure 16.17 Diaphragm wall construction

The excavation is taken out in alternate panels 3–6 m long. A pre-assembled reinforcing cage is lowered into the bentonite-filled trench and then concrete is placed by tremie pipe. The intermediate panels are then constructed in a similar manner. Diaphragm walls are designed as retaining walls using conventional methods for calculating earth pressure (Section 8). However, they cannot usually be designed to act as cantilever walls at the final stage of excavation, and they require to be propped by shores (or held at the top or intermediate levels by ground anchors) as described on p. **16**–22.

Contiguous bored pile walls faced with reinforced concrete can also be used for basements (see Figure 16.50(f)).

WATERPROOFING BASEMENTS

Watertightness of a basement can be obtained either by relying on impervious concrete and leaktight joints, or by providing an impermeable membrane in the form of trowelled-on asphalt tanking or preformed sheathing material. Neither method is entirely reliable.

If complete watertightness is required for functional reasons in a basement it is probable that the asphalt tanking method has a slight advantage compared with relying on the concrete alone, as tanking is a distinct operation carried out by skilled operatives, and the work can be restricted to favourable weather conditions and subjected to intensive supervision; whereas if the concrete alone is to be relied upon for watertightness, the concreting operations proceed in stages over a long construction period, in all weathers, with comparatively unskilled labour, and in congested situations, thus making close supervision difficult at all times.

Asphalt tanking is laid on blinding concrete beneath the basement floor and may be applied either to the exterior of the retaining walls if space is available around the excavations or, in restricted space conditions, it can be applied to a vertical backing wall before constructing the main wall (Figure 16.18). It is useless to apply tanking to the interior of the structural wall as the water pressure will merely force it off. Tanking applied to the exterior of the retaining wall should be protected by a 100 mm thick backing wall (in a manner similar to Figure 16.18) to prevent damage by sharp objects in the backfill materials.

Asphalt tanking is covered by BS 988 and BS 1162 for limestone aggregate and natural rock asphalt aggregate respectively. The tanking should be applied in three coats

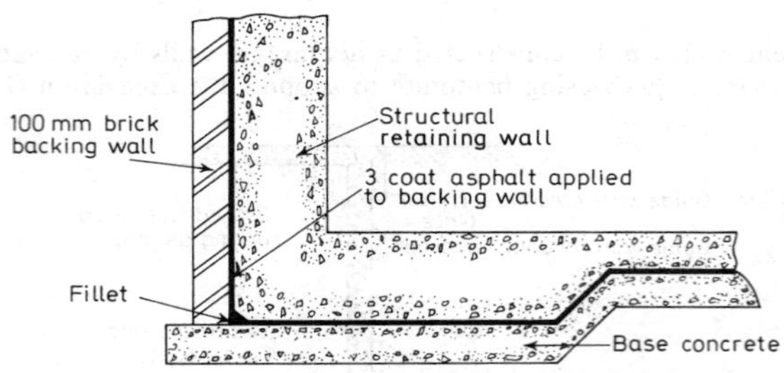

Figure 16.18 Asphalt tanking to basement

to a total thickness of not less than 27 mm for horizontal work and 20 mm for vertical work. Other points of workmanship are covered in CP 102.

Pumps keeping down the ground water level around the excavation should not be shut down until the structural concrete walls have been concreted and have attained their design strength.

CONSTRUCTION OF BASEMENTS

If space around the substructure permits, the most economical method of constructing a basement is to form the excavation with sloping sides, followed by concreting the floor slab and then the retaining walls. If the space is restricted it will be necessary to support the vertical face of the excavation with steel sheet piling (Figure 16.19) or by horizontal timber sheeting in conjunction with vertical soldier piles (Figure 16.20). The sheet piling method is suitable for soft or waterbearing ground where continuous support is necessary and where it is desired to maintain the surrounding ground water table at its normal level to safeguard existing structures. Horizontal sheeting can be used in 'dry' ground conditions, or where drainage towards the excavation can be permitted. In the latter case, hydrostatic pressures do not develop with correspondingly reduced loads to be carried by the bracing system.

The bracing system required to support sheeting to excavations of moderate width (say up to 30 m) can be in the form of horizontal struts and walings restrained against buckling by king piles and vertical cross bracing (Figure 16.21). The struts can be preloaded by jacking to minimise inward movement of the sides. Where wide excavations have to be supported it is preferable to use a system of ground anchors (shown in various stages of construction in conjunction with sheet piling in Figure 16.19) or raking shores (shown in conjunction with horizontal sheeting in Figure 16.20).

Ground anchors have the advantage of providing a clear working space within the excavation and they can conveniently provide a preloading force to minimise inward movement, but there may be problems with existing sewers or other obstructions preventing their installation, also it may be impossible to obtain wayleaves from surrounding property owners. Raking shores obstruct the working space and require substantial bearing blocks at the toe. These may give difficulties with maintaining waterproofing in thin basement slabs.

Inward movement of the sheeted sides of an excavation will take place inevitably owing to relief of lateral pressure on removal of the excavation, the compression of the supporting struts (or stretch and creep of ground anchors) and the thermal movements

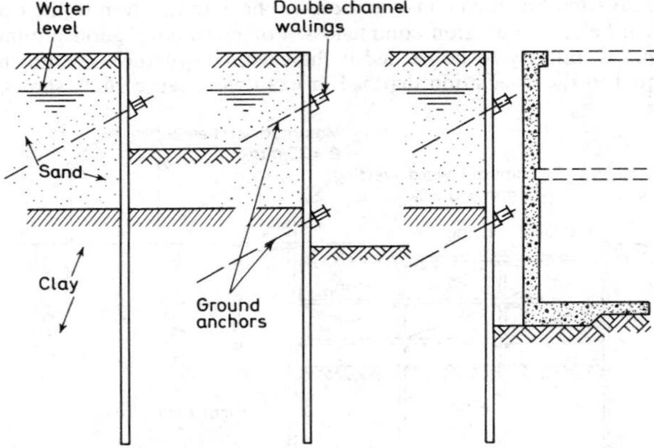

Figure 16.19 Excavation supported by tied-back sheet piling

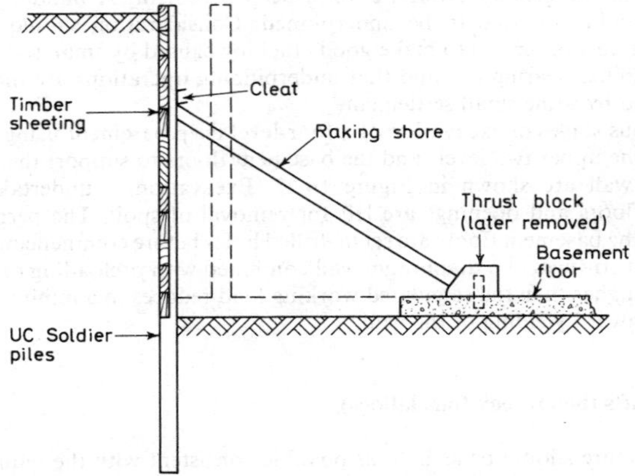

Figure 16.20 Excavation supported by soldier piles and sheeting

of the support system. The inward movement is proportional to the depth of the excavation and its magnitude is governed by the type of soil rather than by the particular support system (assuming the system to be properly designed and carefully executed).

In soft clays and silts, movements of up to $2\frac{1}{2}\%$ of the excavation depth have been observed.[9] In loose sands and gravels various measurements have shown movements in the range of 0.2–0.5% of the excavations depths, while in stiff London clay Cole and Burland[10] observed 0.3% yielding. In soft clays the inward movement is accompanied by a vertical settlement of the same magnitude of the ground surface close to the perimeter of the excavation. The settlement is about half this maximum value at half the excavation depth from the face and falls to a negligible amount at a distance of three or four times the excavation depth from the face. In London clay the observed settlement close to the excavation was only $\frac{1}{2}$ to $\frac{1}{3}$ of the horizontal movement.[10]

If there are existing structures within a distance of three times the excavation depth from the excavation line then consideration will have to be given to the need for underpinning them before excavation commences. For reasonably good ground conditions, underpinning is unlikely to be needed if the existing structures are not nearer than a distance equal to the excavation depth. For example, Figure 16.21 shows the order of

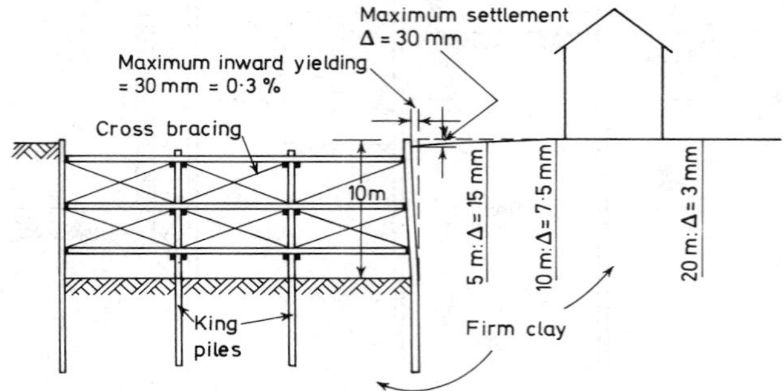

Figure 16.21 Bracing to wide excavation (also showing inward movement)

settlements of the ground around a 10 m deep basement. A building in the position indicated would not need to be underpinned. Consideration should be given to the comparative cost of repairs to make good cracking caused by small settlements and that of underpinning, bearing in mind that underpinning operations are themselves usually accompanied by some small settlement.

The various stages of excavation of a four-level deep basement using ground anchors to support the upper two levels and the basement floors to support the lower levels of a diaphragm wall are shown in Figure 16.22. Excavation is undertaken beneath the completed floors and openings are left for removal of spoil. The permanent columns supporting the basement floors are set in drilled holes before commencing the excavation. The inherent stiffness of a diaphragm wall combined with preloading of ground anchors say to 50% higher than the calculated working load reduces to a minimum (but does not eliminate) inward yielding of the wall.

Buoyancy rafts (hollow box foundations)

The substructure should be as light as possible consistent with the requirement of stiffness. A cellular ('egg box') construction is suitable. This structural form does not normally allow the substructure to be used for any purpose other than its function as a foundation element.

Cellular buoyancy rafts may be designed as caissons (Figure 16.23), which is an economical method of sinking for soft ground conditions, but ground disturbance during sinking can result in some settlement. A buoyancy raft should preferably be constructed within an open excavation. If necessary the cells may be constructed in individual small areas or strips which are subsequently bonded together. By limiting the area of the excavation in this way, the heave and subsequent reconsolidation of a soft clay can be minimised to a marked degree.

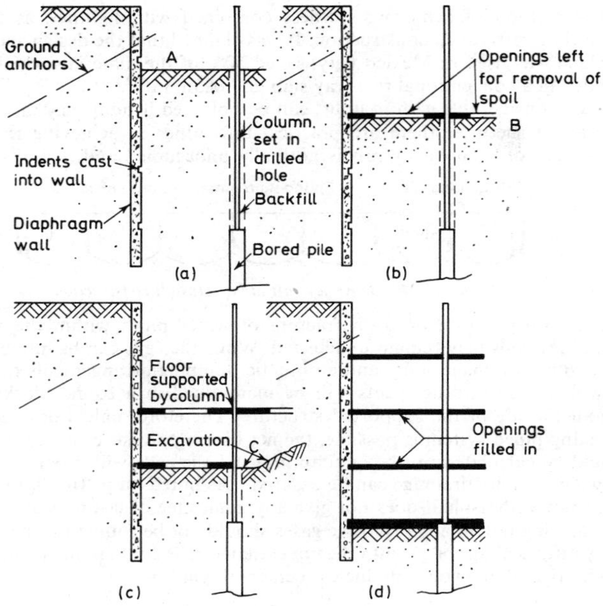

Figure 16.22 Construction of deep basement. (a) Excavation to level A and ground anchors installed; (b) excavation to level B and floor slab cast; (c) excavation to level C and further floor slab cast; (d) completed excavation with all basement floor slabs cast

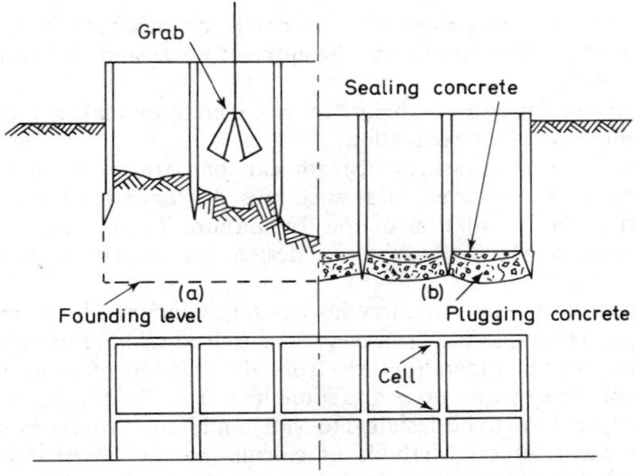

Figure 16.23 Caisson-type cellular buoyancy raft

Another method of achieving stiffness combined with lightness is to design the foundations in barrel shell construction. It was claimed that the design shown in Figure 16.24, which was used in Mexico City, saved 50% of the materials which would have been needed in a conventional two-way slab and beam raft.[11]

Although considerable gain in uplift can be obtained if buoyancy rafts are designed as watertight structures, there are practical difficulties in achieving this. The space within the cells of a buoyancy raft is normally unoccupied and, if leaks occur, either through the substructure or from fracture of water pipes within the structure, the flooding of the cells may remain undetected. While the cells can be interconnected and provided with a drainage sump and automatic pumping arrangements there can be no certainty that these arrangements will be maintained in a sound working condition throughout the life of the supported structure. Therefore, unless drainage by gravity to an existing piped system is possible, the net bearing pressures beneath the buoyancy raft should be calculated on the assumption that the cells will become flooded to the level at which gravity drainage can be assured. As noted on p. **16**–19, the tanking of a buoyancy raft with asphalt does not give any guarantee of lasting watertightness.

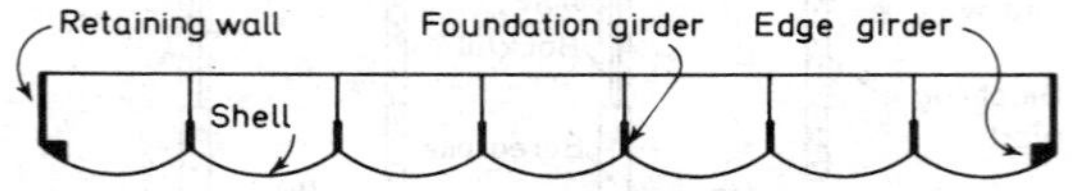

Figure 16.24 Buoyancy raft in barrel vault construction

Pipes carrying potentially explosive gases should not be routed through the cells of a buoyancy raft. Leakage of gas into the unventilated cells could remain undetected with a consequent risk of an explosion due to accidental ignition.

Caisson foundations

GENERAL

The types of caisson foundation are as follows:

A box caisson is closed at the bottom but open to the atmosphere at the top.

An open caisson is open both at the top and bottom.

A compressed air or pneumatic caisson has a working chamber in which air is maintained above atmospheric pressure to prevent the entry of water and soil into the excavation.

A monolith is an open caisson of heavy mass concrete or masonry construction containing one or more wells for excavation.

The allowable bearing pressures beneath caissons are calculated by the methods described in Section 8. However, allowance must be made for the disturbance which may occur during the installation of the foundation. These factors are noted in the following subsections which describe the design and construction methods for the various types.

Caissons are often required to carry horizontal or inclined loads in addition to the vertical loading. As examples, caisson piers to river bridges have to carry lateral loading from wind forces on the superstructure, from the traction of vehicles on the bridge, from river currents, from wave forces and sometimes from floating ice or debris. Caissons in berthing structures have to be designed to withstand impact forces from ships, mooring rope pull, and wave forces. Methods of calculating the bearing pressures beneath eccentrically loaded foundations are described on p. **16**–10. A caisson will be safe against overturning provided that the bearing pressure beneath its edge does not exceed the safe bearing capacity of the foundation material, but it is also necessary to ensure that tilting due to elastic compression and consolidation of the foundation soil or rock does not exceed tolerable limits.

The walls of caissons are frequently subjected to severe stresses during construction. These stresses may arise from launching operations (when caissons are constructed on a slipway and allowed to slide into the water), from wave forces when floating under tow or during sinking, from racking due to uneven support whilst excavating individual cells, from superimposed kentledge, and from the drag effects of skin friction.

Initially lateral pressures on the exernal walls of caissons may be relatively low, corresponding to active pressure of soil loosened by the sinking process. However, with time the loosened soil will reconsolidate, and because the walls may be rigid and unyielding the conditions of earth pressure 'at rest' may develop (the coefficients appropriate to 'active' or 'at rest' earth pressure conditions are stated in Section 8). Where caissons are sunk through stiff over-consolidated clays or shales it may be necessary to cut the excavation larger than the plan dimensions of the foundation. With time the soil will swell to fill the gap and substantial swelling pressures may develop on the external walls.

CP 2004 permits the stresses for temporary work in reinforced concrete or pre-cast components that are only temporarily subjected to loading to be $33\frac{1}{3}\%$ above the working stresses specified in CP 114. However, if permanent members are cast in place, the stresses should be limited to those specified in CP 114 or even lower stresses adopted for work in adverse conditions.

Mild steel used for caisson construction should comply with the requirements of BS 4360. The workmanship and stresses should conform to BS 449 with a permissible increase at the designer's discretion up to 25% for temporary work or temporarily loaded members.

BOX CAISSONS

Box caissons are designed to be floated in water and sunk on to a prepared foundation bed. The stages of sinking are shown in Figure 16.25. The foundation bed is prepared

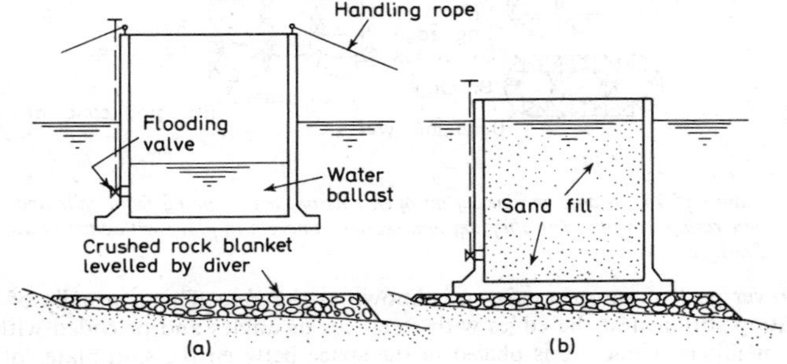

Figure 16.25 Stages in sinking a box caisson. (a) flooding valve opened to admit water ballast; (b) caisson sunk in final position

under water by divers, and the caisson is lowered by opening flooding valves to allow the unit to sink at a controlled rate. Box caissons are suitable for site conditions where the bed can be prepared with little or no excavation below the sea bed or river bed. Thus they are unsuitable for conditions where scour can undermine a shallow foundation. They are also unsuitable for conditions where scour can occur during the final stages of sinking by the action of eddies and currents in the gap between the base of the caisson and the bed material as the gap diminishes. For founding on soft clay or in scouring conditions, box caissons can be sunk on to a piled raft constructed underwater but this method is normally more expensive than adopting an open well caisson.

Box caissons can be of relatively light reinforced concrete construction, since they are not subjected to severe stresses during sinking. Light construction is desirable to give the required freeboard whilst floating. After sinking they can be filled with mass concrete or sand if dead weight is required for the purpose of increasing the resistance to overturning or lateral forces.

OPEN CAISSONS

Open caissons are designed to be sunk by excavating while removing soil beneath them through the open cells. They are designed in such a manner that the dead weight of the caisson together with any kentledge which may be placed upon it exceeds the skin friction of the soil around the walls and the resistance of the soil beneath the bottom (cutting) edges of the walls. To aid sinking, the soil may be excavated from beneath the cutting edges, or kentledge may be placed on the top of the walls to increase the dead weight. The skin friction around the external walls can be reduced considerably by injecting a bentonite slurry above the cutting edge between the walls and the soil. On reaching founding level, mass concrete is placed to plug each cell after which any water in the cells can be pumped out and further concrete placed to form the final seal. The portions of the cells above the sealing plugs can be left empty, or they can be filled with mass concrete, sand, or fresh water depending on the function of the unit and the allowable net bearing pressure. The stages of sinking are shown in Figure 16.26.

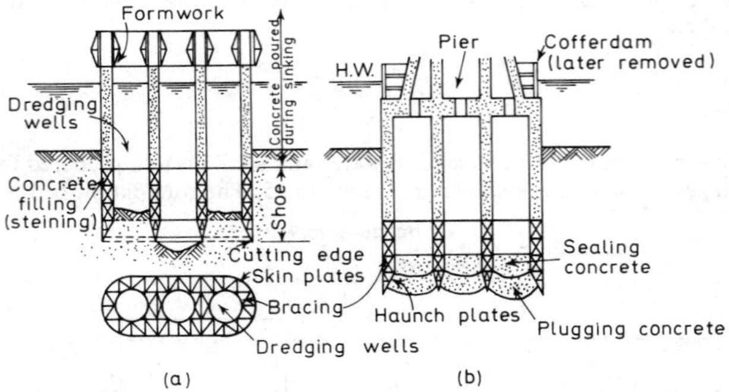

Figure 16.26 Stages in sinking an open caisson. (a) grabbing from cells and concreting in walls; (b) plugging and sealing concrete in place with caisson at final level

The lower part of an open caisson is known as the *shoe*. This is usually of thin mild steel plating stiffened at the edges with steel tees or angles and provided with internal bracing members. Concrete is placed in the space between the skin plates of the shoe to provide ballast for sinking through water and thereafter more concrete and further strakes of skin plating are added to obtain the required downward force to overcome skin friction and the bearing resistance of the soil beneath the cutting edges. While the top of the shoe is still above water level, formwork is assembled and the walls are extended above the shoe in reinforced concrete. The formwork is usually arranged in lifts of about $1\frac{1}{2}$ m, and a 24 hour cycle of operations comprises grabbing to sink $1\frac{1}{2}$ m, erecting steel skin plating or formwork in the walls, placing the concrete, and striking the formwork. Sinking proceeds steadily throughout this cycle. Thick walls are needed for rigidity and to provide dead weight. As well as being reinforced to withstand external earth and hydrostatic pressures, they must resist racking stresses and vertical tension stresses. The latter may occur when the upper part of the caisson is held by skin friction and the lower part tends to fall into the undercut and loosened zone beneath the shoe.

The form of construction, incorporating a shoe fabricated in steel plating, is the traditional method of design, which provides optimum conditions for control of sinking at all stages. However, the introduction of bentonite injection techniques to aid sinking has improved the control conditions making it possible to design caissons entirely in reinforced concrete.

Some typical values used to give a rough guide to skin friction are shown in Table 16.2.

Table 16.2 (AFTER TERZAGHI AND PECK)[12]

Type of soil	*Skin friction* (kN/m^2)
Silt and soft clay	7–30
Very stiff clay	50–200
Loose sand	10–35
Dense sand	30–70
Dense gravel	50–100

The soil is excavated from within the cells and where necessary from below cutting edge level by mechanical grab. In uncemented granular soils the spoil can be removed by an air lift pump. On reaching founding level any kentledge placed on the walls is removed to arrest sinking and mass concrete is quickly placed at and below cutting edge level in the corner cells to provide a bearing on which the caisson comes to rest. The remaining outer cells are then plugged with concrete followed by completion of excavating and plugging of the inner cells. The concrete plugs are placed under water and after the concrete has hardened the cells are pumped out and further sealing concrete is placed.

Accuracy in the positioning of caissons and control of verticality while sinking are necessary. Various methods of achieving these are:

(a) sinking within piled enclosure (Figure 16.27)
(b) sinking between moored pontoons (Figure 16.28)
(c) sinking through a sand island (Figure 16.29)

The choice of method depends on the site conditions, i.e. the depth of water, degrees of exposure, and velocity of sea or river currents. It also depends on the number of caissons to be sunk on any particular project. The cost of an elaborate floating sinking set as shown in Figure 16.28 is justified if spread over a number of sinking sites. Lowering during sinking can be achieved by using suspension links and jacks (Figure 16.27), by lowering from block and tackle (Figure 16.28), by free sinking with the use of guides, or by the controlled expulsion of air from the cells in conjunction with air domes (Figure 16.30).

Open well caissons are best suited to sinking to a moderate depth in soft or loose soils to reach a founding level on stiff or compact material, i.e. through materials which can be readily dredged and are free of obstruction such as boulders, tree trunks or sunken vessels. They are unsuitable for ground containing obstructions which cannot be broken out from beneath the cutting edge, and are unsuitable for sinking on to an irregular rock surface. Problems also arise when founding on soft rocks. Grabbing through water causes softening and breakdown of the rock, making it difficult to judge when a satisfactory bearing stratum has been reached and to clean the rock surface to receive the concrete plug.

Removal of soil from within or below the cells of an open caisson causes quite appreciable loss of ground, i.e. the total volume of soil excavated exceeds the volume displaced by the caisson. Therefore open caissons are unsuitable for sinking close to existing structures.

Some of the difficulties mentioned above can be overcome by providing an open caisson

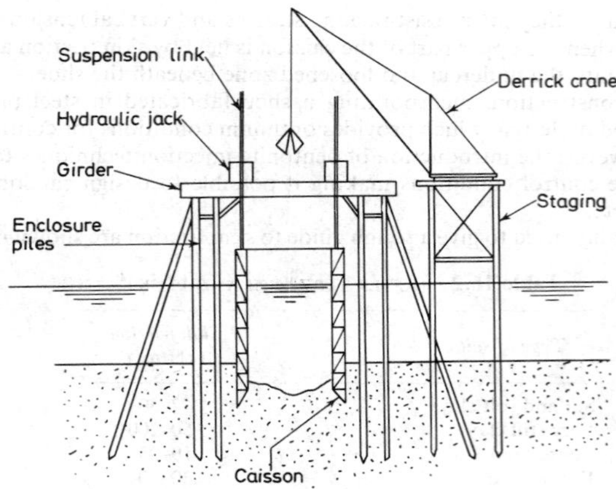

Figure 16.27 Lowering caisson from piled staging

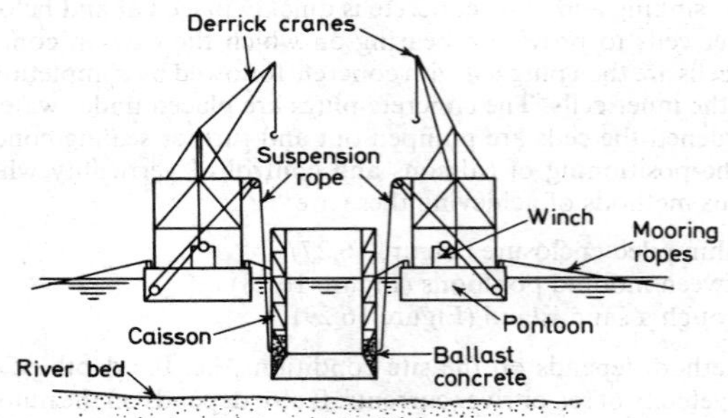

Figure 16.28 Lowering caisson from pontoons

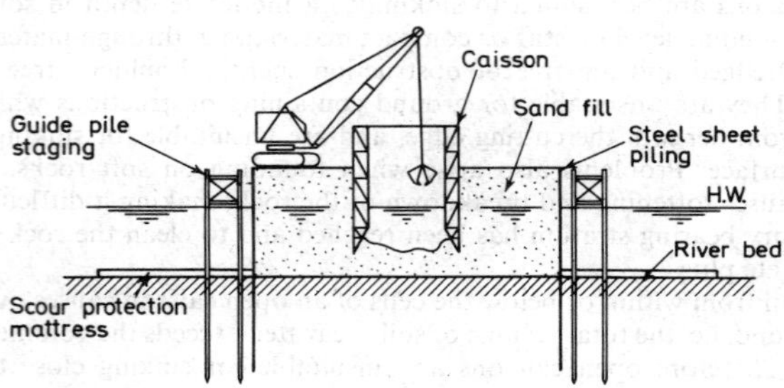

Figure 16.29 Sinking caisson through a sand island

with air domes. These are provided with air locks and are designed to be placed over individual cells as required. Having placed a dome on top of a cell, compressed air is introduced to expel water, after which workmen can enter through an air lock to remove obstructions or to prepare the bottom to receive the sealing concrete. There are limits to the air pressure under which operatives can work in this manner (see p. **16**–31). Air domes provided on all cells can be used as a means of floating an open caisson to the sinking site and for controlling its verticality during sinking by varying the rate of expulsion of air from individual cells. Caissons designed in this way are known as flotation caissons. A design used for the Tagus River Bridge[13] is shown in Figure 16.30. The cutting edge of this caisson was 'tailored' to suit the profile of the rock surface on which the caisson was landed. The domes of flotation caissons are not normally provided with an air lock.

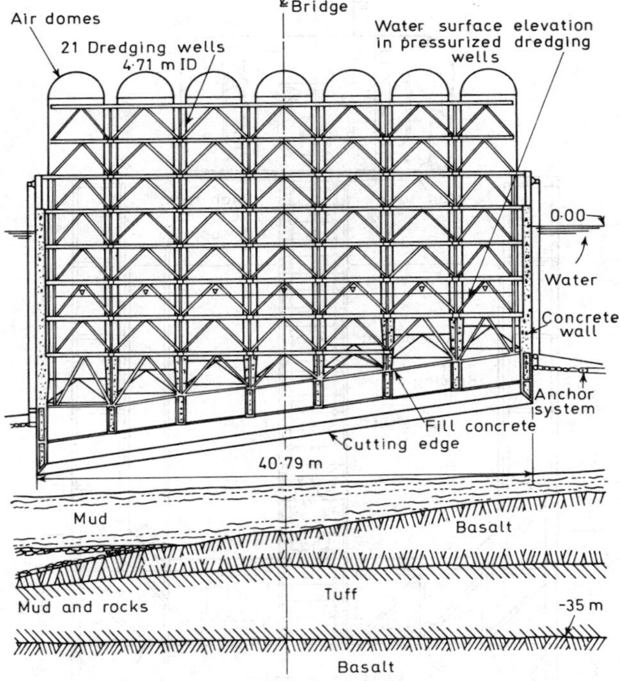

Figure 16.30 Flotation caisson for the Tagus River Bridge (after Riggs[13])

After they have been removed, grabbing proceeds in the normal way for open well caissons.

PNEUMATIC CAISSONS

Pneumatic caissons are designed to be sunk with the assistance of compressed air to obtain a 'dry' working chamber. The general arrangement is shown in Figure 16.31. The caisson consists of a single working chamber surrounded by the shoe with its cutting edge, and a heavy roof. Walls are extended above the shoe in the form of double steel skin plating with mass concrete infilling. The height of the walls depends on the weight required to provide sinking effort and the need to provide freeboard when sinking through water. The airshaft extends from the working chamber to the full height of the caisson and it is surmounted by a combined man lock and muck lock. As their name implies the former is used for access and egress by operatives and the latter for removal of spoil in crane buckets. The man locks must at all times be above the highest tide or river flood levels, with due allowance being made for rapid sinking in soft or loose soils.

Work in pneumatic caissons is regulated by the Factories Act.[15] The regulations require 10 ft^3 (283 litres) of fresh air per minute per person in the working chamber

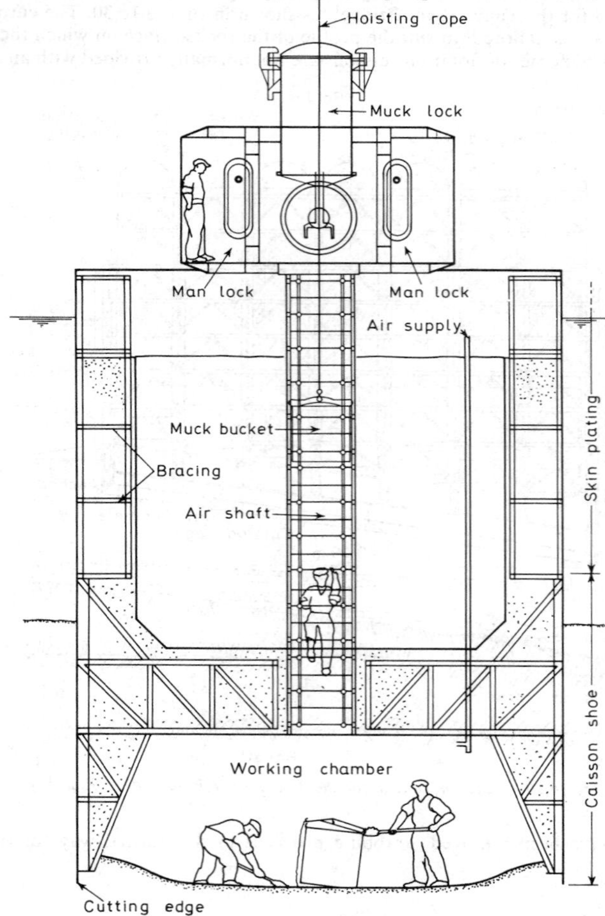

Figure 16.31 Compressed air caisson (after Wilson and Sully[14])

at the pressure in the chamber. The air is supplied from stationary compressors powered by diesel or electric motors. Standby power must be available if the site conditions are such as to endanger life or property in the event of failure of the main supply. To improve working conditions and to reduce the incidence of caisson sickness the air supply should be treated to warm it for working in cold weather and to cool it for hot weather working.

In tropical climates the air should be de-humidified to keep the wet bulb temperature at less than 25 °C. In very permeable ground the escape of air into the soil beneath the working chamber may cause too great a demand on the air supply. This can be reduced by pregrouting the ground with clay, cement or chemicals.

If the dead weight of the caisson together with any added kentledge is insufficient to overcome the skin friction, the effective sinking weight can be temporarily increased by 'blowing down' the caisson. This involves removing the operatives from the working chamber then reducing the air pressure by about one-quarter of the gauge pressure.

On nearing founding level, concrete blocks are placed on the floor of the working chamber and the roof is allowed to come to rest on them. The working chamber is then filled with concrete and the air shaft and airlocks removed.

The pneumatic caisson is suitable for sinking close to existing structures since the excavation is not accompanied by loss of ground. It is also suitable for sinking in ground containing obstructions, and for founding on an irregular rock bed. Pneumatic caissons have the severe limitation that the depth of sinking cannot exceed a level at which the required air pressure to exclude water from the working chamber exceeds the limit at which operatives can work without danger to their health. Generally a pressure of 50 lbf/in^2 (345 kN/m^2) is considered to be a safe maximum but stringent medical precautions and supervision are required at all stages of the work. Generally the high cost of compressed air sinking precludes pneumatic caissons for all but special foundations where no alternatives are feasible or economically possible.

MONOLITHS AND CYLINDERS

Monoliths are open caissons of reinforced concrete or mass concrete construction (Figure 16.32) and are mainly used for quay walls where their heavy weight and massive

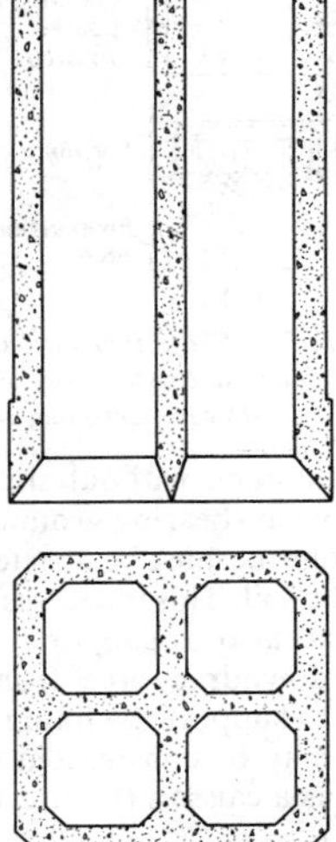

Figure 16.32 Concrete monolith

construction are favourable for resisting the thrust of the filling behind the wall, and for withstanding the impact forces from berthing ships. Because of their weight they are unsuitable for sinking through deep soft deposits. Generally, their design and method of construction follow the same principles as those for open caissons (described on p. **16**–26).

Open caissons of cylindrical form and having a single cell are sometimes referred to as cylinder foundations.

SHAFT FOUNDATIONS

Where deep foundations are required for the heavily loaded columns of a structure it may be desirable to sink the foundation in the form of a lined shaft excavated by hand or by mechanical grab. This type of foundation is similar to the large bored pile as described on p. **16**–43 but its distinguishing characteristic is the construction of the lining in place, taken down stage by stage as the shaft is deepened. The shaft foundation would be selected in cases where the required diameter was larger than the capability of the large-bored pile drilling machine, in ground containing boulders or other obstructions which could prevent machine drilling or caisson sinking, and in localities where specialist pile drilling plant is not available but where labour for hand excavation can be provided from local resources.

Shaft foundations can be of any desired shape but the cylindrical form is the most convenient since internal bracing is not required. The lining can consist of mass concrete placed *in situ* behind formwork (Figure 16.33(a)) or bolted pre-cast concrete, steel or cast-iron segments (Figure 16.33(b)). The *in situ* concrete lining is suitable for relatively

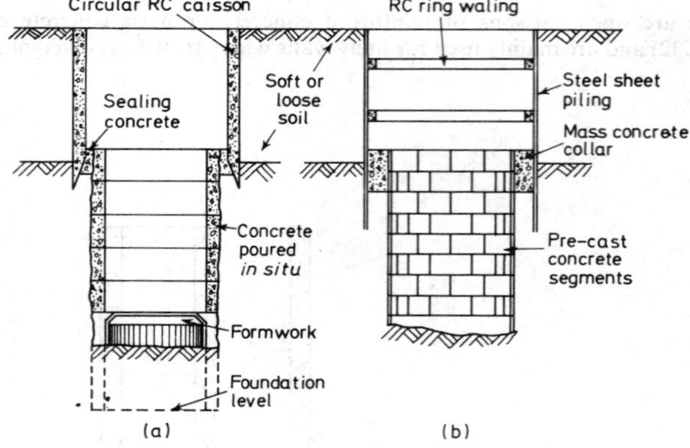

Figure 16.33 Shaft foundations. (a) with mass concrete lining constructed below a caisson; (b) with pre-cast concrete segmental lining constructed below a sheet-piled cofferdam

dry ground which can stand without support for a height of about $1\frac{1}{2}$ m. Segmental lining can be used in water-bearing ground which can stand unsupported for the height of a segment. Cement grout must be injected at intervals into the space between the back of the segments and the soil. This is necessary to prevent excessive flow of water down the back of the lining, and also to support the segments from dropping under their own weight augmented by downdrag forces from the loosened soil. The collar at the top of the shaft is also required to support the lining.

Shaft foundations may be constructed as a second stage after first sinking through soft or loose ground as a caisson (Figure 16.33(a) or at the base of a sheet piled cofferdam (Figure 16.33(b)).

PILED FOUNDATIONS

General descriptions of pile types

There is a large variety of types of pile used for foundation work. The choice depends on the environmental and ground conditions, the presence or absence of ground water, the function of the pile (i.e. whether compression, uplift or lateral loads are to be carried), the desired speed of construction, and consideration of relative cost. The ability of the pile to resist aggressive substances or organisms in the ground or in surrounding water must also be considered.

In CP 2004, piles are grouped into three categories:

(a) Displacement piles: These include all solid piles, including timber and precast concrete and steel or concrete tubes closed at the lower end by a shoe or plug, which may be either left in place or extruded to form an enlarged foot.
(b) Small displacement piles: These include rolled steel sections, open-ended tubes and hollow sections if the ground enters freely during driving and screw piles.
(c) Non-displacement piles: These are formed by boring or other methods of excavation; the borehole may be lined with a casing or tube that is either left in place or extracted as the hole is filled.'

Displacement or small-displacement piles in preformed sections are suitable for open sites where large numbers of piles are required. They can be pre-cast or fabricated by mass production methods and driven at a fast rate by mobile rigs. They are suitable for soft and aggressive soil conditions when the whole material of the pile can be checked for soundness before being driven. Preformed piles are not damaged by the driving of adjacent piles, nor is their installation affected by ground water. They are normally selected for river and marine works where they can be driven through water and in sections suitable for resisting lateral and uplift loads. They can also be driven in very long lengths.

Displacement piles in preformed sections cannot be varied readily in length to suit the varying level of the bearing stratum, but certain types of pre-cast concrete piles can be assembled from short sections jointed to form assemblies of variable length. In hard driving conditions preformed piles may break causing delays when the broken units are withdrawn or replacement piles driven. A worse feature is unseen damage particularly when driving slender units in long lengths which may be deflected from the correct alignment to an extent at which the bending stresses cause fracture of the pile.

When solid pile sections are driven in large groups the resulting displacement of the ground may lift piles already driven from their seating on the bearing stratum, or may damage existing underground structures or services. Problems of ground heave can be overcome or partially overcome in some circumstances by re-driving risen piles, or by inserting the piles in pre-bored holes. Small-displacement piles are advantageous for soil conditions giving rise to ground heave.

Displacement piles suffer a major disadvantage when used in urban areas where the noise and vibration caused by driving them can cause a nuisance to the public and damage to existing structures. Other disadvantages are the inability to drive them in very large diameters, and they cannot be used where the available headroom is insufficient to accommodate the driving rig.

Driven and cast-in-place piles are widely used in the displacement pile group. A tube closed at its lower end by a detachable shoe or by a plug of gravel or dry concrete is driven to the desired penetration. Steel reinforcement is lowered down the tube and the latter is then withdrawn during or after placing the concrete. These types have the advantages that the length can be varied readily to suit variation in the level of the bearing stratum, the closed end excludes ground water, an enlarged base can be formed by hammering out the concrete placed at the toe, the reinforcement is required only for the function of the pile as a foundation element (i.e. not from considerations of lifting and driving as for the pre-cast concrete pile), and the noise and vibrations are not severe when the piles are driven by a drop hammer operating within the drive tube.

Driven and cast-in-place piles may not be suitable for very soft soil conditions where the newly placed concrete can be squeezed inwards as the drive tube is withdrawn causing 'necking' of the pile shaft, nor are they suitable for ground where water is encountered under artesian head which washes out the cement from the unset concrete. Ground heave can damage adjacent piles before the concrete has hardened, and heaved piles cannot easily be redriven. However, this problem can be overcome by driving a number of tubes in a group in advance of placing the concrete. The latter is delayed until pile driving has proceeded to a distance of at least $6\frac{1}{2}$ pile diameters from the one being concreted if small (up to 3 mm) uplift is permitted, or 8 diameters away if negligible (less than 3 mm) uplift must be achieved.[16] The lengths of driven and cast-in-place piles are limited by the ability of the driving rigs to extract the drive tube, and they cannot be installed in very large diameters. They are unsuitable for river or marine works unless specially adapted for extending them through water, and they cannot be driven in situations of low headroom.

Non-displacement piles or bored piles are formed by drilling a borehole to the desired depth, followed by placing a cage of steel reinforcement and then placing concrete. It may be necessary to support the borehole by steel tubing (or casing) which is driven down or allowed to sink under its own weight as the borehole is drilled. Normally the casing is filled completely with easily workable concrete before it is extracted when the concrete slumps outwards to fill the void so formed.

In stiff cohesive soils or soft rocks it is possible to use a rotary tool to form an enlarged base to the piles which greatly increases the end bearing resistance. Alternatively, men can descend the shafts of large-diameter piles to form an enlarged base by hand excavation.

Care is needed in placing concrete in bored piles. In very soft ground there is a tendency to squeezing of the unset concrete, and if water is met under artesian head it may wash out the cement from the unset concrete. If water cannot be excluded from the pile borehole by the casing, no attempt should be made to pump it out before placing concrete. In these circumstances the concrete should be placed under water by tremie pipe. A bottom opening skip should not be used. Breaks in the concrete shafts of bored piles may occur if the concrete is lifted when withdrawing the casing, or if soil falls into the space above the concrete due to premature withdrawal of the casing.

Bored piles have the advantages that their length can be readily varied to suit varying ground conditions, the soil or rock removed during boring can be inspected and if necessary subjected to tests, and very large shaft diameters are possible, with enlarged base diameters up to 6 m. Bored piles can be drilled to any desired depth and in any soil or rock conditions. They can be installed without appreciable noise or vibration in conditions of low headroom, and without risk of ground heave.

Bored piles are unsuitable for obtaining economical skin friction and end bearing values on granular soils because of loosening of these soils by drilling. Boring in soft or loose soils results in loss of ground which may cause excessive settlement of adjacent structures. They are also unsuitable for marine works.

Details of some types of displacement piles

TIMBER PILES

In countries where timber is readily available, timber piles are suitable for light to moderate loadings (up to 300 kN). Softwoods require preservation in creosote in accordance with Group I, Table 4 of CP 98. If this is done they will have a long life below ground water level but are subject to decay above this level. Where possible, pile caps in concrete should be taken down to water level (Figure 16.34(a)). If this is too deep, a composite pile may be installed, the upper part above water level being in pre-cast concrete or concrete cast-in-place jointed to a timber section (Figure 16.34(b)).

To prevent damage to timber piles during driving, the head should be protected by a steel or iron ring, and the toe by a cast-iron shoe (Figure 16.35).

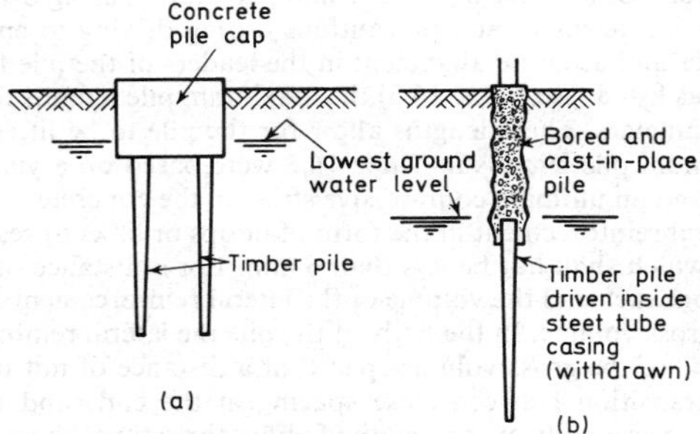

Figure 16.34 Methods of avoiding decay in timber piles

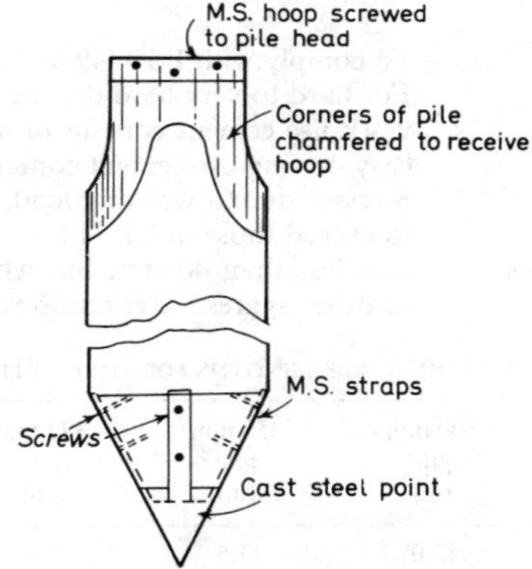

Figure 16.35 Protecting the head and toe of a timber pile

CP 2004 requires the working stresses in compression on a timber pile not to exceed those tabulated in CP 112 for compression parallel to the grain for the species and grade of timber used, due allowances being made for eccentricity of loading, non-verticality of driving, bending stresses due to lateral loads, and reductions in section due to drilling lifting holes or notching the piles. The working stresses of CP 112 may be exceeded while the pile is being driven.

PRE-CAST AND PRESTRESSED CONCRETE PILES

Pre-cast reinforced concrete piles are not economical for general use because a considerable amount of steel reinforcement is needed to withstand bending stresses during lifting and subsequent compressive and tensile stresses during driving. Much of this reinforcement is not required once the pile is in the ground. Pre-cast concrete piles are also liable to damage on handling and during driving in hard ground.

The effect of prestressing of solid or hollow concrete piles in conjunction with high-quality concrete is to produce a unit which should not suffer hair cracks while being lifted or transported and therefore should produce a more durable foundation element than the ordinary pre-cast concrete pile. This is advantageous in aggressive ground conditions. However, prestressed concrete piles are liable to crack during driving, requiring careful detailing of reinforcement, and precautions during driving to ensure concentric blows of the hammer and accurate alignment in the leaders of the pile frame.

Saurin[17] has listed (see Table 16.3) the maximum pile lengths for main reinforcement of various diameters. These lengths allow for the pile to be lifted at the head and toe.

The maximum pile lengths in Table 16.3 were based on a yield stress in the steel of 262 N/mm^2 and an ultimate compressive stress in the concrete of 27.8 N/mm^2. CP 2004 requires lateral reinforcement in the form of hoops or links to resist driving stresses, the diameter of which shall not be less than 6 mm. For a distance of three times the width of the pile from each end the volume of the lateral reinforcement should not be less than 0.6% of the gross volume. In the body of the pile the lateral reinforcement should not be less than 0.2% of the gross volume spaced at a distance of not more than half the pile width. The transition between close spacing at the ends and the maximum spacing should be made gradually over a length of about three times the width. A typical pre-cast concrete pile of solid section designed for fairly easy driving conditions and the minimum transverse reinforcement required by CP 2004 is shown in Figure 16.36. Other CP 2004 recommendations are

Reinforcement	—To comply with BS 4449 and 4461.
Concrete mixes	—For hard to very hard driving conditions and all marine work use cement content of 400 kg/m^3. For normal or easy driving use cement content of 300 kg/m^3.
Concrete design	—Stresses due to working load, handling and driving not to exceed those in CP 114.
Cover to reinforcement	—Not less than 40 mm, but where exposed to sea water or other aggressive conditions not less than 50 mm.

Table 16.3 MAXIMUM PILE LENGTHS FOR GIVEN REINFORCEMENT

Bar diameter for 4 no. bars (mm)	300 mm *pile* (m)	350 mm *pile* (m)	375 mm *pile* (m)	400 mm *pile* (m)
20	12.0	11.5	—	—
25	15.0	14.0	13.75	13.5
32	—	17.5	17.25	16.75
40	—	—	21.0	20.5

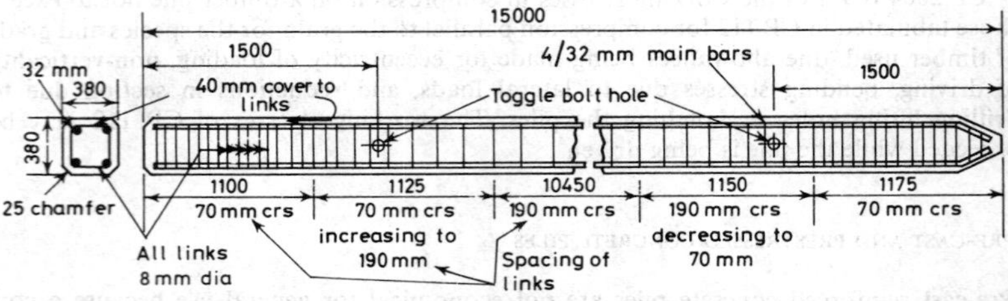

Figure 16.36 Design of pre-cast concrete pile suitable for fairly easy driving conditions

Where piles are driven through hard ground or ground containing obstructions liable to damage the toe of a pile, a cast steel or cast iron shoe should be provided as shown in Figure 16.37(a). For driving on to a sloping hard rock surface a rock point should be provided as shown in Figure 16.37(b) to prevent the toe skidding down the slope. A shoe need not be provided for easy driving in clays and sands when the pile may be terminated as shown in Figure 16.37(c).

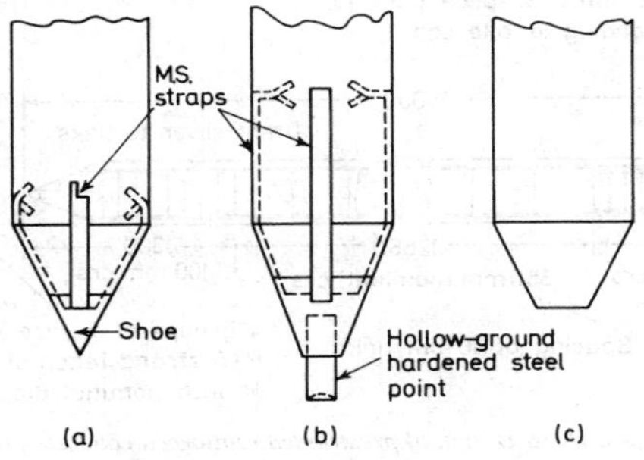

Figure 16.37 Design of toe for pre-cast or pre-stressed concrete pile

The recommendations of CP 2004 for prestressed concrete piles are as follows:

Materials —To be in accordance with CP 115.

Design —Maximum axial stress calculated from 28-day works cube stress less prestress after losses. (The stress should be reduced if the ratio of effective length to radius of gyration exceeds 50).

Static stresses produced by lifting and pitching not to exceed 3.1 N/mm^2 in tension and 13.9 N/mm^2 in compression for 28-day cube strength of 42 N/mm^2. (For 28-day cube strength of 50 N/mm^2, tension stress not to exceed 3.4 N/mm^2 and compression stress not to exceed 17 N/mm^2.)

Prestress —Minimum prestress is related to ratio of weight of hammer to weight of pile thus:

Ratio	0.9	0.8	0.7	0.6
Minimum prestress for normal driving (N/mm^2)	2.0	3.5	5.0	6.0
Minimum prestress for easy driving (N/mm^2)	3.5	4.0	5.0	6.0

The minimum prestress for diesel hammer should be 5.0 N/mm^2

Lateral reinforcement—Mild steel stirrups not less than 6 mm diameter spaced at a pitch of not more than side dimensions less 50 mm. At top and bottom for length of three times side dimension stirrup volume not less than 0.6% of pile volume.

Cover —As for pre-cast concrete piles (see above).

To minimise damage to pile heads during driving, pre-cast concrete or prestressed concrete piles should be driven with timber or plastic packing between the helmet and the hammer. The hammer weight should be roughly equal to the weight of the pile and never less than half its weight. The drop should be 1 to 1.25 m. Particular care is necessary when driving with a diesel hammer when an uncontrollable sharp impact can break the pile if the toe meets a hard layer. Drop hammers or single-acting hammers are preferable for these ground conditions.

A typical pre-stressed concrete pile designed to the above recommendations is shown in Figure 16.38.

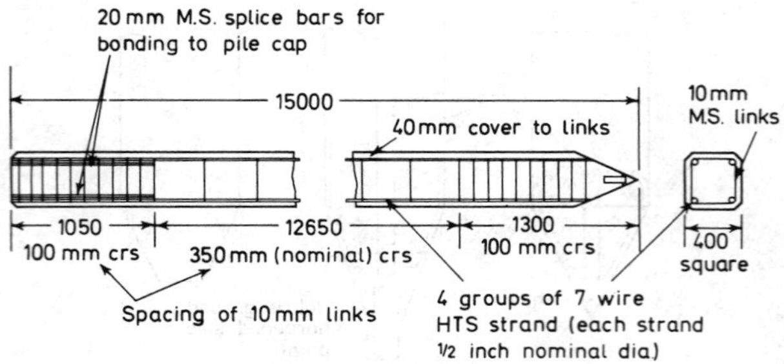

Figure 16.38 Design of prestressed reinforced concrete pile

JOINTED PRECAST CONCRETE PILES

One of the drawbacks of ordinary pre-cast or prestressed concrete piles is that they cannot be readily adjusted in length to suit the varying level of a hard bearing stratum. Where the bearing stratum is shallow a length of pile must be cut off and is wasted. Where it is deep the pile must be lengthened with an inevitable delay in the process of splicing on a new length. This drawback can be overcome by the use of pre-cast concrete piles assembled from short units. Two principal types are available. The West's pile (Figure 16.39(a)) consists of short cylindrical hollow shells made in 368, 444, 508 and 610 mm outside diameters. The shells are threaded on to a steel mandrel which carries a shoe at the lower end. The driving head is designed to allow the full weight of the drop hammer to fall on the mandrel while the shells take a cushioned blow. Shells can be added or taken away from the mandrel to suit the varying penetration depths of the piles. On completion of driving, the mandrel is withdrawn, a reinforcing cage is lowered down the shells and the interior space is filled with concrete. Care is needed with this type of pile in driving through ground containing obstructions. If the mandrel goes out of line there is difficulty in withdrawing it and the shells may be displaced. The shells are also liable to be lifted due to ground heave in firm to stiff clays. Piles driven in groups should be pre-bored for part of their length or the order of driving arranged to minimise ground heave.

The Herkules pile (Figure 16.39(b)) comprises lengths of solid hexagonal section pre-cast units with bayonet-type locking joints which are stronger than the concrete section. The joints are capable of withstanding uplift caused by ground heave. The lengths are manufactured to suit the requirements of the particular job and additional short lengths are locked on if deeper penetrations are required.

STEEL PILES

Steel piles of tubular, box, and H-section have the advantages of being robust and easy to handle and can withstand hard driving. They can be driven in long lengths and have a good resistance to lateral forces and to buckling. They are advantageous for marine work. They can be lengthened by welding on additional lengths as required and cut-off

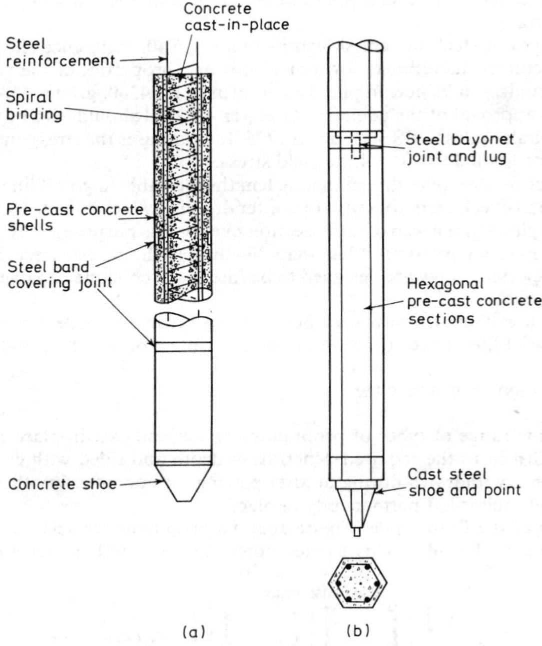

Figure 16.39 Jointed pre-cast concrete piles. (a) the West's shell pile; (b) the Herkules pile

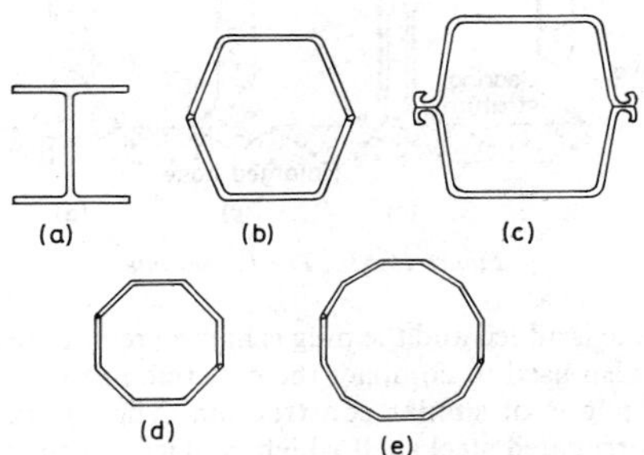

Figure 16.40 Steel bearing piles of various types. (a) universal bearing pile (UBP); (b) Rendhex foundation column; (c) Larssen box pile; (d) Frodingham octagonal pile; (e) Frodingham duodecagonal pile

sections have scrap value. If a small displacement is needed to minimise ground heave the H-section can be used, or tubular piles can be driven with open ends and the soil removed by a drilling rig.

Various types of steel pile are shown in Figure 16.40. Reference should be made to steel manufacturers' handbooks for dimensions and properties of the various sections.

CP 2004 requires mild steel in piles to conform to BS 4360, grades 43A, 50B, or other grades to the approval of the engineer. High-tensile steel should also be to BS 4360 and the steel in tubular piles to BS 3601 or BS 1775. In both cases the stress under the working load should be limited to 30% of the yield stress.

Slender-section steel piles driven in long lengths are liable to go off line during driving. It is desirable to check them for curvature after driving by inclinometer (a small-diameter tube can be welded to the web of an H-section pile for this purpose). If H-piles or unfilled tubular piles have a curvature of less than 360 m they should be rejected. Tubular piles need not be rejected if they are designed to be filled with concrete capable of carrying the full working load.

Steel piles are liable to corrosion but allowance can be made for corrosion losses within the useful life of the structure or special protection can be provided (p. **16**–59).

DRIVEN AND CAST-IN-PLACE PILES

There is a wide range of types of proprietary driven and cast-in-place piles in which a steel tube is driven to the required penetration depth and filled with concrete. In some types the tube is withdrawn during or after placing the concrete. In other types the tube or a light steel shell is left permanently in place.

In the case of the Franki pile (Figure 16.41) a drop hammer acts on a plug of gravel at the bottom of the tube. This carries down the tube and on reaching the bearing

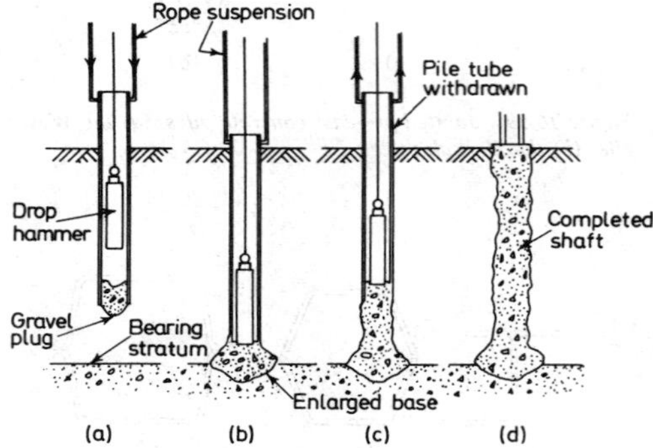

Figure 16.41 The Franki pile

stratum further concrete is added and the plug is hammered out to form an enlarged base. The drop hammer is also used to compact the concrete in the shaft as the tube is withdrawn. The Dowsett pile is of similar construction. The Franki pile can be provided with a light section corrugated steel shell which is placed in the tube before filling with concrete to provide a permanent casing to withstand 'squeezing' ground conditions.

The Simplex, Vibro, Delta, Holmpress and Alpha piles are similar in principle. A steel drive tube (Figure 16.42) is provided with a cast iron shoe and is driven to the required penetration by a drop hammer or diesel hammer acting on top of the tube. A reinforcing cage is then placed in the tube and concrete is poured before or during withdrawal of the tube. The various proprietary types have different arrangements for compacting the concrete.

Driven and cast-in-place piles of the types described above are cast to nominal outside diameters ranging from 330 to 520 mm. Their lengths are limited by the capacity of the rig to pull out the drive tube, but deeper penetration is possible if the upper part of the shaft can be pre-bored to reduce skin friction on the tube.

In the Raymond Standard Taper Pile (Figure 16.43(a)), light gauge tapered steel shells are driven to the required depth on a mandrel. The latter is then withdrawn and the shells are filled with concrete. Straight section shells progressively reducing in diameter (step taper piles) can also be used (Figure 16.43(b)) to drive to depths greater than are possible with the standard taper pile.

Placing concrete in the shells should be delayed until ground heave has ceased when driving these piles in groups. Ground heave can be reduced by preboring.

The BSP cased pile system consists of driving a fairly light spirally welded steel tube either by a hammer on top of the pile or by a drop hammer acting on a plug of concrete

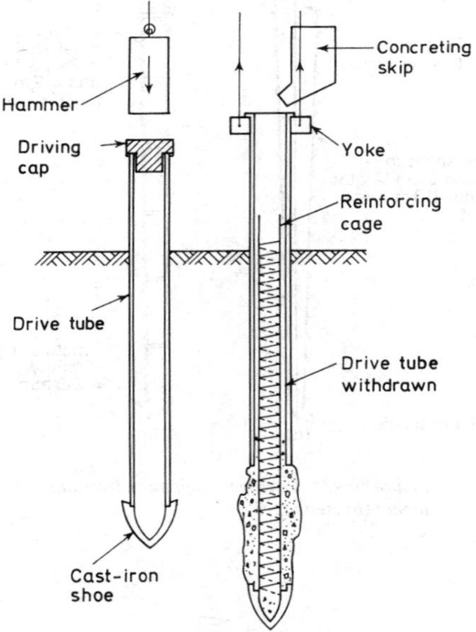

Figure 16.42 Driven and cast-in-place pile

at the bottom of the closed-end pile. On reaching founding level the whole pile is filled with concrete. This type of pile can be used for marine works. Inside tube diameters range from 245 to 508 mm.

CP 2004 requires the concrete of all driven and cast-in-place types to have a cement content of not less than 300 kg/m^3. The average compressive strength under working loads shall not exceed 25% of the specified 28-day works cube strength. Care should be taken to ensure that the volume of concrete placed fills the volume of the soil displaced by the drive tube or the volume of shells left in place. This is a safeguard against caving of the ground while withdrawing the tube or collapse of shells.

SCREW PILES

Screw piles or cylinders (Figure 16.44) consist of a helical blade which is screwed into the ground by applying torque to a solid or hollow circular shaft. Screwcrete cylinders are hollow cylindrical concrete shells threaded on to a mandrel which carries the helical

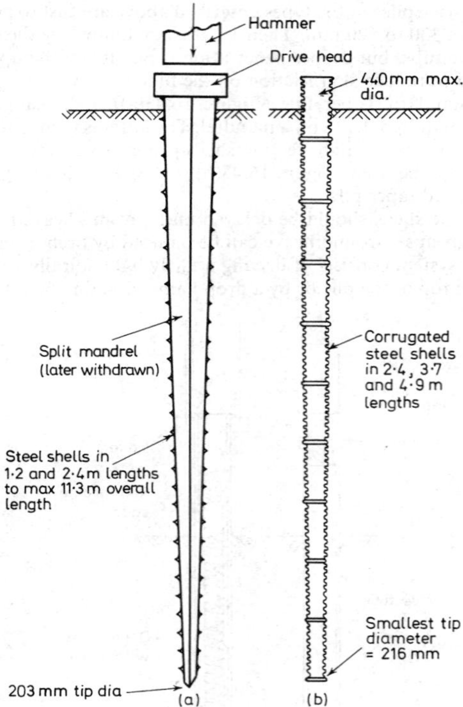

Figure 16.43 Raymond shell piles. (a) standard taper; (b) step taper

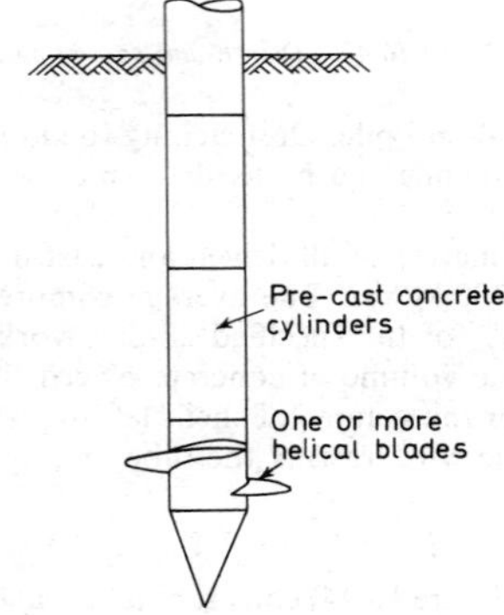

Figure 16.44 Screw cylinder pile

blade at its lower end. On reaching founding level the mandrel is withdrawn and the interior of the shells is filled with concrete. The large-diameter blade (up to 2.8 m) gives the piles a high uplift and compression resistance but they cannot be screwed to any great distance into compact or stiff soils.

Types of non-displacement piles

AUGER DRILLED PILES

If the soil is capable of remaining unsupported for a short time the pile borehole can be drilled by a rotary spiral plate or bucket auger. After drilling to the desired depth a steel casing is lowered into the borehole to prevent later collapse of the sides and to prevent contamination of the concrete from soil or ground water. In stiff cohesive soils or soft rocks it is possible to form a base enlargement by a rotary under-reaming tool or, if the shaft diameter is large enough, the enlarged base can be excavated by hand.

Rotary bucket augers can drill to depths of up to 60 m and can form shaft and base diameters of up to 5 m and 6 m respectively.

In 'squeezing' soils, or in ground contaminated by substances aggressive to concrete, a light steel or plastics tube can be lowered down the pile borehole before placing the concrete to form a permanent casing.

Safety precautions in bored piling work are covered by CP 2011.

PERCUSSION BORED PILES

In ground which collapses during drilling, requiring continuous support by casing, the pile boring is undertaken by baling or grabbing. For small-diameter (up to 600 mm) piles the tripod rig is used to handle the drilling tools and to extract the casing. For large-diameter piles a powered rig which combines a casing oscillator and a winch for handling grabbing and chiselling tools is used to drill to diameters of up to 1.5 m and depths of 50 m or more.

Problems of placing concrete in difficult conditions, e.g. in 'squeezing' ground, can be overcome in special cases by placing concrete under compressed air with the assistance of an airlock on top of the casing (i.e. the Pressure pile) or by placing pre-cast concrete sections in the casing and injecting cement grout to fill the joints between and around the sections while withdrawing the casing (the Prestcore pile).

CONCRETE FOR NON-DISPLACEMENT PILES

The cement content should not be leaner than 300 kg/m^3. The average compressive stress under the working load should not exceed 25% of the specified works cube strength at 28 days. CP 2004 permits a higher allowable stress if the pile has a permanent casing of suitable shape.

The concrete should be easily workable and capable of slumping to fill all voids as the casing is being withdrawn without being lifted by the casing. If a tremie pipe is necessary for placing concrete under water the mix should not be leaner than 400 kg of cement per m^3 of concrete and a slump of 150 mm is suitable.

Raking piles to resist lateral loads

Where lateral forces are large it may be necessary to provide raking piles to carry lateral loading in compression or tension axially along the piles. Arrangements of raking pile foundations for a retaining wall and a berthing structure are shown in Figures 16.45 (a) and (b) respectively.

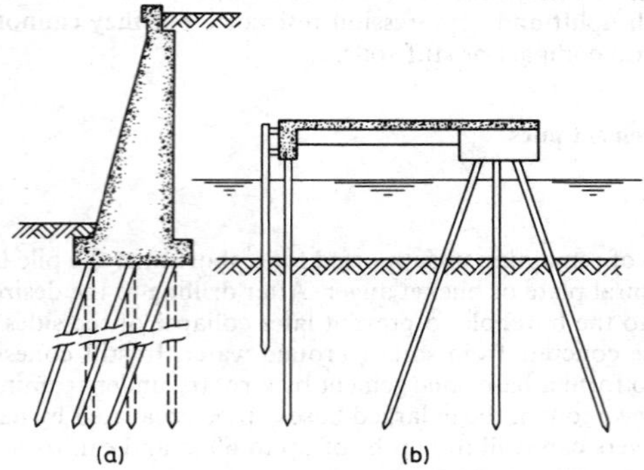

Figure 16.45 Raking piles to resist lateral loads. (a) beneath retaining wall; (b) in marine berthing structure

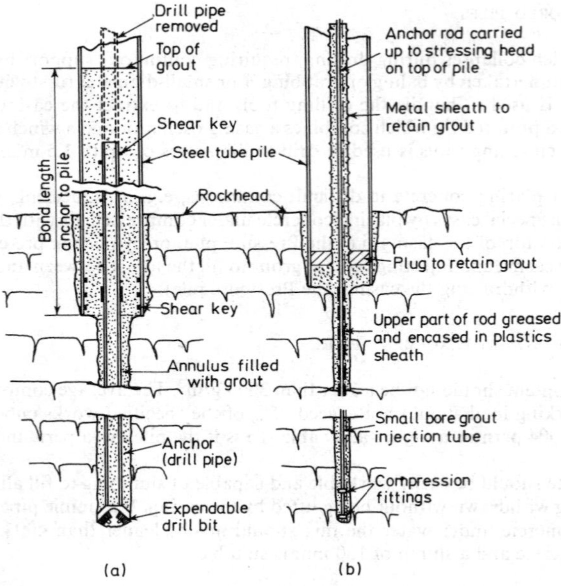

Figure 16.46 Methods of anchoring piles to resist uplift. (a) 'dead' anchors for small to moderate uplift loads; (b) high capacity prestressed anchors

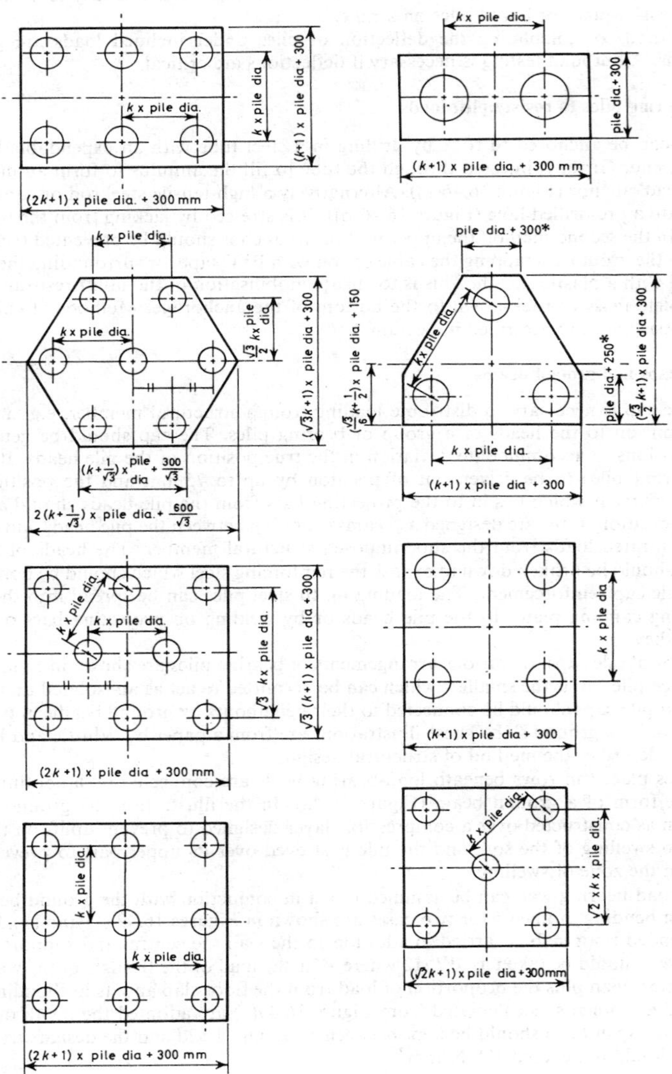

Figure 16.47 Designs for pile caps, k is spacing factor. These dimensions are determined by reinforcement*

Raking piles should not have a rake flatter than 1 in 3 if difficulties in driving are to be avoided, but flatter rakes are possible with short piles. It is not easy to install driven and cast-in-place or bored piles on a rake.

Methods of calculating the deflection of piles under inclined loads are given by Broms.[18], but load testing is necessary if deflections are critical.

Anchoring piles to resist uplift loads

Piles can be anchored to rock by drilling in a steel tube with an expendable bit at its lower end. Grout is injected through the tube to fill the annulus to form an unstressed or 'dead' anchor (Figure 16.46(a)). Alternatively a high-tensile steel rod or cable can be fed into a pre-drilled hole (Figure 16.46(b)). It is stressed by jacking from the top of the pile. In the second method the upper part of the anchor should be prevented from bonding to the grout by wrapping the cable or rod with PVC tape, or surrounding the greased metal with a plastic sheath. This is to ensure mobilisation of the uplift resistance of the complete mass of rock down to the bottom of the anchorage. Methods of calculating this resistance are described in Section 8.

Pile caps and ground beams

A pile cap is necessary to distribute loading from a structural member, e.g. a building column, on to the heads of a group of bearing piles. The cap should be generous in dimensions to accommodate deviation in the true position of the pile heads. It is usual to permit piles to be driven out of position by up to 75 mm and the positioning of reinforcement which ties in to the projecting bars from the pile heads should allow for this deviation. Caps are designed as beams spanning between the pile heads and carrying concentrated loads from the superimposed structural member. The heads of concrete piles should be broken down to expose the reinforcing steel which should be bonded into the pile cap reinforcement. The loading on to steel piles can be spread into the cap by welding capping plates to the pile heads or by welding on projecting bars or lugs as shear keys.

Typical pile caps for various arrangements of bearing piles are shown in Figure 16.47. A three-pile cap is the smallest which can be permitted to act as an isolated unit. Single- or two-pile caps should be connected to their neighbours by ground beams in two directions or by a ground slab. These illustrations are from a paper by Whittle and Beattie[19] which describes the method of structural design.

Piles placed in rows beneath load-bearing walls are connected by a continuous cap in the form of a ground beam (Figure 16.48). In the illustration the ground beam is shown as constructed over a compressible layer designed to prevent uplift on the beam due to swelling of the soil, and the pile is sleeved over its upper part to prevent uplift within the zone of swelling.

A load-bearing wall can be assumed to act in conjuction with the ground beam. The design bending moments for two cases are shown in Figures 16.49 (a) and (b). Where a suspended floor slab is carried in addition to the wall the central and support bending moment should be taken as $WL/12$ where W is the load of the triangle of the wall above the beam span plus the proportion of load from the floor slab and its live loading. If the bending moments as calculated from Figure 16.49(a) are adopted the depth of ground beam-to-span ratio should be kept between 1 : 15 and 1 : 20 and the design stress in the steel should not exceed 108 N/mm^2.

Test loading of piles

Loading tests on piles may be needed at two stages: in the first stage to verify the carrying capacity of the piles either in compression, uplift or lateral loading, and in the second stage to act as a proof load to verify the soundness of workmanship or adequacy of penetration of working piles.

For first-stage testing either the constant rate of penetration test (CRP) or the maintained load (ML) method may be used. The latter is to be preferred if it is desired to obtain information on the deflection of the pile under the working load, or at some multiple of this load.

For proof loading of working piles the maintained load test should be made. It is not usual to apply a load of more than 1.5 times the working load in order to avoid overstressing the pile.

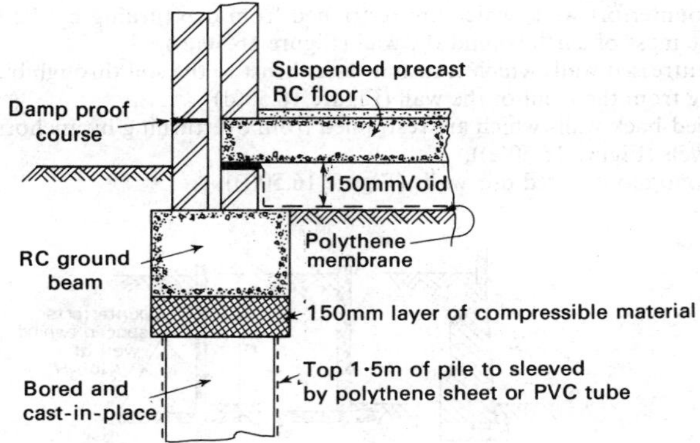

Figure 16.48 Ground beam for piles carrying a load-bearing wall

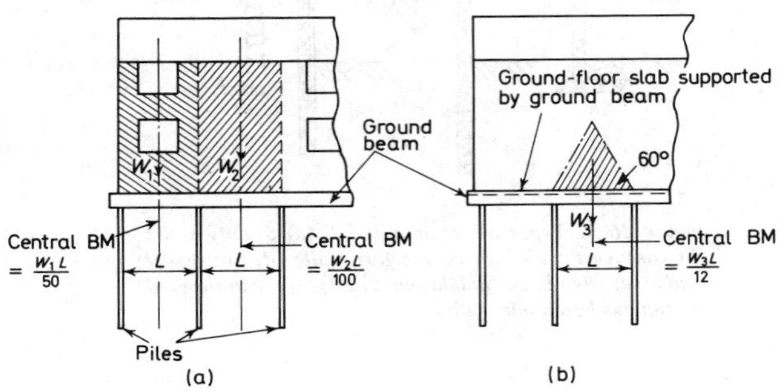

Figure 16.49 Bending moments on pile-supported ground beams

The procedures for the CRP and ML tests are described in CP 2004.

RETAINING WALLS

General

This section covers the design and construction of free-standing or tied-back retaining walls. The design of retaining walls for basements, bridge abutments and wharves is described on p. **16**–18 and in Sections 18 and 24 respectively.

Free-standing or tied-back retaining walls can be grouped for design purposes as follows:

(a) Gravity walls which rely on the mass of the structure to resist overturning (Figure 16.50(a)).
(b) Cantilever walls which rely on the bending strength of the cantilevered slab above the base (Figure 16.50(b)).
(c) Counterfort walls which are restrained from overturning by the force exerted by the mass of earth behind the wall (Figure 16.50(c)).
(d) Buttressed walls which transmit their thrust to the soil through buttresses projecting from the front of the wall (Figure 16.50(d)).
(e) Tied-back walls which are restrained from overturning by anchors at one or more levels (Figure 16.50(e)).
(f) Contiguous bored pile walls (Figure 16.50(f)).

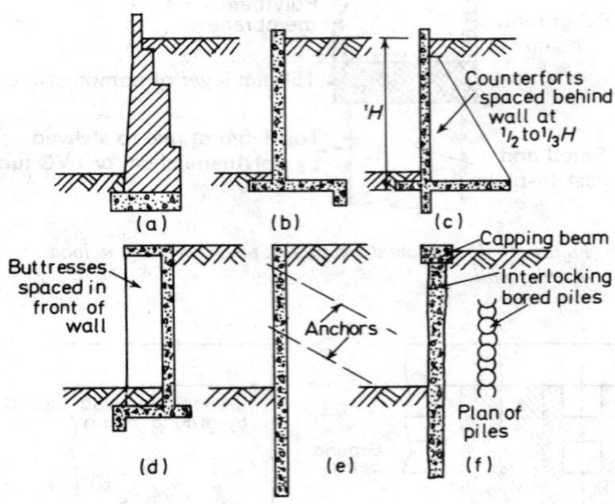

Figure 16.50 Types of retaining wall, (a) gravity wall; (b) cantilever wall; (c) counterfort wall; (d) buttressed wall; (e) tied-back diaphragm wall; (f) cantilevered contiguous bored pile wall

It is assumed that sufficient forward movement of free-standing walls takes place to allow the earth pressure behind the walls to be calculated as the 'active pressure' case (see Section 8). Where the foundation of the wall is at a shallow depth below the lower ground level, the passive resistance to overturning or sliding is neglected since it may be destroyed by trenching in front of the wall at some future time.

The forces acting on a gravity and simple cantilever wall are shown in Figures 16.51 (a) and (b). The force R is the resultant of the active earth pressure P_A and the weight of the wall W and backfill above the wall foundation. The surcharge on the fill behind the wall is allowed for when calculating P_A but is not included in the weight W. To prevent overturning of the wall the resultant R should cut the base of the wall foundation within its middle third, i.e. the eccentricity must not exceed $B/6$.

Having determined the position and magnitude of R, the bearing pressures at the toe and heel of the base are determined as described on p. **16**–10. These should not exceed the allowable bearing pressure of the ground, and the settlement at the toe should be within tolerable limits. Then the resistance to sliding of the base should be determined. If this is inadequate the base should be widened or taken down to a depth where the passive resistance in front of the wall may be safely mobilised (Figure 16.52).

Hydrostatic pressure behind the retaining walls should be avoided by the provision of a drainage layer behind the wall combined with weepholes and a collector drain (as shown in Figure 16.54).

Gravity walls

Typical designs for gravity walls in brickwork, mass concrete, and cribwork, are shown in Figure 16.53 (a) and (b). Walls of these types are economical for retained heights of

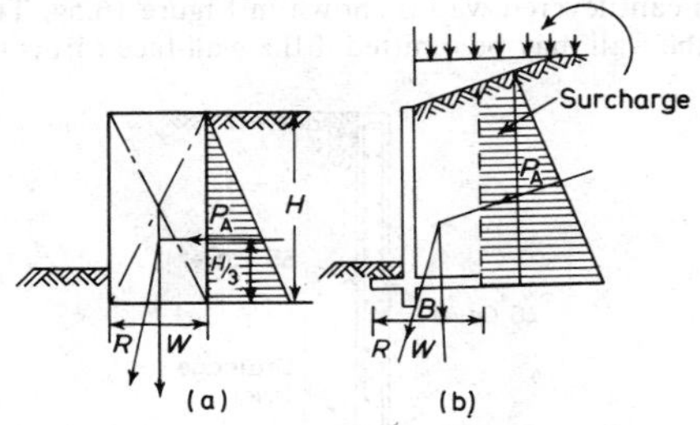

Figure 16.51 Forces acting on a free-standing retaining wall. (a) simple gravity wall; (b) cantilever wall

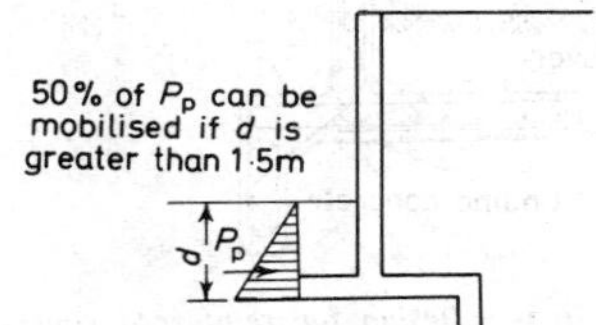

Figure 16.52 Passive resistance at toe of retaining wall

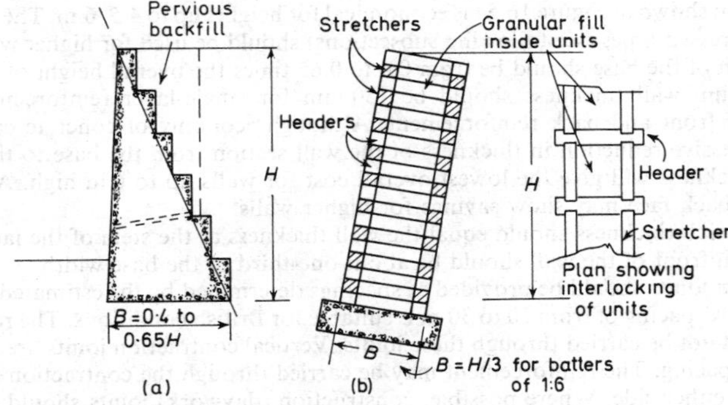

Figure 16.53 Designs for gravity retaining walls. (a) mass concrete; (b) cribwork

up to 2 to 3 m, or up to 5 m for cribwork walls. The width of the base should be about 0.4–0.65 times the overall height. For walls designed to present a 'vertical' appearance the front face should be battered back slightly say to 1 in 24 to allow for the inevitable slight forward rotation. The sloping wall and base (Figure 16.53(b)) provides the best alignment to resist earth pressure. Vertical joints in brick walls should be at 5 m to 18 m and in concrete walls at 20 m centres or at some convenient length for a day's 'pour' of concrete. A preformed joint filler strip in bituminised fibre or PVC may be used.

Cantilevered reinforced concrete walls

A typical design for a cantilevered wall is shown in Figure 16.54. The projection of the base slab in front of the wall may be omitted if the wall face forms the boundary of the

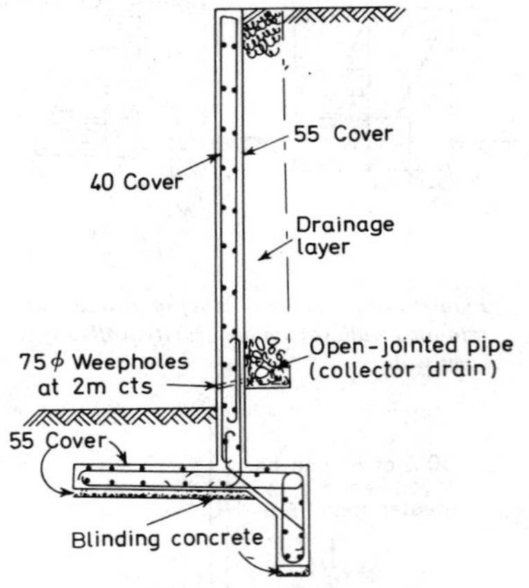

Figure 16.54 Design for reinforced concrete cantilever wall

property but this arrangement should be avoided if at all possible because of the high pressure on the soil at the toe and the consequent risk of excessive forward rotation.

The design shown in Figure 16.54 is economical for heights up to 4.5–6 m. The counterfort or buttressed types (see following subsections) should be used for higher walls.

The width of the base should be from 0.4 to 0.65 times the overall height of the wall. The minimum wall thickness should be 150 mm for single-layer reinforcement and 230 mm for front and back reinforcement. Although economy of concrete can result from progressive reduction in thickness of the wall section from the base to the top, a uniform thickness will give the lowest overall cost for walls up to 6 m high. A sloping or stepped-back face may show savings for higher walls.

The base slab thickness should equal the wall thickness at the stem of the latter. The projection in front of the wall should be about one-third of the base width.

Expansion joints should be provided at spacings determined by the estimated thermal movement. A spacing of from 20 to 30 m is suitable for British conditions. The reinforcement should not be carried through these joints. Vertical contraction joints are required at 5–10 m spacing. The reinforcement may be carried through the contraction joints or stopped on either side. Where possible, construction (daywork) joints should coincide with expansion or contraction joints. The minimum cover to the reinforcing steel, appropriate in each case to the exposure conditions, is shown in Figure 16.54.

Counterfort walls

The wall slab of counterfort retaining walls spans horizontally between the counterforts except for the bottom 1 m which cantilevers from the base slab. The counterforts are designed as T-beams of tapering section, and they are usually spaced at distances of one-third to one-half the height of the wall. The base of the counterfort must be well tied into the base slab. The latter acts as a horizontal beam carrying the surcharge load of the backfill and spanning between counterforts or from back beam to front beam. The counterforts transmit high bearing pressures to the ground at their front ends and may require piled foundations or a stiff front beam to distribute the pressure along the front of the wall.

Buttressed walls

Buttressed walls are economical for walls higher than 6 m designed to be cast against an excavated face, whereas the counterfort wall is more suitable where the ground behind the wall is to be raised by filling. The wall slab spans horizontally between the buttresses except for the bottom 1 m which cantilevers from the base slab. The buttresses act as compression members transmitting loading to the base slab or to piles on weak ground.

Tied-back walls

The stages in constructing a tied-back wall in the form of a diaphragm wall are shown in Figure 16.55. In stage I the wall must be designed to cantilever from the stage I excavation

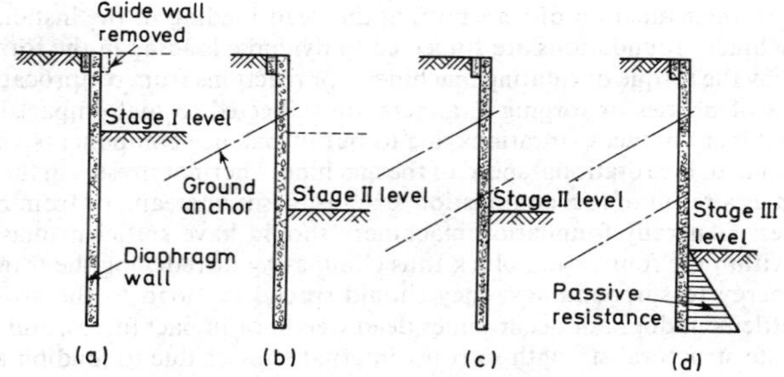

Figure 16.55 Stages in constructing a tied-back diaphragm wall. (a) excavating to first stage in preparation for installing top-level ground anchors; (b) top-level anchors installed, excavation to second stage in preparation for installing bottom-level anchors; (c) bottom-level anchors installed; (d) excavation to third (final) stage

level. For the stage II excavations the wall spans between the anchorage level and the soil at the excavation line, similarly at stage III. At the latter stage the passive resistance of the soil in front of the buried portion must be adequate to prevent the wall moving forward at the toe, and the pressure beneath the base of the wall due to the vertical component of the anchor stress must not exceed the allowable bearing pressure of the soil.

The use of the tied-back wall as a basement retaining wall is described on p. **16**–22.

Contiguous bored pile walls

Retaining walls formed by a continuous line of bored piles can be designed as simple cantilever structures (Figure 16.50(f)) or as tied-back walls. Walls of this type are economical when constructed by rotary auger drilling methods (see p. **16**–43) in self-supporting ground above the water table. In these conditions the piles can be installed merely as abutting units.

In water-bearing cohesionless soils the piles must interlock. If this is not done water and soil will bleed through the gaps causing loss of ground behind the wall. Interlocking is done by drilling and concreting alternate piles then using a chisel to drill in the space between these piles forming a deep groove in each of the latter. The drilled-out space is then filled with concrete to form the continuous wall. Construction in this manner is likely to cost more than the diaphragm wall.

Materials and working stresses

Concrete mixes and the quality of bricks or blocks should be selected to be suitable for the conditions of exposure, attention being paid to frost resistance. Information on the durability of these materials in aggressive conditions is given on pp. **16**–59 and **16**–60. Generally the materials and working stresses for reinforced concrete should be in accordance with CP 114.

MACHINE FOUNDATIONS

General

In addition to their function of transmitting the dead loading of the installation to the ground, machinery foundations are subjected to dynamic loading in the form of thrusts transmitted by the torque of rotating machinery, or reactions from reciprocating engines. Foundations of presses or forging hammers are subjected to high impact loading and rotating machinery induces vibrations due to out-of-balance components vibrating at a frequency equal to the rotational speed of the machine. Thermal stresses in the foundation may be high as a result of fuel combustion, exhaust gases or steam, or from manufacturing processes. Generally foundation machinery should have sufficient mass to absorb vibrations within the foundation block thus eliminating or reducing the transmission of vibration energy to surroundings; they should spread the load to the ground so that excessive settlement does not occur under dead weight or impact forces, and they should have adequate structural strength to resist internal stresses due to loading and thermal movements.

Machinery foundation blocks are frequently required to have large openings or changes of section to accommodate pipework or other components below bedplate level. These openings can induce high stresses in the foundation block due to shrinkage combined with other effects. Abrupt changes of section should be avoided, and openings should be adequately reinforced.

Foundations for vibrating machinery

When the frequency of a foundation block carrying vibrating machinery approaches the natural frequency of the soil, resonance will occur and the amplitude may be such as to cause excessive settlement of the soil beneath the foundation, or beneath other foundations affected by the transmitted wave energy. This is particularly liable to occur with foundations on loose granular soils. Knowing the weight of the machine and its foundation and the vibration characteristics of the soil, it is possible to calculate the resonant frequency of the machine–foundation–soil system. Ideally the frequency of the applied forces should not exceed half of this resonant frequency for most reciprocating machines and should be at least 1.5 times the resonant frequency for machines having frequencies greater than the natural frequency. If the applied frequencies are within this range there is a danger of resonance and excessive amplitude. These criteria are recommended by Converse[20] who describes various mathematical theories for calculating natural frequency and amplitude, and tabulates recommended ratios of foundation weight-to-engine weight for various types of machinery. Generally the aim in design is to provide sufficient mass to absorb as much as possible of the energy within the foundation block and to proportion the block in such a manner that energy waves are reflected within the mass of the block or transmitted downwards rather than transversely to affect adjacent property. In some cases it may be advantageous to mount the foundation block on special mountings such as rubber carpets or rubber–steel sandwich blocks.

Foundations for turbo-generators

The foundation blocks for large turbo-generators are complex structures subjected to periodic reversing movements due to differential heating and cooling of the concrete structures, to moisture movements related to ambient humidity, steam and water leakage, and to dynamic strains within the elastic range. They are also subjected to progressive movements resulting from long-term settlements of the foundation soil and from shrinkage and creep of concrete. These movements may be of sufficient magnitude to cause misalignment of the shafts of the machinery.[21]

FOUNDATIONS IN SPECIAL CONDITIONS

Foundations on fill

If granular fill can be placed in layers with careful control of compaction, the resulting settlement due to the foundation loading and the settlement of the fill under its own weight will be small. Provided that the fill has been placed on a relatively incompressible stratum the settlement of the structure will be little if anything greater than would occur with a foundation on a reasonably stiff or compact natural soil.

However, in most cases of construction on filling, the material has probably not been placed under conditions of controlled compaction but has been loosely end-tipped, and the age of the fill may not be known with certainty. However, it is usually possible to obtain a good indication of the constituents of the fill and its state of compaction from observations in boreholes and trial pits (preferably the latter). From these observations an estimate can be made of the likely remaining settlement due to consolidation of the fill under its own weight and that of the superimposed loading. As a guide to the order of settlement of various types of filling, Table 16.4 shows the observed percentage of settlement of fills settling under their own weight.

Table 16.4 SETTLEMENT OF FILLS

Type of material	*How placed*	*Settlement as % of original thickness*	*Time for 90% completion of settlement* (years)
Rock	Well compacted in layers	$\frac{1}{4}$ to 1	10
Sands and gravels	Well compacted in layers	$\frac{1}{2}$ to 1	5
Shales and mudstone	Fairly well compacted in layers	1 to 2	10
Sand	End-tipped no compaction	2 to 3	5
Sand	Hydraulically allowed to settle through water	3 to 4	5
Sand	Hydraulically, above water level	1 to 2	5
Clay	Hydraulically, above water level	10	10
Domestic refuse	Loose or end-tipped	30	10

For shallow granular fills, strip or pad foundations are suitable for most types of structure. For deep granular fills which have not had special compaction, e.g. on restored opencast mineral workings where the settlement may be of the order of 1 or 2% of the thickness, it will be necessary to use raft foundations for structures which are not very sensitive to differential settlement, or piled foundations for structures for which small settlements must be avoided.

Ordinary shallow foundations can be used on hydraulically placed sand fill where this can consolidate by drainage but not when the fill has been allowed to settle through water. Piled foundations are necessary for structures on hydraulically placed clay fill or on domestic refuse.

Raft or piled foundations can be avoided on loose granular fills if any one of the following ground treatment processes are adopted:

Preloading
Cement injection
Deep vibration

Information on these processes is given on pp. **16**–14, 15.

Where piled foundations are used in fill areas, consolidation of the fill and of any underlying natural compressible soil will cause dragdown forces on the pile shafts which must be added to the working load from the superstructure (Section 8). Where bored piles are used through the fill the dragdown or negative skin friction forces may be very high, and for economy it may be desirable to minimise the dragdown by adoption of slender preformed section (e.g. H-piles) or to surround the pile shaft with a sleeve or a layer of soft bitumen.

Fills consisting of industrial wastes may contain substances which are highly aggressive to buried concrete or steelwork.

Foundations in areas of mining subsidence

GENERAL

Cavities are formed where minerals are extracted from the ground by deep mining or pumping. In time the ground over the cavities will collapse wholly or partially filling the void. This leads to subsidence of the ground surface. Movements of the surface may be large both in a vertical direction and in the form of horizontal ground strains and the foundations of structures require special consideration to accommodate these movements without resulting damage to the superstructure. The majority of foundation problems in the United Kingdom are due to coal mining, but subsidence can occur due to extraction of other minerals such as brine.

In the nineteenth century and earlier, coal was extracted by methods known variously as 'pillar and stall', 'room and pillar', and 'bord and pillar'. The galleries were mined in various directions from the shaft followed by cross galleries leaving rectangular or triangular pillars of coal to support the roof above the workings (Figure 16.56).

The current method of coal mining is by 'longwall' methods in which the coal seam is extracted completely on an advancing face (Figure 16.57). The amount of subsidence at ground level is less than the thickness of coal extracted owing to bulking of the collapsed strata.

The problems of foundations of buildings on old mine workings have been discussed by Price *et al.*[22]

FOUNDATION DESIGN IN AREAS OF PILLAR AND STALL WORKINGS

The risk of collapse depends on the condition of the 'roof' over the workings. Where this consists of weak or broken rock, at some stage collapse will occur and the void so formed will gradually work its way up to the ground surface to form a 'crown hole' (Figure 16.58(a)). If, however, the roof is a massive sandstone it will bridge over the cavity for an unlimited period of years (Figure 16.58(b)). However, the pillars of coal may suffer slow deterioration at an unpredictable rate.

In considering the design of foundations over workings of this type an appraisal is made of the general geological conditions. Where the collapse of overburden strata or

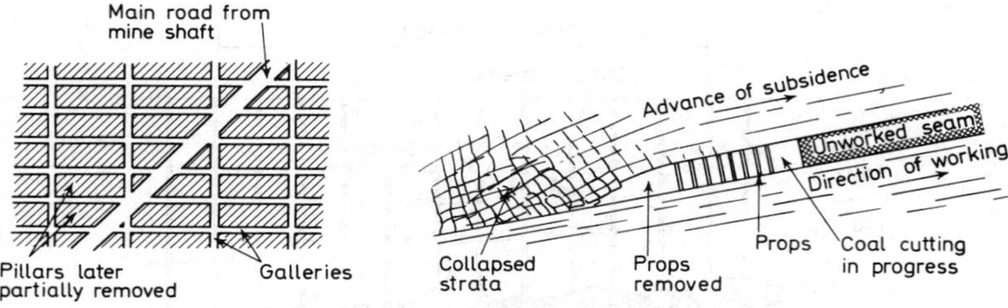

Figure 16.56 'Pillar and stall' mine workings

Figure 16.57 Extraction of coal by longwall method

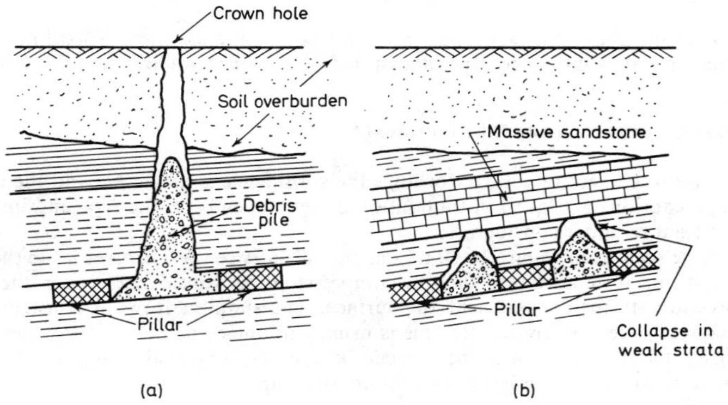

Figure 16.58 Subsidence due to collapse of cavities in mine workings. (a) weak strata over coal seam; (b) strong 'roof' over coal seam

coal pillars could result in severe local surface subsidence, then precautions against these effects must be taken. Methods which may be considered are:

(i) Filling the workings by injection techniques, or
(ii) Constructing piled or deep shaft foundations to a founding level below the workings

Method (i) is used where the workings are at such a depth that method (ii) is uneconomical. No attempt is made to locate individual galleries but the area of the structure is ringed by a double row of injection holes at close spacing. Gravel or a stiff sand–cement grout is fed down these holes to form a barrier in the voids of the worked seam. Holes are then drilled on a nominal grid in the space within the barrier and low-cost materials are fed down these holes to fill all accessible voids. These materials may consist of sand: pulverised fuel ash: water slurry, or a lean sand: pf-ash: cement grout. A schematic drawing of the method is shown in Figure 16.59.

Where deep shaft or piled foundations (method (ii)) are used the shaft is sleeved where it passes through the overburden to prevent transference of load to the foundation in the event of subsidence. The outer lining forming the sleeve must be strong enough to resist lateral movement caused by subsidence. Where structures are to be built on soft compressible soils overlying mine workings, piled foundations bearing on a thin cover

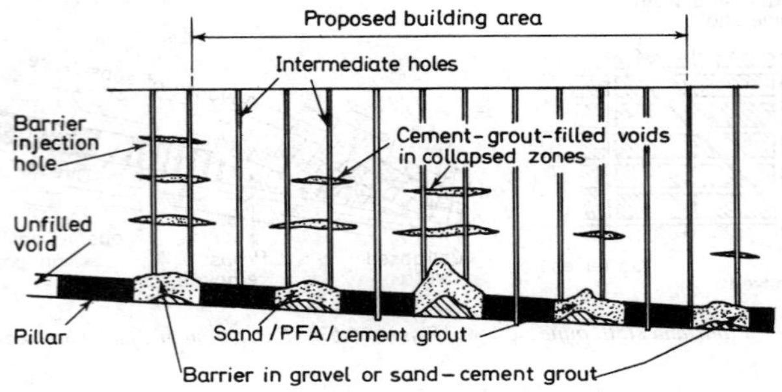

Figure 16.59 Filling of abandoned mineworkings

of rock strata above the workings must not be used since the toe loading from the piles may initiate subsidence. Buoyancy raft foundations should be used (see p. **16**–22).

FOUNDATION DESIGN IN AREAS OF LONGWALL WORKINGS

In the case of current or future workings the subsidence is inevitable and the degree to which precautions are taken in foundation design depends on the type and importance of the structure under consideration.

It will be seen from Figure 16.60 that as the subsidence wave crosses a site the ground surface is first in tension and then in compression. As subsidence ceases, the residual compression strains die out near the surface. The simplest form of construction is a shallow reinforced concrete raft. This is usually adopted for houses for which the cost of repairs due to distortion of the raft can be kept to a reasonable figure.

Points to note in the design of raft foundations are:

(i) The underside of the raft should be flat, i.e. it should not be keyed into the ground.
(ii) A slip membrane is provided beneath the raft to allow ground strains to take place without severe compression or tension forces developing in the substructure.
(iii) Reinforcement is provided in the centre of the slab to resist bending stresses caused either by hogging or sagging.

A raft may not be suitable for heavy structures such as bridges or factories. In these cases, the principle to be adopted is to use bearing pressures as *high* as possible so minimising the foundation area and hence the horizontal tension and compression forces transmitted to the superstructure. If the layout permits, the structure should be supported on only three bases to allow it to tilt without distortion.

Mauntner[23] has analysed the conditions of support of a single base shown in Figure 16.61 as follows:

Maximum pressure on foundation

$$= q_{max} = \frac{4qb}{3(b-2l)} \text{ for cantilevering (Figure 16.61(a))}$$

and

$$q_{max} = \frac{qb}{b-l} \text{ for free support (Figure 16.61(b))}$$

where b is the length of structure in the vertical plane under consideration, q is the uniformly assumed bearing pressure for undisturbed ground and l is the unsupported length for cantilevering or free support.

The value of q_{max} depends on the length which in turn depends on the ratio q_{max}/q. When q_{max} approaches the ultimate bearing capacity of the ground, yielding will occur causing tilting in the cantilver case (Figure 16.61(a)) or more-or-less uniform settlement for the free-support case (Figure 16.61(b)). In both cases the effect of yielding of the

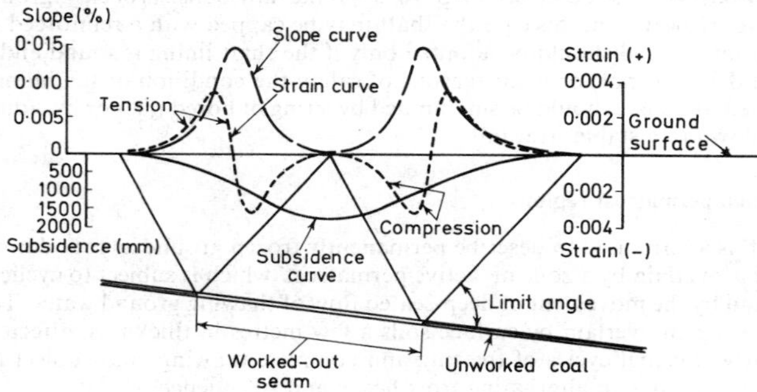

Figure 16.60 Form of subsidence above longwall mineworkings (adapted from NCB Handbook)[24]

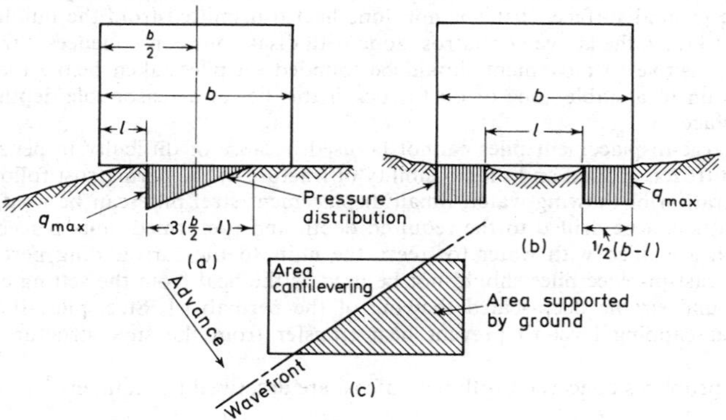

Figure 16.61 Pressure distribution beneath structures on areas of subsidence (after Mauntner)[23]*. (a) structure acting as cantilever; (b) structure acting as a beam; (c) wavefront oblique to structure*

ground will be to reduce l either in the cantilver or beam case and thus to reduce the stresses in the substructure and superstructure. Thus the ratio q_{max}/q should be as small as possible or, in other words, the design bearing pressure should be as close as possible to the ultimate bearing capacity. The latter can be determined by soil mechanics tests or plate bearing tests. Various directions of the advancing wavefront must be analysed (Figure 16.61(c)) to calculate the worst support conditions.

Trenching around a structure can be used to reduce compressive strain but this method is ineffective in countering tension strains.

For general information on mining subsidence and various methods of protection, reference should be made to the *Subsidence Engineer's Handbook* published by the National Coal Board.[24]

FOUNDATIONS ADJACENT TO EXISTING SHAFTS

Foundation problems may arise owing to collapse of deteriorated shaft linings followed by surface subsidence. The type of material used for filling the shaft should be ascertained. If it is granular, the loose material and any cavities can be consolidated by injection of a low-cost grout as described on p. **16**–55. If the infill consists of clay, grouting may be ineffective. However, in this case the shaft may be capped with a reinforced concrete slab. The latter method should be adopted only if the shaft lining is sound and durable over its full depth. If not, or if for reasons of safety the condition of the lining cannot be ascertained, the shaft should be surrounded by a ring of bored piles or by a diaphragm wall taken down to a stable stratum.

Foundations in permafrost regions

'Permafrost' is a term used to describe permanently frozen ground as found in northern latitudes. It is overlain by a zone of 'active permafrost' which is subject to cyclic changes in climate and by the movement of deep-seated flow of thawing ground water. The active permafrost is again overlain by surface soils a few metres in thickness, affected by the normal yearly seasonal cycles of freezing and complete thawing, and subject to severe ground movement due to alternating frost heave and subsidence.

Light structures which are insensitive to settlement may be placed on shallow foundations. The shallow soils subjected to annual freeze–thaw cycles are stripped off and replaced by non-frost-susceptible gravel. The ground floor of the building is supported clear of the ground surface. If this is not done heat transmitted from the building will commence to thaw the active permafrost zone with disastrous consequences. Structures which are sensitive to movement should be founded on piles taken below the active permafrost on to a stable zone or on to rock if this lies at a reasonable depth below ground surface.

Driven large-displacement piles cannot be used because of difficulty in penetration of the hard frozen ground and the possibility of fracture of the permafrost followed by deep penetration of thawing water. Small displacement steel piles can be used. Alternatively holes can be drilled to the required depth and preformed units placed in the holes which are filled with water to freeze the units to the surrounding permafrost. Bored and cast-in-place piles should not be used as the heat from the setting concrete will cause undesirable deep-seated thawing of the permafrost. Steel piles should be insulated at capping level to prevent heat transfer from the superstructure to the permafrost.

General problems concerned with permafrost are described by Rathjens.[25]

THE DURABILITY OF FOUNDATIONS

General

Foundation materials are subjected to attack by aggressive compounds in the soil or ground water, by living organisms and by mechanical abrasion or erosion. The severity of the attack depends on the concentration of aggressive compounds, the level of and fluctuations in the ground water table or the variation in tidal and river levels, and on climatic conditions. Immunity against deterioration of foundations can be provided to a varying degree by protective measures. The protection adopted is usually a compromise between complete protection over the life of the structure, and the cheaper partial protection while accepting the possible need for periodic repairs or renewals. Methods of protection of various foundation materials are described in the following sections.

Timber

Timber piles are liable to fungal decay if they are kept in moist conditions, i.e. above the ground water level. Piles wholly below the water level, if given suitable preservative treatment, can perform satisfactorily for a very long period of years. Properly air-seasoned timber, if kept wholly dry (i.e. moisture content less than 22%), will also remain free of decay for an indefinitely long period. The best form of protection against fungal attack and termites is pressure treatment with coal tar creosote or the copper (chrome) arsenic-type waterborne preservative. Creosote protection should be applied in accordance with BS 913 and the waterborne type to BS 4072.

Timber piles in marine structures are liable to destruction by molluscan and crustacean borers which inhabit saline or brackish waters. Although preservative treatment gives some protection against these organisms, the longest life is given by a timber known to be resistant to their depredations. Greenheart, jarrah, and blue gum are suitable for cold European waters. In other countries, billian in the China seas, turpentine in New South Wales, black cypress and ti-tree in Queensland, spotted gum in Tasmania and teak in India have been found to have some immunity.

Protection can be given by jacketting piles in concrete before driving, or by gunite mortar after installation. Concrete can also be used in land foundations either in composite concrete–timber piles, or in deep pile caps down to ground water level (p. **16**–35).

Metals

Protection can be given to steel piles by impervious coatings of bitumen, coal-tar pitch, or synthetic resins, but these treatments are not effective for piles driven into the ground since the coatings are partially stripped off. It is the normal practice to provide sufficient cross-sectional area of steel to allow for wastage over the useful life of the structure while still leaving enough steel to keep the working stresses within safe limits.

Fancutt and Hudson[26] quote the following rates of corrosion for bare steel:

In soil	0.3 to 3 mils per year (from tests by the National Bureau of Standards in USA; in British soils it was in the lower half of this range. 1 mil = 10^{-3} in)
Normal sea water	0.5 to 5 mils per year
Industrial atmosphere in Britain	3 to 5 mils per year

In severe conditions, i.e. in polluted ground, it may be necessary to adopt a system of cathodic protection. Steel piles in river and marine structures can be protected above the soil line by a good coating of bitumen or coal-tar pitch or pitch resin. However, these coatings are liable to damage by floating objects or barnacle growth and cathodic protection is necessary in marine structures if a long life is desired.

Cast iron has a similar corrosion resistance to mild steel and protective coatings provide the best method of treatment of substructures such as cylinder foundations constructed from cast-iron segments.

Concrete

The principal cause of deterioration of concrete in structures below ground level is attack by sulphates in the soil or ground water. Sulphates occur naturally in some soils and in peats. They occur in sea water at a concentration of about 230 parts per 100 000 which is greatly in excess of the figure regarded as marginal between nonaggressive and aggressive. However, because of the inhibiting effect of the chlorides in sea water the sulphates do not cause an expansive reaction to normal Portland cement concrete if it is of good quality and well compacted. However, it is a good precaution to use sulphate-resisting cement in *reinforced* concrete structures immersed in sea water.

Concentrations of sulphates may be high in industrial wastes, particularly in colliery wastes and some blast-furnace slags. Where fill material contains industrial wastes a full chemical analysis should be made to identify potentially aggressive compounds.

The precautions to be taken to protect concrete substructures are listed in Building Research Station Digest No. 90[27] and also in CP 2004, but these recommendations do not give much consideration to the workability required for the recommended concrete mixes for the particular placing conditions. The author prefers his own table of recommendations.[28]

High-alumina cement and supersulphate cements give protection in acidic ground conditions where the pH values can be as low as 3.5, but high-alumina cement concrete cannot be used where alkalis are present in strong concentrations.

In normal climatic conditions in Great Britain, concrete at a depth greater than 300 mm is unlikely to suffer disintegration due to frost expansion. In severe conditions of exposure a dense concrete mix should be used with a water/cement ratio of less than 0.5. If the ratio is between 0.5 and 0 6 there is a risk of frost attack and above 0.6 the risk becomes progressively greater.

The required cover of steel reinforcement to prevent corrosion of the steel for various exposure conditions is listed in CP 116.

Brickwork

Bricks with a high absorption should be avoided since they are liable to frost disintegration, and they can absorb sulphates or other aggressive substances from the soil or from filling in contact with the brickwork.

In sulphate-bearing soils or ground water the brickwork mortar should be a 1 : 3 cement : sand mix made with sulphate-resisting cement or in severe conditions with supersulphated cement.

Concrete bricks or blocks may be used for foundations if they are in accordance with British Standard 1180. Precautions should be taken against sulphate attack by specifying the type of cement and the quality of concrete to be resistant to the concentration of sulphates as determined by chemical analysis.

EARTH AND ROCKFILL DAMS AND EMBANKMENTS

General design

Earth and rockfill dams and embankments are structures built for the purpose of impounding water or controlling floods. The principal design requirements for an embankment dam are:

(a) The soil or rock on which the dam is built must be capable of sustaining the load from the embankment without general shear failure or excessive settlement.
(b) A barrier must be provided beneath the dam to cut-off the flow of water through pervious strata. Alternatively suitable means must be provided to control such flow to prevent instability due to erosion.

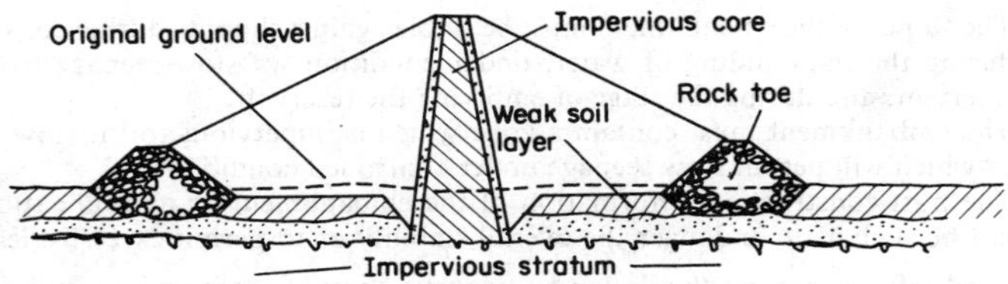

Figure 16.62 Thin-core dam showing use of rock fill to stabilise the toe

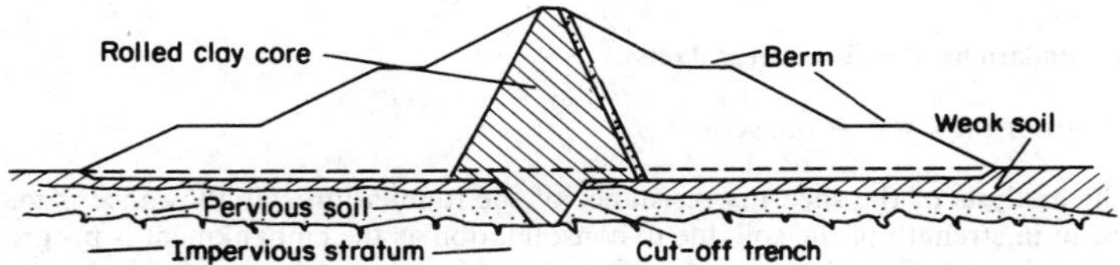

Figure 16.63 Wide-core dam with berms constructed on weak soil layer

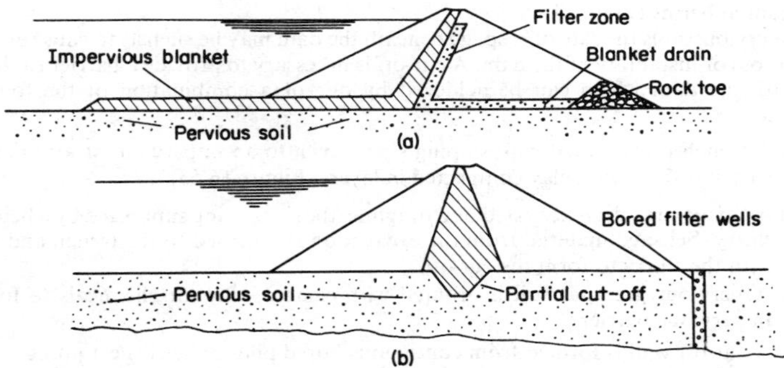

Figure 16.64 (a) Dam with impervious upstream blanket and sloping thin core
(b) Dam with partial cut-off and control of underseepage by bored filter wells

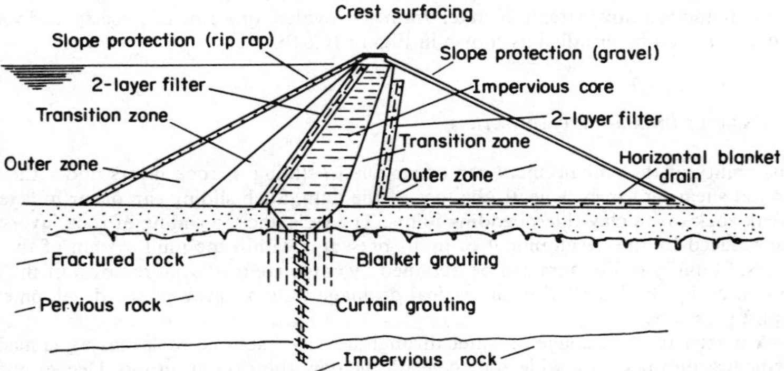

Figure 16.65 Dam of zoned construction showing blanket grouting and curtain grouting

(c) The slopes of the embankment must be stable against slipping during construction, during the impounding of water, under conditions of slow seepage from a full reservoir, and during the stage of emptying the reservoir.
(d) The embankment must contain a zone which is impervious to the flow of water or which will permit slow seepage under controlled conditions.
(e) The post-construction deformation of the embankment or its foundations must not be such as to endanger the overall stability or to permit excessive leakage.

Detailed information on the design and construction of earth and rockfill dams can be found under references 29 and 30.

The foundations of embankment dams

DAMS BUILT ON A SOIL FOUNDATION

In the analysis of the shearing resistance of the foundation soil allowance is made for the gain in strength of the soil due to consolidation as the embankment is progressively raised in height. Complete removal of weak soil layers may be avoided by constructing a substantial rock toe to the slope extending through the weak zone (Figure 16.62). Alternatively, the load from the embankment can be spread over a wide area by providing substantial berms (Figure 16.63).

In pervious soils the rate of seepage beneath the dam may be such as to cause excessive water loss or instability of the dam. A cut-off is necessary to provide a barrier to the flow beneath the dam which can be achieved by one or a combination of the following methods:

(a) A trench is excavated with sloping sides down to an impervious stratum and the trench is filled with clay compacted in layers (Figure 16.63).
(b) A narrow trench is excavated by a dragline, the sides being supported by a bentonite slurry. Selected material from the excavation is returned to the trench and mixed with the slurry to form the cut-off.[31]
(c) A pervious granular soil is grouted with cement, clay, or chemicals to form an impervious barrier.[30]
(d) A cut-off wall is formed from contiguous bored piles or steel sheet piling.

If the water loss from the reservoir is acceptable, controlled seepage can be allowed through pervious soils beneath a dam. The control can be effected by an impervious blanket upstream of the dam (Figure 16.64a) or a partial cut-off (Figure 16.64b). With both methods a downstream blanket drain is provided, or a row of close-spaced vertical filter wells can be installed as shown in Figure 16.64b.

DAMS BUILT ON A ROCK FOUNDATION

Instability of the embankment can occur due to sliding of rock layers under the horizontal shearing forces beneath the toes of the dam. Such sliding can occur in layers of weak rock or on clay-filled bedding joints. The resistance to sliding of rock layers may be reduced by the development of uplift pressures within the joint system of the rock mass. Stability of the dam can be obtained by complete or partial removal of the weak rocks, or by the installation of vertical drainage wells to prevent the development of uplift pressures.

A barrier to the seepage of water through heavily fractured rock masses, is made by cement grouting over a wide zone beneath the dam (blanket grouting). Deeper and less pervious rocks require a cut-off in the form of a grout curtain taken down to a relatively impervious formation (Figure 16.65). Vertical drainage wells, if required, are provided downstream of the grout curtain.

Embankment design for an earthfill dam

GENERAL DESIGN FEATURES

A dam of *homogeneous* construction consists wholly of clay compacted in layers and provided with a downstream drainage blanket to control seepage (Figure 16.66). Where there is insufficient clay for homogeneous construction or where climatic conditions are unfavourable for placing and compacting large volumes of clay a *zoned* construction

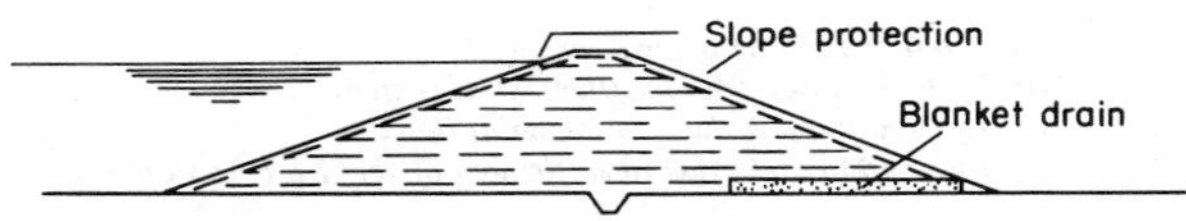

Figure 16.66 Dam of homogeneous construction on clay foundation soil

is used. A zoned construction makes the most economical use of all locally-occurring materials. The principal design features as shown in Figure 16.65 are listed as follows:

(a) An *impervious core.*
(b) *Filter zones* (which may also serve as transition zones) on either side of the core to intercept seepage and to control pore pressures developed during construction.
(c) *Transition zones* between the core and the coarse granular soils or rockfill of the outer zones, or between the core and the downstream filter.
(d) *Outer zones* or shells which form the main support of the interior impervious and filter zones. The coarsest and most stable material is placed in the outer zones. Where rockfill is used the dam may be classed as one of earth-rock design.
(e) A *horizontal drainage blanket* to intercept seepage from the downstream filter zone and to prevent the development of uplift pressures at the interface between the dam and its foundation.
(f) *Slope protection* against wave action on the upstream face and against rain and wind erosion on the downstream face.
(g) *Crest surfacing* which may take the form of a roadway.

THE IMPERVIOUS CORE

A thin core is provided where there is a shortage of suitable clay or where climatic conditions are unfavourable to large-scale placement of a wide clay core. The thin core may be placed vertically as in Figure 16.62 or as a sloping core (Figure 16.64a). The latter has advantages in that construction of the main body of the dam can continue when the weather is unfavourable to the placement of clay, but there are complications in narrow valleys due to the varying geometry of the dam cross-section towards the flanks.

The wide rolled fill core is provided where there is ample clay available at the site, or where only semi-pervious material is available, and where climatic conditions are favourable to large scale placement and compaction of clay fill. Care is necessary in the selection, processing and compaction of the clay to avoid cracking of an impervious core.[33, 34] Transverse cracking may occur due to differential settlement between the centre and ends of the impervious zone (Figure 16.67a). Horizontal cracking may be caused by arching of clay placed on a steep hillside (Figure 16.67b) or when a vertical sided core wall settles relatively to the outer zones (Figure 16.67c). To avoid cracking in the impervious zone the clay should be placed at a moisture content on the wet side of the optimum such that it is in a plastic rather than brittle state after compaction. With a

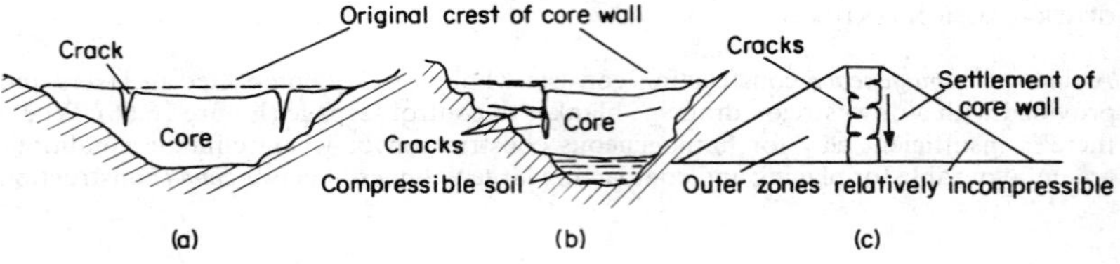

Figure 16.67 Cracking in clay core wells
(a) Vertical cracks due to settlement at centre
(b) Horizontal crack due to arching at flank
(c) Horizontal cracks due to settlement of vertical core wall

properly designed filter zone any cracks which occur in a well-designed and constructed impervious core should be self-healing.

A thin concrete core wall may be provided if no suitable clay is available, but in normal circumstances a concrete core wall is uneconomical compared with a thin clay core.

THE FILTER ZONES

The grading of the material used in a single- or multi-layer filter is determined by the rules described on page **8**–67. The finest filter material is placed in contact with the core or downstream transition zone with the object of preventing migration of soil particles from the core into the filter if water seeps through any cracks which may develop. The need for additional layers depends on the grading of the soil in the transition or outer zones of the dam.

THE TRANSITION ZONES

These usually comprise fill material of finer grading than the material in the outer shells. Sometimes the transition zones may form the outer layers of a filter, or they may be omitted entirely, depending on the material used for the outer zone.

THE OUTER ZONES

The material for the outer zones, sometimes referred to as as shells or shoulders, is selected having regard to the relative economics of providing well-graded granular material or rockfill having a high shearing resistance and permitting steep slopes to the embankment, or it may consist of more readily available clayey fill or random poorly-graded soil or rock from spillway or tunnel excavations associated with the dam, for which flatter slopes may be necessary. Horizontal drainage layers may be required if clayey fill is used in the outer or transition zones. These are used to assist the dissipation of pore water pressures developed during construction. The arrangement of drainage layers used in the Backwater Dam[32] are shown in Figure 16.68.

SLOPE PROTECTION

The upstream face of the dam must be protected from wave action. The protection may take the form of dumped rockfill (rip-rap), hand-packed stone, articulated or monolithic

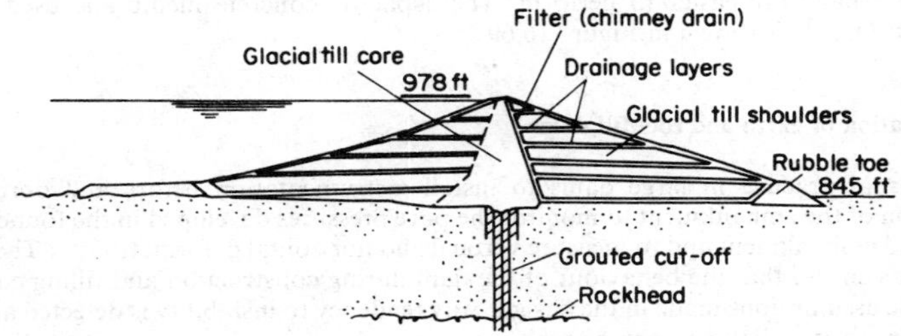

Figure 16.68 Horizontal drainage layers in the Backwater Dam

concrete slabs, or asphaltic concrete. The required size of rockfill depends on the maximum wind velocity and the 'fetch' of the reservoir.[35]

Rockfill or handpitched stone is preferable to monolithic concrete slabs if appreciable settlement is expected. A filter layer is placed beneath the rock. The downstream slope must be protected against wind and rain erosion by grassing, or a layer of gravel or broken stone.

ROCKFILL DAMS

Embankment dams constructed principally of rockfill with only a thin clay core are not strictly definable as 'rockfill dams'. Sherard *et al.*[29] class them as 'earth-rock' dams and confine the term rockfill dam to an embankment constructed wholly of rock, the water-

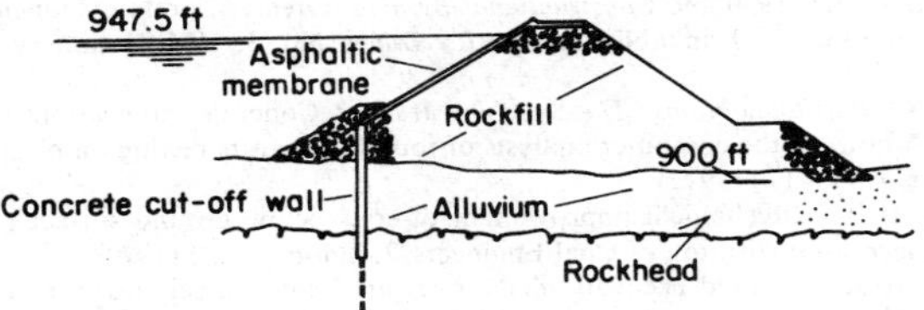

Figure 16.69 Asphaltic concrete membrane for the Dungonnell Dam. Membrane consists of: Double seal coat; 50 mm asphaltic concrete outer layer; 50 mm asphaltic concrete under layer; 125 mm Bitumen-macadam drainage layer; 75 mm asphaltic concrete underseal; Surface dressing on selected rockfill

tightness being provided by an upstream membrane of reinforced concrete, asphaltic concrete, or welded steel plate. An advantage of this type of dam, on sites where river diversion works are costly or impracticable, is their ability to allow a flood to flow over the partly completed structure.[36] Whereas at one time the rockfill was placed by end-tipping in high lifts with sluicing by high pressure water jets to obtain compaction, nowadays the end-tipping method is more usually confined to fill placed below water level (e.g. where a rockfill toe forms the upstream cofferdam). Present day practice is to place the rockfill in layers with a layer thickness up to twice the largest rock size (with a maximum layer thickness of 1.8 m). Placement and compaction of the layers of rockfill gives a better control of construction and lesser deformation than the end-tipped and sluided fill method.

Asphaltic concrete is favoured for the upstream impervious membrane since it can conform to post-construction deformations with less cracking than a concrete membrane, and repairs are easier to perform. The asphaltic concrete membrane used at Dungonnell Dam[37] is shown in Figure 16.69.

Instrumentation of earth and rockfill dams

It is the usual practice in large dams to install instrumentation before and during construction of the embankment to monitor the pore pressures developed in the foundation soil and embankment and to measure vertical and horizontal deformations.[38] These observations ensure that the behaviour of the dam during construction and filling conforms to the assumptions made in the design. Any tendency to instability is detected and the appropriate remedial measures are taken.

Safety requirements

In the United Kingdom dams impounding water in a reservoir of more than 5 m gallons (22 750 m^3) capacity must be designed by an engineer registered under a Home Office panel. The dams must be inspected after construction by a member of the panel at intervals of not more than 10 years.

REFERENCES

1. BJERRUM, L., *Proc. European Conf. on Soil Mechanics and Foundation Engineering,* 2, Deutsche Gesellschaft für Erd und Grundbau E.V., Essen (1963)
2. *The Building Regulations* 1972, HM Stationery Office, London (1972)
3. *The Building Standards (Scotland) (Consolidation) Regulations* 1971, HM Stationery Office, London (1971)
4. *London Building Acts* 1930–39, *Constructional By-laws,* Greater London Council
5. 'House foundations on shrinkable clays', *BRS Digest* No. 3, HM Stationery Office, London (1950)
6. BAKER, A. L. L., *Raft foundations: The Soil Line Method,* Concrete Publications (1937)
7. SAWKO, F., 'A note on the computer analysis of foundation rafts resting on plastic soils', *Struct. Engr,* **50,** No. 4, 171–174 (1972)
8. GREENWOOD, D. A., 'Mechanical improvement of soils below ground surface', *Proc. Conf. on Ground Engineering,* Institution of Civil Engineers, London, 11–22 (1970)
9. 'Measurements at a strutted excavation', Norwegian Geotechnical Institute, *Tech. Rep.,* Nos. 1 to 9 (1962–66)
10. COLE, K. W. and BURLAND, J. B., 'Observation of retaining wall movements associated with a large excavation', *Proc. 5th European Conf. on Soil Mechanics and Foundation Engineering,* Madrid, 1, 445–453 (1972)
11. ENRIQUEZ, R. and FIERRO, A., 'A new project for Mexico City', *Civ. Engng (USA),* 36–38 (June 1963)
12. TERZAGHI, K. and PECK, R. B., *Soil Mechanics in Engineering Practice,* Wiley, New York, 563 (2 edn 1967)
13. RIGGS, L. W., 'Tagus River Bridge—Tower piers', *Civ. Engng (USA),* 41–45 (Feb. 1965)
14. WILSON, W. S. and SULLY, F. W., 'Compressed air caisson foundations', *Instn Civ. Engrs,* Works Construction Paper 13 (1949)
15. Statutory Instrument No. 61: 'Work in compressed air—special regulations 1958', HM Stationery

Office, London (1958). (See also 'A medical code of practice for work in compressed air', CIRIA Rep. 44)
16. COLE, K. W., 'Uplift of piles due to driving displacement', *Civ. Engng and Publ. Works Rev.*, **67**, 788, 263–269 (Mar. 1972)
17. SAURIN, B. F., 'The design of reinforced concrete piles, with special reference to the reinforcement', *J. Instn Civ. Engrs*, **32**, 5, 80–109 (Mar. 1949)
18. BROMS, B., 'The lateral resistance of piles in cohesive soils', *Proc. Am. Soc. Civ. Engrs*, **90**, SM2, 27–63 (Mar. 1964); 'The lateral resistance of piles in cohesionless soils', *Proc. Am. Soc. Civ. Engrs*, **90**, SM3, 123–156 (May 1964); 'Design of laterally loaded piles', *Proc. Am. Soc. Civ. Engrs*, **91**, SM3, 79–99 (May 1965)
19. WHITTLE, R. T. and BEATTIE, D., 'Standard pile caps', *Concrete*, **6**, 1, 34–36 (Jan. 1972); **6**, 2, 29–31 (Feb. 1972)
20. CONVERSE, F. J., 'Foundations subjected to dynamic forces', *Foundation Engineering*, McGraw Hill, London, 769–825 (1962)
21. FITZHERBERT, W. A. and BARNETT, J. H., 'Causes of movement in reinforced turbo-blocks and developments in turbo-block design and construction', *Proc. Instn Civ. Engrs*, **36**, 351–393 (Feb. 1967)
22. PRICE, D. G., MALKIN, A. B. and KNILL, J. L., 'Foundations of multi-storey blocks on the Coal Measures with special reference to old mine workings', *Q. J. Engng Geology*, **1**, 4, 271–322 (June 1969)
23. MAUNTNER, K. W., 'Structures in areas subjected to mining subsidence', *Proc. 2nd Int. Conf. on Soil Mechanics and Foundation Engineering*, **2**, 167–177, Rotterdam, Koninklijk Instituut van Ingenieurs Gravenhage (1948)
24. *Subsidence Engineer's Handbook*, National Coal Board Production Department (1965)
25. RATHJENS, G. W., 'Arctic engineering requires knowledge of permafrost behaviour', *Civ. Engng (USA)*, 645–647 (Nov. 1951)
26. FANCUTT, F. and HUDSON, J. C., 'The choice of protective schemes for structural steelwork', *Proc. Instn Civ. Engrs*, **17**, 405–430 (Dec. 1960)
27. *Concrete in sulphate-bearing soils and ground waters*, Building Research Station Digest No. 90, 2nd series (Feb. 1968, revised 1970)
28. TOMLINSON, M. J., *Foundation Design and Construction*, 2nd edn, Pitman, London, 758–761 (1969)
29. SHERARD, J. L., WOODWARD, R. J., GIZIENSKI, S. F. and CLEVENGER, W. A., *Earth and Earth-Rock Dams*, Wiley, London, 1963
30. Design Criteria for large Dams, *Amer. Soc. of Civil Eng*, 131p (1967)
31. JONES, J. C., 'Deep cut-offs in pervious alluvium combining slurry trenches and grouting, *Trans 9th Int. Cong. on Large Dams*, Istanbul, I, 509–524, 1967
32. GEDDES, W. G. N., ROCKE, G. and SCRIMGEOUR, J., 'The Backwater Dam', *Proc. Inst. of Civil Engrs*, **51**, 433–464 (March 1972)
33. GORDON, J. L. and DUGUID, D. R., 'Experiences with cracking at Duncan Dam', *Proc. 10th Int. Cong. on Large Dams*, Montreal, **I**, 469, 1970
34. KJAERNSLI, B. and TORBLAA, I., 'Leakage through horizontal cracks in the core of Hyttejuvet Dam', *Norwegian Geotechnical Institute Publication No. 80*, **39**, 1968
35. US Bureau of Reclamation, *Earth Manual*, 1st ed. (revised 1963)
36. OLIVER, H., 'Through and overflow rockfill dams—new design techniques', *Proc. Inst. Civil Eng*, **36**, 433–471 (March 1967)
37. POSKITT, F. F., 'The Asphaltic lining of Dungonnell Dam', *Proc. Inst. Civil Eng*, **51**, 567–579 (March 1972)
38. PENMAN, A. D. M., 'Instrumentation for embankment dams subjected to rapid drawdown', *Building Research Station*, Current Paper CP1/72 (Jan. 1972)

LIST OF CODES OF PRACTICE AND STANDARDS PUBLISHED BY THE BRITISH STANDARDS INSTITUTION AND REFERRED TO IN SECTION 16

CP 2:1951*
Earth retaining structures
CP 3:Chapter V part 1 (1967)
Chapter V part 2 (1972) Loading
CP 98:1964
Preservative treatments for constructional timber
CP 101:1972

* Published by the Institution of Structural Engineers

Foundations and substructures for non-industrial buildings of not more than 4 stories
CP 102: 1973
Protection of buildings against water from the ground
CP 112: Part 1 1967: Part 2 1971
The structural use of timber
CP 114: Part 2: 1969
The structural use of reinforced concrete in buildings
CP 115: Part 1 1959: Part 2 1969
The structural use of prestressed concrete in buildings
CP 116:1965
The structural use of precast concrete
CP 2001:1957
Site investigations
CP 2003:1959
Earthworks
CP 2004: 1972
Foundations
CP 2011:1969
Safety precautions in the construction of large diameter bore holes for piling and other purposes

BS 449 Part 1 (1970), Part 2 (1969), Supplement No. 1 (1970), Addendum No. 1 (1970)
The use of structural steel in building
BS 913: 1973
Pressure creosoting of timber
BS 988; 1076; 1097; 1451: 1973
Mastic asphalt for building (limestone aggregate)
BS 1162; 1410; 1418:1966
Mastic asphalt for building (natural rock asphalt aggregate)
BS 1180: 1972
Concrete bricks and fixing bricks
BS 1775: 1964
Steel tubes for mechanical, structural, and general engineering purposes
BS 3601:1962
Steel pipes and tubes for pressure purposes, carbon steel: ordinary duties
BS 4072:1966
Wood preservation by means of waterborne copper/chrome/arsenic compositions
BS 4360: 1972
Weldable structural steels
BS 4449:1969
Hot rolled steel bars for the reinforcement of concrete
BS4461: 1969
Cold worked steel bars for the reinforcement of concrete

17 LOADINGS

LOADINGS 17

17 LOADINGS

T. A. WYATT, Ph.D.
Department of Civil Engineering, Imperial College

LOADING

The process of structural design usually leads to some criterion of acceptability based on comparing the maximum predicted action of loads with an assured value of structural resistance. The assessment of the loading is thus as important as the structural analysis proper, although it has tended in the past to receive much less critical attention. This lack of attention has been fostered by a tendency for design loadings to be specified by clients or by governmental authority in broad terms to a degree of rigidity that leaves little freedom of choice to the designer.

Virtually all structural loadings are subject to some degree to statistical uncertainty; in other words, the maximum load that will act on any given structure during its life cannot be precisely known in advance, even if the probability of exceeding any particular value is known. In conjunction with the statistical uncertainty in the actual strength of any structure, the problem of safety is essentially probabilistic: a satisfactory design is one that limits the chance of occurrence of a load exceeding the actual strength to an acceptably low value (indeed, probably very small indeed, so that both its calculation and assessment of its significance are rather difficult), but does not make this event strictly impossible.

The format advocated by the ISO (International Standards Organisation) to tackle this problem in structural specifications in general follows proposals made by CEB/CIB and suggests definition of a 'characteristic load' based on statistical description of the maximum value of the load to occur during the design lifetime. The characteristic load may be taken as the expected maximum value (i.e. the mean of the values of the maximum load that might be observed in the life of an 'ensemble' of structures of the given type), or preferably as a value having a lower chance of occurrence, such as the expected maximum plus (say) one standard deviation of the variation of the maxima over the ensemble. The characteristic load is then augmented by a partial safety factor to yield a 'design load', to be equated to the 'design strength' which is related on the same basis to the statistics of strength of the postulated ensemble of structures. This format is very useful if the variabilities of loads and strength are not too dissimilar, so that the circumstances that could lead to failure (at, of course, a very low probability) call strongly for the conjunction of abnormally high load with abnormally low strength, at a value lying fairly close to the 'design' value as just defined.

The position reached in the United Kingdom at the time of writing (1973) varies considerably between different types of load and different types of structure. The Code of Practice for functional requirements of buildings includes a chapter (CP 3, Chapter V) on loading, covering most gravitational loadings and also the effect of wind. Only for the latter is an indication given of the relation between value of load and probability of occurrence. The use of this code is frequently made obligatory by building regulations and by-laws. Bridge loadings have in the past been given in the steel bridge specification (BS 153: Part 3A), and will in future be given in the unified code for bridge structures together with comprehensive specification of the corresponding partial safety factors.

Increasing realisation of the complexity of the interaction between loading and structural stressing in determination of structural safety has unfortunately led to increasingly sophisticated specifications for specialised structures, including loading; for the foreseeable future different methods of evaluating wind effects are likely to be called for in building, bridge, chimney, and mast and tower specifications.

The objective of this section is to give some guidance on the fundamental characteristics of various types of loadings. No attempt is made to summarise specifications in detail, nor to give densities of building or other materials such as would normally be found in a data handbook (for example, reference 1), but rather to present background material and to highlight features where a lack of appreciation of fundamental characteristics could lead to misuse of specified values. Shortage of space has prevented discussion of certain difficult specific problems, such as the probability of simultaneous occurrence of high wind loading and ice accretion on slender structures. On the problem of snow loading in general, reference may be made to the French specification (Règles Neige-Vent).

OCCUPANCY LOADS ON BUILDINGS

The floor loadings specified for office or residential buildings have remained little changed for many years and are undoubtedly based as much on experience that the accepted values lead to a satisfactory level of safety as on detailed knowledge of actual loads in service. Two major surveys, covering office[2] and retail[3] premises respectively, are, however, now available. The raw observations of actual weight loadings in the office premises were first used to determine the actual average load over notional 'bays' of various sizes (irrespective of the real structural systems of the floors concerned); the

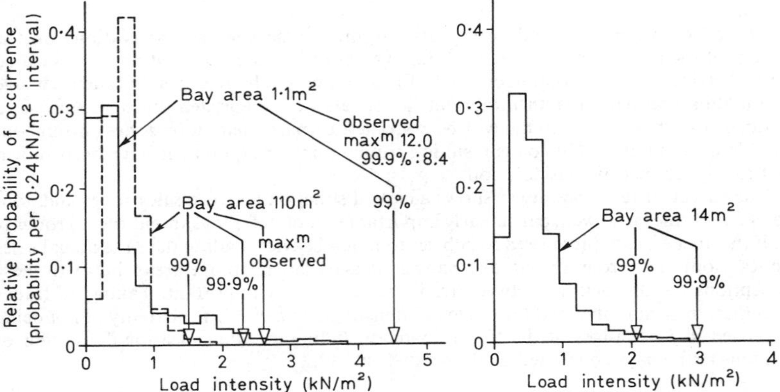

Figure 17.1 Observed load intensities in office buildings (ground floors and basements excepted)

relative frequency of finding a 'bay' to be subject to any given load is shown by Figure 17.1, for three selected sizes of bay. The most remarkable characteristic is the wide range of loads observed, even when the average is taken over quite a large region of floor. The average observed value (including personnel or other 'mobile' loads) was about 0.62 kN/m^2, leaving lowest basement floors out of account, but loads considerably in excess of 2.5 kN/m^2 were observed, even among values averaged over bays of 100 m^2.

Values of load having 99% and 99.9% probability of not being exceeded in the design life (which has been taken as equivalent to 12 complete changes of load such as would occur on a change of occupancy of the premises) for office premises taken from reference 1 are given in Table 17.1.

Table 17.1 AVERAGE LOAD INTENSITIES IN OFFICE PREMISES CORRESPONDING TO 99% AND 99.9%. PROBABILITIES OF NOT BEING EXCEEDED WITH 12 CHANGES OF OCCUPANCY (EXCLUDING GROUND FLOORS AND LOWEST BASEMENTS)

Area of bay (m^2)	1.1	5.2	14	31	111	192
Load for 99% probability (kN/m^2)	9.4	4.3	3.2	2.6	2.15	1.7
Load for 99.9% probability (kN/m^2)	17.4	5.3	4.3	3.5	3.2	2.3

The United Kingdom Code CP3 Chap. V (1967 issue) specified 2.5 kN/m^2 for general office premises, with a moderate reduction permissible when designing beams or further members supporting areas greater than 46 m^2. It is difficult to relate the above results to the standard format, as the sensitivity of the values to the probability considered is such that the usual partial load factors applied to *strengths* would no longer adequately fulfil their role of assuring a consistent level of safety between different structural types or materials. Furthermore, the dependence of this sensitivity on the size of bay suggests that the partial factor applied to the load would have to vary with the area. This problem may prove to be better treated by a more advanced probabilistic specification format.

Broadly, however, simple replacement of the existing code values by the 99% probability values with an appropriate reduction of partial load factor (say 1.25 in place of 1.6) would give a more rational balance of safety against size. It should be noted that a moderate improvement of safety with increasing size is desirable, in view of the likelihood of more serious consequences following the failure of a large bay. The variation of design load with bay size would thus be much more than hitherto accepted. The extent to which the average load intensity on any bay should be modified according to the shape of the influence function for the structural effect under consideration, to allow for local concentrations of high intensity within the bay area, is also discussed in the reference quoted.

The office occupancy load survey also permits some general observations on the nature of the load. The occurrence of high values of loading was commonly associated with shelving, often in conjunction with filing cabinets. The relatively frequent change of building occupation is an important factor, and it is suggested that it is unwise to assume that these heavy items will be restricted to particular floor zones throughout the life of a building. The loads resulting from computer equipment have been shown not to require special consideration.

The survey of retail premises[3] shows a clear distinction between sales zones and non-sales zones; the latter were particularly important in food retailing, amounting to roughly half the area of such premises and subject to much heavier loading than the actual sales zones. Books and ironmongery also showed heavily-laden storage areas, but with these exceptions the distinctions between trades were not very important. Taking all trades together, the result obtained for the load intensity having 99.9% probability of not being exceeded in 14 changes of the 'fixed' loads, including an allowance for the weight of persons in the bays concerned, is shown in Table 17.2.

Table 17.2 AVERAGE LOAD INTENSITIES IN RETAIL PREMISES CORRESPONDING TO 99.9% PROBABILITY OF NOT BEING EXCEEDED WITH 14 CHANGES OF OCCUPANCY IRRESPECTIVE OF TRADE

Area of bay (m^2)	1.1	5.6	15	28
Load on sales areas (kN/m^2)	9.1	5.4	4.0	3.3
Load on non-sales areas (kN/m^2)	18.2	10.8	7.7	6.3

The statistical variability was rather less wide than in the case of office premises, and the reduction of load intensity as the area increases was also less marked; for sales zones there was virtually no further reduction beyond 28 m^2 area, but insufficient evidence was available for larger areas in non-sales zones.

For buildings for special purposes the lessons of the possibility of wide statistical variation should be borne in mind and in particular the possibility of change in use, unless the structural layout imposes positive constraints on the use that would ensure qualified professional consideration being given prior to any major change. One important practical example of such constraint is the multi-storey (or any other roofed) car park, where restricting headroom to approximately 2 m effectively ensures that vehicles heavier than private cars are excluded.

CONTAINERS FOR GRANULAR SOLIDS

The general problem of forces in a body of granular material is more appropriately classified in the field of soil mechanics rather than loading, but some important special factors can occur, particularly in the form of large transient forces or dynamic effects during the discharge of material from bunkers or silos. The terminology is somewhat imprecise, with no strict distinction between these two terms; both can be described as bins (a common usage in America). A hopper may be a container with inclined walls only (i.e. an inverted cone or pyramid), or the section with inclined walls forming the base of a parallel-sided bin.

It is also necessary to distinguish between 'mass flow' and 'funnel flow' when discharging. In mass flow the movement of material towards the outlet is uniform across the cross section with the exception of a fairly localised 'boundary layer' adjacent to the walls, whereas in funnel flow movement is localised in a relatively narrow pipe or core which is replenished from the top. The former behaviour is often called for when storing perishable material to ensure that material is discharged in approximately the same order as it was loaded, but has the disadvantage of being associated with considerable increases during discharge in the loads acting on the walls of deep bins.[4] These increases are imperfectly understood; they appear to be rather inconsistent, and although often referred to as dynamic loads they are not generally true inertial effects.

The most common basis for design of deep bins is the theory of Janssen, the horizontal load p_h at a depth h below the free surface being related to a material parameter k by the equation

$$p_h = \frac{\rho R}{\mu'}(1 - e^{-h/c})$$

where ρ is the density of material, R the ratio of area to perimeter of bin cross section (one quarter of diameter for circular bin), μ' the coefficient of friction of material on wall, and $c = R/k\mu'$.

In Janssen's derivation k is the ratio of the horizontal pressure to the vertical pressure in the active state, given by $k = (1 - \sin\phi)/(1 + \sin\phi)$, ϕ being the angle of internal friction of the material. This angle may in practice be somewhat less than the angle of repose; some values are given in Table 17.3. Further values are given by Reynolds.[1]

Table 17.3 GRANULAR MATERIALS

	Angle of friction ϕ	*Density,* ρ (10^3 kg/m^3)
Gravel	35°–45°	1.6–2.2
Coal	20° (fines) to 40° (washed coal)	0.9
Grain	30°	0.5 (oats) to 0.8 (wheat)
Cement	10°–18°	1.4

The coefficient of friction of these materials on a concrete wall is about 0.5, rather less on a steel wall. Experimental results generally imply a rather lower value of k than given by the above (i.e. smaller loads near the top but asymptotically the same at greater depths). The larger value $k = 0.5$ suggested by Reynolds would seem to include some allowance for the dynamic effects, although the increase of load thus predicted does not entirely conform to the description that follows.

The dynamic effects during discharge may increase the effective horizontal loading by a factor as great as 3, and many failures have been reported as thus caused. According to Jenike[5] (who also gives an excellent bibliography) the explanation is that the condition at rest is approximated by the 'active' state with the major pressure nearly vertical (as in Jenssen's theory), but that on withdrawal of material from below there is vertical expansion producing a 'switch' to a 'flow' condition with the major pressure nearly horizontal. This approximates to the passive pressure state, corresponding to arching across the bin. Once this state is established each 'arch' has only to support its own weight and the horizontal loading is again similar to the Janssen theory, but at the instant of 'switch' the top of the arching region has to give some support to the 'active state' material above, producing a very high horizontal loading locally. The vertical expansion required to cause the switch is very small, so that the switch generally propagates rapidly upwards and the strength provided must at all points cater for the corresponding concentrated load. At the time of writing these theories had not been fully verified and demonstrated. The Russian specification (see reference 4) suggests that the basic value given by the Janssen formula should be doubled over the lower 65% of the height of deep bins to allow for this dynamic effect. The lower 15% of height of circular bins with flat floors are also exempted from the dynamic loading, because of the formation of a dead zone of inert material. The specification of the American Concrete Institute for grain silos requires allowance only where the outlet is markedly eccentric, a condition which can give rise to severe 'ovalisation' loading; qualitative warning is given about dynamic effects in other materials.

Dynamic effects are relatively small in 'funnel flow', but the design features necessary to ensure funnel flow are not fully established. Funnel flow is assured if the depth does not exceed 1.3 to 1.5 diameters, the lower limit applying to grain silos. A perforated tube or a lattice tower placed over the discharge orifice also prevents the type of mass flow that can cause dynamic loading. Projecting circumferential fins on the inside face of the wall are of rather less certain action. The problem of the very fine materials such as cement has been discussed by Leonhardt.[6]

ROAD BRIDGES

Probabilistic considerations are also important in traffic loading on road bridges. For loaded lengths exceeding about 30 m the reasonable design load is significantly less than the maximum weight of vehicles that could possibly occupy the carriageway, because only a very small proportion of traffic consists of vehicles giving the highest load intensity (weight per unit length of carriageway lane) and the probability of more than a few such vehicles consecutively without 'dilution' by lighter vehicles becomes extremely small. The maximum load intensity of vehicles (excluding abnormal load vehicles that only circulate under supervision) is a little over 30 kN/m (24 t four-axle vehicles about 7.2 m long for bulk loads such as heavy liquids or powders), and the British HA specification is based on a maximum of four such vehicles consecutively.[7] When the Motor Vehicle (Construction and Use) Regulations of 1955 were amended in 1964 to permit total weights exceeding 24 t, *minimum* permissible values of wheelbase were specified for such vehicles, in a number of increments up to 9.7 m (corresponding to an overall length of about 12 m) for a 32 t five-axle articulated combination. The *maximum* permitted overall length is 13 m, so that this class gives a loading of about 25 kN/m, with little scope for variation.

The growth of containerised freight traffic has increased the pressures to raise the permitted length and weights of vehicles; 20 m/46 t has been suggested, but such extreme values do not at present seem likely to gain acceptance. The longest experience of operation of 38 t articulated vehicles is to be found in Germany, and a typical vehicle is shown in Figure 17.2, together with existing British vehicles and the idealised 32 t vehicle which is the basis of the American H20–S16 specification.

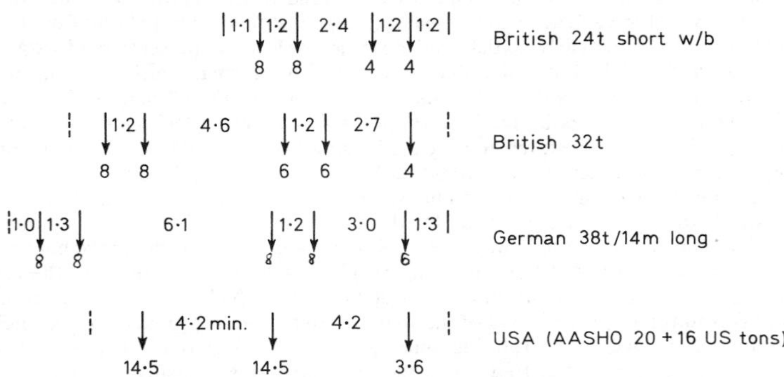

Figure 17.2 Road vehicles. Masses in tonnes. Length and spacing in metres

Only about 2% of goods vehicles in use in the United Kingdom at the time of writing (1973) approached the maximum load intensity; a further 12% gave approximately 0.8 of the maximum intensity, 19% gave 0.6, and the average for goods vehicles was less than 0.4 of the maximum intensity. The average for passenger vehicles (mostly private cars) was only about 0.1 of the maximum intensity. It should be noted, however, that overloading is common.

The total load on any span greater than about 20 m clearly requires that the traffic shall be stationary or extremely slow moving, due to the greatly increased space between vehicles at speed even in the worst cases of closely following traffic. Thus, for such spans impact or dynamic effects have negligible influence on the required design strength, and the actual probability of reaching close to the specified loading (or the likely number of occurrences of, say, 60% of the specified load) depends greatly on the number of occasions when stationary traffic is likely to occupy the span and on the proportion of

Table 17.4 TYPICAL TRAFFIC WEIGHT DISTRIBUTION

	I	*II*	*III*
Inter-city motorway (e.g. M1)—all traffic	10	35	45
slow lane, night	20	65	15
Urban industrial district main route	15	50	35

Column I: percentage of lane occupation by vehicles of weight per unit length exceeding $\frac{3}{4}$ of specification maximum load of 30 kN/m

Column II: percentage of lane occupation by vehicles of weight per unit length between $\frac{1}{4}$ and $\frac{3}{4}$ of specification maximum load

Column III: percentage of lane occupation by vehicles of weight per unit length less than $\frac{1}{4}$ of specification maximum load

heavy commercial vehicles in the traffic. The latter depends not only on the general nature of the route but also on the time of day (being much larger at night on main roads) and on the possibility of segregation of traffic types on multi-lane roads or any other case where the light traffic may opt to avoid mixing with the heavy. Although some progress has been made on probabilistic analysis of these problems,[8] application of this to design will require more data on occurrences of stationary traffic and on the bunching of heavy vehicles. Some indication of the nature of traffic in the United Kingdom is given in Table 17.4.

Further information on vehicle loads can be found in the report[9] on which fatigue loads in the British specification are based, although it should be noted that this is based on studies made in the early sixties, and does not reflect the present trend to greater vehicle size and weight but lower load intensity. The normal traffic loading in that specification is reduced to a uniformly distributed load which is a function of the loaded length only, plus a single point load (more specifically a 'knife edge' line-load across the traffic lane). The loading expressed as a u.d.l. should be a function of the shape of the influence line for the action considered as well as the loaded length, as a 'pointed' influence line is clearly sensitive to a cluster of heavy vehicles, and the addition of the knife-edge load makes some provision for this effect.

The dynamic action of the load is predominantly a question of the movement of the vehicle on its springs (or of the 'unsprung weights' of wheels and axles on the tyres); the dynamic effect of the addition of the weight of the vehicle *per se* arising from the time it takes to travel from the end of the span to somewhere near midspan is negligible, presuming the vehicle to be running smoothly on a smooth road. The excitation is therefore at a vehicle natural frequency; the fundamental is typically about 1.4 Hz for commercial vehicles, which corresponds to a suspension having a static deflection of about 150 mm. Only rarely is the fundamental frequency significantly lower than this, although this may not remain true if the trend continues towards selflevelling suspensions that can have a much larger equivalent static deflection; it might be considered possible for the vehicle to be resonant with a structural frequency within the range from 1.0 to 3.0 Hz, the latter value being limited to older or only part-laden vehicles. The natural frequency of the unsprung masses on the tyres is in the range from 8 to 14 Hz, above the range of basic frequencies of the whole bridge, but possibly significant for deck units.

It is useful to distinguish two classes of possible oscillation of the vehicle which then leads to excitation of the structure; passage over a single severe road surface irregularity leading to a large vehicle motion which is then damped out by the vehicle dampers, or a more random motion caused by the succession of small imperfections in the surface. The former is believed to be the governing factor on most bridges, associated usually with the joint between abutment and bridge. The first pulse (half cycle) of excitation to the bridge can be taken as applied at a distance from the bump equal to the distance travelled by the vehicle in one quarter of the natural period, and subsequent pulses clearly progress across the span but rapidly diminish in amplitude; a vehicle damping of 15% of critical damping can be assumed, so that each half-cycle has an amplitude 0.6 of the preceding one. The amplitude of the first pulse can conservatively be taken from Table 17.5.

Table 17.5 DYNAMIC LOADING CAUSED BY SINGLE MAJOR SURFACE IRREGULARITY

Speed of vehicle (m/s)	10	20	30
$\dfrac{\text{Amplitude of first load pulse}}{\text{Weight of vehicle}}$	0.25	0.4	0.6

Only a small proportion of vehicles (perhaps one in five or one in ten) will have the combination of suspension characteristics that would give the worst conditions for any specific structure, so that the effect on fatigue life is small, even on spans less than 30 m where the fatigue of steel members can be quite significant. The British specification includes an allowance of 25% on the maximum effect of one axle load for dynamic effects; it is difficult to compare this directly with the values in Table 17.5, as the latter is the initial pulse, at which time the vehicle is only just entering the span.

It remains to consider the possible reaction of users to any noticeable oscillation, especially if pedestrians have access to the bridge. A suggested approach is to consider the response of the system to a unit sequence of load pulses corresponding to a vehicle with suspension resonant with the bridge and thus to deduce the magnitude of excitation (measured by the amplitude of the first pulse) necessary to induce a response that would be considered unwelcome by a typical pedestrian; say, an oscillation building rapidly to an acceleration amplitude of 0.1 g, subsequently decaying at the relatively slow natural damping of the bridge. By comparison of the critical excitation with the nature of the expected commercial traffic density and its speed, Table 17.5 will enable an estimate to be made of the proportion of pedestrians using the structure who would regard the motion as unpleasant.

RAILWAY BRIDGES

Train weights and loadings are generally closely under the control of the owner, and a single train can extend to cover almost any loaded length of practical interest. The statistical variability of the estimate of the maximum loading that will occur on any specific structure is thus relatively small, but because trains crossing the bridge at speed may frequently approach close to the maximum weight, dynamic effects are important.

Design loadings may be in the form of an idealised train, specifying axle loadings and spacings (the body of the train apart from the locomotives is commonly taken as a uniformly distributed load); or as an equivalent u.d.l. tabulated as a function of span. In either case, most existing specifications are based on steam locomotive practice and somewhat outdated freight vehicle types, with the weight per unit length of the locomotives about twice that of the trailing vehicles. With modern traction, the disparity between locomotive and trailing load intensities is much less, although when two diesel locomotives run coupled together a very sharp load concentration arises from the two bogies coming adjacent to the coupling. Typical modern rolling stock for European standard-gauge railways is illustrated in Figure 17.3.

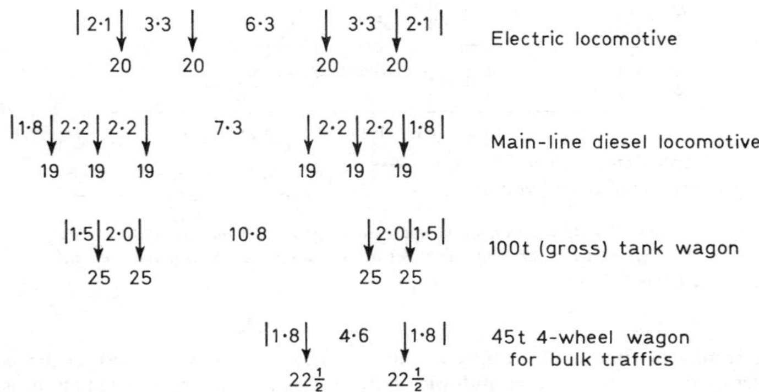

Figure 17.3 Railway vehicles. Masses in tonnes. Length and spacing in metres

Where the equivalent u.d.l. presentation is made, the value is based on the effect of a train on a simple span. Because the locomotives are placed at the end of the train (although this is no longer invariable operating practice) there is a considerable difference between the effects on a strongly skew influence line (such as end shear) and on a symmetric influence line, and the British RB loading gave separate tabulations for moment and for end shear. The RB specification included dynamic effects; an axle-train formulation (RA loading) was also given. Whichever formulation is adopted, unless the client is able to give specific advice, the designer should consider the possibility that the action of modern rolling stock on continuous-beam members (including deck units) of spans between 7 m and 20 m may differ markedly from a specification based on steam traction and simple spans.

The dynamic effects on railway bridges were fully studied in Britain during the twenties, with particular reference to the 'hammer blow' caused by steam locomotives. The Report[10] is an excellent exposition of the factors involved, although great changes have since occurred in the numerical values; bridge natural damping is now generally much lower than then assumed (an unfavourable factor), natural frequencies are generally lower (favourable except for some short-span cases), and excitation is now predominantly by the reaction from the bouncing or other oscillations of the rolling stock on its suspensions.

A recent investigation in which the ratio of the peak measured bending stress to the corresponding value calculated statically from the nominal train weights (i.e. statistical variability in actual weights included with dynamic effects) for a large number of tests on short spans in Britain gave results summarised in Figure 17.4. The present British fatigue spectra are based on application of these factors to the load cycles predicted on a quasistatic basis.[12] Special studies based on the actual traffic envisaged would often be justified for special cases; the number of occurrences of high loadings on short spans is clearly strongly dependent on the number of double-headed trains, and on long spans on trains of liquid or powders in bulk or special mineral traffics. Lines of predominantly passenger traffic, especially suburban or commuter operations, and the possibility of extension of use of powered vehicles distributed through the train to eliminate locomotives (e.g. to 'liner' trains for containerised traffic), clearly also create special cases.

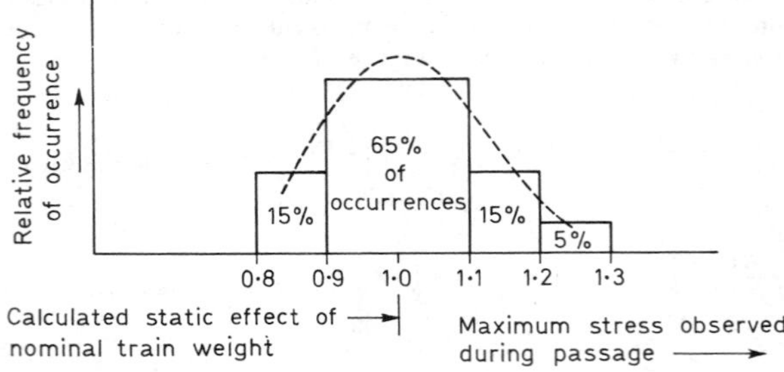

Figure 17.4 Histogram of measurements of maximum stress caused by passage of train (or single vehicle) at speed, compared with the nominal effect of the load

The results in Figure 17.4 imply an impact factor $I = 0.3$, defined as the *addition* to cover dynamic response as a multiple of the static value. More detailed analysis under the research programme of the Union Internationale des Chemins de Fer[11] has led to the formula

$$I = \frac{0.65K}{1-K+K^2}$$

in which the parameter $K = V/2nL$, where V is the train speed, n the lowest natural frequency of the bridge and L the span. Parameter K is the ratio of one half of the bridge natural period to the time taken for the train to cross the span, and thus seems more appropriate to consideration of the effect of the rate of build-up of the load as a whole than to the effect of rough riding or oscillation of the rolling stock. This may become a major factor as speeds increase beyond (say) 60 m/s, and is important in design of structures for tracked air-cushion vehicles, but for present railway structures the parameter K rarely exceeds 0.3.

WIND LOADING

Wind is by its very nature a dynamic loading. It is obvious that the gustiness always noticeable in strong winds will cause a fluctuation of the loading; but in addition to this action, aerodynamic instability of the flow pattern round the structure, or interaction between motion of the structure and the flow, may cause periodic fluctuation of the loading that can result in serious oscillation of the structure. The gust action is most important as regards excitation in the downwind direction and increases rapidly with wind speed, so that this is normally the action governing the 'static' strength required. The instabilities usually cause maximum motion perpendicular to the flow, and may have their most serious effect at relatively moderate wind speeds, so that these may be most important in respect of fatigue damage, comfort of occupants, or in some cases deflection serviceability criteria. The instability excitation is usually only significant on slender structures, but in both cases it is generally true that reducing either the natural frequency or the natural damping markedly increases the risk of serious dynamic response. Current trends in design are thus forcing designers to pay much more attention to these problems.

The basis for calculation of the required strength is the equation

$$p = \tfrac{1}{2}C_p \rho V^2$$

in which p is the pressure on the structure (for example, N/m^2), ρ is the density of air, 1.23 kg/m^3, and V is the wind speed (for example, m/s).

C_p, the 'pressure coefficient', is dependent on the shape of the body. A complete analysis thus requires:

(i) analysis of local meteorological records and extrapolation to determine the strength of wind having a given low probability of being exceeded,
(ii) a 'model' of the gusts, involving definition of their fluctuation in space (area of influence of any one gust) and time (dynamic effects),
(iii) knowledge of the pressure coefficient for the given shape,
(iv) dynamic analysis of the structure to determine the maximum value of stress in any selected structural element.

Meteorological data

The strength of the wind is usually conveniently expressed by its hourly mean value ($\bar{V}$, say), because at the peak of a major storm (very localised tornado phenomena excepted) the gusts can be treated as a 'stochastic' (random) process that is 'stationary' (having constant statistical properties although the instantaneous values at any point are changing) over such a period. Furthermore, in step (iv) the fluctuations in the response prove to be sufficiently rapid that an hour provides a large sample, and the maximum reached in the sample is then relatively insensitive to the actions of chance: thus, to estimate the overall maximum response having a given probability of occurrence, take the hourly mean wind having that probability and multiply the mean response by the 'expected' value (average that would be found from a number of statistically similar samples) of the ratio of the peak gust response to the mean. The probability of a worse condition arising from a particularly adverse low-probability gust action in an hour of lower mean speed can be neglected.

The method usually adopted for extrapolation of meteorological records to predict the wind speed having the selected low probability of occurrence is to take the maximum values from each year of the available record, and fit a Fisher-Tippett Type I extreme value distribution. Shellard[13] has shown the application of this method in the UK. The main difficulty is that a long run of reliable and consistent records is required, and serious distortion can occur if there is a systematic change within the duration of the record, for example due to change or resiting of the anemometer or even to change of its exposure. Davenport[14] and others have made useful progress towards shortening the duration of record that is required by taking account of the whole wind history of the site rather than only the annual extremes.

The records are usually corrected at source to the value applicable at a height of 10 m above ground, and are most often from sites in open terrain. The simplest satisfactory empirical formula for variation with height is the power law

$$\bar{V}_z = \bar{V}_{10}\left(\frac{z}{10}\right)^{\alpha}$$

in which $\bar{V}_z$ is the hourly mean speed at z (m) above ground and α is the wind gradient index, a function of terrain roughness.

Some values of α, together with a factor R to relate the 10 m speed at site to a common basis, and the intensity of turbulence I_{10} to express the gustiness, are given in Table 17.6. By definition

$$\bar{V}_{10} = R\bar{V}_0$$

$$\sigma_{10}(V) = I_{10}\bar{V}_{10}$$

in which $\bar{V}_0$ is the hourly mean speed at 10 m in the typical inland open terrain, and $\sigma_{10}(V)$ is the rms value of gust speed fluctuation at 10 m.

In the past[15, 16] it has been suggested that $\sigma(V)$ is invariant with height, but recent measurements show a decrease with height; a reasonable empirical formula giving values conforming to the present UK Code[17] is

$$\sigma_z(V) = \sigma_{10}(V)\left(\frac{z}{10}\right)^{-0.085}$$

Table 17.6 TERRAIN PARAMETERS

Description of terrain	α	R	I_{10}
Sea coast exposure (up to 5 km from coast if open terrain)	0.14	1.10	0.14
Open country inland, very few trees or other obstacles	0.16	1.0	0.17
Arable farmland (except as qualified above or below)	0.18	0.93	0.19
Suburban areas, woodland, farms with small fields and large hedgerows	0.22	0.80	0.23
Towns having extensive suburbs of 2–4 storeys mixed forest on rough topography	0.28	0.60	0.32

The values in Table 17.6 apply only in strong winds; say $\bar{V}_{10} > 10$ m/s. In principle R should reflect not only the terrain roughness, but also the local topography (ground contours), but in practice this is rarely possible unless anemometer data are available to compare the site, for all wind directions, with local undisturbed conditions. A site on a smooth-contoured hill top may suffer significantly accelerated wind speeds.

The maximum values of gust speeds averaged over 3 seconds or over 15 seconds can be obtained by adding $3.5\sigma_z(V)$ or $2.2\sigma_z(V)$ respectively to the hourly mean, $\bar{V}_z$.

Analysis of gust action

The discussion above has included the strength of gustiness, but gives no model of the variation of gust speeds in space or with time. The simplest approach is simply to consider the correlation of gust speeds at any instant over the area of the structure, which is summarised by the 'integral scales' of gustiness, discussed (for example) in paper 3 of reference 16. The application of this approach can also be found in reference 16, paper 6; it is particularly suited to very slender ('line like') structures and is being applied to derivation of the wind loading specification for bridges in the UK. The value of this approach is greatest for cases where the influence line for the structural action under consideration contains both positive and negative regions, so that the critical gust action depends on the greatest difference between action on the respective regions. Conventional codes of practice do not provide rational means of coping with this problem.

A simplification in the presentation of this approach can be made by noting that the correlations of gust speeds in the directions along-wind and perpendicular to the wind are closely related, and that as the gusts are carried along by the mean wind, the correlations perpendicular to the wind are thus related to the duration of the gust. It may assist appreciation of this point to visualise the gust as the effect of one side of an eddy that is at the same time being carried along by the mean flow, the eddy having a characteristic 'shape'. The shape is found to give a well correlated increase of velocity over a width of about one-half unit (taking the corresponding along-wind dimension as one unit) for gusts at a considerable height above the ground, varying down to well under one-third unit horizontally near the ground. Thus for a mean speed of 25 m/s the along-wind length of a 15 s gust would be 375 m and effective correlation would exist over about 125 m crosswind. For a large prismatic body such as a building, allowance must be made for correlation over both coordinates of the frontal area, and to some degree in the longitudinal direction also, as the pressures on the body are influenced by the flow pattern over some considerable distance. The suggestion of deriving equivalent loading for structures of dimension exceeding 50 m on the 15 s gust used in the UK Code[17] for buildings and structures includes these allowances and may give a somewhat low estimate for slender lattice or 'line-like' structures.

To include the effect of fluctuation with time, recourse is made to power spectrum analysis, and the wind speed is subjected to a form of Fourier analysis, using an integral

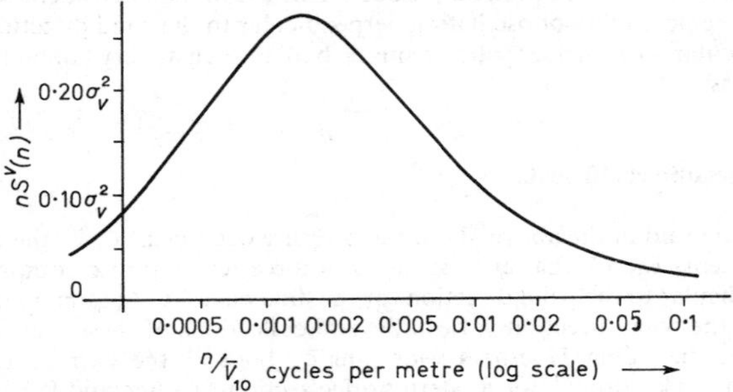

Figure 17.5 Wind gust speed spectrum. Universal nondimensional form for strong winds. S^v, power spectrum of wind; σ_v^2, variance of wind speed due to gustiness; n, frequency (Hz)*; $\bar{V}_{10}$, hourly mean speed at height 10* m

transform in place of the familiar Fourier series. In this way, the speed is represented not as a series of discrete harmonic components each having an identifiable amplitude, but as an integral of infinitesimal components over a continuous range of increments of frequency. This does not imply that identifiable periodicity exists in the wind, but does give a measure of the extent to which a structure of any given natural frequency would pick up excitation. The wind speed spectrum $S^V(n)$ is found to have a universal shape (in the appropriate nondimensional form) for any height or terrain, as shown in Figure 17.5. A power spectrum portrays the distribution of the *square* of the quantity considered with frequency, so the units of $S^V(n)$ are $(m/s)^2/Hz$: to cover the wide frequency range of the natural wind a logarithmic scale of frequency (n, say) is preferred and $nS^V(n)$ is then plotted so that areas on the plot retain their significance as the distribution of the square of the gust speed fluctuations, $nS^V \, d(\log n) = S^V \, dn$. The use of the spectrum has been explained by Davenport[15, 18] and space here permits only to point out that the spectrum can be operated upon by frequency-dependant functions expressing the correlation of the gusts over the structure and the dynamic magnification in terms of response in each natural mode of the structure. The correlation of any infinitesimal frequency component (measured by the 'coherence') follows similar rules to the correlation of the gross gust speed except that the characteristic longitudinal dimension is now the wavelength corresponding to the given frequency, $\bar{V}/n$. The effective crosswind dimension of frequency component n (twice 'lateral scale') is normally about $2\bar{V}/7n$, but is much less than this value in the horizontal direction near the ground.

Inspection of Figure 17.6 and the substitution of dimensions, such as for a typical example $\bar{V} = 25$ m/s, lowest natural frequency $n_1 = 0.5$ Hz, size of structure 50 m, shows that the resonant frequencies lie well on the diminishing upward tail of the gust spectrum and that the 'resonant gust' is small (here $2\bar{V}/7n = 14$ m) compared with the size of the structure. For most cases the dynamic effect is not large and is deemed to be covered by the load factor, but it is clear that the effect is sensitive to the value of frequency in relation to size, as well as to structural damping, and investigation is advisable if either of these factors are suspected to be lower than is usual. If such study is made it is the opinion of the writer that 10–20% trade-off from the load factor is justifiable. A study of this kind (or better) is obligatory in Canada[19] for buildings exceeding 120 m in height. For structures of simple shape and simple dynamic first mode shape, tabulated solutions are available[19, 20] (based on a slightly different empirical form of spectrum due to Davenport), or for slightly more complex cases, a term can be added to the static correlation analysis (paper 6[16]).

The power-spectrum method of analysis is directly applicable to some problems involving deflection or vibration amplitude criteria, although consideration should also be given to the possibility of oscillation perpendicular to the wind direction as discussed below, particularly for slender solid prismatic bodies, such as very tall buildings having a uniform cross section in plan.

Force and pressure coefficients

Little has been said in the above about the pressure coefficient C_p or the corresponding force coefficients C_D, C_L that express the total force resolved into components parallel and perpendicular to the wind direction (or the 'body axes' $C_x C_y C_z$ maintaining constant direction as the wind direction varies). Several collections of these factors can be used for reference, including Hoerners wide-ranging book,[21] the current UK and other national codes, Cowdrey[22] for modern bridge sections, Cohen and Perrin[23] for lattice masts, and the present author[24] for antenna structures. Very large numbers of ad hoc tests have been reported, notably by the National Physical Laboratory (Teddington, Middx, UK).

For prismatic shapes having sharp edges the coefficients are truly independent of wind speed in the practical range, whereas for rounded sections there is generally a sharp change in the flow pattern when the speed reaches a 'critical Reynolds number', which is usually about 10^5: in normal environmental conditions Reynolds number $= 6 \times 10^4$

Vd, where *V* is the speed in m/s and *d* the diameter in m. The essential feature of the change is that the point at which the flow separates sharply from the surface moves towards the back of the body and the drag coefficient decreases; this change will be provoked at a lower Reynolds number by roughness of the body surface near the separation point, and above the critical speed roughness significantly increases the force coefficient. Unfortunately the nature of roughness on most structures (joints, fins left by formwork imperfections, bolt heads) does not accord readily with the ideal 'sand-grain' roughness on which most tests are based, but certainly practical structures are not smooth in the sense that a polished test model is smooth.

For example with $H = 50$ m, $n = 0.5$ Hz, $\bar{V} = 25$ m/s

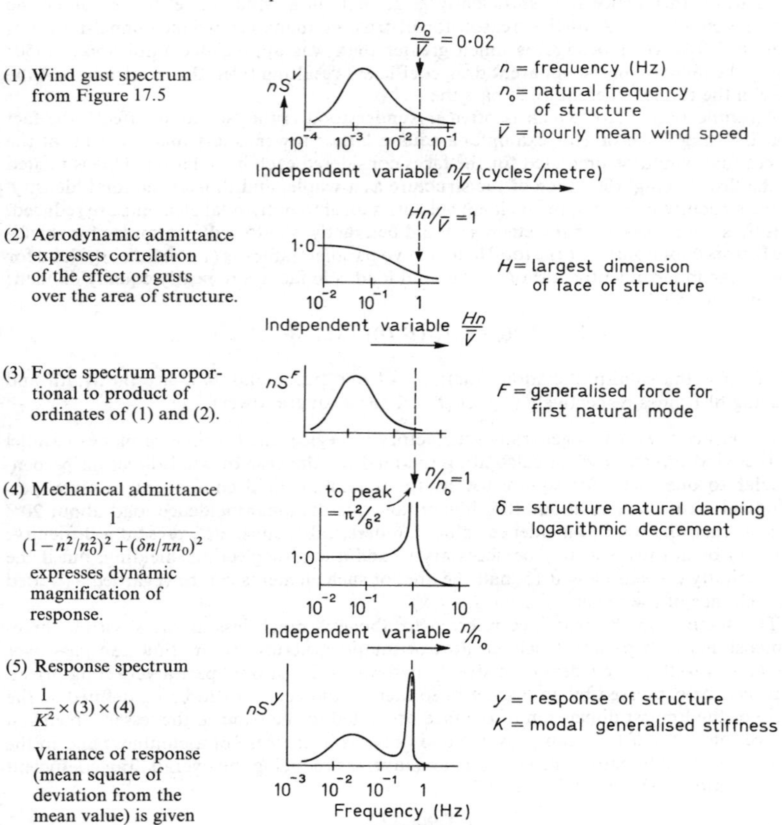

Figure 17.6 Illustration of steps in power spectrum analysis of gust action

Most wind-tunnel tests in the past have been conducted in a smooth flow, with the 'gustiness' reduced as far as possible. Turbulence has deliberately been introduced in relatively few cases, and its effect is by no means fully understood. The rear-face suction ('base pressure') is principally affected and in the case of prisms of substantial 'depth' in the downwind direction (square, or downwind greater than crosswind dimension) the result is to reduce the loading (care is necessary when reading test reports as an *increase* in the absolute value of base pressure corresponds to decrease of load): the substantial reduction in the force coefficients specified for *buildings* in the current UK Code[17] by comparison with earlier editions is in recognition of this. Conversely for prisms of small downwind dimension (approaching a plate) the load may be slightly increased. These effects are unfortunately scale dependent, as they are caused by the small rapid fluctuations corresponding to 'gust wavelengths' smaller than the cross-section of the structure. For small bodies such as members in lattice structures the corresponding components of turbulence in the natural wind may not then be sufficient to cause the reduction of coefficient referred to above, but it is not possible to reproduce the natural turbulence to a sufficiently large scale in a wind tunnel to determine the limiting conditions. A further reason for distrusting many early wind-tunnel results is that the effect of 'blockage' is much greater than was appreciated until about 1950; this is the increase in the apparent drag coefficient resulting from the constriction of the flow in the tunnel where it must pass the test body.

An important factor which is often misunderstood is the 'shielding effect', the fact that the drag force on (for example) a square lattice tower is less than the sum of the forces that would be predicted for the faces considered each in isolation. This is related to the flow 'seeing' the shape of the structure as a whole, and thus as the total 'density' of the structure increases, the incident velocities local to individual elements are reduced. It is thus not necessary that elements should be exactly 'shadowed'. Formulae for shielding factors commonly give the total load on two parallel frames as $(1+\eta)$ times the load for one frame independently, although the total load is in fact more nearly equally divided; for example,[25]

$$\eta = 1.0-1.7(\phi-0.1\sqrt{(B/D)}) \text{ (with } B/12D<\eta<1.0)$$

where ϕ is the solidity ('shadow fraction') of one frame and B, D are the width and spacing of frames respectively (e.g. $B/D = 1$ for a square tower).

For lattice towers it is generally satisfactory to neglect the bracings in planes parallel to the wind direction when calculating the load for the case of wind direction perpendicular to one face. For square towers the maximum load component resolved perpendicular to a face is about 10% higher, and the diagonal-incidence load about 20% higher, than the normal-incidence value. Ladders, cable runs, etc., present a difficulty; for very open towers it may be necessary to add a load neglecting shielding, but if the face solidity exceeds (say) 0.15, half the area of such elements can be regarded as added to each face of the tower.

The suction on the rear face referred to above is much less in cases with a three-dimensional flow pattern (such as short prismatic structures where flow can pass over the end as well as the sides) then strictly two-dimensional flow past a very long prism. This can be expressed by an 'aspect-ratio factor', the aspect ratio being defined as the ratio of the longest dimension of the face presented to the wind to the lesser dimension (twice this value if flow can pass one end only, as in the case of a chimney) and in the absence of specific information this can be taken as reducing the overall drag coefficient by the ratio k_1, shown in Table 17.7.

Table 17.7

Aspect ratio	∞	40	20	10	5	2	1
Reduction factor k_1	1.00	0.90	0.80	0.72	0.67	0.63	0.60

It will be noted that the effect is important even for quite large aspect ratios. A small obstruction to end flow, such as the supporting post of a signboard, is not significant, but a larger obstruction such as the deck of a bridge covering the end of a supporting pier destroys the effect.

Wind-excited oscillation

The flow pattern passing a slender prismatic body commonly has unsteady features, often leading to the regular build-up of vortices in the wake which, on reaching a limiting size, are swept away by the stream. Unless this occurs simultaneously symmetrically on both sides (which is unusual) the result is a periodic fluctuation of the aerodynamic force in the crosswind direction. The frequency at which this process takes place depends on the cross-section shape, and is expressed nondimensionally by the Strouhal Number, S:

$$S = nD/V$$

where n is the frequency (Hz), D the cross-section dimension (usually that perpendicular to flow) and V the flow velocity.

Thus, as the velocity increases, so does the vortex-shedding frequency, and a critical condition is likely to arise when the frequency is resonant with the structure natural frequency N (say). For this reason the velocity may be expressed nondimensionally as the 'reduced velocity', V_R:

$$V_R = V/ND$$

The critical condition is clearly $V_R = 1/S$. For most cross-section shapes giving trouble in practice V_R lies between 5 and 8; 5 for circular or roughly circular cross sections (decagon, etc.), 6 to 8 for rectangular sections, up to higher values for rectangular sections where the downwind dimension is large (e.g. wind perpendicular to the *narrow* face of a slab block) although the excitation then becomes relatively weak. Sufficient data on Strouhal numbers, or critical values of reduced velocity, are available to permit prediction of the critical speed for most cases, but the strength of excitation is much less regular and often requires ad hoc wind-tunnel testing if the critical speed falls within the practical range of speed for the site. The strength of excitation is often sensitive to detail in the cross-section shape, for example details at the leading edge of prismatic bridge cross sections. Test results are conveniently expressed by relating the steady-state amplitude to the damping nondimensional parameter $m\delta_s/\rho D^2$, in which m is the mass per unit length, δ_s the structure damping as a logarithmic decrement, and ρ the density of air.

The problem is often important in the case of circular sections, such as chimneys. Excitation is reduced by presence of the free end and is thus not normally significant for chimneys of height less than eight diameters. The excitation is also Reynolds-Number-dependent; below the critical Reynolds Number the response is typically very steady as long as the wind speed is critical. In the region of the critical Reynolds Number (2×10^5 to 4×10^5) the excitation is very weak. At higher values excitation is again present but is more random, vortices of relatively short length along the body forming and shedding with poor correlation along the length, producing a response in which the natural frequency predominates but the amplitude is continuously modulated. In this region the test results are usually quoted as a root-mean-square (rms) value, either as rms displacement, or the rms amplitude which is $\sqrt{2}$ larger. The random modulation means that allowance must be made for values significantly larger than the rms; it may be assumed that the amplitude has a Rayleigh probability distribution. The available results[26] suggest that if the response rms amplitude exceeds some critical value, typically about 0.015 D, the motion acts considerably to improve the correlation of shedding along the length and the response then increases sharply, to a magnitude that would be unacceptable to most massive structures. The influence of taper of a chimney may be important, clearly tending to reduce the likelihood of well correlated shedding, although quite well correlated shedding can occur below the critical Reynolds Number at a frequency determined by the diameter near the top. The interaction of motion of the structure, of wind turbulence and change of mean, speed with height is complex and little explored above the critical value, with or without taper; power spectrum methods have been proposed.[27]

Methods have been developed to reduce the excitation by aerodynamic means. The addition of helical 'strakes' to chimneys is now quite widely practiced[28] and can effectively eliminate response but at the expense of considerably increased drag. An alternative[28] is the 'perforated shroud', rather more expensive to make, but minimising the drag penalty. Responses in modes other than the fundamental have rarely been reported for massive structures, but should be considered. They have been observed on guyed cylinders and are common on tensioned cables; the singing of telegraph wires is of this type, and has led to the whole class of vortex-induced oscillation being referred to as 'Aeolian'.

The shedding of vortices from a structure can also have very serious effects on other structures nearby downwind. This effect can extend over quite substantial separations,[29] up to at least 10 D. For simple cases, such as pairs of similar chimneys, the ad hoc test technique is straightforward, simulating the upwind element by a fixed model in the wind tunnel. For complex cases, such as slender tower-type buildings in a city-centre environment, very complex procedures are required, including simulation of the general incident turbulence, which can be undertaken only at a very few specialist research centres.

There are several other forms of aerodynamic excitation of oscillation in a crosswind direction. The best known arises from a 'negative lift slope' condition, and is often referred to as 'galloping' from the very large amplitude oscillations suffered by electricity transmission lines when their aerodynamic coefficients are modified by the shape of ice accretions. This behaviour can briefly be explained as follows: downwards motion (say) of a horizontal prismatic structure causes the incidence of a horizontal wind to appear as if inclined upwards when viewed from the structure, and if this causes a *decrease* in the lift force (positive upwards), the motion is reinforced. In most cases this is not the case, but it does occur over a limited range of incidence angle for a square prism (and a range of other rectangular prisms)[30] and over quite a wide range of angle for D-section or pear-shape cross sections as on power lines with ice accretion. Clearly a true circular section is immune. In the simplest case, where a principal elastic plane of the structure coincides with motion perpendicular to the wind (i.e. excitation in that direction causes motion exactly in that direction) and the variation of lift coefficient (C_L) with angle of incidence (α) is closely linear in the region of the mean incidence, the critical wind speed for the onset of oscillation is given by

$$V = \frac{4mN\delta_s}{\rho D_0(\mathrm{d}C_L/\mathrm{d}\alpha)}$$

where D_0 is the dimension used in defining C_L. The other variables are as previously defined.

Galloping is distinguished by the direct proportionality of the critical wind speed to the structural damping. If this speed is exceeded, the amplitude grows to a large value, limited only by the curvature of the C_L verses α relationship. Methods for taking this into account are given in the references quoted. In contrast, vortex-shedding excitation is distinguished by a peak of response at the speed corresponding to $V_R = 1/S$, although the amplitude may increase again to higher values at substantially higher wind speeds in a turbulent wind. In response to vortex shedding, the amplitude (rather than critical speed) is damping dependent, commonly roughly inversely proportional; although there may be a controlling damping value above which oscillation is virtually entirely suppressed, or marking a sharp change in amplitude as related to the effect of motion on correlation of shedding discussed above.

More complicated behaviour, possibly involving the phenomenon of flutter, with or without interaction with the two mechanisms already described, can lead to strong oscillation of sections akin to airfoils (having relatively large dimensions in the plane of the wind), such as slender bridges. An introduction has been given by Selberg[31] and by the present author elsewhere.[32]

EARTHQUAKE EFFECTS

The effect of earthquakes on civil engineering structures is primarily a question of the dynamic response of the structure excited by motion of the ground; in general, it is the horizontal components of ground acceleration that govern, although increasing attention is being paid to the effect of the vertical component of ground motion on such cases as large-span sheds having only small live load effects from other causes. The ground motion is normally assumed to be the same at all points on the foundation of the structure; it is not within the practical power of the engineer to deal with the possibility of major relative movement on some fault-line passing within the foundation, apart from site investigation to minimise the risk of building over an existing or incipient fault where movement can be expected.

The equations of motion of the masses of a structure excited by ground motion can be manipulated to give exactly the same differential equations for the displacements *relative to the ground* (and thus the strains in the structure) as for the case of the structure on a fixed base, subjected to horizontal loads applied to every mass equal to the product of the mass and the ground acceleration. A simple basis for design is thus to express an acceleration as a fraction of the acceleration of gravity (g) and to design for this fraction of the weight of the system, treated as a horizontal loading. Due to the dynamic nature of the problem, however, this equivalent acceleration is not simply equal to the maximum ground acceleration but will depend on the natural frequency (or natural period) of the structure and on the history of the ground motion extending over some time prior to the instant when maximum relative displacement is found to occur. For a given ground motion it is a straightforward matter to solve the equations of motion numerically and record the maximum response; repeating this process for single-degree of freedom structures of varying natural frequency (or period) leads to the *spectrum* of the earthquake. The so-called velocity spectrum, S_v, is the most commonly given form; the equivalent acceleration for design is ωS_v, where ω is the 'circular' natural frequency, rad/s. The spectrum is also dependent on the natural damping of the structure. The maximum of ωS_v typically occurs at a frequency of the order of 3 Hz; the structures of fundamental natural frequency below 3 Hz are progressively relatively less sensitive to earthquakes, although the effect of higher modes may become significant.

Unfortunately, the prediction of a ground motion to form a reasonable design basis for any specific structure is subject to many uncertainties. An earthquake occurs when strain energy gradually built up in the earth's crust is suddenly released by movement on some fault plane. The energy released is measured by the *magnitude* of the earthquake, whereas its effect at some point on the ground is the *intensity* at that point. A rather crude single-parameter measure of intensity is given by scales such as the Modified Mercalli or Rossi-Forel ratings, which are based mainly on an only roughly quantified description of the human sensation or structural damage experienced (or expected).[33] The intensity experienced at a given distance from an earthquake of given magnitude depends greatly on the subsoil or shallow-rock conditions and considerably worse ground motion can be experienced where a thick layer of low-density low-stiffness material overlies heavier, stiffer, material. The duration and frequency content (and thus the shape of the spectrum) can also vary greatly even for cases where the overall intensity rating would be similar; a motion of given intensity recorded close to a low-magnitude shock would be shorter and have higher predominant frequencies by comparison with motion of the same intensity recorded distant from a high-magnitude shock. In the case of energy release from long faults the movements may be progressive along the fault, again leading to considerable differences in duration and frequency content from point to point.

The final factor to be introduced before describing the most useful approaches to design is that experience has shown that for most structures and in most regions where earthquake is a major design consideration, it would be highly uneconomic to base design on an 'elastic' or 'no significant damage' criterion. For most structures the aim must be to prevent major failures causing collapse and loss of life, while making use to the full of the possibility of inelastic structural behaviour resulting in dissipation of energy that is to a substantial degree analogous to increased structural damping. The obvious exceptions to the application of this principle are cases where even moderate damage must be prevented, such as nuclear reactor containment vessels, or buildings housing vital post-disaster services.

The most widely used format for a design code incorporating the factors described above is exemplified by the Unified Building Code of the United States. The total horizontal load (base shear) V is given by

$$V = CK\left(\frac{n^{\frac{1}{3}}}{20}\right)W$$

in which n is the lowest natural frequency, Hz, and W the weight of the structure. The liability of the site to earthquakes (seismicity) is expressed by the coefficient C, normally specified by 'regions', the value unity being applicable in regions of high risk. Considerable efforts are currently being made to take account of more local subsoil conditions by 'microregionalisation'. The factor K expresses the capacity of the structure for inelastic energy dissipation, and varies from 0.67 to 1.33 (1.50 for exceptional cases) with low values for 'brittle' structural forms, high values where ductility is assured by appropriate design. The factor in parenthesis is an approximation to the spectrum ($2\pi n S_v$) for the region of frequencies below about 1 Hz, being a slightly more conservative approximation for very low frequencies than for higher values, because the low values generally apply to very large structures where the consequence of collapse would be especially severe. The low-frequency components are also more sensitive to subsoil fluctuations.

The total force (V) is then distributed over the structure in proportion to the product of the mass and the mode shape function for the first mode (the latter is often approximated by direct proportionality to the height above the ground). It has been noted above that slender tall structures may also show significant higher-mode response, and this is most liable to increase stresses near the top (a so-called 'whiplash' effect); an added proportion of the total load, perhaps 15%, may thus be required to be applied at the highest point.

When it is desired to give more detailed consideration to the behaviour of the structure in the inelastic range, the 'reserve energy' technique is simple to apply and can quickly give very useful guidance and economy in design. To proceed to greater detail requires ad hoc computer step-by-step solution of the response to a given ground motion; this is increasingly commonly done in both USA and Japan, and is general practice in the latter country for buildings exceeding 15 storeys. Two important points must be noted. Firstly, that most of the available ground motion records to input to this procedure were obtained at a substantial distance from a large shock, so that special consideration is necessary for sites in a region whose more localised energy release is typical (producing a higher characteristic frequency in ground motion) as well as sites on soft subsoil (possibility of lower frequencies as well as overall magnification). Secondly, any one record is but one chance example of the superposition of ground wave motions of considerable complexity. Although the broad statistical properties of the ground motion are thus generally representative, the actual net peak response of one specific structure will vary greatly owing to the random factors in this superposition. One technique is to generate artificial ground motion sequences, all having the same broad statistical properties, so that the calculated maximum responses can be averaged (or the value for any given probability of occurrence selected). A somewhat more crude method to make use of a single record is to repeat analysis with a scale factor applied to the mass of the structure to modify the natural frequency. Averaging the responses obtained over a range of (say) $\pm 30\%$ of frequency greatly reduces the probable error due to the random factors.

Finally, it is worth repeating that design to ensure ductility can give much more benefit for a given cost than directly increasing strength. Good design keeps to simple shapes and simple structural forms to reduce the risk of large-scale 'stress concentrations' which would arise, for example, between two wings of a building having different natural frequencies. The conference proceedings that includes reference 34 is strongly recommended for further reading; a modern reference handbook[35] is also available on this subject.

REFERENCES

1. REYNOLDS, C. E., *The Reinforced Concrete Designers Handbook*, CACA
2. MITCHELL, G. R., and WOODGATE, R. W., 'Floor loading in office buildings—the results of a survey', Building Research Station Current Paper 3/71
3. MITCHELL, G. R., and WOODGATE, R. W., 'Floor loading in retail premises—the results of a survey', Building Research Station Current Paper 25/71
4. TURITZIN, A. M., 'Dynamic pressure of granular material in deep bins', *Proc. ASCE*, **89**, ST2 (Apr. 1963)
5. JENIKE, A. W., and JOHANSEN, J. R., 'Bin loads', *Proc. ASCE*, **94**, ST4 (Apr. 1968)
6. LEONHARDT, F., *et al.*, 'The safe design of cement silos', CACA translation No. 94
7. HENDERSON, W., 'British highway bridge loading', *Proc. Instn civ. Engrs*, **3** part 2 (June 1954)
8. STEPHENSON, H., 'Highway bridge live loads related to the laws of chance', *Proc ASCE*, ST4 (July 1957)
9. LEONARD, D. R., 'A traffic loading and its use in the fatigue life of assessment of highway bridges', Transport and Road Research Laboratory, Tech. Note 311 (1968)
10. *Report of the Bridge Stress Committee*, HMSO (1928)
11. 'Dynamic effects in railway bridges' (a report on work of ORE Committee D23), *Rail International* (Jan. 1971)
12. 'Discussion on the basis of the revised fatigue clause for BS 153', *Proc. Instn civ. Engrs*, **27**, (Feb. 1964)
13. SHELLARD, H. C., 'Extreme wind speeds over Great Britain and Northern Ireland', *Met. Mag.*, **87** (1958)
14. DAVENPORT, A. G., 'The dependance of wind loads on meteorological parameters, wind effects on buildings and structures', *Proc. Int. Seminar, Ottawa* 1967, Univ. Toronto Press
15. DAVENPORT, A. G., 'The application of statistical concepts to wind loading of structures', *Proc. Instn civ. Engrs.*, **19** (Aug. 1961)
16. *The modern design of wind-sensitive structures*, CIRIA (1971)
17. 'Wind loads', CP3 chap. V, part 2, *Loading*, British Standards Institution, London (1972)
18. DAVENPORT, A. G., 'The response of slender line-like structures to a gusty wind', *Proc. Instn civ. Engrs*, **23** (Nov. 1962)
19. National Building Code of Canada 1970 and NBC Supplement No. 4, Canadian Structural Design Manual, chap. 1 (1970)
20. DAVENPORT, A. G., 'Gust loading factors', *Proc. ASCE*, **93** ST3 (June)
21. HOERNER, S. F., *Fluid-dynamic drag*, pub. by Author (1965)
22. COWDREY, C. F., 'Time average aerodynamic forces on bridges', *NPL Aero Report* 1327 (1971); continuation NPL Mar. Sci. Report 1–72 (1972)
23. COHEN, E., and PERRIN, H., 'Design of multi-level guyed towers—wind loading', *Proc. Am. Soc. civ. Engrs*, **83** ST5 (Sept. 1957)
24. WYATT, T. A., 'The aerodynamics of shallow paraboloid antennas', *Ann. NY Acad. Sci.*, **116**, 1 (1964)
25. SCRUTON, C., and NEWBERRY, C. W., 'On the estimation of wind loads for building and structural design', *Proc. Instn civ. Engrs*, **25** (June 1963)
26. WOOTTON, L. R., 'The oscillations of large circular stacks in wind', *Proc. Instn civ. Engrs*, **43** (Aug. 1969)
27. VICKERY, B. J., and CLARK, A. W., 'Lift or across-wind response of tapered stacks', *Proc. ASCE*, **98** ST1 (Jan. 1972)
28. WALSHE, D. E., and WOOTTON, L. R., 'Preventing wind-induced oscillations of structures of circular section', *Proc. Instn civ. Engrs*, **47** (Sept. 1970)
29. WHITBREAD, R. E., and WOOTTON, L. R., 'An aerodynamic investigation for tower blocks for Ping Shek Estate, Hong Kong', *NPL Aero Special Rep.* 002 (1967)
30. NOVAK, M., 'Galloping oscillations of prismatic structures', *Proc. ASCE*, EM1 (Feb. 1972)

31. SELBERG, A., 'Oscillation and aerodynamic stability of suspension bridges', *Acta Polytechnica Scandinavia,* Ci 13, Trondheim (1961)
32. WYATT, T. A., 'The effect of wind on slender long-span bridges', *Proc. Symp. on Wind Effects on Buildings & Structures, Loughborough Univ.* (1968)
33. NEUMANN, F., 'Seismic forces on engineering structures', *Proc. ASCE,* **88** ST2 (Apr. 1962)
34. BLUME, J. A., 'Analysis of dynamic earthquake response', ASCE/IABSE Conference, Planning and Design of Tall Buildings (Lehigh) 1972, Paper 1b/6
35. WIEGEL, R. L. (Ed), *Earthquake Engineering,* Prentice Hall (1970)

18 BRIDGES

18 BRIDGES

D. J. LEE, B.Sc., Tech.D.I.C., F.I.C.E., F.I.Struct.E.
B. J. RICHMOND, B.Sc.(Eng.), Ph.D., M.I.C.E., M.I.Struct.E.
Maunsell & Partners

This chapter covers the selection and analysis of bridges and attempts to relate the most frequently used bridging materials—steel and concrete.

As extensive treatment as is possible in the space available is given to box girder analysis as this is important in most modern bridge construction. Information about individual bridges will be found in the bibliography. Reference to these specific examples will assist an understanding of the historical background and the existing state of the art. A good general review of the structural form of bridges is given by Beckett,[2] whilst a sensitive aesthetic assessment is provided by Mock.[1]

Masonry arches and steel trusses have not been dealt with but interesting examples of these types of bridges are contained in References.

The principles developed in this chapter for open or closed sections are applicable to trussed structures if suitable modifications are made to allow for shear behaviour of the truss system.

Thus the authors hope that from a reading of this chapter it is possible to make a preliminary assessment for most modern bridge designs by methods which enable the essential nature of the structural behaviour to be perceived and can be developed to detailed analysis without the necessity of revising the basic principles.

ECONOMICS AND CHOICE OF STRUCTURAL SYSTEM

Cost comparisons which would make it possible to arrive at the most economical choice of material, structural form, span, etc., have been sought for many years by bridge engineers but since the costs of any one bridge depend on the circumstances prevailing at that time, the information is always imprecise. Cost data must be up-to-date and must be sufficiently detailed to allow adjustments to be made for changed circumstances. It is the changes in these factors that lead to new methods of construction and new structural systems; a major change of this kind has been that involving box girders, plate girders and trusses.

A very early steel box girder bridge, the Brittania Bridge,[3] built by Stevenson over the Menai Straits (main spans 140 m or 459 ft, completed in 1850) was very successful and was in regular use for railway trains until recently when it was damaged by fire. Each span was lifted into place in its entirety by hydraulic jacks. The advantages of truss construction were, however, sufficient to convince engineers for the next hundred years that box structures were not economical though plated structures were used in the form of I beams for smaller spans and lighter loads. The steel box girder re-emerged as a structural system for bridges after the second World War although short span multi-cellular bridges in reinforced concrete had been used for short spans in the 1930s. By 1965 a large proportion of structures other than short spans were built as box structures of one form or another. A greater degree of selectivity then began to emerge and open cross sections even for substantial spans were again being used provided no problems of

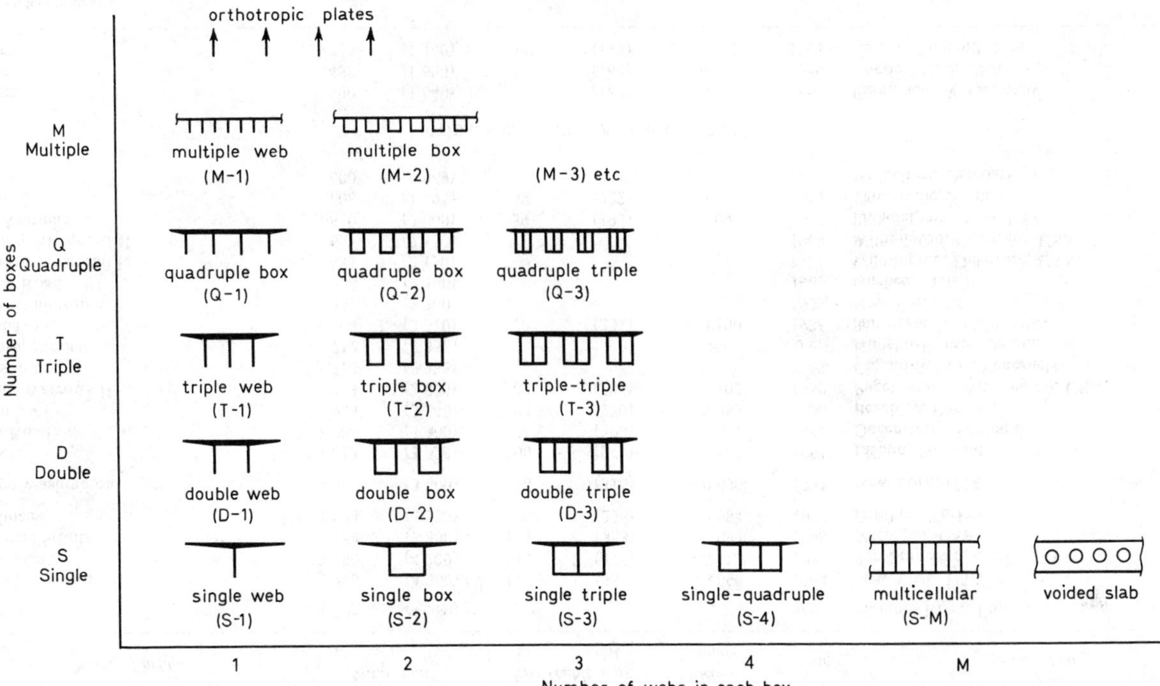

Figure 18.1 Classification of bridge deck cross sections

Table 18.1 THE WORLD'S LEADING SUSPENSION BRIDGES

Name of bridge	*Main span* m	ft	*Rise (cable sag)* m	ft	*Rise/span*	*Year*	*Location*
Humber	1 396	(4 580)				u.c.	Humber River, England
Verrazano Narrows	1 298	(4 260)	117	(385)	−0.090	1964	New York, USA
Golden Gate	1 280	(4 200)	145	(475)	−0.113	1937	San Francisco, Calif. USA
Mackinac Straits	1 158	(3 800)	108	(353)	−0.093	1958	Michigan, USA
Bosphorus	1 074	(3 525)	90	(296)	−0.084	1973	Ortakoy, Turkey
George Washington	1 067	(3 500)	96	(316)	−0.090	1931	New York, USA
Tagus	1 013	(3 323)	106	(349)	−0.105	1966	Lisbon, Portugal
Forth Road	1 006	(3 300)	91	(300)	−0.091	1964	Queensferry, Scotland
Severn	988	(3 240)	82	(270)	−0.083	1966	Beachley, England
Tacoma Narrows II	853	(2 800)	87	(286)	−0,102	1950	Puget Sound, Washington, USA
Angostura	712	(2 336)				1967	Cuidad Bolivar, Venezuela
Kanmon Strait	712	(2 336)				u.c.	Honshu-Kyushu, Japan
Transbay	2 × 704	2 × (2 310)	70	(231)	−0.100	1936	San Franciso, Calif. USA
Bronx–Whitestone	701	(2 300)				1939	New York, USA
Quebec Road	668	(2 190)				1969	Quebec, Canada
Delaware Memorial I	655	(2 150)				1951	Wilmington, Delaware, USA
Delaware Memorial II	655	(2 150)				1968	Wilmington, Delaware, USA
Walt Whitman	610	(2 000)	59	(195)	−0.097	1957	Philadelphia, Penn. USA
Tancarville	608	(1 995)	68	(222)	−0.111	1959	Tancarville, France
Lillebaelt	600	(1 969)				1969	Middelfart, Denmark
OTHER SUSPENSION BRIDGES OF INTEREST							
Emmerich	500	(1 640)	55	(182)	−0.111	1965	Emmerich, W. Germany
Newport	488	(1 600)	55	(182)	−0.114	1969	Rhode Island, USA
Tamar	335	(1 100)	34	(112)	−0.102	1961	Saltash, England

u.c. = under construction

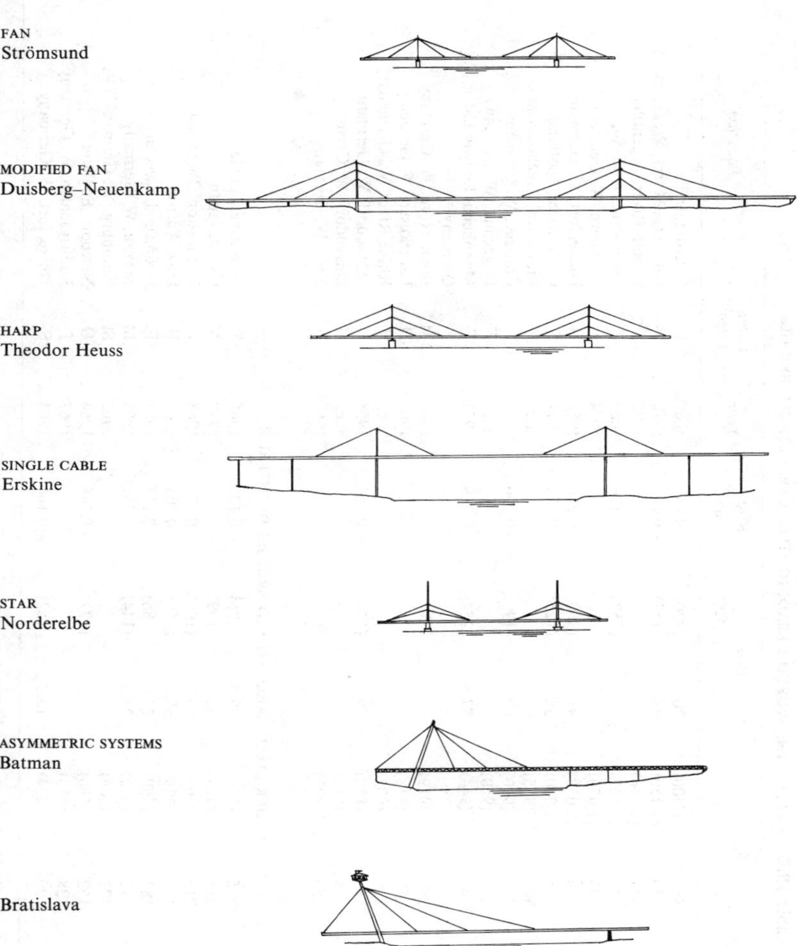

Figure 18.2 Examples of different cable systems. Scale: 1/10 000 approx.

aerodynamic stability were possible. The use of plate girders has been further encouraged by the reaction caused by failures of steel box girder bridges but it seems likely that a balanced view of the merits of various forms of construction will prevail.

Figure 18.1[4] shows the possible cross sections for bridge structures which can include truss systems if the plane of each triangulated panel is represented by either a web or flange member. The significance of box structures in a more general sense now becomes clear. It is the open cross section that is a particular, although important, form of construction, whereas the box system is perhaps a misleadingly simple description of the general range of structures.

The most basic structural dimension for a given span affecting both the least cost and

Table 18.2 SOME OF THE WORLD'S LEADING STEEL CABLE STAYED BRIDGES

Name of bridge	*Span* m	*Span* ft	*Rise* m	*Rise* ft	*Rise/Span*	*Year*	*Cable system*	*Location*
Hooghly	457	(1 500)	70	(230)	−0.153	u.c.	F	Calcutta, India
Duisberg–Neuenkamp	350	(1 150)	48	(157)	−0.137	1970	M	Duisberg–Moers, W. Germany
Zarate–Brazo	340	(1 116)				1972	S	Paraná River, Argentina
West Gate	336	(1 102)	45.7	(150)	−0.136	u.c.	F	Melbourne, Australia
Köhlbrand High Level	325	(1 070)				1974	M	Hamburg, W. Germany
Kniebrücke	320	(1 050)	91.6	(300)	−0.286	1969	H	Düsseldorf, W. Germany
Erskine	305	(1 000)	38	(125)	−0.124	1971	S	Clyde River, Scotland
Bratislava	303	(995)				u.c.	F	Bratislava, Czechoslovakia
Severins	302	(990)	60.5	(199)	−0.200	1959	F	Cologne, W. Germany
Donaubrücke	290	(950)				u.c.	F	Deggendu, W. Germany
Nordbrücke	288	(944)	77.5	(254)	−0.269	1972	F	Mannheim–Ludwigshafen, W. Germany
Friedrich–Ebert–Brücke	280	(919)				1967	M	Bonn–Nord, W. Germany
Leverkusen	280	(918)				1965	H	Leverkusen, W. Germany
Speyer	275	(902)				u.c.	F	Rhine River, W. Germany
Theodor Heuss (Nord)	260	(853)	39	(128)	−0.150	1958	H	Düsseldorf, W. Germany
Oberkassel	258	(845)				u.c.	H	Düsseldorf, W. Germany
Rees	255	(837)				1967	H	Rees, W. Germany
OTHER STEEL CABLE STAYED BRIDGES OF INTEREST								
Wye	235	(770)	29.3	(96)	−0.125	1966	S	Wye River, England
Toyosata–Ohhashi	216	(709)	30.2	(99)	−0.140	1970	M	Osaka, Japan
Batman	215	(705)	59.7	(196)	−0.278	1968	F	Launceston, Tasmania
Donaubrücke	215	(705)	65.5	(215)	−0.305	u.c.	H	Linz, Austria
Strömsundbrücke	183	(599)	27.0	(86)	−0.147	1955	F	Strömsund, Sweden
Maxau	175	(575)	44.0	(144)	−0.251	1966	M	Maxau, W. Germany
Norderelbe	172	(564)				1962	St	Hamburg, W. Germany
George Street (Usk)	152	(500)	37.2	(122)	−0.248	1964	H	Newport, England
Hochstrasse Ludwigshafen	138	(453)				1967	F	Ludwigshafen, W. Germany
Julicherstrasse	99	(324)	15.8	(52)	−0.160	1964	S	Düsseldorf, W. Germany

Cable system: F fan, M modified fan, H harp, S single cable, St star. u.c. = under construction

Table 18.3 SOME OF THE WORLD'S LEADING CONCRETE CABLE STAYED BRIDGES

Name of bridge	*Span* m	*Span* ft	*Rise* m	*Rise* ft	$\frac{Rise}{Span}$	*Year*	*Cable system*	*Location*
Wadi Kuf	282	(925)	54.1	(177)	−0.192	u.c.	S	Beida, Libya
Tiel	267	(876)	46.3	(152)	−0.173	1974	F	Waal River, Netherlands
Corrientes	248	(814)	47.4	(155)	−0.193	u.c.	M	Paraná River, Argentina
Maracaibo	235	(771)	42.5	(139)	−0.180	1962	S	Lake Maracaibo, Venezuela
Polcevara	210	(689)	45.1	(148)	−0.214	1967	S	Genoa, Italy
Magliana	145	(476)	34.0	(111)	−0.234	1967	S	Rome, Italy

u.c. = under construction
Cable system: F fan, M modified fan, H harp, S single cable

Table 18.4 THE WORLD'S LEADING STEEL ARCH BRIDGES

Name of bridge	Span m	Span ft	Rise m	Rise ft	Rise/Span	Year	Location
Bayonne	504	(1 652)	81	(266)	0.161	1931	New York, NY, USA
Sydney Harbour	503	(1 650)	107	(350)	0.212	1932	Sydney, Australia
Fremont*	383	(1 255)				u.c.	Portland, Oregon, USA
Port Mann*	366	(1 200)	76	(250)	0.208	1964	Vancouver, Canada
Thatcher†	344	(1 128)				1962	Balboa, Panama
Laviolette†	335	(1 100)				1967	Trois-Rivières, Canada
Zd'ákov	330	(1 083)	42.5	(139)	0.129	1967	Lake Orlik, Czechoslovakia
Runcorn–Widnes	330	(1 082)	66.4	(218)	0.202	1961	Mersey River, England
Birchenough	329	(1 080)	65.8	(216)	0.200	1935	Sabi River, Rhodesia
Glen Canyon	313	(1 028)				1959	Arizona, USA
Lewiston–Queenston	305	(1 000)	48.4	(159)	0.159	1962	Niagara River, USA–Canada
Hell Gate	298	(977)				1917	New York, NY, USA
OTHER STEEL ARCH BRIDGES OF INTEREST							
Rainbow	289	(950)	45.7	(150)	0.158	1941	Niagara Falls, USA–Canada
Askerofjord	278	(912)	40.5	(133)	0.146	1960	Askerofjord, Sweden
Fehmarnsund*	249	(816)	43.6	(143)	0.175	1963	Fehmarnsund, W. Germany
Adomi (Volta)	245	(805)	57.4	(188)	0.234	1957	Adomi, Ghana
Kaiserlei*	220	(722)				1964	Frankfurt a.M., W. Germany

u.c. = under construction
* Tied arch
† Cantilver arch

Table 18.5 THE WORLD'S LEADING CONCRETE ARCH BRIDGES

Name of bridge	Span		Rise		$\frac{Rise}{Span}$	Year	Location
	m	ft	m	ft			
Gladesville	305	(1 000)	40.8	(134)	0.134	1964	Sydney, Australia
Rio Paraná	290	(952)	53.0	(174)	0.183	1965	Paraná River, Brazil–Paraguay
Arrabida	270	(885)	51.9	(170)	0.192	1963	Portugal
Sandö	264	(866)	40.0	(131)	0.151	1943	Angerman River, Sweden
Shibenik	246	(808)				1967	Krka River, Yugoslavia
Fiumarella	231	(758)	66.1	(217)	0.286	1961	Catanzaro, Italy
Novi Sad	211	(692)				1961	Danube River, Yugoslavia
Linenau	210	(689)				1967	Bregenz, Austria
Van Stadens	200	(656)				1971	Van Stadens Gorge, S. Africa
Esla	192	(631)				1942	Esla River, Spain
Rio das Antas	180	(590)	28.0	(92)	0.156	1953	Brazil
Traneberg	178	(585)	26.2	(86)	0.147	1934	Stockholm, Sweden
Plougastel (Albert Louppe)	173	(567)	33	(108)	0.190	1930	Elorn River, France
Selah Creek	168	(550)				1971	Yakima, Wash., USA
La Roche–Guyon	161	(528)	23.0	(75)	0.143	1934	France
Cowlitz River Bridge	158	(520)					Mossyrock, Wash., USA
Caracas–LaGuaira	152	(498)	39.0	(128)	0.257	1952	Caracas, Venezuela
Puddefjord	145	(492)				1956	Norway
Podolska	145	(492)				1942	Czechoslovakia
OTHER CONCRETE ARCH BRIDGES OF INTEREST							
Revin–Orzy	120	(394)	10.0	(33)	0.083		Meuse River, France
Glemstal	114	(374)	27.1	(89)	0.238		Stuttgart, W. Germany
Slängsboda	111	(364)	12.0	(39)	0.108	1961	Stockholm, Sweden

u.c. = under construction

Table 18.6 THE WORLD'S LEADING CANTILEVER TRUSS BRIDGES

Name of bridge	Span m	Span ft	Year	Location
Quebec Railway	549	(1 800)	1918	Quebec, Canada
Forth Railway	2 × 521	2 × (1 710)	1890	Queensferry, Scotland
Delaware River	501	(1 644)	u.c.	Chester, Penn–Bridgeport, NJ, USA
Greater New Orleans	480	(1 575)	1958	New Orleans, La., USA
Howrah	457	(1 500)	1943	Calcutta, India
Transbay	427	(1 400)	1936	San Francisco, Calif., USA
Baton Rouge	376	(1 235)	1968	Baton Rouge, La., USA
Tappan Zee	369	(1 212)	1955	Tarrytown, NY, USA
Longview	366	(1 200)	1930	Columbia River, Wash., USA
Queensboro	360	(1 182)	1909	New York, USA
I Carquinez Strait	2 × 335	2 × (1 100)	1927	San Francisco, Calif., USA
II Carquinez Strait	2 × 335	2 × (1 100)	1958	San Francisco, Calif., USA
Second Narrows	335	(1 100)	1960	Vancouver, Canada
Jacques Cartier	334	(1 097)	1930	Montreal, Canada
Isaiah D. Hart	332	(1 088)	1967	Jacksonville, Fla., USA
Richmond–San Rafael	2 × 326	2 × (1 070)	1956	San Pablo Bay, Calif., USA
Grace Memorial	320	(1 050)	1929	Cooper River, SC, USA
Newburgh–Beacon	305	(1 000)	1963	Hudson River, NY, USA
OTHER CANTILEVER TRUSS BRIDGES OF INTEREST				
Auckland Harbour	244	(800)	1959	Auckland, New Zealand

u.c. = under construction

Table 18.7 SOME OF THE WORLD'S LEADING STEEL GIRDER BRIDGES

Name of bridge	*Span* m	ft	*Depth* (d) *at midspan* m	*Depth* (d2) *at pier* m	$\frac{d_1+d_2}{Span}$	*Year*	*Type*	*Location*
Niteroi	300	(984)	7.4	12.9	0.068	1974	B	Rio de Janeiro, Brazil
Sava I	261	(856)	4.6	9.8	0.055	1956	P	Belgrade, Yugoslavia
Zoo	259	(850)	4.5	10.0	0.056	1966	B	Cologne, W. Germany
Sava II	250	(820)				1969	B	Belgrade, Yugoslavia
Koblenz	235	(771)				u.c.	B	Rhine River, W. Germany
San Mateo–Hayward	228	(750)	4.6	9.2	0.060	1967	B	Calif., USA
Hochbrücke 'Radar Insel'	221	(727)	5	9.5	0.066	u.c.	P	Nord–Ost see Canal, W. Germany
Moselle	219	(718)				u.c.	B	Moselle Valley, W. Germany
Milford Haven	213	(700)	5.9	5.9	0.055	u.c.	B	Pembroke Dock, Wales
Fourth Danube	210	(689)				1970	B	Vienna, Austria
Düsseldorf–Neuss	206	(676)	3.3	7.8	0.054	1951	B	Düsseldorf, W. Germany
Wiesbaden–Schierstein	205	(673)	4.4	7.4	0.057		P	Rhine River, W. Germany
Europa	198	(650)	7.7	7.7	0.078	1964	B	Sill Valley, Austria
Köln–Deutz	185	(606)				1948	B	Rhine River, W. Germany
Poplar Street	183	(600)	6.2	7.6	0.070	1967	B	St. Louis, Miss., USA
Italia	175	(574)	8.5	8.5	0	1969	B	Lao River, Italy
Avonmouth	174	(570)				1974	B	Gloucs., England
Gemersheim	165	(541)	9.1	5.4	0.058	1971	B	Rhine River, W. Germany
Speyer	163	(535)	3.4	6.40	0.060	1956	B	Rhine River, W. Germany
Concordia	160	(525)	4.9	4.9	0.060	1967	B	Montreal, Canada
New Temerloh	151	(500)	3.7	5.9	0.064	1974	B	Temerloh, Malaysia
OTHER STEEL GIRDER BRIDGES OF INTEREST								
Calcasieu River	137	(450)	2.1	7.0	0.078	1963	P	Louisiana, USA
St. Alban	135	(443)	2.8	9.3	0.062	1955	P	Basel, Switzerland
Amara	82	(269)	3.7	12.1	0.087	1958	B	Tigris River, Iraq

u,c. = under construction
Bridge type: B box girder, P plate girder

Table 18.8 SOME OF THE WORLD'S LEADING CONCRETE GIRDER BRIDGES

Name of bridge	*Span* m	*Span* ft	*Depth* (d) *at midspan* m	*Depth* (d_2) *at pier* m	$\frac{d_1+d_2}{span}$	*Year*	*Type*	*Location*
Urato	230	(754)	4.0	12.5	0.072	1972	C	Shikoku, Japan
Three Sisters	229	(750)				u.c.	C	Potomac River, Washington, DC, USA
Bendorf	208	(682)	4.4	10.4	0.071	1965	C	Bendorf, W. Germany
Gardens Point	183	(600)				u.c.	C	Brisbane, Australia
Brisbane Water	183	(600)				u.c.	C+SS	N.S.W., Australia
Medway	152	(500)	2.2	10.8	0.086	1963	C+SS	Rochester, England
Neckarsulm	151	(496)	4.2	7.4	0.078	1968	C	Neckarsulm, W. Germany
Moscow River	148	(485)				1957	CG	USSR
Amakusa	146	(479)				1966	C	Japan
Kingston	143	(470)	2.4	10.0	0.087	1970	C	Glasgow, Scotland
Victoria	142	(467)				1970	C+SS	Brisbane, Australia
Tocantins	142	(465)				1961	C	Tocantins River, Brazil
Bettingen	140	(459)	3.0	7.0	0.089		C	Main River, W. Germany
Don	139	(455)				1964	C	Rostow, USSR
Pine Valley	137	(450)				u.c.	CG	Calif., USA
Alnö	134	(440)				1964	C	Alnösund, Sweden
Öland	130	(426)				1972	C	Kalmar Sound, Sweden
Schelnicha	128	(420)				1964	C	Moscow, USSR
Rio Colorada	124	(407)				1972	SA	San Jose, Costa Rica
D'Omonita	120	(394)	1.8	6.0	0.065		C+SS	Honduras
Nusle	115	(378)	6.4	6.4	0.111	u.c.	CG	Prague, Czechoslovakia
Bassein Creek	115	(376)				1970	C	Bombay, India
			OTHER CONCRETE GIRDER BRIDGES OF INTEREST					
Worms	114	(375)	2.5	6.5	0.079	1952	C	Rhine River, W. Germany
Koblenz	114	(374)	2.7	7.2	0.087	1954	C	Moselle River, W. Germany
Nötesund	110	(361)	2.2	5.7	0.072	1966	C	Orust, Sweden
Siegtal	105	(344)	5.8	5.8	0.110	1969	CG	Eiserfeld, W. Germany
Chillon Viaduct	104	(341)	2.2	5.6	0.072	1973	CG	Chillon, Switzerland
Narrows	97	(320)	2.2	4.2	0.068	1959	C+SS	Perth, Australia
Oleron	79	(259)	2.5	4.5	0.089	1966	CG	Rochefort, France

u.c. = under construction

the least weight methods of measuring efficiency is the effective lever arm of the structure for resisting bending moments resulting from the vertically acting forces from self weight and imposed loads and vertical components of the support reactions. In bridges which depend on horizontal reactions from the ground this distance is the rise of an arch above its foundations or the dip of a suspension cable between towers. If the supports are at different levels the dip or rise is measured vertically from the chord joining the supports.

The high strength-to-weight ratio of steel wire and favourable price-to-strength ratio results in dip-to-span ratios of 0.1 being suitable for even the longest suspension bridges, see Table 18.1. The shallow cable has a higher tension which improves its capacity for carrying uneven loads without large deflection and increases its natural frequency of vibration. The cost of the cable alone is not, however, sufficient to reach conclusions on economics since the cost of foundations to anchor the cables is substantial and varies with the ground conditions.

The lower strength-to-weight ratios of steel in compression and concrete combined with the destabilising effect of the compressive force of the thrust lead to the rise-to-span ratios being considerably higher on average (see Tables 18.4 and 18.5). Good foundations and the requirements of local topography may lead to reduced ratios, and arches, such as Gladesville,[5] which are in flat country and yet have the roadway running above the arch rib, require a low rise to minimise the cost of approach embankments.

The depth between compression and tension flanges is the lever arm of a simply supported beam structure, such as a truss, plate girder or box girder. If the structure is continuous at both ends then the sum of the depths at the centre span and one of the supports is the lever arm (see Tables 18.7 and 18.8).

The cable-supported bridge can be seen as either a suspension bridge or a continuous beam with the effective depth at the supports equal to the height of the tower. Tables 18.2 and 18.3 show the leading bridges of that type using the rise in the suspension bridge sense as the leading dimension of depth. Figure 18.2 shows the various arrangements of cables that are used.

The choice of span depends on the foundations, depth of water and height of the deck, but in many cases other requirements such as navigation clearances dictate the minimum span. It is usually only shorter spans where, proportionately at any rate, there is considerable variation possible. It has been claimed in the past that at the most economic span of a multi-span structure, the cost of foundations equals the cost of the superstructure less the basic deck structure costs. The assumptions necessary for this to be valid are that the cost of superstructure per unit length should increase linearly with span and that

Table 18.9

Depth	0.5	0.6	0.7	0.75	0.8
Af	0.25 t	0.2 t	0.15 t	0.125 t	0.1 t
Section modulus Z	0.167 t	0.180 t	0.187 t	0.187 5 t	0.187 t

Note: total cross section throughout = 1.0 t

of the substructure should vary inversely with span. The slopes of the respective cost–span curves are then equal and opposite at the point of intersection of the curves provided that any constant cost in both foundations and superstructure are first subtracted. If the cost of the superstructure is assumed to increase proportionately to the square root of the span, however, then the same approach requires that half the superstructure cost should equal the foundation cost. In modern structures it is difficult to separate the costs of the basic deck system from the total of the multi-span structure.

The well known rule that for maximum economy the total area of the flanges of a beam should equal the area of the web is a more useful guide. Table 18.9 shows that for a given web thickness and a total area of cross section of 1.0 t the maximum section modulus is at a depth of 0.75 where the total flange area is one third the web area, but

Table 18.10 PRECAST CONCRETE BRIDGE BEAMS

Type of beam	*Name of beam*	*Classification (as Figure 18.1)*	*Span*	*Beam section*
I	C & CA I section beam	M-1	12–36 m	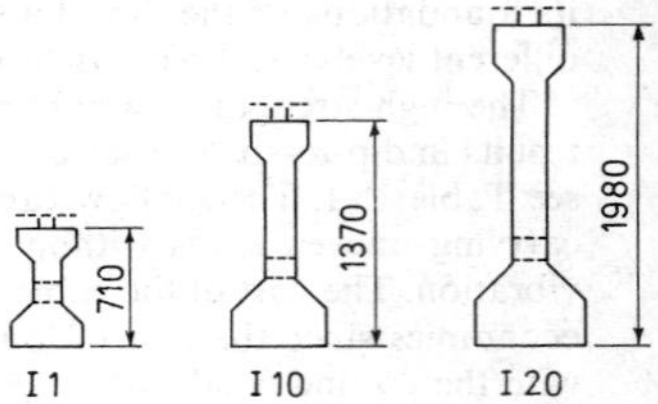 I 1 (710), I 10 (1370), I 20 (1980)
Inverted T	C & CA inverted T beam for spans from 7 m to 16 m	orthotropic slab	7–16 m	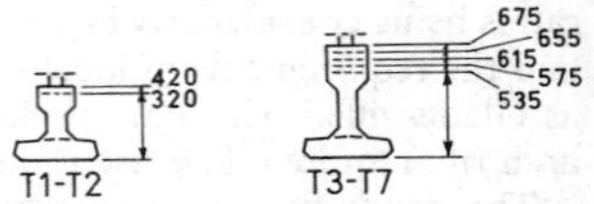 T1-T2 (420, 320), T3-T7 (675, 655, 615, 575, 535)
Inverted T (M range)	MoT/C & CA prestressed inverted T beam for spans from 15 m to 29 m	(a) T beam M-1 (b) Pseudo box S-M	15–29 m	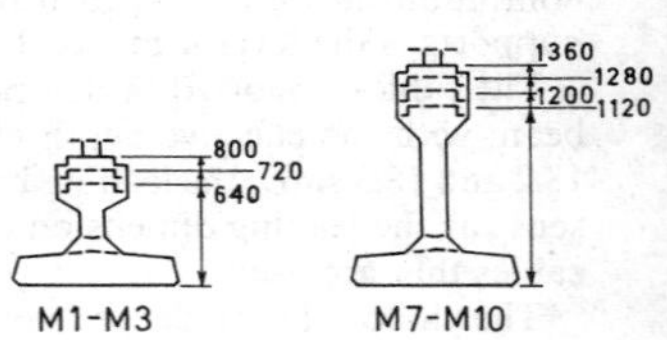 M1-M3 (800, 720, 640), M7-M10 (1360, 1280, 1200, 1120)
Box	C & CA box section beam	S-M	12–36 m	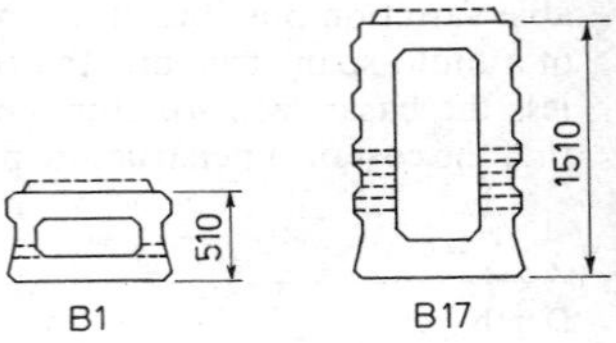 B1 (510), B17 (1510)
Box	top hat	S-M	15–40 m	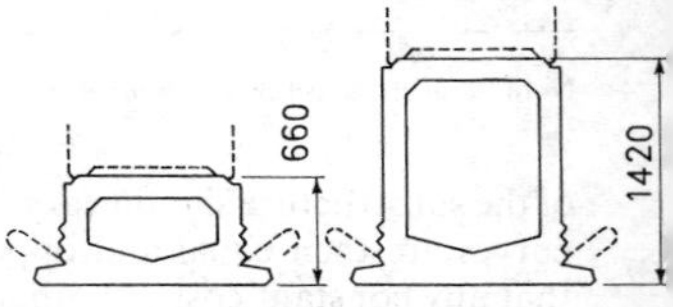 (660), (1420)
U	U beam	M-2	15–36 m	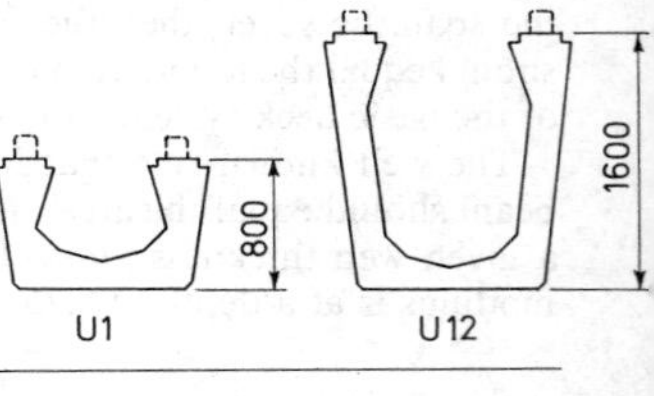 U1 (800), U12 (1600)

Table 18.10 (*continued*)

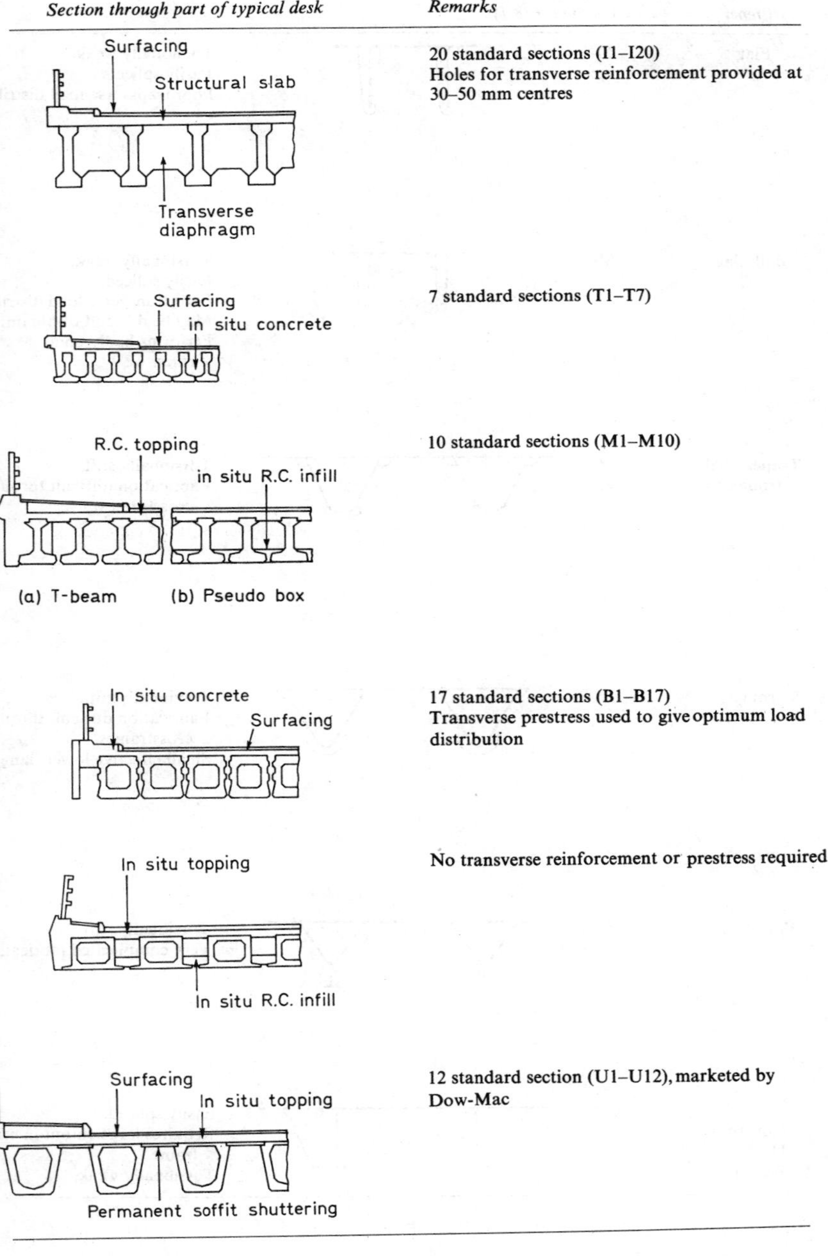

Section through part of typical desk	*Remarks*
	20 standard sections (I1–I20) Holes for transverse reinforcement provided at 30–50 mm centres
	7 standard sections (T1–T7)
	10 standard sections (M1–M10)
	17 standard sections (B1–B17) Transverse prestress used to give optimum load distribution
	No transverse reinforcement or prestress required
	12 standard section (U1–U12), marketed by Dow-Mac

Table 18.11 LONGITUDINAL STIFFENERS FOR ORTHOTROPIC DECKS

Type of stiffener	*Classification (as Figure 18.1)*		*Remarks*
Flat	M-1 etc	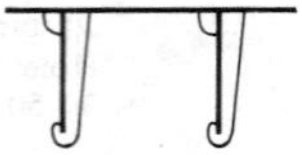	Torsionally weak. Easily spliced. Poor transverse load distributio
Bulb flat	M-1	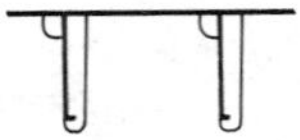	Torsionally weak. Easily spliced. Poor transverse load distributio May be difficult to obtain. Performs badly under Merrison rules.
Trapezoidal trough	M-2 etc	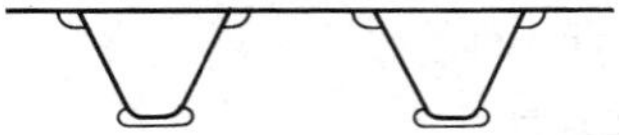	Torsionally stiff. Fabrication difficult through crossframes.
V trough	M-2	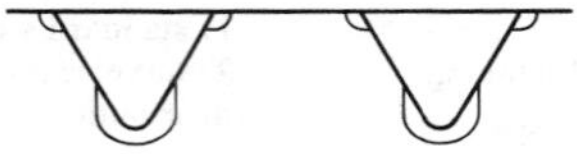	Torsionally stiff. Fabrication difficult through crossframes. Small effective lower flange are
Wineglass	M-2	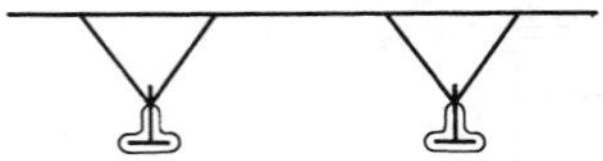	Torsionally stiff. Very complicated fabrication.
(cut from Universal Beam)	M-1	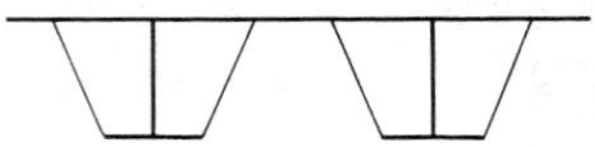	Easily spliced. Requires large cutout in cross-frame. Torsionally weak.

CONCRETE GIRDER BRIDGE

Bettingen Bridge, Frankfurt a/M West Germany (Polensky and Zöllner)

CANTILEVER TRUSS BRIDGE WITH BOX GIRDER WIDENING

Auckland Harbour Bridge, New Zealand (Freeman Fox and Partners)

STEEL GIRDER BRIDGE

Rio–Niteroi Bridge, Brazil Built by Redpath Dorman Long Ltd (a subsidiary of the British Steel Corporation) and the Cleveland Bridge and Engineering Co. Ltd (Trafalgar House Group; in association with Montreal Engenharia SA)

CONCRETE CABLE
STAYED BRIDGE
Tempul Aqueduct, Spain
(Torroja Institute, Madrid)

STEEL TRUSSED CABLE
STAYED BRIDGE
Batman Bridge, Tasmania, Australia (G. Mounsell and Partners)

STEEL ARCH BRIDGE
Fehmarnsund Bridge, West Germany (Beratungsstelle für Stahlverwendung, Düsseldorf)

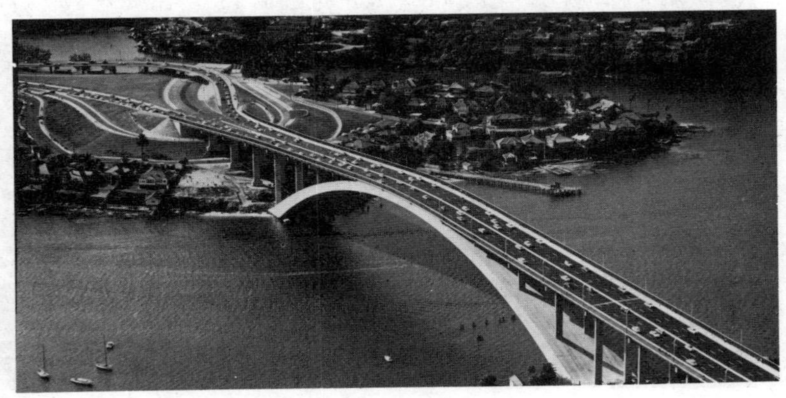

CONCRETE ARCH BRIDGE
Gladesville Bridge, Sydney, Australia (G. Maunsell and Partners)

SUSPENSION BRIDGE
Bosphorus Bridge, Istanbul (Freeman, Fox and Partners)

STEEL BOX CABLE STAYED BRIDGE
Severins Bridge, Cologne (Beratungsstelle für Stahlverwendung, Düsseldorf)

at a depth of 0.5 where the flange and web areas are equal the section modulus is only 11% lower. A shallower beam is usually more economical because a simpler web is then possible provided that the shear force can be carried. Fabrication, transportation and erection are also less costly.

Table 18.10 shows the types of standardised pre-cast concrete beams that are appropriate to various parts of the short-span range. Apart from the cost advantages of standardisation and factory production, which may be offset by higher overheads and transport costs, there are the following advantages:

(a) Estimates of cost more reliable
(b) Speed of construction
(c) No temporary staging required
(d) Sample beams can be tested to demonstrate level of prestress and ultimate strength

In simple right spans the system chosen, apart from span, depends on construction depth limitations, difficulties of access and of course prevailing prices. For example, the top hat beam system[6] is suitable for restricted access and small construction depths. The 'U' beam system[7] is suitable for similar conditions but requires an increased depth. At the greater depth it is more economical. An advantage of torsionally stiff structures of this type, particularly when they are designed to be spaced apart in the transverse direction, is that they can readily be fanned out to support the structures with complex plan forms that are now common.

The standard concrete beams are essentially a series of elements that can be placed across the complete span, requiring only simple shuttering to support the transversely spanning top slab. Diaphragm beams at the supports are required and occasionally intermediate diaphragms may be provided.

Steel beams can be used as an alternative form of construction in the same span range. Either a series of I sections or small box girders can be used.

Pre-cast or prefabricated elements can be made as transverse rather than longitudinal elements and then joined together on site by prestressing in concrete structures or welding or bolting in steel structures. This approach, sometimes known as segmental construction, was used for the structures of Figures 18.4(d), (e), (f), (h), (i), (j) and (k). It was also used for the steel structures of Figures 18.4(l) and (m). The remaining steel structures shown in Figures 18.4(n) to (r) were constructed by a similar process but with the subdivision taken a stage further. Each transverse slice was built up on the end of the cantilevering structure from several stiffened panels.

In situ concrete, reinforced or prestressed, can be used to form complete spans in one operation or else the cantilevering approach can be used. In the latter case the speed of construction is limited by the time required for the concrete to reach a cube strength adequate for the degree of prestress necessary to support the next section of the cantilever and the erection equipment. Segmental methods of construction[8] avoid such delays. In shorter spans, provided that the restrictions on construction depth are not too severe, *in situ* concrete structures can be built economically using the cross section of Figure 18.4(c). The simple cross section[9] was developed to suit the use of formwork which, after supporting a complete span, could be rapidly moved to the next span. The resulting machine is only economical for multi-span structures.

The stiffened steel plates (Table 18.11) are used for deck systems of long-span bridges and movable bridges in order to reduce the self-weight of the structure.

THE CHARACTERISTICS OF BRIDGE STRUCTURES

The following theories have been chosen and developed for their value in demonstrating the principal characteristics of various types of bridge structure. Other methods of calculation, based on finite elements for example, may be more accurate and more economical in certain circumstances. The theories are, however, linked to the main structural properties of the bridge types considered and are meant to assist the process of synthesis necessary before detailed calculations begin. The concepts described are also useful for idealising structures when using computer programs and for the interpretation and checking of the computer output.

Theory of suspension bridges and arch bridges

The basic theory of arch and suspension bridges is the same and the equation derived below for suspension bridges is applicable to arches if a change in sign of H and y is made.

SUSPENSION BRIDGES WITH EXTERNAL ANCHORAGES

The dead load of the cable and stiffening girder is supported by the force per unit length of span produced by the horizontal component of the cable force and the rate of change of slope of the cable:

$$H_g y''(x) + g = 0 \tag{1}$$

where y, etc., are shown in Figure 18.3.

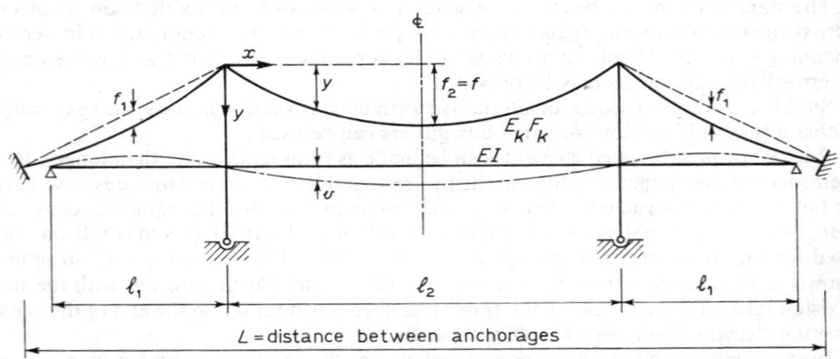

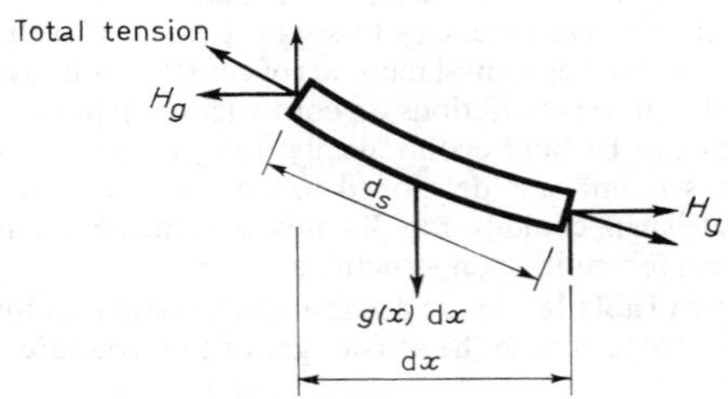

Figure 18.3 Suspension bridge notation

For a parabolic shape of cable corresponding to constant intensity of load across the span l, $y''(x) = -8f/l^2$ and

$$H_g = gl^2/8f \tag{2}$$

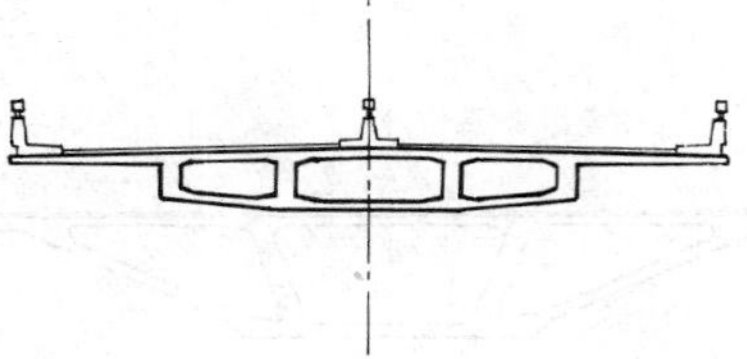

(a) Westway, Section One

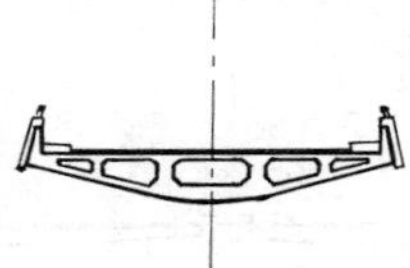

(b) Tunnel relief flyover, Liverpool

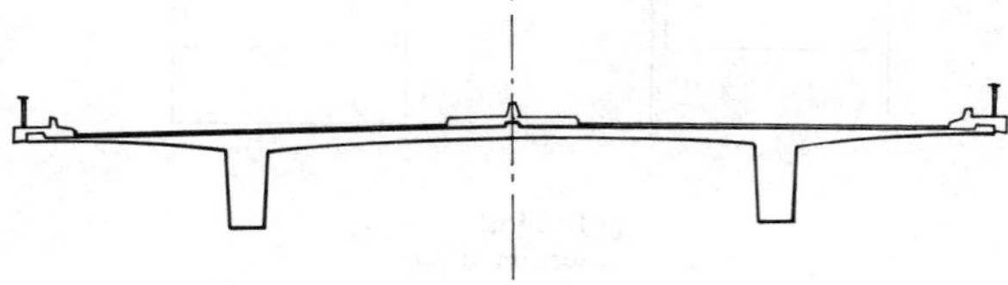

(c) Vorlandbrücke Obereisesheim

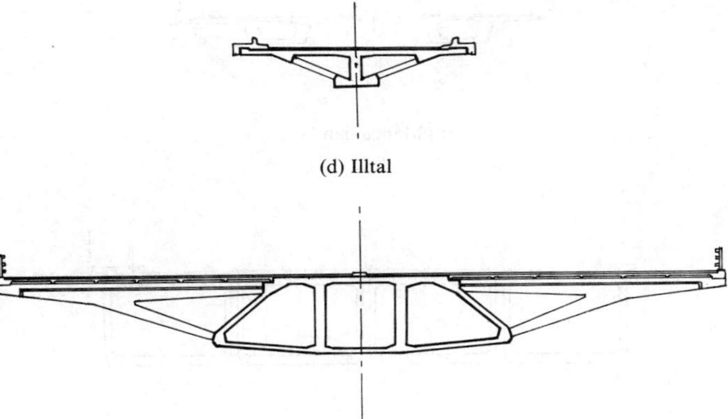

(d) Illtal

(e) West Gate Approach Viaducts

Figure 18.4 Elevated roadways

Figure 18.4 (*continued*)

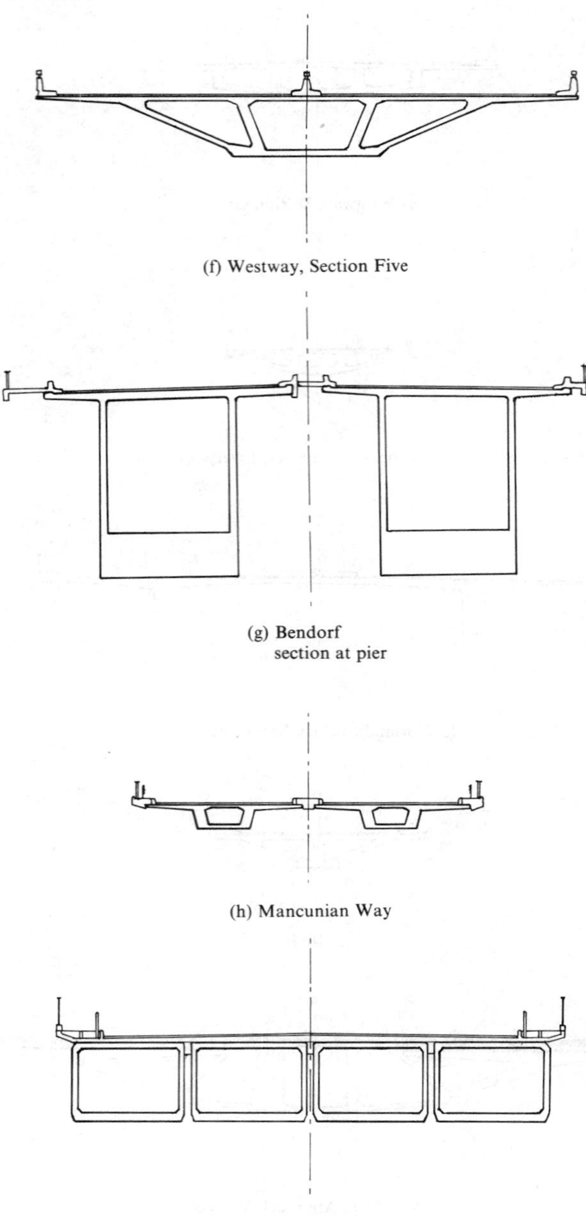

(f) Westway, Section Five

(g) Bendorf
section at pier

(h) Mancunian Way

(i) Gladesville

Figure 18.4 (*continued*)

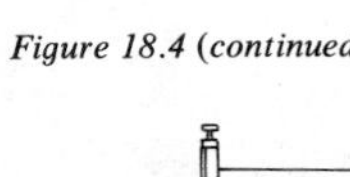

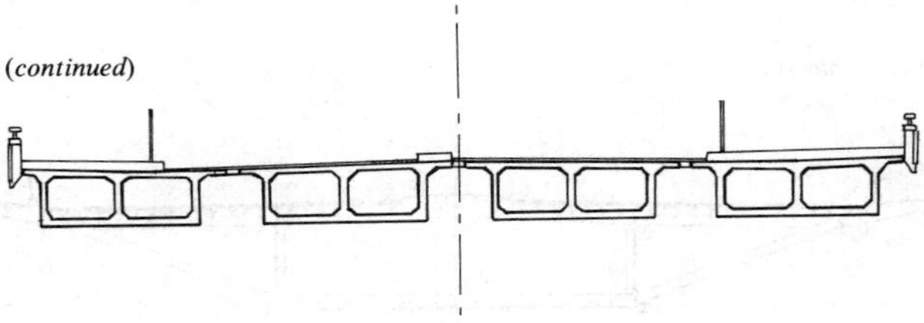

(j) London

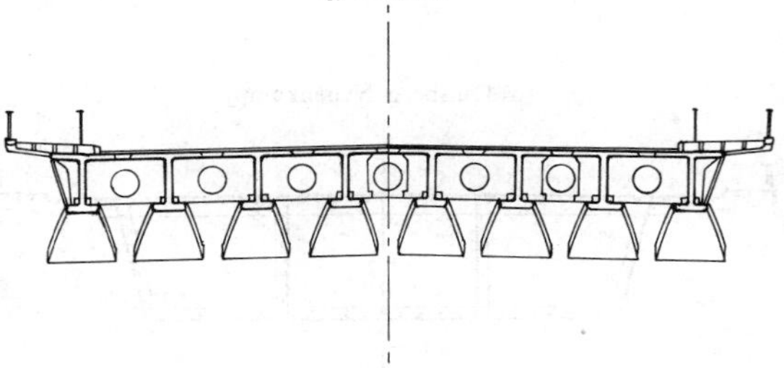

(k) Narrows

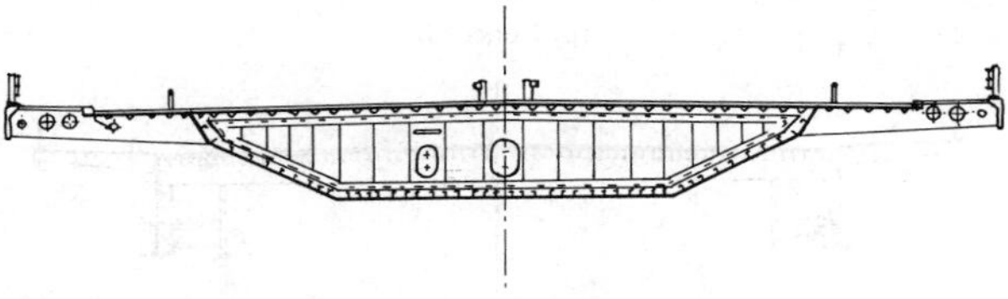

(l) Erskine

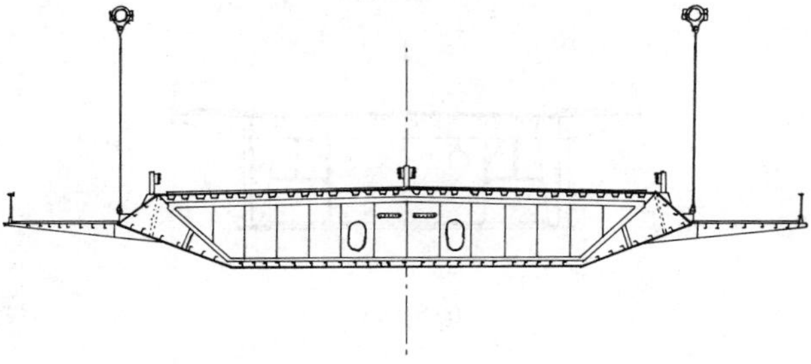

(m) Severn

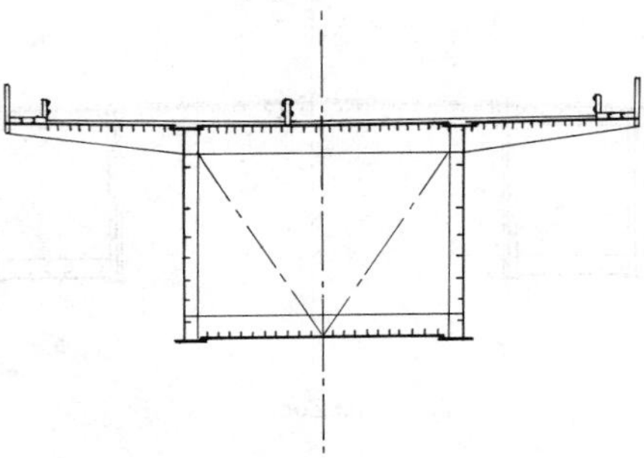

(n) Europa

Figure 18.4 (*continued*)

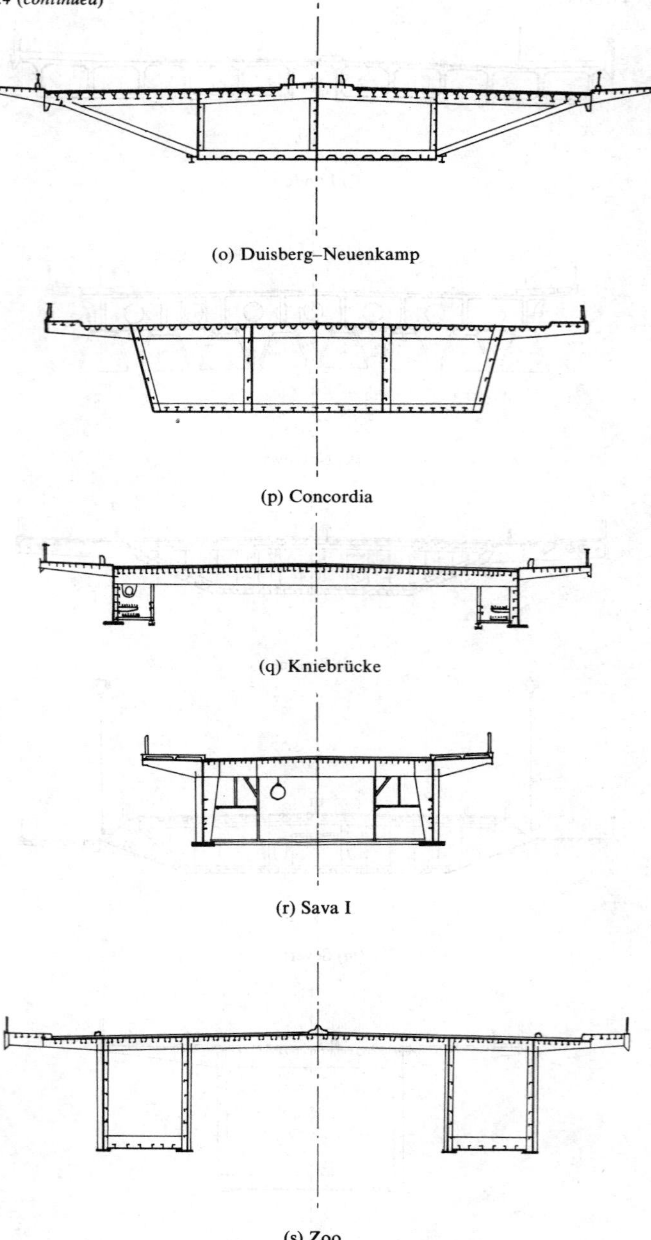

(o) Duisberg–Neuenkamp

(p) Concordia

(q) Kniebrücke

(r) Sava I

(s) Zoo

The cable tension increases under live load $p(x)$ to

$$H = H_g + H_p \tag{3}$$

The increase in support from the cable is $-[Hv''(x) + H_p y''(x)]$ where $v(x)$ is the vertical deflection of the cable and stiffening girder. The stiffening girder contributes a supporting reaction per unit length of $[EIv''(x)]''$ and adding the cable and stiffening girder contributions and equating them to the intensity of the applied load gives

$$[EIv''(x)]'' - Hv''(x) = p(x) + H_p y'' \tag{4}$$

The term $H_p y''$ is added to the live load in order to show that the equation can be represented physically by the substitute structure of Figure 18.5. y'' is $-8f/l^2$ and therefore represents a force in the opposite direction to the live load.

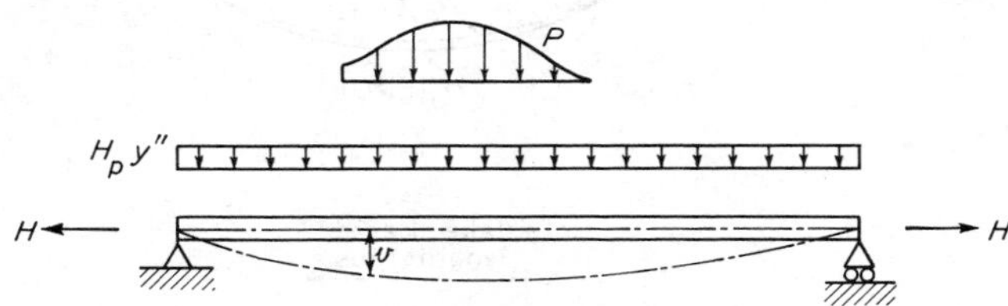

Figure 18.5 Substitute girder

H_p depends on the change in length of the cable and if Δ dx is the horizontal projection of the change in length of an element ds then for fixed anchorages,

$$\int_0^L \Delta\,\mathrm{d}x = 0$$

Integrating along the cable and allowing for a change in temperature of ΔT gives

$$\int_0^L \Delta\,\mathrm{d}x = H_p \frac{L_k}{E_k F_k} \pm \alpha_T \Delta T L_T + y'' \int_0^L v(x)\,\mathrm{d}x = 0 \tag{5}$$

Approximate values of L_k and L_T are:

$$L_k \simeq \left(1 + 8\frac{f^2}{l^2} + \frac{3}{2}\tan^2 v_0\right) + \frac{S_1}{\cos^2 v_1} + \frac{S_2}{\cos^2 v_2}$$

$$L_T \simeq \left(1 + \frac{16}{3}\frac{f^2}{l^2} + \tan^2 v_0\right) + \frac{S_1}{\cos v_1} + \frac{S_2}{\cos v_2} \tag{6}$$

See Figure 18.6.

Equations (4) and (5) must be satisfied simultaneously and although this makes the problem nonlinear, by solving for two assumed values of H the correct value can be satisfactorily determined by interpolation. Each assumed H gives an incorrect solution to equation (5) and, assuming the error varies linearly, the correct value of H can be found. For each assumed value of H the structure behaves as a simple beam and influence lines can be constructed for bending moments, etc., and for $\int_0^L v(x)\,\mathrm{d}x$ Hawranek and Steinhardt suggest that for a particular loading case the bending moment and shear forces be found from both sets of influence lines as well as the $\int_0^L v(x)\,\mathrm{d}x$ values. H is

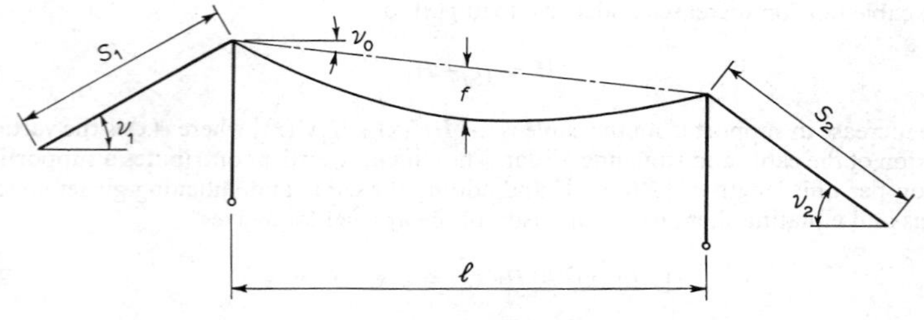

Figure 18.6

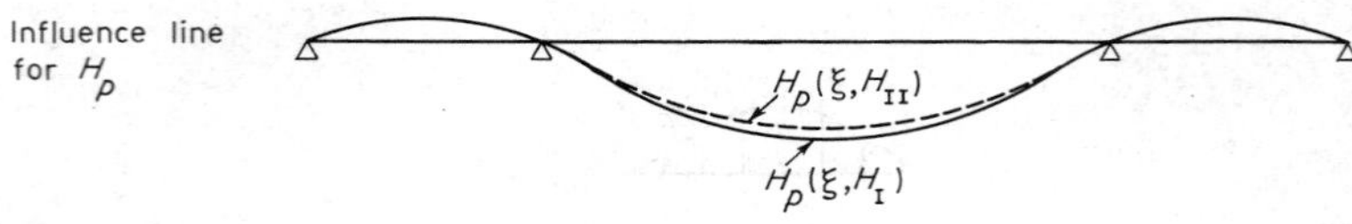

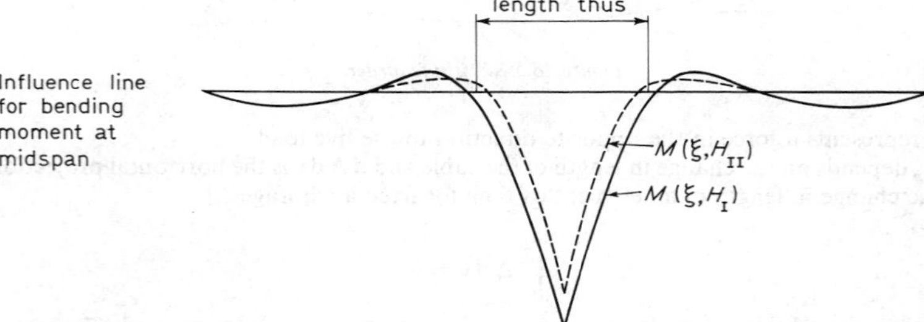

Figure 18.7 Influence lines

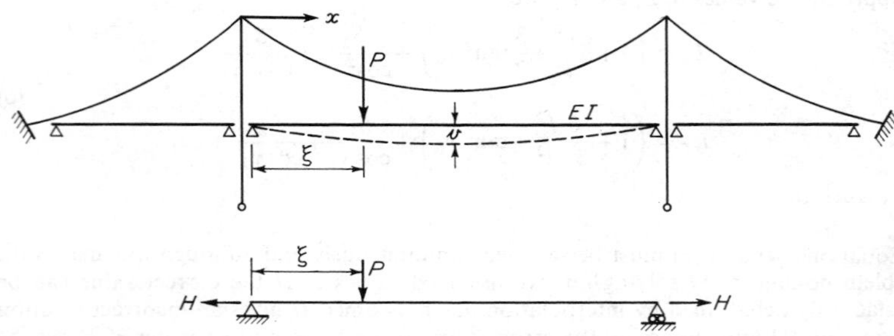

Figure 18.8

(Figures 18.3, 18.5 to 18.8 are reproduced from 'Theorie und Berechnung der Stahlbrücken' published by Springer-Verlag, Berlin[10])

found by interpolation and then the final bending moments and shears are found by interpolating between the two sets of values already found from the influence lines.

Typical results for a continuous stiffening girder are shown in Figure 18.7.

The above treatment follows that given by Hawranek and Steinhardt[10] who also give a comprehensive set of standard solutions for the substitute girder. The result quoted below illustrates the form the solutions take:

Using

$$\mu^2 = H/EI$$

For the load case of Figure 18.8, deflections as a function of x are given by:

$$v(x,\xi) = PG(x,\xi) = P\frac{l}{H}\left[\frac{x}{l}\left(1-\frac{\xi}{l}\right) - \frac{\sinh \mu x \sinh \mu(l-\xi)}{\mu l \sinh \mu l}\right] \quad \text{for } \xi \geqslant x$$

$$v(x,\xi) = PG(x,\xi) = P\frac{l}{H}\left[\frac{\xi}{l}\left(1-\frac{x}{l}\right) - \frac{\sinh \mu \xi \sinh \mu(l-x)}{\mu l \sinh \mu l}\right] \quad \xi \leqslant x \tag{7}$$

$G(x, \xi)$ is known as a Green's function.

And

$$F(\xi) = \int_0^L v(x)\,\mathrm{d}x = P\frac{l^2}{H}\left[\frac{\xi(l-\xi)}{2l^2} - \frac{1}{(\mu l)^2}\left(1-\frac{\cosh \mu(l/2-\xi)}{\cosh \mu(l/2)}\right)\right] \tag{8}$$

Computers can be used to analyse suspension bridges either by following the above approach or by means of standard framework programmes provided that the interaction of axial loads and deflections is allowed for. In other words, the change in geometry of the cable is considered. In some programmes the axial loads must be stated as part of the data in the same way that H is assumed in obtaining a solution to equation (4). In others an interactive process produces the correct axial forces. The structure solved can include the actual system of suspenders, tower properties, etc., or can be a very simple solution of the substitute structure of Figure 18.5.

SELF-ANCHORED SUSPENSION BRIDGES

The horizontal component of the cable tension can be resisted by the stiffening girder which then acts as a laterally loaded compression member between suspenders. The net tension on the structure is therefore zero and the substitute girder has a zero axial load acting on it. The structure is substantially linear in its response to live load whereas the externally anchored bridge has an increasing stiffness with increasing deflection.

ARCH BRIDGES

The design of arches is based on the thrust line following the shape of the arch so that there is either no bending moment or a reduced bending moment in the arch member.

The shape of arch can only satisfy one condition of loading without bending moments being developed. Temperature changes, creep, foundation movements and imperfections must however introduce some bending in all but the three-hinged arch. In a bridge structure, live loading will produce a varying distribution of loading which will introduce bending. Clearly the higher the proportion of dead load the more nearly can the arch be designed to be in pure compression.

The most common shapes are the circular arch, the parabolic arch and more recently the inclined leg frame (Figure 18.9 (a), (b), (c)). Loadings over the whole of (c) can be examined in two stages, which enables a design to be produced before detailed dimensions are known (Figure 18.10).

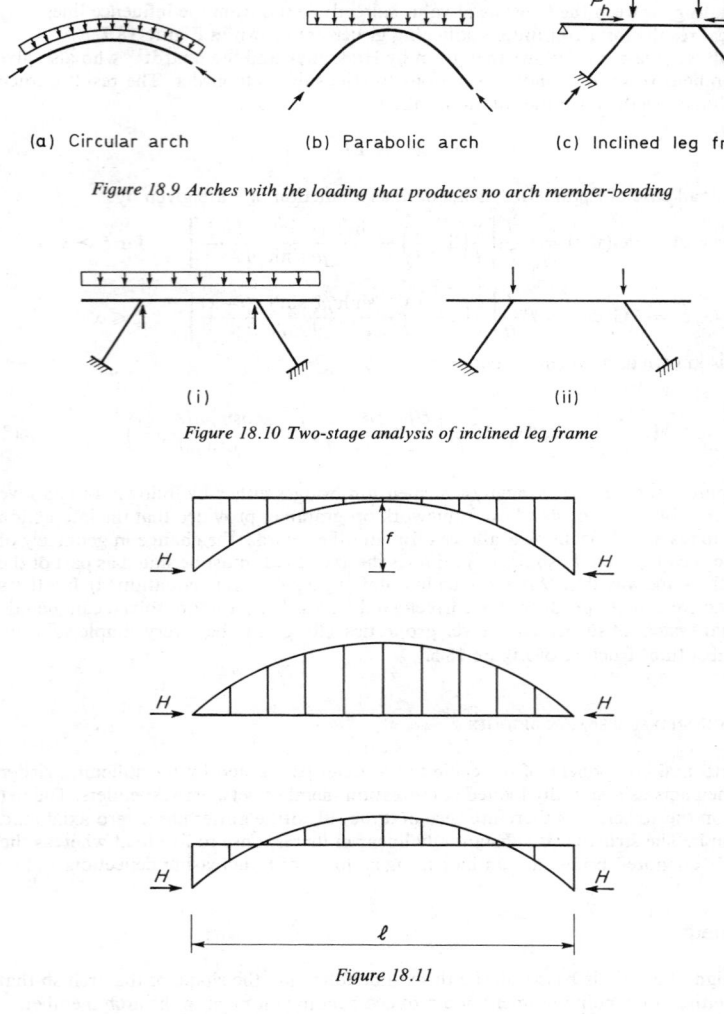

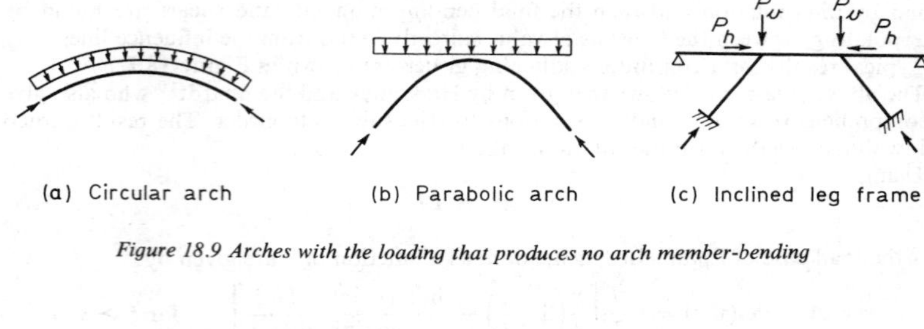

Figure 18.9 Arches with the loading that produces no arch member-bending

(i)

(ii)

Figure 18.10 Two-stage analysis of inclined leg frame

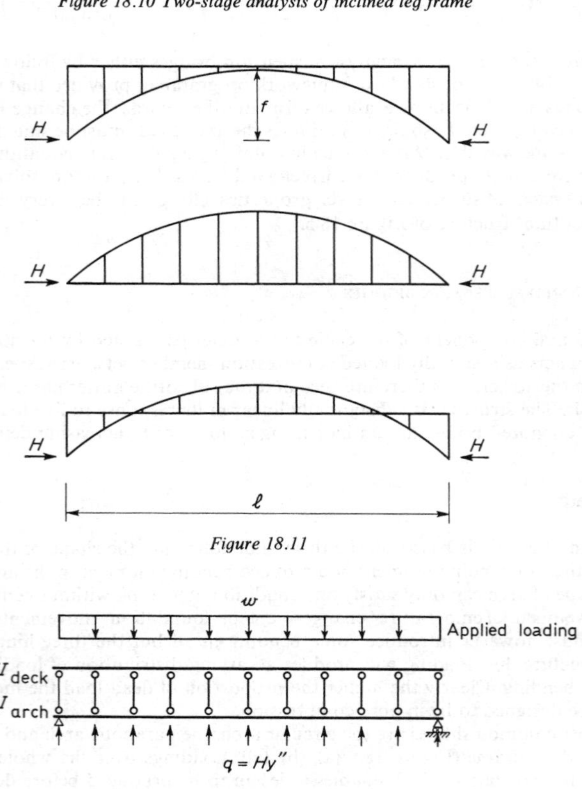

Figure 18.11

Figure 18.12 Application of substitute girder

Arches for bridges frequently have continuous beams supporting the deck as in Figure 18.11.

At the design stage, since the dead load carried by the deck will depend on the construction method and the thrust jacked into the arch, the structure is to some extent determinate but clearly some bending of the deck beams between supports is introduced. The system can be represented as in Figure 18.12 where

$$H \simeq wl^2/8f$$

At the preliminary design stage live loading can be examined by splitting it into symmetrical and antisymmetrical components. (Figure 18.13). The symmetrical system will produce to a first approximation small bending moments and the anti-symmetrical system is equivalent to a simple beam with half the arch span (zero thrust due to opposite effects of load).

The local effects of loading on the deck can always be examined as a beam between columns and the overall behaviour can be seen as that of an arch with a total EI of

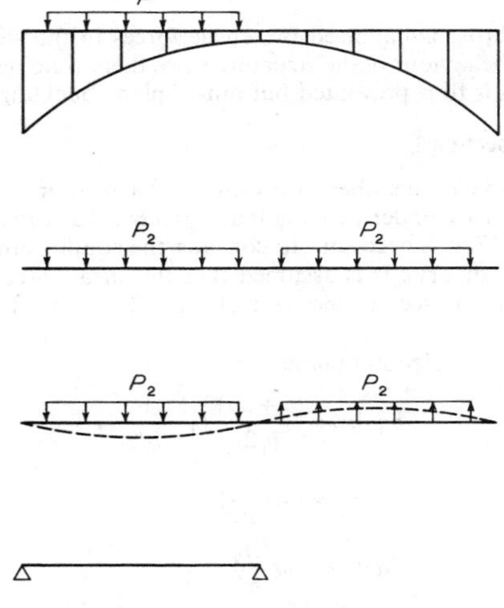

Figure 18.13 Arch live loading

$EI_{deck} + EI_{arch}$. The bending moments produced in the parts will be in proportion to stiffness.

Arches may require correction for deflections since the thrust will magnify these in the same way that a strut is affected by end load. A two-hinged arch with a uniform loading buckles as in Figure 18.14 and to a first approximation the effective length of the strut is $l/2$. The magnification of moments can be found as for a strut using $1/(1 - H/H_{cr})$ as the factor. Critical loads are available for a variety of cases as in Figure 18.15.

Equation (4) for suspension bridges is applicable to arches with changes of sign in

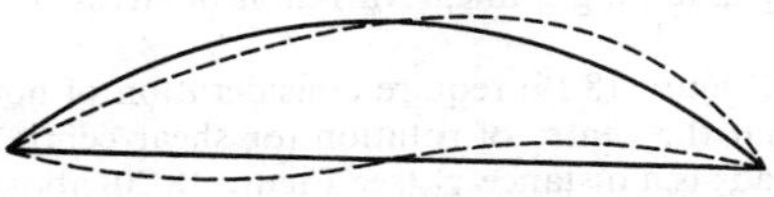

Figure 18.14 Buckling of a two-hinged arch

H and y. The nonlinear effects of deflections can be determined using that equation or the substitute beam in compression instead of tension. Computer programmes can be used to analyse arches in the way already described for suspension bridges but if H/H_{cr} is small it is unnecessary to use programmes which allow for changes of geometry.

Out-of-plane deflections can cause buckling or significant stresses in single arches

f_l	0	0.1	0.2	0.3	0.4	0.5	
For a two hinged arch	39.4	35.6	28.4	19.4	13.7	9.6	$H_{cr} = \frac{kEI}{l^2}$
For a fixed arch	80.8	75.8	63.1	47.9	34.8	24.4	

Figure 18.15 Factors k for determining the critical arch thrust H_{cr} for parabolic arches under constant uniformly distributed load

or arches not braced laterally. This effect can be readily investigated using a grillage programme which allows for the interaction of axial loads and transverse deflections. The plane of the grillage must be considered as the plane of the arch for that purpose.

TIED ARCHES

In the tied arch, the thrust is balanced by tensile forces in the stiffening girder which simplifies the in-plane behaviour of the structure since there is no net thrust on the structure. In-plane buckling is thus prevented but out-of-plane buckling is still possible.

Bridge girders of open section

The cross girder and bracing members are shown as broken lines in Figure 18.16 since they do not affect the girder under twisting loads unless it has torsionally stiff members.

To find q (Figure 18.17) it is necessary to consider the equilibrium equation of the top flange (cf. simple beam theory). It is assumed that the shear force on the top flange is zero as shown and that the second moment of area of the top flange about a vertical axis is I_T.

Taking moments in a horizontal plane,

$$bq\,dx = \frac{(\sigma+d\sigma)}{b/2}I_T - \frac{\sigma}{b/2}I_T$$

$$= d\sigma\frac{2I_T}{b}$$

therefore

$$q\,dx = d\sigma\frac{2I_T}{b^2}$$

This is the same equation as that used for simple beams if their top flange area A is made equal to $2I_T/b^2$. It follows that q can be found by considering an equivalent simple beam with A_T replacing the deck and acted on by shear forces Q (Figure 18.18). Note that Q is not altered by horizontal shears and is therefore the same as in a simple beam loaded by W.

Any twisting load can be referred to the web positions giving the loads to be applied to the effective girder of Figure 18.18. The bending moments produced in the effective girder, acting as a single beam with the span of the actual structure, are applied to the effective girder cross section. The second moment of area I^v_{eff} is used to find the longitudinal stresses between the top and bottom flanges. The remainder of the top flange stresses can be found by assuming a linear variation of stress between the web–flange junctions.

More complex loads (Figure 18.19) require consideration of horizontal forces and it is convenient to note that the centre of rotation (or shear centre) of the cross section under purely twisting loads is a distance y_1 (see Figure 18.20) above the top flange. This can be seen if it is recognised that longitudinal strain distribution and hence curvature results in the ratio of deflections in Figure 18.20 to be $w/v = 2y_1/b$ or $w = 2(y_1/b)v$. Therefore, at a height above deck of y_1 the normal to the midpoint of the deck must cut the vertical axis.

The vertical members of the cross section may be individual boxes or thick-walled concrete webs with significant torsional stiffness in both cases. If transverse bracing is provided, thus preventing the shape of the cross section from changing, the rotation of the boxes will equal that of the deck. This effect can be included to give the following governing equations with the deflections shown in Figure 18.19 which include a vertical displacement of the whole cross section z. I_z and I_y are the usual second moments of area about a horizontal and vertical axis respectively. GK is the sum of the torsional stiffness of the whole cross section.

$$EI_z z^{iv}(x) = P_1(x)$$

$$EI^v_{eff} v^{iv} - GK\frac{2}{b^2}v'' = P_1\frac{e_2}{b} + P_2\frac{(e_1+y_1)}{b} \qquad (9)$$

$$EI_y\left(w^{iv} - \frac{2y_1}{b}v^{iv}\right) = P_2$$

Equations (9) are not put forward for solution as a set of differential equations but as a description of the various mechanisms involved. The second is similar to the suspension bridge equation (4), since $GK(2/b^2)v''$ can be compared with the Hv'' term. There is no term corresponding to $H_p y''$. The equivalent structure is, therefore, a beam of stiffness EI^v_{eff} under axial tension $GK(2/b^2)$.

The deflections and bending moments will be correctly duplicated by such a structure but the stresses do not of course require a direct contribution from the imaginary tensile force.

More general behaviour of suspension bridges and arches

The above treatment of a girder can be extended to include the usual twin cables of a suspension bridge. The positions of the cables are shown in Figure 18.21. Equations (9) become extended into:

$$\text{(a)}\quad EI_z z^{iv}(x) - (H_1+H_2)z''(x) - \underline{(H_1-H_2)\frac{2e}{b}v''(x)} = P_1(x) + (H_{p1}+H_{p2})y''$$

$$\text{(b)}\quad EI^v_{eff} v^{iv}(x) - GK\frac{2}{b^2}v''(x) - (H_1+H_2)\frac{2e^2}{b^2}v''(x) - \underline{(H_1-H_2)\frac{e}{b}z''(x)} \qquad (10)$$

$$= P_1\frac{e_2}{b} + P_2\frac{e_1}{b} + P_2\frac{y_1}{b} + (H_{p1}-H_{p2})y''\frac{e}{b}$$

$$\text{(c)}\quad EI_y\left(w^{iv}(x) - \frac{2y_1}{b}v^{iv}(x)\right) = P_w$$

The terms underlined are fairly small and can be neglected, which leads to (a) and (b) being independent equations in z and v.

Further, if purely torsional loading is assumed, $H_{p1} \simeq H_{p2} = H_p$ and writing $H_1 + H_2 = 2H$ the second equation becomes

$$EI^{eff}_v v^{iv}(x) - \left(GK\frac{2}{b^2} + 2H\frac{2e_2}{b^2}\right)v''(x) = P_1\frac{e_2}{b} + P_2\frac{(e_1+y_1)}{b} + 2H_p\frac{e}{b}y'' \qquad (11)$$

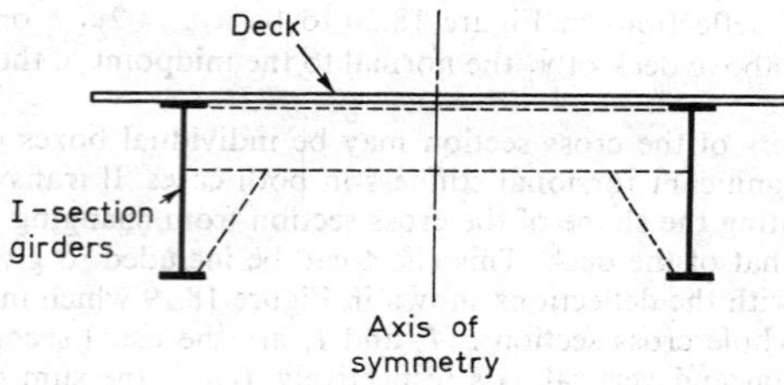

Figure 18.16 Girder of thin walled open section

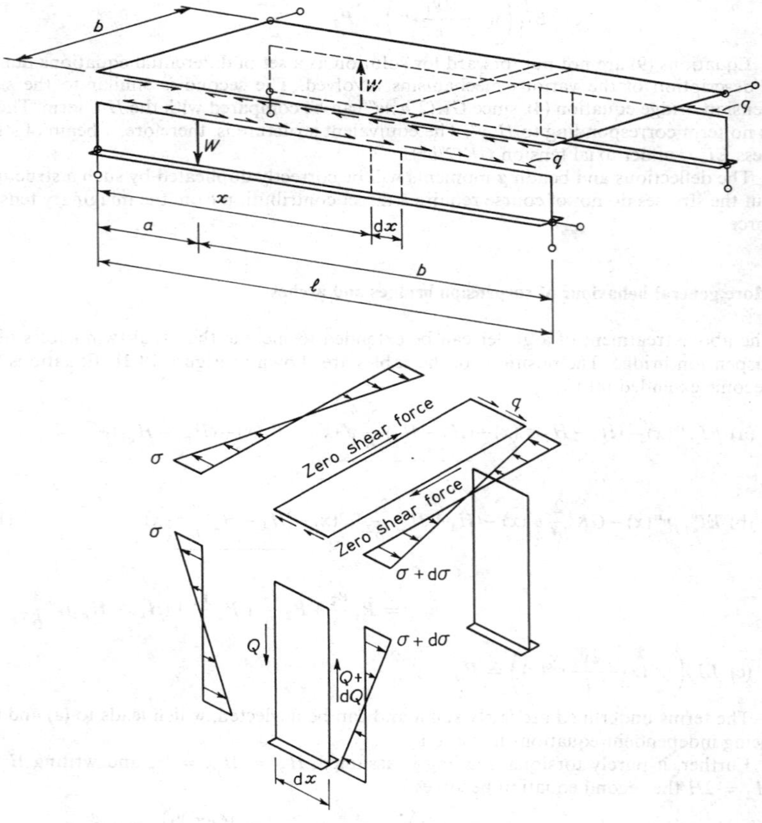

Figure 18.17 Loading and stresses on thin walled open section

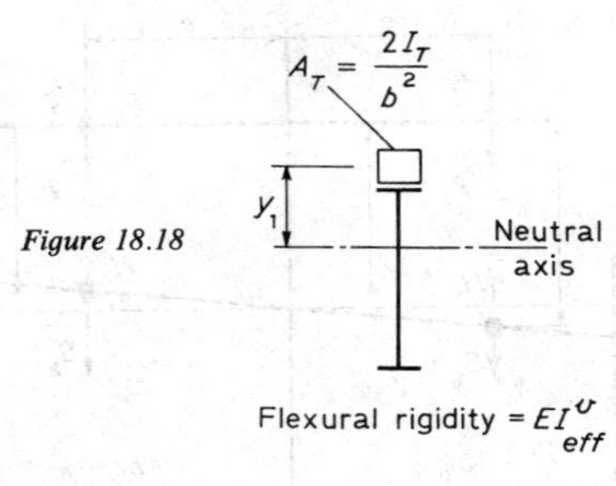

Figure 18.18

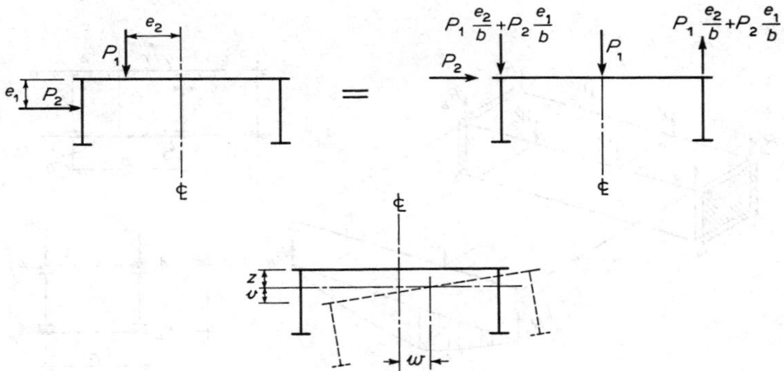

Figure 18.19 Overall loading and displacement

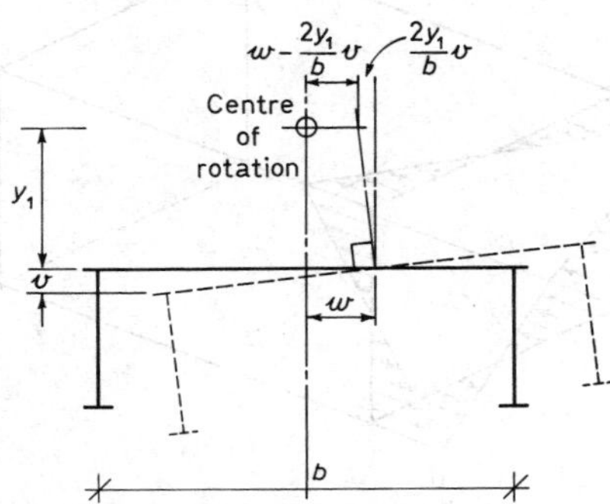

Figure 18.20 Twisting about centre of rotation

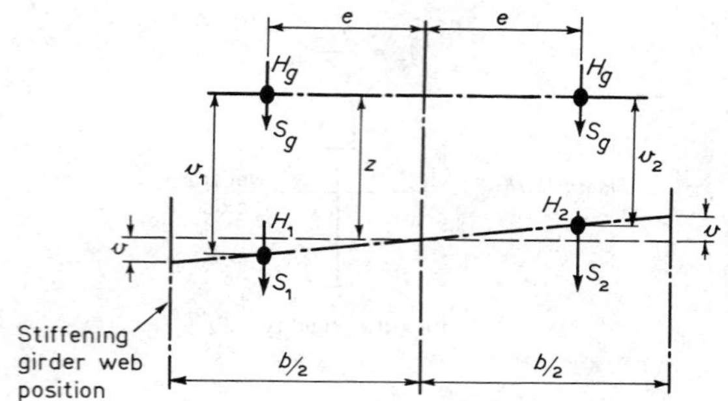

Figure 18.21 Change in cable position

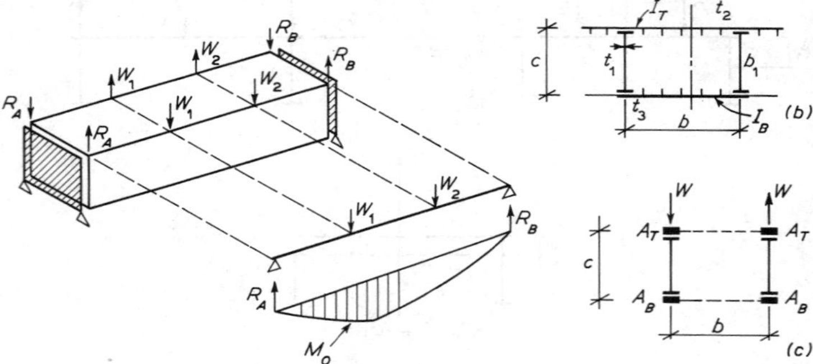

Figure 18.22 Equivalent I *beam*

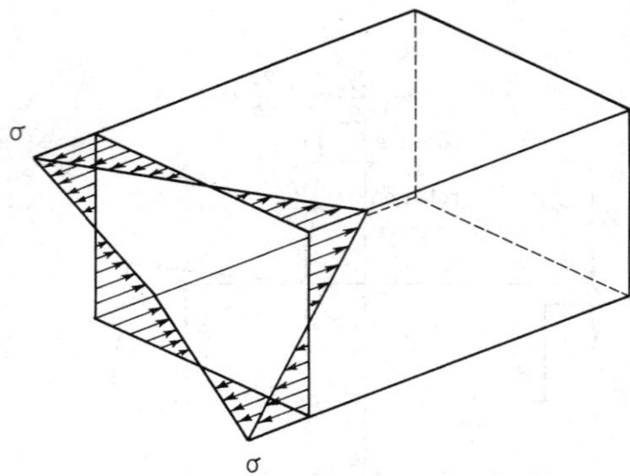

Figure 18.23 Warping stress distribution

For each cable the condition $\int_0^L \Delta\, dx = 0$ must be satisfied leading as before to the correct value of H_p. Equation (11) is clearly of the same form as the ordinary suspension bridge equation for vertical deflections except that EI^v_{eff} replaces EI and $GK(2/b^2)+2H(e^2/b^2)$ replaces H. Therefore, the same results can be used to solve this equation.

The above equations may be used to investigate aerodynamic and other vibrational effects as well as live loading. The above treatment follows that of Hawranek,[8] but the effective beam concept has been used in order to give the equations more physical significance. Hawranek works in terms of a warping function and relates loads to the shear centre. Hawranek and Steinhardt give, however, a number of useful results for the natural frequency of various systems.

Horizontal deflections due to wind may produce a significant lateral component of the cable force but the complete properties of the structure must be known before the effect can be determined, e.g. the tower stiffness influences this type of behaviour. It is not considered in the above equations. For a given system, approximate results can be estimated by simple calculations but rigorous results can be obtained using, for example, a grillage programme which includes the interaction of axial forces and deflections. The plane of the grillage must be assumed to be the plane of the cable for this purpose.

Single cell box girder

The full torsional stiffness of a single cell, $4A^2G/\oint(1/t)\, ds$ is only mobilised if the twisting forces are applied in the cross section in a distribution corresponding to a constant shear flow around the box. A structure with effective diaphragms at the supports only (Figure 18.22(a)) and with hinges at the long edges has no torsional stiffness under the twisting loads shown. They are carried in differential bending. The associated stresses are warping stresses (Figure 18.23). Figure 18.22(a) shows how the warping moments M_0 are found and Figure 18.22(c) shows the effective beams carrying equal and opposite values of M_0. The properties of one of the effective beams can be calculated from the following values:

top flange area $\quad A_T = \dfrac{2I_T}{b_T^2}$

web as in actual box beam

bottom flange area $\quad A_B = \dfrac{2I_B}{b_B^2}$

where I_T and I_B are, respectively, the second moments of area of the top flange assembly and the bottom flange assembly about the vertical axis of symmetry.

The warping stresses in the effective beam are determined as in normal beam theory and the remaining warping stresses in the flanges can be found by assuming a linear variation of stress across the top and bottom flanges.

Complex box structures can be solved most economically by suitable finite-element programmes. Elaborate calculations by alternative means are not justified but in order to understand the behaviour of box structures or for structures where an approximate result shows that further investigation is unnecessary, the following methods are of value.

Boxes with discrete diaphragms

Steel boxes usually have a number of diaphragms formed from a lattice system or solid plate and sometimes unbraced frames. The following influence-coefficient approach is one of several methods of dealing with such structures acted upon by twisting loads.[11, 12]

The structure is rendered statically determinate for twisting loads by releasing the

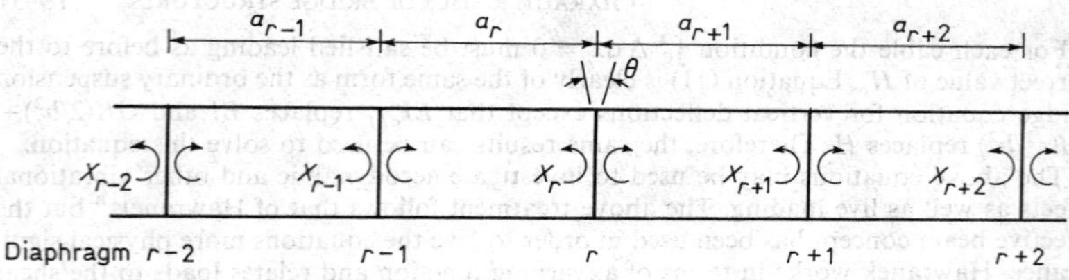

Figure 18.24 Box with discrete diaphragms

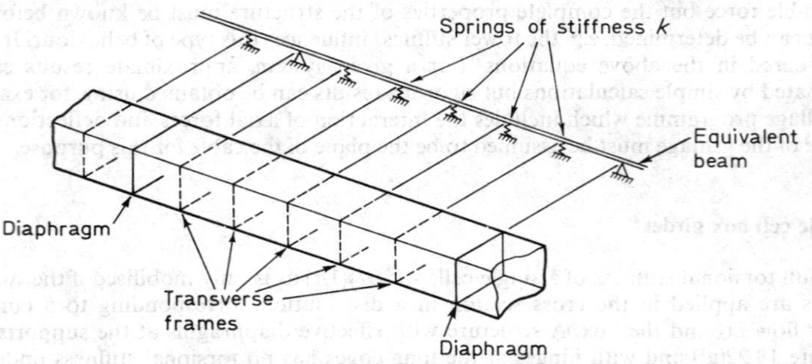

Figure 18.25 Equivalent beam on elastic supports

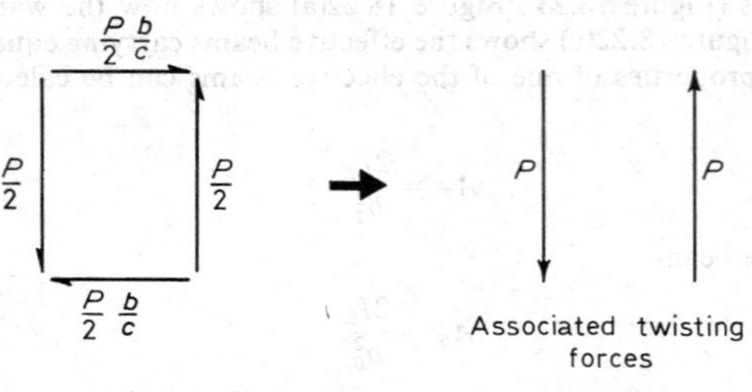

Figure 18.26 Distortional force system

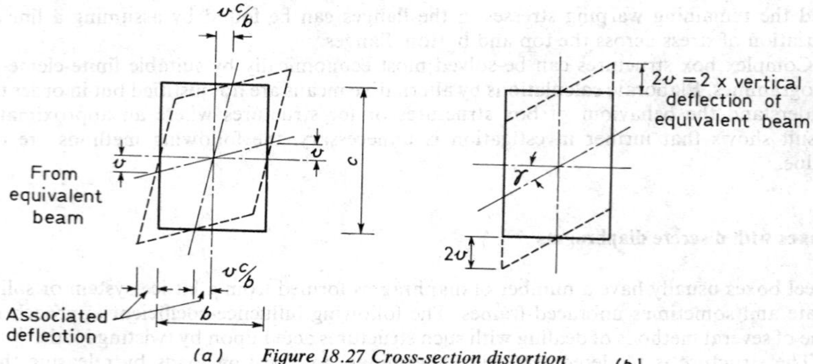

Figure 18.27 Cross-section distortion

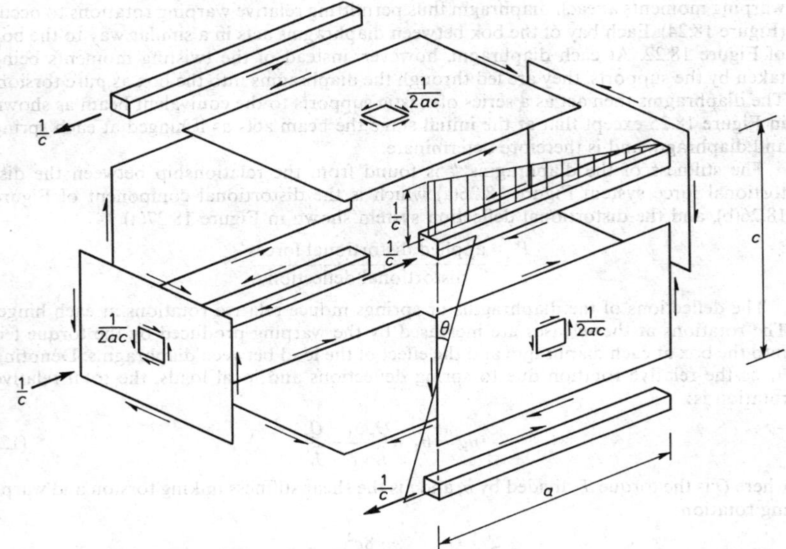

Figure 18.28 Exploded box girder

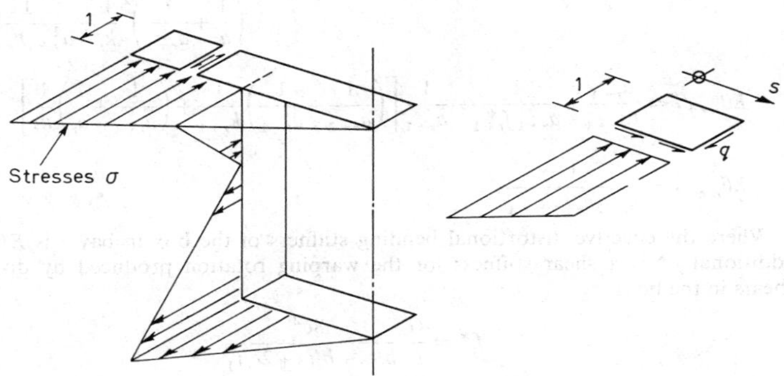

Figure 18.29 Half box

warping moments at each diaphragm thus permitting relative warping rotations to occur (Figure 18.24). Each bay of the box between diaphragms acts in a similar way to the box of Figure 18.22. At each diaphragm, however, instead of the twisting moments being taken by the supports, they are fed through the diaphragms into the box as pure torsion. The diaphragms then act as a series of elastic supports to the equivalent beam as shown in Figure 18.25 except that at the initial stage the beam acts as if hinged at each spring and diaphragm and is therefore determinate.

The stiffness of the diaphragms k is found from the relationship between the distortional force system Figure 18.26(a) which is the distortional component of Figure 18.26(b), and the distortional deflection system shown in Figure 18.27(a).

$$\frac{P}{v} = \frac{\text{applied distortional forces}}{\text{distortional deflection}}$$

The deflections of the diaphragms or springs induce relative rotations at each hinge. The rotations at the releases are increased by the warping produced by the torque fed into the box at each diaphragm and the effect of the load between diaphragms. Denoting θ_1 as the relative rotation due to spring deflections and local loads, the total relative rotation is:

$$\theta_r = \theta_1 + \frac{Q_{r+1}}{\bar{f}_{r+1}} - \frac{Q_r}{\bar{f}_r} \tag{12}$$

where Q is the torque T divided by b, and $\bar{f}$ is the shear stiffness linking torsion and warping rotation:

$$\bar{f} = \frac{G}{E}\,\frac{8c^2}{b/t_3 + b/t_2 - 2c/t_1} \tag{13}$$

r and $r+1$ refer to bays between diaphragms as in Figure 18.24.

The influence coefficients for solving the series of compatibility equations are:

$$E\theta_{r,r-2} = \frac{1}{a_{r-1}a_r k_{r-1}}$$

$$E\theta_{r,r-1} = \frac{a}{6I_r} - \frac{1}{a_r f_r^*} - \frac{1}{a_r}\left[\left(\frac{1}{a_r} + \frac{1}{a_{r-1}}\right)\frac{1}{k_{r-1}} + \left(\frac{1}{a_r} + \frac{1}{a_{r+1}}\right)\frac{1}{k_r}\right]$$

$$E\theta_{r,r} = \frac{a}{3I_r} + \frac{a_{r+1}}{3I_{r+1}} + \frac{1}{a_r f_r^*} + \frac{1}{a_{r+1} f_{r+1}^*} + \frac{1}{a_r^2 k_{r-1}} + \left(\frac{1}{a_r} + \frac{1}{a_{r+1}}\right)^2 \frac{1}{k_r} + \frac{1}{a_{r+1}^2 k_{r+1}}$$

$$E\theta_{r,r+1} = \frac{a_{r+1}}{6I_{r+1}} - \frac{1}{a_{r+1} f_{r+1}^*} - \frac{1}{a_{r+1}}\left[\left(\frac{1}{a_{r+2}} + \frac{1}{a_{r+1}}\right)\frac{1}{k_{r+1}} + \left(\frac{1}{a_{r+1}} + \frac{1}{a_r}\right)\frac{1}{k_r}\right] \tag{14}$$

$$E\theta_{r,r+2} = \frac{1}{a_{r+2}a_{r+1}k_{r+1}}$$

Where the effective distortional bending stiffness of the box in bay r is EI_r and an additional f^* is a shear stiffness for the warping rotation produced by distortional shears in the box:

$$f^* = \frac{G}{E}\,\frac{8c^2}{b/t_3 + b/t_2 + 2c/t_1} \tag{15}$$

The various results given above can be found from concepts of virtual work using the mechanism shown in Figure 18.28 consisting of a series of shear webs and booms of axial stiffness. It can be used to obtain more general results such as those for boxes of trapezoidal cross section.[11, 14]

The warping produced by torsion only results in stresses if there is a change in torsion and therefore incompatible warping. Longitudinal stresses act to remove the lack of continuity. An upper bound estimate of these stresses can be made by assuming that the cross section cannot deform. The warping moments produced by a change in torque $T = Pb$ is:

$$X = \frac{P}{2}\frac{\sqrt{f^* I_\theta}}{f}\exp\left[-\sqrt{(f^*/I_\theta)}x\right] \tag{16}$$

where x is the distance from the cross section at which the change occurs.

The distribution of shear forces associated with warping stresses can be found from the warping stress distribution (Figure 18.29). Starting from the edge of the cantilever and assuming σ is the change in longitudinal stress over a unit length:

$$q = \int_0^s \sigma t \, ds$$

where q is the shear flow τt

The complementary shear to q is on the face of the cross section. The shear at the centre of the top flange can be assumed zero at the first stage of the calculation, which enables a simple shear system to be found. A pure torsional shear flow must be added to remove any component of pure torsion acting on the cross section.

Box beams with continuous diaphragms

In concrete boxes the frame action of the webs and flanges provides a continuous resistance to distortion, consequently special diaphragms are not usually necessary except at disturbances such as bearings and other support points.

Smaller steel box beams which rely on the stiffness of the sides plus the frame action of web and flange stiffness or larger box beams with special frames which leave the interior of the box unobstructed, have similar characteristics to the concrete boxes. Frames are generally flexible compared with braced or plate diaphragms but stiff frames can be made which will have properties that can only be explored fully by a treatment which is suitable for discrete diaphragms. The validity of the following approach for steel boxes can be determined from the half wavelength which should be greater than twice the spacing of the frames.

The effect of twisting loads P applied at the corners of the web on the vertical deflections caused by distortional bending of the box and allowing for the diaphragm action of the cross section is:[13]

$$v = \frac{P}{2\alpha\beta}\mathrm{e}^{-\alpha x}\left\{\beta\left[\frac{\lambda^2}{k}+\frac{1}{2E}\left(\frac{1}{f^*}-\frac{1}{\bar{f}}\right)\right]\cos\beta x+\alpha\left[\frac{\lambda^2}{k}-\frac{1}{2E}\left(\frac{1}{f^*}-\frac{1}{\bar{f}}\right)\right]\sin\beta x\right\} \tag{17}$$

assuming that the distance to a support is infinitely long. EI_θ is the effective distortional bending stiffness of the box, k is the diaphragm stiffness per unit length.
λ is defined by:

$$\lambda^4 = \frac{k}{4EI_\theta}$$

and

$$\alpha = (\lambda^2 + k/4Ef^*)^{\frac{1}{2}}, \qquad \beta = (\lambda^2 - k/4Ef^*)^{\frac{1}{2}}$$

The generalised warping stress resultant is:

$$x = \frac{P}{2\alpha\beta}\mathrm{e}^{-\alpha x}\left[\beta\left(\frac{1}{2}+\frac{\lambda^2 I_\theta}{\bar{f}}\right)\cos\beta x+\alpha\left(-\frac{1}{2}+\frac{\lambda^2 I_\theta}{\bar{f}}\right)\sin\beta x\right] \tag{18}$$

If the properties of the box are within the range $\lambda^2 - (k/4Ef^*) < 0$ then $\beta = (k/4Ef^* - \lambda^2)^{\frac{1}{2}}$ replaces β and hyperbolic functions are used.

The half wavelength is $\pi/2\beta$.

In many cases the effect of shear is not considerable and $\alpha = \beta = \lambda$. The equations then simplify into the standard beam on elastic foundation results. The above equations include, however, the effect of the change in torque due to the twisting loads P and in order to correct the results of beam on elastic foundation theory the warping moment of equation (16) should be added. The correction is likely to be of most significance in boxes which are much wider than their depth.

The above equations are valid for boxes of trapezoidal cross sections if the concepts are generalised as in reference 14.

Box girders with cantilevers

The advantages of box girders over structures of open cross section are sometimes only marginally linked to structural efficiency. However, where a compact structure is to

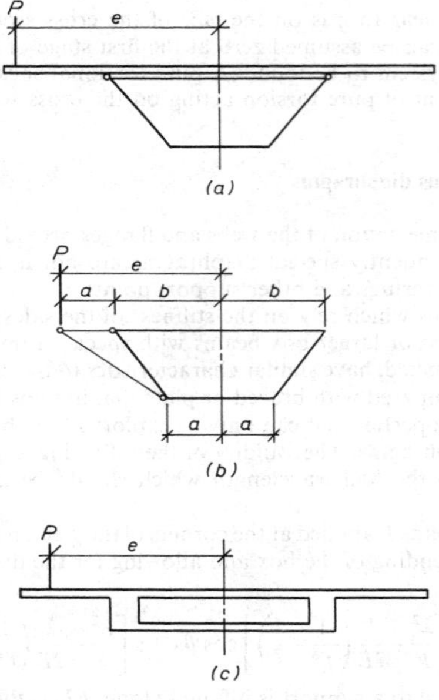

Figure 18.30

carry a much wider deck, producing a large cantilevered section, the torsional stiffness and strength of the box is of primary importance. The importance of the interaction of the cantilever and the box and the magnitude of the stresses is correspondingly great since any loss of torsional strength and stiffness could result in a major increase in the shear stresses and longitudinal stresses with a corresponding reduction in the load factor.

The three types of cantilever in Figure 18.30 show:[15]

(a) Transmission of torque Pe into twisting couple which must be resisted by diaphragm action

(b) Cantilever bracket which, in the position shown, produces horizontal loads which

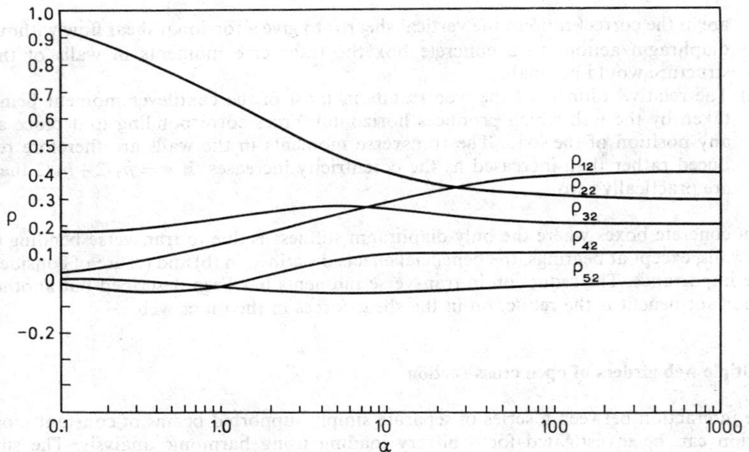

Figure 18.31 Distribution coefficients for a five-beam bridge, beam two loaded

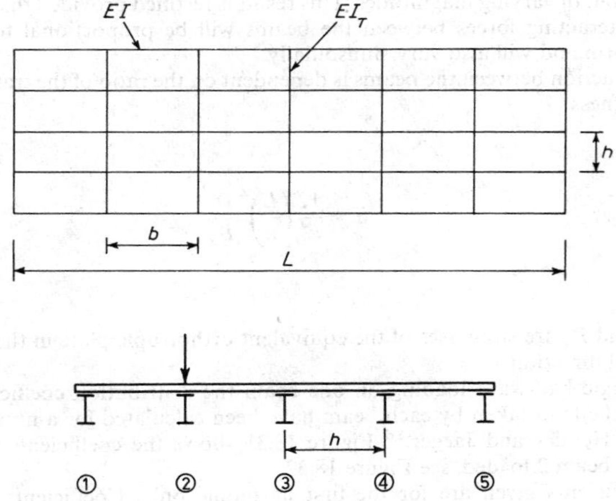

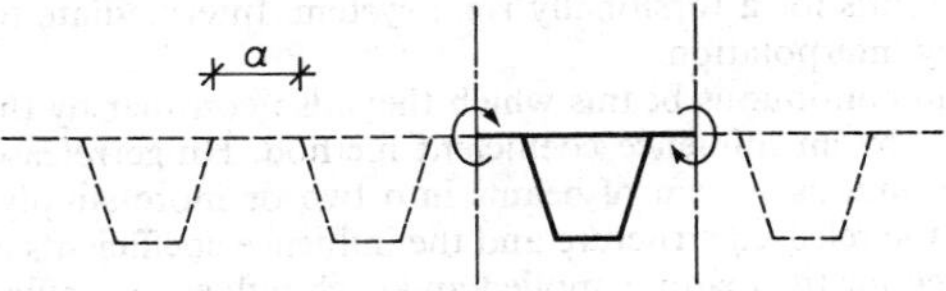

Figure 18.32 Five-beam bridge

(*Figures 18.31 and 18.32 are reproduced from 'The analysis of grid frameworks and related structures' published by Chatto and Windus*[16])

Figure 18.33 Interconnected multiple box girders

are in the correct ratio to the vertical shear P to give a torsional shear flow without diaphragm action. In a concrete box the transverse moments in walls of the structure would be small.

(c) The relative stiffness of the web results in most of the cantilever moment being taken by the web which produces horizontal forces corresponding to a brace at any position of the load. The transverse moments in the walls are therefore reduced rather than increased as the eccentricity increases. If $e = b_T/2 + b_B/2$ they are practically zero

In concrete boxes where the only diaphragm stiffness is due to transverse bending of the walls, except at bearings, the beneficial effects described in (b) and (c) are of considerable importance. The reduction in transverse moments has been described but another important benefit is the reduction in the shear forces in the outer web.

Multiple web girders of open cross-section

The interaction between a series of separate simply supported beams of constant cross section can be investigated for arbitrary loading using harmonic analysis. The sine series is the most suitable approach because any load which varies sinusoidally over the span in a complete number of half waves produces a deflected profile for each beam of the same form but of varying magnitudes. This result is justified provided that it is recognised that the interacting forces between the beams will be proportional to the transverse deflected form and will also vary sinusoidally.

The interaction between the beams is dependent on the ratio of the transverse to longitudinal stiffness:

$$\alpha = \frac{12}{\pi^4}\left(\frac{L}{h}\right)^4 \frac{D_y}{D_x} \qquad (19)$$

where D_y and D_x are stiffnesses of the equivalent orthotropic plate in the transverse and longitudinal direction.

For a single half-wave loading on one beam the distribution coefficients giving the fraction of the load taken by each beam have been calculated for a number of different systems by Hendry and Jaeger.[16] Figure 18.31 shows the coefficients for a five-beam bridge with beam 2 loaded, see Figure 18.32.

The coefficients given are for the first harmonic only. Coefficients for subsequent harmonics can be found by varying α as appropriate for the shorter wavelength. Alternatively if a sufficiently close approximation is given by the first harmonic alone for distribution to the unloaded beams, then the behaviour of the loaded beam is given by its 'free deflection' curve less that which has been distributed to the other beams.

The results in Figure 18.31 are for a system with zero torsional rigidity; Hendry and Jaeger also give results for a torsionally rigid system. Intermediate torsional stiffnesses can be analysed by interpolation.

Fixed ended and continuous beams which they also consider by this method may be more easily solved by an influence coefficient method. Hinge releases at the supports can convert a continuous system of beams into two or more simply supported spans. The behaviour of the released structure and the influence coefficients can be found using the above approach for the loading applied and each influence coefficient.

Grillage programmes enable solutions to be obtained by electronic computer with the advantage that complex geometries can be simulated without difficulty. In both cases it is necessary to estimate the effective top flange unless the more complex form of beam and slab programme using finite elements is used.

Multiple single cell box beams

A series of box beams connected by a top deck and in some cases cross beams and stiffening diaphragms can be analysed by various approaches. The grillage approach using an electronic computer is not necessarily suited to all problems of this type but it is discussed first because in determining the properties of the members the essential mode of behaviour of this form of structure emerges.

Figure 18.33 shows part of a typical cross section which could represent a series of concrete main longitudinal beams spanning between 20 and 60 m or the trough stiffeners on a steel deck system spanning 4 m.

In such systems the interaction between the beams is through the deck slab or deck plate only if no special cross beams, etc., are provided. The magnitude of the interacting forces will be mainly dependant on the overall deflections of the beams and so the distortional stiffness of the individual boxes will be practically equal to the frame stiffness of the sides. The distortional bending or warping stiffness of the boxes may be assumed to be nil except for local wheel loads which will be mentioned later.

Isolating one of the boxes and its share of the deck slab its behaviour can be considered further (Figure 18.34). A unit value of the antisymmetrical component of the vertical

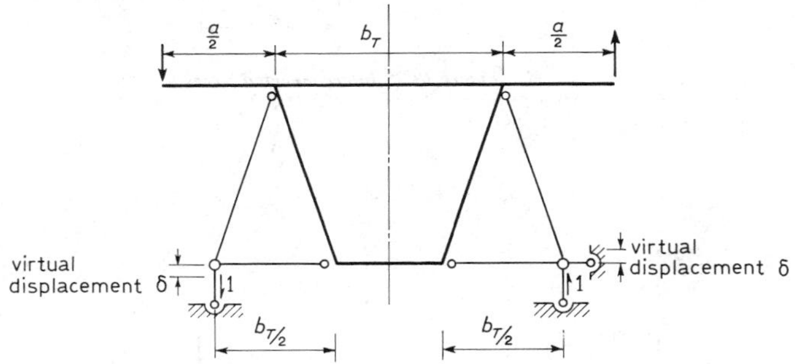

Figure 18.34 Torsional support system for slice of beam to give frame stiffness

shearing forces will act as shown. The twisting effect will be resisted by pure torsion if there is no significant distortional resistance of the box beam except for diaphragm action. The cross section, acting as a frame is loaded by the distortional component of the twisting load. The net effect can be obtained by the device illustrated. The pin jointed bars are placed so as to apply a pure torsional shear flow to the cross section. They also only prevent a pure rotation of the cross section since distortional deflections do not have components in the direction of the restraints. A simple plane frame analysis of the system gives the deflections and hence effective stiffness of a transverse member cantilevering out a distance $(a+b)/2$ from a grillage element with the torsional and flexural stiffnesses of the box. The antisymmetrical moments and symmetrical shears and moments can be applied to find the appropriate stiffnesses and if the simplest form of grillage is used a single compromise value must be chosen. The one based on antisymmetrical shears alone has been found to give results which compared well with a three-dimensional finite element simulation of a series of concrete boxes at two-metre centres.

The effect of local wheel loads on the transverse moments must be added to the above

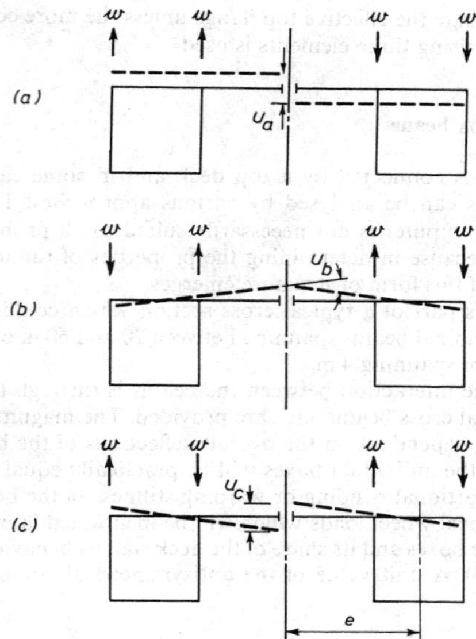

Figure 18.35 Interconnected boxes

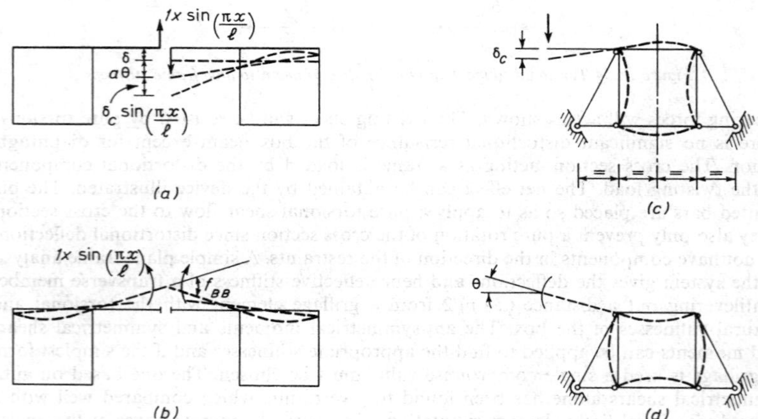

Figure 18.36 Unit loads and couples applied at releases

transverse moments by assuming that the boxes provide rigid supports. The maximum moments in the slab were, in the case mentioned, unaffected by the local slab loading.

Another effect already mentioned is the distortion of the box due to wheel loads applied on one side only. In fact for boxes that are spaced at up to 2 m centres the wheel loads are spread over a width sufficient to make the highest loads fairly symmetrically disposed about an individual box. An allowance can be made, however, by using the single cell theory to calculate stresses which are superimposed on those described above.

A similar approach has been used for steel deck systems, assuming points of contra-flexure halfway between stringers, but instead of treating the structure as a series of discrete beams it is transformed into an orthotropic plate. The transverse flexural stiffness is included in the torsional stiffness of the plate and is, therefore, taken as zero in the plate. The longitudinal flexural stiffness is determined in the usual way. Transversely the deflections of the plate are represented as a sine series in order to solve the plate equation for wheel loading. A large number of terms in the series are required because of poor convergence which, together with the difficulties in obtaining detailed stress values other than longitudinal ones, from the solution, make the method of limited value. Graphs have, however, been produced[17] for the wheel loading used in the USA which are useful for preliminary estimates.

Where a small number of large box girders are used it may be necessary to allow for the various components of the interacting forces more exactly. An example of this is shown in Figure 18.35 in which the nonuniform component of load on two boxes is

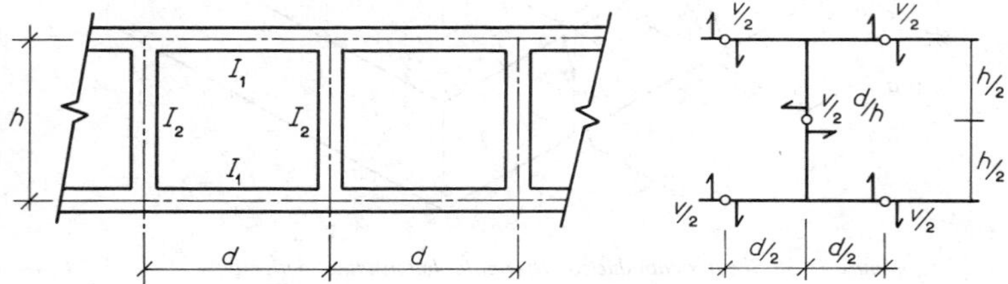

Figure 18.37 Transverse frame

split into three load systems with the properties of either symmetry or antisymmetry. Releasing the forces at the centre of the connecting slab or cross girder produces a lack of compatibility in each case. The influence coefficients are found from the unit forces of Figure 18.36(a) which relate to the compatibility equations including u_a and u_c and Figure 18.36(b) for u_b. δ represents the overall bending deflection of the box beam, $a\theta$ is the deflection produced by torsional rotation and δ_c is the deflection of the cross girder or deflection of the slab. δ_c will be found from Figure 18.36(c) if it is a concrete box of the type already discussed, and θ_c similarly from Figure 18.36(d). Sinusoidally varying forces can be used for boxes without discrete diaphragms except at the supports, otherwise the influence coefficients can be related to individual cross members.

Multi-cellular bridge decks

Bridge structures similar to the top hat beam deck of Table 18.10 which has 115 mm thick webs and no diaphragms between supports, have cross sections which are relatively flexible in transverse shear. The usual grillage or orthotropic plate approach in which shear deformations are neglected is consequently invalid. The following treatment[6] is also relevant to cellular steel decks which may be even more flexible owing to higher web depth-to-thickness ratios.

Transverse shears are carried by the Vierendeel frame action (Figure 18.37) and the flexibility of the frame can be simulated by an equivalent web area of the transverse beams,

$$A_w = \frac{E}{G}\,\frac{12\,h/d}{(dh/2I_1)+(h^2/I_2)} \tag{20}$$

The grillage programme used must include the effects of deflections due to shear strains. It is necessary to differentiate between rotations of initially horizontal and initially vertical lines when shear deformations are considered. In the grillage programme used for the Table 18.10 structure the rotation of vertical lines was the variable used. The flexural

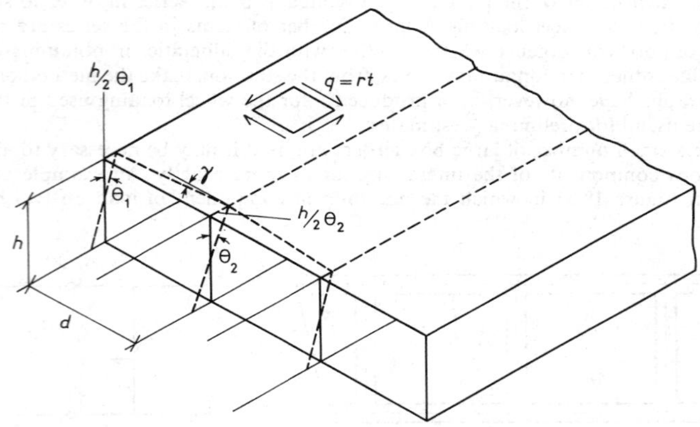

Figure 18.38 Shear strain due to differential longitudinal deflections

parameters are derived in the usual way but the torsional stiffnesses of the grillage members require further consideration.

Symmetrical loading

Loads disposed symmetrically in the transverse sense produce only relatively small transverse movements. True torsion is absent but Figure 18.38 shows that transverse members of grillage will be subjected to twisting which in the actual structure is simply a set of shear strains leading to shear transfer between the beams in a horizontal plane. This is sometimes referred to as shear lag. If the shear lag is small the shear transfer is high and the whole flange will be uniformly stressed. If the transverse members are assigned a stiffness per unit longitudinal distance of $h^2t/2$ this effect will be simulated. The torsional stiffness of the longitudinals is largely immaterial since they do not rotate significantly.

Antisymmetrical loading

Loads which cause twisting produce rotations of the cross section which are the mean of a rotation of a vertical and horizontal line. The vertical line component is given directly by the grillage. The rate of change of rotation of a horizontal line is almost equal to the

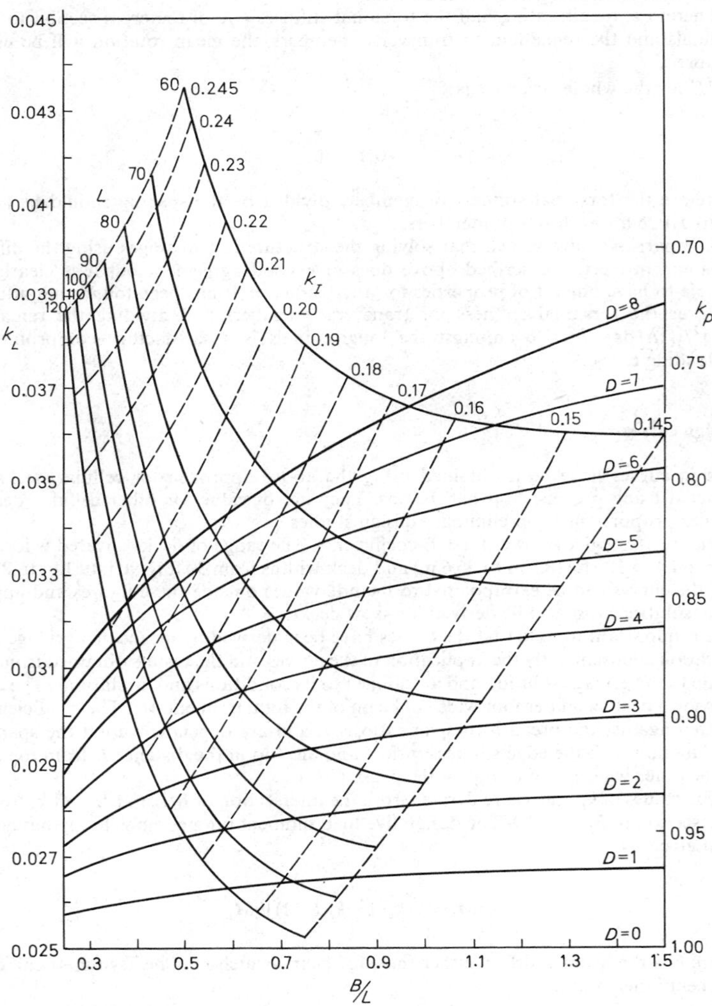

Figure 18.39 Design curves, HB *coefficients*

rate of change of rotation of a vertical line considered in the longitudinal and transverse direction respectively, since the true webs only undergo small shear strains compared with the frame of the cross section.

Therefore, by allocating half the torsional stiffness GK of the cross section to longitudinals and the remainder to transverse members, the mean rotation will be used as required.

GK for the whole structure is:

$$4GA^2 \Big/ \oint \frac{\mathrm{d}s}{t}$$

therefore the torsional stiffness of members divided by the spacing should be $(4GA^2/\oint(1/t)\,\mathrm{d}s)/2b$ for both sets of members.

Comparisons have shown that solving the structure in two stages using the different torsional properties described above does give good agreement but it is clearly preferable to have one set of properties for any load case. It has been found that adopting $h^2t/2$ as the torsional stiffness for transverse members and dividing the remainder, $4GA^2/\oint(1/t)\,\mathrm{d}s-(h^2t/2)b$ amongst the longitudinals is a satisfactory compromise for all loading cases.

Design curves

Design curves have been obtained using the above approach for cellular decks constructed using pre-cast 'top hat' beams. They are of value for other cellular decks of similar proportions for preliminary design studies.

Figure 18.39 gives curves for HB coefficients. The range of decks covered is for spans from 60 ft to 120 ft (18.3 m to 36.6 m) and deck widths from 30 ft to 90 ft (9.1 m to 27.4 m) but the curves can be extrapolated to include values outside these figures and approximate solutions can also be derived for skew decks.

It is important to note that the curves have been derived by means of a grillage representation. Consequently the application of the curves and the results obtained from them relate to the grillage solution and are subject to its conditions and limitations. The curves provide a coefficient per foot width of beam of the total moment M_L. The coefficients are plotted against the breadth to span ratio B/L and are dependent upon the span L in conjunction with the edge stiffness ratio I and also upon the distance D from the centre of the outer wheels to the edge of the deck.

The values of k_L and k_I are derived from the intersection of B/L and L, and k_p from the intersection of B/L and D. The design live load moment for a composite top hat beam is then given by

$$M = k_L k_p[1-k_I(I-1)]bM_L \tag{21}$$

where b is the beam width of either the edge beam, which may be asymmetrical, or the adjacent inner beam.

STRESS CONCENTRATIONS

Sudden changes in loads or in the shape of a structure produce stresses that cannot be calculated by normal beam theory. Concentrated loads such as the reactions at bearings and holes cut out of flanges are obvious examples of such changes but alterations in the direction of flanges, variations in thickness or width may produce significant effects. The introduction of strengthening members such as diaphragms or stiffening around

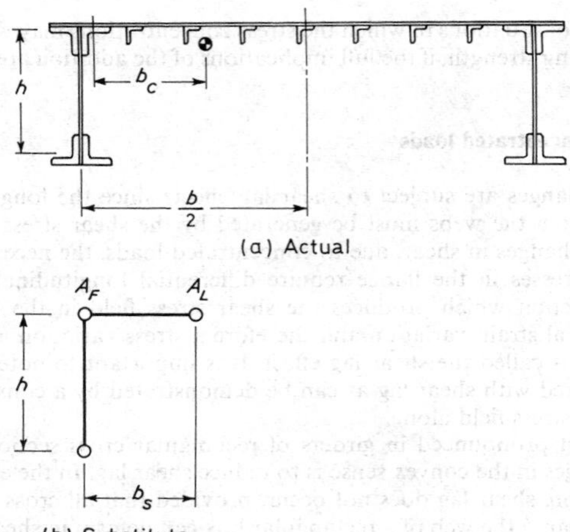

Figure 18.40 Transformation of actual into substitute beam cross section

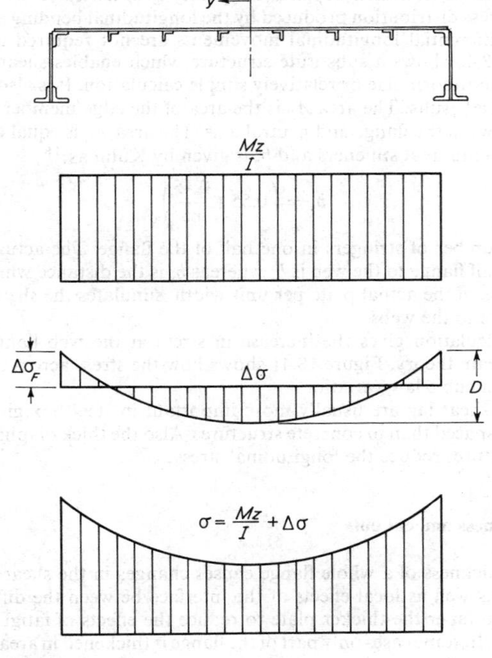

Figure 18.41

(Figures 18.40 and 18.41 are reproduced from 'Stresses in aircraft and shell structures' published by McGraw Hill[18])

holes are examples of situations in which the stress concentrations may weaken the structure instead of adding strength, if the full implications of the addition are not considered.

Shear lag due to concentrated loads

Beams with wide flanges are subject to shear lag effects since the longitudinal stresses at points remote from the webs must be generated by the shear stress field across the flange. At sudden changes in shear, due to concentrated loads, the necessary changes in the longitudinal stresses in the flange require differential longitudinal movements in the transverse direction which produces the shear stress field in the flange. In other words, a longitudinal strain variation and therefore a stress variation across the flange is produced which is called the shear lag effect. It is important to note that transverse stresses are associated with shear lag as can be demonstrated by a consideration of the statics of the shear stress field alone.

Shear lag is most pronounced in girders of rectangular cross section. The effect of cambering the flanges in the convex sense is to reduce shear lag. In the extreme case of a circular cross section, shear lag does not occur, provided that all cross sections remain circular. This is because the web of a rectangular box can deform in shear without shear deformations of the flange being necessary, whereas the circular beam can only deform in shear if all parts of the beam are subjected to shear strains. The distribution of shear strain in the latter case, which depends on purely geometrical considerations, agrees with the shear stress distribution produced by the longitudinal bending stresses of normal beam theory. Differential longitudinal movements are not required and no shear lag occurs. Figure 18.40 shows a substitute structure which enables shear lag effects to be found from standard formulae or relatively simple calculation. It is also useful for evaluating more precise results. The area A_F is the area of the edge member plus one third of the web area between the flange and neutral axis. The area A_L is equal to half the area of the plate plus longitudinal stiffeners and b_s is given by Kuhn as:[18]

$$b_s = \left(0.55 + \frac{0.45}{n^2}\right) \tag{22}$$

where n is the number of stringers in one half of the flange. The actual distance of the centroid of the half flange to the web is b_c whereas b_s is the distance which, together with the shear stiffness of the actual plate per unit width, simulates the shear lag property of the flange relative to the webs.

The above calculation gives the increase in stress at the web flange junction, $\Delta\sigma_F$, above normal beam theory. Figure 18.41 shows how the stress across the flange may be found assuming a cubic law variation.

The effects of shear lag are usually more important in steel box girders as the webs are more widely spaced than in concrete structures. Also the thick diaphragms at bearings in concrete structures reduce the longitudinal stresses.

Changes in thickness and cut outs

The change in thickness of a whole flange causes changes in the shear stresses between web and flange as well as local effects at the interface between the different flanges. In steel it is usual to taper the thicker plate to reduce the effects of fatigue and possibility of brittle fracture. In some cases only part of the flange is thickened in areas of concentrated load such as forces from supporting cables or prestressing cables. The effect of such thickening is to tend to concentrate all flange forces in that part of the flange which must be allowed for by either gradually tapering out the increased area or by carrying the greater thickness through to a more lowly stressed region. A premature end to the reinforcement may overload the connecting unreinforced section.

The reinforcement required for cut outs must be continued or tapered for similar reasons but the need to do so is more obvious. The transverse and shear stresses associated with cut outs are, however, also of considerable importance. In steel structures they are likely to cause buckling, fatigue and brittle fracture problems whereas in concrete structures they cause cracking. It is, therefore, necessary to connect diaphragms, etc., to the structure by much more than nominal amounts of reinforcing steel. It is instructive to note with reference to steel structures that a number of large tankers have experienced local failures at cut outs owing to the effects described above. It seems likely that the extrapolation of design knowledge from smaller structures was not backed up with sufficient research into the complex stress systems produced and the associated buckling phenomenon. Similarly the causes of failure of several steel box girder bridges have been mainly due to stress concentrations due to cut outs in stiffeners, support reactions, and a cut out produced by partially unbolting a main compression flange splice. An earlier failure of a plate girder bridge due to the stress concentrations produced by a flange cover plate completes an unanswerable case for the importance of allowing for stress concentrations in structural design. The basic engineering solution to this problem is to avoid severe stress concentraions and all those mentioned could have been avoided without significant cost or difficulty. Some degree of stress concentration is, however, unavoidable and only by using test data can the distinction be drawn between the acceptable and unsafe forms of structures, structural details, and associated stress levels.

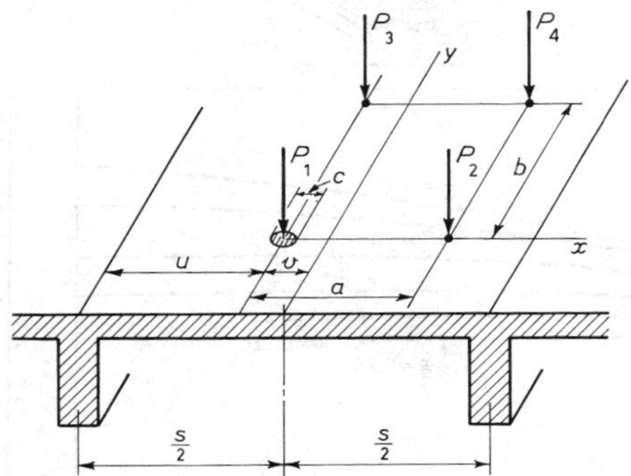

Figure 18.42 Slab supporting wheel loads

The stress levels themselves due to known loads can be found with considerable accuracy using, for example, two- and three-dimensional finite element methods. Kuhn[18] describes ingenious methods for the approximate analysis of shell structures with cut outs as well as the shear lag approach described above. They give valuable insight into the structural behaviour of such systems but are more expensive to use than computer based techniques using finite elements.

CONCRETE DECK SLABS

The usual form of deck system is a concrete slab spanning transversely between longitudinal beams which are often the main members of the bridge. Cross girders can be used to produce a longitudinally spanning slab or the slab can be supported on a series of stringers spanning between cross girders. The simplest system is generally the most economical and only where there are special requirements are stringers and cross girders used. In large steel trusses, for example, the loads must be carried to intersection points requiring more complex systems.

The slab must be designed for three different modes of behaviour:

(a) Local flexure due to the transfer of wheel loads to the adjacent beam members
(b) Flexure due to relative movements of various parts of structure
(c) In-plane stresses due to beam action of main and secondary members of structure

Local stresses can be found by assuming that all supporting members are rigid when evaluating the slab moments and shears due to wheel loading. The remaining effects can then be found by applying loads equal to the reactions of the supporting members to those members. It is important that the latter loads should be statically equivalent to the vehicle loading but the exact spanwise distribution is not usually required.

There are several publications giving influence surfaces for local slab bending moments for several types of support, i.e. simply supported on four sides, cantilever slabs, fixed

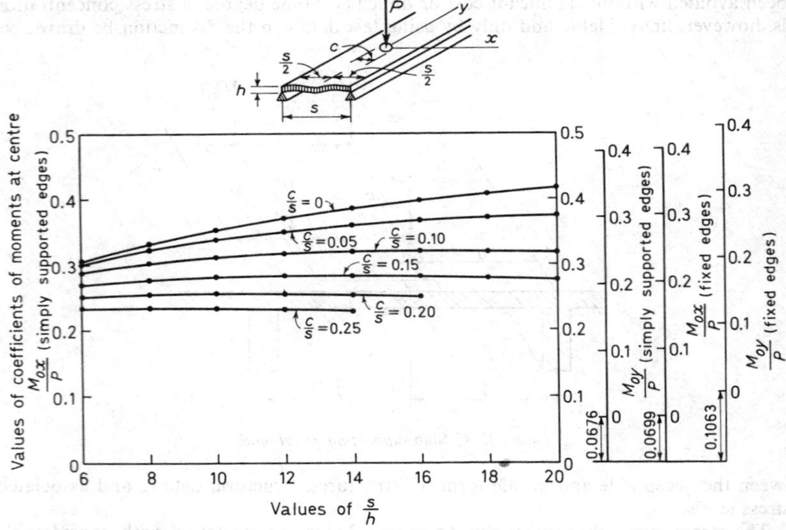

Figure 18.43 Coefficients of bending moments M_{ox} and M_{oy} in directions of x and y respectively, produced at centre of slab by a central load P distributed uniformly over the area of a small circle with diameter c

on four sides.[19, 20] The best known treatment is by Westergaard[21] and was used to derive the equivalent uniformly distributed loads given in the British loading specification for bridges.[22] Westergaard uses Nádai's equations to obtain the bending moment under a wheel load:

$$\left.\begin{matrix} M_x \\ M_y \end{matrix}\right\} = \frac{(1+\mu)P}{4\pi}\left[\ln\left(\frac{4s}{\pi c_1}\cos\frac{\pi v}{s}\right)+\frac{1}{2}\right] \pm \frac{(1-\mu)P}{8\pi} \qquad (23)$$

The meaning of the symbols is given in Figure 18.42 except for c_1 which is the equivalent

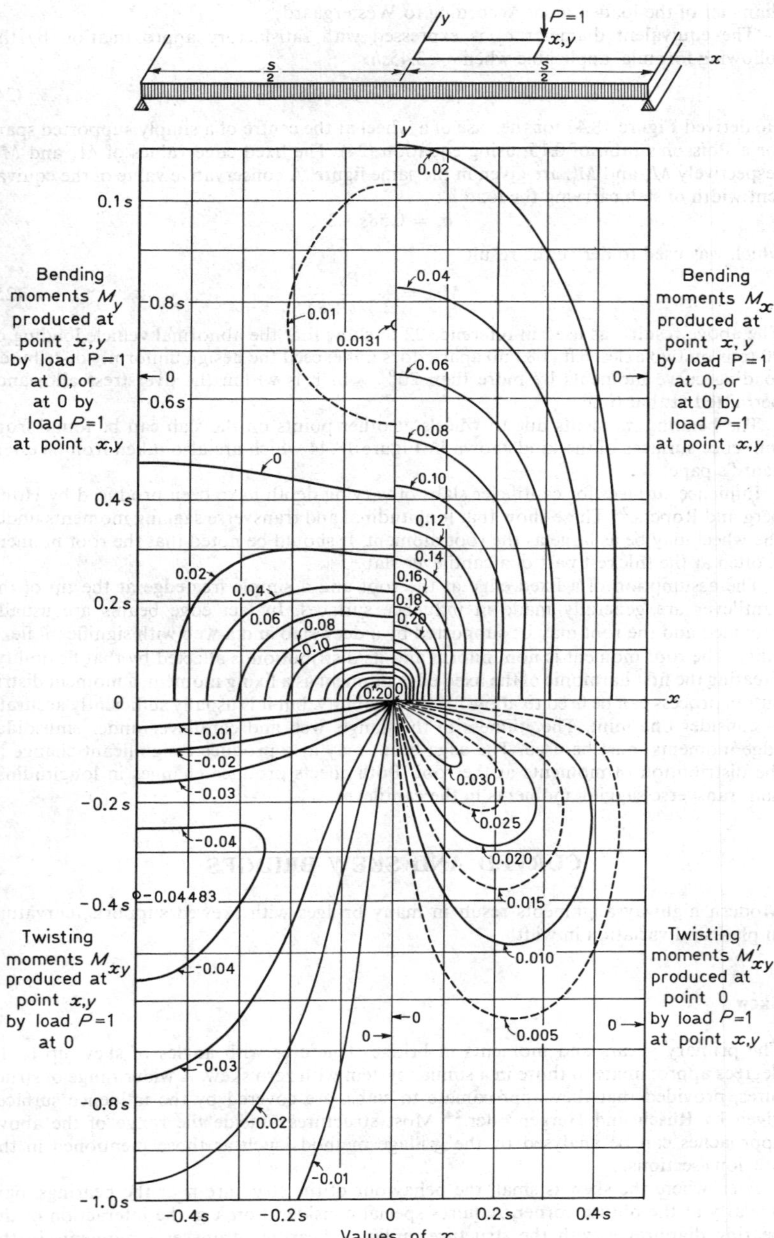

Figure 18.44 Contour lines of surfaces representing moments. Poisson's ratio $\mu = 0.15$

diameter of the loaded area. According to Westergaard:

'The equivalent diameter c_1 is expressed with satisfactory approximation by the following formula, applicable when $c<3.45sh$

$$c_1 = 2[(0.4c^2+h^2)^{\frac{1}{2}}-0.675h]\text{'} \qquad (24)$$

He derived Figure 18.43 for the case of a wheel at the centre of a simply supported span, for a Poisson's ratio of 0.15, using equation (24). The fixed edge values of M_x and M_y, respectively M'_x and M'_y, are given in the same figure. A conservative value of the equivalent width of slab carrying the load is:

$$b_e = 0.58s+2c$$

which was used to derive the result:

$$M_{0x} = \frac{Ps}{2.32s+8c} \qquad (25)$$

The above result was used in reference 22 to show that the abnormal vehicle loading of 90 tons on two axles, 6 ft (1.83 m) apart, does not exceed the design uniformly distributed loading curve moments by more than 20%, which is within the overstress allowance permitted at that time.

The bending moments due to wheels at other points on the slab can be found from influence surfaces of the kind shown in Figure 18.44 which are also taken from Westergaard's paper.

Influence surfaces for cantilever slabs of varying depth have been produced by Homberg and Ropers.[23] These show that longitudinal and transverse sagging moments under the wheel may be as large as the root moment. It should be noted that the root moment is often at the thickest part of a cantilever slab.

The assumption of a fixed edge at the root and a simple free edge at the tip of the cantilever are generally made in influence surfaces. In fact edge beams are usually provided and the root may be supported by a deck slab and a web with significant flexibility. The root moment is nonuniform and its distribution is affected by that flexibility. Treating the first harmonic of the fixed edge moment as a fixing moment, a moment distribution process can be used to allow for the flexibility, but it is usually sufficiently accurate to consider one joint. The stiffness of the flange, web and cantilever under sinusoidal edge moments must be used. The edge beam may also produce a significant change in the distribution of moments at the root. Both effects produce changes in longitudinal and transverse sagging moments in the cantilever.

CURVED AND SKEW BRIDGES

Modern highway alignments result in many bridges with skewed supports, curvature in plan, and variation in width.

Skew

The primary shears and moments in bridge structures with angles of skew up to 15 degrees approximate to those in a similar system with zero skew. A wider range of structures, provided that they approximate to slabs, are covered by the influence surfaces given by Rüsch and Hergenröder.[24] Most structures outside the range of the above approaches can be analysed by the grillage methods such as those mentioned in the previous sections.

Even where the skew is small the behaviour of the structure near the bearings, particularly at the obtuse corner, requires special consideration, e.g. the interaction of the bearing diaphragm with the structure, uplift at bearings, transverse moments in the deck and shear distribution. These effects depend on the detailed form of the structure and the articulation of stiffness of the bearings and piers. In torsionally stiff structures, with twin bearings at both ends of a span, skew may produce significant end fixity with correspondingly high loads on the inner bearings (obtuse corners) and low loads on the others.

Curved in plan

Curved, torsionally stiff, structures supported against torsion at each vertical support can be analysed for the effects of bending continuity as if they were straight beams provided that the following requirements are satisfied:[25]

$$\gamma < 1 \quad \text{and} \quad \alpha < 30^\circ$$
$$\gamma < 5 \quad \text{and} \quad \alpha < 20^\circ$$
$$\gamma < 10 \quad \text{and} \quad \alpha < 15^\circ$$

where $\gamma = EI/GK$ and α is the angle subtended by one span. The resulting bending moments are within 6% of curved beam solutions.

The solution found by that approach only allows for the vertical loads. The twisting moments produced by eccentricity of the loads relative to the shear centre must be considered separately. The vertical loads alone, however, produce torsional moments in the structure and graphs are given in reference 25 which enable these to be found from the continuity moments. The effects of applied torques for the ranges of structures given above can be estimated from considerations of statics and relative stiffnesses. More generally the standard theory of curved beams can be applied to curved bridge structures or grillage programmes can be used.

In all curved structures the longitudinal stresses produce lateral loads due to the curved path they follow. The effect of these loads, including the local loads caused by the curvature of prestressing cables, must always be given careful consideration. The resulting forces are resisted by the frame action of the cross section producing deformations of the cross section. In exceptionally thin-walled concrete box structures or steel box structures without special frames or diaphragms these deformations can result in significant changes in the primary bending moments.[26] In the usual type of box structure, however, it is possible to consider this effect separately and, provided that the forces acting on the cross section allow for curvature, reasonable results can be obtained by assuming the box is straight and using theories such as those described previously.

BUCKLING OF WEB SYSTEMS OF PLATE GIRDERS

There are two basic approaches to the design of web systems:

(i) Elastic critical buckling as in the German specification DIN4114[27]
(ii) Partial tension field as in BS 153[28] and the American Institute of Steel Construction specification[29]

The Merrison design rules, which are expected to form the basis of the new specification for steel box girder bridges are based on a combination of the approaches of (i) and (ii) but with the major innovation that the strength of the stiffeners as well as their stiffness must be considered. They are applicable to plate girders as a special case.

In the German approach the elastic critical stress of the individual plate panels is calculated for the combined in plane stress field. The critical stress system is corrected for nonlinearity by factors depending on the level of the equivalent stress if it is above 80% of the yield. No allowance is made for post-buckling strength other than the use of low safety factors. The methods adopted for stiffeners in webs are the same as the plate panel approach, using appropriate critical stresses. DIN 4114 contains formulae for the critical stresses of simple plate panels and stiffened panels. Further values are given in Klöppel and Möller;[30] In important bridges it is necessary to carry out eigenvalue calculations as the web systems used are unlikely to be covered in any published set of graphs. The lack of straightness of the stiffeners is not allowed for which means that realistic estimates of strength are not made.

The British and American specifications are based on experimental evidence[31, 32] which is particularly abundant in the case of plate girders with vertical stiffeners only. The web plates are assumed to carry shear by a partially developed diagonal tension field after the critical load is reached and in the table of permissible stresses it is assumed that the flange has the properties necessary for that to be possible. In BS 153 the flange is required to have an area which satisfies a stiffness criterion for web restraint and in the AISC specification the diagonal tension field is not assumed to react on the flange. The American tests were, however, all carried out on plate girders with flanges at least twice the thickness of the web and it seems likely that a minimum flange area is necessary in the vicinity of the web to give the restraint implicit in their tests. It may be concluded

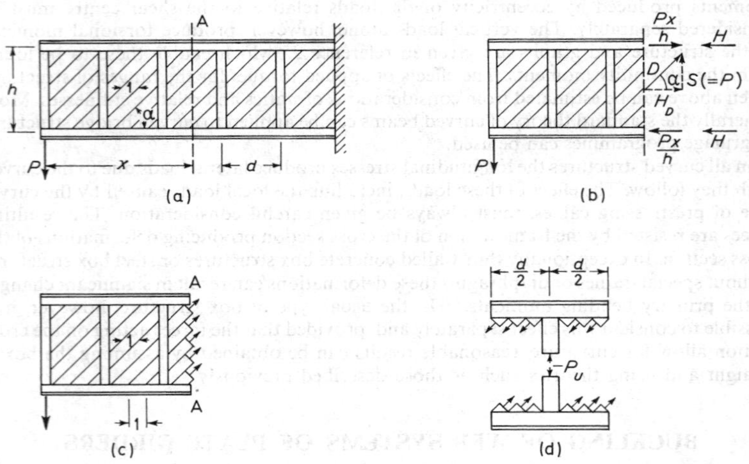

Figure 18.45 Forces in diagonal-tension beam

(*Reproduced from 'Stresses in aircraft and shell structures' published by McGraw Hill*[18])

that the British and American specifications are only applicable to plate girders with relatively thick flanges.

In the German and British specifications horizontal stiffeners can be used to limit the panel size in order to use increased web depth-to-thicknesses ratios. The German approach being general includes any stiffener arrangement whereas BS 153 is limited to two horizontal stiffeners at specified positions on the web. The BS 153 requirements for vertical stiffeners are based on a considerable experimental evidence whereas the horizontal stiffeners are based on critical buckling formulae for bending stresses with relatively little experimental evidence to support them. The stiffeners required for the most commonly used plate girders with depths up to about 2 m require light stiffeners according to most specifications and so in practice heavier ones than are necessary are often used. The uncertainties in areas of all three specifications referred to above make this a wise precaution and the safe use of those specifications requires an appreciation of their theoretical basis and the experimental evidence on which they are based.

Diagonal tension

It is possible to design webs assuming that all the shear force is carried by pure diagonal tension, resulting in the stress system of Figure 18.45. In aircraft structures a modified form of that approach has been used making an allowance for the critical buckling shear capacity of the web between vertical stiffeners. The relatively heavy flange–web angle connections can be designed to act as beams carrying the tension field in the web to the vertical stiffeners. In bridges, however, welding has eliminated these angles, thereby removing a considerable part of the material available for restraining the web. In flanges of thicknesses comparable to that of the web, for example, it is only possible for the flange to restrain the web in a direction longitudinal to the web–flange interface. The effects of such restraint can only be safely evaluated by experiment. Theories have been advanced by Bastler and Rockey which allow for the restraint implicitly by mechanisms which assume respectively

(a) the diagonal tension field is only supported by the web stiffeners
(b) the diagonal tension field is carried by a beam consisting of the flange and part of the web.

In plate girder structures with vertical stiffeners only there is sufficient experimental evidence to show the limits within which the theories can be applied. In more complex structures or where loads are applied in circumstances not covered by the tests, e.g. heavy loads between stiffeners, it is necessary to provide additional material in order to carry the diagonal tension by a clearly defined beam. A beam member is also required to carry the tension field along the vertical edge at the free ends of the web.

Compression flanges

The compression flange requires lateral support in order to prevent lateral buckling and resist wind loading. The most critical condition for the support system is buckling between supported points of contraflexure at the supports as in Figure 18.46. Each bay of the flange acts as if simply supported and if the component of out of straightness of the supports, due to initial eccentricity and deck loads, in an alternating pattern is e, then from statics the load on a support is:

$$R = \frac{(4P/l)e}{1 - P/P_{crit}} \tag{26}$$

where P is the factored axial load in the flange and P_{crit} is the critical buckling value of the support system. In a system with more than four supports,

$$P_{crit} = k/4l \tag{27}$$

for a spring stiffness k.

Lateral buckling of the flange between supports can be examined by normal strut methods allowing for imperfections and including the axial load in the web, provided that the outstand-to-thickness ratio is limited to prevent torsional buckling.

STEEL BOX GIRDERS

Webs

The web system of a box girder is not in principle different from the web of a plate girder but it may be exposed to more severe stress conditions unless the following characteristics of box girders are recognised.

(a) Flanges are more likely to be thin thus giving less restraint to web
(b) Shear lag effect in wide box flanges can produce high longitudinal stresses in web
(c) Uneven shear stress distribution may be produced by bearings under diaphragms
(d) Moment-rotation curves may not have sufficient plateau region to permit re-distribution of stress in continuous structures

In the regions near supports or whenever stress redistribution within the web is required, the use of additional beam material at the flange web interface, as mentioned in the sub-section on plate girder webs, is desirable.

The stiffener requirements are covered in the Merrison design rules by methods that allow for the combined effects of shear, bending and end loading from cross girders on

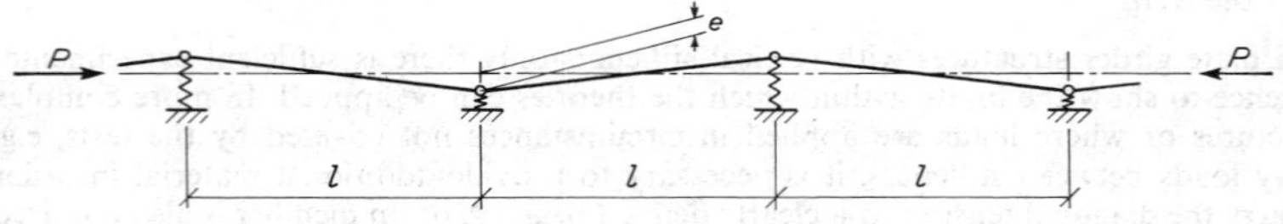

Figure 18.46

the vertical stiffeners. In order to produce robust structures, however, transverse loading from wheel loads, etc., must be carried by means other than the main web plate to the vertical stiffeners. The rules are designed to cover a wide range of problems safely and, in major structures where the advantages of a particular structural system are to be determined, special investigations are necessary. An approach that has been found very useful is based on a computer programme for grillages in which the out-of-plane effects of axial loads are included by a modification to the stiffness matrix of the members. The web system is simulated by grillage members representing the stiffeners and a closely spaced lattice of beams representing the web plate. The longitudinal stresses and shear stresses are represented by appropriate axial loads in the members. The effects of partial diagonal tension can be considered by a suitable choice of axial loads in the diagonals of the lattice members. Imperfections in the plate and stiffeners, interaction with flange stiffeners, and holes in the web can be readily allowed for. Comparisons made with large deflection solutions for webs with vertical stiffeners only showed good agreement. The grillage approach can, of course, be applied to any stiffener system including both horizontal and vertical stiffeners.

Compression flange

The stiffened compression flange can be treated as a series of struts with effective lengths equal to the spacing of the transverse stiffeners. The stress in the outstand is limited by the yield stress provided torsional buckling is avoided and by the local buckling capacity of the plate on the other side. In wider flanges the effects of transverse compression and shear are more likely to require investigation as they may affect overall buckling as well as local buckling.

The transverse stiffeners act in preventing strut buckling in the vertical direction the same way that the cross frames prevent lateral buckling in plate girders (see plate girder flanges, 18–59). Simply supported transverse stiffeners of width b will deflect in a half sine wave with a stiffness $k = \pi^4 EI/b^4$ for vertical loading per unit length. The critical load per unit width of flange, not assuming buckling between transverse stiffeners, is found from equation (27):

$$\frac{4N_{cr}}{l} = k = \frac{\pi^4 EI}{b^4}$$

or

$$N_{cr} = \pi^4 \frac{EI}{b^4}\frac{l}{4} \tag{28}$$

The effects of imperfections, buckling over a longer wavelength and support from torsionally stiff transverse members can be treated as in the strut on elastic support theory, provided the stiffnesses used allow for transverse sinusoidal variations as described above.

DYNAMIC RESPONSE

The stresses produced by the dynamic response of a bridge to a vehicle traversing it are covered by loading specifications. Unusual structures may require a special investigation but the main problem will be that of defining the vehicle size, speed and frequency of occurrence. Damping, though difficult to quantify, is not likely to be important in this type of problem since the main effects are usually short-term ones where damping is of little significance.[33] The prolonged oscillations produced by a series of vehicles acting in phase with each other is highly improbable at high live-load stress levels.

The effect of vibration on the user of a bridge is, potentially, more important since small accelerations, 0.02g, can produce discomfort in pedestrians and occupants of stationary vehicles. Occupants of moving vehicles cannot distinguish bridge movements as they are masked by the normal oscillations of the vehicle. Figure 18.47 from reference 34 shows limits proposed by various workers in the industrial field compared with those found by the Road Research Laboratory from the reactions of people to a footbridge erected in the laboratory, excited to a steady state resonant motion. Short periods of oscillation much greater than the limits in the figure are likely to be acceptable to bridge users if they understand that it is part of the normal behaviour of the structure. The stationary car occupant is a different case since resonance can produce oscillations of increased amplitude if the natural frequency of the vehicle, 1–3 Hz and the bridge are nearly the same. Wyatt has commented, however, that for spans over 30 m.

'. . . an obvious design aim would be to avoid the frequency range of 1–3 Hz. It has, however, also been shown that as a result of 40 years' design progress the economic structure is likely to have a frequency in this range and it is tantamount to throwing away this progress if higher frequencies are specified. Any specification which imposes a limit on frequency is thus strongly to be deplored.'

Wind-excited oscillations are known to be of primary importance in long-span structures of the suspension bridge and cable braced type. What is not clear is the significance of this phenomenon in more modest spans with or without cable supports. Theoretical investigations of the problem must allow for effects of vortex shedding on the particular cross section in question, similarly the changes in lift characteristics with rotation and deflection. Structural and aerodynamic damping are both important and of course the elastic characteristics of the structure must be known.

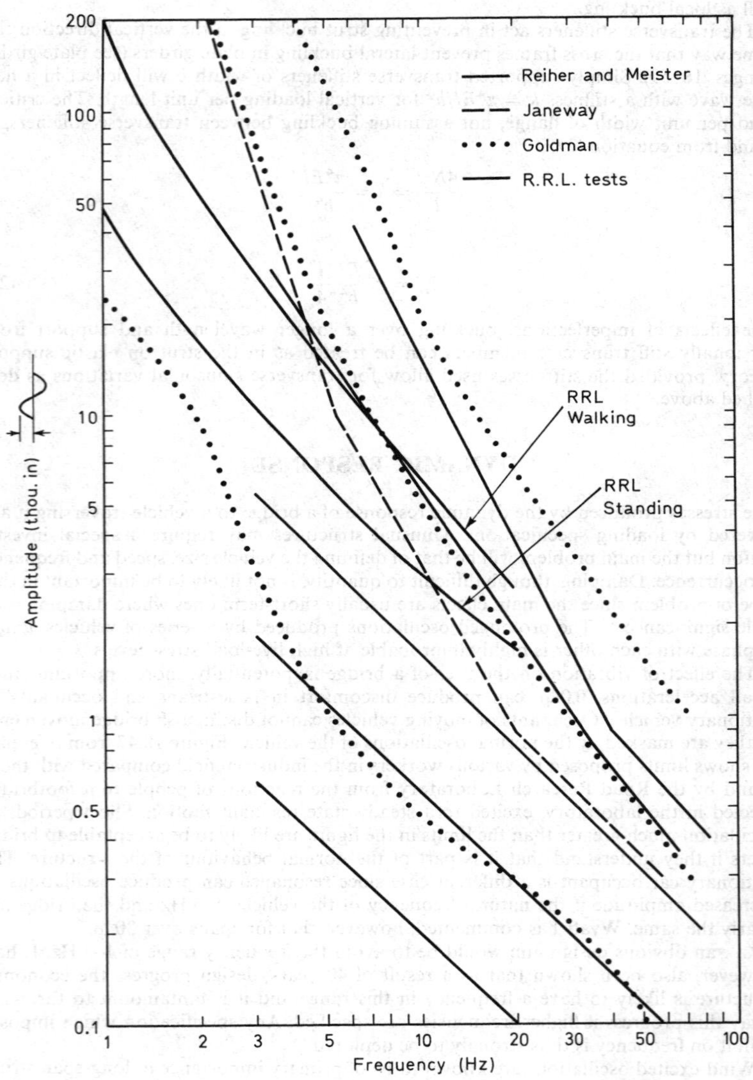

Figure 18.47 Comparison of tolerance levels to vibration

The behaviour of long-span structures is usually determined with the aid of wind tunnel tests on models of segments of the deck and stiffening girder. Spring supports can reproduce to scale the natural frequency calculated for the prototype. The use of wind tunnels will continue to be important for major bridges where the economic advantages of determining the most suitable deck system are considerable.

In the intermediate range of spans the data for various shapes of cross section from wind-tunnel tests are likely to be sufficient in the near future to permit an analytical estimate of the effect of wind without being unduly conservative.

ITEMS REQUIRING SPECIAL CONSIDERATION

This section has considered in general terms the types of bridges used throughout the world, primarily of steel or composite construction or a combination of both types, together with general principles of analysis which enable the preparation of the main members of the design to be correctly established. Space precludes treatment of many interesting aspects of bridge design.

The authors would like to draw attention to several points which, from experience, are likely to give rise to difficulties unless adequate consideration is given to them. Many are obvious but it is a regrettable fact that it is obvious points which give rise to recurring difficulty.

It is essential that adequate time and effort is devoted to the checking of any bridge design by a senior engineer fully experienced in the type of work involved. Much time can be wasted in elaborate analysis either by slide rule or computer when the simplest elementary assessment of some details will demonstrate inadequacies or inaccurate conception of the mode of behaviour. It is important for bridge engineers to have some grasp of three-dimensional actions of structures rather than a preoccupation with cross sections, plans and elevations.

To this end, the engineer is encouraged to make simple models of paper or cardboard so that the implications of what is apparently an obvious decision can be grasped and interpreted. The relationship of frame members in space, the junctions of beams and columns, the implications of running bridge girders into diaphragms and end blocks are typical examples.

Attention to the following check list may perhaps help bridge engineers to reduce the number of errors or difficulties on site due to lack of application in the drawing office.

(a) Check all the detailed items of the bridge structure and not just the main structural members.
(b) Check the basic principles of reinforced concrete design used. The same applies to steelwork details for welding and bolting.
(c) Can the reinforcement be fixed by a man of normal size?
(d) Can the concrete be placed and vibrated properly?
(e) Can welds be made in a suitable sequence and do bolt spanners, etc., foul other completed parts of the construction?
(f) The geometric calculations should be checked. The leading dimensions on the general arrangement drawings cause considerable confusion if they are not correct.
(g) Secondary forces and stresses should not be forgotten. It is better to consider them by elementary methods than to omit taking them into account.
(h) Do not forget constructional tolerances which will cause lack of fit and possibly additional loads and stresses due to eccentricity.
(i) Particular attention should be paid to bearing design and details for drainage and expansion joints. The specifications for these items should not be skimped.
(j) Remember that lateral loads are generated from lack of straightness in tension. These are often forgotten and include things as widely diverse as:
    (i) substantial lateral loading on elements of cellular construction;

(ii) change of direction of thrust members as at portal frame joints;
(iii) haunches on the soffit of a beam;
(iv) overall curvature of an arch;
(v) the behaviour of prestressing cables in plan and elevation imposes a local loading as well as the effect of the line of pressure.

(k) Post-tensioned prestressed concrete anchorages should not only be considered in terms of bursting, splitting and the length of transmission of force for the single anchorages, but in terms of sets or the whole group.
(l) Openings for services and access for maintenance and inspection.
(m) Anchors require careful detailing. Holes, ties, etc., become a stress raiser and they must be detailed to suit.
(n) Temporary openings and drain holes necessary during construction should receive adequate consideration at an early stage, also for the removal of items of plant, jacks, falsework, shuttering, lifting gear, etc.
(o) Engineers sometimes apply normal beam theory without considering whether this is applicable. Many elements act as deep beams, i.e. brackets, corbels, halved joints, anchorages, tie backs. They require application of a specific approach which adequately caters for behaviour which would actually occur in service either in bending or in shear or torsion.
(p) Consideration should be given to the weathering of a structure in terms of orientation appearance and durability.
(q) The designer should consider whether the concept of erection is clear in his mind and what the effect would be of lack of compliance by the contractors.
(r) Are there any unforeseen events which would provoke an unfortunate chain of circumstances such as traffic collision on any parts of the structure, crane booms and dump trucks inadvertently striking part of the bridge during construction, fires, etc?
(s) When inspecting bridges under construction, keep a look out for things which are not in accordance with the intention of the design from a safety point of view.
(t) Is the articulation of the structure as the designer intended?
(u) Is reinforcement in the bottom of cantilevers rather than the top?
(v) Has distribution or secondary reinforcement been omitted by mistake at the detailing stage?

APPENDIX: MOVABLE BRIDGES

The principles associated with the various types of movable bridge are well established but increasing knowledge particularly in the fields of control machinery and materials technology has made new applications possible. The selection of the type of movable bridge is largely governed by the nature of the site. The relative priority of the two thoroughfares, usually a road or railway over a waterway, will help to establish the acceptable closed position headroom and also the speed and conditions under which the bridge will be required to operate.

Where an existing bridge is to be replaced, the condition of the existing foundations will influence the choice of type for the new bridge. For example, it would normally be uneconomic to excavate a tail chamber for a trunnion bascule bridge out of existing heavy concrete foundations.

There is a need for speed during those phases of construction that will obstruct navigation channels. This obstruction can be minimised by finally assembling the bridge as a few large items; the assembly of the leaves of one modern bascule bridge has been done in the bridge open position.

Minimum dead weight has the obvious advantage of reducing the required capacity of the bridge machinery; an orthotropic steel deck is used on most modern bridges

Table 18.12

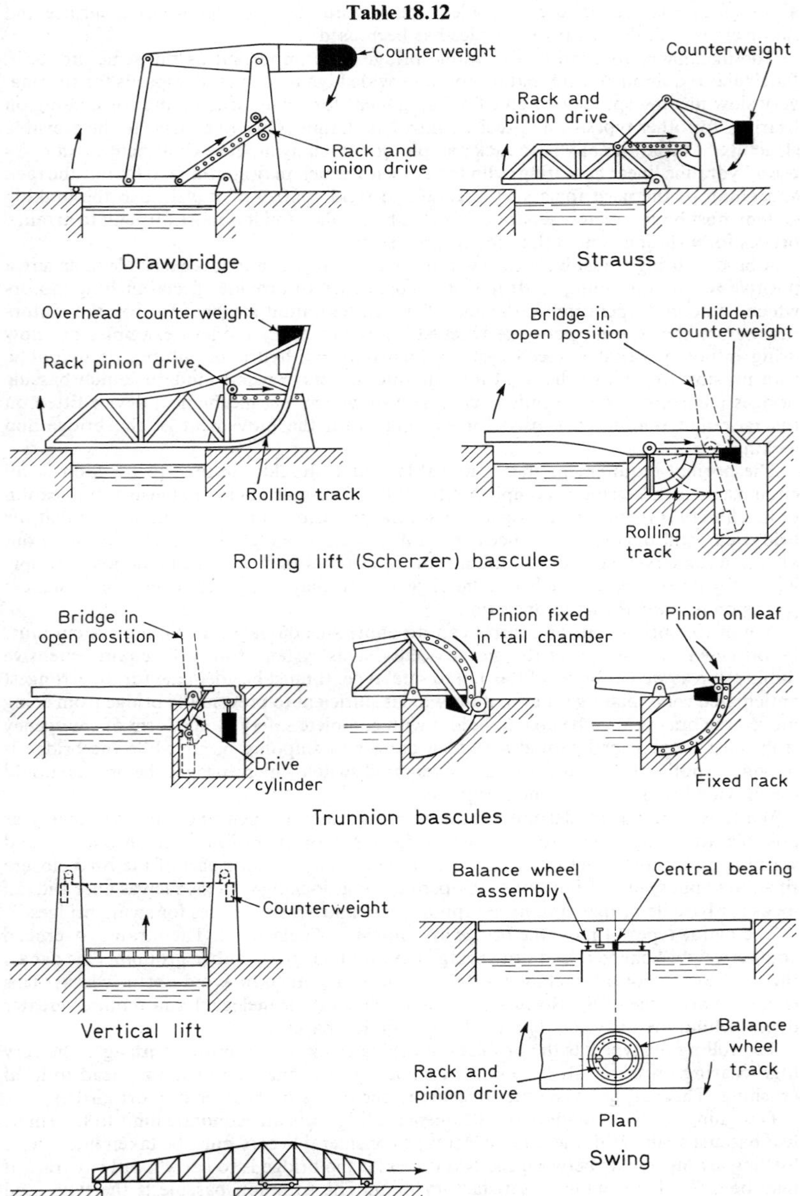
Counterweight
Rack and pinion drive
Drawbridge
Counterweight
Rack and pinion drive
Strauss
Overhead counterweight
Rack pinion drive
Rolling track
Bridge in open position
Hidden counterweight
Rolling track
Rolling lift (Scherzer) bascules
Bridge in open position
Drive cylinder
Pinion fixed in tail chamber
Pinion on leaf
Fixed rack
Trunnion bascules
Counterweight
Vertical lift
Balance wheel assembly
Central bearing
Balance wheel track
Rack and pinion drive
Plan
Swing
Retractable

although care is needed to ensure adequate adhesion between the wearing surface and the steel deck. A flexible p.v.c. surface has been used.

The machinery to operate a movable bridge is as important as the structure itself. Particular requirements are that it should provide high torque at low speeds for starting, very slow inching speeds—to give final alignment for swing bridges and for landing on bearings for other types—and precise control at all times. The final drive to the movable structure has traditionally been rack and pinion but many modern Continental examples use a hydraulic linear actuating cylinder. A factor which may continue to favour the rack and pinion is the need for a very precise and rigid control of travel; also the braking system must be completely separate from the hydraulics and it is more difficult to arrange brakes for a sliding system than for rotary pinions.

A bascule bridge must be landed without violent impact and it is desirable to ensure a positive reaction in doing so. It has been customary to provide special inching motors which come into operation at the end of travel, with limit switches to stop the motors just before the bearing pads are reached. However many modern examples are now using infinitely variable speed systems incorporating hydraulic motors or hydraulic transmission to remove the need for separate motors. Several modern Dutch bascule bridges incorporate the so-called 'snail' feature which is a mechanical device fitted on the rack and pinion drive used for slowing down the movement of the bridge and locking it.

The normal requirement that a movable bridge should continue to function in all circumstances necessitates complete reliability of its machinery. Extensive duplication of machinery is incorporated together with the provision of auxiliary motors so that the bridge can still operate, at reduced speed if necessary, with some of its machinery out of action for servicing or due to failure. In the event of failure of the main power supply a standby power source, such as a diesel generator, may be provided; in some instances provision is made for hand operation.

A movable bridge out of control can do enormous damage; a braking system must be provided to ensure that this never occurs. This system, too, will require extensive duplication as well as being of the 'fail to safe' type. It must be adequate for the strongest anticipated wind loading for even if the wind is sufficient to prevent the bridge from being moved the brakes must be able to hold it with complete safety. Some form of emergency stop must be provided should a sudden danger to shipping occur while the bridge is moving. Another safety measure is an overspeed switch which engages the brakes should the moving bridge speed up unacceptably.

While it is obviously desirable that the bridge should open and close as quickly as possible attention must also be given to the speed of operation of crash barriers and warning lights and hooters as these operations occupy a major part of the road closure time. Machinery must be provided to operate bridge locking systems such as nose and tail locks for bascule bridges and nose support jacks and centre wedges for swing bridges.

The major types of movable bridge are shown in Table 18.12. The Strauss, overhead counterweight Scherzer, and drawbridge have all their parts above ground. This avoids the need to provide a tail chamber, but care must be taken to avoid obtrusiveness. This is particularly true of the Strauss bascule which is rather inelegant and requires greater depth behind the quayside than the fixed trunnion bascule.

The rolling lift type has the problem of rolling tracks deteriorating with age. The very high bearing pressures at the points of contact of the curved rollers may lead to local crushing. This can be overcome by using wider tracks on heavier support girders.

Designing for wind loading on the opened bridge sets an economic limit to the single leaf bascule span. With the double leaf type considerable care must be taken in the nose locking arrangements between the two leaves; if the bridge is to carry a railway track it may be difficult to obtain a satisfactory joint. The trunnion bascule is the type used in many modern examples and may be driven in several ways. The rolling lift bascule gives a wider clearance with the bridge in the open position than a fixed trunnion bridge of the same span, although it therefore requires a greater depth behind the quay.

The leaf of a bascule bridge is normally designed to be sufficiently torsionally rigid to be able to be opened with the drive applied to only one of its main girders, when the other set of machinery is inoperative.

The vertical lift bridge sets a headroom limit and is expensive for narrow crossings. However it can be used for very long spans without the nose locking problems of the double leaf bascule. The lifting machinery may be either at the top of the lifting towers or in the piers; a mechanical or electrical linkage connects the separate motors to ensure synchronised parallel motion of the corners of the lifting section.

The swing bridge often provides the cheapest solution for a given span. It may be of the balanced cantilever type or have a shorter tail span ('bob-tail' type), and may turn on a rim bearing or a central pivot bearing. Its major disadvantage is the need to protect the bridge in the open position. Space must be provided at the quayside to lay the span in the open position; this may not be possible especially where there are adjacent locks requiring a clear quay for handling mooring ropes. The retractable bridge is not often used as it requires a suitable approach to accommodate the span in the open position and heavy rolling or sliding ways.

REFERENCES

1. MOCK, B., *The Architecture of Bridges,* Museum of Modern Art, New York (1949)
2. BECKETT, D., *Bridges,* Hamlyn, London
3. FAIRBAIRN, W., *Conway and Britannia Tubular Bridges* (1849)
4. LEE, D. J., 'The selection of box beam arrangements in bridge design', *Developments in Bridge Design and Construction,* Crosby Lockwood (1971)
5. BAXTER, J. W., GEE, A. F. AND JAMES, H. B., 'Gladesville Bridge', *Proc. Instn Civ. Engrs,* **34** (June 1966)
6. HOOK, D. M. A. AND RICHMOND, B., 'Precast box beams in cellular bridge decks', *Struct. Engr,* **48** (Mar. 1970)
7. CHAPLIN, E. C., *et al.,* 'The development of a precast concrete bridge beam of U-section'
8. LEE, D. L, 'Western Avenue Extension, the Design of Section 5', *Struct. Engr.,* **48** (Mar. 1970)
9. BEYER, E. and THUL, H., 'Hochstrassen', Baton-Verlag CMBH (1967)
10. HAWRANEK, A. and STEINHARDT, O., 'Theorie und Berechning der Stahlbrücken', Springer-Verlag, Berlin (1958)
11. RESINGER, F., 'Der dünnwandige Kastenfrager, Köln', Stahlban-Verlag (1959)
12. RICHMOND, B., 'Twisting of thin-walled box girders', *Proc. Instn civ. Engrs,* **33** (Apr. 1966)
13. RICHMOND, B., 'Trapezoidal boxes with continuous diaphragms', *Proc. Instn civ. Engrs* (Aug. 1969)
14. DALTON, D. C. and RICHMOND B., 'Twisting of thin-walled box girders of trapezoidal cross section', *Proc. Instn civ. Engrs* (Jan. 1968)
15. RICHMOND, B., 'The relationship of box beam theories to bridge design', *Conf. on Developments in Bridge Design and Construction,* Crosby Lockwood (1971)
16. HENDRY, A. W. and JAEGER, L. G., 'The analysis of grid frameworks and related structures', Chatto & Windus, London (1958)
17. *Design manual for orthotropic steel plate deck bridges,* AISC, New York (1963)
18. KUHN, P., *Stresses in aircraft and shell structures,* New York, McGraw-Hill (1956)
19. PUCHER, A., *Influence surfaces of elastic slabs,* Wein Springer-Verlag (1964)
20. KRUG, S and STEIN, P., *Influence surfaces of orthogonal on isotropic plates,* Berlin, Springer (1969)
21. WESTERGAARD, H. M., 'Computation of stresses in bridge slabs due to wheel loads', *Public Roads,* **II,** No. 1 (1930)
22. HENDERSON, W., 'British Highway Bridge Loading', *Proc. Inst civ. Engrs,* **3,** Part II (1954)
23. HOMBERG, H. and ROPERS, W., 'Kragplatten mit verändlicher Dicke', *Beton and Stahlbetonbau* (Mar. 1963)
24. RUSH, H. and HERGERÖDER, A., *Influence surfaces for moments in skew slabs,* Munich (C. & C.A. translation) (1961)
25. GARRET, R. J. and COCHRANE, R. A., 'The analysis of prestressed beams curved in plan with torsional restraint at the supports', *Struct. Engr,* **48** (Mar. 1970)
26. DABROWSKI, R., *Curved thin-walled girders, theory and analysis,* Springer-Verlag, Berlin (C. & C.A. translation) (1968)
27. *Stabilitätsfälle (Knickung, Kippung, Beulung),* DIN 4114 (1961)
28. *Specification for steel girder bridges,* BS 153, British Standards Institution, London (1966)

29. *Specification for the design fabrication and erection of structural steel for buildings,* AISC (1963)

30. KLÖPPEL, K. and MÖLLER, K. H., *Beulwerte ausgesteifter Rechleckplatten,* Wilhelm Ernst, Berlin (1966)

31. KERENSKY, O. A., *et al.,* 'The basis for design of beams and plate girders in the revised British Standard 153', *Proc. Instn civ. Engrs,* **5,** Part III (1956)

32. BASLER, K., 'Strength of plate girders in shear', *I.Struct. Div. Am. Soc. civ. Engrs,* **87,** ST7 (Oct. 1961)

33. BIGGS, J. M., *et al.,* 'Vibration of simple-span highway bridges', *Trans. ASCE,* **124** (1959)

34. LEONARD, D. R., 'Human tolerance levels for bridge vibrations', *Road Research Laboratory Report No. 34* (1966)

BIBLIOGRAPHY

AMMAN, O. H., *et al.,* 'George Washington Bridge', *Trans ASCE,* **97,** 1–442 (1933)

AMMAN, O. H., *et al.,* 'Verrazano Narrows Bridge', *Proc ASCE, J. Constr. Div.,* **92** (CO2), 1–192 (Mar. 1966)

ANDERSON, J. K., 'Runcorn–Widnes Bridge', *Proc. Instn civ. Engrs,* **29,** 535–70 (Nov. 1964)

ANDERSON, J. K., 'Tamar Bridge', *Proc. Instn civ. Engrs,* **31,** 337–60 (Aug. 1965)

ANDERSON, J. K. and BROWN, C. D., 'Design and Construction of the Kingsferry Lifting Bridge, Isle of Sheppey', *Proc. Instn civ. Engrs,* **28,** 449–70 (Aug. 1964)

ANDERSON, J. K., *et al.,* 'Forth Road Bridge', *Proc. Instn civ. Engrs,* **32,** 321–512 (Nov. 1965)

ANDREW, C. E., 'Unusual design problems—Second Tacoma Narrows Bridge', *Proc. ASCE,* **73** (10), 1483–97 (Dec. 1947)

BALDWIN, R. A. and WOOLLEY, C. W., 'Cumberland Basin Bridges, Scheme 2, Construction', *Proc. Instn civ. Engrs,* **33,** 289–312 (Feb. 1966)

BAUR, W., 'Die Durchstichbrücke Neckarsulm', *Beton u. Stahlbetonbau,* **64** (3), 57–61 (Mar. 1969)

BAXTER, J. W., BIRKETT, E. M. and GIFFORD, E. W. H., 'Narrows Bridge, Perth, Western Australia', *Proc. Instn civ. Engrs,* **20,** 39–84 (Sept. 1961)

BAXTER, J. W., LEE, D. J. and HUMPHRIES, E. F., 'Design of Western Avenue Extension (Westway)', *Proc. Instn civ. Engrs,* **50,** 177–218 (Feb. 1972)

BEYER, E., 'Die Kniebrücke in Düsseldorf', *Stahlbau,* **39** (6), 185–9 (June 1970)

BEYER, E. and ERNST, H. J., 'Brücke Jülicher Strasse in Düsseldorf', *Bauingenieur,* **39** (12), 469–77 (Dec.1964)

BEYER, E., GRASSL, H. and WINTERGERST, L., 'Nordbrücke Düsseldorf', *Stahlbau,* **27** (1958, 1–6 (Jan.); 57–62 (Mar.); 103–7 (Apr.); 147–54 (June); 184–8 (July)

BEYER, E. and THUL, H., *Hochstrassen,* Beton-Verlag GMBH (1967)

BIGGART, A., 'The Erection of the Forth Bridge', *The Engineer,* 438–9 (Nov. 25, 1887)

BILL, M. and MAILLART, R., *Bridges and Constructions,* 3rd edn, Architectural Publishers Artemis, Zurich (1969)

BINGHAM, T. G. and LEE, D. J., 'The Mancunian Way elevated road structure', *Proc. Instn civ. Engrs,* **42,** 459–92 (Apr. 1969)

BORELLY, W., 'Die Nordbrücke Mannheim–Ludwigshafen im Bau', *Stahlbau,* **39** (5), 156–7 (May 1970)

BOYNTON, R. M. and RIGGS, L. W., Tagus River Bridge, *Civ. Engng* (New York), **36** (2), 34–45 (Feb. 1966)

BREBBIA, C. A. and CONNOR, J. J., *Fundamentals of Finite Element Techniques,* Butterworth, London (1973)

BROWN, C. D., 'Design and Construction of George Street Bridge over River Usk, at Newport, Monmouthshire', *Proc. Instn civ. Engrs,* **32,** 31–52 (Sept. 1965)

BRUMMER, M. and HANSON, C. W., 'Development and Design of Walt Whitman Bridge', *Proc. ASCE, J.Struct Div.,* **82** (ST4) (July 1956)

CHETTOE, C. S. and HENDERSON, W., 'Masonry Arch Bridges: a Study', *Proc. Instn civ. Engrs,* **7,** 723–62 (Aug. 1957)

COOMBS, A. S. and HINCH, L. W., 'The Heads of the Valleys Road', *Proc. Instn civ. Engrs,* **44,** 89–118 (Sept. 1969)

COVRE, G. and STABILINI, P., 'Steel spans over three bays of the Italia viaduct on the Salerno–Reggio Calabria motorway', *Acier-Stahl-Steel,* **35** (7–8), 307–13 (July–Aug. 1970)

DANIEL, H., 'Die Bundesautobahnbrücke über den Rhein bei Leverkusen', *Stahlbau,* **34** (1965), 33–6 (Feb.); 83–6 (March); 115–9 (Apr.); 153–8 (May); 362–8 (Dec.)

DANIEL, H., 'Die Rheinbrücke Duisberg-Neuenkamp, Planung, Bau wettbewerb und seine Ergebnisse', *Stahlbau,* **40** (7), 193–200 (July 1971)

'Design of the 1675 ft Killvankull steel arch bridge', *Engng News Record,* 873–877 (13 Dec. 1928)

DIAMANT, R. M. E., 'Maracaibo Bridge', *Civ. Engng* (London), **58** (681), 482–4 (Apr. 1963)

ERDE, J. M., 'Lowestoft double-leaf trunnion bascule bridge', *Civ. Engng & P.W. Rev.,* **67** (790), 465–9 (May 1972)

FABER, L., 'Die Europabrücke, Uberbau', *Stahlbau,* **33** (7), 193–9 (July 1964), 'Der Unterbau der Europabrucke', *Bautechnik,* **41** (7), 217–27 (July 1964)

FAIRHURST, W. A. and BEVERIDGE, A., 'Superstructure of Tay Road Bridge', *Struct. Engr,* **43** (3), 75–82 (March 1965)

FAIRHURST, W. A., BEVERIDGE, A. and FARQUHAR, G. F., 'The design and construction of Kingston Bridge and elevated approach roads, Glasgow', *Struct. Engr,* **49** (1), 11–33 (Jan. 1971)

FALTUS, F. and ZEMAN, J., 'Die Bogenbrücke über die Moldau bie Zdakov', *Stahlbau,* **37** (11), 332–9 (Nov. 1968)

FARAGO, B. and CHAN, W. W., 'The Analysis of Steel Decks with Special Reference to Highway Bridges at Amara and Kut in Iraq', *Proc. Instn civ. Engrs,* **16,** 1–32 (May 1960)

FEIDLER, L. L., 'Erection of Lewiston–Queenston Bridge', *Civ. Engng, ASCE* (New York), **32** (11), 50–53 (Nov. 1962)

FEIGE, A., 'The evolution of German Cable-Stayed Bridges and overall survey', *Acier-Stahl-Steel,* **31** (12), 523–32 (Dec. 1966)

FEIGE, A. and IDELBERGER, K., 'Long Span Steel Highway Bridges Today and Tomorrow', *Acier-Stahl-Steel,* **36** (5), 210–22 (May 1971)

FINCH, R. M. and GOLDSTEIN, A., 'Clifton Bridge, Nottingham: Initial Design Studies and model test: Design & Construction', *Proc. Instn civ. Engrs,* **12,** 289–316, 317–52 (Mar. 1959)

FINSTERWALDER, U. and SCHAMBECK, H', 'Die Spannbetonbrücke über den Rhein bei Bendorf, *Beton u. Stahlbetonbau,* **60** (3), 55–62 (Mar. 1965)

FREEMAN, R., Sr., 'Sydney Harbour Bridge, design of structure and foundations', *Proc. Instn civ. Engrs,* **238,** 153–93 (1934)

FREUDENBERG, G., 'Die Doppel klappbrücke (Herrenbrücke) über die Trave in Lubeck', *Stahlbau,* **37** (10), 289–98 (Oct. 1968)

FREUDENBERG, G., 'Die Stahlhochstrasse über den neuen Hauptbahnhof in Ludwigshafen/Rhein', *Stahlbau,* **39** (9), 257–67 (Sept. 1970)

FREUDENBERG, G., 'The world's largest double-leaf bascule bridge over the Bay of Cadiz', *Acier-Stahl-Steel,* **36** (11), 463–72 (Nov. 1971)

FREUDENBERG, G. and ROTHA, O., 'Die Zoobrücke über den Rhein in Köln', *Stahlbau,* **35** (1966), 225–35 (Aug.); 269–77 (Sept.); 337–46 (Nov.)

FREYSSINET, E., MULLER, J. and SHAMA, R., 'Largest Concrete Spans of the Americas', *Civ. Engng* (New York), **23** (3), 41–55 (Mar. 1953)

GEE, A. F., 'Cable Stayed Concrete Bridges', *Developments in Bridge Design & Construction,* Eds K.C. Rockey, J. L. Bannister and H. R. Evans, Crosby Lockwood (1971)

GIFFORD, E. W. H., 'The Development of Long Span Prestressed Concrete Bridges', *Struct. Engr,* **40** (10), 325–35 (Oct. 1962)

GILL, R. J. and DOZZI, S., 'Concordia Orthotropic Bridge; Fabrication and Erection', *Engng J.,* **49** (5) 10–18 (May 1966)

GOWRING, G. I. B. and HARDIE, A., 'Severn Bridge: foundations and substructure', *Proc. Instn civ. Engrs,* **41,** 49–68 (Sept. 1968)

GUTFLEISCH, W. and KRÜGER, U., 'Der Stahlüberbau der Rheinbrücke Gemersheim', *Stahlbau,* **41** (2), 33–40 (Feb. 1972)

GUYON, Y., 'Long-span Prestressed Concrete Bridges constructed by the Freyssinet System', *Proc. Instn civ. Engrs,* **7,** 110–79 (May 1957)

HARDESTY, S., 'The Rainbow Bridge at Niagara Falls', *Civ. Engng, ASCE* (New York), **11** (9), 532–4 (Sept. 1941)

HARTWIG, H. J. and HAFKE, B., 'Die Bogenbrücke über den Askerofjord', *Stahlbau,* **30** (1961), 289–303 (Oct.); 365–77 (Dec.)

HAVEMANN, H. K., ASCHENBERG, H. and FREUDENBERG, G., 'Die Brücke über die Norderelbe im Zuge der Bundesautobahn Südliche Umgehung Hamburg, *Stahlbau,* **32** (1963), 193–8 (July); 240–8 (Aug.); 281–7 (Sept.); 310–7 (Oct.)

HECKEL, R., 'The Fourth Danube Bridge in Vienna—Damage and Repair', *Developments in Bridge Design and Construction,* Eds K. C. Rockey, J. L. Bannister and H. R. Evans, Crosby Lockwood (1971)

HEDIFINE, A. and MANDEL, H. M., 'Design and Construction of Newport Bridge', *Proc. ASCE, J.Struct. Div.,* **97** (ST 11), 2635–51 (Nov. 1971)

HESS, H., 'Die Severinsbrücke Köle', *Stahlbau,* **29** (8), 225–61 (Aug. 1960)

HILTON, N. and HARDENBERG, G., 'Port Mann Bridge, Vancouver, Canada', *Proc. Instn civ. Engrs,* **29,** 677–712 (Dec. 1964)

HOPPE, C., 'Die erste Strassenbrücke über den Panama-Kanal', *Der Bauingenieur* (Berlin) (5), 205–6 (1963)

HUET, M., 'Le Pont de Tancarville', *Soc. Ingénieurs Civ. France Mem.*, **113** (5), 23–42 (May 1960)

HYATT, K. E., 'Severn Bridge; fabrication and erection', *Proc. Instn civ. Engrs,* **41,** 69–104 (Sept. 1968)

'Inverted suspension span is simple and cheap', *Engng News Record,* 27–31 (11 May 1972)

KAROL, J., 'Calcasieu River Bridge', *Welding J.,* **42** (11), 867–70, 877–80 (Nov. 1963)

KERENSKY, O. A., HENDERSON, W. and BROWN, W. C., The Erskine Bridge', *Struct. Engnr,* **50** (4), 147–70 (Apr. 1972)

KERENSKY, O. A. and LITTLE, G., 'Medway Bridge: Design', *Proc. Instn civ. Engrs,* **29,** 19–52 (Sept. 1964)

KIER, M., HANSEN, F. and DUNSTER, J. A., 'Medway Bridge: Construction', *Proc. Instn civ. Engrs,* **29,** 53–100 (Sept. 1964)

KLINGENBERG, W., 'Neubau einer Hängebrücke über den Rhien bei Emmerich', *Bauingenieur,* **37** (7), 237–9 (July 1962)

KONDO, K., KOMATSU, S., INOUE, H. and MATSUKAWA, A., 'Design and Construction of Toyosato–Ohhashi Bridge', *Stahlbau,* **41** (6), 181–9 (June 1972)

LACEY, G. C., BREEN, J. E. and BURNS, N. H., 'State of the Art for Long Span Prestressed Concrete Bridges of Segmental Construction', *J. Prestressed Concrete Inst.,* **16** (5), 53–77 (Sept.–Oct. 1971)

LEE, D. J., 'Prestressed Concrete in Britain, Bridges 1966–70', *Concrete,* **4** (6), 227–48 (June 1970)

LEE, D. J., 'The design of bridges of precast segmental construction', *Concrete Society technical paper,* PCS 10.

LEE, D. J., 'Prestressed Concrete elevated roads in Britain', *Concrete Society technical paper,* PCS 12.

LEE, D. J., *The theory and practice of bearings and expansion joints for bridges,* Cement and Concrete Association (1971)

LEMIEUX, P., KALNAVARNS, E. and MORDNZ, N., 'Trois-Rivières Bridge, design and construction', *The Engineering J.* (Montreal), 16 pp. (Sept. 1968)

LEONHARDT, F., BAUR, W. and TRAH, W., 'Brücke über den Rio Caroni, Venezuela', *Beton u. Stahlbetonbau,* **61** (2). 25–38 (Feb. 1966)

LIPPERT, E., 'Die Bauausführung der Rheinbrücke Bendorf, Los 1', *Beton u. Stahlbetonbau,* **60** (4), 81–92 (Apr. 1965)

MODJESKI and MASTERS, *Greater New Orleans Bridge over Mississippi River,* Final Report to Mississippi River Bridge Authority, 2 vols. (1960)

MURPHY, F., 'Building the world's highest arch span', *Civ. Engng, ASCE* (New York), **29** (2), 86–89 (Feb. 1959)

NEW, D. H., LOWE, J. R. and READ, J., 'The Superstructure of the Tasman Bridge, Hobart', *Struct. Engnr,* **45** (2), 81–90 (Feb. 1967)

O'CONNOR, C., 'Design of Bridge Superstructures', Wiley, New York (1971)

PURCELL, C., 'San Francisco–Oakland Bay Bridge', *Civ. Engng, ASCE,* 183–7 (Apr. 1934)

RADOJKOVIĆ, 'The evolution of welded bridge construction in Jugoslavia', *Acier-Stahl-Steel,* **31** (12), 533–541 (Dec. 1966)

RAWLINSON, SIR JOSEPH and STOTT, P. F., 'The Hammersmith Flyover', *Proc. Instn civ. Engrs,* **23,** 565–600 (Dec. 1962)

'Record prestressed box girder span under way', *Engng News Record,* 14 (24 June 1971)

RICHMOND, B., 'Twisting of thin-walled Box Girders', *Proc. Instn civ. Engrs,* **33,** 659–75 (Apr. 1966)

ROBERTS, SIR GILBERT, 'Severn Bridge: design and contract arrangements, *Proc. Instn civ. Engrs,* **41**, 1–48 (Sept. 1968)

ROBERTS, G., and KERENSKY, O. A., 'Auckland Harbour Bridge—Design', *Proc. Instn civ. Engrs,* **18,** 459–78 (Apr. 1961)

'Rolling falsework arch cuts construction time', *Engng News Record,* 32–33 (26 March 1970)

'San Mateo–Haywood Bridge', *California Highways and Pub. Wks* (Sept., Oct. 1964)

SCHAFER, G., 'The New Highway Bridge over the Sava between Belgrade and Zemun (Yugoslavia)', *Acier-Stahl-Steel,* **22** (5), 213–8 (May 1957)

SCHÖTTGEN, J. and WINTERGERST, L., 'Die Strassenbrücke, über den Rhien bei Maxau', *Stahlbau,* **37** (1968); 1–9 (Jan.), 50–57 (Feb.)

SCHRÖTER, H. J., 'Hängebrücke über den Kleinen Belt in Dänemark', *Stahlbau,* **37** (4), 122–4 (Apr. 1968)

SCHRÖTER, H. J., 'Zwei neue stählerne Hochbrücken in Norddeutschland', *Stahlbau,* **39** (10), 314–6 (Oct. 1970)

SCOTT, P. A. and ROBERTS, G., 'The Volta Bridge', *Proc. Instn civ. Engrs,* **9,** 395–432 (Apr. 1958)

SHIELDS, E. J., 'Poplar Street Bridge—design and fabrication', *Civ. Engng* (New York), **36** (2), 52–5 (Feb. 1966)

SHIRLEY-SMITH, H., *The world's great bridges,* revised edn, Phoenix House, London (1964)

SHIRLEY-SMITH, H. and FREEMAN, R., JR., 'The design and erection of the Birchenough and Otto Beit Bridges, Rhodesia', *Proc. Instn civ. Engrs,* **24,** 171–208 (May 1945)

SMITH, H. S. and PAIN, J. F., 'Auckland Harbour Bridge—Construction', *Proc. Instn civ. Engrs,* **18,** 459–78 (Apr. 1961)

SORGENFREI, O. F., 'Greater New Orleans Bridge Completed', *Civ. Engng* (New York), **28** (6), 60 (June 1958), also **28** (2), 96 (Feb. 1958)

'Spectacular Venezuelan Concrete Arch Bridge', *Engng News Record,* 28–32 (11 Sept. 1952)

STEIN, P. and WILD, H., 'Das Bogentragwerk der Fehmarnsundbrücke', *Stahlbau,* **34** (6), 171–86 (June 1965)

STEINMAN, D. B., GRONQUIST, C. H., JOYCE, W. E. and LONDON, J., 'Mackinac Bridge', *Civ. Engng* (New York), **29** (1) 48–60 (Jan. 1959)

STELLMAN, W. L. O., 'Brücke über den Rio Paraná in Foz do Igaçú Brasilien', *Beton u. Stahlbetonbau,* **61** (6), 145–9 (June 1966)

STRAUSS, J. B., 'The Golden Gate Bridge, Golden Gate Bridge and Highway District' (Jan. 1938)

TALATI, J. B., HOLLOWAY, B. G. R. and CHAPMAN, R. G., 'A twin leaf bascule bridge for Calcutta', *Proc. Instn civ. Engrs,* **48,** 285–302 (Feb. 1971)

'The Batman Bridge, Tasmania', *Building with Steel,* British Constructional Steelwork Association, 69 (Nov. 1969)

'The Quebec Bridge', *The Engineer,* 138–140 (15 Feb. 1918)

THOMA, W. and PERRON, M., 'Pont de Revin-Orzy', *Construction,* **18** (9), 333–7 (Sept. 1963)

THUL, H., 'Brückenbau (Beiträge der Deutschem Gruppe der FIP, "Bedeutende Spannbetonbauten" zum V. Internationalen Spannbeton-Kongress)', *Beton u. Stahlbetonbau,* **61** (5), 97–115 (May 1966)

THUL, H., 'Entwicklungen im Deutschen Schragseilbrückenbau', *Stahlbau,* **41** (6), 161–71 (June 1972)

TIMOSHENKO, S. and WOINOWSKY-KRIEGER, S., *Theory of plates and shells,* 2nd edn, McGraw-Hill (1959)

VAN NESTE, A. J., 'Ten Years of Steel Bridges at Rotterdam', *Acier-Stahl-Steel,* **35** (1970); part 1, 343–8 (July–Aug.); part 2, 388–96 (Sept.)

VAVASOUR, P. and WILSON, J. S., 'Cumberland Basin bridges scheme Planning and Design', *Proc. Instn civ. Engrs,* **33,** 261–88 (Feb. 1966)

VIROLA, J., 'The proposed Ahashi Straits Bridge, Japan, compared with other great suspension bridges', *Acier-Stahl-Steel,* **32** (3), 113–6 (March 1967)

VIROLA, J., 'World's Greatest Suspension Bridges before 1970', *Acier-Stahl-Steel,* **33** (3), 121–8 (March 1968)

VIROLA, J., 'The World's Greatest Cantilever Bridges', *Acier-Stahl-Steel,* **34** (4), 164–70 (Apr. 1969)

VIROLA, J., 'The World's Greatest Steel Arch Bridges', *International Civ. Eng.,* **2** (5), 209–24 (Nov. 1971)

WALTHER, R., 'Spannbandbrücken' *Schweizerische Bauzeitung,* **87** (8), 133–7 (Feb. 1969); English transl., *International Civ. Eng.,* **2** (1), 1–7 (July 1971)

WARD, A. and BATESON, E., 'The New Howrah Bridge, Calcutta, Design of Structure, Foundation and Approaches', *Proc. Instn civ. Engrs,* **28,** 167–236 (May 1947)

WEITZ, F. R., 'Entwicklungstendenzen des Strahlbrückenbaus am Beispiel der Rheinbrücke Wiesbaden–Schierstein', *Stahlbau,* **35** (1966), 289–301 (Oct.), 357–365 (Dec.)

WEST, R. E., 'New Manchester Road bridge in the Port of London', *Proc. Instn civ. Engrs,* **48,** 161–94 (Feb. 1971)

WITTFOHT, H., 'Die Siegtalbrücke Eiserfeld im Zuge der Autobahn Dortmund–Giessen', *Beton u. Stahlbetonbau,* **65** (1), 1–10 (Jan. 1970)

WITTFOHT, H., 'Spannbeton-Kongress 1970 (Bericht), Arbeitssitzung V. Bemerk enswerte Bauwerke-Brücken', *Beton u. Stahlbetonbau,* **66** (2), 25–31 (Feb. 1971)

WITTFOHT, H., BILGER, W. and SCHMERBER, L., 'Neubau der Mainbrücke Bettingen', *Beton u. Stahlbetonbau,* **56,** 85–96, 114–22 (1961)

ZEMAN, J., 'A 1083 ft span steel arch bridge in Czechoslovakia', *Proc. Instn civ. Engrs,* **37,** 609–631 (Aug. 1967)

19 BUILDINGS

BUILDINGS 19

19 BUILDINGS

J. RODIN, B.Sc., F.I.C.E., F.I.Struct.E., M.Cons.E.
Building Design Partnership

The design of the total building including its internal and external environment has traditionally been the responsibility of the architect but this is now so complex a task that, except for the simplest of buildings, a multidisciplinary involvement is necessary whereby engineering, surveying and other specialist skills are integrated with those of the architect to achieve consistent quality throughout the project.

Internal form and environment will be determined by the functional requirements of the occupying organisation, the space needed to meet these functional requirements and the required comfort levels in regard to such items as noise, temperature, humidity and lighting. The external form and environment will be determined by the characteristics of the site and adjacent buildings. Influencing all aspects will be the constraints arising from time and cost, town planning and building regulations.

GENERAL MANAGEMENT

The procedures for handling large-scale building projects as opposed to civil engineering projects are complicated by the larger number of individual professional parties involved and by the large amount of legislation on permissions and approvals. The handling of such projects has been studied by the RIBA and is described in their *Handbook of Architectural Practice and Management*.

The overall procedures for the organisation of building projects are covered in Part 3.220: 'Plan of Work for Design Team Operation'. Table 19.1, taken from this section, shows the twelve discrete stages into which the project can be divided and briefly indicates the contents of each stage and the parties directly involved. Full details of the work required from each of the several professions and contractors at each stage are shown on separate diagrams. For example the detailed breakdown of Stage C, OUTLINE PROPOSALS, is shown in Table 19.2.

Table 19.1 OUTLINE PLAN OF WORK

Stage	*Purpose of work and decisions to be reached*	*Tasks to be done*	*People directly involved*	*Usual terminology*
A. INCEPTION	To prepare general outline of requirements and plan future action.	Set up client organisation for briefing. Consider requirements, appoint architect.	All client interests, architect.	BRIEFING

Stage	Purpose of work and decisions to be reached	Tasks to be done	People directly involved	Usual terminology
B. FEASIBILITY	To provide the client with an appraisal and recommendation in order that he may determine the form in which the project is to proceed, ensuring that it is feasible, functionally, technically and financially.	Carry out studies of user requirements, site conditions, planning, design, and cost, etc., as necessary to reach decisions.	Clients' representatives, architects, engineers, and QS according to nature of project.	
C. OUTLINE PROPOSALS	To determine general approach to layout, design and construction in order to obtain approval of client on outline proposals and accompanying report.	Develop the brief further. Carry out studies on user requirements, technical problems, planning, design and costs, as necessary to reach decisions.	All client interests, architects, engineers, QS and specialists as required.	SKETCH PLANS
D. SCHEME DESIGN	To complete the brief and decide on particular proposals, including planning arrangement appearance, constructional method, outline specification, and cost, and to obtain all approvals.	Final development of the brief, full design of the project by architect, preliminary design by engineers, preparation of cost plan and full explanatory report. Submission of proposals for all approvals.	All client interests, architects, engineers, QS and specialists and all statutory and other approving authorities.	
Brief should not be modified after this point.				
E. DETAIL DESIGN	To obtain final decision on every matter related to design, specification, construction and cost.	Full design of every part and component by collaboration of all concerned. Complete cost checking of designs.	Architects, QS engineers and specialists, contractor (if appointed).	WORKING DRAWINGS
Any further change in location, size, shape, or cost after this time will result in abortive work.				
F. PRODUCTION INFORMATION	To prepare production information and make final detailed decisions to carry out work.	Preparation of final production information, i.e. drawings, schedules and specifications.	Architects, engineers and specialists, contractor (if appointed).	
G. BILLS OF QUANTITIES	To prepare and complete all information and arrangements for obtaining tender.	Preparation of Bills of Quantities and tender documents.	Architects, QS, contractor (if appointed).	
H. TENDER ACTION	Action as recommended in paras. 7–14 inclusive of 'Selective Tendering'*	Action as recommended in paras. 7–14 inclusive of 'Selective Tendering'*	Architects, QS, engineers, contractor, client.	

Stage	Purpose of work and decisions to be reached	Tasks to be done	People directly involved	Usual terminology
J. PROJECT PLANNING	Action in accordance with paras. 5–10 inclusive of 'Project Management'*	Action in accordance with paras. 5–10 inclusive of 'Project Management'*	Contractor, sub-contractors.	SITE OPERATIONS
K. OPERATIONS ON SITE	Action in accordance with paras. 11–14 inclusive of 'Project Management'*	Action in accordance with paras. 11–14 inclusive of 'Project Management'*	Architects, engineers, contractors, sub-contractors, QS, client.	
L. COMPLETION	Action in accordance with paras. 15–18 inclusive of 'Project Management'*	Action in accordance with paras. 15–18 inclusive of 'Project Management'*	Architects, engineers, contractor, QS, client.	
M. FEEDBACK	To analyse the management, construction and performance of the project.	Analysis of job records. Inspections of completed building. Studies of building in use.	Architect engineers, QS, contractor, client.	

*Publication of National joint Consultative Council of Architects, Quantity Surveyors and Builders.

BRIEF

Buildings are either purpose built for a particular user or are speculative. In either case, the first step is to compile an agreed brief setting out the basic requirements of the project covering:

Purpose, function and scope including limitations of cost and time; proposed activities and organisation including numbers and types of people concerned, internal and external service requirements, particular systems such as document retrieval, special functional requirements such as security;
Design factors and required standards covering internal and external environment; spatial requirements, organisational relationships and required groupings affecting layout;
Internal and external traffic and required access for pedestrians, vehicles and materials;
Factors affecting type of construction, expansion, alteration, change of use, life.

Of primary importance is the building use and the associated schedule of basic accommodation including the number and nature of the intended occupants. By adding allowances for circulation, services, plant, toilet and ancillary accommodation, a close assessment of the gross floor area can be made and thereby the size of building determined. By considering the relationships between the different activities the optimum grouping of the spaces provided for them can be analysed in preparation for their translation into a physical plan to suit the particular site.

A user client may have special requirements: most buildings are expected to have a useful life of sixty to a hundred years, but in some cases, a more limited life span may be envisaged dictating a light form of construction which can be easily and cheaply demolished and replaced. Alternatively, a client may require a robust building shell of long life in which internal adaptation can be carried out to suit a later, and perhaps unknown, alternative use. Substantial mechanical and electrical service requirements, as occur in hospitals and some specialist laboratories and factories, may dominate the design leading, perhaps, to the incorporation of near-storey height service floors alternating with the functional floors.

THE SITE

Early site appraisal is vital. Suitability for the purpose intended requires consultation with various planning authorities to confirm zoning and land use definition. Access for vehicles, people and goods must be checked and the availability of public transport and future road or transport links determined. Increasingly, good access to major international air and rail termini or proximity to the national road network is a prerequisite of a site.

Subsoil deficiencies and underground service easements may present difficulties in development. Investigation of old mineral workings (brick clay, salt, sand and gravel extraction for example), coal mines, shafts and wells, should be undertaken; particularly where such work is known to have occurred in the vicinity. Evidence of filling should be investigated and dated. A subsoil survey should be recommended to the client (together with a cost estimate) early in the life of the project to identify the underlying conditions which may ultimately influence the building location, arrangement and cost.

The local climate requires early checking: high wind speeds will involve special stiffening; atmospheric pollution or salt laden coastal winds will require the selection of suitable materials and careful detailing of exposed building elements. Excessive external noise from major roads, railways or airports may necessitate soundproofing in the building or sound screening between the building and the noise source.

Confined city sites introduce problems such as: delivery and storage of building materials and components; the threat of restrictions or stoppages arising from local objection to construction noise; protection of adjoining property which may need underpinning and should be surveyed for dilapidations before work commences on site.

TOWN PLANNING

Almost all construction work is governed by the various Planning Acts. In areas of dense development, particular attention must be given to the three-dimensional aspects of a project to ensure compatibility with adjoining structures. Permitted densities, in terms of plot ratio, people or habitable rooms per acre, are usually allocated to a site as a result of its location. Car parking standards apply to most sites: inner city locations tend to have more restrictive allocations than suburban or rural locations. Height limitations must be determined and compliance with the daylight angles (published by HM Stationery Office) assured.

Certain special areas of interest or quality are now defined as Conservation Areas and particularly stringent requirements are placed upon the design of buildings proposed. Frequently, in such areas, demolition of existing buildings is prohibited or strictly limited and new work must marry with retained structures.

Early and full consultation with the Planning Officer ensures an understanding of the local planning requirements and such items as the possible need for Office Development Permit or Industrial Development Certificate or eligibility for special government grant. The programme of planning committee dates, the sequence of application required and the time scale of decisions should be determined. To establish the acceptability of a site for a particular purpose it is frequently advisable to submit an outline planning application which, if successful, can be followed by a detailed application.

Compliance with the Building Regulations, fire escape requirements and Public Health Acts will need to be checked with various other local government departments. Other points requiring their agreement include: highway access, traffic generation and sight lines; building and improvement lines—constraints on proposed developments;

Table 19.2 STAGE C: OUTLINE PROPOSA

(To determine general approach to layout, design and construction, in order to obt

Col. 1 Client Function	*Col. 2 Architect Management Function*	*Col. 3 Architect Design Function*	*Col. 4 Quantity Surveyor Function*
1. Contribute to meeting: note items on agenda in Col. 8.	1. Organise design team. Call meeting to discuss directive prepared in stage B, action 9 (Col. 2): establish responsibilities, prepare plan of work and timetable for stage C. See Col. 8 for items for agenda for meeting.	1. Contribute to meeting: note items on agenda in Col. 8.	1. Contribute to meeting: note items on agenda in Col. 8.
2. Provide all further information required by architect. Assist as required in all studies carried out by members of design team. Initiate and conclude according to timetable, any studies that are required within own organisations. Make decisions on all matters submitted for decision relevant to stage C.	2. Elicit all information relevant to stage C by questionnaire, discussion, visits, observations, user studies, etc: Initiate studies by consultants and client as required. Maintain and coordinate progress throughout this stage.	2. Carry out studies relevant to stage C, e.g.: (a) study published analyses of similar projects, visit if possible. (b) study circulation and space association problems. (c) try out detail planning solutions and study effect of planning and other controls.	2. Carry out studies relevant to stage C, e.g.: (a) obtain all significant details of client's require ments relevant to cost ar contract information on site problems, etc. (b) re-examine, supplement and confirm cost inform tion assembled in stage F
			3. Outline design implicatio of cost range or cost limit
		3. In consultation with team assimilate information obtained in action 2, and produce diagrammatic analyses, discuss problems.	
		4. Try out various general solutions; discuss with team; modify as necessary, and decide on one general approach. Prepare outline scheme, indicating, e.g. critical dimensions, main space locations and uses and pass to team.	4. Collaborate in preparatio of outline scheme. Prepar quick cost studies of alternative structural and services solutions, and advise on economic aspec of solutions.
		5. Assist QS in preparation of outline cost plan; discuss and decide on cost ranges for main elements, and method of presentation of estimate to client.	5. Confirm cost limit or give firm estimate based upon user requirements and outline designs and proposals. Prepare outline co plan in consultation with team, either from comparison of requirements with analytical costs of previous projects or from approximate quantities based on assumed specification.
	6. Compile dossiers provided by team members on final (or alternative) sketch designs, recording all assumptions, and issue to all members of the team.	6. Contribute to design dossiers, assemble all sketches and note relevant assumptions.	6. Record basis of estimate t contribute to design dossiers.
	7. Prepare report as coordinated version of all members' reports, including fully developed brief.	7. Contribute to preparation of report.	7. Contribute to preparation of report.
8. Receive architect's report; consider, discuss and decide outstanding issues. Give instructions for further action.	8. Present report to client; discuss and obtain decisions and further instructions.		

authoritative approval of the client on the outline proposals and accompanying report.)

Col. 5 *Engineer* *Civil and Structural Functions*	*Col. 6* *Engineer* *Mechanical and Electrical Functions*	*Col. 7* *Contractor (if appointed)* *Function*	*Col. 8* *Remarks*
1. Contribute to meeting: note items on agenda in Col. 8.	1. Contribute to meeting: note items on agenda in Col. 8.	1. Contribute to meeting: note items on agenda in Col. 8.	ITEMS FOR AGENDA FOR MEETING: 1. *State objectives and provide information:* *(a) brief as far as developed.* *(b) site plans and other site data.*
2. Carry out studies relevant to stage C, e.g.: (a) site surveys, soil investigation. (b) complete questionnaires on structural and civil requirements.	2. Carry out initial studies relevant to stage C, e.g.: (a) environmental conditions, user and services requirements, appraise M & E loadings on an area or cube basis. (b) consider possible types of installation and analyse capital and running costs, possible sizes and effects of major services installations, main services supply requirements.	2. Carry out studies relevant to stage C, e.g. visit site and investigate: (a) ground conditions, access and availability of services for construction. (b) local labour situation. (c) local subcontractors and suppliers to assess quality reliability, production potential and price level, etc.	*(c) re-state cost limits or cost range, based on client's brief.* *(d) timetable.* *(e) agree dimensional method.* 2. *Determine priorities.* 3. *Define roles and responsibilities of team members and methods communication and reporting.*
3. Advise architect on, e.g.: (a) types of structure. (b) methods of building. (c) types of foundation. (d) roads, drainage, water supply, etc.	3. Advise architect on design implications of studies made, e.g.: (a) factors which would influence efficiency, and cost of engineering elements, i.e. site utilisation, building aspect and grouping, optimum construction parameters, etc. (b) possible services solutions and ramifications of them. (c) regulations and views of statutory authorities.	3. Advise architect on findings and also on: (a) approximate times for construction of alternative methods. (b) effect of construction times on cost, etc.	4. *Define method of work, tender procedure and contract arrangements.* 5. *Agree drawing techniques.* 6. *Agree systems of cost and engineering checks on design.* 7. *Agree type of bill of quantities.* 8. *Agree check list of actions to be taken.* 9. *Agree programming and progressing techniques.*
4. Collaborate in preparation of outline scheme, prepare notes and sketches, consider alternatives, agree decision on general approach, and record details of alternative plans and assumptions.	4. Collaborate in preparation of outline scheme, check that services decisions remain valid; record details of alternative plans and assumptions.	4. Collaborate in preparation of outline scheme: continue to advise on time and cost implications of alternative designs or methods. Record details of proposals and assumptions.	
5. Provide QS with information for outline cost plan, with sketches on which to base estimate, and agree QS proposals.	5. Provide QS with cost range information for outline cost plan, and agree QS proposals: interpret agreed standards by illustration.	5. Provide QS with information affecting price levels, for outline cost plan and agree QS proposals.	
6. Compile dossier of essential data collected in actions 2 to 5 above.	6. Compile dossier of essential data collected in actions 2 to 5 above.	6. Compile dossier of basic cost information agreed with QS and architect.	The report includes: (a) the brief as far as it has been developed;
7. Contribute to preparation of report.	7. Contribute to preparation of report.	7. Contribute to preparation of report.	(b) an explanation of the major design decisions; and (c) firm estimate with outline cost plan.

street closures—construction and long term benefits, if practical; special licences—for petrol storage, theatres or cinemas, music or dancing, public houses, etc; 'bad neighbour' developments requiring general publication and site notices—e.g. refuse or sewage disposal, slaughterhouse, places of entertainment causing crowd and noise; connection to public sewers—separate, partially separate or combined system for foul and surface drainage, trade effluents; discharge into river or watercourse—specific consent required from local river authority.

PUBLIC UTILITY

Once an outline brief exists and a site is under consideration the various public utility organisations (GPO, gas, electricity, water, sewer authorities) should be consulted to determine the availability of their various services.

FEASIBILITY

The compatibility of brief and site with the external constraints in their varying forms logically leads to the preparation of a feasibility study. This is normally the first design exercise and provides the design team with an opportunity to explore the problem, to propose solutions, to cost the alternatives and to identify options for the client. Presentation of a preferred option with objective data supporting the preference completes the first stage and forms the basis for the final design.

COST

Cost is an important factor at all stages of the design process. Alternative design solutions or materials must be carefully considered to ensure that cost is within budget, that money is allocated in a balanced way to best suit the client's needs and that, throughout the project, good value is obtained for the money spent The most significant decisions affecting cost occur in the concept and outline planning stage.

Of first importance is the economic use of space in the proposed building. Although the basic range of accommodation is fixed, considerable additional space is required for circulation and access, stores, plant rooms and toilet facilities. This additional space, sometimes called 'balance area', can vary considerably according to the layout adopted and should be kept to the minimum by efficient planning of staircases and service ducts, grouping of toilet facilities and a restriction on the area of circulation routes. The economic plan form will also aim at reducing the ratio of external wall area to total floor area thus saving expensive wall materials and reducing heat losses (or gains) and hence minimising the installation and running costs of the heating, ventilation or air conditioning systems. The reduction of storey heights to a minimum will have similar cost benefits.

It is usual to prepare a cost plan for the project in elemental form. Initially it is a cost estimate based on the preferred scheme and structural system together with a specification covering the main building elements. In the long term it forms a cost structure for monitoring the cost effect of changes and the detailed development of the design. The cost plan should state whether it includes for price inflation to tender stage or building completion, or is based upon rates current at date of estimate.

Major elements should be kept in reasonable balance: for example, the use of an expensive cladding material could leave too little money for the remainder of the work resulting in a visually pleasing but operationally unsuccessful building. The cost plan is an excellent means of checking the balance between the different elements of structure, finishes and services though the relative percentages of the overall cost will vary from case to case according to the type of building and its user requirements.

Whilst the capital construction cost of a building is of primary importance other costs will also be significant and could affect design. The annual running cost is one such item and services installations particularly, should be considered in terms of operational as well as initial cost. Similarly, the use of an expensive but hardwearing material may be justified in terms of subsequently reduced outlay on cleaning or maintenance. Discounting techniques and, possibly, tax considerations are necessary to make true cost assessments of such comparisons.

The total cost of a building project will also include expenditure on land, borrowed capital and the fitting out of the completed building; compensation to adjoining owners and other associated costs as well as legal and design consultant's fees and expenses. In some cases, the earlier a development can be occupied the better the cost advantage to the client. The construction method and programme are then significant and may affect the design form. It is often possible to assess the financial advantage of early completion and by comparative financial analysis to justify additional construction cost to shorten the construction period.

AN EXAMPLE

Halifax Building Society: new head office

THE DESIGN, BROAD CONCEPT

There were three main recommendations contained in the Brief. The first was that the general office requirements of the Society should be planned using open office or burolandschaft principles, whilst the Executive and Directors' accommodation should be contained within individual offices with permanent partitions. The intent was to provide the greatest degree of flexibility which would allow for growth and change in the general offices in balance with optimum conditions of privacy and quiet for the people whose work would benefit from such conditions. The second recommendation was that an automatic retrieval system should be used for the storage of deeds and correspondence files. The third recommendation was that the new head office should be completely air conditioned.

These three aspects have strongly influenced the final form of the design which places the whole of the general office accommodation, approximately 5000 m² gross in area, on one floor at third-floor level, thus taking up the total area of the site. This important decision was taken for the following reasons:

> *The organisation of work processes was considered to be best served by a horizontal relationship of working groups or departments. The scale of space provided is compatible with the principles of burolandschaft agreed by the premises committee. Any other solution would entail the breakdown of general office accommodation on to separate floors, and probably the splitting of larger departments on to more than one floor.*
>
> *The third floor is the lowest level on which a complete floor can be made available for the general office. This will become evident as the reasons for placing accommodation on lower floors are explained.*
>
> *At third-floor level it is possible to obtain panoramic views over Halifax as all the neighbouring buildings, except for the Computer Building, are of less height.*

The storage area for the retrieval system is situated below, where it is possible to obtain maximum security. The enclosing sub-basement structure is designed to withstand fire, explosion, or collapse of the superstructure. Security from flooding is also achieved. Furthermore, placing this large volume below ground avoids obstruction in planning the occupied areas and lightens the apparent bulk of this large building.

The recreational facilities for the building were required to be planned in association with the staff restaurants and to be readily accessible after normal working hours. The design

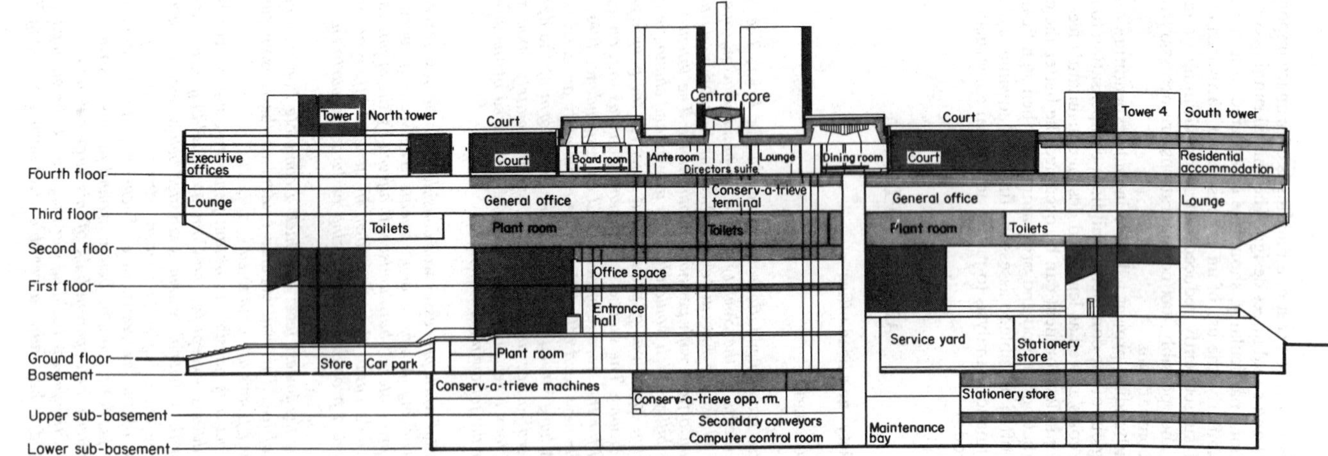

Figure 19.1 The broad concept of the design

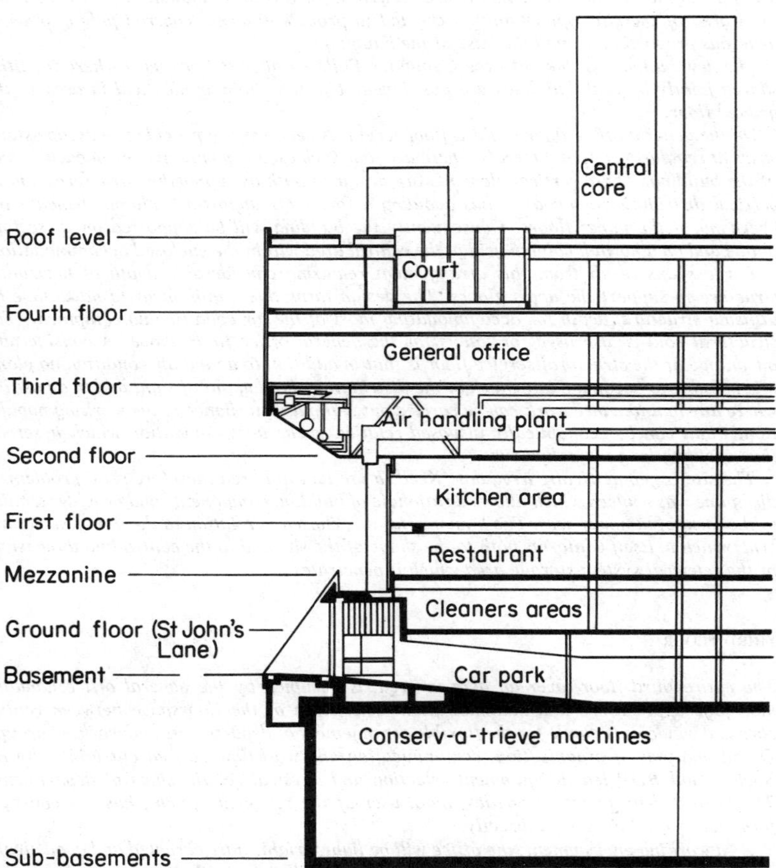

Figure 19.2 Part section of building showing arrangement of accommodation

Sub-basement. Occupied by conserv-a-trieve machines and their ancillary systems in two separate sections for deeds and files. The system can handle up to 4700 references an hour.

Basement. Car park with 60 parking spaces. Kitchen, post and stationery store.

Ground Floor and Mezzanine. Main entrance, reception and centralised cloakroom with recreation suite above. Includes staff restaurants divided into self-service and waitress service. The other end of the site features a large pool.

First Floor. Kitchen area and a bridge link to separate computer building.

Second Floor. All air-handling plant is on this floor together with a major part of the distribution ducting within the structural void.

Third Floor. General office which handles mortgage applications and administration.

Fourth Floor. Directors' offices, suit and restaurant. Accommodation is linked to the twin lift cores by corridors which define landscaped courts. Flats for resident engineer and caretaker are situated around the tower.

groups all the social accommodation for the building around the main entrance hall at ground and mezzanine levels. As the area required for this accommodation is less than the total area of the site, opportunity is created to provide external covered public areas of generous proportion around the base of the building.

The new building is linked to the Computer Building at first-floor level where facilities shared jointly by both buildings are positioned. The first floor is identical in area to the ground floor.

On the general office floor at third-floor level it is necessary to provide fire escape staircases at regularly spaced intervals; there are four such escape positions, one in each corner of the building. These vertical elements are designed both as supporting structures and as vertical distribution ductways, thus avoiding a forest of supporting columns beneath the overhang of the upper floors. Consequently the building will be supported on these four points and on a normal column grid in the central area within the enclosed accommodation.

Large spans result from this arrangement requiring considerable depth of horizontal structure to support the upper floors. The design turns this requirement to advantage by utilising structural depth for accommodating most of the air conditioning equipment. The structural void is also used for toilets for the general office floor above. A considerable advantage of the structural/service floor is that in addition to actual air conditioning plant, it is possible to accommodate much of the distribution ducting out of sight and in a position where any maintenance work can be carried out without disturbance to the working population. Plant can be economically disposed relative to the accommodation which it serves, thus avoiding long main duct runs.

The site is geometrically irregular. Need to use its total area therefore set a problem; a discipline was evolved which allows a rationale of building components and provides a guide to the form of minor spaces. The 'minor grid' was chosen in relation to the major structural grid, which in itself conforms both to the shape of the site and to the controlling dimensions of the retrieval system storage area which it penetrates.

THIRD FLOOR

The entire third floor, over an acre in area, is occupied by the general office handling mortgage applications and administration. The depth of the floorspace between central core and windows is ample for the flexible arrangement of departments and working groups. Detail analyses of organisation, communication and workflow carried out jointly by the Society and BDP led to equipment selection and physical relationships of departments. The largest department, securities, main user of the retrieval system, has the conveyor terminal located to serve it directly.

As a working environment, the office will be light, bright, fully carpeted and acoustically controlled. The degree of screening or openness will vary with the needs of different departments and individuals.

At each corner of the floor, behind the service-stair towers, is a refreshment lounge, giving broad views out for those working deep in the room.

Acknowledgements. John Laing Construction Limited, main contractors; Roneo Vickers Limited, automatic filing installation; Building Design Partnership, architects, quantity surveyors, civil, structural, mechanical and electrical engineers.

INTERNAL ENVIRONMENT

Thermal environment

The required comfort conditions and tolerances are determined by the intended function of the space concerned.

Thermal comfort depends on a complex of inter-related factors: air temperature,

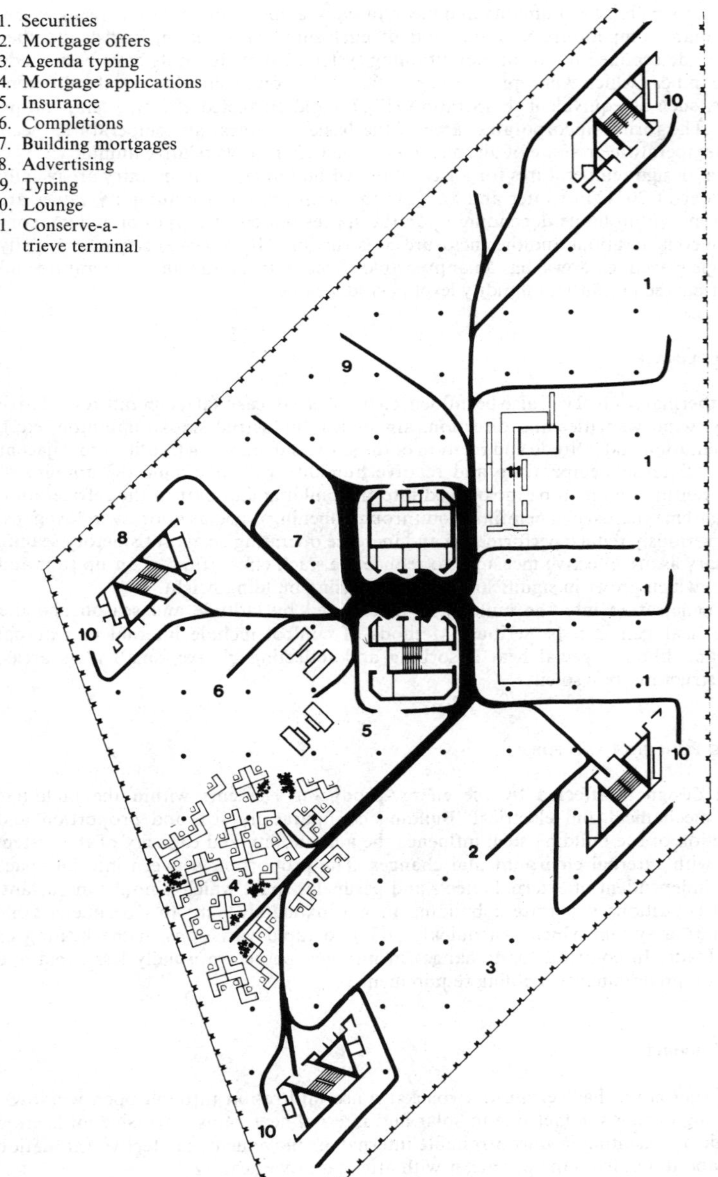

Figure 19.3 Third floor

ventilation rates, relative humidity and mean radiant temperature of the enclosing space. Mean radiant temperature is a function of enclosure construction, whilst all other factors are determined by the air conditioning system. Many attempts have been made to devise indices which will represent in one figure the composite effect of the different variables, such as equivalent temperature (Eq.T.) and corrected effective temperature (C.E.T.). The former incorporates three of the basic variables: air temperature, mean radiant temperature and rate of air movement; the C.E.T. adds relative humidity.

Internal design temperatures for air-conditioned buildings in this country are usually of the order of 20 °C in winter and 22 °C in the summer; relative humidity values are usually kept within limits depending upon the spaces served, the types of system, condensation considerations and the enclosure construction. Glass area and type, especially large single glazed windows, has an appreciable effect on mean radiant temperature and also restricts the permitted humidity level in cold weather.

SITE AND CLIMATE

Internal thermal control will also be influenced by external seasonal temperatures, relative humidity, wind velocities and direction, air quality (industrial smoke pollution, etc.), solar orientation and latitude and relation of the site to surrounding locality and adjacent buildings. External temperature and relative humidity will determine the amount of external ventilation air to be introduced into the building and part of the refrigeration load requirements. Excessive infiltration through openings such as doors, window gaps, etc., can seriously reduce performance and increase operating costs; satisfactory sealing is necessary as are effective measures to reduce the stack effect (flow of air up stair and lift areas) which grows in significance with increasing building height.

Solar penetration into the building is determined by latitude and season and the resulting heat gain can be serious. Methods of control include internal or external louvres and blinds, special heat absorbing and reflecting glasses, small glass areas, shading structure, brie soleil.

BUILDING FUNCTION AND FORM

Thermal design is affected by the energy-producing elements within the building: human, mechanical and electrical. Building configuration, size and proportion and construction of the building shell influence the adaptability and capacity of the system to cope with external environmental changes. The proportion between interior space which is independent of external effects and perimeter space which is not, is important. External conditions penetrate a building to approximately 6 m: this perimeter zone will require a system which can quickly adapt to rapid variations in the heating or cooling loads. In contrast, load changes in interior spaces are usually less rapid and represent a predominantly cooling requirement.

Air conditioning

Natural ventilation has certain drawbacks: noise infiltration through open windows; overheating during summer due to solar and internal heat gains; excessive infiltration of outside air resulting in uncontrollable internal air movement; ineffective ventilation beyond about 5 m from the perimeter with attendant overheating.

Mechanical ventilation solves only a few of these problems. Noise and outside air infiltration are reduced as windows are opened less frequently. During warmer weather, however, overheating and high humidity can occur due to the inherent inability of the system to supply air at the correct thermal condition. This inability is overcome by the inclusion of refrigeration, thereby changing the system from mechanical ventilation to air conditioning.

Air conditioning provides a controlled internal thermal environment which is independent of the external conditions or of any changes in the internal load conditions. Planning and configuration of the building will be influenced by the provision of air conditioning. Deep space can be created with the knowledge that a satisfactory internal thermal environment will be achieved. Similarly, non-opening windows avoid infiltration problems which are accentuated with increased building height.

Moisture control and filtration of the incoming air are integral parts of full air conditioning giving a cleaner, healthier and more comfortable atmosphere compared with ventilation by natural methods. Redecorating costs and absenteeism may be reduced and working efficiency increased.

AIR CONDITIONING SYSTEMS

Many types of air conditioning systems are available and can be classified into three basic groups: 'centralised', 'decentralised' and 'selfcontained' systems; some solutions are combinations from these three.

Centralised systems. (a) Systems where air is processed at a central plant and distributed for use without further treatment:

(i) single duct all-air systems using high, medium or low velocity distribution;
(ii) double duct all-air systems using high, medium or low velocity distribution with local terminal mixing units.

(b) Systems where air is processed at a central plant, but with final heat addition or subtraction at the point of use:

(i) single duct all-air reheat/recool systems, using high, medium or low velocity air distribution with associated heating and/or cooling water distribution;
(ii) perimeter induction air/water systems using high, medium or low velocity primary air distribution with secondary heating and/or cooling water distribution on a two, three or four pipe principle.

Decentralised systems. Systems where a liquid medium is distributed from a central point to units which condition air locally: some such systems also have a supplementary primary air supply from a central plant to the unit or space:

(1) room fan coil unit air/water system with two, three or four pipe water distribution and local outside air connections;
(2) as (1) but with supplementary primary air from central plant;
(3) localised zone air handling unit all-air systems with associated heating/cooling water distributions and with low velocity air distribution to conditioned spaces from the units;
(4) radiant ceiling systems supplied with heating/cooling water distribution and supplemented with separate single duct all-air system.

Selfcontained systems. Systems where selfcontained air conditioners process and supply air at the point of use.

Each system has merits and limitations. The simpler low velocity all-air single duct systems require a large amount of duct space and are not a practical solution where a large number of zones of varying use are to be served. In these cases medium or high velocity distribution would be adopted. Double duct all-air systems mix air from separate hot and cold distribution ducts using ceiling or cill mounted terminal mixing boxes. This system is very adaptable, but the combination of two supply ducts plus a return air duct requires considerable service space, even when using high and medium velocity distribution.

Induction systems are probably the optimum presently available for large buildings incorporating extensive perimeter accommodation. The induction unit discharges primary air supplied from the central plant through high-pressure nozzles and this induces air from the space into the unit which then mixes with the primary air before discharging back to the space; temperature control is achieved by a heating/cooling coil. The basic difference between two-, three- and four-pipe associated water distribution systems is that the latter two can provide, at the point of use, the simultaneous facility for either heating or cooling, while the two-pipe system is restricted at any one time to one or the other.

Fan coil systems incorporate a heating/cooling coil and a circulating fan. Primary air can be ducted direct to the units from a central system or discharged to the space independently, or, alternatively, each unit can draw in air direct from outside. Self-contained packaged air conditioning units are usually restricted to smaller specialised projects.

Radiant heating/cooling ceilings, when used with a supplementary air system, can provide an effective environment although their adaptability to meet rapid fluctuations in heating and cooling loads is limited.

AIR CONDITIONING—DISTRIBUTION AND INTEGRATION

Considerable duct distribution space is required and air outlets and extracts are often incorporated in the detailing of light fittings and suspended ceilings. From the earliest stages, therefore, the air conditioning system should be integrated into the total planning and detail design process of both the building elements and the structure.

Perimeter units can be served from a network of air ducts or water pipes concentrated in zones near the outer wall: within the under-cill or ceiling void for horizontal piping or ducts and within structural column enclosures for vertical distribution. Alternatively, the perimeter area may be served from the central core with ducts and pipes accommodated above a false ceiling, within a structural hollow floor or beneath a raised floor.

In areas where little flexibility for changing use is required, a totally integrated solution using the structure to accommodate air and water distribution may produce some economies including reduced storey height. Where a high degree of flexibility is required as, for example, in open-plan buildings, ceiling distribution on a modular basis for interior zones and cill or ceiling distribution for the perimeter becomes essential and a false ceiling is required, the ceiling space being used to accommodate the ducts and pipes.

Air may be exhausted through light fittings or ceiling grills; the former arrangement cools and hence increases the efficiency of the light source: it also removes excess heat (arising from high light levels) which can be transferred for use elsewhere, e.g. the perimeter area, but is more commonly vented to the exterior. Air is injected via diffusers or slots incorporated in the light fittings or suspended ceiling. Careful design of the outlets coupled with adequate ceiling height, 4 m to 5 m if possible, is necessary to prevent draughts.

There are three basic air supply and exhaust systems, two of which use the ceiling space as one large duct or plenum:

(1) Negative plenum—air is extracted into the plenum through outlets in the false ceiling which are usually part of the light fittings. Air supply is ducted to diffusers or slots incorporated in the ceiling design.
(2) Positive plenum—the plenum is used as the supply duct, air being forced through ports in the false ceiling. Extracted air is ducted from terminals usually incorporated in the light fittings.
(3) Fully ducted—ducts to both supply and extract terminals.

The completely ducted system has fewer thermal problems, but occupies more space and is more expensive. The plenum systems substantially reduce duct requirements, but are less efficient: they also require careful control of temperature to prevent condensation and, sometimes, the incorporation of insulation on the underside of the structural floor to confine the plenum effects to the storey intended.

LOCATION OF AIR CONDITIONING PLANT

Air conditioned buildings require plant room space up to 8 to 10% of the gross area of the building and there are many judgements which affect its location: for example, the air intake must be separated from the air discharge. The top of the building is often the preferred location particularly with all air systems involving recirculated air. There is a reasonable economic limit to the distance served by an individual plant room of about 10 storeys, but in a tall building, intermediate plant rooms would be provided at approximately 20-storey intervals since the plant room can serve both up and down. Basement plant rooms are also common but if basement construction is expensive, because of water or rock, essential requirements such as car parking or deep storage would take precedence over the plant space. Alternatively, height limitations on the superstructure could, on a prime site, make the provision of extra basement space for plant worth while in spite of high cost.

Heating/cooling generation

Arrangements for the heating and cooling generating plant will depend on a number of general and localised factors: availability, suitability, and economic costs associated with the utilisation of fuel and power; resources peculiar to the site; and utilisation of recoverable energy associated with the heating and cooling systems installed within the building.

Fuel and power considerations are complex and include a detailed appraisal of operating and capital costs for various fuel alternatives (coal, gas and oil) and power. Boiler plants incorporating combined dual firing burners suitable for gas (town or natural) and oil can offer attractive capital and operating cost characteristics combined with greater flexibility.

Heat recovery systems have been gaining popularity. A common arrangement is to utilise low grade heat being rejected from refrigeration machines. Another is to transfer heat extracted from the interior of deeply planned areas, which have to be cooled, to spaces requiring a heating load, such as perimeter zones, during winter and certain mid-season periods.

On larger specialised projects, total energy is finding an application. This is based on the concept that the total energy requirements, in all its forms can be provided from a single fuel source. These systems incorporate electrical generation with heat being produced as a by-product. Refrigeration, which can be met by either electricity or heat, is usually a complementary part of such an integrated energy system.

ENVIRONMENTAL ENGINEERING (for example on pp. **19**–10–**19**–11)

Planning and services/structural considerations indicate the need for two main services plant rooms. The optimum positions for these are one at an intermediate floor level immediately below the main office floor and the other at basement level. A further plant room is also necessary at roof level. The intermediate level plant space is chosen because of its close proximity to the majority of the spaces to be provided with air treatment and will accommodate the air-conditioning and ventilation plants. Main heating and cooling plant is located at basement level facilitating the heating interconnection with the Computer Building and also assisting the control of extensive noise and vibration associated with this type of equipment. The plant room at roof level houses cooling towers, lift motor rooms, storage tanks and exhaust ventilation fans.

The supporting towers accommodate stairways and certain secondary services requiring a basement/upper level relationship, such as drainage, pipework services and exhaust ventilation.

Utilisation of the structural/services void at second-floor level had enabled secondary vertical services distribution to be planned at the tower positions and enables vertical distribution to be diversified over the main office floors with the result that horizontal distribution has been localised. Similarly the primary distribution system for serving perimeter spaces of office floors is located within the void, small riser positions being located along the outer walls.

Diversification of the vertical distribution and routing of the perimeter primary distribution have reduced the ceiling void depths at both main office floors resulting in lower floor-to-floor heights. In addition the positioning of a major portion of the distribution in the structural/services void improves maintenance aspects of the services installation. Disturbances for maintenance at office floor levels is minimised and access to services improved. Distribution units, which are normally placed below the window glazing, are located in the structural/services void.

Oil was rejected as fuel owing to the fire risk to the deeds store. Halifax was one of the first industrial areas to be converted to natural gas and tariff rates indicate its suitability. Addition to the present Computer Building heating load will marginally improve the tariff rate as will the use of natural gas as the energy source for the refrigeration cooling plant.

The air-conditioning system is designed to take advantage of the principles of heat recovery. Excess heat resulting from a high level of illumination is to be extracted at source and transferred to areas or systems requiring a heating load. Low-grade heat rejected from refrigeration equipment and exhaust ventilation will be utilised in a similar manner.

Thermal insulation

The object of thermal insulation together with heating is to obtain, irrespective of the prevailing weather conditions, a near constant internal temperature determined by requirements of human comfort and satisfactory conditions for manufacturing processes or storage of goods. Adequate insulation is needed to avoid excessive expenditure on heating plant and fuel. The insulation and heating of buildings for human occupation are normally designed to maintain a temperature of 16 to 21 °C according to use when the outside temperature is 1 °C. Generally good thermal insulating materials are those which entrap air, such as lightweight concrete, wood wool slabs, glass or mineral fibre wool.

Calculation of the thermal transmittance or U value of a wall, floor or roof is carried out by adding together the thermal resistances of the materials and surface coefficients and taking the reciprocal of the answer to provide the thermal transmittance of the composite construction.

$$U = 1/R$$

$$R = R_s1 + R_s2 + \frac{(L1)}{(K1)} + \frac{(L2)}{(K2)} \text{ etc.} + R_a + R_h$$

where R is the total thermal resistance of the structure, R_s1 is the surface resistance of the inner face, R_s2 the surface re tan resistance of the outer face, R_a the resistance of the air cavity if present, R_h the resistance of unit material such as hollow block where resistance per unit thickness does not apply, and, L/K is the resistance of one layer of material or thermal conductivity K and thickness L.

To evaluate the total heat loss from a room or building the thermal transmittance of the walls, windows, floor and ceiling must be calculated and allowance made for the losses involved in heating up the ventilating air and the structure when heating is intermittent.

Structural members penetrating the full thickness of a wall produce 'cold bridges', locally reducing the thermal resistance and internal surface temperatures, with consequent added risk of condensation. In such cases the cold bridge should be reduced in width or eliminated by appropriate insulation.

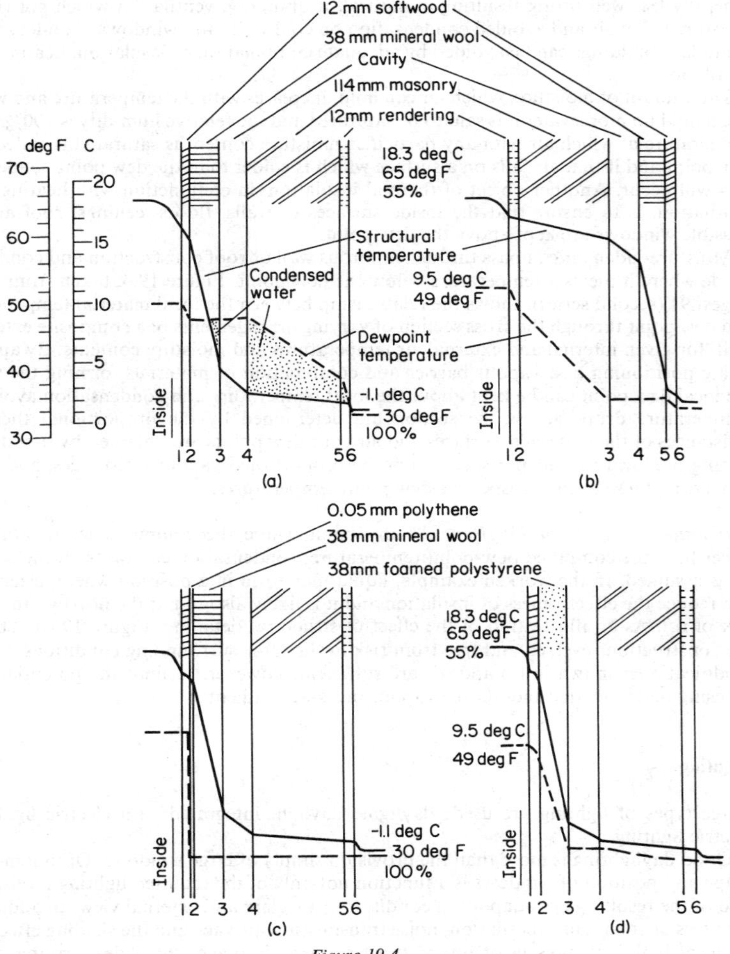

Figure 19.4

Urea formaldehyde foam is sometimes used to fill the cavity in cavity wall construction resulting in an almost 80% reduction in heat loss through the walls. Double glazing, as well as reducing heat loss, has advantages in increasing the temperature on the inside window surface and may improve internal comfort conditions.

Condensation problems have increased due to new methods of building, standards of heating and control of ventilation, and changing family habits which have led to intermittent heating coupled with the generation of more moisture inside the dwelling. Old buildings, particularly domestic ones, usually had open fires and flues and windows were generally less well fitting resulting in natural, if draughty, ventilation which got rid of moisture laden air and avoided condensation on cold walls and windows. Condensation in modern buildings can be avoided by adequate combination of insulation, heating and ventilation.

The amount of moisture, which air can hold, increases with the temperature and when it can hold no more water it is said to be saturated and the relative humidity is 100%. The temperature at which air with any particular moisture content is saturated is called the dew point and if that air falls on a surface which is colder than the dew point condensation will occur. Another object of thermal insulation, in conjunction with heating and ventilation, is to ensure that the inside surfaces of walls, floors, ceilings, roof and if possible windows are kept above the dew point.

Moisture-laden air can pass through a porous wall or roof construction and condense inside where it meets a temperature below the dew point. Figure 19.4, taken from BRS Digest 91 (second series), shows the relationship between the local material temperature and dew point through the cross section of varying arrangements of a composite external wall, for given internal and external air temperatures and moisture contents. By appropriate positioning of a vapour barrier and combination of materials forming the wall, the local dew point can be kept above the local temperature and condensation avoided. Temperature drops across the section are determined by the proportional thermal resistances of the materials, surfaces and air gap; dew points are obtained by first determining the local vapour pressures, from the proportional vapour resistances, and then converting these to their respective dew point temperatures.

Estimation of condensation risk. At any point where the computed temperature is lower than the computed dew-point temperature, condensation can occur in the conditions assumed. In the worked example, liquid may form in a position where, clearly, it can reduce the effectiveness of insulation and it is likely also to put the nearby timber at risk of rot. As an illustration of the effect of structural detailing. Figure 19.4(b) shows the construction reversed and free from risk in the same surrounding conditions. Slight modifications, shown in (c) and (d) are sufficient, however, to limit the potential risk by using materials that modify the vapour pressure gradient.

Lighting

Three types of lighting are used: daylight; daylight integrated with electric lighting; electric lighting.

Good daylighting is more than the provision simply of large windows. Optimum size, shape and position of windows is a function not only of the required lighting levels, but also of the resulting eye adaptation conditions, sky glare and external view. In addition, heat loss or solar gain, ventilation, noise transmission, privacy and the shading effects of adjacent buildings, present or future, must be taken into account. Side-lit rooms often appear badly illuminated because of the contrast between the areas adjacent to and those remote from the windows, even though working illumination levels may be adequate throughout.

When electric lighting costs were high, daylighting appeared cheap and its real cost went unquestioned. The present position is different: modern light sources cost less and are more efficient while the true cost of daylight is recognised in terms of added cost in construction, maintenance, heat loss or gain and, in urban areas, the inefficient use of the available site area. Simultaneously, the expected standards have increased in both quantity and quality and, in many modern buildings, daylighting would not be relied upon as the sole source of light even during periods of good outdoor light.

By introducing electric lighting of a colour to blend with daylight it is possible to provide adequate illumination over the whole working area without a sense of deprivation of daylight. Moreover such arrangements, known as permanent supplementary artificial lighting of interiors (PSLAI), can be applied without visual discomfort over areas much greater than can be lit by daylight alone, irrespective of the prevailing outdoor light—its added cost must be weighed against the direct and indirect costs of higher ceilings and bigger windows, reduced floor space for light wells, or restricted useful depth of rooms.

Quality of the electric light is as important as quantity and design should take into account: brightness and colour patterns; directional lighting where appropriate; control of direct or reflected glare from light sources; colour rendering; prevention of excessive contrast between adjacent areas.

The most common light sources are tungsten lamps and fluorescent tubes with a growing acceptance of mercury fluorescent lamps. Tungsten lamps are common in domestic and decorative installations, but are comparatively inefficient in their light output and are generally uneconomic for the lighting levels required in most modern buildings. Fluorescent tubes are the most commonly used, but can take up a considerable amount of ceiling space. Mercury fluorescent lamps provide similar benefits of efficiency and long life, but more closely approach a point source, permitting greater freedom in ceiling design. The ability to accommodate an economic light fitting will depend upon the planning and structural grids. When these are not appropriate to the light fitting, the lighting system will be expensive in itself and may also cause extra cost in removing the unwanted heat.

The light fittings have to be carefully spaced to provide adequate lighting levels over the whole working plane. Due to the physical discomfort which can be caused by the brightness of the light source, careful attention must be given to the prevention of direct or reflected glare. Glare standards exist for most types of working environments and the glare characteristics of lamp fittings and control diffusers are readily available.

The varying colour qualities and corresponding luminance efficiencies of the available light sources have an important bearing not only on the visual environment, but also on the degree of heating or air conditioning that may be required. Where good colour rendering is preferred, lamps of lower efficiency but better spectral quality should be used. Some of these have colour rendering properties very close to those of daylight and are thus particularly suitable for supplemented installations (PSALI).

In some buildings, the energy for lighting can be a substantial part of the total required for all purposes. Since most of that provided for light appears as heat the possibility exists of using this as a major, and perhaps the only, source of internal heating; alternatively the extra heat load may prove an embarrassment to the air conditioning system. In either case the lighting must be treated as an integral part of the total environmental design.

LIGHTING FOR VARIOUS CATEGORIES OF BUILDING

Speculative offices. Such buildings are generally leased without lighting fittings to avoid inhibiting either the letting pattern or the tenant's partition layout. Where lighting fittings are supplied the preference is often for surface-mounted hot-cathode fluorescent tube units with prismatic light controllers. Lighting levels are currently in the region of 400 lux.

Offices—purpose design. An average level of 1000 lux with fluorescent fittings is now commonplace and the trend is still upwards. At such levels the basic problems are providing suitable light-controllers to meet the limiting glare indices and disposing of the heat load which at a 1000 lux may be of the order of 70 W/m^2. The first may be countered by the use of properly designed plastic or metal controllers. The second calls for air circulation through the fitting resulting in deep ceiling voids and suitably sized holes through the structure for trunking.

Offices—burolandschaft. The above comments apply but with a growing emphasis on quality. One result is an increased interest in high pressure discharge lighting for commercial interiors since they confer the benefits of good colour rendition, excellent modelling, low maintenance and reduced heat load. The depth of fitting is greater than that of a fluorescent fitting and thus calls for a deep ceiling void.

Hospitals. The difficulties of reconciling the lighting needs in wards of patients who may either be lying supine or sitting up in their beds has led to separate systems being installed. In the latter case wall-mounted units are preferred and these are often incorporated into continuous horizontal trunking runs which may contain other services such as oxygen, sound broadcasting, nurse call systems, etc. The former requirement is met by fluorescent fittings generally of the suspended pattern. There are many specialised considerations, such as operating theatres and anaesthetic rooms where totally enclosed, noise-proof fluorescent fittings sealed into the ceiling structure provide general illumination whilst shadowless operating-table lighting fittings incorporating tungsten light sources produce intensities up to 10 000 lux in the operating area.

Housing. Ceiling or wall-mounted tungsten fittings remain prevalent with perhaps fluorescent fittings in kitchens above worktops and the use of table lamps and floor standards. In the stairways and common areas of high rise blocks, fluorescent bulkhead fittings are used and are robust and inexpensive to run.

Schools. Cost considerations usually dictate surface-mounted fluorescent fittings with prismatic light controllers with levels in the region of 600 lux. In rooms where the seating has a fixed orientation, directional fittings may be used.

Industrial buildings. When ceiling heights are below about 4 m, fluorescent fittings are still the most used light source. Above this, high-pressure mercury or sodium discharge lamps in reflector fittings are used with a wide range of distribution curves both symmetrical and asymmetrical. The colour rendition of mercury fluorescent, mercury halide or high-pressure sodium light sources are satisfactory, but care has to be exercised in machine shops because of stroboscopic effects.

Car parks. The majority of multi-deck car parks use bare fluorescent tubes in fittings with moisture-proof lampholders and glassfibre or PVC-coated bodies. In open car parks of the larger variety, increasing use is being made of high mast lighting.

Noise

The control of noise requires consideration of its nature, source and mode of transmission. Typically the main problems are: reduction of noise to an acceptable level for efficient working; and effective noise barriers for privacy. Problems of sound insulation and sound absorption are involved.

Externally the main source of noise is air or road traffic; penetration is reduced by double glazing (cavity preferably not less than 200 mm), minimum window area and heavy wall construction. In extreme cases, windows must be kept permanently closed and the building air conditioned.

Internally, structural walls and floors are generally of sufficient mass to provide effective barriers against airborne sound but impact sound is not reduced by mass alone and a resilient material must be added to provide adequate total sound insulation. The lighter building elements, such as suspended ceilings or demountable partitions, do not provide good sound insulation. Continuity of sound insulation, where it is required, is important: for example, a sound-insulating wall would need to extend through the void above a suspended ceiling, unless the ceiling is itself a good sound insulator.

The use of sound-absorbing surface materials and shapes is effective in reducing the ambient noise level and may be so successful in burolandschaft offices that a degree of manufactured ambient sound may be needed to mask and hence reduce the disturbance from local intermittent noise.

Appropriate planning and detailing of the building is vital to the elimination of noise problems and the establishment of privacy. Wherever possible, areas requiring low noise levels should be divorced from noisy areas such as plant rooms, loading bays and lift motor rooms. Many items of mechanical and electrical equipment produce airborne noise which can pass along air conditioning or ventilation ducts which then require

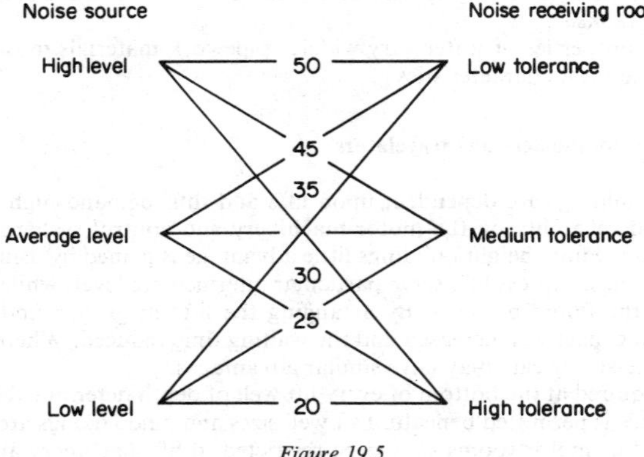

Figure 19.5

silencer units. Equipment located in occupied rooms must be selected with appropriate low noise characteristics; in certain cases, especially on high-pressure systems, secondary silencer units are required. Rotating or reciprocating plant should be isolated from the structure to prevent structure-borne noise or vibrations. The increase in plant noise within buildings is increasingly a factor in modern design, requiring specialist advice.

Rooms with a high level of sound within them do not require such a good standard of insulation from adjoining rooms of similar level, but low-tolerance rooms will require a high standard. Figure 19.5 gives an indication of sound reduction levels for different room tolerances (from *Acoustics, Noise and Building* by D. H. Parkin and H. R. Humphreys).

The sound reduction of dense walls varies with the sound frequency and with the weight of wall. At 550 Hz, the sound reduction is as follows:

Weight (kg/m^2)	3	6	12	25	50	100	200	400	800	1000
Sound reduction (dB)	20	24	28	32	36	40	44	49	54	55

For a cavity wall a reduction value corresponding to the combined weight of the two leaves is used and to this is added the additional assistance provided by the cavity which varies with its width as follows:

Air space (mm)	30	40	50	60	80–100	150	200
Added sound reduction (dB)	6	8	9	10	12	10	6

If the wall contains a door, the equivalent resistance is an intermediate value between those for the wall and door, dependent upon the relative dB values and areas. For a brick wall of say 46 dB and door of 20 dB and the wall 10 times the area of the door, the equivalent sound reduction values (obtained from charts, e.g. Neufert) is 30 dB. The full insulation value is obtained only if all holes, e.g. for services, are sealed; even very small openings such as key holes and open joints represent serious sound leaks and must be avoided if good sound insulation is to be maintained.

Water supply

Most water authorities require storage for 12 or 24 hours' consumption: a significant requirement in terms of both volume and weight. Water stored at the top of the building avoids pumping and is always immediately available for use. When stored at low level, pumping is required to transfer the water to a high-level small storage tank, or to provide a continuous pressurised pump system. Too high a static head may be avoided in tall buildings by sectionalising the system so that a number of floors are served from intermediate storage tanks.

The chemical properties of water vary widely; pipework materials must be carefully chosen to suit the local characteristics.

Lifts, escalators, paternosters and travelators

Many modern buildings are dependent upon lifts and thus demand high standards of performance and reliability on the motor machinery and control systems. Lift speeds vary: for normal medium height buildings little advantage is gained by using high-speed lifts; in tall buildings, express lifts serve particular intermediate levels while slower stopping lifts serve the floors between. By arranging the lifts in groups under combined control, the total capacity is increased and the waiting time reduced. Where plan area is restricted, double-storey cars may have similar advantages.

A lift pit is required at the bottom of every lift well of depth determined by lift speed; no occupied space is permitted beneath. Lift well sizes and guide fixings are given by the manufacturers. Lift motor rooms should be restricted to lift machinery and associated equipment. The lift well enclosure, pit and motor room form part of the building construction and may require particular construction as a 'protected shaft' passing between fire compartments.

Escalators have a much greater carrying capacity, but can only serve between two floors. Their use is mainly in high-flow areas with a limited number of floors. Capacity is varied by width and speed and can exceed 10 000 people per hour.

Paternosters are suitable for continuous heavy traffic through all floors of a building and are permitted in most countries except in blocks of flats. They are comparatively slow and should be accompanied by at least one ordinary lift for sick or handicapped people. All parts must be noncombustible and landings must be enclosed by fire-resisting walls and doors. Maximum capacity for a two-person car is about 600 people per hour.

Travelators are used basically for horizontal movement but increasing use is being made of them on shallow inclines to replace pedestrian ramps. They are used in transport terminals and interchanges.

REGULATIONS AND FIRE REQUIREMENTS

Building Regulations

The Building Regulations now replace the previous system of local authority by-laws; their purpose is to ensure that buildings are healthy and safe for their occupants. They apply to new buildings, alterations or extensions to existing buildings, new or replacement fittings and also to buildings where there is an intended material change in use. Some buildings are exempt, e.g. local authority schools and Crown buildings and some regulations apply only to certain types of building, e.g. the requirements for sound and heat insulation apply only to dwellings.

Any person intending to carry out work governed by the Building Regulations must give notice and deposit plans which must be passed or rejected by the local authority within five weeks (this period may be increased to two months by mutual consent). The local authority must also be given at least 24 hours notice of commencement of work and before covering foundations, drains or private sewers. If in a particular case a Building Regulation requirement would be unreasonable a relaxation or dispensation may be sought: in most cases the local authority will decide the matter, but in certain instances, mainly concerned with structural stability and structural fire precautions and all applications by the local authority itself, reference must be made to the Secretary of State. In the event of a local authority refusing an application for relaxation or dispensation an appeal can be made to the Secretary of State.

The Building Regulations contain three basic types of technical requirement, functional, performance and specific, supported by 'deemed to satisfy' examples of construction which will satisfy the mandatory provisions often by reference to British Standards or British Standards Codes of Practice. The major items covered are:

Materials. Any material conforming to a British Standard and used in accordance with the relevant Code of Practice is deemed to satisfy.

Site preparation and resistance to moisture. Sites must be drained and cleared of vegetation; ground floors must prevent penetration of ground water; hardcore must be free of sulphates or other harmful materials; walls must prevent rising damp and external walls must be weatherproof and damp must be avoided in cavity walls; roofs must keep out rain and snow.

Structural stability. Loading according to the Code of Practice; foundations safe and free from damaging settlement or other movements including safeguard against swelling, shrinking or freezing of subsoil; structure capable of carrying loads to foundations without causing damage to the building; deemed to satisfy Codes and Standards plus schedules specifying sizes for timber joists and rafters and certain load-bearing walls of bricks or blocks; buildings of five or more storeys must resist progressive collapse (see below).

Structural fire precautions. Buildings classified into eight purpose groups with differing fire standards dependent upon use; internal fire spread controlled by division into compartments bounded by walls and floors of specified fire resistance, common examples given; external fire spread between buildings controlled by limiting the area within external walls having less than the necessary fire resistance, depending upon the distance from the site boundary; compartment walls and floors together with party walls must in general be imperforate, but certain openings may be allowed, e.g. for pipes and protected shafts enclosing services or staircases; fire-resistant doors must be selfclosing and satisfy certain tests, special provision for some doors in lift shafts; stairways and landings must in general be noncombustible; fire stops necessary where fire-resisting elements are penetrated and at junctions and in cavities; ceilings and wall surfaces must limit the rate of flame spread; roof construction limitations within stated distances of the site boundary.

Thermal insulation (for dwellings only). Minimum insulation requirements for walls and roofs exposed to the external air with 'deemed to satisfy' provisions.

Sound insulation (for dwellings only). 'Deemed to satisfy' provisions lay down required performance and give specific examples for suitable construction for party walls and floors; procedures for sound measurement given.

In addition to the Building Regulations, local authorities will check compliance with other statutory requirements. In particular the Public Health Act 1936 requires the local authority to be satisfied on:

entrances and exits in certain buildings: buildings concerned are theatres, halls and places of public assembly or worship, restaurants, shops and stores; warehouses; clubs; nonexempt schools.

means of escape from fire: for most types of building.

PROGRESSIVE COLLAPSE

The Building Regulations require buildings of five or more storeys (including basements) to be so constructed that the area of structural collapse consequent upon the failure of any one structural member shall be kept within specified limits, i.e. a local failure should not lead to the progressive collapse of a major part of the building. Regulation D19 requires that the collapse resulting from the removal of any one 'portion' of any one structural member must be localised to the storey containing that member and the storeys next above and next below; or if, for particular members, this limitation cannot be achieved then these members must be capable of resisting at near ultimate stresses an additional load of 34 kN/m^2 from any direction applied directly to itself plus any reaction arising from adjacent construction subjected to the same pressure. Regulation D20 limits the area of structural failure, within each of the three permitted storeys, to the lesser of 70 m^2 or 15% of the storey area. Buildings using large pre-cast concrete panels are deemed to satisfy the requirements if they comply with CP 116: Part 2: 1969. A 'portion' is defined as that part of a structural member situated between adjacent supports or between a support and the extremity of the member; but for walls the 'portion' is limited in length to 2.25 times the wall height.

These Building Regulations apply only to England and Wales, but the provisions in Scotland as to progressive collapse are similar. In London, amendments to the building by-laws make similar provisions, except that the length of a load-bearing flank wall to be considered removed is not limited to 2.25 times its height; there is also an added requirement that party walls between dwellings shall have a minimum weight of 3.4 kN/m^2 or a proved resistance to 7 kN/m^2 of horizontal pressure in either direction.

The Institution of Structural Engineers document RP/68/05 gives the view that fully framed structures in concrete or steel in which at least the columns are continuous can accommodate the unpredictable loads and effects envisaged on the progressive collapse requirements of the Building (5th Amendment) Regulations 1970 (now incorporated in the 1972 Building Regulations), provided that: Building Regulation D8 is satisfied; the structure accords with BS 449 and the concrete CPs 114, 115 and 116 (to be superseded by CP 110); a minimum tying strength of 25 kN/m, at working stresses, is provided in all floors and roof in each of two directions approximately at right angles; and that all floors and roof slabs are sufficiently anchored in the direction of their span to each other or to their supports to resist a horizontal tensile force of 25 kN/m. This view has been accepted and may be used to obtain relaxation of the requirements of D19.

The GLC in their explanatory notes divide structures into three categories: fully continuous, such as reinforced concrete frames designed for continuity at the joints, in which it is assumed that the interaction between components will be sufficient to meet the requirements against progressive collapse; structures with nominal continuity only which require checking to verify that there is adequate horizontal and vertical continuity, namely a peripheral tie of 100 kN at normal stresses, ties in direction of span of 25 kN/m and transversely 12.5 kN/m, vertical ties in reinforced concrete walls and columns of

0.2% and 0.8% respectively of the cross-sectional area; and partially discontinuous structures in which the horizontal or vertical continuity is insufficient to produce the desired degree of interaction between structural members—in these cases limited damage only following the removal of structural members must be proved or the member strengthened to resist the 34 kN/m^2 added force. Debris load must be taken into account, particularly if there is any danger of impact, as must the loss of lateral support when a horizontal member is removed—double-storey columns with staggered joints help.

When considering the resistance of a building to progressive collapse following the removal of a structural member, account may be taken of any mode of action of any existing building component whether it is structural or not. Thus arching action in load-bearing walls can resist major horizontal forces provided that there is sufficient vertical load present and horizontal restraint at top and bottom. Composite action between walls and floors, as deep beams or cantilevers, to bridge gaps beneath, and catenary action, where appropriate details and anchorage exist, can be used to demonstrate the continued stability of the damaged structure . The three-dimensional action of certain structures may make an important contribution. Of greatest importance is the combination of structural plan form and the joints between members—appropriate attention to these aspects during the basic planning of the building can introduce at little, if any, extra cost a general 'robustness' which reduces or eliminates any tendency for local damage to initiate a major collapse. Thus, the incorporation of a spine wall in a load-bearing reinforced concrete cross wall building coupled with joints designed to produce the necessary composite action for three-dimensional bridging action, can ensure virtual immunity from progressive collapse.

On the other hand a load-bearing flank or gable wall requires special consideration if it is not suitably buttressed by a spine wall. It is itself more prone than the internal walls to the effects of horizontal pressure; alternatively, in the event of an adjacent explosion, it may lose the support of the end façade construction as well as lateral support from the floors.

FIRE PROTECTION

Fire protection measures include: structural fire resistance; means for the safe evacuation of the occupants; detection and warning devices; fire extinguishing devices like sprinklers; and means of containing the fire. Fire protection should be considered as an overall system embracing all these measures with the objective of safeguarding life and property.

The Building Regulations classify buildings according to their use and consequent fire risk, specify the fire-resisting requirements for structural and other elements within or bounding a building or compartment, and specify the required flamespread resistance of finishes.

Size is an important factor and the required fire periods vary with the area and volume of the building or compartment. This leads to the division of large buildings into compartments bounded by walls, floors and other components of sufficient resistance to the spread of fire. Separation between buildings is needed to resist the spread of fire and the regulations specify the minimum distance of an external wall from the adjacent site boundary depending upon its size and percentage of 'unprotected' area or, given the boundary distances, the regulations define the minimum fire-resisting qualities of the wall.

Fire-resisting doors or shutters are permitted as part of a separating or compartment wall providing they have the same resistance to fire and spread of flame as the wall and are selfclosing. Fire stops are required to prevent the spread of fire through pipeholes or other gaps penetrating a compartment wall or floor and, within cavities, to stop flame or smoke spreading unseen between compartments. This requirement is often in conflict with ventilation systems: firebreaks in shafts and automatic fire dampers in ducts may be required.

Means of escape from a fire and access for fire fighting require discussions with the local fire authority. Maximum travel distances to reach a place of safety, such as a fire-protected staircase or the open air, are specified in, for example, the GLC Code of Practice 'Means of Escape in Case of Fire' which also specifies the minimum width of escape passages and staircases.

Warning systems can take the form of heat or smoke detectors and can be made to operate automatically. Fire extinguishing devices such as sprinklers and CO_2 systems are sometimes demanded by fire prevention officers in high-risk accommodation and may require underground water tanks and automatic pumps. In shops, the maximum permitted floor area in the Building Regulations may be doubled if a sprinkler system is installed. If brought into use sprinklers can cause considerable damage to the building contents: CO_2 systems prevent this, but must have safety devices to ensure that CO_2 is not injected during periods of occupation. Most buildings incorporate a manual form of fire fighting system such as a wet or dry riser sometimes coupled with an automatically pressurised pumping system, requiring additional water storage tanks.

Fire protection of structural steel has moved towards the use of lightweight encasements: prefabricated plaster; asbestos board with an outer finish in steel; in situ lightweight plaster; pre-cast lightweight concrete sections particularly for industrial applications because of its durability and resistance to damage. The most recent developments are: intumescent paints which expand under heat to form a heat shield providing up to three hours fire resistance depending upon the thickness used; surrounding the section with a thin sheet of steel and filling the gap with a lightweight insulating material, before delivery, or on site; the use of water as a coolant within hollow steel sections.

Fire protection of concrete structures is based upon the provision of adequate thickness of construction and adequate cover to the reinforcement or prestressing tendons. Lightweight concrete has an improved fire resistance because of its better thermal resistance and, if made with artificial lightweight aggregates, is virtually free of spalling during a fire.

The fire period for timber walls and floors is usually half an hour but one hour is required for walls separating dwellings—the junction between the two requires care. Timber can be treated to delay ignition and to inhibit active flaming, but these do not prevent its eventual decomposition at high temperature. By careful selection of timber species and glue, laminated sections give as good a fire resistance as the solid material.

All plastics decompose and are combustible at relatively low temperatures; their use in building therefore has been in conjunction with other materials. Polystyrene for example has excellent thermal resistance, but makes no contribution to fire protection because it melts too soon; on the other hand, it is used as a cavity filling and thermal insulant in walls and floors has not produced any additional fire problems. The use of a steel facing prevents ready combustion of polyurethane foam panels and spread of fire in the cavity. Reinforced polyester with an aerated concrete backing gave satisfactory fire test results for use as cladding panels on high-rise flats. Plastics vary in composition and form and, to establish a safe and national attitude to their use in buildings, performance data in relation to flammability, generation of toxic gases and smoke must be determined and considered carefully in relation to the intended use.

MATERIALS

The principle materials used in buildings are concrete, steel, brick and masonry, and timber: each has its own developing technology described in Sections 11, 12, 14 and 15. Other materials include: aluminium, various alloys, glass, plastics, rubber. When used structurally, the essential properties concern, strength, rigidity, durability and fire resistance. Relevant general properties concern: hardness, thermal characteristics of insulation and expansion, weight, uniformity, appearance and workability. All these may be affected by changing temperature, humidity and weather. The choice of material for a particular building element will be determined by suitability for the intended purpose, cost and availability and compatibility with other materials.

Concrete

As a building and structural material concrete, in itself, is durable and impermeable and with the appropriate choice of cement and aggregates, is resistant to chemical attack; it is readily available and using different cements, aggregates, forms and surface treatment it can provide a wide range of strength, density and appearance. Its mass automatically provides good sound insulation and high thermal capacity.

Lightweight concrete is used in both structural and nonstructural elements and is adopted when weight is at a premium, or to take advantage of its better fire and thermal resistance. Typical applications include: building blocks, screeds, fire casings, pre-cast floor, roof and cladding units, in situ prestressed or reinforced concrete structures. Its reduced stiffness should be allowed for in design.

Normal-weight concrete is used in most other contexts and is generally reinforced with high-tensile or mild-steel bars. Plain concrete is used in situ for mass foundations, gravity retaining walls, screeds and blinding and for some pre-cast elements such as paving, curbs and building blocks. Prestressing is used in pre-cast floor and long-span beam units and occasionally in in situ construction as a means of controlling deflection or reducing member size: it has a special application in hanging and transfer structures particularly when structural depth is limited. Fibre reinforcement (steel, glass, plastic) is used in special situations where increased resistance to impact or reduced concrete thickness is an important advantage; current applications are in pre-cast cladding, pipes, paving slabs and pile shells.

Cladding is a major application of concrete in buildings; generally pre-cast, it may be a nonstructural element or may incorporate a structural function in resisting horizontal or vertical loads or both. Often, such pre-cast cladding incorporates windows or doors and is of sandwich construction including a layer or thermal insulation. A large variety of surface treatments is available; great care is required in the detailing and texturing to ensure satisfactory weathering. Aggregates may be exposed or the surface covered with mosaic or tiles; plane and sculptured profiles are common.

Pre-casting methods include centrifugal, vibratory, pressure, vacuum and extrusion equipment while, on site, concrete is often pumped into position using prefabricated formwork systems and powerful tower cranes are available to erect heavy pre-cast elements. The choice between pre-cast and in situ construction will depend upon such factors as cost, speed, access and availability of labour.

Steel

In one form or another steel is extensively used in practically all buildings. As a structural element it is available in a wide range of section and composition to suit the particular requirements of stress, deflection, corrosion or jointing technique: it is of high strength; in itself, occupies little space and is prefabricated for easy and rapid erection on site; it readily lends itself to extension or alteration. It has two disadvantages, fire and corrosion: several methods exist to overcome its fire sensitivity; various coatings are applied to resist corrosion and some steels (Corten) can be left exposed without treatment. Castellated beams are useful to reduce deflection and to provide holes for the passage or services; hollow sections, of various wall thicknesses, are used in tubular structures, columns, trusses, space frames; combined sections are commonly used and composite action, via shear connectors, with concrete floor construction can be advantageous. Sheet steel applications include roof and wall cladding, floor infilling systems, ducting.

Brick and masonry

Brick and masonry have the advantages of a long heritage of experience and simple construction based on traditional skills; building plant costs are low, but the labour content is high and not always in sufficient supply. The common materials are: bricks and concrete blocks of various types, finishes and strength: and natural or reconstituted stone. The main applications occur in load-bearing walls and piers, particularly in low- and medium-rise buildings, and as cladding. Internally they are used as the inner skin of cavity construction or as partitions and, within limits, may be used to brace framed construction. Reinforcement can be added in the horizontal joints to produce beam action or vertically, in piers, for tying purposes.

In general the massive nature of these materials provides good sound and thermal insulation together with good compressive strength and durability. Movement due to shrinkage, temperature and moisture change, and chemical action must be allowed for and provision made in design against progressive collapse.

Timber

The main advantages are: it is readily available in a wide variety of types and section; it is light in weight; it is easily worked employing traditional skills. Typical applications are: floors, roofs, framing to light buildings, cladding, wall and ceiling construction and in surface finishings. It is often used in temporary buildings and for temporary works and formwork.

Size, form and consistency limitations have been reduced by the introduction of glued laminates and special fastening systems have made larger-scale structures possible. Combustibility, rot and insect infestation can be retarded by chemical impregnation while treatment with steam or ammonia gas introduces flexibility. Temperature and moisture movements remain problems; fire resistance can be developed only to a degree but is, generally, sufficient for domestic purposes.

WALLS, ROOFS AND FINISHES

External finishes, materials and weathering

The design of the external fabric involves a knowledge of the behaviour of the materials and elements of construction and includes consideration of weathering and water-shedding characteristics. External materials must be durable under the influence of climatic extremes and the local environmental conditions including wind velocity and prevailing direction, and whether coastal, urban, industrial or rural. The cost of maintenance, and accessibility for maintenance are also important considerations. The major functional requirements for the external wall include heat and sound insulation and damp proofing.

Design elements range from screws to complete assemblies. Deterioration due to weathering may be aesthetic or functional and may be visible or concealed, and design details should be such as to avoid structural deterioration especially in concealed situations and should be designed bearing in mind the possible colour change or staining effects of weathering.

The weathering characteristics of the main materials used externally are described in detail in 'Weathering and performance of building materials', published by the University of Manchester.

Concrete finishes depend upon the moulds used, the material properties and surface treatment. Blemishes such as surface air holes cannot always be avoided and untreated smooth surfaces generally weather badly, although when, with hard concretes of plain or white mixes, the surface laitance is ground off and sealed, good results can be obtained. Patterning and texture can provide interesting finishes, as with rough board markings, or ribbed surfaces which may be hammered or tooled in various ways. Deep patterning can also be very effective. Exposed aggregate finishes are available from a wide variety of processes and aggregates, and these generally weather well.

Brickwork is traditionally used as external walling and a variety of colours and textures are available in facing bricks. Attention must be given to weathering performance, porosity and freezing effects, efflorescence, sulphate attack, etc.; when associated problems are recognised, solutions are available. Details must be designed to accommodate relative movement of the brickwork and other building elements.

Timber and timber products may be used as cladding, colour changes usually occur on exposure to light and to water, which can also lead to damage such as splitting, warping, and dirt penetration. Various preservation treatments are available, including the use of varnishes, synthetic resinous clear finishes, opaque paints and applied film overlays.

With the use of metals externally, their particular properties as to electrochemical corrosion in relation to weathering and the problem of bimetallic electrolytic action need to be understood. Aluminium, bronze and copper weather well and are used in roofing, cladding, window framing and flashing applications. Lead may be used in sheet form for special roof covering and more frequently for flashings. Zinc provides useful coatings and flashings. Outstanding durability can be obtained with the use of stainless steel and low-alloy steels of good weathering properties are available.

Curtain walling can provide an economic form of cladding and glazing, with advantages of: lightness; thinness (as affecting usable floor area); flexibility of fenestration; speed of erection without external scaffolding. It is provided in two main types: unit assemblies in which selfsupporting panels are prefabricated including glazing and solid infilling with interlock or lap joints and part assemblies in which frame members are erected and the glass and solid sheets added. Weather resistance and connection to the structure must be adequate to meet high local wind pressures and driving rain conditions. Sealing methods include the use of mastics, gaskets, cover tapes and spring strips. Thermal movements may be very large since the panels have low heat capacity and respond rapidly to changes in temperature, giving rise to differential movements between one part of the curtain walling and another, and between the curtain walling and the structure.

Different types of glass are available including: clear, coloured or opaque; heat absorbing, filtering or reflecting; toughened, single or bonded into insulated sandwich construction.

A wide range of plastics for external use is available, with different resistances to ultraviolet light, temperature, water, oxygen, micro-organisms, atmospheric pollution and loading. These include PVC, used for rainwater goods, glassfibre-reinforced polyester resins forming sheet or shell products, polymethyl methacrylate providing transparent sheets of high strength and durability and the phenolic and amino resins for laminated sheets. Polymer films may be applied to other materials such as boards or metals to improve durability.

Floor, ceiling and wall finishes (internal)

Such finishes may be integral with the structure or applied. Type of usage, cost, chemical resistance, aesthetic requirements, fire resistance or 'spread of flame' requirements, maintenance, are some of the factors influencing selection.

Floor finishes integral with concrete rely on good workmanship: power-float finish, use of hardeners, dust inhibitors, waterproofers, or the application of granolithic concrete finishes to 'green' concrete. Timber and metal decking can also be in the 'integral' category. Applied floor finishes can vary from simple sheet or tile materials stuck (or laid loose in some cases) to the structural slab, with or without levelling screeds. Damp-proof membranes or vapour barriers may be required for slabs on ground, depending on the type of finish to be applied and/or the ground water conditions. Screeds may require to be of adequate thickness to allow for the running of service conduits, or thickened, reinforced and isolated by insulation in the case of floors to be heated or used for impact sound insulation.

Integral wall finishes result from the use of controlled shuttering on concrete work, fair-faced brick or blockwork or self-finished plane or profiled sheet materials. Applied wall finishes can be basically divided into wet and dry applications; plastering—by hand or spray—being typical of the former and dry-lining such as plasterboards and proprietary insulation boards, acoustic finishes, are among the final finishes that may be required.

Integral ceiling finishes result from the use of untreated structural soffits such as concrete, metal or timber. Applied finishes may be divided into direct and suspended, the former, as indicated, being the application of wet or dry 'lining' or finishing direct to the structural soffite, such as paint, plaster, sprayed finishes, plasterboards, acoustic insulation boards or tiles. Suspended ceilings can be used to conceal structural members, to provide space for services, to reduce room heights for functional or aesthetic reasons, to provide a grid for flexible layout of partitioning. Such ceilings may be partly or wholly demountable for access to services or may be 'monolithic' or 'fixed', e.g. plaster on expanded metal.

Building Regulations must be referred to when considering internal finishes; requirements for resistance to fire being particularly stringent in areas such as staircases and circulation spaces, but applicable to other areas also and varying according to building use, area and volume.

Roofs

Roofs must keep out the weather, be durable and structurally stable, provide heat insulation in most cases and in certain cases provide light and ventilation. The choice of roof structure will generally be determined by the general form of the building and the activities for which it is designed. Unlike floors there is not usually any restriction on the depth of a roof and this gives a wide flexibility for economical and appropriate solutions. Sometimes the roof structure will be outside the main building.

A roof must carry its own weight together with imposed loads of roof finish and usually insulation, snow, the effects of wind, normally maintenance and often plant. It must resist excessive deflection or distortion which though not leading to collapse may damage decorations and services and if visible can lead to lack of confidence and anxiety in the occupants. In accommodation for sedentary work or living, heat insulation, lighting, ventilation and sound insulation are important.

In general, the spacing of supports should be as close as possible consistent with present or possible future use.

Short-span roofs below 25 ft (7.6 m)—houses, blocks of flats, many multi-storey buildings and some warehouses. On houses, roofs are often traditional in design but with sheet materials allowing a lower pitch the uplift effect of wind is important. Flats and multi-storey buildings are normally roofed with a concrete slab similar to the floor construction.

Medium span roofs 25–80 ft (7.6–24 m)—industrial buildings, warehouses, transit buildings, etc. Here intermediate supports are often a nuisance. Appropriate systems are in situ, pre-cast or prestressed concrete, trusses, lattice girders and portal frames.

Long-span roofs over 80 ft (24 m)—exhibition halls, industrial buildings, leisure buildings, sports stadia and transport buildings. Many of these buildings require roofs which only keep the elements off the occupants. Systems would include lattice girders, space frames, roofs supported by suspended cables, prestressed concrete, arched construction, concrete folded plates and hyperbolic paraboloids.

It is important to note that in the Building Regulations roofs are not designated as an element of structure and generally are not required to be fire resisting. Exceptions occur where another building with windows adjoins the roof and in the London area where buildings coming under section 20 of the London Building Act and the regulations of the larger privately owned estates do require specified fire resistance for roofs.

Roof coverings include slates and tiles for houses, sheet materials flat or profiled, asphalt, felt, new materials based on synthetic rubber, plastics, sprayed on materials and glass. It must always be remembered that provision must be made for roof drainage.

Fire spread is important in relation to roof coverings and is covered by BS 476: 1958: Part 2.

Thermal insulation is often required to conserve heat in the buildings and is covered by the Thermal Insulation (Industrial Buildings) Act 1957 and by the Building Regulations, but it is also important to reduce solar gain and avoid excessive expansion in the structure of the roof sometimes distorting the structural frame and outside walls. A reflective external finish to the roof also assists.

Condensation is a serious problem in roofs and is usually overcome by placing a continuous vapour barrier on the warm side (the face nearest the inside of the building) of the insulation. External venting is provided to the insulation. This is a subject on which a great deal of research has been carried out in recent years and deserves careful study.

Partitions

Partitions divide large areas into individual spaces for specific purposes such as stores, offices, etc., and separate circulation from working or living areas. The type of partition is determined by requirements of acoustic or thermal insulation, security, privacy, fire resistance and flexibility of planning. When the partitions are structural, brick or blockwork or concrete are commonly used: however, partitioning is generally kept separate from the structures.

Commonly used partitions, in increasing weight, are: light framing with infilling of glass or building board, plasterboard dry partition panels, woodwool and compressed straw building slabs, sandwich composite panels, pre-cast autoclaved concrete panels, light to dense blockwork, brickwork and concrete.

STRUCTURE

The design of building structures is an iterative process by which the type, shape, dimensions, materials and location of the various structural elements are initially chosen as a first approximation; loads are then determined and the design developed by a process of adjustment and verification that structural performance will be satisfactory. The structure must also satisfy the functional needs of the building, site factors and the many technical requirements concerned with the safety, health, comfort and convenience of the occupants.

Assessment of structural behaviour must cover:

'SERVICEABILITY LIMIT STATES'

Concerned with acceptable horizontal and vertical deflections and structural cracking and the compatibility of these with the secondary elements supported by the structure, such as partitions, cladding, finishes.

'ULTIMATE LIMIT STATES'

Concerned with the provision of adequate reserves of strength to cater for variations in materials, structural behaviour, loading and consequences of failure. Partial factors are used for this purpose as follows:

γ_m allows for variations in strength and is the product of:

γ_{m1} to take into account the reduction in strength of materials in the structure as a whole, as compared with the control test specimen, and

γ_{m2} to take account of local variations in strength due to other causes, e.g. the construction process.

γ_f allows for variability of loads and load effects and is the product of:

γ_{f1} to take account of variability of loads above the characteristic values used in design, and

γ_{f2} to allow for the reduced probability of combinations of loads, and

γ_{f3} to allow for the adverse effects of inaccuracies in design assumptions, constructional tolerances such as dimensions of cross section, position of steel and eccentricities of loading.

γ_c takes into account the particular behaviour of the structure and its importance in terms of consequential damage, should failure occur. It is the product of γ_{c1} and γ_{c2} where:

γ_{c1} takes account of the nature of the structure and its behaviour at or near collapse (whether brittle and sudden or ductile and preceded by warning) and the extent of collapse resulting from the failure of a particular member (whether partial or complete), and

γ_{c2} takes account of the seriousness of a collapse in terms of its economic consequences and dangers to life and the community.

The values allocated to these partial factors are based upon engineering judgement, experience and, where available, statistical data, as appropriate. The first British Code to incorporate limit state philosophy is CP 110: 1972, *The Structural Use of Concrete,* which includes the two partial factors γ_m and γ_f. Values for the subcomponents (γ_{m1}, γ_{m2}, etc.) are not given, but the quoted global values vary with the circumstances and the load combinations being considered. Thus for the ultimate limit states:

γ_f values	*Dead load*	*Imposed load*	*Wind load*
For dead plus imposed load	1.4	1.6	—
For dead plus wind load	0.9	—	1.4
For dead plus wind load plus imposed load	1.2	1.2	1.2

γ_m values
For steel 1.15
For concrete 1.5

HAZARDS

Building structures may be subjected to such hazards as: impact from aircraft or vehicular traffic: internal or external explosion caused by, for example, gas or petrol vapour or by sabotage; fire; settlement; coarse errors in design, detailing or construction; and special sensitivities, for example as to acceptance of movement or differential movement or as to conditions of elastic instability, not appreciated or allowed for in design.

Except in special circumstances, these hazards cannot be quantified. However, for buildings of five or more storeys, the Building Regulation requirements concerning progressive collapse (see p. **19**–26) provide a general level of protection whereby the stability of a building is not put excessively at risk as a result of local structural damage arising from whatever cause. In cases of known risk the special requirements should be included in the design brief.

Many methods are available for confining the effects of accidental damage to the immediate locality of the incident. These include designing to accept the forces involved, the provision of alternative paths for the loadings, 'fail-safe' and 'back-up' structures. Research has been carried out on partial-stability conditions, whereby the remaining components of the building framework are capable of bridging or stringing over an area of total local damage by beam, catenary or membrane action.

Statutory requirements as to fire resistance and means of escape are devised to ensure continued stability for sufficient time to permit evacuation of the building and fire fighting to protect adjoining property.

STRUCTURAL TESTS

The behaviour of the structure is normally assessed by analytical methods but may also be estimated by tests on prototypes or models or by a combination of analysis and experiment. Prototype testing is sometimes used, for example, in pre-cast concrete construction where the accuracy of the design assumptions may be in question or, in cases of repetitive application, where a better or more reliable understanding or structural behaviour may lead to economy. Model testing may be used to determine internal forces or stresses and in special cases photoelastic analysis may be used to check complex local stress conditions, e.g. around service openings in major structural members.

WIND EFFECTS ON BUILDINGS

The air flow around a particular building is affected by the adjacent land and building complex and by the shape and size of the building itself, the roof type, the position and size of overhangs, the area and location of openings and the direction of the wind. Account needs to be taken of the loading effect of wind turbulence on the building as a whole (and perhaps during construction) and on components; the local wind environment in the vicinity of buildings, the general weather tightness, natural ventilation and the air pollution around buildings also need to be taken into account.

CP3 Chapter V, Part 2: 1967 gives the method to be used for assessing wind loads on individual buildings and takes account of the location of the building, the topography, the ground roughness, the building size, shape and height and a statistical factor related to the probability of given wind speeds occurring during the specified life of the buildings. For groups of buildings, particularly those including tall buildings, the environmental effects are frequently studied by means of model tests in wind tunnels.

Movement

The problem of movement in buildings is not so much the determination of its absolute value, but more that of achieving a compatibility of movement between parts. Without this, cracking and other disturbances are inevitable and are likely to recur even after repair; in such cases, the accurate assessment of movement serves little purpose. On the other hand, the simple recognition that relative movement will occur, coupled with a broad assessment of its significance, are essential first steps in establishing a compatible design in which the actual amount of movement is relatively unimportant.

When significant parts of a building tend to move appreciably relative to each other, they can be separated into independent blocks. Within each block differential movement will occur between elements, e.g. between the structure and partitions, and also between different parts or different materials forming a particular element.

Factors affecting the division of a building into blocks include: differential foundation settlement due to load or soil variations, changes in foundation type or major intervals in construction; longitudinal movements due to changes in temperature, shrinkage or prestress (immediate and long term); abrupt changes in building plan or floor or roof level; abrupt differences in structural stress. Within each block, the most common movement problems are: partitions damaged by floor deflection or relative longitudinal movement; crushing or buckling of cladding and partitions due to relative vertical movement between them and the structure; separation of surfacings due to movement relative to the backing material.

Structural joints include hinge details, to permit rotation; expansion and contraction joints, commonly used with flexible seals; and complete separation, for example, where double columns are introduced. Joints between nonstructural elements allow for expansion or contraction or lozenging by providing appropriate horizontal and vertical gaps between the elements which are sealed with a flexible material or by cover strips secured to one side of the joint. Joints exposed to the weather require special attention in detailing and manufacture and in the choice of sealing materials. The open drain joint coupled with an effective air and water back seal has proved a successful and reliable joint.

Wherever possible, restraints to longitudinal movement should be avoided by appropriate design and location of wind stiffening cores or bracing walls. For example, in rectangular buildings it is frequently appropriate to locate a service core at one end, providing rigidity in two directions, and a cross wall at the other which provides the necessary torsional stiffness but permits free movement in the longitudinal direction. Temperature movements are minimised by keeping the structure within the insulation envelope.

Structural arrangement

A great variety of structural arrangement is used in practice depending upon the planning, functional, aesthetic and economic requirements of the building and site. Even for similar buildings, the relative priorities attached to the individual factors affecting structural decision will vary depending upon the particular circumstances and the views of the client. General rules governing structural arrangement cannot be given, but in most cases the following principles are valid: vertical loads should be transmitted along the shortest and most direct path to the supporting ground; when minimum structural sections are dictated by nonstructural requirements, e.g. sound insulation, they should be deployed to gather extra load; vertical load-bearing elements should be stacked directly over each other; transfer structures, including vertical ties, should be used only when justified; structural layout should be regular to increase repetition of identical building components and improve construction rhythm.

The structural arrangement, construction method and material may be chosen on the basis of some overriding consideration such as the provision for future alteration or extension, passage for services, speed of construction, availability of labour and materials or difficulty of access. Where alternatives are equally appropriate, the choice is generally made by cost comparison, but this must take into account all aspects affecting the total cost of the building including, where appropriate, running and maintenance costs.

The primary structural systems available for spanning vertical loads across space are the catenary, acting in tension; the arch, in compression; and the beam, in bending; the last being of most importance in buildings. Columns and walls are the commonly used members for vertical load support, and on occasion tension members are used to suspend lower work from high-level beam or cantilver construction. Frame structures are of two basic types, those in which horizontal forces are taken by shear walls or bracing and those in which the frames, comprising columns rigidly jointed to beams or slabs, are designed to accept horizontal as well as vertical loading.

Structures relying solely on frame action for stability become increasingly inefficient with height and reach a normal practical limit of about fifteen to eighteen storeys. For tall framed buildings, sway limitations (which must take account of increased lateral displacement due to the action of vertical loads on frames deflected by wind loading) are such that much larger and stiffer members than necessary for vertical loads alone would need to be used.

Resistance to vertical load

COLUMNS AND WALLS

Column and load-bearing wall positions are determined mainly by the building use. Where large clear spans are not necessary and regular and permanent space division is required, as for example in multi-storey flats, load-bearing walls are commonly adopted. Column spacing, in conjunction with the floor construction, will be affected by the available structural depth and the necessary provisions for the passage of mechanical and electrical services. Concrete columns may be of any reasonable shape; in tall buildings the section may be kept constant for construction convenience or reduced at upper levels to save usable floor area. When floor space is particularly valuable, spiral reinforcement or solid steel sections may be used or tension supports may be provided; the latter may be prestressed, in stages, to minimise the required sectional area and eliminate cracking.

FLOORS

In addition to their structural function, floors may need to provide impact and airborne sound insulation, thermal insulation and appropriate fire resistance depending upon their location in the building and the building type.

In situ concrete floors. Reinforced and prestressed concrete flat slab construction has the advantages of: minimum structural depth; adaptability to irregular arrangements of column or walls; does not require a suspended ceiling, but if one is provided gives complete freedom for the passage of services. Solid reinforced concrete construction, 150–250 mm thick, may be used for spans up to about 7 m, or greater if post-tensioned. Punching shear at the columns, midspan deflection and the size and location of openings require special attention. When larger spans or weight reductions are required, the slab may be coffered to provide one- or two-way spanning: standard or purpose-made plastic, steel, or fibreglass moulds are available and can produce a visually attractive self-finished ceiling; alternatively, permanent cavity formers may be left in position. Beam and slab construction, at the expense of greater floor depth and slower construction, has the advantages of: longer spans; ready provision for large opening, e.g. for stairs and lifts; adaptability to varying size and building shape; relatively lightweight. It is most economic when large repetitive areas, or a heavy loading is required. In situ beam and slab construction may be the only valid method of construction for complex shapes and areas.

Pre-cast floors. These have the advantages of speed of erection and quality and accuracy of manufacture; they are economic for medium to large spans particularly where layouts are straightforward and repetitive; they may be used in conjunction with steel or concrete frames in addition to load-bearing wall construction. The largest use of pre-cast flooring is in the form of hollow or solid slabs, reinforced or prestressed, with widths varying normally from 300 mm to 2.7 m and up to 7 or 8 m in the case of large panel construction where such slabs may incorporate openings, ducting and floating screeds. Pre-cast slabs may be designed to act compositely with the supporting concrete or steel beams. For longer spans single or double tee beams, which combine floor slab and beam are available; they are connected by welding, bolting or in situ jointing to provide secondary load distribution and equalise deflection.

Composite floors. These rely on the composite action between an in situ concrete topping and pre-cast concrete soffit elements, which may take the form of pre-cast concrete ribs, planks or slabs incorporating the tension element of the composite slab and having projecting reinforcement to ensure composite action. Alternatively, permanent steel shuttering of hollow construction for the passage of services, or of expanded metal may form the soffite. These floors are easily erected, do not need shuttering and provide the shallow depth of in situ slabs. A further example of composite action is that between concrete floor slabs, in situ or pre-cast, and steel beam or frame construction.

TRANSFER STRUCTURES

It is frequently found in multi-storey construction that the special functions of the lower storeys require an arrangement of columns or bearing walls very different from that required for the efficient support of the superstructure above. A transfer structure is then required to transmit the typical floor column loads to the fewer but larger supports beneath. Often a very substantial structure involving storey height beams is required—this should be taken into account in the early stages of design since it may provide a suitable location for plant. An alternative is to place the transfer structure at roof or intermediate upper floor level and to suspend the lower structure by means of steel or prestressed concrete hangers.

Resistance to horizontal load

In low- to medium-rise buildings, the structural system is designed primarily to resist the vertical loads and is then checked for lateral forces which may be taken by moment resisting frame action, braced frames or shear walls conveniently located around lift shafts or stair wells. Simple shear walls provide the necessary horizontal restraint for 'pin-jointed' frameworks which are often convenient and economic. A minimum of three bracing walls are required so disposed as to provide: resistance in each of two directions at right angles; an overall torsional resistance; minimum restraint to thermal or similar movement.

Tall buildings

With increasing height, resistance to horizontal forces begins to dominate the design and may add substantially to the total cost. In addition to structural safety, sway limitations must be satisfied in terms of horizontal accelerations as well as actual movement. In principle, the lateral resistance may be provided by frames in bending, braced frames or by shear walls, as in lower structures, but the greater magnitude of the forces and movements necessitates a more sophisticated approach. The various systems in use are illustrated in Figure 19.6.

FRAMES IN BENDING

Internal frames are comparatively inefficient and flexible as a result of the planning necessity for a wide spacing of internal columns and limited floor beam depth. On the other hand, exterior frames, formed in the plane of the external wall, may have closely spaced columns connected by deep spandrels. In this way, the entire perimeter of the building may be developed as a major lateral load-resisting system referred to as the 'boxed frame' or 'framed tube'. Subject to 'shear lag' considerations, the building walls act respectively as the webs or flanges of a box section cantilevering from the foundation. To allow for 'shear lag', two channel sections may be considered operative in place of the complete box. Deep spandrels, although advantageous, are not essential and apartment buildings have been constructed in the USA up to 46 storeys in height using part of the adjacent flat slab floor as the beam continuous with the closely spaced external columns.

BRACED FRAMES

Braced frames usually incorporate single or double diagonal braces or K bracing within the beam and column framework and may be used internally, around service cores or in the external wall. When used externally, intermediate transfer structures may be incorporated to transmit a major proportion of the total vertical load to the corner columns. Such transfer structures may also support and hence separate different configurations of internal supports required when the function varies between different vertical zones of the building. Such arrangements have a major impact on the internal planning and external appearance, and have important relevance to the planning and construction of very tall buildings using steelwork. Similar external truss action may be achieved in concrete by blocking out windows to form solid and continuous diagonal members or, for limited height, by using pre-cast cladding forming a multiple diagonal system. When they can be accommodated, internal trusses can bring into action lengths of external wall otherwise rendered ineffective by 'shear lag'. By alternating the plan position of such internal trusses, from storey to storey, the structural span of the floors may be reduced to half of the architectural planning bay, by providing additional hanging supports from the trusses in the storey above.

SHEAR WALLS AND CORES

Shear walls may be internal or external or may surround internal service areas to form cores; their location and dimensioning are major design elements since they seriously impinge on internal planning and may affect external appearance. In the early formative stages of design, quick structural appraisal of alternative shear walls will be required followed by careful design and analysis of the final arrangement.

In office buildings the service core which includes lifts, stairs, ducts and toilets can occupy 20% or more of the total floor area while fire and sound insulation require this area to be bounded by heavy wall construction. These conditions naturally lead to the use of the service core as a major vertical wind brace. However, away from the core area, open office space is generally required and even if partitions are used they would be demountable to allow for future alteration; internal bracing walls are therefore a planning impediment in offices and are generally avoided. External bracing walls, however, are often used, generally in conjunction with internal cores. In housing or hotels, partitions are normally fixed, need to be heavy for sound insulation and are regularly spaced; they therefore provide many convenient locations for internal bracing walls.

When shear walls alone are used, the general structural requirements are: at least three must be provided of which at least two must be parallel and widely spaced, to provide torsional resistance, with the third at right angles; the centroid of the shear walls should be close to the centre of gravity of the loading; walls likely to need very large openings should be avoided if alternatives are available since their stiffness and hence load-resisting contribution will be substantially diminished.

Walls with openings produce a stiffness intermediate between that of the total combined length acting as a monolithic wall and the sum of the stiffnesses of the parts acting separately, depending upon the relative size and location of the openings. Normal analysis assumes that all the shear walls or cores act from a completely rigid foundation such that relative rotation or vertical movement does not occur. Since even small relative movements could seriously invalidate the design, it is important that this assumption is realised in the foundation design or if this is not practical or economic, the shear wall system should be designed to suit. In general, the total horizontal load is distributed between the shear walls in proportion to their relative stiffnesses taking into account any eccentricity of the applied load. The floor system then acts as a horizontal diaphragm equalising horizontal displacement and rotation at each floor level.

Shear walls and cores are almost invariably constructed in concrete and are often slip formed. Pre-cast large panels have been successfully used, particularly in high-rise blocks of flats, with the combined functions of load bearing walls and vertical wind braces. The joints between the panels and the lintels over openings require particularly consideration in the design of such structures.

The use of shear walls or cores is an economic and efficient method for resisting large horizontal forces but, in most cases, deflection limits their use to below 30–40 storeys. However, if the shear wall, and building, are shaped in plan along their length, a vertical shell or folded plate action could be developed permitting greater heights; otherwise a 'boxed frame' or one of the combined systems described below is required to control deflection. Another limitation of shear walls, particularly if lightly reinforced, arises from the possibility of brittle failure which could make them unsuitable for seismic structures. However, by suitable framing around the shear wall, the necessary ductile behaviour can be obtained to absorb the considerable strain energy arising from the earthquake.

COMBINED SYSTEMS

Internal cores may be used in conjunction with external moment-resisting frames so that the substantial overturning resistance of the façade frame is combined with the excellent shear resistance of the core to form a highly efficient total system known as 'tube in tube'. This arrangement still relies upon closely spaced external columns; when widely spaced external columns are required, a beneficial interaction between the core and the external columns can still be obtained by connecting the two with deep stiff beams rigidly connected to the core and located at convenient levels (roof and service floors). In this arrangement, the core continues to take the shear but the overturning resistance of the full building depth is called into play and deflection is reduced. Another advantage of this system is that it can help control the effects of differential expansion or contraction of the external columns.

VERTICAL MOVEMENT

Another aspect distinguishing tall buildings is the need to consider the possibility of differential vertical movements due to temperature or stress and, in the case of concrete structures, those due to creep and shrinkage. The movement is most marked between the internal structure and the external columns particularly if the latter are totally or partially outside the external wall and glazing. It affects most the external cladding and partitions located at right angles to the external wall, as well as any linking structural element.

The effects are best controlled by attempting to achieve uniformity of stress and exposure (including insulation where necessary) and uniform surface/volume ratios for the concrete elements to reduce differential shrinkage or creep. When the problem is particularly severe, the building can be divided into two or more sections by incorporating intermediate transfer structures, in effect producing horizontal expansion or contraction joints. Alternatively the movement can be restrained by stiff beams connecting the external columns to the core. Another approach is to freely permit the movement and incorporate appropriate movement details in the structure, partitions and finishings.

Special structures

Special structures include means of covering or enclosing large areas without internal supports (by means of beam, membrane, tension, skeletal or pneumatic structural action), space frames and tall buildings, each of which has its own particular field of application and specialised technology. They also include cases where the normal assumptions regarding structural action may not apply. For example, deep beams, in which the span-to-depth ratio is small (less than 5:1) behave differently from shallower beams and the normal theory of flexure does not apply. In portal frames, in which the beam spans are large in relation to the column lengths, the distribution of bending moments is highly sensitive to the relative stiffnesses of the members, and the normal assumptions of stiffness of reinforced concrete sections (whether cracked or uncracked) or of steel sections

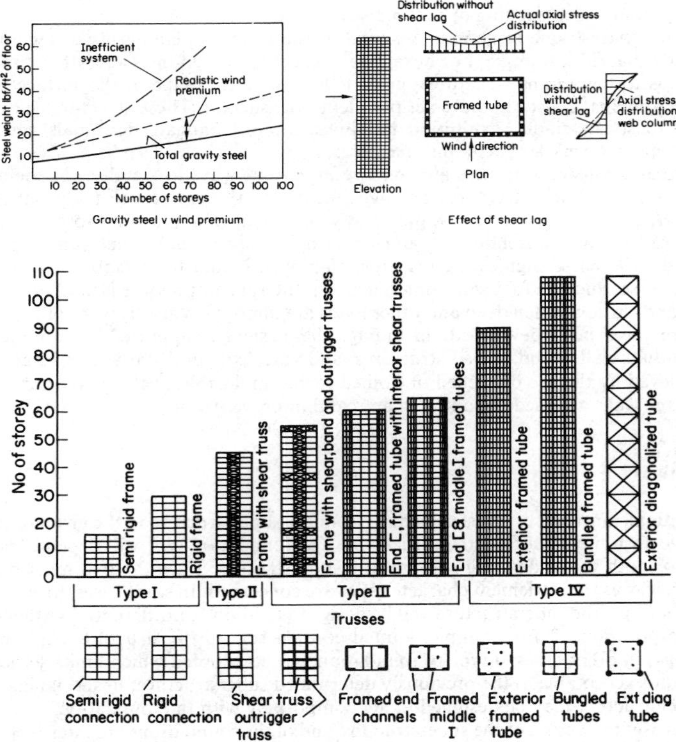

Figure 19.6 (From a Paper by Hal Iyengar: 'Preliminary Design and Optimisation of Steel Building Systems', ASCE/IABLE Conf. on Planning and Design of Tall Buildings, USA, 1972.)

at yield stresses are insufficiently accurate. Similarly the geometry of the system may exaggerate the effects of movement.

Membrane structures involve the use of thin surfaces geometrically arranged to support vertical loading mainly by forces parallel to the surface. They may be folded plates or singly curved, as in barrel vaults and arches, or doubly curved as in domes, hyperbolic paraboloids and other special forms of curved surface. While secondary bending effects are present at right angles to the surface, the main internal forces are parallel to it and it is essential that the surface and boundary conditions are geometrically correct for resisting the loading. The simple example is the dome, in which compressive forces occur within the dome, and ring tensions or external restraints are required at the perimeter. Concrete, timber and sometimes steel are used in forming long-span membrane structures; they use materials efficiently, but their economic viability depends mainly on the workmanship and labour required. Folded plates provide functional and aesthetically interesting roofs, in which normal flexural action occurs; additional considerations of edge support, end shear, buckling and distribution of out-of-balance loading also apply.

Tension structures involving cable-supported sheeting have been used in exhibition buildings and in sports stadia; the structural system involves steel cables acting in catenary from which decking or tenting is supported.

Skeletal space-frame structures are used in the form of plane grids of rectangular, diagonal ('diagrid'), triangular or hexagonal pattern, arches, domes and other structures analogous to membrane structures, in that the geometric shape of the surface is such that the principal resisting forces act parallel to the surface. These structures have been applied to sports stadia, exhibition buildings, aircraft hangars, terminals and other places requiring very large column free areas.

Pneumatic construction uses air pressure in various ways to stabilise the membrane of the building. Such structures are light, economical, easy to erect, dismantle and transport, and have proved to be practically successful in application to housing temporary exhibitions, warehousing, and covering to sports halls and other specialist buildings. The basic engineering principle employed is that the membrane can accept tensile stresses and will fold when not in tension. Internal air pressure is used to maintain membrane tensions when dead and other loads are imposed. Various types of pneumatic structure have been developed, including, the basic air-supported membranes and inflated dual-walled and ribbed structures, and various hybrid types. The larger spans are achieved by the use of arched or domed forms, and cables, cable nets, and internal membrane walls are used to control shapes and improve stiffness.

Foundations

Foundations must safely transfer loading from the superstructure to the ground without excessive absolute or differential settlement. The foundation type will depend upon the nature of both the underlying soil and the superstructure since the two must be compatible as far as the settlement characteristics are concerned. In some cases, to economise in foundations, the superstructure will be designed to allow for differential settlement by the incorporation of, for example, jointed construction in the structure, cladding and finishings; in other cases, a type of foundation will be adopted which limits settlements to amounts acceptable to the previously determined superstructure; in special instances, the superstructure may be designed to act compositely with the foundation.

In all cases the design of the superstructure and of the foundations are interdependent: knowledge of the soil conditions is thus essential at all stages of the design process beginning with at least broad local knowledge prior to land acquisition. Such information may also influence the location of particular buildings on the site.

The risk of mining subsidence must be assessed at an early stage and the appropriate course of action decided: whether to seal and grout the workings, pile through the workings, design the superstructure to accommodate subsidence, or to accept the risk without special action.

The main types of building foundation are individual pad footings, strip footings, rafts, piers and deep footings, deep spread foundations and piles. Basement construction is useful in reducing settlements and in controlling differential settlements between parts of buildings of different height.

Ground floor slabs may be suspended, but normally bear directly on the soil: they are required to span local weaknesses or voids due to settlement and to transfer local loading to an appropriate area of the ground. Their design is largely an empirical process; specification, workmanship, joints and surface treatments are important.

Basement construction and ground floor slabs must resist penetration of water or water vapour, to a degree determined by the building use; provisions may include: land drainage; good quality concrete of suitable thickness and with appropriate workmanship and details particularly at the joints as to be adequately waterproof; and the provision of internal or external waterproof membranes. In some buildings it may be more economic to accept some leakage and provide for discharging it, than to attempt completely watertight construction. The presence of water may give rise to serious uplift problems in basements during or after construction unless adequately provided for in design.

BIBLIOGRAPHY

General design and management

Job Book, RIBA
LONGMORE, J., *Daylight Protractors,* HMSO
NEUFERT, E., *Architects Data,* Crosby Lockwood, London (1970)
Plan of Work, RIBA
SLIVA and FAIRWEATHER, *A. J. Metric Handbook,* The Architectural Press, London (1969)
The Architects Journal, A. J. Information Library

Cost planning and control

Cost Control in Building Design, HMSO (1968)
NISBET, J., *Estimating and Cost Control.*
STONE, P. A., *Building Design Evaluation: Costs-in-Use.*
STONE, P. A., *Building Economy: Design, Production and Organisation.*

Internal environment

'Condensation in Dwellings', Current Paper 31/71, HMSO: Part I—'Design Guide'; Part II—'Remedial Measures'.
FABER, O., and KELL, J. R., *Heating and Air-Conditioning of Buildings,* The Architectural Press, London (1958).
IHVE Guide to Current Practice, The Institution of Heating and Ventilating Engineers (1970).
'Insulation Against External Noise'—1 and 2, Building Research Station Digests 128 and 129 HMSO
KNUDSEN, V. O., and HARRIS, C. H., *Acoustical Designing in Architecture,* Wiley, New York and London (1962)
PARKIN, P. H., and HUMPHREYS, H. R., *Acoustics, Noise and Buildings,* Faber and Faber, London (1958).
'Prevention of Condensation', Building Research Station Digest 91 (1968).
PURKIS, H. J., *Building Physics Acoustics,* Pergamon, London (1966). 'Recommendations for Lighting Building Interiors', IES Code, Illuminating Engineering Society, London (1968).
'Theoretical and Practical Aspects of Thermal Comfort', Current Paper CP14/71. Building Research Establishment.

Regulations

Building Regulations 1972—General Guidance Note, HMSO (1972).
ELDER, A. J., *The Guide to the Building Regulations.* The Architectural Press (1972).
Public Health Act 1936 HMSO
Public Health Act 1961 HMSO
The Building Regulations 1972 HMSO

Progressive collapse

AMER, G. S. T., and JUMAR, S., 'Tests on Assemblies of Large Precast Concrete Panels', Building Research Establishment Current Paper CP20/72.
'Building Regulations 1965, Multi-Storey Framed Construction', Department of the Environment Circular 11/71, HMSO (1971).
WILFORD, M. U., and YU, C. W., 'Catenary Action in Damaged Structures'. Seminar 'The Stability of Pre-cast Concrete Structures', DOE/CIRIA, London (1973)

Collapse of Flats at Ronan Point Canning Town, HMSO (1968).
London Building (Constructional) Amending By-Laws 1970, Notes for Guidance, GLC.
CREASY, L. R., 'Partial Stability', *Struct. Eng,* (Jan. 1972).
Resistance of Buildings to Accidental Damage, Institution of Structural Engineers RP/68/05 (1971).

Tall buildings

COULL, A., and IRWIN, A. W., 'Analysis of Load Distribution in Multi-Storey Shear Wall Structures', *Struct. Eng,* (Aug. 1970)
KHAN, F. R., 'The Future of High Rise Structures', *Progressive Architecture,* (Oct. 1972).
KHAN, F. R., and AMIN, N. R., 'Analysis and design of Framed Tube Structures for Tall Concrete Buildings', *Struct. Eng* (Mar. 1973).
PEARCE, D., and MATTHEWS, D. D., *An appraisal of the Design of Shear Walls in Box Frame Structures,* Department of the Environment, HMSO (1973)
Planning and Design of Tall Buildings, ASCE/IABSE International Conference; Lehigh University, Bethlehem, Pennsylvania, USA (1972).
Tall Buildings: the proceedings of a Symposium on Tall Buildings Edited by A. Coull and E. Stafford Smith, Pergamon, London (1967).
'The Behaviour of Large Panel Structures', CIRIA Report No. 45, London (1973).

Wind

'The Assessment of Wind Loads', Building Research Station Digest 119, HMSO.
'Wind Effects due to Groups of Buildings', Building Research Establishment CP23/70.
'Wind Environment Around Tall Buildings', Building Research Station Digest 141, HMSO.

Special structures

FISCHER, R. E., Ed., *Architectural Engineering—New Structures,* McGraw-Hill (1964)
SIEGAL, C., *Structure and Form in Modern Architecture,* Crosby Lockwood, London (1962)

Miscellaneous

Design Guides on Roofs, Department of the Environment.
NISSEN, H., *Industrialised Building and Modular Design,* Cement and Concrete Association, London (1972).
Principles of Modern Building, Vols 1 and II, HMSO.
Proc. Symp. on *Design for Movement in Buildings,* Concrete Society London (1969)
Proc. Symp. on *Weathering of Concrete,* Concrete Society, London (1971)
Specification—Vols I and II, The Architectural Press.
Weathering and Performance of Building Materials, University of Manchester, Medical and Technical Publishing Co. Ltd (1970)

20 HYDRAULIC STRUCTURES

HYDRAULIC STRUCTURES 20

20 HYDRAULIC STRUCTURES

A. R. THOMAS,
Formerly Consultant, Binnie and Partners

OPEN-CHANNEL STRUCTURES

Basic concepts

THE BERNOULLI THEOREM

Two important theorems in the hydraulics of flow through structures are the Bernoulli and pressure–momentum theorems. The former (see Section 5) expresses conservation of energy, and when applied to straight line flow in an open channel is

$$Z+d+V^2/2g = \text{constant} \tag{1}$$

(see Figure 20.1) where Z is the elevation of the bed, d is the depth of flow above the bed, V the mean velocity, and g the gravitational constant. Where there is head loss, the constant C is replaced by $C-H_L$ where H_L is the head loss between two specified points. Where the flow is curvilinear, depth will vary across the channel and d is a mean value. In cases of significant departure of velocity distribution from uniform, a factor α should be included in the term for velocity head. Under normal conditions of flow in wide uniform channels, $\alpha = 1.02$ for smooth boundaries and higher values for rough. For example, if $n/d^{\frac{1}{6}} = 0.0225$ (where n = Manning's roughness) $\alpha = 1.12$. In order to simplify calculations for ordinary purposes α is often assumed to be unity.

Specific energy of flow is energy related to a given elevation, usually the floor or bed level of the channel, thus specific energy head H is given by

$$H = d+V^2/2g \tag{2}$$

In the case of channels of rectangular cross section, equation (2) can also be expressed as

$$H = d+q^2/2gd^2 \tag{2a}$$

where q is the discharge per unit width of channel, Q/B, where Q is the total discharge and B the width. To derive d from known q and H, Figure 20.2 may be used.

CRITICAL DEPTH

In critical flow $V = (gd)^{\frac{1}{2}}$. It follows from equation (2a) that, in a rectangular channel, critical depth d_c is given by

$$d_c = (q^2/g)^{\frac{1}{3}} = \tfrac{2}{3}H \tag{3}$$

As may be seen from Figure 20.2, critical flow represents the condition of minimum specific energy for a given discharge. In a channel of nonrectangular cross section,

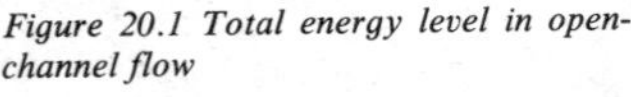

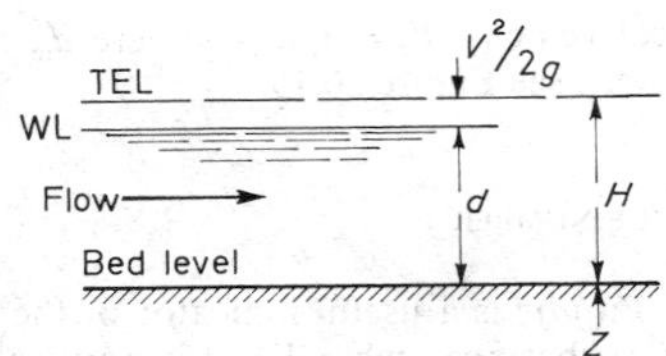

Figure 20.1 Total energy level in open-channel flow

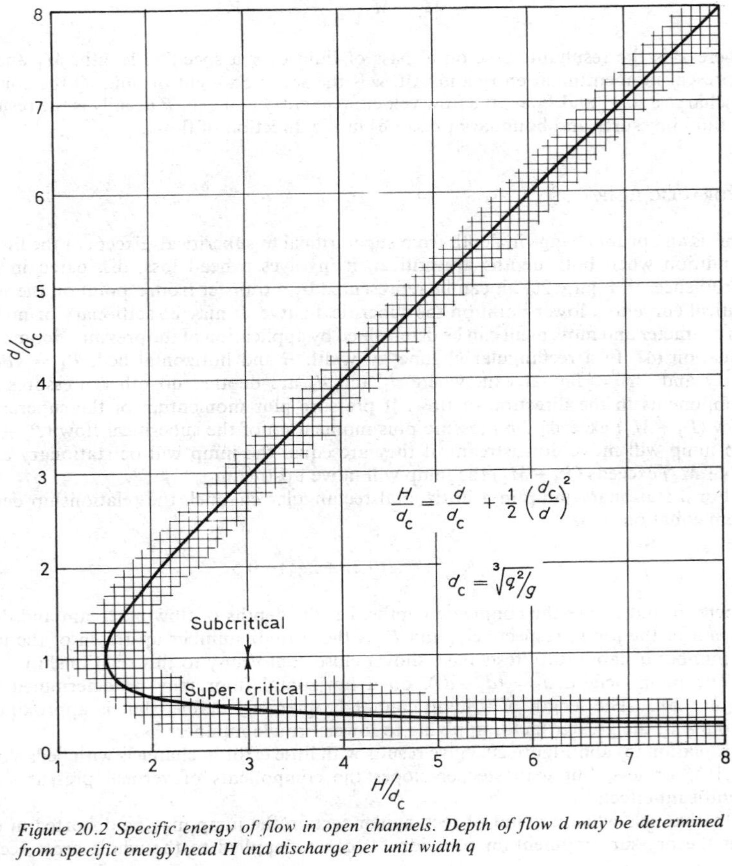

V²/2g
TEL
WL
Flow
d
H
Bed level
Z

8
7
6
5
4
3
2
1
0
d/dc
H/dc = d/dc + 1/2 (dc/d)²
dc = ∛(q²/g)
Subcritical
Super critical
1
2
3
4
5
6
7
8
H/dc

Figure 20.2 Specific energy of flow in open channels. Depth of flow d may be determined from specific energy head H and discharge per unit width q

critical velocity $V_c = (gd_m)^{\frac{1}{2}}$ where d_m is mean depth. For critical depths in circular channels, see Figure 20.19.

FROUDE NUMBER

$F = V/(gd)^{\frac{1}{2}}$ is a useful indicator of the stability of free surface flow. When $F<1$, the flow is subcritical; when $F = 1$ it is critical and when $F>1$, supercritical. As F approaches unity from either direction, the flow becomes unstable and surface waves may develop. Surface undulations may occur in subscritical flow when F exceeds 0.5.

THE PRESSURE–MOMENTUM THEOREM

Unlike the Bernoulli theorem, this applies whether there is head loss or not. It follows from Newton's second law and can be expressed as

$$P = M_2 - M_1 = \frac{w}{g} Q(V_2 - V_1) \tag{4}$$

where P is the resultant force on a mass of fluid over a specified length, M_1 and M_2 represent momentum at entry and exit, w is the specific weight of fluid, Q the constant discharge and V_1 and V_2 are the flow velocities at entry and exit. P usually is the resultant of fluid pressures and boundary pressures in the direction of flow.

HYDRAULIC JUMP

This is an abrupt change in depth from supercritical to subcritical. Except at the limiting condition when both depths are critical, it involves a head loss, dissipated in extra turbulence. In Figure 20.2 it can be represented by a transfer from a point on the supercritical curve to a lower point on the subcritical curve. It may be stationary or moving. Its character and movement can be determined by application of the pressure–momentum equation (4). In a rectangular channel of width B and horizontal bed, $P_1 = \frac{1}{2}Bd_1^2$ at entry and $P_2 = \frac{1}{2}Bd_2^2$ at exit, where d_1 and d_2 are depths; no other pressures have components in the direction of flow. If pressure plus momentum of the supercritical flow $(P_1 + M_1)$ exceeds the pressure plus momentum of the subcritical flow $(P_2 + M_2)$, the jump will move downstream, if they are equal the jump will be stationary and if $(P_2 + M_2)$ exceeds $(P_1 + M_1)$ the jump will move upstream.

For a stationary jump in a horizontal rectangular channel, the relationship derived from equation (4) is

$$\frac{d_2}{d_1} = \sqrt{(0.25 + 2F_1^2)} - 0.5 \tag{5}$$

where d_1 and d_2 are the conjugate depths, i.e. the depths of flow upstream and downstream of the jump, respectively, and F_1 is the Froude number upstream of the jump. A number of laboratory tests have shown close conformity to this relationship.

The jump height, $d_j = (d_2 - d_1)$, on a horizontal floor may be determined from Figure 20.3. The length of a jump cannot be precisely defined but is approximately 5 to $8 \times d_j$, the greater factor applying to lower Froude numbers.[1]

Equation (5) and Figure 20.3 give results with little error in channels with beds sloping at 10% or less, but with steeper slopes the components of vertical pressures have significant effect.

In channels which are not of rectangular section the jump may be distorted in plan, but the pressure–momentum equation (4) can be applied to the whole cross section.

Several methods for calculating the conjugate depths in channels of various shapes are available.[2, 3, 4, 25]

Transitions—subcritical flow

In channels of variable cross section, equation (1) or Figure 20.2 may be used to determine depth of flow, provided changes are sufficiently gradual. In converging flow q and

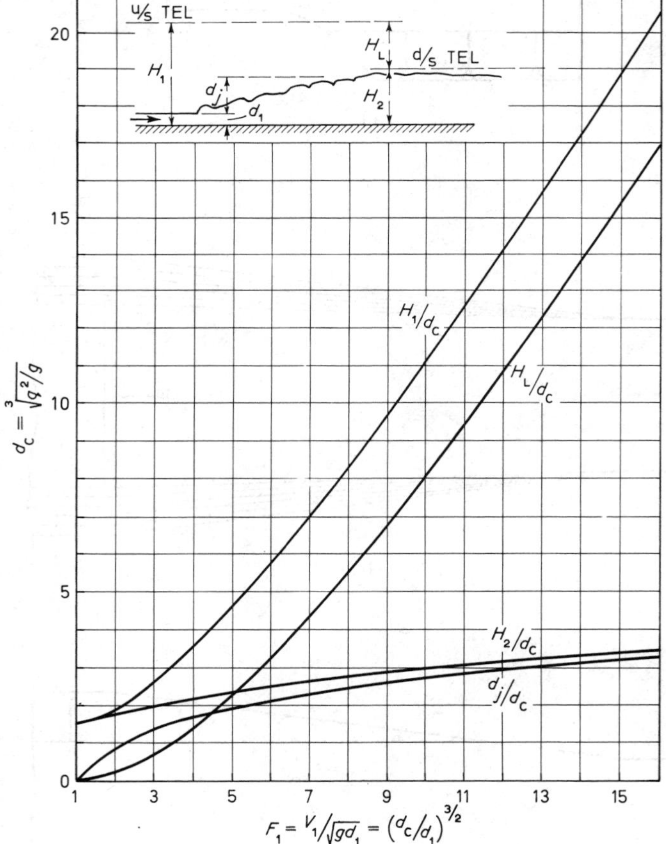

Figure 20.3 Hydraulic jump relationships[23a]

hence d_c increase with the reduction in width. Therefore with constant specific energy H, it is evident from Figure 20.2 that d reduces. As examples, a channel may be contracted at a bridge and allowed to expand downstream, or a gated regulator may have a raised sill. In both cases the surface is depressed in the contraction. Provided the flow remains subscritical the process is reversible in a downstream expansion. If, however, a contraction reduces the depth to the critical value, any further contraction has the effect of raising the upstream head, because critical depth is the minimum depth possible for

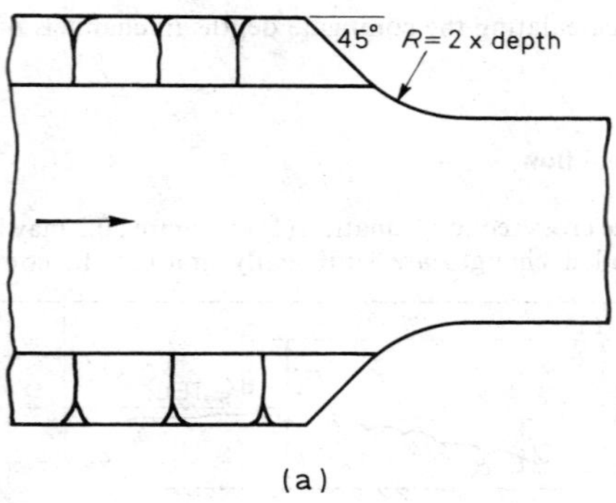

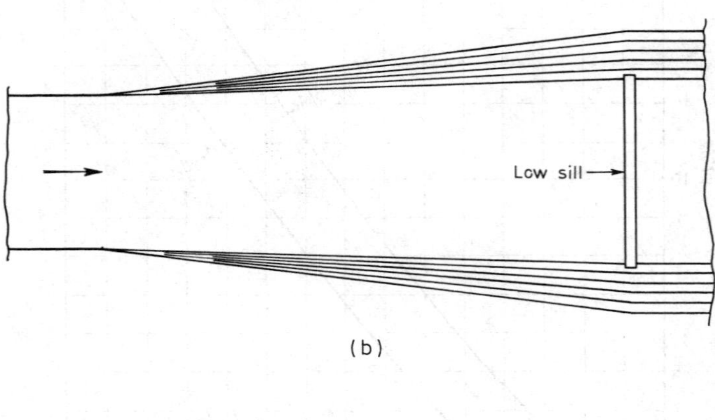

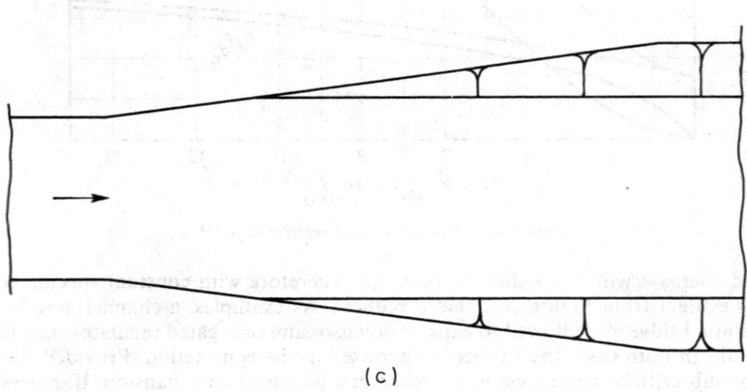

Figure 20.4 Typical transitions for subcritical flow. (a) Contraction from sloping to vertical sides; (b) warped expansion; (c) expansion with vertical sides

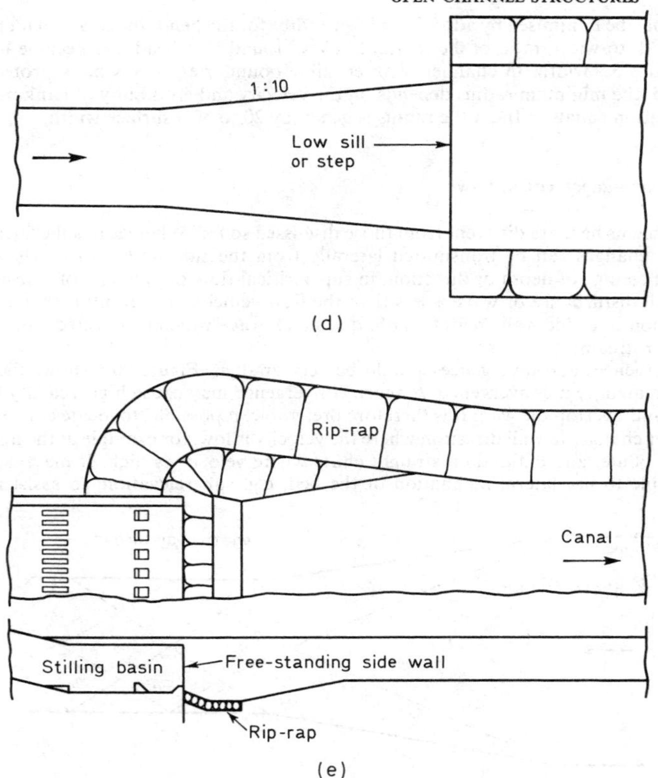

Figure 20.4 (continued) (d) short expansion; (e) example of transition from stilling basin to canal in erodible material

any given specific head—see Figure 20.2. The result is a rise in upstream water level, the excess head generates supercritical flow downstream of the throat, or section of maximum contraction, and is lost in a hydraulic jump where the flow changes back to subcritical. The throat is then acting as a 'control'. If head loss is to be avoided, the Froude number should not be allowed to approach close to unity.

Convergences for subcritical flow may be rapid but external angles in the side walls should be avoided by the use of large-radius curves, as shown in Figures 20.4(a) and (b). Diverging channels in subcritical flow are liable to result in separation of flow from one or both side walls unless expansion is gradual. Side expansions of 1 : 10 are usually satisfactory. Sharper divergences may be followed in some conditions;[5] expansion is assisted by a rising floor, baffle blocks or a raised sill downstream and by a hydraulic jump. The expansion ratio is also a factor—see page **20**–31—where expansions in enclosed flow are discussed. Some examples of diverging transitions are shown in Figures 20.4(c) to (e).

Changes of direction cause head loss because of the secondary flow which distorts the flow pattern; the flow near the bed is deflected more sharply than the surface flow. If the bend is very sharp there may be complete separation at the inner boundary. These effects may be minimised by adopting a large radius for the bend. In rectangular channels with depth to width ratio of 0.6 to 1.2, Shukry[6] found that head loss became minimal with radius 3 × width. In channels with erodible boundaries, unless bank protection is provided, the minimum radius depends on the velocity and erodibility of bank material. On irrigation canals in India the radius is generally 20 to 30 × surface width.

Transitions—supercritical flow

The problems here are different from those discussed so far. Whereas in subcritical flow, pressure changes can be transmitted laterally from the side walls to the whole flow, inducing change of depth or direction, in supercritical flow the velocity of transmission of a small disturbance or wave is less than the flow velocity. The result is that a change in direction of a side wall creates an oblique shock wave which is reflected from side to side downstream.

Convergences and divergences should be very gradual. Figure 20.5 shows the shock waves created by a convergence. A sharp convergence may cause high velocity flow to ride up and overtop the wall. It is therefore preferable, if possible, to locate convergences and other changes in wall direction where the velocity is low, for example at the upstream end of a chute, and maintain a straight chute where velocity is high. It may, however, be possible to use lateral inclination of the bed, e.g. superelevation, to assist in con-

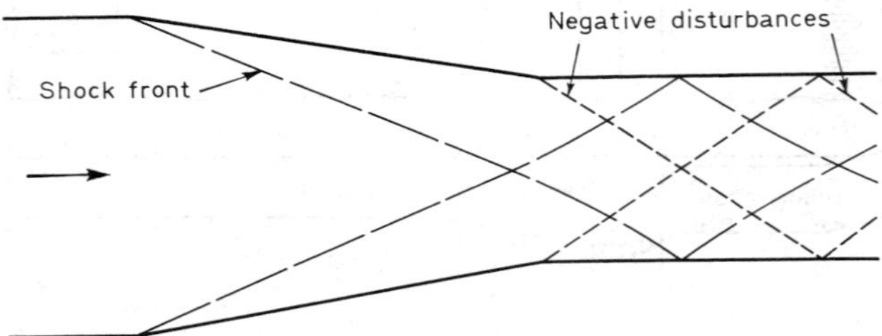

Figure 20.5 Example of shock waves at convergence in supercritical channel. (After Ippen[7])

vergence or divergence. Where shock waves are unavoidable, they will occur in a zigzag pattern for some distance downstream owing to reflection from side walls. The side walls should therefore be high enough to contain them at points of reflection. Sloping side walls, as in trapezoidal channels, are particularly vulnerable. Methods are available for the calculation of pattern and height of shock waves in simple cases and for minimising their effects.[7] Scale models may also be used.

Long-radius bends are preferable to short radius, especially where overtopping is a danger. Changes in direction can be induced by (i) lateral inclination (superelevation) of the bed, (ii) change of side-wall direction and (iii) introduction of divide walls. For (i) the cross slope should be V^2/gr where V is the velocity and r the bend radius; this is the most effective method in wide channels where there is not a large variation of velocity. Gradual approach and exit transitions are needed in which the alignment of the side walls must be matched to the lateral bed slope. (ii) is a method often used, sometimes in combination with (i). Knapp[7] recommends compound curves for the side walls, with radius $2r$ in the approach and exit over a length of $B/\tan\beta$ in each case, where r is the radius of the centreline of the main curve, B is the channel width and $\sin\beta = F$, the Froude number. This arrangement creates counter waves which tend to neutralise the shock waves generated by the main curve, so reducing disturbances downstream.

Method (iii), provision of divide walls, reduces disturbance by shock waves, but may not be acceptable in channels carrying debris. The noses of divide walls in high-velocity flow should be robust to avoid fracture by fluctuating pressures due to local flow separation. It is usually best to adopt long noses at a flat slope.

Large high-velocity flows entrain air—see page **20**–41. Shock waves increase air entrainment and side walls should be high enough to contain the aerated flow.

Supercritical flow on floors which are horizontal or rising may be potentially unstable in that a hydraulic jump may occur, converting the flow to subcritical. If the discharge is variable the flow may be subcritical at low flows and supercritical at high flows, with an intermediate range when either may occur. To determine whether supercritical flow is stable, assume that a jump exists and determine whether it will move upstream or downstream. The subcritical TEL (total energy level) downstream of the jump should be calculated by reference to a possible downstream control. The TEL needed to sustain the jump may be calculated from equations (4) or (5) or derived by obtaining H_2 from Figure 20.3 and adding to the bed level. If the former TEL exceeds the latter, subcritical flow can exist; if not, the jump would move downstream and supercritical flow would be stable at that discharge. Problems concerning the location of a hydraulic jump can be solved in a similar way.

Weirs

Weirs are used to control flow or water levels, or to measure flow. They range from low walls across streams to the spillway crests of high dams.

The basic equation for free flow over weirs is

$$q = c(2g)^{\frac{1}{2}}H^{\frac{3}{2}} \tag{6}$$

where q is the discharge per unit width, c is a discharge coefficient, g the gravitational acceleration and H the total head level upstream above weir crest, normally taken as $d+V^2/2g$, where d is the upstream depth above crest and V is the mean velocity of approach. For convenience in practice, a coefficient C is substituted for $c(2g)^{\frac{1}{2}}$ giving

$$q = CH^{\frac{3}{2}} \tag{6a}$$

and, assuming q to be constant across the weir, $Q = CBH^{\frac{3}{2}}$, where Q is the total discharge and B the width of waterway. Whereas c is nondimensional, the value of C depends on the units used. To convert from English to metric units:

$$C \text{ (ft s units)} = 1.81\ C \text{ (m s units)} \tag{7}$$

In general C varies with head over the weir; only in a few special cases is it constant. There are many weir profiles, each with different characteristics in relation to discharge coefficient and modularity. Weir flow is said to be 'modular' or 'free flow' when it is unaffected by tailwater level. The point at which a rising tailwater begins to affect the upstream head or flow is termed the 'modular limit'. Values of the coefficients of weirs of many different profiles have been published, for example by King and Brater.[8] (See also Section 5.) In this Section, some types in general use are considered.

Sharp-crested weirs formed of metal plates are used for precise measurements of flow. Flow over weirs with narrow crests having rectangular upstream corners is effectively sharp crested, with a coefficient C approximately 1.8, provided the nappe springs clear and is fully vented. In practice such weirs are often not fully vented, resulting in a rather higher, variable coefficient.

Triangular profile weirs have sensibly constant coefficients throughout their modular range; no venting is required and the coefficient is greater than that of a sharp-crested weir. For example, the Crump weir, Figure 20.9, with 1 in 2 upstream and 1 in 5 downstream slope, has a free flow coefficient C of 1.96 and a modular limit (within 1% of discharge) of 0.75 (see *Submerged weirs*, **20**–12). Weirs of this type are widely used for measurements of stream flows.

Trapezoidal profile weirs, with flat upstream and downstream slopes and a narrow horizontal crest, formed by the gate sill, are often used in gated controls and barrages—see for example Figure 20.11. They have a free-flow coefficient which is variable but generally exceeds 1.7 and under drowned conditions the afflux is small.

Broad-crested weirs have horizontal crests wide enough for parallel flow effectively to develop. Control is then at the point of critical depth so that $C = 1.70$. To ensure that this condition applies and C is constant, the upstream edge should be rounded to avoid the formation of a roller above crest level. In practice the value of C is 1% to 3% lower due to friction loss. If the downstream floor falls at a gentle slope—say 1 in 10, the

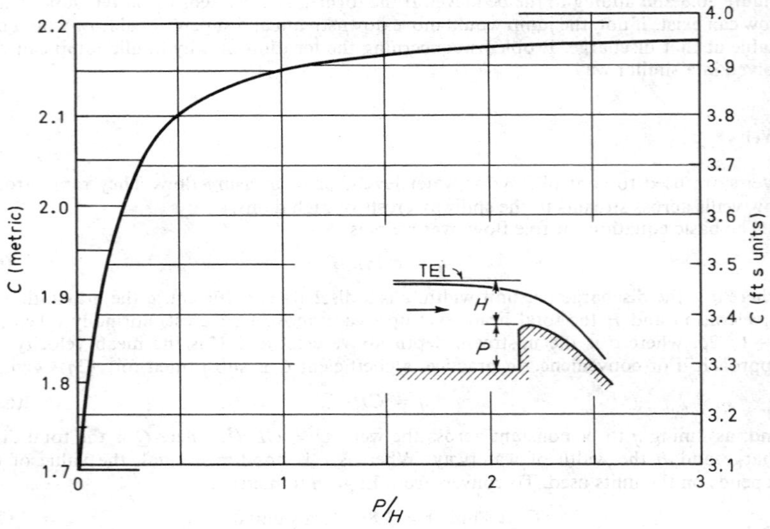

Figure 20.6 Discharge coefficient of free-nappe weirs at design discharge. (Based on USBR (US Bureau of Reclamation) data)

modular limit is between 0.7 and 0.8. Broad-crested weirs have been extensively used for flow measurement and for proportional distribution of flow at dividing points in irrigation systems.

Free-nappe profile weirs, with profile according to the shape of an undernappe of flow over a sharp crested weir (Figure 20.6) have been widely used for overflow spillway crests. The standard profile is one with vertical upstream face and weir height P large compared with head over crest, H. The profile varies with smaller values of P/H and sloping upsteam faces. This profile has the advantages that C is comparatively high for the profile discharge (i.e. the discharge corresponding to the nappe profile used), sub-atmospheric pressures do not develop within the range up to profile discharge, no venting is required and the flow characteristics are well documented and predictable.

The coefficient C of the standard weir at profile discharge is shown in Figure 20.6. By adopting a profile discharge lower than the maximum discharge, a higher coefficient is obtained at flows exceeding profile discharge.[9] Discharge in excess of the profile discharge causes pressures on the face of the weir to fall below atmospheric in the vicinity of the crest where the curvature is sharp. This is usually acceptable provided that the structure is safe against uplift, and a reasonable margin of pressure is allowed above cavitation level to allow for fluctuations. In tests with free-nappe profile weirs the following minimum pressure heads were measured on the crest[10]—see Table 20.1.

Table 20.1

Minimum pressure heads

H/H_0	1.00	1.17	1.33	1.50
h_p/H_0 with no piers	0	−0.20	−0.43	−0.78
At centre of pier bay	+0.04	−0.02	−0.11	−0.36
Along piers	−0.01	−0.21	−0.49	−0.82

H is the actual head above crest level, H_0 the free-nappe head above crest level and h_p the minimum pressure head above weir surface.

Figure 20.7 shows the variation of C with head and with downstream floor level. Table 20.2 provides profile coordinates from which weirs of standard profile and some variations can be designed.

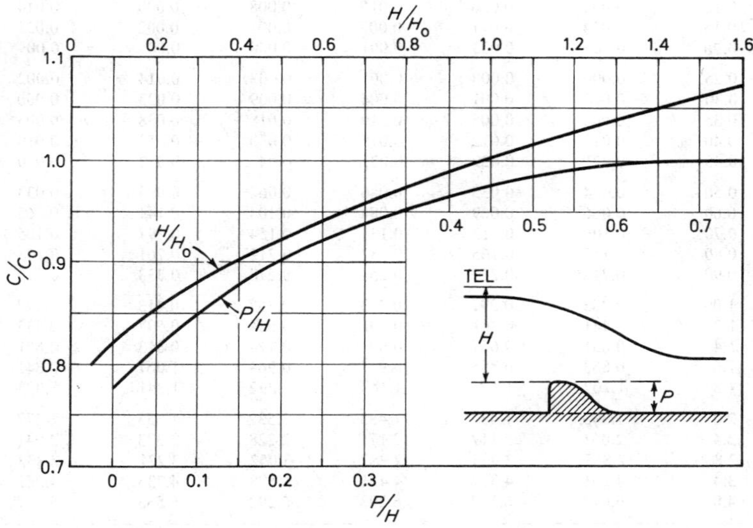

Figure 20.7 Free-nappe profile weirs. Upper curve shows effect of head. H_0 = design head of profile. H = actual head on weir. Lower curve shows effect of high floor level. (Based on USBR data)

Sharp side contractions at the abutments of weirs reduce the discharge capacity locally. They should be curved as in Figures 20.4(a). Piers may have a similar effect, to avoid which spillway piers are often extended upstream, so that the contraction at the pier noses occurs in a region of lower velocity. If a weir of this type is used in a river with a sediment load, the upstream bed level may be expected to be raised by deposits, reducing the effective height P and giving rise to a nonuniform flow distribution. Allowance for these effects is required in the design.

Submerged weirs

The effect of a tail water level above the modular limit is to raise the upstream water level for a given discharge. The degree to which the upstream head or discharge is affected

Table 20.2 PROFILE COORDINATES FOR FREE-NAPPE PROFILE WEIR DESIGN

Horizontal coordinates X/H	Vertical coordinates y/H measured downwards — Upstream weir slope					
	Vertical	*Inclined downstream*				*Inclined upstream*
		$1H:3V$	$2H:3V$	$1H:1V$	$2H:1V$	$1H:3V$
0.0	0.125	0.098	0.066	0.045	0.012	0.159
0.05	0.066	0.051	0.032	0.021	0.002	0.080
0.10	0.033	0.026	0.015	0.008	0.000	0.044
0.15	0.014	0.011	0.005	0.001	0.002	0.022
0.20	0.004	0.003	0.001	0.000	0.007	0.009
0.25	0.000	0.000	0.001	0.003	0.014	0.002
0.30	0.000	0.001	0.004	0.009	0.023	0.000
0.35	0.004	0.005	0.010	0.018	0.038	0.003
0.40	0.011	0.012	0.019	0.030	0.053	0.010
0.45	0.021	0.022	0.031	0.045	0.072	0.020
0.50	0.034	0.035	0.046	0.062	0.093	0.033
0.60	0.066	0.069	0.084	0.104	0.141	0.065
0.70	0.106	0.113	0.131	0.154	0.197	0.105
0.80	0.157	0.165	0.187	0.212	0.261	0.153
0.90	0.216	0.227	0.251	0.278	0.333	0.210
1.00	0.283	0.297	0.323	0.352	0.413	0.277
1.2	0.441	0.461	0.491	0.524	0.597	0.433
1.4	0.631	0.657	0.691	0.728	0.813	0.621
1.6	0.853	0.885	0.923	0.964	1.061	0.841
1.8	1.107	1.145	1.187	1.232	1.341	1.083
2.0	1.393	1.437	1.483	1.532	1.653	1.377
2.4	2.061	2.117	2.171	2.228	2.373	2.041
2.8	2.857	2.925	2.987	3.052	3.221	2.833
3.4	4.291	4.377	4.451	4.528	4.733	4.261
4.0	6.013	6.117	6.203	6.292	6.533	5.977

Derived from lower nappe profile of flow springing from sharp-crested weir.[11] H is total head above designed weir crest. Vertical datum is designed weir crest level. Horizontal datum is upstream edge of crest (i.e. crest of sharp-crested weir).

depends on the weir profile; moreover in certain ranges of submergence the flow pattern is uncertain and may change from diving nappe, which follows downstream weir face, to surface nappe, which separates near the weir crest, a roller developing beneath. Observations of discharge related to upstream and downstream heads or water levels therefore cannot be reagarded as of general application. Nevertheless, good indications can be obtained. Figure 20.8 shows the effect of submergence on standard free-nappe profile weirs[11] and Figure 20.9 the effect on Crump triangular profile weirs.[12]

Measuring weirs

For laboratory and other small-scale measurements, sharp-crested weirs consisting of thin plates in the form of rectangular or V-notch weirs are found convenient. Standard formulae or tables of discharge for these are available.[8] For measurement of larger flows in the field, however, sharp-crested weirs have drawbacks, particularly the need to vent the nappe, the head difference required to ensure modular conditions and the effect of accretion of upstream bed level following the erection of a gauging weir. Current meters have been widely used in the field and still have application to larger rivers, where fixed

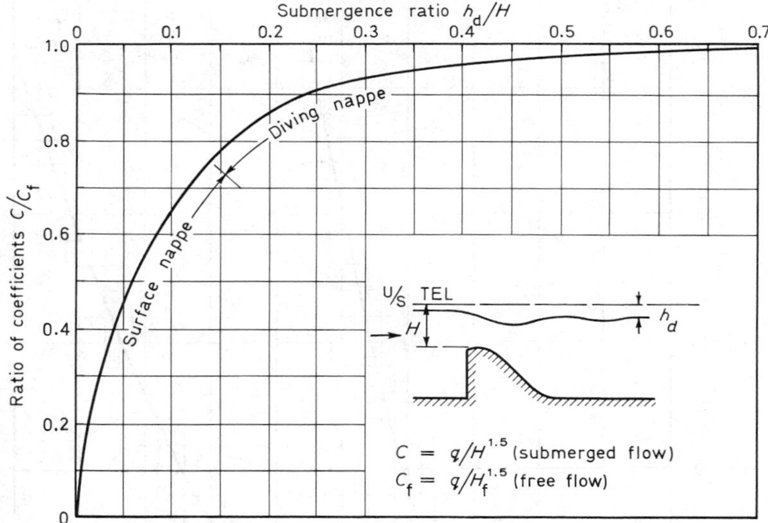

Figure 20.8 Free-nappe profile weirs. Effect of tailwater on discharge coefficient. (Based on USBR data[11])

structures may not be practicable. The latter offer advantages in the case of small rivers, streams and canals in demanding less labour and ability to provide continuous records of flow. Critical depth meters are often used in such cases.

Flow at critical velocity and depth can be induced by providing a raised sill or weir or by contracting the width of flow, or both. Measurement depends on the characteristics of critical flow for which the general weir formula (6) can be used, provided that the crest is high enough, or degree of contraction sufficient, to ensure that the critical flow is not drowned by the tailwater. The throat may be of any suitable shape to obtain the desired characteristics or to match upstream channel shape. For precise measurement, allowance should be made for head losses.[13]

Unless there is already a local drop in level, the introduction of a measuring device will result in a rise in upstream level, though this may be quite small if a device with high modular limit is chosen, or a Crump weir with crest tapping which can be used when drowned by high tailwater level.[12] Where the range of discharge is large and it is desired to obtain an accurate measurement of low flows, a stepped weir may be used, consisting of a short weir at low level for the low flows flanked by longer weirs at a higher level. Alternatively a flat vee weir may be used with crest tapping for submerged conditions.[12]

In the UK, broad-crested weirs with a round nose and Crump weirs have been accepted as standard.[14] In the United States, Parshall measuring flumes have been widely used.[15]

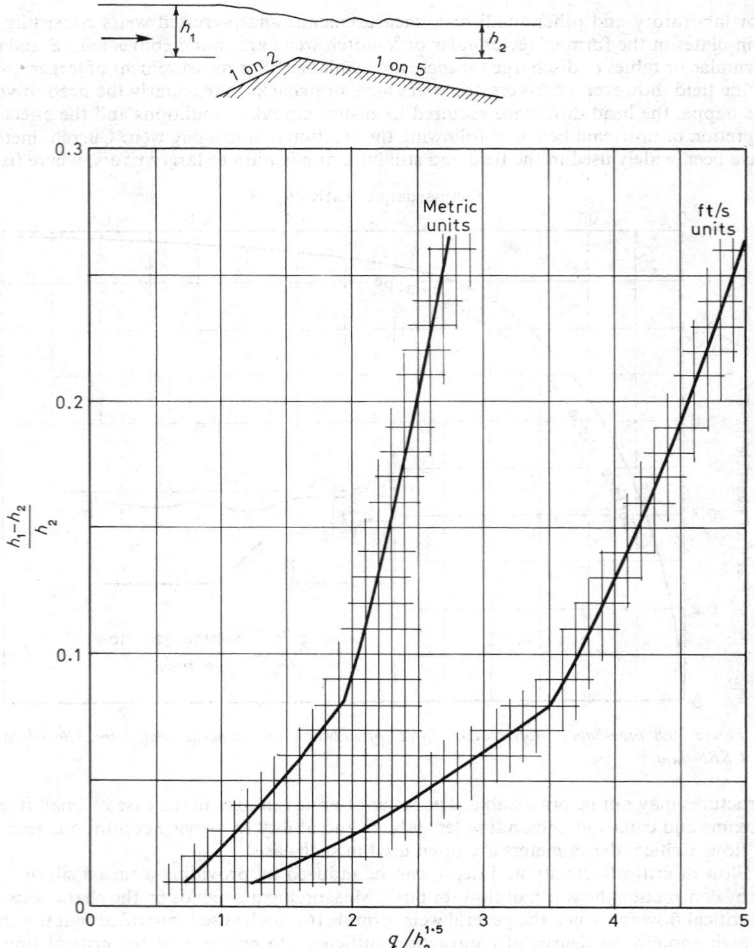

Figure 20.9 Afflux at submerged Crump weirs. (Based on data of White[12])

These were designed with plane surfaces so that they might be easily constructed of wood or concrete. C is not constant but calibration formulae and tables are available. Where the stream to be gauged carries appreciable bed load, a critical-depth flume with a flat or nearly flat bed at the channel bed level is desirable. The bed load can then pass through without excessive accretion upstream, though there may be some at the sides. A measuring flume of this type is shown in Figure 20.10. The degree of contraction sufficient to ensure modular flow can be checked by comparing calculated upstream water levels (using $C = 1.66$) with existing tailwater levels. The broad-crested weir coefficient is applicable, adjusted for head loss upstream of the location of critical depth.[13]

Structures of many other types are used for flow measurement, mostly depending on

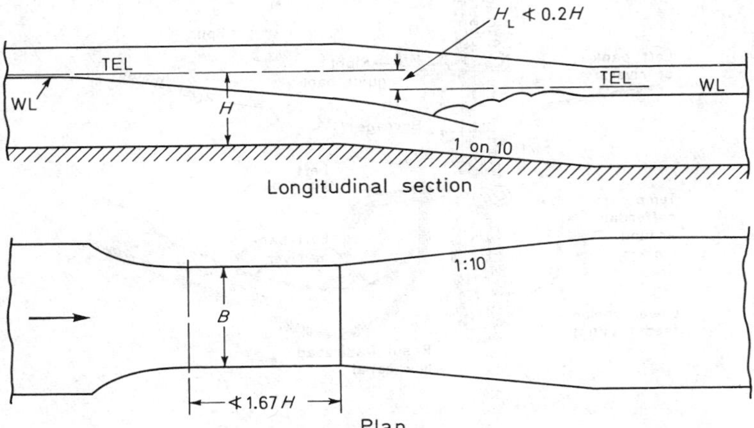

Figure 20.10 Measuring flume with flat floor for debris-laden flow

the critical depth principle or on orifice control, as for example devices on irrigation canal outlets.[16]

Control weirs and barrages

Weirs are used to control the upstream water levels of a river or canal by raising it, for such purposes as diversion of flow into canals, extraction of water by pumping, creating head for hydroelectric power or maintaining a required depth of water for navigation. A fixed weir also raises flood levels, which may not be acceptable. A gated weir, or barrage, however, does not have this drawback if the gate sill is level with the river bed, or on a low weir crest. The gates are kept closed during low flows, maintaining the required upstream water level, but opened as may be necessary to pass floods. The range of water level is thus much less than with a simple weir, and the gates can be operated to maintain constant water level over a wide range of flow. Types of gates are described on pages **20**–47 to **20**–49.

The choice of crest profile depends on the circumstances. For example, a weir with a free-nappe profile is suitable where the crest is to be above the upstream channel bed and there is considerable head difference from upstream to downstream. On the other hand a low crest with flat triangular profile is better suited where, at high rates of flow, the afflux or rise of upstream water level due to the weir must be kept to a minimum.

Control structures in alluvial rivers

Whereas structures in rivers with rocky beds and banks can often be of simple design, with an upstream cut-off wall into the rock and a basin or bucket energy dissipator

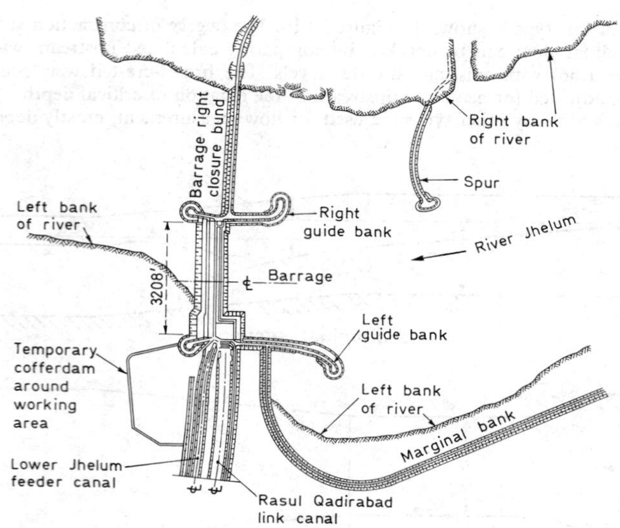

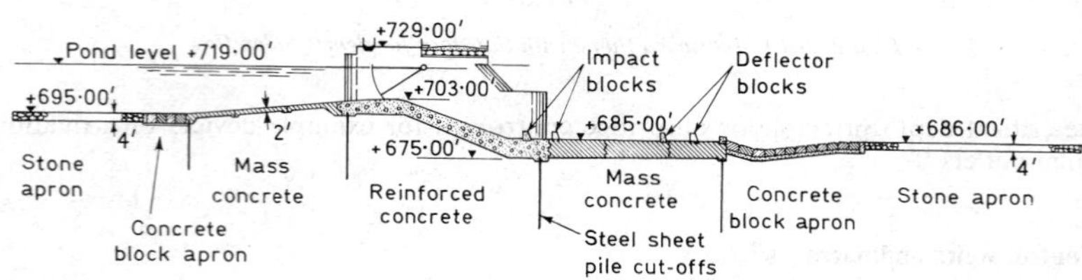

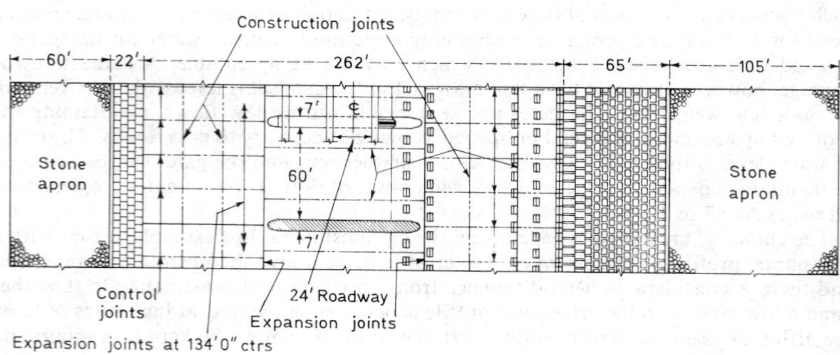

Figure 20.11 Rasul barrage on the River Jhelum, Pakistan. General layout (top), longitudinal section (centre) and part plan (bottom). Note that flow from right in plan (top) and from left in section and plan (bottom). Dimensions in feet
(Consulting Engineers, Coode and Partners)

downstream, the design of control structures in alluvial rivers requires consideration of many other factors.

Firstly the site and orientation of the structure in relation to the river channel pattern is most important and generally should take priority over other considerations. Alluvial rivers without constraint by structures, training works or outcrops of rock or clay, may change course over a period of years, forming new patterns of river channels. The history of a river course is a good guide to such tendencies. The site for a control structure should be a stable one in the long term, i.e. it should remain adequate despite changes in the channel pattern over a number of years, if necessary with the aid of training works. Where a weir or barrage is used for diversion or abstraction of water it is usually desirable to ensure that the quantity of sediment in the water abstracted is a minimum. The best location for the offtake with this in view is generally on the outside of a bend, and the training works should be located to maintain the approach channel accordingly. This consideration applies even where special arrangements are made for sediment exclusion.

A typical barrage forming the headworks of an irrigation canal system on a large river in Pakistan is shown in Figure 20.11. A weir or barrage may occupy only a small part of the width of river channel and flood plain. For example in India and Pakistan it is general practice to make the width of waterways between abutments equal to or rather greater than the width of Lacey régime channel $4.8Q^{\frac{1}{2}}$ where Q is the maximum design discharge in m^3/s.[33] Flanking bunds or embankments are then required extending from the abutments to high ground on either side. Where flood levels are being raised by the control, marginal bunds or flood embankments are often provided extending upstream on each bank. To prevent oblique approach, protect the bunds and avoid outflanking, guide banks are required extending upstream from the abutments—see Figure 20.11. In stable rivers these may be quite short, but where there may be wide swings in the river course they should be approximately equal in length to the width of waterway between them. In addition, in rivers of this type, spurs or groynes may be provided upstream, but these may cause further trouble unless correctly located. Model tests are desirable before construction. Similar measures are used to train alluvial rivers at bridges, The guide banks and spur heads are protected against scour, by rip rap or concrete slabs (see pages **20**–25 to **20**–27).

Low-level sluices provided in the weir or barrage, generally adjoining the canal regulator, have three functions: firstly, they enable the approach to the regulator to be sluiced at intervals so that sediment will deposit there and not enter the canal during normal operation; secondly if kept open during a flood they draw the main stream towards the canal regulator, thus ensuring a deep channel for supplies to be diverted through them at low level during construction. To fulfil these functions the sill should be well below the canal regulator sill level and the sluices should have sufficient capacity to influence flood flow distribution. A divide wall is often provided normal to the weir between undersluices and weir—see Figure 20.11—to enable the canal to draw supplies from a pocket of low velocity water, the undersluices being kept closed, and to facilitate the sluicing operation which is carried out by opening the undersluices with canal regulator gates closed. Sluicing with regulator open may lead to sediment entering the canal; if the canal must operate continuously, control of coarse sediment can be provided by tunnels beneath the level of the canal regulator sill, which draw off the bed load and discharge it downstream.[17]

Downstream of the weir and undersluices, a floor is provided to protect the foundations against scour (Figure 20.11). The drop in water level across the weir or undersluices is accompanied by the formation of a hydraulic jump, except possibly at high flood flows when it may be drowned. A flexible apron of loose stone or concrete blocks is beneficial as an extension to the floor. For design of floor and apron see pages **20**–19 and **20**–25. To allow for nonuniform discharge distribution, the design discharge per unit width of floor should exceed the mean by an allowance depending on the approach conditions. In India and Pakistan a factor of 20% has generally been added for alluvial rivers but in extreme conditions it should be higher—for example where curvature of approach could cause a high concentration.

Permeable foundations

The foundations of a structure on permeable materials require special consideration. Two important requirements are that every part of the structure must be safe against uplift pressures beneath and that underflow or seepage through the permeable materials should be controlled so that there is no failure by 'piping'. Where a continuous impermeable stratum is within reach, underflow can be prevented by a line of sheet piles or a curtain wall intersecting it, or possibly by grouting, but the sealing must be perfect. If, however, the permeable materials are too deep for this treatment, the floor must be safe against uplift pressures exceeding the tailwater level acting on the underside of the structure throughout.

Uplift depends on the hydraulic gradient of flow through the material beneath the work, reducing from the upstream water level to the downstream water level. Its distribution may be considerably affected by the non-uniformity of the materials so a prior investigation of the character of the material, its uniformity and the existence of strata of different permeability is necessary. The floor upstream of a weir or gates is safe against uplift because of the water load above but the downstream floor is particularly vulnerable at times of high upstream and low downstream water levels. Measures to reduce uplift pressures on the downstream floor include the lengthening of the upstream floor and provision of transverse lines of sheet piling upstream or beneath the weir, both serving to lengthen the effective seepage path, and provision of relief drains. Typical protective measures beneath a gated structure are shown in Figure 20.11. The weight of downstream floor and superstructure must exceed the uplift pressure by a factor to allow for any nonuniformity of the material which may not have been represented in the flow net. In some cases allowance is made for the weight of water on the downstream floor but in others this is ignored in case the floor may be dewatered for inspection or repairs.

'Piping' consists of the removal of foundation material by the flow of seepage water. It can occur at the tail end of a structure where the underflow emerges and is a potential cause of undermining and ultimate failure of the structure. It is caused by excessive exit gradient. Theoretically, piping occurs when the exit gradient is 1 : 1 but, in design, much flatter limiting gradients are adopted. Khosla *et al.*[18] proposed the following safety factors:

Shingle (which includes gravel and cobbles)	4 to 5
Coarse sand	5 to 6
Fine sand	6 to 7

It is usual to protect against piping, where the foundation material is granular, by providing coarser filter material to intercept the seepage over its exit area. This is generally covered by loose stone or other protection against scour, but in case this should fail, other measures are needed to reduce the exit gradient. Such measures include the lengthening of the structure and the provision of transverse lines of sheet piling to reduce the overall hydraulic gradient, provision of relief drains and the provision of a curtain wall or line of sheet piling at the tail end of the floor. The last is most important to avoid a locally steep gradient and protect the floor from undermining by scour, but it should not be too deep because it increases uplift beneath the floor. The upstream or central sheet piling should extend laterally into the flanking embankments, and lines of piling are carried around as may be necessary to intersect seepage paths and box in the foundations.

It will be noted that some of the measures serve to protect against both uplift and piping. The lengthening of the upstream floor can be effected by a relatively thin reinforced concrete floor but leakage through it must be avoided, so all joints should be sealed with flexible water stops. Its efficiency is greatly increased by a line of sheet piles at its upstream end, which also serves to protect against undermining there by scour.

To determine the length of floor in a preliminary design, Lane's empirical weighted creep method may be used.[19] The seepage path is assumed to be the sum of the lengths of vertical faces (e.g. each side of sheet piles) plus one-third of the horizontal length. Sloping surfaces steeper than 45° are treated as vertical and those flatter as horizontal. The minimum safe ratio of seepage path to head across the structure ranges from 1.6 in hard clay, through 3 in soft clay or coarse gravel, 5 in coarse sand, 7 in fine sand to 8.5 in silt. Uplift pressure is approximately in proportion to the weighted creep distance from upstream to downstream, but this assumption can lead to error. Uplift may be estimated more precisely by the use of a flow net, the electrical analogue method or mathematically. Khosla *et al.*[18] have provided solutions to a number of standard cases. Luthra and Joglekar[20] have dealt with cases of stratified subsoils. For general design procedures, reference may be made to Haigh,[17] Leliavsky[21] and Foy and Green.[22]

Irrigation canal structures

Canal head regulators are usually located immediately upstream of a weir or barrage in line with the abutments (see Figure 20.11). The principles for design are the same as for a barrage. On alluvial rivers the crest is well above the sill of the undersluices. A stilling basin of sufficient depth, to ensure that the hydraulic jump is retained within it, is essential where the canal bed is erodible and is also generally provided where the canal is lined. Allowance may be necessary for nonuniform distribution due to oblique approach, depending on the velocity upstream.

Where the general ground slope exceeds the design slope of a canal, falls or drop structures are required at intervals to dissipate the excess head and lower the canal to conform to the ground level. Falls are designed in a similar way to weirs, with ungated crest and stilling basin. To reduce cost, the width of waterway is often made less than the width of canal. The upstream contraction presents little difficulty, but the downstream expansion must be gentle to avoid asymmetrical flow downstream (see pages **20**–5 to **20**–7).

Energy dissipation. Stilling basins

At weirs, barrages, sluices, spillways, tunnel outfalls, canal falls and in general where a sharp fall occurs in total energy level, a stilling basin is needed to contain the flow in the region of energy dissipation. This is especially important where the channel bed is erodible. The surplus energy may be dissipated by water spilling into a pool, which may be in bed rock, or lined rip-rap or concrete.

In most cases the energy head to be dissipated is sufficient to create supercritical flow, defined on page **20**–4. A hydraulic jump is then generally the most effective and economical way of dissipating the surplus energy. The object is to provide a stilling basin lined with nonerodible material, usually concrete, deep enough to retain the jump over the whole range of flow conditions and long enough for the eddies generated in the jump to be reduced to an acceptable intensity before reaching the channel downstream. The minimum depth is thus related to the characteristics of the jump while the minimum length is related mainly to the degree of stilling required. Where the channel bed is erodible a greater length of basin is generally required than where it is in rock or concrete lined. In the basin, chute blocks, baffle blocks or piers are often provided to help to stabilise the jump and reduce the length of basin required.

As shown earlier, the stability of a hydraulic jump is expressed by the pressure–momentum equation (5) representing the condition at which the jump is at its limit of stability, i.e. any increase in discharge or upstream head would cause 'sweep-out' or movement of the jump downstream and possibly out of the basin. In the design of stilling basins, however, the quantities which are known are usually the discharge, head drop and tailwater level and it is required to determine the basin floor level. Equation (5) therefore cannot be directly applied, but the maximum acceptable floor level can be easily found with the aid of Figure 20.3. The procedure is to compute upstream and downstream total energy levels (water level + velocity head), compute $H_L = H_1 - H_2$ (see Figure 20.3), compute critical depth d_c by equation (3), compute H_L/d_c, read off H_2/d_c directly beneath H_L/d_c, i.e. for same F_1, and compute H_2. This gives the minimum depth of basin floor beneath tailwater total energy level. It applies to a plain floor and may be reduced by 10% to 20% if chute blocks and/or baffle blocks and end sill are provided. However, it is often the practice to provide the full depth and consider the blocks to provide a safety margin in addition. It is usually necessary to determine minimum basin depth for several discharges throughout the range, because the most severe case is not always with the maximum discharge. When determining q in cases of non-

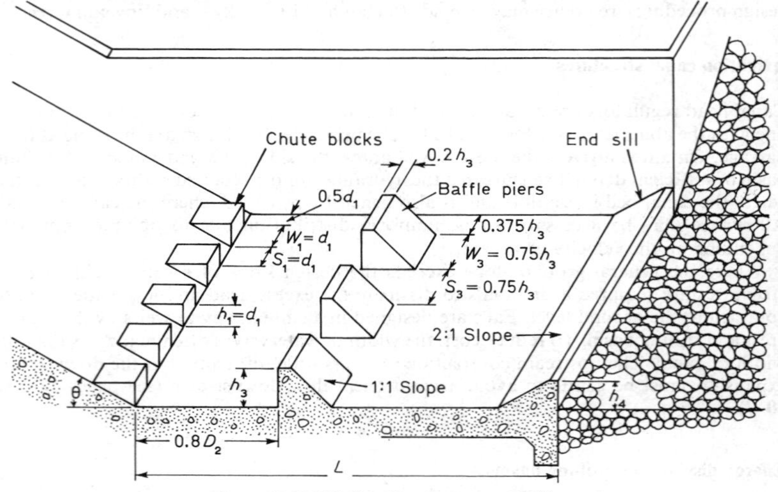

Figure 20.12 USBR stilling basin, type III[24]

uniform distribution across the basin it may be necessary to use a value rather higher than mean $q = Q/B$, where Q is the total discharge and B the width. Tailwater level is clearly of critical importance for the stability of the jump and it is necessary to have a reliable stage discharge curve, with allowance for future changes, as for example due to degradation downstream. The lowest probable levels should be used. In the case of basins for gated spillway releases, where discharge may be increased rapidly over a short period, allowance should be made for low tailwater levels due to time lag.

The length of basin required cannot be defined so precisely. On a plain floor the length of a jump may be 4 or 5 times the depth d_2 in the basin. If residual eddies can be tolerated downstream because the bed is not erodible or is protected by a flexible apron, as in Figure 20.11, a length of 4 d_2 may suffice. Where chute blocks and baffle blocks are provided in such cases a length of 2.5 d_2 is sometimes considered adequate (but see below).

Many standard designs of hydraulic jump stilling basins have been developed from model tests, one of the most comprehensive being that of Bradley and Peterka.[23, 24] Four types of jump were defined according to the Froude number F_1, each with somewhat different characteristics, namely:

F_1 from 1.7 to 2.5 Pre-jump, low energy loss

F_1 from 2.5 to 4.5	Transition, rough pulsating water surface
F_1 from 4.5 to 9.0	Range of good jumps least affected by tailwater variations
F_1 exceeding 9.0	Effective but rough

If F_1 is in the range 2.5 to 4.5 the pulsations are likely to produce surface waves which are propagated downstream. The Froude number is generally determined by other factors, but if there is any choice it is clearly desirable for it to be within the range 4.5 to 9.0. Bradley and Peterka's basin III for F_1 between 4.5 and 9 is shown in Figure 20.12. The dimensions of the chute blocks are made equal to the depth d_1 and those of the baffle blocks range from 1.3 d_1 for $F_1 = 4$ to 3 d_1 for $F_1 = 14$. The height of end sill ranges from 1.2 d_1 for $F_1 = 4$ to 2 d_1 for $F_1 = 14$.

Where F_1 is between 2.5 and 4.5 (basin IV) the chute block height is $2d_1$ and the baffle blocks are omitted. Where F_1 exceeds 9 (basin II) the baffle blocks are omitted and a dentated end sill is recommended. Basins II and IV, having no baffle blocks, are required to be longer than basin III, with floor lengths of approximately $4d_2$. In the case of high head structures, if the velocity much exceeds 15 m/s, chute blocks and baffle blocks are liable to be damaged by cavitation. They can be omitted or protected by steel cladding, as at Mangla Spillway.[52]

Erosion of bed and banks immediately downstream of the stilling basin can be a serious problem, whether the head drop through the structure is great or small—see remarks on transitions, p. **20**–5.

A normal cause of erosion is the residual turbulence from the hydraulic jump. This may scour the bed beneath the level of the basin floor, so a flexible protection such as rip-rap is needed which will adjust its level to the scoured bed downstream of it (see p. **20**–25). When the banks are formed of erodible material they need slope protection to guard against local velocities and wave wash. In the case of weirs and barrages on alluvial rivers the banks are carried downstream a short distance—perhaps equal to a quarter of the width of river channel (see Figure 20.11). A loose stone apron is provided at the toe. In the case of canals where the banks are erodible, the slope protection is continued for a distance in which the surface waves will be reduced and velocity distribution will become normal.

A layout of stilling basin and canal banks which has been found satisfactory is shown in Figure 20.4(e). The gently diverging side walls are free-standing at their downstream ends, where they consequently do not have to serve as high earth-retaining walls; the channel downstream is widened to accommodate the side rollers which will develop and the banks are protected by rip-rap.

In the case of small flows, shorter and simpler structures have been used, for example the straight drop spillway basin of the US Department of Agriculture.[26]

For large flows and high heads, experience has shown that hydraulic jump basins are generally satisfactory. Damage which has occurred has been due mainly to the basin being of inadequate depth, to cavitation where baffle blocks have been exposed to high velocity flow and to abrasion due to loose materials in the basin.[28, 29] In some cases these materials may have remained from river diversion operations but in other cases bed material and even rip-rap has been carried into the basins by backwash. There have also been instances of vibration and shock due to flow instability. In large-scale basins it is especially necessary to guard against flow separation at the side walls, which can be a cause of both these last effects and of back wash.

Bucket energy dissipators

The hydraulic jump stilling basins described above are effective but costly, especially for high discharge concentrations. Where the foundations of the structure are in rock, even an erodible rock, a much higher degree of residual turbulence may be acceptable. In such cases, if the head is sufficient, a flip bucket, also termed 'ski jump' and 'trajectory bucket', provides a much cheaper solution. Chute spillways of this type are described on page 20–40 (see Figure 20.23).

A submerged roller bucket (see Figure 20.13) is suitable over a wide range of Froude numbers. The bucket is placed well below the tailwater level so that a submerged roller forms in the bucket and exit velocities are not excessive. Compared with a hydraulic jump basin, it is deeper but shorter and generally less costly; but the range of tailwater level for satisfactory operation is limited, which precludes its use in some cases. Rules for design have been given by McPherson and Karr[30] and by Bleichley and Peterka[25, 31] who found that slotted buckets were superior to plain buckets.

Terminal structures for pipes and valves

High-velocity jets from pipes and terminal valves have considerable erosive power, even on hard rock. Means of protection include the use of valves which disperse the jet in the air, for example the cone valve, or valves which project the jet some distance, where a plunge pool can be provided, or structures devised to contain the jet and allow most of the energy to be dissipated before discharge into an erodible channel.

Figure 20.14 shows an impact stilling basin developed by the US Bureau of Reclamation (USBR)[25] for pipe and open-channel outlets with discharges up to 10 m^3/s and

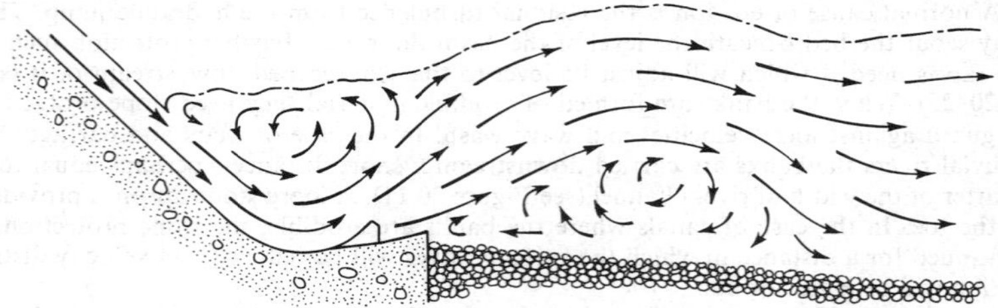

Figure 20.13 Submerged roller bucket—Angostura-type slotted bucket[31]

velocities up to 9 m/s. It may also be considered for terminal valves within the limits stated. The required dimensions may be obtained from Figure 20.14.

A special basin has been developed by the USBR for hollow jet valves.[25, 32] Basins have also been used for cone valves. A very effective energy dissipator for pressure pipe outlets is a vertical well in which the pipe outlet is deeply submerged at a short distance above the bottom. Figure 20.15 shows a USBR type of well. Regulation can be provided by a sleeve valve at the pipe outlet, operated from above (see page 20–55).

Depth of scour at structures

Apart from scour downstream of stilling basins, structures such as bridges, jetties, groynes and constrictions forming obstructions to flow in rivers and channels with erodible beds can give rise to scour due to disturbance of the normal flow pattern. Scour can also be caused by oblique flow at the upstream of control structures such as barrages and regulators. It is generally required to estimate the depth of scour so that adequate protection can be provided or so that the foundations can be located at sufficient depth to avoid the possibility of undermining.

Local scour results from the deflection and hence concentration of flow caused by an obstruction. The depth of scour depends on the shape of the obstruction, its orientation to the flow, the channel cross section and discharge, the character of the erodible bed

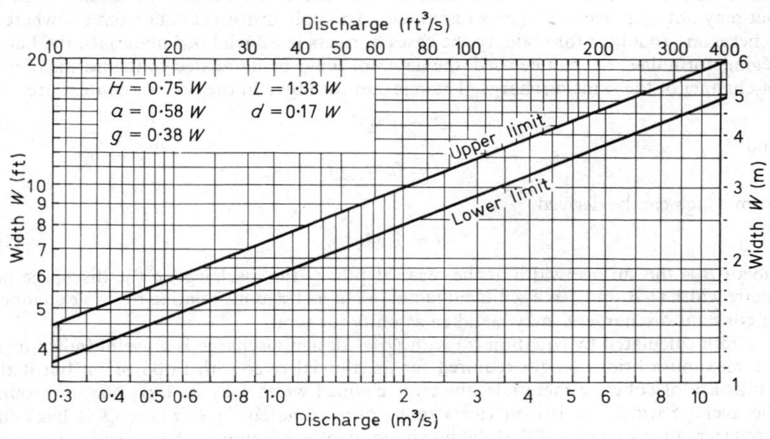

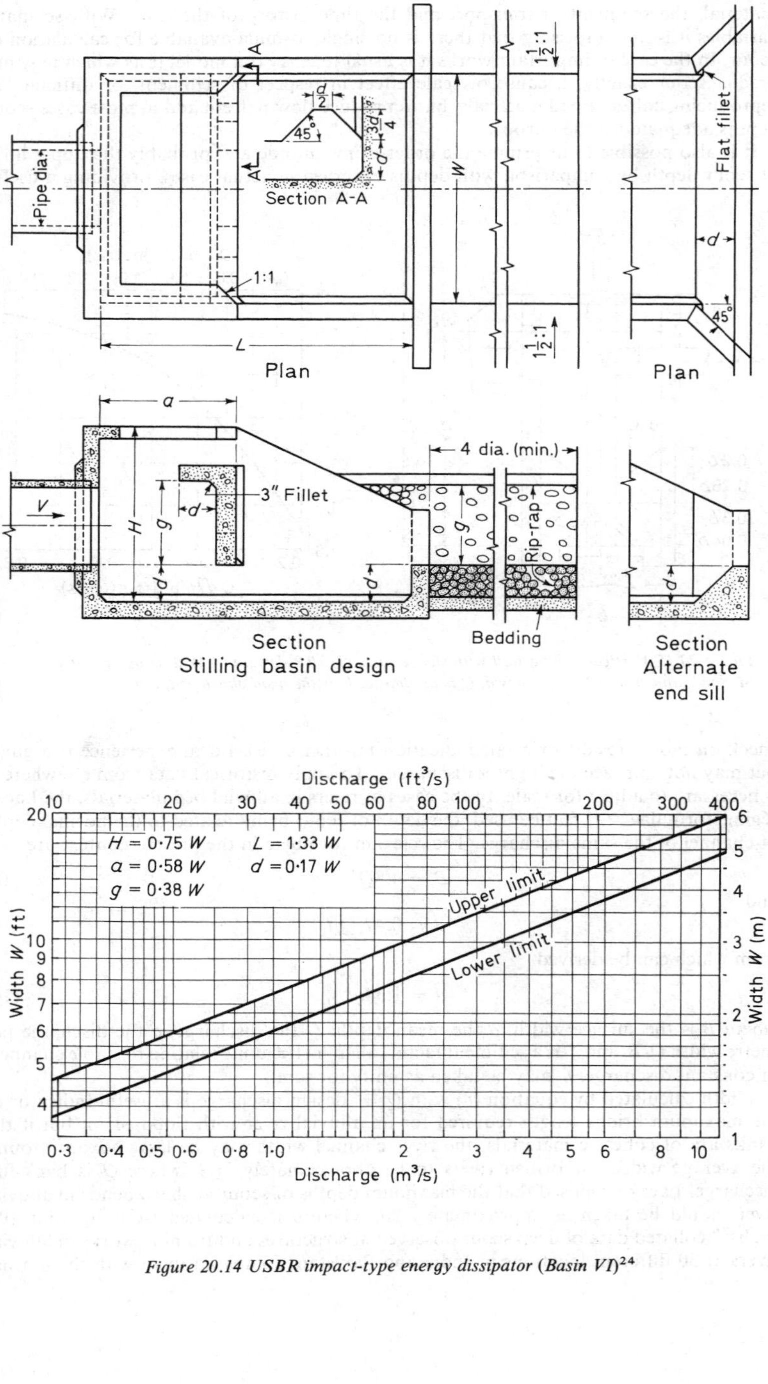

Figure 20.14 USBR impact-type energy dissipator (Basin VI)[24]

material, the sediment in transport and the time history of the flow. With so many variables it is not surprising that there is no single formula available for calculation of scour. In the case of important works it is usual to carry out model tests which may not predict scour exactly, because of scale effect in respect of sediment, or difficulty in reproducing cohesive bed materials, but reproduce flow pattern and in most cases scour depths adequate for the purpose.

It is also possible to determine the order of magnitude and probably the upper limit of scour depth by comparison with depths observed in actual cases, providing a useful

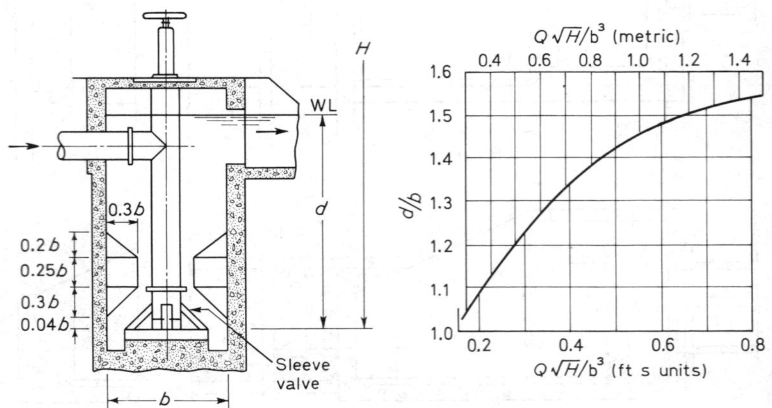

Figure 20.15 Vertical stilling well with sleeve valve (USBR design).[24] *Well is of square section in plan with corner fillets as shown. Q is discharge, H is the head above pedestal*

check on model results or a fair indication in other cases. Local experience is a guide but may not embrace the highest discharges. To apply historical data from elsewhere it is necessary to adjust for scale. In the cases of rivers in alluvial bed materials, the Lacey régime formulae[33, 34] can be used, the depth of scour being related to the normal depth of channel of the same discharge. The relevant formulae in the present context are

$$B = 4.8Q^{\frac{1}{2}} \tag{8}$$

and

$$d = 0.47(Q/f_L)^{\frac{1}{3}} \tag{9}$$

from which can be derived

$$d = 1.34q^{\frac{2}{3}}/f_L^{\frac{1}{3}} \tag{10}$$

where B is the surface width, d the mean depth, Q the discharge, q the discharge per metre width Q/B, and f_L is a sediment factor, all in metric units relating to stable channels of constant discharge. f_L may be taken as unity for sand.

Width calculated by equation (8) with $Q =$ design discharge is a useful indicator of the maximum bridge length required for an alluvial river with flood plain, but if the banks are of cohesive materials, the river channel width may be less; Nixon[35] found the average widths of British rivers to be approximately $3Q^{\frac{1}{2}}$, where Q is bank-full discharge. Lacey proposed that the maximum depths of scour at sharp bends in alluvial rivers could be taken as approximately $2d$, where d is calculated from equation (9). Inglis[36] collated data of deep scour observed at structures and training works in alluvial rivers at 30 different locations in India and Pakistan, compared them with the normal depths indicated by (9) and reached the following conclusions for maximum depth of scour below water level:

At bridge piers, $2d$
At large radius guide banks, $2.75d$
At spurs along river banks, $1.7d$ to $3.8d$, depending on length of spur projection, sharpness of curvature of flow, position and orientation

Here d is Lacey's normal mean depth calculated from equation (9) using estimated peak discharge. It will be appreciated that large flood flows cannot be measured but are estimated, while maximum depths of scour are transient and may in fact have been greater than observed. The scour depths are related to the total rather than the local flow on the grounds that the scour results from the concentration of the whole flow. In the case of braided rivers, allowance could be made for the division of total flow into several channels. Scour depth in rivers in gravel and boulders would be less than indicated but the difference may be small. Scour depths in cohesive materials could be less because of the time required to reach maximum scour.

For the purpose of design of aprons upstream and downstream of barrages in northern India, maximum depth of scour below water level was taken as $1.5d$ at the upstream end of the hard floor and $2d$ at the downstream end of the basin.[18] Here d was calculated from equation (10) using mean q.

Scour at bridge piers has been studied in some detail in models, scour depth being related to discharge per unit width and sometimes expressed as depth below upstream bed level.[37, 38] To apply such relationships, the discharge concentration, which can in the worst case greatly exceed the average, has to be estimated, and the upstream depth then to be calculated for the corresponding flood condition. The latter can be done using equation (10) which is likely to give a conservative value because of time lag and sediment load. It should be checked by use of the appropriate Inglis factor above.

Protection against scour

This generally consists of one of the following materials.

Boulders from the river bed which are generally rounded and therefore less stable than quarried stone of similar weight.

Rip-rap or pitching of quarried stone is widely used. In some cases it is hand packed, especially on side slopes which are expected to remain as placed without settlement, but with increasing use of mechanical equipment it is more often placed in a random manner. This is also preferable in locations where it is expected to settle or move down due to scour. On side slopes the thickness of rip-rap should be sufficient to accommodate the biggest stones without large gaps—at least $1.5 \times$ median stone diameter—and an underlayer or filter of smaller stone is generally required to prevent the base material from being washed out by wave action. In the case of bed protection, surfaces not subject to scour may be treated in the same way, but at transitions from stilling basins and in general where the channel bed may scour beneath the apron level, the volume of rip-rap should be sufficient to protect a slope at the angle of repose of the rip-rap on the bed material (for a sand bed generally 1 on 2) extending from the apron level to the level of anticipated deepest scour. The rip-rap may be laid on a prepared slope or it may be laid in a horizontal apron which it is assumed will settle to a slope when scour occurs. The assumed thickness of the stone on the slope is that sufficient to prevent removal of the fine material beneath, under the particular flow conditions, and in estimation of the volume required a margin should be included to allow for stone carried away in the flow. In some cases a filter layer of smaller material is provided, but its availability and location after settlement of the apron is uncertain.

The size of rip-rap which will remain stable may be estimated from the diagram in Figure 20.16. 60% by weight of the material should be equal to or larger than the size shown.

Derrick stone is stone in blocks too heavy to be placed by bulk handling and which therefore requires individual placing. It is usually placed on an underlayer of graded rip-rap.

Concrete blocks, slabs or units of various shapes. As the density of concrete is less than that of stone, larger and heavier blocks are required than the corresponding stone sizes. Concrete blocks are used in locations where stone of suitable quality and weight is not

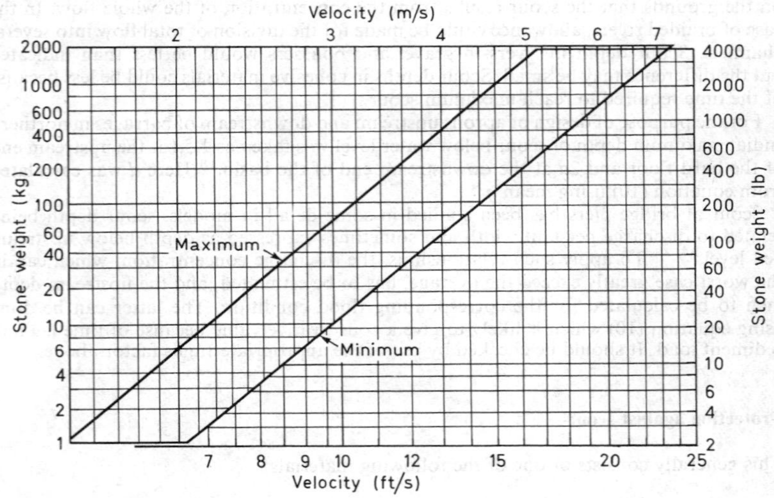

Figure 20.16 Stability of loose rock in flowing water.[10]
Specific gravity of rock 2.65 shown above
Specific gravity of rock weight = weight shown × $1.75(s-1)^3$
Use minimum weight in normal flow and maximum weight in very turbulent flow

available or is too costly. Concrete blocks or slabs on edge, for example 2 m wide × 0.5 m long × 1 m deep, have been used successfully for flexible aprons downstream of barrages in rivers with sand beds. Concrete slabs are used in slope protection but require good compaction of fill beneath to avoid uneven settlement. Concrete units of special shapes have been developed which require less concrete than do concrete cubes for the same duty; some of these are extensively used in coastal protection and can also be used in river channel works.

Gabions, consisting of wire crates containing boulders or broken stone, generally wired together to form an apron, form an economical temporary protection against erosion and have been used in permanent works, though the wire crates may be subject to corrosion. The standard metric size is 2 m × 1 m × 1 m but thinner mattresses are available. They offer great resistance to removal by flow and a gabion apron has considerable flexibility in adjusting to scour, though less than an apron of rip-rap.

Asphalt provides a smooth impervious cover but does not have much flexibility.

Sheets of nylon and other synthetic materials, woven to a fine mesh provide an effective filter layer over sand and have been used to form thin mattresses, with pockets filled with cement grout, for side slope protection.

Brushwood fascine mattresses, a traditional protection consisting generally of willow twigs bound in bundles and formed into longitudinal and lateral layers bound together before it is launched by weighting with stone and sinking into place, is still used and has a considerable life under water.

Vegetation, in particular certain grasses and shrubs, when established above normal water level, can protect a bank against occasional high level wave wash or even shallow overtopping.

For protection of formed banks in cut or fill, any of the above materials would be suitable (see also Section 26) subject to adequate protection against scour of the toe of the bank or the channel bed near it. This may be provided by a line of sheet piling at the toe or by a flexible apron laid horizontally which will subside and protect the underwater slope when scour occurs. Quarried stone rip-rap is usually used for the apron where available. Similar aprons are used at the upstream and downstream ends of barrages, regulators and stilling basins. The stone should be large enough not to be transported by the flow (see Figure 20.16) and the volume of stone should be sufficient to protect the slope to the level of deepest scour. In the case of rigid structures it may be dangerous to rely on loose stone protection; it is generally best to provide foundations at low levels beneath possible scour. If stone or concrete blocks are used to protect existing structures they should be placed as low as possible beneath normal bed level.

ENCLOSED FLOW

Head loss in large conduits and tunnels

Head loss in pipes is dealt with in Section 5. Head loss in large conduits and tunnels may similarly be estimated by the Darcy formula:

$$i = \lambda V^2/2gD = \lambda V^2/8gm \qquad (11)$$

where i is the hydraulic gradient, λ the friction factor, V the mean velocity, D the diameter of circular conduit flowing full, or m the hydraulic radius to be used for part full and noncircular conduits. λ and Manning's n are related by

$$n = \lambda^{\frac{1}{2}} D^{\frac{1}{6}}/10.8 = \lambda^{\frac{1}{2}} m^{\frac{1}{6}}/13.6 \qquad (12)$$

In nearly all actual cases of large conduits the boundary cannot be classed as smooth or rough but falls within the transition region. λ therefore depends on the effective roughness and on the Reynolds number VD/ν or $4Vm/\nu$, where ν is the kinematic viscosity (for values of ν see page **5**–15). Although many types of roughness are composite, for example smooth concrete with projections due to formwork joints, and therefore the equivalent sand roughness concept is not completely representative, it does provide a method of predicting the friction factor, based on recorded experience. In the case of new works this depends on the type of forms used, quality of workmanship and degree to which projections are ground down. Deterioration occurs with age and use. A steel lining may corrode and be roughened by tuberculation, as for pipes. Concrete inverts may be roughened by abrasion during river diversion. There may be deposits due to leaching through joints and cracks in a concrete lining, even vegetation and animal growths, while the deposit of slime by untreated water is commonplace.

Typical values of equivalent sand roughness k for new surfaces, based mainly on Ackers[39] and USBR experience,[40] we give in Table 20.3.

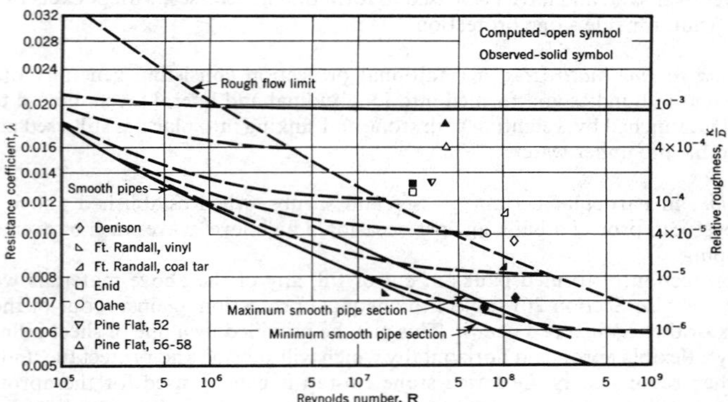

Figure 20.17 Head loss in uniform conduits. Open symbols, computed; solid symbols, observed

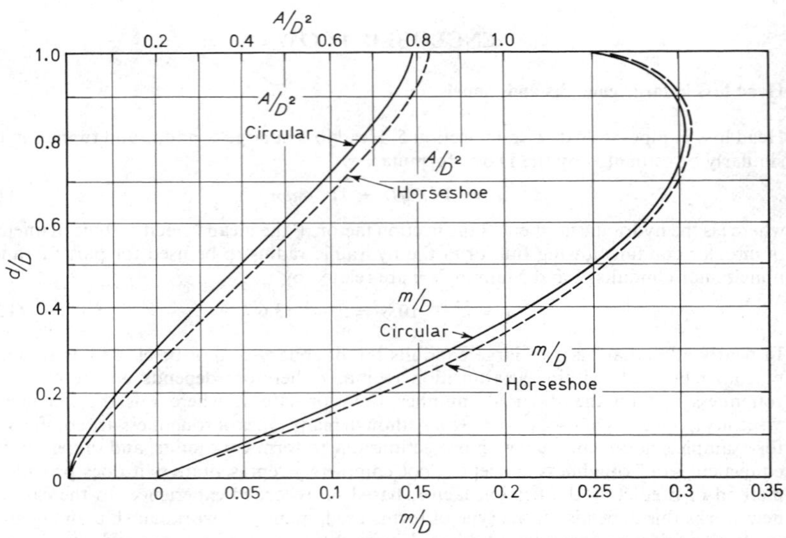

Figure 20.18 Area and hydraulic radius of conduits, part full. For key, see Figure 20.19

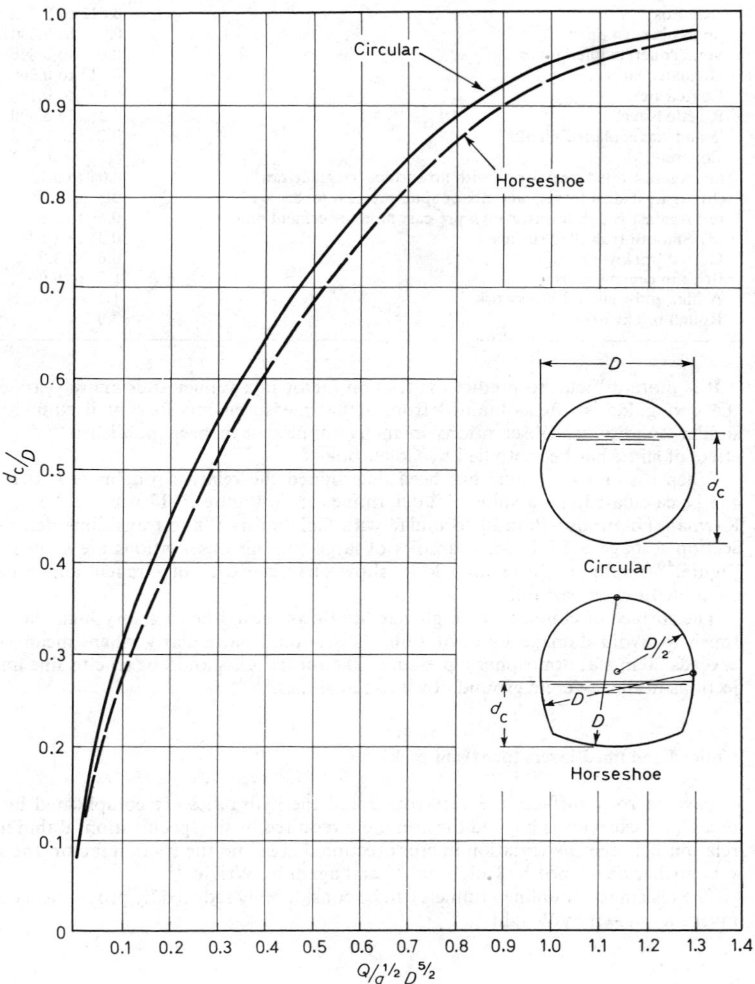

Figure 20.19 Critical depth in circular and horseshoe conduits

Table 20.3

Surface	*k range* (mm)
Asbestos	0.012 to 0.015
Spun bitumen lined	0.0 to 0.030
Spun concrete lined pipes	0.0 to 0.030
Uncoated steel	0.015 to 0.060
Coated steel	0.03 to 0.15
Rivetted steel	0.3 to 6.000
Wood stave, planed planks	0.2 to 1.5
Concrete:	
(a) Against oiled steel forms with no surface irregularities	0.04 to 0.25
(b) Against steel forms, wet mix or spun pre-cast pipes	0.3 to 1.5
(c) Against rough forms, rough pre-cast pipes or cement gun	0.6 to 2.0
(d) Smooth trowelled surface	0.3 to 1.5
Glazed brickwork	0.6 to 3.0
Brick in cement mortar	1.5 to 6.0
Ashlar and well laid brickwork	1.5
Rough brickwork	3.0

It is more difficult to predict the friction factor in a tunnel after many years of use; the best guide is often obtained from actual measurements in similar tunnels under similar conditions. Observations in many tunnels have been published.[40, 41, 42] The effect of slime has been studied by Colebrook.[42]

When a suitable k value has been determined the relative roughness k/D or $k/4m$ can be calculated and a value of λ determined from Figure 20.17 which is based on the Karman–Nikuradse–Prandtl formulae with Colebrook–White transitions described in Section 5, page **5**–15. Some examples of large conduit observations are shown on the Figure.[40,41] Figure 20.18 and 20.19 show characteristics of circular and horseshoe conduits flowing part full.[11]

The surface of conduits for high-velocity flows should be to a very high standard of finish to avoid damage by cavitation. This applies particularly where mean velocity exceeds 20 m/s at atmospheric pressure. The formwork should be true to line and projections in the concrete ground down to flat slopes.[43, 44]

Unlined and lined-invert tunnels in rock

Excavated rock surfaces are very rough and the hydraulics are complicated by 'overbreak', i.e. excavation beyond the minimum required by the specification. Rahm found a relation between the variation in cross-sectional area and the friction factor, the subject was further developed by Colebrook[42] and again by Wright.[45]

The resistance of unlined tunnels can be considerably reduced by providing a concrete invert, as shown by Wright.

Transitions and bends

Transitions may be from circular to noncircular sections or vice versa, or from one circular section to another of different diameter. In conduits for high-velocity flows, transitions are generally gradual to avoid flow separation and possibly cavitation. It is also necessary to adopt moderate rates of expansion if head is to be conserved and instability of flow downstream avoided. Circular sections can be merged into rectangular or horseshoe sections without double curvature and avoiding sharp local divergences. Figure 20.20 shows diagrams from which head loss may be estimated in diffusers of circular section; curves of similar pattern but slightly differing values apply to diffusers of rectangular section.[46] It will be seen that for expansion ratios of 2 or more the head loss may be considerable unless the angle of divergence is small. Where rapid expansion is required, divide walls may be used so that the flow is carried in a number of ducts, each of which is a reasonably efficient diffuser while the overall angle of expansion may

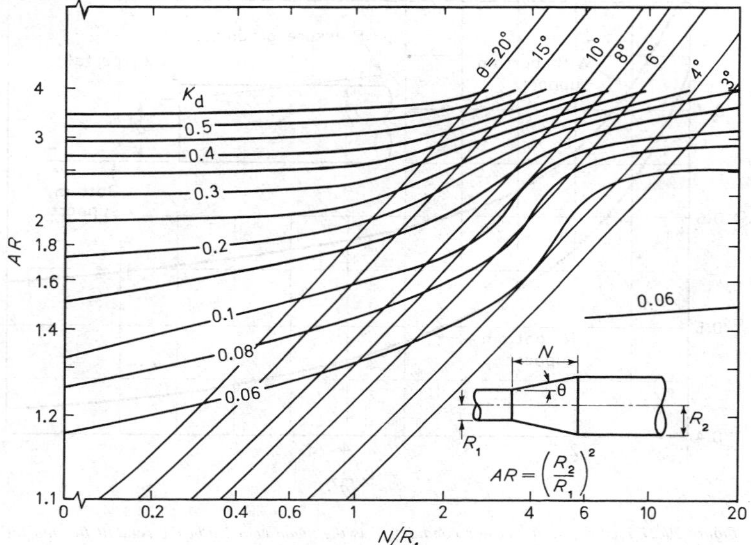

Figure 20.20 Head loss coefficient K_d of conical diffusers with a tailpipe

be as much as 90°. For head loss in a sudden enlargement, see Section 5, page **5**–17; expansions of this type are sometimes used for energy dissipation in closed conduits. Contractions may be more rapid than diffusers, but to avoid head loss due to the formation of a *vena contracta* in the downstream conduit, it is desirable to provide a rounded external angle between the transition and conduit at a radius of at least one-sixth diameter. In high-velocity flow this is an area of potential cavitation and the radius should be greater.

Head losses at bends are similar to those in pipe bends (**5**–18). In large conduits a compromise has often to be reached between the greater head loss of a short radius bend and the greater cost of long radius bend; bend radii of between 1.5 and 3 diameters are often adopted. The flow instability induced by a bend persists for a distance of many diameters downstream and may affect the performance of turbines or pumps.

Bifurcations and manifolds, dividing the flow, for example, for two or more machines, are generally designed with great care to achieve a smooth change of velocity, absence of swirl and minimum head loss. Model tests with air are useful to indicate flow pattern and pressure drop in closed conduit transitions; relatively low pressures are used to avoid compressibility effects.

Transitions leading from subcritical flow in open channels to closed conduit flow may, where the approach velocity is low, be designed on the same principles as apply to intakes from reservoirs—see page **20**–24. Sharp corners lead to the formation of a *vena contracta* and head loss; this may be avoided by providing a rounded or bellmouth entry. With higher approach velocity the transition should be more gradual, with curves of larger radius. If the waterway area reduces towards the conduit, the flow will accelerate, and if it becomes supercritical an hydraulic jump would occur upstream of the closed conduit, resulting in head loss and possibly air entrainment in the conduit. To avoid this the design should be such that the contact between free surface and roof occurs where

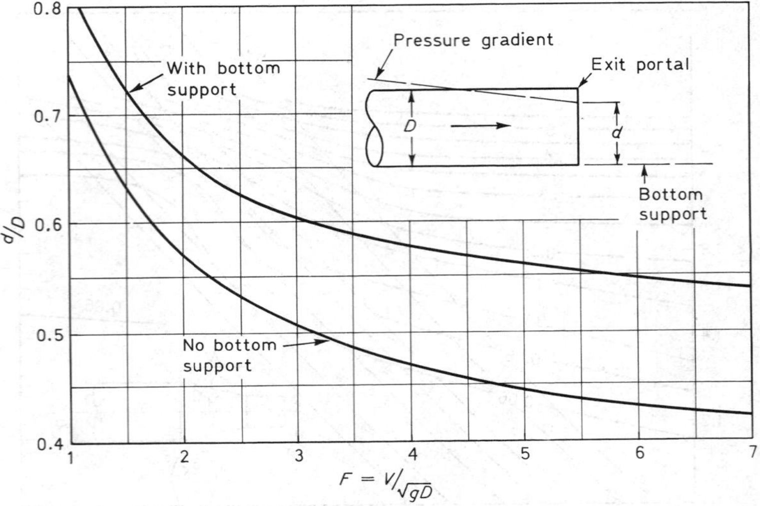

Figure 20.21 Exit depth in circular conduits. V is the mean velocity in the conduit flowing full. (Based on USWES data[10])

flow is subcritical, preferably with Froude number well below unity. Downstream of this point the waterway can be further reduced if desired but should be checked for lower discharges.

If the exit of a conduit is fully or partially submerged, head loss can be reduced by providing a gradual expansion, which can often be extended in the tail channel. If the conduit exit is not submerged, a free surface may develop some distance upstream of the exit portal, even when the conduit is flowing under pressure. The depth of the exit depends on downstream conditions but where tailwater level is low the flow becomes supercritical and the conduit exit acts as a control. The end depth still depends to some extent on the tail channel, particularly whether the flow is supported at the sides and bed, but the end depth may be estimated from Figure 20.21. If the emerging flow is supercritical and downstream flow subcritical, a hydraulic jump will occur and a stilling basin may be needed.

Flow through conduits can be routed and energy gradient plotted by use of the Bernoulli equation (see page **20**–2). Allowance should be made for head loss due to friction, bends, transitions and hydraulic jumps. Subcritical flow is routed in an upstream direction starting from the tail channel or a control, and supercritical flow in a downstream direction, using step methods if necessary. To locate an hydraulic jump the pressure–momentum theorem can be used, taking account of the slope of the conduit and the pressure against the conduit roof if submerged downstream. The method is described by Kalinske and Robertson.[48] Critical depths in circular and horseshoe conduits may be determined from Figure 20.19, and diagrams facilitating computation of jumps in conduits of circular and other cross sections have been published.[2, 3, 4] In cases where hydraulic jumps might occur in closed conduits with undesirable results, due to additional head loss or air (see below), it is recommended that the routing be repeated for several discharges using both high and low values of head loss coefficients and upper and lower limits of tailwater rating curve, to obtain a complete account of the flow.

Air problems

Air entrained upstream, or released from solution due to pressure drop, tends to collect at high points in long conduits and if not released can cause restriction of the water flow passages. Air release valves, often combined with vacuum relief to admit air if pressure falls below atmospheric, are therefore provided at high points. Vents are often provided in horizontal tunnels downstream of junctions where entrained air may enter. Air which has collected beneath the soffit tends to be carried forward by the flow, even against

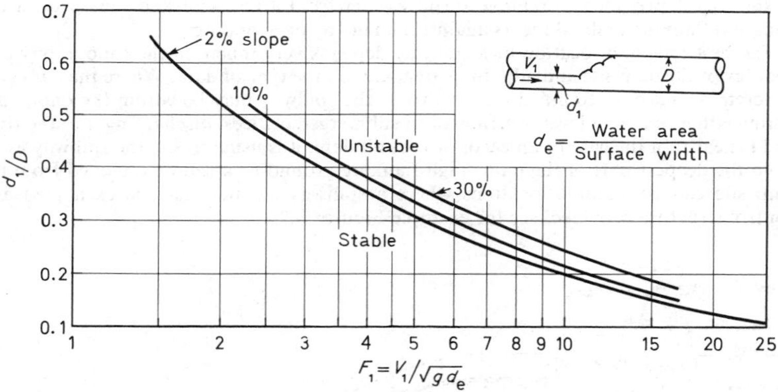

Figure 20.22 Stability of entrained air downstream of hydraulic jumps in circular conduits. (After Kalinske and Robertson[48])

a small gradient, but with a variable flow may move upstream and downstream at different times. At vertical shafts in pressure conduits and at deeply submerged exits the intermittent escape of air produces shock waves due to slap on the soffit as water replaces the air. This effect can be minimised by vents for controlled air release.

Hydraulic jumps entrain air and when a jump in a conduit is in contact with the soffit much of the air is released downstream. Following model tests in a conduit with various slopes by Kalinske and Robertson[48] and others, and several observations at full scale, the US Army Corps of Engineers[10] use the formula

$$\beta = 0.03(F_1 - 1)^{1.6} \tag{13}$$

which gives higher values than found in the model tests to allow for scale effect. Here β is the air/water ratio Q_a/Q_w, $F_1 = V_1/\sqrt{(gd_e)}$, V_1 is the upstream velocity and d_e the effective upstream depth (= water area/surface width). A particular application of these formulae is the estimation of air demand downstream of gates or valves located in closed conduits, where high velocity flow at part openings is transformed to full conduit flow through a jump—see Figure 20.22. Full-scale observations in three different cases showed that with rectangular gate openings peak demand occurred at 60% to 85% opening, with a secondary peak at about 5%.

The air pumped by the jump may be carried downstream by the full bore flow but, if the velocity is insufficient for this, air will collect immediately downstream of the jump and when a quantity of air has accumulated it will 'blow back' through the jump. Figure 20.22 shows the limiting conditions for the air just carried by the flow, as found by Kalinske and Robertson.[48] Sailer[49] compared these curves with conditions in a number of full-scale inverted siphons and found verification in that five cases where blowback had occurred were represented by higher values of $(F_1 - 1)$ than shown by the curves, while others giving no trouble were on or below the curves. With large flows, blowback through the jump is, like 'blowout' at the exit, explosive and potentially dangerous.

SPILLWAYS

Purpose and types

A spillway is provided to remove surplus water from a reservoir and thus protect the dam and flanking embankments against damage by overtopping.

The best type and location of a spillway depends very much on the topography and geology of the dam site and adjoining area, and on the type of dam. Where the dam is of concrete or masonry founded on hard rock, the spillway may be within the dam, consisting either of a high level overflow or of submerged orifices, discharging into the river bed beneath. In the case of an earth or rockfill dam, it is usual to site the spillway away from the deepest part of the dam; high flanking ground or a saddle removed from the dam site can be suitable locations where a spillway channel may be excavated and control structure provided, see for example Figure 20.23.

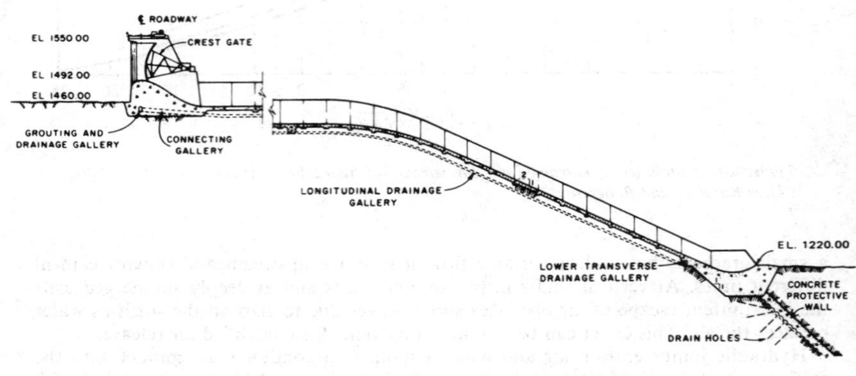

Figure 20.23 Chute spillway, Tarbela dam, Pakistan (Consulting engineers: Tippetts–Abbott–McCarthy–Stratton)

Where the dam is built on a narrow gorge and there is no suitable separate site for the spillway, a side-channel spillway is often adopted.

If control is by a fixed ungated weir, the maximum retention level of the reservoir is the weir crest level; at times of spill the reservoir level rises and sufficient freeboard has to be allowed above maximum water level, which is the level at which the design maximum flood discharge is released. In the case of fully regulated spillways, on the other hand, flood flows can be discharged with reservoir at retention level, which need never be exceeded. For a given dam height, storage can thus be greater but, because there is less flood storage, the spillway capacity also may have to be greater. The gates, however, enable the reservoir to be drawn down in advance of a flood peak, given adequate forecasting. Low-level orifices, having greater capacity than required for purposes of normal supply, have greater capability than has a gated crest overflow in drawing down a reservoir in the event of damage to the dam, an important aspect in areas where earthquake risk is present. But crest overflow weirs have a greater rate of increase of capacity as a reservoir level rises above normal, thus providing additional safety margin.

Cost is a major consideration in the choice between a regulated and an unregulated spillway, but spillways without gates have advantages in respect of reliability, absence of mechanical maintenance problems and no power requirements. They are therefore often adopted at remote sites and also in the case of small dams where the cost of gates would not be justified.

Siphon spillways carry some of the advantages of both gated and ungated spillways. They can be designed to prime and operate to maximum discharge within a small range of reservoir level and they are automatic, with no moving parts.

Another type of spillway, particularly suited for use with earth or rockfill dams is the bellmouth or 'morning glory' spillway, which can be built quite independently from the dam. If the reservoir is for water supply, the bellmouth and shaft are often combined

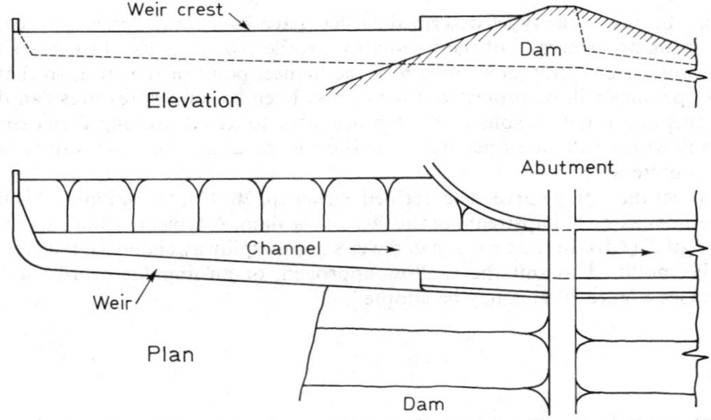

Figure 20.24 Side-channel spillway

in the same structure as a draw-off tower and the low-level tunnel can be used for river diversion during construction.

In many cases it is advantageous to provide more than one spillway. Instead of relying on a single spillway to control all floods up to catastrophic, it may be safer and more economical to provide a main or service spillway, fully regulated and capable of controlling all floods up to perhaps a 20- or 50-year return period, and a secondary or emergency spillway which will bring the combined capacity to the catastrophic level. The capacity of the emergency spillway should be adequate to control normal inflows alone for a reasonable period should the main spillway be damaged.

Channel spillways

The simplest form of spillway, consisting of a channel excavated in rock, is often used for small reservoirs and for emergency spillways. The control is usually provided by a hard sill or weir at the entrance. The channel downstream should be given sufficient slope to ensure that the weir will not be drowned by backwater effect. If the rock is erodible, a curtain wall with bed protection or stilling basin is needed to avoid erosion undermining the downstream end of the weir.

In the case of emergency spillways, a 'fuse plug' is generally provided, consisting of an erodible bank across the channel. Its crest is below the dam crest level but above normal operating level. When overtopped it quickly erodes down to the level of the hard sill, bringing the full discharge capacity of the channel into operation. To avoid excessive draw-down of the reservoir, emergency spillways of this type are preferably wide and shallow. Bed and bank protection are usually minimal but it is essential that there is no risk of erosion progressing upstream and breaching the reservoir rim beneath the level of the hard sill.

A side channel spillway consists of a weir discharging into a parallel channel, as in Figure 20.24 where the weir is aligned on a ground contour normal to the axis of the dam. The channel must be of sufficient width and depth to allow for energy dissipation and the generation of exit flow without drowning the weir. Calculation is based on the momentum principle.[11, 50]

Weirs

These may be any of the types described earlier (page **20**–9) but, where elevated, as in a concrete dam, are generally of the free-nappe profile type. It is usual for gates to close onto the weir face slightly downstream of the highest point of the crest, so that the jet at small openings will be projected downwards. Even so, underpressures can develop, and for this and other reasons it is often desirable to avoid prolonged releases under high heads with small gate openings. Provision of separate sluices or valves for small releases is preferable.

Piers affect the rating curve, as described earlier (p. **20**–11); to minimise this the piers are sometimes extended upstream of the face of the dam. A typical rating curve is shown in Figure 20.27(**AB**). In the case of gated weirs across spillway channels, the crest profile should be modified to suit the shallow approach, or an angular profile, as used in barrages (see Figure 20.11), may be adopted.

Orifices

These are generally of rectangular section and regulated by radial gates—see for example Figure 20.25. Where the head exceeds 10 m, care is required to avoid cavitation damage to the invert; with high heads the curvature should be very slight and the surface finished to a high standard. When the reservoir is drawn down so that the orifices are not fully submerged, flow is of the free-surface weir type. The rating curve of an orifice spillway with gates fully open therefore consists of two main parts, as may be seen from the example in Figure 20.27. The lower part CD relates to weir flow and the upper part EF to submerged orifice flow. Between the two is a transition DE. The discharge for weir flow is $Q = C_1 B H_1^{1.5}$ where Q is the discharge, C_1 is the weir coefficient, B is the width and H_1 is the upstream head above the sill. C_1 is generally a variable.

In the range of orifice flow, $Q = C_2 B \sqrt{(2gH_2)}$ where C_2 is the coefficient of discharge and H_2 is the effective head below reservoir level. If the jet springs clear, so that atmospheric pressure obtains around the whole periphery, H_2 is best measured from the centre of the jet at its exit from the orifice. If the jet emerges on a horizontal floor confined within vertical side walls, H_2 is more correctly measured from the soffit level as representing the effective elevation plus pressure head over the jet at the point of separation from the soffit.

The requirements in design are those of high-pressure outlets, described on page **20**–44. If these are met the coefficient can approach unity. In the case of Mangla spillway[52]

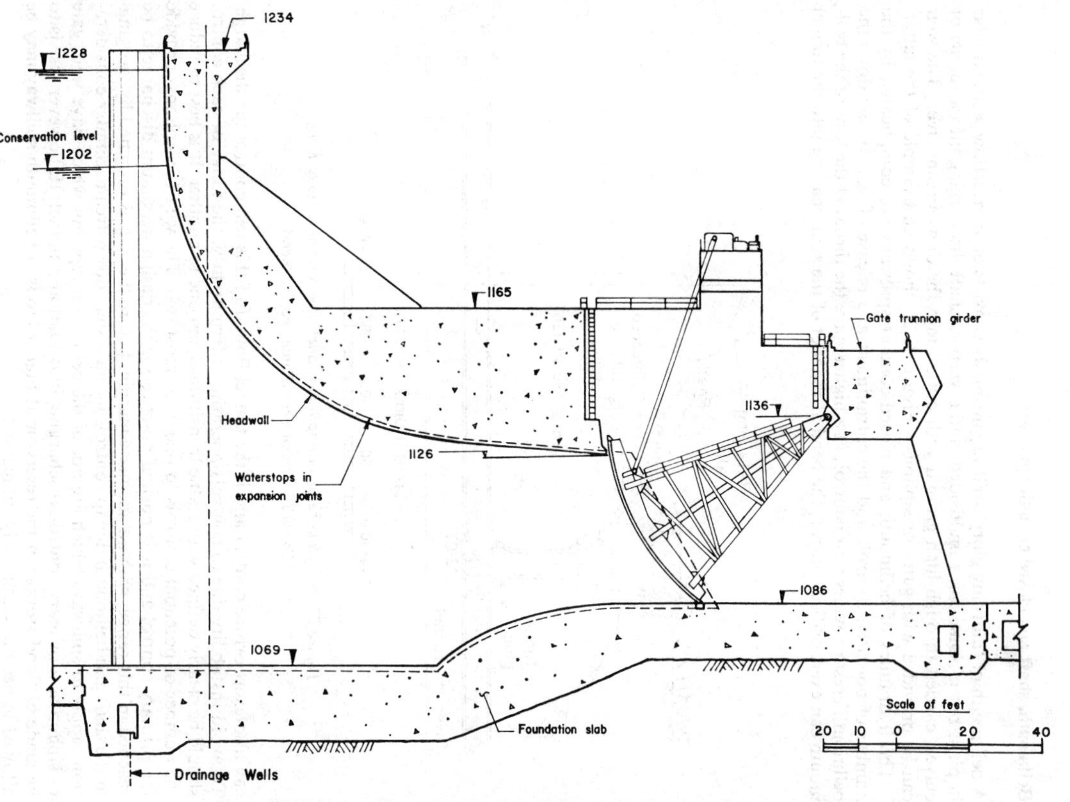

Figure 20.25 Main spillway, Mangla dam, Pakistan. Section through gate structure
(Consulting engineers: Binnie and Partners in association with Hanza Engineering Co. and Preece, Cardew and Rider)

(Figure 20.25) it was approximately 0.95. Model tests are used to indicate pressures on the boundary surfaces and provide rating curves for full and partial gate openings.

Bellmouth, shaft and closed conduit spillways

A bellmouth or 'morning glory' spillway normally consists of an overflow weir, circular in plan, but in some cases multisided, and a vertical shaft discharging into a tunnel or culvert carried through high ground with outfall into the downstream river. The weir can be provided with gates or siphons—Figure 20.26 shows an example of the latter.

The hydraulics of bellmouth and closed-conduit spillways are complicated by the number of potential controls and the entrainment and release of air. At low flows the bellmouth crest provides weir control; at a higher stage the throat at the foot of the bellmouth can exert orifice control; the bend at the foot of the shaft leading into the tunnel

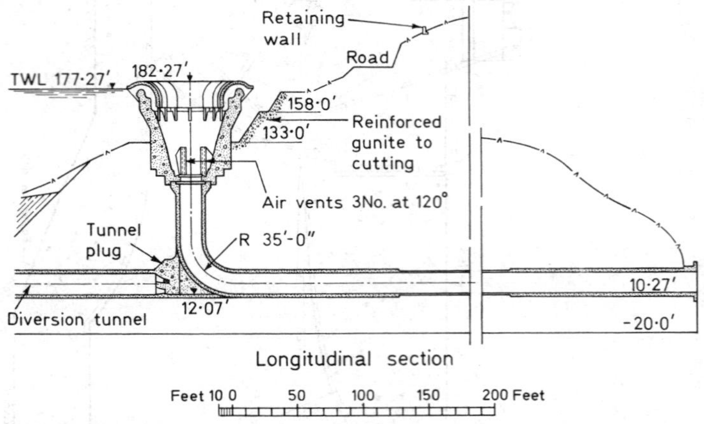

Figure 20.26 Siphon bellmouth spillway, Shek Pik reservoir, Hong Kong (Consulting engineers: Binnie and Partners)

can also exert orifice control and if the tunnel flows full this can control the discharge. To avoid instability due to controls operating intermittently, the range of each control should be clearly defined with stable transitions from one to another. It is best to reduce the number of potential controls to one or at most two. The weir (or siphons) provide the primary control and it is normal practice for the design maximum discharge to be reached in this range, with an adequate margin, before the weir is drowned by 'gorging' in the shaft and bellmouth due to controls in the system downstream. Similar considerations apply to spillways which are not of the bellmouth type but which have weir, gate or siphon as a primary control, discharging into shaft and tunnel. If, however, use is to be made of flood storage in the reservoir at higher levels, a bellmouth spillway may be allowed to become completely submerged.

As in the case of straight weirs, sharper curvature raises the coefficient of discharge. Profiles based on the shape of undernappe of a jet have been designed for weirs circular in plan.[53, 11] In several cases of bellmouth spillways measures were necessary to reduce swirl and prevent vortex formation particularly at the highest flows, when it could greatly reduce discharge capacity. Vortex flow is induced by asymmetrical approach in the reservoir; anti-vortex measures usually consist of crest piers or vanes, or a single divide wall on a diameter extending from below crest level to above maximum reservoir level.[54, 55] When the jets from opposite sides of the shaft intersect, the entrainment of air can result in negative pressures on the walls. Vents may be necessary to avoid instability and vibrations.

A typical rating curve is shown in Figure 20.27. Weir control is represented by the curve AG. At low flows there is a free surface flow in the bend and tunnel but with rising discharge and downstream conduit not flowing full the bend begins to act as an orifice

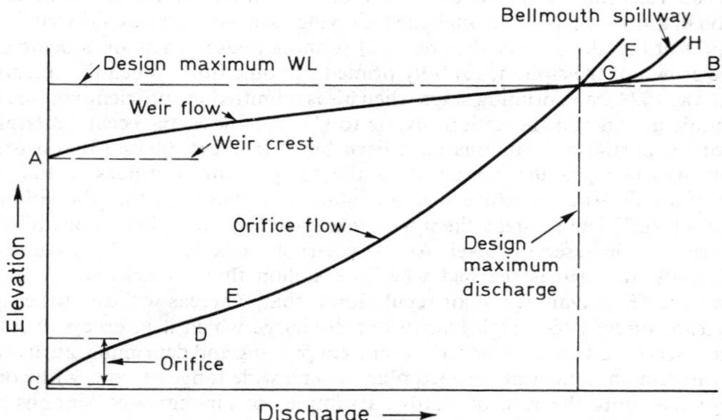

Figure 20.27 Typical rating curves for high-level weir, deep orifice and bellmouth spillways

with water level rapidly rising in the shaft. When it reaches crest level it begins to drown the weir flow. The bellmouth is then said to be 'gorged'. This is represented by the intersection of the two curves at G, above which the bend assumes control of the rate of flow. The rating curve of the spillway is therefore AGH with a short transition at G representing drowned weir control and bend orifice control at H.

If the bend is too sharp, flow from the bend to the conduit is very disturbed; if it is too easy the downstream culvert may flow full; a bend radius of 1.5 to 2 times the diameter is generally satisfactory.

For proper control of flow at the bend and smoother flow in the conduit a deflector may be placed on the inside wall at the upstream of the bend.[56] Where the conduit is used for river diversion during construction, a properly shaped bend can later be formed when the diversion intake is plugged. Unless the conduit is very short, flow with a free surface is desirable with sufficient air space above for entrained air to be released without trouble. Sufficient slope should be provided to ensure that the depth does not exceed the desired limit. As the result of a model study, Mussalli and Carstens[56] recommend upper limits for the proportion of water flow in such conduits, ranging from 97% of the area, when the Froude number is 2, to 50% when it is 8.5. Where the velocity is high enough to entrain air it is desirable to provide an air vent at or near the bend. With high velocities it is also best to avoid bends and other conditions downstream which could cause a hydraulic jump to form in the conduit.

It is evident that the concrete surface at the base of the shafts of bellmouth spillways can be subjected to high impact loads by water, possibly with ice and logs, spilling from a great height. In some cases steel or cast-iron lining has been provided in this area, but from a survey of 16 bellmouth spillways, of which eight with unlined concrete inverts had undergone a fair test, Bradley[55] found no erosion of a serious nature. Dense concrete with smooth surface finish is called for here and in the conduit.

Siphon spillways

Compared with a free-surface weir, flow through a siphon reaches a high rate of discharge per unit width with only a small rise of reservoir level needed to prime the siphon. This permits a higher retention level for a given maximum water level, or alternatively a higher concentration of flow in a restricted width.

Reservoir retention level is equal to siphon crest level. As the reservoir level rises, the action of a siphon passes through the following successive phases: (1) weir flow, when water spills at low depth over the crest; (2) priming phase, when air is being extracted from the crown of the siphon; (3) fully primed siphonic flow. When the reservoir falls, (3) gives way to (4), a depriming stage when air is admitted in sufficient volume to break the siphonic action and the action reverts to (1), weir flow. In recent years many air-regulated (or partialised) siphons have been built. In these, phase (3) consists of two parts: in (a) when priming has occurred the entry of air continues so that the flow consists of an air–water mixture. The air intake is so designed that the volume of air admitted is insufficient to break the siphon (except at low flows) but is controlled by very small variations in reservoir level. As the reservoir rises further the volume of air is reduced until stage (b) is reached when the siphon flows 'blackwater', i.e. with no entrained air. The advantage of air regulation is that, whereas without it the siphon on priming runs directly to a high blackwater discharge which if in excess of inflow will draw the reservoir down and lead to intermittent priming and depriming, an air-regulated siphon will remain in the fully primed phase over a wide range of flow, with continuous discharge matching the rate of inflow. Examples of air-regulated siphons are Eyebrook,[57] Shek Pik[58] and Plover Cove.[59]

Spillway siphons are generally designed to prime automatically when the reservoir has risen to a level such that (a) an upstream air inlet is submerged, (b) the siphon outlet is sealed by a deflected jet or a downstream weir and (c) the flow is sufficient to entrain and remove air from the crown of the siphon. Various priming devices have been used[60] and the priming depth above crest level is in some cases as little as one-sixth of the throat diameter.

The blackwater discharge capacity of a siphon can be expressed as

$$Q = CA\sqrt{(2gH)}$$

where C is a coefficient allowing for head losses, A is the cross-sectional area of the flow at exit and H is the head from upstream reservoir to effective exit level—usually the downstream lip of the hood. The value of C obtained in the Plover Cove siphon model[59] was 0.68. That of the model of Shek Pik bellmouth siphon, where the shape was radial and no sealing weir was provided,[58] was 0.66.

Surface waves in the reservoir are an important factor in siphon design. Model tests showed that despite provision of baffles, waves caused surging in the siphon but the air intakes could be designed to counterbalance the effects of surging and wave wash. The head in siphons is usually limited to about 7 m to avoid cavitation at the crest. Tests on the Plover Cove siphons showed that wave action resulted in transient pressures below average pressures, but a total head of 7.3 m was still feasible. Each case should, however, be examined in the light of the particular conditions obtaining.

Chutes

Chutes may be built into the downstream faces of concrete dams; longer chutes are often provided to convey the flow from side channel or other flanking spillways to the river bed downstream—see Figure 20.23. The gradient is likely to be steep enough to generate high-velocity flow. As lateral changes of direction could result in overtopping of the side walls, any essential changes in the alignment should be made near the control structure where the velocity is relatively low and thereafter the chute should be straight. There should also preferably be no changes of alignment in the side walls where flow is supercritical because diagonal shock waves would be created which might cause overtopping downstream.

High-velocity flow can give rise to high pressure and the most careful precautions are necessary to avoid uplift pressures developing beneath the chute slabs. A high standard of surface finish is called for and the profile should contain only very gradual curvature. Joints between slabs should be keyed and bridged by flexible water stops sealed at intersections. Projections at the joints facing upstream should not be allowed but offsets facing downstream up to 12 mm are often accepted or even specified. Drains are generally provided beneath and parallel to all joints, so that in the event of leakage, uplift pressure cannot build up. For additional protection, chute slabs are often anchored to the foundation rock. Special care is needed where the chute rests on jointed or fissured rock because pressure can be transmitted from leakage at a higher level despite underslab drainage. Chutes at a steep slope are especially vulnerable and may call for deep anchors. Stability should be checked for all possible modes of failure.

In general, high head spillways with chutes should not be used for routine releases of water for supply, because of the risk of cavitation damage with small gate openings, also because the chute may have to be taken out of service in the dry season for repairs, and because low flows can create problems of erosion downstream.

Calculation of depth of flow in chutes may be done by a step method beginning at the top, using curves showing energy against depth for the required discharge per unit width, similar to that of Figure 20.2. The calculation should be performed with two roughness coefficients, one representing a maximum, for use in determining side-wall height, and the other representing a minimum, for use in the design of energy-dissipating works. The height of side walls should include an allowance for bulking due to air entrainment. The US Corps of Engineers[10] provide a design curve based on observed data, with the equation

$$c = 0.436 \log_{10} (S/q^{\frac{1}{5}})+0.971,$$

where c is the ratio of air volume to air-plus-water volume, q the discharge per foot width in m^2/s, and S the sine of the angle of chute inclination.

Energy dissipation

The energy to be dissipated at the outfall from a spillway is very considerable. The means of protection of the dam and other structures from its erosive action depend largely on the rate of discharge and its head, the erodibility of the materials of the river bed and surrounding ground and on the proximity of the dam.

In the case of rivers in alluvium or other easily erodible ground, a stilling basin designed to contain a hydraulic jump is often provided. This may be of the rectangular type or a submerged roller bucket (see sub-sections on pp **20**–20 and **20**–21). The former is generally more efficient but more costly, especially in the case of large structures where deep retaining walls would be required. Where the river bed material is rock a ‘ski jump’, trajectory or ‘flip’ bucket is generally provided. This is elevated above maximum tailwater level, so that the jet trajectory carries the water into a plunge pool some distance from the bucket. Dissipators of this type are generally less costly and are suitable where the bedrock is resistant enough so that erosion does not progress back and endanger the foundations of the dam or other structures. Ski jumps have also been used where the river bed is of alluvium, and the structures are suitably protected against erosion.

The radius of flip buckets is not critical provided it is large relative to the depth of flow.

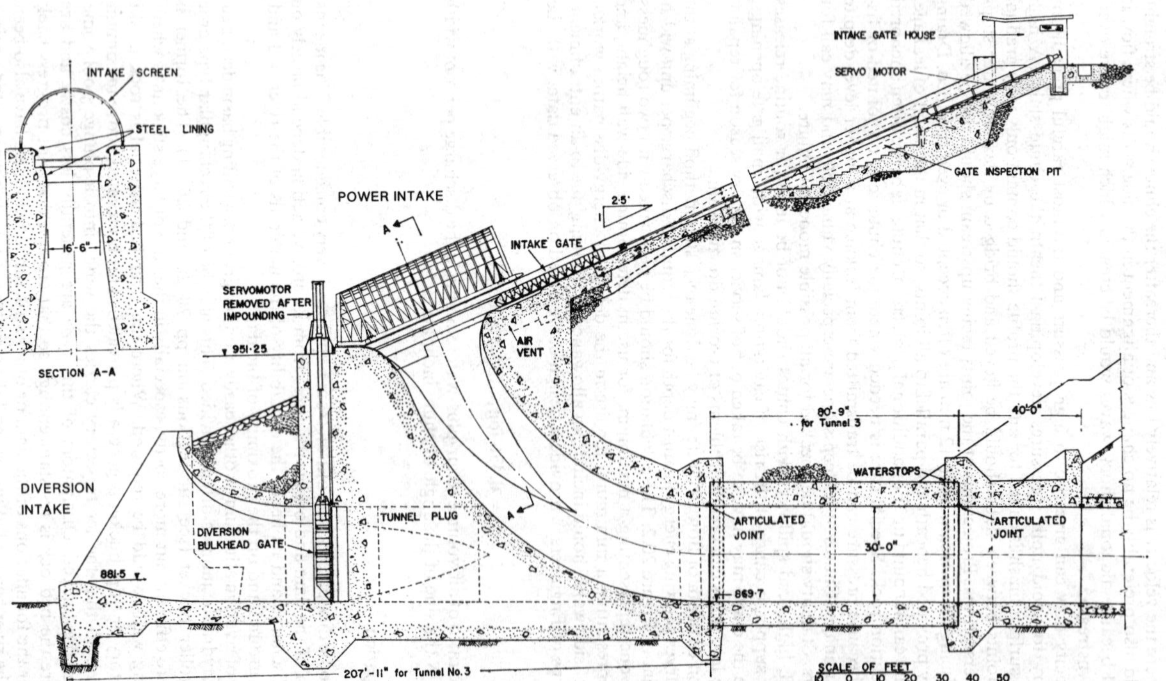

Figure 20.28 Intakes at Mangla dam, Pakistan. Longitudinal sections. (Consulting Engineers: Binnie & Partners in association with Harza Engineering Company and Preece, Cardew & Rider)

The exit angle is important as it determines the throw distance; the angle is generally between 20° and 40° and the theoretical throw distance is given by x in the formula

$$\frac{x}{h_v} = \sin 2\theta + 2\cos\theta\left(\sin^2\theta + \frac{y}{h_v}\right)^{\frac{1}{2}} \tag{14}$$

where h_v is the velocity head at the bucket lip, θ the bucket exit angle, measured from

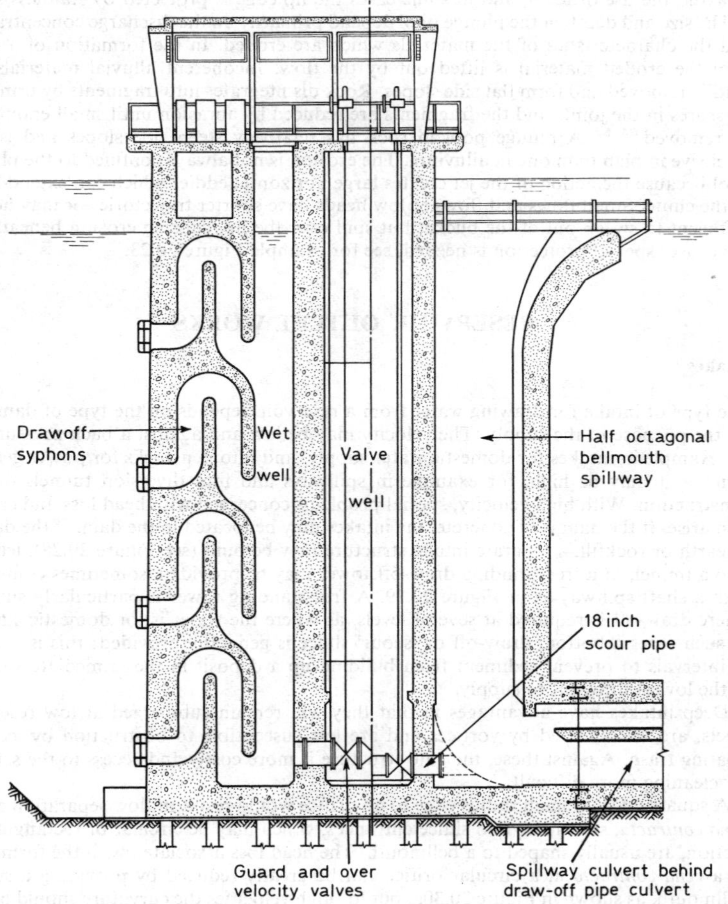

Figure 20.29 Combined draw-off tower and spillway, Seletar reservoir, Singapore. Air-controlled siphons are used instead of valves for draw-off purposes (Consulting engineers: Binnie and Partners, Malaysia)

the horizontal, y the vertical height of bucket lip above tailwater.[10] However, because of air resistance and internal shear the jet tends to break up and diffuse so that the actual trajectory distance may be 10% to 30% less than indicated by the formula. Elevatorski[25] quotes examples of models and full-scale spillways.

Flip buckets are not necessarily of circular profile, nor axisymmetrical. There are several instances of buckets composed of flat deflecting surfaces, some designed to deflect the jet to one side to suit downstream requirements.[61]

In the design of the sidewalls, allowance must be made for the additional lateral pressure due to centrifugal force. This may be calculated by methods of Gumensky,[62] Balloffet[63] or the US Army Corps of Engineers.[10] To avoid cavitation damage it is usual to avoid the use of teeth, and in some cases the lip edge is protected by stainless steel.

The size and depth of the plunge pool depend primarily on the discharge concentration and the characteristics of the materials which are eroded. In the formation of a deep pool the eroded material is lifted out by the flow. Incoherent alluvial materials are readily removed and form flat side slopes. Rock disintegrates into fragments by transient pressures in the joints and the fragments are reduced by abrasion until small enough to be removed.[64, 65] A plunge pool in rock has relatively steep side slopes and is less extensive in plan than one in alluvium. The erosion is not always confined to the plunge pool because the action of the jet creates large horizontal eddies which can extend back to the chute. Small flows and flows at low heads have shorter trajectories or may not be sufficient to sweep out of the bucket but spill over the lip, causing erosion beneath. In such cases special protection is needed, see for example Figure 20.23.

RESERVOIR OUTLET WORKS

Intakes

The type of intake for drawing water from a reservoir depends on the type of dam and on the purpose of the supply. The velocity may be low and against a back pressure, as for example in intakes for domestic water supply, and into penstocks for power generation, or it may be high, for example in spillways and into diversion tunnels during construction. With high velocity, special problems concerned with head loss and cavitation arise. If the dam is of concrete, the intakes may be located in the dam. If the dam is of earth or rockfill, a separate intake structure may be built (see Figure 20.28), leading into a tunnel, or a free-standing draw-off tower may be provided, sometimes combined with a shaft spillway as in Figure 20.29. A free-standing tower is particularly suitable where draw-off is required at several levels, as where the water is for domestic supply. In such cases a bottom draw-off or 'scour' sluice is generally provided; this is opened at intervals to prevent sediment from building up a deposit in the immediate vicinity of the lowest draw-off to supply.

Deep intakes have advantages in that they will remain submerged at low reservoir levels, are less affected by vortices and are less susceptible to obstruction by ice and floating trash. Against these, the gate structure is more costly and access to the screens for cleaning more difficult.

A square edge or small radius edge to an orifice would result in flow separation and a *vena contracta,* so orifices and sluice entrances, which may be circular or rectangular in section, are usually shaped to a bellmouth. The head loss associated with the formation of a *vena contracta* at a circular orifice can be greatly reduced by providing a simple bellmouth, as shown in Figure 20.30a, but for high velocities the curvature should be less to avoid low pressures which might result in cavitation damage. Compound curves of two or more radii and elliptical curves are often suitable profiles. A typical example of an elliptical profile is shown in Figure 20.30b. With this profile the minimum pressure at the boundary is approximately $0.1\,V^2/2g$ below the corresponding pressure in parallel flow in the orifice downstream where V is mean velocity. As this is a mean pressure, lower pressures may occur owing to fluctuations. For very high velocities this may not be acceptable and the profile may be based on the profile of a jet springing from a sharp-edged orifice[66, 67] or may be compound elliptical.[10] In the case of important works, especially with high velocities, intake entry curves are usually tested in hydraulic models.

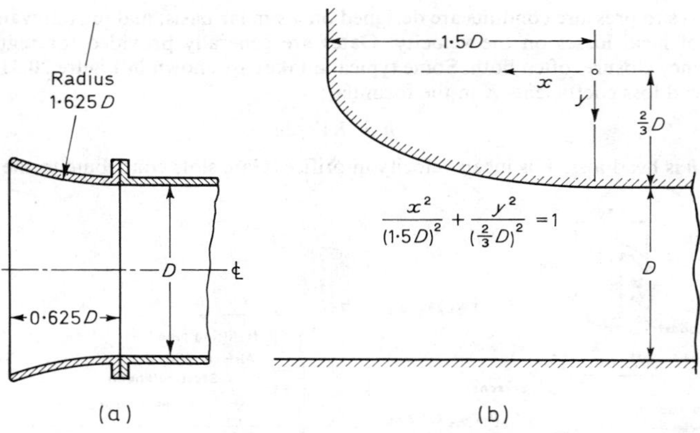

Figure 20.30 (a) Simple bellmouth; (b) elliptical roof profile for conduit intake with parallel sides and horizontal floor[10]

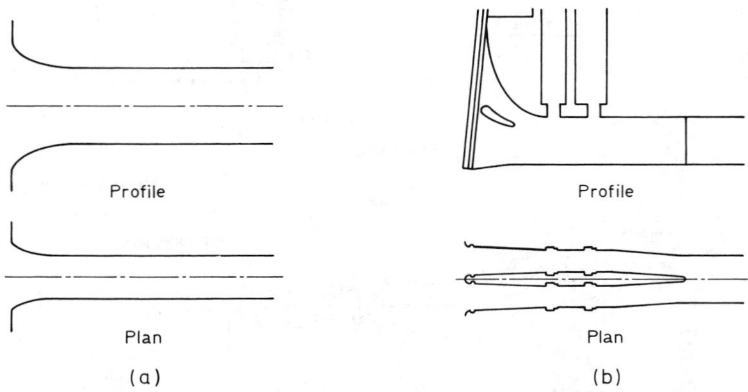

Figure 20.31 Typical intakes to pressure conduits. (a) Single intakes in concrete dams, K = 0.07 to 0.16; (b) double intakes to tunnels at earth dams, K = 0.12 to 0.25

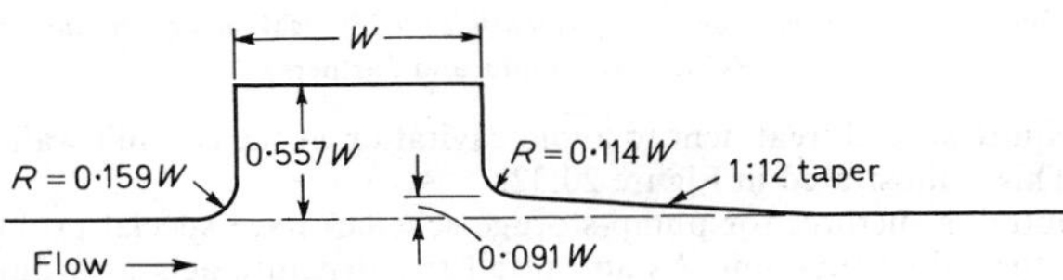

Figure 20.32 Typical gate slot with downstream offset (USWES design[10]*)*

Intakes to pressure conduits are designed on a similar basis, making allowance for the effect of head losses on the velocity. Gates are generally provided for regulation or emergency closure, often both. Some typical intakes are shown in Figure 20.31, together with head loss coefficients K in the formula

$$h = KV^2/2g$$

where h is head loss, V is mean velocity in orifice. Gate slots contribute to the head loss

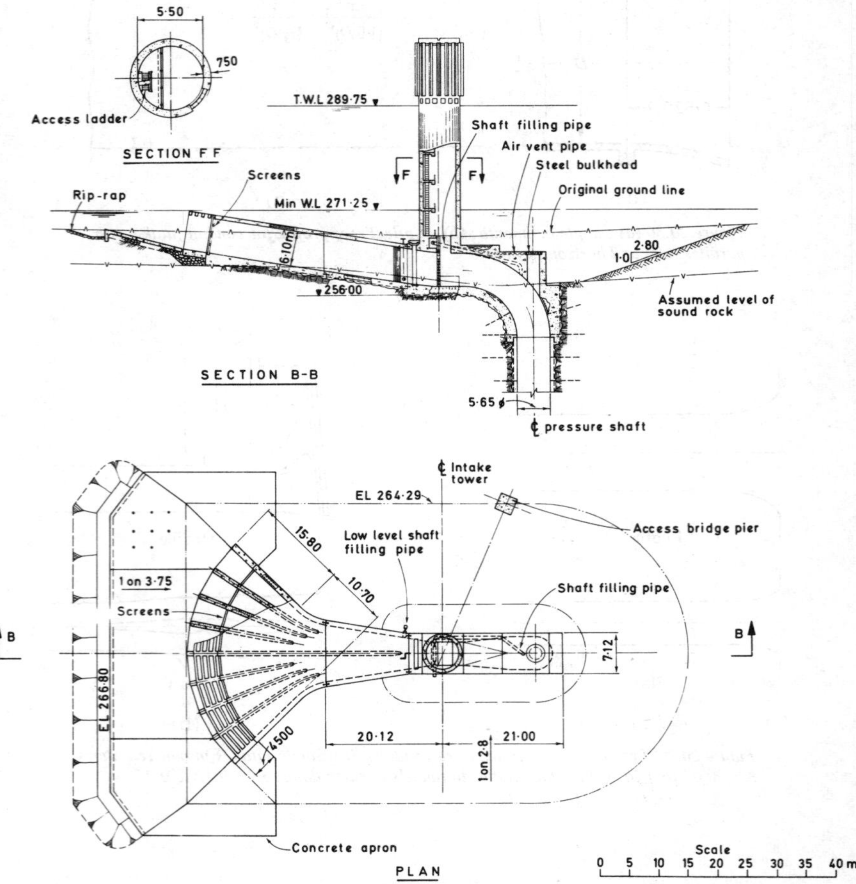

Figure 20.33 Design of upper intake/outfall for 230 MW pump storage scheme
(Courtesy: Binnie and Partners)

and also require special treatment to avoid cavitation at the conduit wall surface downstream.[68] This is illustrated in Figure 20.32.

Intake/outfall structures for pump-storage schemes have special problems in design because of their dual function. As an outfall the structure acts as a diffuser; velocity head has to be converted to static head with minimum head loss, in a system complicated by bends and gate chamber. In consequence of the more stringent requirements of diffuser flow, the structure is usually efficient as an intake but vortices may be a problem, especially at the stage when the reservoir level is approaching its lower limit. Figure 20.33 shows the upper intake designed for Camlough pump storage scheme.[69] The head loss coefficients observed in the model tests were 0.235 for inflow to the shaft and 0.36 for outflow to the reservoir.

Vortices

Though a slight surface swirl may be of no consequence, a vortex with an air core extending to an intake can be harmful in reducing the discharge capacity of the intake, causing gate vibration or resulting in admission of air to pumps or turbines. Any tendency for a vortex to form in a model test should be carefully investigated because vortices form more readily and develop further at full scale.

A free vortex tends to form in accelerating flow towards a region of low pressure, as at a submerged outlet. It is facilitated by boundary geometry consistent with vortex shape, and by an initial circulation in the reservoir, and is more marked the greater the pressure drop to the outlet relative to the depth below surface. Vortex action is reduced by deeper submergence of the intake, by reduced velocity at the intake and by obstructions to the rotation, such as horizontal grids and projecting walls.[70, 71]

Vortices are also a problem in pump sumps where even a slight swirl may affect pump efficiency and more refined measures are needed.[72, 73] Velocity of approach is generally limited to 1 m/s, but eddies can still form at points of separation. Sharp wall angles and regions of dead water should be avoided and expansions should be gradual. Model tests are often needed to determine suitable sump geometry.

Screens

Screens or trash racks are provided at intakes to hydroelectric plants, pumps and water treatment works. Log booms are often placed upstream, but it is generally required to intercept small debris and possibly fish. The spacing of the bars may be 2 to 20 cm depending on the duty. The main requirements in design are that the bars are stiff enough not to vibrate and are arranged for easy cleaning. As a general guide for screen area the mean velocity is usually limited to 0.6 m/s or less. Vibration is avoided if the dimensions of the bars are such that the natural frequency of the bars is higher than the forcing frequency.[74] Screens may be fixed and cleaned by raking or lifted above water for cleaning.[75] For ease of cleaning from above, the vertical bars generally project upstream of the lateral bars.

Gates and valves

Gates are used to control flow in open channels or closed conduits by restricting or closing the water way. They may be required:

(a) for regulating the flow, when they must be capable of operating at any required degree of opening;

(b) for emergency or guard purposes, when they must be capable of closing under any condition of runaway flow which could occur;

(c) as bulkhead gates for closing a conduit for inspection, maintenance or construction works. When they are permanent installations, such gates are generally designed to open and close only under balanced pressures, but when used to close diversion tunnels during construction works, closure may be against a considerable flow. Stop logs are similar in function to bulkhead gates, but are in smaller units handled individually and placed above one another.

The types generally used in outlets from reservoirs and in spillways, barrages and canals are as follows.

Vertical lift gates are supported by guides in slots at the side walls of the conduit. They may open by raising or by lowering; in some cases they are in two or even three parts, each operating independently. They may have sliding contact with the guides or may have wheels (fixed-wheel gates) or a moving train of rollers (Stoney gates). They have seals at the sides and (in orifices or closed conduits) also at the top and generally close onto a steel sill with an inset compressible seal if required. Advantages of vertical lift gates are their simplicity and ease of maintenance; disadvantages are the requirement of slots in the side walls and limitations of loading on axles at very high heads. Sliding gates can be used for high heads but require correspondingly powerful actuators. However, sliding gates have been installed for heads as high as 200 m, carrying a water load exceeding 1000 tons.[76] A disadvantage in the use of vertical lifting crest gates for spillways, barrages and canals is the requirement of a high superstructure. In the case of conduits the hydraulic disadvantage of side slots in high velocity flow has been overcome by the introduction of 'jet flow' gates; in these, narrow side slots are provided but an upstream contraction causes the flow to spring clear of the slots, re-attaching to the side walls downstream. Caterpillar-track-mounted gates are sometimes used for emergency closure, operating on flat bearing faces on the upstream face of a dam. One gate is generally sufficient for a number of orifices, controlled by a mobile gantry from the top.

Inclined lift gates are similar in many respects to vertical lift gates but operate on inclined tracks, see Figure 20.28. They are sometimes used for guard or emergency purposes at intakes in earth or rockfill dams, the track being laid on the upstream face of the dam. An advantage of this gate is its low cost compared with alternatives of a vertical lift gate in a tower in the reservoir or in a gate shaft within the dam. Disadvantages may include the remoteness and inaccessibility of the gate in case of emergency and the long vent shaft.

Radial (or tainter) gates An example may be seen in Figure 20.25. The gate skin is of cylindrical shape and is supported on cross members spanning between two radial arms which rotate on short axles extending from the side walls or piers. The resultant of the water pressures passes through the centreline of the axles creating no moment opposing gate operation; therefore powerful actuators are not required and in the event of power failure quite large gates can be operated manually. Other advantages are simplicity, reliability and low cost. They do not require side slots; side seals are of the sliding type and the gates close onto a steel sill. Where it is required to allow passage of floating debris, the top of the gate may consist of a hinged flap opening downwards. They are widely used for both weir and orifice control in spillways. They are also sometimes used in pressure conduits but need more space than vertical lift gates and problems of access and removal for maintenance have to be considered.

Segmental gates which are rotated to lower them into a recess in the bed are particularly suited for waterways used by ships. Special arrangements are needed to maintain the recess clear of deposits.

Hinged leaf, bascule or flap gates are sometimes used for crest control where water depth is not great. They are hinged at the bottom and may be used for regulation with water spilling over them; they need venting. They have the advantage of allowing floating debris to pass at small openings but, although they can be of curved profile, the weir crest has to be rather wide to provide the recess. On many European rivers, bascule gates are used for regulating upstream water level, operated by hydraulic actuators located below the weir crest. Hinged gates can be made to open automatically by a simple mechanical device when the upstream water level rises to a given height. Gates hinged at the top are also used where the whole assembly is retractable to allow the passage of ships.

Drum and sector gates (see Figure 20.34) These are crest gates which open downwards, retracting into a recess in the crest. They may be hinged on the upstream (drum) or downstream (sector). The upper surface can be shaped to suit the weir profile when fully open. In the examples shown, the gate consists of a watertight vessel controlled by application of headwater pressure beneath; it is sealed at the hinge and gate seat. These gates can be arranged to operate automatically by the upstream water level.

Bear trap gates when raised are in the form of a flat 'A', with upstream and downstream leaves forming the two legs, hinged at the bottom, with seals at the hinges and the apex. The gate is raised by admitting water under pressure from the headwater. When lowered

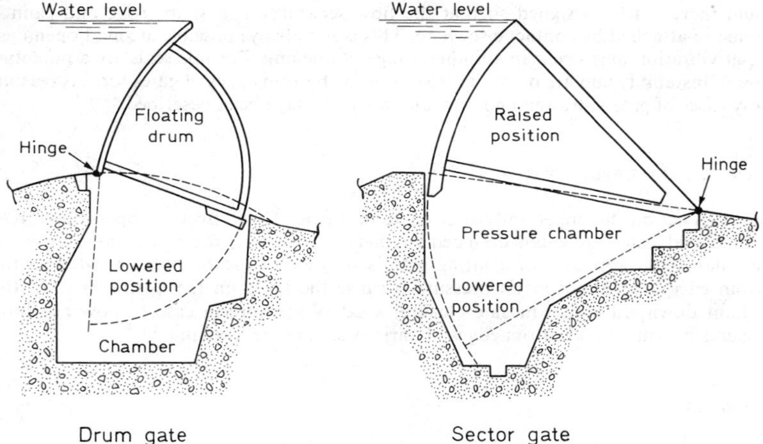

Figure 20.34 Drum and sector gates

the upstream leaf overlaps the downstream leaf and both fold flat. Bear trap gates have for long been used in Europe and the United States for river regulation.

Rolling gates consist of a roller with toothed wheel meshing with an inclined toothed rack at each end. The gate is rotated by a chain and accordingly moves up and down the racks. Roller gates have been used for river regulation.

Cylinder or ring gates have been used as crest gates on bellmouth spillways and for bottom outlets.[76] Some of the former open by being lowered vertically into a recess in the weir crest, controlled by water pressure, others and the bottom outlet gates by being lifted from above. Lateral control of the gate motion is provided by guides.

PARTIAL GATE OPENINGS

These constitute orifices of which three sides are formed by the fixed boundaries and the fourth by the gate lip. When unsubmerged, the jet springs clear from the gate lip forming a *vena contracta* the dimensions of which depend on the shape of the gate skin and lip and on the upstream profile of the sill. It is therefore usual to calibrate gates by model tests. Details of calibration of various gates are available.[10, 76, 78, 79, 80] Where gates are partially submerged downstream, calibration is complicated by the additional variable, and is also less reliable because of possible variation of flow pattern. Submergence may also lead to vibration problems.

VIBRATION

In general, vibration is a result of resonance where the frequency of a pulsating force is equal or nearly equal to the natural frequency of a flexible part of the structure. Gates are liable to vibrate when overtopped and not adequately vented, when significant flow occurs both over and under a gate, when the gate is partially or fully submerged, when the location of flow separation is unstable or there is flow re-attachment. Vibration by the two latter causes may occur at the bottom edge of a lifting or radial gate which should therefore be designed so that the flow separates at a sharp edge and cannot become re-attached by contact elsewhere. This is not always possible at small openings, so that vibration may occur in a limited range of opening. Flexible seals are a potential cause of instability and are often omitted from the bottom edges of gates for this reason. Many cases of gate vibration and remedial methods have been described.[81, 82]

DOWNPULL AND UPTHRUST FORCES

These can act on the upper and lower edges of a gate. They affect the operating forces required and gates are often designed so that the resultant force is of assistance. In particular, if the top edge of a lifting gate is subjected to static pressure whereas the bottom edge is at atmospheric pressure because the lip is on the upstream edge, the resultant downpull assists in gate closure, a safety measure in case of power failure. Pressures measured on bottom edges of various shapes are available.[10, 83]

GATE SEALS

Where a small amount of leakage can be tolerated, as in most works in the open, the seal at the bottom of a lifting or radial gate is usually metal to metal, between the gate edge and a steel sill set flush in the floor. A bottom slot would fill with debris and a projecting rubber seal may vibrate, but if leakage is to be minimal a rubber seal may be inset flush in the floor. Side and top seals can be metal to metal but with close tolerances these may be costly. Flexible rubber seals are therefore frequently used, in the form of strip tightly clamped with small projection, or moulded into bulbous shapes (e.g. music-note type) and arranged to be held in contact by the water pressure. As frictional resistance between metal and rubber seal increases with pressure, brass cladding is often used for sliding seals where the head exceeds 60 m. For very high heads, metal to metal contact may be required.

VALVES

Valves are used to regulate flow or pressure in pipes and conduits or to close them against flow, often at high pressure. A service valve, whether for regulation or closure is generally protected by an upstream gate or guard valve which can be closed against flow to isolate the regulating valve for maintenance or repair, and to prevent leakage if the regulator is not adequately sealed. Valves may be ‘in-line’, i.e. with pipe line upstream and downstream, or ‘terminal’ at the discharge end of a pipe line. Variations in valve design are numerous; only a few types which are normally used in reservoir outlet and hydro-electric power systems are described below. Of these, gate, spherical, butterfly, needle and tube valves are generally used in-line while needle, tube, hollow jet and sleeve valves are used as terminal regulators.

The discharge capacity of a valve may be expressed as

$$Q = CA\sqrt{(2gH_L)}$$

where Q is the discharge, C the coefficient, A the cross-sectional area of the valve at entry and H_L the head loss in the valve. In special cases the coefficient is based on the throat area of the valve. $H_L = KV^2/2g$, where V is mean velocity at entry, therefore $K = 1/C^2$.

Gate or sluice valves in their simplest form (Figure 20.35) consist of a sliding leaf in a valve body with side slots, thus resembling a vertical lift gate with operating rod sealed to contain the pressure. The sealing contact is metal to metal and the leaf is usually wedge

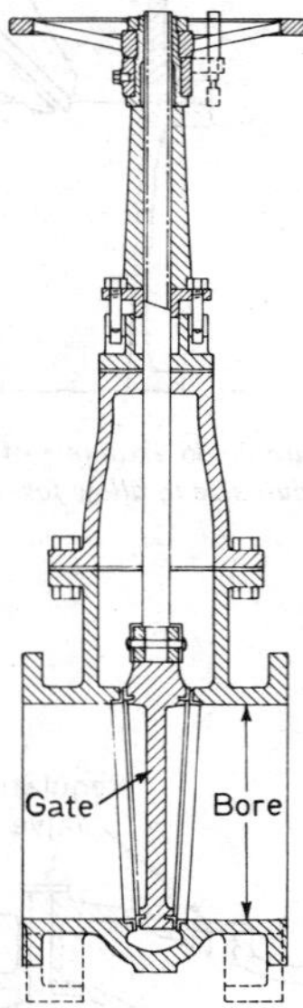

Figure 20.35 Typical gate valve, wedge type (Courtesy: J. Blakeborough and Sons Ltd)

shaped to provide tight sealing when fully closed. Parallel guides prevent vibration at part openings, but gate valves are not suited for regulation except at low or moderate pressures. A bypass is usually provided to balance pressures but a valve for guard or emergency duty may have to close under unbalanced pressure. Advantages of gate valves include the simplicity of design and low head loss when fully open. Disadvantages are the considerable power required to operate under unbalanced heads, cavitation damage to the slots at high velocity and damage by abrasion of the sealing faces if sediment is carried in the flow. The main disadvantages of slots are overcome in the 'ring-follower' valve in which the gate when raised is followed by a cylindrical ring which effectively covers the slots. It requires a cavity for the ring beneath the conduit and is suitable only for guard purposes.

Spherical or rotary plug valves are used for guard and on-off duties. They generally consist of a short length of tube of the same diameter as the conduit and length about the same dimension, which is in line with the conduit in the open position and is rotated

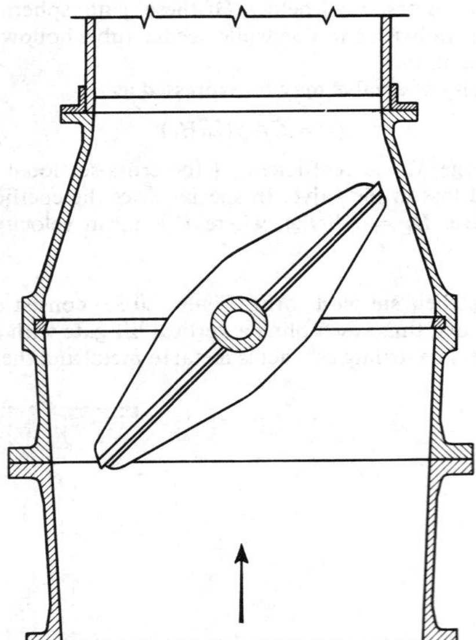

Figure 20.36 Example of butterfly valve. The valve diameter exceeds the conduit size to allow for obstruction of waterway by the blade

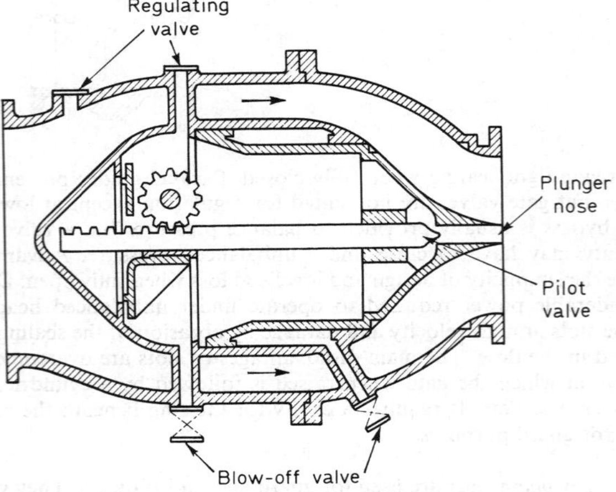

Figure 20.37. Larner–Johnson needle valve, with internal pilot valve control

through 90° to effect a closure. This is enclosed in a body of roughly spherical shape. The great advantage of this valve is that it offers no obstruction to the flow when in the fully open position. Hydraulic characteristics are given by Guins.[84]

Butterfly valves (Figure 20.36) are widely used as guard or isolating valves but under low or medium pressure conditions they can be used for regulation. The blade or disc is mounted on a shaft, or on two stub axles; rotation by hydraulic piston and crank is simple and direct. The obstruction caused by the blade in the fully open position inevitably results in head loss and downstream turbulence. The former is reduced by adopting a slim blade of greater diameter than that of the pipe. However, a slim blade may not have the strength to resist high-pressure loading, so to meet this need, blades in some types of valve consist of two thin parallel discs rigidly connected by structural members parallel to the flow; these have great strength while offering less resistance to the flow than corresponding solid blades. Where the valve is in a terminal position, or is guarding a terminal valve, the pressure of the surrounding fluid may be near atmospheric; this also calls for slim blades to avoid cavitation at high velocities. The resultant torque due to fluid pressure is always acting to close the valve, but for safety when closed it is best for the lower half of the disc (if the axis is horizontal) to close in the direction of flow. Guard valves are often designed to be opened under balanced pressure, for which a bypass valve is provided, but to close against full flow in case of emergency.

Butterfly valves present special sealing problems because of the axles but this problem has been overcome and very low leakage rates have been achieved[77] with rubber seals mounted on disc or body. In some valves the axles are offset to facilitate replacement of the sealing ring. For very high heads rubber is not suitable and metal to metal seals are required.

Hydraulic and torque characteristics of valves with blades of various shapes are available.[84, 85, 86]

Needle valves have long been used for precise flow regulation. They consist of a needle or tapered plunger which moves axially within an orifice forming part of the valve body (see Figure 20.37). The plunger is located by guides and operated by screw or hydraulic pressure. In the Larner–Johnson valve, actuation is by the pressure difference across the valve controlled by a pilot valve in the nose of the plunger. The plunger and body are precisely shaped so that the flow accelerates through the valve, to assist actuation, while avoiding cavitation under operating conditions. The discharge coefficient for a fully open valve is from 0.4 to 0.72 depending on the throat area ratio.

Tube valves which were used on some USBR reservoir outlets resemble needle valves but the part of the needle downstream of the sealing ring is omitted. This reduced cavitation problems but vibration was experienced with some valves. This valve has been satisfactorily operated fully submerged. The discharge coefficient, when fully open, based on valve outlet diameter is about 0.6. An in-line regulator of similar general form is shown in Figure 20.38.[86] This has tubular ports at its discharge end which direct jets towards the centre of the downstream pipe where excess head is dissipated, thus avoiding cavitation damage. The port openings are regulated by an axial movement of the plunger.

Hollow jet valve (Figure 20.39) This was developed by the USBR as successor to needle and tube valves and has been generally satisfactory as a high-pressure terminal regulator. It resembles a needle valve with the downstream half of the needle omitted, while the remaining half advances upstream to seal against a ring on the body. The flow is deflected by the surrounding tubular body and is trajected in the form of a jet with nearly parallel sides and hollow centre. Pressure is admitted to the plunger to assist operation, which is mechanical. When fully open the valve normally has a coefficient of discharge of 0.7 based on the outlet diameter.[87, 88] USBR hollow jet valves are stated to operate satisfactorily when partially submerged up to centre level, but should not be operated fully

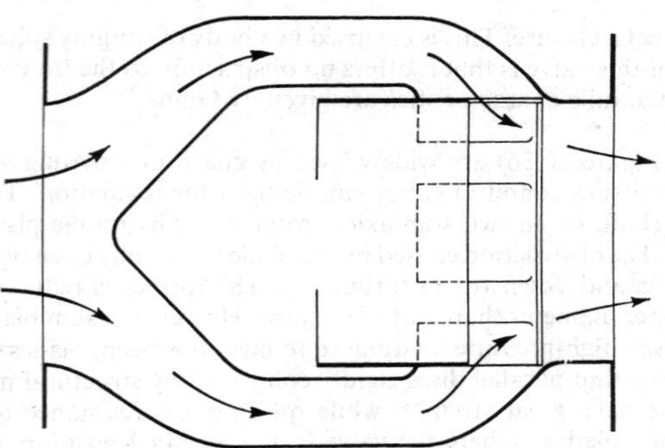

Figure 20.38 In-line regulating valve **(Courtesy: Glenfield and Kennedy Ltd)**

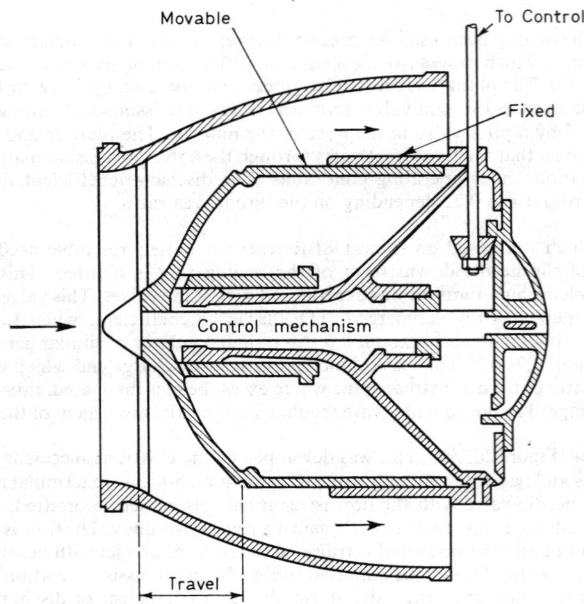

Figure 20.39 Hollow-jet valve

submerged.[77] The trajectory can be calculated approximately by the mechanics of a projectile. Though some aeration and dispersion of the jet occurs the jet fall-out is concentrated in a relatively small area and in some cases erosion is a problem. A stilling basin has been developed for this valve.[25, 32]

Fixed cone-sleeve valves, also known as Howell–Bunger valves (Figure 20.40), are widely used in free discharge terminal applications, including pressure relief for turbines. They have a tubular body on the outside of which is a cylindrical sleeve. This is operated by screws or hydraulic servomotors and retracts to form an opening through which the discharge occurs, deflected outwards by the fixed cone. Discharge coefficient at normal maximum opening is approximately 0.85.[89]

Asymmetry of approach flow may lead to vibration; the valve should not be located too close to a bend in the conduit. A tapered contraction from conduit to valve assists

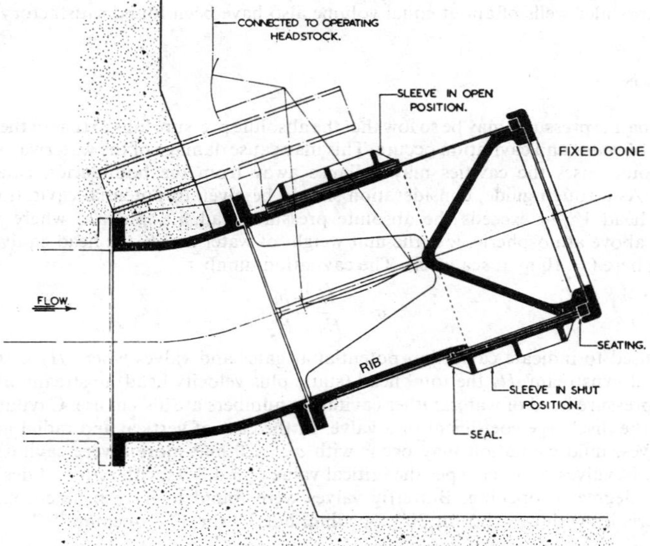

Figure 20.40 Fixed-cone sleeve valve (Courtesy: Glenfield and Kennedy Ltd)

to stabilise the flow. The sharply increasing diameter as the water leaves the valve forces the jet to break up, which is excellent for energy dissipation but sometimes creates problems due to fallout from drifting spray. The limits of the trajectory can be calculated by assuming projectiles at the upper and lower points and two sides of the jet, leaving the valve at velocity equal to $(2gH)^{\frac{1}{2}}$ where H is the pressure plus velocity head in the valve body, and initial trajectory according to the angle of the cone, normally 45° to the axis. Fallout distance is, however, appreciably reduced by air resistance and affected by wind. In some installations a cylindrical hood is provided to restrict the dispersion of the jet, which then becomes tubular in form. To avoid vibration and failure by fatigue this should be rigid and is often of steel with concrete surround. Because of the large air demand, free access of air is essential

The advantages of sleeve valves are simplicity and relatively low cost, low actuating power and the energy-dissipating characteristics of the jet. They have been made in sizes up to 3850 mm diameter and (smaller valves) for heads up to 280 m. Several cases of damage due to vibration have been recorded, particularly fatigue failure of the ribs attaching the fixed cone to the body, but this weakness has been overcome by increasing rigidity and avoiding causes of instability, in particular by fairing the leading edges of the ribs. Valves of this type have been operated fully submerged but trouble has occurred with partial submergence.

Submerged sleeve valves Located in a stilling well, a valve of this type is an excellent terminal regulator. As developed by the National Engineering Laboratory, the valve has an internal sleeve sliding in a perforated cylinder. The perforations result in numerous small jets, which can be stilled in a small chamber, and the perforations can be graded to obtain any desired discharge/stroke characteristic. In some valves large ports are provided with the latter object but causing less obstruction and often the full opening of the sleeve is utilised, without perforations or ports. In this case the discharge coefficient is greater but a larger stilling well is needed. Dimensions of stilling wells providing adequate energy dissipation may be determined from Figure 20.15, but shallower, wider wells of about equal volume also have been found satisfactory.

CAVITATION

Local dynamic pressures may be so low that the absolute pressure falls to that of the vapour pressure of water and cavitation occurs. This may cause damage of the waterway surfaces but in some cases the cavities may collapse away from the boundaries, causing no damage. As a rough guide, consideration should be given to potential cavitation when velocity head $V^2/2g$ exceeds the absolute pressure head, i.e. $p/w+h_a$ where p is the pressure above atmospheric, w is the unit weight of water and h_a the head equivalent to 1 atmosphere (= 10 m at sea level). The cavitation number

$$\sigma = \frac{H_2-h_v}{H_T-H_2} \tag{15}$$

may be used to indicate cavitation potential at gates and valves where H_2 is the static pressure downstream, H_T the total head (static plus velocity head) upstream and h_v the vapour pressure head of water. Other cavitation numbers are also in use. Cavitation also reduces the discharge coefficient of a valve. In the case of vertical and radial gates and gate valves, mild cavitation may occur with $\sigma = 2.0$ and more severe cavitation with $\sigma = 1.0$. In valves of other types the critical value of σ differs with different designs and with the degree of opening. Butterfly valves were found to have incipient cavitation characteristics with $\sigma = 1.5$ for 30% opening, but 3.9 for 80% opening.[86, 90]

AIR DEMAND

Air vents are provided downstream of valves in conduits, to relieve low pressures which develop due to regulation and to avoid cavitation or column separation following valve closure. The rate of air demand of an hydraulic jump is discussed on page **20**–33. During closure of the valve, downstream pressure falls and the water standing in the vent pipe is gradually drained. At the same time the water in the conduit, if flowing full, is decelerated, but if the conduit is long and the valve closes before the vent is admitting air, pressure might fall to cavitation level. In such cases it is necessary to limit the speed of valve closure, especially when approaching final closure. Pressure can be estimated by a step calculation.

REFERENCES

Hydraulic jump

1. BAKHMETEFF, B. A., and MATZKE, A. E., 'The hydraulic jump in terms of dynamic similarity', *Trans. ASCE*, 101, paper 1935, 630 (1936)

2. STEVENS, J. C., 'The hydraulic jump in standard conduits', *Civ. Engng*, New York, **3** (Oct. 1933)
3. MASSEY, B. S., 'Hydraulic jump in trapezoidal channels—an improved method', *Water Power*, **13**, 232 (June 1961)
4. SILVESTER, R., 'Hydraulic jump in all shapes of horizontal channels', *Proc. ASCE*, **90**, HY1, 23 (Jan. 1964)

Open-channel transitions

5. SIMMONS, W. P., 'Transitions for canals and culverts', *Proc. ASCE*, **90**, HY3, 115 (May 1964)
6. SHUKRY, A., 'Flow around bends in an open flume', *Trans. ASCE*, **115**, paper 2411, 751 (1950)
7. IPPEN, A. T., KNAPP, R. T., ROUSE, H., and HSU, E. Y., 'High velocity flow in open channels—a Symposium', *Trans. ASCE*, **116**, paper 2434, 265 (1951)

Weirs and barrages

8. KING, H. W., and BRATER, E. F., *Handbook of Hydraulics*, 5th edn, McGraw-Hill, New York (1963)
9. ROUSE, H., *Fluid mechanics for hydraulic engineers*, McGraw-Hill, New York, 318 (1938)
10. Corps of Engineers, *Hydraulic design criteria*, US Army Engineer Waterways Experiment Station, Vicksburg, Miss. (1952–70)
11. US Dept of the Interior, Bureau of Reclamation, *Design of small dams*, Denver, Colo. (1960)
12. WHITE, W. R., 'The performance of two-dimensional and flat-vee triangular profile weirs', *Proc. Instn Civ. Engrs*, paper 7350S (1971)
13. ACKERS, P., and HARRISON, A. J. M., *Critical-depth flumes for flow measurement in open channels*, Hydraulics Research Station, Wallingford, Hyd. Res., paper 5 (1963)
14. *Methods of measurement of liquid flow in open channels, long base weirs*, BS 3680, Part 4B, British Standards Institution, London (1969)
15. PARSHALL, R. L., *The Parshall measuring flume*, Colorado Agricultural Experimental Station, Fort Collins, Colorado, Bull. 423 (March 1936)
16. THOMAS, C. W., 'World practices in water measurements at turnouts', *Proc. ASCE*, **86**, IR2, paper 2530 (June 1960)
17. HAIGH, F. F., 'The Emerson barrage', *J. Inst. Civ. Engrs*, **2**, paper 5227 (December 1941)
18. KHOSLA, A. N., BOSE, N. K., and MCKENZIE-TAYLOR, E., *Design of weirs on permeable foundations*, Central Board of Irrigation, India, pubn 12 (1936)
19. LANE, E. W., 'Security from under-seepage masonry dams on earth foundations', *Trans. ASCE*, **100**, Paper 1919, 1235 (1935)
20. LUTHRA, S. D. L., and JOGLEKAR, D. V., 'Uplift pressures below hydraulic structures on stratified, permeable foundations', *Proc. Int. Assn. Hyd. Res., Dubrovnik*, Paper 11/1 (1961)
21. LELIAVSKY, S., *Design of dams for percolation and erosion*, Chapman and Hall, London (1935)
22. FOY, SIR T., and GREEN, H. S., 'Barrages and dams on permeable foundations', *Handbook of applied hydraulics*, Eds C. V. Davis and K. E. Sorensen, 3rd edn, Chapt. 17, McGraw-Hill, New York (1969)

Energy dissipation

23. BRADLEY, J. N., and PETERKA, A. J., 'The hydraulic design of stilling basins', *Proc. ASCE*, **83**, HY5, papers 1401–1406 (October 1957)

23a. THOMAS, A. R., Discussion on above, *Proc. ASCE*, **84**, HY2, paper 1616, 34 (April 1958)

24. PETERKA, A. J., *Hydraulic design of stilling basins and energy dissipators*, US Dept. of the Interior, Bureau of Reclamation, Water Resources Eng Mono. 25 (revised 1963)
25. ELEVATORSKI, E. A., *Hydraulic energy dissipators*, McGraw-Hill, New York (1959)
26. DONNELLY, C. A., and BLAISDELL, F. W., 'Straight drop spillway stilling basin', *Proc. ASCE*, **91**, HY3, 101 (May 1965)
27. American Society of Civil Engineers, Task Force on energy dissipators for spillways and outlet works, Progress report (with bibliography), *Proc. ASCE*, HY1, paper 3762, 121 (January 1964)
28. BERRYHILL, R. H., 'Stilling basin experiences of the Corps of Engineers', *Proc. ASCE*, HY3, **83**, paper 1264 (June 1957)
29. BERRYHILL, R. H., 'Experience with prototype energy dissipators', *Proc. ASCE*, **89**, HY3, paper 3521, 181 (May 1963)
30. MCPHERSON, M. B., and KARR, M. H., 'A study of bucket-type energy-dissipator characteristics', *Proc. ASCE*, **83**, HY3, paper 1266 (June 1957)
31. BEICHLEY, G. L., and PETERKA, A. J., 'The hydraulic design of slotted spillway buckets', *Proc. ASCE*, **85**, HY10, paper 2200, 1 (October 1959)
32. BEICHLEY, G. L., and PETERKA, A. J., 'Hydraulic design of hollow-jet valve stilling basin', *Proc. ASCE*, **87**, HY5, paper 2924, 1 (September 1961)

Scour

33. LACEY, G., 'Stable channels in alluvium', *Proc. Instn Civ. Engrs*, **229**, 259 (1929–30)

34. LACEY, G., 'Flow in alluvial channels with sandy mobile beds', *Proc. Instn Civ. Engrs,* **9,** 145 (February 1958)
35. NIXON, M., 'A study of the bank-full discharges of rivers in England and Wales', *Proc. Instn Civ.* Engrs, **12**, 157 (1959) and discussion, **14**, 416 (1959)
36. INGLIS, SIR CLAUDE, The behaviour and control of rivers and canals, Central Waterpower Irrigation and Navigation Research Station, Poona, India, Pt II, 327 (1949)
37. LAURSEN, E. M., 'Scour at bridge crossings', *Trans. ASCE,* **127,** 166 (1962)
38. NEILL, C. R., 'Measurements of bridge scour and bed changes in a flooding sand-bed river', *Proc. Instn Civ. Engrs,* **30,** 415 (February 1965) and discussion, **36,** 397 (February 1967)

Enclosed flow

39. ACKERS, P., 'Resistance of fluids flowing in channels and pipes', *Hydraulics Research Paper No.* 1, HMSO (1958)
40. BRADLEY, J. N., and THOMPSON, L. R., *Friction factors for large conduits flowing full,* US Dept of the Interior, Bureau of Reclamation, Water Resources Engg, Mono. 7 (Revised 1962)
41. American Society of Civil Engineers, Task Force on Flow in Large Conduits, 'Factors influencing flow in large conduits', *Proc. ASCE,* **91,** HY6, paper 4543, 123 (November 1965)
42. COLEBROOK, C. F., 'The flow of water in unlined, lined and partly lined rock tunnels', *Proc. Instn Civ. Engrs,* **11,** 103 (September 1958)
43. KENN, M. J., *Factors influencing the erosion of concrete by cavitation,* Construction Industry Research and Information Association (CIRIA), London, Tech. Note 1 (1968)
44. BALL, J. W., 'Construction finishes and high-velocity flow', *Proc. ASCE,* **89,** CO2, 91 (September 1963)
45. WRIGHT, D. E., *The hydraulic design of unlined and lined-invert rock tunnels,* Construction Industry Research and Information Association (CIRIA), London, Report 29 (1971)
46. MILLER, D. S., *Internal flow. A guide to losses in pipe and duct systems.* Brit. Hydromechanics Res. Assn., Cranfield (1971)
47. RAJARATNAM, N., 'Hydraulic jump in horizontal conduits', *Water Power,* **17,** 80 (February 1965)
48. KALINSKE, A. A., and ROBERTSON, J. M., 'Closed conduit flow', Symposium on 'Entrainment of air in flowing water', *Trans. ASCE,* **108,** paper 2205, 1435 (1943)
49. SAILER, R. E., 'San Diego aqueduct', *Civ. Engng* (New York), 268 (May 1955)

Spillways

50. American Society of Civil Engineers, Task force on hydraulic design of spillways: Bibliography on the hydraulic design of spillways, progress report, *Proc. ASCE,* **89,** HY4, paper 3573, 117 (July 1963)
51. FARNEY, H. S., and MARKUS, A., 'Side-channel spillway design', *Proc. ASCE,* **88,** HY3, paper 3143 (May 1962)
52. BINNIE, G. M., GERRARD, R. T., ELDRIDGE, J. G., KIRMANI, S. S., DAVIS, C. V., DICKINSON, J. C., GWYTHER, J. R., THOMAS, A. R., LITTLE, A. L., CLARK, J. F. F., and SEDDON, B. T., 'Engineering of Mangla', *Proc. Instn Civ. Engrs,* **38,** 343 (November 1967)
53. WAGNER, W. E., 'Morning-glory shaft spillways: Determination of pressure-controlled profiles', *Proc. ASCE,* **80**, 432, (April 1954)
54. BINNIE, G. M., 'Model experiments on bellmouth and siphon bellmouth overflow spillways', *J. Instn Civ. Engrs,* **10,** 65 (November 1938)
55. BRADLEY, J. N., 'Morning-glory shaft spillways: Prototype behaviour', *Proc. ASCE,* **80**, HY, paper 431 (April 1954)
56. MUSSALLI, Y. G. and CARSTENS, M. R., *A study of flow conditions in shaft spillways,* Georgia Institute of Technology, Atlanta, Ga., USA, WRC-0669 (1969)
57. OLIVER, G. C. S., 'Eye Brook reservoir spillway', *J. Instn Water Engrs,* **13,** 205 (May 1959)
58. FELLERMAN, L., *Model tests of Shek Pik Siphon Spillway, Tung Chung Water Scheme;* Brit. Hydromechanics Research Assn, Cranfield, RR841 (1965)
59. Hydraulics Research Station, Wallingford, *Plover Cove reservoir; air regulated siphon spillway,* Rep. No. EX539 (1971)
60. CHARLTON, J. A., *Self priming siphons—an appraisal*; Brit. Hydromechanics Research Assn, Cranfield, SP725 (1962)
61. RHONE, T. J., and PETERKA, A. J., 'Improved tunnel spillway flip buckets'; *Proc. ASCE,* **85,** HY12, paper 2316 (December 1959)
62. GUMENSKY, D. B., 'Design of side walls in chutes and spillways'; *Proc. ASCE,* paper 175 (February 1953)
63. BALLOFFET, A., 'Pressures on spillway flip buckets', *Proc. ASCE,* HY5, paper 2930 (September 1961)
64. GUNKO, F. G., *et al.*, 'Research on the hydraulic regime and local scour of river bed below spillways of high head dams', *Proc. Int. Assn for Hyd. Res., Leningrad,* paper 1. 50 (1965)

65. AKHMEDOV, T. K. L., 'Local erosion of fissured rock at the downstream end of spillways' (trans. from Russian) *Hydrotechnical Const., ASCE*, No. 9, 821 (September 1968)

Reservoir outlet works

66. ROUSE, H., Ed., *Engineering Hydraulics*, Wiley, New York, 32 (1950)
67. JOGLEKAR, D. V., and DAMLE, P. M., 'Cavitation-free sluice outlet design', *Proc. Int. Assn for Hyd. Res.*, paper B2 (1957)
68. BALL, J. W., 'Hydraulic characteristics of gate slots', *Proc. ASCE*, **85**, HY 10, paper 2224, (October 1959)
69. TRUSCOTT, G. F., 'Camlough Pumped Storage Scheme, Part 1, Upper reservoir model studies. Brit. Hydromechanics Research Assn, Cranfield, Report RR, 1159 (1973)
70. DENNY, D. F., and YOUNG, G. A. J., 'The prevention of vortices and swirl at intakes', *Proc. Int. Assn for Hyd. Res.*, paper C1 (1957)
71. ANWAR, H. O., 'Prevention of vortices at intakes', *Water Power*, **20**, 393 (October 1968)
72. FRASER, W. H., and HARRISON, N., 'Hydraulic problems encountered in intake structures of vertical wet-pit pumps and methods leading to their solution', *Trans. ASME*, 643 (May 1953)
73. HATTERSLEY, R. T., 'Hydraulic design of pump intakes', *Proc. ASCE*, **91**, HY2, paper 4276, 223 (March 1965)
74. SELL, L. E., 'Hydroelectric power plant trashrack design', *Proc. ASCE*, **97**, PO1, paper 7819, 115 (January 1971)
75. American Society of Civil Engineers, Committee on Operation and Maintenance of Hydro-electric Generating Stations of the Power Division: 'Design, operation and maintenance of intakes, racks and booms'; *Proc. ASCE*, **85**, PO5, paper 2226, 71 (October 1959)

Gates and values

76. BLEULER, W., 'Sluice gates', *Water Power*, **15**, 460 (November 1963)
77. KOHLER, W. H., and BALL, J. W., 'High pressure outlets, gates and valves', *Handbook of Applied Hydraulics*, C. V. Davis and K. E. Sorensen, 3rd edn, McGraw-Hill, Section 21 (1969)
78. BRADLEY, J. N., 'Rating curves for flow over drum gates', *Trans. ASCE*, **119**, paper 2677 (1954)
79. TOCH, A., 'Discharge characteristics of Tainter gates'; *Proc. ASCE*, **79**, paper 295 (October 1953)
80. ANWAR, H. O., 'Discharge coefficients for control gates', *Water Power*, **16**, 152 (Apr. 1964)
81. SIMMONS, W. P., 'Experiences with flow-induced vibrations', *Proc. ASCE*, **91**, paper 4414, 185 (July 1965)
82. SCHMIDGALL, T., 'Spillway gate vibrations on Arkansas river dams'; *Proc. ASCE*, **98**, paper 8676, 219 (January 1972)
83. COLGATE, D., 'Hydraulic downpull forces on high head gates', *Proc. ASCE*, **85**, paper 2245, 39 (November 1959)
84. GUINS, V. G., 'Flow characteristics of butterfly and spherical valves', *Proc. ASCE*, **94**, HY3, paper 5933, 675 (May 1968)
85. MCPHERSON, M. B., STRAUSSER, H. S., and WILLIAMS, J. C., 'Butterfly valve flow characteristics', *Proc. ASCE*, **83**, HY1, paper 1167 (February 1957)
86. MILLER, E., 'Flow and cavitation characteristics of control valves', *J. Inst. Water Engineers*, **22**, No. 7 (October 1968)
87. THOMAS, C., 'Discharge coefficients for gates and valves', *Proc. ASCE*, **81**, paper 746 (July 1955)
88. LANCASTER, D. M., and DEXTER, R. B., 'Hydraulic characteristics of hollow jet valves', *Proc. ASCE*, **85**, HY11, paper 2263, 53 (November 1959)
89. ELDER, R., and DOUGHERTY, G. B., 'Characteristics of fixed dispersion cone valves', *Proc. ASCE*, **78**, paper 153 (September 1952)
90. TULLIS, J. P., and MARSCHNER, B. W., 'Review of cavitation research on valves'; *Proc. ASCE*, **94**, HY1, paper 5705, 1 (January 1968)

21 HIGHWAYS

21 HIGHWAYS

T. D. WILSON, B.Sc., F.I.C.E., F.I.Struct.E., F.I.Mun.E., F.I.H.E.
Deputy Chief Engineer, Department of the Environment

HIGHWAY ADMINISTRATION UK

Introduction

The Romans were the first to construct roads in Britain but thereafter little was done until the Industrial Revolution in the eighteenth century when turnpike trusts were set up and tolls were levied on road users.

In 1888 main roads were made the responsibility of the County Councils; then in 1909, because of the growth in motor traffic, a Road Board was set up and in 1919 the Ministry of Transport was established and took over this Board's responsibilities. In 1970 this Ministry was absorbed into the Department of the Environment.

In the UK, trunk roads provide the route network for through traffic. For administrative purposes other roads are classified as principal and nonprincipal. See Table 21.1.

Highway authorities

A highway authority exercises powers under highway and traffic legislation and has various public responsibilities and duties. The highway authorities are:

1. For trunk roads—the Secretary of State for the Environment
2. For principal and other roads—Councils of Counties and County Boroughs, with certain exceptions non-County Boroughs and Urban Districts.

Financial arrangements

The financial responsibility for the construction, improvement and maintenance of trunk roads lies with the Secretary of State for the Environment. On nontrunk roads, the Local Highway Authority has this responsibility with certain direct and/or indirect contributions from central government.

Secretary of State's responsibility

The Secretary of State for the Environment's regional responsibilities for roads in England are administered by eight Regional Controllers (Roads and Transportation), by the Department of Environment headquarters for London, and by six Road Construction Units each of which has certain delegated powers; the latter deal with major new road projects of value over £1 million.

Agent authorities

Certain local highway authorities (normally County Councils and large County Boroughs) act as Agents for the Secretary of State through the Regional Controller (Roads and Transportation), on maintenance and other works on trunk roads.

Highway legislation

Highway law for England and Wales was first codified in 1835 and since then there has been much legislation, most of which was consolidated in the Highways Act of 1959. A further Act of 1971 dealt with miscellaneous matters not covered in the 1959 Act. Pratt and MacKenzie's *Law of Highways*[1] is a useful reference and cumulative supplements are issued from time to time between editions. Most highway authorities have

Table 21.1 MILEAGE OF PUBLIC HIGHWAYS IN GREAT BRITAIN—1 APRIL 1971

Administrative area		*Trunk roads*		*Principal Roads*	*Other roads including green lanes*	*Total*
		Motorway	*All-purpose*			
County Boroughs and small Burghs	England	30 (48)	130 (209)	1 875	18 147	20 182
	Scotland	—	22 (35)	371	4 451	4 844
	Wales	7 (11)	21 (34)	90	897	1 015
Greater London		16 (26)	134 (215)	875	6 862	7 887
Counties	England	630 (1 014)	5 152	11 514	113 420	130 716
	Scotland	66 (106)	1 877	4 201	18 505	24 649
	Wales	16 (26)	996	1 390	17 504	19 906
Totals	England	676 (1 088)	5 416	14 264[a]	138 429	158 785
	Scotland	66 (107)	1 899	4 572[b]	22 956	29 493
	Wales	23 (37)	1 017	1 486	18 451	20 977
Total Great Britain		765 (1 232)	8 332	20 322	179 836	209 255

[a]Includes 13 miles of Local Authority Motorways.
[b]Includes 2 miles of Local Authority Motorways
Note: figures in parentheses are approximate values in kilometres (1 mile = 1.61 km)

Table 22.2 EXPENDITURE ON ROADS, GREAT BRITAIN—FINANCIAL YEAR 1970–71

	(Thousand £) *From Central funds*	*From Local Authority funds*	*Total*
New construction and improvement	383 403	107 933	491 336
Maintenance	23 755	137 814	161 569
Cleansing—watering and snow-clearing	2 439	31 373	33 812
Administration	6 035	46 245	52 280
Total expenditure on roads	415 632	323 365	738 997
Road lighting	2 722	37 777	40 499
Car parks	30	31 574	31 604

power to acquire land compulsorily for highway purposes but, if there is objection, the Secretary of State must, in general, confirm the Compulsory Purchase Order. The Highway Acts give powers to protect future routes and lay down statutory procedures which safeguard the rights of individuals affected by proposed roads.

The main planning legislation was re-enacted in the Consolidated Town and Country Planning Act, 1962, with some further amendments in the Acts of 1968 and 1971.

Maps

ORDNANCE SURVEY

The Ordnance Survey (Head Office: Romsey Road, Maybush, Southampton, SO9 4DH) produces maps for official and public use and also offers many other services which are byproducts of its normal work of triangulation, levelling, aerial and field surveys, drawing, printing and publication.

The Ordnance Survey publishes large-scale maps at four scales:

1 : 1250	These are all based on the National Grid	
1 : 2500	These may be either National Grid or County Series	
1 : 10 560	or six inches to one mile. Mostly based on the National Grid	The replacement of the 1 : 10 560 scale by the 1 : 10 000 scale began in 1969.
1 : 10 000	Based on the National Grid	

The 1 : 1250 and 1 : 2500 maps represent the features of the ground to scale but the 1 : 10 560 and 1 : 10 000 maps are generalised.

GEOLOGICAL SURVEY

Geological maps and memoirs issued by the Institute of Geological Sciences (Headquarters: Exhibition Road, South Kensington, London SW7) give information on solid rock formations and drift deposits throughout the UK.

SITE INVESTIGATIONS

British Standard Code of Practice CP 2001, *Site Investigation,*[2] provides a general guide to the planning and execution of site work.

FEASIBILITY STUDIES

This subsection outlines the surveys, studies and other factors which need to be considered in a design for a new road or new road network.

Feasibility studies

Feasibility studies determine the practicable cost-effective solutions and also the priority order for their implementation.

TERMS OF REFERENCE

The terms of reference for a feasibility study should indicate what is known and what is wanted, namely:

- (i) The *study area* boundaries and the *problem*
- (ii) Available information, e.g. (a) planning studies; (b) traffic and/or transportation studies; (c) geological reports
- (iii) Possible solutions to be investigated
- (iv) General planning factors to be taken into account, i.e.
  - (a) Other highway schemes currently under construction or under study for possible inclusion in a future programme
  - (b) Adjacent New Towns or other large-scale development proposals
  - (c) Proposals for new river crossings, e.g. bridges, barrages or tunnels
  - (d) Extension of, or new, industrial areas, hospitals, sports centres, docks or airports

(v) Physical aspects to be considered in making a selection from alternatives, e.g.
  (a) Topography
  (b) Soils and geology
  (c) Meteorology
  (d) Effect on the environment, landscape, etc.
  (e) Effect on land and property
  (f) Effect on neighbouring development
  (g) Effect of noise due to traffic
  (h) Design standards
  (j) Possibility of stage construction
(vi) Traffic aspects including assumptions and techniques for predicting Design Year (usually 15 years after opening)[3] and economic assessment year flows
(vii) Economic evaluation method
(viii) Programme relating to the report on the study.

CONTENT OF REPORT

The Report on the Study, supplemented by line diagrams, maps, etc., should then succinctly cover:

(a) Existing conditions for traffic, etc.
(b) Design Year conditions—supplemented by such additional detailed surveys, etc., as are necessary to justify a recommendation. Flow diagrams are a clear way of demonstrating this, not only to indicate future flows on sections of route but also the turning movements at junctions (see Figure 21.1)
(c) The preferred choice after analysis of the advantages and disadvantages in relation to items (iv), (v), (vi) for each alternative
(d) The comments of statutory bodies, Government Departments, local and regional planning authorities affected by the proposals
(e) Estimates of cost at current prices, subdivided into roadworks, bridgeworks, land costs and any other particular items
(f) Economic Assessment summarising the benefits, savings in travel time, reduction in accidents, reduction in delays at major junctions, extra maintenance costs incurred; other special items may be added in particular cases
(g) Other factors—potential for redevelopment, dereliction clearance, area resuscitation, etc.

SUPPLEMENTARY TRAFFIC STUDIES

Supplementary traffic information may be required to support the report's recommendations, namely:

1. Vehicular and pedestrian traffic counts[4, 5] including daily and seasonal variations
2. Origin and destination surveys[5, 6]
3. Speed and delay studies[4, 6]
4. Accident statistics[4, 6, 7]
5. Parking studies[4] on usage, availability, etc.
6. The estimation of future traffic[5, 6, 8–11]
   —trip generation, trip destination
   —traffic generation and traffic assignment

If the broader aspects of transportation have to be covered then this may require Land Use Transportation Studies.[8, 11, 12] These will mean home interviews as well as a wide range of other statistics so that the future movements of vehicles and people can be assessed and, in this case, 'model split' or the subdivision of traffic into the various modes of transportation has then to be assessed.

Table 21.3 THE VARIATION IN TRAFFIC[a] ON THE ROADS OF GREAT BRITAIN OVER THE DECADE 1960–1970 (Courtesy: Director, Transport and Road Research Laboratory)

	Average daily flow		*Annual motor vehicle miles*					
	1960	*1966*	*1960*	*1966*	*1967*	*1968*	*1969*	*1970*
			(*thousand million*)		*Index numbers* (*1966 = 100*)			
Motorways	...	19 501	0.45	2.85	130	169	177	212
Urban roads								
Trunk	9 007	14 305	5.68	9.00	101	102	107	111
Class I	6 423	9 468	14.80	22.42	104	107	110	116
Class II	3,166	4,884	4.55	7.10	105	106	106	113
Class III	1 427	2 417	2.85	5.01	100	99	104	110
All trunk and classified	4 377	6 698	27.88	43.54	103	105	108	114
Unclassified[b]	661	918	9.42	15.63	...	...	...	...
All urban roads[c]	1 809	2 514	37.30	59.17	...	...	...	...
Rural roads								
Trunk	3 996	5 912	9.57	14.27	101	106	108	112
Class I	2 108	2 938	10.32	14.40	105	113	114	119
Class II	8 404	1 240	4.22	6.18	103	110	117	126
Class III	355	427	5.33	6.76	108	115	117	121
All trunk and classified	1 140	1 479	29.74	41.60	104	110	113	118
Unclassified[b]	113	172	232	3.78	...	...	...	...
All rural roads[c]	659	906	32.06	45.38	...	...	...	...
All roads	1 008	1 456	69.81	107.40	105	110	113	120

[a]Results are not available in terms of the new classification of roads (see Introduction)
[b]Traffic on unclassified roads was counted only in 1960 and 1966
[c]Other than motorways

The variation of traffic over a daily, weekly and annual period is as shown in Figures 21.1, 21.2 and 21.3 respectively.

The effect of turning movements at junctions affects the speeds and delays through junctions. Figure 21.4 is an example of a turning-movement diagram.

OTHER SURVEYS

In regard to the physical aspects which must be considered in the report:

(a) *Topography* The Ordnance Survey maps will need supplementing. Aerial survey techniques are widely used and may serve to highlight geological weaknesses (see references 13 and 14).

(b) *Soils geology* The Institute of Geological Sciences has an extensive aerial and ground survey coverage and this is supplemented by memoirs. The Soil Survey of England and Wales has mapped the quality of agricultural soils. Preliminary soils surveys for a feasibility report are usually supplemented at detailed design. Soil survey costs between 0.3% and 1% of construction costs (see references 15–19).

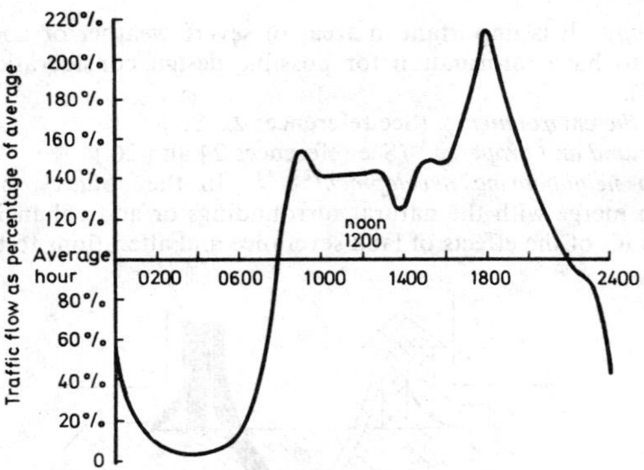

Figure 21.1 Pattern of traffic throughout the day. (National figures for summer)

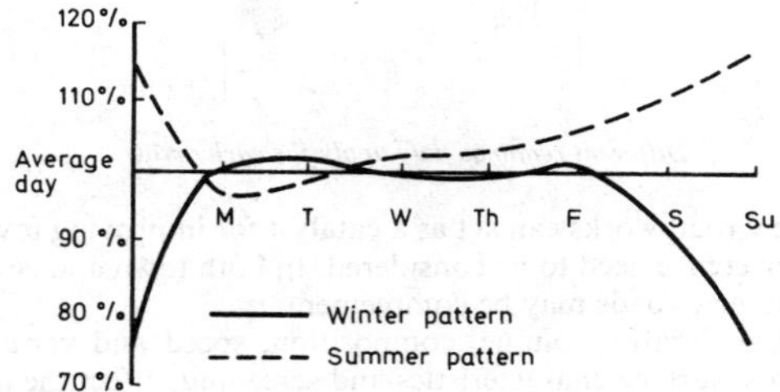

Figure 21.2 Pattern of traffic throughout the week

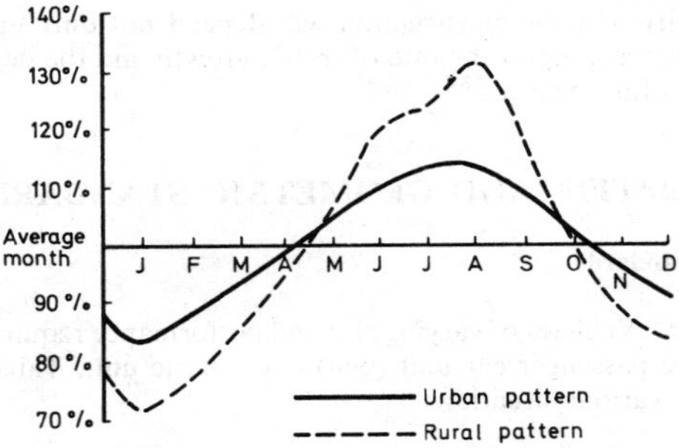

Figure 21.3 Pattern of traffic throughout the year

(c) *Meteorology* It is important in areas of severe weather or unexpected changes, e.g. fog, to have information for possible design consideration (see references 20 and 21).
(d) *Effect on the environment* (See references 22–26.)
(e) *Effect on land and property* (See references 24 and 26.)
(f) *Effect on neighbouring development*[22–27] In the country, new roads should appear to merge with the natural surroundings or add enhancingly to them and take account of the effects of land severance and alterations to the present use. In

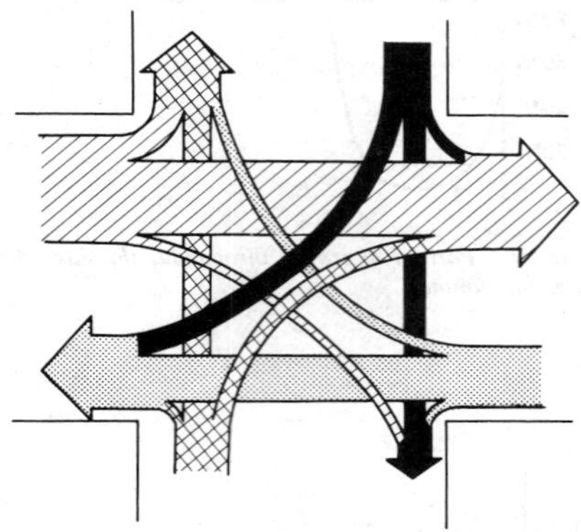

Different readings may apply for each arrow

towns the new road works can act as a catalyst for improving town areas but their effect or interference need to be considered. In both town and country, dereliction clearance and new roads may be complementary.
(g) *Effect of noise* Traffic volume, composition, speed and variability, as well as road gradients, surface characteristics and screening, affect the ability of noise to disturb.[28–38] The Government Committee on the Problem of Noise Final Report (HMSO 1963), gave recommended noise levels in buildings, and the Noise Advisory Council recommended 70 dBA as the maximum permitted level at the building facade. This is equivalent to 55 dBA inside an average building close to a road. (See also references 38a, b.)
(h) Design Standards will vary according to the class of road and are referred to in the next subsection.
(j) The possibility of stage construction will depend not only upon economic considerations but also upon the rate of traffic growth and the degree of interference with traffic at future stages.[39]

TRAFFIC AND GEOMETRIC STANDARDS

Traffic capacity standards

Passenger car units Vehicles of varying size and performance require different amounts of road space. The passenger car unit (pcu) is the basic unit. Table 21.4[40] gives conversion factors for various situations.

Through carriageways Generally in the UK roads are designed for traffic capacities forecast for a 'design year' 15 to 20 years after the road is opened to traffic, and to either 'urban'—for urban roads and for interchanges when morning and evening peak hour flows are used—or 'rural' when the average 7 day 16 hour flow in August (5 weekdays plus Saturday and Sunday) is used. These correspond to design speeds of 80 km/h (50 mph) and 113 km/h (70 mph) respectively.

Table 21.4

Class of vehicle	*Equivalent value* (pcu)			
	Rural standards	*Urban standards*	*Roundabout design*	*Traffic signal design*
Private cars, motor cycle combinations, taxis and light goods up to 1500 kg unladen	1.00	1.00	1.00	1.00
Motorcycles (solo), scooters and mopeds	1.00	0.75	0.75	0.33
Goods, vehicles over 1500 kg unladen and horse drawn vehicles	3.00	2.00	2.80	1.75
Buses, coaches, trolley vehicles	3.00	3.00	2.80	2.25
Pedal cycles	0.50	0.33	0.50	0.20

RURAL ROADS

Table 21.5[40] shows design capacities for high-standard limited-access dual-carriageway roads or motorways. The longer the average journey the higher the level of service which can be considered appropriate.

Table 21.5

Standard (level of service)	*Average journey length* (km)	*Daily capacities (average 16 hour August day)* (pcu)			*Peak hourly capacity per lane* (pcu)
		Dual 2-lane	*Dual 3-lane*	*Dual 4-lane*	
1	Over 40	33 000	50 000	66 000	1 000
2	15 to 40	40 000	60 000	80 000	1 200
3	Under 15	50 000	75 000	100 000	1 500

The width chosen for the route (usually Standard 1) will be considered adequate for the more densely trafficked sections if the latter comply with Standards 2 or 3 as appropriate.

Table 21.6

Carriageway width (m)	*Design speed* (km/h)	*Design capacity* (pcu/day) *Rural standards*
14.60 (dual)	120	66 000
11.00 (dual)	120	50 000
7.30 (dual)	120	33 000
10.00	100	15 000
7.30	100	9 000
7.30	80	9 000
—	60	—

Table 21.6[40] shows recommended capacities of rural roads from dual four-lane carriageways to single two-lane carriageways. These capacities relate approximately to Standard 1.

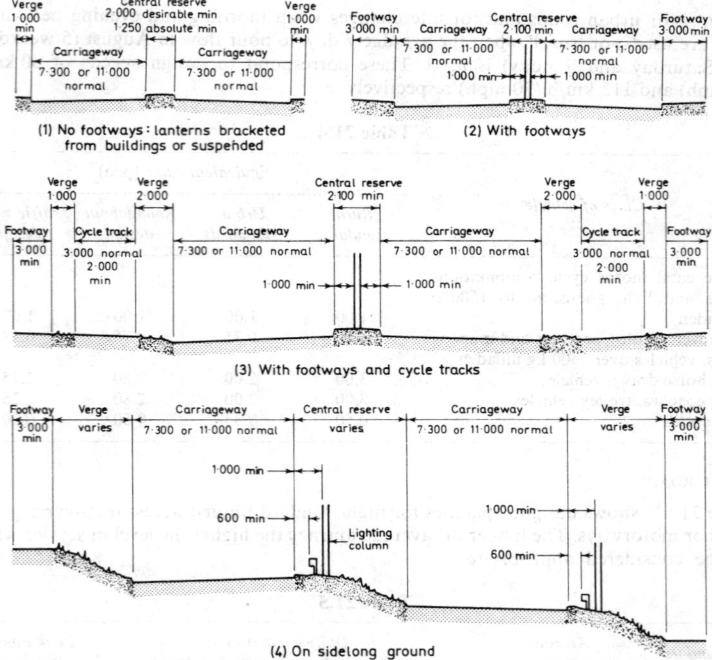

Figure 21.5 Urban all-purpose roads—primary distributors. The clearances shown are suitable for speeds above 48 km/h (30 mph). (Courtesy: HMSO—from 'Roads in Urban Areas'[43])

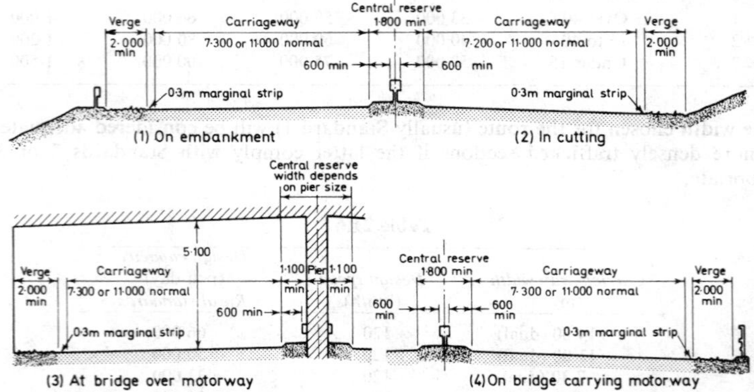

Figure 21.6 Urban motorways. Notes: (1) The central reserve should be bordered by raised kerbs where it is 2.0 m wide or less, or where the face of any safety fence is less than 1.25 m from the adjoining carriageway. Where greater widths are available, flush marginal strips 0.3 m wide may be used instead of kerbs. (2) On lengths of motorway without paved verges the near-side edge of each carriageway should be bordered by raised kerbs and should be at least 0.6 m clear of the face of any safety fence on the verge and at least 1.0 m clear of bridge piers, retaining walls, lighting columns, etc. (3) The clearance between the carriageways and any fixed obstructions on the inside of bends at the sides of the road or on the central reserve should be increased where necessary to ensure the requisite visibility standards. (Courtesy: HMSO—from 'Roads in Urban Areas'[43])

Table 21.7

(a) PRACTICAL CAPACITIES OF TWO-WAY URBAN ROADS

Effective width of carriageway in metres (excluding refuges or central reserve)	*2-lane*			*3-lane*		*4-lane*			*6-lane*			
	6	*6.75*	*7.3*	*9*	*10*	*12*	*13.5*	*14.6*	*18*	*20*	*22*	*Remarks*
Description	*Capacity* (pcu/h) *for BOTH directions of flow*					*Capacity* (pcu/h) *for ONE direction of flow*						
Urban motorway with grade separation and no frontage access								3 000			4 500	Applicable to the highest category of *distributor*
All-purpose road with no frontage access, no standing vehicles permitted and negligible cross-traffic	1 200	1 350	1 500	2 000	2 200	2 000	2 200	2 400	3 000	3 300	3 600	Appropriate for all-purpose *distributors*
All-purpose street with high-capacity junctions and 'No Waiting' restrictions	800	1 000	1 200	1 600	1 800	1 200	1 350	1 500	2 000 2 200	2 250 2 450 for dual carriageways	2 500 2 700	Applicable to those *distributors* and *access roads* where access to development is frequent but capacity is not unduly restricted by junctions
All-purpose street with capacity restricted by waiting vehicles and junctions	300 to 500	450 to 600	600 to 750	900 to 1 100	1 100 to 1 300	800 to 900	900 to 1 000	1 000 to 1 200	1 300 to 1 700	1 500 to 2 000	1 600 to 2 200	Typical of existing roads where waiting vehicles and junctions with heavy cross traffic severely limit capacity

Table 21.7 (continued)

(b) PRACTICAL CAPACITIES OF ONE-WAY URBAN ROADS

Effective width of carriageway in metres (excluding refuges)	*6*	*6.75*	*7.3*	*9*	*10*	*11*	*12*	*13.5*	*14.6*	*Remarks*
Description	*Capacity* (pcu/h)									
Urban motorway with grade separation and no frontage access			3 000			4 500			6 000	Applicable to the highest category of *distributor*
All-purpose road with no frontage access, no standing vehicles and negligible cross-traffic	2 000	2 200	2 400	3 000	3 300	3 600	4 000	4 400	4 800	Appropriate for all-purpose *distributors*
All-purpose street with high-capacity junctions and 'No Waiting' restrictions	1 300	1 450	1 600	2 150	2 400	2 650	3 000	3 350	3 700	Applicable to those *distributors* and *access roads* when access to development is frequent but capacity is not unduly restricted by junctions
All-purpose street with capacity restricted by waiting vehicles and junctions	800	950	1 100	1 650	1 900	2 150	2 500	2 800	3 200	Typical of existing roads where waiting vehicles and junctions with heavy cross traffic severely limit capacity

URBAN ROADS

On urban roads the practical traffic capacities are as shown in Table 21.7.

The width of the carriageway may be varied to accommodate weaving between traffic lanes or to cater for slow-moving vehicles on hills. In the case of high-speed roads, additional emergency-stopping lanes may be provided. The criteria for these are given in *Layout of Roads in Urban Areas*,[41] *Urban Traffic Engineering Techniques*[42] and *Roads in Urban Areas*.[43]

Horizontal geometric standards

In horizontal alignment, transition curves (i.e. having a curvature which varies in a progressive way) may be used between straights and circular arcs to allow time for a driver and vehicle to adapt to the change in curvature and to provide distance to adjust the amount of superelevation or crossfall provided on the road.

The transition curves used are spiral, lemniscate, cubic parabole, or clothoid (used in British Integrated Highway Design Program System for computer evaluation[115]).

SUPERELEVATION

Because of the range of vehicle speeds on any road the *maximum* superelevation is limited to 1 in 24 (desirable maximum) and 1 in 14½ (absolute maximum); coupled with this, drainage considerations require a *minimum* crossfall of 1 in 40.

Transition curves Transition curves should be provided for all curves less than the radii given in Table 21.8[40] at each design speed.

Table 21.8

Design speed (km/h)	*Minimum radius with no transitions* (m)
120 (75 mph)	1 500 (on motorways 3 250 m minimum)
100 (62 mph)	1 350
80 (50 mph)	1 200
60 (37 mph)	600

Setting out tables for a range of speeds and radii have been produced for transition curves.[116]

COMBINATION OF CURVES IN AN ALIGNMENT

Phase 2 of the British Integrated Program system incorporates a program HORAL which allows the designer to define the various elements of an alignment and permits the elimination of straights between curves of opposite hand.

Vertical and horizontal curves should as far as possible be similar in length and in phase with each other. Sharp horizontal curves should not be introduced at the top of crests or the bottom of sags.

Visibility Forward visibility to overtake or to stop if any obstruction on the carriageway is sighted is essential for road safety. On horizontal curves, obstructions on central reservation and on verges should be checked.

On single-carriageway two-way roads, sufficient visibility for safe overtaking is desirable but it should not be less than the minimum stopping distance given in Table 21.9.

Sight distances, both vertical and horizontal, should be measured between points 1.05 m above the carriageway along the centrelines of both the nearside and offside lanes of the carriageway. On dual-carriageway roads, sight distances should be checked on both carriageways, and in this case the minimum stopping distance can be the criterion.

Table 21.9[43] shows recommended visibility for a range of design speeds.

Table 21.9 MINIMUM SIGHT DISTANCES

Design speed (km/h)	*Sight distances*	
	Minimum overtaking distance (single) carriageway (m)	*Minimum stopping distance (single and dual carriageways* (m)
80	360	140
60	270	90
50	225	70
30	135	30

For motorways in the UK, sight distance, both vertical and horizontal, should be checked along the centrelines of both the nearside and the offside lanes and must be at least 300 m (950 ft). Longer sight distances are provided where practicable, particularly at junctions.

For the offside lane the measurement should be made along a chord passing through a front not more than 1.00 m inside the right-hand lane of the opposite carriageway. This gives a curve of minimum radius 1650 m in conjunction with a standard central reserve of 4 m.

The vertical profile

GRADIENTS

A road profile consists of uniform gradients connected by summit and valley curves. The maximum gradients recommended in references 40 and 43 are given in Table 21.10.

Table 21.10

Type of road	*Gradients (%)*	
	Maximum desirable	*Absolute maximum*
Rural motorways	3	4
Urban motorways and primary distributions	4	5
Other rural roads	4	—
Rural slip roads	Up 5 Down 7	—
Urban slip roads	5	8

However, in hilly country, steeper gradients may have to be adopted particularly on the less important routes.

VERTICAL CURVES

At all changes in gradient, vertical curves should be provided. The curvature should be large enough to provide sight distances to allow for safe stopping at design speed. For summits, the stopping sight distance should be measured from points 1.05 m above the carriageway; for sags (or valleys) the stopping sight distance should allow headlamp beams to show up objects on the carriageway. On single-carriageway roads there should be, where practicable, sufficient visibility distance on summits to allow for overtaking. The stopping sight distances and overtaking sight distances for different design speeds are given in Table 21.9.

Preferably

Minimum vertical curve length $L = 175 \times$ (algebraic difference of the two gradients) for summit curve

and

$L = 75 \times$ curve (algebraic difference of the two gradients) for valley or sag curves

Furthermore the minimum length of curve in metres should not be less than half the design speed in km/h (or length in feet never less than three times the design speed in mph), nor less than 70 m on main carriageways.

VERTICAL CURVES—MOTORWAYS

For motorways with 70 mph (120 km/h) design speed, curves should if possible be designed with lengths not less than 300 m (1000 ft) and minimum radii of 18 000 m (60 000 ft) for summits and 5000 m (30 000 ft) for valleys. If there is no alternative to shorter curves and/or smaller radii a 300 m (950 ft) minimum sight distance at summit curves is still required.

Junction design standards

Principles of junction design are fully covered in *Layout of Roads in Rural Areas,*[40] *Roads in Urban Areas*[43] and *Urban Traffic Engineering Techniques.*[42]

PRIORITY (OR UNCONTROLLED) JUNCTIONS

Capacity The frequency and duration of gaps between vehicles in the main road flow governs the number of vehicles able to cross or join from the side road.[44] Figure 21.7[43] indicates the maximum volume of side-road traffic that can cross varying flows on the main road, based upon a single-lane approach and random distribution of main road traffic.

The capacity can be improved by widening the side road near the junction to permit more than one vehicle to merge abreast or by providing a central reservation (4.30 m wide) in the major road to enable cutting of the main road stream to take place in easy movements.

Visibility At priority junctions there should be full visibility at eye level height of 1.05 m above the road level over the area defined by:

(a) A line x metres long measured along the centreline of the side road from the continuation of the nearer edge of the major road carriageway

(b) A line y metres long measured along the nearer edge of the major road carriageway from its intersection with the centreline of the side road

(c) A straight line joining these points, see Table 21.11[43] and Table 21.12.[40]

Table 21.11 VISIBILITY DISTANCES AT PRIORITY JUNCTIONS IN URBAN AREAS

Type of major road	*Speed limit* km/h	*Minimum visibility distance y* m
All-purpose primary distributor	80 (50 mph)	152
	64 (40 mph)	122
District or local distributor	48 (30 mph)	91
Access road	(30 mph)	61

The x distance should normally be 9 m. Some reduction may be reasonable if the side road is lightly trafficked, as may be so for some cul-de-sac and other access roads. On such roads the x distance may be reduced to 4.6 m in urban areas, but the y distance should remain the same. In certain cases, where buildings obstruct the visibility, a 2.1 m x distance may have to be accepted as the absolute minimum.

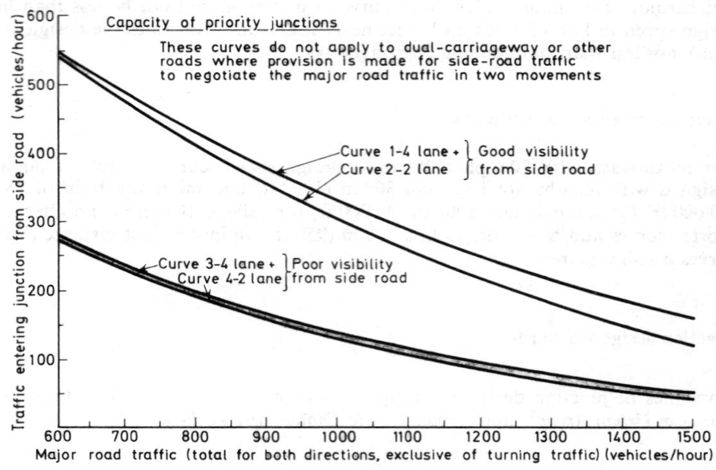

Figure 21.7 Maximum volumes of side-road traffic that can cross varying flows on the main road (Courtesy: HMSO—from 'Roads in Urban Areas'[43])

Table 21.12

Design speed km/h	*Visibility distance at junctions* m	*Acceleration lane length including nose* m	*Nearside deceleration lane length including nose* m	*Right-turn deceleration lane length including taper* m
120 (dual carriageways)	230	400	210	200
100	210	280	160	160
80	180	(210)	(130)	130
60	140	(140)	(110)	110

Note: ( )Speed-change lanes not usually needed except in special circumstances, e.g. where grade separation is provided at a junction.

Speed-change lanes Table 21.12[40] shows the lengths appropriate to major road speeds.

The lengths of acceleration and deceleration lanes may be varied to take account of up-grade or down-grade approaches.

Carriageway radii within junctions The recommended minimum radii without acceleration lane is 10.5 m although a 6 m radius may have to be accepted in residential streets. When an acceleration lane is provided, the minimum radius should be 25 m.

Carriageway width within junctions In appropriate cases extra width should be allowed, to cater for road curvature and for broken-down vehicles. (See Table 21.13.[40])

Table 21.13 WIDTHS OF CARRIAGEWAYS IN JUNCTIONS

Inner radius m	*Design speed* km/h	*Single lane width* m	*Single lane width with space to pass stationary vehicle* m*	*Two-lane width for one or two way traffic* m
10.5	18	5.50	10.30	11.50
15	23	5.20	9.40	10.60
20	27	5.00	8.80	10.00
30	32	4.60	7.90	9.10
40	37	4.50	7.50	8.70
50	41	4.50	7.20	8.40
75	50	4.50	7.00	8.20
100	57	4.50	6.80	8.00
125	62	4.50	6.60	7.80
150	64	4.50	6.40	7.60
Straight		4.50	6.00	7.30

*The excess width in column 4 over column 3 may be constructed to lower standards provided that the surface is either hatched or of a different surface to discourage its use by normal moving traffic.

Traffic-signal-controlled junctions

When priority junctions become overloaded, traffic signals may be able to reduce the number of traffic cuts by imposing time-sharing between conflicting movements and consequently the capacity of a junction is usually improved.

Vehicle-actuated signals cater more precisely for the demands of the moment and consequently reduce delays and provide greater flexibility for dealing with variations in traffic movements.

Coordinated signals enable a sequence of signals to reduce overall delay to traffic. Area traffic control on this basis by computer is now in use.

Roundabouts

General Roundabouts generally have less capacity than traffic signals for the same area of land but circumstances which may favour their provision are:

1. A high proportion of right-turning traffic
2. More than four approach roads
3. Restricted approach widths prohibiting provision of extra lanes
4. Close proximity of other junctions which might cause queuing difficulty with traffic signals
5. A 'Y' junction layout lending itself to roundabout design.

Capacity The overall roundabout capacity is determined by the capacity of individual weave lengths. The following formula gives practical capacity equal to 80% of the absolute capacity.

$$Q_p = \frac{280w(1+e/w)(1-p/3)}{1+w/l}$$

where Q_p is the practical capacity of the weave section (pcu/h), w is the width of the weave section (m) (range 6 to 18 m), e is the average entry width (m) (range of e/w is 0.4 to 1.0), c is the weaving length (m) (18.0–90.0 m range of $w/1$ is 0.12 to 0.4) and p is the proportion of weaving traffic.

For details see references 41 and 43, and Figure 21.8.

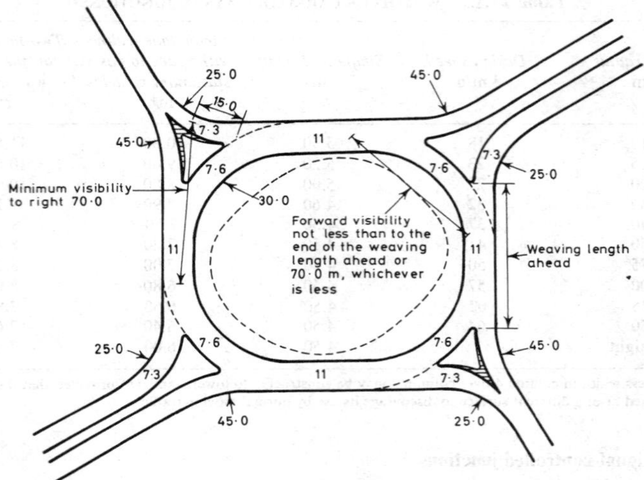

Figure 21.8 Dimensions in metres. (Courtesy: HMSO—from 'Layout of Roads in Urban Areas'[43])

Mini-roundabouts

General Mini-roundabouts act as a series of priority junctions with traffic on the approach roads giving way to traffic within the roundabout. As it is more difficult to influence speeds with these layouts their use is generally best confined to roads subjected to a 64 km/h (40 mph) speed limit or less. The design is based on two principles:

(a) that traffic streams will cut, not weave
(b) the central island takes up valuable space that can profitably be used by road vehicles. The optimum size of the central island is 6 m.

Layout It is essential to ensure that carriageway width on the roundabout should be at least equal to the preceding entry width.

Capacity An approach formula for capacity[46] is given by:

$$Q = K(EW + \sqrt{A})$$

where Q is the total entry volume (pcu/h), EW is sum (m) of basic road widths of all approaches (not half width), A is the area added to the junction by the flared approaches (m^2), K is a factor depending on site conditions (e.g. 3-way junction $K = 80$ pcu/h, 4-way junction $K = 70$ pcu/h, 5-way junction $K = 65$ pcu/h).

At each entry point more precise calculations are necessary in regard to total flows and gap acceptance so as to determine the number of lanes required.

A graphical method[47] has also been put forward as an alternative method for calculating the capacity of mini-roundabouts.

Interchange design standards

General Principles governing the provision of interchanges and their design are covered in *Layout of Roads in Rural Areas*[40] and *Roads in Urban Areas*.[43]

Lane capacity The practical capacity of through lanes at an interchange can normally be considered similar to the lane capacity away from the interchange but on link roads, owing to more adverse gradients and curvature, the lane capacity is less and is usually taken as about 80% of normal lane capacity.

At more complicated interchanges, space-sharing arrangements can often be incorporated with additional direct links for particularly large traffic movements. These links may form a second stage of construction timed to fit in with the growth in traffic, hence the interchange design should be flexible enough to allow for this stage of development.

Layout and location Interchanges should allow all entries and exits to be made on the left via acceleration and deceleration lanes.

MERGING AND DIVERGING SECTIONS

The distance between successive points of convergence or divergence should allow at least six seconds at design speed for drivers to make the appropriate decisions, with a minimum of 183 m between bifurcations. Traffic signs should therefore be an integral part of the design. The lengths of acceleration and deceleration lanes are given in references 40 and 43.

The lengths of acceleration lane required for different conditions are indicated in Table 21.14.[43] They should normally be a direct taper unless, unavoidably, they must be located on bends—in which case a parallel acceleration lane avoiding abrupt entry onto the major road is preferable.

Table 21.14 ACCELERATION AND DECELERATION LANE LENGTHS

Design speed of major road km/h	*Gradient of major road* %	*Acceleration lane length* m	*Deceleration lane length* m
80	4% up	244	82
	level	171	91
	4% down	134	104
64	4% up	155	76 (4% up, level, 4% down)
	level	104	
	4% down	76	
48	4% up	64 (4% up, level, 4% down)	64 (4% up, level, 4% down)
	level		
	4% down		

WEAVING SECTIONS

Wherever possible avoid siting successive entry and exit slip roads so close as to cause hazardous weaving manoeuvres. Where weaving sections are unavoidable, their minimum length for speeds of about 48 km/h are given in reference 43 (table 13.1) and Table 21.15.[43] The desirable minimum is 183 m.

Table 21.15 WEAVING LENGTHS

Weaving volume pcu/h	*Minimum length of weaving section* m
1 000	76
1 500	122
2 000	198
2 500	290
3 000	411
3 500	564

The number of lanes required for minimum weaving sections may be estimated from:

$$N = \frac{W_1 + 3W_2 + F_1 + F_2}{C}$$

where N is the number of lanes, W_1 is the larger weaving flow, W_2 is the smaller weaving flow, F_1 and F_2 are the outer nonweaving streams and C is the normal lane capacity of the major road.

Where a length greater than the minimum can be provided, the value of N can be reduced by replacing $3W_2$ in this formula by

$$\left(\frac{2 \times (\text{length given in Table 21.15})}{\text{actual length}} + 1\right) W_2$$

SLIP ROADS

Design speed The design speed of slip roads should normally be between two-thirds and half that of the more important major road at an interchange. Exceptionally a loop slip road to a lower design speed may be necessary where space is restricted. 24 km/h (15 mph) is an absolute minimum in urban areas.

Width Slip-road carriageways should normally carry one-way traffic only and have widths in accordance with Table 21.13 with allowance for passing a halted vehicle at all points.

Radii The minimum slip road radii and stopping distances for various speeds are given in Table 21.16.[43]

Table 21.16 MINIMUM SLIP-ROAD RADII AND STOPPING DISTANCES

Design speed km/h	*Minimum radius* m	*Minimum stopping sight distance* m
64 (40 mph)	122	91
48 (30 mph)	73	58
40 (25 mph)	52	46
32 (20 mph)	34	34
24 (15 mph)	18	21

Gradients The normal 4% should be considered as a maximum gradient on slip roads carrying a large volume of commercial traffic but otherwise they may be steepened to 5% up-grade and 7% down-grade if necessary.

INTERCHANGE LAYOUT

The choice of layout of an interchange depends on topography, land-use and other restraints but adequate standards must be maintained with the principle that better standards should be provided for the more intense traffic movements.

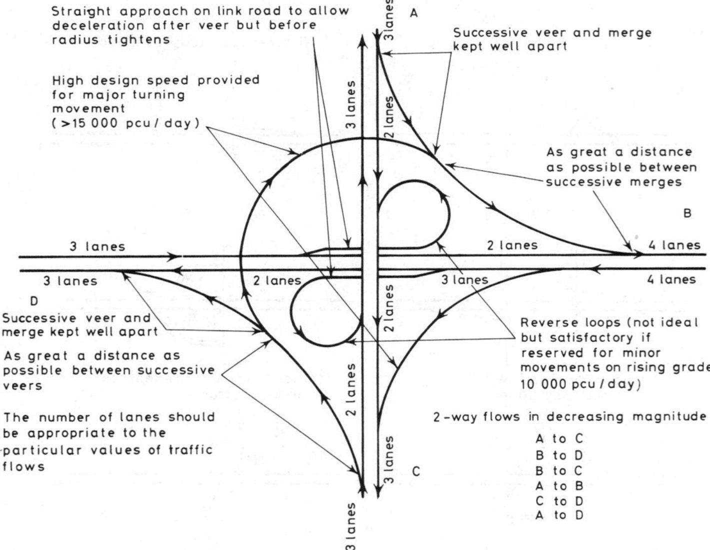

Figure 21.9 Typical motorway/motorway interchange. Note: Flow A to D too small to be provided for. Points to note: 1 No weaving; 2 Points of veer and merge kept to minimum with a maximum of two for any movement; 3 Good visibility of all points of veer and merge; 4 Sharper radii on rising grade where grade assists with deceleration; 5 Smallest turning movements not catered for as these would be difficult to accommodate satisfactorily without much additional expense; 6 Major flow on the right at all points of veer and merge (C to A > C to B + C to D also C to B > C to D and C to B > A to B); 7 Changes in vertical alignment kept to a minimum; 8 Normal motorway standards incorporated for major 'through' movements, lower standards for turns; 9 Separate take-offs from main carriageway where total turning traffic exceeds total through traffic (B to D < B to C + B to A). The major flow keeps to the right at each veer point and enables a single lane reduction to take place at each if required

A typical layout of a motorway-to-motorway interchange embodying the principles discussed above is shown in Figure 21.9.

Types of interchange The application of the principles and standards can lead to many different layouts but between a major route and less important roads there are three basic types: diamond, grade-separated roundabout and partial cloverleaf, all of which permit the traffic conflicts to take place on the minor road. Examples of these layouts are shown in Figure 21.10.

With diamond-interchanges, slip-road exits and entrances should preferably not be located opposite to one another so as to avoid the possibility of vehicles on the exit slip road continuing across the minor road into the access slip road.

Where several local roads intersect or where there is a large proportion of turning vehicles, a grade-separated roundabout is likely to be preferable to a diamond layout, particularly in rural areas where land is more likely to be available. This allows easy

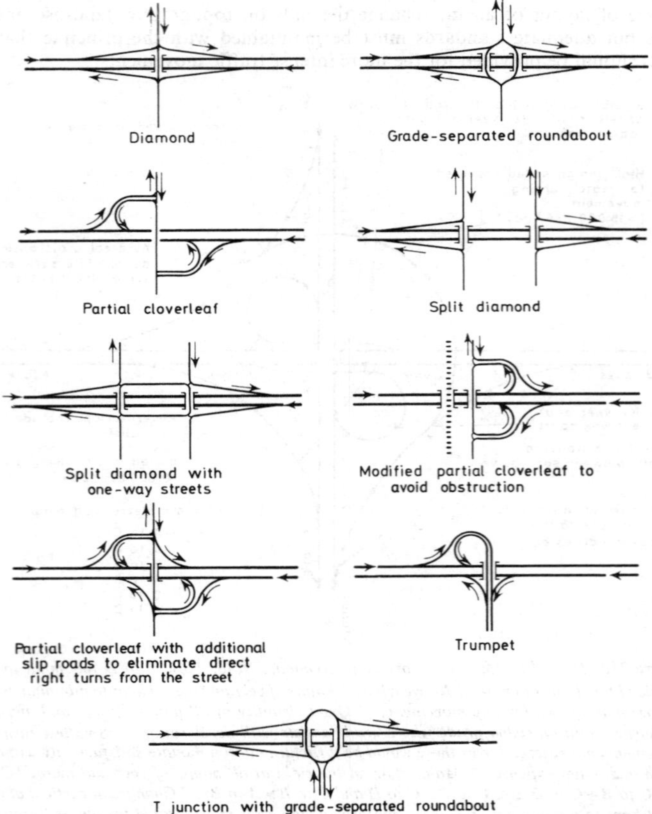

Figure 21.10 Interchange layouts. (Courtesy: HMSO—from 'Roads in Urban Areas'[43])

turning movements and, in urban areas, the use of mini-roundabouts either singly or in pairs might achieve similar capacities with less land area required.

Where particular right-turn movements are high or where site conditions are difficult, part-cloverleaf layouts can be used to advantage. The layout chosen should minimise conflict by keeping as many traffic movements as possible left-turning.

At interchanges between two or more major routes, traffic cuts between major flows should be eliminated so as to allow continuous movement of traffic between one route and the other. The most suitable type of interchange depends upon the configuration of the site, the traffic volumes and distribution and a combination of roundabouts, loops and direct links is the answer. It is preferable, where possible, to change the road layout to simplify the traffic pattern; for example a five-way interchange might be resolved into a three-way and a four-way interchange spaced adequately to cater for weaving traffic between them.

SPACING OF INTERCHANGES

Spacing of interchanges should aim at attracting maximum traffic volumes to the grade-separated route and in rural areas spacing should not normally be less than three miles with an absolute desirable minimum of half a mile in urban areas. Where closer spacing is necessary or weaving volumes are particularly high it is often beneficial to incorporate collector–distributor roads which are connected to the major route at a limited number of points but themselves having several connections to the local road system. These roads, which flank the major road, will usually be designed for a lower speed and consequently the lengths required for speed-change lanes and weaving will be less.

STRUCTURE, SURFACING, DRAINAGE, FENCING, LIGHTING SIGNING AND MARKINGS

Road earthworks—subgrades

MATERIALS

Table 1 from the British Standard Code of Practice on Earthworks[48] lists the characteristics of soils used in earthwork construction. An efficient earthworks operation requires a detailed soil survey, a works programme to suit the soil conditions, the specification and the plant to be used as well as an understanding of the ultimate loading when the road will be complete and in use.

PLANNING

Major roadworks contracts in the UK are normally of two years' duration, hence with the climate in mind a starting time towards the end of the year permits two full seasons for earthworks.

It is normal for proper programming and plant selection to indicate the quantities of materials classed as suitable, unsuitable and rock in each cutting together with the extent and location of soft areas to be removed. Trial embankments may be carried out to confirm techniques of earthworks construction.

SPECIFICATION—EARTHWORKS

Suitability for forming embankments The Ministry of Transport Specification for Roads and Bridgeworks[49] is used for all major roads in the UK and it defines suitable and unsuitable material.

The maximum use of excavated material can be achieved by layering unsuitable material in a filling with another drier material: combinations of wet sand and dry sand, pulverised fuel ash (PFA) and chalk, burnt colliery shale and silt, rock and wet boulder clay, etc.

Compaction The Ministry of Transport specification[49] gives the number of passes of particular compaction plant and the thickness of the layer of soil to be compacted as opposed to the 'Relative Density Specification' which relates the minimum acceptable density compared with the maximum obtainable in the BS compaction test.[50] A method specification implies that there must be sufficient plant to carry out the specified number of passes and that it is always working.

DESIGN

Soil slopes An indication of angles of repose is given in Table 21.17, but maintenance also needs to be considered.

Table 21.17

Material	*Angle of repose* deg	*Weight* kg/m³	*Weight* lb/ft³
Sand, dry	30	1 400–1 600	90–100
moist	35	1 600–1 800	100–110
wet	25	1 800–2 000	110–125
Topsoil, dry	30	1 400–1 600	90–100
moist	45	1 600–1 800	100–110
wet	15	1 800–1 900	110–120
Gravel	40	1 400	90
Hardcore	45	1 600–1 800	100–110
Gravel and sand	25–30	1 600–1 800	100–110
Clay, dry	30	1 900–2 200	120–140
moist	45	1 900–2 600	120–160
wet	15	1 900–2 600	120–160
Mud	—	1 700–1 900	105–120
Ashes	40	600	40

Some soils such as fissured clays deteriorate with time. This may mean comparing the extra cost of building to a flatter slope initially with the anticipated maintenance cost;[51] 1 in 3 or 1 in 4 slope in conjunction with face drainage will normally be a reasonable compromise in such cases.

Rock slopes Rock slopes require a detailed knowledge of the joint pattern and the geology of the site. As with soil slopes a knowledge of the piezometric level is essential to a proper understanding of the problem.[52]

Settlement or instability in embankments This may mean controlling the rate of filling, incorporating a berm in the side slopes or excavating whole or part of any underlying soft material using selected filling materials in or under the bank.

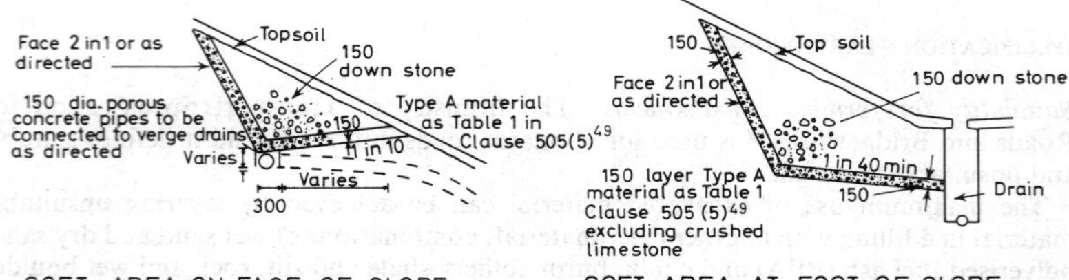

Figure 21.11 Possible treatment of soft areas on cuttings. Dimensions in mm

Soft areas In Figure 21.11 typical treatment of soft areas both on the face of the slope and at the foot of a cutting slope is shown.

Road pavement design

INTRODUCTION—ROAD NOTE 29

Pavement design in the UK is normally based on the recommendations of Road Note 29 *A guide to the structural design of pavements for new roads*[53] and reference must be made to the third edition of the Road Note and the Department of the Environment Specification for Road and Bridgeworks[49] when designing road pavements.

In Road Note 29, traffic is defined in terms of the cumulative number of 8200 kg axles (referred to as standard axles) to be carried during the design life of the road. The equivalent number of standard axles is that number which has the same damaging power as the actual traffic on the road (only the number of commercial vehicles exceeding 1500 kg unladen weight and their axle loadings need be considered—lighter vehicles have no significant effect).

ESTIMATION OF TRAFFIC FOR DESIGN PURPOSES

Initial traffic and growth rate Obtain the traffic data or, in the UK where no estimate of growth rate for commercial vehicles can be made, an average value of 4% per year may be taken.

Design life The structural life chosen will be influenced by the type of road, by its probable use after the end of the design period and whether a flexible or concrete pavement is to be used.

Concrete pavements are normally designed for a life of 40 years; flexible pavements for a shorter life of 20 years and overlay for an increase of life.

Cumulative number of commercial vehicles Road Note 29[53] gives the cumulative number of commercial vehicles carried by each slow lane for design lives of up to 40 years at different rates of growth.

Equivalent cumulative number of standard axles Table 21.20 (Table 2 in Road Note 29[53]) gives, for three classes of road, the conversion factors to be applied to the number of commercial vehicles to obtain the cumulative number of standard axles upon which the design is to be based.

Residential roads and associated development areas Table 21.19 (Table 1 of Road Note 29)[53] shows the initial traffic intensities that should be assumed for the design of such roads.

DESIGN OF PAVEMENT THICKNESS

Given the cumulative number of standard axles the recommended minimum thickness for the various layers of the pavement can be obtained from Figures 6–11 of Road Note 29.[53] The thicknesses of individual layers are intended to be rounded upwards to the next 10 mm intercept. The materials referred to are those specified in the Department of the Environment Specification[49] and any Clause numbers quoted refer to that Specification. The first step, however, is to consider the strength and condition of the subgrade.

Subgrade The overall thickness of the pavement depends mainly on subgrade strength but if the subgrade is susceptible to frost in the UK the overlying thickness of the sub-base base and surfacing must be sufficient to give a total thickness of construction of *not less than* 450 mm. Appendix 1 in Road Note 29[53] gives guidance on the frost susceptibility of soils and road materials.

The strength of the subgrade is assessed on the California Bearing Ratio (CBR) scale—see Table 21.21 (Table 3 of Road Note 29). Wherever practicable the water should be prevented from rising to within 600 mm of the formation level. In other cases the high-water-table condition referred to in Table 21.21 should be used in design.

By using selected filling just below the subbase layer the CBR value of the subgrade can be improved and so the depth of subbase material reduced. Road Note 29 does not cover this aspect but subbase thickness can be reduced to 150 mm if selected fill layers of the following minimum CBR values and thicknesses are used to suit the category of traffic and the strength of the natural subgrade.

Table 21.18 gives a general indication of that which is acceptable in the UK.

Table 21.18

Traffic category (millions of standard axles)	*Minimum CBR value of the selected fill*	*CBR of subgrade (%) (depths in mm)*						
		Under 2	*2*	*3*	*4*	*5*	*6*	*7*
25 and over	15%	650	500	350	260	200	140	100
6 and up to 25	15%	590	440	310	220	170	110	—
Over 2 and up to 6	10%	550	400	270	190	140	—	—

Flexible pavements The term 'flexible pavement' is used not only for pavements with bituminous road bases but also those with bituminous surfacing on cement-bound macadam bases which are often referred to as composite pavements.

(i) Subbase Figure 6 of Road Note 29[53] relates the thickness of subbase required to the CBR value of the subgrade. If the subgrade CBR is less than 2% an extra 150 mm of subbase above that for CBR 2% is required.

The minimum CBR value of the subbase material should be 30% for cumulative total of traffic in excess of 0.5 million standard axles and 20% for lesser traffic. Type 1 subbase (Clause 803),[49] soil cement (Clause 805)[49] and cement-bound granular material (Clause 806)[49] can be assumed to fulfil the CBR requirement of 30% without test.

Where subbase is required the minimum thicknesses that should be laid are 80 mm where the cumulative traffic is less than 0.5 million standard axles and 150 mm where traffic is in excess of that value (see Figure 6 of Road Note 29).

(ii) Road base The recommended minimum thickness of road base using rolled asphalt (Clause 812), dense tarmacadam (Clause 810), dense bitumen macadam (Clause 811), lean concrete (Clause 807), cement-bound granular material (Clause 806), wet-mix macadam (Clause 808), dry-bound macadam (Clause 809) and soil cement (Clause 805) (Clauses refer to reference 49) are given in Figures 7–10 of Road Note 29.[53]

Soil cement and cement-bound granular material are only recommended for a cumulative number of standard axles less than 1.5 million and 5 million respectively.

Where cumulative traffic is over 11 million standard axles and lean concrete, wet-mixed or dry-bound macadam is used for the roadbase, any additional thickness of surfacing indicated in Figures 9 and 10 of Road Note 29 in excess of the minimum requirement of 100 mm may be made up in approved bituminous road base or basecourse material to form a composite roadbase.

The current policy in the UK, however, is not to use wet-mix or dry-bound macadam road base for motorways or heavily trafficked trunk roads.

(iii) Surfacing courses The minimum thickness of surfacing is obtained from Figures 7–10 of Road Note 29[53] and depends upon the type of roadbase material to be used. The materials recommended for the surfacing vary with the cumulative traffic to be carried and details are given in Table 21.22 (Table 4 of Road Note 29) together with the minimum thickness of wearing courses.

Concrete pavements Rigid pavements normally have concrete as the running surface although in some cases a bituminous overlay may be given, e.g. in city streets, where extensive disturbance by statutory undertakers has occurred.

(i) Subbase Three classifications of subgrade: 'weak', 'normal' and 'very stable' related to CBR values are considered for concrete roads and Table 21.23 (Table 5 of Road Note 29)[53] gives the minimum thickness of subbase for each type. If the subgrade is susceptible to frost then the thickness of the subbase should be increased if necessary to ensure a total pavement thickness of not less than 450 mm.

(ii) Concrete slabs Figure 11 of Road Note 29[53] gives the thicknesses required for reinforced and unreinforced concrete slabs in terms of the cumulative number of standard axles to be carried for the three types of subgrade considered in Table 21.23.

On residential roads and similar roads built for light traffic, the pavement may be required to carry comparatively heavy loads during the construction of the surrounding development; alternative design thicknesses are given in Figure 11 of Road Note 29.

(iii) Reinforcement For reinforced concrete road slabs the minimum weight of reinforcement required is given in Figure 12 of Road Note 29.

(iv) Spacing of joints in reinforced concrete slabs The recommended maximum spacing of joints in relation to the weight of reinforcement is shown in Figure 13 of Road Note 29.

Where limestone aggregate is used for the full depth of the slab, the maximum joint spacing may be increased by 20%.

Longitudinal joints should be provided so that the slabs are not more than 4.5 m wide, except where special reinforcement is used (see Road Note 29, paragraphs 53 and 54).

(v) Spacing of joints in unreinforced concrete slabs The maximum spacing of expansion joints recommended is 60 m for slabs of 200 mm or greater thickness and 40 m for slabs of lesser thickness, with intermediate contraction joints at 5 m intervals where aggregates other than limestone are used; where limestone is used these may be increased by 10%.

Expansion joints (as for reinforced slabs) may be replaced by contraction joints in unreinforced concrete roads constructed to full carriageway width in the summer months (21 April to 21 October).

Tied warping joints (Appendix 2 of Road Note 29) may be substituted for some of the sliding contraction joints but not more than three warping joints should be used in succession.

Longitudinal joints should be provided so that the slabs are not more than 4.5 m wide.

(vi) Details of joints are given in Road Note 29.[53]

Pavement construction in rock cuttings

(i) Hard rock Where the rock is of adequate strength, regulate with, say, 50–75 mm of lean concrete and treat as a lean concrete roadbase then add bituminous surfacing in accordance with Figure 9 of Road Note 29.[53]

Table 21.19 (*Table 1 of ref. 53*) COMMERCIAL TRAFFIC FLOWS RECOMMENDED FOR USE IN THE DESIGN OF ROADS IN RESIDENTIAL AND ASSOCIATED DEVELOPMENTS WHEN MORE ACCURATE ASSESSMENTS ARE NOT AVAILABLE

	Type of road	*Estimated traffic flow of commercial vehicles per day (in each direction) at the time of construction*
1	Culs-de-sac and minor residential roads	10
2	Through roads and roads carrying regular bus routes involving up to 25 public service vehicles per day in each direction	75
3	Major through roads carrying regular bus routes involving 25–50 public service vehicles per day in each direction	175
4	Main shopping centre of a large development carrying goods deliveries and main through roads carrying more than 50 public service vehicles per day in each direction	350

Table 21.20 (*Table 2 of ref. 53*) CONVERSION FACTORS TO BE USED TO OBTAIN THE EQUIVALENT NUMBER OF STANDARD AXLES FROM THE NUMBER OF COMMERCIAL VEHICLES

Type of road	*Number of axles per commercial vehicle (see paragraph 14)* (*a*)	*Number of standard axles per commercial axle* (*b*)	*Number of standard axles per commercial vehicle* (*a*) × (*b*)
Motorways and trunk roads designed to carry over 1000 commercial vehicles per day in each direction at the time of construction	2.7	0.4	1.08
Roads designed to carry between 250 and 1000 commercial vehicles per day in each direction at the time of construction	2.4	0.3	0.72
All other public roads	2.25	0.2	0.45

Table 21.21 (*Table 3 of ref. 53*) ESTIMATED LABORATORY CBR VALUES FOR BRITISH SOILS COMPACTED AT THE NATURAL MOISTURE CONTENT

Type of soil	*Plasticity index* %	*CBR (%)*	
		Depth of water-table below formation level	
		More than 600 mm	600 mm *or less*
Heavy clay	70	2	1
	60	2	1.5
	50	2.5	2
	40	3	2
Silty clay	30	5	3
Sandy clay	20	6	4
	10	7	5
Silt	—	2	1
Sand (poorly graded)	nonplastic	20	10
Sand (well graded)	nonplastic	40	15
Well-graded sandy gravel	nonplastic	60	20

Table 21.22 (*Table 4 of ref. 53*) RECOMMENDED BITUMINOUS SURFACINGS FOR NEWLY CONSTRUCTED FLEXIBLE PAVEMENTS (SEE NOTE 1)

Traffic (cumulative number of standard axles)			
Over 11 millions (1)	*2.5–11 millions* (2)	*0.5–2.5 millions* (3)	*Less than 0.5 million* (4)
Wearing course (crushed rock or slag coarse aggregate only) Minimum thickness 40 mm Rolled asphalt to BS 594 (pitch-bitumen binder may be used) (Clause 907)		Wearing course Minimum thickness 20 mm Rolled asphalt to BS 594 (pitch-bitumen binder may be used) (Clause 907) Dense tar surfacing to BTIA Specification (Clause 909) Cold asphalt to BS 1690 (Clause 910) (see note 4) Medium-textured tarmacadam to BS 802 (Clause 913) (to be surface-dressed immediately or as soon as possible—see note 4) Dense bitumen macadam to BS 1621 (Clause 908) (see note 4) Open-textured bitumen macadam to BS 1621 (Clause 912) (see note 4)	Two-course (a) Wearing course—Minimum thickness 20 mm Cold asphalt to BS 1690 (Clause 910) Coated macadam to BS 802, BS 1621, BS 1241 or BS 2040 (Clauses 913, 912 or 908) (see Notes 2 and 4) (b) Basecourse Coated macadam to BS 802, BS 1621, BS 1241 or BS 2040 (Clauses 906 or 905) (see note 2) Single course Rolled asphalt to BS 594 (pitch-bitumen binder may be used) Dense tar surfacing to BTIA Specification (Clause 909) Medium-textured tarmacadam to BS 802 (Clause 913) to be surface-dressed immediately or as soon as possible—see note 4) Dense bitumen macadam to BS 1621 (Clause 908) (see note 4) 60 mm of single-coarse tarmacadam to BS 802 (Clause 906) or BS 1241 (to be surface-dressed immediately or as soon as possible —see note 4) 60 mm of single-course bitumen macadam to BS 1621 (Clause 905) or BS 2040 (see note 4)
Basecourse Minimum thickness 60 mm Rolled asphalt to BS 594 (Clause 902) (see note 2) Dense bitumen macadam or dense tarmacadam (crushed rock or slag only) (Clause 903 or 904)	Basecourse Rolled asphalt to BS 594 (Clause 902) (see note 2) Dense bitumen macadam or dense tarmacadam (Clause 903 or 904) (see note 3)	Basecourse Rolled asphalt to BS 594 (Clause 902) (see note 2) Dense bitumen macadam or dense tarmacadam (Clause 903 or 904) Single-course tarmacadam to BS 802 (Clause 906) or BS 1241 (see notes 2 and 5) Single-course bitumen macadam to BS 1621 (Clause 905) or BS 2040 (see notes 2 and 5)	

Notes:

1 The thicknesses of all layers of bituminous surfacings should be consistent with the appropriate BS spec.

2 When gravel, other than limestone, is used, 2% of Portland cement should be added to the mix and the percentage of fine aggregate reduced accordingly.

3 Gravel tarmacadam is not recommended as a basecourse for roads designed to carry more than 2.5 million st. ax.

4 When the wearing course is neither rolled asphalt nor dense tar surfacing and where it is not intended to apply a surface-dressing immediately to the wearing course, it is essential to seal the construction against the ingress of water by applying a surface dressing either to the roadbase or to the basecourse.

5 Under a wearing course of rolled asphalt or dense tar surfacing the basecourse should consist of rolled asphalt to BS 594 (Clause 902) or of dense coated macadam (Clause 903 or 904).

(ii) Soft rock Where: (a) the surface of the rock is likely to crumble owing to the passage of heavy construction plant, regulation with lean concrete may not be satisfactory and granular subbase may be more practicable; (b) there is soft

Table 21.23 (*Table 5 of ref. 53*) CLASSIFICATION OF SUBGRADES FOR CONCRETE ROADS AND MINIMUM THICKNESSES OF SUBBASE REQUIRED

Type of subgrade	*Definition*	*Minimum thickness of subbase required*
Weak	All subgrades of CBR value 2% or less as defined in Table 21.21	150 mm
Normal	Subgrades other than those defined by the other categories	80 mm
Very stable	All subgrades of CBR value 15% or more as defined in Table 21.21 This category includes undisturbed foundations of old roads	0

Table 21.24 (*Table 6 of ref. 53*) DIMENSIONS FOR SEALING MATERIALS AND GROOVES FOR JOINTS IN CONCRETE ROADS

Type of joint	*Spacing* (m)	*Width of groove* (mm)	*Depth of seal* (mm)
Contraction joint	Under 8	10	20–25
	8–15	15	20–25[a]
	15–20	20	25–30
	Over 20	See note *b*	25–30
Warping joint	All spacings	5	15–20
Expansion joint	All spacings	5 mm greater than thickness of filler	25–30
Longitudinal joint	—	5	20–25

[a]When warping joints are used the spacing applicable is the distance between adjacent sliding joints.
[b]For contraction joint spacings in excess of 20 m the width of groove should be increased by 5 mm for each 5 m in excess of 20 m.

Table 21.25 (*Table 7 of ref. 53*) DIMENSIONS OF DOWEL BARS FOR EXPANSION AND CONTRACTION JOINTS

Slab thickness mm	*Expansion joints*		*Contraction joints*	
	Diameter mm	*Length* mm	*Diameter* mm	*Length* mm
150–180[a]	20	550	12	400
190–230	25	650	20	500
240 and over	32	750	25	600

[a]Dowel bars are not recommended for slabs thinner than 150 mm.

rock, this may soften up and deteriorate if wet and it may be better to regulate with cement-bound materials; (c) rocks are susceptible to frost, the 450 mm minimum thickness of pavement construction will be needed.

Surfacing characteristics

RIDING QUALITY

The Department of the Environment specification for Road and Bridgeworks[49] requires that the longitudinal regularity of the surfaces of wearing courses, base courses and concrete slabs should be within the relevant tolerance—Table 21.26 gives an indication of the expected standards in the UK.

Table 21.26

			Wearing surfaces on flexible and concrete carriageways				*Flexible basecourses wearing surfaces on flexible and concrete hardshoulders and lay-bys*			
	**Irregularity exceeding*	mm	3		6		3		6	
		in	$\frac{1}{8}$		$\frac{1}{4}$		$\frac{1}{8}$		$\frac{1}{4}$	
	Traverse length	m	300	75	300	75	300	75	300	75
		ft	1 000	250	1 000	250	1 000	250	1 000	250
Maximum permitted number of surface irregularities	Pavements designed for more than 2.5×10^6 standard axles		20	9	2	1	40	18	4	2
	Pavements designed for 2.5×10^6 or less standard axles		40	18	4	2	60	27	6	3

*An irregularity is a variation in the profile of the road surface as measured by the rolling straight edge. No irregularity exceeding 10 mm ($\frac{3}{8}$ in) shall be permitted.

SKIDDING RESISTANCE

Texture depth of the surface is of primary importance on high-speed roads, and the resistance to polishing of the surface aggregate is the important factor on lower speed roads.

Surface texture depth A rugous surface, i.e. sufficient texture depth, is essential for high-speed roads. The 'sand patch method' (see Clause 2709 of the Department of the Environment Specification)[49] is used to measure this. For concrete carriageways the average texture depth in the wearing surface should be not less than 0.75 mm. (See Clause 1021 of reference 49.)

For flexible construction, the wearing course surface on motorways and other high-speed roads is normally pre-coated chippings in hot-rolled asphalt. No standard is currently specified, reliance being placed on other specification requirements such as grading and rate of spread of chippings, rolling temperatures, etc., to obtain a satisfactory surface texture; however, 1.5 mm is often recommended.

A satisfactory texture depth can generally be obtained with the rate of spread of $\frac{3}{4}$ in (19 mm) chippings of 90 to 110 sq yd to the ton specified in BS 594[54] provided that the chippings are to a higher range of grading and with a reduced flakiness index, say 20 instead of 35.

Increase in texture depth has a secondary benefit in reducing spray from vehicles. Research has been undertaken on the spray problem using open-textured wearing courses which allow surface water to pass through and flow away at an impervious layer immediately below this.

Polished-stone value For concrete roads no actual requirements for the polished-stone value of aggregates in the top layer of the slab are normally specified although currently the Department of the Environment specification[49] places a restriction on the use of limestone in the top 5 cm because of its tendency to polish under traffic.

For flexible surfacings the required polished-stone value will depend on the type of road or 'site' and it is recommended that it should be not less than the value set out in Table 21.27.

Table 21.27 POLISHED-STONE VALUES OF AGGREGATE FOR BITUMINOUS PAVEMENT COURSES AND SURFACE DRESSINGS

Site		*Minimum laboratory determined polished-stone value for aggregate required for:*	
Category	*Type*	1 Chippings for (i) surface dressing (ii) bedding into: (a) hot-rolled asphalt (b) DTS (c) fine cold asphalt (d) mastic asphalt 2 Wearing courses of: (i) tarmacadam (ii) bitumen macadam (iii) rolled asphalt without chippings bedded in (iv) DTS without chippings bedded in 3 Unsurface dressed basecourses of tarmacadam and bitumen macadam to be used as a running surface for a considerable period	The coarse aggregate component of rolled asphalt and DTS having chippings of higher psv bedded into the surface
A	'Difficult sites', e.g.: (i) Roundabouts and their approaches (ii) Bends of less than 152 m (500 ft) radius on unrestricted roads (iii) Gradients 1 : 20 or steeper and longer than 91 m (100 yds) (iv) Approaches to traffic signals on unrestricted roads	62	45
B	'Average sites', i.e. (i) Motorways and (ii) All other roads including those in urban areas which carry more than 2 000 vehicles per day	59	45
C	'Easy sites', i.e.: Roads not included in Categories A or B but carrying more than 1 000 vehicles per day	45	No requirements

Where polishing of road surfaces has occurred the skid resistance and texture depth may be restored by surface dressing with chippings. Increasing use is being made of synthetic resin binders with artificial polish-resistant road-stones such as calcined bauxite—at dangerous bends, junctions and approaches to traffic signals.

FLEXIBLE SURFACING

Flexible surfacing normally comprises a basecourse and a wearing course but on less heavily trafficked roads a 'single-course' type of material is often used without a separate wearing course.

(i) Basecourse materials The type of material used for a basecourse is selected according to intensity of traffic, whilst the nominal size depends on the thickness. On new roads designed to carry over 2.5 million standard axles, basecourses of hot-rolled asphalt or dense coated macadam are used. These materials are also used for regulating and strengthening existing roads which carry heavy traffic. They have excellent load-spreading properties and an ability to carry the heaviest traffic soon after cooling. The specifications for dense coated macadam in the Department of the Environment specification[49] are as follows:

Clause 903 Dense bitumen macadam for basecourse
Clause 904 Dense tarmacadam for basecourse

These materials are available in 30.75 mm, 28 mm and 20 mm nominal sizes suitable for thicknesses of 65–75 mm, 50–65 mm and 30–50 mm respectively.

These materials are equally suitable for use on less heavily trafficked roads at the lower of the viscosities quoted for either type. *Open-textured basecourse* material to the appropriate British Standard is also used on less heavily trafficked roads but the Department of the Environment specification[49] (Clauses 905 and 906 for bitumen macadam basecourse and tarmacadam basecourse respectively) makes some qualification as to grading and binder.

The grading of aggregate in *'single-course' material* provides a reasonably close surface. Its primary purpose is to strengthen light and medium trafficked roads and to provide an improved running surface. It is customary to seal the surface with coated grit when it is to be overlaid by a bituminous carpet or surface dressed within a limited period. Single-course material is also being increasingly used as a normal basecourse material.

(ii) Wearing-course materials A wide variety of bituminous materials are used for wearing-course construction in thicknesses ranging normally from 20 mm to 37.5 mm. The wearing-course materials in the Department of the Environment specification[49] are:

(a) Rolled asphalt to BS 594 (Clause 907)
(b) Dense tar surfacing (Clause 909)
(c) Open-textured bitumen macadam (Clause 912)
(d) Dense bitumen macadam (Clause 908)
(e) Cold asphalt (Clause 910)
(f) Tarmacadam (Clause 913)

Wearing course adds additional strength to the pavement and it forms an impervious layer over the construction.

When the wearing course does not itself provide an impervious layer as in, for example, the case of open-textured material, it is important to surface dress it so as to protect the lower layers and the subgrade against the ingress of water. Dense-coated macadam or rolled asphalt used in basecourses or roadbase do, however, provide adequate protection to underlying construction.

Wearing-course materials are selected according to traffic intensity. For new construction, see Table 21.22. For maintenance and resurfacing work the range can be more extensive especially if strength and waterproofing are not the primary needs. Fine cold asphalt and bitumen macadam have given satisfactory performances on existing heavily trafficked roads of adequate strength, with qualifications, of course, on the adequacy of their skid resistance retention.

Dense tar surfacing is used for motorway service areas and other vehicle parking areas where resistance to softening by oil droppings is important. The material should be 14 mm nominal size and the coarse aggregate content should be 50%.

SURFACE DRESSING

Surface dressing[55] has three main purposes:

(i) To seal the road surface against the ingress of water
(ii) To resist disintegration
(iii) To renew skid-resistance properties and surface texture.

The type and size of chipping and the viscosity and rate of spread of the binder has to be chosen to suit traffic and road conditions, including the hardness of the road surface. The binder may be tar or cut-back bitumen and reference should be made to Road Note 1[56] or Road Note 38,[57] as appropriate, for full details of the recommendations for surface dressing.

Road drainage

The design of drainage for highways is in accordance with the general principles of drainage in which the following factors are especially relative to roads.

Rainfall Estimates should be based on the Bilham formula[58] which indicates the main rate of rainfall over time for various storm frequencies.

STORM FREQUENCY

Recommended storm frequencies for design capacity estimates:

Culverts—normal conditions 'once in 10 years' But the flooding risk at the inlet for storms of a higher level than once in 10 years should be estimated and where necessary a storm frequency level higher than the norm should be substituted.

Surface water drainage

(i) Normal conditions 'once per year' (pipes assumed flowing full but no surcharge) but should be checked at the level of flow for a 'once in 10 years' storm and adjustments made to ensure that surcharge does not flood the road subbase construction.
(ii) Outfalls from valley in cuttings where damage or risk to the safety of road users arises. Capacity, without surcharge, for a 'once in 10 years' storm required.

Main rivers Widths of channel and bridges must be agreed with the river authority responsible.

In *normal conditions,* a storm frequency not less than *'once in 100 years'* should be used but for a large river see 'Floods in Relation to Reservoir Practice'.[59]

SURFACE WATER RUN-OFF—ESTIMATION

Culverts and watercourses (*rational design method*) The whole of the catchment area draining to the culvert inlet including, where necessary, surface water drainage outfalls from the new roads, should be calculated on an area basis using the formula

$$\text{Run-off} = AIR$$

where run-off is in ft^3/s, A is the area to be drained in acres, I is the percentage of impermeability of the area and R the mean average rate of rainfall over the drainage area in inches per hour during the 'time of concentration'

Surface water drainage of roadworks should be based on TRRL Road Note 35.[60].

Capacity of the system—evaluation The capacity evaluation, velocity, roughness coefficients, etc., should be derived from the tables included in Hydraulics Research Paper No. 4.[61]

It is also prudent to consider existing structures on the watercourses and known flooding records and other local information as a general guide in cases of unchanged circumstances. Culvert sizes and locations must be agreed with the river authority for the watercourse.

PIPE STRENGTHS—DESIGN

Loading (*rigid pipes*) See National Building Studies Special Report No. 37[62] and 'Simplified Tables of External Loads on Buried Pipelines (1970)' published by the Building Research Station.

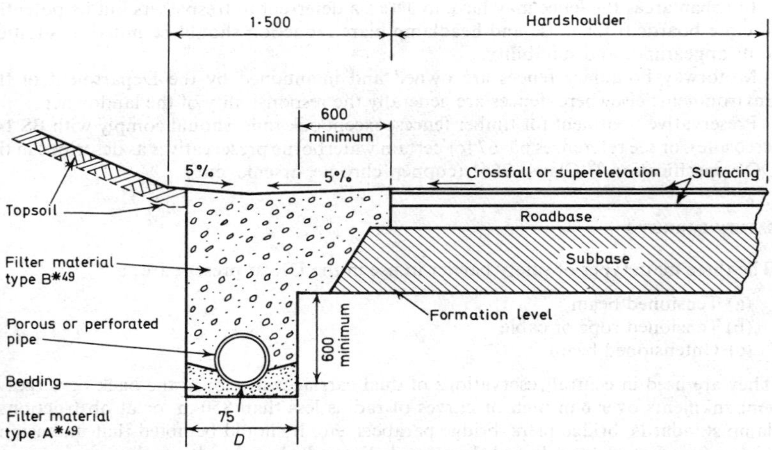

Figure 21.12 Combined surface water/subsoil drain, rural motorway verge. Dimension D = 600 mm (150 mm pipe); 675 mm (225 mm pipe); 750 mm (300 mm pipe). (N.B. Dimensional convention: Where no units are shown, dimensions are in millimetres, or in metres to three decimal places.)

Pipe bedding (*rigid pipes*) Figure 21.12 indicates the granular bedding material and bedfill for rigid pipes. A bedding factor of 1.9 should be used in the designs.

Pipe bedding (*flexible pipes*) Figure 21.12 also illustrates the bedding of flexible pipes suitable for normal ground conditions with a bearing capacity of at least one ton per square foot.

Pipes, manholes, gully frames and gratings Reference should be made to the DOE Specification for Road and Bridgeworks[49] for description of materials and also to TRRL Report LR 277[63] and LR 346.[64]

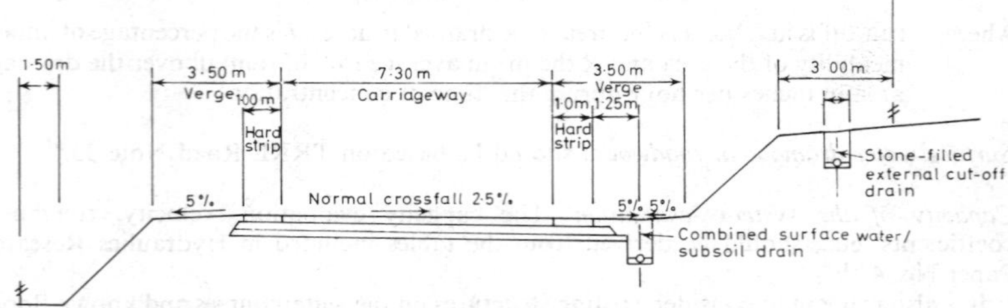

Figure 21.13 Rural all-purpose road. Two-way 7.3 m carriageway

For typical road drainage layouts see Figures 21.12–21.16. (For dimensional convention, see caption to Figure 21.12.)

Road fencing

FENCING

BS 1722, Parts 1 to 11, covers a wide range of fencing types.

In urban areas the fence may have to act as a deterrent to trespassers but its potential (if close boarded) for noise and headlamp glare reduction should be noted in addition to its appearance and suitability.

Motorway boundary fences are owned and maintained by the Department of the Environment; elsewhere, fences are generally the responsibility of the landowner.

Preservative treatment for timber fences, except oak rails, should comply with BS 144 (creosote), or see references 65–67 for certain waterborne preservatives as described in the DOE specification,[49] Clause 2641 (copper, chrome, arsenic, etc.).

SAFETY BARRIERS

The three main types of safety barrier in use in the UK at present are:

(a) Tensioned beam
(b) Tensioned rope or cable
(c) Untensioned beam.

They are used in central reservations of dual carriageways, or at the back of verges on embankments over 6 m high or curves of radius less than 850 m, or at obstructions—lamp standards, bridge piers, bridge parapets, etc. It should be noted that untensioned beams (c) are the least safe and should only be used when the alternatives are impracticable. A British Standard is in course of preparation but the DOE publication[49] provides a specification. See also references 68–72.

Road lighting

The choice of geometry of the system can be made by reference to CP 1004, parts 1 to 9,[73] in which preferred arrangements of column spacing, overhand and minimum light

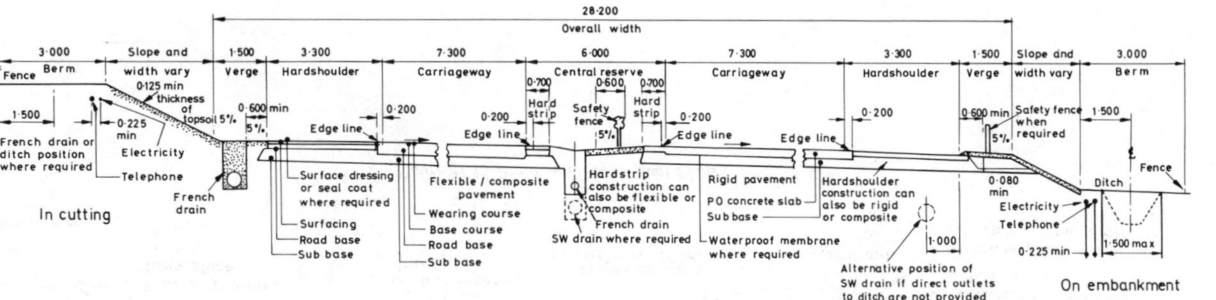

Figure 21.14 Rural motorway—dual 2 lane

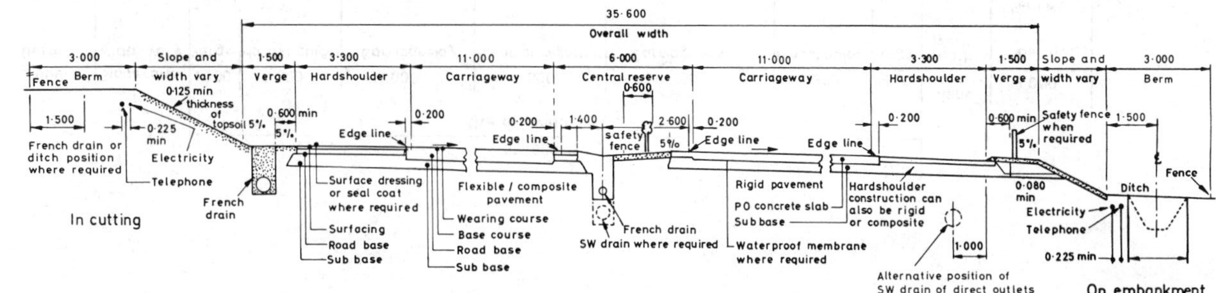

Figure 21.15 Rural motorway—dual 3 lane

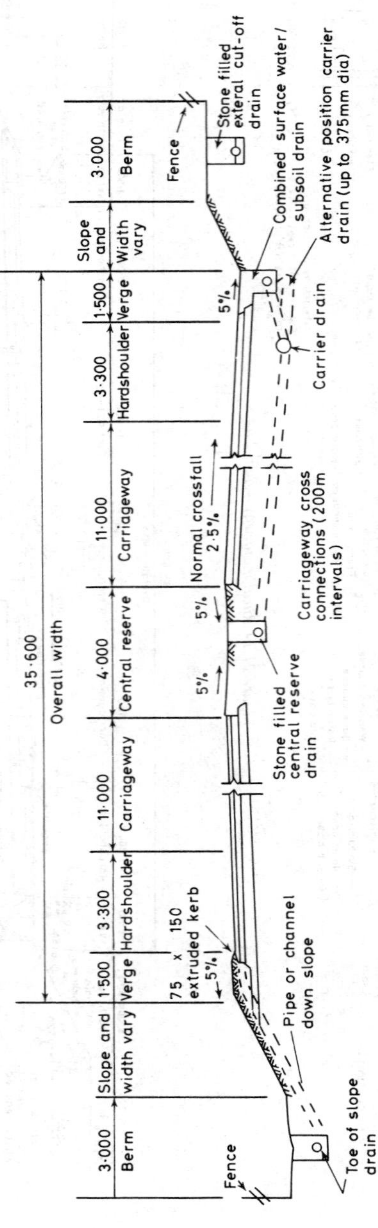

Figure 21.16 Rural motorway—dual 3 lane

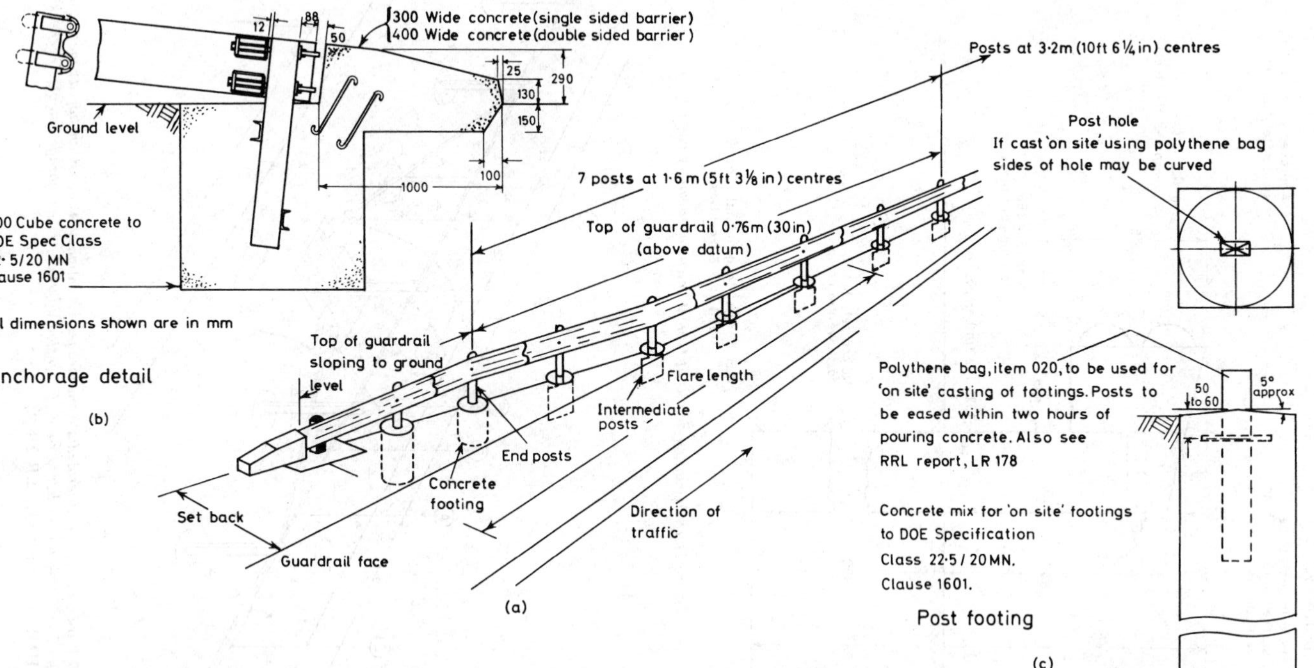

Figure 21.17 Tensioned guardrail safety fence—single-sided flared end. Each post position to be measured from a datum passing through the first post not affected by the flared end. Note: Set back/flare length = 1/10. Tolerances: ±1 mm. (c) *Post footing is* 310 mm *square or* 350 mm *dia., and* 950 mm *deep; post depth below ground level is* 415 mm. (Courtesy: HMSO—from **TRRL LR** 278,[71] amended)

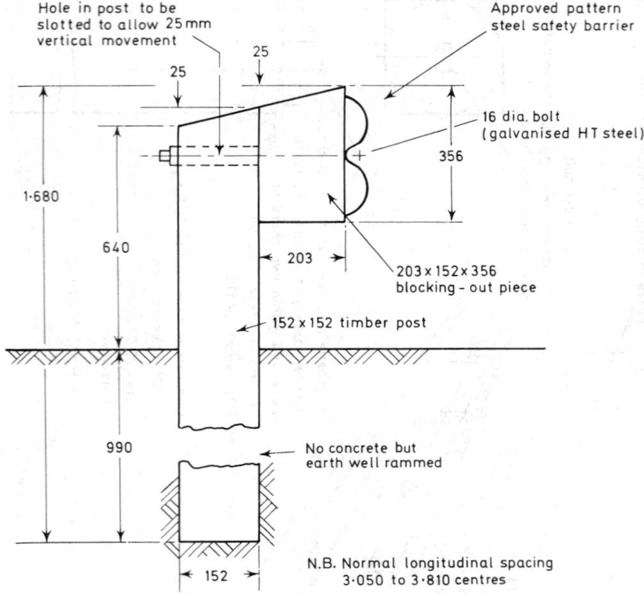

Figure 21.18 Mounting details of typical untensioned safety barrier

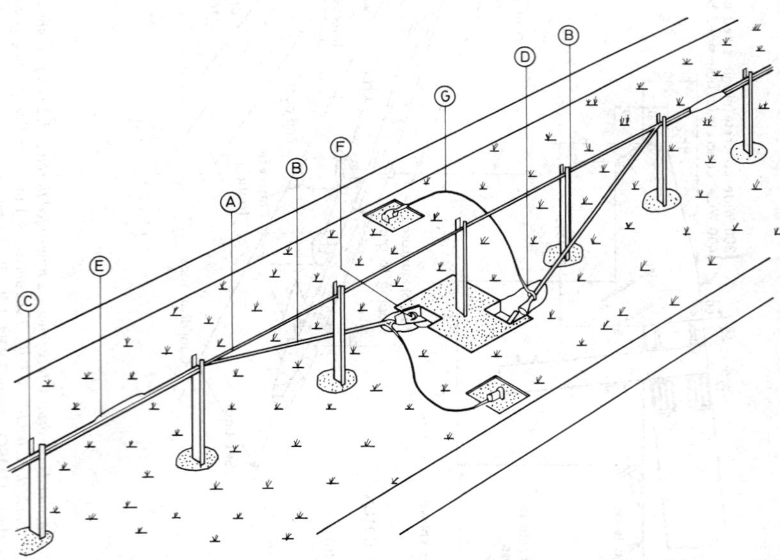

Figure 21.19 Wire rope barrier. Intermediate anchorage. A, continuous rope; B, anchored rope; C, slotted post; D, 'knock-off' link for anchored rope; E, tension adjuster; F, anchorage spigot; G, check wire. (Courtesy: HMSO—from TRRL LR 278[71])

output are specified for a basic dimension of lantern mounting height for all types of roads, junctions, roundabouts, tunnels and underpasses.

Tables 21.28 and 21.29 give details of this basic geometry for all group A lighting for cut-off and semi-cut-off systems respectively. Tables 21.30 and 21.31 are derived from

	Table 21.28			Table 21.29		
	Group A1	*Group A2*	*Group A3*	*Group A1*	*Group A2*	*Group A3*
Mounting height in feet	H	H	H	H	H	H
Max overhang A in feet	0.25 H	0.25 H	0.25 H	0.25 H	0.25 H	0.25 H
Min spacing S in feet	3 H	3.2 H	3.4 H	4 H	4.4 H	4.8 H
Max effective width W for this spacing	0.8 H	0.85 H	0.9 H	H	1.1 H	1.2 H
Min light flux per lantern	8 H^2	7 H^2	6 H^2	10 H^2	9 H^2	8 H^2

Table 21.30 GEOMETRY FOR GROUP A LIGHTING ARRANGEMENTS. CUT-OFF SYSTEM

Arrangement	*Mounting height* H (m)	*Width between kerbs* (m)															
		6	*7*	*8*	*9*	*10*	*11*	*13*	*15*	*17*	*19*	*21*	*23*	*25*	*27*	*29*	*31*
		Design spacing (m)															
Single side	8	26															
	10	35	34	30													
	12		42	42	38	35											
Single central, single carriageway	8					28	28	24									
	10						35	35	32	28							
	12							42	42	41	36						
Staggered	8	26	25	22	19	17	16	13									
	10	32	32	32	30	27	25	21	18	16	14						
	12					38	36	30	26	23	21	19					
Opposite or off-set opposite	10								35	32	29						
	12									42	41	37	34	31	29	27	25
Twin central, dual carriage-way (width between kerbs is per carriageway)	10	35	34	30	27												
	12	42	42	42	38	35	31										

the basic parameters and are based on Group A2 lighting. The majority of Group A lighting designs can achieve an adequate performance by the use of lanterns at 10 m mounting height giving approximately 12 000 lumens in the lower hemisphere.

SPACING ON BENDS

A row of lanterns on the outside of the curve is required when the radius of curvature is less than about 80H. Table 21.32 gives the recommended spacing in such cases, interpolated from Table 17 of CP 1004, Part 2, which is in imperial dimensions.

MAINTENANCE

An essential feature of each lighting installation is that it should be properly maintained. Maintenance includes regular visual inspection and cleaning of lanterns, lamps, refractors, etc., painting, greasing, replacement of worn or broken parts and re-lamping, etc., on a planned basis. Local conditions generally dictate the intervals at which maintenance should be performed.

HIGH MAST LIGHTING

Siting, mounting height, and lumen outputs should be considered initially on the basis of a specified minimum value of horizontal illumination on any part of the carriageway

Table 21.31 GEOMETRY FOR GROUP A LIGHTING ARRANGEMENTS. SEMI-CUT-OFF SYSTEM

		Width between kerbs (m)														
Arrangement	*Mounting height* H (m)	*6*	*7*	*8*	*9*	*10*	*11*	*13*	*15*	*17*	*19*	*21*	*23*	*25*	*27*	*29*
		Design spacing (m)														
Single side	8	35														
	10		47	41												
	12			56	53											
Single central, single carriageway	8					38	38	32								
	10							47	44							
	12									56	50					
Staggered	8	35	35	35	34	31	28									
	10	44	44	44	44	44	44	37	32	28	25					
	12							53	46	41	37	33				
Opposite, or off-set opposite	10							50	50	50	50					
	12									60	60	60	60	56	52	48
Twin central, dual carriageway (width between kerbs is per carriageway)	10		47	41	37											
	12		56	56	53	47	43									

Note: It is assumed that lanterns are mounted over the kerb or within $\pm 0.1\ H$ of the kerb line. When this limit is exceeded the lateral separation between rows of lanterns should be used instead of the width between kerbs.

Table 21.32 SPACINGS FOR LANTERNS ON THE OUTSIDE OF REGULAR CURVES

Radius of curvature (m)	*150*	*200*	*250*	*300*	*350*	*400*	*450*	*500*	*550*	*600*	*650*	*700*
Spacing (8 or 10 m mounting height) (m)	21	25	28	30	33	35	37	39	41	43	—	—
Spacing (12 m mounting height) (m)	25	29	34	36	39	41	44	46	49	51	54	56

surface (which should not be less than 1 lumen/ft^2). However, the design should be modified to incorporate to some extent the principles of normal road lighting to ensure reasonable uniformity of road surface luminance.

The lanterns, which are mounted on carriages capable of being lowered to the ground for maintenance, may number from three to eight. The lamps used can be (a) high-pressure mercury-vapour fluorescent (MBF), or high-pressure mercury vapour with halides (MBI) usually producing an axial (circular) distribution; (b) low pressure sodium vapour (SOX or SLI) which can be arranged for axial or axial asymmetric (elliptic) distribution; (c) high-pressure sodium vapour. 1000 watt MBF lamps are most commonly used: where SOX lamps are used these are usually 2 × 180 watt lamps per lantern. A recent development, the variable geometry lantern, can give a nonaxial asymmetric (bean-shaped) distribution and can be orientated in the horizontal plane to produce the required light distribution from the mast as a whole.

For raising and lowering the lantern carriage a selfsustaining winch is regarded as essential and power assistance is usually provided. It is possible to replace the carriage by a maintenance platform to facilitate painting the mast and servicing the mast top equipment but this needs provision of a safety holding device should the hoisting gear fail.

Road signing

Traffic signs and road markings aid the driver to gain the maximum advantage from the road geometry and must form an integral part of road design.

The regulations concerning signs and road markings are contained in Statutory Instrument No. 1857.[74] Advice on the use of signs and road markings is given in an illustrated manual, the *Traffic Signs Manual*.[75]

On completion of a draft layout of traffic signs the following questions need answering:

Are there too many signs?
Are any signs misleading?
Are all the signs truthful?
Are all the signs located to the best advantage?

Following implementation of a signing scheme, studies should be made of traffic behaviour and appropriate amendments made.

There are three main types of sign:

(a) Informatory, to indicate direction (routes and places) or facilities (e.g. parking)
(b) Warning, to forewarn of road hazards (e.g. bends)
(c) Regulatory, to enforce conditions.

The construction, illumination and siting of signs is dealt with in references 76 and 77.

SITING OF SIGNS

Advance signs must be of sufficient size to allow a driver ample time to read their message and be sited so that a driver has time to carry out the necessary movement or instruction safely.

Experiments were carried out to determine these factors.[78] The results obtained are shown in Tables 21.33 and 21.34.

Table 21.33 ADVANCE DIRECTION SIGNS

Approach speeds of private cars 85 percentiles km/h		*x-height* of lettering* mm	*Distance of sign from intersection* m	*Minimum clear visibility distance of sign* m
Up to and including 48 (30 mph)		100	45	60
Over 48 (30 mph) up to and including 64 (40 mph)		100	45	60
Over 64 (40 mph) up to and including 80 (50 mph)		150	150	105
Over 80 (50 mph) up to and including 97 (60 mph)		200	225	135
Over 97 (60 mph)	(a)	250	300	180
	(b)	300	300	180

Note: *The height of lettering suited to legibility at the speed quoted.
(a) is for high standard all-purpose dual carriageway roads; (b) is for motorways.

Table 21.34 TRIANGULAR WARNING SIGNS

Approach speeds of private cars 85 percentiles km/h	*Height* mm	*Distance of sign from hazard* m	*Minimum clear visibility distance of sign* m
Up to and including 48 (30 mph)	600	45	60
Over 48 (30 mph) up to and including 64 (40 mph)	750 (600)	105	60
Over 64 (40 mph) up to and including 80 (50 mph)	900 (750)	180	75
Over 80 (50 mph) up to and including 97 (60 mph)	1 200 (900)	240	75
Over 97 (60 mph)	1 200 (1 500)	300	105

Road markings

The road markings[74, 75] are used to guide, control, warn or inform drivers. They may be lines, continuous or broken, hatchings or messages. The UK uses white for all lines and markings except those used in connection with waiting regulations, which are yellow.

Road markings for motorways are given in metric units but imperial units are still quoted for all-purpose roads until such time as metric equivalents are released. Details for the many markings can be obtained in the references 74, 75, 78, 79.

ROAD SPECIFICATION AND MATERIAL TESTING

Introduction

The Department of the Environment Specification for Road and Bridge Works, 1969, Fourth Edition (to be referred to as the 'Specification'), used in conjunction with design standards such as Road Note 29,[53] is generally standard for major roadworks in the UK and aims to ensure an adequate but economic quality of construction by prescribing standards of materials, workmanship and finished road representing the best value for money in terms of initial cost and subsequent performance during its design life.

It is now generally incorporated by reference in road and bridgeworks contracts. It was evolved after detailed consultation with governmental, local authority and private contractors' interests aided by research from TRRL.

Application of specification

There are two types of clause in the Specification:

(a) The first schedule lists those clauses which state the general requirements qualified by the phrases 'as described in the Contract' or 'as shown on the drawings' and the additional information must be added before the clause becomes effective.

(b) The second schedule lists clauses which, although effective as they stand, contain such phrases as 'unless otherwise agreed by the Engineer' which gives some freedom of method to suit particular circumstances.

A metric version of the Specification is being prepared to update the current fourth edition of the Specification.

DRAINAGE—SERIES 500

The clauses in this series are intended to apply primarily to the relatively shallow construction which is usually associated with surface water drainage in roadworks. Additional clauses with more stringent specification requirements may be needed for drainage authority works and others of a specialised nature, e.g. work in headings, pipe jacking, deep trenches, etc.

EARTHWORKS—SERIES 600

Plans and longitudinal sections showing the positions and details of boreholes and trial pits should be supplied to tenderers supplemented by all factual information derived from the site investigations.

BRITISH STANDARD SPECIFICATIONS

The Specification indicates most materials used in highways work by reference to the appropriate British Standard, a list of which is given in contract documents.

SUBBASE AND ROADBASE—SERIES 800; FLEXIBLE SURFACING—SERIES 900

The Specification allows the contractor to select the permitted materials that he considers will be the most economical when laid to the thicknesses required by the contract.

Materials testing

TESTING

A specification may be 'end-product' in which only the finished work is tested and may result in rejection of the whole, or may be by 'method specification' in which the method of working is defined, e.g. compaction of earthworks in the DOE Specification. In most cases, however, some blend of the two types of specification is adopted. The cost of testing varies between 0.3 and 1% of the value of a project.

EARTHWORKS

Cuttings Tests are taken primarily to determine whether the excavated materials are suitable for forming embankments. In the case of a cohesive soil this will depend on the height of the embankment but provided that it is below 10 m in height an undrained shear strength of 50 kN/m^2 will be adequate for conventional plant. Suitability is also sometimes judged from the moisture content of the soil, which should be lower than 1.2 to 1.4 times the plastic limit dependent upon the material. BS 1377[80] describes the methods of determining the undrained shear strength, moisture content, plastic limit and other soil characteristics.

California Bearing Ratio (CBR) Tests are taken at formation level to determine the pavement construction depths.[81]

Embankments In 'end-product' specification, compaction is controlled by *in situ* density tests. Field densities are compared with the maximum obtainable in a standard laboratory test and the ratio is expressed as a percentage and termed relative density; BS 1377[80] gives several standards but the most frequently used is the heavy-compaction Test 12. Specifications often require 95% of this standard although 90% may be used if the material is silty and above optimum moisture content, i.e. the moisture content at maximum density for a given amount of compaction.

The sand replacement and core cutter methods (Tests 14B and 14D of BS 1377[80]) for determining densities are generally used although nuclear devices have been used with variable results.

AGGREGATES

The road pavement consists of:

Rigid construction—subbase, concrete slab (reinforced or unreinforced); or
Flexible construction—subbase, base, surfacing
Composite construction—concrete, lean concrete or cement stabilised base with bituminous surfacing.

The aggregates used in these are crushed rock, gravel or slag and a binder of cement or bitumen, etc., except for dry-bound macadam.

Aggregates and binders may be tested separately.[82–84]

CONTROL TESTING OF MIXED MATERIALS

BS 598[85] and Road Note 10[86] describe the various methods of sampling and analysing bituminous materials. Cement-bound materials are not easy to test in the wet state but development towards this is taking place.[87]

ACCEPTANCE TESTING OF THE FINISHED PRODUCT

Concrete is tested in its hardened state,[87] e.g. concrete cubes, but for concrete roads the cylinder splitting test[88] which measures the tensile strength is now used and is more logical since the performance of a concrete road depends on its ability to withstand tensile stresses. As a nondestructive test, ultrasonic testing is being developed.

Fatigue is a factor in bituminous materials of cement-bound bases. Measurements of the deflection of the surface under a standard load gives a guide to the life of the pavement.[89] The French Deflectograph (modified by TRRL) and the Benkelman Beam Test are used; the plate Bearing Test is often used at subbase level.[90]

STATISTCAL ACCEPTABILITY

Statistical methods for sampling and acceptance testing which aim at evening-out the risks between the contractor and the engineer are described in TRRL Report LR 276 for hot-rolled asphalt, and 299, 300, 301 for concrete.

Construction plant (see references 91–96)

Since the capital costs on major road projects may be up to 30% of the value of the scheme, modern road designs and specifications allow a choice of methods and equipment to maximise on plant developments (see Figure 21.20), e.g.

(1) Excavators and compactors
(2) Drainage and elevating plant
(3) Tunnel moles
(4) Wire guidance systems for formation and pavement layer trimming, and for bituminous and concrete pavers
(5) Top soiling and seeding plant
(6) Batch mixing units
(7) Piling, pre-casting and bridge construction techniques.

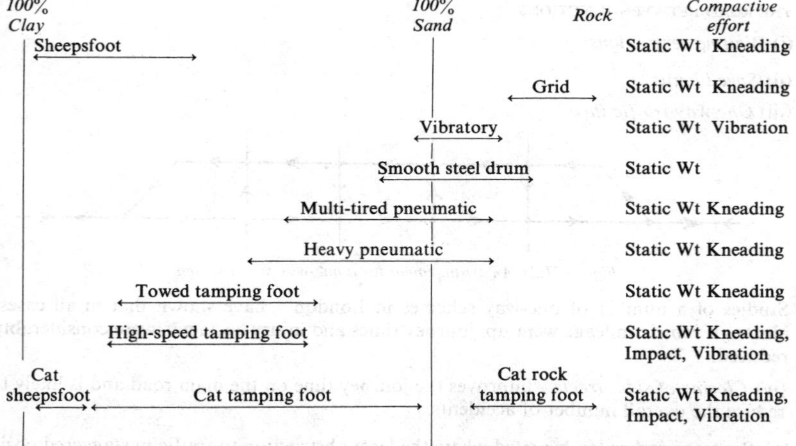

Figure 21.20 Economic zones of application into which each of these basic types of compactor fall. Each type has been positioned in what is considered to be its most effective and economical zone of application, although it is not uncommon to find them working outside of these zones

IMPROVEMENTS AND TRAFFIC MANAGEMENT ON EXISTING ROADS

Improvements on existing roads

The speed, capacity and safety of many roads can be improved by the widening and realignment of existing carriageways or by the application of traffic management techniques. These would be considered before deciding whether or not to build a new road.

Traffic management aims to

(i) increase road capacity

and

(ii) improve road safety

The need to improve capacity can often be assessed from simple observations of traffic congestion and queuing at junctions although speed and delay measurements[97] are sometimes required and also a study of accidents particularly at road junctions where more than half the fatal and serious accidents in built-up areas occur. Typical values for personal injury accident rates in Great Britain are given in Table 21.35.

Table 21.35 ACCIDENT RATES IN 1968 IN PERSONAL INJURY ACCIDENTS PER MILLION VEHICLE MILES

	Class 'A' roads	*Class 'B' roads*	*Class 'C' roads*	*All roads*
Urban	2.96	3.05	3.65	3.21
Rural	1.07	1.27	1.54	1.20
Motorways	—	—	—	0.314

The accident rates in shopping streets can be well over 10.

Typical traffic management measures

THE ROAD BETWEEN JUNCTIONS

(i) *Waiting restrictions*

(ii) *Speed limits*

(iii) *One-way traffic flow*

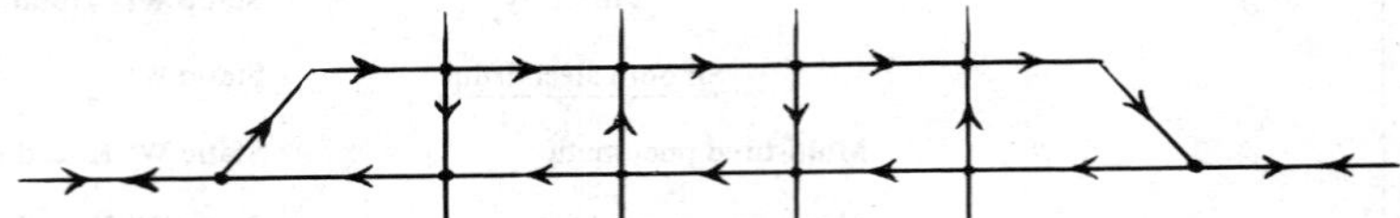

Figure 21.21 An arrangement for a one-way street system

Studies of a number of one-way schemes in London[98] have shown that in all cases, although vehicle mileage went up, journey times and injury accidents were considerably reduced.

(iv) *Closing of side streets* improves the journey time on the main road and is likely to reduce the overall number of accidents.

(v) *Bus stops and/or lay-bys* sited where the least obstruction to traffic in staggered positions so that on leaving the stops buses will move away from each other.

(vi) *Restrictions of heavy vehicles, play streets, shopping precincts*

(vii) *Street lighting*

(viii) *Tidal flow* A greater number of lanes is allocated to the direction of greater flow in peak hours. Gantries carrying secret signs are often used to ensure control of the system at its extremities.

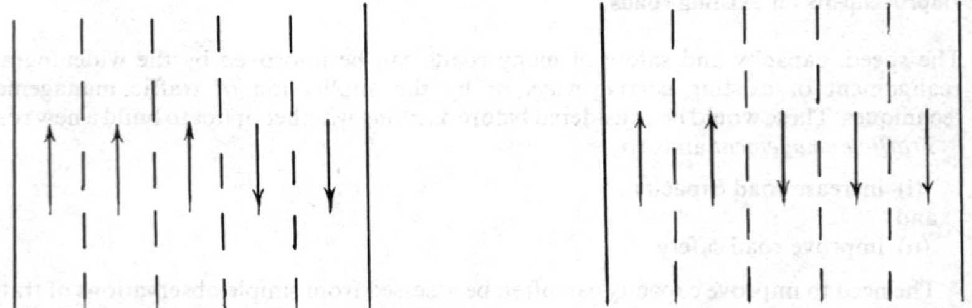

Figure 21.22 Tidal flow on a city street. Left: morning peak—3 lanes inward, 2 lanes outward. Right: evening peak—2 lanes inward, 3 lanes outward

AT JUNCTIONS

(i) *Channelising islands and kerb realignment*
(a) to separate conflicting traffic streams
(b) to assist intersecting or merging at suitable angles
(c) to control vehicle speeds
(d) to encourage drivers to take the correct path
(e) to provide shelter for vehicles about to manoeuvre, e.g. right turns
(f) to facilitate pedestrian crossing movements.

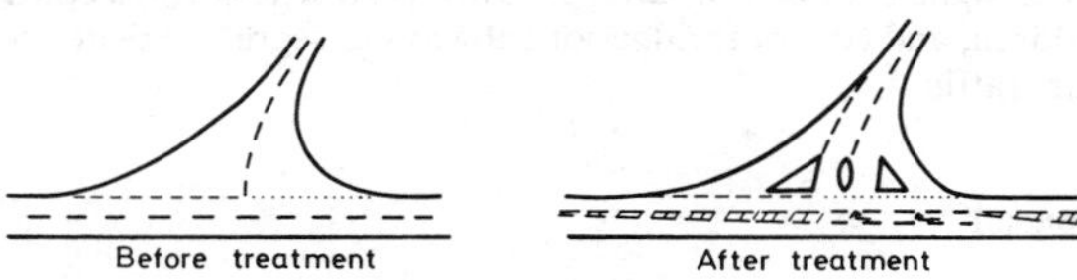

Figure 21.23 Channelising islands

(ii) *Improved visibility*

(iii) *Staggered junctions* To deter straight-over movements at crossroads, right–left turn sequences generally preferable.

(iv) *Right-turn ban*

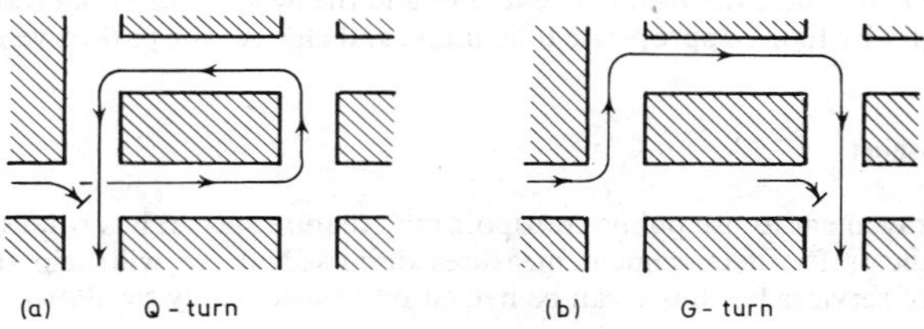

Figure 21.24 Banned right turns

(v) *Roundabouts and traffic signals* Standard roundabouts normally require more land than traffic signals and in city streets the introduction of the mini-roundabout may be apposite.

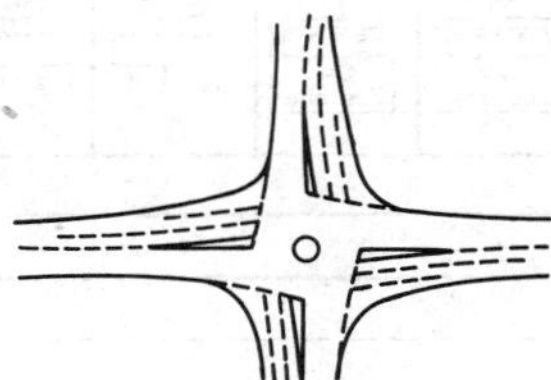

Figure 21.25 Mini-roundabout

Factors which favour the provision of a roundabout are:

(a) high proportion of right-turning traffic
(b) more than four approaches to the junction
(c) narrow approach widths

(vi) *Box junctions* To prohibit drivers from entering the junction unless it is clear.

Pedestrian facilities

Thirteen out of fourteen pedestrian casualties occur on urban roads. A quarter of all casualties on such roads and a third of those killed or seriously injured are pedestrians Many of the traffic management measures above are helpful to pedestrians but the following are specifically for pedestrian safety: (i) guard rails to channel pedestrians to the proper crossing points; (ii) central refuge islands to enable pedestrians to cross in two movements; (iii) uncontrolled crossings (zebra) to give pedestrians priority over vehicles; (iv) pedestrian signal control facilities such as all-red phases, special pedestrian phase and special signals for pedestrians (pelican); (v) traffic-warden controlled crossings mainly for children, and (vi) footbridges or subways to segregate pedestrians completely from vehicular traffic.

Area traffic control[99]

An area traffic control system coordinates traffic signals over an area by means of a central controller, usually a computer. Normally these devices aim to minimise overall journey times within the area.

Parking of vehicles[60]

The total amount of parking space provided should be balanced against the capacity of the streets to handle the traffic it generates and the needs of essential traffic should be given priority by fixing appropriate time limits and charges for parked vehicles.

Public transport

The encouragement to use public transport may require special bus routes or bus lanes. Many of the traffic management measures discussed above will help the speed and reliability of services but buses can be helped more specifically by allowing 'buses only'

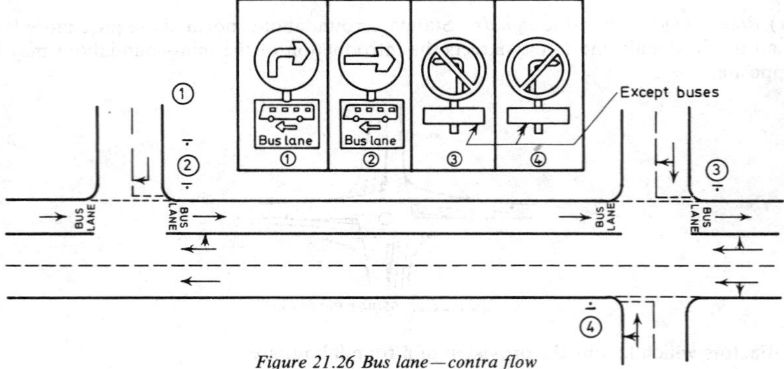

Figure 21.26 Bus lane—contra flow

access to streets otherwise barred to vehicles, 'bus only' right turns at junctions and by providing 'bus only' lanes both with and against (contra flow) the normal direction of traffic flow.

The use of public transport can also be encouraged by parking controls in town centres and the provision of cheap or free car parks near to public transport routes on the outskirts of towns.

Traffic regulations and orders

The legal aspects of traffic regulations are covered in '*Road Traffic Regulation Act 1967*'[101] as amended by the Transport Act 1968[102] and The Local Authorities Traffic Orders (Procedure) (England and Wales) Regulation 1969,[103] The application of the legal requirements and advice on the use of the orders are contained in '*Traffic Management and Parking Manual*'.[104] It may be worth introducing these experimentally to test the efficiency of the proposal.

Environment

The effect of traffic management schemes on the environment may mean localised detriment for overall benefits. The banning of heavy vehicles, limited parking, and the creation of other facilities, e.g. play streets, may serve to reduce any detriment.

Traffic signing and road marking

Traffic signing and road marking is an essential element in traffic management (see '*Traffic Signs Manual*'[105]).

Monitoring of traffic management

'Before' and 'after' studies serve to show the effectiveness of the scheme or need for further adjustment.

An indication of the value of the traffic safety aspect of traffic management schemes has given the following approximate percentages:[97]

Description of change	*Reduction in personal injury accidents found in sample studies* (%)
Staggering of cross roads	60
Application of No Waiting regulations	up to 30
Improvement of road alignment at bends	80
Provision of dual carriageways	30
Improved visibility at junctions	30
Provision of traffic signals	40
Provision of roundabouts	50

HIGHWAY MAINTENANCE

General

The object of maintaining a highway is to preserve the fabric in such a condition that it provides safe passage for all traffic. The Report of the Committee on Highway Maintenance provides proposals for dividing maintenance operations into

(i) Structure
(ii) Aids to movement and safety
(iii) Amenity.

Within those groups it is not easy to ascribe priorities without assessing the type of traffic, its contribution to the community and the advantages to be gained by incurring the expenditure or even the disadvantages of not incurring the expenditure.

In Britian some measure of the first of these is established by the designation of the road:

(i) Motorway and trunk road
(ii) Principal road
(iii) Nonprincipal road
(iv) Nonclassified road

Within these groups the nature of the traffic must be identified, e.g. significant numbers of pedestrians establish a need for safe footpaths by using upstanding kerbs, safety rails, etc.

Staff for maintenance

The type of staff required for maintenance includes those:

(i) for inspection, estimating and programming of work
(ii) for superintending day-to-day operations
(iii) for carrying out the physical work—either by direct labour or by contract.

The officer with control of highways should be a well qualified engineer who can encompass the whole range of operations—provision of new highways and bridges and their physical effects on the country through drainage arrangements, earthworks, etc.,

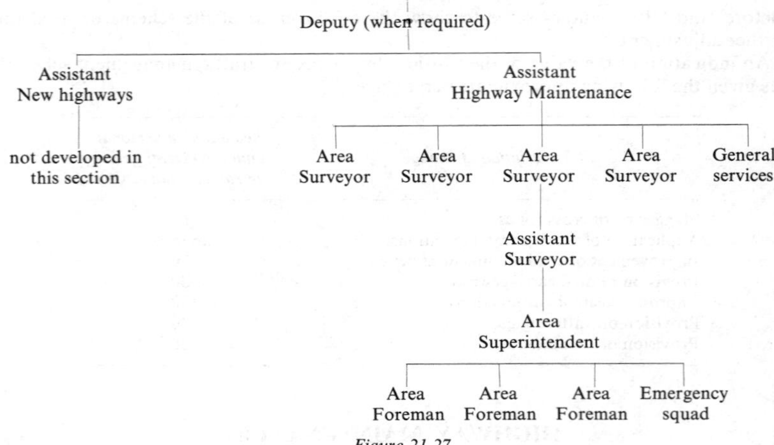

Figure 21.27

the maintenance of the road, and its management in terms of control of traffic, surveillance of operations of public utilities and of security generally. He may require a second-in-command over the whole range of interests but up to about 3000 km (2000 miles) a single chief officer could probably cope.

Figure 21.27 illustrates a typical organisation.

The Marshall Report indicates that about 800 km (500 miles) of road would form a useful area unit; this could be increased significantly if comprehensive support services were readily available.

The qualifications of the Chief Officer have been described. This covers his Deputy as well, and is likely also to apply to the Assistant who will tend to be a younger edition of his seniors. He will have professional qualifications and will be capable also of making contact with road users and industrial interests. If contract arrangements are necessary to cope with the task he should be capable of drawing up the necessary documents.

The Area Surveyor should for preference be qualified and should be able to cope with some of the design problems he will meet from time to time. He will be required to interpret survey data provided as a basis for work programming and to deal with the financial implications'

The qualities looked for in the Area Superintendent are preferably good experience with a tradesman background, an ability to organise labour and an ability to undertake ground surveys for work programming purposes. The number of foremen must depend on the nature of the area and of the men under their control, e.g. rural conditions vary from urban conditions. Many of the inspections mentioned in the following summaries will be carried out by the foremen and by members of their squads. The 'General services' available for maintenance work could include for example

(i) A signs store, traffic signals, etc.
(ii) A plant depot
(iii) A winter maintenance depot and salt stores
(iv) Workshops

The 'Emergency' squad is a pool of skilled labour who can be diverted to special operations.

Assessment of structural needs

A method of assessing the structural needs of highways is set out fully in reference 109. Briefly this describes the equipment and measurements required; a subsequent document will develop a method of assessing the maintenance rating to establish a work priority.

Table 21.36 STRUCTURAL ELEMENTS OF HIGHWAY MAINTENANCE

Element	*Item*	*Correction*	*Remarks*
Carriageway	(1) Condition Shape Irregularity	Reconstruction, resurfacing, surface dressing	Depending on degree of fault
	(2) Strength	Reconstruction, resurfacing	Depending on degree of fault
Surface	(1) Lack of skid resistance only	Surface dressing	This is corrected within other correction operations as above
	(2) Patching	Surface dressing	To maintain a watertight surface
	(3) Defect through utility services		These should be masked during surface operations
	(Road markings are entered in Table 21.37)		
Drainage	(1) Gully emptying		Dependent on local conditions—referred to also under sweeping and cleaning
	(2) General performance	Width of running water in channels	Standards relate to the flow in channels reducing the effective width of carriageways

Table 21.36—*continued*

Element	*Item*	*Correction*	*Remarks*
	(3) Concentrated flow across carriageway	Install extra channels or gulleys	Not to be confused with normal flow across cambered roads
		(See also 'Structures' below.)	
Footways (Urban)	Irregularities related to safe use	Re-lay paving slabs	(1) Suggestions for height of projections are given and also rates of inspection in different areas
		Resurface flexible material	(2) Legal implications must be recognised
(Rural)	Irregularities and surface water related to safe use	Minimum maintenance in rural areas	Rate of inspection given
Kerbing	(1) As drainage feature	Install to delineate drainage channel and support edge of carriageway	
	(2) Edge of footway	Defective kerbing	A normal height of kerb in urban and rural conditions is given as safety feature for pedestrians
		Kerbs sunk to carriageway level or lower	Serves to support edge of carriageway
Structures	(1) Bridges, culverts, walls	Necessity for programming of painting of steelwork	Rate of inspection to be assessed on local conditions
	Safety of road-user is paramount	Deterioration of concrete or other fabric Underwater inspection where necessary	
	(2) Embankments and cuttings; to include ditches where appropriate	Regular inspection for incipient slips or failures Advice to adjoining landowners about structural or drainage defects	Legal implications are important

Table 21.37 AIDS TO MOVEMENT AND SAFETY

Element	*Item*	*Correction*	*Remarks*
Road markings	(1) Advisory markings	Re-make	General recognition should be possible
	(2) Mandatory and statutory markings	Re-make	These markings must lie within legal limits
	(3) Reflecting studs	Loose or ineffective studs should be replaced. A wholesale change might be made at the end of effective life	An inspection before onset of winter conditions is advisable and equally an inspection should be made after winter maintenance operations are complete
Traffic signs and bollards	(1) Illuminated signs and bollards	Regular inspection for (i) light failure (ii) drainage (iii) cleaning (iv) supports and frames	Particular attention to be paid to mandatory signs, e.g. 'Stop' signs, etc.

Table 21.37—*continued*

Element	*Item*	*Correction*	*Remarks*
	(2) Nonilluminated signs	Regular inspection for: (i) drainage (ii) cleaning (iii) supports	See above
	(3) Traffic signals	Regular inspection for (i) light failure (ii) general maintenance and cleaning (iii) phasing (iv) alignment (v) Mechanism (vi) Painting	(1) All highway personnel should report faults wherever discovered (2) Contract maintenance and guarantees to be operated. On-call arrangements to be made
Pedestrian crossings	(1) Beacons (2) Road markings	As for traffic lights (i) Slippery surfaces should be corrected (ii) Obscure markings should be made good	Legal implications are important
Road lighting	(1) Lanterns	Regular inspection for (i) illumination (ii) cleaning	The safety of the road user—driver and pedestrian—is paramount in this connection
	(2) Columns	Conditions to be reported during inspection	
Guard rails and safety fences	Pedestrians and vehicular	To be included in regular inspections. Where risk to public is involved speedy action is necessary. Inspection to cover: (i) condition (ii) painting (iii) cleaning	Note legal implications
Winter maintenance	(1) Precautionary salting	On receipt of frost warning roads should be treated The rate of salt-spread should be 14 g/m^2. Salt in accordance with BS 3247: Part 1: 1970	Treatment should be applied within a limited period after a warning: (a) Rural main roads and motorways—2 hours (b) Other important roads and accesses to emergency services—2 hours (c) Urban main roads—1 hour Crews of salting equipment should be on stand-by duty Neat salt should be used This should be mixed with grit in special circumstances only
	(2) Snow and ice clearance	Use of specialist equipment	Major routes as in (a) above should never become impassable to traffic. This is related to traffic flows which have the effect along with salt of keeping snow from accumulating. Snow ploughs with blades effectively remove slush from main routes; crews

Table 21.37—*continued*

Element	*Item*	*Correction*	*Remarks*
			should be on stand-by duty Roads as in (c) above should not be impassable for longer than one hour. Public transport is a major factor in the clearance of snow Roads in other priorities should generally be cleared in 4 to 6 hours unless conditions are exceptional. Pedestrian ways may require special treatment in town centres Footpath clearance should be confined to busy areas, steep hills, etc.
	(3) Snow fencing		Local knowledge is necessary to establish the siting and timing of snow-fence erection. Care must be taken to clear this with landowners
	(4) Salt storage		Care must be taken to site salt heaps to avoid drainage to local crops, watercourses

(Winter maintenance plant is dealt with later.)

AMENITY ITEMS

Items included under a general heading of the amenity functions of a road nevertheless have a safety element in that grass cutting, lopping of trees, etc., ensures that sight lines at bends and the visibility of traffic signs are maintained.

Amenity functions are summarised in Table 21.38.

Table 21.38 AMENITY ITEMS

Element	*Item*	*Correction*	*Remarks*
Grass cutting	(1) Prevention of obstruction of sight lines (2) Maintenance on certain roads of reasonable pedestrian access (3) Control of noxious weeds	Grass cutting	Standards vary for urban or rural situations: *Rural* (a) Major roads—6 ft (1.8 m) of the verge should be kept below 6 in (15 cm). Elsewhere one or two cuts should be employed to keep grass to 12 in (30 cm) long (b) Minor roads—one cut per year will normally suffice (c) Spraying of grassed areas can be used to control noxious

Table 21.38—*continued*

Element	*Item*	*Correction*	*Remarks*
			weeds. Consideration must be given to width of road, nature of traffic and, not least, the culture of the grass growth *Urban* (a) Major roads—grass should be kept down to 3 in (8 cm) (b) Minor roads—minimum maintenance consistent with the environment
		Use of chemical sprays	These should be used with caution where access for cutting is not available and for growth control generally. Examples are: around sign supports; central reserves with safety barriers; urban walkways
Hedge trimming	Prevention of obstruction of visibility at bends and at traffic signs	Tractor-mounted equipment might be employed	This is not normally a function of the highway authority. They have power to require land-owners to reduce hedge heights and to control trees Legal implications should be noted carefully
Trees	This repeats very much the information under hedge trimming. Trees may call for specialist advice which is often available in the Parks Department of a Local Authority		Legal implications and ownership are important factors
Sweeping and cleaning	(1) Objects and material shed by vehicles	Heavy and dangerous items must be removed by hand This is material which can cause broken wind-screens and mechanical damage	The rates of inspection and activity on this account vary with the weight of traffic, between daily inspection on motorways and weekly visits on less busy roads All highway staff should be aware of the necessity to remove dangerous items whenever they see them
	(2) Vegetation and detritus	This is material which can block drainage systems which can obscure road markings, and cause dirty windscreens	Legal implications are important Again rates of inspection and activity are dictated by traffic but where traffic is heaviest, cleaning and scavenging is difficult. In rural areas carriageway sweeping is not necessary more often than at two-monthly intervals.

Table 21.38—*continued*

Element	*Item*	*Correction*	*Remarks*
			In town centres daily attendance is required, reducing to weekly attention in residential areas.This should include footpaths. The rate of gully emptying is dependent on the build-up of such material In dry weather it may be necessary to top up gulleys with water to permit drainage traps to operate. This can fit in with road-washing operations

WINTER MAINTENANCE PLANT

(i) Specialist salt spreaders
(ii) Salt spreaders—demountable
(iii) Fixed snow plough vehicles (blade)
(iv) Demountable snow blades
(v) Specialist rotary snow ploughs
(vi) Demountable rotary ploughs

The above plant can be used in smaller models for pedestrian precincts.

Maintenance plant

A wide range of plant is used some of which is specialised. The following are examples of the commoner pieces of maintenance plant.

SWEEPER COLLECTORS

Modern vacuum sweeper collectors are mobile selfcontained units. Hopper sizes range from 1 to 4 cubic yards.

GULLY EMPTIERS

Special-purpose vacuum tankers vary in size from 120 gallon capacity trailer mounted or electric pedestrian controller units to the large 1500 gallon vehicle.

INDUSTRIAL TRACTORS

The size varies from 30 h.p. to 65 h.p. with attachments for:

(a) Front loading shovel
(b) Trailer
(c) Grass cutters and hedge trimmers
(d) Verge trimmers
(e) Sweeping brush
(f) Air compressor
(g) Auger

The range of capacities of this equipment is outside the scope of this reference but details are available from the manufacturers.

Summary

The report of the Committee on Highway Maintenance[106] indicates proposed national maintenance standards for the UK—good maintenance extends the life of a road pavement and adds to the convenience and safety of the public using the road. It means:

(a) Day-to-day maintenance to maintain a road in proper condition for the traffic using it. This includes patching, surface dressing, gully emptying, repairs to drainage, kerb and footway maintenance, maintenance of bridges and other structures, embankments and verges, repair and maintenance of traffic lights, carriageway markings, street lighting and street furniture, snow and ice clearance. (See sub-section on winter maintenance.)
(b) Structural work required to enable a road to carry an increased volume or weight of traffic.

Tables 21.36, 21.37 and 21.38 set out the needs in relation to structural works, aids to movement and amenity. Regular inspections and a system to determine priorities for short- and long-term attention, i.e. a maintenance rating system, will be required.

The structural needs of the road including resistance to skidding, surface irregularity, changes in traffic and usage of the road and a 'maintenance' rating[107] for, say, each 500 m length of road require to be derived. Computer programs[108] are being developed to handle maintenance ratings for an extensive road network.

The possibility of skid-resistance checking using the Sideways Force Coefficient Routine Investigation Machine (SCRIM) developed by TRRL—which provides a printout of sideways force coefficient (SFC) at 10 m intervals on a road and can test about 1500 km per year[109]—is being considered in the UK.

The strength of the road structure, and any reinforcement overlay required[110, 111] can be assessed by the Benkelman Beam which measures the deflection of the pavement under a standard wheel load, and consideration is being given to using the deflectrograph originally developed in France[109] which enables speedier measurement of such deflection to be obtained.

MAINTENANCE SYSTEMS

The systems range from the individual or lengthman system, to the collective or 'gang' system, aided as necessary by mechanical equipment and specialised gangs for surface dressing, road markings, lighting, signing, bridge maintenance, etc.

WINTER MAINTENANCE

Setting up a winter maintenance system means determining the standards required then examining the meteorological data for the district so as to establish the period during which the winter maintenance organisation will be required to function. The organisation with its resources and plant will undertake de-icing, gritting, snow clearance and similar work.

MINOR ROADS IN THE UNITED KINGDOM

Minor public roads could be defined as those which are unclassified; about 50% of roads in the UK are of this type.

Minor roads are the capillaries from the main arteries. Without them the quality of life would suffer.

In rural areas

Railway closures have increased the importance and usage of minor roads. The altered pattern of country life has meant larger vehicles for farm deliveries, oil tankers, grain transporters and milk collection. Leisure and tourist traffic seek out the minor roads and by-ways.

Since rural minor roads have derived generally from old tracks in the UK, the biggest problem is to provide adequate pavement strength. Most minor roads rely on ditches for their drainage, and where verges are narrow (Figure 21.28) this is a weakness.

Passing places with gullies and piped surface water (SW) drainage are a cheaper way

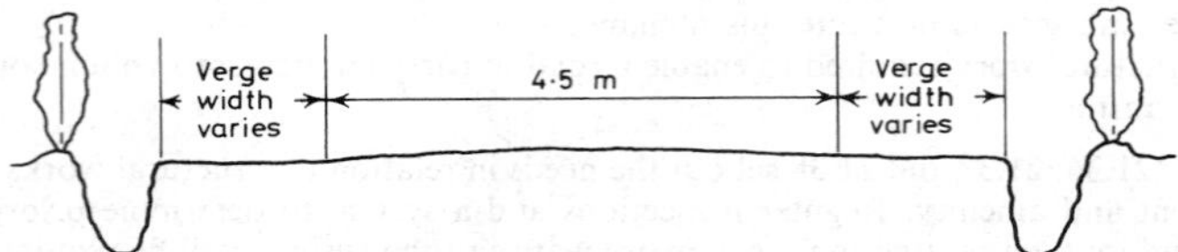

Figure 21.28 Typical minor road in flat country

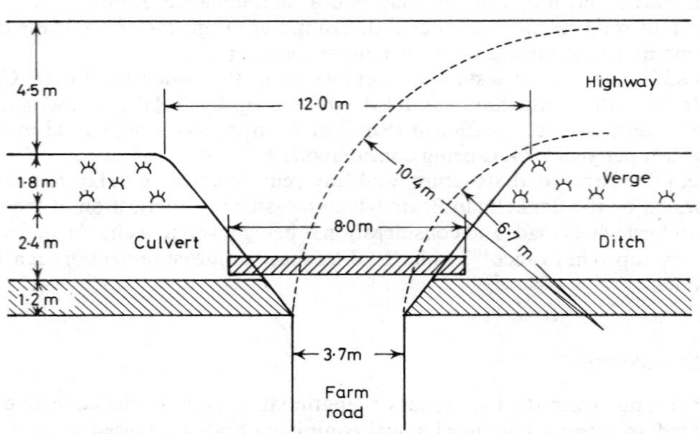

Figure 21.29 Typical farm main entrance

to cater for larger and heavier vehicles than large-scale widening and may be almost as effective.

Adequate visibility which affects safe usage especially at all junctions can very often be achieved at low cost by cutting back hedges and regrading slopes.

In general, road geometry requires 6.71 m minimum radius to cope with articulated vehicles and adequate width of gates—3.66 m where these are on the minor road.

In urban areas

Broadly, there are four main types of urban road:

(i) Primary distributors for long-distance traffic to and from other towns

(ii) District distributors for traffic to residential, business and industrial districts
(iii) Local distributors for traffic to environmental areas (Figure 21.30)
(iv) Access roads which give access from local distributors to buildings, housing estate roads and land within the environmental areas (Figure 21.31).

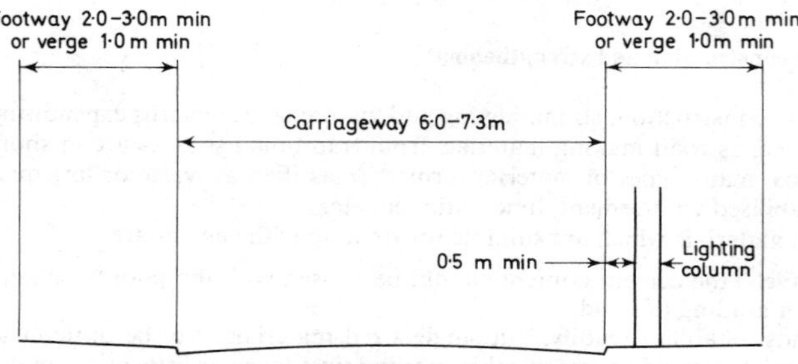

Figure 21.30 Local distributors. Typical cross section

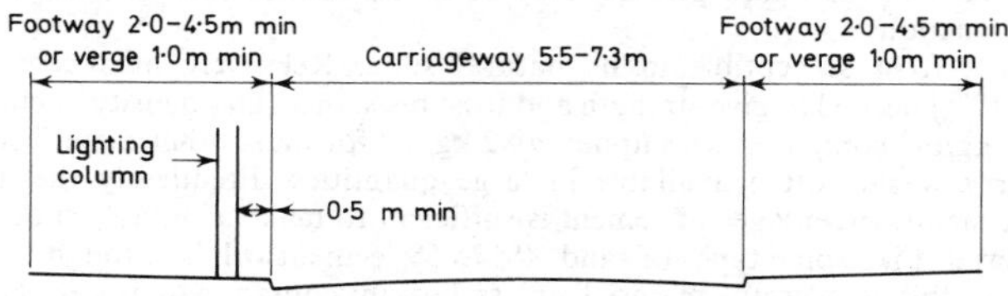

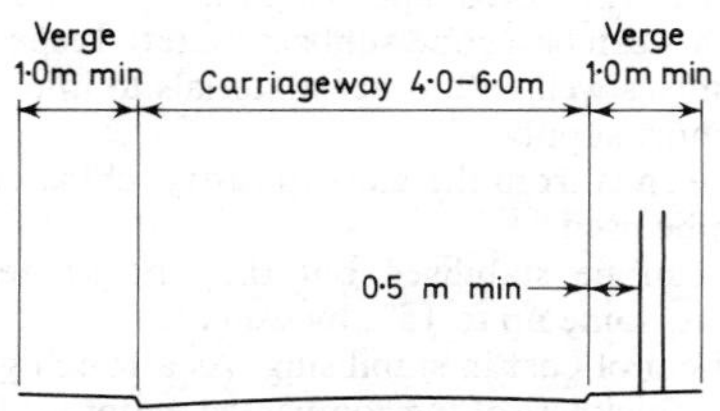

Figure 21.31 Access roads. Typical cross section. (Courtesy: HMSO—from 'Roads in Urban Areas'[43])

In general, access roads (or urban minor roads) are bounded by primary and district distributors and may even be culs-de-sac.

PAVEMENT DESIGN

Carriageway construction should if possible be to the appropriate strength standard as in Road Note 29 (above) for the rest of the road system.

In the case of housing-estate roads:

(1) They should be economic, durable and require little maintenance.
(2) If constructed in advance of the houses the builder should be able to move materials and plant on them in all weathers.

(3) They should possess good skid resistance and a uniform reflecting surface for visibility at night (the latter is particularly apposite to concrete).
(4) If necessary a coloured carriageway should be provided, for architectural or aesthetic reasons.

Pavement construction and strengthening

Low cost in construction, strengthening and maintenance governs expenditure on minor roads, hence, as road-making materials from traditional sources are in short supply in many areas, many types of material formerly classified as waste or low quality can be used if stabilised with cement, lime, bitumen, etc.

Typical materials which are suitable for treating with cement are:

Gravel—the cement content should be chosen with the poorest material or most open grading in mind.

Sands—stabilise readily, but single sized materials may be difficult to compact, in which case add some of other grading sizes (even as little as 5% may do).

Limestone—some are relatively soft and are frost susceptible. When stabilised with cement frost susceptibility may be inhibited. Suitable for subbase and base construction.

Chalk—frost susceptible in its natural state. Relatively high cement content (10–12%) needed to give strength and frost resistance. Dry density of chalk is about 1201 kg/m^3 compared with about 2082 kg/m^3 for most other materials.

Quarry waste—often available in large quantities. Frequently the addition of quite small percentages of cement is sufficient to make a suitable subbase or base material. Like some types of sand, 4% to 5% cement will be enough.

Shale—this is a useful material unless heavily contaminated with sulphates. In Scotland oil shale has been successfully stabilised with the addition of about 6% of cement (M8 and Forth Bridge approaches).

Pulverised Fuel Ash—can be a good subbase material especially if there is a near-by source. PFA also mixes well with other materials to improve a grading deficiency when fines are in short supply.

Industrial wastes—sands from the glass industry, china clay waste, or other hard inert materials can be used.

Soils—almost all can be stabilised but the fine-grained materials need large quantities of cement, some up to 15% by weight.

Cement is the major item of cost in stabilising. As a rough guide each 1% of cement costs about 1p per inch of thickness of the compacted material. The use of poorer quality of materials, wastes, etc., has a dual purpose of permitting better use of resources and also aiding in conservation and clearance of dereliction.

Low-cost roads for a developing country

GENERAL

In most developing countries a new road should be capable of adjustment to meet future needs.

The economic considerations for the solutions of low-cost road construction cannot be dissociated from an appreciation of the risk run when dimensions and characteristics inferior to the usual accepted standards are adopted. The risk itself is dependent strictly on the manner in which the roads are made and are brought into operation.

The benefits of mechanisation are often economically obvious although developing countries will have their own policy on mechanisation versus manual methods. Certain

Table 21.39

Item	*Nature of work*	*Operations*	*Classification*
1	Clearing operation	Preparation of ground	B
2	Earthwork	Excavation	B
		Transportation: small leads	B
		long leads	C
		Compaction	C
3	Soil stabilisation	Borrowing of soil	B
		Pulverising of soil	B
		Adding stabiliser to the soil	B
		Dry mixing	B
		Adding required moisture	B
		Wet mixing	B
		Compaction	C
4	Granular base and surface courses	Laying of aggregate	B
		Adding filler material	B
		Watering	B
		Rolling	C
5	Bituminous pavements		
	(a) Surface dressing	Spraying of bitumen	B
		Spreading of stone chips	B
		Rolling	C
	(b) Open graded premix surfacings	Mixing	C
		Laying	B
		Rolling	C
	(c) Grouted macadam	Laying of aggregate	B
		Spraying of bitumen	B
		Rolling	C
	(d) Asphaltic concrete	Mixing	C
		Laying	C
		Rolling	C
6	Maintenance		
	(a) Earth or gravel roads	Routine maintenance of surfacing	B
		Regrading	C
		Regravelling	B
	(b) Surfaced roads	Patching	A
		Resurfacing	B
	(c) All roads	Routine maintenance of verges	B
		Grassed verges	B
		Side drains and ditches	B
		Culverts	A
7	Quarrying and crushing		
	(a) Large quarrying	Quarrying crushing	C
	(b) Small quarrying	Quarrying crushing	B
	(c) Small-scale crushing of graded stone		C
	(d) Breaking into single size stone		B

Classification
A Work that can be better done by hand than by machine.
B Work that can be effectively done by hand or alternatively by machine.
C Work that can be satisfactorily done only by machine.

constructional operations may not, however, lend themselves to reasonable quality of work by manual methods.

Table 21.39 is a summary of acceptable alternatives.

Development stages The timing of stages will depend on the development of traffic in the particular country, its geography, climate, etc. The stages can be broadly classified as follows:

(i) *Track or unimproved road* usable by motor traffic but frequently affected by water.

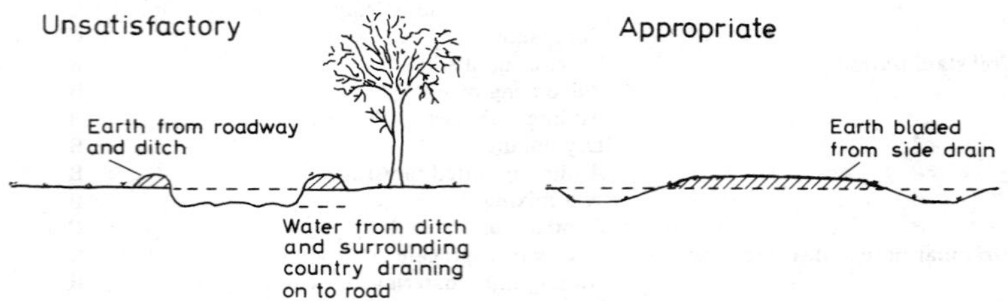

Figure 21.32 Correctly and incorrectly maintained earth track

(ii) *Earth road or improved road* consists of drained roadway with (probably) permanent structures at all watercourses.

(iii) *Paved road surfaced* in black top or concrete. Permanently open to traffic.

Geometric design Good alignment of a new road is essential; poor alignments are usually old tracks opened to motor traffic as the cheapest immediate solution.

ROUTE SELECTION

Route selection should take into consideration, *inter alia*:

(1) Topography and hydrology of area
(2) Nature of soils and availability of materials
(3) Need to serve the population (traffic)
(4) Climate

PAVEMENT DESIGN

A design life of 10 years is usually appropriate with an assessment of what the traffic will be over this period and the growth factor thereafter.

The graph (Figure 21.34) shows the relationship between the number of times a 'standard' axle load is applied and the equivalent axle load. The graph is based on a table of equivalent axles prepared by AASHO* and can be used to determine the effect on the road of vehicles which are heavier than the normal traffic of the area.

For example, in an area where heavy equipment is likely to be transported by road such as oil drilling rigs, power station equipment, transformers, turbines, etc., it would be wise to allow for such extraordinary traffic.

Table 21.40 DESIGN CHARACTERISTICS FOR ROADS IN DIFFERENT TYPES OF TERRAIN

Chacteristic road type	*Terrain*	*Design speed* km/h	*Minimum radius of curvature* m	*Maximum gradient* %	*Maximum length of grade* m	*Formation width (Permanent surfacing and shoulders)* m	*Width of permanent surfacing* m
Primary	Flat to rolling	80–110	190–360	4	None	10–13	6–7.5
W = 50 m	Hilly	55–80	90–190	5–7	600 over 4%	10–13	—
	Mountainous	40–55	50–90	7–9	400 over 6%	8–10	—
Secondary	Flat to rolling	60–80	110–190	5	None	10–12	6–6.8
W = 30 m	Hilly	50–60	75–110	5–7	None	10–12	—
	Mountainous	35–50	35–75	7–9	750 over 6%	8–9	—
Feeder	Flat to rolling	50–60	75–110	7	None	7.5–8	5.5–6
W = 20 m	Hilly	35–50	35–75	7–9	None	7.5–8	—
	Mountainous	25–35	30–35	9–12	1000 over 9%	7.5–8	—

Allows for superelevation of 10%.
W = right of way to allow for construction, maintenance, future widening.

Maintenance and strengthening Because of their inherent weakness low-cost roads must be constantly maintained and more regularly inspected than normal highways since they are gradually called upon to carry greater volumes of traffic than those for which they had been built, even taking into account their lower safety coefficient.

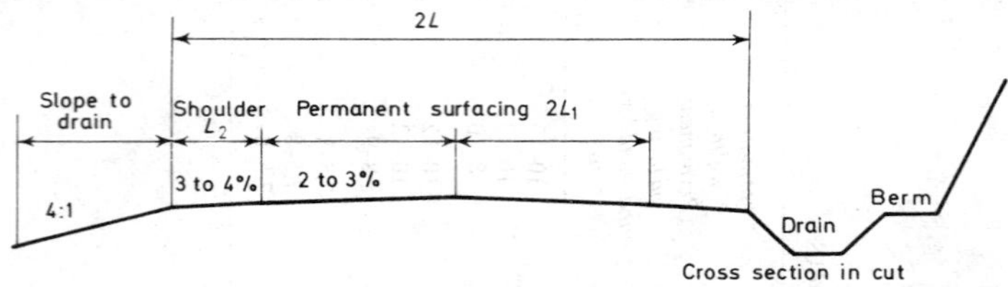

	Flat, rolling or hilly terrain			Mountainous terrain		
	$2L$ (m)	$2L_1$ (m)	L_2 (m)	$2L$ (m)	$2L_1$ (m)	L_2 (m)
Primary roads	10–13	6–7.5	⩾2.0	8–10	6–7.5	⩾1.0
Secondary roads	10–12	6–6.8	⩾2.0	8–9	6–6.8	⩾1.0
Feeder roads	7.5–3.0	5.5–6.0	⩾1.0	7.5–8.0	5.5–6.0	⩾1.0

Figure 21.33 Typical standard cross-section of road. Formation width 2L

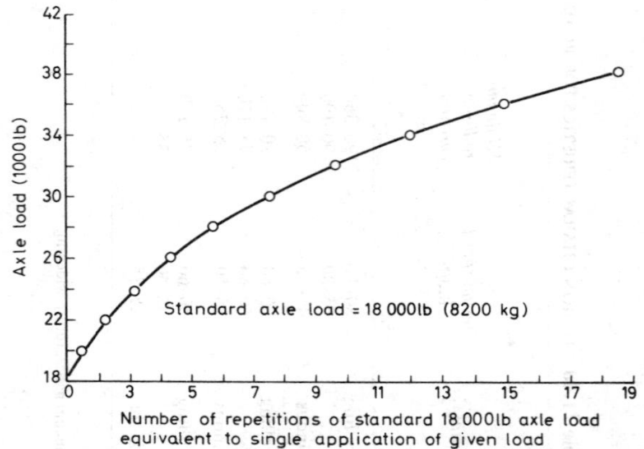

Figure 21.34 Equivalence factor and damaging power of different axle loads. (Based on AASHO data). (1lb = 0.4536 kg)

The measurement of surface deflections by Benkleman Beam remains the most effective method of estimating the residual strength of pavements under standard loading.

STATUTORY UNDERTAKERS' APPARATUS IN THE ROAD

Introduction

Over the years, an intricate network of mains and services have been laid in the public highways in the UK and the operation of this apparatus necessitates frequent disturbance of the highways. The installation of new services and the maintenance of the apparatus and the highway will cause considerable temporary obstructions and even permanent damage to the road structure unless correct reinstatement procedures are adopted.

Legal position

The legal position in Britain regarding apparatus owned by Statutory Undertakers laid in public highways is defined in the Highways Acts 1959, 1969, 1971, the Pipelines Act 1967 and the Public Utilities Street Works Act 1950.

(A) ALL-PURPOSE ROADS

For all-purpose roads the Statutory Bodies have rights under the Highways Acts to install apparatus within the highway boundaries subject to the approval of the Highway Authority for the actual location of the service.

(B) MOTORWAYS AND SPECIAL ROADS

The position regarding motorways is somewhat different to that pertaining to other public highways.

With the exception of the Post Office, Statutory Undertakers have no rights to lay apparatus on, under, or over land along the route of a motorway or special road although permission may be granted under special conditions.

(C) PUBLIC UTILITIES STREET WORKS ACT 1950

This defines the procedure to be adopted wherever works are carried out in a highway. A brief outline of the four parts of the Act and of the general procedure to be adopted by Highway Authorities and Undertakers is as follows:

(i) *Part 1* (Sections 3–14 and 1st, 2nd and 3rd Schedules) protects the highway authorities responsible for streets, etc., when Statutory Undertakers exercise their own statutory powers to open streets, etc.
(ii) *Part 2* protects Statutory Undertakers where their apparatus is affected by road, bridge or transport works.
(iii) *Part 3* is a miscellaneous section and deals with the effect of one Statutory Undertaker on another, road closures, etc.
(iv) *Part 4* covers financial provisions, special application of the Act in London and Scotland, etc.

Location of services

It is of advantage if the location of each individual service relative to apparatus owned by other Statutory Bodies is consistent, as for example in Figure 21.35.[112]

CROSSING OF ROAD WORKS

(i) Services at crossing locations should be placed at heights and depths which ensure the protection of the apparatus and the safety of the traffic.

(ii) Depths should be such as to provide not less than 300 mm of cover below the *formation* (i.e. the underside of the subbase) of the highway pavement.

(iii) Crossings of new roads by overhead cables should provide the clearances from finished level shown in Table 21.41.

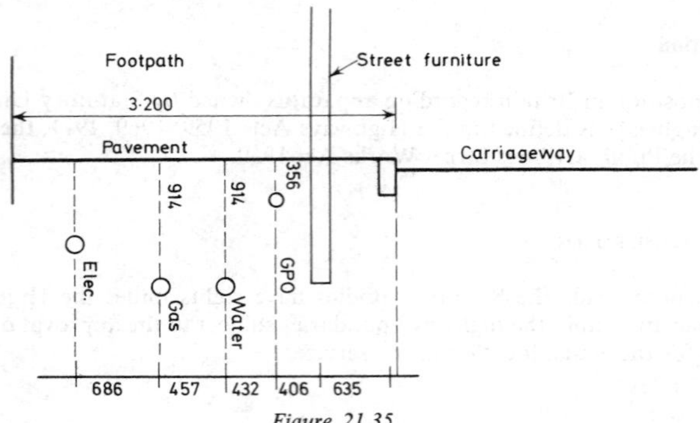

Figure 21.35

Table 21.41

Service	Minimum height above finished road level (m)
Electricity (overhead lines)[113]	
(i) 33 kV and lesser voltages	5.8
(ii) Exceeding 33 kV to 66 kV	6.0
(iii) Exceeding 66 kV to 132 kV	6.7
(iv) Exceeding 132 kV to 275 kV	7.0
(v) Exceeding 275 kV	7.3
Telephone lines	
(i) Motorways, trunk roads, principal roads, main urban roads	6.8
(ii) Minor roads	5.6

(iv) With all apparatus it is desirable that future maintenance may be carried out without necessitating access to the highway. Underground cables across motorways should normally be placed in sleeved ducts with access facilities located outside the highway boundary fences.

Reinstatement

The Statutory Undertaker is obliged to provide interim restoration of openings to the original surface levels to the requirements of the Highway Authority. Permanent restoration is normally to be carried out by the Highway Authority.

Programming

Statutory Undertakers' works and alterations must be completed to a rigid, restrictive and predetermined programme because of their responsibility to consumers. This often means altering existing apparatus prior to the commencement of the Highway Contract works.

Finance

Costs arising from the diversion of existing apparatus owned by Statutory Undertakers which must be altered owing to the roadworks, falls on the Highway Authority, except that the Public Utilities Street Works Act provides for offset payments in respect of:

(1) Betterment—i.e. where different or improved apparatus is requested by the Undertaker.
(2) Deferment of renewal of the apparatus.

ROAD RESEARCH

Research on highway engineering and associated subjects in the UK is focused on the Department of the Environment's Transport and Road Research Laboratory (TRRL) based at Crowthorne (Berkshire), and at Livingston (West Lothian).

The laboratory comprises five divisions (Design, Construction, Traffic, Safety and Administration and Services) with a subdivision into several sections. A number of external Advisory Committees exist to advise on the various aspects of research and these include leading engineers and scientists.

Apart from work carried out at the Laboratory, research is sponsored at Universities and other research establishments, let out as industrial research contracts or conducted in cooperation with industry.

Results of research are issued as TRRL Reports, papers and articles in the scientific press and as 'Road Notes'. Each year's work is summarised in the annual publication 'Road Research' (published by HMSO).

ACKNOWLEDGEMENTS

Acknowledgement is made to the Director of the Transport and Road Research Laboratory, Crowthorne, Berks, for permission to publish extracts from Laboratory Reports and other TRRL publications, and to the Controller, Her Majesty's Stationery Office, for permission to reproduce certain illustrations.

REFERENCES

1. PRATT and MACKENZIE *Law of Highways*, 21st edn, Ed. H. Parrish and Lord de Mauley, Butterworths, London (1967)
2. *Site Investigation*, CP 2001, British Standards Institution, London (1957)
3. *Traffic Prediction for Rural Roads*, HMSO (1968)
4. WELLS, G. R., *Traffic Engineering—An Introduction*, Charles Griffin & Co. (1970)
5. *Research on Road Traffic*, TRRL (1965)
6. *Urban Traffic Engineering Techniques*, HMSO (1965)
7. GOOD, G. E., 'A Gravity Distribution Model', *Traffic Engineering and Control*, **13** (Dec. 1971)
8. BRUTON, M. J., *Introduction to Transportation Planning*, Hutchinson (1970)
9. *Comparative Analysis of Traffic Assignment Techniques with Actual Highway Use*, Highway Research Board, USA (1968)
10. BURRELL, J. E., *Zoning and Network Coding for Traffic Studies*, PTRC Symposium (Feb. 1972)
11. BURRELL, J. E., *Distribution Models and their Calibration*, PTRC Symposium (Feb. 1972)
12. DAVIES E., Ed., *Traffic Engineering Practice*, 2nd edn, Spon London (1968)

13. DUMBLETON, M. J. and WEST, G., *Air Photograph Interpretation for Road Engineers in Britain,* TRRL Rep. LR 369,(1970)
14. Topographical Survey Tender Document, issued by Department of the Environment (Aug. 1970)
15. *See reference* 2
16. WEEKS, A. G. 'Soil Survey Procedures in Highway Design', *Instn Highway Engrs,* **18**, No. 1, 5–11 (Jan. 1971)
17. Ministry of Transport (DOE) Specification for Roads and Bridgeworks, HMSO (1969)
18. Plans of Abandoned Mines—Scales—various, Divisional Plans Record Offices of the National Coal Board
19. *Soil Mechanics for Road Engineers,* TRRL (1952)
20. GAFFNEY, J. A. and HENRY, J. K. M., 'Special Problems Encountered in the Design & Construction of the M62 in West Riding and the M6 in Westmorland', Motorways Conference, Institution of Civil Engineers, London, Paper 9 (Apr. 1971)
21. HOGBIM, L. E. 'Snow—Fences', TRRL Rep. LR 362
22. CROWE, S. *Landscaping of Roads,* London Architectural Press (1960)
23. SPEARING, G. D., and PORTER, M. R., 'Rural Motorways and The Rural Environment', Proc. *Instn civ. Engrs,* Paper 7304 (Oct. 1970)
24. 'Motorways in Britain Today and Tomorrow', *Proc. Conf. Instn civ Engrs,* London (1971)
25. *See reference* 20
26. WOOD, A. A., 'Urban Transportation and the Environment', *J. Instn Highway Engrs,* **19**, No. 5,
27. WILSON, T. D., 'The Engineering of Conservation', *J. Instn Highway Engrs,* **17**, No. 7, 27 (July 1970)
28. MOORE, J. E., *Design for Noise Reduction,* Architectural Press
29. LANGDON, F. J., and SCHOLES, W. E., 'The Traffic Noise Index: A Method of Controlling Noise Nuisance', BRS Current Paper 38/68
30. SCHOLES, W. E., and SARGENT, J. W., 'Designing against Noise from Road Traffic', BRS Current Paper 20/71
31. RICHARDS, E. J., and CROOME, D. J., 'The Problem of Traffic Noise', *Proc. Int. Conf. on Roads in the Landscape* (July 1967)
32. MAEKAWA, Z., 'Noise Reduction by Screens', *Appl. Acoustics,* **1**, No 157 (1968)
33. SCHOLES, W. E., SALVIDGE, A. C., and SARGENT, J. W., 'Field Performance of a Noise Barrier', BRS Current Paper 24/71
34. 'Insulation against External Noise–1' BRS Digest 128 (April 1971)
35. 'Insulation against External Noise–2', BRS Digest 129 (May 1971)
36. 'Double Glazing and Double Windows', BRS Digest 140 (April 1972)
37. Government Committee on the Problem of Noise, Final Report, Command Paper 2056, HMSO (1963)
38. Urban Motorways (Noise), Parliamentary Statement: Peter Walker (Secretary of State for the Environment) Hansard (24 June 1971)

38a. 'New Housing and Road Traffic Noise', Design Bulletin 26, Building Research Station

38b. National Physical Laboratory reports: AC 54, AC 56, AC 57, AC 58

39. TRRL, LR 286 (1969)
40. *Layout of Roads in Rural Areas,* Ministry of Transport, Scottish Development Department, The Welsh Office, HMSO (1968)
41. *Layout of Roads in Urban Areas,* HMSO (1966)
42. *Urban Traffic Engineering Techniques,* Ministry of Transport, Scottish Development Department, HMSO (1965)
43. *Roads in Urban Areas,* Ministry of Transport, Scottish Development Department, The Welsh Office, HMSO (1966)
44. TANNER, J. C., *A Theoretical Analysis of Delays at Uncontrolled Intersections,* Biometrika (1962)
45. WEBSTER, F. V., and NEWBY, R. F., 'Research into the Relative Methods of Roundabouts and Traffic-Signal-Controlled Intersections', *Proc. Instn civ. Engrs,* **27**, 47 (January 1964)
46. 'Junction Designs', *Technical Memorandum H7/71,* Traffic Engineering Division, Department of the Environment (June 1971)
47. BENNETT, R. F., 'The Designs of Roundabouts since the Priority Rule', *J. Instn Highway Engrs,* **18**, No. 9, 13–23 (September 1971)
48. *Earthworks,* CP 2003, British Standards Institutions, London (1959)
49. Ministry of Transport (DOE) Specification for Roads and Bridgeworks, 4th edn, HMSO (1969)

50. *Methods of Testing Soils for Civil Engineering Purposes*, BS1377, British Standards Institution, London (1967)
51. SYMONS, 'The Application of Residual Shear Strength to the Design of Cuttings in Overconsolidated Fissured Clays', Road Research Laboratory Report LR 227
52. HOEK, E., 'Rock Slope Stability in Open Cast Mining', Imperial College Rock Mechanics Progress Report No. 4 (July 1970)
53. 'A guide to the structural design of pavements for new roads', TRRL Road Note 29, 3rd edn, (1970)
54. *Rolled Asphalt (Hot Process)*, BS 594, British Standards Institution, London (1961)
55. 'An experiment comparing the performance of roadstones in surface dressing', TRRL Rep. No. 46,(1966)
56.–57. Road Notes 1 and 38 now replaced by Road Note 39, 'Recommendations for Road Surface Dressing', TRRL (1972)
58. BILHAM, E. G., *Classification of Heavy Falls in Short Periods*, HMSO (1962)
59. *Floods in Relation to Reservoire Practice*, Institution of Civil Engineers (1960)
60. 'A Guide for Engineers to the Design of Storm Sewer Systems', TRRL Road Note 35, (1963)
61. 'Tables for the Design of Storm Drains, Sewers and Pipe-Lines', Hydraulics Research Paper No. 4, Hydraulics Research Station, HMSO (1969)
62. 'Loading Charts for the Design of Buried Rigid Pipelines', National Building Studies–Special Report 37, HMSO (1966)
63. 'The Hydraulic Efficiency and Spacing of BS Road Gulleys', TRRL Rep. LR277,(1969)
64. 'Selection of Materials for Subsurface Drains', TRRL Rep. LR346,(1970)
65. *Glossary of Terms Relating to Timber Preservation*, BS 4261, HMSO (1968)
66. *Coal Tar Creosote for Preservation of Timber*, BS 144, HMSO (1954)
67. *Pressure Creosoting of Timber*, BS 913, HMSO (1954)
68. 'Provisional Specification for a Wire Rope Crash Barrier', TRRL Rep. LR 98,(1968)
69. 'Instructions for Using RRL Post Setting Rigs when Erecting Tensioned Beam Crash Barriers', TRRL Rep. LR 178,(1968)
70. 'Impact Test on a Modified Christiani-Nielson Crash Barrier', TRRL Rep. LR 246,(1969)
71. 'Specification and Installation Procedure for the RRL Tensioned Corrugated Beam Safety Fence', TRRL Rep. LR 278, HMSO (1969)
72. 'A Post Driving Technique for the Erection of Tensioned Beam Crash Barriers', TRRL Rep. LR 338,(1970)
73. *Street Lighting*, CP 1004, parts 1–9, British Standards Institution, London (1963)
74. 'The Traffic Signs Regulations and General Directions 1964', Statutory Instrument No. 1857, HMSO (1964)
75. *Traffic Signs Manual*, Ministry of Transport, HMSO (1965)
76. 'The Traffic Signs (Speed Limits) Regulations and General Directions 1969', Statutory Instrument No 1487, HMSO (1969)
77. *The Construction of Road Traffic Signs and Internally Illuminated Bollards*, BS 873, British Standards Institution, London (1970)
78. *Research on Road Traffic*, TRRL,(1965)
79. *Specification for Road Marking Materials (Superimposed Type)* (at present under revision), BS 3262, part 1, British Standards Institution, London (1970)
80. *See reference* 50
81. *See reference* 53
82. *Methods for Sampling and Testing of Mineral Aggregates, Sands and Fillers*, BS 812, British Standards Institution, London (1967)
83. *Specification for Aggregates from Natural Sources for Concrete*, BS 882, British Standards Institution, London (1965)
84. *Portland Cement (Ordinary and Rapid Hardening)*, BS 12, British Standards Institution, London (1958)
85. *Sampling and Examination of Bituminous Mixtures for Roads and Buildings*, BS 598, British Standards Institution, London (1958)
86. 'Rapid Methods of Analysis for Bituminous Road Materials', TRRL Road Note 10, (1967)
87. *Methods of Testing Concrete*, BS 1881, parts 1–6, British Standards Institution, London (1970–72)
88. *See reference* 17
89. MILLARD, R. S., and LISTER, N. W., 'The Assessment of Maintenance needs for Road Pavements', *Proc. Instn civ. Engrs*, **48**, 223–244(Feb. 1971)

90. *See reference* 19
91. 'Mechanisation for Road and Bridge Construction', Institution of Civil Engineers/Institution of Highway Engineers Conference, London (1972)
92. PARSON, A. W., and BROAD, B. A., 'Belt Conveyors—Feasibility Studies', TRRL Reps LR 336, LR 337 (1970)
93. RUSSELL, W. L., 'A Machine for Spreading Coated Chippings on Rolled Asphalt', *Roads and Road Construction,* **41,** 70–73 (March 1963)
94. GEDDES, S., *Building and Civil Engineering Plant,* Crosby/Lockwood, London (1951)
95. Manufacturers' Recommendations—see relevant technical brochures
96. BURKS, A. E., 'Plant and Methods for Concrete Road Construction', Concrete Society Meeting, London (1967)
97. *Urban Traffic Engineering Techniques,* Ministry of Transport/Scottish Development Dept, HMSO (1965)
98. DUFF, J. T., *Traffic Engineering Practice,* Chapt. V., Ed. E. Davies, Spon, London (1968)
99. Papers on Area Traffic Control, PTRC Symposium on Traffic Management, London, 19–22 October 1971
100. *Parking in Town Centres*, Ministry of Housing and Local Government and Ministry of Transport, HMSO (1965)
101. Road Traffic Regulation Act 1967, Chapter 76, HMSO (1967)
102. Transport Act 1968, HMSO (1968)
103. The Local Authorities Traffic Orders (Procedure) England and Wales Regulations 1969, HMSO (1969)
104. *Traffic Management and Parking Manual*, Ministry of Transport, HMSO (1969)
105. *Traffic Signs Manual*, Ministry of Transport, HMSO (1965)
106. *Report of Committee on Highway Maintenance*, Chap. 4 'Standards' and Appendix (1), 'Proposed Initial Standards of Maintenance, HMSO (1970)
107. *Report of Committee on Highway Maintenance*, Chap. 5 and Appendix (2): 'Maintenance Rating System', HMSO (1970)
108. Planning and Transportation Research and Computer Co Ltd—Seminar Proceedings: Maintenance Rating Schemes, 12 November 1971 (1971)
109. 'Methods of Ascertaining Structure and Maintenance Needs of Highways', TRRL Rep. to be published in 1973
110. *See reference* 89
111. *See reference* 53
112. *Report of the Joint Committee on Location of Underground Services*, Institution of Civil Engineers (First published 1946, revised July 1963)
113. Electricity (Overhead Line) Regulations 1970
114. *Specification for a Maintenance Rating System using Histograms,* TRRL, to be published in 1973
115. *British Integrated Highway Design Program System,* Highway Engineering Computer Branch, Department of Environment, London
116. CRISSWELL, H., *Highway Spirals Superelevation and Vertical Curves,* 3rd edn, Carriers Publishing Co. (1967)

22 AIRPORTS

AIRPORTS 22

22 AIRPORTS

E. V. FINN, C.Eng., F.I.C.E., F.I.Struct.E., F.R.S.H., M.Cons.E.
and
A. D. TOWNEND, B.Sc.(Eng.), A.C.G.I., D.I.C., C.Eng., F.I.C.E.
Sir Frederick Snow & Partners

INTRODUCTION

Rapid developments in airport design have taken place in recent years. This has been made necessary by the considerable increase in the size and weight of the aircraft in use; by the number of aircraft movements, and also by the corresponding increase in the number of passengers and variety of services required at a modern airport.

Large airports are no longer merely landing strips for aircraft, but rather, they must be planned as transportation interchanges. The various surface transport routes for motor cars, buses and perhaps trains, and the air lanes, all converge at the airport. Here, all necessary facilities must be provided for the passengers and their baggage, for cargo, for the parking and servicing of all the types of vehicles involved, as well as for the airport staff. Indeed airports require many of the facilities of a town, and airport design involves many aspects of civil engineering such as design of load-bearing pavements, roads, storm-water drainage, water supply, fire mains, fuel installations, buildings varying from large steel- or concrete-framed terminals, administration blocks, hangars and maintenance bases, to small brick buildings housing radar and other equipment. Electrical and mechanical engineering works are involved with the supply of power, lighting, heating and ventilating to buildings, as well as airfield and approach lighting.

The methods of design of many of these items differ in no way from methods to be adopted where these structures are unconnected with an airport. This section will therefore deal only with those aspects of design particular to airports and airfields.

GENERAL RECOMMENDATIONS AND REQUIREMENTS

Details of international requirements for the layout of airfields are covered in the International Civil Aviation Organisation (ICAO) Standards and Recommended Practices for Aerodromes, Annex 14[1] and this publication is revised periodically. Any aerodrome (airfield or airport) that is not State owned and operated requires a licence to accept a commercial service. The technical and other requirements for the licensing of a site as an aerodrome in the UK are incorporated in Civil Aviation Publication CAP 168: Licensing of Aerodromes[2] published by the Civil Aviation Authority (CAA). In general this conforms with and amplifies the information given in ICAO Annex 14, except for certain modifications which have been found appropriate to aerodromes in the UK.

The following definitions are taken from the above publications.

Aerodrome

Any area of land or water designed, equipped, set apart or commonly used for affording facilities for the landing and departure of aircraft and includes any area or space, whether on the ground, on the roof of a building or elsewhere, which is designed, equipped or set apart for affording facilities for the landing and departure of aircraft capable of descending or climbing vertically, but shall not include any area the use of which for affording facilities for the landing and departure of aircraft has been abandoned and has not been resumed.

Aerodrome beacon

Aeronautical beacon used to indicate the location of an aerodrome.

Aerodrome elevation

The elevation of the highest point of the landing area.

Aerodrome reference point

The designated geographical location of an aerodrome.

Apron

A defined area on a land aerodrome, intended to accommodate aircraft for the purpose of loading or unloading passengers or cargo, refuelling, parking or maintenance.

Barrette

Three or more aeronautical ground lights closely spaced in a transverse line so that from a distance they appear as a short bar of light.

Clearway

A rectangular area at the end of the take-off run available and under the control of the aerodrome licensee, selected or prepared as a suitable area over which an aircraft may make a portion of its initial climb to a specified height.

Crosswind component

The velocity component of the wind measured at or corrected to a height of 10 m above ground level at right angles to the direction of take-off or landing.

Instrument approach runway

A runway intended for the operation of aircraft using nonvisual aids providing at least directional guidance in azimuth adequate for a straight-in approach.

Noninstrument runway

A runway intended for the operation of aircraft using visual approach procedures.

Runway selected basic length

The length selected as a basis for the design of a runway and associated physical characteristics of the land aerodrome. It should be the length that would be required at a level site at sea level in standard atmospheric conditions and in still air to meet the needs of the aircraft for which the runway is provided in order to comply with the relevant maximum cross wind components.

Runway actual length

The basic length of runway corrected to take account of elevation, temperature and humidity.

Shoulder

An area adjacent to the edge of a paved surface so prepared as to provide a transition between the pavement and the adjacent surface for aircraft running off the pavement.

Stopway

A defined rectangular area at the end of the take-off run available, prepared and designated as a suitable area in which an aircraft can be stopped in the case of a discontinued take-off.

Strip

An area of specified dimensions enclosing a runway to provide for the safety of aircraft operations.

Taxiway

A defined path, on a land aerodrome, selected or prepared for use of taxiing aircraft.

Threshold

The beginning of that portion of the runway usable for landing.

In addition to the official definitions given above, other terms are used and some of these, together with explanations, are given below.

Airport reference letters

Reference code letters A, B, C, D or E may be assigned to airports, depending on main runway lengths and other physical requirements. Although reference should be made to ICAO Annex 14[1] or the CAA CAP 168[2] for the detailed standards and recommendations regarding airport layout, the recommendations for lengths, clearances and for the vertical alignment of runways and taxiways given by Annex 14 are summarised in Table 22.1.

Obstruction surface

An imaginary surface which extends over the whole area occupied by the airport and extends beyond its limits is defined. It is necessary to restrict the creation of new objects, and to remove or mark existing objects, whether man-made or naturally occurring, which project above this imaginary surface.

Table 22.1. ANNEX 14: RECOMMENDATIONS (ALL DISTANCES IN METRES)

Runway code letter for longest runway	A	B	C	D	E
Runway selected basic length	2 100 and over	2 099–1 500	1 499–900	899–750	749–600
Minimum width of runway	45	45	30	23	18
Minimum width of taxiways	23	23	15	10	7.5
Distance between any point on the edge of a taxiway and the edge of a runway					
– Instrument runway	150	150	150	—	—
– Other runway	75	73	73	36	29
Distance between any point on the edge of one taxiway and the edge of another taxiway	62	52	43	27	23
Distance between any point on the edge of a taxiway and a fixed obstruction	38	30	26	18	16
Vertical alignment of runways					
Maximum effective slope	1%	1%	1%	2%	2%
Maximum slope	1.25%	1.25%	1.5%	2%	2%
Maximum change between consecutive slopes	1.5%	1.5%	1.5%	2%	2%
Maximum rate of change of slope per 30 m	0.1%	0.1%	0.2%	0.4%	0.4%
Minimum distance between successive points of intersection of vertical curves is the sum of the absolute numerical values of the corresponding grade changes multiplied by the factor given	30 000	30 000	15 000	5 000	5 000
Maximum transverse slope	1.5%	1.5%	1.5%	2%	2%

A plan view of this imaginary surface is shown in Figure 22.1 and its components are:

1 An inner horizontal surface located 45 m above the airport elevation extending to a horizontal distance *a* measured from the aerodrome reference point. (Distances vary according to the code letter of the aerodrome and are given in Table 22.2.)

2 A conical surface with a slope of 5% above the horizontal and with outer limits contained in a horizontal plane located at a height *b* above the inner horizontal surface.

3 Transitional surfaces established for each runway used for landing aircraft. The slope of these surfaces shall be *c*% and the outer limit shall be determined by its intersection with the inner horizontal surface.

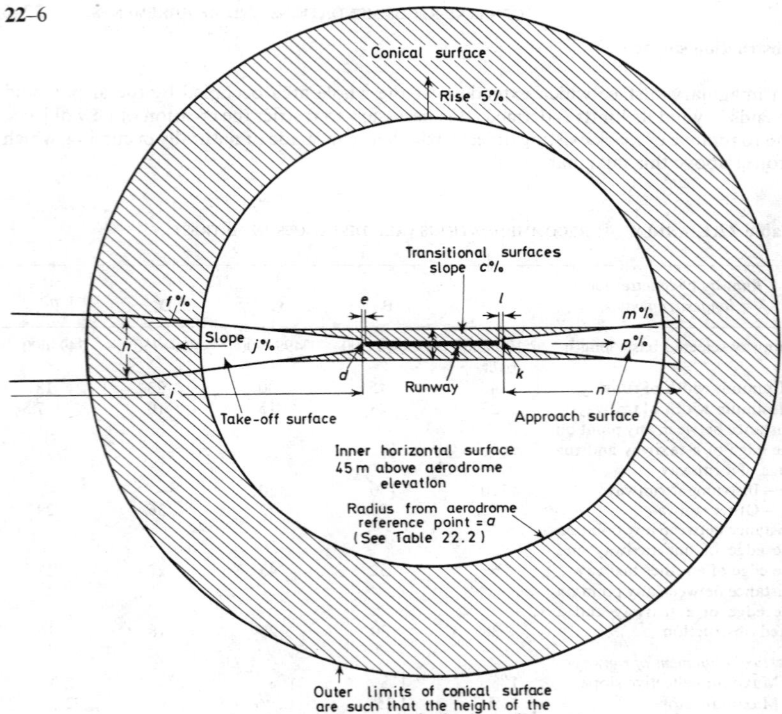

Figure 22.1 Plan view of obstruction surface

Table 22.2. DIMENSIONS (IN METRES) ASSOCIATED WITH INNER HORIZONTAL CONICAL, TRANSITIONAL, TAKE-OFF AND APPROACH SURFACES

Surface	*Dimension*	*Runway code letter*				
		A	B	C	D	E
Inner horizontal	*a*	4 000	4 000	4 000	2 500	2 000
Conical	*b*	100	100	75	55	35
Transitional	*c*	14.3%	14.3%	14.3%	20%	20%
Take-off	*d*	180	180	180	80	60
(main take-off runways only)	*e*	60	60	60	30	30
	f	12.5%	12.5%	12.5%	10%	10%
	h	1 200	1 200	1 200	580	380
	i	15 000	15 000	15 000	2 500	1 600
	j	2%	2%	2%	4%	5%
		Instrument approach	*Other approach surfaces*			
		A, B, C	A, B	C	D	E
Approach	*k*	300	150	150	80	60
	l	60	60	60	30	30
	m	15%	10%	10%	10%	10%
	n	15 000	3 000	3 000	2 500	1 600
	p	2%	2.5%	3.33%	4%	5%
	q	2.5%	—	—	—	—

4 *Take-off surfaces* established for each runway direction used for the take-off of aeroplanes. The limits of the take-off surfaces are determined by an inner edge, two diverging sides and an outer edge, the inner and outer edges being perpendicular to the flight path. The inner edge has a length d and is at the end of the clearway or stopway if provided, or is at a distance e from the end of the runway. Each side diverges at a rate of $f\%$ relative to the flight path centreline until a specified maximum lateral width h is reached, continuing thereafter at that width to the outer edge. The distance between the inner and outer edges, or length of the take-off surface, is i and the surface slopes up at $j\%$ to the horizontal. The figures given in Table 22.2 refer only to the main take-off runway at Code A, B or C aerodromes, and Annex 14 should be referred to for other take-off runways.

5 *Approach surfaces* established for each runway direction used for the landing of aeroplanes. Like the take-off surfaces, the approach surfaces are bounded by an inner edge, two diverging sides, when viewed from the runway, and an outer edge. The inner edge of length k is located at a distance l from the runway threshold. Each side diverges at a rate $m\%$ from the extended centreline of the runway to the outer edge and the length of the approach surface is n. The slope of the surface above the horizontal is $p\%$ although the slope of the outer 3000 m of Instrument Approach Surfaces on code A, B or C aerodromes is at $q\%$.

AIRPORT LOCATION

The factors affecting the choice of a site for an airport are now much more complex and some are interrelated. The impact of an airport of even moderate size on the surrounding communities is such that each factor, whether physical or economic, must be considered carefully and, if possible, it should be quantified, usually in money terms, so that alternatives can be compared. Cost/benefit analysis is a valuable tool in assisting the choice of location.

Some of the factors which are mentioned in the following paragraphs will not directly affect the civil engineer but he should be aware of them.

Passenger catchment area

It is usual to try and define a catchment area for the passengers who are likely to want to use the airport. A major international airport such as Heathrow (London) will attract passengers from many parts of the country by virtue of the variety of services it provides but a regional airport such as Newcastle will have a relatively well defined catchment area.

Where regional airports or smaller ones are concerned, a travel time of 45 min by car is commonly accepted as reasonable. This figure is valid for the UK and similar countries but would require local considerations to be taken into account elsewhere. Mean travel speeds of 80 km/h for motorways; 70 km/h for dual carriageway 'A' class, 55 km/h single carriage 'A' class and 30 km/h for urban areas can be assumed for guidance.

The presence of a competing airport within about $1\frac{1}{2}$ h travel time will modify the catchment area. The plotting of journey time contours and time indifference lines is the basis of the technique used to assess airport demand.

Environmental aspects

An airport has two main effects on the environment through noise and land use.

Aircraft engines have increased their levels of noise emissions with increases in power. In 1972, new certification procedures adopted by ICAO came into force and, as a result, new subsonic jet aircraft have to conform to certain maximum noise limits measured in Effective Perceived Noise Decibels (EPNdB) at particular points on the ground in relation to the aircraft all-up weight.

This certification will probably not begin to have an appreciable effect until about 1980 when the numbers of the present generation of aircraft will have been reduced significantly.

The intrusive effect of aircraft noise is measured by the Noise and Number Index (NNI) which is a concept put forward by the Wilson Committee on Noise in 1963[3] in conjunction with a social survey round Heathrow (London) Airport.

$$\text{NNI} = \Sigma\ \text{PNdB} + 15 \log_{10} \text{N} - 80$$

where NNI is the noise effect at a point on the ground; PNdB is the average peak noise level received at the point from a group of aircraft of similar noise generating level, measured in Perceived Noise Decibels; N is the number of aircraft in that group in an average operating day often taken as 0600–1800 h during the busy three months of the year; Σ is the logarithmic sum of the group NNI values.

A second survey of aircraft noise at Heathrow (London) Airport which investigated more deeply the annoyance caused by aircraft noise, was carried out in 1967.[4] Whilst some slight modifications to the NNI concept were suggested,[5] it is still accepted as a basis for assessing the intrusion of noise at medium and large airports in urban areas.

Land use is affected by an airport and one of the most important restrictions arises from aircraft noise. Some authorities have recommended that construction of various types of buildings should be restricted in relation to the NNI contours, and one example is:

60 NNI and above—no development
50–60 NNI —no major development; infilling only with sound insulation
40–50 NNI —no major development; infilling allowed

In addition, there will be a demand for land for housing the staff and their families, for access roads, ancillary industries, off-airport telecommunications, landing and navigational aids, etc.

The area of land within the airport boundary may vary from a few hundred hectares for a single runway airport to some 3500 ha for a four parallel runway airport.

Economic aspects

There are three main economic aspects taken into account in considering airport location, namely employment, direct spending and the multiplier effect.

The construction and the operation of an airport is a source of employment but its location must also be assessed for its effect on restricting employment in other fields, e.g. effect on local coal mining or removal of local industry originally on the prospective airport site.

The airport is a spending generator in that people are attracted to it and its employees' income is spent in the local area. The airport has a secondary or tertiary effect called the 'multiplier effect' whereby it generates housing, services and industry to provide facilities for the airport employees and their families and this multiplies itself, enlarging the community.

Surface access

Roads are the primary method of access for airports. Instances such as Gatwick Airport where 40–45% (estimated) of all passengers use the train, and Heathrow Airport which will be connected to central London by underground in 1975, are exceptions.

The airport should be integrated with the road network and should not be added on. The design capacities for the roads will be derived from the total numbers of cars for air passengers (see 'Car Parking' in the subsection on 'Design') together with the requirements for airline airport staff and industries associated with the airport. Gradually, other modes of surface transport will come into use on future major airports and they may affect location decisions. These modes include the monorail, hovertrain and other rapid-transit systems. Others such as the 'automatic taxi' and high-speed travelators are liable to affect internal airport design more than the location but they cannot be neglected. Many systems are under continuing investigation and development.

Air traffic control

The siting of an airport in relation to controlled air space is critical. In the UK the National Air Traffic Services (NATS) is responsible for air traffic control, and provides a combined service to both the Civil Aviation Authority and the Ministry of Defence. At some airports NATS provides the air traffic control (ATC) service but at others it provides a licensing and training function to ensure that all controllers operate to similar standards and similar competency.

The UK, parts of the USA—notably the north-east and west coast air corridors—and some parts of Europe are particularly congested by air transport and the introduction of a new airport, even a medium-sized one, could cause problems. In less developed parts of the world, there is not likely to be much difficulty.

Topography

Ideally, an airport should be located on relatively flat ground although it is most important that the natural drainage of the site should be effective. Suitable land must be acquired not only for the area of the airport itself but also for surface access routes and for off-airport facilities such as radio and radar landing aids, navigational aids and approach lighting. The proposed site should not be hemmed in by hills, rivers, main roads, towns or other developments which are likely to hinder any conceivable future expansion of the airport. Topographical maps, aerial photographs and development plans should be obtained and studied so that alternative sites can be assessed in detail.

The vertical alignment of runways is an important item to be considered.

On the one hand, runways should follow the contours of the natural ground as closely as possible, if expensive earth moving is to be avoided. However, gradients have a significant effect on the operation. The slope at any point affects the acceleration or deceleration of an aircraft and the maximum slope recommended in Annex 14 is shown in Table 22.1. The effective slope of a runway is the difference between the maximum and minimum elevation along the runway centreline divided by the runway length, and the maximum recommended effective slope of a runway is also given in Table 22.1. Annex 14 recommends that the actual runway length adopted should be the selected basic length for the runway code letter increased at the rate of 10% for each 1% of effective slope.

Earthworks can be reduced by making several changes in the slope along the length of the runway, but this practice should be avoided if possible. Data giving the maximum change in slope, the maximum rate of change of slope and the minimum distance between successive points of intersection of vertical curves are given in Table 22.1 and reference should be made to Annex 14 for sight distance considerations. The elevation of the airport is important and Annex 14 recommends that the basic length of a runway should be increased at the rate of 7% per 300 m elevation above mean sea level.

Obstructions

Objects projecting above the imaginary surfaces (see p. **22**–5) are obstructions and should be removed, if possible, or marked. The location of an airport may be influenced by potential obstructions, particularly those under the aircraft approaches, and the cost of removing obstructions must be considered when comparing alternative locations or alignments.

The decision as to when an obstruction is serious or not is finally one for the competent authority, the CAA in the UK, but, in general, proposed airports should avoid obstructions, and obstructions should not be added in the vicinity of airports.

Meteorology

The use of an airfield is controlled to a certain extent by the wind. Crosswind components may prevent safe usage of the runway and the direction of the runway should be aligned

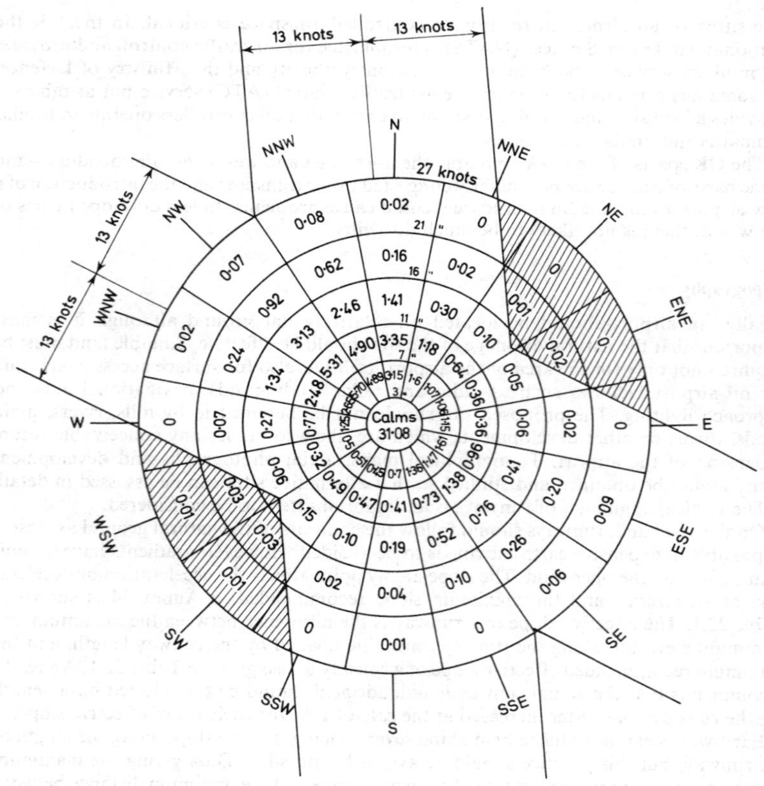

Figure 22.2 Graphical method to determine runway usability

to keep these to a minimum. To do this, a full summary of wind duration, speed and direction is required, taken over a period of years. From this a convenient graphical method of determining runway orientation as devised by Marwick is as follows.

The recorded hours (as a percentage total) of each velocity are plotted in the sectors intercepted between concentric circles representing velocities (Figure 22.2). The runway axis is then drawn in a trial direction through the centre of the circles and two parallel lines representing 13 knots (or any permissible crosswind component) to the same scale as the circles.

All winds falling outside these lines are in excess of the critical for that particular runway direction. Further trial and error establishes the desired pattern.

Alternatively a computer may be employed to follow a similar process in order to establish the percentage usability of an airfield having one or several runways in various orientations.

In a multi-runway layout, the main runway may be set in the direction of the most frequent winds of moderate and greater force and the subsidiary runways are laid in the direction which yields for the whole system the required maximum crosswind component and percentage usability. The present tendency is to aim for a single runway system with high permissible crosswind components.

The prevalence and nature of gusts and air turbulence in the area must be considered separately.

The average daily temperature (over 24 h) for the hottest month of the year is of interest to the designer and it will be necessary to increase the length of the runways where high temperatures are likely (see Annex 14).[1] The elevation of the airport has a like effect, and the basic length of a runway should be increased as described on p. **22–9**.

Other meteorological factors which may need to be assessed for each airport site under consideration are the frequency of occurrence of fog (low visibility) and low cloud base, of icing and the frequency and quantity of snowfall, as well as rainfall. Factors such as electrical disturbances, dust storms or smoke may have to be evaluated at certain locations.

The provision of landing and navigational aids can be directly related to the frequency of adverse conditions, particularly low visibility and low cloud base at different weather minima levels. These levels can also be assessed from meteorological records for up to 10-year periods and preferably not less than 5 years. During exceptional weather conditions at the airport, approaching aircraft will have to be diverted to other airports, and the likelihood of similar conditions closing any or all of these alternative landing places at the same time will need careful consideration.

AIRPORT CONCEPT AND LAYOUT

Concepts

The concepts for airports have changed considerably in the past 10 years, particularly due to the great increase in the number of passengers handled, the increase in size of aircraft and the improved ability to land in strong crosswinds.

The number of air passengers carried throughout the world on scheduled services by airlines of ICAO member states has risen from 111 million in 1961 to 403 million in 1971. The present average annual growth rate is 14.5%.

The size of aircraft has increased greatly, for example, the Boeing 747 is 70.500 m long with a wingspan of 59.700 m and a height of 19.400 m above the ground, whereas 10 years ago the early Boeing 707 was in regular service and had corresponding dimensions of 44.200 m, 39.880 m and 12.700 m.

Modern aircraft are able to land in crosswinds of over 20 knots and this has resulted in a much reduced need for subsidiary runways. This in its turn, has allowed the terminal area to expand in length between the parallel runways.

The airport is designed to meet many requirements but compromise is inevitable since some of the most important present varying degrees of incompatibility. The main factors are:

rapid and efficient handling of passengers;
minimum walking distances;

simple directional guidance for passengers;
maximum runway movement rates;
minimum taxiing times;
rapid aircraft turn-round on the apron.

Modern airports for large and medium aircraft generally follow the pattern of parallel runways with the terminal facilities based on one of two principles, either (a) centralised handling or (b) decentralised handling. In the former, all the facilities such as check-in, baggage handling, customs and immigration, restaurants, bars, concessions, banks, etc., are usually concentrated in one building. At major airports several centralised buildings may be required either split for operation by different airlines or groups of airlines, as at Kennedy, New York, or by route type, as at Heathrow (London), or into arrival and departure, or into domestic and international functions. There is, however, limited airside and landside frontage. Aircraft have to be parked away from the building with access by piers or apron buses and landside car parks tend to involve long walking distance.

With decentralised handling, a series of smaller unit terminals is built, each linked to one or more general-services buildings which contain common facilities such as restaurants, bars, concessions, banks, etc. The unit terminals contain only those facilities which are related directly to passenger and aircraft handling such as check in, baggage handling, lounge, car parking, etc. However, there can be, and is, every variation between these two limiting principles.

Various aspects of planning the airport layout are described in greater detail in the following.

The small airport for light aircraft will almost cetainly have centralised handling facilities and it may well require one or more cross runways owing to the inability of light aircraft to operate in strong crosswinds.

Runways and taxiways

The number of runways at any airport other than one for light aircraft will be determined from the number of aircraft expected in a given period, usually an hour, but it is difficult to give general guidance since the capacity of any runway or runway system depends on a variety of factors such as:

aircraft types and relative numbers of each taxiway system;
ATC techniques;
apron capacity;
landing aids.

There will be significant differences between the capacities under instrument flight rules (IFR) and visual flight rules (VFR) and the IFR capacities will be lower. The major airports handling high rates of commercial air transport movements operate under IFR even in good weather conditions.

As an indication, the capacity of a single runway handling a mixture of air transport and general aviation aircraft will have a capacity of between 12 and 40 (per hour) under IFR conditions depending on the factors noted above. The maximum figure could rise to 50 under VFR.

At busy airports there will certainly have to be a parallel taxiway for the full length of the runway and, at some of the more sophisticated airports, there may be double or even treble parallel taxiways. Intermediate fast turn-off points from the runway may be provided at about the 2/3 points. At the other end of the scale, an airport with only low movement rates may not require a parallel taxiway and backtracking on the runway would be acceptable. Taxiways should lead directly on to the end of the runway so the aircraft are automatically aligned on the centreline.

Terminal area: apron, terminal building and car parking

The terminal area has three main constituents: the aircraft apron, the terminal building and car parking with the associated road system. Their relation to each other will be determined in principle by whether the centralised or decentralised concept is adopted, and on major airports by the method of internal surface transport.

There are other factors which influence the decision such as the pattern of airline operations, the ratio of domestic to international passengers, number of transfer passengers, etc.

CENTRALISED CONCEPT

An example of this is Gatwick Airport London (Figure 22.3) which was opened in 1953 and which is still being expanded progressively. The terminal building contains all the

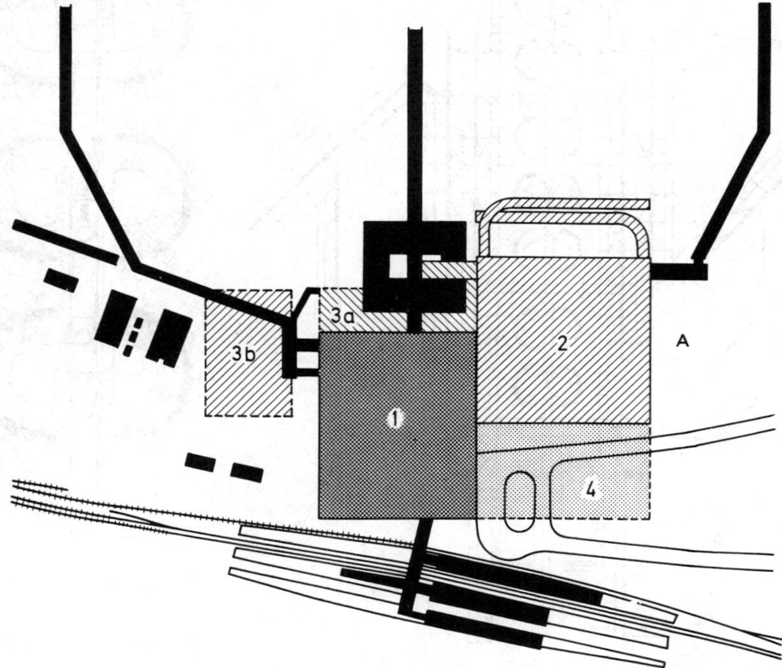

Figure 22.3 (1) existing terminal building; (2) extension under construction; (3a) proposed western extension; (3b) proposed southern building; (4) possible further northern extension; (A) car park

facilities required by the passenger, such as check-in, lounges, customs and immigration, restaurant, bar, concessions, etc. The car parking is concentrated at one side and fingers give access to the aeroplanes. The North East Regional Airport at Newcastle is a smaller example of this concept.

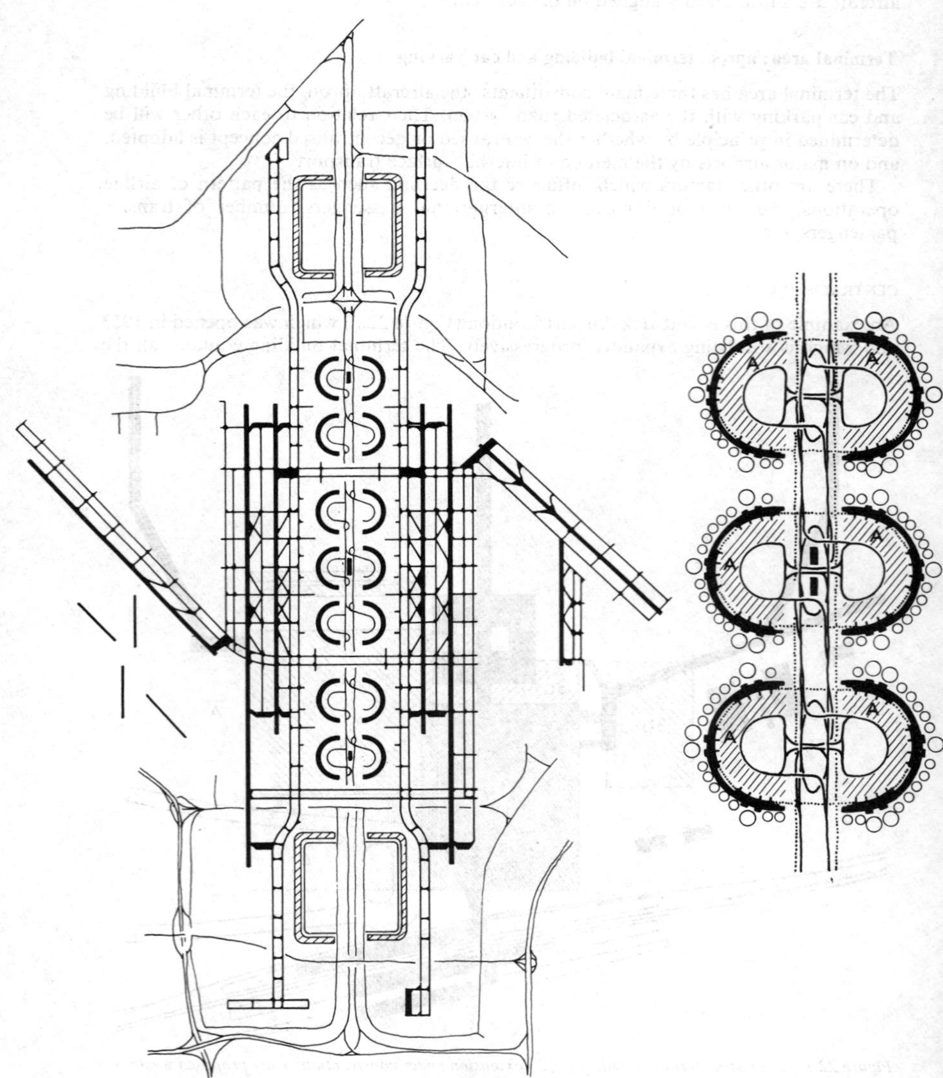

Figure 22.4 Dallas–Fort Worth
(A) car park

DECENTRALISED CONCEPT

There has not yet been an airport built truly to this concept but the proposals for the major airport at Dallas, Forth Worth employ the principle throughout (Figure 22.4).

There are variations on these concepts such as Paris Roissy (Figure 22.5) where satellite terminals containing only waiting lounges are separated from a main terminal building, containing the remaining facilities. There are three main buildings with satellites. Access from the main terminal to the satellites is by tunnel and aerobridges are provided

Figure 22.5 Roissy, Paris. Note 'drive through' parking
(A) car park

from the satellites to the aircraft. In another example of this layout at Houston Intercontinental (Figure 22.6), satellites are connected by covered accesses incorporating travelators.[6] The development of rapid-transit systems for passengers between a main terminal and satellites as at Tampa, Florida (Figure 22.7), or between several terminals, vehicle parks and rail terminus, or linking unit terminals, will fundamentally affect future airport concepts.

Ancillary buildings

It has been customary to collect the remaining airport buildings under this heading but some, such as large hangars, cargo terminals, etc., may be major projects in their own right.

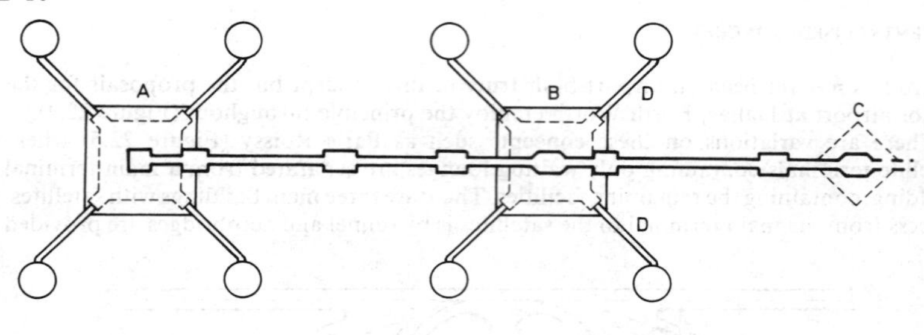

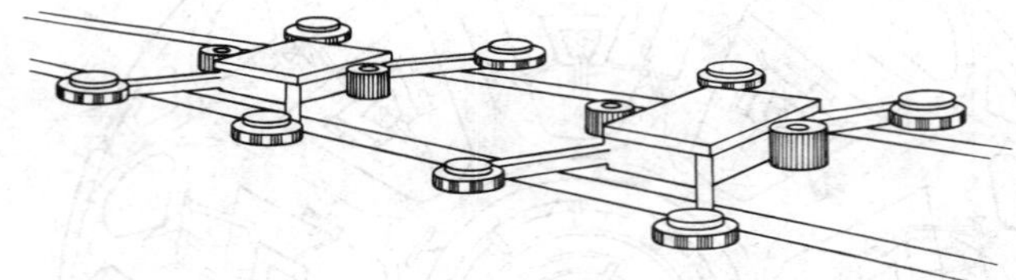

Figure 22.6 Houston Intercontinental. A, B, unit terminals; C, hotel; D, access with travelator

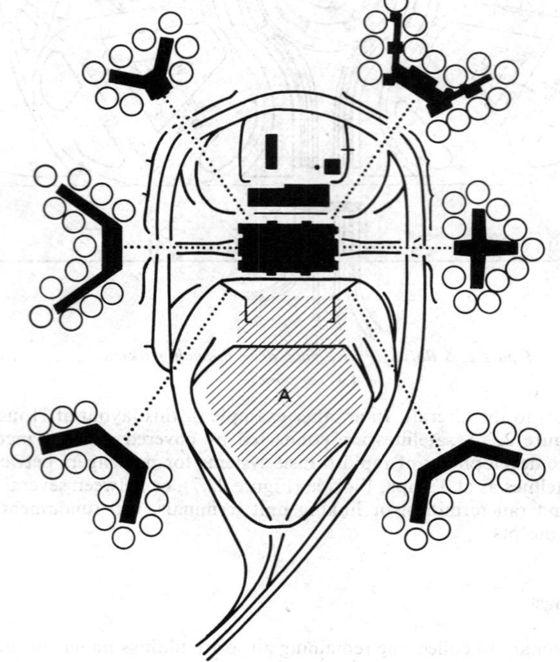

Figure 22.7 Tampa, Florida
(A) car park

CONTROL TOWER

This should give controllers a view of all the runways and is designed round the equipment required for air traffic control. Large areas of false floor for cabling may be needed. The tower is generally a separate building and not a part of the terminal building.

APRON CONTROL

Some airports include an apron control cabin located so that an apron controller can direct aircraft to the apron stands from the taxiways.

AIRCRAFT CATERING BUILDING

This should preferably be located close to the terminal area and is a specialist catering building.

CARGO TERMINAL BUILDING

This facility may be a simple framed building or a sophisticated terminal such as that of BEA/BOAC at Heathrow Airport which comprises transit sheds housing computer-controlled mechanical handling equipment, office blocks, vehicle parking, loading bays and circulation. There are no particular civil engineering requirements.

MAINTENANCE HANGARS

These may range from a simple framed building to a major structure such as the BOAC hangar for the Boeing 747 at Heathrow Airport London.[7] The structure basically is a cladding for the maintenance requirements but considerations of large clear spans and door openings will determine the structural forms.

BUILDINGS FOR ELECTRICAL AND ELECTRONIC EQUIPMENT

These, generally, are simple buildings designed to house particular equipment some of which may require a controlled environment. The manufacturers advise on this point. The buildings for certain navigational aids cannot have ferrous metal above a specified level and the manufacturer's advice should be sought. Others may require special shielding.

Generally speaking there are no particular constructional problems.

Airfield lighting

The extent of the approach and runway lighting provided not only depends on the airport classification but should be compatible with the radio and radar landing aids provided. It generally consists of high-intensity centreline and crossbar lighting for the approach areas, together with runway centreline and edge lighting. Taxiway lighting usually consists of green centreline lights supplemented with blue edges lights at junctions and around the apron areas.

For visual guidance in the angle of descent, Visual Approach Slope Indicators (VASIs) are provided in two groups on each side of the runway.

Recommendations and requirements are given in ICAO Annex 14[1] and include information on airport beacons, obstruction lighting, floodlighting and other aspects. Control systems to provide safety, reliability and fault indication are vital aspects.

Telecommunications and navigational aids

All modern airports incorporate a degree of telecommunication and navigational aid facilities to assist aircraft in airport location, approach and landing. Recommendations and requirements are given in ICAO Annex 10. The following facilities are typical of those provided at the larger airports.

VHF and/or UHF transmitters.
VHF and/or UHF receivers.
Locator beacons.
Instrument Landing Systems (ILS).
Radar (Approach and/or Surveillance).
HF communications.

Meteorological equipment is also provided and used in conjunction with the telecommunications equipment.

The positioning of the various telecommunication and navigational aids is extremely important and must conform to the accepted recommendations in respect of siting and the grading of the surrounding areas. This latter requirement is particularly important in respect of ILS and radar installations.

DESIGN

Traffic forecasts

The capacity of the apron, terminal building and car parks is determined from the traffic forecasts. Such forecasts are normally made on an annual basis and are split into scheduled and charter flights for both domestic and international services.

These annual forecasts are determined by one of two main methods. Firstly, extrapolation of historical data, and secondly, by analysis of such factors as future income levels, regional and national development plans, future population forecasts, tourist potential, etc. This analysis is usually carried out on a computer.

The annual figures are then converted to hourly flows called Standard Busy Rates (SBR). The SBR is defined as that rate which is exceeded 29 times in the year, and has been found to give a reasonable basis for design.

Table 22.3

Annual passenger movements	*SBR/Annual ratio*
100 000	0.002 –0.003
250 000	0.001 –0.002
500 000	0.0007–0.0012
1 million	0.0006–0.0010
2–5 million	0.0004–0.0009

It is essential to obtain an estimate of the short-period flow rates for passengers and aircraft. This can be done by an analysis of monthly, weekly, daily and hourly aircraft movement patterns but unless a lot of relevant data is available an assessment using ratios of standard busy hour passenger rates to annual movements is more likely to be the only suitable method. In general the ratios decrease with increasing annual movements, they also tend to be higher at airports with high proportions of international leisure travel and they tend to be high where one route dominates the schedules as occurs at many small airports and therefore such airports need to be independently considered. Table 22.3 gives an indication of the ranges of ratios.

The passenger SBR is then used to determine the SBR of the aircraft movements estimating the likely mix of aircraft, capacity and load factors, taking future trends into account.

Pavements

Suitable pavements must be provided for aprons and maintenance areas as well as for the runways and taxiways. The general requirements which must be fulfilled by the designs are as follows:

(1) Adequate strength for all aircraft types likely to use the airport (including future aircraft types)
(2) Adequate fatigue strength
(3) Absence of loose particles which could be sucked into jet engines
(4) Resistance to jet blast
(5) Resistance to fuel spillage (particularly on aprons and maintenance areas)
(6) Good surface drainage
(7) Ability to accept temperature movements
(8) Good skid resistance
(9) Good riding surface for comfort in the aircraft
(10) Ease of maintenance

From the operational point of view, it is difficult at busy airports to close down a runway for the purposes of maintenance or for strengthening of the pavement; indeed maintenance work may have to be carried out at night. Pavements designed to fulfil the needs of only the immediate future may prove to be expensive in the long run. A runway may be required to have a life of 20 years or more and the designer must anticipate requirements as far into the future as is possible.

The three types of pavement construction may be grouped as follows:

(a) Rigid.
(b) Composite.
(c) Flexible.

An example of each of the three types is shown in Figure 22.8 which is reproduced from a paper prepared by the Department of the Environment[8] based on that written by Martin and Macrae.[9] The choice of construction depends on the availability and cost of the materials required, and on the operational requirements at any particular location. In practice, there is often a case for using different types of construction at different areas in the same airport.

In the past, it was the practice to design the pavement to resist a single wheel load. The surface area over which this acted depended on the magnitude of the load and the tyre pressures used. Nowadays this design approach is less relevant since, for example, the Boeing 747 Jumbo jet has a maximum all-up weight of 376 tonnes and is supported on 18 wheels. The stresses induced in a pavement depend on the undercarriage configuration, the individual wheel loads and tyre pressures, the type of pavement construction, the thickness of the various layers and the subgrade.

Dawson and Mills[10] give diagrams of the undercarriage configurations of most transport aircraft in use and the configurations, aircraft weights and tyre pressures for most transport aircraft are published by the Department of the Environment.[11] The concept of the Load Classification Number (LCN) has been used to classify aircraft according to their effect on pavements but recently Martin and Macrae have recom-

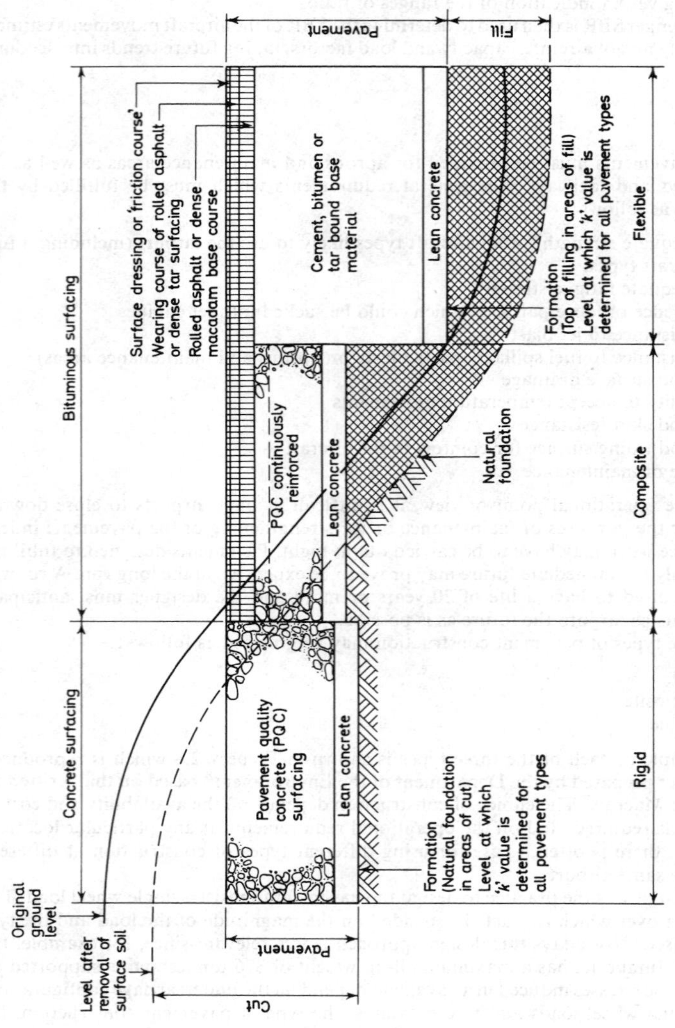

Figure 22.8 Alternative recommended types of aircraft pavements

mended the use of seven Load Classification Groups (LCG) for both aircraft and pavements and this approach is covered in both Department of the Environment documents.[8, 11] For example, the Concorde airliner is classified as LCG II while the small HS125 executive airliner is LCG VII. It is the LCG of the heaviest aircraft likely to use the pavement on repeated occasions which is of interest to the designer.

If the LCG system for aircraft loading is used then the corresponding pavement design chart reproduced from Morton and Macrae[9] should be used to determine the required thickness of each layer in a rigid, composite or flexible pavement (see Figure 22.9). The lowest layer recommended in all cases is 100 mm of dry lean concrete. The modulus of subgrade reaction, k, should be determined by using a 762 mm diameter plate. This chart is simple to use, but it is recognised that the resulting designs may tend to be rather conservative.

Great care should be taken if the selected pavement design in layers or materials is different from that in the paper[9] and it may be advisable to depart from the LCG system, reverting to LCN and using former empirical methods or analytical methods.

Analytical methods of design have often been based on Westergaard's theory. More recently, the flexural stresses in rigid pavements under various configurations of undercarriage loading have been calculated by using a computer programme developed by the Portland Cement Association, and design charts have been published based on results obtained by this method.[12]

Flexible pavements under load have been analysed by using a computer programme named BISTRO,[15] which has been developed by The Shell International Petroleum Co. This method permits the computation of the stresses, strains and displacements at any point in a multilayered elastic system due to any arrangement of uniformly loaded circular areas on the surface: it is necessary to assume values for the elastic modulus and Poisson's ratio for each layer.

The use of these analytical methods does greatly assist the designer to understand how the pavement will behave under load, especially when new materials or types of construction or unusual environments are encountered. The problem associated with their use lies in the necessarily oversimplified assumptions regarding the behaviour of the materials in the pavement and subgrade, and in the choice of values for the material constants.[13, 14]

Neither analytical nor empirical methods therefore can be considered to be entirely satisfactory individually and there is much to commend the use of both approaches when large areas of pavement are being designed.

RIGID PAVEMENTS

A concrete pavement is usually considered as being rigid because the load is spread over a wide area of subgrade by virtue of its inherent flexural strength. The concrete is usually unreinforced and is divided into rectangular bays to restrict the tensile stresses which are induced by a combination of three factors:

(1) Contraction of the slab due to falling temperature and concrete shrinkage. This movement is restricted by the friction between the slab and the subgrade and as a result tensile stresses are induced in the slab.

(2) Warping of the slab due to a temperature gradient through the thickness of the slab. High surface temperatures cause the slab to dome until it is supported mainly at the edges, whilst low surface temperatures cause the corners to curl upwards.

(3) Loading. Slabs are usually most susceptible to loading near their corners which may cause cracks to form across a corner. Acute angles in slabs should therefore be avoided at all costs.

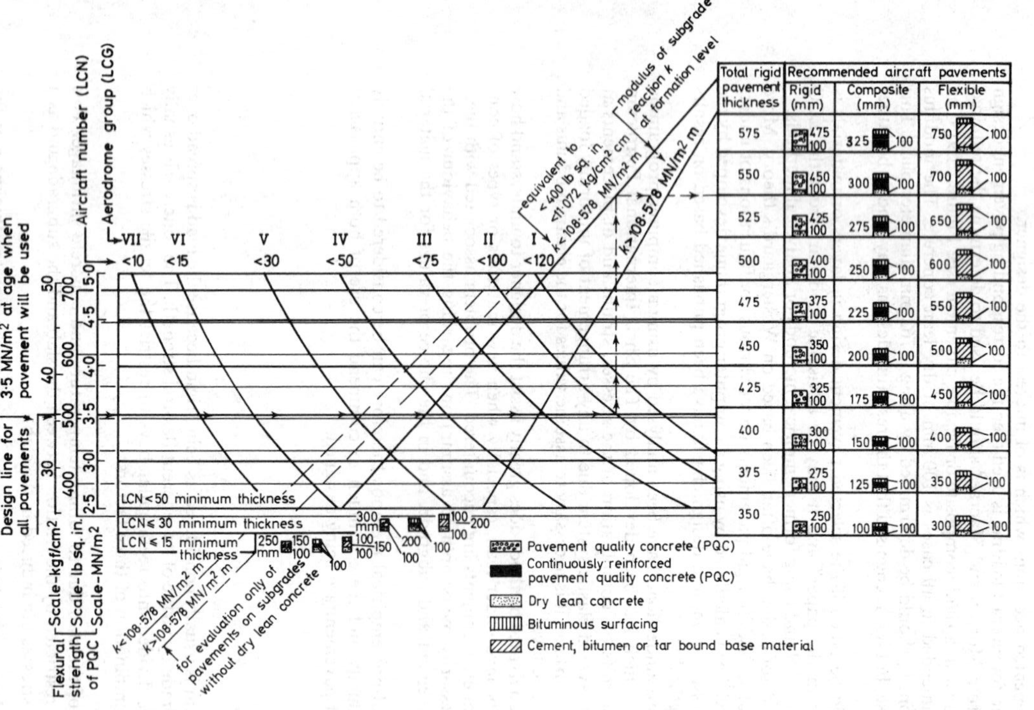

Total rigid pavement thickness	Rigid (mm)	Composite (mm)	Flexible (mm)
575	475 / 100	325 / 100	750 / 100
550	450 / 100	300 / 100	700 / 100
525	425 / 100	275 / 100	650 / 100
500	400 / 100	250 / 100	600 / 100
475	375 / 100	225 / 100	550 / 100
450	350 / 100	200 / 100	500 / 100
425	325 / 100	175 / 100	450 / 100
400	300 / 100	150 / 100	400 / 100
375	275 / 100	125 / 100	350 / 100
350	250 / 100	100 / 100	300 / 100

Figure 22.9 Design and evaluation chart for rigid, composite and flexible aircraft pavements. The example shown is for the centre longitudinal strips of the runways, taxiways and aprons of a LCG 11 aerodrome founded on a formation of k $< 108{\cdot}578$ *MN/m²* *to carry aircraft of LCN≤ 100. Construction could be:*

425 mm of PQC surfacing on 100 mm of lean concrete

or

100 mm of bituminous surfacing on 275 mm of continuous reinforced PQC 100 mm of lean concrete

100 mm of bituminous surfacing on 650 mm of cemented base material on 100 mm of lean concrete

depending on the surface required and the economics of the construction.

The outer strips of the runways, taxiways and aprons for the same aerodrome may be based on LCG III requirements

The bays are separated by contraction joints and the bay size depends on the slab thickness. The maximum bay sizes should be as follows:

Table 22.4 MAXIMUM BAY SIZES OF CONCRETE RUNWAYS

Slab thickness	*Bay size*
150 mm or less	3 m
151–224 mm	3.75 m
225–274 mm	5.25 m
275 mm and over	6 m

The contraction joints may be formed by using crack inducers as shown in Figure 22.10.

Load is usually transferred between adjacent slabs by aggregate interlock, in which case no dowel bars are needed. If the aggregate particles in the concrete are not too hard, however, a more satisfactory solution is achieved by continuous casting of the slab, perhaps employing slip-forming techniques, followed by sawing the joints after the concrete has set. Slots in the surface of concrete pavements, whether preformed or sawn, should be as narrow as possible; they should be filled with a semicompressible material such as hardboard or fibreboard and need not be sealed.

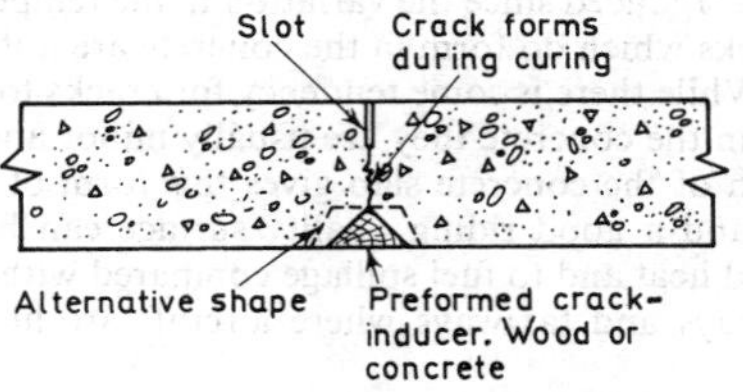

Figure 22.10 Contraction joint

Expansion joints may be provided in thin slabs but may be entirely omitted in slabs more than 250 mm in thickness.

Single-butt construction joints as shown in Figure 22.11 are recommended since those incorporating a joggle are susceptible to cracking. Dowels may be omitted for slabs 275 mm thick and over.

The pavement quality concrete (PQC) used in rigid pavements should be designed on the basis of its flexural strength measured by loading 152 mm × 152 mm test beams, rather than on cube strength. It is the strength of the concrete when it is first loaded which is of importance so that age factors may be taken into account. The aggregate/cement ratio should not exceed 6.3 : 1 and the water/cement ratio should be less than 0.50.

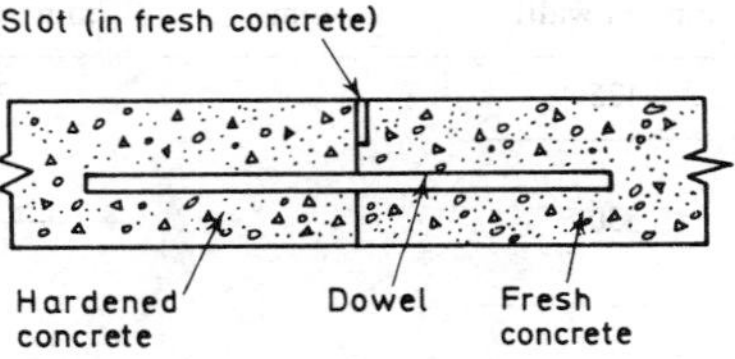

Figure 22.11 Construction joint

The PQC slabs should be placed over a layer of dry lean concrete having an aggregate/cement ratio of 15 : 1 and a minimum cube strength of 5.2 MN/m^2.

In order to improve the skid resistance of the concrete surface, the concrete may be wire combed or small transverse grooves may be cast into the wet concrete surface. It is essential that experiments are carried out to ensure that such treatment is applied at the correct time. Alternatively the hardened concrete may be scored with diamond cutting drums.

Well constructed concrete pavements show little cracking and are resistant to both jet blast and fuel spillage. They do have the disadvantage, however, that, because of the frequent joints, they tend to be somewhat bumpy. They are ideal at runway ends, taxiway junctions, aprons and on maintenance areas where aircraft stand or are slow moving.

Joints can be largely eliminated if prestressed concrete construction is adopted but this form of construction is unlikely to be economic under most conditions.

COMPOSITE PAVEMENTS

In a continuous reinforced concrete pavement, cracking accentuated by exposure to heavy traffic, is likely to develop whatever quantity of reinforcement is incorporated. However, if the continuous reinforced concrete pavement is overlain by bituminous surfacing the cracking is reduced since the variation in the temperature in the concrete is lowered and those cracks which do form in the concrete are not subject to wear and are unlikely to be severe. While there is some tendency for cracks to form in the bituminous surfacing above those in the concrete they are usually minor and can be easily resealed.

The flexural strength of the concrete slab gives this form of construction good load spreading properties, and a good riding quality surface can be obtained. It is not as resistant to jet blast and heat and to fuel spillage compared with the rigid pavement so it is often used on runways and taxiways where aircraft are likely to be moving fairly rapidly.

Pavement quality concrete should be used for the reinforced slab, overlying 100 mm of dry lean concrete. The minimum cross-sectional areas specified for reinforcement for the appropriate concrete slab thickness, as recommended by Martin and Macrae, are given in Table 22.5.

Table 22.5 STEEL REINFORCEMENT IN RIGID PAVEMENTS

Slab thickness mm	*Schedule of reinforcement*			
	Main steel		*Transverse steel*	
	Minimum area mm²/m width	*Spacing limits* mm	*Minimum area* mm²/m width	*Spacing limits* mm
100	425		295	125–175
125				
150	530			
175				
200				
225	635	125–175		
250	740		170	150–225
275				
300				
325	825			
350				

The surfacing, normally 100 mm in thickness, should be rolled Marshall asphalt or dense tar surfacing laid in two courses. This two-course work reduces the tendency to sympathetic cracking in the wearing course over cracks in the underlying slab.

FLEXIBLE PAVEMENTS

A flexible pavement[11, 13] is one which depends on its thickness and elasticity to disperse the load to such an extent that the subgrade is not overstressed. It is made up of a number of layers of granular materials increasing in rigidity and decreasing in flexibility towards the surface. The lower materials may be unbound, or bound with bitumen or cement. The middle layers should be asphalt, bitumen or tar macadam. The surface layers should be impervious and Marshall asphalt or dense tar surfacing specifications are usual. The following factors have to be considered in relation to the design:

(a) The overall depth of pavement must be such that the strength of the subgrade is not exceeded.
(b) The strength of each individual layer of the pavement must be such as to resist the pressure at that level.
(c) The shearing strength of the surfacing and layers beneath must exceed the shear stresses produced by the tyre load.

For very light-duty pavements several layers may be omitted. When dry lean concrete is used on the subgrade to provide a good working surface it must be weak, otherwise cracks which form in this layer are likely to spread upwards towards the surface. An aggregate/cement ratio of 18 : 1 for gravel or 22 : 1 for crushed rock is usually suitable.

Well designed flexible pavements have good riding qualities but some surfaces are susceptible to jet heat and fuel spillage may cause softening of the surface.

Relatively high landing and take-off speeds of modern aircraft, combined with the flat transverse slopes on runways, have led to the problem of aquaplaning. Various devices such as friction coatings on flexible runways and grooving, wire combing or scoring on rigid pavements have been adopted to alleviate this.

Aprons

The engineering design of the apron has been described (see p. **22**–19 *et seq.*). The apron should preferably be constructed in concrete to avoid damage from fuel and hydraulic fluid spillage. Dense tar surfacing (DTS) can be used as a second choice.

The detailed layout of aprons is beyond the scope of this book but Table 22.6, giving typical stand sizes for various groups of jet aircraft, may be used as guidance.

Table 22.6

	Nose-in parking m square	*Self-manoeuvring* m diameter
Air bus	85	100
Long haul	65	90
Medium haul	50	60
Short haul	40	50
General aviation	—	30

The clearances given in Table 22.1 will have to be observed. The number of stands required on the apron is derived from the aircraft Standard Busy Rate. This is a complicated process in the case of major airports and will involve the use of computers for simulation. Smaller airports are treated more empirically and a rough rule is to increase the SBR by 10% and round up to the next whole number.

Terminal buildings

The functions, flow pattern, accommodation, configuration and size of the terminal building or buildings need individual assessment for the factors of influence are many and differ in each case. The use of simulation and computer models to aid design of this most complex of buildings have been developed and are likely to be used for the larger terminals.

Perret[16] gives the following approximate guide for the capacity of a terminal:

(a) 1500 passengers per hour in each direction for every 15 000 m^2 of area available to the public

(b) 1500 passengers per hour each way for every 25 000 m^2 of total terminal (excluding office accommodation)
Reductions of 30–40% could be made in the areas for terminals handling predominantly domestic traffic, conversely the space could be increased drastically if, for example, there were a high proportion of visitors.'

The allocation of areas to the various functions such as check-in desks, lounges, customs and immigration, etc., is also based on the passenger SBR figures but it is necessary to split these down into international and domestic categories, and also to differentiate between terminal and transit passengers. It should be noted that there may be a higher flow rate in a particular category outside the standard busy hour and the worst case must then be used as the design basis.

Perret[16] gives useful data on terminal building design and other publications by the International Air Transport Association[17] and the American Society of Civil Engineers[18] are helpful.

Car parking

The problems arising out of making provision for car parking are among the most difficult facing the airport designer. Generally speaking, the majority of passengers travel to and from airports by car and the areas required to accommodate these cars are very considerable.

There are five main categories of car parking:

1 Kerbside—for setting down and picking up.
2 Short term—say up to 48 h.
3 Long term.
4 Staff—both airport and airline.
5 Spectators.

At large airports, if it is intended to provide long-term parking accommodation on the airport to satisfy the demand, then the accommodation required becomes vast and fundamental to the concept of the airport.

MAINS SERVICES

Surface water drainage

A carefully designed surface water drainage system is a necessary requirement of an airport. Badly drained subgrades have greatly reduced load-bearing capacity and may cause failure in pavement areas. Badly drained surfaces decrease skid resistance on paved areas, and cause breakdown of the surface vegetation on earth areas.

HYDROLOGICAL DATA

As much information as possible must be gathered on rainfall statistics, stream flow, flood levels, water-table fluctuations and locations of subsoil drainage channels in the area of the airport. Sources of information in Britain include the Meteorological Office records,[19] 'The Surface Water Year Book of Great Britain',[20] Local Authorities, local River Boards and River Conservators.

DRAINAGE FROM RUNWAYS, TAXIWAYS AND APRONS

Drainage from paved areas should be by gullies, continuous gratings, or slot drains. Ditches are not used adjacent to construction but may be used to drain the strip and carry away surface water remote from aircraft operations.

DRAINAGE FROM MARGINS

The main requirement for margin drainage is to prevent water ponding, particularly near the edges of construction. This requirement is met by grading the surface to gentle falls and providing French drains in the valleys. Cut-off French drains will be necessary adjacent to construction where the ground slopes towards pavement areas, but care needs to be taken to avoid the possibility of stone ingestion by aircraft engines.

SUBSOIL DRAINAGE

This may be necessary to drain low-lying waterlogged areas, or to keep a fluctuating water table well below subgrade level. Open-jointed porous pipes laid in a 'herringbone', 'parallel' or 'gridiron' system should be used. Depths should be as great as possible, bearing in mind soil permeability and outfall levels, but should generally not be less than 600 mm or greater than 1.200 m below the surface.

Suggested spacings of parallel subdrains are given in table 3 of British Standard Code of Practice 301.[21] See also Section 28.

GENERAL DESIGN PROCEDURE

The surface water drainage system should follow the natural drainage system of the area as closely as possible. Where the runways and taxiway lie across natural drainage channels, these channels must be either diverted or culverted, the choice depending upon relative costs, existing bed gradients, and possible difficulties with regard to the purchase of additional land for diversion channels.

On deciding the gully grating or slot drain layout pattern for the paved areas, the pavement drainage layout is determined so as to give the shortest possible system connecting the inlets to the nearest natural main-drainage channel(s). The calculation of pipe sizes may be done by any of the standard rainfall run-off methods, e.g. Lloyd Davies's using the rainfall-duration-intensity formula applicable to that locality, or the RRL Hydrograph method. Gradients on an airfield will usually be very flat, but the minimum velocity of flow in a pipe should not be allowed to fall below 0.6 m/s.

The margin drainage system of French drains should be designed on the same basis, connecting the system to natural water courses or to the pavement drainage system by catch-pit manholes. The minimum depth of the French drain should be 750 mm, the minimum pipe diameter 100 mm and the minimum gradient 0.15%.

Impermeability coefficients for use on airfields are given in Table 22.7.

Table 22.7 IMPERMEABILITY COEFFICIENTS

Surface	*Impermeability coefficient*
Asphalt pavement	0.80–0.95
Concrete pavement	0.70–0.90
Impervious soil (heavy)	0.40–0.65
Impervious soil with turf	0.30–0.55
Pervious soils	0.05–0.40
Pervious soils with turf	0.00–0.30

The capacity of the subsoil drainage system when used should be based on the difference between the impermeability coefficient for the soil and unity, for a 6–12 h storm, depending on whether the permeability is high or low. In the design it should be noted that French drains will intercept some subsoil water. Minimum pipe size should be 150 mm and minimum gradient 0.15%.

Where existing watercourses and ditches are required to be filled in, open-jointed porous pipes should generally be provided to carry away any ground water percolating into the line of the old water course.

PETROL INTERCEPTORS

On areas where fuel spillage is possible, surface water drainage may have to be routed through fuel interceptors to prevent river pollution, and to localise the fire hazard of burning fuel. Generally, these interceptors must be very large in order to cope with large volumes of fuel spillage but, alternatively, an interceptor of reduced size may be designed so that the first flush of water is routed through the interceptor and, as flow increases above a certain level, bypasses it (see Figure 22.12). If subsoil drainage passes through the interceptor then the design rate of flow at which the bypass comes into operation should be not less than twice the base flow.

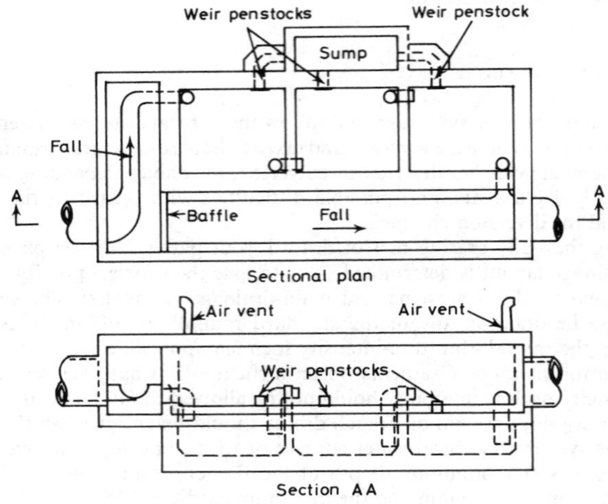

Figure 22.12 Fuel interceptor

STORAGE PONDS

Where the increased storm-water run-off, brought about by the construction of large impervious surfaces, might tend to overload the natural drainage system downstream of the airport, storage ponds controlled by sluices or weirs may be required. For a short high-intensity storm, storage ponds spread the total quantity of storm water over a longer period of run-off, thus reducing the rate of flow at any particular time. By consultation with the local River Board or Local Authority the maximum rate of flow into the water-courses may be determined and, accordingly, the size of the storage pond(s).

Sewerage scheme

The soil sewerage system of an airport should be designed to allow for the large fluctuations that occur not only during a day but from season to season. Ideally the system should be separate, in which case dry-weather flow may be equal to the average water consumption of a peak day (excluding fire-fighting water consumption).

The size of the drain should be designed to carry six times dry-weather flow, and should be checked to ensure that the system will cope with the maximum peak hourly flow. Disposal may be into the existing main sewer or, if not available, a sewage treatment plant will have to be provided. Owing to the necessarily level configuration of airport sites pumping mains may have to be provided. Adequate provision should be made for disposing aircraft sanitation wastes into the airport soil sewer. This may take the form of a small brick building with a sump for receiving aircraft wastes in the floor. This sump should be connected to the soil drainage system through a comminutor. Adequate cold-water flushing facilities should be provided in this building.

Water supply

The water supply requirements must be based on the expected number of staff, passengers and spectators. Suggested consumption figures are:

35–45 litre/day per head for staff;
20–25 litre/day per head for passengers and visitors;
5–10 litre/day per head for spectators.

Additional requirements for fire-fighting purposes will also be required.

The number of visitors and spectators depends on the airport location and the proximity of large centres of population.

Fire fighting, water supply

Water mains supply should be provided at the airport fire station for the filling and washing of the airport fire tenders. Hydrant mains are valuable in the vicinity of airport buildings, but are not required on the airfield in the vicinity of runways or taxiways. If life is to be saved from an aircraft fire, the tenders have to be capable of extinguishing the fire extremely rapidly in a period that is too short for the connecting of hoses to hydrant points.

Electrical mains installation

The extent of the electrical installation on any given airport depends on the facilities to be provided but generally speaking it comprises a high-voltage system supplying a number of substations strategically located to serve the various buildings and technical facilities.

A typical substation would contain high- and medium-voltage switchgear, transformers and, in some cases, stand-by equipment for the essential services.

The civil engineer will need to provide, in addition to the substation, ducting, junctions and inspection pits for cables. Pit covers must be capable of withstanding the potential loading on them, depending on location, and large pits within aircraft paving areas are avoided where possible. The engineer should also realise that there can be interference between contiguous cables carrying a variety of frequencies, from mains electricity to high radio frequencies, and he should take advice regarding cable routing.

ANCILLARY SERVICES

There are several ancillary services associated with the airport and the main ones are described in the following sections.

Aircraft sanitation

The aircraft toilets are emptied into vehicles and the contents are disposed of at aircraft sanitation buildings. These house macerators or comminutors which discharge into the foul drainage system. The buildings require an electrical power supply.

Fuel installation

The supply of fuel to aircraft is normally carried out by the fuel companies who contract for a specified period after tendering. There are two means of distributing fuel to the aircraft aprons:

1 by aircraft refuellers;
2 by a hydrant system.

Aircraft refuellers are usually employed and they range from 2250 l to 82 000 l. The larger is an articulated vehicle with an overall length of 21.500 m, a height of 3.650 m (including radio aerial) a width of 3.200 m, a turning circle of 21.400 m absolute minimum, and a laden weight of 91 t.

Hydrant systems consisting essentially of a distribution network terminating in pits in the apron which contain hose couplings, have been installed at some airports but have, up to now, not been very popular for two main reasons. The first is the inherent inflexibility. The mixes of aircraft at any airport change rapidly and the parking stands on aprons have rarely remained constant for more than two or three years with the result that the hydrant point has frequently been in the wrong position almost as soon as it has been installed. The second is that the fuel companies tenure is normally shorter than the hydrant system life and individual companies have not been prepared to finance the high capital cost. However, with the advent of very large aircraft of enormous fuel consumption, even larger fuel dispensers become less attractive and the hydrant system is likely to become of increasing interest in the future. Nose-in aircraft parking is becoming the standard with jet aircraft; this is tending to prolong the life of fixed apron stand positions and is a factor encouraging the greater use of hydrants.

Both the aircraft refuellers and hydrant systems incorporate safety features which prevent the pumping of fuel if the hosepipe should become disconnected.

G ɔund movement signs

These signs are placed adjacent to taxiways and aprons to direct the pilots. Details are given in Annex 14[1] and CAP 162[2]. In addition, aircraft stand-number signs are provided, either free-standing or fixed to the terminal buildings or piers.

All these signs will require an electrical power supply.

Crash and rescue services

Fire engines and crash tenders are housed in buildings with quick and easy access to the aprons, taxiways and runway. The scale of provision for the UK is given in CAP 168[2] and the requirements are related to the heaviest aircraft in regular operation at the airport.

At some airports where a crash in water is possible, rescue boats should be provided.

Boundary and security fences, including crash access

Airports should be properly fenced and the choice of fence depends on availability and cost. Whilst a 1.2 m fence is adequate over most of the perimeter, security and customs may require a higher fence topped with barbed wire strands in the terminal area separating the landside from the airside. The airside/landside fence will require manned gates at all accesses, which should be kept to a minimum.

The perimeter fence should be provided with a number of frangible gates so that crash and rescue services can get quickly to the scene of crashes which occur outside the boundary.

REFERENCES

1. *International standards and recommended practices for aerodomes,* Annex 14, 6th edn, International Civil Aviation Organisation (ICAO) (Sept. 1971)
2. *Licensing of aerodromes,* CAP 168, Civil Aviation Authority (1971)
3. *Noise—Final Report,* Committee on the Problem of Noise, Chairman: Sir Alan Wilson, Command-2056, HMSO (July 1963)
4. *Second Survey on Aircraft Annoyance around London (Heathrow) Airport,* HMSO (1971)
5. *Aircraft Noise: Should the Noise and Number Index be revised?,* Report of Research Sub-Committee of Noise Advisory Council, HMSO (1972)
6. 'L'Architecture D'Aujourd Hui', *Airports* (June 1971)
7. JOYNER, K. J., *et al.,* 'Structural Aspects of the Boeing 747 Hangar at Heathrow Airport', *Proc. Instn civ. Engrs,* **47,** Paper 7326, 483–513 (Dec. 1970)
8. *Design and Evaluation of Aircraft Pavements,* Department of the Environment (1971)
9. MARTIN, F. R. and MACRAE, A. R., 'Current British pavement design', paper 6, Aircraft Pavement Design Conference, Institution of Civil Engineers (1971)
10. DAWSON, J. L. and MILLS, R. L., 'Undercarriage effects on (a) rigid pavements, (b) flexible pavements', paper 2, Aircraft Pavement Design Conference, Institution of Civil Engineers (1971)
11. *Pavement Classification for Civil and Military aircraft,* Vol. 1, Department of the Environment (1971)
12. BREIHAN, 'Developments in pavement design in the U.S.A.; rigid pavements', paper 3, Aircraft Pavement Design Conference, Institution of Civil Engineers (1971)
13. AHLWIN, R. G., 'Developments in pavement design in the U.S.A.: flexible pavements', paper 4' Aircraft Pavement Design Conference, Institution of Civil Engineers (1971)
14. PEUTZ, M. G. F., VAN KEMPEN, H. P. N. and JONES, A., 'Layered systems under normal surface loads', *Highway Research Record,* 228 (1968)
15. EDWARDS, J. M. and VALKERING, C. P., *Structural design of asphalt pavements for heavy aircraft,* Shell International Petroleum Co. Ltd., London (1970)
16. PERRET, J. D., 'The Capacity of Airports—Planning Considerations', *Proc. Instn civ. Engrs,* paper no. 7372, **50,** 435–450 (Dec. 1971)
17. *Airport Terminal Reference Manual,* 5th edn, I.A.T.A. (1970)
18. *Airport Terminal Facilities,* American Society of Civil Engineers, Airport Operators Council International (1967)
19. *British Rainfall,* Meteorological Office, HMSO (1964)
20. *Surface Water Yearbook of Great Britain,* Water Resources Board, HMSO (1965–66)
21. *Building Drainage,* CP 301, British Standards Institution, London (1971)

23 RAILWAYS

RAILWAYS 23

23 RAILWAYS

D. H. COOMBS, B.Sc., C.Eng., F.I.C.E., M.I.Mech.E.,
Barrister-at-Law, Railway Engineering Consultant,
Formerly Permanent Way Engineer, British Railways Board

EARTHWORKS AND DRAINAGE

The contours of the territory to be crossed by a railway are obviously decisive as to its average gradient but they are also the background to fixing the maximum permissible gradient within the limits of tractive and braking adhesion. The lower the difference between average gradient and maximum gradient, the greater is the practicable train load and the lesser are the deviations from constant speed running. The minimum curvature to be used also determines the differences between the line speed limit and local speed restrictions. Long curves of small radius on heavy gradients may involve derailment hazards for very long and heavy trains, arising from braking or tractive effort surges along the train. Both maximum gradient and minimum curvature have a large effect on the earthworks cost of constructing a railway and because of this the ideal of constant speed running is often subject to heavy qualification. This is particularly the case in mountainous country.

Railway alignment needs to be planned to give a volume balance between excavation in cuttings and tipping in embankment, subject to the material excavated being suitable for tipping to form embankments and subject to minimising the haul of the excavated soil. Recourse to borrow pits for embankments and spoil hauls for cuttings should be minimised.

Soil formations

The side slopes of cuttings or embankments, based on the natural angle of repose of the specific soil involved, are generally made within the following ratios of horizontal to vertical:

Rock	Between 0:1 and 1:1
Gravel	About 0.9:1
Dry sand	About 1.3:1
Compact earth	Between 1:1 and 1.5:1
Well drained clay	Between 1:1 and 1.5:1
Wet clay	Between 2.5:1 and 4:1
Peat	Between 1:1 and 4:1

Steep rock faces, chalk cliffs and boulder strewn hill slopes may present rockfall or chalkfall problems which may need special watchmen or signalling provisions or special fence or apron or tunnel protection. Special precautions may also be required for deep cuttings in wet clay or sections of line subject to wash out inundations.

Where the depth of cutting in weak clay or silt exceeds about 4 m it is desirable, outside of arid locations, that a detailed local soil survey be made to determine at what slope the soil will theoretically be stable. The results of such a survey needs to be interpreted in the light of engineering experience with the stability of slopes in similar soils. The alternatives which may need to be considered may include decreasing the slope, retaining the slope but moving back the upper half to form an intermediate berm (which, whilst being a good protection against a circular slip, may not be very effective in preventing surface or weathering slips) or building drained hardcore buttress drains with a stepped bottom (which gives some protection against both forms of strip). In heavily built-up areas mass concrete retaining or toe walls, or 'cut and cover' or tunnels may be appropriate.

The shear strength of the ground to be tipped to form an embankment must be known as a starting point in design and field investigations to ensure the absence of any sizeable soft pockets are desirable. If such pockets are found laboratory consolidation/time tests are invaluable. The softer the subsoil support on which the embankment is to be built the less steep are the appropriate embankment side slopes, indeed, lightweight fills such as ashes, or subsoil strengthening by excavation, or injection of cement grout, or the installation of vertical sand drains, may need to be considered.

It is important to finish the formation surface at a crossfall of from 1 in 15 to 1 in 35 from the centre of the track or from one side of the other in order to facilitate draining, particularly when the formation is in impervious soil. The provision of a formation waterproofing layer of plastic or bitumen sheeting or of sprayed bitumen is an important aid to keeping the formation surface free from slurry. In cold climates this diminishes the hazard of ice forming pockets in the frozen subsoil resulting in a swelling of the formation by frost heave to produce dangerous twists or humps in the track at a time when correction by ordinary tamping is impossible. If this occurs temporary wooden wedge packings would have to be inserted under sleepers and later removed, to maintain track levels. Where such a waterproofing layer is applied it is usual to cover it with a 75–150 mm thick compacted layer of sand, ashes, stone or slag just prior to placing the ballast. The provision of a protected waterproof membrane diminishes the impaction of ballast into the formation face, keeps the moisture content of the subsoil, cutting or embankment more constant, and so reduces the rate of deterioration in the geometry of the track.

The formation crossfall can be applied either across one or two tracks or from the centre of one or two tracks. In the case of a double track, the latter alternative involves the installation of a drain in the 6-foot or inter-track space and the creation of a stepped ballast profile in the 6-foot or in the inter-track space which, if too marked, may weaken the lateral resistance to buckling of the inside track due to the weak ballast buttressing of the sleeper ends. It is possible that a continuous formation slope under two curved tracks can create clearance difficulties, but these can sometimes be overcome at the design stage by marginally raising the soffit of structures or widening the inter-track space on the curve affected.

Formation is the cut or made-up earth surface on which the permanent way is constructed; its strength and stability under all operating conditions and at all seasons of the year depend upon (i) the reliability of the beds under the earthworks; (ii) the quality of the soil in the earthworks; (iii) its adequate compaction in embankments; (v) the drainage of surface and ground space water in cuts and embankments; and (v) protection of the surfaces of earthworks against washing out and erosion of the soil.

Almost any non-cohesive or draining soil may be freely used in embankments though some caution may be needed in using dusty sand. If clay is used, the lumps should not be too large not only because of the subsequent compaction effort but in order to facilitate the immediate passage of machines and transport vehicles over the tipped material. Top soil with grass and turf should not be used in embankments less than 1 m high or on a transverse slope of more than 1 in 5. Soft clay, chalky and talcous soils are to be avoided if possible. Peat, silt, silty loam, fine sand/silt, and soils containing more than 5% of soluble salts should never be used.

Embankments and flooding

Embankments up to 2 m high on wet swamps and the lower part of higher embankments should be formed of a draining soil. All embankments should be made up in homogeneous horizontal layers of soil but if a layer of draining soil is placed over a layer of non-draining soil it is important that the latter is surfaced to a crossfall from the centre of 1 in 25 to 1 in 10. It is important to avoid any core or pocket of draining soil enclosed with non-draining soil and it is also important that draining soil should not be tipped on a steeply inclined flank of non-draining soil within the embankment. The thickness of successive layers of tipping should be related to the compaction capacity of the machines moving over successive layers. Compaction should of course be progressively checked on site. Where tamping is not likely to be effective, as with dry sand in an arid country, or rock, or rocky earth, an allowance for settlement, which may amount to 5% of the embankment height, needs to be calculated and allowed for.

Embankments across bogs need special hydrogeological investigation, both before the line is routed across the bog and before design and construction methods are selected.

Bogs situated on hillsides or watersheds or on floodplains are generally small, shallow and capable of being drained, but bogs in enclosed basins are often wide, deep and incapable of being drained. In broad terms it is better to avoid bogs especially on flood plains, but if this is not practicable then at least it is desirable to cross the bog at its most shallow and narrow section.

The transverse slope of the hard bottom of the bog is important. It should not exceed:

1 in 10 where the peat has a stable consistency capable of taking an external load in compression without being squeezed out;
1 in 15 for peat of an unstable consistency which is forced out by an external load; and
1 in 20 for a bog filled with water or bog ooze. If the inclines are greater than stated then the alternatives to be considered are levelling the bottom, relocating the line or replacement of the embankment by a trestle construction.

The embankment height above the bog surface should be at least 0.8 m if the peat is completely removed from under the embankment, or at least 1.2 m if it is left in place and in any case 1.0 m above the highest known flood level. Bogs should be drained if economically feasible.

Embankments of 3 m or more above the bog may be built on the bog-peat if it is of stable consistency; under this height some of the peat must be removed in order to ensure adequate compression of the remaining peat. Where the peat is of an unstable consistency the embankment should be founded on the hard bottom where the bog is up to 3 m deep; if it exceeds this depth, the tipped embankment materials should have a minimum depth of 3 m below the surface of the peat that is, the total construction depth of the embankment should be at least 4 m. The slopes of the underwater sides of the embankment should be less than the above water sides and should be within the range 1 : 1.75 and 1 : 2 for coarse granular soils and 1 : 1.3 to 1 : 1.5 for tipped rock or stone. On flood plains it may be expedient to incorporate protective beams of about 2 m width in the sides of the embankments if they are not made of rock tipping to a height of 0.5 m above flood level and where high flow velocities occur such protective beams should be formed of large rock. Alternatively the slope of the embankment sides can be reduced if the water flow velocity is not great or in substitution or in addition stone or concrete slab pitching to protect the sides of the embankment can be considered. Whatever flood protection is decided upon, it should be applied progressively as the embankment is built rather than as a separate and final operation in the construction schedule, if the work cannot be completed with certainty before flooding occurs.

Hillside locations of railway tracks often present special stability problems and therefore necessitate careful soil mechanics and hydrogeological investigations prior to design and construction.

Water bearing strata are a particular hazard and if practicable should be cut through by hillside ditches. If a stable formation bedding is not attainable by normal earthworks construction methods a tunnel or deep founded retaining wall may be a necessary alternative to re-routing the line.

On a stable hillside conventional earthworks design standards can be applied on a transverse slope up to about 1 in 5, provided the turf or other vegetative layer is first removed for slopes down to 1 in 10, or ploughed or scarified for lesser slopes. For slopes between 1 in 5 and 1 in 3 it is necessary to construct horizontal steps of 1–3 m width on which to found the formation embankment after removal of any loose surface soil. The width of the steps needs to be related to the excavator or skimming machines to be used in cutting the steps. On hillsides between 1 in 3 and 1 in 1.5 slope it is desirable to provide a counter bench or berm on the lower side founded on a stepped bed; for slopes steeper than 1 in 1.5 retaining walls are probably inevitable.

Where the railway passes through wastes of very fine sands, i.e. drifting or travelling sands, the main problem is to minimise the erosion of embankments and the filling of cuttings. This can be secured in varying degrees in varying locations by routing the railway round established sand dunes, endeavouring to build the line on embankment, preferably about the height of the existing sand dunes where the wind blows along the track, and in cutting where it blows across the track increasing the width of the land occupied by the railway in order to encourage the local survival of vegetation, planting trees and shrubs to hold the sand, the erection of shields of reeds, brushwood, cane and other straight stemmed grass and covering the embankment slopes with mats of the same material, the spraying of embankment slopes with bitumen to bind the sand and the covering of the slopes of cuttings and embankments with a layer of clay or gravel. In a drifting sand terrain protection against wind erosion is a primary objective; drainage provision and compaction provisions are probably of little consequence in such an environment, but it is of protective value to deposit the ballast immediately the formation is constructed.

Drainage

Water is commonly the greatest enemy of the permanent way and where the natural run-off in cuttings is not adequate to the drainage of track and cess, or where the ground water level in cuttings or on the flat is seasonally sufficiently high to diminish sensibly the foundation strength of the formation, drainage is required.

In some situations unlined ditches may be adequate but for heavy service these may prove difficult to keep clear and to a regular fall; lined channels are more reliable. These are usually formed with precast concrete units but where a large volume of water is involved *in situ* concrete or brickwork may be more appropriate.

If it is desired to lower the water table across the formation pipe drains are more effective and in a deep cutting it is desirable to site the drain, if only one is installed, on the upstream side of the watershed. The pipes can be of corrugated steel, glazed earthenware with spigot and socket joints, pitch fibre or cast iron with spigot and socket joints where the drain needs to cross under the track. Pitch fibre and corrugated steel pipes are often used with water collecting perforations. Glazed earthenware pipes are naturally limited in length and where speed and cheapness of construction is of paramount importance pipes which can be laid in lengths of 2.7 to 3.6 m are advantageous but since most track drains are primarily collecting drains longer lengths than about 1 m need to be perforated.

Channel drains can be laid at very slight gradients as they are readily accessible for cleaning and regrading. Pipe drains must be laid at a self-cleansing gradient. In general it is likely to be a wasteful investment to lay track drains of less than 150 mm diameter and it is sensible to regard 230 mm diameter as the normal minimum size. If possible pipe drains should not be laid at a lesser gradient than 1 in 200 but 1 in 1000 should be taken as an absolute minimum. In the best drainage work the joints at least are supported on a haunching of concrete over the bottom third of the periphery, especially with flat gradients where rapid silting up of the pipe may result from local loss of level.

Normally track drains are open jointed except where they cross under the track at the outfall end. They should preferably be placed in trenches battered at about 1 in 6 when in cohesive soil and back filled with side linings of filter material to keep the silt in the drain water from getting into the pipe, with a main backfull of coarse stone, hardcore or ballast. If necessary a layer of chippings, gravel, sand or clinker ashes may be used as a top cover. In clay soil a track drain may have an effective life of 20 to 70 years. The principal determinants of such life in any given cohesive soil environment are first, avoidance of slurry creation by suitable formation protection and efficient filter protection in the drain construction. The drain will normally follow the gradient of the track. A satisfactory average depth is about 1 m below rail level. The upper end should have at least 150 mm cover to the drain pipe and about 1.2 m below ground level is a reasonable depth limit at the lower end.

Manholes with catchpits, i.e. with manhole inverts about 230 m below the drainpipe invert should be provided every 30 m or so in a cohesive soil to permit access for cleaning the drain by rodding or pulling; these can be built up with precast concrete sections, with *in situ* concrete or with glass fibre prefabricated units. It is normal to cover catchpits or manholes for safety reasons and a cover in the form of an open grid is useful in allowing the state of the drain to be observed without removal of the cover. The manhole is usually finished a few inches above ground or ballast level.

Where drain pipes need to cross under the tracks it is desirable that the pipe should be at least 1 m below rail level or 0.75 m below rail level if protected by concrete at least 150 m thick.

Drainage for the protection of cuts and embankment works during and after their construction needs to be considered and implemented well before such works are commenced to secure safety in construction stability of the earthworks after construction and to avoid checks in the progress of the works. This is especially true of hillside locations, swamps, cuts and embankments on wet soils, major station or other building sites and of locations where soluble soils or rock strata are involved and where slow draining soils are concerned.

Soiling of slopes

It is desirable in the interests of earthwork stability and amenity to soil the sides of slopes with a layer of 10 cm of good top soil and sow it with grass seed or slips of indigenous root propagating plants. The vegetation chosen needs to give good ground cover quickly with little top growth which could require cutting back to prevent seeding or the presence of a fire hazard. In Britain a rye free grass seed mixture with some low growing clover in it is generally suitable.

The plants protect the slopes from erosion by run-off water, from wind erosion, from surface or weeping slips and from large and rapid variations of surface moisture content causing crack formation of a kind associated with circular or deep seated slips, besides saving the track drainage system from rapid silting up. Also if the slopes are not soiled and sown, weed infestation is likely.

If soiling and sowing is not practicable bitumen spraying can be useful in delaying weather erosion of slopes especially where sand or soft clay soils are involved. Where flooding is a hazard, stone or concrete slab pitching on the slopes may be appropriate. Where slopes are soiled and sown provision of watering by spray or trickle to establish a stable growth may be necessary.

BALLAST

The term ballast refers to all the material placed between sleeper and formation and, more specifically, to the material directly supporting the sleepers.

It is usual to cover the formation with a 100–150 mm layer of sand or ashes, or gravel with a sufficient cohesive soil content to bind it. Such a layer is useful to 'blind' a cohesive formation such as clay or silt in order to prevent slurry rising into the main ballast but very fine ash or sand may themselves cause slurry contamination if the formation becomes distorted through overload to cause ponding on the surface. The blinding layer also serves to prevent the main ballast pieces being driven into a soft formation under load pressures. The primary function of the main ballast is to secure stability of the track geometry within acceptable tolerances through the various seasons. In order

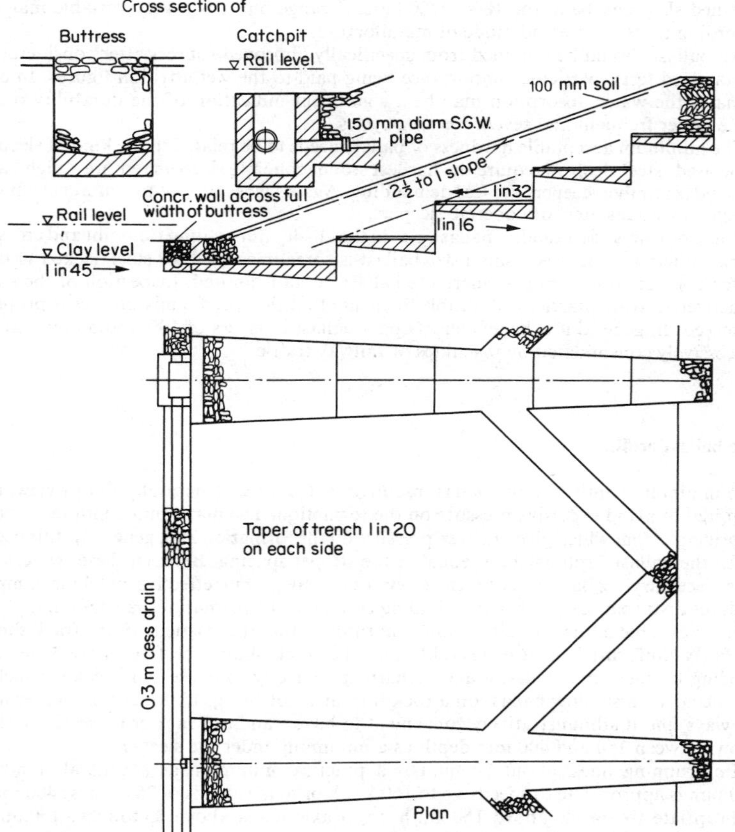

Figure 23.1 Buttress drain

to do this effectively, it needs to have (i) good frictional interlocking characteristics and (ii) a sufficient depth.

To a large extent, the choice of ballast is governed by technical adequacy at minimum cost at the working site. In some locations ashes, sand, gravel or beach ballast may be adequate for light loads at moderate speeds but these materials are variously liable to wind erosion, loss of bearing strength when saturated, water erosion, or lack of frictional interlocking. Crushed stone in a progressive grading between about 20 mm and 60 mm is most generally the standard ballast. Steelworks slag is also widely used where available. A cubical shape is generally preferred to a laminal one but some tolerance in this detail may be necessary to permit locally available stone to be used. Hardness is a primary requirement. A minimum strength of 700 kg/cm^2 in crushing is desirable. Good quality basalt or granite or any other stone in the 1600–4000 kg/cm^2 crushing strength range is a first technical choice.

Carboniferous and silicaceous limestones are likely to be adequate at 1600 kg/cm^2 but some limestones are soft with a crushing strength down to 200 kg/cm^2 when rapid powdering occurs, to the detriment of ballast drainage and also of rolling stock bearings. Crushed slag may be in the 1800–2700 kg/cm^2 range but it is a very variable material according to its source and mode of manufacture.

All ballast should be obtained from specifically approved sources after consideration of standard tests, particular importance being paid to the wet attrition figures. In cold climates the water absorption may be a significant indication of the durability of the stone under frequent and severe freezing cycles.

The minimum acceptable hardness of ballast needs to be related to the kind of sleepers to be used: steel sleepers require the hardest stone; concrete sleepers require a high hardness index; timber sleepers are less demanding. Again, ashes are apt to contain chemicals which may attack steel or concrete sleepers.

Some railways use a smaller ballast, e.g. in the 20–40 mm grading for point and crossing work; others do not. Also, small size ballast is sometimes used for steel decked bridges or for concrete floor bridges where the ballast depth is limited. Inspection of the stone as delivered from quarries is desirable to ensure that the specific mix of sizes is properly observed. In general it is beneficial to apply ballast in layers of 100–150 mm which are successively consolidated by machines or railway traffic.

The ballast profile

The minimum depth of stone ballast required is, from the structural point of view, that required to avoid excessive pressure on the formation. The maximum depth required is, in principle, that which gives an even pressure on the formation and, generally, this occurs when the ballast depth is about equal to the sleeper spacing. Between these two criteria it is necessary to have a sufficient stone depth to permit effective machine tamping without disturbance of either the blinding course or the formation, say 150 mm.

It is desirable in the interests of uniform running that the elasticity of the track should be fairly uniform; it is also desirable that the track should be free of weak sections needing frequent maintenance and of hard spots likely to cause rail breakage such as can occur with shallow ballast on a rock formation or bridge abutment. However most railways find it administratively convenient to have standard track cross sections which show between 150 and 400 mm depth as a minimum under the sleeper.

For running lines, about 200 mm is a practical minimum as a general standard; 300 mm is appropriate to speeds up to 160 km/h or to axle loads of 25 tonnes; 400 mm is appropriate to speeds above 160 km/h or to axle loads above 25 tonnes; 150 mm is appropriate to light railways or sidings.

The depth of ballast can also be related to gross tonnage since the frequency of maintenance tamping and lining is directly related to this. On this approach 230 mm is appropriate to say 15 million tonnes, 300 mm up to say 25 million tonnes per annum and 400 mm up to, say, 40 million tonnes annually.

Where there are significant local variations in soil strength it may be economical and technically appropriate to adjust the total construction depth of ballast and blanket underlay from point to point on the basis of results given by the Californian Bearing Ratio test. Although it does not directly measure soil strength, this is a well tried practical field procedure not needing any elaborate equipment or interpretation.

If practicable, it is of value to consolidate the ballast before the track is laid. This helps to preserve the formation from furrowing and reduces subsequent slacking of the track in service. On the other hand if the ballast is not laid as soon as the formation is prepared it may be distorted by the passage over it of men and machines.

The ends of the sleepers need to be boxed in with shoulder ballast to a minimum width of 150 mm for any trades, 200–250 mm for running lines carrying moderate speed moderate axle load traffic, 300 mm minimum for welded track on the straight and 350 mm for welded track on curves. Generally there is little advantage in extending shoulder ballasting beyond 300–350 mm. In the last decade or so, the practice of raising the shoulder ballast in a slope from about top of sleeper level at the rail to about top of rail level at the shoulder edge has become common on European railways. This practice not only increases the lateral stability of the track but provides a useful reserve of boxing ballast which can be temporarily utilised to make good the boxing ballast when the track is tamped. The angle of the shoulder should be about 55°.

SLEEPERS

Timber sleepers

The traditional track support is the timber sleeper but there is a growing movement toward various forms of concrete track supports. In some countries the baseplated timber sleeper is already more expensive to instal than a baseplateless prestressed concrete sleeper.

In Britain the sleepers and point and crossing timbers have traditionally been of softwood, chiefly Douglas Fir from Canada. Maritime Pine from SW France and Corsica, and Baltic Redwood from Poland and Russia. Home grown fir sleepers have been used to the extent that they have been available. In continental Europe, beech and oak sleepers have been widely used. All softwood sleepers, and most hardwood sleepers which can be impregnated are pressure creosoted by the Bethell or Ruping process before being baseplated. The harder softwood sleepers such as Douglas Fir or Scots Fir and hardwood sleepers need to be incised prior to creosoting. One to three gallons of creosote per sleeper is a normal absorption, according to species. In North and South America a mixture of mineral oil and creosote is used.

A timber sleeper may have a first or running line life of between 6 and 50 years with an average of about 20 years according to traffic loading, weather exposure, the nature and incidence of the indigenous vegetative and insect enemies of timber, the presence or absence of baseplates, the nature of the fastenings, the quality of the ballast and the maintenance and, of course, the species of timber and, significantly, the rate at which it has grown.

The traditional British timber sleeper is 10 in × 5 in × 8 ft 6 in with point and crossing timbers of 12 in × 6 in section. In continental Europe sleepers mainly of oak or beech are used to a depth of 160 mm. Until the last few years British sleeper spacing was 0.76 m or 2112 to the mile. In 1970 this was altered to 2288 to the mile or 26 to the 18 m rail length with provision for 28 on curves and under timber sleepered CWR, i.e. 2464 to the mile. Outside of Britain, 2500 to the mile is a normal sleepering density and 2800 is not exceptional.

The density of sleepering applied represents a compromise between overall sleeper cost and the lessened frequency of maintenance attention and reduced rail bending stress which results therefrom. In addition a decrease in sleeper spacing may be appropriate where the formation is weak or where it is not practicable to increase the total track construction depth by increasing the depth of ballast. Since also increasing the density of sleepering increases both the lateral and vertical resistances to buckling movements of the track, it is general practice to reduce sleeper spacing when laying track intended to carry long welded rails. Where the rail formation is regularly subjected to frost heave distortions there may be a case for deliberately choosing a rather weak rail with complementary close sleeper spacing in order to allow the track to accommodate itself to a frost heave contour without imposing undue bending stresses in the rails.

Steel sleepers

Steel sleepers are widely used in Germany, Switzerland, Greece, in most of the countries of South America, in North, Central and South Africa, and in several Asian countries.

Where atmospheric pollution is low, 50–60 years appears to be a realistic range of average life during which time from 20–30% of the sleeper may be lost in corrosion. Since, however, most steel sleepers fail through cracks from stress concentration points, or the wearing through of the sleeper at the rail seat, sleeper design features, especially those associated with the mounting of rail fastenings, and accumulated gross tonnage carried are also important factors determining service life.

In tropical climates the immunity of the steel sleeper to insect or spore damage is important. Steel sleepers also have a higher scrap value than timber or concrete sleepers; they pack neatly for transport; they have good resistance to lateral or longitudinal movement of the track and give a high consistency of gauge both at installation and in service. They are also repairable by welding or pressing. Nor do they burn or suffer from exposure to dry heat.

On the other hand steel sleepers are more expensive than either timber or concrete sleepers, although in some parts of the world the first cost differential is not very large. Steel sleepers also accelerate the wear degradation of ballast, needing therefore a rather harder stone than is necessary for timber or concrete sleepers. Steel sleepered track can be fine lined both manually and by machine but the manual lining adjustment is more expensive than with other types of sleeper and with both sorts of lining shows a tendency to move back where small slews are involved. For this reason special care in securing a good initial alignment is necessary where steel sleepers are installed. Before the advent of mechanical on-track tamping machines the packing of steel sleepered track was an expensive process. With modern machine tamping however there is no cost differential compared with concrete or wooden sleepers.

The lower individual weight of the steel sleeper lowers the track resistance to vertical buckling with long welded rails. Steel sleepered track is also slightly more noisy than timber or concrete sleepered track and this may be of relevance in heavily built up areas.

The electrical conductivity of the sleeper necessitates the use of insulating rail pads and fastening shoes or insulating washers to permit track circuit signalling and where conductor dust or detritus can accumulate on the track this aspect may present a major problem. In any case since an insulator failure on a steel sleeper is immediately decisive the use of such sleepers entails a considerable amount of preventive maintenance in the inspection and renewal of the insulating components. In addition on electrified tracks, particularly where direct current is used, the ready loss of return current to the ground may interfere with service pipes and mains.

The steel sleeper is susceptible to heavy corrosion loss in marine atmosphere, particularly when exposed to salt spray, in tunnels and at locations where chemically aggressive ground water is present. In general steel sleepers appear to be favoured for slow moderate speed lines; they are not much used in high speed lines.

Ballast practice with steel sleepers varies: some railways use large 40–70 mm ballast; others use small ballast (20–40 mm). The range of fastenings used with steel sleepers is limited.

Most sleepers are made from a steel containing from 0.08 to 0.15% copper to enhance corrosion resistance, and are hot dipped in tar.

Concrete sleepers

The diminishing availability of timber for railway sleepers throughout the world coupled with an escalating price has been accompanied over the last decade by the increasing use of concrete sleepers. Over the same period the financial difficulties of many railways have caused a cutback in sleeper renewals so that an increasing hidden demand has been building up. The logic implicit in this situation is that use of the concrete sleeper will expand progressively.

The practicability of the concrete sleeper was greatly advanced first by the two block reinforced type which needed little capital expenditure on factory equipment; and secondly by the development of the prestressed and post-stressed designs which reduced the steel reinforcement and gave the concrete sleeper a valuable elasticity under load and a greater durability. At the present time the two-block concrete sleeper and the through or monobloc concrete sleeper cost about the same, though in the past the two-block sleeper has been slightly cheaper. The monobloc sleeper appears to be used mainly on fast or heavily trafficked lines; the two-block sleeper mainly on medium speed medium loaded lines.

The British Railways standard F27 prestressed concrete sleeper (Figure 23.2) is a twenty-two wire sleeper 2.51 m long with a depth under the rail of 203 mm. It weighs 301 kg and is normally laid at 26 per 18 m length under 113 lb/yd section welded rail.

Figure 23.2 B.R. F27 prestressed concrete sleeper (Costain Concrete Co. Ltd.)

This type of sleeper has so far been in widespread use for 25 years and based on experience to date, coupled with experience of previously installed reinforced concrete sleepers, it seems probable that the F27 sleeper will have a service of at least 60 years; possibly 80 years. Some administrations have had trouble with structural failures of prestressed concrete sleepers, especially where a desire to minimise first cost has resulted in an inadequate specification. The rate of failure of prestressed concrete sleepers of the F27 type, used mainly at the wide spacing of 750 mm has not at the time of writing exceeded 1 in 16 000 per annum.

Behaviour of the concrete sleeper in a derailment is a factor of importance. With the two block type of sleeper a single pair of derailed wheels can so bend the tie rods between the blocks to cause the gauge to tighten to the extent of causing progressive derailment of following wheels. With the multiple wire prestressed concrete sleeper, general fracturing of the concrete at some location inside the gauge is not likely from experience to produce a gauge tightening in excess of about 3 mm. With timber sleepers on the other hand gauge is usually spread by a heavy axle derailment and there is the ensuing possibility of progressive wheels being derailed by spreading of the gauge.

The two block sleeper is not considered by some to give the stability of track geometry which can be secured with monobloc sleepers due to the tendency of the two blocks to assume a variable mutual inclination. This tendency leads to variable gauge and spalling of the concrete at the point of entry of the encastred ends of the tie bar into the blocks, especially under heavy axle loads on narrow gauge railways.

The damage to concrete sleepers arising from derailment is commonly supposed to be worse in extent and cost than with timber sleepers but experience does not clearly support this assumption. Indeed it appears that short grained hardwood timber sleepers particularly in arid country, will shatter more completely than monobloc prestressed concrete sleepers, though softwood sleepers can perhaps stand more flange impact without needing to be replaced than concrete sleepers. In either case a high level of the boxing ballast will minimise this type of damage.

The need for an adequate inter-rail insulation for track circuiting can be met with most concrete sleepers without technical difficulty or great expense; but the design of the fastenings and their housings are important in this regard. Where the heads of the bolts holding the rail clips are housed in a connecting metal tie bar, the need to maintain the insulating parts to a high standard is as important as with steel sleepers.

FASTENINGS

Rail fastenings are referred to as direct when the fastening holds the rail directly to the sleeper and indirect when it holds the rail to a chair or baseplate which is separately held to the sleeper. In addition, fastenings may be rigid or elastic, depending on whether a spring element is incorporated, and also adjustable or self tensioning.

The square section cut spike was the original fastening used for flat bottom track (Figure 23.4). In Europe it is now regarded as obsolete except for sidings and light narrow

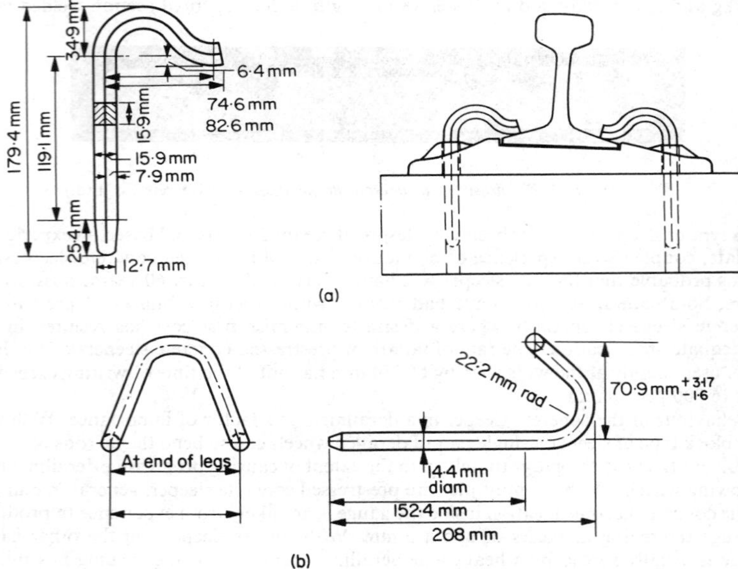

Figure 23.3 Types of spring spike

gauge railways; but in America and on American-built railways in various parts of the world it is still in general use on lines carrying up to 35 tons axle loads. Features of this type of fastening are the need for a high sleepering density (in America 3200 to the mile is regarded as a minimum); continuous replacement of sleepers; the normal use of heavy section rail and rail anchors each side of each sleeper; a short sleeper life and variable track gauge. It is a simple and robust system which lends itself to rapid pioneer development but is not well adapted to high speed welded rail track in countries where the annual temperature range is high, nor to countries where timber and steel are scarce resources.

In Europe the coach screw direct fastening has been preferred and is still extensively used. Both spike and screw have poor gauge holding when used as direct fastenings but the position is improved when baseplates are used, especially when additional screws or spikes are used for indirect fastening, i.e. to secure the baseplate to the sleeper; and it is now general practice to fit baseplates on curved track fastened with spikes or screws. The main advantage of the screw over the cut spike is that it exerts a positive pressure on the rail foot where the spike does not. If therefore holds the rail against rail drive and the loss of expansion gaps. In France and on French-built railways, the screw used with a spring clip—the RN and the RNS clips—is now the general standard.

In Germany and much of central Europe the K fastening of the DB has been the standard fastening since 1926. This consists of a rolled steel baseplate giving a 1 in 40 rail inclination with two heavy ribs forming a channel seating for the rail resting on a 5 mm compressed poplar pad and slotted to take a T-headed bolt each side, inverted rigid U-shaped clips being held by the tee bolts to which heavy spring washers are fitted. Four screws are used to hold the baseplate down to timber sleepers in Germany but in some other countries, e.g. Belgium, only two are used. In Germany spring steel washers are commonly used on the baseplate screws. The K baseplates adaptable to a number of spring clips driven into the rib of the baseplate.

In Britain the Pandrol type self-tensioning clip has been the standard British Rail fastening since 1964 on both concrete and timber sleepers. It is also used extensively in Africa and Australia. Other self-tensioning fastenings which have been used extensively in Britain include the Heyback clip, the Mills 'C' clip and the SHC flat plate clip. The Fist self-tensioning stirrup has been widely used in Sweden and South Africa.

In many parts of the world an increasing shortage and expense of track maintenance labour has induced a growing interest in 'fit-and-forget' type tracks so that self-tensioning fastenings are becoming more popular, especially for continuous welded rail track. Here the importance of fastenings maintaining a constant rail holding pressure is fundamental to the lateral stability of the track, particularly when the ballast profile is below standard. The clamping force holding the sleeper to the rail needs to be enough for the rail to drive the sleeper in the ballast before the rail slips over the sleeper. A nominal clip pressure of 1.6 to 2.0 tonnes per rail is generally adequate to secure this. It is however necessary that any self-tensioning clip should have a large deflection to ensure that its nominal pressure on the toe of the rail is substantially retained throughout its service life in the face of cumulative adverse manufacturing tolerances in rail flange thickness, clip, rail pad, insulator and rail clip housing, together with wear of these elements. A spring clip with 10 mm deflection losing 5 mm of its deflection loses 50% of its toe load where a clip having a designed deflection of 20 mm loses only 25% of its pressure. It is also important that the design of the clip and its mounting taken together should hold accurate gauge by positive and unyielding location of the rail.

RAILS

The flat bottom or Vignole rail is now an almost universal standard so that the obsolescent bull head rail of past British practice need not be developed here.

Generally rail sections or weights are derived from progressive experience, not because of analytical difficulties but because of inadequate quantified knowledge of what conditions of loading and support actually occur.

A simple rough guide to appropriate rail weight in kg/m is to multiply the axle load in metric tons by two. However, it is necessary to make allowance for a speed factor and the following formula given by Schramm does this:

$$\text{Rail weight in kg/m} = 156 - \frac{10\,600}{A\alpha + 67}$$

where

A = static axle load in metric tons;
α = the speed factor.

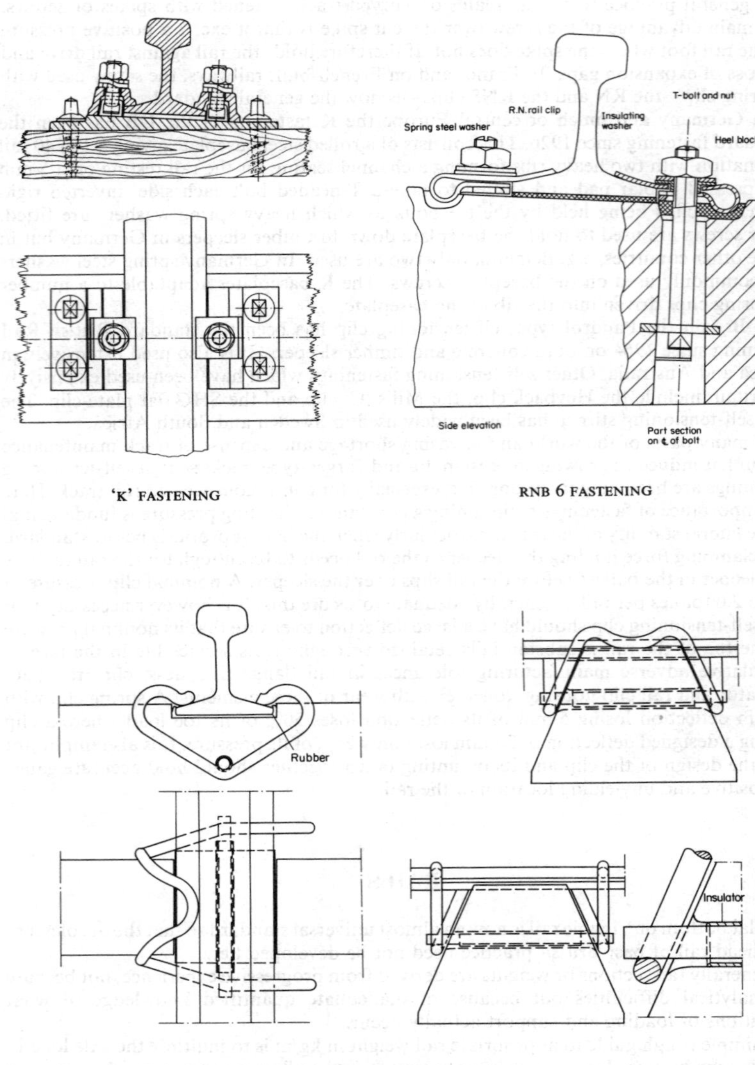

Figure 23.4 Types of rail fastening

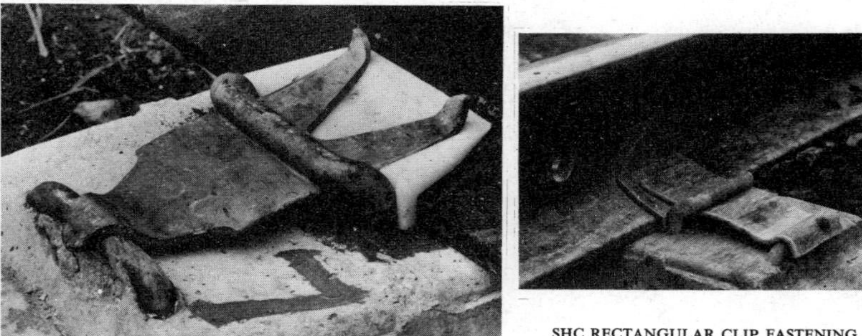

SHC RECTANGULAR CLIP FASTENING

GHC 'VEE' FASTENING

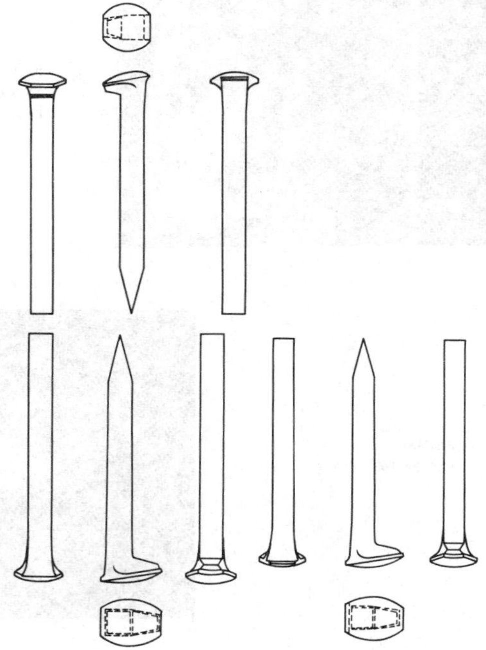

AREA CUT SPIKE FASTENINGS

Figure 23.4 (continued) Types of rail fastening

MILLS
'C' TYPE

PANDROL CLIP
AND LOCKSPIKE
FASTENING

HEYBACK
FASTENING

Figure 23.4 (continued) Types of rail fastening

The speed factor is still under critical review but it can be evaluated by three formula which are widely accepted for practical use:

$$\alpha = 1+\frac{V^2}{30\ 000} \qquad \text{up to 100 km/h}$$

$$\alpha = 1+\frac{4.5V^2}{10^5}-\frac{1.5V^3}{10^7} \qquad \text{up to 140 km/h}$$

$$\alpha = 1.18+\frac{0.706V^3}{10^7} \qquad \text{over 140 km/h}$$

From the above expressions, Table 23.1 has been produced. The figures give an indication of appropriate weights within ±15%.

Table 23.1 APPROXIMATE RAIL WEIGHTS, KG/M

Static axle weight in tons	*Speed in* km/h				
	50	100	140	160	200
15	28	34	36	37	39
20	36	42	44	46	49
25	44	50	52	54	58
30	50	56	59	61	65
35	57	61	65	67	70

British rails are available from 12 to 55 kg/m; in Europe rails up to 75 kg/m are used and up to 77 kg/m in the USA. The present standard in Britain and over a large part of Europe is the 54 kg/m rail, adapted from the earlier British standard 109 lb rail. There is however an increasing use in continental Europe of 60 kg/m rails and in the USSR the 75 kg/m rail. Table 23.1 relates to a representative sleeper spacing of 630 mm. If the spacing is wider than this, and until recently British practice utilised a sleeper spacing of 750 mm, the stress due to bending moment may be up to 9% greater but in any case the occurrence of three loose sleepers can put up rail bending stress by 100% as can a heavy wheel flat at 20 mph. The formulae quoted do not differentiate the sprung and unsprung proportions of the static axle weights indicated. At rail ends a 12 mm dipped joint, can with an unsprung mass of the order of 20% of the static load at a speed of 160 km/h create a dynamic increment of load equal to the static load. Since the great majority of rail breaks occur at rail ends it can be stated that the rail weight selected must take into account the unsprung masses on the heavier axles and also the state of maintenance of joints which can be realised.

The value of the dynamic increment according to the Research Division of British Rail is given by

$$F = 2\alpha V\sqrt{\left(\frac{K \, . \, M_2}{g}\right)}$$

where

2α = dip angle in radians
V = velocity in inches/sec
K = effective track stiffness inches/sec
M_2 = Unsprung weight
g = gravitational constant in inches/in/sec

The weight of rail chosen has an influence on the shear loading of the subsoil. According to Eisenmann increasing the rail weight from 98 lb/yd to 140 lb/yd diminishes shear stress by 20%.

A further consideration affecting wheel load in relation to wheel size and rail quality is that the sub-surface Hertzian stresses arising from the wheel/rail contact are directly related to the square root of the wheel load and inversely as the square root of the wheel radius and that shelling failures are likely when the shear stress at the contact zone exceeds about 50% of the ultimate tensile strength. Because of this high tensile quality rails can take a larger wheelweight/wheel size ratio without shelling.

As rail weights increase, the breadth of the rail foot increases so that changing to a

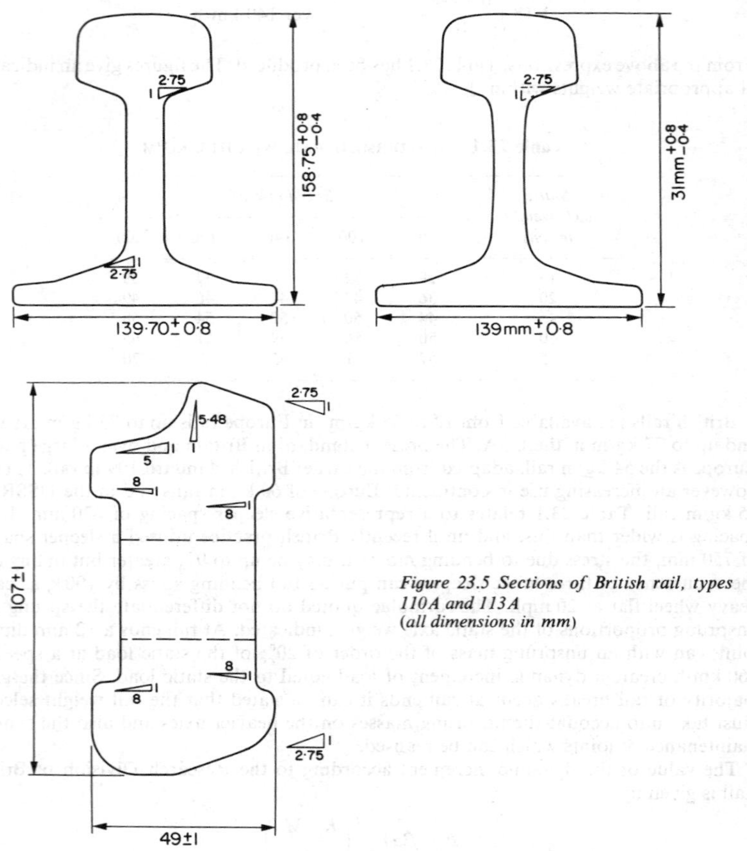

Figure 23.5 Sections of British rail, types 110A and 113A (all dimensions in mm)

heavier weight rail will generally involve changing the baseplates or, where cast-in inserts are used, re-spacing the inserts, unless the new rail is limited to new sleepers. Similarly the administration of permanent way stocks is simplified and the tied up capital reduced if the range of rail sections in use by a railway is limited.

Where rails are to be welded into long rails or continuous welded rails it is also prudent to choose a rail weight which has a margin of strength corresponding to the contraction stresses to which the rail will be subject.

The principal specifications for rails having an international use are Specification

860 and 860–2 of the Union Internationale des Chemins de Fer (UIC) and the specifications of the American Railway Engineering Association. The range of flat bottom rails normally available is from 12 to 77 kg/m.

The most commonly used rail steels are defined in the above standards. The special steel used most generally for rails in the UK is 12–14% manganese steel, which surface hardens rapidly in service but which is expensive and also needs great care in machining if fatigue cracks are to be avoided. It is largely used for points and crossings but since it has a coefficient of expansion several times that of ordinary rail steel and poor electrical conductivity its use is generally limited to isolated sections of junction work.

Rails with slightly higher than normal carbon or manganese contents (see UIC Specification 860–1) are sometimes used on curves to reduce sidewear but generally effective rail lubrication gives a better return. Rails with higher carbon or about 1% chromium are sometimes used where heavy axle loads cause plastic flow of the rail head.

Rail joints are either suspended or supported. The suspended joint is mainly associated with bulkhead tracks. Joints are also square, i.e. opposite each other, or staggered by up to half the rail length. In some administrations rail joints are square on the straight and staggered on the curves. Rail and fishplate sections need to be considered together in selecting a rail section, since if the ratio of the *Ixx* of the fishplate is much below 25% of the *Ixx* of the rail, broken fishplates and battered rail ends are likely to have a high incidence.

Broken and cracked rails

It is important for any railway to keep a classified record of cracked and broken rails and periodically to review the rate at which defective and broken rails are occurring, particularly if speeds, axle loads or gross tonnage on the system are increasing. A standard report form is a basic starting point. In addition it is necessary for any railway system to endeavour to detect cracked rails before they break and for this purpose a system of periodical visual, magnetic or ultrasonic inspection, particularly at rail ends, is essential. The best system technically is an ultrasonic rail flaw detection car with automatic film recording of the signals and automated scanning of the record but the cost of this provision is beyond the scope of most railways outside of the larger national systems although hiring a car of this type may be possible. Alternatively portable battery operated ultrasonic hand sets are available.

CURVED TRACK

The main curves of the railway are nominally of constant radius that is, circular curves; curves made up of two or more circular curves of different radius curving in the same direction are called compound curves. Straights are generally and desirably connected to circular curves by transition curves of progressively varying radius; and adjoining circular curves of different radius are commonly joined together in a similar way if the difference of radii exceeds about 10%. Two adjoining circular curves curving in opposite directions comprise a reverse curve, and here, whatever the radius, the presence of connecting transitions is relatively more important than with circular or compound curves.

For practical purposes the cubic parabola $y = kx^3$ gives a uniform change of curvature between tangent point on the straight and tangent point on the curve, or between the tangent points of two curves, and is the most used form of transition curve. The versines, measured on half-chord overlapping chords along the transition, change with linear uniformity from zero to R although it is usual to smooth out the rate of increase and decrease at the start and end of the transition by putting on about one-sixth of the rate of the first versine at the zero station and reduce the increment at the final transition station to about five-sixths of the half-chord rate of increase. Versines are conveniently measured in millimetres.

The geometrical relation between a circular curve and a transition curve tangenting on to both the straight and the circular curve involves the moving inwards along a diameter normal to the straight of the circular curve by an amount termed the shift. The transition curve bisects the shift at its midpoint measured along the straight tangent line and the offset from that line to the tangent point of the transition and circular curve is four times the shift or eight times the offset at the mid-transition point. It follows that where no transition exists or where it is insufficient in length the institution or extension of a transition can be done only by sharpening the radius of the circular curve concerned or moving the tangent straight away from the curve, though this may be worthwhile since any transition is better than none.

The length of a transition curve is determined primarily by what is judged to be an acceptable rate of change of cant or cant deficiency. For standard gauge plain track a desirable rate may be 35 mm/sec. with a maximum of say 55 mm/sec to secure passenger comfort. In switches and crossings a rate as high as 80 mm/sec may be applied but a good standard of switch and lead design is desirable for this rate of loss or gain of cant or cant deficiency.

Some limiting cant and cant deficiency values observed in British practice on a 4 ft $8\frac{1}{2}$ in gauge are listed below.

Maximum cant:	
on curved track	150 mm
at station platforms	110 mm
gradient	1 in 400
deficiency on plain line, continuous welded rails (CWR)	110 mm
deficiency on plain line, jointed track	90 mm
deficiency on switches and crossings welded into CWR	
on through line	110 mm
on turn out	90 mm
with negative cant on turn out	110 mm
at switch toes	125 mm
deficiency on switches and crossings in jointed track	
on through line	90 mm
on turn out	90 mm
with negative cant on turn out	110 mm
at switch toes	125 mm
Maximum rate of change of cant	
on plain line	55 mm/sec
on switches and crossings	55 mm/sec
Maximum rate of change of cant deficiency	
on plain line	55 mm/sec
on switches and crossings	55 mm/sec
Maximum rate of change of cant deficiency	
on plain line	55 mm/sec
on switches and crossings (inclined design)	55 mm/sec
on switches and crossings (vertical design)	80 mm/sec

The amount of cant applied to a track depends upon consideration of the following factors:

(i) The line speed limit, that is, the maximum speed at which traffic is allowed to run on a line or branch or section of a line or branch. This limit is usually fixed with reference to the value and distribution of permanent speed restrictions on the line or branch, or section thereof, involved.

(ii) The proximity of permanent speed restrictions, junctions, stopping places, etc.

(iii) Track gradients which may cause a reduction in the speed of freight or slow moving passenger trains without having an appreciable effect on the speed of fast trains.

(iv) The relative importance of the various types of traffic.

(v) On reverse curves where the speed is above say 100 km/h the permissible rate of variation of cant should be not more than twice the corresponding maximum rate of variation of curvature, if the straight between the reverse curves is less than about 30 m.

Generally where fast and slow trains run on the same lines an intermediate speed is selected to fix the cant. In this situation it may exceptionally be necessary to limit the cant and therefore the maximum speed to prevent surface damage to the low rail by heavy axles on slow moving freight trains.

Each line of a double line should be separately assessed. In exposed locations subject to high winds it may be desirable to limit cant to below the 150 mm maximum.

Normally cant or cant deficiency is uniformally gained or lost within the length of a transition curve, or, where there is no transition curve, as may occur in switches and crossings, over the length of the virtual transition, which for practical purposes is taken as the shortest distance between the centres of bogies of coaching stock using the line. If the desired cant cannot be put on within this length observing a maximum cant gradient of 1 in 400, the cant loss or gain is continued on to the circular curve. The 1 in 400 cant gradient mentioned relates to axle twist derailment possibilities, especially where four wheeled vehicles are concerned.

The maximum permissible speed on circular curves appropriate to the determination of permanent speed restrictions may, for standard gauge railways be obtained from the following expressions:

$$\text{Equilibrium (or theoretical) cant} = E = 11.82\,\frac{Ve^2}{R}\ \text{mm}$$

$$\text{Equilibrium speed} = Ve = 0.29\sqrt{(RE)}\ \text{km/h}$$

$$\text{Maximum speed} = Vm = 0.29\sqrt{R(E+D)}\ \text{km/h}$$

where

R = radius in m

E = cant, which may be either actual cant or equilibrium cant but in practice the difference is not likely to be significant, though the distinction has to be be kept in mind in certain circumstances

D = maximum allowable cant deficiency in mm.

Desirable lengths of transition curves can be derived from the greater of the two values obtained from:

$$\text{Length} = L = 0.0075EVm \text{ metres}$$

or

$$L = 0.0075DVm \text{ metres}$$

where

E = cant in millimetres

D = deficiency of cant in millimetres

Vm = maximum permissible speed in km/h.

If space is limited the length of the transition may be reduced to two-thirds L subject to a minimum cant or cant deficiency gradient of 1 in 400.

On compound curves to be traversed at a uniform speed the desirable length is obtained from the greater of the two following expressions:

$$L = 0.0075(E1-E2)Vm \text{ metres}$$

or

$$L = 0.0075(D1-D2)Vm \text{ metres}$$

where $E1$ and $D1$ are the cant and cant deficiency conditions for one curve and $E2$ and $D2$ are the similar values for the other curve.

Table 23.2 CURVE FORMULAE

Given	*Sought*	*Formula*	*Given*	*Sought*
E, M	C	$C = 2M\sqrt{\left\{\dfrac{E+M}{E-M}\right\}}$	α, D	L
E, R	C	$C = \dfrac{2R\sqrt{\{E(2R+E)\}}}{R+E}$	C, E	M
E, T	C	$C = \dfrac{2T(T^2-E^2)}{T^2+E^2}$	C, α	M
E, α	C	$C = 2E\dfrac{\sin\frac{1}{2}\alpha}{\text{Ex sec}\frac{1}{2}\alpha}$	E, α	M
M, T	C	$O = C^3-2TC^2+4M^2C+8M^2T$	R, C	M
M, R	C	$C = 2\sqrt{\{M(2R-M)\}}$	R, E	M
M, α	C	$C = 2M\cot\frac{1}{4}\alpha$		
R, T	C	$C = \dfrac{2TR}{\sqrt{(T^2+R^2)}}$	R, T	M
R, α	C	$C = 2R\sin\frac{1}{2}\alpha$	R, α	M
T, α	C	$C = 2T\cos\frac{1}{2}\alpha$	T, C	M
R	D	$\sin\frac{1}{2}D = \dfrac{50}{R}$	T, E	M
α, L	D	$D = \dfrac{100\alpha}{L}$ approx.	T, α	M
C, M	E	$E = M\dfrac{C^2+4M^2}{C^2-4M^2}$	D	R
C, α	E	$E = \frac{1}{2}C\dfrac{\text{ex sec}\frac{1}{2}\alpha}{\sin\frac{1}{2}\alpha}$	C, E	R
M, α	E	$E = \dfrac{M}{\cos\frac{1}{2}\alpha}$	C, M	R
R, C	E	$E = \dfrac{R^2}{\sqrt{\{(R+\frac{1}{2}C)(R-\frac{1}{2}C)\}}}$	C, α	R
R, M	E	$E = \dfrac{RM}{R-M}$	E, α	R
R, T	E	$E = \sqrt{(T^2+R^2)}-R$	M, E	R
R, α	E	$E = R\,\text{ex sec}\frac{1}{2}\alpha$		
T, C	E	$E = T\sqrt{\left\{\dfrac{2T-C}{2T+C}\right\}}$	M, α	R
T, M	E	$O = E^3+E^2M-ET^2+MT^2$	T, C	R
T, α	E	$E = T\tan\frac{1}{4}\alpha$	T, E	R
R, α	E	$E = \dfrac{R1-\cos\frac{1}{2}\alpha}{\cos\frac{1}{2}\alpha}$	T, M	R

D = Degree of curve
R = Radius
α = Ext angle = central angle
L = Length of curve

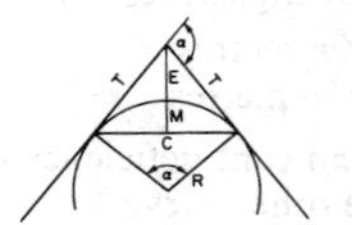

M = Mid-ordinate
T = Tangent
C = Long chord
E = External distance

Table 23.2 (*continued*)

Formula	*Given*	*Sought*	*Formula*
approx	T, α	R	$R = T\cot\frac{1}{2}\alpha$
$+M^2E+\dfrac{MC^2}{4}-\dfrac{C^2E}{4}$	C, E	T	$O = 2T^3-T^2C-2TE^2-CE^2$
an $\frac{1}{4}\alpha$	C, M	T	$T = \dfrac{C(C^2+4M^2)}{2(C^2-4M^2)}$
s $\frac{1}{2}\alpha$	C, α	T	$T = \dfrac{C}{2\cos\frac{1}{2}\alpha}$
$\sqrt{\{(R+\frac{1}{2}C)(R-\frac{1}{2}C)\}}$	E, α	T	$T = E\cot\frac{1}{4}\alpha$
E	M, E	T	$T = E\sqrt{\left\{\dfrac{E+M}{E-M}\right\}}$
$\dfrac{R^2}{\sqrt{(T^2+R^2)}}$	M, α	T	$T = M\dfrac{\tan\frac{1}{2}\alpha}{\text{vers}\,\frac{1}{2}\alpha}$
rs $\frac{1}{2}\alpha$ $\sqrt{\left\{\dfrac{2T-C}{2T+C}\right\}}$	R, C	T	$T = \dfrac{CR}{2\sqrt{\{(R+C/2)(R-C/2)\}}}$
$-E^2)$	R, E	T	$T = \sqrt{\{E(2R+E)\}}$
$+E^2$	R, M	T	$T = \dfrac{R\sqrt{\{M(2R-M)\}}}{R-M}$
t $\frac{1}{2}\alpha$ vers $\frac{1}{2}\alpha$	R, α	T	$T = R\tan\frac{1}{2}\alpha$
D	D, L	α	$\alpha = \dfrac{DL}{100}$ approx
$R^2\left(\dfrac{4E^2-C^2}{8E}\right)-\dfrac{RC^2}{4}-\dfrac{C^2E}{8}$	M, C	α	$\tan\frac{1}{4}\alpha = \dfrac{2M}{C}$
$+(\frac{1}{2}C)^2$ / M	M, E	α	$\cos\frac{1}{2}\alpha = \dfrac{M}{E}$
$\frac{1}{2}\alpha$	R, C	α	$\sin\frac{1}{2}\alpha = \dfrac{C}{2R}$
c $\frac{1}{2}\alpha$	R, E	α	ex sec $\frac{1}{2}\alpha = \dfrac{E}{R}$
M	R, M	α	vers $\frac{1}{2}\alpha = \dfrac{M}{R}$
$\frac{1}{2}\alpha$	R, T	α	$\tan\frac{1}{2}\alpha = \dfrac{T}{R}$
$\dfrac{CT}{\sqrt{\{(2T+C)(2T-C)\}}}$	T, C	α	$\cos\frac{1}{2}\alpha = \dfrac{C}{2T}$
$\dfrac{(T+E)(T-E)}{2E}$	T, E	α	$\tan\frac{1}{4} = \dfrac{E}{T}$
$R^2\left(\dfrac{M^2+T^2}{2M}\right)+RT^2-\frac{1}{2}MT^2$	R, α	L	$L = 0.017\,453\,292\,5\,R\alpha$

Note: ex sec $A = \sec A - 1 = \dfrac{1-\cos A}{\cos A} \notin$ vers $A = 1-\cos A$

Similarly on reverse curves the transition lengths are given by:

$$L = 0.0075(E1+E2)Vm \text{ metres}$$

or

$$L = 0.0075(D1+D2)Vm \text{ metres}$$

It should be noted that in using the above formulae the constants used have reference to a standard gauge, to an assumed height of the centre of gravity of vehicles of about 1.5 m and a subjective passenger comfort assessment of the tolerable rate of change of cant or cant deficiency. To this extent the values are arbitrary rather than absolute and refer to standard gauge. Further qualifications are that the springing of vehicles may be such as to produce excessive lean under cant deficiency running to encroach significantly on side clearances or to diminish passenger comfort, whilst if the centre of gravity of a vehicle is higher or lower than that assumed the maximum permissible speed is affected in inverse ratio, so that certain vehicles may be permitted to run at a higher speed than other vehicles.

The versine or middle ordinate of a chord on to a curve is proportional to the curvature and is the basis of all railway curve alignment, checking and adjustment. Its value as determined from triangular analysis is given by:

$$2R = \frac{(C/2)^2}{V} + V$$

where

R = radius of a curve
C = length of chord on which the versine is measured
V = the versine

but since the value of V is very small in relation to R in railway situations, the final V of the expression may be disregarded so that for both field measurements and calculation purposes:

$$V = \frac{C^2}{8R}$$

In Britain and on the Continent, railway curvature is usually described by the length of the radius measured in metres. The American practice is to describe a curve by the angle subtended at the centre of the curve by a chord of 100 ft. For railway work it is sufficient in transposing degree units into radius units to divide 5730 by the degree of the curve to give the radius in feet, or to divide 1746 by the degree of the curve to give the radius in metres. Thus what is described in American practice as a 10° curve would be described in Europe as a curve of 175 m or 8.7 chains radius.

Main railway curves are initially set out by theodolite generally on the basis that equal chords are subtended by equal angles but informal setting out by an offset from a tangent followed by the use of overlapping chords of convenient length using the versine appropriate to the radius can give a degree of accuracy sufficient for less important curves. A selection of curve formulae covering most calculations for the design or setting out of railway circular curves is set out in Table 23.2.

It is desirable to keep rail joints more or less square, or at a constant stagger where this is preferred, on curves by inserting short rails in the inner rail of curved track. The difference in rail length, D, for standard gauge and for 60 ft or 18.3 m rails is given by:

$$D = \frac{27.45}{\text{Radius in metres}} \quad \text{metres}$$

Alternatively the approximate difference in length can be obtained from the formula $D = \frac{2}{3}$ versine. It is normal to take a standard range of short rails from the makers and for 18.3 m rails, these are 18.25, 18.20 and 18.15 m.

The precise alignment of curves is vital to a smooth ride, stability of the track geometry and minimising track wear. There are several alternative methods of adjusting railway curves as a maintenance operation but all are based on versine measurements and relate, either to a smoothing or averaging approach, in which the differences between a limited number of adjacent versines is adjusted to give a fairly uniform rate of change in transitions or to a fairly uniform value on circular curves; or to a design lining approach in which the whole of the versines of a curve are adjusted as one revision operation to give a precise rate of increase or decrease in transitions and a strictly uniform value on the circular curve portion, as far as clearances from structures etc. will permit. It is generally convenient to use smoothing or local adjustment techniques after a curve has been aligned on a design basis. It is also normal practice to set markers or monuments or pegs in or beside the tracks when design lining is carried out.

Curve realignment can be based on manual measurement of versines or with less accuracy on the versine measurements of a track geometry recording car or of a track lining machine.

The side and interbody clearances shown on the diagram contained in the Ministry's Requirements (see Figure 23.14) relate to straight track and may need to be augmented on curved track for end throw or centre throw of vehicles, which are generally of the same magnitude. (These requirements are currently under revision).

$$\text{Centre throw} = C = \frac{B^2}{8R}$$

$$\text{End throw} = E = \frac{L^2 - B^2}{8R}$$

where

B = wheelbase or bogie centres
L = length of vehicle
R = radius

To this increase of effective body width on curved track must be added a further allowance due to the tilting of the vehicle to the low side on canted track. In Britain the loss of side clearance at the vehicle cornice is about two and a quarter times the actual cant.

It is general practice slightly to widen the track on very sharp curves to allow all vehicles especially locomotives with a long rigid wheelbase to pass round such curves without straining the track. The normal British practice is:

Curves of 10 to 7 chains	6 mm
7 to $5\frac{1}{2}$ chains	12 mm
Less than $5\frac{1}{2}$ chains	19 mm

Appropriate values of gauge widening for a given track radius depend on free play of wheelsets, length of half wheel flange below rail level and rigid wheelbase of the critical vehicle.

In the UK it is the Ministry of Transport's enjoined practice to install check rails on curves of 10 chains or less radius. Check rails are also sometimes installed in Britain on flatter curves to diminish side cutting of the high rail or as a protection against inadvertance or mishap at the bottom of a gradient. Wide flangeway check rails or guard rails may also be installed to protect bridge supports or girders against being struck by derailed vehicles. In the UK this provision is a Ministry of Transport requirement.

WELDED TRACK

About two-thirds of the on-track maintenance work put into jointed permanent way is at or adjoining the rail joints and the great majority of rail failures occur at or near fishplated joints. For these and other reasons the tendency today is to use long welded or continuously welded rails.

This procedure entails the elastic containment of the longitudinal expansion and contraction stresses arising in the rails as a consequence of variation in the rail temperature with reference to the temperature at which the rails were fastened to the sleepers. If, as is usual, the rails are fastened down on installation within a prescribed narrow temperature range situated at or a little above the mean of the annual rail temperature range in a normal or stress free condition, then the amount of tension in winter and compression in summer is limited to what is judged to give an optimum compromise between a cold season hazard of broken rails and a hot season hazard of track buckles.

The installation and maintenance of welded rail track need more technical insight and attention than ordinary track. The lateral strength of the track depends upon three main elements, the *Iyy* of the two rails, the framework stiffness of the assembled rails and sleepers as developed by the fastenings, and the frictional loading of the ballast on the sides, ends and bottom of the sleepers. The resistance of the track to vertical buckling, which generally occurs in combination with, or as a trigger to, lateral buckling, is deter-

Figure 23.6 Rail hydro-stressor. This has a 70 ton pull and 15 in stroke. For pushing action the clamps are turned through 180° (The Permanent Way Equipment Co.)

mined mainly by the total weight of the track and in this context concrete through type sleepers show a considerable advantage. In a fully-ballasted track the ballast representatively accounts for about two-thirds of the total moment of lateral resistance. Loose sleepers and kinks in the alignment are of especial significance to the stability of welded track so that a high standard of maintenance is essential for this sort of permanent way.

Fastening welded rails

The need to fasten the rails in order to get a stress-free rail at about the mean of the annual temperature range would mean, in many parts of the world, a very short season for laying welded rails without artificial assistance. This assistance is normally supplied either by heating the rails by propane-gas travelling heaters or by using hydraulic tensors, so that when fastened down after extension by heat or tensile pull, the rail has a length equivalent to a stress free condition at a temperature between prescribed limits (21–27 °C for UK). In this way it is possible to lay welded track all the year round. The pull required to extend a 54 kg/m rail is of the order of 1.6 tonnes per °C. The tensors have a capacity of about 70 tons.

Welded rails are not normally laid in Europe in curves with a lesser radius than 600 m. Long welded rails are made up by flash butt welding of standard rail lengths into welded rails of 200–400 m length in depot and welded into continuous welded rails by Thermit welds, except in Russia where a transportable flash butt welder is applied to join rails in the track.

It is normal practice in Britain and France to install sliding switches at the ends of long welded rails but in Germany this is not done, reliance being placed on very firm fastening of the welded rails where they connect with jointed track.

The advantages of welded track include an extension of rail lives by about a third, a reduction of on-track maintenance by about half, a dramatic reduction of rail breakages, an increase in running speeds, less damage to the formation, increased sleeper lines, improved ride comfort, a reduction of traction energy of up to 5%, and a reduction in train noise. In effect welded rails, prestressed concrete sleepers and self-tensioning fastenings are complementary to each other in extending the life and reducing the long term overall cost of permanent way.

SWITCHES AND CROSSINGS

All railway switch and crossing work is built up of three basic units: switches, common or acute (angle) crossings, and obtuse (angle) or diamond crossings. Layouts of switches and crossings are illustrated in Figures 23.7 to 23.10.

Switches

The heel of a switch is the point from which it is free to move. Early switch designs had a loose heel formed by a semi-tight fishplated joint but this type is now rarely used outside of sidings. It is now the general practice for the heel of the switch to be formed by a bolted connection through a block between switchrail and stockrail so that the switch joint is well behind the heel. The length of the switch planing on the head of the rail varies between about 0.5 m and 11 m and most railway administrations have a range of two to eight standard switches to meet the requirements of short leads in depots and various speeds of traffic on the running lines.

Originally ordinary or straight planing of switch blades was universally applied but during the last few decades curved planing, in which the switch rail is bent elastically whilst being planed on one side of the head, has become general. In this way the curve running through the lead is continued to the switch tip. A recent practice is to apply a larger radius to the planing of the switch than is applied to the switch between the end of the planing and the heel followed by a sharper radius between the heel and the crossing. In this way a transitioned turnout is formed which permits raising the speed through the turnout, an important point for high speed running.

Crossings

Crossings are described by the angle at the crossing nose. In continental Europe it is common to state this in degrees but over the English speaking world the angle is usually described as 1 in N. However, the angle of a crossing expressed as 1 in N, varies slightly

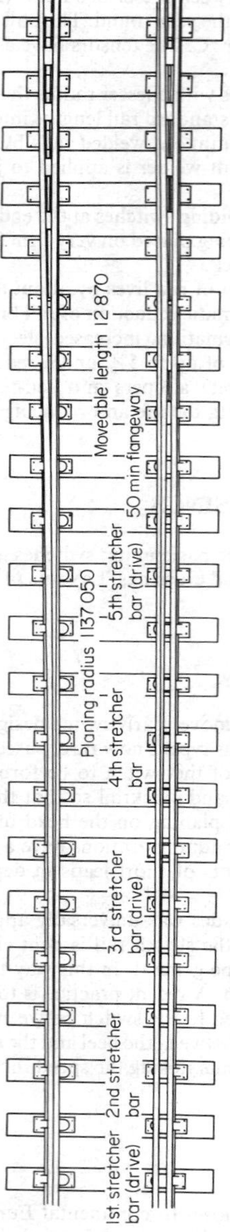

Figure 23.7 Timbered layout of long switches

Figure 23.8 Cast common crossing (Edgar Allen & Co. Ltd.)

Figure 23.9 Part welded crossing in lead on prestressed concrete bearers (British Railways, London Midland Region)

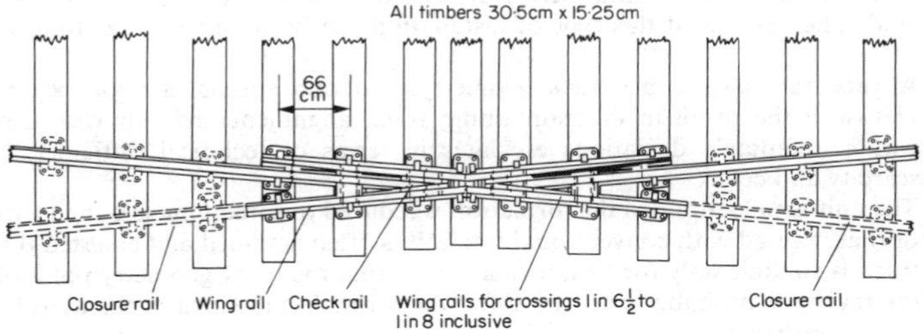

Figure 23.10 Obtuse crossing

according to the measuring convention adopted. Thus if 1 in 8 is obtained by measuring 8 units along the centre line for 1 unit of symmetrical spread normal to the centre line (centre line measure), then measuring 8 units along one leg of the vee for a spread of 1 unit measured at right angles to that leg (right angle measure) will give 1 in 7.969 whilst if 8 units is measured along each leg to give a spread of 1 unit the crossing size will be 1 in 8.016 (isosceles measure). It is therefore necessary for any railway to have a specific mode of measuring crossings which is known to all supplying manufacturers of crossings, or to state the angle in degrees.

In the UK fixed diamond crossings are limited by Ministry requirement to be not flatter than 1 in 8. In continental Europe 1 in 10 is the general limit. Where a line crossing flatter than the allowed limit is necessary the points of the diamond crossings are constructed as switches and have to be set to suit the train movement desired.

There is no official limit on the angle of common crossings but 1 in 28 is about the flattest angle at which the crossing nose has sufficient robustness. Crossings are either made up from rails or are cast in 12–14% manganese steel. Built up crossings have the inherent limitations of bolted assemblies and of recent years it has become general practice in the UK to weld part or whole of the crossing assembly. Latterly welding of the vee section and Huck fastening of the wings has been favoured. In general, it is preferred practice to have cast crossings for high speed or heavily worked lines using built up crossings for less exacting duty. It is also preferred to use cast crossings for angles flatter than about 1 in 20. In the USA it is common practice for the nose section of the crossing to be a manganese casting to which ordinary rail section wings are attached to form what is known as a rail bound cast crossing.

Leads

The design of railway connections is based on standard leads, that is, on specific combinations of switches and crossings forming turnouts of specific length giving specific departure angles. Exceptionally, as at important junctions where space is limited, leads are specially designed on the basis of standard switches with special crossings, e.g. instead of utilising either a 1 in 8 or 1 in 9 crossing, a crossing is made to a fine decimal angle, such as 1 in 8.319, but this is avoided as far as possible in the interests of standardisation. However, reference to a standard text book on permanent way, such as *British Railway Track* or *Railway Permanent Way* by Hepworth and Lee is recommended to civil engineers not familiar with field work.

SLAB TRACK DEVELOPMENT

A number of continuous slab track installations (Figure 23.11) have been put into service in various parts of the world and there is little doubt that this development will continue to expand. The benefits of this type of installation can be summarised as follows:

1. An accurate and stable track geometry. As train speeds increase beyond say 160 km/h the problem of maintaining track alignment and top with certainty within acceptable deviations escalates in terms of technical difficulty, labour scarcity and cost.
2. The unit area loading on the formation is reduced to something less than a quarter of that realised with conventional track. It is often assumed and asserted that slab track is suitable only for firm formations. From the track geometry point of view the reverse is probably true; the softer the formation the greater the need for slab track construction.
3. The day to day maintenance cost of well designed continuous slab track is potentially asymptatic to zero.

4. The requirement of conventional tracks for considerable and recurring engineering possessions to permit the working of on-track liners, tampers and ballast cleaners is eliminated.
5. The hazard of track buckling either vertically or horizontally is eliminated. Since conventional sleepered track has a factor of safety against lateral buckling which characteristically at the high temperature limit is of the order of 1.3 and which demonstrably falls to less than 1.0 when some maintenance or design imperative is transgressed, the installation of continuous slab track is from this aspect most attractive where the annual temperature range exceeds say 70 °C.
6. Since with conventional track the lateral buckling hazard depends on a limited rail temperature rise above a fairly high fastening temperature, the substitution of slab track enables the fastening temperature of continuous welded rails to be substantially lowered so that the tensile stresses in the rails in winter are small. This is obviously favourable to the elimination of rail fractures.
7. Slab track, more especially continuous slab track, offers the possibility of continuous support to the rail which cannot fail in the way that individual sleeper supports can fail. Even if the rail is supported on pads at normal sleeper spacing the support is still more stable than that afforded by a sleeper. For this reason alone there are grounds for reducing rail weights on slab track.

With slab track installations it is difficult and costly to restore the track geometry if the slab subsides in a general way; although concrete road engineering experience indicates that if the slab is well designed, and suitably protected from damage by frost

Figure 23.11 Section of slab track (British Railways, LM Region)

heave its long term stability should not be in much doubt. It is likely to cost 25% to 75% more than conventional high quality CWR track and its construction rate for a new line would be about one-third the rate of laying of conventional track. The possession time for relaying existing lines would probably be at least thirty times that needed for conventional CWR track. In addition the site organisation necessary for the delivery of large quantities of concrete represents a formidible problem where road access is limited.

Prefabricated slab track

The previous remarks relating to continuous or ribbon slab track are also true to a lesser degree about prefabricated unit slab track which has been installed on a trial basis in many parts of the world. The main problems here are to secure a sufficiently accurate compacted bed on which to lay the units and to avoid high edge stresses in the unit slabs. In this connection, the design of load transferring joints between adjacent slabs has not yet been established at a satisfactory standard. A further alternative is an open centre frame without any load transferring connection beyond the running rails. The main attraction of any prefabricated unit is that it is consonant with track renewals on the very short possessions of say 5 to 30 hours which are obtainable on many existing railways without either closure, or long spells of single line working or diversions.

Estimates of the limit of speed at which trains can be run on conventional tracks vary between about 150 and 250 mph. The degree therefore to which future railway speeds are limited by dwindling energy resources may well affect the prospects for slab track railway construction.

TRACK MAINTENANCE

Generally, the greater the strength and inherent stability of the design of the track construction and its foundation, the less demanding is the maintenance task. The stiffness of the rail, the sleeper spacing and the depth of ballast are important factors affecting the cost of maintenance.

The evolution of the automatic levelling, lining and tamping machine in the 1960's has changed the general pattern of track maintenance. The original manual method of maintenance was to drive individual pieces of ballast under the sleeper with a beater pick to obtain the required rail level and firmness of support. A later development pioneered in France, was to jack the sleeper and spread stone chippings over the bearing bed. This system was subsequently refined by measuring the static height the rail needed to be lifted, by calibrated sighting boards or a telescope every few sleepers and, after the dynamic deflection of the rail under traffic had been measured by voidmeters or dansometers and the readings added to the static values, distributing measured canisters of chippings along the sleeper beds, spreading the chippings evenly over a specified distance each side of the rail. The amount of chippings for sleepers between measuring points were interpolated as seemed prudent. This system is called measured shovel packing (MSP).

With experienced track maintenance men, beater packing and MSP give a high quality finish to the track. However, with the present shortage of manual labour partial mechanisation of track packing has been attempted with hand held power tools with either a vibrating blade or a vibrating impact rammer. This approach at present limited to locations where for one reason or another it is not possible to deploy on-track tamping machines.

As a result of the increasing reliance on automatic tamping and lining machines and increasing difficulties of labour recruitment there is a general movement towards substituting mobile staff from towns to working site for the traditional small lineside gangs. Another consequence is the increasing importance of preplanning the major maintenance work up to a year or more ahead.

The principal on-track machines used in addition to the tampers are the ballast cleaner and the track geometry recording car. The latter produces a graphical record of right and left hand top, track gauge variations, twist (the mutual angular deflection between two axles on a rigid frame), right and left hand versines, and an event line showing reference points, e.g. mileposts, bridges, stations, switches and crossings. An electronic analyser is now a normal accessory in the recording car and this makes a

Figure 23.12 Automatic tamping, lining and consolidating machine (Plasser and Theurer)

progressive record of accumulated exceedences or fault points for twist, top and line relating to the chosen unit length of say 0.5 km, when a print out of the parameter totals and gross total is automatic. In addition, a punched tape can be provided to process the results into priority categories to give a programme of selective tamping and/or lining or to schedule future track maintenance programmes.

On-track ballast cleaner

Over the course of time, all ballast degenerates with powdering and contamination with slurry and other debris which, in combination, impedes the natural drainage of the ballast and lowers its elastic support capacity. The ballast gradually assumes a plasticity asymptotic to that of the supporting ground. In this situation, cleaning or replacement of the ballast becomes essential to restore track stability. In general, the only economic or technically feasible way of doing this without gross interference with traffic, or indeed at all, in many cases, is the use of an on-track ballast cleaner (Figure 23.13). A complementary necessity for ballast cleaners arises from the common situation that the formation has become furrowed and distorted to destroy its original drainage plane. When timber sleepers with a characteristic life of 12–20 years was the normal circumstance it was possible to keep ballast condition up to a tolerable standard by the manual cleansing associated with sleeper renewal. Now however that the prestressed concrete sleeper with a prospective life of 60–80 years is increasingly the standard for modern track it is becoming more imperative than previously to ensure that new track is laid on a ballast support which can be left undisturbed for the greater part of a century. Ballast cleaning by on-track machines is therefore an integral part of providing a'fit and forget' type of track which combines with a system of mechanised maintenance operating on very long repetitive cycles.

Organisation of track inspection

Track foot patrols, previously carried out by resident gangers or sub-gangers, who then themselves proceeded to apply minor maintenance attention as necessary with their own gangs, need a more formal system of organisation when mobile labour is substituted. The local distribution and collection of track wallsers, the mode of making track inspection

Figure 23.13 Plasser ballast cleaner at work (Plasser and Theurer)

reports, and the organisation necessary to ensure that they are properly scrutinised and cleared by a track supervisor, need careful attention.

Mechanical maintenance also needs a higher standard of organisation of labour, equipment and track possessions than is essential to manual maintenance and therefore formal training in planning for track supervisory staff is essential to the success of mechanisation and mobile labour. In this, some form of work and method study to determine the necessary frequency and labour and machine content of constituent maintenance operations over a section of track is essential.

In this connection it is common to apportion the tracks on the larger railway systems into a number of categories based on speed and weight of traffic, e.g. primary, secondary and tertiary lines, or A1, A2, B1, B2, etc. where the letter is a speed index and the number is a gross tonnage index. Such categories are a useful guide to the appropriate frequency of track inspection and tamping and lining, quality of materials used in track renewals, maintenance standards, and manpower requirements.

RAILWAY STRUCTURES

In the UK, railway constructions are governed by the code set out in the Railway Construction and Operation Requirements for Passenger Lines and Recommendations for Goods Lines (1963 Reprint) issued by the Ministry of Transport (Department of the Environment). The dimensions, although originally given in Imperial units, have been converted to SI units for the purposes of this section.

Its main requirements affecting structures are given in the following paragraphs.

Stations

It is desirable to avoid constructing a station on, or providing a siding in connection with, a line which is laid upon a gradient steeper than 1 in 260. Where this is unavoidable either in the station or in connecting sidings, further requirements are stipulated.

The lines leading to passenger platforms should be arranged so that the platform roads may be entered in the normal direction of movement without reversing; and so that, with double lines and, generally, with passing places for passenger trains on single lines, each line shall have its own platform face.

Curvature of platform lines, and of station yards generally, is to be avoided as far as possible.

The minimum clear width of any platform throughout its length to be 1.8 m. At important stations the width should be not less than 3.6 m except for short distances at either end in any case of difficulty. The descent at the ends of platforms to be by ramps. Columns for the support of roofs, and other fixed works, to be not less 1.8 m clear from the edge of platforms. A general clear headway of not less than 2.4 m to be provided over platforms. The height of platforms should as a rule be 0.91 m, not less than 0.84 m or more than 0.91 m at permanent stations without special approval.

The edges of the platforms to overhang not less than 0.3 m, and the recess so formed to be kept clear as far as possible of permanent obstruction.

Waiting rooms or shelters, and conveniences, should be provided at junction stations and elsewhere as may be necessary.

Footbridges or subways should be provided for passengers to cross the railway at all exchange and other important stations.

Staircases or inclined approaches forming the main communication to or from platforms to be at no point narrower than at the top and the available width to be in no case contracted by any erection or fixed obstruction below the top.

The steps of staircases should not be less than 280 mm of tread nor more than 180 mm in the rise and midway loadings to be provided where the height exceeds approximately 3 m.

The slope of inclined footways and ramps not to exceed 1 in 8.

Bridges and viaducts

The Requirements in respect of bridges and viaducts apply particularly to new underbridges, more especially to structures under 91 m span, but they should be followed as far as practicable in the reconstruction of existing bridges. For bridges over a railway, the Requirements should be followed where applicable.

It is desirable that bridges and viaducts should, where practicable, be wholly constructed in some form of masonry, brickwork or concrete. The design and construction of steel girder bridges should be governed by the provisions of BS 153.

Cast iron must not be used in any portion of the structure of a bridge carrying a railway except when subject to direct compression only, nor may cast iron columns of small size be used for abutment or pier work in high structures. Structural timberwork to be avoided as a rule, and to be protected from fire.

For all bridges and viaducts under a railway the types of loading specified in Part 3A of BS 153 for Girder Bridges are recommended.

A standard of 18 units represents approximately maximum existing (1959) loading. To allow for a possible increase of axle loads, a standard of 20 units is recommended for

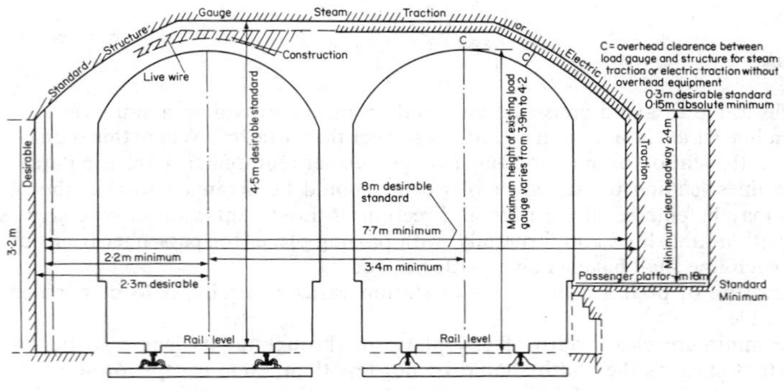

Figure 23.14 M.O.T. structure and load gauge

all important main lines. A lower standard will be accepted for lines restricted to lighter loading.

For highway bridges over a railway, use of the Ministry of Transport Standard Equivalent Loading Curve (1931) is recommended. (It is however to be noted that British Railways also observe their own domestic standards of loading and design).

Items to be taken into account in calculating stresses should follow, as far as they are applicable, the provisions of BS 153, Part 3A, for Girder Bridges.

The permissible stresses are:

MASONRY AND BRICKWORK

Between one-quarter and one-tenth of the ultimate crushing strength of the material.

REINFORCED CONCRETE

British standards to be followed for all materials. Steel reinforcement stress should not exceed 18 000 lb/sq. in or precast work 20 000 lb/sq in. For prestressed concrete, working stresses in the reinforcement may be higher depending on the character of the steel used and its anchorage or adhesion in the concrete.

TIMBER

Working stress not to exceed one-fifth of the ultimate strength.

STEEL

The working stress of structural mild steel to BS 15 should be in accordance with BS 153 for Girder Bridges. The working stress of high tensile structural steel to BS 548 should not exceed 247 N/mm^2. The working stress of steel members embedded in a concrete matrix may be higher than the above provided no account is taken of the strength of the matrix. Cast iron to be normally to BS 321. The working stress in direct compression must not exceed 154.4 N/mm^2.

Certain of the requirements regarding clearances are illustrated in Figure 23.14. Refuges for the safety of men working on the line need to be provided where the un-

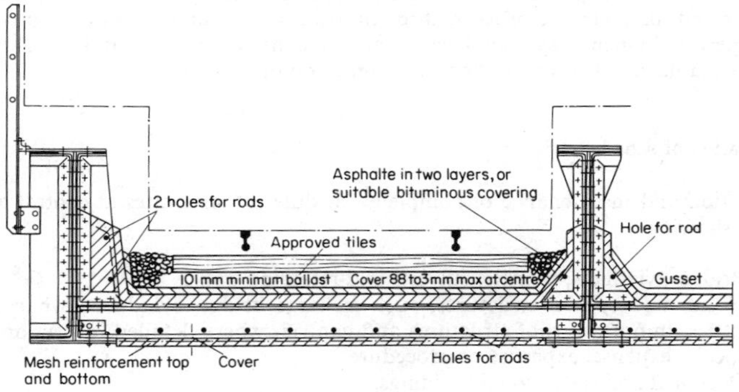

Figure 23.15 Concrete encased steelwork

impeded clearance from running rail is less than 1.5 m in vertical cuttings, retaining walls, tunnels and on long viaducts.

Past practice for steel underline bridges has been to provide longitudinal timbers to carry the rails but this is not now favoured. The general practice today is to provide a steel deck to carry normal ballasted track.

The important design considerations for railway bridges outside of standard specification and Ministry requirements are that the whole of the structure should be resistant to corrosion, that all exposed steel parts should be accessible for inspection and painting, that its maintenance should be economical, that neither its erection or subsequent maintenance should interfere with traffic in any avoidable way. The line possession time for the renewal of a moderate span bridge may be of the order of 36 hours or less on a busy line and the choice of design and mode of dismantling or moving the old and erecting or rolling in the new bridge need to be related to this.

With new construction it is generally preferable that bridges and culverts should be built simultaneously with the construction of the adjoining formation so that the back filling of abutments and wing walls can be completed immediately the bridge or culvert is built as a progressive part of the main formation work. In some cases however it may be expedient to erect temporary bridges from stock materials, with or without a detour round the permanent structure in order to facilitate rail delivery of heavy materials and machines and equipment either for the bridge or culvert itself or for the railway works being constructed beyond it, since the permanent structure usually takes longer to build than a temporary one.

INSPECTION OF STRUCTURES

The large number, wide dispersal and diverse nature of the occupied or operational structures belonging to the railway or looked after by the railway on an agency basis is such that a closely organised system of inspection is vital. The essentials of such a system are as follows.

Responsibilities

Definition of the responsibilities of the various grades of staff involved from chief engineer and his executive officers, through local officers and works staff down to the non-expert permanent way man, because his constant presence about the railway makes him a valuable lay observer of the visual condition of structures.

Preparation of schedules

Preparation and maintenance of complete schedules of structures at central and local offices showing:

(a) Bridges, viaducts, tunnels, culverts and retaining walls.
(b) Nominated special structures, such as large bridges, viaducts, wharves, piers, jetties, important roof structures and gantries where detailed inspection requires special expense, expertise or procedure.
(c) Occupied and operational buildings.

Frequency of inspections

Establishment of appropriate frequencies of inspection and by whom. Superficial examinations by a competent examiner may be fixed between say six months and twelve months according to the nature of the structure. Detailed examination about every twelve months may be appropriate for tunnels, dock, harbour and sea defence works and selected structural parts under water and between about three to six years for bridges, culverts, buildings and frames of large buildings, large roofs and tunnel shafts. Provision needs to be made for variation of frequencies by mutual agreement between central and local officers.

Requirements to be observed

Publication of a code of general and particular requirements to be observed in the examination of bridges, gantries, buildings and other structures. The general requirements may include references to the form of reporting, the special marking of any items in the report likely to affect safety, the noting of hidden parts not examined, changes in

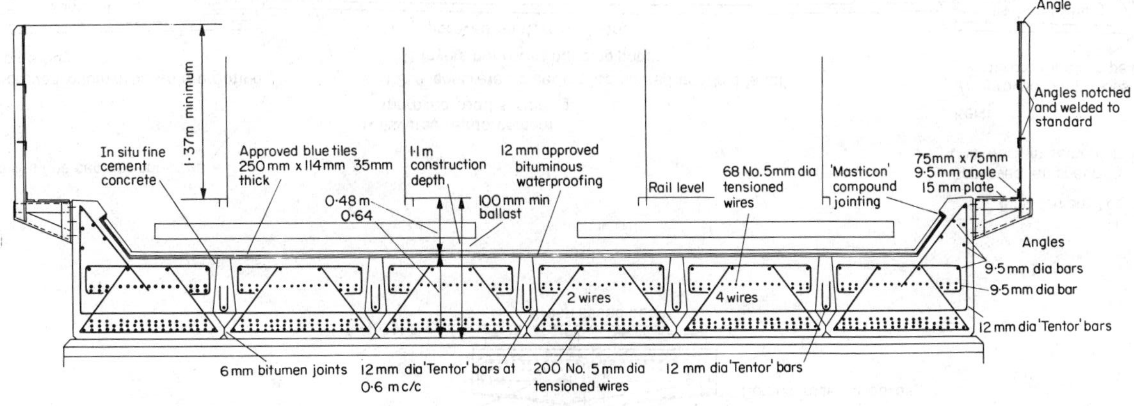

Figure 23.16 Prestressed concrete underbridge

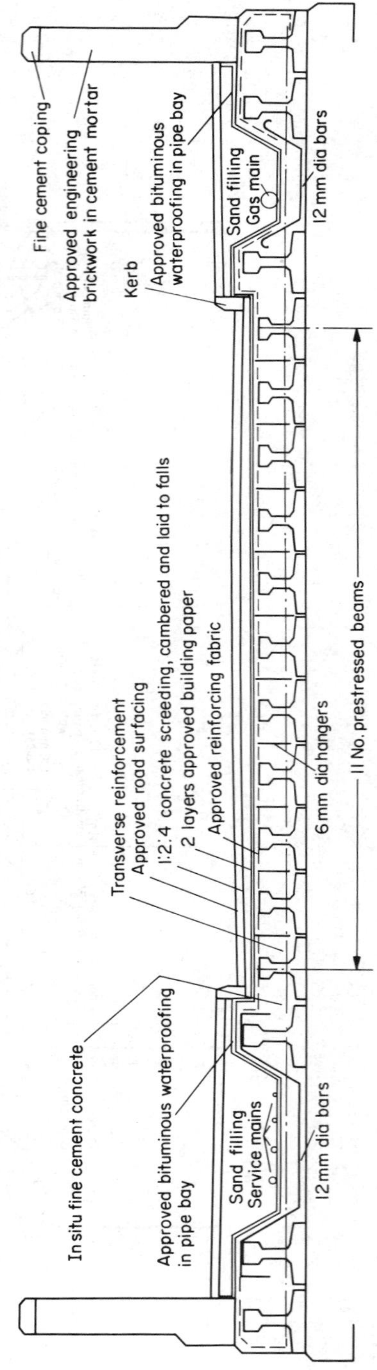

Tentor bars
11 No. H.T.W.
13 No. H.T.W.
Steel pipe
Semi-circular grooves
Prestressed beam
Fine cement coping
Approved engineering brickwork in cement mortar
Approved bituminous waterproofing in pipe bay
Kerb
Sand filling
Gas main
12 mm dia bars
Transverse reinforcement
Approved road surfacing
1:2:4 concrete screeding, cambered and laid to falls
2 layers approved building paper
Approved reinforcing fabric
6 mm dia hangers
11 No. prestressed beams
In situ fine cement concrete
Approved bituminous waterproofing in pipe bay
Sand filling
Service mains
12 mm dia bars

Figure 23.17 Prestressed concrete overbridge

use since the last examination, the provision of explanatory sketches, the reference to a senior of parts requiring further examination, the use of road or rail mounted inspection platforms, the observance of safety standards, the requirements of local, river and police authorities and of regulations affecting electrified lines, the scrutiny prior to the examination of previous reports on the structure, advance arrangements for the removal of hoardings, gutters, laggings, decking, glazing, roof cladding, etc. advance notice of requirements for staging, scaffolding, barges, rescue launches, divers, look-out men, line possessions, switching off of traction current.

The particular requirements may include reference to the need for viewing a structure in daylight if it is to be examined at night, the recording of all metal wastage due to corrosion, a schedule of parts specifically to be examined, the condition of 'tell tales', the plumbing of the vertical alignment of piers, columns, girders, abutments and walls, the surcharging of walls, distortion or deteriorations of sub-assemblies, joints or bearings, the ingress of water and weephole condition. It may also be a requirement to observe structures under moving load, and to observe and record any evidence of settlement in bedstones, bearings, trestles and foundations, the need to check and record evidence of scour adjacent to a pier, wall or abutment, alterations in a river course adjacent to a bridge or viaduct, the noting of joints in rails or longitudinal timbers adversely affecting a structure. Attention should be drawn to any features of track, roadway, drainage or traffic having an adverse effect and the condition of paint or other protective coatings with reference to previous painting date. Notes should be available of precautions to be taken before any suspension bolt is dismantled for inspection and action to be taken when a defect is found. The requirements may cover the specific inspection of bolts and holding attachments to bridges or buildings such as smoke plates, handrails, walkways or external stairs, the condition of longitudinal timbers and troughs carrying the track, the prescence of accumulated rubbish or ballast at girder ends, the condition of drainage outlets, the presence of vegetation, brick or concrete spalling, loose jointing mortar, not in the ends of timber roof members, the condition of air bricks, the security of roof coverings, the condition of roads, paths and fences adjacent to railway buildings, surfaces of platforms, loading docks and the condition of notice or name boards.

Report forms

The issue of standard reporting forms for the examination of bridges and structures, buildings, and tunnels with appropriate schedules of parts specifically to be reported upon.

The essentials of the system are that the inspections are regularly and thoroughly carried out by competent personnel, that all structures are included, that the inspection reports are scrutinised by an appropriate professional engineer and that appropriate action is taken in due time and recorded.

RESEARCH, DEVELOPMENT AND INTERNATIONAL COLLABORATION

The principal railway co-ordinating body in Europe is the UIC (Union Internationale des Chemins de Fer) whose headquarters is in Paris. Its primary purpose is the encouragement and development of standard railway policies and practices throughout Europe. It has a permanent secretariat but it works through committees covering the activities of most railway departments. Twenty-six nations are constituent members.

Its technical research co-ordinating arm is ORE (Office for Research and Experiments) which has its office at Utrecht. The UIC issues specifications and standards for railway practice to which member administrations are expected to adhere. The ORE is represented on all technical committees of UIC and issues reports made by working parties or by specially commissioned persons of academic or professional distinction to member administrations of UIC.

The field and laboratory work of ORE is assigned to member administrations or to universities or technical schools; such work is usually progressed by an appropriate working party composed of delegates from member administrations serviced by a technical secretary from ORE. Distinguished men from industry and universities are sometimes co-opted to serve on specific working parties. UIC and ORE reports are normally issued simultaneously in English, French and German versions, and these languages are used at UIC and ORE meetings; French being the primary procedural tongue.

Another European body which makes a significant contribution to railway technical literature in its periodical Bulletins, is the International Railway Congress (IRC) whose administration is in Brussels.

The principal technical railway co-ordinating organisation in the USA is the American Railway Engineering Association (AREA) of Chicago which issues a periodical journal covering minutes of meetings, reports of special committees and recommendations for specifications and technical procedures. The AREA also sponsors research and development projects which are mainly progressed by the Association of American Railroads (AAR) at its Research Centre at Chicago. The Pan American Railway Congress is a similar but less developed organisation covering the countries of South America. Much valuable railway research and development work is done in Japan and the English version of the monthly Journal of the Permanent Way Society of Japan, founded in 1958 in Tokyo, provides a comprehensive contemporary reflection of this work.

Important permanent way research centres also exist in Prague, Moscow, Belgrade and Johannesburg but the results of their activities is normally limited to domestic distribution. The chief technical and development railway work carried out in the UK is associated with the Research and Development Centre at Derby which has a considerable international reputation. Its reports are generally limited to internal circulation within the BRB but most of its railway research projects are reported through papers and articles for the British engineering institutions, chiefly the Institution of Civil Engineers, the Institution of Electrical Engineers, the Institution of Mechanical Engineers, and the technical press. It also accepts commissions for research from outside the BRB.

The national railways of UK, Canada, France, Germany and Japan are also associated with sponsored railway consultancy organisations which whilst primarily concerned with the techniques of railway organisation also offer technical advice to clients.

Contemporary railway civil engineering practice and developments in the UK and elsewhere are more to be found in papers presented to the railway division of the Institution of Civil Engineers and to the Permanent Way Institution, in the bulletins and journals of ORE, IRC, and the AREA and in the railway technical press of Britain, France, Germany and America, rather than in the few standard textbooks on railway engineering which are necessarily less up to date.

REFERENCES

1. COOMBS, D. H., 'British Railway Track', London (1971)
2. UKRAS, 'British Permanent Way', London (1964)
3. HEPWORTH and LEE, 'Railway Permanent Way', Manchester (1922)
4. SCHRAMM, G., 'Permanent Way Technique and Economy', Darmstadt (1961)
5. Soil Mechanics for Road Engineers, HMSO
6. PASCHOUD and DEL PEDRO, 'Railway Steel Sleepers', Lausanne
7. Railway Construction and Operation Requirements, HMSO (1963)
8. PEARSON, H. M., 'Railway Works Construction', Odhams Press
9. TURTON, F., *Railway Bridge Maintenance*, Hutchinson (1972)
10. DAVEY, 'Tests on Road Bridges', HMSO (1953)

11. SONNEVILLE and BENTOT, 'Elasticity and lateral strength of Railway Track', IRCA Bulletin (March 1955/June 1956)
12. BERRIDGE, P. S. A., 'The Girder Bridge', Robert Maxwell (1969)
13. Railway Engineering and Maintenance Encyclopaedia, Simmons Boardman
14. COUR-PALAIS, 'Railway Points and Crossings', Thurclare Co.
15. SEARLES and IVES, 'Railway Surveying', Wiley
16. BALL, J. D. W., 'Reinforced Concrete Railway Structures', Constable
17. Report of the Bridge Stress Committee, HMSO (1928)
18. Experiments on the Stability of Long Welded Rails, British Transport Commission (1961)
19. PRUD'HOMME and BENTOT, 'The Stability of tracks laid with long welded rails', IRCA Bulletin (July 1969)
20. PERROT, S. W. and BADGER, F. E. C., 'Practice of Railway Surveying and Permanent Way Work', London (1920)
21. CHELTONE, DOVEY and MITCHELL, 'The Strength of Cast Iron Girder Bridges, J. Inst. Civil Eng., **22** 243 (1944)
22. EISENMANN, J., 'Stress Distribution in the Permanent Way due to heavy axle loads and high speeds', AREA Proc. (1969)

BRITISH STANDARDS

BS 9 : 1935
Bullhead railway rails
BS 11 : 1959
Flat bottom railway rails
BS 15 : 1961
Mild steel for general structural purposes
BS 47 : 1959
Steel fishplates for bullhead and flat bottom railway rails
BS 64 : 1946
Steel fishbolts and nuts for railway rails
BS 105 : 1919
Light and heavy bridge-type railway rails
BS 144 : 1954 (amended March 1963)
Coal tar creosote for the preservation of timber
BS 153 Steel Girder Bridges, Part 1 : 1958. Materials, workmanship, protection against atmospheric corrosion; Part 2 : 1958. Weighing, shipping, erection; Part 3A : 1954. Loads; Part 3B : 1958. Stresses; Part 4 : 1958. Design and construction
BS 449 1959. The structural use of steel on buildings (incorporating BS Code of Practice CP 113): and Supplement 1 (1959) Recommendations for design and Addendum 1 (1961) and The use of cold formed steel sections in building.
BS 500 : 1956
Steel railway sleepers for flat bottom rails
BS 751 : 1959
Steel bearing plates for flat bottom railway rails
BS 812 : 1960
Methods for sampling and testing of mineral aggregates, gravels and filters
BS 913 : 1954
Pressure creosoting of timber
BS 968 : 1962
High yield stress (welding quality) structural steel
BS 986 : 1945
Concrete railway sleepers
BS 1377 : 1961
Methods of testing soils for civil engineering purposes
BS 1924 : 1957
Methods of test for stabilised soils
BS 2762 : 1956
Notch ductile steel for general structural purposes

24 HARBOURS AND DOCKS

HARBOURS AND DOCKS 24

24 HARBOURS AND DOCKS

D. H. LITTLE, BSc., M.I.C.E.
Walter C. Andrews & Partners.

Maritime structures need to be simple and robust. Finesse and the finer points of structural design must give way to what might be crudely termed 'thick and heavy' structures; cover to reinforcement should be 50 mm thick; steel members never less than 10 mm thick; galvanising should be of the heaviest weight and fittings such as bollards, guard rails and ladders must all have ample reserve of strength.

Generally there is little precision in maritime work and there is probably no other branch of the profession in which experience and judgement play so important a part. Waves can be analysed on a mathematical basis for ideal conditions but in practice so many variables are possible—speed, direction and duration of wind, slope and nature of sea bed, configuration of coastline—that not much more than broad inferences can be drawn from theoretical considerations. Littoral drifts, currents and siltation may be observed and measured for an existing set of conditions but there is no exact way of calculating the effect on these likely to result from a change in conditions brought about by a proposed new work. These overall limitations apply more strongly to harbours than to docks. Where the works are large enough, hydraulic models should be used.

HARBOURS

Siting and location of harbours

Normally a harbour is the corollary of other activities and the choice of general site is dictated by commercial and economic requirements. In detail, too, the position of entrances may be governed by natural channels and the draft considered necessary in them, but full advantage should be taken of any protection likely to be afforded by other natural features such as outer reefs and nearby headlands. If breakwaters are to be of the rubble mound type, a dominating fact may well be the source of supply of the stone.

Waves

THEORETICAL WAVEFORMS

(See also Section 29.)

Waves are formed by wind blowing over the surface of the water. In deep water waves remain oscillatory; in shallow water they break and become translatory. Theoretical ideal waves can be dealt with on a mathematical basis, while they remain oscillatory.

Referring to Figure 24.1: length λ is the distance from crest to crest; height $2a$ is from crest to trough; period T is the time taken for two successive crests to pass a fixed point; speed C is the apparent rate at which a crest advances; steepness is the ratio H/λ, where H is the wave height ($H = 2a$).

It follows that

$$C = \lambda/T \tag{1}$$

Although single oscillatory waves appear to be moving forward, elements of water at the surface actually only move in a vertical circle with diameter equal to the height H. It can be shown mathematically and demonstrated practically that such surface movement also entails circular motion of elements below the surface but of decreasing diameters according to the formula

$$\text{diameter at depth } D = He \exp(-2\pi D/\lambda) \tag{2}$$

whence at depth $\lambda/10$, diameter of orbit $= H/2$; at depth $\lambda/2$, diameter of orbit $= H/25$;

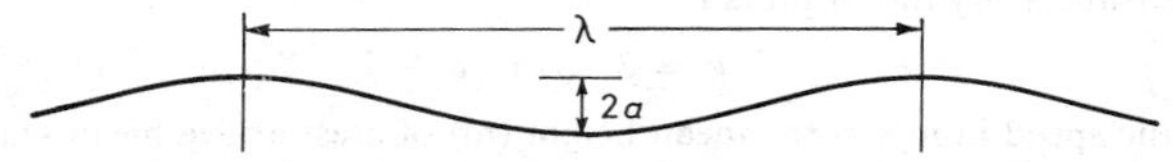

Figure 24.1 Waveform

at depth λ, diameter of orbit $= H/600$ so that beyond depths greater than half a wavelength wave effects are unimportant.

The actual wave shape is a trochoid (the curve traced out by a point on the spoke of a wheel rolled along a horizontal line), the radius of the wheel being $\lambda/2\pi$ and the distance along the spoke from the centre being $H/2$.

From consideration of the centrifugal forces set up by the circular motion it can be shown that

$$\lambda = gT^2/2\pi \tag{3}$$

It will be noted that theoretical relationships between length, period and speed can be established but height is independent of these variables.

FETCH AND WAVE HEIGHT

By assuming that a wind blows with constant strength and direction over an initially calm expanse of water the length (and therefore the speed and period) and the height of the resulting wave can be calculated; these being greater as the length of fetch, the strength of the wind and the time of the blow increase. If the fetch is not long enough the maximum values within the capacity of a given wind will not be reached; similarly if the time of blow is not long enough the maximum values will not be reached, no matter how long the fetch may be. It may be noted that winds less than 10 knots (18.5 km/h) do not set up waves of any significance. From this it follows that Stevenson's well known empirical formula

$$H = 0.36\sqrt{F} \tag{4}$$

where H is the height in m and F the fetch in km, can only be true for very limited conditions. It ignores the time factor but, as is usual for civil engineering requirements, if only the maximum value is wanted this is on the safe side. It also ignores wind strength. For fetches up to 160 km it would appear to be applicable to winds up to 90 km/h (gradient speed) and for fetches between 160 and 800 km the corresponding wind is 110 km/h. So long as these limitations are remembered, Stevenson's formula is useful for preliminary guidance. Much closer assessments, however, can and should be made for final use, but this is a complex process which has to be handled with experienced judgement and should be done by a meteorologist with specialist knowledge of wave forecasting.

WAVES IN SHALLOW WATER AND BREAKING WAVES

When waves approach shallow water their speed and length decrease but only when the depth of water becomes less than half the wavelength. For beach slopes up to 1 : 10 (and possibly 1 : 5) and for water depths less than 3 m, the speed in m/s of all waves is given by the formula

$$V = 3.2\sqrt{d} \tag{5}$$

provided the waves do not break (d = depth in metres). For waves which have broken and become translatory the formula is

$$V = 3.2\sqrt{(h+d)} \tag{6}$$

where V is the speed in m/s, h the mean height (m) of crest above mean sea level and d the depth of water (m) below mean sea level, i.e. $(h+d)$ is the height of crest above sea bed.

In the absence of wind, waves break when the depth of water is one and a half times the height of the breaking waves, i.e. when $d = 1.5H_b$. On entering shallow water all waves tend first to decrease slightly in height and then to increase again to breaking height. Short steep waves increase only back to their original deep water height H before breaking, i.e. $H_b = H$, while long low waves increase to twice their original height, i.e. $H_b = 2H$. A short steep wave has H/L greater than 0.02 and a long low wave has H/L less than 0.01. With a strong on shore wind, however, or a steep beach (i.e. slope greater than 1 : 50) the breaking depth of water may be twice the breaking wave height while with a strong off shore wind or a very gradual slope the ratio may be only 1.

The above variables are (i) the calculated mean wave height, (ii) the actual height, (iii) the breaking height and (iv) the breaking depth; these may be summarised as follows, and tabulated as in Table 24.1.

The deep sea height can be calculated as an ‘equivalent constant wave’ and this calculated height indicated as H. The actual height is, however, affected by interaction

Table 24.1

Calculated mean height	*Actual height varies from*	*Short steep waves*				*Long low waves*			
		Breaking height $H_b = H$	*Breaking depth*			*Breaking height* $H_b = 2H$	*Breaking depth*		
			Offshore wind	*No wind*	*Onshore wind*		*Offshore wind*	*No wind*	*Onshore wind*
H	$\frac{2H}{3}$ to $\frac{4H}{3}$	$\frac{2H}{3}$ to $\frac{4H}{3}$	$\frac{2H}{3}$ to $\frac{4H}{3}$	H to $2H$	$\frac{4H}{3}$ to $\frac{8H}{3}$	$\frac{4H}{3}$ to $\frac{8H}{3}$	$\frac{4H}{3}$ to $\frac{8H}{3}$	$2H$ to $4H$	$\frac{8H}{3}$ to $\frac{16H}{3}$

of waves and may vary between $\frac{2}{3}H$ and $\frac{4}{3}H$. The breaking height H_b for a short steep wave is the same as the actual height. The breaking height H_b for a long low wave is twice its actual height. The breaking depth is H_b for off shore wind or flat beach, $1.5H_b$ for no wind, $2H_b$ for on-shore wind or steep beach (steeper than 1 : 50).

This indicates that the breaking depth can vary between $\frac{2}{3}H$ and $\frac{16}{3}H$, depending on circumstances. It is important that the estimated upper and lower limits should be taken into account when designing any works exposed to wave action.

Since the speed and length of waves decrease as the depth of water decreases it follows that waves approaching a shore obliquely will tend to wheel round until parallel to the shore. If waves actually finished on the shore without breaking they would all end parallel to the shore irrespective of the deep sea angle of approach, but breaking prevents this full completion of the wheeling.

Wind

BEAUFORT WIND STRENGTH SCALE

Wind strength is measured in terms of miles per hour (or km/h). Speed and even direction vary considerably with height. Over the sea, owing to friction, surface speed may only be $\frac{2}{3}$ that at a height of about 600 m, while over land, owing to friction and physical obstructions, the ratio may be $\frac{1}{3}$. The speed at a height of about 600 m, however, can be estimated by meteorologists from isobars and is known as the 'gradient speed'. In predicting wave effects by calculation, a steady gradient speed is always taken for the wind and an effective value for this has to be converted from the probable actual gusty surface wind. This emphasises the need for specialised experience for really accurate forecasting.

Table 24.2 BEAUFORT SCALE FOR WIND

Scale number	*Description*	*Velocity* mile/h	*Noticeable effect at sea*
0	Calm	Less than 1	Sea is mirror smooth
1		1–3	Small wavelets like scales but no foam crests
2	Light	4–7	Waves are short and more pronounced
3		8–12	Crests begin to break. Foam has glassy appearance not as yet white
4	Moderate	13–18	Waves are longer. Many white horses
5	Fresh	19–24	Waves are more pronounced; white foaming crests seen everywhere
6	Strong	25–31	Longer waves form; foaming crests more extensive
7		32–38	Sea heaps up, foam begins to blow in streaks
8		39–46	Waves increase visibly. Foam is blown in dense streaks
	Gale		
9		47–54	Waves increase visibly. Foam is blown in dense streaks
10		55–63	High waves with long overhanging crests; great foam patches
	Strong gale		
11		64–75	Waves so high that ships are hidden in the troughs. Sea covered with streaking foam; air filled with spray
12	Hurricane	Above 75	

1 mile/h = 1.6 km/h

For everyday practical use wind speeds are given on the basis of the Beaufort scale devised by Admiral Beaufort in 1806. In this scale wind values up to 84 mile/h (135 km/h) and over are divided into twelve sections and the wind strength increases in value from 0 to 12. Originally the descriptions were in sailing terms but these have been revised to suit the passing of sail and with practice in its use quite fair estimates can be made. The scale is set out in Table 24.2.

Forces on breakwaters

So far as forces are concerned there are two main types of breakwaters (1) rubble mounds with sloping faces and (2) walls with vertical faces. Waves impinging on rubble mounds must reach a depth at which the waves break and become translatory. With vertical wall breakwaters, the depth at high tide is usually sufficient for the waves to remain oscillatory. At lower states of the tide, breaking conditions may obtain and it is essential to take this into account. Where the water is deep enough to maintain oscillatory conditions, however, a waveform termed the 'clapotis' is set up and is such that against the face of the vertical wall the wave height is doubled and the mean sea level raised. The theory of the clapotis and the forces imposed by it were published by Sainflou[1] and this, together with a recommendation based on it by the *Sixteenth International Congress of Navigation* at Brussels in 1935, enables the forces of oscillatory waves to be assessed with some exactitude.

Forces on breakwaters are therefore best considered under two headings: (a) those due to translatory waves and (b) those due to oscillatory waves.

FORCES CAUSED BY TRANSLATORY WAVES ON VERTICAL FACES

When a wave breaks the water moves forward bodily as a mass and by similarity with a jet of water impinging on a plate the pressure intensity is

$$p = \rho V^2/g \tag{7}$$

Hence if the velocity V of a breaking wave could be known the force set up by it could be calculated. Unfortunately no simple relationship exists between the breaking velocity and the wave height. Wind strength and direction affect it; so does the slope of the beach and the length and steepness of the waves. It is, therefore, only possible to indicate ranges or limits of the effects; thus if d is the depth of water below mean sea level $V = 3.2\sqrt{(h+d)}$ (equation 6) where h is the height of crest above mean sea level.

Both h and d can be expressed in terms of H, the calculated deep sea height.

$$V = 3.2K\sqrt{H} \tag{8}$$

where K is a constant.

Hence

$$\begin{aligned} p = \rho V^2/g &= 100(3.2)^2K^2H \quad \text{kg/m}^2 \\ &= K^2H \quad \text{tonnes/m}^2 \end{aligned} \tag{9}$$

For short steep waves H_b = breaking height = H = deep sea height, thus

$$h = H/2 \tag{10}$$

d may be H_b, $1.5H_b$, or $2H_b$, i.e. H, $1.5H$, or $2H$.

For long low waves $H_b = 2H$:

$$h = H \tag{11}$$

As for short waves d may be H_b, $1.5H_b$, or $2H_b$, i.e. $2H$, $3H$, or $4H$.

For both long and short waves the actual wave height may vary between $\frac{2}{3}H$ and $\frac{4}{3}H$. These variables are set out in Table 24.3.

Civil engineering works are usually concerned with short waves and as beach slopes exceed 1 : 50 as a rule and rubble mound slopes always do a breaking depth of twice the deep sea height is the safest assumption, whence $p = 3.3H$. In order to include for some possibility of long waves it is suggested that the pressure intensity in tonnes/m² be taken as

$$p = 4.0H \tag{12}$$

where H is the mean calculated deep sea height.

Table 24.3 (H = CALCULATED MEAN DEEP SEA HEIGHT; ACTUAL DEEP SEA HEIGHT = $2H/3$ TO $4H/3$)

Variable	*Unit*	*Short waves* ($H_b = H$)						*Long waves* ($H_b = 2H$)					
H_b	m	$2H/3$	$4H/3$	$2H/3$	$4H/3$	$2H/3$	$4H/3$	$4H/3$	$8H/3$	$4H/3$	$8H/3$	$4H/3$	$8H/3$
$h = H_b/2$	m	$H/3$	$2H/3$	$H/3$	$2H/3$	$H/3$	$2H/3$	$2H/3$	$4H/3$	$2H/3$	$4H/3$	$2H/3$	$4H/3$
$d = H_b$, $1.5H_b$ or $2H_b$	m	$2H/3$	$4H/3$	H	$2H$	$4H/3$	$8H/3$	$4H/3$	$8H/3$	$2H$	$4H$	$8H/3$	$16H/3$
$h+d$	m	H	$2H$	$4H/3$	$8H/3$	$5H/3$	$10H/3$	$2H$	$4H$	$8H/3$	$16H/3$	$10H/3$	$20H/3$
$\sqrt{(h+d)}$	m	$\sqrt{H}$	$1.4\sqrt{H}$	$1.2\sqrt{H}$	$1.6\sqrt{H}$	$1.3\sqrt{H}$	$1.8\sqrt{H}$	$1.4\sqrt{H}$	$2\sqrt{H}$	$1.6\sqrt{H}$	$2.3\sqrt{H}$	$1.8\sqrt{H}$	$2.6\sqrt{H}$
$V = 3.2\sqrt{(h+d)}$	m/s	$3.2\sqrt{H}$	$4.5\sqrt{H}$	$3.8\sqrt{H}$	$5.1\sqrt{H}$	$4.2\sqrt{H}$	$5.8\sqrt{H}$	$4.5\sqrt{H}$	$6.4\sqrt{H}$	$5.1\sqrt{H}$	$7.3\sqrt{H}$	$5.8\sqrt{H}$	$8.3\sqrt{H}$
$p = 100V^2/1000$	tonnes/m^2	H	$2.0H$	$1.5H$	$2.6H$	$1.7H$	$3.3H$	$2.0H$	$4.0H$	$2.6H$	$5.3H$	$3.3H$	$6.8H$
p for $H = 6$	tonnes/m^2	6.0	12.0	9.0	17.0	10.2	19.8	12.0	24.0	15.6	3.3	19.8	40.8
	Actual height	$2H/3$	$4H/3$	$2H/3$	$4H/3$	$2H/3$	$4H/3$	$2H/3$	$4H/3$	$2H/3$	$4H/3$	$2H/3$	$4H/3$
		Offshore wind		No wind		Onshore wind		Offshore wind		No wind		Onshore wind	
		$d = H_b$		$d = 1.5H_b$		$d = 2H_b$		$d = H_b$		$d = 1.5H_b$		$d = 2H_b$	

If an actual observed maximum height of wave is adopted instead of a calculated mean deep sea height, then the actual height would be H and not $4H/3$. The corresponding maximum pressure intensity then becomes $p = 2.6H$ for short waves; $p = 3.0H$ for long waves; and a fair figure is $p = 3.0H$.

For a wave height of 8 m p becomes 32.0 tonnes/m^2 which is the value adopted by Luiggi.[2] Other actual recorded[3] pressures do not appear to have exceeded 40.0 tonnes/m^2 and as wave heights are unlikely to exceed 10 m this affords another check on the suggested formula $p = 4.0H$. Distribution of pressure is best assumed to follow the straight line proposed for oscillatory waves at the Brussels Congress in 1935. Referring to Figure 24.2 let AC be a breakwater face of unlimited height; BE is mean sea level and CD the line of the sea bed. Then the pressure diagram is ABCDE such that pressure

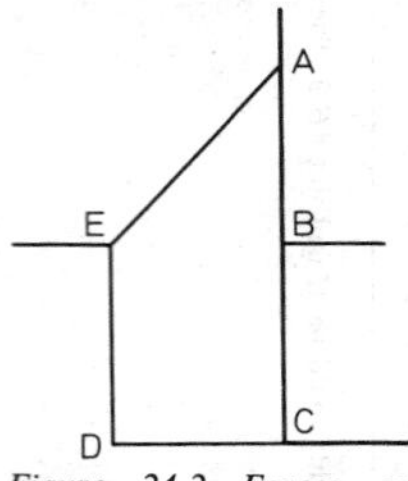

Figure 24.2 Forces on breakwater of unlimited height

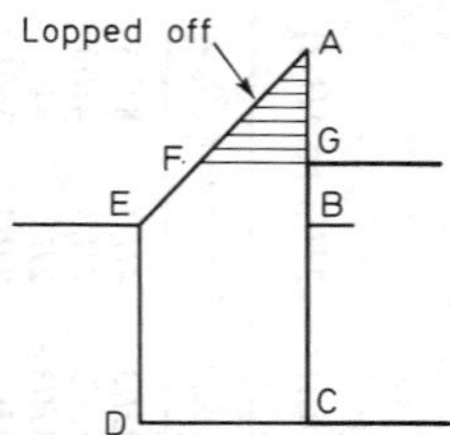

Figure 24.3 Forces on breakwater of limited height

intensity at BE is given by $p = 4H$ and the height AB above the mean sea level at which this pressure reduces to zero is taken as $H/2$. If the actual height of breakwater does not extend to A but stops at G (Figure 24.3) then the top portion of the diagram AGF is 'lopped' off, leaving GBCDEF as the net value. It is not claimed that this way of assessing the effects of breaking waves is conclusive or very satisfactory but until full scale researches are available it can hardly be improved upon.

FORCES CAUSED BY TRANSLATORY WAVES ON SLOPING FACES

When waves strike a sloping surface they must break and when that surface is formed from tipped stones the energy of the waves is destroyed, partly by dashing against the stones and partly by passing over and up them.

Table 24.4 WEIGHT OF BLOCKS IN KG—NATURAL STONE DENSITY 2.5 TON/M^3

Slope α	Wave height (H), m					
	1	*2*	*3*	*4*	*5*	*6*
1:1	∞	∞	∞	∞	∞	∞
1:2	122	980	3 300	7 800	15 250	26 300
1:3	44	352	1 190	2 820	5 500	9 500
1:4	29	228	775	1 830	3 590	6 200
1:5	23	184	620	1 470	2 880	4 968
1:12	15	125	400	1 000	2 000	3 200

2.5 ton/m^3 = 170 lb/ft^3. 1 kg = 2.2 lb. 1000 kg = 1 tonne.

The critical portion of a rubble breakwater is upwards from about 5 m below low water, and an empirical formula for assessing the weight of individual stones over this section is the 'Spanish' or 'Caribbean' formula, thus:

$$P = \frac{NH^3 d}{(\cos\alpha - \sin\alpha)^3 (d-1)^3}$$

where P is the weight of block kg, $N = 15$ for natural quarried stone, 19 for cast blocks, H is the height of the wave at breakwater (m), d the density of the blocks (tonne/m^3), and α is the angle with the horizontal of slope of breakwater face. See Table 24.4. More modern work has been carried out by H. R. S. Wallingford for CERA.[4]

FORCES CAUSED BY OSCILLATORY WAVES ON VERTICAL FACES

When waves impinge on a vertical wall and the depth of water alongside is sufficient to prevent them breaking, pressures can be assessed by the Sainflou theory.[1, 5] Referring to Figure 24.4 this can be briefly stated as follows:

When the reflected wave from a vertical wall is superimposed on the incident wave the wave height H is doubled and the mean line of the resulting clapotis is H_0 above still water level such that

$$H_0 = \frac{\pi H^2}{\lambda} \coth \frac{2\pi D}{\lambda} \tag{13}$$

where λ is the length of wave and D the depth of water below still water level.

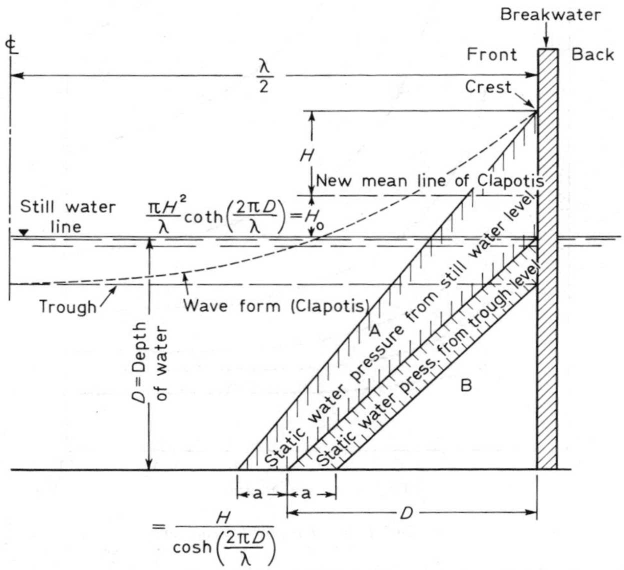

Figure 24.4 Saintflou theory

The crest of the clapotis rises to a maximum value $H + H_0$ above still water level and for the crest in front of the wall the intensity of pressure at any depth D in terms of head of water is given by

$$a = H \cosh \frac{2\pi D}{\lambda} \tag{14}$$

It is to be noted that a is the excess of pressure above that from still water level. Similarly for the trough in front of the wall the still water pressure exceeds the trough pressure by *a*. If the breakwater effect is 100 per cent efficient the water behind it will be dead calm and at still water level. This would be an impossible ideal upper limit while the worst possible lower limit would be to assume that the trough of the wave in the harbour behind the breakwater went to the same level as the clapotis trough in front. Hence, the range of pressure is (where P = resultant thrust on the breakwater):

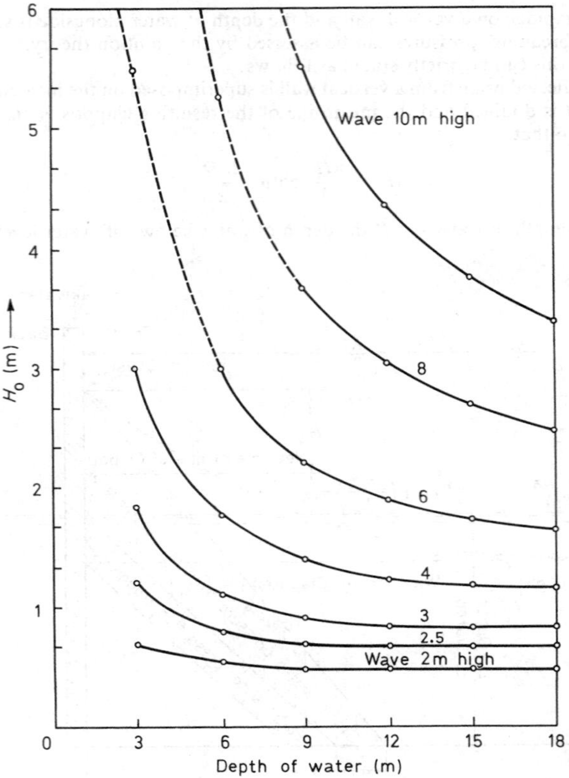

Figure 24.5 Value of H_0 for typical waves

For still water at the back and clapotis crest at the front

$$P = \rho \times \text{area A} \quad (15)$$

For clapotis trough at the back and clapotis crest at the front

$$P = \rho(\text{area A} + \text{area B}) \quad (16)$$

These pressures may be regarded as positive, i.e. acting towards the harbour. When the trough is in front of the wall there is a maximum negative pressure acting in the opposite direction of $\rho \times$ area B. Values of H_0 and a for some typical waves are given in Figures 24.5 and 24.6, the dotted portions of the curves indicating that in those depths of water the waves would almost certainly have broken. At the Brussels Congress of 1935 a

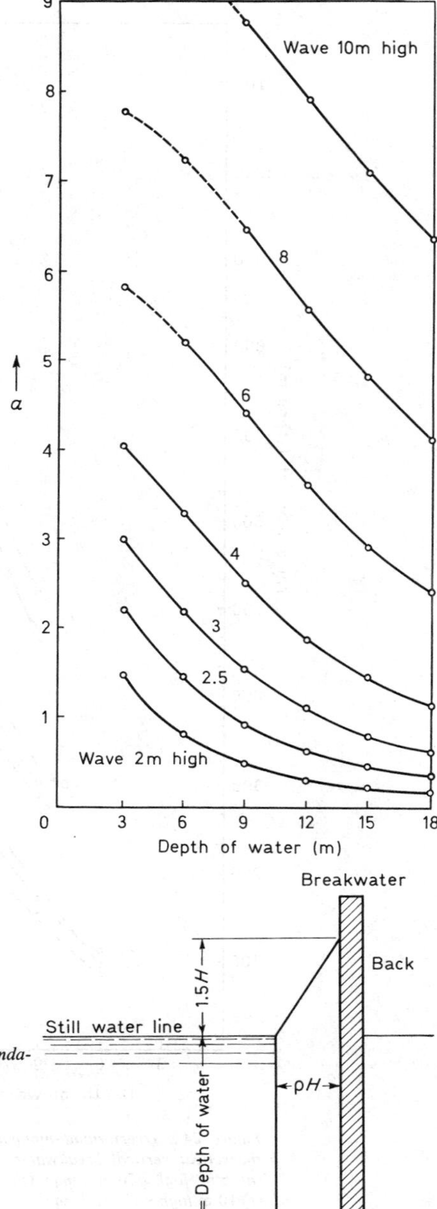

Figure 24.6 Value of a for typical waves

Figure 24.7 Brussels Congress (1935) recommendations for pressure diagram

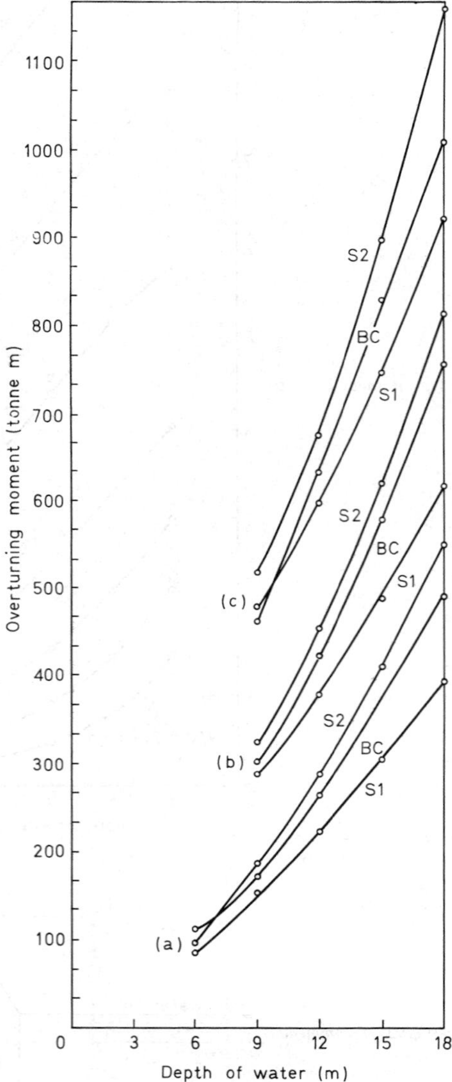

Figure 24.8 Overturning moments for wave acting on theoretical vertical breakwater of unlimited freeboard. (a) 6 m high × 70 m long; (b) 8 m high × 90 m long; (c) 10 m high × 110 m long

simplified pressure diagram as indicated in Figure 24.7 was recommended as a practical equivalent of the Sainflou formula.

In Figure 24.8 are shown the overturning moments for waves 6 m, 8 m and 10 m high acting on a theoretical vertical breakwater of unlimited freeboard. For each wave height the moments have been calculated on the basis of (1) Sainflou theory assuming still water at back S1, (2) Sainflou theory assuming clapotis trough at back S2 and (3) Brussels Congress recommendation BC. Similar curves have also been drawn for wave heights of 2 m, 1.5 m, 3 m and 4 m and in all cases the Brussels Congress curve lies between the two limits of the Sainflou theory indicating that the former is a very sound practical compromise. When the limited freeboard of an actual breakwater is not as

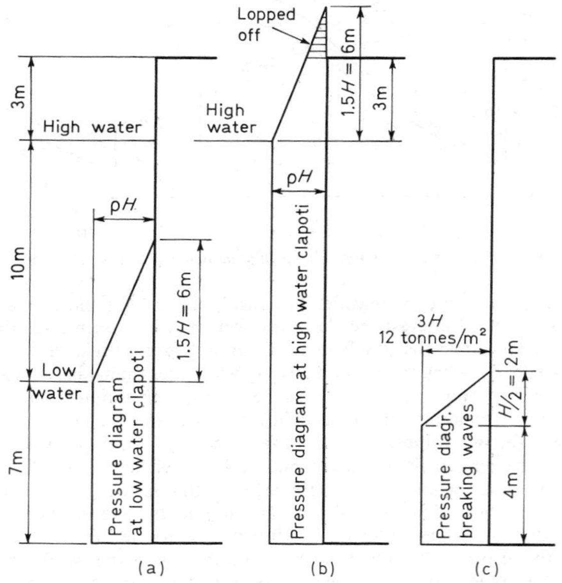

Figure 24.9 Typical pressure diagrams for 20 m *breakwater. (a) Clapotis effect with* 7 m *water alongside; (b) clapotis effect with* 13 m *water alongside; (c) breaking effect with* 4 m *water alongside*

high as the theoretical pressure diagram then the upper portion of the latter is lopped off as in Figure 24.3.

CONDITIONS AT HIGH AND AT LOW WATER

As has been stated, when designing vertical-wall breakwaters the possibility that waves may break at low water must be considered.

Example Assume a site with tidal range of 10 m in which two types of floating caisson breakwaters are to be sunk, one with a height of 20 m and the other 13 m. With each type a minimum freeboard of 3 m at high water is proposed so that at low water the depth alongside one will be 0 to 7 m and the other will always dry out. The maximum wave anticipated is 3 m high by 5 m long—as approaching from the open sea. This is a short steep wave and it will be assumed to increase in height to 4 m and then break.

Thus at low water both sections of breakwater may be assailed by breaking waves but at higher states of the tide the waves will remain oscillatory. Figure 24.9 shows typical pressure diagrams for the 20 m breakwater: (a) clapotis effect with 7 m water alongside, (b) clapotis with 16 m water, and (c) breaking effect with 4 m water. From these and similar diagrams the total thrusts and the total overturning moments can be calculated

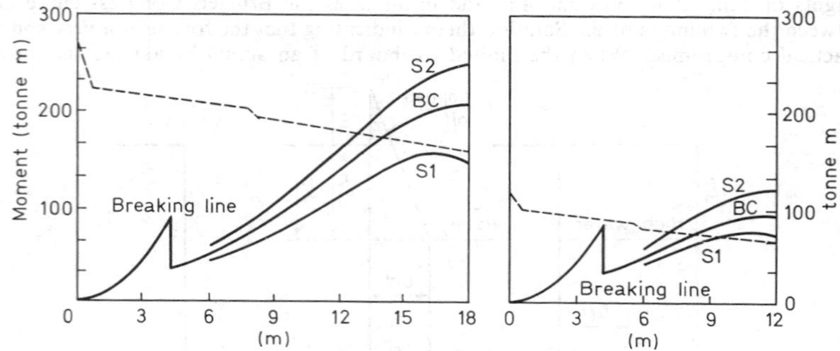

Figures 24.10 and 24.11 Total overturning moment against depth diagrams

for varying depths of water alongside. The results have been plotted in Figure 24.10 to 24.13. As a matter of interest the clapotis effects based on Sainflou's theory are also shown and it will be noted that the Brussels Congress figures for thrust and moment are a mean. Theoretically, when the water depth is just greater than 4 m the wave should not break and for the purposes of the curves the thrusts and moments at that depth are calculated both for breaking and for oscillatory waves and the curve is assumed to be vertical between the two values; while from zero depth to 4 m the curves approximate to straight lines from zero to the breaking values. At water depths of 20 m and 13 m the respective breakwaters would be just submerged with considerable quantities of water passing right over them. The exact thrusts and moments have not been calculated for these depths, the curves merely being continued on smoothly from the values at 16 m and 10 m depths respectively. It will be seen from the curves that while breaking thrusts are similar to the clapotis thrusts, moments from the latter are the greater. This follows, of course, from the fact that while breaking waves give high intensity of pressure the clapotis effect acts over a much greater depth. The example chosen is perhaps unusual on account of the tidal range of 10 m but it serves to illustrate the principles applicable to any conditions.

Types of breakwater

There are three main traditional types of breakwater,[5] vertical sided, Figure 24.14; rubble mounds, Figure 24.15; and composite, i.e. a rubble mound as major foundation carrying a vertical sided superstructure, Figure 24.16. Considerable attention is now being given to floating breakwaters but up to 1973 none had been constructed successfully on an actual functional basis.[6]

VERTICAL-SIDED BREAKWATERS

The earliest form consisted of a piled timber formwork filled with relatively small size rubble and as such is only suited to small depths and minor works. Blockwork with vertical and horizontal joints or with sloping joints ('sliced work') has been used for large scale works,[3] the former being suited to sound foundations and the latter more accommodating to possible settlement on poor foundations. More modern types have consisted of reinforced concrete caissons[7] floated out and filled with sand.

It is important to guard against scour up to a depth of 10 m on the exposed side and a relatively heavy rubble mound may be required for this. Overturning forces at the crest of the wave will obviously be catered for but with the trough of the wave alongside the net force will be outwards away from the harbour and wall panels in reinforced concrete caissons must be designed for this 'negative' condition. In heavy seas masses of water may cascade down from a great height on to the deck of the breakwater and a continuous strong surface is essential to withstand this.

RUBBLE MOUND BREAKWATERS

These may be entirely of rubble with a maximum weight of 10 tonnes for individual stones (but more usually 5 tonnes) ranging down to 1 tonne, 0.1 to 0.25 tonnes, and ordinary small rubble. Seaward slopes at depths lower than about 5 m below low water remain natural for rock at $1\frac{1}{4}$ to 1; from this height up to low water wave action becomes active and slopes must relate to the weight of individual stones.

In lieu of rubble throughout, mounds may consist of rubble to a depth of 5 m below low water and then be completed with 'pell-mell' concrete blocks. The weights of such blocks vary between 20 and 50 tonnes and while individual blocks may be moved at

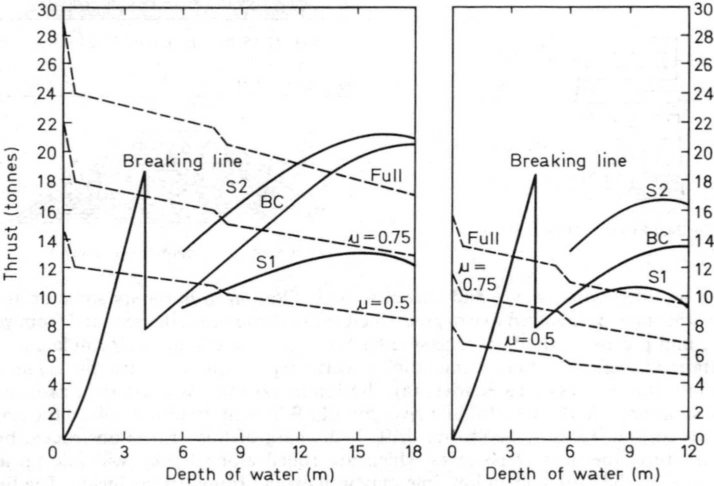

Figures 24.12 and 24.13 Total thrust against depth diagrams

times the blocks as a mass are not flattened down to easy slopes like smaller rubble and the slope in fact remains at about $1\frac{1}{4}$ to 1 just as does the rubble at lower depths.

COMPOSITE BREAKWATERS

Vertical walls may be built on a rubble mound foundation either because the natural sea bed is too weak to take a wall or because the depth of water is too great. In very deep water, of the order of 30 m, the construction of a complete vertical wall becomes impracticable and the dimensions of a complete rubble mound—especially with a large tidal range—would be prohibitive. The compromise consists of a rubble mound at natural slopes of about $1\frac{1}{4}$ to 1 from sea bed to within not less than 5 m of low water, and above this level a vertical wall. Almost invariably the reduction in water depth due to the mound results in translatory waves attacking both mound and wall and the success of these breakwaters depends on adequate design against the major effects of such waves, i.e. the direct impulsive forces on the wall, the dragging and sucking of the undertow on both wall and mound and the 'punching' and 'tearing' of the falling columns of water on the horizontal surface of wall and mound.

Siltation and dredging

SILTATION

Analytical theories have been developed with some success to deal with siltation in inland rivers and canals.[8] Such watercourses are reasonably well defined; a certain degree of control can be exercised over their route and shape and they are not too much

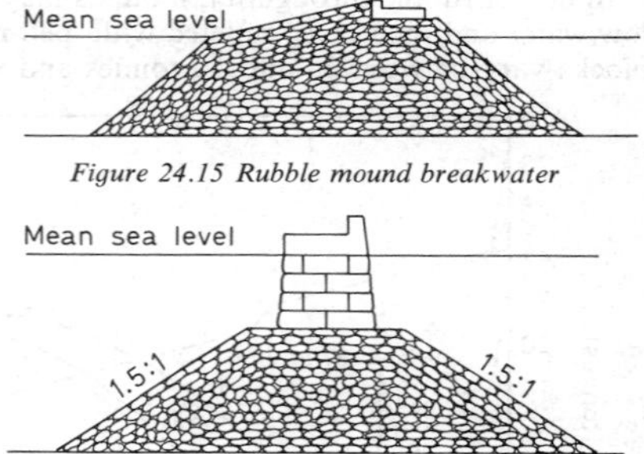

Figure 24.15 Rubble mound breakwater

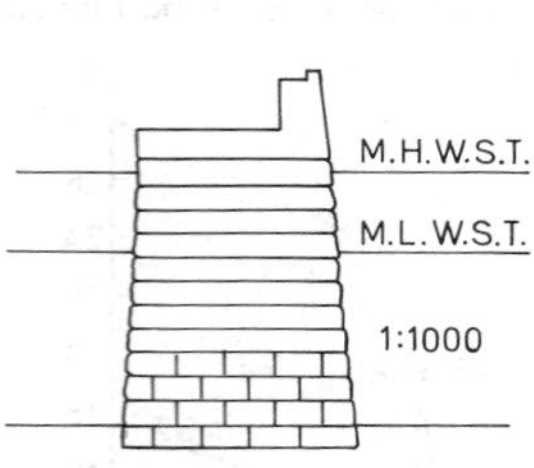

Figure 24.14 Vertical sided breakwater

Figure 24.16 Composite breakwater

affected by tides, wind, waves and currents. With siltation in harbours some or all of these factors may be involved to a degree which varies throughout the year and from year to year, and precise methods of assessing harbour siltation cannot be formulated.

If siltation is defined as the accumulation of earth deposit on the sea bed by any means then it falls into two main categories: (a) siltation in harbours with no river associated and (b) siltation in harbours sited at a river mouth. Where no river is involved the single source of accumulation is the littoral drift. If the path of this drift is obstructed by a solid structure, the heavier particles which are rolled along bodily will pile up and accumulate on the drift side and in time may well extend round to the inside. The finer particles of the drift which, outside the harbour, are kept in suspension by the wave agitation of the sea and transported by currents will, on entering the calmer confines of the harbour, no longer be maintained in suspension but will settle out.

Where a harbour is at a river mouth the material carried down by the river is a second source of siltation; the interaction between this material and the littoral drift makes for further complication, with the added difficulty of the difference in density between fresh and salt water.

With both types of harbour a certain amount of beneficial scour can be provided at the entrances by keeping the tidal compartment large in relation to the entrances and so inducing large tidal currents. The speed of such currents should not exceed 6 km/h

(4 mile/h) otherwise navigation would be affected, but velocities well below this have good scouring effects as the following figures show:

Velocities of 0.6 km/h will move fine sand
Velocities of 0.8 km/h will move coarse sand
Velocities of 1.0 km/h will move fine gravel
Velocities of 1.1 km/h will move gravel the size of beans
Velocities of 2.6 km/h will move shingle 25–50 mm diameter
Velocities of 3.0–4.0 km/h will move shingle 25–75 mm diameter

DREDGING

Harbours which naturally maintain their depth everywhere are extremely rare; almost invariably dredging has to be undertaken. The most common dredging equipment[9, 10] comprises bucket ladder dredgers with attendant dumb or selfhopper barges, such units being able to deal with upwards of one million cubic yards per annum. For confined spaces alongside jetties or inside dock entrances, grab buckets are necessary but these have a much lower output rate. In suitable material suction dredgers have the highest rate of output, as much as one million m^3 per month. Rock dredging can sometimes be done by bucket dredgers if the rock is laminated and the laminations lie at a favourable angle. Where blasting has to be used the time and cost can increase by more than twenty times compared with normal dredging.

Layout and entrances

The main factors involved in the layout of a harbour are waves, currents and siltation; as the considerations based on them frequently conflict with each other, compromise is inevitable. From the immediate point of view of navigation the purpose of a harbour is protection from waves and the biggest wave reduction is effected with the smallest entrance sited remote from the direction of approach of the waves. Approaching a narrow entrance with heavy beam seas, however, is difficult and might even be disastrous. As every harbour must be prepared to serve as a harbour of refuge, i.e. a protection to be sought by vessels during the height of a storm, it is usual to site entrances facing the heaviest seas, thereby improving ease of access at the expense of smoothness within the harbour. But where the harbour is operated on the lighterage basis of unloading cargoes into barges and thence to the shore the reverse may have to be adopted, as at Madras,[2, 5] since such work cannot be carried out in waves much more than 1 m high. Wave height reduction within a harbour is improved with added distance of the entrances from the shore line and more so with increased width parallel to the shore in which the waves passing through can spread themselves. Stevenson's empirical formula for assessing wave height within a harbour is

$$h = H\left[\left(\frac{b}{B}\right)^{\frac{1}{2}} \frac{D^{\frac{1}{4}}}{50}\left(1+\frac{b^{\frac{1}{2}}}{B}\right)\right] \tag{17}$$

where h is the wave height in the harbour at distance D from the entrance; H is the wave height approaching the harbour; b is the width of the entrance; and B the width of the harbour at distance D from the entrance.

Siltation is of primary importance in maintaining a harbour and the layout must take account of this. The windward breakwater accordingly is usually made to extend beyond the line of the opposite leeward one, and the two curve in towards each other. The object of this is to keep the current across the entrance to a minimum and to ensure that the littoral drift does not get trapped by the leeward breakwater. It frequently happens that the exact alignments of the ends of the breakwaters are adjusted on site as the actual consequences of the new works become apparent.

Experiments on models

In view of the many variable natural factors involved in harbour design and the manner in which it is possible for them to react on each other, models[9, 10] would seem to offer almost the only solution to most of the problems. On the whole this is probably the case, but the very diversity of causes and effects which make the use of models almost compulsory also make the model design extremely complex and the results have to be interpreted with much caution.

DOCKS

Jetties

Jetties jut out into the water, usually at right angles to the shore line and, in order to avoid siltation and to maintain an even water flow in the harbour, they are mostly open

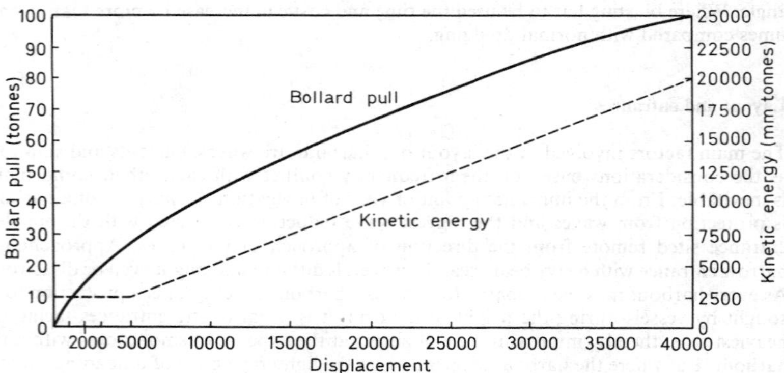

Figure 24.17 Bollard pulls and kinetic energy against displacement

structures. Earth pressures from retained filling do not, therefore, have to be considered and the main loadings are (i) vertical deck loading, (ii) bollard pulls from berthed vessels and (iii) impact blows from berthing vessels.

VERTICAL LOADING

Vertically applied deck loading should never be taken as less than 1000 kg/m^2 unless it is known definitely from specified usage of the structure that it cannot be as much. In addition to the uniform load of 1000 kg/m^2 the knife edge concentrated load of the Ministry of Transport loading curve is included where heavy lorry traffic is anticipated. With pile spacings at 3 to 5 m centres it will be found that a uniform load of 2500 kg/m^2 will result in pile loadings similar to those obtained from the Ministry of Transport loading; this can be used for preliminary calculations. Where extensive stacking of machinery or merchandise may be involved uniform loads of 2500 kg/m^2 or even

5000 kg/m^2 may result. Standard gauge railway tracks will involve standard railway loadings but no allowance for the impact of fast moving traffic need be taken. No general value can be given to cover railway and crane loadings and each case must be dealt with separately using values provided by the makers. In particular, container cranes, because of the weights to be lifted and the large areas covered, can be as severe as shipbuilding cranes.

BOLLARD PULLS

Under stress of accident or bad weather bollard pulls can be high, and it is not uncommon for bollards to be broken or uprooted. In sheltered waters, however, a reasonable upper limit for the largest of normal vessels is 100 tonnes and, except for very small structures, a similar lower limit is 25 tonnes. With bollards spaced at 33 m centres these loads are equivalent to 3 tonnes and $\frac{3}{4}$ tonnes per lineal m and an overall figure between these limits is the best approach for the structure as a whole, bearing in mind the necessity for dealing with the local concentration of load at the bollard. Mammoth tankers and bulk carriers are special cases.[11]

BERTHING IMPACT

Berthing impacts are almost as indefinite as bollard pulls. Unless a full head-on collision occurs—and that might be regarded as an accident against which reasonable provision cannot be made—half the weight of a vessel can be assumed to be effective, i.e. if W is the weight of vessel, V the velocity of berthing impact,

$$\text{kinetic energy to be absorbed} = \tfrac{1}{2}W\frac{V^2}{2g} = \frac{WV^2}{4g} \qquad (18)$$

In the absence of any specific estimate of likely berthing speed a value of 150 mm/s is a good average figure. Larger vessels, however, will be handled more carefully than small ones and the potential damage from a vessel of 500 tonnes may well be as high or higher than that from one of 5000 tonnes. As a guide for preliminary design work Figure 24.17 shows bollard pulls and kinetic energy plotted against displacement of normal ships. The kinetic energy is taken as constant for displacement between 1000 and 5000 tonnes on the assumption that in this range increase in weight will be offset by increased care in handling. Tankers and bulk carriers can be dealt with on a similar basis but special criteria are usually specified.

FENDERS

The absorption of kinetic energies of 500 mm tonnes upwards can only be effected by some form of flexible fendering. For values exceeding 25 000 mm tonnes large concrete clumps or patent bells[12] and tubes[12] have been used, but rubber in compression has now almost completely superseded these. Rubber in shear[13] can be used for values below 100 mm tonnes but small compression blocks are more commonly used now.

Although massive floating fenders are an advantage from the point of view of spreading blows over many vertical fenders, they have the disadvantage of applying the blows as point loads at water level. Hence fender piles have to be strong enough in themselves to distribute the blows in the form of reaction into the ground and into the resilient support at the deck. Where floating fenders are not used and where the vessels to be catered for are 'wall' or vertical sided it can be assumed that blows are applied at deck level, the fender pile being flexible enough in itself to 'deflect' away from any local application of load at an intermediate position.

TYPES OF PILES

Piles are the most common form of support for jetties. Almost invariable they are fixed at the heads either in the deck slab or by bracing below deck level and they are usually fixed by the ground, but at what depth in the ground is not easy to define. Calculations based on certain assumed conditions for ground characteristics indicate[14] that even in poor ground the depth required to develop full fixity is not high; and practical experience confirms this, for even in weak muds of river estuaries, timber piles if broken accidentally usually snap at river bed level. A reasonable assumption for 'fixity depth' is one third the total penetration. From this the gross effective column length follows and to allow for fixity at both ends the net effective length can be taken as 0.8 × the gross length.

Concrete piles (see also page **24**–38) are not strong in bending and if used only as vertical piles they must be well braced to reduce their lengths, and consequently the bending moments set up in them by horizontal loads. Bracing, however, involves tidal or even underwater work; to eliminate this, raking piles are to be preferred. Loads in raking piles are best calculated by the simple method of 'triangle of forces' as in Figure 24.18. More elaborate methods have been evolved but for maritime construction these hardly seem justified. The actual load to be taken by a pile cannot be assessed very

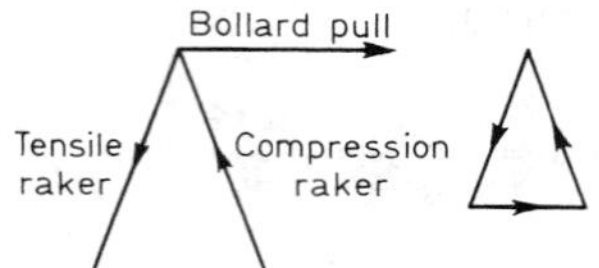

Figure 24.18 Calculation of loads in raking piles

accurately; the effective column length of the pile is only an approximation and with raking piles the actual rate of rake must vary somewhat from the precise outline shown on the drawings.

Timber piles could be driven to a rake but as bracing can be readily carried out (since timber can be so easily worked on site) vertical piles are usually adopted.

Steel piles of box sections or hollow tubes are now most common. They are very strong in bending and can withstand large horizontal forces even with long unbraced lengths. On this account they can often be used in a structure composed entirely of vertical piles, without bracing and without rakers. The horizontal displacement might be high, e.g. 2 or 3 in (6–8 cm), but unless cranes are carried by the jetty such occasional movements could be accepted, and the resilience or capacity for absorbing kinetic energy begins to become appreciable with displacements of this order. Where cranes are involved, however, a movement of more than 6 mm should not be exceeded—mainly because movements on jetties always seem worse than they are and their effects are always exaggerated by observers and users—and rakers or bracing may have to be introduced to achieve this. It may be noted that a movement of 6 mm cannot involve an appreciable quantity of kinetic energy; even if the force set up is 200 tonnes the energy absorbed is only 600 mm tonnes and with any form of piled jetty—concrete or steel piles, heavily braced vertical piles or unbraced raking piles—an independent system of flexible fendering should be provided.

In deep water with current speeds of 5 knots the possibility of oscillation must be taken into account.[15]

DECKING

With timber piles and main beams the final decking would be of 75 mm timber planks but for permanent work with steel or concrete piles the deck would also be of concrete. The most common form is thick slabs—the thickness may be 0.5 to 1.0 m—which are as economical as beam and slab construction in spite of the extra volume of concrete involved. The slabs contain a hidden beam system and, in order to reduce the slab depth, are best reinforced with high-tensile cold-worked bars (60 000 lb/in^2 proof stress and 30 000 lb/in^2 working stress) top and bottom and in both directions. Usually one size of bar is sufficient, 12 mm or 18 mm, the beams being formed by closing the general spacing of bars from 300 mm to 100 mm or 150 mm. Expansion joints are unnecessary; jetties up to 400 m long have been built without them in Great Britain and up to 700 m in the USA.

SAFE LOADS ON PILES

Pile driving for jetties and the assessment of pile loading capacity follow the same technique generally as for land work except that soft estuarial mud is more likely to be encountered and, in that case, the normal dynamic pile driving formulae are usually completely misleading. Test loading is essential.

SCREWED PILES

Screwed piles or even screwed cylinders may be used in poor ground; these too act as friction piles, the large diameter of cylinders and the numbers of screws used merely serving to increase the effective frictional area.[16]

PIERS AND CYLINDERS

When the depth of sea bed becomes so great that piles cannot be used because of buckling tendencies, solid piers or cylinders may have to be used but these forms of construction are invariably more expensive than simple piling. If the foundation level is likely to vary much, cylinders are to be preferred to solid piers. Cylinders are built up of pre-cast reinforced concrete rings, for example, 2m in diameter, 1 m deep and about 100 mm thick.

Cylinders may consist of a group of piles driven well down to a firm foundation and covered from just below sea bed upwards with pre-cast concrete tubes, or the tubes may be carried down to the firm foundation and piles dropped in to serve as reinforcement to the insitu underwater concrete placed in the tubes. In either case both the cylinder and the piles must contribute towards the load carrying capacity but it is usual to base design work on the assumption that only one is fully operative. The capping of cylinders usually consists of deep heavy beams so as to form portal bents in cross sections which carry the longitudinal girders forming the main basis of the decking system. In order to reduce weight from the constructional viewpoint, prestressed concrete will no doubt be used for pre-cast units in such designs. For very heavy jetties, however, deep insitu girders are necessary for the main longitudinal beams, the whole jetty system then consisting of portal construction in both directions.[17]

As an alternative to cylinder portals, solid piers made of cast insitu concrete deposited under water between steel sheet piling as shuttering can be cheaper provided that ground conditions are well defined and a good foundation is available at only a few feet below dredged bed level. Above low water concrete would be deposited in the dry with back shuttering over the sheet piling and the latter can then be removed and re-used. Concrete placed under water against steel sheet piling does not adhere to it at all; on the contrary the laitance formed by the concrete sliding into place against the piling—it cannot be rammed under water—seems to serve as a lubricant when it comes to withdrawing the latter.

JETTIES ON ROCK FOUNDATIONS

The construction of jetties where the surface of the sea bed is rock is bound to be costly and slow. If the rock is a softish shale and steel piles can be driven at least two or three feet into it, then an open piled structure becomes a possibility, but where the rock is so hard that even steel piles cannot be driven then drilling may have to be adopted and piles planted in the drilled holes. Where very favourable conditions obtain for the making and launching of reinforced concrete caissons, these have been used to form solid pier foundations for jetties on rock but normally such caissons are far too expensive.

DOLPHINS

Dolphins usually form a mooring to which vessels can be tied up or they are guiding devices for assisting vessels to negotiate narrow entrances. Occasionally they are built in line, with connecting walkways, when they serve as jetties or wharves and then have to withstand both berthing impact and mooring pulls. Reinforced concrete piles are not well suited to dolphin construction. Mooring loads can only be taken by raking piles and if these may be applied at any angle rakers are necessary in two planes and driving becomes very involved. Neither is concrete suitable where resilience is required for berthing impacts; until recent years most dolphins have been of heavily braced timber construction. During World War II the Americans introduced a timber dolphin into Great Britain consisting of a mass of piles driven touching each other in a circle and securely bound at the top with wire so as to form a structure similar to a huge bundle of faggots. Steel piles, because of their heavy bending strength, can be driven vertically and if adequately fixed at their heads with a heavy reinforced concrete slab form a suitable anchorage capable of withstanding mooring loads in all directions.[14] For purely isolated dolphins used as guides to navigation and liable to be subjected to unusual impact loads, large diameter steel tubes may be a solution but they are expensive.

Wharves

SOLID AND OPEN TYPES OF WHARF

Wharves are berths built parallel to the waterfront and may be of solid construction when they have to hold back the earth of the foreshore or of open construction when the earth of the foreshore slopes down in the form of an embankment. Very deep wharves may have to provide as much as 15 m of water at low tide and with large tidal ranges and allowing for some freeboard above high water the depth from coping to dredged bed may exceed 25 m. With such depths open structures are seldom economical. If the foundation is sound ballast or rock and if good rock is available to form the foreshore embankment then the slope of the latter may only extend over a horizontal distance of 35 m, but even this minimum width is excessively costly for a piled platform. Moreover, in poorer ground the danger of a slip in the embankments of such height becomes an important consideration.

TYPES OF SOLID CONSTRUCTION

Wharves of solid construction may consist of mass (in situ concrete) walls, monoliths, caissons floated into place and sunk, mass walls of pre-cast concrete blocks either with horizontal joints or nearly vertical (i.e. 'slice' work) or sheet piling of concrete or steel. Whatever the construction the problem of assessing earth active pressures and capacity for passive resistance is common to all.

PRESSURES ON WHARF WALLS

Active pressures The classical earth pressure theories of Coulomb, Rankine and Bell are all similar fundamentally except that Bell[18] was the first engineer to appreciate the full significance of the cohesive property of clays. Modern theories developed in the study of soil mechanics have by no means superseded the old. The most significant contributions of soil mechanics are the better assessment, through improved sampling and testing technique, of cohesive strength, and the analysis of overall failures of walls brought about by the shearing of the ground along slip circles.

So far as active earth pressure on wharf walls is concerned the presence of water greatly simplifies the problem. In waterlogged ground the minimum hydrostatic pressure must be that due to sea water, i.e. a fluid of 1000 kg/m³. To this must be added the effect of the earth particles in the water; if these are granular or noncohesive the pressure may be found by Rankine's formula:

$$p = \rho h \frac{(1-\sin\phi)}{(1+\sin\phi)} \tag{19}$$

where p is the horizontal intensity of pressure at depth h, ρ is the density of the material and ϕ the angle of internal friction.

For the best of backfill (e.g. rock) the buoyant weight may be taken as 500 kg/m³ and ϕ as 45° whence

$$p = h \times 500 \frac{(1-\sin 45°)}{(1+\sin 45°)} = h(500 \times 1/6) = 80h \tag{20}$$

Hence waterlogged rock or similar good filling gives a pressure equivalent to a fluid weighing 1080 kg/m³ (i.e. water at 1000 plus buoyant rock at 80).

At the other extreme even if a mud is assumed to have ϕ as zero, so that

$$p = \rho n \frac{(1-\sin\phi)}{(1+\sin\phi)} = \rho h \tag{21}$$

then with the buoyant weight still at 500 kg/m³ the combined effect of water and mud is only equivalent to a fluid weighing 1500 kg/m³, i.e. 1000 for water and 500 for buoyant mud. This of course is an excessive upper limit since it amounts to assuming that the mud acts as a perfect fluid. Probably a 'fluid factor' of 0.9 is more nearly correct; hence the range between buoyant rock and buoyant mud is represented by pressure effects of 1080 to 1350 kg/m³.

Fully buoyant pressure seldom if ever acts right from coping level. A conservative estimate is from mean high water surface but some engineers work to mean sea level. As a preliminary design basis or for a quick check on a given design the assumption of fluid pressure at 1200 kg/m^3 acting right down the back of the wall from coping level will not be far wrong for any site. Taking the full fluid pressure from coping level will usually include sufficient allowance for surcharge. Quite often such an assumption will prove accurate enough for the final design. For a site with a large tidal range, a large clearance between high tide and coping, and good backfilling, the assumption would give too high a value; conversely, for a site with medium tidal range and freeboard and poor mud as backfill right down the back of the wall it might not give a high enough value, especially if surcharge loading at surface level is likely to be high.

Some illustrations of the assessment of total active pressures in this manner are shown in Figure 24.19 in which the chain line is for an equivalent fluid of 1200 kg/m^3 and the

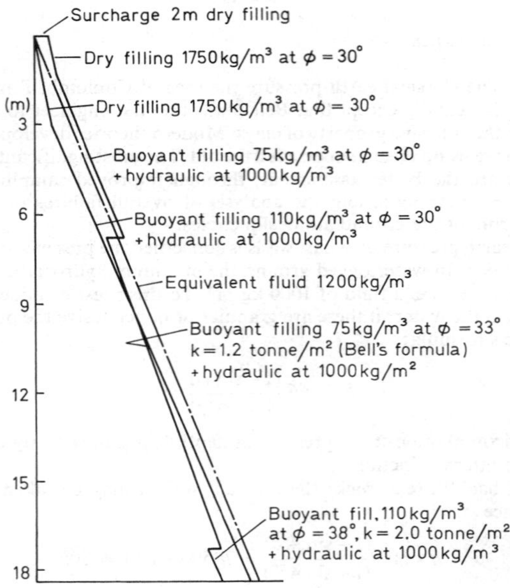

Figure 24.19 Assessment of total active pressure

full lines are arrived at in the more orthodox manner indicated. Pressures on the front of the wall are best kept separate from those on the back.

Passive pressures Passive resistance below dredged bed level is not so easy to deal with as active pressure on the back since the water component on the front can have no passive potential. The earth passive pressure is only a potential too and some movement of the wall, no matter how slight, is essential to bring it into play. Such movement with flexible walls like steel sheet piling can take place locally, i.e. within the ground and without being apparent at coping level but can hardly be local in the case of massive solid walls of concrete; in fact many such walls have moved a matter of inches at coping level and not a few have moved several feet.

By Rankine's formula, passive potential is given by

$$p = \rho h \frac{(1+\sin \phi)}{(1-\sin \phi)} \tag{22}$$

i.e. the fluid multiplier is the reciprocal of that for the active pressure, whence for good ground with $\phi = 45°$

$$p = \rho h \frac{(1+\sin 45°)}{(1-\sin 45°)} = 6\rho h \tag{23}$$

and with buoyant ρ at 500 kg/m^3

$$\rho = 3000h \tag{24}$$

With the water effect at $1000h$ added the gross value becomes

$$p = 3000h+1000h = 4000h \tag{25}$$

Practical experience has proved that for high values of ϕ, Rankine's formula underestimates passive capacity and it is customary to double the value obtained in this way. The water effect, however, must not be doubled, so the absolute maximum for passive resistance becomes

$$p = 2\times 3000h+1000h = 7000h \tag{26}$$

With very low values for ϕ the empirical multiplier of the Rankine formula must not be used since at the extreme of ϕ = zero the material may have become little better than a heavy fluid. In such extreme cases the equivalent passive capacity may be as low as $p = 2000h$, including the water effect.

Muds as poor as this are really useless from the point of view of passive resistance and instances are known where sheet piled walls in such material have failed with less than 3 m difference in level on the active and passive sides of the piling.

To summarise, and using the equivalent fluid analogy: active pressure may vary from $p = 1080h$ to $1350h$ and passive capacity may vary from $p = 2000h$ to $4000h$ for rigid walls to $p = 2000h$ to $8000h$ for flexible walls.

Sliding wedge method for passive pressures As an alternative to the equivalent fluid analogy the sliding wedge method, solved graphically, is best. When the soil constants of ϕ (angle of internal friction) and C (cohesive strength) are known accurately from laboratory tests on undisturbed samples the sliding wedge method is the only logical approach to the problem; it is simple to understand and can be carried out quite quickly and is applicable to all forms of surcharge or reliving platforms. Essentially it is 'trial and error'; sliding of the earth is assumed to take place along a straight line and the problem is treated as one of statics with one body of earth sliding over the other along the assumed inclined plane of rupture. Referring to Figure 24.20(a) let ABD represent the back of a wall with a relieving bank sloping up BE. Then as a first trial assume the earth will fail in sliding down the plane AF. If the earth is cohesive only and has zero angle of internal friction, the only forces on the sliding plane will be W, the weight of earth ABEF; R, the reaction at right angles to AF, since $\phi = 0$; C, the cohesive force $= c$ AF where c is cohesion; and P, the balancing horizontal force on back of wall, P is found by the usual method of parallelograms of forces as in Figure 24.20(b).

Passive resistance is found similarly except that the cohesive force acts in the opposite direction and for noncohesive soils the reaction R is inclined at an angle ϕ to the normal to the plane AF. Different slopes are tried for the plane AF and the maximum value of P for active pressures, or the minimum for passive resistance, is adopted.

SOLID WALLS

The cheapest form of solid gravity wall is mass concrete placed in trench. Where conditions have been such that no blows in the bottom were feared, trenches have been taken out to a depth of 35 m, but in poorer ground and for greater depths monoliths are adopted. These may be of pre-cast blocks or may be cast in situ above ground level as the monolith is sunk. Monoliths as large as 16 m × 16 m with nine cells have been sunk up to 50 m from coping level and in spite of the weight involved the main difficulty, even in the poorest ground, is to get them to go down, blasting and bentonite often having to be resorted to. Gaps up to 3 m must be left between blocks to allow for sinking out of true vertical and the gaps have to be closed after sinking is finished. From the normal design

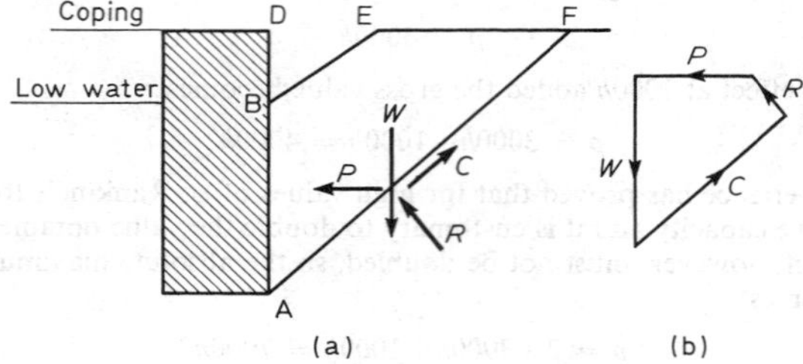

Figure 24.20 Sliding wedge method. (a) Back of wall with relieving bank sloping up BE; *(b) determination of P by triangle of forces*

viewpoint monolith construction is unsatisfactory because the deeper a monolith is taken down the less stable it appears to become. Figure 24.21(a) in which the active and passive pressures are shown in the equivalent fluid form illustrate this. This theoretical difficulty can be offset by taking friction into account as indicated in Figure 24.21(b) (i.e. downwards on the back of the wall; upwards on the front) and by increasing the passive capacity, but the uncertainties attached to assuming these factors are themselves unsatisfactory.

A better way of relieving a load on the back of a wall is by means of a relieving bank carried down to low water so that provided adequate drainage is arranged the tide rises and falls equally behind and in front of the wall and no hydrostatic difference is developed. Referring to Figure 24.21(a) let AD represent the face of a wall 17 m high from coping

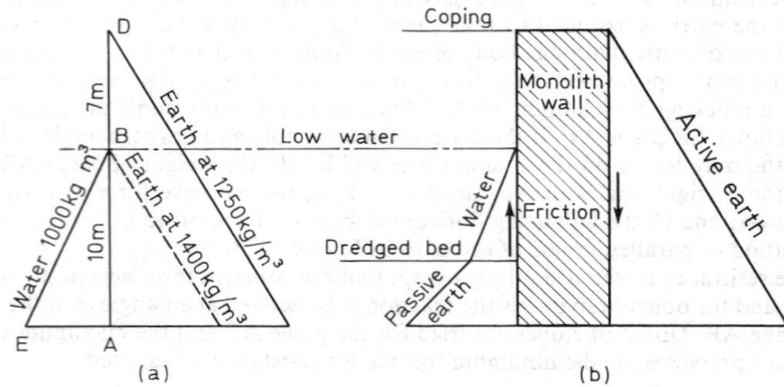

Figure 24.21 Forces on solid walls

D to foundation A with low water B at 6 m down from coping. Then the water pressure on the front of the wall below B is

$$\frac{\rho h^2}{6} = \frac{1000 \times 11 \times 11}{2 \times 1000} = 60 \text{ tonnes}$$

The moment of this pressure about A is

$$\frac{\rho h^3}{6} = \frac{60 \times 10}{3} = 200 \text{ tonne m}$$

Similarly with active fluid pressure on the back of the wall at 1250 kg/m^3 from coping

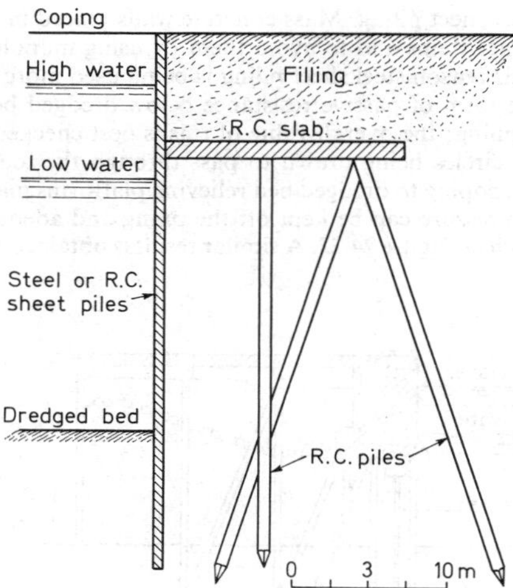

Figure 24.22 Sheet piling

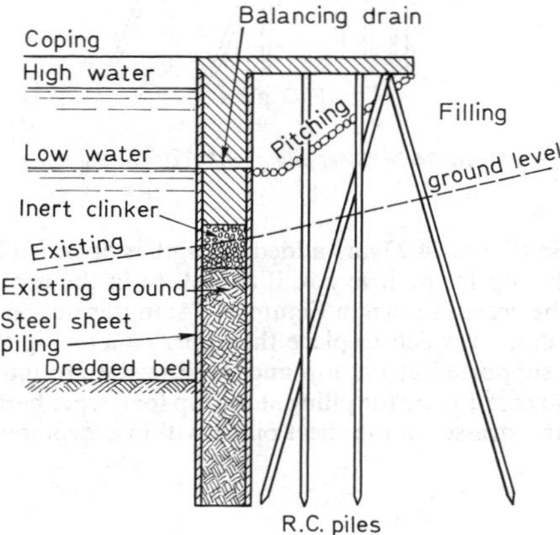

Figure 24.23 Sheet piling with gravity wall

level D the corresponding active values are 180 tonnes and 1020 m tonnes respectively. If a relieving bank can be included so that the active pressure line commences at B then the intensity may increase from 1250 kg/m to (say) 1400 kg/m^3 due to surcharge effect, but the total values become about 90 tonnes for the pressure and 300 tonnes for the overturning moment about A, i.e. the relieving bank eliminates half the pressure and nearly two thirds the moment.

SHEET PILING

This principle can be applied to any form of construction and is being increasingly used in conjunction with steel sheet piling. Mass concrete walls placed in the dry may necessitate extensive temporary dams and to avoid them by using monoliths is usually even more expensive. For this reason steel sheet piling is being used more extensively.

Good support in the form of passive resistance below dredged bed level is the vital requirement for sheet piling; the overall value of this is best checked by the method of 'slip circle' failure, the circles being drawn to pass through the toe of the piling. For heights up to 13 m from coping to dredged bed relieving platforms may not be necessary, but for greater depths pressure can be kept off the piling and adequate top anchorage provided by a design such as Figure 24.22. A similar result is obtained with the top gravity

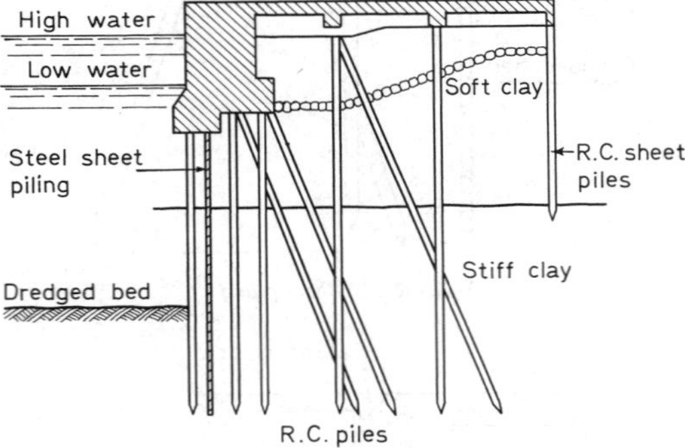

Figure 24.24 Sheet piling with fixity at the top

type of wall shown in Figure 24.23 and added strength is obtained by regarding the sheet piles as 'fixed' at the top in the heavy wall as well as in the ground. Fixity at the top is also achieved in the design shown in Figure 24.24; in this design the sheet piling could be used as a coffer dam in which to place the upper concrete in the dry. The design of sheet piling simply supported at the top and fixed in the ground is fully dealt with in papers[19] and textbooks.[20] Fixing the piling at the top for deeper berths becomes necessary as a means of keeping stresses in the sheet piling within acceptable limits.

CAISSONS

For construction in extensive open water followed by subsequent reclamation, reinforced concrete floating caissons may have to be used but unless special facilities, such as an old floating dock, are available for the launching of the caisson, such work must be very expensive. Fundamentally, too, it is uneconomical to have to build a structure with all the care necessary to enable it to float only as a temporary measure. After sinking, filling is essential to provide mass for stability and this, too, is expensive unless favourable conditions (e.g. a good supply of sand filling) obtain.

CYLINDERS

Cylinders are commonly used for extensions to existing gravity walls when extra depth alongside is required. When the work can be reached from land-based cranes and underwater concreting in the cylinder can be accepted this form of construction is economical. Cylinders can also be used for new work in situations where otherwise a large dam would be needed, each cylinder in effect being its own selfcontained dam. For small depth the backfill can slope down under the deck, the cylinder and the deck then forming in effect a platform over the slope of the bank. But for deep berths such a platform becomes too wide and, to reduce it, sheet piles may be introduced at the back to hold up the fill to a vertical face. The cylinders then have to be braced together with portals to support the piling; the construction requires careful design and good-quality concrete placed in the dry so that costs tend to be high.[17]

Dry or graving docks

Dry or graving docks are structures into which a ship can be taken afloat; the caisson or gate is then shut, the water is pumped out and the ship is left dry on the docking blocks. The dock may be needed only for carrying out minor repairs, inspections, and painting as with most commercial trading ports, or it may be needed for heavy repairs or complete refittings as for most shipbuilding and repair yards. In either instance the users state their requirements as to depths, length and breadth, cranage facilities, general services such as water, compressed air, steam, oil fuel and electricity.

In the design of dry docks[21] attention must be continually paid to constructional problems and, even after a design is finished, the method adopted for carrying out the work may necessitate modifications. A temporary dam is invariably required and while details of this may be best left to the contractor the designer must consider this first and may even make provision for it in the permanent work.

When satisfactorily completed a properly designed graving dock will float in the ground just as a floating dock floats in water. The floor and walls will act together as one unit and very often the total weight of the completed dock filled with water is less than the weight of the earth excavated for its formation, so that foundation pressures generally are less after completion than those that existed before work started. This is borne out by the fact that occurrences of dock floors lifting under hydrostatic pressure are not uncommon, while settlements in dry docks are almost unknown. The support that the walls receive from the floor is vital to their stability and during the construction period great care must be taken to ensure that walls built in trenches are not allowed to get into an overloaded condition before the floor is placed. This danger is offset somewhat by the fact that, during construction, full hydrostatic pressure is unlikely to build up behind the wall but it may be necessary to ensure added safeguards by some preliminary 'open cut'[21] or by limiting the extent to which walls may be built ahead of the floor without support, by back shoring off the 'dumpling' excavation.

WALLS

The front profile of dry-dock walls for modern commercial vessels at trading ports (i.e. where heavy repairs are unlikely) may be almost vertical.[22, 23] In shipbuilding yards where badly damaged vessels have to be catered for, or where major rebuilding of ships is undertaken, the profile will be stepped at intervals to provide altars off which timber raking shores can be strutted.

Full hydrostatic uplift invariably acts under the floor and walls of a dry dock and unless the floor is vented sufficient dead weight must be provided to counteract it. On this account and because of the heavy nature of the work generally, walls are usually of cast in situ mass concrete gravity design. Earth pressures on the back are similar to those for wharf walls, the safest assumption being that, on completion, the ground will be waterlogged up to the mean high water surface. In built up areas where work has to be carried out in full trench a section of the wall may be 'back hung' as shown in Figure 24.25, the back steps following the outlines of the trench timbering. Such walls are heavier than most orthodox sections but where heavy cranes are required the expense is offset by the fact that both tracks for the crane are carried on the wall. Where construction space is not limited a more economical section is shown in Figure 24.26. Horizontal joints should be at 1 to 1.5 m intervals and in order to be watertight should be stepped down 0.3 m to the back and be covered with cement mortar 25 mm thick as the first stage in each concreting operation. Longitudinal joints should be vertical at intervals of 13 to

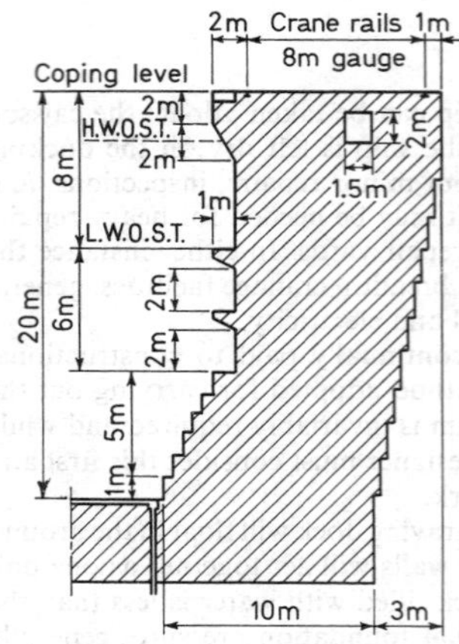

Figure 24.25 Back hanging of wall section

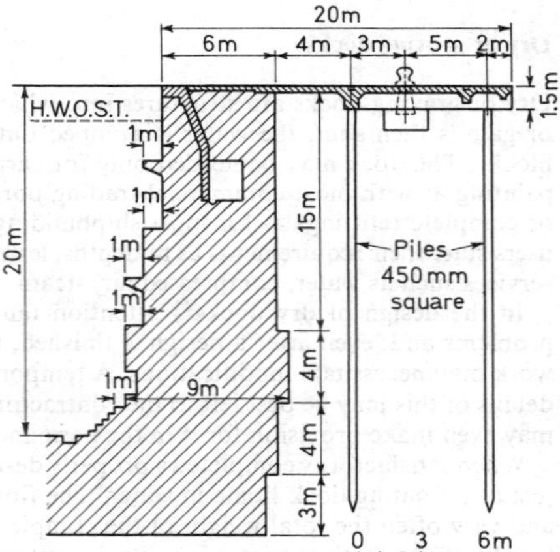

Figure 24.26 Economical wall section

17 m and as they are essentially contraction joints they should be sealed with a bitumen compound.

FLOORS

Floors of dry docks have to withstand hydrostatic uplift when the dock is dry and there is no ship in the dock and they have also to be strong enough to transmit a ship load to the ground without fear of settlement. In good ground (not rock), considerations of uplift usually result in a floor thickness capable of carrying the ship load; in poorer ground, if the principle that a dry dock floats in the ground is accepted, then a section generally as for good ground can be adopted. But if this principle is not accepted piles might be used to support the ship load and the capacity of these for resisting uplift may be taken into account. Without piles the floor acts as a flat arch or restrained beam and as a rough guide to the thickness required the thrusts T (Figure 24.27) may be assumed to act at one foot from the edges and equated to the uplift bending moment thus:

$$\frac{WL^2}{8} = T(R-2) \qquad (27)$$

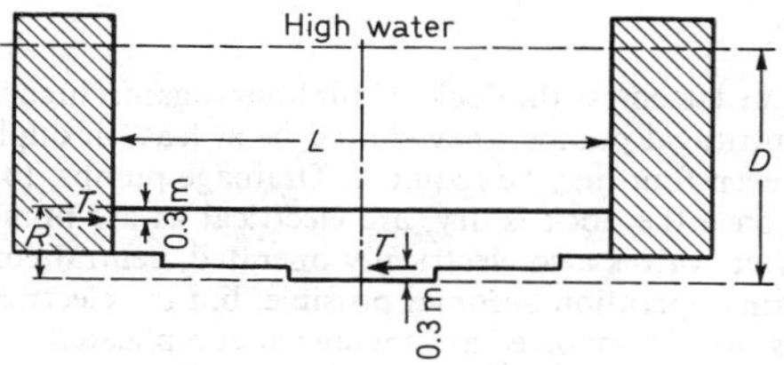

Figure 24.27 Thrusts on floors

where $W = 1000D - 2200R$; 1000 being the weight of salt water and 2200 the weight of concrete.

The thrust T should not result in a stress in the concrete greater than 250 tonnes/m^2.

For construction in rock the total quantity of water likely to require pumping if the floor is not vented is not excessive. As there is no need to distribute the ship loading over a large area it is usual to vent against uplift; the floor thickness then becomes a matter of judgement. Some Continental floors have been put in as thin as 0.5 m but for very heavy ship loads and a softer rock a value of 1–2 m may be preferred. Construction joints should be keyed or sloped as necessary to develop shear strength. They should extend the whole length of the floor and there should be no horizontal joints. The whole floor should be of constant permeability throughout its full depth. Nearly all the troubles with the lifting of dock floors are associated with construction comprising an almost impermeable top thin layer of granite stones or sets overlying a relatively thick and permeable layer of mass concrete.

FACINGS

Granite facings undoubtedly provide the best wearing surface for exposed sections, especially copings, altars and stairs, but are seldom used now because of the cost. For the watertight meeting faces, against which the caissons bear, granite was invariably used in the form of large blocks of 1 to 1.5 m^3 with a face left 10 mm proud for dressing back into one plane to an accuracy of 0.1 mm. To guard against hydrostatic uplift each stone should be anchored down into the mass concrete with round bars up to 50 mm diameter screwed to take nuts and heavy washer plates. Modern practice is, however, to eliminate granite even for meeting faces and to substitute fine concrete placed in situ and rubbed down, if necessary, to a tolerance of ± 2 mm. This construction, however, can only be used in association with a rubber gasket device on the caisson.

CASSIONS AND GATES

Caissons for closing docks are of steel construction[24] whether they are mitre gates, sliding caissons, ship or floating caissons or horizontally pivoted 'box' gates. Timber is sometimes used for mitre gates but usually only for replacement of existing gates.

Timber watertight faces of greenheart are dressed to an accuracy of 0.05 mm but the deflection of the caisson as a whole renders it advisable to make the haunches up to

3 mm proud of the general plane. When rubber gaskets are incorporated in the timber the latter takes the load but as the rubber, squeezing up into a chase in the timber, forms the watertight seal, excessive accuracy in the dressing of the timber is not necessary.

PUMPS AND CULVERTS

Main pumps are designed to empty the dock in four hours against high tide and are usually electrically driven centrifugal pumps; they should be at least in duplicate and for very large docks three or even four may be required. Drainage pumps, to deal with seepage and general leakage once the dock is dry, are electrical as are pumps for a fire main service. With the culvert valves also electrically operated, central control of the whole flooding and dewatering operation becomes possible, but the electrical and mechanical engineering problems thereby involved are somewhat complicated.

Culverts have to be large enough to fill the dock in one hour and the emptying culvert must be large enough to feed water to the pumps without danger of cavitation. When more than one main pump is required the emptying culverts lead into a large sump into which the pump suctions dip, and experience suggests that the average velocity into this sump should not exceed 3 m/s. Velocities in the filling culverts may be as high as 17 m/s and all transitions for changes of direction or size must be very gradual.

DOCKING BLOCKS

Docking blocks along the centreline of the floor together with an outer row on either side of the centreline for the side keels vary considerably. Oak or other suitable hardwood is probably the best, when obtainable, for the main body of the block, with a soft wood capping piece; cast steel and reinforced concrete have been used. A good height above floor level is 1.5 m.

SERVICES AND FITTINGS

Services for a dry dock in a shipyard will be extensive—sewage, fresh water, fire mains, compressed air, electrical power, welding facilities, gas, distilled water and fuel oil—and one or more subways in the main walls will be needed with branches off to serve pits in the coping. Subways should be high enough to walk in upright and of ample width to accommodate bends and junctions for branch pipes. Access manholes should be at intervals of about 50 m.

Fittings such as step irons, ladders, handrails, ring bolts, eye bolts for securing docking blocks and gratings to floor sumps and culvert openings should be generously proportioned to allow for corrosion and should be heavily galvanised.

Wet docks or basins

Wet docks or basins are large areas of water bounded by vertical faced solid walls approached through locks or a simple entrance off the main waterway. Water level is kept almost constant at high water and, unless a lock entrance is provided, entry and exit can only be at times of high water. Within the wet dock vessels tie up to the walls for discharge of cargo or in the case of shipbuilding yards they lie alongside for major refits or internal repairs. In commercial ports, wet docks are by far the most extensive maritime structure: and as a result the word 'dock' in commercial practice has come to stand for wet dock rather than graving dock.

WALLS OF WET DOCKS

Walls for wet docks are similar to those for wharves except that they can never be of open construction. The fact that the water level in front of them remains sensibly at high tide is of prime importance in the problem of stability. Referring to Figure 24.28, DA represents a wall height of 17 m as in Figure 24.21(a), but BA is now 15 m to represent high tide level as compared with 10 m to low tide level. Thus the water pressure on the front of the walls below B is

$$\frac{\rho h^2}{2} = \frac{1000 \times 15 \times 15}{2 \times 1000} = 113 \text{ tonnes}$$

The moment of this pressure about A is

$$\frac{\rho h^3}{6} = \frac{113 \times 15}{3} = 565 \text{ tonne m}$$

These values are two and three times respectively greater than for a similar wharf wall at low tide as indicated by Figure 24.21(a).

A striking example of the importance of extra water support on the front of wet dock walls was afforded during World War II when a Continental port was bombed. The two caissons of a lock entrance to a long narrow wet dock were both hit so that the dock water drained out rapidly with falling tide. Water remained dammed up behind the dock walls which were of the gravity mass concrete type and, at low water, a length of nearly 400 m collapsed in one long line by overturning on its face into the water. Normally, however, it would not be considered economical to design for such an occurrence and

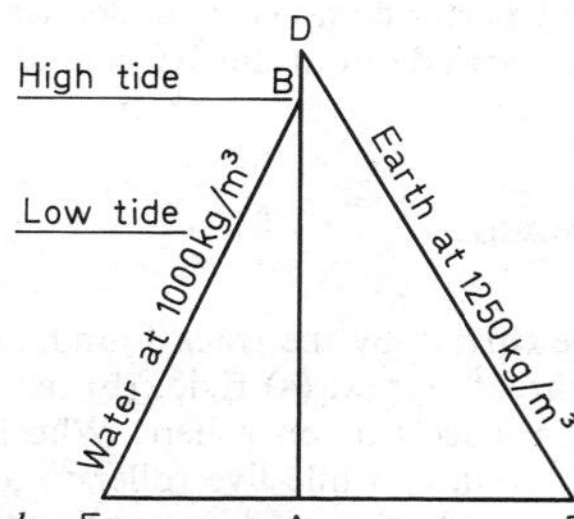

Figure 24.28 Walls of wet dock

advantage would be taken of the additional water support. This has resulted in relatively light mass concrete walls being built in the past which are now only standing with very fine factors of safety. Although the driving of piles close behind old mass walls is not usually likely to be a source of trouble, where wet dock walls are concerned it is advisable to be very cautious. Instances have been known where the driving of pre-cast piles has had to be abandoned in favour of core drilled in situ piles.

Leakages and consequent reduction in water level are to be expected in wet docks and where adjoining dry dock facilities exist provision is often made for 'topping' up the wet dock by the main pumps from the dry dock. It is often sufficient to rely on the external tidal source for maintaining an adequate water depth. The leakages are invariably sufficient to saturate the back filling up to high water level but it may have taken years for the water to accumulate, and once this first accumulation has been pumped away it is often possible to carry out very deep excavation behind walls in comparatively dry conditions,[25] provided that the wall foundations are firm and sound.

Slipways

SLIPWAYS FOR SHIP BUILDING

Slipways may be for the building of ships or for repairs. Building slips are usually arranged so that building operations are carried out above high water and the lower end of the slip is tidal and open to the water. Sometimes, however, temporary dams are provided at half tide level so as to increase the effective working length. Retaining walls, floors, cranage and rail facilities follow normal maritime procedure for such structures.

SLIPWAYS FOR REPAIRS TO SHIPS

Repairing slipways are the more common type and the term slipway or maritime slipway is usually taken to mean this type. Such a slipway consists essentially of a cradle which can be run out on a rail track foundation under water so that vessels can be centred over it; the cradle and vessel are then hauled back up the track, the weight of the vessel gradually coming on to the cradle and both coming right out of the water and clear of high tide. Broadside slipping is sometimes adopted where the value of the foreshore is not high and where the river or sea bed slopes steeply, but longitudinal slopes down the foreshore are by far the most common. Slopes vary between 1 : 10 and 1 : 20 but to make the most of some natural slopes the slipway may sometimes be curved. The rails used are standard flat bottom railway sections from 36 to 40 kg/m and the civil engineering problem is essentially one of providing a crane track foundation, partly in the dry, partly between tides, and partly completely under water. Like all maritime work it is best, wherever possible, carried out in the dry behind a dam.

CRADLES AND LOADING

The loading to be carried by the track foundations is dependent on the type of cradle used. There are three[26] types: (a) rigid, (b) semirigid and (c) collapsible; each may be mounted either on wheels or on rollers. Wheels and bearings are the more common mountings in this country while live rollers[27] are used in the USA. Rigid cradles are single long units and when carried on continuous rollers, with docking blocks packed up towards the after end so as to be in a horizontal line, they impose the least severe loading on the track foundation. Such cradles are essential for vessels much above 2000 tonnes and it is claimed[28] that with them vessels of 10 000 tonnes and over could be handled. These cradles require longer slips; to offset this the semirigid type may be used, consisting of a series of sections closely coupled together, special attention being paid to making the coupling system as simple as possible. Collapsible cradles give the shortest possible length of slip and reduce expensive underwater work to a minimum. These cradles are, in effect, a series of bogies connected by chains and if the actual bogie length is one third the centre spacings the collapsed length can be little more than one third the extended length.

When keel blocks are set at a constant height and, therefore, parallel to the slope of the slipway, the vessel during the slipping operation only bears on the cradle gradually.

At the first point of contact the vessel is completely waterborne, but as the cradle is hauled up the vessel is supported partly in the water and partly at the first point on the cradle, this latter load slowly increasing until the vessel is completely clear of the water and its weight evenly spread over the whole cradle length. This increasing point load is termed the 'sue load' and can only be determined from an exact knowledge of the vessel's displacement and an assumed line of its trim. Any pronounced deviation from the assumed line of trim due to a damaged condition would have to be treated as a special case.

Sue loads can be very high and when confined to a single short bogie of a collapsible cradle they are correspondingly severe on the track and the track foundation. With certain types of vessel the sue effect may be transferred from the first to the second section of the cradle and if the connecting chains are not exactly equal in length (which is difficult to ensure), severe racking loads may be induced in the track beams. These may be offset by connecting the first two sections by rigid links and bracing them together for extra longitudinal stiffness. When the cradle is carried on three lines of track the apparent reduction in loading on each track is dependent on the tracks being at the same level on cross section. With underwater work especially this may not be easy to guarantee and may be upset by unequal settlement of the supports. Most vessels are so built as to impose all the weight on the centre keel and, therefore, most loading on the centre track where three are provided. The question of level supports is not then so important, but some vessels are as heavy along their sides as along their centres, which implies that the loading may be equally distributed over the track supports. If this is relied upon in the design to reduce either pile loads or beam sizes then it is essential with three or more supports that they should all be at the same level. Because of the uncertainty of the ship loading, continuous footings, whenever good ground conditions permit, are preferable to piles and beams since occasional very high loads will probably be distributed safely by a continuous footing but may well cause local failure in a beam or pile.

HAULAGE

Haulage in British practice is mainly by single wire rope, the two ends of the rope returning to the double drum of a winch after being reeved round a system of sheaves on the cradle and on a head block fixed to a concrete foundation in front of the winch. This results in a fully balanced haulage system in which all pulls are equalised. The winch may be driven electrically or by petrol or diesel engine. A warping drum is provided for back hauling the cradle, a wire rope passing from this round a sheave fixed at the lower end of the slipway back into the cradle. Slipping speeds may vary from 1 to 3 m/min and loads to be designed for in the winch foundation, the main hauling head block and the down haul block are governed by the power unit and the pulley system adopted.

TRAVERSING SLIPWAYS

Some slipways have been constructed on which the ship, after being pulled up to the highest point, can be traversed sideways to other berths. To enable this to be done the rail tracks on the upper part of the slipway are carried on a series of separate foundation blocks and in the spaces between the blocks there are bogies carrying similar rail tracks. When the ship is being slipped the rails on the bogies are held in register with those on the blocks, thus forming a continuous track. At the proper point all the cradle wheels are carried by the rails on the bogies. These bogies are themselves carried on rails running at right angles to the slipway. The bogies, cradle and ship can then be traversed sideways to an adjacent berth when the ship is transferred by jacks to the fixed berthing blocks clear of the cradles. The cradle and bogies are then returned to the main slipway line and the cradle lowered to take another ship.

Equipment

RAILWAY TRACKS

Railways are for slow-moving goods traffic mostly in confined areas; usage by loco cranes with wheel loads up to 25 tonnes is common; access to workshops may necessitate sharp curves and frequently the railway and road have to form one common service even away from jetties and wharf walls. For the latter, rail and road must form a flush surface and flat bottomed rails resting on concrete with the road formed in concrete is the best combination. Because of this, flat bottom rails—36 to 40 kg/m are generally standard practice for dock work. On concrete roads the edge of the concrete on the inside of the running track should be protected against spalling by a check bulb angle or old flat-bottomed rail; if cranes with double flanged wheels are to use the track, this protection must be provided on each side. Switches are spring loaded with nonremovable levers all set flush into the concrete.

Curves should not be smaller than 50 m radius if possible, but if necessary a truck crocodile can be forced round a curve of 30 m radius.

HEAVY CRANE TRACKS

Crane tracks are usually carried on piled reinforced concrete beams. Very large travelling cranes may have four sets of wheels with four wheels in each set, i.e. a total of sixteen wheels. Under the worst loading conditions each of the four wheels in any one set might be loaded to 50 tonnes, making a total of 200 tonnes per set. The concrete beams for such loading may be 1.5 m deep by 0.75 m wide and with piles spaced at 3 m centres the excessive stiffness of the beam in relation to its relatively short span becomes a vital consideration. Unless yielding of the supports is taken into account calculated values for pile loads will be too high while bending moments in the beam will be too low and may even be of the wrong sign.[29] With deep concrete beams horizontal surge may be ignored apart from nominal cross beams at (say) 30 m intervals. Very heavy wheel loads are usually carried on double wheels with the flanges placed centrally side by side. This necessitates a track of two flat-bottom rails riveted on to a 20 mm steel base plate, the whole forming what is commonly termed duplex track.

ROADS

Road construction for docks follows normal practice with some extra care paid to the fact that loads will be high but slow moving or even static, as with mobile cranes. The effects of loads from these cranes can be very severe.

SERVICES

Underground services such as water, gas, oil fuel, compressed air and electricity also follow normal practice apart from extra strength at traffic crossings. Cast iron and steel are mainly used for pipes; under certain conditions in waterlogged mud, sulphide-producing bacteria may attack them, leading to a softening of the metal. Services usually have to be taken to service pits at the dockside copings where portable couplings can be effected for supplying the fuel or power to ships. With solid wall construction a subway to accommodate all services can be readily provided and it is a simple matter to form branches to the coping pits. In open jetty construction the provision of services is far from simple and the exact requirements of the user must be known before design is commenced as these requirements may have a fundamental bearing on the design.

BOLLARDS

Bollards should be of cast steel and are of two main types: (a) vertical or mushroom, used mainly for steadying vessels when being manoeuvred and (b) horned or lipped, used for mooring stationary vessels. The latter usually have a thick base plate and are bolted down into the dock wall or jetty deck near the coping; the former may be set into a solid wall or carried in independent foundation blocks well back from the coping line. Capacities vary from 10 to 100 tonnes for lip bollards, but the mushroom types are mainly of the higher capacity. Both should be made on generous lines to allow for excessive corrosion, wear and accident.

FAIRLEADS

Fairleads are a form of fixed 'snatch block' used to guide the rope from capstan to ship with a 'fair lead'. If properly maintained and greased the deeply grooved wheel will revolve on its bush when used.

CAPSTANS

Small capstans may be used for shunting railway wagons, and large ones, up to 30 tonnes capacity, for handling ships. Modern types are mainly electrically drive but compressed air has been used with very good results, the mechanism being simple, robust and not sensitive to exposure and rough usage.

FIXED CRANES

Fixed cranes may vary from small 5 tonne derricks to 250 tonne cantilever or goliath cranes carried on four steel towers or legs, the maximum load down one leg being as much as 500 tonnes. For the heavier loads solid piers are usually favoured, the two front legs as a rule being carried by the dock wall.

SHEDS AND WAREHOUSES

Dockside sheds and warehouses usually involve heavy floor loadings (1.5 tonnes/m^2 is common and 2.5 or even 5.0 tonnes/m^2 may have to be allowed for); wind forces are high and general exposure to salt laden atmosphere is severe, but apart from this the structural problems are similar to other framed buildings.

COALING PLANT

For coal dispatch, conveyor systems are becoming increasingly popular in this country, but various other systems may be used such as (1) tipping from a high level formed by an embankment or (2) from a high level in the form of a mechanical lift or hoist within a steel framed tower or (3) by cranes which lift wagons direct or hoppers filled from wagons. Coal receipt into large stack grounds may be by overhead travelling grabs mounted on high gantries.

The details of most equipment, especially cranes, pumps, capstans, coal conveyors and hoists are essentially for mechanical and electrical engineers but heavy structural and foundation problems may be involved and close collaboration with the civil engineer is essential.

PILES

Piles for land work, as for open sea-water work, are best in pre-cast reinforced concrete or prestressed concrete but, unless ground conditions are very bad, in situ concrete may be used and some systems can be driven to a rake of 1 : 8. Insitu piles are particularly advantageous, too, in old reclaimed areas where many unknown obstructions may be likely. For very long piles, between 30 and 50 m, steel sections are much lighter to handle and being so much stronger withstand lifting stresses very well. A good ratio of driving hammer weight to pile weight is also possible with standard hammers.

For maritime work, piles are usually more than 15 m long. With reinforced concrete piles a minimum cover of 50 mm is essential to protect reinforcement against water and, since long column lengths are generally unavoidable, the usual size for main steel for ordinary reinforced concrete is 15 mm, 32 mm or 40 mm. Long piles are often made of large cross section on the assumption that this provides extra strength against lifting stresses. Actually, if the moment of resistance of a section is calculated by steel beam theory and with a limiting stress of 840 kg/cm^2 to keep tensile cracks to a minimum it

Table 24.5 ORDINARY REINFORCED CONCRETE

Main reinforcement	*Moment of resistance* (kN m) *Steel beam theory* 840 kg/m^2					*Maximum length* (m) *2 pt lifting at $\frac{1}{5}$ points*				
4 at 25 mm	14.9	19.2	21.5	23.4	27.7	18	18	17.5	17	16.5
4 at 32 mm	22.5	29.2	32.5	35.9	42.5	22	22	21.5	21	20
4 at 40 mm	31.1	40.9	45.4	50.0	59.5	26	26	25.5	25	24.5
Pile size (cm)	30 × 30	35 × 35	37 × 37	40 × 40	45 × 45	30 × 30	35 × 35	37 × 37	40 × 40	45 × 45

can be shown that for 2 point lifting (i.e. at $0.2L$ from each end) bending strength for ordinary reinforced concrete does not increase with size quite so much as bending moment due to increased weight, from which it follows that so far as lifting stresses are concerned increased length is best met by decreased cross section.

Prestressed concrete is now being used more extensively for piles for maritime work and Table 24.5 does not apply to these. Because of the usual long column length, applied stresses from the structure are not very high and the basic objection of prestressing a member, whose main purpose is to carry direct compressive loads, does not apply. With prestressed concrete piles cracking is unlikely and for factory piles, at least, the concrete should be of higher quality, and 40 mm cover is sufficient protection for the reinforcement. For bonding into the slab it may be necessary to incorporate mild steel bars of sufficient length to cater for predicted variations in driven pile length.

Construction

DAMS

Keeping water out is one of the most important constructional problems in dock work. When possible, permanent sections of the work should be so designed and programmed as to serve as dams for the remainder, but usually dams have to be temporary. Steel sheet piling of flat sections with high tensile strength in the clutches is used to form selfsupporting cells, the cells being earth filled and the piles supporting the filling by hoop tension.

Sections of the deep type to withstand longitudinal bending are normally used as the watertight seal to heavy embankments formed of whatever material happens to be available, the piling eliminating the need for restricted choice. Where space is too limited for embankments, double rows of piling with tie bars and walings, and earth filled, may be adopted. The design of such sheeting follows normal lines, except that the assessment of the earth pressures is extremely complex. With adequate drainage of the filling the inner sheeting is not loaded on the outside and on the inside has to retain drained material; a reasonable assessment of this is to regard it as an equivalent fluid of 400 kg/m^3. The outer or seaward sheeting has water on the outside and saturated filling on the inside; it would seem reasonable to assume that the net result is a pressure outwards, although some methods of analysis suggest a net pressure inwards. If the outer sheeting was withdrawn the filling would undoubtedly collapse outwards and the safest procedure is to make both the inner and the outer sheeting similar. This means that the latter is also designed for a fluid pressure of 400 kg/m^3 and as the outside water weighs 1000 kg/m it follows that the inside saturated fill is taken as 1400 kg/m^3, though this is somewhat on the high side.

GROUND WATER LOWERING, CHEMICAL CONSOLIDATION AND GROUTING

Purposely designed dams will keep out open water and if taken down deep enough there will be little or no seepage under them. But on some sites artesian water may be present or, for example, in an excavation below the level of an existing dock wall. In such instances groundwater lowering by well points or by deep wells may be practicable, or it may be possible to form a seal by chemical consolidation.

When seepage may be localised but is sufficiently extensive to cause flooding to the same extent as through an open rubble bank, it is probably impossible to drill enough for chemical consolidation and the flow would be too great for groundwater lowering. Plain grouting with cement or a mixture of cement and clay may then prove a solution. The pumping of such grout requires careful handling, but each case has to be treated on its merits and highly specialised technique is not involved.

The author wishes to acknowledge the assistance he has received from the N.M.B. Memo 135/45 of the Hydrographic Department, Admiralty, in connection with general wave theory.

REFERENCES

Harbours

1. SAINFLOU, M. G., *Annales des Ponts et Chaussées,* **4,** 5 (1928)
2. MINIKIN, R. R., *Harbours Engineering,* **158,** 161 (1944)
3. CUNNINGHAM, B., *Harbour Engineering,* London (1928)
4. 'Rip-rap Protection for Slopes subject to Wave Attack', CERA Res. Rep. 4 (May 1966)
5. CAGLI, H. C., 'Italian Docks and Harbours', *J. Instn civ. Engrs,* **1,** 465 (1928)
6. WEBBER, N. B. and HARRIS, R. J. S., *Floating Breakwaters,* Southampton University (1972)
7. STROYER, R., *Concrete Structures in Marine Work,* London (1934)
8. LACEY, G., 'A General Theory of Flow in Alluvium', *J. Inst. Civil Engrs,* **27,** 16 (1946)
9. ALLEN, J., *Scales Models in Hydraulic Engineering,* London (1948)
10. Hydraulics Research Station, Wallingford

Docks

11. *Conference on Tanker and Bulk Carrier Terminals, Inst. Civil Engrs* (Nov. 1969)
12. BAKER, A. L. L., 'Heysham Jetty', *Instn civ. Engrs Maritime Paper, No. 9* (1947)
13. LITTLE, D. H., 'Some Designs for Flexible Fenders', *J. Inst. Civil Engrs,* Pt II, 42 (1953)
14. LITTLE, D. H., 'Some Dolphin Designs', *J. Instn civ. Engrs,* **27,** 48 (1946)
15. SAINSBURY, R. N. and KING, D., 'The Flow Induced Oscillation of Marine Structures', *Proc. Inst. Civil Engrs* (July 1971)

16. MORGAN, H. D., 'The Design of Wharves on Soft Ground', *as above*, **22,** 5 (1944)
17. PATERSON, D. E. and GUTHRIE BROWN, J., 'Sturrock Graving Dock, Capetown; Captain Cook Graving Dock, Sydney', *J. Inst. Civil Engrs,* **28,** 286 (1947)
18. BELL, A. L., 'Clay Pressures', *Proc. Inst. Civil Engrs,* 199 (1915)
19. PACKSHAW, S., 'Earth Pressure and Earth Resistance', *J. Inst. Civil Engrs,* **25,** 233 (1946)
20. LEE, D. H., *Sheet Piling, Coffer dams and Caissons,* LONDON (1946)
21. LITTLE, D. H. and EVANS, E. J., 'Stresses in Docks', *J. Inst. Civil Engrs,* **23,** 75 (1944)
22. BURNS, T. F., 'Design and Construction of No. 8 Dry Dock, North Shields', *Proc. Inst. Civil Engrs* (June 1955)
23. ROSS, K., RENNIE, W. J. H. and COX, P. A., 'The New Dry Dock at Belfast', *Proc. Inst. Civil Engrs* (Feb. 1972)
24. EASTON, F. M., 'Dock Gates', *J. Inst. Struct. Engrs,* **16,** 359 (1938)
25. LITTLE, D. H., 'Widening No. 10 Dock, Devonport', *Inst. Civil Engrs Maritime Paper, No. 6* (1946)
26. MINIKIN, R. R., *Slipways,* Dock and Harbour Authority (1945)
27. GLOVER, W. G., *Marine Slipways,* Dock and Harbour Authority (1946)
28. CRANDELL, J. S., *Railway Dry Docks,* Dock and Harbour Authority (1947)
29. ELSBY, W., 'Continuous Beams on Elastic Supports', *Concrete and Construction,* **381,** 399 (1943)

25 ELECTRICAL POWER SUPPLY

ELECTRICAL POWER SUPPLY 25

FUELS

N. G. EGGLETON, B.Sc., F.I.C.E.
Central Electricity Generating Board

In order to generate electricity in sufficient quantity for public supply, generators are normally driven by turbines which are themselves powered by steam raised from the combustion of coal or oil, or from nuclear fission (thermal energy), or by gas produced from the combustion of oil (also thermal energy) or by water pressure (hydraulic energy).

In England and Wales and the South of Scotland, practically all electricity generation is steam powered, although gas turbines are used to generate a small amount of electricity

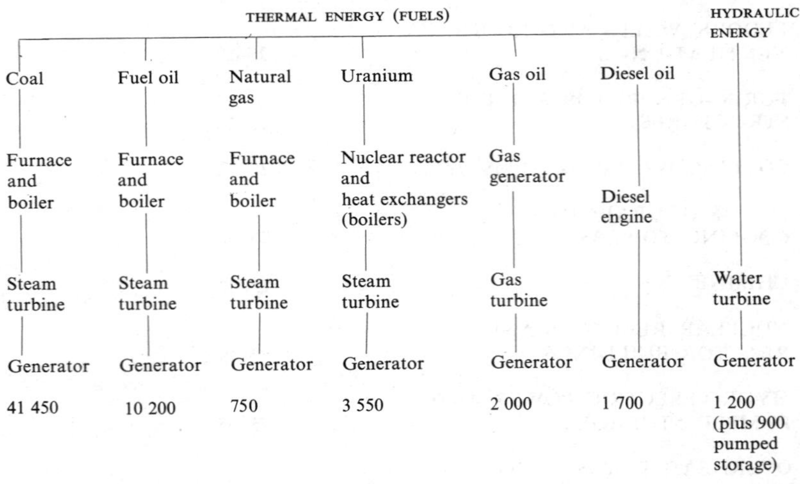

Figure 25.1 The last line of the diagram gives the total installed capacity (megawatts) in 1971 of generating plant in England, Scotland and Wales

at times of peak demand. The generation of hydro-electric power is relatively small and is concentrated in Scotland.

Steam-powered generation

Electricity is produced from steam power by the association of two major plant items of matched output which, together with ancillary plant and services, form a unit:

(i) a boiler for raising steam, heated either by a coal- or oil-burning furnace or by the heat produced by the nuclear fission of radioactive material in a reactor

(ii) a steam turbine coupled to an electricity generator (a turbo-generator) in which the turbine converts the heat and pressure energy of steam into mechanical power to drive the generator

Figure 25.2 shows diagrammatically the plant used to generate electricity from steam power, and the flow and losses of energy through the unit. The energy transmitted from the turbine to the generator is the electrical rating of the unit—typically 500 or 660 MW (megawatts). The efficiency of a steam turbine increases with the drop in steam temperature across the turbine and, to achieve the lowest temperature at exhaust, the steam from the low-pressure stage of the turbine is discharged into condensers carrying water-cooled tubes. The condensate is returned to the boilers for recirculation, while the cooling water which has gained the heat lost by the steam in the condensers is either returned to source for the heat to be dissipated or is returned to cooling towers before recirculation through the condensers.

In Britain the electricity supply industry was nationalised in 1948, and in England and Wales the construction and operation of power stations and the transmission of electricity at high voltage is the responsibility of the Central Electricity Generating Board (CEGB).

Fuel cost and the capital cost of the plant and buildings of a power station are the dominant costs of electrical power. The relative cost of electricity generated from coal, oil and nuclear fuel used in CEGB stations commissioned in the early 1970s is as follows (nuclear = 100%):

	Coal	*Oil*	*Nuclear*
Fuel cost	93	79	25
Capital charges	35	35	71
Operating and transmission and other costs	4	4	4
Total	132%	118%	100%

The operating and transmission costs are approximately the same in each case. The capital costs are approximately the same for coal and oil stations but are much higher for nuclear stations. The cost of oil fuel is slightly lower than of coal, but the cost of nuclear fuel is substantially lowest. In total, the cheapest electricity is produced from nuclear fuel, followed by oil, and with coal producing the most expensive electricity. There is greater potential for future technical improvement in the utilisation of nuclear fuel than in the combustion of oil or coal, so that from technical considerations, nuclear fuel used in future power stations should produce still cheaper electricity under stable cost conditions.

The total fuel consumption for the generation of public supply electricity in England and Wales in 1970–71 was 95.7 million tonnes (coal equivalent) in the following proportions:

Coal	72%
Oil	20.6%
Nuclear Fuel	7.4%

The policy of the nationalised industry is to develop a flexible system capable of responding to the relative costs and availabilities of different fuels. Accordingly, if electricity generation doubles between 1970 and 1980, the industry could, by conversions to oil and by constructing new capacity preponderantly to burn either oil or nuclear fuels, change the consumption by 1980 to approximately equal percentages (in coal equivalent) for each type of fuel. The trend of new construction towards oil and nuclear stations is shown in Table 25.1 by the list of CEGB stations commissioned between 1960 and 1978.

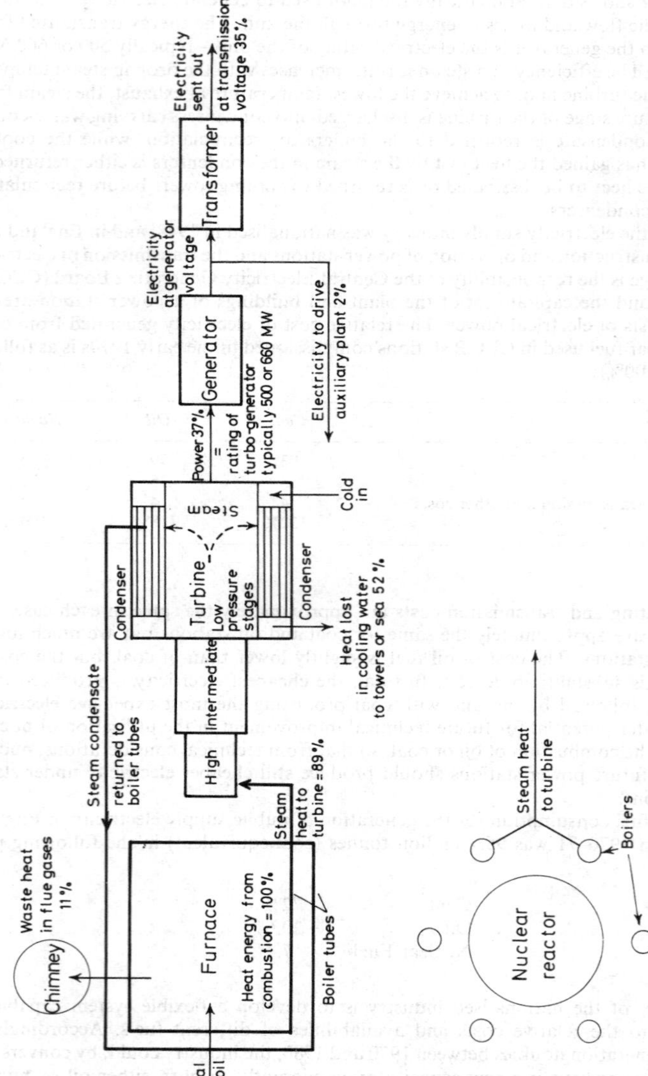

Figure 25.2 Flow of energy through plant used to generate electricity from steam power. Energy shown as percentage of heat in fuel

Table 25.1 FUELS AND NUMBER AND SIZE (MW) OF UNITS IN POWER STATIONS COMMISSIONED BETWEEN 1960 AND 1978

Years of completion	*Stations*	*Fuel*			
		Coal	*Oil*	*Nuclear*	
				Magnox	*AGR*
1960	Bold B (3), Elland (3)	6 × 60			
1960–1963	Belvedere (4), Plymouth B (2) S. Denes (4), Bankside (3) Littlebrook C (4)		17 × 60		
1962	Bradwell			6 × 50 +3 × 20	
1962	Berkeley			4 × 80	
1963	Aberthaw A	6 × 100			
1964	Hunterston A (Scotland)			6 × 60	
1960–1963	Agecroft C (2), Blyth A (4), Drakelow B (4), Staythorpe B (3), Northfleet (6), Padiham (2), Rugeley A (5), Richborough (3), Skelton Grange B (4), Uskmouth B (3)	36 × 120			
1961–1963	Belvedere (2), Bankside (1)		3 × 120		
1962–1963	High Marnham (5), W. Thurrock (2), Willington B (2)	9 × 200			
1965	Hinkley Point A			6 × 100 +3 × 30	
1965	Dungeness A (4), Trawsfynydd (4)			8 × 150	
1965	Thorpe Marsh	2 × 550 compound			
1966	Sizewell			2 × 325	
1966–1967	Blyth B (4), W. Thurrock (3), Drakelow C (2)	9 × 300			
1968	Oldbury			2 × 312	
1968	Tilbury B	2 × 350			
1968–1972	Ferrybridge (4), W. Burton (4), Eggborough (4), Ratcliffe (3), Ironbridge B (2), Aberthaw B (3), Rugeley B (2), Didcot (4), Fiddlers Ferry (4), Cottam (4)	34 × 500			
1971	Wylfa			4 × 330	
1971	Fawley		4 × 500		
1972	Kingsnorth (alternatively coal or oil fired)	(4 × 500) or	(4 × 500)		
1972	Pembroke		4 × 500		
1973	Drax	3 × 660			
1973–1975	Hinkley Point B (2), Hartlepool (2), Hunterston B (2) (Scotland), Heysham (2), Dungeness B (2)				10 × 660
1978	Ince B		2 × 500		
1978	Isle of Grain		5 × 660		

PLANT LAYOUT, BUILDINGS LAYOUT AND STATION SITING

N. G. EGGLETON, B.Sc., F.I.C.E.
Central Electricity Generating Board

For operational efficiency it is desirable to use high-output boilers and turbo-generators and, as Table 25.1 shows, almost exclusively 500 MW and 660 MW units were installed in stations completing between 1968 and 1978. The number of units installed in each station is decided according to the anticipated demand for electricity in the area, or—in the case of coal stations located near coal fields and of nuclear stations—according to the supply required to the national transmission grid; stations normally consist of from two to six identical units of either 500 MW or 660 MW output.

The best arrangement for construction and operation is to have the units placed side by side as shown in Figure 25.3, a plan of a 2000 MW station consisting of four 500 MW oil-fired units. The row of boilers so formed is contained in the boiler house and the row of turbo-generators is contained in the turbine house. The boiler house and turbine house are built parallel and in contact, except for a de-aerator and tank bay interposed between them. There is an auxiliary switchgear bay and generator transformers bay on the side of the turbine house; from the other side of the boiler house come the main gas flues leading through the dust precipitators and the induced draught fans to the chimney. Construction progresses from one end of the power house to the other, and units are commissioned in corresponding sequence. A cross section through the power house is shown in Figure 25.4.

The principal dimension in the layout of the power house is the distance between the centrelines of adjacent boilers, and this (and also the dimension from front to back of the boiler house) is kept to a minimum consistent with providing adequate space for construction, operation and maintenance.

The boiler spacing may be affected by the arrangement of the turbo-generators, which are desirably kept at the same centres as the boilers to give repetition of design.

Turbo-generators may be arranged longitudinally (i.e. with their shafts in line along the turbine house as in Figure 25.3), or transversely (with their shafts parallel and across the turbine house, having the high-pressure cylinder of the turbine nearest to the boiler and the generator farthest from the boiler, as shown diagrammatically in Figure 25.2). The longitudinal arrangement gives the least turbine house width and consequently the least crane span, but the overall length (allowing for rotor withdrawal) of turbo-generators of 500 MW or greater capacity usually exceeds the desired pitch of the boilers, i.e. approximately 60 metres. Hence most stations with sets of 500 MW or over have a transverse arrangement of the turbo-generators; this demands a larger turbine house, but the extra floor space has been found to facilitate greatly the location of auxiliary plant and the dismantling of turbo-generators for maintenance.

Also with the transverse arrangement, the lengths of high-pressure steam pipework between boilers and turbines and the lengths of conductor between generator and external transformer are shorter than with the longitudinal arrangement. On some sites, limitations of the site area may dictate a long and narrow power house, in which case the longitudinal layout of turbo-generators would be used and the boiler spacing increased accordingly.

The boiler-house length and width are therefore determined by the size of the boilers and their spacing, taking into account the arrangement of turbo-generators. The turbine-

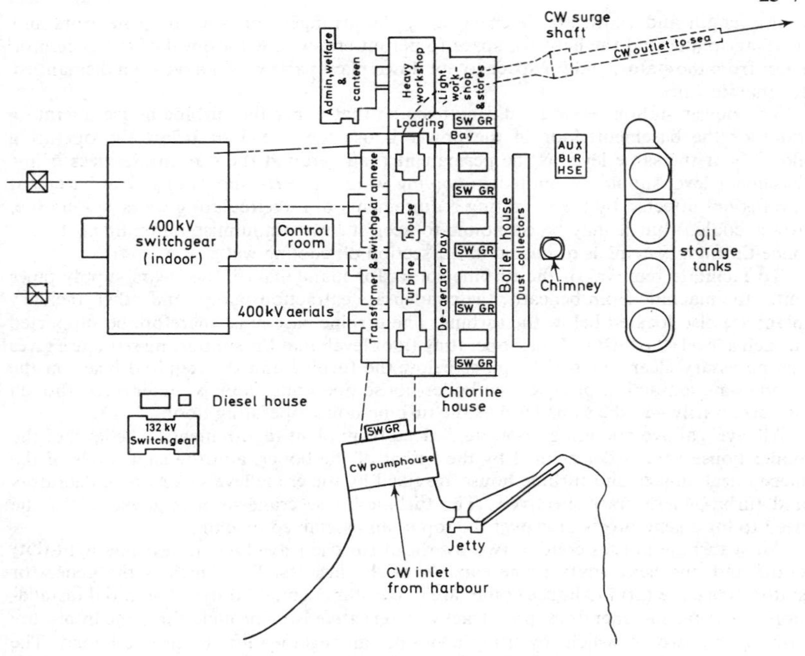

Figure 25.3 Power station, 4 × 500 MW, oil fired (Courtesy: Central Electricity Generating Board)

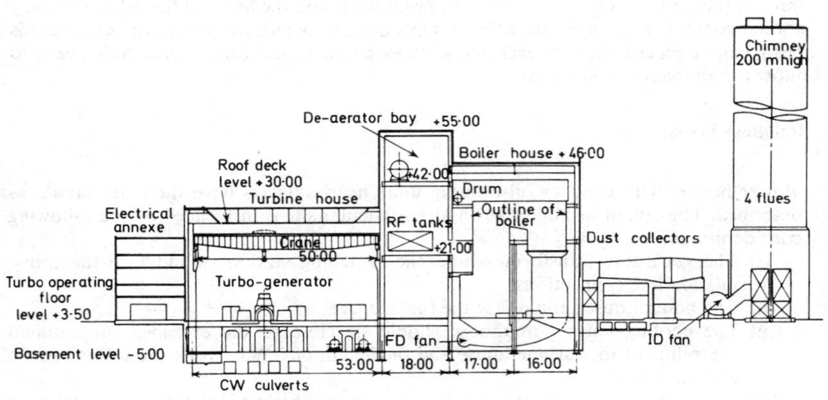

Figure 25.4 Cross section through 4 × 500 MW oil-fired power station (Courtesy: Central Electricity Generating Board)

house length and width are determined by the arrangement of turbo-generators and auxiliary plant, and the need for space to permit end-wise withdrawal of the generator rotor from the stator, and for space to put down plant parts which have been dismantled for maintenance.

The power station levels should be fixed so that either the turbine house operating floor or the basement floor of the power house (i.e. the floor below the operating floor) is at the same level as the permanent roads around the station. Services below basement level should be limited to cooling water culverts and drains. The basement level is not affected by the operating requirements of a recirculating water system (i.e. tower cooled) but it may be economic to keep it low to minimise pumping costs in a once-through system, as discussed in the section on cooling water.

To facilitate removal of the turbine rotors for maintenance, the steam supply pipes enter the machine from beneath; the condensate extraction pumps and other ancillary plant are also located below the turbine. The turbine base must therefore be supported at such a level (the turbine house operating floor level) that the supporting structure gives the necessary clearance for the plant below the turbine and the required head on the condensate extraction pumps. The boiler-house operating floor is preferably—though not necessarily—at the same level as the turbine-house operating floor.

All levels above operating floor are derived from plant requirements. The level of the boiler house roof is determined by the height of the boiler, and the roof levels of the mechanical annexe and turbine house are fixed by the crane levels over the de-aerators and turbo-generators respectively. The turbine-house crane level is governed by the need to lift a generator stator over the top of an assembled machine.

Most turbine houses contain two overhead electric travelling cranes: one light-duty crane and one heavy-duty crane capable of the heaviest lift, which is the generator stator. Since the turbine-house cranes are in constant demand during the initial installation of the turbo-generators, an attractive alternative is to provide three medium-duty cranes, any two of which, by using a beam, can together lift the heaviest load. The availability of three cranes speeds both installation and subsequent maintenance of the plant.

In a transverse arrangement, the length of the turbo-generators determines the crane span and, in the case of 500 MW machines, this span would be about 60 m and the stator weight about 300 tonnes. A single travelling beam for this span and load would be very heavy and to avoid this two cranes may be provided, the beam of one spanning about one third of the turbine house width over the generators, and the beam of the other spanning about two thirds width over the turbines. Only one column for the intermediate crane rails beam can be placed between each set, so these columns and beams tend to be heavy to allow for all loading conditions.

Buildings layout

Power houses with coal- or oil-burning units nearly always have the plant layout as described. The layout of buildings on any particular site is influenced by the following considerations:

(i) The generator transformers and the switching station should face the transmission line connections
(ii) The boiler house should face the fuel storage
(iii) The cooling-water pumphouse should lie close to the extended longitudinal centreline of the turbine house and the length of CW culverts should be kept to a minimum.

These requirements are all met in the layout shown in Figure 25.3. Other siting features to be noted are that the control room is located near to the centre of the turbine house, though detached from it; the administration, welfare and canteen building is convenient to the site entrance and is connected to the power house by a footbridge at operating floor level; and the loading bay and heavy workshops are an extension of the turbine house, thus allowing the turbine house cranes to travel over.

At this coastal station, the fuel oil storage tanks (capacity one week's burn at full output) are supplied by pipeline from a nearby refinery. The circulating water system is once-through, with the length of culverts before and after the condensers being kept to a minimum. The jetty near the CW intake is for handling plant loads of up to 300 tonnes brought by ship, as this is often more economical than strengthening bridges on a road route. The 400 kV transmission switchgear is housed in a steel structure clad with metal sheeting on walls and roof to protect the installation from sea spray.

Station siting

Ideally, thermal power stations are located to give—without harm to local amenity—the minimum combined costs of capital charges, fuel costs, operating costs, and transmission costs. Low capital charges result from good foundation conditions, and good labour availability for site construction. Low transport costs keep down the price of coal or oil, hence coal-fired stations are located near the coalfields, and oil-fired stations near to oil refineries; the tonnage of nuclear fuel used is relatively small and transport costs do not influence the location of nuclear stations. Low transmission costs result from power stations being located near to centres of demand for electricity, although the national transmission grid in Britain permits a wide location of stations.

An important requirement of any power station site is proximity to a supply of river or sea water adequate for operation of the cooling-water system. Within a given area of search, the best site is that which offers the best combination of desirable features—least harm to amenity, low capital costs, low fuel costs, low transmission costs, a route for transmission lines from the site to the national grid, and availability of cooling water. Difficult and costly foundation conditions are often accepted when outweighed by a combination of other desirable features.

An oil-fired power station of 2000 to 4000 MW output using generators of 500 MW or greater capacity, could be contained in the following land areas (1 hectare = 10 000 m^2):

	Area (hectares)
Station buildings	6
Oil storage tanks	4
Cooling towers	18
Switching compound or buildings	4
Surrounding area and approaches	15

A nuclear station of 2000 MW output using Advanced Gas Reactors (AGR) and with direct-cooling-water system (i.e. no cooling towers) would require a total of about 30 hectares of which 6 hectares would be occupied by permanent buildings.

In addition to these areas, an area of about 20 hectares would be needed temporarily during the construction period for offices, hostels and canteens, car and bus parks, material storage areas and working and fabrication areas.

The area containing the permanent buildings and within the permanent stations roads has to be levelled, and made safe against flooding. If the levelled area of the site is above flood level of river or highest tide, special precautions are not required. If part of the area is below flood level, it can be raised on imported fill or protected by a flood bank.

The main differences between the three types of steam power station (coal, oil and nuclear) should be borne in mind when reading the descriptions of aspects of power station construction which follow. In general, the turbo-generators and their support structures, the condensers, cooling-water system and turbine house are the same. Coal boilers and oil boilers are similar, but the boiler houses would differ in the arrangements for storage and handling of fuel and the disposal of ash; nuclear reactors are quite unlike the coal and oil boilers which they replace, except in having the same function of producing steam for the turbines.

POWER HOUSE STEELWORK

N. M. GRIEVE, M.I.Struct.E.
Central Electricity Generating Board

In a 2000 MW station with four 500 MW boiler and turbo-generator units, each boiler measures about 24 m × 46 m in plan and is suspended from girders about 60 m above ground level. Under operating conditions (i.e. filled with water and steam) it weighs about 12 000 tonnes; other heavy items to be supported by the structure are the de-aerator (300 tonnes), the boiler drum (300 tonnes) and two water tanks (660 tonnes each). The boiler is surrounded by access floors at about 10 m intervals of height, and steelwork has proved to give the best structure for housing the boilers and ancillaries and carrying the loads. Generally the support structure for one boiler only need be designed as the remainder will be a repetition of the first.

The turbine-house framework may be of concrete or steel but, because the adjoining boiler house is invariably steel, there is no advantage to be gained from having a concrete-framed turbine house. The weight of the turbo-generators is carried directly to ground on concrete or steel structures (turbo-blocks) and is not carried by the turbine house framework.

The power-house building therefore consists mainly of a boiler house and turbine house whose basic dimensions depend on the number and arrangement of boiler/turbo-generator units. In deciding the plant layout, it is important to have the boiler support columns arranged symmetrically around the boiler.

Design

After the plant layout has been finalised, it is necessary to ascertain if the boiler contractor has any special design requirements or limitations to be imposed on the boiler support structure. This is an important preliminary to design since the columns supporting the boiler will, with their connecting girders and bracing, form the core of the structure, and will provide transverse and longitudinal stability to both turbine and boiler house (see Figure 25.5). This is the reason for having a plant layout which allows a symmetrical arrangement of boiler support columns.

The main factors which depend on requirements generally obtained from the boiler contractor are as follows:

(a) The structure has to carry the wind load and must therefore be of sufficient stiffness to ensure that the combined maximum lateral movements under wind and operating temperatures are within the limits permissible for the high-pressure steam pipes and other connections to the boiler.

(b) The boiler drum of a 500 MW unit is a steel cylinder approximately 30 m long and 2 m in diameter; it weighs 260 tonnes without water. It is fabricated off site and, when the boiler support structure is sufficiently stable, is hoisted into position above the boiler. Large open areas must be left in the floors to enable the boiler drum to pass through. The position and size of these openings must be established at an early date as it is important to design for continuity of the buildings during

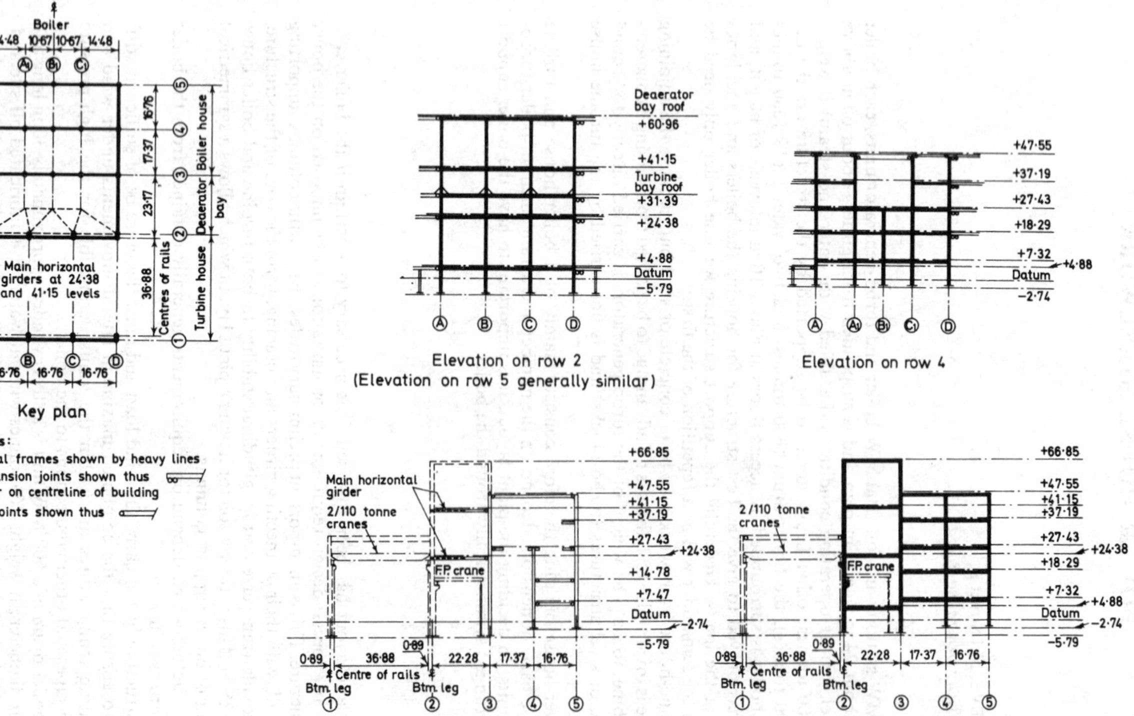

Figure 25.5 Boiler support structure

the drum-hoisting operation. Afterwards the open areas are generally filled in and form part of the flooring system.

(c) The boiler contractor may allow only limited deflection of the beams under important items of plant such as the de-aerator vessel. Excessive differential deflections between the vessel supports can be detrimental to its efficient functioning.

Should the limitations on the horizontal movements of the structure due to wind load and temperature under item (a) be very stringent (say 50 mm in 60 m) then a braced frame system may be necessary to obtain the degree of stiffness required. This type of construction, whilst structurally effective, does not allow full flexibility of choice in the routing through the building of large-diameter pipework and ducting. Should the horizontal deflection requirements be less stringent, then a type of portal frame construction can be considered. This has the advantage of providing clear spaces between the girders and columns which form the frames.

The remaining part of the main building such as the floors, walls and roofs, are of conventional design. The turbine-house roof may require more detailed attention since the span can vary from approximately 30 m to 55 m. Limitations may be required on its deflection to ensure effective roof drainage. The roof girders are sometimes used to erect the heavy-duty cranes.

The structural framework is designed in accordance with BS 449 *The use of structural steel in building*. The members and their connections are required to take full account of the moments and stresses due to all possible combinations of loads and other effects including vertical loadings, wind loads, temperature effects and dynamic effects from cranes and other moving plant.

Superimposed loading

Superimposed loads are generally allowed for in accordance with CP3 Part 1, Chapter V: *Loading—Dead and Imposed Loads*, but all plant loads, loads from tanks, crane wheel loads, etc., should be the known weights of such items. Where accurate weights are not known because structural design has to be ahead of detailed plant design, then safely high assessments are made to enable steelwork design to proceed and estimates to be given for construction of the foundations. Temporary loads such as people, stored materials, movable plant and equipment, plant laid down during erection or maintenance, may be allowed for in various areas in accordance with Table 25.2.

Wind loading

The structure should be designed for wind loading in accordance with CP3, Part 2, Chapter V: *Loading—Wind Loads*. The effects of wind on the partly erected and clad structure as well as on the finished structure should be considered, and the absence of the stabilising effect of heavy items of plant should be taken into account. From tests carried out on models of framed structures in various stages of erection, it has been found that wind conditions can be more severe during the erection period than on the completed structure. It may therefore be necessary during erection to temporarily stiffen certain main connections, and also provide temporary bracing to stabilise the structure against this more severe wind condition.

Turbine-house crane loading

In the design of the crane support columns the transverse and longitudinal forces caused by the movements of the cranes should be taken into consideration as well as the station vertical loading. It is important that the crane support columns should be made stiff enough to reduce horizontal deflection due to roof loading and eccentric loading from the cranes to a minimum. Over-flexibility from the columns has in the past resulted in serious crane crabbing.

Thermal loading

The maximum temperature rise in the boiler house under certain conditions can be in excess of 35 degC and the structure should be designed to accommodate thermal forces

Table 25.2 TEMPORARY LOADING ON FLOORS AND ROOFS

Operating floors in turbine and boiler houses	12.5 kN/m² (but 50 kN/m² in areas used to lay down turbo-generator parts). In the case of a large operating floor where there is little likelihood of the whole area being fully loaded at one time, a reduction of up to 20% of the temporary load can be taken on main beams and columns.
Loading bay in turbine house	Design for transporter with heaviest load
Basement floors in turbine and boiler houses (i.e. floor immediately below operating floor)	25 kN/m²
Cable basement floor	7.5 kN/m²
Gallery floors (walkways only)	5 kN/m²
Gallery floors in large circulation areas in boiler house (e.g. between boilers), de-aerator floors and tank floors	7.5 kN/m². The main beams and columns may be designed for a temporary load of 5 kN/m² but no reduction should be taken in the desi n of secondary beams.
Stairways in turbine and boiler houses	5 kN/m²
Flat metal-deck roof on turbine house	1.5 kN/m² to allow access for maintenance only. Load on main steelwork can be reduced to 0.75 kN/m².
Concrete shell roof on turbine house	0.75 kN/m² to allow access for maintenance only
Flat metal-deck roof on boiler house	1.5 kN/m² for deck and main steelwork. 4 kN/m² for areas which have to take construction or maintenance loads.
Control-room floor	4 kN/m²
Switchgear floors	Design for actual loads (generally between 10 and 25 kN/m²
Instrument workshop and laboratory floors	4 kN/m²
Electrical workshop floors and areas used for light engineering	7.5 kN/m² checked for point loading from heaviest plant load
Stores and heavy-machine shop floors	15 kN/m² checked for point loading from heaviest plant load

resulting from temperature changes of this magnitude. Although the high temperatures usually occur at the top of the boiler house, high differential temperatures have been noted at various levels in localised areas depending on the distance of the particular part of the structure from the boiler casing. Expansion joints are usually necessary to avoid excessive movements and stresses due to thermal forces.

Load combinations

The structure should be designed to cater for the worst combination of loading without exceeding the specified permissible stresses or the specified limits of vertical and horizontal deflection. It may also be considered necessary in the case of portal frame construction to check the beam to column connections for possible 'pattern' loading due to alternate beam spans being loaded temporarily.

Construction

The total tonnage of structural steelwork in the turbine and boiler house of a 2000 MW power station can vary from 25 000 to 30 000 tonnes depending on the plant layout and the height of the building. About 85% of this total tonnage is required for the support of the plant and the remaining 15% is required to provide the building framework.

The main columns in the turbine house are generally of welded box section about 2 m deep by 1.5 m wide with approximately 32 mm-thick wall plates. A central division plate is generally required to reduce the unsupported width of the wall plates. Horizontal diaphragm plates are required with manholes positioned centrally for internal access during erection, and ladders are provided between each diaphragm plate. The span of the turbine house roof truss can be up to 55 m with a depth of 4.5 m.

The main boiler support columns are also generally of welded box section and vary from 0.9 to 1.4 m square with wall plates up to 50 mm thick. The columns are connected together by plate girders from 1.2 to 2.7 m deep with flange plates up to 75 mm thick. Where these members combine to form portal frames the girders are usually haunched to accommodate the large bending moments. Should diagonally braced frame construction be preferred, the bracing usually consists of heavy universal column sections.

The heavy loading and large spans at the ancillary plant levels in the boiler house demand the use of welded plate, box and lattice girders in the construction. Probably the most interesting of these girders are those which form the main boiler suspension steelwork. Some typical forms of suspension steelwork are as follows:

(i) Two principal lattice girders per boiler up to 30 m span and weighing up to 100 tonnes with smaller secondary cross girders.

(ii) Welded box girders varying from 15 to 21 m long and from 3 to 4.5 m deep, the flanges being from 0.9 to 1.2 m wide and up to 75 mm thick. They are stiffened internally by vertical and horizontal stiffeners. Where vertical diaphragm stiffening plates are used they are usually provided with access openings for the steel erectors. The box girders are fabricated in two lengths to suit the erection craneage and are connected together at site by high-strength friction-grip bolts of diameters varying between 19 and 32 mm. The weight of a complete main box girder can be up to 160 tonnes.

(iii) A series of closely spaced plate girders about 27 m span and approximately 4.5 m deep with flanges up to 75 mm thick.

Erection

A typical steelwork erection scheme is shown in Figure 25.6. The average steelwork erection rate is in excess of 860 tonnes per month. This average rate has to be increased considerably during the first twelve months in order to provide the necessary stability to the structure to enable the boiler contractor to install the first boiler drum. A sustained erection rate of 1400 tonnes per month for a period of twelve consecutive months has been achieved. On another station 1040 tonnes per month was achieved for a period of 31 consecutive months.

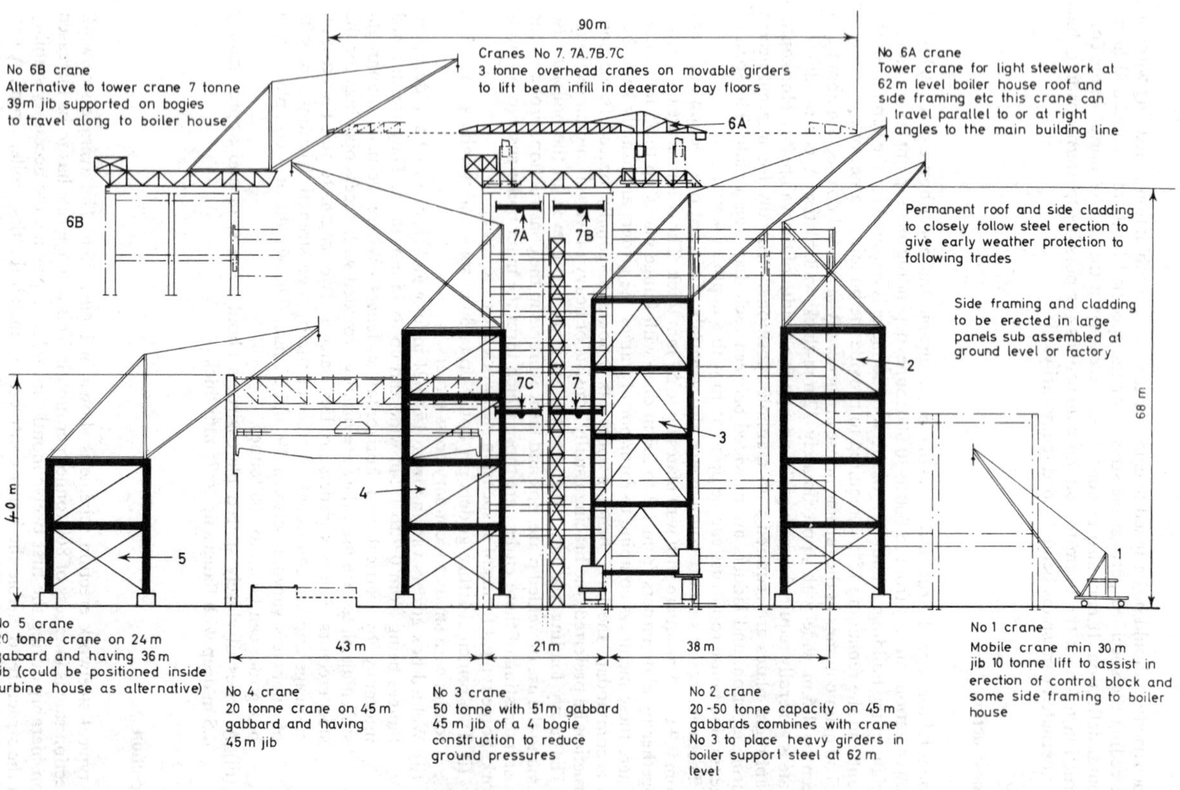

Figure 25.6 Steelwork erection scheme

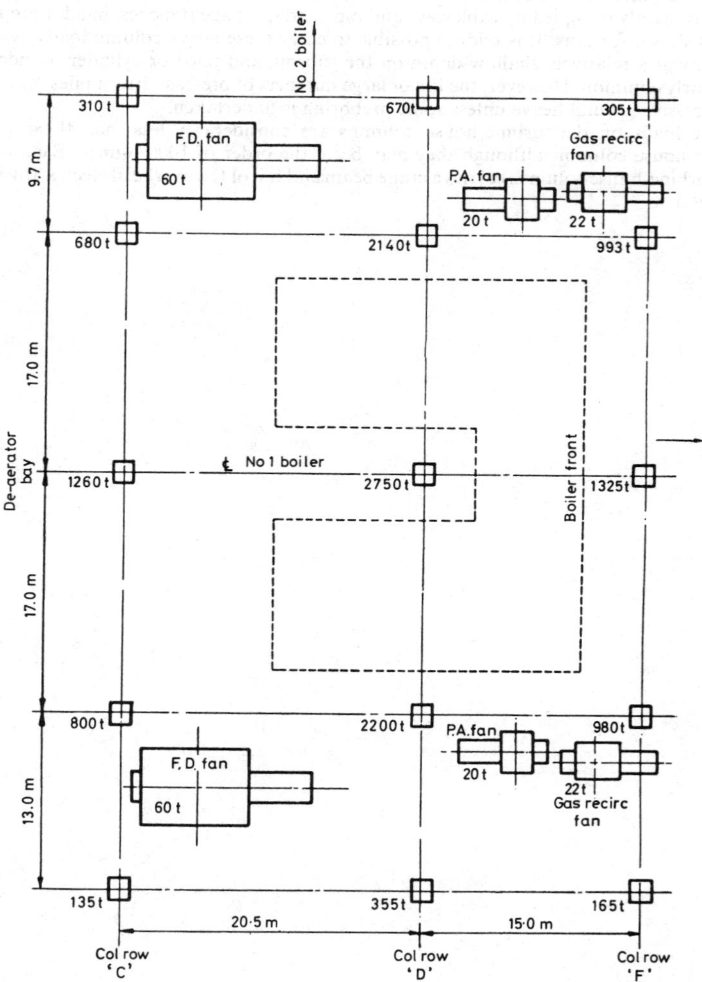

Figure 25.7 Boiler foundation loading plan. t = tonne (Courtesy: Central Electricity Generating Board)

Foundations

The loads resulting from the boiler and supporting steelwork of a 500 MW unit are shown in Figure 25.7. The space between the boiler columns below ground (or basement) level is mainly occupied by cableways and pipe and drainage trenches, in addition to the bases shown for fans. It is seldom possible to carry these heavy column loads by direct bearing at a relatively shallow depth on the subsoil, and piled or cylinder foundations are fairly common. However, the use of large numbers of pre-cast driven piles may result in excessive ground heave unless some preboring is undertaken.

The loads on the turbine-house columns are considerably less than those on the boiler-house columns although they may be of the order of 1000 tonnes. Each row of the turbine-house columns carries a crane beam and rail of the overhead electric travelling crane(s).

FLOORS, WALLS AND ROOFS: VENTILATION

A. MEREDITH, M.I.C.E.
Central Electricity Generating Board

Roofs, external walls and floors must be durable and capable of being erected quickly on the supporting steelwork to give support and protected conditions to allow the installation of plant to proceed with a minimum of delay.

Roofs

Power station roofs are generally flat roofs having a drainage slope not exceeding 5°, built of either reinforced concrete or metal-deck construction.

Flat concrete roofs are suitable for heavy traffic, as on coal- or oil-burning boiler houses, also on annexes, where there is a risk of damage from objects falling from a higher level. They may be formed of pre-cast units or of *in situ* concrete, or of a combination of these, and screeded to falls. They are waterproofed with two 10 mm layers of mastic asphalt (laid on an insulating layer) and covered with a solar reflective layer of 10 mm white chippings. The insulating layer may be fibreboard, or a lightweight cellular screed may be used instead of an ordinary mortar screed; its main purpose is to reduce the differential thermal movement between the concrete and the supporting steel structure.

Flat metal-deck roofs are suitable for low-trafficked roofs, such as for turbine houses, reactor buildings and transmission switch houses; in each case the low self-weight of

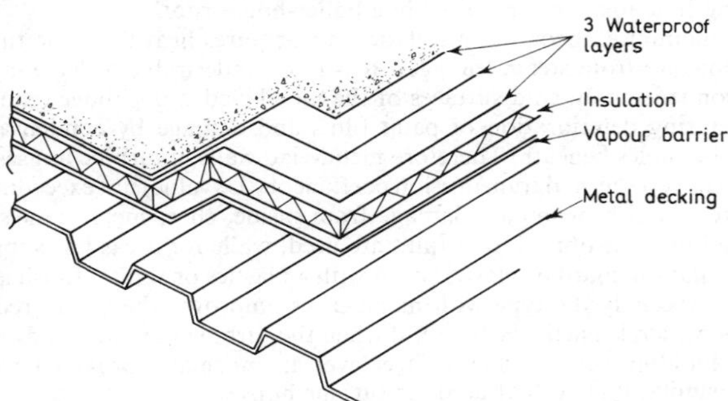

Figure 25.8 Metal-deck roof (Courtesy: Central Electricity Generating Board)

the roof covering is important for the long roof spans of the building; these roofs are of the composite construction shown in Figure 25.8. The metal deck is secured to the roof steelwork and is usually troughed sheet made from aluminium or galvanised steel, plastics-coated or painted; steel decks deflect less than aluminium decks, but when their coating is damaged they corrode more quickly. A vapour barrier of bituminous felt or other impervious sheeting is bonded to the metal deck and this is covered by an insulating layer of fibreboard which is bonded to the vapour barrier and mechanically fastened to the metal deck. The waterproofing layers (usually three) are bituminous felt sheeting bonded with hot bitumen, first on the insulating layer and then on each other, and these are covered by a sun-reflecting layer of 10 mm white chippings. The insulating and waterproofing layers of metal deck roofs are susceptible to accidental damage and heavily trafficked areas should be paved with asbestos cement (or pre-cast concrete) tiles bedded in bitumen. Ponding on the top surface may occur if thermal movement is taken up in local deflections or if due allowance is not made for the deflection of aluminium decking. Properly constructed metal-deck roofs will achieve an FAA fire-resistance rating (Flat Roof—Grade AA) in accordance with BS 476 Part 3.

Walls

The lower part of external walls (the 'plinth' extending from ground level to about 2 m above operating floor level) is usually made of pre-cast concrete panels in both boiler and turbine house to give reasonable resistance to impact damage. The external walls above the plinth are usually of profiled metal fixed to sheeting rails, the profiled cladding being formed from steel sheet finished with a plastics coating, or from aluminium sheet left with a mill finish or coated with plastics. Glazing can be incorporated to give natural lighting and to provide an architectural feature. In the turbine house, a glazing band is usually incorporated between the top of the plinth and the crane rail level to give natural lighting at operating level.

Insulating linings

In a boiler house the air is hot under normal working conditions owing to heat emission from the boilers and it is not necessary to insulate the metal clad walls to reduce loss of heat from the boiler house—indeed, loss of heat by conduction through the cladding is desirable. In a flat metal-deck roof of composite construction, the insulating layer is an essential component to support the waterproofing layers, so it must be retained although its insulating function is not required in a boiler-house roof.

In a turbine house, incoming ventilation air acquires heat from the turbo-generator and also moisture from steam leakages. It is most undesirable to have moisture in this warm air condensing on cold surfaces of the metal-clad walls above crane level and of the roof, causing deterioration of paint films and damage by condensation dripping on to plant or cables beneath. Therefore metal-clad walls in turbine houses are normally insulated to have a thermal transmission coefficient ('U' value) not exceeding 2 W/m^2 °C. Where there is a risk of impact damage from inside, sheet metal panels backed with mineral wool or glass-fibre slab or quilt are used, while for areas less subject to impact damage, insulation board or plasterboard with a plastics or similar finish is used. Insulation lining, especially the type with metal trays, improves the sound reduction factor of the walls, which is particularly useful when the turbine house faces dwelling houses.

Reactor buildings above charge face level are normally provided with insulation linings of a similar type to that used for turbine houses.

In transmission switch houses, the electrical equipment produces very little heat. Insulation is normally provided to reduce the risk of condensation inside the building.

Glazing

Fixed glazing may be incorporated in wall cladding as an architectural feature and to admit natural lighting, although in practice artificial lighting is continuous in power houses. Openable glazing is not recommended because the operating gear is susceptible to damage through misuse or becomes inoperable after long periods without operation or maintenance.

Glazing in the turbine-house roof is not recommended because the cold surfaces induce condensation in cold weather. Glazing in the boiler-house roof is unnecessary.

Floors

Open-grid flooring is used for stairways and galleries and for a large proportion of the floor areas inside the buildings. It has the advantage of low self-weight and also minimises obstruction to the upward flow of air for the removal of heat in boiler houses. Sometimes imposed loadings require that solid suspended floors must be provided in certain locations; it is also necessary to use solid suspended floors in areas where liquid spillage requires to be collected, e.g. at de-aerator level. The floors at or about operating level have to carry operating plant, or dismantled components of plant when maintenance work is being carried out. Reinforced concrete is used extensively for floors of this type and filler joist construction still has a limited application for the floors carrying the highest loadings; pre-cast concrete unit and jack arch floors have the advantage that access is obtained immediately following erection. Where there are many perforations, trimming is easier with *in situ* concrete construction.

Ventilation

Figure 25.9 shows a cross section of a power house where the turbine house adjoins the boiler house and there is no wall between them. Heat is emitted by the boilers (and to a lesser extent by turbo-generators and other plant) and this raises the air temperature inside the buildings. The air is hottest around the boiler casing, so that the column of warm internal air for the height of the boiler house (about 50 m), being of lighter relative density than the cooler external air, causes the boiler house to act as a chimney, with sufficient air movement to ventilate adequately both the boiler house and turbine house if there is no wall between them. Cool air is drawn in through low-level louvres in the external walls of the boiler house and turbine house, and heated air escapes through outlets at roof level: some of the heated air is abstracted from the top of the boiler house by forced draught (FD) fans which feed it as combustion air to the boilers.

The turbine house inlet louvres contribute about one-third of the total air flow but the turbine house plant contributes only one-fifth of the total emitted heat. The temperature rise above outside ambient will therefore be less in the turbine house than in the boiler house. A turbine house which is not open to an adjoining boiler house (e.g. on a nuclear power station) must be mechanically ventilated, since natural ventilation of a turbine house without the chimney effect of the boiler house would demand excessive areas of inlet and outlet louvres to achieve the same limitation of temperature rise.

For a given rate of emission of heat, the air temperature in the power house is related to the rate at which air flows through it. The object in designing the ventilating system, therefore, is to provide inlet and outlet louvres in appropriate locations and of appropriate size to permit an adequate flow of air through the building.

The design sequence is as follows:

(i) Decide on the maximum permissible rise of inside temperature above outside temperature. This is derived from medical considerations of heat stress in personnel and of local climatic conditions; in Britain a maximum rise of 15 degC is suitable.

(ii) Assess the heat emission from plant and the solar gain or conduction loss through the fabric of the building to give the total quantity of heat to be removed by the ventilating air. For a 500 MW unit, about 12 MW of heat is emitted from the

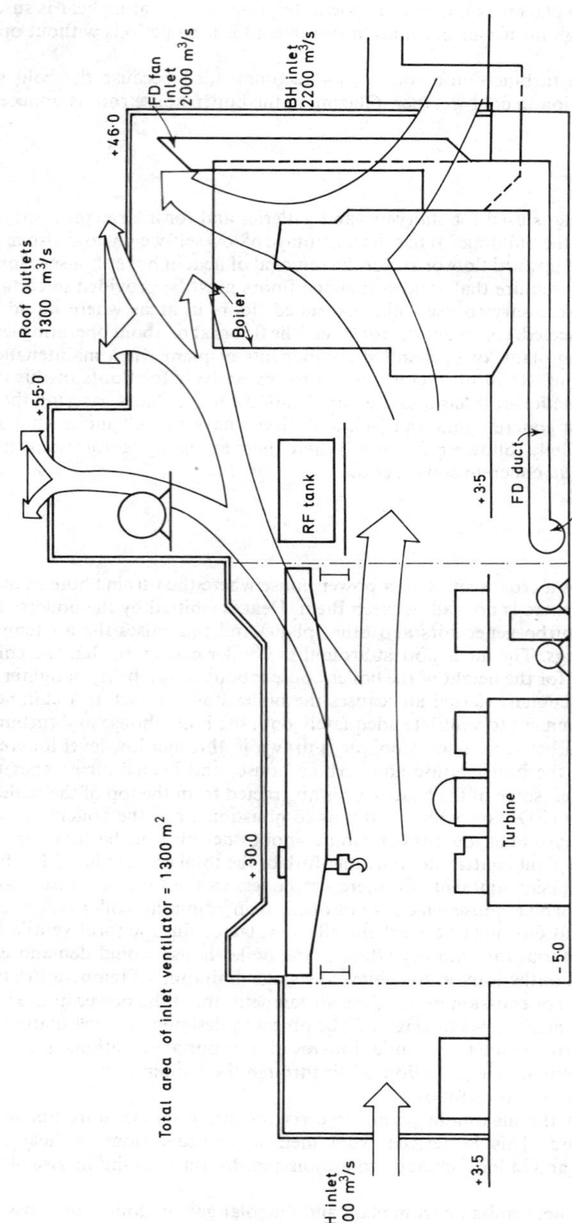

Figure 25.9 Cross section through 4 × 500 MW oil-fired power station showing ventilating air flow pattern

boiler plant and about 3 MW from the turbo-generator. The solar gain or conduction loss is relatively small and has little effect on the subsequent calculations.

(iii) Calculate the total rate of air flow necessary to remove the heat; this gives the required air flow through the inlet ventilators.

$$\text{Quantity of Air} = \frac{\text{Heat to be removed by air}}{\text{Specific Heat of air} \times \text{temperature rise}}$$

For example, for a 2000 MW Station of four 500 MW units allowing 15 deg C temperature rise:

$$Q = \frac{4 \times (12+3) \times 10^6 \text{ J/s}}{1200 \text{ J/m}^3 \text{ deg C} \times 15 \text{ deg C}}$$

$$= 3300 \text{ m}^3/\text{s}$$

(iv) Determine the net area of the inlet ventilators to admit the total air flow. Using a simplified method given in CP3:
Flow of air through ventilators = Net area of ventilators × velocity of air flow

$$\text{i.e.} \quad Q = A \times C_v \times \left(2gh\,\frac{(t_i - t_o)}{T_o}\right)^{\frac{1}{2}}$$

where Q is the flow of air through the inlet ventilators, A is the net area of ventilator openings, C_v the coefficient of velocity of air flow through ventilators, h the height between inlet and outlet ventilators, $t_i - t_o$ the difference between the average inside temperature and the outside temperature (this can be taken as half the permissible temperature rise deg C) and T_o is the absolute temperature of the outside air in K, i.e. $T_o = 273 + t_o$ °C.

Then for $Q = 3300$ m³/s, and assuming $h = 36$ m,

$$t_i - t_o = \frac{15\,°\text{C}}{2} = 7.5\,°\text{C}$$

$$T_o = 273 + 20 = 293 \text{ K}$$

$$C_v = 0.6 \text{ (according to type of ventilator chosen)}$$

$$A = \frac{3300}{0.6 \times (2 \times 9.81 \times 36 \times 7.5/293)^{\frac{1}{2}}}$$

$$= 1300 \text{ m}^2$$

This is the net area of the inlet ventilator openings and also of the outlet ventilators. Louvred ventilators on power stations typically have a net opening area to gross ventilator area ratio of 0.5 so that the gross area of the ventilators is 2600 m².

(v) The outlet ventilators must have the same total $A \times C_v$ value as the inlet ventilators to allow for the temporary condition at the end of shift when combustion ceases in the boilers. Then the FD fans stop running, but the boilers continue to emit heat which can only be removed by discharging the total air flow through the outlet ventilators. Under normal conditions, with the boilers and FD fans operating, the fans may take 2000 m³/s of combustion air, equal to 60% of the total air flow. It should be possible to close an equal percentage of the area of the outlet ventilators, preferably around the FD fan intakes, so that a balance is maintained in the system. If the necessary proportion of the outlet ventilators is not closed, the FD fans may draw air into the boiler house through the outlet ventilators.

In a nuclear power station, the size of the reactor building and the enclosure (by walls and floors) of the space around the reactor containment does not permit adequate natural ventilation of working places. The ventilation required in various parts of the reactor building also depends upon the operations carried out there, and in consequence most of the reactor building is mechanically ventilated. The natural ventilation of the turbine house of a nuclear power station cannot be augmented by the chimney effect of adjoining boiler house: consequently the turbine houses of nuclear stations are mechanically ventilated. In order to avoid large temperature variations, it is usual to provide ducting to introduce the incoming cold air at positions near to heat-emitting surfaces.

TURBO-GENERATOR SUPPORT STRUCTURES

D. L. McKIE, M.I.C.E.
Central Electricity Generating Board

The turbo-generator support structure (the turbo-block) is built up from the basement floor to carry the turbo-generator at operating floor level: it provides sufficient internal space to accommodate plant and services which must be located beneath the turbine. The turbo-block is itself carried on a substantial foundation.

The earlier 500 MW turbo-generators were equipped with a 'bridge' condenser which was located beneath the low-pressure cylinder of the turbine. The turbo-block was traditionally made in reinforced concrete with the bridge condenser placed within it on concrete plinths. This arrangement gave a height of about 12.5 m from basement to operating floor for 500 and 660 MW units. A loading diagram for a 500 MW turbo-generator with a bridge condenser on a reinforced concrete block, is shown in Figure 25.10.

The later 500 MW and larger turbines are equipped with a pannier condenser located on each side of the low-pressure cylinder; the use of structural-steel turbo-generator blocks was introduced to Britain at the same time as pannier condensers. Structural-steel blocks weigh less than reinforced-concrete blocks and give a reduced foundation loading. A 500 MW unit on a steel block and with pannier condensers is shown in Figure 25.11. The reduction in height of the block should be noted.

The advantages of steel blocks compared with concrete blocks are:

(i) The slender dimensions of steel columns permit auxiliary equipment (particularly steam pipes and electrical connections to the generator) to be accommodated more conveniently.
(ii) The improved ventilation maintains the steel columns at a more uniform temperature than concrete columns. There have been many instances of large turbines being put out of alignment owing to temperature variations in the concrete columns at the turbine end.
(iii) The behaviour under load of the steel block is consistent and maintained, whereas concrete blocks are subject to movement due to curing and creep. This has led to some machines needing realignment after initial operation.
(iv) Steel blocks have shown no deterioration by attrition under the turbine sole plates, whereas crumbling of concrete surfaces has been observed occasionally.
(v) It is possible to make economies in construction by integrating the condenser and generator casings into the steel support system.
(vi) Steel blocks can be readily adjusted to eliminate any local resonance.
(vii) The time for erection of a steel block is much less than for a concrete block, although turbo-block erection time is not critical in the overall construction programme.
(viii) The steel block can be designed and supplied by the turbo-generator contractor thereby avoiding any division of responsibility for alignment between turbo-generator contractor and concrete-block contractor.

The disadvantages of steel blocks are:

(i) Their capital cost is considerably greater than that of concrete blocks.

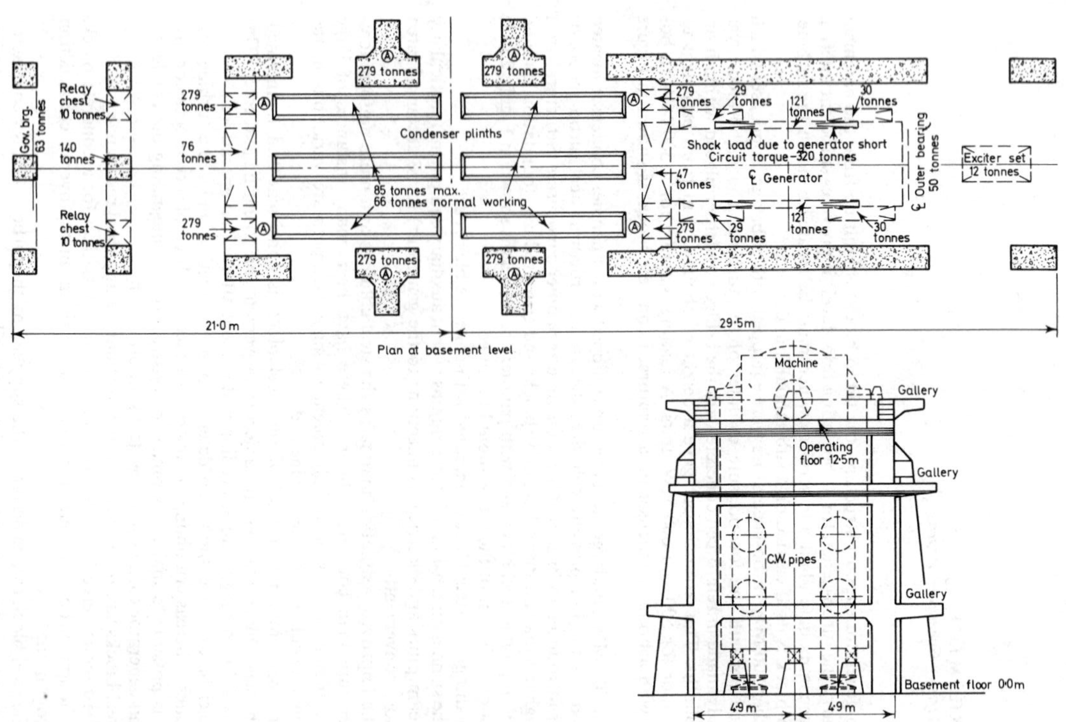

Figure 25.10 Simplified loading plan for 500 MW turbo-generator on concrete block with bridge condenser. (Courtesy: Central Electricity Generating Board)

(ii) The steel columns require an automatic water-spray system for fire protection.
(iii) The design of mountings for instruments and other attachments to the structure requires careful consideration since the absolute amplitude of steel block movement can be greater than that of concrete blocks.

Concrete blocks have not been completely superseded, especially for smaller machines. The outline of the concrete block is determined by the machine manufacturer who provides the block designer with loading conditions and restrictions on deflections at bearings. With regard to deflection, it is an advantage if the bearing supports can be mounted over piers or columns. Certain modifications may be permitted by the manufacturer to facilitate construction, to assist in keeping deflections within the prescribed

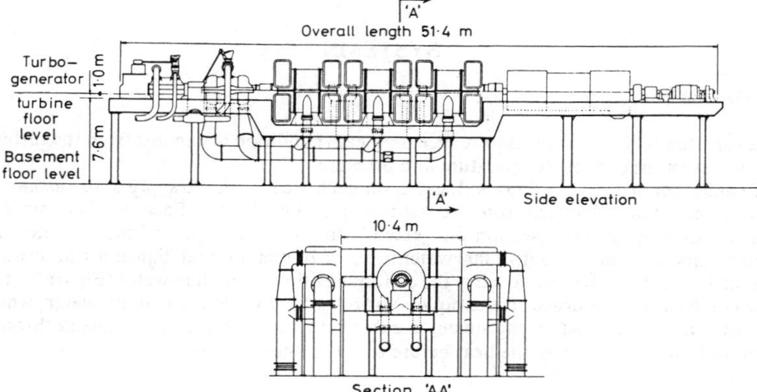

Figure 25.11 Arrangement of 500 MW turbo-generator on steel block with pannier condensers (Courtesy: Central Electricity Generating Board)

limits and to ensure that resonance will not occur at the machine's normal running speed.

Concrete blocks should, where practicable, be built several months ahead of machine erection, in order to allow as much concrete shrinkage as possible to take place beforehand. Although a well placed and good quality concrete is required, a very high strength is not necessary; the mix design should aim at reducing shrinkage and thermal movement. Other important points in the design of concrete blocks are the choice of positions of joints between lifts and the inclusion of adequate reinforcement (not less than 50 kg/m^3) which should be placed in three planes; even though this may not be theoretically necessary.

A soils investigation should be carried out prior to the design of the foundation. Vibration and compaction of the subsoil, with consequent settlement, must be considered. In order to reduce transmission of vibration from the machine, the foundation should be isolated from the surrounding basement floor and, as far as possible, from other structures and services.

Useful guidance for the design of turbo-generator blocks is given in German Standard DIN 4024, 1955[1]. It is anticipated that a British Standard Code of Practice dealing with foundations for reciprocating, rotating and impact machines will be available about 1975.

A programme of measuring and determining the causes of movement in turbo-generator blocks in use has been carried out by the CEGB so that the effect of movements on future (larger) turbo-generators can be assessed and satisfactory support structures can be designed.[2]

COOLING-WATER SYSTEMS

W. G. JONES, C.G.I.A., M.I.C.E.
Central Electricity Generating Board

SYSTEMS

Condensers

The function of a condenser is to reduce the steam condition at exhaust from the turbine to the lowest practicable temperature and pressure.

Pannier condensers are 'boxes' located on each side of the low pressure turbine to receive the steam exhausted from the last row of turbine blades. Each box is about 3 m wide × 8 m deep in cross section and extends for the full length of the low-pressure turbine (about 20 m). About 10 000 water tubes of 20 mm internal diameter run through the full length of each box and carry the continuous flow of cooling water (Figure 25.12). The condensate is returned to the boilers for reheating, while the cooling water, which has gained the heat lost by the steam, is either returned to source or is passed through cooling towers to dissipate the heat before re-use in the condensers.

Once-through and recirculating systems

There are two alternative systems for maintaining the necessary flow of cold water through the condensers:

(i) Direct or once-through cooling (see Figure 25.13(a)) in which cold water is pumped from the sea or a river through the condensers and is then discharged back to the source in such a position that heated water is not taken into the intake again. In the case of cooling water taken from and returned to a lake, recirculation is unavoidable eventually, but the lake is channelled so that the heated water must flow a sufficient distance between discharge and intake for the heat to be dissipated before re-use.

(ii) Recirculating or tower cooling (see Figure 25.13(b)) in which the heated water from the condensers is pumped to distribution channels at high level inside a cooling tower, from which it falls as a spray over a stack of timber or asbestos slats. The cooling tower is designed to produce an upward flow of air so that heat in the descending water droplets is lost by evaporation and by contact with the air. The cooled water is collected in a shallow pond over the base area of the tower and is then recirculated by pumping through the condensers.

DIRECT (ONCE-THROUGH) INSTALLATION (FIGURE 25.13(a))

In passing from intake to outfall in a direct cooling-water (CW) system, the water goes through the following sequence of installations:

Intake. This facilitates flow of water from source to the coarse screens. On rivers or protected coasts it is usually a channel, dredged to sufficient width and depth to carry the

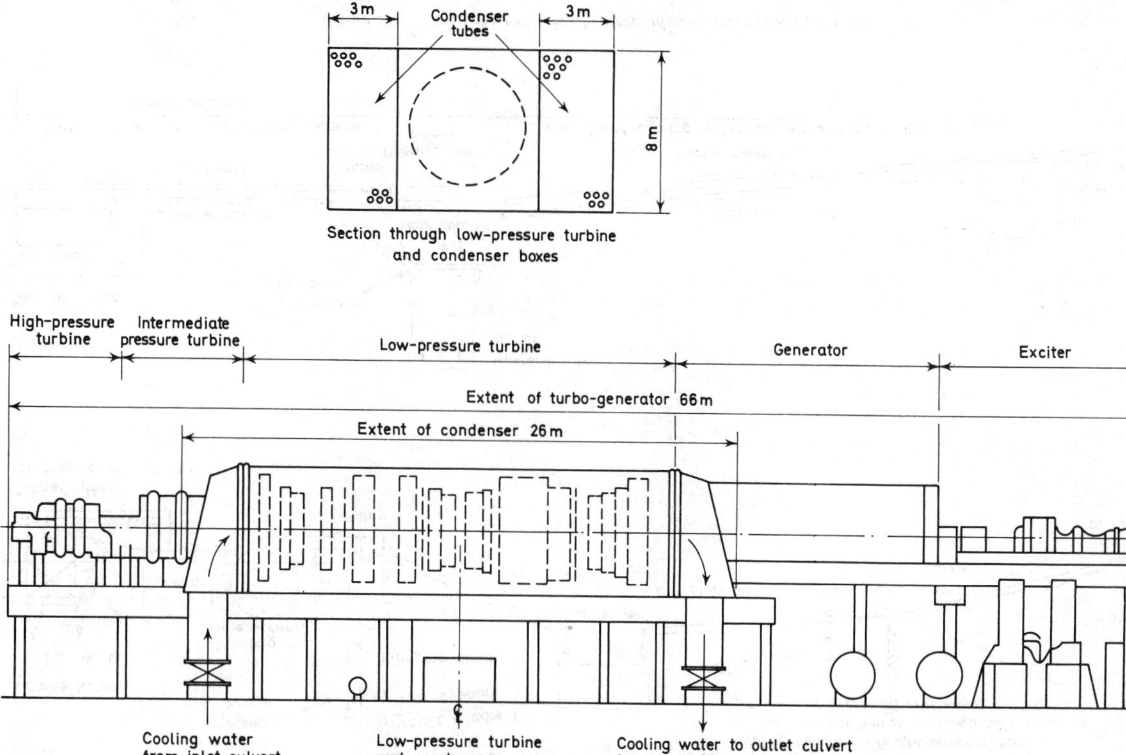

Figure 25.12 Turbo-generator with pannier condensers. Condenser tubes total about 20 000 (tube dia. about 22 mm). Sizes shown are typical for a 500 MW turbo-generator

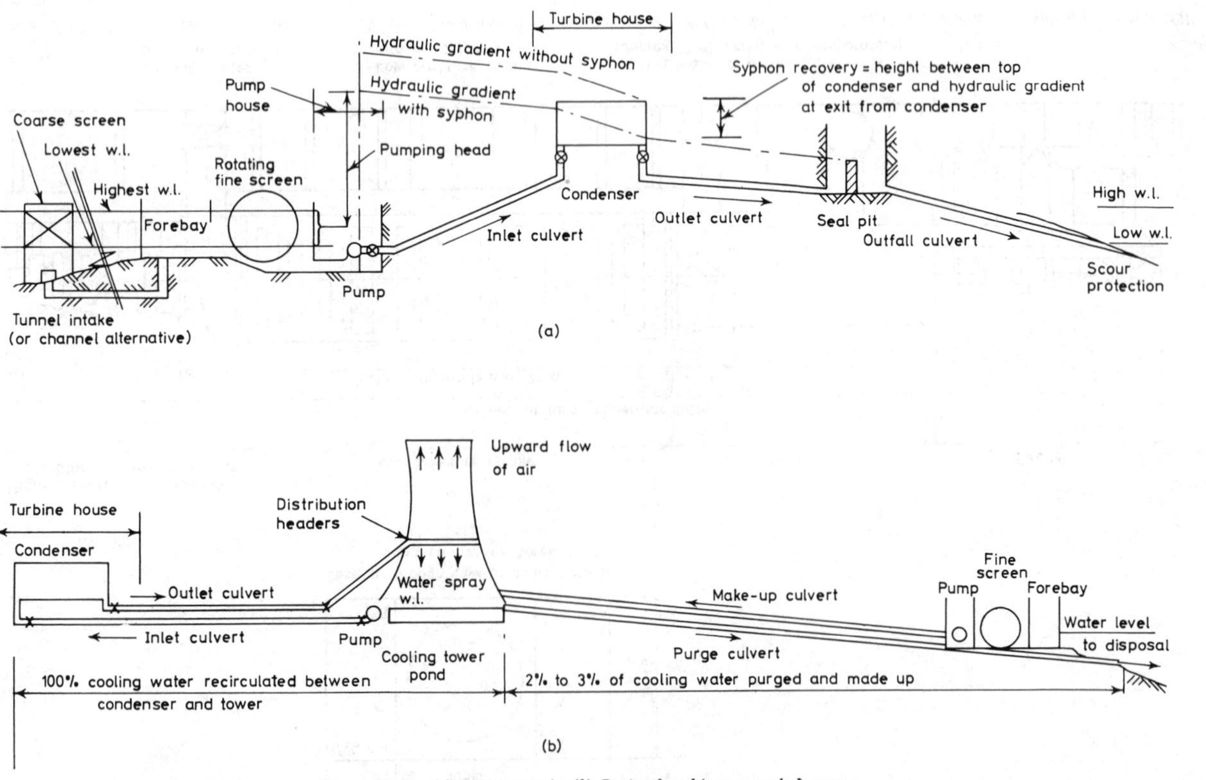

Figure 25.13 (*a*) *Direct system.* (*b*) *Recirculated* (*tower-cooled*) *system*

maximum flow at a velocity at which sand is not carried into the system. On exposed coasts, the intake is usually a concrete-lined tunnel extending from the forebay to a position offshore where the sea bed is below the effect of wave action, so that silt and sand is not drawn in (Figure 25.14). The tunnel terminates in a vertical shaft to the sea bed, the shaft being concrete lined, with a streamlined concrete cill around the shaft entrance at about a metre above bed level. There is usually a permanent headworks structure above the intake shaft to facilitate cleaning of coarse screens around the intake shaft and to facilitate closure of the intake shaft when necessary for maintenance[3].

Coarse screens to exclude tree trunks and other large debris, are located at the entrance to tunnel intakes or at the landward end of channel intakes where water enters the forebay.

The forebay is a large open chamber sited on land to dissipate turbulence of the incoming water and induce good flow conditions to the screen chamber and pumphouse following.

Fine screens (where not installed in the headwork of tunnelled intakes) may be either travelling-band screens or rotating-drum screens. Drum screens are made of a layer of metal mesh 2.5 m wide, mounted circumferentially between a pair of rotating wheels about 15 m in diameter. Two concrete walls shaped as saddles lie under the wheels and divide the screen chamber into three compartments. Unscreened water from the forebay passes to the outer compartments and then into the space between the wheels. It then flows through the circumferential screen into the central compartment beneath the drum. From here the screened water flows through a draught tube to the pump. The rotating-drum carries trash retained inside the circumferential screen to trash removal equipment operating on the underside of the drum at the top. The drum dimensions have to be such that the trash removal equipment is above high tide, and there is sufficient screen area submerged at low tide.

Pumps are located below lowest water level so that they can be started without priming. On a station generating 2000 MW, the CW pumps deal with 240 000 m^3 of water per hour.

A non-return valve is located on the discharge side of the pump to prevent damage to the pump and motor resulting from reversal of flow from the condenser. The valve closes automatically within seconds if the pump is stopped by causes other than normal operating control.

Inlet culverts connect the pumps to the condenser and are subject to the maximum pumping head.

Outlet culverts take the cooling water to the seal pit after discharge from the condenser. They are subjected to little more pressure than the tidal head.

Seal pit: to limit the condenser syphon height. The flow of water through a condenser in a once-through system can be made to operate as a syphon, but the syphonic recovery of hydraulic head has to be limited because of the need to ensure stability of flow conditions. This is done by discharging the down leg of the syphon (i.e. the culvert that carries flow from the condenser) into a pit of water in which the surface is kept—by an overflow weir—at a level calculated to maintain optimum syphonic recovery in the condenser. The culvert discharge is kept wholly submerged to prevent backflow of air into the culvert. Normally, the weir level is not more than nine metres below the top of the condenser.

Outfall culverts take the cooling water from the seal pit to the sea.

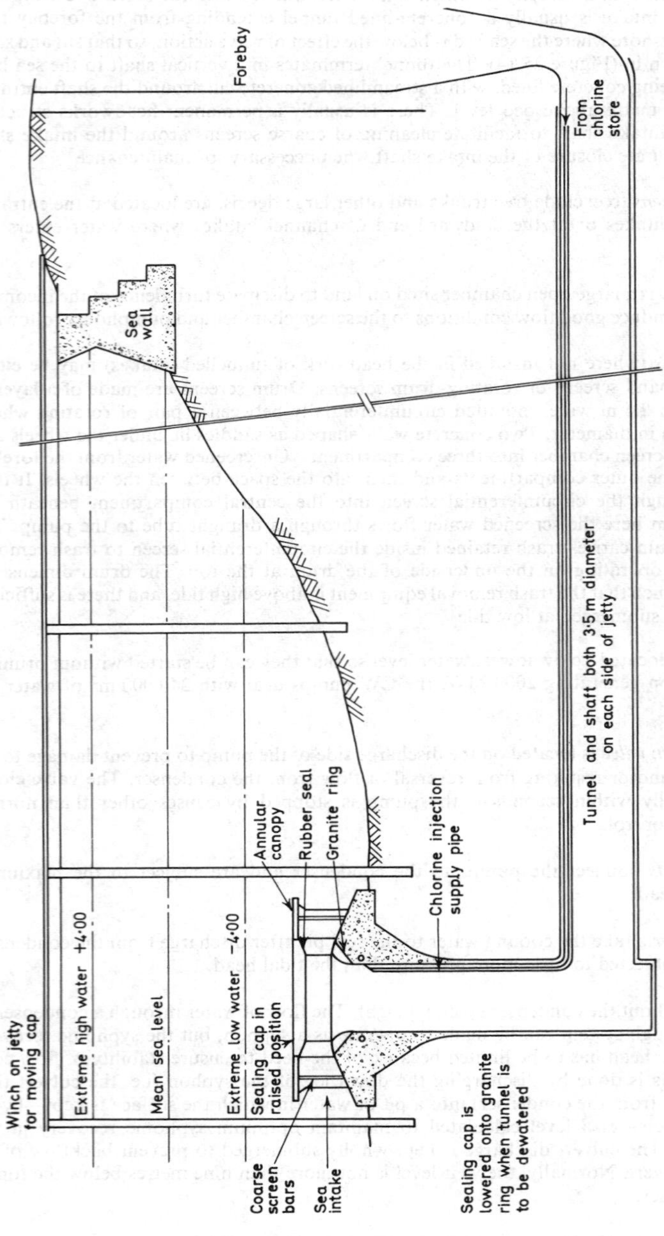

Winch on jetty for moving cap
Extreme high water +4·00
Mean sea level
Extreme low water −4·00
Sealing cap in raised position
Coarse screen bars
Sea intake
Sealing cap is lowered onto granite ring when tunnel is to be dewatered
Annular canopy
Rubber seal
Granite ring
Chlorine injection supply pipe
Sea wall
Forebay
From chlorine store
Tunnel and shaft both 3·5 m diameter on each side of jetty

Figure 25.14 Sea-bed water intake

For a 2000 MW station with four 500 MW turbo-generators, the CW system typically includes:

Intake for 240 000 m^3 water per hour
Coarse screens
Forebay
4 Rotating fine screens, 16 m diameter × 2.5 m wide
4 Pumps each capable of pumping 60 000 m^3 water per hour
4 Inlet (pressure) culverts between pumps and condensers, interconnected for interchange facility in case of pump failure, each capable of carrying a flow of 120 000 m^3 per hour at a velocity of 3 m/s
4 Outlet (low-pressure) culverts between condensers and seal pit
Outfall culvert(s) from seal pit to sea

RECIRCULATING (TOWER COOLED) INSTALLATION (FIGURE 25.13(b))

From the collecting ponds of the cooling towers, the cooled water gravitates along open channels to the pumphouse. It is pumped through the condensers and up to the distributors in the cooling towers about 10 m above ground level. From the distributors it trickles down over the surfaces of the slats stacked in the lower part of the tower, there being cooled by the natural up-draught of air; it is finally collected in the pond for recirculation.

About 1% of the water passing through a tower evaporates and this causes an increase in the dissolved salts content of the recirculated water. The increase in dissolved salts is kept to an acceptable limit by purging—i.e. discharging to waste—about 2% of the circulating water, and both the purged and evaporated quantities are replaced with fresh supplies of water (make-up). Make-up water is drawn through coarse and fine screens from a nearby river and is pumped to one or more cooling tower ponds. From another point in the recirculating system, a gravity pipeline is laid to take purged water back to the river at a point downstream of the intake screens.

For a 2000 MW station with four 500 MW turbo-generators, the CW system typically includes:

8 Natural draught cooling towers about 90 m in diameter at base and 110 m high
4 Circulating pumps, each capable of pumping 60 000 m^3 of water per hour against the head of the distributors
Pressure culverts between pumps and condensers and between condensers and distributors, interconnected for interchange facility in case of pump failure and tower maintenance
Gravity culverts between tower ponds and pumps forebay

and the make-up system includes:

River intake, with coarse and fine screens
Three make-up pumps (giving 50% stand-by) each capable of pumping 5000 m^3 of water per hour against the head of the cooling tower ponds
Purge culvert to return 3300 m^3 of water per hour to river or waste

Comparison of systems

An important difference between tower and one-through systems is in their total pumping head. In a tower system the pumping head is the frictional head plus the static head (the height of the distributors above the pond surface, about 10 m); in a once-through system the pumping head is the height above tide level of the highest point on the hydraulic gradient (see Figure 25.13(a)).

A typical pumping head for a once-through system is 15 m and for a tower cooled system is 24 m. In both systems the quantity of water pumped through the condensers is approximately the same. The excess energy for pumping in the tower system takes 0.7% of the turbo-generator output, or reduces the station overall efficiency by 0.3%.

Generally, the water temperatures in a once-through system are from 5 degC to 10 degC lower than the temperatures in a recirculated system and this gives greater turbine efficiency.

Therefore, once-through systems have the financial advantage, but they are restricted in Britain to coastal or estuarine locations, since the heat rejection in cooling water from large stations (equivalent to about 1.3 × the electrical output) is no longer acceptable in rivers inland.

Effect of system on power-house floor levels

Tower systems are little affected economically by the height of the condensers above the river which is the source of make-up water, and the CW system has no effect on the basement and operating floor levels in the power station.

Pumping costs in a once-through system, however, are increased as the height of the condensers above sea level increases. In cases where it is practicable to lower the floor levels of the turbine house by excavation, without risk of flooding, the resulting saving in pumping costs has to be considered relative to the cost of extra excavation.

CULVERTS

Layout of pumps and culverts

The inlet culverts are interconnected so that flow can be maintained to all condensers in the event of failure of any one pump. Figure 25.15 shows a diagrammatic layout of

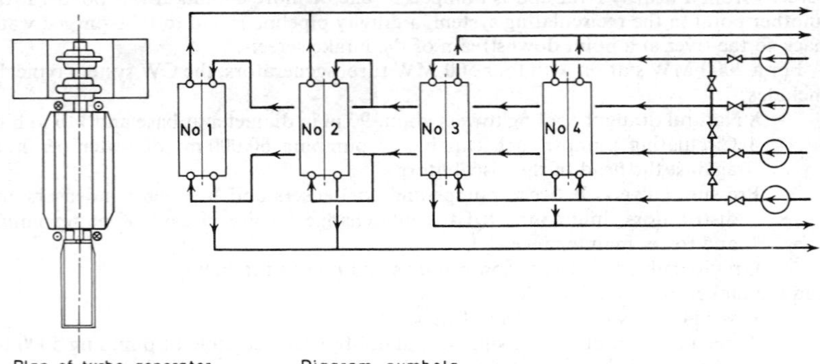

Figure 25.15 Layout of pumps and culverts for transverse arrangement of turbo-generators

the pumps, valves, and inlet and outlet culverts for four turbo-generators with pannier condensers.

Sectional area of culverts

A culvert of small cross section and high velocity tends towards high head loss and high pumping energy cost, but towards low capital installation cost. A larger culvert would have lower velocity and lower pumping energy cost, but higher installation cost.

The economic size of culvert is selected by calculating the cross section of a number of culverts which give velocities of from 3 to 4 m/s in the inlet culvert. For each size the total cost of energy required to pump water through the culvert for the life of the station and the capital cost of installing the culvert are calculated. Energy and installation costs are added together for each size of culvert. The economic size is that which has the lowest combined cost. Culverts are normally about 3 m in diameter for inlet and outlet culverts and from 3 to 4.5 m in diameter for outfall culverts.

Hydraulic pressures

Cooling-water culverts are subjected to the following hydraulic pressures:

(i) Normal running pressures. These are as calculated for the hydraulic gradient.

(ii) Starting pressure (closed-valve pressure). Pumping is started with the valves in the system closed. Each valve is opened in succession when the section before the valve has been filled. This procedure subjects the pressure culverts to the closed valve pressure of the pumps.

(iii) Sudden closure pressures (surge pressures and vacuum conditions). If the pump motor stops through other than normal operational control, the butterfly valve immediately downstream of the pump is designed to close automatically within seconds. This is to prevent damage to the pump and motor through reversal of flow from the condenser. When the valve closes, the water between it and the condenser briefly oscillates and causes alternatively high surge pressures and vacuum conditions in a length of culvert downstream of the closed valve.

Culverts may be designed for the following hydraulic conditions:

Once-through System

Inlet culvert —closed-valve pressure of pump. Also length of culvert downstream of pump (say equal to 4 diameters of culvert) to withstand vacuum

Outlet culvert —Normal working pressure

Outfall culvert—Pressure at entry to culvert

Tower-Cooled System

Inlet culvert —as for once-through system

Outlet culvert —closed valve pressure of pump (because of valve at entry to cooling tower)

Make-up pipelines can be quite long and surge and vacuum conditions can follow the normal shut-down of make-up pumps. The pipelines may therefore be designed for the closed valve head of the pumps, allowing an additional 10% for surge over the first 25% of length from the pumps; the same length should be designed to withstand vacuum conditions. Steel pipe is normally used for make-up pipelines so that the variations of pressure can be absorbed in the elasticity of the steel.

Loading on culverts

Culverts may be subjected to the following types of loading:

(i) Internal water pressure (normal running, closed valve, and surge)

(ii) External air pressure (vacuum consequent on self-closure of valve)

(iii) Handling stresses in steel pipes during installation

(iv) Ground load and superload

(v) Self-weight of culvert and water

(vi) Buoyancy—the weight of the culvert when empty must withstand any upthrust from groundwater

Concrete culverts

Concrete culverts in section may be circular or rectangular with chamfered corners, the latter being preferred for multiple culverts. They are extremely durable, but construction is slow and the open excavations may hinder traffic movement about the station site for long periods.

Present practice is to design for normal working pressure to CP 2007 (concrete not subject to surface cracking) and to check for surge conditions and closed-valve pressures to CP 114 which permits higher concrete stresses than CP 2007. In the design the internal pressures may be reduced by a safely low estimate of permanent external (ground) pressure where this is present. The culvert must also be designed to withstand a safely high external loading when it is empty or subject to vacuum. The test pressure for concrete culverts should be the normal working pressure.

Steel culverts

Steel culverts may be installed in the form of pipes made from steel plates with welded longitudinal and circumferential joints. These steel pipes, with a suitable external coating against corrosion when laid underground, are estimated to have a life of 40 to 60 years which is adequate for the working life of the power station. Steel culvert offers speedier installation than concrete culvert and minimises the duration of open-culvert excavations, and hence restriction of movement on site.

Steel pipe has been installed under the following conditions:

(i) Laid overground with plate thickness not suitable for vacuum conditions
(ii) Laid overground with plate thickness suitable for vacuum conditions
(iii) Laid underground without concrete surround if the external loading (including vacuum conditions when applicable) can be carried by the pipe with lateral earth support; changes of diameter not exceeding 5% can be accepted without harm to the pipe
(iv) Laid underground with complete or partial concrete surround if the external loading exceeds that which can be carried by the pipe with earth support only

If steel culverts were designed only to withstand internal pressures, the plate would be too thin to withstand the handling stresses during installation. There are therefore minimum practical thicknesses of plate from which pipes are made, these being 12 mm plate for pipes from 1.6 to 2.5 m in diameter, then thickening by 2 mm for each 0.5 m increase of diameter to 22 mm plate for 5 m diameter pipe. Such pipes are more than strong enough for internal pressures in cooling-water culverts, but for diameters exceeding 1.6 m they are not suitable to resist vacuum conditions.

For steel pipe culverts subject to vacuum conditions, laid above ground without concrete surround, the plate thickness can be calculated from the formula for buckling:

$$p = \frac{2E}{1-\mu^2}\left(\frac{t}{d}\right)^3$$

where p is the external collapse pressure (taken as 2 atmospheres to include a factor of safety of 2 to allow for manufacturing tolerances = 0.2 N/mm^2), E is the modulus of elasticity (taken as 200 000 N/mm^2), μ is Poisson's ratio (taken as 0.3) and t/d is the ratio of plate thickness to pipe diameter.

From this, t should be not less than $d/130$, giving plate thicknesses varying from 16 mm for 2 m diameter pipe to 40 mm for 5 m diameter pipe. This takes into account only the external load due to vacuum conditions and applies only to steel pipes laid above ground without concrete surround. External loading and ground support must be considered where steel pipes subject to vacuum conditions are laid below ground.

Water velocities and mussel growth

Velocities are kept above the selfcleansing velocity of 0.8 m/s. The minimum velocity for sea water in once-through systems is considered to be 3.25 m/s, since below this velocity mussels can settle on the culvert walls and floor. When established, mussel growth can form a thick rough lining on the walls and floors of the inlet culverts, so restricting water flow. Mussels may also become detached and can cause damage if the shells are carried to the condenser to become lodged in the water tubes. However, it is impossible to maintain a velocity above 3.25 m/s at all times and provision is made to inject chlorine into the incoming cooling water during the spatting season to kill mussels or spat. A similar method is used to reduce algae growth on tower-cooled stations.

NATURAL DRAUGHT COOLING TOWERS

I. W. HANNAH, B.Sc., M.I.C.E.
Central Electricity Generating Board

A natural-draught cooling tower consists essentially of a large reinforced concrete chimney or shell into which air is admitted around the base in such a manner that the induced flow of air intimately mixes with and cools a falling stream of water which has been heated in passing through the turbine condensers. This water is normally distributed uniformly across the area of the shell by a sprinkler system, supported at about 10 m above ground level. Packing in the form of timber or asbestos-cement slats is arranged below the sprinklers to maintain the falling water in either droplet form, or in thin sheet form, in order to expose and maintain as large a surface area as possible to the cooling air (Figure 25.16) The cooled water is finally collected in a concrete pond which covers the base area of the tower, for recirculation to the condensers. Two cooling towers about 90 m in diameter and 110 m high are required to cool the water flowing through the condensers of a 500 MW unit.

The engineering design of a cooling tower thus presents a combination of hydraulic, aerodynamic, thermodynamic and structural problems. Only the last will be considered here.

The pack and sprinkler support structures are normally of pre-cast reinforced concrete to the new BS 'Unified' Code of Practice. Two variations from this code lie in the use of minimal cover to reinforcement in order to minimise the sectional size of the pre-cast members and their impedance to cooling air, and in specifying abnormally tight dimensional tolerances to avoid unacceptable cumulative errors during the erection of the thousands of pre-cast elements used in the pack support structure.

The tower shell

Hyperbolic, or near-hyperbolic shells have been used for cooling towers since 1918, when the shape was adopted to provide stability against the gross foundation settlement which was anticipated by Van Itersen in his design for the Dutch State Mines. This shape has a substantial advantage in economy of both material and surface area over any equivalent cylindrical structure. When the further aesthetic advantage of the conic section is added, it is not surprising that the shape has remained virtually unchallenged since that date. It is now a shape uniquely associated with cooling towers.

The design of the shell presents the only unconventional structural design features in a cooling tower. The only significant loading on a tower shell is that due to wind and in consequence the structure is very light when compared with other structural shells. As the wall thickness is much smaller than the other principal dimensions, the structure tends to resist wind loading predominantly by membrane and tangential shear stress resultants rather than by bending resistance.

Examination of the history of tower design shows that advantage was taken of this characteristic to introduce a drastic simplification of the 8th-order differential equations which represent the approximate linear behaviour of the hyperbolic shell. Neglecting the bending and transverse shear forces allows the basic hyperbolic equations to be integrated along the characteristics which, for a hyperboloid, are straight lines[4]. This method allows the stress resultants and tangential stress to be calculated and the 'membrane' theory can also be used to find the shell displacements. It is noteworthy that the stresses thus calculated can be shown to be independent of Young's modulus, shell thickness and Poisson's ratio.

It is worth noting that later more sophisticated analyses do not normally show any significantly different values of these direct stress resultants and it is arguable that there

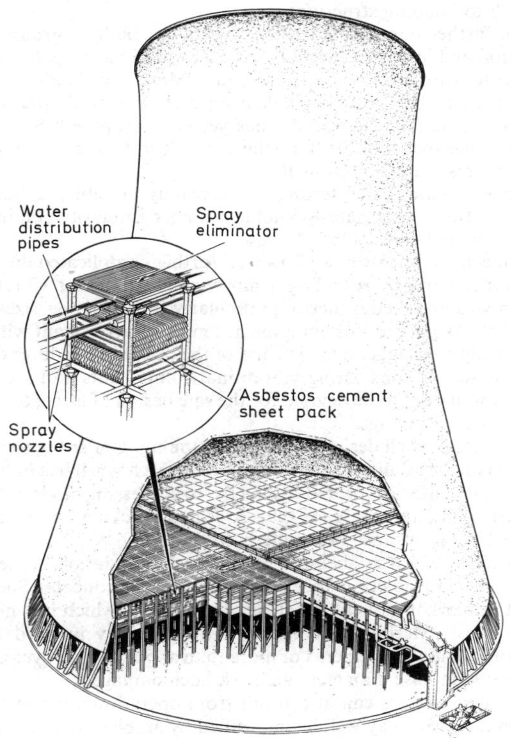

Figure 25.16 Natural-draught concrete cooling tower (Courtesy: Central Electricity Generating Board)

is thus little point in undertaking such complex analyses.

It is generally considered preferable, however, to adopt the full 'bending' analysis, which has become calculable with computer development. Solved originally by Albasiny[5], the problem has since been resolved by most of the modern numerical mathematical techniques.

Several tower design computer programs[6, 7] are now available. As comparable results from several such programs have been shown to be in excellent agreement, the principal ones can be used with confidence.

Wind loading

The estimation of wind loading on cooling towers is much less understood. The collapse of three large towers at Ferrybridge 'C' power station in 1965 showed dramatically that contemporary design values of wind pressure were both inadequate and inapplicable for such wind-sensitive structures[8]. Appreciation of the important relationship between structural geometry and the duration of the wind forces has led to the 5 or 10 seconds maximum mean values of the incident wind speed prediction being almost universally used for subsequent designs. (The recent revision of BS CP3, Chapter V, reflects this change of approach to building structures.)

The situation is further complicated when towers are built in groups. These cause significant distortion and intensification of the basic wind-loading patterns, particularly for towers on the leeward side of such an array. Maximum loadings on individual towers are then found to be highly dependent on certain critical wind directions. No reliable method of calculating these forces has yet been proposed. Safety margins are therefore imposed somewhat arbitrarily by the use of load factors to increase the wind-induced direct-stress resultants in the shell.

Recourse to aerodynamic model testing is not readily possible, as facilities capable of even approaching the appropriate Reynolds Number similarity conditions are rare, inflexible in operation and expensive.

Recent wind-tunnel experiments have, however, further complicated the determination of tower-shell wind loading by revealing a much more dominant vibrational loading characteristic than was expected hitherto, particularly for the larger sizes of tower (i.e. above 110 m height). As part of this component tends to be resonant with the shell, its significance can hardly be exaggerated. The use of the design procedures outlined above is totally inappropriate for considering vibrational components and a completely new design method may well need to be devised for the safe design of larger towers, especially when arranged in close groupings.

Several other aspects of shell design deserve mentioning. Under high-wind conditions the possibility of elastic instability exists. Although research work has indicated that this form of failure is significantly less probable than the overstressing mode in conventionally sized and designed shells, the presence of meridional cracking can cause a dramatic decrease in the buckling resistance of a hyperboloid.

Again the sensitivity of large shells to differential foundation settlement may be marked, in apparent contradiction to Van Itersen's early concept. Such movements cause nonsymmetric horizontal bending effects in the shell, which it is not intrinsically well prepared to resist. Meridional cracking can be readily induced by inadequate foundations and differential settlements of more than about 25 mm over a quadrant of a tower 110 m high + 90 m base diameter can cause such damage.

Cracking in the shell surface can also result from operational thermal and moisture movement transients. These may well be intensified by ambient temperature variations. As above, the effect of such cracking alone is only structurally significant in its consequential effect of rendering the shell more susceptible to buckling or vibrational failure.

Support columns and foundations

The design of foundations and diagonal support columns for tower shells is largely conventional, but any possibility of inducing uplift conditions on the upstream side of larger towers requires most careful consideration. Circumferential continuity of the foundations and/or pond wall is strongly recommended in preference to discrete pad foundations below each support column node.

The transfer of load and horizontal shear from the shell into the support columns is effected through a thickening of the lower shell section, or ring beam. Very high maximum column loads, which may result from elastic analysis of the support complex, are likely to be modified or smoothed out by redistribution in practice, since it is clear that the loads are transient by nature and that the least degree of nonlinearity induced in the columns by momentary overstress will permit such a redistribution. Some allowance for this effect is normally permitted in design.

Figure 25.17 shows a cross-section through a typical foundation for a cooling tower.

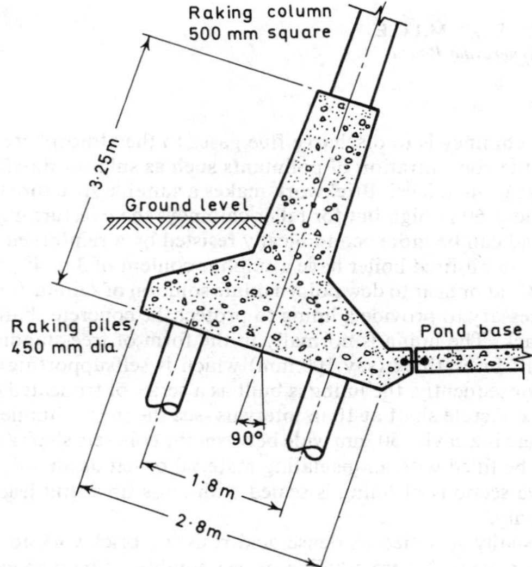

Figure 25.17 Cooling-tower foundation

The concrete ring beam (supported on raking piles) carries a triangulated system of raking support columns, which in turn support the ring beam created by the thickening of the lower section of the tower shell. The water in the pond and the tower packing are carried by the base slab.

CHIMNEYS

K. P. GRUBB, B.Sc.(Eng.), M.I.C.E.
Central Electricity Generating Board

The function of a chimney is to discharge flue gases to the atmosphere at such a height and velocity that the concentration of pollutants such as sulphur dioxide is kept within acceptable limits at ground level. Brickwork makes a suitable structure for free-standing chimneys up to about 60 m high but for taller chimneys the overturning moment due to increased wind load can be more economically resisted by a reinforced concrete shaft.

Flue gases from an oil-fired boiler have a sulphur content of 3 to 4% and the temperature is about 150 °C; at or near to dew point a dilute solution of sulphuric acid is produced. It is therefore necessary to provide a lining to protect the concrete shaft internally from heat and acid attack. The lining is normally in the form of free-standing acid-resisting brickwork one-half brick thick (say 100 mm) which is selfsupporting for a maximum height of 10 m. Consequently, the lining is built as a series of truncated cones carried on corbels inside the concrete shaft at 10 m intervals (see the inner chimney detail at 'A' in Figure 25.18). There is a cavity 50 mm wide between the concrete shaft and the brickwork lining which may be filled with an insulating material or left as an air gap; the junction between successive sections of lining is sealed with glass fibre and lead to exclude flue gases from the cavity.

The lining is usually specified as dense acid-resisting brickwork to BS 3679, laid in potassium silicate mortar. Where alkaline or wet conditions may be experienced (as at the top of the chimney) a synthetic resin should be used instead of mortar to avoid softening of the joints which are normally kept as thin as possible, say 3 to 5 mm. Linings are usually one-half brick thick for flues up to 6 m in diameter, but in the lower levels of the chimney and around gas entry points, a lining one brick thick is provided.

Multi-flue chimneys

The column of hot gases rising inside a chimney continues to rise as a 'plume' without appreciable dispersion for some height after leaving the top of the chimney and so increases the effective height of emission; for example, with a chimney 200 m high the effective height of emission could be 500 m.

From the expression

$$C \propto \frac{Q}{H^2}$$

where C is the concentration of polutant at ground level, Q is the rate of emission of flue gas and H is the effective height of emission, it can be seen that an increase in the effective height of emission has considerable effect in reducing the concentration of pollution at ground level. Research has shown that the plume rise (which governs the effective height of emission) is largely dependent upon the heat content of the plume; therefore for a power station with several boilers (each with its own flue to maintain

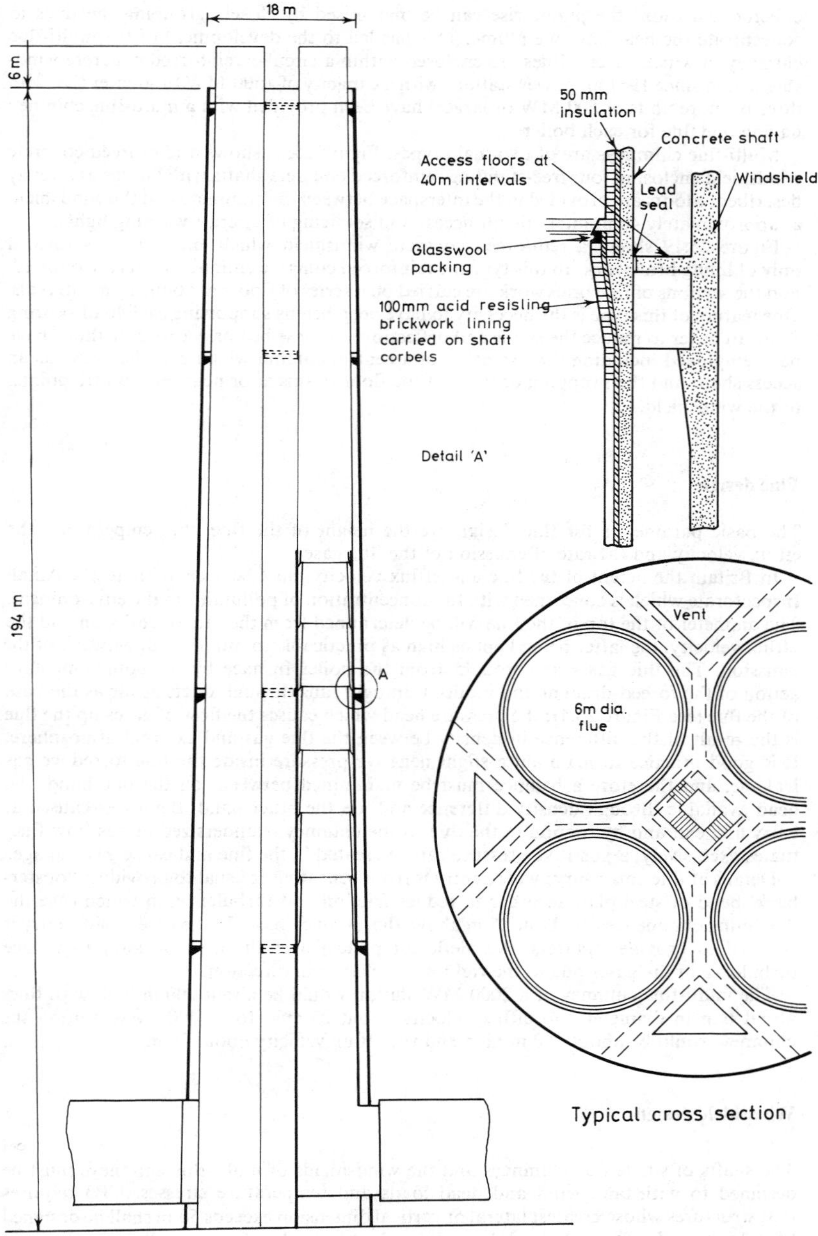

Figure 25.18 200 m multi-flue chimney with free-standing flues

effective emission) the plume rise can be maximised by closely grouping the flues to concentrate the heat into one plume. This has led to the development of the multi-flue chimney in which several flues are enclosed within a circular reinforced concrete windshield, and since 1961 all power stations with a capacity of 1500 MW or greater (i.e. with three or more units of 500 MW or larger) have been provided with a multi-flue chimney having one flue for each boiler.

Multi-flue chimneys are of two main types. Figure 25.18 shows a reinforced concrete windshield enclosing four free-standing reinforced concrete shafts with linings as already described. Floors are provided in the interspace between the chimneys and the windshield at approximately 40 m intervals for access and servicing of aircraft warning lights.

Figure 25.19 shows a reinforced concrete windshield which encloses flues formed only of lining brickwork. In this type the reinforced concrete chimney shafts are omitted, and the sections of flue brickwork are carried on a series of floors at about 10 m intervals. One feature of this type is the necessity to have deep beams supporting each load-bearing floor. In order to reduce the spans and deflections of these beams, various methods have been employed including the provision of a central column which can also serve as an access shaft, and the propping or tying of the floor beams at or near their centre points, to the windshield.

Flue design

The basic parameters for flue design are the height of the flue, the temperature, the efflux velocity and the rate of emission of the flue gases.

In Britain the height of the flue and efflux velocity must be acceptable to the Alkali Inspectorate which is concerned with the concentration of pollutants in the environment. The diameter of the top of the flue will be determined from the rate of emission and the efflux velocity, the latter being kept as high as practicable to minimise downwash of the emission. The flue gases are brought from the boiler furnace by the complementary action of the forced-draught and induced-draught fans through ducts as far as the base of the flue (see Figure 25.4); the pressure head which causes the flow of gases up the flue is the result of the difference in density between the flue gas and external atmosphere. It is good practice to maintain a slight negative pressure inside the flue to reduce gas leakage, and therefore a balance must be maintained between, on the one hand, the head available through density difference and, on the other hand, the losses caused at entry and exit and by friction in the flue. If the chimney is undersized or gas flow fluctuates excessively, a possitive pressure can be created in the flue and cause gas leakage.

Details at flue entry vary; where entry is from beneath it is usual to provide a 'lobster-back' bend of steel plate in order to reduce friction and turbulence, in which case the flue lining commences at about 25 m above the chimney base. In the case of side entry it is usual to provide 'splitters' (i.e. deflector plates) over the area of entry to reduce turbulence in the gases due to the right-angle change in direction.

The multi-flue chimney of a 2000 MW station would be about 200 m high with flues about 6 m in diameter and efflux velocity about 23 m/s; for a 4000 MW station the chimney would be about 260 m high and the efflux velocity about 26 m/s.

Windshield design

The shafts of single-flue chimneys and the windshields of multi-flue chimneys must be designed to withstand wind and dead loads and temperature stresses. CP3 requires that structures whose greatest lateral or vertical dimension exceeds 50 m shall be designed for a 15-second gust wind speed, but it is prudent to apply a factor to allow for dynamic effects in the preliminary design. The basic design, as a cantilever resisting overturning

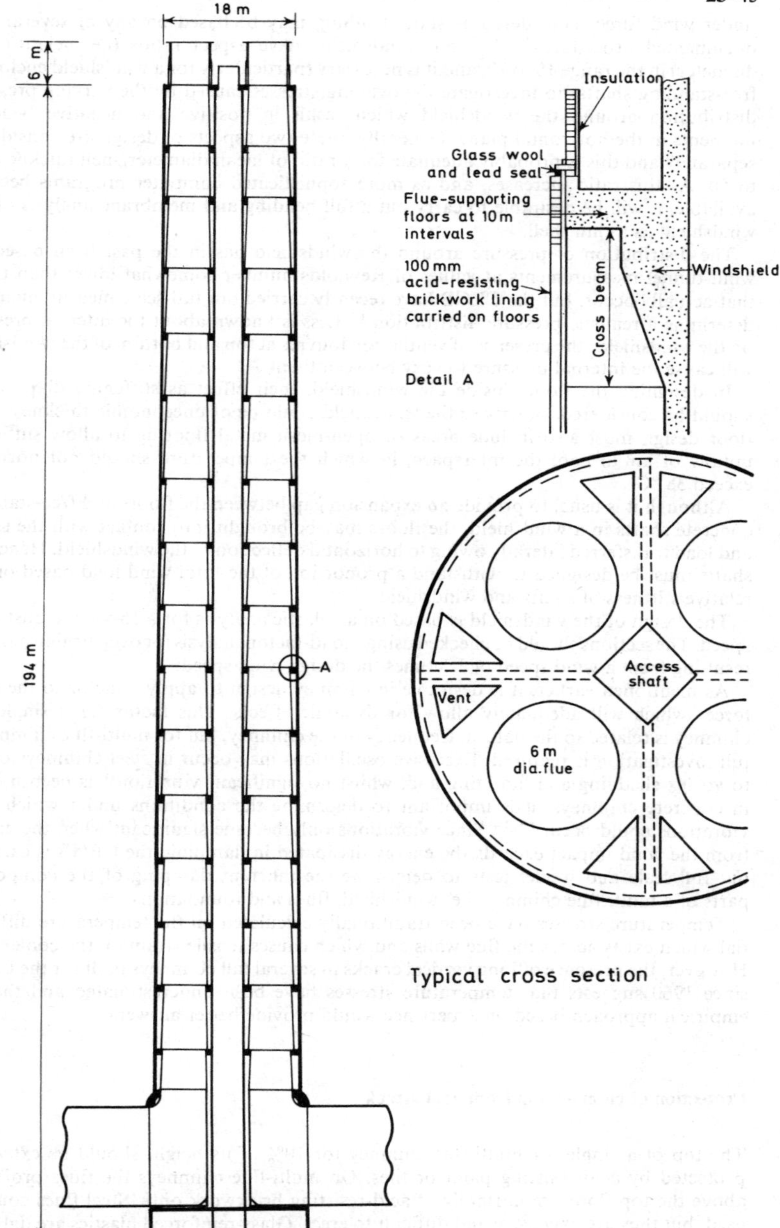

Figure 25.19 200 m multi-flue chimney with flues supported on windshield

under wind forces considered as static loading, may be based on any of several well documented procedures,[9, 10, 11] but windshields have aspect ratios (i.e. height/mean diameter) in the range 10 to 12, and it is necessary (particularly for a windshield enclosing free-standing shafts) to investigate the ovalling stresses caused by the varying pressure distribution around the windshield which result in positive and negative bending moments in the horizontal plane. Generally, these two aspects of design are considered separately and this is probably adequate for a ratio of mean diameter/shell thickness up to 50. As this ratio increases, and as more sophisticated computer programs become available, it will be desirable to carry out a full bending and membrane analysis of the windshield as a thin shell.

The distribution of pressure around the windshield has in the past been based on wind-tunnel measurements at values of Reynolds number somewhat lower than those that actually occur, but the CEGB have recently carried out full-scale measurements to determine a realistic pressure distribution.[12] Less is known about the internal pressure on the windshield; the presence of ventilation louvres at top and bottom of the windshield will cause the internal pressure to vary between them.

In designing the floors inside the windshield, their effect as stiffening diaphragms should be considered, otherwise the windshield could be of uneconomic thickness. The floor design must also include areas of open-mesh metal flooring to allow sufficient upflow of air to cool the interspace, in which the temperature should not normally exceed 38 °C.

Although it is usual to provide an expansion gap between the floors and free-standing concrete shafts in a windshield, the floors may be brought into contact with the shafts and load transferred laterally owing to horizontal deflection of the windshield. Hence the shafts must be designed to withstand a proportion of the total wind load based on the relative stiffness of shafts and windshield.

The design of the windshield is based on an elastic analysis for a 15-second gust wind speed. The sections should be checked using a load-factor analysis for overturning moment resulting from a wind speed of 1.5 times the design wind speed.

As mentioned earlier, it is desirable in chimney design to apply a factor to the wind forces which will adequately allow for dynamic effects. This factor for a single-flue chimney is related to the natural frequency of the chimney, but for multi-flue chimneys a full investigation is required. Excessive oscillations may occur in steel chimneys owing to vortex shedding and buffeting and, whilst no significant vibration has been noticed in concrete chimneys, it is important to determine the conditions under which such vibrations could occur.[13, 14] Since vibrations only become significant when the energy from the wind impact exceeds the energy dissipated in damping, the CEGB is carrying out full-scale and model tests to determine the inherent damping of the component parts of a multi-flue chimney, i.e. windshield, flues and foundations.

Temperature stresses have been traditionally calculated on the temperature differential which exists across the flue walls and which causes tensile strain on the cooler face. However, the presence of long vertical cracks in several tall chimneys built for the CEGB since 1960 suggests that temperature stresses have been underestimated and that an empirical approach based on experience would provide better answers.

Protection of chimney top from acid attack

The top of a single- or multi-flue chimney for 10% of its height should be externally protected by acid-resisting paint or tiles. On multi-flue chimneys the flues projecting above the top floor are normally of acid-resisting brickwork only. Steel flues could be used, but they are expensive and difficult to erect. Glass-reinforced plastics are light and easy to erect but there is insufficient operating experience to justify their general use at present. The top floor may be covered with quarry tiles.

Aircraft warning lights

Aircraft warning lights must be provided in accordance with regulations, usually at 50 m intervals vertically and at the top of the chimney. Three lights at 120° intervals (or four at 90° intervals, for a multi-flue chimney with four flues) are provided at each level.

On multi-flue chimneys, the lights are usually fixed on the outer face of inward-opening doors in the windshield, accessible at the appropriate floor level. On single-flue chimneys, lamps are maintained by steeplejacks and fittings are duplicated.

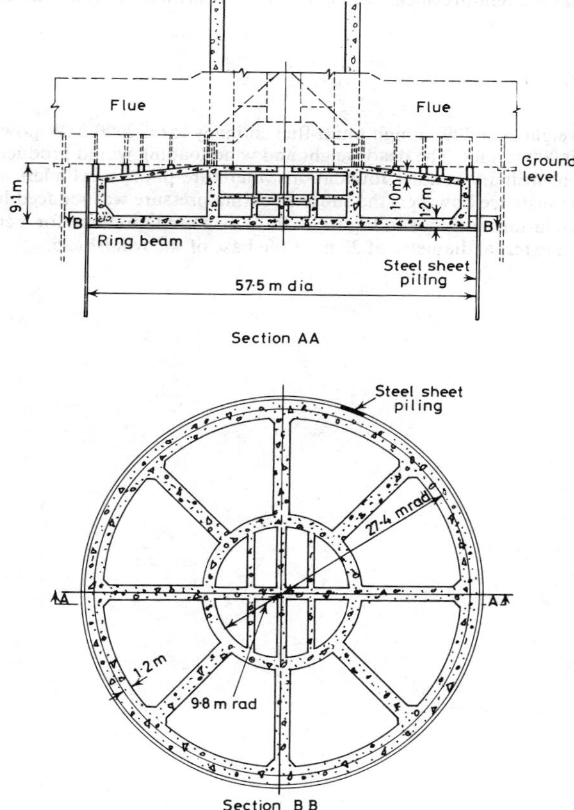

Figure 25.20 Cellular foundation in reinforced concrete, for multi-flue chimney (Courtesy: Central Electricity Generating Board)

Access

Bronze sockets in which steeplejacks may insert ladder fixing hooks should be built in the external face of single-flue chimneys; they are unnecessary on multi-flue chimneys because there is internal access to the top.

Lightning Protection

A lightning protection system is necessary. A coronal band is provided and CP 326 permits the use of steel reinforcement in a concrete structure as down conductors, provided that the reinforcement cage is adequately earthed and tested on completion for continuity.

Foundation

The total weight of a 200 m high multi-flue chimney for a 2000 MW power station approaches 20 000 tonnes. The dead weight and wind loading would produce high bearing pressures on ordinary contact foundations, therefore piles or cylinders are commonly used. In one instance however, the subsoil bearing pressure was reduced by the use of a cellular foundation as shown in Figure 25.20. This foundation was for a chimney 200 m high with an external diameter of 22 m at the base of the windshield.

NUCLEAR REACTORS AND REACTOR BUILDINGS

N. G. EGGLETON, B.Sc., F.I.C.E.
Central Electricity Generating Board

In nuclear power stations, the turbo-generators are the same as in coal or oil-burning stations, but instead of heat being produced from the combustion of coal or oil in the furnace of a boiler, heat is produced by the nuclear fission of a radioactive material in a reactor. In Britain, the reactors which have been built for the public supply industry use a circulating gas coolant for the nuclear fuel and are classified as gas-cooled reactors.

The reactor core consists of nuclear fuel elements housed in vertical channels formed in a graphite moderator which is assembled from accurately machined graphite blocks. The moderator slows down the neutrons radiated from the nuclear fuel in order to enhance the frequency of neutron collision and fission in the mass of nuclear fuel. The reactor is designed to contain sufficient mass of nuclear fuel to give a selfsustaining reaction. The reactor core is enclosed within a pressure vessel made of either steel or prestressed concrete, and heat exchangers ('boilers') are arranged around its circumference; the boilers are located outside steel pressure vessels, but are contained inside concrete vessels or are contained in cavities within the wall of concrete vessels. The reactor shown in Figure 25.21 is contained in a prestressed concrete pressure vessel. The gas coolant is circulated inside the pressure vessel; it flows up the channels in the moderator which house the fuel elements and there it gains heat, then it flows outwards from the core and down through the boilers, where it gives up heat to water and steam in the boiler–turbine steam circuit. It is then recirculated upwards through the core. In order to increase its capacity to carry heat, the gas is kept at high pressure.

The first nuclear power station programme in Britain consisted of nine Magnox stations commissioned between 1962 and 1968, and the second consists of five Advanced Gas-cooled reactor stations to commission between 1972 and 1976 (See Table 25.1). The Magnox reactors use natural uranium fuel elements in metal form, encased in magnesium alloy (magnox) finned containers about 25 mm in diameter and $\frac{1}{2}$ to 1 m long to support the fuel and contain the radioactive products of fission; they are graphite moderated and are cooled by carbon dioxide gas. The metal fuel and magnox cans limit the coolant temperature so that the steam temperatures and pressures reached in the boilers are suitable only for turbo-generators up to 300 MW capacity. The fuel irradiation is 3000 to 4000 MW days per tonne.

The Advanced Gas Reactors (AGR) use enriched uranium (uranium dioxide) fuel. The natural uranium is enriched in a separate process by increasing the proportion of the radioactive isotope present. The fuel elements consist of 36 stainless-steel ribbed tubes containing 14.5 mm diameter uranium dioxide pellets, the tube cluster being contained within a graphite sleeve 190 mm in diameter. Eight such elements each 1 metre long, are linked together by a tie bar to form a fuel stringer, and each channel in the reactor core accommodates one stringer. Oxide fuel and stainless-steel cans permit higher operating temperatures and the steam conditions reached in the boilers (170 kgf/cm^2 and 540 °C) are suitable for 660 MW turbo-generators. The fuel irradiation is about 18 000 MW days per tonne. The moderator in the core remains graphite and the coolant remains carbon dioxide gas, but it is circulated at much higher pressure to increase its heat transfer capacity.

The gas operating pressure in the series of Magnox reactors progressed from about 1000 kN/m^2 to 2000 kN/m^2 and the earlier reactors were contained in steel pressure vessels up to 90 mm thick. As reactors of increasing size were designed and coolant pressures increased, the fabrication of still larger and thicker steel vessels became

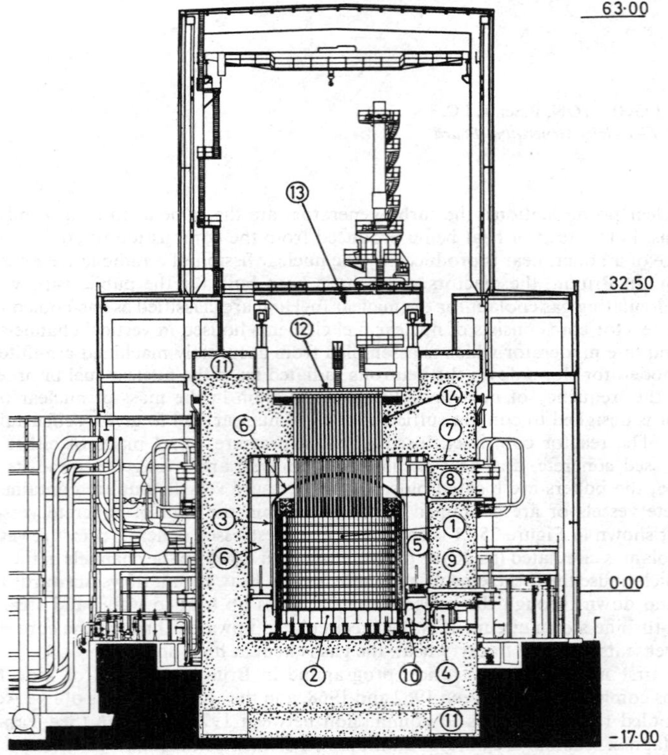

Figure 25.21 Advanced gas-cooled reactor. 1 Reactor core; 2 Supporting grid; 3 Gas baffle (steel cylinder 14 m diameter, without bottom and with torispherical dome); 4 Gas circulators; 5 Boilers; 6 Thermal insulation; 7 Reheat steam penetrations; 8 Main steam penetrations; 9 Boiler feed penetrations; 10 Prestressed concrete pressure vessel (19 m diameter and 19.35 m high internally); 11 Cable-stressing galleries; 12 Charge face level; 13 Fuelling machine; 14 Standpipes (one standpipe above every channel in core for refuelling or control)

impracticable, and prestressed concrete pressure vessels were used in the later Magnox reactors and in all the AGRs. The AGR pressure vessels operate at a coolant pressure of 4000 kN/m^2.

Reactor pressure vessel

The prestressed concrete pressure vessel shown in Figure 25.21 is a vertical cylinder with helical multi-layer post-tensioned cables in the walls, so arranged that no cables are required across the top and bottom slabs. The pattern of cables is shown in Figure 25.22.

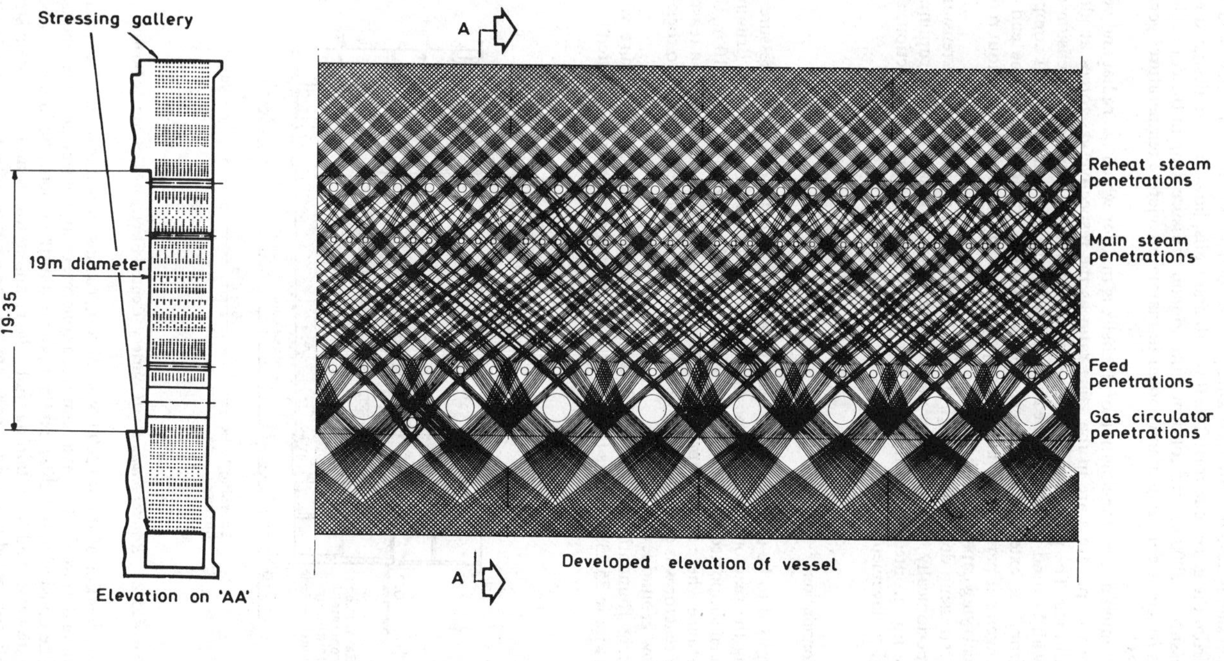

Figure 25.22 Prestressing system of pressure vessel

The bottom slab is designed for a working pressure of 4200 kN/m^2 abs which is the outlet pressure from the gas circulators. The top slab and walls, however, are designed for the lower pressure of 3900 kN/m^2 following the coolant pressure drop in the core. The vessel inner surfaces are insulated and cooled to maintain concrete temperatures generally below 70 °C.

The optimum angle of inclination of the helix is that for which the radial and vertical components of prestress are in the same ratio as the respective gas forces at ultimate load conditions. The prestressing cables are taken up into an annular extension of the cylinder wall beyond the flat slabbed end to such a height that the radial component of load from the extended cables provides sufficient force to restrain the end slabs. This arrangement permits any number of penetrations in the top slab without reducing the load-carrying capacity of the slab.

The prestressing design has a high degree of redundancy and the cable tensions are checked periodically; the pressure vessel is therefore very safe against rupture. The cables are not grouted in and they can be removed individually for inspection, and, if necessary, replacement.

Reactor foundations

The main load to be carried by the reactor building foundations is the pressure vessel containing the reactor core. Because of the load distribution, reactors in Britain have usually been supported by direct bearing on the subsoil at a depth of about 10 m, though in one instance where it was impracticable to spread the load sufficiently at a reasonable depth the reactor was carried on concrete shafts 2.3 m in diameter taken into rock about 40 m below ground level.

A contact foundation for an AGR with a prestressed concrete pressure vessel is shown in Figure 25.23. Here the total weight of the loaded vessel is transmitted to a rock

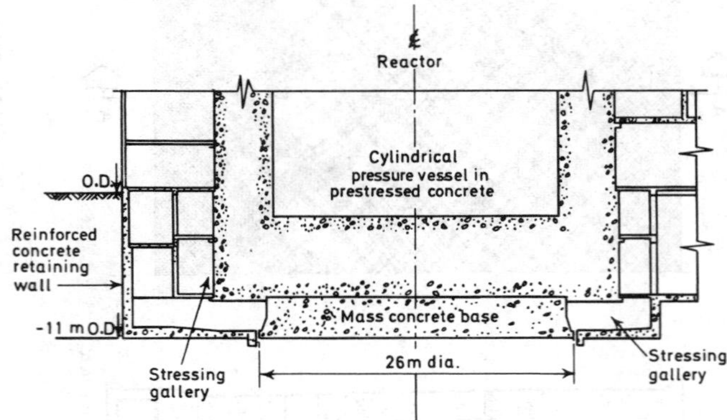

Figure 25.23 Reactor foundation (Courtesy: Central Electricity Generating Board)

stratum at a loading of about 1200 kN/m^2 through a mass concrete base 26 m in diameter. Surrounding the circular base is a reinforced concrete retaining wall, and the beams which support the considerably lighter external portions of the reactor building span between the pressure vessel and the retaining wall. The retaining wall is an independent structure and is only connected to the mass concrete base by a continuous rubber water stop.

Stringent precautions are taken in the design and construction of reactor foundations to ensure that long-term overall settlement and differential settlement (both between buildings and that which results in tilt of the reactor) will be within acceptable limits.

Layout of reactor building

The Magnox and AGR nuclear stations each have two reactors, with one or more turbo-generators to each reactor. As with oil- or coal-burning stations, the layout has developed to an in-line arrangement of the units (reactor and turbo-generators), generally with transverse arrangement of the turbo-generators.

In addition to the pressure vessels and access facilities around them, the reactor building has to accommodate the following principal items.

Services annexe—consisting of offices, stores, laboratories and changing rooms and ablutions for personnel.

Fuelling machine, which traverses the charge face over the reactor core. Operated by remote control, it removes cans or stringers of irradiated fuel from the reactor core and inserts new fuel elements. Also located at the charge-face level are workshops for maintenance of the fuelling machine and other reactor equipment and the decontamination centre.

Shielded block, which contains the new and irradiated fuel-handling equipment.

Active waste disposal. After removal from an AGR core, the fuel element stringers are separated and certain components of the assembly are waste. This scrap has become radioactive whilst in the reactor core, and is disposed of by burial in deep vaults in the reactor building.

Cooling pond in which irradiated fuel elements are stored for about three months after removal from the reactor core. This allows the radioactivity to decay to a level at which they can be transported away from the power station for processing. There is some six metres depth of water in the cooling pond to shield operatives from radiation.

Cooling-pond water and active-effluent treatment plant

Carbon dioxide treatment plant

Ventilation plant. Most of the reactor building is artificially ventilated, because the size of the reactor building leaves many working areas with insufficient natural ventilation and also because many areas must be kept under either forced or plenum ventilation conditions to prevent the possibility of spread of radioactive contamination. Extracted air is also filtered before discharge to outside atmosphere. The ventilation plant occupies a substantial part of the reactor building.

Station control room, which contains the instrumentation and control panels for the operation of the reactors, boilers, turbo-generators, and ancillary plant.

Electrical switchgear

Emergency generators. These are stand-by generators to safeguard against failure of the station supply and of the incoming supply from the public system.

The main difference between the layouts of various reactor buildings is in the location of the foregoing items, particularly of the services annexe. If the reactor spacing is kept to the minimum, the services annexe has to be located between the reactors and the turbo-generators, which results in long high-pressure pipework between boilers and turbines. By increasing the spacing of the reactors and locating the services annexe between them, the turbo-generators can be brought close to the reactors and the length of pipework reduced; this results in an increase in building volume and costs, but is more than offset by the saving on pipework.

HYDRO-ELECTRIC POWER AND PUMPED STORAGE

N. G. EGGLETON, B.Sc., F.I.C.E.
Central Electricity Generating Board

Hydro-electric power

In hydro-electric power plants, turbines are powered by passing the greater part of the water flow of a river through the turbines. The plants fall into two categories according to the absence or use of storage:

(i) Run-of-river plants, which have insignificant reservoir capacity. At any particular time the power available for generation is limited according to the river flow at that time, and there is no reserve to meet high demand for electricity. These plants usually provide continuous generation (base load supply) and their output/capacity ratio (load factor) is high.

(ii) Storage plants, which have a significant reservoir capacity sufficient to enable the water flow through the turbines to be regulated according to the demand for electricity. The load factor of these plants is usually low.

Thus, a run-of-river scheme will have a low dam (or weir) to raise the river level up to the intake works which give controlled conditions for flow of water to the pressure pipes or tunnels leading to the turbines. A storage scheme will have a high dam located in the best position, taking into account the volume impounded, elevation and construction cost. The intake works of a storage scheme are normally incorporated in the dam, or built as a tower either on the face of the dam, or free-standing in the reservoir.

For either run-of-river or storage schemes, the power station housing the turbines and generators is built at a lower level downstream of the dam, so that water flowing through the turbines is returned to the course of the river; the pressure head which operates the turbines is that due to the difference in level between the water surface at the intake and the tail-race level after leaving the turbines. The power station may be located at (or incorporated in) the base of the dam (Figure 25.24(a)) or it may be located at some distance from the dam and connected to the intake by shafts and pressure tunnels (Figure 25.24(b)) or it may be located underground intermediately between the bottom of the shafts and the emergence of the tunnels (Figure 25.24(c)). In the latter case, the tunnel upstream of the turbines are pressure tunnels and the tunnels downstream of the turbines form the tail-race to the river.

In the case of a scheme with a long pressure tunnel leading to the turbines, fluctuations of demand for electricity cause fluctuations in the speed of the turbines and of the flow of water through them, with consequent acceleration or deceleration of the mass of water in the tunnel and with pressure variations. A surge tank is constructed over and near the end of the tunnel to protect it from surges of pressure, and to facilitate the response of the plant to variations of demand.

According to the hydraulic operating head, different types of turbine are used, approximately as follows:

High head (1500–300 m)	Pelton wheels (horizontal or vertical shafts)
Medium head (400–30 m)	Francis turbines (horizontal or vertical shafts)
Low head (40–5 m)	Kaplan turbines (vertical shafts only)

Pumped storage

Demand for electricity varies throughout each 24 hours, being greatest at mid-morning and mid-afternoon and least at night. Normally, generating capacity in excess of demand has to be closed down but, if electricity could be stored in large quantities, some of the more efficient steam plant could continue to generate at night and the stored output could

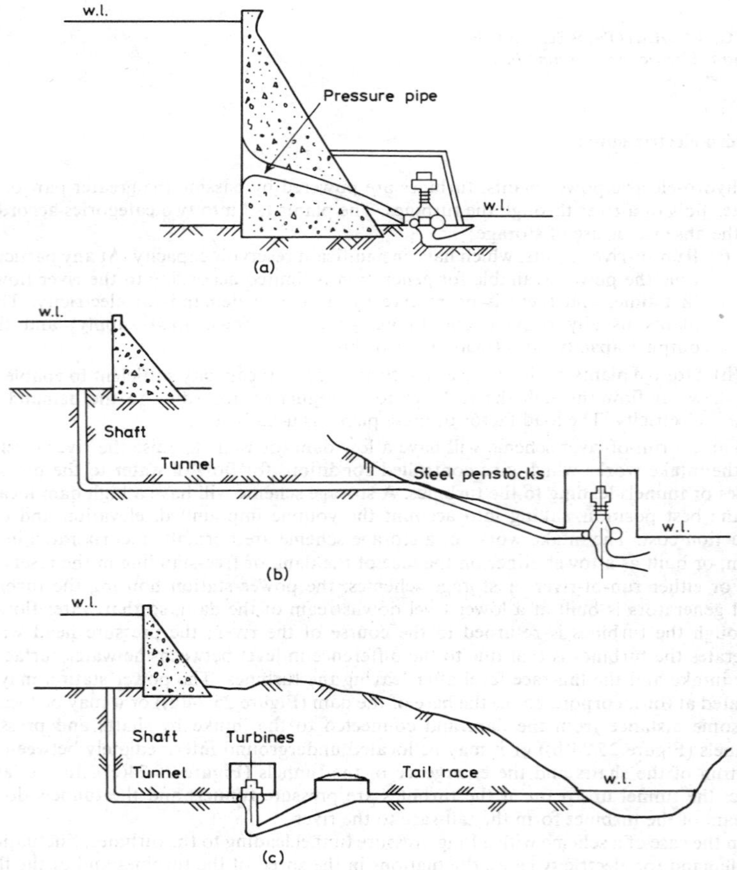

Figure 25.24 Locations for power station in hydro-electric scheme. (a) Low to medium head; (b) medium to high head; (c) high head

be released later to save generating from the least efficient plant during the daytime peaks of demand.

Pumped storage is an adjunct to highly efficient steam plants to achieve the effect of storing electricity. It operates by using their surplus generating capacity at the time of low demand to pump water from a low-level reservoir up to a high-level reservoir, and then at times of high demand using the water stored in the high-level reservoir to drive hydraulic turbines and generate electricity. The scheme is built as a conventional storage-type of hydro-electric installation, except for the following differences:

(a) water after discharge from the turbines is stored in a low-level reservoir
(b) the turbines must be reversible to act as pumps, or the pumps and turbines may be separate machines
(c) the electrical alternators operate as motors to drive the pumps, or as generators when being driven by the turbines

The Ffestiniog pumped storage scheme in North Wales is similar to Figure 25.24(b). There is an upper reservoir of 1.7×10^6 m^3 capacity which, with the available head of approximately 300 m, permits a daily output of 1.2 million kWh by operating the four 75 MW generators for 4 hours daily.

The upper reservoir is formed by a dam across the outlet from a lake. The dam is 275 m long and 36 m high, having a buttress gravity centre section 245 m long, with a solid gravity spillway on the west flank. There are two vertical intake shafts 18 m upstream from the dam, 200 m deep and 4.5 m in diameter inside a concrete lining 600 mm thick. Each shaft bifurcates at the bottom into tunnels approximately 3 m in diameter and 1150 m long. This allows two turbines to operate while two tunnels and a shaft are drained for inspection. For 550 m upstream of the tunnel portal, the tunnels have steel linings surrounded by concrete, and the remaining length of each tunnel has a concrete lining 450 mm thick. The tunnels emerge at a portal chamber about 60 m above the power station. The portal chamber is connected to the power station by steel penstocks 2.5 m in diameter and 200 m long. Each penstock bifurcates before the power station, the upper branch being connected to the turbine inlet and the lower branch to the pump discharge. Surge shafts are not provided over the tunnels because of the high head/length ratio of the scheme, but the turbines are fitted with relief valves.

The lowest reservoir is about 2 km from the upper reservoir and is retained by a solid gravity dam 550 m long and 12 m high. Regulators are installed to control the discharge of the river flow from the lower reservoir.

OVERHEAD TRANSMISSION LINES AND SUPPORTS

(Summarised from second edition)

N. G. EGGLETON, B.Sc., F.I.C.E.
Central Electricity Generating Board

Electricity is distributed at high voltage in transmission lines from power stations through a sequence of substations, in which the voltage is transformed down to the consumer's voltage. Heat is generated in a conductor by the flow of current and the conductor size must be such that the temperature does not rise above the annealing point of the conductor material (approximately 75 °C for hard-drawn copper and aluminium). For a given conductor, the heat generated (which is also the power lost in transmission) is proportional to the voltage, but the power transmitted is proportional to the square of the voltage, which makes transmission at high voltages desirable. In Britain, alternating current (three-phase, 50 Hz) is transmitted at 11, 33, 66, 275, and 400 kV, and higher voltages are under consideration. Conductors may be insulated cables laid underground or may be bare conductors carried at a safe height above ground between towers. The high degree of insulation necessary on underground cables makes them expensive, and overhead conductors have a considerable economic advantage.

The range of overhead conductor sizes normally used is:

Voltage limit	*Conductor size* (equivalent copper area mm^2)
up to 33	16– 161
66	23– 197
132	81– 258
275	twin 113–twin 258
400	twin 258

Design standards

The safety regulations covering the design of high-voltage lines in Britain may be summarised (using approximate SI equivalents) as:

(i) Conductors—minimum factor of safety = 2 (on breaking load) when at −5 °C they have a 10 mm radial thickness of ice and are subjected to an 80 km/h wind on the full projected area of the ice-coated conductor (equivalent to 384 N/m^2).

(ii) Supports are to withstand the longitudinal, transverse and vertical forces imposed by the conductors under the above conditions of loading without damage and without movement in the ground. Wind pressure on supports = 384 N/m^2 on projected area, and with compound structures such as steel towers the pressure on the lee-side members may be taken as one half that on the windward side. Minimum factors of safety under these maximum working loads, calculated on the crippling load of struts and the elastic limit of tension members, are:

Iron or steel	2.5
Wood	3.5
Concrete	3.5

(iii) Minimum height of conductors:

Maximum ac voltage (kV)	66	110	165	exceeding 165
Ground clearance (m) at 50 °C	6.1	6.4	6.7	7.0

Conductors

The three main conductor types are hard-drawn copper (BS 125), stranded aluminium (BS 215 Part 1) and steel-reinforced aluminium (BS 215 Part 2). Steel-reinforced aluminium is used for the majority of extra-high-voltage transmission lines, because the high-strength conductors permit long spans between supports.

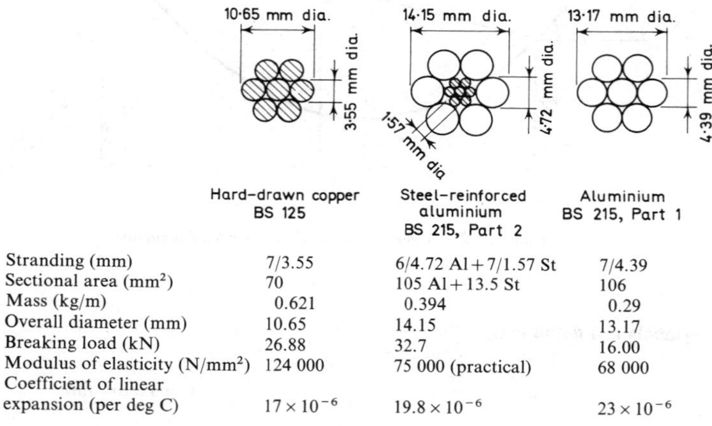

	Hard-drawn copper BS 125	Steel-reinforced aluminium BS 215, Part 2	Aluminium BS 215, Part 1
Stranding (mm)	7/3.55	6/4.72 Al + 7/1.57 St	7/4.39
Sectional area (mm^2)	70	105 Al + 13.5 St	106
Mass (kg/m)	0.621	0.394	0.29
Overall diameter (mm)	10.65	14.15	13.17
Breaking load (kN)	26.88	32.7	16.00
Modulus of elasticity (N/mm^2)	124 000	75 000 (practical)	68 000
Coefficient of linear expansion (per deg C)	17×10^{-6}	19.8×10^{-6}	23×10^{-6}

Figure 25.25 Comparison of three conductors of 70 mm² equivalent copper area

Conductors consist of three or more individual wires stranded together and are categorised according to the details of stranding (see Figure 25.25).

The mechanical characteristics of conductor materials are:

	Weight (kg/mm^2 m)	*Tensile strength* ($N\ mm^2$)	*Coefficient of linear expansion* (per deg C)	*Modulus of elasticity* ($N\ mm^2$)
Copper (hard drawn)	0·008 9	415	17×10^{-6}	124 000
Aluminium (hard drawn)	0.002 7	160	23×10^{-6}	68 000
Steel	0.007 9	1 340	11.5×10^{-6}	200 000

Conductor sags and tensions

Variations in conductor sags and tensions result from changes in temperature and loading, and conductors must be strung so that sags do not exceed that which ground clearance permits, nor tensions exceed the required factors of safety.

The general case of a conductor suspended between supports at different levels is shown in Figure 25.26. Calculations within the practical requirements of accuracy, can, for simplicity, be made on the basis of a parabolic curve, although a catenary curve would be more accurate.

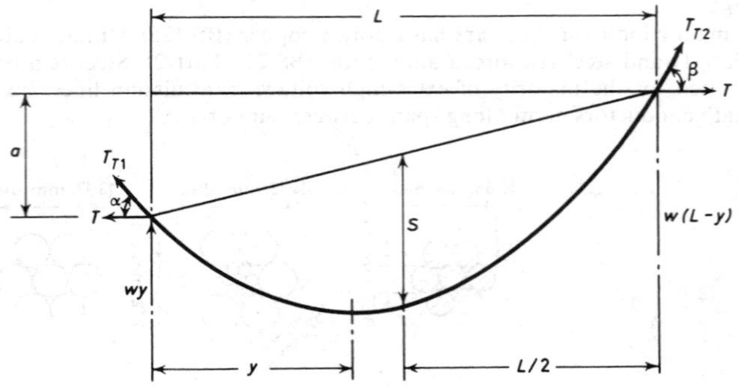

Figure 25.26 General case of the suspended conductor

(a) Symbols and definitions

General

L = span (horizontal distance between supports)

S = sag (distance measured in the direction of the resultant load between the conductor and the midpoint of a straight line joining the two supports)

T = tension (horizontal component of the tension under load w, being uniform throughout any one span)

T_T = tangential tension (actual tension at any given point in a conductor in the direction of the tangent to the curve)

w = resultant load per metre of conductor from vertical load of self-weight + ice (if present) and horizontal wind load

a = difference in level of adjacent supports

y = horizontal distance of lowest point of curve from lower support

t = temperature rise from initial to final conditions

Conductor data

A = actual cross-sectional area of conductors

d = overall diameter of conductor

E = modulus of elasticity of complete conductor.
Virtual modulus for composite conductor of materials a and b

$$= \frac{E_a m + E_b}{m+1} \quad \text{where } m = \frac{\text{area of a}}{\text{area of b}}$$

c = coefficient of linear expansion of complete conductor.
Virtual coefficient for a composite conductor of materials a and b where E and m are as above

$$= \frac{c_a E_a m + c_b E_b}{m E_a + E_b}$$

C_1 = a constant = $\sqrt{EA/24}$

C_2 = a constant = cEA

(b) Formulae

$$S = \frac{wL^2}{8T} \tag{1}$$

$$\text{Length of conductor} = L + \frac{8S^2}{3L} = L + \frac{w^2 L^3}{24T^2} \tag{2}$$

$$\left(\frac{C_1 w_2 L}{T_2}\right)^2 - T_2 = \left(\frac{C_1 w_1 L}{T_1}\right)^2 - T_1 + C_2 t \tag{3}$$

where the suffixes to w_1 T_1 and w_2 T_2 denote

conditions at initial and final temperature respectively.

$$y = \frac{L}{2} - \frac{aT}{wL} \tag{4}$$

Vertical reaction at higher support $= w(L-y)$

Vertical reaction at lower support $= wy$

(c) Example

To determine the maximum sag at 50 °C of a steel-reinforced aluminium conductor of 100 mm² nominal aluminium area (6/4.72 mm aluminium + 7/1.57 mm steel) on a span of 200 m. Factor of safety = 2.0 on breaking load with 10 mm radial thickness of ice at −5 °C and 80 km/h wind.

The sequence of calculations is as follows:

(i) Evaluate the conductor data (see Figure 25.25)

$$A = 105\ \text{mm}^2\ \text{Al} + 13.5\ \text{mm}^2\ \text{Steel} = 118.5\ \text{mm}^2$$

$$d = 14.15\ \text{mm}; \qquad \text{breaking load} = 32.7\ \text{kN}$$

$$C_1 = \left(\frac{EA}{24}\right)^{\frac{1}{2}} = \left(\frac{75\,000 \times 118.5}{24}\right)^{\frac{1}{2}} = 609$$

$$C_2 = cEA = 19.8 \times 10^{-6} \times 75\,000 \times 118.5 = 176$$

$$t = 50\ ^\circ\text{C} - (-5\ ^\circ\text{C}) = 55\ \text{deg C}$$

(ii) Calculate the initial tension T_1 in the cable at −5 °C, the resultant load in the conductor w_1 at −5 °C and w_2 at 50 °C

$$T_1 = \text{initial tension at } -5\ ^\circ\text{C} = \frac{\text{breaking load}}{\text{factor of safety}} = \frac{32.7\ \text{kN}}{2} = 16\,350\ \text{N}$$

$$\text{Horizontal wind load} = 384\left(\frac{20 + 14.15}{1000}\right) = 13.2\ \text{N/m}$$

Weight of ice (at 0.91 g/cm³) on conductor at −5 °C

$$= \frac{0.91}{1000} \times \frac{\pi}{4}\left[\left(\frac{34.15}{10}\right)^2 - \left(\frac{14.15}{10}\right)^2\right] \times 100 = 0.686\ \text{kg/m}$$

$$\begin{aligned} \text{Total weight} &= \text{ice} + \text{self weight} \\ &= 0.686 + 0.394 && = 1.08 \text{ kg/m} \\ \text{hence } w_1 &= [13.2^2 + (9.9 \times 1.08)^2]^{\frac{1}{2}} && = 17 \text{ N/m} \\ w_2 &= \text{load due to self-weight} = 9.9 \times 0.394 = 3.9 \text{ N/m} \end{aligned}$$

(iii) From equation (3) find horizontal tension T_2 at 50 °C

$$\left(\frac{C_1 w_2 L}{T_2}\right)^2 - T_2 = \left(\frac{C_1 w_1 L}{T_1}\right)^2 - T_1 + C_2 t$$

$$\left(\frac{609 \times 3.9 \times 200}{T_2}\right)^2 - T_2 = \left(\frac{609 \times 17 \times 200}{16\,350}\right)^2 - 16\,350 + 176 \times 55$$

$$\left(\frac{475\,000}{T_2}\right)^2 - T_2 = 9430$$

T_2 is solved by trial and error using a slide rule.
In this case $T_2 = 4100$

(iv) From equation (1)

$$\text{Sag } S \text{ at } 50\ ^\circ\text{C} = \frac{w_2 L^2}{8T_2} = \frac{3.9 \times 200^2}{8 \times 4100} = 4.75 \text{ m}$$

Supports

The configuration of supports (but not necessarily the form and material used) depends initially on the electrical requirements of number of circuits, conductor size and type, insulation and clearances, and on the arrangement of conductors and earth-wires. The structural design of supports is related to the imposed loads:

HORIZONTAL TRANSVERSE LOADS (P)

(i) Wind on bare or ice-coated conductors. Calculated on the support 'wind span' = half the sum of the adjacent span lengths (see Figure 25.27) (P_w)

(ii) Wind on supports. For square-lattice structures, wind on the leeward face is taken as half that on the windward face; this shielding factor decreases with rectangular shapes until the full wind is taken on both faces. On cylindrical members, wind pressure is taken on 0.6 of projected area. (P_s)

(iii) Conductor tension at line deviations.

$$\text{Transverse load} = 2T \sin\frac{\theta}{2}$$

(θ is the angle of deviation and T is the maximum conductor tension, see Figure 25.27) (P_a)

HORIZONTAL LONGITUDINAL LOADS (T)

(i) Full conductor tension at line terminals (T)

(ii) Out-of-balance conductor tensions due to broken conductors or earthwires. At supports with suspension insulators, a reduced conductor tension, usually 70 per cent, is allowed for the swing of insulators into the unbroken span. (0.7 T or T)

(iii) Out-of-balance conductor tension at angle or section positions. Only encountered in special cases, e.g. change from single to double earthwires. (T_x)

VERTICAL LOADS (V)

(i) Weight of bare or ice-coated conductors calculated on basis of support 'weight span' (from equation (4)) (V_w)

(ii) Weight of insulators, etc. (V_i)

(iii) Support weight (V_s)

WIND AND ICE LOADS

The relationship between wind velocity (V km/h) and pressure (P N/m^2) may be taken as:

Flat surfaces	$P = 0.1\ V^2$
Round surfaces (e.g. conductors)	$P = 0.06\ V^2$

Figure 25.27 indicates the locations relative to the conductor in which the main types

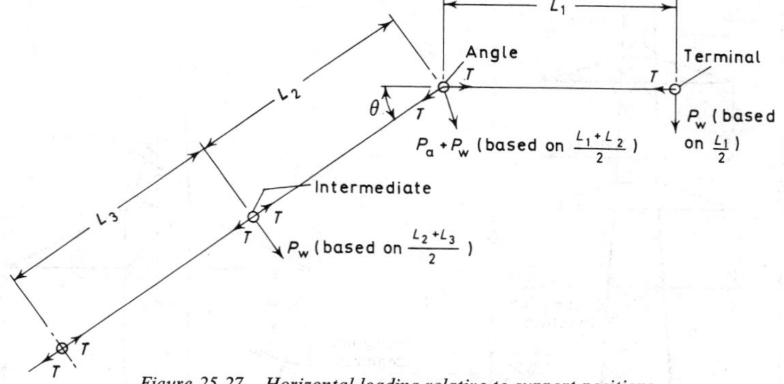

Figure 25.27 Horizontal loading relative to support positions

of support are used, i.e. intermediate, angle and terminal.

The structural form and materials of supports may be classified as:

Single or composite wooden poles
Single or composite reinforced concrete and prestressed concrete poles
Steel tubular poles
Narrow-base towers—lattice structures of rolled steel and tubular sections, with single block foundations, increasingly with the use of guy wires
Broad-base towers —steel lattice structures, with a separate foundation for each leg (Figures 25.28 and 25.33).

Design of broad-base towers

The following description applies chiefly to the design sequence for broad-base towers, but the principle applies equally to other types.

With the electrical requirements resolved, the first step in design of the supports is to decide upon the 'standard span', i.e. the most economic span assuming level ground. Exploratory design is concentrated on the intermediate supports (being the majority) and the following interdependent factors are taken into consideration to determine the general outline and the arrangement and height of the cross arms:

(1) Height to bottom conductor, which is the minimum specified ground clearance, plus the maximum sag of the conductor.

(2) Conductor spacing

(i) Minimum horizontal and vertical spacing to provide adequate midspan clearance dependent upon span length, sag and voltage, as well as factors such as ice shedding overcome by offsetting the conductors (Figure 25.28)

(ii) Minimum live-metal-to-earth clearance, taking into account the maximum swing of suspension insulators related to horizontal (P_w and T) and vertical (V_w and V_i) conductor loading. Figure 25.29 shows a wire clearance diagram;

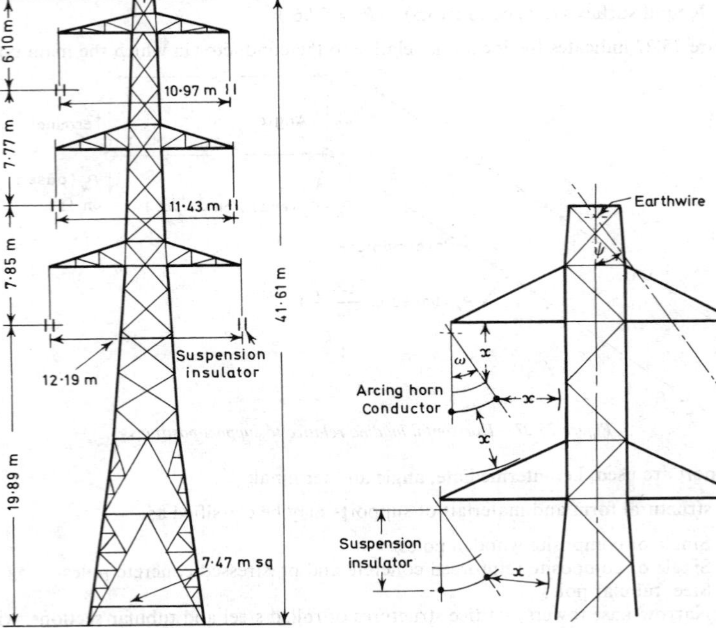

Figure 25.28 Broad-base tower

Figure 25.29 Wire clearance diagram

the live-conductor-to-earthed-support clearances are decided according to the transmission voltage.

(iii) Earthwire spacing. Protection against lightning is obtained by shielding the conductors with an overhead earthwire suitably earthed at the structures to intercept and earth direct lightning strokes. The shielding angle ψ is preferred to be not greater than 30°. The earthwire sag should not exceed that of the conductor and the relative spacing is determined by the shielding angle (Figure 25.29).

When the standard span has been decided, the most economic tower to meet the prescribed conditions can be designed. For the intermediate tower, the basis of loading may be:

wind span = greatest wind load = wind load on standard span + 10%
weight span = greatest weight load = weight load from standard span + 100%
maximum length of span = length of standard span + 40%.

Final design is undertaken graphically by means of stress diagrams, usually on the basis

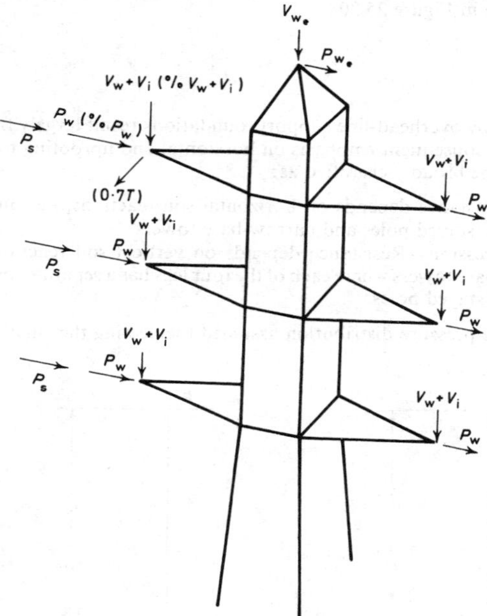

Figure 25.30 Loading diagram for a double-circuit intermediate tower

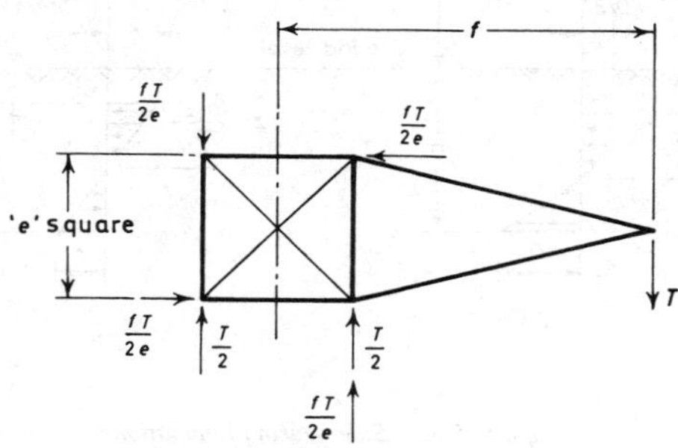

Figure 25.31 Torsion loading

of working loads. The factors of safety (e.g. in Great Britain 2.5 under normal, and 1.5 under broken-wire conditions) are applied when the individual member loads are tabulated. Vertical loads are omitted from the stress diagrams, but at the tabulations are shared equally over the four main legs.

A typical loading diagram shown in Figure 25.30 includes a condition of any one conductor or the earthwire broken (shown in brackets for a top conductor but it must be considered individually at each conductor and earthwire position). Note the proportionate reductions of P_w and V_w for the broken-wire condition and the equal distribution at cross-arm level of the wind load P_s on the tower.

Stability under torsion loads due to broken wires depends on the tower being adequately braced in plan, essentially at cross-arm level but possibly at other levels dependent on particular design features. Figure 25.31 shows the reactions for a tower of square cross section of the type shown in Figure 25.30.

Foundations

The forces to be resisted by overhead-line support foundations result largely from overturning moments, with a consequent emphasis on horizontal and uprooting forces. The types of foundation can be broadly classified as:

(i) Side bearing—Resistance depends on horizontal soil reactions, i.e. single foundations used for unstayed poles and narrow-base towers.
(ii) Uplift and compression—Resistance depends on vertical soil reactions, i.e. in the case of bread-base towers where each of the four legs has a separate foundation, and in the case of stayed poles.

Figure 25.32 shows the pressure distribution assumed (neglecting the small values of

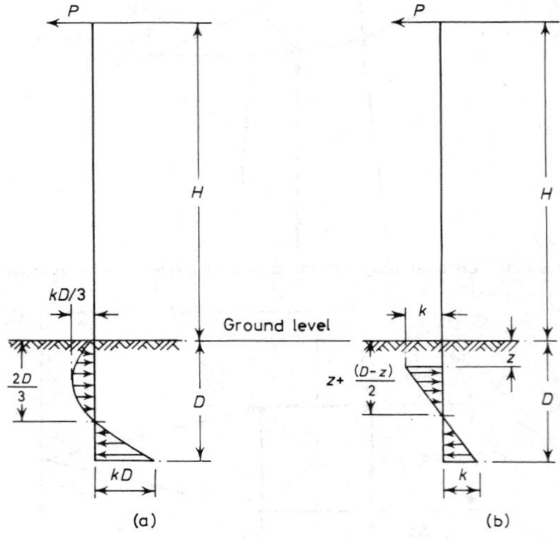

Figure 25.32 Side-bearing foundations

direct horizontal shear) in two formulae used for pole and shallow concrete block side-bearing foundation. In Figure 25.32(a) (parabolic distribution) the pressure developed is based on the horizontal movement relative to the pivotal point and assumes that soil resistance increases proportionately with depth:

$$P\left(H+\frac{2D}{3}\right)=\frac{kbD^3}{12}$$

where k is a constant and b is the breadth of the foundation.

Figure 25.32(b) involves similar assumptions but is based on constant soil resistance:

$$P\left(H+z+\frac{D-z}{2}\right)=\frac{kb(D-z)^2}{6}$$

were z is the amount of topsoil to be neglected, assumed to be at least 300 mm.

Figure 25.33 shows an uplift and compression foundation, each footing consisting of

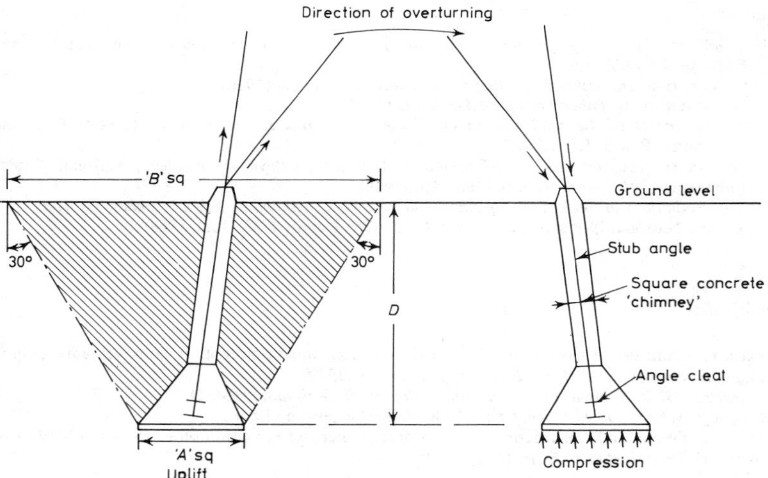

Figure 25.33 Uplift and compression foundations

a shallow concrete pad surmounted by a truncated pyramid and chimney enclosing the stub angle. With an overturning moment, one pair of foundation blocks will tend to be uprooted and the other pair forced downwards. With intermediate towers the loading is largely due to wind and is reversible so that all four footings are identical. Ultimate uplift resistance is calculated on an assumed frustrum of earth above the foundation block.

REFERENCES

Turbo-generator Support Structures

1. *Supporting structures for rotary machines* (especially pier foundations for steam turbines), German Standard DIN 4024 (English translation available)
2. FITZHERBERT, W. A. and BARNETT, J. H., 'Causes of movement in reinforced concrete turbo-blocks and developments in turbo-block design and construction', *Proc. Instn civ. Engrs* (Feb. 1967)

Cooling-Water Systems

3. CHAPMAN, E. K. J., GIBB, F. R. and PUGH, C. E., 'Cooling water intakes at Wylfa power station', *Proc. Instn civ. Engrs* (Feb. 1969)

Natural Draught Cooling Towers

4. MARTIN, D. W. and SCRIVEN, W. E., 'The calculation of membrane stresses in hyperbolic cooling towers', *Proc. Instn civ. Engrs* (Aug. 1961)
5. ALBASINY, E. L., 'Equilibrium stresses in hyperbolic cooling towers', *Int. Symp. on Shell Structures in Engineering Practice, Budapest* (1965)
6. LOVELAND, A., *Metric cooling tower design programme,* Central Electricity Generating Board, MS/C/P3 (Aug. 1970)
7. MERRYLEES, S. H. and WONG, Y. C., *Bending analysis of cooling towers,* Central Electricity Generating Board, RD/C/H247 (Apr. 1969)
8. *Report of the committee of inquiry into collapse of cooling towers at Ferrybridge,* Central Electricity Generating Board (1966)

Chimneys

9. *Specification for the design and construction of reinforced concrete chimneys,* American Concrete Institute, ACI 307–69
10. TAYLOR, C. P. and TURNER, L., *Reinforced concrete chimneys* (1960)
11. FABER and MEAD, *Faber's reinforced concrete* (1961)
12. *Measurements of the wind loading on Fawley generating station chimney,* Central Electricity Generating Board, RD/L/N6/72
13. *Wind-tunnel tests on chimneys of circular section at high Reynold's numbers,* National Physical Laboratory, NPL Aero Report 1266 (Apr. 1968)
14. *The oscillations of model circular stacks due to vortex shedding at Reynold's numbers from* 10^5 *to* 3×10^6, National Physical Laboratory, NPL Aero Report 1267 (June 1968)

Bibliograph:

BAKER, L. H. and MCPHERSON, D. H., 'Cockenzie and Longannet power stations: novel features in the design and construction', *Proc. Instn civ. Engrs* (Aug. 1970)

Central Electricity Generating Board Annual Report and Accounts, HMSO

Modern Power Station Practice, Central Electricity Generating Board

RAE, F. A., 'Design and construction of a reinforced concrete foundation block for a 200 MW turbo-generator', *Proc. Instn civ. Engrs* (May 1962)

26 WATER SUPPLIES

WATER SUPPLIES 26

26 WATER SUPPLIES

R. C. S. WALTERS, B.Sc., F.I.C.E., P.P.I.W.E., F.G.S., and
B. H. ROFE, M.A.(Cantab), F.I.C.E., F.I.W.E., F.G.S.
Rofe, Kennard and Lapworth

MANAGEMENT AND REORGANISATION

The management of water resources in England and Wales up to April 1974 was the responsibility of three groups:

(a) The River Authorities
(b) The Statutory Water Undertakers
(c) The Sewerage and Sewage Treatment Authorities

Early nineteenth century legislation,[1] including the Waterworks Clauses Act of 1847 and the Public Health Acts of 1848, 1863, 1875 and 1878, was largely superseded in 1945 by the Water Act of the Ministry of Health and a Supplementary Act in 1948. The twenty-nine River Authorities in England and Wales were formed under the Water Resources Act 1963 and at the same time the Water Resources Board was formed with the duty of advising the River Authorities in respect of their new responsibilities in connection with water resources.

The Statutory Water Undertakers comprise individual local Authorities, joint Boards and Water Companies numbering approximately 200. They were responsible for the maintenance of a reliable supply of potable water and distributing the same. Sewerage and sewage disposal were wholly local government services and in most of the country they were administered by County Borough or District Councils. In a few areas a Joint Board, or Main Drainage Authority, was responsible for sewage disposal and trunk sewers, whilst the Constituent Authorities were responsible for the local sewers. There were over 1200 sewerage authorities at the present time.

In a White paper issued in December 1971 by the Department of the Environment, the Government set out its proposals for the reorganisation of water supplies in England and Wales and, after consultation which resulted in some modifications, these proposals were embodied in the Water Act 1973. The act provides for the management of all the services required to implement water resources to come under the jurisdiction of ten Regional Water Authorities. These new authorities take over the functions of, and supersede, the River Authorities, Statutory Water Undertakers and Sewerage and Sewage Treatment Authorities; except that the Water Companies are retained to supply water on an agency basis. Sewerage will generally remain the responsibility of the new District Councils, who may also continue to manage sewage treatment and disposal works as an agency to the Water Authority. In many cases the original Water Undertaking area is to be retained as an operational unit for water supply, but elsewhere new units are being formed to manage the operation of both the clean and dirty water ends of the hydrological cycle.

PRESENT CONSUMPTION AND FUTURE ESTIMATED DEMAND

As a requirement of the Water Resources Act the River Authorities must produce a detailed summary of the consumption in their areas during each water year. For the water year ended 30th September, 1970, from the figures quoted in the eighth Annual Report of the Water Resources Board,[3] the total quantity of water abstracted for public water supply was just over 5000 million cubic metres (13.9 million cubic metres/day or 3060 million gallons/day). A further 11 500 million cubic metres was abstracted by the Central Electricity Generating Board and other industry but the majority of this was used for cooling water and returned to source.

In Great Britain the average daily consumption of potable water per head of population varies considerably but the figures given below in Table 26.1 are based on returns in 1967 and are quoted from the Manual of British Water Engineering Practice 1969.

Table 26.1 AVERAGE DOMESTIC CONSUMPTION PER HEAD IN GREAT BRITAIN

Area	*Average*		*Range*	
	gallons	*litres*	*gallons/head/day*	*litres/head/day*
England and Wales	35	159	93 to 16	423 to 73
Scotland	57	258	115 to 29	523 to 132
N. Ireland	44	200	73 to 8	332 to 36

In addition to the differences in average daily consumption during the year there are seasonal and maximum diurnal and hourly consumptions which should be considered for the purposes of detailed design and new works.[4] The maximum daily rate of domestic consumption is generally considered to be $1\frac{3}{4}$ times the mean, and the maximum hourly rate 3 times the mean. Waste is included in the average figure.

In waterworks practice, industrial requirements are often referred to as measured or metered water. In 1969 the average metered consumption in England and Wales was equivalent to 20 gallons per head per day (approximately 90 litres per head per day) but this can of course vary in local areas with heavy industrial demand and low population to as much as 150 gallons (or 700 litres) per head per day and should always be assessed on an individual basis.

Domestic consumption (details)

A typical breakdown of the present consumption and an estimate of future demand was made for a group of six selected Water Undertakings in south-east England by Sharp in 1967.[5] This is shown in Table 26.2.

Table 26.2 BREAKDOWN OF DOMESTIC CONSUMPTION (1967)

Component	*Estimated 1967 average consumption*		*Forecast of possible average consumption in 2000*	
	gallons/head/day	*litres/head/day*	*gallons/head/day*	*litres/head/day*
Drinking and cooking	1	4.5	1	4.5
Dishwashing and cleaning	3	13.5	4	18
Laundry	3	13.5	5	22.5
Personal washing and bathing	10	45.5	13	59
Closet flushing and garbage disposal	11	50	14	63.5
Car washing	—	—	1	4.5
Garden use and recreation	1	4.5	6	27.5
Waste in distribution	5	22.5	8	36.5
Total	34	154	52	236

Industrial consumption (details)

No generalisation can be made as the industrial consumption in each town varies considerably according to the nature of the industry both in quantity and quality requirements. The water is generally required during the working day and this factor must be taken into account in the design of pumps, pipes and reservoirs as it affects the peak rates of flow. As general guidance the following examples are typical:[6, 7]
For brewing, the quantity of water is substantially the amount brewed, but for beer the water is preferably hard; for stout, soft; cider must be made from pure soft water without iron.

Canning is best done with hard water (except for peas), and iron must be less than 0.5 parts per million (ppm) or milligram per litre (mg/litre); anything between two and four gallons of water are required per lb canned (20 to 40 litres per kilogram).

The dyeing industry requires soft, iron-free water, and about 10 000 gallons are required per 1000 lb processed (100 litres per kilogram); mercerising textiles takes 25 000 gallons per 1000 lb (250 litres per kilogram).

Industries such as distilling, ice-making and mineral water making require large amounts of water, plus that for power purposes in steam raising.

Leather requires 8 gallons/lb (80 litres/kg) of raw hide tanned, water rich in sulphates being preferred. Rubber requires 70 litres per kg processed.

Paper or cardboard manufacture requires anything between 13 000 and 80 000 gallons/ton (or 60 to 360 litres/kg).

A ton of soap requires about 500 gallons (2200 litres) of water in its manufacture.

In this country sugar beet takes about one gallon per two lb of sugar (5 litres/kg) used in washing the beet, dissolving the sugar, transporting the material in the factory and in steam raising.

In the heavier industries the following quantities may be taken as approximate, e.g. railways take about 0.1 gallon of water per ton of goods carried per mile (0.22 litres per 1000 kg per km). Cement takes 750 gallons of water per ton (3 litres per kg). Coke might consume 3000 or 4000 gallons of crude water per ton for cooling (13 to 18 litres per kg). Electricity works take 15 gallons (or 67 litres) per kilowatt generated, of crude water, for make up or loss in cooling towers, and $\frac{1}{3}$ gallon (1.5 litres) per kilowatt of fresh water for boilers. Steelworks would consume some 2000 gallons of mostly crude water per ton of steel manufactured (or 9 litres/kg).

The cost of industrial water or metered supplies varies considerably but is generally in the range 10p to 50p per 1000 gallons (4500 litres).

Agricultural requirements

In addition to the human population, allowance must be made in a dairy farming district for the cow population. A cow requires as much as 30–40 gallons (135 to 180 litres) per day and there may be special requirements such as for bottling—(100 gallons (450 litres) of milk means 200 gallons (900 litres) of water); manufacturing 1000 lb (455 kg) dried milk needs 120 gallons (550 litres), 1000 lb (455 kg) alum needs 1000 gallons (4500 litres), 1000 lb (455 kg) cheese needs 170 gallons (750 litres). Where intensive fruit farming is practised a complete network of pipes is required throughout an orchard for treatment and irrigation. Similarly, considerable quantities of water are required to maintain bowling greens, golf courses, race courses and overhead irrigation of crops by rotary sprinklers is on the increase. In the Thames Valley a total of 20 mg/day (90 Ml) has been estimated as the requirement for irrigation for the year 2000. Very high consumptions of the order of 5000 to 10 000 gallons/acre (50 000 to 100 000 l/ha) per day would be required by a market gardener, and for tomatoes under glass approximately 20 000 gallons/acre (200 000 l/ha) per day. Again the watercress industry at certain times of the year consumes very large quantities of water and may require between 0.3

and 0.5 million gallons/acre (2.5 and 5.7 million l/ha) per day. Paradoxical as it may seem, more water may be required in winter to keep watercress from freezing than in summer to keep it from scorching.[8]

Fire protection

Generally, hydrants are spaced not more than 130 m (150 yd) apart. Important buildings may require additional protection, i.e. more than two hydrants within 90 m (100 yd). For less important buildings one hydrant within 140 m (150 yd) may suffice. Hydrants should be 6 m (20 ft) or more away from buildings, are best placed at crossings or corners, and are usually fixed on short 80 mm (3 in) branches from the main which should not be less than 100 mm (4 in). Fire mains should deliver 120 gallons/min (550 litres/min) at each hydrant expected to be in use at the same time (generally two). As pressure in a main to command the highest buildings is generally impracticable, fire engines are used to deliver 260 to 450 gallons/min (1200 to 2000 litres/min) to a height of 50 m (160 ft) through a 24 mm ($\frac{15}{16}$ in) dia. nozzle; the larger fire engines deliver up to 1000 gal/min (4500 litres/min). A residual pressure of 3 m (10 ft) at the ground is desirable to avoid the engine creating a vacuum in the main on the suction side.

In towns the calculation for the distribution of water is based on very general assumptions of the amount of water required at any given moment, and it may not be practicable to design adequately for fire protection if the mains are assumed to be taking the maximum hour's domestic and industrial requirements as well. A good practical arrangement of valves and hydrants based on experience and checked occasionally by some such method as that of Hardy Cross is of more value than any very exact calculations. Nowadays the fire authorities work closely with the water authorities to determine the positions of fire hydrants.

Waste

Some consumption of water by waste is inevitable and few Statutory Undertakers can seriously claim a figure of less than 10%, whilst in some areas where pressures are higher or the mains and services are old or in poor condition, or where efficient waste prevention methods are not applied, the wastage may amount to as much as 50% or more. Waste may be due to a number of factors including: leakage from reservoirs, mains and other works of an Undertaking, and from consumers' pipes and fittings through apertures, fractures, defective joints; or faulty washers and valve seatings; bad design, failure to turn off taps; and in all cases leakage and waste are intensified by unduly high pressures. Waste can be detected by detailed examination of the distribution system or house-to-house inspection, apart from a detailed check on the main reservoirs and aqueducts, etc.

EXAMINATION OF THE SYSTEM

It is best to examine the water system section by section between midnight and 5.00 a.m. and check the night flow by a meter capable of reading small flows and recording them on a chart. A specific test on a 12 mm ($\frac{1}{2}$ in) lead pipe under 3.2 kgf/cm^2 pressure gave a loss of 46 000 litres per day for a 0.6 cm hole, 17 000 litres per day for a 0.3 cm hole and 1600 litres per day for 0.15 cm hole. Tests on newly laid mains often call for a loss not exceeding 1 litre per day per centimetre of diameter per kilometre of length. House-to-house inspections are probably in most cases the most effective way of checking waste. A dripping tap wastes up to 500 litres per day and one running full as much as 10 000 litres per day. The provision by the Water Authority of facilities for the rewashering, renewal and adjustment of taps, and repairs to service pipes at the lowest possible cost, undoubtedly encourages consumers to report leakages promptly and is an overall economy.

TRANSMISSION AND DISTRIBUTION OF WATER

Water may be transmitted under gravity along open or covered channels, through tunnels or through pipes. Open channels are often used for catchwaters, waste-water channels or for river intakes to pumping stations in pumped storage schemes. Nowadays they are not generally used for the transmission of treated water due to the danger of pollution. In some cases canals are adapted as aqueducts for the transmission of water. Some large aqueducts have been constructed with sections of covered channel constructed by 'cut and cover' methods and modern practice is to construct these of plain or reinforced concrete. Where an aqueduct is required to pass through ground appreciably higher than the hydraulic gradient, tunnelling is necessary. The general principles of tunnelling are described elsewhere but for waterworks purposes the tunnels are usually lined, even in rock, partly to ensure that a fall does not block the waterway but also to reduce the friction. The use of pressure tunnels through the centre of a congested city with modern tunnelling methods is now becoming an economically satisfactory alternative to large trunk mains laid near the surface.

Pipes

The major part of water transmission is carried out through pipes and there has been a considerable increase in the numbers of new types of pipes and joints of all sizes in the last few years; lists of those available and approved for loan sanction are available from the Department of the Environment and include spun and cast iron, ductile iron, steel, concrete, asbestos cement, and their range of joints. The lists also include small sizes only, unplasticised PVC (type 1140) and polythene (type 425 and 71 plasticised) pipes with their corresponding joints. Several technical factors affect the final choice of pipe material, including internal pressures, hydraulic and operating conditions, maximum permissible diameters, external and internal corrosion, and any special conditions of laying.

Joints may be classified into three categories, depending upon their capacity for movement, namely rigid, semirigid, and flexible. Rigid joints are those which admit no movement at all and comprise flanged, welded and the now obsolete turned-and-bored joints. The semi-rigid joint is represented by the spigot-and-socket caulked lead joint which has given service for well over a century but is now largely obsolescent. Flexible joints are used where rigidity is undesirable and comprise mainly mechanical and rubber ring joints which permit some degree of deflection at each joint. Amongst the joints included in this category are the Tyton joint for cast iron, spun iron and ductile iron pipes, the Johnson coupling and Fastite Joint for steel pipes and the lock joint for prestressed concrete pipes and the detachable and Widnes joints for asbestos cement pipes. Victaulic joints are frequently used where longitudinal tension is required.

CAST IRON PIPES (GREY IRON AND DUCTILE IRON)

The use of vertically cast iron pipes is now limited to the flanged pipes employed in connection to reservoirs, pumps and treatment plant, the bulk of the iron pipes in waterworks service being spun iron, centrifugally cast in metal or sand moulds. Such pipes may be of grey iron or ductile iron, the latter having the advantage of higher tensile strength and reduced tendency to fracture.

The production of spun iron pipes to imperial sizes has largely ceased and they are now generally available in metric sizes from 80 mm upwards.

Grey iron pipes and fittings of sizes 80–700 mm are covered by BS 4622 and ductile iron pipes and fittings of sizes 80–1200 mm by BS 4772. The former classes B, C, and D for maximum working pressures of 200, 300 and 400 feet head of water, have been replaced, in BS 4622, by classes 1, 2 and 3 which represent (for spun iron pipes with socket and spigot joints) recommended maximum working pressures, inclusive of surge, of 10, 12.5 and 16 bar; maximum working pressures for flanged pipes and fittings in BS 4622 are lower than those for spun iron pipes and where necessary the Standard advises the use of ductile iron or strengthened grey iron fittings. Pressure ratings for ductile iron pipes (class K9) and fittings (class K12) vary with size: 40 bar up to 300 mm, 25 bar for 350–600 mm and 16 bar for 700–1200 mm (BS Code of Practice 2010, Part 3).

The standard length of spun iron socket and spigot pipes to BS 4622 is 5.5 metres, and available joints include Tyton (the most widely used) and mechanical flexible joints of bolted-gland type. The standard length of flanged pipes is 4 m.

The range of pipes likely to be available in future differs in some respects from that of BS 4622; it is believed that the largest British manufacturer's grey iron pipes will cover 80 mm to 600 mm sizes and classes 1 and 3 only (these correspond roughly to the former classes B and C) and ductile iron pipes will be available in the sizes 80 mm to 1200 mm.

Protective coatings, include bitumen sheathing, and the application of centrifugally applied concrete or bitumen lining can be provided where conditions warrant these additional safeguards. Where aggressive soil conditions exist, the pipe may be protected by a tubular polythene sleeve.[9]

(*Note.* 1 = 14.5 lb/in^2 (1.02 kg/cm^2) = 33.4 ft (10.2 m) head of water.)

STEEL PIPES

BS 534 covers the manufacture of steel spigot and socket pipes and specials. Manufacturers of steel pipes are generally able to manufacture special pipes of any reasonable size, thickness or shape to suit customers' requirements. Pipes vary in size from 50 mm up to 1800 mm with wall thicknesses varying from 2.5 mm to approximately 20 mm. They may be jointed by welding with internal sleeve welds only, or internal sleeve welds and external sleeve welds to facilitate testing or butt welds. Alternatively if greater flexibility is required plain ended pipes are used in conjunction with Johnson couplings. The pipes may be protected with bitumen, concrete, or a sheathing of bitumen wrapped in hessian plus bitumen or coated in bitumen plus asbestos sometimes reinforced with woven glass. Where the surrounding ground water is aggressive and the soil has a resistivity of less than 5000 Ω/cm^3 then cathodic protection is required which may be provided either by sacrificial anodes, or by the imposition of a protection current from a direct current source such as an accumulator or transformer rectifier unit.

ASBESTOS PIPES

Asbestos cement pipe[10] is made of a mixture of asbestos and portland cement to form a laminated material of great strength and density. The material is less subject to encrustation in soft water districts and is not affected by electrolytic action. Flexible joints are used exclusively throughout the range of sizes up to 900 mm in diameter for working pressures up to 90 to 122 m head according to size. Special bends, tees and adaptors are not made in asbestos and those of cast iron are generally used for connections to asbestos pipes. BS 486 applies to A-C pressure pipes.

CONCRETE PIPES

Standard concrete pipes of plain or reinforced concrete are made up to a diameter of about 2300 mm and are chiefly used to convey liquids not under pressure. Sizes from

150 mm to 1800 mm are covered by BS 556. The joints are generally of the flexible type such as the Stanton–Cornelius. Prestressed concrete pipes can now be manufactured over a wide range of sizes, varying in diameter from 635 mm to 1800 mm (BS 4625 covers sizes from 400 mm to 1800 mm), and usually having a thin steel shell with a spun concrete interior lining stressed externally by prestressing wire on the outside of the steel shell, the whole then being protected by an outer covering of cement mortar. Working pressures of up to 120 m head of water can easily be obtained. In sizes over 1200 mm, longitudinal prestressing wires are normally employed and the steel cylinder is not used. The lock joints of the simple push-in selfcentering type are completely reliable provided that the manufacturerers' jointing instructions are followed precisely.

ALUMINIUM PIPES

Aluminium pipes[11] are now available up to 700 mm diameter, manufactured by the helical method, but these have not been used to any great extent in water supply. The evidence would so far seem to indicate that this is a material which might be more widely used provided that suitable precautions are taken to protect the material similar to those adopted for steel. The main advantage is in the reduction in weight particularly for overground purposes.

PVC AND FIBREGLASS WRAPPED PIPES

PVC pipes are light and easy to handle, corrosion resistant, and are generally available in sizes up to 400 mm in lengths of approximately 9 m. Larger pipes for waterworks purposes may require to be strengthened and this can be achieved by the use of glass fibre reinforcement. The joints are usually made by a push-on type of rubber ring joint or by a solvent welded joint, the latter only being practical where site conditions permit. It has also proved possible to mole plough long lengths of up to 200 m of this pipe up to a diameter of 300 mm under ground without surface trenching. It should be noted that the coefficient of expansion of PVC is eight times greater than that of steel and considerable movement can take place in long lengths of rigidly jointed pipelines.

In considering the design of the pipeline the external loads generally arise from the weight of the pipe and its contents, the trench filling, superimposed loads including impact from traffic, and from subsidence. The design of pipelines and the strength of the pipes required has been considered empirically and the design method commonly used is that proposed by Marston and Spangler and described by Young and Smith.[12] When a pipeline has to be laid above ground over some obstruction it may either be carried on a pipe bridge or be designed as a selfsupporting arch. Special design and fabrication are necessary in these cases.

Flow in pipes

STREAMLINE FLOW

Osborne Reynolds found by experiment that the average velocity below which streamline flow could be maintained for various diameters of pipes is approximately give by the equation

$$\text{diameter (in)} \times \text{mean velocity (ft/s)} = \tfrac{1}{2} \text{ or less (Imperial units)} \quad (1)$$

or

$$\text{diameter (cm)} \times \text{mean velocity (cm/s)} = \tfrac{1}{24} \text{ or less (metric units)}$$

In a pipe of 12 in diameter, for example, there must be a mean velocity of under 1/24 ft/s, or 720 gallons/h, for maintaining streamline flow; in practice, however, it

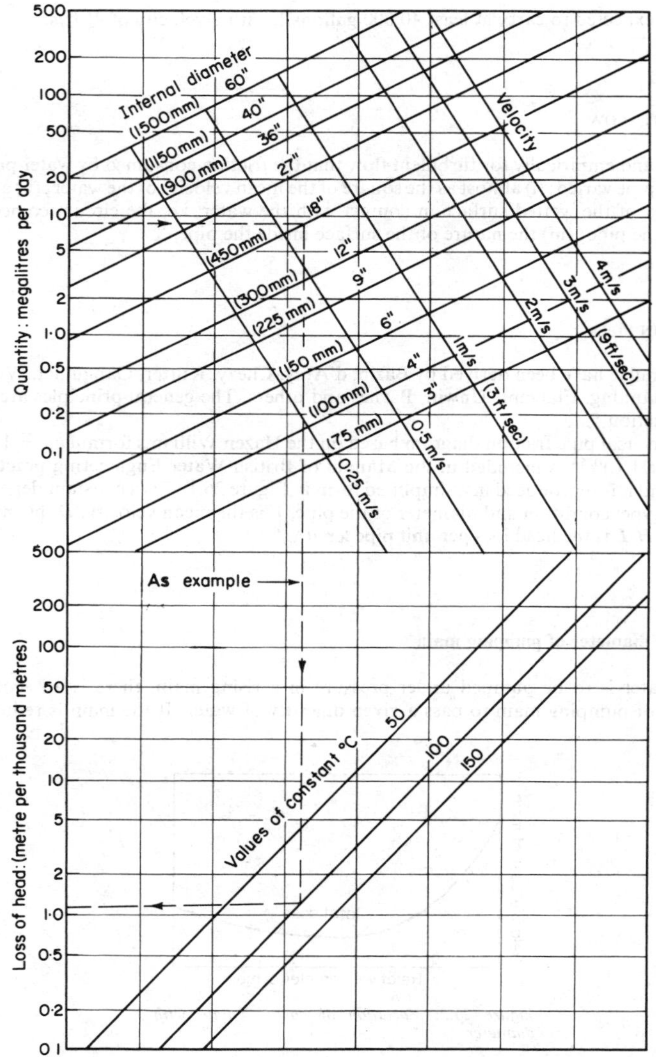

Figure 26.1. Pipe-friction diagram.

Example of use

450 mm pipe is required to carry 9Ml/day. What would be the velocity and head loss? On the top half of the diagram, read across from rate of flow to size-velocity read off diagonal line giving 0.6 m/sec. Strike down vertically to appropriate e value (assume 100). Read across horizontally to left scale

$$\therefore \text{Head loss} = 1.2 \text{ m}/1000 \text{ m}.$$

Multiply by lengths of pipe to get total friction loss.

would be expected to carry at least 40 000 gallons/h with a velocity of $2\frac{1}{4}$ ft/s.

TURBULENT FLOW

Froude found empirically for turbulent flow that the friction consumed by water passing through a pipe varied: (i) almost as the square of the mean velocity of the water; (ii) almost as the area of the wetted surface in contact with the water, i.e. the circumference and length of the pipe; (iii) the nature of the surface inside the pipe.

FRICTION IN PIPES

Basic formulae have been derived by Bazin, d'Arcy, Chezy, Kutter, Ganguillet, Wisbert, Hazen, Manning, Flamant, Unwin, Barnes and others. The general principles are dealt with in Section 5.

The universal pipe friction diagram based on the Hazen Williams formula $v = 1.318C (D/4)^{0.63} (H/L)^{0.54}$ is included in the Manual of British Water Engineering practice as chart D and it is reproduced in a simplified form in Figure 26.1. C is a constant depending upon the type, condition and diameter of the pipe, V is the mean velocity, D the internal diameter, H/L is the head loss per unit pipe length.

Economic diameter of pumping main

Where water is to be pumped under pressure in a rising main, there is an economic diameter of pumping main to pass a given quantity of water. If the main is reduced in

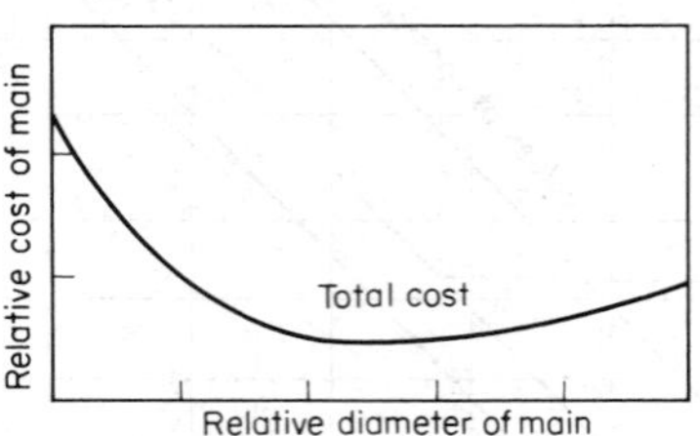

Figure 26.2. Variation of cost of main with diameter.

diameter the cost of the main will be less but the friction will be increased and the cost of pumping allowing also for larger machinery required will be more. The converse also applies. Figure 26.2 shows the combined annual cost of the sinking fund taken at 35 years on the main together with 15 years on the machinery and pumping at a given unit rate of electricity. The curve is particularly instructive in showing that it is generally more economical to err by choosing too large a diameter than too small a diameter as the left-hand side of the curve rises more steeply than the right-hand side. Lea[13] has proved mathematically that the economic diameter lies between 0.535 and 0.675 times

$Q^{\frac{1}{2}}$ (Q being the quantity pumped in ft^3/s). Hence the velocity for economic pumping can be deduced as lying between 2.75 and 4.5 ft/s (0.8 to 1.4 m/s).

Valves

STANDARD GATE OR SLUICE VALVE TO BS 1218

These valves, of $1\frac{1}{2}$ to 12 in (37 to 300 mm) diameter corresponding to cast iron pipes, are to be had for class 1, working pressure 300 ft (90 m) and class 2 for a working pressure of 400 ft (120 m). Valves up to 48 in diameter are available. Valves are usually flanged to enable them to be removed, repaired and re-inserted without disturbing the rest of the pipe lines. The nonrising spindle type has the screw totally enclosed within a casing, is operated by a key, and usually opens by turning anticlockwise although this should be checked by looking at the arrow on the casing. For special purposes there are valves with spigots and sockets, double spigots, Victaulic joints, hand wheels, exposed screw rising spindle types, and anticlockwise opening or a combination of any of these.

Some sluice valves of the larger sizes, such as those situated in valve towers of impounding reservoirs, are often geared to facilitate operation by one man. Sluice valves may be provided with indicators to measure the amount of opening for both rising and nonrising spindles. There are locking devices. The larger sizes are also often provided with small bypasses to relieve pressures on opposite sides of the gate and the sizes of these bypasses are a matter of calculation, but the values in Table 26.3 may be taken as typical.

Table 26.3 DIAMETER OF VALVES AND BYPASSES

Main valve diameter		*Bypass diameter*	
in	mm	in	mm
up to 8	up to 200	$\frac{1}{2}$–$\frac{3}{4}$	10–20
9–12	225–300	1–$1\frac{1}{4}$	25–30
14–21	350–525	2–3	50–75
22–36	550–900	3–4	75–100
36–48	900–1 200	4–6	100–150

BUTTERFLY VALVES

Butterfly valves in accordance with BS 3952 are now extensively used as they are easier to operate than gate valves, smaller in size and generally cheaper. They should not be operated at water velocities of over 16 ft/s (5 m/s) where rubber seatings are included.

REFLUX VALVES

Reflux valves are also known as nonreturn, recoil, retaining, foot and flap valves. These are made up to 60 in (150 cm) diameter or more. There are many types, single door, multiple door, horizontal, vertical and tilting discs.

AIR VALVES

Air valves are put at the highest points of mains and also on flat gradients of under 1 in 500 where the distances are more than half a mile. Those having large orifices let out air in large mains when being filled and those with small orifices are used for letting out air as it accumulates in coming out of the water. The double air valve has one large and one small orifice, a vulcanite ball being used for the large orifice and a rubber ball for the small, $\frac{3}{32}$-in orifice. This is the type most generally in use but there are other modifications, e.g. that with and without an isolating valve to enable the balls to be inspected without emptying the main, or a refined type—kinetic—which prevents the balls slamming shut through a sudden rush of air and water. In this type the air is bypassed around the ball. For large mains a double air valve (isolating, kinetic type) is the most likely to be adopted, as it is capable of inspection and cleaning, which should be done at yearly intervals, and is not liable to slam shut when the mains are filled or refilled.

HYDRANTS

A hydrant consists usually of: (a) an 80 mm (3 in) branch from the main with a duck foot on which rests the screw-down hydrant and stand pipe; (b) a screw-down hydrant and standpipe placed on the main itself; (c) an 80 mm (3 in) pipe and hose attachment.

WASHOUTS

Washouts are usually branches of (say) 80 to 150 mm (3 to 6 in) diameter, leading from the main to a ditch or river with an ordinary sluice valve control. Special branch tees having the invert of the branch coincident with that of the main are made to enable any sediment to be washed out of the main. Flap valves should be put on the ends of the branches.

VALVES FOR SPECIAL PURPOSES

The pressure-reducing valve The ordinary sluice valve, half closed or throttled, is often used for reducing the pressure in a pipeline, but should not be, as special valves of various types are made for the purpose. The common form of pressure-reducing valve provides a constant pressure downstream at less than that upstream; the downstream water presses against a piston loaded with weights, and as the downstream pressure rises it forces the loaded piston upwards, and, through a system of levers, closes the valve. Modifications in this type of valve enable (a) the downstream pressure to vary with the rate of flow through the valve, or (b) a reduction in the head through the valve by a constant amount.

The pressure-retaining valve This type is sometimes called a sustaining valve and is an adaptation of the pressure-reducing valve. It enables a variable inlet pressure to be converted to a constant pressure or to prevent the upstream pressure from falling.

The pressure relief valve This type is also sometimes called a sustaining valve and is an adaptation of the pressure-reducing valve. It may be a loaded spring affair or an adaptation of the pressure-retaining valve by which, for important installations, the times of opening can be regulated.

Flow control valves Such valves are intended for controlling constant flows in pipelines irrespective of pressure, and modifications of them provide for dividing flows into two, or introducing other flows to make up the quantity. Some flow valves have balanced discs electrically or hydraulically operated; other forms, often of the needle type, are designed to close hydraulically in the event of a burst main, or to open and close hydraulically or electrically at stated hours, as in pumping stations.

Cost

The cost of water mains laid complete is of the order of 65 p per cm diameter per metre (£1.50 per inch of diameter per lineal yard) with 10% to be added for valves and fittings (1972), including excavation (cover 4 ft) and backfilling, but excluding road restoration.

Surge

High surge pressures can be caused in pipelines by a sudden alteration in the velocity of flow, such as can occur if a pump stops or a valve is closed suddenly. This situation, commonly called 'water hammer' has been reviewed by Lapworth[14] and a method of graphical analysis has been described by Lupton.[15] It is important to restrict and control the pressure surge as otherwise considerable damage can be caused to pumps and pipelines.

Surge can be controlled in a number of ways including: provision of air-vessels with bypasses, relief valves, surge towers, mechanical surge suppressors, air entry points or increasing the flywheel capacity or inertia of pumps.

The velocity a of a pressure wave of water in ft/s may be calculated from a simplified formula (which includes Young's modulus for the material of the pipe, bulk modulus of water (lbf/in^2) and Poisson's ratio):

$$a = 4600/(1+0.01\ D/e)^{\frac{1}{2}} \quad \text{for steel pipes}$$

or

$$a = 4600/(1+0.02\ D/e)^{\frac{1}{2}} \quad \text{for cast-iron pipes}$$

where D is the diameter of pipe (in), e the thickness of metal of pipe (in).

When $D = 9$ in and $e = 0.6$ in,

$$a = 4600/(1+0.02+15)^{\frac{1}{2}} \quad \text{for cast-iron pipe}$$
$$= 4600/1.15^{\frac{1}{2}} = 4100 \text{ ft/s} = \text{velocity of surge}$$

Thus if a flow of 11 000 gallons/h in a pipe 5000 ft in length at an elevation giving a manometric head of 275 ft, the velocity of 1.1 ft/sec could be attained for carrying the pressure-wave of 4100 ft/s.

As a first approximation the resulting surge pressure = ±(velocity of water in the pipe × velocity of the pressure wave) ÷ acceleration due to gravity, and to obtain the pressure in the pipeline this should be added to (or subtracted from) the total manometric head at any point. Thus:

$$\text{Surge pressure} = \pm(4100/32.2) \times 1.1 = \pm 140 \text{ ft head (43 m)}$$
$$\text{Therefore, net surge head} = 275+140 = 415 \text{ ft (127 m)}$$
$$275-140 = 135 \text{ ft (41.5 m)}$$

For detailed design, one of the graphical methods should be used, or the equivalent computer programs now available. It should be remembered that negative, resulting pressures (or vacuum) can often cause as much damage as excessively high pressures.

MEASUREMENT OF WATER IN STREAMS

V-notch

For the measurement of small flows up to 10 000 gallons/h (50 000 litres/h) a plate with a V-notch is frequently used; the formula, in imperial units, where Q is the flow in ft^3/s ($1\ ft^3/s = 28.3$ litres/s) and H is the head in ft (1 ft = 0.3048 m) is:

(i) For a 90° V-notch, $Q = 2.5\ H^{\frac{5}{2}}$

(ii) For a 60° V-notch, $Q = 1.43\ H^{\frac{5}{2}}$.

The head is measured approximately 3 ft (1 m) upstream of the notch.

Rectangular notch

For larger flows in the open, a wide rectangular notch is suitable.

The formula is $Q = 3.33\ LH^{\frac{3}{2}}$ (units as above with L = length in ft). If the depth of water is measured at the notch, then instead of 3 ft (1 m) upstream, approximately $\frac{1}{2}$ in (12 mm) should be added if the velocity of flow is 2 ft/s and approximately $\frac{3}{4}$ in (18 mm) if the velocity is 3 ft/s for rough calculations.

Table 26.4 DISCHARGES FROM 90° V-NOTCH AND RECTANGULAR WEIRS

H mm	90° *V-notch*	*Rectangular*	*H* in	90° *V-notch* gal/hr	*Rectangular*
12.5	0.03	2.5	0.5	22	621
25	0.14	7.2	1	112	1 756
50	0.80	20.6	2	636	4 968
75	2.2	37.8	3	1 752	9 124
102	4.5	58.2	4	3 594	14 051
127	7.9	81.4	5	6 290	19 637
152	12.5	107.0	6	9 922	25 811
178	18.4	134.8	7	14 564	32 528
203	25.7	164.6	8	20 330	39 724
229	34.5	196.0	9	27 294	47 436
254	44.9	230.0	10	35 530	55 523
279	56.7	265.5	11	44 928	64 060
305	70.9	302.0	12	56 160	73 008

Weirs

For still larger flows, e.g. waterfalls in rivers, overflow weirs in reservoirs, special calibration is usual from models representing the same shape and to some extent the roughness.

Using the basic formula, however, $Q = 3.33\ H^{\frac{3}{2}}$, the following quantities and weights of water flowing over a reservoir crest per foot of its length may be taken as a rough guide (see Table 26.4).

CURRENT METER

The flow in a wide river is calculated by measuring the cross sectional area of the river and the velocity by a current meter at several points across the section. A current meter is an instrument provided with a propeller screw which, when immersed, is turned by the velocity of the water, and the number of turns is a measure of that velocity; the mean of all the velocities multiplied by the cross-sectional area of the water is a measure of the quantity flowing. This method is often supplemented by dilution gauging for small rivers with high turbulence. Methods of dilution gauging are defined in BS 3680 Parts 2A and 2C. Further information on flow measurement is given in Sections 20 and 29.

CRUMP WEIR

The Crump weir[16] has a sharp horizontal crest with a 1 : 2 slope on the upstream side and a 1 : 5 slope on the downstream side. Considerable research has been undertaken into the characteristics of this type of weir, and this is summarised in Water Resources Board Publication TN8.[17] It is capable of measuring high discharges and can function in partially drowned conditions which means that the size of the structure can be kept down, thus avoiding objections by amenity and fishery interests.

Compound types with side section separated by piers can be used to improve discharge of low flows, and a 'flat V' type with the crest sloped in at 1 : 10 also provides for this.

Each type should be individually calibrated for reliable results. The results can be read directly on a staff gauge or recorded on a chart or punch tape for later data processing.

MEASUREMENT OF WATER IN PIPES

For measuring the flow of water within a pipe the most useful apparatus is the Venturi meter, named by the American inventor Herschel after an Italian of the eighteenth century who experimented on the flow of water in tapered pipes. As the heads in the pipe and at the throat vary with the velocity through the pipe the quantity passing through the meter is proportional to the square root of the difference of the pressure at the inlet and at the throat. By pressure pipes, connected to the upstream side of the meter and the throat, the rate of flow is recorded either visibly as in a manometer or transferred to a pen and paper chart. This rate of flow can also be integrated instrumentally to show the total quantity. Friction losses are 1 or 2 ft for good design.

SERVICE RESERVOIRS

An important item in the distribution of water is the covered service reservoir, not to be confused with the open impounding or pumped storage reservoir.
Whereas the function of an impounding reservoir is to store crude water for use in dry months or years, the service reservoir stores the drinking water for immediate use.

The covered service reservoir is an integral part of the distribution system and its object is to balance the daily fluctuations of demand on the one hand and the method of delivery from the source, such as pumping, on the other. All service reservoirs must be at the highest possible level necessary to serve the houses which are to be supplied. If there is no natural ground sufficiently high for the purpose the water must be raised by being perpetually pumped (i.e. boosted), or the reservoir must be elevated and so becomes a water tower. At 1972 prices, the cost of pumping is about 1p per 1000 litres per 100 m lift (1.5p per 1000 gallons/100 ft lift).

Covered service reservoirs

All service reservoirs must be covered to keep the clean water which is put into them from being fouled by exposure to the atmosphere, which encourages algal growths (especially with hard water), by dirt from the air or by vermin from the ground. Such

Table 26.5 COVERED SERVICE RESERVOIRS ON LEVEL GROUND

Capacity million gallons (million litres)	$\frac{1}{2}$ (2.25)	1 (4.5)	$1\frac{1}{2}$ (6.75)	2 (9.0)	$2\frac{1}{2}$ (11.25)	3 (13.5)
Depth ft (m)	12 (3.7)	15 (4.6)	18 (5.5)	21 (6.4)	24 (7.3)	27 (8.2)
Side of square ft (m)	82 (25)	104 (32)	115 (35)	123 (38)	129 (39)	133 (41)
Diameter in ft (m)	92 (28)	117 (36)	130 (40)	139 (42)	144 (44)	149 (46)
Cost 1972 (thousand £)	50	90	125	155	180	200

reservoirs are usually built half in and half out of the ground, the material of excavation being used for banking material for the walls and covering the roof. About 12 in of topsoil is usually placed over the whole expanse of roof to improve the appearance and also to keep the concrete of the roof at an even temperature to prevent expansion and cracking. Walls may be of mass concrete, reinforced concrete, brick, masonry, puddled clay: columns may be of plain concrete, reinforced concrete, brick, steel, masonry. Roofs may be of reinforced concrete, brick arched, or asbestos sheeting on mild steel trusses. Floors may be of plain concrete, reinforced concrete, or bricks on puddled clay. Internally reservoirs may be lined with asphalt. Different circumstances and different persons dictate the choice; mass or reinforced concrete walls, reinforced concrete roof and floors, concrete or coated steel columns being the author's usual choice.

The principles of design follow orthodox practice for masonry, concrete, or steel structures. For primary design purposes the data given in Table 26.5 are applicable.

Water towers

The cost of water towers is balanced against the cost of boosting, i.e. the capital cost of the tower and its maintenance against the capital cost of boosting plant and its maintenance. If a reduction of capital expenditure is of paramount importance, boosting may be chosen, but in many cases greater security is felt with an elevated tank.

Towers are of several designs. For a given capacity the globular form, of steel, seen in the USA, is the most economical in the weight of steel. The cylindrical form with 'dished' bottom is next best. The rectangular form conveniently built up of steel plates is often seen in this country, particularly for industrial uses. The majority of towers for waterworks purposes are of reinforced concrete, consisting of cylindrical structures, either on legs or totally enclosed; they are often lined with asphalt, and firms specialising in reinforced concrete are usually employed for their construction.

A form much favoured is the cylindrical steel tank surrounded with a thin external shell of reinforced concrete having a space of 3 or 4 ft between the steel tank and the shell. The shell lends itself to architectural treatment to harmonise with the surroundings, can always be maintained to present a pleasing appearance, and can often be constructed by a local contractor. The steel tank inside the concrete shell can be inspected externally for leakage, and preserved by painting; this is not so easy with the reinforced concrete tank, particularly if it shows signs of leakage.

The largest single towers in Great Britain contain between $\frac{3}{4}$ and 1 million gallons, and the highest are about 30 m (100 ft). Costs vary considerably according to height and location. A 50 000 gallon (225 000 litre) reinforced concrete tower, 12.2 m (40 ft high could cost £30 000; a 30.5 m (100 ft) £50 000 and a 200 000 gallon (900 000 litre) tower 12.2 m (40 ft) £100 000; a 30.5 m (100 ft), £150 000.

Valves and fittings for service reservoirs

The appurtenances of a reservoir consist of suitable inlet, supply and washout valves, overflow arrangements, roof air inlets, a depth indicator and recorder, and an outlet meter and recorder. Flow through the reservoir is desirable, and to facilitate this the inlet is placed at one end and the outlet at the other. Air ventilators are provided to prevent accumulation of gases and accommodate the rise and fall of the water level. The overflow pipe is usually a standpipe with a bellmouth on top. In nine cases out of ten, and particularly where underground water is the source of supply, the service reservoir should rarely (e.g. 5 or 10 years) need cleaning and, on these occasions, as the operation need not take more than a few hours, the supply can usually be maintained by bypassing perhaps through a small tank or at night; partition walls with all the necessary duplication of pipework are only required for the larger sizes where there is no alternative storage available during the period out of service. Some form of telltale or automatic level recorder is necessary, and there are many types, to inform the operator, at the source, of the water level. Although there may be a meter at the source to regulate the quantity of water flowing into the reservoir it is often desirable to have a meter on the outlet side to ascertain the rate of draw-off hour by hour in the day.

THE UNDERGROUND SCHEME

Development of a new source

The origin of the water for an underground water scheme is rain which has fallen on the surface of the ground and has sunk in; to retrieve this water a well is dug in the fissured formation into which the rain has penetrated. If the strata are not sufficiently open to yield enough water an adit or heading may be driven in the bottom of the well. If the strata are well fissured and yield water readily, one or more boreholes are adopted. Whatever the 'hole' in the ground may be, its function is to accommodate a pump of the right dimensions for the economical pumping of water; this pump being low enough to reach the water when the water level in the well is at its lowest. Occasionally, where the water level never falls or is never likely to fall more than 25 ft (7.6 m) below the surface, a surface pump will do; otherwise the pump must be, and generally is, immersed in the water below the ground. To drive the pump there are the alternatives of oil and electricity; oil is seldom used since the trend of modern waterworks practice in Great Britain today is towards the sole use of electricity. For wells and boreholes, pumps are usually of the submersible type on account of the lower cost and the small buildings required for the pumping station. Whatever the pumping machinery may be, however, for waterworks practice it must be absolutely reliable, and it is desirable to have provision for standby equipment and boreholes.

The capacity of the borehole pumps must be determined by the water requirements of the district and the capacity of the underground works. Some pumps may be left to themselves, others may require men in charge. Pumps may have to pump the whole daily supply of twenty-four hours in eight or sixteen hours to fit in with shifts, or even more during weekdays so as to close down on Sundays. Hence the capacity of the pumping machinery, and any treatment works and pumping mains, if designed for eight hours pumping, must be three times as large as those designed for twenty-four hours pumping, although the total daily or annual quantity pumped to the town is the same. The water, after being lifted by the well pumps to the surface, is pumped on either by the same pump or a modification thereof, or by a surface pump, with or without a balancing reservoir.

Automatic pumping machinery for waterworks pumping is increasing in popularity owing to its improved reliability and saving in cost.

The water, where necessary, may be softened, filtered (rare for underground supplies), treated for removal of iron and/or manganese. The pumping main which must be designed in accordance with the pumping rate may be of asbestos, iron, steel or PVC with various coverings. It is best to leave the pumping main free from tappings for house services and take it direct to the covered service reservoir. The service reservoir from which the water is distributed should hold at least three days supply: the distribution mains from the reservoir should be designed to carry a rate of about three times the average consumption to allow for peak demands.

Geology of source[18, 19]

The source of water for a pumping scheme which depends upon obtaining water from an underground source, in England particularly, rather than in Wales or in Scotland, is based on the following geological formations, in order of merit.

CHALK

This includes all three divisions, Upper, Middle, and Lower, with overlying gravels and crags, occurs in Yorkshire, Lincolnshire and East Anglia. Chalk, which is associated with overlying porous gravels, Thanet Sand (and other Lower London tertiaries) is present in the Home Counties (Buckinghamshire, Berkshire, Surrey, Middlesex, Hertfordshire and Essex), London, Kent, Sussex, Hampshire, Isle of Wight, and with underlying Upper Greensand in Wiltshire and Dorset.

BUNTER AND KEUPER SANDSTONES

These are present in South Lancashire, Cheshire, Yorkshire, Nottinghamshire, Staffordshire, Warwickshire and parts of Somerset and Dorset.

OOLITES

Lower Oolites include somewhat arenaceous deposits in East Yorkshire; the Lincolnshire Limestone of Lincolnshire; thin beds in Northamptonshire and Oxfordshire; the Inferior Oolites and Cotteswold Sands of the Cotswold hills of Gloucestershire and Worcestershire; and the somewhat arenaceous limestones of Somerset and Dorset. Upper Oolites are found in Yorkshire, Oxfordshire and Wiltshire.

LOWER GREENSAND

The chief development of the Lower Greensand is all around the foot of the Chalk beneath the Gault of the North and South Downs, which enclose the Weald of Kent and Sussex, and parts of Surrey and Hampshire. Some of the Lower Greensand provides useful soft water at a depth of over 1000 ft in the Slough area. Other divisions include spreads of Greensand at the northern foot of the Chalk in Bedfordshire and Hertfordshire; and in Lincolnshire.

CARBONIFEROUS

Under this general term may be included the Carboniferous Limestone which collects water in large fissures as in the Mendips of Somerset, Derbyshire and Yorkshire; the Millstone Grit and other grits with small fissures of the Yoredales and Coal Measures of Lancashire, Yorkshire and Wales; grits in the Upper Coal Measures for Coventry, and small supplies in the Culm of Devonshire.

PERMIAN AND MAGNESIAN LIMESTONE

This occurs in the north-eastern counties.

ASHDOWN SAND AND TUNBRIDGE WELLS SAND

These are present in the Kent and Sussex Weald.

OLD RED SANDSTONE

Old Red Sandstone is present in South Wales, the Forest of Dean and Herefordshire.

In addition there are water-bearing gravels in proximity to rivers or other so-called water-bearing formations in juxtaposition with many of the above water-bearing strata.

The following may be taken as a rough indication of the extent and quantities of water that are pumped from the main four formations (see Table 26.6):

Table 26.6

	Exposed surface square miles	*Underground extent* square miles	*Quantity pumped* million gallons per day (1960)
Chalk and Upper Greensand	5 000	7 000	350
New Red Sandstone	1 750	1 250	80
Oolites (Lower and Upper)	2 500	1 000	25
Lower Greensand	1 000	5 000	20

1 square mile = 2.6 square km; 1 million gallons/day = 4.55×10^6 litres/day

Selection of source[20]

The natural factors governing the quantity of water obtainable at any one spot are the direction of flow of the water underground, the general geological arrangement of the strata, the area of strata exposed to the skies, the thickness of formation, the extent of the formation underground, the nature of the fissures and formation and porosity of strata, the amount of rainfall, the evaporation and dissolved salts from the strata.

The factors governing the final choice of site are those enumerated above, with conditions of proximity of buildings, proximity to places to be supplied, cost of development and other engineering considerations, elevation, acquisition of site.

The direction of flow of water underground can be ascertained by plotting contours of underground water levels from the records of water levels, referred to Ordnance Datum, in existing wells. Such contours plotted in outline have now been done for the main formations in England and, when 'read' in conjunction with the area of strata exposed to the skies, give an indication of the amount of water available at the selected site.

The arrangement of the strata such as faults and/or rolls, anticlines, synclines, thinning out, and change in the lithology may profoundly modify the potential yield at any one spot.

As to the proportion of rainfall which is collectable in large areas, this should approximate to the amount measurable in a surface percolation gauge. Thus, with an average rainfall of 600 to 900 mm per year on an average, the surface percolation is from 200 to 300 mm per year. Lapworth[21] has suggested that percolation in the Chalk is 0.9 of the average rainfall less 13.5 in (340 mm) per year. The round figure of 10 in (250 mm) per year is often assumed for a working basis and this is equivalent to 400 000 gallons per day per square mile (or 0.7 million litres per day per square kilometre) of gathering ground, the area being determined from the extent shown by the underground contours which may be assumed to flow (or be drawn) to the selected source.

With regard to natural factors affecting adversely the cost of underground water these are mainly (a) hardness where rain acidified by vegetation dissolves calcium and magnesium from limestone; (b) chlorides, which come from contacts with rock salt as in Cheshire, or from rocks containing highly mineralised water from ancient seas or other sources and which have been protected throughout the ages by a clay layer; (c) iron from acidified rain dissolved from the rocks, ground or peat through which it passes as in the ironstone of the Weald.

The waters for new sources should be analysed chemically much more fully than those already in use, and it may be necessary, for example, to look for such a substance as fluorine which affects teeth.

Man-made factors affecting the quality of water are those of proximity of buildings, sewage, and noxious effluents, e.g. gas liquors on the gathering ground, refuse tips, manure, all objectionable in varying degrees according to their distance from, and the potential fissures underground leading to, the well. The site of the well should, of course. be chosen as near as possible to the place where the water is required; every mile of main adds greatly to the cost, e.g. a 12 in (300 mm) main at a pumping rate of 1 million gallons/day (5 million litres/day) costs £20 000 per kilometre. Among other engineering considerations, the highest underground water level when pumping is significant, because for every 100 ft (30.5 m) in elevation saved about 1.5 p per 1000 gallons (4500 litres) is saved, which for 1 million gallons/day might represent a capital sum of £60 000 For further information see references 18–21.)

Construction of boreholes and adits

The diameter of the boreholes for water are much larger, and the depths are much less, than for oil wells. Boreholes under 12 in (300 mm) diameter are seldom made for a permanent source of public water supply in Great Britain. Steel lining for boreholes is standardised to BS 879 with diameters up to 48 in (1200 mm). Bored wells, however, may be made up to 12 ft (3600 mm) diameter under water without pumping. Wells and adits when made by hand are dug in the dry to rest water level, and by pumping below rest water level.

Boreholes are made by rotary drilling or by percussion; drilling enables exact cores to be obtained, whereas the percussion method pounds up the material. Both methods are used according to the conditions of hardness of the strata and any special requirements as to necessity of cores for the identification of strata. In sandy formations, boreholes are sunk by the mud flush system, the mud keeping back the sand as drilling proceeds through the soft sand; this process enables gravel to be inserted outside the perforated tubes, which is a very satisfactory way of keeping back the sand when the site is brought into commission. A recent method of drilling which has been tried is known as the 'reversed' flow system where water is pumped under pressure into the borehole while being drilled. Sites are often developed by two boreholes so that duplicate machinery may be inserted in each. Often a sandstone site may be developed by several boreholes spread over the site with a small pumping unit in each in order not to pump too large a quantity at any one spot and so draw in sand.

Boreholes in the Chalk are risky, for although the site may be geologically a good one the borehole may quite easily miss water-bearing fissures. A geophysical survey of the site to ascertain where there is least electrical resistance to indicate the place most likely to be the most fissured may be useful to save the cost of a well and adit.

Large wells with adits are seldom economical at present costs but can be adopted where there is an existing station requiring large volumes of water. Unwanted water entering a perforated lined borehole may be eliminated at any horizon by grouting by the rubber sleeve manchette process.

The cost of drilling and testing a lined 24 in (600 mm) diameter borehole over 50 m deep with a neighbouring observation hole would be approximately £150 per metre depth.

Water-well casing

The lining of a borehole may be plain for lining out clay or other non-water-bearing material. Usually the top 50 ft (15 m) is lined out to prevent surface contamination. The water-bearing portion of the borehole, depending on the capability of the rocks to stand up and not collapse, is often unlined. As some boreholes, even those in the Chalk which were thought to be safe, have collapsed it is often considered prudent to line a borehole throughout its depth with perforated or slotted tubes in the water-bearing horizons and plain tubes in the clay or unstable sections.

The tubes in the water-bearing portion of the borehole may be perforated with holes at centres or by slots 3 to 12 mm wide, 150 mm or more long spaced at 100 to 150 mm centres. The British Standard casing (BS 879) for water wells includes lap welded and welded steel tubes. The main points of general difference are in the joints which may be screwed and socketed with (a) V thread screwing, (b) square thread screwing or (c) screw flush butt joints with square form thread parallel screwing.

Testing boreholes and wells

The testing of newly developed sources is often complained of as being costly, but it is extremely necessary, as it is the basis upon which any pumping scheme rests. Some newly constructed holes show stationary pumping levels almost within a few minutes of commencement of the test; whereas in others the water levels are not stabilised for some time, even after three days or more. The frequent stopping and starting of pumping often improves the yield of a newly drilled borehole and the rate of rise at the end of the test gives the measure of the inflow into a well.

Yields may be increased in Chalk and other limestones by treating the boreholes with about 5–10 tons of hydrochloric acid. Fissures, clogged by boring, are cleared.

Specific yield of boreholes

For comparing the yield of one borehole with another it is convenient to compare the specific yield, the quantity pumped divided by the difference of level between rest and stabilised pumping level. Typical specific yields range from 75 million litres per hour per metre (poor) to 750 million litres per hour per metre of lowering (good).

Theoretically the yield of wells varies as the logarithm of their diameters, but the mathematical theory of wells cannot be applied, except broadly, since conditions in the fissuring of strata are too variable and uncertain in practice. English geology does not lend itself to any mathematical treatment which presupposes that every cubic metre of strata over many square kilometres extent and many metres depth is of absolute uniform composition throughout. Measuring devices for the testing of wells include, of course, a flow recorder, or method for measuring the quantity, and a depth recorder for measuring the depth of water; both instruments are provided with charts. The motive power for pumping may be electricity if available on the site, or an oil-driven engine brought to the site. The test must be continuous day and night, and the commonly accepted period for a public water scheme is 14 days of 24 hours nonstop work. Frequently, longer periods are necessary where the water level is not stable or additional information about the aquifer is required.

SURFACE WATER SCHEMES

Development of a new source and requirements[22]

A surface water scheme usually includes the construction of a reservoir to store river water at times of high flows for use in times of low flows in order to give a uniform daily rate during a design drought period consisting of consecutive drought years, with a given probability of occurrence—often taken as once in fifty or a hundred years. Such a scheme basically depends on: (a) the *quantity* of rain falling on the gathering ground and (b) the *evaporation* or loss in quantity after it falls on the gathering ground to the dam; (c) the *storage* of the reservoir consistent with cost and its reliable *yield* and (d) the suitability of the geological conditions for safety of the *submerged area* and the *dam site*. Thus, meteorology, engineering and geology are interdependent for ensuring the safety and cost of a surface water scheme.

The main uses of reservoir conservation are for:

(1) Domestic and industrial water requirements
(2) Hydro-electric generation
(3) Irrigation
(4) Regulation of the flow of a river by increasing dry weather flows and by reducing floods
(5) Recreation (fishing, sailing).

The chief types of reservoirs are:

(1) Impounding reservoirs. So-called because the sides of a valley, with the dam, impound the natural flow of the river
(2) Pumped-storage reservoirs are formed by a dam or bund remote from the river from which they are filled only by pumping
(3) Some impounding reservoirs may also be used partly for pumped storage
(4) Both types could be used for river regulation, i.e. regulating the flow of a river to maintain abstractions lower down.

The chief types of modern dams are:

(1) Gravity concrete dams. Triangular in section where the line of pressure passes through the 'middle third' of the dam. So-called 'gravity' because any cross section could stand without overturning
(2) Modifications of concrete dams, i.e. curved gravity, prestressed, reinforced, thick arch, thin arch, double curvature
(3) Buttress, multiple arch concrete dams
(4) Earth dams with various forms of clay cores which are supported by sands, gravels, soft sedimentary and all other soft rocks. (Various types of construction, e.g. hydraulic fill dams, can be included in this category)
(5) Rockfill dams with impervious cores, or faced with asphaltic concrete.

Types (1), (2) and (3) concrete dams are usually adopted on rock foundations, while (4) and (5) earth rock dams are generally the best solution on soft alluvial or sedimentary deposits.

Geology of source

The chief valleys in Britain where surface waters are impounded are on the following geological formations, which are relatively impermeable:[18]

(i) The Ordovician, Silurian and Old Red Sandstone. The grits and shales of the Ordovician and Silurian in Wales present many developed (e.g. Claerwen, Clywedog) and potential sites in Wales (e.g. Llandegfedd) and in the Lake District, the artificially enlarged lakes, Thirlmere, Haweswater, Crummock.
(ii) The Carboniferous Series, Yoredales, and other grits and shales below the Coal Measures are responsible for most of the sites which have been developed in Yorkshire, Lancashire and Cheshire (Scammonden, Erwood, Lamaload).
(iii) Other clay formations, e.g. the Keuper Marl (Chew) and Forest Marble of Somerset (Sutton Bingham), the Lias Clay of the Midlands (Eyebrook and

Draycote), the Ashdown Sand and Weald Clay of Sussex (Weir Wood and Bough Beech).

(iv) The granites of Dartmoor and Cornwall (e.g. Meldon, Siblyback) and the igneous and metamorphic rocks of Scotland, support the pre-stressed Allt na Lairige, the thin double curvature arch Monar and very many buttress dams for hydro-electric power.

The geographical requirements are that the valley should be wide and flat for the site of the reservoir but narrow at the site of the dam, sufficiently elevated to command the town and large enough to provide an adequate yield.

Economics of storage and yield of reservoirs

The storage of a reservoir is related to the annual run-off of the gathering ground.

If the required daily quantity of water is less than the driest weather run-off and can be acquired, there should be no necessity for a reservoir. If the required quantity is more

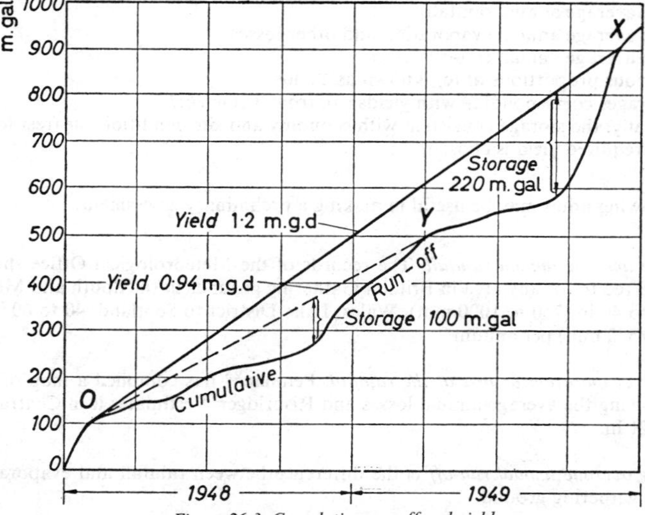

Figure 26.3. Cumulative run-off and yield.

than the daily run-off, the storage is based on a proportion of the annual run-off. Thus if the proportion is under 75% or 80% of the average annual run-off (generally known as the flow of the three driest consecutive years, which has been determined by records of gauging for 35 years), economic reservoirs of reasonable size are assured.

For proportions greater than 80%; the size and cost of the reservoir may be doubled for only a very small increase in yield.

STORAGE AND YIELD FROM RIVER FLOW RECORDS

Where the flow of the river is known over a series of years the storage necessary for different yields can best be calculated from plotting the flows cumulatively as a mass diagram. Figure 26.3 is a typical diagram for a small scheme in which the wavy line OX represents the cumulative run-off from a gathering ground of 1050 acres (425 hectares) for two years. The straight line OX represents the uniform yield of 1.2 million gallons/day (5.5 million litres/day) during 22 months. The vertical distance between the two lines represents the quantity by which the total actual run-off is below the average at any time.

Thus the maximum deficiency during the 22 months is 220 million gallons (100 million litres) which occurs in September 1949 hence the storage required to maintain the guaranteed yield of 1.2 million gallons/day (5.5 million litres/day) is 220 million gallons (1000 million litres).

Similarly, for a smaller yield, i.e. for 0.94 million gallons/day (4.25 million litres/day) represented by the line OY the storage required would be 100 million gallons (450 million litres) based on a period of 14 months.

STORAGE AND YIELD FROM RAINFALL AND EVAPORATION RECORDS

Where the run-off of a stream or river is unknown, the relationship between storage and yield may be assessed for the gathering ground of the dam in the following steps:

(a) The average annual rainfall
(b) The average annual evaporation and other losses
(c) The average annual run-off $(a-b)$
(d) Various proportions of (c), known as Yields
(e) Storages corresponding with yields (d), from Table 26.7.
(f) Finally, the storage consistent with economy and site conditions nearest to giving the required yield is chosen.

The following notes may be useful in making a preliminary assessment.

(a) *The average annual rainfall.* The records of the Meteorological Office should be referred to for any area in Britain. Gathering grounds in the South and Midlands, 30 to 40 in (750 to 1000 mm), Wales, Lake District to Scotland, 40 to 60 in (1000 to 1500 mm) per annum.

(b) *The evaporation or loss of the rainfall.* Penman[23] has compiled a map of Britain showing the average annual losses and Risbridger[24] estimates it in Central Wales at 21 in.

(c) *The average annual run-off* is the difference between rainfall and evaporation on the gathering ground.

(d) *For the yield-storage relation,* reference for a first approximation should be made either to the Deacon diagram[25] or the Lapworth chart,[26] from which Table 26.7, below, gives extracts of the yield–storage relation for various run-offs. (The rainfall less evaporation is also known as the available yield.)

The determination of the size of the reservoir to balance 50% or less of the available yield is difficult as the problem becomes sensitive to the dry weather flow as in the first two examples above and special droughts may 'wreck' the calculations. In connection with the storage and yield problems for rain fall and evaporation data special reference may be made to the Deacon diagram in the thirteenth edition of the *Encyclopaedia Britannica* article on Water Supply and to a paper by Lapworth,[26] but Table 26.7 gives some main connections between storage and yield.

The percentage of run-off taken for the yield and storage depends on the physical conditions of each site and on the daily requirements of each town.

Table 26.7

Average run-off from gathering ground (inches)	*Storages in inches*					
	5	*10*	*15*	*20*	*25*	*30*
			yields in inches			
10	5	7.5	9.0	10.0	—	—
20	10.5	14.0	16.2	18.0	20.0	—
30	15.0	19.7	22.5	24.5	26.5	28.0
40	18.5	24.5	28.5	31.0	33.0	35.0
50	21.0	29.0	33.9	37.0	40.0	42.2
60	23.0	32.3	38.7	43.3	47.0	50.0
70	25.0	35.2	42.9	48.2	53.0	57.0
80	27.0	38.2	46.0	53.0	58.0	63.0

CONVERSIONS: STORAGE: One inch of rainfall on one acre = 22 610 gallons
10 mm of rainfall on one hectare = 100 000 litres
RUN-OFF: One inch per annum on one acre = 62 gallons/day
10 mm per annum on one hectare = 274 litres/day

EXAMPLE: Assume a gathering ground of 1000 acres
Assume average annual rainfall = 70 inches
Assume average annual evaporation = 20 inches
Assume average annual run-off = 50 inches

From Table 26.7, for average annual run-off of 50 in:

Storages of 50-inch run-off	10 inches	20 inches	30 inches
Corresponding yields for 50-inch run-off	29 inches	37 inches	42.2 inches
From conversions:			
Storages			
(gallons/acre)	226 100	452 200	678 000
(gallons/1000 acres)	226×10^6	452×10^6	678×10^6
Yields: m.g.d.	1.8	2.3	2.6
m^3/day	8 000	10 500	11 800

Catchwaters

In order to augment the yield of a reservoir, catchwaters are often resorted to whereby additional water is led into a reservoir from another gathering ground. As it is not economical to design catchwaters to take maximum floods, only up to 90% of the water available is taken. A catchwater may be a tunnel or more often an open channel graded to suit the contour of the land. This is the cheapest form of structure and provided that the length is not too great it is cheaper than a pipe.

The design of an open channel can be based on the Chézy formula, see Section 5.

FORMATION OF RESERVOIRS

Reservoirs may be formed with earthen (or rockfill) dams, concrete dams (gravity, arch, multiple arch, cupola, buttress, reinforced, prestressed), by raising existing lakes or by enclosing with artificial embankments, generally filled by pumping. In addition to the dam forming the reservoir certain ancillary works such as draw-off tower, overflow and diversion works are normally required.

Valve towers

A water supply reservoir usually has a valve tower to contain pairs of valves at different levels, to draw off the water as it rises and falls and to ensure the water at that level is consistent with good quality. Recording instruments are often situated within the valve tower: (a) to record the level of the water below top water level, (b) to record the level of the water when the reservoir is overflowing, (c) to record the quantity of water taken to supply, (d) to record the quantity of water discharged for compensation. A rain gauge is usually placed in the vicinity of the dam.

Floods

The determination of the magnitude of a flood is difficult, but the Institution of Civil Engineers Interim Report[27] produced valuable data based on experience of floods in upland areas.

For many lowland areas, however, the upland floods may safely be halved or even quartered.

The amount of flood is proportional to the size of the gathering ground and typical intensities are as in Table 26.8.

The above are intended to be normal maximum floods but catastrophic floods may have doubled the above values, or more in some areas.

There are various types of overflow weirs: (i) the straight form of weir, either on the crest of the dam, or constructed at right angles to the crest, built upstream on one or both valley sides; (ii) the bellmouth form of overflow weir which consists of a bellmouth and vertical shaft built upstream, the flood water travelling down the shaft into a tunnel which takes the water round the end of the dam or through the embankment itself; (iii) siphons.

Overflow weirs

The function of the overflow is to carry flood water safely over the crest of the dam. The commonly accepted maximum depth of water which should be permitted is 2 to 4 ft, particularly with earth dams. With concrete dams, in the small gathering grounds

Table 26.8 MAGNITUDE OF FLOODS

Area of gathering ground (acres)	*Approximate upland flood anticipated* ft^3/s 1000 acres	*Approximate lowland flood anticipated* ft^3/s 1000 acres
2 000	1 000	300
5 000	700	200
10 000	500	150
15 000	400	125
20 000	350	100
50 000	250	70
100 000	200	50
250 000	140	35

1 ft^3/s = 28.3 litres/s; 1 million gallons/day = 4.54×10^6 litres/day

of Great Britain, ample length can usually be provided so that the depth of water even for a catastrophic flood is very moderate. In 1956 at Eyebrook earth dam, Corby, the water level was raised by 2 ft 6 in thereby increasing the storage by 250 million gallons and a siphon was added which would discharge safely the maximum flow of 4000 ft^3/s working under the same head as before the siphon was erected.

For an earthen embankment, however, as long spillways become expensive, it is useful to estimate the maximum flood as closely as possible in order to keep down cost.

Draw-off and diversion culverts

Tunnels or culverts preferably constructed around the ends of the dam in solid strata are used for diverting streams during the construction of reservoirs, and their dimensions during construction depend upon the best approximation of the magnitude of the flood. In the case of bell-mouth overflows the tunnel is required to carry water which flows over the bellmouth permanently.

Frequently the same tunnel is used to carry water drawn off from the reservoir to supply or for regulation.

Earthen embankments and earth dams

General The adoption of an *earthen* embankment for an impounding reservoir is largely a matter of choice, depending on the geological factors governing the site. If clay predominates at the site of the dam, an earthen embankment in all probability will be adopted; whereas in rocky country a masonry or concrete dam would be more suitable. This generalisation is not rigid, for on the granite of Cornwall in the same valley there is a concrete dam upstream with an earthen embankment downstream. The earthen embankment usually depends upon a puddled or rolled clay core for watertightness, both in the trench below ground and in the body of the embankment above ground, although a concrete-filled cut-off trench is often adopted. The trench is one of the most difficult and expensive portions of the work and its depth in the valley and the length into the hillside cannot usually be determined in detail until the geological structure is completely revealed. Even so there may be faults and water-bearing fissured rock of unknown extent which can only be excavated or cut out at prohibitive cost, and reliance must be placed on pressure grouting under the cut-off trench.

Both upstream and downstream, earthen embankments depend for stability on adequate drainage by rubble and 'selected' material (known as 'filters', specially graded to certain rules) which is placed against the core or laid in layers in the ordinary filling. Broken rock, or material containing a large percentage of broken rock, permits steeper slopes to be adopted; thus for the upstream slope 1 (vertical) in 3 or 4 (horizontal) might suffice, and for that downstream 1 in 2 or 3; whereas, for the ordinary clayey materials so common in this country, 1 in 5 or 6 or more upstream, and 1 in 3 to 5 downstream are common. The guiding principle in such embankments is adequate drainage, as well as the application of the principles of soil mechanics[28, 29] such as tests for shear strength and other properties of clays, and deduced slip planes and stability diagrams.[30] Bearing pressures of the subsoil below the embankment must be known if the weight of the embankment will be sustained without subsidence. (See Section 8.)

COST OF EARTH DAMS

It is useful to be able to estimate the approximate cost of a dam when the shape of the valley at the dam section is known, and Mitchell[31] has suggested the following formula, particularly for comparing several sites when making a preliminary survey of alternative sources:

$$\text{Cost} = \pounds(\underset{(1)}{3.6aLH^2} + \underset{(2)}{0.6bLH^2} + \underset{(3)}{1.83cLD} + \underset{(4)}{16\,500H} + \underset{(5)}{50\,000})$$

where L is the length of crest in metres, H the mean height from ground to crest in metres, i.e. area of cross section of valley divided by length of crest or chord across valley, D the mean depth of cut-off trench in metres.

Item (1) Cost of forming embankment where *a* is rate per cubic metre (£1.95 at 1970 prices)
Item (2) Extra-over cost of rolled clay core where *b* is rate per cubic metre (varying from zero to one-third of *a*)
Item (3) Cost of concrete-filled cut-off trench 1.8 m wide where *c* is rate per cubic metre (£26.00 at 1970 prices)
Item (4) Cost of stream diversion, overflow and valve shaft
Item (5) Cost of miscellaneous items such as pipework and valves, footbridge to valve shaft, reinforcement and steelwork, and recording instruments for water level, overflow and discharge below dam.

SPECIAL PROBLEMS CONCERNING EARTH DAMS[32]

Floods over earth dams For earthen dams and embankments it is important that the 'Freeboard' and distance between top-water level and crest should be adequate: it may depend on the 'Fetch' (the maximum distance the water is impounded at right angles to the dam), the density of trees and vegetation, the elevation of the site, intensity and direction of winds, and, of course a correct assessment of maximum and 'catastrophic' floods. In Britain the usual *freeboard* is 1.5 to 2 m (5 or 6 ft).

(N.B. Some engineers consider the term 'Freeboard' to denote the distance between the level of the maximum flood and the crest).

DESIGN OF EARTH DAMS

Some of the basic factors considered in design are:

(1) To ascertain by site investigation and laboratory testing of the materials, available for constructing the dam, preferably those nearest to the site for economy
(2) To ascertain the conditions and properties of the strata under the embankment to resist sliding or slipping
(3) To analyse the factor of safety over *slipping* for the particular dam section by choosing a slip plane or circle through the probable weakest line of failure in and under the dam. This can now be done by one of several computer programs.

Common slopes of embankments vary between 1 : 1 and 1 : 6.

THIXOTROPIC CLAYS [32a]

Certain Bentonite clays or slurries have been used for sinking dam trenches in soft strata without timbering. Although the process is not well understood, the clays have the effect of remaining liquid when the trench is being dug but form a gel or colloid (slightly heavier than water) when disturbed which has the effect of keeping the walls of the trench from falling in. Recently Bentonite cement mixtures have been used to effect a permanent cut-off seal in porous strata below embankments.

PORE PRESSURE UNDER EARTH DAMS

Where embankments are constructed of, or rest on, cohesive materials containing water within the pores, a pore pressure is set up within the material when loaded either during construction or on the subsequent filling of the reservoir. The increase in pore pressure leads to a reduction in the shear strength of the material and a corresponding reduction in the factor of safety which can lead to failure of the embankment by slipping. To obviate this effect, sloping layers or 'blankets' of drainage material are included in the embankments to enable the pore pressure to be dissipated before it reaches a dangerous level.

A similar effect can be observed by infiltration of ground water and if this is anticipated then relief wells or vertical sand drains should be incorporated under the embankment.

Pore pressure can be measured by installing piezometers, which are ceramic pots sealed in the strata and connected to a gauge by fine tubes.[33]

DEFORMATION OF EARTH DAMS

Under this term is included the causes of the shapeless early nineteenth century embankments often seen before the Safety Provisions Act of 1930.

These include: (i) subsidence of the crest—some have sunk nearly below the overflow top water level; (ii) irregular shape of the embankment due to local slips; (iii) irregular toe lines due to slides.

Although sagging of the crest can be remedied by levelling and adding additional material, and irregular shapes and toes can be regraded and re-aligned, little or nothing is known of what is going on in the strata inside a dam and, since it is dangerous to take such measures without analysing the original cause of the deformation, instruments have been developed to indicate both horizontal and vertical movements.[34]

Horizontal movements can now be measured by vibrations in an electrically stimulated wire (embedded between two concrete blocks) whose tension varies with the horizontal stress.

For vertical movements, the relative displacement can be ascertained by lowering an induction coil down through a tube connecting two plates, or by various types of instruments using the U-tube principle. A fuller exposition on instrumentation in earth dams is given in reference 33.

For schemes including recent earthen embankments see also references 52 to 57.

COMPACTION OF EARTH AND ROCKFILL DAMS

In earth dams it is important to have the impervious clay core well compacted either by heavy rollers or light vibrating rollers operating on layers of 0.25 to 0.5 m. The 'voids' percentage should be kept below 5%. The compaction of the rest of the 'fill' need not be to such a high specification. In a case such as Scammonden Dam (height about 250 ft), an earth and rockfill dam near Huddersfield, a special investigation had to be made to ensure minimum settlement because a six-lane motorway goes over the crest.[34]

For this particular site (consisting of alternating grits, sandstones and shales of the Millstone Grit formation) extensive large-scale experiments revealed that the best compaction depended upon (a) how the rock was quarried, (b) how the material was deposited on the embankment and (c) the best type of compacting machines.

Concrete dams

The adoption of concrete for constructing a dam depends on the topography and geology of the site. The trench, as in the case of an earthen embankment, may be filled either with puddled clay or concrete, depending on the hardness or softness of the strata penetrated and the cost of filling it either with clay or concrete. Below the trench there may be necessity for extensive grouting, to reduce leakage under the dam. The surface for the broad foundation for the superstructure must also be prepared. These foundation works may cost as much as the superstructure seen above ground.

For a straight gravity concrete dam, 100 ft high, the broad foundation would be about 60 ft in width, and if rock or other stable formation is not found reasonably near (say

20 ft below) the surface, considerable expense in foundation work may also be entailed. For a buttress or multiple-arch dam, it would be less, but the strata sustaining the buttresses would have to be stronger. Most dam failures may be attributed to faulty foundations.

Above ground, the *gravity* dam is so-called because any section can stand by itself because of its weight (150 lb/ft^3). Concrete is generally vibrated, and in dam construction particularly, to eliminate air pockets, prevent leakage and increase speed of setting. Shuttering must be especially well constructed to withstand vibration during construction.

Curved-on-plan gravity dams are sometimes substituted for straight dams for aesthetic reasons, but the gravity section of the dam cannot be reduced.[35]

COST OF CONCRETE GRAVITY DAMS[31]

$$\text{Cost} = £(\underset{(1)}{0.375xLH^2} + \underset{(2)}{0.675xLW^2} + \underset{(3)}{0.75yLD_1} + \underset{(4)}{1.83zLD_2} + \underset{(5)}{35\,000})$$

where H is the mean height from broad foundation to crest in metres, i.e. area of cross section of valley divided by length of crest or chord across valley, L is the length of crest in metres, W the width of dam at crest in metres, D_1 the mean depth of broad foundation below ground level in metres, D_2 the mean depth of cut-off trench below broad foundation in metres.

Item (1) Cost of concrete dam where x is rate per cubic metre (£16.00 at 1970 prices)
Item (2) Cost of crest road or footpath
Item (3) Cost of excavating broad foundation where y is rate per cubic metre (£1.30 at 1970 prices)
Item (4) Cost of concrete filled cut-off trench 1.8 m wide where z is rate per cubic metre (£20.00 at 1970 prices)
Item (5) Ancillary works such as pipework and valves, reinforcement and steelwork and recording instruments for water level, overflow and discharge below dam.

Note: Like the corresponding formula for earth dams, this formula is intended for comparing a number of sites when making a preliminary survey of alternative sources.

BUTTRESS AND MULTIPLE-ARCH DAMS

The buttresses of a buttress dam form part of two adjacent halves of two arches (thick) which act as cantilevers and hence if the bearing pressure of one buttress differs relatively from the other, movement may occur at the centre of the arch, where an expansion joint is (or should be) inserted. The internal buttresses of the multiple-arch dam are merely blocks of concrete acting as abutments for supporting the two halves of two rigid (thin) arches. Hence the foundations for the buttresses of the multiple-arch dam must have equal bearing capacity to avoid fracture of the true thin arches; whereas in those for the buttress dam, some inequality is taken care of by the expansion joint between the two cantilever arms of the buttresses.

Thin-arch dam The true thin-arch concrete dam is suitable for the valley which has a good foundation and where the width at the level of the dam crest is not more than three times the proposed maximum height of the dam.

The volume of concrete in an arch dam is about half that in a comparable gravity dam.

Preliminary calculations are directed to finding the thickness t in feet of the dam at any depth in terms of the water pressure in ton/ft^2(P) (i.e. on a foot strip of dam), and the radius R of the upstream face in feet and the compressive strength S in ton/ft^2. If the abutment pressure is greater than the compressive strength of the strata on which the abutment rests, the concrete should be increased in width.

Thick-arch dam The thick-arch dams lie between the thicknesses of the gravity and arch dams. They have been adopted in valleys with chord/height ratios up to five or six. The theory of design involves doubtful assumptions, but nevertheless tests on models seem to confirm that these assumptions are reasonable. The saving in concrete and cost for all arch dams is well worthwhile but the supporting foundation strata must be above suspicion.

Cupola, dome, or double curvature arch dam This type of dam is generally suitable for valleys with a chord height ratio of under three. It is economical in concrete and its strength, for this thickness, is like an egg shell. Calculations are complex but models for testing to destruction are used with success. Foundations must be above reproach.

PRESTRESSED CONCRETE DAM

Prestressed concrete dams have been developed and adopted in recent years for economy of concrete where good foundations are available. The thin concrete structure is anchored by steel cables embedded vertically in the concrete of the structure and with grout inserted in boreholes in hard strata below.

Existing dams have been raised successfully and for this purpose the use of prestressed steel enables the existing structure to be little interfered with beyond drilling vertical boreholes for the prestressed cables to be inserted.[36]

SPECIAL PROBLEMS CONCERNING CONCRETE DAMS[32]

Floods over dams The precise estimate of floods is not so important as those for earth dams but nevertheless overtopping the *crest* should not be permitted not only because of the extra weight of water pressure on the dam but particularly the risk of scouring the strata under the toe. It is true that an overflow of 100 m over the Vaiont dam caused no damage to the dam but the abutments were against hard dolomite limestone.

The interim report of the Institution of Civil Engineers is in general use for the assessment of floods for upland areas.

Rock testing by seismic methods. The velocity of sound through rock may give a valuable indication of its state below the surface, i.e. whether it is faulted, dry, wet, disturbed, open or revealing unsuspected faults or whether the density of concrete foundations is sufficient and the efficiency of grout curtains and contact grouting particularly for concrete dams which are on rock.

Pore pressure and uplift Pore pressure is dangerous under a concrete dam as the pressure is upward and 'lightens' the dam tending to turn it over and make it slide.

In some cases pore pressure leads to leakage at the toe of the dam. In other cases it appears after a few years possibly from some kind of clogging and the pressure has to be reduced either by grouting or alternatively by drilling drains under the dam which, although it increases the leakage, reduces dangerous uplift.

Deformation of dam and strata Strain gauges embedded in the dam give a measure of any untoward trouble going on in a solid concrete dam. They consist of electrically stimulated vibrating wires in which any change in vibration speed indicates change of stress in the dam, indicating degree of movement.

Other indicators of deformation are surface effects due to weather and temperature for which thermometers are inserted in the dam. Although movement of a dam can be ascertained by elaborate surveying equipment, pendulums and inverted pendulums inserted in boreholes inside the dam measure deformation more exactly.

Pendulums for high dams are used especially for showing the movement when the reservoir is filled and empty and if these values are the same or whether they change over the years. The relative movement of the strata with the dam can also be found.

Examples of raised lakes

There are two or three instances of the utilisation of natural lakes (other than Thirlmere and Haweswater which have been developed by high dams), the chief of which is Loch Katrine, for Glasgow, where the natural surface of the lake has been raised 14 ft to draw off 3 ft below, the total supply available being about 70 million gallons/day. Similarly the Crummock Lake for Workington has been raised 2 ft; the draw off pipe is 7 ft 6 in below this level. This is estimated to give a gross supply of 13 million gallons/day. The utilisation of existing lakes raises special methods of tunnelling to draw water from lower existing levels as well as raising the level of water and at the same time coping with storm water.

Pumped storage reservoirs

The largest examples of these reservoirs, constructed in this country on clay and with clay cores and supported by any suitable material nearest the site, are those of the Metropolitan Water Board. These reservoirs are of the order of 60 ft in height, cover many hundreds of acres, and store water from the Thames.

DESALINATION[37]

Desalination should be regarded as a method of treatment to remove impurities, particularly salts, from a saline water.

It has come into use at an increasing rate during the last 20 years; and in certain circumstances can compete with orthodox sources which depend on conventional treatment of water from boreholes, impounding reservoirs and river waters. In 1972 the cost of water from 'traditional' sources was about 15p–25p per 1000 gallons (4500 litres) whereas from desalination plants it may vary between 50p and £1.50 per 1000 gallons (4500 litres). This variation in the cost of desalination arises from:

(a) The degree of salinity to be treated, e.g. sea water (chlorides 35 000 ppm), brackish water (5000 ppm) down to 500 ppm which is tasteless to most palates.
(b) The location and availability and cost of power, heat, transport.
(c) The selection of plant, i.e. (i) multi-stage flash (MSF) and other variations of this distillation plant, (ii) electrodialysis and the somewhat similar reverse osmosis plant, (iii) several other types used on ships and in factories and other special purposes.

Multi-stage flash distillation (MSF) (vacuum separation)

If sea water is evaporated, steam is condensed as pure water leaving solid salt behind, as in the Dead Sea region. In the mechanical MSF process the sea water is pumped through a pipe (sufficiently long to ensure not drawing in sand in rough weather), heated, and passed into a tank under partial or reduced vacuum. As water boils at a lower temperature than normal when at a lower pressure (as on a mountain), fresh water 'flashes off' as steam which is cooled by incoming pipes conveying the sea water and condenses to fresh water. This process is repeated in several stages to increase the efficiency. Many problems arise, apart from the multiplicity of pipes, particularly the elimination of alkaline and calcium sulphate scale which can be controlled by the addition of polysulphide, acid, and lime to increase the pH from 5 to 7. If temperatures could be used above 120 °C the cost could be reduced.

Electrodialysis (membrane-electric separation)

If brackish water is pumped through a tank between two membranes on each side of which is a positive electrode and a negative electrode, the electropositive sodium will go through one membrane to the negative electrode and the electronegative chlorine will go through the other membrane to the positive electrode. The water between the membranes, thus denuded of sodium chloride and other salts, is fresh. The method is only economic for water containing up to about 10 000 mg/litre, and for reductions down to 500 mg/litre, a cost of 25p to 50p per 1000 gallons is incurred.

Operational plants up to 22 million litres/day (4.8 million gallons/day) output have been installed in the Middle East, and the method is now well established.

Reverse osmosis (membrane pressure separation)

'Osmosis' may be envisaged as a natural flow of fresh water into sea water when in contact with each other; whereas 'reverse osmosis' acts when pressure is applied to the brine which, when pushed through a special membrane such as cellulose acetate, causes the fresh water to flow out of the brine for separate use.

Pilot plants producing 22 000 litres/day (5000 gallons/day) are in operation and act in the same range as electrodialysis plant. The main disadvantages are the high operating pressure (about 600 lbf/in^2) and the fine limits involved in production of the membrane. When these problems are solved the process offers great potential.

Freezing

Two freezing processes are being evaluated—vacuum and secondary refrigerant—but neither are yet proven in a full-scale operation.

TREATMENT OF WATER FOR POTABLE SUPPLY[38, 39]

The type of treatment required varies considerably according to the source of supply of the raw water, whether it be a surface water or from underground sources.

Surface waters may be divided into upland and lowland sources. Except in the case of very small supplies, upland waters are usually impounded in the catchment area and are good quality waters, low in dissolved solids and with little organic contamination from the biological point of view, although they are frequently high in organic colour due to deposits of peat and can also contain, particularly in the Pennine area, iron, manganese, and aluminium in solution.

Although impounded water is generally of good quality it can be subject to disturbance due to stratification, thermal turnover if the water is deep, or by surface winds and flash run-off if the water is shallow. In these circumstances there is a marked and often very sudden deterioration in the quality of the water and, although this may be for only a short duration, it must be given full consideration when a treatment plant is being considered.

The increasing demands being made on upland sources, the difficulty of finding suitable reservoir sites and the extremely high cost of trunk main laying has led to a reappraisal and has indicated that in many cases a greater reliable yield can be established by using a reservoir for regulation of the river flow and abstracting direct from the river in its lower reaches, e.g. Clywedog.[40] Similarly lowland pumped storage reservoirs (e.g. Grafham Water, Draycote and Empingham) are filled from low-quality river waters

Therefore, increasing use is made of lowland waters taken from the lower reaches of comparatively slow-moving and turbid rivers. They present far greater problems from the point of view of organic and industrial pollution and the treatment aspect becomes far more complex.

Increasing sophistication in the equipment available for automatic control is leading to consideration of continuous monitoring of raw water qualities for automatic control of the treatment process, but it will be some time before this becomes established practice.

Underground supplies from aquifers such as limestone, chalk, sandstone and greensand are normally biologically pure, but are often very hard and can contain objectionable levels of iron, manganese, sulphates and chlorides, as well as excess CO_2 and H_2S. In some circumstances river gravel can also be used as an underground source, although this is not necessarily of the same degree of organic or biological purity. Although there are large quantities of water in old mine workings, it tends to be very high in dissolved solids, particularly sulphates and chlorides. Boreholes near the coast can also suffer from an infiltration of salinity with resultant brackish water.

Water characteristics

The main characteristics of a raw water which affect treatment processes are summarised in Table 26.9 as follows:

Table 26.9 CHEMICAL CHARACTERISTICS OF RAW WATER

Group	*Main constituents affecting treatment*
Gases	Oxygen, carbon dioxide, hydrogen sulphide, ammonia
Suspended solids	Clays, minerals, siliceous matter, vegetable debris
Organic matter	Organic acids, humus, peat, algae, faecal matter
Dissolved solids	(*a*) *Hardness salts:* *Permanent*—calcium and magnesium sulphates, nitrates and chlorides *Temporary*—calcium and magnesium bicarbonates (*b*) *Nonhardness salts:* sodium sulphates, chlorides, nitrates or bicarbonates

A range of possible treatments for different characteristics of the raw water is summarised in Table 26.10.

tion, precipitation, softening, flocculation, sedimentation, pH control, filtration,

The processes given in Table 26.10 are associated with the appropriate sedimentation and filtration plant and techniques. The full range of treatment for a poor quality lowland river water may include: storage, algal control, aeration, pH control, coagulation, precipitation softening, flocculation, sedimentation, pH control, filtration, chlorination, dechlorination, pH adjustment, and taste control.

The range of chemicals commonly used in treatment processes is summarised in Table 26.11.

Table 26.10 RANGE OF TREATMENTS FOR DIFFERENT CHARACTERISTICS

Characteristics	*Possible treatment*
Gases	Aeration
Dissolved impurities	Precipitation—(oxidation)
Suspended matter	Coagulation, settlement
Colloidal matter	Coagulation
Colour	Coagulation, activated carbon, ozone, chlorine
Odour	Aeration, activated carbon
Taste	Activated carbon, chlorine, chlorine dioxide, ozone
Acidity/Free carbon dioxide	Aeration, control by alkali
Hardness	Lime and/or soda precipitation, or ion exchange methods
Iron and manganese	Aeration, precipitation and filtration with iron removal media
Other metals	Coagulation and precipitation
Salinity (brackish)	Distillation, demineralisation, reverse osmosis, freezing
Oil	Flotation, coagulation
Algae	Straining, copper-sulphate, chlorine, cupri-chloramine
Biological impurities	Storage, chlorine, chloramine, ozone, ultra-violet light
Industrial pollution	Combination of above as required

Table 26.11 CHEMICALS COMMONLY USED IN WATER TREATMENT

Substance	*Formula*	*Purpose*
Activated carbon	C	Taste and odour control
Aluminium sulphate (alum)	$Al_2(SO_4)_3$	Coagulant
Ammonia	NH_3	With chlorine for sterilisation
Ammonium sulphate	$(NH_4)_2SO_4$	A source of ammonia for chloramine
Calcium carbonate (chalk)	$CaCO_3$	A source of bicarbonate alkalinity
Calcium hydroxide (slaked lime)	$Ca(OH)_2$	Softening and pH control
Calcium hypochlorite (bleach powder)	$Ca(OCl)Cl$	Disinfection
Calcium oxide (quick or burnt lime)	CaO	Softening and pH control
Chlorine	Cl_2	Disinfection
Chlorine dioxide	ClO_2	Disinfection, taste and odour removal
Copper sulphate (bluestone)	$CuSO_44H_2O$	Algal control
Ferrous sulphate (copperas)	$FeSO_47H_2O$	Coagulant
Ferric chloride	$FeCl_3$	Coagulant
Ozone	O_3	Disinfection, taste, odour and colour removal
Potassium permanganate	$KMnO_4$	Removal of iron, manganese, algal control
Sodium aluminate	$Na_2Al_2O_4$	Coagulant
Sodium carbonate (soda ash)	Na_2CO_3	Removal of permanent hardness and pH control
Sodium chloride (common salt)	NaCl	Regeneration of zeolites
Sodium hypochlorite (Chloros or Voxsan)	NaOCl	Disinfection
Sulphur dioxide	SO_2	Dechlorination

The basic principles adopted in each stage of these treatment processes are briefly described in the following, with typical examples.

Storage

For a scheme using direct river abstraction a raw-water storage of seven days is recommended to allow for settlement of heavy silt load to even-out any rapid changes in water quality and to allow for rejection of water containing accidental and heavy pollution (e.g. Oxford,[41] Nottingham[42]).

Algae

The growth of severe blooms of algae which would interfere with the treatment process can be inhibited by the use of an algicide (e.g. copper sulphate) and the exclusion of direct sunlight.

Aeration

The level of dissolved gases can be substantially reduced by an aeration system which in the order of ascending efficiency takes the form of cascades, sprays, and induced draft towers. If the water is particularly 'flat' in appearance the level of oxygen can be increased and the appearance of the water enhanced by similar means (e.g. Ardleigh,[43] Oxford[41]).

Coagulation

Any raw water containing colour or finely divided suspended solid needs the addition of a coagulant to neutralise the electrical charges causing dispersion and induce the impurities to coalesce and flocculate. This process is often assisted by slow-speed agitation to increase the collision between the particles. Normally the reagents are delivered by road vehicle and taken into bulk storage at the treatment works. The method of adding the coagulant most usually adopted is the use of positive displacement ram-type metering pumps which can easily be controlled to vary the dose according to the treatment flow and, if required, to the water quality (e.g. Colchester, Swansea).

pH control

For efficient coagulation the pH value is critical and as the pH of the water with the added coagulant is unlikely to be at the required level it is necessary to correct this by the addition of acid or alkali, preferably under automatic control. Variation of pH from the level necessary for optimum coagulation can produce light fluffy and fragile flocs and high residual dissolved coagulant (e.g. Bradford, Londonderry).[44]

Precipitation

In any treatment process which includes coagulation and sedimentation it is possible to precipitate the hardness salts by the addition of lime and/or soda and the precipitated salts are effectively removed in the general system and can in some circumstances increase the efficiency of the treatment although inevitably producing an increase in the volume of sludge to be discharged from the works (e.g. N.E. Lincolnshire,[45] Sheffield[46]).

Mixing

As the volume of reagent is small compared with the volume of water being treated it is critical to ensure that the chemical is fully dispersed into the body of the water and also that the reagents are added in the correct sequence according to the chemical requirements. Reagents can be diluted to ease the mixing problem, this is really two-stage mixing, and are then added in an area of turbulence induced either hydraulically or mechanically: hydraulically in the nappe of a weir or the standing wave of a venturi flume or mechanically by high-speed mixing and sometimes by a pumped recirculating system (e.g. Bristol[53]).

Flocculation

After the reagents have been added and fully mixed it is normal to induce flocculation by passing the water through an area of slow agitation which, again, can be induced either hydraulically or mechanically (e.g. N. Derbyshire).

Sedimentation

In its simplest form, sedimentation is the use of tanks giving a retention time that is long enough for the floc particles to settle and compact into sludge on the bottom of the tank, from which point the solids are discharged for disposal and the settled water is decanted to the following filters. However carefully such tanks are designed the physical retention seldom exceeds 40% of the nominal retention and this has led to the development of the upflow type of treatment unit. After a flocculation zone the water is induced to flow vertically upwards through an area of suspended sludge where the floc particles have a large area of contact which greatly assists in the formation of denser agglomerates. Such tanks are designed so that the sludge can be withdrawn at the rate at which it is forming and provide a stable process that can be controlled over a varying range of duties. Construction can be in either steel or concrete, with the units either square or circular in plan, and as the capacities of Treatment Works increase it is generally more economical to consider a smaller number of circular-type tanks (e.g. Bradford, Derwent Valley.)

Filtration

After coagulation and sedimentation, the water still retains a quantity of suspended matter which is removed by filtration. It should be noted that the filtration process associated with the treatments being described is that known as 'rapid' as distinct from slow-sand filtration which is a biological process completely in itself.

The settled water passes through a layer of comparatively fine and specially graded sand supported on underbeds of graded pebbles with a piped header and lateral underdrain system, or supported on a flat floor with a system of closely spaced nozzles. In either design the clean water is collected from the base of the filter and as the resistance to flow increases, in proportion to the quantity of intercepted matter building up, the filter bed is cleaned; first by expanding the compacted bed, usually by the application of an air scour, which effectively loosens the intercepted impurities which are then flushed out to waste by a reverse flow of water.

The filter bed can be contained equally well in a steel pressure vessel or an open-topped gravity tank and the siting of the plant relative to the hydraulic gradient can determine which method is preferable. Where large flows are being considered the gravity-type filter does not have the same restriction on the size of individual units and it is unusual to follow sedimentation, requiring open-topped type tankwork, by pressure filters (e.g. Lune Valley, West Glamorgan[47]).

Filtration technique is currently going through a period of change with advocates for downward, upward and sideways flow, for deep beds and shallow beds, for single media, multi-media, and multi-layer media.

All these variations have some application, however limited, and provide filtration in the depth of the bed rather than on the surface, with a greatly increased efficiency. Although there is very little, if any, long-term operating experience on some of the designs, the use of a two-layer downflow filter with a top stratum of graded anthracite resting on a shallower layer of conventional sand is a system that has been in use for a number of years in the United Kingdom and in gaining support as it has been proved that filter ratings can be increased and the length of filter runs extended.

Backwashing

Air for expanding and scouring the filter bed is normally provided by electrically driven blowers delivering large volumes of air at relatively low pressure direct to the induction system which is usually the underdrain system used for collecting the filtrate. Wash water is most often provided by direct pumping to the same common induction system (e.g. Wakefield[48]).

Chlorination

After the water has been satisfactorily clarified it is still necessary for potable duties to make sure that it has been fully disinfected and chlorine is the most usual reagent for this duty. It is normally delivered as a liquid under pressure in either cylinders or drums, depending on the quantity required, and in some of the largest installations it is being delivered by bulk tanker and transferred into storage vessels at the treatment plant.
As it is considered prudent to carry a chlorine residual into the reticulation system as a measures of safety this has to be controlled at a level low enough to be unobjectionable to the consumer. Current practice is often to dose above the chlorine demand of the water and to control the residual passing into supply by adding sulphur dioxide to neutralise any excess. Sulphur dioxide is handled as a liquid under pressure in exactly the same way as chlorine and the dosing is normally under automatic control to ensure that the final chlorine residual is maintained at the correct level (e.g. West Surrey).

pH adjustment

Depending on the treatment adopted the final water is unlikely to be at the pH required for distribution purposes and will need correction by the addition of acid or alkali and at the same time it is important to correct any undue corrosive tendencies which may be inherent in the treated water (e.g. Bedford,[49] N. Devon[50]).

Taste control

Taste which is objectionable as far as the consumer is concerned can be present in the raw water or can develop during the treatment process and activated carbon is often used to absorb the elements that are responsible. It can be added as a powder before the sedimentation process and removed with the sludge, directly as a powder or granule on to the filter beds and removed with the washwater, or as a granular filter medium in a separate filtration stage added to the end of the clarification treatment. In the first two applications the carbon is not recovered but if it is used as an additional filtration unit it can either be regenerated on site or returned to the manufacturer for this purpose (e.g. East Surrey,[51] Oxford[41])

Waste products

Whatever the process used for clarification, or precipitation softening, the impurities removed are concentrated in the form of a sludge which under the best operating conditions is unlikely to be less than 95% water. In this form it can be fed to a filter press, or possibly a centrifuge for a softening sludge, to produce a dry solid suitable for mechanical handling and disposal. The filtrate, or centrate, has to be disposed as a liquor.

In a few cases attention is being given to the possibility of treating the sludge with acid to recover the coagulant but this process is not yet proven as commercially viable (e.g. Fylde, Mid Northants).

Sludge disposal in water works is not such a problem as sewage sludge. Local conditions can normally cope with the smaller quantities of waterworks sludge by distributing it on land, quarries, pits, river, sea which may be and generally are available.

Transport of sludge should be in closed containers in hilly districts, otherwise there is loss of sludge from well-filled open-top vehicles!

REFERENCES

1. MCDOWELL, H. R. and CHAMBERLIN, C. F., *Michael and Will on the Law Relating to Water*, 9th edn, London (1950)
2. 'Water Reorganisation in England and Wales', Department of the Environment Circular 92/71, HMSO
3. Water Resources Board 8th Annual Report, HMSO
4. PARKER, P. A. M., *The Control of Water*, London (1949)
5. SHARP, R. G., 'Estimation of future demand on water resources in Britain', *J.I.W.E.*, **21,** 232 (1967)
6. POLLITT, A. A., *Technology of Water*, London (1924)
7. LEVERIN, H. A., *Industrial Waters of Canada*, Department of Mines and Resources (1942)
8. 'Watercress Growing', Ministry of Agriculture Bulletin, No. 136, HMSO (1967)
9. HAYTON, J. G., 'The use of loose polythene sleeving as a form of protection to spun iron water mains against external corrosion', *J.I.W.E.*, **18,** 465 (1964)
10. KLEIN, R. L., 'The use of asbestos-cement pressure pipes in water supply practice', *Water & Water Engineering*, **63,** 356 (1959)
11. 'Development in aluminium pipes for water supply', *Water & Water Engineering*, **66,** 418 (1962)
12. YOUNG, O. C. and SMITH, J. H., 'Simplified tables of external loads on buried pipelines', Building Research station (1970)
13. LEA, F. C., *Hydraulics*, London (1938)
14. LAPWORTH, C. F., 'Surge Control in pipelines', *Trans.I.W.E.*, **49,** 29 (1944)
15. LUPTON, H. R., 'Graphical analysis of pressure surge in pumping systems', *J.I.W.E.*, **7,** 87 (1953)
16. CRUMP, E. S., 'A new method of gauging stream flow', *Proc. Instn civ. Engrs*, Part 1, **1,** 749 (1952)
17. *Crump Weir Design*, TN8, Water Resources Board (1970)
18. Geological survey maps and memoirs
19. WALTERS, R. C. S., 'Hydro-geology, Chalk', *Trans. I. W. E.*, **34,** 79 (1929), 'Oolites', *Trans. I.W.E.*, **41,** 134 (1936)
20. INESON, J., 'Development of Groundwater Resources in England and Wales', *J.I.W.E.*, **24,** 155 (1970)
21. LAPWORTH, C. F., 'Percolation in the Chalk', *J.I.W.E.*, **2,** 97 (1948)
22. ARMSTRONG, R. B. and CLARKE, K. F., 'Water Resource planning in South-East England', *J.I.W.E.*, **26,** 11 (1972)
23. PENMAN, H. L., 'Evaporation over the British Isles', *J.I.W.E.*, **8,** 415 (1954)
24. RISBRIDGER, C. A. and GODFREY, W. H., 'Rainfall, run-off and storage: Elan and Claerwen', *Proc. Instn civ. Engrs*, Part III, **3,** 345 (1954)

25. DEACON, G. F., 'Water Supply', in 13th edn of *Encyclopedia Britannica*
26. LAPWORTH, C. F., 'Reservoir Storage and Yield', *J.I.W.E.*, **3**, 269 (1949)
27. 'Interim Report of the Committee on Floods in relation to Reservoir Practice', Institution of Civil Engineers (1933)
28. TERZAGHI, K. and PECK, R. B., *Soil Mechanics in Engineering Practice*, New York (1967)
29. TAYLOR, D. W., *Fundamentals of Soil Mechanics*, New York (1948)
30. BISHOP, W. A., 'The use of the Slip Circle in the stability analysis of slopes', *Geotechnique*, **5**, 7 (1955)
31. MITCHELL, P. B., 'Reservoir Site Investigation and Economics', *J.I.W.E.*, **5**, 445 (1951)
32. *Proc. Symp. on Grouts and Drilling Muds in Engineering Practice*, London (1963)
32a. WALTERS, R. C. S., *Dam Geology* (Appendices by J. L. Knill), 2nd edn, Butterworths, London (1973)
33. ROFE, B. H. and TYE, P. F., 'Application of Instrumentation to Earth Dams', *J.I.W.E.*, **25**, 137 (1971)
34. WILLIAMS, H. and STOTHARD, J. N., 'Rock Excavation and Specification Trials for the Lancashire–Yorkshire Motorway—Yorkshire (West Riding) Section', *Proc. Instn civ. Engrs*, **37**, 607 and *Discussion*, **38**, 135 (1967)
35. FARRAR, R. E. S., 'Meldon Reservoir', *Civ. Engng and Publ. Wks Rev.*, **67**, 895 (1972)
36. The heightening of Argal Dam for the Falmouth Corporation Water Undertaking', *Water and Water Engineering*, **65**, 537 (1961)
37. SILVER, R. S., Desalination, Central Office of Information UKAEA, HMSO (1967)
38. SKEAT, W. O., Ed., *Manual of British Water Engineering Practice*, 4th edn, Vol. 3 (1969)
39. HOLDEN, W. S., Ed., *Water Treatment and Examination*, London (1970)
40. FORDHAM, A. E., COCHRANE, N. J., KRETSCHMER, J. M. and BAXTER, R. S., 'The Clywedog Reservoir Project', *J.I.W.E.*, **24**, 17 (1970)
41' CARTWRIGHT, F., 'Design of Farmoor Treatment Works, Oxford Corporation Water Department', *J.I.W.E.*, **18**, 381 (1964)
42. ADAMS, R. W., ROBINSON, R. D. and KENNETT, C. A., 'The River Derwent Scheme of the Nottingham Corporation', *J.I.W.E.*, **27**, 15 (1973)
43. 'The Ardleigh Reservoir Scheme in North-East Essex', *Water and Water Engineering*, **76**, 3 (1972)
44. WILCOCK, E. J. and SARD, B. A., 'Design and Operation of the Carmoney Water Treatment Works: Faughan River Scheme—Londonderry R.D.C.', *J.I.W.E.*, **18**, 477 (1964)
45. ASH, R. V., 'The Great Eau Scheme: North-East Lincolnshire Water Board', *J.I.W.E.*, **20**, 435 (1966)
46. EARNSHAW, F., 'Design of the Yorkshire Derwent Headworks', *J.I.W.E.*, **16**, 139 (1962)
47. 'The Usk Reservoir Scheme of the Swansea Corporation', *Water and Water Engineering*, **59**, 377 (1955)
48. COLLINS, P. G. M. and GIBB, O., 'Design and Construction of the Fixby Treatment Works of the Wakefield and District Water Board', *J.I.W.E.*, **18**, 491 (1964)
49. 'New Water Treatment Works of the Borough of Bedford Water Undertaking', *Water and Water Engineering*, **63**, 61 (1959)
50. 'The Meldon Reservoir Scheme of the North Devon Water Board', *Water and Water Engineering*, **76**, 353 (1972)
51. SHINNER, J. S. and DAVISON, A. S., 'The development of Bough Beech as a source of supply (The East Surrey Water Company)', *J.I.W.E.*, **25**, 243 (1971)
52. HALLAS, P. S. and TITFORD, A. R., 'Design and construction of Bough Beech Reservoir', *J.I.W.E.*, **25**, 293 (1971)
53. PICKEN, J. A., 'The Chew Stoke Reservoir Scheme', *J.I.W.E.*, **11**, 33 (1957)
54. WALTERS, R. C. S. and WALTON, R. J. C., 'Water Supply for the Yeovil district (Sutton Bingham Scheme)', *Proc. Instn civ. Engrs*, **8**, 71 (1957)
55. KENNARD, J. and KENNARD, M. F., 'Selset Reservoir', *Proc. Instn civ. Engrs*, **21**, 277 (1962)
56. KENNARD, M. F., 'Balderhead Reservoir', *Civ. Engng and Publ. Wks Rev.*, **58**, 633 (1963)
57. CRANN, H. H., 'Design and Construction of Llyn Celyn', *J.I.W.E.*, **22**, 13 (1968)

27 SEWERAGE AND SEWAGE DISPOSAL

SEWERAGE AND SEWAGE DISPOSAL 27

27 SEWERAGE AND SEWAGE DISPOSAL

L. B. ESCRITT
Civil and public-health engineer

FLOW CALCULATIONS

DISCHARGE OF SEWERS

The discharge of circular sewers in average condition can be calculated by empiric formulae including:

$$Q = 0.000\,35D^{2.62}/I^{0.5} \quad (1)$$

$$V = 26.738D^{0.62}/I^{0.5} \quad (2)$$

where Q is in litres per second, D is the diameter in millimetres, V is in metres per minute and I is the length divided by the fall.

It is now known that the formulae applicable to fully charged pipes or culverts cannot be used with accuracy for open channels or partly filled pipes because of the loss of energy due to waves on the water-to-air surface.[2] The following formula is recommended as suitable for open channels or partly filled pipes constructed to average standards of workmanship:

$$V = 76.253\left(\frac{A}{P+0.5W}\right)^{0.62}\Big/I^{0.5} \quad (3)$$

where V is in metres per *second*, A is the cross section of flow in square metres, P the wetted perimeter in *metres*, W the width of water-to-air surface in metres and I the length divided by the fall.

Formula (3) can be applied to calculate the proportional velocities, as compared to the velocity when flowing full, and on the former can be based the proportional discharges. Table 27.1 is calculated on this basis.

Table 27.1 PROPORTIONAL VELOCITIES AND DISCHARGES FOR PARTLY FILLED CIRCULAR PIPES AND CULVERTS

Proportional depth	*Proportional velocity*	*Proportional area*	*Proportional discharge*
1.0	1.000 0	1.000 0	1.000 0
0.9	1.039 4	0.948 0	0.985 4
0.8	1.018 9	0.857 6	0.873 8
0.7	0.976 5	0.747 7	0.730 2
0.6	0.917 3	0.626 5	0.574 7
0.5	0.842 5	0.500 0	0.421 3
0.4	0.751 7	0.373 5	0.280 8
0.3	0.642 7	0.252 3	0.162 1
0.2	0.510 2	0.142 4	0.072 7
0.1	0.337 3	0.052 0	0.017 5

The above formulae should, on the average, give good accuracy for *all* sizes of pipes, culverts and channels of fair workmanship.

Measurement of flow

Devices for measuring flow of sewage should be such that they do not tend to cause silting or the retention of heavy solids in the channel through which the flow to be recorded passes. The devices themselves should be such that their pipework or flow chambers do not silt, or else should be so arranged that they are easily cleansed. The types of apparatus that can be satisfactorily used for clean water do not all serve well for sewage or sewage sludge without special modifications. Nevertheless, venturi meters and other meters involving small-diameter tubes for recording differences of pressure can be used if provision is made for flushing with clean water. Electric meters with no parts that need cleaning are available.

STANDING WAVE FLUME

For recording flows of sewage generally, the most useful apparatus is the standing wave flume with automatic flow recorder. It is also used for controlling water level with regard to Comminutors and constant-velocity detritus channels (see formula (11)). The standing wave flume can be in the form of a contraction in the channel, the invert of which is practically level; alternatively, the invert can be raised to form a hump which is not so marked as to interfere with the transport of solids. Again, lateral contractions together with a hump may be combined in one flume. A rectangular standing wave flume should be proportioned more or less as follows if it is to record accurately:

Width of throat equals maximum depth of flow $= H$

Length of parallel sides of throat $= 2H$

Radius of upstream contraction (the curve of which is tangential to throat) $= 2\frac{1}{2}H$

If width of channel above and below flume is $2H$, then length of curved contraction in direction of flow $= 1\frac{1}{2}H$

Below the throat, each side slopes outwards at 1 in 6, so that if width of channel below flume is $2H$, length of this portion $= 3H$

The floor of this portion must fall away so that the lowest water level below the flume is 25 mm below the invert of the throat.

The recorder float is in a chamber 0.76 m square connecting with the flume at a distance $3\frac{1}{2}H$ above the commencement of the curved contraction, through a 150 mm square opening.

The straight channel before this point should be at least eight times its width in length.

The loss of head through a flume is about 15 to 20% of the value of H at the time.

The approximate discharge of a standing wave flume is given by the following formula:

$$Q = 1.706\ BH\sqrt{H} \tag{4}$$

where B is the width, (m), H the head (m) and Q the flow (m^3/s).

For other methods of flow measurement see Section 5.

SEWERAGE

SEWERAGE SYSTEMS

Towns are sewered according to three systems:

1. The 'combined system' in which surface water and soil sewage from premises are

collected by one combined system of sewers and delivered to sewage treatment works or outfall.

2. The 'separate system' in which surface water from roads and roofs is delivered by surface water sewers to the nearest natural water-course, and soil sewage is collected by soil sewers and delivered to sewage treatment works or to sea outfall.

3. The 'partially separate system' in which part of the surface water e.g. from roads and front portions of roofs and premises is taken by the surface water sewers, while soil sewage and rainwater from back portions of roofs and backyards are taken by the soil sewers as may be sanctioned by the local authority.

GENERAL REQUIREMENTS

Apart from rare exceptions sewers are designed as if they were open channels, i.e. the hydraulic gradient is taken as being the gradient of the crown of the sewer. Sewers need to be laid to gradients that will ensure velocities of flow sufficient to prevent settlement of solids. Present-day practice is to arrange the gradients so that when the sewers are flowing full the velocity is not less than 0.76 m/s.

Table 27.2 APPROXIMATE MINIMUM GRADIENTS FOR BEST WORKING CONDITIONS IN CIRCULAR SEWERS.* VELOCITY OF FLOW 0.76 m/s

Diameter of sewer nominal (mm)	*Gradient* 1 *in:*
150	174
175	212
200	252
225	291
300	410
375	545
450	685
525	813
600	965
675	1 113
750	1 284
825	1 439
900	1 613
975	1 758
1 050	1 970

* Nearest figures in *Escritts' Tables.*[1]

No sewer or rising main should be less than 100 mm in diameter and public sewers are usually not less than 150 mm in diameter. Sewers are laid in straight lines and to even gradients between manholes, except that sewers of more than 750 mm in diameter may change direction (not grade) between manholes. Manholes on private sewers should not be more than 90 m apart. On public sewers manholes should be placed at all junctions of sewers and generally at not more than 110 m apart. Sewers should be laid with not less than 1.2 m of cover from road level to crown of pipe or not less than 0.9 m of cover where not in roads. Clay pipes, concrete pipes and asbestos-cement pipes should be given concrete protection according to the most recent requirements of the Department of the Environment, plus other concrete protection as may be required according to local conditions.

In design it is usually assumed that the hydraulic gradient is the slope of the top of the sewer. Wherever there are junctions of sewers of equal or different diameters the crown of the upstream sewer must not be lower than that of the downstream sewer, as otherwise the hydraulic gradient is interrupted. Similarly, the downstream sewer must have its invert not higher than that of the upstream sewer, as otherwise the grade of the invert will be interrupted, causing solids to be held back. There must be continuous fall, from the top end of every lateral sewer to the lowest point on the system, at suitable gradients according to the requirements of the flow and to secure self-cleansing conditions.

SOIL SEWERAGE: RATES OF FLOW

The flow in separate soil sewers during dry weather is equal to, or, as a result of infiltration, somewhat greater than, the water supply to the same district. The flow is not constant throughout the day, for it rises from early morning to about twice the average about midday and falls off gradually until after midnight, when the domestic flow is practically nil. For design purposes, the peak domestic flow may be taken as being four times the average dry weather flow. A certain amount of subsoil water finds its ways into sewers, the quantity depending on the age and state of the sewers and the height of the water table relative to the sewers. To allow for infiltration to soil sewers the latter are designed to accommodate, when running full, four to six times the average dry weather flow, according to circumstances. The following are indications of the dry-weather flows to be expected from different classes of towns, including moderate quantities of trade waste:

Small villages: 0.045 to 0.07 m^3 per head of population per day
Moderate size provincial towns: 0.12 to 0.15 m^3 per head of population per day
Average towns: 0.25 m^3 per head of population per day.

SURFACE-WATER AND COMBINED SEWERAGE

Surface-water and combined sewers are designed to accommodate the run-off of rainfall from impervious surfaces. The latter take soil sewage in addition to surface water, but the quantity is comparatively so small that it can be neglected in calculations. Partially separate systems are designed according to local circumstances.

Surface-water sewerage design involves a compromise between extravagance and inadequacy. There are rare occasions when a combination of circumstances results in so great a flow that it could not be accommodated without the construction of sewers at a disproportionate expense; on the other hand, if sewers were not made big enough to prevent a number of floods every year they unquestionably would be unduly small. Practice in Great Britain and abroad has been for surface-water and combined sewers to be made capable of accommodating without surcharge the maximum flow likely to occur in a short period of years and for surcharge to allow for heavier, less frequent storms.

It is a matter of experience that if the sewers are calculated to accommodate the greatest rainfall run-off liable to occur about once a year there will be no flooding provided that some factor of safety is allowed to cover the less frequent, more intense storms. This, in fact, was approximately the British practice for many years. It appears, however, that a better rule is to design on the storm likely to occur once in three years and, at the same time, make closer calculations for the other factors such as impermeability of surfaces: accordingly the once-in-three years storm is considered in the following text. Occasionally calculations have been made on the basis of very rare storms but, while these storms do occur, the areas they cover are limited and it appears that allowance for such storms amounts to extravagance.

Another matter of experience is that, of the various storms liable to occur once in three years, the storm that has the duration equal to the time of concentration of the particular part of the catchment in question is the storm that produces the greatest run-off. Time of concentration is the time taken for the surface water to run from the farthest point of the catchment to the point where flow is desired to be known for the purpose of determining the diameter of sewer. The procedure in design is, therefore,

Table 27.3 RUN-OFF COEFFICIENTS AND INTENSITIES OF STORMS LIABLE TO OCCUR ONCE IN THREE YEARS IN GREAT BRITAIN

Duration of storm (min)	*Run-off coefficient* C	*Intensity* (mm/h) R	*Duration of storm* (min)	*Run-off coefficient* C	*Intensity* (mm/h) R
11	0.406	51.3	55	0.674	17.7
12	0.427	48.5	60	0.682	16.7
13	0.446	46.1	65	0.690	15.8
14	0.463	43.9	70	0.696	15.1
15	0.478	42.0	75	0.701	14.4
16	0.492	40.3	80	0.706	13.8
17	0.504	38.7	85	0.710	13.2
18	0.515	37.3	90	0.715	12.7
19	0.526	36.0	95	0.718	12.2
20	0.535	34.8	100	0.721	11.8
21	0.545	33.7	110	0.727	11.1
22	0.553	32.7	120	0.732	10.4
23	0.560	31.8	130	0.736	9.9
24	0.567	30.8	140	0.740	9.4
25	0.574	30.1	150	0.743	9.0
26	0.581	29.3	160	0.746	8.6
27	0.586	28.6	170	0.749	8.2
28	0.592	27.8	180	0.751	7.9
29	0.597	27.3	190	0.753	7.6
30	0.602	26.6	200	0.755	7.3
32	0.611	25.5	210	0.757	7.1
34	0.620	24.5	220	0.759	6.9
36	0.627	23.6	230	0.760	6.7
38	0.634	22.7	240	0.761	6.5
40	0.640	22.0	250	0.763	6.3
42	0.646	21.2	260	0.764	6.1
44	0.651	20.6	270	0.765	6.0
46	0.656	20.0	280	0.766	5.8
48	0.661	19.4	290	0.767	5.7
50	0.665	18.9	300	0.768	5.6

to make an estimate of time of concentration of the catchment or part of catchment on the basis of which the appropriate intensity of rainfall can be determined from rainfall statistics: this rainfall on the impermeable surface gives the amount of run-off.

In Great Britain, practice has developed over the years from crude rules of thumb. For some time because of inadequate knowledge of rainfall statistics and statistics of changes of impermeability of surfaces during rainfall, the tendency was to develop over-elaborate theory and this resulted in sewers being constructed larger than was economic. Moreover, some of the methods were difficult to comprehend or too laborious and time-wasting and some were mathematically or logically unsound. In view of research it is now possible to use simple methods that can be applied with greater statistical accuracy and speed than was hitherto thought possible. The method recommended here is identical with the best form of the American 'rational' method applied to British rainfall and run-off statistics.

RAINFALL AND RUN-OFF

The magnitudes of storms are different in various parts of the world, for which reason local rainfalls have been studied and suitable formulae based on them. In Great Britain, Bilham's formula[3] is accepted at present for determining the magnitude of short storms:

$$R = \frac{267.7N^{0.2817}}{t^{0.7183}} - \frac{152.4}{t}, \tag{5}$$

where R is millimetres of rainfall per hour, N is the number of years between storms of this magnitude and t is the duration of the storm in minutes.

Workings according to this formula for storms liable to occur once in three years are given in Table 27.3.

The whole of the rainfall that falls on to roofs, roads or other surfaces does not run off because some of the water is evaporated, soaked into materials or collected in puddles, gullies, etc. A maximum of 95% may run off metal, glazed-tile or slate roofs and a maximum of 90% off perfect asphalt road surfaces, whereas as little as 5% may run off park land. However, observations in Great Britain show that a fair estimate of impermeability of surfaces for industrial or housing areas can be made by measuring all paved

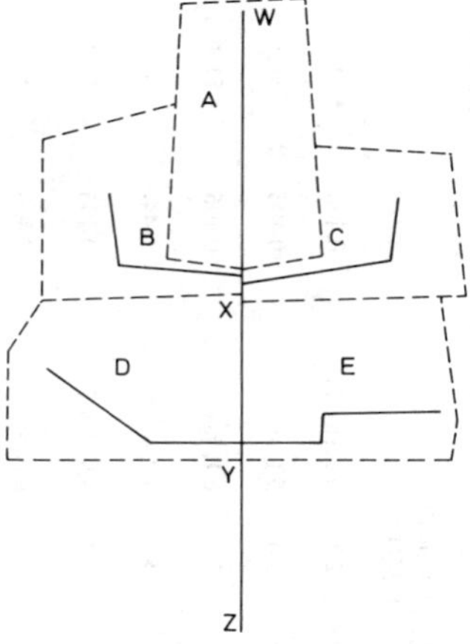

Figure 27.1 Drainage-area diagram

and roofed surfaces and assuming that 80% of the rainfall runs off these surfaces, and, in urban calculations, run-off from unpaved surfaces such as parks and garden land can usually be neglected.

This 80% of paved and roofed surfaces is 'ultimate impermeability', for it is the proportion of rainfall that runs off the surfaces after rainfall has stopped and run-off has come to an end. The actual impermeability, however, changes during rainfall. At first all the rainwater soaks into surfaces, collects into puddles, fills dried-out gullies, etc.

Table 27.4 TYPICAL 'RATIONAL-METHOD' CALCULATION*

Sewer	*Areas contributing*	*Area covered by roofs and pavements* hectares	*Time of concentration* min	*Run-off coefficient*	*Rainfall intensity* mm/h	*Run-off* m^3/min	*Length divided by fall*	*Required diameter of sewer* mm	*Capacity* m^3/min	*Velocity* m/min	*Length* m	*Time of flow* min
W to X	A	2.5	6.5, say 11	0.406	• 51.3	8.7	425	460	9.55	57.5	374	6.5
B lateral	B	3.0	5.1, say 11	0.406	51.3	10.4	530	528	12.0	55.0	280	5.1
C lateral	C	2.7	6.1, say 11	0.406	51.3	9.4	480	460	8.71	52.5	320	6.1
X to Y	A, B, C	8.2	6.5 + 6.5 = 13	0.446	46.1	28.0	600	763	30.2	66.1	430	6.5
D lateral	D	3.2	4.9, say 11	0.406	51.3	11.1	310	460	11.5	69.2	340	4.9
E lateral	E	3.1	6.0, say 11	0.406	51.3	10.8	420	528	13.8	63.1	380	6.0
Y to Z	A, B, C, D, E	14.5	6.5 + 6.5 = 13	0.446	46.1	49.5	585	917	50.1	75.9	295	3.9

* This example took 24 minutes to calculate and 18 minutes to check using *Escritts' Tables of Metric Hydraulic Flow*.[1]

But, once all surfaces are thoroughly wet, the impermeability increases during rainfall. In Table 27.3 run-off coefficients are given suitable for Great Britain and based on an 80% ultimate impermeability for all roofed and paved surfaces: should a different ultimate impermeability be chosen, the figures in the table would need to be modified accordingly.

THE 'RATIONAL' METHOD

In the American 'rational' method the following formula is applied:

$$Q = 0.16\,CRA \tag{6}$$

where Q is the run-off of storm water to the sewers in cubic metres per minute, C is the run-off coefficient, R the rainfall intensity in millimetres per hour and A the roofed and paved part of drainage area in hectares.

Each area or component area is considered according to its particular time of concentration and area of roofed and paved surfaces. First, a guess is made as to time of concentration; a sewer size is calculated on the basis of the above formula and the available gradient and, from the velocity of flow through the sewer as so sized, the time of concentration can be determined. If this time of concentration is longer or shorter than it would have been on the basis of the first guess, the calculation of diameter is repeated until the correct time of concentration and therefore the correct size of sewer has been found.

Table 27.4 is a typical calculation sheet for the area shown in Figure 27.1, according to the 'rational' method. In this calculation the values of run-off coefficient and rainfall intensity have been taken from Table 27.3. It will be noted that there are no values of these for times of concentration less than 11 min: this is because, below approximately 11 min, the run-off coefficient decreases more rapidly than rainfall intensity increases and that, therefore, all times of concentration of less than 11 min are taken as 11 min for the purpose of determining run-off coefficient and rainfall intensity. As will be seen from the table, this greatly reduces the amount of work that has to be done: in fact, all the sewers in many small areas can be designed throughout on the basis of a run-off coefficient of 0.406 and a rainfall intensity of 51.3 mm/h.

When a calculation has been completed in the manner shown in Table 27.4, the engineer may, according to his practice or experience, add a factor of safety and make a moderate increase in the diameters of sewers where this may appear desirable or vice versa.

DISCHARGE OF STORM WATER

At one time it was considered that storm water in excess of about six times dry-weather flow could be discharged from combined sewers to inland water courses without treatment. This practice became an increasing menace to inland waters and, with a view to eliminating it, sections 30 and 31 of the Public Health Act, 1936, required sewage to be purified before discharge to streams, canals, etc., and that local authorities should not create a nuisance. Action to prevent pollution of streams can be taken under Common Law reinforced by these sections. Nevertheless, for some time, engineers continued to design works flagrantly in contravention of these requirements and, consequently, the condition of British rivers and streams became grossly contaminated.

What would appear reasonable practice for the relief of existing overloaded combined sewerage systems would be to provide storm-overflow chambers discharging to storm tanks capable of storing the storm water which could afterwards be passed back to the sewers for delivery to the sewage-treatment works. These tanks should be similar in capacity and design to those provided at the sewage-treatment works. Each case would, however, require consideration on its own merits, bearing in mind the proportion of dilution water in the stream to which the sewage were to be discharged and the restrictions applied by the river authority.

When new sewerage systems are being designed, no storm overflows whatsoever should be provided except at the sewage-treatment works where the storm water should be given an acceptable standard of treatment.

SOAKAWAY SYSTEMS FOR SURFACE WATER

In limestone districts where there are dry valleys, and in many other areas where the subsoil is such that water will soak away, roof water can be discharged to soakaways. A general rule, to be applied with discretion, is that where there are no natural water-courses, surface water sewerage is not applicable and soakaway systems should be used. The discharge of surface water from roads is, however, permissible only where there is no risk to water supplies: where water soaked into the ground finds its ways to wells, surface water from roads treated with tar is not admissible.

Usual practice is to provide large soakaway chambers having storage capacities equal to 15 mm of rainfall over the impervious area served. Soakaways should *not* be filled with rubble as this is costly and wastes capacity.

There are two systems of soakaway installation: first, to connect a group of gullies to each soakaway, the soakaway being so designed to take the flow from them; secondly, to lay 150 mm diameter surface water drains in the road and to place large soakaways at intervals of not more than 110 m in lieu of manholes, an extra-large soakaway being placed at the lowest point on every line of drain. The advantage of this method is that if one soakaway should fail to function it will overflow to the next.

SEWER FLUSHING

When sewers cannot be laid to sufficiently self-cleansing gradients, or when it is expected that the flows in them will not be sufficient to keep them clean at the gradients at which they are laid, means of flushing are provided. At the heads of main sewers having inadequate gradients, automatic flushing tanks are constructed which are usually filled from water authority's mains. The water supply to the flushing tank must discharge into a separate chamber which must be isolated from the flushing tank by a trap so as to prevent contamination of the water main. The flushing tank discharges to the sewer by means of an automatic siphon, the discharge rate of which should be sufficient to at least half fill the sewer but not to surcharge the sewer. It is important that a flushing tank should be of sufficient capacity to flush not only the top end of the sewer but also the lower end. As full investigations have not been made into the tail out of a flush over a length of sewer, several rules of thumb for capacity of flushing tank have been suggested. Probably the most practical is to allow a flushing tank capacity of not less than one tenth of the total cubic capacity of the length of sewer to be flushed. To facilitate the occasional flushing of top ends of branch sewers (where silting is particularly likely to take place because there is little flow), flushing penstocks should be provided in all the topmost manholes. The manholes can then be filled from a hydrant or by means of a tank vehicle and the sewer flushed by opening the penstock.

VENTILATION OF SEWERS

Sewers are ventilated for the purposes of relieving air pressure so as to permit free flow of sewage, and removing gases and vapours which are explosive, poisonous or liable to cause deterioration of structures. Petrol vapour and methane are the principal causes of explosions in sewers. Petrol vapour and sulphuretted hydrogen are rapid poisons. Sulphuretted hydrogen has a destructive effect on Portland cement and therefore on concrete and brickwork. Ventilation is a means of minimising these dangers. Except in very rare instances, sewerage systems are ventilated by natural ventilation only. The flow of air in sewers is due to the wind and the bellows action of increasing and decreasing quantities of sewage in the sewers, convection due to temperature differences and the drag of the flow of sewage. Ventilation takes place between ventilating pipes, any ventilating columns or ventilating manhole covers provided, and also the open ends where sewers discharge to treatment works or outfall. The direction of flow of air cannot be determined as it varies according to the wind, temperature, etc., at the time.

In localities where interceptors are not provided on the drains to premises, the sewers receive adequate ventilation through the soil-and-vent pipes of the buildings. In localities where interceptors are used it becomes necessary to place a ventilating column at the head of every branch sewer and at key positions on the sewerage systems, as may be necessary. Ventilating manhole covers may be used where they will not be liable to cause nuisance. To be effective a ventilating column should have a cross sectional area of not less than one-tenth the cross sectional area of the sewer it serves. It should be carried to above eaves level of any nearby buildings.

The desirability of installing interceptors in the final manholes of drainage systems has long been a matter of controversy. The advantages of omitting interceptors are the greatly improved ventilation of the sewerage system, doing away with the necessity for unsightly columns or objectionable ventilating covers, together with avoidance of the occasional stoppages that are due to silting of the traps themselves. The reasons given for installing interceptors are that they prevent large quantities of sewer gas from flowing up the drains of premises, with the possibility of undesirable results should a faulty drain permit the flow of gas into a building. They are said to interfere with the movement of rats, but observations have proved this to be untrue. Opinion at present is tending towards the use or omission of interceptors according to circumstances, as follows: where drains are old and faulty, or where drains passing under buildings are constructed other than in cast iron, interceptors are desirable; where drainage systems consist entirely of cast iron drains and fittings and building sanitation is of good design and construction, interceptors serve no useful purpose.

Construction of sewers

CONSTRUCTION OF PIPE SEWERS

The vast majority of drains and sewers are constructed of small pipes of up to, say, 225 mm diameter. In order that these shall not be liable to stoppages they must be laid to even gradients and in straight lines, particular care being taken to ensure that there are no impediments in the invert. Unfortunately deterioration of workmanship is very noticeable and, therefore, great care has to be taken to ensure that construction is satisfactory.

Drain pipes can be laid on earth or on concrete. In no case should any pipes be laid on refilled earth, hardcore, ballast or granulated fill, for all of these materials are certain to settle in course of time and this may cause damage to the structure. When pipes are laid on earth the excavation must be made to the exact depth and gradients and, if material is replaced owing to excavation being made too deep, this must be fine concrete or cement mortar. The pipes must be set for level by placing a straight edge in the invert resting in the pipe previously laid, in the one that is being laid and on an accurately adjusted level peg: it is not permissible to decide line and level by measuring to the outside of the barrel or the sockets.

When pipes are being laid on concrete the concrete should be constructed to the true gradient and socket holes cut in the concrete while it is green. The pipes should then be bedded on cement mortar throughout the full length of the barrel and set truly to gradient.

Where pipes are jointed in cement mortar, any cement that has entered the pipe during construction should be removed immediately after each pipe has been laid: it should not be permitted for several pipes to be laid and jointed afterwards.

Materials for small-diameter drains and sewers include vitrified clay pipes, asbestos-cement pipes and cast-iron pipes. Very large quantities of vitrified clay pipes are used. These have sockets for jointing in cement mortar and also with various designs of flexible joints. The advantage of flexible joints is that they remain watertight in spite of movements of the pipe due to temperature changes, etc. This is of importance because it promises that future drains and sewers will be virtually watertight throughout, whereas, in the past, all drains and sewers except those of special construction (e.g. cast iron) developed leaks resulting in contamination of the subsoil and also infiltration adding considerably to the quantity of sewage to be dealt with.

CONSTRUCTION OF LARGE-DIAMETER SEWERS

Large-diameter sewers are constructed of concrete pipes with socket or ogee joints, cast iron pipes, cast iron tubbing, brickwork, mass concrete, mass concrete lined with brickwork, reinforced concrete, pre-cast concrete segments, and steel tubes.

Pre-cast concrete pipes are available up to 1800 nominal mm diamter: cast iron pipes with spigot and socket joints are normally available up to 1200 nominal mm diameter. For larger diameters than mentioned above, pipes of concrete or cast iron can be obtained to special order, but for much larger diameters cast iron tubbing and *in situ* constructional works are to be preferred. Cast-iron tubbing is particularly applicable to the construction of large-diameter sewers in heading with the aid of compressed air. The segments are normally 508 mm long in the direction of the length of the sewer. There is a number of segments of similar form with radial joints, bolted together to form the greater part of the circle, including the invert; a small wedge-shaped key is inserted at the crown, and the two segments, one on each side of this key, have one flange not radial so as to permit the insertion of the crown key.

Most sewers are circular in shape, for the circle is the theoretically ideal form for a pipe or culvert as regards hydraulic properties. Deviation from the circle is sometimes adopted to reduce constructional costs when work is executed in open cut or in trapezoidal heading, and less often to secure the hydraulic advantages of some special form of sewer.

CONSTRUCTION OF MANHOLES AND CHAMBERS

Sewer manholes are chambers constructed to give access to sewers for maintenance purposes. In the case of small-diameter pipe sewers which, in the event of a stoppage, could be cleared by rodding, the chamber should not be less than 1.37 m long and 0.915 m wide and, except where the depth of the sewer below ground level is too shallow, the headroom should not be less than 2 m from the top of the benching to the underside of the roof, in order that the operative may stand upright in comfort.

Manholes are constructed of brickwork, mass concrete, pre-cast concrete pipes and combinations of these materials. Reinforced concrete is rarely justifiable. Cast-iron tubbing is used in special circumstances.

Manhole chambers other than of shallow depth are approached by an access shaft, the most satisfactory proportions of which are 0.7 m by 0.8 m, the latter dimension being measured from the face of the wall which bears the ladder or step-irons. Ladders should be at least 300 mm wide and have 300 mm rung spacings. They are preferable to step irons in nearly all instances. Step irons should also be 300 mm wide: narrow step irons wide enough to take one foot only and staggered from left to right are very commonly used owing to their cheapness, but they are not suitable for other than shallow manholes. They are, however, the usual provision with pre-cast concrete manholes.

The inverts of the manholes should be shaped and graded to the forms and gradients of the inverts of the sewers on which they are placed. From the level of the centre of the sewer the sides of the channels should be brought up vertically to crown level and turned over sharply to form the benchings, which should slope evenly at 1 in 6.

The tops of manholes are corbelled-in or covered by concrete slabs having openings equal to those of the manhole covers, i.e. 0.56 m diameter and never less than 0.51 m diameter.

Manhole covers for general sewerage work are heavy road covers with circular openings and circular plugs, the circular form being the strongest in proportion to the weight of metal. Various types are available, suitable for different forms of road construction, such as bolted airtight, nominally airtight (the usual variety), or ventilating.

Side-entrance manholes. Manholes giving access to large-diameter sewers which can be entered by men for maintenance purposes should be side-entrance manholes, that is, the shaft of the manhole is to one side of the centreline of the sewer and the lower end of the ladder terminates at a platform at one side clear of the sewer and above top water level. From this platform a flight of steps leads to the invert. The purpose of this arrangement is to minimise accidents and to facilitate rescue of casualties. Safety chains or bars should be provided at side-entrance manholes. These should be arranged to stretch across the sewer on the downstream side of each manhole and two should be provided on all very-large-diameter sewers.

INVERTED SIPHONS AND SEWERS ABOVE GROUND LEVEL

When sewage has to be conveyed across a valley, river or canal and the fall is adequate for flow under gravity but the ground or bed level is below the hydraulic gradient, the sewer must be carried above ground level supported on piers or pylons, or else an inverted siphon must be constructed.

Sewers constructed above ground level are usually cast iron or steel pipes and, as far as practicable, are laid in straight lines from end to end so as to minimise the effects of temperature movements. Expansion joints should be inserted at calculated distances, each length of pipe being anchored intermediately between the expansion joints.

Inverted siphons are frequently the only practicable means of crossing a valley under gravity and for that reason are used, although they are considered to be sources of trouble unless designed with care. An inverted siphon is liable to become silted unless the flow through it is frequently sufficiently high to produce a definitely self cleansing velocity. For this reason the siphon should be in most instances of such a diameter that the peak daily flow produces a self cleansing velocity. There should also be sufficient fall to ensure that the maximum wet-weather flow is able to pass through the siphon. In most instances inverted siphons should be provided in duplicate. Inverted siphons on combined sewers need to be multiple siphons consisting of two or more pipes. The smallest pipe should be of such a diameter that it is self cleansing at the peak daily flow. The larger siphons should not come into operation except when the flow is so great that it spills over storm weirs and passes through them. Design should be such that scum is not trapped in the manhole at the upstream end of siphon.

Sewage pumping

When sewage has to be lifted, the types of apparatus employed usually fall under the following classifications: 1 centrifugal pumps, 2 reciprocating pumps, 3 pneumatic ejectors.

CENTRIFUGAL PUMPS

These, usually of the unchokable, comparatively low-efficiency, variety, are used for pumping crude sewage, the impeller being so designed that it will pass a ball of diameter slightly less than that of the delivery and that fibrous material cannot fold over the leading edge of the impeller. At large pumping stations the sewage is screened and may then be passed to ordinary centrifugal pumps of high efficiency. High-efficiency pumps are also used for clarified effluents. Unchokable pumps can be used for pumping sewage sludge.

The specific speed of a centrifugal pump expresses the speed in revolutions per minute of an imaginary pump geometrically similar to that of the actual pump being considered. Various formulae are used in different parts of the world. The metric formula at present in use is:

$$N_s = 3.65nQ^{0.5}/H^{0.75} \tag{7}$$

where N_s is the specific speed, n the revolutions per minute, Q the delivery in cubic metres per second and H the manometric head in metres at maximum efficiency.

RECIPROCATING PUMPS

These are much less used at present than formerly. They are particularly suitable for high and varying heads and they are of high efficiency except at low heads. But when used for pumping sewage they should be protected by screens having clear spaces of not more than 16 mm. Reciprocating pumps for sewage pumping are usually of the ram type, that is, the piston moves through glands displacing water in a vessel: it is not arranged to fit closely in a cylinder. Reciprocating pumps have been used for dealing with difficult sludges, but in most circumstances are now usually ruled out because of their high capital cost.

PNEUMATIC EJECTORS

These have the two advantages of avoiding the necessity for installing screens and of doing away with a below-ground suction well. They can also be installed completely underground in a dry chamber, a pump house not being essential. They have the disadvantages of being very inefficient except when operating against low heads, and of costing more than pumps.

POWER CALCULATIONS

It is advantageous to install centrifugal pumps that have the maximum efficiency at the average working head, not at the maximum working head, and usually power consumption estimates are based on this. The maximum head against which a pump has to deliver is made up of the dead lift in metres measured from the lowest working level in the suction well to the crown of the rising main at its outlet, its highest point or the highest water level in the chamber into which it discharges, whichever is the higher, together with the friction head in the rising main, pipe connections and valves, the friction loss in the pump itself, and the velocity head. The friction head in the rising main can be determined by calculating the hydraulic gradient required to cause the flow and by finding the level above the pumping station to which it would be necessary to deliver to be able to gravitate to the point of outlet through a pipe of equal length and diameter to those of the rising main. Friction losses in bends valves, etc., are usually roughly estimated in terms of equivalent length of pipe, the friction loss in the pump is taken as part of the pump's efficiency, while the velocity head is generally too small to take into account in all except very detailed calculations.

The MECHANICAL load on motor in kilowatts can be found by the following formula:

$$\text{kW} = \frac{QH}{6.118\,3} \times \frac{100}{\text{pump efficiency (\%)}}, \tag{8}$$

where Q is in cubic metres per minute and H is the manometric head in metres.

This value, multiplied by 100 and divided by percentage motor efficiency, gives the electricity demand in kilowatts which, multiplied by 100 and divided by percentage power factor, gives kilovolt amps on which kVA charges are made. Approximate efficiencies and power factors of motors are given in Table 27.5.

AUTOMATIC ELECTRIC PUMPING

Unattended automatic electric sewage pumping stations are very largely used. These should have small suction wells so as to avoid accumulation of sludge and scum: consequently, motors are liable to make very frequent starts and electric gear should be capable of withstanding this. Broadly, 'frequent-duty' starters should always be specified

Table 27.5

kW	*Efficiency* (%)	*Power factor* (%)
7.5	85	88
15.0	87	89
30.0	89	90
60.0	90	91

and the capacity of wells so proportioned that individual motors will not start more frequently than 15 times per hour. It is important that all pump sets should be entirely independent of each other, both mechanically and electrically, so that the failure of one pump cannot have any effect on the operation of the rest. Wherever practicable, there should be two independent supplies of electricity and not dependent on manual operation should one source of current fail.

SEWAGE TREATMENT

AREA OF LAND FOR SEWAGE-TREATMENT WORKS

Sufficient land for construction of works plus adequate space for future extension should always be purchased and reserved, as lack of land can lead to difficulties and heavy expenditure in the future. The following formula gives a fair estimate of the amount of land that should be required for treatment works other than those involving land treatment:

$$A = P^{0.\dot{6}}/74 \tag{9}$$

where: A is the area of sewage works property in hectares and P the head of population.

CHEMISTRY OF SEWAGE

By the strength of a sewage is meant its organic content or the amount of material, both suspended and in solution, that can be oxidised by micro-organisms. Strength is estimated by chemical analysis according to a number of methods described in the Ministry of Housing and Local Government publication *Methods of Chemical Analysis as Applied to Sewage and Sewage Effluents*.[4]

McGowan's formula was for long used in Great Britain for estimating strength of sewage as a prelude to the design of sewage treatment works but, abroad, biochemical oxygen demand (or BOD value) is normally determined for this purpose, and it is now usual for engineers and chemists in Great Britain to neglect McGowan's formula and to rely on BOD value. Biochemical oxygen demand is determined by measuring the quantity of oxygen consumed by a sample of sewage (or of sewage effluent) which is incubated at a constant temperature for a period of (usually) five days. This is known as BOD_5.

BOD load can be expressed in terms of kilogrammes of BOD per head of population per day. The following are typical figures:

	Kilogrammes BOD *per head per day*
Domestic sewage from a housing estate	0.045
General design figure for a mainly domestic area on separate sewerage	0.055
General design figure for a mainly domestic area on combined sewerage	0.066
Typical balanced domestic and industrial area on combined sewerage	0.077

The following is an analysis for a sewage of about average strength:

	mg/litre
Ammoniacal nitrogen	35.3
Albuminoid nitrogen	9.1
Total organic nitrogen	22.5
Oxidised nitrogen	trace
Total nitrogen	58.5
Oxygen absorbed at 27 °C in 4 hr	112.7
$B.O.D._5$	375
Chlorine	91.6
Solids in suspension	294.0

STANDARDS OF EFFLUENTS

The Royal Commission on Sewage Disposal considered that the most reliable index of the nuisance-producing power of a polluted stream is the 5 days BOD value, and suggested that an effluent should be considered satisfactory if it contained not more than 30 mg/litre of suspended matter and had a 5 days BOD value of not more than 20 mg/litre, it being assumed that the river into which discharge was made diluted the effluent by at least eight volumes of river water to one of effluent. It is now accepted that higher standards are necessary in particular circumstances.

When sewage is discharged into the sea, the possibility of pollution of bathing beaches or shellfish beds has to be considered. Float experiments have to be made to determine tidal currents (see *Public Health Engineering Practice, Vol. II*[2] for methods of making tidal experiments). According to the results of these experiments, the sewage is stored in storage tanks or tank sewers for discharge at those states of the tide only when there will be the least possibility of its approaching the foreshore. The treatment given to sewage discharged into the sea varies from no treatment to full treatment (i.e. to Royal Commission standard). In the majority of instances, screening and sedimentation are desirable.

TREATMENT OF STORM WATER

In Great Britain the usual practice has been to provide for full treatment of a flow equal to three times average dry-weather flow from combined and partially separate sewerage systems and to pass any excess of storm water to storm tanks or stand-by tanks. In these tanks the storm water was stored and any excess over their capacity passed over weirs to the river or stream. After the storm the contents of the tanks were decanted for full treatment at the sewage works and then the remaining sludge removed. These tanks had a detention period of 6 h dry-weather flow.

Having studied specific cases the writer is convinced that these practices were inadequate and liable to cause severe pollution of streams. His view is that wherever practicable up to 6 × dry-weather flow should be passed for full treatment and the storm tanks should have a capacity as calculated by the following formula:

$$C = 62Ap^{1.5}N^{0.5}/P^{0.5} \tag{10}$$

where: C is the storage capacity in cubic metres, Ap is the impermeable area in hectares, N the number of years between occurrences of storm (one year is suggested) and P is fives times dry-weather flow in cubic metres per minute.

Sewage-treatment works taking flows from separate soil sewers do not require storm tanks but should be capable of accepting all flows coming to them, say six times dry-weather flow.

Storm tanks may be similar in design to sedimentation tanks except that they have weirs to make them fill in turn so that every small storm does not foul them all; also they have floating-arm draw-offs for decanting their contents to the sewage works. Mechanical arrangements may be provided for sludging.

Processes of sewage treatment

Sewage to be given full treatment is first treated for the removal of suspended solids; secondly, it is oxidised by bacterial action; and finally, it is settled for the purpose of removing any suspended solids remaining, including those produced by aeration. At most sewage works the order of treatment is as follows:

1. Screening (before storm-water separation)
2. Detritus settlement (before storm-water separation)
3. Sedimentation of organic content and finely divided mineral matter.
4. Aeration in percolating filters, activated sludge tanks or channels, or on land
5. Settlement of humus after percolating filter treatment, or of activated sludge after activated sludge treatment.

SCREENING

The purpose of screens is to remove large solids liable to interfere with the components of the treatment works. For most purposes, bar screens with 16 to 19 mm spaces are suitable. At small works, hand-raked screens can be used: these require a submerged area of 1 m^2 per 7000 head of population. They can be placed in small tanks that can serve also for detritus settlement.

At other than small works, mechanically raked screens are required which may be automatically controlled and provided with disintegrators for breaking up the screenings. These screens need to be large enough to pass the maximum rate of flow at the velocity of 1 m/s between the bars but the chamber should be so proportioned that settlement of detritus does not occur.

Comminutors are proprietory mechanisms that serve both purposes of screening and disintegrating the screenings. They may be applied in many circumstances.

Screenings can be:

1. disintegrated and returned to the flow of sewage
2. disintegrated and passed to the sludge disposal plant

3. buried in trenches
4. incinerated (expensive).

DETRITUS SETTLEMENT

Sand, gravel, etc., that otherwise would settle in various sewage or sludge tanks are removed either before or after screening. At very small works the chamber housing the screens will serve for this purpose but at large works either mechanised tanks or constant-velocity channels are used. One in the former class is the Dorr Detritor: this is a flat-bottomed square tank from which detritus is removed by a rotating mechanism and passed to a grit washer. The tanks have a surface area of about 1 m^2 for every 1555 m^3/day maximum rate of flow.

An interesting and efficient method for removing detritus containing little organic material is the constant-velocity detritus channel of parabolic section in which the velocity flow is controlled by a standing-wave flume. Two or more channels are installed in order that one or more may be put out of use for maintenance purposes. The depth of the channels is usually not less than the depth from crown to invert of the incoming sewer; the length should be about thirty times the depth, and the width of top water level is found from the formula:

$$X = 4.92Q/H \qquad (11)$$

where X is the width of the channel at the water level in metres, H is the depth of flow in metres and Q the rate of flow at that depth in cubic metres per second.

The channel is truly parabolic in shape or, for convenience, an approximation to the parabola avoiding the use of curved surfaces, a W section is often used. Downstream of each parabolic channel is a standing-wave flume of rectangular section, the invert of which is level with the invert of the detritus channel. The width of the throat of the flume is determined by formula (4).

Constant-velocity detritus channels of large size may be cleansed during use by dredger or travelling grit pump. The removed grit is comparatively clean, but can be improved by mechanical washing if it is required for any purpose. Detritus channels for small works are mostly easily cleansed by closing the inlet to the channel to be cleansed when the water in the channel runs out through the flume at the lower end, leaving behind the grit which can be dug out by hand labour.

SEDIMENTATION OF SEWAGE

Before treatment of sewage by aeration in percolating filters, in activated sludge works, or on land, strength can be considerably reduced by settlement of suspended solids. Sedimentation, apart from being an economical means of reducing the strength of sewage, amounts almost to a necessity as a prelude to efficient aeration by some methods e.g. percolating filter treatment.

The capacity of sedimentation tanks is not critical. A large change of detention period tends to produce a small modification of percentage of suspended solids settled. On the other hand, attention to detail is important. First, a badly designed inlet causes excessive turbulence in the tank and this can ruin a performance. Secondly, there must be capacity for storage of sludge if some designs of tank are to give good results. Convenience of working must not be overlooked regarding the type of tank and the way in which it is to be used.

Municipal sedimentation tanks should be capable of removing upwards of 70% of the suspended solids in crude sewage. At the same time the BOD content should be reduced by about 42%.

Sedimentation tanks can be classed as cross-flow tanks and upward-flow tanks, although there is a complete range of intermediate designs between cross-flow and upward-flow types. A cross-flow tank is one in which flow is from side to side or centre outwards and an upward-flow tank is one in which the flow enters at a low level and is withdrawn by weirs at the surface. Most present-day tanks have inlets at a moderate distance below the surface and outlet weirs at the surface, and so they do not fall distinctly into any one of the above categories.

While long rectangular tanks with flow from end to end are still constructed in some circumstances, most are now either square hopper-bottomed tanks with central inlets and peripheral weirs, or circular conical or flat-bottomed tanks with central inlets and peripheral weirs, the latter with mechanisms for continuous sludging.

Upward-flow, pyramidal and other semi-upward flow primary tanks should have surface areas of not less than 1 m^2 for every 10 m^3 per day dry-weather flow. A detention period of about 6 h dry-weather flow is now frequently allowed in British practice.

The pyramidal bottoms of square tanks are, in Great Britain, usually sloped at 60° to the horizontal to ensure that the sludge does not hang up. In America steeper slopes are usual. Sludge should be withdrawn at least once a week, and preferably daily, and withdrawal should be slow in order that sewage is not withdrawn with the sludge. An important point to be kept in mind in the design of pyramidal bottomed tanks is that there should be sufficient capacity for sludge storage below the central inlet. A practical rule is to allow for a sludge storage of one week's production and this should be sufficiently below the inlet for the sludge not to be disturbed. The sludge pipes for pyramidal bottomed tanks are normally 150 mm in diameter and are carried up from the bottom of the tank to form a rodding eye above top water level. From this pipe a branch connects at 1 to 1.5 m below top water level to the sluice valve which controls the draw-off and discharges to a manhole on a line of drain that leads to the sludge treatment works.

When practicable all sedimentation tanks are provided with wash-out pipes in order that they may be emptied under gravity. In the case of pyramidal tanks the depth of the pyramid often makes it impracticable for the tank to be completely emptied in this manner.

Circular tanks with flat or comparatively flat bottoms and equipped with mechanical raking mechanism for the continuous removal of sludge are commonly employed at the larger sewage works. There are many designs of automatic raking mechanisms.

SLUDGE TREATMENT AND DISPOSAL

The disposal of sludge is a major problem at sewage works for, even in those circumstances where it is suitable as manure, the quantity tends to exceed the demand and the latter is seasonal, whereas sludge production is virtually constant throughout the year. The problem is now becoming more difficult because of the increasing quantities of toxic substances which tend to render sludge unsuitable for any purpose and call for precautions regarding its disposal and also the future use of any land on to which it has been dumped.

The amount of sludge produced at the average municipal sewage works is about 0.079 kg per head of population per day which, at an average moisture content of 95.5%, gives about 1.76 litres of sludge per head of population per day. This sludge is withdrawn from the primary sedimentation tanks but includes the primary sludge from the sewage plus the humus from percolating-filter schemes or the surplus activated sludge from activated-sludge schemes. For design purposes it is a safe rule to make pumping mains and drains from two and a half to three times the above volume to allow for variations in moisture content.

Means of disposal of sludge include:

(i) discharge into trenches and covering over, or ploughing into land
(ii) distribution by pipe line to nearby farm land for utilisation as fertiliser (where safe)
(iii) drying on drying beds (usually the best method)

(iv) drying with the aid of filter presses or vacuum filters
(v) heat drying
(vi) digestion followed by drying
(vii) heat treatment followed by filter pressing (Porteous process)
(viii) incineration
(ix) dumping at sea (can be very economical where practicable)

Of these methods those most commonly used include some form of dewatering, the chief form of dewatering being drying on sludge-drying beds. Vacuum filters are now improved and are without some of the drawbacks of previous designs. Dumping of sludge at sea is limited to those vicinities where the sludge can be delivered to ships or to sea by pipeline. Disposal by pumping liquid sludge to land is limited to a minority of works and those circumstances where sludge does not contain too high a concentration of any poisonous element.

SLUDGE-DRYING BEDS

Sludge-drying beds are shallow lagoons usually with concrete floors. A layer of clean clinker or other medium is laid to a depth of 300 mm including about 50 mm of fine medium graded from 6 to 12 mm diameter. The beds are under-drained with agricultural tiles or false floors of filter tiles from which the sludge liquor gravitates to a pumping station for return to the flow of crude sewage. Sludge is discharged from the sedimentation tanks to form a layer about 250 mm deep which is then left to dry. As soon as it is dry enough it is dug out of small works or removed by special machines at large works. The area required for sludge-drying beds is in the region of 0.24 m^2 per head of population served. There should be several beds to allow for the cycle of filling, draining and sludge removal.

SLUDGE DIGESTION

The digestion of sludge with the aid of alkaline anaerobic bacteria reduces the solids by converting these into methane and carbon dioxide. After digestion the sludge loses its unpleasant odour, is usually easier to dewater and is reduced in quantity: the number of pathogenic bacteria is reduced and most weed seeds are destroyed. The gas can be collected and used in dual-fuel engines to supply most or all of the electric power required on the works.

Sludge can be digested at day temperature but a detention period of three or preferably four months is required: gas collection is also not economic. If, however, the sludge is heated to about 30 °C (mesophilic digestion) digestion is much more rapid and smaller tank capacities are required. In this process it is usual to have two-stage digestion with heated primary tanks arranged for stirring, mixing and gas collection and unheated secondary tanks arranged for decanting of strata of sludge liquor that form at or near the surface. The primary tanks should have a capacity of about 1 m^2 for every 1.55 kg/day of volatile solids in the sludge or, on the average, 1 m^3 for every 2.14 kg/day of total sludge solids. This usually works out at 0.037 m^3 of tank per head of population or about three weeks' detention period. If the sludge as withdrawn from the sedimentation tanks has a moisture content of more than 95.5%, dewatering tanks should be arranged to come before the primary digestion tanks.

The secondary digestion tanks following heated primary tanks should have a capacity of 0.064 m^3 per head of population served, or about 0.09 m^3 head if a high standard of dewatering is desired. These tanks should have several decanting valves at various levels.

Sludge gas consists of about 67% of methane, about 3% of various gases and 30% carbon dioxide, provided that there is an adequate detention period. If the tanks are too small the proportion of carbon dioxide to methane will be greater and the gas will be useless as a fuel. When the gas is burnt in dual-fuel engines the heat that can be recovered from the exhaust gases and cooling water is usually sufficient for heating the sludge in the primary digestion tanks.

The amount of sludge gas is about 0.96 m^3/kg of solids digested or in the region of 0.0222 m^3 per head of population per day. It has a net calorific value of about 22 343 kJ/m^3 or approximately 1 m^3 of gas will produce 1.86 kW h with average efficiency dual-fuel engines and alternators.

There are several methods of collecting sludge gas including the provision of spiral-guided gasholder roofs having a storage capacity of 4 to 8 h gas production.

Mechanisms for heating and stirring digestion tanks and for decanting water bands are manufactured by Dorr–Oliver Co. Ltd, and Ames Crosta Ltd.

TREATMENT OF SEWAGE ON LAND

In the method of 'land filtration' which is applicable to sites with light porous soils, settled sewage is irrigated into trenches and the treated effluent collected in under-drains which run parallel to, and between the trenches. Where the soil is impermeable, the less effective method of 'broad irrigation' is used, that is, the sewage is made to pass over the surface of the land by means of irrigation trenches which follow the direction of the

Table 27.6

Class of soil and subsoil, and method of working	*Cubic metres of settled sewage per hectare per day*	*Area required for sludge disposal per 1 000 cubic metres of sewage per day* (hectares)
Class I. All kinds of good soil and subsoil, e.g. sandy loam overlying gravel and sand:		
a Filtration with cropping	135	0.445
b Filtration with little cropping	280	0.445
c Surface irrigation with cropping	78	0.445
Class II. Heavy soil overlying clay subsoil:		
Surface irrigation with cropping	56	1.07
Class III. Stiff clayey soil overlying dense clay:		
Surface irrigation with cropping	34	1.78

contours, the land is 'contour ploughed' so as to prevent rapid flow over the surface, and the treated effluent is collected by a series of ditches. At one time it was recommended that land treatment should be used where the land was suitable and of reasonable cost. Now the tendency of British practice is towards obsolescence of land treatment.

Table 27.6 is based on figures in the Fifth Report of the Royal Commission on Sewage Disposal, and gives some indication of the areas of land required.

PERCOLATING-FILTER TREATMENT

Percolating-filter treatment is used in Great Britain at more sites than any other method of sewage treatment. The tendency of practice is to use percolating filters at all new works for rural schemes and for municipalities of moderate size. For very large towns and regional schemes, activated sludge methods are more usual. A percolating filter consists of a bed of coke, clinker, broken stone or other suitable material which will not disintegrate under the action of the weather or of corrosive sewages. The material generally in use at present is of the grade described as 'coarse medium' in the Reports of the Royal Commission, the particles seldom being smaller than 25 mm or larger than 75 mm in diameter, excluding the large material used at the bottom of the bed to assist ventilation. The finer grades of such media are somewhat more efficient on the average than the coarser grades; on the other hand, they are more liable to become choked at or near the surface.

Effluent from sedimentation tanks is discharged over the surface of the percolating filters by mechanical sparges. In the case of circular beds (by far the most common) rotating distributors, which are fairly simple mechanisms, are used. The arms of the distributors are perforated in such a manner that distribution of tank effluent is more or less in direct proportion to the area of bed covered by each part of the distributor. Except at large works, dosing tanks with automatic siphons are required, so that at all times of the day the sewage is passed to the percolating filters at rates of flow adequate to actuate the mechanism which depends for its rotary motion on the reaction of the jets of sewage discharging from its arms. These dosing tanks should have capacities of not less than one minute's discharge of dosing siphon (the siphon being sized to discharge at the maximum rate of flow to filters) or not less than 3 m^3 per 1000 m^2 of filter bed surface.

Rectangular beds, which are used particularly where economy of land is necessary but which otherwise are not quite so economical as circular beds, are served by travelling distributors which move from end to end of each bed, usually actuated by the flow of sewage through the mechanism itself, but sometimes cable-driven and actuated either by water-wheel or power drive.

The floors of percolating filters are usually constructed of 150 mm of mass concrete laid to falls of about 1 in 200. Drainage is assisted by the use of large material of about 100 to 150 mm diameter and under-drains of agricultural tiles discharging to effluent collecting channels. These under-drains and the coarse material also assist in ventilating the beds, adequate ventilation being essential to the process of aeration. Occasionally, false floors of special filter tiles are constructed. The walls of the filters are constructed of brickwork bonded in cement mortar, concrete or reinforced concrete: formerly dry walling of stone or boulder clinker was common.

Percolating filters are now constructed between the limits of about 1.4 m deep and 3 m deep. Excessively shallow beds are expensive to construct and comparatively inefficient. It is now known that there is an optimum rate of flow in the region of 5 m^3 per day per m^2 of bed at which the filter gives its maximum efficiency. Where practicable the depth of the filter should be adjusted to give this figure. Alternatively, alternating double filtration or recirculation may be used.

Percolating filters of normal construction and operated without recirculation of alternating double filtration should be of adequate capacity if they have 7.3 m^3 of medium per kilogramme of BOD per day in the crude sewage. This means that about 0.4 m^3 of medium per head of population would be required to treat a mainly domestic sewage from a separate system, 0.48 m^3 of medium per head would serve a generally domestic area with combined sewerage and 0.56 m^3 would probably serve a balanced domestic and industrial area with combined sewerage.

HUMUS TANKS

The sewage trickling through the bed forms a film of moisture over each particle of stone or clinker. This moisture absorbs oxygen from the air which is utilised by aerobic bacteria in oxidising the organic content of the sewage. These bacteria develop rapidly after the filter has been put into use and coat the particles of the medium. Other organisms develop in the beds. Particularly noticeable are algal and similar growths which are to be found at the surface because they require light. These would rapidly choke the surface and cause ponding, were it not for worms and larvae, particularly of the psychoda fly, which feed on the vegetable matter. Ponding at the surface is most common during the winter months when the worms and larvae are driven down from the surface by the cold. In the spring these return to the surface and, breaking up the vegetable growth, cause a heavy discharge of humus.

Humus is easily settled. On the other hand, it is liable to gas and rise after settlement. For this reason care has to be taken in the maintenance of humus tanks. Humus tanks are a normal provision and are omitted only where the filters are followed by land irrigation, or when they are more a trouble than an advantage, as for example in the case of very small works that treat the sewage from isolated buildings.

Humus tanks are usually given capacities in the region of 4 h dry weather flow. Weir outlets are normal, and scumboards are essential because of the tendency for gassing humus to float.

ALTERNATING DOUBLE FILTRATION AND RECIRCULATION

If settled sewage is dosed on to the surface of percolating filters at too low a rate a proportion of the medium does not become wet and the organic growth does not develop there and, therefore, the filters do not achieve their maximum efficiency. Alternating double filtration is a method in which there are two batteries of filters in series one with another, arranged for each in turn to be the primary stage. Recirculation is the return of part of the filter effluent to mix with the feed. Both of these methods increase the rate of flow per unit of surface area and make the filters more efficient.

THE ACTIVATED-SLUDGE SYSTEM

The activated-sludge system is a method of oxidising the organic content of sewage with the aid of bacteria that are in suspension in the sewage forming flocculi. The bacteria grow in the sewage, are removed by final sedimentation tanks and returned to the inlet end of the aeration tanks in sufficient quantity to maintain the proportion of the suspended solids or 'activated sludge'. The oxygen used in the process comes out of solution and has to be replaced. This is effected by one of two methods: injecting air as fine bubbles (diffused-air system) and aeration by splashing at the surface (surface aeration or mechanical agitation). There are several proprietary varieties of these methods. The most popular in Great Britain are the diffused-air system of Activated Sludge Ltd, and the 'Simplex' surface aeration of Ames Crosta Ltd.

The capacity of the tanks and the power required for aeration depend on the quantity of sewage and its strength. Contrary to previous opinion it has been shown by a large body of experimental results that the degree of aeration achieved in a specified time does not depend directly on the capacity of the aeration tanks only, but on the capacity of the aeration tanks plus that of the final sedimentation tanks. The power input can also be varied within limits without affecting the final results provided that the aggregate volume of the aeration and final sedimentation tanks is altered inversely as the alteration of the power input. In Great Britain it is usual to have large aeration and sedimentation tank capacities but moderate power consumption, whereas in America capacities are smaller but power input greater. Nevertheless, on the average, performance seems to be the same.

DIFFUSED-AIR SYSTEM

The detention period and power input required to achieve the desired final effluent depends not only on the strength of the sewage but also on its particular difficulty of treatment in the individual case. On the average the size of the works could be calculated by the formula:

$$T = 564/A \tag{12}$$

where T is the aggregate detention period of the aeration and final sedimentation tanks in hours, and A is the air supply in cubic metres of free air per kilogramme of BOD in the settled sewage delivered to the aeration tanks.

(The value of 'A' is the average working supply: the maximum design figure should be about 1.67 times higher).

The detention period in Great Britain averages 12 h which is shared between the aeration tanks and the final sedimentation tanks. The final sedimentation tanks seldom have a detention period of less than 4 or more than 6 h. In the aeration tanks and any associated aerated channels is accommodated the remainder of the detention period calculated as necessary for the process, but the aeration tanks must not be too small to house the diffusers. For example, if the diffusers were spaced 0.3 m along parallel air mains laid on the bottom of flat-bottomed tanks at 0.75 m centres, the normal design rate of air flow would be 0.08 m^3/min per square metre of water surface in tank. This would be a reasonable figure for tanks 2.5 m deep. The diffusers that produce the small bubbles in the equipment provided by Activated Sludge Ltd are known as 'dome diffusers'. They are mushroom-shaped ceramic tiles and are designed to be fixed to the tops of the air mains.

Air has to be supplied at the pressure necessary to deliver against the total water depth of the aeration tanks plus friction losses in the system of mains. The mains should be sized larger than in other compressed-air practice so as not to give a loss of head of more than 0.06 m per 100 m length of pipe. The air should preferably be compressed by rotary blowers and have dust removed by electrostatic or other effective air filters.

Activated sludge is recirculated in the quantity of about two-thirds of the dry-weather flow and it should be possible to make recirculation equal to dry-weather flow. Surplus activated sludge averages 3.88 m^3/day per 1000 head of population and has an average moisture content of 99.33%. It is usually returned to the flow of crude sewage.

'SIMPLEX' SURFACE-AERATION SYSTEM

In this process there are rows of vertical uptake tubes in flat-bottomed tanks. At the top of each tube is a mixed-flow centrifugal impeller known as a 'high-intensity cone'. These impellers are individually driven by electric motors. Tank capacities and motor powers are in somewhat similar order to those of the diffused-air system according to local differences. Activated sludge is recirculated in similar proportions but usually pumped whereas low-lift air lifts are used in diffuser-air schemes.

EFFLUENT POLISHING

When a higher standard than a Royal Commission effluent is required the usual method of securing this is to filter the effluent from a normal treatment works. Micro-strainers or rapid sand filters may be used for this purpose.

INDUSTRIAL WASTES

The treatment of any industrial waste is an individual problem and has a literature of its own. Accordingly it is not possible to deal with trade waste treatment in a limited space other than generally, and the reader interested in any particular waste is referred to the textbooks[5, 6] on the subject.

The first method of dealing with industrial wastes, which should always be considered, is reduction of polluting content by salving any useful material. Next, acids, alkalis and harmful chemicals may be neutralised or removed by the addition of suitable reagents or precipitants.

Many organic wastes can be treated like sewage by screening and sedimentation followed by oxidation by percolating filters or activated sludge methods. The percolating filter method of recirculation, which was originated for trade waste purposes, is particularly applicable.

Where a trade waste is treated by sewage-treatment methods it is usually advantageous if it is mixed with an excess of normal sewage. For this reason, apart from the dangers of toxic substances, it is acceptable that such wastes should be discharged to public sewers to be treated by the local authority.

SEWAGE DISPOSAL FOR ISOLATED BUILDINGS

While it is possible for the methods applicable to isolated premises to be sanitary, most private installations seriously suffer from ill design and neglect and cesspools, in particular, generally amount to little more than a pretence of sanitation, involving a serious danger to health. Wherever practicable, main drainage should be installed. Where methods applicable to isolated buildings must be used specialist literature should be studied.[2,7]

DANGERS FROM SEWAGE

There are increasing hazards to public health, livestock and crops due to poisonous elements in sewage sludge, land that has been used for sewage treatment and sewage effluents. Great care must now be taken before making use of any sludge, land formerly used for sewage works or effluent.

REFERENCES

1. ESCRITT, L. B. and V. P., *Escritts' Tables of Metric Hydraulic Flow*, Allen & Unwin, London (1971)
2. ESCRITT, L. B., *Public Health Engineering Practice*, Vol. II, *Sewerage and Sewage Disposal*, MacDonald and Evans, London (1972)
3. BILHAM, E. G., 'Classification of Heavy Falls in Short Periods', *British Rainfall*, 262, HMSO (1935)
4. *Methods of Chemical Analysis as Applied to Sewage and Sewage Effluents*, Ministry of Housing and Local Government, HMSO, London (1956)
5. SOUTHGATE, B. A., *Treatment and Disposal of Industrial Waste Waters*, HMSO, London (1948)
6. RUDOLFS, W., *Industrial Wastes*, Reinhold, New York (1953)
7. ESCRITT, L. B., *Sewers and Sewage Works with Metric Calculations and Formulae*, Allen & Unwin, London (1971)

28 LAND DRAINAGE AND RIVER MAINTENANCE

LAND DRAINAGE AND RIVER MAINTENANCE 28

28 LAND DRAINAGE AND RIVER MAINTENANCE

G. McLEOD, M.Sc., F.Inst.C.E., F.Inst.W.E.
Usk River Authority

INTRODUCTION

Objects of land drainage

The object of land drainage works as understood by the river engineer is to prevent flooding or to benefit land by the removal of surplus water. The benefits of preventing the flooding of built-up areas are obvious although not always easy to evaluate in financial terms. The benefit resulting from improved drainage of agricultural land depends on the type of land and the use to which it is put. By lowering the water table in an area of grassland the productivity may be increased by fifty or one hundred per cent, but a slightly greater lowering may make the land suitable for arable cultivation so that the value of the crop may be several times what it was before the land was drained.

History of land drainage

In England some early land-drainage works were carried out by the Romans, but the earliest known drainage authority was the Lords, Bailiff and Jurats of Romney Marsh, in Kent. Large-scale drainage works were undertaken in the seventeenth century in the Fens and in Hatfield Chase in south Yorkshire, under the supervision of a Dutch engineer Sir Cornelius Vermuyden, who cut straight artificial channels through the great areas of marshland to lead the waters to the tidal estuaries.

Organised drainage

Commissioners of Sewers were appointed to organise the drainage of various areas and drainage boards were set up under local Acts of Parliament. The system was reorganised by the Land Drainage Act of 1930 under which 52 catchment boards were set up covering those parts of the country where it was considered that there was most need of drainage works. The catchment boards were empowered to maintain and improve the 'main rivers' which were designated within their areas by the Ministry of Agriculture and Fisheries, and they had a general power of supervision over the drainage boards within their area, now called 'internal drainage boards'. The catchment boards could submit schemes for the formation of new internal drainage boards, which could be established by Ministerial Order. The catchment boards were concerned with land drainage only.

River Boards

Under the River Boards Act of 1948 the catchment boards were absorbed into the river boards which had additional powers of control of fisheries and pollution. There were only

33 river boards, which with 'Conservancy Catchment Boards' for the Thames and the Lee, covered the whole of England and Wales. Their land drainage powers were still derived from the Land Drainage Act of 1930, with minor amendments. There was a new Land Drainage Act of 1961 which introduced General and Special Drainage Charges to enable revenue to be obtained from agricultural land outside the internal drainage board areas, but few of the river boards took advantage of these powers.

River Authorities

A further change resulted from the Water Resources Act of 1963 which set up 27 river authorities which took over the duties of the river boards but also had extensive powers

Figure 28.1 Areas of river authorities

of control over water abstraction, and a duty to survey the water resources and demands within their areas, and where necessary take action to augment the resources. Distribution of water remained the responsibility of the water undertaking but any new supplies of water they needed and even the abstraction of water from their existing reservoirs required a licence from the river authority. The authorities undertook a great deal of work on the measurement of river flow, rainfall and evaporation for which grants were now available under the guidance of the Water Resources Board which had also been set up under the Act of 1963. River flow and rainfall had been measured on a smaller, but increasing, scale by the catchment boards and river boards, but the work was now organised on a national basis (Figure 28.1).

Regional Water Authorities

The surveys of water resources and demand indicated the immense scale of the works required in the next decades to meet rising demand, and it was found that this work would have to be organised over larger units of area than those of many of the river authorities. It was also felt that because many water supplies in the future will come from abstractions in the lower reaches of rivers, the body responsible for the supply of water

Figure 28.2 Proposed regional water authorities

should also be in control of the treatment of sewage discharge to the rivers. The Water Act recently enacted provides for the setting up of ten new Regional Water Authorities which will take over the land drainage, fisheries, water resources and pollution prevention functions of the river authorities as well as the duties of water distribution and supply and sewage treatment. They will also provide water recreation facilities.

Future organisation

The Regional Water Authorities will have regional land drainage committees and may set up area land drainage committees for parts of the region. The Authorities will carry out works on 'main river' and will have powers to control the operations of other bodies and individuals affecting main river, e.g. the construction of bridges or culverts. They will exercise a general supervision over internal drainage boards and over the maintenance of the principal secondary watercourses in the parts of their area not covered by such boards. It is expected that they will continue to carry out sea-defence works intended to prevent flooding by tidal waters (as distinct from coast-protection works intended to prevent the erosion of higher coastal land) and will, as in the case of former drainage authorities, operate works for irrigation by maintaining suitable water levels in the drainage systems during dry periods.

Finance

Land-drainage works have in the past been financed from several sources. The river boards and river authorities derived their basic land-drainage income from precepts on the county boroughs and county councils and internal drainage boards within their areas. Approved improvement schemes received grants from the Ministry of Agriculture, Fisheries and Food and in some cases contributions were made to the cost of works by persons or corporate bodies deriving benefit. General or special drainage charges could also be levied under the Land Drainage Act 1961. Internal drainage boards derive their income from drainage rates levied on land and other property within their area. Land drainage work will in the first instance be financed as at present by a precept on local authorities but may be transferred to water supply charges at a future date.

DRAINAGE PROBLEMS

Agricultural

Bad drainage of grassland results in a high water table below which the roots of the grass cannot extend and the depth of soil available to the plants is therefore limited. Because of the shallow root system a temporary lowering of the water table during droughts will dry out the surface layers of soil and cause 'burning' of the turf. The high water table makes the land cold and retards growth and nutrients in the waterlogged soil are leached away.

These disadvantages apply with greater force to arable cultivation, especially of root crops. Potatoes will be killed and rot if they are flooded for twenty-four hours and most arable crops require a depth of several feet of well drained soil. One of the difficulties of the drainage engineer is to obtain authoritative guidance on the optimum depth of the water table from field surface level but it may be taken that water level in the ditches should be at least three feet below field level for grassland, or three feet in peat lands to six feet in silts used for arable cultivation. All these margins might be increased in winter conditions when there is no danger of overdraining, provided that this does not cause difficulties where the ditches form the fences between fields.

Urban

In urban areas also it is not sufficient to ensure that water does not cause visible flooding. If land is saturated to surface level, basements, subfloors, pits in garages or for heating boilers will fill with water and floors and foundation walls will be damaged. Visible flooding of buildings, either domestic or industrial will immediately raise an outcry and a demand for remedial measures.

Assessment of damage

In order to be satisfied that the cost of remedial works is justified it is necessary to evaluate the damage caused by bad drainage. In the case of agricultural land it is possible to estimate the increased value of annual production which could be obtained by improved drainage and to capitalise this figure. In urban areas a comprehensive or sample survey may be made to ascertain the number of houses of graded sizes or values liable to flood and to estimate the loss per house of each class likely to result from each flood. From a knowledge of the frequency of flooding (see p. **28**–8) the present value of all the damage likely to occur during the expected life of the remedial works may be arrived at. Similar techniques may be applied to industrial premises but there is more difficulty in assessing the cash value of preventing the flooding of roads or recreational areas, or relieving the anxiety which may be suffered by residents in flood areas even when floods do not actually occur.

Remedial measures

The methods used to deal with flood problems fall into the main classes listed below:

- Catchwater drains
- Channel improvements
- Embanking
- Washlands and detention basins
- Reconstruction or removal of weirs
- Control structures
- Pumping
- Sea-defence works

These methods may be applied singly or in combination. For example, flood levels in a river may be lowered by channel improvements but may still require embankments to contain the highest floods. The methods will be discussed in subsequent paragraphs but it may be advisable first to consider some more general matters.

QUANTITIES OF WATER TO BE DEALT WITH

Measurement

The ideal basis on which to decide on the flood flows for which remedial works will be designed would be a long period of recorded measurement of flow in the river or stream concerned. This is rarely available but a record from a gauge on a neighbouring similar stream or at a different point on the same river may be used with correction for the difference of catchment area or other relevant parameters.

In recent years large numbers of gauging stations have been established in England and Wales and the records from them have been collected and published by the Water Resources Board in successive issues of the *Surface Water Year Book*. In waterworks practice, gauging weirs on streams of moderate size, such as those proposed to be impounded to form reservoirs, were generally sharp-edged rectangular or triangular plate weirs. Their accuracy was dependent on gravel or other debris not being allowed to accumulate behind the plate. On small streams a temporary weir of this type may be quickly installed to measure floods during a single winter as an aid to the assessment of more infrequent major floods.

The most common type of gauging weir is now the 'Crump' weir with a triangular profile with the upstream face at a slope of 2:1 and the downstream face at a slope of 5:1.[1] It has the advantage of being 'modular' until 75% drowned, will allow gravel to pass over it, and is passable by migratory fish to heights up to at least three feet above tail water level. In rivers with a wide variation of flow it is difficult to design a Crump weir which will measure low flows accurately without raising flood levels unduly and compound weirs with a low section for low flows are used. The basic weir formula for the Crump weir is

$$Q = 1.96\, bH^{3/2} \quad \text{(metric units)}$$

or

$$Q = 3.55\, bH^{3/2} \quad \text{(ft s units)}$$

where H is the total head including velocity head.

In terms of the gauged head h,

$$Q = Cv\, 1.96\, bH^{3/2}$$

where Cv can be found from the equation

$$0.196\left(\frac{h}{h+p}\right)^2 Cv^2 - Cv^{2/3} + 1 = 0$$

Tables for the evaluation of Cv are available in *Crump Weir Design* published by the Water Resources Board[2].

A 'flat-V' weir with the crest sloping from the sides to the centre has also been used, but produces a concentration of flow in the centre of the channel downstream which may cause scour.

Early gauging stations on large rivers were mainly velocity–area stations where velocities are measured by current meter at a carefully selected and surveyed cross section. This should be in a straight and uniform length of channel with a level bed free from obstructions. In most cases the current meter is suspended from a cableway with a winch on the bank from which the horizontal and vertical position of the meter can be controlled. Most modern meters are of the helical propeller type. In fast-flowing rivers, sinkers up to 50 kg are required for use with the meters. Water velocities are measured at one, two, or more points on each of a number of verticals spaced across the river.

Dilution methods of gauging depend on measuring the degree of dilution of a salt or other chemical injected into the stream. Where the water is used for abstraction, for fisheries, or for stock, care is obviously needed in the selection of the tracer.

Variation with area

According to Nash and Shaw[3]:

$$\bar{Y} = 0.009\, A^{0.85}\, R^{2.2}$$

where $\bar{Y}$ = mean annual flood in ft^3/s
A = catchment area in square miles
R = mean annual rainfall in inches

If the value of R is assumed to be constant over the whole catchment, the flows derived from records at a gauging station may be applied to parts of the catchment, or to neighbouring similar catchments, by using the relationship

$$Q \propto A^{0.85}$$

Calculation of flood flows

Where no records of flow measurement exist the flood flow for design purposes must be estimated from the characteristics of the catchment. For many years the most favoured method was the 'rational' or 'Lloyd–Davies' method which was based on the assumption that maximum flow would occur as the result of the maximum rainfall intensity to be anticipated within the 'time of concentration', i.e. the time taken for rainfall falling on the furthest part of the catchment to reach the part of the river under consideration. In its simple form this method assumed that:

(a) rainfall intensity was constant during the time of concentration,
(b) rainfall intensity was constant at all parts of the catchment,
(c) that the permeability of the catchment surface was uniform over the period of the time of concentration.

These assumptions could only be justified in respect of small, generally urban, catchments and the method has now been superseded.

Modern methods are based on the concept of the 'unit hydrograph'. The unit hydrograph of duration T is the storm run-off due to unit volume of effective rainfall falling uniformly in time and space on the catchment in time T. The instantaneous unit hydrograph is the limit to which the unit hydrograph tends as the duration T is diminished indefinitely. It is assumed that the hydrograph of the flow produced by other quantities of rainfall falling in periods equal to T will be in the same form as the unit hydrograph but with the ordinates increased or decreased in proportion to the quantities of rainfall. If the hydrographs produced in this way for successive intervals of time are plotted on an appropriate time base the ordinates of them may be added to give the total flow at each moment during the whole period. A synthetic hydrograph for the flood resulting from any sequence of rainfall may thus be produced.

Rainfall frequencies

Bilham's formula is widely used for the estimation of probable storms of a given frequency. It may be doubted if a single formula can be applicable to areas of all sizes in all parts of the country, although the very exceptional storms of 150 mm or more are not confined to any particular area. Bilham's formula is:

$$r = \left[\left(\frac{1.25T}{N} \right)^{1/3.55} - 0.1 \right] \times 25.4$$

where r is the total rainfall in mm, T is the storm duration in hours and N is the probable number of such storms in 10 years.

Flood frequencies

From a study of flood records it is possible by Gumbel's or other methods to establish the frequency of a flood of a given volume and to assess the rate of flow to be expected *on average* once in 10 years, 20 years, or 100 years. It is important to realise that this does not mean that the '10-year flood' will occur at 10-year intervals, but that the probability of such a flood occurring in a given year is 1 in 10.

Most river authorities have been trying to build up maps showing the extent of floods of various frequencies and in many river valleys can state the probability of the flooding of particular areas.

Knowing the average frequency of floods of various magnitudes it is possible to decide, on the basis of the property at risk, what flow should be adopted for design purposes. It may be acceptable for grassland to be flooded for a short period once a year, or arable land once in 10 years, but where residential or industrial property is concerned once in 100 years may be the acceptable risk, or if the flooding would be deep enough to involve serious loss of life the design period may be even longer. It is emphasised that two 100-year floods may occur in one year and what is meant is a flood with a 1% probability.

In some cases it costs little more to design for the 100-year flood than for the 10-year flood, while in others the cost may rise rapidly with the rate of flow provided for.

Flood studies

A study unit has been set up by the National Environmental Research Council to make a study of floods in British rivers and when the report of this study is published it is expected to form the most reliable guide to the assessment of flood flows to be provided against in river improvement schemes.

CHANNEL DESIGN

Flow formulae

The formula most generally used in channel design is that of Manning:

$$V = \frac{1}{N} R^{2/3} S^{1/2}$$

where V is the mean velocity in the channel in m/s, N is a constant equal to Kutter's N, R is the hydraulic mean depth in metres and S is the surface slope (fall/length).

In foot-second units the formula is

$$V = \frac{1.486}{N} R^{2/3} S^{1/2}$$

the value of N remaining unchanged for a particular channel in either system of units. Tables of appropriate values of N are given in various textbooks and in some of these[4] selection is aided by photographs of channels where certain values would be applicable. One method is to measure flows and corresponding surface gradients in the existing river channel during high flows and calculate the value of N from the formula. This can then be used for the calculation of flood levels.

In lowland channels of typical trapezoidal form, newly cut in fine bed material, N may be as low as 0.022 5 and provided that such a channel is well maintained the mature channel may have a value of 0.025 which is a figure commonly used for the design of lowland drainage channels.

A natural river with a gravel bed may have a roughness coefficient of 0.030 to 0.035, and a badly obstructed channel may have values as high as 0.050. The latter figure is not likely to be used in design calculations.

For very smooth concrete pipes or channels N may be as low as 0.012 but in general cases 0.014 may be a safer figure. Where there is surface roughness or deterioration, construction joints or algal growths, i.e. conditions which may be expected some years after construction, the value of 0.017 may be taken. The size of a channel to carry a given flow may therefore be reduced by lining it with concrete and this may be worthwhile where space is restricted or particularly valuable, e.g. through a factory site.

Other formulae

Other formulae which have been used for flood calculations are Bazin's, Kutter's and the more recent Colebrook–White formula

$$V = (32gRS)^{1/2} \lg \left(\frac{k}{14.8R} + \frac{1.255r}{R(32gRS)^{1/2}} \right)$$

where k is the linear projection of the roughness and r is the kinematic viscosity. This

Table 28.1 OPEN-CHANNEL WATER LEVEL CALCULATION

River.. From.. To...............................

Reach	Downstream water level (m)	Assumed rise in reach (m)	Water level at mean section (m)	Mean bed width and side slopes (m)	Mean bed level (m)	Depth d at mean section (m)	Area A of mean section (m^2)	Wetted perimeter P of mean section (m)	Mean hydraulic radius $R = A/P$ (m)	Discharge Q (m^3/s)	Mean velocity $V = Q/A$ (m/s)	Length of reach L (m)
1 U/s face bridge 'A' to d/s face bridge 'B'	3.460	0.026	3.473	13 $1\frac{1}{2}:1$	1.460	2.013	32.3	20.25	1.60	14.0	0.434	400
	Afflux through bridge 'B' = 0.022 m (see p. 28–13).											
2 Bridge 'B' to tributary 'C'	3.507	0.060	3.530	13 $1\frac{1}{2}:1$	1.505	2.025	32.5	20.28	1.601	14.0	0.431	1 000
	Flow in tributary 'C' is 3 m^3/s.											
3 Tributary 'C' to bridge 'D'	3.570	0.06	3.600	13 $1\frac{1}{2}:1$	1.560	2.040	32.8	20.35	1.61	11.0	0.335	1 600
Water level at bridge 'D'	3.570 + 0.062 = 3.632 m											

Notes

1 Downstream water level for first reach (at bridge 'A') must be known or may be derived by starting further downstream from a range of assumed levels. The resultant water level curves will be found to converge to indicate the level at 'A'.

2 The first value assumed for the rise in the reach may be based on the rise in bed level, i.e. depth is assumed to be constant. If the calculated value $(h_f + h_v)$ is significantly different, start again with a new assumed rise near to the first calculated value. Unsuccessful assumptions have been omitted from Table 28.1.

3 Water level at mean section is downstream water level plus half the rise in the reach.

4 Area of cross section $A = d(13 + 1\frac{1}{2}d)$
Wetted perimeter $P = 13 + 3.606d$

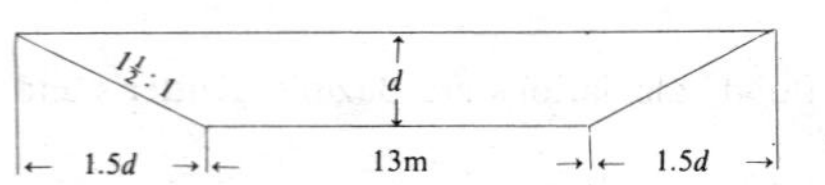

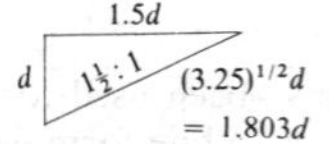

Calculated by .... Date ....

Checked by ....

Coefficient of rugosity N	Loss of head due to friction h_f (m)	Upstream bed width and side slopes (m)	Upstream bed level and water level (m)	Upstream depth (m)	Upstream Q (m^3/s)	Upstream area (m^2)	Velocity head at downstream end of reach h_{v1} (m)	Velocity head at upstream end of reach h_{v2} (m)	Loss of velocity head $h_{v1}-h_{v2} = h_v$ (m)	Total loss of head in reach h_f+h_v (m)
0.025	0.0251	13 $1\frac{1}{2}:1$	1.485 3.485	2.000	14	32.0	0.0099	0.0102	−0.0003	0.0248
ater level upstream = 3.507 (3.460 + 0.0245 + 0.022)										
0.025	0.062	13 $1\frac{1}{2}:1$	1.525 3.567	2.042	14	32.9	0.0102	0.0092	+0.001	0.063
ow in next reach is 14 − 3 = 11 m^3/s										
0.025	0.059	13 $1\frac{1}{2}:1$	1.595 3.630	2.035	11.0	32.65	0.0092	0.0057	+0.0035	0.062

5 Frictional head loss $h_f = \dfrac{N^2V^2L}{R^{4/3}}$

and if $N = 0.025$ $\quad h_f = \dfrac{0.000625\ V^2L}{R^{4/3}}$

6 With such low velocities as in this example the calculation of velocity heads would normally be omitted. Unless there is a change of section or flow, only one value need be calculated for each reach, the value of h_{v1} being the same as that calculated for h_{v2} in the previous reach. $h_{v2} = \dfrac{V_2^2}{2\times 9.81}$ where V_2 is the upstream velocity (Q divided by upstream area).

formula is most easily applied by the use of charts which have been published by the Department of Scientific and Industrial Research.[5]

Manning's formula, however is much more easily used and has given good results in practice. Much work has been done by many investigators working in different conditions on the evaluation of N and the engineer can select a suitable value and can use the formula readily with a slide-rule, calculator or computer.

Design procedure

Having established the design flow, the flow formula, and the average gradient (total fall/length of channel) some trial cross sections can be used to find a width and depth which will give the required flow when running at the known gradient and with uniform flow. These dimensions are adopted as a provisional design and the more precise calculations of flood levels are then made. For this purpose the channel is divided into reaches sufficiently short to make it reasonable to assume that within the reach the surface fall will be the same as if the depth were constant and equal to the mean depth.

Starting from a known or assumed flood level at the downstream end of the first reach a trial figure is taken for the surface fall in the reach. Half this fall plus the downstream level will give the mean surface level and subtracting the mean bed level gives the mean depth for the reach. The corresponding area, wetted perimeter, hydraulic mean radius and velocity can now be calculated and hence the fall in the reach. (See example, Table 28.1) For this purpose it is convenient to transpose Manning's formula to the form:

$$h = \frac{N^2 V^2 l}{R^{4/3}} \quad \text{(metric units)}$$

The calculated fall is now compared with the assumed value and if the difference is significant a new value is assumed nearer to the calculated value and the process repeated until agreement is close. The final calculated fall is added to the downstream level to give the level at the head of the first reach, which is the starting point for the second reach.

In calculations in connection with lowland drains and slow-moving rivers the changes of velocity head between one part of the channel and another are very small and can be neglected, but in fast-flowing rivers the change of velocity head in each reach must be calculated and allowed for. If a constricted length of channel causes an increase of velocity from 2 m/s to 4 m/s this change alone will require a surface fall of 0.61 m in addition to frictional losses.

If the channel is of uniform section and other parameters remain constant through the successive reaches the calculation could be carried out by computer with the advantages of increased speed and accuracy, but in many cases variations in side slopes, afflux at bridges and other changes would make programming difficult.

The calculation is selfcorrecting to some extent because if the fall in one reach is overestimated giving too high a water level at the head of the reach, the depth in the next reach will be increased, giving a flatter gradient and bringing the calculated surface level back towards the correct value.

Having calculated the flood surface level in the channel, embankments and flood walls can be designed, with a suitable freeboard. This will depend on the nature of the land or property protected. Embankments are normally given more freeboard than walls because of the greater risk if they are overtopped and the greater risk of settlement or damage by animals.

Side slopes

Side slopes of channels can be based on those of the existing channel in the same material but should never be steeper than 1:1 (more commonly 2:1). In major channels the selection of safe side slopes should be the result of proper soil mechanics investigation of shear strength and stability. Serious difficulties have arisen on some major schemes due to inadequate investigation of side-slope stability.

Afflux at bridges

On most rivers there will be a number of bridges which form local constrictions of the flow and the loss of energy at these will form a step in the surface gradient. This afflux is dependent on the degree of streamlining of the bridge opening. With perfect streamlining the loss of head at entry would be

$$H-h = \frac{V^2}{2g}-\frac{v^2}{2g}$$

or, in terms of the flow Q,

$$H-h = \frac{Q^2}{2g}\left(\frac{1}{A^2}-\frac{1}{a^2}\right)$$

where H is the velocity head under the bridge, h is the velocity head upstream, V is the velocity under the bridge, v is the approach velocity upstream, A is the cross-sectional area of the bridge opening and a is the cross-sectional area of the channel upstream.

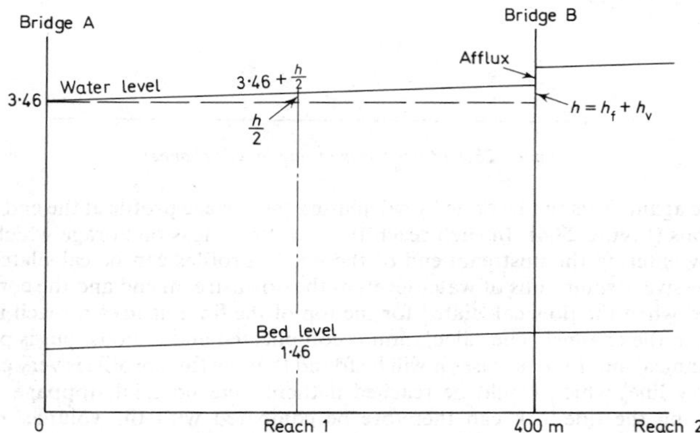

Figure 28.3 Water-level calculation

In practice the loss will be greater by some factor C_a (which may be 1.2 for bridges with square abutments or 1.1 where there is some streamlining). Usually there is no recovery of head downstream so the afflux to be taken into the water level calculations is

$$H-h = C_a\frac{Q^2}{2g}\left(\frac{1}{A^2}-\frac{1}{a^2}\right)$$

Example

Table 28.2 illustrates a tabular form of setting out the water-level calculation for a river channel. (See also Figure 28.3) By the use of a computer or desk-top calculator the length of reaches can be reduced and the fall in each reach determined accurately without the tedious process of 'manual' trial and error.

Unsteady flow

There are many channels where the flow is subject to rapid and predictable variation such as a lowland drain discharging through a sluice into a tidal river. In such a case the critical condition arises when the tidal door has been closed for the maximum period and the water reaching the channel during the period of tide-lock has been added to the water which was running in it when the tidal door closed. The water then reaches the 'maximum ponding level' at the moment when the tidal doors open and the level begins to fall again. The problem is to find out how low the water levels will fall before the tidal

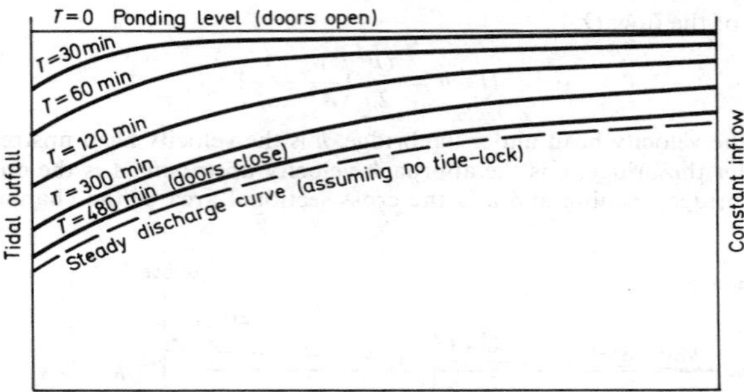

Figure 28.4 Water levels in tide-locked channel

doors close again. This can be done by calculating the surface profile at the end of successive intervals (Figure 28.4). In each reach there will be a loss of storage which will add to the flow entering the upstream end of the reach. Profiles can be calculated starting from successive assumptions of water levels at the downstream end and the correct value is indicated when the flow calculated for the top of the final upstream reach is equal to the inflow to the channel. The calculation is complicated and tedious, but is practicable with mechanical aid. In most cases it will be found that the final profile is very close to the 'steady flow line' which would be reached if there were no tidal stoppage. The total inflow during the tide-lock can therefore be compared with the volumes of storage available above the steady-flow line when the channel is filled to various levels, to establish the ponding level. In important schemes, or where the ponding level is particularly critical, or the tide-lock period is unusually long, the full calculation is advisable.

RIVER RÉGIME

General

In a stable channel, i.e. one which does not erode its bed or banks and does not form shoals or banks of silt, there is a relationship between the width, depth, gradient and river flow. If the width, for example is increased beyond the 'régime width' silting will occur along the banks or as shoals forming islands so as to restore the width to what the river requires and can maintain.

Much work has been done in India, Africa and America to establish empirical formulae expressing the régime width, depth, and gradient in terms of flow and bed material. Most of the work has been done on channels in silt and sand and much of it on irrigation canals where the flow is constant, and there is a pressing need for research into the application of régime theory to natural rivers with gravel beds. Fortunately this research is now being undertaken by the Hydraulics Research Station with the cooperation of several river authorities.

Régime flow

The application of the régime theory to natural rivers raises the problems of what flow should be taken as significant in determining the dimensions of the stable channel. The low flows which go on for most of the time cause little or no change in the channel section, while the maximum flood flows which occur for a few hours at intervals of several years cause rapid but temporary changes which the river will subsequently tend to restore.

Various suggestions have been made as to the 'dominant flow' to be used, but the most favoured suggestion is the 'bank-full flow' which has been quantified as the flow which is exceeded for 0.6% of the time. In many rivers this would correspond to the annual maximum flood. The term 'bank-full flow' is not very satisfactory, as some parts of a river are bank-full more frequently and for longer periods than others. Further research may reveal that in a natural river there is a pattern or combination of flood flows which is more significant than any single value.

Régime formulae

Following the work of Kennedy and Lindley,[6] Lacey[7] put forward the following empirical formulae based on the analysis of data from straight irrigation canals:

$$V = 1.15(fR)^{1/2}$$
$$P = 2.67\sqrt{Q}$$
$$S = 0.000\,55\,f^{5/8}Q^{-1/6}$$
(foot second units)

where V is the mean velocity, R is the hydraulic mean depth, P the wetted perimeter, S the slope of the channel and f is a coefficient depending on the diameter D (in mm) of the sediment transported, such that $f = (2.5D)^{1/2}$.

Later workers used mean depth d and surface water width W instead of R and P and expressed the dimensions of the channel in terms of the flow Q.

Blench[8] introduced a bed factor Fb and a side factor Fs expressing the different characteristics of the bed material and bank material and put forward equations which were made dimensionally consistent by including kinematic viscosity and the acceleration due to gravity. Inglis[9] attempted to take account of bed sediment charge, i.e. the quantity of sediment in motion along the bed, as distinct from suspended solids in the water. He put forward equations relating slope, depth, and breadth to the sediment charge.

There are great practical difficulties in measuring the sediment charge in rivers of any considerable size and Nixon[10] studied data from twenty-nine British rivers and tried to simplify the formulae by eliminating bed load and bed and side factors. This may be an oversimplification because it would imply that a channel of given flow and slope would have a cross section of a certain shape independent of the bed and side material, which is contrary to experience.

The principal equations suggested by various investigators are listed in Table 28.2 and would be applicable to channels in fine materials.

Various investigators have criticised the empirical approach and have tried to derive régime relationships from theoretical studies including such parameters as sediment diameter, specific gravity, viscosity, and channel friction but nothing has yet emerged to replace the empirical formulae resulting from a study of stable channels. Even these have little application to natural rivers with gravel beds and pending the results of current research the design engineer dealing with such cases can probably get most

Table 28.2 RÉGIME EQUATIONS (foot-second units)

Investigator	*Width*	*Depth*	*Velocity*	*Data*
Lacey	$2.58\ Q^{0.5}$	$0.52\ Q^{0.33}$	$0.75\ Q^{0.17}$	Canals in India
Blench	$(Fb\ Q/Fs)^{0.5}$	$(Fs\ Q/Fb)^{0.5}$		American and other rivers
Nixon	$1.67\ Qb^{0.5}$	$0.55\ Qb^{0.33}$	$1.11\ Qb^{0.17}$	British and American rivers
Nash	$1.32\ Qb^{0.54}$	$0.93\ Qb^{0.27}$	$0.61\ Qb^{0.24}$	Statistical analysis of Nixon's data

Qb = bank-full flow

help from a study of the dimensions and slope of stable reaches of the river for which the new or improved channel is required.

Practical applications

In spite of the considerable time since the early work on the régime theory little use has been made of it in the design of new and improved channels in Britain until very recent years, probably due to the difficulties already discussed of the application of the theory to natural rivers and coarse bed materials. Consequently many new channels or river diversions have resulted in erosion or silting which has entailed excessive remedial works or maintenance costs.

A channel may be capable of carrying the required flow with smaller dimensions and a steeper gradient than the régime values but the high velocities generated will increase the width and depth by scour and will reduce the gradient by the deposit of the eroded material at the lower end of the reach. Similarly many channels have been made with excess width to carry the maximum flood flows, but will then reduce their width by the deposit of silt with a narrower meandering deep channel within the main channel. The most common error arising from lack of knowledge of the régime theory is when a river is straightened by cutting across meanders without considering whether the resultant shortening of the course and increase of slope will produce scouring velocities.

CATCHWATER DRAINS

Purpose

In many cases the water from areas of high ground runs down into low-lying areas to create or accentuate drainage problems. The upland area may be greater than the low-land, and also produces a greater run-off per unit area. There is obviously an advantage in diverting this upland water before it runs down to the low level from which it is difficult to discharge, and this may be done by a catchwater drain. In pumping schemes especially, the diversion of the upland water will reduce both the capital cost of the pumping plant and its subsequent running cost.

The catchwater drain is made to follow the edge of the upland and to collect the water of the various streams running down from it. Control structures will be necessary to allow some flow to follow the original course in dry periods to avoid a shortage of water is the lowland area. In flood periods the whole flow will be diverted down the catchwater to a suitable outlet which may be on the coast or a tidal river (Figure 28.5).

Examples

An early example of a catchwater was the Carr Dyke which is said to have been made by the Romans round the edge of the Fens from Lincoln to Peterborough. A later example is the Grand Military Canal, which follows the edge of the high land surrounding Romney Marsh, in Kent, and acts as a catchwater, although it was made for other purposes.

The most notable example in recent years is the catchwater which forms part of the Great Ouse Flood Protection Scheme and leads the flood waters of the Lark, Wissey,

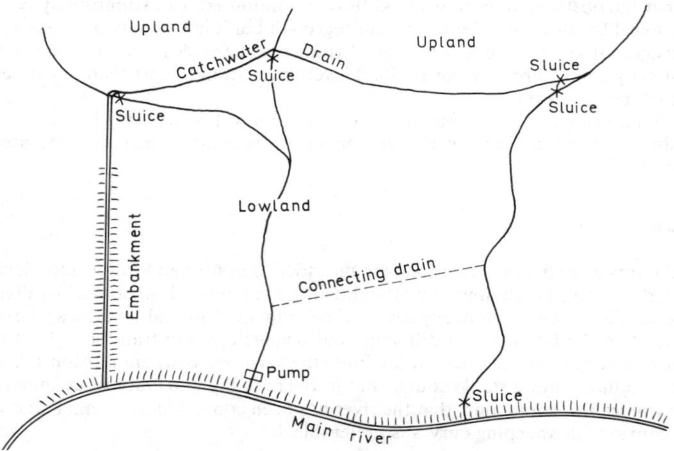

Figure 28.5 Drainage system with catchwater

and Little Ouse round the southern and eastern edge of the Cambridgeshire fens to Denver sluice where the water is discharged into the tidal Ouse or the Cut-off Channel. The flood waters of these rivers formerly meandered across the fens in embanked channels whose embankments were constantly liable to threatened or actual breaches.

Design

The design of catchwaters will follow the lines suggested in the subsection on Channel Design, p. **28**–9. The flow to be carried can be derived from knowledge of flows in the existing watercourses, or calculated from the area of upland drainage to the new channel. In the latter case the possibility of error is greater and an allowance should be made to provide a factor of safety. In some cases the existing channels may continue to carry a part of the flow up to their safe capacity, but where the lowland area is to be pumped the whole upland water flow will probably be diverted.

The route and the gradient will be largely determined by the topography, and having assessed the flood flow the design of the channel is straightforward.

Control structures

These will most commonly be vertically lifting steel gates at the crossings of the original river courses, designed to allow the release of controlled flows down the old river courses. Other structures may be required in particular cases. (See also Section 20.)

CHANNEL IMPROVEMENTS

Pioneer work

A natural river is generally found to be badly obstructed by trees and bushes either growing out into the channels from the sides or fallen into the water where many continue to grow, collecting debris and forming islands. The first stage of improvement in such a case is the cutting and removal of trees and brushwood. In the interests of amenity and to avoid complaints from fishermen of loss of shade for the fish, selected trees should be left singly and in groups along the banks. These can be selected and trimmed in such a way that the obstruction even to flood flows is minimised. Cut stumps may be treated with chemical brushwood killer to prevent regrowth but it will be found necessary to cut back new growth every three or four years. In many cases the clearance of trees and brushwood offers greater improvement of flood levels in relation to cost than any other form of channel improvement.

In lowland drains and slow-moving rivers the removal of weeds will also result in a significant lowering of flood level. This operation is dealt with under 'Maintenance' on page **28**–35.

Alignment

Natural rivers and streams often follow a meandering course and lower flood levels in a particular area can be obtained by straightening the course downstream by diversions across meanders or by a completely new channel. Practical difficulties arise as the original course is often the boundary of different land ownerships and transfer of land may be difficult to arrange. Realignment will eliminate sharp bends where erosion takes place and will produce a more stable course, but in rivers of high amenity value long straight reaches will lead to complaints that the river has been converted to a 'canal' and in such cases a course with sweeping curves is preferable.

Gradients

The flood level at the downstream end of the improvement is fixed by conditions outside the scope of the improvement. Levels at other points on the river course will be fixed by the designer from consideration of what is required to bring flood levels below the level of adjacent property, how far embankments will be used to retain exceptional floods and the relative costs of works required to obtain various maximum flood levels. The last criterion may necessitate the preliminary design of several alternative schemes before a decision can be made.

Having decided on the acceptable flood levels at points along the course of the channel the designer will consider the depth of water required to carry the flow and thus can derive the bed levels and the width. In some cases the possible lowering of the bed level will be limited by sewer crossings, bridge inverts, or other obstacles and the width of channel will be selected to provide the required flow capacity. In other cases the width may be limited by buildings or roads and the depth is the variable factor. In either case the selected channel cross section must comply with régime requirements (see 'River Régime', p. **28**–14) in order to avoid excessive future maintenance works to deal with scour or silting. In lowland areas where water velocities are low, régime considerations may not apply.

Weirs

Where all the requirements cannot be met it may be necessary to undertake special works. For example, where bank erosion is anticipated the width may be artificially controlled by revetments of various types described below. Similarly where realignment results in a gradient steeper than the stable régime gradient, corrective works are needed. The total fall which will produce a stable gradient can be calculated and subtracted from the fall on the desired alignment. The excess fall represented by the difference can then be lost by the provision of weirs. A series of low weirs at intervals will often be preferable to one or two high ones, which produce greater problems of energy dissipation and may cause dangerous scour holes downstream. The design of weirs is dealt with under 'Structures', p. **28**–29. The reduction of gradient by weirs or otherwise will reduce velocities and will call for an increase in channel cross section, generally by an increase in width.

In small channels, or in short lengths of larger channels, bed scour may be prevented by the use of mattresses of wire or plastic mesh filled with stone, or stone pitching, or by concrete. On works of any size these measures are expensive and they are not generally used except in urban areas.

Revetments and linings

Many different materials have been used to protect the banks and beds of watercourses from scour. One of the earliest methods was the use of brushwood faggots or fascines and these are still used in certain conditions. For revetment purposes the brushwood, particularly willow and hawthorn, is made up into tight bundles about one metre long and about 0.3 m diameter. These are laid side by side end-on to the channel, with successive layers stepped back to conform with the bank slope. The faggots are held down by wires to stakes driven through into the ground below, but after a few floods or tides the whole mass is impregnated with silt which holds and preserves the brushwood. Brushwood is also used in longer lengths to build up large mattresses which are floated into position and sunk by loading with stone to protect the river bed or the lower bank slopes. This is a specialised procedure developed by the Dutch, but is carried out in Britain by certain specialist firms.

Stone has also been used since early times for revetment. This may be used as drystone pitching placed by hand or machine on the river bank, according to the size of stone, or it may be grouted solid with cement or bitumen. The latter material allows for minor settlement of the banks. With grouted pitching the provision of weepholes is important to prevent the building up of hydraulic pressure behind the revetment during variations of river level. With drystone the stone should be large enough not to be moved by the river during floods and observation of the largest stones in the river bed will be some guide. Stone used for revetment should be capable of withstanding the action of severe frost when saturated with water.

Where the cost is justified and conditions require it, steel sheet piling is used. This is commonly in the neighbourhood of weirs, locks or sluices, where there is wave action or heavy turbulence. The piling is generally finished at the top edge with a capping beam which is tied back at intervals to anchor piles or blocks placed beyond the zone of soil supported by the piling. Without these anchorages the height of vertical face which can safely be supported by cantilever piles is limited. The section of piling selected is generally one with a relatively high minimum thickness so as to offer a long life. Steel with a copper content is sometimes specified to increase the life.

In fine bed material such as silts, corrugated asbestos cement sheets have been used as an inexpensive form of sheet piling. Some specially made sheets, together with driving helmets, are available, but in some cases standard heavy-gauge roofing sheets have been used. This material is only suitable to support up to one-metre faces where damage by boats, etc., is not expected.

Another form of revetment is the use of gabions or crates of wire mesh filled with stone. These are commonly 2 m × 1 m × 1 m but longer lengths up to 6 m are used. These boxes are subdivided by diaphragms and the cells are filled by carefully packing the stone round the faces and filling the centres in a more random fashion. It is a mistake to think that the gabions can be filled with river bed material tipped in by a mechanical excavator. The gabions can be obtained prefabricated in hexagonal or square mesh, or can be made up on site from ordinary reinforcement fabric (75 mm mesh 5 gauge). The square mesh is of heavier gauge and is easier to fill without distortion. The resultant gabions are less flexible.

The gabions are built up like bricks to the required height, each course being stepped back 0.1 to 0.5 m behind the course below (Figure 28.6). The bottom course should stand on a mattress 0.3 or 0.5 m thick and extending to form an apron 3 to 4 m wide on the river

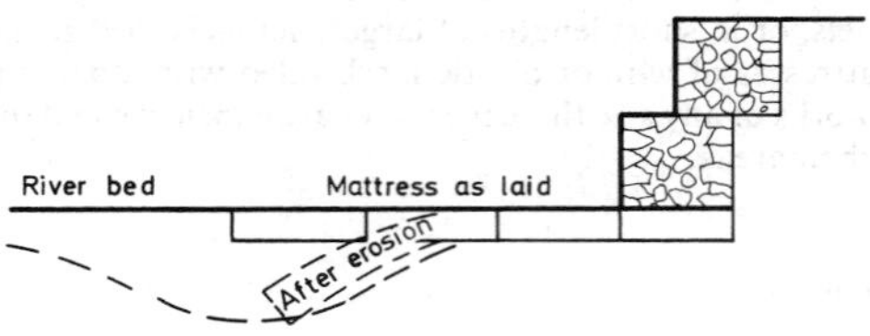

Figure 28.6 Gabion revetment

bed. This should be laid flush with the river bed, and if erosion proceeds it is flexible enough to sag down so as to prevent undermining of the gabion wall. Similar mattresses are sometimes laid on sloping river banks, but the slope must not be too steep and the mattress must be subdivided into small cells to prevent the stone running down to the lower edge.

Some new systems are now available. Plastic grids can be laid and filled with soil and seeded. The plastic reinforces and holds the turf to prevent erosion of slopes exposed to occasional flood flows. On banks more frequently exposed to the action of the water, nylon mattresses can be laid, consisting of a series of pockets running side by side down the bank. These are filled with concrete to form a continuous sheet which can extend down the lower part of the bank slope which is permanently under water. Sandbags filled with dry mixed concrete have been used for the same purpose but cannot be laid accurately under water.

EMBANKMENTS

Purpose

The flood flows of most rivers exceed the capacity of the channels at some time in most years and major floods may be three times the channel capacity. The channel capacity can be increased by clearance, widening and deepening but it may well be found that it would be impracticable, or at least undesirable, so to enlarge the channel as to carry the major floods. In such cases overflow is prevented by the use of embankments or flood walls. The embankments are situated outside the normal river channel and at flood level they provide the necessary increase in the waterway section by providing both extra width and extra depth without overflow. The further back from the edge of the channel they are placed the less height of embankment will be needed and the safer they will be from erosion and slipping of the natural banks, but in urban areas the land required to set back the embankments may not be available. In fact there may not be space for the wide-based embankment at all, and a flood wall, or a 'half-bank' supported by a wall may have to be used (see Figure 28.7).

Materials

Most flood embankments in Britain are of moderate dimensions, not exceeding 4 m in height and many not exceeding 2 m. They are constructed of the best material available locally such as sandy clay, loam silt and even peat, although the latter material is undesirable and gives much trouble. Pure strong clays also cause trouble due to drying out and cracking. River bed material consisting of graded sand and gravel has been used for low banks exposed to short duration floods, and provided with surface cover of soil and turf. Clay cores are not generally included in the design but in some cases have been pro-

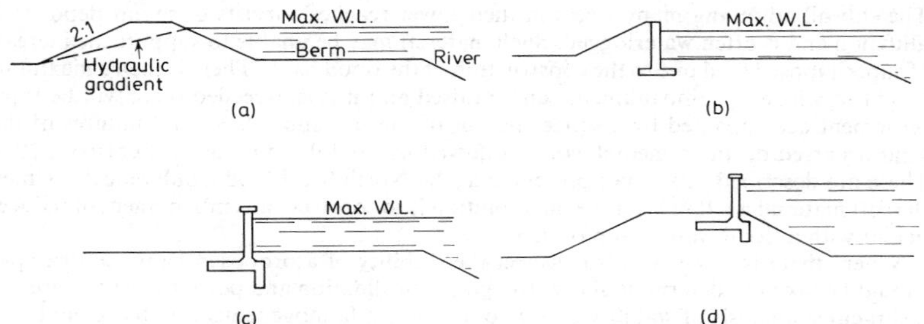

Figure 28.7 (a) Flood embankment; (b) half-bank with wall; (c) flood wall; (d) flood wall on embankment

vided subsequently to cure seepage by excavating a trench along the centre of the embankment and filling with clay. Steel sheet piling has been used in the same way to cure local seepage.

Design

The first stage of design is to establish the bank top levels. These should provide a freeboard above maximum flood level of 0.5 to 0.7 m. This is to allow for:

(1) Possible errors in calculation of flood level
(2) Floods exceeding the design figure
(3) Settlement of the embankment
(4) Formation of low spots by cattle or pedestrians
(5) Damage to the upper layers by burrowing animals
(6) Cracking due to drying.

The bank top width is often about 2 m but may be increased to permit the passage of a tractor along the top with various maintenance equipments. Greater top width will also improve the hydraulic gradient through the embankment and give increased safety.

Side slopes on the river face should not be less than 2:1 and on the landward face may be somewhat flatter. Where space permits and fill is available a 'roll-over' bank is constructed with very flat slopes so that the area may be mown as part of the adjoining field. Arable cultivation is not desirable as the embankment level will gradually be reduced. The planting of trees or shrubs on the embankment should also be discouraged. The ideal surface treatment is good turf grazed by sheep or cut mechanically several times per year.

Hydraulic gradient

The hydraulic gradient through the embankment is the slope of a line from the high water mark on the river side to the landward toe of the embankment and this should not exceed 1:4. In important cases a 'flow-net' through the embankment and its foundation should be calculated to reveal any risk of 'piping' due to excessive rates of seepage. This may indicate the need for a drainage layer of stone and an inverted filter at the landward toe to relieve pore pressures, but these measures are very rarely required or constructed in practice.

Stability

The subsoil adjoining many rivers in their lower reaches consists of recent deposits of alluvium and is often waterlogged. Such material may be unable to support the increase of superimposed load due to the construction of the flood bank. There is then a maximum height to which the embankment can be raised and if it is exceeded there will be rapid settlement accompanied by a squeezing out of the soft subsoil. Several failures of this kind occurred during remedial works followed by tidal flooding of 1st February, 1953. These are described in a paper presented at the North Sea Flood Conference[11]. Unless the soft material is a thin layer the only solution is to construct an embankment of reduced height with a flood wall on top of it.

Where there is any reason to suspect the stability of a proposed bank, soil samples should be taken to determine shear strength, consolidation and pore-water pressure and a slip-circle analysis of stability carried out[12]. For a homogeneous clay bank built on a deep deposit of similar clay having a shear strength τ and density y the critical height is given by $H = 5.5\tau/y$ (foot-pound units). The calculation of factors of safety and stability under high water conditions is described in a paper by Marsland[13].

Construction

The area of the base of the embankment should be stripped of turf and topsoil to a depth of 0.25 m. This may be stockpiled for later use on the surfacing of the bank. The fill should then be deposited and rolled down in layers, generally by bulldozer or tractor shovel. The material is often too soft for the use of conventional rollers. It may be necessary to build the bank initially to a greater thickness and lower level by longitudinal deposit and rolling and to finish by working up the side slopes to the finished level. Finally the embankment will be cased with topsoil and sown with grass seed. The height of the newly completed bank should make allowance for settlement which is likely during the first year at the rate of about one-tenth of the height of the bank, assuming good material, well compacted. Where banks are to be grazed the grass can be a good pasture mixture based on rye grass and clover. In other cases such as public open spaces special short grasses as used in parks and playing fields may be used to reduce maintenance cutting. The advice of a reputable seed merchant or the Ministry of Agriculture's advisory service will be found helpful in the selection of suitable grasses. Rates of application are from 3.5 to 6.3 g/m^2.

Flood walls

Where space or foundation conditions do not permit the construction of embankments, flood walls may be used. A slight reduction in freeboard may be acceptable because the wall will be reliable up to its top level, will not be liable to settlement and will be less liable to damage by overtopping. The wall should be designed as a retaining wall with hydraulic pressure on the riverside face calculated on a level slightly above the wall top to allow for overtopping or wave action. Particular attention is required to the foundation to prevent water passing under the wall and also to construction joints. Although the escape of water through these may not cause flooding it will certainly attract public attention and criticism. To meet amenity requirements the wall may have to be cased in natural or artificial stone or a half bank may be formed behind it which will provide support, seal any leakage, and meet objections to an exposed concrete face.

Where the appearance is acceptable a simple flood wall may be formed by a line of steel sheet piling driven to a suitable depth and finished on the top with a concrete capping beam. This automatically solves the problem of preventing seepage under the wall.

DETENTION BASINS AND WASHLANDS

Storage

The peak discharge in a river or stream can be reduced by temporarily storing some of the flow in a detention basin. The simplest form of this arrangement is where the stream flows through a natural or artificial lake or through a low area where overflow takes

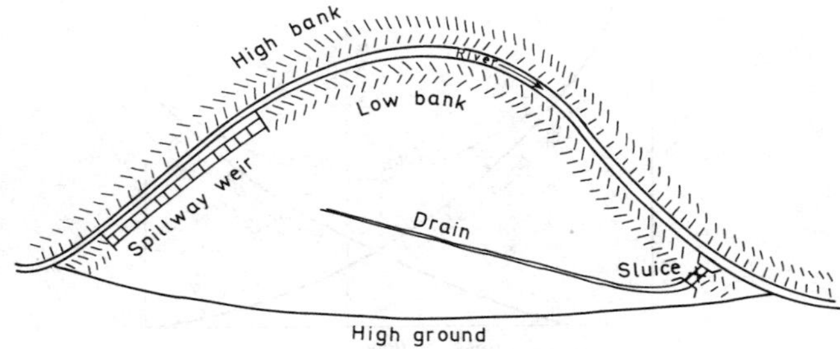

Figure 28.8 Flood storage basin

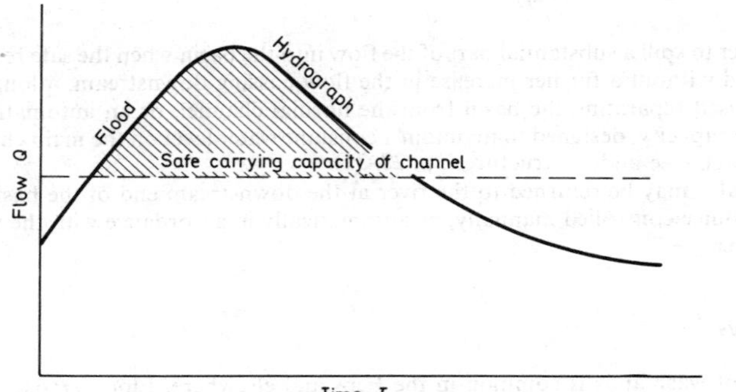

Figure 28.9 Hatched area represents storage volume required

place as soon as the flow increases above normal. But in such a case the storage may be nearly full by the time the maximum flow occurs and will then be ineffective. The flow into the basin should therefore be controlled by banks and weirs so that flows up to the bank-full capacity can pass on down the stream leaving the basin empty. Any surplus flow can then be spilled into the basin over a weir or through a sluice so that flow down the channel is still restricted to a safe value (Figure 28.8). When the peak of the flood has passed the basin can be emptied through an escape sluice back into the river. The following points should be noted:

(1) Flow up to the maximum capacity of the river channel should bypass the basin.
(2) The basin should be empty at the start of the flood.

(3) The capacity should be sufficient to hold the surplus flow in the 'design flood' which may be the maximum flow to be expected in a 5-year, 10-year, 25-year or 100-year interval according to the properties at risk (Figure 28.9).
(4) There should be the possibility of emptying the basin rapidly as soon as river flow subsides, in case of a second flood in quick succession.

Provided that an adequate area of suitable land exists in the river valley the maximum flow passing downstream may be reduced to any desired extent above the normal flow, but the volume of storage required increases in greater proportion than the reduction of residual flow and economy may require a compromise between the provision of storage and channel improvements downstream (Figure 28.10).

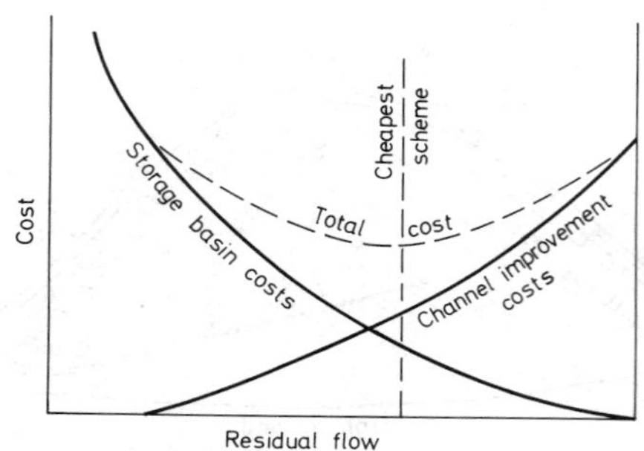

Figure 28.10 Combined channel improvement and storage

In order to spill a substantial part of the flow into the basin when the safe residual flow is reached without a further increase in the flow passing downstream, a long side weir may be used separating the basin from the normal channel, or an automatic sluice of adequate capacity, designed to maintain a constant water level on the main channel side, may be used (see under 'Structures', p. **28**–31).

The water may be returned to the river at the downstream end of the basin through another sluice controlled manually, or automatically in accordance with the water level in the river.

Washlands

The use of washlands is common in the Fens and elsewhere. Flood embankments are constructed well back from the river enclosing an area which may be up to a mile wide and twenty miles long. This area acts both as a detention basin and as a flood channel. Large quantities of water can be temporarily stored and because of the large cross-sectional area of the waterway at flood level, velocities and surface gradients are very small indeed. The area of land involved is considerable and its use is usually restricted to summer grazing or the production of hay or lucerne. If the washland is only required for major floods lesser floods may be excluded by a lower 'cradge' bank alongside the river and the washland may be cultivated, with acceptance of the risk that crops may be lost once in five or ten years. The most notable examples of washlands are Whittlesey Wash on the River Nene below Peterborough, and Welney Wash on the River Great Ouse between Earith and Downham Market. In both cases water is discharged into and out of the washland through control sluices so that the washland is not flooded until its use becomes necessary. The full capacity is then available. At the other end of the scale a wide berm may be excavated on one or both sides of a channel with a level a little above normal flow level, to provide storage and extra flow capacity at flood levels.

PUMPING

General

Pumping is used where the level of the land relative to the river or estuary to which water must be discharged is too low for satisfactory drainage by gravity. Sometimes gravity drainage will avoid surface flooding and permit summer grazing but not intensive cultivation, and the capital and running costs of pumping must be measured against the increased productivity which it can provide.

Capacity required

Very often areas of high land discharge their surface water into the low-lying land and the combined flow is discharged through a single gravity outfall. The high land may have a greater area as well as a higher rate of run-off per unit area, and in such cases it is important to divert as much of the upland water as possible away from the pumps by catchwater drains or embanked high-level carrier drains. When this has been allowed for, the run-off from the remaining area must be assessed. On the flat land, often with peat or silt soils, which generally comprise the pumped areas, normal 'time of concentration' methods of calculation are not applicable, and more empirical methods are used. These are often based on equivalent rainfalls or on flows per unit area. In the Fens in the 1947 floods it was observed that those pumping stations designed to discharge the equivalent of 9 mm of rainfall per day (15.75 ft^3/s per 1000 acres) were able to drain their areas by continuous running whereas many pumps with smaller relative capacities were overloaded. In other parts of the country with higher rainfalls, capacities equivalent to 12 mm of rainfall per day have been provided. The run-off from areas of higher ground or built-up areas must be assessed separately. For agricultural land of the type normally bordering on fen land a figure of 50 ft^3/s per 1000 acres may be used. If possible, gaugings should be made of flood flows in the streams flowing from these areas.

Allowance may have to be made for inflows of water from other sources such as groundwater or motorways. Maximum flows from roads and built-up areas are often calculated for the design of surface-water sewers on the basis of storms lasting only a few minutes. What is important in a normal pumped drainage area is the maximum average run-off over a period of several hours.

Single or multiple pumps

When the total flood run-off has been arrived at the pump capacity can be specified. It is generally economical to pump the whole district at one station but if this involves long deep drains, possibly crossing areas of high or difficult ground, it may be better to divide the flow between two or more stations. Even in a single station it may be decided to use one, two or more pumps. Because of the very low load factor it is not usual to provide stand-by pumps above the maximum required capacity and the use of multiple pumps does provide against the breakdown of a unit. Pumps of varying capacity allow flexibility in running but add to capital costs and complicate automatic running. The most common arrangements are single pumps in small stations, duplicate pumps in medium stations, and three or more in large stations.

Selection of site

The site should be as near the lowest part of the drainage area as possible. It should, if possible, be sited beside the river or estuary to which the water is to be discharged but sometimes this would mean cutting a very deep drain through high ground near the river. The pump may then discharge into a high-level drain connecting to the river, but this will have to be embanked to maximum river flood level or closed by a tidal sluice. In the latter case difficulties may arise due to prolonged high river levels.

Motive power

Historically pumps have been powered by horse mill, windmill, steam and diesel engines but most modern installations are electrical. This provides the considerable advantages of automatic operation and reduced maintenance. In some rural areas failures of supply are not uncommon but these are not often long enough to cause difficulties. The electrical supply also simplifies the provision of lighting and heating in the station.

Modern diesel installations, when used, have vertical, multi-cylinder engines running at relatively low speeds and dowrated to allow for 24-hour running when required. They require auxiliary equipment for starting and gears to reduce the pump speed. Vertical spindle pumps are driven through bevel gears. The installations generally require an attendant.

Pumps

Land-drainage pumps are required to produce very large outputs at low heads, generally between 3 m and 7 m, but occasionally up to 10 m. The output of the pure centrifugal

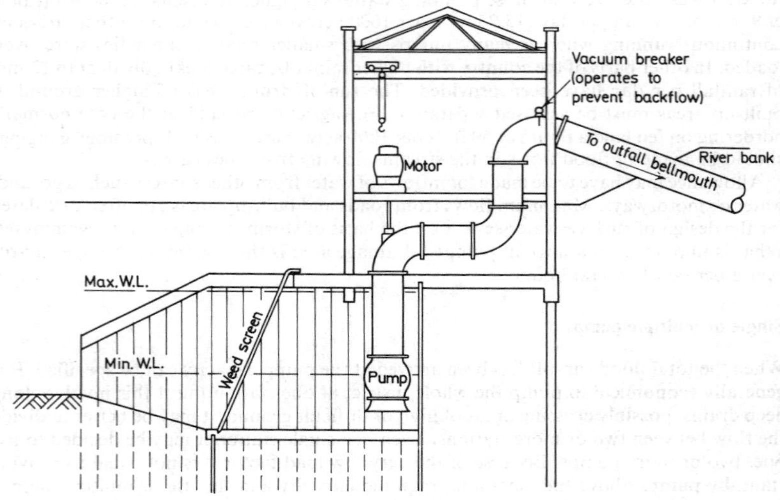

Figure 28.11 Vertical spindle pump

pump falls off with increasing head while the pure axial flow pump requires a high horsepower at high heads. Mixed flow pumps are therefore favoured. Most modern pumps are mounted vertically with the impeller permanently submerged and the motor on a floor above (Figure 28.11). Output will vary with head and the capacity required will be specified to be attained either against the maximum anticipated head, or as an average over the tidal cycle if discharge is to a tidal outfall. The motor or engine should provide 10% more power than is required by the pump against any anticipated combination of intake and discharge levels.

Controls

Electrical installations are started and stopped automatically at preset levels in the sump or intake bay. The sensor may be floats, electrodes, or a pressure recorder, which records the water level in the sump against time, starts and stops the pump at the appropriate levels, and records running times. A cut-out may be required in some cases to prevent the possibility of overflow in the discharge channel. Small installations may have 'direct-on-line' starting but larger motors in rural areas may require controlled starting to avoid excessive load on supply lines during starting. Normal no-volt and overload releases should be provided, and low-capacity heaters to keep the control equipment dry.

Sumps and buildings

Vertical spindle pumps are suspended in sumps the back walls of which are curved to conform with the bell-mouth entry to the pump over an arc of about 90°. This reduces the risk of the formation of destructive vortices which would occur if the pump were suspended in the centre of a large sump. Horizontal spindle pumps, where the impeller is remote from the bell-mouth, do not suffer this trouble.

The main walls of the sump may be of reinforced concrete or steel sheet piling. The entrance to the sump will be provided with a weed screen of sloping steel bars, carefully designed so that it can be raked without the rake catching on stays or spacers. In a few cases automatic screening or raking equipment has been installed. Without this the stations are not truly automatic as frequent clearance of the weed screens is required.

Buildings should be substantial enough to protect the plant from weather and from interference. Good architectural design of the exterior is important, but the main form of the building should be determined by the plant. The building should provide the headroom and floor space required for overhaul of the plant.

Other types

In recent years there have been a number of departures from the conventional vertical-spindle pump arrangement. One is the floating pumping station where the pump and motor are mounted inside a steel pontoon, which floats in an intake bay at the end of the main drain. Intake to the pump is through a grating at the end of the pontoon and discharge is through a pipe with swivelling joints at the pontoon and bank ends. These joints are one of the sources of difficulty, and condensation inside the pontoon is another, but most of the installations are quite successful. The arrangement has the advantage that the equipment rises with water level and cannot be flooded in exceptional conditions such as might arise from the breaching of a tidal embankment or catastrophic flood. Many fixed stations have been put out of action in such conditions when they were most needed, and motors in fixed stations should be kept above maximum flood level.

Another modern development is the outdoor pump where weatherproof motors are used. The vertical-spindle pump is mounted in the sump in the normal way with the motor above it but the only building is a small kiosk to house the electrical control gear. This arrangement is economical and suitable for areas where interference with the plant is unlikely. In high-rainfall areas it may raise problems when maintenance of the plant is required.

Recently some submerged pumps have been used. These are of the borehole type, arranged vertically or horizontally in the bottom of the sump (Figure 28.12). Again the only building is a small kiosk and in this case the pump and motor is free from interference. Effective screening arrangements are important to prevent debris reaching the pump as clearance of blockages would be troublesome. The sump would have to be closed by stop-logs and pumped out.

Archimedean pumps consisting of a rotating spindle carrying a helical blade working in a close-fitting trough have been used. They are practically immune from blockage by

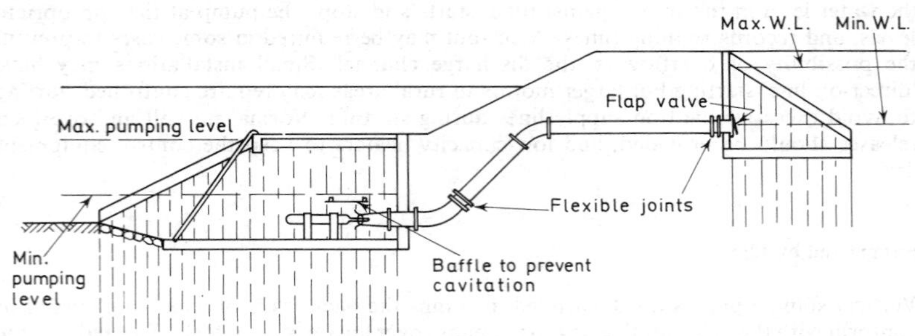

Figure 28.12 Submerged pump

weeds and debris and high efficiency is claimed. The whole of the working parts are on view and any trouble would be immediately visible. They are not suited to wide variations of lift. The trough is formed approximately to size, coated with cement mortar, and given its final shape by rotating the screw with a thin packing attached to give the desired clearance.

Running costs

In an automatic, electrically operated station practically the whole of the running cost will be the charges for electricity. These vary in different areas. 'Off-peak' running may be worthwhile in some cases, but it necessitates considerable storage capacity in the drainage system for the inflow during the 'on-peak' periods, and the resultant variations in water level and intermittent flow are undesirable. In addition the plant will have to have double capacity to discharge the daily inflow in the limited running hours.

Charges are often based on 'maximum demand' and it will be advantageous if this can be the monthly maximum rather than the annual maximum so that advantage can be gained during summer months when it is not necessary to pump.

Sometimes arrangements can be provided for gravity discharge in certain conditions but it is generally found that when the advantages of pumped drainage are experienced there is reluctance to revert to gravity for a minor saving in running cost. Provision for gravity flow may be advisable for emergency use in the event of a prolonged stoppage of the pump.

SEA DEFENCE

Sea defence as carried out by the land drainage engineer is confined to the prevention of flooding of low-lying land by sea water. It is distinct from coast protection work carried out by local authorities under the Coast Protection Act, which comprises the prevention of erosion of higher land by the sea. Sea defence, as thus defined, requires the construction and maintenance of barriers of various kinds to hold back the sea from the low land and also various works to preserve and protect these barriers.

Further information on *methods* of protection is given in Section 29 ('Coastal and Maritime Engineering'). Figures 28.13 and 28.14 show types of protection for sea walls.

STRUCTURES

Retaining walls

The most common structure met with by the river engineer is the riverside retaining wall, supporting merely the river bank or a road or building adjacent to it. The structural design of such a wall is straightforward and is discussed in Sections 11/36, but some special points deserve mention. The wall will be subject to varying water levels and stresses, and reinforcement should be calculated for high-water and low-water conditions. These may produce resultant pressures in opposite directions. After a period of high water the soil behind the wall will be saturated, lowering its shear strength and increasing its weight, and the river may then fall rapidly, removing support from the front of the wall. The wall should be designed for this condition. Weepholes and rubble drains behind the wall should be provided to relieve the hydraulic pressure as soon as possible. The design of the base of the wall should aim to prevent undermining or sliding due to water lubrication. In concrete walls, stresses and cover should be in accordance with the Code of Practice for reinforced concrete structures for the storage of liquids[14].

Weirs

Weirs are constructed for a variety of purposes, the most common being to retain a required water level upstream, to reduce the gradient of an erosive river, or for gauging. Some or all of these purposes may be combined. The purpose will affect the form of crest, but some other features are common to all weirs. The design, which is discussed further in Section 20, must prevent the passage of water under or around the weir and should dissipate the energy of the falling water without causing progressive scour of the bed and sides of the channel downstream.

The method of preventing the escape of water depends on the nature of the subsoil. On extensive sands, as found in some Indian rivers, the base of the weir has to be very long with several cut-off walls constructed down into the sand to lengthen the flow path and to reduce the hydraulic gradients under the weir. In British conditions, foundations are usually on more impermeable material and the length of structure required for hydraulic design will generally be adequate to provide a satisfactory hydraulic gradient especially if steel sheet piling or cut-off walls are incorporated at the edges of the structure to prevent undermining. Where the head retained is more than one metre, a flow-net analysis of pressures under the structure should be made.

Whatever the form of weir it is desirable that the proposed design should be tested if possible by means of a hydraulic model.

Fish passes

Fish passes are often required in conjunction with weirs to allow migratory fish to travel upstream to spawn. Salmon and trout are able to ascend Crump weirs up to about 1 m high but weirs with steeper faces are more difficult and if the water falls on to an apron covered to a shallow depth ascent will be impossible. On weirs with a flat downstream slope but height over 1 m a pass may be formed merely by a timber baulk or low wall running diagonally on the slope to form a channel with a concentration of flow at a

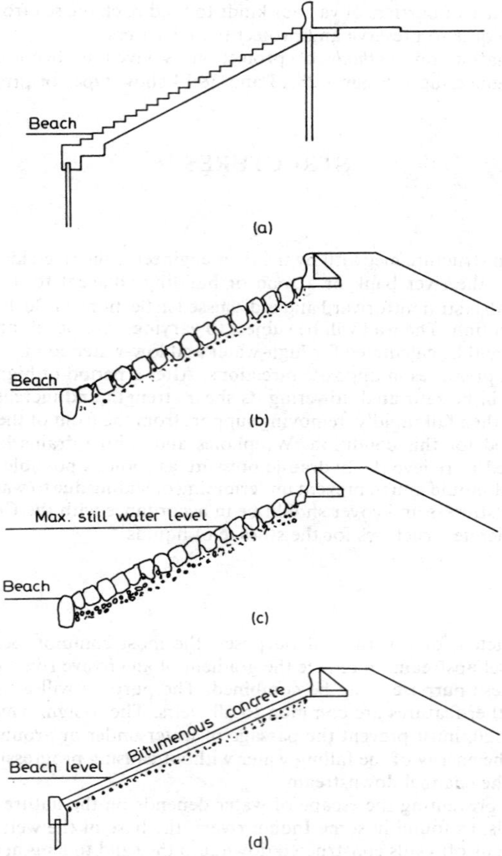

Figure 28.13 Types of protection for sea walls. (a) Stepped concrete; (b) asphalt-grouted stone; (c) stone pitched; (d) bitumen

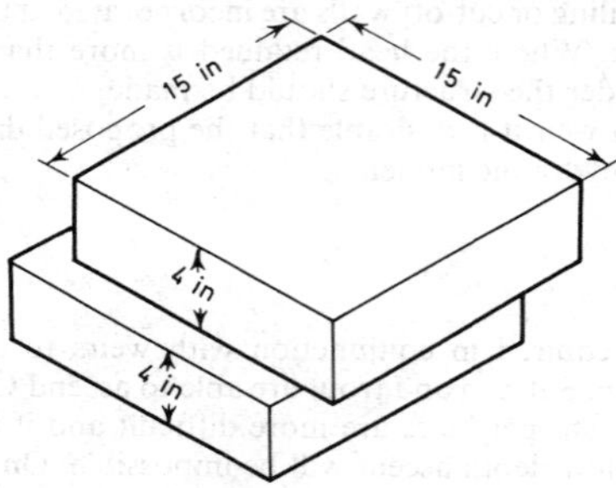

Figure 28.14 Interlocking pre-cast block for sea walls

reduced gradient along its upstream side. Another simple pass in a low weir is simply a square notch 1.0 m to 1.5 m wide in the weir crest to produce a local concentration of flow and reduction of height.

The most usual form of pass for larger weirs is a series of pools connected by submerged or surface openings. The pools should not be less than 3 m long and 2.5 m wide with not less than 1 m depth below the connecting openings. The difference in level between successive pools should not exceed 0.5 m. The connecting openings may be square or streamlined passages through the walls not less than 0.5 m wide and 0.25 m deep. It may be found best to make the openings larger, with slots in the sides to receive boards by which the width and depth can be adjusted to find the best working conditions.

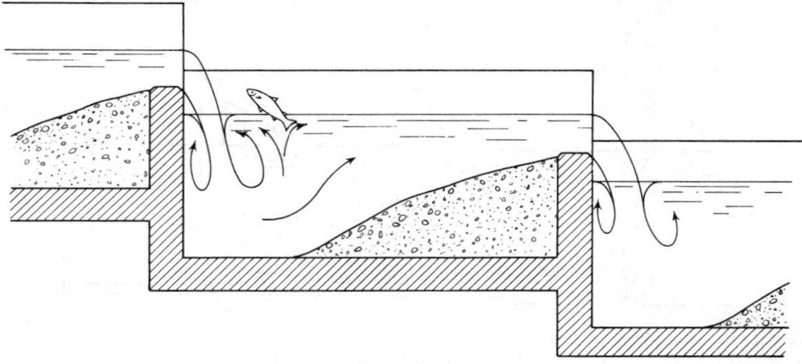

Figure 28.15 Pool-type fish pass

Excessive turbulence and aeration of the water in the pools will discourage the passage of fish, but the upward currents produced round the falling jet of water encourage and assist the fish to jump.[15] If the fall is sufficiently reduced fish will be able to swim up the jet into the next pool. In some passes, including the well known example at Pitlochry, the connections between pools are submerged pipes, but this requires a large flow of water through the pass, which may not always be available. Stuart[15] suggests the form of pool shown in Figure 28.15 which imitates a natural pool. The gravel is shaped by the falling water jet.

Sluices

Water level upstream of a weir rises with increasing flow but a sluice may be designed to produce a constant water level or even to reduce levels at periods of high flow. The most popular types of sluice are vertical lifting, radial, or falling gates (Figure 28.16). (The design of gates and valves is discussed further in Section 20.)

Vertical gates in small sizes may be penstocks in cast iron or fabricated steel sliding in cast-iron frames and operated by screws. Larger gates are of steel, with guide rollers or wheels running on tracks or rails built into a recess in the abutments. Operation is by roller chains or ropes passing over sprockets or drums mounted on a superstructure and often carrying a counterbalance. This may be connected directly to the chains or ropes, or through a pulley arrangement to reduce the travel of the counterweight, although increasing its weight. Very large gates, where the pressure on the wheels would be very great, have been made with rollers carried in a separate frame moving half the travel of the gates so as to produce a free roller-bearing movement. Opening and closing of the gate may be by hand winch or electrical. In the latter case operation may be made automatic by the use of electrodes which cause the gate to open if the water level is more than say 0.10 m above normal, to stop when the level returns to normal, and to close if the level falls 0.10 m below normal. Operation must be slow or limited in duration to avoid 'hunting', i.e. the gate continuing to open and close due to rapid changes of water level produced by the gate itself.

Radial gates are constructed so that the water face is a segment of a cylinder, pivoted about its axis so that all water pressures pass through the axis and there is no hydraulic rotating moment. The gate structure is balanced by counterweights on balance arms. The lower edge of the gate closes on to a sill and is sealed by a strip of rubber or similar material

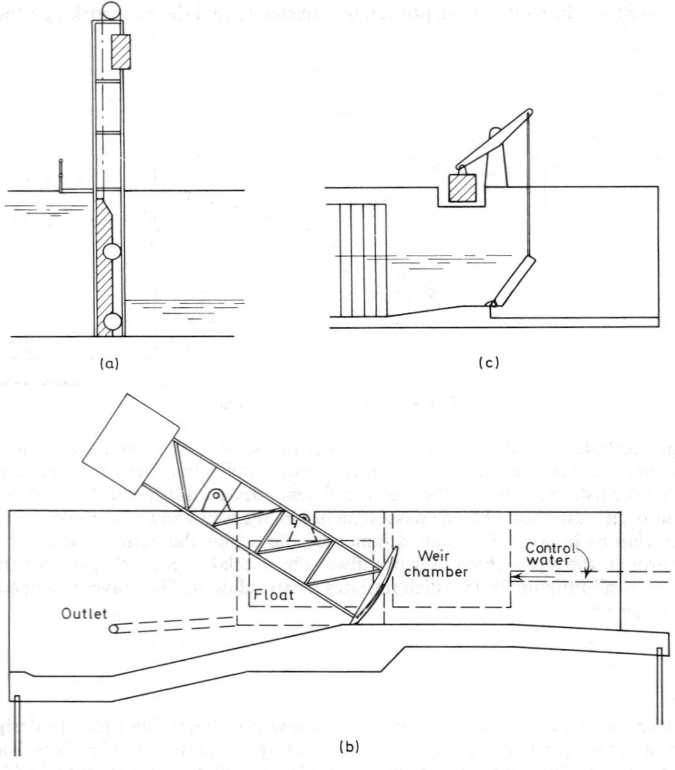

Figure 28.16 (a) Vertical roller sluice gate; (b) radial gate sluice; (c) falling weir sluice

The ends of the gate have leather seals sliding on a metal strip in the abutments. Operation can be by hand or electric winch but more commonly by floats. These are placed in chambers in the abutments or piers. Water from the river upstream passes over an adjustable weir and into the float chamber, from which it escapes at a constant rate through a valve. If the upstream water level rises so does the flow over the weir and water accumulates in the float chamber, causing the gate to open slowly until the water level returns to normal. Similarly if the level falls, flow over the control weir does not balance the escape of water from the float chamber, and the gate closes until normal level is restored. Again the design must be such as to avoid hunting but in practice it has been found possible to control the level of a large river within 12 mm of a set level. In major floods the gate rises completely out of the water and there is a possibility that the downstream water level will take control and prevent the gate from closing when it should. This situation can be avoided by proper design of the control arrangements. This design, and the structural design of the gates, will normally be carried out by firms experienced in the work, and the civil engineer will only need to specify the duty required of the gate and to design the associated work on the foundation, sill and abutments. These call for little comment, except that it should not be overlooked that the horizontal thrust on a large sluice gate is very considerable, and must be taken into account in calculating foundation pressures and resistance to sliding. The possibility of excessive percolation under the structure and of scour below the downstream edge of the apron must be examined. In these respects the structure is similar to a weir.

Radial and vertical gates have the disadvantage that flow takes place under them and floating debris of all kinds collects against them and has to be removed manually. This is avoided by tilting gates which are hinged at the bottom edge and allow the water to pass over them. The gates are lowered by links or chains and may be operated manually or automatically by electricity. The gate when fully open lies flat in a shallow pit formed in the foundation to give maximum discharge, or in some cases still forms a low weir in the lowered position. In rivers carrying coarse sediment there may be difficulty in lowering the gate to bed level, but in practice the water weiring over the gate generally keeps the area immediately downstream clear. This type of sluice has the advantage that any failure of power supply or other operating failure will not allow water levels to fall below the gate level. As an adjustable weir it is particularly suitable for small installations with manual operation.

Tidal outfalls

Tidal outfalls are required where drainage channels discharge through a sea wall or tidal embankment. Their function is to allow discharge at low tide but to prevent the tidal water from flowing back into the drainage system during the high tide. Essentially they consist of a culvert through the tidal embankment with a tidal flap or door, which may

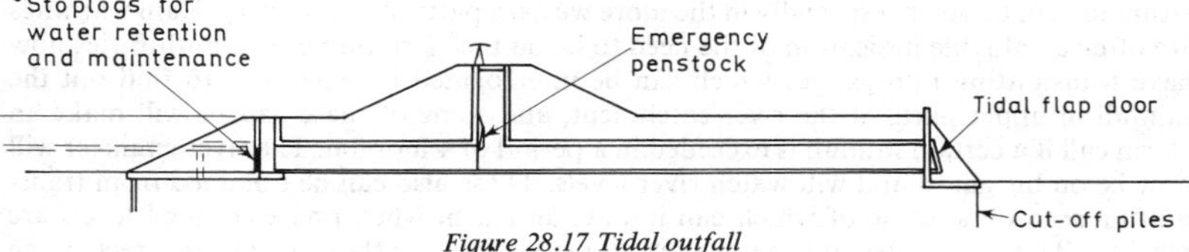

Figure 28.17 Tidal outfall

be at the outer end of the culvert or in a chamber in the embankment. Where beach levels near the sea wall are relatively high the culvert may be extended for some distance out on the foreshore and a door on the outer end could be subject to severe wave action. At the wall itself the door can be sheltered by wing walls and breakwaters, and still greater protection is obtained by the use of a chamber in the embankment. The door is usually circular or square, of cast iron, steel, or plastic, hung from the top by double hinges which allow the door to seat freely and to accommodate small obstructions such as weed or sticks on the seat. There should be sufficient space between the bottom of the door and the apron on to which the water falls to avoid debris such as stones or tin cans from being trapped behind the door and preventing closure. There should also be adequate clearance between the sides of the door and the wing walls or sides of the chamber. When the door is in a chamber it must be mounted on the upstream wall and the chamber itself must be built up above high tide level. It is not advisable for the chamber to be sited inland of the tidal embankment as the intervening culvert will be under pressure at high tides. If the door is some distance out on the foreshore there is the possibility of tidal water breaking into the culvert behind the door and it is advisable to provide a sluice or penstock capable of shutting off the culvert at the tidal embankment or at the inland end.

The outfall has to discharge the whole inflow during the tidal period of 12 h 25 min in a part of that time depending on the level of the outfall and the form of the tidal curve. It must therefore have a bigger capacity than if it could run continuously. This may be calculated by multiplying the amount of inflow by the ratio of the total tide cycle to the discharge period. This will err on the safe side because the rate of discharge will be increased for the first part of the discharge period since the level upstream of the outfall will have built up during the period of tide-lock. Discharge through the door will start as soon as the level on the tidal side falls below the ponded level on the inland side and will increase until the tide falls to the level where free discharge takes place. In the case of a rectangular culvert this will be when the tidal level is at two thirds of the depth of the culvert above the sill. Discharge will then fall slowly with the falling water level inland until the tide returns and the door closes again.

FLOOD WARNING SYSTEMS

Objects

In many areas the complete elimination of flooding, even if possible, would be prohibitively expensive or would require the demolition of too much of the property liable to flood, and the continued risk of some flooding must be accepted. In such cases, or where there will inevitably be delay before flood protection works can be carried out, there is a need for a warning system, so that furniture in houses or stock in shops and factories can be removed from places liable to flood, and livestock may be moved from low-lying land. Most, if not all river authorities, have such systems in operation.

The elaboration of the system will depend on the extent and nature of the areas at risk. The Meteorological Office is experimenting with the use of radar to enable estimates to be made of quantities of rainfall to be expected but at present only very approximate estimates can be given, especially in the more western parts of the country. Rain warnings are often a valuable indication of the need to be on the alert. Some river authorities now have transmitting rain gauges which can be interrogated by telephone to find out the rainfall in upper parts of the river catchment, and some of these gauges will make an alarm call if a certain rainfall is exceeded in a period of 4 h or 8 h. The river engineer will now be on his guard and will watch river levels. These also can be obtained from transmitting recorders, some of which can initiate an alarm when predetermined levels are reached. This alarm message may be telephoned to river authority staff or direct to the police, who will consult with the river authority staff and issue the warning to the public by public address vehicles, or in rural areas by telephone.

Many authorities have arrangements for the manning of an operations room when floods are anticipated. Here data relative to rainfall and river levels are collected and river level predictions are made.

Flood prediction

River levels resulting from a given rainfall or an observed flow at a point upstream can only be predicted on the basis of records of previous floods, so it may take many years to build up a system which will give reasonable accuracy. The effect of rainfall in particular depends on the 'antecedent precipitation', the evaporation and transpiration, and the amount of ground water reaching the river. The last three of these can all be related to the time of year, and the antecedent precipitation index (API) is taken as the sum of the rainfall in preceding days, each day's fall being reduced by a factor $(0.90)^n$ where n is the number of days since the fall occurred. Coaxial diagrams have been used to take account of the rainfall, API and time of year, and to read off predicted flood levels at one or more key points on the river.

A simpler system depends on the study of levels reached at various points along the river in previous floods so that when a certain level is reached at a point in the upper reaches the level to be expected, and the probable time of the flood peak at a flood danger point downstream, can be predicted. The automatic transmitting level gauges can be set to give warnings at levels which may result in the flooding of riverside meadows, and a second warning when built-up areas are threatened. The warnings are best transmitted in words from a tape recording and the persons receiving them must have clear instructions as to action to be taken. In some of the longer rivers where there are years of record it is possible to give a close estimate of the time and height of floods as much as two days in advance.

Short-duration warnings

In short steep rivers and streams the flood rises too quickly for such a system and in some of these a simple float-operated alarm is used which rings a bell or flashes a light in a police station or factory control room when water level approaches flood level. In some cases two levels are signalled so that the first warning gives an alert and the interval before the second warning gives an indication of the rate of rise.

With all warning systems care should be taken to avoid too many false alarms, which will lead to disregard of the warning which is really needed.

MAINTENANCE

General

The maintenance of a river system obviously calls for a wide variety of repairs and replacements to structures, walls, piling, and sluices but most of this is not peculiar to river work except that it may have to be done in the periods when low river levels may be expected. Two common types of river maintenance are the periodic dredging of accumulated mud or silt, and the clearing of trees, brushwood and weeds.

Trees and brushwood

Trees and brushwood encroaching on the channel need to be cleared or cut back every two or three years. Where complete removal is desirable the stumps may be sprayed with a brushwood killer to prevent regrowth. In many rivers the clearance of fallen trees from the channel, especially from bridge openings, is a regular task.

Weeds

In lowland watercourses weeds grow annually and can hold back water to a surprisingly high level. The weeds grow on the bank slopes and, where the water is less than 2 m deep, also on the bed. Removal can be done mechanically or chemically.

Hand cutting has now practically died out and cutting of weeds on the banks is done by various cutters of the mowing or flail types mounted on an arm on a tractor running alongside the channel, or sometimes on a boat. Weeds growing in the water are most commonly cut with horizontal V-shaped reciprocating blades carried by a special launch. Another type of launch carries a band-saw type of cutter working round a vertical U-shape. Both types will cut weeds effectively but difficulties arise in channels with many small bridges and culverts. The weeds after cutting are allowed to drift up against a simple rope boom and removed by hand or mechanical elevator.

Another type of cutter is mounted on the edge of a special bucket on a hydraulic excavator. The cutter is driven by a hydraulic motor and the weeds are lifted out in the bucket. This equipment is suitable for the narrower streams.

Chemical methods

In many areas weeds growing on the banks or emerging from the water are dealt with by chemical spray. Great care is necessary in the choice and application of the chemicals but the method has been used for many years in some districts without trouble. The Ministry of Agriculture, Fisheries and Food approves certain chemicals for this type of use and publishes advice on the subject.[16] Dalapon, in various proprietary formulations is effective against phragmites and other common weeds and is not toxic to fish or livestock. Aminotriazole is also used. Some other chemicals are toxic and can only be used when the watercourse contains no fish and is not used for livestock. No chemicals are in general use at present for destroying submerged weeds and sometimes the clearing of the emergent weeds leads to a greater development of the submerged weed. Chemical methods are particularly useful in watercourses which contain only a shallow depth of water during the summer so that submerged weeds are not a serious problem.

The chemical may be applied from a knapsack sprayer carried by the operator but more commonly from a lance or boom supplied from a tank carried on a tractor-drawn trailer. Spraying from aircraft has been tried but there is difficulty in controlling drift which may damage adjoining crops. This is also a danger with ground methods in windy weather but can be reduced by control of drop size, and care in application. There are sometimes difficulties in getting access to streams because of hay or other crops. Spraying should be done in early summer when there is enough growth to absorb the chemicals but not enough to make penetration difficult or to cause de-oxygenation of the water by large masses of decaying foliage.

FISHERIES AND RECREATION

Fisheries

There are said to be three million anglers in Great Britain and where there are fisheries in rivers, or fisheries may be developed in the future, the river engineer must have regard to them in carrying out his work. Weirs and sluices must allow the passage of migratory fish and channel improvements must not produce long stretches of shallow even flow where fish will not lie. The banks must be such that anglers can get access to the water and in clearance schemes sufficient trees must be left to give shaded areas where fish will lie. Channel regrading may destroy valuable fishing pools and lead to claims for compensation. The engineer will be wise to consult with the river authority fishery officer and with the fishermen concerned before finalising his plans.

Other recreation

There is increasing interest in all forms of water-based recreation and the future water authorities will have greater powers to carry out work for recreational purposes. These may include riverside walks and footbridges, boat-launching ramps and boathouses, the restoration of locks and navigation channels and the like. In some recreational areas very flat bank slopes and artificial beaches may be called for to provide for paddling and bathing, where water quality permits. Boating lakes and model yacht ponds are often constructed by the river. These should be separate from the river channel to avoid problems of currents, silting, and flood damage.

In carrying out river works in general a high standard of appearance will be called for as the rivers will, in the future, be popular centres of recreation of all kinds.

REFERENCES

1. CRUMP, E. S., 'A new method of gauging stream flow with little afflux by means of a submerged weir of triangular profile', *Proc. Instn Civ. Engrs* (Mar. 1952)
2. *Crump Weir Design,* Water Resources Board
3. NASH, J. E. and SHAW, B. L., 'Flood frequency as a function of catchment characteristics'
4. VEN TE CHOW, *Open Channel Hydraulics,* McGraw-Hill
5. 'Charts for the hydraulic design of channels and pipes', *Hydraulic Research Paper No. 2,* HMSO
6. LINDLEY, E. S., 'Regime channels', 1919, *Proc. Punjab Engineering Congr.,* **7**, p. 63
7. LACEY, G., 'Stable channels in alluvium', 1929 min. *Proc. Instn Civ. Engrs,* **229**, p. 259; 'Uniform flow in alluvial rivers and canals', min., *Proc. Instn Civ. Engrs,* **237**, p. 421 (1935)
8. BLENCH, T., 'Hydraulics of canals and rivers', *Civil Engineering Reference Book,* Vol. 2, 2nd edn Butterworths, London (1961)
9. INGLIS, C. C., 'Meanders and their bearing on river training', Institution of Civil Engineers Maritime and Waterways Divs (1947); 'The effect of variations in charge and grade on the slopes and shapes of channels', Int. Assoc. for hydraulic research, 3rd Congress, Grenoble (1949)
10. NIXON, M., 'A study of the bank-full discharges of rivers in England and Wales', *Proc. Instn Civ. Engrs,* **12,** p. 157 (1959)
11. COOLING, L. F. and MARSLAND, A., 'Soil mechanics studies of failures in sea defence banks of Essex and Kent', *Proc. Conf. on North Sea Floods,* Institution of Civil Engineers, London
12. TAYLOR, D. W., *Fundamentals of Soil Mechanics,* Wiley, New York (1948)
13. MARSLAND, A., 'The design and construction of earthen flood banks'. *J. Instn Civ. Engrs,* **11,** No. 3 (May 1957)
14. *Design and construction of reinforced and prestressed concrete structures for the storage of water and other aqueous liquids,* CP 2007, British Standards Institution, London
15. STUART, T. A., *The leaping behaviour of Salmon and Trout at falls and obstructions',* Department of Agriculture and Fisheries for Scotland, HMSO (1962)
16. *Approved products for farmers and growers,* Ministry of Agriculture, Fisheries & Food, HMSO (1973); *Code of Practice for the use of herbicides on weeds in watercourses and lakes,* Ministry of Agriculture, Fisheries & Food, HMSO (1967)

29 COASTAL AND MARITIME ENGINEERING

COASTAL AND MARITIME ENGINEERING 29

29 COASTAL AND MARITIME ENGINEERING

F. L. TERRETT, M. Eng., F.I.C.E.
Lewis & Duvivier

TIDES

Tide-raising forces

The alternate rising and falling of sea level is caused by the attractive forces of the moon and the sun on the rotating earth. The predominant effect, that of the moon, can be explained in a simplified form by omitting in the first place the rotation of the earth and moon about their own axes and considering the relative motion of the two bodies about their common centre of rotation G (Figure 29.1). They revolve about G independently, not as a single rigid body, and points P_1 and P_2 on the earth's surface rotate about G_1 and G_2 in which GG_1 and GG_2 are respectively parallel to CP_1 and CP_2, and P_1G_1 and P_2G_2 are parallel to CG. The attractive force of the moon on a particle of mass m

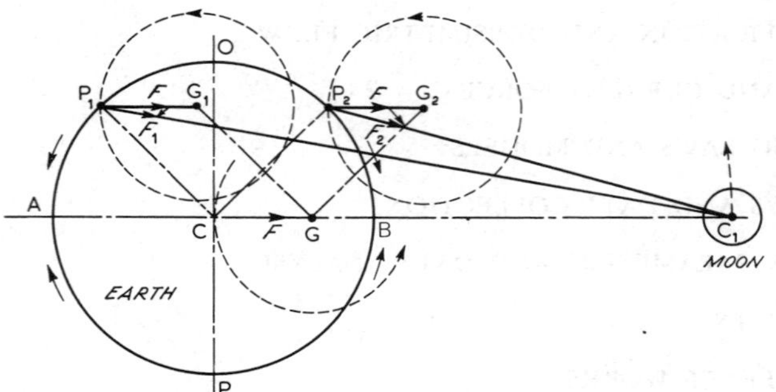

Figure 29.1 Tide-raising forces

at the centre of the earth is gmM_1/L^2 in which M_1 is the mass of the moon, L the distance between the centres of the moon and the earth and g is the gravitational constant. This attractive force is the centripetal force F which restrains the particle in its circular orbit round G.

If particles of water, also of mass m, are to remain in position at points P_1 and P_2 they must also be acted upon by forces F towards their centres of rotation G_1 and G_2. The attraction of the moon on these particles is respectively F_1 which is less than F, and F_2 which is greater than F since P_1C_1 and P_2C_1 are respectively greater and less than L. The vector differences shown in the figure, **F** minus $\mathbf{F}_1$, and $\mathbf{F}_2$ minus **F**, are the tide-raising forces. The vertical components of these forces are small in relation to the earth's gravity and are of little importance; the horizontal components which are towards A and B, respectively, generate the tidal wave. They are zero at points A and B in line with the moon and near points O and P at right angles to AB, and are a maximum midway between these points. Their directions, indicated by the circumferential arrows in the figure, cause two high waters, one directly under the moon and the other on the opposite side of the earth.

The sun produces similar tide-raising forces but of barely half the magnitude of those due to the moon. As the earth rotates the tides are phased with the apparent motion of the moon so that the interval between successive high waters is approximately half the lunar day of about 24 h 50 min.

Tidal variations—effects of declination

Variations in tide level result from the varying positions of the sun and the moon relative to the earth; at times of new and full moon the tide-raising forces of the sun reinforce those of the moon giving spring tides and when the moon is at the first and third quarter the sun's tide-raising forces counteract those of the moon giving neap tides.

In many places there is a marked inequality in the height and range of succeeding tides which is largely due to the angles between the plane of rotation of the earth about its axis and the planes of the orbit of the moon round the earth and of the earth round the sun. These varying angles, which are the declination of the moon and the sun, introduce a diurnal component which combines with the semidiurnal tides. It is possible for one high water to be suppressed altogether and for an inequality in time also to be caused by declination so that the interval from high to low water may not be the same as from low to high water.

Tidal currents—coastal effects—reflection and resonance

In the open oceans the tide generated by the attractive forces of the moon and sun takes the form of a progressive wave in which the associated currents are in the direction of wave propagation below the crest and in the opposite direction in the trough. The maximum current velocities are at the crest and trough, i.e. at high and low water, and zero at half tide on both rising and falling tides.

This simple description of the tidal motion is, however, much altered by many factors, in particular by the shape and disposition of the land masses and the depth of the seas around them. Reflection of the tidal wave from the shores and resonance effects in enclosed or partially enclosed gulfs and straits result in standing oscillations in which the tidal current is zero at high and low water and a maximum at half tide or thereabouts. An example of such a standing oscillation is found in the eastern half of the English Channel where high water between the Isle of Wight and Dover occurs within about 10 minutes at all places along the English and French coasts.

As the tidal wave enters shallow water, in an estuary for example, it is distorted; the speed of propogation is reduced and the wave crest tends to overtake the preceding trough. Thus the time interval from low to high water is reduced and from high to low water increased, and the flood current becomes stronger than the ebb. Also, the height of the tide may increase as the estuary narrows inland.

Coriolis force

The tidal currents are affected by the rotation of the earth, being deflected to the right in the northern hemisphere and to the left in the southern. This is known as the Coriolis effect after the French scientist of that name (1792–1843). In a narrow sea, such as the English Channel, deflection of the flood current to the right is inhibited by the proximity of the shore lines and the Coriolis force leads to higher tides along the French than the English coast.

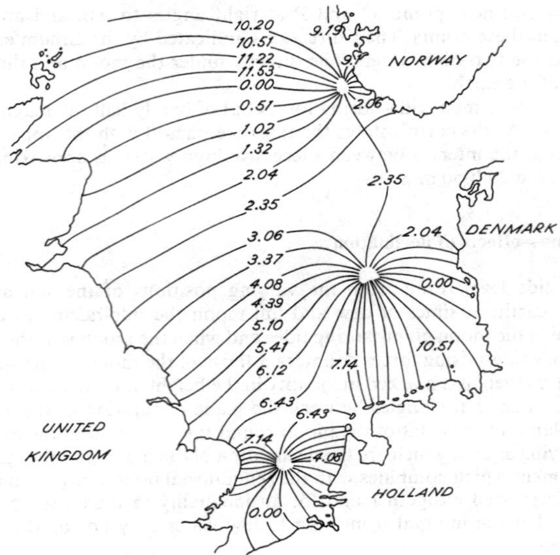

Figure 29.2 Amphidromic points in North Sea

In a more open sea the tidal wave and its associated currents may become rotary about an amphidromic point at which the currents are zero and there is no tidal variation in level. There are three such amphidromic points in the North Sea (Figure 29.2).

In the relatively small areas of water with which coastal engineers are normally concerned the Coriolis effect is not significant, but in studies for larger projects, such as the Delta Scheme in the Netherlands, it cannot be ignored.

Prediction of tides

The astronomical tide-raising forces create semidiurnal and longer frequencies in the tidal cycle: shallow-water effects introduce higher frequencies. The recorded tidal curve at any place can be broken down into these various frequencies and the individual constituents recombined to give tidal predictions. Such predictions for places throughout the world are made by the Institute of Oceanographic Sciences and are published annually by Her Majesty's Stationery Office.

WAVES

General

Coastal and estuarine process are complex and it is seldom possible to find solely analytical solutions to practical problems. A great deal of theoretical work has, however, been carried out and the more important results are given below with notes on their significance and application. For their derivation see reference 1.

In Figure 29.3: T is the wave period (= time interval for motion to recur at a fixed point), c is the velocity of wave propogation or wave celerity, $\eta(x, t)$ is surface elevation at position x and time t, u is the horizontal component of instantaneous velocity of fluid element, v is the vertical component of instantaneous velocity of fluid element, p the instantaneous 'static' pressure, H the wave height ($= 2a$), ρ the density (mass per unit volume) and ν the kinematic viscosity.

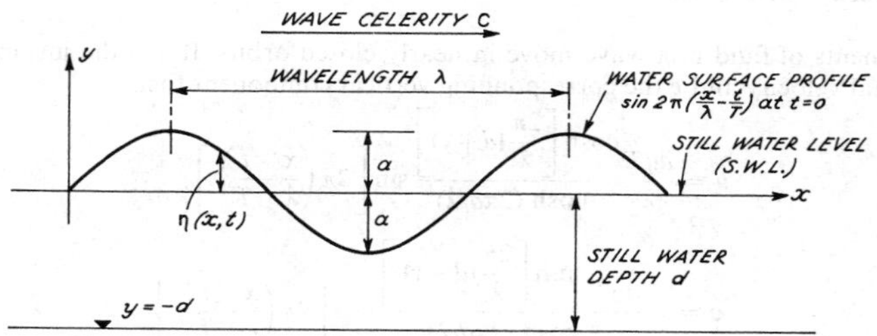

Figure 29.3 Coordinate system

Wave length, celerity and period as functions of depth

$$c^2 = \frac{g\lambda}{2\pi}\tanh\left(\frac{2\pi d}{\lambda}\right) \tag{1}$$

$$\lambda = cT \tag{2}$$

Equation (1), derived from theoretical work by Stokes,[2] is strictly accurate only for waves of small amplitude but the error resulting from its application to practical problems is small and the theory may be used with confidence. When a wave moves from deep to shallow water (or vice versa), c and λ both change while T necessarily remains constant. Published tables and graphs[3, 4] relating the variables are available so that manipulation of the equations is not necessary.

Waves are conveniently classified into types according to the relative depth d/λ as follows:

Shallow-water (long) waves	$d/\lambda = 0$ to $1/20$	$\tanh(2\pi d/\lambda) \simeq 2\pi d/\lambda$
Deep-water (short) waves	$d/\lambda = 1/2$ to ∞	$\tanh(2\pi d/\lambda) \simeq 1$
Intermediate waves	$d/\lambda = 1/20$ to $1/2$	

Using these approximations, equations (1) and (2) become:

(a) *For shallow water*

$$c = (gd)^{\frac{1}{2}} \tag{3}$$

$$\lambda = T(gd)^{\frac{1}{2}} \tag{4}$$

(b) *For deep water*

$$c = gT/2\pi \tag{5}$$

$$\lambda = gT^2/2\pi \tag{6}$$

Within the limits indicated above, the formulae (3) to (6) resulted in errors of less than 1%. For many purposes such precision is unnecessary and the limits may be widened.

Fluid velocity and pressure

The elements of fluid in a wave move in nearly closed orbits. If u is the instantaneous horizontal velocity and v the corresponding vertical component then

$$u = \frac{agT}{\lambda} \frac{\cosh\left[\dfrac{2\pi}{\lambda}(d+y)\right]}{\cosh(2\pi d/\lambda)} \sin\left[2\pi\left(\frac{x}{\lambda}-\frac{t}{T}\right)\right] \tag{7}$$

$$v = -\frac{agT}{\lambda} \frac{\sinh\left[\dfrac{2\pi}{\lambda}(d+y)\right]}{\cosh(2\pi d/\lambda)} \cos\left[2\pi\left(\frac{x}{\lambda}-\frac{t}{T}\right)\right] \tag{8}$$

For any given phase angle (the sine and cosine terms) the velocities diminish with increasing depth, and where $y \geqslant -\lambda/2$ there is no appreciable motion.

If the water depth is less than $\lambda/2$, v becomes zero at the bed and the fluid above the bed is constrained to move in elliptical orbits in which A the major axis and B the minor axis are given by

$$A = 2a \frac{\cosh\left[\dfrac{2\pi}{\lambda}(d+y)\right]}{\sinh(2\pi d/\lambda)} \tag{9}$$

$$B = 2a \frac{\sinh\left[\dfrac{2\pi}{\lambda}(d+y)\right]}{\sinh(2\pi d/\lambda)} \tag{10}$$

For shallow water, as defined on page **29**–5, A and B become:

$$A = a\lambda/\pi d \tag{11}$$

$$B = 2a(d+y)/d \tag{12}$$

For deep water:

$$A = B = 2a \exp(2\pi y/\lambda) \tag{13}$$

These results which are shown schematically in Figure 29.4 are of value in estimating the depths at which sediment may be disturbed by wave action. In practice, however, particularly in the case of shallow water waves of large amplitude, the trajectories of the

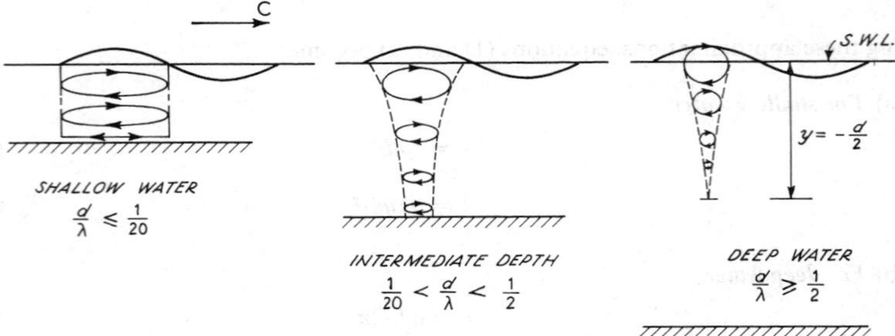

Figure 29.4 Water particle trajectories

water particles are not closed orbits, the velocities in the direction of wave propagation being greater and those in the reverse direction less than predicted by the theory. There is a resultant net movement of fluid in the direction of wave propagation at the surface and, in the case of shallow-water waves, close to the bed also, which in a confined system must be balanced by a return flow, normally at mid-depth.

Pressure fluctuations under a wave also diminish with depth and the instantaneous hydrostatic pressure p is given by

$$p = \rho g \left[a \sin \left[2\pi \left(\frac{x}{\lambda} - \frac{t}{T} \right) \right] \frac{\cosh \left[\frac{2\pi}{\lambda}(d+y) \right]}{\cosh (2\pi d/\lambda)} - y \right] \tag{14}$$

The sine term has limiting values of ± 1 under the crest and trough of the wave, and at the bed where $d = -y$ the maximum and minimum pressures are

$$\rho g \left(d \pm \frac{a}{\cosh (2\pi d/\lambda)} \right) \tag{15}$$

A pressure-sensing instrument located at a moderate depth may thus be used for the measurement of wave heights. It should also be noted that the pressure gradient will impose forces on any solid object within the fluid in addition to the drag forces arising from the fluid velocities u and v.

Superposition

Most of the properties arising from separate wave trains, i.e. surface elevation, particle velocity and instantaneous pressure, may be superposed, provided that both the amplitude of the component trains and the amplitude of the combined trains remain small. For example:

Combined surface elevation at position x and time t $\eta_\tau = \eta_1 + \eta_2 + \eta_3 + \dots \eta_n$

Combined instantaneous horizontal velocity $u_\tau = u_1 + u_2 + u_3 + \dots u_n$

Combined instantaneous vertical velocity $v_\tau = v_1 + v_2 + v_3 + \dots v_n$

The resulting surface elevations can be most readily determined graphically; when considering the interaction of wave trains travelling in different directions the surface profiles at, say, quarter-period intervals should be sketched to give a visual appreciation of the motion. Wave energy is proportional to the square of the wave amplitude a and cannot therefore be superposed.

The mathematically derived results for some special cases are given in (1)–(3) as follows:

(1) Two wave trains of amplitude a_1 and a_2 and the same period moving in the same direction with different phase angles δ_1 and δ_2:

$$\eta_\tau = a_1 \sin \left[2\pi \left(\frac{x}{\lambda} - \frac{t}{T} + \delta_1 \right) \right] + a_2 \sin \left[2\pi \left(\frac{x}{\lambda} - \frac{t}{T} + \delta_2 \right) \right] \tag{16}$$

If the two components are in phase, i.e. $\delta_1 = \delta_2$:

$$\eta_\tau = (a_1 + a_2) \sin \left[2\pi \left(\frac{x}{\lambda} - \frac{t}{T} + \delta_1 \right) \right] \tag{17}$$

If the two components are half a wavelength out of phase, i.e. $\delta_1 - \delta_2 = 1/2$:

$$\eta_\tau = (a_1 - a_2) \sin \left[2\pi \left(\frac{x}{\lambda} - \frac{t}{T} + \delta_1 \right) \right] \tag{18}$$

If the two components differ in phase by quarter of a wavelength, i.e. $\delta_1 - \delta_2 = 1/4$:

$$\eta_\tau = a_1 \sin\left[2\pi\left(\frac{x}{\lambda}-\frac{t}{T}+\delta_1\right)\right] - a_2 \cos\left[2\pi\left(\frac{x}{\lambda}-\frac{t}{T}+\delta_1\right)\right] \tag{19}$$

(2) Two wave trains moving in the same direction but with different periods T_1 and T_2 (wavelengths λ_1 and λ_2):

$$\eta_\tau = a_1 \sin\left[2\pi\left(\frac{x}{\lambda_1}-\frac{t}{T_1}+\delta_1\right)\right] + a_2 \sin\left[2\pi\left(\frac{x}{\lambda_2}-\frac{t}{T_2}+\delta_2\right)\right] \tag{20}$$

Equation (29.20) is not harmonic but it may be periodic. The constants δ_1 and δ_2 are the phase angles at the arbitrary origin where $t = 0$, $x = 0$, but the phase difference will change continuously, and an alternative origin may be found at which $\delta_1 = \delta_2 = 0$ and η_τ is zero. If after an interval of time T, η_τ is again zero and $T = mT_1 = nT_2$ in which m and n are integers, the resultant wave train is periodic, the period T being the smallest value that will satisfy $T = mT_1 = nT_2$.

If $a_1 = a_2$ and η_τ is periodic, nodes will occur at periods T, $2T$, $3T\ldots$, the envelope of the crest being another wave of period T, amplitude $2a_1$ and celerity

$$\frac{\lambda_1\lambda_2}{T_1 T_2}\left(\frac{T_2-T_1}{\lambda_2-\lambda_1}\right)$$

This special case is known as 'pure beat'.

(3) For two progressive waves moving in opposite directions:

$$\eta_\tau = a_1 \sin\left[2\pi\left(\frac{x}{\lambda_1}-\frac{t}{T_1}+\delta_1\right)\right] + a_2 \sin\left[2\pi\left(\frac{x}{\lambda_2}+\frac{t}{T_2}+\delta_2\right)\right] \tag{21}$$

If the origin of coordinates is chosen so that $\delta_1 = 0$ and if in addition $T_1 = T_2(\lambda_1 = \lambda_2)$, then

$$\eta_\tau = a_1 \sin\left[2\pi\left(\frac{x}{\lambda}-\frac{t}{T}\right)\right] + a_2 \cos\delta_2 \sin\left[2\pi\left(\frac{x}{\lambda}+\frac{t}{T}\right)\right]$$
$$+ a_2 \sin\delta_2 \cos\left[2\pi\left(\frac{x}{\lambda}+\frac{t}{T}\right)\right] \tag{22}$$

Equation 29.22 applies to an incoming wave which is totally or partially reflected by a structure such as a breakwater when $a_2 = K_r a_1$ in which K_r is the reflection coefficient. In practice K_r is close to 1 for small waves impinging on a vertical wall.

(a) *Perfect reflection* ($K_r = 1$)

If a wave train is perfectly reflected ($K_r = 1$ and $a_2 = a_1$) by a vertical barrier at $x = x_1$, then

$$\eta_\tau = 2a \sin\left[2\pi\left(\frac{x_1}{\lambda}-\frac{t}{T}\right)\right] \cos\left[\frac{2\pi}{\lambda}(x-x_1)\right] \tag{23}$$

It will be noted that η_τ is the product of two harmonic terms, one a function of x only and the other a function of t only. Thus there are certain times when $\eta_\tau = 0$ for *all* values of x, i.e. the water surface is flat, and certain positions where $\eta_\tau = 0$ for all values of t, i.e. where there is no vertical displacement of the surface at any time; the latter points are called nodes and will be located where

$$\cos\left[\frac{2\pi}{\lambda}(x-x_1)\right] = 0$$

i.e.

$$x_{\text{node}} = x_1 - \frac{(2n+1)\lambda}{4} \tag{24}$$

in which n can have any of the values 0, 1, 2, 3, ... Thus the nodes will occur at $\lambda/4$, $3\lambda/4$, $5\lambda/4$ ... from the barrier. This condition of stationary nodes is known as a 'standing wave' or 'clapotis'. The instantaneous horizontal and vertical water particle velocities in a standing wave are

$$u = \frac{2agT}{\lambda} \frac{\cosh\left[\frac{2\pi}{\lambda}(d+y)\right]}{\cosh(2\pi d/\lambda)} \cos\left[2\pi\left(\frac{x_1}{\lambda}-\frac{t}{T}\right)\right] \sin\left[\frac{2\pi}{\lambda}(x-x_1)\right] \tag{25}$$

$$v = -\frac{2agT}{\lambda} \frac{\sinh\left[\frac{2\pi}{\lambda}(d+y)\right]}{\cosh(2\pi d/\lambda)} \cos\left[2\pi\left(\frac{x_1}{\lambda}-\frac{t}{T}\right)\right] \cos\left[\frac{2\pi}{\lambda}(x-x_1)\right] \tag{26}$$

Since the nodes occur where $\cos[(2\pi/\lambda)(x-x_1)] = 0$ there are horizontal motions only under the nodes and vertical motion only under the antinodes. This result is important to the understanding of the relationship between tidal level and tidal current in coastal waters where the reflection or interaction of the tidal wave may result in a partial or complete standing wave.

The pressure p within a standing wave is given by

$$p = \rho g\left(\eta_\tau \frac{\cosh\left[\frac{2\pi}{\lambda}(d+y)\right]}{\cosh(2\pi d/\lambda)} - y\right) \tag{27}$$

and is hydrostatic under the nodes where $\eta_\tau = 0$ and there is no vertical movement.

(b) *Imperfect reflection* ($K_r < 1$)

If partial reflection takes place at a vertical barrier at $x = x_1$, $a_2 = K_r a_1$,

$$\eta_\tau = a\sin\left[2\pi\left(\frac{x-x_1}{\lambda}-\frac{t}{T}\right)\right] - K_r a \sin\left[2\pi\left(\frac{x-x_1}{\lambda}+\frac{t}{T}\right)\right] \tag{28}$$

The maximum and minim values for this expression occur in the same positions as the antinodes and nodes for the case of perfect reflection, the maximum and minimum amplitudes being

$$a_{\min} = a_1 - a_2$$

$$a_{\max} = a_1 + a_2$$

and

$$K_r = \frac{a_{\max} - a_{\min}}{a_{\max}}$$

Wave trains and wave energy

GROUP CELERITY, C_G

For a pure beat with waves travelling in the same direction the node and hence the wave group between each pair of nodes progress at a celerity of

$$\frac{\lambda_1\lambda_2}{T_1T_2}\left(\frac{T_2-T_1}{\lambda_2-\lambda_1}\right)$$

(see p. 29–8). The group of waves between any pair of nodes may be considered separately from the preceding and succeeding groups and it can be shown that as T_1 approaches T_2 the 'group celerity' C_G becomes

$$\frac{c}{2}\left[1+\frac{4\pi d/\lambda}{\sinh(4\pi d/\lambda)}\right] \tag{29}$$

Thus in deep water the group celerity is half the celerity of the individual waves in the group while in shallow water it approaches the celerity of the individual waves.

In a finite group of waves travelling in otherwise undisturbed water, wave crests will form at the back of the group, travel through it, in deep water at twice the speed of the group but at decreasing relative velocity as the water becomes shallow, and disappear at the front. It is evident that the energy within the wave train travels at the group celerity, not the wave celerity, and that the time taken for waves to reach a location distant from the area in which they have been generated is a function of C_G.

ENERGY

The average potential energy density (average potential energy per unit surface area) which is attributable to the presence of a progressive wave on the free surface is $\rho g a^2/4$; the average kinetic energy density is also $\rho g a^2/4$ and the total average energy density E is given by

$$E = \rho g a^2/2 \tag{30}$$

For a two-component composite wave train with both waves travelling in the same direction the average potential and kinetic energy densities are both $(\rho g/4)(a_1^2+a_2^2)$, and

$$E = \frac{\rho g}{2}(a_1^2+a_2^2) \tag{31}$$

For a standing wave, $E = \rho g a^2$ where a is the amplitude of the incident and reflected waves.

The proportion of the total energy which is carried along with a progressive wave train is given by

$$E\times\tfrac{1}{2}\left(1+\frac{(4\pi d/\lambda)}{\sinh(4\pi d/\lambda)}\right) = E\times\frac{C_G}{c} \tag{32}$$

In deep water this is half of the total energy, while it approaches the total energy in shallow water.

Transformation of waves

When waves travel from deep water into shallow water there will be no reflection of energy if the bed slope does not exceed 1 in 20, and the energy flux across any two planes parallel to the wave crests will remain constant provided no energy is dissipated or generated between the two planes. Using this principle of energy conservation, and allowing for changes in channel width b, or crest length due to wave refraction, changes in wave length, height and celerity can be calculated.

It is usual to refer these parameters to the corresponding values in deep water to which is ascribed the suffix '0', the basic transformation expressions being:

$$\frac{c}{c_0} = \frac{\lambda}{\lambda_0} = \tanh\left(\frac{2\pi d}{\lambda}\right) \tag{33}$$

and

$$\frac{H}{H_0} = \left(\frac{b_0}{b}\right)^{\frac{1}{2}} \times \left[\frac{2\cosh^2(2\pi d/\lambda)}{(4\pi d/\lambda)+\sinh(4\pi d/\lambda)}\right]^{\frac{1}{2}} \tag{34}$$

Change in wave steepness H/λ is obtained by combining equations (29.33) and (29.34).

Tables[3] of the various wave functions are available, from which typical values are given in Table 29.1

Table 29.1 WAVE TRANSFORMATION FUNCTIONS

d/λ_0	1.0	0.5	0.1	0.05	0.01	0.005
$c/c_0 = \lambda/\lambda_0$	1.0	0.99	0.71	0.53	0.25	0.15
C_G/c	0.5	0.52	0.81	0.91	1.0	1.0
H/H_0	0.98	0.92	0.90	1.0	1.1	1.15

It should be noted that at a certain depth, depending upon wave height and length, the wave will start to break and the above relationships then become invalid.

Reflection coefficients

Reflection coefficients for abrupt changes in geometry must normally be determined by experiment but some guidance may be obtained from published results.[5] It should be noted that the reflection coefficient is a function of both the incident wave steepness and the geometry of the solid boundaries.

Dissipation of wave energy

The rate at which energy is dissipated as a wave train travels through deep water is exceedingly small. The resulting reduction of wave amplitude with distance and time can be derived from:

$$a = a_0 e^{-\alpha x} = a_0 e^{-\alpha C_G t} \tag{35}$$

in which the damping modulus α is given by

$$\alpha = \frac{4\nu}{c}\left(\frac{2\pi}{\lambda}\right)^2$$

This expression gives the typical times and distances for the wave height to be reduced to half of its original value as shown in Table 29.2.

Table 29.2 DISTANCE AND TIME FOR 50% REDUCTION IN WAVE HEIGHT

λ	30 m	3 m	0.3 m
t	1 700 h	17 h	10 min
$x = C_G t$	22 000 km	64 km	210 m

For shallow water the theoretical solution is inaccurate owing to turbulence near the bed. Experimental work has shown α to be considerably larger than given above, the best fit to the available data being

$$\alpha = \frac{13.5\pi^{\frac{3}{2}}(T\nu)^{\frac{1}{2}}}{(4\pi d/\lambda)+\sinh(4\pi d/\lambda)} \tag{36}$$

These results cannot be applied to very long waves or intermediate waves where the extent of the turbulence is uncertain. If the bed is permeable there will be an energy loss due to wave-induced flow and the wave heights will be less than predicted.

Finite amplitude theory—breaking of waves

The foregoing results are strictly applicable only to waves of small amplitude. They are, however, sufficiently precise for many purposes and it is usually only necessary to have recourse to the more difficult finite-amplitude theory to obtain an appreciation of the processes which limit the maximum possible height of wave.

It should be noted that waves of finite amplitude have longer, shallower troughs and shorter, steeper crests than the sine wave assumed in small-amplitude theory, and this departure should be taken into account when determining the height of structures above mean sea level and the forces on them.

The wave celerity is insensitive to second- and higher-order effects but waves of finite height travel faster than small waves.

In the finite theory developed by Stokes it is assumed that if the water-particle velocity at the crest of the wave exceeds the celerity of the wave it will 'topple over' or 'spill'. The crest angle determined for this condition in deep water is 120° when the wave steepness $H/\lambda = 1/7$. From this the height of breaking waves H_b in deep water is given by $H_b/gT^2 = 0.0272$ which fits experimental data when $d/T^2 > 1$ m/s². The crest height above mean sea level a_c reaches a maximum value of $0.68H$.

In shallow water, long waves can be looked upon as 'solitary waves' (see below). This applies when $d/T^2 < 0.1$ m/s² when the ratio a_c/H approaches 1 and $H_b/d = 0.78$.

These results are shown in Figure 29.5 in which the curve in the range $0.1 < d/T^2 < 1.0$ has been fitted empirically to the available data.

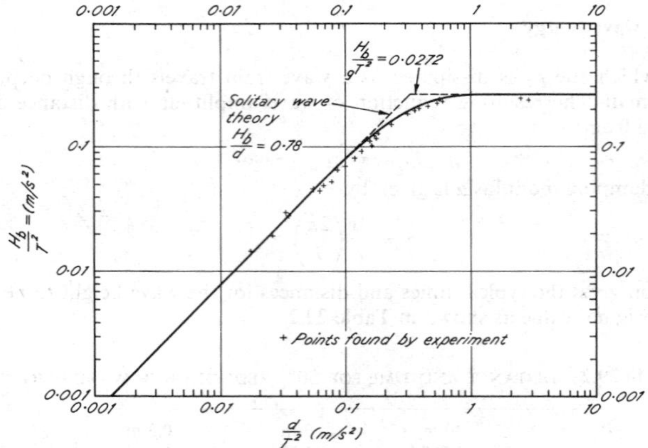

Figure 29.5 Breaking index curve (after Reid and Bretschneider)

The solitary wave

It is possible for a single wave, lying entirely above the still-water level, to be generated; such a wave propagates at constant velocity and is unaltered in form. In nature, waves generated by landslides or earthquakes may approximate to this type and, as already mentioned, long oscillatory waves moving into shallow water.

The surface profile of such a wave is given by the relation

$$\eta = H\left(\operatorname{sech}\left[\left(\frac{3H}{4d^3}\right)^{\frac{1}{2}}(x - ct)\right]\right)^2 \tag{37}$$

and

$$c = [g(H+d)]^{\frac{1}{2}} \tag{38}$$

the origin of x being at the wave crest.

Wave generation

When wind blows across a free water surface at a very low velocity the interface remains perfectly stable and the mirror calm is undisturbed. If the velocity increases slightly, ripples appear; these are capillary waves and have a length of about 17 mm and a period of 0.07 s. With further increase in wind speed the ripples start to grow and become gravity (rather than capillary) waves. Although great interest has been shown in these critical wind speeds, they are of little importance in engineering practice.

Once gravity waves have formed, energy is transferred from the air to the water in three ways:

(i) by shear at the interface
(ii) by pressure differences due to the form resistance of the waves
(iii) by random pressure fluctuations associated with the turbulent air stream

Initially the second of these will be dominant but as the wave length increases, shear at the interface becomes more important. If the wind blows for sufficient time over a long enough 'fetch' the energy input from the wind will become equal to the losses within the wave motion and further growth ceases. This is known as a fully developed sea for which Bretschneider gives

$$\frac{c}{U} = 1.95 \tag{39}$$

and

$$\frac{gH}{U^2} = 0.283 \tag{40}$$

where U is the wind speed. It will be noted that the waves are travelling at nearly twice the wind speed. However, other workers give significantly different results, for the physical processes involved are not sufficiently understood and so-called wave forecasting methods which relate wave height, period and length to wind speed and fetch, cannot therefore be other than semi-empirical. Confirmation of the results is difficult owing to the problems associated with measuring wave properties in deep water and in analysing the results. The approximate time and distance or 'fetch' for which a steady wind has to blow in one direction over deep water, and the corresponding significant wave height as assessed by the '*Wave Spectrum*' method of Pierson *et al.*, for a fully developed sea, and by the '*Significant Wave*' method of Sverdrup, Monk and Bretschneider for a 90% developed sea, are compared below.

Table 29.3 RELATIONSHIP BETWEEN WAVE HEIGHT, WIND SPEED, TIME AND FETCH

Wind speed (m/s)	*Wave spectrum method* (*fully developed sea*)			*Significant wave method* (90% *developed sea*)		
	H (m$^{\frac{1}{3}}$)	*Time* (h)	*Fetch* (km)	H (m$^{\frac{1}{3}}$)	*Time* (h)	*Fetch* (km)
5	0.4	2	18	0.7	17	150
10	2.1	9	125	2.6	33	600
15	6.7	22	500	5.9	50	1 400
20	13.0	40	1 250	10.5	66	2 550
25	22.2	65	2 400	16.2	83	3 850
28	29.2	82	3 500	20.1	92	4 700

The 'significant wave' method gives such large values for time and fetch for a fully developed sea as to be unattainable in reality. For short fetches and high wind speeds the following expressions are derived from the work of Bretschneider:

$$H_{\frac{1}{3}} = 0.024\,(U^2F)^{\frac{1}{3}} \tag{41}$$

$$T_{\frac{1}{3}} = 0.6(U^2F)^{\frac{1}{6}} \tag{42}$$

$$\frac{F_{\text{min}}}{t_{\text{min}}} = 1.26(U^2F)^{\frac{1}{6}} \tag{43}$$

$H_{\frac{1}{3}}$ is the significant wave height in metres (mean of the highest one-third of the waves), $T_{\frac{1}{3}}$ is the significant wave period in seconds (mean period of the highest one-third of the waves), U is the windspeed (m/s), F the fetch length (km), F_{min} the minimum fetch for the wave condition to develop, and, t_{min} the minimum duration for the wave condition to develop.

Since in real storms the wind is not constant in speed or direction nor unlimited in extent, the practical application of these results is complicated.

Wave heights are randomly distributed about some mean value and it is not possible to define a 'highest' wave from a recording of a wave train with a given mean height, or the energy level; it is however possible to determine the probability of a certain height being equalled or exceeded in a given sample. It appears to be generally agreed that in deep water the spectrum of wave heights fits approximately to a Rayleigh distribution and the following relationships have been derived:

$$H_{\frac{1}{3}} = 1.6 \times H_{\text{mean}} \tag{44}$$

$$H_{1/10} = 2.03 \times H_{\text{mean}} \tag{45}$$

$$(H_{1/10} = \text{mean of the highest } 10\% \text{ of the waves})$$

About 1% of the waves will equal or exceed about $2.8H_{\text{mean}}$, about one wave in 10 000 will equal $3.4H_{\text{mean}}$ and about 16% of the waves will exceed $H_{\frac{1}{3}}$; it is usual to measure $H_{\frac{1}{3}}$ when examining wave records. It cannot be assumed that these results for deep water, based on the Rayleigh distribution, can be applied to waves in shallow water for which less prototype data are available.

Wave generation in shallow water

The physical processes involved in the generation of waves in shallow water are the same as those in deep water except that when $d/T^2 < 0.75$ m/s the waves 'feel the bottom' and the growth of the longer-period waves is restricted. If the fetch and duration are unlimited the relationship between windspeed, depth and significant wave height, based on the work of Bretschneider, is as shown in Table 29.4.

Table 29.4 WAVE HEIGHTS IN SHALLOW WATER

Wind speed m/s	*Water depth* (m) 1	2	3	5	7	9
	Wave heights (m)					
10	0.30	0.48	0.63	0.90	1.13	1.35
15	0.38	0.60	0.78	1.14	1.44	1.73
20	0.45	0.71	0.92	1.35	1.72	2.04
25	0.52	0.81	1.07	1.53	1.95	2.32
28	0.56	0.88	1.14	1.64	2.06	2.45

Figure 29.6 gives forecasting curves after Thijsse and Schijf[6] for wave height and length when both fetch and depth are restricted.

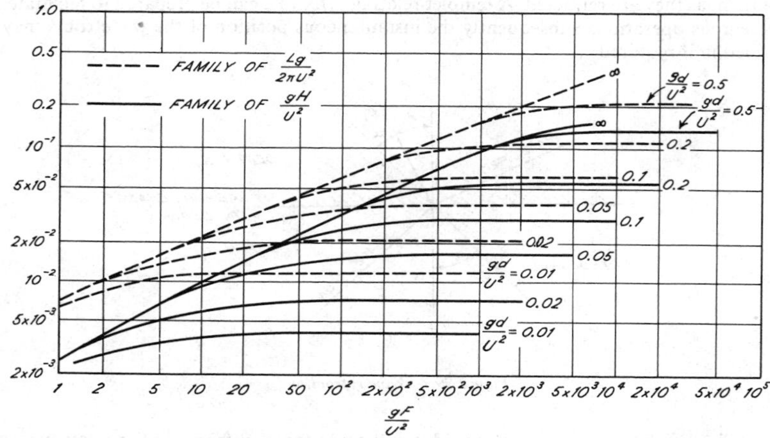

Figure 29.6 Growth of waves in limited depth (after Thysse and Schijf[6])

Wave decay

When waves generated in deep water travel out of the generating area, or when the generating wind abates, they lose their extreme irregularities and diminish in height. The processes which result in this decay of wave height are principally the following:

(i) lateral diffraction of energy
(ii) selective attenuation; the long-period part of the wave spectrum travels faster than the short-period part and the energy is spread out in the direction of wave propagation
(iii) air resistance or directly opposing winds
(iv) viscous damping in the water.

A conclusion derived empirically by the Admiralty in 1942 is that waves lose one-third of their height each time they travel a distance in nautical miles equal to their length in feet. Charts based on the work of Sverdrup, Munk and Bretschneider[7] are available.

Propagation of waves into shallow water—refraction

When waves travel into shallow water ($d<\lambda/2$) their speed diminishes, their form alters and if the wavefronts are long and not parallel to the contours they are refracted and become curved. Since any wave train consists of a number of components of varying length the bottom will thus have a sorting effect, the longer components being affected sooner and therefore to a greater overall extent than the shorter components.

If the direction of wave propagation is represented by rays (orthogonals—see Figure 29.7), refraction of the waves is precisely analogous to the bending of light rays passing from one medium into another of greater density, and Snell's law, $\sin i/\sin r = c_1/c_2$, applies.

If the wave period and depth are known, c_1 and c_2 can be calculated from the relationships given on pages **29**–5 and **29**–10, or more expeditiously from published tables. Starting from the assumed direction of the orthogonals in deep water, it is possible to plot them as they are refracted. A templet relating c_1/c_2 to α can be prepared to facilitate this tedious operation. Subsequently the instantaneous position of the wavefronts may be filled in if required.

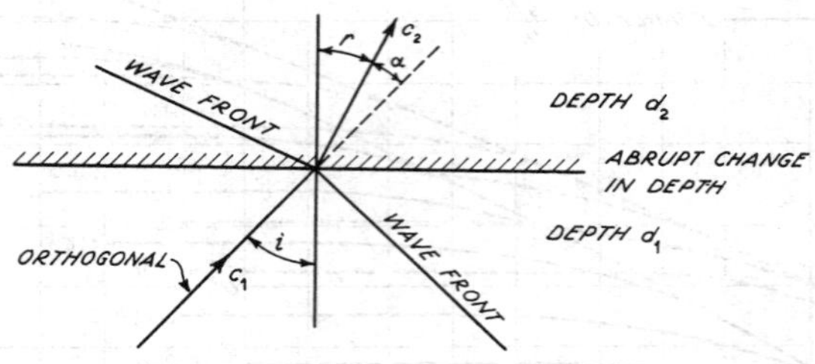

Figure 29.7 Wave refraction

It will be noted that a sea-bottom ridge will cause the wavefronts to become concave and the orthogonals to converge whereas a valley will have the reverse effect.

Wavefronts approaching a shoreline at an angle will be convex with diverging orthogonals.

It is usual, and nearly correct, to assume that no energy crosses the orthogonals so that if the wave height in deep water is known the wave height at any other point in the diagram can be calculated from equation (34).

Wave forecasting

Using the principles set out on pages **29**–13 to **29**–16 it should in theory be possible to compute from the meteorological synoptic charts a wave spectrum for a given return period for any coastal site. The practical application of the various methods available is, however, complex and beyond the scope of this article. For further information the reader is referred to publications dealing specifically with forecasting methods.[1, 7] Moreover, the theoretical and empirical basis for the methods is far from perfect and errors will be introduced at each stage; if at all possible, design decisions should therefore be based on statistical analysis of wave records for the site. It is then doubtful whether consideration of the physical processes generating the waves is necessary except to establish the typicality of the sampling period.

Diffraction

If waves pass the end of a breakwater or through a hole in a barrier they spread into the water behind the obstruction, and like refraction, the phenomena is analogous to the diffraction of light. Mathematical solutions of problems involving diffraction are difficult but have been carried out for a few simple cases, e.g. for a train of uniform low waves in uniform depth passing the end of a breakwater for which the diffraction pattern is shown in Figure 29.8.

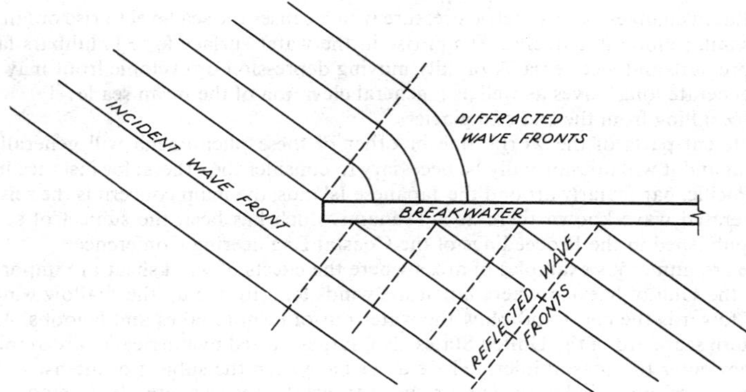

Figure 29.8 Wave diffraction

The application of diffraction theory to harbour design is limited since in a real harbour basin the diffraction pattern will normally be curtailed by the boundaries of the basin and refraction and reflection will occur simultaneously; therefore the study of problems of diffraction usually requires an hydraulic-model investigation.

EXCEPTIONAL WATER LEVELS

Long waves—surge

In addition to the tidal wave with a dominant period of about 12½ hours and wind-generated storm waves which normally have dominant periods between about 4 and 12 seconds, the surface of the sea is frequently disturbed by long waves of intermediate periods or by solitary waves. Although these long waves are normally of very small amplitude and are therefore barely noticeable to the casual observer and difficult to measure on a wave record, they can be of great importance since the ability of waves to penetrate into sheltered bays, harbours and estuaries increases with their length.

Consideration of long-wave activity is therefore necessary when designing a new harbour or harbour basin since such waves cannot be excluded. If the wave period coincides with that a of a mode of oscillation of a harbour basin or of a natural tidal inlet, the disturbance will be amplified and can lead to problems in the mooring of ships; in extreme cases a ship's motion may become sufficiently violent for it break its mooring lines.

The effects of long waves or surges which may have a period of several hours must be taken into account in determining the crest levels of sea walls and flood defences. It should also be noted that such disturbances can cause the sea level to fall below predicted tide levels so that ships may stand or sea water intakes become temporarily exposed. Disturbances of this kind may be generated in a number of ways of which the most important are as follows:

(i) Seismic activity (earthquakes)
(ii) Wind shear on the water surface in shallow seas or lakes which causes not only short-period waves but also an inclination of the water surface
(iii) Swell waves of slightly different periods reaching the locality at the same time from separate storms and producing *surf beats*

(iv) Rapid changes in barometric pressure which causes the sea level to rise or fall by a similar amount, namely 300 mm rise in the water surface for 34 millibars fall in pressure and vice versa. A rapidly moving depression or cyclonic front may thus generate long waves as well as a general elevation of the mean sea level

(v) Ice falling from the end of a glacier.

In different parts of the world, one or other of these phenomena will generally be dominant and it will not normally be necessary to consider the others; for instance in the North Pacific, particularly around the Japanese Islands, the main concern is the seismically generated wave known there as a Tsunami which has been the subject of several papers published in the Proceedings of the Coastal Engineering Conferences.

There are numerous examples of areas where the effects of wind shear are important, notably the Gulf of Mexico where hurricane winds tend to pile up the shallow water of the gulf towards the coast and blow the water out of inshore lakes and lagoons. Along the Eastern seaboard of the United States also, surges caused by hurricanes are amplified when they enter the coastal inlets. These areas have been the subject of intensive study and, since hurricane wind speeds and rates of travel do not vary greatly, attempts have been made to understand the physical processes involved and to predict the effects of future storms of this kind.

In the North Sea, which is very shallow, wind shear again plays a dominant role particularly when combined with a fall in barometric pressure. Here prolonged northerly winds may generate a surge or surges as in 1953 when it reached a height of 2.7 m along the East Anglian coast and exceeded 3.0 m in the Delta area of the Netherlands. According to the Admiralty Tide Tables, depression of the sea level or negative surges of 0.6 to 0.9 m occur several times a year in the southern North Sea; levels 2.1 m below tidal prediction were recorded at Southend in 1967.

In the UK extreme high water levels have been studied on a statistical basis using measured departures from predicted sea levels irrespective of their cause.[8, 9] The duration of a surge at any particular locality is generally no more than a few hours and the probability that the peak of an exceptionally high surge will coincide with a very high astronomical tide is small as the two phenomena are independent; return periods for the highest conceivable water levels are therefore very long, and the choice of crest level for flood protection works requires careful judgment in balancing the cost of the works against the risk of damage and perhaps loss of life if they are overtopped.

Well known examples of enclosed waters in which long-period wind-induced oscillations are set up are Loch Ness with a period of 33 min, Lough Neagh (period 45 min) and the Baltic (period 15 h).

Waves of appreciable height and periods in the range of 20 s to a few minutes, which cause surging in tidal dock basins, have been observed in many places and may be expected in any coastal waters open to the ocean. They are particularly prevalent in the southern hemisphere.

Analytical results which are of help in considering long-wave problems are given below.

Wind set-up

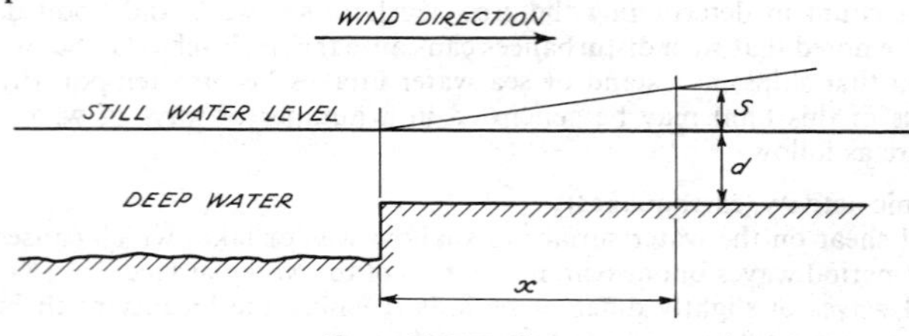

Figure 29.9 Wind set-up

For a wind of constant direction and speed U the wind set-up S above still water level can be determined from the water slope:

$$\frac{dS}{dx} = \frac{KU^2}{g(d+S)} \tag{46}$$

in which the constant K depends on surface stress, which is a function of the wave state and the current structure associated with shear and roughness of the bottom. From a study of Lake Okeechobee in Florida K has been evaluated for enclosed waters as 3.3×10^{-6}.

For an open coastline with the wind perpendicular to the shore (Figure 29.8) an approximation is obtained by assuming constant depth; then equation (29.46) can be solved giving

$$S = d\left[\left(\frac{2KU^2x}{gd^2}+1\right)^{\frac{1}{2}}-1\right] \tag{47}$$

In order to allow for bottom slope it is suggested that $K = 3.0 \times 10^{-6}$ be used instead of 3.3×10^{-6}.

Wave set-up

In addition to wind set-up the breaking of waves on a beach also raises the mean sea level locally and this may be as much as 10 to 20% of the incident wave height. Consequently tide gauges on open beaches will give misleading results while a partly sheltered beach will be subject to littoral currents flowing from the exposed to the sheltered region.

Resonance in harbour basins

Resonant standing wave systems can be demonstrated in a simple manner if an open-topped tank part filled with water is moved to and fro with the correct frequency; if the tank is mounted on rollers and driven by a variable speed drive it can be excited in several different modes, the first or slowest of which occurs when the length of the tank is equal to half the wavelength of the progressive gravity wave which would occur in that depth of water. The tank then contains one 'cell' of a standing wave system with one 'node' at constant level, but with maximum horizontal velocities at its centre, and maximum vertical movement and vertical velocity at its ends.

Resonance in a rectangular basin of length a, width b and depth d will occur when the period T of the varying exciting force coincides with one of the modes of oscillation of the basin, i.e.

$$T = \frac{2}{(gd)^{\frac{1}{2}}}\left[\left(\frac{n}{a}\right)^2+\left(\frac{m}{b}\right)^2\right]^{-\frac{1}{2}} \tag{48}$$

where n and m are integers representing the various modes of oscillation in directions a and b respectively. For oscillation in one direction only:

$$T_a = \frac{2a}{n(gd)^{\frac{1}{2}}} \tag{49}$$

$$T_b = \frac{2b}{m(gd)^{\frac{1}{2}}} \tag{50}$$

Periods of oscillation for circular and elliptical basins can be evolved analytically while numerical solutions may be used for irregular basins.[1]

In a harbour basin this kind of oscillation, often referred to as *ranging* or *scend,* can be initiated by long-wave activity in the approaches to the harbour. If the opening to the basin through which the forcing wave enters is at one end, oscillation in the first mode will be encouraged; if it is at the centre this mode of oscillation will be suppressed, but a second mode oscillation may occur with two nodes at one-quarter of the basin length from either end. The determination of the modes of oscillation for basins of irregular shape, with openings of appreciable width interconnected with one another and the open sea, is complex and requires the use of either a hydraulic model or a mathematical model solved on a computer or a combination of both techniques.

Ranging of moored ships

Vessels within a harbour which is subjected to long-wave resonance are likely to be of much smaller length than the waves, and will respond not only to the horizontal movement of the water which is much greater than the vertical movement, but also to the continually changing slope of the water surface. If the vessel is unrestrained it will accelerate down the slope of the wave in one direction, and then decelerate to rest before accelerating again in the opposite direction as the water slope changes. For a vessel which is restrained by elastic moorings, resonance will occur if $T^2 = 4\pi^2 M/k$ in which M is the 'virtual mass', i.e. the mass of the vessel plus the mass of water associated with the motion, and k the stiffness of the mooring. In practice, resonance may occur if the moorings are moderately stiff. If the moorings are very stiff but with some slack, which is the most usual case, the motion is irregular and may become extremely violent. In either case large forces may be imposed on the moorings and lines may be broken. For further information the reader is referred to Russell.[10]

SEA-BED AND LITTORAL SEDIMENTS

Sources of material

The primary sources of sea bed and littoral sediments are the adjacent land masses, from which the material is derived either by the normal processes of subaerial denudation and transported to the coast by streams and rivers, or from erosion of the coastline under wave attack. There is little evidence of any significant transport of material to the shore from deep water, apart from silt which finds its way into some of the estuaries in the UK,[11, 12] and in some cases sand, e.g. from the Irish Sea into Liverpool and Morecambe Bays. There is some evidence that shingle may be moved from offlying shoals in shallow water on to the shore but as a source of beach-building material the quantities so moved are of little importance.

Modes of transport—currents and waves

Movement of sediment may be caused by currents alone, depending upon grain size and current velocity (Figure 29.10).[13] Where the currents are strong, as is frequently found close inshore at headlands and in the entrances to rivers and tidal inlets, they may have a significant effect on the sea-bed profile and configuration of the shore. Along the greater part of the coast, movement of material is initiated by wave action and the resulting direction of littoral transport is dependent upon the relative strengths and directions of the wave induced and tidal currents and the grain size of the material. Along the foreshore and in the breaker zone, wave action predominates and the material is moved along a zigzag path due to the uprush and backwash of obliquely approaching waves. Within and to seaward of the breaker zone along-shore currents have an increasing effect and as the depth increases may become dominant. It is thus possible for transport outside

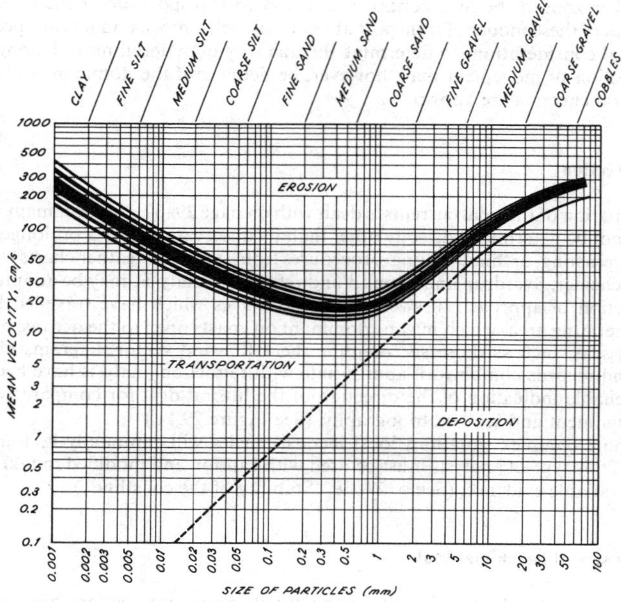

Figure 29.10 Erosion, transportation and deposition curves (after Kuenen[13])

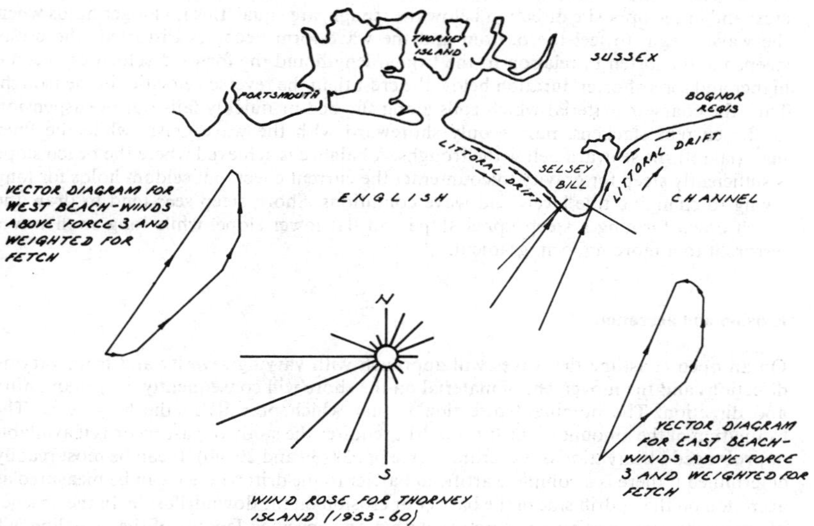

Figure 29.11 South coast of England—Portsmouth to Bognor Regis (after Duvivier[14])

the breaker zone to be in a contrary direction to transport along the foreshore. The mechanics of these modes of transport are not precisely known and it is not possible from theoretical considerations to determine the quantity of material moved along the coast. The direction of movement may, however, be deduced if the dominant wave direction and littoral currents are known.

Wave direction

Determination of the tidal currents is dealt with on page **29**–37. The dominant wave direction cannot be determined directly other than by observation over a prolonged period of time. Forecasting, or 'hind-casting' procedures from synoptic weather charts as developed by Bretscheider, Sverdrup and Munk,[1] and refraction analysis, may be adopted to assess the direction of approach of swell waves (i.e. waves which have travelled beyond the wind generating area) which may be dominant on coasts open to the oceans, but for more enclosed seas, such as those surrounding the UK, wind vector diagrams plotted from local wind records combined if appropriate with refraction analysis have been found to give a reliable indication of the direction of the littoral drift for comparatively regular shore alignment and sea-bed topography. (See Figure 29.11.)

For more complex situations local inconsistencies will commonly be found and the results of this type of analysis must be used with caution and modified to take account of all the available evidence. (See p. **29**–39, 'Stability of the coastline'.)

Effect of size of beach material

It is commonly found that the coarsest material comprising the foreshore and adjacent sea bed is at the crest of the beach, i.e. at the limit of wave uprush, and gradually becomes finer towards and below low water mark. This sorting of beach material is generally ascribed to the varying velocities in the oscillating wave currents. Whereas in deep water the velocity of the water particles, in the direction of wave propagation below the wave crest and in the opposite direction below the trough, are equal, this no longer holds when the waves begin to feel the bottom and the wave form becomes distorted; the crests steepen and shorten in relation to the trough length and the forward velocities become higher and for a shorter duration below the crest than the reverse velocities in the trough. Thus the coarsest material which rolls along the bed or quickly falls out of suspension as the current slackens, moves only shoreward with the wave crests while the finer material returns seaward below the troughs. A balance is achieved where the beach slope is sufficiently steep for gravity to counteract the current effect, but seldom holds for long owing to changing tide levels and wave conditions. Short, steep seas tend to draw the beach down forming a steep upper slope and flat lower slope, while long-swell waves restore it to a more uniform gradient.

Erosion and accretion

On an open coastline the waves will approach with varying severity and from varying directions and the movement of material on the shore will consequently vary in quantity and direction. The dominant direction is that which prevails in the long term. The estimation of the amount of the littoral drift requires the study of past records if available or prolonged observation or experiment (see pp. **29**–39 and **29**–40). It can be most readily determined if there is a complete artificial barrier to the drift where it can be measured as accretion on the updrift side of the barrier or erosion on the downdrift side. In the absence of such a barrier, erosion or accretion within any specified lengths of the coastline will depend upon the balance of transport into and out of the area, and it is to be noted that a stable coastline does not indicate that there is no littoral transport or that the reversals of direction of transport with the varying direction of wave approach, and the direction and strength of wave induced and tidal currents are equal. Apart from the movement of material into the area from the beaches updrift of the length under study and out of the area to the downdrift beaches, account must be taken of other sources of material supply and losses from other causes. Supply to the area will be augmented by material eroded from the back-land and of this the finer material will quickly be washed out to sea and finally deposited in deep water. Attrition will reduce the size of the coarser material until it also may be lost in the same way and there may be direct loss of coarse material in the event of deep water, or deep gullies existing close to the shore. Sand may be lost by wind action resulting in the formation of dunes behind the foreshore. Exceptionally material may find its way on to the beaches from offlying shoals and banks. Estimation of these sources of supply and loss can only be made by comparison of surveys of the shore and sea bed.

STRATIFICATION AND DENSIMETRIC FLOW

Saline wedge in estuaries

The difference in density of fresh and sea water has an important effect on the movement of sediments in estuaries and estuarial dock systems and on the behaviour of effluent discharges and industrial cooling water intakes and outfalls.

In the absence of turbulence, liquids of different densities do not readily mix and if brought together tend to form layers, the less dense rising to the surface and the more dense remaining below. In turbulent flow the density difference required to maintain a stable interface between the two liquids depends upon the degree of turbulence but is commonly quite small; for example the 2.5% difference in density between fresh water and sea water is more than sufficient to inhibit mixing, and the interface in the case of a river or stream discharging into the open sea, unless it does so via a long tidal estuary, is frequently visible as a sharp line on the surface.

If the two liquids of density ρ_1 and ρ_2, which in nature may be salt and fresh water, or silty and clear water, are initially separated by a vertical gate which is then removed the

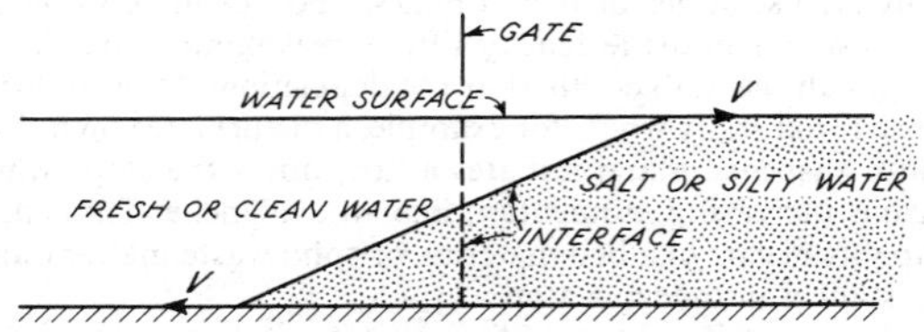

Figure 29.12 Saline wedge in estuaries

more dense salt or silty water will flow in one direction as a wedge under the less dense fresh or clear water while the fresh or clear water flows in the opposite direction forming an upper wedge. The velocity of the interface at the surface and the bed (see Figure 29.12) is given by

$$V = 2/3\left(\frac{\rho_1 - \rho_2}{\rho_1 + \rho_2}gd\right)^{\frac{1}{2}} \tag{51}$$

For salt and fresh water and a depth d = 12 m, which is not unusual in estuarial dock systems, V = 0.8 m/s which is sufficient to erode and transport fine bed material. Examples of this phenomenon leading to a siltation problem in impounded docks are found on the Thames and the Mersey.

In estuaries and tidal rivers the fresh water rises over the sea-water, the interface moving up and down river with the rising and falling tide; as a result the seaward transport of silt, which is mainly in the form of saltating bed load, is arrested or reversed with consequent siltation in the tidal reaches.

Effluent outfalls

Studies and experimental work by Abraham have shown that, for sewage discharged into the North Sea off the Netherlands coast, a dilution of 50 times, which reduces the density difference to 0.05%, is required if the formation of a stable field is to be prevented. Dilutions of this order of magnitude can be achieved within the buoyant plume which rises from an effluent outfall discharging close to the sea bed in moderate depth, and can be calculated from data published by various workers.[15, 16, 17] The dilution depends almost entirely on the ratio of the diameter z of the discharge port to the water depth d, from approximately 10 for $z/d = 10$ to approximately 150 for $z/d = 100$. The initial jet angle and velocity of efflux v are not significant except for very small values of z/d. It should be noted that the densimetric Froude number given by $v[gz(\rho_0-\rho_s)/(\rho_s)]^{-\frac{1}{2}}$, in which ρ_0 and ρ_s are respectively the density of the receiving body of water and effluent, cannot in practice be less than 1; if the selected port area produces a value less than 1, then the port will discharge only 'part full' as an inverted weir, or in the case of a multiport system the flow will be concentrated in only a few of the ports.

A current flowing across the point of discharge will lengthen the plume with a corresponding increase in dilution. Field studies by Agg and Wakeford[18] have shown that a cross current of velocity u increases the dilution by $10 \times u/v$ but this result should not be applied outside the range of the experiments which was small.

Density and turbidity currents

If silt-laden water, brine or some other fluid having a density greater than sea water $\rho_0 = 1.025$ g/cm^3) is discharged on to a sloping sea bed it will flow down the slope as a discreet and coherent stream until the density difference is eliminated either by dilution due to turbulent mixing with the ambient fluid or the deposition of the silt load. Streams of this kind can also occur naturally when, for example, a disturbance on a sloping sea bed puts fine material into suspension and initiates a flow down the slope which scours further material and gains increasing momentum like an avalanche. These density or turbidity currents are capable of carrying bed material, or solid waste matter dumped on a sloping sea bed, long distances into deep water.

Calculation of the velocity of a turbidity current requires an assumption as to the way the current will spread laterally unless it is confined by the sea bed topography and it is necessary also to establish that the current will be stable, which will only be the case on moderate slopes.

The results of the experimental work by Tresakar[19] relates velocity v to a number θ developed by Keulegan for density currents, i.e.

$$\theta = \frac{1}{\mathrm{v}}\left(\frac{vg\,(\rho_s-\rho_0)}{\rho_s}\right)^{-\frac{1}{3}} \tag{52}$$

where ρ_0 is the density of ambient fluid, ρ_s is the density of fluid forming the current, and ν is the kinematic viscosity of fluid forming the current.

θ is a function of slope only and is found by Tesaker[19] to be 0.02 for a slope of 1 in 10 and 0.027 for a slope of 1 in 20. For a theoretical treatment of the subject the reader is referred to Elliston and Turner's paper.[20]

WAVE AND CURRENT FORCES

Forces on a circular cylinder or pile

The most usual approach to the assessment of wave and current forces on piled and braced structures or pipelines in the sea is by the summation of 'drag' and 'intertia' components due respectively to the velocities and accelerations of the water particles in the motion. The horizontal force on an elementary length ds of a cylinder or pipe of diameter D is then

$$\mathrm{d}F = \left(\tfrac{1}{2}C_{\mathrm{d}}\rho Du\bar{u} + C_{\mathrm{m}}\rho\frac{\pi D^2}{4}\frac{\mathrm{d}u}{\mathrm{d}t}\right)\mathrm{d}s \tag{53}$$

in which C_{d} is the drag coefficient, C_{m} the inertia or mass coefficient, and $\bar{u}$ is the absolute value of u.

Using the small-amplitude theory (pp. **29**–5 and **29**–6) the required values of u and du/dt are obtained from equation (7). (Similarly vertical forces can be obtained using v and dv/dt from equation (8).

It is convenient to transfer the origin of the coordinates x and t to the wave crest and to assume that the elementary length ds of the cylinder is located at $x = 0$ and elevation s above the sea bed ($d+y = s$). By substituting from equations (29.1) and (29.2) the equations for velocity and acceleration can be reduced to:

$$u = \frac{\pi H}{T}\frac{\cosh(2\pi s/\lambda)}{\sinh(2\pi d/\lambda)}\cos\left(\frac{2\pi t}{T}\right) \tag{54}$$

and

$$\frac{\mathrm{d}u}{\mathrm{d}t} = \frac{2\pi^2 H}{T^2}\frac{\cosh(2\pi s/\lambda)}{\sinh(2\pi d/\lambda)}\sin\left(\frac{2\pi t}{T}\right) \tag{55}$$

In the case of breaking waves it is suggested that the Solitary Wave[21, 22, 23] theory should be used with graphical and tabulated values of u and du/dt as given by Munk.[24]

For steady flow the theoretical value of $C_{\mathrm{m}} = 2.0$ while C_{d} is related to Reynolds' Number (uD/kinematic viscosity). For orbital flow in waves the coefficients have been determined experimentally on model and full-scale piles.[25, 26] A mean value of 2.5 is reported for C_{m} but it is to be noted that 20% of the results were above 3.5 and 10% above 4.0. Attempts to relate C_{d} to Reynolds' number were unsatisfactory—see Figure 29.13. It has however been demonstrated that more consistent results can be obtained if the data are analysed using the higher-order wave theory; in that case values of $C_{\mathrm{m}} = 2.0$ and $C_{\mathrm{d}} = 0.7$ can be justified. For an approximate assessment of forces using the linear wave theory the values $C_{\mathrm{m}} = 2.5$ and C_{d} from an envelope of the experimental data in Figure 29.13 are recommended.

Forces on sea walls and breakwaters

In an analysis of the forces exerted by waves on a structure it is first necessary to decide if the waves will break against or immediately in front of the structure (see p. **29**–12). Forces due to nonbreaking waves will be essentially hydrostatic whereas breaking waves exert additional pressures due to the dynamic effects of turbulent water motion and entrapped

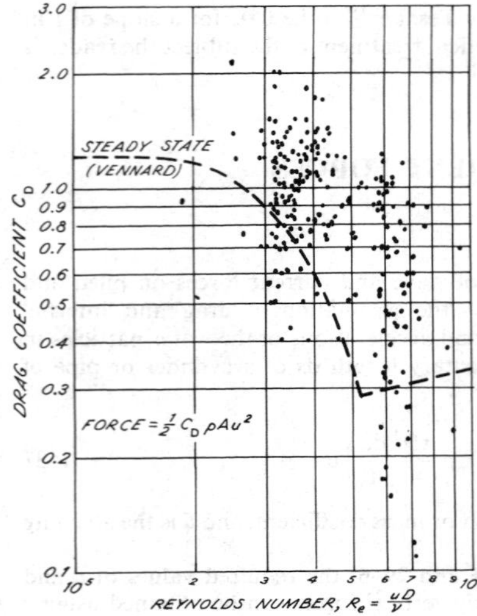

Figure 29.13 Coefficient of drag C_D *as a function of Reynold's number* R_e *for waves higher than* 10 ft (3 m) (After Wiegel, Beehe and Moon, *J. Hydraulics Div*, paper 1199). A *is the projected area,* u *the particle velocity,* D *the pile diameter,* υ *the kinematic viscosity and* ρ *the density of water*

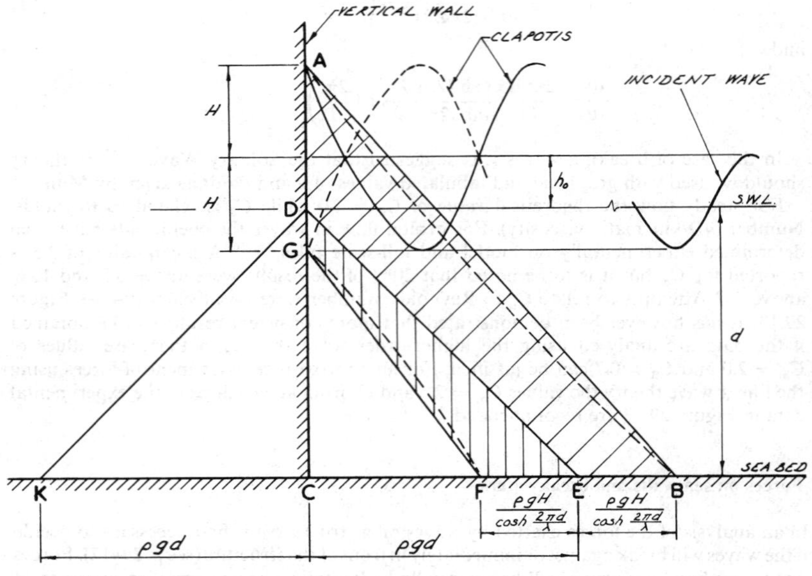

Figure 29.14 Sainflou diagram

air pockets; breakwaters should not be designed for unbroken waves unless the circumstances are exceptional.

NONBREAKING WAVES—SAINFLOU METHOD

With complete reflection from a vertical face a 'standing wave' or 'clapotis' will be set up. (See p. **29**–8, $K_r = 1$). The wave height at the wall becomes twice the incident wave height H, the mean level or orbit centre being a height h_0 above still water level where

$$h_0 = \frac{\pi H^2}{\lambda} \frac{1}{\tanh(2\pi d/\lambda)} \tag{56}$$

The pressure against the face of the wall is given by equation (29.27) (p. **29**–9) and is a maximum at the base of the wall where

$$p_{max} = \rho g \left[\frac{\pm H}{\cosh(2\pi d/\lambda)} + d \right] \tag{57}$$

The variation in pressure against the face of the wall is shown by the broken lines AB and GF in Figure 29.14 when the crest and trough of the wave respectively are against the wall. Bearing in mind the idealised situation upon which this theoretical approach is based it is sufficiently accurate to use the straight lines AB and GF in place of the hyperbolic curves.

If there is still water behind the wall there will be an outward pressure represented by the triangle DKC and the resultant force on the wall is then given by the area ABED in the landward direction when the wave crest is at the wall and by the area DEFG in the seaward direction when the wave trough is at the wall.

If the crest of the wall is less than $(H + h_0)$ above still-water level the usual procedure is to assume the full clapotis but to omit the part of area ABED which lies above the crest level. Clearly this will give a conservative estimate of the total force.

BREAKING WAVES

Evidence from various sources, including analysis of existing structures, suggests that the forces due to breaking waves will be in the range 100 to 500 kN/m^2 but no satisfactory analytical methods have been devised whereby they can be predicted with any great confidence.

The following expression for the maximum dynamic pressure P_{dm}, based on experimental work by Bagnold,[27] has been proposed for the case in which the sea bed shelves steeply (in excess of 1 in 20) from deep water up to the face of a vertical wall.

$$P_{dm} = \pi \rho g H_b \frac{d}{\lambda_0} \left(1 + \frac{d}{d_0} \right) \tag{58}$$

in which H_b is the height of the breaking wave, d the depth at the wall, d_0 the depth away from the wall, and λ_0 the wavelength in depth d_0.

It is suggested that the maximum pressure P_{dm} at still-water level falls off parabolically to zero at the crest and trough of the wave and is added to the hydrostatic pressure as indicated in Figure 29.15.

With still-water pressure behind the structure the resultant pressure is given by the shaded area AFGBED.

An alternative approach is to assume that, immediately before breaking, the water mass in the wave moves forward with the velocity of wave propogation c. Using the approximate relationship for shallow water $c = (gd)^{\frac{1}{2}}$ (equation 3), the dynamic pressure $P_d = \rho c^2 = \rho g d$ which is added to the hydrostatic pressure as shown in Figure 29.16.

With still-water behind the structure the resultant pressure is given by the shaded area AFGHBED.

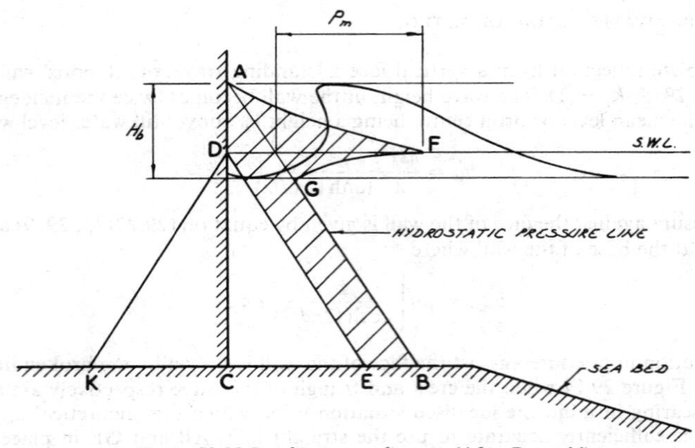

Figure 29.15 Breaking wave diagram (After Bagnold)

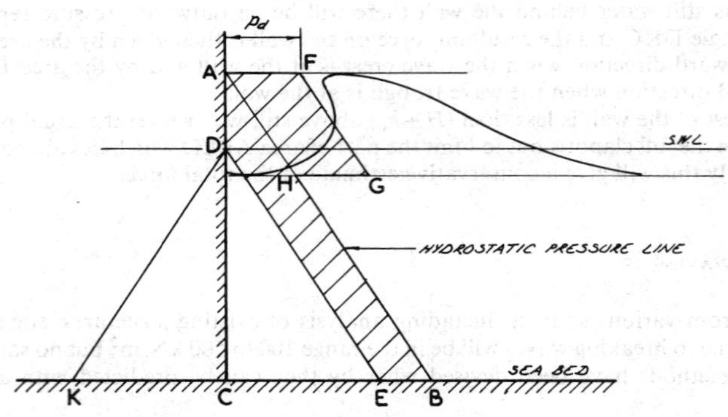

Figure 29.16 Breaking wave diagram—Momentum method

Neither of these methods on its own can be relied upon and in any particular case the results should be checked by comparison with the performance of existing structures in similar situations, supplemented for large and important projects by model testing.

SLOPING WALLS AND BREAKWATERS

The analysis given in the preceding paragraphs can be applied to walls with sloping faces (Figure 29.17).

According to Miche[28] the maximum steepness of waves which are completely reflected is given by

$$\frac{H}{\lambda} = \frac{\cos^2 \beta}{\pi}\left(1 - \frac{2\beta}{\pi}\right)^{\frac{1}{2}} \tag{59}$$

For breaking waves the dynamic pressure is reduced by $\cos^2 \beta$.

Wave forces on sloping rubble-mound breakwaters or sea walls are undoubtedly the most difficult to treat analytically, for the hydrodynamics are almost hopelessly complicated by the highly variable characteristics of such structures. Therefore the calculation

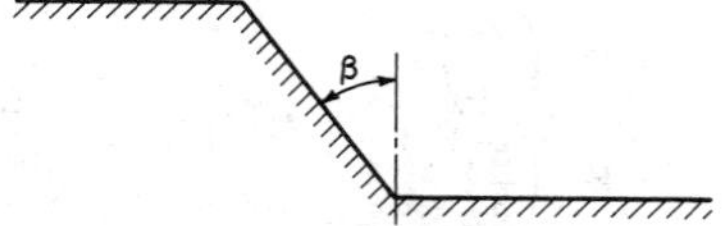

Figure 29.17 Sloping wall

of the required weight of armour unit has been invariably based on model testing. The most used formula is that of Irribarren as later modified by Hudson,[29] namely

$$W_r = \frac{\gamma_r H^3}{K_D(\gamma_r/\gamma_w - 1)^3 \cot \alpha} \tag{60}$$

in which W_r is the weight of armour units, γ_r the specific weight of armour units, γ_w the specific weight of water, H the wave height, γ the angle of breakwater slope to the horizontal, and K_D an empirical coefficient. (See p. **29**–56, 'Rubble-mound breakwaters'.)

SCALING LAWS AND MODELS

General

Although an understanding of the physical processes involved is essential to the solution of most problems in coastal engineering, few are amenable to mathematical analysis alone, and it is therefore necessary to rely on past experience aided where appropriate by model studies. In some cases, for example studies of wave diffraction and refraction, model study results will give an accurate interpretation of the prototype behaviour, but in others, such as studies of siltation in large estuaries, both the operation of the model and the interpretation of the results will require a large measure of judgment based on experience and observation.

Scaling and similarity

According to Buckingham's Π theorem, which provides the most general and elegant approach to the problem of similarity and scaling, if a_0 is a function of n independent dimensional parameters $a_1, a_2, a_3 \ldots a_n$, then the variables can be combined so that the dependent dimensionless group A_0 is a function of $n-3$ independent dimensionless groups $A_1, A_2, A_3 \ldots A_{n-3}$

where

$$A_1 = (a_1^{x_1} \; a_2^{y_1} \; a_3^{z_1})a_4$$

$$A_2 = (a_1^{x_2} \; a_2^{y_2} \; a_3^{z_2})a_5$$

$$A_0 = (a_1^{x_0} \; a_2^{y_0} \; a_3^{z_0})a_0$$

the indices being chosen by trial and error to make the relationships dimensionally correct.

This approach yields directly, and in only a few lines, to many well known results, e.g. the drag coefficient of a smooth object in a fluid stream which is a function of Reynolds number alone.

In general, apart from geometrical similarity, it is necessary for the following parameters to be effectively the same:

$$\text{Froude number} = V/(Lg)^{\frac{1}{2}}$$

$$\text{Reynolds number} = VL/v$$

$$\text{Weber number} = V\Big/\left(\frac{\phi}{\rho L}\right)^{\frac{1}{2}}$$

$$\text{Mach number} = V\Big/\left(\frac{E}{\rho}\right)^{\frac{1}{2}}$$

$$\text{Euler number} = V\Big/\left(\frac{2\Delta p}{\rho}\right)^{\frac{1}{2}}$$

in which V is the velocity (m/s), L the length (m), v the kinematic viscosity (m^2/s), ϕ the surface tension (N/m), E the elastic modulus (N/m^2), ρ the density (kg/m^3), and Δp the pressure difference (N/m^2).

In many cases, similarity in these parameters is only possible if the model is the same size as the prototype or, using a special fluid, nearly the same size. The greatest difficulty occurs with the modelling of the sediments which form the mobile beds of seas and estuaries; the particle size cannot be reduced to scale as the material then becomes clay with entirely different settling characteristics from the sand or silt of the prototype. For this type of model, claims have been made for the use of plastics, wax, boiled wood chippings or other materials with a specific gravity only a little above 1.0 but such materials are expensive and therefore are used infrequently in practice. In other situations also, similarity cannot always be maintained; models of tidal estuaries and rivers for example are almost invariably distorted since, if they are not, either the depth would be so small that the Reynolds number becomes unacceptably low or the model has to be impracticably large. Nevertheless it has been shown repeatedly that satisfactory reproduction of tidal flows, currents and salinity differences can be obtained in such models.

It can be shown by mathematical analysis that a distorted model may be able to reproduce prototype behaviour, or it can be justified in more general terms as follows. If a vertical sided basin is connected to a tidal sea through a long channel of appreciable hydraulic resistance, the tidal response of the basin (provided it is short compared with the length of the tidal wave) depends only on the area of the basin and the hydraulic resistance of the channel, and a deep narrow channel can be made with the same hydraulic properties as a wide shallow one.

Tidal models

Models of estuaries or other large tidal inlets have frequently been built, and they commonly have vertical scales of the order of 1 : 100 and horizontal scales of the order of 1 : 500. When the roughness of the bed has been adjusted with reference to prototype data, they reproduce accurately the tidal levels and currents in the main channels, but not currents across mudflats or other areas which are uncovered at low water. Such models may therefore be used with confidence to predict the effect on currents and tide or surge levels of, for instance, dredging, training works or a barrage in the upper estuary. Salt water intrusion and the related phenomena of dispersion and flushing of wastes (either effluents or warm water from power station discharges) will be less well reproduced but, with careful calibration, correspondence in a particular reach is sufficiently good to justify the use of the model in the case of important projects. The movement and deposition of silt is the least satisfactory aspect of the tidal model, and requires the most elaborate and careful calibration and very extensive prototype data. Since, however, the cost of dredging the approaches to many ports is very large there have been frequent attempts to build and operate models of the type, and from time to time exceptional claims have been made as to the ability of particular mobile bed models to predict changes in the prototype; all such claims should be regarded with some reserve and the report on a model study of this nature should not be expected to be more than a qualified subjective opinion as to whether proposed changes by construction works or in dredging procedures would lead to accretion or erosion.

An estuary model investigation together with the associated field studies will usually require the employment of teams of four or five engineers and assistants in the laboratory and the field for at least a year and the cost will be substantial.

Harbour models

Provided that the model is large enough for the waves to be unaffected by surface tension forces and bed friction, wave disturbance models of harbours with unbroken waves obey Froude's law precisely and the results can be used with confidence. Although the procedure adopted for the operation of a particular model (e.g. whether the input should be an irregular wave train or a series of single period waves, or whether it is sufficient to assume that the response of the harbour will be linear so that only one wave height at sea for each wave period need be considered) is a matter on which opinions differ but the construction and testing of a model in some form is usually justified before embarking on major harbour works. The linear scale adopted is commonly of the order of from 1 : 50 to 1 : 100, the time scale being the square root of the linear scale. Judgment and experience are required in assessing the acceptable wave height at any particular berth within the harbour because of the lack of basic data and difficulty in determining the response of moored vessesl to waves of specified periods and heights.

The response of the harbour to long waves, i.e. those which may cause the harbour basins to resonate in the first or second modes, is difficult to model because the waves are reflected and rereflected from the boundaries of the basin which contains the model. These extraneous reflections can be suppressed by the provision of wide, flat beaches (their width being large compared with the wavelength) along the boundaries of the basin but this results in either an impracticably large basin or an unacceptably small model. For this type of study it is therefore necessary to have recourse to a mathematical model to solve the whole problem or to specify the wave characteristics at the harbour entrance, and then to reproduce them in the model.

Forces on structures

Forces on structures, i.e. breakwaters, piles, pipelines, etc., and on moored ships also obey Froude's Law provided that the scale is large enough to ensure fully developed turbulence and that air entrainment is not important. The first proviso is easily satisfied as model scales of about 1 : 20 are usually not inconvenient, while the second applies only to structures in breaking waves. With these qualifications it is generally accepted that forces measured on model structures may be scaled up and used for design purposes. For rigid structures the forces may be measured by pressure cells, dynamometers or strain gauges recording electronically on paper or magnetic tape. For rock-mound breakwaters and similar structures for which the stability of the individual armour units is the main consideration, observation of the percentage of the units which rock or are displaced during each test run provides the basis for design.

In the past it was customary to use only regular trains of waves in the laboratory, testing with a variety of heights and periods, and from these to produce the necessary design figures. While the procedure is probably sound for unbroken waves, if the waves break in front of the structure, which may be due partly to the reflection of the previous wave from the structure itself, it is necessary to use irregular waves. The irregular wave train may be either a synthetic combination of a number of sine waves or a 'facsimile copy' of real waves recorded at sea; the latter procedure is more convincing and no more difficult in application. Since the magnitude of the force on a structure caused by a breaking wave depends very greatly on the form of the wave and exactly how it breaks, and since even regular waves produce a very irregular pattern of forces, the largest forces from a train of irregular waves will almost certainly not be caused by the highest wave and a statistical approach to the analysis of the results becomes essential.

Overtopping

The crest level required for a breakwater to provide adequate protection, or for a sea wall to be safe from scour on its landward side, requires a knowledge of tide and surge levels and wave heights from which it can be assessed by experience or can be estimated from a model study. The amount of water thrown over the top of a sea wall or the size of the regenerated wave on the harbour side of a breakwater can be measured in a wind–wave flume but, unless the crest level is abnormally low relative to the sea level, air entrainment and water droplet size will be of great importance. These are not susceptible to scaling; the size of the drop of water blown from the surface will be the same in both model and prototype but thereafter the prototype drop will be exposed to a stronger wind and be broken up into fine spray whereas the model drop will fall relatively quickly on to the rear slope of the wall or into the harbour. Overtopping tests should therefore only be used to establish the relative merits of various types of structure and their crest levels and one of known performance should be included as a basis of comparison.

Mathematical models

If the equations of motion of the fluid are known, mathematical models can be constructed in the same way as physical models; the high-speed digital computer makes it practical to handle the equations. Such models can be checked and adjusted in the same way as physical models, e.g. in an effluent disposal problem the model would be calibrated by establishing the coincidence of model and prototype float tracks and the spreading of dye patches.

At the present time, mathematical models have three disadvantages; firstly the average engineer is much less at home with columns of numbers than with the visual and tactile, secondly the effects of small changes are less readily and immediately appreciated than in the physical model, and, thirdly, more experience is required in order to judge the validity of the results.

Littoral processes

The part played by wave action in the movement of sediments can only be modelled if it is secondary to the tidal currents, i.e. only in the circumstances in which waves initiate movement by putting the sediment into suspension but do not control the direction in which it is moved.

Small-scale experiments with waves and beaches have been of value in providing an understanding of littoral processes but these are not 'models' of real beaches and cannot be used for the solution of particular problems.

SURVEYS AND DATA COLLECTION

Sources of information

Hydrographic surveys are costly and every effort should therefore be made to obtain the results of previous surveys and measurements which may be relevant. In the UK, information may be obtained from:

(i) Published Admiralty Charts which also contain information on sea-bed materials and tidal currents, and the unpublished originals which contain more information than the published charts
(ii) Geological and Ordnance Survey maps
(iii) Port Authorities charts and records
(iv) Authorities or engineers responsible for the construction of works in the area
(v) Research Laboratories concerned with coastal processes, e.g. the National Institute of Oceanography, the Hydraulics Research Station, the Fisheries Laboratory and the Water Pollution Research Laboratory.

General

The problems associated with the collection of information for the design of maritime works increases with distance from high-water mark and the degree of exposure of the coastline. Between tides, conventional land surveying techniques are frequently adequate and can be extended a short distance beyond low water mark with the assistance of a diver or chainman in a wading suit, using a marked line of 'corlene', or other fibre which floats and does not shrink when wet, for measuring distances, and a sounding chain or marked pole.

Conventional subsoil investigation procedures are also appropriate for work on the foreshore or in shallow water. A lightweight boring rig can be set up quickly and a useful amount of work done in the few hours the foreshore is uncovered, and the working time can be extended, if the coastline is not too exposed, by using a small scaffold working platform.

Beyond low-water mark and up to two or three miles offshore, floating craft or other specialised equipment are necessary to form a working platform, but conventional optical surveying techniques are still adequate for position fixing.

Beyond the limit of normal visibility it is almost essential to employ electronic position fixing systems. Although the equipment is expensive the instruments can give direct readings or plot coordinated positions and it is thus possible to reduce the staff employed and to continue work by day and night even in poor visibility. Its use may therefore be worthwhile closer inshore for intensive survey work or on the construction of major works.

Instruments and equipment for use offshore are now being developed rapidly under the stimulus of the offshore oil and gas industries, and some of them are coming into use for civil engineering work; for special applications, research laboratories, survey companies and instrument makers should be consulted.

Position fixing

VISUAL METHODS

Position fixing is normally carried out by resection, the required horizontal angles being measured simultaneously by two observers with sextants; accurate work requires practice and trained surveyors should be employed where possible. For greater accuracy, intersection fixing by theodolite observers on shore may be employed but this requires an efficient radio communication system.

In some cases the work can be facilitated by the use of transit lines on shore, which may be lighted for working at night and for greater accuracy may be supplemented with a laser beam.

RADIO METHODS

With a good navigational radar set, distances can be measured to an accuracy of about 1%—which is sufficient for current measurements by float tracking and for large-area preliminary surveys; radar reflectors should be set up in known position on shore and positions fixed by trilateration.

The seas around North Western Europe are covered by the Decca Navigation system comprising a series of synchronised transmitting stations. The signals from pairs of these stations are in phase along the lines of a hyperbolic grid plotted on special charts and any vessel equipped with a special receiver which senses the difference in phase of signals from three stations simultaneously is able to follow its course across the grid as a numerical display. Although not primarily intended for survey work, position fixing by Decca Navigator is a useful alternative to fixing by radar where great accuracy is not required. To within 15 m of true position can be expected with careful calibration but the accuracy varies with the intersection angle of the grid curves and falls off towards the edges of the charts. Also, the system tends to be unreliable close inshore where there are high cliffs and at dawn and dusk. Decca Navigator sets are installed in most vessels of the type normally chartered for survey work.

A special 'Decca' Chain which works on the same principle as the navigation system may be specially installed for a particular project and will give positions to within about 1 m. If only one vessel is employed, a master transmitter may be installed on the ship with two slave stations on shore and the fixing grid lines are then circles.

MARKING POSITIONS AT SEA

Temporary positions at sea are most conveniently marked with small plastic buoys secured to small sinkers. The length of mooring line needs to be greater than the maximum water depth and the movement of the buoy with wind and tidal current therefore increases with depth and tidal range. It may be reduced by using a two-buoy system in which the light surface buoy is secured by a relatively short line to a larger buoy held a short distance below low water level by a taught line to the sinker. In exposed situations, unless very heavy gear is used, which is not always possible or justified, frequent checking of the location of the buoy is necessary to ensure that it has not dragged its mooring. A buoy may be used for fixing further positions either by visual observations or, if equipped with a reflector, by radar. Positions on the sea bed can be marked with 'sonar' devices which transmit sound pulses either continuously or on call.

The level of the sea bed

Isolated soundings can readily be taken with a lead and chain but for a survey of any extent an echo sounder should be used. This instrument consists of a 'transducer' which generates acoustic pulses, focuses them as a beam downwards to the sea bed and picks up the reflected waves. The time taken for the sound wave to travel from the surface to the sea bed and back is recorded and gives a measure of the water depth. If it is wished to resolve small features on the sea bed the pulses must be focused into a very narrow beam and it is then only possible to work under calm conditions when there is little rolling or pitching of the survey boat. For most purposes it is usual to use an instrument giving a beam with a 15° spread which allows work to continue in waves and swell for so long as the other operations, e.g. position fixing and control of the survey boat, are practicable.

Echo sounders installed for navigation purposes, particularly if they record on dry paper, may be used for preliminary surveys but for precise work an instrument designed for survey work should be used. These usually record on paper 250 mm wide, are portable and can be easily installed in the type of vessel normally used for survey work. Echo sounders need to be calibrated frequently: this is done using a reflector plate suspended below the transducer. A watch must be kept for extraneous reflections from objects lying on the bottom or even shoals of fish swimming below the boat; if sediment forming the sea bed is very soft some of the sound wave energy may penetrate and reflections will be recorded both from the bed and from the rock head or other dense stratum below the soft bed.

Having recorded the sea-bed level relative to the surface it is then necessary to relate the sea surface to some fixed datum, for which a tide gauge, either autographic or visually read, has to be established at some convenient point on the shore; predicted tide levels should not be used.

The nature of the sea bed

The Mariner's sounding lead has a recess in its base which is filled with tallow so that small samples of any loose sediment may be picked up from the sea bed; Admiralty Charts give information on the sea bottom obtained in this way. Larger samples can be recovered with a small scoop or spring-loaded grab.

The next stage in an investigation for civil engineering works will be to inspect the bed by diver who will collect samples by hand and probe with a bar or an air or water jet. If visibility is good, which is most likely to be the case in coastal waters following a spell of calm weather and at neap tides, the diver swimming along a tagged line will rapidly collect a great deal of information and a permanent record can be made with underwater photographic or television equipment. If the water is turbid the diver will be working largely by feel and the information obtained will be more restricted and less reliable. If possible an engineer or geologist trained in the use of selfcontained aqualung equipment should be employed for this work. In depths greater than 40 ft (12 m) diving times are restricted and above 100 ft (30 m) decompression procedures must be followed however short the time spent on the bottom (see Section 41).

A more general survey of sea-bed features can be made with 'transit' or 'side-scan' sonar equipment which transmits a fan-shaped acoustic beam very narrow in the horizontal plane. The equipment is towed behind or alongside the survey vessel and scans an area of the sea bed normally up to 500 m wide to one side of the vessel's track so that large area can be covered quickly. The record is made to scale on a wide strip of wet paper. A uniform flat surface will produce little return except directly under the boat, but any upstanding feature, e.g. rock outcrop; wreck, pile stump, etc., will be recorded as a dark mark on the paper with a white shadow behind. The equipment gives good resolution and a clear record of any upstanding features larger than about 300 mm across.

The nature of material below the sea bed

Offshore borings are expensive and may take up a great deal of time. It is therefore important to keep their number to a minimum and to ensure that they are located so as to give the maximum amount of data for the expenditure involved. All possible alternative sources of information should be explored before the borings are undertaken, e.g. by examination of exposures of the strata on shore, from onshore borings and by superficial inspection and sampling of the sea bed (see p. 29–35). If further information is required, a seismic survey can give a great deal of information in a short time. The equipment used is similar in principle to an echo sounder but the pulse energy is very much greater and high-gain amplifiers are required to process the reflected signals. The sound pulses from the transducer are reflected from the sea bed as with an echo sounder, but also penetrate the bed below and are reflected from any interfaces between materials through which the velocity of sound differs. The junction between loose sediment and solid rocks can normally be located with confidence and it is unusual if at least one further reflecting horizon is not recorded. The geological structure of the rock below the sea bed is drawn by the instrument on a continuous roll of paper; it should be noted that it measures (and plots) the travel time of the pulse—the determination of depth to the reflecting horizons requires a knowledge of the velocity of sound through the materials. For a precise calculation of depth, borehole cores are required for direct measurement of the velocity but, for a preliminary interpretation, figures from other sites may be employed, for example:

	Wave velocity (m/s)
Sand	300– 800
Glacial moraine	1 500–2 100
Chalk	2 000–2 600
Lower greensand	1 600–2 000
Gault (clay)	1 800–2 200
Jurassic limestone	1 800–3 600
Carboniferous	2 400–3 600
Ordovician (N. Wales)	2 400–3 600
Cambrian (N. Wales)	4 000–6 500
Limestones	3 500–6 500
Granites	4 600–7 000

Unless the strata are dipping steeply the interpretation of the record at more than twice the water depth below the sea surface is difficult owing to multiple sound reflections from the strata interfaces and the surface.

When the information obtained above has been considered it is usual to prepare tentative designs of the proposed works from which the most suitable position and depth of the offshore borings can be decided.

Three methods for setting up the boring rig are available:

(a) Above the sea surface on a piled platform or jack-up barge
(b) On a barge or selfpropelled craft
(c) On the sea bed.

If methods (a) or (b) are used the drilling procedure and equipment will be the same as would be employed on land. In reasonably sheltered water, e.g. the English Channel, shell and auger boring to moderate depths can be carried out rapidly at reasonable cost from a platform erected over the side of a small coaster or large fishing vessel. As the exposure and/or the depth of the hole increases, and for core boring, a jack-up barge or a boring vessel with a central well is required, or the boring rig can be set up on the sea bed. Fully automatic underwater equipment is available for shallow holes and can be operated in any depth of water; with it a 20 ft core barrel is driven into the sea bed and the rig then recovered by the attendant surface vessel. For deeper holes (up to 300 m), diver-operated rigs can be used but in this case the operating water depth is limited to about 50 m.

If cores are not recovered from a borehole, identification of the strata and information on the permeability and density of the rock can be obtained from *in situ* measurement of the electrical resistivity and the scatter and absorption of gamma radiation by instruments lowered down the hole.

Current measurement

The measurement of currents at a fixed point in the sea is usually carried out with a current meter or by repeatedly timing floats over short distances while the movement and mixing of masses of water over large areas is carried out by tracking floats, dye patches or radioactive tracers.

FLOATS AND TRACERS

Floats less than 1 ft (30 cm) deep are susceptible to wind and wave action and their paths will not be representative of the general current movement; floats made from poles ballasted at their lower end, so as to float vertically, are better but surface effects will still be significant and unpredictable. Floats consisting of a substantial ballasted cruciform drogue attached by a line to a small surface float with a marker stick or flag, or a lamp for night observation, are preferable for most purposes. Under favourable conditions one boat can track several floats at a time.

If information on the turbulent dispersion of a neutrally buoyant effluent is required, measurements can be made of the dilution of indicators such as Rhodamine B dye which can be detected in a fluorimeter at concentrations up to 1 part in $10.^{9}$ Radioactive tracers, e.b. bromine 82, have also been used, and bacteriological tracers, e.g. Serratia, have been suggested. Experiments with tracers in the open sea are expensive and difficult and should not be attempted until as much information as possible has been obtained from float tracking and other sources.

CURRENT METERS

The earliest and simplest type of current meter comprises a free-running propellor with a mechanical system for counting the number of turns and a vane to align it with the current. Later instruments have been fitted with a switch to count the turns electrically while in modern propeller instruments the revolutions are counted electronically without any mechanical switching. Instruments can be arranged to read directly or the information can be transmitted by a cable to a display screen in the survey vessel or on shore if not too distant. Instruments of this latter kind can be coupled to a radio buoy which transmits the data to a receiver on shore where they can be recorded either automatically or manually. Completely selfcontained instruments which record the data on paper tape as a graph or on magnetic tape are now more popular than the transmitting instrument. As it is seldom possible to observe the direction of the current visually, the meter is usually fitted with a compass and a device for transmitting or recording the heading of the instrument. A recent development is the electromagnetic current meter which has no moving parts and is therefore more robust and less prone to damage or malfunctioning if fouled by seaweed or other debris.

Current meters must be secured at a fixed level and not attached directly to a floating buoy or survey boat. The laying and recovery of the meters has to be carried out with great care to ensure that they are not damaged during these operations, and their location must be chosen with due regard to the risk of damage by shipping. It has to be remembered that the recording will be affected by wave action, depending upon the wave height and length and water depth, and on the response of the instrument to variations in the strength and direction of the current.

Waves and tides

The practice of estimating wave heights by eye either from the shore or from a ship is unreliable and data obtained in this way or by observation against a graduated pole, although of value for some purposes, are not suitable for numerical analysis; owing to variability, wave data have to be analysed statistically and a very large number of systematic observations are required.

Instruments for measuring wave height are of three kinds:

(i) Double-integrating accelerometers housed in a moored ship or a buoy designed to rise and fall with the water surface
(ii) Pressure gauges which are held at a fixed level below the surface
(iii) Wave staffs which pass through the sea surface and record its level directly.

Instruments of the first kind have been installed in lightships but it is clear that in this situation they will not record the true waveform and the amplitude may also be distorted by the response of the ship. If installed in a small, shallow-draft buoy with a light mooring, instruments of this kind should produce satisfactory records. Long waves and tides will not be recorded.

Instruments of the second kind can either be completely selfcontained or the pressure cell can be laid on the sea bed and connected by a cable or pressure hose to a recording instrument on shore. The drawback of this type of instrument is that the pressure variation caused by a wave of given height decreases with increasing value of the ratio depth/wave length; since in tidal waters both quantities vary continuously and at any one time waves of various periods will be present, the interpretation of the records is laborious and not always easy. Shorter waves tend to be lost owing to the high attenuation. If the instrument is arranged to record on paper rolls the tidal variation should be compensated within the instrument or the record will be too cramped for satisfactory interpretation. If information on waves with periods of over, say, 15 s is required, hydraulic filters can be inserted between the cell and the measuring instrument to suppress the pressure variations due to the shorter-period storm waves and the record can then be scanned visually for long-wave activity.

The third type of instrument is a recent development and where it can be installed on a fixed structure is to be preferred as a direct, undistorted record is provided. It consists of a single vertical insulated wire connected to a circuit designed to measure changes in its capacitance with variations in the depth of immersion. It thus differs from laboratory wave gauges in which the variation in the resistance between two vertical parallel bare wires partly immersed in the water is continuously measured with an ac bridge circuit. This type of instrument is not suitable for use in the field as the wires must be kept perfectly clean. Wave staffs can be mounted on tall pillar buoys which, if they are correctly designed, barely respond to storm waves; if the buoy does so respond its movement can be measured with a double-integrating accelerometer incorporated in it, and combined with the record from the wave staff.

Owing to the vast amount of data that would otherwise be collected (about 6×10^6 waves per annum) wave recorders do not usually run continuously and an arbitrary sampling period, e.g. 15 min per 2 h, is usually adopted. It would be better to have a trigger circuit arranged to sample waves of increasing height with increasing intensity, e.g. 0–1 m, no measurements; 1–3 m, 5 min/h; over 3 m, continuous recording.

The measurement of very long-period waves, e.g. the tide, presents little difficulty provided short-period waves can be excluded from the record; either the second or third type of wave gauge described above can be adapted for the purpose. The traditional tide gauge consists of a float in a stilling well mechanically connected to a clockwork-driven drum recorder.

Meteorological data

Tide levels are affected by wind and barometric pressure and waves are generated by the wind which also affects the currents; therefore the recording of any wave, tide, or current data should be accompanied by concurrent and immediate antecedent wind and barometric pressure records. This information can frequently be obtained from an existing meteorological station, but failing this a recording anemometer and barograph should be set up. Portable direct-reading cup anemometers should be carried in the survey boat for checking local variations in wind speed during the survey.

Stability of the coastline—beach and sea-bed sediment movement

If the sea bed consists of loose, erodable material evidence will be required as to its long-term stability. The most fruitful sources of information in the UK are the early Ordnance Survey maps and Admiralty Charts.

Other engineering works, e.g. groynes and harbour piers, in the vicinity, should be inspected and enquiries made as to changes they may have caused in the foreshore and sea bed. Conclusions drawn from these observations may need to be supplemented by further soundings and beach-level surveys.

It should be noted that almost complete stability does not preclude the possibility that very large quantities of sediment are passing through an area and the investigation should be extended along the coast in both directions to points where the passage of sediment is barred by topographical features.

Estimates of the quantity of material moving along a coastline can be made by labelling the sediment with radioactive or fluorescent tracers but, since transport rates may vary widely from year to year and one exceptional storm can have as much effect as many months of typical weather, such experiments are of doubtful value unless continued over a long period; experimental groynes or dredged channels may well yield more information for an equivalent cost and effort.

Along coastlines exposed to the open sea, where sediments are moved primarily by wave action as bed load, material finer than sand will be rapidly dispersed and settle in deep water so that measurements of suspended solids are not normally necessary. In sheltered estuaries the situation is quite different and measurements of suspended solids may be required. Provided the concentration is low and the particle size fairly uniform, the observations can be made *in situ* with an optical instrument, but usually it is necessary to collect samples for laboratory analysis. However, an instrument which measures the backscattering of gamma radiation, which does not have the limitations of the optical method, is being developed.

Water properties

Instruments which give direct readings of salinity, temperature and dissolved oxygen, and are quick and easy to operate, are available commercially so that information on these characteristics can readily be obtained while a survey team is in the field.

Domestic sewage can be detected in sea water up to concentrations of about 1 part per million by culturing samples and counting the number of colonies of coliform bacteria which develop. These measurements, which require the services of a bacteriological laboratory, are the best means of judging the degree of pollution from an existing outfall.

Chemical tests for the presence of other pollutants may also be required when considering the effects of industrial outfalls; very small concentrations are difficult to identify but filter-feeding shell fish effectively concentrate many of the more objectionable pollutants and it may be worthwhile collecting them for analysis with the water samples.

DESIGN PARAMETERS AND DATA ANALYSIS

Ground conditions

The inspection, testing and analysis of superficial and subsoil samples of sea bed and beach materials and of the underlying strata following the site investigations (outlined on pages **29**–35 and **29**–36) will follow normal soils and rock testing procedures, including grading of sediments and laboratory tests for compressive strength, internal angle of friction, cohesion, etc., as described in Section 10.

The work may need to be extended, however, to include a study of sea-bed and foreshore stability. For this purpose all available charts and surveys should be plotted or reproduced to a common scale and datum, and carefully scrutinised for changes between one survey and the next. The significance of any observed variation can be determined by superimposing cross sections from successive surveys or by plotting accretion and erosion contours from which changes in successive periods of time in the volume of material within a given area can be calculated.

Waves, tides and currents

The other measurements which may have been made, for instance of wave height and period, tides, currents, salinity, silt content and similar phenomena, will almost always be scattered over a wide range and it will not normally be possible to make direct use of them. In some cases, particularly if the data are of limited extent or quality, their use may be confined to corroborating, or modifying in some particular, decisions based on previous experience. Alternatively the data may be analysed statistically with a view to choosing values for the required design parameters to give a selected low probability of their being exceeded within the estimated lifespan of the works to be constructed. In this case the method of analysis adopted will be limited by the amount of data available but within this limit it has to be remembered that a very elaborate processing procedure will only be justified if techniques are available whereby the results can be rigorously applied in the subsequent design and subjective judgments avoided at a later stage.

As far as possible, therefore, certain decisions should be made before data collection is started, for example:

(i) The type of structure to be built, or range of possible types, or the kind of solution most likely to be adopted for some particular problem

(ii) The methods by which the data will be processed and the designs carried out, which may be by numerical analysis by hand or by computer, by a model study or by a combination of these, and hence the form in which the data are to be recorded. These decisions will be influenced by questions of cost and time and by the availability of computer and model study facilities.

Suppose, for example, a breakwater is being considered; if the design is to be based upon empirical formula, e.g. the Hudson equation for a rubble mound or the Sainflou method for a mass concrete or blockwork breakwater, then a straightforward record of wave height for a relatively short period, say one winter, to enable a value for $H_{\frac{1}{3}}$ or $H_{1/10}$ (see p. **29**–13) to be estimated may be all that is required, with visual observations of the direction of wave approach. On the other hand if some less traditional type of structure is envisaged, e.g. the cellular structure adopted for the Brighton Marina Breakwaters (see Figure 29.29) for which model testing in random waves is necessary (see p. **29**–31) then the wave record should include all wave characteristics, namely, height period and shape, preferably in a form that can be used directly to control the model wave generator, or be simulated by a number of sinusoidal components following Fourier analysis of the prototype wave train, and the measurements made in the model should in turn be in a form which can be analysed statistically to produce the worst combination of wave energy, water level, wind speed and direction of wave approach.

MATERIALS

The maritime environment is among the most aggressive in which civil engineering works are constructed and the question of durability, therefore, is one of the main factors in the choice of materials. In this context four zones can be identified as follows:

(a) The splash zone above high-water level where surfaces may become coated with salts due to alternate wetting by spray and evaporation, and abrasion may be caused by blown sand and shingle thrown up by the sea
(b) The beach zone between high- and low-water marks where severe abrasion by wave driven shingle or gravel may override other considerations
(c) The tidal zone away from a beach where corrosion of metals is likely to be most severe
(d) The submerged zone below low water which is the least aggressive of the four

Most of the common engineering materials can be employed in maritime works, for example, rock, mass concrete, reinforced concrete, timber, iron, steel, special steels, bronzes, plastics and bitumen. Some of these can be used in most situations while others have more limited application as discussed below.

Rock

Most of the harder and more durable types of rock are suitable for any of the zones defined above and may be in the form of masonry or random placed blocks in revetments or rubble mounds. Usually the quality of the stone used is dependent upon what is economically available; there is no recognised testing procedure to determine the suitability of a rock specifically for maritime works but tests developed for other purposes, e.g. for filter media and roadstone and concreting aggregates (BS 1438 and BS 812) are useful for comparative purposes. The rocks most commonly used are of the igneous and metamorphic types but also some of the harder sedimentary rocks. It is to be noted that some of the softer sedimentary rocks may be attacked by piddock or pholas near or below low water mark.

Brickwork

Engineering brickwork is suitable for use in zone (a) and has from time to time been used with successful results in zones (b) and (c). In all instances it is essential to ensure that the joints are completely filled with strong durable mortar.

Concrete[30]

Mass concrete is suitable in all situations, the mix being designed for maximum density and durability rather than high strength; these properties can usually be achieved with concrete having a 28-day compressive strength of 25 N/mm^2. It may be cast *in situ* or in the form of masonry or random rubble blocks. For *in situ* work large pours are desirable and special attention should be paid to construction joints which are vulnerable to attack by the sea. Ordinary Portland cement is usually used in mass concrete unless exposure to abrasion at an early age is anticipated when rapid-hardening Portland cement may be preferred. For very large structures a worthwhile economy may be effected by replacement of up to 25% of the cement by pulverised fly-ash, while replacement to a similar extent with ground slag will improve the durability of the concrete. For concrete placed under water either by tremie or bottom-opening skip it is usual to increase the cement content by about 25%.

Normal reinforced concrete construction is satisfactory in zones (c) and (d) but special precautions may be necessary in zones (a) and (b). A generous cover is required to the reinforcing steel, normally not less than 2 in (5 cm) and additional cover up to 4 in (10 cm) or more where abrasion is likely to occur. If there is a possibility of leeching and evaporation of sea water from the surface of thin sections in the tidal zone, sulphate-resisting cement should be used.

Timber

Timber is widely used in coastal engineering works notably for jetties, slipways, groynes and outfall pipe support trestles. Although subject to various forms of deterioration, an acceptable lifespan can generally be assured by suitable choice of species and/or preservative treatment. Above water level where the timber will be continuously damp, or alternately wet and dry, decay by normal wet or dry rot will be most prevalent; in the beach zone abrasion may be a more important consideration while near and below mean tide level attack by marine borers, of which the Toredo and Limnoria are the most common, will be of greatest concern.

A number of naturally resistant hardwoods are available of which Greenheart is the best known and outstanding as regards strength and resistance to attack by marine borers; others in common use are Jarrah, Ekki, and Opepe. Softwoods need to be treated by impregnation with creosote or one of the proprietary preservatives which are

Table 29.5 TIMBER PROPERTIES

Species	*Specific gravity at 50% moisture content*	*Ultimate stresses parallel to grain* (N/mm^2)			*Modulus of elasticity* N/mm^2
		Bending	*Compression*	*Shear*	
Greenheart	1.32	135	70	9	18 000
Jarrah	1.01	68	36	9	10 000
Ekki	1.32	120	68	16	14 000
Opepe	0.95	90	50	12	12 000
Pitchpine	0.74	54	25	6	10 000
Douglas fir	0.64	53	26	6	11 000

mostly based on copper, chromium or arsenic salts. The usual specification for this is impregnation by the full cell process at a pressure of 1.2 MN/m^2 for a period of 4 h or to refusal; alternatively a retention of 8 kg of preservative salt per cubic metre of timber. Softwoods most commonly used in marine work are Pitchpine and Douglas Fir. These timbers must be incised in order to achieve adequate penetration of the preservative but are outstanding in respect of resistance to abrasion in the beach zone.

The more important properties of the above timbers are given in Table 29.5.

Working stresses will depend upon the quality of the timber and the location and size of blemishes such as knots and shakes but will usually be from one-fifth to one-third of the ultimate stresses given in Table 29.5.

Iron

In the past cast iron, in pipes and jetty or pier columns, was frequently used and has a long life in the marine environment except where subjected to abrasion. Apart from abrasion, deterioration occurs by graphitisation in which the iron corrodes leaving a residue of soft and porous iron oxide and graphite. In modern practice the use of cast iron is limited to pipe specials and valves, spun iron or ductile iron now being more usual for pipes.

Wrought iron is most durable and outstanding for many purposes. It is not now readily obtainable and its use is confined to special small items such as pipe straps and fittings in inaccessible situations.

Steel

On grounds of economy and ease of construction steel in the form of piles and structural sections is frequently employed in maritime structures in spite of its relatively high rate of corrosion in many situations. The normal rate of atmospheric corrosion tends to be accelerated particularly in the tidal and splash zones and precautions are needed to give an acceptable life span. Tests over a long period by the Sea Action Committee of the Institution of Civil Engineers and measurements on existing structures indicate an average loss of thickness of 0.08 mm per annum in sea water. Proctective coatings of tar, bitumastic, or other paints, wrapping with coal tar or synthetic tapes or galvanising will retard commencement of corrosion but coatings may need to be renewed at frequent intervals if they are to give a significant increase in the life of the structure. Below low-water level, coating or wrapping systems combined with cathodic protection by impressed current or sacrificial anodes, as used frequently on steel jetties and submerged pipelines, can prolong the life of the structure indefinitely. These protective measures are seldom economical for steel sheet piling for which it is more usual to make allowance for corrosion in the design; it is usual to assume that a wall of steel sheet piling will have reached the end of its useful life by the time 50% of the metal has been lost by corrosion, when the initial stresses in the material will have doubled. For the sections of piling most commonly used, this criterion gives a life of from 60 to 100 years.

Steel has very poor resistance to abrasion and on shingle beaches may have a life of not more than one-tenth of normal; steel sheet piling in groynes for example is known to have worn through in five years or less in some situations. For these conditions concrete encasement of the vulnerable parts of a steel structure is probably the best form of protection, or cladding with timber which can be renewed from time to time.

Corrosion-resistant metals

Metals in this category such as stainless steel, bronzes and monel metal are expensive and their use is generally confined to small special items.

Of the various types of stainless steel available, austenitic steel containing nickel and chromium has the most suitable strength and corrosion-resistance for use in structural and civil engineering work. It is durable in most situations encountered in maritime engineering with the exception of anaerobic conditions which may occur due to marine growth or below the sea bed, particularly in estuarine muds, and in stagnant conditions where the oxygen supply is low. Under these circumstances, stainless steel is subject, owing to the breakdown of the protective oxide film, to pitting and crevice corrosion, a tendency which is increased in the presence of the chlorides in the sea water. Care is therefore needed to avoid laps and crevices at joints.

Stainless steel containing molybdenum is less prone to pitting and crevice attack, and may also be chosen, but at extra cost, to avoid superficial rust staining where appearance is important.

Monel metal, an alloy of 30% copper and 66% nickel with 4% iron and manganese, combines high strength with maximum resistance to corrosion in all maritime situations.

For use in sea water, bronzes must be zinc free. They are normally used only for such items as valve seatings and trims.

Synthetic materials

This category of materials includes sealing compounds such as polysulphides and epoxies which are used in marine work although some are difficult to apply in damp conditions. A recent development is an epoxy coating which can be applied under water. PVC pipes are suitable as liners in outfall and intake structures or, strengthened with a glass reinforced polyester or epoxy coating, as pipelines for various purposes; polythene is outstanding for pipes as it can be supplied in long lengths, is easily floated or towed into position and is a good material for abrasive conditions. All these materials are corrosion resistant, but expensive, and apart from polythene not very tough; their use is generally restricted to special applications where there are problems with the more traditional materials.

Bitumen

Bitumen[31, 32, 33] is widely used in the form of sand mastic (sand, filler and bitumen) as the binder in asphaltic concrete, as a jointing material for stone or concrete blockwork revetments, as asphaltic carpets above or below water and for grouting rubble revetments, groynes or breakwaters. These uses have largely been developed for coast defence work in the Netherlands but are now finding an increasing use in other countries. This form of construction has the advantage of being sufficiently flexible to accommodate long-term settlement without fracture while being rigid enough to withstand the large but short-term dynamic forces of breaking waves.

In order to achieve good durability, asphaltic concrete requires a higher bitumen content than normal paving mixes but limited so as to resist flow down the slopes on which it is to be laid and to avoid segregation during mixing and handling. Usually 6% to 9% of bitumen of penetration grade 40/50, 60/70 or 80/100 will be used depending upon the situation of the work and climatic conditions. A graded aggregate of maximum size 25 mm is commonly used, although it can be gap graded, with a filler content (usually ground limestone or cement) between 8% and 13%. Mixes of this type require to be compacted and their use is therefore confined to work above water level.

For the jointing of blockwork revetments the filler may be replaced with a fibrous material such as asbestos and a typical composition for the mastic would be: bitumen 40/50 45%, sand 50%, asbestos fibre 5%. The composition for any specific work depends upon the size of the joints, the angle of slope and the range of ambient temperature and should be determined by trial on the basis of the preceding figures which are for a pouring mastic. If the joints are to be trowelled, a typical mix would be: bitumen 40/50 25%, sand 70%, asbestos fibre 5%.

The inert materials in sand mastic (sand and filler) are overfilled with bitumen so that the mixture is pourable when hot and does not require compaction. It can therefore be used both above and below water level. The composition can vary widely depending upon the requirements which may be an asphalt carpet for scour protection as used in the Delta Works in the Netherlands, either placed *in situ* or as prefabricated mattresses reinforced with wire mesh, or as a material for grouting rubble revetments or breakwaters either above or below water level. The materials and their proportions should be chosen to give a viscosity which is low enough at the placing temperature to give adequate spread or penetration but not so high as to allow excessive flow on slopes; it will normally be in the range 10 to 10^2 Ns/m^2. At ambient temperature the viscosity must be high enough to prevent excessive long-term flow and will normally be in the range 10^6 to 10^9 Ns/m^2.

For use above water, a bitumen of penetration grade 20/30 to 80/100 is likely to be chosen, and for underwater work one of the softer grades, e.g. 180/200 or 280/320. A typical mix is: sand 75%, filler 10%, bitumen 15%. For underwater work the temperature of the mastic should be relatively low (160 to 180 °C) so as to reduce heat losses while placing and the production of excessive steam which otherwise could cause the mastic to harden as a spongy mass with a tendency to float.

SEA-DEFENCE WORKS

Sea walls

Sea walls may be required for the protection of land which is being eroded by the sea, for the protection of low-lying land against flooding by the sea, or for the purpose of reclaiming land from the sea. The type of wall to be built in any particular case is very much dependent upon circumstances such as the exposure of the site, the material forming the foundation upon which the wall will be built, whether or not there is a beach in front of the wall and type of beach, and the cost of the work in relation to the value of the property or land to be protected. They thus range from heavy mass concrete or masonry structures founded upon rock or other firm foundation, and frequently used for the protection of urban areas, to light timber breastworks or pitched revetments for the protection of open land.

Within the range of types indicated above, the following are typical examples.

Figure 29.18 shows a cross section of a mass concrete vertical sea wall at Tenby in South Wales; it is founded upon shale bedrock underlying a flat sandy beach. The exposure is not severe, being in a sheltered bay within which there is little littoral drift. The foundations of the wall are protected against erosion of the sand by a concrete apron and toe wall in a trench in the rock, but scour is not a serious risk at this particular site and the fact that waves are reflected from this type of wall has not had any significant effect upon beach conditions.

Walls of this type have been used with varying degrees of success in places where the exposure is severe. They are eminently suitable in two extreme conditions; firstly, where there is no beach, the foreshore on which the wall is founded being exposed bedrock, and at the other extreme where large quantities of beach material in transit with the littoral drift provide an ample supply of material to make good scour which is likely to take place in front of such a wall during storm conditions.

This type of wall may be the first choice in situations where abrasion by wave-driven shingle or gravel is severe, since, whether built in concrete or masonry it can suffer an appreciable loss of material without significant effect upon its stability.

In situations where the supply of beach material with the littoral drift is not high the vertical type of sea wall may aggravate the situation. Under these conditions the less reflective sloping type of wall is more appropriate. An example is shown in Figure 29.19 of a wall at Crosby in Lancashire where the beaches consist of a variable depth of fine sand overlying silts and clays. With gravel or shingle beaches where abrasion may be a problem the stepped work shown in Figure 29.19 is better replaced by a smooth slope which under extreme conditions may have to be paved with granite blocks or other hard-wearing surface.

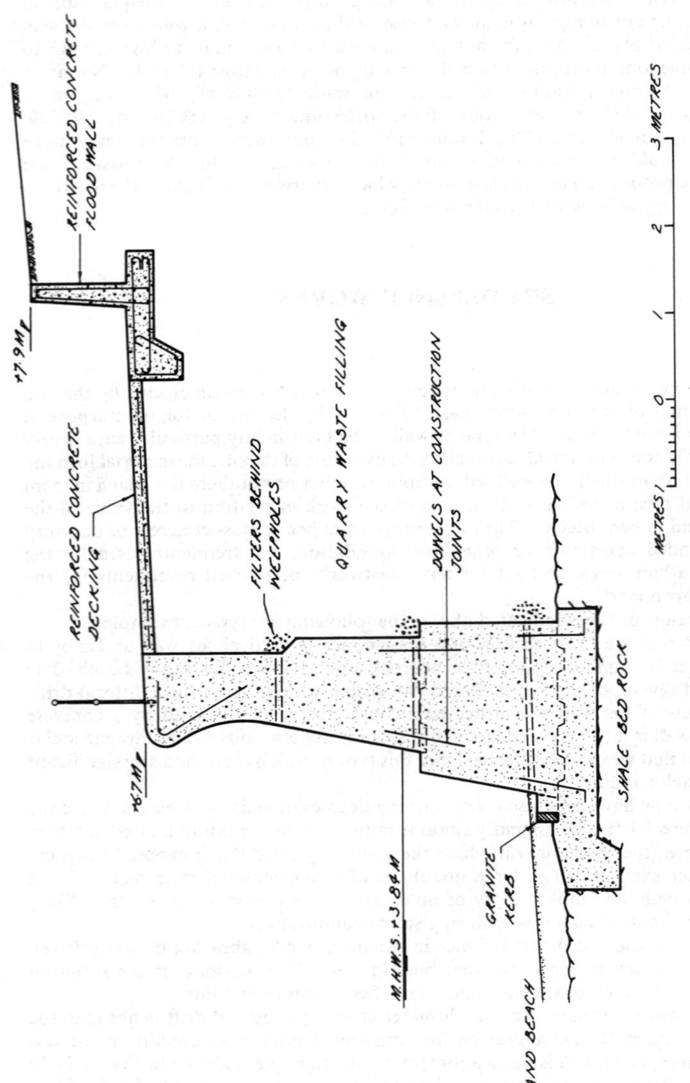

Figure 29.18 Mass concrete sea wall at Tenby, Pembrokeshire

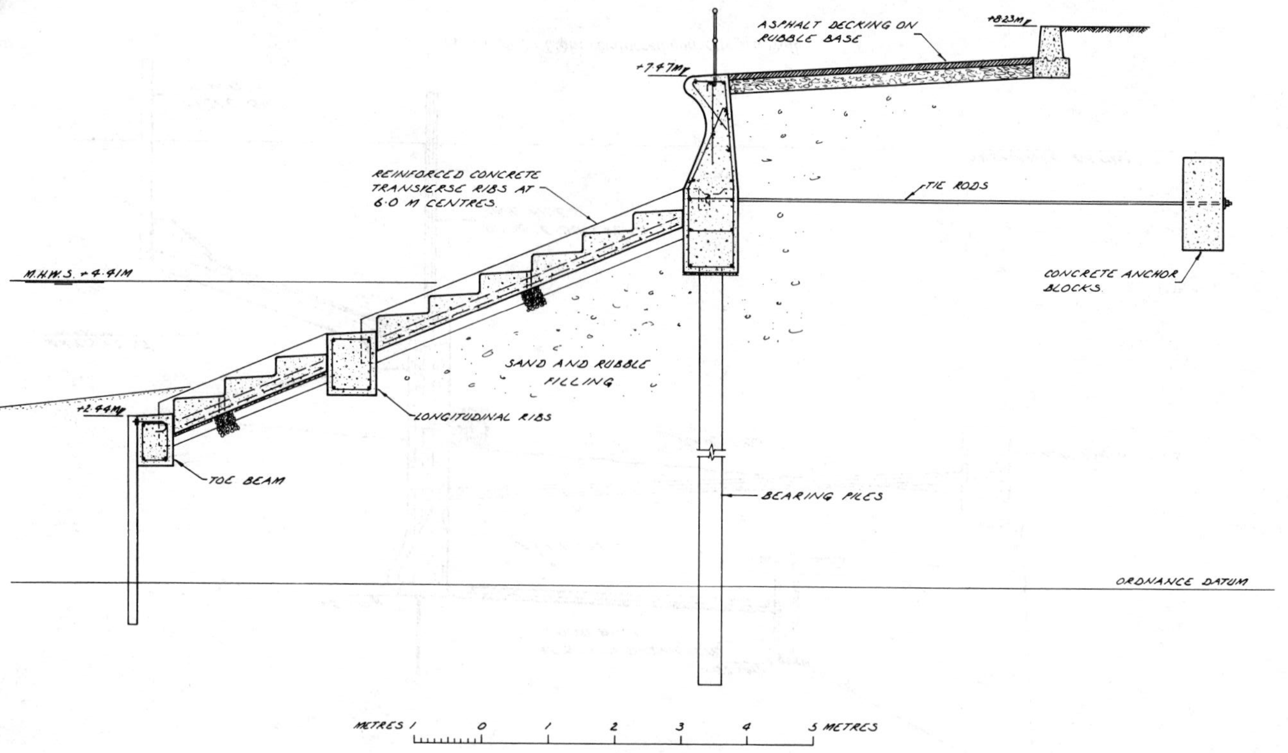

Figure 29.19 Sloping sea wall at Crosby, Lancashire

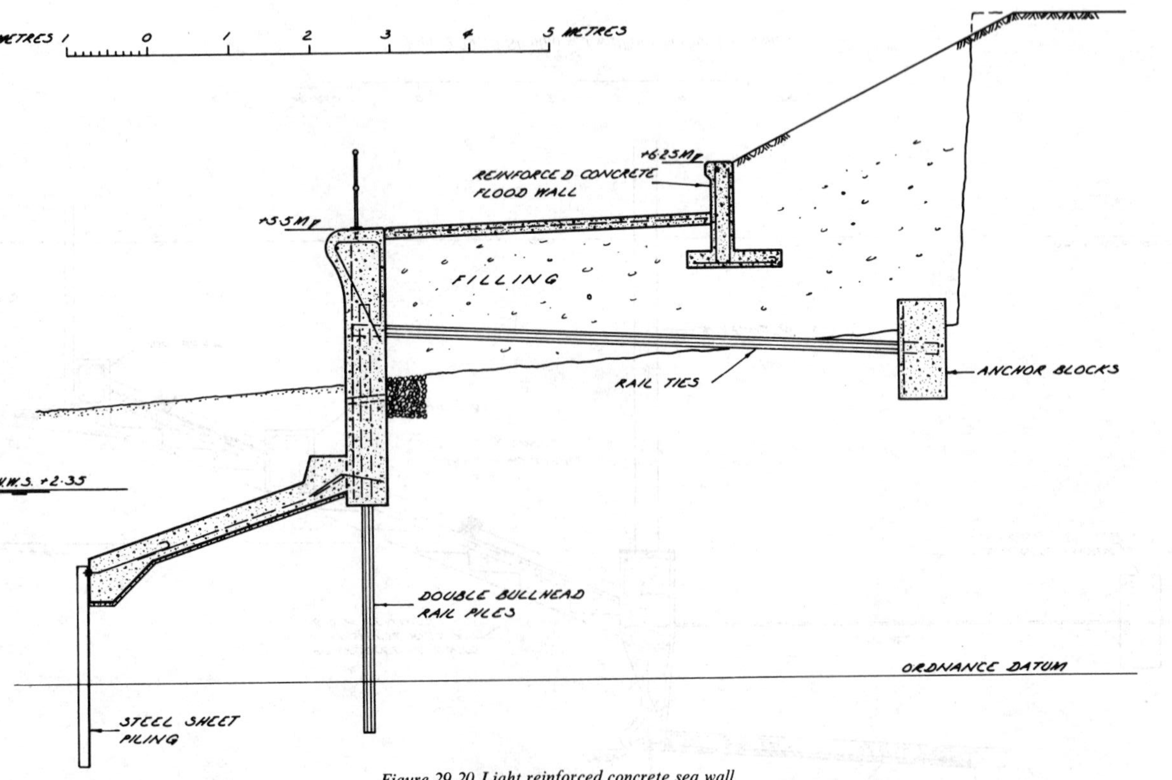

Figure 29.20 Light reinforced concrete sea wall

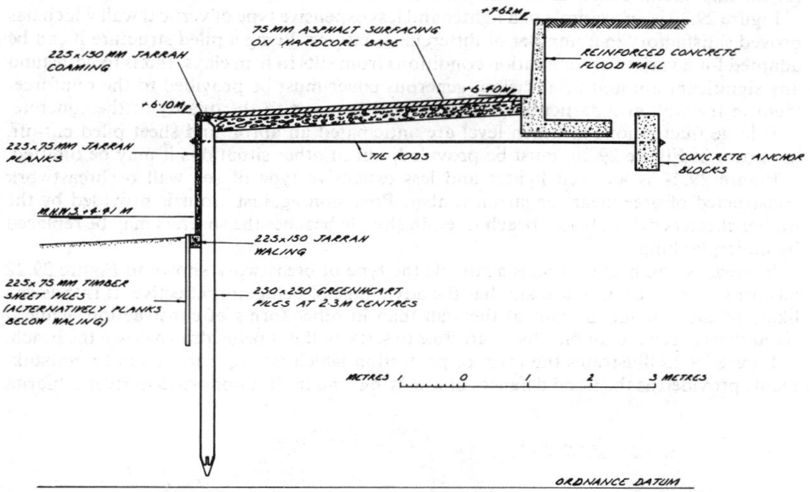

Figure 29.21 Timber breastwork

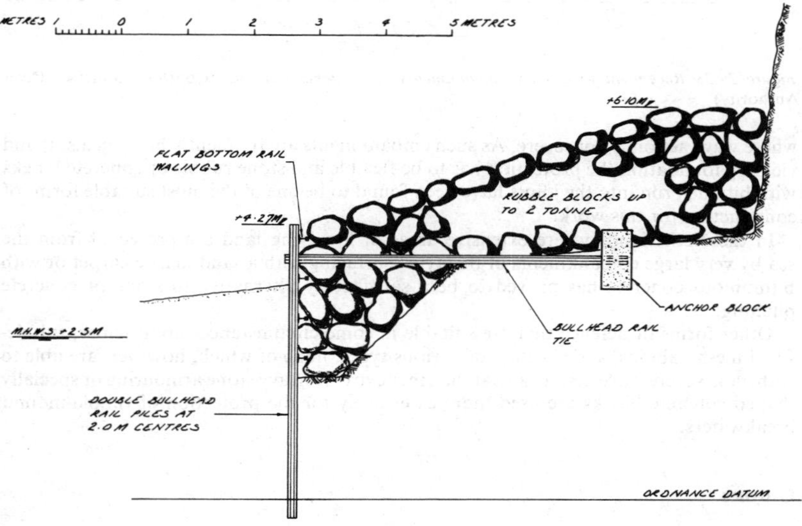

Figure 29.22 Rail pile and rubble breastwork

This form of construction may be chosen where foundation conditions are unsuitable for the high loading associated with the massive vertical type of wall.

Figure 29.20 is an example of a lighter and less expensive type of vertical wall which has proved satisfactory in a number of different situations. Being a piled structure it can be adapted for a variety of foundation conditions from silts to firm clays. If it is to withstand any significant amount of abrasion, generous cover must be provided to the reinforcement in the wall and particular attention must be given to the quality of the concrete.

If large fluctuations in beach level are anticipated an apron and sheet piled cut-off, as shown in Figure 29.20, must be provided, but in other situations it may be omitted.

Figure 29.21 is an even lighter and less expensive type of sea wall or breastwork constructed of greenheart or jarrah timber. Provision against scour is provided by the timber sheeters driven below beach level. In shingle beaches the sheeters may be replaced by underplanking.

In areas where heavy stone is available the type of breastwork shown in Figure 29.22 has proved to be economical and has the advantage of being nonreflective. It is thus less likely to cause scour in front of the wall than in other forms of construction, while if scour does occur the rubble blocks are able to settle without being drawn down the beach.

Figure 29.23 illustrates the form of protection which is frequently given to embankments provided as the flood defences of low-lying land in situations such as river estuaries

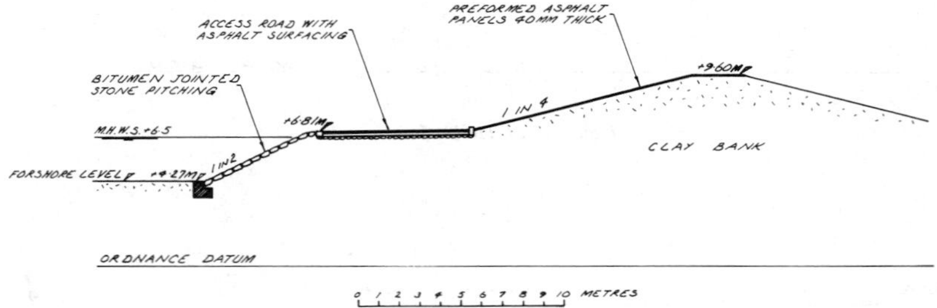

Figure 29.23 Revetment to estuary embankments: Hurditches sea wall (Courtesy Somerset River Authority)

where wave action is not severe. As such embankments are frequently built on a soft and yielding foundation the protection has to be flexible and stone rubble or concrete blocks with bitumen run into the joints have been found to be one of the most suitable forms of construction for this work.

In the Netherlands where extensive areas of low-lying land are protected from the sea by very large embankments of this type, surfacing with a sand mastic carpet or with bituminous concrete has proved to be a satisfactory alternative to stone or concrete pitching.

Other forms of defence that are suitable in some circumstances are fascines, rubble-filled mesh gabions, and sheeting of various types, none of which, however, are able to withstand severe wave action, and at the other extreme heavy stone armouring or specially shaped concrete blocks are used more extensively for the protection of rubble-mound breakwaters.

Groynes

A naturally occurring shingle beach is one of the best forms of defence against the sea but in the long term will only remain so if the supplies of shingle at least equal the losses;

otherwise the beach will be drawn down during storm conditions allowing the waves to attack the land and the strata underlying the beach and continuing or at least intermittent erosion will occur. In this latter situation the building of sea walls or breastworks to protect the land when the beaches are drawn down may not alone provide the solution to the problem as the land itself may be one of the sources from which the beach material has been derived and the construction of sea walls or breastworks will have cut off this part of the supply. Under these circumstances two courses are open to the authorities responsible for coast protection, either to groyne the beach or to recharge it artificially from time to time. It should be noted that groynes alone are unlikely to be the solution to a problem of this kind unless the quantities of material in transit along the coast with the littoral drift are very large, for groynes will not prevent the beaches from being drawn down under storm conditions so that intermittent erosion of the land will continue albeit at a slower rate until the groynes are outflanked at their inner ends when they will cease to serve the purpose for which they were built.

The preceding paragraph was concerned specifically with the groyning of shingle beaches. The same considerations will apply where the beaches are composed of sand and shingle or gravel but where they are composed of sand alone the situation is different. In exposed situations, sand beaches cannot be considered an effective defence against the sea unless losses by wind, waves and currents are continually replenished. If there is an ample supply of fresh sand a natural dune belt may be formed inshore of the beach which if it is wide enough and sufficiently stabilised by vegetation may form an adequate defence against the sea on account of its bulk and depth.

In the absence of a sufficient accumulation of sand by natural means the beaches may be encouraged to build up by the construction of groynes.

Although the physical processes involved in the movement of beach material are not fully understood there is sufficient experience in the control of shingle and gravel beaches

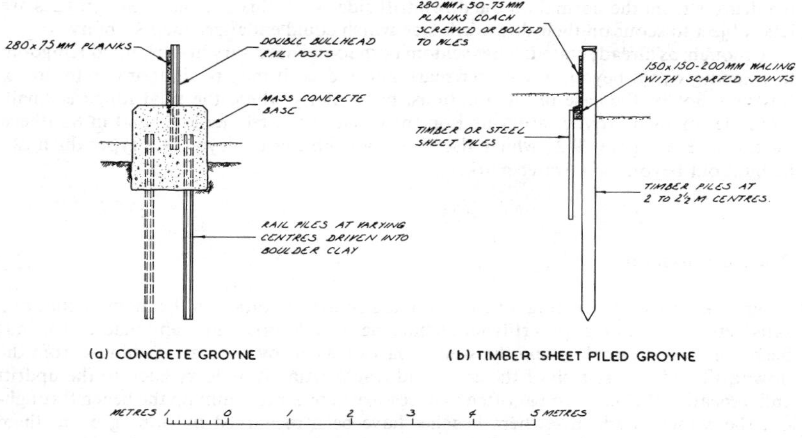

Figure 29.24 Concrete and timber groynes

for systems of groynes to be built with confidence. As with sea walls and breastworks the type of construction adopted in any particular situation will depend upon the exposure of the site, the nature of the beach itself and the foundation conditions. Figure 29.24 shows two typical examples.

As the slopes associated with shingle and gravel beaches are steep, fluctuations in level can be very considerable and in order to control these fluctuations it is usually necessary for the groynes for such beaches to be built at fairly close centres, the actual spacing in any particular case being dependent upon direction of wave approach and volume of littoral drift. It has been found from experience in the United Kingdom that with very few exceptions it is unnecessary to extend the groynes beyond low-water mark and at from 40 m to 70 m centres, which is close enough to avoid undue fluctuations in level; this leads to the provision of approximately 1 m of groyne for every metre of sea front protected.

For the very flat gradients of sand beaches the groynes can be much more widely spaced but need also to be much longer, and again the provision of approximately 1 m length of groyne for every metre of frontage protected is a reasonable guide to the probable requirement.

For gravel beaches it is usually necessary to provide widely spaced groynes across the lower sandy foreshore with short intermediate groynes to control the movement of the coarser material at the top of the beach.

Since groynes can only serve their purpose by interrupting the alongshore movement of beach material, whether the movement is predominantly in one direction or not, a successful system of groynes along any one stretch of the coast must inevitably lead to a reduction in the amount of material available to the downdrift beaches. In general, therefore, a system of groynes should not be made so efficient as to form a complete barrier to the littoral drift. Apart from the expense involved, this is one reason for not extending the groynes beyond low-water mark thus allowing some escape of material round their outer ends. Initially the groynes should be kept low relative to the beach level and only built up gradually as material accumulates. The alignment of the groynes relative to the coastline is also of importance; if the littoral drift is not predominantly in one direction the groynes should be built approximately normal to the coastline but where there is a marked drift in one direction it is better to lay off the groynes up to ten degrees from the normal on the downdrift side. With this alignment the groynes are less subject to scour on their downdrift side which could endanger their stability.

Although, as already stated, it has seldom been found necessary in the United Kingdom to extend groynes beyond low-water mark and indeed it may be undesirable to do so, this may not be the case in all situations, particularly where the tidal range is small. Examples of such groynes are found on the coast of the Netherlands and in Southern Portugal. (See Figure 29.25 which also shows a rubble crib groyne of a type which can be built out beyond low-water mark.)

Beach nourishment

As an alternative to groyning, or as a complementary operation in the maintenance of a satisfactory beach defence, artificial recharging is undertaken in many places. This has been done, for example along the south coast of Kent, by excavating shingle from the downdrift end of a stretch of the coast and transporting it by lorry back to the updrift end, repeating the operation as often as necessary in order to maintain the beach throughout the whole length. Elsewhere beaches have been recharged by tipping on to them quarried stone or gravel excavated from inland gravel pits. Along the coasts of Northumberland and Durham, colliery waste tipped on to the shore as a convenient means of disposal forms a protective beach along many miles of the coastline. Although objection may be raised to a beach of this type from the point of view of amenity, it does form an effective sea defence.

Sand beaches are more difficult to renourish but it has been accomplished by pumping sand from the updrift side of an obstruction across to the downdrift side, and by pumping dredged sand from offshore on to beaches. Attempts to renourish by depositing dredged sand close to the shore from hopper barges have not, however, proved successful.

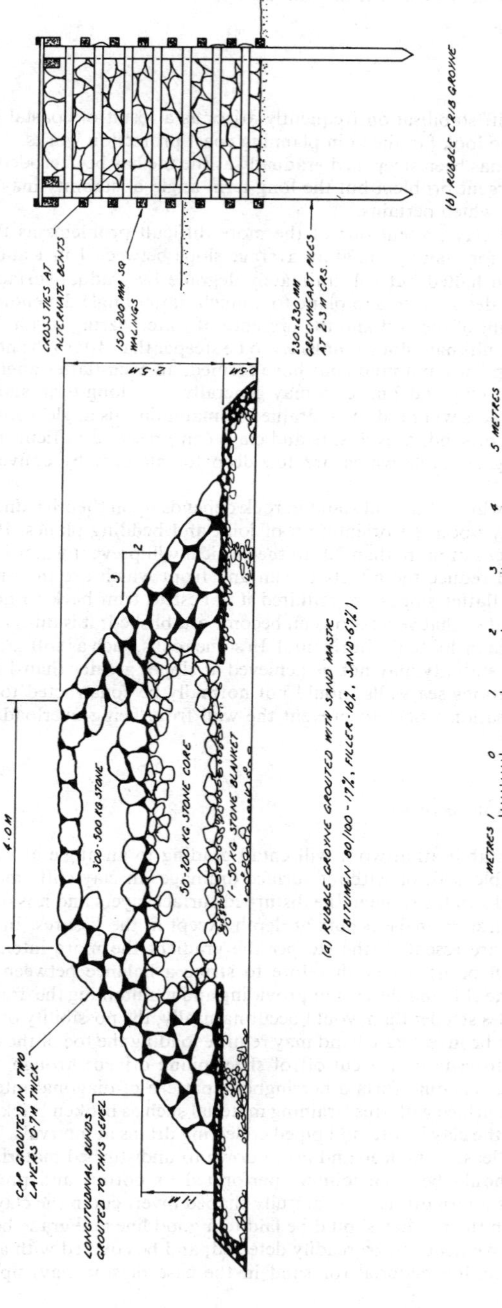

Figure 29.25 Rubble groynes

Cliff stabilisation

GENERAL

The need for cliff stabilisation frequently arises as a result of coastal erosion and it is then necessary to look far ahead in planning coast protection works.

A cliff which has been steepened gradually as the toe has been eroded may not appear to present an urgent problem but the long-term angle of stability may be considerably flatter than that which pertains.

Cliffs in hard clay present one of the more difficult problems as they may remain relatively stable for many years at an average slope between 1 in 1 and 1 in 3 after toe erosion has been halted but will ultimately degrade by gradual surface slipping or by movement on a deep seated slip plane to a much flatter angle depending upon the type of clay, the height of the cliff and the presence of waterbearing strata. In London Clay, for example, the ultimate slope is unlikely to be steeper than 10° to the horizontal (1 in $5\frac{1}{2}$) and, where deep slips and mud runs have formed, an even flatter angle may eventually prevail. Boulder clay and Lias clay may generally have long-term stability at 1 in 2 to 1 in $2\frac{1}{2}$ but in places will need to be drained to maintain this angle against surface creep.

Mixed strata of sand, gravel, silts and clays can present a difficult problem as water may be flowing at levels which are too deep to intercept by conventional drainage systems.

The angle at which cliffs will stand in rock depends upon their hardness and durability and very largely upon the orientation of joint and bedding planes. If they have to be trimmed a batter not more than 20° to the vertical will prevent water from collecting on the surface and reduce the affects of sun and front which are the principal agents of denudation. If flatter slopes are required it is best to trim back to not more than 50° to the horizontal so that vegetation can become established; it is difficult to soil and seed a slope steeper than 40° to the horizontal. In some rocks, such as soft sandstone, marl and chalk, ultimate stability may not be achieved at slopes greater than 1 in $1\frac{1}{2}$.

In such situations sea walls should not normally be constructed to stop the erosion without stabilisation works to prevent the wall from being overloaded by slipping or falling material.

METHODS OF STABILISATION

In most cases stabilisation work will entail grading to an angle at which the material can remain stable with or without surface drainage. In clay cliffs movement of water occurs mainly down the slope in the disturbed surface layers and it is questionable if any significant natural drainage occurs at depth except in the fissures. In these conditions, surface drains are essential, the steeper the gradient the more intensive the drainage required; it will be necessary therefore to strike a balance between the cost of bulk excavation in the cliff and the cost of providing and maintaining the drains. If it is decided to drain at slopes steeper than would occur naturally, the possibility of movement below the drains must be investigated and may require loading the toe of the cliff with material placed on the foreshore or a cut-off of sheet piling driven through any potential slip plane. The drains should form a herringbone pattern of diagonal interceptors brought up close to the surface with free-draining material such as broken brick, rubble or gravel, or with sand if the clay is soft, and piped collecting drains at intervals. The drains should normally be at least $1\frac{1}{4}$ m deep and taken down to undisturbed material where possible.

The pipes should be open jointed, perforated or porous and preferably laid on a concrete bed or alternatively in a carefully shaped invert cut in the clay to prevent water flowing beneath them. They should be laid to a good line and grade between inspection pits, so that movement can be readily detected, and be covered with a gravel filter with coarser free-draining material (or sand in the case of soft clay) up to ground level.

A drainage system of this kind has the additional effect of buttressing the surface layers of the cliff and improving its overall stability.

If filling is required at any places in the cliff face, usually near the toe, it is an advantage to provide a sheet drain of sand or ballast beneath the fill as well as to drain the surface.

In mixed soils with free draining-materials at a low level, drains bored from the foot of the cliff and lined with perforated pipes or vertical sand drains have an application.

For rock cliffs, alternatives to trimming back are rock bolting and grouting. Clay slopes have also been stabilised by grouting.

TOE PROTECTION

The toe protection provided for coastal cliffs need only vary from that provided in other situations to take account of potential instability.

With clay cliffs the slope of the wall should not be significantly greater than the clay slope behind and may require a wide decking or berm, possibly at two levels in order to attain the required height.

Drainage behind the walls will usually be necessary and should be at a low level in order to reduce water pressures in the cliff. As it is difficult to keep the sea out of drains below tide level and the outfalls clear of beach material and seaweed it may be necessary to pump the flow.

If complete cliff stability is not certain, a flexible form of construction as shown in Figures 29.21 and 29.22 is economical to construct and to repair if damaged by cliff falls or movement.

BREAKWATERS

Breakwaters can be designed either to reflect or absorb wave energy but for reasons of construction or economy many are composite structures which reflect waves of certain characteristics or at particular water levels while under other circumstances they cause the waves to break and dissipate their energy.

Vertical-faced structures

The traditional mass concrete, masonry or concrete blockwork breakwater of which there are numerous examples in the UK and elsewhere is of the reflecting type, a typical example being the breakwaters at Dover,[34] one of which is shown in section in Figure 29.26. These structures differ mainly in the method adopted for the construction of the foundations; at Dover this was done by excavating a trench in the chalk sea bed working from a diving bell, and in it the blockwork was placed by goliath crane. Variations of this form of construction are the slicework breakwater, e.g. at Colombo,[35] in which the blocks are able to settle without disrupting the structure, and breakwaters with outer walls of blockwork or masonry and a rubble or concrete hearting. At Newhaven[36] (Figure 29.27) the breakwater is of mass concrete cast *in situ* on a foundation of 100-ton concrete sacks deposited from a special bottom-opening vessel. A similar form of construction at Whitby[37] was carried out within shutters erected on the sea bed from a gantry which was designed to 'walk' ahead of the work.

An alternative form of construction for this type of breakwater is the caisson, constructed partly on shore and partly afloat in a building dock, floated into position and sunk on to a prepared foundation. The breakwater at Helsingborg (Figure 29.28) is a typical example.

The main advantage of caissons is that very large units can be handled and with adequate space for building them progress can be rapid. However, their use is limited by the depth in which they can be floated.

A caisson system which does not have this limitation was used for two breakwaters at Hanstholm[38] (Denmark) and has been adopted for the harbour arms currently being built for the Brighton Marina. The caissons are 12 m diameter reinforced concrete cylindrical units weighing up to 600 tonnes each, and open at their lower ends. They are constructed on shore and transported to the end of the completed part of the breakwater

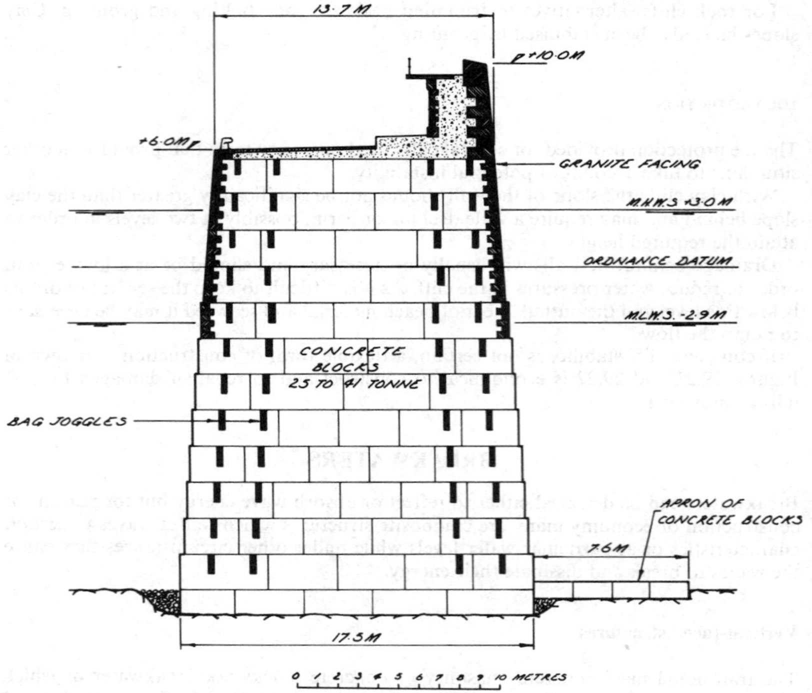

Figure 29.26 Dover harbour west breakwater

on a selfpropelled trolley where they are suspended in position clear of the sea bed by a special crane while tremied concrete is placed in the bottom to form a massive foundation plug. A cross section of the Brighton breakwater is shown in Figure 29.29.

Rubble-mound breakwaters

The rubble mound is the traditional form of energy-absorbing breakwater but there are numerous variations. In most cases quarried rock forms the core of the mound and, when economically available in suitable sizes, has frequently been used for the armour layers and capping. Where large rock is difficult to obtain, concrete armour units of various shapes have been used, among the best known being the Tetrapod, Tribar, Akmon, Stabit and Dolos. For rock armour, K_D values in the Hudson Formula (equation (60)) between 2.0 and 2.8 are suggested, depending upon the angularity of the rock and the position in the breakwater. For concrete units, values of from about 8.0 to 18.0

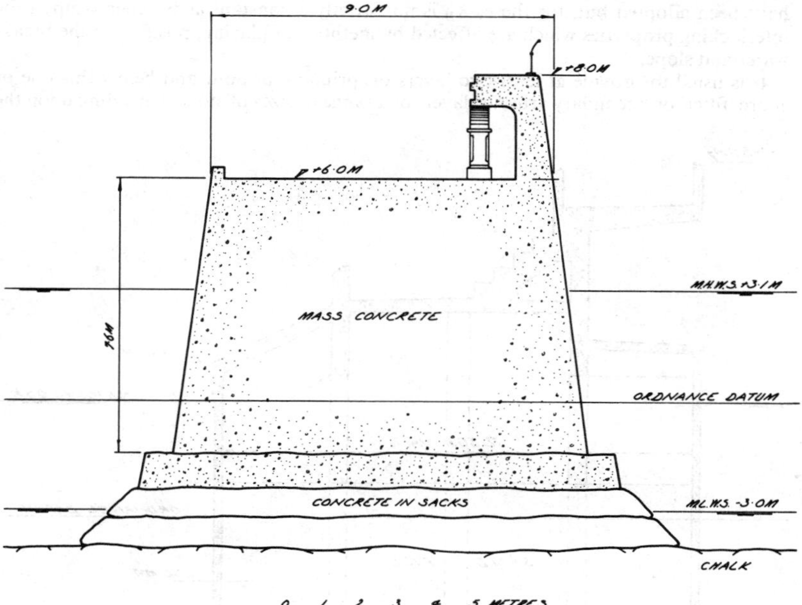

Figure 29.27 West breakwater, Newhaven

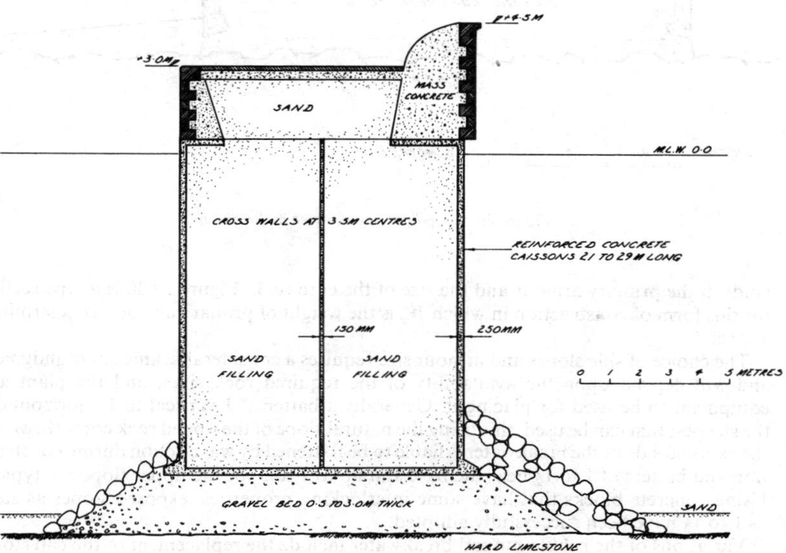

Figure 29.28 Caisson breakwater at Helsingborg, Sweden

have been adopted but, for these, K_D is not strictly a constant as it depends upon the interlocking properties which are affected by methods of placing, position in the breakwater and slope.

It is usual to provide at least two layers of 'primary' armour and below this one or more 'filter' or 'secondary armour' layers of a range of sizes of stone depending upon the

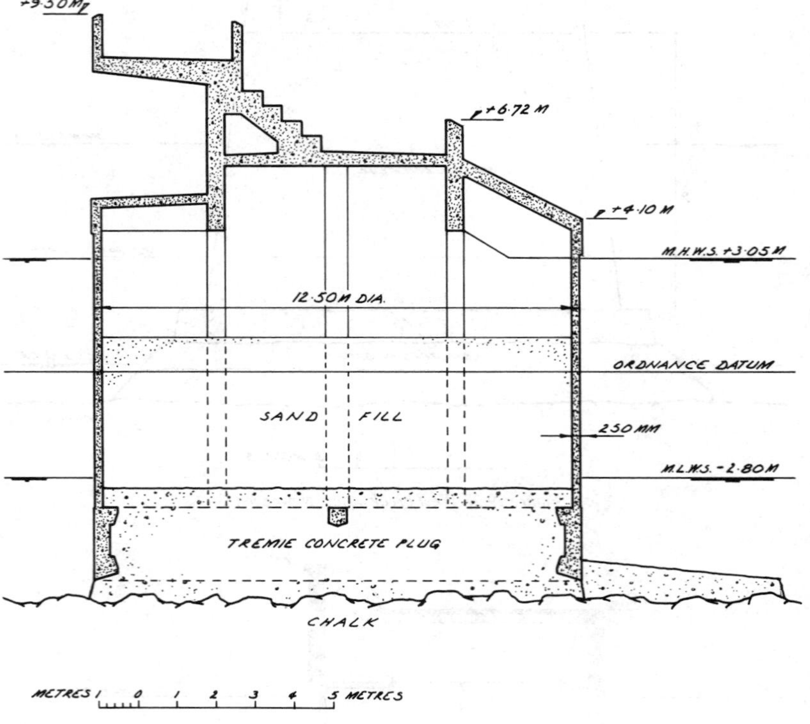

Figure 29.29 Cellular breakwater at Brighton

voids in the primary armour and the size of the core rock. Figure 29.30 is a type section for this form of construction in which W_r is the weight of primary armour as determined above.

The choice of side slopes and armour size requires a considerable amount of judgment and will depend upon the availability of the required rock sizes, and the plant and equipment to be used for placing it. Generally a batter of 1 vertical to $1\frac{1}{4}$ horizontal is the steepest that can be used, this being the natural slope of the tipped rock core. However, the exposed side of the breakwater is liable to be flattened by wave action during construction and batters of 1 to $1\frac{1}{2}$ near the base and up to 1 to 3 for the upper slope are typical. Using concrete blocks that have some interlocking properties, exposed slopes as steep as 1 to $1\frac{1}{3}$ have been successfully adopted.

Variations of the rubble-mound breakwater include the replacement of the cap stones by a concrete deck and parapet, a recent example being the breakwaters at Port Talbot,[39] and the use of a sand core placed between successive pairs of rubble mounts built up as the work proceeds.

At a number of places, where overtopping of the breakwater, and the resulting wave action behind it, can be tolerated, the crest level may be close to or even below water level. The new North Mole at the Hook of Holland is in this category (Figure 29.31.)

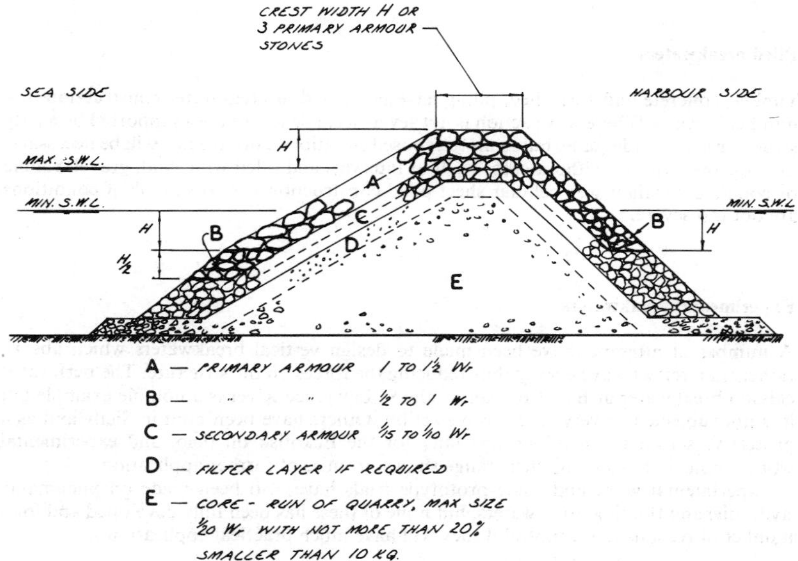

Figure 29.30 Type section for rubble-mound breakwater. A, *primary armour* 1 *to* $1\frac{1}{2}$ W_r; B, *primary armour* $\frac{1}{2}$ *to* 1 W_r; C, *secondary armour* $\frac{1}{15}$ *to* $\frac{1}{10}$ W_r; D, *filter layer if required; E, core—run of quarry, Maximum size* $\frac{1}{20}$ W_r *with not more than* 20% *smaller than* 10 kg

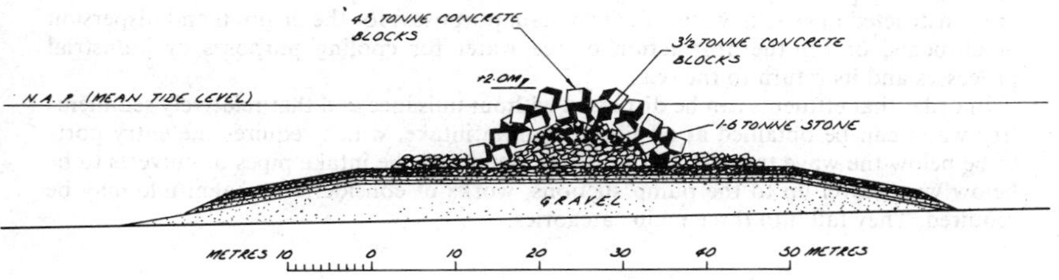

Figure 29.31 North Mole at Hook of Holland

Rockfilled crib breakwaters

A type of breakwater intermediate between the fully reflecting vertical wall and the mainly absorbent rubble mount is the rockfilled crib. It can be constructed with piles and walings, as the crib-groyne in Figure 29.25, or the cribs can be fabricated on shore and floated into position.

Composite construction

Apart from the above distinct types of construction numerous combinations of them have been built and, for very deep water, mass concrete, blockwork and caissons have been founded on rubble mounds as at Casablanca, Bari (Italy), Marseilles and Funchal.

Piled breakwaters

Timber, concrete and steel sheet piling have been used in breakwater construction in a number of ways. Where wave action is not severe a single line of piles supported by a jetty structure may be adequate, but in more exposed situations a double line will be necessary, tied together with or without cross walls at intervals and filled with sand, gravel, rubble or concrete. Straight web cellular sheet piled construction is also suitable if conditions are not too severe.

Experimental breakwaters

A number of attempts have been made to design vertical breakwaters which absorb rather than reflect wave energy thus reducing the forces on the structure. The perforated caisson breakwater at Baie Comeau on the St Lawrence River is a notable example but it is not subjected to very severe wave action. Others have been built in Sicily and as a protective screen to an oil storage tank for the Eckofisk oil field, and experimental work[40] aimed at extending their range shows promise of further application.

Experimental work and some prototype trials have also been made on pneumatic, hydraulic and floating breakwaters, but none of these has been fully developed and for a number of reasons it is doubtful if they will have much practical application..

SEA-WATER INTAKES AND OUTFALLS

Apart from pipes for gas and oil which are outside the scope of this Section, pipelines are constructed in coastal waters for two main purposes: for the disposal and dispersion of effluents, or for the abstraction of sea water for cooling purposes or industrial processes and its return to the sea.

In order that effluents can be dispersed without nuisance and that relatively sediment-free water can be obtained at all times from an intake, which requires the entry ports to be below the wave troughs at lowest low water and the intake pipes or culverts to be below water level up to the pump suctions, works of considerable magnitude may be required. They fall into three main categories.

Jointed pipelines

These include pipes laid and jointed by diver in a trench excavated in the sea bed and backfilled with imported or dredged material, or partially or wholly with concrete in the case of a rocky sea bed, or supported above the sea bed on piled trestles or concrete saddles.

The most usual type of pipe for this form of construction is cast or spun iron with flexible joints, but steel, plastics, concrete and aluminium have all been used. The choice of pipe material should be made for each project having regard to durability and method of installation; the cost of the pipe itself is usually a less important consideration.

The design of the work should take account of methods of construction and the type of equipment to be used, which may be conventional plant working on the shore between tides, from a gantry, from floating craft, or from a jack-up spud platform.

Allowance must be made for possible changes in sea-bed level to ensure that the pipe or its supports are not undermined, and if the pipes are supported above the sea bed they must be designed to resist forces due to wave action and currents (see p. **29**–25).

Pipelines towed or floated into position

Pipes installed by these methods, in which they are fabricated into long lengths on shore and towed into position by tugs or barge-mounted winches, have normally been of steel, with an epoxy, bitumastic, or cement mortar lining, wrapped externally (see p. **29**–43) and coated with concrete or gunite to protect the wrapping from mechanical damage and to make the line just sink when sealed and full of air.

Usually the pipes are joined into long 'strings' by butt welding and if floated and sunk into position the strings are joined under water with flexible couplings. For lines towed into position along the sea bed the strings are butt welded together and the lining and wrappings completed at the joints as the pipes are pulled out from the fabricating site.

For pipelines not exceeding 1 m in diameter, polythene pipe has been used and has an advantage in flexibility and ease of handling. Because they are light in weight polythene pipes must be anchored if they are to be left exposed on the sea bed.

There is some evidence that a pipe left exposed on a sandy sea bed will, at least partially, embed itself naturally, but in exposed situations where the sand level is liable to fluctuate it is necessary to bury the pipe. This is usually done by dredging a trench prior to launching the pipe but ploughs and jetting equipment have been used for this purpose.

Tunnels and shafts

There have been many instances of tunnels being constructed as sea water intakes and outfalls. In soft ground they have been confined to fairly shallow depths allowing the use of compressed air. In rock there is more latitude in the choice of depth and cover, fissured strata being treated ahead of the work by grouting.

The connection of the tunnel to the sea has in some cases been made by flooding the tunnel and breaking through with a single large round of explosive, but more usually by one or more shafts leading to some permanent structure (a tower of caisson) on the sea bed, or drilled from floating plant or a jack-up platform. In the latter case the tunnel may be a culvert, running full, or it may have pipes installed in it.

REFERENCES

1. IPPEN, A. T., *Estuary and Coastline Hydrodynamics,* McGraw-Hill (1966)
2. STOKES, G. C., 'On the Theory of Oscillating Waves', *Trans. Cambridge Phil. Soc.,* **8** and Suppl., *Sci. Papers,* **1** (1847)
3. WIEGEL, R. L., 'Gravity Waves, Tables and Functions', Council on Wave Research, The Engineering Foundation (1954)
4. 'Technical Report No. 4', Beach Erosion Board, US Government Printing Office, Washington
5. BOURODIMOS, E. L., and IPPEN, A. T., 'Wave Reflection and Transmission in Channels of Variable Section, *Proc.* 11*th Conf. on Coastal Engng*, Vol. 1, London (1968)
6. THIJSSE, T. TH, and SCHIJF, J. B., 'Report on Waves', *17th Int. Nav. Congr.,* Section II, Communication 4 (1949)
7. BRETSCHNEIDER, C. L., 'Revised Wave Forecasting Relationships', Proc. 2nd Conf. on Coastal Engng (1952)

8. SUTHONS, C. T., 'Frequency of Occurrence of Abnormally High Sea Levels on the East and South Coasts of England', *Proc. Instn civ. Engrs*, **25** (August 1963)
9. LENNON, G. W., 'A Frequency Investigation of Abnormally High Tidal Levels at Certain West Coast Ports', *Proc. Instn civ. Engrs*, **25** (August 1963)
10. RUSSELL, R. C. H., 'A Study of the Movement of Moored Ships subjected to Wave Action', *Proc. Instn civ. Engrs*, **12** (April 1959)
11. INGLIS, SIR CLAUDE, and KESTNER, F. J. T., 'Changes in the Washes as affected by training walls and reclamation Works'. *Proc. Instn civ. Engrs*, **11**, 435–466 (December 1958)
12. PRICE, W. A., and KENDRICK, M. P., 'Field and model investigation into reasons for siltation in the Mersey Estuary', *Proc. Instn civ. Engrs*, **24** (April 1963)
13. KUENEN, PH. H., *Marine Geology*, Wiley (1950)
14. DUVIVIER, J., 'Selsey Coast Protection Scheme', *Proc. Instn civ. Engrs*, **20** (December 1961)
15. ABRAHAM, G., 'Jet Diffusion in Stagnant Ambient Fluid', Delft. Hyd. Lab., Pub. No. 29 (1963)
16. FRANKEL, R. F., and CUMMINGS, J. D., 'Turbulent Mixing Phenomena of Ocean Outfalls', *Proc. Am. Soc. Civ. Engrs*, 91.S.A. (1965)
17. ANWAR, H. O., 'Behaviour of Buoyant Jet in a Calm Fluid', *Proc. Am. Soc. civ. Engrs*, 95.HY4 (1969)
18. AGG, A. R., and WAKEFORD, A. C., 'Field Studies of Jet Dilution of Sewage at Sea Outfalls', *J.Inst.P.H.E.* (April 1972)
19. TESAKER, E., 'Uniform Turbidity Current Experiments', *Proc. 13th Congr. Int. Assn Hydraulic Res* (1969)
20. ELLISTON, T. H., and TURNER, J. S., 'Turbulent Entrainment in Stratified Flow', *J. Fluid Mech.*, **6** (1959)
21. WIEGEL, R. L., and BEEBE, K. E., 'The design Wave in Shallow Water', *J. Waterways Div., Proc. Am. Soc. Civ. Engrs*, No. W.W.I. (March 1956)
22. WIEGEL, R. L., and SKJEI, R. E., 'Breaking Wave Force Prediction', *J. Waterways and Harbors. Div. Proc. Am. Soc. civ. Engrs*, **84**, No. W.W.2 (March 1958)
23. BRETSCHNEIDER, C. L., 'Selection of Design Wave for Offshore Structures', *J. Waterways and Harbors. Div. Proc. Am. Soc. civ. Engrs*, **84**, No. W.W.2 (March 1958)
24. MUNK, W. H., 'The Solitary Wave Theory and its Application to Surf Problems, Ocean Surface Waves' (*Ann N.Y. Acad. Sci.*), **51** (May 1949)
25. WIEGEL, R. L., BEEBE, K. E., and MOON, J., 'Ocean Wave Forces on Circular Cylindrical Piles', *J. Hydraulics Div. Proc. Am. Soc. civ. Engrs*, **83**, No. HY2 (April 1957)
26. MORISON, J. R., JOHNSON, J. W., and O'BRIEN, M. P., 'Experimental Studies of Forces on Piles', *Proc. 4th Conf. on Coastal Engng* (October 1953)
27. BAGNOLD, R. A., 'Interim Report on Wave Pressure Research', *J. Inst. Civ. Engrs*, **12** (June 1939)
28. MICHE, R., 'La Pouvoir Réflechissant des Ouvrage Maritimes Exposés a l'Action de la Houle', *Ann. Ponts Chauss.*, 121 (1951)
29. HUDSON, R. Y., 'Laboratory Investigation of Rubble Mound Breakwaters', *Waterways Harbors. Div., Proc. Am. Soc. civ. Engrs*, Paper No. 2171 (September 1959)
30. ALLEN, R. T. L., and TERRETT, F. L., 'Durability of Concrete in Coast Protection Works', *Proc. 11th Conf. on Coastal Engng*, London (1968)
31. Van Asbeck, W. F., *Bitumen in Hydraulic Engineering*, Vol. 1, Shell Int. Pet. Co., London (1959); Vol. 2, Elsevier, Amsterdam (1964)
32. VISSER, W., *Coast Protection with Bitumen*, Shell Bitumen Reprint No. 20, Shell International Petroleum Company
33. KERKHOVEN, R. E., *Hydraulic Applications in the Netherlands*, Shell International Petroleum Company Report No. 110 F
34. WILSON, M. F. G., 'Admiralty Harbour, Dover', Min. of *Proc. Instn civ. Engrs*, **209** (1921)
35. KYLE, J., 'Colombo Harbour Works, Ceylon', Min. of *Proc. Instn civ. Engrs*, **87** Part I (1886/87)
36. CAREY, A. E., 'Harbour Improvements at Newhaven, Sussex', Min. of *Proc. Instn civ. Engrs*, **87**, Part I (1886/7)
37. MITCHELL, J., 'Whitby Harbour Improvement', Min. of *Proc. Instn civ. Engng*, **209** (1921)
38. LUNDGREN, H., 'A New Type of Breakwater for Exposed Positions'. *Dock and Harbour* (November 1962)
39. MCGAREY, D. G., and FRAENKEL, P. M., 'Port Talbot Harbour: planning and design'. *Proc. Instn civ. Engrs*, **45** (April 1970)
40. TERRETT, F. L., OSORIO, J. D. C., and LEAN, G. H., 'Model Studies of a Perforated Breakwater', *Proc. 11th Conf. on Coastal Engng*, London (1968)

30 TUNNELLING

TUNNELLING 30

30 TUNNELLING

A. M. MUIR WOOD, M.A., F.I.C.E., F.G.S.
Sir William Halcrow & Partners

Traditions have prevailed in tunnelling longer than in any other branch of civil engineering. This characteristic explains the relatively rapid advances in tunnelling in recent years as the art has at last become increasingly married to technology. Advances in tunnelling usually arise not from research so much as from innovations in methods of design and construction. Monitoring of the results in the field may then follow, supported by research where existing knowledge fails to explain the findings.

Two essential elements to economic tunnelling are:

(i) The tunnel (unless permanently unlined) must be considered as a composite ground–lining structure. Not only does the lining support the ground but the ground in its turn supports the lining.

(ii) The design of the permanent tunnel must be considered in association with the methods of construction. The overall cost of the process requires to be minimised and the finished geometry is only one of many factors.

There are many barriers to a full understanding of the behaviour of the ground around a tunnel: the three-dimensional, time-dependent nature of the problem; the complexity of the stress–strain relations in soft ground; the effects of the initial state of stress, discontinuities and joints upon the behaviour of a rock; the dependence upon the method of excavation; the standard of workmanship; inhomogeneity of the ground.

Full-scale tunnels provide the one reliable laboratory for testing theory against practice. Modelling techniques, with centrifugal models used to overcome limitations of scale, may assist materially in an understanding of the problems.

THE OPTIONS FOR A TUNNEL ROUTE

The ground is the principal determinant of the cost of a tunnel of a given size. For this reason great economic benefits may derive from the capability of selecting a favourable and relatively consistent type of ground for tunnelling. Until the geological structure is known, the object should be to keep the options for a tunnel route as open as possible.

For each type of tunnel there are certain geometrical constraints and other specific factors affecting cost. For a road tunnel, for example, acceptable gradients and curves will be related to the design speed and hence to traffic costs. For a pressure tunnel, on the other hand, there is little direct geometrical constraint and the differential cost of construction in relation to the ground would need to be considered against the capitalised head losses.

A general knowledge of the geological structure will indicate whether or not the most direct route conforms to a favourable geological horizon or whether on the contrary it may encounter unstable ground such as squeezing rock, running sand, major fault zones, decomposed rock, karstic limestone or similar hazards which may only be penetrated at great expense.

Where there is a possibility of adopting an economic method of tunnelling, related specifically to a type of ground with limited variation, there may be a considerable benefit from diverging from the most direct route, in order to situate the tunnel throughout in such ground.

At the earliest stage in planning, such factors should be considered so that the options may be described, systematically tested and reduced as information arises from the first stage of site investigation.

COSTS OF TUNNELLING

Principal factors

Attempts are made periodically to set out tunnelling costs in a systematised form, with costs per unit length of a certain size of tunnel related to a few generalised ground types and to a few other simplified categories of accessibility and tunnel length. Except for specific areas in which the ground can be reliably depended upon, there is no valid way of expressing tunnelling costs on a simple unit cost basis.

From a knowledge of the ground a system of tunnelling may be selected and the costs evaluated on an assumed average rate of progress. The rate of progress may be assessed from experience in similar ground elsewhere, taking account of any innovation in the tunnelling method, and not forgetting the costs of ground treatments or similar ancillary operations. In general, the extent of variability in the cost of tunnelling is increasing for these principle reasons:

(a) Tailor-made tunnel systems to suit a particular type of ground permit increasing economies in construction.
(b) The cost of labour-intensive tunnelling systems adopted for difficult ground or in congested circumstances will naturally reflect the trend of labour costs including incentive payments.
(c) The demand for tunnels in urban development tends to reduce the options available for a tunnel route.

As the result of these factors, at the present time there is at least an order of magnitude between the unit cost of constructing the cheapest and the most expensive tunnel of the same size. Hence, there is an increasing benefit to be derived from undertaking studies appropriate to choosing the most economic expedient in each situation. Table 30.1 sets out the unit costs (inclusive of internal construction and services) of the principal post-war road tunnels in Britain with a brief account of the major factors affecting the cost.

A feature that may be overlooked in comparing the costs of tunnels concerns the means of access during construction. While a shallow urban tunnel or a short tunnel through a hill may be approached directly from the ground surface, long and subaqueous tunnels usually require working shafts and access headings, adding not only to the direct cost of the project but also to the cost of all the consequent tunnelling operations.

Effect of tunnel size

The cost of a given tunnel is specific to its situation and its timing, on account of the varying differences in prices, varying local skills and technical capabilities. There is therefore no simple factor to be applied to the cost of a tunnel in order to determine its hypothetical cost at a different place or time.

Neither is there a simple formula to determine the cost of a tunnel by consideration of another tunnel in the same ground and conditions but of different size. As a simplification, where variation in size does not entail a change in basic techniques, we may consider each factor in construction as entailing a unit cost U expressed as:

$$U = A + Bd + Cd^2$$

where A, B and C are constants and d is the finished diameter. For a highly mechanised

Table **30.1** RECENT BRITISH ROAD TUNNELS

Tunnel	*Dartford*[1]	*Clyde*[2]	*Blackwall*[3]	*Tyne*[4]	*Heathrow Cargo*[5]	*Great Charles Street*[6]	*Mersey*[7a, b]
Period of construction	1956–63	1957–63	1960–67	1961–67	1966–68	1967	1967–(74)
Purpose	River crossing	River crossing	River crossing	River crossing	Link under runways	Urban link under streets	River crossing
Length (m)	1 429	2 × 760	893	1 688	900	2 × 280	2 × 2 244
Internal diameter (m)	8.6	9	8.6	9.5	10.3	12.2 (equivalent with flat invert)	9.6
Internal area (m^2)	58	63	58	71	82	59	72
Means of working access to tunnel	Working shafts and passages	Working shafts and passages	From shield pits sunk as caissons	From 2 shafts and 2 pits with surface approaches	Direct from surface with shield pits	Direct from surface	Direct from surface approach for machine drives: shafts for hand drives
Ground	Chalk, gravel, peat and river mud	Sandstone, shale boulder clay, silt, sand, gravel	Woolwich Beds, London clay, gravel, peat, river mud	Glacial deposits and coal measures	London clay	Bunter sandstone	Bunter sandstone
Variability of ground	Moderate	High	Moderate	High	Very low	Very low	Low

Table 30.1 (*continued*)

Tunnel	*Dartford*[1]	*Clyde*[2]	*Blackwall*[3]	*Tyne*[4]	*Heathrow Cargo*[5]	*Great Charles*[6] *Street*	*Mersey*[7a, b]
Water problems	Considerable	Considerable	Severe	Considerable	Below water table but natural clay seal	Above water table	Significant
Means of excavation	Pilot by shield. Full face: mostly by shield. Spoil disposed as pumped slurry	Hand shield, air tools and blasting	Two pilots and main tunnel by hand shield	Pilot by shield, main tunnel by hand shield and by arch erector and excavator	Hand shield, clay spades	Part by tunnel machine part by tools	Pilot blasted. Full face: tunnel machine (82% of length). Hand driving with arch erector (18% of length)
Aids to excavation	Cement and clay/cement grouting and local clay blanket	Compressed air, ground treatments locally	Multi stage ground treatments, compressed air	Compressed air, ground treatments general	Nil	Nil	Local grouting
Normal rate of driving (m/week)	6–9	3–10	5–6	$3\frac{1}{2}$–$7\frac{1}{2}$	25–31	10	1st tube: 18–21 2nd tube: 30–31
Approximate cost per linear metre of tunnel (excavation and lining as percentage)	£5 320 (82%)	£6 150 (82%)	£7 150 (75%)	£4 400 (70%)	£1 910 (48%)	£1 256 (77%)	1st tube: £4 720 (58%) 2nd tube: (estimated £4 600) (60%)

Information by courtesy of Mott, Hay and Anderson, and Sir William Halcrow and Partners.

system, A will be high, while for a labour-intensive system C will be high. For excavation there will be an appreciable element in spoil disposal costs for which C will predominate while for temporary tunnel supports A and B will be the principal factors.

As the size of tunnel is reduced, the increasing congestion leads to reduced efficiency in working. In consequence there is a size of tunnel for which the costs will be a minimum; the greater the degree of mechanisation, the greater will be the size d_{min} for minimum cost (i.e. B and $C \rightarrow 0$ as $d \rightarrow d_{min}$). For a long length of tunnel in the London clay the minimum cost is obtained for a tunnel diameter of about 100 inches (2.54 m) while for certain machine-driven tunnels in soft rock the optimum diameter has been found to be 10–11 ft (about 3 m), and about 7–8 ft (2 m) for a hand-driven tunnel in hard rock.

SYSTEMATIC SITE INVESTIGATION

Geological data

The scheme for determining the geological conditions should work from the general towards the particular. This will entail a study of geological maps and papers, first on a regional and then on a local basis. In Britain there are normally available sheets at scales of 1 in ≡ 1 mile and 2½ or 6 in ≡ 1 mile with explanatory memoirs, produced by the Institute of Geological Sciences. Where geological maps do not exist, aerial photographs often provide useful information on the geological structure.

Objects

According to the apparent options for the tunnel the scheme of site investigation may then be designed with these main objects:

(a) to test geological data at doubtful points
(b) to explore particular areas of tunnelling difficulty
(c) to obtain information necessary to complement available data on important aspects of geology and geohydrology
(d) to obtain samples for testing and to undertake *in situ* tests in order to establish the suitability of ground for alternative methods of tunnelling
(e) to determine design and construction parameters.

Means

A few large boreholes or adits may be justified for direct examination, *in situ* testing and for subsequent inspection by tendering contractors and others.

There is no general rule on the spacing between boreholes. At one extreme, for sedimentary rocks of a uniform character it may only be necessary to be able to establish general continuity of the geological sequence by identification of marker beds or horizons. At the other extreme, igneous intrusions and metamorphosed rocks may present so complex a pattern as to necessitate a tunnelling method highly tolerant to change, however well the ground may be investigated. A good general rule is to establish during site investigation a set of hypotheses, concerning the geological structure and the properties of the ground, to be tested so that when a conflicting anomaly is indicated by a new borehole its significance is appreciated (i.e. is the benefit of an additional borehole likely to exceed its cost?). Where there is doubt concerning the practicability of adopting a mechanical system of tunnelling, special care is required to ensure exploration of the ground in sufficient detail to determine the feasibility of the scheme.[8]

Geophysical methods of exploration may serve not only to extend the data from individual boreholes in the second and third dimension but also to reveal specific features such as faults and igneous intrusions. Without adequate 'fixes' geophysical results may permit widely different possible interpretations.

TUNNELLING METHODS RELATED TO THE GROUND

Historical background

The history of tunnelling is one of increasing diversification of methods with an increasing capability to explore and to understand the ground.

While Brunel used the first tunnelling shield for the Thames Tunnel in 1825–28, tunnels throughout the nineteenth century continued generally to be constructed by means of one of the traditional methods of excavation and timbered support.[9] Although these are now largely of historical interest only, the English method, widely and successfully used, sometimes in soft ground and in broken jointed rocks where other methods had failed, merits mention.

An essential feature of the English method concerned the use of longitudinal crown bars, supported at the forward end on props and sill and at the rearward end on the last section of completed permanent lining, which might be brickwork or masonry. In this way continuous support was provided to the ground over the tunnel from the time of first excavation and, in principle, the method may be considered as the forerunner of the tunnelling shield.

Shield tunnelling

Shield tunnelling is strongly associated with the name of Greathead. He worked with the first circular shield designed by Barlow for the Tower Subway beneath the River

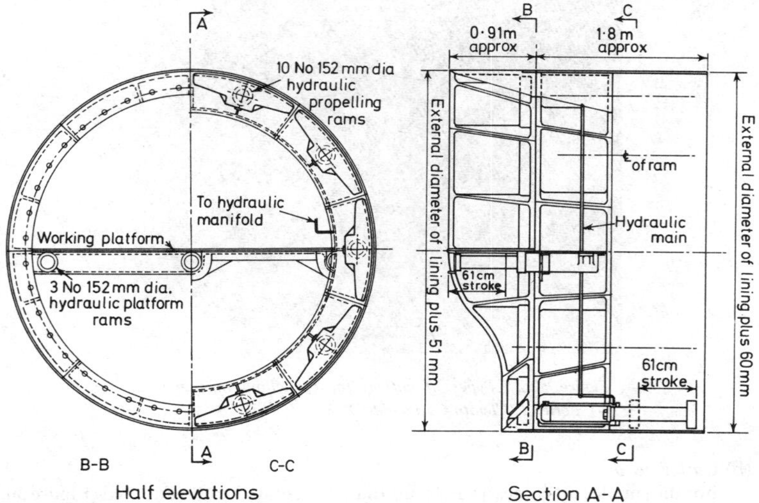

Figure 30.1 Hooded Greathead shield with platform rams suitable for 3.5 m diameter tunnel

Thames in 1869. For the South London Railway in 1886–90 Greathead designed a shield (Figure 30.1) incorporating most of the essential features which have survived to the present day.[10] Greathead not only recognised that a shield reduced the risks in tunnelling in water-bearing ground but he was also one of the few of his time to appreciate that it permitted faster and cheaper tunnelling in good ground.

The first shield with a mechanical cutting head was the Price excavator used in 1897 for the Central London Railway. Numerous types of rotary machine have been developed subsequently;[11] while these generally assume a stable face to cut against, several recent machines have been used in soft clay, silts and sands with a plated face to provide support, incorporating adjustable cutters which rotate or oscillate.[12]

Rock tunnelling machines

Many tunnelling machines for rock have been evolved since 1956, although here again the prototype machine belongs to the last century, usually attributed to Beaumont,[13] used for Channel Tunnel heading in 1881–82 and subsequently for the Mersey Railway Tunnel.[14] There are several features of such machines[15] which merit differentiation:

(a) *Cutters* (see Figure 30.2)
For the softest rock the cutters are fixed picks which chisel the rock out as a succession of grooves. For harder rocks, generally in ascending order of hardness, machines make use of single or multiple disc cutters, toothed cutters, roller cutters or cutters with tungsten carbide insert buttons.

Figure 30.2 Types of cutters for tunnelling machines (a) 'Series 12' Tooth Cutter–MNX;

(b) *Cutter heads*
For the smaller machines a single full-diameter rotary head is adopted (Figure 30.3). As the machine size increases so is there a tendency to introduce planetary cutters to share the work between cutters and reduce the range between minimum and maximum cutter speeds.

(c) *Thrust of machine*
For the softest rock, machines receive purchase by rams thrusting against a gripper ring expanded against the periphery. With a few exceptions the remainder of the

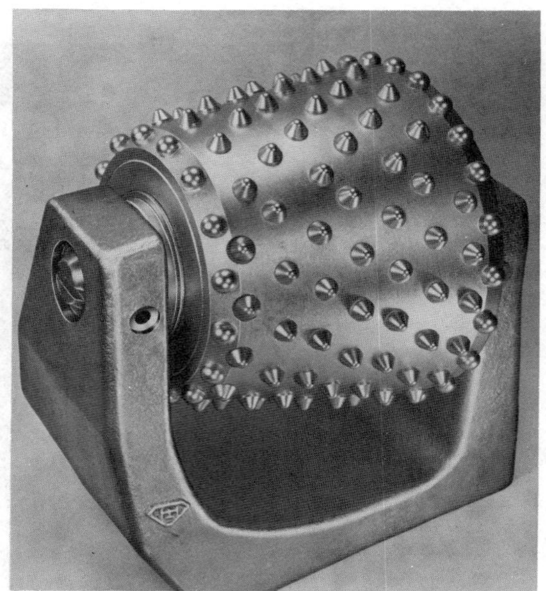

Figure 30.2 (b) 'Series 12' HHIX Cutter;

Figure 30.2 (c) Bolt-on-Disc Cutter, type DGX
(Courtesy: Hughes Tool Co.)

Figure 30.3 Cutter head of hard rock machine (Courtesy: The Robbins Co.)

Figure 30.4 Atlas Copco full face boring machine (Courtesy: Atlas Copco)

machines obtain their forward thrust by means of diametrically opposed thrust pads jacked against the ground. All such machines advance by periodically withdrawing and repositioning the thrust ring or pads. One machine uses a central pilot drill which is firmly anchored into a hole ahead of the face. Not only does this provide a means for pulling the machine forward but it also establishes a firm forward bearing for the cutter head which may be a valuable feature where rock variation in the face causes uneven loading on the head.

There are in addition three other types of machine meriting special mention:

(i) Machines which use planetary heads with picks assembled to form a milling cutter with the axis at a slight angle to the machine axis (Figure 30.4). Excavation is then by undercutting with relatively little axial thrust. Consequently, for soft rock such machines may be track mounted.

(ii) A machine which mounts rotating drum heads, traversing along one or more rotating arms (Figure 30.6). Each head has picks on its face and on its periphery. Thus, the ratio of cutting thrust to forward thrust is low and an additional feature is this machine's capability to excavate noncircular galleries, up to 5.2 m high by 3.7 m wide.

Figure 30.5 Road header in iron ore mine (Courtesy: Anderson Mavor Ltd)

(iii) The road header, a machine developed originally for mining, has a rotary milling head on a telescopic boom, attached to the body of the machine by a universal joint (Figure 30.5). Thus a typical machine may excavate a gallery up to 4.5 m high and 5.8 m wide. The cutter head usually mounts picks in the pattern of a conical scroll.

The selection of a tunnelling machine must take account of the rock types to be encountered along the length of drive. The efficiency of the machine is related to the inherent properties of the rock, principally to strength, the extent of jointing, strain modulus and abrasion. The aim is to minimise the specific energy needed to fragment a rock by keeping the size of particle high and by provoking brittle fracture. The cutter action aims to set up a high local difference in principal stresses and high enough tensile stresses to induce cracking. Each type of cutter and pattern of cutters operates most efficiently in a rock whose properties lie within a limited range. Difficulties may arise

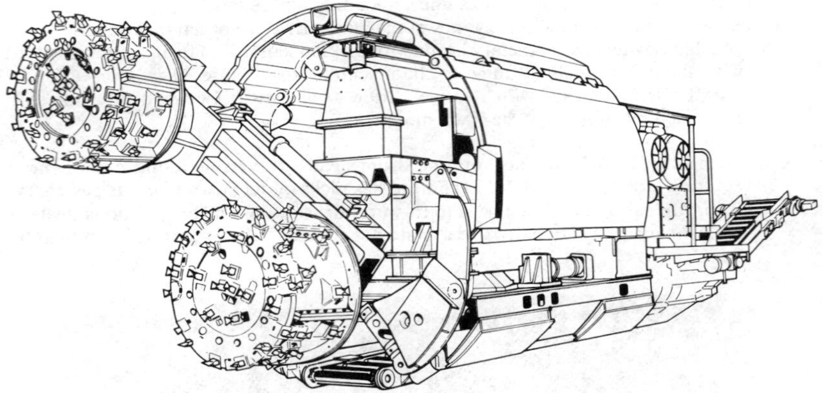

Figure 30.6 Greenside-McAlpine rock tunneller (Courtesy: Sir Robert McAlpine and Sons Ltd)

from several causes; for example from excessive wear of the cutters in hard rock which grooves instead of fragmenting, from inefficient fracture of rock too soft or plastic for the type of cutter, from excessive wear due to overheating in the presence of high content of silica, from excessive bearing loads where rock varies appreciably in the face, from jamming of the head where hard rock is heavily jointed and tends to collapse onto the machine or at the face.[7b]

A simple criterion for the economics of machine excavation concerns the cost of repairing and replacing cutters. In sound soft rock this will be found to be a trivial sum in relation to other costs of excavation. For the hardest rock, the cost will be found to climb to a figure of several pounds per cubic metre, with stoppages every few metres for replacement and this stage represents the present economic limit.

The tolerance of a machine to the full range of rock types to be encountered should be considered in weighing the overall merits and costs of its introduction. Another essential question concerns prediction of the need for temporary support close to the face, a process that presents greater difficulty for the full face machines and for which object certain machines make special provision in their design.[5]

TUNNEL CONSTRUCTION

Drilling and blasting

The traditional scheme of advancing rock tunnels has been by drilling and blasting and this method continues to be generally adopted for short tunnels, hard rock tunnels and for tunnels in variable ground. At the present date, for example, machine tunnelling is unlikely to be economic in shattered rock or in rock of strengths greater than 200 MN/m^2.

The principle behind blasting in a tunnel is to obtain the greatest 'pull' for the minimum explosive charge and for the minimum damage of the rock around the tunnel. Secondary objectives are to fragment the rock adequately and to form a compact stock pile against the face.

The pattern of drill holes is designed to suit the rock and the explosive. Cut holes are arranged towards the centre of the face, usually inclined towards each other in order to remove a cone or wedge. One or more central unloaded holes of larger diameter may be used to assist the cut. The remainder of the holes are drilled parallel to the tunnel axis. Delays of a few milliseconds are used between groups of drill holes, from the cut outwards, so that the excavation is enlarged with the travel of the shock wave.

Considerable effort has been applied to establishing the neatest periphery to the excavation by the trimming holes, which may be charged or uncharged. In the technique of pre-splitting, the trimming holes are fired before the remainder with distributed charges to cause cracking around the periphery between adjacent holes. Another technique which has been used in tunnelling is termed smooth-blasting, whereby the line of trimmer holes is required to coincide with the periphery of the excavation, each being loaded with a reduced distributed charge and fired with a short delay after the remainder. Generally, experience shows that neither method is economic for tunnelling. It may nevertheless be well worth considering means for reducing overbreak by careful control of the spacing, line and charging of the trimmer holes. The geometry of the drills or the drill carriage should be designed to permit the trimming holes to be drilled as parallel as possible to the tunnel axis. Care in these respects may show considerable benefit not only in reduction of direct overbreak but also in the reduced extent of the surrounding zone of cracking and displacement of the rock, with consequent savings in the extent of temporary support.

Sectional drawings of tunnels usually indicate the periphery of the 'minimum section' and the 'payment line' which allows payment for overbreak to be assessed in relation to the volume of excavation and the volume of concrete lining. Occasionally a 'limit line' is also shown, beyond which a leaner concrete mix may be used for filling. Overbreak in a tunnel is frequently expressed as a percentage of sectional area but, without knowledge of the size of tunnel, this designation has little significance. It is more helpful usually to describe overbreak to the arch (and possibly to the invert separately) in terms of average distance beyond the minimum section.

For small tunnels, hand-operated drills are used on telescopic air-legs. For larger tunnels there is usually a wider choice, including ladder drills, light mobile boom-mounted drills or heavier drills mounted on a jumbo. The latter may provide advantage in controlling the drill pattern and with the speed of drilling, also in protection close to the face for other operations; the main disadvantage arises from inflexibility in the event of departure from full-face driving. For the Mont Blanc Tunnel, for example, it was fortunate that a jumbo was used only from the French end, since difficulties encountered along the Italian drive compelled the enlargement from headings over a considerable length of tunnel.[16]

Spoil handling

The handling of spoil from the face cannot be considered separately from the method of excavation. Mechanical shields and tunnelling machines have built-in chain or belt conveyors loading to a hopper or to another conveyor. The same operation is achieved in a drill-and-blast tunnel by means of a mechanical loader, often with composite face shovel and conveyor. The general trend is to use rail wagons for transport for tunnels up to about 7 m diameter and for tunnels worked from vertical shafts and to use dump trucks for large tunnels directly accessible from the surface or for tunnels at a gradient of more than about 2.5%.

Conveyors are also used for dry materials and where access is by inclined shaft. The pulverising of spoil and its discharge by pipe as a slurry has been adopted for suitable soft rock. Frequently the bottle-neck in materials handling is found to occur at the foot of a working shaft and here mining practice has introduced the use of automatic tipping of tunnel wagons into large hoppers from which shaft skips are rapidly loaded. The entire process of excavation and removal of spoil merits considerable study at an early stage as to its adequacy, with contingency plans to overcome foreseeable causes of breakdown. Many expedients are available to handle wagons at the face for the orderly loading of spoil. The 'cherry-picker' traverses the wagon by overhead lift, the California crossing provides a crossover slid along the tracks. The more elaborate sliding platform, advanced in sections by hydraulic jack, may extend for several hundred metres and, among other benefits, may eliminate the need to lay trackwork near to the faces.

Tunnel lining

The method of tunnel lining is essentially related to the nature of the ground and to the scheme of excavation. General purpose tunnel lining has economic application to small tunnels in variable ground. Recent progress and attendant economy have been demonstrated to result from the capability for designing the lining specifically to the condition of the ground and the overall tunnelling system.

The first subdivision in type of lining results from whether or not the need exists for an immediate support at the face. In the United States the common practice in tunnelling in soft ground has been to tunnel by hand, to erect continuous support in timber sets or steel liner plates and subsequently to place an *in situ* concrete lining. In Britain, and generally throughout Europe, shields have been more widely used together with permanent primary segmental linings.

The traditional lining over more than one hundred years has been the ring of bolted cast iron segments built within the protection of the tail of the shield, with the external annulus often grouted with lime or cement. Improvements in site investigation procedure have allowed the development of alternative types of lining which can be adopted in certain restricted types of soft ground.

Reinforced concrete segments[17] have been preferred to cast iron segments for reasons of cost since 1938 except where loading is heavy or where watertightness is an essential object.

The Moscow method of tunnelling utilises reinforced concrete segments in sand with the intention, not realised in practice, of synchronising the filling of the external annulus by cement grout with the advance of the shield.

Another general type of tunnel lining is built in rings of segments immediately behind the shield. Each ring is then expanded directly against the ground with elimination of the procedures of bolting and grouting. Evidently the system can only be used where the ground around the tunnel is selfsupporting over the width of a ring for a short period and thus a certain minimum apparent cohesion of the ground is necessary. Many such techniques have been developed in the London clay, each being applicable to a certain range of tunnel diameter.[18] Two types of lining based on this principle merit mention.

The Donseg lining is created from rings of tapered segments, expansion against the ground being achieved by the process of inserting alternate segments, as longitudinally tapered keys, into the ring by the shield rams (Figure 30.7). This is a highly economic method, limited to tunnels of diameter not exceeding about 3 m, because of the geometry of the lining.

For larger tunnels, the Halcrow lining provides for articulating joints between segments. In this way, a part ring of segments may be assembled clear of the extrados. The insertion and expansion of jacks between special segments cause the ring to expand against the ground, accompanied by relative rotation between adjacent segments (Figure 30.8). A special feature of a lining of this type is that secondary stresses are limited to a low level with consequent savings in the structural thickness of the lining.

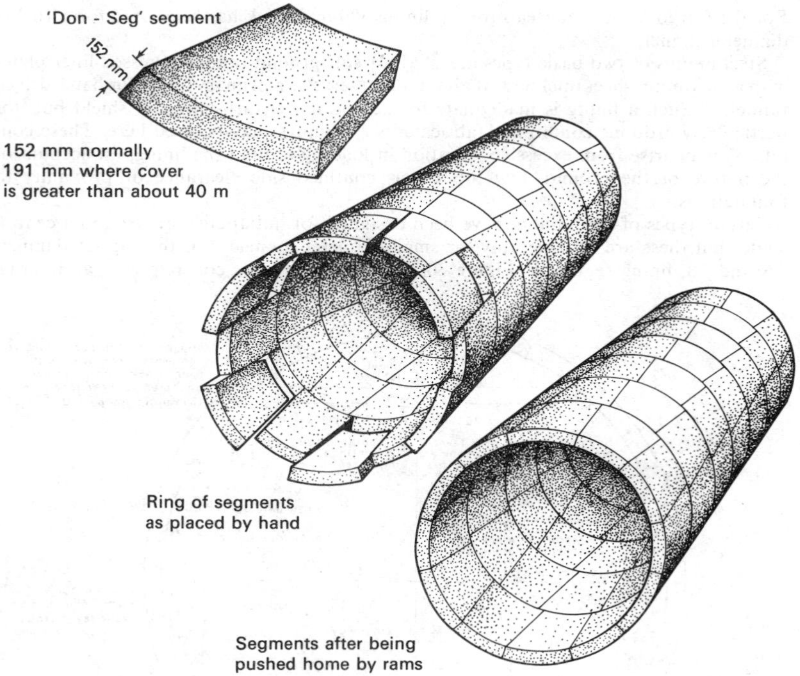

'Don - Seg' segment

152 mm

152 mm normally
191 mm where cover
is greater than about 40 m

Ring of segments
as placed by hand

Segments after being
pushed home by rams

Figure 30.7 Donseg tunnel lining

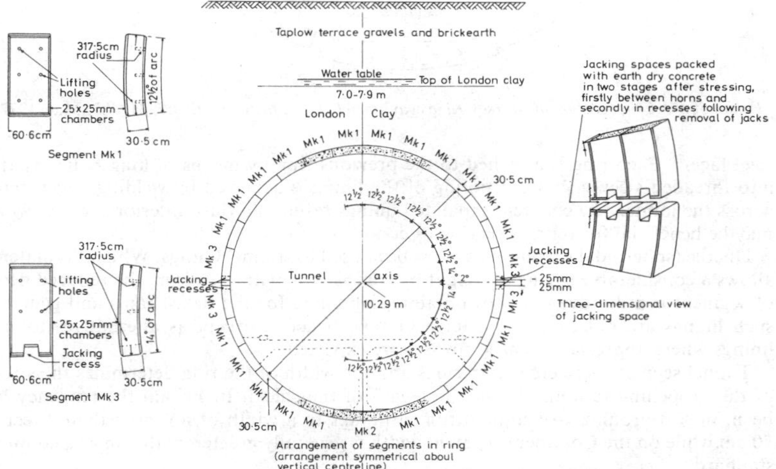

Figure 30.8 Lining for Cargo Tunnel at Heathrow Airport, London

For the Cargo Tunnel at Heathrow[5] a lining 300 mm thick has been used for a 10.3 m diameter tunnel.

Steel linings of two basic types are used for soft-ground tunnels. Pressed liner plates with a maximum sheet thickness of about 8 mm serve as a primary lining for hand-driven tunnels.[19] Such a lining is inadequate for accepting the thrust from a shield but, for particularly arduous conditions, fabricated steel linings may be used here. These conditions may arise from excessive variation in loading around the lining, on account of the nature of the ground, low top cover, confined side clearance or proximity to foundations.

Several types of flush lining have been designed for initial erection around a central spider but these are suitable only for small-diameter tunnels. For the Mersey Tunnels 3A and 3B, lining segments (Figure 30.9) are made in mass concrete with an internal

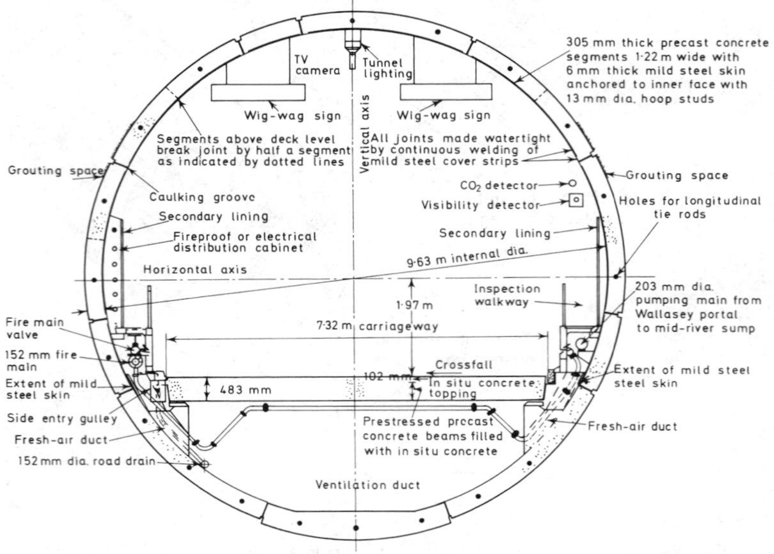

Figure 30.9 Cross section of Mersey Kingsway tunnel (*Proc. Instn civ. Engrs,* **51,** 487, Mar. 1972)

steel face.[7b] Each ring is attached to the previous ring by means of long bolts inserted into threaded sleeves. Waterproofing of the lining is achieved by welding cover plates across the joints. The concrete expanded linings result in a flush interior surface, which may be beneficial for tunnels serving as conduits.

Ductile (spheroidal graphite) iron has been used for tunnel linings. While this material allows a considerable saving in weight by comparison with grey iron, the reduced depth of segment is a disadvantage for obtaining purchase for the thrust rams and generally such linings are not found to offer economic benefits, except as alternatives to steel linings where appreciable tensile stresses are expected.

Tunnel segments are erected in rings and the width of the ring determines the stroke of the propelling ram and hence the length of the shield. In Britain the tendency has been, in soft ground, to maintain tunnel linings to a width of no more than 2 feet or 70 cm while on the Continent segment width is generally greater, with 1 m as a common standard.

Rock tunnels are usually lined *in situ* with concrete placed behind shutters. The lining may be cast in discrete lengths or continuously behind telescopic shutters travelled forward in a retracted mode. Concrete is usually pumped, with placers possibly used for filling the crown. Subsequent contact grouting is usually necessary to fill shrinkage cracks and voids between lining and rock.

Thrust-boring

Thrust boring of tunnels[20] has developed from pipe jacking, whereby lengths of steel pipe are pushed through the ground, from a jacking pit, with the addition of a new length of pipe at the rearward end after each extension of the jack. Thrust-bored tunnels are frequently in the form of reinforced concrete elements. The limiting distance of thrust boring depends upon the ground, the geometry of the tunnel and the capacity of the jacks. This may be extended by the use of an external lubricant such as bentonite or by using intermediate jacking points to control the maximum length to be advanced at a time. For small pipes excavation is often by continuous-flight auger; for larger tunnels, excavation may be by hand or by small mechanical excavator. In unstable ground the pressure at the face may be balanced by keeping the cutting edge buried.

Waterproofing

Waterproofing of rings of tunnel segments has been traditionally by caulking between the joints and by grummets placed beneath the washers at bolts. Usually it is considered uneconomic to attempt to make a tunnel 100% dry. The caulking material has been metallic lead strip or, for less severe conditions, asbestos cement compounds used dry. Polyalkathene is a suitable material for grummets, since it flows under pressure as the bolt is tightened into the space to be sealed. Where water pressures are low, sealing over joints and bolts may be achieved by glass fibre reinforced epoxy resin sheets built up *in situ*. Schemes have been devised for achieving a watertight gasket between segments but these usually provide inadequate tolerance to the dimensional variations. Such a scheme has been used, with apparent success, in conjunction with epoxy resin grouting at the corners.[21]

Temporary support

In rock tunnelling, the permanent lining cannot be considered separately from the scheme of excavation and temporary support. The initial stability of the excavated ground depends not only upon the inherent quality of the rock but also on the method and quality of the excavation process. Generally, mechanical excavation will not only provide a better shaped arch around the tunnel but, more important, also much less disturbance of the surrounding rock. Recent studies have indicated that blasting may cause cracking of the rock up to a diameter outside the tunnel.

The essence of good tunnelling in jointed rock is to provide adequate support to incipiently collapsing rock as soon as possible. The means for achieving this end are directly related to the nature of the rock and its jointing. The situation may be summarised thus:

(a) Where the rock is highly shattered or with frequent open joints, effective support may require the use of heavy arches. These must be provided with adequate foot supports to avoid punching into the invert and must be blocked off the rock sufficiently frequently to avoid excessive bending stresses.[22] Arches of the yielding type,[23] designed originally for colliery support, are now widely used in tunnels in recognition of their ease in erection and the virtual equivalence of their major and minor second moments of area, and hence greatly reduced tendency to distort.

There are sometimes advantages if the support system allows some initial controlled relaxation of the ground so that natural arching may occur; the utmost caution must be observed.

(b) Where the rock is subject to progressive deterioration or to surface weathering, an immediate application of concrete or mortar may provide great benefit. A thin application of pneumatically applied mortar (gunite) or fine concrete (shotcrete)[24] will often serve in this respect, applied preferably to enter open crevices between blocks so that an adequate arch is provided around the tunnel. A somewhat heavier and more expensive version with the same general object may be provided by an initial concrete lining placed against the newly exposed rock, possibly behind perforated steel sheeting supported by arches.[25] There are often great advantages in the reduction of overbreak if support of this nature can be applied so close to the face as to receive benefit of the three-dimensional dome that occurs here. There is also a certain time dependence of the tendency for collapse from a tunnel roof; thus a great deal of the barring down of an unstable tunnel roof can frequently be avoided by immediate support.

(c) The action of rock bolts in supporting the ground around a tunnel depends upon the nature of the jointing.[26, 27] Where joints are open in a 'tension zone', rock bolts will require to be anchored in sound rock beyond. Elsewhere it is generally more effective to use the rock bolts to form a reinforced zone of rock, with the rock bolts serving the purpose of shear reinforcement. Where the rock cleaves predominantly along a single series jointing or bedding planes crossing the tunnel at a low angle, the rock bolts may serve to tie together the laminations to form a continuous slab. Where the pattern is more complex, the bolts may serve to form a reinforced arch. Bolts should be inserted to a regular pattern, recognising that local surface variations of the rock will not be significant over the depth of bolting. The effective action of one bolt depends upon the spacing to its immediate neighbours.

There are many types of rock bolt but these may conveniently be considered in two groups: those which rely upon end anchorage, usually by some method of mechanical expansion of the end of the bolt, and those which are keyed along their length. The latter type are preferably deformed bolts, either set in cement grout or in epoxy resin or similar adhesive; the cement grout may be introduced in an expanded metal cage around the bolt or through a tube or hollow bolt. The end anchorage bolt is the cheaper expedient and is more readily stressed but, in soft or weathered rock the anchoring should be achieved by a resin bonding. The head of the bolt should be fitted with plate washers or a short length of channel to spread the load adequately over the surface of a soft rock. Progressive failure of a jointed rock may be controlled by a wire mesh between bolts acting as a containing cage which is also a useful safety measure. Evidently the effective depth of a bolted rock arch or slab depends upon the bolt size, length and spacing, the length usually requiring, for overall economy, to be twice the spacing or more.

Rock bolting and shotcreting are often used in association, the former providing the major support, the latter controlling surface deterioration without which aid the bolts would be effective for a short time only. Surface cracking of shotcrete provides early warning of continuing movement of the ground.

Advance by full face or by heading

A first consideration in excavating a large tunnel concerns the practicability of full-face excavation. This will depend upon the stability of the rock in relation to the tunnel size and upon the need for any advance heading to explore the ground and to provide an opportunity for undertaking ground treatment ahead of the main excavation. In very large tunnels it may be economic to excavate a top heading first in order to insert supports for the crown and then subsequently to work the invert section as a vertical bench. A variation to such a method may utilise a bottom heading in addition, serving for drainage and for removal of spoil.[28]

In swelling ground (i.e. in rock containing a montmorillonitic clay) the difficulty in support may be roughly expressed as proportional to the area of the tunnel. In consequence there may be considerable benefits in utilising a series of small headings around the periphery of the tunnel in which the permanent lining for the full arch and invert is cast section by section, as for the Straight Creek Tunnel, Colorado.[29]

AIDS TO TUNNELLING

Compressed air

The application of the use of compressed air to soft ground tunnelling is another development associated with Greathead and the South London Railway (1886–90).[10]

In soft clays, compressed air will provide direct support to the ground. In silts and sands the compressed air displaces the greater part of the pore water and causes cohesion between grains of the soil by surface tension. The effect allows running sands to be treated in excavation as a soft rock. Another side effect of compressed air in the ground is to reduce its permeability to the flow of water (by as much as an order of magnitude for silts).

The use of compressed air necessitates a considerable outlay in low-pressure compressors, air coolers, air locks (including a medical lock) and the associated control and monitoring system. Even a momentary loss in air pressure might have fatal consequences and hence the need for a high degree of duplication and stand-by equipment. The working conditions in compressed air owe a great deal to pioneering studies by Professor J. S. Haldane, leading to a set of recommendations by the Institution of Civil Engineers later revised and issued as a set of regulations under the Factory Inspectorate.[30] Comparable standards have been evolved in other countries which undertake tunnelling or caisson work in compressed air.

Compressed air introduces increased direct and indirect costs, the latter arising from the reduction in effective working time and the period spent in 'locking out' which may for instance increase from about $\frac{1}{4}$ hour to $\frac{3}{4}$ hour for a 6-hour shift as the working pressure (measured above atmospheric) rises from 1 to 2 atmospheres. The upper limit, without special air mixtures, is about 3 atmospheres. It is recognised that it is necessary for strict medical supervision to be provided for workmen in compressed air.[31] For many years it has been known that the amount of nitrogen dissolved in the blood is related to the period of exposure to a given pressure so that if the pressure is lowered too rapidly bubbles are formed, particularly at the joints, leading to the condition known as 'the bends'. More recently compressed air has become associated with a more serious complaint, that of bone necrosis, which leaves the victim more-or-less crippled. One reaction to this discovery has been to resort to other forms of aid in order to dispense totally with the use of compressed air. A more reasonable attitude appears to be to discover the causative process and to eliminate the offending factor, since alternatives for compressed air may not only entail high cost but also introduce new hazards. In the past, before bone necrosis was associated with compressed air, medical inspection of workmen passed them fit to work in compressed air without attention being given to any latent defect of the bones or joints, an oversight that should not recur for future compressed air working.

Compressed air has been used for many subaqueous tunnels in soft ground. The problem of balancing the external water pressure increases with the height of the tunnel as well as the size of the tunnel. The depth below the water surface and the texture of the ground will determine the quantity of air required. In coarse sand a rule of thumb for determining the maximum demand has been stated as 24 D^2 cfm (or 7.5 D^2 m^3/min) where D is the tunnel diameter[32] in feet (or metres).

Where the ground comprises clay interbedded with thin layers of sand or silt it has frequently been found that a relatively low ratio between the pressure of air and the external head of water is adequate to provide greatly improved stability to a tunnel face.

In open ground, air losses may be reduced by locally sealing the exposed face, for which purpose bentonite dust has been used. A further problem area arises, in the construction of a segmental tunnel lining behind a tunnelling shield, in the avoidance of collapse of the ground onto the lining immediately behind the tail of the shield. This

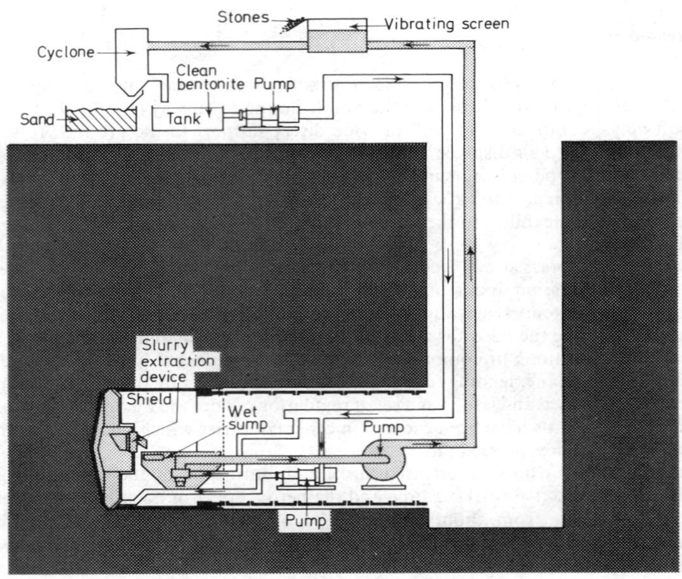

Figure 30.10 Bentonite tunnelling machine (After National Research Development Council)

has been countered by grouting with bentonite through the skin of the shield in order to increase the capability of supporting the ground by compressed air. An alternative method has been to fill the annular space with pea gravel as the shield advances, by no means easy to perform satisfactorily.

The bentonite shield

It is possible, where frequent access is not required to the face of a shield, to provide for ground support by compressed air, water or mud confined to the face of the shield. Air is not however recommended for this purpose. The use of mud in this application offers considerable benefits in permitting an approximately balanced pressure over the full height of the face and in providing a suitable medium for the pumping away of spoil. The first such shield was used in Mexico City,[33] utilising the mud formed from the natural montmorillonitic clay spoil of the area. Subsequently, a true bentonite tunnelling machine has been used successfully in London (Figure 30.10) in sands and gravels. The most advantageous application is in ground of so open a nature as to present difficulty in retaining compressed air, but the method also has promise for finer soils.

Ground treatments

A wide choice of grouting media[34] is now available for consolidating weak or water-bearing ground:

(a) Setting grouts containing cement, bentonite, fly ash and other materials may be selected, at the lowest cost compatible with adequate travelling capability for the dimensions of pores and joints to be filled. Bentonite may also be used on its own as a lubricant for the extrados of the shield skin, for thrust bored tunnelling and for shaft sinking. Bentonite mixtures are thixotropic, i.e. they form a gel in the absence of shearing motion.

(b) Chemical grouts are used in medium to fine sands, single chemical systems having a time-dependent control of setting and two-chemical systems, of which the Joosten is the most familiar process, depending on contact between the two components.

(c) For silty sands, resin grouts may be used, low viscosity grouts being available for permeabilities down to about 10^{-3} cm/s.

Generally the finer the ground and the lower its permeability the more expensive the grouting process. The principle in grouting variable ground is therefore one of working through the available grouts from cements, clays, chemicals and resins as appropriate, so that the cheaper grouts are used to confine the travel and hence the 'take' of the more

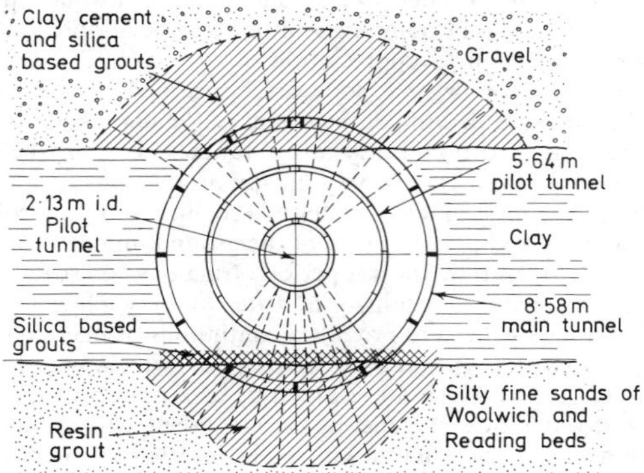

Figure 30.11 Pattern of ground treatment for north end of second Blackwall Tunnel, London (*J. Instn civ. Engrs*, **35**, 19, Oct. 1966—with acknowledgement to the Council of the Institution of Civil Engineers)

expensive grouts (Figure 30.11). In fine material, electrochemical grouting may be used in tunnels in the future; it has already been used for foundations.[35] In this process, electro-osmosis accelerates the rate of penetration of the chemical agent through the ground.

Freezing

Freezing has had a longer history as an aid to sinking mine shafts than for civil engineering applications.[36] It has been used in civil engineering works predominantly for situations of unusual difficulty and for installations using vertical freezing-holes sunk from the surface. In each hole is inserted a U-tube or, alternatively, a composite freezing tube with concentric inner and outer tube through which cold brine is circulated, usually down the inner and up the outer tube. The brine is usually used at a temperature of about −20 °C but this may be reduced to −35 °C.

Freezing was widely used for construction of the Moscow underground in the early 1930s, for vertical shafts and for inclined escalator tunnels. For several lengths of recent sewer tunnel in Germany freezing has been adopted with horizontal freezing holes. A freezing operation with the use of brine usually occupies several weeks after the installation of the tubes and equipment.

Freezing has been used for six shafts for the Ely–Ouse water scheme,[37] each requiring control of groundwater to a depth of 25–65 metres below surface.

> The cost of freezing (1969) was about £560/metre for a 4.5 m i.d. shaft and £740/metre for a 7.5 m i.d. shaft.

A new development has entailed the use of liquid nitrogen as the freezing agent. Since the operating temperature may then be lowered to −150 °C the freezing operation occurs rapidly and the process has frequently been used for penetrating relatively thin bands of water-bearing ground during the sinking of shafts.

Dewatering

Control of water for tunnelling may be achieved by water-table lowering by pumping or by diverting the water as a tunnel is lined. The first requires no further explanation here beyond the observation that pumping continues to be adopted in association with urban tunnelling with inadequate appreciation of the risks of settlement to adjacent buildings, particularly where organic soils are concerned. In certain circumstances recharging wells may be used to control the extent of the depression of the water table.

If the water is permitted to flow freely into a tunnel there may be a risk of ground settlement but in rock tunnels this is only an important consideration in crushed or altered fault zones. While major flow of water will require to be controlled in consideration of pumping capacity and deterioration of the ground, minor flow will only present problems for the lining operation. Over a period of many years several expedients have been devised for the diversion and control of water to allow placing of the lining. One of the successful methods has been to provide a continuous protection of plastic sheeting around the tunnel supported on panels of steel mesh with longitudinal french drains along each side of the invert which are grouted up as a final operation. An alternative arrangement, where water flow is general but not great, will use a quick-set mortar pneumatically applied onto steel mesh with pipes inserted at intervals to concentrate the water flow. The pipes are stopped off on completion of lining or, occasionally, allowed to flow where permanent drainage and pressure relief are intended.

Where water is confined locally to joints it may be adequate to form a stopping of flash-set mortar around a flexible tubular former to provide a drainage path.

GROUND MOVEMENTS

Excavation for a tunnel may give rise to associated ground movements for two principal reasons. These may either be caused by over-excavation, leaving cavities beyond the space occupied by the lined tunnel, or by release of original stresses in the ground, giving rise to elastic or plastic deformation towards the tunnel.

In rock, over-excavation may occur from roof collapses or from failure to line solidly against the ground. Rock falls may develop to domes, arches or chimneys depending upon the nature of the ground and the pattern of initial stresses. High horizontal stresses for example will tend to limit the extent of the cavity. As rock fractures it increases in bulk; in consequence, once a plug is provided at the base of the cavity its upward extent will be limited and may be approximately calculated.

Even small cavities immediately behind the lining are serious in that they may lead to uneven loading on the lining and consequential failure; hence the need for systematic contact grouting. In soft incompetent ground, over-excavation will usually be transmitted in full to the surface approximately to equate to the volume of surface settlement. However, in dense sand the total settlement at the surface will be reduced; in loose sand, settlement may occur as a result of disturbance by tunnelling even in the absence of any over-excavation. It is often impossible in soft ground to subdivide the effects of over-excavation and of changes in the stress pattern, the latter tending to give rise to loss of ground towards the exposed face.

The shape of the 'bowl' of settlement at the surface will usually be influenced by loss of ground along a length of tunnel somewhat greater than its depth below surface, the influence factors being highly dependent upon the geological structure. In homogeneous soft ground and for a tunnel advanced with consistent standards of design, workmanship and progress, a characteristic depression will develop over the tunnel which may be described approximately in terms of the shape of statistical normal distribution curves.[38] Approximately half the total settlement will have occurred immediately above the advancing tunnel face, for tunnels at no great depth.

Tests undertaken during the construction of tunnels in London clay indicate that with increasing depth, there is a greater tendency for loss of ground arising from deformation towards the advancing face. In general, the contribution to loss of ground may be as set out in Table 30.2.

Table 30.2 CONTRIBUTION TO LOSS OF GROUND AROUND A SHIELD-DRIVEN TUNNEL

Nature of ground loss	*Computation*	*Normal limits* (%)
Ground loss at face	$\pi d^2 h/4$	0.1–?
Ground loss behind cutting edge	πdt	0.1–0.5
Ground loss along the shield	$\pi l\, v/8$	0–1
Ground loss behind the tail	$\pi d(d-d_0)/2$	0–4
	($\pi d(d-d_0)/4$ above water table)	0–2

where the loss per unit length is expressed as a percentage of area of tunnel face and d is the diameter of the shield, d_0 is the external diameter of the lining, t is the relief behind the cutting edge, v is the 'look up' of the shield measured as the extent of out of plumb on vertical diameter, l is the length of shield and h is the horizontal movement of ground at the face per unit length of advance of shield.

A special cause for settlement over a tunnel may occur where a shield in soft ground can only be kept to correct line by means of maintaining an appreciable 'look up' on account of a tendency to settle at the cutting edge. This loss may be countered to some extend by grouting above the shield as it advances, with fly ash or similar material.

TUNNEL DESIGN

Stresses around a tunnel

The state of stress in real ground around a full-size tunnel during the course of construction is too complex to analyse fully. A more rewarding process is to idealise the problem to a certain degree and then, by inference and judgement, determine the significance of inadequacies of the conceptual model.

We start by considering the two-dimensional problem of a long unlined circular tunnel pierced instantaneously at great depth in perfectly elastic ground. We can in this instance build up the overall stress pattern around the tunnel by superposition of its constituents.[39] The initial vertical loading will be redistributed and will set up the circumferential and radial principal stresses σ_T and σ_R shown in Figure 30.12 for the vertical and horizontal axes. At the periphery

$$\sigma_R = 0 \tag{1}$$

and at axis and crown level

$$\sigma_T = 3\sigma^* \quad \text{and} \quad \sigma_T = -\sigma^* \tag{2}$$

respectively where σ^* was the initial vertical loading in the ground. A similar set of relationships may be obtained for the horizontal loads $N\sigma^*$. For ground loaded from

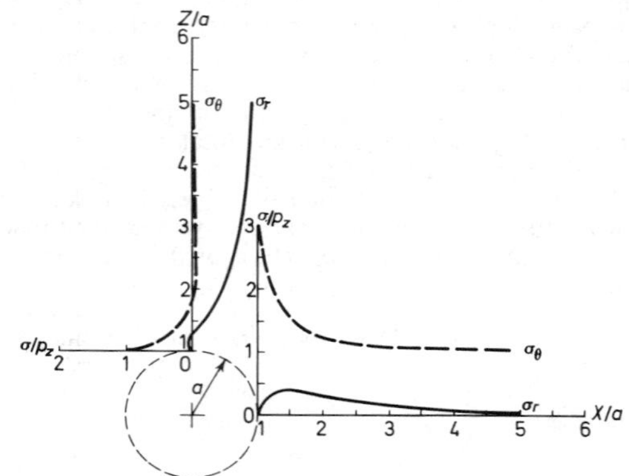

Figure 30.12 Stresses around circular tunnel in elastic ground initially stressed in vertical direction only

above and laterally constrained, it can readily be shown that $N = \nu/(1-\nu)$ where ν is Poisson's ratio. For ground loaded and then subjected to reduction of vertical loading, N may be greater than unity and, indeed, in over-consolidated ground where appreciable surface erosion has occurred N, according to the circumstances, may be 2, 3 or more. Evidently if $N = 1$, equation (2) indicates by superposition that $\sigma_T = 2\sigma^*$ around the periphery.

The most important departures from this simple model may be caused by:

(a) nonelastic behaviour of the ground
(b) limiting ultimate strength of the ground
(c) inability of the ground to accept tension
(d) discontinuities in the ground.

The simplest nonelastic model is that for ground assumed to behave elastically up to certain limiting differences between maximum and minimum principal stresses (generally the stress parallel to the tunnel axis may be considered as intermediate between the other two) and thereafter to deform perfectly plastically. For example a jointed rock might be considered as elastic for stresses lying within the Mohr's envelope with plastic deformation occurring at the limiting shear stress of

$$T = C' + \sigma_N \tan \phi' \tag{3}$$

The stress pattern around a circular tunnel[40] might then be represented as Figure 30.13. It will be noted that full development of the plastic zone will entail appreciable movement of ground into the tunnel and theoretical considerations suggest delay in supporting ground to reduce to a minimum the load on tunnel supports (see p. **30**–18). In most tunnels, the object is to provide support as rapidly as possible and then to consider merits of systems that will yield noncatastrophically at excess loads. A diagram such as Figure 30.13 permits examination of the reduction in plastic movement as a result of increased σ_R at the periphery of the tunnel by means of ground support.

It is suggested that, at low containing stresses, jointed rock fails by riding over irregularities on the joints and that these are increasingly sheared through under higher

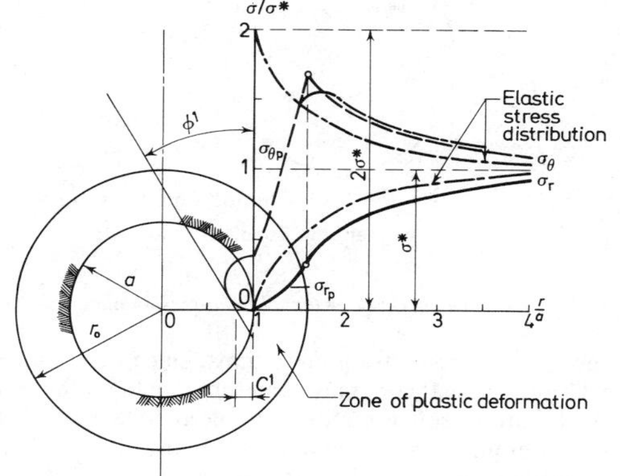

Figure 30.13 State of stresses around a circular tunnel in ground yielding to plastic and elastic strains (N = 1) (After Kastner[40])

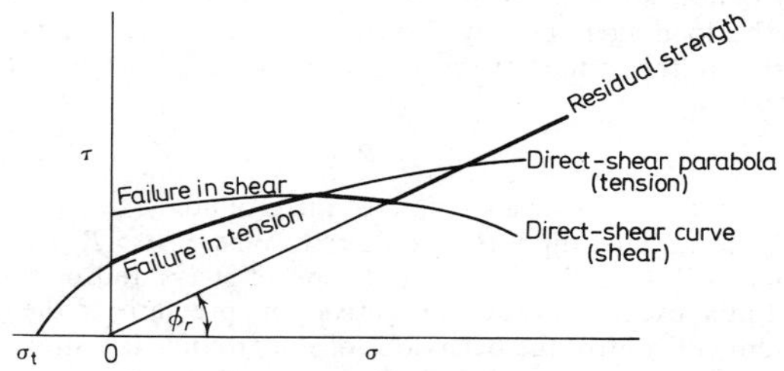

Figure 30.14 Failure limits for rock (After Lajtai[41])

loading.[41] Ultimately the strength of the rock in true plastic yielding may be represented as a purely frictional material, giving a limiting strength line as in Figure 30.14, to be reproduced in an analysis of the failure of the ground around a tunnel.

From equation (2) it is readily seen that $3 > N > \frac{1}{3}$ will cause no tension in elastic ground. Outside these limits, tension zones will occur for a circular tunnel. We can alternatively treat the ground as a no-tension material and consider the ellipse circumscribing the tunnel which will satisfy this condition. Supports must then be provided for the zone of material between the circular tunnel and this ellipse (Figure 30.15) beyond which the ground is considered as intact.

The degree of knowledge of the ground, its homogeneity and the extent of the tunnel will determine the length to which simple analysis, models and numerical methods may appropriately be applied to defining the economic basis for tunnel design. The method of finite elements[42] has almost unlimited potential for solving this class of problem but the cost of the solution increases rapidly with increasing complexity. Great care is

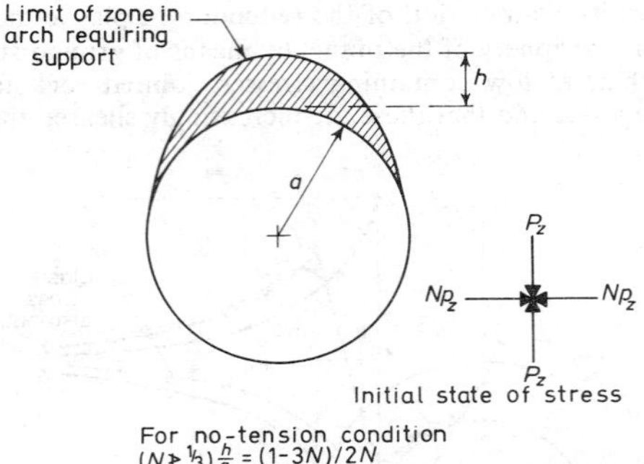

Figure 30.15 Circular tunnel in no-tension material

required for the nonlinear stress–strain conditions, since the result depends upon the loading sequence. Critical state theory may soon throw light on the behaviour of a tunnel at the condition of failure in soft ground.[43] Simple models have yielded much insight on the behaviour of nonuniform or jointed rock[26] and, for qualitative solutions, inexpensive physical models should not be overlooked.

Competence of the ground

The competence of a rock is a measure of its capacity to resist deformation under a given loading. Since the loading is usually directly related to the overburden, it appears helpful, in classifying the ground from the view point of tunnelling, to define a competence factor[44] as

$$F_c = P/\sigma^* \tag{4}$$

where P is unconfined compressive strength of the ground.

Where $F_c < 2$, immediate support is required. Where $10 > F_c > 2$, stability of unsupported ground will depend on the initial state of stress and on stress–strain–time characteristics. Initial excavation causes negative pore pressures in the ground and their rate of dissipation will control the behaviour of the ground. If a simple analysis points to local overstressing of the ground, a decision is required as to the need for support before dissipation can occur or whether advantage may be taken of the development of a plastic zone to reduce maximum level of stressing. Where $F_c > 10$, the ground will be competent and the real problem concerns discontinuities and joints, pre-existing or caused by tunnelling.

Stiffness of the tunnel lining

The tunnel lining and the surrounding ground should be treated as a composite structure when considering states of stress and deformations. A question of first importance concerns the relative stiffness of the lining and the ground it displaces. For elastic conditions for the simplest, 'elliptical', mode of deformation of a circular tunnel,[45] we may express this stiffness ratio as

$$R_s = 3EI/a^3\lambda \tag{5}$$

where E and I relate to lining stiffness, a is tunnel diameter and λ the appropriate ground strain modulus, whereby the share of imposed loading borne by the tunnel lining may be explored.[42] It can readily be shown that for a normal load $p = p_o \cos 2\theta$ between ground and tunnel (i.e. a load varying continuously between maxima and minima on perpendicular axes), the bending moment M in the lining per unit length is

$$M = p_o a^2 R_s/6(1+R_s) \tag{6}$$

and diametral deflection

$$\Delta = 2a^2 M/3EI \tag{7}$$

This approach immediately indicates the wastefulness of designing over-stiff linings which result in unnecessarily high values for R_s. In Britain, in recent years the benefit of articulating linings in soft ground have been appreciated and exploited. Thin flexible linings are also becoming more widely recognised as desirable for tunnels in soft rock.

Tunnel geometry

In the softest ground ($R_s \ll 1$) a circular tunnel is economic; where $R_s \sim 1$ there is greater tolerance in departures towards an elliptical cross section; the economic section, if permitted by use and construction, is elliptical with the major axis parallel to the major initial principal stress.

For rock, the geometry of the tunnel may be related within reasonable limits far more directly to the use and the method of construction; generally, there are benefits in support costs if the ratio of height to width take some account of the initial stress field in the ground. Except in swelling and fragmented rock, the invert may be flat or slightly dished, with possible advantages in construction. The construction options require to be carefully considered before establishing a definitive tunnel section. Especially where a tunnel has no invert lining, the conditions of stress in the ground in the vicinity of sidewall footings require checking. Water pressure, grouting or other treatment may directly influence the loading pattern on the lining, or indirectly affect it by virtue of resulting modification to effective stresses in the ground.

Towards a better understanding

The advance in techniques of tunnelling is a continuing process and the results of monitoring the behaviour of a tunnel by comparison with prediction provide valuable correction to design techniques and assumptions. Great skill is required in interpreting the quantities of data that may be obtained in this way in a comprehensible manner without introducing the risks of oversimplification. Such monitoring will normally be concerned with strains in the ground and with deformations of, and stresses in, the lining.[6] Generally the simplest and most robust instruments should be used to avoid disappointment from incomplete results arising from damage.

REFERENCES

1. KELL, J., 'The Dartford Tunnel', *Proc. Instn Civ. Engrs*, **24,** 359–372 (Mar. 1963)
2. MORGAN, H. D., HASWELL, C. K. and PIRIE, E. S., 'Clyde Tunnel design, construction and tunnel services', *Proc. Instn Civ. Engrs*, **30,** 291–322 (Feb. 1965)

3. KELL, J. AND RIDLEY, G., 'Blackwall Tunnel duplication', *Proc. Instn Civ. Engrs.*, **35**, 253–274 (Oct. 1966)
4. FALKINER, R. H. and TOUGH, S. G., 'The Tyne Tunnel 2: construction of the main tunnel', *Proc. Instn Civ. Engrs.*, **39**, 213–234 (Feb. 1968)
5. MUIRWOOD, A. M. and GIBB, F. R., 'Design and construction of the cargo tunnel at Heathrow Airport, London, *Proc. Instn Civ. Engrs*, **48**, 11–34 (Jan. 1971)
6. LYONS, A. C. and SCOFIELD, J., 'Great Charles Street Road Tunnel', *Tunnels & Tunnelling*, **1**, 23–26 (May/June 1969)
7a. MEGAW, T. M. and BROWN, C. D., 'Mersey Kingsway Tunnel: planning and design', *Proc. Instn Civ. Engrs*, **51**, 479–502 (Mar. 1972)
7b. MCKENZIE, J. C. and DODDS, G. S., 'Mersey Kingsway Tunnel: construction', *Proc. Instn Civ. Engrs.*, **51**, 503–533 (Mar. 1972)
8. GRANGE, A. and WOOD, A. M. MUIR, 'The site investigations for a Channel Tunnel', 1964–65, *Proc. Instn Civ. Engrs*, **45**, 103–123 (Jan. 1970)
9. SZECHY, K., 'The art of Tunnelling (1st edn in English), Akadémiai Kiadó, Budapest, 891 (1966)
10. GREATHEAD, J. H., 'The City and South London Railway; with some remarks upon subaqueous tunnelling by shield and compressed air', *Instn Civ. Engrs Papers on London Underground Railways* 1885–1929, Paper 2873, 39–73 (Nov. 1895)
11. CLARK, J. A. M., HOOK, G. S., LEE, J. J., MASON, P. L. and THOMAS, D. G., 'The Victoria Line—some modern developments in tunnelling construction', *Proc. Instn Civ. Engrs* (suppl.), Paper 7270S, 397–451 (1969)
12. German Soil Mechanics Congr. in Hamburg, *Ground engineering*, 19–20 (Jan. 1969)
13. 'English boring machine', *The Engineer*, **55**, 455 (June 1883)
14. FOX, F., 'The Mersey Railway', *Proc. Instn Civ. Engrs*, **86**, 40–49 (May 1886)
15. NICHOLSON, W. E., 'Big bits drill big', *World Mining:* Part 1, 41–48 (Sept. 1967); Part II, 44–51 (Oct. 1967)
16. SANDSTRÖM, GOSTA E., *The history of tunnelling*, Barrie & Rockliff, 427 (1963)
17. GROVES, G. L., 'Tunnel linings with special reference to a new form of reinforced concrete lining', *J. Instn. Civ. Engrs*, **20**, 29–42 (Mar. 1943)
18. DONOVAN, H. J., 'Modern tunnelling methods', *J. Instn Publ. Health*, LXVIII, Part 2, 103–130 (Apr. 1969)
19. APEL, F., *Tunnel mit Schildvortrieb*, Werner-Verlag, 293 (1968)
20. RICHARDSON, M. A., 'Pipeforcing: An appraisal of 10 years of operation', *Tunnels and Tunnelling*, 215–219 (July 1970)
21. RICHARDSON, C. A., 'Constructing a soft-ground tunnel under Boston Harbour', *Civ. Engng (ASCE)*, 42–45 (Jan. 1961)
22. PROCTOR, R. V. and WHITE, T. L., *Rock tunnelling with steel supports*, Commercial Shearing and Stamping Co., Ohio (1946)
23. CUNLIFFE, J. and JOHNSTON, A. G., 'Roadway support with special reference to yielding arches', *Trans. Instn Min. Engrs*, **117**, 805–818 (1958)
24. ALBERTS, C. and BÄCKSTRÖM, S., 'Instant shotcrete support in rock tunnels', *Tunnels and Tunnelling*, **3**, No. 1, 29–32 (Jan. 1971)
25. WÖHLBIER, H., 'Der Ausbau unterirdischer Hohlräume mit S und A Bleche System', *Bergbauwissenschaften*, **16**, 117–126 (1969)
26. LANG, T. A., 'Theory and practice of rock bolting', *Trans. Am. Inst. Min., Metall. and Petrol. Engrs*, **220**, 333–348 (1962)
27. PENDER, E. B., HOSKING, A. D. and MATTNER, R. H., 'Grouted rock bolts for permanent support of major underground works', *J. Inst. Aust. Civ. Engrs*, **35**, 129–150
28. ANDERSON, D., 'The Construction of the Mersey Tunnel', *J. Instn Civ. Engrs*, **2**, 473–516 (Mar. 1936)
29. GLOVER, E. F. and O'REILLY, M. P., 'Tunnelling in the USA', *Tunnels and Tunnelling*, **3**, 431–437 (Nov.-Dec. 1971)
30. 'The work in compressed air', Special Regulations (Statutory Instruments No. 61), Ministry of Labour (1958)
31. MCCALLUM, R. I. (Ed.), *Decompression of compressed air workers in civil engineering*, Oriel Press, 329 (1967)
32. HEWETT, B. H. M. and JOHANNESSON, S., *Shield and compressed air tunnelling*, McGraw-Hill, 465 (1922)
33. HARRIES, D. A., 'Constructing the deep-level drainage system of Mexico City', *Tunnels and Tunnelling*, **3**, 35–42 (Jan./Feb. 1971)
34. ISCHY, E. and GLOSSOP, R., 'An introduction to alluvial grouting', *Proc. Instn Civ. Engrs*, **21**, 463–465 (Mar. 1962)

35. CARON, C., 'Consolidation des terrains argileux par électro-osmose', *Ann. de l'Inst. Tech. du Bâtiment et des travaux publics,* **285,** 75–91 (Sept. 1971)
36. MUSSCHE, H. E. and WADDINGTON, J. C., 'Applications of the freezing process to civil engineering works', *Inst. Civ. Engng Works Constr.,* Paper 5 (1946)
37. COLLINS, S. P. and DEACON, W. G., 'Shaft sinking by ground freezing: Ely Ouse-Essex Scheme', *Proc. Instn Civ. Engrs,* Paper 7506S (Feb. 1972)
38. MUIRWOOD, A. M., 'Soft ground tunnelling', *Technology and potential of tunnelling,* **1,** 167–174, **11,** 72–75, Johannesburg (1970)
39. TERZAGHI, K. and RICHART, F. E., 'Stresses in rock about cavities', *Geotechnique,* **3,** No. 2, 57–90 (1953)
40. KASTNER, H., 'Statik des Tunnel- und Stollenbaues', Springer-Verlag, 2nd edn, 269 (1971)
41. LAJTAI, E. Z., 'Shear strength of weakness planes in rock', *Inst. J. Rock Mech. Min. Sci.,* **6,** No. 5, 499–515 (Sept. 1969)
42. ZIENKIEWICZ, O. C., VALLIAPPAN, S. and KING, I. P., 'Stress analysis of rock as a "no tension" material', *Geotechnique,* **18,** 56–66 (Mar. 68)
43. SCHOFIELD, A. N. and WROTH, C. P., *Critical state soil mechanics,* 1st edn, McGraw-Hill (1968)
44. MUIRWOOD, A. M., 'Tunnels for roads and motorways', *Q. J. Engng Geol.,* **5,** 111–126 (1972)
45. MORGAN, H. D., 'A contribution to the analysis of stress in a circular tunnel', *Geotechnique,* **11,** 37–46 (1961)

31 CONTRACT MANAGEMENT AND CONTROL

CONTRACT MANAGEMENT AND CONTROL 31

31 CONTRACT MANAGEMENT AND CONTROL

P. H. D. HANCOCK, M.A., F.I.C.E., A.M.B.I.M.
George Wimpey & Co.

INTRODUCTION

How to use this section

The reader is recommended to start by reading up to page **31**–6. However, if he needs to refer to some particular point in a hurry he should turn direct to Figure 31.1 which is an index to the section in the form of a grid showing how the various aspects of contract management interlock.

Contract Management is about getting people to do things and is therefore a less clear-cut subject than most in this book. Any opinions expressed or implied are the author's own.

Essential references

Contract Management is such a large subject that this chapter can only cover the ground by referring frequently to two relevant Institution of Civil Engineers' (ICE) publications:

Civil Engineering Procedure[1]

ICE Conditions of Contract[2]

Civil Engineering Procedure[1] is a concise account of the normal procedures for managing civil engineering work in the UK. This section concentrates more on underlying principles but matters that are well covered in reference 1 are treated very briefly here. The serious reader should equip himself with his own copy of both publications. Neither is expensive.

Of course the reader will encounter other procedures and other forms of contract but the above together with this section should enable the reader to orient himself to whatever system he finds himself required to operate.

Four other very useful general references[3–6] are included in the list at the end of this section.

The purpose of this section

Contract Management is like swimming in that you don't learn much until you get into the water and start splashing about. Some would go further and say that you only start to learn when you are out of your depth! This section is designed to make the painful process of learning by experience quicker and less chancy by giving the learner a framework of concepts which will show him what he can learn from his work and help him to store what he learns in an orderly fashion.

Planning & controlling: Progress (Programming, Progressing)	Planning & controlling: Quality (Specs, Quality control)	Planning & controlling: Safety	Planning & controlling: Economy (Estimating, Tendering)	Planning & controlling: Economy (Costing, Valuations, Financing)	Planning & controlling: Constraints (Certainties, Risks, Insurances)	Organising & activating (Roles, Relationships, Procedures)	Managing resources: Site (Ground, Climate, Access, Services, Temp installns)	Managing resources: Information (Drawings, Specs, BOQ, Contract, Records etc.)	Managing resources: Men (Staff, Labour)	Managing resources: Machines (Plant, Transport, Tools & Tackle)	Managing resources: Materials	Managing resources: Subcontractors	Managing resources: Third parties	Problem solving, Improving, Innovating (Management techniques)
														81 to 84
													80 to 81	81 to 84
												76 to 78		81 to 84
											71 to 73	80		81 to 84
										10	75	80		81 to 84
									65 to 67 69 to 71		45,71,74, 75 to 76	80	67 to 68	81 to 84
								56 to 61	56 to 58, 67,68	65	76	80		81 to 84
							8 to 10, 56	33,52,64	68		73 to 74	79 to 80		81 to 84
						53 to 56		63 to 64	53 to 56 69 to 71		71 to 75	78 to 79	19	81 to 84
					6 to 8, 50 to 53		50 to 52	36 50 to 51	35 50 to 51	50 to 51	50 to 51	50 to 51	50 to 51	81 to 84
				6 to 8, 37 to 42, 46 to 50	50	47		38,47,48, 49,58 to 60	35,41 to 45, 46	35 42 to 44	35,38,42,45 to 46 74 to 75	42,46		81 to 84
			6 to 8, 31 to 33, 36	31,36	35 to 37	33	33	33,37,38, 47,58 to 60	33,35	35	35	35		81 to 84
		6 to 8, 23 to 24, 29	30 to 31	30,31	24 to 30	24 to 26	26	27	26 to 27 45	27	28	28	28 to 29	81 to 84
	6 to 8, 19 to 20		20	20	21	20 to 21	21	21 to 22	21 to 22 45	65	22	22 to 23		81 to 84
6 to 8, 10 to 13			33 to 34	15,33,34, 47		14 to 15	8 to 10 15	15 to 16	16	16	17 to 18	18 to 19	19	81 to 84

Figure 31.1 The contract management process—grid index to this Section. Numbers in boxes are page numbers in the text. To examine all the aspects of a topic, refer to the pages numbered in the relevant heavy box and in all the boxes ranged horizontally and vertically from that box (see pp. **31**–2 *and* 6)

A second purpose is to help the engineer with wide experience in one sector of the industry to understand better the activities and attitudes of those in other sectors.

A simple picture of the total process

Contract Management is the middle phase of a total process which begins when a need first starts to take shape in someone's mind and ends when the completed road, bridge, dam or whatever, becomes part of the landscape.

This process is most conveniently shown by a simple precedence diagram (Figure 31.2) which shows along the top the three parties contractually involved and, down the page, the principle activities in their logical sequence. This diagram is based closely on reference 1 where the reader will find a straightforward description of the roles of Promoter, Engineer and Contractor and of the activities covered by each box.

For the purposes of this section, Contract Management is defined as that part of the total process which starts with the preparation of Contract Documents and ends when the Contractor gets his final payment.

The reader must bear in mind that in real life many of the activities overlap and so the sequence is seldom as simple and clear-cut as it appears in Figure 31.2. For example it is almost universal for the Contractor to be paid in stages as the work proceeds, it is common for design and construction to develop more or less in parallel and it is not unknown for construction to start before a contract has been signed or even before the Promoter has acquired all his land and raised his funds!

Roles may vary

The division of tasks is not always as clear-cut as it appears in Figure 31.2. There are several variants of which five are sufficiently common to merit mention here.

1 *Direct labour:* See reference 1, para. 42, Twort[4] and Lovell[16]
2 *All-in contract:* See reference 1, paras. 41 and 49 and Twort[4]
3 *Engineer a permanent salaried employee of the promoter:* His position under the ICE contract[2] is unaltered but in practice he often finds it difficult to maintain the degree of independence implied by Figure 31.2 and recommended by reference 1, para. 19.
4 *Promoter uses a consulting engineer but fields his own strong technical team:* The motive is sometimes understandable but this arrangement always demands special care to avoid dangerously divided responsibilities. In bad cases both Promoter and Contractor can find themselves having to cope with an engineer who is a two-headed monster.
5 *Management contract:* Under this arrangement, an experienced consultant or contractor takes on much of the Promoter's role and manages his contract for him for a fee, engaging and coordinating the work of design consultants and construction contractors. This mode of working is fairly new in this country and it seems most appropriate for large complex jobs which would otherwise require the Promoter to make a large and continuous contribution to the running of the job. See Frame.[12]

A valid model whatever the contractual arrangements

The reader who is accustomed to some kind of nonstandard project procedure (see previous paragraph) must not be put off by the fact that this section is based on the procedure and uses the terminology set out in Figure 31.2. Whatever may be the contractual

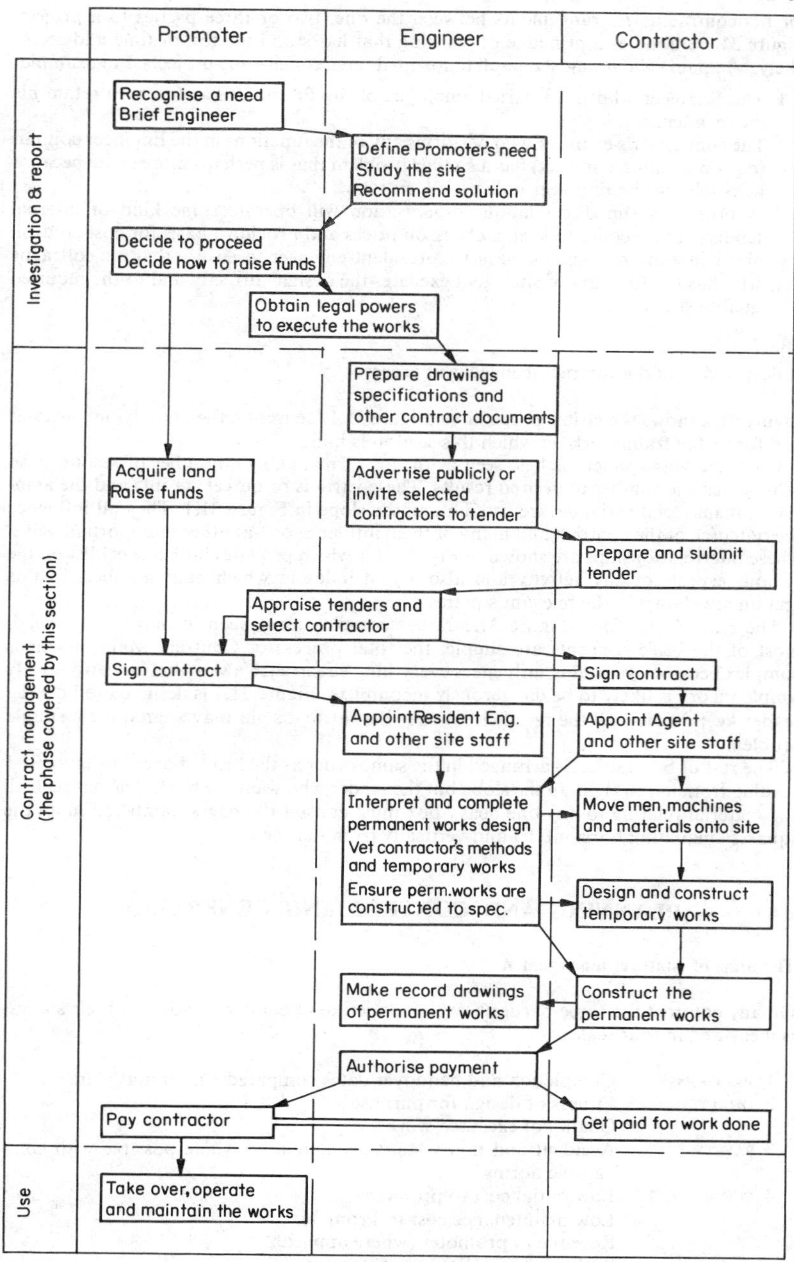

*Figure 31.2 Civil engineering procedure—simplified diagram showing how civil engineering work is normally organised (see p. **31**–4)*

(or noncontractual) arrangements between the one, two or three parties to a project, Figure 31.2 embodies a principle of working that has stood the test of time and seems likely to remain valid for most one-off tailor-made civil engineering projects. For example:

1 The Engineer who is a salaried employee of the Promoter has to try to retain his independence.
2 The contractor's engineer who performs any of the functions in the Engineer column (e.g. on an all-in contract) has a similar problem that is perhaps more acute because he is seldom the Engineer under the Contract.
3 A prudently run direct labour organisation will operate some kind of internal tendering procedure to keep a check on prices and provide a basis for cost control, and will appoint someone akin to a resident engineer to ensure that his colleague who has the function of site agent executes the design correctly and to the required quality standards.

A deeper view of the contract management process

Figure 31.2 shows the visible procedures. Figure 31.1 focuses on the underlying purposes and forms the framework on which this section is hung.

Contract Management is best seen as the art of managing a number of resources so as to achieve a number of desired results. These various resources, results and the associated managerial activities are listed along the slope in Figure 31.1. They all influence the outcome of the contract and many of them influence one-another in important ways. These interrelationships are shown in Figure 31.1 which provides both a checklist on the various aspects of any activity and also a grid index in which page numbers of this section are shown in the relevant squares.

The reader who finds Figure 31.1 rather complex must keep in mind that though most of the basic concepts are simple, the total process of Contract Management is complex because everything influences everything else in such a way that any attractively simple theory is likely to be dangerously incomplete. Figure 31.1 is designed to help the reader keep in view all the ramifications of what at first sight may seem quite a simple problem.

The rest of this section is arranged in the same order as the heavy boxes in Figure 31.1 reading from left to right up the slope but the reader who wants to be sure he has seen all the material relating to any one heavy box must consult the pages numbered in all the squares emanating horizontally and vertically from that box.

PLANNING AND CONTROLLING GENERALLY

The aims of contract management

On any contract the three parties have to cooperate to common ends and their success is measured in four scales.

1 PROGRESS	Completion and handover dates compared with initial plan
2 QUALITY	Fitness of design for purpose Quality of executed work
3 SAFETY	Accident and injury statistics compared where possible with comparable norms
4 ECONOMY	Low initial cost to promoter Low maintenance cost to Promoter Revenue to promoter (where applicable) Profit and cash flow to contractor

Obviously these four objectives conflict, and perfection on any one usually requires unrealistic sacrifices on the other three. Moreover, the various parties often have different ideas of what is the 'right' standard to aim at. Contract Management is the art of achieving that balance between progress, quality, safety and economy which best satisfies the three parties involved. (See for example reference 1, para. 78.)

Planning and controlling to achieve the desired results

For each desired result there are well established planning and control systems with familiar names. These are set out in Table 31.1.

Table 31.1

Desired result	*Planning systems*	*Control systems*
PROGRESS	Programming	Progress control
QUALITY	Drawings Specifications	Quality control
SAFETY	Law on Safety Contract clauses Site instructions	Safety statistics
ECONOMY	Estimating Budgeting Cash flow forecasting Profit forecasts	Cost control Budgetary control Cash flow control Financial control

These various systems have certain features in common which are discussed subsequently.

Notes on each individual system are given in separate paragraphs later in the section and these may be located from the grid index, Figure 31.1.

Planning and control—definitions

The words 'planning' and 'control' unfortunately mean different things to different people. For the purposes of this section their meanings are defined as follows.

Planning means deciding what you want to achieve and how you propose to achieve it. It obviously includes the selection of methods and sequences.

Control means comparing what is with what ought to be. In the European languages other than English, the word means this and no more but unfortunately in English it is also used in the wider sense of 'commanding' or 'influencing' as in 'control of labour'. It is only because the same word can in English mean two such fundamentally different things that a clear definition is needed for this section.

Planning and control necessary and inseparable

No purposeful human activity, however simple, can start without a plan of some sort because as the Irishman said, 'If you don't know where you want to go you aren't very likely to get there'. By the same token, a plan without control is pretty well useless because if you don't know where you are you can't decide how to get where you want to go!

In practice most people will agree that you need a contract plan of some sort but, they will mostly disagree on the amount of effort that should be put into preparing it. It is human nature to hope for the best and let things take their course in the knowledge that a crisis when it comes, will make it much easier than it is in normal times to impose one's will on people. The answer to this is that management by crisis is a fine sharp tool if kept for exceptional situations but if it has to be used every day it soon gets blunt.

So, one object of planning control is to goad people into doing things before they become crises. More haste more cost!

The essence of the planning and control process

Planning and control are inseparable strands of one process which is best summed up in a simple flow chart, Figure 31.3.

This shows that any plan of action contains three essential elements:

1 A statement of what it is desired to achieve and the methods proposed (Box 3.1). See also below.

2 A statement of the difficulties or 'constraints' that are expected and allowed for (Box 3.3). See also pages **31**–50 and 51.

3 Following from 1 and 2, a statement of the resources that will be needed (Box 4). The broad categories of resources appear as main headings in the grid index Figure 31.1.

All this may seem so obvious as to be hardly worth saying. Yet it is surprising how often a plan goes wrong because those who made it have failed to answer properly one or other of the questions in Figure 31.3.

Planning and control a continuous process

Figure 31.3 shows that contract planning and control is a continuous process of providing for the future with wisdom gained from studying the past. At the outset of a contract one has to rely mainly on one's own previous experience and the stored wisdom of one's own organisation, but as soon as the work is under way one starts the process of comparing achievement with plan, and replanning where necessary, and this process never stops until the completed works are handed over and paid for.

If the reader can keep always in mind this continuousness of the planning and control process, he will avoid the pitfall which so often gives 'planning' a bad name. This is the notion that one can, at the outset of an undertaking make a plan in great detail for months or years ahead and then expect things to work out exactly as planned. Nothing could be more wrong. Planning a contract is like planning a journey. One knows where one is trying to go, but there are many details that can only be settled as one goes along. In fact, the whole art of successful planning is to identify and make those decisions that need to be made and to leave those that are better left until nearer the event.

Methods, plant, temporary works and site layout inseparable from planning but covered elsewhere

From Figure 31.3 and the foregoing, it is obvious that planning cannot be separated from the process of deciding construction methods, plant, site layout and temporary

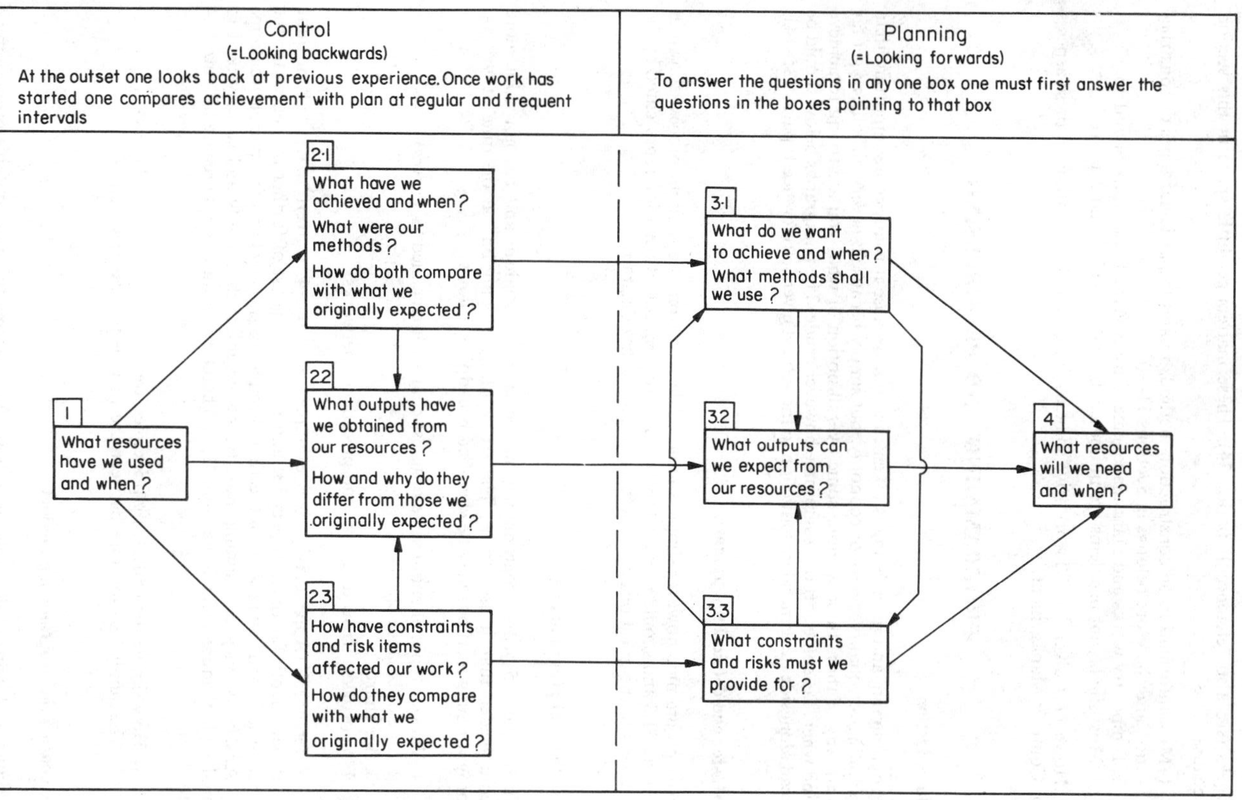
Control
(=Looking backwards)
At the outset one looks back at previous experience. Once work has started one compares achievement with plan at regular and frequent intervals
Planning
(=Looking forwards)
To answer the questions in any one box one must first answer the questions in the boxes pointing to that box
1
What resources have we used and when?
2·1
What have we achieved and when? What were our methods? How do both compare with what we originally expected?
2.2
What outputs have we obtained from our resources? How and why do they differ from those we originally expected?
2.3
How have constraints and risk items affected our work? How do they compare with what we originally expected?
3·1
What do we want to achieve and when? What methods shall we use?
3.2
What outputs can we expect from our resources?
3.3
What constraints and risks must we provide for?
4
What resources will we need and when?

Figure 31.3 The essentials of the planning and control process (*see p.* **31**–8)

works (see also reference 1, para. 134). These matters get little space in this section because:

1 Most information on construction methods relates to particular types of construction and therefore belongs in Sections 11–30
2 Temporary works and plant management are covered in Sections 34 and 35
3 Most site layout problems are intimately bound up with 1 and 2 above

However, Calvert[3] and Twort[4] give some useful general information on these aspects of Contract Management.

PROGRAMMING AND PROGRESSING

Introduction

Programming and progressing is taken first, not because it is any more important than planning for quality, safety or economy, but simply because it is very difficult to get far with any of these three without some overall notion of what you are trying to achieve and when, i.e. some sort of programme however crude. The paragraphs below should be read alongside the general introduction to the subject given in reference 1, paras 135–141.

Master and detailed programmes

Any programme however simple should be derived from the thought process shown in Figure 31.3 and should provide for recording the control information represented by the answers to the questions in the left-hand half of that diagram.

Methods of programming

In the past 15 years programming has become a specialism and has got a lot more complicated than it used to be. The art is still developing but, at the time of writing, programming methods may be classified broadly as follows:

1 Various types of cumulative progress charts (often called 'S' curves)
2 Line of balance charts
3 Bar charts
4 Networks (arrow diagrams and precedence networks)

Reference 1 (Appendix 3) gives examples of 3 and 4. Calvert[3] covers all in quite sufficient depth for the nonspecialist and gives a large list of further references for the specialist. Twort[4] gives some interesting examples of special charts for special situations.

Any book on programming must treat the subject thoroughly and so may tend to make it look more difficult than it really is. Therefore the author has prepared Figure 31.4 to show

(i) How simple the underlying principles are
(ii) The family relationships between the various programme methods.

Choosing the programming method

Figure 31.4 helps to illustrate the criteria for choosing the most suitable programming method for any particular piece of work. These criteria are summarised below. It is

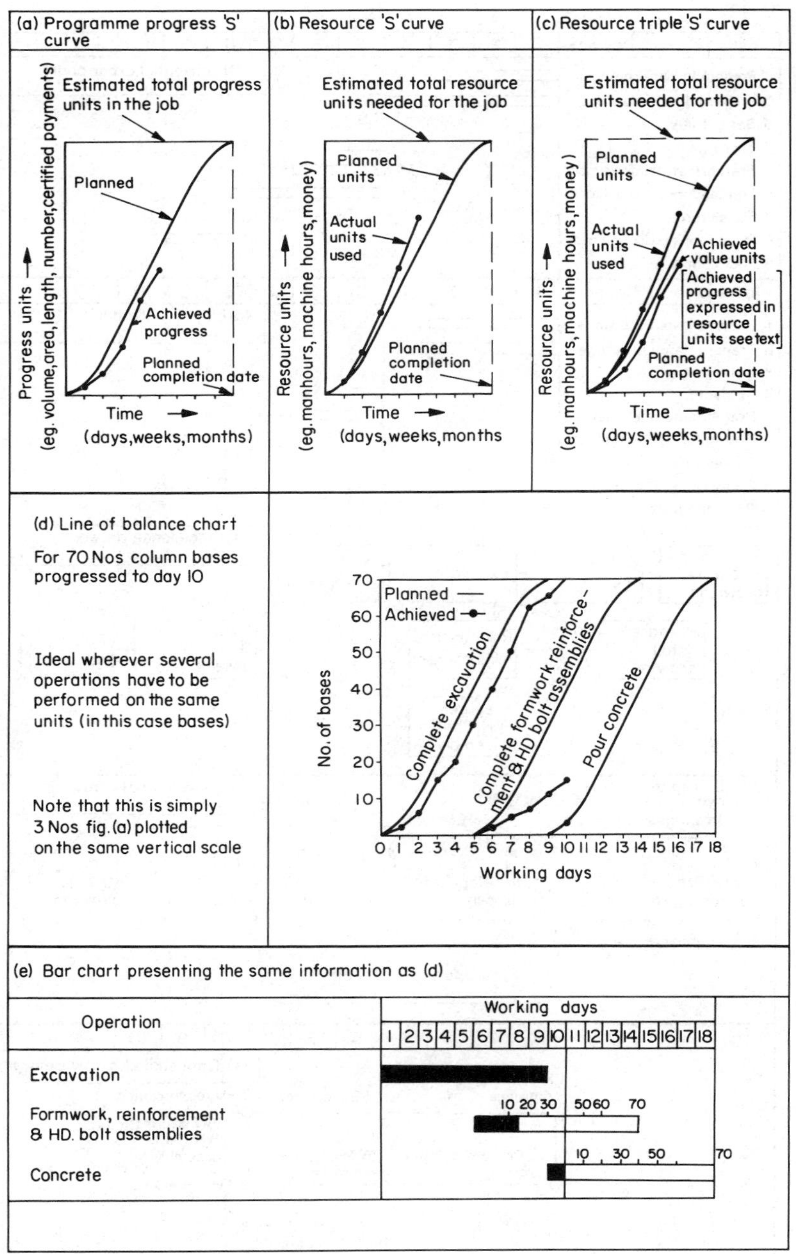

Figure 31.4(a)–(e) Simple programme-progress charts for repetitive work and for other situations where progress can be adequately defined by a single measure unit (see pp. **31**–10–**31**–13)

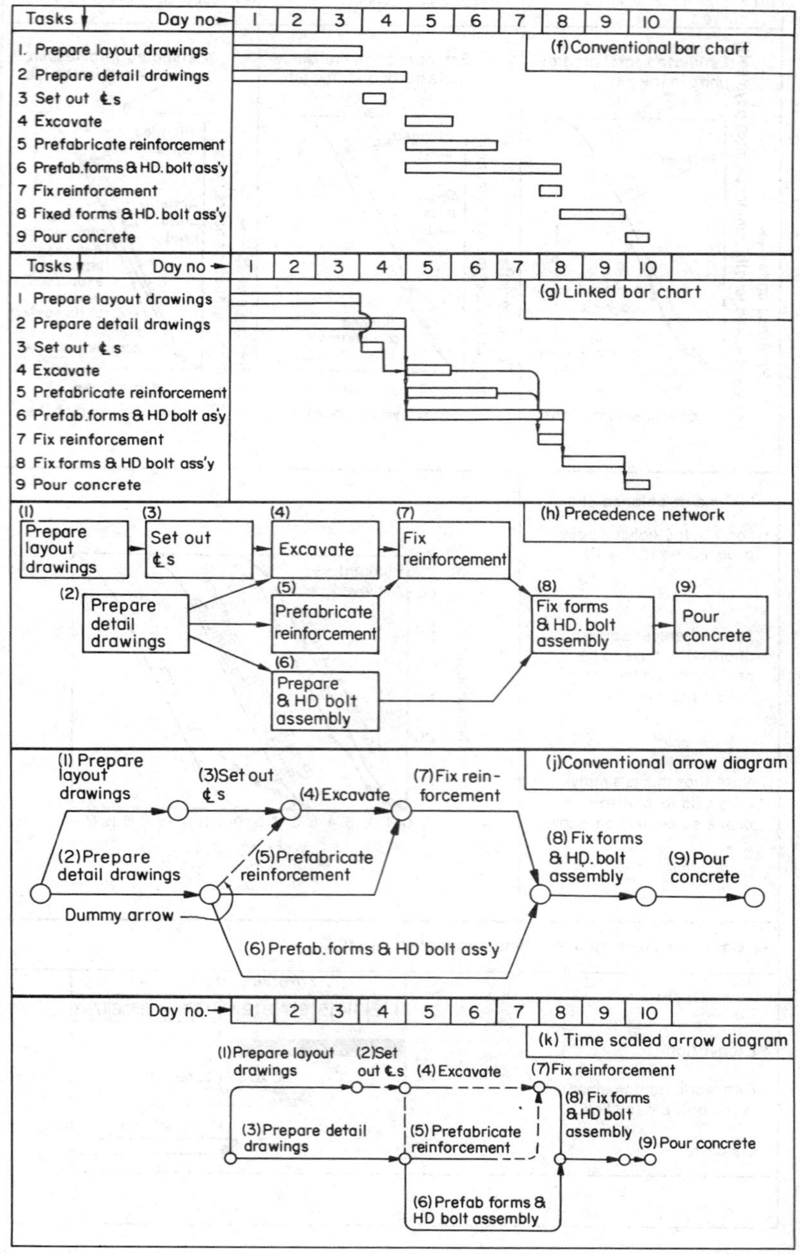

Figure 31.4(f)–(k) More complex programme-progress charts, for one-off work and for other situations where progress cannot be adequately defined by a single measure unit. Note: To illustrate the relationship with Figures 31.4(a)–(e), the above five charts all represent the same single column base which could be one of the 70 numbers in Figures 31.4(d) and (e) (see pp. **31**–10–**31**–13)

always wise to consider one of the simpler methods before choosing one of the more complex ones.

'S' curves (Figures 31.4(a)–31.4(c))

These are the simplest of all programming methods.

The *programme-progress 'S' curve* (Figure 31.4(a)) is so well known that it needs no introduction.

The *resource 'S' curve* (Figure 31.4(b)) is often useful in its own right and can also be used as an indicator of physical progress *provided that* (and this is a big proviso) *one knows from other sources the relationship between resources used and physical progress achieved.* Such situations do occur and where they do, the resource 'S' curve is attractive because it is so very cheap to use. It is nearly always cheaper to record money or man-hours put into a job than it is to measure physical progress.

The *resource triple 'S' curve* (Figure 31.4(c)) is used where one is not confident about the relationship between resources used and physical progress achieved. Progress is measured physically in the normal way and is translated into units of achieved value (i.e. resource units that would have been used if their productivity were as estimated). The chart then gives some information about costs as well as progress.

Note that, if (b) and (c) related to the same piece of work, (b) would be inappropriate because it would give a misleading impression of progress.

S curves are cheap to construct and easy to understand. They can therefore be used in their own right as a simple planning method for simple situations or as a means of summarising in a simple and compelling way the overall status of work that is being planned and controlled by more elaborate methods.

Line of balance (Figure 31.4(d))

Line of balance is the best method for planning and controlling repetitive work where a number of operations has to be performed in a fixed sequence on a number of similar units. It is simply a family of 'S' curves which by definition cannot cross and should stay more or less the planned distance apart so as to maintain the planned flexibility in day-to-day operations.

Work of this type can be programmed by bar chart (see Figure 31.4(e)) but line of balance is superior because it shows trends and history more vividly.

Bar charts

The bar chart remains popular and probably always will because it is familiar, compact and above all, as shown in Figure 31.4 it is the one method that can be used with some success on any type of work. However, just because it is more adaptable, it is nearly always less effective in situations that clearly call for one of the more specialised methods. When in doubt a good compromise to start with is a linked bar chart (Figure 31.4(g)) supported by a summary 'S' curve (Figure 31.4(a), (b), (c)).

Network analysis

Whatever method is finally chosen for detailed planning and control, networks and particularly precedence diagrams (see Figure 31.4(h)) are a wonderful tool for 'back of envelope' planning, for working out initially the logical sequence of operations for any undertaking however large or small. The other advantages of network analysis are well known but many people have had disappointments with it. So, before embarking on a full-scale networking exercise one should ask four questions.

1 *How flexible is the task sequence?* In Figure 31.4 the sequence of tasks on any one base is completely inflexible but when there are 70 bases the task sequence for the whole job should be pretty flexible (because to guard against hold-ups there should always be several bases, ready for the next task to be performed on them). Networking is ideal when the task sequence is inherently inflexible. Where there is some flexibility, networking can still be used but it requires much more effort to keep the programme in line with the work.
2 *How much repetition is there?* Where there is a lot of cyclic or repetitive work a network may be very valuable for analysing one cycle in detail but is almost certain to be unduly laborious if applied to the whole job.
3 *Is speed the only thing that really matters?* Where it is, and resources can be allowed to queue up to do the required tasks, networking is ideal and easy to use. Such situations do occur, but usually only in small parts of jobs. For whole jobs it is more usual to try to arrange things so that there is a steady queue of work for a not-too-rapidly fluctuating force of labour and plant. On such jobs networking can still be used but the necessary 'resource levelling' makes the whole thing more laborious and it is often better to use a network to work out the overall logic and then develop the detailed planning manually so as to give smooth resource-usage curves.
4 *What is the planning budget?* Networking is a fine sharp tool for job planning and control but it is relatively expensive (see p. **31**–15). If insufficient effort is put into it, it is worse than useless. So, if the money for proper networking is not available, choose a simpler method.

Organising programming/progressing work

For every job two decisions have to be made early on and often reviewed subsequently:

1 *How will the planning work be shared between specialist planners and the managers who control the work?* Specialist planners can often give useful part-time help on quite small straightforward jobs but by and large they are most likely to be needed:

   Where the number of managers or departments to be coordinated is large.
   Where the job itself is large and/or complex.
   Where the job, though it may be small, is of a type that is novel to the managers concerned.
   Where the key managers are unusually nonplanning-minded!

2 *If specialist planners are required, should they be based at head office or at the job?* In the author's experience, the planner should keep as close as he can to the centre of gravity of the work he is planning. This often means a lot of travelling.

Both these points may seem obvious to the novice. Experienced managers and planners know how often planning can falter because of problems in finding the right answer to these two organisation questions.

Relations between 'planners' and 'doers'

As programming has become more elaborate and scientific it has increasingly become the province of specialist planners and planning departments. However, separating the planning from the doing always creates the problem of how to maintain a fruitful working relationship between 'planners' and 'doers' (i.e. agent, general foreman, section foreman, etc.). This problem is not unique to construction. Roget's Thesaurus suggests such synonyms for 'planner' as 'meddler', 'fusspot', 'interferer', 'nuisance'. Nor is it a new problem, as the reader will find if he reads Hotspur's indignant report of his encounter with a staff officer after the battle of Holmedon (Henry IV Pt. 1, Act 1, Scene 3, Lines 29–69). The fact is that programming is a service which some managers think they can do without. Therefore, though money seldom changes hands, the good planner has always to 'sell' his services. He does this by:

1 Keeping close to the job and the men doing it. A great part of planning is spotting and influencing significant decisions while they are being made
2 Taking great trouble to present his ideas in a clear, concise and compelling manner
3 Acquiring a reputation for being right on questions of fact. If his facts can be trusted his opinions will be valued, otherwise not
4 Showing he knows what things cost. Programming and estimating are closely interrelated. So are progressing and cost control
5 Accepting that bargaining between managers over dates is the rule not the exception
6 Not trying to programme further ahead or in more detail than the situation warrants. (See p. **31**–8.)

Costs of programming

As programming has got into the hands of specialists, it has become a recognisable cost which people naturally try to keep down. There are two ways of looking at this cost.

1 *Cost versus project cost.* Figure 31.5 summarises such information as the author has been able to glean. It is imperfect but should provide the newcomer with a benchmark to start from.
2 *Cost versus benefit.* There are certain jobs, or, more commonly, certain small parts of jobs, where the penalties of not finishing on time are so great that Figure 31.5 does not apply at all because it is worth spending whatever needs to be spent to ensure that the programming yields its full potential benefit. Examples in the author's own experience are diversion tunnel closures on hydro-electric schemes, and shutdowns for overhauls on process plants.

Programming and the site

Programming is inseparable from the process of deciding construction methods, plant, site layout and temporary works (but see p. **31**–8).

The site itself also imposes some of the stiffest constraints and risks that have to be allowed for in preparing a programme and recording progress (see page **31**–50 and 51).

Programming/progressing and the contract information system

All the elements of the contract information system shown in Figure 31.11 (see p. **31**–57) are relevant to the programming/progressing function.

Man-to-man discussion and *meetings* are essential tools for establishing realistic programmes and for finding out the real reasons why progress is not in line with programme. Minutes of meetings often form an essential part of progress records.
All the elements of the contract information system shown in Figure 31.11 (see p. **31**–57) *Drawings and specifications* are essential for establishing what has to be done. A programme will often need to show when particular drawings are required and a progress record will include records of when drawings and revisions were issued. See also reference 1, paras 39 and 40.

Correspondence should be used sparingly. Discussions and meetings are usually more productive though correspondence is a useful way of putting and recording complex queries and decisions.

Bills of quantities are unfortunately often of only limited use to the programmer because the work breakdown in the bill differs from that needed for programming and progressing the work. The magnitude of this problem varies from job to job but it can produce quite severe problems in reconciling physical (programme) progress with financial (i.e. bill) progress.

Conditions of contract. ICE Clause 14 covers the submission of a programme by the contractor once he has been awarded the job. In the past engineers have not always insisted on such a programme because, if it is prepared properly, it must show the tasks to be performed by the engineer and the promoter and can therefore become, in the eyes of the engineer, a 'two-edged sword'. In fact some engineers refer to critical path analysis as 'critical claims analysis'. (See also reference 1, para 137.)

ICE Clause 14 has been stiffened and much amplified in the fifth edition (1973)[2] and bears careful study. ICE Clauses 41 to 48[2] are also directly relevant to the programming of the work and again have been substantially revised in the 5th edition. For further comments the reader should refer to Marks, Grant and Helson.[5]
and achieved through the job in case of subsequent disputes. (See also p. **31**–48.)

Disputes and claims. These occur even on the best regulated job. The contribution of programming/progressing is to keep accurate and detailed records of what was intended and achieved through the job in case of subsequent disputes. (See also p. **31**–48.)

Labour programming

See reference 1, para. 143.

Good programming plays a vital part in keeping a happy and productive labour force on site, by providing:

1 Reasonably smooth labour curves with not too steep a build-up at the start or run-down at the finish. 'Reasonably smooth' and 'not too steep' have their own meaning on every job and this is something on which management must make a judgement using the best available advice.

2 A work programme whose demand for labour is as immune as it can be made to delays on individual parts of the work. This means a programme in which tasks queue for resources, i.e. wherever possible there are two or three other jobs the plant and men could go on to if the one they are programmed to be on is unexpectedly held up. (See also p. **31**–14.)

It is now more clearly recognised than it used to be that a well planned job and good industrial relations tend to go together.

Plant programming and progressing

See reference 1, para. 144.

As with materials, a vital factor in plant programming is the lead time, i.e. the time it takes to get a piece of plant on to site at the right price once you have decided you need it. Where anything can be got on to site at a day's notice without price penalty the plant programme can be pretty perfunctory. At the other extreme, on overseas jobs where plant has to be shipped round the world, transported over difficult terrain and assembled on site, the plant usage programme must be considered most carefully and plant procurement becomes a vital programme/progressing exercise in its own right. The most normal situation is somewhere in between, where plant can be got to site at short notice but only at a price, i.e. proper programming will save money.

Materials scheduling, programming and progressing

This involves:

1 Making from the drawings, specification and bill of quantities, a list with descriptions and quantities of all the materials that will be required for the works
2 Deciding, from the construction programme, when they will be required
3 Ensuring that they are ordered early enough
4 Ensuring that they are delivered to site by the required dates

Since drawings and quantities usually change as the work progresses, the schedules have to be continuously updated and in order to save effort it is necessary at the outset to:

1 Identify and deal promptly with the items that have long lead times (see below)
2 Separate the bulk materials from the itemised materials

Lead time is the total time needed to obtain quotations, obtain engineer's approval (if needed), place order, manufacture and deliver to site. A good buyer will know the

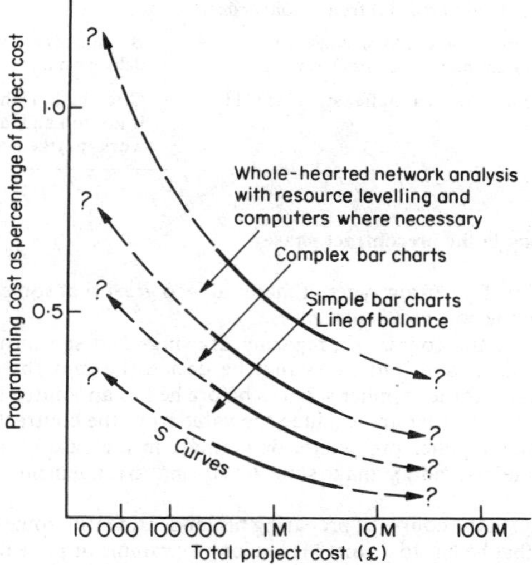

Figure 31.5 Costs of programming—very rough guide to the probable costs of using various methods of programming (see p. **31**–15)

current realistic lead times for the materials he handles regularly. He will also know how much reliance he can place on any particular supplier's promised delivery dates. Suppliers vary enormously in this respect.

Bulk materials include the obvious things like cement, unshaped timber, reinforcing rod, screws, nails and also any individual items of which a large number of identical units is required.

Itemised materials are individual items of which only a very small number is required.

In most cases the distinction between the two is obvious but there are some items that could be 'bulk' on one job and 'specific' on another. For example, if 1 No only of a certain type of valve is required it is obviously a specific item, but if 200 Nos identical are required all over the job they would probably be treated as a bulk item.

The essential differences for scheduling, programming and progressing are summarised in Table 31.2.

Table 31.2

	Bulk materials	*Itemised materials*
Takeoff/Scheduling	For the job as a whole From GA drawings/BoQ and other information which gets more precise as the job develops	For individual pieces of work From working drawings or other exact information
Lead time	May be long, but within the agreed delivery period individual consignments can usually be called forward at fairly short notice	Can often be very long for unexpected small value items
Ordering	Bulk orders placed in good time on the understanding that final quantities may differ (within reason) from initial order	Firm orders for exact quantities
Programming and progressing	By graphs or analogous means to maintain max.–min. stock on site	By individual Order Nos and delivery dates
Criticality of specific consignments	Usually low as a 'buffer stock' is held on site	Often high. If the specific item is not to hand, a large piece of work may be held up

Materials programming in the precontract phase

Sometimes the lead time for certain types of material is so great that some programming has to be done before the contract is placed.

The Engineer may, in the course of preparing drawings and specifications, identify some materials with such long lead times that he decides he must by one means or another, reserve capacity at the supplier's works before he has appointed the contractor. He may place a firm direct order and exclude the value from the contract, he may do it through the nominated supplier procedure, or he may, in the case of a supplier with whom he has a close understanding, make some less formal arrangement which produces the same result.

The Contractor may, in the course of preparing his tender, identify some materials with such long lead times that he has to distort his tender programme or price for supply from a more expensive source that offers earlier delivery or even go back to the engineer to see whether a mutually acceptable solution (e.g. by substitution) can be negotiated. Of course this is one of the many situations where the conscientious contractor risks being outflanked by the carefree one who shuts his eyes to the problem and relies on sorting the thing out in a claim at the end of the day.

Programming and progress control of subcontractor's work

The ideal programme for a subcontractor runs something like this. 'You will be able to start on such and such a date (some weeks or months ahead). You will have the work area entirely to yourself and you must complete by such and such a date' (which allows him more time than he estimates he needs).

This ideal can sometimes be achieved, but more usually the matter is more complicated because, for example:

Start dates cannot be predicted accurately months in advance.
The subcontractor's work interlocks with that of the contractor and of other subcontractors and they often have to work in the same area at the same time.
The overall programme is seldom sufficiently long to allow each subcontractor as much time as he wants plus a bit to spare.

Even when the subcontractor's programme is completely selfcontained (like the ideal sketched above) the prudent contractor will get a programme from the subcontractor before he starts work, and monitor his progress at frequent intervals to make sure he is likely to complete by the promised date. The more complex the interlock between various subcontractors and the contractor's own work, the more vital this programming–progressing function becomes and on jobs with a really large number of interlocking subcontracts, the coordination of subcontractors can become one of the contractor's most demanding tasks. In these situations it is usually necessary to hold programme–progress meetings (possibly as frequently as weekly) of authorised representatives of all the subcontractors whose work interlocks in a particular sector of the job.

In programme discussions with subcontractors, each side has to keep one eye on his rights under the agreed subcontract. However, the main contractor should always act on the basis that however hard he 'screws down' the subcontractor in the legal documents, his best protection against subcontractor delays which would disrupt his own work is to do all in his power to help subcontractors do their job. This may include doing part of their job for them in areas where they may turn out to be weak, such as, for example, programming and progressing.

Programming, progressing and organising third parties

Third parties usually only figure in a programme as constraints, as the source of limitations on access, noise, night working and so on. Sometimes, however, work to be done by third parties (e.g. statutory undertakings or other contractors) form an integral part of the programme for the complete project. Where this situation arises it can bring most tantalising problems of programming, organising and maintaining progress because the contracting parties can apply so little leverage to the actions of third parties. (See for example NEDO[6] para. 2.7.)

QUALITY AND ACCURACY

Planning for quality during the design period

In doing this, the engineer will go through a thought process very much on the lines set out in Figure 31.3.

1 He will form in his mind a somewhat idealised picture of the job he wants to get built (Box 3.1)
2 He will then modify his concept to take account of the relevant constraints (Box 3.3). For example (especially on overseas work) some types of material, craft skill or plant may be unobtainable or so expensive that an alternative must be specified
3 He will then consider the resources he needs (Box 4). These will include drawings, specifications, conditions of contract, resident engineer and inspection staff, site labs., offsite testing facilities, etc.

4 Finally he will draw up some sort of programme related to the construction programme, showing when these various resources will be needed

Planning and controlling to achieve the desired accuracy in setting out the works

In most works, setting out is something well within the normal range of contractor's and engineer's site staff. However, some classes of work require very great precision and this may require special thought by the Engineer (and by the tendering contractors), special specification clauses and perhaps the help of some specialist surveying staff. See also reference 1, para. 174, and Section 33.

Balancing quality against cost

By and large a better quality job costs more initially but lasts longer and costs less to maintain. Achieving the most satisfactory balance requires from the Engineer some of the most important and difficult judgements that he has to make, particularly on works that are intended from the outset to have a finite life (see reference 1, para. 34).

This is really outside the scope of this section but it sometimes impinges on contract management, for example where unforeseen conditions demand major new design decisions during the course of construction.

Organising to achieve the desired quality standards

(See reference 1, paras. 110–112.)

Quality illustrates very well the separate creative and regulative aspects of organisation explored on p. **31**–54 and Figure 31.10.

Table 31.3

Role	*Creative activities*	*Regulative activities*
Engineer	Creating a workmanlike design in the Contract Documents Adapting the design to suit new conditions encountered during the contract Commenting helpfully on the Contractor's proposed construction methods	Ensuring that the Contractor adheres to the drawings and specification
Contractor	Selecting the best construction methods to give the specified quality	Ensuring that materials and workmanship comply with drawings and specification

The creative activities create the conditions for success while the regulative activities ensure that these conditions are maintained.

ICE,[2] Clauses 17, 36–39 and 49 lay responsibility for quality and accuracy squarely on the contractor but here, as in so many other matters of organisation, pre-allocating blame is not always the best way to ensure that the job will be done. If the contract is not interpreted too literally, there is room for creative discussion and agreement between Engineer and Contractor on construction methods and sometimes also on the permanent works design. A problem that often rears its head is 'Should the RE's (resident engineer) staff ensure that the work is correctly constructed or should they merely ensure that the contractor ensures this?' The latter is the contractually correct answer but in reality Engineer and Contractor are equally interested in getting a good job built on time and it is hard to imagine any RE deliberately letting defective work be done and then later rejecting it in order to teach the contractor a lesson. In fact on a happy job the RE's inspectors and inspecting engineers are the contractor's best allies in getting the job done on time and to the required standard but the contractor who thus effectively delegates much of the day-to-day inspection must never forget that the responsibility for quality still rests with him, especially if things go wrong.

Quality and the Engineer/Contractor relationship

Money disputes can be deferred and progress problems do not usually become obvious until the job is well advanced, but quality problems have to be resolved as they arise and so they provide most of the ground on which Engineer and Contractor establish at the outset of the job, the tone of a relationship which ultimately governs the success of every aspect of the contract. Civil engineering work has two inherent characteristics which make it more prone than (say) mechanical engineering to produce quality problems.

1 Civil engineering is essentially the art of moulding natural landscapes and using materials that are either natural or like bricks and cement inherit characteristics from their natural origins. People who work with Nature know that she is not wholly predictable and that one gets the best results by treating her as an ally not an enemy. This means that the Engineer cannot always specify in advance exactly what he will want in every situation that may arise. In other words he has to be a bit 'flexible'. This of course sets the Contractor the problem of deciding how 'flexible' the Engineer will be in any situation that demands flexibility!

2 In some types of work 'good trade practice' can be a quality standard in its own right (see p. **31**–22) but Engineer and Contractor may very well disagree on what is good trade practice on the job in question.

Incidentally, both these factors produce uncertainty and risk in the precontract phase. The tendering contractors try to forecast how 'inflexible' the Engineer will be and the Engineer when examining the tenders tries to assess how far each contractor may try to push him on matters where both sides agree there is room for a bit of flexibility (see also Twort[4]).

The site and quality

Just as the site is a fundamental factor in the design, so is the climate at site (including the prevailing air and water pollution) an essential factor in deciding what quality standards to aim at. Extremes of heat, cold, damp, drought or wind often demand special thought.

The contract and quality

The rights and duties of Engineer and Contractor are spelt out very clearly in ICE Conditions,[2] clauses 17, 36 to 39 and 49.

Influence of drawings and specifications on quality

These are part of the design process which is covered in detail in many other sections in this book. Short notes and references on the contract management aspects are given on pages **31**–58 and 59. Three important points must be kept in mind when writing any specification:

1 It is easy but seldom right to specify 'the best'. It is usually necessary but much more difficult to decide and specify the quality standard which is economically and in other ways appropriate for the job.
2 Because of 1, expressions such as 'to the satisfaction of the Engineer' are sometimes unavoidable. However, the specifier should keep in mind that these expressions always create doubts in the minds of tendering contractors (just what will satisfy this particular Engineer/resident engineer?), doubts which may be reflected in higher tender prices. They may also lead during the contract to disputes which might have been avoided by more careful drafting of the specification.
3 The extent to which 'good trade practice' serves as a satisfactory standard varies enormously with the type of work. It means quite a lot on work like traditional building where well established crafts are using familiar materials in a familiar way. On the other hand, in mechanical and electrical work and any work where unfamiliar materials are being used, it is necessary to write very exact specifications. Much civil engineering work falls somewhere between these two extremes.

Men for control of quality

Quality control is a major part of the job of all the resident engineer's staff but the front line men are usually the inspectors and any specialist testing staff. The duties of inspectors are well set out in reference 1, paras. 102 and 103, and in Twort.[4] Specialist testing staff may be junior engineers who are temporarily specialising or they may belong to outside testing agencies working under contract.

Materials management and quality control

Proper quality control of materials is a major factor in achieving a finished job of the right quality at the right price. The main stages in this control process are:

1 Inspection at works or sources or of samples
2 Correct packing and shipping
3 Inspection on arrival at site
4 Correct storing and handling on site
5 Correct incorporation in the works and acceptance by engineer.

The technical aspects of these five stages are outside the scope of this section. The managerial aspect is that rejection by the engineer at stage 5 produces much more chaos on site than does rejection or corrective action at earlier stages. It is therefore in the interests of both Engineer and Contractor to agree on sensible ways of applying all reasonable checks at stages 1 to 4. This is one of the areas where neither party should try to limit himself too rigidly to the minima spelt out in the contract. (See also Figure 31.14.)

Quality control of subcontractors' work

The Contractor is responsible for the quality of materials and workmanship of all his subcontractors including those nominated by the Engineer. At the very least he has to avoid certifying and paying for subcontractor work which is later condemned by the engineer. His principal ways of maintaining this control are:

1 Inspect materials and workmanship frequently, particularly in the early stages of work by a new subcontractor

2 Refuse to pay for defective work until it is put right.

In these tasks the resident engineer's staff, especially his inspectors can be useful allies, though they have always to be wary of seeming to relieve the contractor of his responsibility. On the other hand, if Engineer-Contractor relations are strained the RE's staff can cause havoc with subcontractors by insisting on the ultimate in quality, without regard to the need for progress and economy, creating in the end a situation which usually rebounds on the Engineer as much as on the Contractor.

SAFETY

Introduction

Compared with most industries, construction has a high accident rate. It is the largest (40%) single contributor to fatal accidents under the Factories Act and perhaps for this reason is among the leaders in accident prevention techniques. The reader should refer to reference 1, para. 169 for some general notes on the problem and what is being done about it.

In this section frequent reference is made to:

The FCEC Supervisor's Safety Booklet[47]
The ICE Report on Safety in Civil Engineering[50]

The FCEC booklet[47] is a mine of useful information and the author believes that every engineer on site should have his own copy.

The reader who thinks safety is a simple or unimportant matter should refer to the three fascinating papers by Merchant, Short and Kinsella[51] (which with discussion run to 90 pages of *Proc ICE*).

Much of the literature on construction safety is heavy with exhortation. The reader will rightly guess that this indicates that there are sometimes rather large gaps between what people do and what they should do about safety. This is because we are looking at a moving object. People's ideas of acceptable safety standards are rising all the time and as long as this continues, precept is bound to stay somewhat ahead of practice.

The principal reasons for the change in the climate of opinion about safety are:

1 Partly due to advances in medical science, people attach a greater value to each human life than they did in the past
2 Resulting from 1 there is a growing feeling that a person injured at work should be adequately compensated for his loss even if it can be proved that the accident was partly his own fault or nobody's fault, i.e. more of the true cost of accidents should be borne by employers and/or government and less by the injured persons
3 Those who have to pay are become more aware of the real cost of accidents
4 Those working full-time on accident prevention are getting more skilful at making people feel they can and should do better.

The following are among the problems that will have to be dealt with before construction safety reaches a standard that satisfies most people most of the time:

1 People in construction tend to be unimpressed by statistics which show the high accident rate in construction compared with manufacturing. They reason that construction, like war, is a dangerous industry that attracts people who like danger
2 More compelling would be statistics compiled on a nationally agreed common footing so as to highlight the high risk areas within the construction industry. Unfortunately there is at present no nationally agreed common footing for industrial safety statistics
3 The present legal framework is overcomplex and, among other things, encourages

the parties involved to think too much about avoiding legal liability and not enough about working safely

4 The old ICE Conditions, so specific on so many matters, was notably unspecific about responsibilities for safety. These are spelt out more clearly in the 5th edition (1973)[2]
5 Under the normal contractual arrangements, the extent of the engineer's responsibility for the safe constructability of his design is not clear-cut. However, it seems now to be accepted that on some types of work he could do better[50]
6 People cannot really regard any level of accidents, however low, as acceptable. So however good one's record there always seems room for improvement.

It seems likely that the climate of opinion and the legal framework on safety will alter rapidly over the next few years, so this section concentrates more on principles than on current legal requirements.

Planning for safety

Safety starts at the job planning stage and for this both Engineer and Contractor need to go through a thought process on the lines of Figure 31.3 but with more specific questions added to boxes 2.3 and 3.3.

2.3 What special hazards have we encountered?
3.3 What special hazards must we provide for?

Many of the *risks* in construction relate to the safety of men, plant and the works and many of these are insurable (see p. **31**–51).

External restraints that relate particularly to safety are notably:

1 The legal framework: Complex in the UK and full of surprises overseas
2 The site (see p. **31**–56)
3 Some third parties (see p. **31**–80)

Safety is a factor in selecting most *resources* (especially plant and some materials) and resources needed specifically for safety will include such items as:

The services of specialist safety staff
The services of some of the third parties listed on page **31**–81
Protective clothing
Safety equipment (e.g. safety nets and harness) for certain operations
Site standing instructions on safety matters
Posters, booklets and other instructional material

These various matters are discussed in the following subsections.

Organising for safety

Safety at present provides a particularly vivid example of the age-old organisational conundrum of how to give operational managers specialist support without reducing their authority and responsibility. (See also Short[53].) *The contractor's agent* is responsible for the safety of all operations (ICE Conditions,[2] clauses 15 and 19) just as he is responsible for progress, cost and quality of the works. He has specialist help in these other areas, so presumably he needs specialist help on safety matters. In fact he is now required by law to have such help.

The safety specialist. Under the Construction Regulations 1961[48] 'firms that employ more than 20 persons in all *at any one time* (not on any particular site) on constructional works, must appoint, in writing, at least one experienced person to be charged with . . .

promoting the safe conduct of the work generally. . . . The safety supervisor need not be a full-time safety officer but his other duties must leave him free to discharge his safety duties with reasonable efficiency'. In practice most firms of any size tend to develop a cadre of full-time expert safety officers who are based on head office and combine a role that is part adviser and part policeman inside the company and key witness for the defence in any complaint against the company.

Obviously this is a difficult role to play well and in fact the key to successful safety organisation is to decide what is the safety officer's role and get it accepted at all levels. In particular the combined role of adviser/policeman to site agents is a tricky one. If the safety officer sees something being done on a site which is wrong and dangerous what should he do?

Should he advise the agent of the sort of accident that might occur and the possible legal and financial consequences?
Should he police the agent by reporting him to higher authority?
Should he have direct authority to overrule the agent and issue orders to have the matter put right?

Of course the answer is that different situations demand different courses of action. The safety officer may have direct authority on certain very specific matters but most of the time he has to exercise his judgement and he becomes accepted as an adviser by acquiring a reputation for sound judgement. However, sound judgement is no use to the safety officer unless he can be confident that on matters that he judges to be really crucial, the people at the top of the company will back him and overrule the agent. This is only right and proper for the agent is the company's agent and if his acts lead to a serious accident he embroils the whole company.

Influence of design and construction methods on safety

About 100 years ago, Sir Benjamin Baker, designer of the Forth Railway Bridge, wrote:

> 'Of the numerous practical considerations to be duly weighed and carefully estimated before the fitness of a design for a long span bridge could be satisfactorily determined, none is more important than those affecting the facility of erection.'

We aren't all building long span bridges but these words are just as valid for many other types of work.

Design and construction methods are in principle inseparable but under the normal contractual system they are assigned to different parties. From the safety angle this arrangement works well enough most of the time in conventional work but it may produce problems when the design is unusual or novel to the contractor.

The recommendations of the ICE Report[50] may be paraphrased as follows:

1 The Engineer should have proper regard for the feasibility of erection of his design and should incorporate any special safety requirements in the specification.
2 The Contractor should thoroughly plan his erection schemes and design his temporary works in detail.
3 The Engineer should check erection schemes and temporary works designs.
4 The Engineer and contractor should cooperate to ensure that the agreed erection scheme is properly executed.

Nobody could argue with any of this, but in practice items 1 and 3 present the Engineer with some thorny problems, e.g.:

—How far can he go in specifying erection methods without sacrificing for his client the chance that some contractor will propose a cheaper, better and equally safe method?
—How stiff can he be on checking erection schemes and temporary works designs

without morally (though not legally) assuming responsibility for the safety of the contractor's operations?

With good Engineer/Contractor relationships these problems can be contained but they are there and if relationships are bad and/or there is a serious accident the problems quickly come to the surface. The ICE Report[50] recommends that the legal and contractual obligations with regard to safety should be clarified.

Safety and the contract documents

A safe job starts with a safe design and specification and this aspect has been discussed in the previous paragraphs.

As regards the *Conditions of Contract* it is a curious fact that until the 5th Edition[2] (1973) the ICE Conditions, though very specific about care of works and insurance (clauses 20–25), said nothing specific about safety. It could be, and presumably was, argued that the Contractor should not need to be told to work safely, but equally one could argue that the conditions of contract spell out what the Employer really cares about. It is obvious from the contract that he cares deeply about quality, workmanship and completion on time and even more deeply about money. If he cares about safety he should say so. This view has prevailed in the 5th Edition[2] (1973) which now contains important references to safety in clauses 8, 15, 16, 19 and 40.

The Association of Consulting Engineers argue convincingly that fee competition would lower the quality of their service. The author believes that safety is an area in which competitive tendering lowers the quality of the contractor's service. The costs of safety and welfare are easy to measure but despite what is said on page **31**–30 the costs of accidents are not always so apparent. Therefore, the Engineer's only really sure way of getting any particular safety and welfare measures that he wants is to spell them out in the contract documents so that no contractor is tempted to cut corners on these matters in order to win the job. (See also p. **31**–25.)

Safety and the site

Some hazards arise from the nature of the site itself. When each new job starts, the safety officer will consider what special hazards if any may arise from such things as:

Bad ground, pits or precipices or cliffs with falling rocks
Buried or overhead cables
Extreme heat or cold
Abnormal winds or tides

This list is by no means complete but gives an idea of the sort of things to look out for.

Safety of men

Safety of men depends partly on proper design of permanent and temporary works and proper choice of construction methods and plant. These are discussed on page **31**–25.

However, a great deal depends on the knowledge and vigilance of every supervisor and every man on site. For guidance on this the reader can do no better than refer to the FCEC booklet.[47] Every engineer on site should get his own copy even if only to learn how to look to his own safety.

Most working rule agreements (see p. **31**–67) say very little about safety but contain various clauses on the closely related matters of welfare and extra money for working in dirty, dangerous or exposed situations. The reader must be warned that the whole area where safety and industrial relations meet is a minefield for the unwary partly because of the present rather unsatisfactory state of the law on compensation for injury (see p. **31**–28). All that can be said here is that accidents are bad for morale and a site which has a lot of silly accidents is most likely to be the one where there are a lot of silly disputes over safety matters.

Safety depends so much on attitudes and there is no doubt that formal off-job training (see FCEC booklet,[47]) short films and propaganda posters, have a cumulative (even if hard to measure) beneficial effect.

Welfare and medical

(See also the FCEC booklet[47] and C.E. Procedure,[1] paras. 171 and 172.)

The following is a checklist of items that commonly need to be considered under this heading:

Protective clothing
Lockers and drying rooms
Shelter for bad weather
Washing and lavatory facilities
Canteens
Hostels
Transport to and from job site
First aid and links with local doctors/hospitals
Medical examinations
Special arrangements for employing disabled persons.

In the UK, minimum standards for many of these matters are laid down by law and/or working rule agreements. Overseas, standards vary enormously and the only thing that is certain is that many of the UK standards will be inappropriate. The good employer at home or abroad finds out what is good local practice and tries to do a bit better but not too much better because his idea of better may seem so strange to the people who are expected to benefit that the effort may be wasted or even counterproductive.

Safety and plant, transport and rigging

All equipment must be:

1 Of safe design and construction
2 Regularly inspected to ensure it is safe to use
3 Used in a safe manner.

Here again the reader should turn in the first place to the FCEC booklet.[47] The Robens report[52] recommends tougher controls on the safe design and construction of equipment.

Rigging/lifting gear deserves a special mention because it is an area where apparently trifling oversights can lead to large and dramatic disasters. (See for example Merchant and Short.[50]) All really major or novel lifts should be calculated and planned in detail beforehand, but many seemingly routine lifting operations also demand an engineering understanding of what is being done. Here the Contractor's engineers, and indeed the

Engineer's engineers, have to exercise considerable judgement when deciding whether to intervene or not.

Materials and safety

Materials management impinges on safety in four main ways:

1 *Safe storage and stacking of ordinary materials:*
10% of accidents are caused by falls of materials. See the FCEC booklet[47] for further details.
2 *Safe storage and handling of dangerous materials:*
(e.g. asbestos, LPG (liquified petroleum gas), explosives, gas cylinders, lead paint, radioactive isotopes). The FCEC booklet[47] gives some information. The supervisor needs to watch out constantly for dangerous materials, especially new or unfamiliar ones.
3 *Special equipment for personal safety:*
e.g. safety harness, safety nets, items connected with diving and compressed air operations, explosive gas detectors, etc.
4 *Protective clothing:*
Hard hats, goggles, gloves, reinforced boots, etc.

Safety and subcontractors and selfemployed persons

Just as competitive tendering encourages main contractors to take short cuts on safety, so does subcontracting encourage subcontractors to take risks, particularly the smaller ones who may be here today and gone tomorrow. The main contractor often needs to watch subcontractors most carefully on safety matters. One might take the view that accidents to subcontractor's men are his own affair but life is seldom as simple as that. If a subcontractor has a serious accident he may seek to put part of the blame on to the main contractor, or the accident itself may injure the main contractor's own men, and damage the works or disrupt progress in a very costly way.

Safety and third parties

The various third parties listed on page **31**–80 (and notably other person's employees) can all affect and/or be affected by accidents on site and the safety officer needs to consider what needs to be done about each of them on each new job as it starts.

The only third party that needs special mention here is the Government whose will is expressed in a mass of legislation covering safety on construction sites. A summary of this legislation is given in the FCEC booklet[47] and more detail on the Construction Regulations is given in the FCEC/NFBTE guide.[48]

The Robens report (1972)[52] recommends sweeping changes to correct the following major defects in current legislation:

1 It is too complex and administered by too many different agencies
2 It encourages those who create risks and those who work with them to rely too much on detailed regulation by external agencies. Instead it should encourage them to take the prime responsibility for safe working
3 It encourages the parties involved to think too much about establishing or avoiding legal liability and this 'adversary' situation often prevents people cooperating as they should to prevent accidents
4 It does not take enough account of the safety of the public (as distinct from employees)

If the law is reformed along these lines, it will bring little short of a revolution in safety matters. In the meantime contractors have to work within the law as it stands, and keep at least one eye on the defects listed above.

Overseas the legal framework needs to be examined very carefully at the outset of a job.

Safety controls and statistics

Safety control in the wider sense of the word, is the responsibility of everyone on the contract. Here, as elsewhere in this section, the word 'control' is used in its narrower sense of 'comparing what is with what ought to be' in order to locate areas where action is needed. Of course there is no accepted view of 'how many accidents there ought to be' but by careful use of statistics a contractor can compare the performance of various sites or groups of sites, and may, in some instances, be able to measure himself against others or against national averages for the industry.

In the UK there are three commonly recognised ways of measuring accident rates

1 *Frequency rate* $= \dfrac{\text{Number of lost time accidents} \times 100\,000}{\text{Total manhours worked}}$

A 'lost-time accident' is one where working time is lost beyond the day or shift when the accident happened. It is not the same as a 'reportable accident' (see 2 below)

2 *Incidence rate* $= \dfrac{\text{Number of reportable accidents} \times 1000}{\text{Number of employees}}$

The Factories Act defines a 'reportable accident' as one in which a man is disabled for more than three days.

3 *Severity rate* $= \dfrac{\text{Manhours lost} \times 100}{\text{Total hours worked}}$

Obviously any one of these figures taken on its own gives an incomplete picture which could be misleading if used blindly. None of them, for example, gives any weight to damage to plant or the works. Furthermore, none of the figures reflects the numbers of people killed or permanently incapacitated. The American National Safety Council attempts to cover this by counting in the severity rate calculation a death or permanently incapacitating injury as 5000 man days lost but it can be argued that unless the data base is very large, one death would so distort the calculated severity rate as to make it almost meaningless as a measure of the ordinary range of accidents. Not surprisingly the Robens Report (1972)[52] recommends that UK statistics should be got on to a commonly agreed footing.

In addition to overall rate calculations the contractor can analyse accidents by, for example:

Types of work

Trades

Effect of productivity incentives

Types of accident

Types of injury

Regions of the UK

Types of weather

Season of year, day of week and hour of day

Age and length of service.

The list is seemingly endless and the contractor's real problem therefore is to decide how much data he can handle and how much of it is likely to be of practical value. A well run safety organisation will review such decisions at regular intervals.

Safety—the costs of accidents

Money is something that everyone understands and those people who realise most clearly how accidents affect their own pockets do most to prevent them. One reads such statements as:

> Industrial accidents lose us many times more man days than do industrial disputes. A recent study in Ontario concluded that the cost of accidents there exceeded by six times the net profit for the whole industry in that province.

But these figures are too generalised to have much impact on individuals on individual sites. The FCEC/NFBTE manual 'Construction Safety'[49] comments as follows:

'The cost of accidents to an employer is made up of many items which vary according to the nature of the accident. Some costs may be covered superficially by insurance (though a portion of the premium is still applicable), others may not even be recognised as legitimate costs because they are hidden in maintenance charges or general overheads. . . .

'Costs of accidents generally can be broken down into

- (a) Wages paid to but not earned by
  - (i) The injured man
  - (ii) Other operatives who stop work from curiosity, from sympathy, or to help
  - (iii) General foreman, trades foreman, timekeeper, etc., in helping investigating, reporting and arranging for continuity of work
- (b) Taking on and instructing replacement labour
- (c) Interruption of the planned sequence of work
- (d) Lower morale of other workers resulting in lower productivity
- (e) Damage to materials and plant
- (f) Plant standing out of use
- (g) Replacing plant
- (h) Administrative incidentals—phone calls, welfare, first aid, etc.
- (i) Initial lower productivity on return to work of the injured man
- (j) Head office costs
- (k) Insurance costs

'It is advisable to produce two separate figures for

- (i) The cost of an injury accident
- (ii) The cost of a damage accident

'It has been calculated that a reportable (three-day) accident can cost on average about £300 (1970) though one single major accident might cost £500 000.'

Insurance premiums are one of the most direct ways in which contractors are made to feel the cost of accidents.

Insurance and safety

Insurance is dealt with in Section 32 so only a few words are needed here.

ICE clauses[2] 22–25 spell out insurance responsibilities in some detail. A great deal of insurance is concerned with the need to compensate people for damage to themselves or their property arising from accidents. In fact insurance premiums are one of the most visible costs of accidents and the enlightened contractor is well aware of how his premiums relate to his accident record. In extreme cases a premium rebate can make the difference between a profit or a loss on the job.

However, in spite of all the insurance, a great deal of the cost of an accident is often still borne by the injured party because as the law stands at present a claim for damages can only succeed if the injury can be proved to be due to negligence of somebody other than the claimant. There are many accidents in which nobody can be proved to have been negligent and others in which several people, including the claimant, are found to have been partly negligent. In such cases the injured party gets little or nothing. There is a growing feeling that this state of affairs is not civilised enough for the 1970s and it is in fact being examined at present by a Royal Commission under Lord Pearson.

ECONOMIC WORKING

Introduction

The process of estimating and controlling costs is simply as in Figure 31.3 with the resource outputs converted to costs and the resource needs converted to sums of money.

Of course, apart from the resources listed in Figure 31.1, money is itself a resource which costs money and this matter is discussed in more detail on pages **31**–49–**31**–50.

Sometimes it is possible to work simply in money so that questions 1, 2.2, 3.2 and 4 (in Figure 31.3) become:

1. How much money did we spend?
2.2 What did parts of the work cost? How and why did they differ from what we originally expected?
3.2 What can we expect parts of the work to cost?
4. How much money will we need?

However, this method is crude and very uncertain when applied to new work in new locations. For most civil engineering work it is necessary sooner or later to go back to first principles and work out what resources (men, machines, materials, etc.) will be needed and how much they will cost.

Estimating by the Engineer

Throughout the period of the project, the Engineer has to do cost estimates for various purposes, the principal four of which are:

1 To predict the total project cost for the promoter at various stages before the contract is let (see C.E. Procedure,[1] para. 34).
2 To compare the cost of alternative designs at various stages before the contract documents are completed
3 To help him fix new rates as provided for in ICE[2] clause 52 (2)
4 To help him adjudicate claims as provided for in ICE[2] clause 52

For *predicting total project cost* and for *costing alternative designs* the Engineer can use published lists of rates or (much preferably) bills priced by contractors for similar complete projects. This method normally gives figures for total project cost that are quite accurate enough at the pretender stage, i.e. they are normally well within the range of tenders subsequently received. For costing alternative designs the method is rather less reliable because, for reasons given on page **31**–37, contractors' priced bills which may differ very little in total, may show quite large differences on important individual rates. In important and difficult cases where he is not too confident about the rates of his disposal, the Engineer will make check estimates of total project cost and costs of alternative designs by methods analogous to those used by contractors (see below).

Estimating for *fixing new rates* for varied work before it starts is not by any means as easy as ICE[2] clause 52 (2) makes it sound. Very often it is a matter for negotiation with the Contractor and the young resident engineer on site will always seek guidance from his superiors before getting too committed. The reader who is interested in seeing what a complex matter it is should refer to the paper by Haswell.[40]

When estimating for *adjudicating claims* after the work is completed the Engineer should be in a better position because if they have done their job properly, the resident engineer's staff will have kept detailed records of the Contractor's activities, with particular reference to anything that may be connected with a possible claim. These records should be highly relevant even if the Contractor has been unable (or unwilling) to forecast the ultimate nature and extent of his claim (and to be fair, it is often very difficult for a contractor to visualise all the effects of a variation or disruption until the work has been completed). See also pages **31**–48 and **31**–60 and Twort.[4]

Estimating by the contractor

Many people outside contracting believe that contractors price their tenders for new work simply by examining their unit costs for recent similar work and making suitable adjustments. This method can be valid on simple repetitive work where one job is very like another, such as road surfacing or drain laying in shallow trenches, but for most civil engineering work it is a formula for disaster because one hardly ever finds two jobs in which the design and construction conditions are sufficiently alike to make recorded costs a sufficient basis for pricing tenders.

It follows that on the great bulk of civil engineering work the contractor has to price each job from first principles, carrying out the following main tasks more or less in the order shown but with a good deal of overlap as each influences all the others.

1 Weigh up the contract documents and the site
2 Prepare construction method statement and programme with labour and plant outputs and requirements
3 Estimate the total cost per working hour of the various types of labour and plant
4 Combine 1 and 2 in labour and plant rates and/or lump sums for the various programmed operations and decide how to place these monies on the bill
5 Obtain quotations for materials from potential suppliers and incorporate these in the bill rates and similarly for work to be subcontracted
6 Price the various overhead items and decide how to place these monies on the bill
7 Prepare a tender summary showing total sums for each of the following:
    Labour
    Plant
    Materials
    Subcontractors
    PC sums (nominated suppliers and subcontractors)
    Promoter's contingencies
    Site overheads
    Head office overheads
8 Decide how much to add for uninsured risks, profit and financing and how to place this money on the bill
9 Complete bill and other tender documents and submit tender

More detail on these tasks is given in the paragraphs below and in C.E. Procedure,[1] paras 70–72 and appendices 1 and 2 which cover pretty comprehensively the items that normally have to be taken into account in preparing a tender.

General procedure for tendering and arranging a contract

The general procedure is set out in C.E. Procedure,[1] paras 64–70 and some aspects are covered in more detail in Marks, Grant and Helson.[5]

Contractors' organisation and manning of estimating and tendering

To the outsider it might seem obvious that a job should be priced by the man who is going to be site agent. In practice this is seldom done because

(a) The prospective agent is fully committed on current work
(b) Several (possibly as many as ten) jobs have to be priced to win one
(c) Getting out good estimates quickly and economically demands the skill which comes from constant practice

Consequently most contractors treat estimating as a specialised activity and employ full-time estimators who are mostly mature men who have had considerable experience on site and elsewhere and have decided to make estimating their career. The estimator leads the tendering team but will get as much advice and help as he can from those people/departments that will be involved if the job is won. He will carry out all the tasks listed on page **31**–32 except number 8 which is a company policy matter normally dealt with at director level.

Estimators are normally based on head office so as to be near other head office-based services (e.g. planning and temporary works design, purchasing, legal, insurance, etc.) and so as to provide a pool of estimating manpower to handle the estimating workload as economically as possible.

Estimating—weighing up the contract documents and the site

The contractor must first study all the contract documents (see p. **31**–58) and visit the site to weigh up ground conditions, climate, access, services and all the other factors which will influence his pricing (see ICE conditions,[2] clause 11). Most estimators take with them on the site visit a comprehensive standard checklist to ensure that they don't forget anything, a notebook for information that is not on the checklist but may later be useful, and a camera. Photos are the ideal way of recording the special character of the site and of storing rapidly a mass of information which is not on the checklist but may turn out to be extremely helpful when one is back in head office working up the prices.

Tender method statement and programme

For the contractor an estimate is the monetary translation of a physical programme of work and jobs are often won by the contractor who devises a better programme and method than his competitors. Therefore, before he can price anything, the contractor has to answer the questions in Figure 31.3 and produce a method statement and programme. It follows that time and timing are essential ingredients of practically every price that the contractor enters in the bill of quantities. However, the bill in its present form takes very little account of this and so provides a fruitful source of contention if work is varied or delayed.

Time/cost tradeoff

It can be shown fairly easily that the contractor's total project cost is related to project duration in the manner shown in Figure 31.6. Though an estimator is unlikely to construct a graph of this type he will (unless he is required to go for speed regardless of cost)

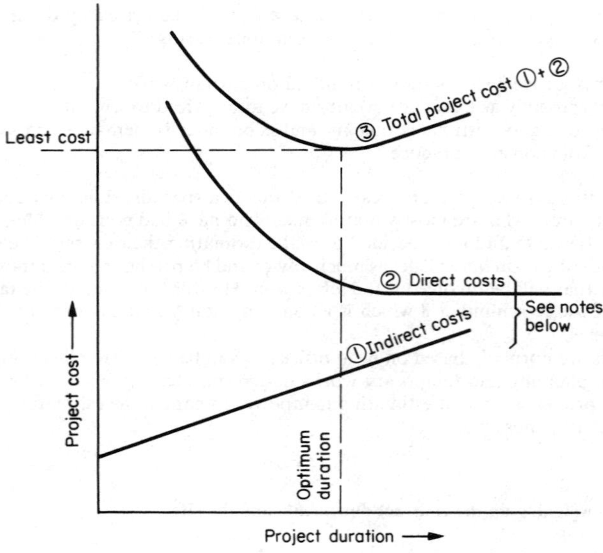

Notes

① *Indirect costs*
(a) Fixed costs: eg. setting up and dismantling site installations
(b) Time-related costs: eg. supervision, plant required for duration of project

② *Direct costs*
eg. most labour and plant, purchase of materials.
As shown in curve ② these all cost more (eg. overtime) if the job has to be accelerated.

Figure 31.6 Project time—contractor's cost relationship

programme the work so as to produce a cost close to the optimum shown in Figure 31.6. With the bill of quantities in its present form, any substantial change in project quantities or duration is likely to push the total cost up the slope one way or the other from the optimum point and so may lead to a claim. For some further details see Marks, Grant and Helson,[5] and Barnes.[31]

Estimating labour, plant and transport costs

Recent research by Barnes and Thompson[31] has shown that most of the difference between contractors' tender prices for the same job arises from differences in estimates of direct labour and plant costs and not, as is commonly supposed (e.g. C.E. procedure,[1] para. 70) from differences in the allowances for risk and profit. It follows that this is an area where much money is easily lost and so the estimator has to deploy his maximum skill and judgement.

Direct labour and plant cost per unit of output
= Manhours per unit of output × total cost of a manhour
+ Machine hours per unit of output × total cost of a machine hour

Manhours and machine hours per unit of output are difficult to predict accurately. There are various books of standard output data but on all but the trivial bill items experienced estimators rightly prefer to use output data they believe from jobs where they know enough about the conditions to be able to judge how to apply the outputs to the work being priced.

The *total cost of a manhour* will include some or all of the items listed under remuneration (p. **31**–44). Unlike outputs, rates per manhour can be established pretty accurately at the time of tendering but at present anyway (1973) are extremely difficult to predict over the duration of the contract. Promoters who require their contractor to carry this risk get correspondingly inflated tenders.

The *total cost of a machine hour* normally includes depreciation, fuel and lubrication, grease, consumables, spares, relevant insurances, and operation and maintenance labour costs. Contractors who own plant normally operate an internal hire system in which plant is hired to the contract at a standard rate which covers some or all of the items listed above (practices differ in different firms). Rates for plant hired from specialist plant hire firms are normally negotiated on the basis set out in the Contractors' Plant Association Schedule of Rates of Hire.[46] This matter is covered more fully in Section 35.

Estimating materials costs

The way in which materials prices are incorporated in the Contractor's tender is shown diagrammatically in Figure 31.14. Just as the Engineer endeavours to place the contract with the lowest reliable contractor, so does the Contractor attempt to obtain his materials from the lowest reliable supplier. In both cases disappointments occur, and the chances of success are greatest when the relationship is a continuing one, not just for the one contract.

On unit rate items there is no point in the Contractor trying to take off the materials quantities with great accuracy because they are very likely to change, often substantially.

On the other hand, for nonvariable lump sum or design and build bids or on any individual bill items where the Contractor bears the risk of increased quantities (e.g. in overbreak concrete on a rock foundation), he has to consider most carefully the quantity to allow for in his pricing. This quantity then becomes an integral part of the estimate data that is fed forward to the contract staff for their cost control.

Subcontractors and estimating

Subcontractors' prices are incorporated in the bid price by the procedure shown in Figure 31.15 and there is nothing further to be added here.

Estimating overheads

The list in C.E. Procedure,[1] Appendix 1 is pretty comprehensive. However, the author treats continengencies (= risks), profit and financing as separate items because they are closely linked and fundamentally different in character from the other types of overhead in the list.

Estimating—allowing for risks and profit

(See also pp. **31**–8, **31**–50, **31**–51)

Risk is a central feature of civil engineering contracting and a large part of the so-called profit markup in a tender is really the Contractor's cover for risks that he has to accept. It follows that the more risks tenderers are asked to accept, the higher will be the offers from experienced contractors. See C.E. Procedure,[1] paras. 54 and 70. Judging the amount to be added to a tender for profit and risk cover is a delicate policy decision which

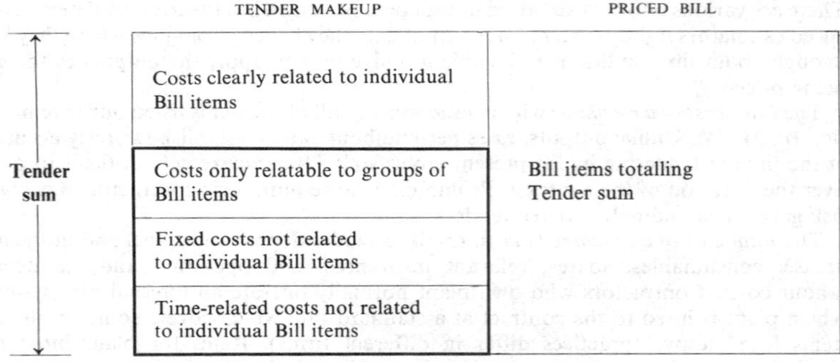

Figure 31.7

is usually taken at director level. A contractor often seeks to reduce his risks by 'qualifying' his tender that is, making it conditional on some amendment to or interpretation of the contract documents. It is common to stipulate that no qualifications to tenders will be allowed. It is equally common for such stipulators to accept qualified tenders! See C.E. Procedure,[1] para. 73 and Twort.[4]

Risks arising from the *resources* to be managed and from the other parties to the contract are discussed on page **31**–50. The *bill* itself also presents risks because, normally, a large proportion of the tender sum has to go on to a small proportion of the bill items. As his pricing of the bill starts to take shape, the estimator will re-examine his makeups for the large value items in the bill and ask himself for each item:

1 Is my pricing of this bill item and particularly the labour and plant outputs as realistic as I can make it?
2 Is the work covered by this bill item likely to be increased/decreased/delayed?

The rate finally entered for each large value item will reflect the contractor's overall assessment of the risks in that item.

Estimating: allowing for financing the work

See pages **31**–39 and **31**–49–**31**–50 and C.E. Procedure,[1] para. 76 and appendix 1, para. 15.

Estimating: allocating the tender sum to the bill items

Materials cost can mostly be allocated directly to individual bill items but much of the rest of the tender sum covers operations which bear no simple relationship to individual bill items. The contractor therefore has to decide how to allocate these sums to the bill items. His problem is best illustrated by a diagram, as in Figure 31.7.

The percentage of the total tender sum made up of items within the heavy box depends so much on the type of work that any figures would be misleading. However, the percentage suggested by this diagram is not fanciful.

The contractor has to spread the monies within the heavy box on to the bill items so as to provide as best he can for the following:

1 Protection against quantity changes. Recent research[31] has shown that on a typical civil engineering contract only 30% of the tender sum covers work whose value survives unchanged in the final account
2 Financing the early stages of the job. See also page **31**–36 and C.E. Procedure,[1] para. 76.

Obviously each contractor makes his own judgement on these matters and the only thing that is certain is that the greater the proportion of the tender sum that falls inside the heavy box, the greater the variation there will be in the rates offered by different contractors for the same bill item and the more money worries there will be for everybody if the work is varied or delayed. See for example Haswell[40] and Stark.[42]

Financial accounting and costing

These two functions have quite different purposes but they are easily confused because they use similar staff and similar procedures which often interlock.

Financial accounting

Is primarily concerned with the historical recording of receipts and payments of cash, amounts owing to or by the business and meeting the requirements of the Law. Though some accounting work is done on site, it is primarily a company (rather than contract) function.

Site accounting is described briefly in C.E. Procedure,[1] paras. 162–164. The engineer who wants simple explanations of such common terms as 'double entry', 'trial balance', 'imprest', discounted cash flow', etc., may consult Allen.[36] Calvert[3] also has a useful short chapter on finance including some aspects of company law and accounting.

Costing and cost control

Costing means calculating what things have cost in order to help managers to do their job better. In fact, in manufacturing it is now treated as one part of the rather grander-sounding function of 'management accounting', but this term does not yet seem to have caught on in construction. Costing is a less precise art than accounting because its prime purpose is to influence future events, not to count candle-ends.

The essential features of any costing system are:

1 *Allocation* of all monies spent to a list of items whose cost it is desired to know. Such items usually have *code numbers* to facilitate allocation and the list of such numbers is usually called a *cost code.*

2 Calculation of the cost of each item at frequent intervals.

Cost control strictly means 'comparing actual costs with what you think they ought to be' (see p. **31**–7). However, it is often also used in the wider sense of 'influencing costs', as will be seen in the next paragraph.

Cost control by the Engineer—general

The one term is used to describe four very different activities:

1 Costing alternative configurations at the investigation/report stage and choosing the one that is economically the most attractive. This is outside the scope of this section but is listed here because it is the essential precursor of Nos 2 and 3. (In fact there are certain parallels between an Engineer's report and a Contractor's tender. Both provide a basis for controlling the cost of the subsequent work.) For further information the reader may refer to C.E. Procedure,[1] para. 34 and to Armstrong.[37]

2 Controlling the estimated cost of the work during the post-report precontract design phase to try to ensure that the value of the contract when placed does not exceed the corresponding figure accepted by the promoter at the report stage. This operation lies at the heart of all engineering design which is the subject of numerous other sections in this book. The Engineer will do check cost estimates (see p. **31**–31) on his design as it develops but his success in controlling the tender sum depends mainly on

(a) The soundness of the design appreciation and cost estimates in his report (see 1 above)
(b) The extent to which he can hold the promoter to the scheme set out in his report
(c) The degree to which his contract documents and his own reputation convince the tendering contractors that his scheme has been properly thought out and will impose no unreasonable risks on the successful tenderer.

3 Controlling the cost of variations and extras during the contract phase to try to ensure that the final account does not exceed the tender sum. This is dealt with under Payment to the Contractor (see p. **31**–48).

4 Controlling his own staff and overhead costs that have to be covered by his fees. This process is analogous to that used by contractors to control labour, staff and associated overhead costs for which see below.

Twort[4] and Marks, Grant and Helson[5] have some useful notes on cost control by the Engineer.

Cost control by the Engineer—savings on materials

Every civil engineering designer regards savings on materials as one of his main criteria of success. In fact, designers who rely solely on priced bills for their cost information tend to err towards overcomplex designs in which much of the saving on materials is cancelled out by extra construction costs arising from the complexity of the work. This problem arises because priced bills in their present form suggest to the unwary that costs vary linearly with quantity and offer little information on the real costs of variety and complexity. The problem is now widely recognised and is treated at some length by Creasy.[33]

Costing by the contractor

Contractors cost their operations for three main purposes which are, in ascending order of difficulty:

1 *To record what work has cost*

e.g. (a) To help in settling the final account
(b) Sometimes on reimbursable work to record the cost to the promoter under subheadings of his choosing.

2 *To help forecast the cost of work*

e.g. (a) To help pre-price variations, extras, etc.
(b) To compare the cost of alternative ways of doing work that is still to be done
(c) To forecast the total cost to complete work that is in progress
(d) To provide data for future tenders.

3 *To help prevent the work costing more than it should*
i.e. to help those supervising the work to run it efficiently.

The problem is that each of these three purposes demands a somewhat different system. Thus a contractor's costing system is normally a compromise designed to serve all three purposes reasonably well and reasonably economically but usually accenting purpose number 3.

The anatomy of costs

The problems of costing are further illuminated by Figure 31.8. This shows the relationship between the headings under which money is spent and the headings under which it is recovered through the bill. The contractor's cost code (Col. C) for a contract is a list of headings under which costs involving more than one type of resource can most conveniently be compared with the estimate. It has to be a compromise between the bill items, the estimate makeup and the way the job is being executed. It usually consists of suitable subdivisions of the four boxes shown in Col C of Figure 31.8. Some items will be lump sums but many will be cost rates per unit of work (hence the term 'unit costs'). The same code can of course be used at levels A and B to classify costs involving only one type of resource.

The different purposes of costing discussed below ideally require data to be collated at different levels in this diagram.

Costing to record what work has cost

Conceptually this is the simplest form of costing. One simply pours all the expenditure through the flow chart (Figure 31.8) to arrive at an end of job total cost for each item in the promoter's code (most probably subdivided into labour, plant, materials, subcontractors and miscellaneous). The practical difficulties arise in formulating and agreeing with the promoter a rational way of allocating to his code the often large sums which fall into boxes 2, 3 and 4 of the contractor's cost code and are often some-

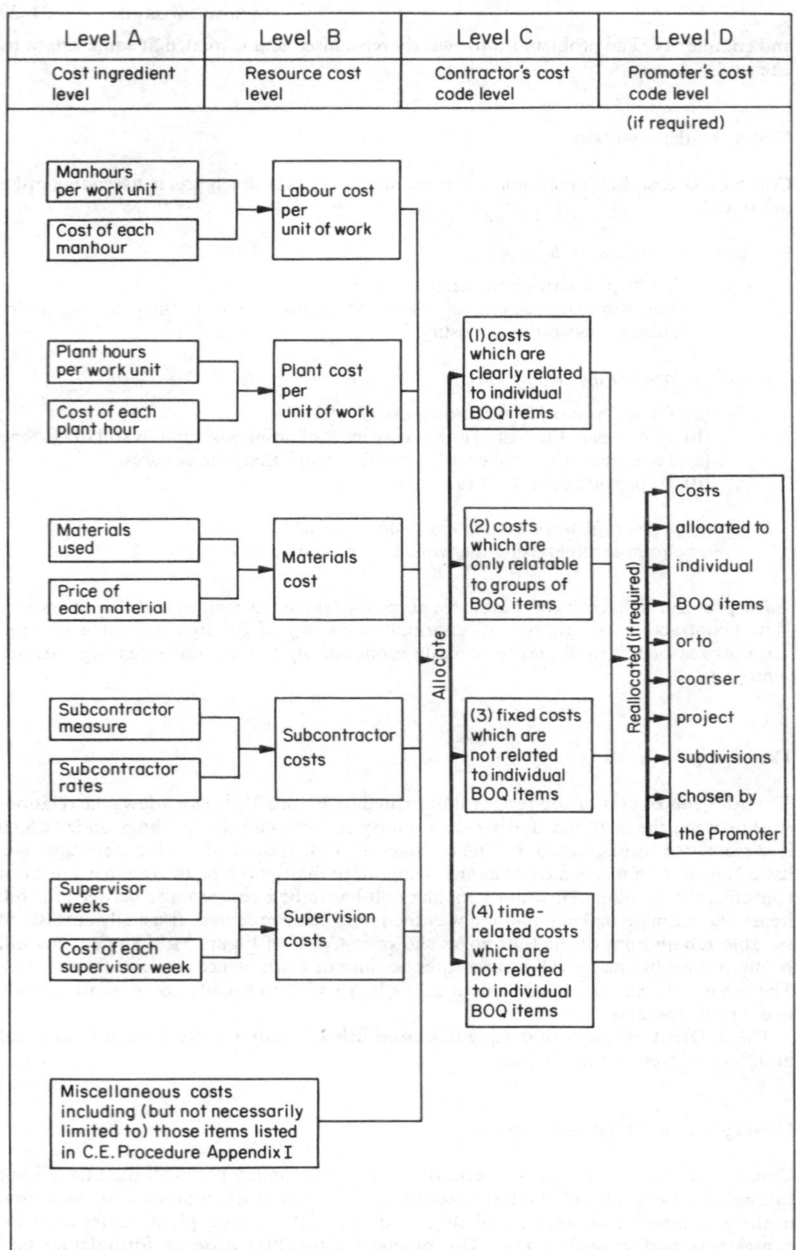

Level A
Cost ingredient level
Level B
Resource cost level
Level C
Contractor's cost code level
Level D
Promoter's cost code level
(if required)
Manhours per work unit
Cost of each manhour
Labour cost per unit of work
Plant hours per work unit
Cost of each plant hour
Plant cost per unit of work
Materials used
Price of each material
Materials cost
Subcontractor measure
Subcontractor rates
Subcontractor costs
Supervisor weeks
Cost of each supervisor week
Supervision costs
Miscellaneous costs including (but not necessarily limited to) those items listed in C.E. Procedure Appendix I
Allocate
(1) costs which are clearly related to individual BOQ items
(2) costs which are only relatable to groups of BOQ items
(3) fixed costs which are not related to individual BOQ items
(4) time-related costs which are not related to individual BOQ items
Reallocated (if required)
Costs allocated to individual BOQ items or to coarser project subdivisions chosen by the Promoter

Figure 31.8 The anatomy of costs (*see p.* **31**–39)

what loosely classed as 'overheads' or 'on-costs'. This is why complete costing to all the bill items is seldom attempted in this country (though it is common in the USA where bills are, by UK standards, ludicrously small).

Costing to help forecast the cost of work to be done

For this the most useful information is the basic ingredient data at level A over a representative period of the contract, together with a close knowledge of the site conditions during that period.

If the data are being used for forecasting the cost of work still to be done on a contract, the person using it will know the contract well and will make almost intuitively the necessary adjustments when future conditions are expected to differ from those prevailing when the data were collected.

If, however, the data are being collected as an aid to pricing some future job as yet unspecified, they are useless to the estimator unless accompanied by:

(a) The final cost record for the job (as a check on the quality of the detailed data at level A)
(b) A very complete statement of the methods and plant used, the site conditions, the rate of progress at different stages of the job and all the other factors that influence cost.

(b) is a fairly tall order on contractor's site staff whose prime objective is to get the job built and so the estimator's best way of getting the figures into context is often to interrogate one or two people who know the job well.

Costing as an aid to job control, i.e. 'cost control'

To the contractor this is the most vital type of costing because it is the only really watertight way of tracking down inefficiency and waste promptly enough to be able to take corrective action. It is also the most difficult because compared with the two types of costing discussed above:

(a) It requires each cost figure to be 'controlled', i.e. compared with a 'value' figure which is what the work should have cost. Such value figures are normally derived from the tender estimate

(b) It requires cost/value comparisons to be prepared at frequent intervals and presented to management within a few days of the end of the period to which the figures refer

(c) Cost control systems rightly make site supervisory staff fear they are being watched and judged by people who only see the figures and don't know the full story. They therefore have a powerful incentive to arrange the costs wherever possible so as to conceal unpleasant facts

The need to compare costs with estimate figures means that the structure of the cost code must be influenced by the structure of the tender estimate and that even so quite a lot of work often has to be done post-tender to recast the estimate figures in a form suited to cost control.

The need for speed means that various shortcuts have to be taken. For example, notional or approximate figures often have to be used for plant hire and materials whose exact invoice cost is not yet known.

The need to take account of the defensive reactions of those whose performance is being watched means that cost control reports have to be checked and 'filtered', e.g.:

(a) Check cost measure against monthly certificate measure
(b) Check recorded costs against corresponding sums in company accounts
(c) Limit the amount of detailed cost information that goes to senior people who know little about the job.

See also page **31**–46.

Contractors' cost control systems

Cost/value comparisons at level A (Figure 31.8) give the most direct and prompt aid to job control.

However, on their own they are seldom sufficient because it is difficult to apply them to the whole of a job and unless the whole job is covered it is difficult to run the necessary checks to ensure that all relevant costs have been included. So a more complex system is needed.

Contractors' cost control systems vary in detail but there seems to be fairly general support for a system on the lines set out in Table 31.4.

The logic behind this seemingly complex system will be apparent from the paragraphs below which deal with each type of expenditure.

Table 31.4

	Cost/value comparisons for job control				
	Weekly		*Monthly*		
Type of expenditure (as Fig. 31.8)	*Level (as Fig. 31.8)*	*Allocated to cost code?*	*Level (as Fig. 31.8)*		*Allocated to cost code?*
	Primary only		*Primary*	*Check*	
Labour	C	Yes	—	B	No
Plant, etc.	Often supplemented by partial data on major items at level A				
Materials	—	—	A	B	No
Subcontractors	—	—	A	B	No
Supervision	As for labour and plant	Yes but no problem as it forms its own code(s)	A	B	No
Miscellaneous	C Some items sometimes	As for supervision	—	B	No

Labour and plant cost control—general

Labour and plant costs are the ones most likely to go astray and so they are the ones that are controlled most frequently (usually weekly) and in most detail.

The principles of labour and plant cost control are shown in Figure 31.9. The divided heavy boxes show the points at which actual may be compared with estimate in order to

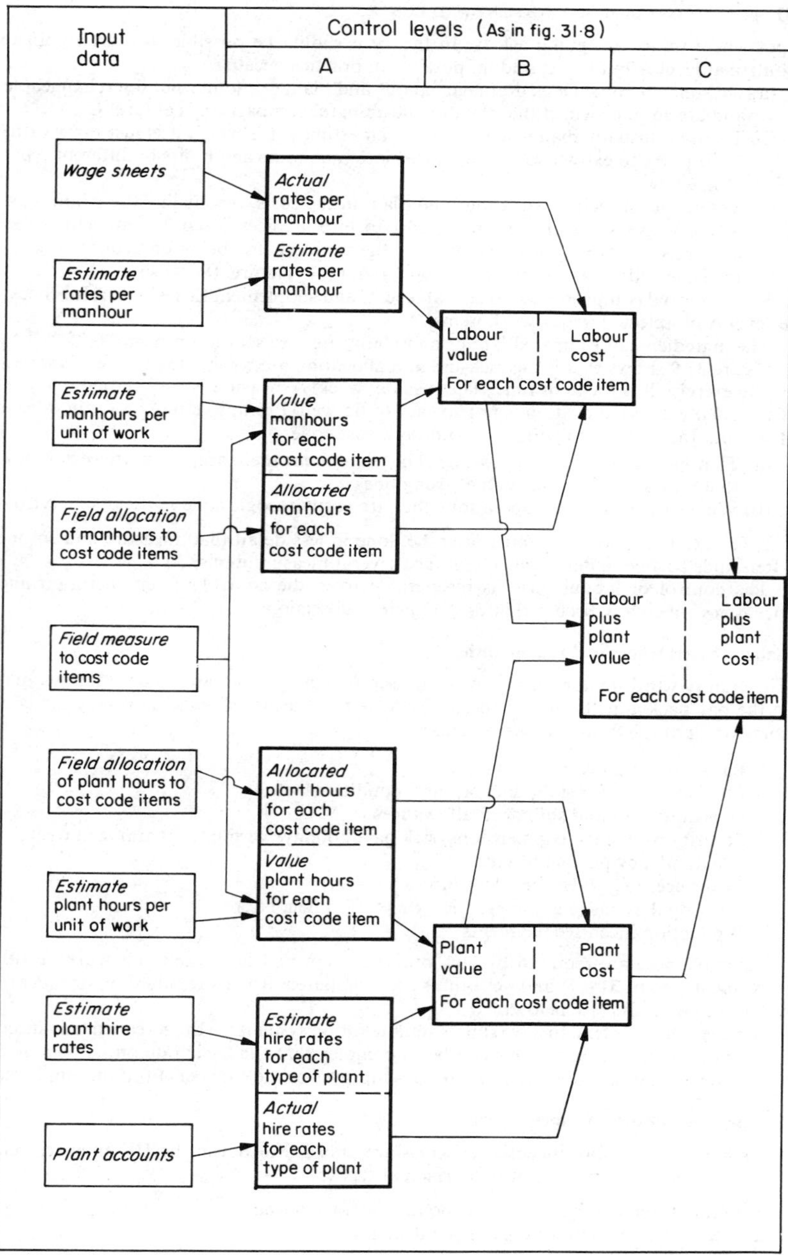

Figure 31.9 Labour and plant cost control. Each heavy box with a central dotted dividing line is a point where a cost control comparison may be and often is made (*see p.* **31**–42)

spot where things are going astray. In theory it should be possible to exert complete control at level A but this is seldom possible in practice because:

(a) On some items, such as drainage, labour and plant are somewhat interchangeable and are so intertwined that the only meaningful comparison is at level C
(b) There are usually many small items in an estimate for which it is just not worthwhile to try to extract values separated into manhours and hours of different types of machine
(c) Yet one must include all labour and plant in the costs, reconciling the labour cost to the wages sheet and the plant cost to hire invoices or equivalent. Otherwise unwanted costs may leak away from the visible items being controlled into an invisible sink so as to present a more favourable picture. (See also p. **31**–41.)

So the normal compromise is to cost at level C and supplement this with careful checks at level A on selected major work items.

The ingredients of estimated labour and plant rates are set out on page **31**–35.

Figure 31.9 shows that if the measure and allocation are wrong, the whole of the rest of the exercise is a waste of time. The (usually weekly) measure is easy though tedious. The (usually daily) allocation is not so easy to do well. Prompt vetting of all timesheets helps, but the best aid to good allocation is a cost code:

(a) That has as few items as possible. The more items there are, the more certain it is that allocators will choose the wrong ones
(b) Whose items describe operations that are easily recognisable by field supervisors

It follows that allocation must often be done in less detail than measure and so one often finds cost code items which embrace several measure items.

The control of labour costs is inseparable from the control of remuneration and incentives which are discussed in the following subsections.

Labour costs: wages and remuneration

The clumsy word 'remuneration' is often used to signify something more than just pay in the pay packet at the end of the week. A 'remuneration package' normally includes some or all of the following ingredients:

Basic rate (usually per hour)
Plus rates (e.g. for skill, danger, dirty conditions, etc.)
Overtime rates and shiftwork allowances
Security payments (e.g. pensions, sick pay, holidays with pay, guaranteed week, redundancy payments, etc.)
Insurances (e.g. NHI, accident insurance, etc.)
Individual extras (e.g. long service, time off in lieu, etc.)
Production incentive payments

Some of these are required by law, others are covered in the relevant working rule agreement (see p. **31**–67) and yet others are negotiated at the site, notably, production incentive payments (see below).

Simple countries tend to have simple remuneration systems. In the so-called 'advanced' countries remuneration systems can be so complex that the basic rate on its own gives very little indication of the true cost to the employer or the true benefit to the employee.

Production incentive payment schemes

On many overseas jobs incentive schemes are unheard of but in the UK they are very common. Such schemes are of two principal types:

1 Bonus schemes—base rate plus bonus for extra output
2 Piece work and labour only subcontracting.

The two are in theory quite different but in practice not so different because they both involve frequent bargaining between employer and men to fix the 'right' rate for the job.

Bonus schemes are described briefly in C.E. Procedure,[1] paras. 153–156 and some notes on the formal framework are given in the relevant working rule agreement. Calvert[3] gives a more detailed history and description of both types of scheme.

The uninitiated suppose that incentive schemes reduce the need for supervision. This may be so on highly repetitive work but usually they require increased effort by supervisors, particularly to ensure a smooth flow of work for the men and to control workmanship, safety and material wastage.

This is a subject on which ideas of what is right and fair vary a lot and change rapidly, a subject on which a little knowledge is a dangerous thing. Wherever an incentive scheme is used it accounts for a substantial and rather unpredictable proportion of the total cost of labour and needs to be handled with great care and skill. The reader who wants a good insight into some of the problems should study a recent short ICE paper by Baird.[45]

Materials cost control—General

Unlike manhours or plant hours, materials can be identified and checked before and after they are used in the works. Therefore there is no point in converting materials into money and allocating the money to the cost code items as is done with labour and plant. Material costs can be controlled entirely at level A (Figure 31.8) by:

(a) *Physical controls:* Checking quantities used against corresponding quantities allowed in estimate and locating/preventing losses and waste. (See below.)
(b) *Financial controls:* Checking prices paid against corresponding prices in estimate. (See below.)

Materials cost control—physical controls

There are five main stages:

1 Ensure that no materials are ordered in excess of those allowed in the estimate without the reasons being known. Of course this is difficult in work where the quantities keep changing but it has to be attempted for important materials. (See also p. **31**–17 and Figure 31.14.)
2 Ensure that materials invoiced by supplier are in fact received on site. (See p. **31**–74.)
3 Handle and store materials on site economically and securely. (See pp. **31**–71 and 74.)
4 Ensure that materials issued from stores are correctly used in the works. (See pp. **31**–74 and 75.)
5 Ensure that surplus materials are gathered up and disposed of advantageously. (See Figure 31.14)

Materials cost control—financial controls

There are three stages:

1 Check order prices against prices used in estimate
2 Check invoice prices against order prices. (See Figure 31.14)
3 Run a regular overall financial control. (See below)

Overall financial control of materials

As an overall check on the efficiency of the whole materials management operation (estimating, purchasing, checking, storing, handling on site, use in the works, inclusion in valuation) it is customary to run a regular (e.g. monthly) contract financial profit and loss account for materials. This is based on the simple formula:

Value of		*Cost of*
Materials in the works		Materials received and paid for
+ Materials in stock on site	−	+ Materials received and not paid for
		− Materials paid for but not received

= Materials profit (or loss if negative)

The necessary cost and value figures may be computed globally for the contract or they may be 'coded', i.e. subdivided into more or less logical groups of similar materials. Such coding makes it easier to track down losses and discrepancies but if too detailed a code is used, miscodings proliferate and defeat the whole object of the exercise.

Control of payments to subcontractors

The contractor should find this pretty easy because the principles and problems closely parallel those involved in the engineer's control of payments to the contractor (see p. **31**–48). The main points that normally need watching are:

(a) Agree measurements promptly excluding defective work until it is rectified
(b) Pay the agreed amount promptly. Subcontractors, especially the small ones, are always short of cash. It is not clever to squeeze a subcontractor too hard. He will allow for it in his price on his next job for you, or worse, he will go broke and disrupt your present job
(c) Recover promptly monies for materials and services supplied to the subcontractor, especially any that were not envisaged when the subcontract was drawn up
(d) Ensure that variations on s/c works are properly covered by Engineer's v.o.
(e) Watch s/c dayworks closely!
(f) Ensure s/c is not holding (and being paid for) excessive material stocks on site
(g) Guard against gross errors by periodically reconciling amounts certified to subcontractors with corresponding amounts certified by Engineer.

Control of supervision and miscellaneous costs

Like materials and subcontractor costs, these costs do not need to be allocated to the work items in the cost code. They are best controlled by comparing actual expenditure with an expenditure programme derived from the estimate and divided into suitable headings which form a separate group of items in the cost code.

Overall financial control of the contract

Pages **31**–42 to 46 describe a system which should provide an infallible early warning of anything going wrong on the contract. However, the people who prepare costs are far from infallible and so prudent contractors run a monthly or quarterly overall financial control on each contract which is based on the simple formula:

Amount certified by Engineer ± adjustments (see following below)	Should equal	Amounts paid through company accounts + Amounts owing to others − Amounts recoverable from others + Allowance for H.O. overheads to date) + Expected profit to date

These may be global figures for the contract or they may be divided into labour, plant, materials, subcontractors, supervision and miscellaneous (as in Figure 31.8) so that when the two sides do not agree it is easier to find out why (e.g. see p. **31**–45, 'Overall financial control of materials). The essential point is that it is a check on the financial health of the contract from information that is quite independent of the contract cost control system, rather like triangulating an important reference point from two separate baselines, and so serves as an audit on the cost control system.

As portrayed here the procedure looks very simple and so it is in principle, but in practice the right-hand side requires a lot of detailed clerical work which is prone to errors, while on the left-hand side the adjustments require the contractor to exercise considerable judgement. These adjustments are of two main types:

1 As explained on pages **31**–37 and **31**–39, there is seldom a simple relationship between the way the contractor spends money and the way he recovers it through the bill, so this relationship has to be assessed each time the control check is done.
2 A realistic judgement has to be made on what will be the final outcome on work done for which rates have still to be negotiated.

Graphs for cost control

It is often useful to present the cost figures for the job, or for certain important parts of it, in graphical form.

The simplest graph, which is equally useful to agent and resident engineer, is a financial progress graph based on the monthly certified payments of the type shown in Figure 31.4a. (See also Twort.[4])

Actual versus budget costs can be plotted as shown in Figure 31.4(b) and this information may be combined with a curve of achieved value to give a triple 'S' curve as shown in Figure 31.4(c). Sometimes, however, even if colours are used, the triple S looks confusing and it is better replaced by three separate graphs each with only two curves on it.

Such simple graphs display the cost history and trends much more compellingly (at least to engineers) than do rows of figures, and can be used on any job. Other more specialised types of graph may be needed to bring out the vital cost facts in special situations.

Organising the cost control function

As will be clear from the relevant paragraphs in the grid index (Figure 31.1), cost control on materials and subcontractors is an integral part of managing the resources and so presents few organisational problems.

On the other hand, labour and plant cost control is usually somewhat separate from the day-to-day management in the field and so can present problems. Cost control is best seen as a service to management and it is useful to think of it in terms of customers and suppliers.

The customers are:

1 The agent (and sub-agents on large sites)
2 The contracts manager (and possibly others at head office)
3 The managing surveyor (see C.E. Procedure,[1] para. 128)
4 The estimators.

Their needs all differ and are sometimes almost impossible to reconcile.
The suppliers are a loose-knit team made up of:

1 Gangers and foremen who may not find it easy to identify the relationship between their work and the cost code items
2 Site engineers who may not understand exactly how their cost measurements are used
3 Cost clerks who may have little technical understanding of the work and may not be encouraged to 'waste time swanning round the site'
4 The agent himself, when he supplies figures to head office.

Different contractors handle these problems in different ways but the author believes that someone like the 'chief production and costing engineer' described in C.E. Procedure,[1] para. 125 is the best person to coordinate the suppliers, keep the cost code up to date and keep the customers satisfied.

Payment to the contractor

This subject is well covered in C.E. Procedure,[1] paras. 76 and 175–189 and in ICE Conditions[2] (especially clauses 60 and 61 but so many other clauses are relevant that they cannot all be listed here). The reader will find much useful further information in Marks, Grant and Helson[5] and Twort.[4] In view of this, the following subsections are confined to some brief remarks on claims, arbitration and litigation.

Disputes and claims

C.E. Procedure,[1] para. 181 discusses the legal meaning of the word 'dispute' and then defines a 'claim' as a '. . . dispute where the Contractor has objected to a decision of the Engineer'.

This definition is widely accepted and so the word 'claim' carries, in many people's minds, an aura of contention which is most unfortunate because the same word is commonly used in a much wider sense to describe any *request* by the contractor for payment for any expenditure or loss due to circumstances which he considers he could not have been expected to foresee or allow for in his tender. As Marks, Grant and Helson[5] put it, a claim under the ICE Conditions[2] 'means a demand for recognition of one's rights which can either be accepted or rejected' (not just one that has already been rejected).

Claims as thus defined most commonly arise from adverse conditions and artificial obstructions (ICE, clause 12) and ordered variations (ICE, clauses 51 and 52). They are discussed briefly in C.E. Procedure,[1] paras. 181–184 and at some length by Marks, Grant and Helson,[5] Twort[4] and in the NEDO Report.[6] Furthermore, each organisation within the industry has its own rich folklore on the subject which is transmitted by word of mouth from senior to junior down the years. The reader who consults the references given above and picks up what he can of the folklore of his own organisation will realise that in the matter of claims things are seldom what they seem and a little knowledge is a dangerous thing.

In the author's opinion the fundamental problem is that a claim of any size is so often a threat to the essential working relationships between those parties involved in a contract. Thus, the Contractor may be reluctant to pursue (or sometimes even to notify) a claim during the course of the work for fear of getting at loggerheads with the Engineer and making all the rest of the work more difficult for everybody. Similarly, the Engineer, particularly if he is the promoter's salaried employee, may feel he needs to demonstrate a certain stiffness about claims in order to retain the promoter's confidence. This in turn tempts the Contractor to present inflated claims so as to give the Engineer the opportunity to make impressive-looking cuts. This may provide an easy way out on a particular contract, but cumulatively it naturally lowers the Engineer's confidence in the veracity of all claims.

The inevitable result of all this is that whatever the contract may say, many claims are left in a woolly state (some call it the 'sin bin') and sorted out after the job is complete. Even then, the settlement that emerges very often depends more on the desire or otherwise of each of the three parties to continue their working relationships on future contracts than it does on the contractual merits of the situation. Couples that want to stay married seldom take their disputes to court! (See also p. 31–59.) The ICE Conditions[2] have now been substantially revised in an attempt to contain this problem.

Arbitration and litigation

The procedure for going to arbitration is outlined in C.E. Procedure,[1] para. 190 and in ICE Conditions,[2] clause 66. The advice most commonly given about arbitration and litigation is 'avoid them if you can', i.e. one should first try every other means of reaching a settlement. However, when all else fails, the two parties can ultimately resolve a dispute by appointing an arbitrator or by taking the matter to the High Court. Both ways are long-winded and expensive. To some extent the aggrieved party can choose between the two and the pros and cons are summarised in Table 31.5.

Table 31.5

Arbitration	*Litigation*
(1) *Slow* Procedure itself is slow. The two parties have to try to agree on an arbitrator. Arbitrator has no power to call for documents or witnesses.	(1) *Relatively quick* Court has powers to call for documents and witnesses.
(2) *Arbitrator understands the subject* Will grasp technical evidence quickly Will spot unsound technical evidence	(2) *Court is composed of laymen* Do not grasp technical evidence quickly May accept spurious technical evidence blindly
(3) *Can be relatively cheap* But winner is not always awarded his costs	(3) *Can be very expensive* Due partly to (2) above.

Arbitration and litigation are matters for the experts and an expert would probably criticise the above table as an oversimplification . The interested reader will find further useful information in Marks, Grant and Helson[5] and Twort[4] but he should never make any irrevocable move in this area without getting expert advice.

Financing and cash flow

For the purposes of this section the following rather simple definitions will serve.

Financing means arranging, usually by borrowing, to have enough money in the bank to pay wages, salaries and other bills when they fall due, which normally means before one can recover the appropriate amounts from others.

Cash flow means inflow and outflow of cash.

A 'favourable cash flow position' means 'there's money in the bank and what's coming in is at least equal to what is going out'.

An 'unfavourable cash flow position' means 'there is serious trouble with the bank manager and bankruptcy may not be too far away'.

Discounted cash flow (*DCF*) is a rational system for adjusting cash flow calculations to take account of the timing of the flows. It is widely used in project appraisal which is outside the scope of this section. It seldom figures as 'DCF' in day-to-day contract management, though every contractor knows and continuously applies the underlying principle which is that a pound paid now is worth more than a pound payable a year hence.

Financing and cash flow are much more widely discussed than they used to be. They are not new problems but problems that have become much more acute because of inflation and high interest rates. The interested reader can find much additional information in Ballard,[38] Parker[41] and NEDO.[6]

Financing a contract

This is touched on in C.E. Procedure,[1] para. 55 and appendix 1. Normally the promoter carries most of the burden of financing the contract but when the contractor is required to raise a substantial portion of the finance, the terms on which he can do this may determine whether or not he wins the job.

Under the normal system of monthly certificates the amount of capital the contractor has to 'invest' in the job is obviously very small (it may be as little as 2–3 months' outgoings on the contract). This is all right as long as things go well but if he gets into difficulties and cannot quickly get additional payment, the amount he has to invest in the contract escalates rapidly, he gets into an 'unfavourable cash flow position' and even if he has the right to some or all of the additional payments, he may go bankrupt before he can establish this right.

Improving the contractor's cash flow

Here the engineer has always had some room for manoeuvre notably under ICE Conditions (4th edition), clause 60 (advances on materials, temporary works and plant). However the matter was recently examined by a NEDO committee[6] who found that there was considerable room for improvement. The problem is not so much that any particular level of financing by the contractor is in itself good or bad, but that the system as it normally operates, makes it very difficult for the contractor to forecast how much finance he will in fact have to find for any particular contract. The contractual position has been much improved in the 5th edition of ICE Conditions,[2] clause 60.

Of course the contractor can seek to help his financing by unbalancing the tender rates (see also p. **31**–37 and Ballard[7] and Stark[42]) but the benefit that can be gained in this way without endangering the final account is very limited indeed.

CONSTRAINTS, RISKS, INSURANCES

Coping with constraints

Contract management is as much a matter of overcoming difficulties as it is of achieving results. These difficulties or constraints form a spectrum ranging from absolute certainties through familiar risks to happenings that no reasonable person could be expected to foresee or provide for. (See also pp. **31**–21, 30, 35 to 37).

Each of the three parties to the project has to provide for constraints as he sees them. The basic principles are shown in Figure 31.3 and the main steps are as follows:

1 *List sources of constraints*
   These will be in two main categories
   (a) The resources to be used
   (b) The other parties to the project
2 *List possible constraints from each source*
   Some typical constraints associated with particular resources are shown in Table 31.6

Constraints imposed by the other parties will include:

(a) Things covered in the contract

(b) Behaviour that can be predicted from the previous behaviour of the party concerned

(c) Behaviour that cannot be predicted confidently

3 *Try to assess the degree of risk in each possible constraint*

4 *Decide how to allow for each constraint in the plan of work*

5 *Decide how best to provide for each identified risk*

6 *As the work progresses, record the actual effect of each constraint* and compare it with what was expected.

Table 31.6 CONSTRAINTS ASSOCIATED WITH PARTICULAR RESOURCES

Resource	*Certainties and near certainties*	*Typical risks*
Site	Ground conditions, climate, access, working space. (CE Procedure[1] para. 30)	Unforeseeable ground conditions Unforeseeable artificial obstructions
Project information Drawings Specifications Instructions	Dependence on design information from suppliers Need to place contract on incomplete information Foreseeable delays in producing post-tender design information	Information not issued when promised Information issued and altered later Information full of discrepancies
Men	Key staff available or not available Type of labour available at site	Unexpectedly low productivity Need to pay unexpectedly high rates to hold men on site Industrial disputes
Plant	Influence of type of plant available on design or construction methods (often minimal, sometimes crucial)	Major accidents to key items Unexpectedly large downtime due to problems arising from e.g.: plant design, spares, climate.
Materials	Is desired material so difficult or expensive that an alternative should be considered? Realistic lead time for each material Import controls or bans (especially on overseas work) Climate requiring special storage for certain materials	Loss or damage in transit Suppliers and shippers fail to meet their promised dates Fire or storms on site Unexpected influence of unfamiliar climate on certain materials
Subcontractors	Influence on design (esp. nominated subcontractor) Influence on method and timing of main contractor's work	Failure to meet programme dates Labour employed on different terms upsets main contractor's labour
Third parties	See pp. **31**–19 and **31**–80	See pp. **31**–19 and **31**–80

Insurance and other ways of providing for risks

The way risks and liabilities are shared between promoter and contractor has an important bearing on the contract price. (See C.E. Procedure,[1] paras. 54, 74, 75) and may even decide the form of contract used. (See C.E. Procedure,[1] para. 52.)

This is a complex matter and it is useful to set out a checklist of sources of risk and ways of providing for them in tabular form as in Table 31.7

Table 31.7 CHECKLIST OF PRINCIPAL SOURCES OF RISK AND WAYS OF PROVIDING FOR THEM

Potential sufferer	*Potential offender*	*How potential sufferer can provide for the risk*
Promoter Engineer Contractor Contractor's Men Plant Suppliers Subcontractors Third parties	Promoter Engineer Contractor Contractor's Men Plant Suppliers Subcontractors Third parties The ground The elements	(i) Make potential offender liable (and exact security where possible) (ii) Get another member of the project team to insure the risk. (iii) Insure the risk himself. (iv) Accept the risk. (v) Make another member of the project team accept the risk even though it is not insurable

Each 'potential sufferer' carries various risks in his dealings with each of the 'potential offenders' and he normally provides for each risk in one of five ways:

1 *Make potential offender liable and exact securities where possible*
This is the obvious way when it can be made to work. Examples are ICE Conditions,[2] clauses 20(i), 47(i) and 49. Securities are touched on in C.E. Procedure,[1] paras. 77 and 89.

2 *Get another member of the project team to insure the risk*
This again is common practice. Examples are ICE Conditions,[2] clauses 21, 22, 23, 24, 25.

3 *Insure the risk himself*
There is a separate section on insurance and so little need be said here. The contractor is contractually obliged to insure some risks and will insure others on his own account. The Engineer will normally have a professional indemnity insurance to cover unforeseen liability to the promoter.

4 *Accept the risk*
For some risks the premium is so high that it seems better to 'take a chance on it' or, more politely 'act as one's own insurer'. The 'excepted risks' in ICE Conditions,[2] clause 20(2) are of this type and are normally accepted by the promoter.

5 *Make another member of the project team accept the risk even though it is not insurable*
In practice this normally arises when a buyer is in a strong enough position to be able to offload the risk on to a seller who feels he must accept it if he wants the work. That a strong buyer is tempted to do this is obvious, but he is not always wise to yield to this temptation. Twort[4] has a very good section on the dangers latent in trying to impose unsuitable risks on a contractor. A particularly well known example is in site investigations where the Engineer may:

(a) Do a perfunctory investigation
(b) Do a careful investigation but keep the data to himself so as to depress the tender price
(c) Do a careful investigation and provide the information to tenderers but disclaim responsibility

In all cases the risk are offloaded on to the Contractor who has to decide how best to provide for them. These practices have now been roundly condemned by NEDO[6] and should diminish in the future.

ORGANISING AND ACTIVATING

Organisation: introduction

Organisation is concerned with the roles of individuals and the working relationships of individuals and groups. It uses subjective and emotive concepts such as 'authority', 'responsibility', 'delegation', 'communication' and so on, and is obviously closely bound up with the management of men. (See pp. **31**–65 to **31**–67).

Unfortunately, the classical theory of organisation structure which has been confidently proclaimed in many books for many years does not seem to tally at all well with the way successful contracts are run! People working in other types of organisation have come to similar conclusions and consequently the search for a more realistic theory of organisation is now in full swing. The reader who is interested in the current flux of new ideas may consult a recent thin paperback by Pugh.[58]

In view of the current unstable state of this art, the author gives here only a very brief resumé of the classical theory together with a few comments and concepts drawn from his own experience.

The classical theory of organisation

To most people the term 'organisation chart' means some sort of pyramid-shaped array of linked boxes, each box containing the name of a person or a function in ascending order of importance with the most important boss sitting in lonely eminence at the top of the pile.

Such charts are of three recognised types, viz. 'military', 'functional' and 'line and staff', and the interpersonal relationships they engender are often further classified as 'direct', 'lateral', 'functional' or 'staff'. Such charts are also categorised as 'shallow' or 'deep', a shallow organisation normally producing wide 'spans of control' and vice versa. Moreover, it is said that a good organisation should have 'clear lines of command' from top to bottom and that each individual should have his 'authority' and 'responsibilities' 'clearly defined'.

These elements of classical organisation theory are described clearly and succinctly by Calvert.[3] They are tacitly accepted by many people who affect to despise all theory and can thus often provide ready-made stumbling blocks when a contract organisation hits a sticky patch.

Problems of applying classical organisation theory to construction contracts

The classical theory is not to be wholly despised but it is an incomplete model of real life on a contract and therefore contains pitfalls for the unwary. For example:

1 A contract is dynamic but a chart is static and so may never catch up with reality
2 A chart tends to isolate people inside their boxes whereas a contract is so full of unpredictables that it requires some overlap of roles, like tiles on a roof, to ensure the job is covered
3 As a picture of the way people work together, the links on a pyramid chart are so incomplete as to be practically useless
4 The pyramid displays in black and white who is above who, whereas much contract

management demands teamwork a bit like that of a football team in attack where whoever has the ball at his feet is for that moment the boss

An alternative concept for contract organisation

The key to successful contract organisation is the simple notion that different types of activity demand different types of relationship between the same people and hence different types of organisation structure.

The author finds it useful to distinguish three main types of activity:

1 Creative
2 Regulative
3 Manning

The meaning of these three terms and the corresponding styles of organisation and interpersonal relationships are set out in Figure 31.10. The key thing to remember about this diagram is that every member of every contract organisation

1 Contributes to one or more creative activities
2 Has to observe a number of established procedures
3 Belongs willy-nilly to a family of expertise

Advantages of task-based organisation models

In contrast with the traditional 'family tree', matrix charts and procedural pyramids as in Figure 31.10 are task-based in the sense that they start with 'what has to be done' rather than 'who is in charge of whom'. This has several advantages:

1 The charts/procedures need only be drawn up for specially difficult or critical tasks as they arise

2 The charts/procedures can be drawn up so as to use each man's strengths and to prop his weaknesses without this being so disturbingly obvious as it is with the normal hierarchy of people

3 The charts/procedures can be drawn up to show very conveniently on one piece of paper the contribution required from people in the separate promoter, engineer and contractor pyramids

4 When shown such charts people do not see themselves as being responsible to Mr X or Mr Y (whose judgement they may not wholly admire) but to the job to be done whose imperatives nobody can deny

5 Following from 4, if people can see clearly what they are trying to achieve, they will know better how to use their initiative when the unexpected arises. Thus detailed control from above can largely be replaced by selfcontrol, which is the key to all successful delegation.

Summing up on organisation

Organisation is full of complexities and pitfalls because it lies at the frontier where man's desire to rationalise does daily battle with his intuitive feeling for what is right and will work with particular people in particular situations. (See also Hancock.[15]) Family trees have to be drawn but the reader must always remember that they conceal more than they reveal about the actual working relationships on a contract.

	Type of activity		
	(1) Creative	(2) Regulative	(3) Manning
(a) Examples of each type of activity (creative/regulative are contrasting pairs)	Winning new work Conceptual design Interpreting design creatively on site Thinking up good construction methods Motivating men to work as a team ie. Getting things done	Spotting snags before taking on a job Checking design calculations Testing concrete cubes Preventing waste theft & fraud on site Timekeeping, disciplinary procedures ie. Preventing & righting wrongs	Getting and keeping good staff Supporting staff in their jobs Developing staff in the long term Removing unsuitable staff Manning the work that needs to be done Holding a good following of good labour
(b) Commonly heard phrases which characterise the the relations between:– People Activities Organisation (Creative/regulative are contrasting pairs	"We can't predict everything "You know the purpose of the exercise" "Use your oun initiative" "Don't just stand aside when you could give valuable help" "Don't stand on ceremony. Talk to anyone as long as it helps to get the job done" "We need willing team work. You will never get it if you are always trying to allocate blame for mistakes" "We bend the organisation to fit our good people"	"We know what always happens if ----" "You know the procedure" "Stick to the book" "You stick to your own job and let other people get on with theirs" "We can't have people bye-passing the proper channels" "We must make individuals responsible otherwise we have no control when things go wrong" "In our organisation people do what they are told, not what they fancy"	"Our staff is our greatest asset" "We must maintain our reputation as a good employer" "We usually try to give misfits a second chance in new surroundings" "When if comes to sacking we try to be firm but fair" "We must keep a lookout for hidden talent" "We must give promising people work that stretches them" "We must back up our people who are in exposed positions"
(c) Type of organisation structure appropriate to the style of relationship	Matrix Key activities (1, 2, 3, 4, 5) × Key people (A, B, C, D) People from various families of expertise contribute in a relatively flexible and informal way to the success of key creative activities. A chart as above can be drawn with each person's contribution described in the relevant square	Procedural pyramids Essentially pyramids of people Makeup of each pyramid is implicit in:– Procedure manuals Books of standing instructions etc which provide for decisions by precedent and for referring new, difficult and disputed decisions upwards towards the top of the pyramid	Families of expertise Conventional Family tree chart All members of pyramid are in the same broad category of expertise eg: Chartered civil engineer Foreman/Ganger Plant Quantity surveyor Admin/accounts etc
Useful analogy	Football team in attack	The long arm of the law	Human family

Figure 31.10 The triple-style organisation. The style of relationship between the same people differs according to the type of activity that produces the relationship (*see pp* **31**–53 and 54).

The author finds it useful to liken the task of structuring a contract organisation to that of trying to maintain a reasonable fit between three slices of cake whose contact faces keep changing.

The resources to be managed

The hundreds of different productive resources needed for a contract are divided into seven main categories which appear as main headings in the grid index, Figure 31.1.

Time and *money* are not included because they are not productive resources but activators of such resources and measures of success. They therefore appear in Figure 31.1 as *'progress'* and *'economy'* along with the other desired results.

THE SITE

This is defined in ICE Conditions,[2] clause 1(n) and important contractor's rights and obligations are covered in clauses 11, 12, 29, 32, 33, 42.

For notes on site layout and temporary works see page **31**–8. Other paragraphs discussing the influence of the site on contract management can be located from the grid index, Figure 31.1.

INFORMATION

The contract information system

The information system is to a contract what the central nervous system is to the human body. Its character largely determines whether the various limbs of the contract team pull together towards a common goal or pull their own separate ways towards confusion, delays, disputes and even litigation. Since information is essentially the means by which people transmit and receive their ideas, wants and fears, it is useful to consider the principal elements of a contract information system on a scale of formality as in Figure 31.11 which also shows the page numbers on which each element is discussed.

In work like civil engineering which is full of unpredictables, the best results are achieved if ordinary day-to-day relations are kept as informal as possible with the more formal means of communication held in reserve for preventing and settling disputes. In fact there is a fairly close correlation between the informal/formal communication spectrum in Figure 31.11 and the creative/regulative organisation spectrum shown in Figure 31.10.

Man-to-man discussion

One often hears the precept 'Don't say it, write it'. This is all right where one is transmitting information or confirming an agreed course of action but it can be a dangerous precept when one is trying to persuade people to do things or provide information. In any situation where one is looking for a response and is not sure what it will be, the author prefers the precept 'If you aren't sure what to say, say it, preferably face to face'.

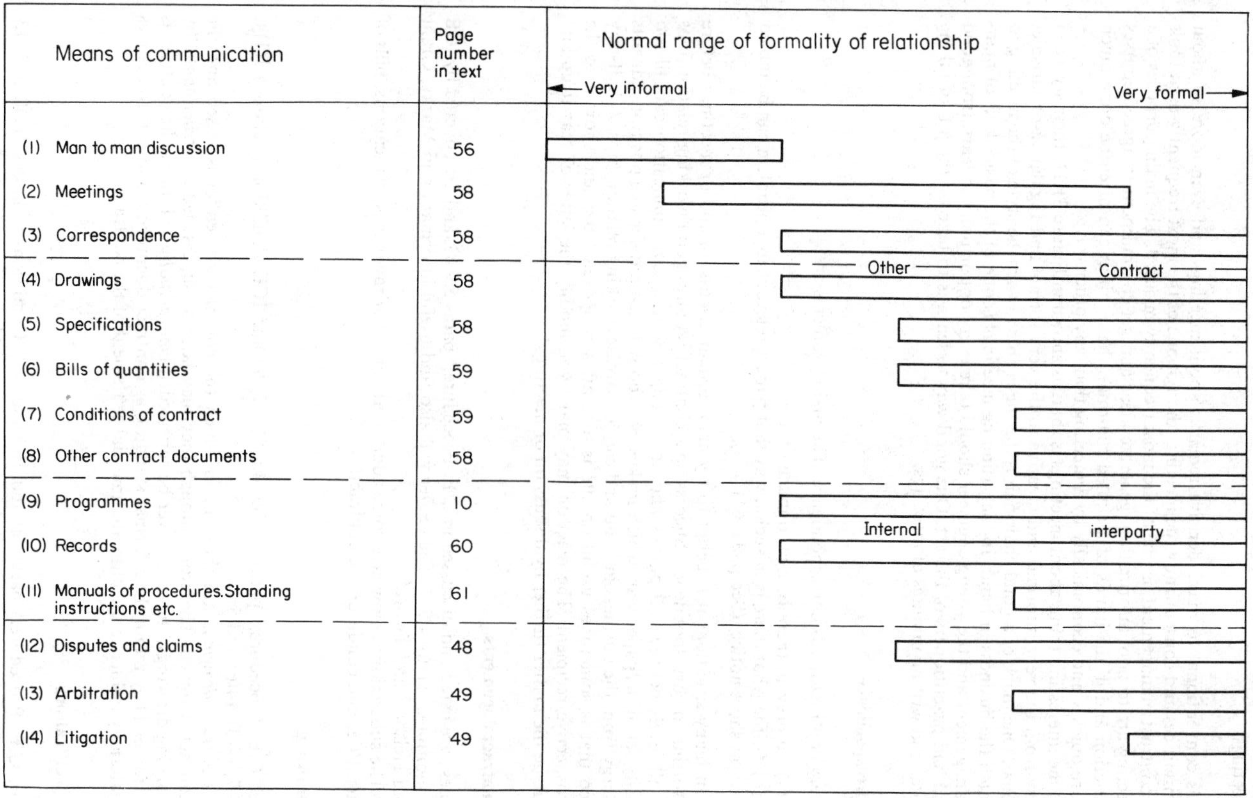

Means of communication	Page number in text	Normal range of formality of relationship (← Very informal … Very formal →)
(1) Man to man discussion	56	
(2) Meetings	58	
(3) Correspondence	58	
Other — Contract		
(4) Drawings	58	
(5) Specifications	58	
(6) Bills of quantities	59	
(7) Conditions of contract	59	
(8) Other contract documents	58	
(9) Programmes	10	
Internal — interparty		
(10) Records	60	
(11) Manuals of procedures. Standing instructions etc.	61	
(12) Disputes and claims	48	
(13) Arbitration	49	
(14) Litigation	49	

Figure 31.11 Contract information: the principal elements on an approximate scale of formality

Meetings

It is conventional to make derisive remarks about meetings and even more so about committees and consequently the meeting, though one of the most frequently used tools of contract management, is one of the most frequently misused. The main purposes of a meeting are to pool information, generate ideas and get commitment to agreed courses of action and in the hands of a skilled chairman these things can often be done much more quickly and economically by a meeting than any other way.

One interesting thing about meetings is that, as shown in Figure 31.11, their style can vary from the very informal and creative to the very formal and regulative or jurisdictional. One mark of a good chairman is that he is able to vary the style of the meeting to match the business in hand. In fact, running meetings is one of the basic skills of man-management and the young engineer should take every opportunity to learn from good and bad practitioners of the art. One good way to learn is to take on the job of taking notes and writing minutes of meetings.

Correspondence

Every contract needs correspondence. The main problems are:

(a) To prevent there being too much of it
(b) To file what there is in such a way that papers relevant to a problem can be tabled at short notice. (See p. **31**–64).

In many situations it is better to try man to man discussion and/or meetings before resorting to correspondence. Success on a contract requires good working relationships at all levels between all parties and an ill-judged written communication can all too easily set off a 'paper war' which makes everybody take up defensive positions, diverts energy from the business in hand and can jeopardise working relationships. A half-way step that is sometimes useful on site is to draft a proposed letter and show it to the prospective recipient. The reaction may make you modify the letter or tear it up and pursue the matter further by discussion or meetings!

Contract documents

These are listed and discussed in C.E. Procedure,[1] paras 56–63, and are covered by ICE Conditions,[2] clauses 5–7. The reader will also find useful information in Marks, Grant and Helson,[5] and Twort.[4]

The subsections below give some additional notes on drawings, specifications, bills of quantities and conditions of contract.

Drawings

See C.E. Procedure,[1] paras 38, 39, 40, 51, 62, 95 and ICE Conditions,[2] clauses 1(1)g, 5–7, 14(3)–14(6) and 51.

Contract drawings are those upon which the contractor's tender is based and the contract is signed. Therefore, though they may later be superseded *constructionally* by working drawings, they can never be superseded *contractually* as long as the contract is in force. They retain their formal standing as contract documents and can sometimes materially influence the ultimate financial settlement of the contract.

Specifications

See C.E. Procedure,[1] paras 60 and 105 and ICE Conditions,[2] clauses 1(1)f, 5–7, 13, 36–40, 51 and, for more detail, Marks, Grant and Helson,[5] and Twort.[4]

C.E. Procedure[1] warns against the danger of conflict between specification clauses and contract provisions. Even if conflict is avoided, a certain amount of overlap is fairly common. For example, on programme and temporary works see C.E. Procedure,[1] para. 60 and ICE Conditions,[2] clause 14.

Bills of quantities

See C.E. Procedure,[1] paras 61 and 149, ICE Conditions,[2] clauses 55–57 and for more detail Marks, Grant and Helson,[5] and Twort.[4]

On work in the UK, the reader is most likely to encounter two 'standard methods of measurement':

1 'The ICE SMM'[29]
2 'The Building SMM'[30]

These are very widely used and even where some other method of measurement is used, their influence is usually evident. They differ in important ways, for example:

1 A civils bill should 'set out in sufficient detail the quantities of work and material necessary . . . without unnecessarily repeating the descriptive matter contained in other documents'. (ICE SMM,[29] clause 6.)
2 A building bill should 'fully describe and accurately represent the works to be executed'. (Building SMM,[30] clause A1.)

In other words a civils bill is always to be read alongside the drawings and other documents whereas a building bill seems often to attempt to combine the function of drawings, specification and bill and is in consequence always more voluminous than a civils bill.

The serious reader should acquire his own copy of both documents[29, 30] and study especially the introductory 'rules', 'general principles', etc.

Both methods of measurement have come under increasing criticism in recent years. See for example Barnes,[31, 32] Creasy,[33] Goldstein[34] and Hansen.[35]

The main criticisms seem to be:

1 The present form of bill causes a lot of unnecessary work because it is so large. Other countries, notably the USA, seem to manage their construction very efficiently with much simpler bills.
2 Large bills with space for a rate for every type of work that could conceivably be needed encourage promoters and engineers to put work out to tender before it is adequately designed. This leads to post-tender design changes which lower the efficiency of the whole industry and are a far greater problem in the UK than in countries like the USA which use simpler bills.
3 Under the unit rate system the contractor is still paid for each piece of work pro rata to its quantity whereas modern construction methods and plant make less and less of his costs directly proportional to quantity. (See also pp. **31**–37 and 39 and Figure 31.7). The result is increasing contention and delay in the settling of final accounts on jobs where work has been varied or delayed. See for example Barnes and Thompson.[31]

At the time of writing (1973) the Civils SMM[29] is being revised and it seems likely that the new version will be less open to the above criticisms. It also seems likely that the Building SMM[30] will soon be revised in the same general direction.

Conditions of contract

The various types of contract and the various factors influencing choice of method are well set out in C.E. Procedure,[1] paras 41–53.

The purpose and philosophy of the conditions of contract are set out in C.E. Procedure,[1] para. 59. The ICE Conditions[2] are an essential reference for this section (see para. 2) but the reader is very likely to encounter two other forms of contract:

1 CCC/WKS/1[21]—the form of contract used by most UK Government Departments
2 The RIBA form of contract[22]—used for contracts that are predominantly building but often contain a substantial proportion of civil engineering work.

The serious reader would be well advised to obtain a copy of each of these documents for his own reference (they are slim documents and not costly). He will encounter other forms of contract, particularly abroad, though many overseas forms of contract[23] closely resemble the ICE Conditions. But whatever situation he is in, he should find that these three documents help him to orient himself to the form of contract in use and to the attitudes of people who are used to forms of contract that are strange to him.

Forms of contract are a most involved subject on which a little knowledge can be a dangerous thing. The serious student should acquire his own copy of Marks, Grant and Helson[5] and should take every opportunity to learn from what happens on the contracts he is on, consulting whenever he can, those on his 'side' who are more expert than him.

What will strike the novice most forcibly are the gaps between what the contract seems to say and what actually happens on site. The primary reason for these gaps is that the contract is designed to contain and regulate disputes, i.e. it provides for relationships at the formal end of the spectrum in Figure 31.11 whereas most day-to-day transactions are concerned with establishing and maintaining the working relationships which are needed to get the job done and are carried on at the informal end of Figure 31.11.

Another reason can be that both Engineer and Contractor tacitly agree that the wording of the contract really ought to be altered. For example, clauses 41 and 51(1) and (2) have been very substantially altered in the 5th edition of ICE Conditions[2] to bring them closer to what really happens, but the wording of the 4th edition was in force for many years.

The reader who finds the relationships between the law, the contract and the real world a bit baffling has only to read some of the recent papers in Proc. ICE (e.g. Abrahamson,[24] Ackroyd,[25] Bell,[26] Haswell,[40] Sharman,[27] Waddington[28]) to find that he is not alone. The paradoxical truth is that though a form of contract is essential the most successful contracts are usually those on which the contract is rarely referred to. Cynics put this the other way round and say 'People only turn to the contract when the bonhomie has evaporated'.

Records

Every document generated by the prosecution of the works, be it a letter, minute of meeting, monthly measurement, purchase order or whatever, automatically becomes part of the contract records. This paragraph deals with those records whose usefulness is not wholly clear at the time but may later become very apparent. Records of this type can help enormously with such things as:

Accidents
Failure or deterioration of completed work
An unforeseen need to do new work adjoining completed work (especially if such work is buried)
Engineer/Contractor disagreements over payments or delays in the construction
Designing and pricing future work.

These 'deferred purpose' records include such things as:

1 *Diaries.* Every engineer on a contract should keep a detailed diary of his own work, however futile this may seem at the time.
2 *Private notes* of verbal instructions or agreements that have not been confirmed in writing in the customary way.

3 *Superseded drawings* must be kept because something done when the drawing was current may two years (and five revisions) later seem very odd or stupid indeed.
4 *As built drawings*. Everybody agrees they are vital but it is surprising how seldom they are done properly and how often even the most respected statutory undertakings are surprised by the odd places in which their buried pipes and cables turn up.

Twort[4] has a very good chapter on 'The Resident Engineer's Office Records' which every contractor's engineer would also do well to read, if only to find out what a competent 'opposing team' is likely to be doing.

Interparty (Figure 31.11) records are obviously more formal than internal ones because behind them lies the implied declaration 'Should there later be a dispute we agree that on such and such a date the state of affairs was so and so'. So each party is on its guard when putting its name to such records. Records for claims come in this category (see C.E. Procedure,[1] para. 182 and ICE Conditions,[2] clause 52(4)). The difficulties that arise in trying to agree such records have in the past been one of the main reasons for claims being left in a woolly state until the work is completed. (See also p. **31**–48).

Manuals of procedures, standing instructions, etc.

Such manuals are best seen as economical and effective devices for pooling and codifying experience and making it available to those engaged in current work. It follows that they have to be constantly updated and so are little use unless they are bound in some sort of loose-leaf binder with nonsequential page numbering so that pages can easily be added, removed or replaced.

Such manuals are of two main types:

1 Internal to one organisation and applying to all its contracts
2 Specific to one contract and usually agreed between the parties to a contract.

In category 1 a contractor will very probably have his own manuals covering some or all of the following (grouped under the same main heads as the grid index Figure 31.1):

Quality	Control and test procedures
Safety	Standing instructions, accident procedures
Money	Accounting, cash handling, costing
Site	Office procedures, security
Engineering information	Procedures for handling drawings, setting out
Men	Wages, bonus, industrial relations, training, staff procedures
Plant	Procedures covering all aspects of obtaining, maintaining and operating
Materials	Purchasing, storing and issuing procedures
Subcontractors	General procedures
Third parties	Important points from relevant legislation.

Category 2 covers anything on which the three parties to the contract need to cooperate and which is not sufficiently provided for in the contract documents. The 'manual' may be anything from a few single sheet standing instructions bound in a file or stuck on notice boards, to a shelf of large volumes covering everything that might conceivably be done on a contract. Such large procedure manuals appear to be most frequently used on large oil and chemical plant contracts where the cooperative activities of promoter, Engineer and Contractor interlock in great detail and need to be clarified for each contract.

The difficulty with any system of manuals is to provide a useful guide and discipline and prevent all the silly mishaps without going so far as to put people in straitjackets in situations when they ought to exert their own initiative. The usefulness of any manual

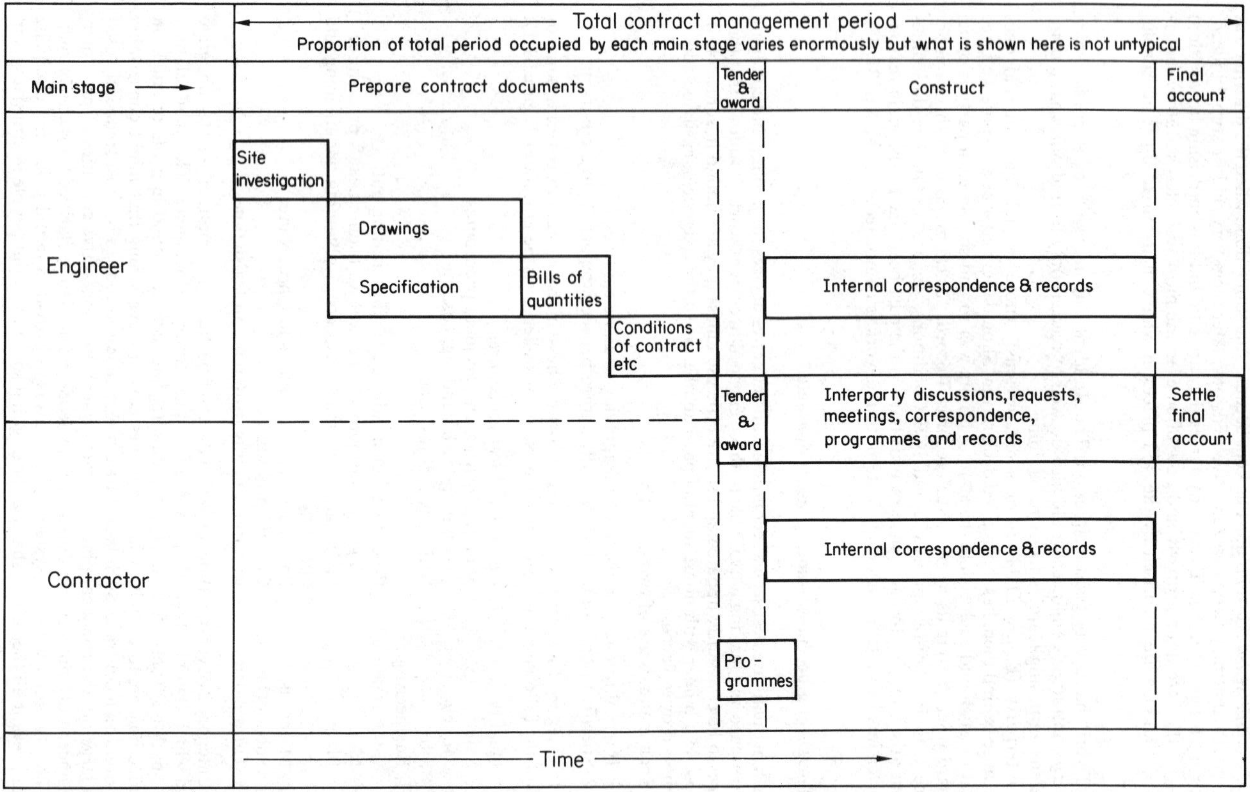

Figure 31.12 Contract information—idealised picture of the timing and sequence in which the main elements are produced (*see p.* **31**–63)

depends directly on the *predictability* of the situation in which it is used. In the terms of Figure 31.10, it is highly relevant to 'regulative' activities but can threaten the success of 'creative' activities.

Organising the flow of project information generally

Figure 31.12 shows an idealised picture of how information for a project is produced, everything in logical sequence with no overlaps between successive stages. (The beginner reading some typical contract documents, could be forgiven for supposing that Figure 31.12 is the way the system really works.)

If the system really worked like this there would be little more to say, but in practice of course many of the operations shown in Figure 31.12 have to overlap each other, notably drawings, many of which often cannot be completed until the construction itself is nearly completed. (See C.E. Procedure,[1] para. 39.) These overlaps produce organisational problems within and between the Engineer's and the Contractor's organisations. The documents and the system as a whole are designed to operate as in Figure 31.12. The system can be bent a bit but if it is bent too far it may break! In fact if the Engineer is put in a position where he cannot produce adequate information at tender stage he is usually well advised to abandon the conventional bill of quantities system and go for some other means of execution. (See C.E. Procedure,[1] paras 41–53.)

Organising drawings

The Engineer's role in preparing and issuing drawings is outlined in C.E. Procedure,[1] paras 38, 39, 40, 51, 62, 95 and in ICE Conditions,[2] clauses 5–7, 14(4)–14(6) and 51, and there are some useful notes in Marks, Grant and Helson.[5]

The *Contractor's* managerial (as distinct from technical) role on site is

1 To record the particulars and date of receipt of each drawing issued by the Engineer and by the Contractor's own organisation and any others involved (e.g. suppliers and specialist subcontractors)
2 To ensure that every drawing goes to the people who need it.

On a job of any size one usually needs a card index divided into job sectors with a card for each drawing showing as a minimum, title, serial No., revision No., size (because it is always easier to find something if you know what size it is), date received and who it was issued to. On a large site it is sometimes worth having a signature book in which recipients sign for drawings received, but this is laborious and is usually only instituted if there is a serious mishap involving work constructed to a superseded drawing and the culprits deny knowledge of the revised drawing.

The best way to keep master sets of contract drawings, working drawings, record drawings and so on is in stick files in which sets of related drawings can be bound as in a book and hung up. The author prefers the kinds that are fairly difficult to dismantle because they discourage the light-fingered borrowers who have lost their own copy and will lose the master copy too if given half a chance.

Organising setting out

Responsibilities are discussed in C.E. Procedure,[1] para. 173 and are clearly defined in ICE Conditions,[2] clause 17. However, to define who is to blame when things go wrong is not the whole of organising to get the work done and, in this matter, as in quality control, there is much scope for creative cooperation and so the reader should refer to page **31**–20.

Organising the filing of correspondence and other records

(See also p. **31**–60.)

Power goes to the man who has the relevant information at his fingertips and such men usually have good filing systems, partly in their brain and partly in cabinets.

Filing is almost entirely a problem of size. If the volume of data is fairly limited, the human brain is the best indexing system known and that is why engineers commonly keep their own files instead of or as well as a central file. When you are looking in a hurry for the correspondence on a certain matter, a central file has to be very good indeed to be better than the engineer who has worked on that particular matter. Nonetheless, a central filing system is essential on a job of any size and the only really foolproof system is the 'double entry' one in which two sets of files are kept:

1 A pair of chronological files, one for incoming and one for outgoing papers. On these files a copy of every letter, report, minute of meeting, etc., is placed in strict date order so that if all else fails, a document known to exist can be located by search and fairly easily if its approximate date is known

2 A set of subject files. The first subdivision can be by subject or by source and the next stage will be subdivision of one within the other so that one gets a matrix of files

			SOURCE					
			Resident engineer	*Head office*	*Minutes of meetings*	*Progress reports*	*Sub-contractors*	etc →
SUBJECT	Earth-works	General						
		Dam						
	Concrete	General						
		Powerhouse						
		etc ↓						

If one has to go to this degree of elaboration one then has the problem of letters, reports, etc., that cover more than one subject. In really desperate cases one then has to get a spare copy of the report or whatever it is and break it up into the subjects of the file. The author has only occasionally had to do this. Twort[4] gives some useful hints on files for the resident engineer's office.

All this is costly and may seem tedious to someone who regards himself as a 'real' engineer. However, no real engineer would dispute that he needs instant access to the drawings, specifications and other contract documents and it is often just as important to be able to call up quickly information that is stored in the contract files.

Information and the site

The Engineer needs detailed information about the site in order to prepare his designs. (See C.E. Procedure (1971), paras 27–30, 38.) But in the past he has tended to withhold this information from tendering contractors or issue it with disclaimers. The NEDO report[6] and the new ICE Conditions,[2] clause 11 should reduce this practice. See also pages **31**–33 and **31**–52.

Influence of available plant on design

The Engineer when designing a job has sometimes to consider what plant may or may not be available and must always be on the lookout for situations where a certain piece of plant will produce a job so much more quickly or economically (or both) that he may need to reconsider the design concept and/or the specification that he originally had in mind. In such cases it is becoming increasingly common for contractors who have access to special pieces of plant to offer alternative designs which are frequently adopted by the engineer. See also Broadbent and Palmer,[9] and Hansen.[35]

MEN

Man-management

'Man-management' covers all those activities to do with assembling, using and caring for human resources. Thus it includes, for example, recruiting, remuneration, discipline, welfare, training, leadership and of course industrial relations, the umbrella under which the practical details of these activities are worked out and put into effect.

Man-management is of course also closely bound up with organisation (see pp. **31**–53 to 56). To distinguish between the two one could say that organisation is the bones of contract management while man-management is its heart and soul.

Man-management must not be confused with 'personnel management' which is a well established but rather misleading term for a range of skills deployed by 'personnel specialists' to help managers to manage their own men. The author does not much like the term 'personnel management' because it can suggest to the uninitiated and the lazy that they can leave the thorny problems of man-management to specialists and blame them when things go wrong. This view may be partly valid in some industries but it is certainly not true in ours. In construction if you cannot manage men you cannot manage anything. (See C.E. Procedure,[1] paras 192–193 and Calvert.[3])

Because it is such an essential day-to-day matter, the climate of ideas on man-management is much influenced by the practical men who affect to despise all theory. What they mean of course is that they despise all theories except their own! The author hopes the next few paragraphs will help the reader to develop his own theory a bit.

Man-management: practices change but principles don't

At the time of writing the industrial relations scene in the UK is changing so rapidly that it is pointless to devote space to the details of particular laws or industrial agreements because they will most probably be obsolete long before this book is generally revised. In fact, in this field, change is the norm as any reader will discover if he takes the trouble to read Terry Coleman's fascinating account of the working conditions of railway navvies in the mid-nineteenth century[60] and compares them with what we now accept as normal and reasonable.

However, though laws, practices and behaviour change rapidly in this country and differ widely in different overseas countries, human nature changes very little if at all

<table>
<tr><td colspan="2" rowspan="2"></td><td colspan="3">Types of activity</td></tr>
<tr><td colspan="2">Maintaining
a contented staff and labour force matched to the needs of the job</td><td>Motivating
people to do the work that needs to be done</td></tr>
<tr><td rowspan="6">1
Factors
that the manager has to understand and take into account</td><td></td><td>Codified</td><td>Tacitly agreed</td><td></td></tr>
<tr><td rowspan="2">At national or industry level</td><td>National laws</td><td>National / Regional / Religious/Tribal customs</td><td rowspan="2">The technical commercial and administrative problems of the work in hand</td></tr>
<tr><td>Industry formal agreements (See 67)</td><td>Industry customs and practices</td></tr>
<tr><td rowspan="2">At working group level</td><td>Declared policies of individual employers and trade unions (See 67)</td><td rowspan="2">Unwritten "club rules" (or "group norms") of the various identifiable working groups involved</td><td rowspan="2">The things that motivate the various identifiable working groups involved</td></tr>
<tr><td>The contract documents (See 58)</td></tr>
<tr><td>At the level of the individual</td><td>Contractual relations between employer and employee</td><td>Unwritten mutual expectations between employer and employee</td><td>The things that motivate the key individuals involved</td></tr>
<tr><td colspan="2" rowspan="7">2
Processes
that are recognised as relevant</td><td colspan="3">Organisation and manning (See 69)</td></tr>
<tr><td>Remuneration</td><td>(See 44)</td><td>Production incentives (See 44)</td></tr>
<tr><td>Safety</td><td>(See 26)</td><td rowspan="3"></td></tr>
<tr><td>Welfare and medical</td><td>(See 27)</td></tr>
<tr><td>Personnel records</td><td>(See Calvert[3])</td></tr>
<tr><td colspan="3">Training and staff development (See 69)</td></tr>
<tr><td colspan="3">Leadership (See 70)</td></tr>
</table>

Figure 31.13 Man-management: the main elements (*see pp.* **31**–67 to 70)

and so it is useful to look at a framework of concepts which will help the young civil engineer to learn something from every man-management situation in which he finds himself.

Figure 31.13—The main elements of man-management and industrial relations

There are two main types of activity:

A. MAINTAINING a contented workforce matched to the needs of the job. This includes preventing/resolving disputes and other forms of breakdown
B. MOTIVATING people to get the job done

Each of course depends upon the other but it is useful to think of them as distinct types of activity if only because they require an understanding of different *factors* (Section 1 of Figure 31.13) and the use of somewhat different *processes* (Section 2 of Figure 31.13).

Most people will accept Figure 31.13 as valid but fewer will agree on what some of the terms really mean. Relevant explanatory pages in this section are referred to on Figure 31.13 and some of these follow immediately below. Where no page is referred to it means either that the meaning of the term is so obvious that no further comment is needed or that it is so subtle and complex that there is not room in this section to discuss it properly and the reader must refer to a specialist text or to his own experience.

Some other paragraphs relevant to man-management can be located from the grid index, Figure 31.1.

Industry formal agreements (Figure 31.13)

In the UK practically all site construction work is governed by some form of 'working rule agreement' (WRA) drawn up between representatives of the employees and trade unions involved (see C.E. Procedure,[1] paras 166–168). The agreements the young civil engineer is mostly likely to come across are:

1 *Civil Engineering Working Rule Agreement*[43]
2 *Building Working Rule Agreement*[44]
3 *Specific Site Agreements*
(Normally only found on large sites involving many different unions (including e.g. mechanical and electrical trades).

The young engineer on site is well advised to obtain for himself a copy of the relevant WRA, if only to inform himself on the very wide range of matters that are covered. One thing he can be sure of is that a competent steward will know the WRA backwards and inside out!

In overseas work, many developed countries have similar rules while in under-developed countries things are much less formalised. The only useful advice is 'find out so as to avoid being caught out'.

Man-management and third parties

In relation to labour, the most significant third parties to be considered are

1 The nation itself, its laws, customs and practices
2 Other employers in the area and particularly other contractors (not subcontractors) if any, on the site.

Item 1 is touched on in Figure 31.13. Item 2 is one that can cause severe problems on some sites, perhaps because on most sites it causes no problems and so people tend to be caught unawares when problems do arise. Obviously, when manning any site, the contractor has to compete with other employers in the area, but the really severe problems usually only arise when several contractors have their own men on the same site but on terms and conditions that are sufficiently different to give grounds for grievances. In difficult multi-contractor sites it is becoming common to negotiate a single site agreement (p. **31**–67) that replaces the normal mixture of WRA and individual employer's practices, and applies to every hourly paid man on the site. To achieve this, the Engineer and/or promoter have to take a more active part in industrial relations than they normally do. Alternatively the promoter may appoint a managing contractor (p. **31**–41).

Interactions between men and design

Competent designers take account of the type of labour that is likely to be available at the site of the works. On most straight civil engineering work this effect is likely to be small but it may be large on some types of overseas work in areas where there is abundant cheap labour that can be used when plant would be used in the UK or where there is a mature local tradition of craftsmanship using local materials in ways that are strange to UK eyes but sensible in their context.

Conversely, the contractor who is faced with a design which requires large numbers of men with skills that are not available in the area, will have to import men or train them on site or both, and the engineer will expect to see this reflected in the tender prices offered by responsible contractors.

Most designers seem less aware of how much influence they can have for good or bad on site morale and industrial relations. Drawings that arrive too late, are altered too often or are full of discrepancies, disrupt the rhythm of the work and lower the men's confidence in the management. A designer who cannot get the right drawings to site at the right time sows the seeds of grievances which will fester and may break out into crippling disputes which are ostensibly about things like bonus, canteens or safety. The author has seen this on several jobs. He has also seen the opposite, where competent design clearly helped to raise morale and improve productivity on site.

Man-management and the contract documents

ICE Conditions,[2] clauses 16, 19, 24, 34 and 35 are relevant. There may also be special contract or specification clauses covering such matters as safety and welfare of the contractor's employees (see p. **31**–26). In any event the documents will spell out what the contractor must provide for the comfort and well-being of the resident engineer and his staff.

Men and the site

The problems of getting and keeping the right number of the right sort of men on the site of the works play a crucial part in the contractor's planning and pricing. The days when navvies would camp on the job are gone for good though some still profess to remember them. On most UK jobs of any size the contractor will be involved in such things as:

Bussing men daily from nearby towns
Lodging allowances and helping men to find lodgings
Purpose-built accommodation on site.

On really big jobs in remote locations, especially overseas, site accommodation can become a major undertaking in itself. The author was on one job where there were, including wives and children, over 10 000 people living in a semi-permanent site town with its own shops, schools, hospital, church, etc. When things reach this scale they are difficult to price accurately in the limited time allowed for tendering and it may pay the promoter to build some accommodation using small local contractors, well before letting the main contract.

Contract organisation and manning

Organisation is discussed in general terms on pages **31**–53 to 58.

The positions and responsibilities of engineer and contractor and their staffs on site are well set out in C.E. Procedure[1] (paras 17–19, 91–128, 157–160 and Figure 31.1). The role of the Engineer's representative is defined in ICE Conditions,[2] clause 2 and that of the Contractor's Agent in clause 15. Twort[4] goes into the matter in greater depth with a good deal of human insight into what really goes on on site, the problems of performing certain roles and of maintaining proper working relationships between resident engineer, agent and their staffs. The interested reader will also find useful information in Baylis,[8] Cairncross,[10] Elsby,[11] Gillot,[13] Greeves,[14] Parsons[18] and Tait.[19]

The paragraphs below only cover certain points which the author feels should be added to what is in the above references.

Manning a contract organisation

Since one always tries to find key men from the parent organisation, it follows that, organisationally, contract manpower divides into two quite distinct categories which are referred to in this section as permanent and transient.

Permanent staff and labour are people drawn from or at least owing a strong allegiance to the parent organisation. They are usually selected by the parent organisation and will look to it for employment when the present contract is complete. It follows that they have a dual loyalty, to the contract organisation and to the parent organisation, which the man in charge of the contract organisation will ignore at his peril because in cases of conflict the parent organisation will usually back their man. It follows that a contract organisation chart is an incomplete picture of reality if it does not show the 'loyalty links' back to the parent organisation over the head of the man nominally in charge of the project

Transient staff and labour are usually selected and recruited by the contract organisation and expect to be paid off when or before the job is completed. Organisationally they are easier to handle because they seldom present the conundrums of dual loyalty but to counterbalance this the problems of selecting, holding and (often) training such people are usually much more severe. Calvert[3] has some useful general notes on recruitment, selection, placement and the associated record-keeping activities. Training is discussed below.

Training and staff development

Ideas on training are at present changing very rapidly. Not so long ago few practising managers saw training as having anything to do with the day-to-day job of contract management except perhaps when as a resident engineer or agent they had to try to find suitable work for an indentured engineer or a few apprentices.

The old ideas stemmed from what the Duke of Edinburgh aptly calls the 'clockwork monkey' theory that one good 'wind' of training at the outset of your career, whether as an engineer, accountant, craftsman or labourer, would keep you going until you retired.

The current rapid technical, industrial and social changes in this country are forcing us to replace this theory by a new one which is that every man needs regular 'rewinds' of training throughout his life and that employers need, in their own interests, to provide much of this training.

This new theory has already led to two types of visible action which directly affect the day-to-day processes of contract management:

1 It is increasingly common on overseas jobs and not unknown in the UK, for contractors to set up schools on site to train men rapidly in skills that are not available locally. In response to this need, contractors are rapidly getting more skilled at training and are sometimes invited to submit evidence of this skill in their tenders (or proposals for negotiated work). Obviously this is going to have a profound effect on the way construction jobs are tackled.

2 For technical and managerial staff, life is tending more and more towards a perpetual 'sandwich course' (mostly hard bread) in which productive work and small layers of relevant training go hand in hand all through life. Of course this is still more vision than fact but the pressure of events is forcing the vision to become fact quite rapidly. This also will have a big influence on the everyday task of staffing contracts.

Leadership

The word 'leadership' is at present rather out of fashion, perhaps because it still conjures up rather obsolete images such as that of the 'born leader' who goes to a famous school and then out to 'darkest Africa' to take up the 'white man's burden', or of the retired major who claims in his job application to be 'good at handling men'. However, it will come back into fashion again and perhaps sooner in contract management than in some other activities because anyone who has worked on a difficult or dangerous contract will know that given the same men and plant different managers achieve very different results and that 'leadership' is really the only single word that explains these differences.

Of course, the fact that leadership is most necessary and most visible in difficult, or dangerous 'crisis' situations produces its own problem in that one always comes across a certain number of leaders who seem to need a supply of crises in order to fulfil themselves and who, if not furnished with crises, may even subconsciously contrive them! Such men are dangerous if left to do things entirely their own way, but they can make really great managers of contracts if they are propped by people of an opposite, highly systematic cast of mind.

In other parts of this section there are sections on systems for planning, progress control, cost control, etc. None of these is any use without leadership in its widest sense of the art of getting people to do willingly more than they can be compelled to do. Much remains to be discovered and some things will never really be understood about this semi-mystic art, but the reader who has understood the ideas developed in Figure 31.13 will have helped himself a little way along the road.

Further reading on man-management

This is a subject on which thousands of books have been written and hundreds of new ones appear every year. Moreover, the theoretical basis of the subject (sometimes referred to as behavioural science) is at present advancing very rapidly. Perhaps in ten or twenty years' time there will be a few recognised standard works but in the meantime the author recommends the interested reader to confine himself to:

(a) Falk,[56] Parkinson[57] and Pugh[58]

(b) Historical works on civil engineering construction[59–65]

(c) The lives of great military commanders who have grappled with the problems of man-management over the centuries. (There is not space to develop the idea here, but the reader who reflects will realise that construction is in many ways more like war than like manufacturing.)

MATERIALS

Introduction

Materials management is the art of

1 Getting materials: of the right QUALITY
in the right QUANTITIES
at the right PRICE
to the right PLACE
at the right TIME

2 Preventing WASTE, and THEFT of materials

For some reason it is a function that seems to be underrated by most engineers. This is surprising because materials account for 30% to 50% of project cost (and more on M and E work) and often cause an even larger percentage of the delays and disruptions that preoccupy site managerial staff. It is obvious therefore that on all types of construction, materials need to be managed with just as much care as is customarily given to the management of men and plant.

Perhaps the function is underrated by engineers because on most run-of-the-mill UK jobs practically everything (at site anyway) can be covered by standard company procedures which are implemented very successfully by storekeepers (who are often ex-labourers), clerks, etc., with very little supervision by senior or highly qualified staff. It is only when he is faced with a large one-off job, especially if it is overseas, where the standard procedures cannot be applied blindly, that the young engineer starts to discover what an elaborate and challenging matter materials management really is.

The main tasks in materials management

Figure 31.14 shows the main tasks in the total process and the people or departments that normally perform them. This diagram looks fairly complex as it is but even so it is only the bare skeleton because in real life one finds all too often that some things are done out of sequence, materials do not arrive when promised, loads do not tally with delivery notes, material quantities in the measured works do not reconcile with quantities issued

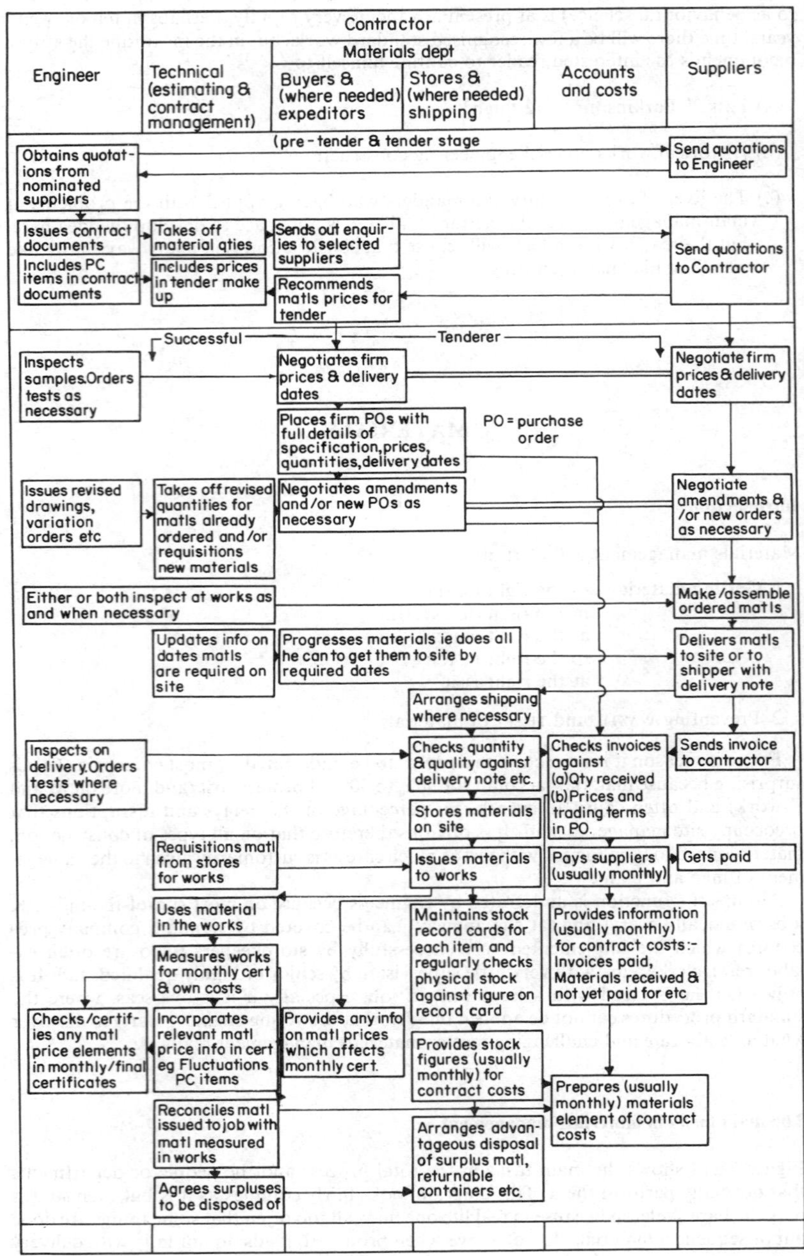
Engineer
Contractor
Technical (estimating & contract management)
Materials dept
Buyers & (where needed) expeditors
Stores & (where needed) shipping
Accounts and costs
Suppliers
(pre - tender & tender stage
Obtains quotations from nominated suppliers
Send quotations to Engineer
Issues contract documents
Takes off material qties
Sends out enquiries to selected suppliers
Send quotations to Contractor
Includes PC items in contract documents
Includes prices in tender make up
Recommends matls prices for tender
Successful
Tenderer
Inspects samples. Orders tests as necessary
Negotiates firm prices & delivery dates
Negotiate firm prices & delivery dates
Places firm POs with full details of specification, prices, quantities, delivery dates
PO = purchase order
Issues revised drawings, variation orders etc
Takes off revised quantities for matls already ordered and /or requisitions new materials
Negotiates amendments and /or new POs as necessary
Negotiate amendments & /or new orders as necessary
Either or both inspect at works as and when necessary
Make /assemble ordered matls
Updates info on dates matls are required on site
Progresses materials ie does all he can to get them to site by required dates
Delivers matls to site or to shipper with delivery note
Arranges shipping where necessary
Inspects on delivery. Orders tests where necessary
Checks quantity & quality against delivery note etc.
Checks invoices against (a) Qty received (b) Prices and trading terms in PO.
Sends invoice to contractor
Stores materials on site
Requisitions matl from stores for works
Issues materials to works
Pays suppliers (usually monthly)
Gets paid
Uses material in the works
Maintains stock record cards for each item and regularly checks physical stock against figure on record card
Provides information (usually monthly) for contract costs:- Invoices paid, Materials received & not yet paid for etc
Measures works for monthly cert & own costs
Checks/certifies any matl price elements in monthly/final certificates
Incorporates relevant matl price info in cert eg Fluctuations PC items
Provides any info on matl prices which affects monthly cert.
Provides stock figures (usually monthly) for contract costs
Prepares (usually monthly) materials element of contract costs
Reconciles matl issued to job with matl measured in works
Arranges advantageous disposal of surplus matl, returnable containers etc.
Agrees surpluses to be disposed of

Figures 31.14 Materials management—outline of main tasks (*see p.* **31**–71)

from stores and so on. If the diagram were complete it would be festooned with loops and subnets for all the various checks and reconciliations that are needed to keep the process running reasonably smoothly and economically. It would thus be far too complex for the purposes of this Section.

Some paragraphs relevant to Figure 31.14 follow below and others can be located from the grid index, Figure 31.1.

Contractors' organisation of the materials function

For organisation purposes the tasks in Figure 31.14 can be summarised thus:

[1] *Engineer*	[2] *Contractor technical*	[3] *Contractor materials dept.*	[4] *Contractor accounts/costs*
Specifies	Requisitions		
Approves		Places orders	Pays suppliers
		Arranges transport to site	Records/reports costs
		Holds in site stores	
		Issues to job	
		Disposes of surpluses	

The contractor's tasks in the columns 2 and 4 fall to well defined units in the organisation and seldom present organisational problems. There is, however, more room for debate on how the tasks in column 3 should be organised. They form successive links in a chain and are so interlocked that ideally they should be under unified command if only to make accountability for delays, breakages, losses, etc., as clear-cut as possible. However, this is often difficult especially on overseas work where the different tasks are performed in different parts of the globe. All that can be said is that the more the tasks in column 3 are carved up, the more effort has to be spent on checks and proddings at the interfaces between the different organisational units involved. (See also C.E. Procedure,[1] para. 161.)

One frequently discussed question is whether purchasing should be done from site or from a permanent base purchasing organisation. Every contractor has his own answer but on balance it seems better to use a permanent purchasing organisation for those materials (usually the great majority) where the contractor can benefit if he places large orders covering several contracts at once and maintains permanent relationships with his trusted suppliers. But see C.E. Procedure,[1] para. 158.

Layout and equipment of stores buildings, yards and compounds

The managerial task here is to weigh up, at the outset, how much attention this matter demands. On a run-of-the-mill UK job, previous experience of competent storekeeping staff should be a sufficient guide as to what is needed. On the other hand, on really large overseas jobs where lead times (see p. **31**–17) are very long, very large buildings and compounds may be needed. In this case, particularly if the site is somewhat cramped, very careful studies may be needed in order to try to predict the variety and volume of materials that will have to be held on site under cover and in the open at various stages of the work.

Even if done with care these predictions will probably be wrong by a fairly large margin so any stores layout should be designed and sited so as to allow for expansion to meet unforeseen needs.

Readers who need to know more about stores layout and equipment may refer to Baily[54] or, for greater detail, to Morrison.[55]

Economic and secure storage of materials

Each type of material must be stored so as to minimise the three principal hazards:

Breakage
Weather damage
Theft

Breakage is the same the world over but *weather damage* problems vary substantially in different climates. In order to control *theft* economically it is essential to know what items are particularly coveted in the area surrounding the job. These can vary from time to time and place to place even in the UK while on an overseas job the list of high theft risk items can be quite surprisingly different from the UK. Apart from the obvious precautions like properly locked and lit storage buildings and compounds, it is useful where possible to keep personnel transport (cars and buses) well away from the work area and in sight of one of the main offices so as to make the export of booty as hazardous as possible.

Checking materials against invoices

The essential steps are shown in Figure 31.14.

The practical complications arise because:

1 It is uneconomic to attempt to check the contents of every crate and the weight of every bulk load before signing for it
2 Invoices seldom match delivery notes one for one
3 In such a highly detailed task, some human error is inevitable.

Economic handling of materials on site

It is often a problem to decide whether to store a particular item or type of material in the stores or adjacent to its final position in the works. Generally speaking big items that are not likely to be pilfered or diverted to some other purpose on the site are best stored within crane reach or hand carrying distance on their final location in the work. However if they can only be put where they will have to be loaded on to a transporter again before use, they may as well be stored in the stores yard under proper supervision.

Ensuring that materials issued from stores are correctly used in the works

Once materials have left the stores to be incorporated in the works, the four principal hazards are:

Wastage (especially of bulk materials like timber, aggregate, cement, etc.)
Accidental damage
Misappropriation
Theft

The best protection against wastage and accidental damage outside the stores area is 'good site housekeeping'. It is only on the dirty untidy sites that you get really serious losses due to such things as aggregate, bricks, (and even timber and steel sections) being 'dozed into the fill'.

Misappropriation is only a problem with certain items like precut pipe and steel sections which can be fairly easily 'modified' by people who need a particular item in a hurry (perhaps because they failed to requisition it at the proper time). A bit of it is inevitable but if it goes unchecked it produces mounting chaos.

The first line of defence against misappropriation, theft and really costly accidental damage is to issue high-risk items to named individuals and, where possible, against specific drawings which are marked up to prevent 'double issues'.

Overall physical checks on site use of materials

It is prudent to run periodic checks on individual high value and high risk items to ensure that:

$$\text{Quantity received} = \text{Quantity in the works} + \text{Quantity in stock}$$

This would be an impossibly large task if applied blindly to every item in use on site and the art of getting value for money in this operation lies in choosing the high value and high risk items, deciding the frequency of the checks, and knowing for each item what is a reasonable percentage to allow for losses, breakages, etc. Materials quantities in the permanent works can be obtained directly from the monthly measurement but quantities used in temporary works and consumables cannot and for this reason these items need to be watched particularly carefully for signs of waste (which can include inadequate re-use, theft or fraud).

Materials management for plant, transport and tools and tackle

Plant, transport and tools and tackle (including such things as scaffolding, steel shuttering, huts, etc.) are normally all hired by the contract from the depot or from outside hire firms.

However, spares and consumables are handled in more or less the same way as materials to be used in the works. In the UK where main dealers and agents are on call and any spare part can normally be got to a site in a few hours, this function may be quite simple and straightforward. But on overseas jobs in remote locations plant spares and consumables can present enormous problems because of the long lead times (see p. **31**–17) and may consume just as much managerial effort as do the materials for the works.

Manning the materials function

See C.E. Procedure[1] paras. 159 and 160.

In view of the importance and complications of materials management, it may seem strange that purchasing staff who handle large sums of money are usually ex-clerks with no special training and that storekeeping staff are often crippled or superannuated labourers. However, it is not so strange when one remembers that only a few generations back, the functions of today's chartered engineers were performed very adequately by unlettered craftsmen. In materials as in engineering there is a steady movement towards greater expertise and qualifications and it is now increasingly common to find materials management in construction in the hands of members of the Institute of Purchasing and Supply.

Materials management and site morale

In the author's experience, materials management has a most potent effect on the morale of the men on site. On a site where work is always being interrupted by materials problems, momentum is lost, the men have time to discuss or dream up grievances and worse still they start to reason that a management which cannot produce the right materials at the right time must be incompetent in other ways as well. Conversely, on a site where the right materials always appear where they are wanted as if by magic, it is easier to build the job than to argue about it and a momentum is built up which can much reduce the need for 'driving' day-to-day supervision. In this matter the contractor is in the firing line both contractually and in the eyes of his men. However, it is in the Engineer's own interest to give the Contractor whatever support he can without upsetting the contractual liabilities.

Project information and the materials function

Obviously the price and availability of materials can fundamentally influence the Engineer's *design* of a job but this is really outside the scope of this Section.

The role of drawings, specifications and bills in the contractor's materials management function is mentioned on pages **31**–17 and **31**–18.

In the ICE Conditions,[2] clauses 36, 37, 39, 53, 54 and 60(1) are particularly relevant and nonstandard *forms of contract* should always be examined most carefully for any special stipulations about purchasing, storage or payment in respect of materials.

Records of orders and deliveries are an inherent part of purchasing and storekeeping systems. Abstracts may be needed:

(a) By Engineer and Contractor when considering any matter connected with delays and/or extra payments

(b) By the Contractor for maintaining his records on the performance of suppliers

As always, the problem is to decide what to abstract and what to destroy at the end of the job.

SUBCONTRACTORS

Introduction

The proportion of work that is sublet varies very widely from job to job but taking the industry as a whole it appears that the average is about 30% of contract value and can be 50% or higher in work which is predominantly building. In any event skill in selecting and controlling subcontractors is vital to the success of most contracts. Twort[4] has a good general chapter which brings to life some of the inherent problems of subcontracting work and the NEDO report[6] shows the trend of current thinking. Marks, Grant and Helson[5] have some useful notes on specific points.

Why subcontract work?

Whatever the law and the contract documents may say, the fact is that subcontracting divides responsibility and dilutes control and so is never to be undertaken lightly. Common reasons for subletting are:

The Engineer often nominates subcontractors before putting the main contract out to tender:

(a) Because he is not confident that the main contractor (whoever he may turn out to be) will command the construction expertise needed for the work in question

(b) Because he relies on the nominated subcontractors for certain specialist design expertise which may include the supply of proprietary materials or components

The main contractor often sublets work:

(a) Because the subcontractor has resources superior to his own and can therefore do the work more cheaply

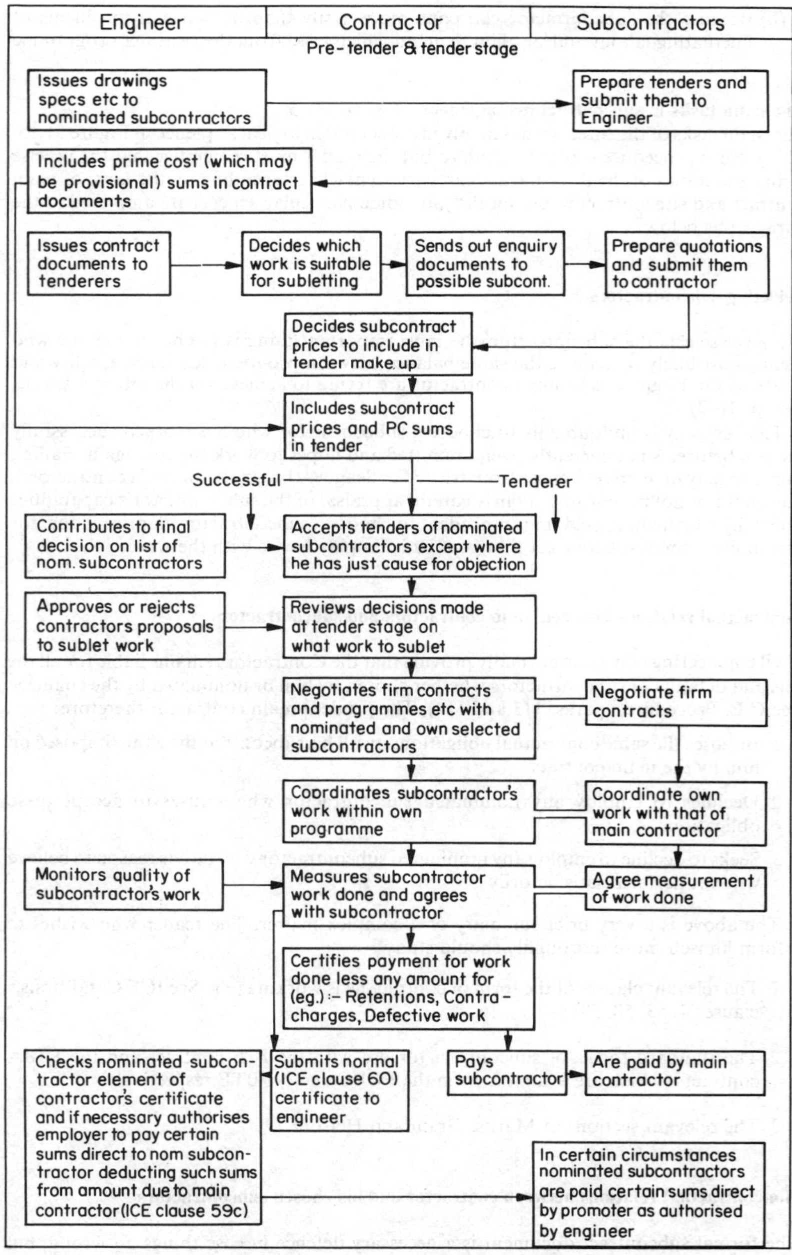
Engineer
Contractor
Subcontractors
Pre-tender & tender stage
Issues drawings specs etc to nominated subcontractors
Prepare tenders and submit them to Engineer
Includes prime cost (which may be provisional) sums in contract documents
Issues contract documents to tenderers
Decides which work is suitable for subletting
Sends out enquiry documents to possible subcont.
Prepare quotations and submit them to contractor
Decides subcontract prices to include in tender make up
Includes subcontract prices and PC sums in tender
Successful
Tenderer
Contributes to final decision on list of nom. subcontractors
Accepts tenders of nominated subcontractors except where he has just cause for objection
Approves or rejects contractors proposals to sublet work
Reviews decisions made at tender stage on what work to sublet
Negotiates firm contracts and programmes etc with nominated and own selected subcontractors
Negotiate firm contracts
Coordinates subcontractor's work within own programme
Coordinate own work with that of main contractor
Monitors quality of subcontractor's work
Measures subcontractor work done and agrees with subcontractor
Agree measurement of work done
Certifies payment for work done less any amount for (eg.) :- Retentions, Contra-charges, Defective work
Checks nominated subcontractor element of contractor's certificate and if necessary authorises employer to pay certain sums direct to nom subcontractor deducting such sums from amount due to main contractor (ICE clause 59c)
Submits normal (ICE clause 60) certificate to engineer
Pays subcontractor
Are paid by main contractor
In certain circumstances nominated subcontractors are paid certain sums direct by promoter as authorised by engineer

Figure 31.15 Subcontract management—outline of main tasks (*see p.* **31**–76)

(b) Because the subcontractor can cope more easily than he can with problems of fluctuating labour and/or plant demand which arise from the contract programme

The main tasks in subcontract management—Figure 31.15

The main tasks of the three parties involved are set out in logical sequence in Figure 31.15. The general procedure is pretty standard but the reader must remember that the precise rights and duties of the three parties can vary somewhat depending on the form of main contract and subcontract in use on the job. Some particular aspects are amplified in the paragraphs below.

Selecting subcontractors

Whoever selects the subcontractor, the most important thing is to choose the one who seems most likely to achieve the same balance between progress, economy, quality and safety as the Engineer and main contractor are trying to achieve on the job as a whole. (See p. **31**–7).

The best way is undoubtedly to choose a subcontractor who has worked successfully for you before, is not currently overcommitted and hopes to work for you again. Failing this, one may take up references from trusted colleagues (Engineers or main contractors). If even this is not practicable, a fairly careful appraisal of the subcontractor's capabilities is usually worthwhile. Seldom is it prudent to choose a subcontractor on price alone, for even quite a small subcontract that goes bad can play havoc with the whole job.

Contractual relations between main contractors and subcontractors

Civil engineering contracts normally provide that the Contractor remains liable for all the acts and defaults of subcontractors whether chosen by him or nominated by the engineer (see C.E. Procedure,[2] paras. 113 and 114). The prudent main contractor therefore:

1 Imposes the same contractual obligations on all his subcontractors as are imposed on him by the main contract

2 Declines to employ any nominated subcontractor who refuses to accept these obligations

3 Seeks to decline to employ any nominated subcontractor who he has reason to believe will not perform satisfactorily

The above is a very brief summary of a complex matter. The reader who wishes to inform himself more thoroughly should study:

1 The relevant clauses of the form of contract he is working (e.g. See ICE Conditions,[2] clauses 4, 53, 58, 59)

2 The standard forms of subcontract for use with the ICE contract and the RIBA contracts (which are available from the FCEC and NFBTE respectively)

3 The relevant sections of Marks, Grant and Helson[5]

Noncontractual relations between contractor and his chosen subcontractors

The formal subcontract document is a necessary defence in case things go wrong, but it does not guarantee that things will go right. In fact the most successful subcontracts are those where people hardly ever look at the documents because they trust each other too well. The main bases of this trust are:

1 Contractor selects his subcontractor carefully
2 Contractor treats his subcontractor in the way that he himself would like to be treated by Engineer
3 Contractor is likely to want to place future work with the same subcontractors.

Noncontractual relations between Contractor, nominated subcontractor and Engineer

Here we often find startling differences between what the contract documents say and what really happens. On paper the nominated subcontractor is bound solely to the main contractor in just the same way as are the Contractor's own chosen subcontractors. In real life, we have a triangular relationship between Contractor, nominated subcontractor and Engineer which can produce problems. For example:

1 Though the subcontractor has a *contract* with the Contractor his prime *loyalty* is usually to the Engineer who has selected him. This can much reduce the main contractor's control of and sense of responsibility for the work of nominated subcontractors. However this problem is mitigated if the nominated subcontractor is one who works regularly with the main contractor and is often directly selected by him
2 Whatever the subcontract documents may say, the nominated subcontractor often in reality accepts responsibility for certain elements of the Engineer's design. This in turn can raise the thorny question of what is the Engineer's responsibility *vis-à-vis* the Contractor, for the sufficiency of the design.

This is difficult territory on which there are no answers that are right for all occasions. The reader who wants to know more about the sort of problems that can arise should refer to Bell[26] and Twort.[4] Many nominated subcontracts go perfectly smoothly but forewarned is forearmed.

Organising subcontractors

Most of what needs saying is covered by the subsections on:

1 Planning and controlling subcontract work (pp. **31**–18, 22, 28, 35, 42, 46, 51).
2 Relationships between the various parts (pp. **31**–76 to 79).

Regular face-to-face discussions, progress meetings, etc., are vital. In dealings with subcontractors perhaps more than anywhere else it is best to keep the paperwork down by following the maxim, 'Say it before you write it unless you are absolutely certain you are right'. In the end, whatever the contract documents may say, successful handling of subcontractors depends on successful day-to-day relationships between key people on site, and, at a deeper level, on Engineer, Contractor and subcontractor understanding how far each depends on the others for his continuing prosperity.

General site information and facilities for subcontractors

The subcontractor has to inform himself on those features of the site which are outside the contractor's control, e.g. climate, ground conditions, public access (road, rail, water, etc.).

He will require to reach agreement with the contractor on such things as:

Access to site: any special arrangements made by contractor
Access within the site at various stages of construction

Arrangements for supply of services, e.g. water, electric power, compressed air (?), steam (?), etc.
Areas available for work, offices, storage, etc.
Arrangements for disposal of debris and noxious wastes

Project information for and from subcontractors

The scope of the subcontractor's work will be defined in the subcontract documents. The subcontractors will probably need other design information as the job progresses and here as elsewhere, the good contractor treats his subcontractors as he himself would wish to be treated by the Engineer.

The situation is more complex when, as is often the case, the Engineer requires certain design information from a nominated subcontractor before he can complete his own working drawings for the Contractor.

Man-management: subcontractors

The main contract and subcontract documents normally contain clauses relating to the terms and conditions of employment of subcontractors' men (e.g. ICE Conditions,[2] clauses 24 and 34). This is not pure altruism. Commonsense and experience show that it is asking for industrial trouble to have on the same site men on very different terms and conditions of employment. On large multicontractor sites it is becomingly increasingly common for the promoter or the main contractor to negotiate a specific site agreement which then applies to all men who work on the site (see p. **31**–67).

Equipment (plant, transport, tools and tackle, etc.): subcontractors

The subcontract documents will regulate the terms on which the subcontractor brings his own equipment on to site (e.g. see ICE Conditions,[2] clause 53(10)) and hires or borrows the contractor's equipment. Obviously it is simpler in theory if the subcontractor is completely selfsufficient but in practice it is often more sensible to arrange for him to use certain items that have to be there anyway. Scaffolding is an obvious example. Compressed air is often another. However, these sharings can never be wholly provided for in the subcontract documents. A bit of give and take is needed to get the job done, but human nature being what it is, most subcontractors need to be closely watched on anything they use without paying for it.

Materials management: subcontractors

For materials that he supplies, the subcontractor needs to make arrangements with the contractor at the outset about areas for storage and arrangements for security. Where the subcontract requires the contractor to supply certain materials as free issue, he needs to take special care to control issues and reconcile materials supplied to materials measured in the work paid for because subcontractors are notoriously profligate with free issue materials.

THIRD PARTIES

The generic term third parties is used in this section to cover all those who may affect or be affected by the design or construction of the works though they have no contractual or commercial links with promoter, engineer or contractor. The term includes such bodies as:

Those who occupy adjacent property (e.g. see ICE Conditions,[2] clause 29)
Those who have rights of way through or close to the site of the works (e.g. see ICE Conditions,[2] clauses 27 and 29)
Other employers in the area and their employees (e.g. see ICE Conditions,[2] clause 31)
Statutory undertakings (see NEDO,[6] paras 2–7)
Central and local government departments (other than the promoter) (e.g. see ICE Conditions,[2] clauses 26, 27, 29, 65, 69)
Police, fire and hospital services
Press, radio and TV

The contractor's duties towards third parties are normally covered in the contract and some relevant clauses are referred to in the list above. The most obvious and universal duty is insurance (see pp. **31**–50 and 51).

However, success on a contract usually depends on doing a bit more than your contractual duty. The above list is not exhaustive and on any proposed project the prudent Engineer and the Contractor make their own lists of third parties who might affect the success of the project and seek to establish with them relationships which take into account that at some point in time they might have a serious influence on the success of the work.

The influence of third parties on certain aspects of contract management is explored further in the paragraphs on the pages listed below:

Programming and progress control	page **31**–19
Safety	page **31**–28
Risks and insurance	page **31**–50
Men	page **31**–67

MANAGEMENT TECHNIQUES

Management techniques for problem solving, improving and innovating

The bulk of this section deals with the day-to-day activities of contract management, i.e. planning and controlling for desired results and organising and managing the necessary resources. Quite distinct from these is a group of activities whose purpose can be summed up as:

1 Putting right things that persistently go wrong
2 Improving things that seem to be going reasonably well
3 Finding new and better ways of doing things

These activities are to the management process what R & D is to a manufacturer's product and it is here that new 'management techniques' constantly proliferate. There are now literally hundreds of such techniques but the only ones that the reader is at all likely to encounter in ordinary contract management are:

Work study
O & M
Network analysis
Statistics and sampling
Operational research (possibly)

Calvert[3] gives a useful introduction to each of these techniques and a good list of further references. The aspects of contract management where these techniques are likely

Table 31.8 MANAGEMENT TECHNIQUES LIKELY TO BE RELEVANT TO CONTRACT MANAGEMENT AND WHERE THEY MIGHT BE USEFUL

Technique / *Aspect of contract management (As in Fig. 31.1)*	*Work study Method study Work measurement Learning curves etc. etc.*	*O & M (Organisation and methods)*	*Network analysis See note 1 below*	*Statistics and sampling*	*Operational research*
OVERALL PLANNING AND METHODS	Const'n methods plant selection temp works site layout		Obviously		
PROGRAMMING AND PROGRESSING	Assessing labour and plant outputs programming repetitive work		Obviously		
QUALITY				Quality control	
SAFETY				Safety controls and statistics	
ESTIMATING AND TENDERING	Estimating labour and plant outputs			Bidding strategy	Bidding strategy
COSTING VALUATIONS FINANCING	Cost investigations		Forecasting and substantiating delays		
CONSTRAINTS RISKS INSURANCES				Assessing risks	
ORGANISING AND ACTIVATING		Organising clerical work procedures form design etc			
SITE				Appraising soils, hydraulic climatic, etc. data supplied	
INFORMATION		As for organising and activating	Programming design work		
MEN	Incentive schemes Improving output esp. on repetitive work				
MACHINES	Appraising new plant Improving outputs Improving repair and main ops.		Streamlining major overhauls on key machines		
MATERIALS	Storing and handling materials	Clerical work in the materials functions	Procurement programmes		Special inventory and stock control problems

Table 31.8—*continued*

Technique *Aspect of contract management (As in Fig. 31.1)*	*Work study Method study Work measurement Learning curves etc. etc.*	*O & M (Organisation and methods)*	*Network analysis See note 1 below*	*Statistics and sampling*	*Operational research*
SUBCONTRACTORS	Helping subcontractors with site production problems		Coordinating subcontractors		
THIRD PARTIES			Impressing on stat. undertakings how their work affects the contract		

Network analysis is included here because it is still regarded by some as a new-fangled management technique. It should perhaps be regarded as just one of a number of programming techniques which are now widely used in the industry. (See p. **31**–10).

to be useful are shown in Table 31.8 in which the headings are the same as in the grid index, Figure 31.1.

Some cautionary remarks about management techniques

All management techniques have certain common features which the reader should keep in mind if he wants to use them.

1 *Innovation*. They start their life as new-fangled notions which practical men view with suspicion. If they take root they eventually become part of the 'commonsense' way of doing things and cease to be regarded as management techniques. Double-entry bookkeeping went through this cycle several hundred years ago and network analysis is going through it at present.
2 *Long timescale*. They demand a longer timescale than most of the activities in contract management. They are therefore most likely to be useful on operations that stay the same for long periods or repeat at regular intervals.
3 *Manning*. They usually require some specialist staff with a more analytical cast of mind than is commonly found in front-line managers. These contrasting types of men sometimes find it difficult to cooperate.
4 *Training*. If a certain technique is to be used, it is vital to teach the front-line men something about it, even if most of the detailed work is to be left in the hands of specialists.
5 *Sales pressure*. Their advocates tend to push new management techniques, like patent medicines, as cure-alls.

Summing up on management techniques

The hazard with management techniques is that they represent the most easily academicised aspect of management and so the serious student of management problems is apt to think they represent all that there is to know about management. Nothing could be further from the truth. Management techniques are essentially vehicles for innovating or at least improving on a long timescale. They tell you practically nothing about the day-to-day problems of contract management which occupy most people most of the time, and that is why they come at the end of this section. Moreover, the reader should always keep in mind that modern management techniques are useless to people who do not understand the most ancient management technique of all: the art of getting, holding and using power.

REFERENCES

Preference has been given to titles that are free or cheap or (like references 1–6) provide in one volume useful backup to many different parts of this Section. The Proceedings of the Institution of Civil Engineers are referred to so frequently that the abbreviation 'Proc ICE' is used throughout.

Essential references

(Without these the reader will make little sense of some sections of this Section. See p. **31**–2.

1. *Civil Engineering Procedure,* 2nd edn, ICE (1971)
2. *'ICE Conditions of Contract'*, 'General conditions of contract and forms of tender, agreement and bond for use in connection with works of civil engineering construction', 5th edn, ICE, Assoc. Cons. Engrs, Fed. Civ. Eng, Contrs (June 1973)

Very useful general references (other than to Proc. ICE)

(These four are referred to many times and the serious reader should acquire his own copy of each.)

3. CALVERT, R. E., *Introduction to Building Management,* 3rd edn, Newnes–Butterworths, London (1970) (See note below)
4. TWORT, A. C., *Civil Engineering Supervision and Management,* 2nd edn, Edward Arnold, London (1972) (See note below)
5. MARKS, R. V., GRANT, A. and HELSON, P. W., *Aspects of Civil Engineering Contract Procedure,* Pergamon (1966).
6. NEDO (National Economic Development Office), 'Contracting in civil engineering since Banwell' (Report of committee, chairman Sir William Harris) HMSO (1968) (Is best read in conjunction with ref. 17)

The main differences between 3 and 4 are not the differences between Building and Civil Engineering but between different ideas of what 'Management' means. Each has gaps. Together they make a strong source.

General references to Proc ICE

*Indicates informal discussion meeting.

7. BALLARD, E. H., 'The control of resources required for the construction of a civil engineering project' (Nov. 1972 and May 1973)
8. BAYLIS, A. L. H., 'The planning and management of the Consulting Engineer's supervisory team at Kainji Dam' (Jan. and Aug. 1972)
9. BROADBENT, B. H. and PALMER, J. E. G., 'How can designers draw on contractors' know-how at all stages?' (Feb. 1967)*
10. CAIRNCROSS, A. A., 'Is the Resident Engineer powerless?' (Apr. 1971)*
11. ELSBY, W. L., 'A resident engineer's duties' (June 1967)*
12. FRAME, A. G., 'Some comments on large project management with particular reference to the Anglesey aluminium smelter construction' (Feb. 1973)*
13. GILLOTT, C. A., TYLER, D. R. and REEVE, J. F., 'Durgapur steelworks project: administration, co-ordination and measurement of the civil engineering and building works' (Nov. 1964 and Jan. 1966)

14. GREEVES, I. S. S., 'The employment of quantity surveyors' (June 1969)*
15. HANCOCK, P. H. D., 'Contractors' administration of large contracts' (Dec. 1967)*
16. LOVELL, S. M., 'Advantages of direct labour in the construction industry' (Feb. 1970)*
17. NEDO 'Contracting in civil engineering since Banwell' (Summary Feb. 1969, Discussion Aug. 1969)
18. PARSONS, G. F., 'Duties of a contractor's agent' (June 1968)*
19. TAIT, W. B., 'The role of the civil engineer in the planning, design and construction of a modern highway' (June 1971)

Project Information—Conditions of Contract

*Indicates informal discussion meeting.
20. ICE CONDITIONS OF CONTRACT (See reference 2)
21. CCC/WKS/1, 'General Conditions of Government Contracts for Building and Civil Engineering Works' HMSO
22. RIBA FORM OF CONTRACT, 'Agreement and Schedule of Conditions of Contract' Joint Contracts Tribunal. Royal Institute of British Architects
23. INTERNATIONAL CONDITIONS OF CONTRACT, 'Conditions of Contract (International) for works of civil engineering construction' (Obtainable from the Federation of Civil Engineering Contractors and others)
24. ABRAHAMSON, M. W., 'Contractual conflicts—or—'First let's kill all the lawyers', *Proc. ICE* (July 1969)*
25. ACKROYD, T. N. W., 'The engineer's responsibility for design under an ICE contract', *Proc. ICE* (June 1968)*
26. BELL, A. T., 'Responsibility for nominated subcontractors and suppliers on civil engineering works', *Proc. ICE* (Nov. 1970)*
27. SHARMAN, F. A. and TURNER, D. A., 'Contractual aspects of piling', *Proc. ICE* (May 1969)*
28. WADDINGTON, J. C., 'Who pays for the unexpected? with particular reference to ICE Conditions of Contract Clauses 11 and 12', *Proc. ICE* (July 1967)*

Project Information—Bills of Quantities

*Indicates informal discussion meeting.
29. ICE S.M.M., 'Standard Method of Measurement of civil engineering quantities' *Instn Civ. Engrs*
30. BUILDING S.M.M., 'Standard Method of Measurement of building works', Royal Institution of Chartered Surveyors and National Federation of Building Trades Employers
31. BARNES, N. M. L. and THOMPSON, P. A., 'Civil engineering bills of quantities' Construction Industry Research and Information Association, Rep. 34 (Sept. 1971)
32. BARNES, N. M. L., 'Civil engineering bills of quantities', *Proc. ICE* (Jan. 1970)
33. CREASY, L. R., 'Economics and engineering organisation', *Proc. ICE* Suppl. (i) (1970)
34. GOLDSTEIN, A., 'Standard methods of measurement of civil engineering work', *Proc. ICE* (Aug. 1965)*
35. HANSEN, F. J., 'Do designers and contractors speak the same language?', *Proc. ICE* (Nov. 1970)*

Finance, estimating, costing, valuations, accounting, etc.

36. ALLEN, A., *Bluff your way in accountancy*, Wolfe Publishing Ltd (1971) (This is a serious and concise booklet with a misleadingly frivolous title)
37. ARMSTRONG, K. G., 'Economic analysis of engineering projects', *Proc. ICE* (Jan. 1965 and Aug. 1966)
38. BALLARD, E. H., See Ref. 7
39. CREASY, L. R., See Ref. 33
40. HASWELL, C. K., 'Rate fixing in Civil Engineering Contracts', *Proc. ICE* (Feb. 1963 and Jan. 1964)
41. PARKER, E. J., 'The planning of project finance', *Proc. ICE* (June and Dec. 1969)
42. STARK, R. M., 'Unbalancing of tenders', *Proc. ICE* (Feb. and Nov. 1969)

Men and machines

43. CIVIL ENGINEERING WORKING RULE AGREEMENT, Copies obtainable from Federation of Civil Engineering Contractors
44. BUILDING WORKING RULE AGREEMENT, Copies obtainable from National Federation of Building Trades Employers (*Note:* There are numerous Regional and Area variations to the National Working Rules)
45. BAIRD, M. ST. C., 'Incentive Schemes: Do they increase productivity?', *Proc. ICE* (June 1971)*
46. CONTRACTORS' PLANT ASSOCIATION, Schedule of Rates of Hire.

(References 3 and 56 also contain relevant material while those interested in comparing man management now with 100 years ago will find interesting material in references 59 and 60.)

Safety

47. FEDERATION OF CIVIL ENGINEERING CONTRACTORS, *Supervisors safety booklet*
48. FEDERATION OF CIVIL ENGINEERING CONTRACTORS, *Guide to the Construction Regulations*
49. FCEC/NFBTE, *Manual on Construction Safety* (Obtainable from sources listed in Ref. 47)
50. INSTITUTION OF CIVIL ENGINEERS, 'Safety in Civil Engineering', *Proc. ICE* (Jan. 1969)
51. MERCHANT, W., SHORT, W. D. and KINSELLA, J., Three papers on various aspects of site safety with discussion, *Proc. ICE* (Mar. and Dec. 1967)
52. ROBENS REPORT—a brief guide, British Safety Council (1972)
53. SHORT, W. D., 'Safety precautions in civil engineering with particular reference to responsibility', *Proc. ICE* (May 1969)*

Materials

Calvert[3] has a chapter on purchasing and Twort[4] some short notes on certain aspects. The author has not found a comprehensive book on materials management specifically for the construction industry. Of the many books on materials management generally, he has found the following two useful:

54. BAILY, P. H., *Purchasing and Supply Management*, 2nd edn, Chapman and Hall (1969)
55. MORISON, A., *Storage and Control of Stock*, 2nd edn, Pitman (1967)

Books on general management

The literature on general management (as distinct from construction management) is so vast and expanding so fast that even the pundits find it hard to separate the wheat from the chaff. The interested reader is recommended to confine himself initially to the first four chapters of Calvert[3] and to three small works that have stood the test of time and are all at present (1973) available as Penguin Paperbacks.

56. FALK, R., *The business of management*
57. PARKINSON, C. N., *Parkinson's Law* (This looks like a frivolous book but it contains more useful management wisdom than many books five times its size)
58. PUGH, D. S., HICKSON, D. J. and HININGS, C. R., *Writers on organisations*

Some relevant historical books

59. BURTON, A., *The Canal Builders*, Eyre Methuen (1972)
60. COLEMAN, T., *The Railway Navvies*, Penguin (1970)
61. HELPS, A., *Life and Labours of Mr Brassey 1805–1870*, Evelyn Adams and Mackay (1969)
62. MIDDLEMAS, R. K., The Master Builders, Hutchinson (1963)
63. ROLT, L. T. C., *George and Robert Stephenson*, Longman
64. ROLT, L. T. C., *Isambard Kingdom Brunel*, Penguin (1970)
65. ROLT, L. T. C., *Thomas Telford*, Longman

32 INSURANCE

INSURANCE 32

32 INSURANCE

J. S. HARVEY, F.C.I.B.
Director of Edward Lumley & Sons Ltd and Edward Lumley (Underwriting Agencies) Ltd, London

THE PURPOSE OF INSURANCE PROTECTION

The general purpose of insurance, expressed in its simplest terms, is to put the Insured, after he has sustained accidental damage to his property from a risk against which he has insured, back into the same position as he was in immediately before the occurrence of the accident.

This is very easy to understand as a general principle; it is also easy to understand in relation to a single building, such as a private dwelling or a small factory, where a realistic value for insurance purposes can be ascertained at the time insurance is arranged, and the extent of damage and/or cost of repairs can be readily expressed in monetary terms. The concept of indemnity for damage from an accidental cause becomes very much more complex when the subject matter of the insurance is the work to be performed in connection with a Construction Contract. On a contract site the value at risk is constantly increasing owing to the work in progress; there can also be large sums of money involved in frequently-changing temporary works which are used to enable the permanent work to be performed. The challenge which this situation presents has been met by the insurance industry by its willingness to provide a comprehensive form of insurance coverage to the engineering community. The cover provided is usually referred to as 'Contractors All Risks Insurance' when applied to civil engineering works and 'Erection All Risks Insurance' when applied to the installation or erection of electrical or mechanical equipment.

On many occasions it is necessary to prepare an insurance policy which is appropriate to both classes of engineering work when the projects concerned are for such things as hydro-electric schemes.

Both classes of insurance have progressed so that at the present time there is a very wide measure of agreement amongst international insurance interests regarding the basic scope of cover which can be granted, although this may be presented in varying forms according to the individual preferences of the various Insurers. The continuing challenge comes in trying to extend, or otherwise modify, the scope of the basic cover, the need for which arises from technological development and the ever-increasing sophistication of integrated projects.

ACCIDENT OR COMMERCIAL RISK?

Much debate takes place as new projects are offered to the insurance market to try to decide where the dividing line is between legitimate insurance and the commercial risks which form part of any trading hazard or progressive development. In such debate it is important to remember that it is not the role of insurance to act as a guarantee that what appears possible in theory, or probable from investigation, can be achieved in practice.

This point can be illustrated by considering the driving of a tunnel. Before driving commences, some degree of investigation of the geology of the intended route and of the general area is undertaken, but however detailed the investigation may be it still only gives partial knowledge regarding the actual conditions that will be encountered during the driving of the tunnel. It is quite conceivable that conditions will be encountered that may lead to abandonment of the project either because of technical problems or because the financial burden of a change in method would be prohibitive to the Principal.

Each major civil engineering project is in many ways a prototype. In mechanical installations new products are being manufactured or new processes are being incorporated and the degree of hazard may be changed very considerably by such modifications. As said above, it is the subject of much and continuing debate to determine as reasonably as possible where the dividing line comes between commercial risk, which has to be borne by the engineering industry as part of its trading risk and development expenditure, and insurable risks which Underwriters can reasonably assess so that with a sufficient volume of business of that type they would be left with a reasonable degree of profit.

REPAIR OR IMPROVEMENT?

It is much to be regretted that all too frequently the finer points regarding division of risk only come to the fore when an accident has occurred. It can happen, and often does happen, that the original method of building or erection on a site is not the most appropriate method to use for the repair work which is necessary following an accident. When such cases arise, it can require detailed negotiations between highly qualified people to establish which costs constitute an increased cost of working, which has to be borne by the Contractor, or his Principal if the Contract Conditions so provide, and which costs are for the account of the Insurers as representing the true and economic cost of repairing the actual damage sustained by the works as originally constructed.

Once again the principles behind these comments can be readily illustrated by considering underground excavations. From time to time it happens that during the excavation of an underground cavern there is a partial collapse of the roof. Investigation may reveal that if longer rock bolts had been used and if these had been spaced less distantly they would have substantially diminished the chances of the fall of rock, and a new pattern of support is introduced into the works of repair and the new pattern is also utilised in the continuing excavation of the balance of the cavern. It depends upon the volume and shape of the void and also upon the ultimate purpose of the cavern as to how the works can be restored to an acceptable condition. It may be filled with concrete or left as a void with a false ceiling constructed to an adequate margin of safety and created in order to obtain the correct dimensions required for the cavern. In such a case there is obviously an admixture of responsibilities for repairing accidental damage and for increased costs of working due to the change to an improved method, and the division of the account can only be decided by negotiation and goodwill on either side.

IS INSURANCE REALLY NECESSARY?

It is a well known fact that the preparation of tenders for large and complex projects is a very costly operation, and this is especially true if the project is located at a great distance from the tenderer's home base. Insurance costs are generally speaking a comparatively small percentage of the ultimate contract price and, in the coordination of all the costings which have to be estimated, the insurance cost is sometimes left to guesswork by the tenderer. It can also happen that after an award a Contractor concentrates so much of his effort on site mobilisation that attention to his insurance obligations under the Contract are deferred until the commencement of the work on site. This may be explained by the unfortunate fact that engineers so often regard insurance as an unnecessary evil and think that to spend too much time on such a subject is to show some degree of lack of confidence in their ability to organise and manage the project in a safe and efficient manner.

How often insurance men have heard comments to the effect that 'nothing could possibly happen; if it could, we would be doing it another way'—yet disasters do occur and, when they do, it is the quality and the aptitude of the insurance coverage which may make all the difference to the Contractor between an overall profit on a project and an overall loss. Any aspect which can have a bearing on the financial stability of a company must be of paramount importance. It is when problems arise on a site that insurance may come into the reckoning and, to make certain that the insurance carried is as effective as possible, it is only prudent to consult an expert in the field of contractors insurance. It is also important to allow adequate time for personnel who are informed about the project to discuss the matter with an insurance expert. This allows the Insurers an opportunity of basing their charges on a proper appraisal of the risks involved, and if there are any points where Insurers are unduly apprehensive these can be discussed and, if necessary, modifications to the Contractors proposals can be mutually agreed. A good understanding between the two parties involved in an insurance contract is also helpful if and when accidents occur because it assists the settlement of claims which are more likely to proceed smoothly and quickly where both parties know the intended limits of their responsibilities, and if contentions do arise, such contentions will be debated in the right environment of informed opinion.

RESPONSIBILITY FROM DESIGN AND TECHNICAL ADVICE

The foregoing remarks are primarily in relation to the works on site performed by a Contractor. Prior to the calling of tenders much effort will have been expended by the Consulting and Design sections of the engineering industry who will have incurred a responsibility for the adequacy of their professional activities. This is another sphere in which the insurance industry becomes involved by granting Professional Indemnity coverage in respect of the legal liabilities incurred by Designers and Consultants, where damages are payable because of errors or omissions. In recent years Professional Indemnity insurance has undergone a very radical change, and what in former years was a comparatively inexpensive form of cover has now become an expensive, albiet necessary, adjunct of any professional activity.

The reasons for this change are manifold. One of the predominant reasons is the greater awareness amongst all sections of the community of the rights which exist, and a greater willingness to commence litigation when it is suspected that some loss or injury has been sustained as the result of negligence by those from whom professional guidance has been sought. There is also an upward trend in the amounts awarded by the Courts by way of damages. In addition, the continual desire to improve existing techniques, or to innovate, brings in its train a greater risk that some element necessary to success will be overlooked. As opposed to Contract Works where the contract value gives a realistic measure of the liability which can be incurred from physical loss or damage, there is no such yardstick by which the potential liabilities from a professional error can be deduced from the fees which are paid for professional services. Another complication arises from the fact that a very long period of time may elapse from the occurrence of the error in the design stage to its manifestation during construction or following completion. It is therefore of paramount importance in connection with the purchase of Professional Indemnity coverage to ensure that cover operates not only for the mistakes which occur during the period of the insurance, but also operates in respect of mistakes made during or prior to the period of insurance and which manifest themselves during the period of insurance.

CAN THE COST OF INSURANCE BE REDUCED?

As in all segments of a competitive industry, the cost of obtaining insurance protection of whatever nature is of considerable importance. This can to some extent be influenced by agreement of an Insured only to seek reimbursement in excess of a predetermined amount. The more important factor in minimising the costs of insurance is to establish a record with the insurance market of safe and efficient working whereby claims, other than those which are obviously beyond the control of the Insured, are reduced to an absolute minimum. In achieving this situation, coordinated research and development, and a wide distribution of the acquired knowledge leading to the standardisation of procedures can be of importance.

If one of the various Standards Institutes has promulgated a specification or Code of Practice, this tends to be regarded as the hallmark certifying the adequacy of the materials and/or procedures being employed; any improvement in standards of safety should give a corresponding diminution to the incidence of accidents and thus an improving experience to Insurers allowing them to reduce their charges for cover.

STRUCTURE OF THE INSURANCE MARKET

The insurance industry consists basically of Brokers and Insurers. The function of the Broker is to give advice and guidance to his client and to negotiate on behalf of his client with Insurers, bringing to bear his knowledge and experience to obtain the most appropriate form of cover applicable to his clients needs at the most economic cost. In this country the legal position of a Broker is that he is an Agent of the Insured. The broking fraternity again breaks down into two basic segments. There are Insurance Brokers who only have direct access to that section of the insurance market which is comprised of Insurance Companies, and there are Lloyd's Registered Brokers who, in addition to having access to the Company market, are authorised to transact business at Lloyd's. In practice, the non-registered Broker has access to the Lloyd's market through the intermediary of a Registered Broker. The Insurance market itself comprises the Insurance companies, Lloyd's and Professional Reinsurance Companies.

Insurance Companies are in a position to accept business either directly from the Insured or through the intermediary of a Broker, whereas the Lloyd's market can only accept business which is channelled to them via a Lloyd's Registered Broker.

The professional Reinsurance Companies have no direct dealings with the insuring public, but form a substantial part of the overall capacity of the insurance market by accepting as reinsurance some part of the liabilities assumed by direct Insurers where this is required. Thus in practice the Insured has a choice of negotiating his insurance requirements directly with an Insurance Company or Companies, or of entrusting the negotiations to a Broker who has access to a much wider section of the insurance market.

Once the decision has been made the negotiation of claims that occur follows the same route, and the Insured will either have to negotiate directly with his Insurers or inform his Broker who would then assist in coordinating all the information necessary for a proper presentation of the facts to Insurers.

BIBLIOGRAPHY

An Examination of the Practice of Parties Negotiating Contracts Requiring Indemnities, Advanced Study Group No. 171, Insurance Institute of London

EAGLESTONE, F. N., *The RIBA Contract and the Insurance Market,* 3rd edn, The Policy Holder Printing and Publishing Co.

HAGART, G. T. N., *Conditions of Contract and Insurance,* Witherby & Co.

HIBBITT, A. J., *RIBA Standard Form of Contract,* Clause 19(2)(a), Chartered Insurance Institute

PIPER, L. J., *Contractors' All Risks and Public Liability Insurance,* Buckley Press

The Underwriting of Contractors' All Risks Policies, Advanced Study Group No. 114 and No. 192, Insurance Institute of London

Table 32.1 MAJOR LOSSES: THE FOLLOWING IS A LIST OF SOME OF THE MAJOR LOSSES SUFFERED BY CONTRACTORS' ALL RISKS UNDERWRITERS IN LONDON IN RECENT YEARS (COURTESY: THE INSURANCE INSTITUTE OF LONDON)

Location	*Type of contract*	*Cause of loss*	*Amount* £
UK	New department store	Fire	185 000
West Africa	Dam construction	Subsidence	200 000
Libya	Harbour construction	Storm damage	300 000
UK	Nuclear power station	Fire	300 000
Greece	Dam construction	Flood	80 000
UK	Power station	Faulty design	80 000
USA	Ammonia plant	Explosion	68 000
Canada	Pulp and paper factory	Mechanical breakdown	300 000
UK	House construction	Faulty workmanship	150 000
UK	Motorway construction	Flood	50 000
UK	Power station	Gale	45 000
Pakistan	Dam construction	Tunnel collapse	500 000
Holland	Tunnel construction	Tunnel collapse	1 000 000
Kenya	Harbour works	Subsidence	400 000
Uruguay	Power station	Flooding	105 000
UK	Erection of turbo set	Breakdown of steam turbine	75 000
Belgium	Erection of air liquefaction plant	Explosion in heat exchanger	75 000
Germany	Hangar construction	Windstorm	85 000
UK	Testing of turbine rotor	Breakdown	100 000
UK	Construction of foundry	Fire	55 000
UK	Bridge construction	Collapse	300 000
UK	Construction of a college	Fire	50 000
UK	Construction of coal conveyor system	Fire	52 500
Korea	Erection of crude oil refinery	Main distillation tower dropped during erection	300 000
Algeria	Construction of 40 in gas pipeline	Storm damage due to torrential rain	125 000
UK	Erection of water turbines	Breakdown	80 000
Libya	Construction of sea pipeline	Storm	300 000
Holland	Erection of sulphuric-acid plant	Loss of catalyst due to overheating	94 000
UK	Turbo alternator	Breakdown	300 000
UK	Boiler economiser	Breakdown	151 000
UK	Boiler	Explosion	248 000
UK	Boiler	Explosion	132 000
Australia	Coal equipment	Fire	300 000
UK	Boiler	Burst under test	405 000
UK	Turbo alternator	Breakdown	200 000
UK	Turbo alternator	Breakdown	150 000
UK	Turbo alternator	Breakdown	300 000
UK	Turbo alternator	Breakdown	200 000
Canada	Turbo alternator	Breakdown	150 000
Japan	Nuclear P.S. boiler	Breakdown	116 000
UK	Heat exchanger	Mishandling	215 000
UK	Bridge	Collapse	135 000
UK	Transformer	Transit	219 000
UK	Reactor buildings	Subsidence	191 000
UK	Reactor building	Fire	255 000
UK	Turbo alternators	Breakdown	928 585
UK	Turbo alternator	Malicious damage	140 000
Sweden	Water tube boiler in chemical plant	Flue gas explosion	135 650
Libya	Construction of breakwater	Storm	187 200
Algeria	Natural gas pipeline	Flood	648 850
Holland	Sweet manufacturing plant	Fire	544 500
Peru	Steel works	Earthquake	312 500

Table 32.1—*continued*

Location	*Type of contract*	*Cause of loss*	*Amount* £
Iraq	Bridge construction	Flood	275 350
Formosa	Pipeline	Tidal wave	115 000
Belgium	Gas turbine at power station	Blade damage jaused by overheating of burner nozzle	109 333
Iraq	Turbine assembly building in power station	Water damage	158 000
Sweden	Water tube boiler at oil refinery	Flue gas explosion	129 375
France	Heat exchanger in nuclear power station	Damage caused by debris from reactor	260 400
Germany	Underground railway	Fire in tunnel	250 000
Pakistan	Link canal	Flood	90 000
Japan	Submarine pipeline	Typhoon	190 230
Japan	Submarine pipeline	Heavy weather	160 960
Italy	Trunk roads	Fire damage to plant	52 940
Belgium	Thermal power station	Coffer dam collapse	56 666
Pakistan	Dam construction	Coffer dam collapse	325 000
Japan	Hydro-electric scheme	Heavy rainfall	69 800
South Africa	Submarine pipeline	Storm	58 260
Iran	Hotel	Fire	77 000
Iran	Fertiliser plant	Fire	53 120
Jamaica	Apartment building	Fire	50 000
Jamaica	Aluminium plant	Accidental damage	145 850
Sicily	Ethylene plant	Accidental damage	55 850
Pakistan	Dam construction	Collapse	200 000
Italy	Tunnel construction	Water penetration	133 300

33 SETTING OUT ON SITE

SETTING OUT ON SITE 33

33 SETTING OUT ON SITE

D. W. QUINION, B.Sc.(Eng.), F.I.C.E., F.I.Struct.E., F.F.B.,
Tarmac Construction

PRINCIPLES

'Setting out' as practised on civil engineering and building sites is the locating of the works to be constructed and checking to ensure that they are dimensionally correct and in their right position. This service is essentially an aid to the labour force and must necessarily be provided in a form that is easy for them to use and understand; the information must be reliable and must be available as and when required. Errors in setting out will in most cases result in remedial works which will be expensive. Whatever lines or levels are provided should be checked to be sure of their accuracy, and they should be provided to the Foreman efficiently so that he can have the necessary confidence in them.

Clause 17 of the Institution of Civil Engineers' Conditions of Contract states:

> 'The Contractor shall be responsible for the true and proper setting out of the Works and for the correctness of the position, levels, dimensions and alignment of all parts of the Works and for the provision of all necessary instruments, appliances and labour in connection therewith. If at any time during the progress of the works any error shall appear or arise in the position, levels, dimensions or alignment of any part of the works the Contractor on being required so to do by the Engineer shall at his own expense rectify such error to the satisfaction of the Engineer, unless such error is based on incorrect data supplied in writing by the Engineer or Engineer's representative, in which case the expense of rectifying same shall be borne by the Employer. The checking of any setting out or of any line or level by the Engineer or the Engineer's representative shall not in any way relieve the Contractor of his responsibility for the correctness thereof and the Contractor shall carefully protect and preserve all bench marks, sight rails, pegs and other things used in setting out works.'

In this Chapter the initials SOE (setting-out engineer) are used to identify whoever is responsible for carrying out the setting out but, in addition to Engineers, this function is performed by Surveyors, Technicians and Foremen. 'The Engineer' is used to define the Client's technical representative.

EQUIPMENT AND GENERAL METHODS

The usual equipment for setting out work comprises a 20-second theodolite complete with legs and optical plumbing, a quick set level complete with legs and level staff. The SOE will usually also have: 30 m steel tape; 1 m folding rule; graduated scale; 30 m fine string line; 500 g plumb bob; triplicate book; club hammer; claw hammer and nails; centre punch; hardened steel point for scribing lines on steel or concrete; cloth for wiping tapes and hands; knife; crayons and waterproof pencil; 0.3 m spirit level.

This additional equipment is usually kept by the Chainman who should have received instruction in the duties required of him. The Chainman is required to take care of the instruments when not in use by the Engineer and to carry these carefully about the site. He needs instruction as to the correct method of holding a level staff, the installation of pegs, profiles and batter rules and the correct use of the tape when measurements are being taken. A well instructed Chainman will greatly ease the job of the SOE, whereas a poorly instructed one can cause errors.

For very accurate work it will be necessary to use a more accurate 1-second theodolite and occasions also arise when specialised equipment such as direct-reading distance measuring instruments and lasers are an advantage. The relatively high cost of these instruments is only occasionally justified unless there is a continuing application on the site.

The reader is referred to Section 6 for more detailed information on the use of surveying instruments and methods, but a number of practical points will be emphasised here.

(i) Theodolites and levels are delicate and easily damaged or strained. They should, therefore, be treated with great care. They should not be erected on potentially slippery surfaces. They should not be left unattended and when not in use should be carefully and correctly replaced in their boxes and the fasteners secured. They should be checked for accuracy and alignment at least once a week and whenever there is any reason for doubt.

(ii) If instruments have to be moved on their legs they should be carried with the legs supported on the shoulder such that the instrument is sitting alongside the head of the bearer in the normal vertical position. When instruments become wet they should be carefully dried by the SOE and they should always be kept clean.

(iii) When taping distances it is usually more accurate to measure from the 1 m mark on the tape, with the end of the tape held clear of the starting position. The Engineer should always make it clear to his Chainman what starting position he requires. Allowances should be made for measuring errors which occur due to slackness in the tape. For example, a 33 m steel tape weighing 0.0219 kg/m with a 5 kg tension will give an inaccuracy of 28.8 mm, and with a 10 kg tension

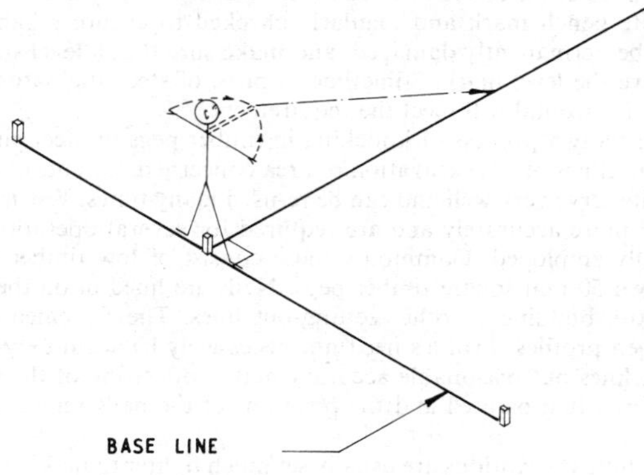

Figure 33.1

an inaccuracy of 7.2 mm. Likewise, correction should be made for measuring on slopes. The measurement of 33 m on a 1-in-50 slope means a 7.5 mm error on the horizontal measurement, and on a 1-in-10 slope would mean an error of 164 mm.

(iv) A 20 s error with a theodolite at 33 m gives an error of 3.2 mm and at 330 m gives 32 mm.

(v) The best method of slope correction where this is important is by taking levels at each end and calculating the correct length which is the taped length along the slope less the sum of

$$\frac{H^2}{2l}+\frac{H^4}{8l^3}$$

where H is the difference in height.

(vi) All setting-out should be completed as a closed traverse and any cumulative errors traced and eliminated by recognised surveying techniques. Unless it is practical, fore and back sight distances should be roughly equal when levelling as this will reduce any levelling inaccuracies if the instrument is in need of adjustment.

(vii) All tapes and bands should be kept clean and lightly oiled to avoid rusting, but not so oily as to pick up dirt.

(viii) The marking out of a right angle without an instrument is quickly achieved by using a 3:4:5 triangle of measurements. Rapid but more approximate results can be obtained by standing over the offset point on the base line with arms out sideways at shoulder height in the line of the base line. As the hands are brought together in front of one they will indicate the line at right angles.

(ix) When establishing a route across difficult country with bushes or other features obscuring required lines, it is often quicker and simpler to locate the positions of markers in clearings where they can be seen and to transfer the lines locally.

(x) When transferring a mark to concrete or steel it is wise to make two temporary marks each side of the required positions and scribe a line between them with a hard steel point. On concrete this can be stencilled in with indelible pencil and on steel the required point can be emphasised with a centre-punch.

TEMPORARY SETTING-OUT POINTS

Temporary Bench Marks (TBMs) are usually established on hopefully immovable features of the site. They can be scribed onto the sides of walls on the tops of foundations or kerbs, or onto piles or bases constructed for the purpose. They must all be levelled in from the main site bench mark and regularly checked to ensure reliability. Finished work should not be permanently damaged, and make sure that a level staff can be held truly vertical above the level mark. Sometimes a piece of steel angle iron, perhaps 1 m long, driven into the ground will meet the requirement.

Setting out is usually a process of knocking in timber pegs or steel pins to mark the extremities or centrelines of the excavation or area concerned. Offcuts of steel reinforcement painted white serve very well and can be re-used many times. When the setting-out lines are required more accurately and are required for several operations then timber profiles are usually employed. Commonly these consist of low timber rails 37 mm × 25 mm fixed to two 50 mm square timber pegs. Nails are lined in on the top to denote the required centre, building or other setting-out lines. The Foremen usually extend string lines between profiles. Profiles need not necessarily be accurately at right angles to the setting-out lines but reasonable accuracy makes offsetting of the line very much easier. The rails are often painted and the positions of the nails referenced on them in pencil.

Profiles for levelling excavations are usually set much tighter to make sighting between them that much easier and heavier timber sections may be necessary.

Setting-out points are usually 50 mm square timber pegs knocked well into the ground and protected with a surround of concrete. Nails locate the true line or intersection point. These points can be scribed directly onto suitable existing concrete or other surfaces. If liable to be damaged they should be clearly marked and where necessary protected by a simple fence or guard.

In special cases it may be justified to erect a small elevated platform above a setting-out point on which to set up the theodolite and gain a clear view not only across the site but in some cases also down into excavations.

Setting-out work has to serve the Foremen and they should be given diagrams clearly indicating how the points and levels relate to the work they have to do and they should be shown the pegs, profiles, etc., from which they will work. Interference with these must not be tolerated.

The Chainman must maintain a suitable stock of timber and steel, preferably cut from salvaged material, which can be used for setting out. He should recover profiles and pegs the Foreman no longer requires. These materials should be painted prior to use. Pegs and profiles are expensive but errors are more expensive and both considerations must not be forgotten.

Colour coding may be necessary where a profile is used with a different length of traveller or boning rod on each side.

A number of mistakes frequently give rise to common errors. It is easy to transfer offset dimensions from drawings to notebook to site and set out bases, etc., on the wrong side of the main setting-out lines. It is easy to give some pegs and markers in offset positions and others on line. It is easy to set up profiles accurately but set the wrong length for the boning rods. Errors of a unit can easily occur in reading tapes and staffs. The Chainman can make simple errors when erecting profiles or holding markers. Simple errors usually arise not from calculation mistakes but from lack of attention; straightforward setting out should always be checked, as well as the apparently more complex. The SOE must always be alert to pegs and profiles which have been inaccurately replaced without his knowledge after being disturbed or damaged.

SITE SURVEY AND PREPARATIONS

Before the commencement of a Contract it is necessary to establish a survey of the site as it currently exists, picking up all natural features and locating the site in relation to established datums such as Local Authority building lines, kerb lines of main roads, or other features that can be regarded as permanent. A principal bench mark should be established on site and this is agreed as a datum with the Engineer. Likewise, basic lines must be agreed for the location and orientation of the works as a whole and about which they will be set out. In cases where there is the possibility of the construction of the works having an effect on adjacent properties as a result of possible ground movements, vibration, etc., it may be necessary to survey and record features of those properties. This may be a recording of levels, inclinations to vertical, positions of cracks. Supporting photographs are valuable in recording the state of such properties.

The SOE now has a basis for proceeding with the setting-out. He is frequently faced with the need to set out the first stages of site construction for an immediate start on the 'access to site date' and the simultaneous need to establish main setting-out lines which may have to last the length of the Contract and be installed with considerable accuracy. Initial construction operations usually consist of site clearance and levelling and approximate setting-out methods can usually enable these operations to commence without delay. In some cases the SOE may not be able to establish the principal datum lines he requires until features of the site such as old buildings, trees, mounds, etc., have been removed. The SOE will establish his principal datum lines in positions where they can be of use to him for as long as possible. He will usually try to establish these clear of buildings and roads. He will try to identify positions for the location of principal points where they can remain undisturbed and free from construction operations. He must decide whether the ground conditions and importance of these datum lines justifies the casting of a concrete block or even a pile, or whether a long peg well driven into the ground with its head protected in concrete will be a sufficient basis for an accurate mark. The SOE will usually find it necessary to prepare a Master Plan indicating his principal setting-out lines, points, and his key bench mark for the site. He can relate his principal setting-out lines to the building lines, centrelines of buildings, roads and principal services required on the site. This information should be used to check the dimensions given on the

Engineer's drawings and copies should be supplied to the Engineer with a request that he confirms that the dimensions are as required. This will also enable the Engineer to satisfy himself on the accuracy of the SOE's setting out. In some cases, to preserve the principal setting-out points, it may be desirable for these to be located right on the site boundaries, or even outside it if permission can be granted by the adjoining land owner. It should be borne in mind that for a large site the Ordnance Survey bench marks around the site will not necessarily correspond within the accuracy with which the site levelling will be done, hence the need to establish a single bench mark on the site for the purpose of the works. From this single key bench mark a number of TBMs will be established and used locally. It is a sensible precaution to check at intervals of not more than one month that these have not moved or been damaged.

SOEs, before commencing their setting out, should discuss with the Foremen the methods to be used in the construction of the works. Foremen will probably require pegs, profiles, batter rules, and other information in locations which will not interfere with the movement of machines, men and materials. They may require offset pegs to be provided by the SOE or may decide to make their own offset measurements. The SOE must determine not only when the setting-out pegs and lines are required for use, so that he can anticipate these dates, but also the accuracy with which the information is required in relation to the purpose for which it will be used. Where considerable accuracy is required it is customary to insert pegs or rails and use nails for the precise position of the line. Lesser accuracy, but greater speed, can frequently be obtained by knocking in steel pins, particularly if the ground is difficult to penetrate.

Checking is all-important and, having established setting-out points and checked that they are in the right position, it is still essential to check the work as it is carried out to ensure that the original setting-out pegs and profiles have not become disturbed during the progress of the work. So, the Engineer must stay in constant contact with the construction operations and provide constant service to the operations. It is wise to check that boning rods are being used properly.

SETTING OUT FOR EXCAVATION AND GRADING WORKS

For these operations a lesser degree of accuracy is needed than for the setting out of foundations and building works. The SOE should bear in mind the likelihood that positions will need to be established and re-established with speed. The initial marking-out of the areas to be excavated and those to be filled will be disturbed when soil stripping takes place. Either long pegs clearly visible from earthmoving machines or smaller pegs with ranging rods to identify them should be used. Attendance will be required by the SOE to provide what is required. As soon as it is practical to do so, lines and profiles should be established around the areas in question and batter rules set up to give guidance on the slopes of cuttings and embankments. It will frequently be necessary as the work proceeds to provide additional profiles and points within the excavation or on the embankments. A typical situation is shown in Figure 33.2.

Where several levels have to be established a colour-coding system on the pegs and profiles should be adopted, and this should be carefully explained to the Foreman and the machine operators, and the Foreman provided with diagrams and explanations from the triplicate book. As the various levels are established, new setting-out points should be provided so that deeper individual foundations and local requirements can be quickly marked out for work to proceed without delay using bulk earth-moving equipment to the best economical advantage if only to remove deposited surplus spoil.

It is important at an early stage to locate the toes of batters and tops of slopes which are curved in plan to ensure that the process of shaping and trimming is carried out quickly and easily the first time. It is better to provide a few pegs or batter rules too many than to provide too little information. It may be necessary in some cases for the Engineer to attend on the excavating machines as they approach formation levels to literally level them in as they proceed. Lasers giving a constant plane of reference can be very useful in this respect. On Motorway and Aerodrome contracts in areas of intersections this attendance by the SOE often saves a lot of secondary grading.

The SOE should take into account whether the bulk excavations are to be taken straight down to formation level or left high to protect the formation until on exposure it can be

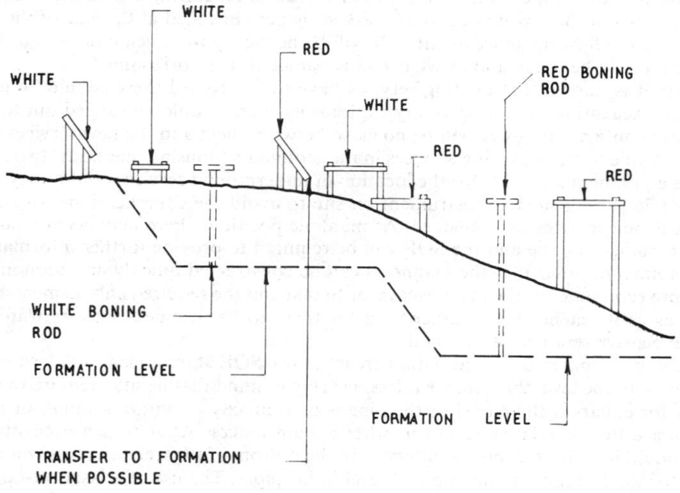

Figure 33.2

immediately blinded or sealed. Likewise with embankments or fill areas, allowance is required for consolidation and settlement. The allowances made in each case, and how the levels given correspond to finished or initial levels, should be made clear in writing to the Foreman.

Where batter rules are set up on varying ground levels to give a continuous finished sloping cut or fill line these can be quickly checked by eye for alignment.

TRENCHING AND PIPELAYING

Pipelines, culverts, service ways and the like are usually tied to specific positions and levels where they enter and leave buildings, pass under roads and intersect with each other. When setting out a particular length of trench or service it is therefore wise first to check between these tie-in points so that if there is any variation from the information shown on the drawings this is dealt with from the outset. It is therefore wise to set out the entire length of trench between consecutive tie-in points and locate the centrelines and essential levels at all junctions, horizontal and vertical bends and manholes. The treating of several sections together in this way will reduce the possibility of errors or late alterations.

In excavating a trench the machine will usually deposit the spoil for backfill on one side of the trench and pipes and other materials will be delivered to the other side. The Foreman will usually require pegs on the centreline of the trench and a specific offset of say 3 m at all key positions. He will require a profile as close to the trench as possible at centres not exceeding 45 m. If these positions are likely to interfere with trench excavation or movement of labour and materials then he may require a further profile offset from the line of the trench. The profiles should be clearly labelled with the length of boning rod or traveller to be used for excavation. The length of a boning rod or traveller should be marked on it. It is not usual to mark out the width of the trench as this will be determined by the bucket of the excavating machine, which would have been selected as the most appropriate, bearing in mind the construction required.

Within a length of pipework between manholes, for example, there may be connections for lead-in pipes from gulleys or other items not requiring a manhole connection. The positions of these will need to be marked by pegs installed at the side of the trench immediately following an excavation. It will be necessary to indicate on which side the connection will be made and at what relative angle to the horizontal.

Where it is known that existing services have to be crossed these should be marked ahead of excavations and if necessary exploratory work should be carried out to locate them and confirm that there will be no clash between them and the new services. Many SOEs are able to trace existing services in the ground by 'dousing' methods. In addition, there is equipment available for the location of underground services, and it may well be worthwhile getting such an instrument on site to avoid the complications which occur when existing services are damaged. At manhole positions there may occur changes in level or changes in line and the SOE will be required to provide further information to the Foreman in order that the manholes can be constructed quickly and economically. It is more economic for the main excavator to take out the required enlargement at these positions as it reaches them rather than for them to be trimmed out afterwards with more expensive removal of the spoil.

Where pipes have to be laid within trenches the SOE should clearly determine with the Foreman the level the latter requires, bearing in mind that he may require to dig out locally for collars if the barrel of the pipe is laid directly on virgin ground, or he may require a different relative level for other circumstances. After trench excavation the SOE should level in steel pins at intervals in the bottom of the trench as this will make it easier for the Foreman to line up and level in his pipes. The use of a laser set-up on the pipe centreline can be a great aid if there will be sufficient use to justify having it on site.

As excavation proceeds the trench should be checked periodically to ensure that it is being excavated to sufficient but not excessive width. With larger-diameter pipes the wrong diameter is easily used and this point should be checked. Care should be taken that cracked or damaged pipes are not used and connections should be temporarily sealed off to maintain cleanliness. The provision of draw wires where specified should not be overlooked.

FOUNDATIONS

Foundations are commonly set out by establishing a series of profiles around the excavations with the location on these profiles of specific setting-out lines notified to the Foreman. Typically these may be as shown in Figure 33.3.

These profiles are usually set just above ground level with the rails horizontal but not necessarily at any particular level. Nails inserted into the profile pick up the particular setting-out lines required and the Foreman can offset these to move from say column centreline to 1 in outside column face or exterior face of the brickwork to suit his requirements. He will normally stretch cord lines or piano wire between the nails and from these he will plumb down using spirit levels or plump bobs. For level purposes he may require level profiles but more usually a series of specific level points can be transferred by the Foreman using straight edges and spirit levels or, more commonly on building contracts, with a water level. Once the Foreman has been provided with setting-out profiles he can usually get by with little further assistance from the SOE other than in checking the various stages of construction. In some cases it may be simpler for the SOE to set up over a setting-out point and to transfer a line directly into an excavation or onto a foundation at a number of points for the operatives' easier use.

Once the foundations have been correctly installed it is comparatively simple to transfer lines and levels up and through the building or structure. On the other hand, inaccuracy in the foundations will be difficult to overcome with later construction, and it is vital that key items such as holding-down bolts for steel frames, reinforcement starter bars for *in situ* concrete construction and pockets for pre-cast concrete columns are in the right position and at the right levels. It is usual to provide some tolerance with these items but it is limited and any alterations which are subsequently required will be costly

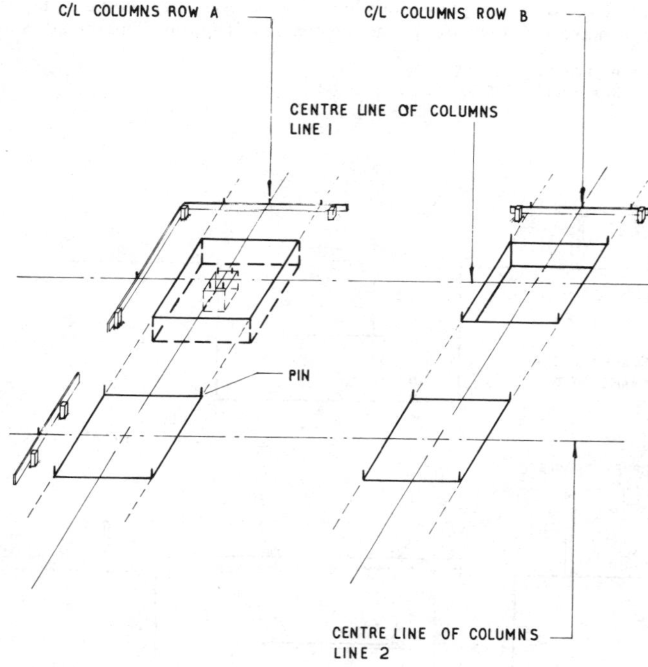

Figure 33.3

and result in delays. The SOE should carefully consider at an early stage whether the tolerances which are likely to be necessary for subsequent stages are realistic in comparison with the cost of remedying any inaccuracies later. To increase the tolerances is also likely to result in some additional cost and a balance must therefore be drawn. Let us consider these three items in more detail.

HOLDING-DOWN BOLTS

These are usually assembled slung from a template which may be made on or off site. The bolts usually have a bottom head surmounted by a washer plate and a sleeve to provide an annular space around the bolts after concreting. The screwed end protrudes through the template to a top nut which by adjustment sets the bolts to required level. It is important that the bolts should hang vertically and the sleeves should enable the bolts to have play after concreting. On many occasions the washer plates are replaced with steel members joining two or more bolts and this helps correct installation. These bolt assemblies have to be supported and, commonly, rigid supports span across the base excavation or shutters. These supports will be subject to deflections and dislodgments and so must be set firmly into position. Where the assembly is difficult to locate and suspend, a frame should be made and concreted to the concrete blinding. The supporting legs may or may not be lost in the construction of the base but an adjustable top portion can be re-used. This method is particularly appropriate if the bolts have to be built solidly without sleeving into the foundations.

When, more conventionally, the bolts have been sleeved they should be tapped with a hammer to make sure they are free as the concrete is setting. The threads should be

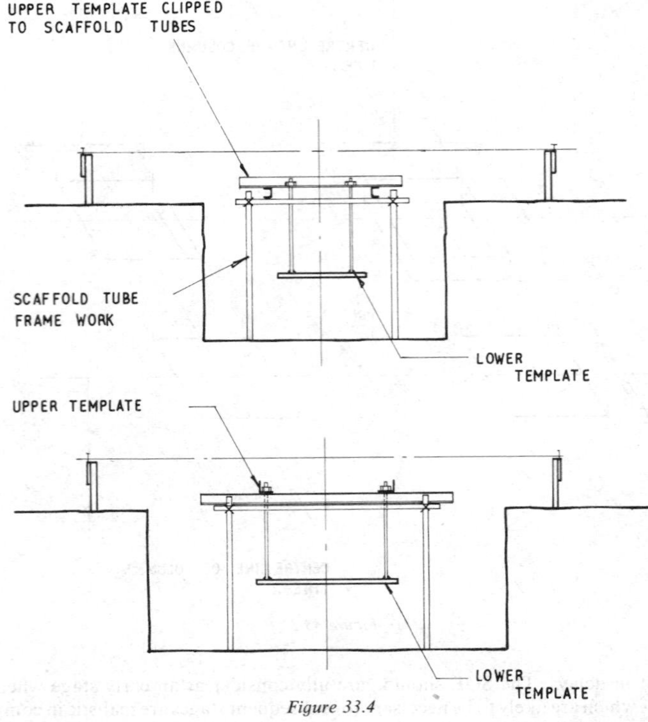

Figure 33.4

cleaned, regreased and wrapped with sacking or similar for protection and the sleeves covered to prevent stones from entering.

The steelwork contractor will usually pack up the base plates of his columns on preset steel shim plates and if the bolts are accurate he will have few problems in lining up the framework. In case his steelwork is all on the large or all on the small size for tolerance it is preferable that he lines up from the centre of the building and does not require a full summation of the tolerances at any place.

REINFORCEMENT STARTER BARS

These are usually assembled with the Column or Wall base reinforcement and tied into the bottom layer. They usually extend one lap length above the height of a small concrete kicker for the columns. They need to be set accurately to provide the correct concrete cover within the column shutter and to lap correctly with the lower end of the column main reinforcement. The starter bars need to be rigidly restrained where they protrude from the top of the column base. Again, it is essential to check before, during and on completion of concreting. Where the kicker is cast integrally with the base it will be checked with the level and positioning of the reinforcement.

COLUMN POCKETS

Where pockets are to be formed for pre-cast concrete or plain rolled steel sections, they should be sized for a reasonable clearance all round as this can finally be very effectively filled with concrete if a slim poker vibrator can be inserted. Such a clearance can be useful for removing debris from the pocket. An excessive clearance on the other hand can affect the size of column base and will increase the temporary wedging and guying used to position the column accurately. A minimum clearance of 50 mm is about right. Again it is important to locate the pocket accurately. The pocket formwork will be subject to an uplift from the fluid concrete which must be resisted. Rather the box should be set low as it is easier to pack up than remove concrete to deepen the pocket. The material from which the pocket is formed should be so arranged that it can be easily removed.

Once column bases have been concreted, it is usual to scribe the centrelines each side of the column position on the concrete. This serves to check that they are correct and is very useful to the erectors of the steel or concrete framework for rapid erection.

PILES AND DIAPHRAGM WALLS

It is customary for the main Contractor to provide centrelines for each base or pile group to a Specialist Piling subcontractor who then locates each pile. It is usually unpractical for all the piles to be positioned from the start as the intensity of operations around any pile position will usually dislodge adjacent markings. The usual tolerance for the centre of the top of the pile is ± 76 mm in position. The top of the pile is located by the protruding starter bars. Although the piling may have been subcontracted, the SOE must still check and ensure that the piles are in the right positions. This is best done when driving or boring has just started and the position of the bore or casing is established.

When a bored pile has been completed it has to be checked. The depth is easily obtained by plumbing with a tape or marked line with weight attached to the lower end. Verticality is frequently a problem with variable ground and when boulders are present. The normal required tolerance is no more than 1 in 80. A protected light on the end of the line will usually give a reasonable indication of alignment verticality by ascertaining how close it comes to the side of the pile at a known depth.

Sheet piling must be checked for verticality in both directions as the work proceeds in order that the clutches will engage when closing a cofferdam and the tolerances in relation to other work are maintained at depth. The cofferdam should be set out to conform with sheet pile dimensions and the final corner closed with the adjacent piles as yet undriven.

Diaphragm walls are formed by digging through concrete-lined guide trenches about 0.9 to 1.8 m deep from ground level depending on the nature of the upper strata. These trenches are formed accurately to a width of 50 mm greater than the width of the digging bucket. It is important that the vertical legs of the guide trench walls are accurately formed and they should be checked periodically to ensure that they do not move under the surcharge of the heavy excavators carrying out the trenching or as the result of ground movements.

TALL BUILDINGS AND STRUCTURES

With multi-storey office blocks and flats it is necessary to control the verticality of the building and the regularity of the storey heights. Accurate foundations are essential with a check at first-storey level to ensure that the framework is proceeding with uniform accuracy. It is difficult to overcome in the superstructure inaccuracies in the foundations. With these structures the cladding is frequently in prefabricated storey-height components and care must be taken to see that the necessary height plus tolerance is available over every storey height. Autoplumbs are usually employed through well openings, to vertically transfer setting-out points. This can also be done with lasers or more cheaply using suspended 15 kg weights on piano wires with the weights suspended in barrels of light oil. Verticality must of course be checked in two directions at right angles to each other. Theodolites set up at ground level can project lines up the face of the building.

With slip-formed concrete structures used for the central cores of tall buildings, for silos, etc., it is essential to carefully check the sliding formwork assembly before sliding commences. The assembly must be truly horizontal and the wall shutters set uniformly inclined to the vertical to avoid any out-of-balance horizontal force. During the sliding the level of the platform must be continuously maintained and kept horizontal. Control is helped by mounting water levels or similar at each jack position and monitoring them in a central control console. A uniform rate of jacking without unnecessary stops will avoid set and snatch conditions. Watch must be kept to identify and correct any rotation of the sliding framework platform.

With steel lattice towers, such as Transmission Towers, a potentially difficult problem is overcome by using large templates. The four lower legs of these towers are inclined to the vertical in both directions and setting out the legs individually to the required tolerance is practical. The excavations for the bases must be marked out individually using steel pins. The lower lengths of the legs which are set into the concrete foundations are then bolted and interconnected into a large fabricated steel template which is set with its top temporary members horizontal and lined up to the transmission line axis or axes.

MARINE STRUCTURES

The main work for the SOE is usually concerned with the initial survey of the area of the works and the contouring of sea bed, river bed or marshy areas. When the area is extensive it is most practical to use aerial photography and echo-sounding methods to obtain the information quickly. When the area is more limited, conventional surveying practices can be adopted.

For dredging works, a relationship must be followed between dredge levels and location. Markers or buoys can be established for position alignment in easy locations and the dredge can usually operate to a chosen depth. In more exposed and offshore locations the problems are more complex but systems exist to deal with them. There are, for example, systems produced by companies such as Decca and Motorola which can simultaneously indicate, relative to two fixed points, the position of the vessel in which the instrument is mounted, and the water depth.

To locate a pile or a structure offshore one usually employs the intersecting-line method from two known points on a base line. Depending on the accuracy required, the SOE may line in the object with two theodolites or establish two pairs of markers for guidance. Depending on the distance apart of the base stations these methods will be more-or-less accurate. When the required location is a considerable distance offshore, more sophisticated equipment such as Deccafix will be used; here, an instrument offshore can be adjusted into a position that is a required distance away from two known base stations where electronic signallers have been sited.

Usually, once an offshore location has been established and centrelines marked, the remaining setting out is simple. It is not always so easy to transfer a level datum from shore to the offshore structure. In most cases there will not be the need for the accuracy used with onshore setting out, and transfer with care by normal surveying methods over long sights may be sufficiently reliable provided that fore and back distances are roughly equal. Astro methods may be necessary over considerable distances offshore.

For the transfer of levels about an offshore structure, a water-level system is very convenient.

When lines or levels need to be transferred to divers working on the sea or river bed, vertical measuring rods are used. These will be difficult to control in flowing-water conditions, but the conditions can be improved by working within a sleeve pipe 1.2 or 1.8 m in diameter into which clear water is introduced under a small pressure head to improve visibility.

TUNNELLING

Working underground is restricting to the SOE as there is usually only a limited access to the works, the works are usually congested, and he usually can only do surveying outside normal working hours which may mean only on Saturday night and Sunday. Because the survey works can only be checked infrequently they must be reliable and firmly established. The most common underground works are tunnels and shafts and often the only access to the former is by the latter and this might involve projecting a 1 mile long tunnel from a 6 m diameter shaft and consequently something less than a 6 m baseline.

Under these conditions theodolites of 1 second accuracy are necessary and much patience is required. It may be necessary for a number of engineers to have to go through the setting-out procedure several times and the mean of their lines used. It may mean that the shafts will have to be watched in case they are moving—in soft-ground conditions they can lean towards large adjacent excavations as the surrounding ground readjusts.

Where considerable accuracy must be maintained, any interference arising from traffic or other vibrations, heat and pollution hazes, and general surrounding activities should be minimal. So even the preliminary ground-level surveying in busy areas is usually carried out under more peaceful conditions at weekends. It is usual to establish across access shafts, or similar structures, the centrelines of the tunnels below. These centrelines are established by means of piano wires suspended down the shafts. The piano wires are wound around screw adjusters at ground level and at their lower ends have heavy 9–15 kg weights in buckets of water or oil. They need to be close to the shaft linings to secure as long a baseline as possible.

At the bottom of the shafts the centreline given by these wires has to be picked up by instruments and established on markers rigidly attached to plates in the crown of the tunnel. A number of surveying methods are employed for this purpose but probably the use of Weisbach's triangle is as reliable as any. Once established the centreline is projected forward whenever the opportunity occurs. An instrument reference point can never be too near to the face particularly if curves are being negotiated. The use of laser reference beams giving the tunnel centreline is now common practice and is a great boon to the tunnel boss.

For negotiating tunnel curves, information is usually provided in the form of offsets at chord lengths with instrumental checks as necessary and possible. It is important to offset on the correct side of the tunnel! Before reaching tangent points it is necessary to work out with the tunnel boss how the tunnel shield will be adjusted to negotiate the curve and whether it can negotiate at the radius required for the finished work. It may be necessary to enlarge the tunnel at these positions using hand methods and such possibilities should be considered from the outset so that the permanent works can, if possible, be adjusted to accommodate the construction practice. Great care is required with horizontal and vertical changes in alignment or level to ensure that the work is accurately set for the new course.

Another important feature to check with tunnels in soft ground is the amount of squat or wander. Using tunnel segments the true diameter may be easily reduced vertically and extended horizontally. It may not be a uniform distortion and considerable difficulty may be experienced with the final tunnel lining to achieve accuracy with the required secondary lining thickness. Measurements of the tunnel segment diameters are kept and checked for every ring both horizontally and vertically.

When the tunnel is being driven under compressed air the centreline has to be transferred through one of the airlocks and this is done by setting up the instrument within the lock, aligning it with the outside door open and then compressing the lock, opening the inner door and transferring the line ahead.

When the tunnel is in free air it may be possible to sink a borehole ahead of the tunnel and pick up a check on the centreline through it.

Larger works underground employ an extension of these methods. When tunnels have to cross or deliberately connect with other services it is wise to locate these intersection points at an early stage. Positions and levels of older services are rarely accurate and adjustments to the new works may be more readily accommodated some distance from the intersection point.

BIBLIOGRAPHY

CLARK, *Plane and Geodetic Surveying*, Vols 1 and 2, Constable

MOTOROLA, 'Aid to dredge positioning', *World Dredging and Marine Constr.* (Apr. 1972)

RICHARDSON and MAYO, *Practical Tunnel Driving*

'Setting out tall buildings', *Proc. Instn Civ. Engrs*, Paper 6719 (July 1964)

SHEPHERD, F. A., *Surveying Problems and Their Solution*, Arnold, London

'The Laser for long distance alignment—a practical assessment', *Proc. Instn Civ. Engrs*, Paper 7524 (May 1972)

34 TEMPORARY WORKS

34 TEMPORARY WORKS

C. J. WILSHERE, B.A., B.A.I., F.I.C.E.
John Laing Design Associates Ltd

THE LEGAL POSITION

In the normal contractual arrangement, the Engineer provides all the information about the permanent structure. However, on many occasions a temporary structure of some type is needed in order to reach the final position. The design and construction of this is a matter solely for the Contractor. In other types of contract things may be somewhat different. For example in a direct labour situation the design of all temporary and permanent works is likely to be in the same hands; or the Engineer may choose to design the temporary works because of their close interaction with the structure. However, in contracts undertaken under the Institution of Civil Engineers' conditions, it is firmly a matter for the Contractor, though the submission of details to the Engineer is normally required. This submission in no way relieves the Contractor of responsibility, but does provide a further check on practicability and safety.

The Engineer has no legal responsibility to the Contractor for approving or 'not objecting to' these drawings. He must be careful not to attract this responsibility to himself unintentionally. But he has a responsibility towards his Client, to ensure that the temporary works will be satisfactory; this means that they must serve this purpose without delaying the work, or cause the Client to be in difficulties because his structure has in some way interfered with others. If the structure takes longer to build through inadequate temporary works design or a failure, the Client suffers.

That very briefly outlines the basic position under English Contract Law. But the position in Common Law is slightly different. Everyone who has a direct supervisory position on the site and who has the ability to make appropriate judgements has a responsibility. Thus, if the Engineer is aware that the temporary works are not all they should be, and some harm befalls, he may well share responsibility with the Contractor should a Court of Law award costs arising out of such an incident. Different forms of contracts in the United Kingdom and elsewhere have comparable provisions.

Other legal requirements arise from problems such as preserving amenities presently enjoyed by neighbours or the public. Requirements are not specifically laid down but follow from this situation.

HM Factory Inspectorate, a part of the Department of Employment, lays down general requirements for health and welfare which will guide the designer, especially in connection with access scaffolding. This is an aspect with which the EEC is presently concerned, and which may lead to standard Common Market rules.

THE TEMPORARY WORKS CONDITION

Design and construction of temporary works comprise a particular case of construction in general but there are certain items which are different. These are:

1. The time for which the structure is in use will be measured not in decades but in months or possibly only hours.

2. Because of this short duration it is easier to predict what loadings would actually have to be carried, which may enable a slightly lower safety factor to be used. Conversely, unless the site is well organised and controlled, unpredicted loads of considerable magnitude can arise.
3. In some cases a collapse of temporary works would be merely a nuisance; in others it will be catastrophic, both to life and property. It may be appropriate to make adjustments to the design parameters depending on circumstances.
4. The available design facilities may be different from the conventional. For example, some temporary works structure may be required at very short notice and the design therefore must be carried out by whoever is available at the time, and checking of a normal level may not be possible.
5. Similarly, the materials may be somewhat different. They may be unusual, and frequently they are not new. This is discussed below under particular materials.
6. Because of the short-term nature of the works, and possible financial advantage, there is a strong tendency to take risks with life and property which would not be tolerated elsewhere.

Classes of construction

Because of these various differences, and the vast range of types of temporary structure that have to be built, there is a variety of classes of construction which may arise. At one extreme, a full design will be made with detailed drawings and a complete specification, which will be supervised very carefully on site. At the other extreme a tradesman will eye something up and decide how to construct it. Thus, there are varying levels of construction and it is desirable that appropriate design stresses are used in each context.

For the efficiently engineered and supervised job, stresses higher than those laid down by the Codes of Practice will be appropriate but, with secondhand material and rule-of-thumb design, stresses considerably lower should be used. This concept of classes is discussed in the Report of the Joint Committee on Falsework,[1] but has wider application in temporary works. Paragraph 3.2 and Table 5.1, reproduced from that document, show the factors proposed for falsework, but the principle has wider application.

Paragraph 3.2 from reference 1:

'*Classes of falsework*

The four classes, described by reference to the degree of specification and detailing, are

Class 1 All members and connections designed and detailed. Working drawings and a full specification for the materials and workmanship provided.

Class 2 All members and major connections designed and detailed. Working drawings and a brief specification of materials and workmanship provided.

Class 3 All main members designed and detailed (but connections not detailed). Working drawings and an outline specification provided.

Class 4 Only a description provided and the falsework built without drawings. The description may be written but will often be only verbal.'

Table 5.1 from reference 1:
Modification factors to allow for the class of falsework

Class of falsework	*Modification factor applied to basic permissible stresses*
1	1.0 to 1.1
2	0.7 to 0.9
3	0.5 to 0.75
4	Not applicable

Limit state design

At the time of writing, considerable interest has been shown in limit state design. However, information to enable calculations to be done is generally not available. But the philosophy of the limit state which involves safety factors more directly related to the various aspects of uncertainty, lends itself admirably to the design of temporary works. This approach is in use already where Engineers adjust stresses and loads.

In the next few years most Codes of Practice will become available in limit state terms.

MATERIALS

Materials common in construction are used and some particular notes are given below. Materials which have already been fabricated to form equipment are described in the next subsection.

Stresses

For virtually all materials in use for temporary works, there is a Code of Practice which lays down the stresses which are appropriate in permanent construction. However, reference to classes in the previous subsection indicates that there are cases where other figures are more appropriate. The table in the Report[1] gives factors ranging from 0.5 to 1.1 according to class.

Steel

Great care is needed to ensure that the chosen grade of steel is used. As there is no permanent system of marking, a basic steel can be inadvertently used in lieu of a higher grade. Steel which has been used is often damaged and is straightened for re-use. Heat treatment or welding can alter properties. Rectification must only be done by those experienced, and almost always the accuracy is limited. Where it is important, as in a strut, a reduction in working stress should be used. Rust-pitted steel should also be used with lower stresses and deeply pitted steel should be discarded. Particular modes of failure with steel joists sometimes overlooked, are instability and web buckling.

STEEL SCAFFOLDING

The scaffold tube in use in this country over the last few years has been exclusively of 4 mm thickness (8 gauge) to BS 1139 (see Table 34.1). However, the tube presently proposed as a European standard is a lighter gauge, though of a higher grade steel. If this tube comes into use in the United Kingdom, it will be necessary to design for it, as it will be a little weaker than existing tube. Alternatively it will be necessary to ensure that only the older, stronger tube is in use in the structure concerned. Table 34.1 gives safe working loads for existing tube, when concentrically loaded. Outside the United Kingdom, tubes of various thicknesses and steel grades are in use.

Timber

Timber available in the United Kingdom is normally graded by the European and American systems which are primarily aesthetic. A small amount of timber is now becoming available graded by strength, either visually or by machine.

Until such time as it is practical to base designs on stress grading, the design stresses used should be based on low-grade timber. Provided that a visual check is made to remove the least suitable timber, a stress grade of 50 is satisfactory; where it is known that the timber is stronger, adjustment should be made accordingly. Timber is a variable material. Where deflection is critical it will be necessary to take the minimum value for Young's Modulus (E). However, on most occasions it will be satisfactory to consider a 'reasonable minimum' E value, which can be twice the minimum.

Timber under site conditions will normally be damp and is unlikely to be dryer than 18%. Designers should use stresses appropriate to wet timber. There are a number of

Table 34.1 SAFE WORKING LOADS FOR CONCENTRICALLY LOADED COLUMNS

Actual length (mm)	*Safe working load* (kN) *for effective lengths*				
	0.7*L*	0.85*L*	1.0*L*	1.5*L*	2.0*L*
600	60.2	58.1	56.0	49.0	42.0
750	57.7	55.1	52.5	43.8	35.3
900	55.3	52.1	49.0	38.7	28.7
1 050	52.8	49.2	45.5	33.6	23.2
1 200	50.4	46.2	42.0	28.7	18.8
1 350	48.0	43.3	38.7	24.5	15.5
1 500	45.5	40.3	35.3	20.9	12.9
1 650	43.1	37.5	31.9	17.9	10.9
1 800	40.6	34.6	28.7	15.5	9.3
1 950	38.3	31.7	25.8	13.5	8.0
2 100	35.9	29.0	23.2	11.8	7.0
2 250	33.6	26.5	20.9	10.5	6.1
2 400	31.3	24.2	18.8	9.3	5.4
2 550	29.0	22.1	17.0	8.3	4.9
2 700	26.9	20.2	15.5	7.5	4.4
2 850	25.0	18.6	14.1	6.8	3.9
3 000	23.2	17.0	12.9	6.1	3.6

48 mm scaffold tube to BS 1139: 1964.

External diameter	48 mm
Thickness of wall	4 mm
Cross sectional area	570 mm²
Weight	4.43 kg/m
Radius of gyration k	15.7 mm
Modulus of cross section z	5 810 mm²
Minimum yield stress	209 N/mm²
Maximum allowable compressive stress	124 N/mm²

Factors for effective length. These must be decided for each case, but the following are generally appropriate:

P	0.7 to 0.85
Q	0.7 to 1.5
R	1.0 to 1.5
S	1.5 to 2.0
T	2.0

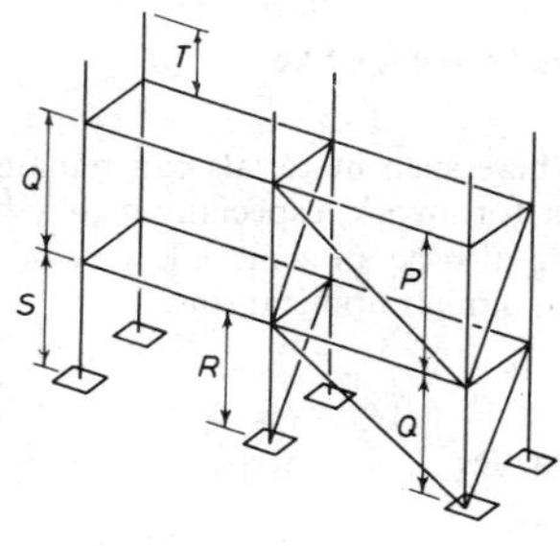

factors in the timber Code of Practice CP112[2] relevant to temporary conditions which enable higher stresses to be used. The Joint Committee's Report on Falsework[1] also discusses this.

PLYWOOD AND MAN-MADE TIMBERS

Waterproof plywood of structural thickness is now available in three main types. Douglas fir from North America, birch from Finland and tropical hardwood from a variety of sources. From structural considerations, there is little to choose between them. But where used as a shutter face, care must be taken to choose appropriately and to apply suitable treatment—see the subsection on Formwork. Large sheets can be obtained to special order with scarfed joints, producing concrete virtually free from joint lines. Data for calculations are available from the manufacturers or suppliers.

Most other materials of this type have a very low E value or are not adequately waterproof.

Aluminium

The lightness of aluminium is countered by its flexibility, and so aluminium has only limited structural applications. As scaffold tube, it is about three times as flexible as ordinary tube and in consequence its use in conjunction with steel tubes must be exercised with care to prevent unexpected distortions. Thus it is desirable to make any one structure of either steel or aluminium. Because of its high scrap value, it is very susceptible to theft.

The ground

In many cases, naturally occurring soil or rock is fundamental to the success of the work. Its properties must be established to enable the design to proceed. However, there are cases where a rigorous, classical analysis will indicate capacities so small as to be unpractical; but for small loads, and short periods, it may still be possible to achieve an acceptable result. Reference may be made to the Joint Committee's Report on Falsework,[1] paragraphs 5.9, 5.10 and 5.12. In all cases, an examination should be made of the ground conditions, at least using a spade. For the situation where the ground must be supported laterally, see the subsection on Temporary Surface Excavations.

Bricks and the like

Where such materials can remain permanently they can form an economical material for formwork, especially if they bring in a trade at that time underutilised. Brickwork can also be an economic way of forming support towers. Concrete, pre-cast or *in situ,* has similar applications.

Plastics

The uses of plastics in temporary works are still few and far between. The difficulties which must be overcome to make it a successful proposition are cost and the low modulus of elasticity. To use plastic effectively, fabricated sections built up in some way to give a large effective depth must be used and this process is expensive in labour. The alternative of thick, solid sections of plastic is considerably more expensive than traditional means, and so it will be some time before this material is in use in large quantities.

There are three main groups: a thermoplastic group in which heat will render the plastic flexible so that it can be shaped again; a thermosetting group in which the plastic once set by the manufacturing process cannot be altered and, thirdly, plastic reinforced with fibres, usually glass fibre. As a structural material, only the third class need be considered, see p. **34**–16. Thermoplastic materials can be used to make textured formwork surfaces and these are discussed by Blake.[3] Small items such as tie cones are very effectively made in plastic, easy stripping from the concrete being the main reason. Bulk with low strength and low weight can be obtained fairly cheaply with expanded plastic such as polystyrene, used for example as a permanent form for a void.

EQUIPMENT

Materials will often be fabricated to form equipment designed especially for the job in hand, but normally it must be written off on the job. There are a number of firms supplying equipment of various kinds useful in Temporary Works, available both for sale or hire. When these are proven items, a saving of time and design cost are immediately available and it is more likely to be economic to write off only a part on the job in question. For example, a set of equipment may be purchased, and the job completed in a given time. Alternatively, by hiring twice as much equipment for a total cost of perhaps half the outlay, the job can be completed in half the time.

It should be noted that there are no standard test methods or factors of safety.

Formwork panels

The face of the concrete will often be formed by panels which may be of steel or of a steel frame with a ply face. This latter is normally a component in a system designed primarily either for walls or soffits. In general these will serve the alternative purpose but not quite so efficiently. These are available commercially, or panels can be purpose made.

Soldiers

The majority of large shutters today rely on soldiers or strongbacks. A number are available to suit most problems: they can be used in other ways, for example as walings.

Centres

Telescopic beams are available in a variety of sizes to support slab formwork. These may present problems of end support owing to the small area in contact with any timber they sit on. Deflections may also need care. In the largest types, spans of 10 m are possible. (See also *Heavy-Duty Girders,* below.)

Form ties

Today form ties in many varieties are available. The simplest approach is a threaded bolt with a cardboard or plastic sleeve; alternatively a plain rod with various proprietary types of friction grip at each end may be used. The form tie will normally be withdrawn and the hole filled. The oldest type of proprietary form tie is the coil tie in which two springs are connected together by rods. The shutter is fixed with special bolts into these springs. On withdrawal of the bolts the holes are made good. Note that threads are not standard and mixing ties and bolts can result in failure.

A comparable group of form ties perform the same function using a she bolt and a threaded rod as the throw-away piece. With this system it is more difficult to obtain accurate spacing or thickness of the wall, and correct location of the portion left in the wall. There are also snap ties in which a complete assembly is cast into the wall and the ends broken off.

Clamps

Clamps have been available for many years for constructing columns. They are also available in a variety of styles for clamping beam shutters.

Props and struts

The ubiquitous adjustable prop in use on all construction sites has recently been the the subject of research.[4] This has led to an appreciation that the published design loads previously used were optimistic in most applications. Fortunately there have been few failures, because props are so seldom loaded to design capacity. Table 34.2 shows the

Table 34.2 SAFE WORKING LOADS (TONNES) ON PROPS

Prop No.	*Height* (m)		
	1.85	3.35	4.85
1 and 2	1.5	1.0	—
3	—	0.9	—
4	—	1.3	0.6

Source: CIRIA Report No. 27

new capacities for these props. This assumes a standard of erection with the tolerances stated below. There are two important workmanship factors:

(i) The prop should not be out of plumb by more than 1 in 60.

(ii) The load should be brought on to the top by an arrangement which is not more than 38 mm from its theoretical central position. On the average site these conditions should be obtained with reasonable supervision.

A standard prop has flat plates at each end. This provides no convenient means of attaching to whatever may be there, to accept a tension load. Push pull props with ends designed specially for this purpose are available, for example, for holding pre-cast units in position until they are secure. They are also useful for aligning formwork.

Unit scaffold

As an alternative to traditional tubular scaffolding, prefabricated unit frames are available. While designed principally for access, they can also be used for soffit purposes.

Typical leg load 4 tonnes.

There are also heavy-duty support scaffolds carrying about half as much again, normally designed to be assembled as towers.

Heavy-duty support equipment

There are a few specialist towers, and Bailey Bridging[5] is available. For even larger loads, Military Trestling is appropriate with a capacity of over 250 tonnes per tower.

Heavy-duty girders

As well as the centres referred to above, Bailey Bridging may be suitable. There are other girders which can be assembled to suit the span.

Piles

The two normally available sheet piles, Larssen and Appleby Frodingham, each come in a variety of types, accompanied by appropriate specials. For data, contact British Steel Corporation. Lighter trench sheeting, some of which is interlocking, is also available usually in lengths up to 5 m. Some identical sections are sold under different names by different suppliers.

For king posts, deadmen and the like, box piles are available. In addition, some foundation piles are suitable.

FORMWORK

Note: BS 4340 : 1968 gives a glossary of formwork terms.

PURPOSE

There are three main aspects. It is important to have formwork which is of the appropriate quality, to produce both satisfactory dimensions and surface appearance. It is necessary that the formwork should be safe and that the risk of damage to people and property should be a minimum. And thirdly it is important that the cost should be as small as possible.

DESIGN DATA

To construct formwork in the most satisfactory way, one must consider the various parameters associated with it. Firstly, loadings. There are two basic cases in formwork, the horizontal and the vertical.

HORIZONTAL LOADING

For the loadings in the horizontal case, reference should be made to CIRIA Report No. 1: *The Pressure of Concrete on Formwork*.[6]

The contents of the data sheet produced by CIRIA from this report are given in the following and in Figure 34.1.

'*The factors to be considered are:*

1 Density of the concrete, Δ (kg/m^3)
2 Workability of the mix, slump (mm)
3 Rate of placing, R (m/h)
4 Method of concrete discharge
5 Concrete temperature t (°C)
6 Vibration (% continuity)
7 Height of lift, H (m)

8 Dimension of the section cast, minimum dimension, d (mm)
9 Reinforcement detail
10 Stiffness of the formwork structure'

P_{max} by calculation

'There are three overall limiting criteria to be considered, the maximum pressure being the least value calculated as follows:

1. (Total height) $\frac{\Delta H}{100}$ kN/m^2

2. (Arching limit) $\left(3R+\frac{d}{10}+15\right)$ kN/m^2 where d does not exceed 500 mm

3. (Stiffening of the concrete) $\left(\frac{\Delta RK}{100}+5\right)$ kN/m^2

where K is a correction factor to allow for concrete temperature and workability derived from the table below.'

Workability	*Concrete temperature* (°C)					
Mean slump (mm)	5	10	15	20	25	30
25	K = 1.45	1.10	0.80	0.60	0.45	0.35
50	1.90	1.45	1.10	0.80	0.60	0.45
75	2.35	1.80	1.35	1.00	0.75	0.55
100	2.75	2.10	1.60	1.15	0.90	0.65

'Where concrete is discharged freely from a height of 2 m or more an addition of 10 kN/m^2 should be allowed for impact. Normal poker vibration is assumed.

An arbitrary ceiling is considered to exist at 150 kN/m^2 for all current practice, due largely to withdrawal of the source of vibration at high rates of placing.' (Courtesy: CIRIA.)

If the full figures indicated are used, formwork will be designed which will be satisfactory in all circumstances. However, there will be a number of cases in which it is over designed, and it may be felt appropriate to reduce the recommended figures by a small amount, with the knowledge that occasionally deflections will be in excess of those intended. Provided that the change in design figures is small, there is little risk, even on these isolated occasions, of actual failure.

Care must be taken in applying this report to judge the conditions under which the concrete will be placed and to ensure that greater forces will not be inadvertently created. For example, the temperature chosen may prove to be optimistic; the concrete may have a greater workability than anticipated; or the rate of pour may be speeded up.

VERTICAL LOADING

The dead load is straightforward to calculate, but care must be taken to establish the type of concrete in use. The loading figures to be applied for live loads are somewhat less certain. Traditionally these have been between 2 and 3.5 kN/m^2, but these relate to placing by barrow or the like. With newer methods of placing such as pumps and crane skips, care must be taken to establish that the figure is appropriate to the case. As far as deflections are concerned, this is unlikely to be a serious problem. For example the impact loadings from a crane skip are of very short duration and it is most probable

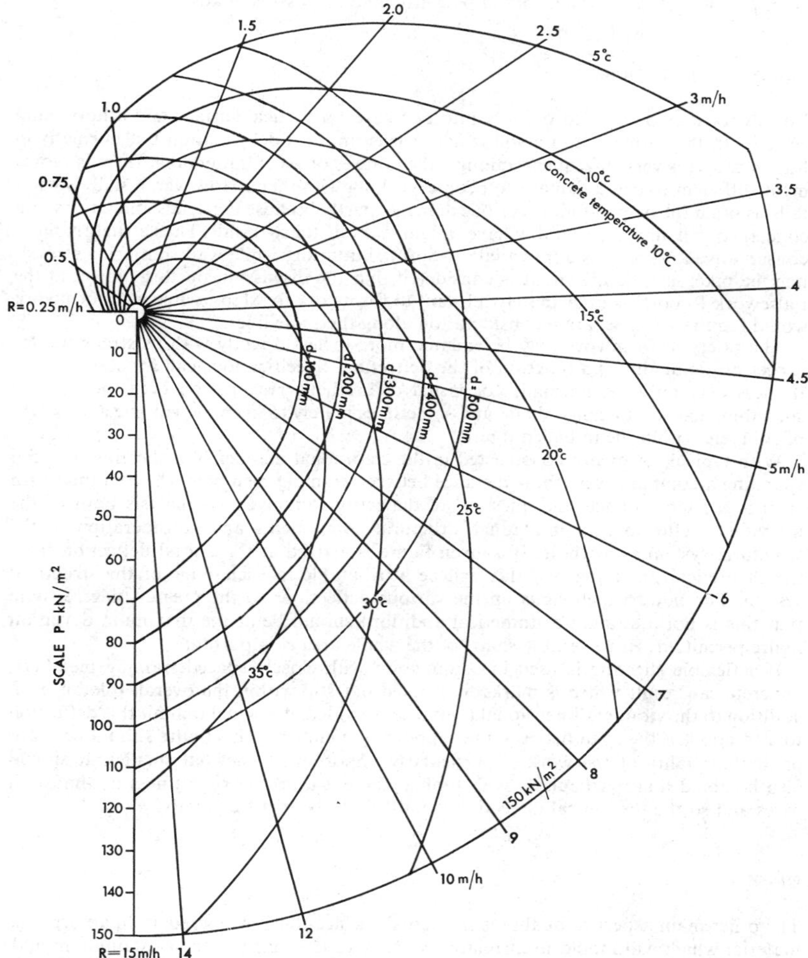

Figure 34.1. P_{max} design diagram. This chart may be used to determine both stiffening and arching limits for most normal practices involving 40 to 50 mm slump concrete. This chart does not include the 10 kN/m² allowance for impact surcharges. See CIRA Report RRI for external vibration. Pressure P is measured radially from point O, along the appropriate 'Rate of placing' line, to the intercept with the curve for concrete temperature or arching limit whichever gives the lesser value. The pressure scale is marked on the radial line representing R = 15 m/h. (Courtesy: CIRIA)

that the formwork will recover, so that any deflections are caused by the dead load only. However, the strength aspect is important.

There may be other environmental loads such as wind, or the possible risk of damage from passing vehicles, which should be considered as possible loads.

DESIGN CONSIDERATIONS

Formwork may be divided broadly into two cases: a vertical surface and a horizontal surface. In the former, failure will result at least in local deformation but normally in little more. It is very infrequent amongst the failures of wall shutters that there is movement sufficient to cause damage to people working with them. However, a slab, when it fails, is often the cause of more serious damage, partly because materials and any person concerned will almost inevitably have a significant distance to fall. Thus a slightly more conservative approach is appropriate to soffit shuttering, while wall shuttering stresses may be taken a little higher. It is considered that the discussion on this subject in the Falsework Report[1] is immediately relevant to formwork for slabs, while wall formwork would require some separate consideration along the same lines.

The safety of formwork may be equated more-or-less directly with its strength. One aspect of the quality is a function of the deflection. Specifications call for standards of flatness over an area, normally for visual or aesthetic reasons, on occasion because something has to fit against them, and sometimes merely to keep the volume and weight of concrete within the intended figure.

With typical formwork constructed in the traditional manner of a sheeting material spanning a comparatively short distance between framing members which in their turn span a greater distance, the question of deflections involves an analysis both of the individual deflections of individual structural components and consideration of the structural system as a whole. If a given figure is arrived at for a total deflection from the theoretical plane, part of this will be attributable to each stage of the structural system. The point which sums up the various deflections to the greatest degree (note that this is not a straight arithmetical addition) should be at the maximum deviation figure permitted. However, the shape of the whole area is important.

If a flexible sheeting is used in conjunction with closely spaced framing members, concrete can result which is markedly rippled but still within the overall tolerance. In addition to the straight dimensional tolerance provided, it is usual to limit this deflection to a fraction of the span between the supporting members. This figure should be 1/270 or, if the quality of the work is particularly important, somewhat smaller. It should also be noted that, particularly with timber, there is a reserve of stiffness in almost all cases and so the theoretical deflection is unlikely to be met frequently.

FINISH

The other main aspect of quality is the actual surface finish. Concrete is an amorphous material which will mould itself relatively easily to the shape of the containing mould, although certain characteristics (sometimes treated as blemishes) will always be present to some degree. The most difficult to eliminate is the blow hole. This is because the air mixed in to the concrete cannot get away but is worked to the edge of the concrete by the compaction and remains at the face of the shutter.

One method which helps to avoid this is to have a shutter surface of absorbent material, although this will produce another characteristic, variation in colour. While concrete has a given colour in given circumstances, these circumstances do not remain sensibly constant. The absorbency of the formwork, differing pressures of the concrete, variations in the actual constituents of the mix from batch to batch and varying compaction will all lead to differently coloured patches on the face of the concrete.

It is difficult to eliminate these completely, though the use of impervious shutters tends to even-up the colour. However, a high gloss finish on the forms such as plastic may create a further problem of dark markings on the concrete. It will thus be clear that it is difficult to have formwork which will give both a very smooth concrete, an evenly coloured concrete and a concrete free from blow holes. The problems of surface finish are discussed at more length by Blake.[3]

While the discussion above relates to a typical piece of concrete against a typical shutter, it is necessary to have joints between one day's concrete and another and between two pieces of formwork which will be dismantled and re-erected on a subsequent occasion. Both of these give rise to variations in the appearance of the concrete. There may be a leakage at the shutter joint, albeit very small, which will locally change the mix and so create a difference of colour. There may be a physical step on such a joint because the formwork was not aligned to the highest possible degree of accuracy. In general these are natural characteristics of the concreting process.

Many designers in accepting this basic difficulty take advantage of it by designing the appearance of the concrete so that these minor disabilities are turned to advantage. For example, along joint lines, a fillet is fixed to the face of the formwork thus creating a groove. If there is a step, or a discolouration, it is inconspicuous by comparison with the grooves, and the arrangement or pattern of these can be designed so that the appearance of the structure is considerably enhanced.

Apart from a straightforward patterning as suggested above, modern plastic materials in particular can provide shaped, textured or patterned concrete, which can provide an alternative way of making the concrete appearance an adjunct to the structure. This is discussed, along with a number of other approaches to surface treatment, in the *Guide to Exposed Concrete Finishes.*[7] The present state of the art does not admit of precise statements on a number of these points. For example, the effect of different mixes and different materials for the mixes will vary the incidence of dark markings and the tendency to leak through a shutter, but there is as yet no known and accepted index to either of these.

THE FORM FACE

To detach the form from the concrete, a release agent is applied to the face before concreting. This is usually an oil, which serves to prevent the concrete physically sticking to the form face. Various types of oil gave different effects and reference should be made to Blake[3] for detailed information. A chemical release agent acts by actual chemical combination with a very thin layer of the outer skin of the concrete to ease stripping, thus reducing damage to shutter face and concrete alike.

Claims have been made that the use of various materials eliminates the need for a release agent; but in all cases a release agent increases the life and makes it easier to strip. Plastic is the most likely material to be acceptable without oil but there is usually a build up of laitance.

For steel, a suitable oil is essential to reduce rust. With timber or plywood, it is often appropriate to paint on a coating to make it more durable. Given the dry conditions needed to ensure its effective adhesion, polyurethane paints give good service, prolonging the life of the face. Many paints react with the alkali in concrete, and so should not be used.

Basic philosophy

The basic design and planning of formwork should take account of the following points.

1 *Strength and stiffness.* The whole structure must be such that there are no weak links, ideally no over-strong sections, and the material content is at a minimum.

2 *Repetition.* In general formwork will be used a number of times, in straightforward designs in a precisely repetitive way. Elsewhere components of the formwork may be re-erected in a different way. As the capital cost is often a considerable part of the total, this always requires careful analysis.
3 *Durability.* As formwork is expected to last, owing to its cost, careful consideration should be given to using materials that are reasonably durable so that the

Figure 34.2. Vertical and horizontal formwork being erected

complete assembly can be used and handled without undue wear.
4 *Stripability.* Many items of formwork have to be set up in re-entrant situations and it is essential to have a design which can be stripped without damaging the formwork and which is not excessively time consuming.
5 *Cost.* The sum of the costs of forming an area of concrete are made up of, firstly, the cost of providing the formwork. This is usually the cost of buying or making the equipment but in some cases it may be hired. The second part of the cost is the labour in erecting, stripping, cleaning and carrying forward to a later use. The cost of any expendable components such as form ties, is the final part. It is essential that the total of these three is kept to a minimum. In today's rising price situation, studying the labour content carefully will best repay effort.

DETAILS

The success of formwork schemes depends on the basic scheme. But failure to cope satisfactorily with any of the details may cause much trouble.

CONSTRUCTION JOINTS

No formwork is required for a horizontal joint but vertical joints in slabs and walls present considerable difficulty. The traditional approach is to use timber or plywood cut in small pieces to surround the projecting reinforcement. This is tedious, though it is not normally necessary to make a very concrete-proof shutter. In many cases a slot at the level of the reinforcement is acceptable. Expanded polystyrene can also be used. An alternative method is to use expanded metal set on a timber frame. This can be used in slabs or walls of considerable depth, and the leakage through the material is very small. If this is carried right out to the edge of the concrete member, rusting on the surface is almost inevitable.

KICKERS

To start a wall shutter or the like from a slab, it is possible merely to stand two shutters on top of the concrete. Two main difficulties arise. The slab is most unlikely to be sufficiently flat to provide a seal at the bottom of the shutter and grout will leak out extensively. In addition, the location of this wall will not be reliable, as the least knock will move the shutters from the position in which they were set up. To get over this problem, it is usual to cast a small height of structure of the same cross section, between 50 and 150 mm high. This can be cast with the slab below, or alternatively may be cast on afterwards, generally enabling greater accuracy to be achieved. Where it is felt more desirable to ensure that the wall above the slab is directly over the wall below, rather than being in the correct theoretical position, a kicker device based on the wall below has considerable advantage. This may be done by making pre-cast blocks which can be set in the top of the shutter below, acting as a spacer to the shutters, and projecting sufficiently high so that when the wall above is to be cast it will act as a spacer at the bottom of that wall. It may be desirable to use this line of blocks to suspend a pair of timbers to form a separate *in situ* kicker. Where slabs are fairly accurate, it is practical to place timbers outside the wall shutter, and nail these down to the slabs.

VOIDS

There are a number of occasions where a completely enclosed empty space is required, and this problem gives scope for considerable ingenuity. Where it is cylindrical, a number of proprietary products are available, in expanded polystyrene, cardboard or sheet metal. It is imperative that adequate steps are taken to tie them down, as the buoyancy is considerable. Where the space is rectangular it may be possible to obtain a suitable cardboard void former; alternatively, light timber may be used. When the void is deep, it will often be sensible to cast it in two lifts. Thus all formwork but the soffit will be recovered. Asbestos cement is a useful soffit material; the makers should be consulted for data on strength. For voids, the design criteria may be relaxed somewhat, as clearly the aesthetic consideration does not apply.

TOP SHUTTERS

It is required in clause F20(a) of the 'Standard Method of Measurement of Building' (published by the Royal Institute of Chartered Surveyors and the National Federation of Building Trade Employers) that surfaces at an angle of more than 15 degrees to the horizontal shall be formed with a shutter. However, this is a somewhat arbitrary requirement, because the various considerations may produce other sensible answers. The factors concerned are the angle of the slope, the concrete mix, the thickness of the slab, the amount of reinforcement and the degree of accuracy of the surface required. In some cases, it is possible to construct a slope at an angle as steep as 60 degrees, and in the case of gunite this can go to beyond 90 degrees. Where a top shutter is felt to be necessary, it is essential that it is both supported off and tied to the lower shutter and designed for the full theoretical hydrostatic pressure. An alternative approach is briefly described in the subsection on *Sliding*, below.

CURVED SHAPES

From a structural point of view the circle is an ideal shape. For example, a circular column form can be designed in pure tension, provided it has enough stiffness for handling. Any kind of curved surface provides a stiffness greater than the equivalent plane one. It is fairly straightforward to use traditional materials for parts of a cylinder, but forming three-dimensional curves is more difficult. Boat-building techniques can be applied to them but, in many cases, the advantage of plastic which can be moulded is paramount. While plastic is a very flexible material, this only matters when it is being used in a situation where bending is important. When used in tension the movement is not of significance. Thus reinforced plastics can be used to good advantage in forming cylindrical shutters and also in forming more complex shapes. Circular column shutters are also available in cardboard and steel.

STRIPPING

The removal of the forms should only take place when the concrete is strong enough. For sections which carry their own weight, the strength required will be considerable. But for other cases, it is merely necessary to resist accidental damage and frost. CIRIA Report 36[8] gives figures which give complete security, though they will delay stripping needlessly in nonfrost conditions, and where the ruling conditions are not virtually identical to a tabulated set. Forms which are left in position will aid curing.

Particular types of formwork

WALLS

In the typical wall it is necessary to have two form surfaces which are held rigidly apart at the required distance. The reaction from the concrete pressure on one side is normally taken by means of form ties through the concrete to the opposite face which is thus balanced. This tie will usually act as a spacer so that the shutter can be set up accurately in the first place. There are a variety of methods within this broad idea. The initial subdivision is based on handling.

It may be that traditional methods are in use in which case it will be man-handled, or alternatively a crane may be available to lift the shutter in relatively large pieces. As an alternative to this, some form of wheels or skids may be provided if the form is moving in a horizontal plane. Where the form is to be crane handled, it is practical to use relatively large structural sections and thereby reduce the number of form ties considerably. This has the effect of reducing labour. On the other hand, where man-handling is essential, the form must be dismantlable into fairly small and light pieces, and it is then more practical to have many more lighter form ties connecting the two forms at many more points, and thus eliminating the need for heavy structural members. Inevitably this leads to a somewhat higher labour content but the capital cost will be less. In addition it is much more versatile, as the components can be built up differently for each successive use. The conditions applicable to the problem in hand will dictate which is the more sensible line of solution.

The actual forms may be of basic materials, or proprietary equipment. This latter is particularly suitable for the piece small approach, but proprietary components can also be assembled into a large form, to be employed as a single unit over a number of uses.

The piece small approach almost inevitably leads to a patterning on the concrete from the individual components that is fairly small. This may be acceptable, or considered desirable, but there are other occasions when large unmarked areas are needed. In these cases the crane approach permits fabricating a form which will remain as a unit with satisfactory joints over a period. Where a wall has to be waterproof, it is very desirable that any form ties have a part cast into the wall and lost to further use. This reduces the risk of leakage which inevitably arises when a number of open holes have to be sealed

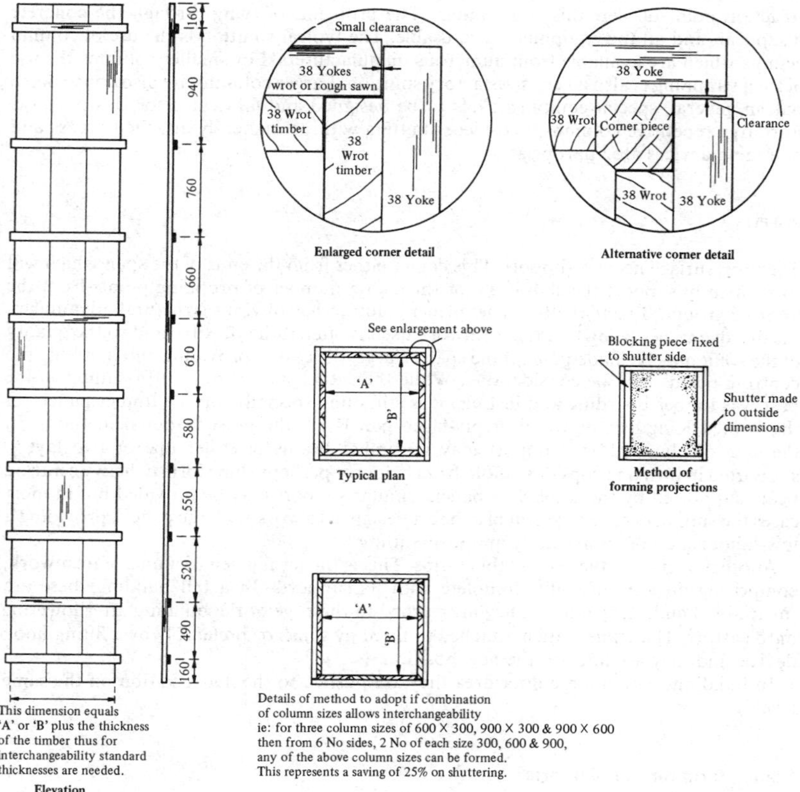

Figure 34.3. Rectangular column shutter using timber sheeting. This arrangement is applicable to columns up to 5.5 m wide. The spacing of yokes is commenced from the top; where this would leave an unsupported length below the lowest yoke of more than 170 mm, an extra yoke is necessary fixed flush at the bottom.

This detail is not suitable for grade 'A' or 'S' work. Column clamps to be handed at alternate levels and if necessary a template used to ensure a square column

after the removal of the forms. If possible, avoid tie-holes and their problems, aesthetic and waterproofing, by tying above or outside the concrete.

COLUMNS

Where forms to a vertical structural member can be tied round the ends, from the construction point of view this is a column. The problems of tying through the concrete disappear, and all the equipment is re-usable. The typical solution is the use of column clamps which are available from numerous manufacturers. For smaller columns the use of steel strapping is also a very satisfactory solution. Larger columns are also constructed but, in general, special equipment has to be designed and fabricated for this purpose. For large repetitive columns, two-piece shutters with permanently attached access and plumbing devices are appropriate.

SOFFITS

The form surface needs a support. This comes either from the ends of the span which will subsequently support the slab itself, or through a number of propping points from the floor or ground directly below. The former solution has obvious structural advantages, the disadvantages being that the formwork itself is often rather flexible and so the quality of the soffit may be inadequate. If the span is large, the cost of providing and handling the centring beams may be considerable. While the actual surface of a soffit shutter in the past was timber boarding and in Europe is still often specially formed timber panels, in the United Kingdom plywood or ply-faced panels are the most favoured methods. In the case of plywood, the support may be timber beams of either one or two layers, supported in turn on props or scaffold from below or perhaps some form of heavy centring beam supported by the walls. For panels, similar support may be provided but in most cases the suppliers have a system of support designed to work with the panels, providing a low-labour method of assembly and dismantling.

Another variant is the use of table forms. This is the term given to a unit of formwork, comprising an area of soffit, complete with its supports. In a tall building these are invariably handled by crane. They are pushed to the edge of the building after lowering onto castors. The crane is then attached to them by slings or preferably by a lifting hook device, and they are lifted to the new position.

In buildings with a large floor area they are pushed to the new position on the same level.

Sliding formwork or slip form

The use of continuously moving formwork is applicable to any structure of constant cross section and appreciable height – 30 m or so. While the traditional application of sliding has been to the strictly constant cross section, tapered structures have been successfully formed and those with changes in cross section at one or more points in their height. The system benefits most from a simple outline and so any such complication must be carefully assessed.

BRIEF DESCRIPTION

Instead of having two shutter panels held together with ties, they are located with a framework over the top of the shutter called a yoke. Fixed to each of these yokes is a jack and this jack operates by climbing a rod or tube through the middle of the wall. Tradi-

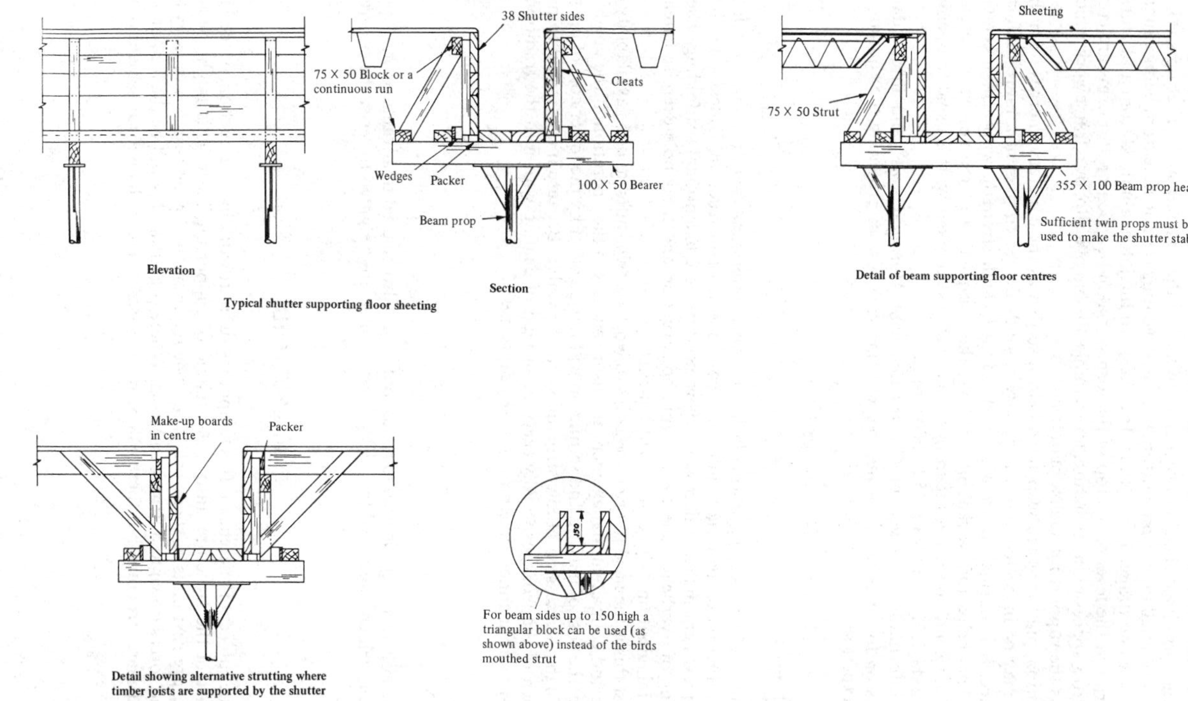

Figure 34.4. Small beam formwork. Where the stripping of the side shutter will not precede the stripping of the soffit, the packers below the beam sides should be omitted. All dimensions in mm

tionally this jack was a screw jack manually operated but this has now almost completely been superseded by a mechanically operated jack, normally from a central hydraulic supply. The form itself should have a smooth face to the concrete and be set up so that there is a very small taper from bottom to top, preventing the concrete from being caught in the shutter and dragged up with it. The access to the job is obtained from a platform or platforms at the level of the top of the formwork, and carried on it. Forms are connected across spaces in the structure, for example, the bin of a silo, by a framework which doubles as the support for such a platform.

In addition, platforms are suspended below, so that as the concrete emerges from the forms, it may be inspected and if necessary smoothed over to present a more uniform appearance. It is normal to use a concrete of fairly high strength, with a high cement content. This helps the workability, so that the compacting of the concrete presents little difficulty, and lubricates the form so that it slides more easily. It is usual to operate 24 hours a day, climbing from 75 to 500 mm/h.

More information is given in *Construction with Moving Forms*[9] and in standard textbooks on formwork.[10,11] A number of firms specialise in providing the equipment and the expertise.

PROS AND CONS

A structure built by sliding normally has no horizontal construction joints. It will be built in a fraction of the time required for conventional construction. Unless some subsequent treatment makes it necessary, there need be no conventional scaffolding. There is a saving of overheads due to the rapid construction and the whole exercise catches people's imagination, with consequent benefits.

Against this is the need to organise gangs to work on two shifts, and the availability of expertise and competent management to ensure success in the operation. If the height is not large, the cost of making and setting up will not be offset by the lower cost of operation, though a succession of low structures may be viable. Because it is essential to preplan much of the work, especially services, instead of dealing with it in the traditional hand-to-mouth manner, time may not permit its use.

VARIANTS

A comparable technique is used in roads, but it can also be used to advantage on slopes. In this case, winches are used to pull up forms, which are weighted to resist the concrete pressure.

FALSEWORK

This is the subject of reference 1. It is temporary support work in construction, needed for some part of the structure until it is capable of supporting itself. This might be wet concrete, pre-cast units or steelwork. The basic requirement for the design is to carry the loads and forces down to a firm foundation. Where practical, this will be the foundation of the ultimate structure, but there will be many cases where such an arrangement is not practical.

Loads

The dead weight which falsework must support is straightforward to calculate. But care must be taken to consider the sequence of loading and to include in the calculations all the other loads which may occur. These are shock loads due to placing the item, the use of machinery on top of the falsework, for example, a dumper truck delivering concrete, or a crane placing the next unit, and other environmental loads such as wind. While the accuracy of loads is known more precisely than is typical in a permanent structure, as the time concerned is so much shorter, the care with which people treat such a structure is very much less, and the unexpected is very much more likely to happen.

Stresses

There are many different circumstances in which falsework may be used. These differ not only in the order of magnitude of the loadings, but in the approach taken to the design and construction. In the ideal case, where the complete design is carried out in full detail, and the site supervision ensures that it is constructed exactly in this manner,

Figure 34.5. A straightforward falsework arrangement

it is possible to use stresses greater than those used for permanent construction. Conversely, where design work is sketchy and supervision not over efficient, it is desirable to use stresses lower than those adopted in permanent work. This approach to design is discussed in the Falsework Report;[1] alternatively, the matter may be dealt with by a method analogous to limit state design. For figures, reference should be made to the appropriate British Standard Specification, or to the Report on Falsework.

Typical construction

Falsework may be constructed out of any of the usual materials as used in a permanent structure. For vertical loads, where any horizontal load is a small part, tubular scaffolding, or the more modern heavy-duty unit scaffolds may be employed to advantage. Where leg loads in excess of 5 tonnes are required, there are other specialised towers available. Bailey Bridging may be used on end, or military trestling may be used. Box piles or other structural-steel sections are appropriate, but their secondhand value is not good.

Other constructions may be made of brickwork or concrete, and on occasion the ground itself can be used. In the past and today in timber-growing countries, wood forms an ideal shoring material. Where the falsework spans horizontally, steel joists are very suitable, and some proprietary adjustable beam units are available in various sizes.

Falsework often comprises a structure set upon a foundation built by a different group of people and supporting in its turn the components of the permanent structure. Where more than one group of people are involved, it is particularly important that the responsibility of each is clearly understood by the other. For example, the foundation for falsework can only be satisfactorily designed when exact information on the loadings is available. This is important as the foundation is often the least satisfactory part of a falsework construction, and to improve its bearing capacity involves considerable expense. Similarly, at the top of the scaffold or falsework, the arrangement in detail of how the load is to be supported is of very considerable importance and is a point where failures have occurred in the past.

HANDLING AND ERECTION OF PRE-CAST UNITS

The advantage of casting concrete other than in its ultimate position is well known, but this leads to the problem of handling it. Where a unit weighs more than 50 kg, and today this may reach 3000 tonnes, there is a need to provide the necessary techniques of moving it into position. The mechanical equipment may include cranes, sticks, gantries, rollers and sliding ways, The range of units will include blocks, beams, columns and walls. The problem, apart from appropriate design, is that the industry has been handling this type of unit for a relatively short time and there is no traditional approach or 'feel' for it. All too often the labour which is asked to deal with this is comparatively inexperienced.

Moving units

Where units can be slid or rolled into position they do not need picking up: jacks underneath can be arranged to do all the vertical movement necessary. The horizontal ways may be bullhead rails with steel balls or well greased timber. But cranes or gantries may well be needed to handle some items. Because the capacity of cranes falls off rapidly as the radius increases, it is sometimes convenient to use two cranes. Effective coordination between two cranes is very difficult, so schemes which involve moving the cranes under load should not be used.

To connect the concrete item to the lifting device requires some form of equipment. Ideally this could consist of slings or a fork-lift device as both of these would be easy to attach to the unit and require no special provision made in the unit itself. But often it is difficult to set the unit down. A traditional approach is to cast loops of reinforcing steel projecting from the top of the concrete and connect to them the hooks of a two- or more-legged sling.

An alternative method would be to put pins of appropriate diameter horizontally through the unit, usual in conjunction with suitable yokes. If a single pin were used a column could be upended ready for placing.

It may be necessary to have some form of anchorage cast into the unit, for example a screwed socket. While this provides a neat appearance in the finished structure, threads are relatively unsatisfactory in this type of work. It is very difficult to keep them clean and thus the thread device which will connect to the crane may not be screwed home properly. The bolt threads wear. Various special devices can be made for lifting items and there are a number of specialised proprietary items available. The width and shape of the unit may well limit the types of fixing which can be placed in it.

Any cast in metal work on a unit to be exposed to the weather is a potential source of rusting or staining. The initial decision should always take this into account and any metal work should be either adequately protected by galvanising or of a nonstaining nature. Alternatively, they could be protected with mortar to prevent the weather causing trouble. Holes in the top of the unit may well fill with water and freeze in cold weather. In some cases units will be split.

Lifting gear

The gear which goes between the unit and the crane is a piece of lifting equipment. Because of this it must be tested to comply with the Regulations[12] and the certificate must be available. The test will normally be an overload depending on the total weight, varying from 10% at a large load to a double load at one tonne. Guidance is given in the Docks Regulations.[13] In addition to the initial test, there is a need for continuing inspection, and the Regulations lay down the minimum. However, badly treated equipment can very quickly become unserviceable and dangerous.

If a unit can be picked up at a central point, it is a convenient arrangement, requiring virtually no lifting gear. But usually units are designed to need support at or near their ends. Similar points of support must be used when they are put down temporarily. If slings are used, the unit will act as a strut. This is clearly advantageous as the cost and weight of a separate spreader beam is eliminated, though it is not always possible. Where it is not very obvious, the top of a unit should be marked. In many cases inverted lifting will cause overstressing or immediate failure.

In the design of all lifting gear, careful consideration must be given to ensuring that all the parts will mate with one another. For example, a large crane will have a large hook which will not go through the ring of size appropriate to the chain sling in use. This will normally be solved if the crane driver has acquired some large shackles but it can prove very embarrassing.

Erection

Many units are completely stable when placed in position. Where tall units are being used, it may be possible to balance them on edge but they could be dangerous. This is a situation which is not obvious, as a concrete unit looks very solid and stable. Thus units should not be released from the lifting gear unless it is certain that they cannot fall. It is normal to use a push-pull prop to give stability. Anchorage will be required on the floor or ground at one end and a fixing in the unit itself. The actual placing of the units will normally be on shims – thin steel plates of various thicknesses to build up the required total thickness. After positioning, the gap will then be filled with a nearly dry mortar packed hard to take the weight of the unit down to its support. In some cases stressing may be used to make vertical units secure but until they have adequate strength in themselves, the prop should not be removed.

Damage

This may occur because crane drivers are not careful. But it is also caused by poor lifting schemes and badly detailed arrangements. Pins in holes must be designed not to impose a load at their extreme ends. Fixings suitable for a direct load must not be used for an angled pull.

The handling of units may involve turning them from a horizontal to a vertical position, as it is often much more convenient to transport them from the casting place to the erection point flat. It will normally be necessary to arrange that the bottom corner about which it turns should be placed in a bed of sand or similar while turning it. It is very difficult for the crane driver to lift so that the heel does not slide as it is raised and this could cause damage to the unit and to anything on which it slides.

ACCESS SCAFFOLDING

In the UK the Regulations[14] made under the Factories Act make two principal requirements for scaffolding: a safe working place and a safe means of access. The main contractor has an overall responsibility for scaffolding but it is also the duty of each employer to satisfy himself that it is in an appropriate condition before he sends his employees onto it. Although the erection may be subcontracted, the responsibility cannot be so simply delegated. A weekly inspection by a competent person is required by the Regulations partly because scaffolding is equipment which all construction operatives feel capable of adapting. Thus the careful examination of it at frequent intervals is highly necessary.

The design and layout of scaffold falls into two headings. It is necessary to have sufficient space on them and sufficient convenient access, so that the operations to be done from them may be carried out quickly. Certain minimum platform widths are laid down in the Regulations referred to above.

It is also necessary to design the scaffold to cope with its own deadload and the liveload of the construction traffic, windloads, and any other environmental loads. While this is frequently done from tables for simple scaffolds, it is necessary on large scaffolds to do calculations. The stability of such a scaffold will almost invariably depend on support from the structure being constructed or maintained. Ties are fixed to it at not more than 7 m spacings in both horizontal and vertical directions. It is essential that where ties have to be taken out temporarily, other ties are put in instead before this happens.

Types

Most scaffolds in the UK are constructed of tube and fittings, although more are now being built with proprietary tubular scaffold, often called unit scaffold. Timber scaffolding was in use for many centuries, and the terminology has largely been derived from it. In the design of scaffolding it is usual to assume that the scaffold has pin joints, and that any stiffness they have does not materially strengthen the scaffold. For this reason diagonal bracing is used on all scaffolds irrespective of size.

An advantage of the unit scaffold is the reduction of labour. It is possible to erect such scaffolds quickly, but the material costs are greater.

Where a building is being constructed of brick, or stone, a putlog scaffold can be used. A single set of standards supports the outer ends of putlogs, whose flattened inner ends rest in the brick joints. But most scaffolds are independent, using two sets of standards to support the decks. These are normally not truly independent scaffold, as they rely on the ties described above.

For maintenance or where the frame of the building is built rapidly—for example in steel—a suspended scaffold is used. While such scaffolds have traditionally been manually operated they are now available in power-operated models and in some cases these are part of the permanent window-cleaning equipment of the building.

Information on scaffolding is given in the Code of Practice,[15] and Table 34.1.

TEMPORARY SURFACE EXCAVATIONS

A great deal of construction work takes place below ground level. This may be done by digging a large hole with sloping sides; by making these sides vertical less excavation is needed—in some cases adjacent buildings prevent the open-cut approach. Where the ground is not good rock, some form of sheeting and a supporting arrangement may be installed to achieve this. Traditionally such sheeting was supported by a maze of struts either raking to the floor of the excavation or spanning across to the other side. Invariably such an arrangement obstructs construction, making it more tedious and expensive. Modern trends in excavation of such holes tend toward providing a completely free working space by the use of anchorage systems in the ground immediately outside the excavation.

Figure 34.6. Ground anchors permit the clean working area shown in this picture

While digging such holes is the contractor's responsibility, and normally it is up to him to propose and to carry out the method, there are many cases where an integration of the ultimate structure with the temporary problem of support to the sides can produce a much better solution. There are many modern examples where either the design *ab initio* has assumed that the structure will form part of the temporary support works as well as becoming the permanent structure in the long term, or where the contractor has decided to adopt this approach and the necessary modifications to the structure have been possible.

Materials

The construction of groundwork supports divides conveniently into three parts:

1 The sheeting.

2 The framing directly supporting individual sheets.
3 The members at right angles to the sheeting which carry the reaction from the earth to some point where it is satisfactorily contained.

Consider first the sheeting. The traditional material is timber planks 37 to 75 mm thick, and this is often the most satisfactory material to use. Additionally today trench sheeting is used as well as steel interlocking piles. For small jobs, timber is preferred, for big ones, sheet piling. Timber is often in short lengths of 900 to 1200 mm fixed vertically. Where the ground is good and machine excavated, it is practical to use boards up to 4 m long. Trench sheets and sheet piles are also fixed vertically, trench sheets being normally in the range 2 to 6 m, whereas piles may be longer, up to 13 m. Where necessary another section will be welded on to create an even longer pile. The limitation of pile size is often a function of the difficulty of driving. Depending on the type of ground encountered and the type of hammer a lighter or heavier section will be necessary.

The framing, either horizontal or vertical members supporting the sheeting, may likewise be of timber, traditionally 225 mm × 225 mm, 300 mm × 300 mm or larger. It may be rolled steel sections, box piles and, on occasions, pre-cast and *in situ* concrete beams. The choice of these depends upon availability, weight, the size of the job and the strength or section modulus required.

In addition to this approach, concrete diaphragm walls are today often used, becoming part of the permanent structure and being both sheeting and framing in one single material. While the diaphragm wall has considerable strength and possible height it is normally constructed in very short lengths of not more than 6 m, whose joints are typically articulated and thus can carry no moment. The support to the diaphragm wall must take this into account. There are more elaborate methods, which permit moment to be carried.

Another approach is to drive a series of soldier piles. If these are placed in prebored holes, the noise of pile driving can be eliminated. Horizontal members are then placed between them as excavation proceeds. While such soldier piles are normally steel, they can also be concrete, for example bored piles. If these are bored to touch each other, a complete wall can be built though normally it will require waling and supports unless it can be designed as a cantilever.

The loads created by the ground are taken through the sheeting and framing to some reaction member. This may be any of the materials which have already been described. In some cases an arch or ring effect is created with a part or whole circle of piles.

A very useful modern development is the ground anchor drilled into either soft ground or rock and providing significant holding power, with no interference whatsoever with subsequent operations. This is very similar to the older method of using a 'dead man', in which an anchorage such as a large block of concrete or a pile or two is put some distance back from the face and a rod ties the sheeting back to it.

Typical problems—a trench

The large number of trench accidents, in which men are killed and hurt through the sides collapsing, is all too well known. The treacherousness of any ground must not be underrated. In a trench the traditional approach with hand digging is to use vertical timbers which are let down rather than driven as a trench is deepened. Individual boards are wedged from the walings to ensure they are tight against the ground. Where the trench is greater than the length of a single plank, a stepping arrangement is required whereby the lower part of the trench is narrower than the upper; alternatively the planks slope outward as they go down. Behind the planks a waling is normally required and between the planks on opposite sides of the trench, struts are installed. When complete this forms an ideal arrangement but the process of reaching this situation during modern mechanical excavation often leaves room for considerable improvement in safety. Various minor differences on this theme involve using trench sheeting as the facing material.

In some cases it is not necessary to close board and individual pairs of boards may be propped apart. It is normally necessary to prop at more than one level, though with short boards of perhaps one metre only a single line of props may be adequate.

As an alternative to this traditional technique, various attempts have been made to devise cages or strutting frames which are put in after the mechanical excavator has operated but before any man descends into the trench. Once started they may be slid forward as required, or lifted out and replaced further along. They enable the timbering gangs to install and make completely safe whatever support arrangement is being provided. There have been several attempts to solve this problem in an acceptable manner, though none have yet achieved widespread acceptance.

Wider excavations

A trench is normally dug to construct a pipeline or duct but it may also be used to construct the retaining wall of the building. Once this is constructed and stable, the remaining earth on the inside of the structure can then be simply dug away. Constructing such a retaining wall in a trench which has many struts is time consuming. It may also present problems of watertightness and quality on the face of the wall.

As an alternative, wider excavations will often be chosen instead of a trench. In this case, sheet piling may be driven before excavation commences, or timbering or sheeting placed as it is in progress. The support necessary to retain these will be provided by a raking shore system to the foot of the excavated hole, or by the ground anchor method mentioned above. The ground anchors will be installed as the excavation proceeds, as soon as it is possible to reach the points on the piling where these are required.

In some cases the traditional king-post solution will be appropriate, where horizontal struts are carried from one side of the excavation to the other but at least at one intermediate point they are supported by a vertical post fixed in the base of the excavation. It is essential to anchor this post down, as otherwise if it rises owing to the pressure horizontally on the struts, the entire cofferdam will collapse. With a diaphragm wall, the same general considerations apply. As this will invariably become part of the structure, it is often appropriate to use the floor beams of the ultimate structure as the strutting holding it in position while the building is constructed.

Loads

The design of works such as this requires consideration of the earth pressures, active and passive, because in many cases the toe at the bottom of the sheeting is a significant part of the support. The superimposed load which exists or may exist adjacent to the excavation must be assessed. For example, there may be an existing building beside it, or it may be possible for heavy lorries to drive along beside it delivering goods to the site. The presence of water must be taken into account. While it is often difficult to make sheet piling waterproof it is equally unrealistic to expect dry conditions on the assumption that the piling will leak. Where there is water such as a river, the situation is clearly simpler, though consideration should be given to any shipping knocking the sheet piles. A large ship cannot be expected to be resisted, but where this is a risk additional fenders may well be used to protect the cofferdam.

Another factor which must be taken into account is temperature change. This question has been raised on many occasions but there are no recorded cases where this has produced any significant problems. Undoubtedly, the temperature range between extremes can be considerable, but this is taken up in elasticity of the struts and of the ground at the ends. Where conditions are extremely cold the earth behind the cofferdam may freeze and create problems.

Newer approaches

Even with the best devised schemes for digging out a hole and then building the structure within it, the time involved is great. If any part of the temporary work can be part of the permanent, time is potentially saved. This can be both the outer face against the earth and the strutting. It is also possible first to construct the piles for a building, designing the upper parts as columns, build the ground floor on these, and then construction proceeds both upwards and downwards simultaneously. In this case, the piles become the basement columns of the permanent building.

Water

Sheet piling is fairly watertight and is quite effective in keeping water out in normal ground conditions. With a small amount of waterproofing it can be satisfactory for temporary works in a river or the sea. As an alternative, a pumped dewatering system may be put in, chemical treatment can be used or, where expense is little object, freezing can be adopted.

In all cases, the pressure from the water must be taken into account in the design.

REFERENCES

1. 'Report of the Joint Committee on Falsework', The Concrete Society and the Institution of Structural Engineers (July 1971)
2. *The Structural Use of Timber*, Part 2, *Metric Units*, CP 112, British Standards Institution, London (1971)
3. BLAKE, L. S., 'Recommendations for the production of high quality concrete surfaces', Cement & Concrete Association, London (1967)
4. BIRCH, N., BOOTH, J. G. and WALKER, M. B. A., 'Effect of site factors on the load capacities of adjustable steel props', Rep. No. 27, CIRIA, London (1971)
5. HATHRELL, MAJOR J. A. E., *The Bailey and Uniflote Handbook*, 2nd edn, Acrow Press, London (1966)
6. KINNEAR, R. G., ACHARYA, D. N. and SADGROVE, B. M., 'Pressure of concrete on formwork', Rep. No. 1, London (Apr. 1965)
7. GAGE, M., *Guide to Exposed Concrete Finishes*, The Architectural Press and The Cement & Concrete Association, London (1970)
8. WEAVER, J. and SADGROVE, B. M., 'Striking times of formwork, tables of curing periods to achieve given strength', Rep. No. 36, CIRIA, London (1971)
9. HUNTER, L. E., *Construction with Moving Forms*, Concrete Publications, London (1951)
10. HURD, M. K., Ed., *Formwork for Concrete*, The American Concrete Inst., Michigan, USA (1963)
11. WYNN, A. E. and MANNING, C. P., *Design and Construction of Formwork for Concrete Structures*, 5th revised edn, Concrete Publications, London (1965)
12. *The Construction (Lifting Operations) Regulations*, 1961 S.I. No. 1581 (1961)
13. *The Docks Regulations*, 1934 S.R.O. No. 279 (1934)
14. *The Construction (Working Places) Regulations*, 1966 S.I. No. 94 (1966)
15. *Metal Scaffolding*: Part 1 (1967), *Common Scaffolds in Steel*, Part 2 (1970), *Suspended Scaffolds*, Part 3 (1972), *Special Scaffold Structures in Steel*, CP 97, British Standards Institution, London

35 SELECTION, PROCUREMENT AND MANAGEMENT OF CONSTRUCTION EQUIPMENT

SELECTION, PROCUREMENT AND MANAGEMENT OF CONSTRUCTION EQUIPMENT 35

35 SELECTION, PROCUREMENT AND MANAGEMENT OF CONSTRUCTION EQUIPMENT

T. MALCOLM, B.Sc., F.I.C.E.
Richard Costain Ltd.
and
R. W. KIMBER, B.Sc., M.I.C.E.
Richard Costain Ltd.

INTRODUCTION

There are two principal, and to some extent antagonistic, aspects of the problem of selection of construction equipment:

(a) Fundamentally technical and concerned with suitability in the performance sense.
(b) Fundamentally economic and commercial and concerned with viability in the business sense.

It is not unusual for these aspects to be the main concern of different groups of individuals within an organisation. The young engineer on a particular construction project is concerned mainly that the equipment should be the 'best' technically to satisfy his method and programme requirements. In justice, it may be said that frequently he has investigated the equipment thoroughly before formulating his requirement but too often, the inexperienced engineer bases his selection on advertisements, sometimes of machines not yet in production.

The plant manager, senior construction manager or director, on the other hand, needs to ensure that the purchase of any new equipment makes sound commercial sense, taking an overall view of the company's policy and operations. Sometimes, therefore, there is a conflict between technical and commercial considerations in the selection of plant and equipment for a particular contract. The object of this section is to show that these attitudes are not irreconcilable and to indicate how the technical and commercial aspects of a purchase may best be approached. It should, incidentally, be obvious that reconciliation must be obtained when both aspects are the concern of one individual, which is usually the case in smaller organisations. It should also be emphasised that a satisfactory solution of the problem of selection cannot be obtained if the technical and commercial pressures do not result in a reasonable compromise.

It is the attempt to divorce them and look blindly at only one aspect which has unfortunate consequences. On the one side this can result in the purchase of an unsuitable item because it is cheap and, on the other, in the purchase of an expensive 'white elephant' which will be used once and rust away thereafter in a plant yard.

TECHNOLOGY OF SELECTION

For many thousands of years man has used tools and implements to supplement the muscular power of his body. The employment of these various primitive implements was usually so obvious that no particular technology was needed to cover their selection and use. With the discovery of natural energy stores in timber, coal, oil, the extension to the use of solar and electrical transformations, and ultimately to the use of the energy of the atomic nucleus, a very different concept has arisen in which machines and tools are designed for a purpose. This has, of course, been associated with the idea of Scientific Method which follows the classical pattern: observed phenomena—hypothesis—trial and proof. This, has been an overriding principle in the sciences and engineering for only the last two hundred years. Over this period, invention has been a considerable part of human activity and it has been accelerating. The tools and equipment of construction have had their fair share of this development.

The reasons for considering a purchase of construction equipment can be variable in time and in relation to the type of organisation. The following paragraphs give some basic categories or combinations which occur in practice.

Need for replacement of an outworn or obsolete machine

This is most likely to occur in a plant hire organisation, or in the holding of a contractor or public authority or in an organisation engaged in a continuous process such as mining. Since the replacement will almost certainly be at least of the same type as the original and may even be the identical machine, there should be no need for a basic assessment of the technical qualities.

An examination of modifications and improvements should be sufficient. There is also the opportunity to look at similar machines of a different manufacture, although there are serious considerations of maintenance familiarity and spares holding, where there is a substantial ownership of the original type.

It is clear, therefore, that in this particular set of circumstances technological considerations are secondary to commercial. This is with the proviso, of course, that the original machine or team has been successful in operation.

The need to supply plant to a large new contract or group of contracts in UK

Availability in the existing holding of plant and the need for replacements, as in the preceeding paragraphs, must be the starting point. This, of course, must be related to method and to output requirements with due consideration of the site conditions. Where plant has not been written down in value to such an extent that its secondhand sale price will not clear the outstanding amounts in the capital ledger, it becomes very difficult to give effect to technical arguments for replacement by new and more efficient equipment.

Certain main information headings may be set down, which must be satisfied before a rational selection may be made on technical grounds. These are as follows:

(a) Scale of the works in terms of value, ground area, volumes of excavation and concrete etc.
(b) Approximate construction period.
(c) Methods envisaged under the sub-headings of excavation, concrete, piling, steelwork erection, paving etc.
(d) Type of plant required and which is available either from existing holding or generally in the market and in time to meet the programme.

The need at estimate and planning stage for precise, accurate and detailed information on site conditions cannot be overstressed. It is suggested that use be made of Code of Practice 2001, 'Site Investigations'.[1]

Although 'Site Investigations' is intended for use by designers and planners it has considerable value in its listing of information which is useful for estimating and for selecting plant. In particular geological information can be of immense value in determining excavation methods and equipment and in problems of availability and processing of concrete aggregates.

Recommendations for purchase must not be based on prejudice, hearsay, out-of-date information or the influence, social or personal, of sales organisations. The staff employed on the plant department side in discussions with estimators, planners and contract management may be qualified in any engineering discipline, civil, mechanical or electrical, but they must be aware of the mechanical, structural and operational qualities of the machines and they must also be thoroughly experienced on site and should be possessed of the imaginative flair which will enable them to see the inevitable working out of any proposed method.

The need to supply plant to a large new overseas contract

This is a somewhat specialised problem which is usually resolved by a decision that the majority of the items will be purchased new for the project, will be operated and maintained by the contract and will be sold at the completion of the project. Costs of shipping back to UK or to another overseas project are often more than the plant is worth at the delivery port. It follows therefore that, initially, free rein may be given to the technologist in selecting the plant which is most suited to the work in terms of method application, size, mechanical quality and reliability, technical back-up from the manufacturer, spares availability, etc. The price of the machine may well be a secondary consideration and so it should be in relation to the ultimate cost.

Overseas contracts are often carried out in remote areas where the contractor is required to be completely self sufficient. It follows that any check or assessment covered in the above paragraphs must be carried out with meticulous care.

The site conditions must be the subject of a most intensive investigation, so that the plant selection may be the best possible, and in order that inevitable consequences may be anticipated and plans may be laid to overcome them. Problems such as extreme heat,

Figure 35.1 Crawler tractor pushing tractor-drawn scraper on motorway project

dust, humidity and undue abrasion must be allowed for, both in selection and in provision of spares and maintenance facilities. A serious failure cannot usually be overcome by telephoning the nearest plant hire organisation for an immediate replacement, and the shipping of replacement equipment may take several months.

The use of new models or new developments should not be considered for overseas work until they have been thoroughly proved on works in UK, W. Europe or USA.

The need to consider purchase of a new model or new development

The advent of a new machine on the market is usually heralded by an extensive advertising campaign. It is not unusual to find that the advertising has considerably anticipated the production availability of the machine and that there are only one or two prototypes available. The manufacturer is, in fact, looking for field testing and proving of his equipment. Even if the charges for the machine are quite nominal, this can be an expensive business for the user, if he gears his method and programme solely to its employment. Reaction to this situation must be conditioned by the reputation of the manufacturer and if involved, his agent.

There are a number of practical actions which may be carried out in the assessment of new machines, once it has been established that the size, output to be anticipated, and

Figure 35.2 Manitovac 4600 special crane handling rock trays to breakwater at Dubai

general characteristics are suitable for a particular duty or for general use in a sequence of similar duties. These actions are as follows:

(a) Sales literature should be regarded as relatively insignificant, and a full technical specification should be demanded. This should enable a competent engineer to make a preliminary assessment of the quality. Important points are: horsepower of engine; weight of machine; sizes of clutch, transmission, brakes, etc; dimensions of tracks, wheels, tyres etc; load distribution and stability; quality of metals; convenience and quality of controls; operator comfort and efficiency; speed and grade capability.

(b) Inevitably even the best manufacturer's specification will be deficient, either by accident or intent, in some item of information which is necessary to a proper appreciation of the machine. This should be requested and should be incorporated in a table showing all the leading data for the machine in question with comparison columns for any similar machines which may be available in the market.

(c) Information should then be obtained, as far as may be possible, on actual performance and preferably from a commercial user rather than one who has been testing a demonstration model.

(d) Demonstrations may be unavoidable but should be accepted only as a last resort. The demonstration model will be new and supertuned and the operator will be a star performer. It is more satisfactory to see the machine in use on a project with an average operator from the contractor's staff. It is usually possible to obtain unbiased opinions from site personnel, if they have not been influenced by high pressure sales talk.

(e) If at the end of these exercises the machine appears to be attractive and the faults which have been discovered appear to be surmountable, an attempt should be made to obtain one on hire. Free loan with option to purchase should be avoided as it imposes some obligation and inhibits the use of the machine to its full capability.

(f) If the machine is ultimately found to be a desirable addition, its purchase must wait on a duty for which it is technically suitable and of a duration to warrant the expenditure. The above procedure is, no doubt, tedious and likely to result in failure to obtain early advantage from a new and revolutionary equipment, but there are very few of these which could be described as immediately successful, and the advantages may well lie in being more cautious.

PLANT HIRE

There is a seasonal variation in public works activity and in the employment of construction plant. This variation becomes more noticeable in a Continental climate such as parts of Europe and the USA. It also results in a complete shut-down in tropical countries which have a heavy seasonal rainfall. However, with the exception of USA, such countries do not have a well developed and organised plant hire market such as is found in the UK. The following remarks, therefore, apply to UK conditions only.

The graph of construction activity (Figure 35.3) is roughly of the following outline and varies with the severity of the winter and the extent of rainfall.

If a contractor or other employer of construction plant can keep all the equipment which he uses occupied for about 75% of the working year he should find it more profitable to own his own machines and to have minimum dealings with the plant hire

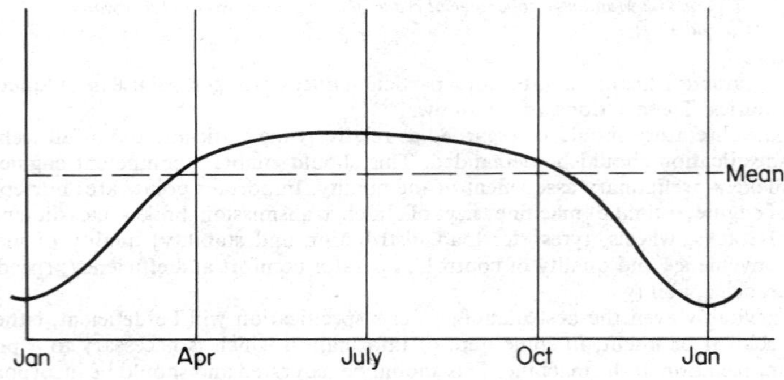

Figure 35.3 Graph showing annual construction activity

market. It is much more probable, however, that in order to cover all his requirements, the average user would need to own machines which could be in use only over a part of the summer season. He is well advised, therefore, to own only sufficient plant to cover requirements below the mean line shown and to hire the balance as required. The plant hire organisations must, of course, suffer to some extent from the same problem and the situation can arise—and has arisen—where certain categories of plant, particularly muck-shifting equipment, are not readily available on hire over the summer months. Nevertheless, with due foresight, it is usually possible to cushion the peak demand by recourse to the hire market.

It should be noted that short term local employments at some distance from a depot are best filled by external hire. The saving in transport costs is usually much more than any hire rate differential.

Special items of plant suffer from a very limited utilisation and are not therefore attractive to plant hire operators. If such an item is unavoidably required there is no alternative to outright purchase. The commercial consequences of this will be discussed later.

Special aspects relating to plant technology

There are some general considerations of the technical aspects of plant acquisition and operation which are worth recording, a summary of these is given below.

(a) A machine cannot be divorced from the duties which it will be required to perform. For example, there are available in the international market a considerable number of crawler tractors of comparable capacity and made by different manufacturers. Adjectives denoting quality cannot sensibly be applied to these machines without regard to their employment. There are perhaps two machines which are very suitable for use on the heaviest civil engineering construction in hard conditions and under the most severe pressure. These are quite rightly labelled 'good' machines, but a much cheaper and lighter tractor is also a 'good' machine when it is employed on light duty work such as on a building site or pulling a roller in easy conditions.
It would be a waste of money to employ a heavy duty machine for such purposes, unless there are over-riding problems of stand-by and standardisation which must be satisfied.

(b) The commonest causes of plant failure are overloading and overspeeding. These result from a mistaken impression that it is a 'good thing' to obtain outputs well in excess of those which could reasonably be expected. The result is excessive breakdown, a reduced life, and an average output, over a long period, which is appreciably less than that which could be obtained by a more sober adherence to manufacturer's recommendations.
A surprising viewpoint on this matter is that these people who would, without qualms, fit an oversize bucket on an excavator or grossly overload a scraper or vehicle would be appalled at any suggestion that they would deliberately overload a crane. Perhaps the shadow of the Factory Inspector or the Law has a bearing, but it should be realised that the excavator or vehicle in these circumstances can be just as lethal as the overloaded crane.
In this connection it should be realised that the less expensive and lighter equipments are much more susceptible to the effects of overloading.

(c) On overseas projects of short duration it may be economic to purchase lightweight equipment with the intention that it will be scrapped at the end of the contract. In any decision of this nature it should be borne in mind that the machines must operate effectively until the last day on which they are needed. On this account it is sometimes necessary to allow extra numbers in the team so that premature failures will not disrupt the programme.

ECONOMIC AND COMMERCIAL ASPECTS OF SELECTION

In discussing technical problems it has not been possible to avoid some mention of the commercial aspects, but the following information is intended to be comprehensive on the economic and commercial side.

There has, over the past 25 years, been a progressive increase in the plant content of the cost of construction works. There is no clear indication of any slackening off in this tendency, although it is possible that political and philosophical arguments against the displacement of labour may have some reversing effect, so long as unemployment remains at a fairly high level. There can be no doubt that the arguments in favour of employment of plant instead of labour are economic rather than social and that they will prevail if the production/consumption impasse cannot be solved in some way which does not require a profit motive and nothing else. Meantime employers and managements must work within the system and must appreciate the necessity of giving as much top level attention to plant problems as they do to the direction and control of the labour force.

While the plant content of cost in a building project is not likely to exceed 15% and is usually somewhat less, the corresponding element of cost in a heavy civil engineering work can be as high as 40%. It is possible for it to be greater than the cost of the labour. It follows that plant control is deserving of as much attention and high quality supervision as does the management of labour, which has always been regarded as the prime concern. There would appear to be a field of occupation in this for qualified and experienced men to take over that part of project management, i.e. the plant, which has been regarded as a secondary issue for civil engineering and building managers and which has frequently been grossly mismanaged.

There are very few purchasers of plant who will start with a clean sheet, i.e. with no legacy of existing plant in usable condition. It follows that the first reaction to any specific requirement must be to look at the holding of plant. If suitable equipment is available the problem is solved. If it is not or if it is in obsolete or poor condition, it must be sold and replaced. This is usually possible if a realistic write-off pattern has been followed, and if utilisation has been good. If the sale does not at least recover the book value an interim loss will be incurred, which can be corrected only by following a realistic commercial pattern from that day forward. If a correct purchase is made, the increased efficiency of operation, and avoidance of undue maintenance costs by intelligent operation should go a long way towards recovery.

It is characteristic of keen and energetic field project managers that they tend to be critical, on grounds of age or obsolescence, of plant supplied to them and at the same time refuse to accept the necessity for hire charges which are levied. If the organisation concerned does not enforce a policy of virtual prohibition of external hire, there will be a tendency on projects to seek short term advantages which can do nothing but make the plant organisation more inefficient. It should be remembered that special items will not be available from external sources and these will of course, be demanded from the plant department.

The nature of the method discussion tends towards a comparison of alternatives, and, apart from the necessary technical comparison, staff employed in the plant organisation must be capable of producing back-up cost figures for each alternative relative to investment, resale, maintenance and repair, transport to and from site and erection and dismantling charges if applicable. There are no short cuts to the production of information of this nature.

Utilisation

The principal factor which influences the purchase of plant for any centralised holding and the setting of appropriate hire rates or equivalent charges to projects, is the anticipated utilisation, i.e. the working hours or chargeable hours which may be expected in any one year or over the life of the machine. The following analysis is designed to show why this factor is of paramount importance.

The basic elements of the cost of operating a machine are as follows:

(a) Capital cost.
(b) Ultimate sale value.
(c) Interest and service charges on the investment.
(d) Cost of insurance, licensing etc.
(e) Cost of maintenance and repair.
(f) Cost of fuel, lubricants, ropes and other consumables.
(g) Cost of operating personnel.

Items (a) and (c), and possibly (d), are continuing costs which do not cease when the machine is idle. Items (e), (f) and (g) are costs which vary directly with the working hours. Items (f) and (g) although they have some influence on selection policy are not normally regarded as part of the plant charges to contracts but are rather site operation costs.

The consequences of failure to assess utilisation properly or of failure to obtain what appeared to be a realistic utilisation can best be illustrated by an example, which at the same time shows the steps which are necessary to establish a hire rate.

Assumptions are made as follows:

Price of machine £10 000
Life 6 years of 1500 working hours, i.e. 9000 total.
Sale value at end £1 000.
Average repairs cost = £1.00 per working hours.

From the above a hire rate per working hour may be obtained as follows:

Price of machine	£10 000
Interest over 6 years	2 700
	£12 700
Deduct re-sale value	1 000
Total capital cost	11 700
Repairs and maintenance	9 000
	£20 700

$$\text{Assessed net hire rate } \frac{£20\ 700}{9\ 000} = £2.30 \text{ per working hour.}$$

If this machine is assumed to obtain or actually does obtain only 1000 hours per year of paid hire i.e. 6000 hours total for its working life, the picture is somewhat altered.

Total capital cost remains at	£11 700
Repairs and maintenance	6 000
	£17 700

$$\text{Assessed net hire rate } \frac{£17\ 700}{6\ 000} = £2.95 \text{ per working hour}$$

This is an increase in the rate of 28% which is much more than any normal tender margin. The difference could not easily be levied on any operating project without showing a serious loss.

If, in setting hire rates or equivalent charges to projects, a misconception, such as the above, on the extent of utilisation is made, i.e. an assessment of 9000 hours against a realisation of 6000, there would be a deficit at the end of 6 years of £3900. This sum would not be available in any fund set up to cover replacement of the machine.

It will, of course, be obvious that in practice exact rounded figures such as the above will not be experienced, but the basic principles are unaffected by this. Appreciation of the true position over a period of time may be affected by very erratic utilisation. For example, a machine which is well used over the first year and appears to be a good purchase, may subsequently lie in a yard for such a time that it is realised much too late that very much heavier charges should have been made in the first year. This accounts for the habit of plant organisations to overcharge during the first year of any machine which does not have an assured utilisation. The special machine, i.e. the one which is never obtainable on hire, is an obvious case for this treatment.

It will be realised, no doubt, that these very firm, not to say rigid, attitudes which are adopted by individuals and groups who are concerned with recommendations and actual purchase of plant, are in fact protective of the overall interest of the construction group. It is unfortunate that, in most instances, the responsibility falls upon the management of a plant holding company or department, and this contributes towards the undoubted ill-feeling, even strife, which exists between contract managements and site personnel and the plant organisation in most of the large construction organisations.

As mentioned previously, this will not be true in a small company where the user is much closer to the purchase problem, but it will be realised that one wrong purchase in a small business will have a much more serious impact than in a large one. The remarks are particularly directed to the attention of the young engineer who cannot understand why he cannot have that perfectly ideal machine which he craves, and at a very ordinary hire rate.

Figure 35.4 Grader trimming sub-base material on motorway project

Comparison of costs

The initial purchase price of a machine is of little ultimate consequence compared with the final cost, which can best be measured in terms of the cost per unit volume of excavation, concrete or whatever may be involved, over a period of years. Apart from the direct costs of breakdown, repairs and maintenance, the costs of disruption, due to breakdown, can be so high that hindsight would encourage the purchase of a machine at twice the original price. It is obvious that no organisation can afford to disregard the initial price, but the accountant who compares three prices and invariably selects the cheapest is a danger to the interests of any construction organisation. An appreciation of the technical facts is essential and one basic inescapable fact is that competing machines never have the same specification or standard of materials and workmanship.

The individual who is over-concerned about purchase discounts and takes great pride in obtaining them is also a menace. The only possible grounds for a special discount are regular purchases of substantial quantities. This obviously allows the manufacturer to produce more efficiently and a discount is legitimate and reasonable. Very frequently such discounts are long established understandings which do not require any negotiation. The purchaser who, for the odd machine, succeeds in 'beating down' the supplier, should realise that the machine was overpriced in the first place, i.e. the discount was added to the price before the negotiation started, particularly if this is known to be a characteristic of the buyer. Where a manufacturer or agent allows himself to be 'beaten down', perhaps because of slow sales, there must be a reaction which is unfavourable to the supply of good service and the ready acceptance of reasonable warranty claims. In this connection it should be mentioned that poor machines are often on quick delivery and that good machines are frequently difficult to obtain. This is a fairly obvious statement, but one which is often obscured by the urgency of a demand.

Replacement

Although the problem of replacement in a general holding or by a plant hire company or an industrial organisation is not acutely one of selection in either the technical or commercial sense, it has associated with it one serious problem which is unavoidable and that is the problem of timing. When is the best time to replace a particular machine?

Very often the decision may be taken in a moment of exasperation and, while this may be regarded as a genuine pressure, it is possible to make some attempt at rationalisation. Little can be done in this direction if reasonably accurate records have not been kept on working hours and repair costs, together with some information on site conditions which have affected these costs.

Although it may be possible to arrive at a policy which can be applied to a group of identical machines it is not likely that this would always be satisfactory. Even with mass produced automobiles it is possible to find considerable variations in quality and reliability. It would be better, therefore, to consider each machine on its own merits with due attention to any conclusion which could safely be applied to a group.

An attempt at rationalising this problem has been made by one of the major international suppliers of construction plant. The basis of their analysis is that the rate of change of the cumulative total cost per working hour is a fair index of the need for a change, particularly if the graph can be extrapolated forward to indicate the future trend. The elements of cost are as follows:

(a) *Capital depreciation.* This must be related to the type, quality and size of the machine and to the duties which it must perform. Due allowance must be made for the inevitable inflation which will occur so that the replacement fund is not badly affected in the course of the years.

As an illustration of the method in respect of this and the other cost elements to follow, a machine is arbitrarily imagined costing £10 000 in 1971 and working 1500 hours per year.

(b) *Interest and insurance charges.* These can be variable, dependent on the scale of the owner's operations, his financial standing and the type of financing and insurance

Table 35.1 DEPRECIATION AND REPLACEMENT COST

	Year end 1	2	3	4	5	6	7	8
Secondhand value %	75	55	40	32	25	20	15	10
Secondhand value	7 500	5 500	4 000	3 200	2 500	2 000	1 500	1 000
Replacement price	10 500	11 000	11 500	12 000	12 500	13 000	13 500	14 000
Deficit	3 000	5 500	7 500	8 800	10 000	11 000	12 000	13 000
Cumulative working hours	1 500	3 000	4 500	6 000	7 500	9 000	10 500	12 000
Cost per cum. hour	2.00	1.84	1.87	1.47	1.33	1.22	1.14	1.08

which is available to him. For the purposes of this exercise it has been assumed that this may be costed at 10% per annum on the average annual employment of capital.

(c) *Cost of repairs and maintenance.* The incidence of these is very much related to the quality of machine, quality of operator and the nature of the duty which it is required to perform. It follows that the tabulated values given below could only by accident agree with an actual experience. An owner must therefore make his own assessment from experience and anticipation of future conditions, and particularly he must try to assess the interim annual costs which are likely to be incurred. This is the factor of greatest significance in determining the best time to get rid of a machine. It must be emphasised

Table 35.2 COST OF INTEREST, INSURANCE ETC

	Year end 1	2	3	4	5	6	7	8
Investment at start of year	10 000	7 500	5 500	4 000	3 200	2 500	2 000	1 500
Annual depreciation	2 500	2 000	1 500	800	700	500	500	500
Investment at year end	7 500	5 500	4 000	3 200	2 500	2 000	1 500	1 000
Average annual investment	8 750	6 500	4 750	3 600	2 850	2 250	1 750	1 250
Interest charge p.a.	875	650	475	360	285	225	175	125
Cumulative charges	875	1 525	2 000	2 360	2 645	2 870	3 045	3 170
Cumulative working hours	1 500	3 000	4 500	6 000	7 500	9 000	10 500	12 000
Cost per cumulative working hour	0.58	0.51	0.45	0.40	0.35	0.32	0.29	0.26

that this part of the exercise cannot be a once for all. Repairs cost records will permit a periodic check with a better prospect of a more accurate forward extra-polation.

The apparently erratic incidence of costs shown below is explained by the fact that heavy costs are incurred at times of major overhaul. In this connection the need for check and assessment is imperative before a major overhaul is put in hand. Otherwise it may be found, after the expenditure has been incurred, that the machine should have been sold and that it must now be retained for a considerably longer period or sold at a loss.

(d) *Breakdowns.* With age it is inevitable that a machine will suffer more frequent and more prolonged breakdown. The increase in the incidence of breakdown will not

Table 35.3 MAINTENANCE AND REPAIR COSTS

	Year end 1	2	3	4	5	6	7	8
Availability %	95	90	80	90	85	80	75	80
Repair costs	350	500	850	1 650	700	1 150	2 800	2 000
Cumulative costs	350	850	1 700	3 350	4 050	5 200	8 000	10 000
Cumulative working hours	1 500	3 000	4 500	6 000	7 500	9 000	10 500	12 000
Cum. costs/working hour	0.23	0.28	0.38	0.56	0.54	0.57	0.76	0.84

be linear because of the local effects of major overhauls which will introduce temporary improvements. Again this is a factor which must be related to the data which an owner should have concerning the machine in question. It is impossible to generalise on the subject of consequential or contingent costs which must vary from project to project and even from time to time on the same project. The results of the following calculation can therefore be no more than a series of index figures to which factors may be applied in the light of information available. The calculation has been based on the cost of hiring an identical machine to replace the one which has broken down. In the case of frequent short term breakdown this would not be possible in any case, but the method is very conservative and gives some idea of the scale of the cost.

(e) *Obsolescence*. This is an obvious selling point in the pressure which is put upon an owner to buy a new machine. Where the old machine is a well-proved, robust and reliable unit, it is wise to have a close look at the reasons which are given for the increased

Table 35.4 COST OF BREAKDOWN

	Year end							
	1	2	3	4	5	6	7	8
Availability %	95	90	85	90	85	80	75	80
Hours lost	75	150	225	150	225	300	375	300
Replacement cost/hr	2.50							
B/D cost	188	375	563	375	563	750	940	750
Cumulative B/D cost	188	563	1 126	1 501	2 064	2 814	3 754	4 504
Cumulative working hours	1 500	3 000	4 500	6 000	7 500	9 000	10 500	12 000
Cost/cumulative hours	0.11	0.19	0.25	0.25	0.28	0.31	0.36	0.38

Table 35.5 OBSOLESCENCE COSTS

	Year end							
	1	2	3	4	5	6	7	8
Obsolescence factor %	—	—	—	15	15	15	20	20
Extra bonus required	—	—	—	225	225	225	300	300
Cost per hour				2.50				
Cost per year	—	—	—	560	560	560	675	675
Cumulative cost	—	—	—	560	1 120	1 680	2 355	3 030
Cumulative working hours	1 500	3 000	4 500	6 000	7 500	9 000	10 500	12 000
Cost/cumulative working hour	—	—	—	0.09	0.15	0.19	0.22	0.25

The various elements detailed above are summarised below.

Table 35.6 CUMULATIVE COST PER WORKING HOUR

	Year end							
	1	2	3	4	5	6	7	8
(a) Depreciation and replacement	2.00	1.84	1.87	1.47	1.33	1.22	1.14	1.08
(b) Interest, insurance, etc.	0.58	0.51	0.45	0.40	0.35	0.32	0.29	0.26
(c) Maintenance and repair	0.23	0.28	0.38	0.56	0.54	0.57	0.76	0.84
(d) Breakdown	0.11	0.19	0.25	0.25	0.28	0.31	0.36	0.38
(e) Obsolescence	—	—	—	0.09	0.15	0.19	0.22	0.25
Total cumulative cost/working hour	2.92	2.82	2.95	2.77	2.65	2.61	2.77	2.81

outputs which are claimed for the current model. If it transpires that the claims are related to greater horsepower obtained by more powerful blowing of the same engine, it may safely be assumed that repair costs and breakdown time are likely to be increased. This must be weighed against the validity of the claim for increased productivity which may turn out to be negative over the working life of the machine. This aspect is not likely to be apparent in any field trial of a brand new demonstration model.

Nevertheless genuine cases of obsolescence must occur and the following method may be used to assess the costs, subject to the proper application of data which should be available. The cost is assessed in terms of the additional hours of the old machine required to give the production of the new.

The following conclusions relate only to this example and are no guide or recommendation whatsoever to action in any other case.

(i) The machine might be exchanged after two years.
(ii) It should certainly not be exchanged immediately after the overhaul in the third year.
(iii) The optimum exchange time appears to be about the end of the sixth year.
(iv) Similarly valuable conclusions may be drawn from such an exercise, carried out with positive data for any particular machine.

There are also certain conclusions in principle which may be drawn from the above table, and which are universally applicable. Most of these are obvious in themselves but a rational integration may be of interest.

(i) Items (a) and (b) always tend toward retention of the machine, provided they are commercially viable.
(ii) Items (c) (d) and (e) tend towards sale of a machine, provided they are not relatively and abnormally low.
(iii) The inter-relationship of (1) and (2) determines the disposal policy.
(iv) If for some good reason an arbitrary write-off period is imposed, it will tend, if very short, towards earlier sale of a machine, and, if very long, towards retention.
(v) Relatively heavy repair costs such as those caused by abnormal conditions or by poor quality in the machine will tend towards early disposal. The converse also applies.
(vi) Absence of serious obsolescence can have a decided influence towards retention.

OUTPUTS OF MACHINES

Excavation

Earthmoving manuals produced by several of the main manufacturers contain basic data on material weights, rolling resistances, traction coefficients, etc. which need not

Figure 35.5 Parallelogram 3-tine ripper mounted on crawler tractor

be repeated here. In order to apply this data properly it is essential that a thorough investigation be made of the project on the lines of CP 2001 'Site Investigations'.[1]

Much of the information called for in this publication is required for design purposes, but it is surprising how relevant it is to plant selection and assessment of operating performance and therefore outputs.

It is essential that users should have an appreciation of the limitations of data contained in manufacturers' earthmoving manuals and handbooks and, even more important, an

Figure 35.6 $1\frac{1}{2}$ cu yd drag hoe or back acter loading hard rock from foundation examination into 15 ton rear dump wagon

appreciation of the value to be placed on manufacturers output claims. These last are invariably optimum and can sometimes be optimistic. They are based on the best possible ground conditions, a first class operator, an unfailing supply of transport and auxiliary plant. The true picture is usually somewhat different.

The usual table of rolling resistances depends, for its effective use, on the user's ability to forecast a ground surface condition in rather vague wording such as 'hard road', 'loose gravel', 'soft clay', etc. Even when a ground condition is being experienced, and is there to be seen and examined, opinions differ as to the value to be set against rolling resistance. This is a field which calls for some extensive research into the relationship between tyre or track penetration and rolling resistance and for the production of some test implement or equipment (preferably simple) which might be used on site at tender or planning stage, and which would reduce the danger of a serious mistake. As an indication of the complexity of the problem, it may be mentioned that liquid mud, hub deep, does not offer serious resistance provided the bottom is hard, and that a tyre penetration of 200 mm into firm plastic clay will result in a very heavy resistance. A dynamometer between a tractor and a towed vehicle will give an accurate assessment, but such an arrangement is not usually convenient for a site visit.

It is not unusual for plant department managements and plant employment advisers to be approached with the question such as 'What is the output of a 22RB or a T.S.24?' The question in this form is meaningless and indicative of lack of experience, and possibly of commonsense, on the part of the questioner. If a fairly comprehensive statement of site conditions and plant disposition can be provided it is possible, from this and from the specified characteristics of the plant, to arrive at an 'optimum' output which is obviously the best which this plant can attain under the basic conditions. This is very far from being the average output to be expected from the machines for purposes of overall planning and estimating.

The methods of calculating optimum output are as follows. The essence of the basic method is that an operating cycle time must be calculated and applied to an assessed capacity.

CALCULATING OPTIMUM OUTPUT

(a) For excavators and the like it is convenient to start the cycle at the dumping action which is nearly instantaneous. The elements are then

(i) slew empty to face and present bucket;
(ii) dig;
(iii) slew laden to dump point;
(iv) dump;
(v) hoisting and lowering if these override slewing times;
(vi) timing of any other relevant movement, such as dragging.

Timings for all these elements can be obtained from a combination of manufacturer's specification details and the physical circumstances. For example the angle of slew has a considerable bearing on the cycle time and planning of the layout should be designed to minimise this. Straight application of the rated slewing speed is not valid, particularly on a small angle, and time must be allowed for acceleration and deceleration. Very generally, on a 90° slew, the cycle times of rope-operated machines range from 25–40 sec and for hydraulic machines, 15–25 sec. If it is anticipated that the digging time may be extended because of hard conditions, consideration should be given to the cost effect of forward preparation of the material by the use of explosives.

An assessment must be made (and in this case there is no substitute for experience) of the fill factor of the bucket. This is the proportion which the bank yardage (measured solid in virgin ground) in the average bucket load bears to the heaped capacity. In

Figure 35.7 Shovel loading special tray-carrying 42T GVW road vehicles at Dubai Quarry

generally cohesive materials, the loose yardage in the bucket tends to approximate to the heaped capacity. This is also usually true of granular materials in shovel or clamshell buckets but in open-ended buckets such as dragline, and, to some extent, drag shovel, loose dry sand or gravel will give a relatively low fill factor. In order to convert loose yardage in the bucket into bank yards information must be available on the solid density and loose density of the material as in the bucket.

From the calculated cycle time and the assessed bucket load an optimum output may be calculated. This will be subject to the application of a number of operational factors which will be detailed later, in order to arrive at an average output for the machine over a prolonged period.

The employment of a bucket which is oversize in relation to the material will result in overload, will increase the cycle time and may well be uneconomical apart from the unnecessary damage which will be done to the plant.

(b) In the case of plant where the essential function is the transport of materials whether on wheels or crawler tracks, information must be available on the anticipated route or routes—analysed into elements of gradient and ground conditions. If the haul length varies appreciably and seasonal variations are expected in ground conditions, separate exercises must be carried out for selected sets of conditions and these must be applied to the appropriate period of the overall programme. Complexities of this nature are not unusual, and any sweeping generalisation is liable to be very misleading. The various elements to be taken into account are as follows.

A Element of distance	B Gradient %	C Rolling resistance %	D Total resistance %	E Maximum speed	F Average speed	G Time

The following paragraphs refer to the table above:

A. As stated previously the laden haul and empty return distances must be split up into gradient lengths and lengths of differing road surface conditions.

B. Gradients should be shown as positive up, and negative down. A rough ground survey is essential.

C. The problem of assessing an anticipated rolling resistance has been mentioned and, failing any better alternative, some reliance must be placed on the tables provided by the principal manufacturers. It should be mentioned, however, that there is always a tendency to underestimate the effects of bad ground conditions. Only on a well made and maintained road can one rely on finding a rolling resistance between 2% and 4%. It should be noted that 2% is generally accepted as the resistance caused by internal friction. Some manufacturers include this part of the resistance in their chart and others do not. The notes attached to the chart should be consulted. A test carried out in UK by dynamometer between a tractor and a roller operating in plastic clay gave a starting resistance of 25% and a moving resistance of 15%, both on level ground. A wheeled vehicle would probably have been rather more affected. It should be borne in mind that the manufacturers who produce information of this nature are mainly concerned with operations in USA, and other places, where work is not expected to proceed in conditions which are fairly normal in UK. It would not be exceptional to have excavation work in this country which must proceed in the face of rolling resistances of 20% either in the cut or on the tip. In this connection it should be borne in mind that traction coefficients fall off seriously in such conditions and a check should be made to ensure that movement is possible. The evil consequences of over loading should not need to be explained in such circumstances.

D. The total resistance is the algebraic sum of B and C. Rolling resistance may be regarded as a small section of gradient which moves forward with the machine.

E. The manufacturer's gradeability chart will provide a maximum speed for the laden or empty vehicle in relation to the total percentage resistance.

F. The manufacturers also provide tables of coefficients to be applied to the maximum speed in order to determine the average speed over a section. These coefficients have been obtained empirically from field trials and they relate to factors such as speed at entry, length of section, accelerations and decelerations etc. They have been found to be reliable by check on results obtained. However they do not take account of factors which are not measurable and which call for the application of judgement. For example a short steep down gradient leading to a bend will never be negotiated at the speed allowed from the maker's chart. There are many similar situations on a haul road lay-out where speed must be assessed in terms of safety.

Even if a haul road surface is fairly hard there can be no assurance that maximum possible speeds can be attained. A very irregular surface will introduce danger and operator discomfort and speeds will inevitably be reduced. There will also be time loss due to breakdown. The calculations so far are based on the assumption that haul roads are maintained in smooth condition, but it is the writer's experience that failure to do so is the single greatest cause of loss of time in the cycle.

G. The time over a section may be obtained from the average speed and the length.

The sum of these elements of time gives the total optimum cycle time. This related to the carrying capacity, which is dealt with on the same lines as for the excavator bucket, gives an optimum output for the machine.

REDUCING FACTORS

As already stated the optimum output is subject to reducing factors which are described below.

(a) *Mechanical reliability of the machine.* This is usually expressed in the form of a percentage availability or of breakdown. It should be borne in mind that a machine cannot break down when it is idle nor is it likely, on the short term, to suffer in idleness conditions leading to breakdown, except in exceptional circumstances such as Arctic cold. It follows that the percentage of availability or breakdown must be determined from working and brokendown hours only.

(b) *Climatic conditions.* The incidence of rainfall, extreme heat, dust, extreme cold etc must have a bearing on output and the probable loss of time must be assessed. See publications on 'Wet Weather Working'.

(c) *Operator capability.* The yardstick is usually the output to be expected from a competent British operator in good average Western Europe conditions. It should be borne in mind that there is a connection between the nationality of the operator and the climatic condition. For example a competent operator from Pakistan on an excavator, without air conditioning, in extreme desert conditions, will produce better results than a good European operator. Such results would not necessarily be as good as those obtained from the European working under good temperate zone conditions. There is obviously room for thought and logical assessment of this factor.

(d) *Nature of materials.* It is essential in works of any size and importance that a full geological and physical specification of the materials should be available. The main points which must be covered are:

(i) Description of rock or alluvium in the state in which it is to be excavated; for example:

Limestone in a cutting or quarry face 15 m high, broken to maximum size of 0.7 m or 750 kg;

or Gravel and sand in pit 8 m deep with water table at 1 m.

It must be kept in mind that construction plant is designed to apply and resist *forces* which are usually proportional to weights and that actual bucket and body volume sizes are no measure at all of the capability of a machine. From data such as above it is possible to arrive at a volume capability such as an allowable bucket size or vehicle body volume which will not overload the machine. A nominal 2 m^3 ($2\frac{1}{2}$ yd^3) excavator can be fitted with a 2.7 m^3 ($3\frac{1}{2}$ yd^3) bucket for digging peat provided there is no danger of the bucket being filled with clay by careless mistake. The reverse situation may arise with very heavy materials. From comprehensive information of this nature it is possible to apply a factor to the nominal output of the machine.

(ii) *Silica and/or Olivine content of the material.* These are the hardest and most abrasive minerals normally encountered in construction works. The presence of undue amounts of silica or quartz or flint or chert, which are all forms of silica, can result in excessive wear and consequent breakdown. Due allowance must be made for this in (a) above. There are also serious cost aspects such as undue tyre or track wear.

(iii) The problem of preparing material for excavation and of the plant required for drilling, blasting, pumping, dewatering, etc. is covered in the appropriate section.

(e) *Associated and ancillary plant.* If a primary producer such as a major excavator is associated with, say, a team of transport vehicles, this team will also suffer from the same variety of efficiency factors. If the best average output of the excavator is to be obtained the number of vehicles must be increased to cover all transport team inefficiencies and cope at all times with the *maximum* possible output of the excavator. Where two or more items of equipment are operating in sequence on a process the efficiency of the whole process will generally be the product of the efficiencies of the individual elements. Some relaxation of this rule may be allowed to cater for some blanket influence such as snow-storm or the like where all machines are out of action at the same time.

(f) *Operator relaxation.* It is impossible for an operator to maintain full efficiency over every minute of a 10-hour shift. Apart from meal and tea breaks and the calls of nature, his attention must be distracted at times. These departures from 100% efficiency can be tolerated and allowed for under the assessment of (c) above where the machine is working more or less in isolation. Where a major unit, or a major assembly of units, is working as part of an extended process sequence the compounding of efficiency losses can become very serious and it is then necessary to carry a sufficient number of spare drivers to ensure that the process is continuous and at high efficiency at least so far as operator attendance and fatigue is concerned. Where such arrangements are intended it is possible to make allowance for a corresponding increase in the overall efficiency of the operation.

HUMAN ERRORS

In conclusion, in the matter of plant outputs it may be said that the cumulative effect of the efficiency factors listed above is to reduce the optimum output by as much as 30 to 40%. This takes no account of human mistakes in organisation which can have an even more serious effect. Examples of these are:

(a) Wrong selection of plant for the duty, perhaps due to inadequate prior investigation.
(b) Encouragement of overloading and overspeeding without regard to inevitable consequences, and possibly related to ill-conceived bonus schemes.
(c) Bad planning of layout.
(d) Bad control of transport resulting in alternate queueing and shortage.
(e) Inadequate organisation and control of servicing and repairs and maintenance.
(f) Inadequate and erratic fuel supplies.
(g) Inadequate vehicle access and lack of provision for recovery and removal of broken down vehicles.
(h) Lack of attention to excavator operating surfaces and to road surfaces, gradients, etc.
(k) Lack of pumps and failure to maintain a natural drainage system.
(e) Ineffective communication and lack of competent supervision.

Concreting

The complexities and intangibles of earthmoving do not apply to the same extent in concreting work. The peak output of the batching and mixing plant should be known accurately, but, if not, it can be obtained by stop watch in a very short test run. If supplies of aggregates, cement and water are adequate the factors which will diminish the peak output are:

(a) *Hard frost.* Due allowance must be made in assessing average outputs for time lost due to frost, and other adverse weather conditions, in relation to the specificational requirements. Improvements may be made by the introduction of a feed water and

Figure 35.8 Batching and mixing plant (20 m per hour) with integral ground bins, drag scraper and elevating weigh hopper operating in UK

aggregate heating system. In this connection it can be advantageous to have an oversize plant which will produce the daily requirement between 10 a.m. and 3 p.m.

(b) *Failure to remove and place concrete at the required rate (for any reason).* This is the prime cause of failure to attain the desired output. The blame, for obvious reasons which reflect on the operating organisation, is usually attributed to the batching and mixing plant. Anything which has been said about gearing up transport and sequential operations to cope with the peak production of the primary unit applies most acutely to concreting operations. If the peak production of the batcher cannot be handled for several hours in the day, the average output required is not likely to be obtained.

(c) *Breakdown of the batching and mixing set-up.* A reasonably well maintained plant should have a mechanical efficiency in excess of 95%. The effects of breakdown, however, at the concrete placing end, can be so serious that quality in the machine and very strict attention to maintenance are matters deserving of the closest attention.

CALCULATING OUTPUTS

Except on extensive concrete paving or very heavy mass concrete pours, where some higher efficiency can be reasonably expected, it would be unwise to supply a batching and mixing plant with a peak output less than 50% greater than the average called for on the heaviest day of concreting in the programme. Failures to obtain output are usually associated with placing in reinforced concrete structures, including thin walls, columns and beams. Such failures are almost entirely due to lack of appreciation of the time necessary to get concrete into the shuttering of slender structures, particularly if cranage is involved.

In very general terms the peak outputs of paddle mixers in batching plants are about 60 m^3 per hour per m^3 of batch capacity. In large tilt or split drum mixers the peak outputs run at about 40 m^3 per hour per m^3 of batch capacity. For mobile mixers of less than 1 m^3 the peak outputs run at about 30 m^3 per hour per m^3 of batch capacity.

The output capability of a concrete paving train is generally determined by the rate of delivery from the batcher, with a limiting factor in the finishing technique, which prevents advance at any greater speeds than 30 or 60 m per hour depending on the number of compactor/finishers employed. With a 8 m slab, 300 mm thick this would give peak requirements of 60 and 120 m^3 per hour. In practice average outputs of about 40 and 80 m^3 per hour are more likely to be attained. A similar calculation may be made relative to any specified slab and any available batching plant and paving train.

There can be output problems with mobile concrete pumps. The makers rated outputs are related to optimum conditions of mix and of distance and take no account of losses due to blockages which are practically unavoidable, although they need not be of very frequent occurrence. It is therefore wise to be conservative and to plan for an average output no greater than 50% of the maker's rating where conditions are expected to be reasonable. If conditions are doubtful some further reduction should be made. In the case of the 150 mm mechanically operated pumps the output is definitely limited to 10–14 m^3 per hour but there is much greater delivery length capability and less probability of blockage. All the above information is, of course, contingent on a continuous supply of concrete to the pump.

Cranage

In connection with cranes it is standard practice for American manufacturers to distinguish between Duty Cycle and Hookwork. By Duty Cycle they mean a regular and continuous process such as grabbing or handling large numbers of trays or units, usually all of the same size and weight. In terms of performance and calculation of output this is indistinguishable from excavator work and anything which has been said about excavators will largely apply, bearing in mind that time must be allowed for attachment and detachment of loads where this is necessary.

Hookwork is the American name for sporadic crane duty as we know it. It involves handling a variety of shapes and weights to varying locations and heights. It often involves holding a load suspended for many minutes. A crane cycle can be calculated from the same basic information as for the excavator but it can only apply to one particular operation. The problem is not usually the speed capability of the crane but rather the organisational capability of the contract management. From experience and particularly from analysis of watt-hour meter recordings from electrically-operated cranes, it may be stated with some confidence that cranes on Hookwork seldom attain 30% of their operational capability and often fall below 15% of the ideal. This is not to say that cranes are usually employed unproductively, but rather that there is seldom enough movement work to keep them fully employed. Failure to have a tightly controlled and well considered cranage programme, particularly with tower cranes, will result in

Figure 35.9 Crawler crane-mounted boom suspended leaders and diesel piling hammer

utilisation figures which are even lower than these quoted above. Apart from the avoidance of the common fault of attempting to satisfy simultaneously competing demands, improvements may be made by attention to time taken to attach and release loads and to unnecessarily prolonged suspension of loads.

General

The previous text relating to outputs indicates the principles which must be applied in an assessment of the output of any item of construction equipment. It is not proposed, therefore, to proceed in detail with all the plant items which have not been mentioned. The following general points may, however, be of interest:

(i) Most plant tends to produce more than is estimated, when it is actually working, and to spend more time in idleness than was thought possible at the estimate stage.

(ii) More output is lost in the end than is gained in the very short term by overloading and overspeeding equipment.

(iii) Tables of alleged 'outputs' for various items of plant are meaningless and dangerous in the hands of those who are not sufficiently experienced. Even ranges of outputs can be so wide as to be ridiculous as reference data. For example, a 0.6 m^3 excavator could be performing well at 8 m^3 per hour in one circumstance and badly at 40 m^3 per hour in another, depending upon the conditions attached to the two duties. Short of experience of a wide variety of conditions, it is unsafe to set off-the-cuff output figures and better to rely on a combination of experience with rational assessment of the particular output problem.

MAINTENANCE AND REPAIR

There is no reputable item of construction equipment which is delivered without a maintenance or service manual and a spare parts schedule. In the service section of the manual there are clear instructions as to each and every cleaning, oiling, greasing and other operation which must be performed at the stated time interval in order to keep the machine in good working order. The majority of responsible site staff are thoroughly aware of this and most of them are adept at finding good reasons why a routine service should be postponed. This is usually associated with some attempt to obtain a short term ephemeral advantage in the way of a record output. Thinking on these lines should be stamped out if the organisation wishes to obtain a long term good average output from machines.

The only circumstances when a service may be deferred is when there is danger of a catastrophe if the machine does not continue at work, and under these circumstances it is always possible to do something to safeguard the machine. If a site supervisor wishes to keep a machine at work, when service is due, in order to prepare for the following day, a common excuse, he has condemned himself on grounds of inability to plan and execute the works.

Transport and easily moveable equipment should be serviced at a central point where fuels, oils etc. can be provided. The more isolated machines should be serviced from a suitably equipped service wagon and fuel bowser.

In operations overseas where conditions, as in the desert, may be very dusty, it is wise to increase the frequency of oil changes and often to at least twice the normal. It is also necessary to ensure that air and oil filters are of the highest quality and that these also are cleaned and/or changed at intervals closer than the maker's recommended frequency.

Cooling of engines, hydraulic systems and the like can be a serious factor in the operation of machines in the tropics. Apart from the necessary changes in specification it is vital that these systems be kept clean both in the coolant and air passages and that they are always in first class condition. Machines should be shaded from direct sunlight which is often sufficient to cause overheating particularly with items such as compressors.

Proximity of repair depots will determine the need for extensive site repair facilities on contracts in UK. It is not usually necessary to install elaborate repair facilities on home contracts, but overseas in remote places on a sizeable contract it is often necessary to be completely self sufficient. This could involve the construction and equipment of the following workshops:

(a) Heavy plant; (b) Light plant; (c) Machine; (d) Transport; (e) Electrical; (f) Welding; (g) Generator house; (h) Compressed air; (i) Painting.

It is not unusual for such an assembly to be considerably more elaborate than a contractor would have in his main plant yard in UK. For example, it is not uncommon for it to be necessary to have a large lathe which a contractor would not consider in UK because he can obtain machining work when required on the occasional very large piece, at some local commercial machine shop.

GENERAL CONCLUSIONS

It will be realised from what has been said previously that the selection, procurement, and management of construction equipment is not a separate business, only slightly related to construction work, and to be carried out by an alien race who must not on any account be made aware of the mysteries, technical and commercial, of the construction process. It is, in fact, an integral and increasingly important part of that process and deserves much more attention than it usually obtains from project management teams. Some confusion, no doubt arises from a mistaken impression that plant companies and departments should have one primary concern only and that is the swift repair of machines which have broken down. While it would be foolish to deny the importance of this function, it must be obvious that someone must be concerned about the factors which made the breakdown inevitable and with the future avoidance of similar occurrences. These factors may well extend back to the day when the machine was purchased and through its employment and possible misuse combined with attendant maintenance failures. It is surely conceivable that a project manager could improve performance on a succession of projects by close attention to these aspects in addition to clamouring for an overnight repair of damaged machines.

The overhaul and repair sections of plant departments can never avoid criticism. Their failures are patently obvious; on the other hand, site failures are frequently buried. It follows that a category of competent men must be available and must have the power to insist on the proper application of techniques of control and management of equipment.

All the photographs in this section are reproduced by permission of Richard Costain & Co. Ltd.

REFERENCE

1. CP 2001: 1957 *Site Investigations,* British Standards Institution

BIBLIOGRAPHY

CERA Report No. 3, *The effect of wet weather on construction of earthworks*
DOE Building Research Establishment, *Guide to Concrete Pumping*
General Motor Corporation, Terex Division, *Production and cost of materials movement with earth moving equipment*
LEWIS, W. A., WASS, F. H. and COWLEY, H. R., 'Road Rollers', *Proc. Inst.Mech.E.* (1966–67)
MORGAN, B. V., 'Off Highway Haulage', *The Quarry Manager's Journal* (Jan.–March 1969)
NICHOLS, H. L. JR., *Moving the earth,* McGraw Hill Publishing Co., London
'Paving—how to get the best results', reprint from Construction Methods and Equipment, McGraw Hill Publications
PAXTON, J. M. (consulting ed.), *Manual of civil engineering plant and equipment,* Applied Science Publishers
STUBBS, FRANK W., JR., *Handbook of heavy construction,* McGraw Hill Publishing Co.
The Caterpillar Tractor Co., *Caterpillar performance handbook*
The Caterpillar Tractor Co., *Fundamentals of earthmoving*
The Caterpillar Tractor Co., *Handbook of ripping*
The Clark Equipment Co., *Michigan wheel tractors in earthmoving*
The Hyster Co., *Guide to Hyster Compaction*
See also various papers on 'The compaction of fill material' and 'Concrete paving' which are published by the Road Research Laboratory

36 CONCRETE CONSTRUCTION

CONCRETE CONSTRUCTION 36

36 CONCRETE CONSTRUCTION

J. M. FISHER, B.Sc., F.I.C.E., F.I.H.E.
John Laing Research & Development Ltd

INTRODUCTION

Concrete is one of the principal construction materials in use throughout the world and it is likely that well over half the total cement produced is used as a component of concrete. In Britain, the rate of cement production is currently 17 million tons per annum, of which it is assessed that about 75% is used in concrete of various classes. Other major uses of cement are in mortars and renderings of many types, tiles and asbestos cement. Assuming that on average concrete will contain about 270 kg of cement per cubic metre then the total concrete production will be of the order of 50 million cubic metres per year.

Some indication of the distribution of concrete production as between sites which produce their own concrete—most heavy civil engineering works using large volumes of concrete will probably fall into this category—ready-mixed plants and pre-cast concrete works, is given by the fact that 40% of all cement produced is used by ready-mixed facilities, whilst 20% goes to pre-cast concrete manufacturers.

In the United States, a higher proportion of the total cement produced—about 60%—is supplied to ready-mixed concrete plants. It is likely that the trend in Britain will be towards the more widespread use of ready-mixed concrete for an increasingly wide range of civil engineering works. However, there will always be a heavy demand for site-mixed concrete as opposed to that from permanent plants, even though concrete production might be let as a subcontract to ready-mixed operators equipped with mobile batching plants. The increasing proportion of the total requirements for concrete aggregate which is now under the control of ready-mixed concrete operators might accelerate this trend.

CONCRETE PRODUCTION

Storage of aggregates

Except in those cases where the components of concrete are readily available at short notice from stockpiles at pits and quarries, ground or bin storage of at least several hours supply of all materials is called for. In some cases, where concreting operations are concentrated into a relatively small proportion of the total contract time—much shorter in fact than the time necessary to assemble the required total quantity of aggregates, as for instance in concrete road construction, the building of large stockpiles adjacent to the batching plant will normally be required.

The equipment required for building and drawing from these large ground stockpiles is expensive and could add 20–25 p per cubic metre to the cost of appregates at the batching plant. Thus the planning of site operations must be directed towards achieving a sound economic balance between the rate of consumption and the rate of supply; it is essential to avoid ground storage of excessive amounts of materials at batching plants and to rely, as far as is possible, on current production from a number of pits and quarries.

Because the nature of the aggregates used in concrete has such a marked effect on its properties in the unhardened state, it is rarely possible to use both quarry- and pit-washed aggregates in combination unless the necessary steps are taken to ensure constant proportions of the two types.

For concretes where the standards of control required are high, provision must be made in the vicinity of batching plants for storing at least two sizes of coarse aggregates and, generally, two days supply of fine aggregate. This latter requirement stems from the fact that fine aggregates, which are normally washed, can contain up to 15% water which will drain away at a fairly rapid rate, allowing the moisture content to stabilise at a much lower figure over a period of perhaps 24 hours.

Large stockpiles should be so constructed that segregation of the larger fractions from graded materials, which could cause embarrassing variations in the overall grading of the concrete aggregate, is avoided to the greatest extent possible. Where large stockpiles of coarse aggregates are to be built, consideration should be given to the use of controlled tipping of lorry loads.

Segregation of the coarser fractions of sands from the finer is not normally a major problem since the moisture content of the whole will be sufficient to prevent particle separation. It is always desirable and will often prove economic to provide adequate paved areas, laid to falls, round the aggregate stockpiles so that access to and drawing from them does not lead to contamination by dirt from the site. It could happen that the loss of aggregate into the ground beneath the stockpile exceeds the cost of providing sufficient hardstanding for aggregate storage.

This paved area will preferably extend well outside the range of any stockpile reclaiming devices and give clean access to and room for manoeuvre round the stockpile. However, it will not normally be necessary to pave right up to the batching plant since with most reclaiming units there will be a certain amount of dead storage which is not removed until concreting operations are virtually complete.

A requirement of any binning arrangements made round batching plants is that their walls should be built high enough to avoid overspilling of one grade of material into another; it is also necessary to ensure that overfilling, which leads to spillage and mixing of different grades of aggregate, does not take place.

It is quite often required that concretes be made from aggregates with special properties, e.g. lightweight (for low-density concretes used for better thermal insulation and lower structural weight) or heavyweight (used for such purposes as radiation shielding to nuclear reactors). Rather than complicate binning arrangement round batching plants used for the major part of the concrete it may be found advisable to use a separate batching/mixing plant of sufficient capacity to meet the requirements of the special concrete.

The sizes of stockpiles and the rates at which materials will need to be taken from them are matters which will need to be given proper consideration by job planners and site management. Amonst the types of equipment most frequently seen operating in Britain are:

(a) ground hoppers fed with materials drawn from stockpiles by, for example, forward-loading shovels and transferred thence by conveyor belts to short-term storage bins above the batching plants
(b) drag shovels, drag lines or grabs mounted atop stockpiles located at the batching plant, feeding into short-term storage bins or pulling materials into live storage areas whence they feed by gravity into weighing equipment

With high-capacity plants these storage bins will, of necessity, hold materials for only a small number of batches, hence the need for adequate ground storage.

Storage of cement

It is generally required that cement be stored after grinding in high-capacity silos at the works. This is done so that the quality of the cement can be assessed for compliance with the appropriate British Standard before delivery and also to ensure, so far as is possible, that the high temperatures attained during the grinding process can be to some extent dissipated before delivery to the site.

The consequences of using physically hot cement as opposed to cooler cement are not thought to be serious, but nevertheless provision must sometimes be made for ensuring that cement temperatures are limited. In some instances where large masses of concrete are to be placed and where overheating due to the exothermic reaction of the cement must be reduced so far as possible, it may be required that the cement be circulated through coolers or kept in circulation through a number of storage units.

Storage of large quantities of cement might be called for to smooth out irregularities in delivery when construction sites are remote from cement works. The present trend towards the movement of bulk supplies of cement from mills to works by special cement trains rather than by road delivery offers the opportunity for short-term site bulk storage in the special high-capacity wagons now coming into use. However, it will generally be necessary for at least part of the journey to be made by road and facilities for pneumatic transfer from rail direct to road vehicles and then into short-term storage at the batching plant, will have to be provided.

On site, overhead cement storage silos of capacities up to about 150 tonnes, either singly or in interconnected groups are quite widely used. These are generally charged pneumatically by dry-air blowers capable of lifting cement to a height of 30 m or so above the ground, mounted on transport vehicles.

A current trend in cement storage is to use high-capacity horizontal silos mounted on road wheels; these are supported on jacks whilst in use as the main storage. From them cement is drawn to charge small-capacity silos immediately above the plant, whence it is fed into the batching plant.

These silos are intended only for storage on site; they are not suitable for hauling cement since they only comply with weight requirements on public highways when unloaded.

Some cement companies are willing to provide such storage facilities on sites and their availability should be investigated before making any firm commitment for contractor ownership.

Water storage

On average, each cubic metre of concrete produced will require between 100 and 140 litres (24 and 30 gallons) of water. In addition, large quantities will be used round the plant at the end of a concreting session for the thorough cleaning down of the whole of the mixing plant and the lorries, skips, pumps or other devices used for the transport of the wet concrete.

These large quantities of water can sometimes be drawn from a Water Authority's mains, but quite often some local source, such as a stream, will have to be used. Permission to extract water from a stream will generally have to be sought from the local River Board.

Overhead storage of water in steel tanks to give a sufficient head to supply a concrete plant can be quite expensive, particularly when drawing from mains is restricted to night-time only and adequate storage has to be provided for all day-time operations. In view of this high cost of water storage it is becoming increasingly common to excavate a hole of sufficient capacity close to the batching plant and to provide a waterproof lining, generally 1000 gauge polythene sheeting, held round the top edge by embedment in concrete or some other means. From this, stored water can be drawn by pump to supply a small header tank above the batching plant. When this mode of storage is used every precaution must be taken to avoid damage to the lining since repair will be difficult.

The large amounts of water used for cleaning down cannot be discharged into any local water course or sewer without first allowing the cement and aggregates to settle out to such an extent that the effluent is acceptable to the Authority. This calls for the building and keeping clean of a comprehensive system of settling tanks from which clear water can be decanted at the end of the line. Provision can be made to recirculate this water into the supply system, but it is doubtful, except where water is very expensive, if recovery is a practical proposition.

Batching concrete

The gauging of concrete to give mixes either of specified proportions or to meet strengths or other requirements is carried out in batching plants using either volume or weight as the unit of measurement. In almost all cases, batching plants incorporate a facility for mixing the concrete. However, some ready-mixed concrete plants and, in parts of

Figure 36.1 Production of concrete for motorway base. Note: large aggregate stockpiles; ground storage of cement in wheeled bulk silos; small overhead cement storage; ground water storage (in background; continuous proportioning and mixing of 4 No. aggregates and cement

the world, plants supplying concrete for paving operations, have a batching facility only, mixing being carried out in mobile mixers on site.

Although there is evidence that given proper control over operations, particularly with regard to the measurement of cement, volume batching of concrete can give high standards of quality control, this system has been almost completely superseded by weight batching. In these plants it is usual to weigh the various aggregates cumulatively in one hopper whilst the cement, any bulky additive such as pulverised fuel ash (PFA) and water will be measured separately.

Water can be batched into a concrete mix either by weight or volume, but with the increasing tendency to use fully atuomatic batching/mixing plants in which the moisture content of the fine aggregate is monitored continuously and its batch weight adjusted accordingly, there is rather more emphasis on weight batching.

A small header tank is generally provided and this in turn is kept continuously charged direct from the supply mains, or by pump when the mains pressure is inadequate or supplies have to be drawn from ground storage.

ADMIXTURES

The use of admixtures to modify properties of unhardened and hardened concrete in one way or another is becoming increasingly common practice in the construction industry. Generally, these materials are added in very small amounts in relation to the size of a batch and it is usual to measure them by volume and feed them into the mixing water supply line from gauge tanks to the mixer.

When using liquid admixtures it is essential to maintain an adequate supply and to ensure that each batch of concrete has its proper dosage added at the correct time. This calls for the full interlocking of water and admixture supplies so that underdosing or overdosing cannot take place.

When PFA is used as an additive then weight batching of the rather higher proportion used is normally required. This calls for a separate overhead supply hopper which can be charged pneumatically from a bulk tanker carrying supplies of ash from a power station. This will normally be dry ash from precipitators.

Mixing concrete

To achieve the full potential strength of a concrete mix it is most important that there should be a proper dispersal of the various constituents within each element of concrete. The speed at which this dispersal takes place will depend upon a number of factors amongst which are:

(a) type of mixer and its speed of rotation
(b) size of charge put into the mixer in relation to the volume of the mixer drum
(c) degree of wear on paddles and blades
(d) order of charging materials

Various types and sizes of mixer are available and the following are commonly used in British practice:

(i) rotating-drum mixers:
  (a) tilting drum
  (b) nontilting drum, including reversing drum
(ii) split-drum mixers
(iii) pan and annular-ring mixers
(iv) trough mixers
(v) continuous mixers

Figure 36.2 shows these mixers in general outline.

Types (i), (ii) and (iii) are commonly described as 'free fall' mixers since their action is derived from the falling within the drum of elements of concrete materials lifted from the bottom towards the top by a series of blades. In types (iv) and (v) the mixing action is more vigorous and it is claimed that this both improves the efficiency of mixing and increases the speed at which a sufficiently high degree of uniformity can be attained.

The largest-capacity batch mixer of any type used to date in British practice has been a 6 m^3 tilting-drum type with an hourly throughput of about 200 m^3. Sizes of the various types available range from a few litres to about 3 m^3 but there are, as noted, some exceptionally large mixers.

ROTATING-DRUM MIXERS

In type (i) mixers which are normally rotated at speeds up to about 20 rpm, mixing is achieved by carrying the ingredients from the bottom of the mixer to the top by a series of paddles of differing form mounted inside the drum. As the paddles approach the top of the mixer, materials are spilled from them and fall to the bottom of the mixer whence they are again lifted towards the top. The speed of the mixer is important in that a slow rotation extends the mixing time while too fast a rate will reduce the efficiency by tending to carry materials over.

A type (i)(a) tiling-drum mixer is charged at the open end with the axis of rotation of the mixer inclined upwards at an angle of about 45°. Mixing takes place whilst the drum

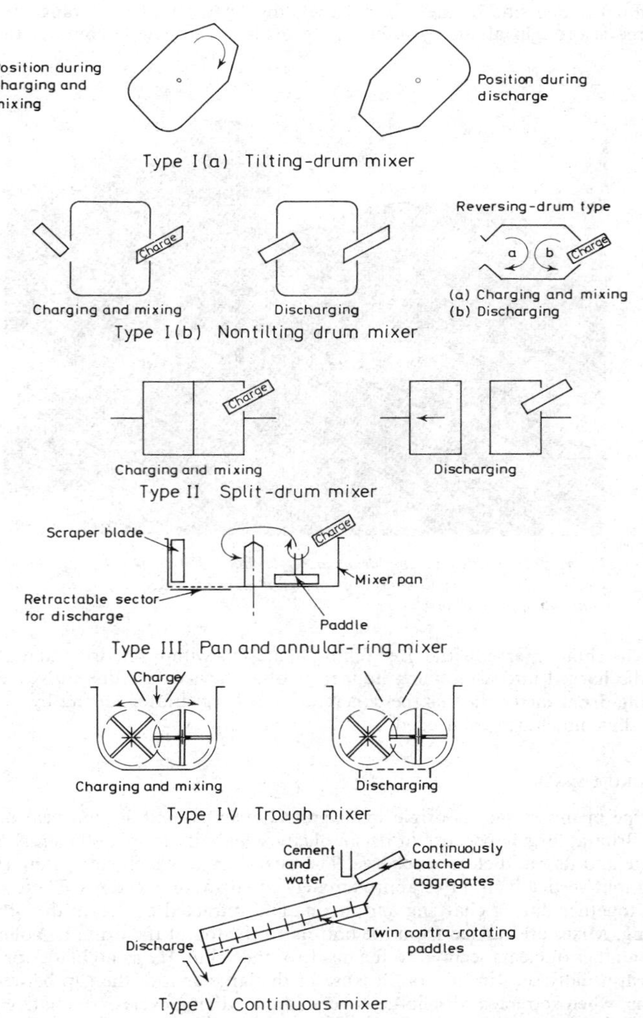

Figure 36.2 Concrete mixers (Courtesy: John Laing and Sons Ltd.)

is in this attitude. Discharge is accomplished by moving the axis of rotation through an angle of about 180°. Depression of the axis below the horizontal is carefully controlled to avoid too high a rate of discharge.

With nontilting drum mixers of type (i)(b), charging is via a retractable or nonretractable shute at one side of the mixer, depending on the loading arrangements, whilst discharge is brought about by inserting an inclined retractable chute at the opposite side. This chute intercepts the 'free falling' materials within the drum and causes them to be discharged into a receiving hopper or other device. An alternative design is the reversing-drum mixer. In this the concrete is discharged after mixing by reversing the drum; thus, no chutes are needed.

Figure 36.3 Small reversing-drum mixer (Winget 400R). Note: drag shovel loading; charge and discharge at opposite ends of mixer drum; means of controlling point of discharge

SPLIT-DRUM MIXERS

This type of mixer had its origin in Belgium but has found a good deal of favour in Great Britain, largely because of its simplicity, its ability to mix efficiently all types of concrete and its rapid clean discharge. The mixer drum, which rotates on a horizontal axis, is split vertically into two approximately equal volume sections. These sections are closed together during charging and mixing and retracted one from the other for discharging. Mixes are carried from the bottom to the top of the drum by cohesion and a small number of cleats secured to the inside of the drum. It has no blades or paddles of the form usually seen in mixers. Because of the large area of the gap between the two sections, when retracted, discharge is very rapid. Mixing cycles are relatively short, in particular because of this rapid discharge, and their efficiency is said to be quite high.

PAN AND ANNULAR-RING MIXERS

To speed up the mixing cycles and at the same time achieve a higher degree of uniformity of the mixed concrete a series of different types of pan mixer have come to be much more widely used in recent years. Mixers of this type, generally in the smaller sizes, have been used in pre-cast concrete works for many years and have proved very efficient. Because a satisfactory degree of uniformity of the mixed concrete can be achieved with this type of mixer in a good deal shorter time than with type (i) their use is now extending to heavy civil engineering work and there has been a steady increase in the sizes of batches which can be mixed in them.

Since the concrete contained in the pans of these mixers is moved round the pan by a series of paddles whose action and speed varies with the design adopted, they have come to be known as 'forced-action' mixers as opposed to the 'free-falling' type described at (i), (ii) and (iii) above. The rotating paddles which mix the materials in the pan have varying forms and action and are driven from either above or below the pan. In some types the paddles rotate on their own axis as well as round the mixer pan and, with these particularly, a high degree of uniformity of the mixed concrete is claimed after only a very short period of mixing.

Because of the large diameter of these mixers, which for efficient mixing will have an average of about 150–250 mm depth of concrete over the area of the pan, they are seldom made with capacities greater than 2 m^3. A mixer of this capacity will have a pan diameter of about 3 m.

All mixers of this type are quickly discharged by retracting a section of the bottom of the pan to allow concrete to be swept from the pan into a receiver. This feature results

Figure 36.4 Medium-size (30 m^3/h) plant (Benford P.B. 40). Note: aggregate stockpiles; overhead cement storage; ground batching; annular ring mixer

in their being mounted quite high in a batching/mixing plant. Each different make of this general type of mixer is claimed to have advantages over its rivals, but it is probable that the final choice will depend more upon such matters as the service afforded by the maker and delivery times. Wear and tear in pan mixers will generally prove to be higher than in a 'free fall' type but wear-resistant metals are coming to be more widely used.

Power consumption is somewhat higher than for 'free-fall' mixers, but this is compensated for by the higher hourly output in relation to the size of the mixer and its accompanying batching arrangements.

TROUGH MIXERS

This very heavy and robust type of forced-action mixer is more widely used in the production of asphalt and coated materials than for concrete though a number of them have been built into batching/mixing plants in recent years. They are often favoured

Figure 36.5 Large-size (110 m^3/h) *plant.* (*Courtesy: Stothert & Pitt.*) *Note: elevator charging of single batch overhead aggregate storage hopper; overhead and vertical silo storage for 2 No. cements; 1 No. 100p* (*2.25* m^3) *trough mixer feeding either* (*a*) *3 No. concrete pumps or* (*b*) *agitator trucks* (*via conveyor belting*) *as required*

by ready-mixed concrete suppliers and several of their plants incorporate mixers of this type. They can be used for mixing in much larger batches than can pan mixers since the effective depth of the concrete in them can be greater.

Mixers of this type may have a single or two contra-rotating shafts carrying blades which are so shaped and disposed as to move the constituents of the mix longitudinally along the axis of the mixer as well as round it. Mixing efficiency for short mixing cycles and hourly outputs in relation to the batch sizes are high for all types of concrete.

CONTINUOUS MIXERS

Continuous mixers proportioning by volume were used for the production of the major part of the very large amount of concrete used in airfield construction during the second world war and in the early post-war years. The generally low quality of much of the concrete produced by this type of machine could have been attributed to both the poor quality of the materials used and the absence of adequate control over the proportioning of all materials, including water, fed into the rotating-mixer drum. However, given reasonably good concrete materials and standards of control over performance, they could be made capable of producing a satisfactory quality of concrete at high rates of output and relatively low cost. More recent developments in machines of this type, which include proportioning of all the individual materials over a series of weight feeders together with the interlocks which close down proportioning and mixing operations on the malfunctioning of any one of the feeders, have resulted in there now being available batching/mixing plants which are capable of outputs approaching 300 m^3 per hour. The continuous flow of accurately proportioned materials through an inclined-trough type of mixer with high-speed contra-rotating paddles ensures very effective mixing with a relatively low power consumption in both proportioning and mixing units of the plant. The rate of feed of the materials can be varied over wide ranges and the retention time in the mixer can be varied by changing the angle of tilt. All types of concrete are quite effectively mixed since the feeders batch constituent materials in their correct proportions on to a continuous moving collecting belt. This ensures that each element of concrete passed through the mixer is correctly proportioned.

In many applications of batching and mixing plants—concrete road construction is one—a very obvious requirement of such a plant is that the time occupied in erecting, dismantling, transporting and re-erecting in a new location, and any heavy craneage required for this, should be reduced to a minimum. For this reason a number of manufacturers are building into their plants either selferecting facilities or features which will result in low costs and short downtimes for these operations. The use of harnessed electrical wiring assembled by no more involved a process than inserting heavy-duty plugs into sockets is a further aid to cost reduction.

For ease of transport it is sought to proportion units of the plant so that they are within acceptable loading gauges on public roads and can be moved without being broken down into smaller units.

Sizes of batching/mixing plants

A decision which will have to be made by a contractor at a very early stage in a contract is that concerning the size of plant to be installed. No doubt early planning of a job whilst preparing a tender will have given a clear indication of what the concreting programme is likely to be and to have set guidelines as to the number, capacity and siting of the batching/mixing plant needed to meet it.

In general it is considered preferable to concentrate concrete production at one central location in the area of maximum demand rather than at a number of dispersed points. This simplifies the problem of supply of materials and of storage. However, over sites to which access can be gained at a number of points it might be thought preferable to locate a number of smaller plants in strategic areas. The early provision of a network of substantially built temporary and permanent site roads will thus be of the greatest value. A further alternative is to provide more than one concrete production unit of smaller size at a central point. Although it will inevitably cost a good deal more to provide a number of small plants with a combined capacity sufficient to meet peak demands than it will one large one, the ability to continue operations when one section is out of commission is particularly attractive. A further point to be considered is that high-capacity single plants producing concrete in large batches need means for carrying these heavy loads around a site and the provision of adequate means for handling the concrete into position.

Mixing efficiency

In order that concrete should meet the requirements of a specification and be in accordance with mix design developed in the laboratory it is necessary that mixing should continue for such length of time that all the materials are uniformly dispersed throughout a batch. This time will vary with the intensity of the mixing action. Thus it is considered that the 'forced-action' types of mixer are able to achieve a higher degree of uniformity in a shorter time than can 'free-fall' mixers. However, for the reasons given, the capacity of 'forced-action' mixers is limited; hence the largest mixers are usually of the 'free-fall' type.

The time of a mixing cycle is governed by the speed at which the operations of charging, mixing and discharging can be carried out.

Of these three components of the mixing cycle the lengthiest is likely to be the second, though mixing is, to some extent, being carried out throughout the three.

Some specifications require that there should be a minimum lapsed time between completion of charging and discharging, this time being dependent upon the measured time required to achieve the degree of uniformity within the mix suggested in BS 3963[1]. It should be noted that BS 3963 specifies a test to be carried out on prototype mixers and gives a guide to the assessment of their performance, but does not define what constitutes satsifactory performance. The test is a complex one needing a good deal of equipment and testing expertise; it is not considered suitable for adoption as a site test.

American practice some years ago had been to lengthen mixing times as batch sizes increased, for instance, whilst $1\frac{1}{2}$ minutes would be required for a 2 cubic yard batch, that for a 6 cubic yard batch would be 3 minutes. Observation of the operation of a number of large tilting-drum mixers has indicated that these mixing times are not insisted on, yet, so far as can be judged, the uniformity of the concrete has not suffered from this curtailment.

READY-MIXED CONCRETE

About 40% of all the cement produced in Great Britain is used in ready-mixed concrete plants which are, by general definition, located off-site. But it would be difficult to assign any proportion of this amount to 'building' and 'civil engineering'. However it would appear that a higher proportion of the concrete used in building work is ready mixed as compared with that in civil engineering works.

A big advantage arising from the use of ready-mixed concrete on civil engineering works is that deliveries can be effected to locations close to the point of use, since the unit price quoted will normally be to site and, provided that a reasonable standard of access road to various points is available, this will include anywhere on the site. Thus, the means of transporting large quantities of mixed concrete from a central plant to points on the site are provided by the ready-mixed concrete operator.

Furthermore, the production of concrete in large centrally located plants should lead to a greater degree of uniformity, which will be increased by agitation of the mixed concrete during transit to the site.

However, ready-mixed concrete is not always as uniform as it should be; and it is generally a good deal more expensive than site-produced concrete.

A contractor using large quantities of concrete on civil engineering work will have to draw up a comprehensive list of points for and against the use of ready-mixed concrete, amongst which will be the requirements of the specification in regard to time between mixing and placing, and in the end he must use his judgement as to which offers, overall, the better proposition.

In this connection it is to be noted that ready-mixed concrete is not used on many major civil engineering works except as the occasion demands, to supplement the amount available from site batching plants when particularly large pours or high placing rates are called for. It is also widely used for the early construction of temporary works before central production units can be put into operation by the contractor.

On some road construction sites both in Britain and abroad, ready-mixed concrete operators have entered into subcontracts with main contractors to supply the aggregates and cement, mix the concrete and distribute it over the site in agitator trucks or other vehicles in the same way that specialist 'black top' contractors do with their bituminous material. Whether or not this will prove attractive will depend upon a number of factors, not the least the question of whether or not a main contractor can himself find the necessary amounts of aggregates in reasonable proximity to the site.

Ready-mixed concrete is supplied to jobs in a number of different ways depending largely on the preference of the operators, but perhaps on the terms of the particular specification in use.

Probably the most widely used, since it dispenses with the need for special mixing plant, is to dry batch all the materials in two or three operations into the truck mixer whilst the latter is turning at about 10–15 rpm below the batcher. On completion of charging, water is added as required and mixing continues at the plant for at least 100 revolutions (10–15 minutes); then mixing continues at about 1–2 rpm until arrival on site. A further 'dry' method is to batch in the same way but to charge the water required into a tank carried on the mixer. Only on arrival at site is the calculated amount of water added to the mix which then continues for a minimum of 100 revolutions (10–15 minutes) before discharge can commence.

In these two methods, mixing is carried out in the truck-mounted drum, which is driven at high or low speed as appropriate from either a power take-off or a donkey engine. The third method involves somewhat higher capital costs for the plant since this will incorporate a high-capacity mixer capable of completely charging a truck mixer or agitator trucks in one, two or three batches as necessary. The concrete is fully mixed at the plant and charged into the truck mixer whilst the latter is rotating at high speed. After complete charging, the concrete is 'kept alive' and prevented from settling and compacting in the drum by rotating at 1–2 rpm en route. Reserves of water for washing out after discharge are always carried by ready-mixed concrete trucks and it is with the aim of stopping the misuse of this water for wetting down the concrete to facilitate discharge and handling that operators are now stipulating that none of this reserve shall be used other than for washing down, otherwise any quality guarantee will be invalidated.

DISTRIBUTION OF CONCRETE

General observations

Whatever the scale of the work, the problem of distributing concrete around a site will arise and a contractor has a wide range of equipment available from which to make his selection. The value of using ready-mixed concrete, with distribution to a number of points already taken into account by the supplier in fixing his charges has been mentioned previously. However, on the assumption that concrete will be produced at a large central batching/mixing plant or a number of smaller plants on the site, methods for distributing around a site will need to be considered.

Where concrete is delivered from a mixing plant to the work by one mode only—for example truck mixer—and chuted direct into the work, this can be regarded as primary distribution only. Where it is conveyed for part of the distance by one means and for the most by another—as for instance an agitator truck, transferring concrete into a pump hopper, thence into the work—this can be regarded as primary and secondary distribution.

Generally, primary distribution will be by means of wheeled transport of one kind or another, but other methods for moving concrete in bulk over fairly long distances—as for instance pumps and conveyor strands—are coming to be used more widely as their design, versatility and standards of reliability are improved.

One problem which sometimes arises, particularly in warm weather or in hot climates is that changes in workability can take place during transfer and occasionally cause difficulties in placing. However, the use is becoming widespread of plasticisers and set retarders in mixes as means by which a consistency suited to proper placing can be maintained for quite long periods, certainly within the time to initial set of most cements.

In the following sections all the commonly used methods for distribution and handling are briefly described. However, as is so often the case, the personal choice of job planners based on their own experience and evidence from previous works, and of course, availability of suitable equipment, will play a major part in determining the precise form that primary and secondary distribution are to take.

Amongst the methods of distribution which will be considered here are:

(1) wheeled transport—for mainly horizontal movement (H)
(2) hoists—for vertical movement (V) only
(3) cranes of different types for H and V
(4) concrete pumps for H and V
(5) pneumatic methods for H and V
(6) conveyors mainly H, but also V
(7) monorail (H)
(8) cableways for H and V

Wheeled transport

LORRIES

The cheapest mode of transport for concrete is undoubtedly the tipping lorry but in general the volume carried and the way in which discharge takes place makes them unsuitable for most purposes. However, they are widely used in paving operations and as flat-bottom transporters for concrete hoppers.

Widely used forms of wheeled transport used for both primary and secondary distribution are dumpers of a range of different types and sizes, flat-bottom tipping

Figure 36.6 1-ton turntable dumper capable of depositing concrete over arc of 180° (Courtesy: Benford Ltd)

lorries (mainly for paving operations) and lorry- or trailer-mounted concrete hoppers. Lorries fitted with conveyor-belt bottoms or ejector plates (as in scrapers used for earth moving) are becoming available and could prove to be most valuable to a contractor in many functions.

DUMPERS

Small, hand-propelled dumpers—wheelbarrows and prams—still have their occasional use on even major works. Mechanisation of a simple form improves the rate at which concrete can be distributed by this means, but it should be noted that the increase in volume carried, which mechanisation allows, does call for rather more sophisticated means of access than just a series of planks serving for a barrow run.

Mechanical dumpers are supplied in a range of different sizes, generally geared to multiples of the batch size of the mixer they are serving. They are often plagued by the fact that the more workable concretes which are readily discharged from them are very easily spilled unless the haul roads along which they are used are of a reasonably good standard and straight. Clean discharge of the less workable concretes, which are not so readily spilled, is more difficult to achieve and may entail a good deal of stripping out. They also have the disadvantage that concrete is literally dumped from them in one mass; this can pose a number of problems such as the effect of sudden heavy impact loading on formwork and the displacement of reinforcing steel.

To overcome these problems a range of dumpers whose rate of discharge can be controlled hydraulically by the dumper driver adjusting the angle of tilt of the body, has been introduced. A further variant of the simple end-tipping dumper is that which allows rotation of the body to feed concrete over an arc of about 180°.

When used for loading a device for secondary distribution, such as crane skips of various forms, a high discharge level for the concrete is a marked advantage. Equipment of this type is now available with a discharge height of up to 2 m and can be used for filling a large range of skips and hoist buckets.

LORRY- AND TRAILER-MOUNTED TRANSPORTERS

For the distribution of concrete from the larger sizes of batching plant—say upwards of 25 m^3 per hour—it is necessary to consider units which are capable of carrying several cubic metres at one time. This implies the use of a chassis of not less than 5 tonnes carrying capacity. High discharge trucks from which discharge is assisted and controlled by the angle of tilt of the body and by a hydraulically driven paddle which propels the concrete towards the outlet, are available and widely used on sites.

Open-top agitator trucks of capacities up to 3 m^3, with facilities for agitating the whole truck load of concrete whilst in transit, were in favour a few years ago but in view of their high capital and operating costs they are coming to be much less frequently used.

On the other hand, a good deal of concrete is now moved around sites from central mixing plants by truck mixers. By this means, mixing of the concrete is continued whilst in transit, in just the same way as ready-mixed concrete would be. Since this method of distribution on sites is coming to be more widely adopted, the relatively high cost of ready-mixed concrete must be thought to be offset in part by the undoubted advantages which it affords.

In those cases where concrete is moved into its final position via crane skips of varying capacity, such as for instance in dam construction, it is common practice to load the skips themselves at the mixer and to transport these on towed trailer bodies to the pick-up point closest to the concrete's final position. These trailer bodies will often be fitted with cradles into which the crane skips can be easily located, since fully loaded skips will have a high centre of gravity and might well prove unstable on the trailers if difficult ground conditions have to be negotiated.

A number of different crane skips and trailer combinations will be available for consideration in the planning and execution of any particular works. Here again, job experience and examination of the advantages and disadvantages of works of a similar nature carried out in the recent past is probably the best guide to the system to be finally chosen.

Hoists

Hoists of various types are used solely for vertical movement of concrete and despite the competition of a vast range of different crane types, they still play an important role in building and civil engineering construction. They range from the rudimentary platform hoist capable of lifting two or three barrows or a pram full of concrete over a short distance to an automatic hoist of up to about 2 tonnes capacity with skip discharge into receiving hoppers at pre-set levels up to about 200 m. Where concrete in substantial volumes has to be lifted to a considerable height, over which travelling times are likely to be extended, the use of twin hoists is sometimes called for. Apart from giving better continuity of flow of concrete such an arrangement can ensure that work will not cease in the event of breakdown of a single unit.

Whilst most hoists currently in use call for the erection of a substantial tower which, for adequate stability may need to be tied in to the structure being erected at frequent intervals, there is for some applications, such as concrete chimney construction, a trend towards the use of rope-guided hoisting systems. Here the bottom works including hoist winches, etc., and the headgear are connected only by tensioned guide ropes. Since these guide ropes pass through eyelets on the outside of the hoist cage they serve to constrain the hoist cage, within close limits, to a vertical path.

Unlike the system which uses an extendable tower which can, within certain limits imposed by stability, be built up in advance of the structure, the guide-rope method requires that these ropes be extended as the work is raised. Thus ground drum storage of sufficient guide rope to reach the full height of the structure must be provided and so must the means for maintaining an adequate tension in the full length guide ropes.

Building operations generally call for the vertical transportation of personnel as well as the materials of construction. Major movement of personnel will generally be confined to fixed periods but materials of all types will be required throughout the day. For this reason some hoist systems have the facility for on and off loading of special wheeled concrete skips as required.

The design and operation of all loading devices are subject to rigorous regulations to ensure safety in operation and it is incumbent on site management to ensure that these are enforced and that detailed inspection of the whole of the mechanism is carried out at frequent intervals.

Cranes

Handling materials by means of cranes and skips is probably one of the oldest construction techniques known to the industry despite the many alternative and perhaps more sophisticated methods currently available. It would be unwise for any contractor to price work on the basis of using recently developed handling methods without giving full consideration to the use of well tried equipment such as the crane and skip.

The range of types of crane available to the construction industry is a wide one, consisting as it does of:

(a) derricks
(b) tower cranes
(c) cranes mounted on crawler base machines
(d) lorry-mounted cranes and wheeled cranes
(e) hydraulic cranes

The lifting capacities of each type also cover a wide range. Categories (d) and (e) have probably seen the greatest number of very recent advances but since their introduction some years ago both (b) and (c) have seen considerable changes in design and stepping-up of capacity. Advances have been made too in derrick design but basically, apart from changes mainly concerned with prime movers, they have remained unchanged for decades. The steam crane is now all but a museum piece; however, its ruggedness, simplicity and general freedom from breakdown still assures its place where electric power cannot be provided at reasonable cost nor is the standard of maintenance available likely to be adequate for diesel power.

DERRICKS

These fall into three basic types:

(1) stiff leg derricks
(2) guyed derricks
(3) mono-tower derricks

and all these three have their frequent uses on civil engineering works. However, type (2) are mainly used for special work such as the handling of very heavy, indivisible steelwork loads. They would rarely, if ever, be used for handling concrete.

Stiff leg derricks with carrying capacities up to about 15 tonnes and jib lengths to 45 m are, like all other lifting devices, restricted to loads much below their maxima when their radius of action approaches the maximum. For example, a derrick as above will only handle the maximum load at a radius of less than 30 m; above this figure it will be much reduced. The stays restrict the effective working arc to rather less than 270°, but this restriction rarely rules out their use, particularly when their range can be increased by mounting them on three bogie trucks set on two parallel pairs of rails.

The range of working height can be increased by mounting the lying legs, with their sole plates, on gabbards, which can again be either fixed or on rail-mounted bogies. Ground conditions are always an important factor in considering the stability of derrick cranes and they become of even greater significance when travelling derricks, mounted on tall gabbards, are called for. The need for soundly constructed rail tracks, laid to close tolerances, cannot be overstressed.

Where full-circle operation of a derrick is needed, then the mono-tower crane can be the answer, but such a machine will be restricted in working area swept by the jib from a fixed point. Mono-towers can be made to travel, but problems of the stability of a single moving tower makes this version much less attractive.

TOWER CRANES

Tower cranes were not widely used as an aid to building and civil engineering work in Britain until 1953 but they were coming to be widely used on the continent, particularly in France, for many years before this. Tower cranes, which can be mounted on fixed or movable towers are of two basic types: luffing jib and saddle jib. Both of them can be mounted on tracked base machines, which can be selfstabilising, or on lorries fitted with hydraulically actuated outriggers whereby the verticality of the tower can be assured.

With both the luffing-jib and saddle-jib types of machines, load-carrying capacity at increasing radius is greatly restricted and manufacturers' specifications should be studied before making a selection. Where very tall towers are to be used, these must be tied in to the erected structure at intervals of about 20 m. When the radius of action of a tower crane is increased by mounting the tower on rail-borne bogies, the same attention to accuracy in track laying—and of course its maintenance—(as for derricks on gabbards) is essential.

Lorry- and crawler-base mounted tower cranes have generally lower lifting capacities than have the static and rail-mounted types, but they fulfil a useful purpose. However, their high capital cost may restrict their use to those sites where a high degree of manoeuvrability from one working place to another, coupled with a reasonable speed of movement, is of greater significance than is high load-carrying capacity.

Restrictions need to be placed on the operation of any lifting device during high winds and this is particularly the case with tower cranes; this might be considered a serious handicap to their use. However, examination of meteorological and downtime records for the majority of sites shows availability to be generally upwards of 90%. The location of the work will of course have some bearing on this availability factor.

CRAWLER-BASE-MOUNTED CRANES

The attractiveness of this type of machine is that craneage is only one of a range of functions which the base machine can perform. Within a short space of time and with the appropriate equipment, it can be re-rigged as a face shovel or a dragline. However, to ensure stability when ground conditions are difficult, it may be necessary to use a machine with wider and longer tracks than would be needed for excavating functions alone.

The range of base machine available ensures their versatility as means of handling a range of capacities of concrete buckets, and their fast rate of slewing gives good output figures in this activity. The versatility of cranes of this type is much increased when the main jib is supplemented by a fly jib since their working range when operating close into a structure is thereby greatly increased.

The higher capacity mobile cranes are normally those available on crawler bases but there seems to be a certain amount of rivalry between makers of both crawler-mounted and lorry-mounted cranes to achieve the accolade of highest capacity.

LORRY-MOUNTED AND WHEEL-MOUNTED CRANES

The former is an independent selfcontained crane unit mounted on a multi-wheeled, multi-axle chassis of appropriate size which can be moved under its own power along public roads. The motive power and controls of the crane are completely separated from those of the lorry chassis on which it is mounted.

A wheel-mounted crane is one in which both the travelling of the crane over the ground and the various motions of the crane unit are controlled from one central point.

Figure 36.7 Hydraulic crane handling roll-over skip (Courtesy: Coles Cranes Ltd)

Because of their rather low ground speed these units are normally to be found working in such sections of construction sites as pre-casting yards, plant yards, etc.; they are moved from site to site on low-loading lorries.

The larger sizes of lorry-mounted crane with extended or heavy-duty jibs will normally be moved in at least two units, one carrying the crane mechanism and the other, sections of the jib and fly jib to make up the required mast length.

With all but the smallest loads for which these crane are used, it is necessary to use the in-built hydraulically actuated outriggers to achieve the necessary degree of stability and to relieve the lorry axles of the crane burden.

HYDRAULIC CRANES

The rapid advances which have taken place in the design of hydraulically actuated cranes with their multi-section telescopic jibs, have possibly outstripped those of almost any other type of lifting device during recent years. Their adaptability seems to be endless and new versions, and the uses to which they can be put, are encountered with astonishing frequency.

In this type of crane the only nonhydraulically actuated function is that of hoisting, but even here hydraulic motors can be used for the cable-drum drive. Slewing, luffing, jib extension and outrigger control are all performed by hydraulic cylinders and rams alone or in combination with a form of cable drive. The speed at which the requirements for a particular type of operation can be met make hydraulic cranes one of the most favoured items of equipment at the commencement of work on many sites.

Concrete skips and buckets

Whilst cranes of the various types already described can be and are used for a multitude of purposes on all construction sites, their role in the handling of concrete is that of moving containers charged in one way or another from point to point on a site. These containers are known variously as skips or buckets and there are several different designs to meet differing loading and placing conditions. In size, they range from capacities of about 400 litres to 9 m^3 which latter size has been used abroad in concrete dam construction, mainly in conjunction with cableways. Skips are of two basic types: the roll-over and the constant attitude. The first named is normally charged with concrete whilst lying on the ground in a horizontal position, close to the mixer or concrete transporter; it assumes a vertical position when hoisted by a lifting device. Concrete is released through the skip discharge in a controlled flow by means of a simple flap, actuated by a lever which is locked in position during transit on the crane hook to the point of deposit. These skips can be fitted with variously shaped outlets and deflectors which permit their being used to good advantage in filling columns and walls.

The design of a skip which will give clean discharge of concrete of a range of consistencies without recourse to hammering and rodding has been the objective of all manufacturers of this type of equipment but it is probable that the answer to the problem of clean discharge lies just as much with the user as with the designer of the equipment. Cleanliness and freedom from build up of concrete in either its wet or hardened state are essential if discharge problems are to be avoided.

Constant-attitude skips are charged with concrete whilst in the same attitude as they will be during transport and discharge. Generally they are of larger capacity than the roll-over skips, but there is no well defined range of sizes used for one type or the other.

In the larger sizes, where the weight of concrete above the outlet might be quite substantial, some form of mechanically operated device will be needed to open the gate for discharge. Simple geared clamshell gates are sometimes used at the lower end of the size range, but they tend to be rather slow in operation. For a faster and more positive gate action, pneumatically or hydraulically actuated rams are built into the structure of the bucket. Such mechanisms add substantially to skip weights.

For air operation, it is essential to have a pressure air line available at the point of discharge and of course, time is spent in coupling up and building up sufficient pressure to actuate the mechanism. Recently developed built-in hydraulically actuated gate-opening devices whose sources of power are hydraulic accumulators charged during hoisting, are said to give very effective control over the discharge of concrete, whether or not the whole batch is to be deposited in one place.

It is clear that the self-weight of a concrete skip or bucket is a factor of much importance to the user since the effective rate of operation of a lifting device will be controlled, in part at least, by this. For this reason, manufacturers have from time to time experimented with different materials for their construction. Very light weight has been achieved by fabricating in glass-reinforced plastic but because of inability to withstand the inevitable rough usage on sites, these skips have been far from successful. Significantly better concrete weight to total weight ratios are achieved by using magnesium in the manufacture of buckets, but only at very considerable additional capital cost. It should be noted that because of the chemical reaction between concrete and aluminium neither this metal nor its alloys are suitable for this purpose. Manufacturers of concrete handling equipment have, during recent years, acquired a good deal of expertise in devising different pieces of plant to suit particular purposes and may well be able to help a contractor who has some special concreting operation to carry out. Underwater concrete is such an operation; this might be more extensively used with the development of off-shore oil fields.

Concrete pumping

The distribution of wet concrete from point to point on a construction site by the expedient of moving it along a pipeline has long been the ambition of contractors in many parts of the world. First references to the developing and testing of equipment to do this are to be found in American literature, but little success seems to have attended these first trials. However, Dutch interest in the problem, following some German experimental work, resulted in the design of equipment which showed considerable promise and worldwide licences to exploit the idea were granted to no more than four commercial interests. The Concrete Pump Company was established in the UK for the purpose of producing and developing a market for concrete pumps but the first to be set to work were imported models. Home production started in 1932 and by 1939 upwards of 70 machines had been built, many for export.

In these first pumps, the concrete was moved down the pipeline by a piston, actuated mechanically by a diesel engine or electric motor. The energy-consuming process of accelerating a column of concrete in the pipeline, allowing it to come to rest whilst a new charge was fed into the pump cylinder and then re-accelerating the whole, soon came to be recognised as a marked disadvantage of the process. Better continuity of movement along the line was achieved by the German device, developed independently by Torkret and Schwing, of using twin cylinders, one charging whilst the other was discharging. Torkret machines used water as the operating medium, bleeding a small proportion away at each stroke to lubricate the pump cylinder. Schwing adopted oil as the operating medium, but of necessity used water as cylinder lubrication.

The upsurge of interest in the use of pumps for the distribution of concrete has resulted in there being a wide choice of equipment available from both the longer established manufacturers and from newcomers. All makes of pump are claimed to have some advantages over others, but users do not necessarily find the advantages claimed and tend to develop an affinity for the make which has given them good service.

The basic differences between various makes of pump are in the actuating medium—oil or water—and in the type of valve used—gate or flapper. So far as can be seen, all types of pump, when operated in the correct manner with suitable concretes, are capable of satisfactory performance but some of the plant servicing departments associated with contractors appear to have their own preferences.

Most of the early single-cylinder ram pumps were of 150 mm bore and suited to pumping concretes with aggregate up to about 38 mm maximum size, at rates approaching 12–15 m^3 per hour, dependent upon the extent to which the pump cylinder was fully charged at each stroke. However, smaller—75 mm and 100 mm—bore pumps were also available for pumping concrete with aggregate up to about 19 mm at rates of around 6–8 m^3 per hour.

Concurrently with the development of hydraulically actuated pumps in Germany and elsewhere, Challenge–Cook Brothers in America introduced an entirely new concept of a concrete pump which they designated the Squeez-Crete pump. In this machine a short length of flexible but highly abrasion-resistant 100 m diameter circular tube in the form of a U is charged with concrete from a hopper, this charging being assisted by maintaining a high vacuum in the surrounding chamber. Concrete is expressed from the flexible tube by rollers which, as they rotate on an axis and round the U from inlet to outlet, depress the tube, pushing the contained concrete forward towards the outlet. Since the rotating of the rollers along the U length of flexible tube is continuous, the very desirable objective of continous flow along the flexible pipe and hence the delivery line, is achieved.

Although 150 mm, 200 mm and even larger diameter pumps are in use, by far the majority of concrete is pumped through pipelines of 75 mm and 100 mm diameter. One of the main reasons for the use of the smaller sizes is that lengths of large-diameter pipe charged with concrete are very heavy and difficult to move on site. Thus, where large-diameter lines are used they will generally be associated with a semi-permanent pump and pipeline installation so that the advantage of the favourable area/wetted perimeter ratio of the large diameter line can be fully exploited. For flexibility of movement around construction sites, the smaller lorry-mounted and thus fully mobile equipment with pumping mains mounted on hydraulically actuated articulated booms is much favoured.

Generally, small-bore pumping equipment is used to handle concrete supplied either from ready-mix plants via truck mixers or concrete mixed on site and distributed in

Figure 36.8 Concreting a bridge deck with two No. concrete pumps fed from truck mixers (ready-mixed concrete). (Courtesy: John Lang and Son Ltd)

agitator trucks. By this means, the flow of concrete from the carrying unit into the re-mixer and feed hopper of the pump can be accurately controlled so as to keep the head of concrete approximately constant.

Where bigger pumping units are used they often form an integral part of a combined batching/mixing/distribution unit where facilities may be provided for alternative methods of distribution as desired.

As would be expected, the effort required to pump concrete vertically is a good deal greater than that for horizontal movement—in ratios varying between 10:1 and 6:1 according to the make. In addition extra effort is required to negotiate bends, which, when they are unavoidably incorporated into a pipeline should be of large radius.

When pumping at their maximum range, which is usually claimed as being from 250 to 500 m horizontally and up to 80 m vertically, the output from all pumps tends to fall away, sometimes quite seriously.

It is generally preferred to use small-bore lines for vertical and larger ones for horizontal transport. However, it is not considered advisable to change diameter along a length of pipe. Where a long horizontal movement of concrete is to be followed by considerable vertical movement it might be thought advisable to transfer into a smaller-bore pipeline, through a supplementary pump.

In view of these limitations in the scope of pumping operations, the scale of the work involved should be carefully studied in advance to determine what is likely to be the most appropriate equipment. The layout of pumping points in relation to reception points should also be arranged to keep distances as short as possible.

When concreting operations include the use of expensive pumping equipment, adequate planning to ensure that utilisation is as high as possible is essential. This implies that an aim should always be to have a sufficiency of work available to fully use semi-permanent installations on construction sites or to exploit the capabilities of hired-in pumps to the maximum.

Pneumatic placing of concrete

The use of compressed air to displace concrete from a container vessel along a pipeline to a concreting point was, despite the problems which followed the methods' introduction, quite a popular technique some years ago, especially for such work as tunnel lining. The new Woodhead railway tunnel built in the late 1940s was lined with concrete 600 mm thick, by means of a pneumatic concrete placer. The process, though at one time quite widely used as being rather better than some other methods of placing concrete, was never regarded with any great enthusiasm and the increasing availability and effectiveness of concrete pumps has led to a marked decline in popularity, except in some cases where it might have advantages—tunnel lining is still one of these.

When using either pumps or pneumatic placers for transporting concrete the need for careful control in maintaining the quality of a properly designed mix is paramount. It can be claimed that either pumps or placers are capable of performing satisfactorily with quite a wide range of mixes provided that the basic requirements of the concrete are met.

Conveyor strands

Conveyor strands are very widely used for moving a vast range of materials cheaply and effectively and so it is natural that there should be interest in using them for moving concrete. The throughput of a conveyor system in relation to the power used in moving it is probably more favourable than with any other method of distribution. This is probably as true with concrete as with any other materials, but there are problems associated with conveying concrete which do not apply with other materials, for example the tendency for coarse aggregates in the wetter mixes to separate out from the matrix and the need for elaborate and effectively maintained belt-cleaning equipment. There is, in practical terms, no limit to the speed at which a belt carrying concrete can be run so that quite narrow ones achieve high rates of delivery.

Whilst concrete for pumping has, of necessity, to have high workability, concrete transported on a conveyor strand can have a wide range of slump values, though it is likely that the wetter mixes could cause more problems than the drier ones. There are examples of concrete having been conveyed over long distances, but it should be borne in mind that they are generally open to the weather and that rain and strong sunshine even over a short time can affect the properties of the wet concrete.

By far the widest application of conveyor strands to the movement of wet concrete from one point to another has been in America. There, the fundamental requirements of a conveyor system—a high degree of flexibility and lightness of individual sections—seem to have been met and a high standard of acceptability achieved.

Perhaps the reasons leading to this extensive use of conveyors in America are to be found in the wide use of ready-mixed concrete and the more ready acceptance by engineers of very large pours, laid without movement joints, in one continuous operation. This latter trend is being noted in Britain but it seems likely that we shall maintain a rather more conservative attitude in regard to bay sizes. It is an essential feature of effective conveyor-belt operation that arrangements for belt scraping should ensure

Figure 36.9 Cableway with constant-attitude concrete bucket used in dam construction (Courtesy: J. M. Henderson & Co. Ltd)

complete emptying of the belt, down to the rubber, at each discharge point, whether from one belt to another or into the work. This implies the use of vulcanised rubber joints only and the immediate replacement or repair of any section of belting which is damaged for any reason. As mentioned previously, high-speed movement of wet concrete can bring about a tendency to segregation of the coarse aggregate from the matrix; to avoid this, each discharge point should be fitted with a hooded funnel within which the separated elements of the mix can be recombined before discharge on to the next section of belt or into the work.

Evenness of flow on to the conveyor is an essential to effective operation and a form of belt feeder to give this uniformity is a worthwhile investment, even though it is another piece of equipment to be maintained.

In some classes of work the final length of belting is a short one to give discharge over a fairly limited area without of necessity moving the whole system. In others, scraper blades are used to sweep a belt clean of concrete at any point along its length and then to direct it into the work. Yet another method is to use a pivoted conveyor on to which concrete can be deposited at any point along its length. This equipment gives a very wide range of placing facilities by the movement of one belt only of a system.

Monorail

Monorail transporters are still to be seen around civil engineering sites and often prove to be the only feasible method of transferring small volume of concrete from a mixing station to the works.

The simple monorail track can be laid on load-spreading planks over difficult ground and the automatic tractor trailer unit capable of carrying about 0.5 m^3 of concrete can be shuttled between the two end points to give a reasonably fast rate of delivery, depending of course on the distance to be travelled. Concrete tipped out of the 'jubilee'-type skip onto a banker board is usually set into position by hand.

Cableways

Cableways, working singly or in pairs, have been one of the principal methods used for placing concrete in many of the world's major dams across valleys of various profiles. To give good coverage of the plan area of a dam, cableways are often set out with a head mast fixed in one position but capable of being pivoted to an angle of about 10° from the vertical. The tail mast is normally mounted on a rail-borne carriage which moves over an arc. The stability of both head and tail masts need special consideration in the light of the load to be carried—concrete and containing skip.

To ensure a steady discharge of concrete from the skip and thus gradual return of the cableway to its unloaded bucket condition, special devices for controlling the rate of discharge are essential. These have been discussed.

Cableways will normally be controlled by an operator from the headmast position, acting under telephone guidance from loading and unloading points; discharge of concrete will be controlled locally at deposit points.

PLACING AND COMPACTING CONCRETE

Placing

Wet concrete is set into the position in which it is to harden with the aid of crane skips, pump lines, conveyor belts, etc., and the primary objective of placing techniques should be to avoid the need for extensive subsequent movement from the point of deposit to its final position. This will normally be within shutters of one form or another.

In the majority of work, the first pours of concrete will need to be set on or against excavated or filled ground and, in order to avoid contamination of the structural concrete, a thin veneer—up to 100 mm thick—of rather lower quality blinding concrete is laid over the formation. Besides preventing contamination, the blinding concrete, which can be set to reasonably accurate levels, can be used as a working platform for the erection of reinforcement where required. When the risk of contamination is low, for instance where concrete is to be built up on a rock formation, reasonably effective cleaning of the rock surface should be carried out but over-insistence on a high degree of cleanliness should be avoided. It is most unlikely that any structural design will include the need for a substantial degree of bond between the formation and the structure.

Wherever it is allowable, the cheapest way of filling concrete into deep excavations is via inclined chutes so positioned as to take concrete direct from a transporter into the work. Purpose-made chuting of light weight but adequate stiffness, achieved by having a narrow but deep section, and supported as necessary, should always be used and generally it should be at an angle greater than about 30° to the horizontal. The range over which concrete can be chuted is sometimes increased by raising the transporter on a specially built movable platform, if necessary, approached by a ramp.

Frequent re-siting of the chutes to avoid too great a build-up of concrete at the discharge point should be aimed at—hence the need for light weight. Alternatively a number of chutes and loading points can be arranged around the work and used in suitable sequence.

Where heavily laden concrete transporters are brought close in to the sides of an excavation to give maximum range for chuting operations the ability of the excavation support to sustain the heavy surcharge should be checked.

When concrete is to be filled into excavations in which water is rising, it will often be necessary to conduct this water through channels around the periphery of the work to low points or sumps from which it can be raised clear of the work. In some instances it will be necessary to install a series of pipes to raise the water with the work prior to grouting and sealing off the pipe when the work has been raised above the water level.

Concrete carried out in series of lifts should have the top surface of each prepared by removing laitance whilst the concrete is still relatively unhardened yet not at risk from the action of water jets or other device used in surface preparation.

With reinforced work, it is particularly important to remove all scraps of tying wire and other debris from joint planes by means of compressed air/water lances or other devices such as suction pipes prior to concreting lest unsightly joints additionally marred by rust stains, should develop.

Concrete should be fed into shutters by a method which will give uniform distribution along a section in layers some 350–500 mm thick, each layer being placed and compacted before a succeeding layer is placed. By this means, the forming of unsightly segregation planes in the concrete as coarse aggregate separates from the matrix will be avoided. There will also be a reduced risk of displaced reinforcement resulting from unbalanced local concrete pressures. As successive layers are placed, each should be properly merged with the preceding layer by shallow penetration of the compacting device, generally an immersion vibrator.

It is becoming common practice to pour relatively thin sections of concrete walling and columns in which the heat gain resulting from placing large masses of concrete will not present a serious problem, in heights up to 10–12 m. When pours of this height are adopted, then it becomes wholly impracticable to use often recommended devices such as full-depth trunking through which to place the concrete, but it may be found advantageous to use special deflector plates, either on the discharge from the concrete skip or at the top of the shutter to direct the bulk of the concrete vertically downwards. There is always a risk that chuting concrete directly into walls between surface reinforcements will result in a certain amount of aggregate separation. This should be of little consequence provided the concrete is designed to be as cohesive as possible and to contain a slight excess of sand.

When walls are poured in deep lifts there is always a risk of excessive water gain at the top surface due to the expressing of water from concrete in the lower levels under the high pressures existing there. This water gain, which can often result in a lower-quality concrete marred by less completely closed surfaces and perhaps colour change, is best countered by using somewhat drier batches of concrete as work approaches the top of the lift.

When more massive sections are to be poured and there is the risk of excessive heat gain, then it will, almost without exception, be necessary to restrict the total height of a pour to some 2 m, irrespective of the use of low-heat cement and/or cement extenders such as ground slag and PFA. A further requirement might be that the area of a section should be carried up in a series of stepped layers, each not exceeding 350–500 mm. This matter is referred to later in considering concrete dam construction.

Joggles set into the top of a concrete lift ostensibly to give a key for succeeding layers are a potent cause of trouble at horizontal construction joints since they are difficult to clean properly and hold water in excess; they should therefore not be used. Instead, joint planes should be slightly crowned to shed any water used in washing down prior to concreting.

Whether or not to use a cement grout or a cement/sand mortar of creamy consistency at joint planes has long been a subject of controversy. Some engineers prefer to use a thin layer of mortar over the area of a joint prior to concreting; others prefer to have the surface dampened but to use no mortar. If mortar is used it is difficult to ensure that it is only thinly spread and worked well into the top of the previous lift; the excessive amounts which may accumulate can give rise to an undesirable amount of shrinkage at the joint plane and perhaps slight colour differences. It is probably wiser to omit mortar—except where a thin (6–12 mm) layer can be properly brushed into the hardened concrete of the previous lift—and to rely on complete compaction of the lowest layer of concrete to give the desired quality of construction joint with as high a degree of bond between the two as it is possible to obtain.

Joints in water-retaining structures

The extravagant use of water bars of different types in concrete construction below ground is perhaps indicative of engineers' and architects' lack of confidence that joints can be made to resist the passage of water without their use. They may be desirable and perhaps necessary where water pressures are high but in many instances it has been found that joints incorporating water bars have tended to show traces of water seepage, sometimes severe, whereas those without such a device are much less prone to trouble.

When such faults are investigated it is generally found that the mode of installation has been faulty rather than that the water bar has been inadequate. They must be installed in properly designed movement joints built precisely in accordance with the drawings; construction methods must be such as to ensure that this is done. Horizontal joints in particular, call for a method which will ensure that correct positioning of the bar is maintained; in vertical joints they must be so secured in position that they remain sensibly normal to the joint plane.

Where joints are built in accordance with known good practice it will rarely be necessary to install water bars in horizontal joints when the effective head is less than 5–6 m; they may, however, be needed in vertical joints as indicated. Joints should be detailed as shown in Figure 36.10. To avoid the need for water bars in the vertical joints of, say, a basement perimeter wall it will be preferable either to cast *in situ* or pre-cast sections some 5–6 m long leaving 1 m gaps between sections. These gaps will be concrete with as dry a mix as can be effectively placed after an interval of more than about three weeks. An effective flexible seal built on to the pressure side of the wall should then assure water tightness.

Underwater concreting

In many civil engineering works it will be necessary to place concrete in situations where it will be under water either all or most of the time. Offshore work between tidal limits are instances in which concrete may have to be placed during a short period of slack water and protected from the effects of scour very soon after placing and whilst still

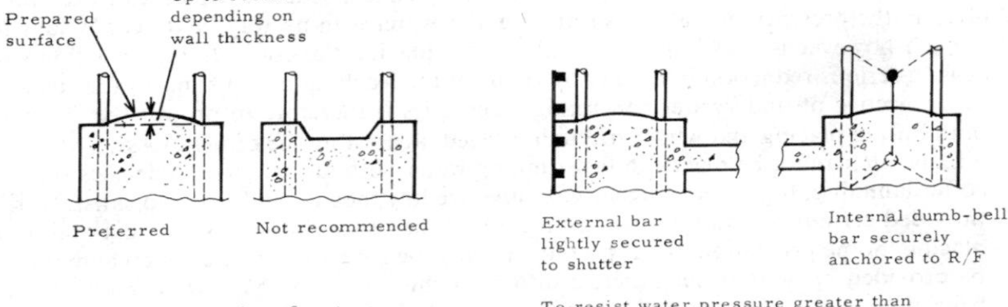

1) HORIZONTAL CONSTRUCTION JOINTS

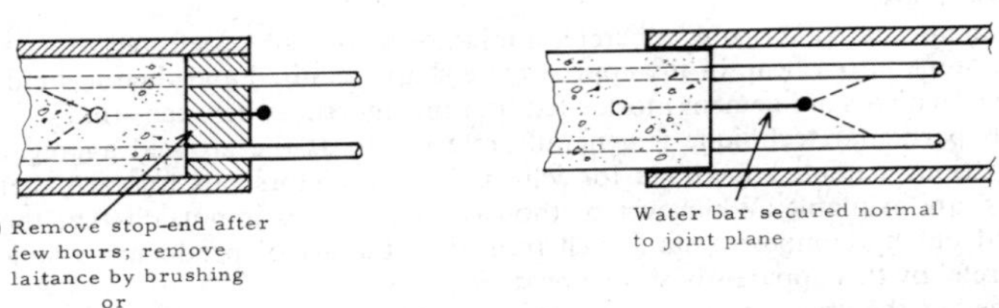

2) VERTICAL JOINT TO RESIST WATER PRESSURE IN CONTINUOUS CONSTRUCTION

External face

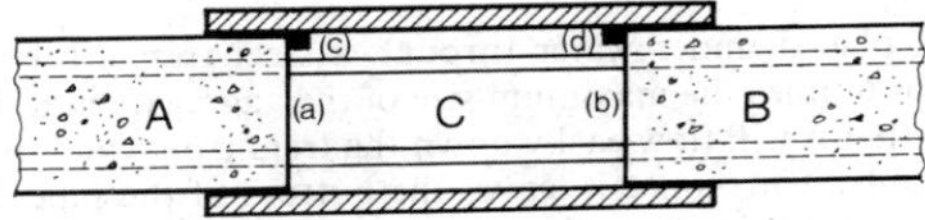

1) Prepare (a) and (b) as in continuous construction
2) Interval between A-B and C not less than 21 days
3) Seal vertical joints at (c) and (d) under favourable conditions

3) VERTICAL JOINT FOR NON-CONTINUOUS CONSTRUCTION PRESSURES UP TO 5 METRES WATER BARS INSTALLED ABOVE THIS PRESSURE

Figure 36.10 Details of construction joints.

unhardened. Other work carried out in tidal and nontidal waters will call for the placing of concrete in parts of structures which are permanently under water. This section concerns itself with this type of work.

Since the hardening of concrete is a purely chemical reaction which can only take place in the presence of water, its setting under water is in no way inhibited. What can happen however is that improper methods for placing the concrete into position can cause a serious reduction in quality by virtue of the leaching-out of some of the cement. The objective of underwater concreting techniques is therefore to protect the concrete both during placing and whilst still unhardened, against undue cement loss.

Concrete cannot be placed in fast-running water without recourse to devices such as cofferdamming, but where current velocities are low then it can be placed satisfactorily provided certain precautions are taken both in the proportioning of the mix and in its placing. Some protection against scour or too wide spreading of the concrete mass can be provided by setting the concrete into steel shutters or perhaps walls built up with bagged concrete. Shutters will normally be used if reasonably accurate concrete levels and shapes are to be achieved by rough screeding carried out under water by divers. Where the shape of the mass of concrete is of no particular concern its only function being to provide a firm base from which the structure proper can be erected, then shuttering is often dispensed with and the concrete mass is allowed to adopt its own angle of repose.

As an alternative to using concrete, it might sometimes be preferred to set single-size coarse stone into a heap of the approximate shape required and then to bind the mass together with a sand/cement mortar fed into the interstices between stones.

This particular technique is generally referred to as the grouted aggregate process and various patented methods for which the proprietors put forward their various claims, are available. Whichever method is selected it is important that the work be carried out by competent staff, well trained in the art of producing a solid mass of 'concrete' by this apparently simple process.

Whether the choice of an underwater concreting technique falls on the use of conventional concrete or on a grouting method, the clearing of unsuitable material from beneath the structure will be essential. It can be carried out by means of a diver-directed grab, suctionpipe or air lift pump. In those situations where resilting will rapidly take place, concreting should be started as soon as possible after the base has been cleared.

METHODS FOR PLACING CONCRETE

Concrete may be placed under water through a steel tremie tube which will have a diameter some 5 or 6 times the maximum size of the aggregate in it. It will preferably be very workable so that it will flow easily down the tremie pipe and over a large area—up to a radius of about 2.5 m—once it leaves the bottom of the pipe. Where bigger areas than those are required it is normally considered better practice to use more than one pipe or to repeat the whole tremie concreting process at a number of points rather than attempt to cover a large area by moving a single pipe from place to place during one operation.

In operation, the tremie tube, which will often be made of a number of sections to allow easy adjustment of its outlet height above the base of the work, will have a receiving hopper at the top. This hopper, together with the tremie tube, will preferably be slung from a crane or overhead structure and fed with concrete from either a crane skip or a pump.

At the start of operations, the tremie pipe is set hard down on the base of the work, water rising up the pipe. A travelling plug of one of a variety of types is set into the outlet from the concrete hopper. As concrete is poured into the hopper it forces the plug down, displacing water from the tremie and preventing the concrete from falling directly through the water. When the plug has been driven down to the bottom of the tube, the tremie is lifted slightly whereupon the weight of concrete finally forces the plug from the pipe, allowing concrete to well out in all directions, including upwards. Further concrete passing down the tremie is encouraged to flow outwards by slightly raising the pipe, but without allowing its end to lift clear of the mass of concrete.

When an area has been brought to the required height, the tremie should be cleared and moved to a new location, where the sequence of operations will be repeated.

For concreting in deep underwater lifts, which will require reducing tremie lengths, provision must be made for supporting the lower lengths of tremie tube whilst adjustments are made.

An alternative to the use of a tremie tube is a special type of bottom-opening skip. This will be fitted with a canvas cover to protect the concrete during lowering through water and with an enveloping metal shroud which will restrain a sudden surge of wet concrete as the flap-type doors are unlatched, only allowing it to flow as the skip is raised slowly off the bottom or away from previously laid concrete.

Generally speaking, skip placing, because of its intermittent nature, will be technically less satisfactory than tremie placing. Also because of the high cost of such skips and the relatively slower rate of placing, the economics of the method may be less favourable than for tremie work. However, where tolerably accurate surface levels without recourse to heavy underwater screeding are required, work may well be easier with skips. Whatever method is adopted, a consistent supply of concrete which has a sufficiently high cement content to cater for the inevitable loss through leaching is essential to the successful completion of this type of work.

Grouted aggregate method. A useful material for this class of work will be graded 76–38 mm and have not more than 10% fines; this is often available as rejects from concrete aggregate processing screens. It should be free from any clay and dusty coatings.

Setting stone into position on a prepared base and grouting should be carried out as quickly as possible, lest silt be deposited over the mass of aggregate or other forms of contamination, such as algae growth which would prevent adequate bond between matrix and stone, should develop. An advantage of using this technique in flowing water is that in passing through the stone mass its velocity will be lowered; this will reduce the likelihood of serious cement leaching.

When using this method of underwater construction the base of the structure should be cleared of any deposits of silt, with an air lift, immediately prior to laying a 76 mm thick bed of sand or pea gravel. Grout pipes are set vertically into the area to be concreted at intervals before stone is tipped round them. These pipes are subsequently coupled through flexible hosing to the grout pumps.

As soon as stone setting is completed—no compaction of any kind will be required—a cement grout of creamy consistency is pumped in turn into the grout pipes and followed with a $1:1\frac{1}{2}$ cement/sand mortar, again of an easy flowing creamy consistency. Grouting will normally start at the lowest point in the mass and as this area is seen to be filled, then pipes can be moved further into the mass of stone until such time as the stone and matrix become a solid mass. When grout tubes are lifted in the work to ease pump pressure, care should be taken to ensure that they are always kept at a level slightly below the grout plane. Underwater work carried out by this method alone is not capable of producing a fair surface, but in combination with normal tremie- or skip-placed concrete it will make a first-class base.

Compacting concrete

The relationship between the degree of compaction of concrete and its compressive strength has been indicated in a previous section and the need for achieving a densely compacted mix will be apparent. What is not sufficiently appreciated is that even highly workable concrete mixes need to have some work done on them in order that they should have an adequate standard of compaction.

With the more workable concretes which are now widely used, the effort required to achieve a satisfactorily high degree of compaction is minimal. Hand punning and, where feasible, foot puddling, is admirably suited to the compaction of such concretes. It is also to be noted that the steady slight up and down movement of the surface of concrete so treated against formwork caused by insertion and withdrawing of a punner tends to reduce the incidence of unsightly air bubbles at shuttered surfaces. However, under present conditions it is difficult to gain acceptance of the idea that compaction can be satisfactorily carried out other than by the use of shutter vibrators, or, generally to be preferred, immersion (poker) vibrators. This being so it is essential to ensure that vibrators are used properly. Though neither type of vibrator should be used as a means of moving concrete other than downwards, i.e. compacting it, some movement horizontally will inevitably follow their use.

Concrete having been set in the shutters as described in continuous layers about 350–500 mm thick along the length of a section, it should be compacted by feeding the vibrator vertically down through the depth of the layer and not more than about 70 mm into the next lower layer. When it is clear that all air has been sensibly expelled from the area being compacted the vibrator should be slowly withdrawn and plunged into the next section of the work, this process being repeated along the length of the section.

The proper compaction of concrete into highly reinforced zones such as the ends of prestressed concrete beams has always been a problem. The concentration of reinforcing bars and hoop steel round cable anchorages is sometimes such that it is virtually impossible to insert even the smallest-diameter poker vibrator. In such cases external vibrators securely fixed to stiff shutters should be used and moved up the work as concreting proceeds. Honeycombed concrete in the areas of highest steel amounts will be avoided if concrete of a suitable workability is filled into the shutter at a very slow rate only, so that visual inspection will be able to reveal areas of inadequate compaction.

In some situations where suitable lifting devices can be made available, it might be considered to pre-cast the anchorage block on end so that immersion vibrators can be used in the direction of the main reinforcing steel and ducting and not across them. Proper treatment of the top surface of these end blocks, as described, should ensure their proper bonding into the beams with no risk of loss of structural strength of the whole.

Curing concrete

In order to achieve full potential strength and to reduce the amount of drying shrinkage and moisture movement to the lowest possible levels it is necessary to carry out the 'curing' process. This connotes a method whereby the amount of water in the newly placed concrete, which is always higher than that required to fully hydrate the cement present, will be retained over a period of at least several days.

Curing large flat areas such as road slabs and monoliths in plain and reinforced concrete work presents no great problem since, as is indicated in later sections, the former can be effectively cured with an impermeable membrane sprayed onto the surface after finishing and the latter with hessian or similar material maintained in a damp condition by means of water sprinklers or occasional drenching with water.

Vertical construction, however, presents a problem which is more difficult to solve. With shuttered vertical faces, the evaporation of moisture from the concrete is effectively barred by the shutters themselves but the economics of construction will generally require that these be used as frequently as possible; this involves removal for re-erection as soon as the concrete has achieved a strength adequate for selfsupport. At this stage, the concrete is liable to lose moisture to an extent which can result in a good deal of surface crazing and shrinkage cracking. Particularly is this the case if the concrete attains a temperature within the mass which is a good deal higher than the ambient.

Thus for vertical construction, curing of some form should be undertaken as soon as possible after the shutters have been removed. This will preferably consist in draping sections with hessian which can be kept damp by frequent spraying with water. Alternatively a spray bar of plastic tubing perforated at intervals with small-diameter holes can be laid along the top of the concrete and connected to a supply point. It is unnecessary to use an excessive amount of water in this process and in fact it is not advisable since underfoot conditions can thereby be made more unsatisfactory. Polythene sheeting is often used as a barrier to the movement of moisture from concrete but unless it is properly secured to the concrete members with adequate ties it may well be less effective than making no attempt at curing whatsoever, since it encourages the more rapid circulation of air around the member.

In some instances contractors have sought to cure vertical surfaces by spraying with an impermeable membrane as used in road construction. This method is not considered to be a satisfactory one since few of the membranes will disintegrate and fall away from the structure after a short time; they will thus prejudice the appearance of the concrete. In many cases their use will severely reduce the bond between concrete and applied finishes such as plaster and cement renderings. When it is considered desirable for the sake of expedience to cure concrete surfaces which are to receive finishings with a membraneous compound, then all traces should be removed by heavy wire brushing or, as has sometimes proved necessary, a more drastic treatment of sand blasting or bush hammering.

The major part of the concreting operations on civil engineering works will be covered by the production, handling, placing and compaction methods outlined in the preceding subsections. However, there are some works where guidance as to the construction methods available to a contractor is perhaps desirable. Road building, which requires the use of highly specialised equipment different from that used in any other form of construction, is clearly one calling for detailed description; this follows in the next subsection.

CONSTRUCTION OF CONCRETE ROADS AND AIRFIELDS

General observations

In Great Britain, only a very small proportion of the total length of motorway and major road construction in the past 25 years has had concrete as both the load-bearing element and the running surface: indeed the proportion has rarely been as high as 10% in any year, and the average has been nearer 5%. Currently (1973) the figure is running at about 25%. This contrasts with the situation in many European countries where the proportion is close to 50%, and with the United States where it is rather over 50%.

Although the amount of concrete road construction has been small, there has been a steady trend towards improvements in the quality of concrete roads in the post-war years. Because of the general adequacy of design, their structural performance has rarely been called into question; riding quality has nearly always been the factor giving rise to adverse comment.

Records of surface profiles of roads laid in recent years indicate that there is now little to choose between the best of the bituminous work and the best of the concrete, but there seems to be some intangible factor which, in the view of many road users, makes a ride on a concrete road inferior to that on asphalt.

Machine-laid concrete roads, which have all but replaced hand-laid work throughout the world, are constructed by two different methods. The first uses forms both to contain the unhardened concrete slab and to support rails on which the road building machines—spreaders, compactors and finishers—are mounted. In Britain this is sometimes referred to as conventional construction; it is often favoured where reinforced slabs are required by the specification. The second method is slip form paving and, as the term implies, the formwork in which the unhardened concrete assumes the required shape of the road slab moves forward as an integral part of the paving machines, leaving the concrete unsupported after only a very short period of time.

In Britain and Europe, conventional construction has held sway for many years, but slip form paving is beginning to make its mark. It is believed that the present Department of the Environment specification option which allows the use of unreinforced but multi-jointed concrete slabs in lieu of reinforced slabs with many fewer joints will encourage this trend though it is likely that the longer established methods will have their strong supporters for many years. In America, slip form paving is relentlessly taking over from conventional work and it seems likely that before long there will be little other than this type of concrete road paving work there.

A high proportion of all the runways, taxiways and loading and parking areas for heavy transport aircraft throughout the world are built of concrete of varying thicknesses, reinforced and unreinforced as the design requires. The same methods as are used for building roads can be used for runways, etc., but as thicknesses are often a good deal greater, construction problems may be somewhat different. For example, slip form paving which is successfully used in road construction may not be so attractive a method for building thicker runway slabs because of the greater risk of serious edge slumping. The greater thicknesses involved may also require that compaction from the surface be supplemented by immerion vibrators.

Conventional construction

This class of work is described under the following headings:

(i) forms and form setting
(ii) trimming of base and laying sliding membrane
(iii) setting of expansion and dummy joint assemblies
(iv) concrete production and transport
(v) spreading concrete
(vi) laying reinforcement—where required
(vii) compacting and finishing concrete
(viii) joint forming and sealing
(ix) texturing and curing of road surfaces

FORMS AND FORM SETTING

All formwork for machine-laid concrete roads should be made from steel plate of at least 5 mm gauge and have an adequate number of stiffeners. They should be fitted with heavy-gauge rails on which construction machines, weighing up to about 12 tonnes when loaded, can be run—and firmly fixed to a concrete base about 100 mm thick at least 3 points along each 3.05 m length. This base is preferably laid with a wire-guided datum laying machine or similar, capable of laying a strip of good quality concrete as wide as the form base, to which the form work can be secured. Normally, laying these base strips to high standards of accuracy should ensure that the forms and rails on which the machines are to run are accurately positioned. Since finishing machines are required to be of the articulated floating beam type (see (vii) on finishing) the accuracy of the top edge of the formwork is not of great importance, but it is often an advantage to have a square, rather than a rounded edge to the formwork. When the forms available are shallower than the slab thickness they can still be used since the concrete base on which they are laid can be any thickness greater than about 100 mm. To achieve a balance between adequate rigidity and lowest cost, form depths will generally be about 175–200 mm.

In those instances where the design of the road calls for the use of a lean concrete base it might be considered preferable to dispense with the concrete form base and use full-depth forms (generally slab depth less 12 mm) laid direct on the lean concrete which will extend beyond the edge of the slabs. This, however, is a matter for cost study by the contractor. It also implies laying the base concrete to a high standard of accuracy.

It should be borne in mind that very high loads are imposed on formwork when spreaders are loaded by side-discharge trucks. If these are used, care should be taken to ensure adequate strength of the form/base combination so that there is no deflection under these impact loads.

TRIMMING OF BASE AND LAYING SLIDING MEMBRANE

Bases to concrete slabs should be accurately laid to ensure that slab thicknesses are correct. Where the bases are of a granular nature they can be set slightly low and trimmed upwards with fine material to be compacted by roller. If the stronger lean concrete bases are required then they should be laid as accurately as possible with no positive tolerance. An accurately laid base will result in lower shrinkage and temperature stresses developing in the slab than would otherwise be the case. These stresses are sometimes required to be reduced by laying a membrane of polythene or heavy gauge kraft paper. A membrane is perhaps not quite so important when a granular base, trimmed with fines, is used.

SETTING OF EXPANSION AND DUMMY JOINT ASSEMBLIES

Under British conditions, which have now fallen into line with long-adopted continental and American practice, expansion joints are not required in concrete roads laid between 21 April and 21 October. It is a natural assumption that contractors will hope to avoid laying concrete in the period October to April, and so expansion joints will not normally be required. However, a full complement of contraction and warping joints is required in the designs. It is essential to have these extensively prefabricated so that setting up at the correct spacing can be carried out expeditiously.

In recent years a dowel bar setting device has become available. This machine provides the means for accurately positioning bars at joints and forcing them down to their correct position within the road slab by vibratory means. The performance of this machine is said to be highly satisfactory as regards both the accuracy with which bars are placed in position and in cost comparisons with methods used hitherto. It should be noted that this device is not thought to be suitable for building dowelled expansion joints.

CONCRETE PRODUCTION AND TRANSPORT

Concrete for road construction can be produced by any of the methods previously described. The method of construction will dictate the means of transport of concrete but it should be noted that high-capacity end-tipping lorries are likely to result in the lowest overall transport costs. Where it is elected to use box spreaders it has generally been considered necessary to use high-discharge side-tipping lorries whose capacity will be no higher than that of the spreader they are loading. However, various devices have been used by contractors to enable them to use the more economical end-tipper lorries for this purpose.

A matter of some importance when operating high-discharge side-tipping lorries is that, in order to achieve rapid and clean discharge, the load must be quite high above the ground. Concrete laden lorries might be rather unstable at speed unless haul roads are well maintained.

The exposed surface of a lorry load of concrete can be noticeably dried out or wetted by exposure to weather during long hauls and a cover should be provided with each lorry so that it is available for use as required.

The overall rate of paving will depend upon a number of factors, including the programmed time for paving with due allowance for unfavourable weather, and the rate at which supplies of aggregate and cement can be made available, supplemented as necessary by materials drawn from stockpiles accumulated during nonpaving work.

The size of batching and mixing plant required will be governed by the expected rate of forward progress, slab thickness and width of road to be built.

Various alternative combinations of lane widths are allowable and often the overall design, on motorways at least, is simplified to the extent that the surfacing of hard shoulders can be concrete rather than a contrasting bituminous material. Where the design and specification of the road structure allows, it will often be found that construction costs can be reduced by paving in equal widths so that batching, mixing and transporting concrete is carried out at the same tempo throughout the work, with constant width placing and finishing equipment.

The prime requisite for high-quality paving work has been found to be a steady rather than a high rate of progress, but there may be some advantages, so far as quality is concerned, in aiming at a higher rather than lower overall speed of construction. What has been found from past experience is that the irregularity index (q) for the road surface is closely linked with the average rate of construction, low q values being associated with high average rates of uninterrupted progress and high values with sporadic working. This implies that the quality of work will be improved if preventive maintenance is carried out diligently to reduce breakdown time of plant to the minimum; the effect of this on the economics of concrete laying will be clear.

SPREADING CONCRETE

The importance of proper spreading of concrete as a factor contributing to the production of a good riding quality of a concrete road with the minimum effort in finishing cannot be overstated. Ideally the equipment used should be capable of so spreading concrete that its loose density as spread over the formation does not vary by more than about 35 kg/m^3 from point to point.

Concrete is almost universally spread by means of hopper spreaders fed from side-tipping lorries of various designs. The outlet from the hopper may be controlled by a helmet gate or by contraction to a narrow opening through which concrete is passed by vibration. With some models, concrete is tipped into the hopper whilst this is in a roughly horizontal position with the outlet remote from the point of loading. For spreading, the hopper is brought into the erect position and only when this is done does concrete flow from the outlet.

The density of concrete as spread is, to some extent, governed by the head of concrete in the hopper and by passing a loaded hopper with open outlet over laid concrete; it is for these reasons that the uniformity of spreading concrete might not be so good as the ideal suggested.

Certain hopper spreaders can be made to function with the hoppers mounted square to the line of the road as opposed to along the line of the road. They can then be loaded by means of end-tipping lorries backed up the formation; concrete is spread in lanes along the line of the road rather than in bands transverse to it.

This latter system involves the use of a joint assembly which can be very readily and securely set into position on the formation without interfering with the movement of concrete lorries to and from the spreader. It should also be noted that these lorries are liable to cause damage to the formation and any sliding membrane laid on it.

LAYING REINFORCEMENT

When reinforcement is used in road slabs it usually takes the form of welded mesh and is set at about 60 mm below the running surface. Three methods for laying are commonly adopted:

(a) Concrete to the level of the reinforcement is laid and compacted. This operation is normally carried out by the first of two spreaders and compacting beams. After reinforcement has been laid as required, surface concrete is laid and compacted by the second pair of machines. Note that the second spreader will have only about $\frac{1}{3}$ to $\frac{1}{4}$ of the throughput of the first.

As an alternative to this method the reinforcing mesh is laid on the uncompacted lower concrete and depressed into its surface by the beam of the compacting machine before laying the surfacing.

(b) The reinforcement is laid in advance of concreting operations, supported at the correct level on closely spaced stirrups set on the formation, the rate of stirrupping

Figure 36.11 Conventional reinforced concrete road construction. Photograph shows from foreground: concrete edge strip carrying rails; sliding membrane and joint assembly; spreading base concrete; laying reinforcement; compacting base concrete through reinforcement; spreading surface concrete; compacting surface concrete; finishing surface with diagonal finishing; testuring and curing (tenting in background)

being about $1/m^2$. When this method is adopted, the spacing of the main longitudinal bars should be adequate to allow the free passage of concrete through them—a minimum of 100 mm is suggested. Various proprietary systems are available for this purpose and they can be used for either mesh or bar reinforcement.

(c) Loose concrete is spread to the depth required for the full slab thickness. Reinforcement is then laid over the area. Prior to compaction in depth the mesh is depressed into the loose concrete to the required depth by vibrating tines. Besides forcing the mesh into the concrete, these vibrating tines make a contribution to the compaction of the concrete but compaction in depth is carried out from the surface only by compacting beams. Whilst this method is fairly widely used in America it has been far from an unqualified success in Britain.

It should be noted that with method (a) it is allowable to use any quality of concrete in the base (provided of course that it complies with the specification) irrespective of the nature of the aggregate; it need not be air entrained. The surfacing concrete, which in Britain and many other countries must be air entrained, can then be laid about 60 mm thick and be made of such materials as will cause the concrete to comply with skid-resistance requirements. Often this smaller volume of concrete for top layer, with air entrainment and a selected aggregate, will be produced in a smaller mixing plant than the rest of the concrete.

With methods (b) and (c) all concrete must be of the same type complying with requirements regarding strength, air entrainment and skid resistance of abraded samples.

In methods (a) and (c) it is usual to lay out the reinforcing mats along the line of the road against the position they will occupy in the slab. As an alternative they can be off-loaded from the supplier's vehicle in bulk on to a wheeled mesh cart which straddles the slab and is either self-propelled or can be towed by one of the leading road-building machines. Mesh is dragged from the pile on the cart as it moves forward with the road slab. Another idea used in America but not yet in Britain is to preload lorry lengths of reinforcing bar onto a cart and to feed these through a device which ensures proper spacing across the width of the road slab. Certainly a good deal of ingenuity is often shown by contractors in their pursuit of the least expensive means for providing specified amounts of reinforcement in roads.

It is claimed that method (a) which requires the final spreading and separate compaction of only 60 mm of surfacing concrete above the base concrete and reinforcement can lead to easier final finishing to the required surface tolerances. This is undoubtedly true in theory and in fact some of the best work has been carried out by this method. On the other hand good work has been carried out when the whole depth of the slab has been compacted and finished as one layer. No doubt selection of a method to be used will depend upon the contractor's estimate of the relative costs and his degree of confidence in being able to achieve the desired results by either one method or the other.

COMPACTING AND FINISHING CONCRETE

Compacting and surface finishing of concrete were at one time carried out by the same machine but, currently, compaction and partial finishing is undertaken by one or two machines of the same type, depending upon whether compaction is carried out to the full depth of the slab or in two distinct layers below and above reinforcement as described. Final finishing is carried out by a further machine which imparts very little compactive effort to the concrete but strikes the surface to a true profile by a to-and-fro screeding action across the slab.

Where two-layer work is adopted, the leading compacting machine will be used to impart a beam finish to the base layer, prior to laying the reinforcement. The surface layer will be compacted and partially finished by the second compacting machine which may have a floating oscillating finishing beam carried on an articulated chassis. Final truing up of the surface will be done with an articulated finishing machine with an angled oscillating single-or double-acting beam. The beam is mounted at an angle of about 50–60° to the line of the road and is of particular value in that it can encourage the quick removal of any excess of 'fat' gathered in front of the beam during final screeding.

In some reinforced road construction work, transverse joints are formed by vibratory or other means in the wet concrete immediately behind the second finisher and kept open prior to sealing by a removable insert of one form or another. Here, an angled beam has a marked advantage over a square beam in that it advances steadily from one end of the joint to the other and does not exert a disturbing pressure on any but a short length of the insert at any one time; this ensures that the joint insert remains in its correct position. Furthermore the angling of the beam extends the wheelbase of the machine considerably and it has been found that this tends to much reduce the 'yaw' of the machine when the resistance to forward movement is greater on one side than the other.

For greatest effectiveness a smoothing beam should be so heavy that there is no tendency to ride over rather than plane off high spots in the concrete. It is considered that for maximum efficiency a beam should weigh about 150 kg/m.

JOINT FORMING AND SEALING

Transverse joints. As noted previously, expansion joints are unlikely to be used in concrete roads except where, inadvertently, construction has been delayed or advanced so that work is in progress during the period mid-October to mid-April. However, contraction and/or warping joints are necessary and these will be provided with the specified number and sizes of load-transfer bars. Weakening of the slab at these contraction joints to ensure cracking there rather than elsewhere within the slab is provided by a wood or plastic fillet secured to the base and by a groove sawn or formed from the surface directly above the fillet. The reduction in slab thickness by fillet and groove should amount to at least 20% and preferably 33% to ensure adequate weakening.

Joint grooves can be formed in the unhardened concrete by means of vibrating blades or wobbly wheels, either of which will displace sufficient concrete to allow a temporary strip of adequate depth and width to be inserted and held secure in the concrete whilst the disturbed surface is being retrued. This refinishing of the surface is most effectively carried out by the angled finishing machine as described.

As an alternative to forming grooves in the unhardened concrete, they may be sawn when the concrete is sufficiently hard to allow this being done without disruption of the concrete along the joint line.

Concrete saws are power driven and the cutting blades are tipped with various grades of silicon carbide or diamonds. The ease or otherwise of cutting concrete depends much more on the kind of aggregate used than on its strength. Limestone concretes are by far the easiest to cut, next in order are other types of crushed rock whilst hardest are quartzites and flints. The last named are particularly difficult and expensive to cut but development in saw type, plant and techniques give constant improvements in the rate of cutting.

A difficulty that often arises in cutting partially hardened concrete by saw is that cracking may be induced at the surface before the concrete is hard enough to be sawn. Limestone concretes crack less readily than other types and can be sawn fairly soon after hardening; they are therefore much to be preferred, provided the skid resistance of the concrete can be made to meet specified requirements. Of course limestone concrete is likely to be more expensive in many parts of the country than is concrete made with the locally produced but much harder cutting quartzite and flint. Careful study of records of previous work, particularly in regard to costing, will be well worth while when considering whether to saw or form joints but on balance it is likely that forming will be the better proposition.

Longitudinal joints. Longitudinal joints may be either full depth at a slab edge or part depth formed by a bottom fillet and a surface groove in the centre of a slab. These surface grooves can be built into the unhardened concrete behind a surface-finishing machine by means of an attachment thereto which displaces concrete and at the same time, inserts a preformed sealing strip. As with the forming of transverse joints, there is some disturbance of the surface when this insertion is carried out; this is best corrected by using the angled finisher.

Since concrete inevitably shrinks away from the preformed sealer its efficiency is somewhat in doubt. But also in doubt is the real need for a completely efficient longitudinal joint seal.

Full-depth longitudinal joints are made against the formwork and will need tie bars across them, as specified. The ties are most readily positioned by cranking to 90° and laying one arm against the form for later recovery and bending into the adjacent lane. These joints are best sealed by sticking a preformed sealer on to the top of the hardened slab just prior to concreting and adjointing slab.

The adequate sealing of joints in concrete paving is a problem which, even now, does not appear to have been solved; perhaps it is insoluble. It is open to question whether or not sealing against water penetration is necessary when the base, as is generally the case, has been built in such a way as to make it much less susceptible to damage from seeping water. However, there is a need to avoid spalling at joints caused by intrusion of stones and perhaps grit in sealing grooves. If joint edges are left completely unsupported, there is also a risk that traffic will, in time, break away some section of joint edge.

TEXTURING AND CUTTING OF ROAD SURFACES

Recent research into the high-speed skid resistance of various types and textures of concrete surfaces has indicated quite clearly that a deeply textured one, which causes discontinuities in the surface water film in wet weather, markedly improves this characteristic.

It is now the practice on all concrete roads which will be used by high-speed traffic to score the surface, after final finishing, with brushes which impart this texture in the thin layer of surface mortar found on all of them. The result of this scoring is often to produce a drumming similar to that caused by running over heavily surface-dressed black-top roads.

Mechanically operated brushes to texture surfaces are now available and are suited to running on formwork, where rail-mounted equipment is used or on the base or formation when slabs have been laid by slip form pavers. Also available is a grooving machine which consists of a series of vibrating blades at irregular spacings. These make incisions into the unhardened concrete. It is claimed that the wholly random spacing of the grooves formed by this machine does much to reduce the noise nuisance which is sometimes induced where too regular and heavy texturing is a feature of the surface.

Adequate curing of concrete road slabs is essential, particularly in periods of low relative humidity, if damaging cracking and other surface defects such as widespread crazing are to be avoided.

Currently, practice throughout the world is to spray the concrete immediately after finishing with a solution of resin in a volatile carrier, or some other liquid, which will retain a high proportion of the water in the concrete for the vital first few days yet will not remain to reduce the skid resistance of the surface, once it is used by traffic. Where hot sunshine prevails for much of the day, a metallic or white reflecting pigment added to the membrane will reduce heat absorption into the slab.

British requirements in respect of curing are currently that the first two or three hours production of concrete slabs should be protected by tenting which supplements the curing effect of the membrane. This does not pose many problems where rail-mounted equipment is used; but it becomes impracticable when slabs are laid at high speed by slip form paving techniques—in fact it is most uncommon to see any form of tenting used when slip form paving methods are adopted, except a small amount carried on or close to the machine, for emergency use only.

Slip form paving

A method of building concrete roads without the use of prefixed formwork—slip form paving—has been developed, mainly in the United States over the past 25 years or so and has made a major impact on construction techniques, not only in America, but over the rest of the world where concrete roads find favour. A high proportion of the total length of concrete road currently being built in America is laid by slip form pavers. One such machine, brought into Britain in 1965, led to a period of intense study and development, not only of the machine but of the techniques associated with its use. It has also led to the design and development of a British counterpart which is claimed to have marked advantages.

Slip form pavers can accurately lay concrete of a wide variety of qualities at high speed; it would thus appear that a stage has been reached where a small number of machines would be capable of carrying out all the concrete road construction likely to be seen in Britain in the near future. Their prospects will depend largely upon their ability to provide a quality of road surface which is acceptable to the motoring public at a competitive cost—and there is evidence that they can do this.

In America, substantial savings in cost of pavement construction as compared with those associated with the use of form- or rail-mounted equipment are claimed for slip

Figure 36.12 Slip form paving with G & Z machine. Unreinforced concrete slab on stabilised base. Transverse joints sawn in hardened concrete of left-hand carriageway after 10–24 hr

form paving. They are said to amount to 10–15% of labour and machine costs but this implies a much smaller saving in overall cost since materials will generally account for 75–80% of the total. Since costs of road building are so much influenced by the extent of plant utilisation, a good throughput of work, i.e. high plant utilisation of slip form pavers, could result in worthwhile savings in comparison with previous methods.

Slip form paving is essentially concerned with laying, on a prepared base, a section of road slab in plastic concrete, within a moving form. The sides of the section are defined by formwork which is rigidly mounted on the machine itself at the limit of width of the slab, whilst its upper surface is formed either by a truly flat 'conforming' plate extending to the full width of the slab over a length of several feet or by oscillating screeds which strike off the surface of the concrete at a controlled level. Some machines use both.

The side forms may vary in length between about 4 and 9 m with different machines. Since they generally operate at forward speeds in excess of 1.5 m per minute, the concrete slab is supported at its vertical edges for a matter of a few minutes only, in contrast with the normal practice of maintaining support by formwork for several hours before striking.

Slip form pavers are normally positioned as to line and level by means of electro-hydraulic equipment. This in turn is controlled by sensors which detect the line and level of wires or cords supported well outside but parallel to and above one or both sides of the slab. Where one wire only is used it is accurately set up to control the position of the nearest slab edge, that of the remote edge being governed by a cross-levelling device on the machine. The locating of this is dictated by the position of the machine in the road, which governs the sense and degree of cross fall.

The concrete for slip form pavers may be deposited on the formation from end tipping trucks or discharged into a spreading device in front of the machine. It is then struck off approximately level across the slab by various means, for example shuttle spreaders, auger screws and transverse paddles. Compaction to its maximum density by a battery of immersion vibrators, supplemented as necessary by surface vibrators, follows. As the machine moves forward, a conforming plate or other levelling device is passed over the highly fluidised concrete and causes it to take up the shape of the slab. When a conforming plate is used, the amount of concrete passed below the plate will depend to some extent on the head of concrete in front; thus the level at its rear may be subject to slight variations. The action of oscillating screeds on the other hand, is to strike off excess concrete or to make good deficiencies with concrete carried in front of them. For this reason there may be some technical advantage in using a machine with oscillating screeds rather than a long conforming plate.

It is often preferred to lay concrete road slabs on a very accurately prepared base of cement- or lime-stabilised material, asphalt, sand–asphalt, lean concrete, etc., which will be adequately strong to carry the weight of the paving equipment. When this is done, the paver works at a fixed height above the base tending, by virtue of its long track base, to smooth out irregularities thereon. Direction has to be controlled by means of a line sensor on the machine but apart from this all levels of the finished road slab are dictated by those of the base on which it is being built.

Slip form pavers are best suited to laying unreinforced concrete slabs built without joints but subdivided—when the concrete has been laid a few hours and hardened sufficiently to resist damage—into short lengths by gang saws. However, by deploying other machines such as spreaders and mesh depressors, it is possible to build slabs which are, to all intents and purposes, the equivalent of our reinforced concrete slabs. When this type of road is built, the basic simplicity of slip form paving is lost and the train of equipment becomes not unlike that used for rail-mounted work as previously described. The capital cost of the machines employed is much higher than for rail-mounted equipment but its potential throughput is also much higher.

In America the final running surface behind a slip form paver is often trued up to a very regular profile by a tube finisher. This simple device consists of a 200 mm diameter tube of length about 1½ times the slab width, mounted on a wheeled chassis which straddles the newly laid concrete slab. The machine moves quite rapidly with the tube just in contact with the concrete surface, firstly in a forward direction with the tube at 45° to the line of the road, then backwards to the completed work, with the tube swung through an angle of 90°. This is repeated until the tube no longer collects variable amounts of surface mortar. When this stage has been reached such minor irregularities as are left behind the slip form paver have been uniformly dispersed over the surface of the slab.

It should be noted that the effective use of these machines relies on the presence of a substantial layer of mortary material at the surface. They should be so manipulated that no deep impressions are formed in the concrete when the tube is lowered on to the surface; nor should they be used to such an extent that an adequate depth of mortar to take the required surface texturing does not remain on the surface.

The profiles of concrete slabs finished by either a slip form paver or a combination of slip form paver and tube finisher are generally very uniform and hence should adequately comply with specification requirements.

So far tube finishers have rarely been used in Europe but there seems no technical objection to their use, provided of course that the mortary layer on which they operate is adequately resistant to the effects of frost- and ice-removing salts. The mortar is ideally suited to taking the favoured heavy brush texture which improves skid resistance, but again there may be some doubt as to the ability of such surfaces to retain texturing under heavy traffic. The answer would seem to be in proper mix design to ensure the correct amount of mortar, good control over all stages of production of the concrete and avoiding its excessive vibration.

Although there do not appear to be any insuperable technical problems associated with the successful operation of slip form pavers there are those of logistics. These are mainly that the machines operate most economically with high throughputs of concrete. Under normal conditions in Britain this necessarily involves stockpiling large amounts of concrete aggregates and perhaps special arrangements for the supply of corresponding amounts of cement. Building and drawing from stockpiles are expensive operations and significant reductions in construction costs have to be achieved to effect the desired savings in overall costs.

Despite some inevitable misgivings regarding the adequacy and cost of slip formed road work it seems that the method's future is assured—always provided that the market for concrete roads is adequate to sustain the use of such costly equipment.

OTHER FORMS OF CONCRETE CONSTRUCTION

Dam construction

Circumstances arise from time to time in the construction of reservoirs for either water supply or hydro-electric schemes when it is advantageous to build a concrete dam rather than an earthen or rock-fill embankment.

The design and specification for the dam will give details of the lengths of each section of the dam, the depth of each lift and the interval which must lapse between the concreting of successive lifts; it will also place a limitation on the depth of concrete which can be poured in the lifts and the time interval between adjacent monoliths.

Section lengths are generally of the order of 15–18 m and joint planes will normally be at 1.50–2.00 m.

Where it is intended that several monoliths in the same area be brought up together, this can be done provided an adequate gap is left between each for subsequent infilling. Gaps of about 2 m have been used and filled some time after the main lengths of the wall, when major drying and thermal shrinkage of the main blocks has taken place. Proper provision for water bars must be left in each side of this gap so that the method involves somewhat heavier expenditure on joint preparation and sealing.

A number of factors have to be taken into account in deciding the method of construction to be adopted.

Firstly will be the duration of the contract and the likely time to be spent on preparation of the foundations before dam construction can begin. Another major factor will be the profile of the valley across which it will be built. In Britain, the sites on which long and high dams can be built are very few so that a contractor is usually faced with building either a small number of high monoliths or a larger number of low ones.

The required number and dimensions of the monoliths will determine the method by which concrete will best be handled into place. Where a long, low dam is required to be built across a shallow valley, it might be considered to carry the concrete from a central mixing plant in crane skips mounted on flat-bottomed lorries or trailer units, thence into the work via a crawler or other type of crane. Flat-bottomed bogies carried on a narrow-gauge rail track along a low-level gantry might also be considered. To build a shorter but higher dam across a narrow steep-sided valley it might be thought preferable to handle concrete in crane skips via a series of derrick cranes mounted on temporary concrete pillars. These could well be sited on the upstream side of the dam, since the greater volume of concrete will be there.

All possible ways of handling concrete are discussed in a previous chapter and the contractor, in preparing his scheme for the work, will make decisions on the rate of concreting and the means by which it is to be placed in position.

It is common practice to build the main mass of a dam wall with a low cement content and hence fairly low-strength hearting concrete, but to use a higher quality having greater durability at the upstream and downstream faces. This facing must be placed within a short period of placing the hearting concrete so that there shall be a complete bond between the two.

Probably the easiest method is to build up the hearting in a 400–500 mm lift over the whole area to within about 400–500 mm of the dam faces—or whatever thickness surfacing concrete is called for in the design. The facing is then filled in to the same depth between hearting and the shuttered upstream and downstream faces. The richer concrete is not normally required to be poured against the shuttering to transverse joints.

Rock suitable for good-quality concrete aggregates is often available at the site of work and it is generally quarried as required to make concrete on site. Since sections are normally large and there is a need to keep cement contents low, aggregate up to 150 mm maximum is often used. The production of lean and sufficiently workable mixes for low-strength hearting concrete is facilitated by using large-size aggregate. However the use of plums or displacers is not economic—nor is it good practice.

In some instances where only poorer qualities of rock—as regards their suitability as aggregate—are to be found at site it might be necessary to import the better quality material for exposed concrete but to use the inferior aggregate for the mass of low strength concrete in the hearting. Proper mix design and placing techniques are quite capable of producing concretes satisfactory for this work from the most unlikely materials,

A disfiguring feature of many dams in the past has been the tendency to seep water through transverse joints and, more rarely, along horizontal joint planes. The former faults can be avoided by proper detailing of joints to facilitate their building according to the intended design. With horizontal joints it is essential to avoid the presence of laitence, downward joggles and deep indentations which will hold water. Laitence can be removed by air/water blast when the concrete is hardened sufficiently to avoid damage. Treatment of concrete with water sprayed on to hessian strips will ensure its proper curing and avoid any shrinkage cracks which might contribute to failure; it will also keep the whole area clean for subsequent operations.

Tunnel linings

In situ concrete linings may be called for in hard-ground tunnels to give support to rock which will deteriorate in the course of time or to improve hydraulic characteristics.

The dimensions of the tunnel—length, diameter, number of access points and other factors—will need to be taken into consideration in deciding the method to be used and the order in which work is to be carried out. Mention has already been made of the use of pneumatic placers but concrete pumping is coming to be more widely favoured. A variety of methods for getting the concrete to the working face have been used in the past and new ideas coming from time to time are adequately described in technical literature.

Mass plain and reinforced concrete sections

Concrete will be placed into sections of this type, which will most frequently be found in power stations and other heavy work, by various combinations of methods already described. The sizes of bay to be concreted will be dictated by the output of the batching plant available for the particular operation in hand. Owing to the complexity of shuttering work in reinforced concrete it might be found advantageous to restrict the depth of pour so as to increase the area to be concreted at any one time—shuttering costs are then likely to be rather lower.

For unreinforced foundations where there is no restraint to shrinkage by steel reinforcement it will generally be found necessary to restrict areas of concrete cast in one pour to avoid shrinkage cracking. Shuttering will be less difficult in these circumstances.

It will probably be necessary, in order to avoid excessively high temperature rises and subsequent shrinkage cracking, to restrict the depth of concrete lift to about 1–1½ m. Whatever criterion for bay size and lift heights is adopted, it is considered essential to carry out effective water spray curing for as long as possible. This applies whether or not such devices as the use of low-heat cement or PFA additive are required by the specification.

Vertical construction with sliding formwork

For many years, certain types of vertical construction in concrete, for example materials storage silos and the service cores of tall buildings, both of which have either a constant cross section throughout their height or only a small number of variations in wall thickness, have been carried out by using formwork which was moved continuously upwards as the concrete was placed within the forms. The infrequent changes in plan have been accomplished by altering the dimensions of the moving shutter as required; this has involved completion of a section and re-erection of one or both faces of the shutter before work recommenced.

More recent developments have seen modifications to the original conception of moving formwork to allow slight and gradual changes in cross section as the work proceeded upwards. The construction of reinforced concrete chimneys and the erection of tall towers serving as supports for TV aerials and amenity buildings are examples of work in which both external dimensions and wall thickness have decreased as the height above ground increased.

Sliding formwork, though now widely used, particularly by companies who have become specialists through mastering the techniques involved, is not a new art, there being reference to such work as long as 60 years ago. However, as the use of the method has become more widespread so have improvements been made in the mode of operation.

Basically the method consists in continuously raising formwork of the correct plan dimensions into which concrete with the designed amount of reinforcement is placed in a series of narrow bands up to 150–200 mm depth, proceeding over the whole plan area. The formwork to the inner and outer faces is connected at close intervals by a series of straddling yokes, each having a device by means of which the yoke and the attached shutter can be moved upwards in relation to jacking rods at each point. These jacking rods are carried from bottom to top of the structure and are located in circular cavities formed within the walls by tubes which surround the rods below the jacking points.

Apart from the external and internal shutters and the jacking systems and their control, it is necessary to provide a working platform which will generally cover the whole plan area of the structure. From this men can operate and on it can be stored such materials as reinforcement and blocking-out pieces for door openings, etc.

Concrete will normally be raised by a hoist or tower crane and distributed around the outside of the structure by hand barrows, light skips, monorail or other devices considered appropriate.

An essential feature of a sliding form is a platform connected with and below the inner and outer faces of the main shutter, from which operatives can carry out work to impart a sufficiently high standard of surface to the concrete emerging from the shutter. Occasionally, when faults have developed in the work, these will be corrected from the same platforms.

In the earliest examples of sliding formwork, raising of the shutter with reference to the jacking rods was carried out by hand-actuated screw jacks. However the specially designed jacks used now are almost universally hydraulically operated from a central control panel. The mode of operation is basically that the jaws on the jacks grip the jacking rod firmly whilst the body of the jacks, attached to the yoke, are moved upwards in about 12 mm intervals, the rate of travel varying between 150 and 500 mm per hour according to circumstances.

The forming of the narrow cavity round the jacking rods ensures that whilst adequate support is given against buckling under vertical loading as they are lengthened, the rods can be recovered for re-use after completion of the slide.

To ensure success of sliding operations it is essential to provide a high standard of control over the quality of the concrete used and the rate of sliding so that when it

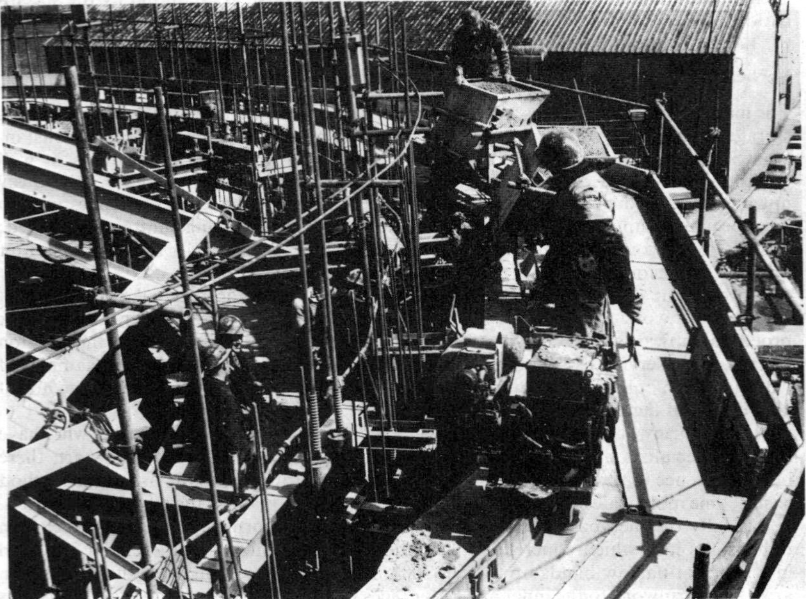

Figure 36.13 Continuous vertical construction of circular silos with sliding formwork. Note: sliding form; jacking points; working platform; monorail transporter distributing hoisted concrete (*Courtesy: John Laing and Son Ltd*)

emerges from the shutter the concrete is capable of selfsupport without slumping. The concrete must also be amenable to surface floating, etc., to remove minor blemishes.

Sliding operations involve 24-hour working, often under very variable weather conditions, and a high standard of job organisation is necessary to ensure that there is continuity of all operations involved. Breakdown of equipment which could result in long delays and perhaps in the extreme, the abandonment of a slide, is best guarded against by either duplication of vital items or constant survey to reduce the risk of untimely failure.

PRE-CAST CONCRETE

It is proposed to deal with only one or two aspects of the subject of pre-cast concrete as used in civil engineering work, although the arrival on the construction scene of equipment capable of handling large indivisible loads has, in recent years, given a new emphasis to work of this nature.

Pre-casting of concrete is widely practised in all branches of civil engineering but perhaps the most spectacular is in maritime work. For example, units weighing thousands of tons to be linked together to form submerged vehicle tunnels across narrow waters are frequently built in docks and made temporarily bouyant by adding bulkheads. They are then floated out to their permanent location where a number are strung together on a prepared base below the sea bed, to make a complete tunnel. This can be appropriately described as pre-cast concrete work. On the other hand, units such as for example, the Ekofisk storage tank for North-Sea oil and the Kish Bank and Royal Sovereign light towers all of which were built—in major part at least—on land before floating out to their permanent location, would not be classed as pre-cast work since they are independent units. Works of this magnitude will not be considered.

Bridges

Many concrete structures can be built either *in situ* or by using a number of pre-cast units which when assembled together, often by *in situ* work, will form an equivalent structure; here, the emphasis will be on bridge work.

Design studies carried out by the engineer and influenced in large measure by his past experience will indicate which method is the more likely to result in lower cost, simplicity and speed of building. The findings of these studies will be incorporated in the contract designs.

When prestressed concrete beams of various types are a feature of the design then the contractor may have to consider either buying-in or making within his own organisation. He will rarely consider setting up equipment to produce long-span box-section beams designed for production by the fully bonded (long-line) system because of the high capital cost involved; but where the beams incorporate their own inbuilt anchorages for prestressing tendons it is often open to consideration whether the beams should be factory made and hauled to site by road and rail or built on the job.

In some instances, for instance where very heavy long-span beams, which could not be brought to the site because of road or rail restrictions, are required, there will be no alternative to site casting. For the smaller, readily transportable, handleable units, such factors as the cost of preparation of suitable casting beds, the cost of concrete production and of providing the high degree of supervision over production need to be considered. Quite often, bridges are built in locations where access is difficult even for small units. Here the alternatives of building on site or waiting until better access can be provided for brought-in beams and the cranes for handling them need to be considered.

Pre-cast concrete units weighing up to about 130 tonnes have formed a substantial part of such recent overhead urban road works as the Mancunian Way in Manchester and West Way and the Five Ways interchange on the western and northern approaches to London. The pre-cast units involved have been made both in casting yards on site and, where of manageable proportions, brought in. These schemes have been well documented in recent literature.

Tunnel works

Not perhaps so much in the public eye as schemes such as the three quoted have been those in which the roadways in bored two-lane vehicular tunnels have been pre-cast and set in position somewhat below the axis of the tunnel. In this arrangement the large area below the road is used primarily for ventilation but also for services of all kinds.

Tunnel road deck units will normally be formed in lengths of up to 6–8 m in pre-casting yards at one or both ends of the tunnel, according to the number required and the time available between the completion of driving and lining and opening to traffic.

A high standard of dimensional accuracy and surface finish is called for, particularly if the upper surface is to form the running surface of the road without recourse to an applied bituminous or further concrete finish; additionally, accurate and sufficient bedding of the units is called for to ensure good performance.

Where ground conditions are appropriate, pre-cast concrete units may be used for lining tunnels. These units may be either solid sections which are stressed into contact with the ground (often man-handled units in the smaller diameter tunnels), or ribbed sections, both of which are put into position with mechanical erectors. Pre-cast tunnel linings are used as widely as is possible because of their low cost as compared with that of cast-iron tubbing.

Cladding panels

Pre-cast units made of lightweight and perhaps heavily air entrained and patterned concrete are sometimes used as architectural features on the facades of buildings. The use of concretes of this type allows of the erection of large panels whose weight is within the capacity of cranes used around a site or hired in for the particular purpose.

REFERENCES

1. *Method for testing the performance of batch type concrete mixers,* BS 3963, British Standards Institution, London (1965)

BIBLIOGRAPHY

BINGHAM, T. G. and LEE, D. J., 'The Mancunian Way elevated road structure', *Proc. Instn civ. Engrs,* London (Apr. 1969)

BRMCA Guide. British Ready Mixed Concrete Association, Ashford, Middlesex (Oct. 1971)

Concrete Manual, 7th edn, US Department of the Interior, Bureau of Reclamation, Denver, Colorado, USA (1963)

'Concrete Pumping', Building Research Station Digest No. 133 (Sept. 1971)

Concrete Roads, Ministry of Transport, Road Research Laboratory, HMSO (1955)

FALKINER, R. H. and TOUGH, S. G., 'Tyne Tunnel 2', *Proc. Instn civ. Engrs,* London (Feb. 1968)

GILLIS, L. R. and SPICKELMIRE, L. S., 'Slip form paving in the United States', Bulletin No. 263, American Road Builders Association (1967)

Report of Committee No. 207 on Large Dams, Journal of the American Concrete Institute, p. 273 (Apr. 1970)

ILLINGWORTH, J. R., *Movement and Distribution of Concrete,* McGraw-Hill (May 1972)

KELL, JASPAR, 'Dartford Tunnel', *Proc. Instn civ. Engrs,* London (Mar. 1963)

KEMPSTER, E., 'Pumpable Concrete', Publication No. CP 39/69, Building Research Station

MUIR WOOD, A. M. and GOBB, F. R. S., 'Cargo Tunnel at Heathrow Airport', *Proc. Instn civ. Engrs,* London (Jan. 1971)

MURDOCK, L. J. and BLACKLEDGE, G. F., *Concrete Materials and Practice,* 4th edn, Arnold (1968)

NUNDY, F. S., 'Construction of Western Avenue Extension', *Proc. Instn civ. Engrs,* London (Feb. 1972)

ORCHARD, D. F., *Concrete Technology,* Vol. 2, *Practice,* C. R. Books London (1962)

'Pumping Ready Mix Concrete', Technical Bulletin TB/1, British Ready Mixed Concrete Association (Mar. 1968)

SHARP, D. R., *Concrete in Highway Engineering,* Pergamon, Oxford (1970)

STEIN, J. and DONALDSON, P. K., 'Techniques and formwork for continuous vertical construction', Concrete Society Limited, London (Oct. 1966)

'Underwater Concreting', Technical Report No. TRCS 3, Concrete Society Limited, London (1971)

37 HEAVY WELDED STRUCTURAL FABRICATION

HEAVY WELDED STRUCTURAL FABRICATION

37

37 HEAVY WELDED STRUCTURAL FABRICATION

J. L. PRATT, B.Sc.(Eng.), C.Eng., M.I.E.E., F.Weld.I.
Research Manager,
Braithwaite & Co. Engineers Ltd.

The vast majority of fabrications in steel are now welded and it is rare to see a new fabrication that is joined by any other method. The steel used is generally to BS 4360 *Specification for weldable structural steels*; the revised 1972 edition now includes the weathering steels and the whole range cover steels with yield stresses ranging from 230 to 450 N/mm^2.

WELDING PROCESSES

Figure 37.1 shows the welding processes most commonly used in steel fabrications; in all cases an arc is struck between the electrode or electrode wire and the workpiece resulting in a high arc temperature which melts off the electrode and deposits it in the joint which has to be made. The Manual Metal Arc (MMA)[1] is the most common process and the electrode is deposited manually with the operator controlling the direction of the weld and the build up of the weld metal. The flux extruded around the core wire when melted by the heat of the arc provides a gaseous shroud which protects the molten pool and arc from atmospheric contamination and controls the weld metal reactions; it can also be the vehicle for supplying certain alloying constituents to the weld metal. The fused slag around the deposited weld metal also helps to form the weld bead shape. There are several types of electrode coverings which function in different capacities and are classified in BS 639 *Specification of covered electrodes for manual metal arc welding of mild steel and medium tensile steel.* Those more particularly used in fabrications are of the classes 1, 2, 3 and 6.

The CO_2 shielded arcs with bare wire of cored wire, can be of the semi-automatic or automatic type.[1, 2, 3] The semi-automatic process utilises a power source, a wire drive unit, incorporating the necessary control units, and a 'gun' which is held by the operator and manipulated manually; the wire is driven through a flexible tube to the gun and a suitable designed nozzle concentric to the gun orifice supplies the CO_2 gas to the arc. The automatic process usually has a heavier 'gun' or head with the wire (also known as electrode wire or feed wire) fed directly through the gun without the intervening flexible tube; the whole apparatus travels automatically for longitudinal welds or may be stationary for circular fabrications. Higher welding currents and deposition rates are generally used with subsequent water cooling of the head being necessary. The weld metal and arc is protected from the atmosphere by the CO_2 shroud but a bare wire must contain deoxidisers such as silicon, manganese and sometimes with aluminium; these are necessary to prevent some oxidising processes which occur within the arc atmosphere. A cored wire has the flux enfolded within the electrode wire as typified in the cross section shown in Figure 37.1; it may be used in semi or full automatic processes. The necessary deoxidants are carried in the flux which may also be the vehicle for additional alloys to be added to the weld; the flux allows for higher welding currents than that in solid or bare wire CO_2 shielded welds, with the slag allowing better bead shapes to be obtained, and is generally more tolerant to rusty plate conditions which

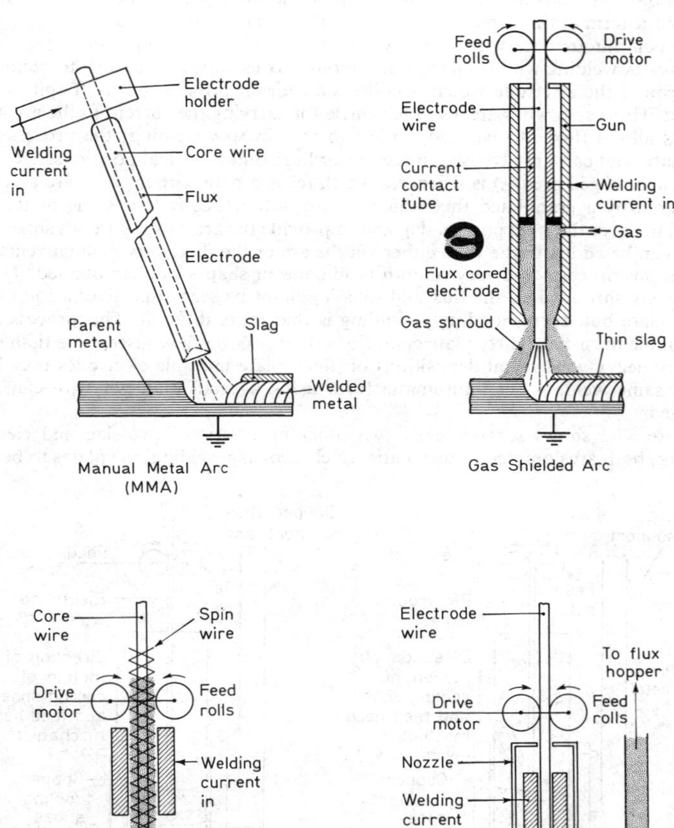

Figure 37.1 Common welding processes

could otherwise lead to porosity. Because of the higher currents used this electrode or cored wire is limited to welding in the flat or horizontal–vertical positions.

Other shielding gases may be used in the foregoing such as helium, argon, argon–oxygen or argon–carbon dioxide, the cheapest and most popular being carbon dioxide; the generic term of this process using any of all these gases is metal arc inert gas (MIG).

The continuous core wire as shown in Figure 37.1 is a continuous core wire like that in MMA welding with similarly continuous flux extruded around it; to maintain the flux around the electrode which is coiled on a drum prior to burn-off spin wires are formed. These spin wires are also the vehicle for carrying the current to the arc and this process allows fluxes to be used similar to that in MAA welding; the process is fully automatic and can only be used in the flat or horizontal–vertical positions.

The submerged arc (SA) is a process which feeds a bare wire into the arc and the arc is covered by a granulated flux which is also automatically fed; some of the flux is melted to cover the weld pool as slag and to provide the arc with a gaseous shield. Again alloys can be added to the weld either via the arc or the flux; very high currents can be used in this process[2] and very smooth bead contour shapes can be obtained. The arc is completely shrouded by the flux and thus it cannot be seen; this gives a total absence of arc glare but, correspondingly, guiding is that more difficult. This process is more susceptible to rusty or dirty plate conditions than MMA but less susceptible than MIG and for very heavy weld metal depositions on thick plate multiple electrodes may be used in the same weld. A semi-automatic form using a small diameter wire can also be obtained.

Figure 37.2 shows schematically two other processes, electro-slag and electro-gas welding; both are completely automatic. In electro-slag welding, the plates to be welded

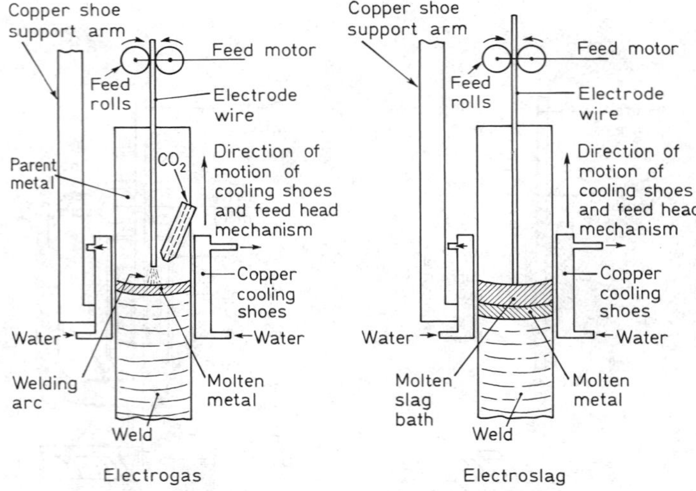

Figure 37.2 Electro-gas and electro-slag processes

are mounted vertically with the edges of the plate square or unprepared; watercooled copper shoes are mounted either side of the weld seam to contain the molten metal. An arc is struck on the starting block with a little granulated flux added to the weld pool; as the wire or electrode burns off the temperature of the slag bath increases and the slag becomes electrically conducting; from then on the electrode protrudes into the bath, the

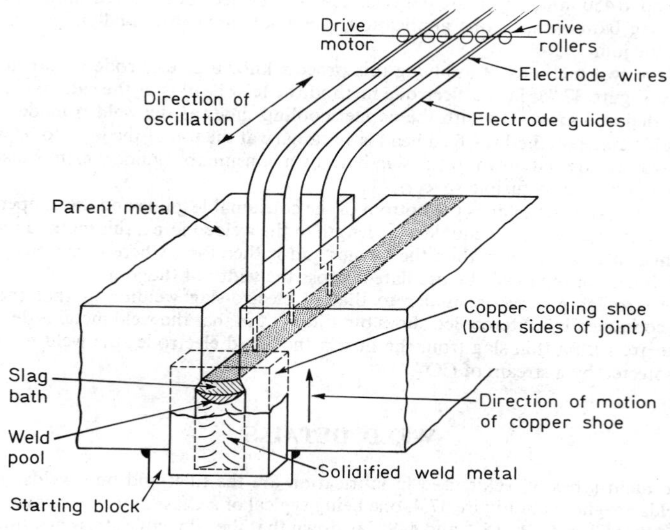

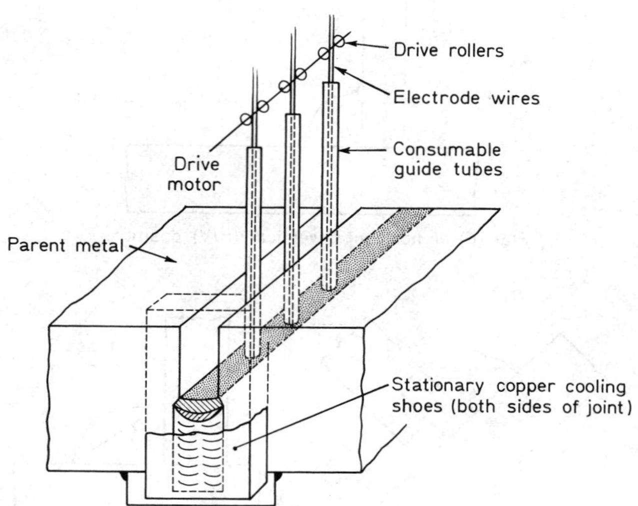

Figure 37.3 Electro-slag welding process

arc extinguishes, and the wire metals off due to the I^2R heating of the current. It is thus not an arc process but a continuous cast process used on plate thicknesses over 25 mm and certainly up to 450 mm; for the greater thicknesses three electrodes are fed simultaneously into the slag bath with, in one application, the whole assembly oscillating across the width of the joint.

There are two methods of applying this process known as electrode or consumable guide (see Figure 37.3); in the electrode method the feed head is at the side of the plate being welded and moves up with the copper cooling shoes as the weld is made. In the consumable guide method the feed head is stationary at the top of the joint to be welded and the wire (s) are fed down to the slag bath by a consumable guide which is insulated from the workpiece by fusible spacers.

As the wire (s) melts so does the bottom of the consumable guide and the copper shoes can be stationary of a length equal to the length of the welded joint; this method requires less sophisticated machinery than the former and is therefore, where it can be applied, cheaper. It cannot obviously be oscillated across the width of the joint.

The electro-gas process is similar to that of electro-slag welding in that the weld metal is contained by watercooled shoes but different in that the weld metal is deposited by a true arc with a thin slag from the flux in the cored electrode; the weld metal and arc is protected by a stream of CO_2.

WELD DETAILS

The two main types of welds used in fabrication are the fillet and butt welds. Typical fillet welds are shown in Figure 37.4, one being typical of a Class 2 or 3 electrode deposit and the other Class 6; BS 153[4] and 449[5] lay down that the allowable stress in a fillet weld

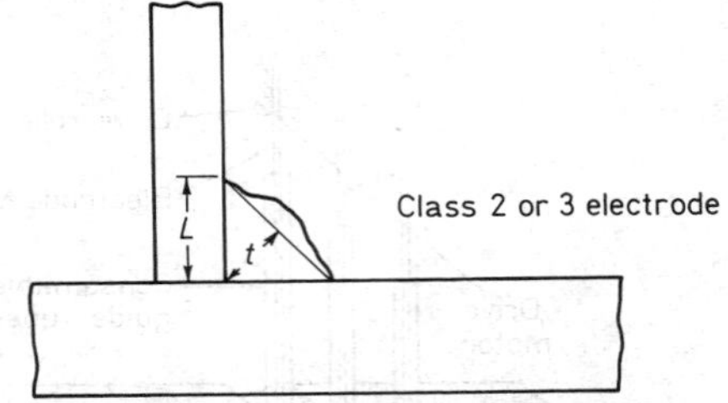

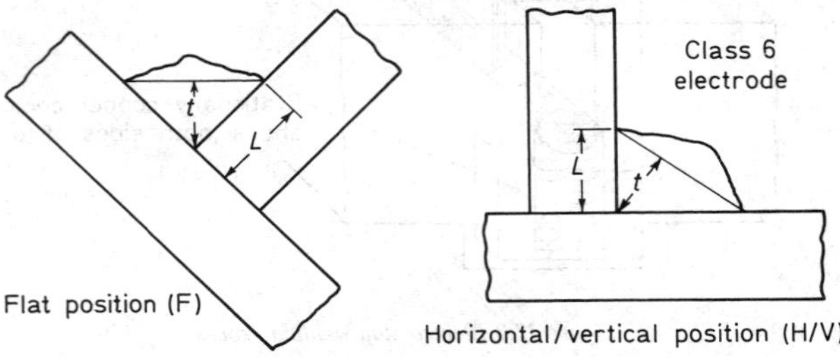

Figure 37.4 Fillet welds. Note: Minimum length of both legs to be measured for L. Note symmetry of leg lengths with Class 6 electrodes in the H/V *position*

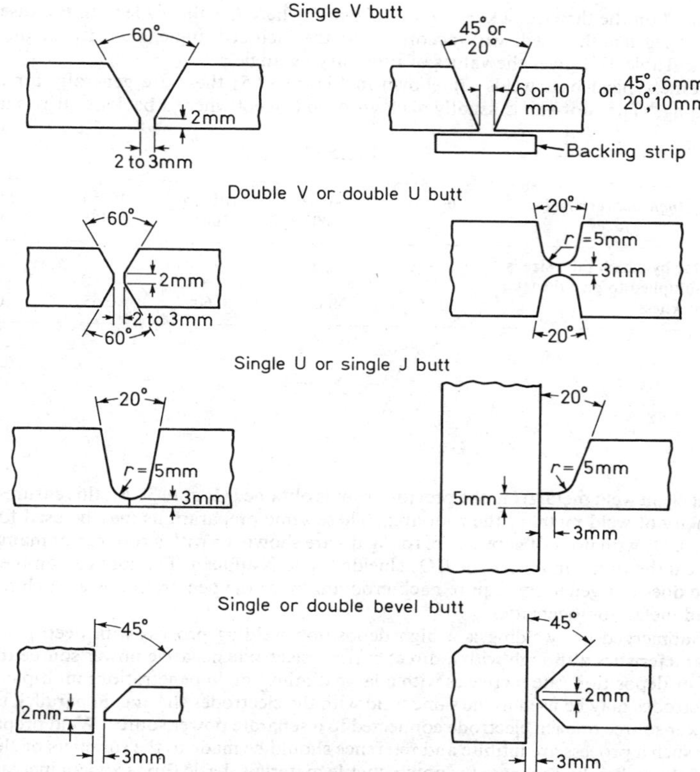

Figure 37.5 Typical butt welds. Note: Angles and dimensions of root gaps and root faces may be altered to suit welding technique and position of weld, the above being suitable for flat position welding. Welding is carried out from both sides of all preparations except where a backing strip is employed. To achieve complete penetration, back gouging (back grooving) may be employed

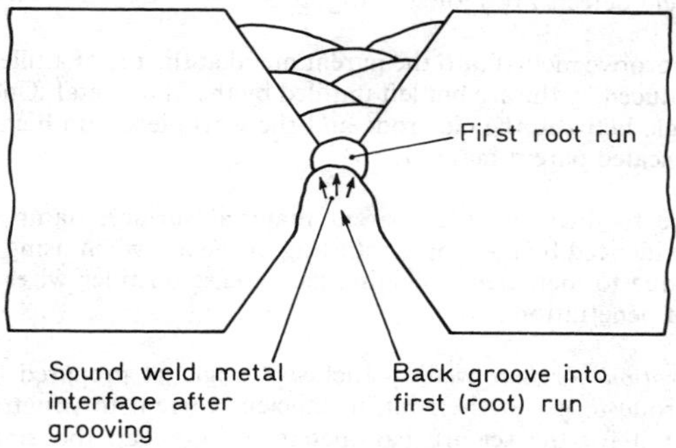

Figure 37.6 Butt weld back groove

is based on the throat thickness, t, or '$0.70 \times L$' where L is the leg length. Because t for a stated leg length L will vary according to the included angle, γ', between the fusion faces. Table 37.1 gives the values of t for varying angle γ.

Some typical butt welds are shown in Figure 37.5; these are generally for manual welding.[4] The root-run is usually back grooved (except where a backing strip is used) so

Table 37.1

Angle between fusion faces	60° *to* 90°	91° *to* 100°	101° *to* 106°	107° *to* 113°	114° *to* 120°
Factor by which fillet size is multiplied to give throat thickness	0.70	0.65	0.60	0.55	0.50

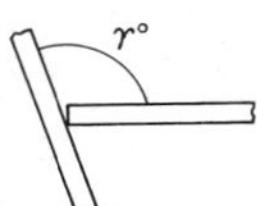

that clean weld metal from the previous root is obtained (Figure 37.6); this ensures homogeneity of weld metal at the root area. These same preparations may be used for cored wire CO_2 with no root gaps where root gaps are shown, or with a root run of manual weld to seal the root gap before the CO_2 shielded weld is applied. The root run on the second side does not generally require back grooving since the penetration is enough to ensure weld metal homogeneity.

Submerged arc welding is a high deposition welding process with deep penetration characteristics although with a direct current electrode negative power source the burn-off or deposition rate increases with a large diminution in penetration; multiple wires or electrodes may be used in the same weld with the electrodes sharing, in parallel, the same power source or each electrode connected to a separate power source. Weld preparations for such a process are infinite and reference should be made to the suppliers of electrodes and fluxes for their advice; for notch ductile materials, basic fluxes and an increase in the number of runs may be necessary.

WELD DEFECTS

Some typical weld defects are shown in Figure 37.7.

'Undercut'. A groove melted into the parent metal at the toe of a fillet weld or root of a butt weld—produced by the arc but left unfilled by the filler metal. Undercut may be due to incorrect angle between the electrode and the workpiece, too high an arc voltage or travel speed or scaled parent material.

'Porosity'. Due to dirty or rusty parent material surface, damp consumables, arc instability (as evidenced by the stop or starting of the arc when using MMA), gas entrapment from air due to inefficient shielding gas, grease on filler wires. When linear can denote 'Lack of penetration'.

'Lack of penetration' for butt welds. Inclusive angle of prepared faces too small to allow the electrode to get at the root, insufficient current to penetrate the landing or landing too small for the set arc parameters or root gap too small to allow more penetration.

'Lack of fusion'. Incorrect manipulation and angle of electrode to ensure side wall fusion, or root fusion.

'Slag inclusions'. Nonmetallic solid material entrapped between runs of weld metal or between weld metal and parent metal. Due to inefficient clearing of the slag between

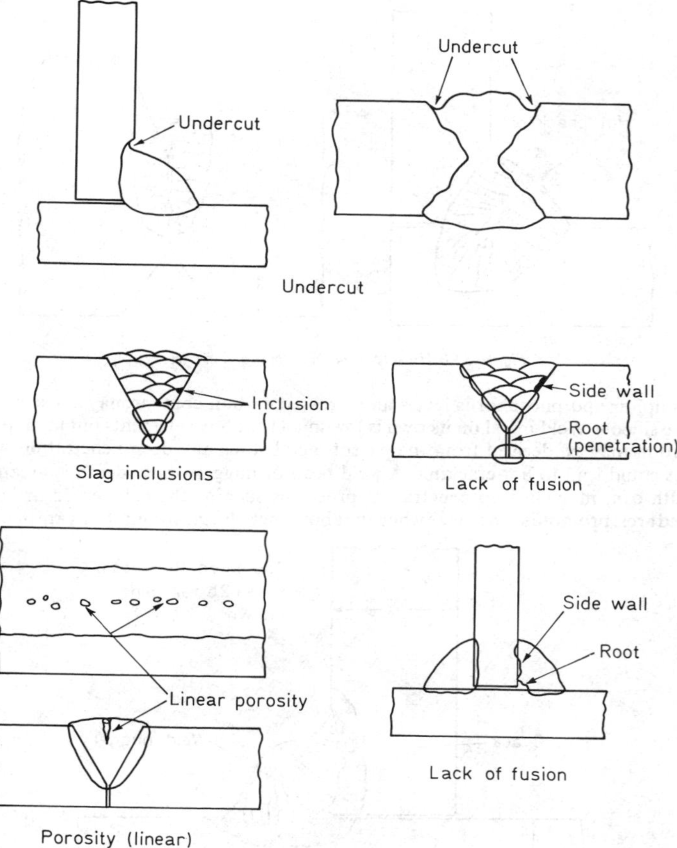

Figure 37.7 Some weld defects

each run which in turn may be due to wrong weld parameters giving the wrong shaped interpass bead and positional requirements.

'Spatter'. Small metallic particles ejected from the weld area and forming on the parent material adjacent to the weld. Spatter varies with the arc process and within that particular process may increase or decrease with differing are parameters.

'Hot-cracking' (solidification cracking). Cracks appearing in the central region of the weld (Figure 37.8) where segregation of sulphur and phosphorous, the lowest melting

point constituents of the weld metal, occurs; thus at temperatures in the region of the solidus thin films of the liquid segregate occur along the grain boundaries (intergranular). The weld metal may thus become susceptible to cracking because of the high shrinkage stresses generated during the cooling of the weld metal. The effect of sulphur may be reduced by obtaining a higher manganese to sulphur ratio in the weld metal whilst at the

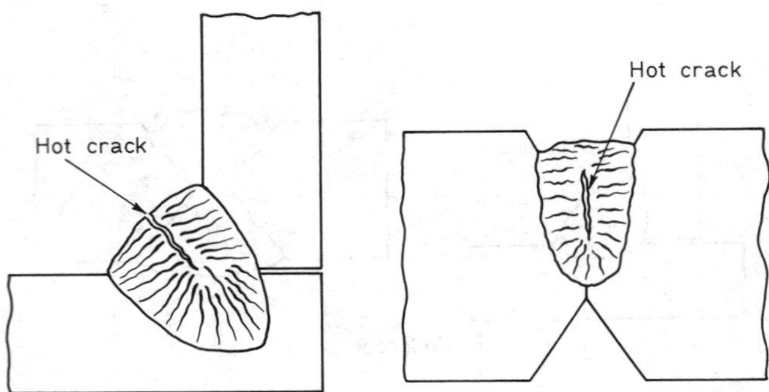

Figure 37.8 Hot-cracking in submerged arc welds

same sulphur and phosphorus levels an increased carbon content may cause cracking; likewise silicon. Weld metal on its own is low in all the above elements but high 'pick-up' or dilution may be derived from the parent metal; thus any deep penetration welding process could lead to hot-cracking. A weld bead or nugget whose depth is greater than its width can, in such deep penetration processes such as the submerged arc or CO_2 shielded arcs, promote such cracks when the above metallurgical conditions are marginally

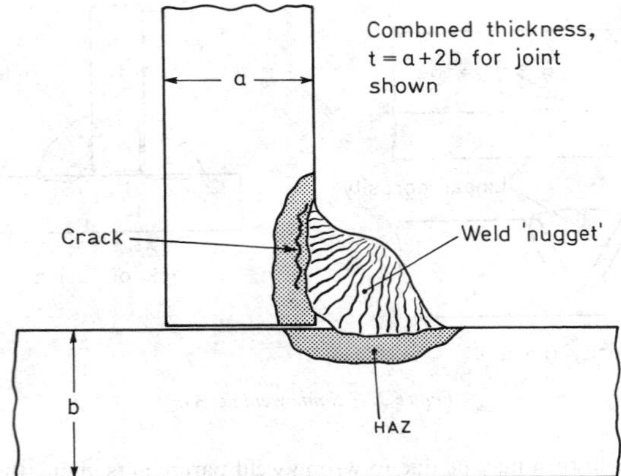

Figure 37.9 HAZ *crack*

operative. In this case a wider preparation or the use of more than one run of weld with lower current values would reduce the dilution factor and the depth-to-width ratio, to decrease the risk of cracking.

'*Cold cracking* (*underbead or HAZ or hydrogen-induced cracking*). The heat-affected zone (HAZ) of a weld is that (generally) narrow zone in the parent metal adjacent to the weld bead affected by the heat input of the weld and whose microstructure and physical properties might be affected by that heat. This zone is rapidly cooled by the mass of the surrounding parent metal and if this cooling rate is high enough a hardened (martensitic) microstructure may be formed. Cracking may develop in this hardened structure, see Figure 37.9, owing to (a) the alloy content of the parent material increasing, (b) high cooling rate, (c) restraint and therefore higher residual stresses resulting from the weld contraction, (d) stresses within the microstructure due to the transformation to a hardened structure, (e) the presence of absorbed hydrogen from the weld diffusing into the HAZ when that weld cools and contributing to the creation of micro fissures, (f) for fillet welds, where the fit-up is bad with root gaps.

In (a) the presence of alloys in increasing amounts increases the hardenability in the HAZ and their effect can be related to that of carbon by the following carbon equivalent formula

$$\text{Carbon equivalent}\,\% = \text{C}\%+\frac{\text{Mn}}{6}\%+\frac{\text{Cr}+\text{Mo}+\text{V}}{5}\%+\frac{\text{Ni}+\text{Cu}}{15}\%$$

Thus any increase of carbon equivalent due to the increase in any of the above alloys will increase the hardenability of the steel. This formula only applies to those steels in BS 4360.

In (b) the cooling rate is assessed partly by the combined thickness t of the joint being welded (Figure 37.9) and partly by the heat input from the weld and any given preheat. The total heat input from any arc may be expressed as

$$H\,(\text{joules/mm}) = \frac{\text{arc voltage} \times \text{current (amps)}\; XT}{L}$$

where T is the time in seconds to deposit L mm of weld.

Lower t and higher H lead to a lower cooling rate in the HAZ with a less hardenable microstructure.

In (c) the restraint increases with the stiffness of the components making the joint. In (e) the hydrogen content can be reduced by using a low-hydrogen process, e.g. hydrogen-controlled (Class 6) MMA electrodes, solid wire CO_2 process, etc. The MMA electrode may have to be baked to reduce its hydrogen content to the lowest level possible; in submerged arc welding the flux would have to be dry and preferably of the fused rather than agglomerate flux. With all automatic wires or electrodes no wire drawing compound contaminates should be present; CO_2 shielded arc processes with solid wire could prove to give weld metals with the lowest hydrogen content.

Preheat curves necessary for combined thicknesses and size of weld deposit (and thus heat input) are given in BS 2642 *General requirements for the arc welding of steel to BS 968 and similar steels*; these preheat requirements are used for BS 4360 Grade 50 steels. Mild steel, or BS 4360 Grades 43, are welded to BS 1856 *General requirements for the metal arc welding of mild steel.*

Both these standards will be shortly replaced by a new standard which will give preheat requirements for all steels to BS 4360; Baker[6] *et al.*[7] have contributed to the study of welding procedures for carbon manganese steels.

Preheat, when applied, reduces the rate of the cooling of the weld and allows more hydrogen to be evolved by the weld metal to the surrounding atmosphere; therefore to be effective it must be applied to the correct temperature and over a sufficient width of the plate. BS 2642 indicates that the width of the preheat zone on each side of the weld should be at least three times the thickness of the plate preheated for all material up to 50 mm thick. In practice the temperature is measured by using thermo indicating crayons or paints, the former melting and the latter changing colour when the correct temperature is achieved, and to make certain that the heat has penetrated the full thickness it is customary to heat the far side of the plate with the temperature indicator on the near side; or by heating the near side until the required temperature is indicated on the same side one minute after the heat source is removed. Although the heats applied are generally low (on the average 100 °C), the wide area over which they are used can lead to more distortion than that of the weld itself; it is therefore preferable to use a higher heat input weld source to reduce the preheat required. It is also more economical.

DISTORTION

Distortion due to welding is dependent on the heat input from the weld; such heat is concentrated in a narrow zone around the weld area. The subsequent contraction of the heated weld metal and parent metal produce undue stresses in the fabricated part; if

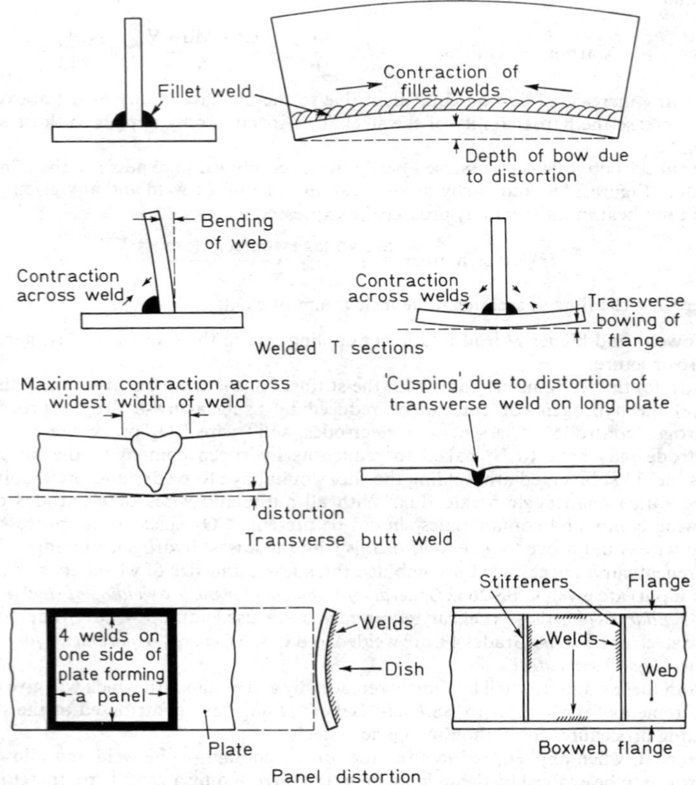

Figure 37.10

unrestrained the fabrication will distort and, if restrained against distortion, residual stresses up to the yield point of the material may occur. The parts being welded may in themselves have residual stresses due to their shape and size and thus their manufacture; these stresses or some of these stresses may be relieved or increased by the local welding heat and thus their distortion due to welding may be difficult to predict. Metals with differing expansion coefficients, thermal conductivities and physical properties will produce different distortion levels with the same weld heat input.

Figure 37.10 shows distortions for typical welds. Figure 37.11 shows joint preparations, welding procedures and some typical plate presetting to compensate for weld distortions. For the heavier type of fabrication it is generally better to fabricate all subsections

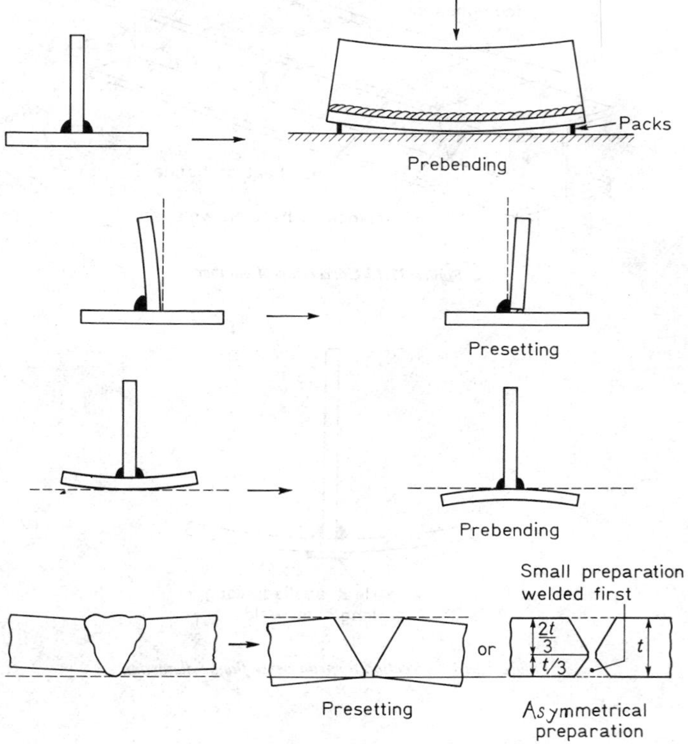

Figure 37.11 Methods to reduce distortion

prior to incorporating them into the main structure but to control the increased distortions for thin walled constructions it may be preferable to assembly and tack the whole assembly to give a much stiffer structure more able to withstand distortion.

CORRECTION OF DISTORTION

For a dished plate (for example, a dish resulting from an area of plate welded all round the periphery of that area on one side of the plate only) the amount of dishing resulting from such a weld depends on the heat input of the weld and the thickness of the plate. To flatten such an area, spot heat from a heating torch can be applied in several places within the dished area on the outside (convex side) of the bulge; this will increase the amount of dishing on heating but on contracting that side will shrink and reduce the

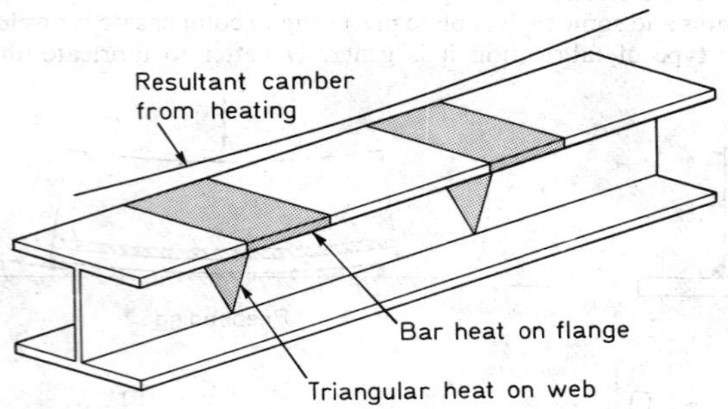

Figure 37.12 Correction of camber

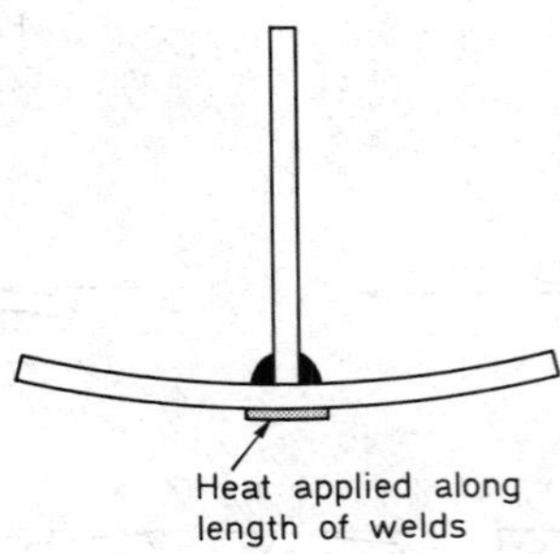

Figure 37.13 Correction for transverse flange distortion

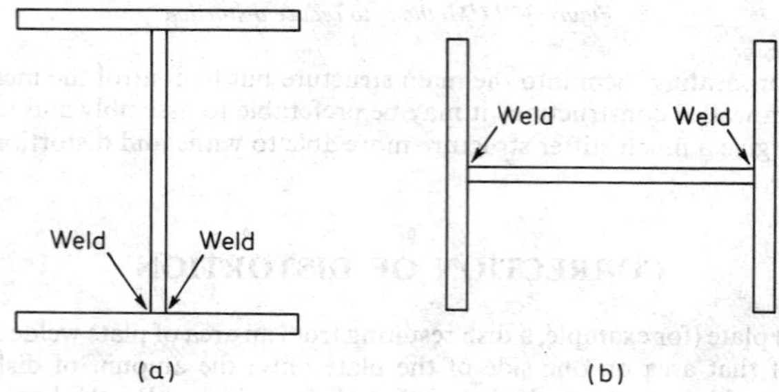

Figure 37.14 Plate girder welds

dish. Heat can be up to red heat (600–650 °C) but does depend on the thickness of the plate; for very thin plate the applied heat may heat both sides to an equal temperature resulting in equal contractions on both sides of the plate with no decrease in the dish.

Triangular heating on the web and bar heating on the flange of a plate girder or section (Figure 37.12) will increase the camber and can also be applied, within certain limits, to box sections. It is important to note that heating the flange and not the web may shrink the flange and increase the camber of the girder but the web, not being heated, cannot shrink to accommodate the increased camber and may therefore buckle.

Angular bending of a flange plate due to the two fillet welds attaching the web to the flange may be corrected by heating in a straight line (Figure 37.13). The effect of introducing heat to shrink areas and to introduce distortions to reduce others, must of a necessity induce stresses into the fabrication; the effect of these stresses and the subsequent increased load on some welds must be carefully watched and if necessary those welds increased in size to accommodate the increased load. All welds or any other form of localised high heats give high residual tensile stresses local to that heat; these stresses in turn generate compressive stresses outside those tensile stress areas. Stress relieving (at about 600 to 650 °C) may relieve the structure of any induced stress but in turn must lead to increased or different distortions to accommodate the subsequent movement of the structure.

ASSEMBLY

'Plate girders.' These may be welded as in Figure 37.14 (a) or (b) by MMA or automatic welding. Tack welding to hold the assembly together must conform to the requirements of BS 1856 and BS 2642, with minimum root gaps; large root gaps may lead to HAZ cracks as described previously or 'burn through' when using high current density automatic welds. For girders with top and bottom flanges of differing thicknesses or with top-flange-to-web and bottom-flange-to-web welds of different sizes, different shrinkages may occur in each flange and hence after the camber of the girder; in such cases it may be necessary to induce triangular heating as described above or to deliberately increase the camber if the web plate is cut to camber in the preparatory stage. For thin flanges it may be found necessary to prebend the plates as shown in Figure 37.11.

For crane girders it may be necessary to make full penetration welds (Figure 37.15) for the web to top flange welds (BS 153). When using automatic welds such as submerged arc, care must be taken that hot cracking does not occur; this can happen when trying to achieve penetration and the dilution of the weld metal by the parent metal is high. To reduce dilution back grooving may be used (but is difficult in this situation) or the web preparation made wider; for submerged arc and CO_2 shielded welds of high current density the physical size of the weld nugget is important. When the weld nugget has a width w and a depth p, hot-cracking may occur if the ratio of p to w is greater than unity. Hot-cracking invariably occurs on the second side of the joint to be welded since the first weld has made rigid or constrained the web–flange assembly and it should be noted that hot-cracking may be contained below the surface of the weld and thus not visible (Figure 37.8).

'Box girders.' These are invariably assembled on one flange as the base fabrication plate and must lie perfectly flat on the assembly stallage or a twist in the box may result; all diaphragms are then placed in position after being subassembled and the two webs then tack welded to diaphragms and flange. As much internal welding as possible is then made before the fourth of closing flange plate is placed in position and tack welded; the four longitudinal web-to-flange welds are then made.

The same comments about differing flange thicknesses or web-to-flange welds in plate girders can apply to box girders.

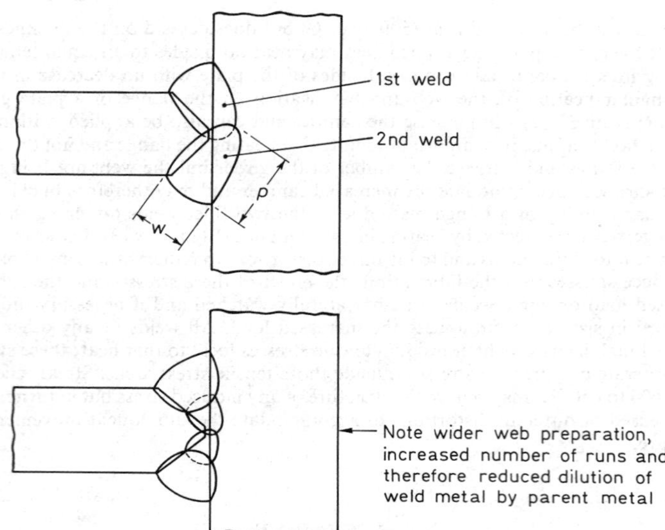

Figure 37.15 Full penetration butt weld, submerged arc

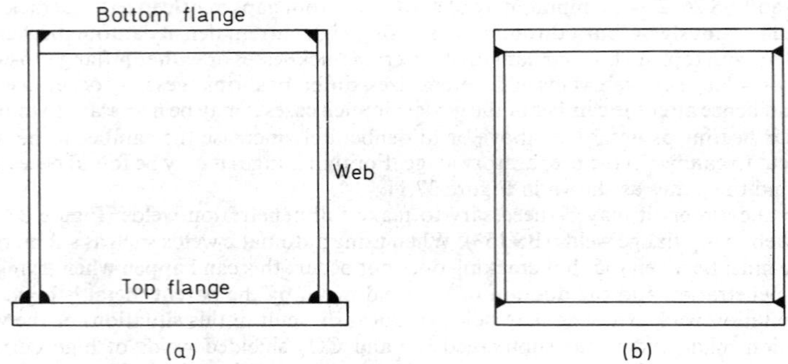

Figure 37.16 Box-girder assembly

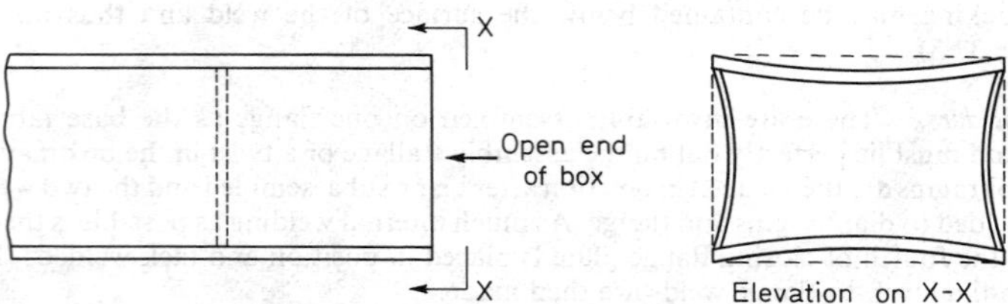

Figure 37.17 Box open-end distortion

The choice of flange-to-web longitudinal weld detail may be dictated by the camber and or curvature required in the box. For a large box where it may be difficult to rotate during fabrication (a) in Figure 37.16 may be preferable; where there is camber, (b) is easier to assemble with the flanges outside the webs than (a) where the box closing flange coming inside the webs can only sit on the diaphragms unless backing strips on the webs are installed between the diaphragms to maintain the closing flange profile.

Where boxes are to be jointed on site the open ends of adjacent boxes should be stiffened if there are no diaphragms close to the open ends to hold those ends to the required square or trapezoidal profile; for all such ends they would tend, without such stiffening, to have inward bows on all flanges and webs although the four corners are dimensionally correct Figure 37.17.

In boxes all stiffeners are subassembled on the webs and flanges before the main assembly is completed; to keep all resultant distortion, shrinkage, etc., to a minimum it is better that intermittent welding be used on such items if the design requirements can be met with such welds. Stiffeners may be of the bulb, angle or T type; the latter two may present difficulties for blast or other type of cleaning after welding and also for the subsequent painting.

STUD WELDING

Stud welding shear connectors on the top flange for bridge girders for composite concrete decks is now a widespread practice; the diameters of the studs usually range from 12 to 25 mm and generally vary from 100 to 150 mm in length although 250 mm studs have been welded. The form of the stud is shown in Figure 37.18 and the head of the stud

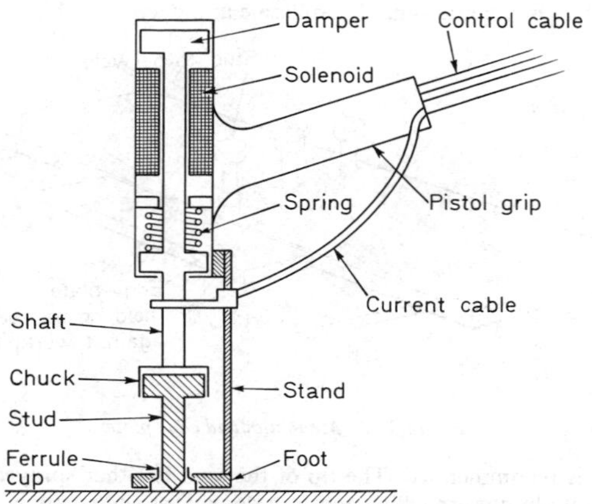

Figure 37.18 Stud welding gun

fixes into a gripping chuck in the operator's gun which in turn is placed vertical over the spot to be welded and which rests firmly on a three point support on the steel surface. When the trigger is pulled an electronic timing device controls the following sequence:

> The chuck is lifted about 3 mm by an electromagnet and a pilot arc is formed about the tapered point (Figure 37.19) which then develops into the main arc, the main arc current being drawn usually from a drooping characteristic transformer–rectifier power source. This arc melts the end of the stud with a resultant melted

area on the workpiece and after a preset time the solenoid is de-energised and the stud is plunged by a return spring on to the workpiece while the arc current is still flowing.

The stud when correctly welded should be of a correct length after welding with a formed upset fillet around it with no undercut; such undercut may be due to incorrect welding parameters or arc blow and when present can lead to easy fracture of the stud from the

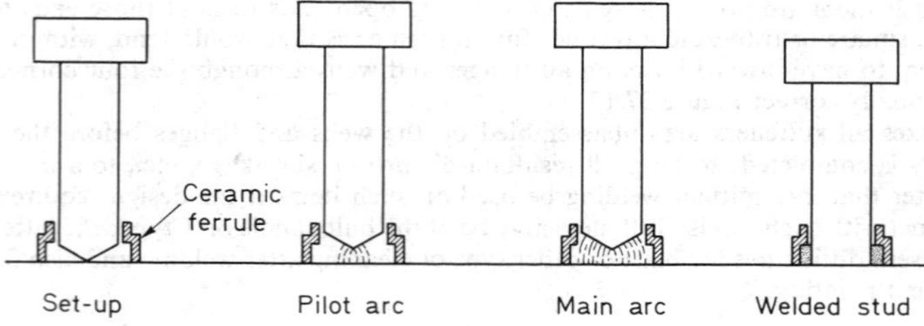

Figure 37.19 Sheer-connector stud

plate surface. Arc blow because of the high, though transient, currents used may be prevalent when the studs are placed near to the edge of the plate; in such cases an edge plate to extend the magnetic field of the current in the main plate may be utilised, see Figure 37.20.

Studs greater than 22 mm ϕ are difficult to weld, leading to erratic arcs and sometimes unsound welds; the state of the plate surface on which the studs are being welded and should be free of all oily contaminants, millscale and deep rust. A light surface grinding

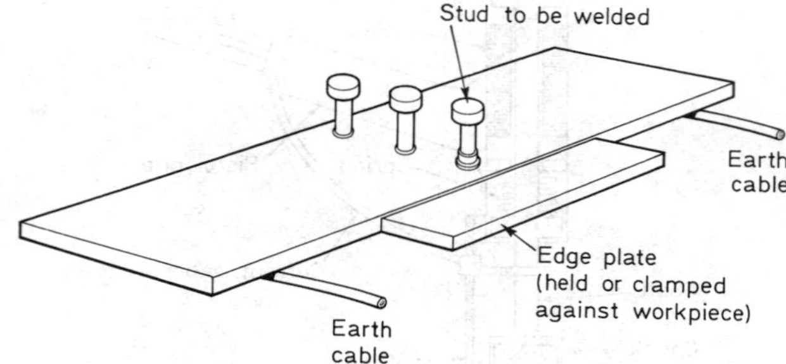

Figure 37.20 Magnetic field edge plate

in the stud area is recommended. The tip of the stud is either sprayed with aluminium or holds a 'slug' of aluminium which acts as a deoxidant when vaporised in the arc; it is important that this deoxidant is not damaged.

One test sometimes employed to ensure the soundness of the steel weld is to bend some to an angle of 30 degrees and to hit or 'ring' the others with a hammer.

TESTING

'Methods.' These may include nondestructive testing methods such as X or gamma radiography, ultrasonics, dye penetrant or magnetic particle testing. Radiography is used almost exclusively on butt welds and ultrasonics on butt and some fillet welds; site welds are invariably tested by ultrasonics and/or gamma radiography employing in

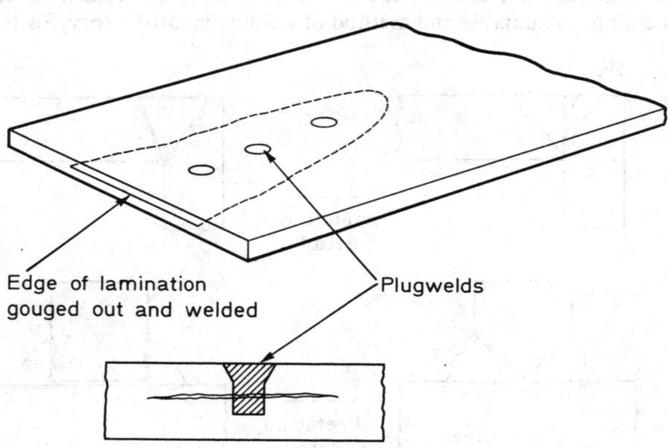

Figure 37.21 Repair to lamination

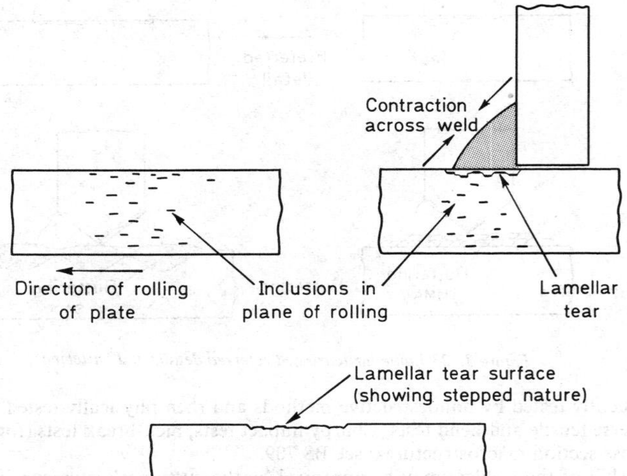

Figure 37.22 Lamellar tearing

general iridium as the source of gamma rays. The standard and scope of testing required is usually determined by the customer and should be ascertained at the enquiry stage.

Before most contracts commence, some welding procedures may have to be approved by the customer; that is, a weld joint simulating the thickness, preparation, etc., of an actual weld configuration used in the fabrication must be welded to prove that the proposed welding consumable and method of welding is satisfactory. Such welds may be

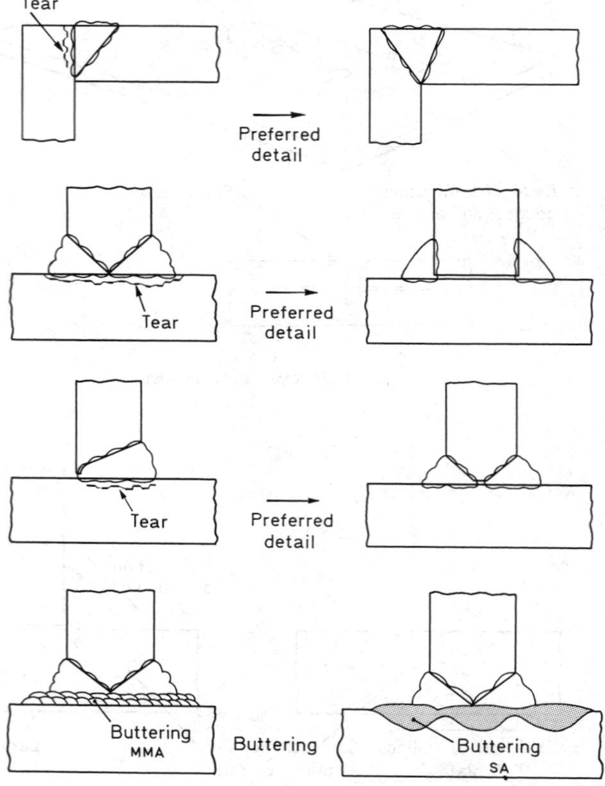

Figure 37.23 Lamellar tearing. Preferred details and buttering

subsequently tested by nondestructive methods and then physically tested by means of transverse tensile and bend tests, Charpy impact tests, nick-break tests (for fillet welds) and cross-section macrostructures; see BS 709.

The skill of the welders may be approved by the customer's own specific test or by BS 4872 *Approval testing of welders when welding procedure approval is not required* or any other subsequent standard with any appropriate nondestructive test requirements.

'Laminations.' Where a plate is laminated or where a plate must be tested for laminations before being incorporated in a fabrication, ultrasonics is the only method by which any such lamination may be detected and the extent measured. Material may be supplied to a standard of lamination testing by the steelmakers; the details of such standards and

the appropriate costs may be obtained from the steel supplier. The effect of any lamination on the stability of the structure must be referred to the designer; for example the effect of a lamination in a compression member is generally more severe than one in a tension member. It is probably true to say that most structures can tolerate a fairly large degree of lamination in a member before repair is required; the repair of such a lamination is shown in Figure 37.21.

'Lamellar tearing.' This is a result of nonmetallic inclusions in the steel in the plane of rolling merging into a tear due to the stress imparted by the weld (Figure 37.22) or other derived stresses normal to the plane of inclusions. Because these inclusions are very small and scattered throughout the thickness of the material they are not detected radiographically, and, up to now, although detectable by ultrasonic inspection, they cannot be quantified to assess potential cracking. The tear, when it occurs, is of a ductile nature, fibrous and stepped or ragged as shown schematically in Figure 37.22, the steps resulting from the inclusions in different planes being joined to form the tear; the presence of such inclusions decreases the ability of the material to withstand extension under a load applied across the thickness of the plate. One destructive method of assessing the material for tearing is to machine a small transverse tensile test piece and to measure its reduction in area at failure of applied tensile load.

Joint details that can promote tearing are shown in Figure 37.23; generally the welds must be large to cause contraction across the welds on cooling to create sufficient tensile stresses across the plate to produce a tear in a susceptible material. For fillet welds to generate such a tear the weld in almost every occasion would have to be greater than 12 mm, although many materials can tolerate much bigger welds, and any details which have both large weld and imposed load stresses across their thickness should be avoided; some preferential details are shown in Figure 37.23. On some suspect material it may be helpful to reduce the risk of tearing by buttering the weld fusion face with MMA or submerged arc welds as also shown.

REFERENCES

1. FLINTHAM, E., *Manual, semi-automatic and automatic welding,* British Oxygen Co. Limited, Hammersmith House, London, W6 (1966)
2. HOULDCROFT, P. T., *Welding Processes,* Cambridge University Press (1967)
3. SMITH, A. A., *CO_2 Welding,* Welding Institute, Abington, Cambridge
4. *Specification for Steel Girder Bridges,* BS153, British Standards Institution, London
5. *The use of structural steel in building,* BS 449, British Standards Institution, London
6. BAKER, R. G., WATKINSON, F. and NEWMAN, R. P., 'The metallurgical implication of welding practice as related to low alloy steels', *Proc. 2nd Commonwealth Welding Conf.,* 1966, Institute of Welding, London, 125–131
7. BAILEY, N., 'The establishment of safe welding procedures for steels', *Am. Welding J.,* Research Suppl. (Apr. 1972)

38 STEELWORK ERECTION

STEELWORK ERECTION 38

38 STEELWORK ERECTION

W. H. ARCH, B.Sc., C. Eng., M.I.C.E.
Redpath Dorman Long Ltd

INTRODUCTION

It is not entirely by chance that the French refer to a bridge as a work of art. For it is indeed a very close liaison between science, and the art of its application, that makes for success in bridge building. This is particularly true for bridges, but is also applicable—though perhaps to a lesser degree—in other types of structures.

A structure, of any type, is essentially designed for its completed condition, and the act of achieving its successful completion is that of coping with logistics of its components, with the partially erected frame, and with the people who will be involved in its erection.

Many variable factors affect the choice of erection method which will be used in any particular case, but on the other hand there are a number of common elements connected with any erection plan which do not vary from job to job.

For any project to be successful it is essential that the designer of the structure has clearly in mind how he visualises the job will be built. It may well be that the final scheme may vary from that which was originally envisaged due to modifications and improvements, but at least the original conception would have been based on a completely thought-out, fully integrated design.

The problems posed by the logistics of a job, the control of materials flow, and also of the organisation of the manpower who will build the structure, are common.

Two examples of design points which the original conception should be able to accommodate, immediately spring to mind. The effects of temporary points of support and possible reversal of stresses during bridge building, and secondly the orientation of column webs and flanges to enable the closing beams to be erected in the final stages of completion of a multi-storey frame.

EFFECTS OF THE SITE ON A PROJECT

The shape and location of the site on which a project is to be built will have had its effect on the basic design of the structure to be erected.

For a tall building for example the size of the site, as well as the eventual use of the building will have determined the column centres, the height or number of storeys will have determined the section of the column and, therefore, the weight. In a bridge the configuration of the area to be spanned together with the ground conditions will have determined the spans. Only in uncomplicated cases will the economics of the actual design of the spans themselves be the sole determining factor in the choice of span—or indeed of the method of construction.

Thus the site through its bearing capacity, or through existing structures or services, has a profound effect on the design of the structure to be erected upon it. That the design of the structure is affected by site conditions is thus clear, but the effect those same conditions have on the erection method to be adopted is even more profound.

A restriction on access for the materials, or the presence of underground services can both affect the placing of cranes and thus the whole approach to the erection plan.

Time on a site costs money and a good scheme should, therefore, aim to reduce as far as is practicable the number of manhours required on that site by the maximum use of prefabrication and subassembly off site. If this is taken to extremes then the components

Figure 38.1 This view of the deck of the Forth Road Bridge during erection shows the changes in profile which can occur during the erection process. Arrangements must be made to accommodate and follow such changes in the crane support arrangements. The connections in the structure itself must also accommodate such changes and be designed accordingly

requiring to be lifted can be too large to transport or too heavy to lift. There is thus an optimum balance for any one job, and in major construction projects it is not often that what is right for one job is also right for another.

The ground conditions will have been fully evaluated before the foundations of the structure are designed. It is of equal importance that the ground conditions under temporary foundations are properly evaluated. In a bridge erection scheme, for example, the whole stability of the project can depend upon the adequacy of a temporary pier and, therefore, of the foundations under that pier. Figure 38.2 shows a bridge span in just such a condition where everything is dependent on the temporary arrangements until the relative safety of the next pier position is attained.

All temporary loads—even of such a transient nature as those under the outriggers

Figure 38.2 Mobile cranes are here being used to subassemble portions of a bridge structure preparatory to the assembled girders being rolled out across the river. The level working area and blocked outriggers can be seen. The Scotch derrick in the background will be used to infill the secondary steel work later in the erection process

of a mobile crane—are of importance. Only by full attention to such matters can the safety of the men working on the job be safeguarded, and the safety of the structure itself be assured against collapse.

THE EFFECT OF PLANT ON DESIGN

Everyone is familiar with the contractor's plant yard. Items go there from one job, to be refurbished, tested as required, often given a coat of paint and there they lie ready to be sent out onto the next job.

The choice of erection plant which is to be used on the next job will, therefore, mean that, for economic reasons, existing tackle will be used where possible. There is, therefore, a need for plant which is as universally useful as possible. Erection cranes with long travel bogies at centres which can be varied from job to job are just one example. At the other end of the scale are the special 'one-off' erection devices built to suit the peculiarities of a particular job. The chances of all the peculiarities repeating exactly on another job are remote, and, therefore, the 'one-off' device is only used on the really big jobs where the very high cost can be written off against that job.

Whatever the type of plant which is being used to erect the steel the economics of that job will be vastly affected by the variance between the actual average weight of each piece to be lifted and the capacity of the crane at each relevant radius. Thus in the ideal situation the weight of every piece should equal the capacities given by the safe-load indicator on the crane. Clearly this is an ideal unlikely to be experienced, but a job on

Figure 38.3 The mobile crane is a standard piece of equipment for bridge erection. Points to note are the proper provision and blocking of the outriggers and the use of the minimum radius possible for handling the lifts

which a conscious effort has been made in that direction will be a more economic job to build.

The choice of plant and the location of that plant during the various stages of building will have a significant effect on the stress generated in the structure during erection. Very careful thought must, therefore, be given to the need for temporary material in additional supports and to the need for additional permanent material to transmit the transient additional erection loads, stresses and deflections.

The use of models can be an invaluable help in the positioning of plant for the best use to be made of the capacity available. The increase in the availability of computer facilities is enabling the optimisation of erection methods, and the resultant requirements in terms of additional material, manpower and plant, to be achieved more easily. In addition a large number of alternative possibilities can be considered.

The temporary supports and erection cranes are arranged to safeguard the safety of the structure during erection. It is of equal importance that full consideration is given to the many working places where men will have to work during that erection. The provision of safe means of access to these working places will have the dual benefit of improving the safety of the individual, and at the same time of speeding the flow of the work. Careful thought must be given to the use, and location of safety nets since a badly used net can be a hindrance to the job and can indeed be a safety hazard rather than providing the protection intended.

TOLERANCES

There are four main areas in which the question of tolerances affects the engineer building a bridge. These are the accuracy of the geometry of the bridge in the horizontal and vertical planes, and then the accuracy of the matching of the components at a splice and lastly, but by no means least, the control of local deformation at a splice, particularly on a welded detail.

Site setting out

The establishment of the centreline and main pier positions is the first essential in any bridge construction job, and upon the correctness of this depends very largely the whole of the project. The accuracy of the initial survey and the standard of marking which is possible in the very early stages naturally leads to a larger tolerance being acceptable at that stage than could be entertained as acceptable when the setting of the bearings is being undertaken later in the construction cycle. At all stages of construction, errors in a transverse direction to the centreline of the bridge should be considerably less than those referring to the overall distance between abutments. Levels should be able to be controlled to a very close tolerance, and agreements in the second place of decimals should be possible using accurate instruments properly handled.

The variation of conditions and the distances involved make it difficult to quantify what is an acceptable approximation for the location of a point on the plan of a bridge, but an oval giving plus or minus 3 mm normal to the bridge centreline and plus or minus 12 mm along the centreline should be attainable when resort has to be made to triangulation for the setting of bearings. Later in the construction when sight lines are shorter, and conditions are generally better for the marking and preservation of lines, considerable improvements in these figures should be possible.

Camber and vertical curvature

Where a bridge is being built using large prefabricated components there is always a possibility that local dimensional variations and twists can occur across the section. Measures must be adopted to accommodate these effects so that force is not necessary to bring the bridge into contact with its bearings. In some cases it may be necessary to provide a temporary support on the centreline of the bridge to allow the structure to take up its natural shape before setting the bearings to suit. If this is necessary care has to be taken to ensure that the required vertical curvature and camber are maintained as closely as possible. It is important to ensure that controllable errors are not allowed to remain uncorrected and that the adjustment to the bearings is not used merely as a means of compensation for careless work.

The use of machine levels and adjusting screws for levelling the bearings will enable extreme accuracy to be attained in setting the upper bearing surfaces of the bearings. There should be a small air gap between the upper and lower halves of the bearing during final setting to ensure an even distribution of load across the bearing before the load is finally transferred.

Great care must be taken to ensure that the bearing is correctly placed below any diaphragm or stiffener in the structure above.

There should be no need to emphasise the effects which can be caused by loads being applied at points not designed to receive them, or of loads eccentric to their designed position.

The specification of the bridge will lay down the precise tolerances permitted for position and level of the bearings in relation to the structure they are supporting. In multi-span continuous girders the tolerances are tighter than those for simply supported members so far as level is concerned. The position of bearings should be generally such that they are aligned within ± 5 mm of the centreline of diaphragms and stiffeners shown on the drawing.

The procedures in the fabricating shop, and the shop assembly of adjacent sections which the specification may have called for, will ensure that these components will recreate the camber called for when they are erected on the bearings we have been discussing.

It is essential to ensure that the methods used for handling, stacking and subsequent erection do not cause permanent distortion of the components. Means of protection should be provided not only to avoid damage to the slings used, but also to prevent damage to the component being lifted or stacked. Incorrect handling can cause deformations and it can also damage any protective treatment which has been previously applied.

Tolerances across site butts

The jigs and shop erection procedures used to ensure accuracy of shop fabrication will have minimised the errors in positioning of stiffeners, etc., and will have kept distortions within practicable limits. The responsibility of those on site is then to reassemble the components and erect them into place.

The tolerances allowable in the butt joints between the components are dictated by the need to limit the eccentricity of adjoining members and the secondary stresses those eccentricities would produce. The nature of the stresses, compression or tension, to which the members will be subjected in service, affects the acceptable limits. The compression condition requires a more severe limitation than does a connection loaded in tension. A temporary or transient loading condition, say, during erection, can of course reverse the permanent condition and this must be taken into account when the connections are being made.

If a butt in the flange plate of a girder is to be made by welding then care must be taken to limit the amount of 'cusping' that can occur when the butt is completed.

At the time of writing, a set of interim design rules governing the general basis of design and method of erection of steel box-girder bridges is in preparation, and this document will also contain a set of fabrication tolerances for the plate panels and for the butt joints associated with them.

The tolerances of both alignment at butts, and of straightness, will be based on measurement taken over a gauge length dependent on the spacing of stiffeners and the amounts of allowable deviations will in general be based on the thickness of the plate or flange being measured. Reference has already been made to the relaxation that is possible in connections subject only to tension loading.

The Committee under the chairmanship of Dr A. W. Merrison which has been drafting their proposed rules covering these tolerances has had the benefit of advice from the fabrication industry in addition to evidence provided from academic sources.

A committee of the British Standards Institution is also working on the latest revision of BS 153 covering bridgework and this committee is also considering the question of tolerances in the light of the latest shop fabrication techniques and research results.

There is clearly a balance between the maintenance of practical limits on the accuracy of workmanship and the achievement of very high standards approaching perfection. The fabrication industry has always been keenly aware of the need to maintain a high standard of workmanship in its products whilst at the same time ensuring that it remains competitive in the market places of the world.

THE INTERACTION OF DESIGN AND ERECTION FACTORS

In the subsection starting on page **38–4** consideration was given to the choice of plant which would be used in a particular project. The weight of the pieces to be lifted is clearly of paramount importance in this consideration.

The weight of a component is a function of its size. Piece weights, therefore, affect the location of the splices connecting the components. In addition there are limitations,

Figure 38.4 A site subassembled column section for a high-rise building being erected. Site subassembly enables more easily transportable components to be built up into economic units for erection. In this case also moment connections could be more easily welded up in optimum conditions and the stub–beam connection reduced to an easily jigged and readily accessible connection

and associated regulations, which limit the size of component that can be transported. Thus the site and any limitations imposed by its access roads can affect the design just as much as the capacity of the crane on the job. All these points must be borne in mind in addition to the purely design considerations which must affect the locations of splices. Particularly where shipping to overseas destinations is involved, the 'nesting' of components must be considered during the design phase in order to minimise the volume required to be occupied by a given weight of material. The cost penalties for shipping a bulky lightweight component are very heavy.

The restrictions referred to, where size must be limited, can result in components which are uneconomical to erect. In these cases consideration should be given to designing to enable subassembly or prefabrication to be carried out at site before the final lifting of the component into place. An example of this in the tall-building field is shown in Figure 38.4. In this example the floor beam stubs have been welded to the column shaft after arrival on site, but before erection. There were a number of advantages in this procedure. The plain components would 'nest' well for transport thus reducing the cost of transportation. The depth of the floor girder sections could be increased at the columns in order to be able better to transmit the wind loads, and lastly the splices remaining to be made up in the air are in a much more accessible location. The location of the splices between sections of columns is limited both by the capacity of the crane available for the erection, and also by the climbing cycle which has been decided upon. Three floor levels in one piece is normal, with a reduction towards the lower sections where heavier scantlings will be used.

FOUNDATIONS AND TEMPORARY SUPPORTS

The design of the permanent foundations for a structure will have been considered in very great detail when the structure was in the design office. From the point of view of the erector, however, the temporary condition must be examined. This applies more particularly to bridge building, but the principle applies in any erection plan. Not only can transient loadings and moments which will be applied to foundations during erection exceed very considerably those which can occur after the structure is complete, but temporary foundations may also be required.

Where temporary foundations are required, as much consideration should be given to their design and construction as their importance to the safety of the erection scheme dictates. The main foundations will not have been designed without reference being made to the results of test borings. Where loads are significant or the effects of settlement on the safety of the structure could be dangerous, then a trial boring should be put down local to the temporary foundation to ensure that adequate precautions are taken.

The additional loads that must be considered when designing temporary foundations should include those that arise from the incompleteness of the structure that they are supporting. The effects of wind loading and aerodynamic instability must be considered. The weights of items of plant, plus the loads they are carrying and any resultant uplift and dynamic loads due to movement, can affect the design.

Particularly in bridgework there will be temporary piers to place on the foundations discussed earlier, and also temporary extensions to the permanent piers. These are commonly constructed of standard components which can be stored and re-used on future work. In the event that specially designed and fabricated temporary supports are not being used, a check calculation should be made to ensure the adequacy of the standard components planned for use.

The points at which it is required to make connections between the permanent structure and the temporary erection structures should be given careful consideration. They must be incorporated in the original design conception. They must be adequate to carry the various loads and stresses to which they will be subjected during the progress of the work. Means of erecting the temporary structures themselves must be provided, and last but not least, provision must be made for the removal of the temporary structures and the making good—or the incorporation as an architectural feature of—the pockets or other connection points provided.

In buildings the temporary structures most commonly used are again props or columns used to support girder structures at intermediate points during the construction, for example, of large lattice girders spanning distances too great to permit erection of the component in one subassembly. It is important here to check the adequacy of the beam carrying the prop, or of the tower column length being temporarily extended. It is not uncommon where temporary erection loads have to be accommodated, for permanent additional material to be incorporated to provide the additional carrying capacity required.

The detailed design of temporary steelwork should take due account of the effects of deflection of the supports. Provision will also have to be provided for unloading the member, and for its removal from the structure after the completion of the member being supported.

THE PARTIALLY COMPLETED STRUCTURE

We have considered the need for temporary supports during the erection of a structure, and these considerations have dealt principally with the need to shorten cantilevers during

Figure 38.5 In the erection of an arch bridge, temporary supports are required to assist in cantilevering the two halves of the arch out from the abutments. Transient loads, deformations and stress reversals which occur during the erection process must all be carefully calculated for all erection stages and constant checks made of the actual behaviour of the structure as it grows out of the crown.

bridge erection or to prop long girders during assembly in a building.

However, all structures are designed to have a minimum amount of redundancy and it follows therefore that until an erection procedure is complete the structure is at risk. It is to eliminate this risk and to take due account of the changing stress patterns that will arise during an erection procedure that must exercise the engineer concerned with those erection procedures. It has already been stressed that an erection scheme should be borne in mind when the original design is being made, since only by that means can a feasible, safe and well regulated design be produced.

If a lattice girder structure is to be built out by cantilevering from a pier it is clear that the stresses in the members will be reversed until the next pier is reached and the girder spans from pier to pier. Similarly only when all the spans are completed will any continuity assumed by the designer be achieved.

Where the lattice girder is supported by piers or, in the case of an arch rib may be supported from above by temporary cables (see Figure 38.5), very special stress patterns can develop. It may be necessary to carry out a number of case studies at progressive stages of erection in order to ensure that the most critical condition for each member and for each connection has been considered and adequately dealt with.

A building designed to derive its rigidity from shear walls or from the composite action of the floor slab will be unstable until these features are completed. Temporary means of providing this rigidity must be arranged.

The location of these temporary members, and their stiffness relative to adjacent members of the partially completed frame must be carefully considered. It must be possible to construct the permanent members without removing the temporary load-carrying members, and to be able to remove the temporary members afterwards.

The permanent structure must be capable of absorbing the loads induced by the temporary members. These can in many cases frame into locations not intended to carry those loads when the frame is completed. Cases of eccentricity need special attention in view of the secondary stresses which can be induced.

STOCKYARDS AND TRANSPORT

We have seen that prefabrication enables manhours to be expended away from the site, and thus to reduce the time that a site is occupied to the absolute minimum. It is this factor that enables a steel frame to be put up so quickly, and to provide an instant support for the follow-up trades.

Inevitably, components must then be transported from the factory to the site. Nothing can ever be perfectly controlled, and a stockyard is therefore necessary to absorb the differences between materials arriving on site, and those actually required from day to day for erection. The effects of delays in the transport system must be cushioned, otherwise expensive erection equipment and manpower would be kept idle.

It has been very truly said that a good stockyard control can make a successful job. The reverse is certainly true: if components and fasteners are not sent out to the erection front as required there will be delays.

The stockyard then is the reflection of the adequacy of the management control on the job and this will be touched on in a later section.

Incoming material must be correctly recorded, and located in order that it can be relocated and transported out when required without resorting to double handling. It is often convenient to colour code the material to conform either to types or area of the project. Craneage will be provided to handle the weights of components involved, with the same rules applying as were discussed in the second section. A light crane with large coverage to handle the light components, and all the heavy components under the heavy crane. A Goliath type of crane is often used for stockyard duties since it occupies little ground area for its tracks and can easily cover a large area without diminution of capacity.

Adequate roads—and rail tracks where required—must be provided to give all-weather service. Axle loads of the vehicles using a stockyard are heavy and maintenance time on broken roads is time lost for erection.

Vehicles bringing material into a stockyard will be the normal form of fixed axle or semi-trailer type of transport common on the public roads. Only in exceptional cases will it be necessary to bring components onto site on special transport vehicles requiring police escort. There are limitations imposed by law on both the weight and dimensions of loads that can be moved on the public highway and a knowledge of those restrictions is essential. Additionally many sites are located in places where the road layout, or low bridges, impose restrictions and a study of these must be undertaken before the component gets stuck and the site is disrupted.

On-site transport is not restricted in the same way, and the size of load is similarly subject to different controls. Site subassembly can lead to awkward loads, and the safe loading and fixing of these on the site transport is important.

Pole trailers are commonly used because only one tractor may be needed to work a number of them. Loading or unloading of one trailer can take place whilst another is on the move.

The proper control and distribution of fasteners is important. Bolts should be re-bagged into 'sets' required for particular connections or areas. This saves time at the splice, and also cuts down the waste of bolts when an excessive numer are issued by the store to a location. The proper storage of welding electrodes, their issue and conditioning while being held at the point of use are also functions often delegated to the stockyard stores control function.

MANPOWER AND SAFETY

None of the operations discussed in this article can be carried out without an adequate and sufficiently skilled number of men.

Steel erection is a task that is best learned by experience. Formal training can teach the basic skills of rigging, scaffolding, slinging, burning, chipping, riveting, crane driving, etc., but only experience can weld all these skills together and produce a good all-round erector. It is the man who provides his experience, but it is up to the employer—the management—to provide him with adequate tools to carry out his work. These tools are not only the cranes and the spanners, but also the working platforms that give him a safe place in which to work.

Figure 38.6 shows a man working on an unsafe platform. Not only is it dangerous for the person working on such an unsafe platform, but the safety of the whole job could be jeopardised. The provision of proper boxes to contain loose tools, and good house-keeping in the handling of small components in general, are conducive to a smooth running and safe job.

A large number of regulations control the manner of ordering of work on a construction site, and these, together with summaries of them are available from H.M. Stationery Offices and from RoSPA. These cover such major items as the testing of plant, etc., but it is seldom the big things that give trouble. Attention paid to the odd drift left lying in a walkway or the piece of handrailing someone else has taken down for access repays itself many times over in the smooth, safe and economical running of the job.

CONSTRUCTION MANAGEMENT

The construction manager is a man whose function is becoming increasingly important both on this side of the Atlantic and in North America where many management techniques were developed.

The basic object of management on a construction site is to have the right component at the right place at the right time with an adequate piece of equipment and men to handle it, and to repeat this over and again to complete the structure on time and within a cost target.

To this basic requirement must be added the coordination of a number of contractors, each with his own problems and targets.

There are a number of contractual forms within which construction work can be carried out, but one important feature of all these is that the roles of the parties to the contract are clearly defined. It is essential that responsibilities are clearly defined in this way since we have seen how closely the original design affects the erection process and vice versa. These interconnecting factors must be appreciated by all the parties or disputes, and perhaps tragedies, can result.

There are a variety of ways in which a prospective owner can have the management of the construction of his project organised. One factor common to all methods, however,

Figure 38.6 An example of an unsafe working platform. To ask a man to work under such conditions endangers not only the man on the platform but also the safety of the other men with whom he is working, and the safety of the whole job could be endangered

is that the earlier in the time scale all the parties who are to be involved can add their particular expertise to the planning of that project, then the sooner the project can be started, and the sooner it can be completed. The attainment of this ideal situation is made easier with some forms of contract and management organisation than with others.

In what is being presented here as the ideal situation a project team is brought together at the inception of the planning of the project. This team will comprise those who will be directly responsible for each of the phases of the project, from the design, not only of the frame, but of all the associated services, right through to the site erection functions, at the other end of the planning and construction time scale.

The other end of the spectrum is the job where all the functions are carried out in watertight compartments, the designs being prepared in a number of engineers' offices, and the contractors tendering in isolation. With this system no meaningful discussions can take place between those who either are, or will be, most intimately involved with the project until so late a time in the construction process that their contribution can only be limited.

SUMMARY

The more complex a project becomes and the more services are involved—a tall building is an excellent example—the greater the need there is for early consultation and planning. In a tall building where the lower floors can be approaching occupancy while those above are having services installed and those at the top are still being framed the whole complex construction pattern is seen in microcosm.

Accesses for foundation construction must integrate with the demolition of what was there before. Material arriving on site requires close control and schedules if traffic chaos and a hopelessly congested site are to be avoided.

The positioning of cranes and the stability of the partially completed frame must be very carefully thought through.

The safety of those working on the structure and of those working and perhaps also living in the vicinity of the new structure must be constantly considered.

Only by the cooperation of everybody who will be involved in the project, be it bridge or building, can the best design be made, the best construction method adopted and the earliest completion date be obtained.

All these factors—design, method of construction and construction time—must be optimised if the owner is to be able to make use of his new facility at the earliest possible time and so be able to generate revenue from his investment.

BIBLIOGRAPHY

BARRON, T., *Erection of Constructional Steelwork,* London, Iliffe (1963)

CONSTRADO, *Steel Designers Manual,* 4th edn, London, Crosby Lockwood (1972)

LEECH, L. V., *Structural Steelwork for Students,* London, Butterworths (1972)

O'CONNOR, C., *Design of Bridge Superstructures,* Wiley-Interscience (1971)

WARD, F. C., BRYANT, E. G. and POUND, R. P., Simply supported bridges in composite construction BCSA, London (1970)

Journals

Acier-Stahl-Steel, Monthly: controlled circulation, published in English and five other languages, address: Constrado, Albany House, Petty France, London SW1

Building with Steel, Quarterly: controlled circulation, British Steel Corporation, address: 33 Grosvenor Place, London SW1 (Tel. 01-235-1212)

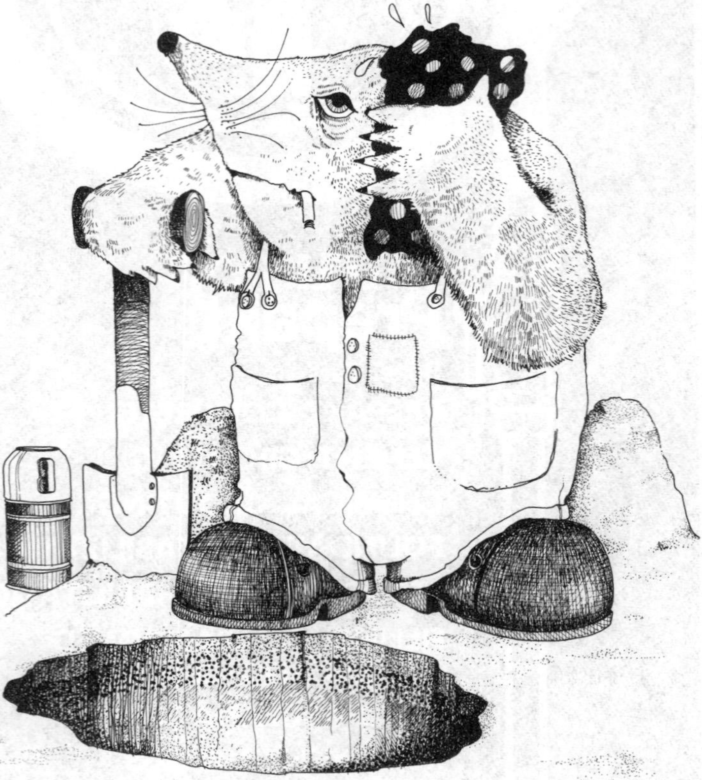

39 BURIED PIPELINE AND SEWER CONSTRUCTION

BURIED PIPELINE AND SEWER CONSTRUCTION 39

39 BURIED PIPELINE AND SEWER CONSTRUCTION

M. A. RICHARDSON, B.Eng., M.I.C.E.,
Rees Construction Ltd.

The survival of modern man is dependent upon the existence of a complex of subterranean pipelines within his environment. Water supply, power and the movement of liquid wastes are the prerequisites of any advanced civilisation. The construction of these complexes must be planned and completed before a thriving community can be established, as was realised by the early civilisations. The Romans and other advanced early civilisations devoted considerable effort to solving the problem of transporting water by gravity aqueducts and pipelines to their settlements and to ensuring that waste water and sewerage was similarly removed to remote areas for disposal.

Later civilisations were less well planned. Communities were established where water was readily available. Sewage disposal was left to the individual. As late as the end of the nineteenth century and still in some areas today, the main sewer was and is the street outside the door.

With the advent of the nineteenth-century Industrial Revolution the increasing demand for domestic and industrial water supply and liquid waste disposal led to the realisation that potable water supplies and sewage disposal were a vital requirement for expansion and prosperity.

CLASSIFICATION

Buried pipelines are readily classified into small and large diameter, shallow or deep.

Small diameters are normally regarded as 100 cm and less and shallow pipelines are regarded as being less than 3 m below the ground surface.

A further classification depends upon whether the pipeline flow is under pressure or partially full under gravity.

Buried pipelines are installed either in trench or by non-surface-disruptive tunnel methods. Small- and large-diameter pipelines are constructed in trench sometimes at depths in excess of 10 m. Recent improvements in tunnelling techniques mean that fewer small-diameter pipes are constructed in trench at depths in excess of five metres in built-up areas where surface disruption is inconvenient or expensive. Deep trench excavation is mainly confined to open-space site location where sheet piling or large-width trenches can be excavated without inconvenience.

Most new pipelines are required within existing conurbations. This creates special requirements on construction methods to give the minimum disturbance to the surrounding environment. Modern towns and cities cannot always be disrupted by deep trenchwork. Modern pipeline technology increasingly involves the use of tunnel methods of one form or another.

USES AND USERS

Pipeline construction is usually commissioned by commercial or public interest such as the water supply and sewage authorities. They are usually faced with the greatest problems of construction as the pipeline is most likely to be required within congested urban areas.

Main sewers are placed at greatest depth as they outfall into the sewage works or treatment plant, at the lowest end of the disposal process. These pipelines present the more difficult construction problems.

The water mains, supplying from areas of high elevation, can be laid at relatively shallow depth. Oil and gas pipelines are invariably pressurised and are shallow.

The route of a pipeline is a compromise between the minimum distance from origin to source of the flow and the physical conditions existing between the origin and source above and below ground. The shortest geographical distance would logically give the most economic pipeline. If, in travelling the shortest distance, it passes beneath surface obstructions or is laid at great depth, it may be cheaper to travel around the obstruction, or miss the area of great depth, the increased length being offset in cost by shallower construction work.

Unless surface dislocation is impossible, shallow pipeline construction work is simple and straightforward and presents no great theoretical or operational problems other than the important consideration of high live loads induced in the pipeline during and after installation.

COSTS

The cost of alternative methods of construction determine the final route of the pipeline and its method of construction.

The final in-ground cost of a subsurface pipeline is the sum of the cost of the component operations in construction and the permanent materials required.

> Total cost = total labour expenditure + total plant machine expenditure + total cost of pipe and associated permanent materials

For any particular diameter and type of pipeline laid in trench the total labour and plant machinery cost is approximately an exponential function of the depth.

The cost of the pipe and other permanent materials is not significantly affected by the depth. The cost of a pipeline laid by nondisruptive methods is largely unaffected by the depth.

The variation in cost of trench and tunnel construction is shown qualitatively graphically in Figures 39.1 and 39.2.

In Figure 39.1 the intersection of Line A with Lines B and C indicates the rate of construction at which tunnel- and trench-constructed pipelines have the same cost.

Figure 39.2 indicates the variation in cost with increasing depth of construction for tunnel and trench. Position D is the depth at which tunnels and trench have the same cost. The depth at which it is economic to construct by nondisruptive tunnel is decreasing as a result of improvements in tunnelling techniques.

TYPES OF PIPELINE MATERIAL

Many varieties of pipe are manufactured, each type being manufactured for specific technical or economic reasons. See Table 39.1.

TRENCH CONSTRUCTION

Deep-trench construction is skilled work and must only be undertaken by operatives fully aware of the dangers inherent in below-ground operations. The method of construction and laying of pipes in trench involves due consideration of the above-ground environment circumstances and the subsurface conditions likely to be encountered.

Where space is available at ground level and the substrata are firm and dry, wide battered trenches can be excavated quickly using large excavations. The trench sides are battered at the angle of natural repose of the substrata but should never be steeper than 2 in 1. The trench sides should be terraced where they exceed 2 m in depth.

Collapses may occur to battered-trench work through failure to reduce batter when the material being excavated deteriorates.

When space at ground level is at a premium or where the trench depth would give an overall excavation of excessive width, vertically sided trenches are excavated using timber or steel sheet piling suitably supported by walings and struts. Trenches of up to 4 m depth

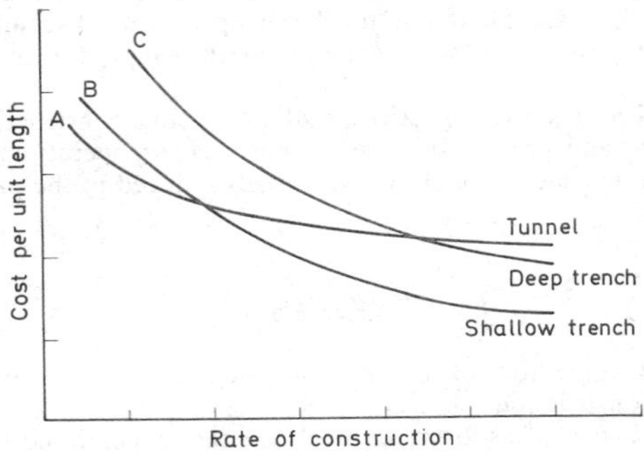

Figure 39.1 Relative cost analysis: tunnel- and trench-constructed pipelines

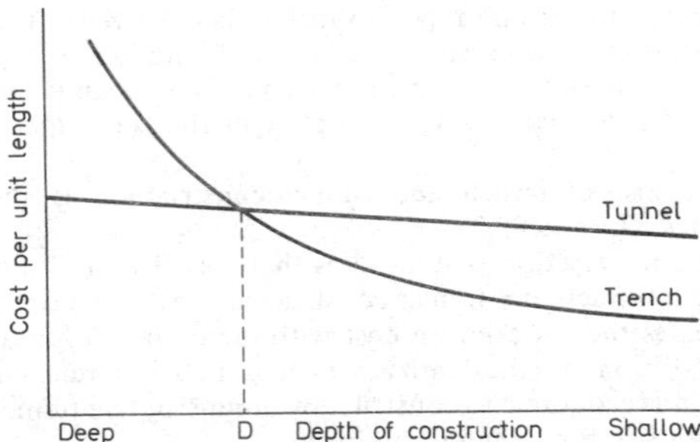

Figure 39.2 Cost variation analysis between tunnel- and trench-constructed pipelines

can be supported by single stages of support. The method of installation will depend upon the firmness of the material being excavated. It is good practice, even in very firm ground and at shallow depth, to give support to a vertical trench side as soon as it has been excavated.

Figure 39.3 shows the correct method of supporting trench sides using timber or light steel sheeting and timber framing suitable for depths up to 4 m.

Figure 39.4 shows the method of trench construction for depths over 4 m.

Figure 39.5 shows the sequence of trench progression where the ground is firm.

Figure 39.6 shows the method of progression where the ground is noncohesive, e.g. sand and gravel.

Under no circumstances should men venture into a trench excavation without support to the trench sides. The support must be extended progressively as the trench is excavated.

Table 39.1

Material	*Range of diameters* (cm)	*Application*	*Remarks*
Clay salt glazed	10–80	Sewers at depths up to 10 m	Now available with flexible joints
Concrete	15–60	Unreinforced:Small-diameter sewers	Substitute for Clay Salt Glazed
	30–350	Reinforced: Large-diameter sewers	May be laid at shallow depth where heavy line loading exists
	36–140	Prestressed: Large-diameter sewers and pressure mains	Reinforced by helical steel winding on outside of pipe barrel
Asbestos cement	15–140	Sewers and pressure mains	Manufactured with asbestos fibres laid circumferentially giving high strength-to-weight ratio
GRP/concrete	30–140	Sewers and low-pressure mains	Manufactured with outer wrapping of glass-reinforced polyester resin
GR Polyester resin	36–120	Sewers and pressure mains	Chemically stable and have exceptional strength-to-weight ratio
Extruded plastic u PVC	8–90	Sewers and low-pressure mains	Available in long lengths, chemically stable
Polyethylene (Polythene)	8–100	Sewers and low-pressure mains	Chemically stable, jointed by heat welding
Steel	8–150	Sewers and pressure mains	Normally require internal and external anti-corrosion protection
Cast iron	8–150	Sewers and pressure mains. Tidal outfalls for sewers	Liable to breakage where ground movement is likely
Wrought iron	8–120	Tidal outfalls for sewers	Has better flexibility than Cast Iron. Not readily available
Ductile iron	8–120	Sewers, pressure mains and tidal outfalls	Largely superseding other ferrous types of pipes. It has good flexibility and noncorrosive characteristics

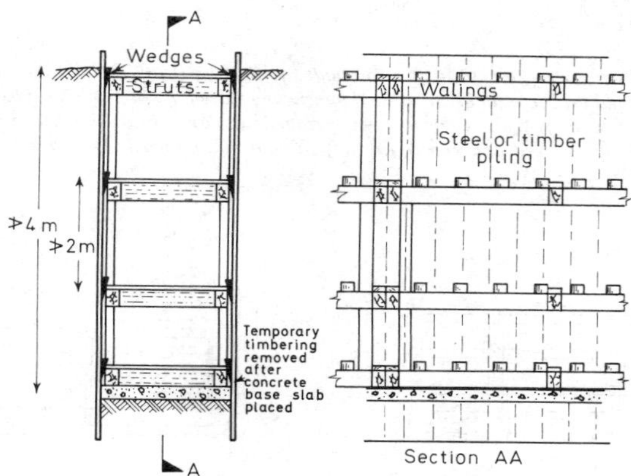

Figure 39.3 Traditional single-stage timbered trench excavation

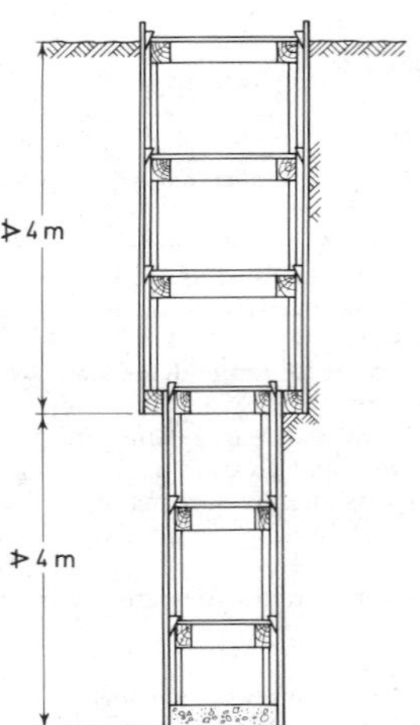

Figure 39.4 Traditional two-stage timbered trench

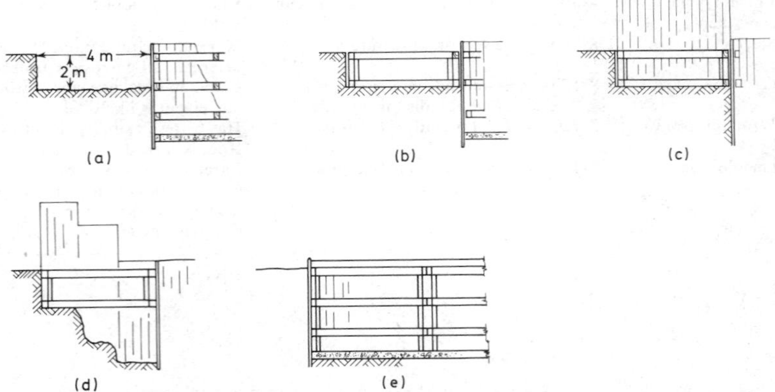

Figure 39.5 Trench progression in firm ground. (a) Section (b) excavated to a depth of 2 m; (b) first and second timber walings, struts and puncheons placed in position; (c) side-support sheets inserted between walings and trench face and wedged; (d) excavation depth increased. Support sheet progressively lowered and re-wedged; (e) third frame inserted, excavation completed and concrete placed

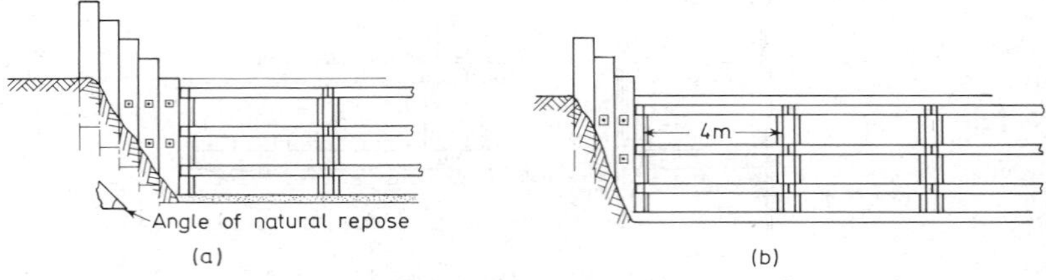

Figure 39.6 Trench progression in loose ground

The most difficult materials in which to trench are compressible silts and warps or sand and gravels where the natural water table is above the bottom of the final formation level. With silts and warps, heavy sheet steel piling will be driven to its full depth, support toe-in of the piles being provided to prevent the trench bottom rising up into the excavation. In very soft silts, the toe-in may need to be at least equal to the depth of the trench itself.

The assessment and design of trench support is determined by a multitude of factors and is not readily calculable. Much theoretical work has been done, notably by Terzaghi and Peck,[1] on the forces imposed on trench support systems.

A trench is a transient structure quickly excavated and back-filled. It is better to design the support system such that the sides of the excavation are stressed by the supports,

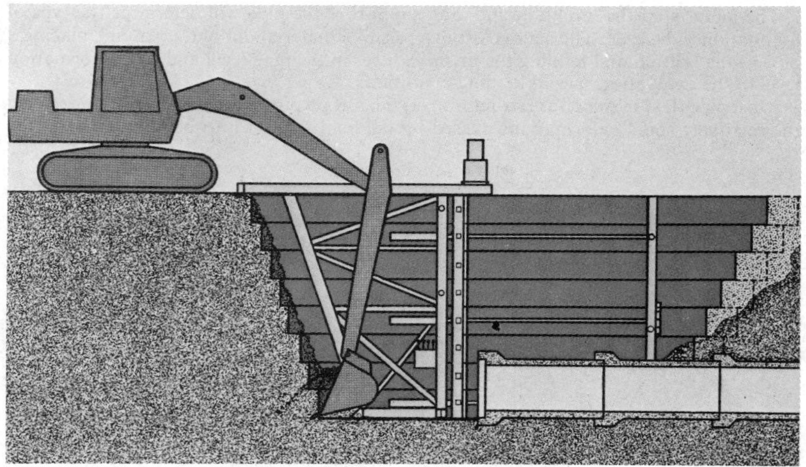

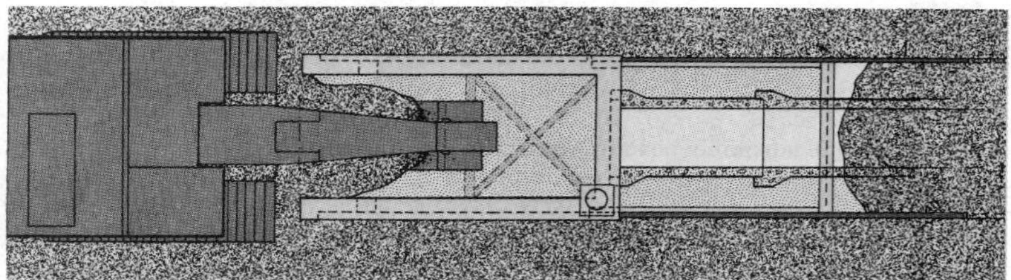

Figure 39.7 Semi-mechanised trench-support system

ensuring that the ground is held back under pressure and minimising or eliminating surface movement adjacent to the excavation.[2]

A recent development in deep-trench excavation has been the introduction of the semi-mechanised selfprogressing trenching machine which, in effect, is a sliding steel cofferdam adapted to progress through the ground, the trench being continuously excavated at the leading edge. As the excavation progresses, horizontal blades are jacked forward in sequence to support the trench sides, pipe-laying and back-filling taking place within the protection of the blades.

Wet sands and gravels may also require heavy steel piling driven to considerable depth prior to excavation. Where sand overlays impermeable strata such as rock or clay, the piles can be driven down and into this strata sealing off the water from the final excavation.[3]

Alternatively, wet sands and gravels may be dewatered by suitable well-point or deep-well techniques.

The well-points and deep-well system can be operated only in sands and gravels which readily yield the ground water.[4]

Well-point operation is effective at depths of trench up to 7 m above which deep-well dewatering may be necessary.

When trench excavation has been completed, the pipe is normally laid in a prepared foundation, surrounded with concrete or granular material and the trench back-filled; the temporary supports are then removed and the surface reinstated.

The material at the level of foundation support of the pipe will determine the type of foundation to be used. The act of cutting a prism of material out of the ground, placing a pipe in the bottom and refilling the prism induces in the pipe itself and on the formation level of the excavation special loading conditions.[5, 6]

Most pipes laid in trench are structurally rigid. A rigid pipe resists loading by converting the relatively small deformations caused by the load upon it into bending stress in the

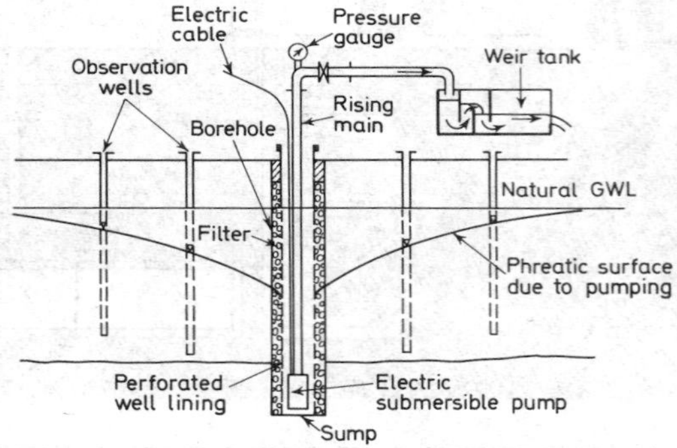

Figure 39.8 Typical deep-well

pipe wall, sufficient to resist the load, without large deformation of the pipe. It behaves as a continuous beam.

The position of maximum bending stress in the rigid pipe wall occurs at the crown and invert and at the spring level. When subjected to its design load, a reinforced concrete pipe, for example, will deform slightly vertically; the tensile stresses on the internal face at the crown and invert and on the external face at springing will cause hairline cracking in the concrete. The strength of the pipe is determined by the load which the pipe will support vertically, producing a crack of predetermined width. British practice allows the crack width in the concrete to be 0.01 in (0.254 mm) at proof load.[7]

A flexible pipe wall, such as thin steel, cannot sustain large bending moments. It can, however, sustain large deformation without damage. Stability is maintained by utilising the passive resistance of the material at the side of the trench generated by a relatively large increase in horizontal diameter and associated decrease in vertical diameter caused by the vertical earth loading.

A flexible-walled pipe, laid in trench, must be provided with compacted material around the sides in order that the passive pressures of the undisturbed material in the trench side can be fully utilised to maintain the stability of the lining. Recent applications of flexible pipe laid in trench have been carried out using 100 cm diameter high-density polyethylene laid at depths up to 13 m.

Table 39.2

Type	*Range of diameters* (cm)	*Application*	*Remarks*
MINING			
Open face without shield			
Timber heading	30–130	Sewers and pressure pipes	Skilled operation. Expensive. Useful where subsurface obstruction anticipated. Pipe laid within heading and back-filled
Steel heading (liner plate)	30–250	Sewers and pressure pipes	Utilises pressed-steel plates with flanges on all four edges for bolting to adjacent plates. Requires secondary lining
Smooth bore. Concrete segmental	130–300	Sewers	Multi-unreinforced segmental block. Segmental ring erected using former. No secondary lining required. May be used as sleeve for pressure pipe
Open face with shield			
Bolted concrete segmental	130–300	Sewers and pressure pipelines	Used in weak ground and where compressed air operation is required. Secondary lining or infill blocks required
Smooth bore. Concrete segmental	130–300	Sewers	Shield used in noncohesive or weakly cohesive ground for safety
Bolted. Smooth bore. Concrete segmental	130–300	Sewers	Same as bolted segmental but requires no secondary lining for sewer use
Smooth bore. Triple segments (Mini Tunnel)	100–130	Sewers	Shield provides security for miner operative in small size. May be used as sleeve for pressure pipes or unlined as a sewer
Hydraulic jacking	90–130	Sewers	Normally only suitable for restricted lengths. Expensive. Uses concrete or steel pipes. Requires extensive temporary work and not always accurate in alignment
MECHANICAL			
Auger boring	15–120	Sewers and pressure pipelines	Suitable for restricted length. Uses concrete or steel pipes. Susceptible to subsurface obstructions and expensive. Requires extensive temporary works and not always accurate in alignment
Percussive auger	15–38	Sewers and pressure pipelines	Useful for short small-diameter crossings of railway lines, canals, etc.

Trench back-filling is a subject all to often neglected. Construction specifications usually require that back-fill material is replaced in thin layers and thoroughly compacted. Granular materials compact well. Cohesive materials are less easy to compact and it may be necessary to import granular material to prevent long-term settlement of the trench fill.

NONDISRUPTIVE CONSTRUCTION

Trench excavation other than at shallow depth or in areas where suitable working space is available, is inconvenient and by its very nature disruptive to the surface and the local environment. Obstructions have to be negotiated where the surface cannot always be disturbed. Common examples are railway lines, canals, buildings, etc. Various nondisruptive techniques are used. These are referred to as tunnelling methods. Tunnelling is any method resulting in the construction of a pipeline without disturbing the ground surface.

Methods include mining, hydraulic jacking of concrete or steel units, and remote rotary mechanical or percussive auger. See Table 39.2.

Timber heading

The oldest soft-ground mining method is the traditional timber heading. A rectangular or square excavation is made through the ground, continual support being given to the ground by timber frames and boards as the work progresses.

On completion of the heading, pipes are laid within to the level and gradient required

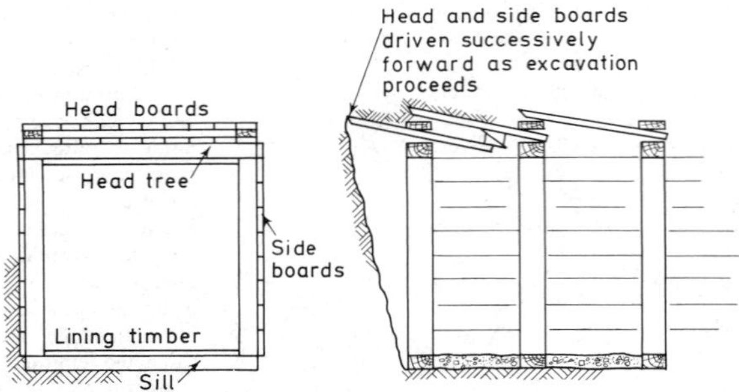

Figure 39.9 Fully timbered heading

then surrounded with concrete or other suitable material. Timber heading can be used for pipelines of up to 130 cm diameter.

The construction of a fully timbered heading is a skilled operation particularly where ground conditions are poor with water present. A reduction in the availability of skilled labour and the rising cost of timber has made the technique expensive and, except in special circumstances, it has fallen into disuse.

Steel liner plate

In firm ground a tunnel can be driven using a steel lining. Steel plates formed to the outer profile of the tunnel excavation are bolted together on longitudinal and circumferential joints. The plates are fabricated of sufficient section modulus to resist the earth's pressure, the completed ring behaving as a flexible pipe during the construction period.

As the work progresses, liquid grout is pumped through preformed holes in the segments to ensure even distribution of the earth's pressure into the ring.

Within the completed liner plate tunnel a secondary lining is formed of *in situ* concrete placed behind a moving shutter or by sliding in a concrete pipe.

Line plate tunnels can also be lined in brickwork or pre-cast concrete lining segments.

Liner plate has been largely superseded by the smooth-bore segmental tunnel and is not as widely used in the UK as it is in the USA.

Bolted concrete segmental block

It has been traditional in the past to regard segmental tunnels as rigid. Early segments for tunnels were fabricated in cast or wrought iron and provided with flanges longitudinally and circumferentially. These were bolted together as the work proceeded and the space between the outside of the tunnel and the excavated earth was filled with gravel or injected grout.

When used for a sewer, the segmental ring was lined with brickwork or *in situ* concrete forming a smooth, rigid structure. During the Second World War, tunnel segments

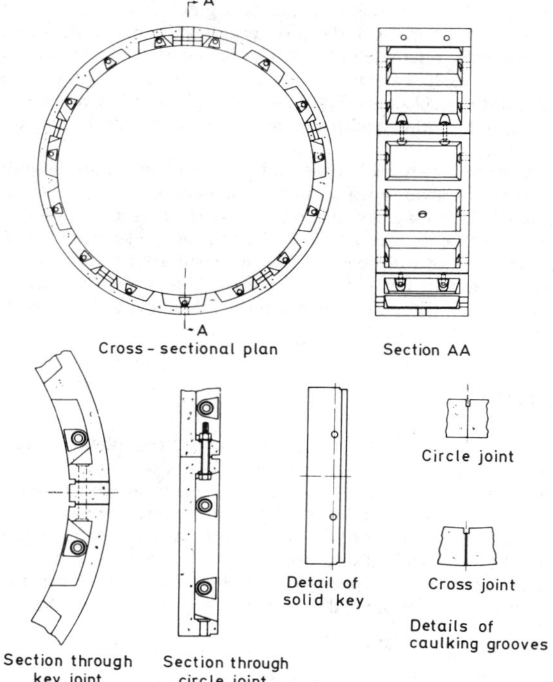

Figure 39.10 Bolted concrete segmental ring

were fabricated in reinforced concrete because of the shortage of cast iron. The concrete segment proved generally as efficient as iron; however, the rigidity of the completed bolted ring was in some doubt. Many tunnels showed cracking at the longitudinal bolted joints, particularly where the earth was soft or was disturbed during the excavation operation.

In these instances, the bolted segmental tunnel lining was behaving as a flexible lining.[8, 9] Bolted segment tunnels are convenient to erect and are particularly useful when used with compressed air.

They are relatively expensive and require to be provided with a secondary lining when used for the transport of water or sewerage. Bolted concrete segments are available from 130 cm diameter upwards.

Smooth-bore bolted segmental block

This type of lining has been recently developed to provide a relatively rigid tunnel lining without the need to line the segments internally on completion. It can be used to advantage in difficult or soft materials encountered during a tunnel drive, where a tunnel shield was not originally installed.

Smooth-bore boltless segmental

The appearance of cracking in bolted concrete segments showed that the lining was behaving as a flexible structure. A lining was developed in which the longitudinal joints are unbolted, being male to female in form. The segments are erected by bolting a steel shoe to each segment, each segment being temporarily connected to its neighbour by coincidental holes in the steel shoes. The space between the segments and the excavated earth is filled with grout or other packing material. The shoes are then dismantled and the process repeated.

Joints between segments are sealed with a bituminous tape and the joint is provided with a caulking groove if ground water infiltration occurs.

The lining has the advantage of cheapness and requires no secondary lining. It is important that the segments are used only in firm ground when installed without a tunnel shield as any downward movement of the earth above the lining will produce a premature deformation which cannot be corrected. This type of lining is available from 130 cm diameter upwards. When installed using a tunnelling shield, it is important that the tunnel shield is steered accurately.

Triple segmental block

Generally, segmental tunnel linings in steel or concrete are difficult to erect in diameters less than 130 cm because of the restriction in working space.

A recent development is an integrated system known as the Mini Tunnel. It is available from 100 cm to 130 cm diameter and a smaller pipeline may be installed within the Mini Tunnel should this be required for hydraulic reasons. Alternatively, prefabricated invert blocks can be used to provide self-cleansing velocity.

In favourable circumstances it is possible, with this system, to construct small-diameter sewers at shallow depth economically, with the added benefits which nondisruptive methods show.

The tunnel is constructed of three identical unreinforced concrete segments; being of three segments, a circular build is ensured and the ring requires no independent support during the erection period.

Each segment ring is built within the rear section of a shield. The shield is driven forward by hydraulic rams, the power being supplied by aerohydraulic pumps. The shield is selfcontained and requires a supply of low-pressure compressed air for its operation.

The shield cuts an overbreak to the segment rings. This overbreak is filled with graded gravel, automatically injected through an orifice in the rear of the shield tail. The gravel packs around the segment ring, giving support to the ground as the shield progresses.

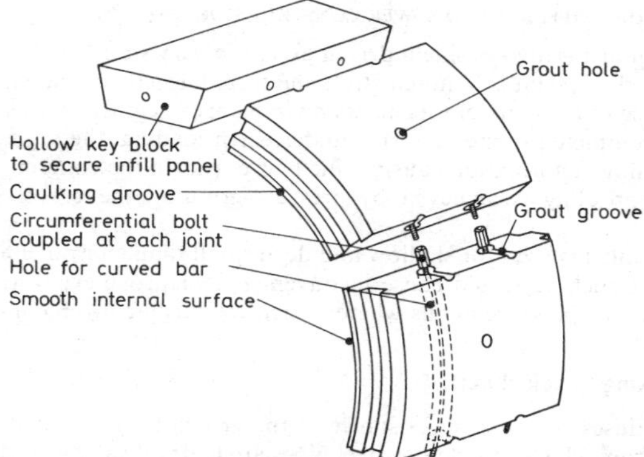

Figure 39.11 Bolted smooth-bore concrete segmental ring (Courtesy: Kinnear Moodie Ltd)

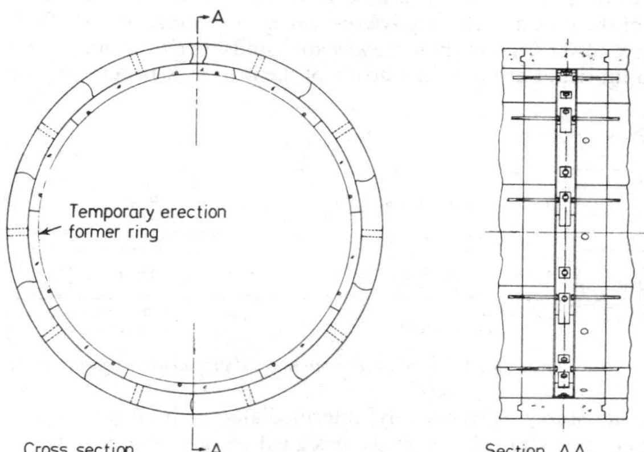

Figure 39.12 Unbolted smooth-bore concrete segmental ring (Courtesy: Kinnear Moodie Ltd)

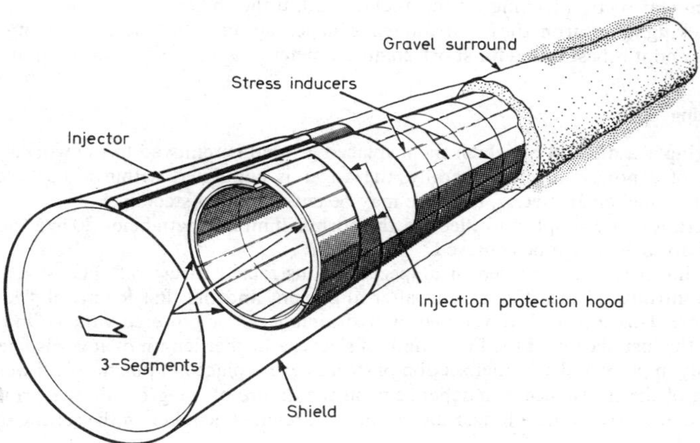

Figure 39.13 Mini Tunnel system of construction (Courtesy: William F. Rees Ltd)

The Mini Tunnel is always shield driven except in very stable rock.

A feature of the system is V notch stress inducers formed coincidentally in the inside and outside face of the concrete segments dividing each segment into subsegments.

When the completed tunnel deforms under earth loading, the segment ring behaves as a flexible lining, deformation causing fine hair cracks in the concrete.

Final waterproofing is achieved by pressure-grouting the gravel surround to the tunnel.

The Mini Tunnel is used for shallow and deep small-diameter pipelines where surface disruption by trenching would cause inconvenience or prove expensive. It can also be used as a sleeve for pressure mains where trench work is inconvenient or impracticable.

Hydraulic jacking (pipejacking)

Pipejacking utilises the passive resistance of the ground at the rear of a working shaft as an anchor from which concrete or steel pipes are hydraulically forced into the ground. The leading edge of the pipeline is fitted with a steel or concrete cutting edge. Material in front of the pipe is excavated by miners as with normal tunnel shield operations.

The maximum length which can be driven is a function of the friction generated on the outside of the pipe and the passive resistance of the anchor wall. Generally, cohesive materials have lower friction than gravels and sands. Friction can be reduced or limited by placing a lubricant around the outside of the pipe as it progresses, a typical material

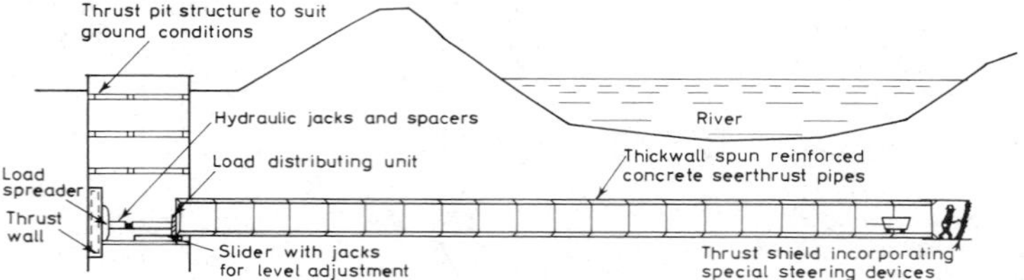

Figure 39.14 Section through normal pipejacking operation

being bentonite slurry. Alternatively, intermediate jacking stations can be introduced into the pipeline at strategic intervals, designed to push the pipeline independently in sections.

The section of pipeline ahead of the intermediate jacking station jacks off the rear section, the rear section jacking off the anchor wall in the shaft.

Pipejacking suffers from the disadvantage of inaccuracy in all but the firmest of ground conditions and it is best suited for short, small-diameter crossings of railways, canals, etc.

Auger boring

Auger boring is a method by which the displaced ground is removed to the working pit by means of a power auger. Generally, the auger is contained within a steel sleeve, although for smaller diameters, the bore may be cut without a steel liner.

The operations are simple and effective, the normal limit in length being 20 m although lengths of up to 30 m can be achieved.

The boring machine is placed in a specially constructed thrust pit. The auger and sleeve are introduced into the ground after alignment and the first length of sleeve is driven home. Thrust force is developed by hydraulic jacks or powered winch. On completion of the installation of the first length of sleeve, a further length of auger is fixed to that already in place and a further section of steel sleeve is placed in position and welded.

Steering of the thrust sleeve is dependent on the nature of the ground. The principle utilised is that of extending a long, rigid column. Care must be taken in the setting of the first steel sleeve and higher accuracies are achieved using long thrust pits allowing maximum length of accurately aligned sleeve.

The machine has diesel hydraulic or electric motive power. A steel main frame carries a sliding subframe propelled by hydraulic rams or winch on which is mounted a manually controlled power unit.

The engine drive is transmitted through a gearbox to the auger. The auger is of an alloy steel comprising spiral flight vanes welded to a tubular shank. The auger heads, of many different designs, generally excavate 10 cm ahead of the lead edge of the steel sleeve and cut an overbreak to the diameter. The overbreak cut minimises the effect of skin friction on the sleeve. The system is inherently expensive, not wholly accurate and has been largely replaced by the pipejacking and percussive auger system.

Percussive auger

It is frequently necessary to lay small-diameter pressure mains and sewers through natural and artificial obstructions. This can be achieved simply and at low cost, using an adaption of the piling hammer.

An outer steel casing artillery shell in shape, contains a heavy reciprocating ram driven by compressed air. The air-driven ram strikes the inside of the nose of the tool and drives the tool forward. A steel or plastic sleeve is fed in behind the tool producing the required pipeline. The accuracy of the final pipeline depends upon the accuracy of the original set-up of the tool and the homogeneity of the material through which it is driving. Unfortunately, the machine may deflect downwards with obvious consequences.

Specialist companies operate this type of machine which can be used successfully at up to 30 m drive length.

PIPE JOINTS

Consideration must be given to the long-term serviceability of new pipeline construction, its durability and ability to withstand a changing above-ground environment and long-term longitudinal movement in the vertical plane.

A pipeline that is incapable of vertical deformation along its length will crack. Rigid pipes must be installed to allow movement at the pipe joint as long-term movement will probably take place.

Previous practice has been to surround rigid pipes with *in situ* concrete in an attempt to strengthen the pipe. Recent work on pipe loading and wall stresses shows that better load distribution and even stressing is gained by using granular surround. Also, longitudinal long-term movement can be more readily accommodated by the pipe joints.

Where deformation of the pipeline in the vertical plane is likely, flexible jointed pipes are used and pipe manufacturers normally issue details and specifications of the types in use. The success of a flexible joint depends upon its airtightness and structural behaviour under deflection. Most engineers require air or water tests on pipelines up to 90 cm diameter after construction.

The usual cause of a test failure is indifferent joining or damage caused to the pipe spigot or socket during handling. Lifting hairpins should always be used when handling and lifting pipes and are usually available from the manufacturers.

FORCES ON BURIED PIPELINES

Structural failure of buried pipelines is usually a result of a misunderstanding of the loads likely to arise during, or as a result of, the method of construction.

Much investigation has been carried out by Spangler[10] into the pipe-wall stresses induced by earth and overburden loads and the principles of design have been well established.

It must be realised that for rigid pipelines, particularly when constructed in reinforced concrete, the structural considerations may be outweighed by the necessity of the material of the pipeline to withstand the long-term effects of chemical action from ground water, carried liquid and erosion effects.

The final in-ground cost of a pipeline or sewer may not be significantly affected by using a better quality of pipe or pipe material than is required by purely theoretical structural considerations.

The forces acting upon a flexible pipeline are not yet fully understood. If the earth pressures acting upon the pipe wall were wholly radial, then the wall would be in a purely compressive stress state. A converse case is a water pipe under internal pressure, when the wall is in a state of pure tension.

A steel or ductile iron pipe, laid in trench, will be subjected to a different set of force conditions during and after construction than a flexibly lined tunnel. The trench-laid pipe will initially be laid on a granular bedding, ideally over a length of 180° of arc. Thereafter, the back-fill material will exert immediate vertical force under gravity together with the forces caused by back-filling. During back-filling, the trench support will be withdrawn and the active pressure of the trench sides will start to exert its influence on the final load pattern and may take many years to reach equilibrium, particularly in cohesive ground.

Precise design methods and final load computations are not yet accurately calculable and, in view of the variation in environmental, geological and soil parameters, it is unlikely that they ever will be.

It has been shown that steel pipes and unbolted segmental concrete tunnels are virtually uncrushable under any fill or distributed nonpunching superloads when the cover is at least equal to the pipe diameter.

With the increasing use of the flexibly lined segmental tunnel, it is of interest to consider the nature of the forces acting upon it and to compute the likely magnitude of stress produced.

Being constructed by nonsurface disruptive means, the tunnel lining is not subjected to the residual weight of the prism of earth above it since this has not been disturbed. Instead, the lining is subjected to directed earth forces which may be approximately radial. If the tunnel is constructed in very soft, wet material, the forces will tend to the ideally radial load condition that would exist if the tunnel were submerged in water.

In this condition, the requirement of the lining to support purely vertical loading no longer exists since the radial pressures induce direct circumferential compressive stress in the lining.

Unlike the pipe laid in trench, the foundation strength of the soil beneath the tunnel does not arise as a design consideration. The foundation to a tunnel laid in very soft silt, for example, will be the material immediately above the lining, since the tunnel will be subjected to hydrostatic uplift even when running full of liquid.

Modern smooth-bore tunnel segments are made in unreinforced concrete. Any deformation in the circular shape is accommodated by movement at knuckle joints between the segments. The deformation takes place because of an imbalance of radial pressures around the segment ring and, in deforming, the lining automatically compensates for the imbalance until the pressure becomes truly radial.

Normally, the vertical diameter will decrease with a corresponding increase in the horizontal diameter, exciting the passive ground pressure at the tunnel sides.

Tunnel lining stresses (normal overburden)

Consider Figure 39.15 – a segment ring, installed at a depth H below ground.

Marston's formula, modified to include for cohesion, becomes:

$$W_t = C_t B_t (wB_t - 2c)$$

where W_t is the load on the ring (kg/m run), w is the unit density of the soil (kg/m^3), B_t is the width of the tunnel including the filled overbreak, c is the cohesive strength of the soil (kg/m^2), C_t is the load coefficient, a function of the depth of the tunnel to the width B_t and the coefficient of internal friction of the soil.

The values of C_t for various ratios of H/B_t are shown in Figure 39.18.

For high values of H/B_t the coefficient C_t tends to a limit which is a function of the angle of internal friction of the material and its ratio of lateral to vertical pressure distribution.

For tunnels laid at great depth, the load on the lining can be calculated using the limiting value of C_t.

The segment ring supports the vertical soil load by the direct concrete stress induced in the lining at the horizontal diameter, as with a cylindrical vessel subjected to internal pressure, i.e. $C_tB_t(wB_t - 2c) = 2tp_{cd}$, where

t is the wall thickness of the lining and p_{cd} is the direct compressive stress in the lining.

As the value of the coefficient of cohesion is the subject of some speculation where a pipeline may extend for several miles, it is not unreasonable to discount its contribution

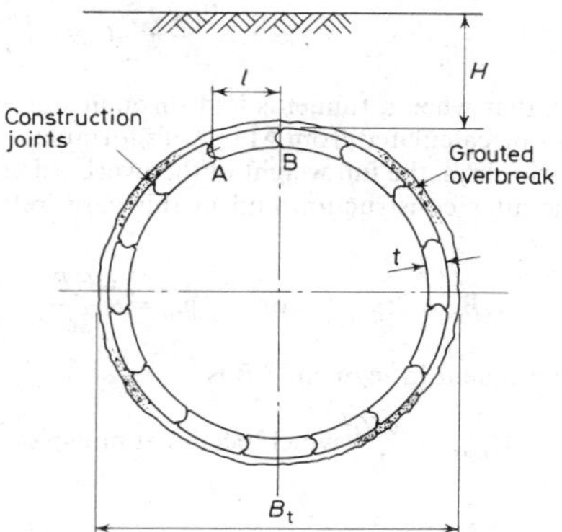

Figure 39.15 Typical concrete segment ring at depth H below ground

to any reduction of soil pressure on the lining as variations in soil strata and water table level can significantly affect its value. In any case, laboratory test results should always be regarded with some conservatism.

This being so, we have:

$$C_tB_t^2w = 2tp_{cd}$$

or

$$p_{cd} = \frac{C_tB_t^2w}{2t} \tag{1}$$

In addition to the direct stress caused by the vertical soil load, bending stresses are created by the soil pressure on each subsegment. These stresses may, for the purposes of simple analysis, be regarded as uniformly distributed load on a simply supported beam.

Again, discounting cohesion, the uniformly distributed load across the segment AB is

$$\frac{C_tB_t^2w}{B_t} = C_tB_tw$$

Therefore

$$M_{ab} = C_t B_t w \frac{l^2}{8}$$

where l is the segment length.

Since the maximum moment (at midspan) is given by

$$M_{max} = p_{cb} z$$

where z (the modulus of section) is $t^2/6$ and p_{cb} is the outer fibre bending stress. We have

$$p_{cb} = \frac{6}{t^2} C_t B_t w \frac{l^2}{8}$$

$$= \frac{3}{4} C_t B_t w \left(\frac{l}{t}\right)^2$$

$$p(\text{max/min}) = p_{cd} \pm p_{cd}$$

$$= C_t \frac{B_t^2 w}{2t} \pm \frac{3}{4} C_t B_t w \left(\frac{l}{t}\right)^2 \quad (2)$$

It is certain, however, that when a tunnel is laid through soil which tends to squeeze or swell, the load cannot be calculated from Marston's formula.

In the case of the plastic clays, the full weight of the overburden is likely to rest on the tunnel lining some time after construction and in this case (referring again to Figure 39.15) we have:

$$wHB_t = 2tp_{cd} \quad \text{or} \quad p_{cd} = \frac{wHB_t}{2t} \quad (3)$$

The maximum bending moment in segment AB is

$$M_{max} = \frac{wHl^2}{8} \text{ which occurs at midspan}$$

so

$$p_{cb} = \frac{3wHl^2}{4t^2} \quad (4)$$

$$p\,(\text{max/min}) = \frac{wHB_t}{2t} \pm \frac{3wHl^2}{4t^2} \quad (5)$$

Superloads

In addition to normal overburden soil pressures, sewers are usually subjected to concentrated and distributed live load, foundations or other distributed superloads.

In Figure 39.17 the single point load P (for example a lorry wheel) causes a superimposed concentrated load W_{sc} where

$$W_{sc} = \frac{C_s PF}{L} \quad (6)$$

in which W_{sc} is the load on the tunnel in kg per unit length, F is the impact factor, C_s is the load coefficient which is a function of $Bt/2H$ and $L/2H$, and L is the effective length of the conduit in metres and can normally be taken as one metre.

Values of C_s can be obtained from Table 39.3 which gives the influence coefficients for the integration of Boussinesq's equation.

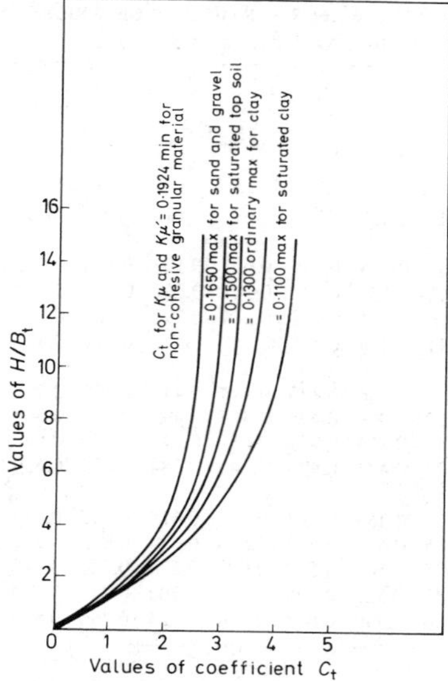

Figure 39.16 Coefficient C_t for tunnels in undisturbed soil

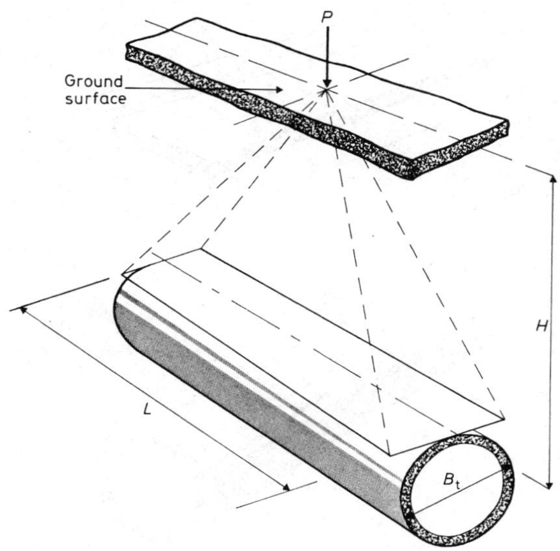

Figure 39.17 Point load centred over tunnel

Table 39.3 LOAD COEFFICIENT, C_s FOR VERTICALLY CENTRED CONCENTRATED AND DISTRIBUTED SUPERIMPOSED LOADS

$\frac{D}{2H}$ or $\frac{B_t}{2H}$	$\frac{M}{2H}$ or $\frac{L}{2H}$													
	0.1	0.2	0.3	0.4	0.5	0.6	0.7	0.8	0.9	1.0	1.2	1.5	2.0	5.0
0.1	0.019	0.037	0.053	0.067	0.079	0.089	0.097	0.103	0.108	0.112	0.117	0.121	0.124	0.128
0.2	0.037	0.072	0.103	0.131	0.155	0.174	0.189	0.202	0.211	0.219	0.229	0.238	0.244	0.248
0.3	0.053	0.103	0.149	0.190	0.224	0.252	0.274	0.292	0.306	0.318	0.333	0.345	0.355	0.360
0.4	0.067	0.131	0.190	0.241	0.284	0.320	0.349	0.373	0.391	0.405	0.425	0.440	0.454	0.460
0.5	0.079	0.155	0.224	0.284	0.336	0.379	0.414	0.441	0.463	0.481	0.505	0.525	0.540	0.548
0.6	0.089	0.174	0.252	0.320	0.379	0.428	0.467	0.499	0.524	0.544	0.572	0.596	0.613	0.624
0.7	0.097	0.189	0.274	0.349	0.414	0.467	0.511	0.546	0.584	0.597	0.628	0.650	0.674	0.688
0.8	0.103	0.202	0.292	0.373	0.441	0.499	0.546	0.584	0.615	0.639	0.674	0.703	0.725	0.740
0.9	0.108	0.211	0.306	0.391	0.463	0.524	0.574	0.615	0.647	0.673	0.711	0.742	0.766	0.784
1.0	0.112	0.219	0.318	0.405	0.481	0.544	0.596	0.639	0.673	0.701	0.740	0.774	0.800	0.816
1.2	0.117	0.229	0.333	0.425	0.505	0.572	0.628	0.674	0.711	0.740	0.783	0.820	0.849	0.868
1.5	0.121	0.238	0.345	0.440	0.525	0.596	0.650	0.703	0.742	0.774	0.820	0.861	0.894	0.916
2.0	0.124	0.244	0.355	0.454	0.540	0.613	0.674	0.725	0.766	0.800	0.849	0.894	0.930	0.956

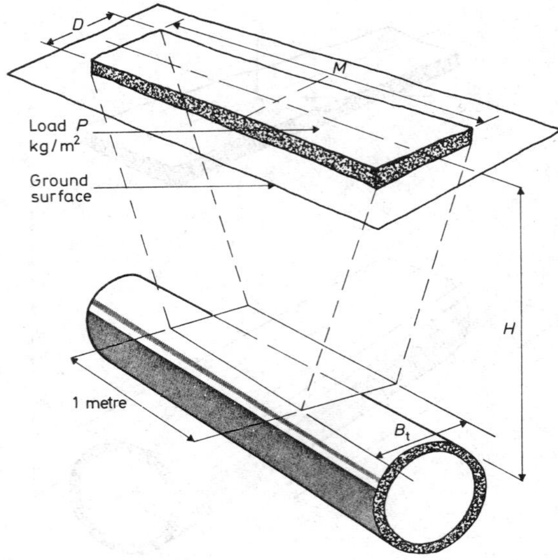

Figure 39.18 Distributed load centred over tunnel

The impact factor is a measure of the dynamic load caused by the movement of the point load across the tunnel. Typical values are:

	F
Roadways including motorways	1.50
Railways	2.00
Airfield runways	1.25

Figure 39.18 shows the condition where the tunnel, in addition to the overburden soil loads, is subjected to a uniformly distributed load of P kg/m^2 at a depth H and the load on the tunnel is given by:

$$W_{sd} = C_s P F B_t \tag{7}$$

Where W_{sd} is the load (kg per unit length of tunnel), P the distributed load (kg/m^2) and D and M are the width and length over which the load is distributed.

Values of C_s can again be read from Figure 39.17 and tunnel lining stresses calculated accordingly.

SUMMARY

The Engineer spends a considerable amount of time in determining the diameters and lengths that he requires in dealing with the drainage problem with which he is faced. Population density and rate of expansion, run-off and porosity factors, storm conditions and times of concentration are only a few of the variable functions which need to be considered in the design.

It is equally important that the method of construction is considered with similar integrity. The technical and financial considerations are interrelated: an economic method of construction which is technically suited to the purpose must be sought. The specification must be drawn up for what it is intended—a specific of the method and quality of work required.

If any construction method is to be specified in principle, then it must be specified in detail. The equipment for construction, necessity for ground stabilisation and all other factors must be catalogued and clarified if the job is to be well done.

REFERENCES

1. TERZAGHI and PECK, *Soil Mechanics in Engineering Practice*, 2nd edn, Wiley, New York (1967)
2. TOMLINSON, M. J., *Ground Engineering—Lateral Support of Deep Excavations*, Institution of Civil Engineers (June 1970)
3. *Larssen Steel Sheet Piling*, 5th edn, South Durham Steel & Iron Co. Ltd (Nov. 1962)
4. CASHMAN, P. M. and HAWS, E. T., *Ground Engineering—Control of Ground Water by Water Lowering*, Institution of Civil Engineers (16th June 1970)
5. CLARKE, N. W. B., *Buried Pipelines*, MacLaren & Sons, London (1968)
6. YOUNG, O. C. and SMITH, J. H., *Simplified Tables of External Loads on Buried Pipelines*, Building Research Station—HMSO (1970)
7. *Concrete Cylindrical Pipes & Fittings*, British Standards Institution, BS 556 (1966)
8. WARD and CHAPLIN, Proc. 4th *Int. Conf. on Soil Mechanics and Foundation Engineering*, Butterworths, London (1957)
9. WARD, W. H., *Yielding of the Ground and the Structural Behaviour of Linings of Different Flexibility in a Tunnel of London Clay*, Building Research Station, CP 34/70 (Oct. 1970)
10. SPANGLER, M. G., 'The structural design of flexible pipe culverts', *Iowa Engng Exp. Station Bull.*, 153 (1941)

1974
THE QUEEN'S AWARD
TO INDUSTRY

40 DREDGING

DREDGING 40

40 DREDGING

J. H. SARGENT, C.Eng., F.I.C.E., F.G.S.
Dredging Investigations Ltd.

It is believed that dredging works originated in Egypt at about 4000 BC: canals were excavated using massive labour forces and the simplest of tools.

In the sixteenth and seventeenth centuries dredging was undertaken using the 'bag-and-spoon' method. During this period the purpose of the dredging was often both to improve navigation and to obtain material to be utilised as ballast for outgoing ships.

An important historical point in the development of dredging plant was reached in about 1590 with the invention in Holland of the 'mud mill'. This machine superficially resembled the modern bucket dredger but it had a chain revolving in the opposite direction and the mud was scooped up a chute which was suspended below the moving chain.

The introduction of the steam engine and the development of the centrifugal dredge pump in the nineteenth century marked a dramatic step forward in the field of dredging work and led directly to the development of modern dredger designs. Although continuous improvement in the design of present-day 'traditional' dredging plant is taking place it is likely that the next major step will be to place dredging units on the sea bottom beneath the area of wave disturbance, the excavated material to be fed through pipeline systems to floating units or directly to shore installations.

DREDGING PLANT AND TECHNIQUES

General remarks

For convenience, dredging plant has been subdivided into (a) mechanically operated dredgers and (b) hydraulically operated dredgers.

The description of plant and techniques is related primarily to the use and application of plant from a civil engineering viewpoint, and does not deal exhaustively with the mechanical design or performance of the equipment.

The aspects of cost and output of dredged materials have only been considered in a very general manner. These aspects will vary over a very wide range since they are dependent on a number of factors which cannot be considered in the present discussion and must vary from project to project. Such factors will include the physical conditions imposed by the site and its boundaries, the working hours and tides available, the water depths existing prior to dredging, the proximity to authorised dumping or disposal areas, the water and sea conditions (allied to the exposure of the site), the availability and skill of local staff, tradesmen and labour, the proximity of the project to navigation and interference from other traffic, the political attitudes (particularly in the undeveloped countries), etc.

Over and above these factors will also be the state of the market in the dredging industry at the time of tender and contract discussion and agreement. Each project must be viewed separately and costed for its own particular set of circumstances.

A general table to assist in considering the type of dredging plant to be employed under a particular set of operating and geological conditions has been prepared by the author. (Sargent, J. H., 'Feasibility Studies for Dredging Projects', *Proc. North Sea Spectrum Conf.*, 1970, Thomas Reed Publications, London, 1971.)

Mechanically operated dredgers

BUCKET DREDGING

The basic bucket dredger design comprises a chain of continuously revolving buckets moving along a steel ladder inclined into the sea or river bed, excavated material being carried to the top of the ladder where it is gravity-discharged into a chute (see Figure 40.6). The majority of bucket dredgers are dumb (i.e. requiring towage to and from site) and discharge through the chute into a selfpropelled hopper or dumb barge for disposal. Some dredgers exist as selfloading hoppers and, in this case, may be selfpropelled.

Since the bucket dredger needs positioning in its working area and in particular requires a reaction when excavating into the face of material to be dredged, the vessel is usually held by several moorings, the headline being of utmost importance to provide the reaction into the digging face. In practice, the headline may extend for several thousand feet for fully effective operation. Short headlines can be used, but result in loss of efficiency.

The bucket dredger is commonly of poor stability when moving or under towage from site to site and consequently towage can be slow and expensive. Depending on its size the dredger will normally require light tug or workboat assistance when on site, both to run lines and to help move the dredger from area to area.

As with most dredging craft based on pontoon designs, the average bucket dredger is susceptible to heavy swell and may suffer badly in swell conditions in excess of about one metre. An important feature in bucket dredging work may be the presence of hoppers or barges alongside, which can also cause serious problems under swell conditions owing to the differential movement between the craft with consequent danger to both craft and men.

The advantage of the heavy bucket dredger as a dredging tool is its ability to dredge a wide variety of materials ranging from soft alluvium to soft rock. The lighter units can tackle a range of soil types but will be restricted in respect of rock and the heavier clay soils.

Foreign material (such as scrap iron, wires, etc.) can also be removed by the bucket dredger, although production (compared with dredging 'normal' soils) may suffer seriously.

Bucket types can be changed to suit the ground conditions: it is common to use a heavy bucket in soil, and a bucket fitted with strengthened lip and fitted with digging teeth when working in soft rock. Bucket capacities vary from 0.5 to 1.0 cubic metres, using bucket chain speeds at 18 to 30 buckets per minute depending on the material and the conditions.

Dredging depths vary with the size of plant but commonly extend to 20 metres or so with extensions to greater depths.

Working problems may occur when using bucket dredgers owing to the noise of the operation. Disturbance is particularly common where work is in progress at night and the site is close to residential property, and even though mechanical design improvements have been made over recent years the noise situation can still be serious.

Outside the civil engineering industry the bucket dredger (or ladder dredger as it is known in the USA) is much used as a basic tool in mining for rare earths and precious metals. Perhaps best known is the extensive use of large bucket dredgers for tin mining in the Far East.

GRAB DREDGING

This type of plant is nearest to normal civil engineering equipment employing a prime mover mounted on a pontoon or vessel loading either into attendant hoppers or barges, or into its own hopper if the vessel is selfcontained. Craft of this type are often self-propelled.

Grab dredger sizes can vary over a wide range from very small units with small cranes mounted on pontoons or barges (such as the small hydraulic cranes used in narrow barges for canal dredging) to large machines capable of carrying grabs of 10 cubic metres or greater.

Simple grab dredgers can be built easily and quickly by running a crawler-mounted excavator onto a pontoon, although in this case production may suffer owing to the instability of the marine design unless work is being carried out in very quiet waters.

Small and medium-sized grab dredgers have found much favour in harbour and ports

Figure 40.1 Pontoon-mounted grab dredger (Courtesy: M.B. Dredging Company Ltd.)

works because of the limited capital expenditure required, their relative ease of maintenance and their particular ability to dredge close to quay walls, jetties and in basins without damage. Further, the quantity of 'foreign' material such as scrap iron, wire, tyres, etc., within port basins is often very considerable and a grab can handle much of this material with reasonable ease. Commonly occurring harbour materials such as silts or silts/clays can also be removed easily with standard grabs making the unit particularly adaptable to the port authority.

Dredging depths will be limited by the size of the available winch drum with a practical limitation imposed by the hoisting speed of the grab and the effect of sea or river currents on the grab when working at or near maximum depths. Working depths up to 25 metres or so below water level are common and greater depths have been undertaken, sometimes for purposes other than navigational dredging such as in Hong Kong Plover Cove earth dam cut-off where dredging was undertaken by a special cantilever grab dredger to considerable depths.

Grab dredgers are able to operate in a swell condition, but production is affected badly over about 1.5 metres swell, and under all conditions the profiling of the bottom must be carefully supervised. Single grab dredging requires a fairly simple pattern of attack working forward after dredging through an arc. With several cranes mounted on the unit the dredger must be moved diagonally to obtain optimum coverage.

Variations on the grab dredger which rely on mechanical power for excavation include:

(i) Back actors (or back-hoes)
(ii) Face shovels or draglines
(iii) Dippers.

The first two have a limited application generally except for specific jobs. The dipper is still in use but has never gained general favour owing to its relatively low production characteristic and limited depth of digging. The dipper dredger can however be valuable when dredging clays in a firm or stiff condition, or soft and friable rock. The tool has also been usefully used for the demolishing of old stone defence works, jetties, locks and similar 'heavy' jobs, which other dredging plant could not handle easily. Dipper dredge plant is normally held by spudded arrangement, often using two forward spuds to pin up the hull, and stern spuds for walking and advancing the dredger.

Hydraulically operated dredgers

This class of dredger comprises the most widely used unit in the world's dredging fleets. Basically the design comprises a centrifugal dredge pump installed in a pontoon or hull with a suctionpipe lowered on a ladder to the bed material. Where the bed material is loose and granular in nature, such as a running sand, it will be drawn under suction directly through the pipe and dredge pump into a discharge line. If the bed material is compact or dense, such as a firm clay or cemented sand, it is necessary to fit a rotating cutter onto the end of the suction line. The cutter will be turned by a separate cutter motor which is commonly mounted at the head of the suction line ladder, although for deep dredging this may be placed close to the suctionhead.

The hydraulic technique of dredging is often claimed to be highly efficient because it combines excavating and transporting waterborne dredged material in one continuous operation. Improvements over the last few years have occurred primarily through better pump design, and dredging pumps now represent a compromise with high efficiency against low power consumption matched against reliability and assessability.

A problem with the technique is inherent in the movement of abrasive materials such as sand, through a pipeline system, since this is the basis for extremely high wear and tear and pump liners and pipes must be designed to stand this potent mixture.

Attention has also been given recently to the use of water jets at or close to the suction mouth to assist in the breakup of *in situ* material and its movement into a suctionpipe. Suctionheads comprise a wide variety of shapes and designs often linked to the dredge manufacturer's or contractor's own individual beliefs and theories.

Suction dredgers have evolved into several classes of plant which for simplicity may be listed as:

(i) Plain suction dredgers
(ii) Cutter suction dredgers
(iii) Trailer suction dredgers
(iv) Pump-ashore units.

PLAIN SUCTION DREDGERS

Plain suction dredgers may be defined as having a plain 'end' to the suction tube, the suctionhead being buried into the material to be excavated. As noted previously, this unit is much used where there is free-running granular material that is required for transportation to reclamation sites as fill, as occurs on an extensive scale in Holland.

After dredging, the material may be passed through side arms into barges moored alongside the dredger or pumped directly through a pipeline system, part of which will usually be floating, to shore.

Some application has been made with this type of dredger for the removal of soft, cohesive materials such as silts or silt/clays in river and estuarine conditions. Although the dredger is useful in these conditions, problems will arise if a suitable dumping area is not available nearby for the disposal of the dredged material which will normally be unsuitable for use on a reclamation site.

The plain suction dredger design is usually based on a dumb pontoon which requires two stern spuds to allow walking when moving forward.

A variation of this type of dredger, developed in the USA, is the 'dustpan' dredger based on a suctionhead which is often up to hull width, fitted with high velocity water jets and capable of dredging loose alluvial materials at very high volumes but with limited discharge pumping. In this case the dredger is often selfpropelled and has been employed to keep open river navigation, pumping sands or alluvial soft silts/clays away from the navigable channel into a nearby area outside the channel where the discharged material is removed or spread thinly by erosion currents.

CUTTER SUCTION DREDGERS

On a global basis, this type of plant is probably the most widely used dredging unit. Dredger sizes range widely from small, two-man designs with 150 mm suctionpipe diameters and limited power, to very large units with suctionpipe diameters over 1 metre and installed horse power in excess of 10 000.

Basically, the cutter suction dredger is pontoon- or hull-built and normally dumb, requiring towage and assistance on site (as for the majority of plain suction dredgers

Figure 40.2 Large cutter suction dredger (Courtesy: Westminster Dredging Company Ltd)

and bucket dredgers). A qualifying feature is the incorporation of a cutter around the suctionpipe mouth powered to turn into the excavation face. The cutter in this case lies with its axis of rotation parallel and in line with the cutter ladder.

When working the unit usually has a spud driven to the river or sea bed mounted in the stern and the vessel pivots about the spud pulled by side anchors in order that the cutter head can be hauled across the working face. A second stern spud known as the 'walking spud' is employed to move the dredger forward. In the majority of cutter suction dredgers, the side lines are led through sheaves fixed to the suction ladder and run to anchors along or just above the sea or river bed. Side booms are often fitted which allow the facility of lifting and moving anchors forward without the necessity to use a separate workboat.

Depending on the structural strength of the design (in particular the weight of the ladder and the available power of the cutter coupled to the power availability of the dredging pump or pumps) the cutter suction dredger can excavate and remove most normal soils and soft rock formations. When considering large units it may be considered as a rough guide that the softer sedimentary rocks such as chalk, marl, coral, weathered sandstones, etc., are within the scope of direct excavation by the cutter suction dredger.

After excavation, the material can either be pumped through a pipeline system (in which case good material will be used as reclamation fill and poor material placed into a shore deposit area) or pumped through the dredge pump and loaded into hoppers or barges which lie alongside. The discharge distance from the dredger will be limited through

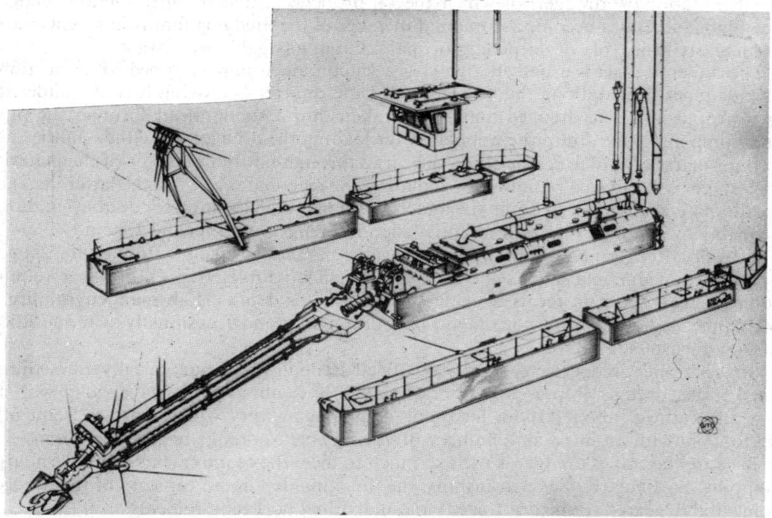

Figure 40.3 Dismantleable cutter suction dredger (Courtesy: I. H. C. Beaver N.V.)

the pipeline system by the installed pump power unless separate booster units are used. The cutter suction dredger is subject to limitations on working when swell is in excess of about 1 metre and in addition a floating pipeline is particularly susceptible to both swell and strong currents and exposure in offshore locations can be dangerous. For optimum production, cutter suction dredgers and floating pipeline systems should be in relatively sheltered waters.

Working depths vary considerably with size of plant but are commonly up to 20 m or so with the larger units. Much research and development is in hand in this respect and greater depths can be achieved.

The design of cutter is also a wide subject with innumerable individual ideas and opinions. Of particular importance is that when working in hard and abrasive materials it is essential that cutter replacement can be effected at high speed, and in order to assist with this problem many rock cutters have replaceable teeth.

Modern cutter suction dredgers are both expensive and relatively sophisticated. In order to work at optimum efficiency, it is necessary to provide adequate control and instrumentation, and to do this it is essential to provide a system to optimise output by coordinating the functions of vacuum (on the dredging pump), cutter torque and the swing speed of the cutter ladder via the side winches.

Of particular interest for small project work is the dismantleable cutter suction dredger typified by a standard range from a major dredger manufacturer with dismantleable units with installed horse powers ranging from 150 to 1500. Although these units can be broken down into component parts for road, rail or water transport, the size and weight of the individual pontoons with the larger units should not be underestimated since, as an example a central pontoon of one of the larger dismantleable dredgers measures 14 m × 2.5 m × 1.8 m

HOPPER SUCTION DREDGERS

In this category of 'suction dredging' plant two types are best known: the trailer dredger and the stationary dredger, of which the selfpropelled trailer hopper suction dredger (see Figure 40.7) has become a dominant member of the dredging family in recent years, particularly in its role of deepening or maintaining navigable waterways.

The trailer dredger is usually designed as a selfcontained ship equipped with a suction-pipe or pipes trailed along the bottom whilst the dredger is moving forward under its own propulsion. The dredged material is taken into a suctionhead through the pipe and pump and passed into the hopper. After loading the dredger will either (in the case of poor material) sail to sea to dump the cargo through bottom doors or (if the material is suitable) will sail to a pump-out installation for reclamation fill. In the latter instance an alternative method is to dump the material in a prepared location for double handling, probably by stationary suction or cutter suction dredger and pipeline system.

A trailer dredger operates in exactly the same manner as a ship without wires, spuds or other impedimenta and is therefore very manoeuvrable. It does however require room to work both in terms of waterway and water depth. High manoeuvrability is often provided by the provision of twin-propulsion units and occasionally by the addition of bow thrusters.

Sizes of trailer dredgers vary greatly as with all dredging plant, but typically the medium-sized trailer dredger in present-day practice will have about 2500 m^3 capacity with an overall length of about 110 m. Dredging depths again vary widely but are frequently up to 23 m with an increasing number of vessels able to dredge in excess of 30 m.

Because the trailer dredger is used so much to meet the requirements of modern large shipping, particularly tanker transport, the dredging depth and capacity of the dredger is inevitably geared to the draft needs and individual port requirements.

Close control of a sophisticated and costly ship such as the modern trailer dredger is vital and the control systems for both the suction pipe and operation of the sand pump are usually centred on the bridge together with the normal controls for propulsion and other navigational equipment. Further automation and dredging aids now provided include the use of concentration, production and loading meters together with multichannel recorders.

A limited number of trailer dredgers have pump-ashore facilities. To effect this operation, a suction line is installed in the hopper well together with an upper set of doors plus the necessary discharge lines. Designs provide either for the sand pump to be connected directly to a shore line or for an additional pressure pump to be included which may be driven by one of the trailer's propulsion engines.

Variations have been built or adapted to meet specific circumstances with trailer dredgers including a combination cutter and trailer dredger. Although attractive on paper, the high cost of such units has not made them generally competitive.

The trailer dredger has a high production characteristic in soft or loose alluvial soils but it is more difficult for the trailer dredger to excavate stiff or hard clays, cemented sands or similar material. The work can be undertaken with adapted dragheads, but lower productions have normally to be accepted. Very fine, loose material such as silt can be easily loaded, but because of its poor settlement characteristic in the hopper it is usual to take only part loads during each cycle.

A dominant and important feature of the trailer dredger when considering the dredging requirements of a new project is its ability to operate in exposed locations, often in swell in excess of several metres and in wind, weather and sea conditions totally beyond the capacity of other plant, such as bucket dredgers, plain and cutter suction dredgers and grabs.

The stationary hopper dredger is of the same basic design as the trailer dredger but but does not move from the working location while dredging and often has the suction-pipe placed in a forward direction. Vessels of this type have, for example, been much employed for the dredging of offshore sand and gravel deposits.

Offloading of the material will be in the same manner as the trailer dredger with the exception that, for sand and gravel extraction operations, the cargo may often be re-handled directly from the hopper by a grab to stock piles or to screening and washing plant ashore. In some instances screening may be carried out on the vessel with reject material pumped overboard, where this is permitted by local legislation.

PUMP-ASHORE PLANT

Pump-ashore units are normally used in conjunction with other dredging plant, the purpose of the installation being to empty the filled barge or hopper by suction means and to transport the dredged material ashore to reclamation area or stockpile.

In essence the technique provides an alternative in the cycle of the total dredging operation and avoids the use of lengthy floating pipeline systems (in the case of suction or cutter suction dredgers) and is applicable in particular where the distance from dredging site to reclamation site is too great or is uneconomic for direct pumping.

ANCILLARY PLANT AND EQUIPMENT

In conjunction with the major items of dredging plant discussed in outline so far, waterborne ancillary plant and craft is essential, including barges, tugs, workboats, survey launches and small marine craft.

Figure 40.4 Split dump barge (Courtesy: M.B. Dredging Company Ltd)

Of prime importance are dumb barges and hoppers, used extensively with bucket dredgers and other plant for carrying dredged material to dump grounds or for reclamation fill.

The size of project and quantities of dredged material will often dictate the capacity and type of craft to be used; selfpropelled hoppers are often used where large quantities and long sailing distances are involved.

Of particular interest since 1960, a development in hopper design has included the 'split' barge. In this case the body of the barge is divided longitudinally along the centreline of the vessel; the two buoyant halves are hinged with one hinge forward at deck level and the second aft. Opening of the barge is usually effected by rams through a hydraulic cylinder; in some designs the barge closes automatically under the action of gravity and the inbuilt buoyancy of the design.

An advantage of this type of hopper is its value in dumping clays with a high adhesive factor (such as boulder clays) which frequently 'arch' in a normal hopper and may be difficult to dislodge.

A useful although comparatively minor feature of split barges is the facility also to break the barge into two sections for transport by rail, flat-bed trailer or water-going vessel. Assembly can then be undertaken both in the dry and in water.

RECLAMATION WORKS

A major contribution of the dredging industry is the use of dredging plant to enable land reclamation (in the sense of land building rather than improvement) to take place. Particular examples of this are the immense reclamation schemes in Holland, including the recent works for Europoort and the projected schemes such as Foulness/Maplin Sands in the UK.

Reclamation methods using dredging plant as the prime mover may be subdivided into reclamation by polder method or reclamation using pumped fill.

In the case of the former (polder) method, normal practice is to construct dykes around a water area and artificially to drain the area. The resulting dry land (frequently below sea or groundwater level) is in Holland termed a 'polder' which gives its name to the method.

In this instance it will be necessary to maintain the water levels inside the polder by pumping through a series of interconnecting drainage ditches and systems.

A major problem where land is open to the sea in tidal areas will be the dyke closure, where the final gap will pass large quantities of tidal flow at increasing velocities.

In the case of the second method, the general land level will be raised by pumping 'in' soil dredged from elsewhere. It is normal practice to elevate the surface above High Water mark for safety, and to obviate artificial pumping.

When considering the economics of the two methods, a major cost of polder reclamation will be the dyke construction. Large areas will therefore often be attractive since the boundary dyke will be low in cost in relation to the reclaimed area.

In considering sandfill reclamation, a significant expense will be the availability of suitable local fill material. Ideally the fill should be easily dredgeable, close to the site and have good pumping characteristics with low wear and tear and high compaction and load-bearing characteristics. In Holland where reclamation is extensive the predominant soil (by UK classification) is a fine sand which meets several (though not all) of these requirements. In the UK an exceedingly wide range of soils will be encountered and a balance will have to be obtained in deciding the choice.

Fill by dredging methods normally means that the material is pumped hydraulically through a pipeline system with water as the transport medium. Reclamation works using dredging methods therefore involve very large quantities of water which must either be drained away or recirculated within the system.

From a geotechnical viewpoint, materials such as sand and gravels will make excellent earthworks fill, but granular materials with increasing particle size (i.e. from sand grading upwards) become increasingly expensive to pump.

Soft clays and silts are easily dredged and pumped but are unsatisfactory fill materials and are usually avoided when possible. However, use has been made of such material, both in Holland and the UK, where deposit grounds are husbanded and redeveloped for agriculture.

In reclamation works, clays with a firm or stiff cohesion characteristic usually 'ball'

Figure 40.5 Aerial view of a typical reclamation project using a cutter suction dredger with short length floating pipeline (Courtesy: Westminster Dredging Company Ltd)

if passed through a dredge pump and, although they may form satisfactory fills, care is needed to analyse and predict the behaviour of such materials.

Prior to reclamation projects, site investigation into the virgin land conditions which will become the subsoil below the dredged fill is important, especially to provide the basic data for the engineering design of substructures.

SURVEYS AND INVESTIGATIONS

Setting-out dredging works

Stationary plant such as bucket dredgers, suction and cutter suction dredgers and grabs can normally be positioned by relatively simple hydrographic survey methods, using beacons, shore markers and sextant observations. Laser systems are also employed and development is in progress with such techniques to integrate control of both the position and depth of dredging automatically.

Trailer dredgers when working within harbours or river areas may utilise similar survey procedures but, when working on long approach channels, particularly offshore, require more sophisticated positioning systems such as radio-positioning techniques. Automation, utilising a system to couple both position-fixing and the navigation system of the ship, has also been successfully used and is particularly applicable when dredging trenches for offshore outfalls, as for the dredging of the trench for a sewer outfall for the Municipality of the Hague, Holland in 1967.

Hydrographic surveys and geotechnical investigations

A vital feature of the assessment of dredging conditions to enable plant selection to be made involves the early hydrographic survey and investigation work.

In order to calculate quantities of material for excavation, accurate hydrographic surveys are required. Additionally, during the dredging operations, surveys may need to be carried out to assess interim payments, particularly if these are made on an *in situ* quantity basis. After the completion of dredging work a particularly accurate survey will be needed to check that the specified depth has been reached, side slopes produced and to agree the final quantities for payment.

Investigations to check the nature and occurrence of the materials to be dredged are within the field of marine geotechnical investigations and will include boring in soils and drilling in rocks together with the recovery of samples or cores for inspection, logging and testing. Laboratory work on the recovered samples and cores is invaluable in determining the properties of the materials and provides essential information for the selection of plant and the costing of the project.

The application of geophysics during the site investigation is proving increasingly useful, provided that sufficient correlation information, possibly in the form of boreholes, is available and is used.

Of particular importance is the delineation of rockhead for a dredging contract. The difference in production by a specific dredger in soil and the same dredger in rock is often so very considerable that the unexpected occurrence of rock on a project will be costly to all concerned! In many cases the dredger mobilised for a soil job may be totally unsuitable for rock work and the mobilisation of a new item of plant can be very expensive and time wasting.

At the present time, it is usual to apply soil and rock mechanics practice to the evaluation of the materials but an increasing value is the recovery of as much *in situ* data as possible, especially with respect to investigations into rock conditions.

The need to estimate whether any form of pretreatment (such as drilling and blasting) is needed or not is particularly relevant to rock. At present the softer sedimentary rocks (soft sandstones, limestones, corals, etc.) can frequently be dredged direct whilst the unweathered igneous and metamorphic rocks will require pretreatment before removal by dredging plant.

In many dredging projects it will be necessary to determine the hydraulic conditions at the dredging site, the dumping area and the project site (especially if reclamation works are to be undertaken). In some cases the construction of a hydraulic model will be needed in order to study the influence of the various hydraulic conditions. A study of current patterns will be very important, in particular to check that navigation problems are not caused, siltation is kept to a minimum and wave action is reduced. Information on the latter is also vital to determine the method of protection to be afforded to dykes and embankments.

If conditions are not too complex, it may be possible to consider the hydraulic conditions using a mathematical model without having to meet the expense of the construction of physical models.

Figure 40.6 Modern large-capacity bucket dredger (Courtesy: Westminster Dredging Company Ltd)

Figure 40.7 Trailer hopper suction dredger

BIBLIOGRAPHY

HUSTON, J., *Hydraulic Dredging*, Cornell Maritime Press, Cambridge, Maryland (1970)

Proc. 22nd Int. Navigation Congr., Paris, 1969, Section II, Subject 2, General Secretariat of PIANC, Brussels (1969)

Proc. Symp. on Dredging, Institution of Civil Engineers, London (Oct. 1961)

Proc. WODCON (World Dredging Conference), Palos Verdes Estates, California (1968, etc.)

41 UNDERWATER WORK

UNDERWATER WORK 41

41 UNDERWATER WORK

Lt-Cdr H. WARDLE, R.D.
Managing Director, Strongwork Diving (International) Ltd

INTRODUCTION

Underwater work is taken to be that requiring underwater activity by divers, submersibles or tools from above the surface. The operation of submersibles is a highly specialised field which is at present confined to offshore works associated with work on pipelines and cables laid in open water. Surface operated 'tools' include dredging and drilling operations (see Section 40), pile driving, etc., which are essentially surface-controlled activities extending underwater. The aim of this Section is to discuss some of the problems associated with the utilisation of divers by the construction industry.

DIVER EMPLOYMENT

The utilisation of divers has increased considerably in the last two decades owing to man's increasing encroachment into the sea. Supertankers have been constructed which, when laden, have a draught of over 30 m, with a consequent increase in the depth of water in which berths and moorings have to be constructed. With increasing world population the disposal of waste by the construction of long sea effluent outfalls has developed and land reclamation schemes are being carried out in many parts of the world. By far the largest development has been that of the production of oil and gas from wells many miles offshore in water of increasing depth. This latter development has led to the evolvement of the oil field diver who may be required to work in 200 m of water or deeper.

TYPES OF DIVER ACTIVITY

Deep diving

This is carried out by divers who have been trained in the use of deep-diving systems, comprising a diving bell in which men are lowered to the work location and from which they operate, supplied by an umbilical hose. The diving bell can be 'locked on' to a compression chamber so that the divers can be transferred under pressure to complete the decompression time. This is the period required for the gradual reduction of pressure to allow controlled release of the inert gas absorbed in the diver's blood stream. The inert gas escapes, via the lungs, as the gas reverts back from a liquid to a gaseous state.

This is known as a TUP (Transfer Under Pressure) System and is normally used in depths greater than 50 m, where it is necessary to replace air with a synthetic gas mixture, normally an oxy-helium mixture. The nitrogen in air under pressure has a narcotic effect on the diver which is not experienced when using a gas with a low atomic weight such as helium.

The use of deep divers is virtually confined to specialised work within the oil industry.

Air diving

By far the greater proportion of commercial diving is carried out using compressed air as the breathing medium. Most underwater construction work on which divers may be employed is in less than 60 m. It is estimated that nine dives in ten are carried out in 10 m of water or less. In this shallow depth range there are no problems in decompressing the diver observing that, in general, it is safe to reduce the absolute pressure to which the diver is exposed by half, irrespective of the time he has been under pressure, i.e. a diver at 10 m (2 atmospheres absolute) can return safely to surface (1 atmosphere absolute).

As the working depth increases so does the time taken to decompress, and the longer the diver stays under pressure the longer is the decompression time. A diver can safely carry out short dives without decompression to approximately 40 m. Long exposures to pressure or a series of short-duration dives without proper decompression can cause serious decompression sickness (bends). Where operations take place in water deeper than 20 m a compression chamber should be available on site for the treatment of bends.

The operation of divers

As ninety per cent of all diving is carried out on work in less than 60 m using air as the breathing gas, this Section deals mainly with this type of operation. Some factors apply equally to the deep diver but, should specific information be required on deep work, reference should be made to one of the international companies providing deep-diving services.

DIVING EQUIPMENT

The following is a summary of diving apparatus in general commercial use for air diving.

Standard diving apparatus

The standard diving apparatus consists of a heavy tinned copper helmet and corselet (attachment between helmet and the heavy twill diving dress), lead-soled boots and weights. Air is supplied via an armoured air hose from the surface to the helmet and escapes from the helmet through a relief valve. The pressure in the helmet is related to the ambient (water) pressure.

The apparatus is used for relatively static heavy work. Although very comfortable for the diver when on the bottom, it is cumbersome and difficult to work in under swell conditions when mobility and vision are restricted.

Its use is now limited to 'old hands' in the diving profession, but it still has a place for localised heavy work and in protecting the diver from cold or pollution. It is sometimes referred to as Helmet or Hard Hat diving gear.

The aqualung

The aqualung normally consists of twin high pressure cylinders supplying air via a manifold and demand valve (reducer), which automatically adjusts the air breathed by the diver to ambient pressure. This automatic adjustment is achieved by the water pressure acting on a diaphragm which, in turn, operates a tilt valve supplying air to the diver. Thus air is provided to the diver at the correct pressure and 'on demand', the action of breathing operating the valve. The diver wears a 'frog suit', fins, mask and a releasable weight belt. This equipment gives a very high degree of mobility but low endurance and is widely used for survey work.

Surface demand diving apparatus

The principle of this apparatus is the same as the aqualung except that the air is piped from the surface to the diver. It is used where a combination of endurance and mobility

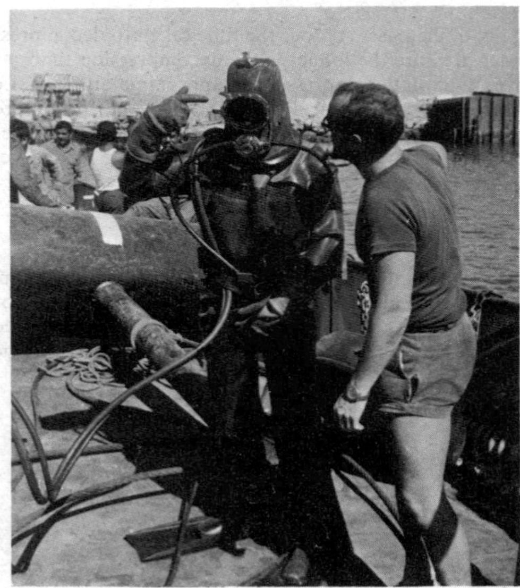

Figure 41.1 A diving suit resistant to hydrocarbons, developed by BP for use in a natural-gas-field environment. The main feature of the suit is a system of exchange breathing. Exhaust gases leave the diver through a buoyed line, clear of the gas-field environment, to avoid formation of a combustible mixture

is required. Some aqualungs have a dual capability, with the diver normally supplied from the surface but carrying 'emergency' air.

The fibreglass helmet

This equipment is an attempt to provide the diver with the comfort and good communications of the standard helmet while retaining the mobility of the frogman. The principle of operation is the same as the standard helmet but with the fibreglass helmet attached directly to a frogman-type suit. The volume of water displaced by this apparatus is less than the standard gear and therefore the weight required to 'sink' the diver is less.

AIR SUPPLY SYSTEMS

Standard diving apparatus

The helmet divers require approximately 1.5 ft^3 of free air per atmosphere, e.g. 3 ft^3 at 10 m, 6 at 20 m etc. The output of the compressor should be sufficient to provide air for a stand-by diver to go to the assistance of the working diver. The output of compressors drops with wear, it is therefore suggested that a safety factor of 50% be applied when calculating required output. A receiver should be in the line of sufficient capacity to allow the divers to surface safely in the event of compressor failure. When working in depths greater than 30 m the standard 100 lbf/in^2 compressor is insufficient to overcome the water pressure and give the diver a sufficient gas flow; a 200 lbf/in^2 compressor is recommended.

The air flow is calculated on the basis of ensuring sufficient ventilation to prevent the diver suffering from carbon dioxide poisoning; the correct air supply is of vital importance.

Surface demand diving apparatus

This apparatus requires gas at pressure at the demand valve. The average volume of air consumed by the diver is one cubic foot per minute; generally, however, the demand valve requires 65 lbf/in^2 above ambient pressure before it will supply at the correct rate to the diver. Therefore, although the volume required is less than that for the helmet diver, pressure is more critical. The 100 lbf/in^2 compressor is only suitable to a depth of approximately 20 m.

The aqualung

HP compressors operating between 2000 and 4000 lbf/in^2 are used. For major surveys, or other works where aqualungs are used extensively, high-pressure cylinder banks are used and the aqualung cylinders are charged by decanting from the banks.

Compressors generally

All must supply air suitable for breathing purposes. Tool or similar compressors should on no account be used.

DIVER QUALIFICATIONS

Transferring construction work underwater increases the degree of difficulty. Clearly it is not reasonable to expect any diver in the underwater environment to carry out all the tasks associated with construction to the same standard of a tradesman. The employer of diving services for simple recovery work has little problem. For complicated work he can either replan the underwater section so it is basic assembly work for the divers or, if this is not possible, employ a specialist diving contractor with engineering back-up. At present there is no formal grading of divers in the relatively small number of men employed in this profession. The larger diving companies do grade divers and the position should be regularised within the next two years. This grading is related to the training and experience of the diver, positively related to diving, e.g. diving plant and equipment, decompression, underwater tools etc. It is not practical to give divers full tradesmen's gradings and training in the many fields in which they are employed.

THE IMPORTANCE OF AN ACCURATE UNDERWATER SURVEY

To prepare a scheme for remedial works on structures underwater it is vital to obtain full and complete information on the state of the structure. One fact is worth a dozen theories. Underwater surveys cost very much more than those on the surface and engineers are often tempted to go for relatively cheap theories rather than establishing basic facts. This of course is putting the case too simply. Those charged with protecting the public, country or corporation purse may authorise expenditure on work which shows a tangible return, such as a new footbridge. When all that is received is a paper report the questions often asked are: 'Was this high expenditure necessary?' 'Why do you have to employ others?' Underwater surveying is a highly specialised and pains-taking task, but it is a vital prerequisite to remedial or underwater construction works.

One practical example is a comprehensive but expensive survey resulting in a cheap and completely successful repair being carried out to the clapping face and cill of a major dock gate which was losing water. On the other hand, inspections of vital bridge foundations have been known to be let to two-man diving contractors, on a competitive-tender basis, at a price which could only be described as ridiculous. All that could be achieved in a case like this is a classical 'yard arm' clearer. 'The foundations were inspected by … on ….'

Probably the worst combination of all is the effect that a well meaning executive produces in trying to get value for money for his employer from the diving contractors employed to carry out a survey, e.g. 'whilst you are there you can fix this,' 'I want all the diving team to be working divers', and so on. The end product is that nothing is done properly. The diving contractor must of course carry out the wishes of his employer, but it cannot be emphasised too strongly that surveying is a job on its own requiring a special approach. Accurate records are vital and, basically, the surveying supervisor must be on the surface, diving only to clarify specific points. Work and surveying are not 'mixed' on the surface nor should they be underwater.

SIZE OF DIVING TEAM

Assuming the diver is properly trained and has been provided with the correct diving apparatus we now consider his actual employment. Statutory regulations require facilities for sending a second diver to assist the first if in difficulties. Someone must 'attend' the 'stand-by' diver so, by inference, the minimum number of divers employed on any operation, from purely diving considerations, is three. The divers will be employed to carry out some specific task requiring surveying gear, tools, etc. and they may also have to provide a boat from which to operate. The minimum number for a diving team is therefore three.

FACTORS THAT AFFECT THE DIVER'S WORK

Effect of tidal flow

The most common limiting factor which adversely affects a diver's work is the velocity of water flow. A flow of one knot has roughly the same effect on a diver as that of an 80 km/h wind force acting on a man on land. The maximum tidal flow in which a diver can work is, not surprisingly, one knot. Limited work can be carried out in about

1.2 knots but, if the flow does not reduce below this figure, physical protection is required to enable the diver to work.

Refraction

A diver looking, as he does, through air, glass and water observes an apparently larger and closer object and is, therefore, liable to give wrong dimensions by observation. A measuring unit consisting of a straight edge with graduations of known size is invaluable. A spirit level may be used to establish level.

Poor visibility

Underwater visibility is generally inversely proportional to the density of population. In rivers which run through highly populated or industrialised areas, the visibility is usually very poor. The converse usually applies in sparsely populated areas. The same crude ruling applies in coastal areas near river estuaries, where the outflow of polluted rivers affects areas many miles offshore. The further one moves into the open sea the better the prospect of good visibility. Apart from the pollution caused by man, sand from the sea bed brought into suspension by storms can reduce visibility to a few inches even well offshore. After a few days' good weather this can change to give a visibility in excess of 30 m. High-candle-power lights only illuminate the particles in suspension so the diver sees myriads of tiny bright lights reflected on the particles. Special low-candle-power lights are used for close inspection in poor visibility.

In areas of permanent low visibility, tactile measuring devices are invaluable. In this case the diver's fingers are his eyes.

As a crude guide, where daylight will not penetrate, underwater floodlights are useless. The effective use of floodlights is primarily limited to night work or in intakes, etc., where natural light is blocked out.

The underwater season

Underwater plant life flourishes at approximately the same time as that on land. In the summer months weeds and other marine growth are at their most prolific particularly in shallow coastal waters. Inspection of outfalls and other structures is therefore best carried out in early Spring.

Fatigue

Breathing underwater involves appreciable effort. The muscles of the rib cage which draws air into the lungs have to work harder to ventilate the lungs with the denser air. With the demand-valve system, some effort is also required to activate the tilt valve. The effort required to swim underwater is heavy but trials have shown that a diver walking in standard gear uses approximately the same effort as the underwater swimmer. Two hours is considered the maximum time a diver can work efficiently using a demand-valve breathing system. Provided that he is working in one area, four hours' work can be carried out in helmet gear.

Sickness

Of necessity, divers are required to have a high degree of physical fitness and, generally, are very rarely ill. Working in cold water and experiencing temperature changes, however, expose the diver to the common cold. This can be serious for the diver since the presence of mucus in the eustachian tubes can prevent the diver clearing his ears owing to the blockage preventing pressure balancing of the ear drum from the inside; so the ear drum is forced inwards. Forcing can cause damage or infection of the ear. Consequently, divers should not dive with a head cold.

PREPARATION FOR UNDERWATER WORK

General

Recognising the difficulties under which the diver has to work, great care is necessary in briefing the diving team. For example, before diving commences on a bridge survey, basic information should first be recorded, e.g.: a fixed datum point above water level which, if required, can be tied to ordinance datum at a later date; if the river is tidal, fix a temporary tide board from which direct readings can be taken during the survey;

Figure 41.2 A diver carrying out inspection of the caissons and bracing legs of 'Sea Quest'

establish chainages along face of piers and abutments; establish river-bed profile in relation to water level and therefore datum.

The above basic information enables a drawing to be produced showing an elevation with dimensions of the structure being surveyed. The diver then proceeds with his inspection, relating his findings to the chainage position and the water level or bed level in recording defects located.

When working in tidal rivers and estuaries reference should be made to the Admiralty Tide Tables to establish the time of high and low water predicted for nearest Primary or Secondary Port. It should be rememberd that on western seaboards strong westerly winds delay the ebbing tide and cause higher tides than those predicted. The converse applies with strong easterly winds.

Offshore

When working offshore, divers can normally work from a vessel in conditions in which vessels can moor and work. The sea states generated by a force 5 wind (Beaufort Scale) normally preclude operations unless land protection is provided as with an offshore wind. In most coastal areas, diving can only take place during high- and low-water slack. Tidal information is provided for fixed locations on large-scale charts with tabulated times, velocity and directions given with times related to the time of high water at a primary port in the Admiralty Tide Tables. It is unlikely that work will be in a tabulated location so interpolation is necessary between the nearest locations on which information is available. A good knowledge of chart work and tide tables is essential for the offshore diver.

The following two important factors should be remembered when assessing tidal speed and direction. The tabulated information refers only to surface tides which affect shipping; on the sea bed the tidal flow could be in the opposite direction to that on the surface. Tabulated information can only give an indication of what may be expected; some adjustment will need to be established during the first day's work. The second factor is that the direction of tidal flow moves around through 180° during the change of the tide. Thus a ship at single anchor may start off over the site of work and swing steadily away as the tide slackens, just when the diver is experiencing the best conditions for working on the bottom.

CONSTRUCTION WORK UNDERWATER—PREPLANNING

The work the diver carries out underwater is normally on some man-made assembly, usually of steel or concrete. The unit or units have been designed to carry out a specific function underwater. As a general rule such units are designed in the same way as similar structures erected on land and little consideration is given to the problems associated with underwater assembly. Construction underwater often takes a long time and it is common to find vital bolts missing simply because the tolerances were too tight or access so difficult that the diver could not insert or set up the bolt.

When planning construction work to be carried out by divers the following factors should be taken into account:

(a) Visibility varies and can be virtually nil, therefore units should be designed in such a way that the diver can work by touch.
(b) Diver's vision is distorted. Fine assembly is difficult. Therefore adopt wider tolerances.
(c) The diver and the units he is working on are affected by tidal flow. Simplify the means of connection to the previously placed works.
(d) Water flow accelerates around mass units placed underwater. The velocity depends on the size of the unit and, of course, on the normal flow. Consider means of protecting the diver.
(e) Lowering heavy units into accurate position from surface floating craft is difficult. Divers can normally handle underwater units that weigh approximately 50 kg on the surface. Movement to location by divers using controlled buoyancy is practical for units up to approximately 250 kg (surface weight) but it must be remembered that the tidal flow acts on the buoyancy unit as well as the unit being placed. A decision on the method of placement obviously is an essential prerequisite of design work.

(f) A diver is virtually weightless underwater and therefore he can apply little downward force. Providing he can establish a footing he can lift to his full strength, but remember that some of his energies may be taken up in combating the tide. The one way in which a diver can positively exert his full force is with his arms and shoulders. Using one hand as a reaction point, he can pull with the other. When preparing the detailed design the use of this motion should be exploited. Modifying the design of concrete blocks to accommodate diver assembly increases design and casting costs, but these are more than offset in increased output from the divers.

(g) A diver's 'feel' when setting up bolts underwater is somewhat blunted. In breathing compressed air he is, in effect, breathing enriched air due to the higher partial pressure of oxygen, which can act as a stimulant. He is liable therefore to apply too much force; therefore the use of a torque wrench is suggested.

(h) In many cases the diver needs one hand to hold himself in position against the tide and so, effectively, he has only one hand free for working. Where a large number of bolts have to be set up, provision should be made to lock the bolt heads.

USE OF TOOLS BY DIVERS

It is difficult for the diver, with the many restraints on his performance, to use tools underwater effectively. He has a number of tools which he may use effectively but careful planning is necessary to ensure their proper utilisation. For example, if it is desired to fix concrete saddles to a rock bed by rock bolts it is sensible to cast the lower section of the saddle block so that it forms a template for the drilling operation, i.e. so that the hole is of sufficient size to accept the drill bit, and the block thickness can be related to drill steel lengths to give the correct penetration. As a general principle, the use of tools should be kept to the minimum. If it is possible, by careful design, to construct underwater by a combination of interlocking concrete blocks and/or units which can be bolted together, so that the diver has a repetitive Meccano-like assembly to carry out, this offers the best solution.

Pneumatic tools

These are commonly used by divers. Generally, all surface air diving tools can be used effectively to about 30 m depth. It will be obvious that extra care is necessary to maintain the tools. Common failings are caused by the lack of consideration of the following basic facts.

(a) Pneumatic tools usually require an effective pressure 'at the tool' of approximately 65 lbf/in^2. At 30 m the back pressure of sea water is approximately 45 lbf/in^2; the minimum pressure required therefore is 110 lbf/in^2.

(b) The diver is normally a long way from the compressor. The use of normal $\frac{3}{4}$ in tool hose causes a large pressure/restriction drop in the line. This must be corrected by the use of a larger-bore hose, the introduction of a 'pig' (receiver) in the line nearer the diver or a combination of both (which is recommended).

(c) The use of an exhaust hose to the surface to remove back pressure appears attractive. The extra restriction to flow and the drag on the exhaust hose removes this apparent advantage. Generally, direct exhaust is used, although a short buoyant length of hose has some advantage in carrying the exhaust air bubbles away from the diver.

Hydraulic tools

These are available but the logistic problems of special pumps limit their use to specialised tasks.

Underwater cutting

There are three basic types:

(a) *The oxy-hydrogen torch* which is little used commercially. Divers need regular practice.

(b) *Oxy arc equipment* Consists of a 'gun' which holds a hollow carbon or steel rod through which oxygen is supplied. Power to the rod is provided by a welding generator. An arc is struck with the rod, as for welding, with the oxygen jet removing molten metal in addition to providing fuel.

(c) *Thermal arc* Similar in principle to the thermic lance but with flexible plastic hose in which metal 'fuel' is moulded. The hose is expensive and the oxygen consumption is high. A quick if crude cutting system which is attractive in that no generator is needed.

Cutting using explosives

The cutting of steel plate or tubular section using explosives is limited in construction work owing to the possibility of damage to adjacent structures. For offshore work in deep or exposed waters, specially designed shaped charges have been developed. These use the minimum weight of explosive to cut through plate or tube of different thicknesses. They are expensive but worth considering where, for example, divers' 'down time' is limited by depth or tide.

Underwater welding

Underwater welding has been carried out for many years although, as yet, the quality of workmanship which can be achieved is lower than that on land. It is suitable for emergency repairs or, with some additional method of providing strength (e.g. a steel sleeve filled with epoxy resin after welding), for permanent works. Habitat welding, i.e. in a 'house' underwater, has been carried out but this is very expensive.

Concreting underwater

Placing concrete underwater by tremie tube has been practised for many years. The difficulties in handling the tremie tube and pouring concrete in such a way as to avoid passing the mix through water, with the consequent loss of cement, are well known. The present-day efficiency of the concrete pump allows direct placement.

Grouting underwater

Grout in mass has been used in lieu of concrete where pumping over a long distance has been required. It is particularly attractive in 'tying together' pre-cast units, interlocked or bolted together, or, perhaps, where both the paths and connections for supplying and controlling the grout flow can be built in. The problems of grouting the foundations of structures in water are much the same as on the surface, i.e. the possible excessive loss of grout through coarse gravel or similar easy paths must be guarded against.

Underwater bolt-firing gun

These are capable of fixing bolts in bricks, steel or concrete. It is not normally practical to fix bolts in rock owing to its variable consistency.

Nondestructive testing

Techniques are similar to those on the surface. Naturally time is required to prepare and clean surfaces where, for example, metal thickness readings are required. The probe may be carried by the diver with a coaxial cable to the instrument. In this way the divers are controlled by telephone from the surface where readings are taken and recorded. Potential difference can be measured in the same way. Encapsulated instruments are available so that a technically trained diver may take his own reading. Where practical, however, surface control is recommended.

Air lifting

Air lifting provides an effective means of excavating material from inside cofferdams when the materials may be deposited outside the cofferdam. Controlling materials ejected from an air lift requires a large, relatively low, platform and, therefore, where materials must be removed from site, normal dredger techniques should be used. The simplest form of air lift consists of a 150–300 mm diameter length of pvc pipe with an air connection inserted approximately 500 mm from one end with a control valve. The diver locates the hose end in the cofferdam to be excavated and opens up the air supply. The air travelling up the tube expands as it ascends, drawing water and materials to the surface.

With an adequate air supply a 10 m long by 200 mm diameter air lift will easily lift 150 mm stones approximately 2 m above the surface. A diver can operate and move such an air lift manually, like a vacuum cleaner, to clear an area.

A pvc air lift can be operated entirely underwater where the materials removed from, say, a circular cofferdam on the sea bed can be scattered broadcast on the surrounding sea floor area. In a water depth of 25 m an air lift 10 m long will work quite effectively although the discharge point is at least 15 m underwater. At this depth the air is compressed and this does not have its full lifting capacity; a greater volume flow of air is therefore required to deal with a given quantity of spoil.

Underwater television

Underwater television is regularly used for instruction and control in offshore work. As previously indicated, 90% of diving work is carried out in shallow water and the greater proportion of this work is carried out in water where the visibility is poor. Television is therefore suitable only for limited localised inspection. However, manufacturers are developing new and more sensitive systems with complex lighting which improves the performance. This development, together with the gradual cleaning up of our rivers, should increase the usage of television for routine shallow underwater inspection. The use of television is justifiable where major repairs are considered necessary to an important structure. A videotape recording taken in the best conditions, which the engineer can study at any time, has obvious attraction.

Underwater photography

Again, owing to poor visibility in closed shallow water, photography is seldom used other than offshore.

Production of drawings

For structural surveys it is clearly preferable to use engineers or draughtsmen who can dive to produce adequate drawings illustrating findings.

UNDERWATER ENGINEERING

Divers can, and do, carry out some remarkable tasks underwater. Such works are, however, invariably the results of sound practical engineering and a proper appreciation of the maritime problems associated with underwater working. Divers need to be used intelligently as part of the construction team.

It is not sensible to ask the diver to carry out heavy manual work where this can be avoided. Fatigue reduces the mental faculties of the diver in the same way as other human beings. Diving is hard work which, to some extent, accounts for loss of memory associated with the job in hand.

Having established a scheme, stick to it and aim to standardise assembly so that the diver can establish his working techniques.

It is said that, given time, the sea will 'eat' anything in it. It is certainly true that the power of water should never be underestimated. The Victorian engineer's principle of good, strong and durable construction is not a bad example to follow for underwater work.

BIBLIOGRAPHY

UK government publications (available from HMSO, London)

Admiralty Diving Manual, revised edn (March 1972)

'Air Ministry Specification AIR 0133', as covered by Exemption Order 41 of The Gas Cylinders (Conveyance) Regulations SR and 0 679 (1931)

'Home Office Specification HOAL 1 for seamless aluminium alloy cylinders for the conveyance of compressed gases'

'Home Office Specification HOAL 2 for the construction of seamless aluminium alloy cylinders for the conveyance of compressed gases'

'Home Office Specification S for seamless alloy steel cylinders for the conveyance of compressed gases'

'Home Office Specification T for seamless alloy steel cylinders for the conveyance of compressed gases'

'The Diving Operations Special Regulations 1960,' Statutory Instruments 1960, No. 688—'Factories'

British Standards (published by the British Standards Institution, London)

BS 399 : 1930 (as amended February 1954)

High carbon steel seamless cylinders for the storage and transport of permanent gases

BS 400 : 1931 (as amended March 1942)

Low carbon steel seamless cylinders for the storage and transport of permanent gases

BS 1045 : 1945 (as amended September 1945)
Manganese steel gas cylinders for atmospheric gases
BS 1319 : 1955 (as amended September 1957)
Medical gas cylinders and anaesthetic apparatus
BS 3595: 1969 (as amended October 1970)
Life saving jackets
BS 4001 :
Part 1 : 1966 *Compressed air open circuit type*
Part 2 : 1967 *Standard diving equipment*

Overseas publications

Classification des engins d'explorations des océans', *Provisional Document* NI 146 BM1, Bureau Veritas (October 1971)

Guide for the classification of manned submersibles, American Bureau of Shipping, New York (1968)

Safety and operational guidelines for undersea vehicles, Marine Technology Society, Washington, DC (1968)

US Navy diving manual, Navy Department, Washington, DC (July 1963)

Other publications

BENNETT, P. B., and ELLIOTT, D. H., *The physiology and medicine of diving and compressed air*, Baillière, Tindall and Cassell, London (1969)

DAVIS, R. H., *Deep diving and submarine operations*, 7th edn, St Catherine Press, London (1962)

The British Sub Aqua Club Diving Manual, 7th edn, London (1972)

The principles of safe diving practice, CIRIA Underwater Engineering Group, London (1972)

42 DEMOLITION

DEMOLITION 42

42 DEMOLITION

R. G. PRICE
Goodman Price Ltd.

INTRODUCTION

Demolition was originally in the hands of relatively unskilled contractors whose sole objective was to salvage anything possible quite irrespective of how long it took to do this. Nowadays, there are a number of recognised contractors most of whom are members of the National Federation of Demolition Contractors. These contractors are highly skilled, mechanised and equipped to undertake the work required in the minimum period of time. As time has elapsed, so the types of structures have altered considerably and this is an accelerating process which will develop exactly on the lines on which civil engineering and building methods of construction are developing. It is, therefore, becoming increasingly important to select a particular project with great care especially where the work is of a difficult nature or where large contracts and short time schedules are concerned. A list of specialist contractors can be obtained from the National Federation of Demolition Contractors.[1] In the case of large complexes in urban areas, one should be especially selective and choose one of those firms who are conversant with the specialist methods necessary to carry out such work with speed and who have the necessary management and technical staff to deal with the difficult problems involved.

Due to the speed with which contracts are now usually required to be carried out, the pattern has changed from the days when demolition was done by hand, apart from the exceptional cases where it would be dangerous to do otherwise. The modern methods are: crane and swinging ball, which is a perfectly safe method in the hands of an experienced contractor and equally experienced crane operator; demolition by engineered deliberate collapse which, within the limits laid down in CP 94,[2] is a perfectly viable method and one which can reduce the risk to life and limb of the men working on the site rather than otherwise; and the more modern and ever increasingly used method of demolition by explosives.

All the above methods are only applicable under certain conditions and it is advisable to call into consultation an experienced demolition contractor at the planning stage to ascertain which method is preferable in a particular case. Some of the bigger firms have their own civil engineers on their staff who are experienced in this particular aspect of their profession and can give advice on methods, on shoring designs etc. Consultation at an early stage is particularly desirable in the case of modern structures incorporating prestressed ungrouted beams etc. and one of the valuable applications of controlled explosives is in this connection. There is no known method of destressing a beam of this type and the removal of superimposed loads will only induce hogging. However, preparatory work can be carried out in safety by placing explosives within the beam and then by firing the charge from a remote point the beam is detensioned in complete safety.

There is, of course, a large element of chance or gamble about estimating for a tender due to the impossibility of knowing the construction of many of the subjects for demolition. This element is reduced in ratio to the experience of the estimator; experienced estimators can identify the type of construction from the appearance of a building and its age. In architecture and structural work, fashion is extremely rigid.

It is, however, very desirable from the demolition contractor's viewpoint that he should be given original construction drawings wherever possible. Normally, the specification calls for demolition down to lowest solid floor level but sometimes the contractor is asked to include for taking out the concrete floors, slabs and foundations. This is manifestly unfair and should never be included as it is necessary to have a crystal ball to price it. The excavation contractor is given a schedule of rates for just such work and the demolition contractor should be treated similarly.

One of the difficulties of demolition at the present time is the increasing tendency for there to be no market for the 'hardcore', which is a byproduct of the trade. It is necessary to make sure that the contractor chosen has accessibility to the necessary transport and tips for the material if the contract is to be completed in time.

SALVAGE

One of the main items of salvage is ferrous scrap which is mainly graded as steel, wrought iron, and cast iron. These are roughly sorted out on site or, where time is of the utmost importance, may be carted away completely intermingled and sorted out at leisure elsewhere. Finally it has to be sorted again into heavy, medium and light cast iron, constructional steel and light steel and piping, and the steel and wrought iron have to be separated and cut into furnace length (1.5 m long) and the cast iron has to be broken down similarly. Mostly this material is disposed of to the major scrap steel merchants who hold orders from the smelters and from foreign countries for its supply.

The steel is cut by oxy-acetylene cutting gear and lifted down or dropped in as large units as possible, and often flame cut again on the ground before loading. With large sections of steel, it may be considered more economic to use the thermic lance (mentioned later in this chapter) for cutting it down. This tool is also very useful for cutting up very thick cast iron. A plentiful supply of oxygen gas is needed for this, and it must be ascertained previously that this is readily available.

MODERN DEMOLITION PROCEDURES

The time honoured methods of breaking out concrete still apply for both cheapness and efficiency. These are drilling by pneumatic drill and diamond points, hydraulic bursting, and the demolition ball where this can be safely used and where the consequent vibration does not preclude it. Hydraulic bursting is probably the best and cheapest method for mass concrete in large quantities and where the resultant debris can be loaded and transported in large pieces. There are also applications for lasers, ultrasonic vibrations, and freezing. Where it is not possible due to noise or inaccessibility to use the three main methods above, explosives or the thermic lancing method[3] may be used.

All these methods will be dealt with in detail later in this section.

Explosives

At present explosives techniques produce the economic and satisfactory answer where other methods may fail on account of noise, costly equipment, insufficient development, inaccessibility to heavy machinery, and the very high cost of materials used; notwithstanding instances where large debris has to be worked down to manageable sizes by a secondary process. The controlled application of explosives enables a trained operator to place a precalculated and tailored charge of a suitable explosive into a piece of material which, when fired, will shatter that section of material into readily manageable pieces without scattering the resultant debris over the site. It can also be used in close proximity to valuable and sensitive machinery, etc.

Figure 42.1 Demolition and site clearance for St. Thomas's Hospital, London (Goodman Price Ltd.)

Figure 42.2 Demolition of Cannon Street (SR) Railway Station, London (Goodman Price Ltd.)

Advantages to be obtained by the explosive technique, when performed by skilled operators, are as follows.

The operation of drilling the holes in which to place the charge by conventional rock drill which, though creating a considerable airborne noise, strangely enough is less offensive than the noise of a jack hammer. On this merit alone the technique has been successfully used in city areas, occupied offices and hotels. Where the charge has been calculated to optimum effect with minimum throw out, the energy is almost entirely expended in the work done and there is very little surplus energy to create dangerous shockwaves in other parts of a structure. Under these conditions, airborne shock waves can be discounted.

With the aid of explosives, a given amount of work can be carried out by fewer men with correspondingly lighter equipment than would be necessary by more conventional methods and can be used in situations relatively inaccessible to other methods of working, particularly when considered against the demolition ball or thermic lancing. As a general guide, any homogeneous mass greater than 300 mm thickness can be economically worked by explosives when considered against any other known method. The greater the thickness of the section to be removed, the more economical becomes the explosive method when related to cubic measure.

In demolition and civil engineering works, 80% of blasting work is performed with nitro glycerine/ammonium nitrate mixtures manufactured under various trade names. These explosives exert an instantaneous pressure of nominally 360 m. tonnes per kg weight at a velocity of 2500–3000 m/s, although the rate of burning is controlled by the diameter of the cartridge. To fragment a structure, holes are drilled at centres and depths relative to the amount of work to be performed and the degree of fragmentation required.

It is foolish to isolate blasting costs and try and relate them to a single alternative method. The total cost of an operation must be considered including the participation of other trades, machinery standing time, size of machinery required, rate of progress, noise nuisance to neighbouring properties, the maximum size of debris that can be handled. When all of these factors have been analysed honestly the balance must come down in the favour of explosives.

A careful study of the object to be shattered should be made from drawings and by a physical examination, so far as this is possible on site. The latter is most important since the object rarely conforms to the drawings other than in general principle. Points to look for are pour levels in concrete, bonds in brickwork, deterioration from chemical attack, movement due to settlement or weakening of supports, degree and nature of reinforcement where it is exposed in reinforced concrete structures.

Having decided on the method and the degree of fragmentation required, the holes may be drilled. Generally the spacing of holes is not greater than the thickness of the material being worked and the burden in front of the hole is usually 80% to 90% of the hole spacing. Depth of hole can be influenced by construction joints, total thickness of the material, type of manpower/machinery that will be effecting removal of debris, working space available above the drill, etc.

Where reinforcement is present, spacing of bars, diameter, laps and the amount of material that can safely be fractured at any one time must be additionally considered.

Having drilled a pattern of holes it may be necessary to deck load them with two or three small charges instead of one large base charge. This decision is once again influenced by construction joints, nature of material and degree of fragmentation desired.

TYPES OF CHARGES AND EXPLOSIVES

On closed sites such as are normally found with demolition contracts and small civil engineering works, the charges are fired electrically with either a magneto type or dynamo condenser type exploder, these machines are capable of firing a number of detonators at any one time by the simple pressing of a button or turn of a handle. Electrical firing is favoured because control of the situation is exercised right up to the instant of firing; should any factor alter on site, right up to the last second, the matter can be dealt with—a situation which does not exist when lighted fuses are used.

Figure 42.3 Demolishing a chimney with explosives at the Thames Reach Barrier scheme (top) Placing and stemming charges of gelignite (bottom) Detonation of charge (G.L.C.)

Figure 42.3 (continued) Demolition completed (G.L.C.)

Other types of explosives[4] used in civil engineering work are:

Gelatines. For underwater work, plaster shooting, cutting metals.
Dynamites. Limited underwater work, cutting metals, rock blasting.
Gelignite and ammonium nitrate mixtures. Maids of all work! Demolitions, rock blasting.
Ammonium nitrates. Rock blasting and roadworks.
Slurry explosives. Rock blasting and roadworks.
A/N diesel oil. Quarry work.
Black powder. Quarry work for masonry stone.

It must be stressed that the foregoing information is in very general terms—'good practice'. There is no substitute for practical experience and the expertise of the blaster is his empirical judgement based on experience within the perimeter of his technical knowledge and mental capabilities.

Thermic lancing process

On occasions, the site of operations is so situated that on no account can noise or vibration be permitted. Here it is advisable to consider breaking out the concrete by thermic lancing which is particularly suitable for the reinforced concrete used in modern designs of building.

Whilst thermic lancing is generally ruled out from a cost point of view when more traditional methods can be employed, it is invaluable where noise abatement is paramount, for example, hospital extensions and alterations, dense commercial developments, and close proximity in accessible positions to highly sensitive plant and machinery.

The cost factor is put into a much more competitive perspective where there are slim concrete sections with a resultant high proportion of reinforcement in relation to the cross sectional area. In this case it is possible to obtain a really good 'burn', increasing production to a rate where the cost is much more reasonable. Here again, the contractors who specialise in this method will usually offer a preplanning advisory service.

Water jet process

Another method which is still in the development stage is cutting of concrete by water jet.[5] This has all the advantages of silence and speed, but of course, in urban developments there is the problem of disposal and control of the resulting water. This method

Figure 42.4 Thermic lancing process (Goodman Price Ltd.)

would appear to be of limited use where the concrete is reinforced. It has been used quite extensively, particularly in Canada where it has been employed for cutting rock and stone in quarries, and cuttings on motorway projects. It could no doubt be used similarly in this country, but one would think its use in demolition would be circumscribed and limited.

Concrete breakers

A new type of hydraulic concrete breaker has now been evolved which will give very much quieter operation. This type of machine is now out of the experimental stage and is being produced and sold successfully in the 1 and 2 breaker size. These have the great advantage for demolition of tall buildings of possessing a low power weight ratio and also of considerably improved silence in operation. The degree of 'punch' at point of contact is also an improvement on previous methods.

There are are now several giant sized breakers for concrete on the market, and these are very successful. They are mounted on the end of the jib of an hydraulic excavator which holds the hammer and point down and on to the concrete to be broken. Either a compressor supplies air direct to the hammer or the excavator hydraulic pressure is applied to it.

These tools have been thoroughly tested, and are a very valuable asset when the large machines servicing them can be brought to the site of operation. They are capable of breaking out concrete at great speed. The air-operated machines are about as noisy as the piling hammers from which they were developed, and whilst the hydraulic ones are quieter, both types cause some vibration, which is not always acceptable.

THE BS CODE OF PRACTICE

About three years ago, the Federation approached the BSI and asked them to assist in compiling a code of practice for the industry. A committee was formed comprising representatives from all the relevant professional bodies, H.M. Inspectorate of Factories, the Greater London Council, and five members of the Federation. Due to the considerable amount of work involved the task took longer than the members originally envisaged but eventually CP 94 : 1971[2] was produced. This has been recognised as a publication of the foremost importance and meets the full approval of the Factories Inspectorate. CP 94 should be carefully studied by all those contemplating, and participating in, demolition. The code deals in detail with the various aspects inherent in demolition work, it includes views on recommended methods of demolition, the use of materials and plant and offers advice on statutory requirements and safety precautions. It details protective precautions prior to and during demolition operations which fall under three headings:

(1) Precautions specifically aimed to safeguard personnel on site.
(2) Precautions for safety of persons not connected with the demolition, including members of the public.
(3) Precautions necessary for the protection of property likely to be affected by demolition operations.

The code also defines methods of demolition including, hand demolition; mechanical demolition by pusher arm, demolition ball, and wire rope pulling; explosives; demolition by hydraulic burster and thermic lance. It recommends methods of demolition suitable for particular types of building such as houses, large buildings, bridges, arches, independent chimneys, steel and concrete structures, spires, pylons and masts, and procedures for the excavation and removal of petroleum tanks.

Adoption and implementation of this code of practice should eliminate the non-professional methods of demolition. The code provides the necessary guide lines to local authorities to compile more comprehensive and detailed specifications for demolition work, administer closer supervision over such contracts and provide the public with a degree of safety hitherto not given on the majority of contracts. All too often the onus for safety is placed squarely on the demolition contractor. Specifications rarely mention specific precautions to be observed, what scaffolding and hoardings should be provided, or the method of demolition to be adopted. Such details are invariably covered by the phrase 'the successful contractor will provide all necessary safety precautions'.

On a specific contract compliance with such a clause can vary from the provision of £2000 worth of scaffolding and hoardings, plus the inconvenience and expense of having to work within such confines, to non-provision of scaffolding and thus rendering buildings dangerous, perhaps to the extent that roads have to be temporarily closed whilst demolition is in progress. It could be said that such a code of practice is long overdue, and it is certain that every efficient and competent contractor will welcome its introduction into the industry.

Preliminary procedures, contract and specification

Where a formal contract is entered into it is the normal practice to use the standard form of RIBA Contract.[6, 7] This contract was formulated for the building construction side of the industry and it is not really suitable for demolition work. When the contract comes to the contractor for signing, there are often more inapplicable paragraphs deleted than those that have been left. The National Federation of Demolition Contractors has a form of contract which it offers as a preferable alternative.

The specification should be kept as simple as possible whilst clearly defining the work required to be done. Work which it is not possible for the contractor to price accurately should be the subject of a 'Provisional Sum' and an attached 'Bill of Rates' for this work. These items include shoring, waterproof rendering, works of reinstatement and making good which cannot be seen or assessed at the time of tendering and the diverting of site services or their cutting off by the Local Authority and Statutory Undertakings.

Many things, which are unforeseen at the time of tendering, can occur after the demolition starts and the fabric of the adjoining buildings is disclosed. For this reason allowance must be made for the need for considerable revision of the programme. The more investigation that is done by the client and his architect prior to the tender documents being released, the less will be the subsequent claims for extra payments and the more realistic will be the time quoted for completion. Party Wall Awards are seldom agreed until after the contract is started. It is, of course, often impossible to finalise these until the building to be demolished is opened up and the face of the party wall exposed, but much preliminary negotiation could be done long before this.

The specification should point out very clearly the levels down to which the buildings are to be demolished, whether cleared out or filled and if any 'earth dumplings' at higher than general level are to be removed; this should be clearly stated and approximate quantities (subject to remeasurement later) should be given.

In cities, the question of any necessary diversion of public services should be high on the list of priorities at the pre-planning discussions. The siting of these should be discussed, the Statutory Undertakers contacted, and if possible, work started or even finished before demolition begins.

Insurance

Another extremely important facet is the question of insurance cover and the professional advisers to the client should make certain that this is adequate for all their requirements. Third party cover should be for not less than £100 000 but most large contractors carry cover of £1 000 000 for any one accident or series of accidents. Workmen's compensation insurance should be unlimited, but of course, this form of insurance is now covered by a statutory regulation. It is also desirable to make sure the contract is covered fully for the contractor's plant and lorries, etc. used on site. It should be a condition of contract that the contractor's insurance policies are to be submitted before the work is started. It is getting increasingly difficult to obtain the necessary cover, especially where the proposer has a bad record for accidents. Statistics show that the accident record of members of the National Federation of Demolition Contractors is considerably better than the national average.

PROTECTIVE MEASURES

We now come to the question of safety on site. Before any contract is placed it should be preceded by a site survey which should be comprehensive and cover the provision of fans, screens and scaffolds for the protection of the public and also methods to protect surrounding buildings from danger of collapse due to withdrawal of support or undermining of foundations. The former is merely a matter of providing suitable protection and the experienced contractor will know exactly what is required, but the question of undermining needs thorough investigation both before and during demolition. It is often not possible to investigate satisfactorily beforehand and it is necessary to have full co-operation between the engineer and the contractor so that the construction of the adjoining buildings can be ascertained as the demolition progresses. This often involves opening up the floors from top to bottom so that walls can be plumbed and thicknesses measured.

Shoring

Support to adjoining buildings can be maintained by leaving up portions of the building to be demolished as buttresses, but this is not always possible and shoring may be necessary.

Shoring may be of three types, but is normally by raking shores or by flying shores in cases where the ground area is required to be kept absolutely clear and where the distance from readily available support is not too far. The larger firms of demolition contractors have their own civil engineers on their staff who are competent to advise on and design shoring suitable for each occasion. It is desirable to have shoring properly designed although, as yet, it is a subject often neglected in the training of an engineer.

The whole operation should be resolved by the consulting engineers and the contractor's engineers. Firstly, the party wall must be exposed and carefully examined to determine the exact degree of support which has to be substituted for the withdrawal of that of the existing buildings, and it must be ascertained whether the floors of the adjoining buildings span on to or are parallel to the party wall. If the front of the building being left standing is not securely bonded into the party wall, shoring or wall ties may have to be erected to take care of this. If it does not interfere with the work proposed on site, buttressing lengths of existing cross walls may be left, but if these must be removed entirely the party walls may have to be made good after their removal. This applies equally to chimney breasts.

The shores must, of course, be designed to be adequate for the anticipated load and the needles must be fixed in good solid brickwork immediately below the horizontal line of the floors on the reverse face of the party wall. The shore base may, with advantage to later operations, be founded at the lowest level of the excavation for the proposed new building. The shore hole has to be sited to miss the new concrete foundations, but also in a position where the shore being erected is bearing for its full height on a good sound

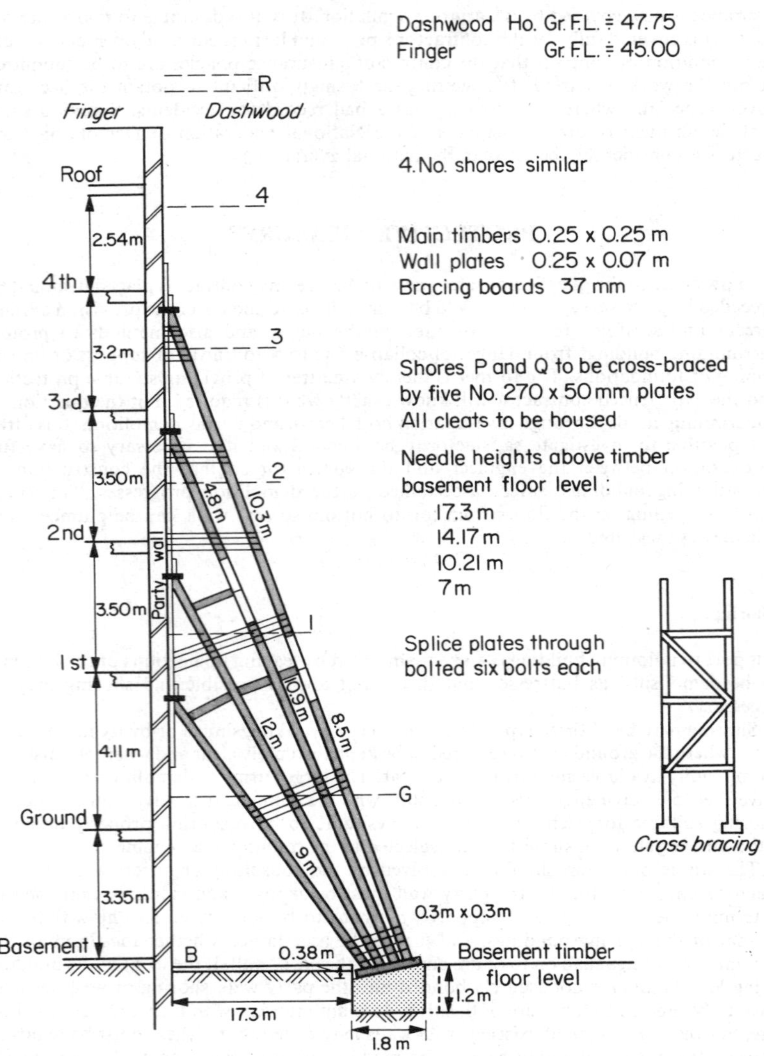

Figure 42.5 Diagram illustrating a typical shoring installation

wall. These conditions sometimes are very difficult to reconcile. The ultimate responsibility for the design must be upon the consulting engineers and for workmanship upon the contractor.

Shores may be erected in timber struts, solid or laminated, or in steel tube. The steel tube is more quickly and easily erected, but timber shoring is more positive and not subject to contraction and expansion on temperature changes. Great care has to be taken with the anchorage of tubular shoring.

Hoardings

Other types of carpentry which come within the purview of the demolition contractor are the provision of hoardings which essentially must be very substantially strutted to prevent collapse under wind pressure. The demolition contractor is also responsible for the provision of temporary roofs where parts of a building may be left occupied. Here again the major contractors carry permanent staffs of men skilled in this work.

Weatherproofing

The contractor should be instructed as to whether this is to be fixed to the party walls of the site. It can take the form of cement rendering (18 mm thick in two coats), bituminous felt sheeting, heavy polyethylene sheathing, or even a bituminous or silicone paint coating if the wall surface is in very sound condition. In the case of the various types of waterproof sheeting it is very necessary to ensure that the contractor has securely fixed it to the wall with *cross* battening at about 1 m centres horizontally and vertically. Random vertical battening is useless, as the wind will quickly strip the wall.

Safety

The safety of workmen and of the general public is fortunately, much more safeguarded in these days, but still more needs to be done in the education of both classes. Workmen do not always foresee things that are likely to contribute to their own downfall (and in this context the word can be used in its literal sense). Contractors must have a safety officer constantly checking for open traps, failure to board up windows, doorways giving access to lift shafts and externally, after these fittings have been removed. It is often true that the more experienced the workman, and however much care he takes when he is actually doing his skilled job, he grows careless about small points which have led to accidents in the past.

The general public are much too venturesome, not appreciating the dangers inherent and will wander over a site whilst it is being demolished, or out of hours when it has been left. This is particularly true of small boys who are intrigued by the ruins and will play on them at weekends undermining free standing walls or falling through dangerous floors. It is impossible always to leave the work in a safe state, and therefore, the safety angle must take the form of denying access by seeing that all doors and windows are secured and, at a later stage in the contract, that proper hoardings, soundly constructed, are erected. The latter must be designed to resist any likely wind pressures, as otherwise they can be a danger in themselves. The general public must also be protected by the erection of scaffolding fans to protect them from debris which may accidentally fall outside the working area. The recommendations of CP 94[2] also give the guide lines as to the safe methods of working according to the height of the buildings and their position in relation to streets and other property. Valuable guidance has also been published by the Department of Employment.[8]

SURVEY OF DEMOLITION METHODS

Hand

This refers to any form of demolition where mechanical means or principles are not employed. There are many forms of hand demolition. Methods vary from where the brickwork is painstakingly hacked away by the use of a mattock, to the equally painstaking and slow dismantling by hand-held pneumatic tools of a reinforced concrete

Figure 42.6 Hand demolition by pick-axe (P. A. Reuter and Goodman Price Ltd.)

building (so situated that it is too dangerous to use any other methods). This type of building, or a steel framed concrete clad structure, can be burnt down by the thermic lance method. During hand demolition of brick structures, short cuts may be taken in certain circumstances by working the walls down so that small masses of brickwork, such as the piers, may be dropped safely in one piece.

The disadvantages of hand demolition are fairly obvious, i.e. slow progress and high cost, and it is only used where the site conditions preclude any other method. Examples of these are demolition around and above any existing railway station or adjoining a busy highway.

Mechanical demolition by pusher arm

The pusher arm is a specially adapted crowd arm on the jib of an hydraulic excavator and is used to push sections of walling progressively into the site to be picked up by the

Figure 42.7 Demolition of building by swinging ball method (Goodman Price Ltd.)

loading machine. The extent of its use is of course controlled by the size and power of the excavator and the length of the jib which limits the height at which it can be utilised.

Mechanical demolition by deliberate collapse

This method can only be used where the contractor has plenty of free ground around the subject. The collapse can be engineered by removing, either mechanically or by explosives, key structural members, thus rendering the building unstable. This method should only be employed by thoroughly experienced personnel, as otherwise, if miscalculated, the structure may only be rendered precarious and will constitute a hazard to life and limb. This must also be considered if the building is not entirely free standing or is only part of a bigger structure which might consequently be rendered a hazard in its turn.

The use of explosives has already been covered in detail earlier in this section.

Crane and ball method

One of the most popular methods of demolition is by the crane with a breaking ball swinging from the jib. Its popularity is mainly because it is economic in labour requirements and is probably the swiftest method of demolition that can be employed. Its disadvantages are that it should not be attempted unless there is an area of 'free ground' around the point of operation and it should only be used in conjunction with a crane operator, thoroughly skilled in its use. The ball should, wherever possible, only be swung in line parallel to the length of the jib and should be secured by a running rope through the fairlead of the crane and attached to the appropriate drum on which it can be wound in or payed out to keep constant control of the ball. 'Side swinging' can cause buckling of the jib of the crane, particularly if the operator misses his target and is very inducive of accidents. This method is recommended to be used only where other methods are not satisfactory and very great care taken to avoid the consequences.

Hydraulic burster

Where one is presented with mass concrete in considerable quantity, undoubtedly the cheapest method of breaking it out is by the use of the hydraulic burster. A row of holes of 75 mm diameter is drilled at a predetermined distance back from the edge of the concrete mass and terminated again at a selected point by another row of holes at right

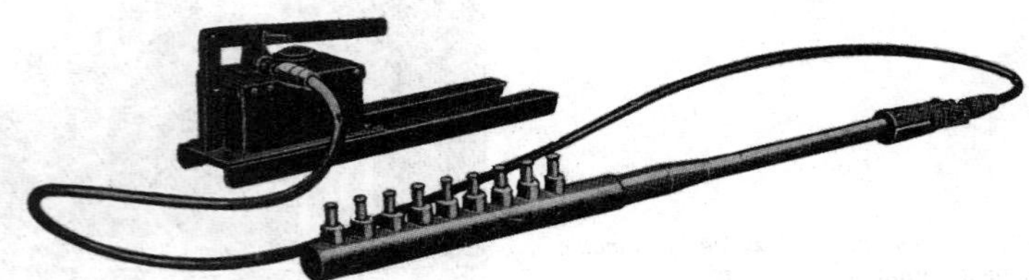

Figure 42.8 Single burster equipment with manual pump (Gullick Ltd.)

angles, and finishing near the edge. The burster cartridge (or more) is inserted progressively but not consecutively along the row of holes, and by the action of hydraulic oil pumped into it, the burster cartridge head forces out the plungers in an horizontal plane, thus exerting pressure and cracking the concrete.

The disadvantage of this method is that it is a hand operation, and therefore, slow, and also the concrete is broken off in very large pieces, and has to be handled and disposed of. It can, of course, be broken up further as a second operation. One advantage of the operation is that the only noise comes from the drilling of the holes, which for special applications could be diamond core drilled. It is not specially satisfactory on reinforced concrete unless the steel content is light.

Thermic lance

The thermic lance[3] consists of a long length of mild steel tube which is packed tightly with Swedish iron rods and the end heated with an acetylene torch. Oxygen is then passed through the tube under pressure and ignited. The concrete is heated until molten and then kept in this state by an increased flow of oxygen and progressively melted away. In doing this the rod-packed lance is burnt away and has to be replaced very frequently.

This method has the great advantage of being the quietest known in that there is no need to use compressed air in any form for drilling. Concrete in any quantity can be melted away, and thus, columns, beams, etc., can be cut, as is steel, by an oxyacetylene cutter. The method enables it to be used in inaccessible places where large machines cannot be used, and large beams etc, can be cut and lifted out by a tower crane or similar method.

The disadvantages are firstly, it is very costly, secondly, is really successful only if there is a fairly large amount of steel in the concrete which helps to get a good 'burn', and thirdly, it produces quantities of dense smoke. The expense can of course, be compensated for by the fact that in a densely populated area, work can carry on during hours when noiser methods would be prohibited.

BS RECOMMENDATIONS FOR THE DEMOLITION OF VARIOUS STRUCTURES

Jack arches

Probably the majority of accidents in the industry have been caused by demolition of jack arch floors and filler joist floors. The former should be demolished in strips parallel to the span of the rings of the arches, i.e. the demolition should proceed in the same direction as that in which the main floor beams run. Sometimes, tie rods are incorporated in the structure and the whole of the arches included in the floor must be demolished before any tie rods are cut.

Filler joist floors

With filler joist floors, a similar method should be followed but dropping each bay by releasing its respective filler joist. The operatives doing this should be standing securely on a temporary decking supported by the main floor beams.

Bridges

In bridges of any type, the first consideration is to remove all dead load possible without interfering with its structural members. In the case of masonry arches, the work should be undertaken in reverse order to that in which it was built, as in the case of most other structures. The support afforded by one arch to another must be taken into account, and if the work is to proceed progressively it may be necessary to erect shoring on the piers and abutments, or centring to the arches. Only contractors well versed in this type of work should be employed, and also it is advisable that an experienced bridge engineer should be consulted.

Figure 42.9(a) Railway bridge demolition. Note the use of sleepers to protect the permanent way (Goodman Price Ltd.)

Special consideration should be given to the additional problems attendant on the demolition of skew-built bridges. In a confined area, a steel or iron bridge should be lightened to the maximum possible extent without rendering it unstable, and the remainder should be supported by suitable craneage before being cut and lowered. If there is sufficient room, the bridge may be collapsed by the use of explosives and in such a case, this is the most convenient and economic method.

Chimneys

Chimneys are a special item in the demolition contractor's portfolio, and are a very interesting and often very spectacular one. The method, of course, depends chiefly on the situation of the chimney which may be closely surrounded by factory or other buildings which are to stay. In such case, the chimney will have to be pulled down piecemeal by hand, the debris being allowed to fall inside the chimney. A doorway is cut through

Figure 42.9(b) Railway bridge demolition. In this instance, the bridge span was small and the majority of debris was collected directly in wagons placed underneath (National Federation of Demolition Contractors)

the side of the chimney at the bottom, and the debris taken out through this. It must be cleared out often and regularly as, otherwise, the pressure exerted on the walls of the chimney may cause its premature collapse. Also the material may 'arch' inside the chimney, and therefore, the foreman has to make sure that this has not happened before sending his men in to load. If it has 'arched', the key will have to be broken before men are allowed underneath.

Normally, a scaffold is built outside the chimney to afford the men demolishing it access to the top and may be of tower type or cover more of the perimeter, even up to 360°. Great care has to be taken to avoid accidents when breaking the cast iron (or sometimes steel) ring which is usually found at the top, and this can be a very difficult handling problem. In certain cases, a scaffold may be erected inside the chimney and the material tumbled down outside if there is plenty of room or else put down in chutes.

Where the chimney is free standing, and a clear avenue of at least one and a half times its height is available, it may be felled. This can be done by gradually chipping away at a predetermined area at the bottom and cutting out a panel of brickwork until the chimney falls due to removal of support. This is a job to be tackled by only very expert personnel, and an experienced contractor. It is then possible to predict accurately the line of fall within a matter of a few feet.

The other and more usual method nowadays is to drill and place explosive charges scientifically in the lower part of the chimney and fire these charges simultaneously so that part of the bottom wall is blown out which by again removing support allows the chimney to fall on a predetermined line. Here again it is emphasised that this is a very skilled operation and there are only a few firms in the UK who are qualified to do this work.

The preceding remarks apply equally whether the chimneys are of brick or reinforced concrete construction, except where post-tensioned cables are concerned. In this case the advice of a chartered engineer should always be sought. In hand demolition of reinforced concrete chimneys, the concrete should be divided into small panels by cutting chases to expose the reinforcement which should then be burned off, the section being supported meanwhile. The lining must be worked down at all times level with the outer skin.

Pylons and masts

Pylons and masts may, on confined sites only, be dismantled piecemeal by hand, and the resulting steel sections carefully lowered unless there is sufficient free space around the base to allow of their being dropped.

If felling is contemplated, there must be a free area one-and-a-quarter times the height of the mast. The legs against the direction of fall should be partially severed and then the rear legs cut through. Prior to this, a steel wire rope should be securely fixed to the top of the mast or pylon and tightened but not strained. When the rear legs are cut it can safely be pulled over. The winch and all personnel must be in a safe area.

Petroleum, oil or gas tanks

The residual gases remaining in these tanks exist for very long periods which one might not readily credit. The *only* safe method when it is proposed to use flame cutting to dispose of the tanks is to render them completely inert. Water should not be used in such a case, but only if the work is going to be transported whole.

The best inhibitor is nitrogen gas which should be passed through the tank after sealing all openings, except the ingress and egress points. Pressure should not be allowed to build up in the tank.

Figure 42.10(a) Demolition of cooling tower at Courtaulds Aintree Works. Demolition contractors: Swinnerton & Miller (Courtaulds Ltd.)

Figure 42.10(b) Demolition of cooling tower (continued)

Figure 42.10(c) Demolition of cooling tower (continued)

Another method is to charge the tank with crystals of solid carbon dioxide 'dry ice' and leave for at least twenty-four hours. Yet another method is to fill the tank completely with suitable foam so that this excludes all gas pockets and to continue feeding foam into the tank whilst cutting. It is advisable to dispose of the tank by any means other than flame cutting, if possible.

FUTURE DEVELOPMENTS

There has been little or no planned educational facility in the demolition industry except for the experience gained gradually on site by the entrant to the industry who, joining as a labourer, graduated through the various grades of cleaner, demolition worker, topman, ganger and up to foreman. This, to a large extent, is still applicable.

As a result of recent action it is hoped that a scheme will be evolved whereby architects, engineers and construction personnel will be able to have a period of training in demolition contractors' offices and on site, as part of their course, and that a week's or fortnight's course for demolition personnel will be arranged at Polytechnics and Universities to acquaint them with the problems of the construction and planning side of the industry. This suggestion has already received the whole-hearted support of the Council of the National Federation of Demolition Contractors.

It is a matter of great concern that many problems of the construction methods employed in modern buildings must be thoroughly understood by the subsequent demolition contractor. It is imperative, therefore, in both his interest and that of the general public, that some system be inaugurated whereby detailed drawings of every construction are micro-filmed and kept permanently in the appropriate local authorities' records or some such other permanent store. This fact should be recorded on a plate in an approved and similar position in all buildings, so that the methods of construction used are available to the demolition contractor when the life cycle of the building is exhausted. The contractor may thus know whether any prestressed beams are pre-tensioned or post-tensioned and whether the building is stable under various conditions and, particularly, whether the retaining walls are self-supporting or rely on parts of the structure for their strength. It is very satisfactory that this point which the writer has been pressing for many years is now becoming recognised as being one of importance and it is hoped that further development will ensue.

The magnitude of the catastrophe which could result from demolition of some modern buildings without full knowledge of their construction is very real and can readily be imagined. It is important for architects and engineers to realise that a building will not last for ever and that its eventual demolition should be taken into account when designing and building new structures. If this is not done it is quite conceivable that the cost of demolishing certain buildings could easily exceed the cost of their construction.

REFERENCES

1. National Federation of Demolition Contractors, 2 Bankcart Avenue, Leicester LE2 2DB
2. CP 94 : 1971, *Demolition*, British Standards Institution
3. The Kaybore Thermic Lancing Co. Ltd., 'Demolition—Thermic Lancing', *The Civil Engineering & Public Works Review,* **67,** 389–91 (April 1972)
4. Nobel's Explosives Co. Ltd., 'A Century of Explosives Manufacture', *Civil Engineering & Public Works Review,* **67,** 390–391 (April 1972)
5. 'High Pressure Jets to blast rock bands', *New Civil Engineer,* **46,** 32–35 (21 June 1973)
6. Specification—Architectural Press, 'Demolition—Site Work', *Specification Journal,* **1,** 10–15 (1973)
7. Specification—Architectural Press, 'Demolition—Specification Clauses', *Specification Journal,* **1,** 16–18 (1973)
8. Department of Employment, 'Health and Safety at Work. 6E Safety on construction works: Demolition', HMSO, London (1973)

BIBLIOGRAPHY

AKAM, E. A., 'Demolition—Erecting a framework; Results of a BRE Survey', *Civil Engineering & Public Works Review,* **67,** 377–381 (April 1972)

PRICE, R. G., 'Demolition, the Art and the Science', *Civil Engineering & Public Works Review,* **67,** 383–385 (April 1972)

SWINNERTON, H. A., 'Demolition, an ancient form of energy adapted to modern practice', *Civil Engineering & Public Works Review,* **67,** 387–389 (April 1972)

GRAINER, W. E. and HANCOCK, T. C., 'Discussion: Whitgift Centre—Croydon', *The Structural Engineer,* **49,** 353–357 (March 1971)

APPENDIX: UNITS, SYMBOLS AND CONSTANTS

APPENDIX: UNITS, SYMBOLS AND CONSTANTS

APPENDIX: UNITS, SYMBOLS AND CONSTANTS

METRICATION

G. R. DARBY, C.Eng., M.I.Mech.E.
Secretary, Metric Steering Committee, C.E.G.B.
and
A. PARRISH, M.B.E., C.Eng., M.I.Mech.E.
Consultant, Formerly with I.C.I. Ltd.

THE SYSTÈME INTERNATIONAL D'UNITES

Introduction

SI is the accepted abbreviation for Système International d'Unités (International System of Units) the modern form of the metric system agreed at an international conference in 1960. This system has been adopted by the ISO[19] and the IEC[18] and its use is recommended wherever the metric system is applied. It is already in the process of being adopted in the legislation of twenty-three countries. The indications are that SI Units will supersede the units of existing metric systems and of all systems based on Imperial Units. The SI is now being adopted throughout most of the world and is likely to remain the primary world system of units of measurement for a very long time.

SI units and the rules for their application are contained in ISO Resolution R1000[10] and a BIPM[15] informatory document 'SI–Le Système International d'Unités'. An abridged version of the former is available as BSI publication PD 5686[6]. BS 3763[5] incorporates information from the BIPM document including matters which deal with units outside the International System which are recognised by the CIPM[17] for use in conjunction with it. The BIPM document is based on resolutions of the CGPM[16] or decisions of the CIPM.

Basic SI units

SI comprises seven basic units from which a wide range of quantities can be derived in the form of products and quotients of these units which are:

Quantity	*Name of unit*	*Unit symbol*
Length	metre	m
Mass	kilogram	kg
Time	second	s
Electric current	ampere	A
*Thermodynamic temperature	kelvin	K
Luminous intensity	candela	cd
Amount of substance	mole	mol

* *Note:* Temperature difference is commonly expressed in degrees Celsius instead of degrees kelvin. The unit of the temperature interval for these scales is the same; 0 K = −273·15°C; 273·15°K = 0°C.

The definition of these units as given in BSI/PD 5686[6] are as follows.

Metre (m). The metre is the length equal to 1 650 763·73 wavelengths in vacuum of the radiation corresponding to the transition between the levels 2 p_{10} and 5 d_5 of the krypton-86 atom. (11th CGPM[16] (1960) Resolution 6).

Kilogramme (*kg*). The kilogramme is the unit of mass; it is equal to the mass of the international prototype of the kilogramme. (1st and 3rd CGPM[16] 1889 and 1901).

Second (*s*). The second is the duration of 9 192 631 770 periods of the radiation corresponding to the transition between the two hyperfine levels of the ground state of the caesium-133 atom. (13th CGPM[16] (1967) Resolution 1).

Ampere (*A*). The ampere is that constant current which, if maintained in two straight parallel conductors of infinite length, of negligible circular cross-section, and placed 1 metre apart in vacuum, would produce between these conductors a force equal to 2×10^{-7} newton per metre of length. (CIPM[17] (1947) Resolution 2 approved by the 9th CGPM[16] (1948).)

Kelvin (*K*). The kelvin, unit of thermodynamic temperature, is the fraction 1/273·16 of the thermodynamic temperature of the triple point of water. (13th CGPM[16] (1967) Resolution 4.)

Candela (*cd*). The candela is the luminous intensity in the perpendicular direction, of a surface of 1/600 000 square metre of a black body at the temperature of freezing platinum under a pressure of 101 325 newtons per square metre. (13th CGPM[16] (1967) Resolution 5.)

The supplementary base units are defined in 'The International System of Units' as follows.

Plane angle (*radian*). The angle subtended at the centre of a circle of radius 1 m by an arc of length 1 m along the circumference.

Solid angle (*steradian*). The solid angle subtended at the centre of a sphere of radius 1 m by an area of 1 m^2 on the surface.

The proposed seventh unit, the *mol* corresponding to the quantity 'amount of substance' is recommended by IUPAP[21], IUPAC[20] and ISO/TC12[22] but needs to be endorsed by the CPGM[16].

The *mole* (symbol '*mol*') is defined as an amount of substance of a system which contains as many elementary units as these are carbon atoms in 0·012 kg (exactly) of the pure nuclide 12C.

The elementary unit must be specified and may be an atom, a molecule, an ion, an electron, a proton etc. or a group of such entities according to a stated formula (58th CIPM[17] (1969) Recommendation 1).

Derived units

SI is a rationalised and coherent system because for any one physical quantity it admits of only one measurement unit with its entire structure derived from no more than seven arbitrarily defined basic units. It is coherent because the derived units are always the products or quotients of two or more of these basic units. Thus the SI unit for velocity is m/s (metre per second) and for acceleration is m/s^2 (metre per second every second).

Special names as shown in Table A.1(a) have been given to some derived units as an aid to communication.

Table A.1(b) shows the relationship of some of the quantities.

Although SI is complete in itself, certain non-SI units are recognised for use in conjunction with it where for traditional, commercial or practical purposes it is difficult to discard them.

For example it is impracticable to disregard the minute (in SI – 60 seconds) and the hour (in SI – 3600 seconds) which are non-coherent units.

Gravitational and absolute systems

There may be some difficulty in understanding the difference between SI and the Metric Technical System of units which has been used principally in Europe. The main difference is that whilst mass is expressed in kg in both systems, weight (representing a force) is

Table A.1(a) SOME DERIVED UNITS HAVING SPECIAL NAMES

Physical quantity	*SI unit*	*Unit symbol*
Force	newton	$N = kg\ m/s^2$
Work, energy, quantity of heat	joule	$J = N\ m = kg\ m^2/s^2$
Power	watt	$W = J/s = kg\ m^2/s^3$
Electric charge	coulomb	$C = A\ s$
Electric potential	volt	$V = W/A = kg\ m^2/As^3$
Electric capacitance	farad	$F = A\ s/V = A^2s^4/kg\ m^2$
Electric resistance	ohm	$\Omega = V/A = kg\ m^2/A^2s^3$
Frequency	hertz	$Hz = \frac{1}{s}$
Magnetic flux	weber	$Wb = kg\ m^2/A\ s^2$
Magnetic flux density	tesla	$T = Wb/m^2 = kg/A\ s^2$
Inductance	henry	$H = kg\ m^2/A^2s^2$
Lummons flux	lumen	$lm = cd\ sr$*
Illumination	lux	$lx = lm/m^2$

* *Note*: One steradian (sr) is the solid angle which, having its vertex at the centre of a sphere, cuts off an area of the surface of the sphere equal to that of a square with sides of length equal to the radius of the sphere.

The SI unit of electric dipole moment (A s m) is usually expressed as a coulomb metre (C m).

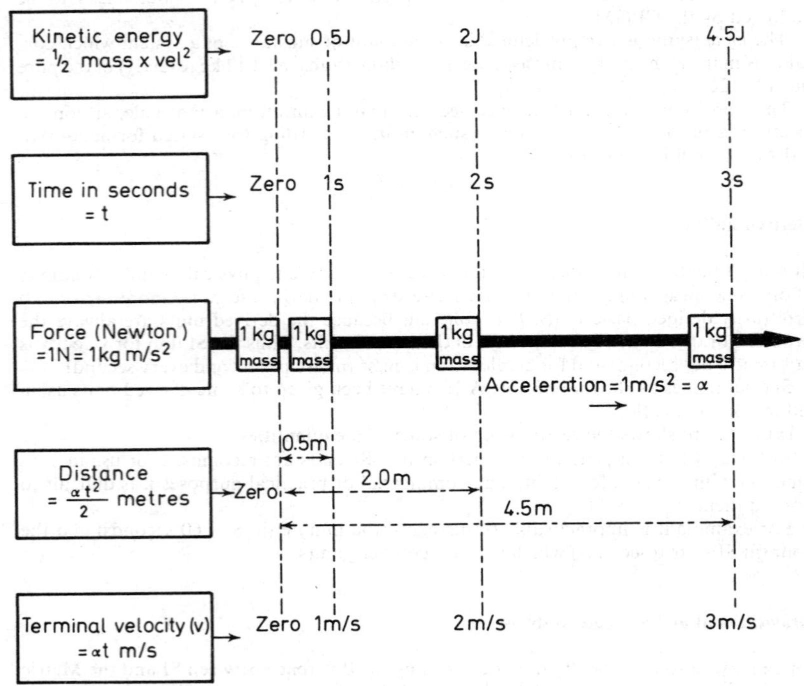

Figure A.1 Absolute Unit of Force (SI)

Table A.1(b) SI UNITS—FAMILY TREE

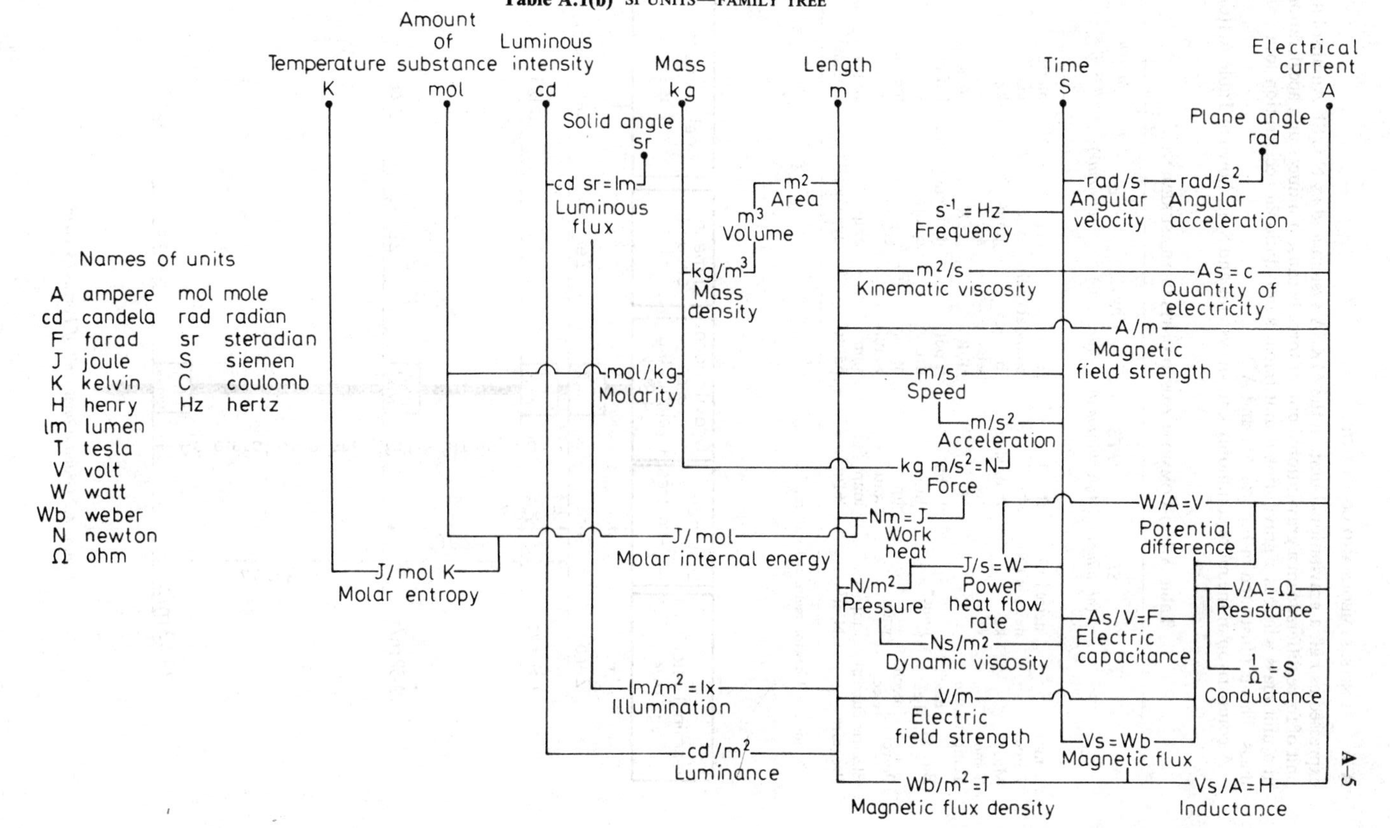

expressed as kgf, a gravitational unit, in the MKSA system and as N is SI. An absolute unit of force differs from a gravitational unit of force because it induces unit acceleration in a unit mass whereas a gravitational unit imparts gravitational acceleration to a unit mass. This is illustrated in Figures A.1 and A.2.

A comparison of the more commonly known systems and SI is shown in Table A.1(c).

Table A.1(c) COMMONLY USED UNITS OF MEASUREMENT

		SI (absolute)	*FPS (gravitational)*	*FPS (absolute)*	*cgs (absolute)*	*Metric technical units*
Length		metre (m)	ft	ft	cm	metre
Force		newton (N)	lbf	poundal (pdl)	dyne	kgf
Mass		kg	lb or slug	lb	gram	kg
Time		s	sec	sec	sec	sec
Temperature		°C K	°F	°F R	°C K	°C K
Energy	mech.	joule*	ft lbf	ft pdl	dyne cm = erg	kgf m
	heat		Btu	Btu	calorie	k cal.
Power	mech.	watt	hp	hp	erg's	metric hp
	elec.		watt	watt		watt
Electric current		amp	amp	amp	amp	amp
Pressure		N/m^2	lbf/ft^2	pdl/ft^2	$dyne/cm^2$	kgf/cm^2

* 1 joule = 1 newton metre or 1 watt second.

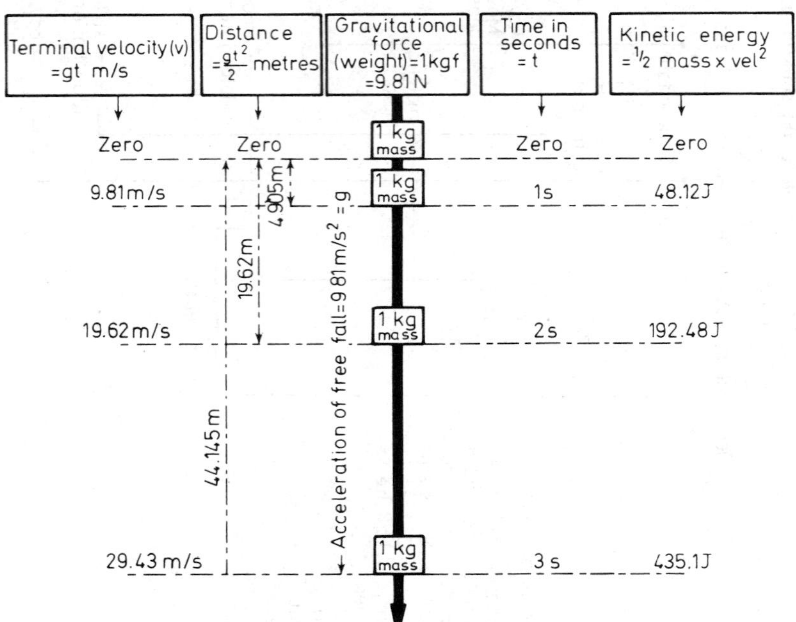

Figure A.2 Gravitational Unit of Forces (Metric)

It should be noted in particular how all energy and power whether from a mechanical, electrical or heat source share a common derived unit in the SI.

Indices

The metric system is a decimal system in which calculations using the numeral 10, multiplied or divided by itself a number of times, is used. Thus quantities containing many zeros, either before or after the decimal sign appear and these can lead to errors. The calculations can be simplified if the index rules are applied.

The number of times by which 10 is multiplied or divided by itself is referred to as the index of the power of 10. These indices may be whole, fractional, or zero, positive or negative and conform to the following rules:

$$10^m = 10 \times 10 \times 10 \times \ldots \text{ to } m \text{ factors} \qquad (1)$$
$$\text{e.g. } 10^3 = 10 \times 10 \times 10 = 1000$$

$$10^m \times 10^n = [10 \times 10 \times \ldots \text{ to } m \text{ factors}] \times 10 \times 10 \times \ldots \text{ to } [n \text{ factors}] \qquad (2)$$
$$= 10 \times 10 \times \ldots \text{ to } (m+n) \text{ factors}$$
$$= 10^{m+n}$$
$$\text{e.g. } 10^2 \times 10^3 = 10^{(2+3)} = 10^5$$

$$10^m \times 10^n \times 10^p \times \ldots = 10^{(m+n+p)} \qquad (3)$$
$$\text{e.g. } 10^1 \times 10^2 \times 10^3 = 10^{(1+2+3)} = 10^6 = 1\,000\,000$$

$$\frac{10^m}{10^n} = 10^{(m-n)} \text{ when } m>n \text{ and } 10^{(n-m)} \text{ when } m<n \qquad (4)$$

$$\text{e.g. } \frac{10^3}{10^2} = 10^{(3-2)} = 10^1$$
$$\text{and } \frac{10^2}{10^3} = \frac{1}{10^{(3-2)}} = \frac{1}{10^1} = 10^{-1} = 0{\cdot}1$$

$$(10^m)^n = 10^{mn} = (10^n)^m \qquad (5)$$
$$= 10^m \times 10^m \times 10^m \times \ldots \text{ to } n \text{ factors}$$
$$\text{or } 10^n \times 10^n \times 10^n \times \ldots \text{ to } m \text{ factors}$$
$$\text{e.g. } (10^2)^3 = 10^{(2\times3)} \text{ or } 10^{(3\times2)} = 10^6 = 1\,000\,000$$

$$10^\circ = 1 \qquad (6)$$

$$10^{-n} = \frac{1}{10^n} \qquad (7)$$

$$\text{e.g. } 10^{-3} = \frac{1}{10^3} = \frac{1}{1000} = 0{\cdot}001$$

$$(10a)^m = 10^m a^m \qquad (8)$$
$$\text{e.g. } (10\times6)^2 = 10^2 \times 6^2 = 3600$$
$$\text{and } (10\times6)^{-2} = \frac{1}{10^2\times6^2} = \frac{1}{3600} = 3{\cdot}6\times10^{-3}$$
$$= 0{\cdot}000\,277$$

$$\left(\frac{10^x}{10^y}\right)^m = \frac{10^{xm}}{10^{ym}} \qquad (9)$$
$$\text{e.g.} \left(\frac{10^3}{10^5}\right)^2 = \frac{10^6}{10^{10}} = 10^{(6-10)} = 10^{-4} = 0{\cdot}0001$$

$$10^{m/n} = \sqrt[n]{(10^m)} \qquad (10)$$
$$\text{e.g. } 10^{4/2} = \sqrt[2]{(10^4)} = 10^2 = 100$$

Example. To convert 1 horsepower to watts

$$1\ \text{hp} = 550\ \text{ft lbf/s} = 5{\cdot}5 \times 10^2\ \text{ft lbf/s}$$

$$1\ \text{ft} = \frac{12 \times 25{\cdot}4\ \text{m}}{1000} = 3{\cdot}048 \times 10^{-1}\ \text{m}$$

$$1\ \text{lbf} = 0{\cdot}4536\ \text{kgf} = 4{\cdot}536 \times 10^{-1}\ \text{kgf}$$

$$1\ \text{kgf} = 9{\cdot}81\ \text{N}$$

$$1\ \text{W} = 1\ \text{N m/s} = 1\ \text{J/s}$$

$$\therefore 1\ \text{hp} = [5{\cdot}5 \times 10^2] \times [3{\cdot}048 \times 10^{-1}] \times [4{\cdot}536 \times 10^{-1}]\ \text{kgf m/s}$$

$$= 5{\cdot}5 \times 3{\cdot}048 \times 4{\cdot}536 \times 9{\cdot}81\ \text{N m/s}$$

$$= 745{\cdot}70\ \text{W}$$

Table A.1(d) shows the more commonly met terms expressed in coherent units.

Table A.1(d) THE USE OF INDICES—MULTIPLES RAISED TO A POWER SHOWN IN TERMS OF COHERENT UNIT

Term	*Process*	*Coherent unit*
1 mm^3	$(10^{-3}\ m)^3$	$10^{-9}\ m^3$
1 μs^{-1}	$(10^{-6}\ s)^{-1}$	$\frac{1}{10^{-6}\ s}$
1 mm^2/s	$(10^{-3}\ m)^2/s$	$10^{-6}\ m^2/s$
1 N/mm^2	$N/(10^{-3}\ m)^2$	$N/10^{-6}\ m^2 =$ $10^6 N/m^2$
1 mm^4	$(10^{-3}\ m)^4$	$10^{-12}\ m^2$
1 mm^2	$(10^{-3}\ m)^2$	$10^{-6}\ m^2$
1 mm^3	$(10^{-3}\ m)^3$	$10^{-9}\ m^3$

Expressing magnitudes of SI units

To express magnitudes of a unit, decimal multiples and submultiples are formed using the prefixes shown in Table A.1(e). This method of expressing magnitudes ensures complete adherence to a decimal system.

Table A.1(e) THE INTERNATIONALLY AGREED MULTIPLES AND SUB-MULTIPLES

Factor by which the unit is multiplied		*Prefix*	*Symbol*	*Common everday samples*
One million million (billion)	10^{12}	tera	T	
One thousand million	10^9	giga	G	gigahertz (GHz)
One million	10^6	mega	M	megawatt (MW)
One thousand	10^3	kilo	k	kilometre (km)
One hundred	10^2	hecto*	h	
Ten	10^1	deca*	da	decagramme (dag)
UNITY	1			
One tenth	10^{-1}	deci*	d	decimetre (dm)
One hundredth	10^{-2}	centi*	c	centimetre (cm)
One thousandth	10^{-3}	milli	m	milligramme (mg)
One millionth	10^{-6}	micro	μ	microsecond (μs)
One thousand millionth	10^{-9}	nano	n	nanosecond (ns)
One million millionth (one millionth)	10^{-12}	pico	p	picofarad (pF)
One thousand million millionth	10^{-15}	femto	f	
One million million millionth	10^{-18}	atto	a	

* To be avoided wherever possible

Examples of use – Length

1 millimetre (1 mm)		= $\frac{1}{1000}$ metre		$= 10^{-3}$ m
10 millimetres	(10 mm)	= 1 centimetre	(1 cm)	$= 10^{-2}$ m
10 centimetres	(10 cm)	= 1 decimetre	(1 dm)	$= 10^{-1}$ m
10 decimetres	(10 dm)	= 1 metre	(1 m)	$= 10^{0}$ m
10 metres	(10 m)	= 1 decametre	(1 dam)	$= 10^{1}$ m
10 decametres	(10 dam)	= 1 hectometre	(1 hm)	$= 10^{2}$ m
10 hectometres	(10 hm)	= 1 kilometre	(1 km)	$= 10^{3}$ m

Rules for use of SI units and the decimal multiples and sub-multiples

The following rules are based on guide rules laid down in ISO R1000[10].

1. The SI units are preferred but it is impracticable to limit usage to these, therefore their decimal multiples and sub-multiples are also required. (For example, it is cumbersome to measure road distances or the breadth of a human hair in metres.)
2. In order to avoid errors in calculations it is preferable to use coherent units. Therefore, it is strongly recommended that in calculations only SI units themselves are used and not their decimal multiples and sub-multiples. (Example: Use $N/m^2 \times 10^6$ not MN/m^2 or N/mm^2 in a calculation.)
3. The use of prefixes representing 10 raised to a power which is a multiple of 3 is especially recommended. (Example: For length, km ... m ... mm ... μm. Thus hm; dam; dm; cm are non-preferred.)
4. When expressing a quantity by a numerical value of a unit it is helpful to use quantities resulting in numerical values between 0 and 1000. Examples:
   $12\ kN = 12 \times 10^3$ N instead of 12 000 N
   $3{\cdot}94$ mm $= 3{\cdot}94 \times 10^{-3}$ m instead of 0·00394 m
   $14{\cdot}01\ kN/m^2 = 14{\cdot}01 \times 10^3\ N/m^2$ instead of $14010\ N/m^2$
5. Compound prefixes are not used. (Example: Write nm not mμm.) Where, however a name has been given to a product or a quotient of a basic SI unit, for example the bar ($10^5\ N/m^2$) it is correct practice to apply the prefix to the name e.g. millibar (10^{-3} bar).
6. In forming decimal multiples and sub-multiples of a derived SI unit preferably only one prefix is used. The prefix should be attached to the unit in the numerator. (Example: MW/m^2 not W/mm^2.) The exception is stress where BSI recommend the use of N/mm^2.
7. Multiplying prefixes are printed immediately adjacent to the SI unit symbol with which they are associated. The multiplication of symbols is usually indicated by leaving a small gap between them. (Example: mN = millinewton. If written as m N this would indicate metre newton.)

SI QUANTITIES AND UNITS

The units shown in Tables A.2(d) to A.2(k) generally obey the above guide rules. However the BSI expect that initially a practical attitude will prevail in the interpretation of these recommendations in the United Kingdom, particularly as certain countries still use Metric Technical Units which include such units as kgf/cm^2. These units may appear for some time alongside SI Units in BS publications.

Tables A.2(a) to A.2(k) contain a list of SI units and a selection of recommended decimal multiples and sub-multiples together with other units or names of units which may be used. These are based on recommendations given in ISO Recommendation R 1000[10] and BSI publication PD 5686[6].

Explanation of Tables

Tables A.2(a) to A.2(k) are divided into 8 columns.

Column 1 indicates the physical quantity name. For physical symbols to be used (e.g. for basic formulae...... M for mass; W for weight, g for gravity) reference should be made to ISO R 31[9] and the latter part of this Section dealing with Symbols and Abbreviations.

Columns 2 and 3 indicate the name and symbol of the SI Unit.

Column 4 indicates the recommended multiple to be used by Industry. The consistent use of 10^n where n is ± 3 or multiples thereof should be noted.

Column 5 indicates multiples of SI units which are listed as non-preferred, but nevertheless will occasionally be met for practical reasons. Although a choice of multiples is shown in the Tables for given quantities, it is recommended that a limited selection should be made within the range stated to suit the engineering discipline.

Column 6 indicates these units other than SI which cannot be omitted for social, traditional or commercial reasons.

Column 7 indicates the value of the units shown in Columns 5 and 6 in terms of the SI Unit, Column 4.

Column 8 gives conversion factors from Imperial units to SI and centimetre-gramme-second (cgs) units to SI together with special remarks applicable to the quantity under discussion. The conversion factors marked thus (+) are exact. Detailed conversion tables for certain selected quantities are given in Tables A.4(a) to A.4(r). Conversion factors are based on BS 350 Parts 1, 2[3] and PD 6203.[7]

Explanatory notes to Tables A.2(a) to A.2(k)

Notes on particular units and expressions follow each of the Tables. The notes reflect the concensus of opinion of British Industry and should be checked from time to time for further developments. National policy statements on such units are promulgated through the BSI News.

Table A.2(a) SPACE AND TIME

Quantity	*Units for common use* — *SI units* — *Name*	*Symbol*	*Decimal multiples and sub-multiples* — *BSI recommended*	*Others*	*Units other than SI*	*Equivalent value in terms of basic unit*	*Conversion factors and remarks*
(1)	*(2)*	*(3)*	*(4)*	*(5)*	*(6)*	*(7)*	*(8)*
Plane Angle	radian	rad				1 rad	1° = 0·017 453 3 rad
					degree (°)	Π/180 rad	1′ = 2·908 88 $\times 10^{-4}$ rad
					minute of arc (′)	Π/10 800 rad	1″ = 4·848 14 $\times 10^{-6}$ rad
(1)					second of arc (″)	Π/648 000 rad	1 right angle = Π/2 rad = 90°
Solid Angle	steradian	sr				1 sr	
Length			km			10^3 m	1 inch = 25·4 mm$^+$
	metre	m				1 m	1 foot = 0·3048 m$^+$
				cm		10^{-2} m	1 yard = 0·9144 m$^+$
			mm			10^{-3} m	1 mile = 1·609 344 km$^+$
			μm			10^{-6} m	1 UK nautical mile = 1·853 18 km
			nm			10^{-9} m	1 angstrom (Å) = 10^{-10} metre
(2)							or 10^{-1} nm
Area			km^2			10^6 m^2	1 sq in = 645·16 mm^2 +
							= 6·4516 cm^2 +
			ha			10^4 m^2	1 sq ft = 0·092 903 m^2
			a			10^2 m^2	1 sq yd = 0·836 127 m^2
	square metre	m^2				1 m^2	1 acre = 0·404 686 ha
				cm^2		10^{-4} m^2	1 sq mile = 258·999 ha
			mm^2			10^{-6} m^2	ha = hectare a = are
Volume	cubic metre	m^3				1 m^3	1 in^3 = 16·3871 cm^3
				dm^3 (litre)		10^{-3} m^3	1 ft^3 = 0·028 3168 m^3
				cm^3 (mlitre)		10^{-6} m^3	1 yd^3 = 0·764 555 m^3
(3)			mm^3			10^{-9} m^3	1 UK fl. oz = 28·4131 mlitre
							1 gal = 4·5461 litre

Table A.2(a)—*continued*

Quantity	*Units for common use*: *SI units*: *Name*	*Symbol*	*Decimal multiples and sub-multiples*: *BSI recommended*	*Others*	*Units other than SI*	*Equivalent value in terms of basic unit*	*Conversion factors and remarks*
(*1*)	(*2*)	(*3*)	(*4*)	(*5*)	(*6*)	(*7*)	(*8*)
					day, hour, minute	7·344 Ms, 3600 s, 60 s	
Time	second	s				1 s	
			ms			10^{-3} s	
			μs			10^{-6} s	
			ns			10^{-9} s	
Angular Velocity	radian per second	rad/s				1 rad/s	1 degree/s = 0·017 4533 rad/s
					rev/min	0·104 720 rad/s	
					rev/s	6·283 19 rad/s	
Velocity	metre per second	m/s				1 m/s	1 km/h = 0·277 778 m/s
					km/h	1/3·6 m/s	1 ft/s = 0·3048 m/s+
					knot (kn)	0·514 444 m/s	1 mile/h = 0·447 04 m/s+or 1·609 34 km/h
							1 UK knot = 0·514 773 m/s
Acceleration	metre per second squared	m/s^2				1 m/s^2	1 ft/s^2 = 0·3048 m/s^2+
							1 in/s^2 = 0·0254 m/s^2+
Angular Acceleration	radian per second squared	rad/s^2				1 rad/s^2	

NOTES ON IMPLEMENTATION OF PARTICULAR UNITS AND EXPRESSIONS

Table A.2(a) SPACE AND TIME

1. *Plane angle*. The Sumerian division of the circle in 360° (hence degrees) is retained for geometry although dynamicists use the radian.
2. *Length—The centimetre*. In the many engineering disciplines the use of the centimetre is non-preferred. It is not recommended for engineering drawings in BS 308[2], but has been adopted as the basic unit of measurement by primary schools and for commercial purposes. Sometimes the centimetre raised to a power (e.g. cm^2; cm^3; cm^4) is used to maintain a sensible range of numerical values in front of the unit. An instance of this concerns steel sections where the moduli of sections and moment of section will be given in steel tables in terms of cm^3 and cm^4 respectively.

Where accuracy to the nearest millimetre is unwarranted, the centimetres can be used to imply a less precise dimension.

3. *Volume and capacity—The litre*. Before 1964 the 1901 litre was equal to 1·000 028 dm^3 At the XII International CGPM[16] meeting on units (1964) the litre was redefined to equate exactly to 1 cubic decimetre. The same conference agreed that the litre should not be used to express the results of precise measurements, so as to make sure that where high precision was involved (say greater than 1 part in 20 000) the possibility of confusion between the former (1901) litre and the new (1964) litre would be eliminated. The UK legal definition will probably be revised accordingly.

It is recommended that the results of precise measurements of volume should be given only in terms of m^3, dm^3, cm^3, mm^3 etc. even though the millilitre (ex cm^3) and litre (ex dm^3) will still be used for operational and commercial purposes.

Because of the possible confusion of the symbol for the litre 'l' with the figure '1', it is strongly recommended that the unit name be spelt in full.

Centilitre is sometimes used for arbitrary quantities implying a greater degree of tolerance.

Table A.2(b) PERIODIC AND RELATED PHENOMENA

Quantity	Units for common use					Equivalent value in terms of basic unit	Conversion factors and remarks
	SI units				Units other than SI		
	Name	Symbol	Decimal multiples and sub-multiples: BSI recommended	Decimal multiples and sub-multiples: Others			
(1)	*(2)*	*(3)*	*(4)*	*(5)*	*(6)*	*(7)*	*(8)*
Frequency			GHz			10^9 Hz	1 c/s (or c.p.s.) = 1 Hz
			MHz			10^6 Hz	
			kHz			10^3 Hz	
	hertz	Hz				1 Hz	
Rotational Frequency	reciprocal second	s^{-1}		rev/s rev/min		1 s^{-1}	
Wavelength	metre	m				1 m	
				cm		10^{-2} m	

NOTES ON TABLE A.2(B)

1. *Rotational frequency (rev/min).* The quantity rev/min is favoured for rotating machinery in place of the rev/s or the SI unit rad/s.

Table A.2(c) MECHANICS

Quantity (1)	Name (2)	Symbol (3)	BSI recommended (4)	Others (5)	Units other than SI (6)	Equivalent value in terms of basic unit (7)	Conversion factors and remarks (8)
	Units for common use						
	SI units		*Decimal multiples and sub-multiples*				
Mass			Mg		tonne	10^3 kg	1 ton = 1016·05 kg or 1·016 05 tonne
	kilogramme	kg				1 kg	1 cwt = 50·8023 kg
			g			10^{-3} kg	1 lb = 0·453 592 37 kg+
(1)			mg			10^{-6} kg	1 oz = 28·3495 g (avoir)
(2)			μg			10^{-9} kg	
Mass density	kilogramme per cubic metre	kg/m³	Mg/m³	g/cm³	g/mlitre; kg/litre	10^3 kg/m³	1 lb/ft³ = 16·0185 kg/m³
				kg/dm³	g/litre	1 kg/m³	1 lb/in³ = 27·6799 g/cm³
Specific volume	cubic metre per kilogramme	m³/kg				1 m³/kg	1 ft³/lb = 0·062 428 m³/kg
					litre/kg	10^{-3} m³/kg	1 cm³/g = 10^{-3} m³/kg+
Momentum	kilogramme metre per second	kg m/s				1 kg m/s	1 lb ft/s = 0·138 255 kg m/s 1 g cm/s = 10^{-5} kg m/s+
Angular momentum	kilogramme square metre per second	kg m²/s				1 kg m²/s	1 lb ft²/s = 0·042 1401 kg m²/s 1 g cm²/s = 10^{-7} kg m²/s+
Moment of inertia	kilogramme square metre	kg m²				1 kg m²	1 lb ft² = 0·042 140 1 kg m² 1 g cm² = 10^{-7} kg m²+
			MN			10^6 N	1 tonf = 9·964 02 kN
			kN			10^3 N	1 lbf = 4·448 22 N
Force	newton	N				1 N	1 ozf = 0·278 014 N
			mN			10^{-3} N	1 pdl = 0·138 255 N
			μN			10^{-6} N	1 dyne = 10^{-5} N+
							1 kgf or kilopond = 9·806 65 N+
Weight					See Note (1)		

Table A.2(c)—*continued*

Quantity	*Units for common use*					*Equivalent value in terms of basic unit*	*Conversion factors and remarks*
	SI units				*Units other than SI*		
	Name	*Symbol*	*Decimal multiples and sub-multiples* BSI recommended	*Others*			
(1)	*(2)*	*(3)*	*(4)*	*(5)*	*(6)*	*(7)*	*(8)*
Moment of force (torque)	newton metre	N m	 MN m kN m μN m	GN m daN m		10^9 N m 10^6 N m 10^3 N m 10 N m 1 N m 10^{-6} N m	1 tonf ft = 3·037 03 kN m 1 lbf ft = 1·355 82 N m 1 pdl ft = 0·042 140 1 N m 1 lbf in = 0·112 985 N m 1 dyne cm = 10^{-7} N m+ 1 kgf m = 9·806 65 N m+
Mass per unit length	kilogramme per metre	kg/m			 tonne/km	1 kg/m 1 kg/m	1 ton/1000 yds = 1·111 16 kg/m 1 ton/mile = 0·631 342 kg/m 1 lb/in = 17·8580 kg/m 1 lb/ft = 1·488 16 kg/m 1 lb/yd = 0·496 055 kg/m
Mass per unit area	kilogramme per square metre	kg/m^2			 kg/ha	1 kg/m^2 10^{-4} kg/m^2	1 lb/acre = 1·120 85 kg/ha 1 ton/sq. mile = 3·922 98 kg/ha 1 lb/1000 ft^2 = 4·882 43 kg/m^2
Mass rate of flow	kilogramme per second	kg/s			 kg/h	1 kg/s 3600 kg/s	1 lb/s = 0·453 592 kg/s 1 lb/h = 1·259 98 × 10^{-4} kg/s 1 UK ton/h = 0·282 235 kg/s or 1·016 05 tonne/h
Volume rate of flow	cubic metre per second	m^3/s			 litre/s litre/min	1 m^3/s 10^{-3} m^3/s $\frac{10^{-3}\ \text{m}^3/\text{s}}{60}$	1 ft^3/s (cusec) = 28·3168 litre/s 1 gal/s = 4·546 09 litre/s
Mass flow rate per unit area	**kilogramme per** square metre second	kg/m^{2}s				1 kg/m^{2}s	1 lb/ft^{2}h = 1·356 23 × 10^{-} 3 kg/m^{2}s

Pressure and stress			GN/m^2			10^9 N/m^2	1 lbf/in^2 = 6·894 76 kN/m^2 = 68·9476 mbar
				daN/mm^2	hbar	10^7 N/m^2	1 torr = 1·333 22 mbar
			MN/m^2	N/mm^2		10^6 N/m^2	1 in Hg = 33·863 9 mbar
				N/cm^2	bar	10^5 N/m^2	1 in W.G. = 2·490 89 mbar
	newton per square metre (pascal)	N/m^2 (Pa)	kN/m^2			10^3 N/m^2	1 kgf/cm^2 = 0·980 665 bar = 98·0665 kN/m^2
					mbar	10^2 N/m^2	1 $tonf/in^2$ = 15·4443 N/mm^2
						1 N/m^2	1 pieze = 10^3 N/m^2
			mN/m^2			10^{-3} N/m^2	1 std atmosphere = 1013·25+ mbar+ = 1·033 23 kgf/cm^2 = 14·695 9 lbf/in^2 = 760 torr+ = 29·921 3 in Hg
(3)			$\mu N/m^2$			10^{-6} N/m^2	
Second moment of area	metre to the power of four	m^4				1 m^4	1 in^4 = 41·6231 cm^4
				cm^4		10^{-8} m^4	1 ft^4 = 863 097 cm^4
			mm^4			10^{-12} m^4	
Section modulus	cubic metre	m^3				1 m^3	1 in^3 = 16·3871 cm^3
				cm^3		10^{-6} m^3	
Dynamic viscosity	newton second per square metre	$N\ s/m^2$				1 $N\ s/m^2$	1 lbf s/ft^2 = 47·8803 $N\ s/m^2$ or 47 880·3 cP
					P (poise)	10^{-1} $N\ s/m^2$	
(4)			$mN\ s/m^2$		cP (centipoise)	10^{-3} $N\ s/m^2$	1 pdl s/ft^2 or lb/ft s } = 1·488 16 $N\ s/m^2$ or 1488·16 cP
Kinematic viscosity	square metre per second	m^2/s				1 m^2/s	1 ft^2/h = 2·580 64 × 10^{-5} m^2/s+ or 25·806 cSt+
					St (stokes)	10^{-4} m^2/s	
(4)			mm^2/s		cSt (centistokes)	10^{-6} m^2/s	1 ft^2/s = 0·092 903 m^2/s or 9·2903 × 10^4 cSt
Surface tension	newton per metre	N/m				1 N/m	1 lbf/ft = 14·5939 N/m
			mN/m			10^{-3} N/m	1 dyne/cm = 10^{-3} N/m
			GJ			10^9 J	
			MJ			10^6 J	1 ft lbf = 1·355 82 J
			kJ			10^3 J	1 erg = 10^{-7} J
Energy, work	joule	J (= N m)				1 J	1 kgf m = 9·806 65 J+
					MW h	3·6 × 10^9 J	1 ft pdl = 0·042 140 1 J
					kW h	3·6 × 10^6 J	1 hp h = 2·684 52 × 10^6 J
					eV (electron volt)	(1·602 10 ± ·00007) × 10^{-19} J	1 kcal = 4186·8 J+
(5)							1 Btu = 1055·06 J

Table A.2(c) —*continued*

Quantity	*Units for common use*					*Equivalent value in terms of basic unit*	*Conversion factors and remarks*
	SI units				*Units other than SI*		
	Name	*Symbol*	*Decimal multiples and sub-multiples*				
			BSI recommended	*Others*			
(1)	*(2)*	*(3)*	*(4)*	*(5)*	*(6)*	*(7)*	*(8)*
			GW			10^9 W	1 hp = 745·7 W
			MW			10^6 W	1 ft lbf/s = 1·355 82 W
			kW			10^3 W	1 metric hp = 735·499 W
Power	watt	W (= J/s)				1 W	1 kgf m/s = 9·806 65 W+
			mW			10^{-3} W	1 erg/s = 10^{-7} W
			μW			10^{-6} W	
				daJ/cm^2		10^3 J/m^2	
				J/cm^2		10^4 J/m^2	
Impact strength	joule per square metre		kJ/m^2			10^3 J/m^2	
		J/m^2				1 J/m^2	
Fuel consumption				litre/km			1 gal/mile = 2·825 litre/km
				km/litre			1 mile/gal = 0·354 km/litre
Specific fuel consumption	kilogramme per joule	kg/J				1 kg/J	1 lb/hp h = 0·168 97 kg/MJ
				kg/MJ		10^{-6} kg/J	
					kg/kW h		
	cubic metre per joule	m^3/J				1 m^3/J	1 pint/hp h = 0·211 68 litre/MJ
				litre/MJ		10^{-9} m^3/J	
					m^3/kW h		

NOTES ON TABLE A.2(C)

1. *Mass (Mass by Weight) (kilogramme).* BSI is considering the spelling of the SI unit for mass so as to replace 'kilogramme' by 'kilogram'.

Confusion sometimes arises over the measuring of the terms 'mass' and 'weight'. Commonly, and in many branches of engineering, it has been the custom to refer to quantities of mass as weights, e.g. weight of coal in kilogrammes.

Weight, however, is dependent upon the gravitational force acting upon the mass. Thus for a mass (M), weight (W) = Mg where g is the local acceleration due to gravity which varies slightly from point to point on the earth's surface. For practical purposes an approximated figure of 9·81 or 9·807 metres per second squared (m/s^2) is used for 'g'.

The force unit in SI is the newton (N) and by using consistent units becomes the force applied to unit mass (kg) to impart unit acceleration (m/s^2) to the mass (as distinct from gravitational acceleration which equals 9·806 65 m/s^2).

Thus it can be more readily understood by comparing the SI system with other systems for mass, weights and measures as shown in Table A.2(d).

Table A.2(d) SYSTEMS OF WEIGHTS AND MEASURES

Quantity	*Foot pound second*	*Metric, technical*	*SI*
Mass	1 lb	1 kg	1 kg
Length	1 ft	1 m	1 m
Force	1 lbf	1 kgf	1 N
Definition of Force	$\text{lbf} = \dfrac{\text{lb} \times \text{ft/s}^2}{g}$	$\text{kgf} = \dfrac{\text{kg} \times \text{m/s}^2}{g}$	$\text{N} = \text{kg} \times \text{m/s}^2$.
Definition of Weight (gravitational force)	1 lbf per 1 lb	1 kgf per 1 kg	9·806 65 N per 1 kg

Consider a load of a mass M kg which is to be lifted up by a crane from a shop floor. This mass (M kg) will be referred to as the weight in kilogrammes so the crane hook will be marked with a safe working load in kilogrammes. The designer however is concerned with the forces exerted by that load on the lifting appliances which will be in newtons. Thus although the crane picks up a load of say 1000 kg the designer will calculate in terms of a force of 9806·65 newtons (equals 9·806 65 kN) assuming that the load is lifted with an acceleration of 1 m/s^2 and this is the minimum force needed to overcome the gravitational pull. It should be noted that if the load is snatched, that is, lifted with greater acceleration, the force becomes proportionally greater and allowances should be made for these larger forces in the design. If, however, the load of 1000 kg is pushed horizontally along the floor, and if the friction is ignored, the force required to move it with an acceleration of 1 m/s^2 is reduced to 1000 N (equals 1 kN) (see also Fig. A.1 and A.2).

2. *Mass (Tonne/Megagramme).* The tonne and kilogramme are generally accepted as replacement units for ton and pounds. In particular most lifting equipment already marked in tons can be considered as adequate for lifting the same number of tonnes because of the small excess (1·6%) of the ton over the tonne.

However in soil mechanics, the megagramme (Mg) rather than the tonne is recommended. This accords with the recommendation of the Institution of Civil Engineers[14] because with large masses involved in work on soil mechanics, confusion between the ton and the tonne could prove very expensive.

3. *Metric Units for Pressure and Stress.* The SI derived unit for force per unit area is the newton per square metre (N/m^2), referred to as the pascal (Pa) and this unit with suitable multiples is favoured as the unit for stress. There are differences of opinion regarding the unit for pressure, but BSI has recommended that, although some flexibility will have to be allowed in the expression of pressure values, the following practice should be adopted[8]:

(a) For the statement of stress property use, without deviation, N/m^2 and appropriate multiples of it e.g. MN/m^2 or this, if preferred, expressed as N/mm^2 or, if essential for non-metallic materials, kN/m^2.
(b) For pressure statements use either N/m^2 (and suitable multiples and sub-multiples of it) or bar or mbar. In such cases the conversion 1 bar = 10^5 N/m^2 will always be quoted for reference.
(c) The unit of pressure should be the same in all related publications issued under the authority of any one Industry Standards Committee. In cases where Industry Standards Committees have no strong views, the secretaries concerned will seek to secure conformity with the practice of closely related committees.
(d) Pressures or pressure differences measured by manometer tube may often conveniently be expressed as a height of a column of fluid, the nature of the fluid being stated. Such readings must be converted to terms of N/m^2 if they are to be used in calculations of flow, etc. On the other hand, manometers are sometimes used merely as indicators that a prescribed operating condition has been met. Judgement is therefore required as to when it can be of advantage to use mmH_2O, mmHg etc. or when it is of advantage to calibrate and read manometers in a suitable multiple of N/m^2 or in mbar. It is understood that manometers calibrated in mbar are becoming increasingly available and it is recommended that pressures expressed as a height of a column of fluid should progressively give place to a suitable multiple of the SI unit or to the millibar.
(e) It has been internationally recommended that pressure units themselves should not be modified to indicate whether the pressure value is 'absolute' (i.e. above zero) or 'gauge' (i.e. above atmospheric pressure). If, therefore, the context leaves any doubt as to which is meant, the work 'pressure' must be qualified appropriately.
e.g. '. . . at a gauge pressure of 12·5 bar'
or '. . . at a gauge pressure of 1·25 MN/m^2'
or '. . . at an absolute pressure of 2·34 bar'
or '. . . at an absolute pressure of 234 kN/m^2'

Table 1.2(e) illustrates some of these practices. Note that this table works on a gauge pressure basis (atmosphere = 0), thus vacuum is shown measured in negative millibars. This continues the custom of associating the higher numerical readings with greater vacuum.

Table A.2(e) GAUGE VACUUM AND PRESSURE

	Vacuum			*Gauge pressure*			
Ins. Hg	*30″ Hg*	*20″ Hg*	*10″ Hg*	*10 lbf/in²*	*100 lbf/ in²*	*1000 lbf/in²*	*2000 lbf/in²*
kN/m^2	−101·3	−67·73	−33·86	68·94	689·4	6894	13789·5
bar	−1·01325 (−1013·25 *mbar*)	−0·6773 (−677·3 *mbar*)	−0·3386 (−338·6 *mbar*)	0·6894	6·894	68·94	137·895
MN/m^2	−0·1013	−0·0677	−0·0338	0·0689	0·6894	6·894	13·7895

Notwithstanding previous practices of referring to pump performances in terms of pressure, the BSI recommends that the pump total head should be specified in linear measure (metres). The symbol N/m^2 will be recognised internationally as the Pascal (Pa).

4. *Viscosity* (*Centipoise*; *Centistokes*). The recognised derived SI units for Dynamic and Kinematic Viscosity are N s/m^2 and m^2/s respectively. However the existing units, centipose (cP) and centistokes (cSt) are so well established internationally, particularly for oils, that the operational use of these units will continue.

5. *Energy* (*MJ/kW h*). The choice of a suitable commercial energy unit common to all energy producing concerns has still to be resolved. The SI unit is the joule and its multiples.

However electrical interests favour the adoption of the kW h (3·6 MJ = 1 kW h).

The following are the probable commercial fuel quantities:

Coal	–	tonne
Electricity	–	kW h
Gas	–	100 MJ (the 'new' therm)
Oil	–	litre; m^3 kg; tonne

6. *Hardness Values* (*kgf*). The ISO Technical Committee TC17[23] has agreed to retain the hardness unit kgf to express the load applied by the indenter and this will ensure that most hardness testing machines and empiracally based formulae are not made obsolete. Thus the Rockwell, Vickers and Brinell hardness numbers will be retained. This number is arbitrary and dimensionless and is dependent upon the resistance offered by the material under test to a definite load.

7. *Concentration.* Concentration should preferably be expressed on a mass/mass basis (i.e. kg/kg; mg/mg; mg/kg) or a volume/volume basis (i.e. m^3/m^3; litre/m^3; millilitre/m^3). It may also be expressed in parts per million (ppm) or as a percentage 'by mass' or 'by volume' respectively.

8. '*ph*' *Scale.* This is a number based on the logarithm, to the base 10 of the reciprocal of the concentration of hydrogen ions in aqueous solution. It is used as a method of expressing small differences in the acidity or alkalinity of nearly neutral solutions in biological and electrolytic processes.

This number will remain unchanged.

Table A.2(f) HEAT

Quantity	*Units for common use*					*Equivalent value in terms of basic unit*	*Conversion factors and remarks*
	SI units				*Units other than SI*		
	Name	*Symbol*	*Decimal multiples and sub-multiples* BSI recommended	*Others*			
(*1*)	(*2*)	(*3*)	(*4*)	(*5*)	(*6*)	(*7*)	(*8*)
Absolute temperature (1)	kelvin	K				1 K	K = °C+273·15 K = 1·8°R (Rankine)
Customary temperature (2)					°C (Celsius)		°C = 5/9 (°F−32)
Temperature interval	kelvin	K			°C		1°C = 1 K = 1·8°F (alternative form 1 deg C = 1 deg K = 1·8 deg F)
Temperature coefficient (linear or volumetric)		1/K			1/°C	1/K	
Heat, quantity of heat, internal energy, enthalpy	joule	J	GJ MJ kJ mJ			10^9 J 10^6 J 10^3 J 1 J 10^{-3} J	1 Btu = 1055·06 J 1 cal (IT) = 4·1868 J+ 1 CHU = 1899·2 J 1 kW h = 3·6 MJ 1 therm = 105·506 MJ 1 erg = 10^{-7} J
Heat flow rate	watt	W	kW			10^3 W 1 W	1 Btu/h = 0·293 071 W 1 kcal/h = 1·163 W+ 1 cal/s = 4·1868 W+ 1 frigorie = 4·186 W
Density of heat flow rate	watt per square metre	W/m^2	MW/m^2 kW/m^2	W/mm^2		10^6 W/m^2 10^3 W/m^2 1 W/m^2	1 Btu/ft^2 h = 3·154 59 W/m^2 1 cal/cm^2 s = 41 868 W/m^2 1 CHU/ft^2 h = 5·678 W/m^2

Thermal conductivity	watt per metre kelvin	W/m K			W/m °C		1 Btu/ft h °F = 1·730 73 W/m °C 1 kcal/m h °C = 1·163 W/m °C+ known as 'k' value
Coefficient of heat transfer	watt per square metre	W/m^2 K	kW/m^2 K	W/cm^2 K	W/m^2 °C	10^4 W/m^2 K 10^3 W/m^2 K 1 W/m^2 K	1 Btu/ft^2 h °F = 5·678 26 W/m^2 °C 1 cal/cm^2 s °C = 41 868 W/m^2+ 1 kcal/m^2 h °C = 1·163 W/m^2 °C+ known as 'U' value
Heat capacity	joule per kelvin	J/K	kJ/K		J/°C	10^3 J/K 1 J/K	1 Btu/deg R = 1899·11 J/°C 1 cal/°C = 4·1868 J/K+
Specific heat capacity (1)	joule per kilogramme kelvin	J/kg K	kJ/kg K		kW h/kg °C kJ/kg °C J/kg °C	10^3 J/kg K 1 J/kg K	1 Btu/lb °F = 4·1868 kJ/kg °C+ 1 cal/g °C = 4·1868 J/kg °C+
Entropy	joule per kelvin	J/K	kJ/K			10^3 J/K 1 J/K	1 Btu/°R = 1899·11 J/K
Specific entropy	joule per kilogramme per kelvin	J/kg K	kJ/kg K	J/g K		10^3 J/kg K 1 J/kg K	1 Btu/lb °F = 4·1868 kJ/kg K+ 1 cal/g K = 4·1868 kJ/kg K+
Specific energy	joule per kilogramme	J/kg	MJ/kg kJ/kg	J/g		10^6 J/kg 10^3 J/kg 1 J/kg	1 Btu/lb = 2·326 kJ/kg+ 1 cal/g = 4·1868 kJ/kg+
Specific enthalpy, specific latent heat	joule per kilogramme	J/kg	MJ/kg kJ/kg			10^6 J/kg 10^3 J/kg 1 J/kg	1 Btu/lb = 2·326 kJ/kg+
Specific heat content (i) Mass basis	joule per kilogramme	J/kg	MJ/kg kJ/kg			10^6 J/kg 10^3 J/kg 1 J/kg	1 kcal/kg = 4·1868 kJ/kg+ 1 Btu/lb = 2·326 kJ/kg+ 1 CHU/lb = 4·186 816 kJ/kg 1 therm/ton = 103·84 kJ/kg

Table A.2(f)—*continued*

	Units for common use						
	SI units						
Quantity	*Name*	*Symbol*	*Decimal multiples and sub-multiples BSI recommended*	*Others*	*Units other than SI*	*Equivalent value in terms of basic unit*	*Conversion factors and remarks*
(1)	*(2)*	*(3)*	*(4)*	*(5)*	*(6)*	*(7)*	*(8)*
(ii) Volume basis	joule per cubic metre	J/m^3	kJ/m^3			10^3 J/m^3 1 J/m^3	1 Btu/gal = 0·232 08 kJ/litre 1 therm/UK gal = 23·208 GJ/m^3 1 cal/cm^3 = 4·1868 MJ/m^3 1 $kcal/m^3$ = 4·1868 kJ/m^3
(2)							
Heat release rate	watt per cubic metre	W/m^3	kW/m^3			1 W/m^3 10^3 W/m^3	1 Btu/ft^2 = 37·2589 kW/m^3 1 cal/cm^3 h = 1·163 kW/m^3

NOTES ON TABLE A.2(F)

1. *Temperature* (*K*; °*C*). The basic unit of thermodynamic temperature is the kelvin. The unit for customary or operational temperature is the °C. The °C is now known as Celsius instead of Centigrade because the latter name can be confused with $\frac{1}{100}$ of a right angle which, in some continental countries, is known as a grade. Thus all temperature readings will be in °C, K (K = °C+273·15) is used for temperatures involved in heat calculations which are equated to absolute zero (i.e. 0 K).

Table A.2(f) shows heat quantities expressed in both K and °C. Where the choice is given the following quantities are those which, for practical purposes, will be measured in customary temperature i.e. °C.

Linear expansion coefficient	$°C^{-1} = \frac{1}{°C}$
Volumetric expansion coefficient	$°C^{-1} = \frac{1}{°C}$
Thermal conductivity	W/m °C
Thermal resistivity	m °C/W
Thermal resistance	m^2 °C/W
Heat capacity	kJ/°C
Specific heat capacity	kJ/kg °C

Originally the basic heat unit was defined in terms of the heat required to raise a unit mass of water through one degree interval of temperature and specific heats were compared with water, taken as unity. The SI joule is not dependent upon the above definition so that the term 'specific heat' is no longer applicable. The word capacity has been added to distinguish the old and new terms so that, for example:

Specific heat capacity of water =	$4{\cdot}2 \times 10^{-3}$ J/kg K
Specific heat—volume basis	MJ/m^3 °C
Specific heat—mass basis	kJ/kg °C
Heat transfer coefficient	W/m^2 °C

2. *Specific heat content* (*ex calorific value*). This quantity (previously known as CV) is used for establishing the heat content of fuels. The term 'calorific' is now a misnomer as calories are replaced by joule, thus the quantities should be known as specific heat content or 'joulerific value'.

To give manageable numbers the following magnitudes are preferred:

Mass	Volume
kJ/kg	MJ/m^3

Table A.2(g) ELECTRICITY AND MAGNETISM

Quantity	Units for common use: SI units: Name	Symbol	Decimal multiples and sub-multiples: BSI recommended	Others	Units other than SI	Equivalent value in terms of basic unit	Conversion factors and remarks
(1)	(2)	(3)	(4)	(5)	(6)	(7)	(8)
Electric current			kA			10^3 A	1 emu = 10 A
	ampere	A				1 A	1 esu = $1/3 \times 10^{-9}$ A
			mA			10^{-3} A	
			μA			10^{-6} A	
			nA			10^{-9} A	
			pA			10^{-12} A	
Electric charge			kC			10^3 C	1 Ah = 3600 C
	coulomb	C				1 C	
			μC			10^{-6} C	C = A s
			nC			10^{-9} C	
			pC			10^{-12} C	
Charge density			MC/m^3	C/cm^3		10^6 C/m^3	1 emu = 10^7 C/m^3
	coulomb per cubic metre		kC/m^3			10^3 c/m^3	1 esu = $1/3 \times 10^{-3}$ C/m^3
		C/m^3				1 C/m^3	
Surface density of charge	coulomb per square metre		MC/m^2	C/mm^2		10^6 C/m^2	
				C/cm^2		10^4 C/m^2	
			kC/m^2			10^3 C/m^2	
		C/m^2				1 C/m^2	
Electric field strength	coulomb per square metre		MV/m			10^6 V/m	
			kV/m	V/mm		10^3 V/m	
				V/cm		10^2 V/m	
	volt per metre	V/m				1 V/m	
			mV/m			10^{-3} V/m	
			μV/m			10^{-6} V/m	

			MV		10^6 V	
Electric potential			kV		10^3 V	
	volt	V			1 V	
			mV		10^{-3} V	
			μV		10^{-6} V	
				C/cm^2	10^4 m^2	
Displacement			kC/m^2		10^3 m^2	C/m^2 = A s/m^2
	coulomb per square metre	C/m^2			1 m^2	
			MC		10^6 C	
Electric flux			kC		10^3 C	
	coulomb	C			1 C	
			mC		10^{-3} C	
Capacitance	farad	F			1 F	
			mF		10^{-3} F	
			μF		10^{-6} F	F = A s/V = C/V
			nF		10^{-9} F	
			pF		10^{-12} F	
Permittivity	farad per metre	F/m			1 F/m	
			μF/m		10^{-6} F/m	$\varepsilon_0 = 8{\cdot}854 \times 10^{-12}$ F/m
			mF/m		10^{-3} F/m	
			pF/m		10^{-12} F/m	
			MC/m^2		10^6 C/m^2	
				C/cm^2	10^4 C/m^2	
Electric polarisation	coulomb per square metre	C/m^2	kC/m^2		10^3 C/m^2	
					1 C/m^2	
Electric dipole moment	coulomb metre	C m			1 C m	
			MA/m^2	A/mm^2	10^{-6} A/m^2	
				A/cm^2	10^{-4} A/m^2	
Current density	ampere per square metre	A/m^2	kA/m^2		10^{-3} A/m^2	
					1 A/m^2	
Linear current density			kA/m	A/mm	10^3 A/m	
				A/cm	10^2 A/m	
	ampere per metre	A/m			1 A/m	

Table A.2(g)—*continued*

Quantity	*Units for common use*					*Equivalent value in terms of basic unit*	*Conversion factors and remarks*
	SI units				*Units other than SI*		
	Name	*Symbol*	*Decimal multiples and sub-multiples BSI recommended*	*Others*			
(1)	*(2)*	*(3)*	*(4)*	*(5)*	*(6)*	*(7)*	*(8)*
Magnetic field strength			kA/m	A/mm		10^3 A/m	1 oersted = $10^3/4\ \Pi$ A/m
				A/cm		10^2 A/cm	
	ampere per metre	A/m				1 A/m	
Magnetic potential difference			kA			10^3 A	
		A				1 A	1 gilbert = $10/4\ \Pi$ A
			mA			10^{-3} A	
Magnetic flux density	tesla	T				1 T	Wb/m^2 = T
			mT			10^{-3} T	1 gauss = 10^{-4} T
			μT			10^{-6} T	
			nT			10^{-9} T	
Magnetic flux	weber	Wb				1 Wb	V s = Wb
			mWb			10^{-3} Wb	1 maxwell = 10^{-8} Wb
Magnetic vector potential			kWb/m	Wb/mm		10^3 Wb/m	
	weber per metre	Wb/m				1 Wb/m	1 maxwell/cm = 10^{-6} Wb/m
Mutual inductance, self inductance	henry	H				1 H	
			mH			10^{-3} H	H = V s/A
			μH			10^{-6} H	
			nH			10^{-9} H	
			pH			10^{-12} H	
Permeability	henry per metre	H/m				1 H/m	$\mu_0 = 4\ \Pi \times 10^{-7}$ H/m
			μH/m			10^{-6} H/m	
			nH/m			10^{-9} H/m	

	metre						
Magnetisation			kA/m	A/mm		10^3 A/m	1 oersted = $10^3/4\,\Pi$ A/m
	ampere per metre	A/m				1 A/m	
Magnetic polarisation	tesla	T				1 T	1 gauss = 10^{-4} T
			mT			10^{-3} T	
Magnetic dipole moment	newton square metre per ampere	N m^2/A				1 N m^2/A	
			GΩ			10^9Ω	
			MΩ			10^6Ω	
			kΩ			10^3Ω	
Resistance	ohm	Ω				1Ω	
			mΩ			10^{-3}Ω	
			μΩ			10^{-6}Ω	
Conductance	reciprocal ohm				kS	10^3 1/Ω or 10^3 S	S = mho = Siemen
		1/Ω			S	1 1/Ω or 1 S	
					mS	10^{-3} 1/Ω or 10^{-3} S	
					μS	10^{-6} 1/Ω or 10^{-6} S	
Resistivity			GΩ m			10^9Ω m	
			MΩ m			10^6Ω m	
			kΩ m			10^3Ω m	
	ohm metre	Ωm				1Ω m	
			mΩ m			10^{-3}Ω m	
			μΩ m			10^{-6}Ω m	
				μΩ cm		10^{-8}Ω m	
(2)			nΩ m			10^{-9}Ω m	
					MS/m	10^6 1/Ω m or S/m	
Conductivity	reciprocal ohm metre	1/Ω m			kS/m	10^3 1/Ω m or S/m	
					S/m	1 1/Ω m or S/m	
Reluctance	reciprocal henry	1/H				1 1/H	
Permeance	henry	H				1 H	

Table A.2(g)—*continued*

Quantity	*Units for common use*					*Equivalent value in terms of basic unit*	*Conversion factors and remarks*
	SI units				*Units other than SI*		
	Name	*Symbol*	*Decimal multiples and sub-multiples BSI recommended*	*Others*			
(1)	(2)	(3)	(4)	(5)	(6)	(7)	(8)
Impedance			MΩ			$10^6\Omega$	
Reactance			kΩ			$10^3\Omega$	
	ohm	Ω				1Ω	
			mΩ			$10^{-3}\Omega$	
Conductance	reciprocal				kS	10^3 1/Ω or S	
	ohm	1/Ω			S	1 1/Ω or S	
					mS	10^{-3} 1/Ω or S	
					μS	10^{-6} 1/Ω or S	
Active power			TW			10^{12} W	
			GW			10^9 W	
			MW			10^6 W	
			kW			10^3 W	
	watt	W				1 W	
			mW			10^{-3} W	
			μW			10^{-6} W	
			nW			10^{-9} W	
Apparent power			TVA			10^{12} VA	
			GVA			10^9 VA	
			MVA			10^6 VA	
			kVA			10^3 VA	
	volt ampere	VA				1 VA	
			mVA			10^{-3} VA	
			μVA			10^{-6} VA	
			nVA			10^{-9} VA	

Reactive power				T var	10^{12} var	
				G var	10^{9} var	
				k var	10^{3} var	
				var	1 var	
				m var	10^{-3} var	
				μ var	10^{-6} var	
				n var	10^{-9} var	
Electric stress	volt per metre	V/m	kV/mm		1 V/m	1 kV/in = 0·039 370 1 kV/mm or 1 V/mil = 39·370 1 kV/m

NOTES ON TABLE A.2(G)

1. *General note.* SI is an extension of the Giorgi rationalised metre – kilogramme – second – ampere (MKSA) system which has been the preferred system of electrical units in use since 1935, thus engineers should be familiar with most of the units.

Changes which are introduced in SI apart from revised definitions for certain base units are:

(a) The tesla (T) as the unit for magnetic flux density.

(b) The siemen (S) for the reciprocal ohm. This has been adopted by IEC[18] and ISO[19] but has not yet been approved by CGPM[16].

(c) A name is required for the unit to describe periodic phenomena, i.e. the hertz (Hz) for cycles per second.

In electricity and magnetism the SI units align with the MKSA rationalised form of the equations between quantities. Thus the magnetic space constant or permeability of free space constant is $4\Pi \times 10^{-7}$ henry per metre and the electric space constant or permittivity of free space constant is 8.854×10^{-12} farad/metre.

2. *Resistivity.* It is the present practice for chemists to work in micro ohms/cm and existing practice is to use dionic instruments etc. calculated in this unit. It will be some time before instruments are calibrated in micro ohm/metre.

Table A.2(h) LIGHT

Quantity	Units for common use					Equivalent value in terms of basic unit	Conversion factors and remarks
	SI units				Units other than SI		
	Name	Symbol	Decimal multiples and sub-multiples BSI recommended	Others			
(1)	(2)	(3)	(4)	(5)	(6)	(7)	(8)
Luminous intensity	candela	cd				1 cd	
Luminous flux	lumen	lm				1 lm	lm = cd sr (candela steradian)
Illumination	lux	lx				1 lx	lx = lm/m^2
				1 m/cm^2		10^4 lx	
Luminance	candela per square metre	cd/m^2				1 cd/m^2	stilb = 1 cd/cm^2 apostilb = $1/\pi$ cd/m^2 1 cd/in^2 = 1550 cd/m^2 1 foot lambert = 3426 cd/m^2 1 lambert = 3183 cd/m^2

Quantities and units of light

The following definitions are based on the International Lighting Vocabulary and on BS 233.

Luminous flux (symbol ϕ)	The light emitted by a source such as a lamp, or received by a surface, irrespective of direction.
Lumen (abbreviation lm)	The SI unit of luminous flux used in describing the total light emitted by a source or received by a surface. (A 100 watt incandescent lamp emits about 1200 lumens.)
Illumination	The process of lighting an object.
Illumination value (symbol E)	The luminous flux incident on a surface, per unit area. The term 'illuminance' has been proposed for international usage. In the Code, the colloquial terms 'illumination' and 'illumination level' have also been used as in current British practice.
Lux (abbreviation lx)	The SI unit of illumination value; it is equal to one lumen per square metre.
Lumen per square foot (abbreviation lm/ft^2)	A non-metric unit of illumination value, equal to 10·76 lux. (Previously called the foot-candle, a term still used in the USA.)
Service value of illumination	The mean value of illumination throughout the life of and installation and averaged over the working area.
Initial value of illumination	The mean value of illumination averaged over the working area before depreciation has started, i.e. when the lamps and fittings are new and clean and when the room is freshly decorated.
Mean spherical illumination (scalar illumination)	The average illumination over the surface of a small sphere centred at a given point; more precisely, it is the flux incident on the surface of the sphere divided by the area of the sphere. The term 'scalar' illumination (symbol Es) has been proposed. The unit of scalar illumination is the lux: care is needed to avoid confusing the unit with the illumination on a plane which is measured in the same unit.
Illumination vector	A term used to describe the flow of light. It has both magnitude and direction. The magnitude is defined as the maximum difference in the value of illumination at diametrically opposed surface elements of a small sphere centred at the point under consideration. The direction of the vector is that of the diameter joining the brighter to the darker element. The ratio of the magnitude of the illumination vector to the scalar illumination has been proposed as an index of modelling.
Luminous intensity (symbol I)	The quantity which describes the illuminating power of a source in a particular direction. More precisely it is the luminous flux emitted within a very narrow cone containing that direction divided by the solid angle of the cone.
Candela (abbreviation cd)	The SI unit of luminous intensity. The term 'candle power' designates a luminous intensitv expressed in candelas.

Table A.2(j) SOUND

Quantity	Units for common use					Equivalent value in terms of basic unit	Conversion factors and remarks
	SI units				Units other than SI		
	Name	Symbol	Decimal multiples and sub-multiples: BSI recommended	Decimal multiples and sub-multiples: Others			
(1)	(2)	(3)	(4)	(5)	(6)	(7)	(8)
Sound intensity	watt per square metre	W/m^2				1 W/m^2	1 erg s^{-1} cm^{-2} = 10^{-3} W/m^2
Sound intensity (logarithmic)					decibel bel	1/10 bel 1 bel	$20 \log_{10}\left(\frac{P}{P_0}\right)$ decibel (dB) where P = measured sound pressure and P_0 = reference sound pressure of 2×10^{-5} N/m^2 (ref. BS. 4196)
Loundness					phon	1 phon	
Attenuation					neper per metre	np/m	

Table A.2(k) PHYSICAL CHEMISTRY AND MOLECULAR PHYSICS

Quantity	Units for common use: SI units: Name	Symbol	Decimal multiples and sub-multiples: BSI recommended	Others	Units other than SI	Equivalent value in terms of basic unit	Conversion factors and remarks
(1)	(2)	(3)	(4)	(5)	(6)	(7)	(8)
Amount of substance					kmol	10^3 mol	1 lb mol = 0·453 592 37 k mol+
					mol	1 mol	
Molar volume					m^3/mol	1 m^3/mol	
Molar mass					kg/mol	1 kg/mol	
					g/mol	10^{-3} kg/mol	
Molar internal energy					J/mol	1 J/mol	1 erg/mol = 10^{-7} J/mol
					J/kmol	10^{-3} J/mol	
Molar heat capacity					J/mol K; J mol °C	1 J/mol K	1 erg mol^{-1} $°C^{-1}$ = 10^{-7} J mol^{-1} K^{-1}
					J/kmol K; Jkmol °C	10^{-3} J/mol K	Universal gas constant 8·3143 kJ/k mol K
Molar entropy					J/mol K	1 J/mol K	
Molality					kmol/l	10^6 mol/m^3	
					kmol/m^3	10^3 mol/m^3	
					mol/l	10^3 mol/m^3	
					mol/dm^3	10^3 mol/m^3	
					mol/m^3	1 mol/m^3	
					kmol/kg	10^3 mol/kg	
					mol/kg	1 mol/kg	

Table A.2(k)—*continued*

Quantity	*Units for common use*					*Equivalent value in terms of basic unit*	*Conversion factors and remarks*
	SI units				*Units other than SI*		
	Name	*Symbol*	*Decimal multiples and sub-multiples BSI recommended*	*Others*			
(1)	*(2)*	*(3)*	*(4)*	*(5)*	*(6)*	*(7)*	*(8)*
Diffusion coefficient	square metre per second	m^2/s				1 m^2/s	
Thermal diffusion coefficient	square metre per second	m^2/s				1 m^2/s	
Molar flow rate					kmol/h mol/h	10^3 mol/h 1 mol/h	1 lb mol/h = 0·453 592 37 kmol/h

Nuclear engineering

It has been the practice to use special units with their individual names for evaluating and comparing results. These units are usually formed by multiplying a unit from the cgs or SI system by a number which matches a value derived from the result of some natural phenomenon. The adoption of SI both nationally and internationally has created the opportunity to examine the practice of using special units in the nuclear industry, with the object of eliminating as many as possible and using the pure system instead.

As an aid to this ISO draft Recommendations 838[11] and 839[12] have been published, giving a list of quantities with special names, the SI unit and the alternative cgs unit. It is expected that as SI is increasingly adopted and absorbed, those units based on cgs will go out of use. The values of these special units illustrate the fact that a change from them to SI would not be as revolutionary as might be supposed. Examples of these values together with the SI units which replace them are shown in Table A.2(1).

Table A.2(1) NUCLEAR ENGINEERING

Special unit			*SI Replacement*
Name		*Value*	
Angstrom	(Å)	10^{-10}m	m
Barn	(b)	10^{-28}m^2	m^2
Curie	(Ci)	$3{\cdot}7 \times 10^{10}$s^{-1}	s^{-1}
Electronvolt	(eV)	$(1{\cdot}602\ 10 \pm 0{\cdot}000\ 07) \times 10^{-19}$J	J
Röntgen	(R)	$2{\cdot}58 \times 10^{-4}$C/kg	C/kg

APPLICATION OF UNITS

Calculations

In applying SI units to calculations care should be taken to see that the formulae and references are applicable to an absolute system of units.

Much of the data in existing literature applies to the imperial foot-pound-second system which is a gravitational system.

The following guide rules may prove useful in calculations or design work.

1. Dynamic formulae written W/g will become M (in kilogrammes); conversely gravitational formulae written in W will become Mg (newtons).
2. To avoid errors convert all quantities to consistent units at the outset i.e. keep all quantities in unity multiplied by relevant powers of 10.
3. Use conversion factors based on BS 350[3].

In formulae which need to be transposed, take care that the constants are consistent with the units to be used in SI. Formulae which are expressions of natural laws are always valid and constants are usually independent of units.

Empirical formula should be carefully checked as the constants invariably depend on the units used in the expression.

Consider the deflection (d) of a cantilevered beam with a load (W) supported at a distance (L)

$$d = \frac{WL^3}{3EI}$$

where E = modulus of elasticity for the material
and I = moment of inertia of the section

This expression is valid in any system provided the units are coherent.

These values may be in metric data sheets expressed in the following magnitudes; W in N; kN: L in m; E in N/mm²; GN/m²; I in cm⁴; d in mm.

In SI the above formula is arranged thus:

$$d(\mathrm{m}) = \frac{W(\mathrm{N})L^3(\mathrm{m}^3)}{3E(\mathrm{N/m}^2)I(\mathrm{m}^4)}$$

$$\therefore\ (\mathrm{m}) = \frac{(\mathrm{N})(\mathrm{m}^3)}{(\mathrm{N})(\mathrm{m}^4)} \times (\mathrm{m}^2) = \mathrm{m}^{(5-4)} = \mathrm{m}$$

Thus, for example, taking:

$$\begin{aligned} W &= 1\ \text{tonf} \\ L &= 10\ \text{ft} \\ E &= 30 \times 10^6\ \text{lbf/in}^2 \\ I &= 69{\cdot}2\ \text{in}^4 \end{aligned}$$

(It should be noted that the units for W and L are inconsistent with the units for E and I).

Converting to SI units:

$$\begin{aligned} W &= 1\ \text{tonf} = 1016\ \text{kg} = 1016 \times 9{\cdot}8605\ \text{N} = 9964{\cdot}02\ \text{N} \\ L &= 10\ \text{ft} = 3{\cdot}048\ \text{m} \\ E &= 30 \times 10^6/\text{lbf/in}^2 = 30 \times 10^6 \times 6{\cdot}895 \times 10^3\ \text{N/m}^2 \\ &= 206{\cdot}85 \times 10^9\ \text{N/m}^2 \\ &= 206{\cdot}85\ \text{GN/m}^2 \\ I &= 69{\cdot}2\ \text{in}^4 = 69{\cdot}2 \times 0{\cdot}254^4\text{m}^4 \\ &= 69{\cdot}2 \times (2{\cdot}54 \times 10^{-2})^4 \\ &= 69{\cdot}2 \times 41{\cdot}61 \times 10^{-8} \\ &= 287{\cdot}94 \times 10^{-7}\text{m}^4 \end{aligned}$$

Putting in these values for the calculations:

$$d = \frac{[9{\cdot}964 \times 10^3] \times [3{\cdot}048^3]}{3 \times [206{\cdot}85 \times 10^9] \times [28{\cdot}794 \times 10^{-6}]}\ \text{m}$$

$$\frac{9{\cdot}964 \times 3{\cdot}048^3}{3 \times 206{\cdot}85 \times 28{\cdot}794}\ \text{m}$$

$$d = 0{\cdot}0158\ \text{m}$$

$$d = 15{\cdot}8\ \text{mm}$$

Note how the powers of 10 readily cancel out and how the I value has been brought to the power 10^{-6} to obey the rule of 10^n where $n = \pm 3$.

The need for this might be questioned but if 10^{-7} had been used the cancelling out would not be so easily carried out.

Sometimes the conversion of formulae is not so straightforward and in such cases it is suggested that the new metric formulae be proved by comparing the metric calculations with the results obtained from the imperial formula using imperial units which should be converted to metric units when resolved. When using imperial formulae containing constants which apply to several units, the calculation should be made using accurate conversion factors and rounding to the required degree of precision as a final step.

Conversion of existing imperial terms

If it is necessary to convert existing imperial terms to a metric equivalent, care should be taken to ensure that the converted value implies the same degree of accuracy. The conversion factor must convey the same order of precision as the original value.

Thus to translate 1 in as 25·4 mm or 1000 ft as 304·8 m conveys a tolerance which in most cases would be too precise.

Table A.3(a) METRIC TO IMPERIAL CONVERSION FACTORS

SI units	*British units*
SPACE AND TIME	
Length:	
1 μm (micron)	$= 39{\cdot}37 \times 10^{-6}$ in
1 mm	= 0·039 370 1 in
1 cm	= 0·393 701 in
1 m	= 3·280 84 ft
1 m	= 1·093 61 yd
1 km	= 0·621 371 mile
Area:	
1 mm^2	$= 1{\cdot}550 \times 10^{-3}$ in^2
1 cm^2	= 0·1550 in^2
1 m^2	= 10·7639 ft^2
1 m^2	= 1·195 99 yd
1 ha	= 2·471 05 acre
Volume:	
1 mm^3	$= 61{\cdot}0237 \times 10^{-6}$ in^3
1 cm^3	$= 61{\cdot}0237 \times 10^{-3}$ in^3
1 m^3	= 35·3147 ft^3
1 m^3	= 1·307 95 yd^3
Capacity:	
$10^6 m^3$	$= 219{\cdot}969 \times 10^6$ gal
1 m^3	= 219·969 gal
1 litre (1)	= 0·219 969 gal = 1·759 80 pint
Capacity flow:	
$10^3/m^3/s$	$= 791{\cdot}9 \times 10^6$ gal/h
1 m^3/s	$= 13{\cdot}20 \times 10^3$ gal/min
1 litre/s	= 13·20 gal/min
1 m^3/kW h	= 219·969 gal/kW h
1 m^3/s	= 35·3147 ft^3/s (cusecs)
1 litre/s	$= 0{\cdot}588\ 58 \times 10^{-3}$ ft^3/min (cfm)
Velocity:	
1 m/s	= 3·280 84 ft/s = 2·236 94 mile
1 km/h	= 0·621 371 mile/h
Acceleration:	
1 m/s^2	= 3·280 84 ft/s^2
MECHANICS	
Mass:	
1 g	= 0·035 274 oz
1 kg	= 2·204 62 lb
1 t	= 0·984 207 ton = 19·6841 cwt
Mass flow:	
1 kg/s	= 2·204 62 lb/s = 7·936 64 klb/h
Mass density:	
1 kg/m^3	= 0·062 428 lb/ft^3
1 kg/litre	= 10·022 119 lb/gal
Mass per unit length:	
1 kg/m	= 0·671 969 lb/ft = 2·015 91 lb/yd
Mass per unit area:	
1 kg/m^2	= 0·204 816 lb/ft^2
Specific volume:	
1 m^3/kg	= 16·0185 ft^3/lb
1 litre/tonne	= 0·223 495 gal/ton
Momentum:	
1 kg m/s	= 7·233 01 lbft/s
Angular momentum:	
1 kg m^2/s	= 23·7304 lbft^2/s

Table A.3(a)—*continued*

SI units	*British units*
Moment of inertia:	
1 kg m^2	= 23·7304 lbft^2
Force:	
1 N	= 0·224 809 lbf
Weight (force) per unit length:	
1 N/m	= 0·068 521 8 lbf/ft = 0·205 566 lbf/yd
Moment of force (or torque):	
1 Nm	= 0·737 562 lbf/ft
Weight (force) per unit area:	
1 N/m^2	= 0·020 885 lbf/ft^2
Pressure:	
1 N/m^2	= 1·450 38 × 10^{-4} lbf/in^2
1 bar	= 14·5038 lbf/in^2
1 bar	= 0·986 923 atmosphere
1 mbar	= 0·401 463 in H_2O
	= 0·029 53 in Hg
Stress:	
1 N/mm^2	= 6·474 90 × 10^{-2} tonf/in^2
1 MN/m^2	= 6·474 90 × 10^{-2} tonf/in^2
1 hbar	= 0·647 490 tonf/in^2
Second moment of area:	
1 cm^4	= 0·024 025 in^4
Section modulus:	
1 m^3	= 61 023·7 in^3
1 cm^3	= 0·061 023 7 in^3
Kinematic viscosity:	
1 m^2/s	= 10·762 75 ft^2/s = 10^6 cSt
1 cSt	= 0·038 75 ft^2/h
Energy, work:	
1 J	= 0·737 562 ft lbf
1 MJ	= 0·3725 hph
1 MJ	= 0·277 78 kW h
Power:	
1 W	= 0·737 562 ft lbf/s
1 kW	= 1·3410 hp = 737·562 ft lbf/s
Fluid mass:	
(Ordinary) 1 kg/s	= 2·204 62 lb/s = 7936·64 lb/h
(Velocity) 1 kg/m^2 s	= 0·204 815 lb/ft^{2}s
HEAT	
Temperature:	
(Interval) 1 degK	= 9/5 deg R (Rankine)
1 degC	= 9/5 deg F
(Coefficient) 1 degR^{-1}	= 1 deg F^{-1} = 5/9 deg C
1 degC^{-1}	= 5/9 deg F^{-1}
Quantity of heat:	
1 J	= 9·478 17 × 10^{-4} Btu
1 J	= 0·238 846 cal
1 kJ	= 947·817 Btu
1 GJ	= 947·817 × 10^3 Btu
1 kJ	= 526·565 CHU
1 GJ	= 526·565 × 10^3 CHU
1 GJ	= 9·478 17 therm
Heat flow rate:	
1 W(J/s)	= 3·412 14 Btu/h
1 W/m^2	= 0·316 998 Btu/ft^{2}h
Thermal conductivity:	
1 W/m°C	= 6·933 47 Btu in/ft^2 h °F

Table A.3(a)—*continued*

SI units	*British units*
Coefficient and heat transfer:	
1 W/m²°C	= 0·176 110 Btu/ft² h °F
Heat capacity:	
1 J/°C	= $0{\cdot}526\ 57 \times 10^{-3}$ Btu/°R
Specific heat capacity:	
1 J/g°C	= 0·238 846 Btu/lb °F
1 kJ/kg°C	= 0·238 846 Btu/lb °F
Entropy:	
1 J/K	= $0{\cdot}526\ 57 \times 10^{-3}$ Btu/°R
Specific Entropy:	
1 J/kg degC	= $0{\cdot}238\ 846 \times 10^{-3}$ Btu/lb °F
1 J/kg degK	= $0{\cdot}238\ 846 \times 10^{-3}$ Btu/lb °R
Specific energy/Specific latent heat:	
1 J/g	= 0·429 923 Btu/lb
1 J/kg	= $0{\cdot}429\ 923 \times 10^{-3}$ Btu/lb
Calorific value:	
1 kJ/kg	= 0·429 923 Btu/lb
1 kJ/kg	= 0·773 861 4 CHU/lb
1 J/m³	= $0{\cdot}026\ 839\ 2 \times 10^{-3}$ Btu/ft³
1 kJ/m³	= 0·026 839 2 Btu/ft³
1 kg/litre	= 4·308 86 Btu/gal
1 kJ/kg	= 0·009 630 2 therm/ton
ELECTRICITY	
Permeability:	
1 H/m	= $10^7/4\ \Pi\ \mu o$
Magnetic flux density:	
1 tesla	= 10^4 gauss = 1 Wb/m²
Conductivity:	
1 mho	= 1 reciprocal ohm
1 Siemen	= 1 reciprocal ohm
Electric stress:	
1 kV/mm	= 25·4 kV/in
1 kV/m	= 0·0254 kV/in

Table A.3(b) UNIVERSAL CONSTANTS IN SI UNITS

The digits in parentheses following each quoted value represent the standard deviation error in the final digits of the quoted value as computed on the criterion of internal consistency. The unified scale of atomic weights is used throughout ($^{12}C = 12$). C = coulomb; G = gauss; Hz = hertz; J = joule; N = newton; T = tesla; u = unified nuclidic mass unit; W = watt; Wb = weber. For result multiply the numerical value by the SI unit.

Constant	*Symbol*	*Numerical value*	*SI unit*
Speed of light in vacuum	c	2·997 925(1)	10^{8} m s^{-1}
Gravitational constant	G	6·670(5)*	10^{-11} N m^{2} kg^{-2}
Elementary charge	e	1·602 10(2)	10^{-19} C
Avogadro constant	N_A	6·022 52(9)	10^{26} kmol^{-1}
Mass unit	u	1·660 43(2)	10^{-27} kg
Electron rest mass	m_e	9·109 08(13)	10^{-31} kg
		5·485 97(3)	10^{-4} u
Proton rest mass	m_p	1·672 52(3)	10^{-27} kg
		1·007 276 63(8)	u
Neutron rest mass	m_n	1·67 482(3)	10^{-27} kg
		1·008 6654(4)	u
Faraday constant	F	9·648 70(5)	10^{4} C mol^{-1}
Planck constant	h	6·625 59(16)	10^{-34} J s
	$h/2\pi$	1·054 494(25)	10^{-34} J s
Fine-structure constant	α	7·297 20(3)	10^{-3}
	$1/\alpha$	137·0388(6)	
Charge-to-mass ratio for electron	e/m_e	1·758 796(6)	10^{11} C kg^{-1}
Quantum of magnetic flux	hc/e	4·135 56(4)	10^{-11} Wb
Rydberg constant	R_∞	1·097 3731(1)	10^{7} m^{-1}
Bohr radius	a_0	5·291 67(2)	10^{-11} m
Compton wavelength of electron	h/m_ec	2·426 21(2)	10^{-12} m
	$\lambda c/2\pi$	3·861 44(3)	10^{-13} m
Electron radius	$e^2/m_ec^2 = r_e$	2·817 77(4)	10^{-15} m
Thomsen cross section	$8\eta r_e^2/3$	6·6516(2)	10^{-29} m^{2}
Compton wavelength of proton	$\lambda c, p$	1·321 398(13)	10^{-15} m
	$\lambda c, p/2\pi$	2·103 07(2)	10^{-16} m
Gyromagnetic ratio of proton	γ	2·675 192(7)	10^{8} rad s^{-1} T^{-1}
	$\gamma/2\pi$	4·257 70(1)	10^{7} Hz T^{-1}
(Uncorrected for diamagnetism	γ'	2·675 123(7)	10^{8} rad s^{-1} T^{-1}
H_2O)	$\gamma'/2\pi$	4·257 59(1)	10^{7} Hz T^{-1}
Bohr magneton	μB	9·2732(2)	10^{-24} J T^{-1}
Nuclear magneton	μN	5·050 50(13)	10^{-27} J T^{-1}
Proton moment	μ_p	1·410 49(4)	10^{-26} J T^{-1}
	$\mu_p/\mu N$	2·792 76(2)	
(Uncorrected for diamagnetism in H_2O sample)		2·792 68(2)	
Gas constant	R_0	8·314 34(35)	J deg^{-1} mol^{-1}
Boltzmann constant	k	1.380 54(6)	10^{-23} J deg^{-1}
First radiation constant ($2\eta hc^2$)	c_1	3·741 50(9)	10^{-16} W m^{2}
Second radiation constant (hc/k)	c_2	1·438 79(6)	10^{-2} m deg
Stefan-Boltzmann constant	σ	5·6697(10)	10^{-8} W m^{-2} deg^{-4}

* The universal gravitational constant is not, and cannot in our present state of knowledge, be expressed in terms of other fundamental constants. The value given here is a direct determination by P. R. Heyl and P. Chrzanowski, J. Res. Natl. Bur. Std. (U.S.) 29, 1 (1942).

The above values are extracts from 'Review of Modern Physics' Vol. 37 No. 4 October 1965 published by the American Institute of Physics.

CONVERSION TABLES

Description of Tables A.4(a) to (r)

Although conversion factors are given in Tables A.2(a) to A.2(k), it is sometimes desirable to refer to a conversion table for the more readily used engineering units. Such conversioning tables from Imperial to SI units are shown in Tables A.4(a) to (r) as follows:

Table	*Description*
A.4(a)	Length – fractions of an inch to millimetres
A.4(b)	Length – feet and inches to metres
A.4(c)	Length – yards to metres
A.4(d)	Length – miles to kilometres
A.4(e)	Area – square feet to square metres
A.4(f)	Area – square yards to square metres
A.4(g)	Volume – cubic inches to cubic centimetres
A.4(h)	Volume – cubic feet to cubic metres
A.4(j)	Volume – cubic yards to cubic metres
A.4(k)	Capacity – gallons to litres
A.4(l)	Mass (Mass by weight) – pounds to kilogrammes
A.4(m)	Mass (Mass by weight) – tons to tonne
A.4(n)	Mass (Mass by weight) – cwt. qr. to tonne
A.4(p)	Pressure – lbf/in^2 to bar
A.4(q)	Stress – $tonf/in^2$ to N/mm^2
A.4(r)	Stress – kgf/mm^2 to N/mm^2

Reading of tables

It will be noted that each table is shown in two parts i.e. main table and auxiliary table. This is to enable all whole number quantities within the range of the table to be expressed in SI units.

Consider the following example from Table A.4(b), feet and inches to metres. To convert 83 ft 7 in to metres.

From Table A.4(b); main table	3 ft 7 in =	1·092 m
auxiliary table	80 ft =	24·384 m
	83 ft 7 in =	25·476 m

For other detailed conversions see BS 350[3] Part 2 and Supplement No. 1 (BSI/PD 6203)[7].

Table A.4(a) LENGTH: FRACTIONS OF AN INCH TO MILLIMETRES
(Correct to the nearest millimetre)

in	0	1	2	3	4	5	6	7	8	9	10	11
	mm											
—	—	25	51	76	102	127	152	178	203	229	254	279
$\frac{1}{16}$	2	27	52	78	103	129	154	179	205	230	256	281
$\frac{1}{8}$	3	29	54	79	105	130	156	181	206	232	257	283
$\frac{3}{16}$	5	30	56	81	106	132	157	183	208	233	259	284
$\frac{1}{4}$	6	32	57	83	108	133	159	184	210	235	260	286
$\frac{5}{16}$	8	33	59	84	110	135	160	186	211	237	262	287
$\frac{3}{8}$	10	35	60	86	111	137	162	187	213	238	264	289
$\frac{7}{16}$	11	37	62	87	113	138	164	189	214	240	265	291
$\frac{1}{2}$	13	38	64	89	114	140	165	191	216	241	267	292
$\frac{9}{16}$	14	40	65	90	116	141	167	192	217	243	268	294
$\frac{5}{8}$	16	41	67	92	117	143	168	194	219	244	270	295
$\frac{11}{16}$	17	43	68	94	119	144	170	195	221	246	271	297
$\frac{3}{4}$	19	44	70	95	121	146	171	197	222	248	273	298
$\frac{13}{16}$	21	46	71	97	122	148	173	198	224	249	275	300
$\frac{7}{8}$	22	48	73	98	124	149	175	200	225	251	276	302
$\frac{15}{16}$	24	49	75	100	125	151	176	202	227	252	278	303

AUXILIARY TABLE

ft	1	2	3	4	5	6	7	8	9
m	0·305	0·610	0·914	1·219	1·524	1·829	2·134	2·438	2·743

Table A.4(b) LENGTH: FEET AND INCHES TO METRES (Correct to the nearest mm)

ft	*in* 0	*1*	*2*	*3*	*4*	*5*	*6*	*7*	*8*	*9*	*10*	*11*
	m											
0	—	0·025	0·051	0·076	0·102	0·127	0·152	0·178	0·203	0·229	0·254	0·279
1	0·305	0·330	0·356	0·381	0·406	0·432	0·457	0·483	0·508	0·533	0·559	0·584
2	0·610	0·635	0·660	0·686	0·711	0·737	0·762	0·787	0·813	0·838	0·864	0·889
3	0·914	0·940	0·965	0·991	1·016	1·041	1·067	1·092	1·118	1·143	1·168	1·194
4	1·219	1·245	1·270	1·295	1·321	1·346	1·372	1·397	1·422	1·448	1·473	1·499
5	1·524	1·549	1·575	1·600	1·626	1·651	1·676	1·702	1·727	1·753	1·778	1·803
6	1·829	1·854	1·880	1·905	1·930	1·956	1·981	2·007	2·032	2·057	2·083	2·108
7	2·134	2·159	2·184	2·210	2·235	2·261	2·286	2·311	2·337	2·362	2·388	2·413
8	2·438	2·464	2·489	2·515	2·540	2·565	2·591	2·616	2·642	2·667	2·692	2·718
9	2·743	2·769	2·794	2·819	2·845	2·870	2·896	2·921	2·946	2·972	2·997	3·023
10	3·048	3·073	3·099	3·124	3·150	3·175	3·200	3·226	3·251	3·277	3·302	3·327
11	3·353	3·378	3·404	3·429	3·454	3·480	3·505	3·531	3·556	3·581	3·607	3·632
12	3·658	3·683	3·708	3·734	3·759	3·785	3·810	3·835	3·861	3·886	3·912	3·937
13	3·962	3·988	4·013	4·039	4·064	4·089	4·115	4·140	4·166	4·191	4·216	4·242
14	4·267	4·293	4·318	4·343	4·369	4·394	4·420	4·445	4·470	4·496	4·521	4·547
15	4·572	4·597	4·623	4·648	4·674	4·699	4·724	4·750	4·775	4·801	4·826	4·851
16	4·877	4·902	4·928	4·953	4·978	5·004	5·029	5·055	5·080	5·105	5·131	5·156
17	5·182	5·207	5·232	5·258	5·283	5·309	5·334	5·359	5·385	5·410	5·436	5·461
18	5·486	5·512	5·537	5·563	5·588	5·613	5·639	5·664	5·690	5·715	5·740	5·766
19	5·791	5·817	5·842	5·867	5·893	5·918	5·944	5·969	5·994	6·020	6·045	6·071
20	6·096	—	—	—	—	—	—	—	—	—	—	—

Table A.4(b)—*continued*

AUXILIARY TABLE

ft	*30*	*40*	*50*	*60*	*70*	*80*	*90*	*100*	*150*	*200*
m	9·144	12·192	15·240	18·288	21·336	24·384	27·432	30·480	45·720	60·960

Table A.4(c) LENGTH: YARDS TO METRES (Correct to the nearest 0·01 m)

yd	*0*	*10*	*20*	*30*	*40*	*50*	*60*	*70*	*80*	*90*
					m					
0	—	9·14	18·29	27·43	36·58	45·72	54·86	64·01	73·15	82·30
100	91·44	100·58	109·73	118·87	128·02	137·16	146·30	155·45	164·59	173·74
200	182·88	192·02	201·17	210·31	219·46	228·60	237·74	246·89	256·03	265·18
300	274·32	283·46	292·61	301·75	310·90	320·04	329·18	338·33	347·47	356·62
400	365·76	374·90	384·05	393·19	402·34	411·48	420·62	429·77	438·91	448·06
500	457·20	466·34	475·49	484·63	493·78	502·92	512·06	521·21	530·35	539·50
600	548·64	557·78	566·93	576·07	585·22	594·36	603·50	612·65	621·79	630·94
700	640·08	649·22	658·37	667·51	676·66	685·80	694·94	704·09	713·23	722·38
800	731·52	740·66	749·81	758·95	768·10	777·24	786·38	795·53	804·67	813·82
900	822·96	832·10	891·25	850·39	859·34	868·68	877·82	886·97	896·11	905·26
1000	914·40	—	—	—	—	—	—	—	—	—

AUXILIARY TABLE

yd	*1*	*2*	*3*	*4*	*5*	*6*	*7*	*8*	*9*
m	0·914	1·829	2·743	3·658	4·572	5·486	6·401	7·315	8·230

Table A.4(d) LENGTH: MILES TO KILOMETRES (Correct to nearest 10 m)

miles	*0*	*1*	*2*	*3*	*4*	*5*	*6*	*7*	*8*	*9*
					km					
0	—	1·61	3·22	4·83	6·44	8·05	9·66	11·27	12·87	14·48
10	16·09	17·70	19·31	20·92	22·53	24·14	25·75	27·36	28·97	30·58
20	32·19	33·80	35·41	37·01	38·62	40·23	41·84	43·45	45·06	46·67
30	48·28	49·89	51·50	53·11	54·72	56·33	57·94	59·55	61·16	62·76
40	64·37	65·98	67·59	69·20	70·81	72·42	74·03	75·64	77·25	78·86
50	80·47	82·08	83·69	85·30	86·90	88·51	90·12	91·73	93·34	94·95
60	96·56	98·17	99·78	101·39	103·00	104·61	106·22	107·83	109·44	111·05
70	112·65	114·26	115·87	117·48	119·09	120·70	122·31	123·92	125·53	127·14
80	128·75	130·36	131·97	133·58	135·19	136·79	138·40	140·01	141·62	143·23
90	144·84	146·45	148·06	149·67	151·28	152·89	154·50	156·11	157·72	159·33
100	160·94	—	—	—	—	—	—	—	—	—

AUXILIARY TABLES

miles	0·1	0·2	0·3	0·4	0·5	0·6	0·7	0·8	0·9
km	0·161	0·322	0·483	0·644	0·805	0·966	1·127	1·288	1·448

furlongs	1	2	3	4	5	6	7
km	0·201	0·402	0·604	0·804	1·006	1·207	1·408

Table A.4(e) AREA: SQUARE FEET TO SQUARE METRES (Correct to the nearest 0·01 m^2)

ft^2	0	10	20	30	40	50	60	70	80	90
	m^2									
0	—	0·93	1·86	2·79	3·72	4·65	5·57	6·50	7·43	8·36
100	9·29	10·22	11·15	12·08	13·01	13·94	14·86	15·79	16·72	17·65
200	18·58	19·51	20·44	21·37	22·30	23·23	24·15	25·08	26·01	26·94
300	27·87	28·80	29·73	30·66	31·59	32·52	33·45	34·37	35·30	36·23
400	37·16	38·09	39·02	39·95	40·88	41·81	42·74	43·66	44·59	45·52
500	46·45	47·38	48·31	49·24	50·17	51·10	52·03	52·95	53·88	54·81
600	55·74	56·67	57·60	58·53	59·46	60·39	61·32	62·25	63·17	64·10
700	65·03	65·96	66·89	67·82	68·75	69·68	70·61	71·54	72·46	73·39
800	74·32	72·25	76·18	77·11	78·04	78·97	79·90	80·83	81·76	82·68
900	83·61	84·54	85·47	86·40	87·33	88·26	89·19	90·12	91·05	91·97
1000	92·90	—	—	—	—	—	—	—	—	—

AUXILIARY TABLES

ft^2	1	2	3	4	5	6	7	8	9
m^2	0·09	0·19	0·28	0·37	0·46	0·56	0·65	0·74	0·84

Table A.4(f) AREA: SQUARE YARDS TO SQUARE METRES (Correct to the nearest 0·01 m^2)

yd^2	0	10	20	30	40	50	60	70	80	90
	m^2									
0	—	8·36	16·72	25·08	33·45	41·81	50·17	58·53	66·89	75·25
100	83·61	91·97	100·34	108·70	117·06	125·42	133·78	142·14	150·50	158·86
200	167·23	175·59	183·95	192·31	200·67	209·03	217·39	225·75	234·12	242·48
300	250·84	259·20	267·56	275·92	284·28	292·65	301·01	309·37	317·73	326·09
400	334·45	342·81	351·17	359·54	367·90	376·26	384·62	392·98	401·34	409·70
500	418·06	426·43	434·79	443·15	451·51	459·87	468·23	476·59	484·95	493·32
600	501·68	510·04	518·40	526·76	535·12	543·48	551·84	560·21	568·57	576·93
700	585·29	593·65	602·01	610·37	618·73	627·10	635·46	643·82	652·18	660·54
800	668·90	677·26	685·62	693·99	702·35	710·71	719·07	727·43	735·79	744·15
900	753·52	760·88	769·37	777·60	785·96	794·32	802·68	811·04	819·41	827·77
1000	836·13	—	—	—	—	—	—	—	—	—

AUXILIARY TABLE

yd^2	1	2	3	4	5	6	7	8	9
m^2	0·84	1·67	2·51	3·34	4·18	5·02	5·85	6·69	7·53

Table A.4(g) VOLUME: CUBIC INCHES TO CUBIC CENTIMETRES OR MILLILITRES (Correct to the nearest cm^3)

in^3	*0*	*10*	*20*	*30*	*40*	*50*	*60*	*70*	*80*	*90*
					cm^3					
0	—	164	328	492	655	819	983	1 147	1 311	1 475
100	1 639	1 803	1 966	2 130	2 294	2 458	2 622	2 786	2 950	3 114
200	3 277	3 441	3 605	3 769	3 933	4 097	4 261	4 425	4 588	4 752
300	4 916	5 080	5 244	5 408	5 572	5 735	5 899	6 063	6 227	6 391
400	6 555	6 719	6 883	7 046	7 210	7 374	7 538	7 702	7 866	8 030
500	8 194	8 357	8 521	8 685	8 849	9 013	9 177	9 341	9 505	9 668
600	9 832	9 996	10 160	10 324	10 488	10 652	10 816	10 979	11 143	11 307
700	11 471	11 635	11 799	11 963	12 126	12 290	12 454	12 618	12 782	12 946
800	13 110	13 274	13 437	13 601	13 765	13 929	14 093	14 257	14 421	14 585
900	14 748	14 912	15 076	15 240	15 404	15 568	15 732	15 896	16 059	16 223
1000	16 387	—	—	—	—	—	—	—	—	—

AUXILIARY TABLE

in^3	0	1	2	3	4	5	6	7	8	9
ml.	—	16	33	49	66	82	98	115	131	147

Table A.4(h) VOLUME: CUBIC FEET TO CUBIC METRES (Correct to the nearest 0·01 m^3)

ft^3	*0*	*10*	*20*	*30*	*40*	*50*	*60*	*70*	*80*	*90*
					m^3					
0	—	0·28	0·57	0·85	1·13	1·42	1·70	1·98	2·27	2·55
100	2·83	3·11	3·40	3·68	3·96	4·25	4·53	4·81	5·10	5·38
200	5·66	5·95	6·23	6·51	6·80	7·08	7·36	7·65	7·93	8·21
300	8·50	8·78	9·06	9·34	9·63	9·91	10·19	10·48	10·76	11·04
400	11·33	11·61	11·89	12·18	12·46	12·74	13·03	13·31	13·59	13·88
500	14·16	14·44	14·72	15·01	15·29	15·57	15·86	16·14	16·42	16·71
600	16·99	17·27	17·56	17·84	18·12	18·41	18·69	18·97	19·26	19·54
700	19·82	20·11	20·39	20·67	20·95	21·24	21·52	21·80	22·09	22·37
800	22·65	22·94	23·22	23·50	23·79	24·07	24·35	24·64	24·92	25·20
900	25·49	25·77	26·05	26·33	26·62	26·90	27·18	27·47	27·75	28·03
1000	28·32	—	—	—	—	—	—	—	—	—

AUXILIARY TABLE

ft^3	0	1	2	3	4	5	6	7	8	9
m^3	—	0·03	0·06	0·08	0·11	0·14	0·17	0·20	0·23	0·25

Table A.4(j) VOLUME: CUBIC YARDS TO CUBIC METRES (Correct to the nearest 0·01 m³)

yd^3	0	1	2	3	4	5	6	7	8	9
					m^3					
0	—	0·76	1·53	2·29	3·06	3·82	4·59	5·35	6·12	6·88
10	7·65	8·41	9·17	9·94	10·70	11·47	12·23	13·00	13·76	14·53
20	15·29	16·06	16·82	17·58	18·35	19·11	19·88	20·64	21·41	22·17
30	22·94	23·70	24·47	25·23	25·99	26·76	27·52	28·29	29·05	29·82
40	30·58	31·35	32·11	32·88	33·64	34·41	35·17	35·93	36·70	37·46
50	38·23	38·99	39·76	40·52	41·29	42·05	42·82	43·58	44·34	45·11
60	45·87	46·64	47·40	48·17	48·93	49·70	50·46	51·23	51·99	52·75
70	53·52	54·28	55·05	55·81	56·58	57·34	58·11	58·87	59·64	60·40
80	61·16	61·93	62·69	63·46	64·22	64·99	65·75	66·52	67·28	68·05
90	68·81	69·57	70·34	71·10	71·87	72·63	73·40	74·16	74·93	75·69
100	75·46	—	—	—	—	—	—	—	—	—

Table A.4(k) CAPACITY: GALLONS TO LITRES (Correct to the nearest 10 ml)

gal	0	1	2	3	4	5	6	7	8	9
					litres					
0	—	4·55	9·09	13·64	18·18	22·73	27·28	31·82	36·37	40·91
10	45·46	50·01	54·55	59·10	63·65	68·19	72·74	77·28	81·83	86·38
20	90·92	95·47	100·01	104·56	109·11	113·65	118·20	122·74	127·29	131·84
30	136·38	140·93	145·48	150·02	154·57	159·11	163·66	168·21	172·75	177·30
40	181·84	186·39	190·94	195·48	200·03	204·57	209·12	213·67	218·21	222·76
50	227·31	231·85	236·40	240·94	245·49	250·04	254·58	259·13	263·67	268·22
60	273·77	277·31	281·86	286·40	290·95	295·50	300·04	304·59	309·13	313·68
70	318·23	322·77	327·32	331·87	336·41	340·96	345·50	350·05	354·60	359·14
80	363·69	368·23	372·78	377·33	381·87	386·42	390·96	395·51	400·06	404·60
90	409·15	413·69	418·24	422·79	427·33	431·88	436·43	440·97	445·52	450·06
100	454·61	—	—	—	—	—	—	—	—	—

AUXILIARY TABLE

Pint	1	2	3	4	5	6	7
litre	0·568	1·136	1·705	2·273	2·841	3·410	3·978

Table A.4(l) MASS: POUNDS TO KILOGRAMMES (Correct to nearest gramme or 0·001 kg)

Pounds	0	1	2	3	4	5	6	7	8	9
					kilogrammes					
0	—	0·454	0·907	1·361	1·814	2·268	2·722	3·175	3·629	4·082
10	4·536	4·990	5·443	5·897	6·350	6·804	7·257	7·711	8·165	8·618
20	9·072	9·525	9·979	10·433	10·886	11·340	11·793	12·247	12·701	13·154
30	13·608	14·061	14·515	14·969	15·422	15·876	16·329	16·783	17·237	17·690
40	18·144	18·597	19·051	19·505	19·958	20·412	20·865	21·319	21·772	22·226
50	22·680	23·133	23·587	24·040	24·494	24·948	25·401	25·855	26·308	26·762
60	27·216	27·669	28·123	28·576	29·030	29·484	29·937	30·391	30·844	31·298
70	31·752	32·205	32·659	33·112	33·566	24·019	34·473	34·927	35·380	35·834
80	36·287	36·741	37·195	37·648	38·102	38·555	39·009	39·463	39·916	40·370
90	40·823	41·277	41·731	42·184	42·638	43·091	43·549	43·999	44·452	44·906
100	45·359	—	—	—	—	—	—	—	—	—

Table A.4(m) MASS: TON TO TONNE

Tons	*0*	*1*	*2*	*3*	*4*	*5*	*6*	*7*	*8*	*9*
0	—	1·0160	2·0321	3·0481	4·0642	5·0802	6·0963	7·1123	8·1284	9·1444
10	10·1605	11·1765	12·1926	13·2086	14·2247	15·2407	16·2568	17·2728	18·2888	19·3049
20	20·3209	21·3370	22·3530	23·3691	24·3851	25·4012	26·4172	27·4333	28·4493	29·4654
30	30·4814	31·4975	32·5135	33·5295	34·5456	35·5616	36·5777	37·5937	38·6098	39·6258
40	40·6419	41·6579	42·6740	43·6900	44·7061	45·7221	46·7382	47·7542	48·7703	49·7865
50	50·8023	51·8184	52·8344	53·8505	54·8665	55·8826	56·8986	57·9147	58·9307	59·9468
60	60·9628	61·9789	62·9949	64·0110	65·0270	66·0430	67·0591	68·0751	69·0912	70·1072
70	71·1233	72·1393	73·1554	74·1714	75·1875	76·2035	77·2196	78·2356	79·2517	80·2677
80	81·2838	82·2998	83·3158	84·3319	85·3479	86·3640	87·3800	88·3961	89·4121	90·4282
90	91·4442	92·4603	93·4763	94·4924	95·5084	96·5245	97·5405	98·5566	99·5726	100·589
100	101·605	—	—	—	—	—	—	—	—	—

Table A.4(n) MASS: CWT; QTR TO TONNE

qtr	*0*	*1*	*2*	*3*	*qtr*	*0*	*1*	*2*	*3*
cwt		*tonne*			*cwt*		*tonne*		
0	—	0·0127	0·0254	0·0381	10	0·5080	0·5207	0·5334	0·5461
1	0·0508	0·0635	0·0762	0·0889	11	0·3588	0·5715	0·5842	0·5969
2	0·1016	0·1143	0·1270	0·1397	12	0·6096	0·6223	0·6350	0·6477
3	0·1524	0·1651	0·1778	0·1905	13	0·6604	0·6731	0·6858	0·6985
4	0·2032	0·2159	0·2286	0·2413	14	0·7112	0·7239	0·7366	0·7493
5	0·2540	0·2667	0·2794	0·2921	15	0·7620	0·7747	0·7874	0·8001
6	0·3048	0·3175	0·3302	0·3429	16	0·8128	0·8255	0·8382	0·8509
7	0·3556	0·3683	0·3810	0·3937	17	0·8636	0·8763	0·8890	0·9017
8	0·4064	0·4191	0·4318	0·4445	18	0·9144	0·9271	0·9398	0·9525
9	0·4572	0·4699	0·4826	0·4953	19	0·9652	0·9779	0·9906	1·0033

Table A.4(p) PRESSURE: LB/IN^{2}F TO 10^5N/m^2 (bar)

lb/in^{2}f	*0*	*1*	*2*	*3*	*4*	*5*	*6*	*7*	*8*	*9*
0	—	0·069	0·138	0·207	0·276	0·345	0·414	0·483	0·552	0·621
10	0·689	0·758	0·827	0·896	0·965	1·034	1·103	1·172	1·241	1·310
20	1·379	1·448	1·517	1·586	1·654	1·724	1·793	1·862	1·931	1·999
30	2·068	2·137	2·206	2·275	2·344	2·413	2·482	2·551	2·620	2·689
40	2·758	2·827	2·895	2·965	3·034	3·103	3·172	3·241	3·309	3·378
50	3·447	3·516	3·585	3·654	3·723	3·792	3·861	3·930	3·999	4·068
60	4·137	4·206	4·275	4·344	4·413	4·442	4·551	4·619	4·688	4·757
70	4·826	4·895	4·964	5·033	5·102	5·171	5·240	5·309	5·378	5·447
80	5·516	5·585	5·654	5·723	5·792	5·861	5·929	5·998	6·067	6·136
90	6·205	6·274	6·343	6·412	6·481	6·550	6·619	6·688	6·757	6·826
100	6·895	—	—	—	—	—	—	—	—	—

AUXILIARY TABLE

lb/in^{2}f	*150*	*200*	*250*	*300*	*350*	*400*	*450*	*500*	*550*	*600*
10^5N/m^2 (bar)	10·342	13·79	17·24	20·68	24·13	27·58	31·03	34·47	37·92	41·37
lb/in^{2}f	650	700	750	800	850	900	950	1 000	—	—
10^5N/m^2 (bar)	44·82	48·26	51·71	55·16	58·61	62·05	65·50	68·95	—	—

To convert from lb/in^{2}f to kN/m^2 multiply above factors by 100; thus for 15lb/in^{2}f = 1·034 bar = 103·4 kN/m^2

Table A.4(q) STRESS: TONS/IN²F TO MN/m² (N/mm²)

ton/ in²f	0	1	2	3	4	5	6	7	8	9
0	—	15·44	30·89	46·33	61·78	77·22	92·67	108·11	123·55	139·00
10	154·44	169·89	185·33	200·78	216·22	231·66	247·11	262·55	278·00	293·44
20	308·89	324·33	339·77	355·22	370·66	386·11	401·55	417·00	432·44	447·88
30	463·33	478·77	494·22	509·66	525·11	540·55	555·99	571·44	586·88	602·33
40	617·77	633·22	648·66	664·10	679·55	694·99	710·44	725·88	741·32	756·77
50	772·21	787·66	803·10	818·55	833·99	849·43	864·88	880·32	895·77	911·21
60	926·66	942·10	957·55	972·99	988·43	1003·88	1019·32	1034·77	1050·21	1065·65
70	1081·10	1096·54	1111·99	1127·43	1142·87	1158·32	1173·77	1189·21	1204·65	1220·10
80	1235·54	1250·98	1266·43	1281·87	1297·32	1312·76	1328·21	1343·65	1359·09	1374·54
90	1389·98	1405·43	1420·87	1436·32	1451·76	1467·21	1482·65	1498·09	1513·54	1528·98
100	1544·43	—	—	—	—	—	—	—	—	—

Table A.4(r) STRESS: kgf/mm² TO MN/m² (N/mm²)

kgf/ mm²	0	1	2	3	4	5	6	7	8	9
0	0·00	9·81	19·61	29·42	39·23	40·03	58·84	68·65	78·45	88·26
10	98·07	107·87	117·68	127·49	137·29	147·10	156·91	166·71	176·52	186·33
20	196·13	205·94	215·75	225·55	235·36	245·17	254·97	264·78	274·59	284·39
30	294·20	304·01	313·81	323·62	333·43	343·23	353·04	362·85	372·65	382·46
40	392·27	402·07	411·88	421·69	431·49	441·30	451·11	460·91	470·72	480·53
50	490·33	500·14	509·95	519·75	529·56	539·37	549·17	558·98	568·79	578·59
60	588·40	598·21	608·01	617·82	627·63	637·43	647·24	657·05	666·85	676·66
70	686·47	696·27	706·08	715·89	725·69	735·50	745·31	755·11	764·92	774·73
80	784·53	794·34	804·15	813·95	823·76	833·57	843·37	853·18	862·99	872·79
90	882·60	892·41	902·21	912·02	921·83	931·63	941·44	951·25	961·05	970·86
100	980·67	—	—	—	—	—	—	—	—	—

AUXILIARY TABLE

kgf/ mm²	150	200	250	300	350	400	450	500	550	600
MN/ m²	1471·00	1961·33	2451·66	2942·00	3432·33	3922·66	4412·99	4903·33	5393·66	5883·99

BIBLIOGRAPHY

1. BS 233
   Glossary of terms used in illumination and photometry; BSI (1953)
2. BS 308
   Engineering drawing practice; BSI (1964)
3. BS 350
   Conversion factors and tables, Parts 1 and 2; BSI (1959 and 1962)
4. BS 2856
   Precise conversion of inch and metric sizes engineering drawings; BSI (1957)
5. BS 3763
   International System (SI) Units; BSI (1964)
6. PD 5686
   The use of SI Units; BSI (1969)
7. PD 6203
   Supplement No. 1 to BS 350, Additional tables for SI conversions; BSI (1967)
8. FIELDEN, G. B. R. Metric units for pressures and stresses; BSI News January 3 (1971)
9. ISO R.31
   Basic quantities and units of the International System of Units (SI units) 2nd edition, International Standards Organisation, Geneva (1965)
10. ISO R.1000
    Rules for the use of units of the International System of Units and a selection of the decimal multiples and sub-multiples of the SI units, International Standards Organisation, Geneva (1969)
11. ISO DR.838
    Quantities and units of atomic and nuclear physics, International Standards Organisation, Geneva (1965)
12. ISO DR.839
    Quantities and units of nuclear reactors and ionizing radiations, International Standards Organisation, Geneva (1965)
13. IEC 50(45)
    International Lighting Vocabulary, 3rd edition, Commission Internationale de L'Eclairage and International Electrotechnical Commission (1958)
14. Addendum to 1967 Standard method of measurement of civil engineering quantities, Institution of Civil Engineers, London (1968)

ABBREVIATIONS IN THE TEXT REFER TO THE FOLLOWING BODIES:

15. BIPM
    The Bureau International des Poids et mesures
16. CGPM
    The Conference Generale des Poids et mesures
17. CIPM
    The Comité International des Poids et mesures
18. IEC
    The International Electrotechnical Commission
19. ISO
    The International Standards Organisation
20. IUPAC
    The International Union of Pure and Applied Chemistry
21. IUPAP
    The International Union of Pure and Applied Physics
22. ISO/TC 12
    ISO Committee, Quantities, Units, Symbols
23. ISO/TC 17
    ISO Committee, Steel

GENERAL REFERENCE PUBLICATIONS

These are additional to those which are referred to in the text.

PD 6031
Use of the metric system in the construction industry; BSI (1968)

PD 6249
Dimensional co-ordination in building. Estimates of timing for BSI work; BSI (1967)
PD 6432
Recommendations for the co-ordination of dimensions in building, arrangements of building components and assemblies within functional groups; BSI (1969)
PD 6444
Recommendations for the co-ordination of dimensions in building. Basic spaces for structures, external envelope and internal sub-division; BSI (1969)
BS 3643
Part 1. ISO metric screw thread. Thread data and standard thread series; BSI (1963)
BS 4318
Recommendations for preferred metric basis sizes for engineering; BSI (1968)
BS 4391
Recommendations for metric basis sizes for metal wire, sheet and strip; BSI (1969)
BS 4500
ISO limits and fits; BSI (1969)

ANDERTON, P. and BIGG, P. H., *National Physical Laboratory Changing to the Metric System. Conversion Factors Symbols and Definitions,* HMSO London (2nd edition 1967)

PARRISH, A., *SI Conversion Charts for Imperial and Metric Quantities,* Iliffe, London (1969)

BSI, *Readimetric,* British Standards Institution, London

MULLIN, J. W., SI Units in Chemical Engineering, The Chemical Engineer, September 1967

Ministry of Public Building and Works, Metrication in the Construction Industry; No. 1 Metrication in Practice; No. 2 Calculations in SI Units Structural, Civil, Heating & Ventilating; No. 3 Craftsman's Pocket Book, HMSO London (1970)

Ministry of Public Building and Works, Going Metric in the Construction Industry; Part 1 Why and When, HMSO London (1967); Part 2 Dimensional Co-ordination, HMSO London (1968)

Public Health Act 1961. The Building Regulations 1965, Metric Equivalents of Dimensions, (Ministry of Housing and Local Government and Welsh Office) HMSO London (1968)

Public Health Act 1961. The Building Regulations 1965, Metric Values. Consultative Proposals, (Ministry of Housing and Local Government and Welsh Office) HMSO London (1969)

FAIRWEATHER, L. and SWILA, JAN. A., *Architects Journal Metric Handbook,* The Architectural Press Limited (1969)

Standard Method of Measurement of Building Works, The Royal Institution of Chartered Surveyors and The National Federation of Building Trades Employers, 5th edition, Metric, July 1968

SI Units for the Compressed Air Industry, Fluid Power International, September 1970

Change to Metric. Reference Manual, The Institution of Heating & Ventilating Engineers 1970

Metrication Guide, Scientific Instrument Manufacturers' Association of Great Britain 1970

Manual on Metrication, British Plastics Federation, June 1969

Metric Units with reference to Water, Sewerage and Related Subjects, Report of Working Party, Ministry of Housing and Local Government, HMSO London (1965)

HADDOCK, A., Going Metric in the UK Petroleum Industry, Journal of Institute of Petroleum No. 548 Vol. 56, March 1970

SYMBOLS AND ABBREVIATIONS

G. R. DARBY, C.Eng., M.I.Mech.E.
Secretary, Metric Steering Committee, C.E.G.B.

Table A.5 QUANTITIES AND UNITS OF PERIODIC AND RELATED PHENOMENA
(Based on ISO recommendation R31)

Symbol	*Quantity*
T	periodic time
$\tau, (T)$	time constant of an
f, ν	exponentially varying quantity frequency
η	rotational frequency
ω	angular frequency
λ	wave length
$\sigma(\tilde{\nu})$	wave number
k	circular wave number
$\log e\ (A_1/A_2)')$	natural logarithm of the ratio of two amplitudes
	ten times the common logarithm of the ratio of two powers
δ	damping coefficient
Λ	logarithmic decrement
α	attenuation co-efficient
β	phase co-efficient
γ	propagation co-efficient

Table A.6 QUANTITIES AND UNITS OF MECHANICS
(Based on ISO recommendation R31)

Symbol	*Quantity*
m	mass
e, ρ	density (mass density)
d	relative density
v	specific volume
p	momentum
b, p_O, p_θ	moment of momentum (angular momentum)
I, J	moment of inertia (dynamic moment of inertia)
F	force
$G(P, W)$	weight
γ	specific weight (weight density)
M	moment of force
M	bending moment
T	torque, moment of a couple
p	pressure
σ	normal stress
τ	shear stress
e, ε	linear strain (relative elongation)

Table A.6—*continued*

Symbol	*Quantity*
γ	shear strain (shear angle)
$\Theta\theta$	volume strain (bulk strain)
μ, ν	Poisson's ratio Poisson's number
E	Young's modulus (modulus of elasticity)
G	shear modulus (modulus of rigidity)
K	bulk modulus (modulus of compression)
x, κ	compressibility (bulk compressibility)
I, I_a	second moment of area (second axial moment of area)
I_p, J	second polar moment of area
$Z, W \left(\frac{I}{v}\right)$	section modulus
$\mu(f)$	co-efficient of friction (factor of friction)
$\eta(\mu)$	viscosity (dynamic viscosity)
γ	kinematic viscosity
$\sigma(\gamma)$	surface tension
A, W	work
E, W	energy
Ep, U, V, Φ	potential energy
E_k, K, T	kinetic energy
p	power

Table A.7 SYMBOLS FOR QUANTITIES AND UNITS OF HEAT
(Based on ISO recommendation R31)

Symbol	*Quantity*
T, Θ	thermodynamic temperature absolute temperature
$t, 0\Theta$	customary temperature
α, λ	linear expansion co-efficient
α, β, γ	cubic expansion co-efficient
β	pressure coefficient
Q	heat, quantity of heat
$\Phi(q)$	heat flow rate
$q(\phi)$	density of heat flow rate
$\lambda(k)$	thermal conductivity
h, k, U, α	co-efficient of heat transfer
$a(\alpha, x, k)$	thermal diffusivity
C	heat capacity
c	specific heat capacity
cp	specific heat capacity at constant pressure
cp	specific heat capacity at constant volume
γ, x, k	ratio of the specific heat capacities
S	entropy
s	specific entropy
$U(E)$	internal energy
$H(I)$	enthalpy
F	free energy
G	Gibbs function
$u(e)$	specific internal energy
$h(i)$	specific enthalpy
f	specific free energy
g	specific Gibbs function
L	latent heat
l	specific latent heat

Table A.8 SYMBOLS FOR QUANTITIES AND UNITS OF ACOUSTICS
(Based on ISO recommendation R31)

Symbol	*Quantity*
T	period, periodic time
f, v	frequency, frequency interval
ω	angular frequency, circular frequency
λ	wavelength
k	circular wave number
ρ	density (mass density)
Ps	static pressure
p	(instantaneous) sound pressure
ε, (x)	(instantaneous) sound particle displacement
u, v	(instantaneous) sound particle velocity
a	(instantaneous) sound particle acceleration
q, U	(instantaneous) volume velocity
c	velocity of sound
E	sound energy density
P, (N, W)	sound energy flux, sound power
I, J	sound intensity
Z_s, (W)	specific acoustic impedance
Z_a, (Z)	acoustic impedance
Z_m, (w)	mechanical impedance
L_p, (L_N, L_w)	sound power level
L_p, (L)	sound pressure level
δ	damping co-efficient
Λ	logarithmic decrement
α	attenuation co-efficient
β	phase co-efficient
γ	propagation co-efficient
δ	dissipation co-efficient
r, τ	reflection co-efficient
γ	transmission co-efficient
α, (α_a)	acoustic absorption co-efficient
R	sound reduction index sound transmission loss
A	equivalent absorption area of a surface or object
T	reverberation time
L_N, (Λ)	loudness level
N	loudness

Table A.9 SYMBOLS FOR QUANTITIES AND UNITS OF ELECTRICITY AND MAGNETISM
(Based on ISO recommendation R31)

Symbol	*Quantity*
I	electric current
Q	electric charge, quantity of electricity
e	volume density of charge, charge density
σ	surface density of charge
E, (K)	electric field strength
V, ϕ	electric potential
U, (V)	potential difference, tension
E	electromotive force
D	displacement (rationalised displacement)
D'	non-rationalised displacement
Ψ	electric flux, flux of displacement (flux of rationalised displacement)
Ψ'	flux of non-rationalised displacement
C	capacitance
ε	permittivity

Table A.9—*continued*

Symbol	*Quantity*
ε_0	permittivity of vacuum
ε'	non-rationalised permittivity
ε'_0	non-rationalised permittivity of vacuum
ε_r	relative permittivity
χ_e	electric susceptibility
χ'_e	non-rationalised electric susceptibility
P	electric polarisation
$p, (p_e)$	electric dipole moment
$J, (S)$	current density
$A, (\alpha)$	linear current density
H	magnetic field strength
H'	non-rationalised magnetic field strength
U_m	magnetic potential difference
F, F_m	magnetomotive force
B	magnetic flux density, magnetic induction
Φ	magnetic flux
A	magnetic vector potential
L	self inductance
M, L_{12}	mutual inductance
$k, (x, k)$	coupling co-efficient
σ	leakage co-efficient
μ	permeability
μ_0	permeability of vacuum
μ'	non-rationalised permeability
μ'_0	non-rationalised permeability of vacuum
μ_r	relative permeability
x, k	magnetic susceptibility
x', k'	non-rationalised magnetic susceptibility
m	electromagnetic moment (magnetic moment)
H_i, M	magnetisation
B_i, J	magnetic polarisation
J'	non-rationalised magnetic polarisation
ω	electromagnetic energy density
S	Poynting vector
c	velocity of propagation of electromagnetic waves in vacuo
R	resistance (to direct current)
G	conductance (to direct current)
e	resistivity
γ, σ	conductivity
R, R_m	reluctance
$A, (P)$	permeance
N	number of turns in winding
m	number of phases
p	number of pairs of poles
ϕ	phase displacement
Z	impedance (complex impedance)
$[Z]$	modulus of impedance (impedance)
X	reactance
R	resistance
Q	quality factor
Y	admittance (complex admittance)
$[Y]$	modulus of admittance (admittance)
B	susceptance
G	conductance
P	active power
$S, (P_s)$	apparent power
$Q, (P_q)$	reactive power

Table A.10 MATHEMATICAL SIGNS AND SYMBOLS FOR USE IN TECHNOLOGY
(Based on ISO recommendation R31)

Sign, symbol	*Quantity*
$=$	equal to
$\neq$ $\neq$	not equal to
$\equiv$	identically equal to
Δ	corresponds to
$\approx$	approximately equal to
$\rightarrow$	approaches
$\simeq$	asymptotically equal to
$\sim$	proportional to
∞	infinity
$<$	smaller than
$>$	larger than
$\leq$ $\leqslant$ $\leqq$	smaller than or equal to
$\geq$ $\geqslant$ $\geqq$	larger than or equal to
$\ll$	much smaller than
$\gg$	much larger than
$+$	plus
$-$	minus
$\cdot$ $\times$	multiplied by
$\frac{a}{b}$ a/b	a divided by b
$\lvert a \rvert$	magnitude of a
a^n	a raised to the power n
$a^{\frac{1}{2}}$ $a^{1/2}$ $\sqrt{a}$ $\sqrt{a}$	square root of a
$a^{1/n}$ $a^{\frac{1}{n}}$ $\sqrt[n]{a}$ $\sqrt[n]{a}$	n'th root of a
$\bar{a}$ $\langle a \rangle$	mean value of a
$p!$	factorial p, $1 \times 2 \times 3 \times \ldots \times p$
$\binom{n}{p}$	binomial co-efficient, $\frac{n(n-1)\ldots(n-p+1)}{1 \times 2 \times 3 \times \ldots \times p}$
Σ	sum
Π	product
$f(x)$ $f(x)$	function f (of f) of the variable x
$[f(x)]_a^b$ $f(x)/_a^b$	$f(b)-f(a)$
$\lim_{x \to a} f(x)$; $\lim_{x \to a} f(x)$	the limit to which $f(x)$ tends as x approaches a
Δx	delta x = finite increment of x
δx	delta x = variation of x
$\frac{df}{dx}$; df/dx; $f'(x)$	differential co-efficient of $f(x)$ with respect to x
$\frac{d^n f}{dx^n}$; $f^{(n)}(x)$	differential co-efficient of order n of $f(x)$
$\frac{\partial f(x, y, \ldots)}{\partial x}$; $\left(\frac{\partial f}{\partial x}\right)_{y\ldots}$	partial differential co-efficient of $f(x, y, \ldots)$ with respect to x, when $y, \ldots$ are held constant
df	the total differential of f
$\int f(x)dx$	indefinite integral of $f(x)$ with respect to x
$\int_a^b f(x)dx$; $\int_a^b f(x)dx$	definite integral of $f(x)$ from $x = a$ to $x = b$
e	base of natural logarithms
e^x; $\exp x$	e raised to the power x
$\log_a x$	logarithm to the base a of x
$\ln x$; $\log_e x$	natural logarithm (Napierian logarithm) of x
$\lg x$; $\log x$; $\log_{10} x$	common (Briggsian) logarithm of x
$\mathrm{lb}\, x$; $\log_2 x$	binary logarithm of x
$\sin x$	sine of x
$\cos x$	cosine of x

Table A.10—*continued*

Symbol	*Quantity*
tan x; tg x	tangent of x
cot x; ctg x	cotangent of x
sec x	secant of x
cosec x	cosecant of x
arcsin x	arc sine of x
arccos x	arc cosine of x
arctan x, arctg x	arc tangent of x
arccot x, arcctg x	arc cotangent of x
arcsec x	arc secant of x
arccosec x	arc cosecant of x
sinh x	hyperbolic sine of x
cosh x	hyperbolic cosine of x
tanh x	hyperbolic tangent of x
coth x	hyperbolic cotangent of x
sech x	hyperbolic secant of x
cosech x	hyperbolic cosecant of x
arsinh x	inverse hyperbolic sine of x
arcosh x	inverse hyperbolic cosine of x
artanh x	inverse hyperbolic tangent of x
arcoth x	inverse hyperbolic cotangent of x
arsech x	inverse hyperbolic secant of x
arcosech x	inverse hyperbolic cosecant of x
i, j	imaginary unity, $1^2 = -1$
$Re\ z$	real part of z
$Im\ z$	imaginary part of z
$\lvert z \rvert$	modulus of z
arg z	argument of z
z^*	conjugate of x, complex conjugate of z
$\tilde{A}$	transpose of matrix A
A*	complex conjugate matrix of matrix A
A†	Hermitian conjugate matrix of matrix A
Aa	vector
$\lvert \mathbf{A} \rvert$, A	magnitude of vector
$\mathbf{A} \cdot \mathbf{B}$	scalar product
$\mathbf{A} \times \mathbf{B}$, $\mathbf{A} \wedge \mathbf{B}$	vector product
$\mathbf{\nabla}$	differential vector operator
$\nabla\phi$, grad ϕ	gradient of ϕ
$\nabla \cdot \mathbf{A}$, div **A**	divergence of **A**
$\nabla \times \mathbf{A}$, $\nabla \wedge \mathbf{A}$, curl **A**, rot **A**	curl of **A**
$\nabla^2\phi$, $\Delta\phi$	Laplacion of ϕ

Table A.11 ABBREVIATIONS OF COMMON UNITS

a	are; year
Å	ångstrom
A	ampere
asb	apostilb
AU	astronomical unit
AT	assay ton
b	barn
bar	bar

Table A.11—*continued*

Bi	Biot (unit of current in electromagnetic CGS system)
Btu BthU	British thermal unit
c	curie
C	coulomb
°C	degree Celsius
cal	calorie
cc	cubic centimetre
cd	candela
CHU	Centigrade heat unit
Ci	curie
cl	centilitre
cm	centimetre
CM	carat
cP	centipoise
c/s	cycle per second
cSt	centistoke
ct	carat
cu. cm.	cubic centimetre
cu ft	cubic foot
cu in	cubic inch
cusec	cubic foot per second
cwt	hundredweight
d	day
dB	decibel
dm	decimetre
dwt	pennyweight
dyn	dyne
e unit E unit	X-ray doseage
erg	erg
eV	electronvolt
f	force
F	farad
°F	degree Fahrenheit
fc	foot candle
ft	foot
ft L	foot Lambert
ft lb	foot pound
g	gramme
G	gauss
gal	gallon
Gb	gilbert
g cal	gramme calorie
gl	gill
gm	gramme
g.p.m.	gallons per minute
g.p.s.	gallons per second
gr	grain
Gs	gauss
h	hour
H	henry
ha	hectare
hp	horse power
hp hr	horse power hour
Hz	hertz
in	inch
in Hg	inch of mercury

Table A.11—*continued*

J	joule
K	kelvin
kc	kilocycle
k cal	kilocalorie
kc/s	kilocycle per second
kg	kilogramme
kgf	kilogramme force
km	kilometre
kn, kt	knot
kV	kilovolt
kVA	kilovolt ampere
kW	kilowatt
kW h	kilowatt hour
L	lambert
l	litre
lb	pound
lbf	pound force
lea	league
lm	lumen
ly	light year
lx	lux
m	metre
m	molality molal concentration
M	molar concentration
mA	milliampere
mbar	millibar
mcps	mega cycles per second
MEV	mega electron volt
mF	millifarad
micron	length -10^{-6} metre pressure -10^{-3} mm Hg
mil	angular $\doteq \frac{1}{1000}$ rt. angle length $\frac{1}{1000}$ inch volume – millilitre
min	minute (time)
mks	metre kilogramme second
ml	millilitre
mL	millilambert
mm	millimetre
mm fd	micromicrofarad
mm Hg	millimetre of mercury
mmm	millimicrons
mol	mole (amount of substance)
mpg	miles per gallon
mpm	metres per minute
m/s, mps	metres per second
mt	metric ton
mV	millivolt
mW h	megawatt hour
Mx	maxwell
N	newton
n. mile, nm	nautical mile
Np	neper
nt	nit

Table A.11—*continued*

ntm	net ton mile
n unit	neutron dose
Oe	oersted
oz	ounze (avoirdupois)
oz. t	ounce (troy)
p	perch
P	poise
P	phon
Pa	pascal
pc	parsec
pdl	poundal
ph	phot
psi	pounds per square inch
pwt	pennyweight
q } ql }	quintal
qts	quart
r } R }	Röntgen
R	Réaumier
°R	degree Rankine
rad	radian
rpm	revolutions per minute
rps	revolutions per second
s	second (time)
S	Siemen
S } St }	stokes
sb	stilb
sn	sthéne
sr	steradian
T	tesla
t	tonne
th	thermie
V	volt
VA	volt ampere
W	watt
Wb	weber
yd	yard

INDEX

INDEX TO ADVERTISERS

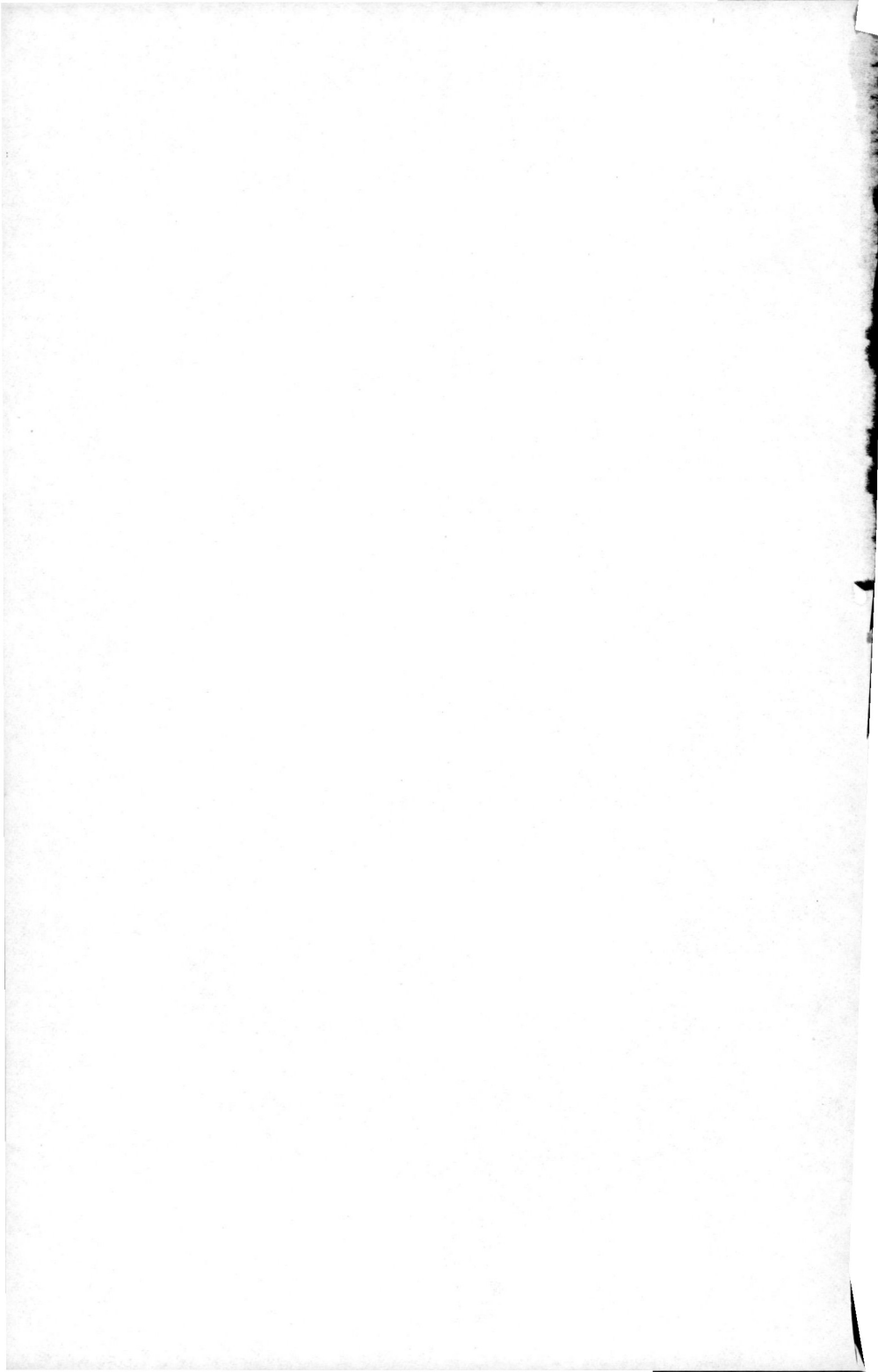